D0147032

LIFE: THE SCIENCE OF BIOLOGY

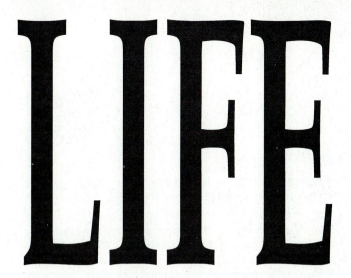

LIFE

THE SCIENCE OF BIOLOGY

THIRD EDITION

William K. Purves

*Harvey Mudd College
Claremont, California*

Gordon H. Orians

*The University of Washington
Seattle, Washington*

H. Craig Heller

*Stanford University
Stanford, California*

SINAUER ASSOCIATES, INC.

W. H. FREEMAN AND COMPANY

THE COVER

American bison (*Bison bison*) meander among the geysers in Yellowstone National Park in Wyoming.

The American bison, or buffalo, is the largest mammal in North America. Before Europeans colonized the continent, as many as 60 million bison ranged from Alberta to Mexico and from western New York to the Rocky Mountains. By 1895, due to relentless overhunting for their meat and hides, fewer than 1,000 bison existed. At the turn of the century, however, conservationists in Canada and the United States began to work to forestall the extinction of the bison by creating wildlife preserves. Today virtually all of North America's 50,000 bison live in national parks and refuges. Chapter 49 discusses ways in which the discipline of conservation biology is attacking the pressures toward extinction that threaten thousands of species throughout the world today. Photograph © Erwin and Peggy Bauer.

THE FRONTISPIECE

The Yellowstone River in Paradise Valley, Montana, at the northern end of the Greater Yellowstone Ecosystem. Photograph by Todd Wilkinson. From *Conservation Biology* 5, September 1991, page 334. Reprinted by permission of the Society for Conservation Biology and Blackwell Scientific Publications, Inc.

LIFE: THE SCIENCE OF BIOLOGY, Third Edition

Copyright © 1992 by Sinauer Associates, Inc.
All rights reserved. This book may not be reproduced
in whole or in part without permission.

Address editorial correspondence to Sinauer Associates, Inc.,
Sunderland, Mass. 01375.

Address orders to W. H. Freeman and Co., Distribution Center,
4419 West 1980 South, Salt Lake City, Utah 84104.

Library of Congress Cataloging-in-Publication Data

Purves, William K. (William Kirkwood), 1934–
　　Life, the science of biology / William K. Purves, Gordon H.
　Orians, H. Craig Heller.
　　　　p.　cm.
　　Includes bibliographical references and index.
　　ISBN 0-7167-2276-3
　　1. Biology.　I. Orians, Gordon H.　II. Heller, H. Craig.
　III. Title.
　QH308.2.P87　1992
　574—dc20
　　　　　　　　　　　　　　　91-19845
　　　　　　　　　　　　　　　CIP

Book and cover design by *Rodelinde Graphic Design*

Printed in U.S.A.

432

To Jean, Betty, and Renu

ABOUT THE AUTHORS

William K. Purves

Bill Purves is Stuart Mudd Professor of Biology as well as founder and chair of the Department of Biology at Harvey Mudd College in Claremont, California. He received his Ph.D. from Yale University in 1959 under Arthur Galston. A Fellow of the American Association for the Advancement of Science, Professor Purves has served as head of the Life Sciences Group at the University of Connecticut, Storrs, and as chair of the Department of Biological Sciences, University of California, Santa Barbara, where he won the Harold J. Plous Award for teaching excellence. His research interests focus on the chemical and physical regulation of plant growth and flowering.

Professor Purves has taught introductory biology each year for the last thirty years and considers the course the most interesting and important of his professional activities. "I can't imagine a year without teaching it," he says. In describing his teaching philosophy, Purves states, "Students learn biological concepts much more rapidly and effectively if they understand where the concepts come from—what the experimental and conceptual background is. 'Facts' by themselves can be boring or incomprehensible, but they become exciting if given a context."

Gordon H. Orians

Gordon Orians is Professor of Zoology and Environmental Studies at the University of Washington. He received his Ph.D. from the University of California, Berkeley, in 1960 under Frank Pitelka. Professor Orians was recently elected to the National Academy of Sciences (1989) and to the American Academy of Arts and Sciences (1991). He is currently serving as President of the Organization for Tropical Studies. In 1976 he was awarded the Brewster Medal by the American Ornithologists' Union.

Professor Orians is a leading authority in ecology and evolution, with research interests in behavioral ecology, plant–herbivore interactions, community structure, and environmental aesthetics. Like the other authors, he draws from his research to bring an added dimension to his teaching and writing. "Teachers who understand research can more easily communicate the excitement researchers feel as they discover new things and the processes involved," Orians says. "All three authors of LIFE, Third Edition, have spent considerable time doing research and we have tried throughout the book to show where our current understanding of biology has come from."

H. Craig Heller

Craig Heller is Bing Professor of Human Biology and chair of the Program in Human Biology at Stanford University, and is a popular lecturer on animal and human physiology. He received his Ph.D. from Yale University in 1970 and did postdoctoral research at the Scripps Institution of Oceanography on brain regulation of body temperatures in mammals. He has continued this research since coming to Stanford in 1972, studying a variety of phenomena ranging from hibernating squirrels to sleeping college students to diving seals to meditating yogis. Professor Heller is a Fellow of the American Association for the Advancement of Science, a member of the editorial board of the *American Journal of Physiology*, and a recipient of the Walter J. Gores Award for Excellence in Teaching.

"A first course in biology requires the student to learn more new words than a first course in a foreign language," Heller says. "The secret to teaching and learning biology is to focus on central and overarching concepts. Once you grasp the concept of how something works, you have a framework on which the facts and vocabulary fall into place. Conceptual understanding also helps you relate what you learn to the real world."

PREFACE

A crucial attribute of a good textbook is its ability to explain things so that students can understand and appreciate them. Our overriding goal has been to explain biological concepts and processes clearly and thoroughly—a goal that guides our classroom teaching as well. Throughout the book we have been at great pains to develop each topic until it has been explained or illustrated sufficiently to meet the needs of both the student who requires extra help and the student who aspires to a greater degree of understanding. At the same time, we have tried not to beat topics to death. Material that is inherently so clear that simple descriptions suffice is treated succinctly.

Another key attribute of a good textbook is that it makes its subject interesting. Biologists do what we do because we find it exciting; a textbook must convey that excitement to the student. We have worked hard to focus on major concepts, explain them as simply and clearly as possible, and then seal them in the mind of the student with unforgettable examples.

As textbook authors and as teachers we want to help our students in every way possible. In the next section ("To the Student") we offer some helpful advice we'd like students to consider as they begin their study of biology. This advice has helped many of our own students. We have also added self-quizzes at the ends of chapters, augmenting study questions of the sort used in the first two editions.

In preparing this third edition of our book we have taken a number of important steps, notably the production of an entirely new art program, the addition of a third author, the addition of four all-new chapters, and a substantial reorganization of the book as a whole.

NEW ART PROGRAM. J/B Woolsey Associates have done a superb job in creating the new art, making this the most immediately evident change from the previous edition. The new art is not just attractive, it has also been designed to have maximum pedagogical value. Hundreds of illustrations have been rethought, redrawn, or newly conceived, with results that will be enormously helpful to students using the book.

When we saw the first color proofs of Figure 24.1 (p. 519), we knew that our art program had truly been revitalized. We had wanted to give students a better sense of what Earth was like in the distant past, so we asked artist John Woolsey to consult with paleobotanist Hermann Pfefferkorn and produce this beautiful and scientifically plausible painting of a Carboniferous forest.

Other examples of effective new art abound. Patrick Lane illustrated the concepts of codominance and multiple alleles with fine paintings of mutant clover leaves (pp. 215–216). Neither introductory biology nor plant physiology texts have had useful illustrations of mineral deficiency systems in plants—look at our brand-new Figure 30.2 on page 681 to see our solution to this problem. The artists have also done dozens of beautiful new illustrations of molecular biological topics, such as the membrane "spread" on pp. 92–93, cell–cell interactions in the immune system (p. 374), and levels of DNA packing (p. 190). The upgrading of the art program applies to all parts of the book. Some more of our favorites are the classification of the moss rose and the blackburnian warbler on pp. 442–443; sponge structure and function on p. 553; and the series of drawings on the generation and propagation of action potentials by nerve impulses on pp. 827, 828, and 831.

NEW AUTHOR. Biology is the broadest of the scientific disciplines, and its diversity of subject matter poses a tremendous challenge to authors as we try to write knowledgeably on all topics. To improve this edition, we have added Craig Heller, who brings great expertise to the chapters on animal physiology.

NEW CHAPTERS. The third edition has four completely new chapters. Rapid progress in molecular biology has necessitated a rearrangement of topics and the addition of a new chapter on recombinant DNA technology (Chapter 14). The chapter that precedes it, on regulation of eukaryotic gene expression, is almost entirely new. It, like Chapter 14, focuses on experimental methods. Chapter 14 covers and illustrates the major techniques of recombinant DNA technology, including cloning, cDNA, electrophoresis, blotting, restriction mapping, chromosome walking, DNA sequencing, the polymerase chain reaction, and more.

The material on plant structure has been completely reorganized, expanded, and modernized to produce a chapter called "The Flowering Plant Body" (Chapter 28). This chapter introduces the whole topic of plant biology.

In order to treat more adequately the animal kingdom, and to support the evolutionary thrust to our treatment of the kingdoms, we have expanded the animal kingdom section to two chapters, one on the protostomes and one on the deuterostomes. This expanded treatment allows us to examine more fully the factors that have influenced the evolution of animal species.

Because species are being lost from Earth at a high rate due to human activities, the media abound with discussions of the "biodiversity crisis." To provide students with a better understanding of why species are being lost, which species are most vulnerable to extinction, and how the science of biology can help us to solve environmental problems and preserve the biological heritage of the planet, we have added a new chapter on conservation biology (Chapter 49).

NEW ORGANIZATION. We have reorganized the book fairly drastically. The most important change was to move forward the main treatment of evolution so that it precedes the chapters on the kingdoms of life. This new arrangement enhances our ability to emphasize evolutionary trends as we discuss the kingdoms.

In addition, we moved the discussion of the kingdoms of life ahead of the chapters on plant and animal physiology. Therefore, when we illustrate the physiological chapters with specific and diverse examples, the student will already have been introduced to those organisms. We have also moved the chapter "Patterns in the Evolution of Life" (Chapter 28) so that it follows the chapters on biological diversity. Thus, when we describe the broad features of the evolutionary record, we can talk about changes in abundance and diversity of groups of organisms that are by then familiar to the reader.

We have reorganized the material on plants and protists and have gathered all life cycle treatments into the corresponding kingdom chapters. Previously some of this material had been contained in the chapter on plant reproduction. The new arrangement tightens the organization of the kingdom chapters. (We have also changed our definition of the kingdom Protista, adopting the "definition by exclusion": Protists are those eukaryotes, unicellular or multicellular, that do not belong in the animal, plant, or fungus kingdoms.)

Current work in animal development and in immunology is informed by advances in genetics and in recombinant DNA technology. Accordingly, we have moved the material on animal development (Chapter 15) and immunology (Chapter 16) into Part Two of the book, "Information and Heredity." This forms a natural grouping of the chapters leading up to and employing molecular biology and its techniques.

The chapters on animal physiology have been completely reorganized and rewritten. The central theme of this section is homeostasis, the maintenance of stability in the internal environment of the body. For this reason, the order of the chapters differs from the usual sequence in which the nervous system comes at the end as the paragon of complexity and integration. We begin animal physiology with the endocrine and nervous systems because of their crucial roles in controlling and regulating all the other physiological systems. Questions about control and regulation are some of the most interesting aspects of physiology, but before one can deal with those questions, one must know something about hormonal and neural mechanisms. Each chapter emphasizes humans because of the relevance of physiology to our own health and well-being. However, physiological systems are also products of evolution, so we present a sampling of their diversity by selecting particularly interesting examples from throughout the animal kingdom to illustrate physiological principles. Frequently an extreme adaptation of an unusual species is just the evolutionary experiment we need to clarify a more general physiological concept.

Before writing this third edition, we commissioned formal reviews of major parts of the second edition. The reviewers were thorough and outspoken, and their many suggestions influenced the plan of the new edition. We are indebted to those six reviewers: Robert K. Colwell (University of Connecticut), Lewis Feldman (University of California, Berkeley), Abraham Flexer (University of Colorado), Douglas J. Futuyma (State University of New York at Stony Brook), Patricia Jones (Stanford University), and Patricia C. Seawell (Palo Alto, California).

As with the first two editions, many of our colleagues reviewed chapters—or entire sections—of the book in manuscript. Those colleagues are listed at right. All the reviewers were helpful, thoughtful, and clearly dedicated to the success of this book, and we appreciate their contributions. We particularly thank Paulette Bierzychudek, Gary J. Brusca, Gregory Capelli, Merrill B. Hille, Richard F. Olivo, William Moody, Irv Tallan, and David Vleck. They stand out among an outstanding group of reviewers because they provided explicit recommendations for extensive improvements. And we thank you, the reader, in advance, for any suggestions and criticisms that you choose to offer as you use this edition.

Perhaps our single greatest debt is to our developmental editor, Elmarie Hutchinson, who performed miracles of clarification, reorganization, and revitalization. We coined the verb "to Elmarie" to describe her contributions. Her remarkable editorial skills are augmented by her technical expertise as a biologist. Best of all, she has a keen sense of what is needed to make material accessible to our student audience.

We are indebted to John Woolsey, Patrick Lane, and the whole crew at J/B Woolsey Associates for their outstanding revision of the art program. Carl May, of Biological Photo Service and Terraphotographics, again directed the photography program; he was ably assisted by Travis Amos, photo editor at W. H. Freeman, and by Jane Potter. Janice Holabird's beautiful layouts and Joe Vesely's production expertise enhanced Rodelinde Albrecht's handsome design. Norma Roche provided a deft copyediting of the manuscript. Carol Wigg coordinated the whole process, kept track of what went where, and made us toe the line.

Andy Sinauer and Linda Chaput, presidents of Sinauer Associates and of W. H. Freeman, respectively, have pooled the resources of those two outstanding publishing companies to make this new edition possible. This synergism has enabled us to produce a book of the highest quality. We are indebted to Andy and Linda for their enthusiastic support and valuable guidance throughout the years of writing, reviewing, rewriting, illustrating, editing, reillustrating, rewriting, and production of this book.

October 1991

William K. Purves
Claremont, California

Gordon H. Orians
Seattle, Washington

H. Craig Heller
Stanford, California

REVIEWERS OF THE THIRD EDITION

Kraig Adler, Cornell University
Barbara L. Bentley, SUNY, Stony Brook
Paulette Bierzychudek, Pomona College
Andrew R. Blaustein, Oregon State University
Richard Briggs, Smith College
Gary J. Brusca, Humboldt State University
Ruth Buskirk, University of Texas, Austin
Gregory Capelli, College of William and Mary
Russell D. Fernald, Stanford University
Robert Fogel, University of Michigan
William J. Grimes, University of Arizona
Paul G. Harrison, University of British Columbia
Merrill B. Hille, University of Washington
James P. Holland, Indiana University
Kent E. Holsinger, University of Connecticut
John Jaenike, University of Rochester
Allan Larson, Washington University
Willam Moody, University of Washington
David O. Norris, University of Colorado, Boulder
Barry M. O'Connor, University of Michigan
Richard F. Olivo, Smith College
Judith A. Owen, Haverford College
Jeffrey Pommerville, Glendale Community College
David M. Prescott, University of Colorado, Boulder
Francis E. Putz, University of Florida
Christopher G. Reed, Dartmouth College
Robert E. Ricklefs, University of Pennsylvania
Fred R. Rickson, Oregon State University
René R. Roth, University of Western Ontario
Carol H. Sibley, University of Washington
Irv Tallan, University of Toronto
Frank Trainor, University of Connecticut
Laurie J. Vitt, UCLA
David Vleck, University of Arizona
Steven A. Williams, Smith College
Edward O. Wilson, Harvard University
Fred Wilt, UC, Berkeley
David S. Woodruff, UC, San Diego

TO THE STUDENT

Welcome to the study of life! In our student days—and ever since—we have enjoyed studying the fascinating and fast-changing field of biology, and we hope that you will, too.

There are a few things you can do to help you get the most from this book and from your course. For openers, read the book actively—don't just read passively, but do things that force you to *think* as you read. If we pose questions, stop and think about them. If a passage reminds you of something that has gone before, think about that, or even check back to refresh your memory. *Ask questions* of the text as you go. Do you understand what is being said? Does it relate to something you already know? Is it supported by experimental or other evidence? Does that evidence convince you? How does this passage fit into the chapter as a whole? *Annotate* the book—write down comments in the margins about things you don't understand, or about how one part relates to another, or even when you find an idea particularly interesting. The point of doing these things is that they will help you learn. People remember things they think about much better than they remember things they have read passively. Highlighting is passive; copying is drudge work; questioning and commenting are active and well worthwhile.

Use the self-quizzes and study questions at the end of each chapter. The self-quizzes are meant to help you remember some of the more detailed material and to help you sort out the information we have laid before you. Answers to all self-quizzes are in the Appendix. The study questions, on the other hand, are often fairly open-ended and are intended to cause you to reflect on the material.

Each chapter has a summary to help you quickly review the high points of what you have read. A way to review the material in slightly more detail after reading the chapter is to go back and look at the boldfaced terms. The boldfacing will probably be more useful on a second reading than on the first.

Two parts of a textbook that are, unfortunately, often underused or even ignored are the glossary and the index. In fact, both can help you a great deal.

When you are uncertain of the meaning of a term, check the glossary first—there are almost 1,500 definitions tucked into it. If you don't find a term in the glossary, or if you want a more thorough discussion of the term, use the index to find where it's discussed.

What if you'd like to pursue some of the topics in greater detail? At the end of each chapter there is a short, annotated list of supplemental readings. We have tried to choose readings from books and magazines, especially *Scientific American*, that should be available in your college library.

Most students occasionally have difficulty in courses, including biology courses. If you find that you are slipping behind in the course, or if a particular topic is giving you an unreasonable amount of trouble, here are some useful steps you might take. First, the basics: attend class, take careful lecture notes, and read the textbook assignments. Second, note that one of the most important roles of studying is to discover what you don't know, so that you can do something about it. Use the index, the glossary, the chapter summaries, and the text itself to try to answer any questions you have and to help you organize the material. Make a habit of looking over your lecture notes within 24 hours of when you take them—find out right away what points are unclear, and get them straightened out in your mind. If this doesn't do the trick, get help! Other students are often a good source of help, especially since they are dealing with the material at the same level as you are. Study groups can be very useful, as long as the participants are all committed to learning the material. Tutors are almost always helpful and useful, as are faculty members. The main thing is to get help when you need it. It is *not* a good idea to be strong and silent and drift into a low grade.

But don't make the grade the point of this or any other course. You are in college to learn, to pursue interesting subjects, and to enjoy the subjects you are pursuing. We hope you'll enjoy the pursuit of biology.

Bill Purves   Gordon Orians   Craig Heller

CONTENTS IN BRIEF

CONTENTS

**PART ONE:
THE CELL**

PART TWO:
INFORMATION
AND HEREDITY

PART THREE: EVOLUTIONARY PROCESSES

**PART FOUR:
THE EVOLUTION
OF DIVERSITY**

PART FIVE:
THE BIOLOGY
OF FLOWERING
PLANTS

PART SEVEN: ECOLOGY AND BIOGEOGRAPHY

1.1 Cool but Not Too Cold
These monarch butterflies are wintering in the highlands of Mexico. This
large roost is one of the few places with suitable winter temperatures.

1

The Science of Biology

A monarch butterfly extracts nectar from a flower, using some of the energy from the nectar to power its flight and some of it to produce eggs. It lays these eggs on a milkweed plant, the prime food source for the caterpillars that will hatch from the eggs. After feeding, growing, and molting, each caterpillar eventually changes into a pupa, within which it is reorganized, and emerges as an adult butterfly. At the end of the northern hemisphere's summer, surviving adult butterflies migrate south to traditional wintering areas in California and the mountains of Mexico (Figure 1.1). To be suitable, a wintering site must have cool temperatures because the butterflies, which do not feed during the winter, must survive on stored food reserves that are used up more slowly at cool temperatures. But if temperatures are too cold, the insects freeze. Very few places reliably provide winter temperatures within the necessary narrow range. Those butterflies that survive the winter migrate north in spring to initiate another annual reproductive cycle. We have no difficulty recognizing these colorful butterflies, and their pupae and caterpillars, as living organisms.

But what is life? Defining life turns out to be more difficult than one might suppose. There is no simple description that sets living organisms apart from nonliving matter. Life is best characterized and distinguished from nonliving systems by the processes that living organisms carry out. Monarch butterflies illustrate them, but we must give these processes closer attention if we wish to develop a more profound understanding of life.

CHARACTERISTICS OF LIVING ORGANISMS

Two key processes characterize all living systems. One of these is **metabolism**, the sum of all the chemical reactions taking place in an organism. Some of these reactions consume energy; others release energy. Reactions that consume energy can occur in living organisms only because they are coupled to other reactions that release it.

All living organisms depend on external sources of energy to fuel their chemical reactions. Some organisms—called **autotrophs**, which means self-feeders—obtain their energy from sunlight or, in a few cases, from the conversion of some very simple mineral substances (Figure 1.2). The remaining organisms, called **heterotrophs** (other-feeders), obtain energy from foods—complex chemical substances in the environment—and metabolize those substances to release energy and to make the chemical building blocks for synthesizing other substances. Heterotrophs obtain their energy directly or indirectly from autotrophs (Figure 1.3).

As a result of their metabolic activities, organisms may increase the number of molecules of which they are composed—that is, they grow. **Growth** is characteristic of all organisms. Some, such as single-celled organisms, only double their original size, whereas others, such as elephants and redwood trees, increase in size more than a thousandfold over their life span. Some nonliving systems also grow, but the ways in which they do so and the patterns produced by their growth are not as complex or variable as those of living organisms.

Reproduction is the second major process characteristic of living organisms. They replicate themselves, but with variation (Figure 1.4). Single-celled organisms reproduce by dividing into two daughter cells. Some organisms reproduce by budding off small portions of their bodies to form new individuals. Most large organisms reproduce by means of special cells produced specifically for that purpose. All the information necessary to form a new individual is transmitted via these cells. Why organisms usually produce offspring that are not identical copies of themselves will form the focus of Chapters 9 and 11. For the present we need only note that *reproducing*

1

1.2 Exposed to the Sun
Autotrophs usually present large surfaces to the sun, the source of their energy. These ferns, horsetails, and grasses are growing in Alaska.

with change makes possible the evolution of life and has produced one of the most distinctive features of life: **adaptation**. When we say that an organism is adapted to its environment, we mean that it possesses characteristics that enhance its survival and reproductive success in that particular environment.

Living organisms are adjusted to their environments in remarkably subtle ways. The wings of birds are adapted for various kinds of flight (Figure 1.5). Camouflaged animals blend into their backgrounds and are difficult for visually hunting predators to locate. Early efforts to explain adaptation invoked some "purpose" or "foresight" and did not provide testable hypotheses. Modern evolutionary biology came into being with the formulation of hypotheses that could be tested by scientific methods. Nearly a century and a half ago, Charles Darwin and Alfred Russel Wallace proposed the first scientifically testable theory about adaptation: evolution by natural selection (Chapter 18).

Adaptation is a uniquely biological phenomenon. It does not make sense to ask about the function of the law of gravity or the adaptive significance of the relationships among the temperature, pressure, and volume of a sample of a gas. These features of the nonliving world are explained in purely mechanistic terms. Biology, the scientific study of living organisms, is concerned with both mechanistic explanations and with the study of function. Studies of

(a)

(b)

1.3 Food from Many Sources
(a) This black-tailed deer feeds directly on plants. (b) The spotted hyena eats other animals. These hyenas did not kill their prey, however, but are eating what remained after a lion ate its fill.

1.4 Offspring Differ from Their Parents
These nursing puppies are all members of a single litter, produced by the white female and fathered by a single male. Genetic variability among the offspring of individual parents is the norm, although it is usually less obvious than the striking differences in the colors of these puppies.

wings, even purely descriptive ones, are strongly influenced by thoughts about function. It is, in fact, difficult to describe a wing without referring to its function. Relating structure to function will be one of the major themes of this book.

HIERARCHICAL ORGANIZATION OF BIOLOGY

Biologists are interested in processes ranging from the structure of simple molecules to the interactions within ecosystems made up of hundreds of species and their physical environment. The subject matter of biology can be visualized as a hierarchy in which objects studied at one level are the building blocks of units studied at higher levels. Thus cells are composed of molecules, tissues are made up of cells, and organs are groups of different types of tissues. Individual multicellular organisms are composed of many different organs; organisms, in turn, form populations and communities. At all levels in the hierarchy, biologists study how the units interact with and adjust to one another.

Some of the most brilliant recent triumphs of biological research have occurred in areas, such as molecular biology and biophysics, that are closely related to physics and chemistry. However, when biologists study molecules they ask different types of questions than chemists do because the study of life is not the study of chemicals in isolation. Biologists study chemical structures and reactions to discover how they are integrated to produce the complex and adaptive metabolism of living organisms. Similarly, at the higher levels in the hierarchy, biologists study how organisms interact with and adjust to one another to form social systems, populations, and ecological communities.

Molecules, Cells, and Tissues

The smallest entities studied by biologists are **molecules**, the basic structural units of compounds (Chapter 2). The principal chemical compounds constituting all living organisms are proteins, nucleic acids, lipids, and carbohydrates (Chapter 3). Within living organisms many molecules are banded together to form complex aggregates, such as membranes (Chapter 5). These aggregates have properties that are essential to their functioning within the organism.

1.5 Wings Adapted for Flight
Most birds use their wings for flight, but different birds fly in different ways. (a) The long, narrow wings of an albatross allow it to sustain a gliding flight for long distances above the ocean, where it searches for fish. (b) The action of a hummingbird's wings allows it to hover in front of flowers while it extracts their nectar.

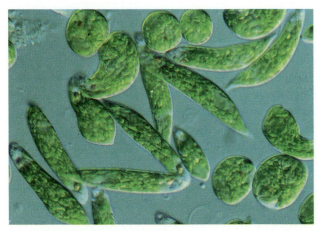

1.6 The Basic Unit of Life
Each of the cells in this picture is a complete, individual organism of *Euglena gracilis*. A single cell can be complex enough to perform all the functions an organism needs to survive and reproduce.

the energy of sunlight. Cells contain **organelles**, complex, membrane-bounded aggregates that perform specific vital functions. The cells of bacteria, which are simpler than those of most other organisms, lack most kinds of membrane-bounded organelles, but some bacterial cells do contain functioning internal membrane systems and are bounded by external membranes.

In organisms consisting of many cells, cells of specific types are organized to form **tissues** (Figure 1.7). Two familiar tissues are epidermis, a layer of flattened cells on the outer surface of many organisms, and mesophyll, the concentrations of specialized cells that capture the sun's energy in plants. Tissues are organized into **organs** such as hearts, livers, roots, and leaves—structures composed of more than one tissue type.

These properties appear only when the aggregates are formed; that is, the isolated molecules do not exhibit them. **Emergent properties**—properties that are exhibited by a system and not by its individual components—are characteristic of all levels of biological organization.

The **cell** is the fundamental unit of life because it is the simplest unit capable of independent existence and reproduction (Figure 1.6), and because all living organisms are composed of cells. Cells have many features in common (Chapter 4), but they come in a wide variety of types, many of which are adapted for specific functions, such as contraction, secretion, storage and transmittal of information, or capturing

Organisms, Species, and Ecosystems

Each creature—each tree, each bacterium, each frog, each person—is an organism. The individual organism is the central unit of study for those biologists concerned with how individuals function and how they maintain themselves in ever-changing environments. Individuals are born, reproduce, and die, and the patterns of their births and deaths determine the directions of evolutionary change, as will be explained in Chapter 18.

Living organisms are highly variable and they come in an enormous diversity of forms to which we may apply such names as robins, garter snakes, howler monkeys, and maple trees. Biologists call the different forms of living organisms **species**; but, as we

1.7 A Leaf Has Several Tissues
The specific cell types in this magnified water lily leaf are gathered into tissues. The cells of the epidermis control the exchange of materials between the leaf its environment. The mesophyll tissue captures solar energy to power photosynthesis; mesophyll cells occur in two different-appearing forms. The dark red structure in the center of the photo is a sclereid; sclereid cells provide strength and support to plants. Plant tissues are discussed in detail in Chapter 28.

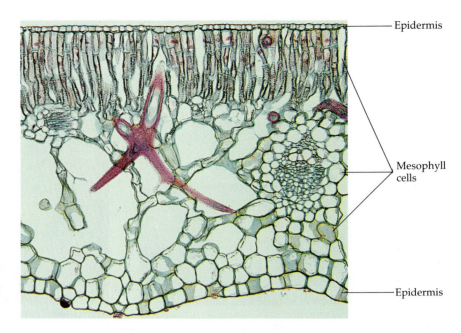

will see, species are not easy to define (Chapter 19). For our present purposes, a species can be defined as a group of similar organisms capable of interbreeding with each other but not with other organisms. The fact that living organisms are organized into species is of profound importance for the structure and functioning of nature. Individuals of many different species typically live together and interact with one another to form **ecological communities** (all the different species in a particular area) and **ecosystems** (communities interrelated through the exchange of energy and matter) (Figure 1.8). Together, Earth's ecosystems form the **biosphere**.

Emergent Properties

Each level of biological organization has properties not found at lower levels. Cells and individual organisms exhibit properties that regulate their functioning within strict limits. The molecules of which cells and organisms are composed lack those properties. Individuals are born and they die, but an individual does not have a birth rate and a death rate. A population does. Populations also have age distributions, densities, and distribution patterns. Ecological communities may be described in terms of the number of species in them and the relative abundances of those species. Ecosystems are characterized by such phenomena as energy flow and nutrient cycles.

Properties such as birth rates and species richness that appear at higher levels of organization are sometimes called, as mentioned earlier, emergent properties. Such properties do not violate principles that operate at lower levels of organization. However, emergent properties usually cannot be detected or even suspected from a study at lower levels. The level of organization at which a biologist works determines the kinds of questions that are most appropriate to ask and to attempt to answer.

Suppose, for example, a biologist walks along the edge of a pond and startles a frog that jumps into the water (Figure 1.9). An obvious question to ask is: Why did the frog jump into the water? A biochemist would answer that question in terms of the molecular mechanisms underlying muscular contraction, pointing out that the contraction results from an interaction between the actin and myosin of the muscles with ATP. A physiologist might answer that the muscles in the frog's legs contracted because they were stimulated by motor nerves that synapse with the muscles. Those nerves, in turn, fired as a result of stimuli initially received by the eyes of the frog. A developmental biologist might answer in terms of the embryological development of the neuromuscular wiring that underlies those physiological responses. An ecologist would suggest that the frog jumped to

escape from a potential predator, moving from an exposed position to a hidden one. Finally, an evolutionist would offer some suggestions as to why frogs hop instead of walking and why they hide in water rather than in some protected site on land.

Is any one of these answers more basic or more important or more correct or true? Not really. All of them are essential parts of the full explanation of the jumping of frogs. This richness of answers to an apparently simple question makes biology a complex science, but it also makes biology an exciting field. In this book, we begin with approaches to biology that concentrate on events and processes at molecular and cellular levels and end with processes at the levels of ecological communities and ecosystems. This is only one of a number of possible ways of organizing a text. We use it because understanding the underlying biochemistry and cell biology is so important for interpreting events at higher levels of organization.

THE METHODS OF SCIENCE

Science is a uniquely human activity (on this planet at least). No other animal practices science. Science, contrary to much popular opinion, is not merely a collection of facts about the world. Science is also a process, a creative form of human behavior. It is a set of ways of discovering things about the universe that has proved to be very powerful.

Scientists employ a variety of methods in attempting to understand the structure and functioning of the universe and its components. They must use diverse methods because different scientific disciplines study such different subjects. The study of atoms, molecules, glaciers, climate, trees, or mammals, to name only a few of the objects of scientific study, requires methods appropriate to each. Also, within disciplines very different questions can be asked about the same subject. And, not surprisingly, because they are people, scientists differ. They differ in the kinds of questions they are comfortable asking, in their skills with particular methods, and in the interpretation of their results. They all agree, however, that the methods must be honest. Ends do not justify means in science. Therefore, scientists worldwide often accept each other's measurements and data even though they may disagree on the conclusions to be drawn from those facts. Nonetheless, experiments yielding surprising or especially important data are often repeated by other scientists to verify or confirm the results before they are widely accepted.

The search for patterns in nature is the essence of science (Figure 1.10). Recognition of patterns is the

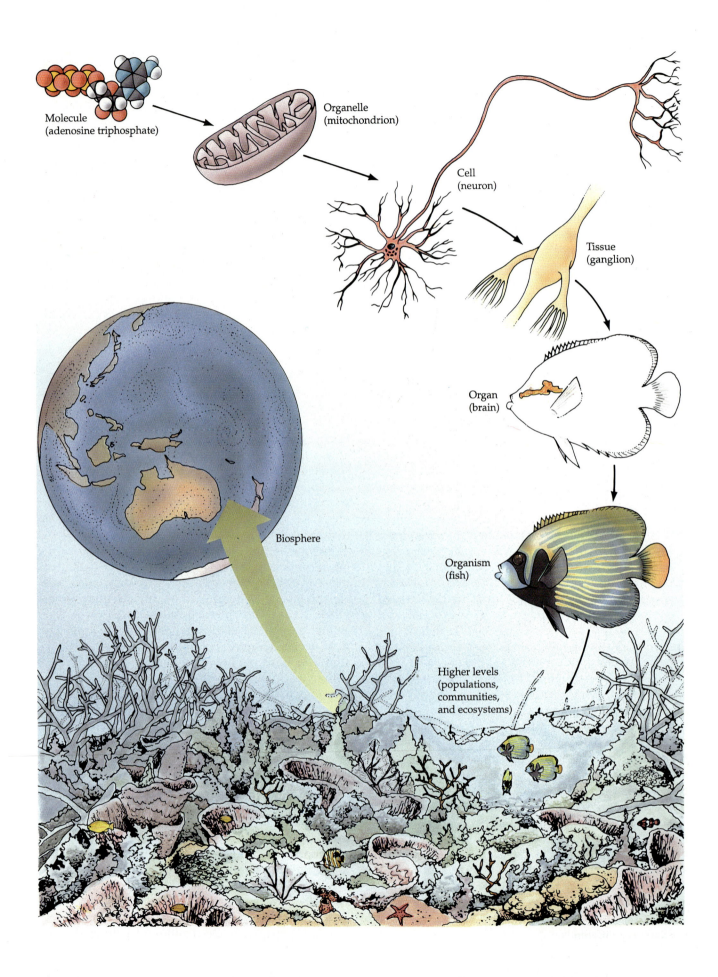

Molecule
(adenosine triphosphate)

Organelle
(mitochondrion)

Cell
(neuron)

Tissue
(ganglion)

Organ
(brain)

Organism
(fish)

Higher levels
(populations,
communities,
and ecosystems)

Biosphere

◄ **1.8   From Molecules to the Biosphere**
The fish—an organism—is a pivotal unit in its coral reef ecosystem. The fish's molecules are organized into cells, tissues, and organs, all integrated in such a way that the fish can extract food from its environment, avoid its predators, and reproduce. The coral reef community is united to other ecosystems by the exchange of energy and materials that is occurring continuously within our biosphere.

foundation for the responses of all organisms. Not surprisingly, the human brain is programmed to seek patterns, which may be one reason why science is such a compelling form of behavior. This desire to find patterns is, however, a two-edged sword. In our eagerness to find them, we may think that we detect patterns that really are not there or we may give scientific meanings to patterns that do not warrant them.

As examples of our desire to detect patterns, consider the delight that people find in recognizing the shape of some person or animal in rocky outcrops. Even though it is obvious that the appearance of, say, the likeness of a human face on a rock is simply an incidental outcome of erosional processes, our interest in the pattern remains strong. Similarly, the constellations of stars in the sky are patterns that do not reveal anything about basic stellar processes. Nonetheless, recognition of the constellations did have functional value because the positions of those stars could be used as navigational aids, and because the movements of stars aided in the measurement of time. The combination of the genuine usefulness of constellations and the limited knowledge our ancestors had about stellar processes fostered an interest in constellations that is still prevalent today, even though we now know that the patterns are incidental byproducts of the movements of galaxies.

1.9   Why Did the Frog Jump?
Scientists from different disciplines might answer this question very differently.

1.10   Pattern in Variation
These beetles differ from one another in many ways, but they all have three main body sections and six legs. You can probably detect several other common themes in the body shapes, color patterns, and length of the insects' appendages.

The Hypothetico-Deductive Method

One basic procedure underlies most of science despite the great variability of subjects studied and methods of investigation. The **hypothetico-deductive method** has four stages: (1) making observations, (2) forming hypotheses, (3) making predictions from those hypotheses, and (4) testing those predictions. Testing predictions, in turn, generates new observations, so the cycle continues indefinitely. A single scientist may carry out all four stages, but some scientists specialize in parts of the process. Indeed, in some fields, like physics, most practitioners are either theoreticians who specialize in developing hypotheses, or experimentalists who specialize in testing hypotheses. This is because the skills needed for each of the two are so complex and so different. Biologists are seldom as specialized as this; it is not uncommon for a single person to carry an investigation sequentially through all four stages of the hypothetico-deductive method many times.

Scientific procedures are easier to grasp if we use a specific example. Biologists have long known that some caterpillars, like those of the emperor gum moth, are conspicuously colored. Others, like those of the peppered moth, are cryptically colored, that is, they blend in with their backgrounds (Figure 1.11). Observations have also revealed that conspicuously colored caterpillars often live in groups but are seldom attacked by birds. These initial observations were used to develop hypotheses, make predictions, and devise tests of those predictions. In other words, they were used to carry out all the stages of hypothetico-deductive science. Let us examine how this was done.

GENERATING A HYPOTHESIS. The conspicuous appearance of some caterpillars, together with the observation that potential predators usually avoid them, suggested to some biologists that the bright color patterns of these caterpillars signal to potential predators that the caterpillars are toxic or distasteful. A related hypothesis is that cryptic caterpillars—inconspicuous to the point of being hidden by their surroundings—are good to eat (palatable) and that their coloration reduces the chance that predators will discover and eat them.

Notice that these hypotheses depend on certain assumptions or on previous knowledge. We must know or have strong reasons to believe that predators such as birds have color vision and that predators can learn about the qualities of their prey by encountering and tasting them. If these assumptions are uncertain, they may be tested first before the hypotheses about toxicity and palatability are tested.

TESTING A HYPOTHESIS. The hypothesis that brightly colored caterpillars are unpalatable was tested by presenting captive blue jays with both monarch butterfly caterpillars and cryptically colored caterpillars. The blue jays were deprived of food long enough to make

them hungry, so they readily attacked the caterpillars. Ingesting even part of one monarch caterpillar caused a blue jay to vomit. Because the birds were housed individually, the experimenters knew which ones had previously tasted monarchs and which ones had not. They found that a single experience with a monarch caterpillar was enough to cause a blue jay to reject all other monarch caterpillars presented to it. In nature, monarch caterpillars live in groups, so a predator readily learns to avoid all group members after having tasted one. Cryptically colored caterpillars, on the other hand, were readily attacked and eaten, and the birds continued to eat additional individuals without showing any signs of sickness or discontent.

Hypothetico-deductive science uses a variety of methods to test hypotheses. Among these are laboratory and field experiments, carefully focused observations, and comparative analysis. Each method has its strengths and weaknesses. The key feature of **experimentation** is the control of most factors that might affect a result so that the influence of those factors that do vary can be seen more clearly. The advantage of working in a laboratory is that control of the environment is easier. Field experiments are more difficult because it is usually impossible to control more than a small part of the total environment. The conditions under which the experiments with blue jays were run allowed the investigators to reject alternative explanations. Their results, for example, could not have been due to lack of hunger on the part of the birds or their failure to see the caterpillars.

Nonetheless, field experiments have one important advantage over laboratory experiments. Their

1.11 Caterpillars Can Be Easy or Hard to See
(a) Many caterpillars are cryptically colored and blend into their surroundings, like this larva of a peppered moth; it resembles a small green twig. (b) This larva of an emperor gum moth, with its conspicuous spines, tubercles, and bright legs, contrasts strikingly with its leafy environment.

(a)

(b)

(a)

1.12 Research Is Essential in Biology

Research and experimentation in biology are carried out in the field and in laboratories. (a) Biologists who study the canopies of rainforest trees use special climbing equipment that allows them to collect data and carry out vital studies in the field. (b) Some scientists study the properties of potentially dangerous chemicals or other substances that must be kept isolated and protected from contamination. A great deal of laboratory testing and experimentation must precede field experiments with such substances.

(b)

results are more readily applicable to what happens where the organisms actually live and evolve. Just because an organism does something in the laboratory does not mean that it behaves the same way in nature. A laboratory experiment demonstrates the *potential* for the organism to act in a certain way in nature, but it does not demonstrate that it will or does act that way. Because we usually wish to explain nature, not the behavior of organisms in the laboratory, combinations of laboratory and field experiments are needed to explain most patterns in biology at organismic and higher levels (Figure 1.12).

By watching blue jays in nature it would be possible to learn the same things revealed by the laboratory experiments, but it would be much more difficult. To observe birds actually encountering monarch caterpillars would require long hours of observation. Also, the investigator would be unlikely to know the histories of the birds being observed. Therefore, if some birds were seen attacking monarchs while others avoided them, at least two expla-

nations would be plausible. One would be that some jays like to eat monarchs whereas others do not. The other would be that none of them like monarchs once they have tasted them, but some had not yet tasted a monarch and had not learned to reject them. Eventually a persistent field observer could detect learning and subsequent rejection of monarch caterpillars, but this would be a very inefficient way of finding out about the palatability of caterpillars. Therefore, although **observational analysis** can be used to test hypotheses, it usually takes much longer to get convincing evidence.

It is inconvenient or even impossible to perform controlled experiments for some problems. Observational analysis is the only available method in those cases. The power of observational analysis is shown by fields such as astronomy, where experiments are rarely possible. This has not prevented astronomers from learning a great deal about their subject.

Comparative analysis is a method unique to the study of life. It is designed to help us understand why organisms have evolved the traits they possess. The comparative method tries to determine the importance of particular traits by comparing different kinds of organisms that do and do not have the trait. For example, the hypothesis that caterpillars with conspicuous color patterns are unpalatable cannot be confirmed in general by demonstrating that monarch butterfly caterpillars are unpalatable to blue jays. Perhaps something is unusual about monarchs and blue jays. The hypothesis can be confirmed only by showing that *most* conspicuous caterpillars are unpalatable, whereas *most* cryptic ones are palatable. Exceptions do not invalidate the hypothesis, because only its most extreme form would assert that *all* conspicuously colored caterpillars are unpalatable to *all* predators. This extreme form of the hypothesis really would state that mortality caused by predators with color vision is the *only* factor that influences the color patterns of caterpillars. Because it is rare for complex traits to be influenced by only one factor to the exclusion of all others, most hypotheses about adaptation are less extreme. This does not make them any less interesting, however. If we determine, for example, that 90 percent of conspicuous caterpillars are unpalatable to their major predators, we have learned that predators are very important agents favoring the evolution of color patterns in caterpillars. Only if there were no difference at all between the percentages of conspicuous and cryptic caterpillars that are unpalatable could we conclude that the hypothesis does not help us understand anything about the color patterns of caterpillars.

ACCEPTING HYPOTHESES. Scientists may differ among themselves about the adequacy of the evidence in support of a particular hypothesis. Also, different scientists may interpret the same set of observations in different ways. Nonetheless, at any given moment, there are many hypotheses that are generally accepted as true or false by most scientists in the field. Others are regarded as not yet convincingly confirmed or rejected. The history of science also shows us that generally accepted hypotheses are frequently overturned by newer discoveries. Sometimes hypotheses that have been convincingly rejected are resurrected by new discoveries. We can neither prove nor reject any hypothesis with absolute certainty. Nonetheless, the features of the hypothetico-deductive method allow us to achieve meaningful understanding of how the world works.

A single validation of a hypothesis rarely leads to its widespread acceptance. Also, rarely does a single contrary result lead to the abandonment of a hypothesis. That is because negative results can be obtained for a variety of reasons, of which incorrectness of the hypothesis is only one. For example, the error may reside in making incorrect predictions from a correct hypothesis. A negative result can also be obtained because of poor experimental design, or because an inappropriate organism, one that does not fit the assumptions of the hypothesis, is chosen for the test. For example, a blind predator, or one that uses primarily its sense of smell, would not be appropriate for testing hypotheses about the color of caterpillars.

A general textbook like this one is based on hypotheses and observations that are generally accepted, but not all of them are supported by equal amounts of evidence. When possible we illustrate hypotheses with observations and experiments that support them, but we cannot, because of space constraints, detail all the evidence. Therefore, you should remember that all these statements of "fact"

involve mixtures of observations, predictions, and interpretations.

Experimentation and Ethics

It is essential to experiment with plants, animals, fungi, and microorganisms to obtain answers to many of the questions posed by biologists. To study the antipredator adaptations of caterpillars, the investigators had to keep jays in cages, make them hungry by depriving them of food, and then feed them caterpillars. This resulted in the deaths of some caterpillars and temporary stress for some of the jays. Determining which chemicals make the caterpillars toxic required the deaths of still other caterpillars.

No amount of observation without intervention could possibly substitute for experimental manipulations. This does not mean, however, that scientists are insensitive to the welfare of the organisms with which they work, any more than the fact that most of us are not vegetarians means that we are not sensitive to the welfare of the animals raised and killed to provide our food. Most scientists who work with animals are continually alert to find ways of getting answers with the smallest number of experimental subjects and in ways that cause the least pain and suffering to them.

SIZE SCALES

Multicellular organisms are composed of many types of cells, each specialized to perform different functions. Among the benefits of multicellularity are improved protection, the ability to adopt a wide variety of shapes, and the ability to increase greatly in size. But why can a unicellular organism not achieve great size and enjoy some of the accompanying benefits? A major reason relates to the changes in the surface-to-volume ratio of an object as it increases in size. A cubical cell 100 micrometers (μm) on an edge has a volume and mass 1,000 times those of a cell 10 μm on an edge, but has a surface area only 100 times greater than the smaller cell (Figure 1.13). The larger a cell is, the more chemical reactions it can carry out in a certain amount of time; or, as a biologist might say, the rate of metabolism of a cell is a function of its *volume*. As a cell metabolizes, it needs to exchange materials and heat with its environment. A cell's rate of exchange of nutrients and waste products with its environment is a function of its *surface area*. Therefore, as a cell grows larger, there is a growing mismatch between metabolic demands for exchange and ability to service those demands. This mismatch limits the practical size of a cell. Some accommodation can be made by increasing surface area through specialized foldings of the external cell membrane, but substantial increases in overall sizes of organisms require the development of structures that transport food, oxygen, and waste materials. Multicellularity, with specialized cells to provide these functions, is the solution to this problem.

The need to dissipate the heat generated by the chemical reactions within organisms means that larger animals have lower metabolic rates than smaller animals. An elephant requires more food per day than a mouse does, but gram for gram the mouse requires more. If an elephant had the same metabolic rate as a mouse, it could not dissipate the heat produced by metabolism fast enough to avoid cooking itself! Conversely, a larger animal does not cool off as rapidly as a small one. Therefore, a larger lizard can heat its body in a warm area and then forage for a longer time in a cool one than a small lizard can. For this reason, some biologists think that large dinosaurs warmed up and cooled off so slowly that

1.14 Proportions Change with Size
Elephants have proportionally much more massive legs than the slender impalas. An animal the size of an elephant with legs the shape of an impala's would collapse under its own weight.

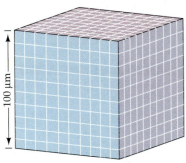

Surface area = 60,000 μm^2
Volume = 1,000,000 μm^3
Surface-to-volume ratio = 0.06

100 μm

10 μm

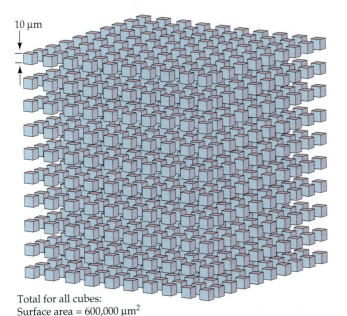

Total for all cubes:
Surface area = 600,000 μm^2
Volume = 1,000,000 μm^3
Surface-to-volume ratio = 0.6

1.13 Small Objects Have Relatively Large Surface Areas
Although a single cube that is 100 μm on an edge has the same volume as 1,000 cubes with edges of 10 μm each, the small cubes have 10 times more surface area.

they effectively maintained relatively high, constant body temperatures throughout much of the year.

Other properties of organisms also change with size. For example, the weight of an animal is related to its volume and is proportional to the cube of its linear dimensions. However, its weight must be supported by its legs. The strength of bones is proportional to their cross-sectional area, which in turn is proportional to the square of their lengths. Therefore, an increase in size must be accompanied by a proportionally greater increase in leg diameter. Delicate, slender legs are characteristic of lightweight impalas, not of heavy elephants (Figure 1.14).

Many ecological interactions among organisms are strongly affected by size. Size determines the food an organism can use, where it can hide, what it can mimic, the area it requires to obtain its food, and its abundance. Not surprisingly, ecologists pay a great deal of attention to the distributions of sizes among organisms in different ecological communities.

TIME SCALES

Modern science requires us to measure time spans both longer and shorter than those we can perceive accurately with our unaided senses. In biology, the longer time spans have caused the greatest conceptual difficulties. The inability of people to believe that Earth is a very old planet delayed the recognition of evolutionary change for a long time.

The most widespread technique for measuring long time spans is the use of naturally radioactive materials as clocks. Because different radioactive materials in Earth's crust decay at specific rates, like the ticking of a clock, it is possible to date the ages of materials deposited as long as several billion years ago. Relative ages of materials can also be assessed by indirect methods such as the vertical positions of rock deposits in relation to one another. This is possible because younger rocks lie on top of older ones unless the rocks have been subjected to dramatic deformations, which are usually evident. By studying the remains of living things found in different layers and correlating their distributions across many sites, it is possible to determine the relative ages of different deposits even if absolute ages are not known.

Biologists working at different levels of organization usually think about and study problems at different time scales (Figure 1.15). Biochemists and physiologists are primarily concerned with **physiological time**, the time required for chemical reactions and physiological changes within an organism. These times range from fractions of seconds to periods of a day or a year. Studies of such physiological processes as aging may require observations extending over the lifetimes of organisms, which may range up to centuries for long-lived plants.

Studies of populations may extend over many generations of the organisms. Generations range from less than an hour in some bacteria to centuries in some plants, but for any group of organisms **ecological time**, the time required for changes in the sizes and distributions of populations (the main focus of many ecological studies), is much longer than physiological time.

The study of changes in the genetic constitution of populations requires us to think in terms of **microevolutionary time**. Significant genetic changes usually take many generations, but when environmental conditions change abruptly they can happen quite rapidly. Most direct studies of evolutionary change have been of microevolutionary changes be-

cause those are the only ones that can be measured directly. Organisms change more substantially over spans of **macroevolutionary time** covering thousands of generations or more. Many macroevolutionary changes are, of necessity, measured indirectly. How we came to recognize the great age of Earth and to study evolutionary changes is a fascinating part of the history of biology.

MAJOR ORGANIZING CONCEPTS IN BIOLOGY

Knowledge about living organisms can be organized in many ways, and we will use several in this book. In some chapters we will look at different groups of

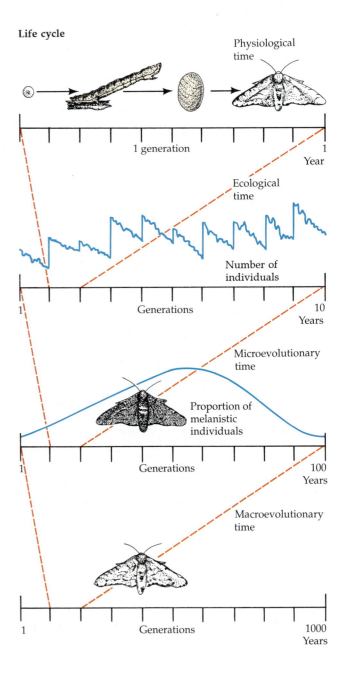

1.15 Biological Time Scales
Biologists use different time scales in their investigations. This study uses a moth that has a single generation each year; its complete life cycle, from hatching to its death shortly after reproducing, takes approximately a year for each individual. Physiological time is the time required for various chemical and physiological changes within an individual's life cycle. Ecological, microevolutionary, and macroevolutionary time scales are progressively longer, encompassing changes to many individuals and populations.

organisms and show how their structures, activities, and adaptations are related to their particular lifestyles. Other chapters focus on major processes carried out by living organisms, such as digestion, respiration, movement, responsiveness, and reproduction. At still other times, our focus is on the mechanisms of evolutionary change. Throughout these discussions we will be guided by several major organizing concepts.

The first of these concepts is the belief that *all properties of living organisms can be explained in physical and chemical terms*. Living organisms appear to be triumphs of organic chemistry. The most complex biological activities, including the mysterious richness of human emotions, are probably manifestations of underlying physico-chemical systems. This belief does not imply that we will be able to understand all complex biological phenomena simply through the study of chemistry. Rather, it implies that biological phenomena must conform to the laws of physics and chemistry. Not too many years ago the idea that the properties of living organisms were the result of physical and chemical interactions was vigorously debated, and lengthy books were written attacking and defending it. Today, however, most biologists accept the chemical basis of life and are attempting to identify the various physical and chemical processes underlying specific biological phenomena. Some basic chemistry and biochemistry is presented in Chapters 2 and 3.

A second key organizing concept comes out of one of the major intellectual triumphs of modern physics: the development and refinement of the notion of energy. *Living organisms can be viewed as systems capable of taking in energy from their environments and converting it to biologically useful forms* (Chapters 7 and 8). The notion of energy is highly abstract. Energy is weightless and occupies no space, yet it exists in a large number of forms. There are experimentally derived formulas for calculating the equivalence of these forms. Energy can never be created or destroyed but only converted from one form to another. But if energy is weightless and occupies no space, how can we measure it and use it in meaningful ways? Actually, we measure energy by its *effects upon matter*, which *can* be weighed and measured. The most useful definition of energy for our purposes is "the capacity to do work," that is, the capacity to change the measurable properties of some particle of matter.

A third major organizing concept is that *genetic information encodes and transmits information between generations*. The discovery that the individuals of the next generation are not preformed within the bodies of individuals in the previous generation, but that information specifying their characteristics is transmitted in the form of a genetic code, is one of the great triumphs of modern biology. The idea that the information necessary to direct the development of complex individuals was carried in certain molecules was so difficult to comprehend that belief in preformation persisted into the present century. The discovery of the molecules that contain the information is even more recent. These exciting developments will be detailed in Chapters 9, 10, and 11.

A fourth major organizing concept in biology is the **cell theory**. Biologists have yet to find an exception to the rule that *all living organisms are composed of cells* and that the cell is, therefore, the basic building block of life. Cells differ greatly in size and in the complexity of their internal structures, but there is a unity to all cells that enables us to discuss many of the properties of living organisms in terms of the way cells perform work and organize themselves.

Magnifying devices were in existence hundreds of years before anyone thought to look carefully at living organisms with them. It was not until 1665 that the Englishman Robert Hooke noticed that cork, wood, and other plant tissues are made up of small, regularly shaped cavities surrounded by walls (Figure 1.16). Hooke called these cavities *cells*. Living, single-celled organisms were first observed a few years later by the Dutch naturalist Anton van Leeuwenhoek, who used a simple microscope of his own design. Neither of these men appreciated the full significance of their observations. The first strong statement that

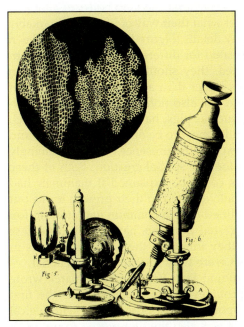

1.16 Hooke's Microscope and What He Saw
Reproduced from a plate in his book *Micrographia*, the diagram shows Robert Hooke's favorite microscope with its source of illumination. In the circle are Hooke's drawings of the empty plant cells found in the bark of the cork oak.

all living organisms consist of cells was made by the German physiologist Theodor Schwann in 1839. In 1858 the German physician Rudolf Virchow suggested that *all cells come from pre-existing cells.* Generally accepted proof of this assertion was provided through experiments by the French chemist and microbiologist Louis Pasteur between 1859 and 1861. Since then the cell theory, as summarized by these two statements, has been a basic tenet in biology.

The final major organizing concept of biology is that *evolution by natural selection results in adaptation.* The mechanisms of evolution will be treated in detail in Part Three, but you will need some knowledge of those mechanisms in order to understand material in the intervening chapters. Fortunately, the basic principles of evolution are simple, even though the details are complex.

EVOLUTIONARY CONCEPTS

Scientific discussions of evolution date back more than 200 years, but little progress was made until a reasonable and testable mechanism was suggested independently in 1858 by the British naturalists Charles Darwin and Alfred Russel Wallace. In the 1760s, the French naturalist Buffon had written his *Natural History of Animals,* which contained a clear statement of the possibility of evolution. Buffon originally believed that all organisms had been specially created for different ways of life, but as he studied animals he observed that the limb bones of all mammals, no matter what their way of life, are remarkably similar in many details (Figure 1.17). If they had been specifically created for different ways of locomotion, Buffon reasoned, they should have been built upon different plans rather than all being modifications of a common one. He also noticed that the legs of certain animals, such as pigs, have toes that never touch the ground and appear to be of no use. Buffon found it difficult to explain the presence of these seemingly useless small toes by special creation. Both of these troubling facts could be explained if mammals had not been specially created in their present forms but had been modified from a common ancestor. Buffon therefore suggested that pigs have two functionless toes because they inherited them from ancestors in which the toes were fully formed and functional.

Buffon's student Jean Baptiste de Lamarck wrote extensively about evolution. Lamarck was the first person to support the idea of evolution with logical arguments and was also the first person to put forth a hypothesis concerning the mechanisms of evolutionary change. He suggested that living organisms have the ability to change gradually over many generations by the inheritance of structures that have become larger and more highly developed as a result of continued use or, conversely, have diminished in

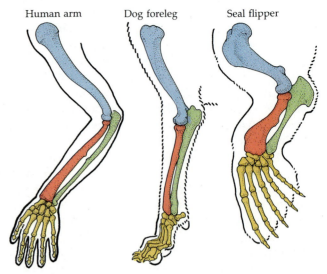

1.17 Mammalian Limbs Have Similar Bones
Mammalian forelimbs may serve different purposes—humans use theirs for manipulating objects, dog forelimbs are for walking on, and seals swim with theirs—but the number and type of their bones are similar. In this diagram, bones of the same type are shown in the same color.

size as a result of disuse. For example, he suggested that water birds extend their toes while swimming, stretching the skin between them. This stretched condition, he thought, can be inherited by the offspring, who will further stretch their skin during their lifetimes and will also pass this condition along to their offspring. According to Lamarck, birds with webbed feet will thereby evolve over a number of generations (Figure 1.18). He explained many other examples of adaptations in a similar way and showed how many domestic plants and animals have departed from the forms of their wild ancestors. We do not now believe that most evolutionary changes have been produced

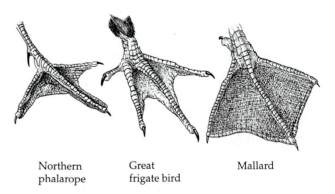

Northern phalarope   Great frigate bird   Mallard

1.18 Partially to Fully Webbed
All of the birds illustrated here stretch their feet while swimming, but their offspring's feet will not be affected by this stretching. The mechanisms that produced this range of webbing from partial to full will be discussed extensively in later chapters.

1.19 Many Organisms Have High Reproductive Rates

If all the offspring in the eggs laid by frogs such as this one grew to adulthood, the world's frog population would be overwhelming. However, many of these eggs will not survive. A rate of reproduction as high as this frog's is usually accompanied by a high mortality rate of eggs and young.

by the mechanisms proposed by Lamarck; we will look at our current understanding of the mechanisms of evolution in Chapter 18.

Lamarck's ideas, however, deserved more attention than they received from his contemporaries, all of whom believed in a static, young universe. By 1859 the climate of opinion had changed enough for the theory of evolution, as proposed by Darwin and Wallace, to receive serious consideration—and to transform biology. By then geologists had shown that Earth has undergone changes over millions of years, and people were willing to think in terms of longer time spans. When a well-documented and thoroughly scientific argument for evolution was presented in the latter half of the nineteenth century, it had a favorable reception.

Charles Darwin and Evolution

Like most great ideas, the Darwin–Wallace theory of natural selection, as presented in Darwin's book *The Origin of Species*, is remarkable for its simplicity. Darwin, who developed the theory in greater detail than Wallace, began with two familiar facts: (1) The individuals of most kinds of organisms are not identical, and (2) offspring tend to resemble their parents. These familiar facts acquired new significance for Darwin when he realized that, given the reproductive rates of most organisms, there must be a heavy mortality (a high death rate) in nature. Without high death rates, even the most slowly reproducing forms would quickly reach enormous population sizes. Darwin suggested—and this is the key point in his theory—that slight variations among individuals significantly affect the chance that a given individual will survive and reproduce. He called this differential reproductive success of varying individuals *natural*

selection, probably because he was deeply interested in the *artificial selection* practices of animal and plant breeders.

Thus Darwin's theory of evolution by natural selection was based on two simple facts (the existence of variability and the existence of similarities between parents and offspring) and one inference (namely, that the variations significantly affect their possessors' probability of survival and reproduction). Many of Darwin's observations on the nature of variability came from domesticated plants and animals. Darwin himself was a pigeon fancier and raised many different breeds. He saw close parallels between artificial selection by breeders and selection in nature.

Darwin recognized that the reproductive rates of organisms are so high that they would result in enormous population increases if all the offspring survived (Figure 1.19). Therefore, Darwin reasoned, mortality must increase as population density increases and competition for space, food, shelter, and other environmental necessities becomes severe, and predation and disease become more prevalent. From this base, Darwin argued his case for natural selection as follows:

How can it be doubted, from the struggle each individual has to obtain subsistence, that any minute variation in structure, habits or instincts, adapting that individual better to the new conditions, would tell upon its vigour and health? In the struggle it would have a better chance of surviving; and those of its offspring which inherited the variation, be it ever so slight, would have a better chance. Yearly more are bred than can survive; the smallest gain in the balance, in the long run, must tell on which death shall fall, and which shall survive. Let this work of selection on the one hand, and death on the other, go on for a thousand generations, who will pretend to affirm that it would produce no effect, when we remember what, in a few years animal breeders effected in cattle, and . . . in sheep, by the identical principle of selection?

That statement, written by Darwin more than 100 years ago, still stands as a good expression of the idea. We now have a much better understanding of the effects of very slight variations, and there is now a wealth of experimental evidence about natural selection, accumulated by several generations of biologists. Nonetheless, the main outlines of the theory stand much as they were proposed in 1858. The main advances since then have involved the development of a rigorous genetic theory of evolutionary changes, the recognition that individuals affect future generations not only through their own offspring but also by helping the survival of relatives who contain the same genes as a result of descent from a common ancestor, increased knowledge of other mechanisms of evolutionary change, and the development of methods by which aspects of evolutionary theory can be subjected to a variety of rigorous tests. These advances, as well as more details of evolutionary theory, will be discussed throughout the book and will be covered in detail in Part Three.

The Importance of a World View

Biologists develop hypotheses, devise tests, and modify their ideas in the light of observational and experimental results; but these activities are carried out within a broader framework. All of us, whether we are scientists or not, operate within the framework of a general world view, which is sometimes called a **paradigm**. The paradigm determines which problems are interesting and which are not. It strongly influences the responses we make to information that seems to run counter to it.

Biology began a major paradigm shift a little over a century ago with the general acceptance of Darwin's theory of evolution by natural selection. The changeover has taken a long time because it required abandoning many components of a different world view. The pre-Darwinian world was thought to be a young one in which living organisms had been created in essentially their current forms. The Darwinian world is an ancient one in which both Earth and its inhabitants have been evolving from forms very different from the ones they now have. It is a world in which we would not recognize most former living organisms if we were transported far back in time, nor organisms of the future if we were transported far forward in time. Acceptance of this paradigm involves not only acceptance of the processes of evolution; it also involves accepting the view that the living world is constantly evolving but without any future "goals." This book is a discussion of the Darwinian paradigm, of how the science of biology appears from that perspective, and of the kind of problems that seem legitimate to biologists who accept this general view.

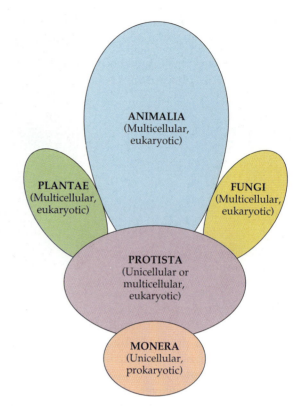

1.20 The Five Kingdoms
Biologists divide living organisms into these five major groups. We will use these classifications when we talk about organisms throughout this book.

THE FIVE KINGDOMS

Perhaps as many as 30 million species of organisms inhabit Earth today. Many times that number lived in the past but are now extinct. To classify this extraordinary diversity of living things, biologists have devised systems that reflect the evolutionary history of life. The details of the system most commonly used will be presented in Chapter 20, but because some of its key terms will be used in the intervening chapters, it is necessary to introduce the broad categories here.

Biologists using this system group living organisms into five large categories called **kingdoms** (Figure 1.20). The kingdoms are based on cellular structure and the mode of nutrition of the organisms. The kingdom **Monera** is composed of the bacteria: Single-celled organisms with distinctive cells referred to as **prokaryotic** (prenuclear) because they lack a nucleus and some of the other internal structures found in the cells of members of other kingdoms (Figure 1.21). Monerans are exceedingly abundant and are found virtually anywhere on Earth where life can exist. Some are critical components of the biogeochemical cycles essential to all life.

Over a billion years ago, some monerans invaded the cells of other monerans. Over time, this relation-

(a)

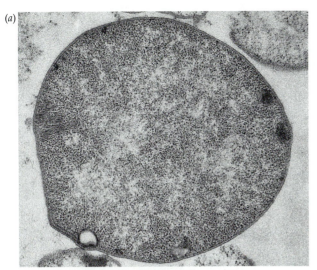

(b)

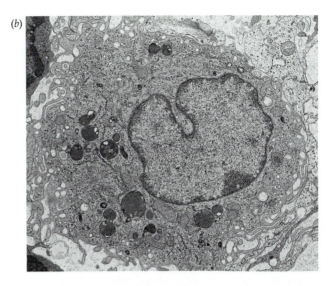

1.21 Cells Differ in Their Structure
(a) The prokaryotic cells of bacteria in the kingdom Monera, such as this archaebacterium, do not have a nucleus; they also lack other internal structures that are
found in the cells of organisms in the other four kingdoms. (b) Unlike the bacterium in (a), this eukaryotic cell
—a type of white blood cell—has a distinct nucleus (the
large, indented structure) and numerous other highly organized and specialized internal compartments. Although
they appear the same size here, this eukaryotic cell is actually about 8 times larger in diameter than this prokaryotic cell. Chapter 4 describes the structures of both
kinds of cells.

ship between hosts and parasites gave rise to the
kingdom **Protista**: Single-celled organisms whose
cells are structurally more complex than the cells of
the monerans. Cells of protists, and of members of
the other remaining kingdoms, are **eukaryotic** (truly
nuclear).

The remaining three kingdoms, consisting primarily of multicellular organisms, are all believed to have
arisen from ancestral protists. The kingdom **Fungi**
includes the molds, mushrooms, yeasts, and other
similar organisms. Fungi absorb food substances
from their surroundings and digest them within their
cells. Many are important as decomposers of the dead
bodies of other organisms. Most members of the
kingdom **Plantae** (plants) convert light energy to the
energy of chemical bonds by the process called photosynthesis. The biological molecules they synthesize
are the primary food for nearly all other living organisms. The kingdom **Animalia** (animals) consists of
organisms that digest food outside their cells and
then absorb the products. Animals depend on other
forms of life for most of their materials and energy.

Sometimes organisms are referred to as "primitive" or "advanced," or "lower" and "higher" forms.
We will avoid these terms because they imply that
some organisms "work better" than others. The
abundance of monerans—the simplest of living
organisms—readily demonstrates that organisms in
all these groups are highly successful. It is useful,
however, to ask why organisms of such different
complexities, sizes, shapes, and ways of life are all
successful. Indeed, providing answers to that problem is one of the major objectives of this book.

QUESTIONS BEYOND SCIENCE

There are questions that modern science cannot answer. Modern medicine has been very successful in
combating many of the diseases that have plagued
humankind over the centuries, but new threats repeatedly arise. The most important recent one is the
rapid spread of Acquired Immune Deficiency Syndrome (AIDS). AIDS already affects a significant proportion of the human population in many countries,
and both the geographic range of the disease and the
number of people infected are steadily increasing.
Scientists are busily investigating the modes of transmission of the disease, how the AIDS virus acts
within the human body, and possible methods of
combating the virus. However, science cannot deal
with the hypothesis that people who contract AIDS
are being punished for their behavior. There are no
scientific tests of that hypothesis. Therefore, scientists cannot, as scientists, provide evidence in support of or against that view. As human beings, scientists hold opinions about such beliefs. However, it
is not to be expected that they will agree or that their
opinions will differ from those held by, say, artists,
lawyers, or theologians.

Science is a powerful way of finding out things
about the world, but it is not appropriately applied
to all questions. In this book we will show the power
of science as applied to many interesting questions
about living organisms. It is important to remember
that there are many other questions about living organisms that we will not raise because they are not
scientific questions.

FOR STUDY

1. Why is it so important in science that tests capable of rejecting a hypothesis be designed and performed?

2. Some philosophers and practitioners of science believe that it is impossible to prove any scientific hypothesis—that we can only fail to find a cause to reject it. Evaluate this view. Can you think of a reason why we can be more certain about rejecting a hypothesis than we can about accepting it?

3. One hypothesis about the conspicuous coloration of caterpillars was described in this chapter and some tests were mentioned. Suggest some other plausible hypotheses for conspicuous coloration in these animals. Develop some critical tests of one of these alternatives.

4. The concept of adaptation means that the features of organisms evolved because they improved the chances that their possessors would survive and reproduce. However, there is no evidence that any evolutionary mechanisms have foresight or that organisms can anticipate future conditions. What, then, do biologists mean when they say, for example, that wings are "for flying?"

5. Consider a single-celled organism. Explain why it is not feasible for this organism to grow to a size of 10 centimeters in diameter. Cover the following topics: (1) surface-to-volume ratio; (2) transport of nutrients; (3) gas exchange; (4) excretion; and (5) support.

READINGS

Darwin, C. 1859. *The Origin of Species by Means of Naural Selection*. John Murray, London. The book that set the world to thinking about evolution; still well worth reading. Many reprinted versions are available.

Irvine, W. 1955. *Apes, Angels, and Victorians*. McGraw-Hill, New York. A delightful account of the reactions of English society to the theory of evolution by means of natural selection.

Kuhn, T. S. 1970. *The Structure of Scientific Revolutions*, 2nd Edition. University of Chicago Press, Chicago. A widely discussed book that developed a view of science as a succession of paradigms.

Margulis, L. and K. V. Schwartz. 1987. *Five Kingdoms: An Illustrated Guide to the Phyla of Life on Earth*, 2nd Edition. W. H. Freeman, New York. A good introduction to the five kingdoms, sometimes at odds with other views of the subject. Excellent examples and illustrations.

Mayr, E. 1982. *The Growth of Biological Thought: Diversity, Evolution, and Inheritance*. Harvard University Press, Cambridge, MA. A monumental synthesis of the development of ideas and concepts in biology, from the early beginnings to the present, by one of the leading contributors to evolutionary biology.

Young, J. Z. 1951. *Doubt and Certainty in Science*. Clarendon Press, Oxford. A book about how scientists develop confidence in their theories.

PART ONE
The Cell

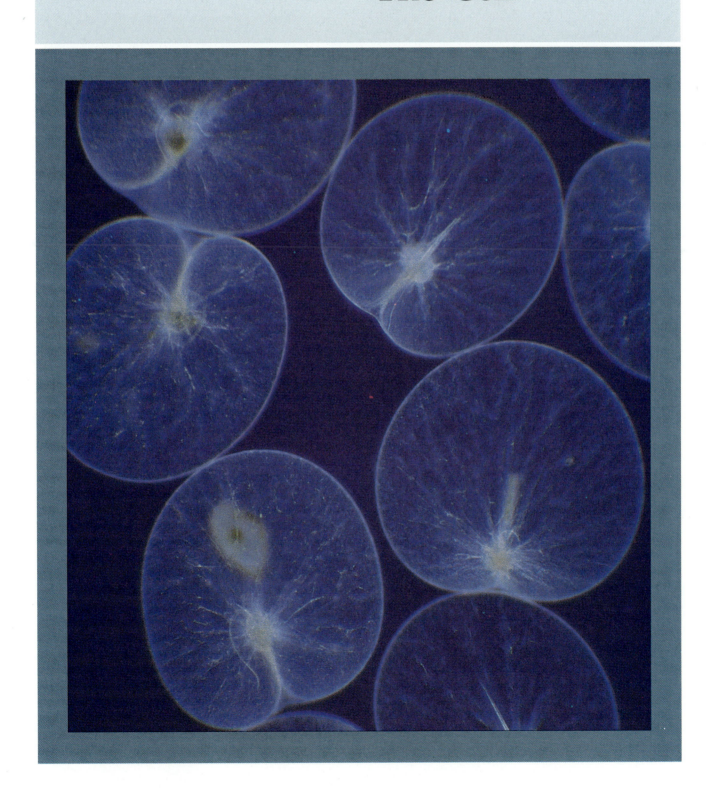

2

Small Molecules

PREVIEW: Organisms and the world around them are composed of chemical substances. All substances are made up of atoms, and atoms can combine to form molecules. Particles within atoms determine the chemical and biological behavior of atoms or molecules. Molecules and atoms undergo chemical reactions of many kinds, some of them producing compounds of importance to living organisms. The activities of cells and organisms arise from chemical reactions within them.

This chapter deals with electrons, protons, neutrons, atoms, elements, isotopes, molecules, electron shells, water, ions, acidity, covalent bonds, ionic bonds, polarity, hydrogen bonds, weak bonds, and simple organic compounds.

The composition of living things is both chemically simple and chemically complex. More than 99 percent of the living matter of all organisms is composed of just four chemical elements—carbon, hydrogen, oxygen, and nitrogen—although there are 92 or more different chemical elements in the world. (A chemical element is a substance that cannot be decomposed into simpler substances by any chemical reaction.) More than half the weight of most living things, including human beings, is made up of water—a single chemical substance composed of only two elements, hydrogen and oxygen. On the other hand, the body of a human being contains hundreds of thousands of chemically distinct proteins. Each of these proteins is made within living cells under the direction of another substance: DNA.

The chemistry of Earth's environment supported and constrained the origin and evolution of life on our planet, and life thrives and evolves today on terms specified by the nonliving world around it. Living things are "fit" for their environment—or, as the Harvard biologist Lawrence J. Henderson pointed out in 1912, the chemistry of our environment is very fit for life. Similarly, the chemistry of life is specialized for the functions of life, such as growth, movement, and communication.

The reduction of biological questions to chemical ones has led to many of the greatest successes of biology. In this chapter and the next we will discuss some of the basic concepts of chemistry and introduce the major groups of chemical substances important to life.

ATOMS

All matter, living and nonliving, is composed of **atoms** (Figure 2.1). More than a million million atoms could fit in a single layer over the period at the end of this sentence. Each atom consists of a dense, positively charged **nucleus**, around which one or more electrons move. The nucleus contains one or more protons and may contain one or more neutrons. Electrons, protons, and neutrons are not indivisible particles. Rather, each has a substructure; however, that level of organization (the world of quarks) has no known consequences for biology.

Scientists have defined the weight, or mass, of a proton as a standard unit: the atomic mass unit, or amu. The amu is also referred to as the dalton (named for English scientist John Dalton, who studied atoms two centuries ago). A single proton or neutron weighs one dalton. A dalton is 1.7×10^{-24} gram (0.0000000000000000000000017 g), whereas the mass of an electron is 9×10^{-28} g (0.0005 dalton). Because they weigh so much less than protons and neutrons, electrons contribute negligibly to the mass of an atom.

The positive electric charge on a proton is defined as a unit of charge. The electric charge of an electron is equal and opposite to that of a proton. Thus the charge of a proton is +1 unit, that of an electron is −1, and that of a neutron is 0. The neutron, as its name suggests, is electrically neutral. The number of protons in an atom equals the number of electrons, so the atom itself is electrically neutral.

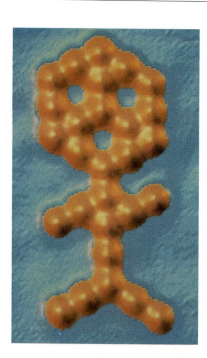

2.1 Atoms
This cartoon figure is one of the world's smallest "drawings." Its images of atoms were created using an advanced instrument called a scanning tunneling microscope. Each gold-colored peak consists of one carbon atom and one oxygen atom; there are 28 peaks in all. It would take more than 20,000 such figures, connected hand-to-hand, to equal the diameter of a single human hair.

ELEMENTS

An **element** is a substance that is composed of only one kind of atom. The element hydrogen consists only of hydrogen atoms; the element iron consists only of iron atoms. An element cannot be decomposed chemically to simpler substances.

A large number of different elements occur in nature, and a growing number of other elements have been produced by physicists using cyclotrons and other particle accelerators. Some of the natural elements, such as silver, gold, and thulium, are extremely rare; others, such as hydrogen, nitrogen, and oxygen, are abundant on this planet. Table 2.1 compares the proportions of some representative ele-

TABLE 2.1
Abundance of Some Chemical Elements

ATOMIC NUMBER	SYM-BOL	ELEMENT	ATOMIC WEIGHT	ABUNDANCE, AS % OF			ATOMIC NUMBER	SYM-BOL	ELEMENT	ATOMIC WEIGHT	ABUNDANCE, AS % OF EARTH'S CRUST
				UNI-VERSE	EARTH'S CRUST	HUMAN BODY					
1	H	Hydrogen	1.008	87	0.14	9.5	28	Ni	Nickel	58.71	0.01
2	He	Helium	4.003	12			29	Cu	Copper	63.54	0.01
3	Li	Lithium	6.939		0.01		30	Zn	Zinc	65.37	0.01
4	Be	Beryllium	9.012				31	Ga	Gallium	69.72	
5	B	Boron	10.81				32	Ge	Germanium	72.59	
6	C	Carbon	12.01	0.03	0.03	18.5	33	As	Arsenic	74.92	
7	N	Nitrogen	14.01	0.01		3.3	34	Se	Selenium	78.96	
8	O	Oxygen	16.00	0.06	46.6	65.0	35	Br	Bromine	79.91	
9	F	Fluorine	19.00		0.03		36	Kr	Krypton	83.80	
10	Ne	Neon	20.18	0.02			37	Rb	Rubidium	85.47	0.03
11	Na	Sodium	22.99		2.83	0.2	38	Sr	Strontium	87.62	0.03
12	Mg	Magnesium	24.31		2.09	0.1	39	Y	Yttrium	88.91	
13	Al	Aluminum	26.98		8.13		40	Zr	Zirconium	91.22	0.02
14	Si	Silicon	28.09		27.7		41	Nb	Niobium	92.91	
15	P	Phosphorus	30.97		0.12	1.0	42	Mo	Molybdenum	95.94	
16	S	Sulfur	32.06		0.05	0.3	43	Tc	Technetium	(97)	
17	Cl	Chlorine	35.45		0.03	0.2	44	Ru	Ruthenium	101.1	
18	A	Argon	39.95				45	Rh	Rhodium	102.9	
19	K	Potassium	39.10		2.59	0.4	46	Pd	Palladium	106.4	
20	Ca	Calcium	40.08		3.63	1.5	47	Ag	Silver	107.9	
21	Sc	Scandium	44.96				48	Cd	Cadmium	112.4	
22	Ti	Titanium	47.90		0.44		49	In	Indium	114.8	
23	V	Vanadium	50.94		0.02		50	Sn	Tin	118.7	
24	Cr	Chromium	52.00		0.02		51	Sb	Antimony	121.8	
25	Mn	Manganese	54.94		0.1		52	Te	Tellurium	127.6	
26	Fe	Iron	55.85		5.0		53	I	Iodine	126.9	
27	Co	Cobalt	58.93				54	Xe	Xenon	131.3	

This list contains only the first 54 of the 92 natural elements. Those elements that are found in the human body are divided into three categories on the basis of abundance; ■, most abundant; ▨, 0.1–3.3 percent; □, trace (less than 0.01 percent).

ments in the human body, in Earth's crust, and in the universe as a whole. A human body consists primarily of the elements hydrogen, carbon, oxygen, nitrogen, phosphorus, and sulfur, along with smaller amounts of several other elements.

How Elements Differ

The chemical properties of an element are determined by the number of electrons its atoms contain. The number of electrons equals the number of protons. This number is called the **atomic number**. An atom of hydrogen contains 1 proton in its nucleus; helium contains 2; carbon, 6; plutonium, 94. Therefore, their atomic numbers are, respectively, 1, 2, 6, and 94.

All atoms except hydrogen have one or more neutrons in their nuclei. The nucleus of a helium atom contains two protons and two neutrons; oxygen has eight protons and eight neutrons. The **mass number** equals the sum of the number of protons and neutrons in its nucleus—electrons are ignored in the mass number because their mass is infinitesimal in comparison with that of a neutron or proton. Helium, therefore, has a mass number of 4 and oxygen a mass number of 16. The mass number may be thought of as the weight of the atom, in daltons.

We represent each element by a one- or two-letter symbol. Thus H = hydrogen, He = helium, O = oxygen, and so forth. Some symbols are not obvious: Fe (from Latin *ferrum*) = iron, Na (Latin *natrium*) = sodium, and W (German *Wolfram*) = tungsten (a metal used for light-bulb filaments). Table 2.1 lists the symbols for 54 of the 92 natural elements. Sometimes we specify the atomic and mass numbers of an element. Then the atomic number is written to the lower left of the symbol, and the mass number to the upper left. In this notation, atoms of hydrogen, carbon, and oxygen are written as $^{1}_{1}H$, $^{12}_{6}C$, and $^{16}_{8}O$.

Isotopes

Until now we have referred to hydrogen and oxygen as if they each were known in only one form. To be more precise, we should have said "the common form of" oxygen or hydrogen because not all atoms of the same element have the same mass number. The common form of hydrogen is ^{1}H, but about 1 out of every 6,500 hydrogen atoms on Earth has a neutron as well as a proton in its nucleus and is thus ^{2}H, called **deuterium**. Furthermore, it is possible to create ^{3}H, **tritium**, which has *two* neutrons and a proton in its nucleus. All three kinds of hydrogen atoms have the atomic number 1, because all three have just one proton. Deuterium, tritium, and common hydrogen have virtually identical chemical and biological properties, although deuterium is twice and tritium three times as heavy as ^{1}H. Such multiple

^{1}H	^{2}H	^{3}H	^{12}C	^{14}C
Hydrogen: 1 proton	Deuterium: 1 proton, 1 neutron	Tritium: 1 proton, 2 neutrons	Carbon-12: 6 protons, 6 neutrons	Carbon-14: 6 protons, 8 neutrons

2.2 Isotopes
The three diagrams on the left are all of hydrogen atoms (each has a single proton), but they show the different numbers of neutrons characteristic of the isotopes common hydrogen, deuterium, and tritium. The six protons in each of the two rightmost atoms identify them as carbon, but the common isotope of carbon has six neutrons, whereas ^{14}C has eight.

forms of a single element are referred to as **isotopes** of the element (Figure 2.2). (The prefix *iso-*, encountered in many technical terms, means "same.")

Many elements exist in several isotopic forms in nature. For example, the natural isotopes of carbon are ^{12}C, ^{13}C, and ^{14}C. Unlike the hydrogen isotopes, the isotopes of most other elements do not have distinct names but rather are written in the form above, and are spoken of as "carbon-12", "carbon-13", and "carbon-14." The vast majority of carbon atoms are ^{12}C, but about 1.1 percent are ^{13}C, and a tiny fraction are ^{14}C. An element's **atomic weight** is the average of the mass numbers of a representative sample of atoms of the element, with all isotopes in their normal proportions. For example, the atomic weight of carbon is 12.011.

Some isotopes, such as tritium and ^{14}C, are **radioisotopes**. That is, they are radioactive, spontaneously giving off energy or subatomic particles that can be detected by various counting devices or imaging techniques such as those described in Box 2.A. In this process of radioactive decay, the original atom is transformed into another type, usually of another element. Thus, for example, $^{14}_{6}C$ is converted to $^{14}_{7}N$ —carbon becomes nitrogen by the emission of an electron, as a neutron becomes a proton. Other isotopes, such as deuterium, are stable (nonradioactive) and can be detected only by virtue of their different mass. Measurement of mass is more difficult than measurement of radioactivity and usually requires the use of an expensive instrument called a mass spectrometer. In biological research, radioisotopes are employed as tracers of biochemical reactions. Some commonly used radioisotopes are tritium, ^{14}C, and ^{32}P (phosphorus-32). Heavy water, containing deuterium, is also useful in studying biochemical reactions.

The decay of any radioisotope is regular—in successive, equal periods of time, the same *fraction* of the remaining radioactive material decays. The rate of decay of an isotope is expressed as its **half-life**.

BOX 2.A

Detecting Radioactivity

Biologists detect and measure radioactivity in various ways. **Autoradiography** is a technique for localizing a radioactive signal, using X-ray film to take a picture. In the photograph below, a specific radioactive substance was added to a fruit fly embryo. The radioactive substance stuck only to regions of the embryo that play a role in organizing body segments in the adult fly. The embryo was washed and then pressed against X-ray film, and left in the dark for several days. Wherever a radioactive decay event occurred, the film was exposed. When the film was developed, silver grains —seen in the autoradiograph as intense black dots—appeared over the parts of the embryo that contained the radioactive substance. Thus biologists are able to study the development of different parts of the fruit fly embryo.

Liquid scintillation counting is the method most frequently used to measure the amount of radioactivity in a sample. The sample is added to a solution containing a substance that emits light (scintillates) when it absorbs the products of radioactive decay. The liquid scintillation counter is a machine that, like a light meter, measures the amount of light. The amount of light detected corresponds directly to the amount of radioactive decay in the sample. The liquid scintillation counter shown here accepts hundreds of samples, each in its radiation-sensitive "cocktail," and lowers each in turn into a darkened well where its emitted light is measured.

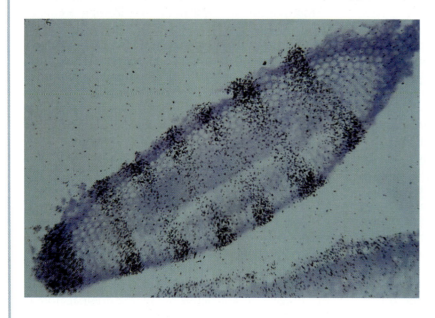

For example, in 14.3 days, one-half of any sample of ^{32}P decays. In the next 14.3 days, one-half of the remaining half decays, leaving only one-quarter of the original sample of ^{32}P, and so on. Thus the half-life of ^{32}P is 14.3 days. Tritium has a half-life of 12.3 years, whereas that of ^{14}C is about 5,700 years. This regularity of decay allows us to use the radioactive isotopes present in nature to determine the ages of ancient bones, rocks, and other materials.

THE BEHAVIOR OF ELECTRONS

The part of the atom of greatest interest to chemists and biologists is the electron. Each element has a characteristic number of electrons in each of its atoms. The number of electrons determines the chemical properties of an atom. All **chemical reactions** consist of changes at the electronic level—that is, reactions are exchanges of electrons, or changes in the sharing of electrons between atoms.

It is impossible to say where a given electron in an atom is at any given time. Instead, we are limited to describing a certain volume of space within the atom where the electron is likely to be found. That particular region in space within which the electron is to be found at least 90 percent of the time is the electron's **orbital** (Figure 2.3). An electron spins like a top—or like Earth on its axis—and, like a top, may spin in one of two directions: clockwise or counter-

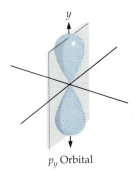

s Orbital

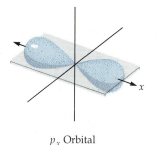

p$_y$ Orbital

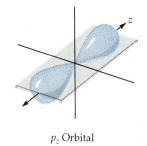

p$_x$ Orbital

p$_z$ Orbital

2.3 Electron Orbitals

Orbitals, indicated by dots, are the regions around an atom's nucleus where electrons are most likely to be found. Two electrons form the spherical *s* orbital. The next two electrons form a larger, spherical *s* orbital (not shown). The next six electrons fill three dumbbell-shaped *p* orbitals, one pair of electrons per orbital. The *p* orbitals are oriented on the *x*, *y*, and *z* axes through a point in the center of the atom. Additional electrons form the still more complicated *d* and *f* orbitals.

clockwise. In an atom, a given orbital can be occupied by at most two electrons, which must spin in opposite directions. Thus any atom larger than helium (atomic number = 2) must have electrons in two or more orbitals. As shown in Figure 2.3, the different orbitals have characteristic forms.

The orbitals are grouped in such a way as to constitute a series of **shells** around the nucleus. The innermost shell, called the K shell, consists of only

one orbital, an *s* orbital. This orbital is filled first, with the electrons of lowest energy. It is filled first because *atoms and molecules are most stable when they have the least energy*. Hydrogen ($_1$H) has one K-shell electron; helium ($_2$He) has two; all other atoms have two K-shell electrons and electrons in other shells as well. The L shell is made up of four orbitals (an *s* orbital and three *p* orbitals) and hence can hold up to eight electrons (Figure 2.4). The M, N, O, P, and Q shells have different numbers of orbitals.

In any atom, it is the outermost shell of electrons that determines what the atom can do chemically. When the outer shell is full, the atom is very stable. Some elements—such as helium, neon, and argon—have full outer shells and are thus chemically nonreactive, or inert. Other elements are reactive in various degrees—they are, in a sense, seeking ways to fill their outer shells with electrons by combining with other atoms.

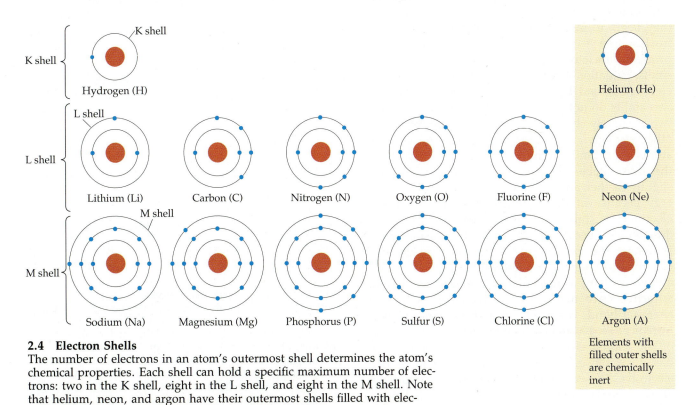

2.4 Electron Shells

The number of electrons in an atom's outermost shell determines the atom's chemical properties. Each shell can hold a specific maximum number of electrons: two in the K shell, eight in the L shell, and eight in the M shell. Note that helium, neon, and argon have their outermost shells filled with electrons. This situation means that atoms of these elements are almost completely inert and almost never react chemically with other atoms.

Elements with filled outer shells are chemically inert

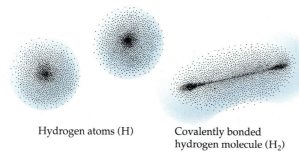

Hydrogen atoms (H)   Covalently bonded
                         hydrogen molecule (H₂)

2.5   Formation of a Covalent Bond
Two hydrogen atoms (H) combine to form a hydrogen molecule (H₂), which is most stable when both electrons are attracted to both protons but the two protons are not so close together as to repel each other strongly. The two electrons are shown here as clouds because electrons cannot be located more precisely. The overlapping of electron orbitals in the K shells of the two hydrogen atoms forms the covalent bond.

CHEMICAL BONDS

Let us see how two atoms can cooperate to give each atom a full outer shell of electrons. In the process the two atoms become joined by a chemical bond. **Molecules** consist of two or more atoms linked by chemical bonds.

Covalent Bonds

A hydrogen atom has but one electron in its only shell. Picture two hydrogen atoms, initially far apart but coming closer and closer to one another, until they begin to interact. The negatively charged electron of atom A is attracted by the positively charged proton in nucleus B, as well as by its own nucleus; a similar situation holds for electron B. So, the two electrons spend time between the two nuclei. The two atoms do not get *too* close together, because the two positively charged nuclei would then repel each other strongly. At a certain distance, the coupled atoms have a minimum amount of energy, so this is the most stable arrangement. (Pulling the atoms slightly farther apart would require an input of energy because of the "gluing" effect of the shared electrons; pushing them closer together would require energy because of the mutual repulsion of the protons.) The two hydrogen nuclei share the two electrons completely. A **chemical bond** joins the two atoms, forming a molecule. A chemical bond is an attractive force that links two atoms. This type of chemical bond, consisting of a shared pair of electrons, is called a **covalent bond** (Figure 2.5). Because of the shared electrons, each hydrogen atom now has, in a sense, a full outer shell containing two electrons. The covalently bonded pair of hydrogen atoms, with completely filled outer shells, is less reactive than the individual atoms, which have incomplete K shells.

A carbon atom has a total of six electrons: two in its (full) K shell and four in its (outer) L shell. Because the L shell has the capacity to hold eight electrons, this atom can share electrons with up to four other atoms. Thus it can form four covalent bonds. For example, an atom of carbon reacts with four hydrogen atoms, forming a substance called methane. The carbon atom of methane has eight electrons in its (full) L shell, and each of the hydrogen atoms has a full K shell, thanks to the sharing. Thus methane is held together by four covalent bonds, each bond being a pair of electrons shared between carbon and one of the hydrogens (Figure 2.6). Bonds in which a single pair of electrons is shared are called **single bonds** (Figure 2.7). When four electrons are shared, the link is a **double bond**. Two oxygen atoms joined by a double bond make up a molecule of oxygen gas. **Triple bonds** are rare in biological molecules, but there is one in nitrogen gas, in which six electrons are shared by two nitrogen atoms. Nitrogen gas is the chief component of the air we breathe.

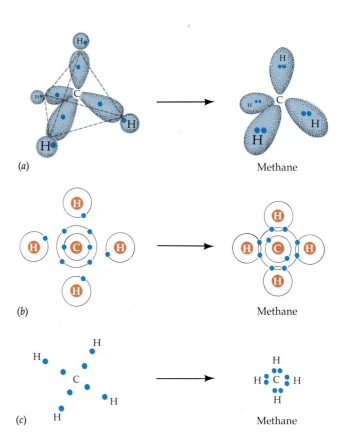

(a)   Methane

(b)   Methane

(c)   Methane

2.6   Covalent Bonding with Carbon
Carbon can complete its outer shell by sharing the electrons of four hydrogen atoms: Each hydrogen shares one of the four outer-shell electrons of the carbon atom, forming methane. The drawings present three methods for showing the formation of the covalent bonds in methane. (a) Overlapping of the electron orbitals of H and C. (b) Filling the outer shells of H and C by sharing electrons. (c) Sharing of electron pairs between H and C. Note that in (a) and (c) only the electrons from the initially unfilled shells are depicted.

H . C : H = H—C—H

Methane

Single covalent bond

O :: O = O ═ O

Oxygen gas

Double bond

N ::: N = N ≡ N

Nitrogen gas

Triple bond

2.7 Single, Double, and Triple Bonds
The single covalent bonds in methane (CH_4), the double bond in oxygen gas (O_2), and the triple bond in nitrogen gas (N_2) can be shown by dots representing electron pairs (left) or by solid lines (right).

The Covalent Bonds of Different Elements

The atoms of a given element tend to form a specific number of covalent bonds with other atoms. This number is based on the number of electrons in the atom's outer shell. Carbon, as discussed above, tends to form four covalent bonds, oxygen two, and hydrogen one. Learning the few numbers given in Table 2.2 will help you keep track of some of the common substances that will be mentioned repeatedly.

The Harvard biologist George Wald once suggested that one reason why the elements hydrogen, oxygen, nitrogen, and carbon are so important to living things is that their atoms are the smallest ones that can fill their outer electron shells by gaining one, two, three, and four electrons, respectively. Larger atoms form covalent bonds with their L-shell or M-shell electrons. These electrons are "screened" from the positively charged nucleus by the K-shell electrons. Hence their L-shell and M-shell electrons are held less tightly by the nucleus and form less stable

covalent bonds. Hydrogen, carbon, oxygen, and nitrogen can form the most stable covalent bonds. In addition, carbon, oxygen, and nitrogen are among the very few elements that can form double or triple bonds (two others are sulfur and phosphorus, which are also essential to living organisms). Wald pointed out that because of double bonding, carbon and oxygen can combine to form carbon dioxide ($O═C═O$), a water-soluble gas readily taken up by plants. On the other hand, silicon (an element otherwise very similar chemically to carbon) can form only single bonds with oxygen.

Ions and Ionic Bonds

When dissolved in water, many substances ionize: Their molecules break apart into charged particles, or **ions**. Hydrochloric acid (HCl) is a good example. HCl is composed of hydrogen and chlorine and is a gas at room temperature. When dissolved in water, HCl separates into hydrogen ions (H^+) and chloride ions (Cl^-; Figure 2.8). When an atom or group of atoms gains or loses one or more electrons that atom or group of atoms becomes an ion; therefore, an ion is not electrically neutral. Ions with one or more positive charges are called **cations**. Ions with one or more negative charges are called **anions**.

In the HCl molecule, the hydrogen and the chlorine atoms share a pair of electrons (one from each)

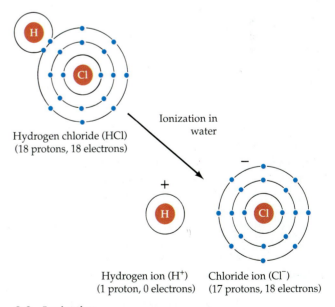

Hydrogen chloride (HCl)
(18 protons, 18 electrons)

Ionization in water

+

−

Hydrogen ion (H$^+$)
(1 proton, 0 electrons)

Chloride ion (Cl$^-$)
(17 protons, 18 electrons)

2.8 Ionization
Hydrogen chloride (hydrochloric acid) ionizes in water. Following ionization, the chlorine atom retains *both* the shared electrons from the covalent bond and becomes a chloride ion (Cl^-). This ion is negatively charged because it contains one more electron than it does protons. The proton of the hydrogen atom, no longer balanced electrically by an electron, becomes a positively charged hydrogen ion (H^+).

TABLE **2.2** Bonding Capabilities of Biologically Important Elements	

ELEMENT	NUMBER OF COVALENT BONDS
Hydrogen (H)	1
Oxygen (O)	2
Sulfur (S)	2
Nitrogen (N)	3
Carbon (C)	4

that form a single covalent bond. When dissolved in water, the hydrogen and the chlorine atoms separate; but the chlorine atom keeps *both* the originally shared electrons. The Cl^- ion has one more electron than elemental chlorine (Cl). The added electron gives the outer, L shell a full, stable load of eight electrons. The H^+ ion has a single positive charge because an electron has been lost—actually, H^+ is just a lonely proton. It is stable because it has no incomplete electron shells.

Some elements form ions with multiple charges by losing or gaining more than one electron to form stable shells. Examples are Ca^{2+} (calcium ion; the calcium atom has lost *two* electrons), Mg^{2+} (magnesium ion), and Al^{3+} (aluminum ion). Also, *groups* of atoms may form ions: NH_4^+ (ammonium ion), SO_4^{2-} (sulfate ion), PO_4^{3-} (phosphate ion). Two biologically important elements each yield more than one stable ion: iron yields Fe^{2+} (ferrous ion) and Fe^{3+} (ferric ion), and copper yields Cu^+ (cuprous ion) and Cu^{2+} (cupric ion).

Oppositely charged ions attract one another. If enough ions are present, or if the solvent (water) evaporates, crystals can form. Solid table salt consists of ions of sodium (Na^+) and chloride (Cl^-) in a highly ordered crystalline array (Figure 2.9). The array is held together by **ionic bonds**, which are chemical bonds in which the attractive force is the electrical attraction between cations and anions. NaCl dissolves readily in water, as does the covalently bonded HCl. Actually, there is no sharp dividing line between covalent bonds and ionic bonds. In some covalent bonds, the electrons are shared equally between the bond partners; but, as will be shown in a later section, some other pairs of atoms share electrons unequally. An ionic bond is simply a case in which one of the partners has the "shared" electron pair *all* the time. In solution, an ionic bond is less than one-tenth as strong as a covalent bond that shares electrons equally, so the ionic bond is much more readily broken.

Weaker Interactions

In addition to covalent and ionic bonds, there are other, weaker types of interaction that are important for biological molecules and structures. We can describe hydrogen bonds and the other weak interactions only after we have introduced further chemical principles upon which they depend.

MOLECULES

Two or more atoms linked by chemical bonds make up a molecule. A substance whose molecules contain more than one kind of atom is called a **compound**. Most biological substances are compounds. A substance (such as oxygen gas) that contains only one kind of atom is called an **elemental substance**.

The **molecular formula** of a compound or an elemental substance shows how many atoms of each element are present in the molecule. This number is written to the lower right of the symbol. For example, the molecular formula for methane is CH_4 (each molecule contains one carbon atom and four hydrogen atoms), that for oxygen gas is O_2, and the molecular formula for sucrose (table sugar) is $C_{12}H_{22}O_{11}$. The hormone insulin is represented by the molecular formula $C_{254}H_{377}N_{65}O_{76}S_6$! Molecular formulas are incomplete descriptions of molecules in that they do not tell us anything about which atoms are linked to which. **Structural formulas** show the atoms and bonds explicitly, as in Figures 2.6 and 2.7.

Molecular Weight and the Mole

Just as each element has an atomic weight, so each compound has a **molecular weight**, which is simply the sum of the atomic weights of the atoms in the molecule. The atomic weights of hydrogen, carbon, and oxygen are, respectively, 1.008, 12.011, and 16.000. Thus the molecular weight of water (H_2O) is $(2 \times 1.008) + 16.000 = 18.016$, or about 18. What is the molecular weight of sucrose ($C_{12}H_{22}O_{11}$)? You can calculate this and find that the answer is approximately 342. By remembering the molecular weights of a few representative biological compounds, you can picture the relative sizes of the molecules that interact with one another (Figure 2.10).

Suppose that we want to compare the effects of sodium chloride (NaCl), potassium chloride (KCl), and lithium chloride (LiCl) on a particular biological process. You might at first think that we could simply

Chloride ions
(Cl^-)

Sodium ions
(Na^+)

2.9   Ionic Bonding
A crystal of sodium chloride (NaCl) is held together by ionic bonds between the sodium cations (Na^+) and the chloride anions (Cl^-).

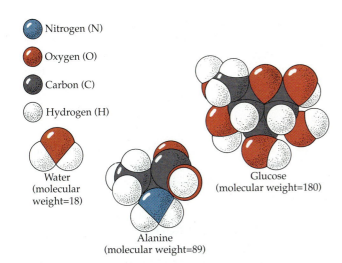

Water
(molecular
weight=18)

Alanine
(molecular weight=89)

Glucose
(molecular weight=180)

- Nitrogen (N)
- Oxygen (O)
- Carbon (C)
- Hydrogen (H)

2.10 Molecular Weight and Size
The relative sizes of three common molecules and their molecular weights. Water is the solvent in which many biological reactions take place; alanine is one of the building blocks of proteins; glucose is a sugar, an important "food" substance in most cells.

give, say, 2 grams (g) of NaCl to one set of subjects, 2 g of KCl to another, and 2 g of LiCl to the third. However, because of the differing molecular weights of NaCl, KCl, and LiCl, the 2-g samples would contain different numbers of molecules of these substances, so the comparison would not be legitimate. Instead, we want to give equal numbers of *molecules* of each substance so that we may compare the activity of one molecule of one substance with that of one molecule of another. However, the weight of a single molecule of sodium chloride is 10^{-22} g, hardly a workable quantity. Individual molecules can be neither weighed nor counted. Instead, we deal in terms of **moles** (also known as gram molecular weights). *One mole of any substance is an amount whose weight in grams is numerically equal to the molecular weight of the substance.* Potassium chloride has a molecular weight of 74.55, so a mole of KCl weighs 74.55 g; a mole of NaCl weighs 58.45 g, and a mole of LiCl, 42.40 g. A mole of one substance contains the same number of molecules as does a mole of any other substance. That number, known as **Avogadro's number**, is 6.023×10^{23} molecules per mole. The concept of the mole is one of the most useful ideas in laboratory chemistry and biology because it enables us to work easily with known numbers of molecules.

CHEMICAL REACTIONS

When atoms combine or change bonding partners, we say that a **chemical reaction** has occurred. As an example, consider the flame of a propane kitchen stove or water heater. When propane (C_3H_8) reacts

with oxygen gas (O_2), the carbon atoms become bonded to oxygen atoms instead of to hydrogen atoms, and the hydrogen atoms become bonded to oxygen instead of carbon. This is shown in Figure 2.11, which uses the convention of representing a chemical reaction by an arrow. The total number of covalent bonds does not change when propane reacts with oxygen gas, but atoms are exchanged between molecules. We call this group transfer. Whenever covalent bonds are formed or broken, there is group transfer.

In the reaction of propane and oxygen, a great deal of energy is released, as evidenced by the heat of the flame and its blue light. In general, chemical reactions are accompanied by changes in energy: Energy may be given off to the environment, as in the reaction of propane with oxygen, or energy may be taken up from the environment (some substances will react only after being heated, for example).

We can measure the energy associated with chemical bonds. Work must be done to break a bond, and that work, or energy, can be expressed in calories. Group transfer, the changing of bond partners, usually results in products with total bond energies that are different from those of the reacting substances; these energies can also be expressed in calories. A calorie is the amount of heat energy needed to raise the temperature of 1 g of pure water (which contains no other substance) from 14.5°C to 15.5°C. The nutritionist's Calorie, which biologists call a kilocalorie, is equal to 1,000 of the heat energy calories. Although defined in terms of heat, the calorie serves as a measure of any form of energy—mechanical, electrical, or chemical.

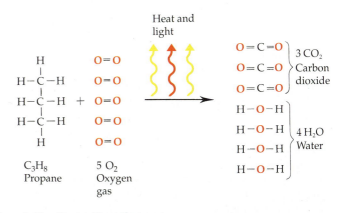

2.11 Group Transfer
The reaction of propane (C_3H_8) and oxygen gas (O_2). The C, H, and O atoms gain new bonding partners during the reaction, and there is a change in energy as indicated by the heat and light of the flame. Covalent bonds are indicated with solid lines; note that one molecule of propane reacts with five molecules of oxygen gas to give three molecules of carbon dioxide and four molecules of water. The total number of covalent bonds is the same after the reaction as before.

WATER

One of the simplest compounds, yet one of the most biologically important and chemically interesting, is water (H_2O). It was in water that life originated on this planet; water covers three-fourths of present-day Earth, and from 45 to more than 95 percent of the weight of any active organism consists of water. Everybody has experienced the biological imperative of a raging thirst. Some organisms live out their lives in water. No organism can remain biologically active without water.

Water is the most effective solvent known—that is, more kinds of substances will dissolve in it than in any other liquid. The chemical reactions of interest to biologists take place in solution, many of them in watery or **aqueous solutions** (although many other reactions occur in solutions in which fatty substances are the solvent). Water itself takes part in a number of important reactions.

Water has a number of properties not shared by its close chemical relatives (such as hydrogen sulfide, H_2S, a foul-smelling gas poisonous to humans). Water can exist in three different physical states—solid, liquid, and gas—at temperatures commonly found on this planet (Figure 2.12). Its solid state, ice, is less dense than its liquid form, which is why ice floats in water. Suppose ice sank in water, as almost all other solids do in their corresponding liquids. Ponds and lakes would then freeze from the bottom up, becoming solid blocks of ice in winter and killing most of the organisms living in them. Once the whole pond had frozen, its temperature could drop well below the freezing point of water. In fact, however, ice floats and forms a protective insulating layer at the top of a pond, reducing heat flow to the cold air above. So fish, plants, and other organisms in the pond can make it through the winter without having to endure subfreezing temperatures. Unless the entire pond freezes, there will be a liquid portion no colder than 0°C, the freezing point of pure water.

As water changes from liquid into its gaseous state, vapor, it takes up an unusually large amount of heat. This is why sweating is a useful cooling device for humans—heat from your body is lost as the water in sweat evaporates. Importantly, it also takes a relatively large amount of heat to raise the temperature of water. The temperature of a given quantity of water is raised only 1°C by an amount of heat that would increase the temperature of the same quantity of ethyl alcohol by 2°C, or of chloroform by 4°C. This important phenomenon contributes to the surprising constancy of the temperature of the oceans and other large bodies of water through the seasons of the year. This constancy is useful to the organisms living in lakes and oceans, for it means that they need not adapt to great variations in temperature. In addition, the relative constancy of water temperature helps to minimize variations in atmospheric temperature throughout the planet.

Water ionizes, but only to a limited extent. In a somewhat simplified form, the ionization of water may be represented as

$$H_2O \rightarrow H^+ + OH^-$$

H^+ is, of course, a hydrogen ion; the OH^- is known as a **hydroxide ion**. (Actually, only about one water molecule in 500 million is ionized at any one time.)

ACIDS, BASES, pH, AND BUFFERS

In *pure* water, the concentration of H^+ ions exactly equals that of OH^- ions, and this "solution" is said to be **neutral**. Now suppose we add some HCl (hydrochloric acid). The HCl ionizes, releasing H^+ ions, so now there are more H^+ than OH^- ions present. Such a solution is acidic. A basic, or alkaline, solution is one in which there are more OH^- than H^+ ions. A basic solution can be made by, for example, adding sodium hydroxide (NaOH), which ionizes to yield OH^- and Na^+ ions.

A compound that can *release* H^+ ions in solution is called an **acid**. Examples are HCl and sulfuric acid (H_2SO_4), one molecule of which may ionize to yield two H^+ ions and one SO_4^{2-} ion. Biological compounds such as acetic acid and pyruvic acid, which contain —COOH (the **carboxyl group**), are also acids, because —COOH → —COO$^-$ + H^+. Compounds that can *accept* H^+ ions are called **bases**. These include bicarbonate ion (HCO_3^-), which can accept an H^+ ion and become carbonic acid (H_2CO_3), and ammonia (NH_3), which can accept an H^+ ion and become an ammonium ion (NH_4^+), and many others.

Note that, although —COOH is an acid, —COO$^-$ is a base, because —COO$^-$ + H^+ → —COOH. Acids

2.12 Water: Solid and Liquid
The ice of McBride Glacier is gradually receding as it melts and joins the sea along the Alaskan coast.

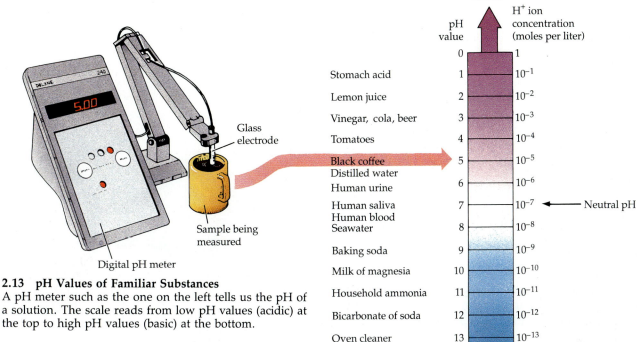

2.13 pH Values of Familiar Substances
A pH meter such as the one on the left tells us the pH of a solution. The scale reads from low pH values (acidic) at the top to high pH values (basic) at the bottom.

and bases exist as pairs, such as —COOH and —COO⁻, because any acid becomes a base when it releases a proton, and any base becomes an acid when it gains a proton.

You may have noticed that the last two reactions discussed were the opposites of each other. The reaction of —COOH to —COO⁻ and H⁺ is a reversible reaction and may be expressed as —COOH ⇌ —COO⁻ + H⁺. A **reversible** reaction is one that can proceed in either direction—"left to right" or "right to left"—depending on the relative starting concentrations of reacting substances and products. In principle, *all* chemical reactions are reversible. Some consequences of this reversibility will be discussed in Chapter 6.

The terms aci*dic* and ba*sic* refer only to *solutions*. How acidic or basic a solution is depends on the relative concentrations of H⁺ and OH⁻ ions in it. *Acid* and *base* refer to *compounds* and *ions*. A compound or ion that is an acid can donate H⁺ ions; one that is a base can accept H⁺ ions.

How do we specify how acidic or basic a solution is? To understand this, you need first to note what the H⁺ ion concentrations of a few contrasting solutions are. In pure water, the H⁺ concentration is 10^{-7} mole per liter (mol/L; also written as 10^{-7} **molar** or 10^{-7} *M*). In 1 *M* hydrochloric acid, the H⁺ concentration is 1 *M*; and in 1 *M* sodium hydroxide, the H⁺ concentration is 10^{-14} *M*. With its values ranging so widely—from more than 1.0 *M* to less than 10^{-14} *M* —the H⁺ concentration itself is an inconvenient quantity to deal with. It is easier to work with the logarithm of the concentration, because logarithms compress the range.

How acidic or basic a solution is is indicated by its pH (a term derived from *potential of Hydrogen*). The pH value is defined as the negative logarithm of the hydrogen ion concentration in mol/L. In chemical notation, molar concentration is often indicated by putting brackets around the symbol for a substance: [H⁺] = the molar concentration of H⁺. We can now write the equation

$$pH = -\log_{10}[H^+]$$

With the H⁺ concentration of pure water being 10^{-7} *M*, its pH is $-\log(10^{-7}) = -(-7)$, or 7. A smaller negative logarithm means a larger number; a lower pH means a higher H⁺ concentration. In 1 *M* HCl, the H⁺ concentration is 1 *M*, so the pH is the negative logarithm of 1 ($-\log 10^0$), or 0. The pH of 1 *M* NaOH is the negative logarithm of 10^{-14}, or 14. When its pH is lower than 7.0, a solution is acidic; it contains more H⁺ than OH⁻ ions. A solution with a pH of 7.0 is neutral, and basic solutions have pH values greater than 7.0. Because the pH scale is a logarithmic scale, the values are exponential: A solution with a pH of 5 is 10 times as acidic as one with a pH of 6 (it has 10 times as great a hydrogen ion concentration); a solution with a pH of 4 is 100 times more acidic than one with a pH of 6. The pH values of a number of common substances are shown in Figure 2.13.

The proper functioning of an organism depends upon tight control of the chemical constituents of its cells. In particular, the pH of each compartment within a cell must be well controlled. The effect of pH on biochemical reactions is discussed briefly in Chapter 6. The control of pH is made possible in part by the presence of **buffers**, which are systems that maintain a relatively constant pH even when substantial amounts of acid or base are added. A buffer is a mixture of an acid that does not ionize completely in water and its corresponding base—for example, carbonic acid (H_2CO_3) and sodium bicarbonate ($NaHCO_3$). If acid is added to this buffer, the added H^+ ions combine with bicarbonate ions to produce more carbonic acid, using up some of the H^+ ions (the Na^+ ions do not participate in the reaction):

$$HCO_3^- + H^+ \rightarrow H_2CO_3$$

If base is added, some of the carbonic acid ionizes to produce bicarbonate ions and more H^+, which counteracts some of the added base. In this way, the buffer minimizes the effects of added acid or base on the pH. Buffers illustrate the reversibility of chemical reactions: The addition of acid drives the reaction in one direction, whereas addition of base drives it in the other direction.

POLARITY

In some molecules the electric charge is not distributed evenly in the covalent bonds. Such molecules are called **polar molecules**. Water is an important example of a polar molecule. In the O—H covalent bonds, the shared electrons are more strongly drawn to the oxygen nucleus, which has eight protons, than to the hydrogen nuclei, which have only one positive charge each. Because of this tendency of the electrons, the hydrogen atoms represent slightly positive regions of the water molecule. In addition, the two

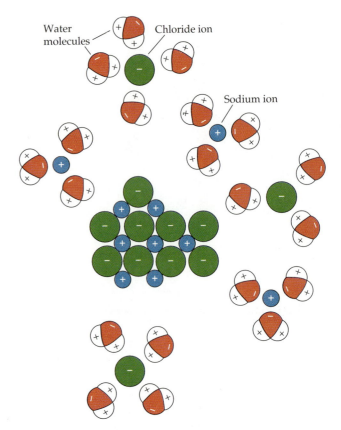

2.15 Ionic Substances in Water
Because water molecules are polar, they cluster around either cations or anions. This "shielding" action of the water reduces the tendency of dissolved ions to reassociate with one another. Sodium and chloride ions are shown surrounded by water molecules in this schematic representation of a solution. The negative oxygen end of the water molecule is attracted to the sodium cation, whereas the positive hydrogen atoms in a water molecule are attracted to a chloride anion.

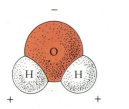 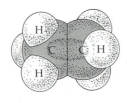

(a) Water molecule   (b) Ethane molecule

2.14 Polarity of Molecules
(a) There is slightly more positive charge on the side of a water molecule near the hydrogen atoms, and slightly more negative charge on the other side. (b) There is an even distribution of electrons over the surface of the symmetrical ethane molecule, and nothing to provide an excess of positive charge at any point; it is therefore nonpolar.

hydrogen atoms of water do not lie on directly opposite sides of the oxygen atom; rather, they are separated by an angle of 104.5 degrees. Because the electrons are drawn away from the hydrogen nuclei, the electron cloud is most dense in the opposite region, which therefore has a slightly negative charge (Figure 2.14a). In contrast, molecules whose electric charge is evenly balanced from one end of the molecule to the other are called **nonpolar molecules**; ethane is an example of a nonpolar molecule (Figure 2.14b).

The polarity of water molecules has many important consequences. For example, water's excellence as a solvent owes much to its polarity. Substances such as sodium chloride dissolve easily in water because the Na^+ and Cl^- ions become hydrated, that is, surrounded by water molecules (Figure 2.15). Because the ions are largely shielded by the water molecules from interacting with one another, they are prevented from dropping back out of solution as solid particles of NaCl.

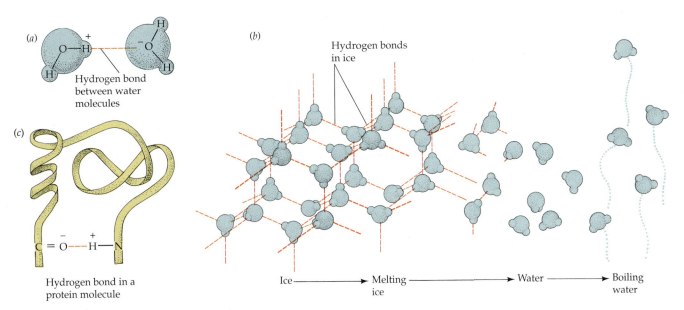

(a)

Hydrogen bond
between water
molecules

(b)

Hydrogen bonds
in ice

(c)

Hydrogen bond in a
protein molecule

Ice ———→ Melting ———————→ Water ———→ Boiling
           ice   water

2.16 Hydrogen Bonding and Ice
Hydrogen bonds are indicated here with dashed red lines. (a) Hydrogen bonds form between water molecules and between other chemical groups. Hydrogen, when covalently bonded to another atom that has a greater affinity for the shared electrons (carbon or nitrogen, for example), has a slight positive charge. (b) The lattice structure of ice, which consists of water molecules held together by hydrogen bonds. This regular array of molecules collapses when ice melts. (c) The specific folding of many large molecules, such as proteins, is maintained in part by hydrogen bonds between different parts of the molecules.

Hydrogen Bonds

Water molecules attract one another because of their polarity. The negative region of one water molecule attracts the positive regions of others, so many water molecules may become loosely attracted to one another, giving substantial order to a puddle of water. The attraction between a slight positive charge on a hydrogen atom and a slight negative charge on a nearby atom is only about one-tenth (or less) as strong as a covalent bond, but it is strong enough to deserve a name: **hydrogen bond**. It is also strong enough to be very important in biology, because hydrogen bonds can form between some chemical groups as well as between water molecules (Figure 2.16). Hydrogen bonding plays major roles in determining the shapes of the giant molecules (proteins and DNA; Chapter 3) and in conserving and decoding genetic information (Chapter 11). Also, the solubility in water of such compounds as sugars depends upon the formation of hydrogen bonds between hydroxyl (—OH) groups on the sugars and the oxygen atoms of water. Hydrogen bonding accounts, too, for most of the unusual properties of water mentioned earlier.

Hydrogen bonding among water molecules gives the liquid a high **surface tension**. This tension creates a surface that is almost like an invisible "skin," and is so strong that some insects can literally walk on water (Figure 2.17). Hydrogen bonding of water molecules to other kinds of molecules gives rise to **capillary action**—the rising of water and watery solutions in narrow tubes. It is in this way that water creeps up through minute spaces in the soil and becomes available to the roots of plants. The ability of water to be pulled up through conducting tissues up to the tops of trees as tall as 100 m (Chapter 29) is also a result of hydrogen bonding among water molecules.

Interactions Between Nonpolar Molecules

There are other types of weak interactions between *non*polar molecules (Table 2.3). When uncharged molecules, or parts of molecules, come so close to

2.17 Surface Tension
A water strider "skates" along, supported by the surface tension of the water that is its home.

TABLE 2.3 Chemical Bonds			
TYPE OF BOND	**BASIS OF BONDING**	**ENERGY**	**BOND LENGTH**
Covalent bond	Sharing of electron pairs	50–110 kcal/mol[a]	<0.2 nm
Ionic bond	Attraction of opposite charges	3–7 kcal/mol	0.28 nm (optimal)
Hydrogen bond	Sharing of H atom	3–7 kcal/mol	0.26–0.31 nm (between atoms that share H)
van der Waals bond	Interaction of electron clouds	~1 kcal/mol	0.24–0.4 nm

[a] kcal/mol = kilocalories per mole; for other abbreviations of units of measurement see the inside front cover.

one another that their electron clouds virtually touch, the electrons of one molecule are weakly attracted by the nuclei of the atoms in the other molecule. The force of this attraction is greater than the repulsive force between the electron clouds. The attractive force is called a **van der Waals interaction**. Although van der Waals interactions are only one-fourth to one-third as strong as hydrogen bonds, they still contribute to the maintenance of the specific structures of large molecules.

Another type of weak attraction, comparable in strength to van der Waals interactions, is the **hydrophobic interaction**. When highly nonpolar molecules, or parts of molecules, come together in the presence of water, they associate with one another in such a way as to minimize their exposure to the water. This is seen when molecules of oil in water minimize the area of oil–water contact by aggregating into droplets. This configuration requires the least energy to maintain; work must be done for the contact area between the oil and water to increase.

SOME SIMPLE ORGANIC COMPOUNDS AND FUNCTIONAL GROUPS

Organic compounds are made of molecules that contain the element carbon. A number of classes of organic compounds are important constituents of organisms, or are produced by them. The simplest class is the **hydrocarbons**, compounds composed of only hydrogen and carbon atoms. Examples of hydrocarbons include methane (CH_4), ethane (CH_3—CH_3), propane (CH_3—CH_2—CH_3), as well as ethylene (CH_2=CH_2) (Figure 2.18). Methane, ethane, and propane are called **saturated hydrocarbons** because they contain no carbon–carbon double bonds and are thus saturated with hydrogens. Ethylene, in contrast, is **unsaturated** and could add more hydrogen to the carbon atoms connected by the double bond, thus becoming the saturated hydrocarbon ethane:

$$H_2C=CH_2 + H_2 \rightarrow H_3C—CH_3$$

Notice that we may write the formula for ethylene as CH_2=CH_2, which makes it easy to note that two identical parts are covalently bonded together, or as H_2C=CH_2, which reminds us that it is the two carbon atoms that share the double bond.

Gasolines are hydrocarbons with 6–10 carbon atoms arranged in a chain; a typical gasoline is octane, with 8 carbon atoms. Motor oils have 12–20 carbon atoms, and waxy semisolids called paraffin waxes are longer-chain hydrocarbons. Polyethylene plastic is a large hydrocarbon, with chains thousands of carbon atoms long. Animal and plant fats have long hydrocarbon chains (Chapter 3). Hydrocarbons as a family are flammable, oily, and immiscible (they do not mix) with water. Things that dissolve in hydrocarbons ordinarily do not dissolve in water, and vice versa.

When a hydrogen atom on a hydrocarbon is replaced by a **hydroxyl group** (—OH), the compound becomes an **alcohol**. Perhaps the most familiar example is **ethanol** (ethyl alcohol; Figure 2.19). Small alcohols like ethanol are soluble in water, but larger alcohols are insoluble in water because of their lengthy hydrocarbon chains.

Both hydroxyl and carbonyl groups are constituents of sugars (Chapter 3). The **carbonyl group** has a central carbon atom with a double bond to an oxygen atom (Figure 2.19). If one of the other two bonds on the carbon atom is to a hydrogen atom, the compound with the carbonyl group is an **aldehyde**; otherwise it is a **ketone**.

Molecules containing one or more carboxyl (—COOH) groups are, as we have seen, acids because of the tendency of the carboxyl group to ionize. **Amines**, on the other hand, are organic bases. These compounds possess an **amino group** (—NH_2), which has a tendency to react with H^+ to give the positively charged —NH_3^+ group. This H^+-accepting character accounts for the classification of amines as bases.

The **amino acids** are an important class of compounds that possess both a carboxyl group and an amino group—and both these groups are attached to the same carbon atom, which is called the α (alpha)

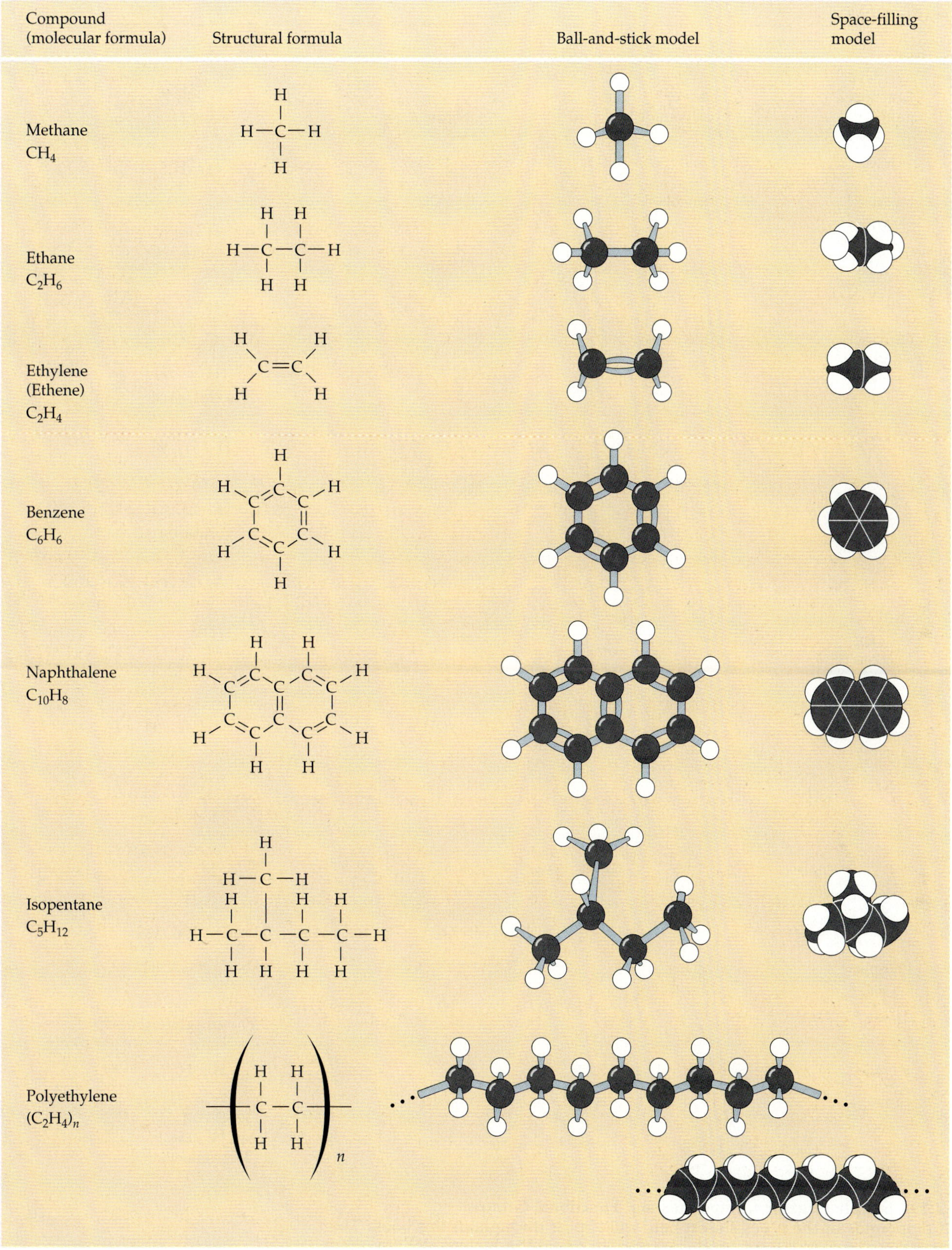

Compound (molecular formula)	Structural formula	Ball-and-stick model	Space-filling model

Methane
CH$_4$

$$\begin{array}{c} H \\ | \\ H-C-H \\ | \\ H \end{array}$$

Ethane
C$_2$H$_6$

$$\begin{array}{c} H \quad H \\ | \quad | \\ H-C-C-H \\ | \quad | \\ H \quad H \end{array}$$

Ethylene
(Ethene)
C$_2$H$_4$

$$\begin{array}{c} H \qquad H \\ \diagdown \quad \diagup \\ C=C \\ \diagup \quad \diagdown \\ H \qquad H \end{array}$$

Benzene
C$_6$H$_6$

Naphthalene
C$_{10}$H$_8$

Isopentane
C$_5$H$_{12}$

Polyethylene
(C$_2$H$_4$)$_n$

2.18 Hydrocarbons

Compare the sizes and structures of these hydrocarbons, noting which are saturated and which unsaturated. The molecules in the figure are represented in four different ways; two representations emphasize the three-dimensional structures. Of these three-dimensional depictions, the ball-and-stick models focus on bond angles and space-filling models focus on the overall shape. You will see many space-filling models in the following chapters. These are the most realistic representations of the appearance of molecules. Atoms of different elements are represented by "knobs" of different sizes and shapes. The individual knobs have radii proportional to the radii of the atoms they represent, and they are constructed to reflect accurately the lengths and angles of molecular bonds. Carbon atoms are shown in black or gray, hydrogen in white, oxygen in red, nitrogen in blue, and phosphorus and sulfur in yellow. Such space-filling models are particularly valuable in helping us to understand how molecules interact with one another, both in particular reactions and in forming structures such as membranes.

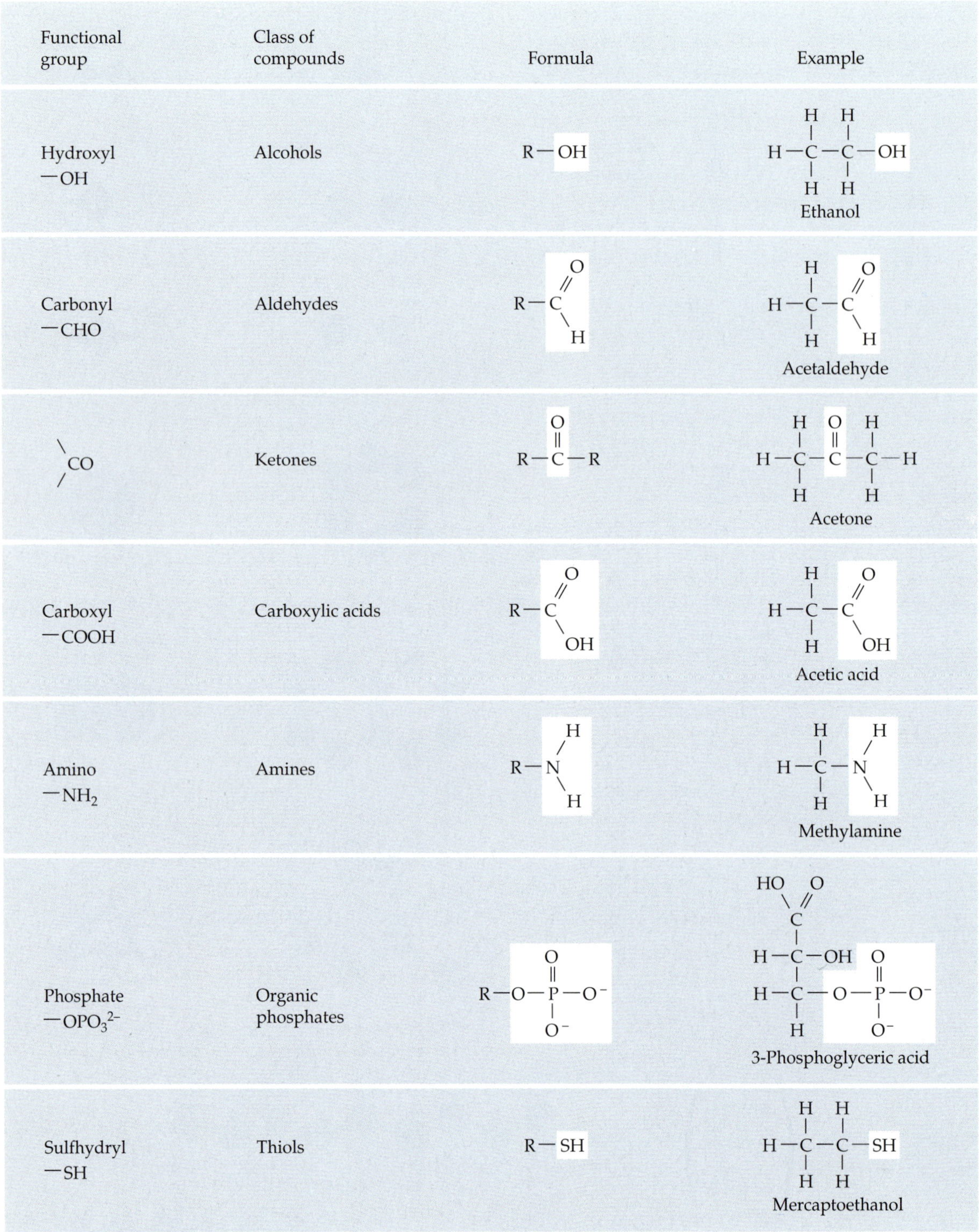

Functional group	Class of compounds	Formula	Example
Hydroxyl —OH	Alcohols	R—OH	$\begin{array}{ccc} H & H \\ | & | \\ H-C-C-OH \\ | & | \\ H & H \end{array}$ Ethanol
Carbonyl —CHO	Aldehydes	$R-C\overset{\displaystyle O}{\underset{\displaystyle H}{\diagdown}}$	$H-\underset{\underset{H}{|}}{\overset{\overset{H}{|}}{C}}-C\overset{\displaystyle O}{\underset{\displaystyle H}{\diagdown}}$ Acetaldehyde
$\diagup$CO$\diagdown$	Ketones	$R-\overset{\displaystyle O}{\overset{\|}{C}}-R$	$H-\underset{\underset{H}{|}}{\overset{\overset{H}{|}}{C}}-\overset{\displaystyle O}{\overset{\|}{C}}-\underset{\underset{H}{|}}{\overset{\overset{H}{|}}{C}}-H$ Acetone
Carboxyl —COOH	Carboxylic acids	$R-C\overset{\displaystyle O}{\underset{\displaystyle OH}{\diagdown}}$	$H-\underset{\underset{H}{|}}{\overset{\overset{H}{|}}{C}}-C\overset{\displaystyle O}{\underset{\displaystyle OH}{\diagdown}}$ Acetic acid
Amino —NH$_2$	Amines	$R-N\overset{\diagup H}{\underset{\diagdown H}{}}$	$H-\underset{\underset{H}{|}}{\overset{\overset{H}{|}}{C}}-N\overset{\diagup H}{\underset{\diagdown H}{}}$ Methylamine
Phosphate —OPO$_3^{2-}$	Organic phosphates	$R-O-\underset{\underset{O^-}{|}}{\overset{\overset{O}{\|}}{P}}-O^-$	$\begin{array}{c} HO\diagdown\;\diagup\kern-0.3em/O \\ C \\ | \\ H-C-OH \quad O \\ | \qquad\qquad \| \\ H-C-O-P-O^- \\ | \qquad\qquad | \\ H \qquad\qquad O^- \end{array}$ 3-Phosphoglyceric acid
Sulfhydryl —SH	Thiols	R—SH	$\begin{array}{ccc} H & H \\ | & | \\ H-C-C-SH \\ | & | \\ H & H \end{array}$ Mercaptoethanol

2.19 Simple Organic Compounds and Functional Groups
Compounds of the types shown here will appear throughout this book. The functional groups are the most common ones found in biologically important molecules.

carbon. Also attached to the α carbon atom are a hydrogen atom and a side chain (Figure 2.20a). Twenty different amino acids constitute the building blocks of the giant protein molecules of living things. Each of the amino acids has a different side chain

(Chapter 3) that gives it its distinctive chemical properties. Because they possess both carboxyl and amino groups, amino acids are simultaneously acids and bases. At the pH values commonly found in cells, both the carboxyl and the amino groups are ionized

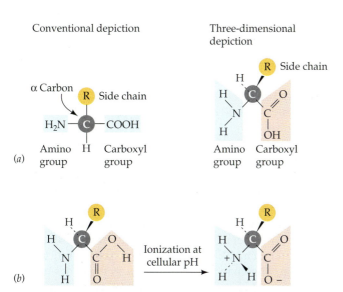

Conventional depiction   Three-dimensional depiction

2.20 Amino Acids
(a) The general structure of an amino acid. The side chain attached to the α carbon differs from one amino acid to another. (b) At pH values found in living cells, both the carboxyl group and the amino group of an amino acid are ionized.

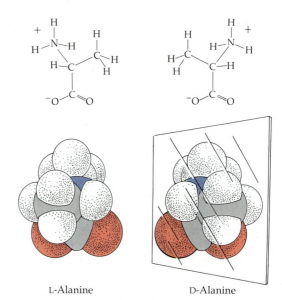

L-Alanine   D-Alanine

2.22 Isomers of Alanine
Structural formulas and space-filling models of the D- and L- forms of the amino acid alanine. Only L-alanine (on the left) is commonly found in living things. (Refer to Figure 2.18 for an explanation of the colors in the space-filling models.)

(Figure 2.20b). The carboxyl group has lost a proton, and the amino group has gained one.

Isomers are compounds with the same chemical formula but different arrangements of the atoms. Whenever a carbon atom has four *different* atoms or groups attached to it, there are two different ways of making the attachments, each the mirror image of the other—any seemingly different arrangement reduces to one of these two if the molecule is simply rotated. Such a carbon atom is called an **asymmetric**

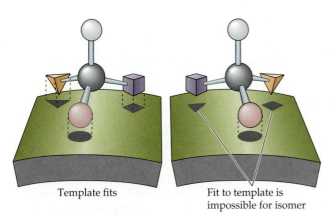

Template fits   Fit to template is impossible for isomer

2.21 Optical Isomers
Optical isomers are mirror images of one another. They result when four different groups are attached to a single carbon atom (the dark gray sphere in the center). If a template is laid out to match the groups on one carbon atom, there is no way the groups on the mirror-image isomer can be rotated to fit the same template.

carbon, and the pair of compounds are called **optical isomers** of each other (Figure 2.21). Your right and your left hand are optical isomers. Just as a glove is specific for a particular hand, so some biochemical molecules can interact with a specific optical isomer of a compound but are unable to "fit" the other. The α carbon in an amino acid is an asymmetric carbon; hence, amino acids can exist in two isomeric forms called D- and L-amino acids (Figure 2.22). D- and L- are abbreviations for *dextro-* and *levorotatory*, referring to the directions (right or left) in which solutions of these compounds rotate the plane of polarized light; they refer to the "handedness" of a molecule with one or more asymmetric carbons. Only L-amino acids are commonly found in most proteins of living things.

Two other functional groups should be introduced here. The **sulfhydryl group** (—SH; Figure 2.19) is important in protein structure (Chapter 3) and in biochemical reactions (Chapter 7). The **phosphate group** (—OPO_3^{2-}) participates in many crucial reactions in which energy is transferred (Chapters 7, 8, 11, 38). Phosphate groups are exchanged between sugars and many other compounds.

The compounds discussed in this chapter include some of the more common ones found in organisms. Between the small molecules and the world of the living stands another level, that of the giant molecules called macromolecules. These large molecules —the proteins, lipids, carbohydrates, and nucleic acids—are the subject of the next chapter.

SUMMARY

Organisms consist primarily of carbon, hydrogen, oxygen, and nitrogen. Sulfur, phosphorus, and several other elements are also essential to living things. Every element is represented by two or more isotopes—elemental forms differing in the number of neutrons in their nuclei.

Two or more atoms linked by chemical bonds (attractive forces) to form a stable structure constitute a molecule. When two atoms share a pair of electrons, the atoms are said to be joined by a covalent bond. Carbon atoms form four covalent bonds, oxygen two, and hydrogen one. Most biological molecules are formed by covalent bonding. Covalent compounds may undergo group transfer (the exchange of covalent bonding partners with one another). This is an important type of chemical reaction in organisms.

Many substances form ions (electrically charged particles) when they dissolve in water. Ions of opposite charge attract each other and in some cases form compounds by ionic bonding. Other weak bonds (hydrogen bonds, van der Waals interactions, and hydrophobic interactions) also help form large molecules and help molecules aggregate into larger structures.

Water, the most abundant substance in organisms, has numerous biologically significant properties. Its polarity makes it an exceptionally effective solvent.

An acid is a substance that can release one or more protons (hydrogen ions). A base can accept one or more protons. An acidic solution is one in which the pH is lower than 7 (the hydrogen ion concentration is greater than 10^{-7} M), whereas a basic solution has a pH greater than 7. A buffer is a system that resists changes in pH upon the addition of acid or base.

SELF-QUIZ

1. The atomic number of an element:
   a. equals the number of neutrons in an atom.
   b. equals the number of protons in an atom.
   c. equals the number of protons minus the number of electrons.
   d. equals the number of neutrons plus the number of protons.
   e. depends on which isotope one is talking about.

2. The atomic weight of an element:
   a. equals the number of neutrons in an atom.
   b. equals the number of protons in an atom.
   c. equals the number of electrons in an atom.
   d. equals the number of neutrons plus the number of protons.
   e. depends on the relative abundances of its isotopes.

3. Which of the following statements about all the isotopes of an element is *not* true?
   a. They have the same atomic number.
   b. They have the same number of protons.
   c. They have the same number of neutrons.
   d. They have the same number of electrons.
   e. They have identical chemical properties.

4. Which of the following statements about a covalent bond is *not* true?
   a. It is stronger than a hydrogen bond.
   b. One can form between atoms of the same element.
   c. Only a single covalent bond can form between two atoms.
   d. It results from the sharing of two electrons by two atoms.
   e. One can form between atoms of different elements.

5. Hydrophobic interactions:
   a. Are stronger than hydrogen bonds.
   b. Are stronger than covalent bonds.
   c. Can hold two ions together.
   d. Can hold two nonpolar molecules together.
   e. Are responsible for the surface tension of water.

6. Which of the following statements about water is *not* true?
   a. It releases a large amount of heat in turning from liquid into vapor.
   b. Its solid form is less dense than its liquid form.
   c. It is the most effective solvent known.
   d. It is typically the most abundant substance in an active organism.
   e. It takes part in some important chemical reactions.

7. A solution with a pH of 9:
   a. is acidic.
   b. is more basic than a solution with a pH of 10.
   c. has ten times the hydrogen ion concentration of a solution with pH 10.
   d. has a hydrogen ion concentration of 9 molar.
   e. has a hydroxide ion concentration of 9 molar.

8. Which of the following compounds is an alcohol?
   a. O_2
   b. $CH_3CH_2CH_2OH$
   c. CH_3COOH
   d. C_3H_8
   e. CH_3COCH_3

9. Which of the following statements about the carboxyl group is *not* true?
   a. It has the chemical formula —COOH.
   b. It is an acidic group.
   c. It can ionize.
   d. It is found in amino acids.
   e. It has an atomic weight of 45.

10. Which of the following statements about amino acids is *not* true?
    a. They are the building blocks of proteins.
    b. They contain carboxyl groups.
    c. They contain amino groups.
    d. They do not ionize.
    e. They have both L- and D-isomers.

FOR STUDY

1. Lithium is the element with atomic number = 3. Draw the structures of the Li atom and of the Li^+ ion.

2. Draw the structure of a pair of water molecules held together by a hydrogen bond. Your drawing should also indicate the covalent bonds in the molecules.

3. The molecular weight of sodium chloride (NaCl) is 58.45. How many grams of NaCl are there in one liter of a 0.1-molar NaCl solution? How many in 0.5 liter of a 0.5-molar NaCl solution?

4. The side chain of the amino acid alanine is —CH_3 (see Figure 2.20). Draw the structures of the two optical isomers of alanine. The side chain of the amino acid glycine is simply a hydrogen atom (—H). Are there two optical isomers of glycine? Explain.

READINGS

Breed, A., T. Rodella and R. Basmajian. 1982. *Through the Molecular Maze*. William Kaufmann, Los Altos, CA. A short, inexpensive guide to the rudiments of chemical concepts and terminology needed by students in introductory courses on the life sciences.

Henderson, L. J. 1958. *The Fitness of the Environment*. Beacon Press, Boston. An essay written in 1912 about physical properties of water and carbon dioxide in relation to life. With a thought-provoking introduction.

Kotz, J. C. and K. F. Purcell. 1987. *Chemistry and Chemical Reactivity*. Saunders, Philadelphia. A well illustrated modern textbook of general chemistry.

McQuarrie, D. A. and P. A. Rock. 1991. *General Chemistry*, 3rd Edition. W. H. Freeman, New York. A first-rate textbook, beautifully illustrated.

Mertz, W. 1981. "The Essential Trace Elements." *Science*, vol. 213, pages 1332–1338. This article reviews the roles of more than a dozen elements needed in small amounts by animals if they are to function normally.

3

Large Molecules

PREVIEW: Living things are made up of many substances, especially lipids, carbohydrates, proteins, and nucleic acids. Some lipids are energy-storing "fuels," some form membranes, and others serve as chemical messengers or as trappers of energy. Carbohydrates function as strengthening elements, as fuels, and in other ways. The diverse functions of proteins include accelerating chemical reactions, defending the animal body against microorganisms, and providing support and protection. Nucleic acids store, transmit, and interpret hereditary information.

This chapter deals with the structures and functions of lipids (triglycerides, phospholipids, steroids, and carotenoids), carbohydrates (monosaccharides, oligosaccharides, and polysaccharides), amino acids, proteins, nucleotides, DNA, and RNA.

The **macromolecules**—giant molecules, or aggregates of molecules, with molecular weights in excess of 1,000 daltons—perform many essential functions in organisms. As we will see, these functions arise directly from the structures of the molecules. Some of the macromolecules fold into globular forms with surface features that enable them to recognize and interact with certain other molecules. Other macromolecules form long, fibrous systems that provide strength and rigidity to parts of an organism; still others contract and allow the organism to move itself. Some macromolecules aggregate to form structures that determine what materials enter or leave the compartments within an organism. The largest of the molecules are all **polymers**: molecules that are made by the combination of many smaller molecules. The small molecules that are a polymer's subunits are called **monomers**. An **oligomer** contains only a few monomers.

There is a flow of *information* among the various classes of macromolecules. The source of the information is DNA (deoxyribonucleic acid), the genetic material. Within the structure of DNA molecules lies the necessary information to dictate the structures of the many different proteins in an organism. Transmitting the information in DNA to proteins is the task of various types of RNAs (ribonucleic acids). Some of these proteins (the enzymes) act to accelerate chemical reactions in the cell. In this chapter, we take a brief look at the major classes of macromolecules in order to see how their structures relate to their functions; in later chapters we will return to the topics raised here and develop them in greater detail.

FROM MONOMERS TO POLYMERS

The largest molecules in living things—polysaccharides, proteins, and nucleic acids—are polymers built from simpler monomers. These polymerization reactions belong to a class of reactions called **condensations** or **dehydrations**, which are of the general type

$$A—H + B—OH \rightarrow A—B + H_2O$$

(A—H is a molecule consisting of a hydrogen atom attached to another part, A; B—OH is a molecule consisting of an —OH group attached to another part, B.) The product A—B is formed along with a molecule of water; the atoms of water are derived from the reactants, with one hydrogen atom coming from one reactant, and an oxygen atom and the other hydrogen atom from the other reactant. **Reactants** are the molecules undergoing a chemical reaction.

The actual polymerization reactions that produce the different kinds of macromolecules differ in detail. In all cases, energy must be added to the system for polymers to form. Other kinds of specific molecules participate; their function is to activate the reactants—to provide the necessary energy for the reactions to be carried out. Large molecules are assembled through the repeated condensations of activated monomers.

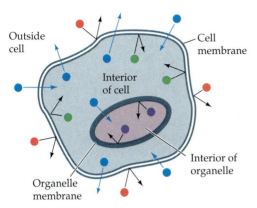

3.1 Lipid-Containing Membranes
Lipid-containing membranes separate the cell from its environment; they also separate the contents of some subcellular organelles from the rest of the cell. Materials that do not dissolve in lipids are generally unable to pass through the membranes from one region to another. Molecules that are lipid-soluble (bright blue symbols in this representation) move through membranes with relative ease.

LIPIDS

The **lipids** are a diverse group of compounds that are insoluble in water but are readily soluble in organic (carbon-based) solvents such as ether. They release large amounts of energy when they break down. Each of these properties is significant in the biology of these compounds. Because lipids do not dissolve in water and water does not dissolve in lipids, a mixture of water and lipids forms two distinct layers. Also, many biological materials that are soluble in water are much less soluble in lipids.

Suppose that you must design compartments, separated from each other and from their environment by barriers that limit the passage of materials. Given the properties of lipids, an effective way to accomplish this would be to use lipid-containing membranes to separate the compartments (Figure 3.1). This is, in fact, the system that has evolved in nature. Molecular traffic within an organism or into and out of its compartments is strictly limited by the solubility properties of the lipid portion of the surrounding membrane. Compounds that dissolve readily in lipids can move rapidly through biological membranes; but compounds that are insoluble in lipids are prevented from passing, or must be transported across the membrane by specific proteins, as will be described in Chapter 5.

The role of lipids in energy storage relates to the topics of oxidation and reduction, which will be described in Chapter 6 when we discuss the processing of energy. For now, suffice it to say that the lipids are marvelous storehouses for energy. Many animal species deposit fat (= lipid) droplets in their bodies as a means for storing energy—as you know, an excess of food results in fat deposition (Figure 3.2). Some plant species, such as olives, avocados, sesame seeds, and castor beans, have substantial amounts of lipids in their seeds or fruits that serve as energy reserves for the next generation.

Triglycerides

One important group of lipids is the **triglycerides**, also known as *simple lipids*. Triglycerides that are solid at room temperature are called **fats**; those that are liquid at this temperature are called **oils**. The triglycerides are composed of two types of building blocks: **fatty acids** and **glycerol**. Fatty acids are carboxylic acids with long hydrocarbon tails. A typical fatty acid found in animal fats is palmitic acid, $C_{15}H_{31}COOH$ (Figure 3.3). Another example is stearic acid, $C_{17}H_{35}COOH$, which has two more carbon atoms and four more hydrogen atoms. These are both **saturated fatty acids** because their hydrocarbon tails contain no double bonds. Another common fatty acid, oleic acid (Figure 3.3), is **unsaturated**. Notice the double bond near the middle of the hydrocarbon chain in oleic acid, causing a kink in the molecule. Other fatty acids, such as linoleic acid, have more than one carbon–carbon double bond and are thus **polyunsaturated**. These molecules have multiple kinks. Unsaturated and polyunsaturated fatty acids can accept hydrogen atoms—that is, they can become hydrogenated. The addition of two hydrogen atoms across the double bond of oleic acid, for example, would produce stearic acid.

Three fatty acid molecules combined with a molecule of glycerol give a molecule of a triglyceride (Figure 3.4). The three fatty acids in one triglyceride molecule are not always the same length, nor are they necessarily all either saturated or unsaturated. The kinks associated with double bonds are impor-

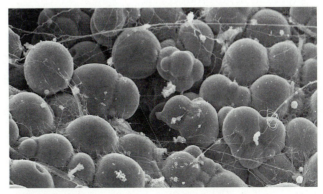

3.2 Fat Cells
These cells served as stores of energy for a mouse.

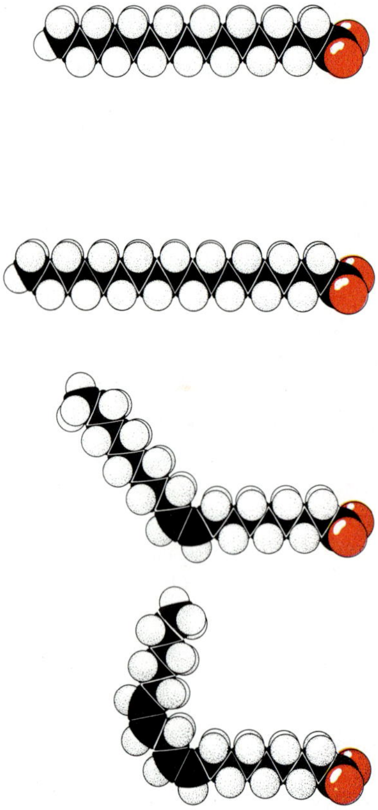

(a) **Palmitic acid**

$$CH_3-CH_2-(CH_2)_{12}-CH_2-\overset{\overset{\displaystyle O}{\|}}{C}-OH$$

(b) **Stearic acid**

$$CH_3-CH_2-(CH_2)_{14}-CH_2-\overset{\overset{\displaystyle O}{\|}}{C}-OH$$

(c) **Oleic acid**

$$CH_3-CH_2-(CH_2)_5-CH_2-CH=CH-CH_2-(CH_2)_5-CH_2-\overset{\overset{\displaystyle O}{\|}}{C}-OH$$

(d) **Linoleic acid**

$$CH_3-CH_2-(CH_2)_2\ CH_2\ CH=CH-CH_2\ CH=CH\ CH_2-(CH_2)_5-CH_2-\overset{\overset{\displaystyle O}{\|}}{C}-OH$$

3.3 Fatty Acids

(a) The absence of double bonds between carbon atoms in the chain means that palmitic acid is a saturated fatty acid; the straight-chain configuration in the model of the molecule is characteristic of saturated fatty acids. *(b)* Stearic acid has two more carbons and four more hydrogens than palmitic acid and is also saturated. *(c)* Oleic acid has a double bond between two carbons in the chain and is therefore unsaturated. *(d)* With two double bonds in its chain, linoleic acid is polyunsaturated.

tant in determining the fluidity and melting point of a lipid. Triglycerides with short or unsaturated chains are usually oily liquids, whereas those with long and saturated chains are waxy solids. Animal fats such as lard and tallow are usually solids with long-chain, saturated or singly unsaturated fatty acids. In these fats, hydrocarbon chain lengths range between 10 and 20 carbon atoms. The triglycerides of plants tend to be more unsaturated, oily liquids. Natural peanut butter, for example, contains a great deal of oil. Peanut butter manufacturers often hydrogenate their product in order to reduce the double bonds and give a saturated, solid product.

Phospholipids

A triglyceride consists of glycerol with three fatty acids bound to it. Having certain phosphorus-containing compounds bound in the place of one of the fatty acids defines a class of substances known as **phospholipids** (Figure 3.5). Many phospholipids are important constituents of biological membranes. If you examine the structure of phospholipids closely, you will find it easy to understand how they are oriented in membranes. The phosphorus-containing portion of the phospholipid molecule carries one or more electric charges, so this portion is **hydrophilic**

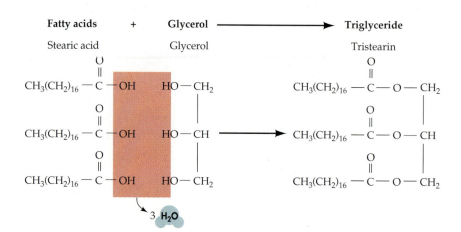

3.4 A Triglyceride and Its Components

Tristearin is a triglyceride composed of glycerol and three molecules of the fatty acid stearic acid. The synthesis of a triglyceride from glycerol and three fatty acids is an example of a condensation. Condensations result in the release of water molecules. (In living things the reaction is more complex, but the end result is as shown here.)

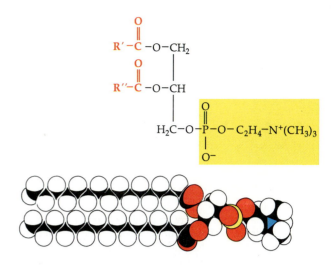

$$CH_3-CH_2-(CH_2)_8-CH_2-\overset{\overset{O}{\|}}{C}-O-CH_2$$

$$CH_3-CH_2-(CH_2)_{10}-CH_2-\overset{\overset{O}{\|}}{C}-O-CH$$

$$H_2C-O-\overset{\overset{O}{\|}}{\underset{\underset{O^-}{|}}{P}}-O^-$$

(a) **Phosphatidate**

$$R'-\overset{\overset{O}{\|}}{C}-O-CH_2$$

$$R''-\overset{\overset{O}{\|}}{C}-O-CH$$

$$H_2C-O-\overset{\overset{O}{\|}}{\underset{\underset{O^-}{|}}{P}}-O-C_2H_4-N^+(CH_3)_3$$

(b) **Phosphatidyl choline (a lecithin)**

3.5 Some Phospholipids

A phospholipid consists of glycerol combined with two molecules of fatty acid and a molecule containing phosphorus. Examples of phosphorus-containing molecules are phosphoric acid, as in the yellow-shaded area of (a), and phosphocholine (yellow shaded area of b). Phospholipids that contain phosphocholine are called lecithins. (c) Cephalins are formed by the addition of both phosphoric acid and ethanolamine, as included in the yellow region. In these diagrams, R' and R'' stand for "residue" and represent any fatty-acid hydrocarbon chains. These chains, shown in red letters, are nonpolar, whereas the shaded phosphorus-containing portions are electrically charged. Other types of phospholipids exist.

$$R'-\overset{\overset{O}{\|}}{C}-O-CH_2$$

$$R''-\overset{\overset{O}{\|}}{C}-O-CH$$

$$H_2C-O-\overset{\overset{O}{\|}}{\underset{\underset{O^-}{|}}{P}}-O-C_2H_4-NH_3^+$$

(c) **Phosphatidyl ethanolamine (a cephalin)**

(water-loving; remember that water is a polar molecule). The two fatty acid regions, however, are **hydrophobic** (water-fearing). Thus in a biological membrane, phospholipids line up in such a way that the nonpolar, hydrophobic "tails" pack tightly together to form the interior of the membrane, and the phosphorus-containing "heads" face outward (some to one side of the membrane and some to the other), where they interact with water, which is excluded from the interior of the membrane (Figure 3.6). The phospholipids form a bilayer, that is, a sheet two molecules thick. Biological membranes and their many important functions will be the subject of Chapter 5. For now, we emphasize that the dark lines of Figure 3.1 represent membranes that are composed of phospholipid bilayers as depicted in Figure 3.6.

Other lipids

The lipids we have considered thus far (phospholipids and triglycerides) are chemically similar. The term *lipid*, however, defines compounds not on the basis of structural similarity, but in terms of their solubility. Remember that lipids are insoluble in water but readily soluble in organic solvents such

3.6 Phospholipids in Biological Membranes
The nonpolar hydrocarbon (fatty-acid) chains gather together in the interior of the phospholipid bilayer by hydrophobic associations. The polar, phosphorus-containing hydrophilic heads of the molecules face outwards toward either side of the membrane. This structure will be shown in more detail in Chapter 5.

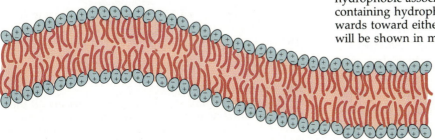

Hydrophilic "head"

Hydrophobic fatty-acid tails

Hydrophilic "head"

Phospholipid bilayer of biological membrane

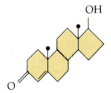

β-Carotene

Vitamin A

3.7 Carotenoids
Carotenoids are shown here in a shorthand chemical notation in which a carbon atom is present at each junction in the rings and at each bend of the chains. Each black dot corresponds to a methyl (—CH₃) group. β-Carotene is symmetrical around the central (green) double bond; the ends of the molecule on either side of the double bond are the same, although the ends are rotated 180 degrees with respect to one another. Two vitamin A molecules are produced by splitting β-carotene in the middle.

as ether, chloroform, or benzene. Some other compounds with these properties (and hence classifiable as lipids) are the carotenoids and the steroids.

The **carotenoids** are a family of light-absorbing pigments found in both plants and animals (Figure 3.7). Beta-carotene (β-carotene) is one of the pigments used to trap light energy in leaves to power the process of photosynthesis (Chapter 8). It is β-carotene that causes plants to grow toward or away from light (a behavior called phototropism, discussed in Chapter 32). In humans, a molecule of β-carotene can be broken down into two vitamin A molecules, from which we make the pigment rhodopsin that is required for vision (Chapter 37). Carotenoids are responsible for the color of carrots, tomatoes, pumpkins, egg yolks, and butter.

The **steroids** are a family of organic compounds based on a multiple ring structure in which the rings share carbons (Figure 3.8). Some steroids are important constituents of membranes. Others are among the hormones, chemical signals that carry messages from one part of the body to another (Chapter 34). Testosterone (Figure 3.8) is a steroid hormone that regulates sexual development in male vertebrates (animals with backbones), and the chemically similar estrogens play a similar role in females. Cortisone is one of a family of hormones that play a wide variety of regulatory roles in the digestion of carbohydrates and proteins, salt and water balance, and sexual development. Vitamin D is a steroid that regulates the absorption of calcium from the intestines. It is necessary for the proper deposition of calcium in bones; a deficiency of vitamin D leads to rickets, a bone-softening disease. Vitamin D is produced in human skin when certain other steroids are irradiated with sunlight or ultraviolet light.

Cholesterol (also shown in Figure 3.8) is synthesized in the liver. In all cells except those of bacteria, cholesterol stiffens membranes. It is also the starting material for making testosterone and several other steroid hormones and for the bile salts that help to get fats into solution so they can be digested. Cholesterol is absorbed from foods such as milk, butter, and animal fats. When there is too much cholesterol in the blood, it is deposited in the arteries (along with other substances), a condition that may lead to arteriosclerosis and heart attack.

Chemically the lipids are quite varied, as you can see by glancing back at Figures 3.3 through 3.8. Their diversity matches the variety of their functions in living things: energy storage, digestion, membrane structure, bone formation, vision, and chemical signaling. Most lipids can be synthesized in the bodies of animals; the synthesis and storage of fats is an important means of locking energy away until it is needed. The few lipids that cannot be synthesized

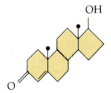

(a) Testosterone

(b) Cortisone

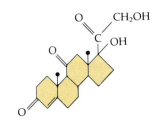

(c) Vitamin D

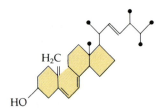

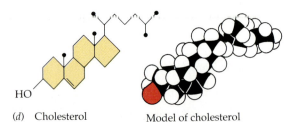

(d) Cholesterol   Model of cholesterol

3.8 Examples of Steroids
Among the important steroids in vertebrates are (a) the male sex hormone testosterone, (b) the hormone cortisone, (c) vitamin D, and (d) cholesterol. All of these steroids have a similar ring structure (orange).

must be obtained in small amounts from the diet. For humans, the diet must include three particular unsaturated fatty acids and the fat-soluble vitamins: A, D, E, and K.

CARBOHYDRATES

Carbohydrates are a diverse group of compounds with molecular weights ranging from less than 100 to hundreds of thousands. They fall into three categories: the **monosaccharides**, or *simple sugars*, which are monomers; the **oligosaccharides**, made up of a few monosaccharides linked together; and the **polysaccharides**, polymeric carbohydrates that include starches, glycogen, cellulose, and many other important biological materials. (*Mono-* means "single," *oligo-* means "few," and *poly-* means "many"; *saccharide* means "sugar.") There is no clear dividing line between a large oligosaccharide and a small polysaccharide, for these are simply terms of convenience used to separate "classes" within what is really a continuum of compounds of various sizes. All share a general formula of approximately $C_nH_{2m}O_m$; that is, there are twice as many hydrogen as oxygen atoms, and the number of carbon atoms is not always the same as the number of oxygen atoms.

Monosaccharides

All living cells contain **glucose**, $C_6H_{12}O_6$, a monosaccharide. It is produced in green plants by photosynthesis (Chapter 8), and it is also obtained by the digestion of certain polysaccharides. In cells it is metabolized to yield energy in the process of cellular respiration (Chapter 7). Glucose exists in both straight-chain and ring forms, in equilibrium with each other (Figure 3.9). There are two distinct ring forms of glucose (α- and β-glucose). These differ in the placement of the —H and —OH groups attached to a particular carbon atom in the molecule (see the carbon atom identified as carbon 1 in Figure 3.9; the numbering convention shown there will be used throughout this book). α- and β-glucose are chemically and physically distinct substances, but they constantly interconvert in aqueous solution.

A number of other simple sugars are illustrated in Figure 3.10. Many monosaccharides have the same formula as glucose, $C_6H_{12}O_6$, including **fructose** ("fruit sugar"), mannose, and galactose. These compounds are all isomers of each other—they are composed of the same kinds and numbers of atoms, but the atoms are combined differently and yield different arrangements such as those shown in Figure 3.10. The six-carbon sugars are referred to collectively as **hexoses**. There are also a number of five-carbon sugars, called **pentoses**. Some pentoses are found primarily in the cell walls of plants, as are several of the hexoses. Two pentoses are of particular importance:

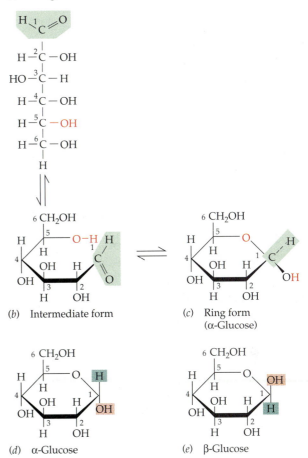

(a) Straight-chain form

(b) Intermediate form

(c) Ring form (α-Glucose)

(d) α-Glucose

(e) β-Glucose

3.9 Forms of Glucose
Glucose exists in several interconverting chemical forms when dissolved in water. The straight-chain form (a) has an aldehyde group at carbon 1 (shaded in green). A reaction between the aldehyde group and the hydroxyl group at carbon 5 (b) gives rise to one of the ring forms (c). The ring form is usually represented as in (d), where the darker lines at the bottom imply that that edge of the molecule extends toward you and the upper edge extends back into the page. Depending on the orientation of the aldehyde group at carbon 1 when the ring closes, either of two rapidly and spontaneously interconverting forms of glucose, α-glucose (d) or β-glucose (e) is formed. α-Glucose and β-glucose differ only at carbon position 1.

ribose and deoxyribose (Figure 3.10), which form part of the backbones of RNA and of DNA, respectively. Ribose and deoxyribose differ by one oxygen atom associated with one of the carbon atoms, carbon 2.

Disaccharides

Larger carbohydrates are made by the combination of two or more monosaccharide molecules. The monosaccharides may be covalently coupled to form specific oligosaccharides and polysaccharides. The smallest oligosaccharides are the disaccharides and the trisaccharides, which are made up of two and three

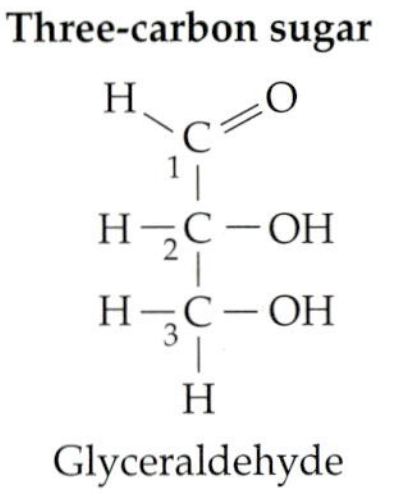

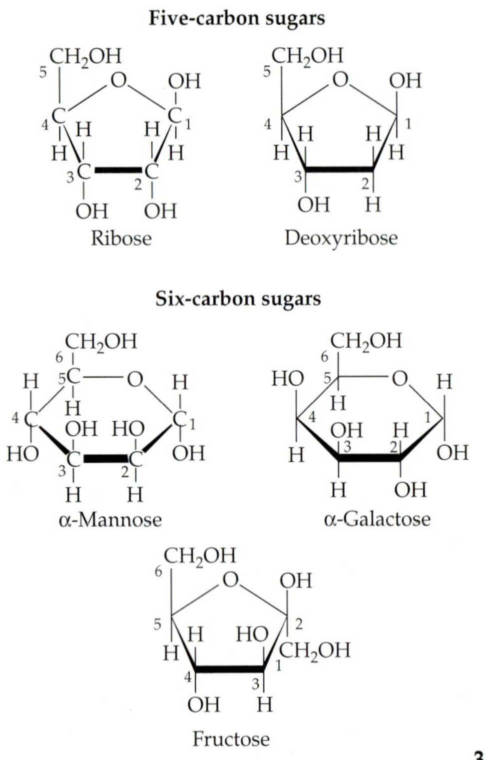

Three-carbon sugar

$$H-C=O$$

H—$\overset{1}{\text{C}}$—OH

H—$\overset{2}{\text{C}}$—OH

$\overset{3}{|}$

H

Glyceraldehyde

Five-carbon sugars

Ribose

Deoxyribose

Six-carbon sugars

α-Mannose

α-Galactose

Fructose

3.10 Monosaccharides

The three-carbon sugar (triose) glyceraldehyde has the formula $C_3H_6O_3$; it is shown in the common straight-chain form. The pentoses, including ribose and deoxyribose, each have five carbons. The three hexoses (six-carbon sugars) shown here all have the formula $C_6H_{12}O_6$, but they are chemically and biologically distinct from one another.

simple sugars, respectively. If one glucose molecule combines with another, as shown in Figure 3.11, the disaccharide product must be one of two types: α-linked or β-linked, depending on whether it is α-glucose or β-glucose that reacts. An α linkage with carbon 4 of a second glucose molecule gives us **maltose**, whereas a β linkage gives **cellobiose**. Both maltose and cellobiose are disaccharides; both have the formula $C_{12}H_{22}O_{11}$; both are composed of two glucose molecules (minus one molecule of water), but they are different compounds—they are recognized by different enzymes and undergo different chemical reactions. Two other commonly occurring disaccharides are **sucrose** and **lactose** (Figure 3.11). Sucrose (common table sugar; also $C_{12}H_{22}O_{11}$) is made from one molecule of glucose and one of fructose. Lactose (milk sugar) consists of glucose and galactose.

Polysaccharides

As we saw in Figure 3.11, maltose consists of two glucose units connected by an α-linkage. Imagine a trisaccharide (three glucose units), a tetrasaccharide

3.11 Disaccharides

A disaccharide is composed of two monosaccharides. As shown in the reaction at the top left (a simplified version of the actual reaction in nature), maltose is produced when an α-1,4 linkage forms between two glucose molecules, while in cellobiose (bottom left) the two glucoses are linked β-1,4. Lactose (bottom right) is made by a β linkage between carbon 1 of galactose and carbon 4 of glucose. In sucrose (top right), carbon 1 of glucose is joined by an α-1,2 linkage to carbon 2 of fructose.

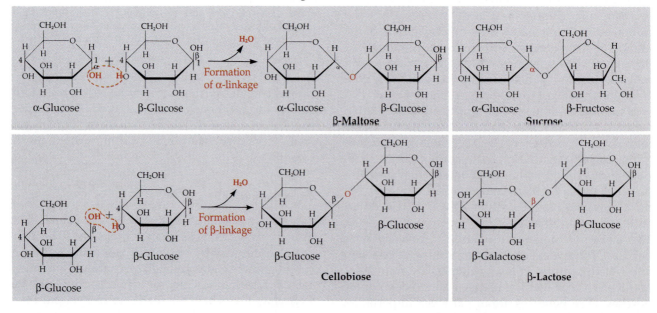

(four glucose units), and finally a giant polysaccharide consisting of hundreds or thousands of glucose units, each connected to the next by an α linkage from carbon 1 of one unit to carbon 4 of the next. This polymer is **starch**, an important storage compound in the plant kingdom (Figure 3.12*b*).

Similarly, there is a giant polysaccharide made up solely of glucose but with the individual units connected by β linkages. This is **cellulose**, the predominant component of plant cell walls (Figure 3.12*a*). Both starch and cellulose are composed of nothing but glucose (when depolymerized, they yield only glucose), yet their biological functions and chemical and physical properties are entirely different. Enzymes that digest one will not affect the other at all. Aggregated starch forms a shapeless solid that crumbles readily, whereas cellulose is largely crystalline and has an impressive ability to withstand longitudinal pulling without breaking. Starch is primarily a storage compound in plants, holding in reserve energy and carbon that can be made available upon digestion. Cellulose is a key structural element in plant cell walls; for example, it provides much of the strength of wood. Humans have enzymes for the digestion of starch but not for the digestion of cellulose. The enzymes of many bacteria, fungi, and snails readily digest cellulose.

Cotton is more than 90 percent cellulose and is a familiar example of this polysaccharide and its properties. Cellulose is the standard building material for woody stalks, fibers, and all types of cell walls in plants. These rigid structures owe almost all their physical strength to cellulose, which is their toughest component and which usually makes up more than one-fourth of the plant cell wall. Cellulose is by far the most common organic compound on this planet, accounting for more than half the carbon present in plant life.

Starch is not actually a single chemical substance; rather, the term denotes a large family of giant molecules of broadly similar structure. All starches are polymers of glucose with α linkages. All are large, but some are enormous, containing tens of thousands of glucose units. An important variable is the degree of branching: Many starches have highly branched chains (Figure 3.12*b*). The starches that store glucose in plants are called **amylose** and are not highly branched. The highly branched polysaccharide that stores glucose in animals is **glycogen** (Figure 3.12*c*). Animals use glycogen to store energy in liver and muscle.

What do we mean when we say that starch and glycogen are storage compounds for energy? Very simply, these compounds can readily be depolymerized to yield glucose monomers. Glucose, in turn, can be further digested, or metabolized—that is, it can undergo chemical reactions—to yield energy for cellular work. Alternatively, glucose can be metabo-

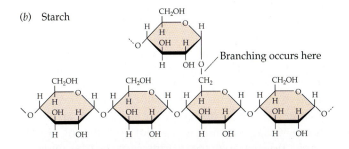

(a) Cellulose

Glucose monomer

Hydrogen bonding to other cellulose molecules can occur at these points

(b) Starch

Branching occurs here

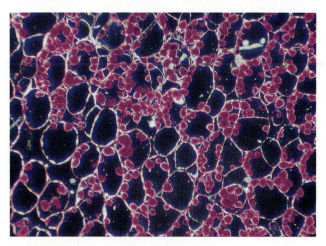

(c) Branched and unbranched polysaccharides

Glucose monomer

Unbranched starch molecule

Highly branched glycogen molecule

3.12 Representative Carbohydrates
(a) Cellulose is an unbranched polymer of glucose; hydrogen bonding to other cellulose molecules can occur, as indicated here with dashed lines. Many adjacent cellulose molecules form the cellulose fibrils in plant cells. *(b)* In starch, branching may occur at the position indicated. In the micrograph, a red dye stains the starch grains in sweet potato cells. *(c)* Glycogen differs from starch in plants only in being more extensively branched.

lized so that its carbon atoms are rearranged to form the skeletons of other compounds. Thus glycogen and starch are storage depots for carbon atoms as well as for energy. Each is chemically stable but readily mobilized by digestion and further metabolism.

Derivative Carbohydrates

Derivative carbohydrates deviate from the general formula $C_nH_{2m}O_m$ by containing other elements. Figure 3.13 shows a sugar phosphate, amino sugars, and chitin as examples. A number of **sugar phosphates**, such as fructose 1,6-bisphosphate, are important intermediates in cellular respiration (Chapter 7) and photosynthesis (Chapter 8). Sugar phosphates have phosphate groups attached to one or more —OH groups of the parent sugar. The two **amino sugars** shown in the figure, glucosamine and galactosamine, have an amino group in place of an —OH group. Galactosamine is a major component of cartilage, the material that forms caps on the ends of bones and stiffens the protruding parts of the ears and nose. The polymer chitin is made from a derivative of glucosamine. Chitin is the principal structural polysaccharide in the skeletons of insects and their

relatives such as crabs and lobsters, as well as in the cell walls of fungi. Fungi and insects (and their relatives) constitute more than 80 percent of the species ever described, and chitin is another of the most abundant substances on Earth.

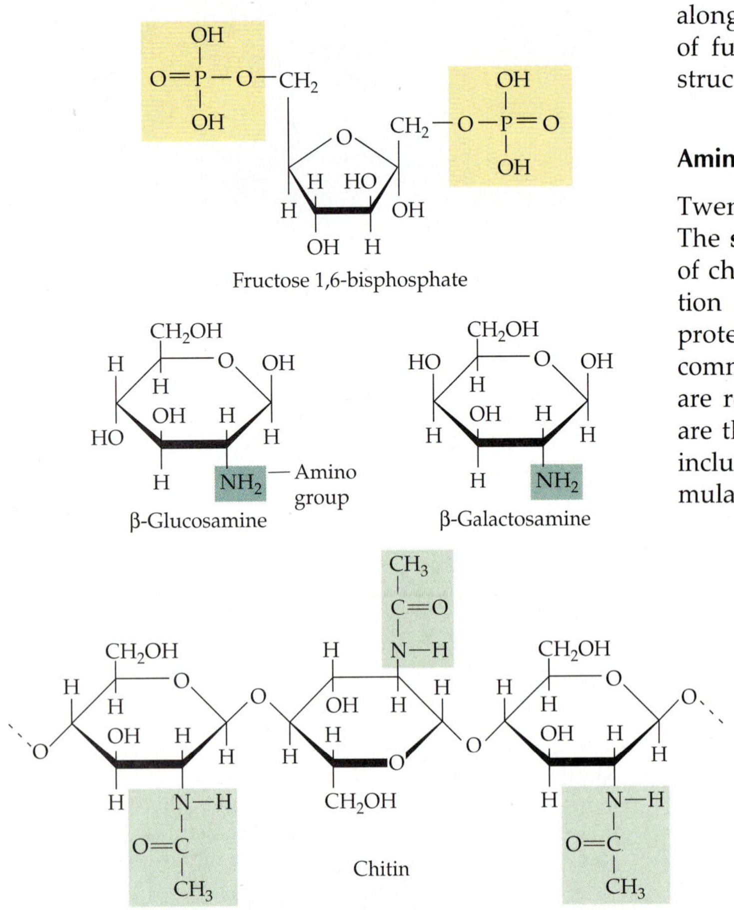

PROTEINS

In Chapter 2 we considered the **amino acids**. These are the monomers from which a fascinating set of polymers are formed—the **proteins**. The proteins account for many of the mechanical elements of living things, from parts of subcellular membranes to skin, bones, and tendons. In vertebrates, other proteins, the immunoglobulins (including the antibodies), form a major line of defense against foreign organisms. The specialized molecules needed to bring about all biochemical reactions are a major class of proteins called enzymes. Our every movement results from the contraction and relaxation of muscles, resulting in turn from the delicately regulated sliding of particular proteins in muscle cells past one another. Still other proteins act as adjustable channels through which sodium ions (Na^+), potassium ions (K^+), and other ions are passed from one side of a nerve-cell membrane to the other, resulting in phenomena such as the transmission of electric signals along a nerve. To understand this stunning variety of functions, we must first see and appreciate the structure of these molecules.

Amino Acids

Twenty different amino acids are found in proteins. The **side chains** of amino acids show a wide variety of chemical properties. Side chains control the function of a protein—they are the reactive groups in proteins. Despite their importance, side chains are commonly left out of structural formulas, where they are represented simply by "R" (for "residue"); they are thus sometimes called R groups. Side chains are included, highlighted in color, in the structural formulas in Table 3.1.

3.13 Derivative Carbohydrates
Fructose 1,6-bisphosphate is a sugar phosphate; the numbers in its name refer to the bonding of the phosphate groups (shaded in color) to the number 1 and number 6 carbon atoms of the sugar. The amino groups (—NH₂) on the amino sugars β-glucosamine and β-galactosamine are also shown in color; recall that the β refers to the position of the —OH group on the number 1 carbon. Chitin is a polymer of N-acetylglucosamine; N-acetyl groups are shown in color.

TABLE 3.1
Twenty amino acids found in proteins

A. Amino acids with hydrophobic R groups

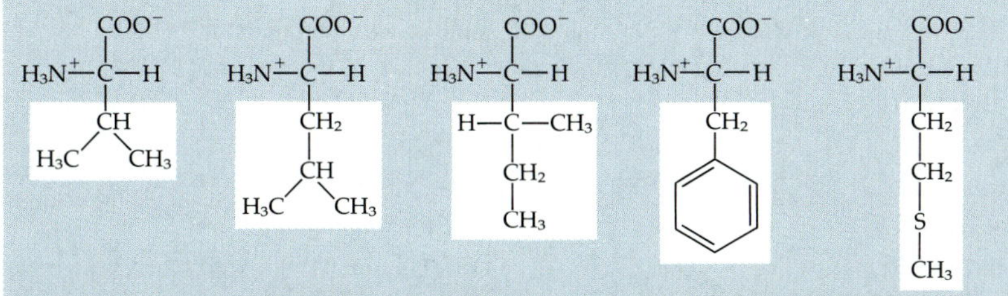

Valine
(Val)

Leucine
(Leu)

Isoleucine
(Ile)

Phenylalanine
(Phe)

Methionine
(Met)

B. Amino acids with hydrophilic groups

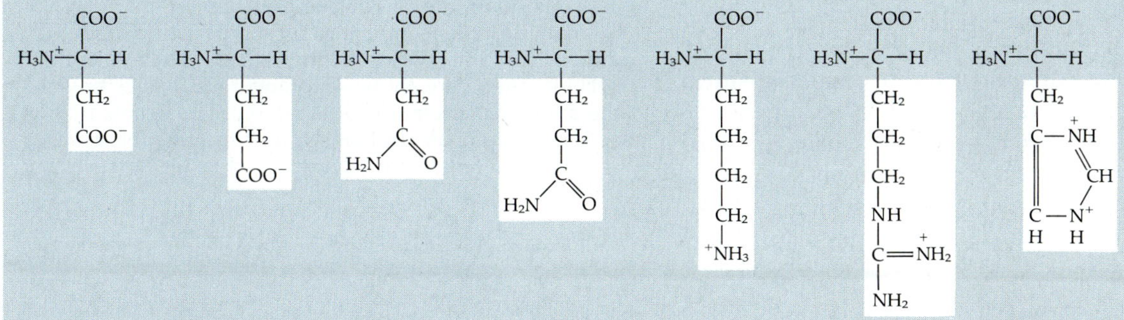

Aspartic acid
(Asp)

Glutamic acid
(Glu)

Asparagine
(Asn)

Glutamine
(Gln)

Lysine
(Lys)

Arginine
(Arg)

Histidine
(His)

C. Amino acids that occur both on the surface and in the interior of proteins

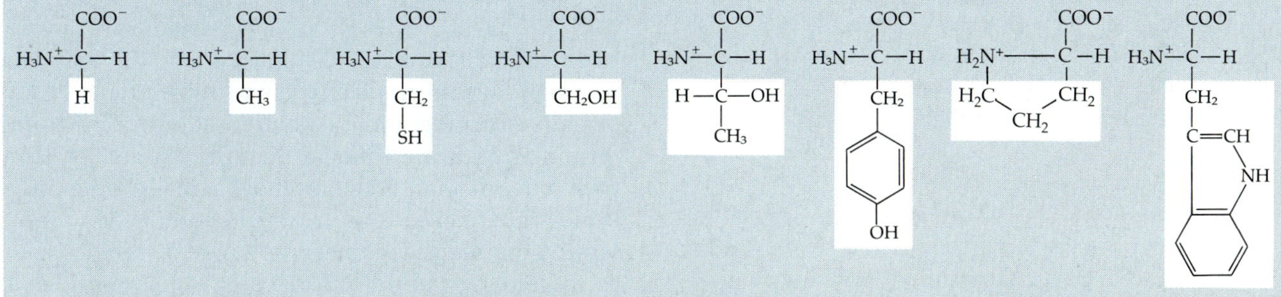

Glycine
(Gly)

Alanine
(Ala)

Cysteine
(Cys)

Serine
(Ser)

Threonine
(Thr)

Tyrosine
(Tyr)

Proline
(Pro)

Tryptophan
(Trp)

The order of a protein's amino acids determines how it folds into a three-dimensional configuration; we will discuss this folding later. One useful classification of amino acids is based on whether they are usually found on the *inside* of a folded protein molecule, on the protein's *surface*, or in *both* places. The side chains of valine, leucine, isoleucine, phenylalanine, and methionine are all either hydrocarbons or very close relatives thereof, so these side chains are hydrophobic (Table 3.1A). They form hydrophobic interactions with other nonpolar molecules. In enzymes, for example, which operate in aqueous solutions, the hydrophobic side chains tend to fold into the interior of the molecule. (These side chains are the protein analogues of the hydrocarbon tails of phospholipids; Figure 3.6.) Other proteins function in other strongly hydrophobic environments such as the interior of a biological membrane. Proteins embedded in membranes have the hydrophobic side chains of their amino acids on the *exterior* of the

49

molecule, where they interact with hydrophobic portions of lipids; they may have hydrophilic side chains on the inside.

Seven amino acids have strongly hydrophilic side chains, so their behavior is just the opposite of that of the hydrophobic chains. The hydrophilic chains are either electrically charged or so polar that they associate readily with water molecules (Table 3.1B). In aqueous solution they tend to orient toward the outside of the molecule. At the pH typically found in cells (around pH 7), aspartic and glutamic acids have negative charges; lysine, arginine, and histidine have positive charges; and asparagine and glutamine are neutral but polar.

Eight amino acids of a third class can be either on the surface of a molecule or in its interior with equal ease (Table 3.1C). The side chains of these eight are moderately polar but are uncharged. Some of them —serine, tyrosine, threonine, and tryptophan—form hydrogen bonds whenever they lie in the interior of a protein molecule. Two cysteine side chains can lose hydrogen atoms so that their sulfur atoms are joined by a covalent bond in a **disulfide bridge** (Figure 3.14). Hydrogen bonds and disulfide bridges help deter-

3.15 Formation of a Peptide Linkage
Two amino acids combine to form a peptide linkage; a molecule of water is lost in the process. The atoms that become linked are shown in orange; the side chains on the amino acids are not shown but are designated by R, as is conventional. (In living things the reaction is substantially more complex, but the end result is as shown here.)

mine how a protein chain folds. The glycine side chain is just a hydrogen atom; thus glycines may fit into tight corners in the interior of a protein molecule, where a larger side chain could not fit.

Peptide Linkages

In the polymerization of amino acids, the carboxyl group of one amino acid reacts with the amino group of another, undergoing a condensation reaction and forming a **peptide linkage**. Figure 3.15 gives a simplified description of the reaction; actually, other molecules must activate the reactants, and there are intermediate steps. A linear polymer of amino acids connected by peptide linkages is a **polypeptide**. A protein is made up of one or more polypeptides. At one end of the polypeptide molecule there is a free amino group, and at the other end a free carboxyl group; the other amino and carboxyl groups are bound in peptide linkages. Thus there is a directionality to a protein; the dipeptide glycine alanine, in which glycine has the free amino group, differs from alanine–glycine, in which alanine has the free amino group.

In the peptide linkage, the C=O oxygen carries a slight negative charge, whereas the N—H hydrogen is slightly positive. This asymmetry of charge favors hydrogen bonding (Chapter 2) within the protein molecule itself and with other molecules, contributing to both the structure and the function of many proteins.

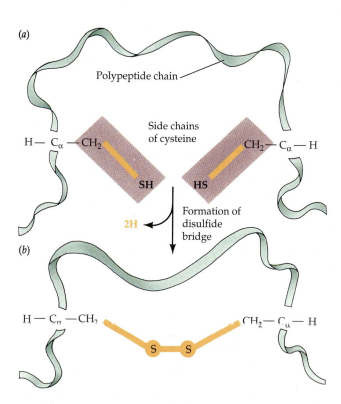

3.14 Formation of a Disulfide Bridge
The SH groups on two cysteine side chains (a) in a chain of amino acids can react to form a linkage between the two sulfur atoms (b). Such disulfide bridges are important in maintaining the proper three-dimensional shapes of protein molecules.

LEVELS OF PROTEIN STRUCTURE

Primary Structure

Protein structure is elegant and complex—so complex that it is described as consisting of several levels. A polypeptide is a linear polymer, or unbranched chain of amino acids (Figure 3.16). The precise sequence of these building blocks in a polypeptide is called its **primary structure** and is dictated by the precise sequence of the monomers (which are called nucleotides) in a linear segment of a DNA molecule. The elucidation of this relationship between DNA primary structure and protein primary structure was one of the triumphs of molecular biology, and it will be described in Chapter 11.

The theoretical number of possible different proteins is enormous. As there are 20 different amino acids, there are $20 \times 20 = 400$ distinct dipeptides and $20 \times 20 \times 20 = 8000$ different tripeptides. Imagine this process of multiplying by 20 extended to even a small protein made up of 100 monomers—there could be 20^{100} of these small proteins. Each of the

(a) Branched polymer

(b) Unbranched linear polymer

(c) Unbranched, linked linear polymers

3.16 Branched versus Linear Polymers
(a) Some biological polymers, such as the carbohydrate glycogen (Figure 3.12c) are highly branched, as depicted in the generalized molecule here. Proteins, however, are unbranched *(b)*, although the chains of amino acids may be linked together, as shown in *(c)*.

different possible proteins has its own distinctive primary structure.

The higher levels of protein structure—from local coiling and folding to the overall shape of the entire molecule—all derive from the primary structure. By presenting side chains of differing character (hydrophilic or hydrophobic, for example) in a specific and unique order, the precise sequence of amino acids in a given protein determines the ways in which the polypeptide chain can twist and fold. By twisting and folding, each specific protein adopts a specific structure that distinguishes it from every other protein.

Secondary Structure

Although the overall structure of each kind of protein is absolutely distinctive, above the primary level there are aspects of structure—such as regular coiling of the polypeptide chain—that may be shared by many proteins. Such shared features are referred to as **secondary structure**. One type of secondary structure, the **α helix** (alpha helix), is a right-handed coil "threaded" in the same direction as a standard wood screw. A twisting of the polypeptide chain about its axis often allows the formation of hydrogen bonds between amino acids four monomers apart along the chain, as shown in Figure 3.17. When this pattern of hydrogen bonding is established repeatedly over a segment of the protein, it stabilizes the twisted form, resulting in an α helix. The ability to form an α helix depends upon the primary structure: Certain amino acids have side chains that distort the coil or otherwise prevent the formation of hydrogen bonds. The α helix is a form of secondary structure that appears in many molecules.

It is particularly evident in a class of fibrous structural proteins called keratins. These include most of the protective tissues found in animals, such as fingernails and claws, skin, hair, and wool. Hair can be stretched because this requires breaking only hydrogen bonds in an α helix, rather than breaking covalent bonds; when the tension is released, both the helix and the hydrogen bonds reform. The contractile machinery of muscle is made of two protein components, actin and myosin (Chapter 38), with the myosin arranged in an α-helical secondary structure.

Silk is an example of a protein with another type of secondary structure, the **β-pleated sheet** (Figure 3.18). Here the protein chains are almost completely extended and are bound into sheets by hydrogen bonds connecting one chain to another. In many proteins, regions of β-pleated sheet are formed by bonding between different parts of the same polypeptide chain.

A third type of secondary structure, the triple helix, is found in collagen (Figure 3.19). This important protein, found in cartilage, tendons, the underlayers of the skin, and the cornea of the eye, consists of

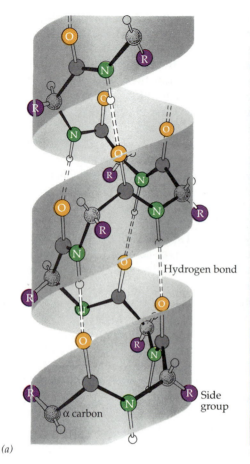

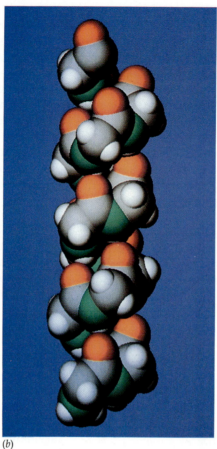

(a)

(b)

3.17   The α Helix

The α helix is an important form of secondary structure in many proteins. (a) A ball-and-stick drawing of an α helix. The right-handed coil of the helix can be seen by following the black bonds in the backbone of the polypeptide. The atoms in the relatively rigid plane of the peptide linkages between amino acid residues are in color, and the hydrogen bonds that stabilize the helix are drawn as dashed lines. (b) The actual positions of the atoms in relationship to one another are realistically rendered in this computer drawing of the α helix. Side chains of the amino acids have been omitted. Carbons are shown in gray, oxygens in red, hydrogens in white, and nitrogens in green. (See Figure 3.20c for the appearance of several lengths of an α helix in a complete protein molecule.)

three polypeptide chains twisted around one another like the strands of a cable. Hydrogen bonds connect the chains, resulting in a structure that is strong, rigid, and unstretchable. The tail of a rat, under the skin, is almost pure collagen.

Tertiary Structure

The α helices and β-pleated sheets sometimes predominate throughout the bulk of a protein molecule, as in the examples cited above. More frequently, however, these secondary structures are found in only limited portions of the molecule and thus do not dominate its overall shape, or **tertiary structure**. A complete description of the tertiary structure specifies the location of every atom in the molecule in

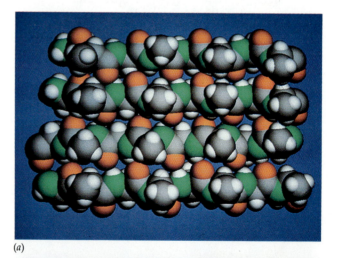

(a)

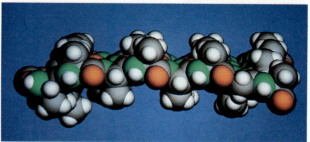

(b)

3.18   The β-Pleated Sheet

(a) This computer drawing shows a localized β-pleated sheet area of a protein. We see four parallel strands of the polypeptide running horizontally in the figure. These strands are joined by hydrogen bonds, forming a sheet. Atoms are colored by the conventions of Figure 3.17. (b) The same material shown in (a), but viewed from the bottom edge to emphasize the "pleats" in the sheet.

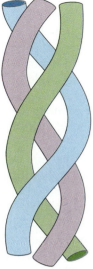

3.19 Collagen

(a) The secondary structure of the protein collagen consists of a triple helix of polypeptide chains in this pattern. Such a triple helix is called tropocollagen. (b) A number of triple helices of tropocollagen join in parallel fashion to create a strong, flexible collagen fibril; several collagen fibrils are shown here, magnified about 20,000 times. The spacing between black bands corresponds to the length of a single tropocollagen molecule.

(a) Secondary structure of collagen

(b) Collagen fibril

three-dimensional space, in relation to all the other atoms. The tertiary structure of the protein lysozyme is shown in various representations in Figure 3.20. Bear in mind that this tertiary structure and the secondary structure emphasized in Figure 3.20c derive entirely from the protein's primary structure. If lysozyme is heated carefully, causing the tertiary structure to break down, the protein will return to its normal tertiary structure when it is cooled. The only

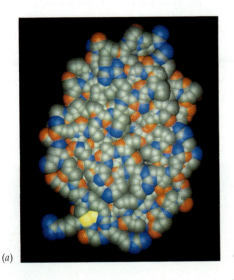

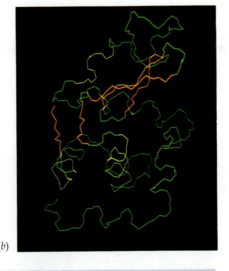

(a)

(b)

3.20 Four Representations of Lysozyme

The tertiary structure of the enzyme lysozyme is shown in four different representations that emphasize different aspects of the structure. All are similarly oriented, although (c) and (d) are slightly rotated relative to the others. (a) This computer-generated, space-filling representation gives the most realistic impression of lysozyme's tertiary structure, which is densely packed. (b) This computer-drawn representation emphasizes the backbone of the folded polypeptide. Regions in green have α-helical secondary structure; those in red constitute a β-pleated sheet. (c) The ribbon representation also emphasizes the secondary structure of certain parts of the molecule; the coiled regions are α-helices and the β-pleated sheet is indicated by the arrows. (d) The "sausage" representation gives a crude picture of the folding of the polypeptide chain. Note the position (shown in yellow) of the active site—the part of the enzyme molecule that binds reactant molecules. From its position in (d), you can infer the position of the active site in the other three representations.

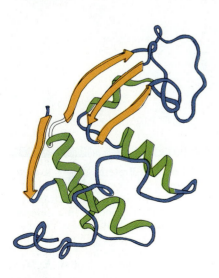

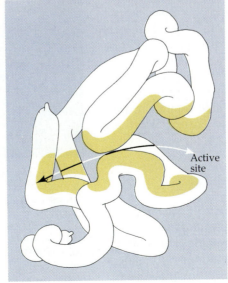

Active site

(c)

(d)

3.21 Tertiary Structure of a Protein
Tertiary structure—the exact three-dimensional folding of a protein molecule—is illustrated for the protein myoglobin. The individual atoms are not shown, nor are the individual amino acids, which form the coiled polypeptide chain. The chain is α helical throughout most of its length. The purple shading shows the overall tertiary configuration of the molecule. The red structure in the upper part of the drawing is an iron-containing heme group.

brings oxygen from the lungs to the tissues and delivers it to myoglobin for storage, illustrates quaternary structure (Figure 3.22). Hydrophobic interactions, hydrogen bonds, and ionic bonds hold four polypeptide chains together to make a hemoglobin molecule. There are two each of two kinds of polypeptides. As the molecule takes up or releases oxygen, its four subunits shift their relative positions slightly, changing the quaternary structure (the relative arrangement of the subunits). Ionic bonds are broken, exposing buried side chains that enhance the binding of molecular oxygen. Each subunit of hemoglobin is folded like a myoglobin molecule, suggesting that both hemoglobin and myoglobin are evolutionary descendants of the same oxygen-binding ancestral protein. But on the surfaces where hemoglobin's subunits come in contact with each other—regions that on myoglobin are exposed to aqueous surroundings—hemoglobin has hydrophobic side chains where myoglobin has hydrophilic ones. Again, the chemical nature of side chains on individual amino acids, as specified by DNA, determines how the molecule folds and packs in three dimensions. Proteins having more than one polypeptide chain function only after they form appropriate quaternary structures.

The four levels of protein structure are summarized in Figure 3.23.

information needed to specify the unique shape of the lysozyme molecule is the information contained in its primary structure.

Myoglobin is an important protein (Figure 3.21). Its function—to store oxygen in certain animal tissues—will be discussed in Chapter 39. Myoglobin has 153 amino acids in its single polypeptide chain; there are no disulfide bridges, and the molecule is unusual in that it consists almost entirely of α helices. (Most proteins do not show a distinct secondary structure over such a large fraction of the molecule.) The eight helices in myoglobin form a pocket that encloses a **heme group**: an iron-containing ring structure that binds O_2. Hydrophobic side chains on the inner sides of the helices help to ensure that the helices fold against one another correctly as the molecule is formed.

Quaternary Structure

Myoglobin and most other proteins are made from a single polypeptide chain. Some proteins, however, are made from two or more polypeptide chains, and the configuration of those two or more polypeptide chains in the protein molecule is the **quaternary structure** of the protein. Hemoglobin, a protein that

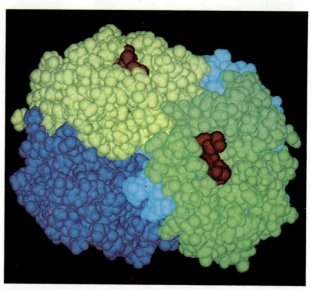

3.22 Quaternary Structure of a Protein
Hemoglobin consists of four folded polypeptide subunits that assemble themselves into the quaternary structure shown here. Each of the subunits is shown in a different color in this computer drawing. Note the heme groups, which are red. The quaternary assemblage includes subunits with tertiary folding, the subunits each contain α-helical regions of secondary structure, and all these levels of structural organization are ultimately dependent on the primary sequence of amino acids in the polypeptides.

NUCLEIC ACIDS

The proteins of today exist because of the structures and activities of various **nucleic acids**. One group of these, the **DNAs**, or **deoxyribonucleic acids**, are giant polymers that carry the instructions for making proteins; another, the **RNAs**, or **ribonucleic acids**, interpret and carry out the instructions coded in the DNAs.

Nucleic acids form from monomers called **nucleotides**, each of which consists of a simple pentose sugar, a phosphate group, and another portion called a nitrogenous (nitrogen-containing) base (Figure 3.24). Molecules consisting of a pentose sugar and a nitrogenous base, but no phosphate group, are called nucleo*sides*. In DNA, the sugar is deoxyribose, which differs from the ribose found in RNA by one oxygen atom (Figure 3.10). As shown in Figure 3.25, the "backbones" of both polymers consist of alternating sugars and phosphates; the bases, which are attached to the sugars, project from the chain. Most RNA molecules are single stranded: Each molecule consists

3.23 Four Levels of Protein Structure
The four levels of protein structure—primary, secondary, tertiary, and quaternary—are summarized here. Secondary, tertiary, and quaternary structure all arise from the primary structure of the protein.

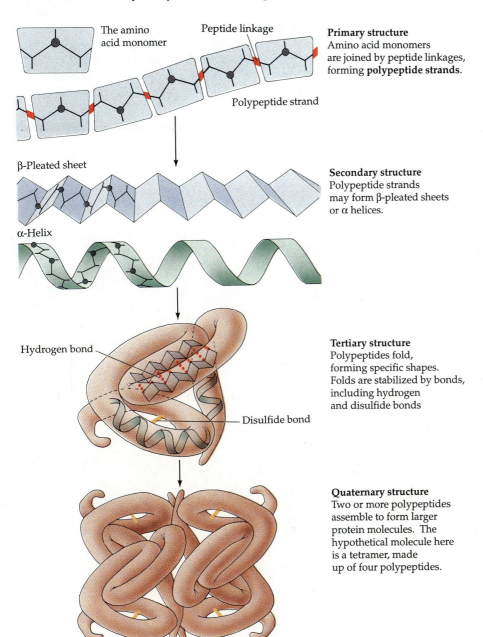

The amino acid monomer

Peptide linkage

Polypeptide strand

Primary structure
Amino acid monomers are joined by peptide linkages, forming **polypeptide strands**.

β-Pleated sheet

α-Helix

Secondary structure
Polypeptide strands may form β-pleated sheets or α helices.

Hydrogen bond

Disulfide bond

Tertiary structure
Polypeptides fold, forming specific shapes. Folds are stabilized by bonds, including hydrogen and disulfide bonds

Quaternary structure
Two or more polypeptides assemble to form larger protein molecules. The hypothetical molecule here is a tetramer, made up of four polypeptides.

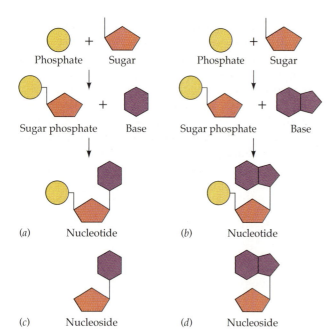

3.24 Components of a Nucleotide
A sugar and a phosphate form a sugar phosphate (either ribose phosphate or deoxyribose phosphate). In nucleotide synthesis, a nitrogen-containing base is then built on the sugar phosphate in several steps not depicted here, forming the complete nucleotide monomer. The four bases fall into two categories, as indicated here by their shapes in (a) and (b). A nucleoside consists of a sugar (not a sugar phosphate) and a base (c and d).

of one polynucleotide chain. DNA, however, is usually double stranded, with two polynucleotide chains being held together by hydrogen bonding between their nitrogenous bases.

Only four nitrogenous bases—and thus only four nucleotides—are found in DNA. The DNA bases are adenine, cytosine, guanine, and thymine. A key to understanding the structures and functions of nucleic acids is the principle of **complementary base pairing**: Particular bases only pair with certain other bases. In DNA, wherever one strand carries an adenine, the other must carry thymine at the corresponding point. Wherever one chain has a cytosine, the other has guanine. The base pairing rules for DNA and RNA are shown in Table 3.2. The pairing scheme maximizes hydrogen bonding between the two strands of DNA. Because one of the larger **purine** bases (adenine or guanine) always pairs with one of the smaller **pyrimidine** bases (thymine or cytosine), all base pairs are the same size. Complementary base pairing between the two strands of the DNA molecule makes it possible to copy DNA molecules very faithfully (Chapter 11). As we will see, base pairing

3.25 RNA versus DNA
A ribonucleic acid (RNA) is made up of a ribose sugar–phosphate backbone with a nitrogenous base attached to each sugar, as shown on the left. On the right is a portion of a double-stranded deoxyribonucleic acid (DNA); it consists of two deoxyribose sugar–phosphate backbones, with the bases attached to the sugars between the strands. The molecule is held together by hydrogen bonds between opposite bases on the two strands. For the bases, A = adenine, T = thymine, G = guanine, C = cytosine, and U = uracil. Note that RNA contains U where DNA contains T.

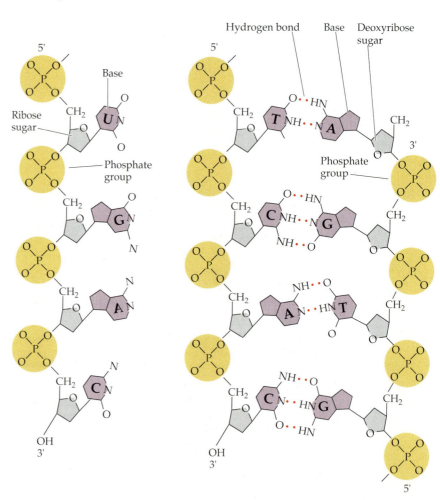

TABLE 3.2 Base Pairing Rules for the Nucleic Acids				

IN DNA:	WHEN RNA AND DNA INTERACT:		WHEN RNA PAIRS WITH RNA:
	RNA	DNA	
A pairs with T	A pairs with T		A pairs with U
T pairs with A	U pairs with A		U pairs with A
G pairs with C	G pairs with C		G pairs with C
C pairs with G	C pairs with G		C pairs with G

with which enzymes react calls for corresponding diversity in the structure of the enzymes themselves. DNAs are similar and uniform and are read by simple machinery; proteins are diverse and interact with a diversity of other compounds.

Throughout this chapter we have treated the classes of macromolecules as if each were completely separate from the others. In fact, certain macromolecules of different classes attach to one another to form covalently bonded products. Many proteins

is also important in understanding RNA functions, and it contributes in part to the structures of certain RNA molecules.

Ribonucleic acids are also composed of four different nucleotides, but the nucleotides differ from those of DNA. The **ribonucleotides** contain ribose rather than deoxyribose, and one of their four bases is different from that in DNA. The four principal bases in RNA are adenine, cytosine, guanine, and uracil (instead of thymine). Although RNA is generally single stranded, complementary associations between nucleotides are important in the formation of new RNA strands, in determining the shapes of some RNA molecules, and in associations between RNA molecules during protein synthesis. Guanine and cytosine pair as in DNA, and uracil pairs with adenine. Adenine in an RNA strand can pair with either uracil (in an RNA strand) or thymine (in a DNA strand).

We will say more about the structure of DNA in Chapter 11. For now, suffice it to say that the three-dimensional appearance of DNA—which corresponds to the tertiary structure of a protein—is strikingly regular. The segment shown in Figure 3.26 could be from virtually any DNA molecule. Through hydrogen bonding, the two complementary polynucleotide strands pair and twist to form a **double helix**. How regular this seems in comparison with the complex and varied structures of proteins! However, this structural difference makes sense in terms of the functions of these two classes of compounds. DNA, on the one hand, is a purely informational molecule. *The information carried by DNA resides simply in the sequence of bases carried in its chains.* This is, in a sense, like the tape of a tape recorder. This message must be read easily and reliably. A uniform molecule like DNA can be interpreted by standard molecular machinery that can read any molecule of DNA—just as a tape player can play any tape of the right size. Proteins, on the other hand, have good reason to differ so greatly. In particular, enzymes must each recognize their own specific "target" molecules. They do this by having a specific three-dimensional form that can match at least a portion of the surface of their targets. Structural diversity in the molecules

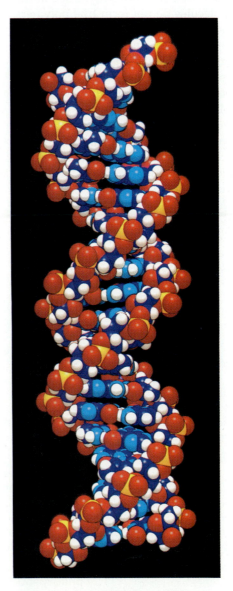

3.26 The Double Helix of DNA
The backbones of the two strands in a DNA molecule are coiled in a double helix. The double helix is demonstrated with a computer drawing showing the atoms in a length of 20 base pairs. Allow your eye to follow the yellow phosphorus atoms and their attached red oxygen atoms in the two helical backbones. The paired bases are stacked in the center of the coil and should become apparent if you concentrate on the light blue nitrogen atoms and the dark blue carbon atoms. The smaller white atoms are hydrogens.

have attached carbohydrates that play roles in the functioning of the whole molecule, and other proteins have attached lipids.

SUMMARY

Many important biological molecules are polymers: They are built from monomers. The monomers combine (polymerize) by undergoing condensations.

The characteristic structures of the lipids differ from the other molecular structures discussed in this chapter. Lipid molecules do not form true polymers. However, all lipid molecules have large hydrophobic regions. As a result, lipids present in any aqueous system tend to aggregate—like oil in water. Triglycerides may accumulate in animal cells, forming enormous fat droplets that may later function as fuel.

The phospholipids tend to aggregate to form a continuous bilayer, with their hydrophilic regions exposed to the surrounding aqueous environment. This arrangement accounts for the formation of cellular membranes. Other lipid molecules, such as cholesterol or the carotenoids, may be inserted into the phospholipid bilayer of the membrane, as may protein molecules with hydrophobic surfaces.

The monosaccharides have two principal functions: They serve as fuel, and as building blocks for polysaccharides—large polymers formed by conden-

sation reactions. The functions of polysaccharides are varied. Starch and glycogen are storage compounds in plants and animals, respectively. The individual molecules of cellulose join other cellulose molecules to form tight crystalline structures, binding the chains into a supramolecular aggregate with great tensile strength. Cellulose is the principal component of the plant cell wall.

Proteins and nucleic acids also form by condensation reactions. Protein molecules have particular tertiary structures as a result of the differing structures and properties of the side chains of the 20 amino acids that are their monomers. The three-dimensional conformation (tertiary structure) of a protein arises spontaneously, on the basis of its specific primary structure (its amino acid sequence). The differences in tertiary structure are the basis for the great specificity of such proteins as the enzymes and immunoglobulins. Local regularities such as α helices and β-pleated sheets are forms of secondary structure. Proteins consisting of more than one polypeptide chain have specific spatial arrangements of these subunits, called quaternary structure.

In contrast to the proteins, the even larger DNA molecules are composed of only four different monomers joined by complementary base pairing. DNA molecules have a very uniform structure, but an enormous information content is based on the sequence of monomers within the polymer.

SELF-QUIZ

1. All lipids:
   a. are triglycerides.
   b. are polar.
   c. are hydrophilic.
   d. are polymers.
   e. are more soluble in nonpolar solvents than in water.

2. Which of the following is *not* a lipid?
   a. A steroid
   b. A fat
   c. A triglyceride
   d. A biological membrane
   e. A carotenoid

3. All carbohydrates:
   a. are polymers.
   b. are simple sugars.
   c. consist of one or more simple sugars.
   d. are found in biological membranes.
   e. are more soluble in nonpolar solvents than in water.

4. Which of the following is *not* a carbohydrate?
   a. Glucose
   b. Starch
   c. Cellulose
   d. Hemoglobin
   e. Deoxyribose

5. All proteins:
   a. are enzymes.
   b. consist of one or more polypeptides.
   c. are amino acids.
   d. have quaternary structures.
   e. have prosthetic groups.

6. Which statement is *not* true of the primary structure of a protein?
   a. It may be branched.
   b. It is determined by the structure of the corresponding DNA.
   c. It is unique to that protein.
   d. It determines the tertiary structure of the protein.
   e. It is the sequence of amino acids in the protein.

7. The amino acid leucine (Table 3.1):
   a. is found in all proteins.
   b. cannot form peptide linkages.
   c. is likely to appear in that part of a membrane protein that lies within the phospholipid bilayer.
   d. is likely to appear in that part of a membrane protein that lies outside the phospholipid bilayer.
   e. is identical to the amino acid lysine.

8. The quaternary structure of a protein:
   a. consists of four subunits— hence the name *quaternary*.
   b. is unrelated to the function of the protein.
   c. may be α, β, or γ.
   d. depends on covalent bonding among the subunits.
   e. depends on the primary structures of the subunits.

9. All nucleic acids:
   a. are polymers of nucleotides.
   b. are polymers of amino acids.
   c. are double-stranded.
   d. are double-helical in structure.
   e. contain deoxyribose.

10. Which statement is *not* true of condensation reactions?
    a. Protein synthesis results from condensation reactions.
    b. Polysaccharide synthesis results from condensation reactions.
    c. Nucleic acid synthesis results from condensation reactions.
    d. Condensation reactions consume water as a reactant.
    e. Different condensation reactions produce the different kinds of macromolecules.

FOR STUDY

1. Phospholipids make up a major part of every biological membrane; cellulose is the major constituent of the cell walls of plants. How do the chemical structures and physical properties of phospholipids and cellulose relate to their functions in cells?

2. Suppose that, in a given protein, one lysine is replaced by aspartic acid (Table 3.1). Is this a change in primary or secondary structure? How might it result in a change in tertiary structure? In quaternary structure?

3. If there are 20 different amino acids commonly found in proteins, how many different dipeptides are there? How many different tripeptides? How many different polypeptides composed of 200 amino acid subunits? If there are four different nitrogenous bases commonly found in RNA, how many different dinucleotides are there? How many different trinucleotides? How many different single-stranded RNAs composed of 200 nucleotides?

4. Contrast the structures of hemoglobin, a DNA molecule, and a protein that spans a biological membrane.

READINGS

Armstrong, F. B. 1982. *Biochemistry*, 2nd Edition. Oxford University Press, New York. Perhaps the most approachable undergraduate text on biological molecules for students with diverse backgrounds.

Doolittle, R. F. 1985. "Proteins." *Scientific American*, October. A strikingly illustrated treatment of protein structure and evolution.

Stryer, L. 1988. *Biochemistry*, 3rd Edition. W. H. Freeman, New York. A relatively advanced but beautiful reference on the subjects of this chapter; outstanding illustrations, concise descriptions, clear prose.

Voet, D. and J. G. Voet. 1990. *Biochemistry*. John Wiley & Sons, New York. A fine advanced textbook, with outstanding illustrations.

4

Organization
of the Cell

PREVIEW: Bacterial cells are membrane-bounded units with little internal compartmentalization. Organisms in the other four kingdoms have more complex cells. Their cells contain membrane-bounded subsystems specialized for particular functions. These functions include the storage of hereditary information; the capture and processing of energy; digestion; nucleic acid and protein synthesis; the secretion of large molecules; the division of the nucleus and cell itself; and the protection of the cell from its own toxic substances. In addition, these intricate cells have complex molecules that aid in movement of the cell.

This chapter deals with the structure of bacterial and eukaryotic cells. These will be considered both as total systems and in terms of their components, including membranes, the nucleus and nuclear envelope, ribosomes, endoplasmic reticulum, Golgi apparatus, mitochondria, chloroplasts, lysosomes, vacuoles, microbodies, microtubules, microfilaments, and intermediate filaments.

The basic unit of organization in living things is the **cell**. All organisms are composed of cells, and all cells come from preexisting cells—these two statements constitute the **cell theory** (Chapter 1). Even viruses, which are not cells themselves, are entirely dependent on the presence and chemical machinery of cells for their reproduction. A cell may arise either by the division of one cell into two or by the fusion of two cells into one (Chapter 9). This situation raises an important question: Where did the first cells come from? This question is the main topic of Chapter 17.

Some cells are free-living organisms in their own right. Others are parts of a multicellular organism (Figure 4.1) In general, each cell in an organism is **totipotent**: it contains all the genetic information required to generate that entire organism. Plants can be used to demonstrate this totipotency of cells. Many plants reproduce by means of seeds, but can sometimes bypass this form of reproduction. With some species it is possible to take a cutting consisting of a bit of stem and a leaf or two, put it in soil and care for it, and end up with an entire plant. Going a step further, one may isolate a piece of tobacco pith, for example, place it in a container with suitable nutrients, and see it proliferate into a tumorlike mass of tissue (Figure 4.2). Treatment with natural substances that control plant growth will induce this mass to form roots and a leafy shoot and thus become an entire plant. But this is not all. From such masses of tobacco cells, one may isolate single cells, and even these may be induced to develop into intact, mature tobacco plants.

However, if we try to go to a level of structure below that of an entire cell, we come to the end of the line. Subcellular structures such as nuclei and chloroplasts may be isolated from cells and induced to carry out their normal functions; but they can never be induced to regenerate whole cells, let alone an entire plant. The inability of even the nucleus to produce any sort of life form demonstrates the essence of a whole cell as a basic unit of function and reproduction.

COMMON CHARACTERISTICS OF CELLS

Cells are tiny: Most have a volume of from 1 to 1,000 μm^3. The eggs of some birds are enormous exceptions, to be sure, and individual cells of several types of algae are large enough to be viewed with the unaided eye. Neurons (nerve cells) have volumes that fit within the "normal" range, but they often have fine projections that may extend for meters, carrying signals from one part of a large animal to another. In spite of these special cases, we may gen-

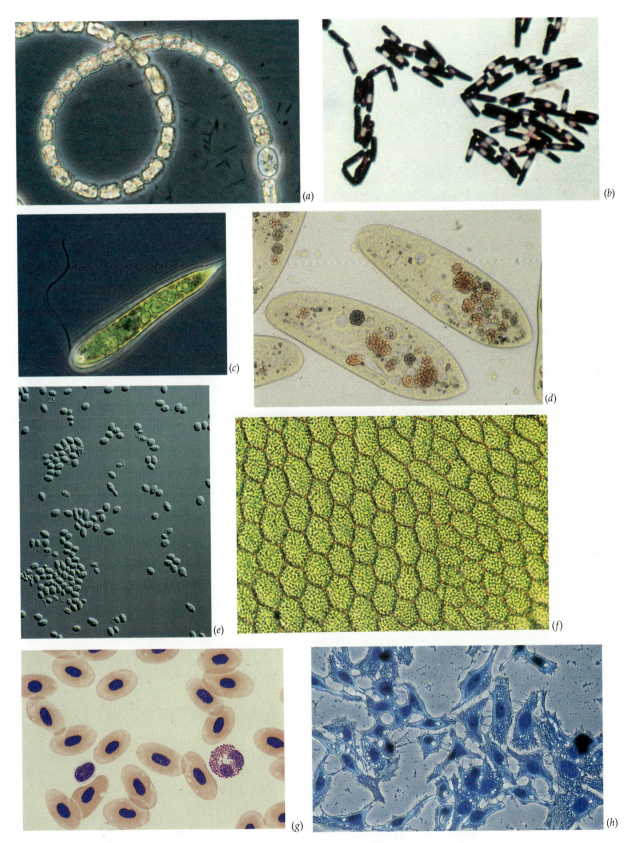

4.1 Cells

Cells come in many shapes and sizes. Here we see *(a)* a filamentous cyano-bacterium; *(b)* a common bacterium; *(c) Euglena*, a plantlike protist; *(d) Parame-cium*, an animallike protist; *(e)* brewer's yeast, a fungus; *(f)* leaf cells of a moss, packed with green, photosynthetic chloroplasts; *(g)* blood cells of a frog; and *(h)* mammalian cells grown in culture in the laboratory.

4.2 Totipotency
Many cells have complete libraries of hereditary information. Here shoots with leaves form from white, tumorlike masses of tobacco pith tissue. The pith cells are totipotent; they contain all the information needed to direct the development of all the cell types in the shoot.

eralize and say that cells are very small objects (Box 4.A). (There is a table of measurement units and their symbols inside the front cover of this book. You may want to refer to it several times as you read this chapter. What is the size range 1–1000 μm^3? If we assume that many cells are roughly spherical, this range of volumes corresponds with a range of diameters of about 1.2–12 μm.)

The cell's activities depend on specific component structures and the organization of these structures into a coordinated whole. A cell must do many things in order to survive: It must obtain energy from its environment; it must be selective as to what materials it allows to enter and leave; all cells must interpret and use the information contained in their DNA; and the chemical reactions essential for life must be kept

BOX 4.A

The Sizes of Things

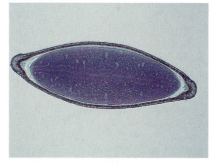 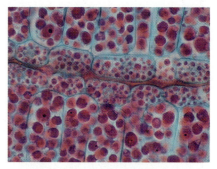

On the left, a pumpkin seed is shown magnified 10 times. When the same seed is magnified 1,000 times, we see the starch grains in the seed's cells as red balls.

Biologists study objects of very different sizes, ranging from molecules with diameters of less than 1 nm to organisms several meters in length. You need a sense of the sizes of things in order to appreciate their functions and the interactions of their parts. The sizes of several important

biological objects are compared in the diagram, which also indicates the methods by which they are usually viewed. To help you develop a sense of sizes we will sometimes identify

the magnifications of figures, as in the pictures above, taken with a light microscope. However, you will need to remember the approximate sizes of cells and their parts.

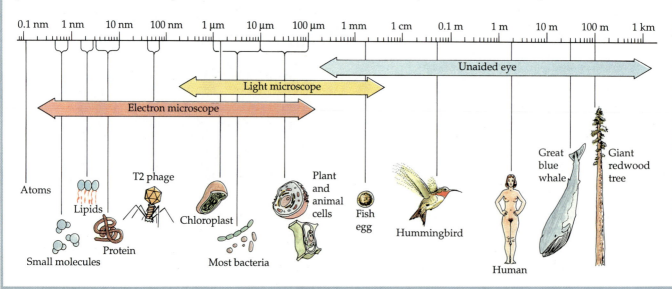

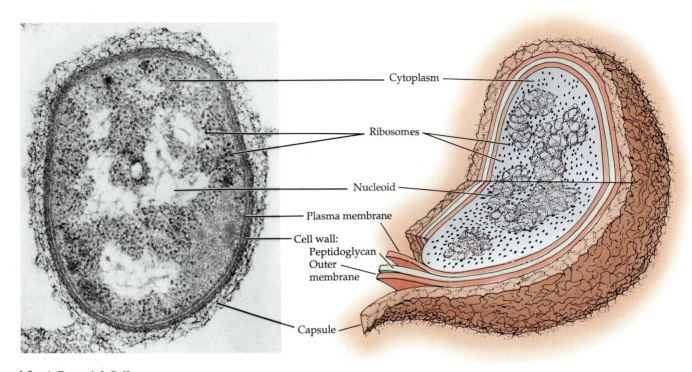

4.3   A Bacterial Cell
An electron micrograph of the bacterium *Pseudomonas aeruginosa*, magnified about 80,000 times. Compare this prokaryotic cell with the eukaryotic cells in Figures 4.9 and 4.10.

from interfering with one another. We will now consider the makeup of cells in terms of the structures and functions of their component parts.

Among all the kinds of cells there are two distinct general arrangements, with only a few intermediate forms in evidence. One general arrangement, usually the simpler, is the **prokaryotic cell**, characteristic of the kingdom Monera (the bacteria). Organisms in the Monera are often referred to as prokaryotes. Their single cells lack nuclear compartments and membrane-bounded internal compartments. The rest of the living world has **eukaryotic cells**—cells that contain true, membrane-bounded nuclei. Eukaryotic cells have other internal compartments that, like the nucleus, are surrounded by membranes. Organisms with this type of cell are known as eukaryotes. Both prokaryotes and eukaryotes have prospered through many hundreds of millions of years of evolution, and both are great "success stories."

PROKARYOTIC CELLS

Let us have a look at the prokaryotic cell (Figure 4.3), typical of the cells found in the Monera. Prokaryotic cells exhibit great variety of internal structure, and, as we will see in Chapter 21, they come in many sizes and shapes.

Features Shared by All Prokaryotic Cells

All prokaryotic cells have, without exception, three things. The first is a **plasma membrane** that separates the cell from its environment and regulates the inward and outward traffic of material. The second is a relatively clear area, as seen under the electron microscope, containing the hereditary material (DNA) of the cell. This region is called the **nucleoid**. Each cell has at least one nucleoid, and some cells may contain more than one. The remainder of the material within the cell is called the **cytoplasm**. At high magnification it is seen to be full of minute, roughly spherical structures called **ribosomes**—the third of the components found in all prokaryotic cells. Ribosomes are approximately 15–20 nm in diameter and consist of three molecules of RNA and about 50 different protein molecules. Their function is to coordinate the synthesis of proteins (Chapter 11). The remainder of the cytoplasm is a complicated solution containing many kinds of enzymes and the other chemical constituents of the cell.

Thus the basics of the prokaryotic cell are a plasma membrane, one or more nucleoids, and a cytoplasm containing thousands of ribosomes and the many molecules that take part in the chemical reactions of a living cell. Structurally, a prokaryotic cell is relatively simple, but functionally it is exceedingly complex. Enzymes in prokaryotic cells direct literally thousands of chemical reactions, with the cell's DNA serving as the molecular memory that allows successive generations of a given cell to be very much like one another.

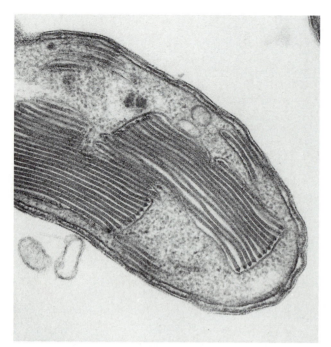

4.4 Photosynthetic Prokaryotes
Photosynthetic membranes fold into "stacks" inside a bacterial cell; such organized collections of internal membranes contradict the mistaken notion that bacteria are nothing more than tiny bags of molecules.

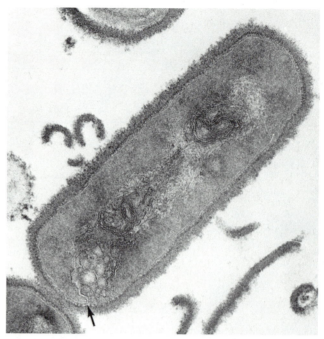

4.5 A Mesosome
Convoluted membranes of a large mesosome extend throughout this bacterial cell. At the lower left (arrow) you can see that the membrane of the mesosome is continuous with the plasma membrane.

Other Features of Prokaryotic Cells

Although the foregoing is a reasonable description of the simplest prokaryotic cells, most prokaryotes have at least a few more structural complexities. Most, for example, have a **cell wall** outside the plasma membrane (Figure 4.3). By its rigidity, the wall lends support to the cell and determines its shape. The cell wall of a bacterium consists mostly of a substance called **peptidoglycan**: a polymer of amino sugars, cross-linked to form a single molecule around the entire cell! Outside the cell wall there is often a layer of slime (composed mostly of polysaccharide), referred to as a **capsule**. The capsules of some bacteria may serve to protect them from attack by white blood cells within the bodies of animals they infect. The capsule provides protection against drying of the cell, and in some cases it may trap other cells for attack by the bacterium. Many prokaryotes produce no capsule at all, and even those that do have capsules will not die if they lose them.

The cyanobacteria (also called blue-greens; Figure 4.1a) and some other bacteria carry on photosynthesis, by which means they convert the energy of sunlight to chemical energy to produce food and to drive other energy-requiring reactions. In these photosynthetic prokaryotes, the plasma membrane folds into the cytoplasm—often very extensively—to form an internal membrane system containing chlorophyll and other compounds needed for photosynthesis

(Figure 4.4). Other bacteria possess different sorts of membranous structures called mesosomes, which may function in cell division or in various energy-releasing reactions (Figure 4.5). Both the photosynthetic membrane systems and the mesosomes are formed by infolding of the plasma membrane. They remain attached to the plasma membrane, and never form the free-floating, isolated, membranous organelles that are characteristic of eukaryotic cells. The fluid portion of a cell's cytoplasm, in which ribosomes and membranous structures are found, is called the **cytosol**.

Some prokaryotes are able to swim about by means of appendages called **flagella** (Figure 4.6a). A single flagellum, made of a protein called flagellin, looks something like a tiny corkscrew. It spins about its axis like a propeller, driving the bacterium along. Ring structures anchor the flagellum to the plasma membrane and, in some bacteria, to the outer membrane of the cell wall (Figure 4.6b). The fact that flagella actually cause the motion of the cell can be shown by removing the flagella; when this is done, the cells no longer move. If the tip of a flagellum becomes attached to an immovable object, the spinning of the flagellum causes the entire cell to rotate.

Along with flagella, **pili** project from the surface of some groups of bacteria (Figure 4.6c). Shorter than flagella, these threadlike structures seem to help bacteria adhere to one another during mating and also to animal cells.

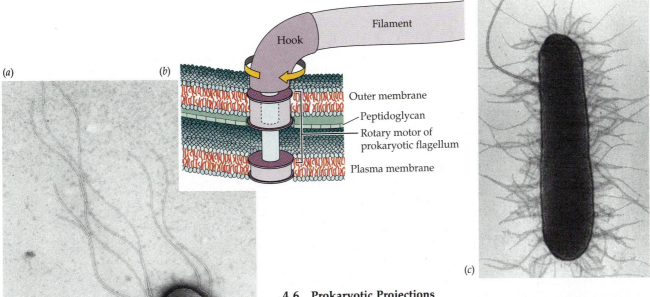

(a)

(b)

Filament

Hook

Outer membrane
Peptidoglycan
Rotary motor of
prokaryotic flagellum
Plasma membrane

(c)

4.6 Prokaryotic Projections

(a) Wavy, whiplike flagella used in locomotion extend from this bacterium. *(b)* The complexity of the basal ends of these tiny structures is shown in a diagram. The mechanism by which bacterial flagella rotate is still under investigation. *(c)* The small, hairlike pili bristling from the surface of this cell of *Escherichia coli* aid in adhesion to other cells.

Kinds of Prokaryotes

Bacteria are often categorized by the shapes of their cells. Spherical cells are **cocci**; rod-shaped cells are **bacilli**; and helical cells, sometimes coiled like corkscrews, are **spirilla** and **spirochetes** (Figure 4.7). Although prokaryotes are individual cells, many types are usually seen in chains, small clusters, or even colonies with hundreds of individuals.

Many prokaryotes have been used extensively in biological research. The most familiar bacterium of all is **Escherichia coli** (Box 4.B).

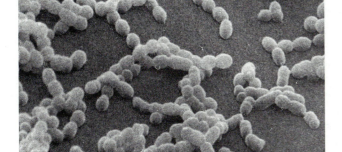

(a)

(b)

4.7 Bacterial Shapes and Growth Forms

(a) These spherical cocci of an acid-producing bacterium grow on tooth enamel and cause decay. *(b)* Rod-shaped bacteria. *(c)* These spirochetes (*Treponema pallidum*) cause syphilis in humans.

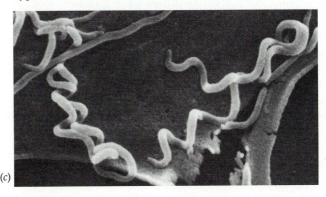

(c)

BOX 4.B

The Best-Known Prokaryote: *Escherichia coli*

Without a doubt the best understood of all living creatures is a humble bacterium living in our intestinal tract: *Escherichia coli*—or, as it is commonly known, *E. coli*. This little rod-shaped bacterium (shown in the figure) is about 2 μm in length and 0.8 μm in diameter, giving it a volume of about 1 μm^3 and a weight of approximately 10^{-12} g (one-millionth of one-millionth of a gram). Thus it is about 100 times larger than the smallest living cells, the mycoplasmas (Chapter 22). Within its tiny body, *E. coli* contains from one to four identical molecules of DNA and 15,000–30,000 ribosomes. Immediately outside the plasma membrane is a cell wall about 10 nm thick, and projecting from the cell are flagella

and pili. The flagella gather into a bundle and push the bacterium at a speed that if magnified to human dimensions would correspond to 30 miles per hour! Every second or so, the bundle of flagella separates and reforms, causing the cell to change its direction. An *E. coli* cell consists of approximately 70 percent water, 15 percent protein, 1 percent DNA, 6 percent RNA, 3 percent carbohydrate, 2 percent lipid, and 1 percent simple ions such as K$^+$ (potassium ions), as well as small amounts of other substances. The genetic material of *E. coli* consists of approximately 1/500 as much DNA as is contained in a single cell of a human being. Nonetheless, as relatively simple as it is, each prokaryotic cell of

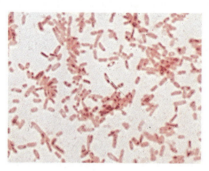

Cells of *E. coli* appear as red rods in this light micrograph of a stained preparation.

E. coli makes thousands of specific proteins.

Several features make this bacterium a favorable subject for biological experimentation. As noted, it is very small. Under the best conditions, it can divide once every 20 minutes, whereas most animal cells require about a day to go through a division cycle. Because of this rapid division, immense populations of *E. coli* can be grown very quickly. One cell can become 8 in an hour, 512 in 3 hours, over a billion in 10 hours, and more than 10^{21} in a day (in principle, and with unlimited food and space). Its nutritional requirements are simple: just water, some mineral ions, and an energy source such as glucose. Unlike some bacteria, most varieties of *E. coli* do not present a great health hazard and so can be grown without extensive precautions. Many genetic strains with different, known characteristics are readily available. As a result of these and other advantages, *E. coli* has been used in countless investigations of genetics, biochemistry, and other areas of biology. It is also used extensively in research on recombinant DNA ("genetic engineering"), a topic considered in detail in Chapter 14.

PROBING THE SUBCELLULAR WORLD: MICROSCOPY

Many significant advances in our knowledge of prokaryotic and eukaryotic cellular structure have depended upon the **resolution**, or **resolving power**, of the instruments available for magnifying tiny objects. We define the resolving power of a lens or microscope as the smallest distance separating two objects so that they can be seen as two distinct things rather than as a single one. For example, most humans can see two fine parallel lines as distinct markings if they are separated by at least 0.1 millimeter (0.1 mm); if they are drawn closer together, we see them as a single line. Thus, the resolving power of the human eye is about 0.1 mm, which is the approximate diameter of the human egg. To see anything smaller than this, we must use some form of microscope.

The **light microscope** (Figure 4.8*a*)—the kind of

microscope you are probably using in the laboratory —made the study of cells possible. In its contemporary form, the light microscope has a resolving power of about 200 nm (0.2 μm, 0.0002 mm), so it gives a useful view of cells and can reveal features of some of the subcellular organelles. Today, half a century after the invention of the electron microscope, the light microscope remains an important tool for the biologist. Many of the illustrations in this book are photographs taken through the light microscope; they are called photomicrographs. The light microscope has its limitations, however. The principal limitation is its 200-nm resolving power. This cannot be improved by adding more lenses, or by taking photomicrographs and then enlarging them. Such enlargements can be made, but they do not increase the **resolution**; as the images become larger, they simply become fuzzier.

Figures 4.3, 4.9, and 4.10 show many cellular

structures that are far too small to be resolved with the light microscope. Ribosomes, for example, being 20 nm or less in diameter, cannot be resolved as individual objects under the light microscope. On the other hand, ribosomes are readily resolved with the **electron microscope** (Figure 4.8*b*). An electron microscope uses powerful magnets as lenses to focus an electron beam, much as the light microscope employs glass lenses to focus a beam of light. The resulting images are called electron micrographs, many of which are seen in this and later chapters. The resolving power of modern electron microscopes is about 0.2 nm, but no biological specimen has yet been seen in such detail. One reason is that the energy of the electron beam at that power is so great that it destroys biological molecules before they can be seen. Because of this and other technical limitations, most electron micrographs resolve detail no finer than 2 nm, and even the best micrographs rarely resolve detail as fine as 1 nm. This corresponds to a resolving power about 100,000 times finer than that of the human eye.

There are two types of electron microscopy. In **transmission electron microscopy**, whose results can be seen in Figures 4.4 and 4.5, electrons pass *through* a sample. It is used to examine thin slices of objects —like the sections shaved off a material and placed on the slides used with a light microscope, but far thinner.

In **scanning electron microscopy**, electrons are directed at the surface of the sample, where they cause other electrons to be emitted; the scanning electron microscope focuses these secondary electrons on a viewing screen. Scanning electron microscopy reveals the *surface* structures of three-dimensional objects, such as the bacteria shown in Figures 4.7*a*, *b*, and *c*. Scanning electron micrographs are usually at a somewhat lower magnification than transmission electron micrographs, because scanning electron microscopy has a resolving power no better than 10 nm.

You might think that with such a resolving power the electron microscope would be used for all microscopic studies, but this is not so. For some applications it would be sheer overkill, like using a magnifying glass to get an overall view of an elephant. A more important limitation is that biological samples have to be killed and dehydrated before they can be examined with an electron microscope. Light microscopy allows us to observe *living* cells.

Samples for transmission electron microscopy have to be thinly sliced. Samples are also often sliced before light-microscope examination. To get a reasonable three-dimensional view of large cells or tissues with a microscope, one looks at many successive slices, rather like examining successive slices of Swiss cheese to "see" one of the holes.

THE EUKARYOTIC CELL

The vast majority of living species, including all animals, plants, fungi, and protists, have cells that are structurally more complex than those of the prokaryotes. A glance at an electron micrograph of almost any eukaryotic cell quickly reveals the most prominent differences (Figures 4.9 and 4.10). Eukaryotic cells are full of membranous structures of wondrous diversity. One or two membranes enclose many of the structures, which are distinct—externally and internally—and which carry on particular biochemical functions. These structures are neatly packaged subsystems, with membranes to control their functions and to regulate what gets in and out. Some of the subsystems are like little factories that make specific products. Others are like power plants that take energy in one form and convert it to a more useful form (Chapter 6). These membranous subsystems, as

(a)

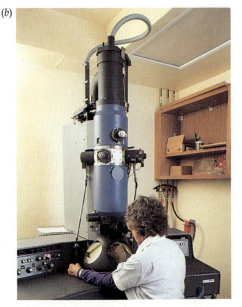

(b)

4.8 Microscopes
(a) A research-quality light microscope. A camera is mounted at the top of the instrument in order to make photomicrographs. (b) A transmission electron microscope. The magnets that focus the electron beam are in the tall cylinder.

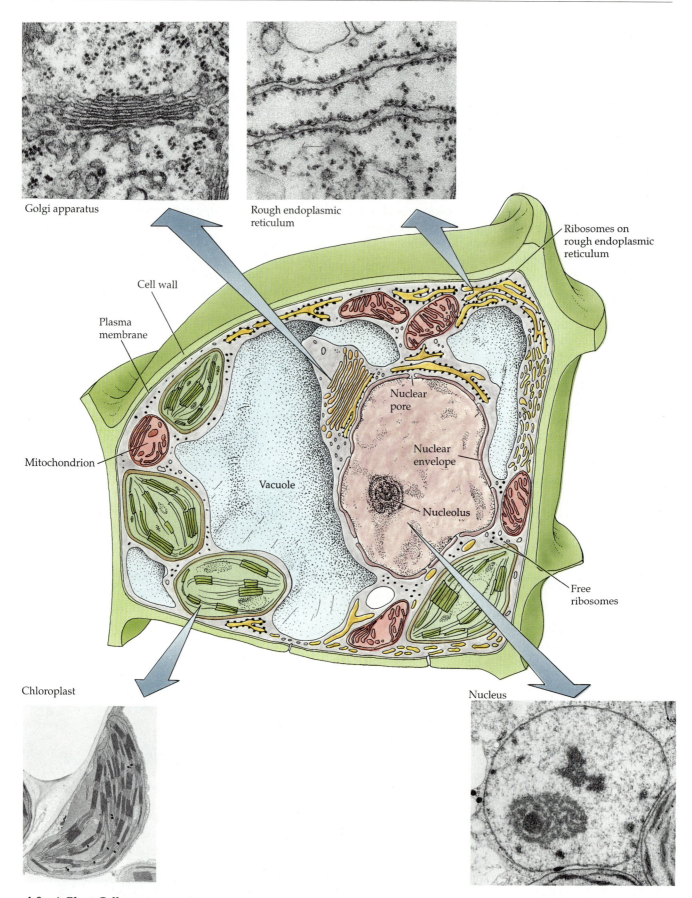

Golgi apparatus

Rough endoplasmic reticulum

Ribosomes on rough endoplasmic reticulum

Cell wall

Plasma membrane

Nuclear pore

Nuclear envelope

Mitochondrion

Vacuole

Nucleolus

Free ribosomes

Chloroplast

Nucleus

4.9 A Plant Cell
This drawing is based on an electron micrograph of a photosynthetic cell from a leaf. Several of the major structures are shown in detail in the electron micrographs.

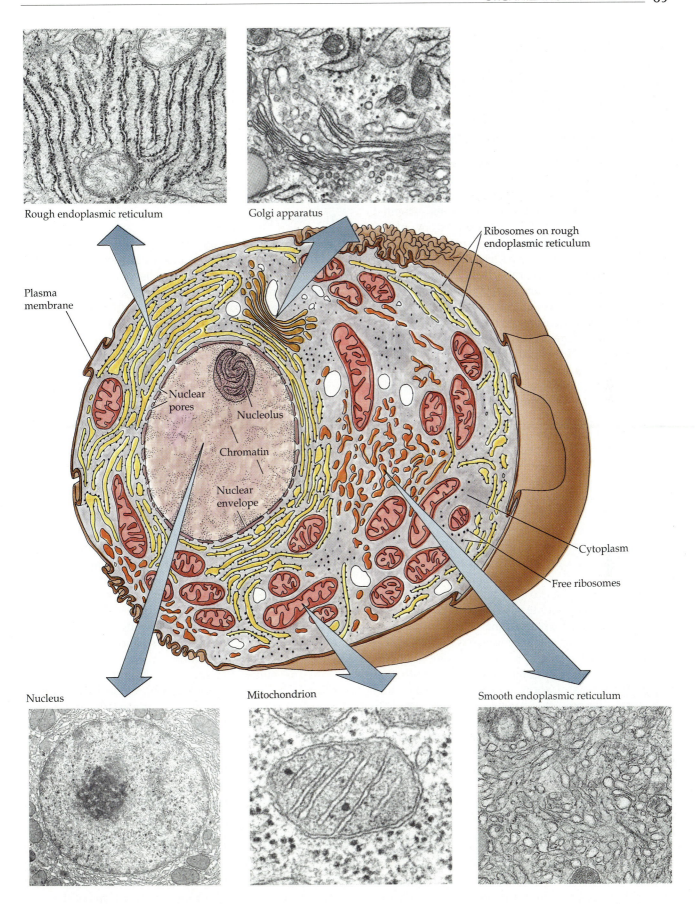

Rough endoplasmic reticulum

Golgi apparatus

Ribosomes on rough
endoplasmic reticulum

Plasma
membrane

Nuclear
pores

Nucleolus

Chromatin

Nuclear
envelope

Cytoplasm

Free ribosomes

Nucleus

Mitochondrion

Smooth endoplasmic reticulum

4.10   An Animal Cell
Drawing of a typical animal cell and details of individual structures.

well as other structures (such as ribosomes) that lack membranes but possess distinctive shapes and functions, are called **organelles**. Like prokaryotic cells, eukaryotic cells have a plasma membrane, cytoplasm, and ribosomes.

Roles of Membranes in Eukaryotic Cells

In 1952, the first people to look at reasonably clear electron micrographs of eukaryotic cells were stunned by the complexity of what they saw. Based on chemical and biological observations, scientists had expected to find cells surrounded by plasma membranes, even though these structures were not resolved with the light microscope. It was also known that cells teemed with organelles, and it was suspected that at least some of these organelles were bounded by membranes. However, it is doubtful that anyone expected membranes to be as profuse in the eukaryotic cell as they actually are. What are all those membranes for? How do they function? What is their structure—or structures, if all membranes are not alike? These questions will be dealt with in detail in Chapter 5, but a few of the most basic ideas are worth considering here.

Most generally, biological membranes regulate molecular traffic from one side of the membrane to the other. The hydrophobic interior of the membrane serves as a barrier to the passage of many materials, especially more polar materials that are readily soluble in water (see Figure 3.6). Other materials are transported through the membrane with the help of highly specific protein molecules. As discussed later in this chapter, the plasma membrane of many kinds of cells can infold and form compartments called vesicles in the cell so as to trap a portion of the cell's environment, as if taking a bite out of it.

Membranes participate in many activities besides transport. They serve as staging areas for interactions between cells. For example, immunologically active white blood cells recognize and interact with their targets by means of specific protein molecules built into their plasma membranes (Chapter 16). The proper development and organization of multicellular animals depends upon recognition between cells, which is mediated by the plasma membrane (Chapter 15). Many intracellular membranes carry the components responsible for energy transformations in cells. Chlorophyll and other substances necessary for energy-capturing photosynthesis are bound in a specific way to membranes in chloroplasts, one type of organelle. The electron carriers that help transform food energy into a form the cell can use are organized as parts of the inner membrane of mitochondria, another organelle type. In many respects, a discussion of eukaryotic cells is a discussion of membranes that are specialized for various cellular activities.

INFORMATION-PROCESSING ORGANELLES

Living things depend on a supply of accurate, appropriate information. Information is *stored* as the sequence of bases in DNA molecules. In eukaryotic cells, the bulk of the DNA resides in the nucleus. Information is *translated*, from the language of DNA into the language of proteins, on the surfaces of the ribosomes.

The Nucleus

The **nucleus** is typically the largest of the organelles (Figures 4.9 and 4.10). It has a diameter of approximately 5 μm in most animal cells. The possession of a membrane-bounded nucleus is the defining property of the eukaryotic cell. (Remember that in prokaryotes there is no membrane separating the nucleoid from the surrounding cytoplasm.) As viewed under the electron microscope, a eukaryotic nucleus is seen to be surrounded by *two* membranes separated by a few tens of nanometers. The **nuclear envelope**, as this pair of membranes is called, is punctuated by **nuclear pores** approximately 9 nm in diameter (Figure 4.11). Each pore is surrounded by eight large protein granules arranged in an octagon where the inner and outer membranes merge. Nuclear pores allow RNA and certain proteins to pass through the envelope to enter or leave the nucleus. The outer membrane of the nuclear envelope sometimes folds outward into the cytoplasm and is continuous with the network called the endoplasmic reticulum (discussed later in the chapter). The endoplasmic reticulum and, to a lesser extent, the outer surface of the outer membrane of the nuclear envelope often carry great numbers of ribosomes. There are no ribosomes on the inner surface of the outer membrane, or on either surface of the nuclear envelope's inner membrane.

In the eukaryotic nucleus, DNA combines with proteins to form a fibrous complex called **chromatin**. The structure of chromatin, with emphasis on the relationship between DNA and protein, will be considered in Chapter 9. Throughout most of the life cycle of the cell, the chromatin exists as exceedingly long, fine threads that are so tangled that they cannot be clearly seen with any microscope. However, when the nucleus is about to divide (that is, to undergo mitosis or meiosis; Chapter 9), the chromatin condenses and coils tightly to form a precise number of readily visible objects called **chromosomes** (Figure 4.12). The chromosomes are the bearers of hereditary instructions; their DNA carries the information required to carry out the synthetic functions of the cell and to endow the cell's descendants with the same instructions.

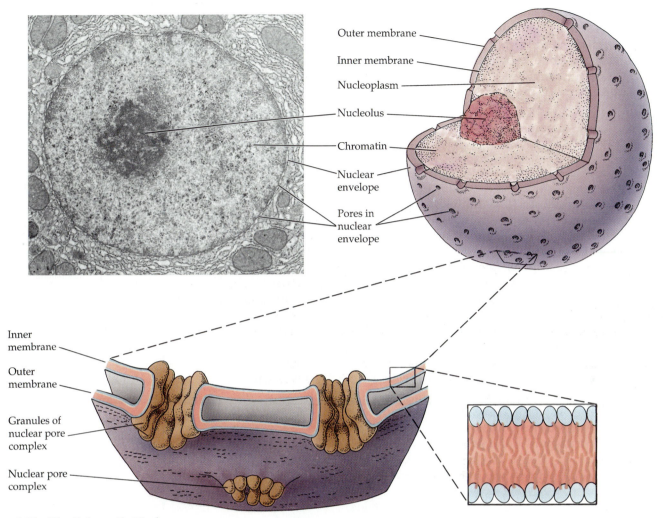

Outer membrane

Inner membrane

Nucleoplasm

Nucleolus

Chromatin

Nuclear envelope

Pores in nuclear envelope

Inner membrane

Outer membrane

Granules of nuclear pore complex

Nuclear pore complex

4.11 The Eukaryotic Nucleus

(Top) The nucleus of an animal cell, viewed by transmission electron microscopy. Notice the appearance of the double-membraned nuclear envelope, the nucleolus, and other common features of animal cell nuclei labeled in the drawing at the right of the micrograph. *(Bottom)* At the left is a diagrammatic representation of nuclear pore complexes in the nuclear envelope. Each complex has a diameter of about 10 nm; eight protein granules surround each pore, leaving a space through which water-soluble molecules can pass between the nucleus and the cytoplasm. The callout on the right shows that each membrane of the nuclear envelope consists of a phospholipid bilayer.

Visible in the nuclei of eukaryotes during most of the nuclear cycle are dense, roughly spherical bodies called **nucleoli** (Figure 4.11). Taken together, the nucleoli contain from 10 to 20 percent of all the RNA in a cell. Ribosomes are assembled in the nucleolus. RNA and protein molecules move into the nucleus and then into the nucleoli. They then combine to form the cell's ribosomes. The ribosomes then move out of the nucleus. Each nucleus must have at least one nucleolus, and those of some species have several. The exact number of nucleoli in its cells is characteristic of a species.

The fluid in which the chromosomes and nucleoli float is called **nucleoplasm**. This fluid is a suspension

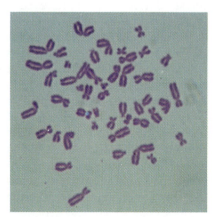

4.12 Chromosomes

The chromosome complement in a normal human cell; you should be able to distinguish 46 chromosomes. At certain times the chromatin condenses and stained chromosomes become readily visible under the light microscope, as shown here.

of various particles, fibers, proteins, and other compounds. (Whereas solid particles dissolve in a solution, they disperse but remain solid in a suspension.) Recall that the fluid portion of the cytoplasm, in which the various organelles, particles, and fibers are suspended, is called the cytosol.

Nucleus and Cytoplasm

The great majority of the cell's DNA resides in the nucleus. The DNA is a repository for the information needed to make most of the macromolecules of the cell. These macromolecules contribute to the activities of the nucleus and of the cytoplasm.

The relationship between the nucleus and the cytoplasm of eukaryotes is well illustrated by the results of experiments performed with giant single-celled algae of the genus *Acetabularia*. Cells of *Acetabularia* reach lengths of a few centimeters and can readily be picked up and handled. The single cell of this organism is just large enough to be easy to use for dissecting and grafting experiments. The cells consist of three principal regions: the cap, the stalk, and the rhizoids (Figure 4.13). The rhizoids anchor the little organism in its watery environment. The nucleus is within the rhizoid region throughout most of the life of the cell. Rhizoids have no ribosomes, and most of the cell's cytoplasm is in the stalk.

If the cap is cut off a cell of *Acetabularia* with a sharp razor blade, a new cap forms over a period of several days, synthesized by the cytoplasm in the stalk. The cell must synthesize many specific proteins and lipids to regenerate its missing cap. If this new cap is removed, still another is formed, and so forth. Grafting experiments can be done with *Acetabularia* cells of two different species to show the origin of the information for the cap structures (Figure 4.13). An *Acetabularia mediterranea* cap looks like an umbrella, whereas an *Acetabularia crenulata* cap looks more like a bunch of bananas. If we cut up the two *Acetabularia* and graft together the rhizoids from *A. mediterranea* and the stalk from *A. crenulata*, a new cap will be formed from the cytoplasm of the *A. crenulata* stalk. Which type will it be? The first new cap to be formed is of intermediate appearance. If we cut off this cap and wait for another to appear, the next cap looks like a typical *A. mediterranea* cap, even though it is made from the cytoplasm of the *A. crenulata* stalk. By more refined experiments, we can show that it is the nucleus—which happens to lie in the rhizoids in both species—that provides the in-

4.13 Domination by the Nucleus
Grafting experiments with the "giant" protist *Acetabularia* pointed to the regulatory activity of the nucleus. The nucleus-containing rhizoids determined the type of cap produced by regenerating *Acetabularia*.

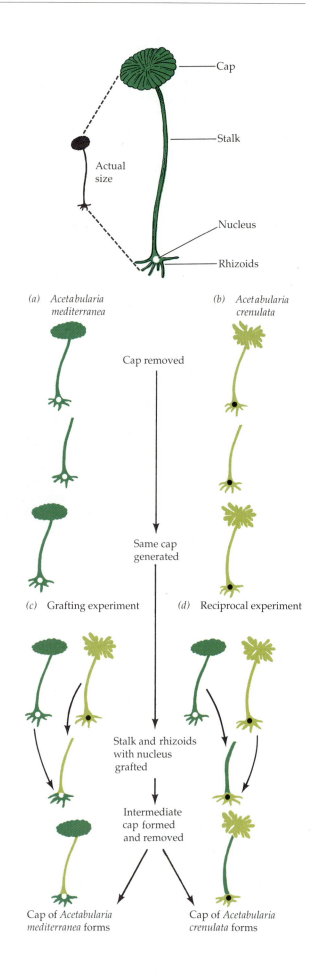

structions for how a new cap is to be made. So the nucleus is the storehouse of information for the cell, even when the cell has been put together by grafting.

Can we be sure that we have interpreted the grafting experiment correctly? Perhaps *A. mediterranea* just happens to be somehow "dominant" over *A. crenulata*. We can test this possibility by turning the experiment around—by doing a reciprocal experiment. To do that, we combine *A. crenulata* rhizoids with an *A. mediterranea* stalk. Again, a cap of intermediate form is made first and is cut away; all subsequent caps are now of the *A. crenulata* type. Now we are more confident of our conclusion: The nucleus controls what the cytoplasm builds.

Ribosomes

In both eukaryotic and prokaryotic cells, ribosomes fill the need for a site where a crucial cellular activity—protein synthesis—can take place. Ribosomes reside in three places in eukaryotic cells: free in the cytoplasm; attached to the surface of endoplasmic reticulum, as will be described later in this chapter; and in the energy-processing organelles discussed in the next section. In each of these places the ribosomes provide the site where proteins are synthesized under the direction of nucleic acids (Chapter 11).

The ribosomes of prokaryotes and of eukaryotes are similar in that both consist of two different-sized subunits. However, eukaryotic ribosomes are somewhat larger. The structure of the prokaryotic ribosome is better understood, and it is illustrated in Figure 4.14. Chemically, ribosomes consist of a kind of RNA to and around which are bound more than 50 kinds of different protein molecules. The ribosome temporarily binds two other kinds of RNA molecules (one large and one small) as it translates hereditary

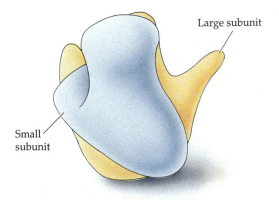

4.14 Ribosome Structure
A ribosome consists of a large subunit and a small subunit; these subunits come together only when participating in protein synthesis. Ribosomes do not have membranes.

information into the structures of the cell's structural and regulatory proteins.

ENERGY-PROCESSING ORGANELLES

In addition to information, cells require energy. Eukaryotic cells have organelles for obtaining energy from food molecules, and some cells of plants and of some protists have organelles in which the energy of sunlight is captured. In prokaryotic cells energy is transformed on the plasma membrane, on membrane infoldings, and in the cytosol.

Mitochondria

Utilization of "fuels" for the eukaryotic cell begins in the cytosol, where chemicals from food substances become molecules that are then taken up by organelles called **mitochondria** (singular: mitochondrion). Mitochondria function primarily to capture the energy from food substances in a form the cell can use. In mitochondria, energy-rich substances from the cytosol are oxidized—that is, electrons are removed from them, as explained in Chapter 7. Some of the energy available from these electrons is used to make a substance—ATP—that stores the energy in two special chemical bonds. The stored energy may be used either immediately or later to perform various kinds of work for the cell. The utilization of food in the mitochondria, with the associated formation of ATP, is called cellular respiration.

Typical mitochondria are small: somewhat less than 1.5 μm in diameter and approximately 2–8 μm in length, which means they are about the size of many bacteria. Mitochondria are visible with a light microscope, but virtually nothing was known of their structure until they were examined with the electron microscope. Electron micrographs show that they have an outer membrane that is smooth and unfolded. Immediately inside this is an inner membrane that folds inward at many points, giving it a much greater surface area than that of the outer membrane (Figure 4.15). In animal cells these folds tend to be quite regular, giving rise to shelflike structures called **cristae**. The mitochondria of plants also have cristae, but plant cristae tend to be much less regular in size and structure, and their inner membranes form both shelves and tubes. Special techniques and very high magnification have been used to show that the inner mitochondrial membrane contains large protein structures now known to participate in cellular respiration (Chapter 7). The region enclosed by the inner membrane is referred to as the **mitochondrial matrix**. Within the matrix one finds some ribosomes and DNA that serve to make some of the proteins needed for the synthesis of mitochondria.

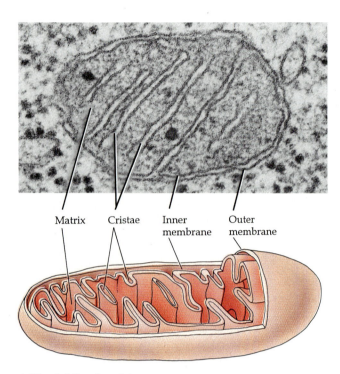

Matrix Cristae Inner membrane Outer membrane

4.15 A Mitochondrion
A mitochondrion as seen by electron microscopy *(top)*.
The drawing shows the mitochondrion's surface cut away
to expose internal structures.

Mitochondria are found in virtually all eukaryotes,
except for a few microscopic forms that live in environments without oxygen. The number of mitochondria in a cell ranges from one contorted giant in some
unicellular protists to a few hundred thousand in
large egg cells. An average human liver cell contains
more than a thousand mitochondria. In all cases, the
primary function of mitochondria is cellular respiration, by which usable energy is derived from food
materials and stored in ATP. The ATP is exported
into the cytosol, where most of it is used; some also
goes into the nucleus. Those cells that require the
most chemical energy tend to have more mitochondria per unit volume. In Chapter 7 we will see how
different parts of the mitochondrion work together
in the respiratory process.

Plastids

Although nuclei, ribosomes, and mitochondria are
found in essentially all eukaryotic cells, one class of
organelles—the **plastids**—is produced only in plants
and certain protists. The most familiar of the plastids
is the **chloroplast**, which is the site of photosynthesis
and contains all the chlorophyll in the plant or protist
cell (Figure 4.16).
Photosynthesis (Chapter 8) is the process by which
light energy is converted into the energy of chemical

bonds; the molecules formed in photosynthesis provide food for the plant itself and for other organisms.
The chloroplast has a number of other metabolic
functions besides photosynthesis. For example, it
plays an important part in making nitrogen available
to the rest of the plant.
Like the mitochondrion, the chloroplast is surrounded by two membranes. However, both membranes are unfolded and surround the organelle as a
smooth, closely fitting, double layer. Arising from
the inner membrane is a series of discrete internal
membranes, whose structure and arrangement vary
from one group of photosynthetic organisms to another. As an introduction, we concentrate on the
chloroplasts of the flowering plants. Even these show
some variation, but the pattern shown in Figure 4.17
is reasonably typical.
The most characteristic feature observed in electron micrographs of chloroplasts is a number of structures that look like stacks of pancakes. These stacks
are called **grana** (singular: granum), and they consist
of a series of flat, closely packed, circular sacs called
thylakoids. Each thylakoid is a single membrane
composed of the usual membrane components
(phospholipids and proteins) to which have been
added chlorophyll and other substances needed for
photosynthetic energy trapping and food production.
All the cell's chlorophyll is contained in the thylakoid
membranes. Thylakoids of one granum may be connected, as shown in Figure 4.17, to those of other
grana, making the interior of the chloroplast a highly
developed network of membranes. The fluid in
which the grana are suspended is referred to as
stroma. Like the mitochondrial matrix, the chloroplast stroma contains ribosomes and DNA. These
ribosomes and this DNA provide some—but only
some—of the proteins of which the chloroplast is
made.

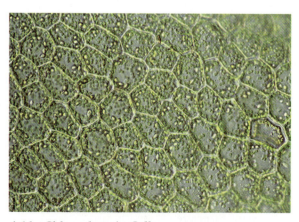

4.16 Chloroplasts in Cells
Each cell in this liverwort "leaf" contains dozens of
chloroplasts, the green organelles that carry on
photosynthesis.

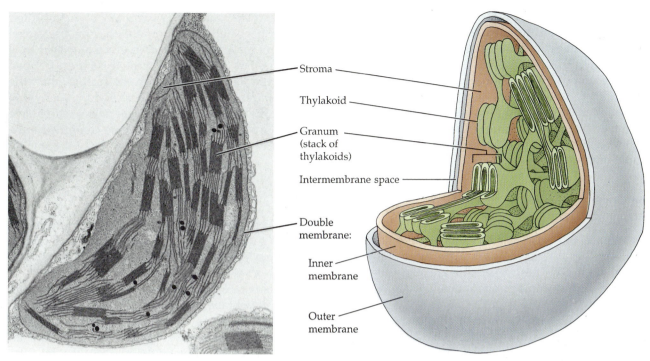

Stroma

Thylakoid

Granum
(stack of
thylakoids)

Intermembrane space

Double
membrane:

Inner
membrane

Outer
membrane

4.17 The Chloroplast

The electron micrograph on the left shows a section of a chloroplast from a leaf of corn. Note the stacks of membranes, called grana, and the membranous connections between the grana, forming an extensive network of photosynthetic membranes in the organelle. Only a thin layer of cytoplasm surrounds the chloroplast in this mature cell. The drawing on the right emphasizes the relationships among the parts of a chloroplast.

The chloroplasts give the leaves their green color. If one makes a thin slice of a leaf and looks at it under the microscope, one quickly discovers that most of the leaf is quite colorless—the only green to be seen is contained in the numerous chloroplasts within its cells. Not all plant cells contain chloroplasts. Most roots, for example, are colorless (or at least not green). This is just as well, because it would be a waste of energy and materials for the plant to provide cells that reside in the dark with chloroplasts, for photosynthesis requires light.

Although animal cells cannot *produce* their own chloroplasts, some animal cells do *contain* functional chloroplasts. These organelles are taken up either as free chloroplasts derived from plants eaten as food, or as bound chloroplasts contained within unicellular algae actually living within the animal tissues. The green color of corals and certain common sea anemones (Figure 4.18) results from chloroplasts in algae that live within the animals. The animals derive some of their nutrition from photosynthesis by these "guest" chloroplasts.

Chloroplasts are not the only kinds of plastids found in plants. The red color of a ripe tomato results from the presence of legions of plastids called **chromoplasts**. Just as chloroplasts derive their color from chlorophyll, chromoplasts are red, orange, or yellow because of the pigments (called carotenoids; Chapter 3) that they contain. The chromoplasts have no known chemical function in the cell, but it is likely that the colors they give to some petals and fruits

4.18 Anemone–Alga Symbiosis

This giant sea anemone owes its green color to the chloroplasts in a unicellular alga that lives and carries on photosynthesis within the the tissues of the anemone. (There is also a green pigment, not related to chlorophyll, in the outer cell layer of the anemone; but it contributes little to the intense color of the animal in this photograph.)

help to attract animals that assist in pollination or seed dispersal. On the other hand, there is no apparent advantage in a carrot root being colored orange. Other types of plastids called **leucoplasts** serve as storage depots for starch and fats. All plastid types are related to one another. Chromoplasts, for example, are formed from chloroplasts by a loss of chlorophyll and some change in internal structure. All plastids develop from **proplastids**, which are very simple in structure.

The Origins of Plastids, Mitochondria, and the Eukaryotes

In the past, biologists tried to grow chloroplasts or mitochondria in culture, outside the cells that they normally inhabit. These organelles are about the size of bacteria, they contain ribosomes and DNA, and they divide within the cell—might they not be treated like little cells in their own right? Although all such efforts at organelle culture failed because the organelles depend on the cell's nucleus and cytoplasm for some parts, the experiments helped nurture thoughts about another important question: How did the eukaryotic cell with its organelles arise in the first place? As we have seen, prokaryotic cells are generally much simpler in structure than eukaryotic cells because prokaryotes lack membrane-bounded organelles. Prokaryotic fossils can be found in sediments well over 3 billion years old, whereas the earliest known eukaryotic fossils date back to only 1.4 billion years ago. It is generally agreed that eukaryotes evolved from prokaryotes. But how?

One suggestion that has alternately been popular and scorned, over and over again, for many years is the **endosymbiotic theory** of the origin of mitochondria and chloroplasts. An important current champion of and contributor to this theory is Lynn Margulis of the University of Massachusetts, Amherst, who proposed the following idea. Picture a time, well over a billion years ago, when only prokaryotes inhabited Earth. Some of them got their food by absorbing it directly from the environment, others were

photosynthetic, and still others fed by eating their prokaryotic neighbors. Suppose that an occasional small, photosynthetic prokaryote was ingested by a larger one but did not get digested, so it sat trapped within the larger cell. Suppose further that the smaller prokaryote survived there and that it divided at about the same rate as the larger one, so successive generations of the larger prokaryote continued to be inhabited (or infected) by the offspring of the smaller one. We would call this endosymbiosis: "living within" another cell or organism, as certain algae live within sea anemones (Figure 4.18).

Could the little green prokaryote "eaten" by the larger prokaryote have been the first "chloroplast"? A present-day chloroplast is surrounded by a double membrane. Such a structure might have arisen when, in the process of engulfing the photosynthetic cell, the membrane of the larger cell stretched around the plasma membrane of the smaller cell, giving it in effect a double membrane (Figure 4.19). Recall, too, that chloroplasts contain ribosomes and DNA. That also fits the endosymbiotic theory, because the chloroplast is proposed to have arisen from an engulfed prokaryote. Another supporting point concerns the sizes of ribosomes. Ribosomes in the eukaryotic cytosol are larger than those of prokaryotes, whereas ribosomes in the chloroplast are similar in size to those of prokaryotes. Similar arguments can be made for the proposition that mitochondria represent the descendants of respiring prokaryotes engulfed by, and ultimately endosymbiotic with, larger prokaryotes. Also, there are striking similarities between some functions of bacterial plasma membranes and mitochondrial inner membranes, and also between the primary structures of certain bacterial and mitochondrial enzymes (Chapter 3). Finally, a few modern cells do contain other, smaller cells as endosymbionts, suggesting that the endosymbiotic theory is a plausible one.

A spectacular example of endosymbiosis is found in the guts of certain Australian termites. We think of a termite as digesting wood; yet, strictly speaking, it cannot. Much of the digestive chemistry is accom-

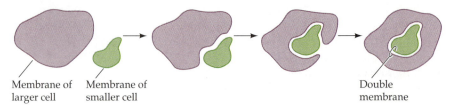

Membrane of larger cell   Membrane of smaller cell

Double membrane

4.19   Creation of a Double Membrane
Double membranes, such as those surrounding chloroplasts and mitochondria, might have been created when a larger cell (purple) engulfed a smaller cell (green) but did not digest the smaller cell. Thus the double membrane would consist of the portion of the larger cell's plasma membrane that enclosed the smaller cell and the plasma membrane of the smaller cell.

plished by an endosymbiotic protist, *Mixotricha paradoxa*, that lives in the termite's gut. But that is far from the whole story. *Mixotricha* itself harbors an amazing colony of endosymbionts. It swims around within the termite gut, apparently propelled by multitudes of flagella. Closer examination shows that although there are a few true flagella in a tuft at one end of the organism, the hundreds of others are not flagella at all—they are long, motile bacteria (spirochetes) that are attached at regular intervals to the plasma membrane of the *Mixotricha* cell and that beat just like real flagella. Also covering the surface of the protist, organized in a precise pattern, are other, smaller bacteria. *And* inside the *Mixotricha* are numerous bacteria of a third species, which are thought to help with the digestion of the tiny wood particles ingested by the protist—which, in turn, obtains them from the gut of the termite that is carrying this strange menagerie around inside itself. Perhaps termites are not all that special. If we carried colonies of *Mixotricha* in our digestive tracts, we could eat wood, too. We *do* carry colonies of *Escherichia coli* that aid in the digestion of our food, as discussed in Chapter 41.

It must be emphasized that mitochondria and chloroplasts are not enough to make a prokaryote into a eukaryote. We must still account for the origin of the nuclear envelope, as well as for other important structures, including those responsible for nuclear division. Thus far, the endosymbiotic theory is incomplete, although suggestions have been made for its extension to deal with the origin of other eukaryotic organelles. Is the endosymbiotic theory true? Certainly it has not yet been proved. A number of compelling objections to the theory have been raised, among them the fact that the DNA responsible for the synthesis of most of the enzymes in chloroplasts and mitochondria resides in the nucleus. However the matter may ultimately be resolved, the endosymbiotic theory is a good example of creative biological thinking; it gives us a useful perspective on the structures, functions, and origins of the mitochondria and chloroplasts.

THE ENDOMEMBRANE SYSTEM

Much of the volume of a eukaryotic cell is taken up by extensive membrane systems that play numerous roles in the life of the cell. These membrane systems are closely interrelated, and they arise from one another. They are referred to collectively as the **endomembrane system**.

Endoplasmic Reticulum

Running here and there throughout the cytoplasm, branching and rejoining, is a network of tubes and flattened sacs called the **endoplasmic reticulum**, or ER. As we have noted, electron micrographs often show this membrane system to be continuous with the outer membrane of the nuclear envelope. Parts of the ER are liberally sprinkled with ribosomes, which are attached to the outer faces of the flattened sacs. Because of their appearance in the electron microscope (Figure 4.20), these regions are called rough ER. The attached ribosomes are sites for the synthesis of proteins that function outside the cytosol, that is, proteins that are to be exported from the cell, incorporated into membranes, or moved into organelles of the endomembrane system. Proteins that are to remain within the cytosol or to move into mitochondria and chloroplasts are synthesized on "free" ribosomes, that is, ones that are not attached to the ER.

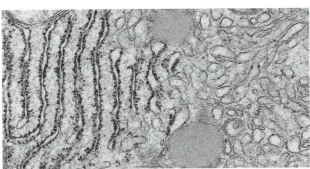

Ribosomes   Membranes

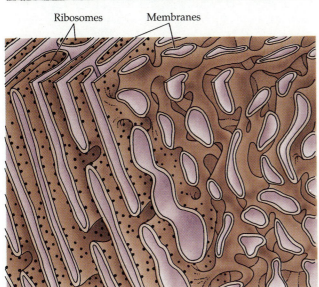

4.20 Endoplasmic Reticulum
The photograph and drawing show rough endoplasmic reticulum on the left and smooth endoplasmic reticulum on the right. Note the absence of ribosomes on the smooth ER.

Other parts of the endoplasmic reticulum lack ribosomes, and are referred to as smooth ER. Smooth ER acts to modify proteins synthesized by the rough ER; both rough ER and smooth ER are shown in Figure 4.20. Another major role of the smooth ER, at least in the livers of vertebrates, is the breakdown of barbiturates and some other drugs. A number of important enzymes are associated with the membranes of the ER. Some of the steps in lipid synthesis are carried out by ER enzymes, as is at least one step in sex hormone synthesis.

As you might guess from its function, the amount of ER in a cell is related to how busily the cell is making proteins for export. Cells that are synthesizing a lot of proteins—glandular cells (Chapter 34) or the immune system's plasma cells (Chapter 16), for example—may be heavily packed with ER, whereas others with less work to do (such as food-storing cells) contain very little ER.

The Golgi Apparatus

In 1898 the Italian microscopist Camillo Golgi reported the discovery of a delicate structure located close to the nuclei of nerve cells. Unfortunately, his technique for staining this structure was tricky and often failed to work, so the **Golgi apparatus** was regarded by most biologists as a figment of Golgi's imagination. However, work with the electron microscope in the late 1950s showed clearly that these structures do exist—and not just in nerve cells, but in most eukaryotic cells.

Its appearance varies in details from species to species, but the Golgi apparatus always consists of a number of flattened sacs lying together like a stack of saucers (Figure 4.21). These stacks may exist as individual units scattered throughout the cytoplasm (as in plants, protists, fungi, and many invertebrate animals), or a few such stacks may form the more complex Golgi bodies of vertebrate cells. The bottom saucer, or forming face, lies nearest the nucleus or a patch of rough ER; the top saucer, or maturing face, lies closest to the surface of the cell. Vesicles from the rough ER travel to and merge with the forming face of the Golgi apparatus. Other small vesicles move between the flattened sacs of the Golgi apparatus, transporting proteins. Associated with the sacs, particularly those toward the maturing face, are numerous tiny vesicles that pinch off from the sacs and then move away, sometimes to merge with each other and finally to merge with the plasma membrane or with other organelles, where they release their contents in a process called exocytosis (see next section). This behavior of the Golgi apparatus is a key to its important cellular functions. The Golgi apparatus serves as a depot in which some of the proteins synthesized on the rough ER are stored, chemically modified, and packaged for delivery either to the environment outside the cell or to other organelles of the cell.

(a)

Vesicles leaving maturing face

Maturing face

Sacs

Golgi apparatus

Forming face

Transport vesicles

Flow of material

Rough endoplasmic reticulum

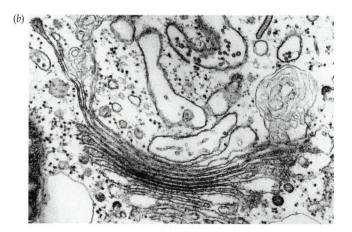

(b)

4.21 Golgi Apparatus
(a) Idealized diagram of a Golgi apparatus. Vesicles from the endoplasmic reticulum deliver substances to the apparatus by fusing with the forming face at the bottom of the stack, and vesicles with substances for export or transport to elsewhere in the cell leave the apparatus by budding off the maturing face. *(b)* A Golgi apparatus in an alga.

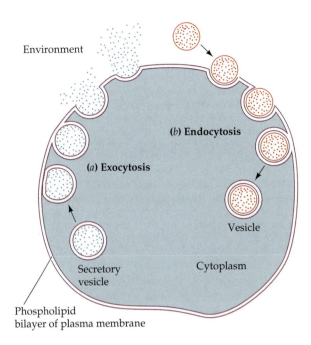

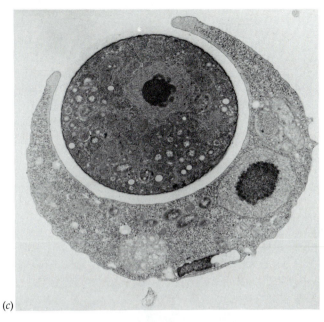

(c)

4.22 Exocytosis, Endocytosis, and Phagocytosis
(a) In exocytosis, a membrane-bounded vesicle containing substances to be exported from the cell fuses with the plasma membrane—first one layer of the phospholipid bilayers of the two membranes, then the other. When this fusion is complete, the contents of the vesicle scatter, and the vesicle membrane becomes part of the plasma membrane. (b) In endocytosis, on the other hand, the plasma membrane surrounds a part of the exterior environment and the whole unit buds off to the interior as a membrane-surrounded vesicle. (c) Phagocytosis (cell eating) is a form of endocytosis. The micrograph shows a small amoeba engulfing another protist.

Exocytosis and Endocytosis

Certain organelles such as the Golgi apparatus secrete materials, made in the ER, to the environment; others *acquire* materials by enveloping a portion of the environment. In eukaryotic cells there is an ongoing traffic—inward and outward—of membrane-bounded "packages," with the active participation of the plasma membrane and the endomembrane system. Macromolecules for export are contained in membranous vesicles that move toward the exterior of the cell and ultimately meet the plasma membrane. The phospholipid regions of the two membranes merge, and an opening to the outside of the cell develops. The contents of the vesicle are released to the environment, and the vesicle membrane is smoothly incorporated into the plasma membrane. This entire process—export of material and transformation of the membrane—is called **exocytosis** (Figure 4.22a).

Materials may be brought into the cell by a related process known as **endocytosis**. The cell surface folds to make a small pocket that is lined by the plasma membrane. The folding increases until the pocket seals off, forming a vesicle whose contents are materials from the environment. This vesicle, enclosed in membrane taken from the plasma membrane, separates from the cell surface and migrates to the interior (Figure 4.22b). Usually the vesicle fuses with a lysosome (as discussed in the next section) and its contents are digested.

Endocytosis is a blanket term for several processes that are sometimes given more specialized names. In particular, a distinction is often made between **pin-**ocytosis (cell drinking), in which tiny, liquid-containing vesicles are formed, and **phagocytosis** (cell eating), in which particles or even entire cells may be trapped in large vesicles (Figure 4.22c). Phagocytosis constitutes an important part of our immune system for defense against foreign cells (Chapter 16). Many kinds of cells are able to engulf food materials from their environment, forming vesicles surrounded by pieces of the plasma membrane.

Lysosomes

Some vesicles from the Golgi apparatus develop into **lysosomes**. These organelles are found in animals, many protists, fungi, and a few plants. They are somewhat smaller than mitochondria, are surrounded by a single membrane, and have a densely staining but usually featureless interior. Lysosomes transport 40 or more different digestive enzymes that accelerate the breakdown of proteins, polysaccharides, nucleic acids, and lipids.

Primary lysosomes form from some of the vesicles pinched off from the Golgi apparatus. One of their functions is to digest materials taken up by endocy-

(a)

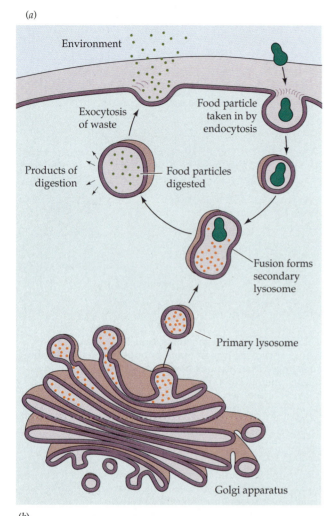

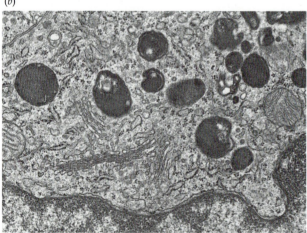

(b)

4.23   Lysosomes

(a) After pinching off from the Golgi apparatus, a primary lysosome, which contains enzymes, becomes a secondary lysosome by fusing with a food-containing vesicle. The enzymes digest the food in the secondary lysosome. Some products of digestion cross the membrane to be used in the cell. Wastes are eventually expelled to the environment by exocytosis. *(b)* The dark compartments in the upper half of this electron micrograph are secondary lysosomes with material being digested; at the bottom is a portion of the nucleus, and above it are some of the flattened membrane sacs of the Golgi apparatus. This is a picture of part of a specialized cell in a rat. One of the roles of such a cell in the body is to take up and digest foreign objects such as disease-causing bacteria, and it carries out this function using the enzymes packaged in its lysosomes. The enzymes must not leak into the rest of the cell.

cytoplasm. The products of digestion then pass through the membrane of the lysosome and are used by the rest of the cell. The "used" secondary lysosome now moves to the plasma membrane, fuses with it, and releases the remaining, undigested contents to the environment by exocytosis. This compartmentalization into lysosomes is obviously a very good arrangement. The digestive enzymes of the lysosome would be highly destructive if they were released into the cell, where they would attack the contents of the cytosol and the other organelles. Instead, enzymes are sealed in the lysosome, from which they cannot escape. The raw materials for their action are brought to them tightly packaged in a vesicle that fuses with the lysosome; the useful products of digestion leak out; finally, the enzymes along with the unusable products are thrown out of the cell.

Why are the lysosomes themselves not destroyed by the enzymes? It would seem that lysosomes also should be subject to attack by the digestive enzymes, yet they survive and function. Despite much interest in this problem, no generally acceptable solution has yet been proposed.

Clearly, the consequences of digestive enzymes escaping from the lysosomes can be severe. At times, though, such digestive activity can be appropriate, as during the development of a frog from a tadpole. The tadpole has a fleshy tail, whereas the mature frog has none. How does the tail disappear? Part of the job is accomplished by the breakdown of lysosomes within the tail cells of the tadpole, releasing enzymes that digest the cells themselves.

tosis (Figure 4.23*a*). When a food-containing vesicle formed by endocytosis fuses with a primary lysosome, a secondary lysosome is formed (Figure 4.23*b*). The effect of this fusion is rather like releasing hungry foxes into a chicken coop. The engulfed particles are quickly digested by enzymes within the secondary lysosome. The activity of the enzymes is enhanced by the mild acidity of the lysosome's interior, where the pH (Chapter 2) is lower than in the surrounding

OTHER ORGANELLES

Microbodies are formed by the pinching off of vesicles from rough endoplasmic reticulum. These small organelles, 0.2–1.7 μm in diameter, are seen under the electron microscope to have a single membrane and a granular interior (Figure 4.24). They are found at one time or another in at least some cells of just about every species of eukaryote. Some microbodies, called **peroxisomes**, house reactions in which toxic peroxides (such as hydrogen peroxide, H_2O_2) are formed as unavoidable side products of chemical reactions. Subsequently, the peroxides are safely broken down within the peroxisomes without mixing with other parts of the cell. Both plant and animal cells have peroxisomes, but another type of microbody, the **glyoxysome**, is found only in plants. Glyoxysomes, which are most prominent in young plants, are the site at which stored lipids are converted into carbohydrates.

You should think of organelles as self-contained compartments or "factories" for specific processes; but you must also remember that materials are constantly being shuttled from one type of organelle to another. For example, there is a considerable traffic of compounds among the organelles in a process called photorespiration (Chapter 8). The process of photorespiration begins in the chloroplasts. One intermediate product is transported from the chloroplasts to the peroxisomes for further chemical changes. Some of the products of peroxisome action are then passed to the mitochondria, whereas others are returned to the chloroplast for still further changes.

In many eukaryotic cells, but particularly in those of plants and protists, there are structures that look quite empty under the electron microscope. They are called **vacuoles** (Figure 4.25). Each is surrounded by a single membrane. Within the vacuole is an aqueous solution containing many dissolved substances.

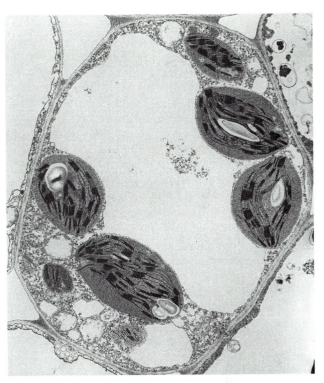

4.25 Vacuoles
The large central vacuole is typical of mature plant cells. Smaller vacuoles are visible toward the two ends of the cell.

For all their structural simplicity, vacuoles play a variety of crucial roles in the lives of cells. For example, consider one of the problems a plant faces. Just like animals and other organisms, plants produce a number of by-products that would be toxic to the organism if not set aside. Animals possess various excretory mechanisms for getting rid of such wastes, but plants are not equipped in the same way. Although plants manage to secrete some wastes to their environment, many compounds must simply be stored within the cells. The solution to this storage problem is to put these wastes in a vacuole. The vacuolar membrane can keep them from getting at the rest of the cell and producing toxic reactions. The vacuoles of many plants store large amounts of chemicals that are either poisonous or distasteful to herbivores (plant-eating animals); this deters animals from eating the plants.

Many plant cells have enormous vacuoles that take up more than 90 percent of the total cell volume and grow as the cell grows. Vacuoles are by no means a waste of space, for the dissolved substances in the vacuole, working together with the vacuolar membrane, provide the turgor, or stiffness, of the cell, which in turn provides support for the structure of nonwoody plants.

Some unicellular protists, as well as sponges and some of the other more ancient invertebrate animals,

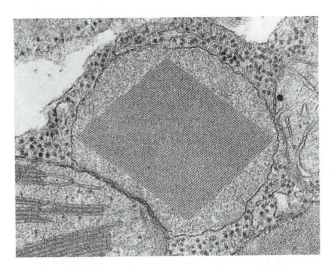

4.24 A Microbody
A diamond-shaped crystal, composed of an enzyme, almost entirely fills this rounded microbody in a leaf cell. The microbody is pressed against a chloroplast, seen at lower left.

obtain nutrients by endocytosis. Particles trapped from the environment in this way end up in vesicles called **food vacuoles**.

Many freshwater-dwelling protists have a highly specialized form of vacuole, called a **contractile vacuole**. Its function is to rid the cell of excess water that rushes in because of the imbalance in salt concentration between the relatively salty interior of the cell and its freshwater environment. Contractile vacuoles are discussed in Chapter 22.

Vacuoles can even play a role in the sex life of plants. Some of the pigments (especially the blue and pink ones) in petals and fruits (and sometimes in leaves) are contained in vacuoles. These pigments—the anthocyanins—serve as cues to encourage animals to visit flowers and thus aid in pollination, or to eat fruits and thus aid in seed dispersal.

THE SKELETON WITHIN

The cytoplasm of eukaryotic cells is divided into numerous compartments by membranes, as described above. But even the cytosol (the space and material inside the plasma membrane but outside the mem-

brane-bounded organelles) is not a simple aqueous solution. In this compartment of the cell there is a set of fibers—the cytoskeleton—that contributes to the cell's shape and physical texture. The fibers are also used to build "motors" that help a cell to move. At least three components of the cytoskeleton are visible in electron micrographs: microtubules, microfilaments, and intermediate filaments.

Microtubules are long, hollow, unbranched cylinders about 25 nm in diameter, with lengths up to several micrometers. They are made up of subunits of a protein called **tubulin** (Figure 4.26). In plants, microtubules help control the arrangement of the fibrous components of the cell wall. Electron micrographs of plants frequently show microtubules lying next to the cell wall, and disruption of the cell's microtubules leads to a disordered arrangement of newly synthesized fibers in the cell wall. In animal cells, microtubules are often found in the parts of the cell that are changing shape, but the mechanisms by which the microtubules might function are not yet known. In some specialized cells, such as nerve cells, the cytoplasm contains microtubules running parallel to the length of long cellular projections. Here the microtubules contribute to the mechanical stability of the projections and serve as tracks along which protein-laden vesicles are moved by a process requiring energy from ATP.

Many eukaryotic cells possess whiplike appendages, the **cilia** and **flagella** (Figure 4.27a). These organelles push or pull the cell through its aqueous environment or promote movement of the surrounding liquid over the surface of the cell. Cilia and flagella are identical in internal structure, but differ in their relative lengths and their patterns of beating. Longer appendages, usually single or in pairs, are called flagella; these propagate waves of bending from one end to the other in snakelike undulation. The shorter appendages, usually present in great numbers, are called cilia; they beat stiffly in one direction and recover flexibly in the other direction (like an oar) so the recovery stroke does not undo the work of the power stroke.

Eukaryotic flagella and cilia are built from specialized microtubules. In cross section, a typical cilium or eukaryotic flagellum is seen to be covered by the

(a)

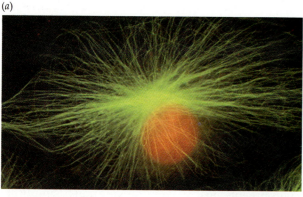

Tubulin
(two subunits)

Assembly
of
microtubule

Breakdown of
microtubule

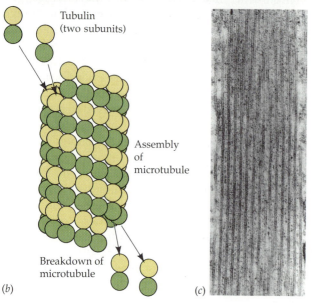

(b)

(c)

4.26 Microtubules
(a) The human cell seen in part here is about to divide. The colors in this micrograph are due to fluorescent stains. Microtubules show up as green, and the DNA-containing chromosomes as orange. These microtubules will participate in distributing the chromosomes to the two daughter cells. (b) The long coil of subunits in a microtubule (much shortened in this idealized drawing) is formed or disassembled by adding or subtracting tubulin molecules in sequence. (c) Parallel microtubules stiffen this cytoplasmic projection of a protist.

plasma membrane and to contain what is usually called a **"9 + 2"** arrangement of microtubules (Figure 4.27b). As you can see from the figure, this nickname is somewhat misleading: There are actually nine fused *pairs* of microtubules, called **doublets**, forming a cylinder, and one pair of unfused microtubules running up the center. The motion of cilia and flagella results from these microtubules sliding past one another, using energy from ATP to drive the process (Chapter 38).

Some prokaryotes, too, have flagella. However, prokaryotic flagella are very different from the 9 + 2 arrangement of the eukaryotic flagellum. There is no structural or evolutionary relationship between the flagella of prokaryotes and those of eukaryotes. In prokaryotes, flagella are much simpler in construction (Figure 4.6), and the entire prokaryotic flagellum has a smaller diameter than that of a single microtubule. Prokaryotic flagella rotate, whereas eukaryotic flagella beat in a wavelike motion.

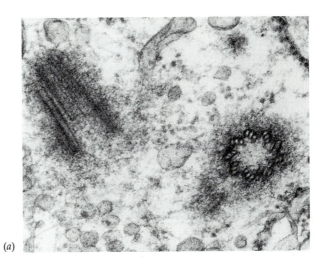

(a)

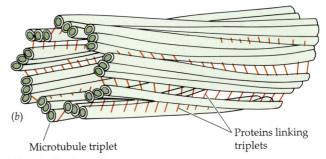

(b)

Microtubule triplet

Proteins linking triplets

4.28 Centrioles
(a) A pair of centrioles, at right angles to each other, are seen in this thin section of a cell. Nine sets of three microtubules are evident in the centriole, seen in cross section. (b) The diagram emphasizes the three-dimensional structure of a centriole.

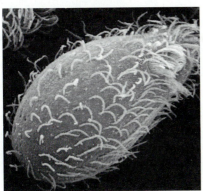

(a)

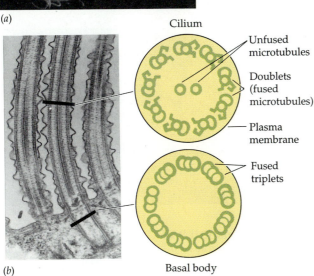

(b)

Cilium

Unfused microtubules

Doublets (fused microtubules)

Plasma membrane

Fused triplets

Basal body

4.27 Cilia
(a) Cilia cover the surface of this protist, *Tetrahymena pyriformis*. The cilia propel the cell through its watery environment. (b) Longitudinal section through three cilia on a cell of *Tetrahymena*. The basic structures seen in cross sections of a cilium and its basal body are shown in the diagrams.

An organelle called a **basal body** is found at the base of every eukaryotic flagellum or cilium (Figure 4.27b). The nine doublets extend into the basal body. In the basal body, each doublet is accompanied by another microtubule, making nine sets of *three* microtubules. The central, unfused microtubules of the cilium or flagellum do not extend into the basal body.

Centrioles are organelles that are virtually identical to basal bodies. Centrioles are found in all eukaryotes except those that never produce cells with cilia or flagella—these exceptions are the flowering plants, the pines and their relatives, and some protists. Under the light microscope, a centriole looks like a small, featureless particle; but the electron microscope reveals that it is made up of a precise bundle of microtubules, arranged as nine sets of three fused microtubules (Figure 4.28). Centrioles help organize the microtubules in cells about to undergo division (Chapter 9).

Microfilaments are another common type of fiber in the cytosol. Microfilaments are assembled from a protein called actin, often in combination with other proteins (Figure 4.29). Each individual microfilament is 9 nm in diameter and several micrometers in

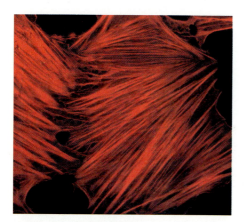

4.29 Microfilaments
The actin microfilaments in this micrograph have been stained so that they fluoresce a brilliant red-orange. Notice their parallel orientation. The microfilaments are in two cells, of a specialized type called fibroblasts, that were grown in culture from embryonic rat tissue. Fibroblasts in tissue culture migrate readily, although slowly; and the actin microfilaments are responsible for this locomotion.

length. Microfilaments may be single or in bundles and networks. Often they are attached to the plasma membrane. Microfilaments help the cell to contract; many types of motion within the cell require their participation. They take part in changes in cell shape (including cell length, as in the contraction of muscles), in the streaming of cytoplasm (a flowing movement observed in some cells), in movements of organelles and particles, and in "pinching" movements such as those that separate the daughter cells after an animal cell has undergone nuclear division.

Filaments of another type, the **intermediate filaments**, play more static roles in stabilizing cell structure and resisting tension. One such role may be maintaining the position of the nucleus in the cell. Intermediate filaments are composed of fibrous proteins similar to those that make up hair and skin. In cells these proteins are organized into tough, ropelike assemblages about 8–10 nm in diameter.

THE SKELETON WITHOUT

The eukaryotic **cell wall** is a semirigid structure found outside the plasma membrane of plants (Figure 4.30a), fungi, and some protists. (There is no comparable structure in animal cells.) The cell wall is made up primarily of polysaccharides. It provides support for the cell, limits the cell's volume, and in some instances may serve to restrict the flow of water into and out of the cell. In plants, modifications of the cell wall are important in determining the specific function of the cell.

Although it might seem that plant cells are com-

pletely isolated from one another by their cell walls, this is not the case. Plant cells are connected by numerous channels, or **plasmodesmata**, that extend through the walls of adjoining cells; a strand of cytoplasm about 4 nm in diameter runs through most plasmodesmata (Figure 4.30b). The diameter of a plasmodesma is 20–40 nm. A plasmodesma allows relatively free passage of many molecules—not only is there no cell wall across the hole, but there is also no plasma membrane.

EUKARYOTES, PROKARYOTES, AND VIRUSES

Let us now review the major differences between the cells of eukaryotes and those of prokaryotes. In contrast to prokaryotes, which have nucleoids, eukaryotes have a true, membrane-enveloped nucleus and, unlike the prokaryotes, eukaryotes have other mem-

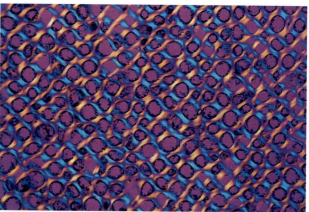

(a)

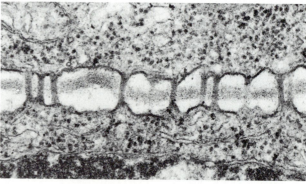

(b)

4.30 The Plant Cell Wall
(a) The brilliant gold and blue structures in this polarizing micrograph of seed tissue are the cell walls. Cell walls make up a substantial fraction of the tissue. (Compare with Hooke's drawing of cork in Figure 1.16.) (b) Plasmodesmata are the gray channels crossing the cell walls (light bands) between the cytoplasm of the cell above and the cell below. This electron micrograph shows portions of two cells from the root tip of timothy grass. Thin sections through plant cell walls around entire cells are shown in Figure 4.25.

brane-bounded organelles that allow various cellular activities to be concentrated in specialized compartments. Eukaryotic cells, but not prokaryotic ones, can perform endocytosis and exocytosis. Eukaryotes also have many specialized molecules, such as tubulin and actin, that are used for movement and are not found in prokaryotes. The cell walls of prokaryotes differ structurally and chemically from those of eukaryotes.

Do such differences mean that eukaryotes are "more advanced," "higher," or "more successful" than prokaryotes? Not at all! Every surviving species is the product of eons of natural selection and is superbly adapted to its environmental niche. Each species has characteristics that enable it to live where and how it does, and to compete successfully against other species. Both eukaryotic and prokaryotic cells are marvels of systematic organization to achieve complex function.

Where do the viruses fit into this picture? They are simpler in structure than the bacteria—so much simpler that they cannot be called cells. Viruses lack ribosomes and must use the ribosomes of a host cell (prokaryotic or eukaryotic) to synthesize the proteins they need. The hereditary material of a virus enters the host cell and there subverts the host's metabolic machinery to make new viruses. Except for a few complex forms, viruses are simply packets of hereditary material wrapped in coats of protein; they in no way demonstrate the life processes of independent cells. The protein coat serves to protect the hereditary material and to introduce it into a cell of a susceptible host. Many animal viruses do, however, acquire a membrane around their protein coat as they bud out through the host cell's plasma membrane. The parasitic mode of existence has served viruses well. Viruses, like prokaryotes and eukaryotes, are enormously successful. A few common viral shapes are illustrated in Figure 4.31, and viruses are treated in detail in Chapter 21.

FRACTIONATING THE EUKARYOTIC CELL: ISOLATING ORGANELLES

During the early days of cell biology in the nineteenth century, virtually all that one could do with the organelles—and only the largest ones at that—was look at them with a microscope as they sat within the cell. Later it became possible to view cells at a higher magnification—with the electron microscope—and to isolate and purify substantial quantities of specific organelles, a process called **cell fractionation**. After obtaining organelles in this way, scientists may study their biochemical activities and physiological functions. Suppose a scientist wanted 1 g of chloroplasts. This would be more than 10 billion individual chloroplasts of average size. It would take forever to pick 10 billion chloroplasts out of cells a few at a time by microsurgical techniques. So before scientists could work with isolated organelles, two feats had to be accomplished: The organelles had to be removed from cells, and the various kinds of organelles had to be separated from one another.

4.31 Viruses
(a) The T4 bacteriophage is a virus that reproduces in cells of the bacterium *Escherichia coli*. *(b)* The tobacco mosaic virus (TMV). *(c)* A spiked, 20-faced protein coat is characteristic of this adenovirus.

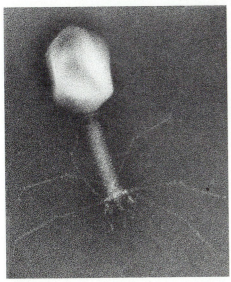
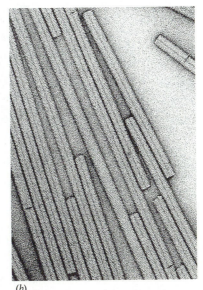
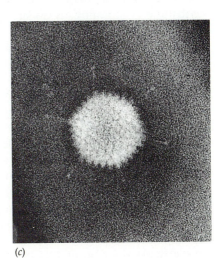

(a)   *(b)*   *(c)*

Rupturing the Cell

The first step in cell fractionation is opening up the cell; during this process, one must be careful not to burst the organelles. A variety of methods can be used to break cells open, most of which employ strong shearing forces. The simplest methods use an old-fashioned mortar and pestle or a hand-operated glass homogenizer, and squeeze and shear cells between two tightly fitting, counter-rotating, ground-glass surfaces. Cells of some tissues can be opened by rapid chopping with a razor blade against a glass plate. A variety of motor-driven homogenizers or blenders are commonly used.

These methods rupture the cells by tearing their plasma membranes and (if present) cell walls. The rupturing of cells is conducted in a solution with a substantial concentration of solutes, which prevents the organelles from bursting. In a more dilute solution or pure water, the organelles would take up water and explode, as described in Chapter 5.

Separating the Organelles

By techniques of the kind just described, one may reduce biological tissue to a crude suspension of mixed organelles, unbroken cells, and debris. But how can the components of such a suspension be separated from one another? The methods of choice are two types of centrifugation. The **centrifuge** is a laboratory instrument that can spin materials extremely rapidly about a fixed axis, which causes the particles in suspension to fall or rise more rapidly than they would under only the force of gravity (Figure 4.32). Different classes of organelles sediment (settle out of a suspension) at different rates in a centrifuge. Factors determining the rate of sedimentation include the organelle's size (radius) and density (weight per unit volume). An organelle that is *more* dense than the liquid in which it is suspended settles toward the bottom of the centrifuge tube, the sedimentation rate being faster the more rapidly the centrifuge is spinning. If an organelle is *less* dense than the liquid, it floats toward the surface; this is why oil floats on water and rises to the top if a mixture of oil and water is shaken and then allowed to separate.

Equilibrium centrifugation can be used to separate two or more types of organelles that differ in density (Figure 4.33). First a centrifuge tube is filled with liquid whose density varies from the top to the bottom of the tube. This could be done, for example, by first putting in a small amount of a highly concentrated (and hence very dense) sugar solution—say, 60 percent sucrose. Next a small amount of a 50 percent sucrose solution is layered on, then 40 percent, and so on. Such a **density gradient** is usually so constructed by an automatic device that the gra-

dient is smooth rather than erratic and jumpy. After the gradient is established, some of the mixture of, say, two organelles to be separated is carefully layered on and the tube is centrifuged. Both populations of organelles sediment into the gradient as long as they are more dense than the surrounding liquid. Once an organelle reaches that part of the gradient where its density equals that of the liquid, it stops sedimenting—it has reached buoyant equilibrium. If it were to be pushed farther down, it would float back up to this point. If the two kinds of organelles have different average densities, they form separate bands. After the tube is removed from the centrifuge, the organelles can be collected separately by the careful use of a pipette or by punching a hole in the bottom of the tube and collecting samples as the liquid slowly drips out.

A second approach, called **differential centrifugation**, depends on differences in either the radius or the density of the particles being centrifuged. No density gradient is used, and the liquid in the tube is usually of low density. A mixture containing various organelles is centrifuged briefly at low speed (and hence low relative centrifugal force). The largest and densest particles sediment out, forming a pellet in the bottom of the tube; this pellet is left behind when the remaining liquid and its suspended contents (together called the supernatant) are poured off. The liquid and its contents are next spun at a higher

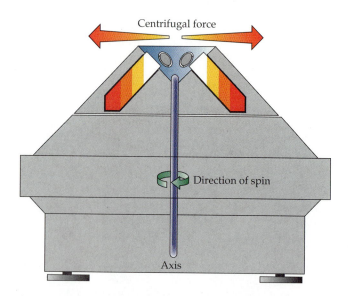

4.32 Centrifugation
Extremely rapid rotation of liquids in centrifuge tubes around an axis produces a force in the tubes analogous to that of gravity, but much more powerful. Suspended particles, denser than the liquid in a centrifuge tube, separate (sink) rapidly. Following centrifugation, the fluid in the tube (called the supernatant) can be decanted (poured off), leaving a pellet of dense particles in the bottom of the tube.

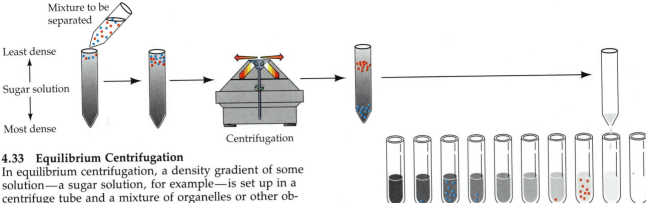

4.33 Equilibrium Centrifugation
In equilibrium centrifugation, a density gradient of some solution—a sugar solution, for example—is set up in a centrifuge tube and a mixture of organelles or other objects to be separated is carefully layered on top. As the tube is spun in the centrifuge, the organelles rise or sink to a level at which their density matches that of the liquid. Ultimately, the organelles collect in bands in the tube. The bands may be separated into different tubes by punching a hole in the bottom of the centrifuge tube and letting out small amounts of liquid into each of a series of tubes.

speed and for a longer time, causing other organelles to sediment to the bottom. By repeating this procedure with ever-increasing centrifugal force, one can separate out many organelles. Nuclei sediment into a pellet even when the centrifuge is spun slowly. The tiny ribosomes, on the other hand, require considerably longer centrifugation at extremely high speeds before they will sediment—in most rotors, the spinning rate must be about 40,000 revolutions per minute, giving forces as great as 100,000 times the force of gravity.

Partially purified organelles are obtained by either equilibrium or differential centrifugation. Further purification is achieved by repeating the centrifugation routine. The purity may be determined by examining the final sample under the microscope—light or electron—or by testing the chemical activities of the sample in comparison with the known behavior of various organelles. We actually *can* get a 1-g sample of chloroplasts (or other organelles) this way and study them to our heart's content.

Early work with isolated organelles focused on their chemical composition and then on the biochemical reactions that take place within them. Gradually we have come to have an extensive, but still partial, knowledge of the activities of the different organelles (as discussed in this and the next few chapters) and hence of the cell itself.

SUMMARY

The prokaryotic cell lacks membrane-bounded organelles. Its DNA resides within a region called the nucleoid and is not combined with protein. These cells also possess ribosomes and a plasma membrane. Most prokaryotes have one or more of the following: a cell wall, a capsule, internal membranes developing from the plasma membrane, and one or more prokaryotic flagella.

Eukaryotic cells, comprising all cells in the other four kingdoms of organisms, have a multiplicity of organelles. The nucleus, surrounded by a double-membraned envelope, contains the genetic information of the cell as well as one or more nucleoli, which function in ribosome synthesis. The nuclear DNA combines with protein to form chromatin. Mitochondria are organelles that process energy, as do chloroplasts in photosynthetic organisms. Mitochondria and chloroplasts have double membranes, complex internal membranous structures, and small amounts of their own DNA and ribosomes. Mitochondria and chloroplasts may have evolved through the incorporation of smaller prokaryotic cells into larger ones.

The nuclear envelope is continuous with the endoplasmic reticulum, a complex network of membranes studded in places with ribosomes. Ribosomes are the sites of protein synthesis, and the endoplasmic reticulum functions in the synthesis of proteins for export or for placement in vesicles or membranes. It also contains a variety of enzymes that carry out other specialized functions. The Golgi apparatus packages and modifies proteins for delivery to the outside of the cell or to other organelles. Lysosomes, which are not found in plants, are centers of digestive activity. Certain other metabolic activities are performed by the microbodies (peroxisomes and glyoxysomes). Large vacuoles are found in many plant cells, and food vacuoles and contractile vacuoles are found in some protists.

The cytoskeleton comprises microtubules, microfilaments and intermediate filaments. Cilia, flagella, and centrioles are composed primarily of microtubules. The cells of plants, fungi, and some protists are surrounded by a cell wall.

SELF-QUIZ

1. Which statement is true of both prokaryotic and eukaryotic cells?
   a. They contain ribosomes.
   b. They have peptidoglycan cell walls.
   c. They contain membrane-bounded organelles.
   d. They contain true nuclei.
   e. Their flagella have the "9 plus 2" structure.

2. Which statement is *not* true of the nuclear envelope?
   a. It is continuous with the endoplasmic reticulum.
   b. The nuclear envelope has pores.
   c. The nuclear envelope consists of two membranes.
   d. RNA and some proteins pass in and out of the nucleus.
   e. The inner membrane bears ribosomes.

3. Which statement is *not* true of mitochondria?
   a. Their inner membrane folds to form cristae.
   b. They are usually one micrometer or less in diameter.
   c. They are green because of the chlorophyll they contain.
   d. Energy-rich substances from the cytosol are oxidized in them.
   e. Much ATP is synthesized in them.

4. Which statement is true of plastids?
   a. They are found in prokaryotes.
   b. They are surrounded by a single membrane.
   c. They are the sites of cellular respiration.
   d. They are found in fungi.
   e. They are of several types, with different functions.

5. Which statement is *not* true of the endoplasmic reticulum?
   a. It is of two types: rough and smooth.
   b. It is a network of tubes and flattened sacs.
   c. It is found in all living cells.
   d. Some of it is sprinkled with ribosomes.
   e. Other parts of it modify proteins.

6. The Golgi apparatus:
   a. is found only in animals.
   b. is found in prokaryotes.
   c. is the appendage that moves a cell around in its environment.
   d. is a site of rapid ATP production.
   e. packages and modifies proteins.

7. Which of the following organelles is *not* surrounded by one or more membranes?
   a. Ribosome
   b. Chloroplast
   c. Mitochondrion
   d. Microbody
   e. Vacuole

8. Eukaryotic flagella:
   a. are composed of a protein called flagellin.
   b. rotate like propellers.
   c. cause the cell to contract.
   d. have the same internal structure as cilia.
   e. cause the movement of chromosomes.

9. Microfilaments:
   a. are found in prokaryotes.
   b. are composed of actin.
   c. provide the motive force for cilia and flagella.
   d. make up the spindle that aids movement of chromosomes.
   e. maintain the position of the nucleus in the cell.

10. Which statement is *not* true of the plant cell wall?
    a. Its chief chemical component is cellulose.
    b. It may contain suberin.
    c. It may contain lignin.
    d. It completely isolates adjacent cells from one another.
    e. It is semirigid.

FOR STUDY

1. Which organelles and other structures are found in both plant and animal cells? Which are found in plant but not animal cells? Which in animal but not plant cells? Discuss, in relation to the activities of plants and animals.

2. Through how many membranes would a molecule have to pass in going from the interior of a chloroplast to the interior of a mitochondrion? from the interior of a lysosome to the outside of a cell? from one ribosome to another?

3. How does the possession of double membranes by chloroplasts and mitochondria relate to the endosymbiotic theory of the origins of these organelles? What other evidence supports the theory?

4. What sorts of cells and subcellular structures would you choose to examine by transmission electron microscopy? by scanning electron microscopy? by light microscopy? What are the advantages and disadvantages of each of these modes of microscopy?

5. Some organelles that cannot be separated from one another by equilibrium centrifugation can be separated by differential centrifugation. Some other organelles that cannot be separated from one another by differential centrifugation can be separated by equilibrium centrifugation. Explain these observations.

READINGS

Alberts, B., D. Bray, J. Lewis, M. Raff, K. Roberts and J. D. Watson. 1989. *Molecular Biology of the Cell*, 2nd Edition. Garland Publishing, New York. An outstanding book in which to pursue the topics of this chapter in greater detail; authoritative treatment of modern cell biology and its experimental basis.

Allen, R. D. "The Microtubule as an Intracellular Engine." 1987. *Scientific American*, February. How microtubules cause two-way transport of materials in cells.

Brandt, W. H. 1975. *The Student's Guide to Optical Microscopes*. William Kaufmann, Los Altos, CA. A short, programmed guide for those interested in learning how to use a light microscope.

Darnell, J., H. Lodish and D. Baltimore. 1990. *Molecular Cell Biology*, 2nd Edition. Scientific American Books, New York. Another excellent middle-level book; fine illustrations.

De Duve, C. 1975. "Exploring Cells with a Centrifuge." *Science*, vol. 189, pages 186–194. A Nobel laureate discusses the uses of centrifugation in studies of cells.

Fawcett, D. W. 1981. *The Cell*, 2nd Edition. Saunders, Philadelphia. Beautiful electron micrographs of subcellular structures in animal cells.

Howells, M. R., J. Kirz and D. Sayre. 1991. "X-Ray Microscopes." *Scientific American*, February. Novel methods of microscopy afford striking improvements in resolution.

Margulis, L. 1981. *Symbiosis in Cell Evolution*. W. H. Freeman, San Francisco. An authoritative and thought-provoking reference on the origin and evolution of eukaryotic cells by a leading student of the problem.

Rothman, J. E. 1985. "The Compartmental Organization of the Golgi Apparatus." *Scientific American*, September. Structure and function in the Golgi apparatus.

Weber, K. and M. Osborn. 1985. "The Molecules of the Cell Matrix." *Scientific American*, October. A clear treatment of microfilaments, intermediate filaments, tubulin, and the ways in which they are studied.

5

Membranes

PREVIEW: Every cell and most organelles are surrounded by membranes. Membranes control movement of material from one compartment to another, recognize specific chemical substances and other cells, hold groups of molecules in place, and participate in phenomena such as the transmission of signals along nerves. Membranes are composed of lipids, proteins, and carbohydrates, each performing specific functions. Molecules and ions may move across membranes in various ways, one of which requires an input of energy. Membranes within cells are constantly being formed and moved about.

This chapter deals with the structures and functions of membrane lipids, proteins, and carbohydrates; junctions between membranes; several kinds of transport across membranes, including simple diffusion, osmosis, facilitated diffusion, and active transport; energy transformations; surface receptors; and the formation and continuity of membranes.

Poised between every cell and its environment is a filmy sheet so thin that it can be seen only with the aid of an electron microscope. Similar sheets are also found within the cytoplasm of many cells; in eukaryotes these sheets surround many of the organelles, dividing the cell into numerous compartments. We refer, of course, to the plasma membrane and other **membranes**, all of which are thin, pliable bilayers of phospholipids with embedded proteins. (The phospholipid bilayer was introduced in Chapter 3, and membranous organelles were described in Chapter 4.) Membranes perform a sweeping array of vital functions and are themselves constantly undergoing change. As we consider membranes, the relationship between physical structure and biological activity is particularly obvious; this relationship is evident at all levels, from the overall shape of the membrane down to its individual chemical components.

Many functions of biological membranes result from their **selective permeability**—some materials move through them more readily than others. The regulation of movement of materials among cells and among compartments within cells is one of the most important functions of a membrane, so we include in this chapter a general discussion of such regulation. The story begins with the fundamental process of diffusion and proceeds to ways in which particular substances may move through the membrane from cell to cell, across internal membranes within the cell, or between a cell and its environment. Long-range transport through many cells in the body of an organism will be taken up in Chapters 29 and 39.

MEMBRANE STRUCTURE AND COMPOSITION

In talking about the chemical makeup and physical organization of a biological membrane, we must consider three classes of biochemical compounds. Each is related to at least one important aspect of membrane function. The three classes are lipids, proteins, and carbohydrates. The lipids serve as an effective barrier to the passage of most materials between the inside and the outside of a cell or organelle; the lipids also account for much of the physical integrity of the membrane. The membrane lipids are present as a double layer that constitutes the continuous portion of the membrane. In this lipid "lake," there "floats" a variety of proteins—the second class of membrane compounds.

Some proteins reach from one side of the membrane to the other, whereas others reside primarily on one side of the membrane. The proteins are responsible for many of the specific tasks performed by membranes. Certain membrane proteins allow some materials to pass through the membrane that cannot pass through the pure lipid bilayer. Other proteins receive chemical signals from the cell's external environment and respond by regulating certain processes within the cell. Some proteins accelerate chemical reactions on the membrane surface, and still others function as sentries that recognize specific molecules that impinge upon them.

The carbohydrates, the third class of membrane compounds, have fewer roles than the proteins, but they, too, are crucial in recognizing specific molecules. The carbohydrates are attached to either lipid or protein molecules. Most carbohydrates are found on the outside of the plasma membrane, where they protrude into the environment, away from the cell.

Generalized membrane architecture is shown in Figure 5.1. The two sides of the membrane are not identical—in fact, the membrane is decidedly asymmetric. Even the lipids on the inward-facing half of the membrane are different from those on the outward-facing half.

Membrane Lipids

The great majority of the lipids in biological membranes are **phospholipids**. Recall that some compounds are hydrophilic ("water loving") and that others are hydrophobic ("water fearing"). Phospholipids are both: They have hydrophilic regions and hydrophobic regions. The large, nonpolar fatty-acid parts of the phospholipid molecule associate easily with other fatty materials but do not dissolve in water. The phosphorus-containing region of the phospholipid is electrically charged and hence very hydrophilic. As a consequence, one way for phospholipids and water to coexist is for the phospholipids to form a double layer with the fatty acids of the two layers pointing toward each other and the polar regions facing the outside. The resulting lipid bilayer separates two aqueous regions (Figure 5.2). Artificial membranes with the same two-layered arrangement are easily made in the laboratory.

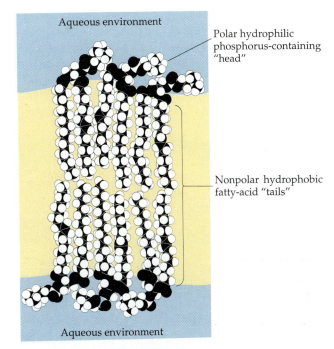

5.2 The Phospholipid Bilayer
The six phospholipid molecules shown here represent a small section of a membrane bilayer. The charged, hydrophilic heads of the molecules are oriented toward the surfaces of the membrane, with the hydrophobic fatty-acid tails mingling with one another in the interior.

Both artificial and natural membranes form continuous sheets. Because of the tendency of the fatty acids to associate with one another and exclude water, small holes or rips in a membrane seal themselves spontaneously. This helps membranes fuse during endocytosis, exocytosis, and cell fusion.

The phospholipid portion of the membrane stabilizes the entire structure. At the same time, it makes the membrane somewhat fluid—about as fluid as lightweight machine oil—so there is the possibility of lateral movement of materials within the membrane. As we will see, some membrane proteins are relatively free to migrate about, and individual phospholipid molecules may also "travel." A given phospholipid molecule in the plasma membrane of a bacterium may get from one end of the bacterium to the other in a little over a second. On the other hand, it is *not* common for a phospholipid molecule in one half of the bilayer to flop over to the other side and trade places with another phospholipid molecule. For this to happen, the polar part of each molecule would have to move through the hydrophobic interior of the membrane. Since phospholipid flip-flops are rare, the two halves of the bilayer may be quite different.

There are important and fundamental similarities among all biological membranes, but different membranes may be quite different in their detailed compositions—even within the same cell. One big area of difference is in lipid composition. The proportions of different types of lipids vary from one

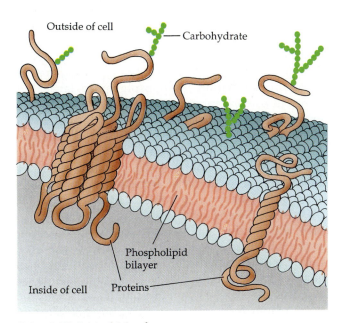

5.1 A Biological Membrane
The general molecular makeup of biological membranes, sometimes called the "fluid mosaic model," is shown in the drawing. A phospholipid bilayer constitutes the continuous portion of the membrane; proteins are embedded in the bilayer, and carbohydrates may be attached to proteins or phospholipids.

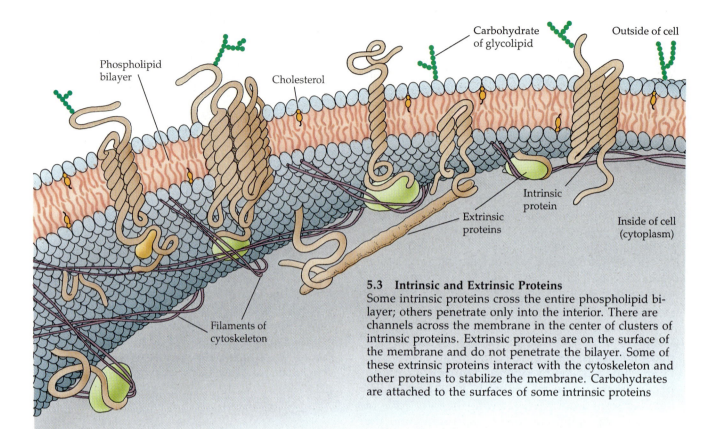

5.3 Intrinsic and Extrinsic Proteins
Some intrinsic proteins cross the entire phospholipid bilayer; others penetrate only into the interior. There are channels across the membrane in the center of clusters of intrinsic proteins. Extrinsic proteins are on the surface of the membrane and do not penetrate the bilayer. Some of these extrinsic proteins interact with the cytoskeleton and other proteins to stabilize the membrane. Carbohydrates are attached to the surfaces of some intrinsic proteins

type of membrane to another. For example, 25 percent of the lipid in many membranes is cholesterol, but some membranes have no cholesterol at all. In a membrane, cholesterol is always next to an unsaturated fatty acid, and its polar region extends into the surrounding aqueous layer (Figure 5.3). Cholesterol plays an important role in determining the fluidity of the membrane. At low concentrations, cholesterol molecules interrupt the hydrophobic interactions of other lipids and thus increase the fluidity of the membrane. Lipids constitute a major fraction of all membranes, and they always form the continuous matrix into which the other chemical components become inserted.

Membrane Proteins

The protein components of biological membranes are either **intrinsic proteins** or **extrinsic proteins** (Figure 5.3). Intrinsic proteins penetrate the phospholipid bilayer; many extend from one side of the membrane to the other. Extrinsic proteins are entirely outside the bilayer; they are attached to the rest of the membrane by weak (noncovalent) bonding with the exposed parts of the intrinsic proteins or, perhaps, with the hydrophilic parts of phospholipid molecules. These two types of proteins play different roles in membrane function.

The membranes of the various organelles differ sharply in their protein composition; quite different

biochemical reactions (many of them requiring membrane-bound enzymes) occur in different organelles. In many of the important reactions of cellular respiration and photosynthesis, membrane-bound enzymes carry electrons from a donor to an acceptor molecule. Accordingly, both mitochondria and chloroplasts have highly specialized internal membranes, and these differ markedly (Figure 5.4).

Many membrane proteins are relatively free to move around within the phospholipid bilayer. Evidence of this migration is dramatically illustrated by experiments using the technique of cell fusion. In the laboratory, specially treated cells of two different species, such as human and mouse, can be fused so that one continuous membrane surrounds the combined cytoplasm and both nuclei. Initially the experimenter can tell by the protein content which part of the plasma membrane came from which species. However, the membrane proteins of the two cells migrate in the joint membrane until, after about 40 minutes, they are uniformly dispersed (Figure 5.5). But there is also good evidence that at least some membrane proteins are *not* free to migrate at will—that they are to an extent held in place. These proteins are "anchored" by microfilaments, as described later in this chapter. Microtubules may also play a role.

Membrane proteins are asymmetrically distributed in the membrane. Many intrinsic proteins extend completely through the membrane and have specific parts of their primary structures on one side of the

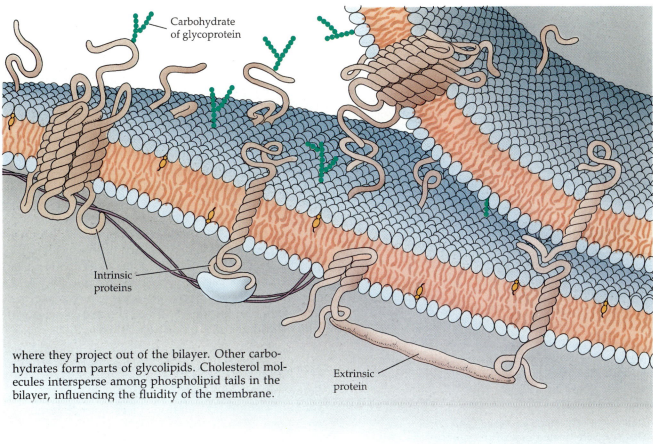

Carbohydrate
of glycoprotein

Intrinsic
proteins

Extrinsic
protein

where they project out of the bilayer. Other carbohydrates form parts of glycolipids. Cholesterol molecules intersperse among phospholipid tails in the bilayer, influencing the fluidity of the membrane.

membrane, specific parts within the membrane, and specific parts on the other side of the membrane. All the other membrane proteins (intrinsic and extrinsic) are localized on one side of the membrane or the other, but not both.

There is an additional form of asymmetry in the distribution of proteins within some plasma membranes. The molecules of one particular type of pro-

5.4 Proteins in Specialized Membranes
(a) The outer membrane has been fractured away from this mitochondrion in a muscle cell, exposing the inner membrane. The image has been magnified about 65,000 times. The particles giving the inner membrane a grainy appearance are proteins necessary for cellular respiration; the indentations are places where cristae extend into the matrix of the mitochondrion; the inner surfaces of several cristae are exposed in the lobe of the mitochondrion extending to the right. (b) Several thylakoids from a spinach chloroplast are seen here at a magnification of about 70,000 times. The distinct protein particles embedded in the membranes are necessary for photosynthesis.

(a)

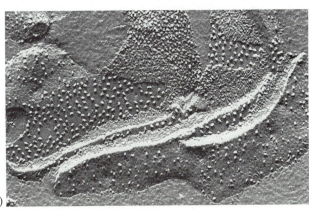

(b)

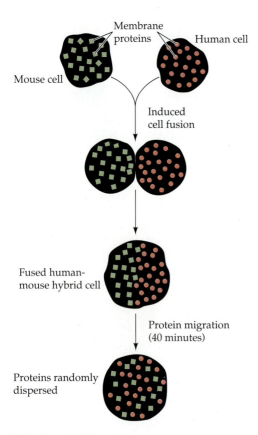

5.5 Protein Migration on Fused Cells
When specially treated mouse and human cells are joined, one continuous membrane surrounds the fused hybrid cell. Distinguished by different fluorescent dyes, many of the mouse and human membrane proteins co-mingle over time, demonstrating the mobility of some proteins in the fluid phospholipid bilayer.

tein may be confined to one part of the cell surface, rather than scattered evenly about. For example, in certain muscle cells, the membrane protein that receives the chemical signal from nerve cells is normally found only where a nerve cell meets the muscle cell. None of this protein is found elsewhere on the muscle cell plasma membrane. However, if the nerve is severed, the protein molecules may later be found evenly distributed over the entire membrane. If the nerve regenerates its attachment to the muscle, then the protein is once again limited to the junction area.

What determines whether a particular membrane protein is intrinsic or extrinsic? If it is intrinsic, what controls whether it reaches all the way through the membrane or is limited to one side? What keeps it in the bilayer, and what determines just how far in it reaches? All these questions are answered in terms of the tertiary structure of the protein (Chapter 3). Recall that the side chains of the various amino acids in a protein differ chemically. What matters here is that some of the side chains are hydrophilic and others hydrophobic. When the polypeptide chain folds into the final tertiary structure of the protein,

large patches of the protein's surface may be predominantly hydrophobic (Figure 5.6). If one end of a folded protein is hydrophilic and the other hydrophobic, it will be an intrinsic protein, sticking out of one side of the membrane. Many intrinsic proteins that reach from one side of the membrane to the other have hydrophobic α-helical regions large enough to penetrate the entire depth of the phospholipid bilayer. They also have hydrophilic ends that protrude into the aqueous environments on either side of the membrane. One such protein, the mysterious glycophorin, is described in Box 5.A. Proteins like this resist being removed from the membrane. If the hydrophobic surface of such a protein is withdrawn part way out of the phospholipid bilayer, it is repelled by the aqueous environment. If an intrinsic protein is pushed farther into the membrane, its hydrophilic end is pushed back by the hydrophobic fatty-acid region of the lipids. Thus such a protein may migrate laterally in the membrane sheet, but it may not push through the membrane or pop out of it.

Membrane Carbohydrates

Some membranes, including all plasma membranes, contain a significant amount of carbohydrate along with the lipids and proteins. For example, the plasma membrane of the red blood cell consists by weight of approximately 40 percent lipid, 52 percent protein, and 8 percent carbohydrate. Some of the membrane carbohydrate binds to lipids to form **glycolipids**. The function of glycolipids is poorly understood, but they may be quite important—there is an indication that the structures of glycolipids change when a cell becomes cancerous. Glycolipids may function in communication between cells.

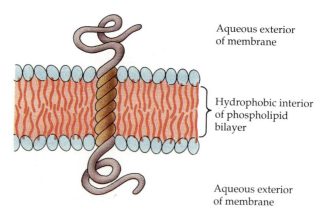

5.6 Surfaces of Intrinsic Proteins
In its folded tertiary structure, an intrinsic membrane protein has both hydrophobic and hydrophilic surfaces. There are hydrophilic surfaces (purple) comprising hydrophilic side chains on amino acids in the polypeptide chain; hydrophobic side chains (brown) on the protein's surface result in a hydrophobic region that buries itself among the fatty acid tails in the interior of the phospholipid bilayer.

BOX 5.A

Glycophorin: A Mystery Protein

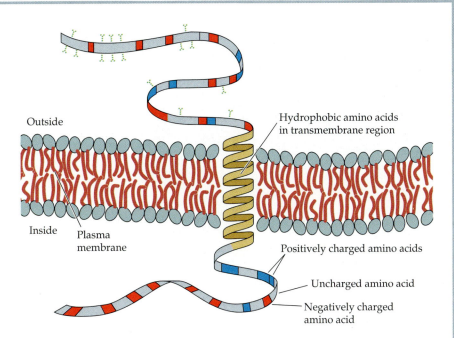

Glycophorin is one of the two most abundant proteins in the plasma membrane of human red blood cells. No cells other than red blood cells have glycophorin, but each red blood cell has over half a million molecules of this protein. Glycophorin carries the great majority of the carbohydrates of the red blood cell membrane. It is an intrinsic protein, passing clear through the membrane and extending to both sides as shown in the figure. The largest part extends outside the cell, where it binds carbohydrates (indicated in green) at 16 of the amino acids. The part of the glycophorin molecule that extends into the interior of the cell contains a number of amino acids with positively or negatively charged side

chains. The part of the molecule that is contained in the lipid bilayer consists entirely of amino acids with hydrophobic side chains (these amino acids are shown in yellow).

Surely a protein present in such abundance and specific to a single cell type must play an important role in the cell. However, the role of glycophorin in red blood cells is totally unknown as this is written. Not only

has no function for glycophorin been discovered, but red blood cells that are devoid of glycophorin seem to function perfectly normally. Owing to a mutation, some people have red blood cells without glycophorin, but these people show no sign of disease. We look forward to the day when we can say what all that glycophorin in the membranes of red blood cells is for.

Most of the carbohydrate in membranes is bound to proteins, forming **glycoproteins**. These bound carbohydrates are oligosaccharide chains, usually not exceeding 15 monosaccharide units in length. The oligosaccharide chains are added to the membrane proteins inside the endoplasmic reticulum and are modified in the Golgi apparatus.

Although they are relatively small and made up from only nine building blocks—the monosaccharides—the carbohydrates linked to membrane proteins and lipids are exceedingly diverse. To understand how this can be, recall that monosaccharides may join to form branched oligomers (Figure 3.16). The possibility of different branching patterns greatly increases the diversity that could be achieved by the monosaccharide sequence of a carbohydrate alone. Also, monosaccharides may link together at any of several different carbons. All in all, membrane carbohydrates have great specificity and diversity. This structural diversity is important in all sorts of reactions at the cell surface in which different membrane carbohydrates each recognize and react with

specific foreign substances. All plasma membrane carbohydrates are on the *outside* of the plasma membrane, as befits their role as recognition sites for foreign substances and cells; none face into the cell.

MICROSCOPIC VIEWS OF BIOLOGICAL MEMBRANES

The plasma membrane and the membranes within cells are too thin to be resolved with the light microscope. However, the electron microscope offers a variety of useful ways to examine membranes. In early work, very thin slices of tissue were made with diamond knives; the resulting sections were examined by transmission electron microscopy. The cuts were very clean, and membranes were often seen in cross section in electron micrographs of the slices (Figure 5.7). A more detailed understanding of membrane architecture had to await methods for seeing surface views and for actually revealing the hydrophobic interior of the membrane.

5.7 Thin Section of a Plasma Membrane
A small portion of the edge of a red blood cell, magnified about 300,000 times. The plasma membrane's phospholipid bilayer appears as two dark lines separated by a light region.

The first successes were achieved by a technique known as **freeze fracture**. The tissue to be examined is frozen solid and then *broken* rather than cut. To visualize the consequences of this fracture, picture a chocolate bar with almonds. If the bar is cut carefully with a very sharp knife, the cut will pass cleanly through the nuts as well as the chocolate. If, on the other hand, the bar is simply broken, the break will pass around any almonds in its path and reveal them where they protrude from the chocolate. Similarly, in the freeze-fracture technique, the break tends to pass around membrane-encased organelles and to pass *between* the two halves of the phospholipid bilayer but *around* the proteins within it. When the exposed surfaces of the fractured bilayer are examined with the electron microscope, they appear

5.8 Freeze Fracturing and Etching
(a) When frozen cells are chipped with a knife, various faces of membranes may be exposed. In addition to the outer and inner surfaces of the entire membrane, the interior of the phospholipid bilayer is often seen. Intrinsic proteins project from the interior fracture faces of each layer of the phospholipid bilayer. (b) Freeze-etching exposes more of the surfaces of membranes and particles in freeze-fractured material by evaporating water (ice) away from the structures. (c) Shadowing with platinum aids in the visualization of surface irregularities by creating contrasting shadows and highlights, as in this freeze-etched preparation of part of a photosynthetic protist. At the lower left is a Golgi body, at the upper right a chloroplast, and at the upper left an inside view of the nuclear envelope, dotted with pores. Note the abundant vesicles between the nucleus and the forming face of the Golgi body. Examples of electron micrographs of freeze fractured preparations are found in Figures 5.4 and 5.9 (top photo).

"bumpy"—the intrinsic proteins are revealed (Figure 5.8a). Further clarification of membrane structure is obtained by the method of **freeze etching**. The freeze-fractured sample is kept cold and put under a high vacuum for a minute or so, allowing some of the frozen water to evaporate, wherever it is exposed to the vacuum. As the water evaporates, it reveals more of the texture of the membrane by uncovering surfaces that were covered by ice (Figure 5.8b). With both freeze-fractured and freeze-etched preparations, contrast is enhanced by **shadowing** with platinum. The metal is "sprayed" on from an angle so that

(a) Freeze fracturing. Frozen cells are chipped with a knife, exposing membrane faces and inner surfaces.

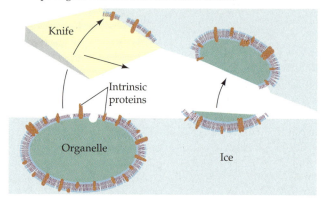

(b) Freeze etching exposes more surfaces by evaporating water away from structures.

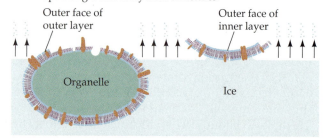

(c) Freeze etched sample shadowed with a heavy metal reveals cellular structures in electron micrograph.

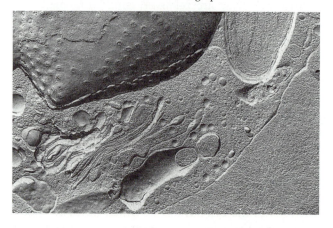

bumps and dips on the surface give shadow patterns that make them easier to see and to interpret (Figure 5.8c).

Freeze fracture and freeze etching were instrumental in formulating our current picture of membrane structure. In particular, they made it clear that the membrane is not just a continuous bilayer of phospholipids with the proteins spread on its surface. Rather, many of the proteins are embedded in the bilayer to various depths.

WHERE ANIMAL CELLS MEET

Electron microscopic studies also revealed that animal cells are surrounded by an **extracellular matrix** consisting of a fine meshwork of polysaccharides permeated by fibrous proteins. The most abundant of the proteins in the extracellular matrix is collagen, whose triple-helical structure was described in Chapter 3. One major function of the matrix is simply to hold the cells together as tissues. Additionally, it helps to direct the ways in which cells interact with one another and to orient the individual cells. This is actually a reciprocal relationship, because the cells themselves secrete the extracellular matrix and establish its orientation. Thus the organizational pattern of a tissue, once established, tends to be maintained as the tissue grows—the first cells orient the matrix, which in turn orients the new cells, and so forth.

Individual cells also come in direct physical contact with one another and may form special links, called **junctions**. The junctions fall into three categories—desmosomes, tight junctions, and gap junctions—depending on their structure and function.

Desmosomes

Desmosomes simply cause neighboring cells to adhere tightly to one another. For example, epithelial cells (cells lining a body cavity or an exterior body surface) are held firmly together by desmosomes. Some desmosomes act like spot welds or rivets at individual points, while others form continuous "belts" around the outer ends of adjacent epithelial cells. Desmosomes are easily recognized in electron micrographs by the characteristic cytoplasmic plaque that appears within each of the "welded" cells (Figure 5.9). The plaques are associated with dense networks of keratin fibers that extend into each of the cells and connect the cells through the space between the two plasma membranes. (Keratin is a fibrous protein that is classed as an intermediate filament; see Chapter 3. It makes up the bulk of our fingernails and hair.) Abundant desmosomes hold adjacent epithelial cells together, strengthening the tissue.

Tight Junctions

Desmosomes cause cells to adhere strongly to one another, even though there is a gap between the plasma membranes. **Tight junctions**, on the other hand, bind so closely that materials cannot move between the joined cells. In fact, there is no space at all between cells at a tight junction. Tight junctions result from the mutual binding of strands of specific proteins in the plasma membranes of the two cells, forming belts that virtually fuse the two membranes (Figure 5.9). Epithelial cells are so extensively linked by tight junctions between all adjacent cells that substances on one side of an epithelium (flat tissue composed of epithelial cells) cannot seep through to the other side. Thus, for example, the contents of the gut (digestive tract) cannot seep between the epithelial cells of the gut lining—the contents can pass through the epithelium only if they are specifically taken up and processed by the epithelial cells and then released into the blood. The selective permeability of the plasma membranes assures that food substances from the digestive tract pass through the epithelial cells to the bloodstream, while many unwanted substances are unable to pass through the membrane. Tight junctions help direct the transport of materials in the body.

Gap Junctions

Adjacent cells in some animal tissues communicate through the third type of junction. A **gap junction** is a gap of 2.7 nm between the plasma membranes of the two cells spanned by many pipelike channels 1.5 nm in internal diameter (Figure 5.9). The channels, called connexons, are made up of a specific protein. Connexons provide a cytoplasmic connection from one cell to the other, through which chemical substances or electric signals may pass.

In Chapters 36 and 38 we will see that the muscle cells of the vertebrate heart, many smooth muscles, and some nerve cells are connected by gap junctions, allowing the direct passage of an electric signal—the nervous impulse—from one cell to the next. Gap junctions are also important in normal embryonic development, for they appear at a specific developmental stage. If animal tissues are experimentally disrupted to dissociate the cells, the cells quickly form new gap junctions as they reassociate. In fact, cells isolated from one species of vertebrate readily form gap junctions with cells from other vertebrate species. Cancer cells, however, never develop gap junctions; this presumably means they do not communicate with other cells as normal cells do.

The differences among the three types of junctions are illustrated in Figure 5.9. Desmosomes allow cells to *adhere* strongly to one another. Tight junctions

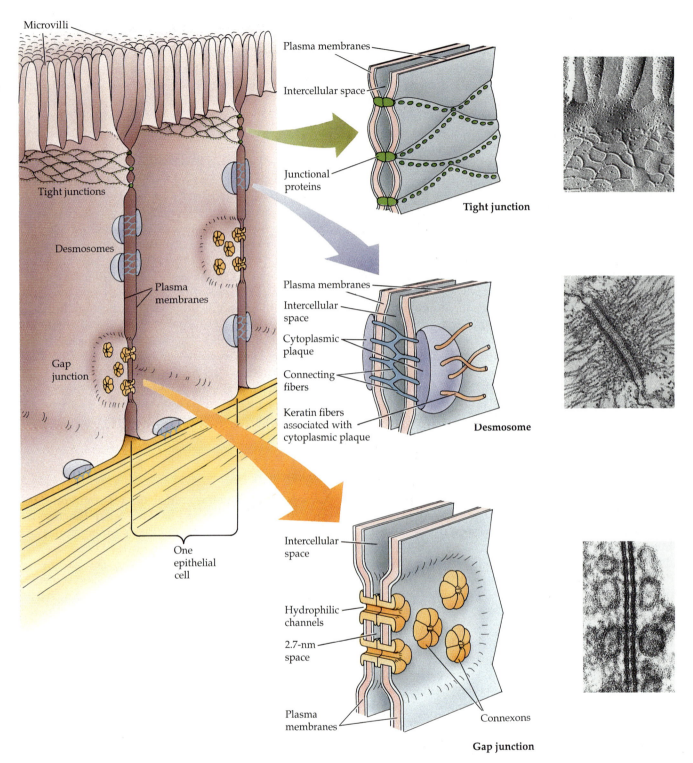

Microvilli

Tight junctions

Desmosomes

Plasma membranes

Gap junction

One epithelial cell

Plasma membranes

Intercellular space

Junctional proteins

Tight junction

Plasma membranes
Intercellular space
Cytoplasmic plaque
Connecting fibers
Keratin fibers associated with cytoplasmic plaque

Desmosome

Intercellular space

Hydrophilic channels

2.7-nm space

Plasma membranes

Connexons

Gap junction

5.9 Junctions between Epithelial Cells

Desmosomes, tight junctions, and gap junctions hold epithelial cells together. Desmosomes link adjacent cells but permit materials to move between them. Connecting fibers bind the neighboring cells to each other; cytoplasmic plaques anchor these fibers in both cells. Within each cell, a meshwork of keratin fibers extends away from the cytoplasmic plaque. Although the two cells are held firmly together, there is a 24-nm space between the two plasma membranes. Tight junctions bar the movement of dissolved materials through the space between epithelial cells. There is no intercellular space where a tight junction occurs. The junctional proteins of adjacent cells bind tightly to fuse the plasma membranes. Long rows of junctional proteins form a complex "lacework," the tight junction. Adjacent cells communicate through gap junctions, where dissolved molecules and electrical signals may pass from one cell to the other through channels. Each channel is made of two connexons. A connexon reaches through the phospholipid bilayer of the membrane and extends into the cytoplasm on one side and into the external environment on the other side, where it abuts a connexon on an adjacent cell. The name "gap junction" refers to the 2.7-nm gap between the two plasma membranes.

serve as *barriers* to the passage of molecules through the space between cells. Gap junctions provide channels for chemical and electric *communication* between the cells on the opposite sides of the junctions.

DIFFUSION

How do molecules move in an aqueous environment? Before we discuss further the movements of molecules across membranes, it is important to consider this.

Nothing in this world is ever absolutely at rest. Everything is in motion, though the motions may be very small. The constant jiggling of molecules and ions in solution increases as the temperature rises. An immediate consequence of the random jiggling is that all the components of a solution tend eventually to become evenly distributed throughout the system. If, for example, a drop of ink is allowed to fall into a container of water, the pigment molecules of the ink will move about at random, spreading through the system until at last the concentration of pigment—and thus the intensity of color—is exactly the same in every drop of liquid in the container. A solution in which the particles are uniformly distributed is said to be at equilibrium. This process of random movement toward a state of equilibrium is called **diffusion**.

There are a few important things to know about diffusion. The first is that the motion of each individual particle is absolutely random, even though the *net* movement of particles is directional until equilibrium is reached. Diffusion is this net movement—always in the direction *from greater* concentration *to lesser* concentration (Figure 5.10). The second important thing is that in a complex solution, the diffusion of each substance is independent of that of the other substances. A third consideration is that one may quantify diffusion, as shown in Box 5.B.

How fast substances diffuse depends on a number of factors, including the diameter of the molecules or ions, the temperature, the electric charge, if any, of the diffusing material, and the **concentration gradient** in the system. The concentration gradient is the change in concentration with distance in a given direction. The greater the concentration gradient, the more rapidly substances diffuse.

Within cells, where distances are very short, distribution of solutes by diffusion is rapid. Small molecules and ions may move from one end of an organelle to another in a fraction of a millisecond, or from the center of a cell to its surface in not much more time than that (Box 5.B). On the other hand, the usefulness of diffusion as a transport mechanism falls off drastically as distances become greater. Diffusion over a centimeter may take an hour or more; over meters it takes years (it is assumed throughout

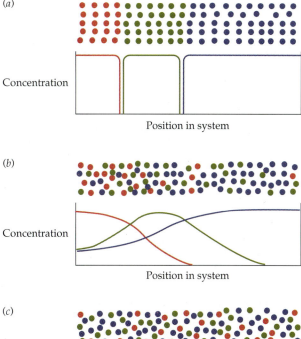

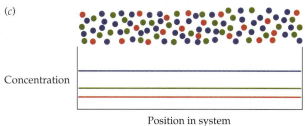

5.10 Diffusion
(a) Initially, each of three dissolved substances is highly concentrated in part of the space in which they have been combined, and absent from the rest of it. (b) As the substances mix, each substance is at a higher concentration—its peak on the graph—in part of the space but is also present elsewhere. (c) At equilibrium, all three substances are uniformly distributed throughout the space, as shown by the straight horizontal lines.

this discussion that the fluid is not stirred or moved in any other way). Diffusion would not be enough to distribute materials over the length of the human body, but within our cells or across layers of one or two cells it is rapid enough to distribute small molecules and ions almost instantaneously.

CROSSING THE MEMBRANE BARRIER

In principle, *all* substances diffuse, although the rates of diffusion vary. In a solution without barriers, all the solutes diffuse at rates determined by their physical properties, as described in Box 5.B, and in directions determined by the concentration gradient of each solute. If a barrier in the form of a biological membrane is introduced, the movement of the different solutes can be affected significantly. Some solutes move fairly readily through the membrane,

BOX 5.B

The Rapidity of Diffusion Within Cells

D and Molecular Weight for Some Biologically Important Substances

SUBSTANCE[a]	D (cm^2/sec $\times$ 10^6)	MOLECULAR WEIGHT
Oxygen gas	19.8	32
Acetylcholine	5.6	182
Sucrose	2.4	342
Serum albumin	0.7	69,000

[a]Acetylcholine carries nerve impulses from cell to cell; sucrose is a common food substance; and serum albumin is a typical protein.

It is possible to talk in quantitative terms about diffusion, even though the process is based on random events. Albert Einstein developed the general interpretation leading to the simple equations given here. These are useful for estimating the distance a molecule moves in some very short time t. The first equation applies in a situation in which diffusion is one-dimensional, as, for example, across the space between two cells. The equation is

$$d^2 = 2Dt$$

where d is the (very small) distance moved, t is the time, and D is what is called the **diffusion coefficient** characteristic of the particular substance. (See the table for typical values of D.) To see how this equation is used, we can calculate the time required for a molecule of acetylcholine to travel across the space between one nerve cell and the next. (Acetylcholine transmits nerve impulses by traveling between nerve cells.) The distance between the two cells is about 20 nm, or 2×10^{-6} cm. From the table we see that the diffusion coefficient for acetylcholine is 5.6. Substituting these values for d and D in the equation, we may solve for t, getting a value of 0.4 microseconds. As you can see, diffusion can be very rapid indeed when distances are very small.

We need a somewhat different equation when we are dealing with diffusion throughout a three-dimensional space. After a compound is produced in a particular organelle, for example, it may diffuse in all directions throughout the cell. Einstein's equation for this is

$$d^2 = 6Dt$$

Try solving this equation for t, using any of the compounds in the table, and using different values for d, ranging from very small ones such as that in the example above to large distances in centimeters and meters. This will give you a feeling for the circumstances under which diffusion is—or is not—likely to be of importance as a transport mechanism.

whereas others are effectively prevented from crossing it. Molecules that can move through the phospholipid barrier diffuse from one compartment to the other until their concentrations are equal on both sides of the membrane. Molecules that cannot cross the membrane diffuse only within their own compartments, so their concentrations remain different on the two sides of the membrane.

Things move through biological membranes in three ways. In **simple diffusion**, small, nonpolar molecules pass through the lipid bilayer of the membrane. Equilibrium is reached when the concentrations of the diffusing substance are identical on both sides of the membrane. Individual molecules are still passing through the membrane at equilibrium, but equal numbers of molecules are moving in each direction, so there is no change in concentration.

A second mechanism is also diffusion—movement down a concentration gradient to produce equal concentrations of solute on the two sides of a membrane. The difference here is that the solute molecules do not diffuse through the membrane on their own. Rather, they combine with **carrier molecules** in the membrane. The carriers, by a mechanism that is still not understood, allow the solute molecules to pass to the other side. This mechanism is **facilitated diffusion**, sometimes called carrier-mediated diffusion.

Both simple diffusion and facilitated diffusion permit the passage of a solute across a membrane down a concentration gradient, that is, from the side of higher concentration to the side of lower concentration. Neither of these mechanisms allows for the transport of a solute *against* a concentration gradient, that is, from the side of the membrane where the concentration is lower to the side where it is higher. Exactly such a phenomenon, however, is of extreme importance to living things, and it is called **active** transport. Like facilitated diffusion, active transport relies on carrier molecules. Unlike facilitated diffusion, this active process by which ions or molecules are moved against their own concentration difference requires a great deal of energy. This input of energy is provided by ATP obtained through the process of cellular respiration (Chapter 7).

In all three mechanisms of movement through membranes, the *rate* of movement depends on the concentration difference across the membrane. In simple diffusion, the net rate of movement is directly proportional to the concentration difference across the membrane (Figure 5.11a). In facilitated diffusion, the rate of movement also increases with the difference in solute concentration across the membrane,

but a point is reached at which further increases in concentration difference are not accompanied by an increased rate (Figure 5.11a). The facilitated diffusion system is said to be saturated at this high concentration. If there are only so many carrier molecules per unit area of membrane, then the rate of movement reaches a maximum when all the carrier molecules are fully engaged in moving solute molecules. In other words, at high solute concentration differences across the membrane, there are not enough carrier molecules free at a given moment to handle all the solute molecules. Like facilitated diffusion, the rate of active transport stops increasing at high solute concentrations (Figure 5.11b).

Simple Diffusion Through a Membrane

For most substances, the ability to pass through biological membranes depends on how soluble they are in lipids. The more lipid-soluble the compound, the more rapidly it diffuses through biological membranes (Figure 5.12). This statement holds true over a wide range of molecular weights. Only certain ions and the smallest of molecules seem to deviate from this rule; materials such as water, K^+, and Cl^- pass through membranes much more rapidly than their solubilities in lipid would predict. We can understand these observations in terms of the chemical structure of the membrane.

5.11 Crossing Biological Membranes
(a) Both simple and facilitated diffusion equalize concentrations of a solute across a membrane. In simple diffusion the rate of movement of a solute across a membrane is directly proportional to the difference in the concentrations of the solute on the two sides of the membrane, as the top graph shows. At equilibrium the concentration of solute inside the membrane equals that outside. With facilitated diffusion, equal concentrations are also reached, but a protein carrier in the membrane allows the rate of solute crossing to be greater. This rate reaches a maximum when all carriers are saturated with solute. (b) Active transport employs energy and can move solutes against a concentration gradient. The rate of movement is similar to that of facilitated diffusion and reaches a maximum when membrane carriers are saturated, but the final concentrations of solute on either side of the membrane can be quite different due to the expenditure of energy.

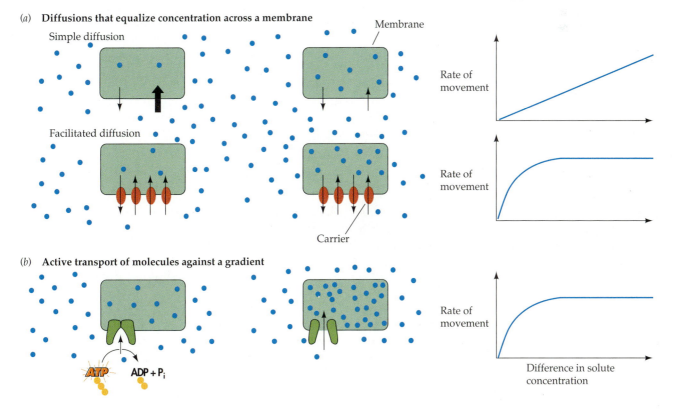

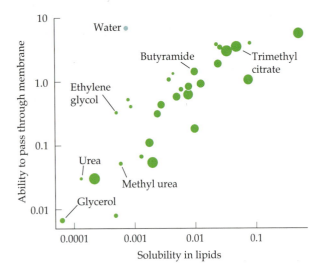

5.12 Membrane Permeability
Substances tend to cross membranes in relation to their solubility in lipids, as shown by the data points in the graph. The sizes of the points correspond roughly to the sizes of various molecules studied; the assortment of molecules of all sizes along the curve indicates that size alone is not an important factor for permeability. For molecules (such as water) that do not cross in relationship to their solubility in lipids, other conditions must be postulated.

Recall that a key feature of membrane architecture is the double layer of phospholipid molecules that forms the framework of the membrane (see Figure 5.2). The inner portion of the double layer consists of the fatty acid chains of the phospholipids, along with cholesterol and other highly hydrophobic, nonpolar materials. When a hydrophilic molecule or ion moves into such a hydrophobic region, it is "rejected" by the lipid layer and forced back again. Such a molecule seldom enters the hydrophobic region; it enters only when energy is available to push it in. On the other hand, a molecule that is itself hydrophobic, and hence soluble in lipids, enters the membrane readily and is thus able to pass through it.

This accounts for most of the information in Figure 5.12, but it does leave the problem of how water itself can move so rapidly through biological membranes —water is not hydrophobic enough to account for its flow in and out of many cells. The diffusion of water

into and out of cells is still under debate because many workers feel that the rapid movement of water through membranes can be explained in several ways.

We must also account for the rapid movement of certain ions through biological membranes. They pass through pores, or **channels**, in the membranes of all eukaryotic cells. There are specific channels for potassium, sodium, calcium, and chloride ions, allowing these ions to diffuse through membranes in spite of their hydrophilic character. Interestingly, these channels are **gated**: They can be opened or closed in response to specific chemical or electric signals. We will see in Chapter 36 how gated channels function in the development of the nervous impulse.

Simple diffusion, then, accounts for much of the passage of lipid-soluble molecules, water, and certain ions in and out of cells and their organelles.

5.13 Membrane Transport Proteins
Substances pass through biological membranes by many mechanisms, as illustrated here. Most of the mechanisms —all except for simple diffusion through the phospholipid bilayer, shown at the far left—involve proteins acting as channels or carriers.

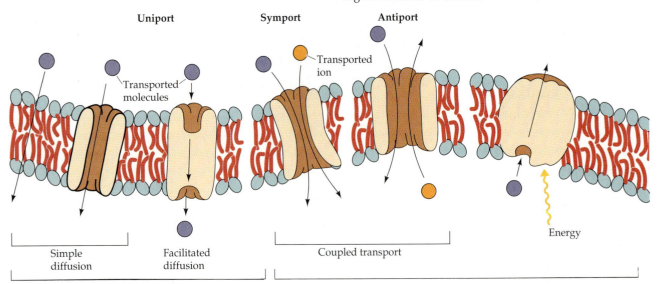

Membrane Transport Proteins

Both the channels through which certain ions diffuse and the carriers for facilitated diffusion and active transport are **membrane transport proteins**, intrinsic proteins that reach from one side of the membrane to the other. Different kinds of membrane transport proteins allow specific substances to pass through the membrane in various ways (Figure 5.13). The channels, for example, have an aqueous region through which ions can diffuse.

Membrane transport proteins may be classified further. **Uniports** transport a single type of solute. **Coupled transport systems** transport two or more different solutes, but neither solute can be moved by a coupled transport system unless the other solute is also present. If the two coupled solutes are transported in the same direction, the system is a **symport**. If they are transported in opposite directions, the system is an **antiport**. One example of an antiport is an anion channel that is abundant in the plasma membrane of red blood cells. Each red blood cell contains about 1 million of these channels, which allow the exchange of bicarbonate and chloride ions. This particular exchange is important in transporting carbon dioxide, which is present in the bloodstream as bicarbonate ions, from working tissues where carbon dioxide is produced to the lungs where it is released (Figure 5.14).

Membrane transport proteins are very specific in their structures. A given membrane transport protein will carry only one particular solute, or solutes that have very similar structures. The great diversity of protein structures allows this high specificity for different transported solutes, just as it allows an enzyme to be highly specific for accelerating a particular chemical reaction.

Facilitated Diffusion

Most biochemical molecules are too hydrophilic to enter the phospholipid bilayer and too large to move through membranes the way water does. Thus they are prevented from passing through the membrane —unless they can interact with a carrier of suitable specificity. As just mentioned, the carriers are transmembrane transport proteins. Where these proteins contact the phospholipid bilayer, their surfaces are hydrophobic, but within them is a hydrophilic region through which the diffusing material passes. As the

solute passes through the membrane, the carrier protein undergoes a change in tertiary or quaternary structure.

In facilitated diffusion, the carrier proteins enable the solutes to pass in *both* directions. The net movement is toward the side where the solute concentration is lowest, simply because on the side where the concentration is greater the carriers encounter more solute molecules to transport.

Active Transport

Things are different in active transport. For one thing, the transported substance may be, and usually is, moved from a region of low concentration to one of higher concentration—that is, it is transported against a concentration difference. This is an "uphill" process that requires an input of energy (Chapter 6). In contrast with the carriers for facilitated diffusion, active-transport carriers operate in one direction only.

There are two basic types of active transport. **Primary active transport** requires the direct participation of adenosine triphosphate (ATP), an energy-storing compound found in all cells. In primary active transport, energy released from ATP drives the movement of specific ions against a concentration difference. As an example of primary active transport, we can compare the concentrations of potassium ions (K^+) and

5.14 An Anion Channel

Red blood cell membranes contain anion channels through which chloride and bicarbonate ions are exchanged. This is important in the transport of carbon dioxide to the lungs. One end of the channel protein binds an extrinsic protein inside the red blood cell.

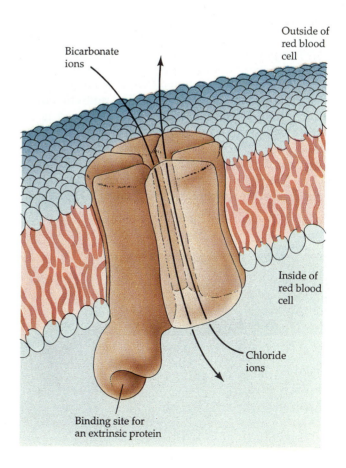

Bicarbonate ions

Outside of red blood cell

Inside of red blood cell

Chloride ions

Binding site for an extrinsic protein

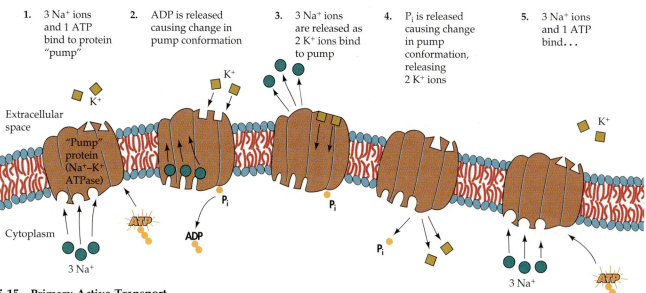

1. $3 Na^+$ ions and 1 ATP bind to protein "pump"

2. ADP is released causing change in pump conformation

3. $3 Na^+$ ions are released as $2 K^+$ ions bind to pump

4. P_i is released causing change in pump conformation, releasing $2 K^+$ ions

5. $3 Na^+$ ions and 1 ATP bind...

Extracellular space

"Pump" protein (Na^+–K^+ ATPase)

Cytoplasm

K^+

P_i

P_i

P_i

ATP

ADP

$3 Na^+$

$3 Na^+$

ATP

5.15 Primary Active Transport

A membrane protein actively pumps Na^+ out of the cell and K^+ into the cell. This movement against the concentration differences is powered by energy from ATP. For each molecule of ATP used, $2 K^+$ are pumped in and $3 Na^+$ are pumped out. The transport protein molecule—the pump—extends clear through the phospholipid bilayer of the membrane; thus the highly hydrophilic Na^+ and K^+ ions need not interact with the hydrophobic center of the membrane. This is an example of primary active transport.

sodium ions (Na^+) inside a nerve cell and in the fluid bathing the nerve (Table 5.1). The K^+ concentration is much higher inside the cell, whereas the Na^+ concentration is much higher outside. In spite of this, the nerve cells continue to pump Na^+ *out* and K^+ *in*, against these concentration differences, ensuring that the differences are maintained. The **sodium–potassium pump** is found in all animal cells and is an intrinsic membrane glycoprotein. It repeatedly breaks down one molecule of ATP to ADP, brings two K^+ ions into the cell, and exports three Na^+ ions (Figure 5.15). The Na–K pump is thus an antiport. There are pumps for the transport of several other ions, but

only cations are transported directly by pumps in primary active transport. The transport of other solutes is achieved by secondary active transport.

Secondary active transport, in contrast to primary active transport, does not use ATP directly; rather, the transport of the solute is tightly coupled to the difference in ion concentration established by primary active transport. The movement of particular solutes, such as sugars and amino acids, is regulated by coupled transport systems that move these specific solutes against their concentration difference, using energy "regained" by letting Na^+ or other ions move *with* their concentration difference (Figure 5.16). Some of the coupled transport systems for secondary active transport are symports, and some are antiports. Putting the two forms of active transport together, we see that energy from ATP is used in one example of primary active transport to establish concentration differences of potassium and sodium ions; the movement of some sodium ions in the opposite direction provides energy for the secondary active transport of the sugar glucose. Other secondary active transporters are used for the uptake of amino acids and other solutes.

Osmosis

For some years there was disagreement about whether water only diffused through biological membranes or was sometimes actively transported as well. It is now clear that water moves through membranes only by **osmosis**, the movement of a solvent through a membrane in accordance with the laws of diffusion. This is a process in which no metabolic energy is expended, and it can be understood in terms of a very few principles, which we will develop here using two simple examples.

TABLE 5.1 Concentration of Major Ions Inside and Outside the Nerve Cell of a Squid		
	CONCENTRATION (MOLAR)	
ION	**IN NEURON**	**IN BLOOD**
K^+	0.400	0.020
Na^+	0.050	0.440
Cl^-	0.120	0.560

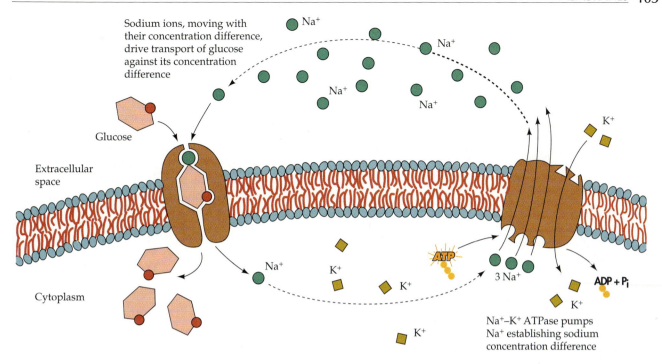

Sodium ions, moving with their concentration difference, drive transport of glucose against its concentration difference

Na⁺

Glucose

Extracellular space

Cytoplasm

Na⁺

K⁺

ATP

3 Na⁺

ADP + Pᵢ

K⁺

Na⁺–K⁺ ATPase pumps Na⁺ establishing sodium concentration difference

5.16 Secondary Active Transport
The Na^+ concentration difference established by primary active transport powers the secondary active transport of glucose and some other substances. Glucose moves through the membrane against its concentration difference, accompanied by Na^+ ions that are moving with their concentration difference (left). The driving force for secondary active transport is the Na^+ concentration difference resulting from ATP-driven primary active transport (right).

Red blood cells are normally suspended in a fluid called plasma, a liquid containing salts, proteins, and other solutes. If a drop of blood is examined under the light microscope, the red cells are seen to have their familiar shape. If pure water is then added to the drop of blood, the cells quickly swell and burst (Figure 5.17). Similarly, if slightly wilted lettuce is put in pure water, it soon becomes crisp; by weighing it before and after, we can show that it has taken up water (Figure 5.18). If, on the other hand, the red blood cells or crisp lettuce leaves are placed in a relatively concentrated solution of salt or sugar, the leaves become limp and the red blood cells pucker and shrink.

These and other observations show that solute concentration is the principal factor in what is called the **osmotic potential** of a solution. The greater the solute concentration, the more negative the osmotic potential of the solution. Pure water has nothing dissolved in it, and its osmotic potential equals zero. Other things being equal, if two unlike solutions are separated by a differentially permeable membrane (one that allows water to pass through but not solutes), osmosis—solvent movement—proceeds toward the solution with the more negative osmotic potential.

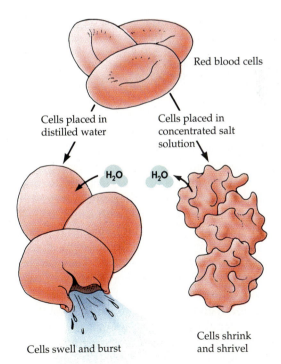

Red blood cells

Cells placed in distilled water

Cells placed in concentrated salt solution

H_2O

H_2O

Cells swell and burst

Cells shrink and shrivel

5.17 Osmosis and Cell Shape
A mammalian red blood cell suspended in plasma has a biconcave shape—indented on both sides. If the cell is placed in distilled water, it swells and bursts due to the entry of water by osmosis. If the cell is placed in a solution in which salts are more concentrated than in plasma, water leaves the cell by osmosis, causing it to shrivel.

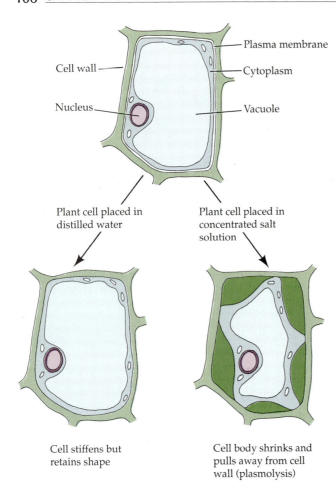

Plasma membrane

Cell wall

Cytoplasm

Nucleus

Vacuole

Plant cell placed in distilled water

Plant cell placed in concentrated salt solution

Cell stiffens but retains shape

Cell body shrinks and pulls away from cell wall (plasmolysis)

5.18 Osmosis and Cells with Walls
Water enters the vacuole of a plant cell placed in distilled water, but the cell retains its shape due to the rigid cell wall. When a plant cell is placed in a concentrated salt solution, the cell loses water to its surroundings; the cell wall retains its shape, but the plasma membrane shrinks away from the inside of the wall as the cell's volume is reduced.

If two solutions have identical osmotic potentials, even though their chemical compositions may be very different, they are said to be **isotonic** to one another. If they are not isotonic, then solution A with a more negative osmotic potential (with a higher concentration of solutes) is said to be **hypertonic** to solution B; solution B is **hypotonic** to solution A. All three of these terms are strictly relative. They can be used only in *comparing* the osmotic potentials of two solutions; no solution can be called "hypertonic," for example, except in comparison with another solution that has a less negative osmotic potential.

In animals, the direction of osmosis is determined by osmotic potentials. A red blood cell takes up water from a solution that is hypotonic to the cell's contents. The cell bursts because its delicate plasma membrane cannot resist the swelling of the cell. The integrity of red blood cells and other blood cells is absolutely dependent upon the maintenance of a constant osmotic potential in the plasma they are

suspended in, because the plasma must be isotonic with the cells if the cells are not to burst or shrink. In contrast, the cells of plants, monerans, fungi, and some protists have cell walls that limit the volume of the cells and keep them from bursting. Cells with sturdy cell walls take up a limited amount of water and, in so doing, build up a pressure against the cell wall that prevents further water from entering. Thus in cells with walls, osmosis is regulated not only by osmotic potentials but also by an opposed **pressure potential**. Osmotic phenomena in plants are discussed in Chapter 29.

SOME OTHER ACTIVITIES OF MEMBRANES

Membranes and Energy Transformations

Certain biological membranes are specialized to process energy—to convert it from light energy to the energy of chemical bonds (Chapter 8), trapping energy released in oxidation–reduction reactions (Chapter 7), and other vital activities. Why should membranes be involved in these activities? The answer is in two parts: structural organization and the separation of electric charges.

Structural organization: Many processes in cells require various substances and take place step by step, with the products of one step being the reactants for the next step. If the necessary chemical substances for these reactions are all moving about at random, only chance collisions will bring them together and the processes will go forward slowly, if at all. If, on the other hand, the different substances are bound to a membrane (and especially if they are arranged in an orderly fashion), the product of one reaction can be released in close proximity to where it is needed for the next step in the pathway, and so forth—a virtual assembly line is established. In this sense, the membrane is a pegboard for the orderly attachment of specific proteins and other molecules.

Separation of charges: A biological membrane can act rather like an electric battery. Work can be obtained from a battery by letting electrons flow from one of its terminals to the other by way of some device, like a motor or a light bulb, that uses the electric current. Something similar takes place in both photosynthesis and cellular respiration. These processes will be discussed in more detail in Chapters 7 and 8. Briefly, because of the limited permeability of the membranes in mitochondria and chloroplasts and because of the activities of certain electron carriers in those membranes, it is possible to establish a substantial gradient of both electric charge and pH across them. When these gradients are discharged by letting electric charge flow back through the membrane, this flow of charge can be used to do work or to form the energy-rich bonds of ATP. Without a membrane to allow the separation of charge, these reactions could not proceed.

Both membrane structure and the ability to separate electric charges relate to the properties of the two bulk components of membranes: lipids and proteins. The pegboard effect of the membrane's structure comes from the ability of the phospholipid bilayer to hold certain proteins in a defined plane so that they do not diffuse freely throughout the cell. Certain membrane proteins create the separation of charges, which is then maintained by the insulating effect of the phospholipid bilayer.

Receptors on the Membrane Surface

Membrane proteins and carbohydrates recognize and bind a variety of things to the outer surface of the plasma membrane. Antibodies (Chapter 16) recognize target cells by virtue of specific proteins or carbohydrates on the surfaces of those cells. Viruses may begin their attack on their intended host cells by attaching to carbohydrates on the host surface. Many hormones, including insulin, are recognized by membrane proteins that serve as receptors (Chapter 34). As already mentioned, many nerve cells pass information to other nerve cells or to muscle fibers by means of a substance called acetylcholine (Chapter 36). Acetylcholine activates these cells by attaching to a membrane protein—the acetylcholine receptor (Figure 5.19)—and changing the permeability of that membrane to ions. One of the most important classes of receptors consists of those that bind substances, called growth factors, that regulate cell reproduction (Chapter 15).

Most animal cells have a mechanism, called **receptor-mediated endocytosis**, that acts to capture specific macromolecules from the cell's environment. The uptake is similar to endocytosis as described in Chapter 4, except that in receptor-mediated endocytosis parts of the plasma membrane contain specific receptor protein molecules. These parts of the membrane containing receptor molecules are called coated pits because the inner surfaces of the membrane are coated with several other, fibrous proteins, of which the best known is called **clathrin**. When a receptor protein binds the appropriate macromolecule from the environment, the associated coated pit invaginates and forms a **coated vesicle** strengthened and stabilized by clathrin molecules (Figure 5.20). Before the coated vesicle fuses with a lysosome or other membranous structure, it is uncoated. Because of its specificity for particular macromolecules, receptor-mediated endocytosis provides a more rapid and efficient method of taking up specific molecules, compared with simple endocytosis.

Receptor proteins must be intrinsic membrane proteins that span the entire thickness of the plasma

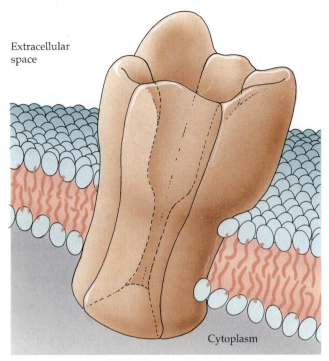

5.19 The Acetylcholine Receptor Protein
The acetylcholine receptor extends through the phospholipid bilayer and protrudes well beyond the bilayer into the extracellular space. Note the channel through the interior of the receptor protein (dashed lines).

membrane. Receptor carbohydrates are bound to such proteins or to phospholipid molecules. When a hormone, virus, or other "visitor," called a **ligand**, binds to its specific receptor, changes occur within the cell, on the cytoplasmic side of the membrane. Generally the visitor does not even cross the membrane to enter the cell. How can such a visitor produce an effect inside the cell if it does not actually enter it? It seems that the receptor protein undergoes structural changes. The portion of the protein that is on the inside is altered, as is the way the protein functions in the cytoplasm.

The specificity of receptors (both proteins and carbohydrates) resides in their particular tertiary structures, that is, in their shapes in three dimensions. Some portion of the receptor protein or carbohydrate fits—hand in glove, as it were—the hormone, growth factor, a part of a virus, or whatever other ligand it is meant to recognize. We will deal in detail with this sort of specificity when we talk about enzymes in the next chapter, and it will arise again in Chapter 16 in a discussion of immune systems.

Cell Adhesion Molecules

As an animal embryo develops, cells move about and become associated with other specific cell types. This behavior is mediated by specific membrane proteins,

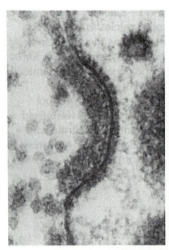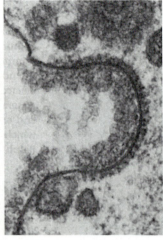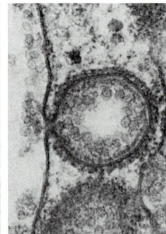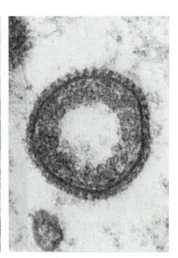

5.20 Coated Vesicle Formation
The micrograph at the left shows a coated pit, its inner surface coated with clathrin and other proteins. The remainder of the sequence illustrates the development of a coated vesicle (right) from the coated pit.

the **cell adhesion molecules**. One type of cell adhesion molecule is responsible for the organization of large numbers of individual nerve cells into nerve cell bundles; other cell adhesion molecules play roles in the organization of other tissues. Groups of cells that are about to migrate within the embryo lose their specific cell adhesion molecules; the cells regain the cell adhesion molecules before reorganizing into a tissue in their new location.

Other Activities of Membranes

We have emphasized some of the more prominent functions of membranes: the compartmentation of cells, the regulation of traffic between compartments, the active pumping of solutes, the mediation of energy-trapping and energy-releasing reactions, the recognition of materials at the cell surface, and the adhesive organization of cells into tissues. However, membranes have additional functions. As discussed in Chapter 4, the membrane of rough endoplasmic reticulum serves as a site for ribosome attachment. Newly formed proteins are passed from the ribosomes through the membrane and into the interior of the endoplasmic reticulum for delivery to other parts of the cell. In nerve cells, as we will see in Chapter 36, the plasma membrane is the conductor of the nerve impulse from one end of the cell to the other. The membranes of muscle cells, some eggs, and other cells are also electrically excitable. Numerous other biological activities and properties discussed in the chapters to follow are integrally associated with membranes.

MEMBRANE INTEGRITY UNDER STRESS

Red blood cells are among the most fragile-seeming of all cells, yet they survive repeated compression and deformation as they squeeze through the finest of capillaries. This surprising resilience of the red blood cell membrane results from the presence of a group of membrane proteins, both intrinsic and extrinsic (Figure 5.21). On the cytoplasmic surface of the plasma membrane, the extrinsic protein spectrin forms a meshwork of microfibrils that supports the membrane. Spectrin is anchored to the membrane at many points by another extrinsic protein, ankyrin, that binds both spectrin and an anion transporter that is an intrinsic membrane protein. This anion transporter exchanges Cl^- for HCO_3^- as referred to earlier in this chapter.

Genetic defects in spectrin and others of these proteins result in abnormal red blood cells and thus in various diseases. Mice with hemolytic anemia have spherical, fragile red blood cells. Their red blood cells have very little spectrin, and the cells take on a normal shape if provided with spectrin.

MEMBRANE FORMATION AND CONTINUITY

Membranes are dynamic structures in the sense that they are the site of numerous physiological and biochemical processes. In addition, we now know they are dynamic structures in the further sense of being constantly formed, transformed from one type to another, and broken down. In eukaryotes, phospholipids are synthesized within the sacs of the rough endoplasmic reticulum and rapidly distributed to membranes throughout the cell. Membrane proteins

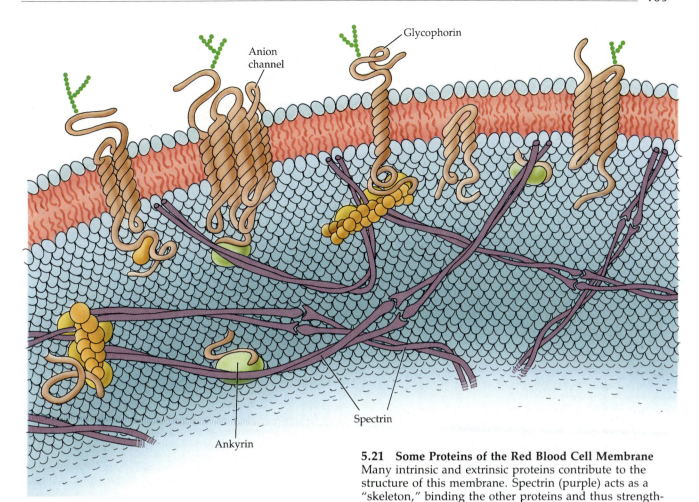

5.21 Some Proteins of the Red Blood Cell Membrane
Many intrinsic and extrinsic proteins contribute to the structure of this membrane. Spectrin (purple) acts as a "skeleton," binding the other proteins and thus strengthening the membrane of this highly flexible cell. Spectrin does not bind the membrane directly but connects by way of linker proteins such as ankyrin (yellow), which in turn binds the anion channel.

are inserted into the sacs of the rough endoplasmic reticulum as they are formed on ribosomes. Sugars may be added to the proteins while they are in the endoplasmic reticulum. Next the proteins are found in the Golgi apparatus, where some have other carbohydrates added to them. The proteins then travel in Golgi-derived vesicles to the plasma membrane and are incorporated into it (Chapter 4).

Functioning membrane itself seems to move about within the eukaryotic cell. For example, portions of the rough endoplasmic reticulum bud away from the endoplasmic reticulum and join the forming faces of the Golgi apparatus (Chapter 4). Rapidly—often in less than an hour—these segments of membrane find themselves in the maturing faces of the apparatus, from which they bud away to join the plasma membrane. Bits of membrane are constantly merging with the plasma membrane in the process of exocytosis, but this is largely balanced by the removal of membrane in endocytosis. In sum, there is a steady flux of membranes as well as membrane components (Figure 5.22).

Given this constant interconversion of membranes, we might expect all subcellular membranes

to be chemically identical. As you already know, this is not the case, for there are major chemical differences among the membranes of even a single cell. Apparently membranes are changed chemically when they form parts of certain organelles. In the Golgi apparatus, for example, the membranes of the forming face are closely similar to those of the endoplasmic reticulum, but the maturing-face membranes are more similar in composition to the plasma membrane. Ceaselessly moving, constantly carrying out functions vital to the life of the cell, biological membranes certainly are not the static, stodgy structures they once were thought to be.

SUMMARY

The continuous matrix of all membranes within cells consists of a double layer of phospholipid molecules, in which cholesterol may be included. The middle of

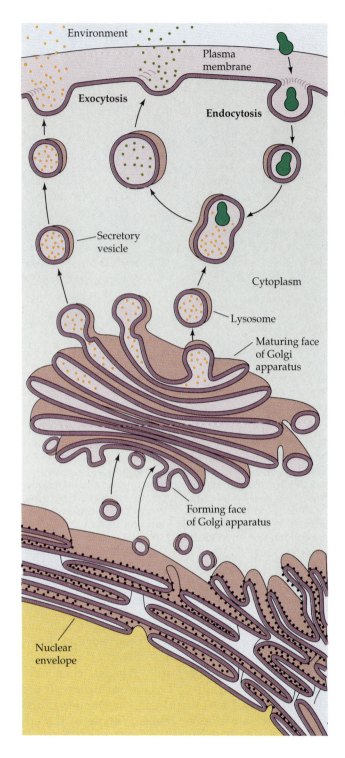

Environment

Plasma membrane

Exocytosis

Endocytosis

Secretory vesicle

Cytoplasm

Lysosome

Maturing face of Golgi apparatus

Forming face of Golgi apparatus

Nuclear envelope

5.22 Membrane Continuity in Cells

Arrows trace how membranes form, move, and fuse in cells. New stretches of membranes may be generated at certain locations, such as the outer membrane of the nuclear envelope. The new membrane may then become incorporated into either the rough or smooth endoplasmic reticulum. Membrane-bounded packets separate from the endoplasmic reticulum, travel, and fuse with the forming face of the Golgi apparatus. The vesicles budding from the maturing face on the opposite side of the Golgi apparatus are also membrane-bounded. The vesicles may remain inside the cell in the form of organelles such as lysosomes, or they may fuse with the plasma membrane, delivering their contents to the exterior of the cell (exocytosis) and adding their membranes to the plasma membrane. Membrane is subtracted from the plasma membrane in the process of endocytosis.

intrinsic and extrinsic. Some membrane proteins are carriers that allow specific molecules or ions to pass through the membranes. Other roles of membrane proteins include the recognition of other cells or molecules, the regulation of specific chemical reactions, and the transmission of information from one side of the membrane to the other.

On the extracellular surface of the plasma membrane there may be a variety of carbohydrates, some attached to lipids and others to proteins. Membrane carbohydrates function in recognition reactions.

Where animal cells meet, they form junctions linking their plasma membranes: desmosomes for adherence, tight junctions that limit flow in spaces between cells, and gap junctions for communication.

Solutes that can pass through a membrane do so by simple diffusion, facilitated diffusion, or active transport. In simple diffusion, the solute must be soluble either in the hydrophobic interior of the membrane or in the hydrophilic interior of a specific ion channel. Facilitated diffusion and active transport require specific carrier proteins. Only active transport can move solutes against a concentration difference. Primary active transport directly uses ATP to pump sodium and other ions; in secondary active transport, energy derived from the flow of sodium ions back into the cell is used to drive solutes through the membrane.

Water moves through membranes only by osmosis. In animal cells, water moves across the plasma membrane toward that side where the osmotic potential is more negative.

Membranes contain many receptor proteins that are important for a variety of physiological processes. Membranes within cells are continually being formed and moved to other parts of the cell.

the membrane is highly hydrophobic, while both exterior surfaces are hydrophilic. Membrane lipids restrict the rates at which most solutes can pass from one side to the other.

Membranes also include a variety of proteins, both

6

Energy, Enzymes, and Catalysis

PREVIEW: Biochemical reactions in organisms are accompanied by changes in energy, which determine the extent to which the reactions proceed. By lowering an energy barrier between the reactants and the products, proteins called enzymes increase reaction rates. Enzymes can function as they do because of their chemical structures. Enzymes themselves are subject to regulation, allowing biochemical reactions to proceed in an orderly fashion.

This chapter deals with metabolism, the first and second laws of thermodynamics, chemical equilibrium, reaction rates, energy of activation, catalysis, enzyme structure and function, prosthetic groups, enzyme inhibition, allosteric control, feedback inhibition, and the effects of temperature and pH on enzymes.

To sustain the processes of life, a typical cell carries out thousands of biochemical reactions each second. The sum of all biochemical reactions constitutes metabolism. What is the purpose of these reactions—of metabolism? Metabolic reactions convert raw materials, obtained from the environment, into the building blocks of proteins and other compounds unique to organisms (Chapter 3). Living things must maintain themselves, replacing lost materials with new ones; they also grow and reproduce, two more activities requiring the continued formation of macromolecules. Many biochemical reactions, including those that synthesize macromolecules, proceed at a negligible pace unless a source of **energy** is provided. (Energy can be defined as the ability to do work.) The energy for some of these reactions is obtained by the breakdown of certain other molecules.

Every cell is separated from its environment by a plasma membrane. Eukaryotic cells are divided into compartments by other membrane systems (Chapter 4). Some substances move through the membranes by active transport, an energy-requiring process in which the substances are moved against a concentration difference (Chapter 5). Active transport plays a key role in numerous vital functions. In animals, these functions include the initiation and transmission of nerve impulses and the contraction of muscles; in plants, the transport of food molecules and the control of the uptake of carbon dioxide for photosynthesis depend on active transport, and thus require energy.

In sum, life depends upon energy—obtaining it from the environment and using it to accomplish the reactions that set living things apart from the nonliving. This chapter builds toward an understanding of the relationship between energy and biochemical reactions and an understanding of the catalysis of reactions by enzymes.

ENERGY AND THE LAWS OF THERMODYNAMICS

Energy is commonly defined as the capacity to do work. It has also been defined as "heat or anything that can be transformed into heat." Energy comes in many forms: heat, light, electric, mechanical, chemical, nuclear, and others. Matter itself is a form of energy—in an atomic explosion (or in a nuclear power plant), a small amount of matter is converted into enormous amounts of energy. When all forms of energy are accounted for, the total amount of energy in the universe is unchanging: Energy can neither be created nor be destroyed. This is the **first law of thermodynamics**. The first law holds for the universe as a whole or for any **closed system** within the universe. A closed system is one that is not exchanging energy with its surroundings.

The various forms of energy are interconvertible (Figure 6.1). In solar batteries, light energy is converted into electric energy. Electric energy can be converted into light, heat, motion, and other forms of energy. Green plants convert light into chemical energy; in muscles, chemical energy is transformed into the energy of motion. Heat, too, can be used to do work (think of a steam engine, for example), but here we run into limitations. The conversion of any other form of energy into heat is not fully reversible; that is, not all the heat can be converted back into the other forms of energy. Biological, chemical, and physical processes are often accompanied by the production of heat, not all of which can be made to do work.

Unusable heat is associated with an increase in disorder. Chemical changes, physical changes, biological processes, and anything else you can think of all tend toward disorder, or randomness. A crystal of sodium chloride is highly ordered. It will dissolve spontaneously in water to form a more random solution of sodium chloride. A sodium chloride solution, however, will not spontaneously reorder itself into a crystal of salt and pure water. Has your room become more or less orderly since the last time you expended energy to straighten it up? Disorder can be discussed in quantitative terms; its measure is a quantity called **entropy**. Greater entropy implies greater disorder in any system. Not all energy conversion processes result in the same ability to do work. Various amounts of useful energy may be lost as the original forms are converted to unusable forms of energy—the heat associated with disorder. In the universe as a whole, or in any closed system, the amount of entropy increases; this is the principle of degradation of energy, also known as the **second law of thermodynamics**. Other ways of stating the second law will appear as the chapter continues.

THERMODYNAMICS AND CHEMICAL EQUILIBRIUM

Chemical Equilibria

In principle, all chemical reactions can run both forward and backward. That is, if compound A can be converted into compound B (A → B), then B can in principle be converted into A (B → A), although at given concentrations of A and B only one of these directions will be favored. One can think of the overall reaction as being the result of competition between forward and reverse reactions. Increasing the concentration of the reactants (A) speeds up the forward reaction, and increasing the concentration of the products (B) favors the reverse (Figure 6.2). At some point the forward and reverse reactions take place at the same rate; in the example in Figure 6.2, this occurs when there are rather few reactant molecules and very many product molecules. At this point no further change in the system is observable, although individual molecules are still forming and breaking apart. This balance between forward and reverse reactions is known as **chemical equilibrium**.

In the example in Figure 6.2, equilibrium lies "to the right" (toward product B) for the reaction A → B. In such a case, in which a reaction goes more than halfway to completion, we say that the reaction A → B is a **spontaneous reaction**, while the reaction B → A is not. A spontaneous reaction is one that, given enough time, goes largely to completion by itself, that is, without the addition of energy. In fact, as spontaneous reactions proceed, they release energy. If a reaction runs spontaneously in one direction (from reactant A to product B, for example), then the reverse reaction (from B to A) will not and cannot go to completion without a steady supply of energy to drive it. For example, starch slowly but spontaneously breaks down in water, producing the disaccharide maltose (Chapter 3). In contrast, maltose does not form starch spontaneously.

Similar considerations apply to entire chemical pathways. The complete oxidation of glucose in the processes of fermentation and cellular respiration is spontaneous, converting energy to a form that may be used to drive a variety of cellular reactions. On the other hand, the synthesis of glucose from carbon dioxide and water is a nonspontaneous reaction that must be driven by energy from the absorption of light in photosynthesis. (The important processes of fermentation, cellular respiration, and photosynthesis will be discussed at length in Chapters 7 and 8.)

Spontaneity has nothing to do with time. A reaction may be extremely slow, yet still be spontaneous.

6.1 Biological Energy Transformations
The leaf traps light energy and produces food by photosynthesis. The sawfly larva obtains its energy by eating the leaf. The larva's movements—including chewing—represent mechanical energy obtained by transforming the chemical energy of food.

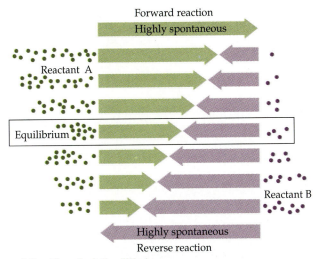

6.2 Chemical Equilibrium
When the concentration of reactant A is high in the spontaneous reaction represented here *(top)*, the reaction is overwhelmingly in the forward direction, toward product B; the reverse is true when the concentration of B is high *(bottom)*. Both forward and reverse reactions proceed toward chemical equilibrium (at which time the rates of forward and reverse reactions are equal).

The burning of a newspaper and the slow browning of pages in old library files are both spontaneous processes, but they have different time scales.

A spontaneous reaction is one that is moving toward equilibrium, using up reactants and making products. Given enough time, every spontaneous process eventually reaches equilibrium concentrations of reactants and products. Cells *prevent* the attainment of equilibrium as long as they live because they utilize a continuous input of energy. In this way, cells keep their chemical composition different from that of the environment around them.

The Equilibrium Constant

The specific point of equilibrium of each reaction is described by an equilibrium constant that is defined in terms of the concentrations of reactants and products. Consider the following example. Every living cell contains glucose 1-phosphate, and its conversion to glucose 6-phosphate is a common event in cells. Later in this chapter, we explain how enzymes work; it is enough here to say that a particular enzyme accelerates this conversion. In our example, this enzyme is added to a solution of glucose 1-phosphate. Let the initial concentration of glucose 1-phosphate be 0.02 *M* (0.02 molar; see Chapter 2). The reaction proceeds until equilibrium is reached, and at that point the reverse reaction to glucose 1-phosphate proceeds at the same rate as the forward reaction to glucose 6-phosphate (Figure 6.3). As the reaction comes to equilibrium, the final concentration of the product, glucose 6-phosphate, rises from 0 to 0.019

M, while the glucose 1-phosphate concentration has fallen to 0.001 *M*. At equilibrium, then, the forward reaction has gone 95 percent of the way to completion—the forward reaction is a spontaneous reaction. This result is obtained every time the experiment is run under the same conditions, namely, at 25°C and pH 7. The **equilibrium constant**, K_{eq}, is defined as the ratio of the concentrations of product and reactant *at equilibrium*:

$$K_{eq} = \frac{[\text{product}]}{[\text{reactant}]}$$

where the brackets indicate concentrations in, for example, moles per liter. In our example,

$$K_{eq} = \frac{[\text{glucose 6-phosphate}]}{[\text{glucose 1-phosphate}]} = \frac{0.019}{0.001} = 19$$

Suppose that we start the experiment, instead, with only the product—with, say, 0.02 *M* glucose 6-phosphate and no glucose 1-phosphate. A reaction occurs, with equilibrium being reached at 0.001 *M* glucose 1-phosphate and 0.019 *M* glucose 6-phosphate, the same as before. Regardless of the starting proportions of glucose 1-phosphate and glucose 6-phosphate, the equilibrium concentrations are defined by K_{eq}—that is, the equilibrium constant *is a constant* for a given chemical reaction as long as other conditions remain unchanged.

In general, for a reaction of the type A $\rightleftharpoons$ B, such as the glucose phosphate example, the equilibrium constant K_{eq} is defined by the equation

$$K_{eq} = \frac{[B]}{[A]}$$

where [B] and [A] are the concentrations of product B and reactant A, in moles per liter, *at equilibrium*. By convention, the reaction products are shown in the numerator and the reactants in the denominator. For the somewhat more complex reaction C + D $\rightleftharpoons$ E + F, the equilibrium constant is

$$K_{eq} = \frac{[E][F]}{[C][D]}$$

For the ionization of acetic acid in water at 25°C,

$$CH_3COOH \rightarrow CH_3COO^- + H^+$$

$$K_{eq} = \frac{[CH_3COO^-][H^+]}{[CH_3COOH]} = 2 \times 10^{-5}\ M$$

which says that the ionization of acetic acid is quite limited—less than one-half of one percent of the acetic acid molecules are ionized, and the ionization is not a spontaneous reaction. Acetic acid is a weak acid. When a strong acid such as hydrochloric acid dissolves in water, virtually all the molecules ionize.

A few more examples of equilibrium constants are

6.3 Concentration at Equilibrium

An enzyme speeds the conversion of glucose 1-phosphate to glucose 6-phosphate; the reverse reaction is boosted by the same enzyme if the starting material is glucose 6-phosphate. (The phosphate groups, highlighted in orange, are numbered according to the carbon to which they are bonded.) At 25°C and a pH of 7, there will always be 95 percent glucose 6-phosphate (purple) and 5 percent glucose 1-phosphate (green) at equilibrium, no matter what the starting percentages of the two compounds.

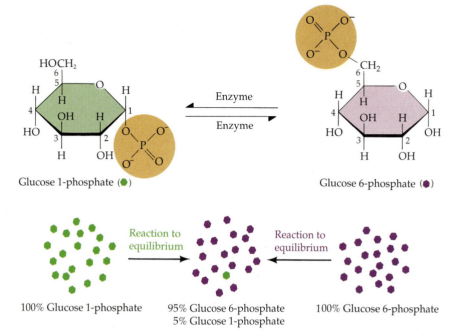

Glucose 1-phosphate (●)

Glucose 6-phosphate (●)

100% Glucose 1-phosphate

95% Glucose 6-phosphate
5% Glucose 1-phosphate

100% Glucose 6-phosphate

given in Table 6.1. Notice that the values vary widely. A high value of K_{eq} means that the reaction goes far toward completion; a very low one means that the reaction scarcely goes at all—in fact, the reverse reaction predominates.

Free Energy and Equilibria

What determines the point of chemical equilibrium? What distinguishes a spontaneous reaction from one that can proceed only with a considerable input of energy? The second question answers the first. A spontaneous reaction—one that goes far toward completion—is one that gives off a great deal of **free**

energy—energy that can be used to do work. Such a reaction is said to be **exergonic**; it releases free energy (Figure 6.4). A reaction with a tiny equilibrium constant can only be made to go by the addition of free energy; it requires energy and is thus **endergonic**. Free energy is symbolized by G (for "Gibbs free energy," named after the nineteenth century Yale thermodynamicist Josiah Willard Gibbs). It cannot be measured absolutely; however, the *change* in free energy, ΔG, of a reaction can be determined readily. It is related directly to the value of K_{eq} for the reaction. Values of ΔG are given in Table 6.1.

In the universe as a whole, or in any closed system, the quantity of free energy is always decreasing

---

TABLE 6.1
Equilibrium Constants and Standard Free Energy Changes of Selected Reactions

REACTION[a]	EQUILIBRIUM CONSTANT	STANDARD FREE ENERGY CHANGE (KCAL/MOL)
Acetic acid + H_2O → acetate + H_3O^+	0.00002	+6.3
Malate → fumarate + H_2O	0.28	+0.75
Fructose 6-phosphate → glucose 6-phosphate	2.0	−0.4
Glucose 1-phosphate → glucose 6-phosphate	19	−1.7
Glucose 6-phosphate + H_2O → glucose + phosphate	260	−3.3
Sucrose + H_2O → glucose + fructose	140,000	−7.0

[a]The reactions are arranged from top to bottom in order of increasing tendency to go to completion as written ("go to the right").

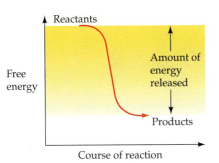

(a) **Exergonic reaction**
(spontaneous; energy-releasing)

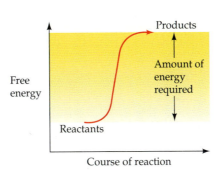

(b) **Endergonic reaction**
(not spontaneous; energy-requiring)

**6.4 Exergonic and Endergonic
Reactions**

In an exergonic reaction *(a)*, energy is released as reactants with a relatively high energy content form products with a lower amount of energy. Energy is required for an endergonic reaction *(b)* as reactants with a low energy content are boosted to products at a higher energy level.

and entropy is always increasing. (This is another way of stating the second law of thermodynamics.) Free energy tends always to a minimum. In a spontaneous reaction, the reactants possess more free energy than do the products; and ΔG is a negative number in an exergonic reaction, such as glucose 1-phosphate $\rightarrow$ glucose 6-phosphate, in which $\Delta G = -1.7$ kilocalories per mole (kcal/mol). Recall that equilibrium for this reaction has a product-to-reactant ratio of 19:1. A reaction that goes nearly to completion ("to the right" as written) has a large, negative ΔG, indicating that it releases a large amount of free energy. A large, positive ΔG means that the reaction hardly proceeds at all as written. If the products are present, such a reaction runs backward ("to the left") to near completion. A ΔG value near zero is characteristic of a readily reversible reaction: Reactants and products have almost the same free energies.

Free Energy, Heat, and Entropy

We have introduced free energy as a quantity related to chemical equilibrium; equilibrium is attained when free energy is at a minimum. Next we recognize that ΔG measures the useful chemical energy obtainable from a reaction. As an exergonic reaction proceeds, free energy is released and may be used to do chemical work such as driving an endergonic reaction. The cell harvests free energy from exergonic reactions such as the oxidation of foodstuffs, or from sunlight, and uses the free energy to drive vital endergonic reactions (such as those of photosynthesis).

Free energy is related to two other forms of energy: heat and a form concerned with the entropy of the system. Both forms can be discussed in the context of spontaneous reactions. There is a tendency toward increasing disorder, or entropy, as we have already noted. In most cases, heat is released by spontaneous reactions. The combustion of propane is an example —in a stove the release of heat is obvious.

Let us now see how free energy, heat, and the energy associated with entropy are related to one

another. In any chemical reaction or physical process, there may be changes in each of these three forms of energy. Entropy, the measure of disorder, is expressed in kilocalories per degree (kcal/deg). To relate entropy to energy, we multiply entropy by the temperature at which the reaction occurs to obtain energy in kilocalories. For this calculation we use the Kelvin temperature scale (the absolute temperature; in Kelvin units, 0 K is equivalent to about $-273°C$). As an example, the increase in entropy as 1 mole of ice melts to form water is 5.26 kcal/deg. The energy lost to disorder when the mole of ice melts at the freezing point (0°C or 273 K) is 5.26 kcal/deg $\times$ 273 K = 1436 kcal.

For any reaction we define the change in free energy in terms of the changes in heat and entropy, as follows:

$$\Delta G = \Delta H - T\Delta S$$

where ΔG is the change in free energy, ΔH the change in heat, ΔS the change in entropy, and T the absolute temperature at which the reaction is occurring. The $T\Delta S$ term shows that the effect of a change in disorder is greater at high temperatures than at low. (For a given value of ΔS, multiplying it by a larger number—that is, a higher temperature—results in a larger energy change.)

As an example illustrating the relative importance of these factors, 673 kcal of heat is given off in the combustion of 1 mol of glucose, and the disorder rises by 0.0433 kcal/deg. At 25°C (which equals 298 K), $T\Delta S = 298(0.0433) = 12.9$ kcal, which can be rounded off to 13 kcal. Both the heat and the disorder terms contribute to the spontaneity of combustion of glucose. The change in free energy in the complete oxidation of 1 mol of glucose is

$$\Delta G = -673 \text{ kcal} - 13 \text{ kcal} = -686 \text{ kcal}$$

Both the change in heat and the change in entropy contribute to the overall change in free energy, and the change in free energy determines the equilibrium constant for the reaction. In this example, although the entropy factor (13 kcal/mol) is far smaller than the heat factor (673 kcal/mol), it does contribute to the total free-energy change. Some other spontaneous reactions have large negative ΔG values because of the entropy term.

REACTION RATES

A knowledge of the overall change in free energy (ΔG) for a reaction tells us where the equilibrium of the reaction lies. The more negative ΔG is, the further the reaction proceeds toward completion as written. However, ΔG does not tell us anything about the **rate of a reaction**—the speed at which the reaction proceeds toward equilibrium. A key to understanding reaction rates lies in the fact that there is an *energy barrier* between reactants and products. Think about a butane lighter. The burning of the fuel (that is, the reaction of butane with oxygen to release carbon dioxide and water vapor) is obviously exergonic—once started, the reaction goes to completion, which comes when all the butane has been burned. Burning butane liberates free energy as light and heat. Accordingly, you might expect this reaction to proceed at once. However, if you simply allow butane to flow and mix with air, nothing happens. To start the burning of butane, you have to provide a spark.

The function of the spark is to provide a bit of heat energy to activate the butane–oxygen mixture. In general, reactions go only after they are pushed over the energy barrier by such bits of energy, which are called **activation energy**, as shown graphically in Figure 6.5. Another illustration of energy in reactions is given in Figure 6.6a, showing a ball partway up a hill. This ball has more free energy than a ball at the bottom of the hill, because somebody expended energy carrying or throwing it up the hill. This energy is then released as the ball rolls down. As the ball

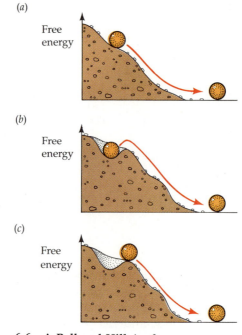

6.6 A Ball-and-Hill Analogy
(a) A ball on the side of a hill corresponds to the reactants in a spontaneous, exergonic reaction; the ball at the bottom of the hill represents the products formed after the "reactant ball" rolls down the hill with a release of free energy. *(b)* If a ball on the side of a hill is in a depression, it will require a bit of energy (activation energy) to get it out of the depression so it can roll down to the bottom of the hill. *(c)* Here the reactant ball on the hill has received an input of activation energy and is poised (activated) for a spontaneous journey down to the product level, releasing free energy as it goes.

rolls down the hill, the reaction is exergonic—the ball is losing energy. For the ball to roll back up the hill would require energy; this would be an endergonic reaction, and it will not happen spontaneously. Suppose now that the ball on the hillside is in a little depression (Figure 6.6b). Rolling down the hill is still an exergonic process, but to start the ball rolling, a small amount of activation energy must be exerted to roll it up out of the depression. In a chemical reaction, the energy barrier—the "hump" over which reactants must "roll" before they can proceed spontaneously to form products—is energy needed to change reactants into intermediate molecular forms called transition-state species. Transition-state species have higher free energies than either the products or the reactants. A transition-state species corresponds to a ball that has just been rolled up from the depression (Figure 6.6c). The activation energy needed to start a reaction is eventually recovered during the ensuing "downhill" phase of the reaction, so the overall drop in free energy, ΔG, is unaffected (Figure 6.5). But the higher the energy barrier to a reaction, the slower the reaction.

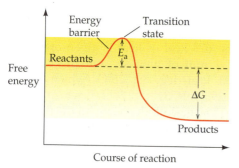

6.5 The Energy Barrier
Even for an energy-releasing reaction, reactants require an initial input of energy to get the reaction started. This activation energy (E_a) allows the reactants to surmount an energy barrier, at which point the reaction may proceed spontaneously with the release of free energy (ΔG).

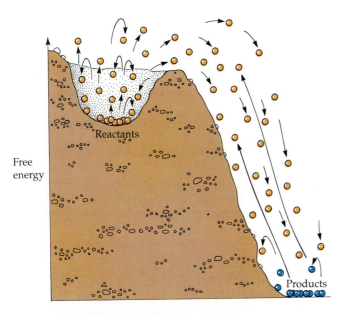

6.7 Energy Hump and Reaction Rate
Extending the ball-and-hill analogy of Figure 6.6, we see that some reactant molecules have sufficient activation energy to boost them over the energy hump; others do not achieve the necessary height and fall back into the depression, remaining there as reactants. If the hump forming the depression were lower, more reactant molecules would have enough energy to overcome it, and the reaction would go at a higher rate.

In any situation, some molecules are going to have more energy than others, that is, there is a mixture of reactant molecules with various energies (Figure 6.7). A minority of the molecules in the mixture may have enough energy to get over the activation "hump" and enter the transition state, and some of these react to yield products. A reaction with a low activation energy goes more rapidly because more of the reactant molecules have sufficient energy to overcome a low barrier. When activation energy is high, the reaction will not go to any significant extent unless more energy is provided, usually as heat. If the system is heated, the reactant molecules all become more energetic: More of them have energy in excess of the required activation energy, so the reaction speeds up.

The idea of an energy barrier is illustrated by the common biochemical reaction of our saliva acting on the starch in our food (Figure 6.8a). Starch, as we saw in Chapter 3, is a polymer consisting of many glucose monomers. The reaction in saliva is one between starch and water. The water cleaves the starch polymer into oligomers by breaking some of the bonds connecting the glucose units. However, a solution of starch in pure water is quite stable because the activation energy of the reaction is high. (When we say "stable" we mean that a reaction does not proceed at all, or proceeds so slowly that it would take a very long time to detect any change.) For the reaction to proceed, water and starch molecules must first collide, then the bonds connecting the glucose units must stretch and break. New bonds must then

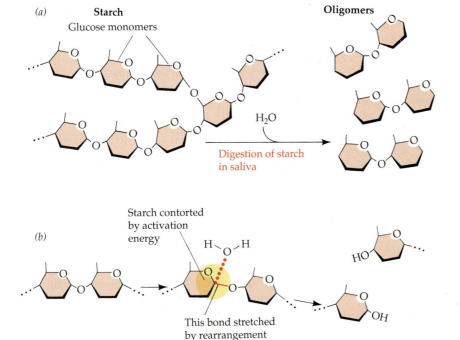

6.8 Breakdown of Starch
(a) In saliva, long starch molecules are digested into oligomers. (b) For starch to break down, an activated intermediate stage must form, as seen in the center here. In the yellow area, activation energy contorts the starch molecule, stressing an otherwise stable bond and preparing the molecule for reaction with water. On the right, the reaction has been completed, with the breaking of the stressed bond and the addition of atoms from a water molecule.

form between the broken ends and either H or OH derived from water. As shown in Figure 6.8b, there is an intermediate stage in which the bond between two glucose units is longer than normal and new bonds are in the process of forming. This is the transition-state species; it would be at the top of the energy barrier in a diagram like Figure 6.5. What is the difference between saliva and water that causes starch to break down in saliva but not in water? The difference will be explained in a moment.

Rate Constants

The rate of any chemical reaction is directly proportional to the concentration of the reactants. The rate —in, for example, micromoles per minute—simply equals the reactant concentration times a **rate constant**, k, related to activation energy. At any given temperature, each specific reaction has its own characteristic rate constant. For the simple case,

$$A \xrightarrow{k} B$$

the rate of formation of product B is equal to k times the concentration of A, or $r_B = k[A]$, where r_B is the rate of formation of B. The rate constant is affected by temperature: As the temperature increases, so does the rate constant k. The increased temperature makes the reactant molecules more energetic, so that more of them can spill over the E_a hump, and the reaction speeds up.

When starch is simply dissolved in water, the rate of its conversion to smaller carbohydrate molecules is extremely slow, as we have already noted. Even when the starch solution is heated to the boiling point, essentially none of the water and starch molecules gain enough energy to exceed the activation energy for the reaction. Nonetheless, we know that starch *does* get digested at a significant rate in the presence of saliva. (That is fortunate, for we would starve to death if all our digestive processes took place as slowly as the spontaneous breakdown of starch dissolved in pure water.) How can a reaction such as that shown in Figure 6.8 be speeded up by the presence of something such as saliva? Speeding up reaction rates is the function of catalysts.

ENZYMES AND SELECTIVE CATALYSIS

A **catalyst** does not cause anything to take place that would not take place eventually without it, and it does not become part of the products. A catalyst merely lowers the activation energy of the reaction, allowing equilibrium to be approached at a faster rate. Most nonbiological catalysts are reasonably nonspecific. Platinum black, for example, catalyzes virtually any reaction in which molecular hydrogen is a

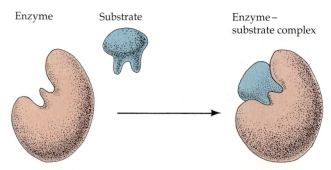

6.9 Enzyme and Substrate
An enzyme is a protein catalyst with an active site capable of binding a particular type of substrate molecule.

reactant because it weakens the bond between the atoms in the H_2 molecule. In contrast, most biological catalysts—**enzymes**—are highly specific. An enzyme usually catalyzes only a single chemical reaction or, at most, a very few closely related reactions. There is an enzyme in saliva that catalyzes the breakdown of starch.

The molecules that are acted upon catalytically are called the enzyme's **substrates**. The substrate molecules bind themselves to the surface of the enzyme at the enzyme's **active site**, where catalysis takes place. The binding of a substrate to the active site of an enzyme forms an **enzyme–substrate complex** held together by one or more means such as hydrogen bonding, ionic attraction, or covalent bonding (Figure 6.9). The enzyme–substrate complex may form product and free enzyme:

$$E + S \rightleftharpoons E{\cdot}S \rightleftharpoons E + P$$

where E is the enzyme, S is the substrate, P is the product, and E·S the enzyme–substrate complex. An enzyme present in saliva acts on the substrate, starch, forming oligosaccharides—the product. Note that E, the free enzyme, is in the same chemical form at the end of the reaction as at the beginning. It can be changed chemically during the reaction, but it is restored to its initial form by the end of the reaction.

The specificity of an enzyme is based upon the exact structure of its active site. The tertiary structure of the enzyme lysozyme is shown in Figure 6.10. Lysozyme is an enzyme that protects the animals that produce it by destroying invading bacteria; it does this by cleaving certain polysaccharide chains in the cell walls of bacteria. Lysozyme is found in tears and other bodily secretions, and it is particularly abundant in the whites of bird eggs. In Figure 6.10, the active site of lysozyme appears as a large indentation filled with the substrate (shown in green). The substrate fits precisely into the active site, whereas other molecules—with different shapes or different chemical groups on their surfaces—cannot form a complex with the enzyme.

Once the enzyme–substrate complex (E·S) has

formed, the enzyme sometimes undergoes a change in shape (Figure 6.11). Such a change results in an **induced fit** between the enzyme and the substrate, improving the alignment of the substrate with the active site. The enzyme α-amylase, which is present in saliva and specific for digesting starch to oligosaccharides, undergoes such a change (Figure 6.8).

When an enzyme lowers the energy of activation, both the forward and the reverse reaction are speeded up, so the enzyme-catalyzed reaction proceeds toward equilibrium more rapidly than the uncatalyzed one. The final equilibrium, it must be emphasized once again, is the same with or without the enzyme. Adding an enzyme to a reaction does not change the difference in free energy (ΔG) between the reactants and the products; it changes only the activation energy and the rate constants (Figure 6.12).

Substrate Concentration and Reaction Rate

For an uncatalyzed reaction of the type A → B, the rate of the reaction is directly proportional to the concentration of A. Let us now look at a graph of an enzyme-catalyzed reaction. For most enzymes, the plot looks like the one in Figure 6.13. The reaction rate increases as the substrate concentration increases, gradually reaching a constant maximum rate.

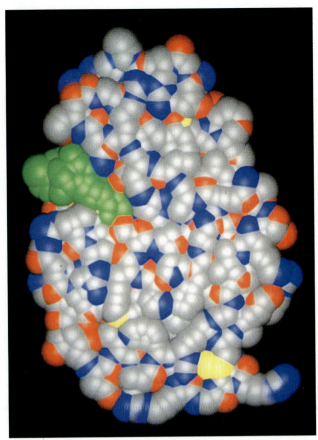

6.10 Tertiary Structure of Lysozyme
A substrate, shown in green, is bound to a lysozyme molecule. In this computer-generated drawing of the enzyme, carbons are gray, oxygens red, sulfurs yellow, and nitrogens blue; hydrogen atoms have been omitted. By attaching in a precise fit to the polysaccharide substrate, lysozyme is able to stress particular bonds, allowing the usually stable polymer to be broken. Lysozyme produced by animals destroys some invading bacteria by cleaving polysaccharides in the bacterial cell wall.

6.11 Induced Fit
Some enzymes change shape slightly when substrate binds. This change is referred to as induced fit. (a) The enzyme hexokinase; the deep cleft on the left side of the molecule divides the overall structure into upper (darker shading) and lower lobes and contains the site where the substrate, glucose, binds. (Atoms of glucose are shown in red.) (b) Hexokinase with glucose bound in the active site. Note that the upper and lower lobes of the enzyme have been induced to come together, closing the cleft.

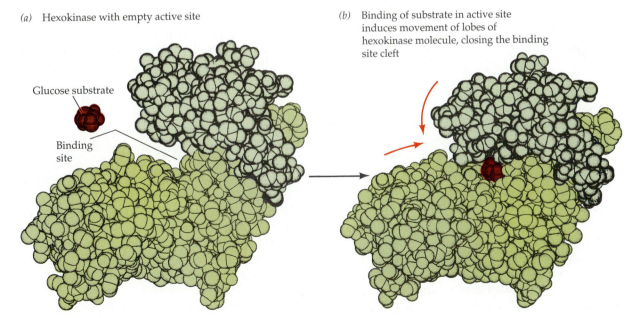

(a) Hexokinase with empty active site

Glucose substrate

Binding site

(b) Binding of substrate in active site induces movement of lobes of hexokinase molecule, closing the binding site cleft

6.12 Enzymes and Energy of Activation

An enzyme-catalyzed reaction has a lower energy of activation (E_a) than does the uncatalyzed reaction, but there is no difference in ΔG.

Reaction without enzyme

E_a

An uncatalyzed reaction requires a higher activation energy than does a catalyzed reaction

E_a

Free energy

Reactants

Reaction with enzyme

ΔG

There is no difference in ΔG between catalyzed and uncatalyzed reactions

Products

Course of reaction

Then, further increases in the substrate concentration will not increase the rate. This is easy to understand if we keep in mind that the concentration of the enzyme is usually much lower than that of the substrate. We are seeing a saturation phenomenon just like those discussed for facilitated diffusion and active transport across membranes (see Figure 5.11). As more substrate is added, more of the enzyme molecules are tied up as enzyme–substrate complexes. Once *all* the enzyme molecules are bound to substrate molecules, nothing is gained by adding yet more substrate—there are no enzyme molecules left to act as catalysts.

The study of the rates of enzyme-catalyzed reactions is called *enzyme kinetics*. As we will see later in this chapter, some graphs of rate versus substrate concentration are quite different from Figure 6.13. Such graphs tell us a great deal about the nature of the enzyme-catalyzed reaction.

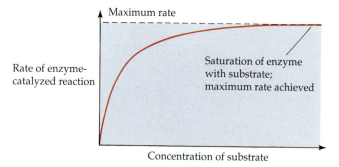

Maximum rate

Saturation of enzyme with substrate; maximum rate achieved

Rate of enzyme-catalyzed reaction

Concentration of substrate

6.13 Enzymes and Reaction Rate

For most enzyme-catalyzed reactions, the reaction rate increases with an increase in the concentration of the enzyme's substrate until it levels off at some maximum rate. At this maximum rate, all enzyme molecules are busy with substrate molecules; because of this, a further increase in substrate concentration does not affect the reaction rate.

Coupling of Reactions

Some of the most important reactions in living organisms are not spontaneous. Even so, they proceed because there are specific enzymes that **couple** them with other reactions that *are* spontaneous. Let us examine a pair of coupled reactions that occur in mitochondria and are catalyzed by the enzyme succinate dehydrogenase (Figure 6.14). The first reaction, which converts succinate to fumarate, is highly spontaneous, with a large drop in free energy. When this reaction takes place in a mitochondrion, the two hydrogen atoms that are removed from succinate are transferred to a molecule of a carrier substance, FAD. The second reaction, the hydrogenation of FAD to $FADH_2$, is nonspontaneous and requires a large input of free energy.

The enzyme succinate dehydrogenase couples the exergonic reaction to the endergonic one by ensuring that hydrogen atoms liberated by succinate are used to make $FADH_2$. One site on the enzyme surface binds succinate; a nearby second site binds FAD. Every time a succinate ion reacts with succinate dehydrogenase, much of the free energy that is released by this highly exergonic process is immediately trapped and used to synthesize $FADH_2$. The $FADH_2$ acts as a **carrier** of the hydrogen and the chemical free energy until another enzyme couples the exergonic dehydrogenation of $FADH_2$ with the endergonic hydrogenation of some other compound (Chapter 7).

In Chapter 5 we introduced other examples of coupled reactions (see Figure 5.15). In animals the sodium–potassium pump (for primary active transport) is an enzyme that couples the exergonic breakdown of ATP to the endergonic pumping of Na^+ and K^+ against their concentration differences. In secondary active transport, another protein couples the exergonic influx of Na^+ to the endergonic influx of glucose. We will see in Chapter 38 how the contractile

Succinate

$$\begin{array}{c} COO^- \\ | \\ H-C-H \\ | \\ H-C-H \\ | \\ COO^- \end{array}$$

Fumarate

$$\begin{array}{c} COO^- \\ | \\ C-H \\ \| \\ H-C \\ | \\ COO^- \end{array}$$

Exergonic dehydrogenation of succinate
(ΔG is negative)

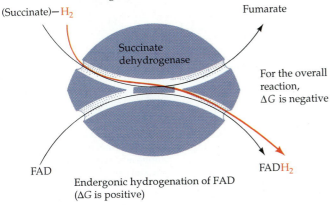

(Succinate)—H₂

Succinate
dehydrogenase

Fumarate

For the overall
reaction,
ΔG is negative

FAD

FADH₂

Endergonic hydrogenation of FAD
(ΔG is positive)

6.14 Coupled Reactions

Because the reaction that converts succinate to fumarate is exergonic, it can drive an endergonic reaction. The enzyme succinate dehydrogenase, which catalyzes the conversion, also couples the nonspontaneous change of FAD to FADH₂ to the conversion of succinate, transferring two hydrogen atoms with their electrons from succinate to FAD in the process. Coupled reactions are the major means of carrying out energy-requiring reactions in biological systems.

MOLECULAR STRUCTURE OF ENZYMES

Until the 1960s, biochemists knew little about the behavior of enzymes at the molecular level. It was generally agreed that the substrates of enzymes bind to an active site on the surface of the enzyme molecule, but the actual structure of an active site was not understood. The remarkable ability of an enzyme to select exactly the right substrate was explained by the assumption that the binding of the substrate to the site depends on a precise interlocking of molecular shapes. In 1894 the great German chemist Emil Fischer compared the fit between an enzyme and substrate to that of a lock and key. Fischer's model persisted for more than half a century with only indirect evidence to support it.

proteins of muscle couple the exergonic breakdown of ATP to the performance of mechanical work against a load. In metabolic pathways, there is another type of coupling, in which successive enzyme-catalyzed steps share compounds. A reaction A + B ⇌ C may be endergonic but still proceed rapidly if the next step C + D ⇌ E is so exergonic that the overall reaction (A + B + D ⇌ E) is exergonic.

The first direct evidence came in 1965, when David Phillips and his colleagues at the Royal Institution in London succeeded in crystallizing the enzyme lysozyme and, using the techniques of X-ray crystallography, determined its structure (Figure 6.15). Since then, the structures of several dozen other enzymes have been determined by such X-ray diffraction studies, and computers are now programmed to draw proteins from X-ray crystallographic data. This work has revealed a great deal about how the enzyme molecule is designed, how it works, and how it is controlled. Small enzymes consist of a single folded polypeptide chain; large enzymes may contain several, often identical, polypeptide chains. Frequently the active site contains a metal ion that enhances the reaction; or there may be a small, nonprotein molecule attached to the active site. These metals and small molecules are called prosthetic groups and will be discussed later in this chapter.

6.15 X-Ray Studies of Proteins

X-ray crystallographic studies of crystallized proteins and nucleic acids produce diffraction patterns that, upon analysis, lead to a visualization of the structure of a molecule; computer drawings such as Figure 6.10 and those in Chapter 3 were created from data provided by X-ray diffraction patterns. Here X-ray crystallographers align a crystal in a powerful instrument they recently designed.

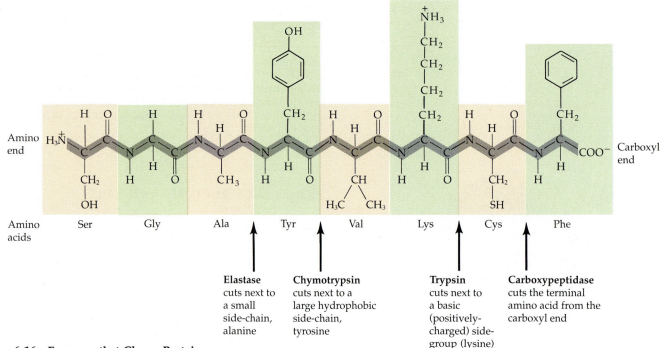

6.16 Enzymes that Cleave Proteins
Various enzymes cleave polypeptide chains at specific places, as defined by side chains on adjacent amino acids. The hypothetical polypeptide in this figure is eight amino acids long; its backbone is shaded in gray and the names of amino acids are indicated at the bottom of the figure. Bonds cleaved by four different enzymes are indicated with arrows, and the relevant side chains on adjacent amino acids are shown in color.

Structures and Actions of Protein-Digesting Enzymes

The X-ray studies suggest that most enzymes behave in similar ways. This can be illustrated by comparing the structures and functions of four enzymes that allow animals to digest proteins: carboxypeptidase, chymotrypsin, trypsin, and elastase. All four break the peptide linkages that connect amino acids in polypeptide chains, but each attacks only a very specific linkage. Carboxypeptidase snips one amino acid at a time from the carboxyl (COOH) end of a chain; the other three enzymes cleave chains at particular places in the middle, as shown in Figure 6.16.

The explanation for this fastidious specificity, which underlies the entire biochemistry of living organisms, lies in the architecture of the enzyme molecule. As Fischer suggested, the structure of the active site is molded to fit the substrate molecule. The binding of the substrate to the active site of the enzyme depends on the same forces that maintain the folded tertiary structure of the enzyme itself: hydrogen bonds, the electrostatic attraction and repulsion of charged chemical groups, and the interaction of hydrophobic groups. Substrates may also be covalently bonded to enzymes. (These forces were described in Chapter 2.)

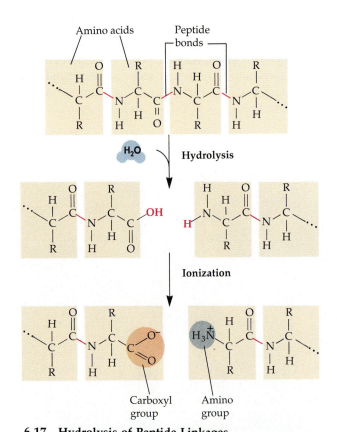

6.17 Hydrolysis of Peptide Linkages
Digestive enzymes such as the four introduced in Figure 6.16 hydrolyze peptide bonds. In the hydrolysis of a peptide, a water molecule donates an —OH group to the carbon atom in the peptide linkage, forming a carboxyl group. The H from water goes to the nitrogen atom, creating an amino group. Both carboxyl and amino groups then ionize.

Protein-digesting enzymes **hydrolyze** a peptide bond—that is, they break the polypeptide chain by adding a water molecule across the peptide bond (Figure 6.17). Carboxypeptidase lowers the energy barrier for this reaction, thus sharply increasing the reaction rate. Because of the specificity of the enzyme, not every peptide linkage is hydrolyzed at this rapid rate, but only those carboxyl-terminal bonds next to bulky hydrophobic side chains that fit comfortably into a hydrophobic pocket in the active site of carboxypeptidase. The other protein-digesting enzymes have different active sites that bind different side chains on the substrate.

Prosthetic Groups and Coenzymes

Although some enzymes consist entirely of one or more polypeptide chains, others possess a tightly bound nonprotein portion called a **prosthetic group**. This may be a single metal ion, a metal ion contained in a small organic molecule, or a **coenzyme**, a complex organic molecule required in some way for the action of one or more enzymes (Figure 6.18). Not all coenzymes are bound as a prosthetic group, however; some are separate and move from enzyme molecule to enzyme molecule.

Some coenzymes assist the catalytic activities of enzymes by accepting or donating electrons or hydrogen atoms (Chapter 7). Other coenzymes alter the structure of a substrate in such a way that the substrate becomes more reactive. In animals, coenzymes of these two types often are produced from vitamins in the diet. Another group of coenzymes transfers phosphate groups from molecule to molecule along with a great deal of free energy. Metal ions attached to enzyme proteins generally function by binding the substrate to the enzyme or by withdrawing electrons from the substrate (Figure 6.19).

REGULATION OF ENZYME ACTIVITY

Various substances, some occurring naturally in cells and other, artificial ones, being applied by scientists or physicians, act upon enzymes to increase or decrease the rates of enzyme-catalyzed reactions. The naturally occurring ones regulate metabolism; the artificial ones are used either to treat disease or to study how enzymes work. Some substances, called **inhibitors**, that inhibit enzyme-catalyzed reactions produce irreversible effects. Other inhibitors produce reversible effects, that is, these inhibitors can become unbound. Enzymes consisting of multiple subunits are subject to another type of control called allosteric regulation. We will discuss all these types of regulation and conclude this section with a consideration of the effects of pH and temperature on enzyme activity.

Irreversible Inhibition

Some inhibitors can irreversibly modify certain side chains at the active sites of enzymes, ruining the enzymes by destroying their capacity to function as

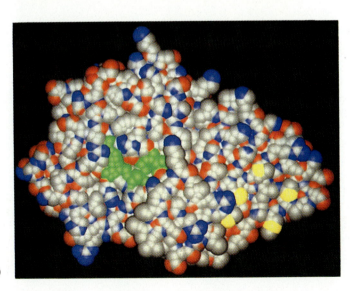

(a)

(b)

6.18 Coenzymes
(a) Biotin covalently bonds to any of several different enzymes that carry a carboxyl group and donate it to another molecule. Because it bonds to its enzyme, this coenzyme is a prosthetic group. (b) The coenzyme nicotinamide adenine dinucleotide, shown in green, attaches to the enzyme glyceraldehyde 3-phosphate dehydrogenase. This shows more realistically than the representation in (a) the relative sizes of enzyme and coenzyme. Hydrogen atoms have been omitted in this drawing.

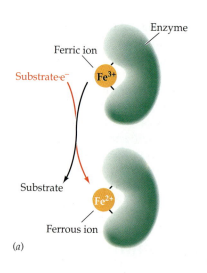

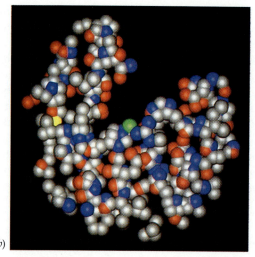

(a)

(b)

6.19 Metal Ions as Prosthetic Groups

(a) A ferric ion attached to an enzyme as a prosthetic group may withdraw an electron from the enzyme's substrate. In this reaction the substrate loses the electron and the ferric ion (Fe^{3+}) becomes a ferrous ion (Fe^{2+}) as it acquires the electron. (b) The enzyme thermolysin, with a zinc ion, shown in green, bound as a prosthetic group. Close to the zinc ion is part of a carboxyl group of a glutamic acid. This carboxyl group and the zinc ion collaborate in binding the substrate. The side chain of the glutamic acid having this carboxyl group is thus the part of the enzyme molecule most important for catalysis.

sential for the orderly propagation of impulses from one nerve cell to another (Chapter 36). Because of their effect on acetylcholinesterase, DFP and other similar compounds are classified as nerve gases.

Reversible Inhibition

Not all inhibitory action is irreversible. Some inhibitor molecules are similar enough to a particular enzyme's natural substrate to bind to the active site, yet different enough that the enzyme catalyzes no chemical reaction. When such a molecule is bound, the natural substrate cannot enter the active site; thus the intruder effectively wastes the enzyme's time, inhibiting the expected catalytic reaction. These molecular "dogs in the manger" are called **competitive inhibitors**. Because they block the enzyme by binding without chemically altering the enzyme (Figure 6.21), the blockage is reversible. A competitive inhibitor may become unbound, leaving the active site unchanged. If enough of the natural substrate molecules are present, they can compete successfully with the inhibitor for empty active sites.

catalysts. An example of such an **irreversible inhibitor** is diisopropylphosphorofluoridate, or DFP (Figure 6.20). DFP reacts with a hydroxyl group belonging to the amino acid serine (see Table 3.3) at an enzyme's active site, preventing the use of this side chain in the catalytic mechanism.

DFP is an irreversible inhibitor for the protein-digesting enzyme trypsin and for many other enzymes whose active sites contain serine. Among these is acetylcholinesterase, an enzyme that is es-

6.20 Irreversible Inhibition

DFP knocks out the function of the digestive enzyme trypsin by bonding covalently to a side chain of the amino acid serine in the active site. The reaction leaves trypsin unable to act on the specific peptide linkages described in Figure 6.16.

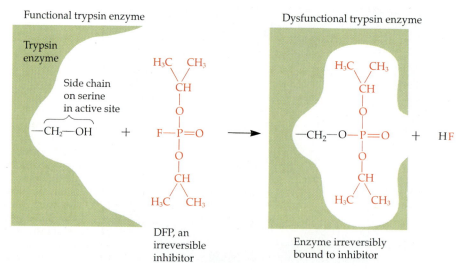

Functional trypsin enzyme

Trypsin enzyme

Side chain on serine in active site

$-CH_2-OH$ +

H_3C CH_3
CH
O
$F-P=O$
O
CH
H_3C CH_3

DFP, an irreversible inhibitor

Dysfunctional trypsin enzyme

H_3C CH_3
CH
O
$-CH_2-O-P=O$
O
CH
H_3C CH_3

+ HF

Enzyme irreversibly bound to inhibitor

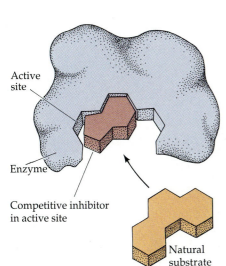

6.21 Competitive Inhibition

A competitive inhibitor binds to an enzyme's active site, thus preventing the natural substrate from entering the site. The binding of a competitive inhibitor does not permanently knock out the function of an enzyme molecule. A substrate molecule may become bound to an active site that was formerly blocked by an inhibitor. Substrate and inhibitor molecules compete for available active sites.

As an illustration of a competitive inhibitor, consider the enzyme succinate dehydrogenase. Recall that this enzyme, found in all mitochondria, removes two hydrogen atoms from a compound called succinate to produce another compound called fumarate; it transfers the hydrogens to another molecule as shown at the top of Figure 6.22. The other molecules shown in the figure are competitive inhibitors of succinate dehydrogenase. They are similar enough to succinate so that the enzyme is fooled into binding them; but having done this, the enzyme can do nothing more with them, because the inhibitors are the wrong size and shape, or have key chemical groups in the wrong places. The enzyme molecule cannot bind a succinate molecule until the inhibitor molecule has moved out of the active site. Dissociation of the inhibitor does occur, because binding of a competitive inhibitor is reversible, *as is binding of the substrate.* For example, when the competitive inhibitor malonate is added to a solution containing succinate and succinate dehydrogenase, the reaction of succinate to fumarate is slowed (as the graph in Figure 6.22 shows). However, the effect of the malonate can be overcome if enough succinate is added. The relative concentrations of substrate and inhibitor determine which is more likely to bind to the active site.

Other inhibitors do not react specifically with the active site; these are called **noncompetitive inhibitors**. Noncompetitive inhibitors bind to the enzyme

(a)

Substrates

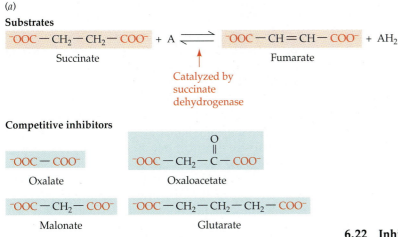

Competitive inhibitors

(b)

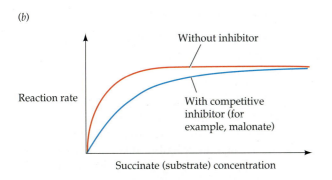

6.22 Inhibition of Succinate Dehydrogenase

Competitive inhibitors slow the reversible conversion of succinate to fumarate catalyzed by succinate dehydrogenase. *(a)* A series of molecules of increasing length—oxalate, malonate, oxaloacetate, and glutarate—compete with succinate for active sites of the succinate dehydrogenase. The similarity that fits them all to the same active site is the presence of two negatively charged carboxyl groups at the ends of the molecule. *(b)* In the absence of a competitive inhibitor, the enzyme increases reaction rate more than it does when one is present. At high concentrations of succinate, however, succinate competes successfully with malonate for the enzyme's active site and overcomes the inhibitory action of malonate; consequently, the same maximum reaction rate is reached at high substrate concentrations.

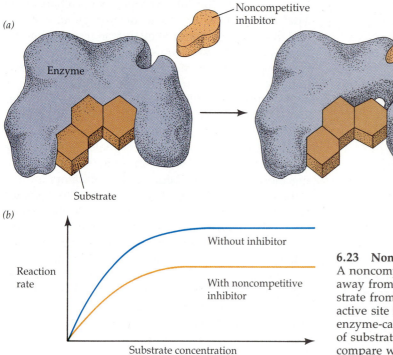

(a)

Noncompetitive inhibitor

Enzyme

Substrate

(b)

Reaction rate

Without inhibitor

With noncompetitive inhibitor

Substrate concentration

6.23 Noncompetitive Inhibition

A noncompetitive inhibitor binds to an enzyme at a site away from the active site. (a) It may not prevent the substrate from binding to the active site, but it modifies the active site in a way that reduces the rate constant for the enzyme-catalyzed reaction. (b) Even high concentrations of substrate do not overcome the effect of the inhibitor; compare with the graph in Figure 6.22b.

at a site away from the active site, but their binding causes a conformational change in the protein that alters the active site (Figure 6.23). The active site still binds substrate molecules, but the rate of product formation is reduced. Noncompetitive inhibitors can become unbound, so their effects are reversible. However, because noncompetitive inhibitors do not bind to the active site, their effects cannot be overcome completely by adding an excess of substrate.

Allosteric Enzymes

The enzymes we have discussed so far are individual polypeptides, although some have associated prosthetic groups. However, many important enzymes, like other proteins, are much larger and more complex. These complex enzymes have quaternary structures (see Chapter 3) consisting of two or more polypeptide subunits, each with a molecular weight in the tens of thousands.

The activity of these complex enzymes is controlled by molecules that may have no similarity either to the reactants or to the products of the reaction being catalyzed. These molecules, called **effectors**, operate by binding to a site on the enzyme other than the active site. Binding at this **allosteric site** can enhance or diminish the reactivity at the active site, so effectors can be inhibitors or activators. Because of the dissimilarity between the effector and the substrate, this phenomenon is called allostery, meaning "different shape." Enzymes subject to allosteric control are called **allosteric enzymes**; all have two or more subunits.

Allosteric enzymes differ greatly from enzymes that consist of a single subunit in the pattern by which they change reaction rates when the substrate concentration is low. Graphs of the rates plotted against the concentration of substrate show this. For an enzyme with a single subunit, the plot looks like Figure 6.13 (repeated in Figure 6.24a). The reaction rate first increases very sharply with increasing substrate concentration, then tapers off to a constant maximum rate as the supply of enzyme becomes saturated with substrate. The plot for an allosteric enzyme is radically different (Figure 6.24b), with a sigmoidal (S-shaped) appearance. The increase in rate with increasing substrate concentration is slight at low substrate concentrations, but there is a range over which the reaction rate is extremely sensitive to relatively small changes in the substrate concentration. Because of this sensitivity, allosteric enzymes are important in fine-tuning the activities of a cell. We can understand this behavior in terms of the structure of an allosteric enzyme.

Mechanism of Allosteric Effects

An allosteric enzyme consists of at least two kinds of subunits. One kind, the catalytic subunit, possesses an active site that binds the enzyme's substrate. The other kind of subunit, the regulatory subunit, has one or more allosteric sites that bind specific effector molecules. A molecule of an allosteric enzyme usually consists of two or more catalytic subunits and two or more regulatory subunits. An allosteric enzyme exists in two or more distinct forms with dif-

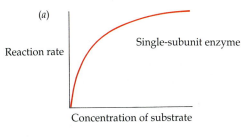

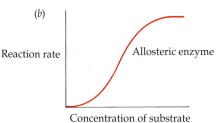

6.24 Allostery and Reaction Rate
(a) As substrate concentration increases, rates of reactions catalyzed by single-subunit enzymes increase—until all enzyme molecules are working. *(b)* The rates of reactions catalyzed by allosteric enzymes rise slowly at low concentrations of substrate, rise steeply for a small range of concentrations, then level off at a maximum where substrate saturates enzyme. Where the curve is steepest, a very small difference in concentration of substrate has a large effect on reaction rate and, consequently, on the activities of a cell.

ferent catalytic efficiencies, and these forms are in equilibrium with each other. In the simple cases we will examine, the **active form** has full catalytic activity, whereas the **inactive form** is totally without activity. In the active form, the active sites on the catalytic subunits are able to bind substrate and convert it to product. In the inactive form, the active sites have been distorted in such a way that they cannot bind substrate; however, the allosteric sites are able to bind an effector, which we will consider in this example to be an inhibitor. The regulatory subunits of the active form of the enzyme have deformed allosteric sites and cannot bind effector. When neither substrate nor inhibitor is present, the active and inactive forms are rapidly converting back and forth and are in equilibrium, with the equilibrium constant being characteristic of the given enzyme.

Now consider what happens when inhibitor or substrate is added to the enzyme solution (Figure 6.25). If substrate is present, some of the substrate binds to the active sites of active enzyme molecules;

6.25 Allosteric Regulation of Enzymes
The hypothetical enzyme shown here has four subunits, two catalytic and the other two regulatory. In the absence of other molecules, the active and inactive forms of the enzyme are in equilibrium, as shown at the top. When the enzyme is in its active form, either or both active sites on the catalytic subunits can accept substrate and catalyze a reaction. In the enzyme's inactive configuration the allosteric sites on the regulatory subunits can accept an inhibitor; the active sites of the inactive form of the enzyme are contorted in such a way that they cannot accept substrate molecules.

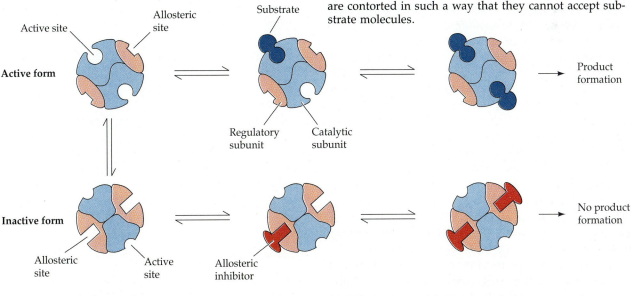

1. In the absence of substrate or inhibitor, active and inactive forms are in equilibrium.

2. In the presence of substrate, active form is stabilized and more active enzyme is formed, restoring equilibrium.

   In the absence of substrate, addition of allosteric inhibitor produces reciprocal effects.

3. Once a site is filled with a ligand, either substrate or inhibitor, binding with a second ligand of the same type is favored.

while the enzyme–substrate complex exists, those enzyme molecules cannot be converted to the inactive form. The presence of a substrate molecule in either active site prevents the enzyme molecule from being converted to the inactive form. However, inactive molecules are being converted to active ones at the same rate as before, so we now have an increase in the concentration of active enzyme as a result of the presence of substrate! Not only that, but because each active enzyme molecule (in this example) has two active sites, each can bind two substrate molecules and simultaneously catalyze reactions of both of them. This explains the upward curvature at the lower left of a plot of the reaction rate versus substrate concentration for an allosteric enzyme: Increasing the substrate concentration increases the availability of active enzyme and of active sites and hence rapidly accelerates the reaction rate.

On the other hand, the addition of an allosteric inhibitor *decreases* the concentration of active enzyme and thus inhibits the reaction. The inhibitor binds to the allosteric site of the *inactive* form of the enzyme, preventing the conversion of the inactive to the active form; but the conversion of the active to the inactive enzyme is not affected by the inhibitor. The overall effect is to decrease the concentration of the active form and thus inhibit the enzymatic reaction. An allosteric activator works in a similar way, except that it binds to the regulatory subunit of the active form of the enzyme and holds it in the active configuration. Note that allosteric inhibitors and activators do not modify the structure of the enzyme; rather, they interfere with its conversion to another form with which it is normally in equilibrium.

This mechanism of allosteric inhibition and activation was proposed by the French molecular biologist Jacques Monod and his colleagues in 1965. Another of Monod's many contributions to biology is described in Chapter 12.

Control of Metabolism Through Allosteric Effects

Metabolism is the totality of the biochemical reactions in a living thing. These reactions proceed down **metabolic pathways**, sequences of enzyme-catalyzed reactions so ordered that the product of one reaction is the substrate for the next. Some pathways synthesize, step-by-step, the important chemical building blocks from which macromolecules are built, others trap energy from the environment, and still others have functions different from these. Some metabolic pathways are branched, with one or more of the intermediate substances being used by more than one enzyme and thus being metabolized through separate branches.

One of the most challenging problems in the study of enzymology was finding a molecular explanation

for the behavior of **regulatory enzymes**—the chemical switches that catalyze reactions at the branching points where two or more metabolic pathways diverge. The end product of a branch pathway may damp the initial step in that branch pathway, reducing the formation of the end product. This is the principle of **negative feedback**, also seen, for example, in thermostats on furnaces (Figure 6.26). (Negative feedback will be a matter of central concern in Chapters 33 and 34.) The *end product* of a particular pathway typically is an allosteric inhibitor of the regulatory enzyme catalyzing the first **committed step** in its own synthesis, that is, the earliest step in the branched pathway that leads only to the synthesis of that end product and no other. The committed steps in metabolic pathways are particularly effective points for feedback control. For instance, inhibition of the B-to-C step in Figure 6.27 shunts all the reactants over into the other branch of the pathway, whereas inhibition of the C-to-D reaction, one step later, would lead only to a possibly harmful and certainly wasteful buildup of substance C. Such wasteful accumulation of unneeded intermediates is by and large avoided in living things.

When two different end products, produced by different branches of a pathway, are both present in excessive concentrations, they may have a concerted effect and act together to inhibit an earlier branch-point enzyme, that is, one that catalyzes the committed step for formation of these two products (see the H-to-L step in Figure 6.27). **Concerted feedback inhibition** like this results in further efficiency: In-

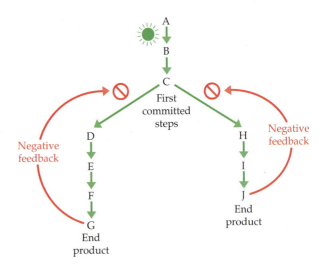

6.26   Feedback in Metabolic Pathways
The reactions C to D and C to H are the first committed steps in the branch pathways leading to end products G and J, respectively. These end products can block the first committed steps by acting as allosteric inhibitors of the enzymes catalyzing them. Thus if the level of product G builds up beyond what the cell needs, its production can be slowed without slowing the production of J, and vice versa.

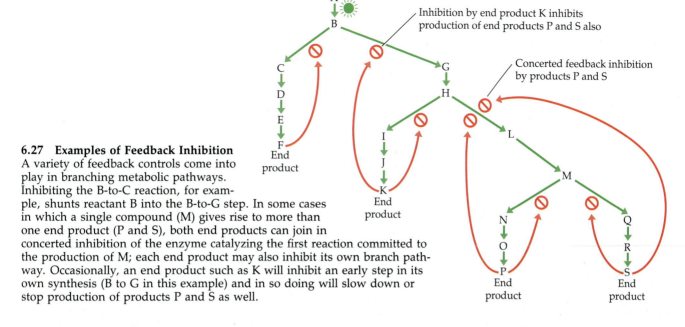

6.27 Examples of Feedback Inhibition
A variety of feedback controls come into play in branching metabolic pathways. Inhibiting the B-to-C reaction, for example, shunts reactant B into the B-to-G step. In some cases in which a single compound (M) gives rise to more than one end product (P and S), both end products can join in concerted inhibition of the enzyme catalyzing the first reaction committed to the production of M; each end product may also inhibit its own branch pathway. Occasionally, an end product such as K will inhibit an early step in its own synthesis (B to G in this example) and in so doing will slow down or stop production of products P and S as well.

termediates L and M do not build up as they would if only the steps to N and Q were inhibited by their individual end products. Concerted feedback inhibition requires that the enzyme have two allosteric sites, both of which must be bound to inhibitors to stop the enzyme's activity.

Occasionally "design flaws" obstruct regulated pathways of this sort. Once in a while, an end product such as K in Figure 6.27 may inhibit not only the step from H to I, but the earlier step from B to G as well. Thus an excess of K shuts down the production of end products P and S as well as its own production. Such cases have been observed in the laboratory. For example, if a culture of *Escherichia coli* is given an oversupply of threonine, it becomes deficient in methionine (another end product) and needs an outside supply of methionine in order to grow normally. Evidently, excess threonine inhibits a regulatory enzyme that affects the methionine branch pathway as well as its own synthesis. It is not surprising that such flaws occasionally appear in metabolic pathways. Biological control networks are the result of long evolution and natural selection. Because *E. coli* do not encounter large overdoses of threonine in their natural environment, which is the vertebrate intestine—or in fact, in any place except the laboratory of a curious biochemist—the flaw in the scheme would pose no disadvantage for the bacterium.

Allosteric regulation is very effective. It allows rapid adjustment to short-term changes in metabolism or in the environment. The activities of enzyme molecules are adjusted by their interactions with small molecules, the end products. It would be a further gain if enzyme *production* as well enzyme *activity* were regulated. If a particular enzyme is not needed, might it not be a good idea simply to stop making it until it is needed? This is indeed the case, and the regulation of enzyme synthesis plays an important role in controlling development and metabolism. This topic will be discussed in Chapters 12 and 14.

Sensitivity of Enzymes to the Environment

Enzymes enable cells to perform reactions under mild conditions, unlike the extremes of temperature and pH employed by chemists in the laboratory. Enzymes themselves are extremely sensitive to changes in the medium around them. For example, the rates of most enzyme-catalyzed reactions depend on the pH of the medium in which they occur. The enzyme is most active at a particular pH, with activity decreasing as the solution is made more acidic or more basic (Figure 6.28). Several factors contribute to this effect. One is the ionization of carboxyl, amino, and other groups on either the substrate or the enzyme. Carboxyl groups (—COOH) ionize to become negatively charged carboxylate ions (—COO⁻) in neutral or basic solutions. Similarly, amino groups (—NH₂) accept H⁺ ions in neutral or acidic solutions, becoming positively charged —NH₃⁺ ions. This means, for example, that in a neutral solution a molecule with an amino group is attracted electrically to another molecule that has a carboxyl group because both groups are ionized and they have opposite charges. The attraction does not occur in acidic solution (in which the carboxyl group does not ionize) or in basic solution (in which the amino group is not ionized). Different enzymes function best at different pH values. Evolution has matched enzymes to their environment—for example, digestive enzymes that act in the stomach (Chapter 41) work best at the very low pH values that prevail in the stomach after a meal.

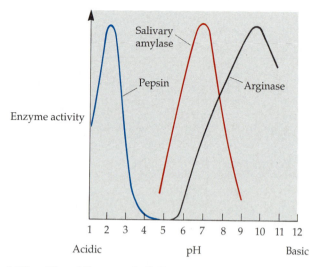

6.28   pH and Enzyme Activity
Most enzymes achieve maximum efficiency at a particular pH, as shown by the peaks of the activity curves for three enzymes.

Temperature also has a profound effect on enzyme activity (Figure 6.29). At low temperatures, warming increases the rate of an enzyme-catalyzed reaction because at higher temperatures a greater fraction of the reactant molecules have enough energy to provide the activation energy of the reaction. However, temperatures that are too high inactivate the enzyme, because at high temperatures the enzyme molecules vibrate and twist so rapidly that some of their non-covalent bonds break. The heat destroys their tertiary structure, and the enzyme molecules lose their activity. Enzymes become permanently inactivated, or **denatured**, at certain temperatures. Some enzymes denature at temperatures only slightly above that of the human body, but a very few others are stable even at the boiling point of water. The graph of enzyme activity versus temperature peaks at the **optimal temperature** for the enzyme. Above the optimal temperature, inactivation of enzyme molecules predominates.

Organisms adapt to changes in the environment. One of the ways they do this is based on groups of enzymes that catalyze the same reaction but have differing physical properties. Enzymes in such a group are called **isozymes**. Isozymes may be chemically similar to one another (made from different combinations of the same subunits, for example) or totally unrelated. Within a given set, different isozymes may have different optimal temperatures. An example is the enzyme acetylcholinesterase in the rainbow trout. If a rainbow trout is transferred from relatively warm water to near-freezing water (2°C), the fish produces an isozyme of acetylcholinesterase that is different from the acetylcholinesterase produced at the higher temperature. The new isozyme has a lower optimal temperature than does the previously formed one, which helps the fish to perform normally in the colder water.

ENZYMES, RIBOZYMES, AND ABZYMES

In the next two chapters we will see how enzymes that catalyze the reactions of two crucial sets of pathways—cellular respiration and photosynthesis — provide cells and organisms with the energy they need to live, grow, and reproduce. Enzymes will appear again and again as we continue our examination of the living world. However, enzymes may not have been the first catalytic macromolecules to evolve! As we will see in Chapter 13, the first biological catalysts may have been RNA molecules. Catalytic RNAs, or **ribozymes**, still function today. We will also learn, in Chapter 16, that some antibodies have modest catalytic activity. Catalytic antibodies, or **abzymes**, have potential as "designer catalysts."

SUMMARY

The various forms of energy (light, motion, electric, chemical bond, heat, and so forth) can be interconverted. The quantity of energy in the universe or in any closed system remains constant; this is the first law of thermodynamics. In the universe or in any closed system, the quantity of free energy decreases and the quantity of entropy (a measure of disorder) increases; this is the second law of thermodynamics.

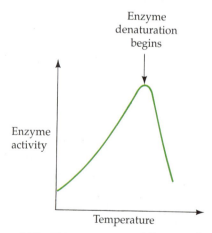

6.29   Temperature and Enzyme Activity
The green curve indicates that a given enzyme is most active at a particular temperature. Organisms are limited to temperatures suitable for enzyme activity, but some organisms have duplicate, differing enzymes that catalyze the same reactions, which permits them to survive and function over a broader range of temperatures and pH than organisms that have only single enzymes for the corresponding reactions.

Every chemical reaction is characterized by an equilibrium constant, K_{eq}, which is related to the change in free energy, ΔG, that accompanies the reaction. Each reaction is also characterized by a rate constant, k, which is related to the energy of activation. Catalysts speed the attainment of equilibrium by lowering activation energy. Catalysts are restored, unchanged, at the end of the reaction.

Reactions in organisms are catalyzed by enzymes. Enzymes are proteins, but some also have nonprotein prosthetic groups that participate in binding the substrate or in the catalytic process. Substrate molecules bind to the active site of the enzyme. The structure of the active site makes an enzyme specific for a single reaction, or for a limited class of reactions. Competitive inhibitors are ones that, because of close chemical similarity to the substrate, bind to the active site. Noncompetitive inhibitors bind elsewhere.

Metabolic pathways are subject to regulation. An important type of control is feedback regulation. The end product of a biosynthetic pathway is an allosteric inhibitor of the enzyme catalyzing the first committed step in the pathway. The rate of a metabolic reaction is affected by the concentrations of substrates and products and by the temperature and the pH of the medium. Some reactions are catalyzed by sets of isozymes, with different isozymes being effective under different circumstances.

SELF-QUIZ

1. Which statement about energy is incorrect?
   a. It can neither be created nor be destroyed.
   b. It is the capacity to do work.
   c. All its conversions are fully reversible.
   d. In the universe as a whole, the amount of free energy decreases.
   e. In the universe as a whole, the amount of entropy increases.

2. Which statement about thermodynamics is incorrect?
   a. Free energy is given off in an exergonic reaction.
   b. Free energy can be used to do work.
   c. A spontaneous reaction is exergonic.
   d. Free energy tends always to a minimum.
   e. Entropy tends always to a minimum.

3. In a chemical reaction,
   a. the rate depends on the equilibrium constant.
   b. the rate depends on the activation energy.
   c. the entropy change depends on the activation energy.
   d. the activation energy depends on the equilibrium constant.
   e. the change in free energy depends on the activation energy.

4. Which statement is *not* true of enzymes?
   a. They consist of proteins, with or without a nonprotein part.
   b. They change the rate constant of the catalyzed reaction.
   c. They change the equilibrium constant of the catalyzed reaction.
   d. They are sensitive to heat.
   e. They are sensitive to pH.

5. The active site of an enzyme:
   a. never changes shape.
   b. forms no chemical bonds with substrates.
   c. determines, by its structure, the specificity of the enzyme.
   d. looks like a lump projecting from the surface of the enzyme.
   e. changes the equilibrium constant of the reaction.

6. A prosthetic group:
   a. is a tightly bound nonprotein part of an enzyme.
   b. is composed of protein.
   c. does not participate in chemical reactions.
   d. is present in all enzymes.
   e. is an artificial enzyme.

7. The rate of an enzyme-catalyzed reaction:
   a. is constant under all conditions.
   b. decreases with an increase in substrate concentration.
   c. cannot be measured.
   d. depends on the equilibrium constant.
   e. can be reduced by inhibitors.

8. Which statement is *not* true of enzyme inhibitors?
   a. A competitive inhibitor binds the active site of the enzyme.
   b. An allosteric inhibitor binds a site on the active form of the enzyme.
   c. A noncompetitive inhibitor binds elsewhere than the active site.
   d. Noncompetitive inhibition cannot be completely overcome by adding more substrate.
   e. Competitive inhibition can be completely overcome by adding more substrate.

9. Which statement is *not* true of feedback inhibition of enzymes?
   a. It is exerted through allosteric effects.
   b. It is directed at the enzyme catalyzing the first committed step in a branch of a pathway.
   c. Concerted feedback inhibition is based on two or more end products.
   d. It acts very slowly.
   e. It is an example of negative feedback.

10. Which statement is *not* true of temperature effects?
    a. Raising the temperature may reduce the activity of an enzyme.
    b. Raising the temperature may increase the activity of an enzyme.
    c. Raising the temperature may denature an enzyme.
    d. Some enzymes are stable at the boiling point of water.
    e. The isozymes of an enzyme have the same optimal temperature.

FOR STUDY

1. How is it possible for endergonic reactions to occur in organisms?

2. Consider two proteins; one is an enzyme dissolved in the cytosol, the other is an ion channel in a membrane. Contrast the structures of the two proteins, indicating at least two important differences.

3. Plot free energy versus course of a reaction for an endergonic reaction and for an exergonic reaction. Include the activation energy in both plots. Label E_a and ΔG in both graphs.

4. Consider an enzyme that is subject to allosteric regulation. If a competitive inhibitor (not an allosteric inhibitor) is added to a solution of such an enzyme, the ratio of enzyme molecules in the active form to those in the inactive form will *increase*. Explain this observation.

READINGS

Dickerson, R. E. and I. Geis. 1969. *The Structure and Action of Proteins*. W. A. Benjamin, Menlo Park, CA. The structure of enzymes is presented as high art in this volume.

Karplus, M. and J. A. MacCammon. 1986. "The Dynamics of Proteins." *Scientific American*, April. This article will correct any misconception of proteins as rigid molecules; it describes the constant, rapid changes in local shape that underly the functioning of proteins.

Koshland, D. E., Jr. 1973. "Protein Shape and Biological Control." *Scientific American*, October. The ability of proteins to change shape in specific circumstances underlines the control and coordination of biological processes.

Morowitz, H. J. 1978. *Foundations of Bioenergetics*. Academic Press, New York. An excellent advanced text on thermodynamics in biology.

Newsholme, E. A. and C. Start. 1973. *Regulation of Metabolism*. John Wiley & Sons, New York. A rigorous treatment of regulation of enzyme activity, with emphasis on feedback control.

Stryer, L. 1988. *Biochemistry*, 3rd Edition. W. H. Freeman, New York. Good discussion of the structure of proteins.

7

Pathways That Release Energy in Cells

PREVIEW: Cells extract energy from food materials by using pathways made up of a large number of small chemical steps. One pathway—glycolysis—is found in virtually all groups of organisms. In the absence of oxygen, some cells perform fermentation to obtain their energy. Respiratory pathways (the citric acid cycle and the respiratory chain) extract much more energy from food substances than fermentation does; however, respiration requires oxygen and cannot be performed by all types of organisms. The pathways of cellular energy metabolism are regulated by a system of allosteric feedback controls.

This chapter deals with anaerobic and aerobic energy metabolism, ATP, NAD, glycolysis, cellular respiration, the citric acid cycle, the respiratory chain, oxidative phosphorylation, the chemiosmotic mechanism, fermentation, and allosteric control.

Living things require a continual flow of matter and energy: matter to reproduce and grow, build new molecules, maintain appropriate concentrations of key compounds, and replace those structures that have worn out; energy to assemble the molecules, perform biological work, and stem the tide of entropy. Matter is obtained from the environment, and energy is drawn from four physiological processes: photosynthesis, glycolysis, fermentation, and cellular respiration (Figure 7.1). These are the biochemical and biophysical processes that power the machinery of life.

Photosynthesis will be the topic of Chapter 8. In this chapter, we are concerned with the extraction of energy from food molecules. Depending on the type of organism extracting the energy and on the nature of the environment—whether the environment contains oxygen gas and is thereby **aerobic** or lacks oxygen gas and is **anaerobic**—either cellular respiration or fermentation is used. Glycolysis takes place in either case, as it constitutes part of the fermentation pathway and also comprises the steps before cellular respiration (Figure 7.1).

The combined operation of glycolysis and cellular respiration is the biological equivalent of burning the sugar glucose (Chapter 3). When glucose is burned with a match, it yields carbon dioxide, water, and energy in the form of heat and light. In cells, glucose is broken down to the same products, but much of the energy is trapped in the energy-storage compound **adenosine triphosphate (ATP).** In this chapter we show how cells use enzymes to "burn" glucose and harness the released energy as ATP.

The key events in the complete biological "combustion" of glucose take place as coupled reactions (Figure 7.2). To understand these events we must first learn the biochemistry of ATP, which is the cell's principal compound for short-term energy storage.

ADENOSINE TRIPHOSPHATE

The structure of ATP is shown in Figure 7.3. All living cells rely on the ATP molecule for the short-term storage of energy, in effect using ATP as a sort of energy currency. Many different enzymes can catalyze the breakdown of ATP; this breakdown yields **adenosine diphosphate** (ADP) and an inorganic phosphate ion. This breakdown is an exergonic reaction, yielding approximately 12 kcal of free energy per mole of ATP under biological conditions—enough energy to drive typical endergonic reactions

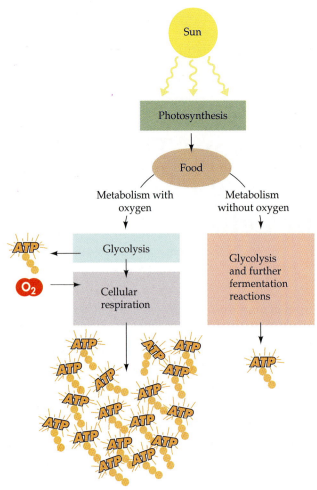

ATP = Energy for work

7.1 Energy for Life

Organisms obtain energy through four types of metabolism: photosynthesis, fermentation, cellular respiration, and glycolysis. Photosynthesis uses light energy to synthesize food compounds. Fermentation releases energy from these food compounds and is particularly important to cells where oxygen is low or depleted. Cellular respiration uses oxygen and releases a far greater amount of energy from a given amount of food than fermentation. Glycolysis is a pathway in which food molecules are converted to a compound that may be used in cellular respiration; glycolysis also serves as the first part of fermentation.

in the cell. ADP, which possesses less free energy than ATP, can combine with a phosphate ion to make a new molecule of ATP if enough energy is provided by some exergonic reaction. (The formation of ATP from ADP and a phosphate ion is endergonic and consumes as much free energy as is released by the breakdown of ATP.)

Many different enzyme-catalyzed reactions in the cell can provide the energy to convert ADP to ATP. The most significant of these in eukaryotes, however, are the reactions of cellular respiration, in which the maximum amount of energy is released from food

molecules and trapped as the stored energy of ATP. A summary of the function of ATP as an energy-storage compound is given in Figure 7.4, which shows the coupling of an exergonic reaction with the formation of ATP, and the subsequent coupling of an endergonic reaction with the splitting of ATP to ADP and a phosphate ion. By this means, energy released in some spontaneous, exergonic reaction may be captured in ATP for use at a later time or in another part of the cell. The phenomenon of bioluminescence provides a visually dramatic example of the use of ATP (Box 7.A).

An active cell requires millions of molecules of ATP per second to drive its biochemical machinery. Even so, it diverts some of its ATP into synthesizing long-term energy-storage compounds. Plants synthesize starch, a long-chain polymer of glucose (Chapter 3) for this purpose, and sometimes synthesize fats. Animals store energy in glycogen (another glucose polymer) and fats. Of course, any large molecule synthesized by the cell—a protein, for example—is a storehouse of energy, but energy storage may not be its primary function. ATP can be considered as the

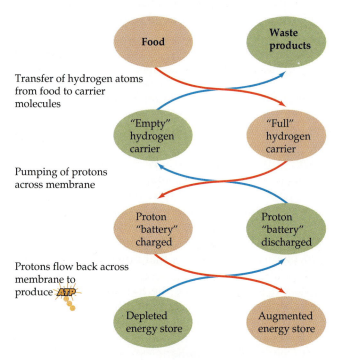

7.2 Coupled Reactions in the "Combustion" of Food

The "combustion" of glucose and other food compounds removes hydrogen atoms, which are transferred to carrier molecules in a coupled reaction. The subsequent removal of hydrogen atoms from the carriers is coupled with the pumping of protons across a membrane, effectively "charging a battery." Finally, the "battery" is "discharged" as protons flow back across the membrane, releasing energy that is used in yet another coupled reaction to form the energy-storing compound ATP. The red arrows emphasize the flow of energy through the coupled reactions.

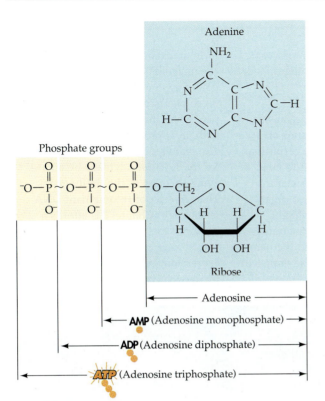

7.3 Structure of ATP

An ATP molecule consists of adenine bonded to ribose (a sugar) with three phosphate groups bonded to another carbon on the ribose. The bonds between phosphate groups that provide energy for cellular processes are shown in color. The compounds ADP and AMP are basically the same as ATP but have one and two fewer phosphate groups, respectively. Adenosine consists of adenine bonded to ribose, without phosphate groups attached.

circulating currency of energy exchange in living organisms, and starches and fats as the savings accounts. When animals need energy, they draw on their deposits of fat and carbohydrates. They break these deposits down into carbon dioxide and water, forming ATP from the energy released in the process. Similarly, plants draw on their stored fats or on deposits of starch, which they convert to glucose; plants then break the glucose down to carbon dioxide and water while forming ATP.

The Energy Content of ATP

To understand the major features of the respiratory pathways, you need to have a feeling for why such large changes in free energy accompany the formation and hydrolysis of ATP. ATP is a **phosphate ester** whose hydrolysis releases a somewhat greater amount of free energy than that for most other esters. An **ester** is an organic compound produced by the reaction of an alcohol with an acid.

For the hydrolysis of ATP to ADP and phosphate, ΔG is about 10 kcal/mol at the temperature, pH, and

substrate concentrations typical of living cells. The hydrolysis of most other phosphate esters produces considerably less than half as much free energy as the hydrolysis of ATP. ATP can therefore transfer phosphate groups to other compounds and "prime" them for later chemical reactions. Part of the unusually large free energy of hydrolysis of ATP comes from the large number of negative charges near each other on its neighboring phosphate groups. When ATP is hydrolyzed, the charges are spread over two molecules, ADP and inorganic phosphate (and hence can get far apart), and these products are thus more stable. The hydrolysis of ADP to adenosine monophosphate (AMP) and inorganic phosphate liberates an even slightly greater amount of free energy than does that of ATP to ADP, although this energy is not often harnessed for work. Hydrolyzing the last phosphate group (converting AMP to adenosine) does not spread out the negative charges any further, so the free-energy change is low, similar to that for any other ester hydrolysis. Because of their larger free energies of hydrolysis, the first and second bonds broken in ATP are sometimes called high-energy bonds, although this refers to the energy of hydrolysis and not to any intrinsic energy of the bond itself. The high-energy bond is sometimes symbolized by $\sim$, and ATP can be written A—R—P$\sim$P$\sim$P, where P represents an entire phosphate group. The phosphate ion, HPO_4^{2-}, is often abbreviated by P_i, meaning inorganic phosphate.

THE TRANSFER OF HYDROGEN ATOMS AND ELECTRONS

Certain pathways of energy metabolism release hydrogen atoms that must be captured and passed on to other reactions. The transfer of either electrons or hydrogen atoms is an oxidation–reduction reaction,

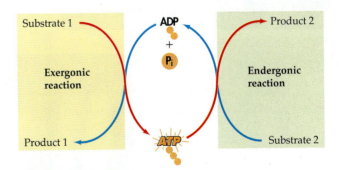

7.4 Formation and Use of ATP

Energy released by an exergonic reaction can be used to combine ADP and an inorganic phosphate ion (P_i) into a molecule of ATP. This ATP may then drive an endergonic reaction, splitting into ADP and P_i in the process.

BOX 7.A

Some Organisms Use ATP to Make Light

Bioluminescence—the production of light by living organisms—always requires ATP as an energy source. Although we know that fireflies use their bioluminescence to communicate between the sexes, we are in the dark as to the role of light in most bioluminescent species. Bioluminescence has evolved independently in many kinds of organisms, from bacteria through vertebrates. A number of fungi are bioluminescent; an ex-

ample is the mushroom *Omphalolotus illudens*, seen in *(a)* as photographed by its own luminescence in a five-hour time exposure.

Many bioluminescent organisms live within the cells or tissues of other organisms, with the result that the host organisms appear to emit light. *(b)* Bioluminescent bacteria populate the kidney-shaped organ below the eye of this "flashlight" fish. *(c)* Another type of bioluminescent bacterium lives within the nematode worms that infest the tissues of these caterpillars.

In an example of bioluminescence "engineered" by scientists, the insertion of a gene from a firefly into tobacco plants produced a dazzling display when the plants were watered with an appropriate substrate *(d)*. The plant's ATP provides the energy for the reaction catalyzed by an enzyme encoded in the firefly gene. This triumph of recombinant DNA technology (Chapter 14) affords a powerful tool for studying such diverse topics as development, gene expression, cellular energetics, and the movement of proteins within cells.

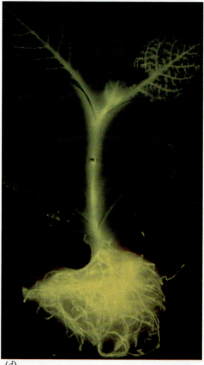

(b)

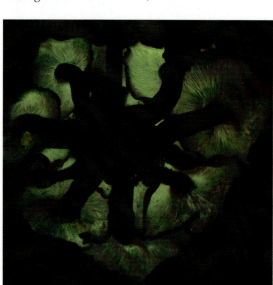

(a)

(c)

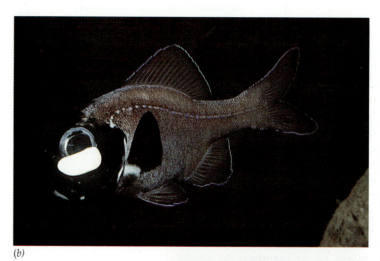

(d)

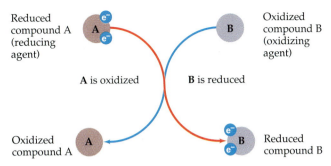

7.5 Oxidation and Reduction
As compound A is oxidized, compound B is reduced. In the process A loses electrons and B gains electrons. In this redox reaction, A is the reducing agent because it donates electrons and B is an oxidizing agent because it accepts the electrons.

or **redox reaction**. The gain of one or more electrons by an atom, ion, or molecule is called **reduction**. The loss of one or more electrons is called **oxidation**. Oxidation and reduction are always defined in terms of traffic in *electrons*. However, it is also appropriate to use these terms when hydrogen atoms (not hydrogen ions) or oxygen atoms are gained or lost because transfers of hydrogen and oxygen atoms involve transfers of electrons. Oxidation is the loss of electrons or the loss of hydrogen atoms or the gain of oxygen atoms; the reverse of any of these is a reduction. Thus the loss of hydrogen atoms by a molecule means the molecule becomes oxidized.

These two processes—oxidation and reduction—always occur together: As one material is oxidized, the electrons it loses are transferred to another material, reducing it (Figure 7.5). Although there cannot be an oxidation without an accompanying reduction, we can talk about the individual "half-reactions," such as the oxidation of one type of iron ion (the ferrous ion, Fe^{2+}) to another (ferric ion, Fe^{3+}). It is convenient for us to talk about such a half-reaction because that oxidation may be coupled to a variety of different reduction reactions—but there must be *some* reduction (the other half-reaction) or there would be no oxidation.

In a redox reaction, we call one of the reactants (the one that becomes reduced) an **oxidizing agent** and the other, which becomes oxidized, a **reducing agent**. An oxidizing agent is something that can accept electrons or hydrogen atoms or can donate oxygen atoms. The oxidizing agent oxidizes the reducing agent, and in the process it becomes reduced (Figure 7.5). Energy is transferred in the reaction, with energy originally present in the reducing agent becoming associated with the reduced product. The oxidative half reactions are highly exergonic, with large negative ΔG values. At the same time, the reductive half reactions are endergonic, with large positive ΔG values. Taking the coupled redox reactions

as a whole, ΔG is always negative as long as the reaction is spontaneous. However, the degree to which the overall reaction is exergonic may be slight or enormous, depending upon the relative ΔG values of the two half reactions. As we will see, some of the key reactions of cellular respiration are highly exergonic redox reactions.

Oxidizing and Reducing Agents in Cells

At some very early stage in evolution, organisms began to form reducing agents and oxidizing agents. The use of certain of these agents—the ones whose redox reactions have suitable values of ΔG—was favored by natural selection as a system for the orderly exchange of electrons, analogous to the use of the ATP–ADP system for the orderly transfer of energy. For example, in Chapter 6 we briefly mentioned FAD as an acceptor of hydrogens in connection with the respiratory conversion of succinate to fumarate; FAD is an oxidizing agent, and $FADH_2$ is a reducing agent.

The most common of the electron banking systems is based on the compound **nicotinamide adenine dinucleotide**, or NAD (Figure 7.6), which exists in two chemically distinct forms: one oxidized (NAD^+) and the other reduced ($NADH + H^+$). The sole function of NAD is to carry hydrogen atoms and free energy from compounds being oxidized and to give up hydrogen atoms and free energy to compounds being reduced (Figure 7.7). The reduction

$$NAD^+ + 2(H) \rightarrow NADH + H^+$$

is accompanied by a free energy increase of 52.4 kcal/mol if oxygen gas is the final oxidizing agent:

$$NADH + H^+ + \frac{1}{2}O_2 \rightarrow NAD^+ + H_2O$$

$$\Delta G = -52.4 \text{ kcal}$$

(Note the representation of the oxidizing agent as "$\frac{1}{2}O_2$" instead of "O." This is to emphasize that it is oxygen gas, O_2, that takes part in the reaction.) In the same way that ATP can be thought of as a means of packaging free energy in bundles of about 10 kcal/mol, NAD can be thought of as a means of packaging <50-kcal/mol bundles.

Various energy carriers are chemically related to one another. One half of NAD looks very much like a molecule of ATP (see Figure 7.3), and it is easy to imagine that several energy carriers constituted from adenine, ribose, phosphates, and other groups evolved over time from a common (and less efficient) precursor molecule. The structures of NAD and some other carrier molecules include compounds that we humans need but cannot synthesize for ourselves; these are classified as vitamins. Nicotinamide, which is part of NAD, forms directly from nicotinic acid, or niacin, a member of the vitamin B complex. Another member of this same vitamin complex is riboflavin,

NAD+ $\quad\quad\quad\quad$ **NADH** + H^+

+ 2H
Reduction
Oxidation

7.6 Nicotinamide Adenine Dinucleotide
Nicotinamide adenine dinucleotide (NAD) exists in two distinct forms: NAD^+, the oxidized form, and NADH + H^+, the reduced form. As the portion of the NAD^+ molecule shown at the top is reduced, it acquires a hydrogen atom and loses its charge. When the same portion of the corresponding NADH molecule is oxidized it loses the hydrogen atom and acquires a charge on its nitrogen atom. NADH + H^+ is a major carrier of hydrogen atoms and free energy in the cell.

which is part of a carrier that we will encounter frequently called **flavin adenine dinucleotide** (FAD). We need only small amounts of vitamins because these carrier molecules are recycled through the metabolic machinery and need to be replaced only slowly. Vitamins are discussed more fully in Chapter 41.

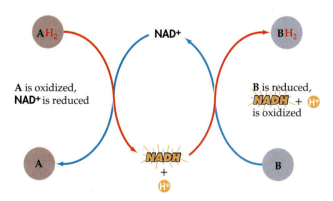

7.7 NAD as Energy Carrier
As compound AH_2 is oxidized, it releases its hydrogen atoms with their electrons to NAD^+, reducing NAD^+ to NADH + H^+. Elsewhere NADH + H^+ may reduce compound B to BH_2, at which time NADH + H^+ is oxidized to NAD^+. Thanks to its ability to carry hydrogen atoms and free energy, NAD is a major and universal energy intermediary in cells.

AN OVERVIEW OF THE
RELEASE OF ENERGY FROM GLUCOSE

Before proceeding, let us divide the energy-extracting processes into pathways that we can consider one at a time (Figure 7.8). As it happens, these pathways tend to be physically separated in the cell; they evolved separately, and perhaps at different times.

The pathway leading up to respiration is the near-universal process of **glycolysis**. This was probably the first energy-releasing pathway to evolve—or, if any earlier pathway existed, it has disappeared from the face of the Earth. Today virtually all living cells use glycolysis, even the most evolutionarily ancient. It is a pathway in which glucose is metabolized to **pyruvic acid** (pyruvate). It contains an oxidative step in which an electron carrier, NAD^+, becomes reduced, acquiring electrons. In addition, a *net* yield of two molecules of ATP is obtained for each molecule of glucose that is processed through glycolysis. The major products of glycolysis are ATP (which the cell will use to drive endergonic reactions), pyruvate, and the two electrons acquired by NAD. Both the pyruvate and the electrons must be processed further.

The process of **cellular respiration** comprises two pathways: the citric acid cycle and the respiratory chain. In eukaryotes in the presence of oxygen and in some bacteria, the pyruvate from glycolysis is oxidized in a cyclical series of respiratory reactions known as the **citric acid cycle** (also called the Krebs cycle or the tricarboxylic acid cycle). In eukaryotes, the reactions of the citric acid cycle are catalyzed by enzymes present in the liquid matrix inside the mitochondrion. Various steps release the carbon atoms of pyruvate (originally the carbon atoms of glucose)

7.8 Energy-Releasing Processes in Cells
The energy-releasing pathways begin with glycolysis, the conversion of glucose to pyruvate. *(a)* Fermentation proceeds in some cells in the absence of oxygen, forming compounds such as lactic acid or ethanol and releasing some ATP. *(b)* When oxygen is present, pyruvate from glycolysis can enter the citric acid cycle in many cells. Compounds from glycolysis and the citric acid cycle are oxidized in the respiratory chain to produce ATP. The pathway from pyruvate to ATP is referred to as cellular respiration. The individual processes appear in full size in Figures 7.10 and 7.12.

(a) In the absence of oxygen, fermentation produces 2 ATPs in its glycolysis reactions for each glucose molecule. Electrons, which in the presence of oxygen would be used to obtain more ATP, are used in lactic acid or ethanol fermentation

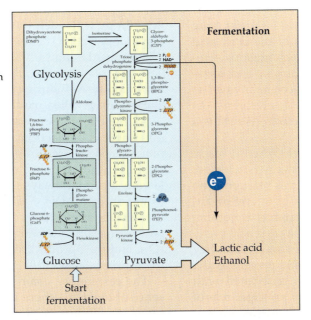

(b) In the presence of oxygen, glucose is further metabolized in the citric acid cycle and the respiratory chain, yielding great quantities of ATP

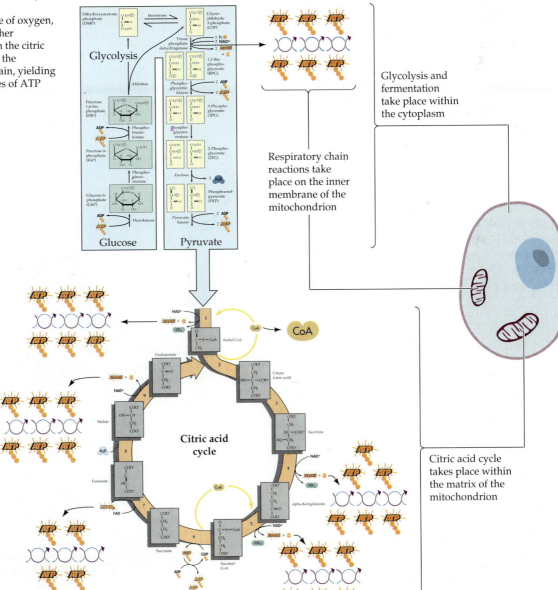

Glycolysis and fermentation take place within the cytoplasm

Respiratory chain reactions take place on the inner membrane of the mitochondrion

Citric acid cycle takes place within the matrix of the mitochondrion

as carbon dioxide (CO_2) molecules, and transfer more electrons to carriers. The products of the citric acid cycle are, then, carbon dioxide, which must be eliminated in some way from the organism, and many more stored electrons (along with accompanying hydrogen nuclei) than are produced in glycolysis. As we are about to see, more stored electrons means a greater ultimate harvest of ATP.

Hydrogen is an outstanding fuel. When it reacts with oxygen, a great deal of free energy is released; better still, the "waste" product of this reaction is no problem either to the environment or to any organism that produces it—that waste product is nothing but water. In both glycolysis and the citric acid cycle, hydrogen atoms are acquired by the molecules they reduce. Most of the energy originally present as the covalent bonds of glucose is now associated with reduced NAD (NADH + H$^+$). The principal role of the **respiratory chain** (Figure 7.8) is to release that energy from the reduced NAD in such a way that it may be used to form ATP. This pathway is a series of successive redox reactions in which hydrogen atoms—or, in the later steps, electrons derived from hydrogen atoms—are passed from one type of membrane carrier to another and finally are allowed to react with oxygen gas and produce water. In eukaryotes, these carriers (and the associated enzymes) are bound to the folds of the inner mitochondrial membranes, which are called **cristae** (Figure 7.9 and Chapter 4). In both prokaryotes and eukaryotes, free energy drops with each transfer of electrons along the respiratory chain. The released energy is used to form ATP from ADP and P$_i$. This is the way in which the vast majority of the ATP in animals is formed. The formation of ATP during the operation of the respiratory chain is called **oxidative phosphorylation**.

As the energy is released, the reduced NAD and other agents of electron transfer are oxidized. They may then be reused in glycolysis and the citric acid cycle, steadily draining off hydrogen atoms and allowing those pathways to continue to operate. This oxidation of NADH + H$^+$ is thus another function of the respiratory chain. The inputs to the respiratory chain are stored hydrogen atoms and oxygen gas (O_2), and the outputs are water and stored energy in the form of ATP.

The respiratory chain cannot function in the absence of oxygen—there is no "bucket" at the end to catch the hydrogen atoms. If we are deprived of oxygen for too long, we die. Without oxygen, the carriers in the cristae of our mitochondria are unable to jettison the hydrogen atoms or electrons bound to them. Soon there are no oxidized carriers available; when that happens, glycolysis and the citric acid cycle stop. With no glycolysis, no citric acid cycle, and no respiratory chain activity, we animals have an insufficient supply of ATP. Without ATP, our cells cannot maintain their structure and metabolism, and we die.

Interestingly, in humans it is the nervous system,

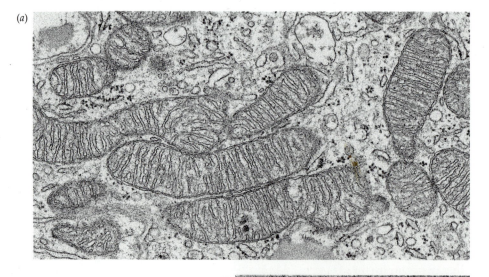

(a)

(b)

7.9 Cellular Powerhouses
(a) Numerous cristae, arising from the inner of the two mitochondrial membranes, reach into the matrix of these mitochondria. The cristae are the sites of the ATP-producing reactions of cellular respiration. (b) A high-magnification view of the inner mitochondrial membrane. Spherical "knobs" project into the mitochondrial matrix; these protein "knobs" catalyze the synthesis of ATP.

including the brain, that dies first. Our muscles have an alternative way to deal with the problem of getting rid of the hydrogen atoms produced during glycolysis: The hydrogens are handed right back to the end product of glycolysis (pyruvic acid) and lactic acid is formed. Because the hydrogen atoms are removed, the electron carriers that held them are oxidized and can be used again to process more glucose. In the absence of oxygen, this continues (even at an increased rate) without the activity of the citric acid cycle or the respiratory chain. Thus ATP continues to be produced, and the cells that have in their cytosol the enzymatic machinery to carry this reaction on are thereby enabled to function for a time in the absence of oxygen. Eventually, however, the concentration of lactic acid in muscles reaches a toxic level. This anaerobic production of ATP, which releases only a small part of the of energy for organisms that require oxygen, is called fermentation. For some organisms that live without oxygen, it is the sole pathway that can combine with glycolysis to release energy. We will examine fermentation in more detail later in this chapter.

GLYCOLYSIS

The metabolic pathway leading up to respiratory metabolism consists of the ten reactions of **glycolysis**, in which a molecule of the six-carbon sugar glucose is gradually converted into two molecules of the three-carbon compound pyruvic acid (Figure 7.10). These reactions are accompanied by the *net* formation of two molecules of ATP and by the reduction of two molecules of NAD^+ to two molecules of NADH + H^+. In other words, ready energy is located in ATP, and four hydrogen atoms are passed on in a reducing agent. The fate of the pyruvic acid depends on the kind of cell carrying out glycolysis and on whether the environment is aerobic or anaerobic. The fate of the NADH + H^+, too, is variable. In most cases, it will be oxidized through the respiratory chain to yield water and NAD^+—a chain of reactions that results in the formation of much more ATP (three molecules of ATP per molecule of NADH + H^+). In fermentation, however, NADH + H^+ is reoxidized to NAD^+ either by pyruvic acid itself or by one of its metabolites, with no further storage of free energy. In either case, glycolysis may be regarded as a series of *preparatory* reactions, to be followed by either the citric acid cycle or the remainder of fermentation.

Within the glycolytic pathway itself, the first five reactions may be viewed as "pump-priming." Each of the five reactions is, in fact, endergonic, taking up free energy; the cell is actually *spending* free energy rather than gaining it during the early reactions of glycolysis. As we see in Figure 7.10, two molecules of ATP are invested to attach two phosphate groups to the sugar and raise its free energy by about 15 kcal/mol (Figure 7.11). We will see in a moment that these phosphate groups will be used to make new molecules of ATP. Although both of the first steps of glycolysis use ATP as one of the substrates, each is catalyzed by a different, specific enzyme. Such specificity is characteristic of enzyme-catalyzed reactions. The enzyme hexokinase catalyzes the first reaction, in which glucose receives a phosphate group from ATP (Figure 7.10). *Kinase* is the generic name for any enzyme that catalyzes the transfer of a phosphate group from ATP to another substrate. In the second reaction, the six-membered glucose ring is rearranged to a five-membered fructose ring; then the enzyme phosphofructokinase adds a second phosphate (taken from another ATP) to the sugar ring (reaction 3). Then the sugar ring is opened, and the six-carbon sugar bisphosphate is cleaved to give two different three-carbon sugar phosphates (reaction 4). One of these sugar phosphates (dihydroxyacetone phosphate) is converted into a second molecule of the other (glyceraldehyde 3-phosphate).

By this time, the halfway point in glycolysis, the following things have happened: Two molecules of ATP have been *used* in the priming reactions, and the glucose molecule has been converted into two molecules of a three-carbon sugar phosphate. By this point, no ATP has been gained, and nothing has been oxidized; in short, it looks as if we are going determinedly backward.

Now, however, the pump is primed and things begin to happen rapidly. In what follows, remember that each step actually occurs twice for each glucose molecule going through glycolysis; that is because each glucose molecule has by now been split into two molecules of three-carbon sugar phosphate, each of which goes through the remaining steps of glycolysis. Each now undergoes a two-step reaction (reaction 6 in Figure 7.10) catalyzed by the enzyme glyceraldehyde 3-phosphate dehydrogenase. The end product of that reaction is 1,3-bisphosphoglycerate (or 1,3-bisphosphoglyceric acid). A phosphate ion has been snatched from the surroundings (but not, this time, from ATP) and tacked onto the three-carbon compound. A glance at Figure 7.11 shows that this reaction is accompanied by an enormous drop in free energy—over 100 kcal per mole of glucose is released in this extremely exergonic reaction. What has happened here? Why the big energy change? The conversion of a sugar to an acid

$$R-\overset{\overset{\textstyle O}{\|}}{C}-H + (O) \rightarrow R-\overset{\overset{\textstyle O}{\|}}{C}-OH$$

is an oxidation and, as you know, oxidations are very

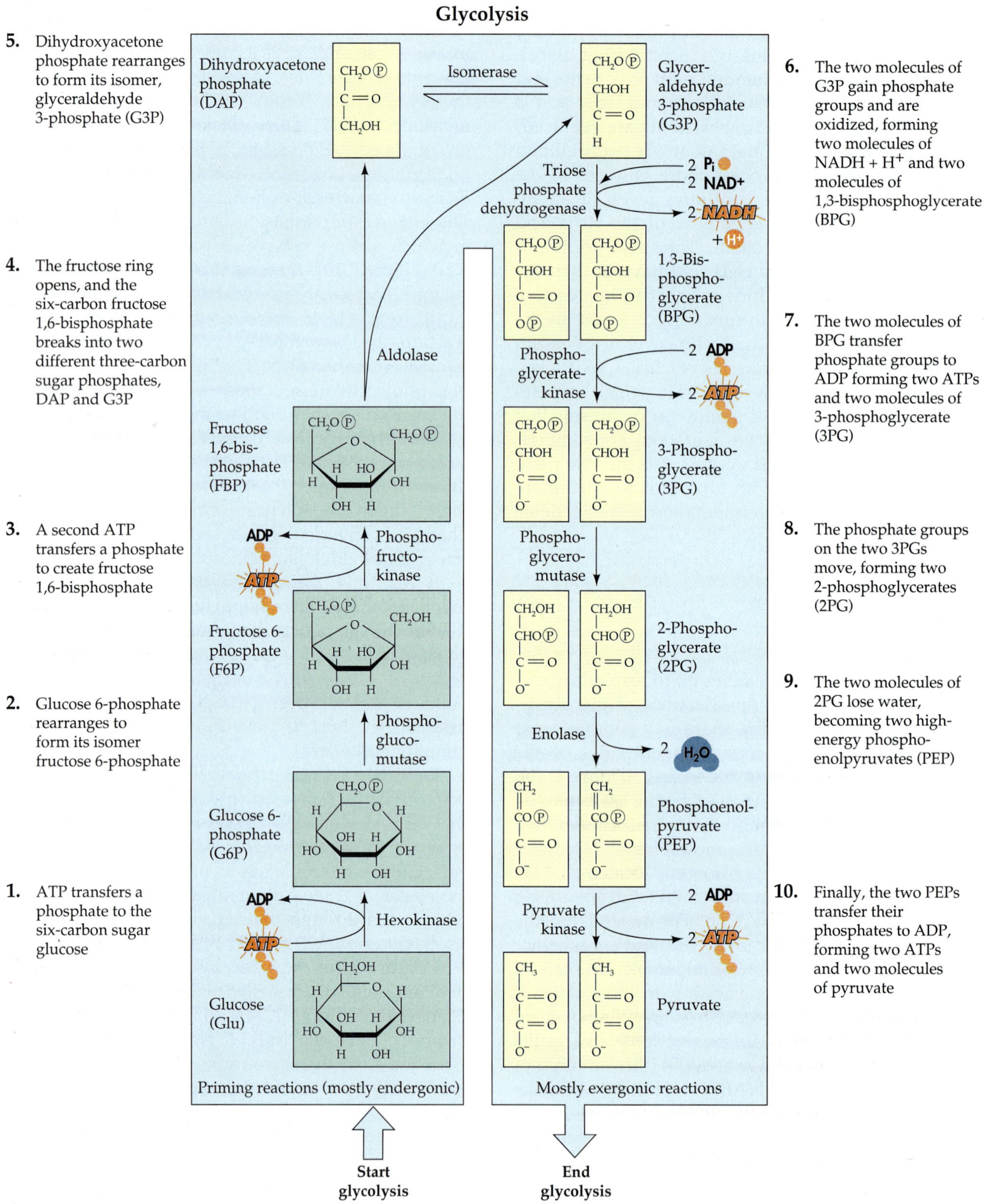

Glycolysis

5. Dihydroxyacetone phosphate rearranges to form its isomer, glyceraldehyde 3-phosphate (G3P)

Dihydroxyacetone phosphate (DAP)

CH_2O℗
$C = O$
CH_2OH

Isomerase

CH_2O℗
$CHOH$
$C = O$
H

Glyceraldehyde 3-phosphate (G3P)

6. The two molecules of G3P gain phosphate groups and are oxidized, forming two molecules of NADH + H⁺ and two molecules of 1,3-bisphosphoglycerate (BPG)

Triose phosphate dehydrogenase

2 Pi
2 NAD⁺
2 NADH
+ H⁺

4. The fructose ring opens, and the six-carbon fructose 1,6-bisphosphate breaks into two different three-carbon sugar phosphates, DAP and G3P

Aldolase

CH_2O℗
$CHOH$
$C = O$
O℗

CH_2O℗
$CHOH$
$C = O$
O℗

1,3-Bis-phospho-glycerate (BPG)

Phospho-glycerate-kinase

2 ADP
2 ATP

7. The two molecules of BPG transfer phosphate groups to ADP forming two ATPs and two molecules of 3-phosphoglycerate (3PG)

Fructose 1,6-bis-phosphate (FBP)

CH_2O℗ CH_2O℗
O
H HO
H OH
OH H

CH_2O℗
$CHOH$
$C = O$
O^-

CH_2O℗
$CHOH$
$C = O$
O^-

3-Phospho-glycerate (3PG)

Phospho-fructo-kinase

ADP
ATP

Phospho-glycero-mutase

8. The phosphate groups on the two 3PGs move, forming two 2-phosphoglycerates (2PG)

Fructose 6-phosphate (F6P)

CH_2O℗ CH_2OH
O
H HO
H OH
OH H

CH_2OH
CHO℗
$C = O$
O^-

CH_2OH
CHO℗
$C = O$
O^-

2-Phospho-glycerate (2PG)

2. Glucose 6-phosphate rearranges to form its isomer fructose 6-phosphate

Phospho-gluco-mutase

Enolase

2 H₂O

9. The two molecules of 2PG lose water, becoming two high-energy phospho-enolpyruvates (PEP)

Glucose 6-phosphate (G6P)

CH_2O℗
H O H
H
OH H
HO OH
H OH

CH_2
CO℗
$C = O$
O^-

CH_2
CO℗
$C = O$
O^-

Phosphoenol-pyruvate (PEP)

1. ATP transfers a phosphate to the six-carbon sugar glucose

Hexokinase

ADP
ATP

Pyruvate kinase

2 ADP
2 ATP

10. Finally, the two PEPs transfer their phosphates to ADP, forming two ATPs and two molecules of pyruvate

Glucose (Glu)

CH_2OH
H O H
H
OH H
HO OH
H OH

CH_3
$C = O$
$C = O$
O^-

CH_3
$C = O$
$C = O$
O^-

Pyruvate

Priming reactions (mostly endergonic)

Mostly exergonic reactions

Start glycolysis

End glycolysis

7.10 Glycolysis

The reactions of glycolysis convert a six-carbon molecule of glucose to two three-carbon molecules of pyruvic acid (pyruvate). Each numbered reaction in the pathway is catalyzed by a specific enzyme. Along the way, some ATP is produced (reactions 7 and 10), and two molecules of NAD⁺ are reduced to 2NADH + 2H⁺ (reaction 6).

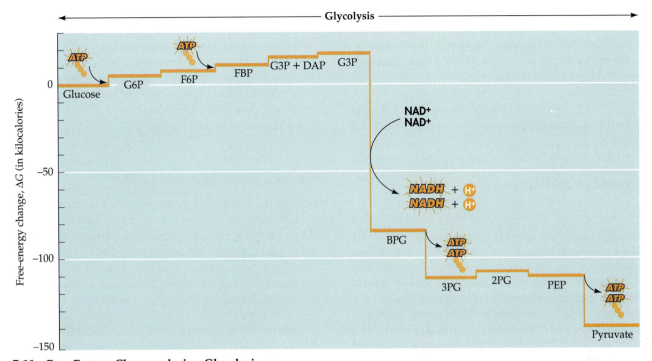

7.11 Free-Energy Changes during Glycolysis
Each reaction of glycolysis changes the free energy available, as shown by the differing energy levels of the series of reactants and products from glucose to pyruvate. Note at the upper left that the investment of each of two ATPs in the priming reactions raises the free-energy content of the sugar. Key energy-releasing reactions account for the largest drops in free energy. Notable among these are the drop from glyceraldehyde 3-phosphate (G3P) to 1,3-bisphosphoglycerate (BPG), where NADH + H$^+$ is formed, and the two later reactions (lower right) that form ATPs. The quantities of NADH + H$^+$ and ATP in the diagram are those for the passage of one molecule of glucose through the reactions of glycolysis—in all 2 ATP and 2 NADH + H$^+$ are released.

exergonic. (Note that here we do *not* write ½O$_2$ because in this case oxygen gas does not participate in the reaction.) The formation of the phosphate ester

$$R-\overset{\overset{\displaystyle O}{\|}}{C}-OH + HPO_4^{2-} \rightarrow R-\overset{\overset{\displaystyle O}{\|}}{C}-O-\overset{\overset{\displaystyle O}{\|}}{\underset{\underset{\displaystyle O^-}{|}}{P}}-O^- + H_2O$$

is slightly endergonic, but not nearly enough to offset the drop from the oxidation.

If this big energy drop were simply the loss of heat, glycolysis would be an extremely inefficient process for providing useful energy to the cell. However, this energy is not lost but is used to make two molecules of NADH + H$^+$ from the two molecules of NAD$^+$. This stored energy is regained later—either in the respiratory chain, by the formation of ATP, or else in the last step of fermentation when pyruvate or its product is reduced and the two molecules of NADH + H$^+$ are restored once again to NAD$^+$. This cycling of NAD is necessary to keep glycolysis going; if all the NAD$^+$ is converted to NADH + H$^+$, glycolysis comes to a halt.

The remaining steps of glycolysis in Figure 7.10

are simpler. The two phosphate groups of 1,3-bisphosphoglycerate are transferred, one at a time, to molecules of ADP, with a rearrangement in between. As a result, over 20 kcal of free energy is stored in ATP for every mole of 1,3-bisphosphoglycerate broken down. Finally, we are left with pyruvic acid—2 mol for each mole of glucose that entered glycolysis.

A review of the reactions shows us that at the beginning of glycolysis two molecules of ATP are used per molecule of glucose, but that ultimately four are produced (two for each of the two 1,3-bisphosphoglycerates)—a net gain of two ATP molecules and two NADH + H$^+$. Under anaerobic conditions, the total usable energy yield from the metabolism of glucose is usually two ATP. The NADH + H$^+$ is rapidly recycled to NAD$^+$ by fermentation for reuse by the glyceraldehyde 3-phosphate dehydrogenase of glycolysis. In the presence of oxygen, on the other hand, eukaryotes and some bacteria are able to reap far more energy by the further metabolism of pyruvate and by reoxidizing the reduced NAD of glycolysis through the respiratory chain.

(By now, you might be wondering why we are using words like *pyruvate* and *pyruvic acid* interchangeably. At pH values commonly found in cells, the ionized form—pyruvate—is present rather than the acid—pyruvic acid. Similarly, all carboxylic acids are present as ions (the *−ate* forms) at these pHs. Thus on grounds of chemical accuracy and simplicity, it is better to name the negative ion than the acid. However, custom often prevails over accuracy, and the acids are often named instead; for example, nobody seems to want to change the name *citric acid cycle* to the apparently more correct form *citrate cycle*.)

THE CITRIC ACID CYCLE

Next we consider an important pathway that takes pyruvate—the end product of glycolysis—as its starting point. Figure 7.11 shows that the metabolism of glucose to pyruvate is accompanied by a drop in free energy of about 140 kcal/mol. About one-third of this energy is captured in the formation of ATP and reduced NAD. Further free energy for biological work can be gained by oxidizing the pyruvate. The citric acid cycle takes pyruvate and breaks it down to CO_2, using the hydrogen atoms to reduce carrier molecules and to pass chemical free energy to those carriers. The reduced carriers are later oxidized in the respiratory chain, which we will discuss presently; and an enormous amount of free energy is transferred from the reduced carriers to ATP in the process. The principal inputs to the citric acid cycle are pyruvic acid, water, and oxidized electron carriers; the principal outputs are carbon dioxide and reduced electron carriers:

$$\text{pyruvic acid } (C_3H_4O_3) + 3\ H_2O + 5\ \text{carrier} \rightarrow$$
$$3\ CO_2 + 5\ \text{carrier} \cdot (2H)$$

The reactions of the citric acid cycle are shown in Figure 7.12, and the energy changes in it and in glycolysis are diagrammed in Figure 7.13. In step 1 of the citric acid cycle, pyruvate is oxidized (yielding useful free energy) and converted to an activated form of acetic acid (CH_3COOH) called **acetyl coenzyme A**. Then acetyl coenzyme A, with two carbon atoms in its acetate group, reacts with a four-carbon acid (oxaloacetate) to form the six-carbon compound citric acid (citrate). The remainder of the cycle consists of a series of enzyme-catalyzed reactions in which citric acid is degraded, leading to the release of two of the carbons as CO_2, to the production of useful free energy from redox reactions, and to the production of a new four-carbon molecule of oxaloacetate from the other four carbons. This new oxaloacetate can react with a second acetyl coenzyme A, producing a second molecule of citrate, and so forth. Acetyl coenzyme A is coming into the cycle from pyruvate, CO_2 is going out, the rest of the compounds in the cycle are being used and replaced, and, as we will see in a moment, energy from redox

reactions is being stored. For now, though, let us concentrate on seeing how the citric acid cycle is maintained in a **steady state**, that is, with material entering and leaving and with intermediate compounds like succinate and malate *turning over constantly* but *without changing concentration*. The concept of steady state is an important one, and you should make sure it is clear to you.

Reactions of the Citric Acid Cycle

During the citric acid cycle as a whole, starting with a single molecule of pyruvate, three carbons are removed as CO_2, and five pairs of hydrogen atoms per pyruvate molecule are used to reduce carrier molecules, with the simultaneous storage of energy. The energy-removing reactions, which are a major reason for the existence of the cycle, are labeled 1, 4, 5, 7, and 9 in Figure 7.12.

Reaction 1 is an interesting one in which several things happen. Its product, acetyl coenzyme A, is 7.5 kcal/mol higher in energy than simple acetate. Acetyl coenzyme A can donate acetate to acceptors such as oxaloacetate much as ATP can donate phosphate to various acceptors. There are three steps in reaction 1: pyruvate is oxidized to acetate with the release of CO_2; part of the energy from this oxidation is saved by reducing NAD^+ to $NADH + H^+$; and some of the remaining energy is temporarily stored by combining the acetate with coenzyme A. An analogous three-step reaction occurs in glycolysis when glyceraldehyde 3-phosphate is converted to 1,3-bisphosphoglycerate (Figure 7.10). In that reaction, a sugar is oxidized to an acid, some of the energy released by oxidation is stored in $NADH + H^+$, and some of the remaining energy is preserved in a second phosphate bond in the molecule. A good metabolic idea is likely to be seen more than once, and in fact we will see this one yet again in a later step in the citric acid cycle. As you might guess, such a complex set of steps as those found in the reaction from pyruvate to acetyl coenzyme A requires more than one type of catalytic protein. This compound reaction is catalyzed by the *pyruvate dehydrogenase complex*, which consists of 72 subunits—24 each of three different kinds of protein, for a total molecular weight of 4.6 million. This is indeed an impressive example of biological organization.

The energy temporarily stored in acetyl coenzyme A helps to drive the reaction with oxaloacetate to make citrate. When this happens, the coenzyme molecule falls away to be recycled and bound to another acetate by the pyruvate dehydrogenase complex. Citrate is rearranged to isocitrate; and in reaction 4, a CO_2 molecule and two hydrogen atoms are removed in converting isocitrate to α-ketoglutarate. As Figure

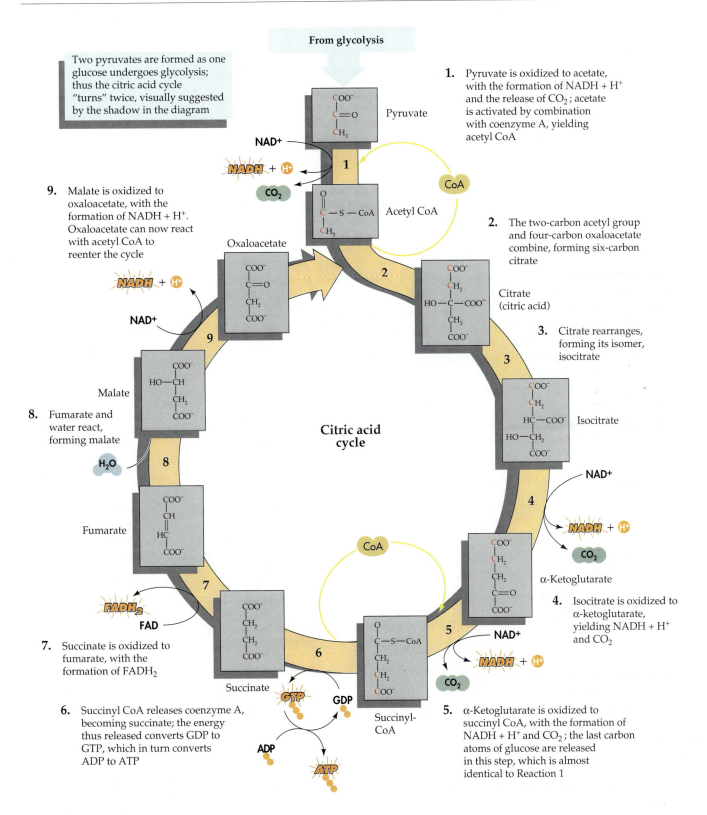

From glycolysis

Two pyruvates are formed as one glucose undergoes glycolysis; thus the citric acid cycle "turns" twice, visually suggested by the shadow in the diagram

1. Pyruvate is oxidized to acetate, with the formation of NADH + H$^+$ and the release of CO_2; acetate is activated by combination with coenzyme A, yielding acetyl CoA

Pyruvate

NAD$^+$

NADH + H$^+$

CO_2

1

Acetyl CoA

CoA

9. Malate is oxidized to oxaloacetate, with the formation of NADH + H$^+$. Oxaloacetate can now react with acetyl CoA to reenter the cycle

2. The two-carbon acetyl group and four-carbon oxaloacetate combine, forming six-carbon citrate

Oxaloacetate

NADH + H$^+$

NAD$^+$

9

2

Citrate (citric acid)

3. Citrate rearranges, forming its isomer, isocitrate

Malate

3

Isocitrate

Citric acid cycle

8. Fumarate and water react, forming malate

NAD$^+$

4

NADH + H$^+$

CO_2

H_2O

8

α-Ketoglutarate

Fumarate

CoA

7

4. Isocitrate is oxidized to α-ketoglutarate, yielding NADH + H$^+$ and CO_2

FADH$_2$

FAD

5

NAD$^+$

7. Succinate is oxidized to fumarate, with the formation of FADH$_2$

6

NADH + H$^+$

CO_2

Succinate

Succinyl-CoA

5. α-Ketoglutarate is oxidized to succinyl CoA, with the formation of NADH + H$^+$ and CO_2; the last carbon atoms of glucose are released in this step, which is almost identical to Reaction 1

6. Succinyl CoA releases coenzyme A, becoming succinate; the energy thus released converts GDP to GTP, which in turn converts ADP to ATP

GTP

GDP

ADP

ATP

7.12 The Citric Acid Cycle
In cellular respiration the three-carbon pyruvate from glycolysis is converted to the two-carbon acetyl coenzyme A (reaction 1). Acetyl coenzyme A reacts with four-carbon oxaloacetate to form a six-carbon molecule of citric acid (citrate). During the remainder of the cycle back to oxaloacetate, two molecules of CO_2 are released. The major overall effect of the cycle is to pass electrons to the carrier molecule NAD. The two carbons entering the cycle from acetyl coenzyme A are traced with colored circles through reaction 5, after which they may be at either end of the molecule (note the symmetry of succinate and fumarate).

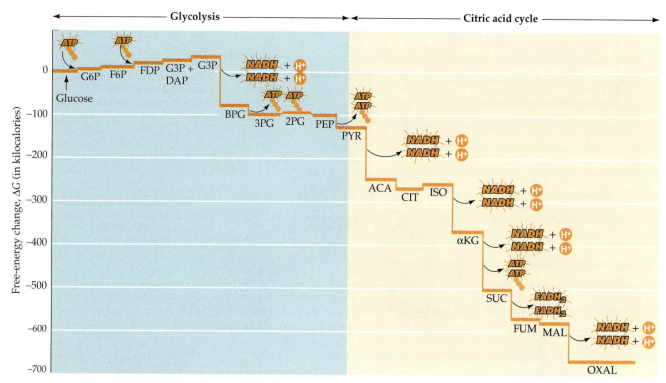

7.13 Free-Energy Change during Glycolysis and Cellular Respiration

Much more free energy is released in the citric acid cycle than in glycolysis. The free energy changes in glycolysis (left-hand side) and the citric acid cycle (right-hand side) are shown. (We saw the left-hand side at a different scale in Figure 7.11.) Electron carriers (NAD in glycolysis; NAD and FAD in the citric acid cycle) are reduced and ATP is generated in reactions coupled to reactions producing major drops in free energy.

7.13 indicates, this reaction produces a large drop in free energy. The released energy is stored in $NADH + H^+$ and can be recovered later in the respiratory chain, when the $NADH + H^+$ is reoxidized.

Reaction 5 of the citric acid cycle is a complex one, like the oxidation of pyruvate to acetyl coenzyme A. Analogous to that reaction, the α-ketoglutarate molecule is oxidized to succinate, CO_2 is given off, some of the oxidation energy is stored in $NADH + H^+$, and some is preserved temporarily by combining succinate with coenzyme A. This temporarily stored energy is saved in reaction 6, in which guanosine triphosphate (GTP) is first made and then used to make ATP. A smaller amount of free energy is released in reaction 7, when two hydrogens are transferred to an enzyme containing FAD (an oxidizing agent similar to NAD^+ that we discussed earlier in this chapter); one more NAD^+ reduction occurs after a molecular rearrangement. The oxaloacetate that is left over after all these reactions is ready to combine with another acetyl coenzyme A molecule and go around the cycle again. Bear in mind that the citric acid cycle operates twice for each glucose molecule that enters glycolysis.

THE RESPIRATORY CHAIN

The roles of NAD^+ and FAD in glycolysis and cellular respiration are critical: Without a suitable oxidizing agent such as one of these ready to be reduced and act as an electron carrier, the oxidative steps of glycolysis and the citric acid cycle could not occur. We can picture the reaction of substrate and oxidizing agent as follows:

$$substrate \cdot H_2 \diagdown \diagup NAD^+$$
$$Substrate \diagup \diagdown NADH + H^+$$

with the hydrogen being passed from the originally reduced substrate (such as malate in the citric acid cycle) to the oxidizing agent NAD^+. We see that the presence of the oxidizing agent is critical, because without it the substrates could not be oxidized, and there would be no respiratory metabolism.

But what about all that NADH? If this reaction continues, it would appear that all the cell's NAD^+ would become reduced, leaving no NAD^+ to act as an oxidizing agent. Fortunately, there is something in most cells—a specific oxidizing agent—that can reoxidize the $NADH + H^+$. This agent is a carrier called ubiquinone (Q). Q acts as follows to oxidize $NADH + H^+$:

$$NADH + H^+ \diagdown \diagup Q$$
$$NAD^+ \diagup \diagdown QH_2$$

Fine! NAD$^+$ is once again available, so glycolysis and the citric acid cycle may continue. But will we not run out of oxidized Q now? No, because there is another carrier, **cytochrome c**, a small protein that can reoxidize the QH$_2$. Does this **respiratory chain** have an end somewhere? Yes—cytochrome c is reoxidized by molecular oxygen, the final oxidizing agent:

cyt c (red) ⟶ ½ O$_2$

cyt c (ox) ⟵ H$_2$O

This is very satisfactory indeed. In most places where life is found on Earth today, oxygen gas is abundant, so there is no worry about running out of oxidizing agent. In addition, the "waste product" of the respiratory chain—water—is nontoxic and presents no disposal problem. The two hydrogens in water, by the way, may be thought of as being the hydrogens that were originally abstracted from some substrate back in the citric acid cycle or glycolysis. They have been handed on from one carrier to another and finally used to reduce molecular oxygen, reoxidizing cytochrome c and allowing the various respiratory pathways to continue.

The respiratory chain is more complicated than we have just indicated, for the chain also contains three large protein complexes through which electrons are passed, as shown in Figure 7.14. Between NADH + H$^+$ and Q lies **NADH-Q reductase**, a complex of 25 polypeptide subunits, with a total molecular weight of 850,000. **Cytochrome reductase**, with 9 subunits and a molecular weight of 250,000, lies between Q and cytochrome c. **Cytochrome oxidase**, with 8 subunits and a molecular weight of 160,000, lies between cytochrome c and oxygen. Different subunits within each of the complexes bear different electron carriers, so electrons are transported *within* each complex. All the components of the respiratory chain are proteins, or attached to proteins, except for Q, which is a smaller molecule.

Why should the respiratory chain have so many links? Why, for example, do we not just use the following single step?

NADH + H$^+$ + ½O$_2$ → NAD$^+$ + H$_2$O

Would this not accomplish the same thing, and more efficiently? To begin with, there is no enzyme that will catalyze the direct oxidation of NADH by oxygen. More fundamentally, this would be an untamable reaction. It would be terrifically exergonic—rather like setting off a stick of dynamite in the cell. There is no biochemical way to harvest that burst of energy in an efficient way and put it to physiological use (there is no metabolic reaction so endergonic as to consume a significant fraction of that energy in a single step). Instead, evolution has led to the lengthy chain found today: a *series* of reactions, each releasing a smaller, relatively manageable amount of energy. Electron transport within each of the three protein complexes results in the formation of ATP. Thus the vast energy supply originally contained in glucose and other foods is finally tucked into the cellular energy currency that is ATP. For each pair of hydrogen atoms passed along the respiratory chain from NADH + H$^+$ to oxygen, three molecules of ATP are formed.

The several carriers of the respiratory chain (including those contained in the three protein complexes) differ as to how they change upon reduction. NAD$^+$, for example, accepts one proton and two electrons, leaving the proton from the other hydrogen atom to float free: NADH + H$^+$. Others, including Q, bind both protons and both electrons in becoming, for example, QH$_2$. The remainder of the chain, however, is only an electron-transport process. As electrons are passed from Q to cytochrome c, the protons wander free into solution and are brought back into the pathway only at the very end of the chain. The cytochromes contain iron atoms that in

7.14 The Oxidation of NADH + H$^+$
Oxidation of NADH + H$^+$ by the respiratory chain produces a great deal of ATP. As traced by the bold arrows, electrons from NADH + H$^+$ are passed through a series of carrier molecules in the inner mitochondrial membrane (or the plasma membrane of an aerobic prokaryote), releasing enough energy to produce ATP from ADP + P$_i$ along the way. The carriers gain free energy and become reduced as electrons are passed to them and release free energy when they are oxidized, passing the electrons to the next carrier in the chain.

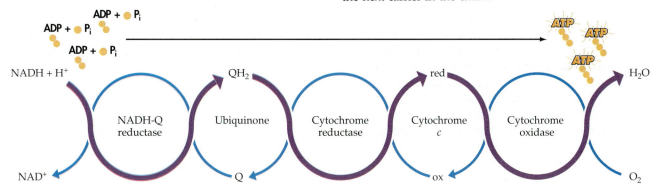

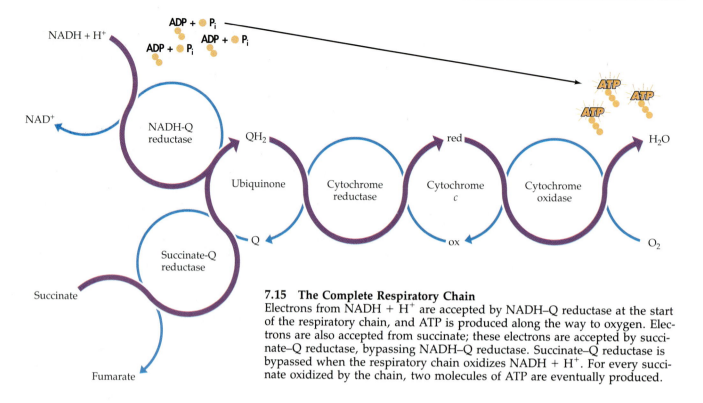

7.15   The Complete Respiratory Chain
Electrons from NADH + H⁺ are accepted by NADH–Q reductase at the start of the respiratory chain, and ATP is produced along the way to oxygen. Electrons are also accepted from succinate; these electrons are accepted by succinate–Q reductase, bypassing NADH–Q reductase. Succinate–Q reductase is bypassed when the respiratory chain oxidizes NADH + H⁺. For every succinate oxidized by the chain, two molecules of ATP are eventually produced.

their oxidized states are Fe^{3+} and in their reduced states Fe^{2+}. The iron atoms are held in place by **heme groups** like that found in hemoglobin (Chapter 3).

Electrons pour into the pool of Q molecules from the NADH + H⁺ pathway; they can also come from another source: the succinate-to-fumarate reaction of the citric acid cycle (reaction 7). Another protein complex, **succinate-Q reductase**, links the oxidation of succinate to the reduction of Q (Figure 7.15). The enzyme that constitutes the first part of succinate-Q reductase has attached to it an FAD carrier molecule, which is reduced by succinate to $FADH_2$. In a later step, hydrogen atoms are given to the Q molecules. No ATP is generated in the succinate-to-Q branch. Hence the pathway from the oxidation of succinate forms only two ATP molecules compared with the three obtained when NAD^+ is the first oxidizing agent.

Oxidative Phosphorylation and Mitochondrial Structure

For many years, biochemists and molecular biologists struggled to understand how the operation of the respiratory chain caused oxidative phosphorylation —the formation of ATP in the mitochondrion. The problem was solved in 1961, when the British biochemist Peter Mitchell proposed the **chemiosmotic theory**. This elegant model illustrates once again the intimate relationship between structure and function in biology, so let us begin by reviewing the placement of the various components of respiratory metabolism within the cell.

The reactions of glycolysis are older than those of the citric acid cycle, having evolved before the most ancient of today's prokaryotes. It is not surprising, therefore, that the enzymes for glycolysis are found free-floating in the cytosol of the cell or bound to the cytoskeleton. They are not enclosed within any organelle and, in fact, are even found in most cells that *lack* organelles. In contrast, the enzymes of the citric acid cycle and the respiratory chain are isolated in the mitochondria. (Some aerobic bacteria also carry out these reactions. Although they lack mitochondria, they do have membrane systems with which these enzymes are closely associated.) A typical mitochondrion from a mammalian cell is shown in Figure 7.16. It has a relatively smooth outer membrane and an inner membrane that is folded back and forth deep into the interior of the organelle, so that the inner membrane has an enormous surface area in relation to the volume that it encloses. That enclosed volume is filled with a protein-rich fluid, the mitochondrial matrix. The enzymes of the citric acid cycle are dissolved in the mitochondrial matrix, with three exceptions: Succinate dehydrogenase, which catalyzes reaction 7 of Figure 7.12, and the two enormous complexes that catalyze reactions 1 and 5. These enzymes are buried in the inner membrane (Box 7.B). The carriers and enzymes of the respiratory chain (other than cytochrome *c*) are also embedded in the inner mitochondrial membrane. Cytochrome *c* is an extrinsic protein (Chapter 5) and lies in the space between the inner and outer mitochondrial membranes, loosely attached to the inner membrane. Ubiquinone, a small, nonprotein molecule, is free to

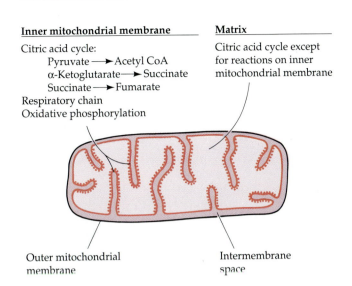

Inner mitochondrial membrane

Citric acid cycle:
     Pyruvate → Acetyl CoA
     α-Ketoglutarate → Succinate
     Succinate → Fumarate
Respiratory chain
Oxidative phosphorylation

Matrix

Citric acid cycle except
for reactions on inner
mitochondrial membrane

Outer mitochondrial
membrane

Intermembrane
space

7.16 Reactions in the Mitochondrion

Most of the important reactions of cellular respiration in eukaryotic cells take place in the mitochondrion's matrix or in its inner membrane.

move within the hydrophobic interior of the phospholipid bilayer of the inner membrane.

Mitchell proposed, and then showed, that operation of the respiratory chain results in the transport of hydrogen ions, against their concentration difference, through the inner membrane of the mitochondrion from inside to outside ("outside" being the space between the two mitochondrial membranes). This movement of H^+ appears to result from the particular location of the various respiratory chain

BOX 7.B

Dissecting the Mitochondrion

We have said that certain enzymes are contained in the mitochondrial matrix, whereas others are embedded in mitochondrial membranes. How do you suppose this was learned? Think for a minute how that question might be approached. Then read the rest of this paragraph, pausing after each step to see whether you can pick up the thread and guess what comes next. To begin with, you will need to use a centrifuge. As described in the closing pages of Chapter 4, the centrifuge can be used to isolate a sample of mitochondria for study. Once a relatively pure sample of mitochondria has been obtained, the matrix must be separated from the surrounding membranes. Think about osmosis (Chapter 6); think about lysis in a hypotonic solution (swelling until a membrane-bounded structure bursts). Yes! Transfer the mitochondria into a hypotonic solution (maybe even distilled water), from which water will rush into the

mitochondria, causing them to burst. If you are clever and do this in a rather small volume of solution, you now have a suspension consisting of the water, the mitochondrial matrix, and fragments of the membranes. What next? Think first.

Centrifuge this suspension. The membranes sink to the bottom of the centrifuge tube, forming a pellet; the matrix remains in solution. Pour off the solution into a second tube; keep the membranous pellet in the first.

Now all that remains is to see which enzyme activities are in each tube. If a particular enzyme (malate dehydrogenase from the citric acid cycle, for example) is found only in the second tube (the tube containing only solution), we may reasonably conclude that it was originally present in the matrix. If another enzyme, such as succinate dehydrogenase, is found only in the first tube (which contains the pellet), it likely is contained in the membrane.

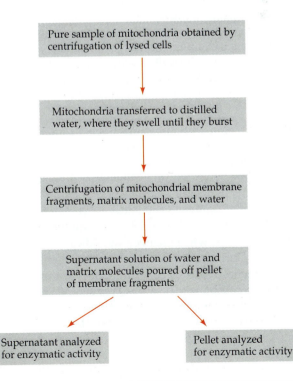

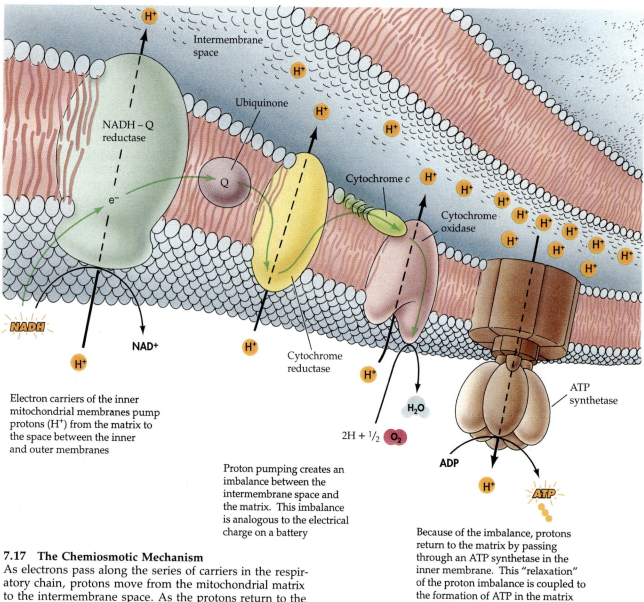

Electron carriers of the inner
mitochondrial membranes pump
protons (H⁺) from the matrix to
the space between the inner
and outer membranes

Proton pumping creates an
imbalance between the
intermembrane space and
the matrix. This imbalance
is analogous to the electrical
charge on a battery

Because of the imbalance, protons
return to the matrix by passing
through an ATP synthetase in the
inner membrane. This "relaxation"
of the proton imbalance is coupled to
the formation of ATP in the matrix

7.17   The Chemiosmotic Mechanism
As electrons pass along the series of carriers in the respiratory chain, protons move from the mitochondrial matrix
to the intermembrane space. As the protons return to the
matrix through an ATP synthetase, ATP is formed.

intermediates in the membrane, and it acts in effect
to charge a "battery" by establishing and maintaining
a difference in pH across the inner membrane. Because of the charge on the proton, this transport also
causes a difference in electric charge across the membrane, further contributing to the battery effect. The
actual mechanisms by which these protons are transported are not yet known. However, as indicated in
Figure 7.17, the protons travel through the membrane in conjunction with electron transport within
the three protein complexes (NADH-Q reductase, cytochrome reductase, and cytochrome oxidase).

It can also be seen in Figure 7.17 that in the inner
mitochondrial membrane there is an enzyme (an ATP
synthetase) that allows the flow of protons through
the membrane and catalyzes the production of ATP

from ADP and P_i. This enzyme is oriented perpen-
dicularly to the surface of the membrane and possesses a specific channel, embedded in the inner mitochondrial membrane, through which the excess
protons on the outside of the membrane may flow
back into the matrix. The ATP synthetase part of this
enzyme sticks out of the inner mitochondrial membrane as a large knob that can be seen in electron
micrographs. The ATP synthetase is postulated to
require unusually acidic conditions in order to produce ATP, and these conditions are achieved by the
passage of protons through the channel.

To summarize the chemiosmotic mechanism: The
flow of hydrogens through the respiratory chain results in a transfer of protons from the inside to the
outside of the inner mitochondrial membrane, lead-

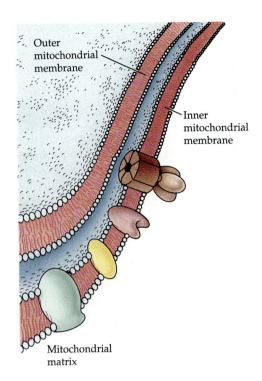

Outer mitochondrial membrane

Inner mitochondrial membrane

Mitochondrial matrix

ing to an accumulation of protons on the outside. By the laws of diffusion (Chapter 5), these excess protons tend to move back spontaneously into the matrix, which they can do only by passing through the channel-like ATP synthetase molecules. In so doing, they provide the acidic conditions necessary for ATP production. Also, as protons diffuse away from an area of their high concentration, energy is released.

According to the chemiosmotic model, one would expect that the mitochondrion could be "fooled" into making more ATP by the following clever trick. A sample of isolated mitochondria is maintained in a solution at pH 8 (slightly basic) until it is fully adjusted. Then suddenly the mitochondria are transferred into a second solution at pH 4 (fairly acidic) containing ADP and P_i. This leads to an excess of protons on the outside of the inner membrane, from whence they should be able to proceed through the proposed ATP synthetase channels, causing a burst of ATP production. That is exactly what is observed, and this acid-induced ATP production by isolated mitochondria stands as one of the stronger pieces of evidence favoring the chemiosmotic theory as the explanation for how oxidative phosphorylation proceeds in cells.

FERMENTATION

Suppose that the supply of oxygen to a respiring cell is cut off, perhaps by drowning or by extreme exertion, leading to an insufficient supply of O_2 in the cell. Glancing back at Figure 7.15, we can see that the first consequence is an inability to reoxidize cytochrome *c*, so all of that compound is soon in the reduced form. Once this happens, there is no oxidizing agent to reoxidize QH_2, and soon all the Q is in the reduced form. So it goes, until the entire respiratory chain is reduced. By this point, there remains no NAD^+ and no oxidized FAD; therefore, the oxidative steps in glycolysis and the citric acid cycle stop. If it is a cell that has no other way to obtain energy from its food, it will die.

If, however, it is one of those cells—such as a muscle cell—that has the necessary enzymes, it will switch to **fermentation**. This process has two defining characteristics. First, a fermentative reaction is one in which $NADH + H^+$ is used to reduce pyruvate or one of its metabolites. This has the important consequence of oxidizing the NADH, regenerating NAD^+. Once the cell has some NAD^+, it can carry some more glucose through glycolysis (that is, through the early steps of fermentation). The amount of NAD^+ obtained from the fermentative step is just enough to take a comparable amount of glucose through glycolysis, with none left over to carry the pyruvate on into the citric acid cycle. Instead, this newly produced pyruvate is also fermented, producing more NAD^+ to oxidize more glucose, and so forth. And now we see the second characteristic of fermentation: By allowing glycolysis to continue, fermentation allows a sustained production of ATP—to be sure, only that ATP obtained from glycolysis and not the much greater yield obtainable with the citric acid cycle and the respiratory chain, but enough to keep the cell going.

In fact, when cells capable of fermentation become anaerobic, the rate of glycolysis speeds up tenfold or even more. Thus a substantial rate of ATP production is maintained, although the efficiency in terms of ATP molecules per glucose molecule is greatly reduced. Some bacteria of the genus *Clostridium*, while growing anaerobically in the presence of glucose, grow and multiply as rapidly as the fastest-growing aerobic bacteria. This rapid growth is made possible by the fact that the *Clostridium* bacteria are running the glycolytic reactions much more rapidly than the aerobes do.

Figure 7.18 shows a particular form of fermentation, namely, that in which the product is lactic acid (lactate). Lactic acid fermentation can occur in our muscle cells. On the other hand, our neurons (nerve cells), lacking the enzyme that reduces pyruvate to lactate, are incapable of carrying on fermentation. Thus our brains are destroyed rapidly in the absence of oxygen.

In other kinds of organisms, different forms of fermentation are observed. Certain yeasts and many plant cells carry on a different process—**alcoholic**

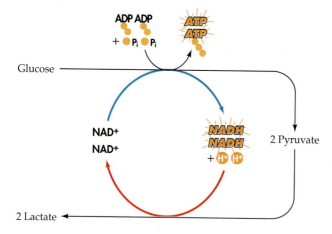

7.18 Lactic Acid Fermentation
Glycolysis produces pyruvate from glucose, as well as ATP and NADH + H$^+$. In lactic acid fermentation, pyruvate is then reduced to lactic acid (lactate) using NADH + H$^+$ as the reducing agent. This type of fermentation is common to the cells of many animals and microorganisms.

fermentation—illustrated in Figure 7.19. In these cells, carbon dioxide is removed from pyruvate, leaving the compound acetaldehyde. This acetaldehyde is reduced by NADH + H$^+$ to produce ethyl alcohol (ethanol). Remember that recycling NAD allows the fermenting cell to produce ATP by glycolysis.

As noted before, some organisms carry on no energy metabolism other than fermentation. Some of these are confined to totally anaerobic environments, whereas others can carry on fermentation in the presence of oxygen. And there are a number of bacteria that carry on cellular respiration—not fermentation—without using oxygen gas as an electron acceptor.

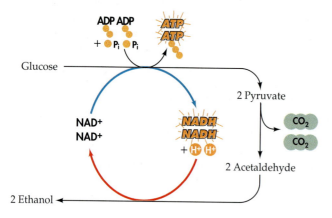

7.19 Alcoholic Fermentation
In alcoholic fermentation, as in lactic acid fermentation, glucose is converted to pyruvate during glycolysis, with the release of NADH + H$^+$ and ATP. In this type of fermentation, however, pyruvate is converted to acetaldehyde, with the release of CO_2. Using the NADH + H$^+$ as a reducing agent, acetaldehyde is reduced to ethanol. Many yeasts carry out these reactions, which are the basis for the brewing industry.

Instead, these bacteria use nitrate ions (NO_3^-) to oxidize their cytochromes, the nitrate being reduced to nitrite ions (NO_2^-) in the process.

COMPARATIVE ENERGY YIELDS

The total yield of stored energy from fermentation is two molecules of ATP per molecule of glucose oxidized. The maximum yield that can be obtained from glycolysis followed by complete aerobic respiration of a molecule of glucose is much greater—about 36 molecules of ATP. Study Figure 7.20 to review where those ATP molecules come from. Why is so much more ATP produced by aerobic respiration? Because carriers (mostly NAD$^+$) are reduced in the citric acid cycle and then oxidized by the respiratory chain, with the concomitant production of ATP by the chemiosmotic mechanism. In an aerobic environment, a species capable of this type of metabolism is going to be at an advantage (in terms of energy availability per glucose molecule) over one that is limited to fermentation.

If glucose is simply burned, the reaction is

$$\text{Glucose} + 6O_2 \rightarrow 6CO_2 + 6H_2O - 686 \text{ kcal/mol}$$

with the 686 kcal of energy all being released as heat and light. The complete biological "combustion"—respiration—of glucose is describable by the same overall reaction, with the key difference that 36 molecules of ATP are formed for each molecule of glucose used. If we count each mole of ATP as storing 12 kcal (the actual amount varies as a function of the concentrations of ATP, ADP, and P_i in the cell), then 36 × 12 = 432 kcal are stored for later use to drive nonspontaneous reactions, instead of being lost as heat. Under these conditions, the efficiency of energy trapping is 432/686 = 0.63, or 63 percent.

CONNECTIONS WITH OTHER PATHWAYS

The respiratory pathways do not operate in isolation from the rest of metabolism. Rather, there is an interchange, with traffic flowing in both directions. For one thing, materials other than glucose can serve as the starting materials for respiratory ATP production. Other monosaccharides may be used, after first being converted to glucose. Polymers such as starch and glycogen are **digested** (that is, hydrolyzed) to glucose and subsequently metabolized to yield ATP. Fats are first digested to yield glycerol and fatty acids (Chapter 3); the glycerol is then readily converted to glyceraldehyde 3-phosphate (an intermediate in glycolysis), and the fatty acids are broken down to form acetyl coenzyme A (an early intermediate in the citric acid cycle).

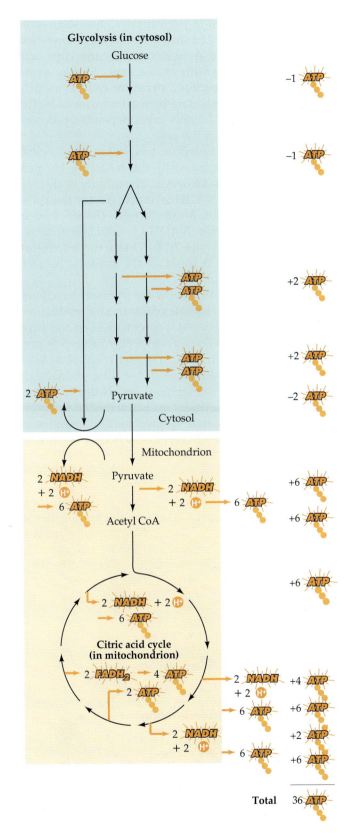

7.20 Energy Yields from Glycolysis and Cellular Respiration

Glycolysis yields two molecules of ATP for every glucose molecule entering the pathway. The ensuing citric acid cycle and respiratory chain produce an additional 34 ATP molecules for every glucose molecule. The source of most of these ATP molecules is the oxidation of reduced carriers (produced in glycolysis and the citric acid cycle) by the respiratory chain. We get three molecules of ATP for each NAD^+ regenerated by the respiratory chain and two molecules of ATP for each FAD. Thus the total gross yield of ATP from one molecule of glucose taken through glycolysis and respiration is 38. However, we must subtract two from that gross, for a net yield of 36 ATP. This is because the inner mitochondrial membrane is impermeable to NADH, and a "toll" of one ATP must be paid for each NADH (produced in glycolysis) that is shuttled into the mitochondrial matrix. The 36 molecules of ATP from the oxidation of glucose through glycolysis combined with cellular respiration still far exceeds the net of two molecules of ATP from fermentation.

needed strictly for the citric acid cycle and ATP formation. When there is more than enough starting material—food, such as glucose—then the cell can divert some acetyl coenzyme A to fatty acid production and some glyceraldehyde 3-phosphate to glycerol formation. The result of this diversion is that we may form fats and add them to the baggage we are carrying around.

Some of the citric acid cycle intermediates are used in the synthesis of various important cellular constituents. Succinyl coenzyme A is a starting point for chlorophyll synthesis, and α-ketoglutarate is a key starting material for amino acid (and, hence, protein) production. Other amino acids are formed from oxaloacetate. (Still other amino acids derive from pyruvate, which is not an intermediate in the cycle.) Acetyl coenzyme A has numerous fates: Other than its role in fatty acid production, it is a building block for numerous pigments, plant growth substances, rubber, and the steroid hormones of animals. The list continues, but this gives you some idea.

The ubiquitous destinies of acetyl coenzyme A also bring a problem to mind: If too many molecules of citric acid cycle intermediates are withdrawn from the citric acid cycle for use in other pathways, the oxaloacetate concentration could ultimately be lowered so much that there would no longer be enough to react with incoming acetyl coenzyme A to keep the citric acid cycle going. This problem is avoided by a number of "replenishing" reactions, which bring in material from other parts of metabolism or which bypass some of the steps of the citric acid cycle, keeping more atoms in the pathway. One such reaction bypasses the first step of the citric acid cycle, in which pyruvate is converted to acetyl coenzyme A with the loss of a carbon atom as CO_2. In the alternative reaction, pyruvate *combines* with a CO_2

Each of these reactions also operates in reverse. Thus, in the synthesis of fats, fatty acids form from acetyl coenzyme A and glycerol forms from glyceraldehyde 3-phosphate. When does this occur? It must be at a time when the cell has an adequate energy supply, otherwise the acetyl coenzyme A would be

molecule, forming oxaloacetate, the four-carbon substance at the end of the cycle. Thus the pool of citric acid cycle intermediates is increased, making up for materials that are lost from other parts of the cycle. We will refer to this combination again in the next section, where the subject is the regulation of respiratory metabolism.

FEEDBACK REGULATION

As you now know, fermentation produces two molecules of ATP for every glucose molecule used whereas passing the pyruvate on from glycolysis to the citric acid cycle and respiratory chain yields 36 ATPs per glucose molecule. Thus an aerobically respiring organism obtains 18 times as much energy per mole of glucose oxidized. In other words, when a yeast cell switches from aerobic respiration to anaerobic fermentation under conditions of low oxygen, it must use glucose 18 times as fast to obtain the same amount of energy. But as soon as aerobic respiration begins again in yeast, glycolysis slows down. Only as much glucose is used as is needed for energy production under the existing conditions, anaerobic or aerobic. This is the **Pasteur effect**, named for its discoverer, Louis Pasteur. What is the mechanism that slows down glycolysis when the respiratory chain begins to operate?

The mechanism by which glycolysis, the citric acid cycle, and the respiratory chain are regulated is **allosteric control** of the enzymes (Chapter 6). Some products of later reactions, if they are in oversupply, can suppress the action of enzymes that catalyze early reactions. On the other hand, an excess of the products of one branch of a synthetic chain can speed up reactions in another branch and divert raw materials away from its own synthesis (Figure 7.21). These negative and positive feedback control mechanisms are used at many points in the energy-extracting processes and are summarized in Figure 7.22.

The main control point in glycolysis is the conversion of fructose 6-phosphate to fructose 1,6-bisphosphate by the enzyme phosphofructokinase. This enzyme is allosterically inhibited by ATP and activated by ADP or AMP (Figure 7.22). The enzyme is also inhibited by citrate, for reasons that will become clear shortly. As long as fermentation proceeds, yielding a relatively small amount of ATP, phosphofructokinase operates at full efficiency. But when aerobic respiration begins producing ATP 18 times as rapidly as before, the excess ATP allosterically inhibits the conversion of fructose 6-phosphate, and the rate of glucose utilization drops.

Pyruvate stands at a key position in the network diagrammed in Figure 7.22. In fermentation it is reduced to lactate, which can either be returned as pyruvate or be used to resynthesize glucose for storage. Under aerobic conditions, pyruvate is converted to acetyl coenzyme A, which enters the citric acid cycle by combining with oxaloacetate. Finally, as we noted in the last section, pyruvate can be used to produce more oxaloacetate by reaction with CO_2. The pathway pyruvate actually takes depends upon conditions and needs in the cell.

In the pyruvate-to-lactate conversion of fermentation, pyruvate is reduced by the $NADH + H^+$ produced in glycolysis. However, the affinity of the respiratory chain for NADH is much greater than that of the enzyme that forms lactate. (Recall that in the respiratory chain NADH-Q reductase oxidizes $NADH + H^+$; see Figure 7.14.) Thus if the respiratory chain is operating, it steals all the available NADH, turning fermentation off.

At a second control point for pyruvate reactions, acetyl coenzyme A regulates oxaloacetate production. If enough oxaloacetate is present to keep the citric acid cycle going as fast as acetyl coenzyme A is produced, the concentration of acetyl coenzyme A remains low. If for some reason too little oxaloacetate is available, acetyl coenzyme A builds up, activating the enzyme that produces oxaloacetate and restoring the level of oxaloacetate needed for the operation of the citric acid cycle. Acetyl coenzyme A is an **allosteric activator** for the reaction.

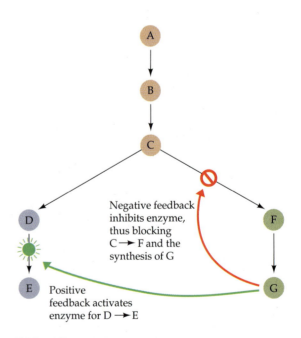

7.21 Allosteric Regulation
Compound G inhibits the enzyme for the conversion of C to F, blocking that reaction and ultimately its own synthesis, demonstrating negative feedback by allosteric regulation. Compound G also provides positive feedback to the enzyme catalyzing the step from D to E, changing that enzyme to a form that will catalyze the reaction.

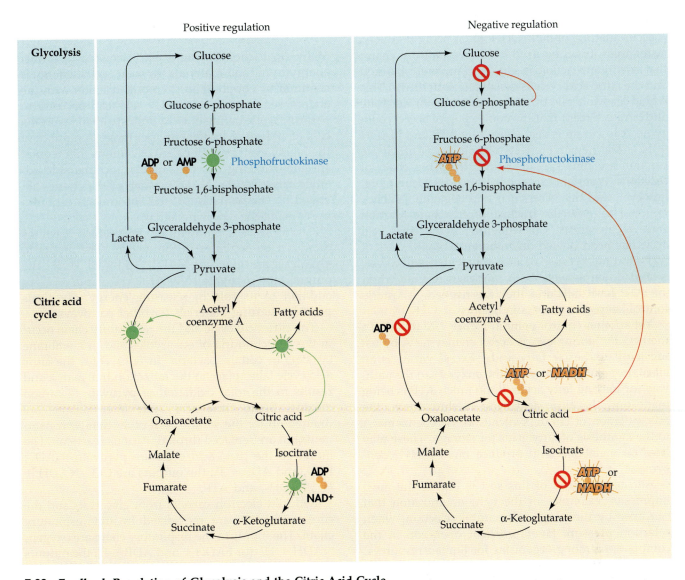

7.22 Feedback Regulation of Glycolysis and the Citric Acid Cycle
Positive and negative allosteric regulation control glycolysis and the citric acid cycle. Negative feedback blocks reactions, as indicated by the red-and-white "stop" signs. Positive feedback boosts reactions, as indicated by the green "go" signs. Here we can see that citric acid is much like compound G in Figure 7.21; it provides negative feedback on the enzyme catalyzing the third step in glycolysis, but also provides positive feedback for the synthesis of fatty acids from acetyl coenzyme A. Note that feedback controls glycolysis and the citric acid cycle at crucial early steps in the pathways; this increases the efficiency of the pathways and prevents buildup of excessive intermediates. Note also that the compounds inhibiting or activating enzymes are often the energy-carrying compounds themselves—ATP, ADP, NAD$^+$, NADH, and so forth. If too much ATP accumulates, for example, ATP inhibits a key reaction and thus slows down ATP production; if much ATP has been consumed, resulting in the formation of ADP and P$_i$, the ADP activates enzymes in the pathway to stimulate the production of more ATP.

Thus the concentration of acetyl coenzyme A determines the balance point between two competing reactions: one that uses oxaloacetate in turning the citric acid cycle and the other that makes more oxaloacetate if it is in short supply. ADP is an opposing **allosteric inhibitor** of this same oxaloacetate-producing enzyme. If the cell is low in ATP, it is high in

ADP; by inhibiting the oxaloacetate-producing enzyme, ADP directs more pyruvate to become citrate, causing the citric acid cycle and respiratory chain to operate more rapidly and form more ATP.

The main control point for the citric acid cycle is the conversion of isocitrate to α-ketoglutarate. ATP and NADH are feedback inhibitors of this reaction, and ADP and NAD^+ are activators. If too much ATP is accumulating, or if $NADH + H^+$ is being produced faster than it can be used by the respiratory chain, the isocitrate reaction is almost completely blocked and the citric acid cycle is essentially shut down. This would lead to a pile-up of large amounts of isocitrate and citrate, except that the conversion of acetyl coenzyme A to citrate is also slowed by ATP and $NADH + H^+$. The negative effects of halting the isocitrate reaction are thus spread backward up the chain of reactions. A certain excess of citrate does accumulate, however, and this excess acts as a negative feedback inhibitor to slow the fructose 6-phosphate reaction early in glycolysis. Consequently, if the citric acid cycle has been slowed down because of an excess of ATP (and not because of a lack of oxygen), glycolysis is shut down as well. Both processes resume when the ATP level falls and they are needed. Allosteric control keeps the process in balance.

Yet another control point in Figure 7.22 involves a method for storing excess acetyl coenzyme A by using it to synthesize fatty acids. Excess citrate is an allosteric activator for one of the enzymes in the pathway for making fatty acids. If too much ATP is being made and the citric acid cycle is shut down, the accumulation of citrate switches acetyl coenzyme A to the synthesis of fatty acids for storage. These may later be metabolized to produce more acetyl coenzyme A.

Allosteric control of this sort is one of the most impressive examples of the tight organization that can arise by the process of natural selection, when selection pressure favors efficient operation in the competition among organisms for limited resources. Each of the feedback controls regulates a part or various parts of the energy-releasing pathways and keeps them operating in harmony and balance. It is unnecessary (and therefore inefficient and disadvantageous) to run the glycolytic mechanism too fast if it is supplemented by the more energy-efficient processes of the citric acid cycle and respiratory chain. In terms of energy production, it is wasteful to produce more acetyl coenzyme A if there is insufficient oxaloacetate to handle it in the citric acid cycle. It is also senseless to shunt too much pyruvate into making oxaloacetate and to neglect production of the fuel acetyl coenzyme A because a two-carbon molecule of acetyl coenzyme A must react with a four-carbon molecule of oxaloacetate to produce the six-carbon citrate—the step that keeps the citric acid cycle turning. Allosteric control maintains the proper balance among the uses of pyruvate by being sensitive to shortages or oversupplies of acetyl coenzyme A. All the other allosteric feedback controls help make the system more efficient and hence contribute to the success of the species that carries them.

SUMMARY

Many small chemical steps act in concert to extract energy from food materials. In some steps, energy is captured by coupling an exergonic reaction with the endergonic formation of ATP. ATP can be used to perform work because its hydrolysis liberates in excess of 10 kcal of free energy per mole. A number of the highly exergonic steps are oxidations that require specific oxidizing agents, notably, nicotinamide adenine dinucleotide (NAD^+). Much of the energy liberated by the oxidation of the substrate is captured in reducing the oxidizing agent.

Glycolysis is a pathway of preparatory reactions catalyzed by enzymes in the cytosol. The inputs to glycolysis are glucose, NAD^+, and $ADP + P_i$; the outputs are pyruvate, $NADH + H^+$, and ATP. In addition to its role in releasing energy, glycolysis provides starting materials for other pathways.

The citric acid cycle and the respiratory chain occur in the mitochondria of eukaryotes and in membrane systems in certain bacteria. In eukaryotes the citric acid cycle takes place in the mitochondrial matrix and results in the total oxidation of pyruvate to carbon dioxide. Some citric acid cycle intermediates are drained off for the synthesis of other cellular constituents but are replaced through various reactions. The inputs to the citric acid cycle are pyruvate, NAD^+, FAD, and $ADP + P_i$; the outputs are CO_2, $NADH + H^+$, $FADH_2$, and ATP.

Reduced electron carriers from glycolysis and the citric acid cycle are reoxidized by the respiratory chain. The inputs to the respiratory chain are oxygen, $NADH + H^+$ (or $FADH_2$), and $ADP + P_i$; the outputs are ATP, NAD^+ (or FAD), and water. For each molecule of $NADH + H^+$ processed by the respiratory chain, three molecules of ATP are formed. As explained by the chemiosmotic model, oxidative phosphorylation (the formation of ATP in the mitochondrion) proceeds by the pumping of protons through the inner membrane during the operation of the respiratory chain.

Many species derive their energy supply from fermentation. The function of fermentation is to oxidize the $NADH + H^+$ produced in glycolysis, ensuring a continued supply of ATP by allowing glycolysis to continue. Fermentation yields 2 molecules of ATP per molecule of glucose utilized. In contrast, glycolysis combined with the citric acid cycle and the respiratory chain yields up to 36 molecules of ATP per molecule of glucose.

A remarkable series of allosteric feedback controls regulates the web of reactions that constitute energy metabolism. The end result is the efficient distribution and storage of free energy, ultimately derived from glucose and other food materials.

SELF-QUIZ

1. Which statement about adenosine triphosphate is *not* true?
   a. It is formed only under aerobic conditions.
   b. It is used as an energy currency by all cells.
   c. Its formation from ADP and phosphate is an endergonic reaction.
   d. ATP provides the energy for many different biochemical reactions.
   e. Some ATP is used to drive the synthesis of storage compounds.

2. Oxidation and reduction:
   a. entail the gain or loss of proteins.
   b. are defined as the loss of electrons.
   c. are both endergonic reactions.
   d. always occur together.
   e. proceed only under aerobic conditions.

3. NAD^+:
   a. is a kind of organelle.
   b. is a protein.
   c. is an oxidizing agent.
   d. is a reducing agent.
   e. is formed only under aerobic conditions.

4. Glycolysis:
   a. takes place in the mitochondrion.
   b. produces no ATP.
   c. has no connection with the respiratory chain.
   d. is the same thing as fermentation.
   e. reduces two molecules of NAD for every glucose molecule processed.

5. Fermentation:
   a. takes place in the mitochondrion.
   b. takes place in all animal cells.
   c. does not require O_2.
   d. requires lactic acid.
   e. prevents glycolysis from taking place.

6. Which statement is *not* true of pyruvate?
   a. It is the end product of glycolysis.
   b. It gets reduced during fermentation.
   c. It feeds into the citric acid cycle.
   d. It is a protein.
   e. It contains three carbon atoms.

7. The citric acid cycle:
   a. takes place in the mitochondrion.
   b. produces no ATP.
   c. has no connection with the respiratory chain.
   d. is the same thing as fermentation.
   e. reduces two molecules of NAD for every glucose molecule processed.

8. Which statement is *not* true of the respiratory chain?
   a. It takes place in the mitochondrion.
   b. It uses O_2 as an oxidizing agent.
   c. It produces ATP.
   d. It regenerates oxidizing agents for glycolysis and the citric acid cycle.
   e. It operates simultaneously with fermentation.

9. Which statement is *not* true of the chemiosmotic mechanism?
   a. Protons are pumped across a membrane.
   b. Protons return through the membrane by way of a channel protein.
   c. ATP is required for the protons to return.
   d. Proton pumping is associated with the respiratory chain.
   e. The membrane in question is the inner mitochondrial membrane.

10. Which statement is *not* true of oxidative phosphorylation?
   a. It is the formation of ATP during the operation of the respiratory chain.
   b. It is brought about by the chemiosmotic mechanism.
   c. It requires aerobic conditions.
   d. In eukaryotes, it takes place in mitochondria.
   e. Its functions can be served equally well by fermentation.

FOR STUDY

1. Trace the sequence of chemical changes that occurs in mammalian brain tissue when the oxygen supply is cut off. (The first change is that the cytochrome oxidase system becomes totally reduced, since electrons can still flow from cytochrome *c* but there is no oxygen to accept electrons from cytochrome oxidase. What are the remaining steps?)

2. Trace the sequence of chemical changes that occurs in mammalian muscle tissue when the oxygen supply is cut off. (The first change is exactly the same as that in Study Question 1.)

3. Some cells that use the citric acid cycle and the respiratory chain can also thrive by using fermentation under anaerobic conditions. Given the lower yield of ATP (per molecule of glucose) in fermentation, why can these cells function so efficiently under anaerobic conditions?

4. Describe the mechanisms by which the rates of glycolysis and of aerobic respiration are kept in balance with one another.

READINGS

Alberts, B., D. Bray, J. Lewis, M. Raff, K. Roberts and J. D. Watson. 1989. *Molecular Biology of the Cell*, 2nd Edition. Garland Publishing, New York. Chapter 7 develops the themes introduced in this chapter; Chapter 3 is also useful as an introduction.

Hinkle, P. C. and R. E. McCarty. 1978. "How Cells Make ATP." *Scientific American*, March. Discussion of the chemiosmotic mechanism, in which ATP is formed by protons passing back through a membrane after having been pumped out by the respiratory chain.

Stryer, L. 1988. *Biochemistry*, 3rd Edition. W. H. Freeman, New York. Though more advanced than this chapter, the section on glycolysis and respiration is straightforward and does not demand an advanced knowledge of chemistry.

Voet, D. and J. G. Voet. 1990. *Biochemistry*. John Wiley & Sons, New York. Another general textbook with a full discussion of energy, enzymes, and catalysis. Outstanding illustrations.

8

Photosynthesis

PREVIEW: Photosynthesis is the main route by which free energy in the environment is made available to the living world; it is also the principal source of reduced carbon compounds. Light energy, trapped by chlorophyll and associated pigments, is used to initiate a series of reactions that result in the formation of ATP and a reducing agent. ATP and the reducing agent are used to reduce a compound produced when carbon dioxide is captured from the environment. Certain plants use extra initial steps in photosynthetic metabolism to trap carbon dioxide more efficiently.

This chapter deals with light, pigments, absorption and action spectra, cyclic and noncyclic photophosphorylation, the Calvin–Benson cycle, C_3 and C_4 photosynthesis, and photorespiration.

We are creatures of the sun. Its light is the source—direct or indirect—of the free energy that powers life on Earth. Our own dependence on the sun, although absolute, is indirect. Like the other animals, the fungi, many protists, and most monerans, we depend upon a ready supply of partially reduced, carbon-containing compounds as a food source. From such compounds we get all the free energy that keeps us alive and functioning. From them, too, we obtain the carbon atoms used in every organic molecule in our bodies. In a word, we are **heterotrophs**: We need to feed upon something else. In a world suddenly populated exclusively by heterotrophs, all life would grind to an end as the food supply disappeared.

In fact, our world owes the continued existence of life to the presence of **autotrophs**—organisms that do not need previously formed organic substances from their environment. For autotrophs, an energy source (such as light) and an inorganic carbon source (such as carbon dioxide gas) are adequate. From these simple ingredients, autotrophs make the reduced carbon compounds from which their bodies are built and their food needs met. By feeding on autotrophs, the heterotrophs of the world meet their needs. The principal autotrophs are **photosynthetic organisms** that use visible light as their energy source. From light, carbon dioxide, and water, they begin the chemistry that sustains almost the entire biosphere. **Photosynthesis** is the transformation of light energy to chemical energy by living things. Organisms that conduct photosynthesis—plants and photosynthetic protists and monerans—stand at the gateway to the living world, at the interface where inorganic becomes organic, where nonlife becomes life. The worldwide extent of photosynthetic activity is stunning: Each year, tens of billions of tons of carbon atoms are taken from carbon dioxide and incorporated into molecules of sugars, amino acids, and other compounds.

EARLY STUDIES OF PHOTOSYNTHESIS

The broad outlines of photosynthesis were visible in the eighteenth and early nineteenth centuries, as it became clear that photosynthesis uses three principal ingredients—water, carbon dioxide, and light—and produces not only food but also oxygen gas. Scientists learned that water comes primarily from the soil (for plants living on land) and must travel from the roots to the leaves; that carbon dioxide is taken in from the atmosphere through tiny apertures, called **stomata**, in the leaves (Figure 8.1); and that light is absolutely necessary for the production of the oxygen and food. The last of the important early discoveries, made during the first decade of the nineteenth century, was the recognition that carbon dioxide uptake and oxygen release are closely related and that both depend upon light action. By 1804 it was possible to summarize photosynthesis in plants as

$$CO_2 + H_2O + \text{light energy} \rightarrow \text{sugars} + O_2$$

It was almost a century and a half before it was possible to determine whether the oxygen released

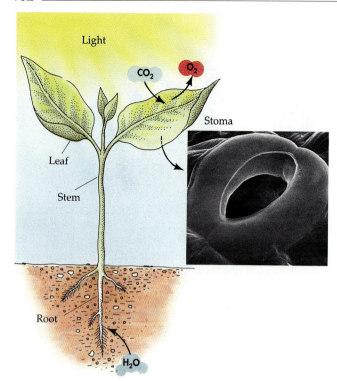

8.1 Ingredients for Photosynthesis
A typical terrestrial plant uses light from the sun, water from the soil, and carbon dioxide from the atmosphere to form organic compounds by the process of photosynthesis. The photograph shows a stoma (plural: stomata) in a cucumber leaf; CO_2 enters the leaf through such openings.

during photosynthesis comes from the carbon dioxide or from the water. The direct demonstration depended on one of the first uses of an isotopic tracer (Chapter 2) in biological research. In the experiments, two groups of green plants were allowed to carry on photosynthesis (Figure 8.2). Plants in the first group were supplied with H_2O containing the heavy-oxygen isotope ^{18}O and with CO_2 containing only the common isotope ^{16}O; plants in the second group were supplied with CO_2 labeled with ^{18}O and water containing only the common isotope. Oxygen gas was collected from each group of plants. It was found that O_2 containing ^{18}O was produced in abundance by the plants given ^{18}O-labeled water but not by plants given labeled CO_2. With this information in hand, and taking into account the number of CO_2 molecules needed to form a simple sugar such as glucose, we may now rewrite the overall equation for photosynthesis as

$$6CO_2 + 12H_2O \rightarrow C_6H_{12}O_6 + 6O_2 + 6H_2O$$

Water appears on both sides of the equation because water is both used as a reactant (the twelve molecules on the left) and released as a product (the six new ones on the right). Note that this equation is essentially the reverse of the overall equation for cellular respiration, which was given in Chapter 7.

The oxygen that is released is in one sense a waste product; in another, it is vital to all oxygen-requiring organisms. The photosynthetic production of oxygen by green plants is an important source of atmospheric oxygen, which most organisms—including plants themselves—require in order to complete their respiratory chains and thus obtain the energy to live.

THE PATHWAYS OF PHOTOSYNTHESIS

It gradually became obvious that the overall photosynthetic reaction just shown cannot proceed in a single step. There is no precedent in all of chemistry for such a complex reaction being a single step. Rather, there must be a whole series of simpler steps. By the middle of the twentieth century, it was clear that photosynthesis comprises two pathways, one of which, driven by light, produces ATP and a reducing agent and the other of which uses ATP and the reducing agent to produce sugar.

Just as NAD (nicotinamide adenine dinucleotide; see Chapter 7) bridges the pathways of cellular respiration, a very similar compound bridges the two pathways of photosynthesis. This electron carrier is **nicotinamide adenine dinucleotide phosphate**, or **NADP**. NADP is virtually identical to NAD, differing from it only in the possession of another phosphate group attached to the ribose portion of the molecule. NAD participates in metabolic breakdown reactions and energy transfers, whereas NADP participates in synthetic reactions requiring energy and reducing

Water and carbon dioxide provided	Photosynthesis	Oxygen released
$H_2O, C\,^{18}O_2$		O_2
$H_2\,^{18}O, CO_2$		$^{18}O_2$

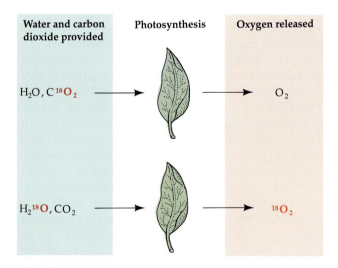

8.2 Oxygen Released during Photosynthesis
Experimenters gave some plants isotope-labeled carbon dioxide, $C^{18}O_2$ (top); they gave other plants isotope-labeled water, $H_2{}^{18}O$ (bottom). Because only plants in the bottom group release isotope-labeled oxygen gas, $^{18}O_2$, we know that water is the source of the oxygen that is liberated during photosynthesis.

power. Like NAD, NADP exists in two forms. One (NADP$^+$) is an oxidizing agent; the other (NADPH + H$^+$) is a reducing agent (Figure 8.3).

One of the photosynthetic pathways uses light and water to produce ATP, NADPH + H$^+$, and O$_2$. This pathway is called **photophosphorylation**, or, loosely, the "light reactions." ATP and NADPH + H$^+$ are carriers of reducing power because reduction is always an endergonic process requiring both energy and electrons.

The reducing power from photophosphorylation is used in the second pathway, which is devoted to the trapping of CO$_2$ and the reduction of the resulting acid to sugar (discussed later in this chapter). These sugar-producing reactions constitute the Calvin–Benson cycle, also known as the photosynthetic carbon reduction cycle or the "dark reactions" (Figure 8.4). Both pathways reside within the chloroplast, but in different parts of that organelle. *Both* processes stop in the dark because, although only photophosphorylation contains steps directly requiring light, the rate of each pathway is dependent upon that of the other. They are tied together by the exchange of ATP and ADP and of NADP$^+$ and NADPH.

Photophosphorylation and the Calvin–Benson cycle function in plant cells along with glycolysis and the pathways of cellular respiration. Plant cells are eukaryotic, they have mitochondria, and the metabolic pathways by which they carry on glycolysis and cellular respiration are the very pathways described in Chapter 7. Respiration proceeds in both the light and the dark in plants, as in other organisms; but plants make their own glucose, unlike heterotrophs.

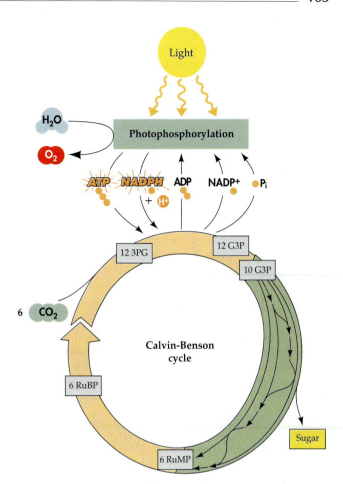

8.4 An Overview of Photosynthesis
Light energy and water are used in photophosphorylation to produce ATP, NADPH + H$^+$, and, in many organisms, O$_2$. CO$_2$ and the ATP plus NADPH + H$^+$ are used in the Calvin–Benson cycle to produce sugars and other food molecules.

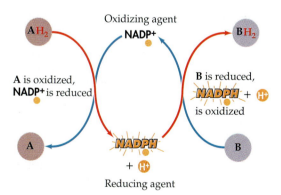

8.3 NADP$^+$ and NADPH + H$^+$
NADP$^+$ is the oxidized form of nicotinamide adenine dinucleotide phosphate, and NADPH + H$^+$ is the reduced form. When NADP$^+$, an oxidizing agent, reacts with a reduced molecule (AH$_2$), AH$_2$ becomes oxidized and NADP$^+$ becomes reduced to NADPH + H$^+$. In turn, NADPH + H$^+$, a reducing agent, may react with an oxidized molecule such as B, becoming oxidized to NADP$^+$ and reducing B to BH$_2$. Thus, NADPH + H$^+$ is an intermediary for energy and reducing power. The red arrows trace the path of electrons in the reactions.

LIGHT AND PIGMENTS

Light plays a marvelous variety of roles in the living world. In photosynthesis, it serves as a source of *energy*; in most other light-related phenomena, it is involved in the transmission of *information*. Many of these phenomena will be described in later chapters. In them, we will find that light can be modulated in many ways to carry information: Its *brightness* may be varied, as may its *color*, and it may be presented for various *durations*, whether continuously or in short, long, or variable periods. Some of the material to be covered will be more meaningful if we first learn to deal in a quantitative way with the brightness, color, and energy content of light.

Basic Physics of Light

First, we note that light *is* a form of energy. Like other forms of radiant energy, it comes in discrete packets, called **quanta**. A quantum of light is sometimes called a photon. Light also behaves as if it were propagated in waves. The **wavelength** is the distance from the peak of one wave to the peak of the next (Figure 8.5). Different colors result from different wavelengths. Light and other forms of radiant energy are **electromagnetic radiation**. The others are cosmic rays, gamma rays, X rays, ultraviolet radiation, infrared radiation, and radio waves. We have listed these forms of radiation here in order of increasing wavelength and of decreasing energy per quantum. Visible light fits into this scheme between ultraviolet and infrared radiation (Figure 8.6). In the past, considerable attention has been devoted to the apparent paradox of light being simultaneously a wave phenomenon and a particle phenomenon, but this is nothing to be concerned about here.

One of the universal constants of nature is the speed of light in a vacuum: 3×10^{10} centimeters per second (186,000 miles per second), symbolized as c. In air, glass, water, and other media, light travels slightly more slowly. Let us consider light as a long train of waves moving in a straight line and see what the train would look like to a stationary observer. Successive peaks of the waves pass the observer with a uniform **frequency** determined by the wavelength and the speed of light. The exact relationship is $v = c/\lambda$, where v (the Greek letter *nu*) is the frequency; c, as just mentioned, is the speed of light; and λ (Greek *lambda*) is the wavelength. Often v is expressed in hertz (Hz), c in centimeters per second (cm/sec), and λ in centimeters. Another frequently used unit of wavelength is the nanometer (nm); one nanometer equals 10^{-9} meter or 10^{-7} centimeter (see table inside back cover).

Our species perceives light as having distinctive colors, for reasons to be discussed in Chapter 37. The colors relate to the wavelengths of the light, as shown in Figure 8.6. Most of us can see electromagnetic radiation with wavelengths from 400 nm to 700 nm. At 400 nm we are at the blue end of the visible spectrum, whereas 700 nm is the red end. Wave-

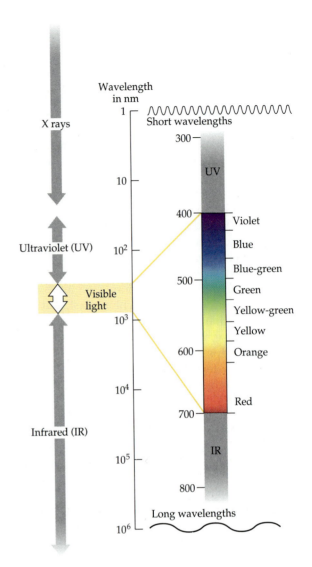

8.6 The Electromagnetic Spectrum

Wavelengths of electromagnetic radiation can be arranged on a scale called the electromagnetic spectrum. A portion of the spectrum in the vicinity of light visible to humans is represented here. Visible light comprises wavelengths between about 400 and 700 nanometers (nm), although not everyone can see over this entire range. Ultraviolet radiation extends from the short-wavelength end of the visible spectrum, and infrared radiation is at the long-wavelength end.

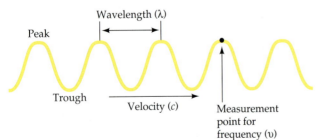

8.5 Wavelike Properties of Light

Light has many features that are best described in terms of waves. The wavelength, λ, is the distance between the peaks of successive waves; and the frequency, v, is the number of peaks passing an observation point in a second. The velocity with which the train of waves moves is c.

lengths in the range from about 100 to 400 nm are in the ultraviolet; those immediately above 700 are referred to as the infrared.

The energy E contained in a single quantum (or photon) is directly proportional to its frequency. The constant of proportionality h is named Planck's constant after Max Planck, who first introduced the concept of the quantum. We may then write $E = hv$, where v is the frequency in Hz; combining this with the equation relating λ, v, and c, we see that $E = hc/\lambda$. Thus shorter wavelengths mean greater energies. A photon of red light of wavelength 660 nm has

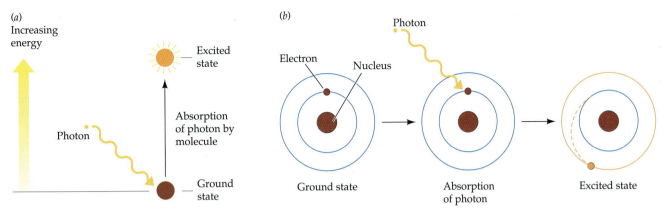

(a) Increasing energy

Photon

Absorption of photon by molecule

Excited state

Ground state

(b) Photon

Electron   Nucleus

Ground state   Absorption of photon   Excited state

8.7 Energy Levels in a Molecule
(a) When a molecule, initially in the ground state, absorbs a photon, the molecule is raised to an excited state possessing more energy. The increase in energy of the molecule equals the energy of the photon absorbed. *(b)* The absorption of the photon "boosts" one of the molecule's electrons to an orbital farther from the nucleus.

less energy than a photon of blue light at, say, 430 nm; an ultraviolet quantum of wavelength 284 nm is much more energetic than either of these. For any light-driven biological process, such as photosynthesis, a quantum can be active only if it consists of enough energy to perform the work required.

The brightness, or **intensity**, of light at a given point is the amount of energy falling on a defined area—such as 1 cm^2—per second. This is usually expressed in energy units (such as calories) per square centimeter per second, but pure light of a single wavelength may also be expressed in terms of photons per square centimeter per second.

Pigments

When a photon meets a molecule, one of three things takes place. The photon may be reflected (bounced off the molecule), or it may be transmitted, simply passing through the molecule. Neither of these causes any change in the molecule, and neither has any biological consequences. The third possibility is that the photon may be *absorbed* by the molecule. In this case, the photon simply disappears. Its energy, of course, cannot disappear, because energy is neither created nor destroyed (Chapter 6). The energy of the absorbed photon is taken over by the molecule, raising the molecule from a **ground state** of lower energy to an **excited state** of higher energy. The difference in energy between this excited state and the ground state is precisely equal to the energy of the absorbed photon. The increase in energy boosts one of the electrons in the molecule into an orbital (Chapter 2) farther from its nucleus; in a sense, this electron is less firmly held by the molecule (Figure 8.7). We will see the chemical consequence of this later in this chapter.

All molecules absorb electromagnetic radiation of

various wavelengths; the specific wavelengths absorbed are characteristic of the particular molecule. However, not all molecules can absorb electromagnetic radiation having wavelengths in the *visible* region. Those that can are called **pigments**. When a beam of white light (that is, light containing visible light of all wavelengths) falls on a pigment, certain wavelengths of the light are absorbed. The remaining wavelengths are reflected or transmitted, so the pigment appears to us to be colored. If, for example, a pigment absorbs both blue and red light, as does the pigment chlorophyll, what we see is the remaining light—primarily, green (Figure 8.8). The fact that chlorophyll absorbs in both the blue and the red region of the spectrum indicates that it has two excited states of differing energy levels, both close enough to the ground state to be reached with the energy of quanta of visible light.

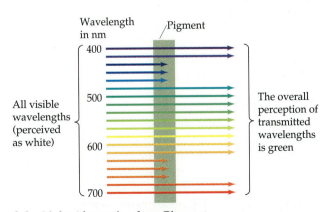

Wavelength in nm   Pigment

400

500

600

700

All visible wavelengths (perceived as white)

The overall perception of transmitted wavelengths is green

8.8 Light Absorption by a Pigment
A pigment (vertical bar) absorbs photons from specific wavelengths of visible light. We see the wavelengths that are not absorbed as the characteristic color of the pigment.

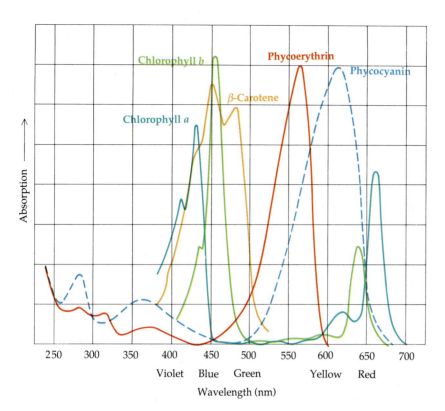

250   300   350   400   450   500   550   600   650   700

Violet   Blue   Green   Yellow   Red

Wavelength (nm)

8.9   Absorption Spectra of Photosynthetic Pigments
Several pigments participate in photosynthesis. Because the absorption peaks for the pigments are at different wavelengths, you can see that photosynthesis uses most of the visible spectrum. Notice how much of the visible spectrum would go to waste if chlorophyll *a* were the only pigment absorbing light for photosynthesis.

Absorption Spectra and Action Spectra

A given kind of molecule can occupy only particular energy levels, that is, it can absorb quanta of only certain specific energies, or wavelengths. If we plot the degree to which a given compound absorbs light as a function of the wavelength of the light, the result is an **absorption spectrum** (Figure 8.9). Absorption spectra are good "fingerprints" of compounds; sometimes an absorption spectrum contains sufficient information to enable us to identify an unknown compound. The fact that the peaks in an absorption spectrum are smoothly rounded, rather than sharp spikes, tells us that a given excited state is not an extremely narrow range of energies. Rather, it consists of a substantial family of energy sublevels, differing by tiny increments of energy much smaller than those contained in a photon of visible light. As molecules move from one of these sublevels to another they absorb or release small amounts of heat (Figure 8.10).

We may also plot the *biological effectiveness* of light —the magnitude of the effect of light on a particular activity such as photosynthesis—as a function of wavelength. The resulting graph is an **action spectrum**. Figure 8.11 is the action spectrum for photosynthesis in the freshwater plant *Anacharis*. As you can see, all wavelengths of visible light are at least somewhat effective in causing photosynthesis, although some are more effective than others. Because light must be absorbed in order to produce a chemical

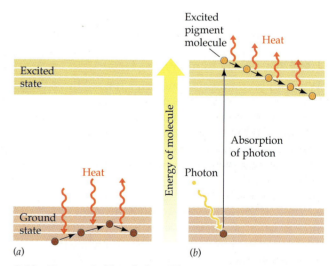

8.10   Energy Sublevels in a Pigment
(a) The ground state of a pigment consists of several sublevels of slightly different energy. When a molecule in the ground state rises from one sublevel to the next, it absorbs a tiny amount of heat; when the molecule falls to the next lower sublevel, it gives off heat. A molecule may be raised from any of these sublevels to an excited state. (b) All excited states also consist of energy sublevels. After reaching an excited state by absorbing a photon, a pigment molecule may give off minute amounts of heat as it drops from one sublevel to the next within the excited state.

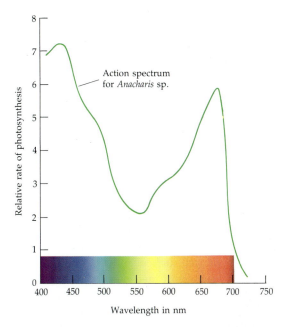

8.11 Action Spectrum of Photosynthesis
An action spectrum plots the biological effect of wavelengths of radiation against the wavelength. Here the rate of photosynthesis in the water plant *Anacharis* is plotted against wavelengths of visible light. As you can see by looking straight down from the peaks of the curve, wavelengths in the blue and red regions of the visible spectrum cause the highest rates of photosynthesis.

or biological effect, action spectra are very helpful in determining what pigment or pigments are used in a particular photobiological process such as photosynthesis. That is, we should be able to find which pigment or pigments have absorption spectra that match the action spectrum of the process.

The Photosynthetic Pigments

A number of pigments are important in biological reactions, and we will discuss them as they appear in the book. Here we discuss pigments that play roles in photosynthesis, that is, pigments found in leaves and in other parts of photosynthetic organisms. Of these, the most important are the **chlorophylls**. Chlorophylls are of universal occurrence in the plant kingdom, in photosynthetic protists, and in virtually all photosynthetic bacteria (with the exception of the halobacteria, which are discussed in Box 8.A). A mutant individual lacking chlorophyll is unable to perform photosynthesis and will starve to death. In green plants, two chlorophylls predominate, **chlorophyll a** and **chlorophyll b**, which differ only slightly in structure. Both have a complex ring structure of a type referred to as **porphyrin**, a lengthy hydrocarbon "tail," and a central magnesium atom

(Figure 8.12). (In Chapter 7 we learned about another porphyrin, heme, which is found in hemoglobin and the cytochromes.)

The chlorophylls absorb light near both ends of the visible spectrum, that is, in the blue and the red ends, as we saw in Figure 8.9. Were these the only chloroplast pigments in photosynthesis, much of the visible spectrum would go unused. However, all photosynthetic organisms possess **accessory pigments** that absorb photons intermediate in energy between the red and the blue and then transfer a portion of the energy to chlorophyll to use in photosynthesis. Among these accessory pigments are **carotenoids** such as β-carotene (Chapter 3); the carotenoids absorb photons in the blue and blue-green wavelengths and appear rich yellow in color. The **phycobilins** (phycocyanin and phycoerythrin), which are found in red algae and cyanobacteria (and contribute to their colors), absorb variously in the

8.12 The Structure of Chlorophyll
A molecule of chlorophyll consists of a porphyrin ring structure with a central magnesium atom and a hydrocarbon "tail," shown extending below the ring. Chlorophyll a and chlorophyll b differ only in the groups attached to the position designated with an R, for "residue," as shown at the top.

BOX 8.A

Photosynthesis in the Halobacteria

From time to time, we discover that some group of organisms conducts its metabolic affairs in ways that previously were totally unexpected. In 1971, for example, biologists found that certain genera of bacteria carry out a form of photosynthesis in which chlorophyll is not involved. The **halobacteria** generally live in extremely salty environments, where the salt concentration is much higher than in the oceans. Under anaerobic conditions, the halobacteria synthes-

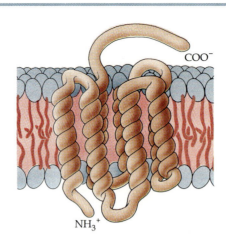

ize **retinal**, which is a carotenoid pigment also found in the vertebrate eye, where it plays a key role in vision (Chapter 37). Retinal, combined with a protein to form **bacteriorhodopsin**, is located in the plasma membranes of the halobacteria; there it absorbs light, resulting in the pumping of protons out of the cell. By a chemiosmotic mechanism of the

sort described in this and the previous chapter, ATP is formed. This ATP serves as the immediate energy source for the metabolism of the bacteria. Presumably, it was sheer evolutionary accident that led to the utilization of retinal in two such different processes as vision and photophosphorylation in organisms so widely separated in ancestry.

Bacteriorhodopsin is incorporated into patches of the cell surface of the halobacteria. These purple membrane patches make up as much as half of the cell surface, and about three-quarters of their mass consists of bacteriorhodopsin. This protein is about 4.5 nm in length, and it is organized into seven helical regions roughly perpendicular to the plane of the plasma membrane. When light is absorbed by bacteriorhodopsin, protons are transported through the membrane.

---

yellow-green, the yellow, and the orange (Figure 8.9). Such accessory pigments, in collaboration with the chlorophylls, constitute an energy-absorbing "antenna" covering much of the visible spectrum.

PHOTOPHOSPHORYLATION

The Activation of Chlorophyll: A "Light Reaction"

The absorption of a photon excites a pigment molecule as was shown in Figure 8.7. The excited state usually is not maintained for very long. One means of returning to the ground state is by **fluorescence**. In this process, the boosted electron falls back from its higher orbital to the original, lower one. This is, of course, accompanied by a loss of energy, which is given off as another photon (Figure 8.13). The molecule absorbs one photon and within approximately 10^{-9} seconds fluorescently emits another. If energy is simply absorbed and then rapidly returned as a quantum of light, there can be no chemical or biological consequences.

In order for biological work to be done, something must happen to transfer energy in some other way during the billionth of a second before a photon can be emitted fluorescently. Photosynthesis conserves energy by using the excited chlorophyll molecule as

a reducing agent. Ground state chlorophyll (symbolized as Chl) is not much of a reducing agent, but excited chlorophyll (Chl*) is a good one. The reducing capability of Chl* is readily understood if we recall that in an excited molecule, one of the electrons is zipping about in an orbital farther from its nucleus than it was before. Less tightly held, this electron can be passed on in a redox reaction to some oxidizing agent. Thus Chl* (but not Chl) can react with an oxidizing agent A in a reaction like this:

$$Chl^* + A \rightarrow Chl^+ + A^-$$

This, then, is the first biochemical consequence of light absorption by chlorophyll in the chloroplast: The chlorophyll becomes a reducing agent and participates in a redox reaction that would not have occurred in the dark. The further adventures of that electron (the one passed from chlorophyll to A) produce ATP and a stable reducing agent (NADPH), both of which are required in the Calvin–Benson cycle. Much of what we know about these reactions stems from the pioneering work of Daniel Arnon of the University of California, Berkeley.

Cyclic Photophosphorylation: ATP Biosynthesis

The formation of ATP from light energy proceeds as shown in Figure 8.14. To begin with, chlorophyll is in the ground state. It absorbs a photon and becomes

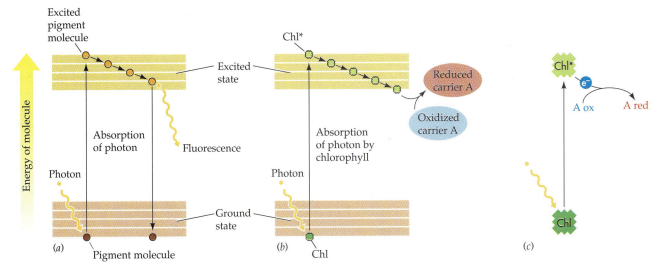

8.13 The Activation of Pigments
(a) When a pigment molecule absorbs a photon, boosting an electron to a higher orbital, the pigment moves to an excited state from its ground state. The photon activates the pigment molecule. Although the molecule may then pass from one energy sublevel to the next, it spends very little time in the excited state. An excited molecule may return to the ground state by fluorescence, in which the electron falls back to its original lower orbital and a pho-

ton is emitted. (b) When a chlorophyll molecule becomes excited it may become a reducing agent; the electron boosted to a higher orbital may be passed to an oxidized electron carrier (A), reducing the carrier molecule. Thus, much of the energy of the excited state is preserved rather than lost in fluorescence. (c) This diagram presents exactly the same information as does that in (b); this will illustrate the conventions to be used in Figures 8.14 and 8.16.

the reducing agent Chl*. The Chl* then reacts with oxidized ferredoxin (Fd) to produce reduced Fd. That reaction is a spontaneous one, that is, it is exergonic, releasing free energy. Fd_{red} is a good enough reducing agent to pass its added electron on to a second oxidizing agent, the plastoquinone complex (pQ), in the reaction $Fd_{red} + pQ_{ox} \rightarrow Fd_{ox} + pQ_{red}$ (another exergonic reaction). Similarly, pQ_{red} passes the electron on to a cytochrome complex, and so forth. This is a series of redox reactions, each exergonic, and one

of them so exergonic that the released free energy can be used to form ATP. Remember, now, that when Chl* passed its electron on to Fd, we were left with a molecule of positively charged Chl^+ (having lost that electron, the chlorophyll has one unbalanced positive charge). In due course, the Chl^+ meets a reducing agent that donates an electron, converting it back to uncharged Chl. That reducing agent is the last member of the electron transport chain in Figure 8.14: plastocyanin. By the time the electron (passed from Chl* and on through the redox chain) comes back to Chl^+ and reduces it, all the energy from the

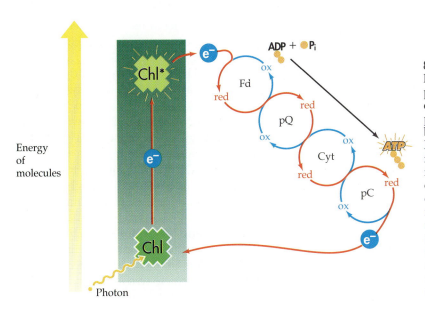

8.14 Cyclic Photophosphorylation
In cyclic photophosphorylation, chlorophyll molecules absorb photons and become excited. Excited chlorophylls (Chl*) pass electrons to an oxidizing agent, Fd, becoming positively charged and reducing Fd. Reduced Fd then reduces pQ, and so forth, in the cascade of redox reactions from Fd through pC in the diagram. The electron transport results in the formation of ATP from ADP and P_i. At the end of the redox chain, the last reduced electron carrier (pC_{red} in the diagram) passes electrons to electron-deficient chlorophylls, returning them to a ground state (Chl) ready to absorb another photon. Abbreviations: Fd, ferredoxin; pQ, plastoquinone; Cyt, cytochrome; pC, plastocyanin.

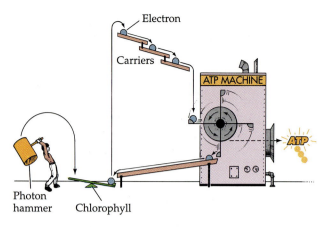

8.15 Cyclic Photophosphorylation Cycles Electrons
Compare this cartoon analog with Figure 8.14.

original photon has been released. In each of the redox reactions, some free energy is lost, until all the original amount has been converted to heat *except* for that used to form ATP. This chain of reactions is called **cyclic photophosphorylation.**

The following analogy may make the concept of cyclic photophosphorylation clearer to you. A person swings a sledgehammer, hitting a lever that throws a heavy ball into the air (like a "ring the bell" game). The ball lands on a ramp, rolling down it and dropping to a second ramp; from there it falls onto a wheel, causing the wheel to turn. When the ball falls from the wheel, it falls to another ramp, rolls down it, and finally falls back onto the lever where it started. If the turning wheel is used to do work (to run a machine), then we can see that this system transforms the original work of the person swinging the hammer into work done by the machine attached to the wheel. The energy of the falling sledgehammer is analogous to the photon in photophosphorylation, the lever to chlorophyll, the ball to an electron, the ramps to the electron carriers, and the turning of the wheel to the formation of ATP (Figure 8.15).

Noncyclic Photophosphorylation: Formation of ATP and NADPH

Cyclic photophosphorylation efficiently traps light energy and converts it into the stored energy of ATP. It may be performed in pretty much the same way in all photosynthetic organisms, although there is currently some debate as to whether green plants use cyclic photophosphorylation. However that debate may turn out, cyclic photophosphorylation is not sufficient by itself to meet the needs of the Calvin–Benson cycle. ATP is needed for two specific reactions in the Calvin–Benson cycle, and one of those reactions also requires a reducing agent: NADPH + H^+. A crucial evolutionary advance was the appear-

ance of a modified form of photophosphorylation, one that uses light energy not only to form ATP but also to form NADPH + H^+. The scheme that evolved is called **noncyclic photophosphorylation**.

In cyclic photophosphorylation, the electron that performs all the reductions makes (in principle) a complete journey through the cycle, and all the reduced carriers become reoxidized. In noncyclic photophosphorylation, electrons from another source replenish chlorophyll molecules that have given up electrons. Instead of electrons returning to chlorophyll molecules as in cyclic photophosphorylation, they are transferred to an oxidizing agent and ultimately transferred to $NADP^+$, thus reducing it to NADPH + H^+. The original source of the electrons is a plentiful one: water. As the electrons are passed from water to, ultimately, NADP, they go through a series of electron carriers. One of the intermediate steps is a highly exergonic one, and some of the free energy released in that step is used to form ATP.

Noncyclic photophosphorylation requires the participation of two distinct molecules of chlorophyll—actually, two separate sets of chlorophyll molecules. One of these sets, called Photosystem I, is used to make a reducing agent strong enough to reduce $NADP^+$ to NADPH + H^+. Photosystem II, the other set of chlorophyll molecules, takes electrons from water and passes them up to the series of redox carriers involved in the conversion of ADP + P_i to ATP. To keep noncyclic photophosphorylation going, both Photosystems I and II must constantly be absorbing light, thereby boosting electrons to higher orbitals from which they may be captured by specific oxidizing agents (Figure 8.16).

We can follow the noncyclic pathway from water to NADP. Photosystem II absorbs photons, sending electrons to an oxidizing agent and itself becoming oxidized to Chl^+. Water ionizes, forming H^+ ions (which we will see again before this paragraph ends) and OH^- ions. The OH^- ions react to produce O_2, more H^+ ions, and electrons. The electrons are passed to Chl^+ of Photosystem II, reducing it once again to Chl, which can absorb photons, and so on. The electron donated by Photosystem II to its oxidizing agent is passed, as in cyclic photophosphorylation, through a series of exergonic redox reactions, and at one point ATP is formed. Photosystem I absorbs photons, becoming excited to Chl^* and then reducing its own oxidizing agent while being oxidized to Chl^+. That Chl^+ is returned to the ground state by accepting electrons passed through the chain from Photosystem II. Now Photosystem I is accounted for, and we must consider only the electrons from Photosystem I and the protons from the original ionization of water at the beginning of the scheme. These are used in the last step of noncyclic photophosphorylation, in which two electrons and two protons (from two operations of the noncyclic scheme) are used to reduce a molecule of $NADP^+$ to $NADPH + H^+$.

In sum, noncyclic photophosphorylation uses two molecules of water, four photons (two each absorbed by Photosystems I and II), one molecule each of $NADP^+$ and ADP, and one P_i ion; from them it returns one molecule of water and produces one molecule each of $NADPH + H^+$ and ATP, and one-half molecule of oxygen (Figure 8.16). A substantial fraction of the light energy absorbed is lost as heat in the several steps, but another significant fraction is trapped in ATP and $NADPH + H^+$.

The two main pathways of photosynthesis—photophosphorylation and the Calvin–Benson cycle—are intimately linked. The Calvin–Benson cycle does not run in the dark because it requires $NADPH + H^+$ and ATP, both of which are produced in photophosphorylation, which is driven by light energy. What may be less obvious is that interruption of the Calvin–Benson cycle by cutting off the supply of CO_2 quickly stops photophosphorylation as well. Just as the Calvin–Benson cycle depends upon a supply of ATP and $NADPH + H^+$, so photophosphorylation requires ADP and $NADP^+$—and these are products of the Calvin–Benson cycle as was shown in Figure 8.4.

ATP Formation in the Chloroplast

In Chapter 7 we considered the **chemiosmotic mechanism** for ATP formation in the mitochondrion. The chemiosmotic mechanism also operates in photophosphorylation. As in oxidative phosphorylation, there is a transport of protons (H^+ ions) across a membrane, resulting in a difference in pH and in electric charge across the membrane. In the mitochondrion, proton pumping is from the matrix, across the internal membrane, and into the space between the inner and outer mitochondrial membranes (Figure 7.17). In the chloroplast, the direction is reversed. The electron carriers of photophosphorylation are located in the thylakoid membranes (see Figure 4.17), and they are so placed as to produce a movement of protons into the interior of the thylakoid, so the inside becomes acidic with respect to the outside. This leads to the passive movement of protons back out of the thylakoid, through protein channels in the membrane. The proteins are ATP synthetases, which are enzymes that catalyze the formation of ATP and which are activated by the movement of H^+ ions through the channels, just as in mitochondria (Figure 8.17).

8.16 Noncyclic Photophosphorylation

In noncyclic photophosphorylation, Photosystems I and II —each containing chlorophyll—operate to keep the process going. ATP is produced by the redox chain between Photosystems II and I, $NADPH + H^+$ is produced by the redox reactions of Photosystem I, and O_2 is produced as a by-product of the breakdown of water. Abbreviations: I, pheophytin-I; Fd, ferredoxin; pQ, plastoquinone; Cyt, cytochrome; pC, plastocyanin.

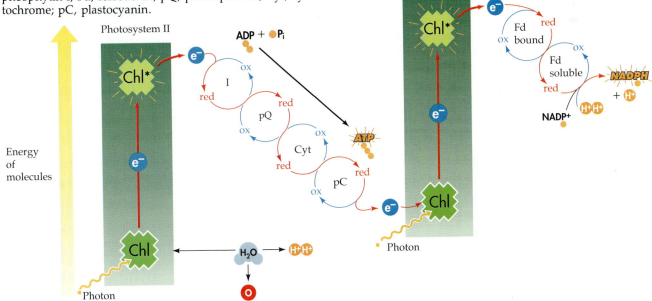

Light absorbed by photosystem II provides energy needed to pass electrons from water to pheophytin I. Further transport of these electrons leads to the generation of ATP. The breakdown of water also yields oxygen gas and protons.

Light absorbed by photosystem I provides energy needed to pass electrons from reduced plastocyanin (ultimately from water via photosystem II) to bound ferredoxin. These electrons, ultimately derived from the breakdown of water, produce $NADPH + H^+$ from $NADP^+$.

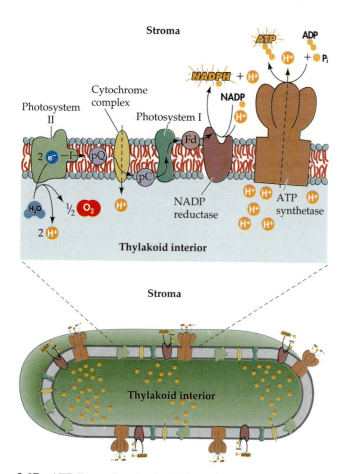

8.17 ATP Formation in the Chloroplast
Light-driven pumping of protons (H⁺) across the thylakoid membrane causes the interior of the thylakoid to become more acidic than the outside. The protons then return to the stroma of the chloroplast through channels in proteins. The proteins are enzymes (ATP synthetase) that catalyze the formation of ATP from ADP and P_i, and they are activated to do so by the passage of the protons on their way out to the stroma. Compare this chemiosmotic mechanism with the activities of the inner mitochondrial membrane shown in Figure 7.17.

Andre Jagendorf (of Cornell University) and Ernest Uribe (now at Washington State University) tested this chemiosmotic model in the following way. Chloroplasts were isolated from spinach leaves and then kept in the dark, so that there would be no light energy to drive the production of ATP. They were first placed in a solution with low pH, so that by diffusion the stroma and, ultimately, the thylakoids became somewhat acidic (Figure 8.18). Then they were transferred to a solution that had a higher pH, so that the interiors of the thylakoids were more acidic than the outsides—mimicking the situation created by light-driven pumping of protons into the interiors. This immediately resulted in the formation of ATP, even though no light was available to serve as the energy source. This is precisely the result predicted by the chemiosmotic model. A very similar experiment, using mitochondria, pH changes, and ATP formation, was described in Chapter 7. (Another chemiosmotic mechanism was described in Box 8.A.)

THE CALVIN–BENSON CYCLE

Only after World War II was there real progress in understanding the second main pathway of photosynthesis, primarily because satisfactory experimental techniques had not been developed before then. A group of scientists at the University of California, Berkeley, led by Melvin Calvin and including Andrew Benson and James Bassham, broke the problem wide open. The problem, as set by the Berkeley group, was to learn the biochemical steps intervening between the uptake of CO_2 and the appearance of the first complex carbohydrates in the chloroplast. Its solution depended on three advances in technique. These three advances were the discovery and the

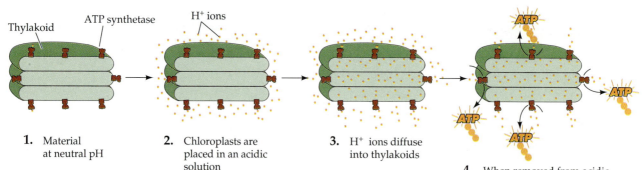

1. Material at neutral pH

2. Chloroplasts are placed in an acidic solution

3. H⁺ ions diffuse into thylakoids

4. When removed from acidic solution, leakage of H⁺ out of thylakoids is coupled with ATP formation

8.18 Artificially Induced ATP Formation
When isolated chloroplasts are kept in a mildly acidic solution in the dark, protons (H⁺) diffuse into the thylakoids, causing the interiors of the thylakoids to become somewhat acidic. When these chloroplasts are transferred to a solution with a higher pH, the interiors of the thylakoids are suddenly acidic with respect to the solution. ATP formation begins immediately as protons leak out of the thylakoids, a result predicted by the chemiosmotic model described in Figure 8.17.

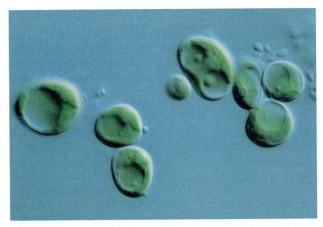

8.19 Algae Used in Experiments on Photosynthesis
The alga *Chlorella pyrenoidosa* was one of several used in
studies by Calvin's group; this image is magnified about
1,600 times. Single-celled algae were chosen over leafy
plants because their enzymes could be inactivated very
quickly by ethanol, which penetrated the cell walls rapidly. This allowed the researchers to stop the reactions in
the cells at a chosen time after treating the cells with radioactive $^{14}CO_2$ in dissolved form. Algae could be grown
continuously in cultures, providing a ready supply of research material with little variation.

availability of a radioactive carbon isotope, ^{14}C; a
rapid technique—paper partition chromatography—
for the separation of individual compounds from
complex solutions; and a technique, called autoradiography, for locating colorless but radioactive
compounds on a paper chromatogram. All three advances are described in Box 8.B.

Armed with these tools (^{14}C, paper partition chromatography, and autoradiography), the Berkeley
group set out to investigate the metabolism of CO_2
by photosynthesizing organisms. Most of their work
was done with unicellular aquatic algae such as the
green alga *Chlorella* (Figure 8.19). Algae were grown
in dense suspensions in a flattened flask (called a
"lollipop" because of its shape) between two bright
lights, ensuring a rapid rate of photosynthesis (Figure
8.20*a*). To start an experiment, a solution containing
dissolved $^{14}CO_2$ was suddenly squirted into the lollipop. At a carefully measured time after this squirt,
a sample of the culture was rapidly drained out into
a container of boiling ethanol. The ethanol performed
two functions: It killed the algae, stopping photosynthesis immediately, and it extracted the ^{14}C-contain-
ing metabolic intermediates of $^{14}CO_2$ from the algae
(along with many other compounds). A sample of
this ethanolic extract was spotted on filter paper for
paper chromatography followed by autoradiography.
Typical results of a 30-second exposure to $^{14}CO_2$ are
shown in Figure 8.20*b*. As you can see, many biochemical reactions had taken place during that short
interval.

The First Stable Product of Carbon Dioxide Fixation

In 30 seconds of continuous exposure to $^{14}CO_2$, many
different products were formed in Calvin's lollipop.
In order to determine which of them was formed
first, it was necessary to repeat the experiment several times, using ever-shorter exposures. Even after
exposures of less than two seconds, half a dozen or
more labeled compounds appeared in the autoradi-
ographs. However, one was produced most rapidly
and in greatest abundance: **3-phosphoglycerate**
(**3PG**, also called 3-phosphoglyceric acid, or PGA),
which we have already encountered as an intermediate in glycolysis. 3PG is the first stable product of
CO_2 fixation. The Berkeley group isolated the individual carbon atoms from this 3PG and found that
the carbon of the carboxyl group was much more
intensely radioactive than the other two carbon atoms
(Figure 8.21). (A single atom either is or is not radioactive. What we mean by "more intensely radioactive" is that in a *population* of molecules, the fraction
of radioactive carboxyl carbons was greater than the
fractions of radioactive carbons in the other two carbon positions.) From this the Berkeley group drew
two important conclusions. First, the heavy labeling
of the carboxyl carbon showed that it is obtained
directly from CO_2. The existence of label in the other
carbon atoms led to the second conclusion: Some
kind of *cyclical* process is involved, a process by
which 3PG is made by adding CO_2 to another compound that is itself produced from photosynthetic
3PG (Figure 8.22).

What Is the Carbon Dioxide Acceptor?

What is this compound, obtained from the further
metabolism of 3PG, that binds with CO_2 to make
more 3PG? Given the structure of 3PG, it was reasonable to expect that the mysterious CO_2 acceptor
would be a *two*-carbon compound of some kind. This
would react with CO_2 and become a *three*-carbon
compound, with the CO_2 becoming a carboxyl group;
the product would be 3PG. If this idea were correct,
it should have been possible for the Berkeley group
to find, on their chromatograms, a compound with
only two carbon atoms, both of them being radioactive after a lollipop experiment. They did *not* find
such a compound. If you think about it now, you
will realize that this made the problem a great deal
more difficult. The "obvious" answer was wrong; the
CO_2 acceptor is not a two-carbon compound. Where
would one go from here?

It was at this point that the recognition of a photosynthetic *cycle* became useful. Consider a tentative
cycle of the sort shown in Figure 8.22*a*: CO_2 reacts
with "X," the CO_2 acceptor, to produce 3PG. From
the 3PG, photosynthetic organisms make products

BOX 8.B

Tools that Cracked the Calvin–Benson Cycle

Calvin's group needed a way to keep track of the carbon atom from CO_2 as it became incorporated into other compounds. This need was met as a by-product of the program to produce the first atomic bomb; as a result of this program, the radioactive carbon isotope ^{14}C was made available to scientists. It was then possible to prepare samples of CO_2 in which some of the carbon atoms were ^{14}C rather than the stable isotope ^{12}C

found in nature. Any compound incorporating the radioactive material would then also be radioactive.

The second need was for an improved method of separating complicated mixtures into their individual components. Living things contain thousands of different chemical components; if we want to study just one of them, it is usually necessary to separate it from all the thousands of others. A powerful new separation tool, called **paper partition chromatography**, became available shortly before the Berkeley group began its work. In many applications, paper chromatography works as follows (figure below, part a). A drop of a solution containing the compounds to be separated is placed near one end of a strip of filter paper. The paper is then lowered into a container with a suitable mixture of organic solvents (liquids such as chloroform and alcohols) until the end of the paper is submerged. The solvent works its way up through the paper by capil-

lary action. As the solvent climbs, the materials in the original mixture are carried up the paper as well. For now, let us say that the compounds in the mixture are pigments, so that we can see what happens to them. By the time the solvent has moved several centimeters up the paper, one usually sees that the mixture of pigments has separated, so the different pigments can be seen as colored spots at various distances up the paper. When such experiments are repeated, it is found that a given compound always travels the same *relative* distance (figure below, part b). If we take the distance (from the starting spot) moved by a particular substance and divide it by the distance (from the same point) moved by the solvent, that ratio is referred to as the R_F ("front ratio") value of that substance under those conditions of temperature and solvent composition. The R_F value of a component can then be used to help identify it.

Paper partition chromatography

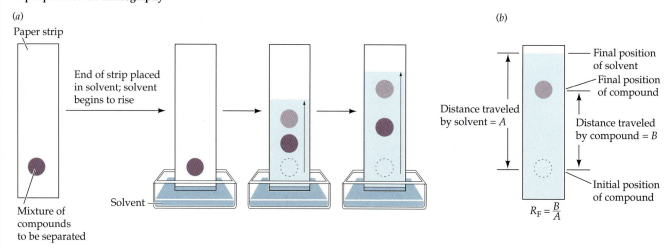

(a)
Paper strip

End of strip placed in solvent; solvent begins to rise

Solvent

Mixture of compounds to be separated

(b)

Distance traveled by solvent = A

Final position of solvent
Final position of compound

Distance traveled by compound = B

Initial position of compound

$$R_F = \frac{B}{A}$$

(things like glucose) and more X; the latter can react with another molecule of CO_2 and keep the cycle going. But what is X, and what are the other intermediates in that cycle? It was observed that 3PG was the only acid phosphate produced in significant amount, whereas there were many kinds of sugar phosphates on the chromatograms. On this basis, the Berkeley group guessed that the first thing to happen to 3PG is its conversion to a three-carbon sugar phos-

phate (glyceraldehyde 3-phosphate, which we will represent as G3P). Such a reaction is a reduction, and reductions are highly endergonic (Chapter 7). The Berkeley group proposed that this reduction would, then, require ATP (Figure 8.22b). At this point in the theorizing, an extremely clever suggestion was made: If Figure 8.22b is an accurate model of what occurs in the chloroplast, then it should be possible to regulate this cycle in the laboratory by simple means.

Two-dimensional chromatography

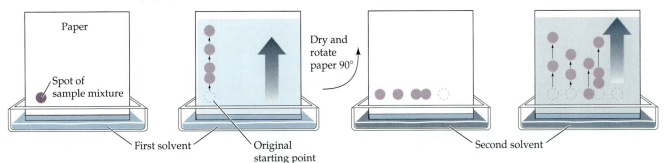

Particular groups of compounds may be difficult to separate even by a technique like paper partition chromatography. It sometimes helps, in such cases, to try **two-dimensional chromatography**. In this technique, the sample mixture is applied as a single spot near one corner of a square of filter paper. Paper chromatography is run as before, using a particular solvent system and achieving a partial separation of the mixture along one edge of the paper. The paper is then dried and rotated through a quarter of a turn so that the row of partially separated materials is now along the bottom edge, and chromatography is run again with a different solvent. If the second

solvent has markedly different properties from those of the first, it will separate some of the things that had failed to separate in the first solvent. The result is what looks like a scattering of materials; in fact, the R_F values in each of the two solvents are consistent, so the precise pattern is repeatable time after time.

The third need was for a method for locating tiny amounts of radioactive material on paper chromatograms (the sheets on which chromatography has been performed). The method developed is called **autoradiography** and works as follows. The chromatogram, which has spots of ^{14}C-containing materials at unknown places on it, is taken into a darkroom and covered either with a type of

photographic film or with a liquid photographic emulsion. This is kept in the dark for a suitable length of time, during which the radioactive decay of ^{14}C results in the release of particles that expose the film in essentially the same way as does light. Later the film is developed, bringing out dark spots (composed of silver grains in the film; see Box 2.A). Because the particles released by the radioactive decay of many radioactive compounds travel only extremely short distances (the ^{14}C particle travels about 10 μm), these dark spots on the film lie right over the places on the chromatogram where there were accumulations of ^{14}C-containing materials, and the R_F values of those materials can thus be determined.

Autoradiography

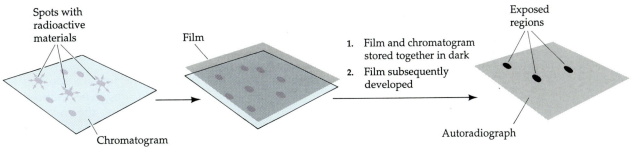

First, turning the light off should specifically block the step from 3PG to G3P, the sugar phosphate. The necessary ATP and NADPH + H$^+$ can only be produced from an input of energy; in photosynthesis, the energy source is light. Therefore ATP should be made in the chloroplast only when the light is on. Second, one can easily block the reaction from X to 3PG by cutting off the supply of CO_2 to the lollipop (Figure 8.22c).

Let us suppose that photosynthesis is proceeding at a steady pace, so the concentrations of 3PG and the CO_2 acceptor X in the cells are constant. The lights are on, of course, and there is plenty of CO_2. Suddenly the investigator turns off the lights, thus blocking the cycle as proposed in Figure 8.22c: What change in 3PG concentration immediately occurs? What change in the concentration of X immediately occurs? *Stop.* Think about this before you read on.

(a)

8.20 A Lollipop and Its Products
(a) The lollipop used in experiments on photosynthesis. The thin flask was filled with a suspension of algae and illuminated from both sides. The supply of CO_2 bubbling through the culture in the lollipop could be shut off, at which time $^{14}CO_2$ was injected. After a designated period of seconds or minutes, the contents were drained into boiling ethanol to stop the reactions in the algal cells. *(b)* A chromatogram showing the products of algal photosynthesis after a 30-second exposure to $^{14}CO_2$. The dark spots are compounds containing ^{14}C—all formed in the 30 seconds following injection of $^{14}CO_2$. The spot labeled PGA (an older notation) corresponds to the position of 3-phosphoglycerate (3PG), a compound we will discuss in the next section.

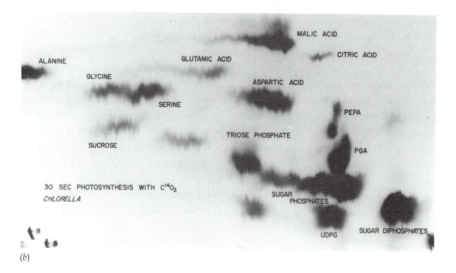

(b)

When the light is turned off, no more ATP is made. Without ATP, the step from 3PG to the three-carbon sugar cannot occur. Therefore, 3PG is no longer being used up. However, there is nothing to stop the formation of 3PG from incoming CO_2 and X, as long is there is any X around. Therefore, the immediate consequence of turning off the light is an *increase* in the level of 3PG. On the other hand, X continues to be used up, because its reaction with CO_2 does not depend on light or ATP; but the formation of X slows down because 3PG can no longer be reduced and ultimately form new X. Therefore, the concentration of X *decreases* (Figure 8.23). That is, these are the changes that we expect to see *if* our model (Figure 8.22c) is correct. Similar reasoning should convince you that when the CO_2 supply is cut off (with the lights on), the concentration of X rises and that of 3PG falls. With no CO_2 available, X is no longer used

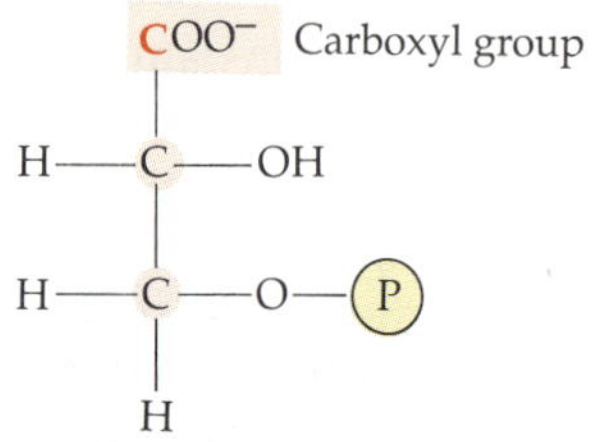

8.21 3-Phosphoglycerate
In experiments in which an algal sample from a lollipop was killed with ethanol only a few seconds after the introduction of $^{14}CO_2$, it was found that 3-phosphoglycerate (3PG) had picked up ^{14}C more rapidly and in greater amounts than the other labeled compounds found after this short time. In most 3PG molecules the ^{14}C was in the carboxyl group, as indicated by the red symbol; smaller numbers of 3PG molecules had the label in the other two carbon atoms. The heavy labeling in the carbon of the carboxyl group indicated the carbon was obtained directly from $^{14}CO_2$.

COO⁻ Carboxyl group

H—C—OH

H—C—O—(P)

H

3-Phosphoglycerate

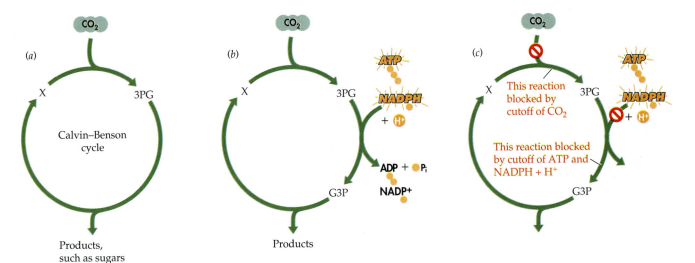

8.22   The Cyclic Nature of Photosynthesis
(a) Once it became apparent that CO_2 combines with some molecule to form 3PG and that a cyclical process is involved, a pathway could be devised in which a molecule of compound "X" combines with CO_2 to form 3PG and in which further reactions regenerate X. (b) Ensuing speculation and experimentation led to the proposition that 3PG was reduced to a sugar phosphate, glyceraldehyde 3-phosphate (G3P); this endergonic reaction, a reduction, would require ATP, as shown here. (c) A cutoff of CO_2 blocks the formation of 3PG from compound X and CO_2; a cutoff of ATP (by turning off the light to the lollipop) blocks the formation of G3P from 3PG.

up, and 3PG is no longer formed; but 3PG can still be reduced to G3P.

Having reached these conclusions, the Berkeley group proceeded to study the effects of changes in light intensity and CO_2 supply on the concentrations of all of the major radioactive compounds found in

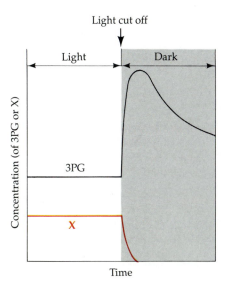

8.23   Changes in the Dark
When the light to a lollipop is turned off (gray area), 3PG builds up as compound X is combined with CO_2. The concentration of X, on the other hand, falls when the light is turned off. X is not replenished because 3PG is not converted to G3P to continue the cycle (light is necessary to provide the ATP for the endergonic 3PG → G3P reaction).

the lollipop experiments. The first thing they noticed was that only one compound underwent the concentration changes proposed for 3PG—and that was 3PG itself. This showed that their model was likely to be accurate. Would any compound behave in the way predicted for the mysterious compound X, the CO_2 acceptor? Yes—and only one, a *five*-carbon sugar phosphate called **ribulose bisphosphate**, or **RuBP** (Figure 8.24). It seemed, then, that there must be a reaction in which CO_2, with its single carbon, combines with the five-carbon RuBP to give *two* molecules of the three-carbon 3PG.

When a reaction like this is proposed, one particular kind of evidence is much desired: One wants to find an *enzyme* that catalyzes the proposed reaction. Indeed, such an enzyme was soon discovered, and from more than one source. The enzyme, RuBP carboxylase (now commonly called **rubisco**), was found in the algae studied by the Berkeley group and also in the leaves of spinach. Best of all, studies of spinach revealed that rubisco is found in only one part of the cell—the chloroplast—and that is exactly where one should find an enzyme concerned with photosynthesis. It was concluded, then, that RuBP is the CO_2 acceptor; it is the previously unknown compound X.

Filling the Gaps in the Calvin–Benson Cycle

Having discovered the first product of CO_2 fixation (3PG) as well as the CO_2 acceptor (RuBP), the Berkeley group proceeded to elucidate the remaining reactions of the cycle. The Calvin–Benson cycle in-

8.24   Ribulose Bisphosphate
Ribulose bisphosphate (RuBP) is the CO_2-accepting compound X in Figures 8.22 and 8.23. The combination of CO_2 and RuBP forms a reaction intermediate, which then splits into two molecules of 3PG. The fate of the carbon atom in CO_2 is traced in red.

cludes some relatively complicated steps between G3P and RuBP; among the intermediates are sugar phosphates with four, five, six, and seven carbon atoms. All the proposed intermediates have been found in chloroplasts, as have all the necessary enzymes. It was also found that ATP is needed at one more point in the Calvin–Benson cycle: in the step producing RuBP from ribulose monophosphate (RuMP). This makes it even easier to understand the effect of turning off the light on the concentration of RuBP (X in Figure 8.23).

For your current purposes it will suffice to learn the material in Figure 8.25, which summarizes the key features of the Calvin–Benson cycle. The reactions of this pathway take place in the chloroplast, with most of the enzymes being dissolved in the

8.25   The Calvin–Benson Cycle
The Calvin–Benson cycle, sometimes called the "dark reactions" of photosynthesis, picks up CO_2 and turns out glucose and other organic molecules that contain the carbon and energy necessary for the myriad processes of life. This diagram presents only the key steps; the values given are those necessary to make one molecule of glucose, which requires six "turns" of the cycle.

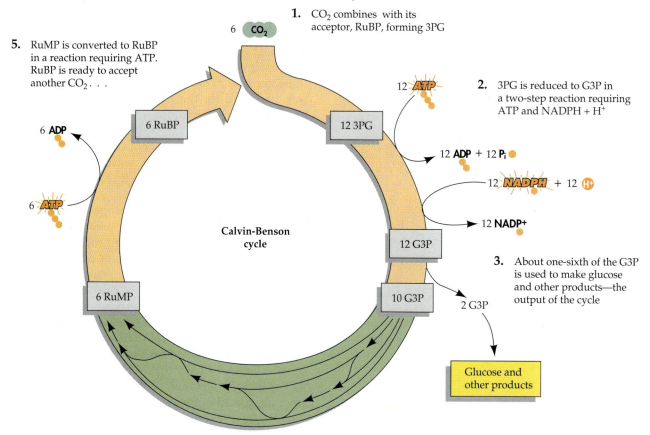

1. CO_2 combines with its acceptor, RuBP, forming 3PG

5. RuMP is converted to RuBP in a reaction requiring ATP. RuBP is ready to accept another CO_2...

2. 3PG is reduced to G3P in a two-step reaction requiring ATP and NADPH + H⁺

3. About one-sixth of the G3P is used to make glucose and other products—the output of the cycle

4. The remaining five-sixths of the G3P is processed in the complex reactions of the "sugar shuffle," resulting in RuMP

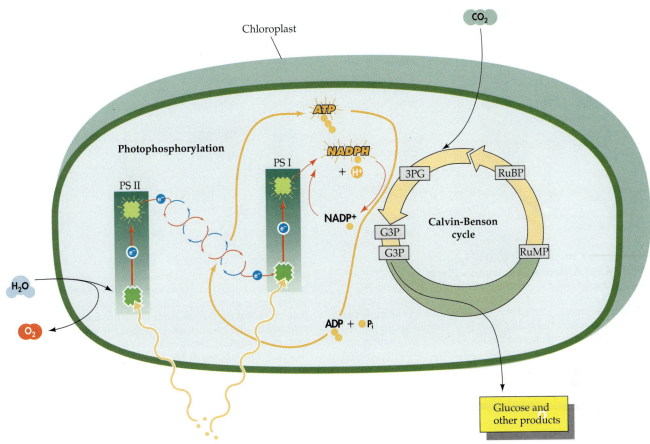

Chloroplast

Photophosphorylation

PS II

PS I

ATP

NADPH

+ H+

NADP+

ADP + Pi

H₂O

O₂

CO₂

3PG

G3P

G3P

RuBP

RuMP

Calvin-Benson cycle

Glucose and other products

8.26   An Overview of the Photosynthetic Reactions

stroma (Chapter 4); rubisco, the enzyme catalyzing the formation of 3PG from RuBP and CO_2, is found in the thylakoid membranes. 3PG is reduced, in a reaction requiring ATP as well as hydrogens provided by NADPH + H$^+$, to a three-carbon sugar phosphate (G3P). There then ensues a complex sequence of reactions with two principal outcomes: the formation of more RuBP, and the release of products such as glucose. The production of one molecule of glucose ($C_6H_{12}O_6$) requires the Calvin–Benson cycle to operate six times on successive CO_2 molecules. A general summary of photosynthesis is given in Figure 8.26.

Glucose produced in photosynthesis is subsequently used to make other compounds besides sugars. The carbon of glucose is incorporated also into amino acids, lipids, and the building blocks of the nucleic acids. The products of the Calvin–Benson cycle are of crucial importance to the entire biosphere, for it is those molecules that serve as the food for all of life. Their covalent bonds represent the total energy yield from the harvesting of light by plants. Most of this stored energy is released by the plants themselves in their own cellular respiration. However, much plant matter ends up being consumed by animals, in which glycolysis and cellular respiration provide free energy from the plant matter for use in animal cell metabolism.

ALTERNATE MODES OF CARBON DIOXIDE FIXATION

From the work of the Berkeley group, it was expected that the exposure of a plant to both light and $^{14}CO_2$ would always lead to the appearance of 3-phospho[^{14}C]glycerate as the first labeled product of CO_2 fixation. Thus it was a surprise to discover that in sugarcane such treatment leads to the formation of four-carbon acids as the first ^{14}C-containing products. Subsequently it was shown that numerous plants follow this pattern. These—the C_4 plants—perform the normal Calvin–Benson cycle; but they add a step that acts as a "CO_2 pump" to increase the rate of photosynthesis even at low levels of CO_2 within the leaf.

In general, C_4 plants live in environments where water is relatively unavailable at times. To prevent excessive water loss, the leaves keep their stomata closed much of the time. This leads to a depletion, by photosynthesis, of CO_2 within the leaf. However, because their leaves contain the enzyme phosphoenolpyruvate carboxylase, or **PEP carboxylase**, C_4 plants have a means of compensating for this depletion. PEP carboxylase catalyzes the reaction of phosphoenolpyruvate (PEP), a three-carbon acid, with

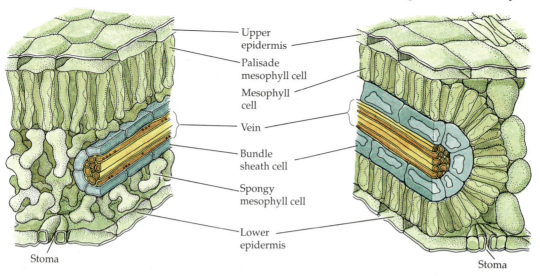

Upper
epidermis

Palisade
mesophyll cell

Mesophyll
cell

Vein

Bundle
sheath cell

Spongy
mesophyll cell

Lower
epidermis

Stoma

Stoma

8.27 Leaf Anatomy of C₃ and C₄ Plants
(a) In the leaf of a C₃ plant, the bundle sheath cells surrounding the vascular elements of a vein are relatively small; the upper part of the leaf is filled with upright palisade mesophyll cells; and loosely arranged spongy mesophyll cells allow gases to circulate in the lower layers within the leaf. Both mesophyll layers carry on photosynthesis. (b) In a C₄ plant, the bundle sheath cells are usu-

ally larger and contain prominent chloroplasts toward their outer edges; uniform mesophyll cells surround the entire vascular bundle. This arrangement facilitates the incorporation of carbon from CO_2 into four-carbon compounds by the mesophyll cells and the passage of these carbon-containing compounds to the bundle sheath cells, where the reactions of the Calvin–Benson cycle take place. The mesophyll serves as a "CO_2 pump."

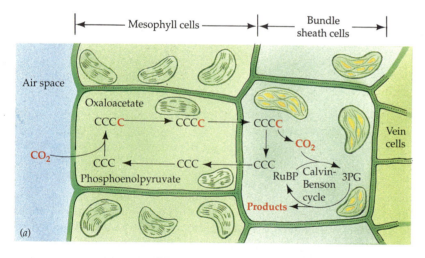

Mesophyll cells

Bundle
sheath cells

Air space

Oxaloacetate

CCC**C** → CCC**C** → CCC**C**

CO_2

Vein
cells

CO_2

CCC ← CCC ← CCC

Phosphoenolpyruvate

RuBP   Calvin–
Benson
cycle   3PG

Products

(a)

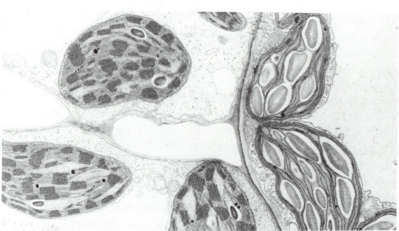

(b)

8.28 C₄ Photosynthesis
(a) In C₄ photosynthesis, mesophyll cells in the leaf pick up CO_2 and incorporate the carbon atom into four-carbon compounds. The four-carbon compounds diffuse into adjacent bundle sheath cells, where the four-carbon molecules are decarboxylated, releasing CO_2. The enzyme rubisco picks up this CO_2, and the usual Calvin–Benson cycle of C₃ photosynthesis ensues. (b) Portions of two mesophyll cells (left) and a single bundle sheath cell (right) from the leaf of a corn plant. Corn is a C₄ plant. Note the numerous grana and few starch grains in the chloroplasts of the mesophyll cells; in the chloroplasts of the bundle sheath cell, where the Calvin–Benson cycle forms the products of photosynthesis, there are numerous large, oval starch granules but very few membranes organized into grana. (See Figure 4.17 to review the structure of chloroplasts.)

CO_2 to yield the four-carbon compound oxaloacetate as the first product of CO_2 fixation. PEP carboxylase has a much greater affinity for CO_2 than does rubisco, so C_4 plants can trap CO_2 even when that gas is present in a much reduced concentration. (You may recall that PEP is a late intermediate in glycolysis (Figure 7.10) and that oxaloacetate is the last intermediate in the citric acid cycle (Figure 7.12). Evolution has led to the use of certain compounds, such as these, in a number of different ways in living things.)

The C_4 plants generally have a specialized leaf anatomy with two classes of photosynthetic cells, each having its own distinctive type of chloroplasts. The cells are arranged as in Figure 8.27, with a **mesophyll** layer of photosynthetic cells surrounding an inner **bundle sheath** layer that is also photosynthetic. In the mesophyll cells are chloroplasts with numerous grana; they trap CO_2 and use it to form various four-carbon acids from oxaloacetate, as well as certain other metabolites. These products diffuse from the mesophyll layer into the bundle sheath (Figure 8.28). There the four-carbon acids are decarboxylated to release CO_2, which is recaptured by rubisco and used in the Calvin–Benson cycle of C_3 **photosynthesis**. The chloroplasts in the bundle sheath lack well-developed grana but typically have substantial starch grains deposited in them because they, rather than the mesophyll chloroplasts, are the sites where sugars are finally formed and starches are stored.

What this system does is, in effect, to "pump" CO_2 from a region where its concentration is low (the intercellular spaces within the leaf) to one where it is relatively more abundant (the bundle sheath layer). PEP carboxylase in the mesophyll chloroplasts can,

by its great affinity, take up CO_2 when rubisco cannot; the temporary products (four-carbon acids) are then loaded into the bundle sheath, allowing the release of sufficient CO_2 to keep the rubisco in the bundle sheath busy.

A related but distinguishable system functions in certain other plants that face frequent water shortages. Most of these are members of the family Crassulaceae, which includes the ice plants and some other succulent plants (Figure 8.29). Because of the way in which their stomata are regulated (Chapter 29), they have access to atmospheric CO_2 only at night; by day their stomata are closed and no CO_2 can enter the leaf. These plants trap the CO_2 at night, using PEP carboxylase; this allows them to store great quantities of CO_2 in the form of carboxyl groups of four-carbon acids. By day, behind closed stomata, the CO_2 is released within the leaves; it is recaptured by rubisco, and photosynthesis then proceeds by way of the Calvin–Benson cycle. The difference between this system and C_4 photosynthesis is that here the PEP comes from the respiratory breakdown of sugars; in C_4 photosynthesis, PEP is produced photosynthetically in a light-requiring reaction.

PHOTORESPIRATION

The enzyme rubisco is not entirely specific for CO_2 as a substrate, and it can catalyze a reaction of RuBP with oxygen. (*Rubisco* stands for *ribulose bisphosphate carboxylase/oxygenase.*) One of the products is glycolate, a two-carbon compound that leaves the chloroplast and diffuses into organelles called **microbodies** (Chapter 4). In the microbody, glycolate is oxidized (in an oxygen-requiring reaction); later the product undergoes reactions leading to the release of CO_2 (Figure 8.30). The rate of the overall process (from RuBP and O_2 to the release of CO_2) is roughly proportional to the light intensity. Because of the dependence on light and because of the uptake of oxygen and release of carbon dioxide, this process is called **photorespiration**. You can see that it interferes with photosynthesis; in fact, it apparently reverses it—but without resulting in ATP formation as does cellular respiration. The role of photorespiration in the life of the plant is unknown, and it may simply result from an unfortunate lack of specificity of rubisco. With this in mind, many scientists are attempting to develop a gene that codes for a more specific form of rubisco, and to insert that gene into crop plants. (See Chapter 14 for a discussion of recombinant DNA technology.) At the same time, it seems odd that rubisco, the most abundant single protein in the living world, should apparently function less than optimally.

Interestingly, the C_4 plants appear to indulge in

8.29 A Crassulacean
This rock dudleya, a member of the family Crassulaceae, is adapted to its desert habitat in several ways, including its reversed stomatal cycle and its use of PEP carboxylase to trap CO_2 at night. Here it shares a rock with numerous lichens.

(a)

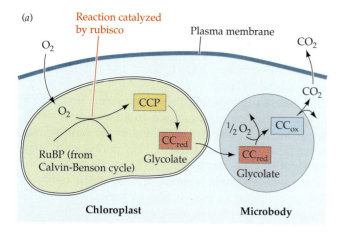

Chloroplast · Microbody

(b)

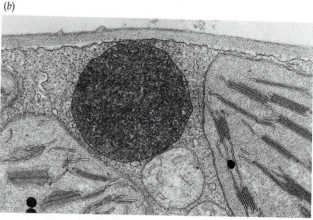

8.30 Photorespiration

(a) In the chloroplast, the enzyme rubisco may catalyze a reaction between RuBP and O_2; further reactions lead to the reduced two-carbon compound glycolate. Glycolate leaves the chloroplast and diffuses into a microbody, where it is oxidized. Later reactions lead to the formation of CO_2, which diffuses out of the cell. With the pickup of O_2 and the release of CO_2, the overall result resembles respiration; and because the process occurs in rough proportion to light intensity, it has been dubbed "photorespiration." (b) The dark, round object is a microbody in a mesophyll cell of a tobacco leaf; microbodies are the organelles in which glycolate from chloroplasts is oxidized (among other reactions). Portions of adjacent chloroplasts are seen to the lower left and far right, a mitochondrion is to the immediate lower right, and the plasma membrane and cell wall are above the microbody in this view.

little photorespiration, in contrast with the C_3 plants, which may photorespire away a substantial fraction of the CO_2 initially fixed in photosynthesis. Seemingly, the C_4 plants have minimized photorespiration in order to maximize the efficiency of their photosynthesis. A comparison of C_3 and C_4 photosynthesis is provided in Table 8.1.

PHOTOSYNTHESIS AND CELLULAR RESPIRATION

In plants, respiration takes place both in the light and in the dark, whereas photosynthesis takes place only in the light. The reactions of glycolysis occur in the cytosol, those of respiration in the mitochondria, and those of photosynthesis in the chloroplasts. Thus photosynthesis and respiration can proceed simultaneously but in different organelles.

For a plant to live, its photosynthesis must exceed its respiration, giving it a net gain of carbon dioxide and energy from the environment. Accordingly, the plant world plus photosynthetic bacteria and protists export food—and oxygen—to the animal kingdom and to all other nonphotosynthetic organisms (with the exception of a few types of bacteria). Animals require both food and oxygen, and they return carbon dioxide that may be used by plants in photosynthesis. Thus there are natural cycles of both carbon dioxide and oxygen.

There are important points of similarity between photosynthesis and respiration. In eukaryotes, both processes reside in specialized organelles that have complex systems of internal membranes. ATP synthesis in both processes relies on the chemiosmotic mechanism, involving the pumping of protons through a membrane. Another key feature of both

TABLE 8.1
Comparison of Photosynthesis in C_3 and C_4 Plants

VARIABLE	C_3 PLANTS	C_4 PLANTS
Perform Calvin–Benson cycle	Yes	Yes
Primary CO_2 acceptor	RuBP	PEP
CO_2-fixing enzyme	Rubisco (RuBP carboxylase)	PEP carboxylase
First product of CO_2 fixation	3PG	Oxaloacetate
Affinity of carboxylase for CO_2	Moderate	High
Leaf anatomy: photosynthetic cells	Mesophyll	Mesophyll + bundle sheath
Classes of chloroplasts	One	Two
Photorespiration	Extensive	Minimal

respiration and photosynthesis is electron transport, that is, the passing of electrons from carrier to carrier in a series of exergonic redox reactions. In respiration, the carriers receive electrons from high-energy food molecules and pass them ultimately to oxygen, forming water. On the other hand, photosynthesis requires an input of light energy to make chlorophyll a reducing agent strong enough to initiate the transfer of electrons. In photosynthesis, water is the source of the electrons, and oxygen is released from water in a very early step. The electrons from water end up in NADPH and, finally, in food molecules.

SUMMARY

The energy for virtually all biological work derives ultimately from the sun, by way of photosynthesis. The energy of a photon is related to its wavelength: $E = hc/\lambda$. When a photon is absorbed by a molecule, the molecule becomes excited. Excited chlorophyll (Chl*) is a better reducing agent than ground-state chlorophyll (Chl). Accessory pigments pass their absorbed energy to chlorophyll.

In photophosphorylation, Chl* passes an electron and energy to an oxidizing agent and thence through a series of carriers, and some of the energy thus released is used to form ATP. ATP is the sole output of cyclic photophosphorylation, while noncyclic photophosphorylation produces both ATP and NADPH + H$^+$. Noncyclic electron flow begins with H$_2$O, proceeds through two chlorophyll-containing photosystems, and ends with NADPH. ATP formation in the chloroplast involves a chemiosmotic mechanism.

Carbon dioxide is incorporated into organic compounds in the Calvin–Benson cycle. The CO_2 reacts with RuBP in the presence of the enzyme rubisco, forming two molecules of 3PG. The reduction of 3PG to a sugar phosphate requires ATP and NADPH + H$^+$, the products of photophosphorylation. The sugar phosphate is metabolized to regenerate RuBP and yield sugars and other compounds.

The C$_4$ plants augment the Calvin–Benson cycle with further reactions that enable the plants to do photosynthesis even at CO_2 levels so low that C$_3$ plants could no longer fix CO_2. Photorespiration, which is common in C$_3$ plants but rare in C$_4$ plants, is a seemingly wasteful process that results from non-specificity of rubisco.

SELF-QUIZ

1. Which statement about light is *not* true?
   a. Its velocity in a vacuum is constant.
   b. It is a form of energy.
   c. The energy of a quantum is directly proportional to its wavelength.
   d. A quantum of blue light has more energy than one of red light.
   e. Different colors correspond to different frequencies.

2. Which statement about light is true?
   a. An absorption spectrum is a plot of biological effectiveness versus wavelength.
   b. An absorption spectrum may be a good means of identifying a pigment.
   c. Light need not be absorbed to produce a biological effect.
   d. A given kind of molecule can occupy any energy level.
   e. A pigment loses energy as it absorbs a photon.

3. Which statement is *not* true of chlorophyll?
   a. Chlorophylls absorb light near both ends of the visible spectrum.
   b. Chlorophylls can accept energy from other pigments, such as carotenoids.
   c. Excited chlorophyll can either reduce another substance or fluoresce.
   d. Excited chlorophyll (Chl*) is an oxidizing agent.
   e. Chlorophylls contain magnesium.

4. In cyclic photophosphorylation:
   a. oxygen gas is released.
   b. ATP is formed.
   c. water donates electrons and protons.
   d. NADPH + H$^+$ is formed.
   e. CO_2 reacts with RuBP.

5. Which does *not* happen in non-cyclic photophosphorylation?
   a. Oxygen gas is released.
   b. ATP is formed.
   c. Water donates electrons and protons.
   d. NADPH + H$^+$ is formed.
   e. CO_2 reacts with RuBP.

6. In the chloroplast:
   a. light leads to the pumping of protons out of the thylakoids.
   b. ATP is formed when protons are pumped into the thylakoids.
   c. light causes the stroma to become acidic to the thylakoids.
   d. protons return passively to the stroma through protein channels.
   e. proton pumping requires ATP.

7. Which is *not* true of the Calvin–Benson cycle?
   a. CO_2 reacts with RuBP to form 3PG.
   b. RuBP is formed by the metabolism of 3PG.
   c. ATP and NADPH + H$^+$ are formed when 3PG is reduced.
   d. The concentration of 3PG rises if the light is switched off.
   e. Rubisco catalyzes the reaction of CO_2 and RuBP.

8. In C_4 photosynthesis:
   a. 3PG is the first product of CO_2 fixation.
   b. rubisco catalyzes the first step in the pathway.
   c. four-carbon acids are formed by PEP carboxylase in the bundle sheath.
   d. photosynthesis continues at lower CO_2 levels than in C_3 plants.
   e. CO_2 released from RuBP is transferred to PEP.

9. C_4 photosynthesis and the acid metabolism in Crassulaceae differ in that:
   a. only C_4 photosynthesis uses PEP carboxylase.
   b. CO_2 is trapped by night in Crassulaceae and by day in C_4 photosynthesis.
   c. four-carbon acids are formed only in C_4 photosynthesis.
   d. only Crassulaceae commonly grow in dry or salty environments.
   e. only C_4 photosynthesis helps conserve water.

10. Photorespiration:
    a. is performed only by C_4 plants.
    b. includes reactions carried out in microbodies.
    c. increases the yield of photosynthesis.
    d. is catalyzed by PEP carboxylase.
    e. is independent of light intensity.

FOR STUDY

1. Both photophosphorylation and the Calvin–Benson cycle stop when the light is turned off. Which specific reaction stops first? Which stops next? Continue answering the question "Which stops next?" until you have explained why both pathways have stopped.

2. In what principal ways are the reactions of photophosphorylation similar to the respiratory chain and oxidative phosphorylation discussed in Chapter 7? Differentiate between cyclic and noncyclic photophosphorylation in terms of (1) the products and (2) the source of electrons for reduction of oxidized chlorophyll.

3. The development of what three experimental techniques made it possible to elucidate the Calvin–Benson cycle? How were those techniques used?

4. If water that is labeled with ^{18}O is added to a suspension of photosynthesizing chloroplasts, which of the following compounds will first become labeled with ^{18}O: ATP, NADPH, O_2, or 3PG? If water labeled with ^3H is added to a suspension of photosynthesizing chloroplasts, which of those compounds will first become radioactive? If CO_2 labeled with ^{14}C is added to a suspension of photosynthesizing chloroplasts, which of those compounds will first become radioactive?

READINGS

Alberts, B., D. Bray, J. Lewis, M. Raff, K. Roberts and J. D. Watson. 1989. *Molecular Biology of the Cell*, 2nd Edition. Garland Publishing, New York. Chapter 7 on energy conversion contains a good discussion of photosynthesis.

Bjorkman, O. and J. Berry. 1973. "High-Efficiency Photosynthesis." *Scientific American*, October. C_4 photosynthesis and what it means to the plants in which it is found.

Clayton, R. K. 1980. *Photosynthesis*. Cambridge University Press, New York. An advanced general treatment of photosynthesis by a prominent photobiologist.

Govindjee and W. J. Coleman. 1990. "How Plants Make Oxygen." *Scientific American*, February. A "clock" in photosystem II that splits water into oxygen gas, protons, and electrons.

Hall, D. O. and K. K. Rao. 1987. *Photosynthesis*, 4th Edition. Edward Arnold, New York. An intermediate-level treatment of all the major topics in photosynthesis and excellent bibliography, all in one hundred pages.

Stryer, L. 1988. *Biochemistry*, 3rd Edition. W. H. Freeman, New York. Chapter 22 gives an advanced but clear treatment of topics in photosynthesis.

Voet, D. and J. G. Voet. 1990. *Biochemistry*. John Wiley & Sons, New York. The discussion of photosynthesis is in Chapter 22.

Weinberg, C. J. and R. H. Williams. 1990. "Energy from the Sun." *Scientific American*, September. Photosynthesis and biomass technology are considered, along with other solar-derived technologies such as wind and solar–thermal, as sources of energy for industrial and other uses.

Youvan, D. C. and B. L. Marrs. 1987. "Molecular Mechanisms of Photosynthesis." *Scientific American*, June. A difficult but interesting article on events in the first fraction of a millisecond of photosynthesis in a bacterium. Part of the article is better read after reading Part 2 of this book.

PART TWO

Information and Heredity

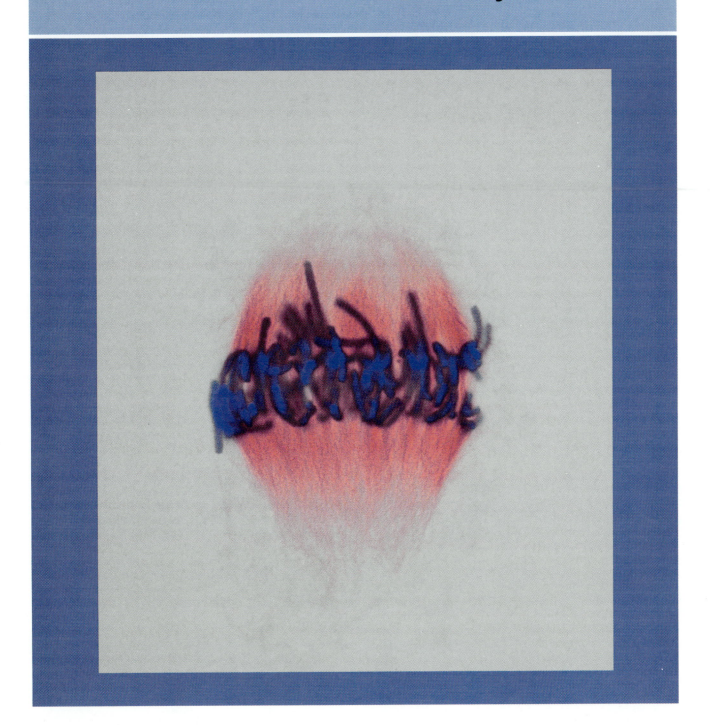

9

Chromosomes and Cell Division

PREVIEW: There are two mechanisms by which cell nuclei may divide: mitosis and meiosis. Mitosis yields two daughter nuclei that are identical to each other and to the parent nucleus in chromosome number and genetic composition. Meiosis yields four daughter nuclei, each with half as many chromosomes as the parent nucleus, and differing in genetic composition. Meiosis constitutes part of the sexual life cycle. During fertilization there is fusion of meiotic products: an egg from the female parent and a sperm from the male parent. This fusion restores the full chromosome number. Chromosomes are composed of DNA, RNA, and proteins.

This chapter deals with nuclei, chromosomes, the phases of mitosis and meiosis, chromatids, centromeres, kinetochores, histones, nucleosomes, spindle fibers, the cell cycle, and ploidy.

Every cell is ruled by its DNA—its genetic material. That DNA is also responsible for the future cells produced when the cell divides, and the cells produced in further generations. The right amounts and kinds of DNA must be provided to new cells and new organisms. Life requires that some new cells be exactly like the cells giving rise to them. A fundamental mechanism used throughout the eukaryotic kingdoms ensures that this happens. Equally fundamental, however, is a mechanism that permits cells to give rise to new cells in which the amount of DNA is precisely reduced. This means that organisms using this mechanism can give rise to new organisms that are *not* exact copies of themselves.

An adult human has over one hundred trillion (10^{14}) cells. As a fertilized human egg—a single cell—develops into a university student, its nucleus gives rise to over one hundred trillion nuclei, each of which contains basically the same genetic information as did the fertilized egg. The intricate mechanism by which the genetic material in the nucleus is first copied and then so partitioned that each of the two daughter nuclei gets one complete copy of the genetic information is called **mitosis**.

The second mechanism for nuclear division is **meiosis**. Because of this mechanism, offspring of the same parents differ. Meiosis produces four daughter nuclei, each with only half the genetic information contained in the original cell, and each differing from the others in the exact information contained. When organisms reproduce sexually, pairs of such cells combine. Each sexual union of cells may produce new genetic combinations.

The division of a eukaryotic cell consists of two steps: first the division of the nucleus, then the division of the cytoplasm. Between divisions—that is, for most of its life—a eukaryotic cell is in a condition called **interphase**. In this chapter we will learn the details of mitosis, meiosis, and interphase, as well as their biological consequences, which are of the utmost importance for heredity, development, and evolution.

NUCLEI AND CHROMOSOMES

All human cells, other than eggs and sperm, contain *two* full sets of genetic information, one full set from the mother of the person and one from the father. Eggs and sperm, however, contain only a single set; any particular egg or sperm contains some information from the mother and some from the father. The genetic information consists of a group of molecules of DNA housed in **chromosomes** in the nucleus. The nucleus appears relatively featureless, however, during most of the life of a cell, and the chromosomes cannot be seen through a microscope (Figure 9.1).

9.1 Nuclei

(a) The bright object in the middle of this spinach cell is a nucleus as resolved through a light microscope. The smaller round spot visible inside the nucleus is a nucleolus, the site of ribosome assembly. *(b)* This view of part of an animal cell required an electron microscope. The two membranes of the nuclear envelope are distinct, as are a number of pores in the envelope. The nucleolus is at the lower right; the other dark material inside the nucleus is chromatin.

(a)

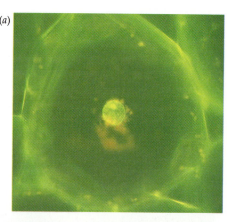

(b)

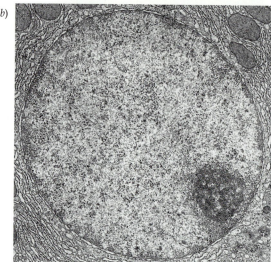

A gigantic, linear, double-stranded molecule of DNA complexed with many proteins is the basic unit of the eukaryotic chromosome. During many stages of a cell's life cycle, each chromosome in a eukaryotic cell contains only one such DNA molecule. At other times, however, the DNA molecule doubles; the chromosome then comprises two joined **chromatids**, each made up of one DNA molecule complexed with proteins. The joining is in a specific, small region of the chromosome called the **centromere** (Figure 9.2). As we will see in the sections that follow, centromeres function in the directed movement of chromosomes when nuclei divide. A body that has a single centromere, whether it contains one or two DNA molecules, is properly called a *chromosome*.

The complex of DNA and proteins in a eukaryotic chromosome is referred to as **chromatin**. It is the DNA of the chromosome that carries the genetic information, with the other components serving to organize the chromosome physically and to regulate the activities of the DNA.

9.2 Chromosomes and Chromatin

(a) A human chromosome in which the centromere is seen as a "pinched-in" region. At the stage of the cell cycle captured here, the chromosome consists of two chromatids lying side by side. Individual fibers of supercoiled chromatin are distinct around the periphery of the chromatids. *(b)* A diagrammatic representation of the chromosome. Kinetochores will be described later in this chapter.

(a)

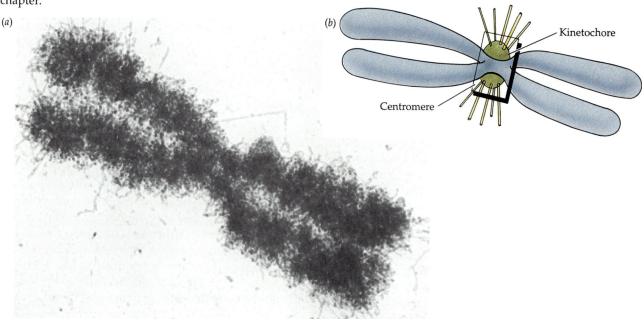

(b)

Kinetochore

Centromere

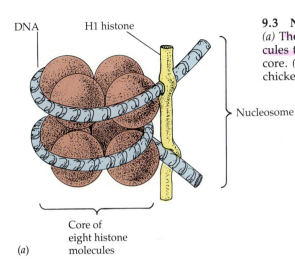

DNA | H1 histone

Nucleosome

Core of
eight histone
molecules

(a)

9.3 Nucleosomes

(a) The DNA double helix coils around a central core of eight histone molecules to make a nucleosome. Another histone (H1) clamps the DNA to the core. *(b)* Nucleosomes of the chromatin fibers from the red blood cell of a chicken look like "beads," and the "threads" connecting them are DNA.

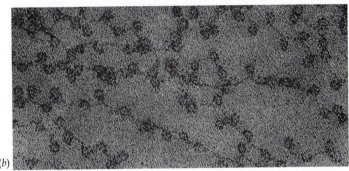

(b)

Organization of Chromatin

Chromatin changes dramatically during mitosis and meiosis. During interphase, the chromatin of a chromosome is strung out so thinly that the chromosome cannot be seen clearly as a defined body under the light microscope; but during most of mitosis and meiosis, the chromatin is coiled and compacted to a high degree, so that the chromosome appears as a bulky object (Figure 9.2). This alternation of forms relates to what the chromatin is doing during interphase and division. Before each mitosis the genetic material is duplicated. Remember that mitosis separates this duplicated genetic material into two new nuclei. This separation is easier to accomplish if the DNA is neatly arranged in compact units rather than being tangled up like a plate of spaghetti. During interphase, however, the DNA must direct the growth and other activities of the cell. As we will see in Chapters 11 and 13, DNA does this by interacting with enzymes, which requires its unwinding to expose the genetic information.

Proteins are closely associated with chromosomal DNA. Chromosomes contain large quantities of five classes of proteins, all of a type known as **histones**. Fairly small as protein molecules go, histones have a positive charge at pH levels found in the cell. The positive charge is a result of their particular amino acid compositions. Histone molecules join together to produce complexes around which the DNA is wound. Two molecules of each of four of the histone classes unite to form a core or spool so shaped that the DNA molecule fits snugly in a coil around it (Figure 9.3*a*). It seems that the fifth class of histone (H1) fits on the outside of the DNA, perhaps "clamping" it to the histone core and helping to pack adjacent DNA–histone complexes. There is strong evidence to show that the chromatin consists of a great number of beadlike units connected by a DNA thread (Figure 9.3*b*). The beads, called **nucleosomes**, are the

complexes we have just described: DNA wound around a histone core and held there by a molecule of histone H1. A chromatid has a single DNA molecule running through vast numbers of nucleosomes. One proposed structure for part of a mitotic chromosome with many nucleosomes is shown in Figure 9.4. During both mitosis and meiosis, the chromatin

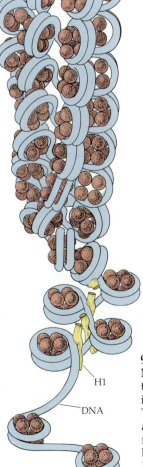

H1

DNA

9.4 Supercoiled Chromatin

Nucleosomes are seen toward the bottom here and are packed into a coil extending upward. This coil would then twist into another, larger coil—and so forth to produce chromatin fibers such as those shown in Figure 9.2*a*. See also Figure 9.5.

becomes ever more coiled and condensed, with further folding of the chromatin continuing up to the time at which chromosomes separate (Figure 9.5). A diverse group of acidic proteins are also present in small quantities in chromosomes. The roles of these proteins will be considered in Chapter 13.

Less is known about the organization of interphase chromatin. We do know, however, that it has nucleosomes, and that they are spaced at the same intervals as in supercoiled chromatin. We know that DNA replicates and directs synthesis of RNA while

remaining associated with histone molecules. However, there is also some evidence that the structure of the nucleosomes may change during some cellular activities. These changes are the subject of current research.

THE CELL CYCLE

A cell lives and functions until it either divides or dies—or, if it is a sex cell, until it fuses with another

9.5 Levels of DNA Packing

Schematic diagram of how DNA is "packed" into a metaphase chromosome.

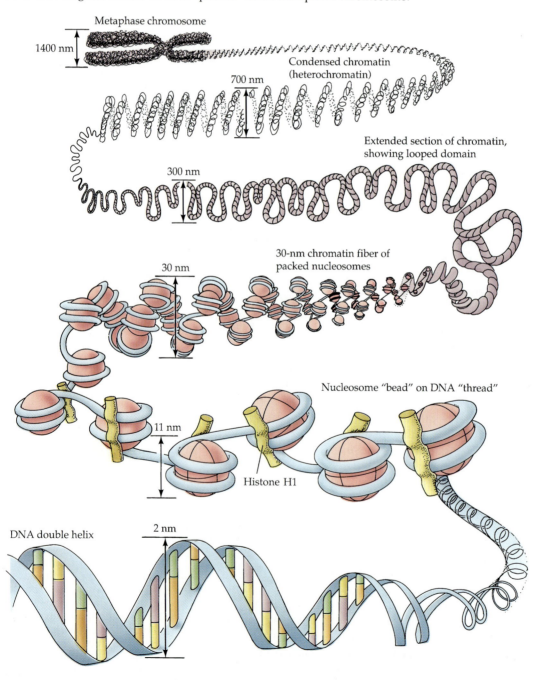

Metaphase chromosome

1400 nm

Condensed chromatin (heterochromatin)

700 nm

Extended section of chromatin, showing looped domain

300 nm

30-nm chromatin fiber of packed nucleosomes

30 nm

Nucleosome "bead" on DNA "thread"

11 nm

Histone H1

DNA double helix

2 nm

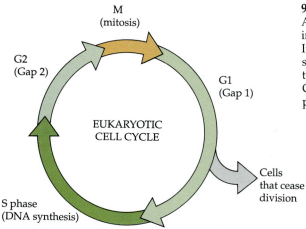

M
(mitosis)

G2
(Gap 2)

G1
(Gap 1)

EUKARYOTIC
CELL CYCLE

Cells
that cease
division

S phase
(DNA synthesis)

9.6 The Cell Cycle

A cell's life history can be represented as a cycle consisting of a short period of mitosis and a longer interphase. In dividing cells, interphase has three subphases. DNA is synthesized only for a short period (the S phase) between two times (phases G1 and G2) when no DNA is made. Cells that do not divide are usually arrested in the G1 phase.

sex cell. Some cells, such as red blood cells, muscle cells, and nerve cells, are incapable of division once they are fully formed; cells of certain other types rarely divide. However, most cells have some probability of dividing, and some are specialized for rapid division. Thus for many kinds of cells we may speak of a **cell cycle** that has mitosis as one phase and interphase as the other (Figure 9.6). A given cell lives for one turn of the cycle and becomes two cells. For life as a whole, the cycle repeats again and again as a constant source of renewal. The cell cycle, even for tissues engaged in rapid growth, consists mainly of the time spent in interphase. Examination of any collection of cells, such as a root tip or a slice of liver, reveals that most of the cells are in interphase most of the time. Only a small percentage of the cells are engaged in mitosis at any given moment; under favorable conditions, this can be confirmed by watching a single cell through its entire cell cycle.

The cell's DNA replicates during a specific portion of interphase called the **S phase** of the cell cycle, the S referring to *synthesis*. There is a significant gap in time between the S phase and the onset of mitosis, referred to as Gap 2 or **G2**. Another gap—**G1**—separates the end of mitosis and the onset of the next S phase. If a cell is not going to divide again, it will remain in G1 for weeks or even for many years until it dies—it seemingly will not waste effort replicating its genetic material. (There are some exceptions in which cells that will not divide synthesize DNA, but the continuation of the G1 phase is the rule in the vast majority of nondividing cells.)

Although the phases of the cell cycle differ in some biochemical activities, most proteins are formed throughout all subphases of interphase. But the histone proteins that we have just been discussing are synthesized primarily during the S phase of the cell cycle, at the same time that DNA is being synthesized. While DNA replicates in the nucleus, histones are synthesized in the cytoplasm; the new histones enter the nucleus through the pores in the nuclear envelope and then combine with the DNA, forming nucleosomes.

The relative durations of the various phases of the cell cycle vary greatly from species to species and from tissue to tissue, but there is a regularity in the cell cycle in a given tissue. What triggers the transitions from phase to phase? Scientists are still studying this, currently by genetic techniques.

MITOSIS

Shortly after the formulation of the cell theory, it became apparent that cell division was preceded by the division of the nucleus. The German botanist Eduard Strasburger first described the division of the nucleus in plants. Five years later, in 1880, the German zoologist Walther Flemming gave a more detailed description of this division in animal cells.

Mitosis is a mechanism by which a single nucleus gives rise to two nuclei that are genetically identical to each other and to the parent nucleus. Mitosis is a process of continuous change. Resist the temptation to think of it as a series of photographic slides in which one scene is replaced directly by another, distinctly different one. Rather, it is like a movie showing continuous action. In spite of this, it is convenient to look at mitosis as a series of important frames selected at intervals from the movie.

Let us begin with interphase, when the nucleus is between divisions (Figure 9.7). We see the nuclear envelope, the nucleoli, and a barely discernible tangle of chromatin. Just before mitosis there is also a pair of **centrosomes** lying near the nucleus. The role of the centrosomes will be described in the next paragraph. In many organisms, each centrosome contains a pair of centrioles. In seed plants and some other organisms, however, no centrioles are present. Where present, each of these pairs of centrioles consists of one "parent" and one smaller "daughter" centriole at right angles to it (Figure 9.8).

Mitosis

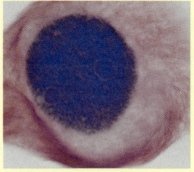

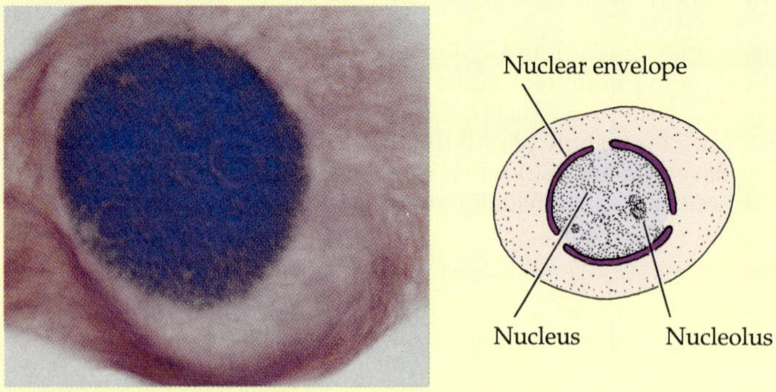

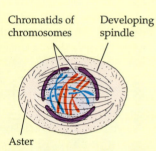

Nuclear envelope

Nucleus   Nucleolus

Interphase:
The nucleus replicates its DNA and centrosomes

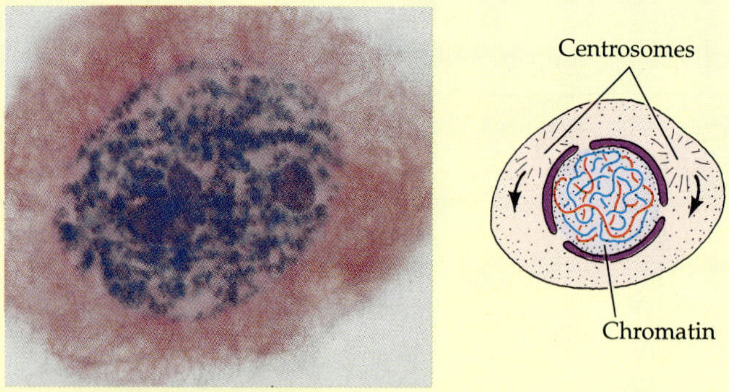

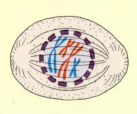

Centrosomes

Chromatin

Interphase-prophase transition:
The chromatin begins to coil

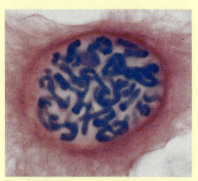

Chromatids of chromosomes   Developing spindle

Aster

Prophase:
The chromatin continues to coil and supercoil, making the chromatin more and more compact. The chromosomes consist of identical, paired chromatids

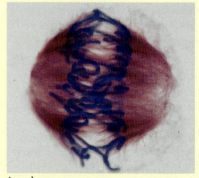

Prometaphase:
Nuclear envelope breaks down. Kinetochore microtubules appear and interact with the polar microtubules of the spindle, resulting in movement of the chromosomes

Equatorial (metaphase) plate

Metaphase:
The duplicated centromere regions connecting paired chromatids become aligned in a plane at the cell's equator

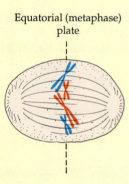

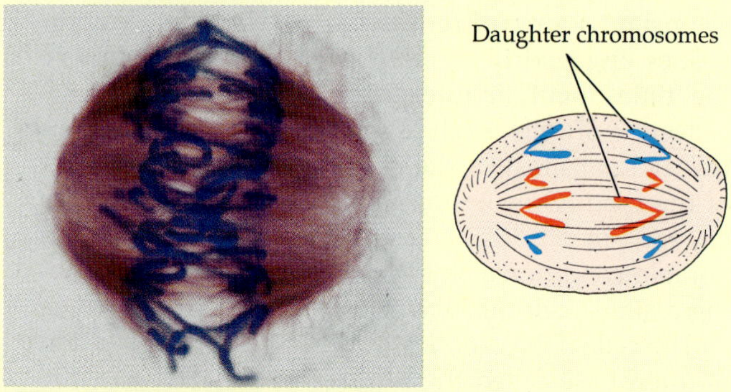

Daughter chromosomes

Anaphase:
Each centromere divides and the new chromosomes (each derived from one member of one of the sets of paired chromatids) begin to move toward the poles

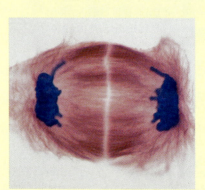

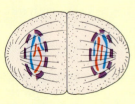

Telophase:
The separating chromosomes reach the poles. Telophase passes into the next interphase as the nuclear envelopes and nucleoli reform and the chromatin becomes diffuse

9.7 Mitosis

Mitosis results in two nuclei, genetically identical to one another and to the nucleus from which they formed. These photomicrographs are of plant nuclei, which lack centrioles and asters. The red dye stains microtubules and thus the spindle; the blue dye stains the chromosomes. In plants, the first steps toward division of the cell itself cause changes in the telophase cell that disrupt staining, causing the white line seen in the photomicrograph. The diagrams are of corresponding phases in animal cells, in order to introduce other structures. In the diagrams, the chromosomes are stylized to emphasize the fates of the individual chromatids.

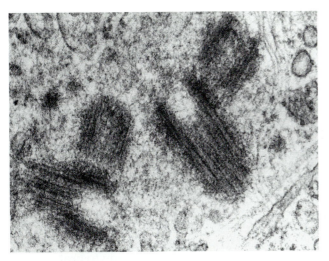

9.8   Centrioles
At a right angle to each "parent" centriole in this animal cell is a "daughter" about one-half its mature length. One parent–daughter pair will migrate to one side of the nucleus early in nuclear division while the other migrates to the opposite side. A centriole consists mostly of microtubules.

Development of the Chromosomes and Spindle

Gradually the appearance of the nucleus changes as the cell enters **prophase**, the beginning of mitosis. The nucleolar material disperses. The centrosomes, with or without pairs of centrioles, move away from each other toward opposite ends of the cell. Each centrosome then serves as a **mitotic center** that organizes microtubules. In animal cells, some of the microtubules point away from the nuclear region and form starlike groupings called **asters**. Other microtubules, called **polar microtubules**, run between the mitotic centers and make up the developing **spindle** (Figure 9.9). The spindle is actually two *half spindles*: Each polar microtubule runs from one mitotic center to the middle of the spindle, where it overlaps with polar microtubules of the other half spindle (Figure 9.9*b*). The polar microtubules are unstable and constantly form and fall apart until they contact polar microtubules from the other half spindle, at which point they become less unstable.

The chromatin also changes during prophase. Originally present as extremely long, thin fibers, it now takes on a more orderly form as a result of coiling, supercoiling, and compacting, as you will

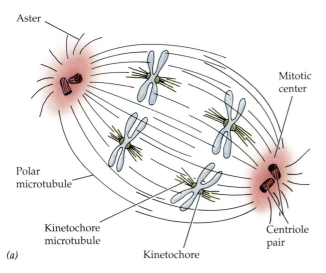

(a)

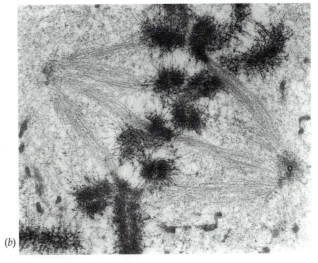

(b)

(c)

9.9   The Mitotic Spindle
(a) A mitotic spindle consists of numerous organized microtubules. Polar microtubules emanate from each pole of the spindle apparatus. Kinetochore microtubules attach both to the kinetochores associated with the centromeres of the chromosomes and to polar microtubules. *(b)* Polar microtubules extend from the poles in this electron micrograph. The large dark objects in the middle are chromosomes. *(c)* Kinetochore microtubules extend down from the top of this electron micrograph to the kinetochore, which is seen as a dark, three-layered "plate."

recall from Figure 9.5. At this level of magnification and at this stage of the nuclear cycle, each chromosome is seen to consist of two chromatids held tightly together over much of their length. The two chromatids of a single chromosome are identical in structure, chemistry, and the hereditary information they carry because one chromatid, formed during the S phase of the previous interphase, is a replica of the other. Within the region of tight binding of the chromatids lies the centromere (Figure 9.2), which must be present in order for the chromatids to become associated with microtubules of the spindle. Very late in prophase, specialized three-layered structures called **kinetochores** develop in the centromere region, one on each chromatid.

Dancing Chromosomes

The end of prophase—which is the beginning of **prometaphase**—is marked by the onset of directed chromosomal movement. Another event in prometaphase is the sudden disintegration of the nuclear envelope into membranous sacs, which takes only 20–30 seconds. Groups of microtubules, called **kinetochore microtubules**, associate with the kinetochores (Figure 9.9c). Some of the polar microtubules attach to the kinetochore microtubules and become stable. The polar microtubules pull, so that the kinetochore and its attached chromosome move toward one of the poles. The polar microtubules break down, and others from the same or the opposite pole attach to the kinetochore microtubules. Thus each chromosome may be pulled around seemingly aimlessly during prometaphase—until, randomly, the kinetochore of one chromatid is connected to microtubules from one pole while the other kinetochore is connected to microtubules from the other pole. When this happens, the microtubules cease falling apart, perhaps because of the tension established by the opposed pulls from the two poles. The polar microtubules pull in such a way that the kinetochores approach a region halfway between the ends of the spindle (Figure 9.7). This region, which may be thought of as an invisible plane perpendicular to the long axis of the spindle, is called the **equatorial plate**, or metaphase plate.

With the arrival of all the kinetochores at the equatorial plate, the cell has reached **metaphase**. The condensation of chromatin that began with prophase continues until the end of metaphase. At this time, the centromeres divide. Metaphase is usually brief, passing directly into **anaphase**, which is the phase in which the chromatids of each chromosome are pulled apart and drawn to the opposite ends of the spindle (Figure 9.7). As the new **daughter chromosomes**—the former chromatids, each containing one double-stranded DNA molecule—move toward the opposite poles of the spindle, it is easy to see that the motion is caused by the microtubules "tugging" at the kinetochore of each daughter chromosome. As the kinetochores are pulled apart, the arms of the chromosomes drag along passively. The mechanism of the tugging is not fully understood. The microtubules are probably shortening, and there may be sliding of the kinetochore microtubules along the polar microtubules. Also during anaphase, the poles of the spindle are often pushed farther apart by the action of some of the polar microtubules, thus contributing to the separation of the daughter chromosomes. The polar microtubules contain dynein, a protein also associated with the microtubules of cilia and flagella (Chapter 4). Presumably, then, the movement of the poles is produced in a manner similar to that in which eukaryotic flagella and cilia are made to beat. Amazingly little energy is expended in moving a chromosome during anaphase. The hydrolysis of 20 ATP molecules is enough to move a chromosome from the equatorial plate to the pole.

The End of Mitosis

As chromosomal movement ceases at the end of anaphase, the cell enters **telophase** (Figure 9.7). Two *identical* collections of chromosomes, which carry identical sets of hereditary instructions, are at the opposite ends of the spindle, which begins to break down, as do the asters if present. A new nuclear envelope forms around each group of chromosomes. The chromosomes begin to uncoil, continuing until all one can see is the diffuse tangle of chromatin characteristic of interphase. The nucleolus or nucleoli reappear at specific sites on specific chromosomes. When these changes are complete, telophase—and mitosis—is at an end, and each of the daughter nuclei enters another interphase. In that interphase, the DNA duplicates and new chromatids form, so that each chromosome consists of two chromatids. The duplication of DNA is a major topic and will be discussed in Chapter 11. Centrioles, if present, replicate during interphase: The two paired centrioles first separate, and then each acts as a "parent" for the formation of a new "daughter" centriole at right angles to it (Figure 9.8).

Mitosis is beautifully precise. Its result is the formation of two nuclei *identical to each other* and to the parent nucleus in chromosomal makeup—and hence in genetic constitution.

Cytokinesis

Mitosis refers only to the division of the nucleus; it is not always immediately accompanied by the division of the rest of the cell, **cytokinesis**. Generally, however, cytokinesis follows immediately upon mitosis. Division of animal cells is usually accomplished by a furrowing of the membrane, as if an invisible thread

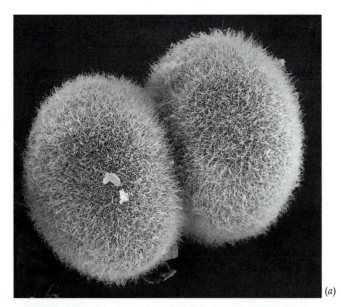

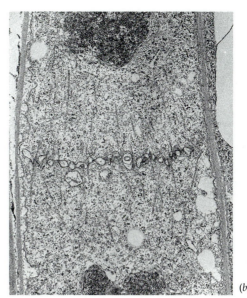

(a)

(b)

9.10 Cytokinesis
(a) A sea urchin egg has just completed cytokinesis at the end of the first division in its development into an embryo. The division furrow has completely separated the cytoplasm of one daughter cell from the other, although their surfaces remain in contact. Tiny, hairlike microvilli cover the surfaces of both cells. (b) The horizontal row of vesicles in this dividing plant cell in late telophase will join to form a cell plate between the cell above and the cell below. Microtubules run between the vesicles and are seen above and below in the cytoplasm.

was tightening between the two parts (Figure 9.10a). The "invisible" threads are microfilaments (Chapter 4) of actin and myosin, two proteins that interact here to produce a contraction, just as they do in muscles (Chapter 38).

The presence of a cell wall in plants requires a different approach to the division of the cell. As the spindle breaks down after mitosis, membranous vesicles derived from Golgi apparatuses appear in the region of the equatorial plate roughly midway between the two daughter nuclei and, with the help of microtubules, begin to lay down a **cell plate**, that is, the beginning of a new cell wall (Figure 9.10b).

Following cytokinesis, both daughter cells contain all the components of a complete cell. The precise distribution of chromosomes is ensured by mitosis. Organelles such as ribosomes, mitochondria, and chloroplasts need not be distributed equally between daughter cells, as long as many of each are present in both cells; accordingly, there is no mechanism comparable to mitosis to provide for their equal allocation to daughter cells. Although it was once thought that centrioles, where present, serve to organize the mitotic spindle, scientists now speculate that the association of the centrioles with mitotic centers simply ensures that centrioles, like chromosomes, are distributed precisely to the daughter cells.

ASEXUAL AND SEXUAL REPRODUCTION

Through the repeated operation of the cell cycle, a single cell can give rise to a vast number of others. The cell could be a unicellular organism reproducing with each cycle, or a cell that divides to produce a multicellular organism. The multicellular organism, in turn, may be able to reproduce itself by releasing one or more of its cells, derived from mitosis and cytokinesis, as a spore *or* by having a multicellular piece break away and grow on its own (Figure 9.11). Either of these processes would be an example of **asexual reproduction**—sometimes called vegetative reproduction. This mode of reproduction is based on mitotic division of the nucleus and, accordingly, produces offspring that are genetically identical with the parent. Asexual reproduction is a rapid and effective means of making new individuals, and it is widely practiced in nature.

The principal drawback of asexual reproduction is its very uniformity, leading to the production of a **clone** of genetically identical progeny. While the clone may be well adapted to its existing environment, it may be at great risk should conditions change. In contrast, parents that produce offspring that differ genetically are more successful when the environment varies unpredictably in time and space, because at least some of their genetically diverse offspring may be individuals able to meet the different challenges of a changing environment.

Diversity is fostered by **sexual reproduction**. There are two key steps in a sexual life cycle (Figure 9.12). The first is the combination of genetic information from two separate cells—usually, at least among animals, contributed by two separate parents. Each parent provides a sex cell, or **gamete**. Each gamete is

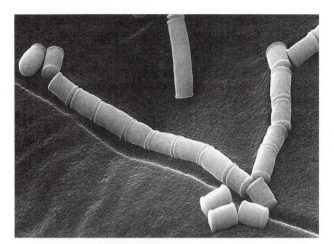

9.11 Asexual Reproduction
These "spools" are asexual spores formed by a fungus. Each spore contains a nucleus produced by a mitotic division; a spore and the fungal body that will grow from it following germination are the same genetically as the parent that fragmented to produce the spores. The general shape of the parent can be guessed from spores still in contact end-to-end.

haploid, meaning that it contains a single set of chromosomes (the number of chromosomes in such a single set is denoted by n). The two gametes—often identifiable as a female **egg** and a male **sperm**—fuse to produce a single cell, the **zygote**, or fertilized egg. This fusion is called **fertilization**. Its consequence, the zygote, contains genetic information from both gametes and, hence, from both parents. A further consequence is that the zygote has *two* sets of chromosomes; it is said to be **diploid**, or $2n$. In many species, including all animals, the zygote proceeds to develop by mitotic divisions into a multicellular adult. Because the zygotic nucleus is diploid, all the body cells produced by mitosis are also $2n$.

What happens at the next round of sexual reproduction? If the gametes were produced by mitosis from a diploid parent, then they too would be diploid. Thus after fusion of two diploid ($2n$) gametes, the next-generation zygote would be $4n$. This is not a tenable situation because subsequent generations would contain more and more chromosomes. There must be a *reduction* step in the sexual life cycle, that

9.12 The Essence of Sexual Reproduction
Fertilization and meiosis alternate in sexual reproduction. Haploid cells or organisms alternate with diploid cells or organisms. A zygote may differentiate into a germ cell, or it may form a multicellular organism that eventually produces germ cells. Whatever their origins, germ cells form haploid cells. A haploid cell may (1) differentiate into a gamete, or (2) it may form a multicellular organism that eventually produces gametes. Different organisms follow different paths around this cycle, and each variation is called a life cycle.

is, a special type of nuclear division that reduces the chromosome number from diploid to haploid. This form of division in sexually reproducing organisms is meiosis.

Meiosis in animal cells leads directly to the production of haploid gametes. However, in plants and some fungi, meiosis gives rise to haploid **spores**, which undergo mitosis, producing multicellular haploid bodies (find the correct path in Figure 9.12). Particular cells in these haploid bodies ultimately give rise, by mitosis, to haploid gametes, and the life cycle continues. Details of some of these life cycles are considered in Chapters 23–25. The simplest possible sexual life cycle is one in which two haploid gametes fuse to give one diploid ($2n$) zygote and this zygote immediately undergoes meiosis, yielding a new set of haploid (n) gametes. Embellishments on this scheme consist mainly of the addition of mitotic divisions leading to multicellularity in the haploid phase, the diploid phase, or both.

In any case, the essence of sexual reproduction is the selection of half of a parent's diploid chromosome set to make a haploid gamete, followed by the fusion of two such haploid gametes to produce a diploid cell containing genetic information from the two gametes. Both these steps contribute to a shuffling of genetic information in the population, so that usu-

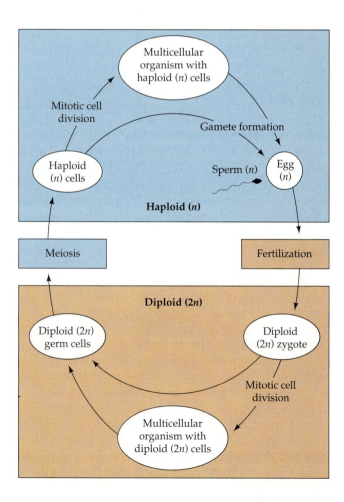

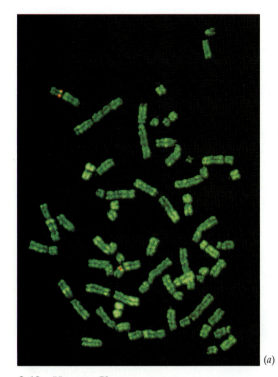

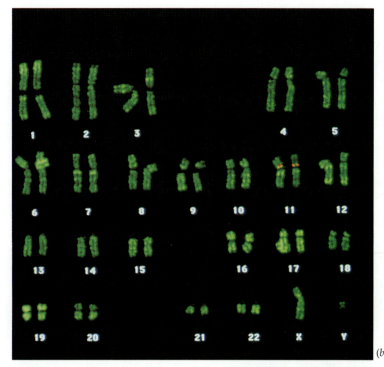

9.13 Human Chromosomes

Chromosomes of a human male. At the left, the chromosomes are spread out because immersion in hypotonic solution bloated and then ruptured the cell, which was in metaphase of mitosis. You should be able to count 46 chromosomes. At the right, a karyotype has been arranged from the metaphase spread of chromosomes. There are 23 pairs of homologous chromosomes, including a pair of sex chromosomes (XY). The chromosomes appear "striped," and the centromeres (which appear as constrictions) occupy characteristic positions on the different chromosomes. You can see that the lengths, banding patterns, and centromere positions are the same on the two members of a homologous pair (except for XY), aiding in distinguishing the pair among all the chromosomes in a metaphase display.

ally no two individuals have exactly the same genetic constitution. This is the opposite of the situation with asexual reproduction. The diversity provided by sexual reproduction has presented enormous opportunities for evolution. Although both asexual and sexual modes of reproduction have been present for billions of years, there are many more species of sexually reproducing organisms than of asexually reproducing organisms.

Ploidy and the Karyotype

When nuclei are in metaphase of mitosis, the centromeres are spread out on the equatorial plate and it is often possible to count and characterize the individual chromosomes. This is a relatively simple process in some organisms, thanks to techniques that can capture cells in metaphase and spread out the chromosomes. A photograph of the entire set of chromosomes can then be made, and the images of the individual chromosomes can be cut out and pasted together in an orderly arrangement (Figure 9.13). Such a rearranged photograph reveals the number, forms, and types of chromosomes in a cell, all of which constitute its **karyotype**. The individual chro-

mosomes can be recognized by their lengths, the positions of their centromeres, and characteristic banding when they are stained and observed at high magnification. When the cell is diploid, the karyotype consists of *pairs* of chromosomes—23 pairs for a total of 46 chromosomes in our species, and greater or smaller numbers of pairs in other diploid species (Table 9.1). In each recognizable pair of chromosomes, one chromosome comes from one parent and one from the other. The members of such a **homologous pair** are identical in size and appearance (with the exception of so-called sex chromosomes in some species; Chapter 10), and the two chromosomes (the homologues) of a homologous pair bear corresponding, though generally not identical, types of genetic information.

Haploid cells contain only one of the homologues from each pair of chromosomes. Thus when haploid gametes fuse in fertilization, the resulting diploid zygote ends up with two homologues of each type. There are circumstances under which triploid ($3n$), tetraploid ($4n$), and higher-order polyploid nuclei are formed. Each of these **ploidy levels** represents an increase in the number of complete sets of chromosomes present.

Meiosis I

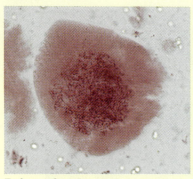
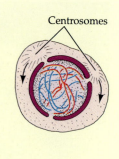

Nuclear envelope

Chromatin

Early prophase I:
The chromatin begins to condense following interphase

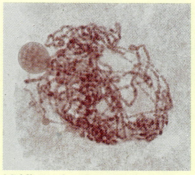
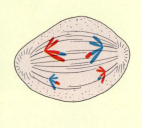

Centrosomes

Middle prophase I:
Synapsis produces bivalents, and chromosomes shorten

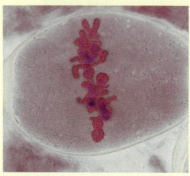
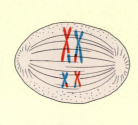

Metaphase I:
The bivalents line up on the equatorial (metaphase) plate

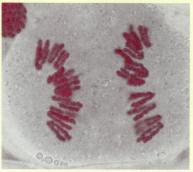
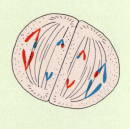

Anaphase I:
The two homologous chromosomes (each with two chromatids) of each bivalent move to opposite poles of the cell

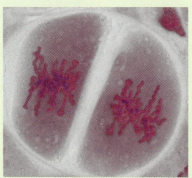
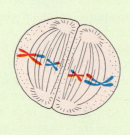

Metaphase II:
Kinetochores of the paired chromatids line up across the equator of each cell

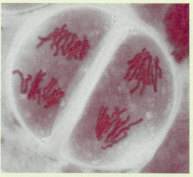
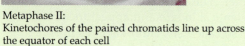

Anaphase II:
The chromatids of the chromosomes finally separate, becoming chromosomes in their own right, and are pulled to opposite poles

MEIOSIS

Meiosis is the mechanism by which the diploid number of chromosomes is reduced to the haploid number for sexual reproduction. To understand the process and its specific details, it is useful to keep in mind the overall functions of meiosis: (1) to reduce the chromosome number from diploid to haploid, (2)

to ensure that each of the four products has a complete set of chromosomes, and (3) to promote genetic diversity among the products. It is helpful to pay particular attention to the fact that the DNA is replicated only once during meiosis, although there are two divisions. Two unique features characterize the first meiotic division, **meiosis I**. The first feature is that homologous chromosomes pair along their entire

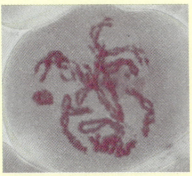
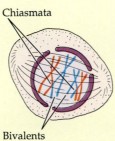

Chiasmata

Bivalents

Middle prophase I:
Chiasmata become evident

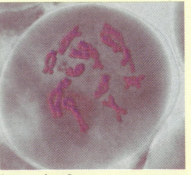
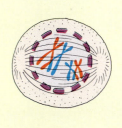

Late prophase I:
Coiling and shortening of the chromosomes continue

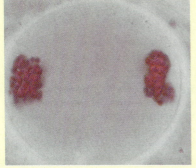
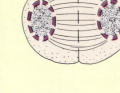

Telophase I:
The chromosomes gather into nuclei, and the original cell divides

Meiosis II

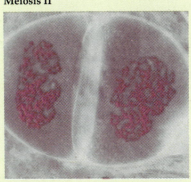
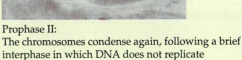

Prophase II:
The chromosomes condense again, following a brief interphase in which DNA does not replicate

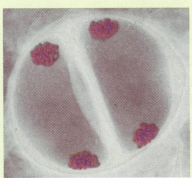
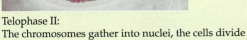

Telophase II:
The chromosomes gather into nuclei, the cells divide

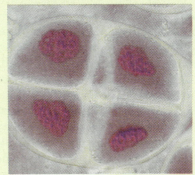

Products of meiosis:
Each of the four cells has a nucleus with a haploid number of chromosomes. Each of the four cells shown here will now develop into a pollen grain

lengths, a process called **synapsis**. This lasts from prophase to the end of metaphase of meiosis I. The second key feature is that homologous chromosomes separate during meiosis I. The individual chromosomes, each consisting of two joined chromatids, remain intact until the end of the metaphase of **meiosis II**, the second meiotic division. Let us now see how this works; you can follow along in Figure 9.14.

9.14 Stages of Meiosis

In meiosis, two sets of chromosomes are divided among four cells, each of which then has half as many chromosomes as the original cell. This happens as a result of two successive nuclear divisions. The photomicrographs shown here are of meiosis in the male reproductive organ of a lily. As in Figure 9.7, the diagrams are of meiosis in an animal.

TABLE 9.1
Numbers of Pairs of Chromosomes in Different Species of Plants and Animals

COMMON NAME	SPECIES	NUMBER OF CHROMOSOME PAIRS
Mosquito	*Culex pipiens*	3
Housefly	*Musca domestica*	6
Garden onion	*Allium cepa*	8
Toad	*Bufo americanus*	11
Rice	*Oryza sativa*	12
Frog	*Rana pipiens*	13
Alligator	*Alligator mississipiensis*	16
Cat	*Felis domesticus*	19
House mouse	*Mus musculus*	20
Rhesus monkey	*Macaca mulatta*	21
Wheat	*Triticum aestivum*	21
Human	*Homo sapiens*	23
Potato	*Solanum tuberosum*	24
Cattle	*Bos taurus*	30
Donkey	*Equus asinus*	31
Horse	*Equus caballus*	32
Dog	*Canis familiaris*	39
Chicken	*Gallus domesticus*	≈39
Carp	*Cyprinus carpio*	52

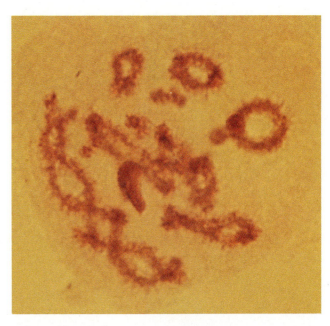

9.15 Chiasmata
Chiasmata—locations where segments of chromatids are being exchanged—are evident in several of these bivalents. Look for X-shaped attachments between chromatids; these attachments are chiasmata, some of which are seen near the middles of chromatids, and others near the ends.

The First Meiotic Division

Meiosis I is preceded by an interphase during which each chromosome is replicated, so that each chromosome then consists of two sister chromatids. There ensues a long **prophase I**, marked by a number of important changes. Very early in prophase I, the homologous chromosomes synapse; they are already tightly joined as soon as they can be clearly seen under the light microscope. Each joined pair of homologous chromosomes is known as a **bivalent**. Throughout prophase I and metaphase I, the chromatin continues to coil and compact progressively, so that the chromosomes appear ever thicker and smoother.

Partway through prophase I, the homologous chromosomes seem to *repel* each other, especially near the centromeres; but the bivalents are held together by regions in which there are physical attachments of some of the homologous chromosomes to each other (Figure 9.15). The regions having these attachments take on an X-shaped appearance and are called **chiasmata** (singular, **chiasma**, meaning "cross" in Greek). In fact, a chiasma reflects an exchange of material between chromatids on homologous chromosomes—what geneticists call crossing over (Figure 9.16). We shall have a great deal to say about crossing over and its genetic consequences in coming chapters. The chromosomes actually exchange material shortly after synapsis begins, but the chiasmata

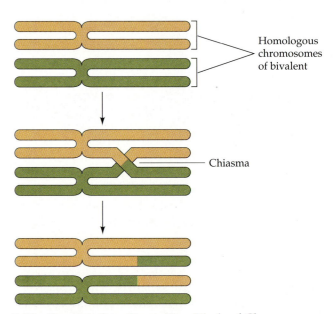

9.16 Crossing Over Forms New Kinds of Chromosomes
Early in prophase I, it often happens that two chromatids of different homologues cross over, break, and rejoin, so that each has some DNA from the other homologue. The products of crossing over are recombinant chromatids, which can have important genetic and evolutionary consequences.

do not become visible until later, during the period of repulsion of homologues.

Prophase I is followed by **prometaphase I**, during which the nuclear envelope and the nucleoli disappear. A spindle is formed, and microtubules become attached to the kinetochores of the chromosomes. In meiosis I, there is only one kinetochore per chromosome, not one per chromatid as in mitosis.

By **metaphase I** the kinetochores have become connected to the poles, all the bivalents have moved to the equatorial plate, and the bivalents are about to be pulled apart. Up to this point, they have been held together by chiasmata; it is this connection that provides the tension needed to stabilize the polar microtubules of the spindle. They separate in **anaphase I** when individual chromosomes, each still consisting of *two* chromatids, are pulled to the poles, one homologous chromosome of a pair going to one pole and the other homologue going to the opposite pole (Figure 9.14). (Note that this differs from the separation of *chromatids* during mitotic anaphase.) Each of the two daughter nuclei from this division contains only one set of chromosomes, compared to the two sets of chromosomes that were present in the original diploid nucleus. Each of these chromosomes, however, has twice the mass of a chromosome at the end of a mitotic division—because it consists of two chromatids rather than just one.

In some species, but not in others, there is a **telophase I**, with the reappearance of nuclear envelopes and so forth. When there is a telophase I, it is followed by an **interkinesis** phase similar to mitotic interphase. During interkinesis there is some, but not complete, uncoiling of the chromatin. There is no replication of the genetic material because each chromosome already consists of two chromatids. In contrast to mitotic interphase, the sister chromatids are generally not genetically identical, because crossing over in prophase I has scrambled the original chromatids to some degree.

The Second Meiotic Division

Meiosis II is similar to mitosis. The chromosomes in each nucleus produced by meiosis I line up at new equatorial plates in metaphase II; the chromatids, each having a centromere, separate; and new daughter chromosomes (consisting now of single chromatids) move to the poles in anaphase II (Figure 9.14). The only major differences between meiosis II and mitosis are these (Figure 9.17): First, DNA replicates before mitosis but not before meiosis II. Second, in mitosis, the chromatids making up a given chromosome are identical, whereas in meiosis II they may differ over part of their length as a result of crossing over in prophase of meiosis I. And third, the number of chromosomes on the equatorial plate of each of the two nuclei is n in meiosis II rather than $2n$ as in

the single mitotic nucleus. The final result of meiosis is four nuclei: Each is haploid, each has a single full set of chromosomes, and each set differs from the others in exact genetic composition. The differences, to repeat a very important point, result from crossing over during prophase I and from the separation of maternal and paternal chromosomes during anaphase I.

Synapsis, Reduction, and Diversity

Let us consider the consequences of synapsis and the separation of homologous chromosomes (Figure 9.17). In *mitosis*, each chromosome behaves independently; its two chromatids are sent to opposite poles at anaphase. If we start a mitotic division with x chromosomes, we end up with x in each daughter nucleus (i.e., each chromosome consisting—at this point—of one chromatid, or double-stranded molecule of DNA). In *meiosis*, things are very different. Synapsis organizes things so that chromosomes of maternal origin are paired with their paternal homologues. Then their separation during meiotic anaphase I assures that each pole receives one chromosome member from each pair of homologous chromosomes. (Remember that each chromosome still consists of *two* chromatids.) For example, at the end of meiosis I in humans, each daughter nucleus contains 23 out of the original 46 chromosomes—one member of each homologous pair. In this way, the chromosome number is decreased from diploid to haploid; in this way, too, meiosis I guarantees that each daughter nucleus gets a full set of chromosomes, for it must get one of each pair of homologous chromosomes.

Diversity among the products of meiosis I is achieved in two ways. First, synapsis during prophase I allows the maternal chromosome to interact with the paternal one; if there is crossing over, the recombinant chromatids contain some genetic material from each. Second, it is a matter of pure chance which member of a pair of chromosomes goes to which daughter cell at anaphase I. If there are two pairs of chromosomes in the diploid parent nucleus, a particular daughter nucleus could get paternal chromosome 1 and maternal chromosome 2, or paternal 2 and maternal 1, or both maternals, or both paternals. It all depends on the random way in which the bivalents line up at metaphase I. Note that of the four possible chromosome combinations just described, two produce daughter nuclei that are essentially the same as one of the parental types (except for any material exchanged by crossing over). You can see that the probability of getting back the original parental combinations decreases rapidly as the number of chromosome pairs increases; most species of diploid organisms do, indeed, have more than two pairs.

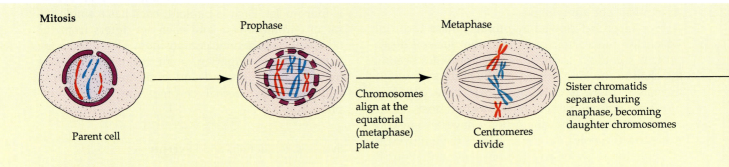

Mitosis

Parent cell → Prophase → *Chromosomes align at the equatorial (metaphase) plate* → Metaphase → *Centromeres divide* → *Sister chromatids separate during anaphase, becoming daughter chromosomes*

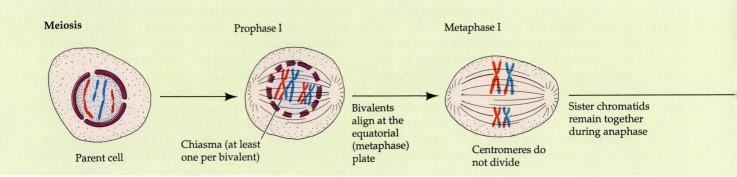

Meiosis

Parent cell → Prophase I — *Chiasma (at least one per bivalent)* → *Bivalents align at the equatorial (metaphase) plate* → Metaphase I — *Centromeres do not divide* → *Sister chromatids remain together during anaphase*

Meiotic Errors and Their Consequences

Occasionally a pair of homologous chromosomes fails to separate during the first meiotic division, or sister chromatids fail to separate during meiosis II or mitosis. This phenomenon is called **nondisjunction**, and it results in the production of **aneuploid** cells. Aneuploidy is a condition in which one or more chromosomes or pieces of chromosomes are either lacking or present in excess. If, for example, the chromosome 21 pair fails to separate during the formation of a human egg, so that both members of the pair go to one pole during anaphase I, then the resulting egg contains either two copies of chromosome 21 or none at all. If an egg with two of these chromosomes is fertilized by a normal sperm, the resulting zygote and infant has three copies of the chromosome—it is **trisomic** for chromosome 21 (Figure 9.18). As a result of carrying the extra chromosome 21, such a child demonstrates the symptoms of Down syndrome: impaired intelligence, characteristic abnormalities of the hands, tongue, and eyelids, and an increased susceptibility to diseases such as leukemia.

Other abnormal events may also lead to aneuploidy. A piece of a chromosome may break away and become attached to another chromosome, in a process called **translocation**. For example, a large part of one chromosome 21 may be translocated to another chromosome. Individuals who inherit this translocated piece along with two normal chromosomes 21 have Down syndrome.

A number of other human disorders also result from particular chromosomal abnormalities. Sex chromosome aneuploidy causes such disorders as Turner syndrome and Klinefelter syndrome, discussed in Chapter 10 in connection with sex determination. Deletion of a portion of chromosome 5 results in cri-du-chat syndrome, so named because the afflicted infant's cry sounds like that of a cat. This syndrome includes severe mental retardation.

Trisomies (and the corresponding monosomies) are surprisingly common in human zygotes, but only those embryos trisomic for chromosome 21 have a high probability of surviving to birth. Trisomies for chromosomes 13, 15, and 18 greatly reduce probabilities of surviving to birth and all lead to death before the age of one year; trisomies and monosomies for other chromosomes are lethal to the embryo. Perhaps four-fifths of all pregnancies self-terminate during the first two months, largely because of such trisomies and monosomies.

MITOSIS, MEIOSIS, AND PLOIDY

Both diploid and haploid nuclei of many organisms divide by mitosis. Multicellular diploid and multicellular haploid individuals form from single-celled beginnings by mitotic divisions. In diploid organisms, mitosis may proceed even when a chromosome from one of the haploid sets is missing or when there is an extra copy of one of the chromosomes (as in Down

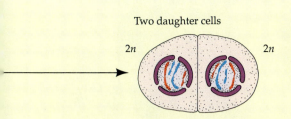

Two daughter cells

2n   2n

9.17   Mitosis and Meiosis Compared
Mitosis is a mechanism for constancy; the parent nucleus produces two identical daughter nuclei. Meiosis is a mechanism for diversity; the parent nucleus produces four daughter nuclei, each different from the parent nucleus and from its sister nuclei. The distinctive features of meiosis are synapsis and the failure of the centromeres to divide at the end of metaphase I.

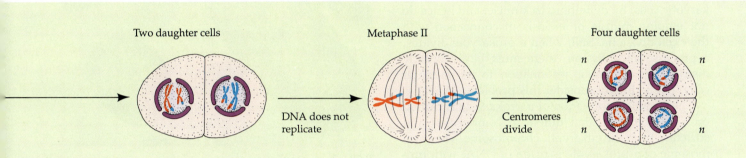

Two daughter cells   Metaphase II   Four daughter cells

n   n

DNA does not   Centromeres
replicate   divide   n   n

syndrome). If through accident or (in some organisms) design, the nucleus has one or more extra full sets of chromosomes, that is, if it is triploid (3n), tetraploid (4n), or of still higher order, this in itself does not prevent mitosis. In mitosis, each chromosome behaves independently of the others.

In meiosis, in contrast, chromosomes synapse to begin division. If an individual chromosome (let alone a full set) has no homologue, then anaphase I cannot send representatives of that chromosome to both poles. A diploid nucleus can undergo a normal meiosis; a haploid one cannot. In a tetraploid nucleus

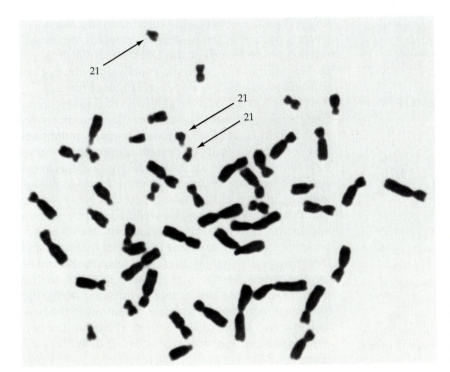

21

21
21

9.18   Chromosomes Showing Down Syndrome
The three copies of chromosome 21 in this spread of chromosomes from a cell in metaphase indicate the person from whom the cell was taken has Down syndrome. The number 21 chromosomes are labeled; you should be able to count 47 chromosomes in the spread.

there is an even number of each kind of chromosome, so all can join to become bivalents, but a triploid nucleus cannot undergo normal meiosis because one-third of the chromosomes would lack partners. This has important consequences for the fertility of triploid, tetraploid, and other chromosomally unusual organisms, which may be produced by plant breeding or by natural accidents.

CELL DIVISION IN PROKARYOTES

In this chapter we have considered the structures and events of the division of the *eukaryotic* nucleus. Prokaryotic cells, by definition, lack nuclei and hence do not employ mitosis or meiosis in connection with their cell divisions. Still, when a prokaryote divides, a process called **fission**, there must be an orderly distribution of genetic information to its daughter cells. Let us briefly consider this problem.

The genetic information of a prokaryote is carried on a chromosome that differs in composition and structure from the eukaryotic chromosome. The chromosome of a prokaryote is made of DNA, with protein components being only temporarily bound and not part of the long-term structure. In the bacterium *Escherichia coli*, the main chromosome is a single circular molecule of DNA about 1.6 million nm (1.6 mm)

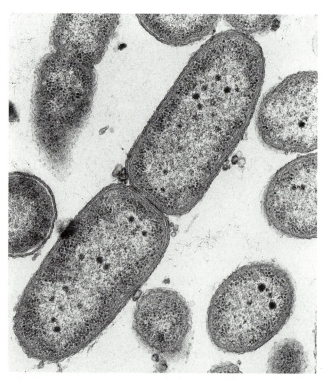

9.20 Bacterial Fission
These two cells of the bacterium *Pseudomonas aeruginosa* have almost completed fission. Plasma membranes have completely formed—separating the cytoplasm of one cell from that of the other—and only a small gap of cell wall remains to be completed. Each cell contains a complete chromosome in the light-toned nucleoid seen in the center of the cells.

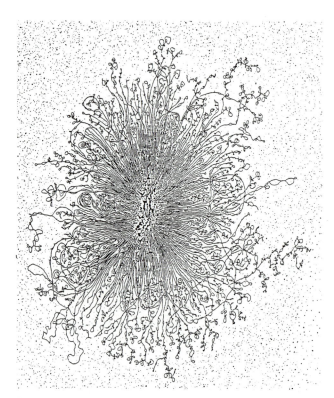

9.19 Circular Chromosomes
The long, looping fibers of DNA from a cell of the bacterium *Escherichia coli* are all part of one continuous circular chromosome.

long (Figure 9.19). The molecule is half a million times longer than it is thick. The bacterium itself is about 1 μm (1,000 nm) in diameter and about 4 μm long. Thus the space into which the long thread of DNA is packed in the bacterial nucleoid is very small relative to the length of the DNA molecule. It is not surprising that the molecule usually appears in electron micrographs as a hopeless tangle of fibers.

When bacterial cells are gently lysed (broken open, Chapter 5) to release their contents, the chromosome, as shown in Figure 9.19, sometimes comes untangled and spreads out to its full length. Several techniques have shown that the bacterial chromosome is a closed circle rather than the linear structure found in eukaryotes. Circular chromosomes are probably to be found in all prokaryotes, as well as in some viruses and in the chloroplasts and mitochondria of eukaryotic cells. The chromosome is attached to the plasma membrane; and when the new DNA molecule is formed from the old one, it, too, attaches to the membrane. As the cell elongates during growth, the two attachment points separate so that when the new wall and membrane material form at fission, the two chromosomes are included in separate daughter cells (Figure 9.20).

SUMMARY

Eukaryotic chromosomes are composed of DNA, some RNA, and proteins, including five histones and a large, diverse group of other proteins. A single chromatid contains one continuous DNA molecule, which at regular intervals wraps around aggregates of histones to form nucleosomes.

Mitosis is a nuclear division that forms two daughter nuclei identical to each other and to the original nucleus in chromosomal makeup and genetic constitution. Mitosis alternates with interphase in the cell cycle. Interphase includes a synthesis phase, in which the genetic and chromosomal material replicates between two gap phases.

During the prophase of mitosis, the chromatin becomes highly compacted into chromosomes, which attach to microtubules of the developing spindle apparatus during prometaphase. The centromeres gather at the equatorial plate in metaphase. Then the kinetochores separate, allowing the daughter chromosomes to move to opposite poles in anaphase. Shortening and movement of the spindle fibers brings about the chromosomal movement before metaphase and during anaphase. In telophase, the new daughter chromosomes begin to uncoil for the next interphase.

Meiosis is a pair of nuclear divisions with no intervening synthesis phase. It results in four haploid daughter nuclei that usually differ among themselves in genetic makeup. During the first meiotic division, homologous chromosomes synapse; the centromeres do not divide at anaphase I. The second meiotic division is almost identical to the mitosis of a haploid cell, except that the chromatids of a given chromosome differ somewhat in genetic composition as a result of crossing over in prophase I.

Mitosis and meiosis are nuclear divisions. Division of the rest of the cell is called cytokinesis. In most animal cells this is a pinching-off motion; in plant cells it involves the formation of a new cell wall.

In prokaryotes the chromosome replicates, and both old and new chromosomes are attached to the plasma membrane. Growth of the wall and membrane separate the chromosome attachment points, and the cell divides between the two attachment points.

In asexual reproduction, the offspring are genetically identical to the parent. The sexual life cycle includes fertilization and, at another stage, meiosis; in consequence, there is much genetic variation among the progeny.

SELF-QUIZ

1. Which of the following statements is *not* true of eukaryotic chromosomes?
   a. They sometimes consist of two chromatids.
   b. They sometimes consist of a single chromatid.
   c. They normally possess a single centromere.
   d. They consist of chromatin.
   e. They are always clearly visible as defined bodies under the light microscope.

2. Nucleosomes:
   a. are made of chromosomes.
   b. consist entirely of DNA.
   c. consist of DNA wound around a histone core.
   d. are present only during mitosis.
   e. are present only during interphase.

3. Which of the following statements is *not* true of the cell cycle?
   a. It consists of mitosis and interphase.
   b. The cell's DNA replicates during G1.
   c. A cell can remain in G1 for weeks or much longer.
   d. Most proteins are formed throughout all subphases of interphase.
   e. Histones are synthesized primarily during the S phase.

4. Which of the following statements is *not* true of mitosis?
   a. A single nucleus gives rise to two identical daughter nuclei.
   b. The daughter nuclei are genetically identical to the parent nucleus.
   c. The centromeres divide at the onset of anaphase.
   d. Homologous chromosomes synapse in prophase.
   e. Mitotic centers organize the microtubules of the spindle fibers.

5. Which of the following statements is true of cytokinesis?
   a. A cell plate is formed in cytokinesis of animal cells.
   b. Furrowing of the membrane initiates cytokinesis in plant cells.
   c. Cytokinesis generally follows immediately upon mitosis.
   d. Actin and myosin are important in cytokinesis in plant cells.
   e. Cytokinesis is the division of the nucleus.

6. In sexual reproduction:
   a. gametes are usually haploid.
   b. gametes are usually diploid.
   c. the zygote is usually haploid.
   d. the chromosome number is reduced during mitosis.
   e. spores are formed during fertilization.

7. In meiosis:
   a. meiosis II reduces the chromosome number from diploid to haploid.
   b. DNA replicates between meiosis I and II.
   c. the chromatids making up a chromosome in meiosis II are identical.
   d. each chromosome in prophase I consists of four chromatids.
   e. homologous chromosomes are separated from one another in anaphase I.

8. In meiosis:
   a. a single nucleus gives rise to two identical daughter nuclei.
   b. the daughter nuclei are genetically identical to the parent nucleus.
   c. the centromeres divide at the onset of anaphase I.
   d. homologous chromosomes synapse in prophase I.
   e. no spindle forms.

9. Which of the following statements is *not* true of aneuploidy?
   a. Aneuploidy results from chromosomal nondisjunction.
   b. Aneuploidy does not happen in humans.
   c. An individual with an extra chromosome is trisomic.
   d. Trisomies are common in human zygotes.
   e. A piece of one chromosome may translocate to another chromosome.

10. In prokaryotes:
   a. there are no meiotic divisions.
   b. mitosis proceeds as in eukaryotes.
   c. the genetic information is not carried in chromosomes.
   d. the chromosomes are identical to those of eukaryotes.
   e. cell division follows division of the nucleus.

FOR STUDY

1. How does a nucleus in the G2 phase of the cell cycle differ from one in the G1 phase?

2. What is a chromatid? When does a chromatid become a chromosome?

3. Compare and contrast mitosis (and subsequent cytokinesis) in animals and plants.

4. Suggest two ways in which one might, with the help of a microscope, determine the relative durations of the various phases of mitosis.

5. Contrast mitotic prophase and prophase I of meiosis. Contrast mitotic anaphase and anaphase I of meiosis.

READINGS

All introductory genetics texts contain chapters on meiosis and how this process distributes genetic information.

Alberts, B., D. Bray, J. Lewis, M. Raff, K. Roberts and J. D. Watson. 1989. *Molecular Biology of the Cell*, 2nd Edition. Garland Publishing, New York. An outstanding book in which to pursue the topics of this chapter in greater detail. Chapters 9, 13, and 15 have definitive modern treatments of the nucleus, mitosis, the cell cycle, meiosis, and more.

Mazia, D. 1961. "How Cells Divide." *Scientific American*, September. A classical description of mitosis by a leading researcher of cell division.

Mazia, D. 1974. "The Cell Cycle." *Scientific American*, January. This article discusses the four major stages of the cell cycle.

Mitchison, J. M. 1972. *The Biology of the Cell Cycle*. Cambridge University Press, New York. One of the few basic texts available on cell division and the cell cycle.

Prescott, D. M. 1976. *Reproduction of Eukaryotic Cells*. Academic Press, New York. A monographic treatment.

Russell, P. J. 1986. *Genetics*. Little, Brown, Boston. A good general treatment of genetics.

Sloboda, R. D. 1980. "The Role of Microtubules in Cell Structure and Cell Division." *American Scientist*, May/June. Includes a discussion of the spindle apparatus.

10

Mendelian Genetics and Beyond

PREVIEW: Gregor Mendel accurately observed patterns of inheritance and proposed a mechanism to account for some of the patterns. Individual traits are determined by genes. Given information about the genes carried by each parent, one may deduce the genetic makeup possible for the offspring, and vice versa. Various kinds of offspring appear in proportions that can be predicted from Mendel's laws. Other biologists showed that patterns of inheritance depend on the chromosomal locations of the genes under study.

This chapter deals with genes, chromosomes, monohybrid and dihybrid crosses, dominant and recessive traits, incomplete dominance, alleles, genotypes, phenotypes, homozygotes, heterozygotes, test crosses, segregation, assortment, linkage, crossing over, sex determination, sex linkage, cytogenetics, polygenes, and the interaction of genes.

We often use the term **Mendelian genetics** to refer to the most basic patterns of inheritance in sexually reproducing organisms with more than one chromosome (and with orderly meiosis). The term honors the Austrian monk Gregor Johann Mendel (1822–1884), the person who first made rigorous, quantitative observations of the patterns of inheritance and proposed plausible mechanisms to explain them. When Mendel began his work with the garden pea in his monastery garden (Figure 10.1), little was known about the sex lives of plants or the consequences of sexual reproduction for the inheritance of traits.

Some useful observations had been made in the late eighteenth century by the German botanist Joseph Gottlieb Kölreuter. He controlled the sexual reproduction of plants that interested him by cross-pollinating them. He attempted a great number of crosses between plants, produced many **hybrids** (the offspring of genetically different parents), and learned a great deal about the process of pollination. In some instances he confirmed the common observation that hybrids are intermediate between their parents with respect to obvious traits such as size, coloration, and flower shape; more important, he found and emphasized that in some cases the hybrids are *not* intermediate but closely resemble just one of the parents. In addition, he studied the offspring from **reciprocal crosses**. These are crosses made in

both directions; that is, in one set of crosses, males with one form of a trait, which we may represent by *A*, are crossed with females with another form of this same trait, *a*, while in a complementary set of crosses *a* males and *A* females serve as the parents. In an example of reciprocal crosses of plants, pollen (which carries the sperm) from a plant with trait *A* is placed on the female organ—from which the sperm can travel to the eggs—of a plant with trait *a* in one set of crosses. In the reciprocal cross, pollen from *a* plants is placed on the female organs of *A* plants. (In many plant species the same individuals have both male and female reproductive organs; each plant may then reproduce as a male, as a female, or as both—which makes such plants excellent material for genetic studies.) In Kölreuter's experience, reciprocal crosses always gave identical results.

MENDEL'S DISCOVERIES

This was essentially the state of knowledge when Mendel began his work. In one sense, the time was ripe for his discoveries, for it had recently been shown that one female gamete combines with one male gamete to bring about fertilization. On the other hand, the role of the chromosomes as bearers of genetic information was unknown, and mitosis and meiosis were yet to be discovered. Mendel himself

10.1 Mendel's Garden
In this monastery garden plot Gregor Mendel performed the experiments that led to his explanation of the patterns of heredity. Today the garden is planted with red and white begonias to illustrate Mendel's laws of genetics, although his classic work was done with peas.

was well qualified to make the big step forward. Although in 1850 he had failed an examination for a teaching certificate in natural science, he later undertook intensive studies in physics, chemistry, mathematics, and various aspects of biology at the University of Vienna. It is probable that his work in physics and mathematics led to his applying experimental and quantitative methods to the study of heredity—and these were the key ingredients in his success.

Mendel worked out the basic principles of the heredity of plants and animals over a period of about nine years, the work culminating in a public lecture in 1865 and a detailed written account in 1866. However, his theory was not accepted. In fact, it was ignored. Perhaps the chief difficulty was that the physical basis of his theory was not understood until the discovery of meiosis, some years later. The most prominent biologists at the time Mendel published simply were not in the habit of thinking in mathematical terms, even the simple ones used by Mendel. His paper on "plant hybridization" appeared in a journal that was received by 120 libraries, and he sent reprinted copies (of which he had obtained 40) to several distinguished scholars. We know that at least one of these scholars died years later without even having opened the pages of the Mendel reprint. Whatever the reasons, Mendel's pioneering paper had no discernible influence on the scientific world for more than 30 years.

Then in 1900, Mendel's discoveries burst into prominence as the result of independent experimentation by the Dutchman Hugo de Vries, the German Carl Correns, and the Austrian Erich von Tschermak. Each of them carried out crossing experiments and obtained quantitative data about the progeny; each published his principal findings in 1900; each cited Mendel's 1866 paper. At last the time was ripe for biologists to appreciate the significance of what these

four geneticists had discovered—largely because meiosis had by then been described. Let us now consider the general way in which Mendel went about his experimentation.

Mendel's Strategy

Mendel chose the garden pea for his most important work because of its ease of cultivation, the feasibility of controlled pollination, and the availability of varieties with differing traits. He controlled pollination by moving pollen from one plant to another; because he did this, he knew the parentage of the offspring in his experiments. The peas Mendel studied naturally self-pollinate if they are untouched—that is, the female organs of flowers receive pollen from the male organs of the same flowers—and he made use of this in some of his experiments.

Mendel began by examining numerous varieties of peas in order to see what heritable traits might be suitable for study. A suitable trait would be one that was "true-breeding." For example, peas with white flowers, when crossed with one another, would have to give rise only to progeny with white flowers, and tall plants bred to tall plants had to give only tall progeny, or they were not considered true-breeding. The suitable traits were also ones that had well-defined, contrasting alternatives that could be obtained in true-breeding form, such as purple flowers versus white flowers. For the bulk of his work, Mendel concentrated on seven pairs of contrasting traits, including purple versus white flowers, spherical versus dented seeds, tall versus dwarf stems, and yellow versus green seeds (Figure 10.2). Before performing a given cross, he made sure that each potential "parent" was from a true-breeding strain.

Mendel then placed pollen collected from one parental strain onto the stigma, or female organ, of a flower of the other parent. These plants were the parental generation, or **P**. In due course, seeds formed, and each seed was planted. The resulting new plants constituted the first filial generation, or F_1. Mendel and his assistants examined each F_1 plant to see which traits it bore and then recorded the *number* of F_1 plants expressing each trait. In some experiments the F_1 plants were allowed to self-pollinate and thus produce a second filial, or F_2, generation. Again, each F_2 plant was characterized and counted. Mendel performed other crosses in which the F_2 was produced by crossing F_1 hybrids with one of the true-breeding parental strains. Always, each type of progeny was counted and scored. This attention to quantitative detail was a unique advance in experimental biology; it allowed Mendel to make numerical comparisons and ultimately to develop a hypothesis, or model—a proposed explanation for the numbers he observed. In sum, Mendel devised a well-organized plan of research, pursued it faithfully

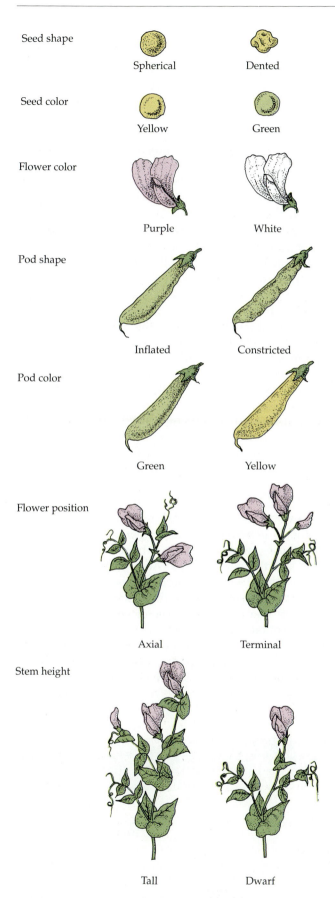

10.2 Inherited Traits Studied by Mendel
Mendel's work on the genetics of peas focused on these seven pairs of traits. He isolated each as a true-breeding trait before he began his studies of various crosses.

and carefully, recorded great amounts of quantitative data, and analyzed the numbers he recorded so as to explain the relative proportions of the different kinds of progeny. His 1866 paper stands to this day as a model of clarity. Let us look at some of his results and the conclusions to which they led.

Experiment 1

"Experiment 1" in Mendel's paper was a **monohybrid cross**, that is, one in which the parents differed for a single trait. He took pollen from plants of a true-breeding strain with dented seeds and placed it on the stigmas of flowers of a true-breeding, spherical-seeded strain. He also performed the reciprocal cross, placing pollen from the spherical-seeded strain on the stigmas of flowers of the dented-seeded strain. In both cases, all the seeds that were produced were spherical—it was as if the dented trait had disappeared completely (Figure 10.3). The following spring Mendel grew out 253 F_1 plants from these spherical seeds, each of which was allowed to self-pollinate to produce F_2 seeds. In all, there were 7,324 F_2 seeds, of which 5,474 were spherical and 1,850 dented.

Mendel observed that the spherical seed trait was **dominant**, being expressed over the dented seed trait, which he called **recessive**. In each of the other six pairs of traits studied by Mendel, one proved to be dominant over the other. When he crossed plants differing for any of these traits, only one of each pair of traits was evident in the F_1 generation. However,

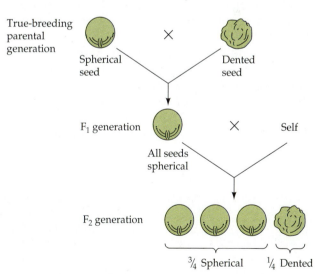

10.3 Mendel's Experiment 1
Mendel crossed plants of a true-breeding, spherical-seeded variety with plants of a true-breeding variety having dented seeds. All the seeds in the resulting F_1 generation were spherical. Mendel then grew plants from these spherical seeds and let them self-pollinate to form a second filial (F_2) generation. About ¼ of the F_2 seeds were dented and ¾ were spherical. The pattern was the same regardless of which parent plant contributed the pollen in the parental generation.

the trait that was not seen in the F_1 reappeared in the F_2. Most important, the ratio of the two traits in the F_2 was always the same: approximately 3:1, that is, ¾ of the F_2 showed the dominant trait and ¼ showed the recessive trait (Table 10.1). In his Experiment 1, the ratio was 5474:1850 = 2.96:1. Reciprocal crosses in the parental generation gave similar outcomes in the F_2.

Terminology for Mendelian Genetics

By themselves the results from Experiment 1 disproved the widely believed theory that inheritance is a "blending" phenomenon. According to the blending theory, Mendel's F_1 seeds should have had an appearance intermediate between those of the two parents, being only slightly dented. Furthermore, the blending theory offers no explanation for the reappearance of the dented trait in the F_2 seeds after its apparent absence in the F_1 seeds. Instead Mendel proposed a **particulate theory**, in which the hereditary carriers are present as discrete units that retain their integrity in the presence of other units. As Mendel saw it, each pea has two such units for each character, one derived from each parent. Each gamete contains one unit, and the resulting zygote (and each cell of the adult that develops from it) contains two. The "unit" is now called a **gene**. Mendel reasoned that in his Experiment 1, the spherical-seeded parent had a pair of genes of the same type, which we shall call S, and the parent with dented seeds had two s genes. The SS parent produced gametes each containing a single S, and the ss parent produced gametes each with a single s. Each member of the F_1 generation had an S from one parent and an s from the other, and an F_1 could thus be described as Ss. We say that S is dominant over s because s is not evident when both genes are present.

The physical appearance of a character is its **phenotype**, which Mendel correctly supposed to be the result of the **genotype**, or genetic constitution, of the organism showing the phenotype. In Experiment 1 we are dealing with two phenotypes (spherical seeds and dented seeds) and three genotypes: The dented-seed phenotype is produced only by the genotype ss, whereas the spherical-seed phenotype may be produced by either SS or Ss. The different forms of a gene (S and s in this case) are called **alleles** of one another. Individuals that breed true for a character contain two copies of the same allele. For example, a strain of true-breeding peas with dented seeds must have the genotype ss—if S were present, the plants would produce spherical seeds. We say individuals that produce dented seeds are **homozygous** for the allele s, meaning that they have two copies of the same allele. Some peas with spherical seeds are homozygous; they are the ones with the genotype SS. However, other spherical-seeded plants are **heterozygous**, having two different alleles of the gene in question; these plants have the genotype Ss. To illustrate these terms with a more complex example, one in which there are three gene pairs, an individual with the genotype $AABbcc$ is homozygous for two genes—in having two A alleles and two c alleles—but heterozygous for the gene with alleles B and b. An individual that is homozygous for a character is sometimes called a homozygote; a heterozygote is heterozygous for the character in question.

Segregation of Alleles

How does Mendel's model explain the composition of the F_2 generation in his Experiment 1? Consider first the F_1, which has the spherical-seeded phenotype and the genotype Ss. According to the model, when any F_1 individual produces gametes, the alleles **segregate**, or separate, so that each gamete receives only *one* member of the pair of genes. Half the gametes contain the S allele and half the s allele. The random combination of these gametes produces the F_2 generation (Figure 10.4). Three different F_2 genotypes are possible: SS, Ss (which is the same thing as

TABLE 10.1
Mendel's Results from Monohybrid Crosses

P			F_2			
DOMINANT	×	RECESSIVE	DOMINANT	RECESSIVE	TOTAL	RATIO
Spherical	×	Dented seeds	5,474	1,850	7,324	2.96:1
Yellow	×	Green seeds	6,022	2,001	8,023	3.01:1
Purple	×	White flowers	705	224	929	3.15:1
Inflated	×	Constricted pods	882	299	1,181	2.95:1
Green	×	Yellow pods	428	152	580	2.82:1
Axial	×	Terminal flowers	651	207	858	3.14:1
Tall	×	Dwarf stems	787	277	1,064	2.84:1

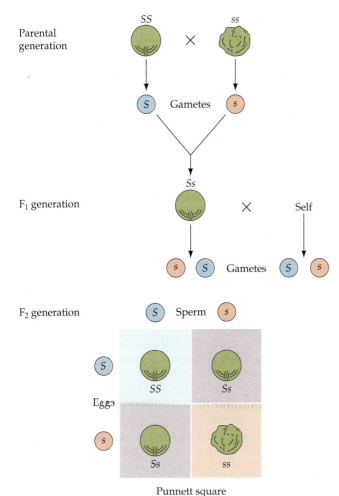

Parental generation

SS × *ss*

S Gametes *s*

F₁ generation

Ss × Self

s *S* Gametes *S* *s*

F₂ generation

S Sperm *s*

Eggs

S / *SS* / *Ss*

s / *Ss* / *ss*

Punnett square

10.4 Mendel's Explanation of Experiment 1
Mendel imagined that heredity depends on factors from each parent, and these factors do not blend in the offspring. The drawing shows a modern version of Mendel's explanation of the experiment in Figure 10.3. A parent homozygous for the allele for spherical seeds is crossed with a parent homozygous for the allele for dented seeds. Each parent makes gametes of only one kind, either *S* or *s*, and these combine at fertilization to form plants that all have the genotype *Ss* and the spherical-seeded phenotype. When the F₁ plants self-pollinate they produce two kinds of eggs, *S* and *s*, and the same two types of male sex cells. These combine randomly in four different ways to form F₂ plants, as shown in the box at the bottom of the figure. Three of these combinations produce genotypes that determine the spherical-seeded phenotype and one produces the genotype for the dented-seed phenotype so that the observed ratio is 3:1. The box in which the F₂ plants are displayed, called a Punnett square, is a convenient device for keeping track of all the ways gametes can combine at fertilization.

sS), and *ss*. Our quantitative way of looking at things may lead us to wonder about what proportions of these genotypes we might expect to observe in the F₂ progeny. The expected frequencies of these three genotypes in our example may be determined in either of two ways: by using the "Punnett square" devised in 1905 by the British geneticist Reginald

Crundall Punnett or by using simple probability calculations. The Punnett square is illustrated in Figure 10.4 and other figures in this chapter, and probabilities are discussed in Box 10.A. By either method (they amount to the same thing), it becomes apparent that self-pollination of the F₁ genotype *Ss* will give the three F₂ genotypes in the expected ratio 1 *SS*:2 *Ss*:1 *ss*. Because *S* is dominant and *s* recessive, the ratio of *phenotypes* is 3 spherical (*SS* and *Ss*) to 1 dented (*ss*), just as observed by Mendel.

Mendel did not live to see his theory placed on a sound physical footing based on chromosomes and DNA. Genes are now known to be portions of the DNA molecules in chromosomes. More specifically, a gene is a portion of the DNA that resides at a particular **locus**, or site, within the chromosome and that encodes a particular function. Remember that the cells in a multicellular organism have the same genotype because they are all derived by mitosis from a single cell, the zygote (Chapter 9). Each diploid cell has two homologous chromosomes of each type, and therefore has two alleles at each locus. Consistent with Mendel's model, if the two homologous chromosomes have copies of the same allele at a given locus, the cell and the organism are homozygous at that locus; if the homologues have differing alleles, the cell and the organism are heterozygous at that locus. Because meiosis reduces the number of chromosomes per cell, each gamete contains only one member of each homologous pair of chromosomes and, hence, only one allele at any given locus. If you visualize *S* and *s* as occupying specific, homologous sites on a pair of homologous chromosomes in an F₁ individual, you will see how they would be inherited through successive generations, by way of meiosis and the random fusion of gametes.

On the basis of monohybrid crosses such as that of Experiment 1, Mendel proposed his first law, called the law of segregation, which says that alleles segregate from one another during the formation of gametes.

The Test Cross

Mendel's theory was fully adequate to explain the ratios of phenotypes observed in F₁ and F₂ generations obtained from crosses of differing, true-breeding strains. To be regarded as fully satisfactory, the theory must also be able to predict—accurately—the outcome of other kinds of experiments. One such challenge was posed by Mendel himself. According to his theory, ⅔ of the F₂ spherical seeds from his Experiment 1 should be heterozygous, each carrying both *S* and *s* alleles. Therefore, if all the spherical seeds were allowed to grow into F₂ adults, and these adults were self-pollinated, then ⅔ of those F₂ plants would be heterozygous, and would thus produce seeds of which about ¾ would be spherical and ¼

BOX 10.A

Elements of Probability

Many people find it easiest to solve genetics problems by means of probability calculations, perhaps because the basic underlying considerations are a familiar part of daily life. When we flip a coin, for example, we expect it to have an equal probability of landing "heads" or "tails." When we roll an honest die, we expect to have equal chances of getting any of the numbers from 1 to 6. We properly bet more money on a coin's giving at least one heads in a pair of tosses than we do on it coming up heads in a single toss—yet if we are even slightly sophisticated, we recognize that on a given toss, the probability of heads is independent of what happened on all the previous tosses. (For

an honest coin, a run of 10 straight heads implies nothing about the next toss. No "law of averages" increases the likelihood that the next toss will come up tails, and no "momentum" makes an eleventh occurrence of heads any more likely. On the eleventh toss, the odds are still 50:50.)

The basic conventions of probability are simple: If an event is absolutely certain to happen, its probability is 1. If it cannot happen, its probability is 0. Otherwise, its probability lies between 0 and 1. A coin toss results in heads half the time, and the probability of heads is ½— as is the probability of tails. If *two* coins (a penny and a dime, say) are tossed, each acts independently of the other. What, then, is the probability of both coins coming up heads? Half the time, the penny comes up heads; of that fraction, half the time the dime also comes up heads. Therefore, the joint probability of two heads is half of one-half, or ½ × ½ = ¼. To find the joint probability of *independent* events, *multiply* the probabilities of the individual events.

To apply this to the monohybrid cross we have been discussing, we

need only deal with gamete formation and random fertilization. A homozygote can produce only one type of gamete, so, for example, an *SS* individual has a probability equal to 1 of producing gametes with the genotype *S*. The heterozygote *Ss* produces *S* gametes with a probability of ½, and *s* gametes with a probability of ½ as well. Consider, now, the F₂ progeny of the cross of Figure 10.4. They are obtained by self-pollinating F₁ hybrids of genotype *Ss*. The probability that an F₂ plant is *SS* must be ½ × ½ = ¼—there is a 50:50 chance of the sperm's being *S*, and this is independent of the 50:50 chance of the egg's being *S*. Similarly, the probability of *ss* offspring is ½ × ½ = ¼. The probability of getting *S* from the sperm and *s* from the egg is also ¼, but the same genotype can also result from *s* in the sperm and *S* in the egg, with a probability of ¼. Thus the probability that an F₂ plant is a heterozygote is ¼ + ¼ = ½. All three of the genotypes are expected in the ratio ¼ *SS*:½ *Ss*:¼ *ss*—hence the 1:2:1 ratio of genotypes and the 3:1 ratio of phenotypes seen in Figure 10.4.

---

dented. The other ⅓ of the F₂ plants, being *SS* homozygotes, would produce only spherical seeds (Figure 10.5). This, in fact, is what Mendel observed.

Another way to test whether a given individual showing a dominant trait is homozygous or heterozygous is by means of a **test cross**. In a test cross, the individual in question is crossed with an individual known to be homozygous for the recessive trait—an easy individual to identify because its phenotype is the recessive one. For the gene that we have been considering, the recessive homozygote for the test cross is *ss*. The individual being tested may be described initially as *S*− because we do not yet know the identity of the second allele. If the individual being tested is homozygous dominant (here, *SS*), all offspring of the test cross will be *Ss* and show the dominant character (spherical seeds). If, however, the tested individual is heterozygous (*Ss*), then approximately ½ of the offspring of the test cross will show the dominant trait; but the other ½ will be homozygous recessive (Figure 10.6). These are exactly

the results that are obtained; thus Mendel's model predicts accurately the results of such test crosses.

Independent Assortment of Alleles

What happens if a cross is made between two parents that differ at two or more loci? When a double heterozygote (for example, *AaBb*) makes gametes, do the alleles of maternal origin go together to one gamete and those of paternal origin to another gamete? Are new associations formed at the time of gamete formation? To answer these questions Mendel performed a series of **dihybrid crosses**: crosses made between parents differing for two independent traits.

In the dihybrid crosses Mendel used peas that differed for two characters of the seeds: One true-breeding strain produced only spherical, yellow seeds and the other strain produced only dented, green ones. Plants of the first strain can be designated *SSYY*, indicating that they are homozygous both for the *S* allele at the seed-shape locus and for the *Y*

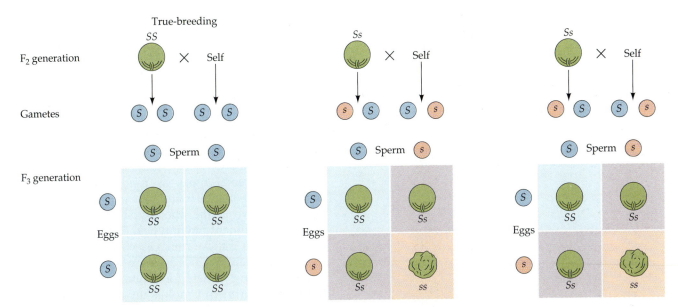

10.5 An F₃ Generation

According to Mendel's ideas, the spherical-seeded F₂ plants in Figure 10.4 are of two kinds. One-third of them (the *SS* homozygotes), upon self-pollination, will produce only spherical-seeded offspring. Two-thirds of them (the *Ss* heterozygotes) will, like the original F₁ plants in Figure 10.3, produce spherical- and dented-seeded plants in a 3:1 ratio.

allele at the seed-color locus. The second doubly homozygous strain is *ssyy*. The doubly heterozygous offspring from a cross between these two strains would be expected to be uniformly *SsYy*. Because the *S* and *Y* alleles are dominant, these F₁ seeds would all be yellow and spherical.

There are two ways in which these doubly heterozygous plants might produce gametes, as Mendel saw it—remember that he had never heard of chromosomes, let alone of meiosis. First, the alleles maintained the associations they had in the original parents, then only two types of gametes would be produced: *SY* and *sy*. Second, the F₂ progeny resulting from self-pollination of the F₁ would consist of three times as many plants bearing spherical, yellow seeds as ones with dented, green seeds. Were such results to be obtained, there would be no reason to suppose that seed shape and seed color were really regulated by two different genes, because spherical seeds would always be yellow, and dented ones green.

The other possibility is that the segregation of *S* from *s* is *independent* of the segregation of *Y* from *y* during the production of gametes. In this case, four kinds of gametes would be produced, and in equal numbers: *SY*, *Sy*, *sY*, and *sy*. When these gametes combined at random, they would produce an F₂ of nine different genotypes (Figure 10.7). The progeny can have any of three possible genotypes for shape (*SS*, *Ss*, or *ss*) and any of three for color (*YY*, *Yy*, or

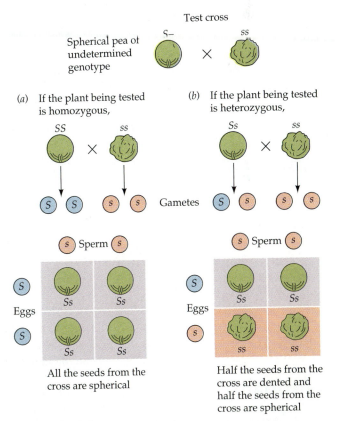

10.6 A Test Cross

A plant with a dominant phenotype may be homozygous or heterozygous. Its genotype can be deduced by crossing it with a homozygous recessive plant—by making a test cross. (*a*) If all the progeny of the test cross show the dominant phenotype, the plant being tested must have been homozygous for the dominant allele. (*b*) If half the progeny of the test cross have the dominant phenotype and half have the recessive phenotype, the plant in question must have been heterozygous.

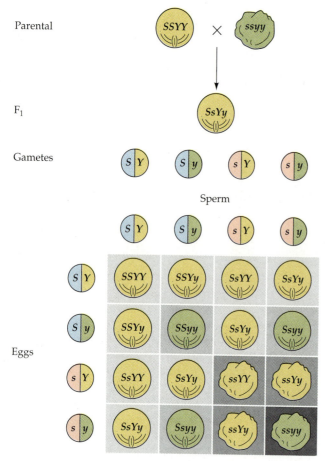

Parental

F₁

Gametes

Sperm

Eggs

10.7 Independent Assortment
The upper edges of this Punnett square show the four kinds of gametes formed by plants heterozygous for two genes. Random combination of these gametes produces equal numbers of the 4 × 4 = 16 combinations displayed in the boxes. Because *S* and *Y* are dominant over *s* and *y*, respectively, these genotypes determine four phenotypes (shown by the differently colored boxes) in the ratio of 9:3:3:1.

yy). These nine genotypes would produce just four phenotypes (spherical, yellow; spherical, green; dented, yellow; dented, green). By using either a Punnett square or simple probability calculations (Figure 10.7 and Box 10.B), we can show that these four phenotypes would be expected to occur in a ratio of 9:3:3:1.

Mendel's actual results were as predicted by the second model. Four different phenotypes appeared in a ratio of about 9:3:3:1 in the F₂, rather than only the two parental types as predicted by the first model. The parental traits appeared in new combinations in two of the phenotypic classes; these are called **recombinant phenotypes**. Accordingly, he formulated what is now known as Mendel's second law: Alleles of different genes **assort independently** of one another during gamete formation. This law of independent assortment is not as universal as the law of segregation. It applies to genes that lie on separate chromosomes, but not to those that lie on the same chromosome. It is, however, correct to say that *chromosomes* assort independently during the formation of gametes.

GENETICS AFTER MENDEL: ALLELES AND THEIR INTERACTIONS

Incomplete Dominance and Codominance

With some genes, dominance of one allele over another is not observed. Instead, the heterozygotes show an intermediate phenotype superficially like that predicted by the old "blending" theory of inheritance. For example, if a true-breeding red snapdragon is crossed with a true-breeding white one, all the F₁ are pink (Figure 10.8). That this is still explainable in terms of Mendelian genes, rather than of a blending theory, is readily demonstrated by a further cross.

BOX 10.B

Probabilities in the Dihybrid Cross

F₂ plants in the example we have been discussing express four phenotypes. The proportions of these phenotypes are easily determined by probabilities. The probability of a seed's being yellow is ¾ (Box 10.A); by the same reasoning, the probability of a seed's being spherical is also ¾. The two traits are determined by separate genes and are independent of one another, so the joint probability of a seed being both yellow and spherical is ¾ × ¾ = ⁹⁄₁₆. For the dented, yellow members of the F₂, the probability of yellow is again ¾; the probability of dented seeds is ½ × ½ = ¼. The joint probability of a seed being both yellow and dented is, then, ¾ × ¼ = ³⁄₁₆. The same probability applies, for analogous reasons, to the spherical, green F₂ seeds. Finally, the probability of F₂ seeds being both dented and green must be ¼ × ¼ = ¹⁄₁₆. Looking at all four phenotypes, we see that they are expected in the ratio of 9:3:3:1.

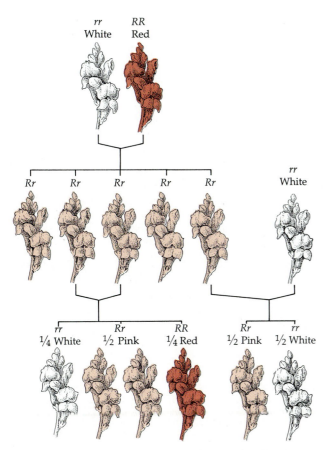

10.8 Incomplete Dominance Follows Mendel's Laws
Being heterozygous makes snapdragon flowers pink because the allele for red flowers is incompletely dominant to the allele for white ones. When true-breeding red and white parents cross, all plants in the F_1 generation are pink. When these F_1 plants self-pollinate, they produce F_2 offspring that are white, pink, and red in a ratio of 1:2:1. A test cross, diagrammed at the right, confirms that pink snapdragons are heterozygous; see Figure 10.6 for the reasoning.

If one of these pink F_1 snapdragons is crossed with a true-breeding white one, the blending theory predicts that all the offspring would be a still-lighter pink. In fact, approximately ½ of the offspring are white and ½ the same pink as the original F_1. Suppose now that the F_1 pink snapdragons are self-pollinated. The resulting F_2 are distributed in a ratio of 1 red:2 pink:1 white. Clearly the hereditary particles—the genes—have not blended, but are readily sorted out in their original forms.

All these results are readily understood in terms of the Mendelian model. The only change required is the recognition that, in cases like this, the heterozygotes show a phenotype intermediate between those of the two homozygotes. Genes code for the production of specific proteins, many of which are enzymes. Different alleles at a locus code for alternative forms of a protein that differ in structure and, when the protein is an enzyme, often have different degrees of catalytic activity. In the snapdragon ex-

ample, one allele codes for an enzyme that catalyzes a reaction leading to the formation of a red pigment in the flowers. The alternative allele codes for an altered protein lacking catalytic activity for pigment production. Plants homozygous for this alternative allele cannot synthesize red pigment, and their flowers are white. Heterozygous plants, with only one allele for the functional enzyme, produce just enough red pigment so that their flowers are pink. Homozygous plants with two alleles for the functional enzyme produce more red pigment and they have red flowers.

When the heterozygous phenotype is intermediate, as in this example, we say that the gene is governed by **incomplete dominance**. Examples of incomplete dominance are more common than those of complete dominance. It is an unusual feature of Mendel's report that all seven of the examples he described are characterized by complete dominance. In order for dominance to be complete, a single copy of the dominant allele must produce enough of its protein product to give the maximum phenotypic response. For example, just one copy of the dominant allele T at one of the loci studied by Mendel leads to the production of enough of a growth-promoting chemical so that the Tt heterozygotes are as tall as the homozygous dominant plants (TT)—the second copy of T causes no further growth of the stem. The homozygous recessive plants (tt) are much shorter because the allele t does not lead to the production of the growth promoter.

Sometimes two alleles at a locus produce different phenotypic effects, *both* of which appear in heterozygotes (Figure 10.9). This phenomenon is called **codominance**.

Pleiotropy

A single allele may have more than one distinguishable phenotypic effect; that is, the allele may be **pleiotropic**. The most familiar example of pleiotropy is the

10.9 Codominance in White Clover
White clover leaves have characteristic patterns of "chevrons" and colored areas, all genetically determined. The leaves on the left are homozygotes of genotype V^hV^h; those on the right are homozygotes of genotype V^fV^f. V^hV^f heterozygotes show both the chevron that is characteristic of V^h and the colored area that is characteristic of V^f; the V^h and V^f alleles are codominant.

allele responsible for the coloration pattern (light body, darker extremities) of Siamese cats, discussed later in this chapter. The same allele is also responsible for the characteristic crossed eyes of Siamese cats. Although these effects would appear to be unrelated, both result from the same protein produced under the influence of that allele.

The Origin of Alleles: Mutation

Why are there different alleles of a gene? The reason different alleles exist is that *any* gene is subject to the process of **mutation**, which means that it can be changed to some *stable, heritable* new form. In other words, an allele can mutate to become a different allele. One particular allele of a gene may be defined as **wild-type**, or standard, because it is present in most individuals and gives rise to an expected trait or phenotype. Other forms of that same gene, often

TABLE 10.2
Multiple Alleles for Eye Color in *Drosophila melanogaster*

GENOTYPE	PHENOTYPE	DEGREE OF PIGMENTATION OF THE EYE
w^+w^+	wild-type (dull red)	0.6800
$w^{col}w^{col}$	colored	0.1636
$w^{sat}w^{sat}$	satsuma	0.1404
w^ww^{col}		0.1114
$w^{co}w^{co}$	coral	0.0798
w^ww^w	wine	0.0650
$w^{a3}w^{a3}$	apricot-3	0.0632
$w^{ch}w^{ch}$	cherry	0.0410
w^ew^e	eosin	0.0324
$w^{bl}w^{bl}$	blood	0.0310
w^aw^a	apricot	0.0197
w^tw^t	tinged	0.0062
ww	white	0.0044

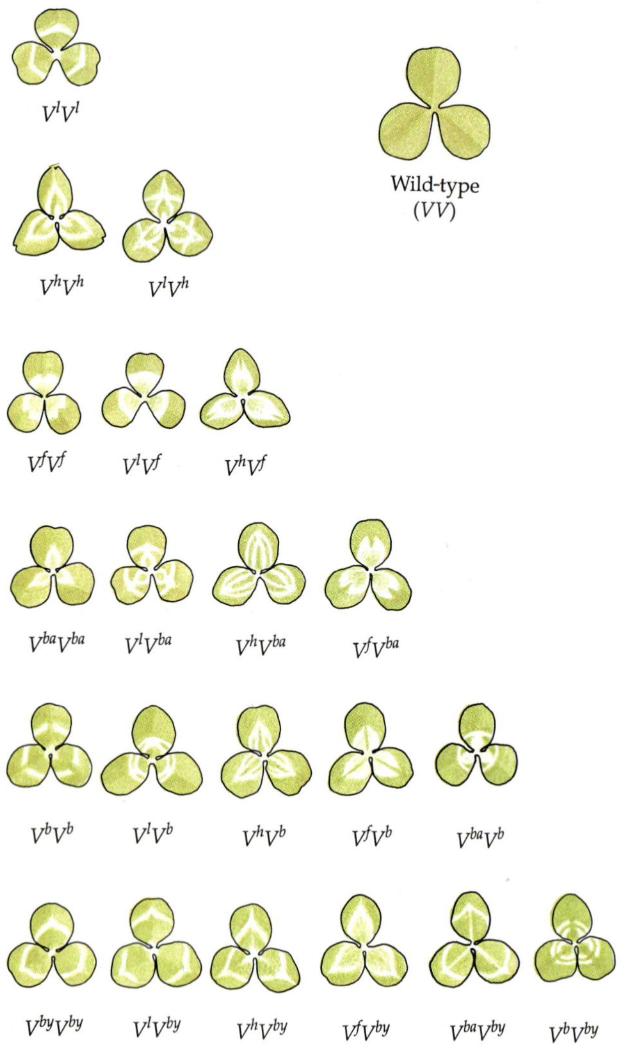

called mutant alleles, may alter the function of the gene somewhat and may produce a different phenotype. The wild-type and mutant alleles reside at the same locus and are inherited according to the rules set forth by Mendel.

Multiple Alleles

Mutation, to be discussed in Chapter 11, is a random process; different copies of the same gene may be changed in a number of different ways, depending upon how and exactly where the DNA changes. This implies that there may be more than two alleles of a given gene in a group of individuals. (Any one individual has only two alleles, of course.) In fact, there are many examples of such **multiple alleles**. Some clover leaves are plain green, while others have chevrons of other colors on their leaves. Seven alleles at a locus control the pattern of chevrons on the leaves of white clover (Figure 10.10). In the fruit fly *Drosophila melanogaster*, a large number of alleles at a locus affect eye color by determining the amount of pigment produced (Table 10.2). The exact color of the fly's eyes depends on which two alleles are inherited.

The ABO blood group system in humans is determined by a set of three alleles (I^A, I^B, and i) at one locus. Different combinations of these alleles in different people produce four different blood types, or phenotypes: A, B, AB, and O (Table 10.3). Early attempts at blood transfusion—made before these blood types were understood—often killed the patient. Around the turn of the century, the Austrian scientist Karl Landsteiner mixed blood cells and serum (which is blood from which cells have been removed) from different individuals. He found that

10.10 Multiple Alleles in White Clover
Seven alleles at the same locus determine the pattern of chevrons and colored areas on white clover leaves. Many of these alleles show codominance in heterozygotes.

TABLE 10.3
The ABO Blood System

BLOOD TYPE	GENOTYPE	REACTION WITH ANTI-A SERUM	REACTION WITH ANTI-B SERUM	TYPE OF DONOR BLOOD ACCEPTED
A	$I^A I^A$ or $I^A I^O$	Clumping of red blood cells	No clumping	A or O
B	$I^B I^B$ or $I^B I^O$	No clumping	Clumping of red blood cells	B or O
AB	$I^A I^B$	Clumping of red blood cells	Clumping of red blood cells	A, B, AB, or O
O	$I^O I^O$	No clumping	No clumping	O

only certain combinations of blood types are compatible. In other combinations, the red blood cells form clumps because of the presence in the serum of specific proteins, called antibodies (Chapter 15), that react with foreign, or non-"self," cells and macromolecules (Figure 10.11). When transfusions are given, it is best to have a perfect matchup of blood types in donor and patient. For example, if the patient has red blood cells of type A, then the blood donor should have type A cells as well. Certain combinations other than perfect matchups are usually also successful, as indicated in Table 10.3.

The Rh factor, so named because it was first found in rhesus monkeys, is another substance on the surface of red blood cells. In most human populations, almost 100 percent of the individuals have the Rh factor, and their blood is said to be Rh^+ (Rh-positive). Among Caucasians, however, only 83 percent are Rh^+; the others lack the Rh factor, and their blood is called Rh^- (Rh-negative). Like the A and B blood types, the Rh factor is genetically determined. A single locus with at least eight multiple alleles is responsible. Certain dominant alleles cause the production of the Rh factor, and Rh^- individuals are homozygous recessives.

Yet another system of multiple alleles, in a scallop, is illustrated in Study Question 1 at the end of the chapter. The question of how differing alleles may be maintained in a population through time will be examined in Chapter 19.

FOCUS ON CHROMOSOMES

Linkage

In the immediate aftermath of the rediscovery of Mendel's laws, it was considered that the second law—independent assortment—was of general applicability. However, some investigators, including Punnett (the inventor of the square), began to observe strange deviations from the expected 9:3:3:1 ratio in some dihybrid crosses. In particular, they observed an apparent excess of parental phenotypes and a shortage of recombinant phenotypes in some of the F₂s. Suppose that the original cross was between the genotypes *AABB* and *aabb*. If alleles at the *A* locus assorted independently of alleles at the *B* locus, the F₂ should consist of 9/16 individuals with the dominant phenotypes for *A* and *B*, 3/16 individuals dominant only for *A* (*A*–*bb*), 3/16 individuals dominant only for *B* (*aaB*–), and 1/16 double recessive homozygotes (*aabb*). What Punnett and others observed, instead, were large excesses of *aabb* over the 1/16 expected.

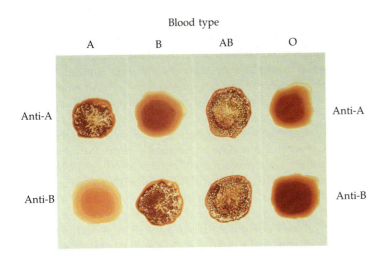

Blood type

A B AB O

Anti-A

Anti-B

10.11 ABO Blood Reactions
Cells of blood types A, B, AB, and O were mixed with anti-A or anti-B antibodies. When antibodies react with red blood cells, they cause them to clump (speckled appearance in photograph); cells that do not react with antibody remain evenly dispersed. Note that anti-A reacts with A and AB cells but not with B or O. Which blood types do anti-B antibodies react with? Note by looking down the columns that each of the types, when mixed separately with anti-A and with anti-B, gives a unique pair of results; this is the basic method by which blood is typed.

These results become understandable when we assume that the two loci are on the *same chromosome*, that is, linked together. After all, since the number of genes in a cell far exceeds the number of chromosomes, each chromosome must contain many genes. Suppose, now, that the *A* and *B* loci are on the same chromosome (Figure 10.12). To remind ourselves that the genes are linked in this fashion, let us write the genotypes differently: One parent is $\overline{AB}\,\overline{AB}$ and the other $\overline{ab}\,\overline{ab}$. The former produces gametes that are of one type, $\overline{AB}$; and the latter produces $\overline{ab}$ gametes. Thus the genotype of the F_1 is $\overline{AB}\,\overline{ab}$. Now, the key difference between dihybrid crosses with linkage and those without is in the formation of gametes by the F_1. Without linkage, as we have seen, four types of gametes are produced in equal frequency (*AB*, *Ab*, *aB*, and *ab*). *With* linkage, however, most of the gametes must be either $\overline{AB}$ or $\overline{ab}$ because the two loci are physically "tied together" on the same chromosome—they are part of the same DNA molecule. (As we will see, a few $\overline{Ab}$ and $\overline{aB}$ gametes will appear also, as a result of crossing over between loci; see Chapter 9.)

The full set of loci found on a given chromosome constitutes a **linkage group**. The number of linkage groups in a species, determined by experiments such as the dihybrid cross described here, should equal the number of homologous chromosome pairs, determined by microscopic examination of nuclei undergoing meiosis or mitosis.

Sex Determination

In Kölreuter's experience, and later in Mendel's, reciprocal crosses apparently always gave identical results. This is because in diploid organisms, chromosomes come in pairs. One member of each chromosome pair derives from each parent; it does not matter whether, for example, a dominant allele was contributed by the mother or the father. However, this is not always the case—sometimes parental origin does matter. To understand certain types of inheritance, we must consider the ways in which sex is determined in different species.

In maize, a plant much studied by geneticists, every diploid adult has both male and female structures. These two types of tissue are genetically identical, just as roots and leaves are genetically identical. Plants such as maize, and animals such as earthworms, which produce both male and female gametes in the same organism, are said to be **monoecious** (from the Greek for "single house"). Some plants, such as date palms and oak trees, and most animals are **dioecious**, that is, some of the individuals can produce only male gametes and the others can produce only female gametes. In most dioecious organisms, sex is determined by differences in the chromosomes; but such determination operates in a bewildering variety of ways (Figure 10.13). The sex of a honeybee, for example, depends on whether it develops from a fertilized or an unfertilized egg. A fertilized egg is diploid, and it gives rise to a female bee—either a worker or a queen, depending on the diet during larval life. An unfertilized egg is haploid and gives rise to a male drone.

In many other animals, including ourselves, sex is determined either by a single **sex chromosome** or by a pair of them. Both males and females have two copies of each of the rest of the chromosomes. In animals that have sex chromsomes, those chromosomes that are present in equal numbers in both sexes are called **autosomes**. The one that is present in different numbers is called the **X chromosome**. For example, female grasshoppers have two X chromosomes, whereas males have only one. These females form eggs containing one copy of each autosome and one X chromosome. The males form two types of sperm. Half contain an X chromosome and one copy of each autosome, and the other half contain only autosomes. This is a natural consequence of meiosis in the two sexes. In females, the two X chromosomes synapse in prophase I; one goes to each of the daughter nuclei from meiosis I. Males have but a single X

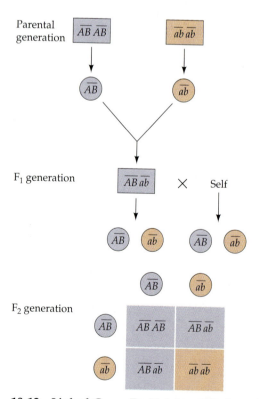

10.12   Linked Genes Do Not Assort Independently
When two genes are on the same chromosome (as indicated by the line above the letters), more of the F_2 offspring from a dihybrid cross have parental combinations (and fewer have nonparental combinations) than predicted by Mendel's laws. In the cross illustrated, genes *A* and *B* are so closely linked that they always segregate together, as if they were a single gene.

Parental generation

F₁ generation   Self

F₂ generation

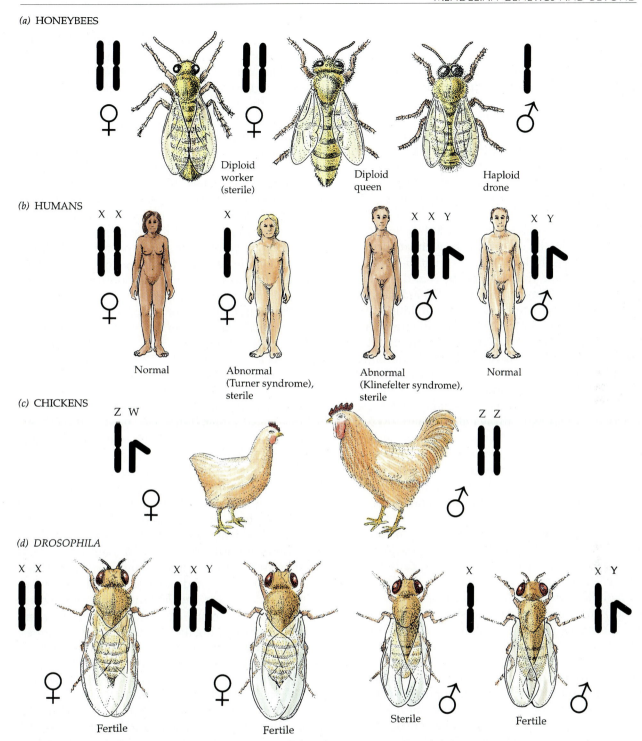

(a) HONEYBEES

Diploid worker (sterile)

Diploid queen

Haploid drone

(b) HUMANS

X X

Normal

X

Abnormal (Turner syndrome), sterile

X X Y

Abnormal (Klinefelter syndrome), sterile

X Y

Normal

(c) CHICKENS

Z W

Z Z

(d) DROSOPHILA

X X

Fertile

X X Y

Fertile

X

Sterile

X Y

Fertile

10.13 Sex Determination in Various Animals
(a) In honeybees, fertilized eggs develop into females and unfertilized eggs develop into haploid males. In other animals, sex is determined by special sex chromosomes. *(b)* Normal human females carry two X chromosomes; normal males carry one X and one Y chromosome. Persons who have some other number of sex chromosomes may develop abnormally. *(c)* In birds, it is the males that carry two identical sex chromosomes (ZZ) and females that have differing ones (ZW). *(d)* *Drosophila* females have two X chromosomes and may also have a Y chromosome; males have an X chromosome and, if they are fertile, a Y chromosome.

chromosome, so there is no synapsis. Thus half the sperm end up containing an X chromosome, but the others get none. Female grasshoppers are described as being XX (ignoring the autosomes) and males as XO (pronounced "ex-oh"). When an X-bearing sperm fertilizes an egg, the zygote is XX and develops into a female. When a sperm without an X fertilizes an egg, the zygote is XO and develops into a male. This chromosomal mechanism ensures that the two sexes are produced in approximately equal numbers. No such mechanism for numerical equality exists in the diploid–haploid system of bees.

In humans and other mammals, females have two X chromosomes and males have one (Figure 10.13). However, males also have a kind of chromosome that is not found in females: the **Y chromosome**. Females may be represented as XX and males as XY. The males produce two kinds of gametes, each having a complete set of autosomes, but differing with respect to their sex chromosomes: Half the gametes carry an X chromosome and the rest carry a Y. When an X-bearing sperm fertilizes an egg, the resulting XX zygote is female; when a Y-bearing sperm fertilizes an egg, the XY zygote is male.

There are some subtle but important differences that show up clearly in mammals with abnormal chromosomal constitutions. These conditions tell us something about the functions of the X and Y chromosomes. In both humans and mice, XO individuals sometimes appear. In humans, XO individuals are females who are moderately physically abnormal but mentally normal and are almost always sterile. The XO condition in humans is called Turner syndrome. In mice, XO individuals are fertile females that are virtually normal. XXY individuals also arise. XXY humans (a condition known as Klinefelter syndrome) are decidedly abnormal, always sterile, and always males. In brief, male sex in humans is determined by the presence of the Y chromosome. In our species, the Y chromosome carries the genes that determine maleness; therefore the absence of Y leads to femaleness, while the presence of Y has a definite masculinizing effect.

The Y chromosome functions in a different manner in the fruit fly *Drosophila melanogaster* (Figure 10.13). Superficially, *Drosophila* follows the same pattern as mammals: females are XX and males XY. However, XO individuals are *males* (rather than females as in mammals) and almost always indistinguishable from normal XY males except that they are sterile. XXY *Drosophila* are normal, fertile females. In *Drosophila*, sex is strictly determined by the ratio of X chromosomes to autosome sets. If there is one X chromosome for each set of autosomes, the individual is a female; if there is only one X chromosome for the two sets of autosomes, the individual is a male. The Y chromosome plays no sex-determining role in *Drosophila*; it is needed only for male fertility.

In birds, moths, and butterflies, *males* are XX and *females* are XY. To avoid confusion, this is usually expressed as ZZ (male) and ZW (female). In these organisms, it is the female that produces two types of gametes. Thus the egg determines the sex of the offspring, rather than the sperm as in humans and fruit flies.

Sex Linkage

How does the existence of sex chromosomes affect patterns of inheritance? In *Drosophila* and in humans,

the Y chromosome carries few known genes, whereas a substantial number of genes affecting a great variety of traits are carried on the X chromosome. This leads to an important deviation from the usual Mendelian ratios for the inheritance of genes located on the X chromosome. Any such gene is present in two copies in females, but in only one copy in males. Therefore, females may be heterozygous for genes that are on the X chromosome, but males will always be **hemizygous** for these genes, having only one of each. It is useful here—as in many instances—to think of loci whose alleles govern easily observable phenotypes as **markers** of the chromosomes on which they are located. Reciprocal crosses of parents differing for markers on the sex chromosomes do not give identical results; this is a sharp deviation from the inheritance of markers on autosomes.

The first and still one of the best examples of **sex-linked inheritance**—inheritance of traits governed by loci on the sex chromosomes—is that of eye color in *Drosophila* (Figure 10.14). The wild-type eye color of these flies is red, and the eye-color locus is on the X chromosome. In 1910, Thomas Hunt Morgan discovered a mutation that causes white eyes. When homozygous red-eyed females were crossed with (hemizygous) white-eyed males, all the sons and daughters had red eyes, because red is dominant over white and all the progeny had inherited a wild-type X chromosome from their mothers. However, in the reciprocal cross, in which a white-eyed female was mated to a red-eyed male, all the sons were white-eyed and all the daughters red-eyed. The sons from the reciprocal cross inherited their only X chromosome from their white-eyed mother; the Y inherited from the father did not carry any gene for eye color. The daughters, on the other hand, got a chromosome with the white allele from their mother and a chromosome bearing the red allele from their father; they were therefore red-eyed heterozygotes. If these same heterozygous females were mated in turn to red-eyed males, half their sons had white eyes, but all their daughters had red eyes.

A number of loci are located on the human X chromosome, and their alleles are inherited in exactly the same way as those for white eyes in *Drosophila*. A good example is hemophilia, a hereditary disorder caused by homozygosity for a mutant recessive allele and characterized by the failure of blood to clot properly; victims suffer from excessive and often fatal bleeding. A hemophilic man married to a homozygous normal woman will not produce any hemophilic children (Figure 10.15). The sons inherit a single, normal X from their mother and will neither have the disease nor transmit it to their children. The daughters get an X chromosome bearing a normal allele from their mother and one bearing the allele for hemophilia from their father. Because hemophilia is recessive, the daughters will not be hemophilic. They

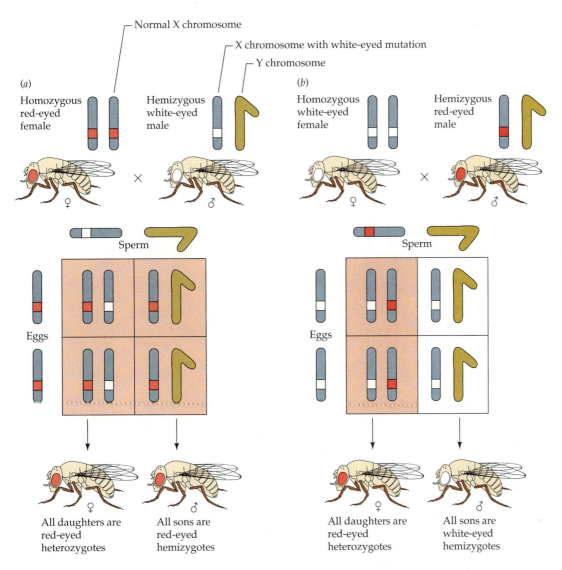

10.14 Sex-Linked Inheritance

Alleles on sex chromosomes are inherited in predictable patterns. In *Drosophila*, the locus for eye color is on the X chromosome. The wild-type allele for red eye color is dominant to the mutant allele for white eyes. Because females carry two X chromosomes, they may be homozygous for either allele, or heterozygous. Males, however, are hemizygous: They carry only one X chromosome, inherited from their mothers. (a) A homozygous red-eyed female mated with a hemizygous, white-eyed male produces heterozygous, red-eyed daughters and hemizygous sons that are red-eyed because their only X chromosome, which they got from her, carries the allele for red eyes. (b) A homozygous, white-eyed female mated with a red-eyed male also produces heterozygous, red-eyed daughters, but her sons are white-eyed because their X chromosome carries the allele for white eyes.

will, however, be heterozygous **carriers**. Such carriers of an X-linked trait will transmit the disease to half their sons and the carrier role to half their daughters. What parental genotypes would produce a female hemophiliac? Her father would have to be a hemophiliac, and her mother a carrier. Because hemophilia is quite rare, two such people are unlikely to meet. Moreover, until recently hemophilic males rarely survived long enough to reproduce. One might expect hemophilic females to be extremely rare, and in fact very few have ever been found.

The small human Y chromosome carries very few loci. Among them are the maleness determiners, whose existence was suggested by the phenotypes of the XO and XXY individuals described above. The pattern of inheritance of Y-linked alleles should be easy for you to work out.

Mendelian Ratios Are Averages, Not Absolutes

You have now been introduced to the basic Mendelian ratios: 3:1, 1:1, 9:3:3:1; you will figure out others as you do homework or test problems. It is essential to remember that these represent highest

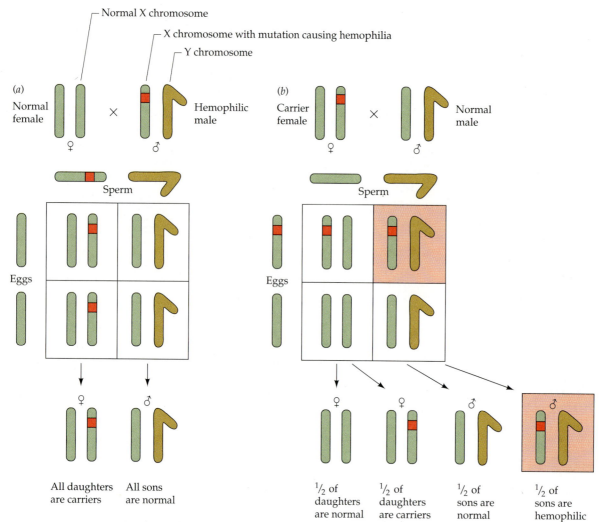

10.15 Inheritance of Hemophilia

Hemophilia, a sex-linked trait in humans, is inherited in the way white eye color is in *Drosophila*. Human males, like male fruit flies, are hemizygous, inheriting their X chromosomes from their mothers. Females carry two chromosomes, inheriting one X chromosome from their fathers and one from their mothers. (a) A son whose mother has normal alleles on both X chromosomes cannot inherit hemophilia even if his father has the allele for hemophilia. His sister is a heterozygous carrier if their father is hemophilic. (b) If a heterozygous woman marries a normal man, half of their children, on the average, inherit the mother's X chromosome that bears the hemophilia allele. Of the children receiving this allele for hemophilia, half, on the average, inherit their father's normal X chromosome and are females and carriers; half inherit their father's Y chromosome and are hemophilic males.

probabilities, not invariant rules. The X–Y system of sex determination in our species results in roughly equal numbers of males and females in a substantial population, but you know that a given family of four children may not consist of two girls and two boys. It is not unusual for four children to be of the same sex; in fact, one family in eight who have four children will have all boys or all girls.

When we are trying to understand the genetic basis for the results of a cross, it is important to know how much deviation from a predicted ratio is within reasonable expectations. A number of statistical methods have been devised for telling the investi-

gator whether the observed deviation from an expected ratio can be attributed to chance variation or whether it is large enough to suggest that the observed ratio is in fact discordant with the predicted ratio. These methods take a number of factors into consideration, but one of the most important is **sample size**. If we expect to find a 3:1 ratio between two phenotypes and we look at a sample of only 8 progeny, we should not be surprised to find 7 individuals of one phenotype and 1 of the other (rather than the expected 6 and 2). It would be surprising, however, in a sample of 80 to find 70 of one phenotype and 10 of the other—you would question whether that really represented a 3:1 ratio. In designing experiments in genetics—and in quantitative biology in general—it is desirable to keep sample sizes large so that the data are easier to evaluate with confidence.

Special Organisms for Special Studies

Many kinds of organisms, both prokaryotic and eukaryotic, have been used in genetic investigations. However, a few species have attracted particular attention because of one advantage or another. Gregor Mendel did his most famous work with the garden pea, but the best-studied higher plant is maize, or corn (*Zea mays*). Originally, this was due in part to the great agricultural importance of maize. Maize has been examined so thoroughly that highly detailed **genetic maps** locating particular genes on each of the chromosomes are available. The first animal to be studied in great detail was the fruit fly, *Drosophila melanogaster* (Figure 10.14). Its small size, ease of cultivation, and short generation time made it attractive. Thomas Hunt Morgan and his students established it as a highly useful laboratory organism in Columbia University's famous "fly room," where such phenomena as sex linkage were discovered. *Drosophila* remains extremely important in studies of chromosome structure, population genetics, the genetics of development, and the genetics of behavior.

There was a great period in genetics in which the focus was the common salmon-colored bread mold *Neurospora crassa*. It was used in a number of historic experiments (Chapter 11). The products of meiosis in *Neurospora* are organized in an unusual way that makes it easy to visualize the results of crossing over.

Meiosis in *Neurospora*

The life cycle of *Neurospora*, like those of other fungi, is complex; it will be considered in detail in Chapter 23. *Neurospora* grows from haploid spores that divide mitotically to produce a feltlike mat of long strands called hyphae. Eventually, as the end result of a sexual process, haploid nuclei from two individuals unite to produce zygotes in a specialized fruiting structure. As soon as a diploid nucleus is formed, it undergoes meiosis. Thus the zygote itself constitutes a greatly reduced diploid generation.

Meiosis in *Neurospora* is a tidy process that packages all of the nuclei produced by the divisions of a single zygote in a long, thin sac called an **ascus** (Figure 10.16). The four haploid nuclei then divide once again by mitosis. The eight nuclei produced by this sequence of events are incorporated into eight spores, all neatly lined up within the ascus. Because the ascus is so thin, the nuclei cannot pass one another as the divisions proceed, so the pairs of spores can easily be identified with the meiotic division that produced them. This makes *Neurospora* an especially

10.16 *Neurospora* Asci
A rosette of asci of *Neurospora crassa*, resulting from a cross of a spore-color mutant with a wild-type strain. Each ascus contains eight spores; their arrangements reflect different segregation patterns.

useful organism in which to examine segregation, assortment, and recombination of genetic markers. Accordingly, we shall use it to illustrate the material in the next section.

Recombination in Eukaryotes

We have already seen that each chromosome contains many loci and that all the loci on one chromosome are thus linked to each other. If homologous chromosomes did not undergo crossing over when paired (Chapter 9), the markers on a given chromosome would all segregate together as a unit. A geneticist would have no way of knowing that linked loci are actually different loci. Mendel's second law (independent assortment of alleles of different loci) would apply only to loci on different chromosomes.

The actual situation is more complex and therefore more interesting. Markers located at different places on the same chromosome do sometimes separate from one another as the result of crossing over. Geneticists use **recombination frequencies** (the observed frequencies in the offspring of marker combinations different from those of the parents) to generate genetic maps that indicate the actual arrangement of markers along the chromosome (Box 10.C).

Genetic markers on the same chromosome pair recombine by **crossing over** (Chapter 9), which results from the physical exchange of corresponding genetic segments between two homologous chromosomes during prophase I of meiosis. That is, recombination occurs at the stage when homologous chromosomes are paired. Recall that the DNA has duplicated by this stage, and each chromosome consists of two chromatids. Thus crossing over occurs at

BOX 10.C

Gene Mapping in Eukaryotes

Neurospora is an excellent organism for illustrating the principles of mapping. In mapping experiments, we do not make use of the orderly packaging of spores in asci; rather, spores are collected at random and examined. Remember that this organism is haploid for most of its life cycle; haploid genotypes thus characterize strains. Suppose that we cross a strain of genotype *AB* with another strain of genotype *ab*. We then determine the **frequency of recombination** between the two markers as follows. Let us say that spores of genotype *AB* make up 40 percent of the total, *ab* 40 percent, *Ab* 10 percent, and *aB* 10 percent. Of all these spores, 40 + 40 = 80 percent are of the parental genotypes, 10 + 10 = 20 percent are recombinant. The frequency of re-

combination between the two markers is thus 20 percent.

To determine the linear sequence and spacing of markers on the chromosome, one may perform a **three-factor cross**. If *Neurospora* strains that are *ABC* and *abc* are crossed, the following spore genotypes might be observed:

1. *ABC*	38.0	percent
2. *abc*	40.2	percent
3. *Abc*	7.2	percent
4. *aBC*	6.6	percent
5. *ABc*	3.1	percent
6. *abC*	3.7	percent
7. *AbC*	0.5	percent
8. *aBc*	0.7	percent

Classes 1 and 2 are parental types. Classes 3 and 4 are single recombinants between *A* and *B*. Classes 5 and 6 are single recombinants between *B* and *C*. Note that all four of the classes 3 through 6 are also recombinants between *A* and *C*. Classes 7 and 8 are recombinant between *A* and *B* and between *B* and *C*, but not between *A* and *C*; they represent *double* crossover types. Note that 7 and 8 are the least frequent classes, which is what one

would expect the double crossover types to be. The two crossovers cancel each other in recombining *A* and *C*, leaving those two markers in the parental arrangement relative to each other.

To compute map distances, we add the crossovers in all the classes that are recombinant between a given pair of markers. The "distance" between *A* and *B* is the sum of classes 3, 4, 7, and 8: 7.2 + 6.6 + 0.5 + 0.7 = 15.0 percent recombination. The distance between *B* and *C* (8.0 percent recombination) is obtained by summing classes 5, 6, 7, and 8. The distance between *A* and *C* is the sum of classes 3 through 6 plus two times class 7 plus two times class 8, because each of these last two classes contains two crossovers between *A* and *C*; so the distance between *A* and *C* = 7.2 + 6.6 + 3.1 + 3.7 + 2(0.5) + 2(0.7) = 23 percent recombination. We can thus draw a map locating the three markers, showing their map "distances," as follows:

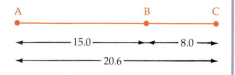

---

the four-strand stage. The exchange event at any point on the length of the chromosome involves only two of the four chromatids, one from each member of the chromosome pair (Figure 10.17). The lengths of chromosome are exchanged reciprocally, so both chromatids involved in crossing over are recombinant (that is, each chromatid contains genes from *both* parents); and no genes are created or destroyed. The points at which the chromatids break in the exchange seem to correspond perfectly (at the level of base pairs in the DNA), so that the amount of material donated by a chromatid exactly equals the amount it receives.

At any point, only two chromatids in a bivalent participate in crossing over; but other crossovers may occur at other points. These other crossovers may involve the same pair of chromatids or any other possible pair that includes one member from each of the homologous chromosomes. The precise arrangement of spores in the ascus of *Neurospora* makes this an ideal organism in which to study the details of

double and multiple crossovers (Figure 10.18). The probability that there will be more than one crossover in the segment between two particular markers depends on the distance between them—the greater the distance, the more likely are crossover events.

Cytogenetics

By making experimental crosses and calculating the recombination frequencies, geneticists can show that certain genes are associated in a linkage group, in a specific order (Box 10.C). Such a linkage group is a logical abstraction. To what extent does it actually correspond with the physical structure of a chromosome as seen under the microscope? To establish a relationship between a genetic linkage group and a chromosome, the cytogeneticist, who studies the microscopic appearance of chromosomes in relation to genetics, tries to find an individual in whom the normal linkage relationships are changed. The cytogeneticist then examines that individual's cells under

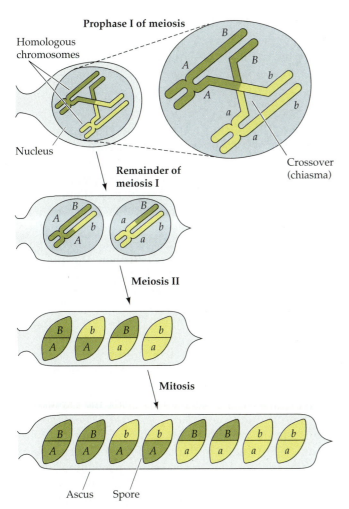

10.17 Crossing Over in *Neurospora*

Neurospora retains the products of meiosis in a single package: The ascus contains the spores. This diagram shows only one of *Neurospora*'s several chromosome pairs, beginning with the diploid nucleus during prophase I of meiosis. A single crossover between genes *A* and *B* forms two recombinant chromatids. In this particular ascus, at the end of meiosis II the nuclei with recombinant chromosomes (*Ab* and *aB*) lie in the middle of the ascus, while those with parental chromosomes (*AB* and *ab*) lie at the ends. A mitotic division follows meiosis II, increasing the number of spores in the ascus to eight. Note that the recombinant chromosomes remain at the middle of the ascus.

the microscope, looking for a corresponding visible change in one or more chromosomes.

In the tissues of most species, the chromosomes are too small for an observer to see any except the most gross changes. One exception is the giant chromosomes in the salivary glands of the larvae of *Drosophila*. Called **polytene chromosomes**, they have replicated their DNA many times without cytokinesis, so that many copies of each DNA molecule lie side by side to form thick, snakelike structures that can be seen clearly even with a low-magnification lens (Figure 10.19). Condensed thickenings, or chromomeres, along these chromosomes are paired with

sister chromomeres on the parallel strands of the polytene chromosome so that the chromosomes appear to have a pattern of transverse bands. (These bands, seen in interphase in polytene chromosomes, are not the same as the bands seen by special staining of chromosomes in mitosis.) Each polytene chromosome has a characteristic band pattern. The two homologous polytene chromosomes are closely synapsed, causing the two chromosomes to look like a single structure. The patterns of bands—their thickness, spacing, sharpness or diffuseness, and so on—are so characteristic for each chromosome that an experienced cytogeneticist can often tell at a glance if the order or position of a group of bands has been changed.

One type of chromosomal change is an **inversion**. Inversions can be detected by genetic mapping as a change in the linkage relations of markers on a chromosome. If the normal order of the markers is *ABCDEFGHI*, an inversion may have the order *ABCD**GFE**HI*. When the polytene chromosomes of such individuals are examined, the order of a group of bands in one of the chromosomes also appears reversed. A linkage group with an inversion in it can be correlated with a chromosome whose bands are inverted, and the geneticist can infer that certain genes are located in certain regions of the visibly altered chromosome.

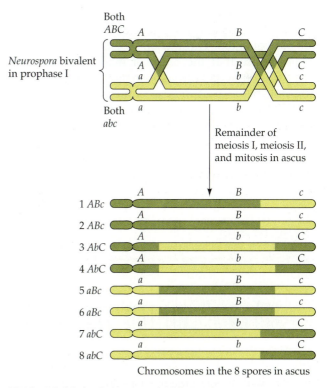

10.18 Multiple Crossovers in *Neurospora*

Even multiple crossovers are faithfully recorded in the order of spores in a *Neurospora* ascus. In this example, as a result of three crossovers during prophase, all eight spores carry recombinant chromosomes.

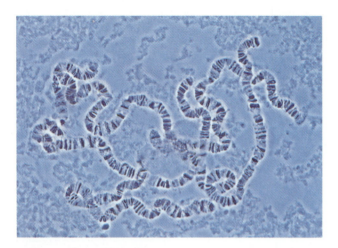

10.19 Polytene Chromosomes
A complete set of banded polytene chromosomes in a cell from the salivary gland of a fruit fly larva. Each long chromosome is actually two homologous chromosomes closely synapsed. The banding pattern along each chromosome is highly specific, allowing researchers to spot differences among individuals.

Other chromosomal changes, such as the **deletion**, or loss, of genetic material, can be useful in correlating the recombination map with the physical chromosome. Where there is genetic evidence of a deletion, a small group of bands (or even a single band) often is missing from one of the chromosomes. The study of a great many chromosomal changes in *Drosophila* has yielded the kind of picture shown in Figure 10.20. This shows genetic and cytological maps of part of the X chromosome of *Drosophila* side by side. Notice that the order of markers deduced by the two methods is in agreement, but the distances are not. Because of the exceedingly complex folding of DNA in a eukaryotic chromosome, one can never be sure that the microscopically observed length of a segment of chromosome is a good reflection of the length of DNA in that segment. Nor is genetic-map distance necessarily a reliable reflection of the length

of DNA. Despite these limitations, the combination of cytogenetics and recombination analysis has been very successful in probing the composition of chromosomes of eukaryotes.

INTERACTIONS OF GENES WITH OTHER GENES AND WITH THE ENVIRONMENT

Thus far we have treated the phenotype of an organism, with respect to a given trait, as a simple result of its genotype; and we have implied that a single trait results from the alleles of a single locus. In fact, several loci may interact to determine a trait's phenotype. To complicate things further, the physical environment may interact with the genetic constitution of an individual in determining the phenotype.

Epistasis

When a particular trait is the result of a series of chemical reactions, each controlled by a different locus, the gene that acts at the earliest step in the series may, in one of its allelic forms, *mask* the expression of one or all of the other loci. This phenomenon of **epistasis**, in which one gene alters the effect of another, is illustrated in the determination of coat color in mice (Figure 10.21). The wild-type color is agouti, a grayish pattern resulting from banding of the individual hairs with a variety of colors. The colors are affected by several loci. The dominant allele *B* determines agouti, whereas the homozygous recessive genotype *bb* results in a black coat. Another, unlinked locus affects an early step in the formation of hair pigments. The dominant allele *A* at this locus allows normal color development, but *aa* blocks all pigment production and results in an all-white albino. As a result, *aa* is said to be epistatic over the *B* locus. Whether the genotype is *BB*, *Bb*, or *bb*, the result is an albino if the alleles of the other locus are both *a*.

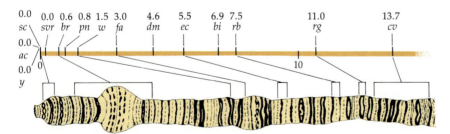

10.20 Map of *Drosophila* Chromosome
One end of a polytene X chromosome from a fruit fly larva. The italic letters indicate marker genes; the numbers above them are distances along the genetic map (top), as determined by recombination studies. The positions on the chromosome (cytological map, shown below) of 8 of the 14 markers have been determined, and their order agrees with the genetic map. Use the lines running between the maps to notice how different distances on the two maps are.

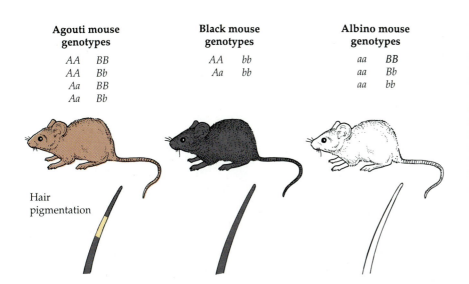

Agouti mouse
genotypes

AA	BB
AA	Bb
Aa	BB
Aa	Bb

Black mouse
genotypes

AA	bb
Aa	bb

Albino mouse
genotypes

aa	BB
aa	Bb
aa	bb

Hair
pigmentation

10.21 Genes May Interact

Two loci epistatically affect the expression of each other in determining coat color in mice. Individuals that have at least one dominant allele at each locus are phenotypically agouti. Individuals with genotype *aa* are albino regardless of their genotype for the other locus, because the *aa* genotype blocks all pigment production. Individuals with *bb* genotypes are black unless they also are *aa* (which makes them albino).

If a mouse with genotype *AABB* and thus the agouti phenotype is crossed with an albino of genotype *aabb*, the F₁ is *AaBb* and of the agouti phenotype. If the F₁ mice are crossed with each other to produce an F₂, the epistasis of *aa* will result in an expected phenotypic ratio of 9 agouti:3 black:4 albino. Can you show why this is so? The underlying ratio is the usual 9:3:3:1 for a dihybrid cross with unlinked genes, but be sure to look closely at each genotype and watch out for epistasis.

Epistasis can also work in both directions between two genes, as first observed by William Bateson and Punnett. They performed a cross between two sweet pea plants (not the edible peas studied by Mendel). Each parent had white flowers. To the astonishment of Bateson and Punnett, the F₁ all had purple flowers! The F₂, obtained by self-pollinating the F₁, were in the ratio of 9 purple:7 white. This looks like a modification of the standard 9:3:3:1 ratio, with the latter three groups being lumped into one. Let us try, as did Bateson and Punnett, to figure it out. First, because this looks like a dihybrid cross ratio, we assume that two different loci are involved. We recall that both dominant alleles are present in ⁹⁄₁₆ of the offspring of a dihybrid cross, and we notice that this is the proportion of the F₂ that are purple. We therefore decide that each individual in this group has at least one copy of each dominant allele and write *A−B−* as their genotype. Because only ⁹⁄₁₆ are purple, it must be that having a dominant allele for only one of the genes will *not* produce a color. *A−bb* and *aaB−* fail to give purple flowers. Let us see how we may represent the whole experiment: If the original parents were *AAbb* and *aaBB*, both would have been white, as observed; all the F₁ would accordingly have been *AaBb* and purple, also as observed. You should now work out the genotypes of the F₂ and then convert them to phenotypes, remembering that all the genotypes give white flowers *except* the ones that are

A−B−. If you do this carefully, you will obtain the observed 9:7 ratio. We may say that *aa* is epistatic to *B* and *bb* is epistatic to *A*, in that both of these doubly recessive genotypes alter the expression of the dominant allele at the other locus, thereby determining that the phenotype will be white. Another way to describe this kind of situation is to say that the two loci are **complementary**. Complementary loci are mutually dependent, the expression of each being dependent upon the alleles of the other.

The epistatic action of complementary genes may be explained as follows. The dominant alleles *A* and *B* in this example code for the production of enzymes that catalyze two separate reactions in the production of a purple pigment. In order for the pigment to be produced, both reactions must take place. If a plant is homozygous for either *a* or *b*, the corresponding reaction will not occur, no purple pigment will form, and the flowers will be white.

Quantitative Inheritance and Environmental Effects

Individual heritable traits are often found to be controlled by many genes, each contributing to the final outcome. As a result, variation in such traits is **continuous** rather than, as in the examples we have been considering, **discontinuous**. In the experiment of Bateson and Punnett, the sweet pea flowers were either white or purple. But many traits that are under genetic control, such as height and other aspects of size, and skin color, vary continuously. We may usefully think of these as being controlled by multiple **polygenes**—loci whose alleles increase or decrease the observed character (Figure 10.22). Polygenes affecting a particular quantitative trait are commonly found on many different chromosomes. One of Mendel's wise decisions was to deal only with discontinuous variation, which is relatively simple. Had he,

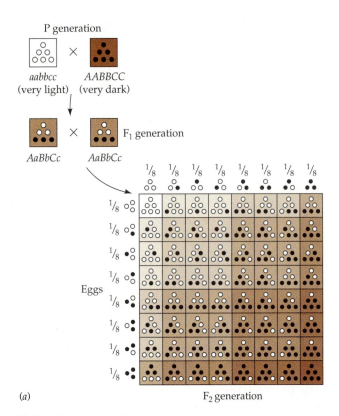

(a)   F₂ generation

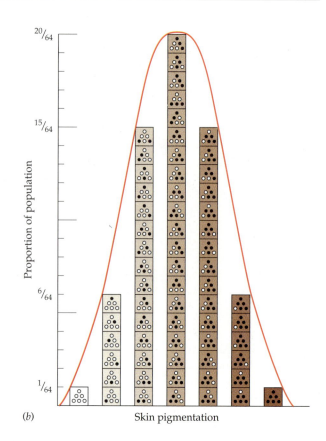

(b)   Skin pigmentation

10.22 Polygenic Inheritance

A small group of genes contributes to the degree of pigmentation in human skin. A model based on three genes is illustrated here. The alleles *A*, *B*, and *C* contribute dark pigment to the skin, but the alleles *a*, *b*, and *c* do not. The more *A*, *B*, and *C* alleles an individual possesses, the darker that person's skin will be. The pattern of inheritance of the alleles is shown in (*a*). The black circles represent the alleles *A*, *B*, and *C*; the white circles represent the alleles *a*, *b*, and *c*. The frequencies of the phenotypes in (*a*) are graphed in (*b*). A couple with intermediate pigmentation (*AaBbCc*, for example) is unlikely to have children with either very light or very dark skin.

like Charles Darwin and others, concentrated on continuous variation, we might still not know the basic rules of heredity!

Humans differ with respect to the amount of a dark pigment, melanin, in the skin. There is great variation in the amount of melanin among different people, but much of this variation is determined by alleles at just four (possibly three) loci. None of the alleles at these loci demonstrates dominance. Of course, skin color is not entirely determined by the genotype, since exposure to sunlight can cause the production of more melanin—that is, tanning.

Such environmental variables as light, temperature, and nutrition can sharply affect the translation of a genotype into a phenotype. A familiar example is the Siamese cat (Figure 10.23). This handsome animal normally has darker fur on its ears, nose, paws, and tail than on the rest of its body. These darkened parts are ones that have a somewhat lower temperature. A few simple experiments show that the Siamese cat has a genotype that results in dark fur, but

10.23 Environmental Effects on Phenotype

This Siamese cat has dark fur on its extremities, where temperature is below the general body temperature. By removing light fur and chilling the area, dark fur can be induced to grow—a clear example of the effect of environment on the phenotype of the cat.

only at temperatures somewhat below the general body temperature. If some dark fur is removed from the tail and the cat is kept at higher-than-usual temperatures, the new fur that grows in is light. Conversely, removal of light fur from the back, followed by local chilling of the area, causes the spot to fill with dark fur.

Genotype and environment interact to determine the phenotype of an organism. For a given population, we may speak of the proportion of individuals with a given genotype that actually show the expected phenotype. This proportion is called the **penetrance** of the genotype. The environment may also affect the **expressivity** of the genotype, that is, the *degree* to which it is expressed. For an example of environmental effects on expressivity, consider Siamese cats that are kept indoors and outdoors in different climates.

The analysis of quantitative inheritance is complicated by uncertainty over how much of the observed variation is due to the environment and how much is due to the effects of the several polygenes. A useful approach to this difficulty is the study of identical twins—individuals that are genetically identical because they developed from a single zygote. Any differences between such twins must be attributed to environmental differences.

It is clear that the phenotype of an organism depends on its total genetic makeup and on its environment. Some of the interactive effects will become more obvious when we focus, in Chapter 12, on the regulation of gene expression.

MAPPING HUMAN CHROMOSOMES

Mapping genes in humans is of special interest because of its potential to improve our ability to diagnose and treat human genetic diseases (Box 10.D). Striking technical advances, and an emerging willingness to pay a great deal of money to invent and implement them, are giving rise to the Human Genome Project, to be discussed in Chapter 14. To set the stage for the Human Genome Project, let us con-

BOX 10.D

Human Genetic Disorders

Our species is subject to more than 150 known heritable metabolic defects, most involving a failure to produce necessary enzymes or other proteins. These tend to be recessive traits, with the recessive alleles being unable to code for the normal proteins. It has been estimated that each of us is heterozygous for 30 or more such recessive disorders. The following are some of the better known genetic disorders.

Cystic fibrosis, caused by an autosomal recessive allele, is a major killer of children; most of its victims die during their early to middle twenties. It involves severe respiratory difficulties, resulting from thick mucus in the lungs, as well as liver, pancreatic, and digestive failure.

About 1 of every 20 people in the United States is heterozygous for the recessive allele for cystic fibrosis, with persons of northern European ancestry being at greatest risk.

Tay–Sachs disease, another example of autosomal recessive inheritance, is commonest among Ashkenazic Jews, occurring in the United States once in every 3,600 births in this group but only about $\frac{1}{100}$ as commonly among the remainder of the American population. Tay–Sachs disease results from a failure to produce a specific enzyme, hexosaminidase A, with resulting degeneration of the nervous system. Children with Tay–Sachs disease appear normal for their first few months, but they usually lose their eyesight by about the age of one year, and they rarely survive more than five years.

Like Tay–Sachs victims, infants with **phenylketonuria** seem normal at birth. Homozygotes for the autosomal recessive allele cannot metabolize the amino acid phenylalanine. The principal consequence is severe mental retardation, and untreated victims usually do not live more than 30 years. Most states require screen-

ing of newborn infants for phenylketonuria, and a diet low in phenylalanine allows normal development.

Sickle-cell anemia, the molecular basis of which is discussed elsewhere in this book, is also an autosomal recessive trait. It is most common among people whose ancestors came from tropical areas or the Mediterranean. About 1 percent of North American blacks have the disease, and about 14 percent are heterozygous for the trait. In homozygotes an abnormal hemoglobin is produced, and this leads to defects in the red blood cells.

Childhood pseudohypertrophic muscular dystrophy is an X-linked recessive trait that appears in the first three years of life. Muscular deterioration is progressive, and victims generally die in their twenties.

Hemophilia, in which the blood does not clot normally, is an X-linked recessive trait. Some hemophiliacs are at risk of death from even minor cuts.

Among the many other hereditary disorders some are mild handicaps at worst, such as red–green color blindness.

sider where things stood only yesterday. Two formidable barriers have obstructed any program for mapping human genes. The first is the long generation time in humans, coupled with the relatively small number of offspring generally produced by a couple. The second barrier is the moral unacceptability of performing human breeding experiments analogous to the controlled breeding programs that were so productive in mapping the genes of *Drosophila melanogaster*, maize, and—as we will see in Chapter 12— bacteria and viruses.

Because of the distinctive character of sex-linked inheritance, the first human genes to be associated with a particular chromosome were all ones found on the X chromosome, beginning with the discovery in 1911 that a gene for color blindness is located there. It was another 59 years before a human gene was located on an autosome.

Pedigree Analysis

Until recently the mapping of human genes relied on existing data from family trees and mostly amounted to assigning a locus to a sex chromosome or autosome. Such a **pedigree analysis** examines the pattern of transmission of a particular genetic characteristic over two or more generations. For fairly obvious reasons, human pedigrees include data on relatively few individuals and few generations. However, useful results may sometimes be obtained by comparing pedigrees of different families, some of whose members have the hereditary trait of interest. Such pedigrees serve as our closest acceptable approximation to the breeding experiments performed with plants and experimental animals.

Some of the symbols commonly used in pedigree charts are shown in Figure 10.24. Figure 10.25 shows some pedigrees; examine them before consulting the legend to see whether you can determine the pattern of inheritance, such as autosomal recessive or sex-linked dominant, demonstrated by each.

Pedigree analysis has also been used to test for the linkage of pairs of traits. By noting the relationships of individuals affected by one, the other, or both traits in a family, one may determine whether the traits are linked or unlinked and even obtain a rough estimate of the map distance between linked loci.

Recent advances in technique have made it possible to map some human genes more efficiently than can be done with pedigree analysis. One of the new techniques is **somatic-cell genetics**, the analysis of hybrid cells containing chromosomes of two different species. The other is **gene cloning**, the production of large numbers of copies of a single gene or group of genes, to be discussed in Chapter 14.

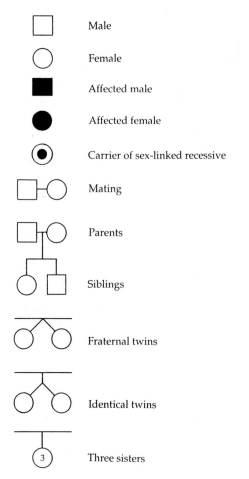

10.24   Symbols for Pedigree Charts
These symbols are used to characterize individuals and relationships in pedigree charts.

Somatic-Cell Genetics

From a rare event, scientists recently developed a new technique. If somatic cells—cells other than gametes—isolated from two animal species are cultured together on a suitable medium, once in a while two cells may fuse to form a *hybrid* cell containing two nuclei, one from each species. Scientists have been able to increase the probability of such hybrid cell fusion by adding certain chemicals, such as polyethylene glycol, to the culture medium.

Once hybrid cells are formed, the two nuclei may fuse to create a nucleus containing the chromosomes of both species. Mitosis in hybrid cells is sometimes abnormal, and during successive mitoses the chromosomes of one of the species are progressively lost. As a result, some of the daughter cells have nuclei containing all the chromosomes of one species and only one or a few chromosomes of the second species. In mouse–human hybrid cells, it is most often the human chromosomes that are lost. At this point, you may be able to guess how mouse–human hybrid

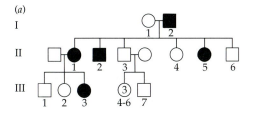

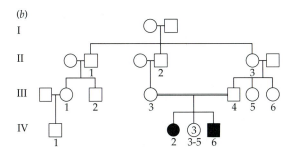

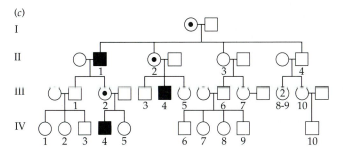

10.25 Patterns of Inheritance
Roman numerals identify generations; the numbers identify individuals within each generation. *(a)* Inheritance of an autosomal dominant trait. Individual I-2 is a male heterozygous for the trait; if he were homozygous, all of the progeny in generation II would show the trait. None of the affected individuals in generations II or III can be homozygotes, because each has received a normal recessive allele from one parent. *(b)* Inheritance of an autosomal recessive trait. If the trait were dominant rather than recessive, then either III-3 or III-4 would have to show the trait in order for IV-2 and IV-6 to have inherited it. III-3 and III-4 are first cousins—their mating is represented by the double horizontal line. There is no indication of sex linkage. *(c)* Inheritance of a sex-linked recessive trait. The affected individuals II-1, III-4, and IV-4 are all males. The carriers are all females. This is the pattern of inheritance for hemophilia (see Figure 10.15).

cells can be used to identify certain genes as being located on particular chromosomes. Researchers can observe several such strains of hybrid cells and can test them for two things: (1) their ability to form the products of a particular gene and (2) their possession of different human chromosomes (based on the characteristic appearances of the individual chromosomes; see Figure 9.13). If enough cells are examined,

a given gene product will be found in all cells that have a particular chromosome and in no cells that lack that chromosome. By means of this technique, a few hundred human genes have been successfully mapped to specific chromosomes. This effort has been enhanced by the availability of data from pedigree analysis. Once a gene has been localized on a particular chromosome by somatic-cell genetic techniques, then it is known that any other genes linked to that one (as discovered by pedigree analysis) must reside on the same chromosome.

Another technique of somatic-cell genetics is chromosome-mediated gene transfer. This differs from the somatic-cell hybridization just described in that instead of combining cells from two species, one combines cells of a recipient species (such as a mouse) with *chromosomes* isolated from metaphase cells of a donor species (such as *Homo sapiens*). When recipient cells are incubated with isolated chromosomes, only some parts of some chromosomes are taken up by the cells, and only a small fraction of that which is taken up gets associated with the host chromosomes of the mouse and is thus made stable. The chances that two genes of the donor will both be stabilized are extremely slight unless the genes are closely linked. By observing the frequency of two loci being transferred together, biologists can determine the map distance between them.

NON-MENDELIAN INHERITANCE

The essence of Mendelian inheritance is that information carried on chromosomes is partitioned with great precision during meiosis. But in eukaryotic cells there are other self-reproducing entities besides the nuclear chromosomes. Mitochondria and chloroplasts carry some genetic information in small circular chromosomes (Chapter 4). The DNA of these organelles is subject to mutation just as is the DNA in the chromosomes of the nucleus, so we may speak of alleles of nonnuclear genes. These genes are not inherited in the same way as nuclear chromosomal genes because the eggs of most species contain large amounts of cytoplasm, but the sperm contain hardly any. Generally speaking, all the mitochondria in a zygote come from the cytoplasm of its mother's egg, even though half the zygote's nuclear chromosomes come from its father. In plant zygotes all the chloroplasts come from the maternal cytoplasm. Hence any particle that is inherited through the cytoplasm is sometimes said to be **maternally inherited**. In such cases, reciprocal crosses give quite different results.

Certain patterns of cellular architecture are inherited in a way that may not involve DNA—let alone

Mendelian genetics—at all. This sort of "pattern inheritance" seems to be especially important in protists such as *Paramecium*; but perhaps we merely notice it more in these protists because they are large, complex cells with distinctive external anatomy. One of the most bizarre examples of pattern inheritance is found in the freshwater protist *Difflugia corona*, which was studied by H. S. Jennings in 1937. Jennings' work was summarized by David Nanney, writing in the journal *Science*:

> The organism constructs a shell by cementing sand grains together with a cellular secretion. The ventral surface of the shell possesses an opening, the "mouth," through which the cell communicates with the outside world. The edges of the openings are surrounded by a symmetrical array of "teeth." Jennings observed that the numbers of teeth varied among individuals, and he explored the question of their heredity by the only means available; he isolated individuals, allowed clones to develop, and inquired into clonal uniformity. Tooth number remained constant within a clone; differences in tooth number were hereditary. He was not able to conduct a breeding analysis, but he noted that, when the cell body divided, one of the daughter cells was extruded naked through the mouth and, while still in contact with its sister, began to construct its own sand castle, beginning in the region of contact [Figure 10.26]. The new mouth structures were therefore constructed in direct contact with structures of the old mouth. This observation suggested to him that the old mouth might serve as a template to guide the organization of the new one—that new teeth were initiated in the interstices between the old teeth. This curious speculation might have remained just that had Jennings not tried his hand at oral surgery. He broke out denticles with a glass needle and examined the consequences of mutilating the parental template. Modified parents produced modified progeny, and new lineages were established with new tooth numbers. A few generations were required for symmetry to be achieved, but once that had been accomplished the tooth number stabilized and a new hereditary state was achieved. After considering these studies, one of my colleagues concluded ruefully that genetic specificity may be based on two different structural foundations—nucleic acid and sand. In view of the notorious instability of structures built upon sand, this conclusion is peculiarly disturbing.

SUMMARY

In monohybrid crosses involving alleles that exhibit complete dominance, F_1 progeny show only the dominant trait. In the F_2 produced by crossing or self-fertilizing F_1 individuals, both parental phenotypes reappear, in the approximate ratio 3 dominant:1 recessive trait. The F_2 consists of three genotypes in the ratio 1 homozygous dominant:2 heterozygotes:1 homozygous recessive. In a test cross, a heterozygote gives two types of progeny in the ratio 1:1; the homozygous dominant gives only the dominant phenotype.

Alleles at most loci do not show complete dominance. A monohybrid cross with incomplete dominance results in a 1:2:1 phenotype ratio in the F_2. Some loci are represented by multiple alleles. Different alleles arise by mutation.

In a dihybrid cross with unlinked genes, alleles of the two loci assort independently at meiosis. If A is dominant to a and B is dominant to b, a cross $AABB \times aabb$ results in an F_2 generation with four phenotypes: $9\ A-B-:3\ A-bb:3\ aaB-:1\ aabb$. If the markers in a dihybrid cross are linked, there is a much higher proportion of parental phenotypes in the F_2 and fewer recombinant individuals.

Crossing over between chromatids produces recombination. From the frequency of recombination, taken as a measure of the distance between markers, mapping techniques can determine the linear order of markers in a linkage group.

The presence or absence of certain chromosomes determines sex in many organisms. Female bees are diploid, and males haploid. In grasshoppers and some other insects, XX is female and XO male. In humans and *Drosophila*, XX determines female, and XY male; the reverse is true in birds, butterflies, and moths. Extra sex chromosomes, or missing ones, may produce abnormalities.

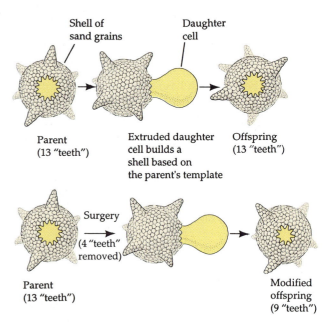

10.26 Non-Mendelian Inheritance in a Protist
Because *Difflugia corona* constructs its shell using its parent's shell as a template, modifications to the number of "teeth" on the shell will be inherited by offspring, although not by Mendelian mechanisms.

Traits governed by loci on the X or the Y chromosome are not inherited in the same ratios as those observed for autosomal markers. Reciprocal crosses for sex-linked markers give nonidentical results.

Human genetics has advanced rapidly since the development of somatic-cell hybridization and chromosome-mediated gene transfer. Pedigree analysis continues to be a useful technique.

The expression of some genes is modified by other genes. A particular recessive homozygous genotype (or a dominant allele) may be epistatic to other loci. Many traits are controlled by the interaction of multiple genes. Both the penetrance and the expressivity of a genotype may be influenced by the environment. Some traits are determined by DNA contained in the mitochondria or chloroplasts. They are inherited only by way of the cytoplasm of the egg, and reciprocal crosses do not give identical results.

SELF-QUIZ

1. Which statement is *not* true for Mendel's cross of *TT* with *tt* peas?
   a. Each parent can produce only one type of gamete.
   b. F_1 individuals produce gametes of two types, each gamete being *T* or *t*.
   c. Three genotypes are observed in the F_2 generation.
   d. Three phenotypes are observed in the F_2 generation.
   e. This is an example of a monohybrid cross.

2. The phenotype of an individual:
   a. depends at least in part on the genotype.
   b. is either homozygous or heterozygous.
   c. determines the genotype.
   d. is the genetic constitution of the organism.
   e. is either monohybrid or dihybrid.

3. Which of the following statements is *not* true of alleles?
   a. They are different forms of the same gene.
   b. There may be several alleles at one locus.
   c. One allele may be dominant over another.
   d. Alleles may show incomplete dominance.
   e. Alleles occupy different loci on the same chromosome.

4. Which statement is *not* true of an individual homozygous for an allele?
   a. Each of its cells possesses two copies of that allele.

   b. Each of its gametes contains one copy of that allele.
   c. It is true breeding with respect to that allele.
   d. Its parents were necessarily homozygous for that allele.
   e. It can pass that allele on to its offspring.

5. Which of the following statements is *not* true of a test cross?
   a. It tests whether an unknown individual is homozygous or heterozygous.
   b. The test individual is crossed with a homozygous recessive individual.
   c. If the test individual is heterozygous, the progeny will have a 1:1 ratio.
   d. If the test individual is homozygous, the progeny will have a 3:1 ratio.
   e. Test cross results are consistent with Mendel's model of inheritance.

6. Linked genes:
   a. must be immediately adjacent to one another on a chromosome.
   b. have alleles that assort independently of one another.
   c. never show crossing over.
   d. are on the same chromosome.
   e. always have multiple alleles.

7. In the F_2 generation of a dihybrid cross:
   a. four phenotypes appear in the ratio 9:3:3:1 if the loci are linked.
   b. four phenotypes appear in the ratio 9:3:3:1 if the loci are unlinked.

   c. two phenotypes appear in the ratio 3:1 if the loci are unlinked.
   d. three phenotypes appear in the ratio 1:2:1 if the loci are unlinked.
   e. two phenotypes appear in the ratio 1:1 whether or not the loci are linked.

8. The sex of a honeybee is determined by:
   a. ploidy, the male being haploid.
   b. X and Y chromosomes, the male being XY.
   c. X and Y chromosomes, the male being XX.
   d. the number of X chromosomes, the male being XO.
   e. Z and W chromosomes, the male being ZZ.

9. In epistasis:
   a. nothing changes from generation to generation.
   b. one gene alters the effect of another.
   c. a portion of a chromosome is deleted.
   d. a portion of a chromosome is inverted.
   e. the behavior of two genes is entirely independent.

10. Individual heritable traits:
   a. are always determined by dominant and recessive alleles.
   b. always vary discontinuously.
   c. can sometimes be controlled by many genes.
   d. were first studied in this century.
   e. do not exist outside the laboratory.

FOR STUDY

1. The photograph shows the shells of 15 bay scallops, *Argopecten irradians*. These scallops are hermaphroditic, so that a single individual can reproduce sexually, as did the pea plants of the F_1 generation in Mendel's experiments. Three colors are evident: yellow; orange; and black and white. The color-determining locus has three alleles. The top row shows a yellow scallop and a representative sample of its offspring, the middle row shows a black-and-white scallop and its offspring, and the bottom row shows an orange scallop and its offspring. Assign a suitable symbol to each of the three alleles participating in color control, then determine the genotype of each of the three parent individuals and tell what you can about the genotypes of the different offspring. Explain your results carefully.

2. Utilizing the Punnett squares below, show that for typical dominant and recessive autosomal traits, it does not matter which parent contributes the dominant allele and which the recessive allele. Cross true-breeding tall plants (*TT*) with true-breeding dwarf plants (*tt*).

Tall Female	× Dwarf Male

Dwarf Female	× Tall Male

Male gametes

Female gametes		

Male gametes

Female gametes		

3. Show diagrammatically what occurs when one selfs the F_1 offspring of the cross in Question 2.

Male gametes

Female gametes

4. A new student of genetics suspected that a particular recessive trait (dumpy wings) was sex-linked. One single mating between a fly having dumpy wings (*dp*; female) and a fly with wild-type wings (*Dp*; male) produced 3 dumpy-winged females and 2 wild-type males. On the basis of these data, is the trait sex-linked or autosomal? What were the genotypes of the parents? Explain how these conclusions can be reached on the basis of so few data.

5. The sex of fishes is determined by the same X–Y system as in humans and *Drosophila*. An allele of the *maculatus* locus on the Y chromosome of the fish *Lebistes* causes a pigmented spot to appear on the dorsal fin. A male fish with a spotted dorsal fin is mated with a female fish with an unspotted fin. Describe the phenotypes of the F_1 and the F_2 from this cross.

6. In *Drosophila melanogaster*, the recessive allele *p*, when homozygous, determines pink eyes. *Pp* or *PP* results in wild-type eye color. Another gene, on another chromosome, has a recessive allele, *sw*, that produces short wings when homozygous. Consider a cross between females of genotype *PPSwSw* and males of genotype *ppswsw*. Describe the phenotypes and genotypes of the F_1 generation and of the F_2 generation produced by allowing the F_1 to mate with one another.

7. On the same chromosome of *Drosophila melanogaster* that carries the *p* (pink eyes) locus, there is another locus that affects the wings. Homozygous recessives, *byby*, have blistery wings, while the dominant allele *By* produces wild-type wings. The *p* and *by* loci are very close together on the chromosome; that is, the two loci are tightly linked. In answering these questions, assume that no crossing-over occurs.

a. For the cross *PPByBy* × *ppbyby*, give the phenotypes and genotypes of the F_1 and of the F_2 produced by interbreeding of the F_1.

b. For the cross *PPbyby* × *ppByBy*, give the phenotypes and genotypes of the F_1 and of the F_2.

c. For the cross of Question 7b, what further phenotype(s) would appear in the F_2 generation if crossing over occurred?

d. Draw a nucleus undergoing meiosis, at the stage in which the crossing over (Question 7c) occurred. In which generation (P, F_1, or F_2) did this crossing over take place?

8. Consider the following cross of *Drosophila melanogaster* with alleles as described in Question 6. Males with genotype *Ppswsw* are crossed with females of genotype *ppSwsw*. Describe the phenotypes and genotypes of the next generation.

9. In the Blue Andalusian fowl, a single pair of alleles controls the color of the feathers. Three colors are observed: "blue," black, and "splashed white." Crosses among these three types yield the following results:

Parents	Progeny
Black × blue	Blue and black (1:1)
Black × splashed white	Blue
Blue × splashed white	Blue and splashed white (1:1)
Black × black	Black
Splashed white × splashed white	Splashed white

a. What progeny would result from the cross blue × blue?

b. If you wanted to sell eggs, all of which would yield blue fowl, how should you proceed?

10. In *Drosophila melanogaster*, white (*w*), eosin (*w^e*), and wild-type red (*w^+*) are multiple alleles of a single locus for eye color. This locus is on the X chromosome. An eosin-eyed female is crossed with a male with wild-type eyes. All the female progeny are red-eyed; half the male offspring have eosin (pale orange) eyes, and half have white eyes.

a. What is the order of dominance of these alleles?

b. What are the genotypes of the parents and progeny?

11. Two people with normal vision have two sons, one color-blind and one with normal vision. If the couple also has daughters, what proportion of them will have normal vision? Explain.

12. A mouse with an agouti coat is mated with an albino mouse of genotype *aabb*. Half the offspring are albino, one-quarter are black, and one-quarter are agouti. What are the genotypes of the agouti parent and of the various kinds of offspring? (Hint: see the section "From Genotype to Phenotype.")

13. Sweet peas (genotype *aaBB*) with white flowers are crossed with sweet peas with purple flowers. Of the progeny, half have purple flowers and half have white flowers. What can you say about the genotype of the purple-flowered parent? (Hint: this is another problem dealing with epistasis.)

READINGS

Mange, A. P. and E. J. Mange. 1990. *Genetics: Human Aspects*, 2nd Edition. Sinauer Associates, Sunderland, MA. Genetics, especially chromosomal inheritance, can be studied using humans as examples; this book does so at an introductory level.

McKusick, V. A. 1981. "The Anatomy of the Human Genome." *Hospital Practice*, April. An article on the increasingly detailed maps of human chromosomes. Explains how the maps are created and what they mean to an understanding of human genetic diseases.

Russell, P. J. 1986. *Genetics*. Little Brown, Boston. A well-balanced treatment of a broad range of topics in genetics. Highly recommended.

Sapienza, C. 1990. "Parental Imprinting of Genes." *Scientific American*, October. When reciprocal crosses aren't equivalent.

Stern, C. and E. R. Sherwood (Eds.). 1966. *The Origin of Genetics: A Mendel Source Book*. W. H. Freeman, New York. A collection of the writings of researchers at the dawn of the science of genetics, including translations of Mendel's papers and letters. The last two articles discuss the likelihood that Mendel fudged his data.

Strickberger, M. W. 1986. *Genetics*, 3rd Edition. Macmillan, New York. Good treatment of the topics covered in this chapter; disappointing in molecular areas.

Sturtevant, A. H. and G. W. Beadle. 1962. *An Introduction to Genetics*. Dover, New York. First published by W. B. Saunders in 1939. Though old, this text holds up as a fine introduction to formal chromosome genetics.

Suzuki, D. T, A. J. F. Griffiths, J. H. Miller and R. C. Lewontin. 1989. *An Introduction to Genetic Analysis*, 4th Edition. W. H. Freeman, New York. An excellent textbook of modern genetics. Chapters 2 and 3 are particularly relevant to this chapter, but Chapters 4 and 5 are also useful.

11

Nucleic Acids as the Genetic Material

PREVIEW: The information for producing specific proteins is encoded in DNA, the genetic material. DNA consists of two polymer strands, and the relationship between the strands is determined by rules of complementarity of base pairing. These rules are important in the mechanisms by which DNA is copied and by which DNA codes for the production of an RNA transcript, which carries genetic information to the sites of protein synthesis. Complementarity also plays a role in the interaction of different kinds of RNA participating in the translation of the genetic information into the sequence of amino acids in a protein molecule. DNA codes for RNA, and RNA codes for protein.

This chapter deals with the one-gene, one-polypeptide theory, bacterial transformation, DNA structure, base-pairing rules, RNA structure, semiconservative replication, the central dogma of molecular biology, messenger RNA, transfer RNA, transcription, translation, codons, anticodons, activating enzymes, ribosomes, RNA viruses, the genetic code, and mutations.

Gregor Mendel described the basic patterns of inheritance in plants and animals, and he devised a powerful explanation for the mechanisms underlying these patterns. The second of these accomplishments is even more impressive than the first because Mendel had no way of knowing the physical basis for his proposed mechanisms. He never knew what a gene is, in chemical terms, nor did he know about the behavior of chromosomes. Hence he could not have known how genes are copied between generations or how new alleles arise. He could not have had the slightest inkling of how a gene works, that is, how the genotype produces a phenotype.

The elucidation of these matters has constituted one of the greatest triumphs of twentieth-century biology. Geneticists, biochemists, biophysicists, and molecular biologists have contributed to the solution. In this and the next two chapters, we will try to show you how the main questions of molecular genetics have been studied.

WHAT DOES THE GENE CONTROL?

There are many steps between genotype and phenotype. Genes cannot, all by themselves, directly produce a phenotypic result such as the color of an eye or a flower, the shape of a seed, or a cleft chin, any more than a thermostat can heat a house. What are the steps intervening between the genes on the chromosomes and the phenotype of the organism? The first hints came early in this century from the work of the Scottish physician Archibald Garrod. Alkaptonuria is a hereditary disease in which the patient's urine turns black when exposed to air. Recognizing that this showed the biochemistry of the affected individual to be different from that of other persons, Garrod suggested in 1908 that alkaptonuria and some other hereditary diseases are consequences of "inborn errors of metabolism." He proposed that the dark urine might be due to a defect in an enzyme. From the pattern of its inheritance, Garrod reasoned that alkaptonuria affects individuals who are homozygous for the recessive allele of a gene that in normal individuals codes for an enzyme needed for the metabolism of the amino acid tyrosine. His proposals were the first plausible approach to the problem of gene expression. However, like Mendel's explanation of inheritance in the garden pea, Garrod's proposals were too advanced for their time and sat almost unappreciated for more than 30 years.

Garrod's ideas were confirmed and extended in a series of experiments performed by George W. Beadle and Edward L. Tatum at the California Institute of Technology in the 1940s, using the bread mold *Neurospora crassa* (Chapter 10). *Neurospora* can be grown on a simple, completely defined medium (that is, one in which all the ingredients are known) containing inorganic ions, a simple source of nitrogen (such as ammonium chloride), an organic source of energy and carbon (such as glucose), and a single vitamin (biotin, Chapter 41). From this **minimal medium**, the enzymes of wild-type *Neurospora* can catalyze the metabolic reactions needed to make all the chemical constituents of its cells. Beadle and Tatum reasoned that mutations might alter the enzymes so that they could no longer do their jobs. In that case, mutants of *Neurospora* might be found that would be unable to make certain cellular compounds and thus would grow only on media to which those compounds were added. Mutants of this type have since been called **auxotrophs** (increased eaters) as opposed to the wild-type **prototrophs** (original eaters) that constituted the original *Neurospora* population. Prototrophs grow on minimal medium, whereas auxotrophs require specific additional nutrients.

Beadle and Tatum irradiated cells of *Neurospora* with X rays to increase the frequency of mutations and then isolated a number of nutritional mutants. These auxotrophs grew on a complex medium enriched with amino acids (the monomers of proteins), purines and pyrimidines (the nitrogenous bases of nucleic acids), and vitamins, but did not grow on the minimal medium that supported the growth of the wild-type strain. Beadle and Tatum then tested these mutants to discover the simplest nutritional supplements that would support their growth (Figure 11.1). Among their collection of mutants were individual strains that required some specific amino acid or vitamin. In almost every case, the nutritional requirement turned out to be simple; only a single compound had to be added to the minimal medium to support the growth of any given mutant. This supported the idea that mutations do, in fact, have simple effects—and, perhaps, that each mutation causes a defect in only one enzyme in the metabolic pathway leading to the synthesis of the required nutrient.

The auxotrophs identified in this way could be divided into classes on the basis of the nutritional supplements that would support their growth. For example, all mutants that did not grow on minimal medium but grew on minimal medium supplemented with the amino acid arginine were classified as *arg* mutants. Other sets of mutants were found that required adenine, or proline, or vitamin B_1, and so forth. Within a group of mutants with the same growth requirement, mapping studies established that some of the individual mutations were at different loci on a chromosome or on different chromosomes. This indicated that different genes can govern a common biosynthetic pathway. For example, Beadle and Tatum found no fewer than 15 different *arg* mutants. These were then grown in the presence of various suspected intermediates in the synthetic metabolic pathway for arginine. Different mutants were able to grow on different intermediates as well as on arginine-supplemented medium (Figure 11.2).

One Gene, One Polypeptide

The growth of various mutants on different intermediates helped to determine the biochemical steps by which various amino acids and other compounds are synthesized in *Neurospora*. Much more important, however, this work led to the formulation of the one-gene, one-enzyme theory. According to this theory, the function of a gene is to control the production of a single, specific enzyme. This proposal strongly influenced the subsequent development of the sciences of genetics and molecular biology. Garrod had pointed in the same direction over three decades before, but only now were other scientists prepared to act on the suggestion.

The one-gene, one-enzyme hypothesis was refined in the mid-1950s as a result of work on sickle-cell anemia. This serious disease results from a recessive allele that, when homozygous, results in defective red blood cells. Where oxygen is abundant, as in the lungs, the cells are normal in structure and function. However, at the low oxygen levels characteristic of working muscles, the red blood cells collapse into a shape that has been described as sickle-like (Figure 11.3). Linus Pauling, at the California Institute of Technology, speculated that the disease results from a defect in hemoglobin, a protein that carries oxygen and that literally fills red blood cells. Human hemoglobin is a tetramer, consisting of two each of two different polypeptide chains. Vernon Ingram, working at Cambridge University in England, showed that one of the two kinds of polypeptides differs by one amino acid between normal and sickle-cell hemoglobin. Ingram's results suggested the more satisfying **one-gene, one-polypeptide theory**: The function of a gene is to control the production of a single, specific polypeptide. Much later, it was discovered that some genes code for forms of RNA that do not get translated into polypeptides.

WHAT IS THE GENE?

For the first half of this century, it was widely assumed that the hereditary material was protein, for it was well known that the proteins have impressive chemical diversity and that some of them—notably enzymes and antibodies—show great specificity. In contrast, nucleic acids were known to have only a few components and seemed too simple to carry the complex information expected in the genetic material. The recognition that the gene is not a protein, but

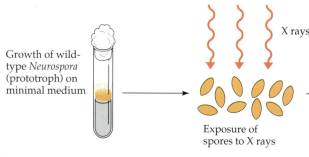

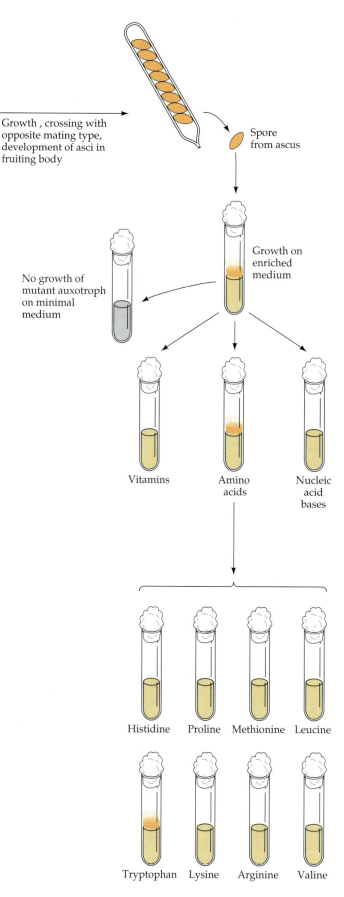

Growth of wild-type *Neurospora* (prototroph) on minimal medium

X rays

Exposure of spores to X rays

Growth, crossing with opposite mating type, development of asci in fruiting body

Spore from ascus

11.1 Detecting Mutations in *Neurospora*

Spores from a wild-type culture are X-rayed, then placed on enriched medium where the survivors, including mutants, form individual colonies. The haploid offspring (spores) are collected. When individually placed on an enriched medium, each spore forms a fungal mat that contains only one kind of haploid nucleus. Part of each mat is placed on minimal medium to test whether it is a mutant auxotroph. The auxotroph's nutritional requirement (in this case, tryptophan) is determined by finding which substance supports its growth. Because each spore is haploid, by the end of the procedure the phenotype reveals the genotype—an important reason for using *Neurospora*.

No growth of mutant auxotroph on minimal medium

Growth on enriched medium

Vitamins   Amino acids   Nucleic acid bases

Histidine   Proline   Methionine   Leucine

Tryptophan   Lysine   Arginine   Valine

Minimal medium supplemented with specific amino acids

rather deoxyribonucleic acid, or **DNA**, was slow in coming and depended on the interaction of several types of research.

The Transforming Principle

Often in the history of biology, research on some specific topic has—with or without answering the question originally under investigation—contributed richly to another, apparently unrelated area. Such a case is the work of Frederick Griffith, an English physician. In the 1920s Griffith was studying the disease-causing behavior of the bacterium *Streptococcus pneumoniae*, or pneumococcus, one of the agents that produce pneumonia in humans. He had identified two strains of pneumococcus, designated S and R because the former produces shiny, smooth (S) colonies when grown in the laboratory, whereas the colonies of the latter are rough (R) in appearance. When the S strain was injected into mice, they died within a day, and the hearts of the dead mice were found to be teeming with the deadly bacteria. When the R strain was injected instead, the mice did not become diseased. In other words, the S strain is virulent and the R strain is nonvirulent. This difference was eventually shown to be due to a difference in the chemical makeup of the bacterial surface. The S strain has a polysaccharide capsule that protects the bacterium from the defense mechanisms of the host (Chapter 16). The R strain lacks this capsule and can be inactivated by a mouse's defenses. In hopes of developing a vaccine against pneumonia, Griffith inoculated other mice with heat-killed pneumococci.

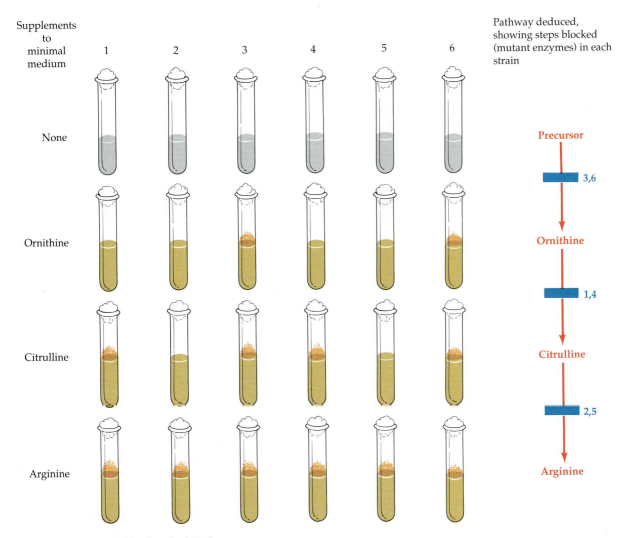

Supplements to minimal medium

Pathway deduced, showing steps blocked (mutant enzymes) in each strain

None

Ornithine

Citrulline

Arginine

Precursor

3,6

Ornithine

1,4

Citrulline

2,5

Arginine

11.2 Dissecting a Biochemical Pathway

You have just isolated six arginine-requiring auxotrophic mutants of *Neurospora*. You know that ornithine and citrulline are chemically related to arginine and think they may be intermediates in its synthesis. Let's use these six mutants to deduce the biosynthetic pathway to arginine. All six strains grow when supplied with arginine (fourth row of cultures) but not on unsupplemented minimal media (top row). Strains 2 and 5 grow only on media supplemented with arginine. What does this suggest? Because these strains cannot grow on either ornithine or citrulline, their mutations interfere with the synthesis of arginine itself—the final step. Strains 1 and 4 grow when supplied with citrulline but not ornithine. What does this suggest? Their mutations block the synthesis of citrulline but permit the conversion of citrulline to arginine. Thus part of the pathway must be citrulline → arginine. Note that strains 1 and 4 do not grow when supplemented with ornithine, suggesting that if ornithine is an intermediate, then it must occur before this genetic block. What can you infer from the behavior of strains 3 and 6? We leave that for you to decide. The simplest pathway consistent with all these observations is the one to the right of the illustration. The horizontal bars show the reactions blocked by the mutation in each strain.

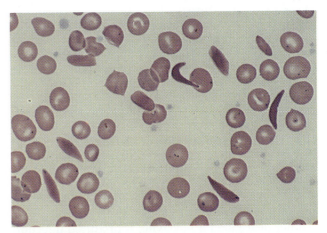

11.3 Sickled Blood Cells

Most of these human red blood cells are normal; note the flattened, roughly circular shape and the concave centers. Some of the cells, recognizable by their shape, are sickled. The sickled cells cannot carry a normal amount of oxygen and nutrient supply to various tissues. The change in shape results from a single amino acid substitution in one of the polypeptides of the protein hemoglobin.

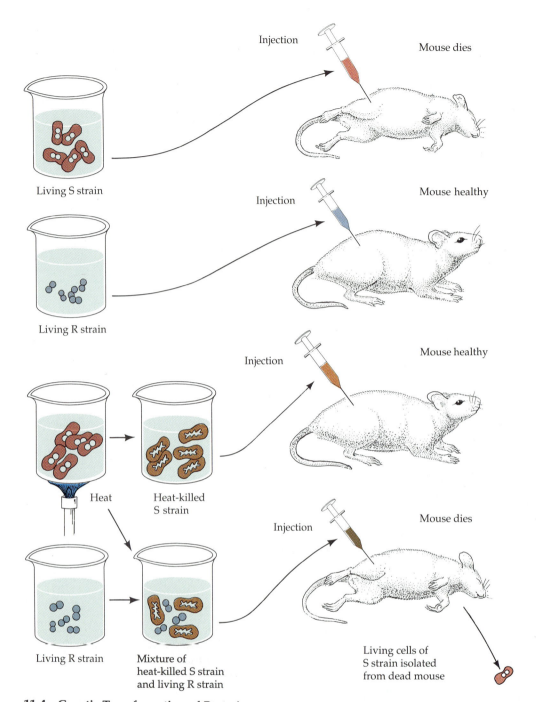

Injection — Mouse dies

Living S strain

Injection — Mouse healthy

Living R strain

Injection — Mouse healthy

Heat — Heat-killed S strain

Injection — Mouse dies

Living R strain — Mixture of heat-killed S strain and living R strain

Living cells of S strain isolated from dead mouse

11.4 Genetic Transformation of Bacteria
Mice injected with living cells of S strain pneumococcus die. Mice injected with living cells of R strain remain healthy, as do those injected with heat-killed S cells. Mice injected with a mixture of living R cells and heat-killed S cells die. Surprisingly, their blood contains living S cells.

Neither heat-killed S nor living R produced infection (Figure 11.4).

Next Griffith inoculated mice with a mixture of living R and heat-killed S bacteria. To his astonishment, all these mice died of pneumonia. When he examined blood from the hearts of these mice, he found it full of living bacteria, many of them belonging to the virulent S strain! He concluded that some of the inoculated R bacteria had been transformed in the presence of the dead S pneumococci.

Did transformation of the bacteria depend upon something done by the mouse? No. It was soon shown that the same transformation occurs when living R and heat-killed S bacteria are simply incubated together in a test tube. Next it was discovered that R cells can be transformed by a cell-free extract (containing all the chemicals but no intact cells) from heat-killed S cells. This demonstrated that some substance—called at the time a chemical **transforming principle**—from the dead S pneumococci can cause a permanent change in the affected R cells (remember that great numbers of *living* bacteria of the S type were always found in the dead mice). The transforming principle thus carried heritable information; and it could be thought of as genetic material. We now know that virtually any genetic trait in pneumococcus or in several other types of bacteria can be passed, by way of transforming principles, from one bacterium to another.

The Transforming Principle is DNA

A crucial step in the history of biology was the identification of the transforming principle, accomplished over a period of several years by Oswald T. Avery and his colleagues at what is now Rockefeller University. They treated samples of the transforming principle in a variety of ways to destroy different types of substances—proteins, nucleic acids, carbohydrates, lipids—and then tested the treated samples to see whether they retained transforming activity. Always the answer was the same: If the DNA in the sample was destroyed, transforming activity was lost; everything else was dispensable. As a final step, Avery, with Colin MacLeod and MacLyn McCarty, isolated virtually pure DNA from a sample of pneumococcal transforming principle and showed that it was highly active in causing bacterial transformation. Their work, published in 1944, is a milestone in establishing that DNA is the genetic material in cells; it did not, however, quickly receive the attention it deserved, and scientists in other laboratories continued their attempts to identify the hereditary material.

The Genetic Material of a Virus

A report in 1952 by Alfred D. Hershey and Martha Chase of the Carnegie Laboratory of Genetics had a much greater immediate impact than did Avery's paper. The "Hershey–Chase experiment" was carried out with a virus that infects bacteria. This virus, which is called T2 bacteriophage, consists of a DNA core packed within a protein coat (Figure 11.5a). A T2 attacks a bacterial cell; about 20 minutes later the bacterial cell lyses, releasing 200 to 1,000 new T2s. The idea behind Hershey and Chase's work was to trace the two components—protein and DNA—during the life cycle of the virus by labeling each with a specific radioactive tracer. Virtually all proteins contain some sulfur (in the amino acids cysteine and methionine), an element not present in DNA, so the proteins could be labeled with ^{35}S. The deoxyribose–phosphate "backbone" of DNA, on the other hand, is rich in phosphorus (see Chapter 3), which is not present in proteins, and could be uniquely tagged with ^{32}P. One batch of T2 was grown in a bacterial culture in the presence of ^{32}P, so that all the viral DNA was labeled with ^{32}P (Figure 11.5b). Similarly,

the proteins of another batch of T2 were labeled with ^{35}S.

In separate experiments, radioactive viruses containing either ^{32}P or ^{35}S were combined with bacteria (Figure 11.5c). After a few minutes, the mixtures were swirled vigorously in a kitchen blender, which stripped the virus coats away but did not burst the bacteria. The bacteria were then separated from the virus coats in a centrifuge. It was found that over three-fourths of the ^{35}S had separated from the bacteria, and that two-thirds or more of the ^{32}P had settled to the bottom of the centrifuge tube with the bacteria. Although the numbers were not as clear-cut as one might like, this suggested that the DNA was transferred to the bacteria while the protein remained outside. Confirmation came when other batches of bacteria and labeled T2 were incubated together for longer periods, so that a progeny generation of viruses could be collected. When this was done, Hershey and Chase found that the T2 progeny contained less than 1 percent of the original ^{35}S but about one-third of the original ^{32}P— and thus, presumably, one-third of the DNA. Because DNA was carried over from virus generation to generation, whereas protein was not, it followed that the hereditary information of the viruses is contained in the DNA. That is, the DNA contains the information for the replication of the virus as well as for the protein coat (Chapter 12). The Hershey–Chase experiment convinced most scientists that DNA is indeed the carrier of hereditary information.

NUCLEIC ACID STRUCTURE

Once it was clearly established that the genetic material is DNA, it became crucially important to learn just what the DNA molecule looks like. In its structure one could hope to find clues as to how the molecule is replicated between nuclear divisions and how it causes the synthesis of specific proteins. Both these expectations were fulfilled.

Evidence From X-Ray Crystallography and Biochemistry

The discovery of the structure of DNA resulted from many types of experimental evidence and theoretical considerations. The most crucial "hard" evidence was obtained by X-ray crystallography. The positions of atoms in a crystalline substance can be inferred by the pattern of diffraction of X rays passed through the crystal, but even today this is not an easy task when the substance is of enormous molecular weight. In the early 1950s, even a highly talented X-ray crystallographer could (and did) look at the best available images from DNA preparations and fail to see what they meant. Nonetheless, the whole at-

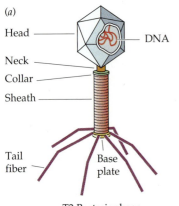

(a)

Head

DNA

Neck

Collar

Sheath

Tail
fiber

Base
plate

T2 Bacteriophage

11.5 The Hershey–Chase Experiment

(a) The external protein structures of the bacteriophage T2. Within the head is a strand of DNA. A virus is made of the two materials that were, at the time of this experiment, leading candidates for the genetic material. (b) Viruses were grown on bacteria containing radioactive ^{32}P or ^{35}S to label their DNA or their protein coats. (c) Labeled viruses were separately mixed with nonradioactive bacteria. Later the mixtures were agitated in a blender to shear the viruses off the cells, then centrifuged to concentrate the cells at the bottom and the viruses in the supernatant liquid. Radioactive ^{32}P from the DNA of the labeled viruses entered the bacteria and appeared largely in the progeny viruses. Radioactive ^{35}S from the proteins of the labeled viruses remained largely in the supernatant fluid, little entering the bacteria. Hershey and Chase concluded that virus DNA entering the bacteria carries genetic information for forming new viruses and must therefore be the genetic material.

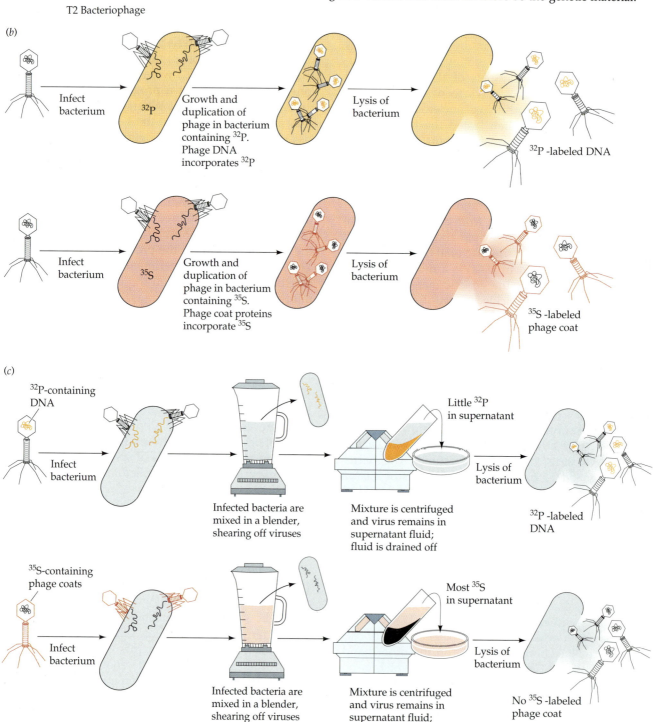

(b)

Infect bacterium → Growth and duplication of phage in bacterium containing ^{32}P. Phage DNA incorporates ^{32}P → Lysis of bacterium → ^{32}P -labeled DNA

Infect bacterium → Growth and duplication of phage in bacterium containing ^{35}S. Phage coat proteins incorporate ^{35}S → Lysis of bacterium → ^{35}S -labeled phage coat

(c)

^{32}P-containing DNA → Infect bacterium → Infected bacteria are mixed in a blender, shearing off viruses → Mixture is centrifuged and virus remains in supernatant fluid; fluid is drained off → Little ^{32}P in supernatant → Lysis of bacterium → ^{32}P -labeled DNA

^{35}S-containing phage coats → Infect bacterium → Infected bacteria are mixed in a blender, shearing off viruses → Mixture is centrifuged and virus remains in supernatant fluid; fluid is drained off → Most ^{35}S in supernatant → Lysis of bacterium → No ^{35}S -labeled phage coat

tempt to characterize DNA would have been impossible without the crystallographs that had been prepared by the English chemist Rosalind Franklin and, earlier, by another English crystallographer, W. T. Astbury (Figure 11.6). Franklin's work, in turn, depended upon the success of the English biophysicist Maurice Wilkins in preparing very uniformly oriented DNA fibers, which made far better samples for crystallography than had previous samples.

Information on the chemical composition of DNA was also important. Biochemists knew the chemical structures of the four monomers, or nucleotides, of DNA and knew how one nucleotide was joined to another to form a polynucleotide chain. A nucleotide of DNA consists of a molecule of the sugar deoxyribose, a phosphate group, and a nitrogen-containing base (see Figures 3.24 and 3.25). The only differences between the four nucleotides of DNA lie in their nitrogenous bases: the purines **adenine** and **guanine**, and the pyrimidines **cytosine** and **thymine**. In 1950 Erwin Chargaff at Columbia University reported observations of major importance. He and his colleagues had found that DNA from many different species—and from different sources within a single organism—exhibits interesting regularities. In *any* DNA the following rules hold: The amount of adenine equals the amount of thymine, and the amount of guanine equals the amount of cytosine. As a result, the amount of combined purines equals that of the combined pyrimidines. The structure of DNA could scarcely have been worked out without this information, yet its significance to the problem was missed for at least three years.

Watson, Crick, and the Double Helix

Another ingredient of the solution of the puzzle was the technique of model building, that is, the assembling of three-dimensional representations of possible molecular structures. This technique, originally exploited in structural studies by the American chemist Linus Pauling, was used by the English physicist Francis Crick and the American geneticist James D. Watson, then both at the Cavendish Laboratory of Cambridge University. Watson and Crick attempted to put together all that had so far been learned about DNA structure into a single, coherent model. Astbury's and Franklin's crystallographs convinced Watson and Crick that the DNA molecule is **helical** (cylindrically spiral) and also gave them the values of certain distances within the helix. The results of density measurements and model building suggested that there are two polynucleotide chains in the molecule. The modeling studies also led to the conclusion that the two chains in DNA run in opposite directions, that is, that they are **antiparallel**.

Crick and Watson attempted a number of models. Late in February of 1953, they built the one that

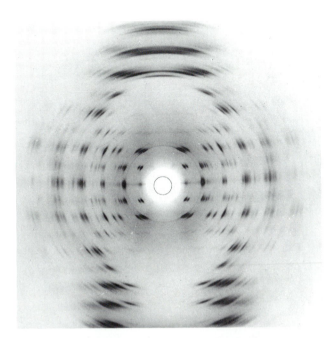

11.6 DNA Seen Using X-Ray Crystallography
An X-ray diffraction pattern of DNA. Analysis of such patterns was crucial in deciphering the structure of the DNA molecule. The pattern seen here was provided by the laboratory of Maurice Wilkins at King's College in London.

established the general structure of DNA (Figure 11.7). There have been minor amendments to their first published structure, but the principal features have remained unchanged.

Key Elements of DNA Structure

Four main points summarize the molecular architecture of DNA: The molecule is (1) a double-stranded helix, (2) of uniform diameter, (3) twisting to the right (that is, twisting in the same direction as the threads on most screws), with (4) the two strands running in opposite directions. As you know, the two strands are polynucleotide chains. The sugar–phosphate "backbones" of the chains coil around the outside of the helix, with the nitrogenous bases pointing toward the center. The two chains are held together by hydrogen bonding between specifically paired bases. As implied by Chargaff's studies, adenine (A) pairs with thymine (T) by forming two hydrogen bonds, and guanine (G) pairs with cytosine (C) by forming three hydrogen bonds. Because the A–T and G–C pairs, like rungs of a ladder, are of equal length and fit identically into the double helix, the diameter of the helix is uniform. A pair of purines would cause a swelling in the molecule and a pair of pyrimidines would cause a constriction, but DNA never includes such pairs. You can see the first two points about DNA's architecture in Figure 11.7: Two strands join to form a double helix of uniform diameter. Before

(a)

(b)

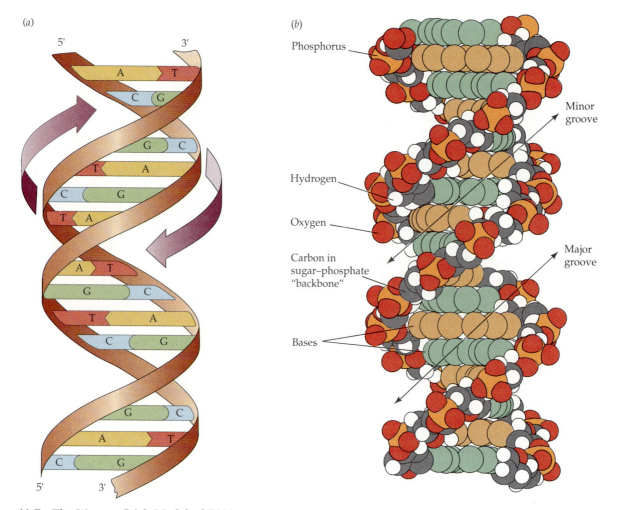

11.7 The Watson–Crick Model of DNA
(a) Watson and Crick proposed that DNA is a double-helical molecule. The brown bands represent the two sugar–phosphate chains, with pairs of bases forming horizontal connections between the chains. The two chains run in opposite directions. (b) Biochemists can now pinpoint the position of every atom in a DNA macromolecule. To see that the essential features of the original Watson–Crick model have been verified, follow with your eyes the double helical ribbons of sugar and phosphate groups and note the horizontal rungs of the bases.

leaving Figure 11.7, notice that two grooves, one broad (the major groove) and one narrow (the minor groove), spiral around the outside of the molecule.

What do we mean when we say that DNA twists to the right and that the two strands run in opposite directions? We mean that the DNA twists in the same way as the threads on most screws. The direction of a polynucleotide can be defined by looking at the linkages between adjacent nucleotides. (These linkages are called phosphodiester bonds.) To avoid confusion with the carbon and nitrogen atoms in the ring structure of the bases which are numbered 1, 2, 3, . . . , a prime is placed after a number that refers to a carbon atom of a sugar: 1′, 2′, 3′, ... In the sugar–phosphate backbone of DNA, the phosphate groups connect to the 3′ carbon of one deoxyribose molecule and the 5′ carbon of the next, linking suc-

cessive sugars together (Figure 11.8). The two ends of a polynucleotide differ from one another. Polynucleotides have a free (not connecting to another nucleotide) 5′ phosphate group at one end—the 5′ end—and a free 3′ —OH group at the other end—the 3′ end—just as polypeptides have a free amino group at one end and a free carboxyl group at the other end. The 5′ end of one strand in a DNA double helix is paired with the 3′ end of the other strand, and vice versa; that is, the strands run in opposite directions.

Alternative Structures for DNA

Thus far we have spoken of DNA as if it has an invariant shape: a right-handed double helix, with two grooves of unequal width spiraling up its side.

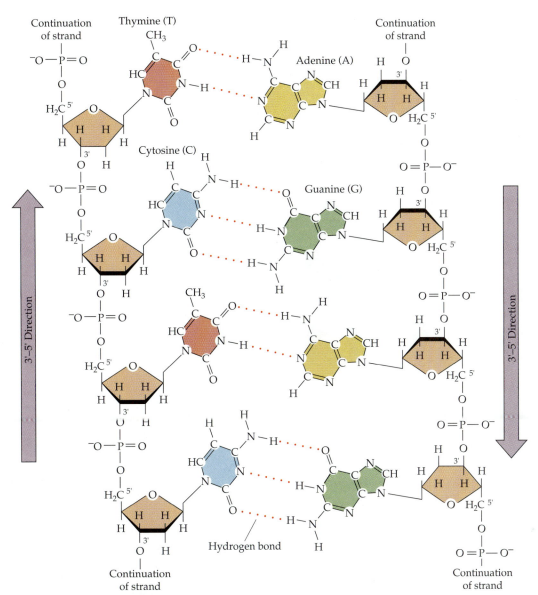

11.8 Complementary Pairing in DNA
In the sugar–phosphate backbone of DNA, each phosphate group links the 3' carbon of one sugar to the 5' carbon of the next sugar along the backbone. This asymmetry gives each DNA strand a 5' "head" and a 3' "tail." Complementary strands line up head-to-tail. Pairs of complementary bases form hydrogen bonds that hold the two strands of a DNA double helix together. T–A pairs form two hydrogen bonds and G–C pairs form three hydrogen bonds.

This is the structure of **B-DNA**, the form in which most DNA appears in living cells. Over the past 30 years, a number of relatively minor variations on this theme have been discovered. In 1979, however, a strikingly different structure was observed in some samples of DNA (Figure 11.9). These **Z-DNA** molecules twist to the *left* rather than to the right, and they have only a single groove. The sugar–phosphate backbones of Z-DNA follow a zigzag course (hence the Z) rather than the smooth spiral of B-DNA backbones. The structure is a left-handed double helix. It

was first observed in small, synthetic DNA molecules with alternating purine and pyrimidine bases on each strand. However, it now seems that short stretches of Z-DNA appear naturally in the DNA of living organisms, for example, in the chromosomes of the fruit fly *Drosophila melanogaster*.

We do not yet know how or whether Z-DNA plays a role distinct from that of B-DNA. It has been suggested that the sharply different shape of Z-DNA makes it recognizable to one or more proteins that may play regulatory roles, but that is mere speculation as this book is written.

Structure of Ribonucleic Acid

When we discuss how DNA functions, we will need to talk about various kinds of ribonucleic acid (**RNA**). RNA is a polynucleotide that is similar to DNA in many ways (see Figure 3.25). However, there are

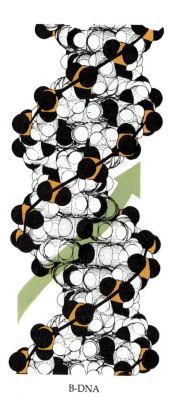

Z-DNA   B-DNA

11.9 The Z Form of DNA

Z-DNA twists to the left rather than to the right as does B-DNA. The difference between the two forms is evident if you compare the phosphate backbones (orange atoms) on each model; the zig-zag ("Z") pattern of the black line connecting the atoms is clearly different from the smooth spiral in B-DNA.

three important differences between the two kinds of nucleic acids. First, RNA generally consists of only a single polynucleotide strand; thus Chargaff's equalities (G = C, A = T) are true only for DNA and not for RNA. Second, the sugar molecule found in ribonucleotides is ribose, rather than deoxyribose as in DNA. Last, three of the nitrogenous bases found in ribonucleotides are identical with those bases found in deoxyribonucleotides—adenine, guanine, and cytosine—but the fourth is uracil (U), which is similar to thymine but lacks the methyl (—CH_3) group. Although RNA is usually single-stranded, it may contain internal complementary base pairing, as we will see in the discussion of transfer RNA later in this chapter. When RNA does pair, uracil is like thymine in that it is complementary to adenine in base pairing.

Implications of the Double-Helical Structure of DNA

The structure of DNA, as first set forth by Watson and Crick, had implications for its biological role. First, the molecule is, in a sense, boring—it runs on and on, nucleotide pair after nucleotide pair, with no kinks and no bulges. There is evidently only one way in which such a molecule could possibly carry and convey information: The information must lie in the linear sequence of the nitrogenous bases.

An implication of the **complementary pairing** of the bases, A with T and G with C, was pointed out by Crick and Watson in their original publication in

the journal *Nature*: "It has not escaped our notice that the specific pairing we have postulated immediately suggests a possible copying mechanism for the genetic material." Each strand of the double helix is complementary to the other—at each point, the base on one strand is complementary with the base on the other strand. If the spirals unwound, each strand could serve as a guide for the synthesis of a new one; the single strands of the parent double helix could each produce a new double-stranded molecule identical to the original. It also appeared that all the information needed to construct new DNA molecules would already be present in the old molecule, because that information is represented by the sequence of bases.

The structure of DNA also suggested an explanation for some mutations: They might simply be changes in the linear sequence of nucleotide pairs. In sum, *the three major properties of genes—function, replication, and mutation—could all be accounted for in terms of the structure of DNA.*

REPLICATION OF THE DNA MOLECULE

Watson and Crick's prediction that the DNA molecule contains all the information needed for its own replication was borne out just three years later in work by Arthur Kornberg at Washington University in St. Louis. He showed that DNA can replicate in a test tube with no cells present. What is required is a mixture containing a specific enzyme, which he called DNA polymerase, and a mixture of the four precursors: the nucleoside triphosphates deoxy-ATP, deoxy-CTP, deoxy-GTP, and deoxy-TTP (Figure 11.10). If any one of the four is omitted from the reaction mixture, DNA does not replicate itself. The requirement that all four nucleoside triphosphates be available suggested that the sequence of the newly formed DNA is specific, not random. The fact that the reaction will not occur without a small amount of intact DNA confirms that the sequence is specific. The intact DNA serves as a template for the reaction—a guide to the exact placement of nucleotides in the new strand. Where there is a T in the template, there must be an A in the new strand, and so forth.

Actually, the model for DNA replication suggested in the original paper by Watson and Crick is only one of three possible ways in which the double helix might be replicated. Gunther Stent and Max Delbrück recognized this, and proposed three possible mechanisms (Figure 11.11). In conservative replication, the original double helix would somehow serve as a template but would either be reconstituted or, perhaps, never unwind at all. Thus the new molecule would contain none of the atoms of the original one. Dispersive replication would include a fragmentation of

the original molecule, the functioning of these fragments as templates, and the assembly of two molecules, each containing both old and new parts, perhaps at random. Finally, there was **semiconservative replication** (proposed by Watson and Crick), in which the original two strands would separate and each would function as the template for a new partner. Each molecule produced by semiconservative replication would consist of one old and one new strand.

Demonstration of Semiconservative Replication

Semiconservative replication was shown to be the correct model in a clever experiment by Matthew Meselson and Franklin Stahl at the California Institute of Technology in 1957. They devised a simple way to tell old strands from new ones. The key was to use a "heavy" isotope of nitrogen. Heavy nitrogen (^{15}N) is a rare, nonradioactive isotope that makes molecules more dense than chemically identical molecules containing the common isotope ^{14}N. To study DNA of different densities (that is, DNA containing ^{15}N versus DNA containing ^{14}N), Meselson, Stahl, and Jerry Vinograd invented a new type of centrifugation procedure. At a certain molarity, a solution of cesium chloride (CsCl) has a density very close to

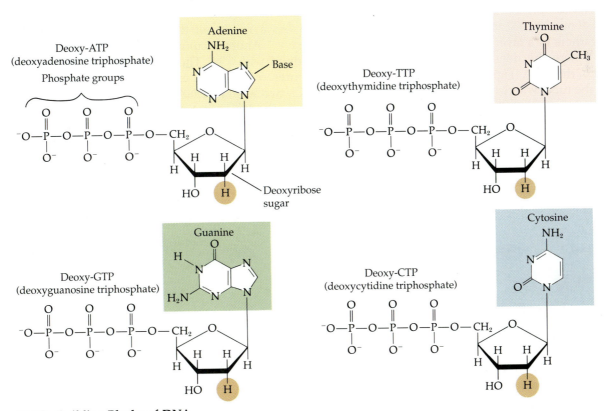

11.10 Building Blocks of DNA
The four deoxyribonucleoside triphosphates that form DNA differ only in their nitrogenous bases.

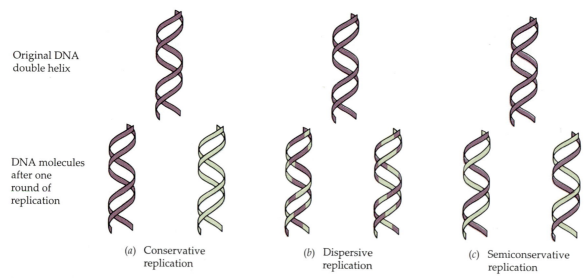

Original DNA
double helix

DNA molecules
after one
round of
replication

(a) Conservative
replication

(b) Dispersive
replication

(c) Semiconservative
replication

11.11 Proposed Models for DNA Replication
Three possible ways for DNA molecules to replicate; all obey the base pairing rules. The two original strands of DNA are shown here in purple, newly synthesized DNA in green. (a) Conservative replication would preserve the original molecule and generate an entirely new molecule. (b) Dispersive replication would produce two molecules with old and new DNA interspersed along each strand. (c) Semiconservative replication would also produce molecules with both old and new DNA, but each molecule would contain one old strand and one new one.

that of DNA; at high gravitational forces produced in an ultracentrifuge, cesium ions actually sediment to some extent, thus establishing a density gradient. When a DNA sample is dissolved in CsCl and centrifuged at about 100,000 times the force of gravity, the DNA gathers in a layer in the centrifuge tube at a position where the density of the CsCl solution equals that of the DNA (Figure 11.12). Any DNA that is lower in the tube, where the density is greater than its own, must rise; DNA that is in a region of lower density must sink.

Once they could measure DNA densities in this way, Meselson and Stahl could begin experimenting. They grew a culture of *Escherichia coli* for 17 generations on a medium in which the nitrogen source (ammonium chloride) was made with ^{15}N instead of ^{14}N. As a result, all the DNA in the bacteria was "heavy." Another culture was grown, as usual, on medium with ^{14}N. They extracted DNA from both cultures. When these extracts were centrifuged with CsCl, two separate DNA bands formed, showing that this method for separating DNA samples of slightly different densities actually worked.

Another culture was grown on ^{15}N medium and

Tubes of DNA and
CsCl before centrifugation

Same tubes after a brief
period of centrifugation

Same tubes much later, after
equilibrium has been reached

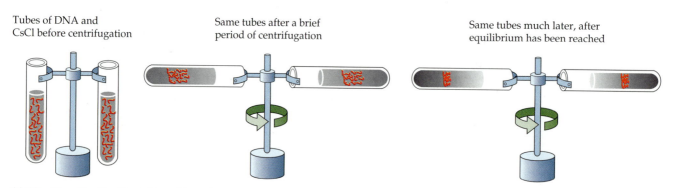

11.12 Density Gradient Centrifugation
When a solution of cesium chloride is centrifuged at extremely high speed, the cesium ions tend to sink slightly, forming a gradient along the tube. Another substance in the tube will float at the point where its density matches that of the gradient. In this illustration, the red DNA molecules aggregate in a single band.

then transferred to normal ^{14}N medium. At various times after the transfer, Meselson and Stahl collected some of the bacteria and extracted DNA from them. Figure 11.13 shows that with time, as DNA was duplicated and the cells divided, the density of the DNA changed. Initially, the DNA was uniformly labeled with ^{15}N and hence was relatively dense. After one generation, when the DNA had been duplicated once, all the extracted double-stranded DNA was of an intermediate density. After two generations, there were two equally large DNA bands: one of low density and one of intermediate density. In subsequent generations, the amount of intermediate-density DNA remained stable, but the amount of low-density DNA increased steadily.

These data can be explained by the semiconservative model of DNA replication. The high-density DNA had two ^{15}N strands, the intermediate-density DNA had one ^{15}N and one ^{14}N strand, and the low-density DNA had two ^{14}N strands. In the first round of DNA replication, the strands of the double helix, both heavy with ^{15}N, separated; during the process of separation, each acted as the template for a second strand, which contained only ^{14}N and hence was less dense. Each double helix then consisted of one ^{15}N and one ^{14}N strand and was of intermediate density. In the second replication, the ^{14}N-containing strands directed the synthesis of partners with ^{14}N, and the ^{15}N strands got new ^{14}N partners.

Were DNA to replicate by the other models, quite different results would have been obtained. Under the conservative model, there would have been two bands (one for heavy ^{15}N–^{15}N and the other for light ^{14}N–^{14}N) after one generation, and no DNA of intermediate density would have formed at any time. In dispersive replication, the first round of replication would have produced DNA of intermediate density, but the density of all the DNA would have decreased after each subsequent replication. In fact, however, the results obtained were the ones predicted by the semiconservative model. The crucial observation was that "half-heavy," intermediate-density (^{15}N–^{14}N) DNA did appear and continued to appear in subsequent generations.

Replicating an Antiparallel Double Helix

How is semiconservative DNA replication accomplished? We now know that there are different kinds of DNA polymerase, with different functions, and that the replication process is an intricate one. Kornberg's basic observations, including the need for a DNA template and for a mixture of nucleoside triphosphates, still hold. He had also shown that nucleotides are always added to the growing chain at the same end: the **3' end**, the end on which the DNA strand has a free hydroxyl group on the 3' carbon of its terminal deoxyribose (Figures 11.8 and 11.14). This hydroxyl group reacts with a phosphate group on the 5' carbon of the deoxyribose of a deoxynucleoside triphosphate (Figure 11.10), and thus the chain grows. Bonds linking the phosphate groups of the

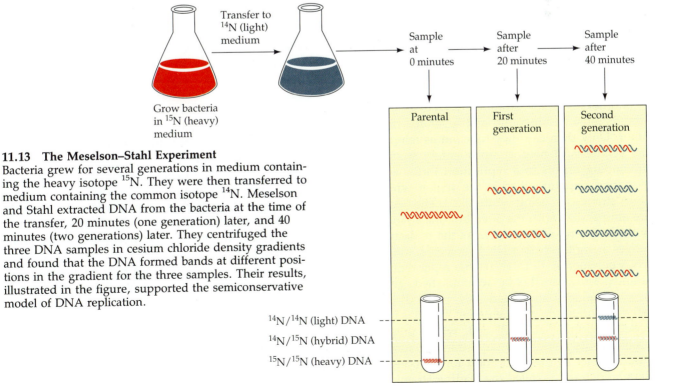

11.13 The Meselson–Stahl Experiment
Bacteria grew for several generations in medium containing the heavy isotope ^{15}N. They were then transferred to medium containing the common isotope ^{14}N. Meselson and Stahl extracted DNA from the bacteria at the time of the transfer, 20 minutes (one generation) later, and 40 minutes (two generations) later. They centrifuged the three DNA samples in cesium chloride density gradients and found that the DNA formed bands at different positions in the gradient for the three samples. Their results, illustrated in the figure, supported the semiconservative model of DNA replication.

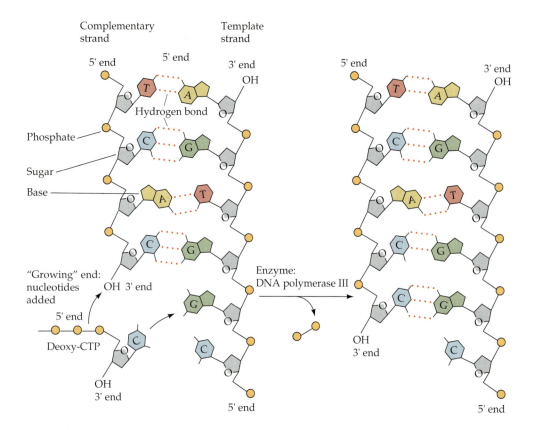

11.14   Growth of a Replicating DNA Strand
A DNA strand, with its 3' end at the top and its 5' end at the bottom, is the template for the synthesis of the complementary strand at the far left. The new strand has its 5' end at the top and its 3' at the bottom—it is antiparallel to the template strand. DNA polymerase III adds the next deoxyribonucleotide, with the base C, at the free —OH group at the 3' end of the growing chain. At the right are the same template and the growing strand, now one base longer.

deoxynucleoside triphosphate break and thereby release the energy for this reaction. Two of the phosphate groups diffuse away, and one becomes part of the sugar–phosphate backbone of the growing DNA molecule.

In order for double-stranded DNA to replicate, the strands must be unwound and separated from each other. Only then can they function as templates for the synthesis of new, complementary strands. The unwinding results in a **replication fork**—a moving Y-shaped structure that is the region where new DNA strands are being synthesized. The two strands are antiparallel, with the 3' end of one strand paired with the 5' end of the other. As the replication fork moves along the parent DNA molecule, the enzyme DNA polymerase III catalyzes the replication of both strands. How can this be accomplished, given that new nucleotides are added only at the 3' end of a polynucleotide chain? One parent strand is being exposed beginning at its 3' end, which presents no problem—its complementary strand is synthesized continuously as the replication fork proceeds. This daughter strand is called the **leading strand**.

The other daughter strand, called the **lagging strand**, is produced discontinuously, in spurts (100–200 nucleotides at a time in eukaryotes; 1,000–2,000 at a time in prokaryotes). These discontinuous stretches are synthesized by adding the 5' to the 3' end, just as the leading strand is, but in the opposite direction with respect to the replication fork. These stretches of new DNA for the lagging strand are called **Okazaki fragments** after their discoverer, the Japanese biochemist Reiji Okazaki. While the leading strand grows continuously "forward," the lagging strand grows in shorter, "backward" stretches with gaps between them (Figure 11.15). The gaps between the Okazaki fragments are then filled in by DNA polymerase III.

Working together, two DNA polymerases and several other proteins (Box 11.A) do the complex job of DNA synthesis with a speed and accuracy that are almost unimaginable. In *E. coli*, the complex makes new DNA at a rate in excess of 1,000 base pairs per second and makes mistakes in fewer than one base in 10^8–10^{12}.

DNA replication begins at just one point, the **ori-**

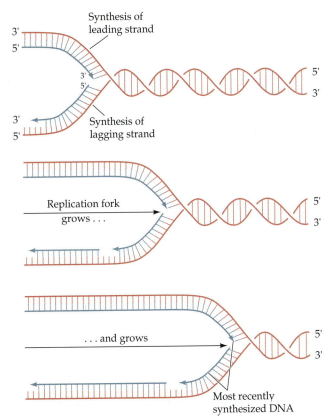

11.15 DNA Replication Fork
Both new DNA strands are synthesized in the 5′ to 3′ direction, even though their template strands are antiparallel. The leading strand is synthesized continuously, but the lagging strand is synthesized in short pieces (Okazaki fragments) that are joined later; thus the replication fork is asymmetrical. Abbreviated here, eukaryotic Okazaki fragments are hundreds of nucleotides long, and prokaryotic fragments are thousands.

gin of replication, on a bacterial chromosome. Each chromosome of a eukaryote has many specific origins of replication. Replication proceeds in both directions from an origin of replication (Figure 11.16).

FROM DNA TO PROTEIN

How does DNA function? We have learned about its structure and how it replicates. Next we need to know what it *does*. Beadle and Tatum demonstrated that genes are responsible for the production of proteins. But how does a gene specify the formation of a protein?

The Central Dogma of Molecular Biology

The **central dogma** of molecular biology is one of the most important concepts to have emerged to explain how genes make polypeptide chains. The central dogma is, simply, the idea that DNA codes for the production of RNA, RNA codes for the production

of protein, and protein does *not* code for the production of protein, RNA, or DNA (Figure 11.17). In Crick's words, "once 'information' has passed into protein *it cannot get out again.*"

Crick contributed two key ideas to the further development of the central dogma. The first solved a difficult problem: How could one explain the relationship between a specific nucleotide sequence (in DNA) and a specific amino acid sequence (in protein)? There is no chemical affinity of nucleotides for amino acids. Crick made a clever suggestion: He proposed that there is an adaptor molecule that carries a specific amino acid at one end and that recognizes some sequence of nucleotides with its other end. In due course—and without being aware of Crick's "adaptor hypothesis"—other molecular biologists found and characterized the adaptor molecules. These are small RNAs called transfer RNAs, or **tRNA**s. They recognize the genetic message *and* simultaneously carry specific amino acids, thus translating the language of DNA into the language of proteins.

Another problem was this: How does the genetic information get from the nucleus to the cytoplasm? The great bulk of the DNA of a eukaryotic cell is confined to the nucleus, but protein synthesis is carried on in the cytoplasm. Crick, together with the South African geneticist Sydney Brenner and the French molecular biologist François Jacob, tackled this problem with another key idea, the "messenger hypothesis." According to the messenger hypothesis,

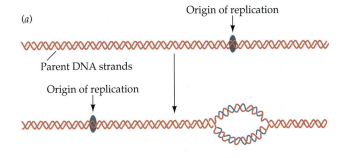

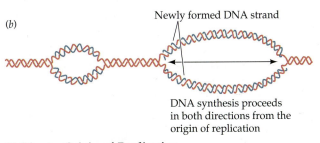

11.16 An Origin of Replication
DNA replication begins at a replication site, of which there are many in every eukaryotic chromosome. (*a*) The parent strands separate at the origin of replication. (*b*) Replication proceeds in both directions from the origin of replication.

BOX 11.A

Collaboration of Proteins at the Replication Fork

The replication of a DNA molecule is an amazingly complex process. We have already described the general problem of synthesizing both a leading and a lagging strand at the same time—but this is only the most obvious difficulty. DNA replication includes all the following steps: (1) unwinding the two parent strands, (2) providing a primer for the synthesis of a new strand, (3) elongation of each of the daughter strands, (4) filling in the gaps between the Okazaki fragments of the lagging strand, (5) connecting the completed Okazaki fragments, and (6) editing the newly synthesized strands for accuracy of replication. Each of these steps requires one or more specific proteins, many of which are enzymes.

The unwinding of the double helix is mediated by two related enzymes called **helicases**, one of which attaches to each of the parent DNA strands. Energy to separate the strands comes from the hydrolysis of ATP. The separated strands would tend to fold back on themselves because of occasional short regions of complementarity, but this is prevented by the attachment of **single-stranded DNA-binding proteins** to each of the separated DNA strands. These binding proteins hold the single strands in a configuration that binds readily to DNA polymerase III. A pair of DNA polymerase III molecules at the replication fork catalyzes the elongation of the leading and lagging strands. However, the discontinuous production of the lagging strand results in a repeated need for a new primer to start the synthesis of the next Okazaki fragment. The primer is a short single strand of RNA, rather than of DNA; it is formed, complementary to the template DNA strand, by an enzyme called a **primase**, which is one of several polypeptides bound together in an aggregate called a **primosome**. DNA polymerase III extends the primer. DNA polymerase I later acts

to replace the RNA primer segments with DNA segments. Finally, each newly completed Okazaki fragment is linked to the completed portion of the lagging strand in a reaction catalyzed by **DNA ligase**, another enzyme.

Besides catalyzing elongation of the leading and lagging strands, DNA polymerase III plays another crucial role in DNA replication: It checks the accuracy of its own work. After it adds a monomer to a strand, it tests the new base pair to see that it is complementary to the nucleotide in the template strand. If an incorrect nucleotide has been inserted, the DNA polymerase excises the erroneously selected nucleotide and tries again. As a result, DNA replication is startlingly faithful, with an error rate on the order of one wrong nucleotide per billion—even though the error rate before the proofreading process is on the order of one wrong nucleotide per 10,000.

Even in this box we have simplified the description of DNA replication. In *E. coli*, more than 30 polypeptides participate. The largest of the proteins, DNA polymerase III, has a molecular weight of 760,000 and consists of seven or eight polypeptide subunits.

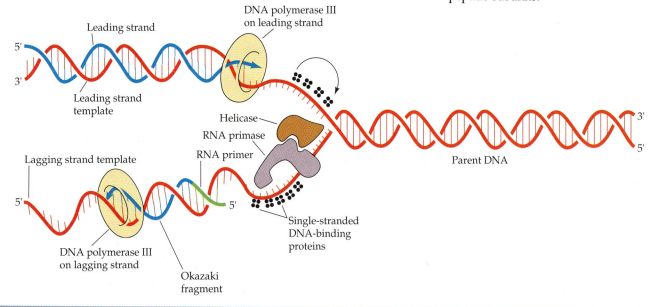

Leading strand

DNA polymerase III on leading strand

5'

3'

Leading strand template

Lagging strand template

Helicase

RNA primase

RNA primer

Parent DNA

3'

5'

Single-stranded DNA-binding proteins

5'

DNA polymerase III on lagging strand

Okazaki fragment

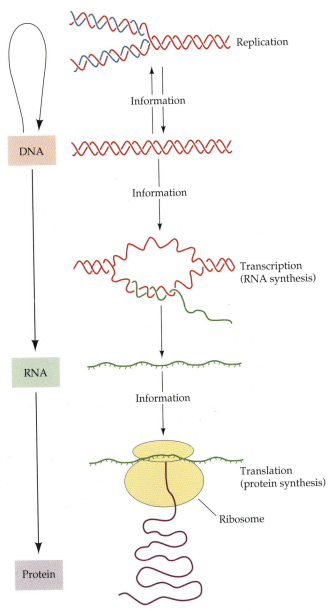

11.17 The Central Dogma
Information coded in the sequence of base pairs in DNA is passed to new molecules of DNA and to molecules of RNA. Information in RNA is passed to proteins.

a specific type of RNA molecule forms as a complementary copy of one strand of the gene. It contains the information from that gene, so there should be as many different messengers as there are different genes. This messenger RNA, or **mRNA**, then travels from the nucleus to the cytoplasm. There it serves as a template on which the tRNA "adaptors" line up to bring amino acids in the proper order into a growing polypeptide chain in the process called **translation**.

Summarizing the main features of the central dogma, the messenger hypothesis, and the adaptor hypothesis, we may say that a given gene is transcribed to produce a messenger RNA complementary

to one of the DNA strands, and that transfer RNA molecules translate the sequence of bases in the mRNA into the appropriate sequence of amino acids.

Transcription

The formation of a specific RNA under the control of a specific DNA is called **transcription**; it requires the enzyme **RNA polymerase**, the appropriate ribonucleoside triphosphates (ATP, GTP, CTP, and *U*TP), and the DNA template. Only *one* of the DNA strands—the **template strand**—is transcribed. The other, complementary DNA strand—the **coding strand**—remains untranscribed. The DNA molecule must partially unwind, as in DNA replication, to expose the bases on the template strand that will be transcribed. RNA polymerase catalyzes the continuous transcription of DNA in only one direction, as was shown in an ingenious experiment using a nucleoside triphosphate (cordycepin triphosphate) that has an unusual nucleoside portion (3'-deoxyadenosine, or cordycepin, Figure 11.18). Cordycepin triphosphate lacks a hydroxyl group at the 3' position of its sugar and thus only its 5' end can join a growing RNA strand; its 3' end cannot. Other, normal nucleotides can attach at either their 3' or 5' position. Now, if mRNA grew by adding nucleotides to its 5' end, cordycepin triphosphate could not form a covalent bond with it. Thus, the addition of this compound would have no effect on transcription. However, cordycepin actually does strongly inhibit mRNA formation. This makes sense only if the 3' position of the RNA molecule is the growing point. Cordycepin triphosphate is mistaken by RNA polymerase

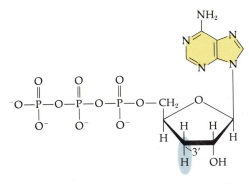

Cordycepin triphosphate

11.18 Cordycepin Triphosphate
This molecule is very similar to adenosine triphosphate as you can see from a comparison with Figure 7.3, but it lacks an —OH group at its 3' carbon end (blue oval). Its incorporation into mRNA blocks further elongation of the strand because there is no attachment site for the next nucleotide. Experiments with this analog of adenosine triphosphate helped to establish that mRNAs grow from the 5' end to the 3' end.

for normal adenosine triphosphate, and its 5' carbon is attached to the 3' end of the growing chain. Once attached, though, it prevents the chain from reacting with another nucleoside triphosphate, because it has no hydroxyl group on its 3' end. Therefore, once cordycepin phosphate has joined a chain, no other nucleotides can be added, mRNA elongation ceases, and only short strands of mRNA with terminal cordycepin molecules can be recovered. This experiment clearly demonstrates that mRNA grows from the 5' end to the 3' end.

When DNA polymerase catalyzes the replication of DNA, the two strands of the parent molecule are unwound, and each strand becomes paired with a new strand. In transcription, the DNA is unwound, but it must then be *rewound*. The DNA strand that is to be transcribed must be partly separated from its partner so that it may serve as a template for mRNA synthesis; but the RNA transcript peels away as it is formed, allowing the DNA that has already been transcribed to be rewound (Figure 11.19).

Transcription of individual genes begins at well-characterized, specific regions called **initiation sites**, which tell the RNA polymerase where to attach and which strand to copy. Similarly, particular base sequences in the DNA specify the termination of tran-

scription. The transcription of DNA is under precise control, so that particular genes are transcribed in some cells, at some times, whereas other genes are transcribed at other times. This intriguing regulatory process is discussed in Chapter 12.

It is not just mRNA that is produced in transcription. The same process is used in the synthesis of tRNA and of ribosomal RNA, or **rRNA**, which constitutes a major fraction of the ribosome; these other forms of ribonucleic acid are coded for by specific regions of the DNA. In prokaryotes, most of the DNA acts as a template for the production of mRNA, tRNA, or rRNA. The situation in eukaryotes is more complicated, as will be explained later in this chapter and in Chapter 13.

Transfer RNA

The genetic information transcribed in an mRNA molecule can be thought of as a series of three-letter "words," that is, each sequence of three nucleotides (the "letters") along the chain specifies a particular amino acid. The "word" is called a **codon**. Let us now see how that codon is related to the amino acid for which it codes. As predicted by Crick, the relation is by way of an adaptor—a specific kind of tRNA. For

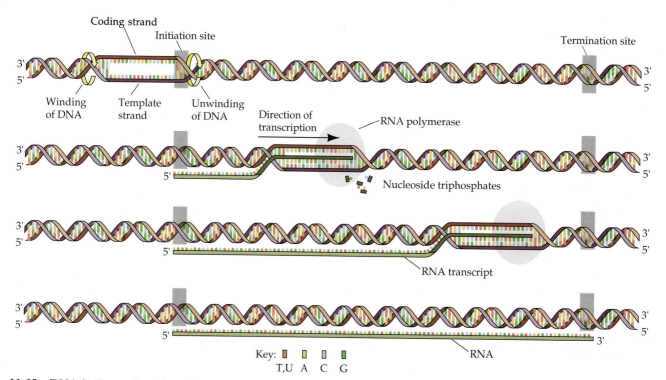

Key: ▌ ▌ ▌ ▌
T,U A C G

11.19 DNA is Transcribed into RNA
The DNA double helix unwinds to give RNA polymerase, moving in the 5' → 3' direction, access to the nucleotide sequence. As the growing RNA transcript is released from the template, the two DNA strands rewind. RNA transcripts are made from only one strand of a DNA double helix at a time. The base-pairing rules are similar to those for DNA: adenine with uracil and guanine with cytosine. The RNA polymerase, really much larger than shown here, would actually cover about 50 base pairs.

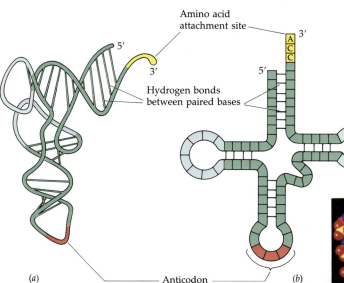

Amino acid attachment site

5′

3′

Hydrogen bonds between paired bases

5′

A
C
C

3′

(a)

Anticodon

(b)

(c)

11.20 Transfer RNA, Crick's "Adaptors"

Molecules of tRNA "read" the genetic code. The three-dimensional shape in (a), brought about by the four regions of base pairing, is diagrammed for clarity in (b). Notice that the region of tRNA that binds to the amino acid is far from the anticodon that interacts with the mRNA. (c) The actual three-dimensional structure of a tRNA is best seen in this computer-generated space-filling representation.

each of the 20 amino acids, there is at least one specific tRNA molecule. (The existence of more than one tRNA species for a given amino acid will be discussed along with degeneracy of the genetic code later in this chapter.)

Compared with most mRNA molecules, tRNA molecules are rather small, consisting of about 75–80 nucleotides. Robert Holley of Cornell University was the first to work out the complete nucleotide sequence of a particular tRNA species. He noticed that several regions, apparently separate when the molecule was viewed as stretched out, had complementary base sequences. Thus these regions could come together by folding and then could stabilize the fold with complementary base pairing. Some years later it became possible to determine the three-dimensional structures of tRNAs, and it was found that this pairing does occur, giving all tRNAs certain shapes in common. At one end of every tRNA molecule is a site to which the amino acid attaches. At the opposite end is a group of three bases, called the **anticodon**, that constitutes the point of contact with mRNA (Figure 11.20). Each tRNA species has its own distinctive anticodon, allowing it to unite by complementary base pairing with one codon. This is the key to the specificity of translation.

The overall three-dimensional shape of the tRNAs plays an important role in allowing them to combine specifically with the binding sites on ribosomes. In all, the structure of tRNA molecules relates clearly to their functions: They carry amino acids, they associate with mRNA molecules, and they interact with ribosomes.

How does a tRNA molecule combine with the correct amino acid? A family of **activating enzymes**, more formally known as aminoacyl–tRNA synthetases, accomplishes this. Each activating enzyme is specific for one amino acid. It must also find its ap-

propriate tRNA, which it does by recognizing short sequences of bases on the tRNA, away from the anticodon. The enzyme reacts first with a molecule of amino acid and a molecule of ATP, producing a high-energy AMP–amino acid that remains bound to the enzyme (Figure 11.21). The high energy results from the breaking of the bonds in the ATP. The enzyme then catalyzes a shifting of the amino acid from the AMP to the 3′ terminal nucleotide of the tRNA, where it is held by a relatively high energy bond. The activating enzyme finally releases this **charged tRNA** (tRNA with its attached amino acid); it can then charge another tRNA molecule. The high-energy bond in the charged tRNA provides the energy for the synthesis of a peptide bond that occurs during translation.

The Ribosome

Ribosomes are required for translation. Each ribosome consists of two subunits, a large, or heavy one and a small, or light one (Figure 11.22a). In eukaryotes, the large subunit consists of three different molecules of rRNA (ribosomal RNA) and about 45 different specific protein molecules, arranged in a precise pattern. The small subunit in eukaryotes consists of one rRNA molecule and 33 different ribo-

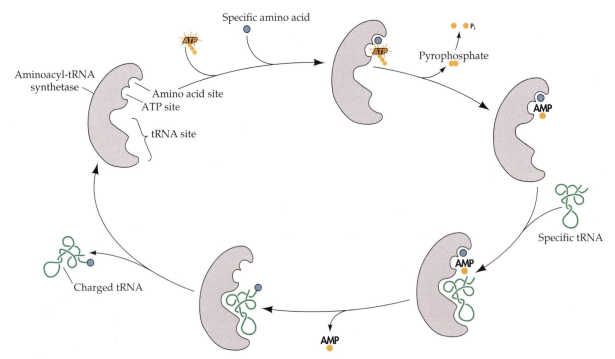

11.21 Charging a Transfer RNA Molecule
An activating enzyme, aminoacyl-tRNA synthetase, has a three-part active site that recognizes three smaller molecules: a specific amino acid, ATP, and a specific tRNA. The enzyme activates the amino acid, catalyzing a reaction with ATP in which high-energy AMP-amino acid is formed. The enzyme then catalyzes a reaction of the activated amino acid with the appropriate tRNA, producing a charged tRNA molecule.

somal protein molecules. The ribosomes of prokaryotes are somewhat smaller than those of eukaryotes. Mitochondria and chloroplasts also contain ribosomes, and in some cases these are even smaller than those of prokaryotes. When not active in the translation of mRNA, the ribosomes actually exist as separated subunits.

Each ribosome has two tRNA-binding sites that participate in translation. The ribosome also binds to the mRNA that it is translating.

Translation

The translation of mRNA begins with the formation of an **initiation complex**, which consists of a ribosomal light subunit bound to the starting point on an mRNA chain and to a charged tRNA bearing the first amino acid (Figure 11.22*b*). The charged tRNA binds to the appropriate point on the mRNA by complementary base pairing between the anticodon of the tRNA and the first codon of the mRNA. Hydrogen bonds form, linking the codon and the anticodon. The heavy subunit of the ribosome then joins the complex. There are two RNA-binding sites on the ribosome: the A site (which accepts a tRNA molecule bearing one amino acid) and the P site (which will carry a tRNA molecule bearing a growing polypeptide chain).

The first charged tRNA now lies in the P site of the ribosome, and the A site is over the second codon (Figure 11.22*c*). The appropriate charged tRNA enters the open A site, its anticodon complementary to the second codon of the mRNA. The first amino acid joins the amino acid on the second tRNA (located at the A site), the peptide linkage forming in such a way that the first amino acid is the N-terminus of the new protein (Chapter 3), while the second amino acid remains attached to its tRNA by its carboxyl group (—COOH). The first tRNA, having released its amino acid, dissociates from the complex, returning to the cytosol to become charged with another amino acid of the same kind. The second tRNA, now bearing a dipeptide, shifts to the P site of the ribosome, which moves up the mRNA by another codon. The process continues, with the latest charged tRNA entering the open A site, picking up the growing polypeptide chain from the one in the P site, and then moving to the newly vacated P site. This continues until an "end-chain" codon enters the A site and terminates translation (Figure 11.22*d*). (The end-chain codons will be briefly described in the section on the genetic code.) The newly synthesized protein separates from the ribosome. The N-terminus of the new protein is the amino acid corresponding to the first codon on the mRNA; the C-terminus is the last amino acid to join the chain.

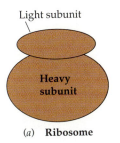

(a) **Ribosome**

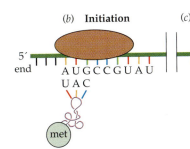

(b) **Initiation**

(c) **Elongation (translation)**

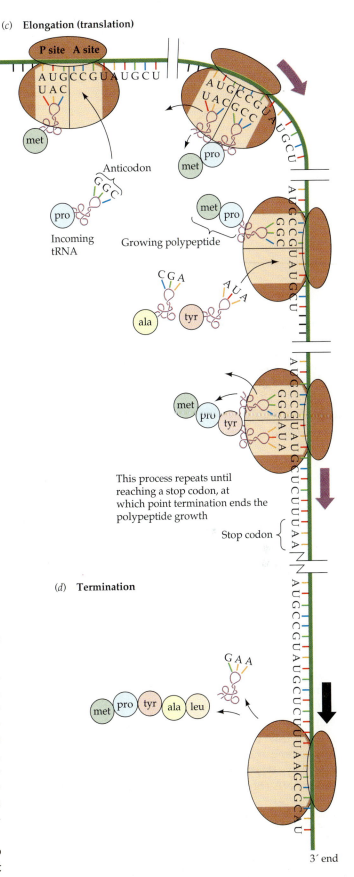

(d) **Termination**

This process repeats until reaching a stop codon, at which point termination ends the polypeptide growth

Stop codon

3′ end

11.22 Translation of Genetic Information

(a) Each ribosome consists of a light and a heavy subunit, which are separate when they are not in use. (b) An initiation complex forms when a light ribosomal subunit joins to the starting point on a molecule of mRNA, to which a tRNA and its amino acid have already attached. Note that the anticodon bases on the tRNA are complementary to the bases of the start codon on the mRNA. After the initiation complex forms, the addition of the heavy subunit completes the ribosome. (c) Manufacture of a polypeptide chain is based on "reading" the base sequence in mRNA at two ribosomal sites, A and P. When translation begins, the first tRNA is in place in the P site. At this time the second codon on the mRNA is exposed in the A site; a tRNA molecule with a complementary anticodon then pairs with the codon in the A site. The amino acid on the first tRNA bonds to the amino acid on the second tRNA. The first tRNA, having released its amino acid, dissociates from the ribosome. The ribosome then moves along the mRNA to the third codon, transferring the second tRNA and its two attached amino acids to the P site and exposing the third codon in the A site. A tRNA with an anticodon complementary to the third codon on the mRNA enters the A site, the two amino acids on the second tRNA in the P site are transferred to the amino acid on the third tRNA in the A site, and so on. The ribosome proceeds, "reading" mRNA codons and adding amino acids to the chain, until an "end chain" signal is reached on the mRNA. (d) Now the completed polypeptide, final tRNA, and ribosomal subunits all dissociate from the mRNA (termination).

Several ribosomes can work at once at translating a single mRNA molecule to produce a number of molecules of the protein almost simultaneously. As soon as the first ribosome has moved far enough from the initiation point, a second initiation complex can be formed, and then a third, and so on. The first ribosome to initiate translation is the first to finish translating the message and be released. The assemblage of a thread of mRNA with its beadlike ribosomes and their growing polypeptide chains is called a polyribosome, or **polysome**. Cells that are actively synthesizing proteins contain large numbers of polysomes and relatively fewer free ribosomes or ribosomal subunits. Figure 11.23 shows a polysome in action.

A given ribosome is not specifically adapted to produce just one kind of protein. It was once thought that there were specific ribosomes for each of the kinds of proteins produced in a cell, but that is not the case. A ribosome can combine with any mRNA and any tRNAs and thus can be used to make dif-

(a)

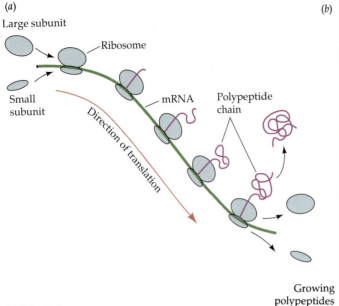

Large subunit

Small subunit

Ribosome

Direction of translation

mRNA

Polypeptide chain

Growing polypeptides

(b)

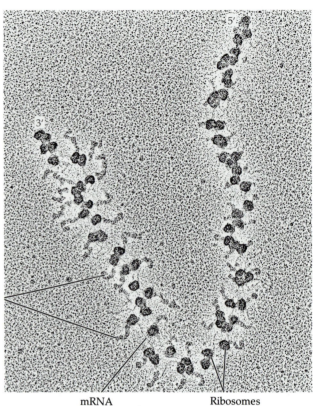

mRNA   Ribosomes

11.23 Polysomes
(a) A polysome consists of ribosomes and their growing polypeptide chains moving in single file along an mRNA molecule. The arrangement increases the efficiency of polypeptide synthesis in the cell, inasmuch as several ribosomes may translate the information in an mRNA simultaneously. (b) In this electron micrograph, the strand of mRNA can be seen between the individual ribosomes at the left.

ferent polypeptide products. The mRNA contains the information that specifies the polypeptide sequence. The ribosome is simply the molecular machine that accomplishes the task. Its structure allows it to hold the mRNA and tRNAs in the right positions, thus allowing the growing polypeptide to be efficiently assembled.

We have simplified this account of translation in at least two ways. First, we have not made specific mention of the proteins and small molecules that play roles in the polypeptide elongation and release processes. Second, we have described protein synthesis as it occurs in prokaryotes. In eukaryotes, things are more complex (Chapter 13). In particular, several steps come between transcription and translation.

In at least one virus certain genes are overlapping. Thus some mRNAs can be translated in more than one way, depending on where the ribosome binds the mRNA (Box 11.B).

The Role of the Endoplasmic Reticulum

As you learned in Chapter 4, one important difference between prokaryotes and eukaryotes is that eukaryotic cells are composed of many individual compartments. This presents a problem in protein synthesis: How are particular proteins targeted to the correct site—electron transport chain components to

the mitochondria, histones to the nucleus, and so forth? The answer is only partly known.

Proteins that are to remain soluble within the cell are synthesized on "free" ribosomes, that is, ones that are not attached to the endoplasmic reticulum (ER). On the other hand, those proteins that are to become parts of membranes, or are to be exported from the cell, or are to end up in lysosomes or peroxisomes, are generally synthesized on the ribosomes of the rough ER. The presence or absence of a specific sequence of amino acids, the **signal peptide**, determines whether a given protein will be made on the rough ER or on free ribosomes.

All protein synthesis *begins* on free ribosomes. The first few amino acids of a membrane protein or a protein destined for export serve as a signal peptide, which attaches to a signal recognition particle composed of protein and RNA. This attachment blocks further protein synthesis until the ribosome can become attached to a specific receptor protein in the membrane of the ER (Figure 11.24). The receptor protein serves as a channel through which the growing polypeptide is extruded, either into the membrane itself or into the interior of the ER, as synthesis continues. An enzyme within the ER interior then causes the removal of the "signal" part of the new protein. Thus the newly formed protein is either built into the membrane or retained within the ER rather than in the cytosol. From the ER it can be transported

BOX 11.B

Making the Most of Your DNA

A virus named φX174, one of the smallest of the bacteriophages, is made up of a few kinds of protein molecules plus a small, circular molecule of DNA. This viral DNA must code for nine kinds of protein, including some proteins that are not part of the mature phage, but are needed during viral replication or release. We know the lower limit for the length of a DNA base sequence that can code for a protein. Scientists puzzled over φX174 because its total DNA content (amounting to about 5,400 nucleotides) appears to be too low by 10–15 percent to code for all nine proteins. This problem was re-

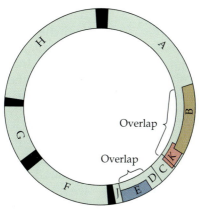

The circle represents the chromosome of the φX174 virus; the letters indicate the proteins coded for by the genes in the colored regions. Note that some regions coding for one protein also include the information for other proteins; for example, the region coding for protein A also codes for protein B and part of protein K. (The few short stretches of the chromosome that do not code for proteins are shown in black.)

solved in a most unexpected way in 1977 when the English biochemist Frederick Sanger and his colleagues reported the complete structure, nucleotide by nucleotide, of the DNA

molecule of φX174. With the nucleotide map spread out before them, and with their knowlege of the primary structures of the proteins, they were able to discover that the genes coding for two of the proteins are embedded within two of the other genes, as indicated in the figure. Putting it another way, the same stretch of DNA can participate in coding for two entirely different proteins. The mRNA transcribed from such a "shared" stretch of DNA also contains codes for both proteins. Which protein is produced depends on where translation begins, that is, on which of two initiation sites becomes bound to a ribosome. When you read about frame-shift mutations later in this chapter, you might recall the story of φX174 and marvel that such an improbable thing should ever have come about. The phenomenon of overlapping genes does not appear to be common; it may have evolved in φX174 because of the limitation on the size of the DNA molecule imposed by the small protein coat.

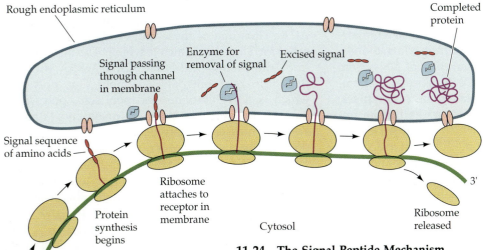

11.24 The Signal Peptide Mechanism
The synthesis of a protein to be exported from the cell begins free in the cytoplasm, as subunits and mRNA come together to form a ribosome. However, the first few amino acids of the protein to be linked together are a "signal" for attachment of the ribosome to a receptor protein in the membrane of the endoplasmic reticulum. As protein synthesis proceeds on the membrane, the polypeptide passes through a channel in the membrane to the interior of the ER and an enzyme clips the signal from the end of the chain. When the protein is complete, it is released to the interior of the ER and the subunits of the ribosome separate from the membrane and detach from one another. The new protein is now functionally outside the cytosol.

to its appropriate location—to other cellular membranes or to the outside of the cell—without mixing with other molecules in the cytoplasm.

RNA Viruses and the Central Dogma

According to the central dogma of molecular biology, DNA codes for RNA and RNA codes for protein. However, there are some variations on the central dogma. Many viruses, such as the tobacco mosaic virus, have RNA rather than DNA as their nucleic acid. Heinz Fraenkel-Conrat of the University of California at Berkeley managed to separate the protein and RNA fractions of the tobacco mosaic virus and then recombine them to obtain active virus particles. When he took RNA from one mutant strain of this virus and combined it with protein from another, the resulting viruses replicated to produce more virus particles like the first, RNA-donating, strain. Thus he showed that RNA is the genetic material of the tobacco mosaic virus. RNA itself is the template for the synthesis of the next generation of viral RNA and viral proteins. In this virus, DNA is left out of the information flow.

In 1964, Howard Temin of the University of Wisconsin studied an RNA virus that causes a cancer in chickens known as Rous sarcoma. The Rous sarcoma virus enters a chicken cell and subsequently causes the cell to make a DNA "transcript" of the viral RNA, the reverse of the usual process. The afflicted cell does not burst, but it changes permanently in shape, metabolism, and growth habit. The new DNA becomes part of the hereditary apparatus of the infected chicken cell. Later, Temin and others showed that the virus carries an enzyme for the manufacture of DNA, using viral RNA as the informational template. The enzyme was named **reverse transcriptase**, inasmuch as it performs the transcription of DNA from RNA rather than RNA from DNA. Viruses that employ reverse transcriptase are known as **retroviruses**; one of the most studied of these is the human immunodeficiency virus (HIV), which causes AIDS. The central dogma requires slight modification to account for the flow of information in retroviruses and their hosts. However, it is still true that information does not flow from protein back to the nucleic acids.

The Genetic Code

Which mRNA codons translate into which amino acids? Molecular biologists broke this genetic code in the early 1960s. The problem seemed formidable at the outset: How could a sequence composed of only four bases be translated into one made up of 20 or so different amino acids? It was not yet possible to determine the base sequence in a nucleic acid, so scientists could not simply compare the primary structure (the amino acid sequence) of a protein with the base sequence of the appropriate DNA or mRNA molecule. However, there were ways of getting partial information about the code even though nucleic acid chemistry was not yet very far advanced.

Marshall W. Nirenberg and J. H. Matthaei, at the National Institutes of Health, made the first breakthrough when they realized that they could use a very simple, artificial polymer instead of a complex, natural mRNA, as a messenger and see what the artificial messenger coded for. In 1961, they published the first of their papers on polypeptide synthesis directed by artificial mRNAs. Nirenberg had prepared an artificial mRNA in which all the bases were uracil; the molecule was called poly U. When poly U was added to a reaction mixture containing ribosomes, amino acids, activating enzymes, tRNAs, and other factors, a polypeptide formed (Figure 11.25a). This polypeptide contained only one kind of amino acid: phenylalanine (Phe). Poly U coded for poly Phe! Accordingly, it appeared that UUU was the mRNA code word—the codon—for phenylalanine. Following up on this success, Nirenberg and Matthaei easily showed that CCC codes for proline and AAA for lysine. (Poly G presented some chemical problems and was not initially tested.) Those were three of the easiest of the codons; different approaches were required to work out the rest.

Har Gobind Khorana, then at the University of Wisconsin, painstakingly synthesized some artificial mRNAs such as poly CA, or CACACA . . . , and poly CAA, or CAACAACAA . . . (Figures 11.25b and 11.25c). Khorana found that poly CA codes for a polypeptide consisting of threonine (Thr) and histidine (His), in alternation (His–Thr–His–Thr. . .). There are two possible codons in poly CA, CAC and ACA. Thus one of these must code for His and the other for Thr—but which is which? The answer came from the results with poly CAA, which produces three different polypeptides: poly Thr; poly Gln, which is polyglutamine; and poly Asn, polyasparagine. To understand this, we must know that an artificial messenger can be read beginning at any point in the chain; there is no specific initiator region. Thus poly CAA can be read as a polymer of CAA, of ACA, or of AAC. Comparing the results of the poly CA and poly CAA experiments, you should be able to figure out which code word is for Thr and which for His.

Nirenberg made a further important, simplifying discovery that led to the decoding of the remaining words in the code book in 1964 and 1965. He found that simple "mRNAs" only three monomers in length, each amounting to a codon, can bind to ribosomes and that the resulting complex can then cause the binding of the corresponding charged tRNA. Thus, for example, simple UUU causes phenylalanyl–tRNA charged with phenylalanine to bind to the ribosome. Complete deciphering of the code

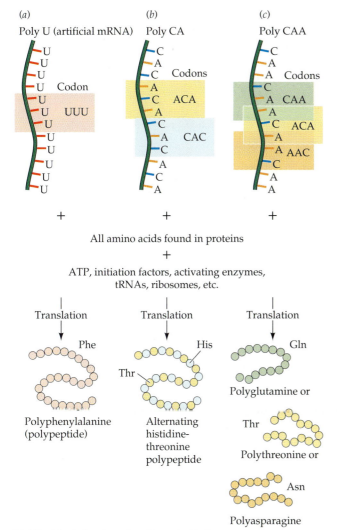

11.25 Deciphering the Genetic Code
The experiments shown here helped to reveal how information is coded in mRNA. (a) A synthetic mRNA containing only uracil is translated in a test tube into a polypeptide containing only the amino acid phenylalanine. This suggests that the RNA code for phenylalanine contains only uracil; we now know it is UUU. (b) The synthetic mRNA . . .CACACA. . . is translated into a polypeptide in which the amino acids threonine and histidine alternate. This suggests ACA and CAC are the codons. (c) Poly CAA, or . . .CAACAACAA. . ., can be read in three different ways: in units of CAA, ACA, or AAC, each of which is translated into a different polypeptide containing only one kind of amino acid.

translation stops and the polypeptide is released from the ribosome–mRNA–tRNA complex. AUG, which codes for methionine, is also the **start codon**, which acts as the initiation signal for transcription. Still, 61 codons are far more than enough to code for 20 amino acids—and indeed there are repeats. Thus we say that the code is **degenerate**, which means that an amino acid may be represented by more than one codon. The degeneracy is not evenly divided among the amino acids. Two of them, methionine and tryptophan, are represented by only one codon, whereas leucine is represented by six different codons.

The term *degeneracy* should not be confused with *ambiguity*, a word used to refer to a situation in which a single codon could specify either of two (or more) different amino acids. Degeneracy in the code means that there is more than one unequivocal way to say, "Put leucine here." Ambiguity would mean that, for a given codon, there would be doubt whether to put in leucine or something else. The genetic code is *not* ambiguous.

It was gratifying to learn that the code appears to be virtually universal, apparently applying to all the species on our planet. It seems that the code must be an ancient one that has been maintained intact through the evolution of the living things around us. There is only one known exception: Within mitochondria the code differs slightly but detectably from that in prokaryotes and elsewhere in eukaryotic cells. It is not even quite the same in the mitochondria of different eukaryotes. The significance of this is not yet clear.

You should remember that the codons shown in Figure 11.26 refer to the mRNA. The master words on the DNA strand that was transcribed to produce the mRNA are complementary to these codons, so, for example, AAA in the template DNA strand corresponds to phenylalanine. Does this code really work? The final proof of the pudding was obtained by synthesizing artificial DNA of known sequence, introducing it into bacteria, and getting them to produce the specific protein coded for by that DNA. We can now program bacteria at will to synthesize proteins no organism ever made before.

book was relatively simple then. To find the "translation" of a codon, Nirenberg could use a sample of that codon as an artificial mRNA and see which amino acid became bound.

The complete genetic code is shown in Figure 11.26. Notice that there are many more RNA codons than there are kinds of amino acids in proteins. Combinations of the four "letters" give 64 different three-letter codons, yet these determine only 20 amino acids. Three of the codons (UAA, UAG, UGA) are end-chain codons, or **chain terminators**; when the translation machinery reaches one of these codons,

MUTATIONS

From what we have said thus far, you might infer that DNA replication, transcription, and translation all proceed without error. In fact, errors do occur, though infrequently, in all three. In particular, errors in the replication of DNA during the production of the gametes are crucial to evolution. If there were no **mutations**—heritable changes in the genetic information—there would be no evolution. Minute changes in the genetic material often lead to easily

Second letter

	U	C	A	G	
U	UUU UUC Phenyl-alanine / UUA UUG Leucine	UCU UCC UCA UCG Serine	UAU UAC Tyrosine / UAA Stop codon / UAG Stop codon	UGU UGC Cysteine / UGA Stop codon / UGG Tryptophan	U C A G
C	CUU CUC CUA CUG Leucine	CCU CCC CCA CCG Proline	CAU CAC Histidine / CAA CAG Glutamine	CGU CGC CGA CGG Arginine	U C A G
A	AUU AUC Isoleucine / AUA / AUG Methionine; initiation codon	ACU ACC ACA ACG Threonine	AAU AAC Asparagine / AAA AAG Lysine	AGU AGC Serine / AGA AGG Arginine	U C A G
G	GUU GUC GUA GUG Valine	GCU GCC GCA GCG Alanine	GAU GAC Aspartic acid / GAA GAG Glutamic acid	GGU GGC GGA GGG Glycine	U C A G

First letter (left side) · Third letter (right side)

11.26   The Universal Genetic Code
Genetic information is encoded in three-letter units—codons—made up of the bases U, C, A, and G. All 64 mRNA codons are shown, along with the amino acid each codes for. To decode a codon, find its first letter in the left column, then read across the top to its second letter, then read down the right column to its third letter. In the corresponding box is the amino acid the codon specifies. For example, AUG specifies the amino acid methionine, and GUA specifies valine. Notice that the number of codons for an amino acid may be one (methionine and tryptophan), two (lysine, tyrosine, and others), four (threonine, alanine, and others), or six (leucine). Three codons (UAA, UAG, and UGA) are stop signals. AUG is the start signal in addition to coding for methionine.

observable changes in the outward form and function of the individual. The detection of a mutation depends on our ability to observe its phenotypic effects. Some effects of mutation are obvious in humans—dwarfism, for instance, or the presence of more than five fingers on each hand. A mutant genotype may be almost equally obvious in a microorganism, for example, if it results in a change in color or in nutritional requirements, as we have discussed for *Neurospora*. Other mutations may be virtually unobservable. In our species, for example, there is a mutation that drastically lowers the level of an enzyme called glucose 6-phosphate dehydrogenase, which is present in many tissues, including red blood cells. The red blood cells of a person carrying the mutant gene are abnormally sensitive to an antimalarial drug called primaquine; when such people are treated with this drug, their red blood cells rupture, causing se-

rious medical problems. People with the normal allele have no such problem with primaquine; before it came into use, no one was aware that such a mutation existed. Similarly, distinguishing a mutant bacterium from a normal one may be a very subtle matter, dependent on what tools are available.

Some mutations are detectable only under certain restrictive conditions, and are not detectable under other, permissive conditions. Organisms carrying such mutations are referred to as conditional mutants. Many conditional mutants are temperature-sensitive, unable to grow at some restrictive temperature, such as 37°C, but able to grow normally at a lower, permissive temperature, such as 30°C. The mutant allele in such an organism may code for an enzyme with an unstable tertiary structure that is altered at the restrictive temperature.

All mutations are alterations in the nucleotide sequence in DNA. We divide mutations into two categories based on the extent of the alteration. **Point mutations** are mutations of single genes from one allele to another brought about by small alterations in the sequence or number of nucleotides—even as small as the substitution of one nucleotide for another. **Chromosomal mutations**, introduced in Chapter 10, are more extensive alterations. Chromosomal mutations may change the position or direction of a DNA segment without actually removing any genetic information, or they may cause a segment of DNA to be irretrievably lost. Both point mutations and chromosomal mutations are heritable.

Point Mutations

Many point mutations consist of the substitution of one base for another in the DNA and, hence, in the mRNA. Often base-substitution mutations change the genetic message so that one amino acid substitutes for another in the protein. Such a **missense mutation** may sometimes cause the protein to be completely nonfunctional, but often the effect is only to reduce its functional efficiency. On rare occasions the functional efficiency is even improved. Individuals carrying missense mutations may survive even though the affected protein is essential to life.

Nonsense mutations, another type of base-substitution mutation, are more often disruptive than are missense mutations. A nonsense mutation is one in which the base substitution results in the formation of a chain-terminator codon such as UAG in the mRNA product (Figure 11.26). A nonsense mutation results in a shortened protein product, since translation does not proceed beyond the point where the mutation occurred.

Not all point mutations are base substitutions. Single base pairs may be inserted into or deleted from DNA. Such mutations are known as **frame-shift mutations**, because they interfere with the decoding of the genetic message by throwing it out of register. Codons may conveniently be thought of as three-letter words, each corresponding to a particular amino acid. Translation proceeds codon by codon; if a base is added to the message or subtracted from it, translation proceeds perfectly until it comes to the one-base insertion or deletion. From that point on, the three-letter words in the message are one letter out of register. In other words, such mutations shift the "reading frame" of the genetic message. Frame-shift mutations almost always lead to the production of completely nonfunctional proteins, especially if they are near the beginning or middle of a gene.

To see how a frame-shift mutation works, consider the following example. Let part of the coding strand of DNA read from 3' to 5' as follows:

3'—AGATACGTGCTGCAT—5'

This is transcribed to yield an mRNA with the following sequence from 5' to 3':

5'—UCUAUGCACGACGUA—3'

which can be divided up into the codons

. . . UCU AUG CAC GAC GUA . . .

which, as you can determine from Figure 11.26, translates to the following amino-acid sequence (from N-terminus to C-terminus):

. . . serine methionine histidine aspartic acid valine . . .

Now suppose that, through a deletion, the DNA strand loses the fifth base in the sequence above, an A, so that it reads as follows:

3'—AGATCGTGCTGCAT—5'

This is transcribed to yield the following mRNA sequence:

5'—UCUAGCACGACGUA—3'

or

. . . UCU AGC ACG ACG UA . . .

which, in turn, is translated to the amino-acid sequence

. . . serine serine threonine threonine . . .

It is small wonder that a frame-shift mutation is so disruptive. An organism carrying such a mutant gene can survive only if the gene product affected is not an essential part of the cellular machinery or if the organism also carries another copy of the gene in its normal form.

A number of chemicals can induce mutations. Among these **mutagens** are base analogues: purines or pyrimidines not found in natural DNA but enough like the natural bases that they can be incorporated into DNA. Base analogues are mutagenic presumably because they are more likely than the natural DNA bases to mispair. For example, 5-bromouracil (Figure 11.27) is very similar to thymine, and it is easily

Thymine   5-Bromouracil   Cytosine

11.27 Mutagens May Cause Point Mutations
Enzymes that replicate DNA cannot distinguish between 5-bromouracil and thymine because the two molecules are so similar in shape. Once incorporated in a DNA strand, 5-bromouracil may rearrange to a form that resembles cytosine and can pair with guanine. In this way, 5-bromouracil causes mutations and is a potent chemical mutagen.

incorporated into DNA in the place of thymine. But the abnormal base is much more likely than thymine to engage in an abnormal pairing with guanine, and it is therefore an inducer of A–T to G–C and G–C to A–T mutations. Certain other mutagens, such as hydroxylamine or nitrous acid, directly alter the structure of the natural bases in DNA, changing them to other bases that tend to mispair. The acridine dyes are mutagens that induce frame-shift mutations.

Chromosomal Mutations

Genetic strands can break and rejoin, with gross disruption of the sequence of genetic information, to cause chromosomal mutations (Figure 11.28). **Deletions** remove part of the genetic material. Like frameshift point mutations, they cause death unless they affect unnecessary genes or are masked by the presence, in the same cell, of normal copies of the deleted genes. One mechanism by which such mutations might occur is easy to imagine: Two breaks might occur in a DNA molecule and the two end pieces might rejoin, leaving out the DNA between the breaks.

Another mechanism by which deletion mutations might arise would lead simultaneously to the production of a second kind of chromosomal mutation: a **duplication**. This would come about if the sister molecules being produced during DNA duplication were to break at different positions and then reconnect to the wrong partners. One of the two molecules produced by this mechanism would lack a segment of DNA, and the other would have two tandem copies of the information that was deleted from the first (see the section on origin of new genes).

Breaking and rejoining the genetic strands can also lead to **inversions**—the removal of a segment of DNA and its reinsertion in the same location but "flipped" end for end so that it runs in the opposite direction. If the breaks leading to an inversion occur within a segment of DNA that codes for a protein, the resulting protein will be drastically altered and almost certainly nonfunctional.

The fourth class of chromosomal mutations, called **translocations,** result when a segment of DNA breaks, moves from one position in a chromosome, and inserts somewhere else. Initially, there is no change in the total genetic material, other than the arrangement of the genes. However, translocations can make synapsis in meiosis difficult and thus sometimes lead to aneuploidy (see the discussion of Down syndrome in Chapter 10).

The Frequency of Mutations

All mutations are rare events. The observed frequencies of mutations are different for different organisms and for different genes within a given organism. Usually the frequency of mutation is much lower than 1

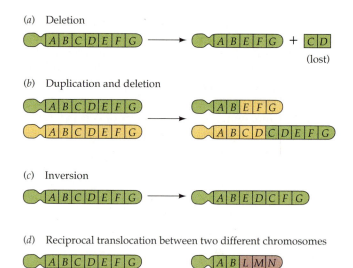

(a) Deletion

(b) Duplication and deletion

(c) Inversion

(d) Reciprocal translocation between two different chromosomes

11.28 Chromosomal Mutations
By breaking during replication, chromosomes may mutate. Genes are represented by letters on the colored chromosomes. (a) Deletions result when a chromosome breaks in two places and a segment is lost when the breaks heal. (b) Duplications and deletions result in mutated chromosomes containing more or fewer copies of normal genes, as the deleted segments from one gene appear in another. (c) An inversion is the reinsertion of a broken part in reverse order. (d) A translocation is the attachment of part of one chromosome to another chromosome. In reciprocal translocations, two chromosomes exchange pieces.

mutation per 10^4 genes per DNA duplication, and sometimes the frequency is as low as 1 mutation per 10^9 genes per duplication. The majority of mutations are point mutations in which one nucleotide is substituted for another during the synthesis of a new DNA strand.

THE ORIGIN OF NEW GENES

The great majority of mutations are harmful to the organism that carries them, and other mutations are neutral (with no effect on fitness). But once in a while a mutation improves the organism's adaptation to its ecological niche or becomes favorable when environmental variables change. Duplication mutations may be the source of "extra" genes. The more complex creatures living on earth often have more DNA and therefore more genes than the simplest creatures. Humans, for example, have 1,000 times more genetic material than bacteria. How do new genes arise? If whole genes are sometimes duplicated by the mechanism described in the previous section, the bearer of the duplication would have a surplus of genetic information that might be turned to good use. Subsequent mutations in one of the two copies of the gene might not have an adverse effect on survival because the other copy of the gene would continue to turn out functional protein. The extra gene might mutate over and over again without ill effect, because its function would be "covered" by the original copy. If the random accumulation of mutations in the extra gene should lead to production of some useful protein (for example, an enzyme with an altered specificity for the substrates it will bind, allowing it to catalyze different—but related— reactions), natural selection would tend to perpetuate its existence. New copies of genes also arise through the activity of transposable elements, to be discussed in the next two chapters.

SUMMARY

Work on *Neurospora* led to the one-gene, one-enzyme theory. Subsequent work, especially that on sickle-cell anemia, led to a revised version: the one-gene, one-polypeptide theory.

Hereditary information can be transferred from dead bacteria to genetically different live bacteria by transformation. The transforming principle is DNA. DNA is the hereditary material in everything except the RNA viruses.

DNA consists of two polynucleotide chains forming a double helix. The chains are held together by hydrogen bonding between their nitrogenous bases. Base pairing is complementary: A–T, T–A, G–C, C–

G. B-DNA has a right-handed helical structure with two grooves. RNA, in contrast to DNA, is usually a single polynucleotide chain. When DNA and RNA associate, the rules of complementarity hold.

DNA replication, catalyzed by DNA polymerases, is semiconservative. Although the DNA strands are antiparallel, both strands are replicated in the 5' to 3' direction, so that the lagging strand must be replicated discontinuously.

The central dogma of molecular biology is that DNA codes for RNA and RNA codes for protein. The dogma must be modified for some RNA viruses; in them, viral RNA sometimes serves as a template for DNA synthesis in the virus's host cell through the activity of a reverse transcriptase.

Transcription is the formation of an RNA molecule with a base sequence complementary to that of one strand of the DNA. In translation, the information transcribed into mRNA is used to make a specific polypeptide from a collection of amino acids. A ribosome binds to the mRNA and also has two specific binding sites for charged tRNAs. Amino acids are brought to the ribosome by specific tRNA molecules and are attached sequentially to the growing polypeptide. The genetic code is degenerate, in that most amino acids are specified by more than one codon. Three codons serve as "end-chain" signals that terminate translation.

Mutations are infrequent. Mutagens increase the frequency of mutation. Point mutations change only a few or even single nucleotide pairs. Chromosomal mutations affect larger stretches of a DNA molecule.

SELF-QUIZ

1. Griffith's studies of *Streptococcus pneumoniae*:
   a. proved that DNA is the genetic material of bacteria.
   b. proved that DNA is the genetic material of bacteriophage.
   c. demonstrated the phenomenon of bacterial transformation.
   d. proved that bacteria reproduce sexually.
   e. proved that protein is not the genetic material.

2. In the Hershey–Chase experiment:
   a. DNA from parent bacteriophage appeared in progeny bacteriophage.
   b. most of the phage DNA never entered the bacteria.
   c. over three-fourths of the phage protein apeared in progeny phage.
   d. DNA was labeled with radioactive sulfur.
   e. DNA formed the coat of the bacteriophage.

3. Which statement about complementary base pairing is *not* true?
   a. It plays a role in DNA replication.
   b. In DNA, T pairs with A.
   c. Purines pair with purines, and pyrimidines pair with pyrimidines.
   d. In DNA, A pairs with T.
   e. The base pairs are of equal length.

4. In semiconservative replication of DNA:
   a. the original double helix remains intact and a new double helix forms.
   b. the strands of the double helix separate and act as templates for new strands.
   c. polymerization is catalyzed by RNA polymerase.
   d. polymerization is catalyzed by a double-helical enzyme.
   e. DNA is synthesized from amino acids.

5. Which of the following does *not* occur during DNA replication?
   a. Unwinding of the parent double helix
   b. Formation of short pieces that are united by DNA ligase
   c. Complementary base pairing
   d. Use of a primer
   e. Polymerization in the direction from 3' to 5'

6. Transcription:
   a. produces only mRNA.
   b. requires ribosomes.
   c. requires tRNAs.
   d. produces RNA growing from the 5' to the 3' end.
   e. occurs only in eukaryotes.

7. Which statement is *not* true of translation?
   a. It is RNA-directed polypeptide synthesis.
   b. An mRNA molecule can be translated by only one ribosome at a time.

c. The same genetic code is in effect in all organisms.

d. Any ribosome can be used in the translation of any mRNA.

e. There are both "start" and "stop" codons.

8. Which statement is false?

a. Transfer RNA functions in translation.

b. Ribosomal RNA functions in translation.

c. RNAs are produced in transcription.

d. Messenger RNAs are produced on ribosomes.

e. DNA codes for mRNA, tRNA, and rRNA.

9. The genetic code:

a. is different for prokaryotes and eukaryotes.

b. has changed during the course of recent evolution.

c. has 64 codons that code for amino acids.

d. is degenerate.

e. is ambiguous.

10. If a mutation results in the codon UAG appearing where there had been UGG:

a. this is a nonsense mutation.

b. this is a missense mutation.

c. this is a frame-shift mutation.

d. this is a large-scale mutation.

e. this is unlikely to have a significant effect.

FOR STUDY

1. The genetic code is described as degenerate. What does this mean? How is it possible that a point mutation, consisting of a replacement of a single nitrogenous base in DNA by a different one, might not result in an error in protein production?

2. Suppose that Meselson and Stahl had continued their experiment on DNA replication for another ten bacterial generations. Would there still have been any ^{14}N–^{15}N DNA present? Would it still have appeared in the centrifuge tube? Explain.

3. Look back at Khorana's experiment with poly CA and poly CAA, in which the codons for histidine and threonine were determined. Using the genetic code (Figure 11.26) as a guide, deduce what results Khorana would have obtained had he used poly UG and poly UGG as artificial messengers. In fact, very few such artificial messengers would have given useful results. For an example of what could happen, consider poly CG and poly CGG. Using poly CG as the messenger, a mixed polypeptide of arginine and alanine (−Arg−Ala−Arg−Ala−) would be obained; poly CGG gives three polypeptides: polyarginine, polyalanine, and polyglycine. Can any codons be determined from only these data? Explain.

4. What causes transcription to start? to stop? What causes translation to start? to stop?

READINGS

Felsenfeld, G. 1985. "DNA." *Scientific American*, October. A well-illustrated description of DNA structure and function.

Judson, H. F. 1979. *The Eighth Day of Creation*. Simon and Schuster, New York. A sparkling history of molecular biology, with the best available description of the events surrounding the discovery of the structure of DNA.

Radman, M. and R. Wagner. 1988. "The High Fidelity of DNA Duplication." *Scientific American*, August. How error avoidance and error correction work. Why don't they work even better?

Smith, M. 1979. "The First Complete Nucleotide Sequencing of an Organism's DNA." *American Scientist*, vol. 67, pages 57–67. How nucleotide sequences in DNA are worked out and the interesting discovery of overlapping genes in the bacteriophage φX174.

Stent, G. S. and R. Calendar. 1978. *Molecular Genetics*, 2nd Edition. W. H. Freeman, New York. A brilliant technical and historical introduction to molecular genetics and the role of DNA.

Suzuki, D. T, A. J. F. Griffiths, J. H. Miller and R. C. Lewontin. 1989. *An Introduction to Genetic Analysis*, 4th Edition. W. H. Freeman, New York. An excellent textbook of modern genetics. Chapters 11, 12 and 13 are particularly relevant to this chapter.

Watson, J. D. 1968. *The Double Helix*. Atheneum, New York. A captivating and, to some, infuriating book in which Watson describes the events leading to the discovery of DNA structure.

Watson, J. D., N. H. Hopkins, J. W. Roberts, J. A. Steitz, and A. M. Weiner. *Molecular Biology of the Gene*, 4th Edition. Benjamin/Cummings, Menlo Park, CA. See especially Chapters 3, 9, 10, and 12–15 of Volume I.

12

Molecular Genetics of Prokaryotes

PREVIEW: Bacteria and viruses take part in sexual processes, and their genes mutate and recombine. Recombination in bacteria occurs in many ways: one involves the transfer of genetic information from dead bacteria to living ones (transformation), and another uses viruses as carriers of bacterial genes from one cell to another (transduction). "Jumping genes" make copies of themselves that appear elsewhere on the chromosome. Some genes serve regulatory functions rather than coding for cytoplasmic proteins. Besides its chromosome, a bacterium may harbor other smaller DNA molecules that also carry genetic information.

This chapter deals with conjugation, the isolation of mutants, mapping, sexduction, lysogeny, transduction, episomes, plasmids, transposable elements, and operons.

Molecular biology has had many triumphs, several of which were described in Chapter 11. How did this new discipline make such rapid advances? A key to some of the early advances was the utilization of the simplest available biological systems: bacteria and bacteriophages (bacterial viruses). Molecular biologists working with bacteria and viruses in the 1950s and 1960s discovered most of the principles described in Chapter 11; it is doubtful whether we would know these principles yet if work had been limited to garden peas, *Neurospora*, corn, and fruit flies. A typical bacterium contains about $\frac{1}{1000}$ as much DNA as a single human cell, and a typical bacteriophage contains about $\frac{1}{100}$ as much DNA as a bacterium. Data on large numbers of organisms can easily be obtained from prokaryotes, but not from most eukaryotes. A single milliliter of medium can contain more than 10^9 *Escherichia coli* cells or 10^{11} bacteriophage particles and cost less than a penny. In addition, a culture of *E. coli* can be grown under conditions that allow it to double every 20 minutes. By contrast, 10^9 mice would cost more than 10^9 dollars and would require a cage that would cover about 3 square miles, and growth of a generation of mice takes about 3 months instead of 20 minutes.

Of course, bacteria and viruses cannot be of use to geneticists and molecular biologists unless they *have* some form of genetics—and, preferably, some form of sex life. That bacteria mutate was finally demonstrated by Salvador Luria and Max Delbrück in 1943. Three years later, Joshua and Esther Leder-

berg and Edward Tatum proved that some bacteria can engage in a form of sexual reproduction—at least in the sense of the transfer of genetic information from one individual to another. These advances, together with the ease of growing and handling bacteria and their viruses, permitted the explosion of molecular biology that came shortly thereafter—you have read about some of these discoveries in Chapter 11. The relative simplicity of these biological systems contributed immeasurably to the elucidation of the nature of the genetic material, the replication of DNA, and the mechanisms of gene expression. Later these same systems were the first subjects of recombinant-DNA technology (Chapter 14). Questions of interest to all biologists continue to be studied in prokaryotes.

MUTATIONS IN BACTERIA AND BACTERIOPHAGES

When *E. coli*—or any other bacterial species—multiplies, a single cell divides to produce two identical offspring. That single cell gives rise to a clone—a population of genetically identical individuals. As long as conditions remain favorable, a population of *E. coli* can double every 20 minutes. *E. coli* may conveniently be grown on the surface of a solid medium containing a sugar, minerals, a nitrogen source such as ammonium chloride, and a solidifying agent such

BOX 12.A

Counting a Population of Bacteria

For many experiments with bacteria and viruses, one of the basic necessities is knowing how many individuals one is dealing with. It is impractical to count these organisms one at a time, so indirect methods must be employed. For bacteria, a standard method begins with spreading a known volume of culture onto the surface of a sterile, solid medium contained in a petri plate. As long as the medium is of the right type, each bacterium placed on the surface divides; its progeny in turn divide; and a colony visible to the naked eye soon appears. Under good conditions, each bacterium gives rise to a separate colony, so a count of colonies directly measures the number of bacteria that were plated.

For this to work, the right number of bacteria must be put on the plate. If there are more than 200–300 colonies on a plate, it may be difficult to count them accurately. If there are fewer than approximately half a dozen colonies, one has to worry about statistical variation. (And if a set of 10 plates produces a total of only one colony, you really have no idea how many bacteria were in the original sample—except that the number was *small*.) To avoid problems like these, one usually prepares a number of plates with different dilutions of the original culture. This is done by the technique of **serial dilution**. One-tenth milliliter of the original sample is added to 0.9 ml of sterile liquid medium in a tube, diluting the culture by a factor of 10. From this diluted sample, 0.1 ml is added to another 0.9 ml of sterile medium in another tube, giving a 100-fold dilution of the original culture. This may be continued through several steps. Then 0.1 ml of the bacterial suspension from each dilution tube is plated onto a separate petri plate. (The term *suspension* is used rather than *solution*, because the bacteria do not dissolve in the medium.) When the colonies are large enough to count, one might obtain results like those in the table. Plate 6, in which the sample had been diluted by a factor of 10^6, showed 26 colonies from the 0.1 ml that had been plated; that is, that dilution had a concentration of 26 bacteria per 0.1 ml, or 260 bacteria per milliliter. From this we

PLATE NO.	DILUTION FACTOR	NO. OF COLONIES
1	10	(too many)
2	10^2	(too many)
3	10^3	(too many)
4	10^4	(too many)
5	10^5	240
6	10^6	26
7	10^7	3
8	10^8	0

Counting Bacteria[a]

[a]Sample data from plates of serially diluted cultures. For each dilution, 0.1 ml was plated on nutrient agar.

deduce that the original culture contained $260 \times 10^6 = 2.6 \times 10^8$ cells/ml. Note that this corresponds reasonably well with the better estimate of 2.4×10^8 cells/ml derived from Plate 5 and the decidedly approximate 3×10^8 cells/ml obtained from Plate 7. The results can be rendered more precise by preparing two or three plates with each dilution and taking the average.

A set of serial dilutions in which a culture of unknown concentration is diluted by successive factors of 10. The tube from which Plate 6 is prepared contains a 10^{-6} dilution of the original culture. In other words, the original culture is diluted by a factor of 10^6, or one million-fold. If 0.1 ml of this suspension yields 26 colonies, we deduce that the original culture contains (26 colonies/0.1 ml) $\times 10^6 = 2.6 \times 10^8$ cells per milliliter.

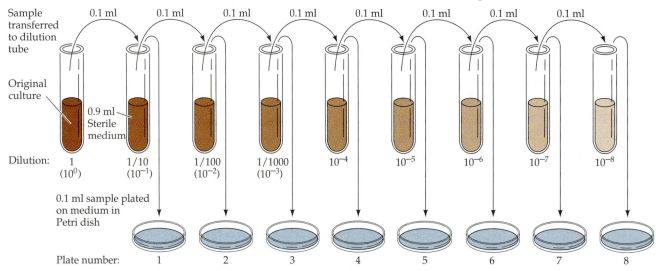

Sample transferred to dilution tube — 0.1 ml (×8)

Original culture — 0.9 ml Sterile medium

Dilution: 1 (10^0), 1/10 (10^{-1}), 1/100 (10^{-2}), 1/1000 (10^{-3}), 10^{-4}, 10^{-5}, 10^{-6}, 10^{-7}, 10^{-8}

0.1 ml sample plated on medium in Petri dish

Plate number: 1 2 3 4 5 6 7 8

as agar (Box 12.A). If bacterial cells are placed on such a surface, under aseptic conditions so that there are no competing microorganisms, each bacterium gives rise to a small, rapidly growing colony. If enough cells (10^7–10^8) are used, the resulting 10^7–10^8 colonies grow until they merge, forming what is called a bacterial "lawn." If the culture begins with a pure type, such as the strain called *E. coli* K, the entire lawn will be of the same type.

We can do an experiment to demonstrate that bacterial mutants arise. Let us mix a large sample of *E. coli* K with a suspension of a bacteriophage, such as the one called T4, and pour the mixture over growth medium in a petri plate. Wherever a virus finds a bacterial cell, it attaches to it, infects it, and eventually causes it to burst, killing the bacterium and releasing a large number of new viruses. These, in turn, attack neighboring cells. Soon visible **plaques**, or circular clearings, begin to appear in the lawn wherever the viruses have killed bacteria (Figure 12.1). A plaque is a circular clearing in the lawn, caused by the virus-induced bursting, or lysis, of bacteria. The plaques grow and grow. However, as you look over several such plates, here and there you find a bacterial colony growing in the midst of a plaque, in spite of the surrounding hordes of viruses. Each of these colonies has arisen from a mutant bacterium resistant to the virus. We call these bacteria *E. coli* K/4 (pronounced "K-bar-4") because they are resistant to phage T4. One can show that resistance is a heritable trait, because such bacteria give rise to colonies of cells that are similarly resistant to T4.

The phages also mutate. If you prepare plates inoculated simultaneously with *E. coli* K/4 and phage T4, you expect to see no plaques because K/4 is T4-resistant. Occasional plaques *are* found, however. These plaques must arise from mutations. But what has mutated, the bacteria or the viruses? Give this question a moment's thought. It is the T4 that have mutated this time. A back mutation of one *E. coli* K/4 back to the wild-type K would not result in plaque formation because only that single cell would be infected by the phage and burst; the neighboring bacteria would still be resistant and would still form an even lawn. However, a mutant phage T4 can infect a K/4 cell and, through its progeny phages, lead to the formation of a plaque. Such a phage is designated T4*h* because it has changed as regards its host—the type of bacterial cell it can infect. The mutation of the phage, like that of the bacterium, is heritable, as seen by the ability of the progeny phages to lyse the K/4 and form a growing plaque.

This simple series of events illustrates the basic ways in which bacteria and bacteriophages are grown in the laboratory, as well as the consequences of certain mutations. What we have seen, by the way, is an example of evolution on a small scale. *E. coli* K

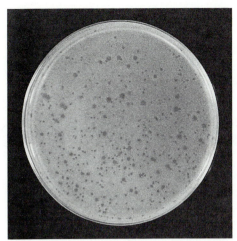

12.1 Lawns and Plaques
This laboratory plate contains a lawn of *Escherichia coli* with numerous plaques (dark areas) caused by bacteriophage. The plaques form where bacterial cells are lysed by the phage.

mutate to K/4 at a low rate all the time in nature, even as bacteriophage T4 mutate to T4*h*. The mutants normally do not take over the entire population but exist in low frequency as members of the bacterial or phage population. However, when the environment favors one genotype in a population over others, the proportions of the different genotypes in the population change. Here, for example, T4 kills *E. coli* K, but not K/4, so K/4 soon becomes predominant.

BACTERIAL CONJUGATION

The existence and heritability of mutations in bacteria and their viruses made these microbes attractive subjects for the investigation of genetic questions. However, if their reproduction were solely asexual, they would not be useful for genetic analysis. Some form of exchange of genetic information between individuals is necessary. Luckily, that does occur. Genetic recombination—a sexual phenomenon—is a rare event in *E. coli*; it was demonstrated by the Lederbergs and Tatum in 1946. They used two nutrient-requiring, or auxotrophic, strains of *E. coli* K12 as parents. Strain I requires the amino acid methionine and the vitamin biotin for growth. We can describe the genotype of this strain as *met⁻bio⁻*. Strain II requires neither of these substances but cannot grow without the amino acids threonine and leucine. Considering all four factors, strain I is *met⁻ bio⁻ thr⁺leu⁺* and strain II *met⁺bio⁺thr⁻leu⁻*. The two mutant strains were mixed and cultured together for several hours on a medium supplemented with methionine, biotin, threonine, and leucine, so that both could grow. The bacteria were then removed from the medium by

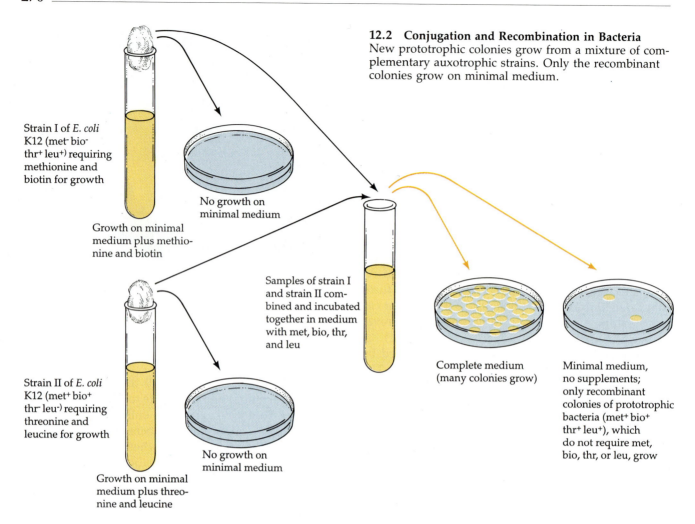

12.2 Conjugation and Recombination in Bacteria
New prototrophic colonies grow from a mixture of complementary auxotrophic strains. Only the recombinant colonies grow on minimal medium.

Strain I of *E. coli* K12 (met⁻ bio⁻ thr⁺ leu⁺) requiring methionine and biotin for growth

No growth on minimal medium

Growth on minimal medium plus methionine and biotin

Strain II of *E. coli* K12 (met⁺ bio⁺ thr⁻ leu⁻) requiring threonine and leucine for growth

No growth on minimal medium

Growth on minimal medium plus threonine and leucine

Samples of strain I and strain II combined and incubated together in medium with met, bio, thr, and leu

Complete medium (many colonies grow)

Minimal medium, no supplements; only recombinant colonies of prototrophic bacteria (met⁺ bio⁺ thr⁺ leu⁺), which do not require met, bio, thr, or leu, grow

centrifugation, washed, and transferred to minimal medium. Neither parental strain could grow on this medium because of their nutritional requirements. However, a few colonies *did* appear on the plates. Because they grew, they must have consisted of bacteria that were *met⁺ bio⁺ thr⁺ leu⁺* (Figure 12.2). These colonies appeared at a rate of approximately 1 for every 10 million cells put on the plates.

In terms of what you have learned thus far, there would seem to be at least three possible explanations for the appearance of those prototrophic, wild-type colonies. One would be mutation. This hypothesis was rejected on the following grounds: First, the observed colonies would have had to arise from *double* mutations because each parent started with two defective alleles. Given the range of single mutation frequencies, one would expect such double mutations to occur in at most 1 of every 10^{12} cells—100,000 times less frequently than actually observed. Neither parental strain, when grown alone under the same conditions, was ever observed to give rise spontaneously to wild-type colonies.

A second possibility was that the Lederbergs and Tatum had simply observed transformation, the incorporation of genetic material from dead bacteria

into live ones, as first described by Griffith (Chapter 11). This possibility was eliminated by Bernard D. Davis of the U. S. Public Health Service. He conducted his experiment with a U-shaped tube, which had its two arms separated by a very fine filter. The pores in the filter were large enough for molecules such as proteins and nucleic acids to pass through, but small enough to prevent the passage of bacteria. He placed a culture of strain I in one arm and a culture of strain II in the other. Then he applied alternating pressure and suction so that the growth medium was flushed back and forth. The flushing mixed the solutions from the two arms, but it did *not* mix the bacteria. Transforming principle from dead bacteria would have been able to pass through the filter. However, Davis found *no* wild-type bacteria on either side of the filter (Figure 12.3). The phenomenon observed by the Lederbergs and Tatum clearly requires that cells of the two strains come into physical contact with one another.

The third possibility was that bacteria had **conjugated** in pairs (Figure 12.4), allowing their genetic material to mix and recombine so that the prototrophic colonies resulted from cells containing *met⁺* and *bio⁺* alleles from strain II and *thr⁺* and *leu⁺* from

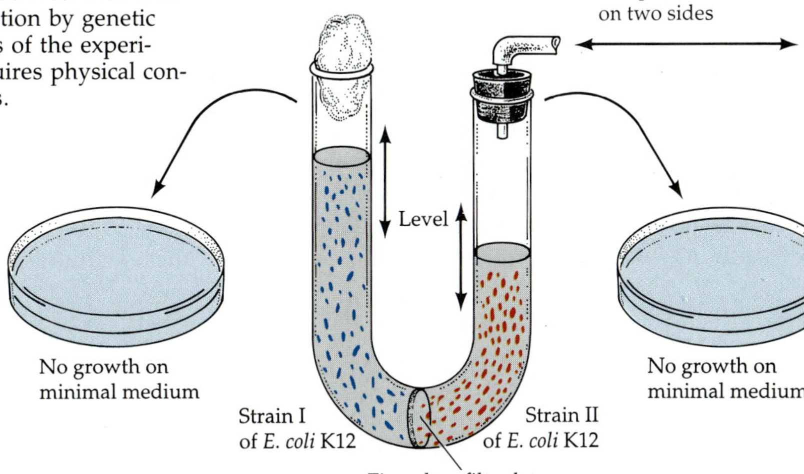

12.3 The Davis U-Tube Experiment

Because no prototrophic recombinant was recovered from either side, the possibility of transformation by genetic material alone was ruled out; the results of the experiment suggested that recombination requires physical contact between cells of the parental strains.

Alternating vacuum and pressure mixes solutions on two sides

No growth on minimal medium

No growth on minimal medium

Strain I of *E. coli* K12

Strain II of *E. coli* K12

Level

Fine glass filter lets molecules (e.g., DNA) pass but not bacteria

strain I. This model was confirmed by other experiments that showed that two cells of differing genotype mated, and one cell—the recipient—received DNA that included the two wild-type alleles for the loci in the recipient. Recombination created a genotype with four wild-type alleles. We will see presently that this recombination resulted not from a simple fusion of cells and subsequent segregation, but from a fundamentally different process.

Isolating Specific Bacterial Mutants

Throughout this chapter we will be talking about experiments that use bacteria and phage with various specific genotypes, as we have just done in the con-

jugation experiment. How can one obtain a strain with a particular genotype, such as *met⁻ bio⁻ thr⁺ leu⁺*?

To isolate a new strain of bacteria carrying a particular mutation—let's say *met⁻*—one starts with a strain carrying the wild-type allele for which mutations are desired. One then subjects these bacteria to procedures that increase the mutation rate, such as irradiation with ultraviolet or X rays, or the addition of some chemical mutagen. Now the search begins. First, one lets the bacteria in the culture grow and increase their numbers by keeping them in a medium that includes the compound that will be needed by the desired mutant strain (in our example, the me-

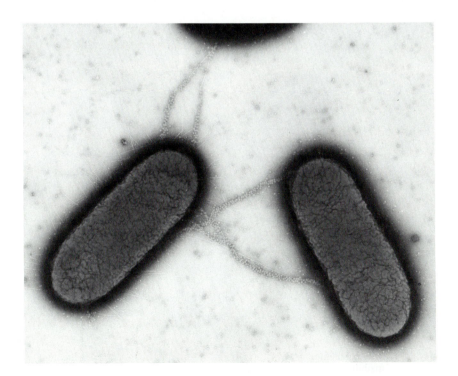

12.4 Conjugating Bacteria

The "male" cell of *E. coli* on the left is connected to two "female" cells by thin tubes called F-pili. In this instance, two F-pili are in contact with each female cell. The tiny "beads" on the pili are a bacteriophage that attaches specifically to F-pili, aiding in visualization of the pili. After cells are joined by F-pili, they are drawn into closer contact, and DNA is transferred from one cell to the other.

dium must include methionine). The overwhelming majority of the cells in the culture are unchanged and will still be wild type; these wild-type cells must be eliminated. There is more than one way to do this, but we will describe one invented by Davis. He knew that the antibiotic penicillin kills only growing bacteria. Therefore, his method was to take a mixed culture of many wild-type and a few mutant bacteria, put it in medium *lacking* the nutrient for which the desired mutants were auxotrophic (again, methio-

nine in our example), and add penicillin. Those cells that do not need methionine grow rapidly—and commit "penicillin suicide." Because they grow, they die. The desired mutants, on the other hand, fail to grow (because the needed nutrient is unavailable) and so avoid damage by the penicillin. They are next transferred to a petri plate that contains medium lacking penicillin but containing methionine, so that they may grow and form colonies (Figure 12.5a). A particular methionine mutant can then be chosen and used

12.5 Isolating and Identifying Auxotrophic Mutants

(a) Penicillin kills growing wild-type cells, but the nongrowing methionine-requiring auxotrophs survive. (b) Because the velvet "prints" an exact replica, colonies present on the master plate but missing from the replica are recognized as auxotrophs. (c) The colony shown in green on the master plate grows only on the plate containing both methionine and biotin (middle) and is therefore met⁻bio⁻. Can you identify the nutritional requirements of each colony on the master plate?

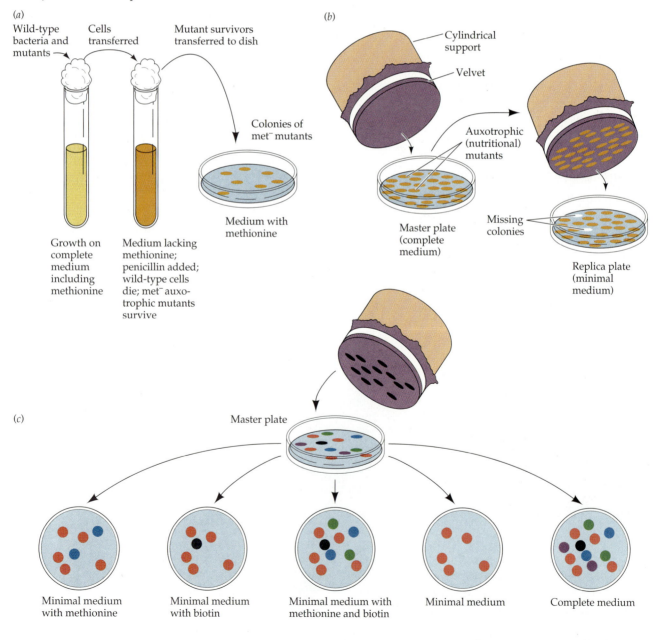

as a parental strain for selecting a *met⁻bio⁻(thr⁺leu⁺)* double mutant by the same general approach as was used to select the single (*met⁻*) mutant. In this case, of course, methionine must be present in the medium at all times, and the absence of biotin is used to protect the *met⁻bio⁻* double mutants from penicillin-induced death. By selecting for the second mutation, one obtains the desired double mutant (*met⁻bio⁻*).

How does one identify the progeny of various recombinant types after performing a cross between two strains of bacteria? One way is by **replica plating**, a technique invented by Joshua and Esther Lederberg (Figure 12.5*b*). Let us see how this can be used to identify bacteria of a particular genotype, say *met⁻bio⁻*, from a mixed population obtained by crossing two strains (one *met⁻bio⁺*, the other *met⁺bio⁻*). A small sample of about 0.1 milliliters of a mixed suspension, presumably including the desired genotype as well as others, is spread on a plate with complete medium and allowed to produce colonies. A sterilized piece of velvet, mounted on a cylindrical support that fits easily into a petri plate, is now pressed gently against the medium. Its fuzzy surface picks up substantial numbers of bacteria from each of the colonies. The velvet is next pressed against the sterile surfaces of new plates containing different kinds of media. In each of these replicate plates, some of the bacteria on the velvet stick to the agar medium—in the same positions relative to one another that they occupied on the original "master" plate (Figure 12.5*b*). In our example, we might use five different replicate plates: one with minimal medium, one with methionine added, one with biotin added, one with both methionine and biotin, and, finally, one with complete medium. Wild-type colonies on the master plate would give growing replicate colonies on each of these plates. What about single mutants (*met⁻bio⁺* and *met⁺bio⁻*)? And what about the desired double mutants? Think about these before studying Figure 12.5*c*. As you can see, replica plating is a powerful means for characterizing mutant colonies.

The Bacterial Fertility Factor

Genetic recombination during conjugation in *E. coli* is a one-way process from a donor to a recipient. English microbiologist William Hayes characterized many strains and found that each strain is either recipient or donor, which can also be called female and male.

The female bacterium itself becomes male after conjugation—in bacteria, maleness is an infectious venereal disease! Hayes also found that a strain of male bacteria gives rise to occasional mutants that no longer function as males—but can now act as females. Hayes rationalized these observations by proposing that maleness in bacteria is due to the presence of a fertility factor, called **F**. Males possess the

factor and are F⁺; females, lacking the factor, are F⁻. In a cross of F⁺ × F⁻, a copy of the F factor is transferred to the female, thus rendering it F⁺, while the original male remains F⁺ (Figure 12.6). The F factor is an extra piece of DNA that can replicate itself and persist in the cell population as if it were a second chromosome independent of the normal bacterial chromosome. Males can change into females simply by losing the F factor through mutation. Genes on the F factor direct a number of processes, among which is the formation on the surface of the male bacterium of long, thin, hairlike projections called **F-pili** (singular, *pilus* = hair). These are tubes with ends that attach to the surface of female cells (see Figure 12.4). Initial contact is made by an F-pilus, and subsequently a mating contact is made that allows DNA to be transferred.

Transfer of Male Genetic Elements

The discovery of genetic recombination by conjugation opened the possibility of mapping the genetic

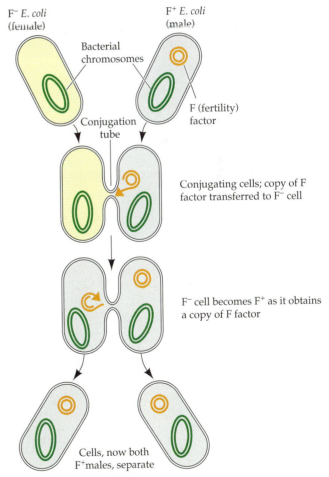

F⁻ *E. coli* (female)

F⁺ *E. coli* (male)

Bacterial chromosomes

Conjugation tube

F (fertility) factor

Conjugating cells; copy of F factor transferred to F⁻ cell

F⁻ cell becomes F⁺ as it obtains a copy of F factor

Cells, now both F⁺males, separate

12.6 Infectious Fertility in *E. coli*
During conjugation, the F⁻ recipient cell receives a copy of the F factor—an extra piece of DNA—from the donor cell by way of a connecting tube (the conjugation tube) and becomes F⁺.

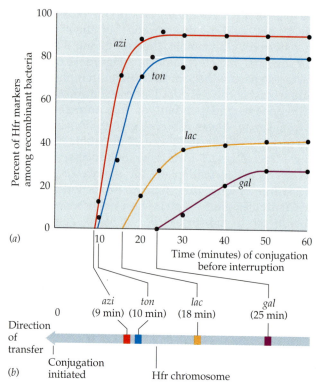

(a)

(b)

azi (9 min) ton (10 min) lac (18 min) gal (25 min)

0

Direction of transfer

Conjugation initiated

Hfr chromosome

12.7 Chromosome Maps from Interrupted Matings
Chromosome maps of *E. coli* were constructed by interrupting conjugating cells at various times and counting recombinants from the matings. (*a*) The map in (*b*) was constructed from data plotted on this graph. In the particular Hfr strain used, recombinants carrying the *azi* allele were first detected from matings interrupted 10 minutes after mixing, as were recombinants carrying the *ton* allele. By extrapolation from other points on the graph, it was determined that the *azi* gene was transferred at 9 minutes, the *ton* gene at 10 minutes. Timings for the *lac* and *gal* genes were established in the same way. (*b*) The map units are in time, in minutes, rather than in recombination frequencies.

most dramatic experiments. They mixed Hfr and F⁻ bacteria at high concentration to initiate conjugation; at various times thereafter they diluted samples of the mixture and agitated them in a kitchen blender for two minutes—this treatment separates conjugating bacteria but does not damage them otherwise. The number of Hfr markers passed to the females depended upon the length of time allowed for conjugation before it was interrupted—the longer the conjugation, the more markers were transferred (Figure 12.7*a*). The markers always entered in a particular order from any particular Hfr strain. The Hfr almost never transferred the F factor itself.

Jacob and Wollman immediately recognized that this interrupted mating technique provided a simple way to map the chromosome. They prepared different mutant strains and crossed pairs of strains; then they interrupted successive samples from the crosses at intervals with a blender. They found that markers are transferred in a particular sequence. The length of mating time required before a particular marker is transferred so that it can appear in recombinant progeny is, then, a measure of its location on the chromosome (Figure 12.7*b*).

Hfr mutants of different origins have different genetic maps, as seen in Table 12.1. However, if you examine the table, you may be able to spot a regularity in the different maps. Jacob and Wollman noticed that whereas different markers enter first in different Hfr strains, the maps are always consistent in that a marker B that lies between markers A and C always does so in any Hfr map. That is, the starting points vary, but the *order* of genes remains constant: Even when genes are in a reversed order, B is still

material of bacteria. However, early attempts at mapping were complicated by the fact that very few recombinant offspring arose from F⁺ × F⁻ crosses. This made it difficult to obtain reliable quantitative data. However, mutant male strains were found that acted as high-frequency donors of genetic information. These were called **Hfr** mutants, for *H*igh *f*requency of *r*ecombination. Hayes found that Hfr males, unlike his ordinary F⁺ males, did not generally transfer their F factor to the female. Also, they transferred only certain markers with high frequency, transferring other markers no more frequently than ordinary F⁺ males did. We know now that in Hfr strains the F factor is actually incorporated into the bacterium's chromosome. Work in 1955 by the French biologists Elie Wollman and François Jacob provided an explanation for Hayes' observations.

Jacob and Wollman showed that the markers from an Hfr male enter the female one at a time. They used the technique of **interrupted mating** in their

TABLE 12.1 Sequences of Markers Transferred by Various Hfr Strains[a]					
ORDER OF ENTRY	**Hfr H**	**Hfr 1**	**Hfr 2**	**Hfr 3**	**Hfr 4**
1	T	L	pro	ade	B₁
2	L	T	T₁	lac	ilu
3	azi	B₁	azi	pro	mal
4	T₁	ilu	L	T₁	trp
5	pro	mal	T	azi	gal
6	lac	trp	B₁	L	ade
7	ade	gal	ilu	T	lac
8	gal	ade	mal	B₁	pro
9	trp	lac	trp	ilu	T₁
10	mal	pro	gal	mal	azi
11	ilu	T₁	ade	trp	L
12	B₁	azi	lac	gal	T

[a]Each column gives data from a specific Hfr strain. Markers at top of column entered first, those at bottom entered last.

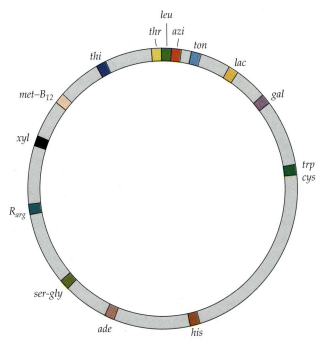

12.8 A Map of *E. coli*'s Circular Chromosome
This map of the *E. coli* chromosome summarizes data from interrupted conjugation experiments (Figure 12.7) with many Hfr strains (Table 12.1). It shows the relative positions of a number of loci. The three-letter abbreviation for each gene is derived from its phenotype: *leu*cine-requiring, *azi*de-resistant, *xyl*ose-utilizing, and so on. This early version of the chromosome map shows only a few genes. By now more than half of the estimated 3,000 *E. coli* genes have been mapped.

between A and C. The simplest conclusion is that the bacterial chromosome is *circular* (Figure 12.8). If you break the circle in different places and convert it into a linear form, you can see what maps are generated.

From these and other experiments, Jacob and Wollman concluded that (1) the *E. coli* chromosome is circular, (2) Hfr males have the F factor incorporated into their chromosome, (3) the location of the insertion varies, giving rise to different Hfr strains, (4) the inserted F factor marks the point at which the chromosome "opens" as conjugation begins, and (5) one end—always the same one—of the now opened chromosome then leads the way into the female. The piece of chromosome continues to move through the mating contact until mating is interrupted naturally or otherwise. At the very tail end of the opened chromosome lies that portion of the F factor that determines maleness.

It was later shown that what moves from the Hfr to the F⁻ is not actually the entire Hfr chromosome. Transfer is initiated by a nick of one strand within the F factor. The nicked strand at the 5′ end begins to unravel from the chromosome and moves to the F⁻, as shown in Figure 12.9. Meanwhile, the transferred strand is replaced by DNA synthesis at the 3′ end of the nick, using the intact circular strand as

the template. Thus the male still contains a complete set of DNA sequences even after donating a fair amount of DNA to the F⁻. The DNA strand that has entered the F⁻ now replicates, becoming double-stranded. Markers on this piece of DNA will not give rise to recombinant bacteria unless they become incorporated into the F⁻ chromosome by crossing over (Figure 12.10). About half the transferred Hfr markers get incorporated in this way, the others being lost as the cell divides.

Sexduction

Sometimes the F factor of an Hfr male separates from the chromosome. In the separation process, the F factor may carry with it a bit of the chromosome. Any

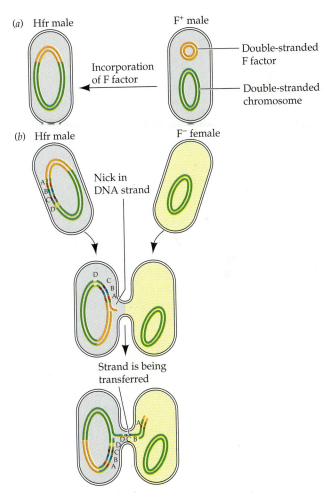

12.9 Origin and Behavior of Hfr Strains
(*a*) When the F factor of an F⁺ cell is incorporated into its chromosome, the cell becomes an Hfr male. (*b*) During conjugation, the male's chromosome opens within the inserted F factor; and one strand of the DNA double helix is transferred to the recipient cell. Because most of the F factor is the last DNA to be transferred, the recipient cell becomes a male only if conjugal transfer of the complete chromosome is achieved. DNA replicates in both donor and recipient strands to make new double-stranded molecules after transfer.

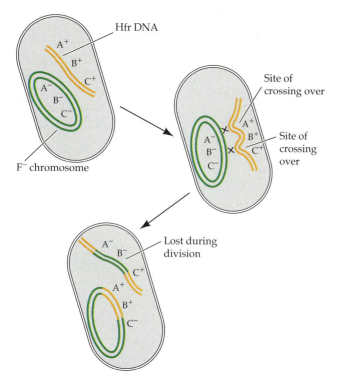

12.10 Recombination Following Conjugation
DNA from an Hfr donor may become incorporated into the recipient cell's chromosome through crossing over. As diagrammed here, only part of the A⁺B⁺C⁺ donor chromosome was incorporated—the part containing A⁺ and B⁺. The resulting A⁺B⁺C⁻ recombinant segment becomes a permanent part of the recipient genotype, and the reciprocal A⁻B⁻C⁺ segment is lost.

chromosomal markers thus captured by the F factor are transferred to the F⁻ recipient when conjugation occurs. This process in which chromosomal markers are carried by the autonomous F factor into the F⁻ is called **sexduction**. The modified F factor is called an F′ (F-prime) factor (Figure 12.11).

An F or F′ factor, like the bacterial chromosome, is a circular DNA molecule. Chromosomal markers carried by the F′ factor are allelic to genes on the main chromosome in the recipient cell. Thus cells harboring an F′ factor may contain more than one allele of a particular gene. Such cells may be used to study dominant–recessive relationships for genes present on F′ factors in these normally haploid bacteria.

BACTERIOPHAGES

Recombination in Phages

Genetic recombination in phages was demonstrated in 1946, the same year the Lederbergs and Tatum revealed the sex life of the bacterium. Here too, mutant characters were needed to serve as markers. Some markers in phages affect the formation of plaques (Figure 12.1), and we have already noted that

mutations can change the ability of a phage to infect certain hosts. To understand the basis for genetic recombination in phages, we must recall that phages reproduce by injecting their genetic material into a host bacterium, which then supports the synthesis of a large number of progeny phages.

Alfred Hershey and Raquel Rotman, at Washington University (St. Louis), performed a series of experiments in which *E. coli* were simultaneously infected by *two* different mutant strains of the bacteriophage T2 (Figure 12.12). In their first experiment, one of the phage strains was genotypically h^+r and the other was hr^+ (we need not worry here about what the actual phenotypes were; just note that h^+ and h are alleles at one locus and r^+ and r are alleles at another locus). The addition of these phages to a culture of *E. coli* would be expected to produce substantial numbers of phages of both parental types. In fact, Hershey and Rotman found not only the parental types but also significant numbers of phages of genotypes h^+r^+ and hr, that is, recombinant phages. As more markers in such phage crosses were studied, a map began to take form. In due course, it was

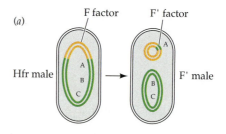

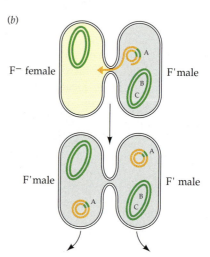

12.11 Sexduction by F′ Factors
(a) The F factor in an Hfr cell sometimes leaves the cell's chromosome and may carry with it some chromosomal genes, indicated here by the marker A, leaving the chromosome deficient in those genes. The factor is then called an F′ factor, and the cell carrying it, an F′ cell. (b) A recipient that conjugates with an F′ strain receives a copy of the DNA of the F′ factor, including the chromosomal genes; the recipient becomes an F′ cell.

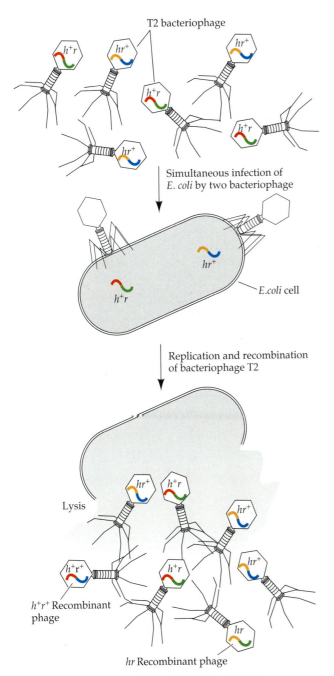

12.12 Genetic Recombination in Bacteriophage
Strains of a bacteriophage can be crossed by infecting a culture of susceptible bacteria with both strains simultaneously.

learned that the phage has a single chromosome and that its map is circular.

Lysogeny and the Disappearing Phages

Thus far, we have treated bacteriophages as if their life cycle was always **lytic**, that is, a cycle in which a phage infects a bacterial cell, the phage replicates, the cell lyses, and many new phages are released to renew the cycle. However, there are many cases in which the phage–bacterium relationship is more like a symbiosis. With some bacteria and some phages, infection does not invariably result in lysis of the bacteria. The phages seem to disappear from the culture, leaving the bacteria immune to further attack by the same strain of phage. However, if bacteria are carefully isolated from such a culture and allowed to grow in fresh medium, some free phages can be detected, showing that they are present in the bacteria. Bacteria harboring such an infection are called **lysogenic**. This name arises from the fact that the combination of lysogenic bacteria with other bacteria sensitive to the phage results in lysis of the sensitive cells.

To see where the free phages come from in a culture of lysogenic bacteria, the French microbiologist Andre Lwoff performed the following delicate experiment with the rather large bacterial species *Bacillus megaterium*. From a lysogenic culture of *B. megaterium*, he isolated a single bacterial cell, mounting it in a drop of medium on a microscope slide. He watched patiently until the cell divided and then removed one of the daughter cells with a micropipette. This cell was transferred to an agar medium to see whether its offspring would be lysogenic. Meanwhile he kept his eye on the daughter cell still under the microscope. When it divided, he again farmed out one of its daughters to solid medium while retaining the other on the microscope slide. He repeated this a total of 19 times, finding that each of the cells transferred to solid medium gave rise to a lysogenic colony. At each transfer, he also sampled the original microdrop to see whether it contained any free phages. It did not. However, in repetitions of this experiment, he sometimes would see the *B. megaterium* burst while it was under the microscope. Whenever this happened, the drop of medium was found to be teeming with free phages.

From such experiments it was evident that the lysogenic bacteria contain a noninfective entity, which Lwoff called a **prophage**. The prophage could remain quiet within bacteria through many cell divisions. Occasionally a lysogenic cell would be **induced**, somehow, and lyse the bacterium, releasing a large number of free phages, which could then infect other bacteria and renew the life cycle. Lwoff learned that ultraviolet radiation is a potent inducer of the production of free phages. Work by many investigators established finally that the prophage is, in fact, a molecule of phage DNA that has been incorporated into the bacterial chromosome. This corresponds exactly with the incorporation of the bacterial F factor into the chromosome that gives rise to an Hfr male. Just as the F factor sometimes leaves the chromosome, so may the prophage—whereupon the phage DNA is activated to multiply rapidly, make many new phages, and lyse the bacterium. The lytic and lysogenic cycles are contrasted in Figure 12.13.

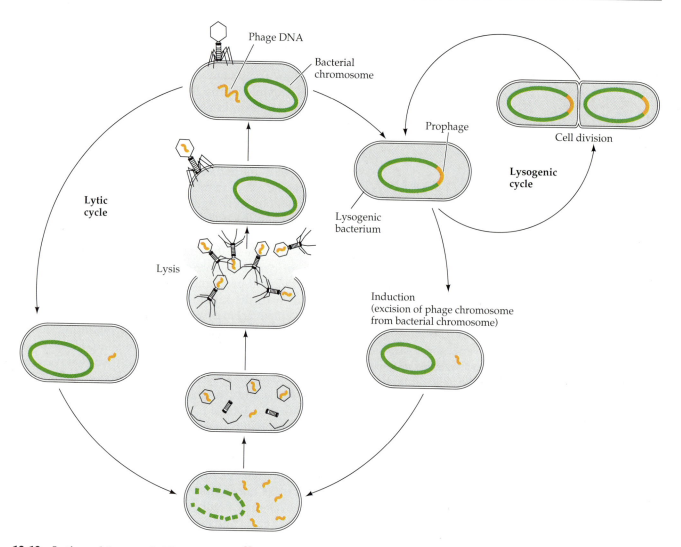

12.13 Lytic and Lysogenic Phage Cycles
Phage DNA (orange) injected into a host bacterium may be incorporated into the host's chromosome (lysogenic cycle) or remain free (lytic cycle). In the lytic cycle, the phage DNA replicates and directs the formation of many new phages, which lyse the host cell (center). Progeny phages may infect other host cells, lysing them and repeating the lytic cycle until all cells in the culture are lysed. In the lysogenic cycle, the phage DNA becomes integrated into the host's chromosome as a prophage. A prophage is replicated as part of the host's chromosome but does not direct the formation of new phages. When a prophage is induced to separate from the host's chromosome, it again becomes a lytic phage, replicating its nucleic acid, forming new phages, and lysing the host cells.

Transduction

If the prophage can escape from the chromosome, then we might expect that on occasion bacterial markers might be taken along by the departing phage DNA (Figure 12.14a). The resulting phages might then introduce these markers into other bacteria that they infect, resulting in genetic recombination in the bacteria. This phenomenon was discovered in 1956 by Joshua Lederberg and was called **restricted transduction**. Transducing phages, which carry bacterial markers, cause newly infected bacteria to become lysogenic. In restricted transduction, only those chro-mosomal genes adjacent to the site of attachment of the prophage may be taken along with the phage DNA.

A related phenomenon, **general transduction**, differs from restricted transduction in that it results from the incorporation of part of the *bacterial* chromosome, *without* the prophage, into a phage coat (Figure 12.14b). The resulting particle, even though it lacks any phage genes, can infect another bacterium, squirting in the piece of DNA from its former host. The bacterium thus infected does *not* become lysogenic, nor does it form new phages and burst as in the lytic cycle—it has not really been infected by a

(a) Restricted transduction

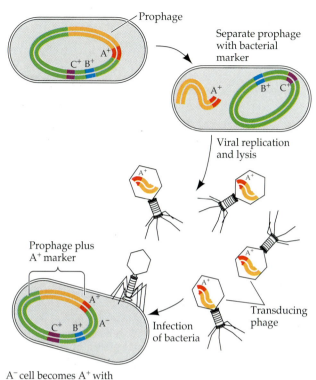

Prophage

Separate prophage with bacterial marker

Viral replication and lysis

Prophage plus A⁺ marker

Transducing phage

Infection of bacteria

A⁻ cell becomes A⁺ with incorporation of marker carried by transducing phage

12.14 Transduction

(a) In restricted transduction, a prophage picks up adjacent bacterial markers as it separates from the bacterial chromosome. Markers from the bacterial chromosome may then be introduced into other bacterial cells that the resulting phages infect. In the diagram an A⁻ bacterial cell—one that is not producing the gene product associated with marker A—becomes A⁺ when a transducing phage introduces the marker into the recipient bacterium's chromosome. (b) In general transduction, parts of the bacterial chromosome are incorporated into phage coats without being accompanied by phage DNA. When these phages infect other bacterial cells, the pieces of chromosome are injected and may become part of the new bacterium's DNA by recombination.

phage. Instead, it now contains a piece of foreign bacterial DNA, rather as if it had conjugated with an Hfr cell. If crossing over takes place between the host chromosome and the transduced DNA, the transfer of markers is completed. In contrast with restricted transduction, general transduction can move any part of the bacterial chromosome. There is no limitation on what chromosomal markers might become enclosed in a phage coat. The phage coat is big enough to house several neighboring bacterial genes; therefore, general transduction serves as another powerful technique for mapping the bacterial chromosome—with viral assistance.

(b) General transduction

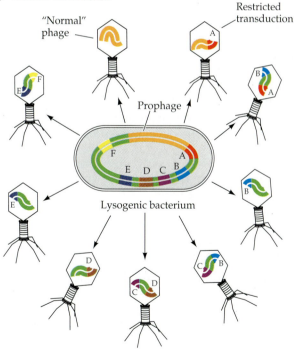

"Normal" phage

Restricted transduction

Prophage

Lysogenic bacterium

EPISOMES AND PLASMIDS

The F factor and viral prophages are examples of **episomes**, by which we mean nonessential genetic elements that can exist in either of two states: independently replicating within a cell or integrated into the main chromosome. They cannot arise by mutation but must be obtained from outside the bacterium. They are, in fact, obtained by infection, whether by a virus or by another bacterium. As we have seen, episomes may be used as vehicles for transferring genetic markers from one bacterium to another. Other nonessential genetic elements, which exist only as free, independently replicating circles of DNA that cannot be incorporated into the bacterial chromosome, are called **plasmids**. (An episome is simply a plasmid that has the possibility of becoming part of the chromosome.)

Resistance factors, or **R factors**, are important plasmids. The R factors first came to the attention of biologists in 1957 during a dysentery epidemic in Japan, when it was discovered that some strains of the dysentery bacterium *Shigella* were resistant to several antibiotics. Researchers found that resistance to the entire spectrum of antibiotics could be transferred by conjugation even when no markers on the main chromosome were transferred. Also, F⁻ cells could serve as donors, indicating that the genes for antibiotic resistance were not carried by the F factor. Eventually it was shown that the genes were carried on plasmids—called the R factors. Each of these carries one or more genes conferring resistance to particular antibiotics. As far as biologists can determine, R factors appeared long before antibiotics were discovered and used, but they seem to have become more abundant in modern times. Can you propose a hypothesis to explain why R factors might be more widespread now than in the past?

Plasmids are not required by a bacterium, because bacteria lacking plasmids still survive. For a particular kind of plasmid to be maintained within a population of bacteria, it must have an origin of replication (Chapter 11). That is, it must be a **replicon**, capable of independent replication, so that it divides at roughly the same rate as the bacterium. Otherwise, it is simply diluted out of the population.

TRANSPOSABLE ELEMENTS

As we have seen, plasmids, episomes, and even phage coats can serve as vehicles for transporting genes from one bacterial cell to another. Another type of "gene transport" within the individual cell relies on segments of chromosomal or plasmid DNA called **transposable elements**. Copies of transposable elements can be inserted at other points in the same or other DNA molecules, often producing multiple physiological effects resulting from the disruption of the genes into which the transposable elements are inserted (Figure 12.15).

The first transposable elements to be discovered in prokaryotes were large pieces of DNA, typically 1,000–2,000 base pairs long, found in many places in the *E. coli* chromosome. Such a sequence can replicate independently of the rest of the chromosome and insert the copy at other, seemingly random places in the chromosome—hence the name *transposable element*. The genes encoding the enzymes necessary for this insertion are found within the transposable element itself. Many transposable elements discovered later were longer (about 5,000 base pairs) and carried one or more additional genes. These elements are called **transposons**.

Although transposable elements were initially referred to as jumping genes, the frequency of such moves is usually very low. Transposition is closely regulated—random insertion would often lead to the inactivation of an essential gene and the death of the cell. The process by which transposable elements move is complex and incompletely understood. One step is the cutting of the DNA at the new site. This is done by an enzyme, transposase, encoded by the transposable element. As shown in Figure 12.16, the enzyme makes a staggered cut, and the copy of the

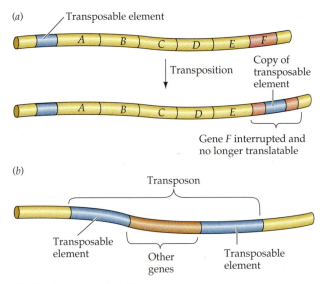

12.15 Transposable Elements
(a) If a transposable element appears in the middle of another gene, that gene can no longer be transcribed to yield an appropriate mRNA; thus the interrupted gene cannot function. (b) A transposon consists of two transposable elements flanking another gene or genes; the entire transposon is copied and inserted as a unit.

transposable element is inserted between the ends. When the gaps are repaired by DNA polymerase, a short duplicated sequence of chromosomal DNA is created at the two junction points.

Transposable elements have contributed to the evolution of plasmids. The plasmids called R factors originally gained their genes for antibiotic resistance through the activity of transposable elements; one piece of evidence for this is that each resistance gene in an R factor is part of a transposon. Transposons on the F factor and on the bacterial chromosome interact to direct the insertion of the F factor into the chromosome in the development of an Hfr male.

CONTROL OF TRANSCRIPTION IN PROKARYOTES

Let us now transfer our attention to the question of how the activities of genes are *regulated*. As a normal inhabitant of the human gut, *Escherichia coli* has to adjust to sudden changes in its chemical environment. Its host may present it with one foodstuff one hour and another the next. For example, the bacteria may suddenly be deluged with milk, the main carbohydrate of which is lactose. This sugar is a β-galactoside—a disaccharide containing galactose β-linked to glucose (Chapter 3). Before lactose can be of any use to the bacteria, it must first be taken into their cells by an enzyme called β-galactoside permease. Then it must be hydrolyzed to glucose and galactose by another enzyme called β-galactosidase.

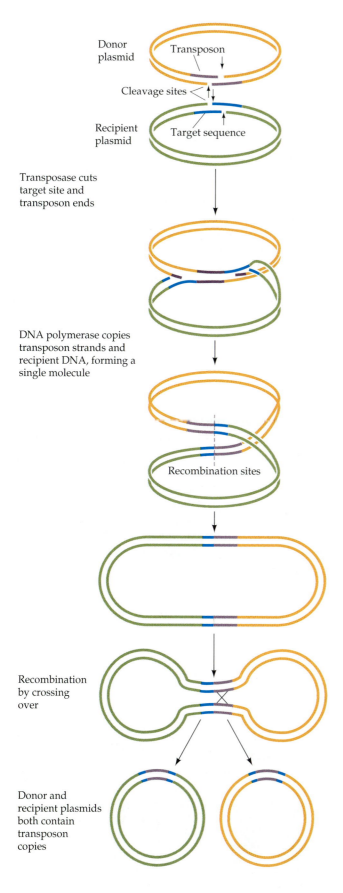

Donor plasmid

Transposon

Cleavage sites

Recipient plasmid

Target sequence

Transposase cuts target site and transposon ends

DNA polymerase copies transposon strands and recipient DNA, forming a single molecule

Recombination sites

Recombination by crossing over

Donor and recipient plasmids both contain transposon copies

12.16 Transposable Element Insertion

The enzyme transposase makes cuts in both the transposable element and the target site of the target DNA. These cuts are staggered; that is, the two strands of the DNA molecules are not cut at the same points. Crossing over occurs between the transposable element and the target DNA, and new DNA strands are formed. At this point the two original DNA molecules are joined in a single molecule. The combined molecule is then separated into two molecules, *both* containing transposable elements, by the enzyme resolvase.

A third enzyme, called thiogalactoside transacetylase, is also required for lactose metabolism. When *E. coli* is grown in a medium that does not contain lactose or other β-galactosides, the level of all three of these enzymes within the bacterial cell is extremely low. If the environment changes so that lactose is the predominant sugar and very little glucose is present, the synthesis of all three of these enzymes begins promptly and their levels may rise more than 1,000-fold.

Compounds that evoke the synthesis of an enzyme (as does lactose in this example) are called **inducers**. The enzymes that are evoked are called **inducible enzymes**, whereas enzymes that are made all the time at a constant rate are called **constitutive enzymes**. If the lactose is removed from the medium, the synthesis of the three enzymes stops almost immediately. The enzyme molecules that have already been formed do not disappear; they are merely diluted during subsequent growth and reproduction until their concentration falls to the original low level within each bacterium.

The blueprints for the synthesis of these three enzymes are called **structural genes**, indicating that they specify the primary structure (that is, the amino acid sequence) of a protein molecule. When Jacob, Wollman, and Monod mapped the particular structural loci coding for enzymes that metabolize lactose, they discovered that all three lie close together in a region that covers only about 1 percent of the *E. coli* chromosome.

It is no coincidence that these three genes lie next to one another. The information from them is transcribed into a single, continuous molecule of mRNA. Such a molecule, containing transcripts of more than one gene, is called a **polycistronic messenger**. This particular polycistronic messenger governs the synthesis of all three enzymes. Either all the enzymes are made or none of them is, depending on whether their common message—their mRNA—is present in the cell.

Processing a Polycistronic Messenger

How can a single mRNA molecule make three different polypeptides? The answer is that the polycistronic mRNA contains punctuation marks that specify the ending of one polypeptide chain and the beginning of the next. A molecule of tRNA is always attached to a growing polypeptide chain, but a finished molecule of protein does not contain any tRNA. This indicates that the last step in prokaryotic protein synthesis must involve not only the termination of the polypeptide chain but also the removal of the terminal tRNA. The termination signal is encoded in the mRNA. Three codons of the genetic code (UAA, UAG, and UGA) mean "terminate translation," and one of them must be present at the end of each structural gene. It is easy to see how a polycistronic messenger can give rise to one polypeptide for each structural gene. A ribosome begins at one end of the message, translates until it comes to the termination signal of the first structural gene transcript, and then releases the first polypeptide (Figure 12.17). However, the ribosome may remain bound and start translating the second structural gene at the next initiation site and, when it finishes, release the second polypeptide, and so on. When the ribosome has translated all the structural gene transcripts on the mRNA, it is released.

Promoters

Some genes are transcribed more often than others. In Chapter 11 it was stated that RNA polymerase attaches to DNA and starts transcribing, but nothing was mentioned about where it attaches. The polymerase does not attach itself randomly; special regions for attachment are built into the DNA molecule. These regions are called **promoters**. There is one promoter for each structural gene or set of structural

genes to be transcribed into mRNA. Promoters serve as a punctuation, telling the RNA polymerase where to start and which strand of DNA to read. A promoter and one or more structural genes are enough to specify the synthesis of an mRNA molecule.

Not all promoters are identical. One may bind RNA polymerase very effectively and therefore trigger frequent transcription of its structural genes; in other words, it competes effectively for the available RNA polymerase. Another promoter may bind the polymerase poorly, and its structural genes are rarely transcribed. The efficiency of the promoter sets a limit on how often each structural gene can be transcribed. An enzyme that is needed in large amounts is encoded by a structural gene whose promoter is efficient, but the synthesis of an enzyme that is needed only in tiny amounts is controlled by an inefficient promoter.

Operons

What about the enzymes, such as those that metabolize lactose, that the bacterium needs in large amounts at some times but not at all at others? The corresponding genes must contain a very efficient promoter so that the maximum rate of mRNA synthesis can be high. There must also be a way to shut down mRNA synthesis when the enzymes are not needed. One solution that has evolved in prokaryotes is to allow an obstacle to be placed between the promoter and its structural genes. The obstacle is a special sort of protein molecule called a **repressor**, which can attach to the DNA molecule at a site called the **operator**. When the repressor (the protein) is bound

12.17   Translation of Polycistronic mRNA

A polycistronic mRNA codes for several polypeptides, one for each of the several loci from which the mRNA was transcribed. A ribosome begins translating the polycistronic messenger at a common initiation site, then moves along the first locus to form the first polypeptide. When the ribosome reaches the first locus's termination signal, the first polypeptide is released and the ribosome continues along the mRNA, repeating the process twice. After it passes the final termination signal, the ribosome detaches from the messenger.

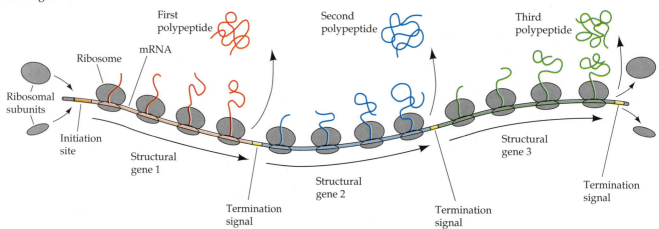

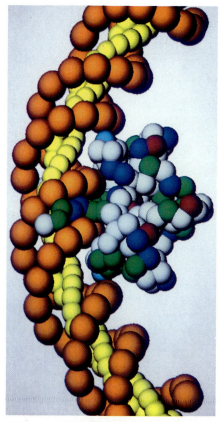

12.18 Repressor and Operator
A repressor protein binds the corresponding operator and thus blocks transcription. In this computer-generated image, the yellow and orange spheres denote parts of the DNA molecule, of which the operator is a part. A part of the repressor has already bound to the minor groove of the operator region of the DNA, and the lower part is about to bind to the major groove. The amino acids of the repressor are colored according to the following conventions: pale blue, hydrophobic; green, hydrophilic; red, positive charge; blue, negative charge.

to the operator (the DNA) it blocks the transcription of mRNA (Figure 12.18). When the repressor is not attached to the operator, messenger synthesis proceeds rapidly. A controllable unit of transcription of this sort is called an **operon**.

An operon always consists of a binding site on the DNA molecule (the promoter) for RNA polymerase, a binding site (the operator) for a specific repressor, and one or more structural genes. How is the operon controlled? The key lies in the repressor and its binding to the operator. The repressor is able to bind not only to its specific operator but also to inducers—in the case of the *lac* operon, lactose and certain other β-galactosides. Binding of the inducer changes the shape of the repressor (by allosteric modification, as explained in Chapter 6), virtually destroying the repressor's affinity for the operator. For example, when lactose (an inducer) is added to a culture of *E. coli*, the inducer enters the cell and promptly combines with its repressor, changing its shape and causing it

to detach from the operator. RNA polymerase can then bind to the promoter and start transcribing the structural genes of the *lac* operon (Figure 12.20). The mRNA transcribed from these genes is translated by ribosomes, which synthesize the enzyme products of the operon. If the inducer molecules bound to the repressor dissociate (which happens as the concentration of lactose drops), the repressor quickly becomes bound to the operator, and transcription of the *lac* operon stops. The mRNA that is already present is degraded over a period of a few minutes, and translation rapidly comes to an end. It is the inducer that regulates the binding of the repressor to the operator—and it is the inducer that gets metabolized by the enzyme products of the operon.

Repressor proteins are coded for by **regulatory genes**. The one that codes for the repressor of the *lac* operon is called the *i* gene (for "inducibility"). The *i* gene happens to lie close to the operon that it controls, but many other regulatory genes are distant from their operons. Like all genes, the *i* gene itself has a promoter, which can be designated p_i. It is a very inefficient promoter, allowing the production of just enough mRNA to synthesize about ten molecules of repressor per cell per generation. There is no operator between p_i and the *i* gene. Therefore, the repressor of the *lac* operon is constitutive, that is, it is made at a constant rate not subject to environmental control. Figure 12.19 shows the sequence of the regulatory gene and the *lac* operon, and Figure 12.20 outlines how the *lac* operon is regulated.

The operon model, proposed in 1961 by Jacob and Monod, implies that the unregulated condition of the *lac* operon is one of being turned *on* and that control is exerted by a regulatory protein—the repressor—that turns the operon *off*. Thus the control mechanism is *negative*. A second important implication of this model is that some genes, such as *i*, produce proteins whose sole function is to regulate the expression of other genes, and that certain other DNA sequences (namely, operators and promoters) do not code for any proteins whatsoever. Promoters are not even transcribed.

Repressible Systems

Inducible systems such as the one for lactose metabolism are of adaptive value to a bacterium that must switch on enzyme synthesis in response to the presence of an outside agent such as lactose. It is equally valuable to a bacterium to be able to switch off the synthesis of certain enzymes in response to an outside agent. For example, if the amino acid tryptophan is present in the medium in ample concentration, it is advantageous to be able to stop making the enzymes for tryptophan synthesis. When the formation of an enzyme is turned off in response to such a biochemical cue, the enzyme is said to be **repressible**.

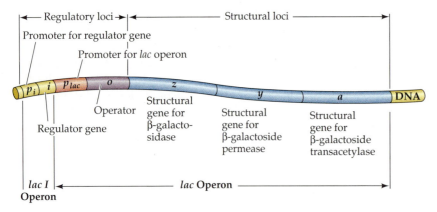

Regulatory loci | Structural loci

Promoter for regulator gene

Promoter for *lac* operon

p_i | i | p_{lac} | o | z | y | a | **DNA**

Operator

Structural gene for β-galacto-sidase

Structural gene for β-galactoside permease

Structural gene for β-galactoside transacetylase

Regulator gene

lac I Operon

lac Operon

12.19 The *lac* Operons of *E. coli*

Think of this stretch of DNA as including two main segments. One segment consists of the *lac* operon itself; the other segment, the *lac I* operon, regulates the transcription of the structural genes of the *lac* operon. The *lac* operon includes a promoter, an operator, and the three structural genes. The *lac I* operon includes its own promoter and a structural gene that codes for the repressor of the *lac* operon. Note that although *i* is a structural locus, it is also a locus that helps regulate the transcription of the main operon's structural genes.

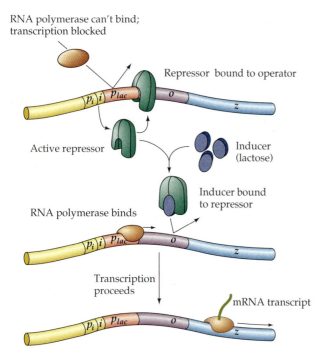

RNA polymerase can't bind; transcription blocked

Repressor bound to operator

p_i | i | p_{lac} | o | z

Active repressor

Inducer (lactose)

Inducer bound to repressor

RNA polymerase binds

p_i | i | p_{lac} | o | z

Transcription proceeds

mRNA transcript

p_i | i | p_{lac} | o | z

12.20 Induction of the *lac* Operon

An inducer alters the shape of a repressor protein and permits RNA polymerase to transcribe the operon's structural genes into mRNA. In an *E. coli* cell growing in the absence of lactose, the repressor protein coded for by gene *i* binds to the operator, preventing RNA polymerase from transcribing the structural genes. When lactose—the inducer—is added, it binds to the repressor, altering the repressor's shape so as to eliminate binding to the operator. As long as the operator remains free of repressor, RNA polymerase that recognizes the promoter can transcribe the operon.

Monod realized that repressible systems, such as the one for tryptophan synthesis, could work by mechanisms similar to those of inducible systems such as the *lac* operon. In repressible systems, the repressor cannot shut off its operon unless it first unites with a **corepressor**, which may be either the nutrient itself (tryptophan in this case) or an analogue of it. If the nutrient is absent, the operon is transcribed at a maximum rate. If it is present, the operon is turned off (Figure 12.21). Note carefully the small but significant difference between inducible and repressible systems. In inducible systems, a compound from the medium (the inducer) interacts with the regulatory-gene product (the repressor), rendering it *incapable* of binding to the operator and thus incapable of blocking transcription. In repressible systems, a compound from the medium (the corepressor) interacts with the regulatory-gene product to make it *capable* of binding the operator and blocking transcription. Although the effects of the external compounds are exactly opposite, the systems as a whole are strikingly similar.

The repressible system is another example of *negative* control of transcription. In both the inducible lactose system and the repressible tryptophan system, the function of the regulatory macromolecule is to prevent transcription. Let us next consider an example of a *positive* control system.

Catabolite Repression

Operons may be regulated by catabolites—molecules produced by catabolic (degradative) reactions. A great many catabolic enzymes function directly or indirectly to furnish the cell with energy. Some of

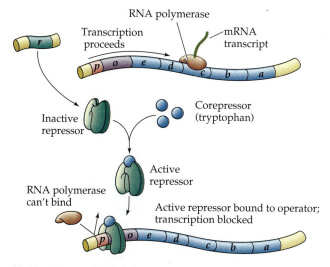

12.21 Repression of the Tryptophan Operon

The tryptophan operon consists of regulatory loci (*p* and *o*) and structural genes (*e, d, c, b, a*); its regulator gene (*r*) is in another part of the genome. A cell growing in the absence of tryptophan produces an *inactive* repressor that does not bind to the operator. This permits RNA polymerase to transcribe the operon's structural genes into mRNAs that are translated into enzymes of the tryptophan pathway. When tryptophan is present, it interacts with the inactive repressor, converting it into an *active repressor*, which does bind to the operator. This blocks RNA polymerase from transcribing the structural genes and prevents the synthesis of the enzymes of the tryptophan pathway. Because tryptophan activates an otherwise inactive repressor, it is called a corepressor.

them produce glucose from various carbohydrates; others catabolize certain amino acids. When glucose or some other excellent carbon source is abundant in the medium, the bacterial cell diminishes or abolishes the synthesis of these catabolic enzymes. This type of transcriptional control is called **catabolite repression**, but its mechanism is entirely different from the kind of operator–repressor mechanism that controls the lactose operon. Catabolite repression is a positive control process; and it relies on promoters, not operators. The promoters bind RNA polymerase in a series of steps (Figure 12.22). First, a special protein (abbreviated **CRP**, for *c*AMP *r*eceptor *p*rotein) binds a low-molecular-weight compound called adenosine 3',5'-cyclic monophosphate (cAMP); next, the CRP–cAMP complex binds close to the binding site of the RNA polymerase and enhances the binding of the polymerase by 50-fold. Glucose causes catabolite repression by lowering the concentration of cAMP, although the mechanism of this process remains unknown. When the glucose concentration falls, the concentration of cAMP rises again and the promoters of many genes responsible for sugar metabolism are activated. cAMP is also present in eukaryotes, where it is involved in the action of many hormones (Chapter 34) and in the regulation of transcription of genes.

As we will see in the next chapter, the regulation of gene expression in eukaryotes is more intricate than in prokaryotes.

12.22 Catabolite Repression

The operon's promoter binds RNA polymerase very poorly, so there is very little transcription of the structural genes, which code for enzymes that catabolize energy sources other than glucose. When supplies of glucose are low, a receptor protein (CRP) interacts with cAMP to form a complex that binds to the promoter. RNA polymerase binds to this promoter complex much more efficiently and transcribes the structural genes. A cell that contains ample glucose and does not require energy from other sources also contains little cAMP and little CRP–cAMP; in such a cell, the structural genes are not transcribed and the enzymes that catabolize the alternative energy sources are not formed.

SUMMARY

Studies of bacteria and their viruses have increased our knowledge of genetics and molecular biology. Mutant bacteria may be isolated for experimental purposes in various ways, including plating on selective media, penicillin suicide, and replica plating. Genetic recombination in bacteria may follow transformation, conjugation (including sexduction), or transduction.

All known bacteriophages have lytic life cycles; some strains also have lysogenic cycles, with the phage DNA maintained as a prophage incorporated into the bacterial chromosome. The prophage may carry bacterial genes when it separates from the chromosome, leading to restricted transduction.

Plasmids (such as R factors) are dispensable, circular DNA molecules that exist within the bacterial cell independent of the chromosome. Episomes, such as the F factor and prophages, are plasmids that can also be incorporated into the chromosome.

Transposable elements are segments of DNA that can cause copies of themselves to be inserted elsewhere in the chromosome. Other genes can be inactivated by insertion of transposable elements.

Transposable elements have played roles in the evolution of plasmids.

Much of the genetic material of bacteria is organized into operons. An operon consists of one or more structural genes, an operator, and a promoter. Transcription of the structural genes is prevented when the operator is bound by a repressor. The operon model explains repressible systems, such as the tryptophan system, as well as inducible ones like the lactose operon.

Positive-control mechanisms include catabolite repression. In this system, binding of the CRP–cAMP complex to the promoter region greatly enhances binding of RNA polymerase. Glucose causes catabolite repression by reducing the concentration of cAMP.

SELF-QUIZ

1. In bacterial conjugation:
   a. each cell donates DNA to the other.
   b. a bacteriophage carries DNA between bacterial cells.
   c. one partner possesses a fertility factor.
   d. the two parent bacteria merge like sperm and egg.
   e. all the progeny are recombinant.

2. Which statement is *not* true of the bacterial fertility factor?
   a. It is a plasmid.
   b. It confers "maleness" on the cell in which it resides.
   c. It can be transferred to a female cell, making it male.
   d. It has thin projections called F-pili.
   e. It can become part of the bacterial chromosome.

3. Hfr mutants
   a. are female bacteria that are highly efficient recipients of genes.
   b. rarely transfer all the markers on the chromosome.
   c. keep their F factor separate from the chromosome at all times.
   d. are unable to conjugate with other bacteria.
   e. transfer markers in random order.

4. Lysogenic bacteria:
   a. lack a prophage.
   b. are accompanied by some free phage when growing in culture.
   c. lyse immediately.
   d. cannot release their phage.
   e. are susceptible to further attack by the same strain of phage.

5. Which statement is *not* true of transduction?
   a. The viral DNA is an episome.
   b. Transduction is a useful technique for mapping a bacterial chromosome.
   c. In restricted transduction, the newly infected cell becomes lysogenic.
   d. Transduction results in genetic recombination.
   e. To carry bacterial markers, the viral coat must contain viral DNA.

6. Plasmids:
   a. are circular protein molecules.
   b. are required by bacteria.
   c. are tiny bacteria.
   d. may confer resistance to antibiotics.
   e. are a form of transposable element.

7. Which statement is *not* true of transposable elements?
   a. A transposable element can be copied to another DNA molecule.
   b. A transposable element can be copied to the same DNA molecule.
   c. A transposable element is typically 100–500 base pairs long.
   d. A transposable element may be part of a plasmid.
   e. A transposable element encodes the enzyme transposase.

8. In an inducible operon:
   a. an outside agent switches on enzyme synthesis.
   b. a corepressor unites with the repressor.
   c. an inducer affects the rate at which repressor is made.
   d. the regulatory gene lacks a promoter.
   e. the control mechanism is positive.

9. The promoter:
   a. is the region that binds the repressor.
   b. is the region that binds RNA polymerase.
   c. is the gene that codes for the repressor.
   d. is a structural gene.
   e. is an operon.

10. Catabolite repression:
    a. produces many catabolites.
    b. requires ribosomes.
    c. operates by an operator–repressor mechanism.
    d. is a form of positive control of transcription.
    e. relies on operators.

FOR STUDY

1. Viruses sometimes carry DNA from one cell to another by transduction. In particular instances, a segment of bacterial DNA may be incorporated into a phage protein coat without any phage DNA. These particles can infect a new host. Would the new host become lysogenic if the phage originally came from a lysogenic host? Why or why not?

2. For studies of metabolism in a particular species of bacteria, you need to isolate mutant strains: You want a histidine auxotroph (a strain, *his*⁻, that cannot synthesize the amino acid histidine) and a tryptophan auxotroph (*trp*⁻, unable to synthesize the amino acid tryptophan). After irradiating a culture of the bacteria with ultraviolet light to increase the mutation rate, you expect to find some *his*⁻ and *trp*⁻ auxotrophs in the culture. Describe all the steps you would take in order to increase the percentages of *his*⁻ and *trp*⁻ auxotrophs, using the "penicillin suicide" technique.

3. You are provided with two strains of *Escherichia coli*. One, an Hfr strain, carries the markers A^+, B^+, and C^+; and it is sensitive to streptomycin. The other is an F^- strain, resistant to streptomycin, and it carries the markers A^-, B^-, and C^-. You mix the two cultures. After 20, 30, and 40 minutes you take samples of the mixed culture and swirl them vigorously in a blender. Next you add streptomycin to the swirled cultures. You examine surviving bacteria by replica plating. Some of the bacteria from the 20-minute sample are B^+; in the 30-minute sample there are both B^+ and C^+ bacteria; but A^+ bacteria are found only in the 40-minute sample. What can you say about the arrangement of the A, B, and C loci on the bacterial chromosome? Explain your answer fully.

4. You have isolated three strains of *E. coli*, which you name I, II, and III. You attempt to cross these strains, and you find that recombinant progeny are obtained when I and II are mixed or when II and III are mixed, but not when I and III are mixed. By diluting a suspension of II and plating it out on solid medium, you isolate a number of separate clones. You find that almost all these clones can conjugate with strain I to produce recombinant offspring. However, one of the clones derived from strain II lacks the ability to conjugate with strain I. Characterize strains I, II, and III, and the nonconjugating clone of strain II, in terms of the fertility (F) factor.

5. In the lactose operon of *E. coli*, repressor molecules are coded for by the regulator gene. The repressor molecules are made in very small quantities and at a constant rate per cell. Would you surmise that the promoter for these repressor molecules is efficient or inefficient? Is synthesis of the repressor constitutive, or is it under environmental control?

6. A key characteristic of a repressible enzyme system is that the repressor molecule must react with a corepressor (typically, the end product of a pathway) before it can combine with the operator of an operon to shut the operon off. How is this different from an inducible enzyme system?

READINGS

Darnell, J. E., Jr. 1985. "RNA." *Scientific American*, October. Discusses aspects of transcription and of gene regulation in prokaryotes and eukaryotes.

Judson, H. F. 1979. *The Eighth Day of Creation*. Simon and Schuster, New York. A constantly fascinating history of molecular biology, with much attention to the regulation of gene expression. For a lay audience.

Nomura, M. 1984. "The Control of Ribosome Synthesis." *Scientific American*, February. How ribosomes are assembled; the roles of operons in regulating the rate of ribosome production in bacteria.

Stent, G. S. and R. Calendar. 1978. *Molecular Genetics*, 2nd Edition. W. H. Freeman, New York. Technical and historical information charmingly presented.

Suzuki, D. T, A. J. F. Griffiths, J. H. Miller and R. C. Lewontin. 1989. *An Introduction to Genetic Analysis*, 4th Edition. W. H. Freeman, New York. An up-to-date revision of one of the field's classic textbooks.

13

Gene Expression in Eukaryotes

PREVIEW: Many eukaryotic genes contain far more DNA than is needed to code for their protein products. Most eukaryotic genes include DNA that does not code for the amino acids of a protein. Moreover, there are many copies of certain genes. Some eukaryotic "genes" may have no function at all. Other genes regulate processes such as cell division, and mutations in these genes have been implicated in causing cancer. Gene expression in eukaryotes can be regulated at the transcriptional and translational levels. Most recent advances in understanding how the genome is regulated and organized have come through the application of recombinant DNA technology.

This chapter deals with nucleic acid hybridization, exons, introns, repetitive DNA, transposable elements, gene families, pseudogenes, splicing, RNA processing, oncogenes, retroviruses, and the regulation of eukaryotic gene expression.

Many eukaryotes are multicellular. Multicellularity requires a division of labor among the cells; therefore a multicellular organism must have a variety of types of cells, containing different proteins and capable of performing different specialized functions. There are at least 200 different cell types in the human body, differing in a few major proteins and many minor proteins. During the development of the multicellular body, different genes are expressed at different times, or in different specific tissues (Chapter 15). How can this be so—how are eukaryotic genes turned on and off?

Before we can address that question, we must answer another one: What are eukaryotic genes like—are they the same as prokaryotic genes? Eukaryotes have enormous amounts of DNA compared with prokaryotes; typically, mammals have on the order of 1,000 times as much DNA per cell as does *Escherichia coli*. Eukaryotes also differ from prokaryotes in the organization of their chromosomes: The eukaryotic chromosome is organized into nucleosomes (Chapter 9). Eukaryotic genes themselves tend to differ from those of most prokaryotes in having stretches of DNA that are not expressed in polypeptide products. That is, most eukaryotic genes are "split" by the presence of noncoding DNA in the midst of the coding sequences. The expression pattern of eukaryotic genes must ultimately control the phenomena of cellular differentiation and development; such a complex task requires many different levels of regulation, including changes in gene structure, transcription, and translation.

As our understanding of the eukaryotic genome grows, molecular biology contributes ever-increasing explanatory power for many other biological disciplines, including immunology (Chapter 16), endocrinology (Chapter 34), and neurobiology (Chapter 36). One of the sources of this success has been the application of recombinant DNA technology (Chapter 14) and related techniques. We will begin our consideration of the eukaryotic genome by explaining one of the most important and powerful of these techniques and showing what it has revealed about the molecular genetics of eukaryotes.

HYBRIDIZATION OF NUCLEIC ACIDS

Nucleic acid hybridization has been crucial in studies of the molecular biology of eukaryotes. This technique depends on the association, through complementary base pairing (Chapter 11), of single-stranded nucleic acids. If we carefully heat a sample of DNA, the hydrogen bonds forming the base pairs are destroyed and the two strands of each double helix separate—we say that the DNA **denatures** (Figure 13.1a). If the temperature is then lowered slowly, the complementary strands join again, or **reanneal**, to

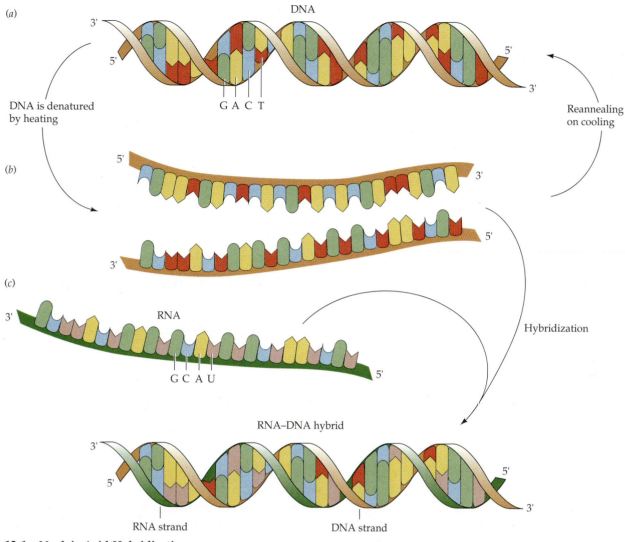

(a)

DNA

3'

5'

DNA is denatured
by heating

G A C T

Reannealing
on cooling

(b)

5'

3'

3'

5'

(c)

3'

RNA

G C A U

5'

Hybridization

RNA–DNA hybrid

3'

5'

5'

3'

RNA strand

DNA strand

13.1 Nucleic Acid Hybridization
(a) Upon being carefully heated, the two polynucleotide strands of a DNA
molecule denature (separate). (b) Complementary base pairing favors the
reannealing of the strands when they cool, reconstituting double-stranded
DNA molecules identical to the original sample. (c) If, however, an RNA transcript is added to the denatured DNA, the RNA competes for the coding
DNA strand, forming double-stranded DNA–RNA hybrid molecules.

form double-stranded DNA, with each base pair
obeying the A–T, G–C pairing rules (Figure 13.1b).
To make this procedure work efficiently, one must
enzymatically or mechanically "cut" the DNA into
short segments a few hundred bases long, and the
temperature and salt concentration in the test solution must be carefully regulated.

Suppose, now, that we heat-denature a sample of
DNA and then combine it with a sample of RNA that
has been transcribed from part of that DNA. That
RNA is complementary to only the one strand of
DNA that coded for it, that is, the coding strand
(Figure 13.1c). As the temperature is lowered, the
RNA transcript and the other DNA strand compete
to attach to the coding strand. The reannealing of the

DNA strands can be avoided by first immobilizing
the denatured DNA on a nitrocellulose filter so that
the separated DNA strands cannot come together
and RNA binding is favored. (The immobilized DNA
is still accessible for hybridization with nucleic acids
in solution.) The reannealing of DNA strands can
also be reduced by outnumbering the DNA strands
with RNA strands.

EUKARYOTIC GENE STRUCTURE

Now we are ready to consider the eukaryotic gene
and how it differs from the prokaryotic gene. The
structure of the eukaryotic gene has been at least

partially understood through the use of hybridization techniques. There are a number of ways to examine the relationship between genes and their RNA transcripts, most involving hybridization. One approach is as follows: One obtains a sample of DNA, denatures it, and then adds cytoplasmic mRNA. One obtains from the hybridization uniform, double-stranded DNA–mRNA hybrid structures associated with single-stranded, looped structures. The loops are the displaced, noncoding DNA strands. In fact, it was discovered in 1977 that the double-stranded hybrid regions are studded with both single- and double-stranded loops. What could this mean?

The mRNA is a faithful transcript of the information required for protein synthesis. Moreover, all the mRNA is bound, through complementary base pairing, with the appropriate region of the single-stranded DNA. However, there is some DNA, *in the middle of the gene*, that is not represented in the mRNA. That is, some of the information in the gene does not end up in the mRNA that encodes the protein product. In fact, most (but not all) vertebrate genes and many other eukaryotic genes contain such *int*ervening sequences, or **introns**—segments of DNA that are transcribed but which do not encode any part of the polypeptide product of the gene. And not all the transcribed RNA gets into the cytoplasm to be mRNA. Those parts of the gene that *are* represented in the mature mRNA product are called **exons** because they are *ex*pressed regions of the gene (Figures 13.2 and 13.3).

Introns do not scramble the sequence that actually codes for a polypeptide, they just reside in the middle of it. The base sequence of the exons, taken in order, is exactly complementary with that of the mature mRNA product. Accordingly, there is no scrambling of the coding sequence for the protein by introns, just a separation. Exons and introns are found in all groups of eukaryotes and even in a few prokaryotes. We are still seeking to understand the significance of introns to the organisms that possess them. Later in this chapter we will describe the posttranscriptional events that remove the transcripts of introns.

REPETITIVE DNA IN EUKARYOTES

Let us see what other things we can learn by using nucleic acid hybridization. If the complementary strands of DNA are separated and allowed to reanneal, we can observe interesting differences in the time course of renaturation. In eukaryotes—but not in prokaryotes—some parts of the genome anneal only very slowly, whereas other segments quickly find partners. Why should one DNA sequence anneal quickly and another slowly? The difference lies in the fact that there are multiple copies of some, but not all, stretches of DNA. If a particular single strand of DNA has, say, a few hundred complementary segments with which it can anneal, it is able to find a partner much more rapidly than one for which only a single acceptable partner exists.

On the basis of this reannealing technique, called liquid hybridization, three classes of eukaryotic DNA have been recognized. The class that reanneals most slowly consists of single-copy sequences—genes

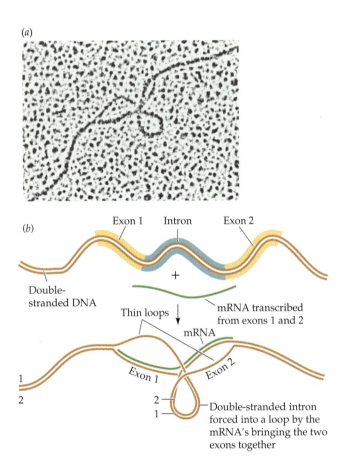

(a)

(b)

Exon 1   Intron   Exon 2

Double-stranded DNA

+

mRNA transcribed from exons 1 and 2

Thin loops

mRNA

Exon 1   Exon 2

1
2

2
1   Double-stranded intron forced into a loop by the mRNA's bringing the two exons together

(c) With no introns:

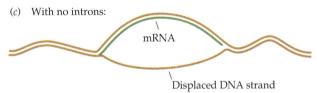

mRNA

Displaced DNA strand

13.2 Evidence for Extra DNA in the Eukaryotic Gene
Mouse DNA was partially denatured and mixed with mRNA transcribed from one of the genes in the DNA. Examination of the resulting mixture by electron microscopy (a) revealed thick nucleic acid, bearing thick loops (here one points downward) and thinner loops. (b) A diagrammatic interpretation. The mRNA hybridized with the coding DNA strand, forming double strands that appear thick in comparison with the loops made by the unpaired, noncoding strand of DNA. The thick loop consists of double-stranded DNA that reannealed because it has no counterpart in the mRNA. It is in the middle of the gene. We see that the mRNA is shorter than the DNA from which it is derived. (c) Hybridization pattern observed when no introns are present, as in prokaryotic DNA.

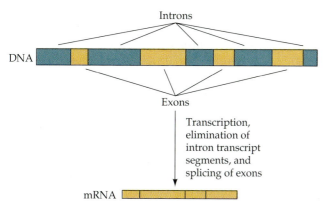

Introns

DNA

Exons

Transcription,
elimination of
intron transcript
segments, and
splicing of exons

mRNA

13.3 Exons and Introns
Many eukaryotic genes contain introns—stretches of
DNA whose transcripts are absent from the mature
mRNA product. Stretches of DNA whose transcripts do
appear in the mature mRNA are called exons.

that, like prokaryotic genes, have only one copy in each **genome**, or haploid set of chromosomes. Single-copy sequences code for most of the enzymes and structural proteins in eukaryotes. Other single-copy sequences form long spacers between successive genes. The class of DNA sequences that reanneals most rapidly consists of highly repetitive DNA. This fraction varies widely from species to species and may make up a third or more of the genome. Although there are half a million copies per genome of some of these segments, their function is not understood. Much of the highly repetitive DNA is located near the centromeres (Chapter 9) and may help maintain the integrity of chromosomes during mitosis and meiosis. Possibly the large number of identical DNA sequences at a centromere is related to the attachment of multiple spindle fibers to each chromosome. DNA of this class does not appear to be transcribed. The class of eukaryotic DNA that reanneals at rates between the two extremes is moderately repetitive DNA, which is present in a few hundred to 10,000 copies per genome. Some of these moderately repetitive genes may be important in the regulation of development, and some are duplicate genes, such as those for rRNA. Transposable elements are moderately repetitive genes with special properties.

TRANSPOSABLE ELEMENTS IN EUKARYOTES

Eukaryotes, like prokaryotes, have **transposable elements**—segments of DNA that insert themselves into different parts of chromosomes and hence have been called "jumping genes" (Chapter 12). It was in maize, a eukaryote, that Barbara McClintock (Cold Spring Harbor Laboratory) obtained the first evidence for transposable elements.

The mechanism for copying transposable elements in at least some eukaryotes differs from the mechanism in prokaryotes in requiring an RNA intermediate. Recall that in prokaryotes the DNA of the transposable elements can simply copy itself. The transposable elements of both prokaryotes and eukaryotes are always parts of chromosomes; unlike plasmids, they do not function as independent pieces of DNA.

The roles of transposable elements in the life of a cell are largely unknown. Although the protein products of some eukaryotic transposable elements have been identified, the products' cellular functions have not been determined. It may be that transposable elements are, in effect, parasites that simply replicate themselves; on the other hand, they may play important roles in the survival of the organisms in whose chromosomes they reside. It is known that transposable elements can act as mutators—that is, by jumping into genes, they can eliminate or change the functions of those genes (Chapter 12). Transposable elements could have several other important functions.

Transposable Elements and Evolution

Transposable elements seem to have played at least one important part in evolutionary history. In Chapter 4 we introduced the endosymbiotic theory of the origin of chloroplasts and mitochondria—the proposal that chloroplasts and mitochondria are the descendants of once free-living prokaryotes. These organelles possess DNA, and some parts of the organelles are encoded by genes on this extranuclear DNA. Other parts of the organelles are coded for by nuclear genes, a finding that might appear to weaken the endosymbiotic theory. However, it has recently been shown that in the course of evolution, genes of some organisms have been transposed to the nuclei from both chloroplasts and mitochondria. Thus the insertion of transposable elements, and the subsequent loss of the originals of these genes from the chloroplasts and mitochondria, can be invoked to counter this argument against the endosymbiotic theory. A finding that had been seen as evidence against the endosymbiotic theory can now be used as evidence *for* the theory.

Transposable elements can bring about deletions, insertions, transpositions, and inversions (Chapter 10); they also may be a source of duplications, in which multiple copies of genes are created. Genes may be inactivated by the insertion of transposable elements, and other genes may be placed in new relative positions, affecting their transcription. For some genes, transposon insertions constitute more than 99 percent of all mutations.

How did the transposable elements themselves arise? Their source is still unknown; but there are interesting hints at a relationship between **retrovi-**

ruses (tumor viruses that use reverse transcriptase to transcribe their RNA to DNA) and the transposable elements of eukaryotes, as we will see when we discuss cancer-causing genes later in this chapter. It is possible that some transposable elements arose from retroviruses.

GENE DUPLICATION AND GENE FAMILIES

Although some genes are represented by thousands of copies in a single diploid cell, as we have seen, others have just two or a few copies. The processes of duplication are various, and include the action of transposable elements. Another source of gene duplication is unequal crossing over, in which mispaired chromosomes cross over in such a way as to put both copies of a gene on the same chromosome.

Once more than one copy of a gene is present, there is a possibility that the two copies will evolve differently. One copy must retain the original function, or the organism may not survive. However, as long as one copy does this, the others may change slightly, extensively, or not at all. A set of genes that is derived in this way is called a **gene family**, and it may reside on different chromosomes, or the individual members may be bunched tightly on a single chromosome.

An evolutionarily ancient gene family found in vertebrates is the globin family. Globins are proteins required for the binding and transport of oxygen; some of them are components of hemoglobin, the oxygen-carrying pigment of red blood cells, and another is found in myoglobin, a related protein that binds and stores oxygen within muscle fibers (Chapter 39). Each molecule of hemoglobin is a tetramer. The four globin polypeptides that make up the tetramer in adult humans are two of one type (α) and two of another (β). The hemoglobin tetramers of human fetuses also contain two types of globin polypeptides, both of which differ from those found in adult hemoglobin; still other α-like and β-like polypeptides are found in the earliest embryonic stages. *All* these globin polypeptides are coded for by globin genes that are descended from a single ancestral globin gene. This has been ascertained from similarities in the amino acid sequences of the polypeptides and from the locations of introns within the genes. The

globin genes, like other gene families, differ more in their introns than in their exons; the exons are probably conserved with little variation because their gene products perform essential functions, whereas those of the introns do not.

The genes for human α-like globins lie in a tight cluster on one of our chromosomes, and the genes for β-like globins lie in another cluster (Figure 13.4). In both clusters there are additional stretches of DNA that are closely similar in base sequence to the globin genes but that are not expressed. Such apparently nonfunctional genes are called **pseudogenes**. How could they have been discovered, if they have no function? They were discovered by recombinant DNA technology—they were located by the fact that they hybridize to a significant extent with adult globin mRNA (as do all other members of the globin gene family).

It is likely that most pseudogenes are duplicate genes that changed so much during evolution that they no longer function. These changes may include inactivated promoters, nonsense mutations in exons, or the inability to remove the transcript of an intron. Some pseudogenes, however, did not arise by gene duplication; they were derived by reverse transcription of the mRNA. Such "processed pseudogenes" lack introns and are found away from the rest of the gene cluster.

Whatever functions the pseudogenes had in the past (when they were not pseudogenes) are now performed by other genes in the family. In many gene families, pseudogenes outnumber the functional genes, often by several fold. Several members of the human globin gene family remain active, serving at different times in development. There are other gene families in which different functional members are active in different tissues or under different conditions. The gene families for rRNA, tRNAs, and histones are examples in which the repetition is used to meet the demand for large amounts of product.

RNA PROCESSING IN EUKARYOTES

When a eukaryotic gene is transcribed, the RNA product contains introns and exons alike. How do we get from this product to an mRNA, and from there to the protein encoded by the gene? The orig-

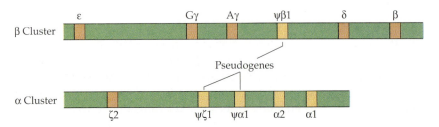

13.4 Gene Families and Pseudogenes
The human α-like and β-like globin gene families are organized into clusters. Each gene cluster includes both functional genes and pseudogenes; pseudogenes are prefixed by the Greek letter psi (ψ).

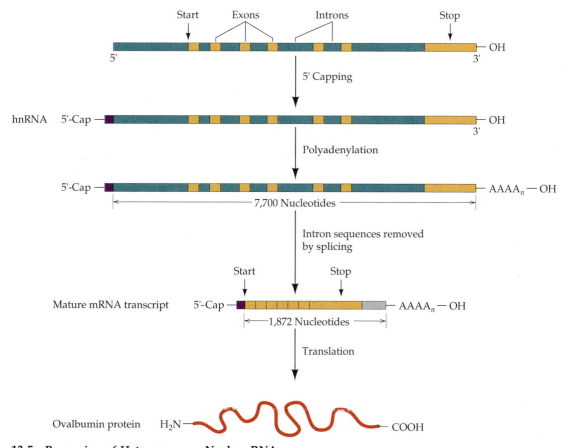

13.5 Processing of Heterogeneous Nuclear RNA
The hnRNA of the ovalbumin gene is capped at the 5' end and has poly A added at the 3' end. It is 7,700 nucleotides long and contains seven introns and eight exons. Splicing removes over three-quarters of the nucleotides and results in a mature mRNA that can be translated to yield ovalbumin.

inal product of transcription of a eukaryotic gene is a heterogeneous nuclear RNA, or **hnRNA**, so called because of the great range of sizes of RNAs of this class. As we will now see, much must be done to the hnRNA to produce a mature mRNA that is ready to be translated.

Capping and Tailing of RNA

Among the first steps in the processing of an hnRNA molecule is the addition of a **cap**—a modified molecule of guanosine triphosphate—at the 5' end of the hnRNA. This modified G cap is retained during the processing of mRNA and facilitates the binding of the mRNA to a ribosome for translation (Chapter 11); it may also help protect the mRNA from degradation.

At the other end of the hnRNA, the 3' end, a string of 100–200 adenine nucleotides, called a **poly A tail,** is added. (These tails constitute 5–20 percent of the length of the mature mRNA molecules that are produced from the hnRNA.) Neither the modified G cap nor the poly A tail is coded for in the DNA; both are added as part of the early processing of hnRNA. The poly A tail is thought to protect the modified hnRNA

and the mRNA from degradation; some evidence suggests that the poly A is needed for translation as well. Somewhat less than one-third of the hnRNA molecules in mammalian cells have poly A tails, and about 70 percent of the resulting mRNA molecules have the tails. We do not yet know why some RNAs get poly A tails and others do not. However, for those hnRNAs that do get poly A tails, the poly A is essential in order for the hnRNA to mature into mRNA.

Splicing of RNA

The next step in the processing of eukaryotic RNA is the removal of the regions coded for by introns in the DNA. If these RNA regions, which are also called introns, were not removed, a nonfunctional mRNA would be produced—or an improper protein. Intron removal is accomplished by a process called **RNA splicing.** Figure 13.5 illustrates how processing of the hnRNA transcript of the 7,700-base-pair gene for ovalbumin, the major protein in egg whites, turns out a mature mRNA of only 1,872 nucleotides. Capping, tailing, the removal of seven introns, and the splicing of eight exons do the job.

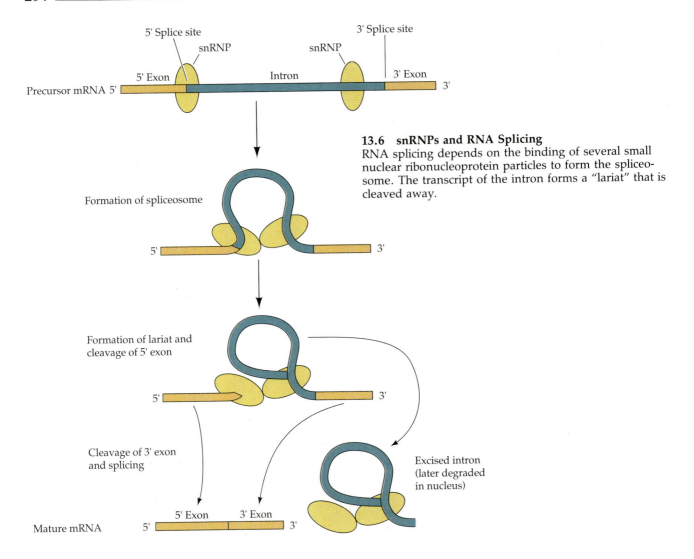

13.6 snRNPs and RNA Splicing
RNA splicing depends on the binding of several small nuclear ribonucleoprotein particles to form the spliceosome. The transcript of the intron forms a "lariat" that is cleaved away.

In the mRNA splicing reaction, a loop is formed so that the intron is extruded and the adjacent exons are brought together. How might the linking of the exons come about? At the boundaries between introns and exons there are **consensus sequences**—short stretches of DNA that appear, with little variation, in many different genes. Eukaryotic nuclei contain small nuclear RNA, or **snRNA**, molecules that contain regions complementary to the consensus sequences. To accomplish the splicing, an snRNA combines with a set of proteins, producing a small nuclear ribonucleoprotein particle, or **snRNP**. One of the snRNPs recognizes and binds the consensus sequence at one end of the intron; a second, different snRNP recognizes and binds the consensus sequence at the other end of the intron; and other snRNPs recognize and bind a sequence in the intron itself. Together, six different snRNPs constitute a "splicing machine" called a spliceosome. As shown in Figure 13.6, the spliceosome joins the exons and releases the introns.

The splicing mechanisms differ among RNA classes: mRNAs, rRNAs, and tRNAs are all spliced in different ways. Molecular biologists were startled to learn in 1981 that in the protist *Tetrahymena thermophila*, the RNA precursor of rRNA can catalyze the splicing of its own intron (Figure 13.7). That is, the RNA, in the absence of any protein, is catalytic. This discovery was made by Thomas Cech of the University of Colorado. Another case of a catalytic RNA (not involving RNA splicing) has been discovered, in *E. coli*. The existence of RNAs with catalytic powers may help explain evolution at the dawn of life, as will be considered in Chapter 17.

The Stability of mRNAs

RNA leaves the nucleus by way of the pores in the nuclear envelope (Figure 4.11). The transport of RNA from the nucleus to the cytoplasm is mediated by carrier proteins, but the details of this mechanism are not yet known. Mature mRNA in the cytoplasm may be relatively stable, lasting for hours or days.

The situation is different in prokaryotes. In prokaryotes, an mRNA molecule usually lasts for only a few minutes following transcription—its life is so

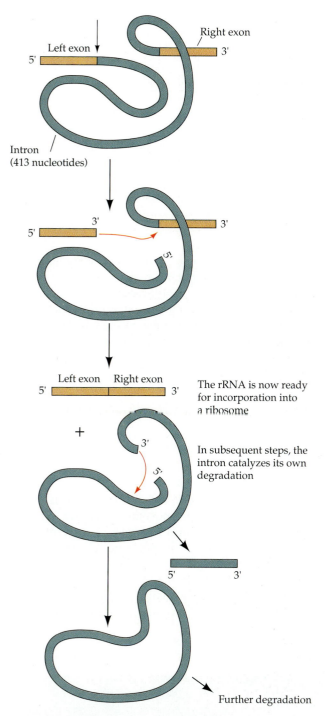

Intron
(413 nucleotides)

The rRNA is now ready for incorporation into a ribosome

In subsequent steps, the intron catalyzes its own degradation

Further degradation

13.7   rRNA Catalyzes Its Own Splicing
In *Tetrahymena thermophila*, ribosomal RNA is spliced after it is transcribed from DNA. The splicing is catalyzed by the RNA itself, without the contribution of any protein. The splicing proceeds in several steps. Note that this is one of a very few examples of catalysis by RNA.

short that translation begins before the mRNA is completely formed. As we will see, different eukaryotic RNAs may differ from one another in their stability. The stability of mRNAs plays an important role in the development of animals and plants.

REGULATION OF GENE EXPRESSION IN EUKARYOTES

The expression—the transcription and translation—of eukaryotic genes is regulated. This regulation is the basis for the phenomenon of **differentiation**, by which different cells become structurally and biochemically specialized for different functions. This subject is of central importance to the study of development (Chapters 15 and 32). The modes of regulation of eukaryotic gene expression are many. In a few cases, gene regulation depends on *changes in the DNA itself*—genes are actually rearranged on the chromosomes. Such rearrangements underlie sex changes in yeast (Box 13.A) and some of the diversity of antibodies (Chapter 16). *Transcription* in eukaryotes is subject to complex mechanisms of regulation, which have a surprising variety. Given the complexity of the eukaryotic gene and of RNA processing, it should not be surprising that RNA processing is also a regulated step, which is known as *posttranscriptional control. Translation* may be regulated by a variety of means. Even the polypeptide products of translation may require further processing before they become biologically active, so we speak as well of *posttranslational control* of the expression of some genes.

Let us begin with some cases in which differentiation is under *transcriptional* control. There are at least three ways in which transcription may be controlled in eukaryotes: large numbers of genes may be inactivated, specific genes may be amplified, or specific genes may be transcribed selectively.

Transcriptional Control: Gene Inactivation

At the close of mitosis or meiosis, chromosomes uncoil, but not completely (Chapter 9). That portion of the chromatin that is diffuse and nonstaining during interphase is referred to as **euchromatin**, while **heterochromatin** retains its coiling and continues to be stainable by the dyes that are used to stain mitotic chromosomes. Heterochromatin generally is inactive in RNA synthesis, that is, it is not transcribed.

Mammals that have two—or more—X chromosomes provide a striking example of heterochromatin and gene inactivation. The normal female mammal has two X chromosomes, and the normal male has one X and one Y. The Y chromosome has few, if any, identifiable genes that are also present on the X chromosome, and the Y appears to be transcriptionally inactive in most cells. Hence there is a 100 percent difference between females and males in the dosage of X-chromosome genes. Why is this not a case of aneuploidy (Chapter 9) involving a rather large chromosome? Aneuploidy for an autosome of comparable size is invariably lethal. Why then is not one sex or the other grossly deformed or completely inviable?

The answer was found in 1961 by Mary Lyon and

BOX 13.A

Cassettes and the Mating Type of Yeasts

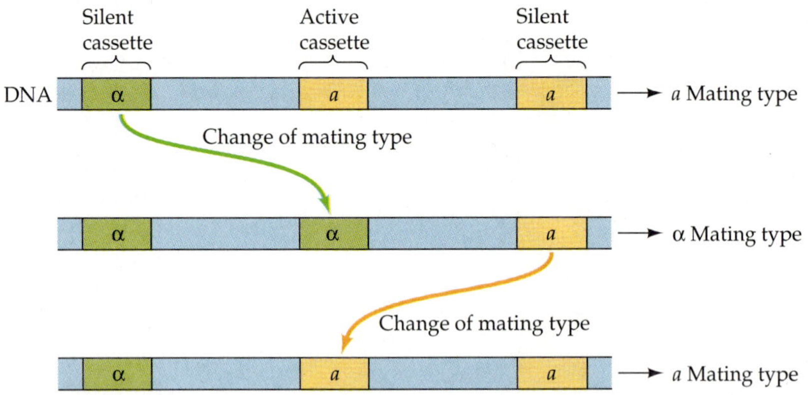

Silent cassette | Active cassette | Silent cassette

DNA → α | a | a → *a* Mating type

Change of mating type

α | α | a → α Mating type

Change of mating type

α | a | a → *a* Mating type

Yeasts have two mating types, *a* and α (alpha). In some yeasts, the mating type may change with almost every generation of cells. The mating type is determined by a locus—the mating-type locus—on one of the chromosomes. If the mating-type locus contains the *a* allele, the cell is of mating type *a*; if the locus contains the α allele, the cell is of mating type α. How does the mating type change so rapidly? The mechanism has been

Replacement of the allele in the mating-type locus with a copy of an unexpressed allele at one of the "cassette" sites changes the mating type if that cassette differs from the one being replaced.

likened to a cassette recorder, with the mating-type locus being the deck into which a cassette is inserted and "played." The yeast cell keeps two cassettes—unexpressed copies of the *a* and α alleles—at other loci on the chromosome that bears the mating-

type locus, as shown in the figure. From time to time the allele in the mating-type locus is removed, and one of the two unexpressed loci is copied to the mating-type locus. If the newly inserted cassette differs from the previous occupant of the mating-type locus, the mating type changes, because yeasts are unicellular, and they are haploid throughout most of their life cycle.

Liane Russell, working independently. Lyon suggested that one of the X chromosomes in each cell of a normal female mammal is inactivated early in embryonic life and remains inactive ever after. The choice of which X in any pair of X chromosomes will remain active is random; because the female embryo already consists of tens or hundreds of cells by the time the choice is made, virtually all female mammals contain patches of tissue in which one or the other X is active. This interpretation is supported by genetic, biochemical, and cytological evidence. Interphase cells of normal female mammals have a single, stainable nuclear body called a **Barr body** (after its discoverer) that is not present in males. The Barr body represents the inactive X chromosome, condensed into heterochromatin (Figure 13.8). The cells of women who have only one X chromosome, like those of normal men, contain *no* Barr bodies. Other women, who have a chromosomal constitution of XXX, have cells with *two* Barr bodies; there are even XXXXY males who have three Barr bodies in each interphase cell. We may thus infer that the cells of each individual, male or female, contain a *single* active X chromosome, so that the dosage of expressed X-chromosome genes is constant and is the same in both sexes.

In individual chromosomes, the presence of limited regions of transcriptionally active euchromatin

may sometimes be observed. This activity is most obvious in polytene chromosomes (Chapter 10), the giant chromosomes found in insect salivary glands. In some preparations of polytene chromosomes, **puffs** are visible (Figure 13.9). These are regions of maximally extended chromatin, whose DNA is being transcribed. In salivary gland preparations, it has been clearly seen that different regions of the chro-

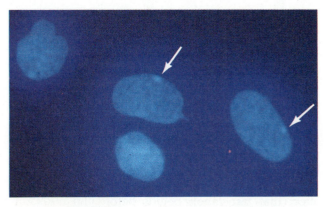

13.8 The Barr Body
The spots marked with arrows in these nuclei of human female cells are Barr bodies. A Barr body represents the condensed, inactive member of the pair of X chromosomes in the cell.

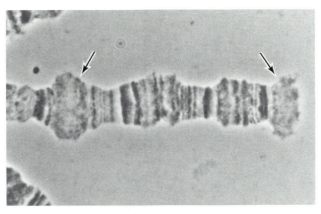

13.9 Chromosome Puffs
Arrows point to two puffs in this chromosome from a salivary gland cell of an insect larva. One of the puffs is at the end of the chromosome. Puffs represent regions where DNA is being transcribed to RNA; as the development of a larva proceeds, different regions of a chromosome puff when different gene products are needed.

mosomes are puffed at different times during development—a clear demonstration that gene activation shifts as the cell develops.

Heterochromatization is not the only means for inactivating genes. Another mechanism is a chemical modification that adds methyl groups to some cytosine residues in certain genes. This process is called **DNA methylation**. In humans and chickens, the DNA coding for globin synthesis is unmethylated in developing red blood cells. However, in cells that will not need to produce globin, the cytosine residues of the globin genes are highly methylated. Because the genes are methylated, they are not transcribed, and no hemoglobin is produced.

Transcriptional Control: Amplification of Genes

One obvious way for a cell to make more of one enzyme or RNA product than another cell does is to have more copies of the appropriate gene and to transcribe them all. Such gene duplication results in more DNA per cell than there would be if there were only one copy of the gene. As we have already seen, there is a class of eukaryotic DNA composed of such multiple gene copies.

The genes that code for histones, the proteins that interact with DNA in producing nucleosomes (Chapter 9), are present in great numbers of copies—perhaps 500 per cell in sea urchins and tens to hundreds per cell in mammals and *Drosophila*. It is not surprising that each human cell contains over 2,000 genes coding for tRNA synthesis. Because 61 kinds of specific tRNA are required in protein synthesis, there are many different tRNA genes, but even so there are many copies of each. In addition, every eukaryotic cell contains in the nucleolar region of the nucleus (Chapter 4) many copies of the DNA that codes

for rRNA. The multiple copies are arranged in a series, one after another; this is called a tandemly repetitive region.

In most cells, the tandemly repetitive region contains enough gene copies to produce rRNA as fast as it is needed. However, in some cells, even these multiple genes are apparently insufficient. The germ cells in amphibians that are destined to become eggs store large amounts of ribosomes for the early development of the embryo. During egg formation, each of these cells—the oocytes—multiplies the number of copies of rRNA genes until there are about a thousand nucleoli, containing more than a million rRNA genes, floating free in its nucleoplasm. Each of these nucleoli consists of DNA with repeating segments coding for rRNA, and each is transcribed to furnish the tremendous amount of rRNA that is stored in a mature oocyte for use by the embryo. This amplification ensures that the oocyte cytoplasm has enough rRNA to sustain rapid development during the entire period from fertilization to the formation of the gastrula (Chapter 15), a structure consisting of hundreds of thousands of cells. This process of actually creating more genes of one kind to enhance transcription is called **gene amplification**. In frog oocytes, each haploid set of chromosomes contains about 500 copies of the rRNA genes before gene amplification; afterward, there are nearly a million. Evidently no genes in these cells except those responsible for rRNA synthesis are amplified. It is thought that genes are amplified in the oocytes of many animals and, under certain conditions, in vascular plants. (The polytene chromosomes of insects represent a form of gene amplification—but not *selective* gene amplification.)

Figure 13.10 shows several active rRNA genes from an amphibian nucleolus. The axial strand, or connecting thread, is nuclear DNA that codes for rRNA. The fuzzy-looking triangles attached to the DNA are composed of many strands of rRNA in the making. Each partial rRNA molecule is attached to the DNA by an RNA polymerase molecule. Many polymerases can transcribe the DNA at the same time. The apex of each triangle is the point at which RNA synthesis starts, so the RNA strands protruding from this region are very short. As transcription proceeds along the DNA, the RNA strands become longer and longer as more and more of the DNA coding for rRNA is read. Many of these fuzzy triangles are repeated along the DNA of the nucleolar organizer, showing us that the rRNA-coding DNA sequence itself is repetitive. Notice, too, that between the triangles there is quite a bit of silent "spacer" DNA that does not seem to be transcribed. The function of these spacers is unknown, but there is evidence suggesting that they serve as a "loading zone" for the RNA-polymerase proteins that transcribe the genes.

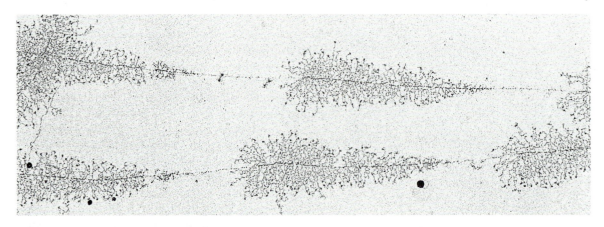

13.10 Transcription in the Nucleolus
Elongating strands of rRNA transcripts form arrowhead-shaped regions, each centered on a strand of DNA that codes for the rRNA.

Transcriptional Control: Selective Gene Transcription

In fact, very few genes become amplified. A much more common pattern of transcriptional control is the specific switching on, or off, of transcription of individual single-copy genes. In eukaryotes, such selective transcription of particular genes can be mediated by steroid hormones (Chapter 34). In insects with polytene chromosomes, these hormones stimulate the formation of chromosome puffs, illustrated in Figure 13.9. The insect hormone ecdysone has three different kinds of specific effects on particular genes. Transcription of some genes, as indicated by puff formation, begins within 4 hours after treatment with ecdysone, while others first form puffs several hours later—the later puffs seem to depend on both ecdysone and the protein products of earlier puffs. In addition, certain other genes *stop* producing puffs when ecdysone is present. Thus a complex repertoire of transcriptional events in differentiation is under hormonal control.

Transcriptional control in prokaryotes relies on operons (Chapter 12) that are subject to either negative or positive regulation. Eukaryotic genes do not have operons, and transcriptional control in eukaryotes is almost always *positive* regulation. In the negative regulation systems of prokaryotes, genes are expressed unless they are turned *off* by proteins. The positively regulated eukaryotic gene is turned *on* by proteins. Eukaryotes have much more DNA than do prokaryotes, so there is a greater danger that regulatory proteins will bind to inappropriate sites. This danger is allayed by requiring binding at *multiple* sites to initiate the transcription of a single gene. As a result, the **promoter** for a eukaryotic gene is more complex than a prokaryotic promoter (Chapter 12). In both eukaryotes and prokaryotes, the promoter is the stretch of DNA to which RNA polymerase binds to initiate transcription.

Prokaryotes have only one kind of RNA polymerase, but there are *three* RNA polymerases in eukaryotes. RNA polymerase I transcribes the DNA that encodes rRNA; not surprisingly, this is the most abundant RNA polymerase in a eukaryotic cell. RNA polymerase II transcribes the structural genes that encode mRNAs, and it thus has the greatest diversity of products. RNA polymerase III transcribes the DNA that encodes tRNAs and some other small RNA species. The rest of our discussion will focus on RNA polymerase II.

RNA polymerase II cannot simply bind to the chromosome and initiate transcription. Rather, it can bind and act only after various regulatory proteins, or **transcription factors**, have assembled on the chromosome. Each transcription factor recognizes and binds to a particular sequence of the DNA. One such sequence is the **TATA box**, an eight-base-pair sequence consisting only of T–A pairs, found in many eukaryotic promoters about 25 base pairs before the starting point for transcription. One of the transcription factors, TFIID, binds to the TATA box. Next, TFIIA and TFIIB bind to other sequences on the promoter. Then RNA polymerase II joins the growing transcription complex, which is completed by the addition of TFIIE (Figure 13.11). Still other DNA sequences, each requiring a different transcription factor, precede the TATA box on the chromosome (Figure 13.12). Transcription begins only after all the sequences have bound their transcription factors. The various DNA sequences and their transcription factors have two principal roles in the regulation of transcription: The TATA box helps specify the exact starting point for transcription, and the others determine the efficiency of the promoter.

Some sequences, such as the TATA box, are common to the promoters of many genes and are recognized by transcription factors found in all the cells of an organism. Other sequences are specific to only

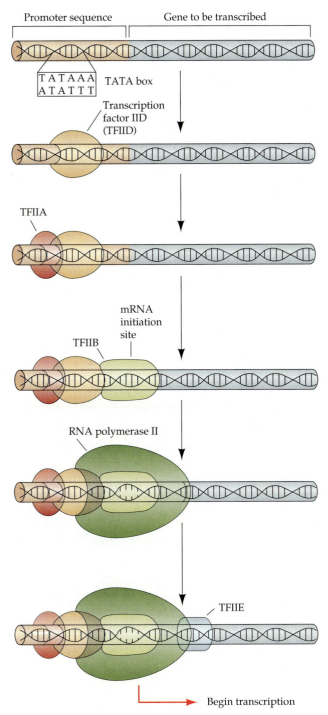

13.11 Forming a Transcription Complex
Interactions among the TATA box, four transcription factors, and RNA polymerase II lead to the formation of the transcription complex.

a few genes and are recognized by transcription factors found only in certain tissues; these play important parts in differentiation.

Although promoters are close to the genes whose transcription they regulate, another important type of regulation is brought about by **enhancers**, DNA sequences that can act at greater distances from the regulated genes. In fact, enhancers may lie on either side of the genes they regulate. Enhancers stimulate specific promoters and thus enhance transcription of specific genes.

Translational Control

Why should eukaryotes need to regulate translation? Each of their mRNAs, unlike those of prokaryotes (Chapter 12) codes for only one polypeptide, so the production of a given polypeptide might seem to be adequately regulated by mechanisms acting on transcription. In fact, translation of some genes is also regulated in eukaryotes. This level of control is a very rapid one and acts closest to the actual formation of the polypeptide product. Let us next consider some examples of translational control of differentiation.

Several different mechanisms provide translational control. As mammals prepare to lactate—to produce milk—a hormone, prolactin, acts on the mammary gland as the final trigger for milk production. The primary effect of this hormone is a dramatic increase in the translation of mRNA for casein, a major milk protein. Prolactin increases the longevity of casein mRNA by a factor of 25, allowing it to be translated that many more times than it would be in the absence of the hormone.

A second mechanism of translational control is employed in the oocytes of sea urchins and certain moths. As already noted, most mRNAs become capped with a modified G unit during RNA processing. Uncapped messages are not translated. Stored mRNA in a tobacco hornworm oocyte, for example, has the G portion of the cap, but the G has not been modified; hence these uncapped messages cannot be translated. When a tobacco hornworm egg is fertilized, the uncapped message is modified to complete the cap, and the mRNAs can then be translated.

An elegant set of controls ensures that hemoglobin is synthesized efficiently and in appropriate quantity

13.12 DNA Modules in the Thymidine Kinase Promoter
The promoter for the gene that encodes the enzyme thymidine kinase is typical of eukaryotic promoters. It contains the TATA box and three other DNA sequences, two of them identical (GC) but oriented in opposite directions. Transcription factors bind to the sequences.

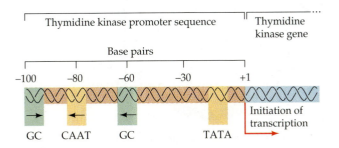

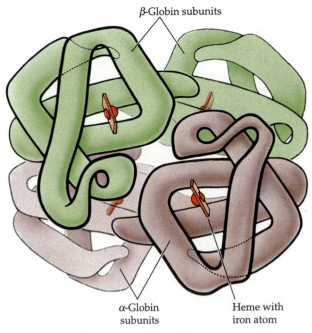

β-Globin subunits

α-Globin subunits

Heme with iron atom

13.13 Hemoglobin Consists of Three Kinds of Molecules
The hemoglobin molecule is made of two α-globin (purple) and two β-globin (green) polypeptides, and each of these globins contains a heme molecule (red). Production of these three compounds must be coordinated for metabolic efficiency.

in developing red blood cells. Hemoglobin is a moderately complex molecule, consisting of two α-globin chains, two β-globin chains (both types of globins are proteins), and four smaller heme molecules, one for each globin chain (Figure 13.13). Thus the complete hemoglobin molecule consists of three distinct kinds of components, in a ratio of 2:2:4. It is essential that the components be synthesized in just this ratio because significant deviations cause severe illnesses. Three separate mechanisms maintain the ratio. First, any excess of heme results in feedback inhibition (Chapter 6) of heme synthesis, thus reducing the imbalance. Secondly, any excess heme increases the translation of globin messengers, further ensuring a suitable balance of heme and globins. Finally, an appropriate ratio of α-globin to β-globin chains is also brought about through control of translation of the two globin mRNAs.

Posttranslational Control

We have considered how steps in cell differentiation may be regulated by the control of transcription, of RNA processing, and of translation. The story does not end here, for the expression of some genes may still be modified *after* translation has taken place. Here, briefly, are a few types of **posttranslational control**.

(1) Some proteins are specifically *inactivated* shortly after their formation by specific degradation. (2) Others, such as insulin and certain other hormones, are not produced in an active form by translation, but rather are made active by later chemical modification of the products of translation (Figure 13.14). (3) Some proteins have to be inserted into particular compartments of the cell, such as mitochondria, and others must be directed through the endoplasmic reticulum and inserted into the plasma membrane. Such proteins must be modified first. Proteins destined to

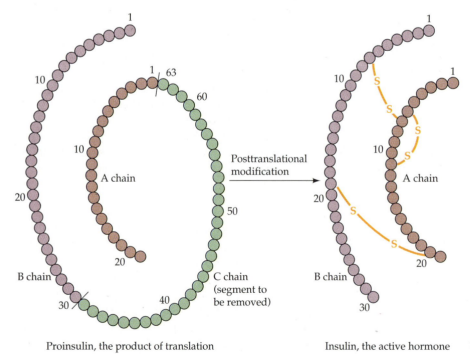

Posttranslational modification

A chain

B chain

C chain (segment to be removed)

Proinsulin, the product of translation

A chain

B chain

Insulin, the active hormone

13.14 Posttranslational Events in Insulin Synthesis
The product of translation is a larger molecule, proinsulin, that is inactive as a hormone. After translation, part of proinsulin is removed, leaving the active polypeptide hormone, insulin.

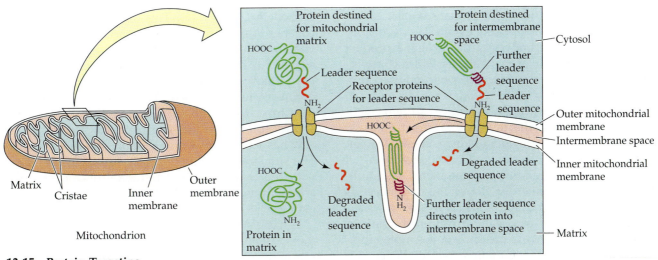

13.15 Protein Targeting
Proteins that pass through membranes are synthesized with leader sequences at the end that carries a free amino group. A single leader sequence allows a protein to pass through both membranes of a mitochondrion. A protein that will be located in the intermembrane space must have a further specific sequence between the main leader sequence and the remainder of the protein.

associate with or pass through particular membranes contain **leader sequences** of amino acids at their N-terminal ends (Figure 13.15). A given leader sequence is recognized by a specific recognition protein in the appropriate membrane, allowing transit through that membrane. Proteases then remove the leader sequence, leaving the final, active protein as a product. We will see that the insertion of glycoprotein molecules into the plasma membrane helps to regulate the adhesion of cells during embryonic development (Chapter 15). (4) Finally, some proteins are inactive until they are incorporated into larger, compound structures. Tubulin and actin form microtubules and microfilaments, respectively (Chapter 4), but the proteins do not become active until the structures have formed. Microtubules and microfilaments are used to change the shapes of cells and to allow the cells to move. Both cell-shape changes and cell movement are important in the differentiation of tissues and organs during embryonic development.

With this brief overview of posttranslational control, we complete our consideration of molecular events in gene expression. In Chapter 15 we will turn to specific examples of differentiation and the control of gene expression during development. We conclude this chapter with a look at what can happen when gene expression is improperly regulated.

GENES, VIRUSES, AND CANCER

An animal consists of legions of cells of many types. The cells divide at different times and at different rates; some do not divide at all. Cell division is tightly orchestrated in a healthy animal.

How is cell division regulated in the various parts of the animal body? Part of the answer lies in a group of proteins called **growth factors**, which circulate in the blood and trigger the normal division of cells. Why does a growth factor act only on certain cells—its target cells? What keeps it from triggering division in any cell it happens to meet? Each growth factor can be bound only by a unique receptor protein, and the receptor proteins for growth factors are embedded in the plasma membranes of their target cells. Upon binding its growth factor, a receptor protein becomes active as an enzyme, catalyzing reactions that participate in cell division. These reactions are not triggered unless the growth factor is present.

What if the system of growth factors and receptor proteins fails? Then cells divide at inappropriate times, and masses of dividing cells—tumors—may form. If the tumors spread throughout the body, the animal has cancer. Recent research has shown that many cancers originate in the activities of cancer-producing alleles, or **oncogenes**, that arise *from the normal alleles that encode growth factors and receptor proteins*. The normal alleles are called **proto-oncogenes**. This name, focusing on a relationship with abnormality and cancer, is in a sense misleading, because proto-oncogenes are absolutely essential to the normal development of the cell. It is the mutant form, the oncogene, that causes cancer. The general biology of cancer will be discussed in Chapter 16.

How do oncogenes produce cancer? The protein products of several oncogenes have been identified, and they fall into three classes. One class mimics the growth factors normally produced by the corresponding proto-oncogenes. In normal cells, the production of growth factor is regulated; but in cancerous cells, the production of proteins similar to the growth factors is unchecked by normal regulatory mechanisms. A second class of oncogene product mimics the receptor proteins for growth factors. Inserted in the plasma membrane, these modified proteins are constantly active as enzymes. In contrast, the normal proto-oncogene products are enzymatically active only in the presence of growth factors. The third class of oncogene products mimics the elements in a pathway leading to DNA replication—yet again, the oncogene leads to unregulated cell multiplication. Thus, while division of the normal cell is regulated, that of the cancerous cell is not.

Proto-oncogenes can lead to cancers by any of three mechanisms. One mechanism is simple mutation—a single point mutation can cause certain proto-oncogenes to become oncogenes and induce tumor formation. A second mechanism is overproduction: multiple copies of the proto-oncogene may form, or normal mechanisms of gene regulation may fail. In the third mechanism the proto-oncogene moves to a new chromosomal site near the promoter of a very active gene, so that the proto-oncogene is constantly transcribed.

Many cancers in mammals and birds are triggered by retroviruses. We have noted that these viruses have RNA, not DNA, as their genetic material. To replicate within cells of their hosts, they must have their RNA copied to DNA by reverse transcription (Figure 13.16). This is accomplished by an enzyme, reverse transcriptase, that is coded for by a viral gene. The DNA reverse transcript of viral RNA is inserted into a host chromosome, and then may trigger the transformation of the host cell into a cancerous cell.

Retroviruses that are highly cancer-producing, or oncogenic, contain, in addition to the three genes necessary for their own reproduction, an oncogene that causes cancerous growth and behavior in the infected cell. Even retroviruses that lack oncogenes can cause tumors in animals, but this induction typically requires many months. In such cases, it has been discovered that the viral DNA has become inserted at a locus very close to a proto-oncogene in the host chromosome. This suggests that cancer formation by oncogenes depends upon the induction of gene activity by the retroviral sequences. The promoters of viral oncogenes are strong and cause frequent transcription.

Origins of Oncogenic Viruses and of Transposable Elements

It is highly likely that the oncogenes of retroviruses arose from animal proto-oncogenes by the incorporation of an RNA transcript of a proto-oncogene into a virus coat along with part of the viral RNA genome (a process similar to transduction, which was de-

scribed in Chapter 12). This origin of retroviral oncogenes is supported by the observation that a retroviral oncogene is similar in base sequence to the suspected "parental" proto-oncogene, except that the proto-oncogene has introns, while the retroviral oncogene is uninterrupted. The oncogene differs further from the proto-oncogene in that it has been shortened at either the N-terminal or C-terminal end and has experienced one or more point mutations. The similarity with the proto-oncogene is consistent with the hypothesis that retroviral oncogenes have arisen from spliced RNA transcripts of proto-oncogenes.

It has also been suggested that some (but not all) transposable elements may have had their origin as retroviruses that became immobilized. This suggestion is based on DNA-sequencing studies; there are strong homologies (similarities) between the base sequences of the ends of certain retroviruses and of certain transposable elements, indicating a possible common origin.

SUMMARY

The most striking difference between most eukaryotic genes and prokaryotic genes is the presence of introns in the genes of eukaryotes. Most eukaryotic genes are present as single copies, but a few other DNA sequences are present in hundreds to hundreds of thousands of copies. The function of most repetitive DNA remains unknown.

Eukaryotes, like prokaryotes, have transposable elements. Transposable elements can cause chromosomal changes such as insertions, deletions, and duplications, and they can modify the transcription of other genes.

Duplication of genes in the course of evolution has given rise to families of genes such as the globin family. Different members of such families may have different functions. Some of them are pseudogenes—altered genes or reverse transcripts that have become incorporated into chromosomal DNA. Because they lack promoters (and introns), pseudogenes cannot be transcribed.

Both exons and introns of eukaryotic genes are transcribed, but the transcripts of introns are later removed by RNA splicing. To become a mature mRNA, an hnRNA undergoes extensive processing, including capping, the addition of a poly A tail, and RNA splicing; snRNPs participate in the splicing for mRNAs. The mature mRNA is typically more stable in eukaryotes than it is in prokaryotes. Long-lived mRNAs play roles in development.

The expression of eukaryotic genes is regulated at several levels: transcription, RNA processing, translation, and posttranslational processing of the poly-

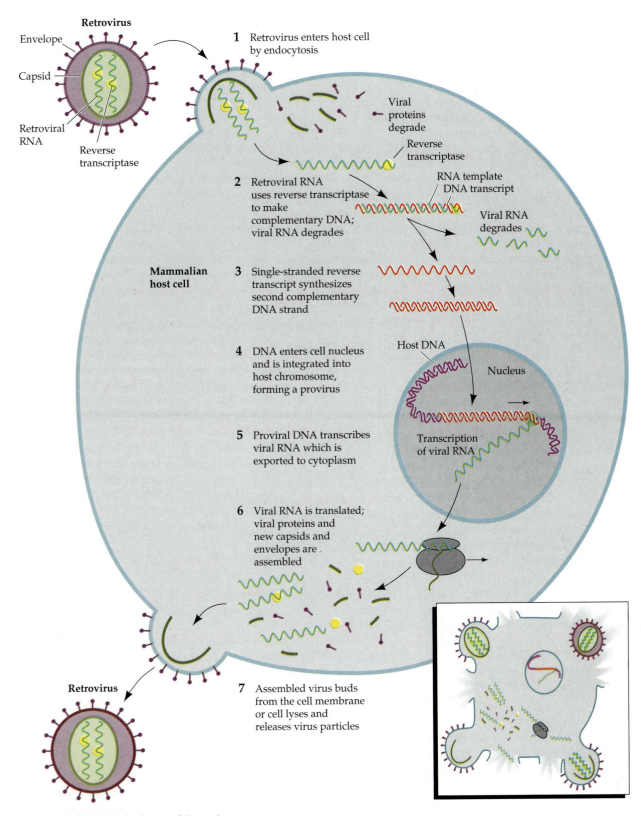

Retrovirus

Envelope

Capsid

Retroviral RNA

Reverse transcriptase

1 Retrovirus enters host cell by endocytosis

Viral proteins degrade

Reverse transcriptase

2 Retroviral RNA uses reverse transcriptase to make complementary DNA; viral RNA degrades

RNA template DNA transcript

Viral RNA degrades

Mammalian host cell

3 Single-stranded reverse transcript synthesizes second complementary DNA strand

Host DNA

4 DNA enters cell nucleus and is integrated into host chromosome, forming a provirus

Nucleus

5 Proviral DNA transcribes viral RNA which is exported to cytoplasm

Transcription of viral RNA

6 Viral RNA is translated; viral proteins and new capsids and envelopes are assembled

Retrovirus

7 Assembled virus buds from the cell membrane or cell lyses and releases virus particles

13.16 Retroviral Transcription and Insertion
The genetic material of a retrovirus is RNA. Upon infection, the viral RNA serves as a template and produces DNA by reverse transcription. The double-stranded DNA transcript becomes integrated into a chromosome of the mammalian host cell. Sometimes infected cells lyse (inset).

peptide product. These effects play important roles in cell differentiation in multicellular eukaryotes.

Regulation of transcription in eukaryotes is generally positive; transcription factors recognize and bind to DNA sequences such as the TATA box, leading to the binding of an RNA polymerase. Eukaryotes have three different RNA polymerases, each transcribing a different type of gene.

Cell division in animals is regulated in part by the growth factors and receptor proteins encoded by proto-oncogenes. Mutation of a proto-oncogene to an oncogene can lead to uncontrolled cell proliferation—cancer. Related oncogenes can be released into cells by retroviruses. Viral and animal oncogenes, retroviruses, and transposable elements seem to share common evolutionary roots.

SELF-QUIZ

1. Which statement is *not* true of nucleic acid hybridization?
   a. It depends upon complementary base pairing.
   b. A DNA strand can hybridize with another DNA strand.
   c. An RNA strand can hybridize with a DNA strand.
   d. A polypeptide can hybridize with a DNA strand.
   e. Double-stranded DNA denatures at high temperature.

2. Which statement is *not* true of introns?
   a. The name is short for "intervening sequence."
   b. They do not encode any part of the polypeptide product of the gene.
   c. They are found in all vertebrate genes.
   d. They are transcribed.
   e. They interrupt, but do not scramble, the coding sequence for a polypeptide.

3. In regard to repetitive DNA:
   a. much highly repetitive DNA lies near the centromeres.
   b. highly repetitive DNA reanneals most slowly.
   c. highly repetitive DNA is transcribed often and rapidly.
   d. single-copy DNA is rare in eukaryotes.
   e. transposable elements are single-copy genes.

4. Capping of hnRNA:
   a. is at the 3' end of the hnRNA.
   b. facilitates binding of the mRNA to a ribosome.
   c. is coded for in the DNA.
   d. prevents its translation.
   e. consists of the addition of a poly A tail.

5. Which statement is *not* true of RNA splicing?
   a. It is directed by consensus sequences between introns and exons.
   b. It is performed by small nuclear ribonucleoprotein particles.
   c. There are different splicing mechanisms for mRNAs, rRNAs, and tRNAs.
   d. RNA splicing is directed by consensus sequences.
   e. RNA splicing lengthens the RNA molecule.

6. Which genes are *not* commonly present in many copies in eukaryotes?
   a. Those that encode histones.
   b. Those that encode mRNAs.
   c. Those that encode tRNAs.
   d. Those that encode rRNA.
   e. Those that are present in a tandemly repetitive region.

7. Which statement is *not* true of selective transcription in eukaryotes?
   a. Different classes of RNA polymerase transcribe different parts of the genome.

   b. Transcription requires transcription factors.
   c. Genes are transcribed in groups called operons.
   d. Control is almost always by positive regulation.
   e. The promoter is more complex than in prokaryotes.

8. Transcription factors in eukaryotes:
   a. consist of DNA.
   b. consist of RNA.
   c. include such sequences as the TATA box.
   d. allow the binding of RNA polymerase to the promoter.
   e. cause operons to be transcribed.

9. Translational control:
   a. is not observed in eukaryotes.
   b. is a slower form of regulation than is transcriptional control.
   c. occurs by only one mechanism.
   d. requires that mRNAs be uncapped.
   e. ensures that hemoglobin is synthesized in appropriate quantity.

10. Which statement is *not* true of proto-oncogenes?
   a. They probably arose from retroviruses.
   b. They possess introns.
   c. They perform essential functions in the normal cell.
   d. They can give rise to oncogenes.
   e. They encode growth factors or receptor proteins.

FOR STUDY

1. In rats a gene 1440 base pairs in length codes for an enzyme made up of 192 amino acid units. Discuss this apparent discrepancy.

2. Describe the steps in the production of mature, translatable mRNA from a eukaryotic gene that contains introns.

3. How can nucleic acid hybridization techniques be used to determine whether a gene possesses introns?

4. Describe the origin and development of gene families such as the globin family.

5. Prepare a list of the possible ways in which transcription and translation may be regulated in eukaryotes. Contrast this with the situation in prokaryotes.

6. Discuss the possible relationships among oncogenes, transposable elements, and retroviruses.

READINGS

Cech, T. R. 1986. "RNA as an Enzyme." *Scientific American*, November. The exciting discovery of the catalytic activity of certain RNAs and its roles in the molecular biology of eukaryotes. These findings help us to understand the origin of life.

Chambon, P. 1981. "Split Genes." *Scientific American*, May. Introns and exons—their origin and how they are dealt with.

Croce, C. M. and G. Klein. 1985. "Chromosome Translocations and Human Cancer." *Scientific American*, March. A clear treatment of oncogenes and their activation.

Donelson, J. E. and M. J. Turner. 1985. "How the Trypanosome Changes Its Coat." *Scientific American*, February. Trypanosomes evade the host's immune system by constantly switching on new genes that code for different surface antigens.

Hunter, T. 1984. "The Proteins of Oncogenes." *Scientific American*, August. A detailed treatment of several specific oncogenes and their products.

Miller, J. A. 1990. "Genes That Protect Against Cancer." *BioScience*, vol. 40, pages 563-566. The roles of antioncogenes, also known as tumor-suppressor genes.

Steitz, J. A. 1988. "Snurps." *Scientific American*, June. How spliceosomes remove intron transcripts.

Suzuki, D. T., A. J. F. Griffiths, J. H. Miller and R. C. Lewontin. 1989. *An Introduction to Genetic Analysis*, 4th Edition. W. H. Freeman, New York. An up-to-date revision of one of the field's classic textbooks.

Varmus, H. 1987. "Reverse Transcription." *Scientific American*, September. Reverse transcription is not confined to the retroviruses. This article describes reverse transcription in eukaryotes and underlines its likely relevance to the emergence of DNA as the genetic material.

Weinberg, R. A. 1988. "Finding the Anti-Oncogene." *Scientific American*, September. Isolation of a gene that prevents a cell from proliferating out of control.

14

Recombinant DNA Technology

PREVIEW: Foreign DNA can be introduced into a cell, in effect converting the cell and its progeny into biochemical factories for the production of specific proteins—this is one form of recombinant DNA technology. Most recent advances in understanding how the genes of eukaryotes are regulated and organized have come through the application of recombinant DNA technology. Recombinant DNA is being used increasingly in agriculture and medicine.

This chapter deals with restriction enzymes, sticky ends, plasmids, vectors, cloning, gene libraries, complementary DNA, Southern blotting, restriction maps, chromosome walking, DNA sequencing, the polymerase chain reaction, genome projects, and agricultural biotechnology.

Techniques from the fields of molecular biology, microbial genetics, and biochemistry have been brought together to create what has come to be called **recombinant DNA technology** and has been popularized as "genetic engineering" or "cloning." By recombinant DNA, we mean *DNA made up of connected segments from mixed sources*—perhaps from two different species or, perhaps, a combination of natural and synthetic DNA. An organism containing recombinant DNA integrated into its own genetic material is called a **transgenic** organism.

An example of combining DNA from different species would be the insertion of a gene from a human into the DNA of a yeast. This could convert the yeast cells into factories for the protein product of the human gene. An early example of combining natural and synthetic DNA was the use of bacteria to produce the human hormone somatostatin. The sequence of this 14-amino acid polypeptide was known. Biologists at the City of Hope Medical Center and at the University of California, San Francisco, proceeded to synthesize, from chemicals off the shelf, a completely artificial stretch of DNA, part of which was designed to code for the amino acid sequence of somatostatin. This DNA was first inserted into a plasmid that had been isolated from a bacterial cell. The scientists then introduced the plasmid into a culture of *Escherichia coli*, where it replicated, producing multiple copies of the recombinant plasmid in each bacterial cell. These

E. coli could then be induced to synthesize human somatostatin. Somatostatin may be useful in treating pancreatitis (a disease of the pancreas), acromegaly (abnormal enlargement of bones in the hands, feet, and face), and insulin-dependent diabetes. A short time ago it could not be considered for medical use because there was almost none of it available from natural sources. Now the possibility of large-scale production of somatostatin by recombinant DNA technology opens new vistas for treatment of these diseases.

Recombinant DNA technology, as a tool, has revolutionized much of experimental biology. Beyond that, this tool will find increasing uses in agriculture, medicine, and other areas of the chemical industry, as well as in combatting environmental pollution. In this chapter we will consider the basic techniques of this technology and some of its applications in the laboratory and beyond.

THE PILLARS OF RECOMBINANT DNA TECHNOLOGY

Recombinant DNA technology depends on the properties of certain enzymes and of DNA itself. It is based on the realization that chemical reactions used in living cells for one purpose may be applied in the laboratory for other, novel purposes. Naturally occurring enzymes that cleave DNA, help it grow, and repair it are numerous and diverse, and many of them are now used in the laboratory to manipulate and combine DNA molecules from different sources.

The nucleic acid base-pairing rules (Chapter 11) underlie many of the fundamental processes of molecular biology. The mechanisms of DNA replication, transcription, and translation all rely on complementary base pairing. Similarly, the key techniques of recombinant DNA technology—sequencing, splicing, locating, and identifying DNA fragments—all make use of the complementary pairing of A with T (or U) and of G with C.

CLEAVING AND SPLICING DNA

Our first examples of how scientists use enzymes and complementary base pairing creatively will be the basic operations of **cleaving** and **splicing** DNA. The enzymes are used in the laboratory to achieve different overall purposes than they would in the living cell.

Restriction Endonucleases

Organisms—even bacteria—must have mechanisms to deal with their enemies. As we saw in Chapter 12, bacteria are attacked by bacteriophages that inject their genetic material into their hosts. Eventually the phage genetic material may be replicated by the enzyme systems of the host. Some bacteria defend themselves against such invasions by producing enzymes, called **restriction endonucleases**, that cleave double-stranded DNA molecules—such as those injected by many phages—into smaller, noninfectious fragments (Figure 14.1). There are many such enzymes, each of which cleaves DNA at a specific site defined by a *sequence of bases* and called a **recognition site**. The DNA of the bacterial host is not cleaved by its own restriction endonucleases because of the activity of specific methylases, enzymes that add methyl (—CH₃) groups to certain of the bases within the recognition sites. The methylation of the bases makes the recognition sites unrecognizable to the restriction endonucleases, and thus prevents cleavage of the host DNA, but the unmethylated phage DNA is efficiently degraded.

A specific sequence of bases defines each recognition site. For example, the restriction endonuclease called *Eco* RI (named after its source, *E. coli*) cuts DNA only where it encounters this sequence in the double helix:

...G–A–A–T–T–C...

...C–T–T–A–A–G...

Other restriction endonucleases recognize different

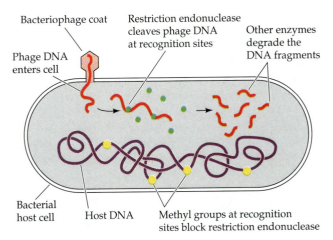

14.1 Bacterium Fights Phage with Restriction Endonucleases
Bacteria produce restriction endonucleases that break up phage DNA. Other enzymes protect bacterial DNA from the cell's own restriction endonucleases by methylating the DNA.

base sequences. The sequence recognized by *Eco* RI occurs on the average about once in 4,000 base pairs; that is, roughly once per four prokaryote genes. Thus this restriction endonuclease can chop a typical high-molecular-weight sample of DNA into pieces containing, on the average, just a few genes. Remember that "on the average" does not mean that it cuts at regular intervals along all stretches of DNA. The *Eco* RI recognition sequence does not occur even once in the 40,000-base-pair sequence of the DNA of phage T7—a fact that is nice for T7 because *E. coli* is its host. Fortunately for *E. coli*, other phages have not evolved in the same way as T7.

Different restriction enzymes that recognize different cleavage sites will cut the same sample into different discrete pieces of DNA. Several hundred restriction endonucleases have been extracted from various bacteria, and many are available for recombinant DNA research. Thus, it is an easy matter to cut a sample of DNA in many different, specific places; and we can use restriction endonucleases as "knives" for genetic "surgery."

Sticky Ends and DNA Splicing

A crucially important property of some restriction endonucleases is that they make staggered cuts in the DNA rather than cutting both strands at the same point. For example, *Eco* RI cuts as shown at the top of Figure 14.2 (note that the cut is within the recognition sequence given above). After the two cuts are made, the two strands are held together by only four base pairs. The hydrogen bonds of those base pairs are too weak to persist at warm temperatures (room temperature or above) and the pieces separate. As a result, there are single-stranded tails at the site of

(a)

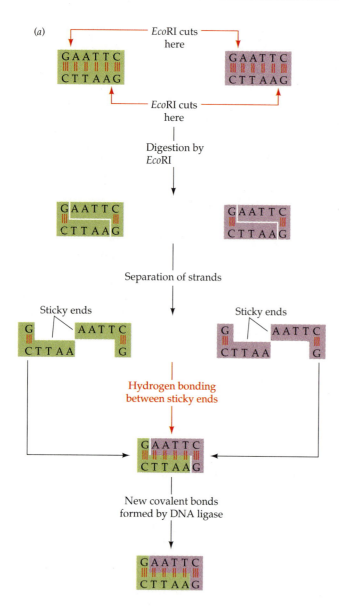

*Eco*RI cuts here

*Eco*RI cuts here

Digestion by *Eco*RI

Separation of strands

Sticky ends

Sticky ends

Hydrogen bonding between sticky ends

New covalent bonds formed by DNA ligase

(b) DNA-*Eco*RI complex

DNA

*Eco*RI protein

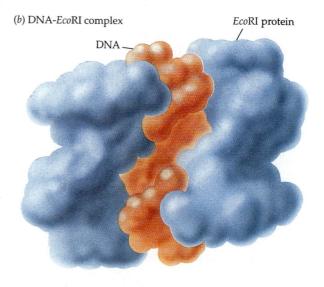

14.2 Cutting and Splicing DNA
(a) Some restriction endonucleases sever one strand of the double helix at one point and the other strand at a different point. The separated pieces have single-stranded "sticky ends" capable of combining with complementary single strands. Newly joined pieces are stabilized by the action of DNA ligase. (b) The binding of the restriction endonuclease *Eco*RI to its recognition site on a DNA molecule.

each cut. These tails are called **sticky ends** because they have a specific base sequence that is capable of binding (by complementary base pairing, at low temperature) with complementary sticky ends (Figure 14.2).

After a piece of DNA has been cut by a restriction endonuclease, it is possible for the complementary sticky ends to rejoin. The original ends can join, or an end may pair with another fragment. If more than one recognition site for a given restriction endonuclease is present in a sample, the enzyme can make a number of cut pieces, all with the *same* sequences in their sticky ends. When the temperature is lowered, the pieces reassociate at random. At the lowered temperature base pairs are more stable, and four base pairs may hold the two pieces of DNA together. The new associations are unstable, however, because they are maintained by only a few hydrogen bonds. The joined sticky ends can be permanently united

by a second enzyme, **DNA ligase**, which makes the joining very stable. The usual function of DNA ligase in the cell, as briefly mentioned in Chapter 11, is to unite the Okazaki fragments of the lagging strand during DNA replication (Box A in Chapter 11). DNA ligase also mends breaks in polynucleotide chains and thus helps in DNA repair.

A piece of DNA can be inserted into a plasmid as shown in Figure 14.3, if both the plasmid and the source for the DNA piece contain recognition sites for the same restriction endonuclease. The DNA to be inserted is cleaved from within its molecule by cutting both ends with a particular restriction endonuclease. The circular plasmid is cleaved with the same endonuclease, transforming it into a linear molecule with sticky ends. The sticky ends of the piece of DNA join the sticky ends of the cleaved plasmid, and DNA ligase seals the joining, regenerating a circular plasmid. The plasmid now contains the inserted DNA— in the middle of what used to be a recognition site for the restriction endonuclease. But notice that the insertion has created two new recognition sites, one on either side of the inserted piece.

CLONING GENES

A typical aim of recombinant DNA work is to obtain many copies of a particular gene. This is accomplished by making transgenic cells containing the

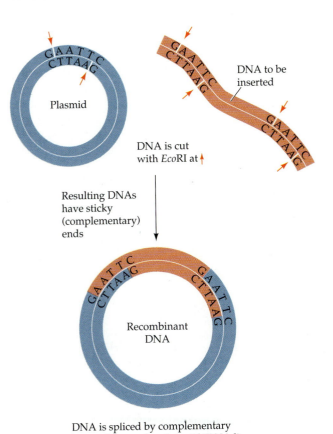

DNA to be inserted

DNA is cut with *Eco*RI at ↑

Resulting DNAs have sticky (complementary) ends

Recombinant DNA

DNA is spliced by complementary base pairing and sealed with DNA ligase

14.3 Insertion of a DNA Sample into a Plasmid
A piece of DNA can be inserted into a plasmid as long as both the circular plasmid and the DNA's source contain recognition sites for the same restriction endonuclease.

desired gene, and allowing the transgenic cells to multiply. Foreign genes, natural or synthetic, must be inserted into suitable **cloning vectors**—viruses or plasmids—before they can be successfully established in a bacterial or yeast host. A stretch of DNA that is introduced by itself into a cell does not get replicated, and it will eventually be degraded, whereas a cloning vector (and any gene contained in it) *does* get replicated.

Cloning Vectors

Plasmids were the first cloning vectors to be employed. Many different plasmids have been isolated from bacterial and eukaryotic sources. For a given experiment, one selects a plasmid with certain specific characteristics. First, the plasmid must be a **replicon**—it must have an origin of replication (Chapter 11)—otherwise it will not be replicated when the host cell divides. Second, it should carry one or more genes conferring particular properties such as resistance to specific antibiotics—it could, for example, be an R factor (Chapter 12). As we shall see in the next section, these properties are used for

selection purposes. Third, the plasmid ideally should have *one* recognition site where the restriction endonuclease will cut it. If it has no such site, it cannot be opened for the insertion of the new genes; if it has two or more, the restriction endonuclease and DNA ligase may form many diverse products rather than just one. With a single site, there is a good probability of achieving the desired insertion (Figure 14.4).

Nowadays most recombinant DNA research uses viral cloning vectors. The viruses used as cloning vectors can, like plasmids, have genes from other sources inserted into their DNA or RNA. Viral cloning vectors are commonly used for work on eukaryotic DNA. The resulting recombinant viral chromosome must still be able to fit into the virus coat. Thus a nonessential piece of the chromosome may be "edited out" before addition of the foreign gene to keep the new recombinant small enough to be packaged into the viral coat for delivery to the cell.

Insertion of Vectors into Host Cells

In transformation, as you will recall from Chapters 11 and 12, a bacterium contacts and takes up isolated DNA from another bacterium. This phenomenon provides the method commonly used for introducing plasmid vectors into bacteria. The bacteria and the solution containing the plasmids are mixed together. Under appropriate conditions, including the presence of a high concentration of calcium ions, some of the bacteria take up the plasmid DNA.

Some viral cloning vectors can infect host cells directly. No special tricks are needed in order to get the recombinant DNA from these vectors into the host cells.

Transfection of eukaryotic cells—the uptake, incorporation, and expression of recombinant DNA—often includes other steps. The cell walls must be removed before DNA can enter plant or fungal cells. The walls can be digested by appropriate enzymes, such as an enzyme found in snail guts that the snail uses to digest its meals. The resulting plant or fungal cell, without a wall, is called a **protoplast** (Figure 14.5). Plant protoplasts are used for transfection with plasmids, for cell–cell hybridization studies, and for other purposes. The most commonly used eukaryotic hosts for recombinant DNA experiments are yeasts, which are, as you know, fungi (Box 14.A). Increasingly, cells of multicellular eukaryotes are being used.

DNA may be inserted into cells in other, more drastic ways. In one method, **microprojectiles**—tiny, high-velocity particles of tungsten, coated with the DNA to be used for transfection—are shot into cells. In a second method, **electroporation**, cells are exposed to rapid pulses of high-voltage current, which temporarily renders the plasma membrane permea-

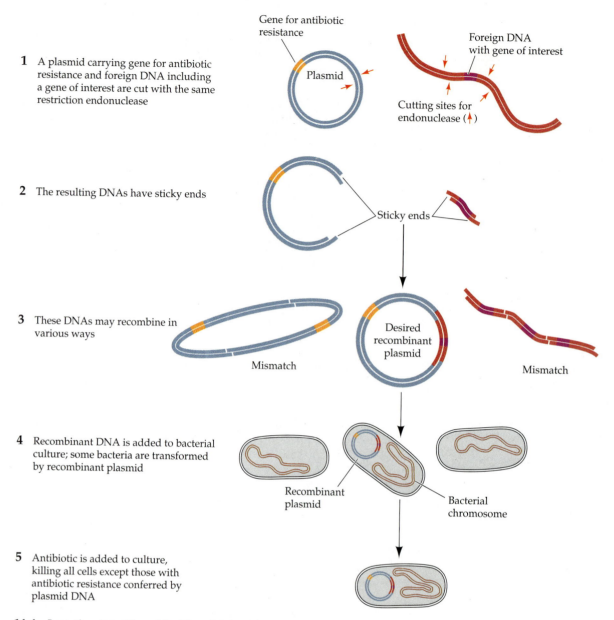

1 A plasmid carrying gene for antibiotic resistance and foreign DNA including a gene of interest are cut with the same restriction endonuclease

Gene for antibiotic resistance

Plasmid

Foreign DNA with gene of interest

Cutting sites for endonuclease (↑)

2 The resulting DNAs have sticky ends

Sticky ends

3 These DNAs may recombine in various ways

Mismatch

Desired recombinant plasmid

Mismatch

4 Recombinant DNA is added to bacterial culture; some bacteria are transformed by recombinant plasmid

Recombinant plasmid

Bacterial chromosome

5 Antibiotic is added to culture, killing all cells except those with antibiotic resistance conferred by plasmid DNA

14.4 Insertion into Plasmids Gives Multiple Products
A plasmid carrying genes for antibiotic resistance and foreign DNA that includes a gene of interest are separately cut with the same restriction endonuclease. The same endonuclease must open the plasmid and also free the gene of interest. The opened plasmid and freed gene are mixed and treated with DNA ligase (Figure 14.2) to form recombinant DNA of various sizes and compositions. Bacteria are transformed with this mixture of recombinant DNA, then exposed to antibiotics that kill all cells except those with antibiotic resistance encoded by the plasmid DNA.

ble to macromolecules in the medium. DNA can be directly injected into mammalian cells, but this is more labor-intensive than other methods, as each recipient cell must be handled individually (Figure 14.6).

DNA may also be coated in various ways to allow it to pass through plasma membranes. It can be complexed with lipids, or it can be enclosed in natural or artificial plasma membranes, which are then fused to

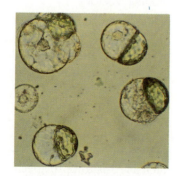

14.5 Plant Protoplasts
Removal of the cell wall of a plant leaves a naked protoplast.

BOX 14.A

Yeasts in Recombinant DNA Technology

The early successes of recombinant DNA technology were achieved with bacteria as hosts. However, bacteria are not ideal organisms for the processing and study of eukaryote genes. By now, cells of a number of eukaryotes have been used in studies of molecular biology and recombinant DNA technology.

Yeasts are now the most commonly used organisms for such studies. The best-studied yeast, *Saccharomyces cerevisiae* (baker's or brewer's yeast), is rapidly becoming the eukaryote best understood at the molecular level, as *Escherichia coli* is the best-understood prokaryote. Unlike most other eukaryotes, yeasts are unicellular rather than multicellular. However, yeasts are typical eukaryotes in most respects, including their genetic mechanisms. At the same time, they share three of the greatest advantages of *E. coli* for work in molecular biology. Because they are tiny and unicellular, they are easy to grow in vast numbers in small volumes of medium. They multiply rapidly—almost as rapidly as *E. coli*. (The division time of a typical culture of *E. coli* is in the range 20–60 minutes and that of *S. cerevisiae* 2–8 hours.) Finally, while mammals have on the order of 1,000 times more DNA per cell than *E. coli*, a haploid cell of *S. cerevisiae* is much more manageable than a mammalian cell— the yeast has only 3½ times more DNA than an *E. coli* cell. The DNA of *S. cerevisiae* is organized into 17 chromosomes (in the haploid phase) that can be separated in the laboratory using a new technique.

Foreign genes can be carried in a yeast cell in three different ways. In one way, the recombinant DNA is present in one or a few copies of a plasmid into which a centromere (cloned from a yeast chromosome) and a replication site have been inserted. The centromere (Chapter 9) makes the plasmids stable, as the mitotic spindle can deliver them to the daughter nuclei. In the second, the recombinant DNA is present in many copies of a plasmid without a centromere—these plasmids are unstable, being passed on to daughter yeast cells at random. In the third, the recombinant DNA is incorporated into a yeast chromosome.

Yeasts are particularly well suited for studies of the relationships between genotype and phenotype. Analysis of mutants, gene cloning, cDNA production, and DNA sequencing can be combined to allow us to understand specific genes, the proteins for which they code, and the functions that are performed by the proteins.

---

the target cells. The nonbiological methods described here are reasonably efficient, resulting in transfection of 1 percent or more of the potential recipient cells. Higher yields are obtained with some of the biological methods.

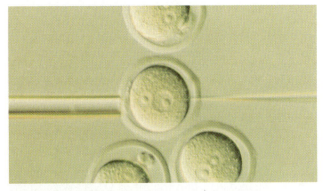

14.6 Direct Injection of DNA into a Nucleus
The micropipette at the right of the nucleus injects foreign DNA into a mouse embryo—a fertilized egg in which the small nucleus to the left is from the egg and the larger one to the right is from the sperm. The injected DNA is entering the sperm nucleus. The pipette on the left holds the cell steady during the "operation."

Selection of Transgenic Cells

Following interaction with a preparation of plasmid vectors, a population of host cells is heterogeneous, as only a small percentage of the cells have taken up the plasmid. One now adds an antibiotic to the suspension—the same antibiotic for which the plasmid carries resistance. Thus all the cells that did not take up the plasmid are destroyed! At this point, the cell population is still heterogeneous. Although the surviving cells all contain the plasmid, only a few of the plasmids contain the DNA sequence we wish to clone. Thus further selection steps are necessary.

Here is one elegant approach to this problem—let us follow it from the start of the experiment. We select, as our cloning vehicle, a plasmid that contains *two* antibiotic resistance genes, for different antibiotics. Each of these genes plays an important part in what follows. *One* of the resistance genes must contain the recognition sequence for the restriction endonuclease being used to cut the insert DNA. We add the endonuclease to the plasmid preparation, cutting the plasmid within the first resistance gene and leaving sticky ends. We then add the insert DNA, which has been cut with the same endonu-

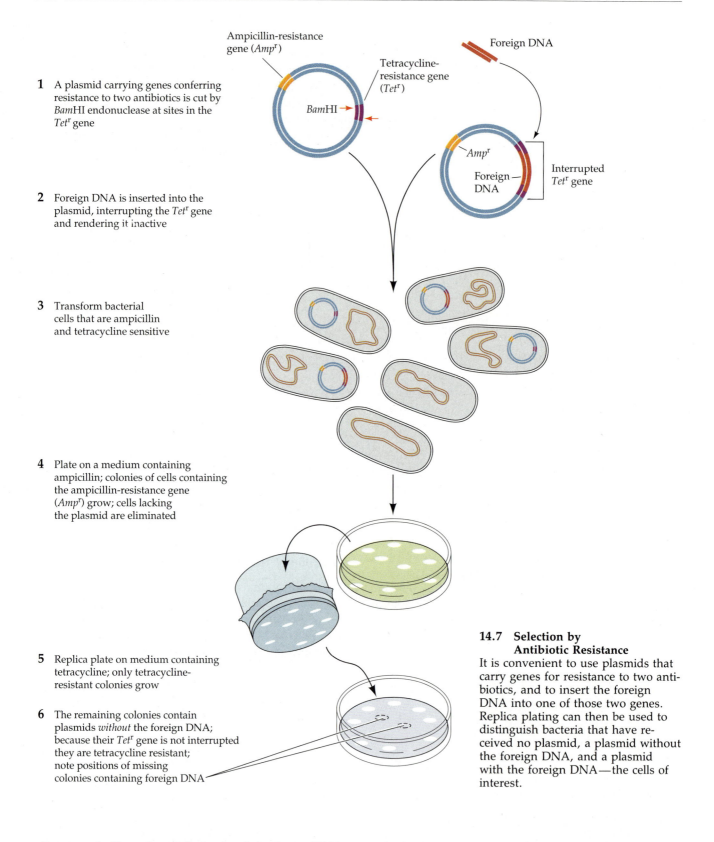

1 A plasmid carrying genes conferring resistance to two antibiotics is cut by *Bam*HI endonuclease at sites in the *Tet*ʳ gene

Ampicillin-resistance gene (*Amp*ʳ)

Tetracycline-resistance gene (*Tet*ʳ)

*Bam*HI

Foreign DNA

*Amp*ʳ

Foreign DNA

Interrupted *Tet*ʳ gene

2 Foreign DNA is inserted into the plasmid, interrupting the *Tet*ʳ gene and rendering it inactive

3 Transform bacterial cells that are ampicillin and tetracycline sensitive

4 Plate on a medium containing ampicillin; colonies of cells containing the ampicillin-resistance gene (*Amp*ʳ) grow; cells lacking the plasmid are eliminated

5 Replica plate on medium containing tetracycline; only tetracycline-resistant colonies grow

6 The remaining colonies contain plasmids *without* the foreign DNA; because their *Tet*ʳ gene is not interrupted they are tetracycline resistant; note positions of missing colonies containing foreign DNA

14.7   Selection by Antibiotic Resistance
It is convenient to use plasmids that carry genes for resistance to two antibiotics, and to insert the foreign DNA into one of those two genes. Replica plating can then be used to distinguish bacteria that have received no plasmid, a plasmid without the foreign DNA, and a plasmid with the foreign DNA—the cells of interest.

cleave, and allow the sticky ends of the insert DNA and the plasmids to recombine. This gives us a heterogeneous population of plasmids, some containing the insert DNA. Combination of the plasmid population with host cells leads to transformation, giving a heterogeneous population of bacteria, some containing a plasmid. Which cells contain plasmids?

Which plasmids contain the insert DNA? We add the antibiotic to which the *other* gene confers resistance. As just described, this treatment kills all the bacteria that do not contain the plasmid. What happens if we now add the other antibiotic to the plasmid-containing cells? Cells containing the plasmid but not the insert DNA are resistant to both antibiotics, since they have both resistance genes. But cells containing the plasmid *with* the insert DNA are sensitive to the second antibiotic—insertion of the DNA into that resistance gene has inactivated the gene by interrupting its base sequence. Therefore, we can isolate the desired transgenic cells by replica plating (Figure 14.7, and see Chapter 12).

Expression of Cloned Genes

A further refinement is sometimes employed; in fact, it was used in the original work on bacterially produced somatostatin, mentioned above. To allow precise control of the transcription of the inserted gene for somatostatin, the investigators arranged for the plasmid vector to include a copy of the *E. coli lac* operon, and they inserted the somatostatin gene into the β-galactosidase structural gene of that operon (Chapter 12). After putting the modified plasmid into bacteria and cloning them, the scientists could cause the production of human somatostatin by adding lactose to the growth medium (Figure 14.8). In the absence of lactose or another inducer, no somatostatin was formed.

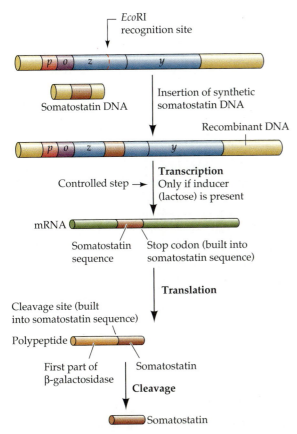

14.8 The *lac* Operon as a Tool with Recombinant DNA
One can control the expression of a foreign gene in a transgenic bacterium if that gene is inserted into a structural gene of the *lac* operon. The gene is expressed only in the presence of an inducer such as lactose.

SOURCES OF GENES FOR CLONING

There are three principal sources of the genes or DNA fragments used in recombinant DNA work. One source is pieces of chromosomes inserted into viruses or plasmids, which are usually maintained as *gene libraries*. A second source is *complementary DNA*, obtained by reverse transcription from specific RNAs. The third source is laboratory synthesis of specific polynucleotide sequences.

Gene Libraries and Shotgunning

DNA from a desired source can be isolated and broken into small fragments, usually by restriction endonucleases. These fragments can then be incorporated into bacteriophage DNA or plasmids that are, in turn, cloned in bacteria. This technique, called **shotgunning**, results in a random collection of clones called a **gene library** (Figure 14.9). Each clone carries a specific fragment of the original DNA, and the library as a whole carries all of the original DNA. The clone or clones carrying a particular gene can be detected in various ways.

Complementary DNA

If a specific RNA, such as a particular mRNA, is available, one can make a complementary DNA, or **cDNA**—DNA made by using RNA as the template. Although the starting amount of specific mRNA is usually small, it can be used repeatedly as the template for the production of large amounts of cDNA; thus cDNA production is a way to *amplify* the amount of material available for hybridization or sequencing studies.

The formation of specific cDNA is illustrated in Figure 14.10. As we have seen, most eukaryotic mRNAs have a string of adenine, or A, residues at their 3′ end (Chapter 13). One allows a sample of mRNA to hybridize with molecules consisting of a string of T residues (oligo-dT). After the annealing, the oligo-dT can serve as a primer and the mRNA can serve as a template for the enzyme **reverse transcriptase**—the enzyme that RNA viruses use to synthesize DNA from RNA templates in host cells (Chapter 11). If the primer and reverse transcriptase are given a source of DNA precursors (the four deoxyribonucleoside triphosphates), they will synthesize a strand of cDNA complementary to the mRNA. After the cDNA strand is removed from the mRNA (by increasing the pH of the solution, thus denaturing the cDNA–mRNA hybrids and degrading the mRNA), it can be used for cloning, hybridization, or other experiments. To clone cDNA by insertion into a DNA vector, the single-stranded cDNA must first be converted into double-stranded DNA, using DNA polymerase.

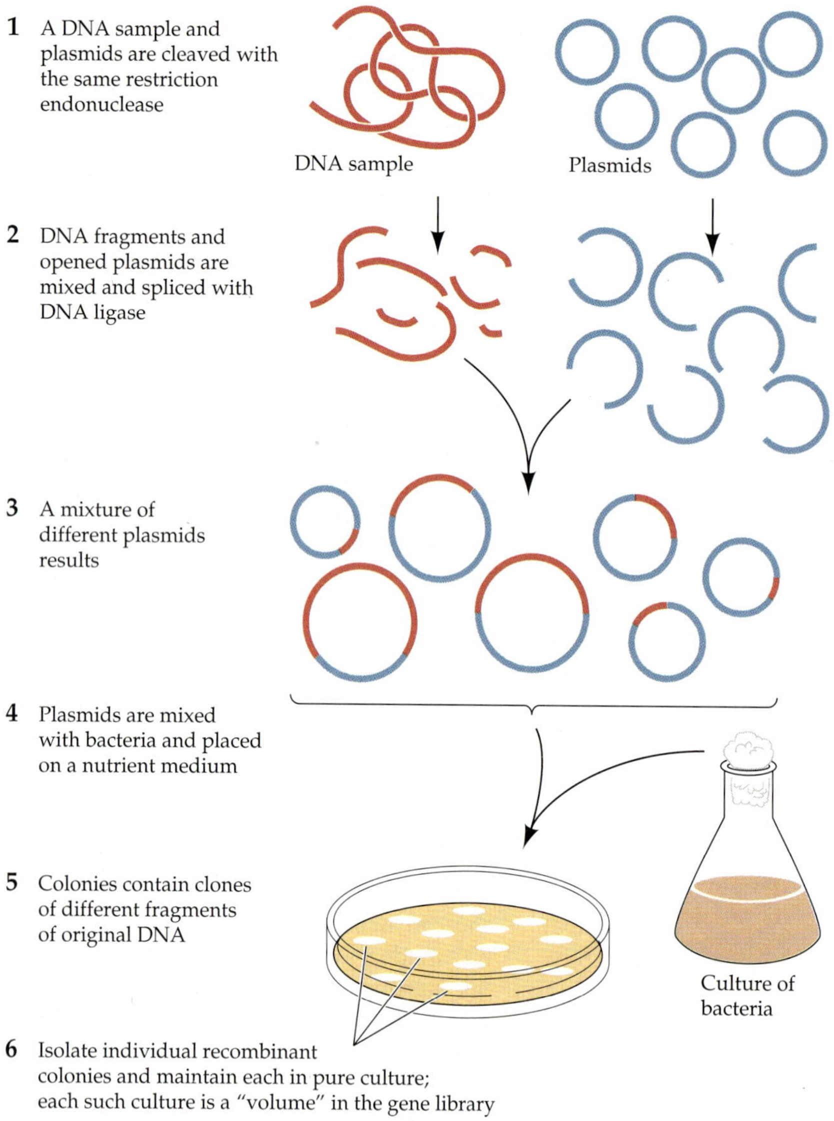

1 A DNA sample and plasmids are cleaved with the same restriction endonuclease

DNA sample   Plasmids

2 DNA fragments and opened plasmids are mixed and spliced with DNA ligase

3 A mixture of different plasmids results

4 Plasmids are mixed with bacteria and placed on a nutrient medium

5 Colonies contain clones of different fragments of original DNA

Culture of bacteria

6 Isolate individual recombinant colonies and maintain each in pure culture; each such culture is a "volume" in the gene library

14.9 Construction of a Gene Library
The DNA to be studied is cleaved into many fragments and the fragments inserted into copies of a suitable phage vector or plasmid. Bacteria are transformed by the vectors, and each transgenic bacterium gives rise to a colony containing part of the original DNA sample. It is possible to get a complete gene library by this "shotgun" method.

Complementary DNAs can be incorporated into bacteriophage vectors; a collection of this type is called a **cDNA library**. In such cDNA libraries, one is dealing only with *expressed* DNA, that is, DNA that is represented by mRNAs. By comparing cDNA libraries from different tissues or different stages of development, one can gain insight into which genes are being turned on or off. In contrast, the advantages of *gene* libraries center on the fact that a gene library *does* include DNA that is not transcribed and DNA that is processed out of the mRNAs (including introns, promoters, and regulatory signals such as the TATA box). "Chromosome walking," to be discussed below, can be done only with gene libraries, not with cDNA libraries.

Synthetic DNA

Another source of DNA for cloning is direct chemical synthesis, which works well for small DNA molecules. This approach was used by the group that first cloned a DNA sequence coding for somatostatin. Commercially made instruments for automated DNA synthesis are now available.

There are a few important considerations in the design of a small, synthetic "gene." First, we must know the amino acid sequence of the desired polypeptide product. Using the genetic "code book" (Figure 11.26), we can figure out the appropriate base sequence for the synthetic gene. What else might we add to this synthetic DNA? How can we assure that translation begins and ends at the right places? We can add codons for initiation and termination of translation. We will need to insert the synthetic gene into a cloning vector—how can we prepare for that? We can add, to the ends of our synthetic gene, appropriate recognition sequences for the restriction endonuclease that we will use to splice the synthetic DNA into the cloning vector. Other refinements are also possible.

In addition to the synthesis of small polypeptides such as somatostatin, there have been two principal uses of synthetic DNA thus far. One use is as "probes" of the sort used to detect specific DNA fragments, to be discussed in the next section. The other use is in "directed" mutagenesis—the production of specific alterations in DNA regions with known sequences.

EXPLORING DNA ORGANIZATION

Recombinant DNA techniques are widely used in studies of the organization of DNA, including whole chromosomes. Such studies may be conducted at many levels, from the examination of the sequence of genes on a chromosome down to the examination of the sequence of bases in a DNA fragment.

Separation of Intact Chromosomes

Human chromosomes can be separated by an elegant automated technique (Figure 14.11). A suspension of chromosomes is stained with two fluorescent dyes and then passed at high speed through a fine tube. Each chromosome takes up the two stains in a particular ratio and thus fluoresces in a distinctive way. Laser beams of two different wavelengths, each absorbed by one of the dyes, are directed at the stream, and the fluorescence of the chromosomes is analyzed. When an appropriately fluorescing chromosome passes the observation point, the drop in which it is contained is given a tiny electric charge. As the drop falls away from the nozzle of the apparatus, it is attracted by a charged plate and falls into a tube that catches it while other, uncharged drops fall straight down into another tube. The system can be

adjusted so that any one of the 22 autosomes, or the X or Y chromosome, is collected—at rates in the hundreds per second.

Separation and Purification of DNA Fragments

DNA fragments differing in length can be separated by **gel electrophoresis** (Figure 14.12). This technique depends on the electric charge of DNA fragments, which results from the ionization of the phosphate groups that connect successive sugars in the polynucleotide chain, forming negatively charged phosphate ions. A mixture of DNA fragments is placed on a porous gel and an electric field is applied across the gel. The negatively charged DNA fragments move through the field toward the positively charged electrode, the smallest fragments moving the fastest and therefore traveling the farthest. The separated materials can then be removed from the gel in their pure form. Different samples may be "run" side by side in different "lanes" in the electric field. DNA fragments of known length or molecular weight are often run next to experimental mixtures to provide a size reference.

Electrophoresis finds many applications in molecular biology, biochemistry, and cell biology. All sorts of molecules, particularly proteins and nucleic acids, can be separated by virtue of their differing sizes and electric charges.

Detection of Specific DNA Fragments

Electrophoresis separates DNA fragments of different sizes but does not itself show which of the separated fragments contains a particular piece of DNA. Detec-

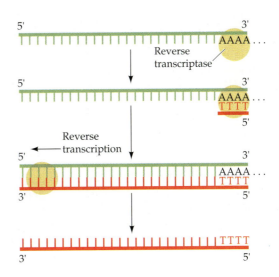

1 mRNA template with 3' poly A tail and reverse transcriptase enzyme

2 A short oligo-dT primer is added and allowed to hybridize with the poly A tail

3 Reverse transcriptase synthesizes cDNA using the mRNA template and deoxyribonucleoside triphosphate substrates, creating a DNA–RNA hybrid

4 When synthesis is completed, mRNA is degraded by a base (NaOH), leaving single-stranded cDNA

14.10 Synthesis of Complementary DNA
Synthesis of single-stranded cDNA requires the enzyme reverse transcriptase, an mRNA template with a poly A tail on the 3' end, a short chain of oligo-dT, and the four deoxyribonucleoside triphosphates. The oligo-dT is added

to the mRNA and allowed to hybridize with its poly A tail. Reverse transcriptase is added; it uses the oligo-dT as a primer, the mRNA as a template, and the deoxyribonucleoside triphosphates as substrates to synthesize DNA.

14.11   Automated Separation of Chromosomes

This apparatus rapidly separates specific human chromosomes from a mixture flowing through the sample chamber. Separation is based on relative staining of the different chromosomes with two fluorescent dyes. Droplets containing the desired chromosomes are diverted into different collecting tubes.

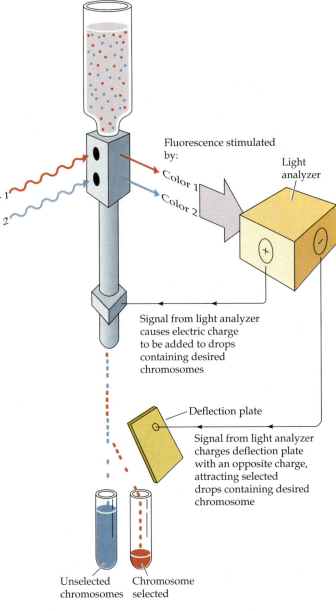

Fluorescence stimulated by:

Color 1

Color 2

Light analyzer

Laser color 1

Laser color 2

Signal from light analyzer causes electric charge to be added to drops containing desired chromosomes

Deflection plate

Signal from light analyzer charges deflection plate with an opposite charge, attracting selected drops containing desired chromosome

Unselected chromosomes

Chromosome selected

tion of a particular piece rests on complementary base pairing with a suitable **probe**—a strand of DNA or RNA known to have the base sequence complementary to the piece of DNA sought. For example, a specific mRNA can be used as a probe to locate the gene from which it was transcribed. However, hybridization experiments with the probe cannot be done in the gel; first the DNA must be transferred to a nitrocellulose filter that binds and immobilizes single-stranded DNA.

A technique called **Southern blotting** is often used in such a search for a specific DNA fragment (Figure 14.13). A mixture of DNA fragments is separated by electrophoresis. The pH of the electrophoresis gel is then made basic, a common procedure that breaks hydrogen bonds and thus separates the strands of the DNA fragments. The gel is then "blotted" with a sheet of nitrocellulose, as shown in the figure, so that some of the DNA is transferred to the sheet. Heating "fixes" the DNA so that it is immobilized on the nitrocellulose in its original position. Now the probe (mRNA or cDNA that has been labeled, radioactively or otherwise) is poured over the sheet. When probe molecules meet DNA strands to which they are complementary, they are trapped by complementary base pairing, forming double-stranded nucleic acid molecules. These trapped molecules remain attached to the sheet, while the other probe molecules are washed away. The positions of the desired fragments on the sheet may then be detected by finding the bound, labeled probe.

Southern blotting was named after its inventor, Scottish molecular biologist E. M. Southern. Related techniques have been named *Northern blotting* (RNA blotting) and *Western blotting* (protein blotting).

Autoradiography

Radioactive materials release energy or subatomic particles that will darken a photographic emulsion. This is the basis of autoradiography, a technique illustrated in Box 2.A and described in Box 8.B. It can

be combined with some of the techniques just described to help scientists locate particular DNA fragments. We will consider next a technique in which autoradiography plays an important part.

Localization of Genes on Chromosomes

Radioactive probes (RNA or cDNA) can be used to determine on which chromosome a gene or other specific DNA segment lies and to give an indication of the location of the gene on that chromosome (Figure 14.14). A microscope slide is prepared with cells in metaphase of nuclear division (Chapter 9). The slide is treated with weak base to separate the DNA strands slightly—break hydrogen bonds and expose single-stranded DNA for hydrogen bonding—and the probe is poured onto the slide. The probe binds only to the DNA sequence to which it is complemen-

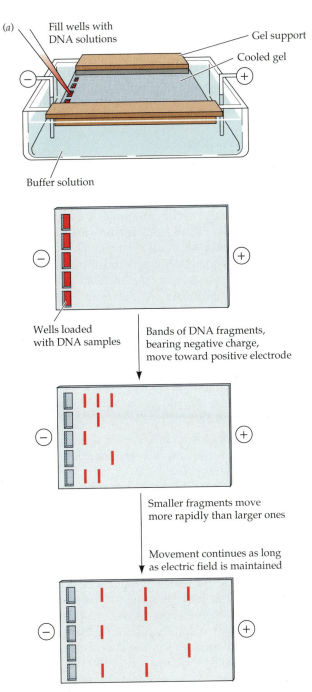

(a) Fill wells with DNA solutions — Gel support — Cooled gel

Buffer solution

Wells loaded with DNA samples

Bands of DNA fragments, bearing negative charge, move toward positive electrode

Smaller fragments move more rapidly than larger ones

Movement continues as long as electric field is maintained

14.12   Gel Electrophoresis
(a) One version of a setup for separating nucleic acid fragments or other macromolecules by gel electrophoresis. Samples are pipetted into wells in a horizontally placed gel slab, and an electric field is applied. DNA fragments move at different rates determined by their sizes. *(b)* Examining a stained slab under ultraviolet light. The pink bands contain the separated DNA fragments.

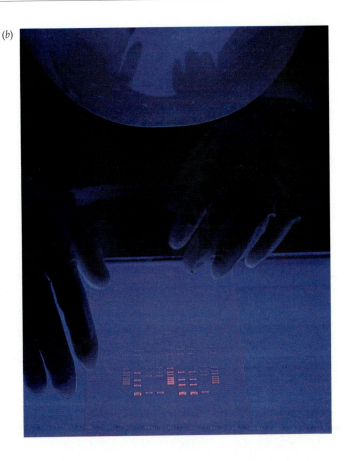

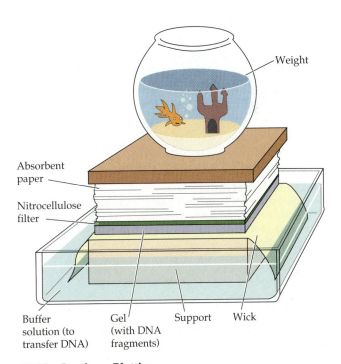

Weight

Absorbent paper

Nitrocellulose filter

Buffer solution (to transfer DNA) — Gel (with DNA fragments) — Support — Wick

14.13   Southern Blotting
A buffer solution moves through a wick, an electrophoresis gel, a nitrocellulose filter, and into a stack of absorbent paper that acts as a blotter. As the buffer moves upward, it transfers DNA strands from the gel to the nitrocellulose filter, which immobilizes the DNA strands. By pouring a suitable probe over the filter with its fixed DNA, investigators can locate those DNA strands of particular interest.

tary, and the unbound probe is washed away. The slide is stained to reveal the bands on the chromosomes, and autoradiography shows where the probe has bound in relation to the banding pattern. This technique is called **in situ hybridization** (*in situ* means "in place").

1 Chromosome preparations are treated with NaOH to denature the strands and expose DNA bases

2 Radioactively-labeled DNA or RNA probe is added

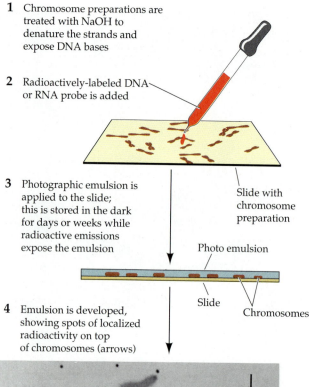

Slide with chromosome preparation

3 Photographic emulsion is applied to the slide; this is stored in the dark for days or weeks while radioactive emissions expose the emulsion

Photo emulsion

Slide

Chromosomes

4 Emulsion is developed, showing spots of localized radioactivity on top of chromosomes (arrows)

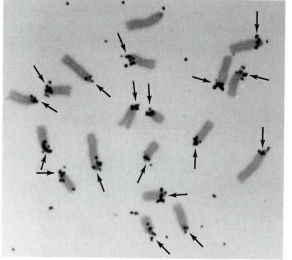

14.14 In Situ Hybridization
A radioactive RNA or DNA probe locates specific genes on specific chromosomes. Recognition of individual chromosomes is made possible by staining that reveals banding patterns.

Restriction Mapping

One of the things molecular biologists were most anxious to do was to determine the actual sequence of bases in samples of DNA—gene fragments, genes, and chromosomes. First, however, it was necessary to obtain a preliminary, less detailed representation of the structure of a DNA molecule. Restriction endonucleases, with their highly specific recognition sequences, can be used to subdivide a DNA molecule into fragments that can then be ordered to form a **restriction map**. Sometimes the base sequences of the individual fragments themselves can be determined, a step on the way to sequencing the entire molecule. An example of restriction mapping is shown in Figure 14.15.

Chromosome Walking

Chromosomes are too large for direct DNA sequencing, and they are too large to clone using viruses or plasmids as vectors. A chromosome is, instead, cleaved into a large number of smaller fragments for individual study. The chromosome can be treated with one or more restriction endonucleases under conditions such that cleavage is incomplete, so individual molecules in the sample are cleaved in different places and the different fragments overlap one another. After we determine the base sequence in each fragment, we need to know the order of the fragments themselves if we are to know the base sequence of the whole chromosome. How does the investigator know the order of those smaller fragments in the original chromosome?

An easy and elegant approach, called **chromosome walking**, is illustrated in Figure 14.16. Two samples of the original DNA are cleaved with different restriction endonucleases, and the fragments from each sample are cloned, creating two gene libraries. A clone from the first library is made single-stranded and used as a probe of the second gene library—the probe hybridizes with a fragment only if the fragment contains a base sequence complementary to the probe on one of its strands. If hybridization occurs, then the other strand of the hybridizing fragment must contain the same sequence as the probe—that fragment overlaps the probe fragment. The cloned fragment from the second library is then used as a probe of the *first* library to identify a third overlapping fragment, the third fragment is used as a probe of the *second* library, and so forth. Chromosome walking, by identifying successive overlapping fragments, reveals the order of the fragments in the original DNA.

Sequencing DNA

How can we determine the base sequence for a fragment of DNA? This problem appeared intractable until the discovery of restriction endonucleases. By using various restriction enzymes, one could cut a sample of DNA into multiple fragments. By using hybridization techniques, biologists could determine which genes are associated with which fragments, so that a partial map of the DNA sample could be made. They still faced the problem of determining the base sequences within these smaller pieces. In the mid-

1970s, the British biochemist Frederick Sanger, who in 1953 had determined the first protein primary structure; Allan Maxam and Walter Gilbert of Harvard University; and others devised techniques for the rapid sequencing of DNA. A number of methods are in current use. One of Sanger's methods is described in Box 14.B.

Why should one want to determine the base sequence of a DNA sample? If one can "read" the primary structure of a gene—this amounts to determining the exact chemical structure of the gene— then knowledge of the base sequence of the DNA, along with knowledge of the genetic code, enables us to determine the primary structure of the protein product. In fact, it is now usually easier to determine the primary structure of a protein by analyzing the corresponding DNA than by analyzing the protein directly.

Most important, knowing base sequences may help us determine how regulatory sites (such as promoters) function. Also, closely related genes may have quite different functions, and sequencing helps us understand how they differ. And we can use DNA sequencing to identify precise molecular defects, as in genetic diseases. Gene cloning coupled with DNA sequencing has provided important information about the insertion of transposable elements in eukaryotes as well as in prokaryotes. DNA sequencing can provide information of interest to evolutionary biology as well. By comparing base sequences of homologous genes from different organisms, we can extend our knowledge of the evolutionary relationships among species.

GENE COPIES BY THE BILLION

Biologists often want to obtain particular pieces of DNA—a particular gene, for example—in quantity sufficient for biochemical studies. A powerful technique, the **polymerase chain reaction**, has made this relatively easy. First described in 1984, the polymerase chain reaction has become one of the most widely used techniques in molecular biology. It can produce billions of copies of a single piece of a DNA molecule—the target sequence—in a few hours.

The polymerase chain reaction is a cyclic process in which the following sequence of steps is repeated

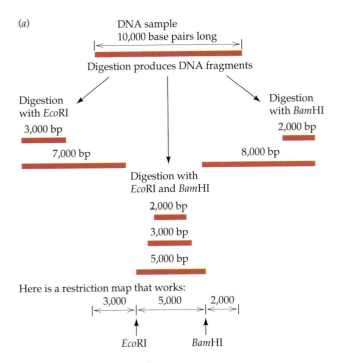

(a)

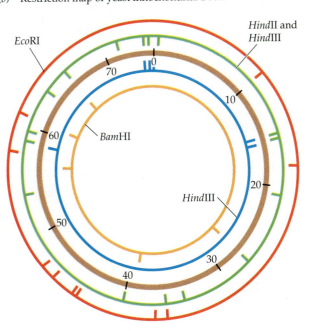

(b) Restriction map of yeast mitochondrial DNA

14.15 Restriction Mapping
(a) Try to see how these pieces, cut from the sample DNA by two restriction endonucleases with different recognition sites, can be pieced together to show the positions of the endonuclease recognition sites in the original DNA. The numbers represent lengths of the DNA fragments in thousands of base pairs (bp). The endonuclease *Bam*HI cleaves the sample into two pieces with lengths 2,000 and 8,000 base pairs; we infer that the *Bam*HI recognition site is about 2,000 base pairs from one end of the original DNA molecule. Where is the *Eco*RI recognition site in relation to the *Bam*HI site and the ends of the sample? (b) The restriction map of DNA isolated from yeast mitochondria. Cutting sites for four restriction endonucleases are shown. The scale is marked in units of thousands of DNA base pairs. Envision the experimental data that gave rise to this map.

1 Restriction endonucleases cleave DNA

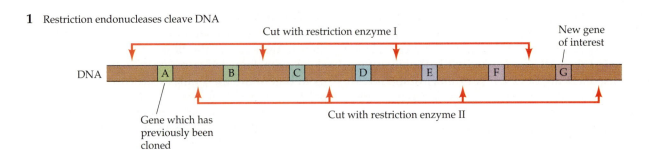

2 Gene libraries from the
two enzymes contain overlapping fragments

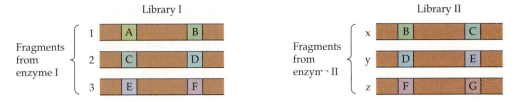

3 Probe libraries
with A and subsequent probes, "walking" down the DNA

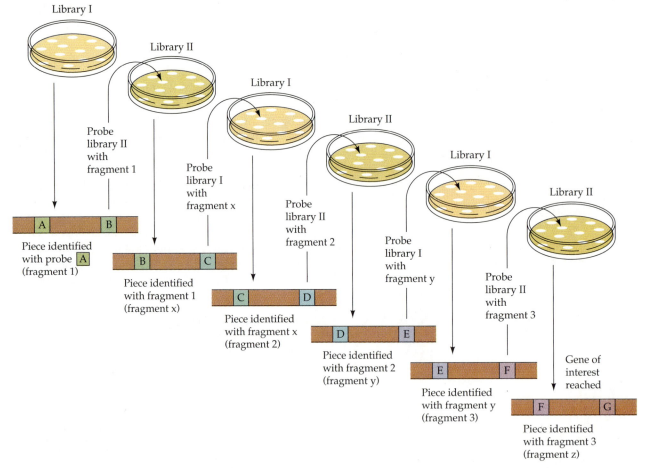

4 DNA is mapped from fragments identified

14.16 Chromosome Walking

The "volumes" in a gene library (Figure 14.9) can be put in the proper order—that is, their order in the original DNA molecule—by chromosome walking.

over and over again. Double-stranded DNA is denatured into single strands, primers for DNA synthesis are added to the 3' ends of the target DNA sequence on the separated strands, and DNA polymerase catalyzes the production of new complementary strands—thus doubling the amount of the DNA sequence in the reaction mixture and leaving the new DNA in the double-stranded state (Figure 14.17). Each cycle takes only one to a few minutes.

To use the polymerase chain reaction, a scientist must know a sequence of about two dozen bases at the 3' end of the target sequence on each DNA strand. Knowing that sequence, the scientist can make a suitable probe, complementary to the sequence, to use as a primer for DNA synthesis.

Separation of the strands of a DNA molecule requires temperatures around 98°C, as we saw in Chapter 13. Therefore, it is necessary to heat the reaction mixture in each cycle of the polymerase chain reaction. This was a problem at first, because the heating destroyed the DNA polymerase that catalyzed the polymerase chain reaction. Now, however, the polymerase chain reaction is run with a temperature-resistant DNA polymerase from the bacterium *Thermus aquaticus*, which lives in hot springs. This polymerase survives the high temperatures used to separate the DNA strands.

Amplification of DNA samples by the polymerase chain reaction has found many uses, ranging from the diagnosis of hereditary diseases in human fetuses

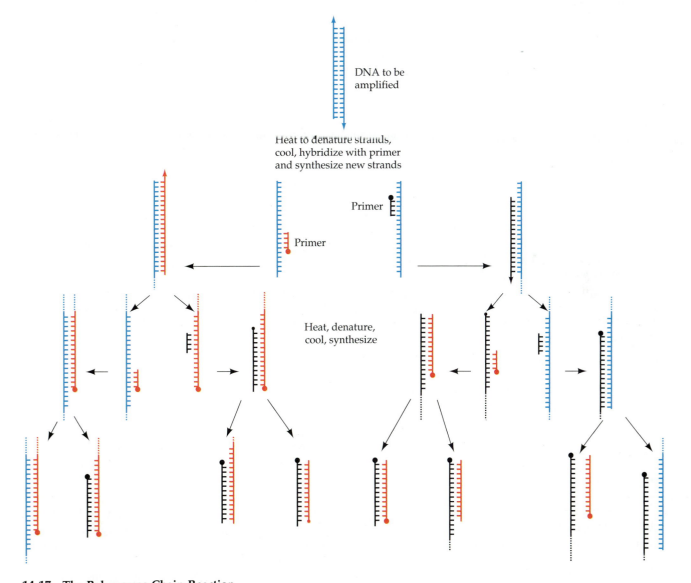

14.17 The Polymerase Chain Reaction
Because it is used so often and in so many laboratories, this technique was named "molecule of the year" by the journal *Science* in 1990. In the presence of appropriate primers, a heat-resistant DNA polymerase copies both strands of the desired DNA sequence; by alternating heating (to separate DNA strands) and synthesis phases, a technician can rapidly make enormous numbers of copies.

BOX 14.B

Determining the Base Sequence of DNA

One of the simplest, most rapid, and most accurate methods for sequencing DNA samples was developed by Frederick Sanger in 1977. This method has similarities to the experiment described in Chapter 11 in which the compound cordycepin triphosphate is used to interrupt RNA synthesis; this experiment demonstrates that RNA synthesis, like DNA synthesis, proceeds from the 5' to the 3' end of the molecule. The nucleoside triphosphates normally used as substrates for DNA synthesis are the 2'-deoxyribonucleoside triphosphates (dNTPs). In the Sanger technique, one also makes use of 2',3'-dideoxyribonucleoside triphosphates (ddNTPs, shown in the figure below). Suppose that a DNA strand is being synthesized, by the addition of one dNTP after another. Now, if a ddNTP should be picked up instead,

it joins the growing chain; but, lacking a free hydroxyl group at C3, it cannot accept a further dNTP. Thus synthesis stops at the point where the ddNTP is inserted—like cordycepin triphosphate, the ddNTP functions to terminate the chain.

In this Sanger technique for sequencing DNA, single-stranded DNA to be sequenced is divided into four samples. To each of the four samples, one adds DNA polymerase (to synthesize the complementary strand), a primer, the four dNTPs (dATP, dGTP, dCTP, and dTTP, one of which is radioactively labeled so that new DNA strands can be detected by their radioactivity), and small amounts of a *different* one of the four ddNTPs (ddATP, ddGTP, ddCTP, or ddTTP). Each sample soon contains a DNA mixture made up of the unknown single strand and shorter complementary strands; the complementary strands are of various lengths, but all are radioactive. In the batch containing ddATP, for example, each time a T is reached on the template strand, the growing complementary strand adds, at random, either dATP or ddATP (figure on facing page). If ddATP is added, chain growth terminates at that point. Depending on chance, some of the replicating strands grow to greater lengths than others before coming to

a ddATP stopping point. Then, within each of the four reaction mixtures, the strands are separated. The strands can then be displayed by gel electrophoresis. Each reaction mixture is run in a different lane on the gel. Because it is known which ddNTP was used in each reaction mixture, it is accordingly known which base is last in the strands of each set. By comparing the migration of terminated fragments from each of the mixtures, one can determine the exact sequence of bases in the original DNA sample, as shown by the colors in the figure. We learn from the gel loaded with the mixture containing ddATP, for example, that there are two Ts at the 5' end of the DNA to be sequenced and one T about in the middle.

More recently, investigators have made use of a simpler technique, in which each of the four ddNTPs has an attached fluorescent molecule, different for each ddNTP. All four ddNTPs are added to the same reaction mixture, containing the four dNTPs, the DNA to be sequenced, primer, and DNA polymerase. No radioactive tracer is needed. Each DNA fragment has its own unique length and a color determined by the ddNTP that terminates it. The DNA sequence can be determined by eye or—better—automatically, by machine.

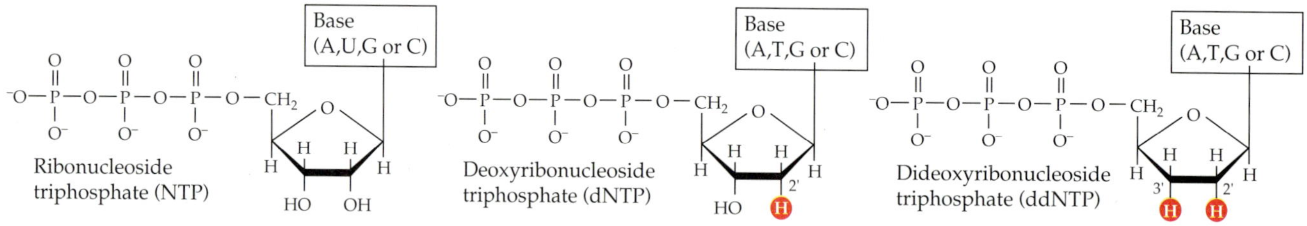

Ribonucleoside triphosphate (NTP)

Deoxyribonucleoside triphosphate (dNTP)

Dideoxyribonucleoside triphosphate (ddNTP)

to the identification of semen samples in criminological investigations and the study of ancient DNA in frozen samples from the last Ice Age.

PROSPECTS

The techniques of recombinant DNA technology have become standard tools for investigators of the molecular biology of both prokaryotes and eukar-

yotes and in both the pharmaceutical industry and the clinical laboratory. They are being used to turn selected strains of bacterial, yeast, plant, and animal cells into factories for the production of important polypeptides. Thus we can supply patients with gene products they cannot make for themselves; surely this will be a major tool of medicine well into the twenty-first century.

In the early days of recombinant DNA research, a number of concerns about safety were expressed, and

1. Single-stranded DNA sequence to be determined. Add samples of this unknown DNA to 4 different reaction mixtures.

2. Add primer, DNA polymerase, 4 dNTPs, and *one* ddNTP—ddATP (Ⓐ) in the case illustrated here. DNA synthesis begins. If dATP is picked up from the mixture, it continues. If ddATP is picked up, synthesis stops. A series of fragments of different lengths is made, each ending with a ddATP.

    The process is repeated with other ddNTPs to deduce the entire sequence.

3. The resulting strands of various lengths are separated by gel electrophoresis which can detect length differences as short as one base.

4. The sequence of the newly synthesized strand of DNA can now be deduced, and converted to the sequence of the template strand.

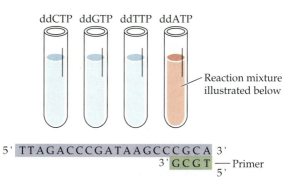

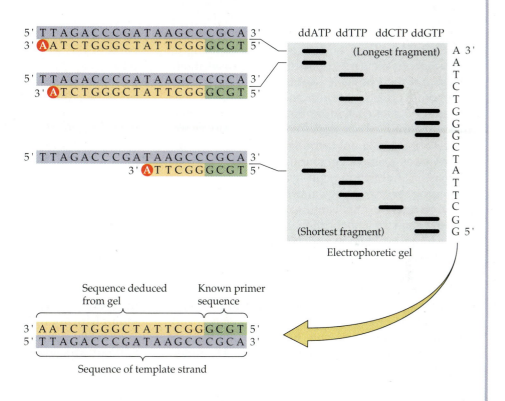

Electrophoretic gel

Sequence deduced from gel | Known primer sequence

Sequence of template strand

at one point there was a moratorium on the research. However, the worst of the fears have abated, and it is widely agreed that the work is safe as long as certain reasonable precautions are observed. These precautions, set forth as guidelines by the U.S. National Institutes of Health, include precisely defined containment provisions (methods for keeping experimental organisms from escaping from the laboratory) and rules for handling different biological materials. Most experiments are thought to pose no hazard. Some are considered risky and, accordingly, are conducted under more restrictive conditions. When the guidelines were first issued, they specifically restricted recombinant DNA work with the genes of cancer viruses. The guidelines have been gradually relaxed over the years since they were first set forth, as various concerns have been put to rest. Relaxation of the guidelines has led to major advances in our understanding of the cancer-producing oncogenes (Chapter 13).

Genome Projects

Our species is subject to hundreds of hereditary diseases, and a new one is discovered every two or three days. It has been estimated that each of us is heterozygous for no fewer than 30 recessive genetic diseases. Recombinant DNA technology affords new methods—often the first methods available—for combating these diseases. A number of **genome projects** are under way (the genome of an organism is all the genetic information the organism contains). These are major efforts to *map* and *sequence* some or all of the human genome and of the genomes of other species. Mapping precedes sequencing and in itself yields much valuable information. The development of effective gene therapies will depend on the mapping and characterization of the gene responsible for a disease, and this provides a major impetus for supporting human genome projects. Genes responsible for cystic fibrosis and other diseases have already been located and can now be studied—enabling us to learn about the gene products, the diseases themselves, and their possible treatment.

Genome projects require the gathering of vast amounts of data, including the locations of genes, their functions, their base sequences, and more. Processing—and simply storing—so much information requires massive computing resources, and the analysis of it all will depend heavily on advances in computing theory. Biology and computer science will be full partners in this venture. While computer science develops new algorithms and parallel processing techniques, biology and biotechnology will develop new physical and biochemical techniques for DNA manipulation and analysis.

The return on this investment will be magnificent, including a revolution in the diagnosis and treatment of hereditary diseases. Even if we sequence every base in the human genome, however, this will not tell us what all the DNA sequences *do*. An understanding of genetic regulation in eukaryotes requires knowledge of much more than base sequences. However, without the mapping and sequencing data from genome projects, these greater problems will remain unsolved.

Plant Agricultural Biotechnology

Most humans depend on a few crop plants for essential calories—rice, wheat, maize, sorghum, and several others, such as potatoes, that dominate locally. Because only a few genetic strains of each of these crops are planted in a given year, the food supply of hundreds of millions of people is in constant peril from mutations in disease-causing organisms that could overcome plant resistance. Equally important threats are changes in weather patterns, or political and economic disruptions that affect supplies of essential agricultural chemicals, fertilizers, or fossil fuels.

Other than the overriding need to contain population growth, the greatest potential for improving human welfare with modern biological technology lies in the search for economically feasible crop plants that are higher yielding; more nutritious; disease resistant; drought, salt, and pollution tolerant; and otherwise able to meet the challenges of our overextended planetary resources.

Plant breeding is a form of "genetic engineering" that has been used for a long time, and that has been practiced intensively in the twentieth century. By the judicious crossing of existing strains of plants, crop yields have been increasing about 1 percent per year in this century. Among the greatest triumphs of the plant breeders were the hybridization of corn (in the 1930s) and the "Green Revolution" (in the 1950s and 1960s, during which improved strains of wheat and rice were used to increase food production in many parts of the world). Plant breeding will continue to play a key role in the development of agriculture. However, it has three significant limitations. First, plant breeding is a slow process, and its projects require many acres of land. Second, plant breeding is nonspecific in that the entire genomes of the parents participate in a cross. Thus, unwanted genes may appear in the progeny along with the desirable ones for which the breeder is selecting. Third, the breeder can work only with strains (and, more rarely, species) that can interbreed with one another.

Addition of recombinant DNA technology to the tools of the breeder addresses each of those three limitations. First, recombinant DNA techniques are often applied to many millions of independent cells at a time, all within a single flask, thus eliminating the space problem; and the cells, unlike whole plants, multiply many times a day. Second, individual genes can readily be transferred from one plant to another without dragging along other, undesired genes. Third, there is no requirement that genes be transferred between closely related plants; in fact, a plant may be given genes from any living thing.

For both conventional plant breeding and recombinant DNA work, there will be a continuing and growing need for a source of suitable genes, sometimes referred to as **germ plasm**, for introduction into existing species. An important goal of conservationists and biologists is to maintain an adequate global supply of germ plasm, both as seeds in repositories and as plants in nature, to ensure genetic diversity in nature and a continuing supply of tools for the breeder and the biologist. Each lost species is a lost treasury of germ plasm. Recombinant DNA itself, perhaps as gene libraries, could serve as a gene repository for germ plasm.

The cloning vector commonly used in recombinant DNA work with plants is a plasmid found in *Agro-*

bacterium tumefaciens, a bacterium. *A. tumefaciens* is a pathogen, causing the disease crown gall, which is characterized by large tumors (Figure 14.18*a*). The bacterium contains a large plasmid, called **Ti** (for *Tumor inducing*). Part of the Ti plasmid is a transposon (Chapter 12) that produces copies of itself in the chromosomes of infected plant cells. That transposon is the key to gene cloning in plants. The gene to be cloned is inserted into the transposon in a Ti plasmid, the plasmid is inserted (by transformation) into *A. tumefaciens*, the bacteria are used to infect the plant, and the gene is copied into the plant's chromosomes along with the rest of the transposon (Figure 14.18*c*). As the recombinant DNA is found only in tumor cells, this might seem like a nonheritable change in the plant's genome. However, tumor cells can be isolated and grown in culture, eventually giving rise to a complete, new, normal plant (Figure 14.19). Each of this plant's cells is transgenic.

Transgenic bacteria will also play growing roles in agriculture. Let us consider but two examples. Nitrogen fixation, the conversion of atmospheric nitrogen gas to ammonium ions usable as a nitrogen source for plants, is crucial to agriculture—and to life on Earth (Chapter 30). Much research is being done on the modification of both the nitrogen-fixing bacteria and the plants that harbor some of them. As a second example, genetically altered bacteria have already been produced that, when present on plant surfaces, prevent frost formation (Figure 14.20).

(a)

(b)

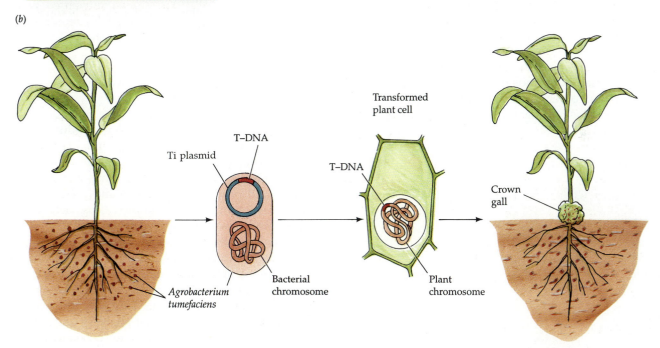

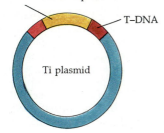

(c) Gene to be cloned is inserted into T-DNA of a modified Ti plasmid

14.18 Crown Gall and *Agrobacterium tumefaciens*
(a) *Agrobacterium tumefaciens* causes tumors—crown gall—on infected plants, such as this geranium. New shoots are forming within the gall, which lies at the base of the geranium stem. (b) *A. tumefaciens* contains the Ti plasmid which, in turn, contains a transposon called T-DNA. Copying of the transposon to a chromosome of an infected plant is a key step in the development of the disease. (c) A scientist can clone a desired gene by inserting the gene into the transposon of a modified Ti plasmid, transforming *A. tumefaciens* with the plasmid, and infecting plants with the bacteria.

14.19 Multiplying a Clone
This rosette of transgenic cotton plantlets is developing from a culture of crown gall tumor tissue. The plantlets contain a gene they picked up from the *Agrobacterium tumefaciens* that caused the tumor.

Recombinant DNA Technology and the Environment

With disturbing frequency, great quantities of oil have been accidentally released from tankers into surrounding waters, causing severe damage to sensitive environments (Figure 14.21). Genetically engineered bacteria, equipped with DNA that codes for enzymes that cleave hydrocarbons and other constituents of oil, are being developed to combat such spills. Several biotechnology companies are producing these and other bacterial strains modified to deal with other types of organic pollutants, such as sewage and dioxins.

14.20 Fighting Frost with Bacteria
Strawberry blossoms and the bacterium *Pseudomonas syringae* in water at −2.8°C. The tube on the left contains normal *P. syringae* and is frozen solid. The *P. syringae* in the tube on the right, however, lack a protein on their surface that provides a "template" for water molecules to align as ice (this is due to deletion of the gene that codes for the protein). Such bacteria sprayed on crops may provide protection from freezing.

14.21 An Environmental Disaster
A tugboat places a floating boom in an effort to contain part of the oil slick spreading from the tanker Exxon *Valdez*. More than 10 million gallons of oil spilled in this 1989 accident.

At the other end of the petroleum industry, genetically engineered bacteria will find uses in the production of chemicals to enhance yields from oil drilling. Oil yield in the United States continues to decline, and products of recombinant DNA technology represent an approach to increasing yields from existing and newly discovered reservoirs.

Yeasts and other microorganisms can concentrate metals such as nickel, gold, and plutonium from dilute solutions. Thus they can be used both in combating pollution and in increasing the recovery of important metals from waste. Genetic manipulation of bacteria and yeast for metal recovery and for the separation of metals from their ores is being actively pursued.

Release of transgenic organisms into the environment in large numbers must be done with care. Generally, bacteria prepared for release into the soil, lakes, or oceans will be designed to do their work effectively but not to persist long in the environment. In spite of such care, debates on the safety and merits of this kind of work will continue—we must always consider environmental impact and health concerns as well as its technological and economic advantages.

SUMMARY

Recombinant DNA technology pervades much of modern biology. In later chapters you will encounter many examples of insights gained by such techniques. The new "biotechnology" is a product of basic science, which discovered the enzymes, the plasmids and viruses, and the methods of separating and identifying specific DNA, RNA, and proteins. At first, practical applications seemed to be only a by-product of the search for knowledge. Now the applications have become important in their own right and are producing by-products that contribute in turn to basic science.

In recombinant DNA technology, foreign DNA is inserted into a suitable vector such as a virus or plasmid with the assistance of restriction endonu-cleases and other enzymes. The modified vectors are inserted into suitable hosts for cloning. Bacteria are widely used as hosts; however, eukaryotic hosts, particularly yeasts, are increasingly used for transfection with eukaryotic genes. DNA for cloning is obtained from gene libraries, cDNA production, or direct chemical synthesis.

DNA fragments are commonly separated by gel electrophoresis and then identified by Southern blotting. DNA fragments and chromosomes may be analyzed by restriction mapping, chromosome walking, and DNA sequencing.

These and other recombinant DNA techniques are becoming ever more useful in basic research, the commercial production of previously rare polypeptides, medicine, agriculture, the petroleum industry, and other applications.

SELF-QUIZ

1. Restriction endonucleases:
   a. play no role in bacteria.
   b. cleave single-stranded DNA molecules.
   c. cleave DNA at highly specific recognition sites.
   d. are inserted into bacteria by bacteriophages.
   e. add methyl groups to specific DNA base sequences.

2. "Sticky ends":
   a. are double-stranded ends of DNA fragments.
   b. are complementary to other specific sticky ends.
   c. rejoin best at elevated temperatures.
   d. are removed by restriction endonucleases.
   e. are identical for all restriction endonucleases.

3. Which statement is *not* true of DNA ligase?
   a. It is an enzyme.
   b. It is a normal constituent of cells.
   c. It can unite sticky ends in recombinant DNA work.
   d. It functions in the normal replication of DNA.
   e. It mends breaks in polypeptide chains.

4. Which feature is undesirable in a plasmid for cloning a gene?
   a. Possession of an origin of replication.
   b. Possession of genes conferring resistance to antibiotics.
   c. Possession of recognition sites for multiple restriction endonucleases.
   d. Possession of multiple recognition sites for the endonuclease to be used.
   e. Possession of genes other than the one to be cloned.

5. Transfection can be accomplished by:
   a. using high-velocity particles of tungsten coated with DNA.
   b. Southern blotting.
   c. gel electrophoresis.
   d. the polymerase chain reaction.
   e. heating the material to denature it.

6. Complementary DNA:
   a. is produced from ribonucleoside triphosphates.
   b. is produced using oligo-dU.
   c. is produced by reverse transcription.
   d. requires no primer.
   e. requires no template.

7. Southern blotting:
   a. is used to detect a specific DNA fragment.
   b. is used to detect a specific RNA fragment.
   c. is used to detect a specific polypeptide fragment.
   d. is a technique for separating nucleic acid fragments.
   e. is used to separate chromosomes.

8. Restriction mapping:
   a. is a useful tool for separating DNA fragments.
   b. is a useful tool for subdividing a DNA molecule into manageable fragments.
   c. cannot be used on prokaryotic DNA.
   d. can be used to produce large quantities of specific DNA.
   e. is an expensive and controversial procedure.

9. The polymerase chain reaction:
   a. is a method for sequencing DNA.
   b. is used to detect a specific DNA fragment.
   c. is used to produce large quantities of specific DNA.
   d. is used to map genes.
   e. is used to transcribe specific genes.

10. Genome projects:
    a. are a matter for the distant future.
    b. are all focused on the human genome.
    c. will tell us what all our genes do.
    d. have already yielded medically useful results.
    e. will probably all be carried out in one carefully chosen university.

FOR STUDY

1. Using examples from this chapter, describe how molecular biologists have found new uses in recombinant DNA technology for enzymes produced by bacteria.

2. Make a thorough list of the phenomena and techniques discussed in this chapter that depend on complementary base pairing.

3. You have attempted to insert a copy of a particular gene into a plasmid, specifically placing your gene in the middle of a gene conferring resistance to the antibiotic streptomycin. The plasmid also has a gene conferring resistance to the antibiotic aureomycin. You have transformed bacteria (sensitive to both antibiotics) with your plasmid suspension. Describe the procedures you would use to select those bacteria that have taken up the plasmid. What additional steps would be required to select those bacteria that have taken up copies of the plasmid that contained your gene?

4. Discuss (a) what you see as important positive features of a human genome project and (b) what you consider to be to be negative features of such a project.

READINGS

Lawn, R. M. and G. A. Vehar. 1986. "The Molecular Genetics of Hemophilia." *Scientific American*, March. Use of recombinant DNA technology and bacteria to produce a blood-clotting protein that may save the lives of hemophiliacs.

Lerner, R. A. and A. Tramontano. 1988. "Catalytic Antibodies." *Scientific American*, March. A powerful new tool, combining the talents of enzymes and antibodies. You might save this one for when you have read Chapter 16.

Murray, A. W. and J. W. Szostak. 1987. "Artificial Chromosomes." *Scientific American*, November. Tools for cloning human genes in yeast and for the investigation of chromosomal behavior during mitosis and meiosis.

Neufeld, P. J. and N. Colman. 1990. "When Science Takes the Witness Stand." *Scientific American*, May. Ethical issues in the use of DNA evidence.

Patterson, D. 1987. "The Causes of Down Syndrome." *Scientific American*, August. Identification and mapping of genes responsible for the commonest cause of mental retardation.

Suzuki, D. T., A. J. F. Griffiths, J. H. Miller and R. C. Lewontin. 1989. *An Introduction to Genetic Analysis*, 4th Edition. W. H. Freeman, New York.

Verma, I. M. 1990. "Gene Therapy." *Scientific American*, November. Introducing healthy alleles to correct heritable disorders.

Watson, J. D., J. Tooze and D. T. Kurtz. 1983. *Recombinant DNA: A Short Course*. W. H. Freeman, New York. Begins at the elementary level but ends up presenting a great deal of molecular biology. A short, intense book, but quite readable.

Weinberg,, R. A. 1985. "The Molecules of Life. " *Scientific American*, October. Introductory chapter to a special issue on the molecules of life; gives a good overview of the role of recombinant DNA technology in various aspects of molecular biology.

Weintraub, H. M. 1990. "Antisense RNA and DNA." *Scientific American*, January. Deactivation of specific genes—a powerful research tool, perhaps some day a medical tool as well.

White, R. and J.-M. Lalouel. 1988. "Chromosome Mapping with DNA Markers." *Scientific American*, February. Describes the powerful tool known as restriction-fragment length polymorphism.

15

Animal Development

PREVIEW: An egg fertilized by a sperm forms a zygote. The zygote then undergoes a period of rapid cell division, during which daughter cells are produced with little or no accompanying cell growth. These divisions are accompanied by cell differentiation and morphogenetic changes in masses of cells. Cells differentiate because different genes are expressed at different times in different cells; thus the cells become specialized for different functions. Morphogenesis results in specialized tissues and organs and depends upon the actions of particular genes in particular groups of cells. Although overall growth may eventually cease, the animal continues to develop until it dies.

This chapter deals with cleavage, blastulation, gastrulation, embryonic germ-layer formation, cellular movements, growth, metamorphosis, prospective potency, prospective fates, differentiation, totipotency, polarity, cytoplasmic segregation, induction, imaginal disks, homeotic mutations, the homeobox, and positional information.

The zygote, a single cell resulting from the union of sperm and egg, gives rise ultimately to all the cells of the adult body—more than a hundred trillion (10^{14}) cells of diverse sorts in each individual of our species. Many intricate steps make up the **development** of the body.

Development is a process of progressive change that continues not just throughout the growth of the embryo but rather until the death of the animal. During development an organism successively takes on the forms of the several stages of its life cycle. An **embryo** is an animal (or plant) in an early state of development. Sometimes the embryo is contained within a protective structure such as an eggshell or a uterus. An embryo does not actively feed because it obtains its food directly or indirectly (by way of the egg, for example) from its mother. A **larva** is an immature form of an animal, feeding independently and differing in appearance from the adult. Examples of larvae include the tadpole of a frog and the caterpillar stage of a butterfly.

Developmental changes take place at all levels of organization—from molecule to organism. Much of the business of development is transacted at the level of nucleic acid and protein synthesis; but we see many of the consequences at the level of the cell, particularly in the phenomenon of **differentiation**, by which cells come to differ in their structure and physiological function. Differentiation requires selective gene expression; what we know about differentiation has derived in large part from recombinant DNA technology, described in Chapter 14. At a higher level of organization, **morphogenesis** establishes the particular shape or arrangement of tissues by the movement of masses of cells within the developing organism; the death of specific groups of cells (which, for example, accounts for a tadpole's losing its tail during metamorphosis into an adult frog); the unequal growth of different parts of the body; and **pattern formation**—the organization of differentiated tissues into specific structures. Another aspect of morphogenesis is **growth**, an increase in volume or mass resulting from cell multiplication and cell expansion. In this chapter we will begin with descriptions of normal morphogenesis, proceed with a discussion of experimental studies of differentiation, and conclude with some historical and modern studies of pattern formation. Throughout this chapter, reference will be made to various types of animals often studied by developmental biologists. Some of these animals are shown in Figure 15.1.

CLEAVAGE

Becoming Multicellular

When an egg is fertilized, the resulting zygote nucleus is activated, and DNA replication and mitosis commence. The activation of the nucleus begins the process of **cleavage**, in which the zygote divides. Cleavage gives rise, over a period of hours, to hundreds of cells, and eventually to thousands. The

15.1 The Cast of Characters
Some of the animals scientists use extensively in research on developmental biology are sea urchins, frogs, and chicks. Here we see each of these as an adult or young animal, along with its embryo or its larva.

divisions of all the cells in most embryos are initially synchronous at each round; first there is a single cell, then there are 2, 4, 8, 16, and so forth.

The patterns of cleavage, including the arrangement of the daughter cells, or **blastomeres**, depend on a number of factors but especially on the distribution of yolk. The sea urchin, for example, has a small egg (150 μm in diameter) with its yolk uniformly distributed. In such an egg, the blastomeres separate completely from one another as they are formed; division proceeds nearly simultaneously from both ends, or poles, of the cell (Figure 15.2*a*).

A frog egg is rather larger (0.5–1 mm), with the yolk concentrated at one pole. After fertilization, there is a substantial redistribution of the cytoplasmic contents of the zygote, including the movement of some pigmented material. As a result, a **gray crescent** forms on one side of the zygote. As we will see, this region is of great significance in later development. Cell division in a frog zygote is complete, but it begins at the pole away from the yolk. This pole is called the **animal pole**; the pole with the yolk is called the **vegetal pole**. The plane of the first cleavage passes through the animal pole, through the middle of the gray crescent on one side of the zygote, and through the site of sperm entry on the other (Figure 15.2*b*).

In a large, extremely yolky egg, such as that of a bird, the situation is markedly different. Yolk occupies the bulk of the cell, with the yolk-poor cytoplasm being confined to the surface or to one end of the cell. Cell division in such zygotes is incomplete—following the mitoses, the nuclei are separated completely, but daughter cells are not completely separated by cell membranes. The embryo develops initially as a disk-shaped mass on top of the yolk (Figure 15.2*c*).

In some animals, the eventual fates of the blastomeres may be fixed as early as the first cell division. In other animals, the parts of the adult body that will be derived from each blastomere are fixed after the second or third division. If one or more blastomeres of such embryos are lost or moved, a corresponding portion of the adult body is not produced. By an analogy, this developmental pattern is called **mosaic development**, with each blastomere contributing a specific set of "tiles" to the final "mosaic" of the larva. **Regulative development**, in contrast, characterizes sea urchins and vertebrates (Figure 15.1; and see Chapter 26). In the embryos of these animals, the loss of some cells during cleavage does not affect development—the remaining cells compensate for the loss. In humans, for example, the separation of the two cells after the zygote divides, and the subsequent development of each, produces identical twins. However, there is a point even in regulative development beyond which a loss of cells does affect the outcome.

Formation of the Blastula

In spite of the differences just discussed, a number of features of cleavage are common to most animals. First, mitosis is always rapid during this stage. In some cases, the blastomeres divide even more rapidly than do bacteria. Second, in most species there is little or no change of overall volume during cleavage. This means, of course, that the blastomeres get smaller and smaller with each division, inasmuch as the ball of cells remains approximately the same size as the fertilized egg. Also, the ratio of nuclear volume to cytoplasmic volume in the embryo increases steadily throughout cleavage, because the original supply of cytoplasm (from the egg) is being shared by ever more nuclei. Finally, during cleavage there is little or no synthesis of new products in the cell—except for DNA and chromosomal proteins.

Throughout cleavage, the embryo retains approximately the same overall external spherical form. In the animals we have been discussing—sea urchins, frogs, and birds—the stage of development called cleavage ends with the formation of a hollow structure, the **blastula** (Figure 15.3). Its cavity, the **blastocoel**, forms because new blastomeres tend to move to the periphery of the other cells, leaving a fluid-filled space in the center. The blastomeres themselves constitute a sheet of cells, the **blastoderm**. In sea urchins the blastoderm is only one cell thick, whereas in other animals, such as frogs, it may be several cells thick. At the blastula stage, the embryo consists of a cavity, the blastocoel, surrounded by a sheet of cells, the blastoderm.

The exact shape of the blastula varies, depending in large part on the size and yolk content of the egg. A frog blastula is a hollow ball, whereas a bird blas-

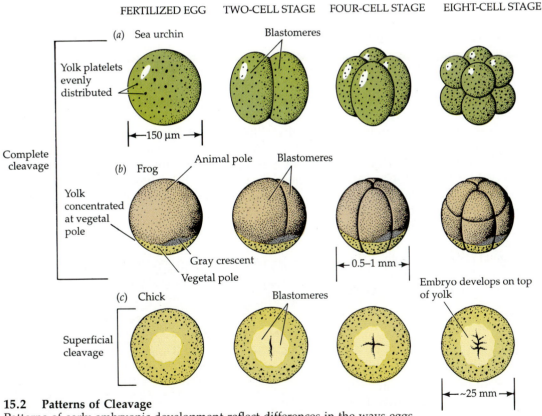

FERTILIZED EGG   TWO-CELL STAGE   FOUR-CELL STAGE   EIGHT-CELL STAGE

(a) Sea urchin

Blastomeres

Yolk platelets evenly distributed

←150 μm→

Complete cleavage

(b) Frog

Animal pole   Blastomeres

Yolk concentrated at vegetal pole

Gray crescent

Vegetal pole

←0.5–1 mm→

Embryo develops on top of yolk

(c) Chick

Blastomeres

Superficial cleavage

←~25 mm→

15.2 Patterns of Cleavage

Patterns of early embryonic development reflect differences in the ways eggs are organized. (a) A sea urchin egg is small, with little yolk (colored dots). The first two cell divisions are complete and the planes of cleavage are both longitudinal; the third division is also complete, but its plane is transverse to those of the prior cleavages. In an eight-cell sea urchin embryo, the cells are approximately the same size and arranged in two layers. (b) A frog egg is larger, with yolk concentrated at the vegetal pole. All cell divisions are complete: the first two begin near the animal pole; the third (far right) is transverse, in a plane near the animal pole. As a result, the cells of the eight-cell stage differ in size, in the amount of yolk they contain, and in their proximity to the gray crescent. (c) A chicken egg is mainly yolk with only a small mass of cytoplasm. The first few divisions of the cytoplasm (seen here from above) do not extend through the yolk; they yield a thin, flat embryo on the surface.

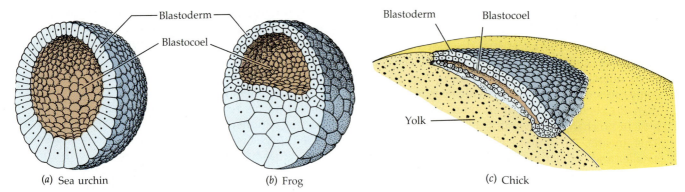

(a) Sea urchin   (b) Frog   (c) Chick

15.3 Blastulas Reflect Patterns of Cleavage
From the beginnings shown in Figure 15.2, continued cell divisions produce hollow blastulas, cut open here to show the blastocoel. (a) In sea urchins and organisms with similar eggs, the blastoderm is a single sheet of cells and the blastocoel they enclose is spherical. (b) In frogs, the blastoderm is many cells thick near the vegetal pole and the blastocoel has a flattened floor. (c) The blastocoel of birds is a lens-shaped cavity defined by a thin blastoderm layer below and a slightly thicker one above.

tula is a disk-shaped structure at the animal pole of the egg.

Thus during a few hours of cleavage, zygotes from small eggs are transformed into hollow blastulas consisting of a large number (from thousands to tens of thousands) of cells. Zygotes from large, yolk-filled eggs develop into disk-shaped sheets with underlying blastocoelic spaces. The developing blastula is a dynamic structure; each division of blastomeres results in changed contacts among the cells. Cleavage itself is thought to be controlled by surface interactions among the blastomeres. Within the blastula, the contents of the embryo are distributed much as they were in the original zygote—the yolk is still more concentrated near the vegetal pole, for example. It is only in subsequent stages of development that massive rearrangements of cells and materials begin within the embryo.

GASTRULATION

A major embryonic redistribution of cells begins with the development of the blastula into a **gastrula**. During gastrula formation—the process of **gastrulation**—some cells move from the surface of the embryo into the interior, resulting in a *two-* or *three*-layered embryo. In sea urchins, the gastrula looks like a punched-in, soft tennis ball; part of the former blastoderm pushes inward, or **invaginates**, and forms an inner germ layer, the **endoderm**, pressed against an outer germ layer, the **ectoderm**. In a few animals, a two-layered embryo consisting of ectoderm and endoderm develops into a two-layered adult with little cellular diversity. Adults of most species, however,

develop from a three-layered embryo: The third germ layer, the **mesoderm**, forms during gastrulation. In all animals, cells from surface layers move to the interior during gastrulation.

Despite the variation in detail, a number of general features are common to the gastrulation of all animals. The rate of mitosis is much slower than during cleavage, and the total volume of the embryo changes little, if at all. There are massive movements of cells, giving rise to adjacent internal and external tissues —ectoderm, endoderm, and mesoderm (Table 15.1). Cells that come to lie together as a result of these movements interact with each other, and new, differentiated cell types result. As development proceeds, different genes are activated in different cells, so that different gene products are formed.

Gastrulation in sea urchins (Figure 15.4) is somewhat complex, in that the invagination of the blastoderm is preceded by a type of movement called **ingression**. In this process, small groups of cells sep-

TABLE 15.1	
Fates of Embryonic Germ Layers in Vertebrates[a]	
GERM LAYER	**FATE**
Ectoderm	Brain and nervous system; lens of eye; inner ear; lining of mouth and of nasal canal; epidermis of skin; hair and nails; sweat glands, oil glands, milk secretory glands
Mesoderm	Skeletal system: bones, cartilage, notochord; gonads; muscle; outer coverings of internal organs; dermis of skin; circulatory system: heart, blood vessels, blood cells; kidneys
Endoderm	Inner linings of: gut, respiratory tract (including lungs), liver, pancreas, thyroid, and urinary bladder

[a]The final structures are complex, containing cells from more than one germ layer. Interactions among tissues are usually important in determining the composition and structure of an organ.

arate from the blastoderm at the vegetal pole and migrate into the blastocoel. Cells at the vegetal pole of a sea urchin blastula become columnar, flattening this end of the embryo slightly. Some of these cells begin to bulge into the blastocoel, eventually breaking free and wandering into the cavity. These wanderers are called **primary mesenchyme** cells; they become part of the mesoderm. Next, the vegetal pole invaginates; the columnar cells become wedge-shaped and buckle inward to produce the endoderm. The ectoderm remains on the outside. (It is interesting that pieces of the vegetal end of a sea urchin blastula, studied in isolation, will spontaneously form small invaginations. Isolated pieces from the animal end will not. This shows that the ability to invaginate is a predetermined characteristic of vegetal pole cells.)

Invagination during gastrulation forms a new cavity, the **archenteron**; this cavity opens to the exterior through the **blastopore**. During invagination in sea urchin embryos, more cells move into the blastocoel from the tip of the invaginating archenteron in a second round of ingression. These **secondary mesenchyme** cells form fine extensions, called filopodia, that extend through the blastocoel and attach to the future oral cavity (mouth) on the animal half as well as to the archenteron. Changes in the shape of the endoderm cells cause further invagination of the archenteron; contraction of the filopodia pulls the archenteron yet further, until the tip of the archenteron reaches the area of the future mouth. Ultimately, the secondary mesenchyme cells become part of the mesoderm. The migration of the primary and secondary mesenchyme cells is guided in part by fibers of a protein, fibronectin, laid down by ectodermal cells at the roof of the blastocoel. The filopodia of the mesenchyme cells become enmeshed in the tangled fibronectin and thus can pull the cells along by contracting.

The archenteron becomes the digestive cavity of the embryo, and its opening to the outside—the blastopore—becomes the anus of the sea urchin. The mouth develops later, from a perforation at the animal pole. Animals such as sea urchins and vertebrates, in which the mouth develops at a distance from the blastopore, are called deuterostomes; other animals, such as earthworms and insects, in which the mouth develops from the blastopore and the anus is formed by a later perforation, are called protostomes. The deuterostome–protostome distinction is one of the bases we use in classifying animals (Chapters 25 and 26).

In frogs and chickens, substantial quantities of yolk complicate the formation of the three germ layers. Cells at one side of a frog embryo change shape and initiate the beginning of invagination. Movement of cells starts just below the center of the gray crescent near the center of the embryo. This first site of

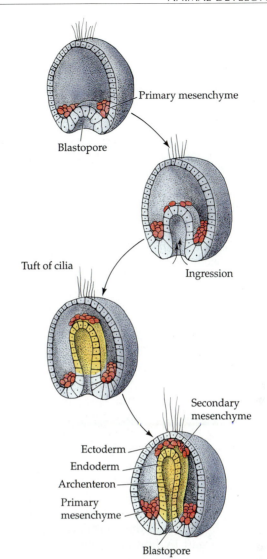

15.4 Gastrulation in the Sea Urchin
During gastrulation, cells in the blastula move to new positions and form the three germ layers of the gastrula (indicated throughout this chapter in pink, yellow, and blue). From these germ layers will develop all adult tissues. First, cells at the blastula's vegetal pole (at the bottom of the blastula in these drawings) move into the blastocoel, forming an opening called the blastopore. As the opening deepens and cells migrate through the blastopore and toward the animal pole (at the top), a new cavity—the archenteron—develops. Cells that line the archenteron become the endoderm layer (yellow), while cells that remain outside the blastopore become the ectoderm layer (blue). At the beginning of gastrulation, groups of primary mesenchyme cells (pink) form near the developing blastopore. Later, groups of secondary mesenchyme cells extend away from the blind tip of the archenteron and connect with the animal pole. Migrating groups of mesenchyme cells eventually form a complete, third germ layer, the mesoderm.

invagination marks the **dorsal lip** of the blastopore. Then two things happen. The cells at the animal end of the embryo begin to increase their surface area and to expand, a process that continues throughout gastrulation. As this expansion at the animal pole pro-

15.5 Gastrulation in the Frog

The developmental sequence begun in Figures 15.2*b* and 15.3*b* continues with gastrulation, which begins *(a)* when cells just below the center of the gray crescent invaginate to form the dorsal lip of the future blastopore. *(b)* Cells at the animal pole (top in these drawings) spread out, pushing surface cells below them toward and across the dorsal lip of the blastopore. Those surface cells flow into the interior of the embryo where they form the endoderm and mesoderm. *(c)* This involution creates the archenteron and obliterates the blastocoel. The dorsal lip of the blastopore forms a circle, with cells on both its dorsal and ventral surfaces; the yolk plug is visible through the blastopore. Continued development of the gastrula *(d* and *e)* gives rise to a notochord derived from mesoderm and to the beginnings of the nervous system (green) derived largely from ectoderm. Figure 15.7 shows the subsequent development of this embryo.

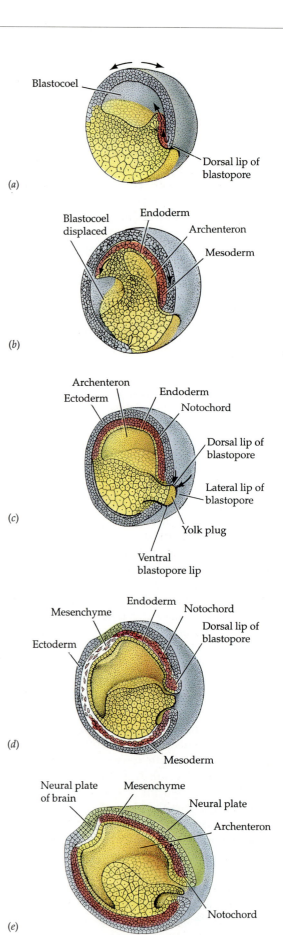

ceeds, cells from the animal half of the embryo surface turn inward at the dorsal lip of the blastopore and "flow" into the interior of the embryo (Figure 15.5). This inward turning of an expanding sheet of cells is called **involution**. The archenteron forms as involution proceeds; and the dorsal lip of the blastopore spreads, eventually forming a complete circle with cells involuting at both its dorsal (upper) and ventral (lower) lips. Within the circular blastopore, a "yolk plug" can be seen. The cells involuting over the lip of the blastopore give rise to both endodermal and mesodermal layers. The formation of the mesoderm is complex, involving both multiplication and migration of cells. Migration of the mesodermal precursors is guided in part by fibronectin fibers, as in sea urchin embryos.

A portion of the mesoderm at the roof of the archenteron differentiates to form the **notochord**, a stiff, supportive rod. Later in the frog's embryonic development, the supportive function of the notochord is taken over by the vertebrae. Box 15.A tells how a biologist figured out how frog gastrulation proceeds.

In bird eggs, with their massive yolk content, gastrulation must proceed by different means. Starting with the disk-shaped blastula that forms from such eggs (Figure 15.2*c*), massive cell movements begin. Surface cells lying on either side of the embryo migrate toward the center line and then forward, forming a depression called the **primitive streak** that is analogous to the blastopore (Figure 15.6). As migrating cells reach the primitive streak, they involute and ingress through it, forming internal sheets that will become the mesoderm and endoderm. In this manner, the gut-forming cells of the endoderm are brought inside the embryo, the skin- and nerve-forming cells of the ectoderm remain external, and the mesoderm, which forms most of the organs, is brought between them. Interactions among the germ layers determine the formation of specific tissues (Table 15.1).

BOX 15.A

Tracking Migrating Cells: Fate Maps

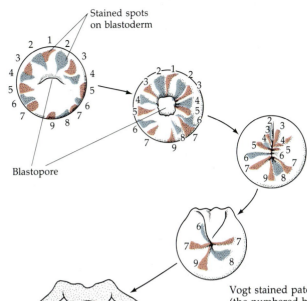

Stained spots on blastoderm

Blastopore

Section through embryo

How was it possible to work out the complex details of gastrulation, which we have covered only superficially? In a classic paper published in 1929, the German embryologist W. Vogt reported the results of experiments in which he made marks on frog blastulas with various dyes. These nontoxic dyes stained the cells where the marks were made and remained within those cells and their descendants; very little dye diffused into neighboring cells. By using dyes of different colors, Vogt could simultaneously stain different parts of the blastoderm and see where they ended up (see figure). During the early stages of gastrulation, the stained regions could be seen to move toward the region where the dorsal lip of the blastopore would form. Then, with the appearance of the dorsal lip, the stained cells involuted into the interior of the embryo. When the embryo was sliced open

later, the stained regions could be relocated, as shown in the figure. By extensive work of this type, Vogt was able to work out fate maps for the blastula and gastrula of the frog. These showed in some detail just what parts of the later embryo, or even the adult, would develop from specific portions of the blastoderm.

More sophisticated methods for marking embryonic cells and their descendants are now in common use. One approach is to inject fluorescent substances into specific cells of the embryo; these substances can

Vogt stained patches of frog blastoderm (the numbered blue and red patches), enabling him to follow their movements and fates. In this embryo (with the animal pole at 1, vegetal pole at 9), cells from the blastula's upper surface are first to involute through the blastopore, leaving cells from the blastula's lower surface to form the gastrula's surface. By cutting gastrulas open, Vogt could follow the fates of cells that involuted early.

be traced later with a fluorescence microscope. In another approach, a specific enzyme is injected into cells. Some cell generations later, the developing embryo or larva can be treated with the substrate of the enzyme. The substrate is converted to its product only in those cells that contain the enzyme—those cells that were derived from the cells receiving the initial injection.

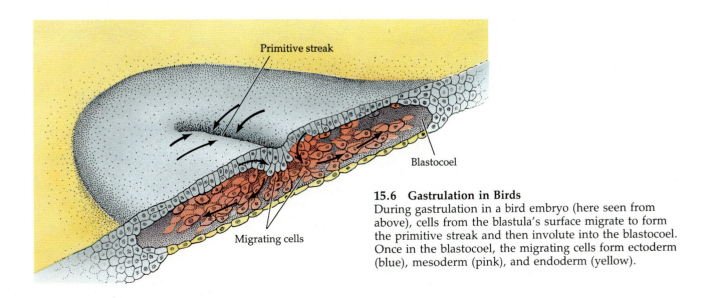

Primitive streak

Blastocoel

Migrating cells

15.6 Gastrulation in Birds

During gastrulation in a bird embryo (here seen from above), cells from the blastula's surface migrate to form the primitive streak and then involute into the blastocoel. Once in the blastocoel, the migrating cells form ectoderm (blue), mesoderm (pink), and endoderm (yellow).

ORGAN FORMATION: THE NERVOUS SYSTEM

As an animal develops from gastrula to adult, many specialized organs and organ systems are formed; developmental patterns differ from one organ to another. One organ system is the nervous system, which processes information obtained from the environment and from other parts of the body. The nervous system of a frog provides a good example of the development of an organ system from the gastrula. The early stages of the formation of the nervous system demonstrate a very common type of morphogenesis: The formation of an internal tubular system from an ectodermal embryonic layer that initially was an external sheet of cells.

As we have seen, the late gastrula of a frog already has three primary tissues—the embryonic germ layers: ectoderm, endoderm, and mesoderm. At this stage, the embryo has anterior (front) and posterior

(rear) ends, the blastopore being at the extreme rear. Now a second axis of symmetry appears, as the embryo develops distinct left and right sides during the process of **neurulation**. On the dorsal side of the embryo (as seen in cross section in Figure 15.7), the ectoderm begins to thicken, forming a flattened **neural plate**. Along the margins of this plate, neural folds appear and gradually thicken. In the center of the plate, a neural groove forms, and deepens as the folds begin to roll toward each other. As this is going on, the embryo as a whole elongates along its anterior-to-posterior axis as a result of the lengthening of the notochord. The neural folds continue to roll toward one another, ultimately touching to form a narrow, hollow cylinder; and this cylinder, the **neural tube**, becomes detached from the overlying ectoderm of the embryonic surface. Thus cells that once were part of a surface sheet are now incorporated into an internal tube. As hinted at in the phases shown in

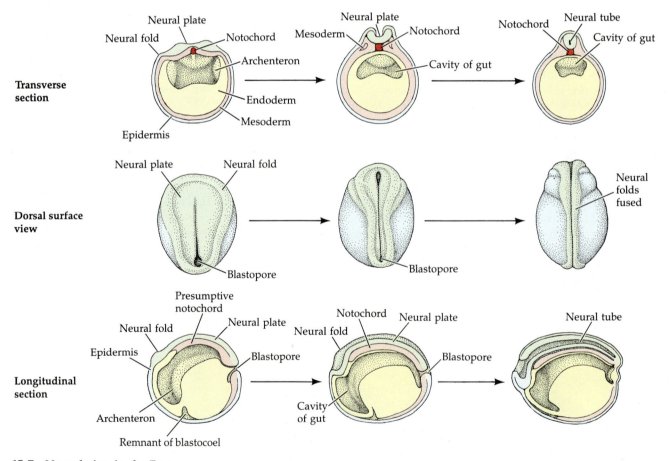

15.7 Neurulation in the Frog
Continuing from Figure 15.5, these drawings outline the development of the frog's neural tube. (*a*) A slice through a late gastrula. The three germ layers are well defined, as is the neural plate, which forms from ectoderm above the notochord. (*b*) As the edges of the neural plate move upward and grow toward one another, the center of the plate sinks, forming the neural groove. (*c*) When the edges of the neural plate grow together and fuse, a hollow cylinder forms and detaches from the ectoderm to become the neural tube. This kind of process, in which a sheet of ectoderm from the surface becomes a tube inside the embryo, is repeated in many developmental contexts.

Figure 15.7, the cellular movements that form the neural tube are complex. First, microtubules in the ectodermal cells reorient, causing the cells to change shape from cuboidal to columnar and to form the neural plate. The action of actin microfilaments on the outer edge of the neural plate causes the cells to move and form the neural tube.

After the neural tube forms, its anterior end, which will develop into the brain, becomes divided into several distinct compartments. These subdivisions develop differently, with the one farthest anterior ballooning out and eventually becoming the cerebrum (the chief coordination center of the nervous system in mammals; Chapter 36).

POSTEMBRYONIC DEVELOPMENT

Thus far we have considered only **embryology**, the study of the early stages of animal development. In most species the embryonic form must undergo many further changes to become an adult. There is usually a substantial amount of cellular division and cellular differentiation, and there are always morphogenetic movements. Growth (irreversible increase in size) is often extensive after feeding begins. In most animals, growth throughout the life of the individual follows what may be described as an S-shaped curve as illustrated in Figure 15.8a. An initial period of slow growth is typically followed by a long phase of rapid growth, with growth slowing markedly at some stage. Details vary considerably among animal groups, however. In many groups, growth continues until the organism dies—lobsters, for example, do this. In humans, overall growth ceases sometime after puberty; and we stay at a more or less constant size throughout most of our adult life (Figure 15.8c). Nonetheless, cellular division continues at a rapid pace throughout our lives, replacing cells such as those sloughed off by our skin (more than 1 gram a day) and intestinal lining as well as the millions of blood cells turned over each minute (Chapter 16).

In considering the lobster, we see another departure from the pattern of Figure 15.8a, because the growth of a lobster is *discontinuous*. Because of its rigid external skeleton, the individual must molt (shed its skeleton) in order to grow (Chapter 38). Accordingly, a lobster grows in spurts during molts (Figure 15.8b).

Larval Development and Metamorphosis

Many animals go through a larval stage in development. A tadpole must change dramatically to become a four-legged adult frog, a radical rearrangement of structures called **metamorphosis**. Most larvae differ strikingly from adults of the same species.

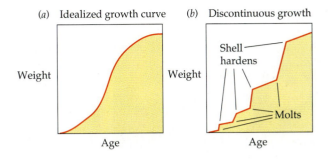

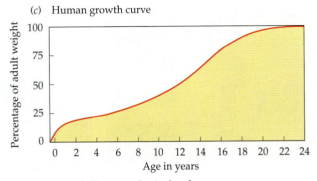

(c) Human growth curve

15.8 Growth Patterns in Animals

All types of animals increase in weight as they grow older; but the quickening and slowing of their growth rates differ, as these growth curves show. The steeper the curve, the faster the growth rate. (a) Idealized S-shaped growth curve characteristic of many species. There is some growth at every age, but the maximum growth rate is during the middle of the life cycle. (b) Discontinuous growth characteristic of hard-shelled animals that must molt to grow. A growth spurt follows each molt, before the new shell hardens. (c) A variant of the S-shaped curve shows the growth pattern for humans. A decade of gradual, preadolescent growth precedes a growth spurt during puberty; within a few years, the final adult weight is approached.

During metamorphosis, new adult structures must be formed and old larval ones lost. It takes many changes to go from tadpole to frog. The gut is shortened, corresponding to the transition from a vegetarian tadpole to a carnivorous adult. The brain is remodeled to allow binocular vision, corresponding to the transition from a tadpole that is a prey organism to a frog that is a predator. Limbs appear and the tail disappears, corresponding to the transition from a swimming tadpole to a jumping frog. The loss of old tissues and old parts takes place by what may be called programmed cell death; that is, certain cells are destined to be destroyed. Cell death plays a part not only in metamorphosis, but also in the embryonic development of most species, thus contributing in a general way to the development of form in animals.

The overall pattern of development in butterflies, moths, and many other insects may be familiar to you (Figure 15.9). From a fertilized egg there develops a creeping larva that feeds voraciously, growing through a series of molts (the larval stages between molts are called **instars**). The newly hatched larva

(a) (b)

(c) (d) (e)

15.9 Complete Metamorphosis in a Moth
The comet-tail moth from Madagascar provides a beautiful example of the mode of insect development called complete metamorphosis. (a) Eggs and first instar larvae (one in the process of hatching). (b) Third instar. (c) Fifth instar. (d) Pupa (removed from cocoon). The sweeping disposal of old tissues and the development of adult structures from imaginal discs in the pupa leads to the designation "complete." (e) Adult male moth, approximately 30 minutes after emergence from the pupa. Its long tails are not yet fully expanded.

has about 10,000 cells that make up the bulk of the body and contribute to its growth. Another 1,000 or so cells, in 19 clusters called **imaginal discs**, remain undifferentiated throughout larval growth. The final instar stops feeding and then surrounds itself with a cocoon and transforms into a **pupa**. In the pupa, tremendous changes take place (Figure 15.10). Some larval cells die, others are reprogrammed to make different products characteristic of the adult, and the imaginal discs differentiate into new adult structures. Such a major revision between larva and adult is referred to as **complete metamorphosis**. In sharp contrast is the **gradual metamorphosis** characteristic of other insects, including grasshoppers and cockroaches (Figure 15.11). In gradual metamorphosis, the instars between molts are known as nymphs (or, when aquatic, naiads) and resemble miniature adults in many physical features.

Imaginal Discs

The fates of the imaginal discs are determined long before metamorphosis. If an imaginal disc is transplanted from one larva to another, it still develops into the same type of organ (a wing, for example, or an antenna) that it would have if left undisturbed;

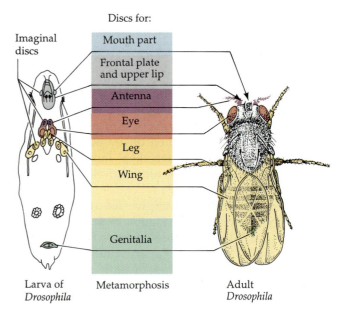

Discs for:

Imaginal discs

Mouth part

Frontal plate and upper lip

Antenna

Eye

Leg

Wing

Genitalia

Larva of *Drosophila*   Metamorphosis   Adult *Drosophila*

15.10 Complete Metamorphosis in Fruit Flies
In insects such as fruit flies and butterflies, the embryo hatches from its egg into a soft-bodied larva that feeds for some time before entering the pupal stage. In the pupa, nearly all the larval tissues die and are resorbed, providing building blocks for subsequent development. The remaining larval tissues are specialized imaginal discs, which proliferate and differentiate to form the organs of the adult insect.

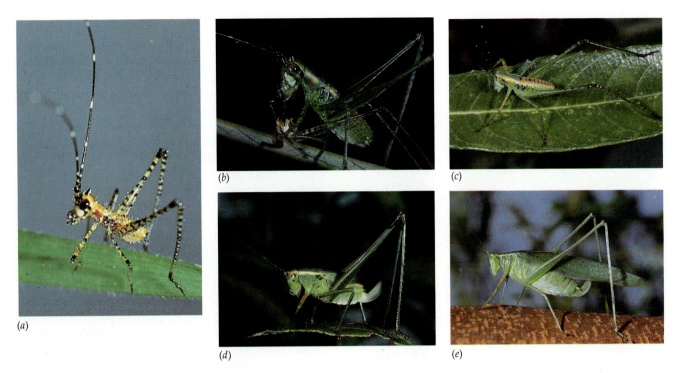

15.11 Gradual Metamorphosis in Katydids
In insects such as water bugs, locusts, and grasshoppers, the embryo hatches
into an immature nymph that resembles the adult of the species but has neither wings nor reproductive organs. During successive molts, the nymph
gradually changes into a mature adult through a series of instars, as shown
by this fork-tailed bush katydid. (*a*) Newly hatched nymph, the first instar.
(*b*) Third instar, eating its own newly shed cuticle. (*c*) Fifth instar; a developing wing bud is visible. (*d*) Sixth instar. (*e*) Adult.

and that organ is formed wherever the imaginal disc
is placed in the host body.

If an imaginal disc is transplanted into an adult
insect, it remains undifferentiated (because the hormonal signal for its development is lacking); but the
cells of the disc will continue to divide within the
new host. Later these transplanted disc cells may be
transplanted to other adult insects or to larvae. If
returned to a larva, the disc cells will *almost* always
develop into the adult organ for which they were
originally determined. However, an occasional imaginal disc will transdetermine in the course of a series
of transplants; that is, it will develop into an organ
other than that normally expected. Transdetermina-
tion shows that imaginal discs have not lost the genes
that they do not normally express, as a disc *may*
express them to produce a different organ.

PROSPECTIVE FATE
AND PROSPECTIVE POTENCY

Staining studies of the sort described in Box 15.A can
determine which adult structures develop from certain parts of the blastula and early gastrula. For in-

stance, the shaded area of the frog blastoderm shown
in Figure 15.12 is destined to become the skin of the
tadpole larva. Does this fate result from a special
property of those particular cells, or does it simply
result from their physical location? To investigate this
question, we can cut out a piece of tissue from that
region and transplant it to another place on an early
gastrula. If we move it to the place indicated in Figure
15.12*a*, it develops instead as neural tissue, ultimately becoming part of the brain of the host embryo.
If, instead, we transplant it somewhat closer to the
blastopore in a region destined to become mesoderm,
it indeed becomes mesoderm; it may then give rise,
for example, to muscle and notochord tissue, as
shown in Figure 15.12*b*. The **prospective potency** of
blastoderm cells, that is, their range of possible development, is great—greater than their **prospective
fate**, which in this example is limited to the formation
of skin.

Does developing embryonic tissue retain its great
prospective potency? Generally speaking, no—the
potency of cells becomes restricted fairly early. Tissue
taken from a late gastrula has a prospective potency
identical with its prospective fate. If taken from a
region fated to develop into brain, for example, late
gastrula tissue becomes brain tissue even if trans-

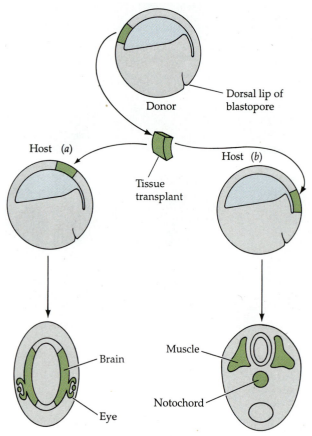

15.12 Developmental Potency in Early Gastrulas
Tissues in early gastrulas retain the capacity to develop in many alternate ways, as demonstrated in these transplantation experiments with frogs. Tissue destined to become part of a tadpole's skin is cut from an early gastrula and transplanted to another gastrula (the host). (a) When transplanted to a region of the host destined to become brain tissue, the donor tissue also develops into brain tissue. (b) Similarly, when transplanted to a mesodermal region of the host, transplanted tissue becomes mesoderm and ultimately muscle and notochord. In each case, the actual fate of the prospective skin is determined by nearby tissues in the host embryo. This wide prospective potency of early gastrulas is lost by the late gastrula stage.

planted to parts of an early gastrula destined to become other structures. The tissue of the late gastrula is said to be **determined**: Its fate has been sealed, regardless of its surroundings. By the late gastrula stage, a specific developmental pattern has been set in motion.

DIFFERENTIATION

Because the cells of a multicellular organism arise by mitotic divisions of a single-celled zygote, they are genetically identical. (A dramatic exception to this rule is found in the immune system, which we will discuss in Chapter 16.) Barring mutation, each cell is like each of the others in its hereditary endowment,

yet the adult organism is made of many distinct, differentiated types of cells. This apparent paradox results from the fact that the expression of the genome is closely regulated.

The genome has been likened to a library full of instruction books. At different times in the life of a cell, different instruction books are taken off the shelves and read; likewise, in the different cells of a multicellular organism, different books are read. At a given time, certain genes are being expressed; that is, the polypeptides for which they code are being synthesized. Other genes, at the same time, are held in check. By switching particular genes on, the cell controls not only the kinds of enzymes that it produces, but also the amounts. Both qualitative and quantitative control of specific protein synthesis are crucial to the proper functioning of the cell.

The zygote is **totipotent**, which means that it has the ability to give rise to every type of cell in the adult body. Its genetic "library" is complete, having instructions for all the structures and functions that will arise throughout the entire life cycle. At some later point in the development of animals (and probably to a lesser extent in plants), the cellular descendants of the zygote lose their totipotency and become determined. Once its prospective fate is achieved, a cell is said to have **differentiated**. Whereas all zygotes are totipotent, most of their descendants become determined and, finally, differentiated. The mechanisms of differentiation relate primarily to changes in the transcription and translation of genetic information (Chapter 13).

Is Differentiation Irreversible?

An early suggestion concerning the mechanisms of differentiation was that the cell nucleus undergoes irreversible genetic changes in the course of development. It was proposed that chromosomal material may be lost, or that some of it may be irreversibly inactivated.

Differentiation is clearly irreversible in certain types of cells. The mammalian red blood cell, which loses its nucleus during development, is an example. Another is the tracheid, a water-conducting cell in vascular plants. The development of a tracheid culminates in the death of the cell, leaving only the pitted cell walls that were formed while the cell was alive (Chapter 28). In these two extreme cases, the irreversibility of differentiation can be explained by the absence of a nucleus. However, it is harder to generalize about mature cells that retain functional nuclei. Most biologists tend to think of plant differentiation as reversible and of animal differentiation as irreversible; but this is not a hard and fast rule. A lobster can regenerate a missing claw, but a cat cannot regenerate a missing paw. Why is differentiation reversible in some cells but not in others? At some stage of development do changes within the nucleus permanently commit a cell to specialization?

At the Institute of Cancer Research in Philadelphia in the 1950s, Robert Briggs and Thomas J. King performed a series of experiments to see whether genetic material was preserved or was permanently inactivated or lost during normal development. To find out whether the nuclei of frog blastulas had lost the ability to do what the zygote nucleus could do, they carried out a series of meticulous transplantations. They first took an unfertilized egg and removed its nucleus (thus forming what is called an enucleated egg). Then, with a very fine glass tube, they punctured a cell of a blastula and drew up part of its contents, including the nucleus. This was then injected into the enucleated egg, and the egg was activated. More than 80 percent of these operations resulted in the formation, from the egg and its new nucleus, of a normal blastula; of these blastulas, more than half developed into normal tadpoles and, ultimately, adult frogs. Clearly, no information has been lost from the nucleus by the time the blastula has formed. On the other hand, Briggs and King found that when the nuclei were derived from older embryonic stages, fewer larvae developed (Figure 15.13).

This work was carried further by John B. Gurdon and his associates at Oxford University, who did similar transplants, using nuclei from gastrulas, swimming tadpoles, and even adults. In a few cases the Gurdon group achieved successes of the following sort. Nuclei from differentiated adult cells were transplanted into enucleated eggs, the eggs were raised to the blastula stage, nuclei were isolated in turn from these blastulas, and these nuclei were used for further transplants. An occasional nucleus obtained by serial transplants of this sort, when placed in an enucleated egg, was able to direct development to a tadpole stage, complete with brain, gut, blood, heart, and other parts. Work of this sort convinced many developmental biologists that the loss of particular genes is not the cause of differentiation, and that genes no longer expressed in certain cells are still present and can be expressed if they are placed in a "younger" environment such as an enucleated egg.

DETERMINATION BY CYTOPLASMIC SEGREGATION

How does determination come about? A number of mechanisms are involved, even in a single animal. Most of the mechanisms fall into two categories, the first based on the segregation of cytoplasmic components into separate cells or parts of cells, and the second on the influences of one part of the embryo on another. We will consider cytoplasmic segregation first.

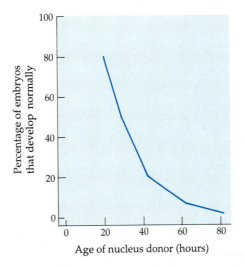

15.13 Loss of Nuclear Potency with Age
Nuclei from older frog embryos transplanted into enucleated eggs are less likely to direct successful development of an embryo than are younger nuclei.

Polarity in the Egg and Zygote

Polarity, the difference of one end from the other, is obvious in development. Our heads are distinct from our feet, and the distal ends of our arms (wrists and fingers) differ from the proximal ends (shoulders) of our arms. Polarity develops early, even in the egg itself. Yolk may be distributed asymmetrically in the egg and the embryo, and other chemical substances may be confined to specific parts of the cell or may be more concentrated at one pole than at the other. In many animals, the original polar distribution of materials in the egg's cytoplasm changes as a result of fertilization, so that a new polar distribution is seen in the zygote. As cleavage proceeds, the resulting blastomeres contain unequal amounts of the materials that were unequally distributed in the zygote. As we learned from the work of Briggs and King and of Gurdon, cell nuclei do not always undergo irreversible changes during early development; thus we can explain some embryological events on the basis of the cytoplasmic differences in blastomeres.

Even as apparently simple a structure as a sea urchin egg shows polarity that can be traced by pigment granules or by the readily dyed animal pole. As the gastrula forms (Figure 15.4), one can see a slight difference in blastomere size, with cells in the vegetal half being somewhat larger. A striking difference between blastomeres can be demonstrated well before that, however. The Swedish biologist Sven Hörstadius showed in the 1930s that the development of sea urchin embryos that have been divided in half at the eight-celled stage depends upon how the separation is performed (Figure 15.14). If the embryo is split into "left" and "right" halves, with each half containing cells from both the animal and the

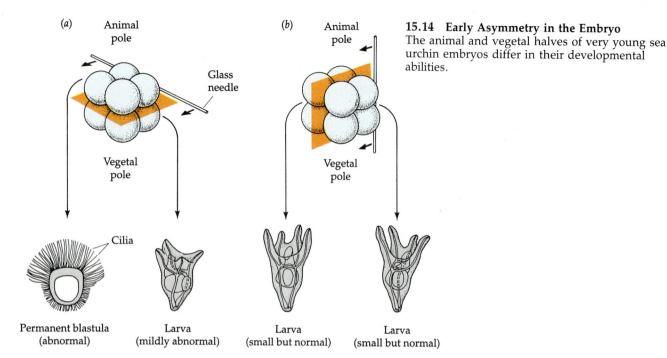

(a) Animal pole

Glass needle

Vegetal pole

Cilia

Permanent blastula (abnormal)

Larva (mildly abnormal)

(b) Animal pole

Vegetal pole

Larva (small but normal)

Larva (small but normal)

15.14 Early Asymmetry in the Embryo
The animal and vegetal halves of very young sea urchin embryos differ in their developmental abilities.

vegetal pole, normal-shaped but dwarfed larvae develop from the halves. If, however, the cut is made so as to separate the four cells at the animal pole from the four at the vegetal pole, the result is different. The animal half develops into an abnormal blastula with large cilia at one end but cannot proceed to form a larva, whereas the vegetal half develops into a small, but almost normal, larva with an expanded gut. For fully normal development, factors from both the animal and vegetal halves of the embryo are necessary. Hörstadius showed that this unequal division of material between the animal and vegetal halves is already present in the unfertilized egg (Box 15.B).

Fertilization of the frog egg leads to a relocalization of some components of the cytoplasm, resulting in —among other things—the formation of the gray crescent. In the 1930s, Hans Spemann found that the gray-crescent region contains materials essential for normal embryonic development (Figure 15.15). In normal cleavage, the first cell division divides the gray crescent equally between the daughter cells. These two blastomeres, if separated, will give rise to normal embryos—this is regulative development as

15.15 Developmental Importance of Asymmetry
The asymmetric distribution of materials in the egg establishes the egg's polarity and determines its developmental architecture. (a) When the plane of the first cleavage divides a frog's egg so that each blastomere receives half the gray crescent, each experimentally separated blastomere forms a normal embryo. (b) When one blastomere receives the entire gray crescent, it forms a normal embryo; the paired blastomere that receives no material from the gray crescent gives rise to a mass of undifferentiated cells.

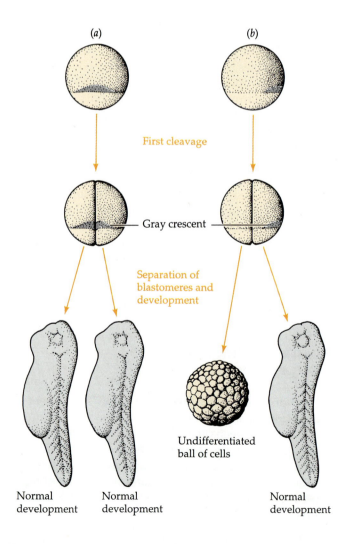

(a)

(b)

First cleavage

Gray crescent

Separation of blastomeres and development

Normal development

Normal development

Undifferentiated ball of cells

Normal development

BOX 15.B

Polarity in the Sea Urchin Egg

We have seen how Hörstadius demonstrated the asymmetric distribution of materials in the eight-celled embryo of the sea urchin. To show that this asymmetry was already present in the cytoplasm of the unfertilized egg, he did some careful microsurgery. With a fine glass needle, he cut unfertilized eggs in half, either along a plane running from the animal to the vegetal pole or through the equator (see figure). A cut through the equator resulted in separate animal and vegetal half-eggs.

To understand what follows, you must know that it is not necessary for such an egg fragment to have a diploid nucleus to undergo cell division and develop. If a fragment *does* contain the egg nucleus, that nucleus fuses with the sperm nucleus following fertilization, giving a diploid zygote. However, fertilization of an enucleated fragment with a sperm gives a "zygote" capable of division and development even though it is haploid; that is, even though it contains only the sperm nucleus.

When Hörstadius divided an unfertilized egg into "left" and "right" halves with a cut passing through both animal and vegetal poles and fertilized each, each developed into a normal, but dwarfed, larva—just as with the "left" and "right" halves of

When a sea urchin egg is divided into animal and vegetal halves (merogones) and the halves are then fertilized by sperm, the animal half forms an abnormal, permanent blastula with an enlarged tuft of cilia. The vegetal half usually develops into a larva that is nearly normal. When an egg is divided on a plane running through the animal and vegetal poles so that both merogones contain equal amounts of animal and vegetal cytoplasm, both halves develop into small, seemingly normal larvae after fertilization.

the eight-celled embryo. However, when he cut an egg into animal and vegetal halves, the vegetal half often gave a small but normal larva, while the animal half produced the same type of abnormal blastula as did the animal half of the eight-celled embryo. This animal-pole blastula could not undergo gastrulation or develop further. Thus the asymmetric distribution of essential materials is already clearly established in the unfertilized sea urchin egg. Normal development can proceed to completion only if the embryo (or embryo fragment) contains the necessary cytoplasmic factors.

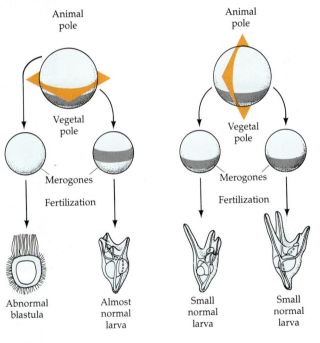

Animal pole

Vegetal pole

Merogones

Fertilization

Abnormal blastula

Almost normal larva

Animal pole

Vegetal pole

Merogones

Fertilization

Small normal larva

Small normal larva

described earlier. In the laboratory, the zygote can be forced to divide in such a way that one of the blastomeres contains the entire gray crescent and the other none of this material. The half with the gray crescent develops normally, but the half lacking the gray crescent region forms only an unorganized cellular mass. These experiments by Hörstadius and by Spemann established that the unequal distribution of materials in the egg cytoplasm plays a role in directing embryonic development.

Cytoplasmic Factors in Polarity in *Drosophila*

In the eggs and larvae of the fruit fly *Drosophila melanogaster*, polarity is based on the distribution of

more than a dozen mRNA and protein species. These **cytoplasmic determinants** are products of specific genes in the mother insect and are distributed to the eggs. They determine the dorsoventral (top–bottom) and anteroposterior (front–rear) axes of the embryo. They were discovered because of the striking appearance of the mutant larvae produced when they are abnormally distributed. For example, larvae formed by females homozygous for the *bicaudal* allele consist solely of two hind ends, joined at the middle (Figure 15.16).

We know that cytoplasmic determinants specify these axes from the results of experiments in which cytoplasm was transferred from one egg to another. Females homozygous for the *bicoid* allele produce lar-

(a) Wild-type

Thorax   Abdomen

Head

(b) *bicaudal*

15.16 The *bicaudal* Mutation
The anteroposterior axis of *Drosophila* larvae arises
from the interaction of several cytoplasmic determinants. *(a)* A larva produced by a wild-type female.
(b) A larva produced by a female homozygous for
the *bicaudal* allele. The larva has no front end or
middle—it consists of two hind ends.

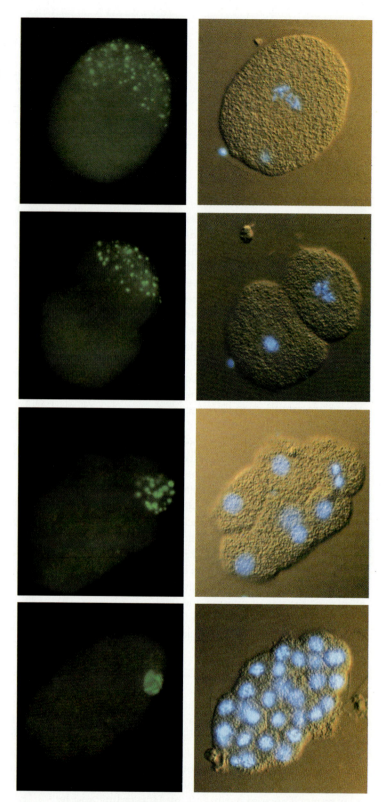

15.17 Distribution of Germ-Line Granules in *Caenorhabditis*
As the embryo of *Caenorhabditis elegans* develops, the germ-line
granules move to the posterior end of the embryo. Eventually the
granules are confined to that cell that gives rise to gametes. The
germ-line granules are the bright spots in the micrographs in the
left column; for comparison, the nuclei are stained blue in the
corresponding micrographs in the right column.

vae with no head and no thorax. However, if eggs of homozygous *bicoid* females are inoculated with cytoplasm from the anterior region of a wild-type egg, the treated eggs develop into normal larvae—with heads developing from the part of the egg that receives the wild-type cytoplasm. Removal of 5 percent or more of the cytoplasm from the anterior of a wild-type egg results in an abnormal larva that looks like a *bicoid* larva. Another allele, *nanos*, plays a comparable role in the development of the posterior end of the larva. Eggs from homozygous *nanos* females develop into larvae with missing abdominal segments; inoculation with cytoplasm from the posterior region of wild-type eggs allows normal development in eggs from *nanos* females.

Germ-Line Granules in *Caenorhabditis*

Various cytoplasmic determinants play roles in the development of the tiny nematode worm *Caenorhabditis elegans*, an animal that will be discussed further at the end of this chapter. **Germ-line granules** are cytoplasmic determinants whose positions in the zygote and embryo are determined by the action of microfilaments. Before the zygote divides, the germ-line granules collect at the posterior end of the cell, and all the granules appear in only one of the first two blastomeres (Figure 15.17). The germ-line granules continue to be precisely distributed during the early cell divisions, ending up in only those cells that will eventually give rise to eggs and sperm.

DETERMINATION BY EMBRYONIC INDUCTION

Induction and the Organizer

Numerous experiments have clearly established that the fates of particular tissues are determined by interactions with other specific tissues in the embryo.

A classic demonstration was performed in 1924 by Hilde Mangold and Hans Spemann, working with newt embryos. The experimental work began with Mangold's transplanting a piece of the dorsal lip of the blastopore from an early gastrula of a lightly pigmented species to a particular place on the surface of an early gastrula of a heavily pigmented species, as shown in Figure 15.18. She could follow the fate of the transplanted piece because it was lighter in color than the surrounding host tissue. As predicted from the known fate of the surrounding tissue—to become mesoderm—the graft itself developed into principally mesodermal products. However, an unexpected and striking thing happened as well: In the region of the graft, an extra neural plate appeared, containing neural folds made from host tissue. The procedure was repeated many times, and in some instances the extra neural plate continued to develop into a secondary embryo attached to the main one! The grafted dorsal lip had induced the nearby host tissue to develop along far different lines than it would have followed without the graft.

Spemann called the dorsal lip of the blastopore the organizer. Not only does this tissue have the fate of becoming part of the notochordal mesoderm, but, in addition, it induces any ectoderm in contact with it to organize into a neural tube. With the formation of the neural tube the principal axes of the embryo—the anteroposterior axis and the dorsoventral axis—are formed; hence the term embryonic organizer.

In the developing embryo there are many instances of **induction**, in which one tissue induces an adjacent tissue to follow a particular line of development. (Note that this embryonic induction is a completely different phenomenon from the induction of enzyme synthesis, discussed in Chapter 12.)

The formation of the lens in frog eyes is a classic example of induction. Lens formation proceeds as follows (Figure 15.19): The developing forebrain bulges out at both sides to form the **optic vesicles**,

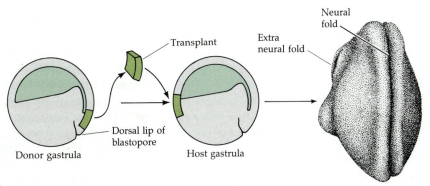

15.18 Embryonic Induction
Mangold transplanted the dorsal lip of the blastopore to a gastrula with different pigmentation so she could distinguish graft tissue from host tissue. She expected the graft to become mesodermal tissue and it did. Unexpectedly, she saw that tissue of the host near the graft developed into a second neural plate. Mangold concluded that the dorsal lip induced nearby host ectoderm to alter its development.

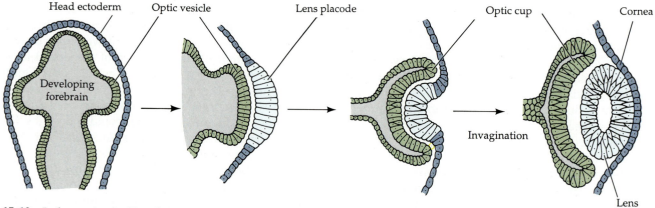

15.19 Inducers in the Vertebrate Eye
The vertebrate eye develops as inducers take their turns. In frogs, for example, the optic vesicle induces overlying ectoderm to form placode tissue that, in turn, induces the formation of an optic cup. The optic cup then induces the placode to invaginate and form the eye's cornea and lens.

and these expand until they come in contact with the ectoderm of the head. The head ectoderm in the region of contact with the optic vesicles thickens, forming a **lens placode**; the lens placode invaginates, folds over on itself, and ultimately detaches from the surface to produce a structure that will develop into the lens. If the growing optic vesicle is cut away, no lens forms in the head region from which the optic vesicle has been removed; if, instead, an impermeable barrier is placed between the optic vesicle and the ectoderm, no lens is induced. These observations suggested that the normal development of ectoderm into a lens depends upon a signal—an **inducer**—obtained by contact with an optic vesicle.

Under some circumstances, implanting an optic vesicle under head ectoderm leads to the formation of a lens placode when the optic vesicle comes into contact with the overlying ectoderm. This experiment works only if the overlying ectoderm is **competent**; not just any ectoderm will do. Induction of the lens requires *two* inductive steps. In the first step, mesenchyme in the developing head induces competence in the surface ectoderm—it "biases" the ectoderm toward development as a lens placode. In the second step, the optic vesicle finalizes matters by inducing the competent ectoderm to become the lens placode.

The interaction of tissues in eye development is a two-way street: There is a "dialogue" between the developing optic vesicle and the ectoderm. The lens determines the size of the **optic cup** that forms from the optic vesicle. If ectoderm from a species of frog with small eyes is grafted over the optic vesicle of one with large eyes, both lens and optic cup are of intermediate size. The lens also induces the ectoderm over it to develop into a cornea.

Induction triggers a sequence of gene expression in the surrounding cells. Tissues do not induce themselves; rather, different tissues interact and induce each other.

One of the most resistant problems in the history of developmental biology has been that of determining the specific chemical nature of the inducers. In some cases, specific diffusible proteins may be involved—the earliest inducer to act in frog gastrulas appears to be a growth factor (Chapter 13). However, in other cases, insoluble extracellular materials such as collagen and other proteins may be involved in induction. Generally speaking, induction is a phenomenon confined to embryonic tissues; however, it continues in the production of certain white blood cells in the adult immune system.

Instructive versus Permissive Induction

There are many different patterns of induction in the embryo at different times. Limbs, hair, feathers, and teeth form as mesoderm and ectoderm interact, inducing each other. Lungs, liver, and pancreas develop in part as a result of inductive interactions between mesoderm and endoderm. Different parts of the mesoderm interact in forming parts of the reproductive and urinary systems. These few examples only scratch the surface of the variety of inductive interactions.

Are the inductive interactions between tissues **instructive** or **permissive**? That is, does the source of the inducer determine which of several paths the target tissue is to follow (instructive induction) or does the inducer simply trigger the target tissue to undergo a developmental pattern that is intrinsic to it and waiting to be started (permissive induction)? Some inductive interactions are instructive and others are permissive; let us consider an example of instructive induction.

In the chick embryo, some cells of the ectoderm of the wing develop into wing feathers, some in the ectoderm of the thigh develop into thigh feathers, and some in the leg develop into the scales and claws of the feet. In each case, the underlying mesoderm induces the appropriate ectodermal derivative to form; it is the source of the mesoderm, not of the ectoderm, that determines what develops. Wing ectoderm, which normally gives rise to wing feathers, can be induced by the appropriate mesoderm to form thigh feathers and even scales and claws. Evidently, the inducing mesoderm is sending *specific* instructions, that is, specific chemical inducers that are characteristic of the particular position of the mesoderm. The mesoderm transmits the positional information, and the induction is instructive.

AGGREGATION OF EMBRYONIC CELLS

How can differences in the cytoplasm of blastomeres lead to the cellular migrations of gastrulation? The answer to this question is not yet known, but the basis for our current thinking is found in a series of papers by Johannes Holtfreter, published about 1940. Holtfreter studied the behavior of isolated cells, isolated tissues, embryonic fragments, and whole frog gastrulas. In one experiment he found that when an isolated dorsal lip of a blastopore is placed on a fragment of endoderm, the lip tissue burrows into the endodermal fragment, forming a small invagination comparable to the beginning of gastrula formation (Figure 15.20a). Thus inherent properties of the dorsal lip cells underlie the beginning of invagination and involution in frog gastrulation.

In another series of experiments, Holtfreter separated cells of various embryonic tissues by exposing them to solutions of high pH. When he returned the pH of the solution to a natural physiological level, the dissociated cells reaggregated to form tissues. If the dissociated cells were of different types, the aggregation was followed by segregation of the different types of cells, such that like cells associated with like. Amazingly, in the final product, the tissues ended up in their normal positions relative to one another. For example, when free cells from a region destined to become epidermis (the outer layer of the skin) were intermingled with cells from the neural plate, the cells sorted themselves out and migrated until the epidermal cells were on the outside and the nervous tissues on the inside (Figure 15.20b). When

15.20 Cellular Affinities in Gastrulas

Studies of isolated gastrula tissues shed light on processes of gastrulation. (a) When a bit of dorsal lip of the blastopore (blue) is placed on isolated endodermal tissue (yellow), the dorsal lip sinks, leaving an indentation similar to the start of a blastopore. (b) Mixtures of cells origi-nally destined to become epidermis (blue) or nervous tissue (green) are dissociated and mixed. Not only do the cells reaggregate into a mass, they sort themselves out so that the epidermal cells form a layer that envelops the nervous tissue.

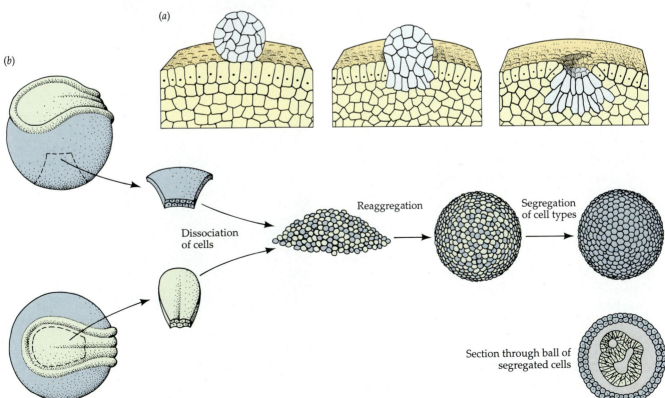

tissues destined to become ectoderm, endoderm, and mesoderm were disaggregated and combined, migration led to the placement of ectoderm on the outside and endoderm on the inside, with mesoderm in between. Again, the behavior of isolated cells and tissues of specific types matched the normal behavior of those tissues in a developing embryo. As Holtfreter put it, there apparently are **cellular affinities** such that cells of the same type preferentially associate with one another. These inherent association properties may underlie the distinctions between the germ layers.

PATTERN FORMATION

One of the outstanding problems of developmental biology is that of **pattern formation**, the development of organs consisting of differentiated cells and tissues in ordered arrangements. The differentiation of cells is beginning to be understood in terms of molecular events, but what of the organization of multitudes of cells into the numerous specific body parts, such as a shoulder blade or a tear duct? It would seem that the imaginal discs of insects should be good models for pattern formation, since they develop in the course of metamorphosis from masses of undifferentiated cells into complex organs. In fact, recent work on the imaginal discs of *Drosophila melanogaster* has led to important insights.

Establishing Body Segmentation

Pattern formation in insects and many other animals consists of the formation of a highly modular body composed of different kinds of modules, such as segments. Complex interactions of different sets of genes determine pattern formation. *Drosophila*, like all insects, has a segmented body. In contrast with segmented worms such as earthworms, the segments of an insect's body differ significantly from one another. The *Drosophila* larva consists of a head, three different thoracic segments, eight abdominal segments, and a genital segment at the posterior end.

The organization of the *Drosophila* larva is determined by a remarkable series of events, the sequential activation of key genes. The overall framework is laid down initially, in the form of anteroposterior and dorsoventral axes, by the activity of the genes that produce the cytoplasmic determinants referred to earlier in this chapter (Figure 15.16). As we saw, mutations in those genes result in the duplication or deletion of body parts such as heads and tails. Next, the number and polarity of the larval segments are determined by the activities of several more genes, the **segmentation genes**. Three classes of segmentation genes participate, one after the other. First, **gap genes** organize large areas along the anteroposterior

axis (Figure 15.21). Mutations in gap genes result in the omission of several larval segments. Second, **pair-rule genes** divide the embryo into two-segment units. Mutations in pair-rule genes result in embryos missing every other segment. Third, **segment-polarity genes** determine the anteroposterior organization of the segments themselves. Mutations in segment-polarity genes result in segments in which some posterior structures are replaced by reversed (mirror image) anterior structures. The three classes of segmentation genes regulate finer and finer details of the segmentation pattern.

Finally, after the basic pattern of segmentation has been established by the segmentation genes, the appropriate structures of the different segments are specified by the activation of **homeotic genes**. These genes tell each segment what to become.

Homeotic Mutations

The key ingredient in understanding these processes was the discovery of dramatic mutations, called **homeotic mutations**, that modify homeotic genes and thus the imaginal discs and the resulting adult body parts. Instead of a normal body part, the insect with a homeotic mutation has another part characteristic

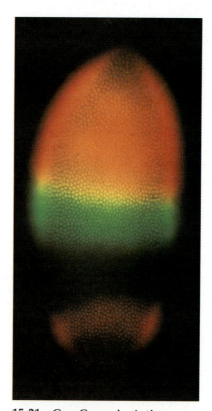

15.21 Gap Genes in Action
Interactions of proteins encoded in gap genes define domains of the larval body in *Drosophila*. In this larva *hunchback* (orange) and *Krüppel* (green) proteins overlap, forming a boundary (yellow) between two domains.

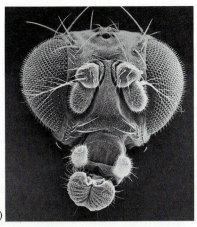
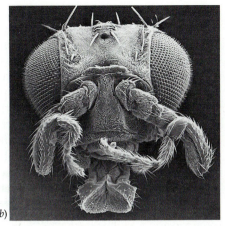

(a) (b)

15.22 A Homeotic Mutation
Scanning electron micrographs of the heads of two *Drosophila melanogaster*.
(a) A wild-type fly, with normal antennae. (b) An *antennapedia* mutant, with
roughly normal legs in the positions usually occupied by antennae. This abnormality results from a homeotic mutation—a drastic mutation causing one
structure to develop in the place of another.

of another body segment. Two bizarre examples are
the *antennapedia* mutant, in which legs grow in the
place of antennae (Figure 15.22), and the *ophthalmoptera* mutant, in which wings grow in the place of
eyes. The homeotic genes determine the differences
between the imaginal discs. The loci of these genes
fall into a few tight clusters, the best characterized of
which is referred to as the **bithorax complex**. The
eight or more genes of the bithorax complex control
the development of the abdomen and posterior
thorax of the fly. Development of the head and anterior thorax is controlled by another cluster, the **antennapedia complex**. The functions of the two complexes interact substantially, and the range of effects
of the antennapedia complex is not completely
known.

Because many mutations in the bithorax complex
are so severe that they prevent development past the
early larval stages, they can be studied only in larvae.
In larvae in which the entire bithorax complex has
been deleted, the third thoracic segment and seven
abdominal segments all develop as second thoracic
segments (Figure 15.23). Another bithorax mutation
leads to the development of larvae in which head,
thoracic, and abdominal segments all differentiate as
normal eighth abdominal segments. In wild-type
Drosophila the second thoracic segment gives rise to
wings and legs in the adult, while the third thoracic
segment produces a pair of legs and a pair of small,
winglike structures called halteres. One group of mutations in the bithorax complex causes the third thoracic segment of the adult to develop exactly like the
second, so that the resulting fly has two pairs of
normal wings and no halteres (Figure 15.24).

As the genotype with the entire bithorax complex
deleted produces a larva that, although highly abnormal, has the normal number of segments, the
bithorax complex clearly does not determine the
number of segments. It is the segmentation genes
that determine the number and polarity of segments.

The Homeobox

Walter Gehring and his associates William McGinnis
and Michael Levine, working in Switzerland, and
Thomas Kaufman, at Indiana University, in the early

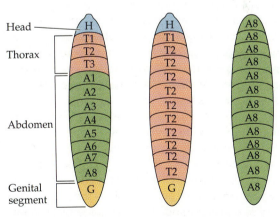

15.23 *bithorax* Mutations in *Drosophila* Larvae
Genotype at loci of the bithorax complex governs how
segments develop in *Drosophila* larvae. The wild-type
larva shown on the left has a head (H), three thoracic
segments (T1-T3) and eight abdominal segments (A1-A8),
and one genital segment (G). The larva in the center lacks
the entire bithorax complex. It has a normal head and
first thoracic segment, but the other segments have developed like normal second thoracic segments. The larva
on the right lacks part of the bithorax complex, specifically the extra sex comb gene; all its segments are like
normal eighth abdominal segments.

15.24 *bithorax* **Mutation in Adult** *Drosophila*
The third thoracic segment of this adult fruit fly developed as if it were a second thoracic segment because of a mutation in its bithorax complex.

1980s undertook a study of the antennapedia complex, using the techniques of recombinant DNA technology. They set out to isolate and clone the *antennapedia* (*Antp*) gene, a member of the antennapedia complex. As part of this study, they prepared a cDNA (Chapter 14) complementary to *Antp* mRNA to be used in locating the *Antp* gene. To the surprise of the investigators, the *Antp* cDNA hybridized with both the *Antp* gene and a nearby segmentation gene (the *fushi tarazu* gene, *ftz*, from the Japanese for "too few segments"). Clearly, the *Antp* and *ftz* genes have a region of close similarity, because part of each gene hybridizes with the same cDNA. Further hybridization studies demonstrated that the same shared stretch of DNA is also found in the bithorax complex, in the *bicoid* gene of the bithorax complex, in some other parts of the *Drosophila* genome, and in genes in other insect species. In fact, this important sequence of DNA, called the **homeobox**, has now been shown to be part of a few genes of all animals that have segmented bodies, including ourselves. The homeobox is also present in the DNA of tomatoes and sea urchins, so we must not assume that its only activities relate to body segmentation.

What is the significance of the homeobox, this small sequence of DNA, 180 base pairs long, that is nearly ubiquitous in living things? The homeobox codes for a 60-amino acid region—the homeodomain—of the protein coded for by the gene containing the homeobox. These proteins remain in the nucleus and bind to DNA, regulating the transcription of other genes. A computerized search of the published sequences of numerous DNA species revealed a similarity between the homeobox and parts of certain regulatory genes in yeast—genes that produce proteins that also bind to specific DNA sequences. Some genes with homeoboxes are expressed only at certain times and in certain tissues as development proceeds.

Let us briefly review pattern formation in *Drosophila*. In the egg, certain genes establish anteroposterior and dorsoventral axes. Later, segmentation genes determine the number and polarity of segments that form; then homeotic genes, largely concentrated in the antennapedia and bithorax complexes, regulate other genes that act in the imaginal discs to determine how the different segments develop. Some of the regulatory effects of the segmentation and homeotic genes probably derive from the homeobox.

What are we to make of the presence of the homeobox in such diverse species as humans, fruit flies, frogs, and tomatoes—and of its presence in several genes in the same organism? It suggests that both the antennapedia complex and the bithorax complex may have arisen from a single ancestral gene. Further, it implies that a single gene in some ancient organism may have been the evolutionary progenitor of what is now a widespread controlling system for development.

Positional Information in Developing Limbs

Multiple factors collaborate in regulating pattern formation. In addition to the genetic controls just described, **positional information**—information about where one group of cells lies in relation to others—plays a role. In the 1970s, the English developmental biologist Lewis Wolpert developed a theory of positional information, based on his studies of developing limb buds in chick embryos.

Chick wing buds are first evident in 3-day-old embryos as a pair of bulges on the surface, consisting of mesoderm covered by a thin layer of ectoderm (Figure 15.25a). The ectoderm has a thickening, the **apical ectodermal ridge**. Over the next week, a wing bud develops into the intricate structure shown in Figure 15.25b, consisting of skin, bone, muscle, tendon, and other tissues. Consider a wing in terms of three coordinate axes: The anteroposterior axis runs from the front to the rear of the chick, the proximodistal axis runs from the base of the limb to its tip, and the dorsoventral axis from top to bottom. Each of these axes has a corresponding type of positional information. We will now consider the two axes about which the most is known.

Pattern formation along the anteroposterior axis is controlled at least in part by a particular part of the wing bud, the zone of polarizing activity, or **ZPA**, that lies on the posterior margin of the bud. It appears that different parts of the limb normally develop at *specific distances from the ZPA*. This hypothesis is supported by the results of grafting experiments such as those summarized in Figure 15.26, in which a ZPA from one bud is grafted onto another bud that still has its own ZPA. The donor ZPA can be placed in different positions on different hosts. If the extra ZPA is grafted on the anterior margin, opposite from

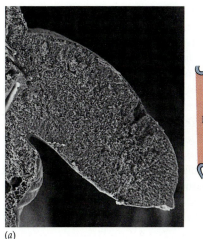

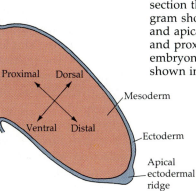

15.25 **Early Development of the Chick Wing**
(a) At the age of 3 days, a chick embryo has two wing buds. The scanning electron micrograph shows a cross section through one bud in a 4-day-old embryo; the diagram shows the positions of the mesoderm, ectoderm, and apical ectodermal ridge as well as the dorsoventral and proximodistal axes. (b) By the age of 9.5 days, the embryonic chick wings have developed to the stage shown in this X-ray photograph, with a full set of bones.

(a)

(b)

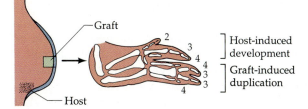

the host ZPA, the distal part of the wing is duplicated, with two complete mirror image units being formed (Figure 15.26a). If the extra ZPA is placed somewhat closer to the host ZPA, incomplete mirror image units appear—one of the digits is missing from each of the units (Figure 15.26b), as if the missing units are of a type that can only form if they are more than some minimum distance from a ZPA. In the third case (Figure 15.26c), with the two ZPAs close together, there is room for some duplication between them, but there is also room enough to the anterior side so that a complete, nearly normal unit forms.

How might the ZPA produce its effects? The effects appear to result from the activity of a substance called retinoic acid. ZPAs produce retinoic acid, and the effect of an extra ZPA can be mimicked by the application of retinoic acid in the place of a grafted ZPA. Target cells in the limb bud have a retinoic acid–binding protein on their surfaces. How the binding of retinoic acid to the protein leads to specific developmental events remains unknown.

A different type of positional effect accounts for pattern formation along the proximodistal axis. The primary source of the positional information is the apical ectodermal ridge. If this ridge is removed from a very young bud, only the bone of the upper wing

15.26 **Positional Information and the ZPA**
A zone of polarizing activity (ZPA) is located on the posterior margin of the chick wing bud as indicated by "Host ZPA." In each of the experiments diagrammed here, the ZPA from one bud was grafted onto another wing bud which still had its own ZPA. (a) ZPA grafted on the anterior margin causes mirror-image duplication of the distal part of the wing-duplication of digits 4, 3, and 2. (b) A ZPA grafted closer to the host ZPA causes duplication of digits 4 and 3, but no digit 2 develops. (c) A ZPA grafted still closer to the host ZPA allows a nearly normal set of digits 2, 3, and 4 to develop on the anterior side but also results in extensive duplication of digits.

Zone of polarizing activity determines the
anteroposterior axis of development

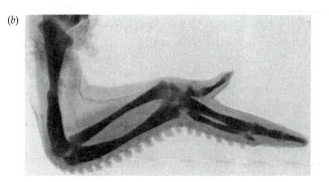

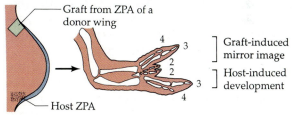

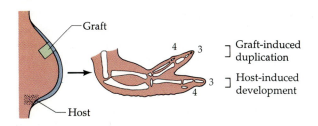

Apical ectodermal ridge determines the
proximodistal axis development

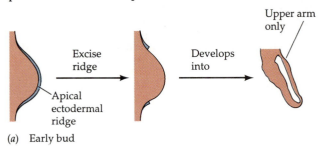

(a)   Early bud

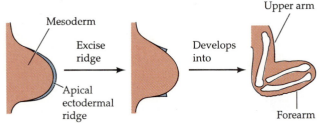

(b)   Late bud

15.27   Effects of the Apical Ectodermal Ridge
To determine the role of the apical ectodermal ridge, biologists removed it from chick wing buds at two different embryonic stages. *(a)* Removal from a very young bud prevents development of the "forearm" and "hand" (compare Figure 15.25*b*). *(b)* Removal at a later stage allows development of the "upper arm" and "forearm" but prevents "hand" development.

develops (Figure 15.27). If the apical ectodermal ridge is removed somewhat later, then the first two segments of the wing are formed; if the apical ectodermal ridge is not removed, a complete wing develops.

As the limb elongates, the different parts of the mesoderm form different parts of the limb. It is believed that the development of the mesodermal cells depends on the length of time they spend close to the apical ectodermal ridge. Those mesodermal cells that spend the least time near the apical ectodermal ridge form the proximal structures; those that spend the most time near the apical ectodermal ridge give rise to the more distal structures. This hypothesis is supported by the grafting experiments in Figure 15.28.

A COMPLETE CATALOG OF DEVELOPMENT IN A TINY WORM

We have by now considered the major aspects of animal development, including events at the molecular, cellular, and organ levels. Is it possible to carry through a detailed analysis, cell by cell, from a zygote to an adult multicellular animal? In fact, just such a

program was initiated in 1963 by a group headed by Sydney Brenner, in Cambridge, England. The object of this study was a 1-millimeter-long nematode, or roundworm, *Caenorhabditis elegans* (Figure 15.29).

The adult worm contains just 959 cells, and it develops from a single cell to the 959-cell stage in just 3.5 days. The relative transparency of the body made it possible, by painstaking observation, to determine the exact history of each of the cells—that is, to determine the fates of all of the developing cells, division by division (Figure 15.30). In addition, it was possible, through examination of many hundreds of electron micrographs, to map out all 8,000 of the connections linking the 302 nerve cells in the adult

Period of mesodermal association with AER
determines **what structures will develop**

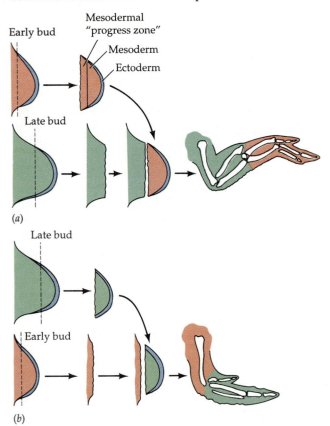

(a)

(b)

15.28   The Apical Ectodermal Ridge Gradually Alters the Fate of the Underlying Mesoderm
(a) Grafting of the tip of an early chick wing bud to the stump of a late bud (obtained by removing its tip) results in a doubling of the "forearm" region (compare Figure 15.25*b*). Both the stump and the grafted tip contain mesoderm that has spent an intermediate amount of time near the apical ectodermal ridge. *(b)* The reciprocal experiment —grafting the tip of a late bud to the stump of an early bud—causes complete omission of the "forearm;" the resulting wing has only an "upper arm" and a "hand." In this experiment, neither the stump nor the grafted tip contains mesoderm that has spent an intermediate amount of time near the apical ectodermal ridge.

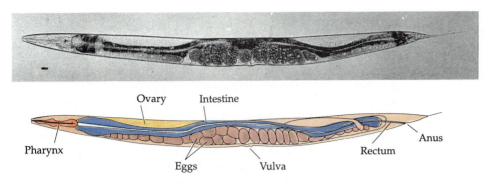

15.29 A Nematode Model of Development
Visibility of internal structures and relative anatomical simplicity make the transparent, colorless nematode worm *Caenorhabditis elegans* a useful organism for following cellular development. It has been possible to trace all divisions from a single cell to the 959 cells found in the fully developed adult.

animal. These nerve connections are almost identical from one individual to another (except for mutants), but there are some small differences in such things as cell shape and connections. Still, this "developmental noise" is slight.

The original hope—indeed the expectation—was that this study would lead to the identification of neat rules and clear mechanisms that would generate the adult form from the simple early stages of embryonic development. In fact, no such thing occurred. Rather, the sequence of cell divisions shows no obvious logic (although the sequence is the same

in every individual). The symmetry of the body of *C. elegans* is not arrived at by symmetrical cell divisions, and body parts are assembled from a strange mixture of sources. Many of the cells die—always the same ones—as part of the worm's development. Brenner himself has said that "there is hardly a shorter way of giving a rule for what goes on than just describing what there is." This apparent lack of rules may be in some ways discouraging; on the other hand, it may give a hint about the nature of complexity in other systems such as the human brain.

Genetic studies of *C. elegans* have been productive.

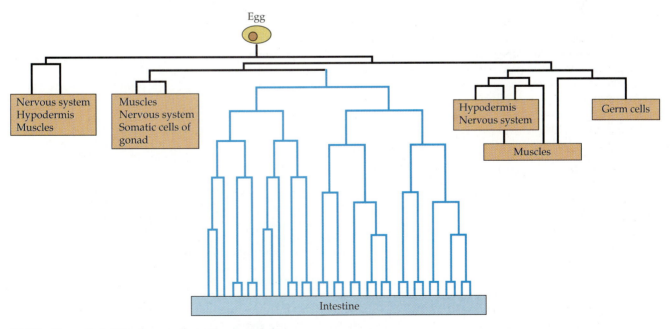

15.30 *Caenorhabditis elegans*, from Egg to Adult
Each fork in this tree represents the mitotic division of a cell. Here we focus on those cells that give rise to the intestine. At the 8-cell embryonic stage, a single cell is identifiable as the source of all subsequent intestinal cells.

Many mutations have already been characterized, including some in genes that control when particular cell lines become determined. These genes are called chronogenes. In some organisms, like *C. elegans*, chronogenes are crucial to the timing of events in development. However, most animals depend upon a different mechanism to ensure that their parts develop properly. Whereas the development of *C. elegans* depends upon a rigidly predetermined sequence of events in particular cells and their descendants, the parts of most animals develop from whatever cells happen to be in particular positions in the embryo at certain key stages.

SUMMARY

A zygote undergoes repeated cell divisions during cleavage, giving rise to a multicellular blastula. The major differentiation of tissue types follows during gastrulation. Gastrulation is different in different animals—cells move by various combinations of these basic mechanisms: expansion of regions on the embryo, invagination, ingression, and involution.

Following gastrulation, specialized organs and organ systems form. Subsequent development may include passage through a larval stage before achievement of the normal adult form. When the adult differs dramatically from the larva, the metamorphosis encompasses both the loss of old parts and the formation of new ones. In insects that undergo complete metamorphosis, the adult insect is formed by the growth of imaginal discs, groups of cells that remain undifferentiated in the larva.

The different parts of the early embryo have different prospective fates. Depending upon the species, the prospective potency of a tissue may differ from its prospective fate. The prospective potency of a cell or tissue may be narrowed by the presence of hormones, by the distribution of cytoplasmic determinants, or by interactions with nearby cells and tissues. In the frog embryo, the dorsal lip of the blastopore functions as an embryonic organizer, determining the fates of cells in its vicinity. More generally, several parts of an embryo may induce specific changes in neighboring tissues.

Pattern formation is based on both genetic and positional information. In insects segmentation is controlled by several genes, including the homeotic genes that determine the differences among the segments. Many of these genes share a particular stretch of DNA, the homeobox. All segmented animals have genes containing the homeobox. The homeobox codes for a homeodomain that regulates the expression of other, specific genes. The orderly development of bird limbs requires positional information, with some structures developing at specific distances from organizing centers.

SELF-QUIZ

1. Which statement is *not* true of cleavage?
   a. The blastomeres are produced by mitosis.
   b. Cleavage patterns depend in part on the distribution of yolk.
   c. The first few divisions are often synchronous.
   d. The first cleavage results in a cell toward the animal pole and one toward the vegetal pole.
   e. In extremely yolky eggs, daughter cells are not completely separated.

2. A blastula:
   a. is a solid ball of cells.
   b. is surrounded by a sheet of cells called the blastocoel.
   c. develops over a few hours of cleavage.
   d. has a lower ratio of nuclear to cytoplasmic volume than did the zygote.
   e. is much larger than the fertilized egg.

3. Gastrulation:
   a. is identical in all animal species.
   b. always results from ingression of cells at the vegetal pole.
   c. always produces a primitive streak.
   d. always results in a roughly spherical gastrula.
   e. is a process in which a 2- or 3-layered embryo is formed.

4. Which statement is *not* true of the dorsal lip of the amphibian blastopore?
   a. It is the first site of invagination as the blastula is forming.
   b. Cells turn in here and flow into the interior of the embryo.
   c. It serves as the embryonic organizer.
   d. It spreads, eventually forming a complete circle.
   e. Cells involuting here give rise to both endoderm and mesoderm.

5. Which statement is *not* true of cytoplasmic determinants in *Drosophila*?
   a. They specify the dorsoventral and anteroposterior axes of the embryo.
   b. Their positions in the embryo are determined by microfilament action.
   c. They are products of specific genes in the mother insect.
   d. They often produce striking effects in larvae.
   e. They have been studied by the transfer of cytoplasm from egg to egg.

6. The organizer of an amphibian embryo:
   a. is a homeotic mutation.
   b. is the homeobox.
   c. induces adjacent ectoderm to organize into a neural tube.
   d. is the ventral lip of the blastopore.
   e. is a product of imaginal discs.

7. Which statement is *not* true of embryonic induction?
   a. One tissue induces an adjacent tissue to develop in a certain way.
   b. Induction triggers a sequence of gene expression in target cells.
   c. Induction may be either instructive or permissive.
   d. A tissue may induce itself.
   e. The chemical identification of specific inducers has been difficult.

8. In establishing body segmentation in *Drosophila* larvae:
   a. the first steps are specified by homeotic genes.
   b. mutations in pair-rule genes result in embryos missing every other segment.
   c. mutations in gap genes result in the insertion of extra segments.
   d. segment-polarity genes determine the dorsoventral axes of segments.
   e. segmentation is the same as in earthworms.

9. Homeotic mutations:
   a. are often so severe that they can be studied only in larvae.
   b. cause subtle changes in the forms of larvae or adults.
   c. occur only in prokaryotes.
   d. do not affect the animal's DNA.
   e. are confined to the apical ectodermal ridge.

10. Which statement is *not* true of the homeobox?
    a. It is transcribed and translated.
    b. It is found only in animals.
    c. Proteins containing the homeodomain bind to DNA.
    d. It is a stretch of DNA shared by many genes.
    e. Its activities often relate to body segmentation.

FOR STUDY

1. Discuss the differences in—and the reasons for the differences in—the gastrulas of a sea urchin, a frog, and a chicken.

2. Consider a cell in the lens of a frog eye. Trace its developmental history back to a particular part of the zygote. Describe all the ways in which other cells or tissues have interacted with it.

3. During development, the prospective potency of a tissue becomes ever more limited, until, in the normal course of events, the prospective potency is the same as the original prospective fate. On the basis of what you have learned in this chapter and in Chapter 13, discuss possible mechanisms for the progressive limitation of the prospective potency.

4. How was it possible for biologists to obtain such a complete accounting of all the cells in the roundworm *Caenorhabditis elegans*? Why can't we reason directly from studies of *C. elegans* to comparable problems in our own species?

READINGS

De Robertis, E. M., G. Oliver and C. V. E. Wright. 1990. "Homeobox Genes and the Vertebrate Body Plan." *Scientific American*, July. A fascinating description of the multiple roles of homeobox genes at different stages of embryonic development.

Gehring, W. J. 1985. "The Molecular Basis of Development." *Scientific American*, October. A lucid account of homeotic mutations and the homeobox.

Gilbert, S. F. 1991. *Developmental Biology*, 3rd Edition. Sinauer Associates, Sunderland. An exceptionally well-balanced treatment of developmental biology, covering both molecular/cellular concepts and embryology. Gives a feeling for the history of the discipline.

Goodman, C. S. and M. J. Bastiani. 1984. "How Embryonic Nerve Cells Recognize One Another." *Scientific American*, December. How do brains develop their specific "wiring"? This article describes experimental methods for dealing with this question.

Holliday, R. 1989. "A Different Kind of Inheritance." *Scientific American*, June. Genes can be methylated in one cell, and the pattern of methylation is inherited by the products of cell division. Thus a pattern of gene activity is transmitted from one cell generation to the next.

Hynes, R. O. 1986. "Fibronectins." *Scientific American*, June. Proteins that guide migrating cells during development; their possible role in cancer.

Stent, G. S. and D. A. Weisblat. 1982. "The Development of a Simple Nervous System." *Scientific American*, January. Details of a specific pattern of development in the leech.

Wolpert, L. 1978. "Pattern Formation in Biological Development." *Scientific American*, October. Positional information helps to determine the fates of cells.

16

Immune Systems and Disease

PREVIEW: The immune system is a highly specific defense system that recognizes, eliminates, and remembers foreign macromolecules and cells. When functioning normally, it can distinguish between self and nonself materials. Several kinds of specialized white blood cells interact in bringing about the immune response. The great specificity of the system is the result of the rearrangement of the genetic material that encodes specific receptor molecules.

This chapter deals with specific and nonspecific defense systems (including phagocytes and the complement system), humoral and cellular immune responses, antigens, antibodies, T-cell receptors, immunological memory, clonal selection, immunological tolerance, antigen processing, antibody synthesis, the major histocompatibility complex, interleukins, AIDS, and cancer.

In the last chapter we considered the development of a zygote into an adult animal. Does development cease at this point, other than for gametogenesis? No. In fact, an animal develops throughout its life, and cell differentiation continues to produce new, highly specialized cells at a great rate as long as the animal lives. The products of this ongoing differentiation include several types of white blood cells that play essential roles in the *protection* of the animal. If parts of this protective system fail, the organism is doomed to disease and, commonly, early death. Such failure is the hallmark of AIDS, the acquired immune deficiency syndrome.

Within a single multicellular organism, the differentiation of cells results from differing expression of the same genotype in different cells. In animals, not only do the individual's own cells differ from one another in structure and function, but their cell surfaces also differ from those of the cells of other organisms, and even from similar cells in another organism of the same species. Each animal uses this cell-surface difference to distinguish **self** from **nonself**, that is, to react in one way to its own cells and macromolecules and in other ways to cells or macromolecules from intruding organisms such as bacteria or fungi.

The environment is alive with biological threats, ranging from large carnivores to potentially lethal microorganisms—the disease-producing organisms or **pathogens**. To cope with such threats, all living organisms are equipped with defense mechanisms. Responses to predators include behaviors such as hiding and counterattacking as well as the evolution of morphological and physiological traits that render the prey harder to attack. Predator–prey interactions will be dealt with fully in Chapter 46. Here we are concerned primarily with the specific and nonspecific defenses of organisms, especially vertebrates, against microscopic pathogens and foreign macromolecules.

NONSPECIFIC DEFENSE MECHANISMS AGAINST PATHOGENS

Organisms possess an array of nonspecific mechanisms that serve as the first line of defense against invaders. Consider the challenges faced by a potential pathogen (such as a bacterium, virus, fungus, protist, or animal parasite) as it approaches the body of an animal. There are four hurdles the pathogen must overcome: It must arrive at the body surface of a potential host, get into the host, multiply inside the host, and finally, prepare to infect the next host. Failure in any of these four steps ends the reproductive career of a pathogenic organism.

The first step—arrival—may be by way of airborne droplets (as from a sneeze), food or drink, an animal that bites, direct contact with an infected individual, or contact with some pathogen-carrying object in the environment. Many of the most massive improvements in public health have come with the control of sewage or with campaigns against insects, ticks, and other animals that carry pathogens; these measures prevent disease-causing organisms from reaching us. "Getting into the host" means different things to different pathogens. Some simply multiply on the surface of a mucous membrane in the throat or intestine, whereas others get into and multiply within these surface cells, and still others pass into deeper tissues of the body or into the circulatory system.

A primary defense against invasion is provided by skin. Healthy, unbroken skin is rarely penetrated by bacteria. There are numerous other defenses against the invasion of our bodies by the parasitic organisms we meet. One of these is our **normal flora**, that is, the bacteria and fungi that live and reproduce in great number on our surfaces without causing disease (Figure 16.1). These natural occupants compete with pathogens for locations and nutrients, and some of them produce inhibitory compounds toxic to potential pathogens.

The one type of healthy surface that *is* penetrable

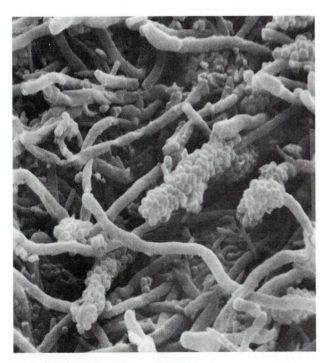

16.1 Normal Flora Gone Rampant
The human mouth harbors a wide variety of microorganisms, most of which cause no damage under normal conditions. When prokaryotic cells accumulate on the surfaces of teeth, the result is called plaque. The presence of plaque contributes to tooth decay. This electron micrograph shows human dental plaque three days after a person stopped brushing.

by bacteria is the mucous membrane, a type of mucus-secreting tissue found in parts of the visual, respiratory, digestive, and urogenital systems. However, these areas of the body have other defense mechanisms to discourage penetration by pathogens. Our tears, nasal secretions, and saliva possess an enzyme called **lysozyme** that attacks the cell walls of many bacteria. Mucus in our noses traps most of the microorganisms in the air we breathe, and most of those that get past this filter end up trapped in mucus deeper in the respiratory tract. They are removed from the respiratory tract by the beating of cilia in the respiratory passageway, which moves a sheet of mucus and the debris it contains up toward the nose and mouth. Another effective means of removing microorganisms from the respiratory tract is the sneeze reflex.

The gastrointestinal tract, too, has a number of defensive strategies. The stomach is inhospitable to most bacteria because of the hydrochloric acid that is secreted into it. (Similarly, the vagina is too acidic for many pathogens.) In the small intestine, the lining is virtually impermeable to bacteria, and some pathogens are killed by bile salts secreted into this part of the gut. The large intestine harbors many bacteria, which multiply freely, however, these are usually removed rather quickly with the feces. (The functioning of the digestive system is described fully in Chapter 41.)

If Pathogens Evade the First Defenses

Microorganisms that manage to penetrate the surface cells of an animal's body encounter still other defenses. These defenses fall into two categories: nonspecific defenses and specific ones, the latter being the immune system. One of the nonspecific mechanisms is a "battle" for iron that takes place between some pathogens and the host. Both require iron for their metabolism, but they utilize it in different forms. They produce different substances to trap the iron, and competition may be intense. If iron is in limited supply, the pathogen seldom wins the battle. The animal host also contains various antimicrobial proteins in its tissues, blood, and lymph.

An extremely important nonspecific defense against pathogens that get past the outer perimeter is afforded by white blood cells called **phagocytes**. Some of these cells adhere to certain tissues; others travel freely in the circulatory system. Pathogenic microorganisms become attached to the membrane of a phagocyte, which ingests the microorganisms by endocytosis (Figure 16.2). Once inside an endocytic vesicle in a phagocyte, pathogens are normally destroyed by enzymes from lysosomes that fuse with the vesicle (Figure 4.23). A single phagocyte can ingest 5–25 bacteria before it dies from the accumula-

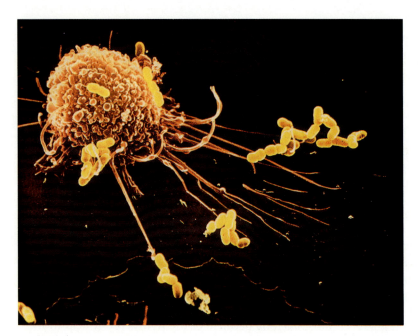

16.2 A Phagocyte and Its Bacterial Prey
Some bacteria (appearing yellow in this artificially colored scanning electron micrograph) have become attached to the surface of a phagocyte in the human bloodstream. Many of these bacteria will be taken into the phagocyte and destroyed before they can multiply and damage the human host. The long protuberances of the phagocyte probably help it move and adhere to other cells.

tion of toxic breakdown products. Even when phagocytes do not destroy all the invaders, they usually reduce the number of pathogens to the point where other defense systems can finish the job. So important is the role of the phagocytes that if their functioning is impaired by disease, the animal usually soon dies of infections.

There is a class of small white blood cells, known as **natural killer cells**, that can destroy some tumor cells and some normal cells that are infected by a virus. It is possible that the natural killer cells seek out cancer cells that may appear in the body.

Another important defense is **inflammation**. The body employs this characteristic, highly generalized response in dealing with infections, mechanical injuries, and burns. The first known written description of inflammation was given in about 30 A.D. by the Roman scholar Celsus: *"Rubor et tumor, cum calore et dolore"* (redness and swelling, with heat and pain). The redness and heat result from dilation of blood vessels in the area. The blood capillaries become leaky, so that some blood plasma (described later in this chapter) and phagocytes escape into the tissue, causing swelling. The pain results both from increased pressure (from the leakage) and from the action of certain leaked enzymes. Certain of the plasma proteins and the phagocytes are responsible for most of the healing aspects of inflammation. The heat may play a role, too, if it raises the temperature beyond that at which the pathogen that triggered the inflammation can multiply effectively. Inflammation is a response to an array of products released from damaged host cells (Chapter 34); hence it is a general, nonspecific response to many different stimuli.

Viral Diseases and Interferon

If one has a viral disease, such as influenza, one is unlikely to develop another viral disease at the same time. An apparent explanation for this phenomenon was provided in 1957 by Alick Isaacs and Jean Lindemann of the National Institute for Medical Research in London. They found that inoculating chick embryo cells with influenza virus causes the cells to produce small amounts of a substance called **interferon** that increases the resistance of neighboring cells to infection by influenza *or other* viruses. Interferons have been found in many vertebrates and are one of the body's lines of nonspecific defense against viral infection.

Interferons differ from species to species, and each species produces at least three different interferons. All interferons are glycoproteins (proteins with a carbohydrate component) consisting of about 160 amino acid units. They bind to receptors in the membranes of cells and inhibit the ability of the viruses to replicate. Interferons have been the subject of intensive research because of their possible applications in medicine.

Nonspecific Defense Mechanisms of Plants

Plants, too, have a variety of mechanisms, both mechanical and chemical, by which they resist or even actively oppose infection by pathogens. The outer surfaces of plants are protected by tissues such as the epidermis or cork. If pathogens get past these barriers, then differences between the defense systems of plants and animals become apparent. Animals generally *repair* tissues that have been infected—they heal, through appropriate developmental pathways. Plants, on the other hand, do not make repairs. Instead, they develop in ways that seal off the damaged tissue, so that the rest of the plant does not become infected. In trees the sealing is accomplished by the production of new wood different in orientation and chemical composition from the previously deposited wood. Some of the new cells also contain substances that resist the growth of microorganisms and hence tend to protect the rest of the plant.

The healing mechanism just described is primarily mechanical. Many plants have chemical defenses as well. Certain fungi, when they infect one of these plants, stimulate the host to produce substances called **phytoalexins**. Phytoalexins are toxic to fungi, and some have limited antibacterial activity as well. Their antifungal activity is nonspecific, that is, the phytoalexins can destroy many species of fungi in addition to the one that originally triggered their production. Physical injuries, viral infections, and even certain chemical compounds can also induce the production of phytoalexins.

SPECIFIC DEFENSES AGAINST INVADERS

Our nonspecific defenses are numerous and effective, but some invaders elude the nonspecific defenses and must be dealt with by defenses targeted against specific threats. This specific destruction of nonself materials is an important function of an animal's **immune system**. The immune system recognizes and attacks specific invaders, such as bacteria and viruses. Once a specific response has been raised against a particular type of nonself cell or macromolecule, the immune system can usually respond more rapidly and powerfully if it encounters the same threat in the future. Thus the functions of the vertebrate immune system are to *recognize*, *selectively eliminate*, and *remember* foreign invaders. Based on the number of cells it contains, the immune system is the largest organ in the body.

If an animal has a defective immune system, it can die from infection by even "harmless" bacteria. Some microorganisms routinely carried in or on an animal's body without harm are potentially pathogenic, and will cause disease if the host's immune system is stressed in some way.

There are two forms of immune response in the vertebrate body. The **cellular immune response**, carried out by cells that recognize nonself cells and molecules, acts against fungi, foreign tissue, multicellular parasites, and viral infections that have become established within cells. The **humoral immune response**, carried out by protein molecules present in the blood, acts primarily against bacteria and against

viruses that have not yet invaded the cells (the name comes from the Latin *humor* = "fluid"). There is considerable overlap in the functions of these two systems, and we will see that there is much in common between their mechanisms. For example, the cells that produce the humoral and cellular immune responses are closely similar.

Cells of the Immune System

The immune system is made up of cells that travel in the blood and lymph. Blood is a fluid tissue. About 55–65 percent of a human's blood is the yellowish liquid matrix called **plasma**; the remainder consists of red blood cells, white blood cells, or **leukocytes**, and platelets. Plasma is mostly water but contains many other important substances. When the substances in plasma clot, the fluid that remains is the **serum**.

As we will see in Chapter 40, some of the circulating components of blood are returned to the heart not by veins but by the lymphatic system (Figure 16.3). **Lymph**, a blood filtrate that accumulates in the spaces outside the blood capillaries, contains water, solutes, and leukocytes that have left the capillaries, but no red blood cells. The lymph is collected in ducts and routed back into a blood vessel near the heart.

Leukocytes are larger and far less numerous than red blood cells in the blood (Figure 16.4). They have nuclei and are colorless. They move like amoebas by extending **pseudopods** and can enter the tissues by squeezing through junctions between the cells making up the walls of blood capillaries. The number of leukocytes in the blood and lymph may rise sharply during infection, providing a useful clue for detecting an infection.

The most abundant of the several types of leukocytes are the phagocytes, which play such an important role in the nonspecific defenses of the body, and the **lymphocytes**. Lymphocytes play predominant roles in the immune systems. Two groups of lymphocytes, the **B cells and T cells**, originate from cells in the bone marrow. The precursors of T cells migrate to an organ of the lymphoid system called the thymus and develop their unique properties there, becoming mature T cells. The B cells migrate from the bone marrow to the peripheral lymphoid system, circulating in the blood and lymph vessels and passing through the lymph nodes and peripheral lymphoid organs such as the spleen. The B and T cells are indistinguishable under the light microscope, but they have quite different functions in immune responses. Both B and T cells give rise to cells that develop in special ways as part of the immune response. Among the cells developing from B cells are the plasma cells, whose role will be described briefly in the next section.

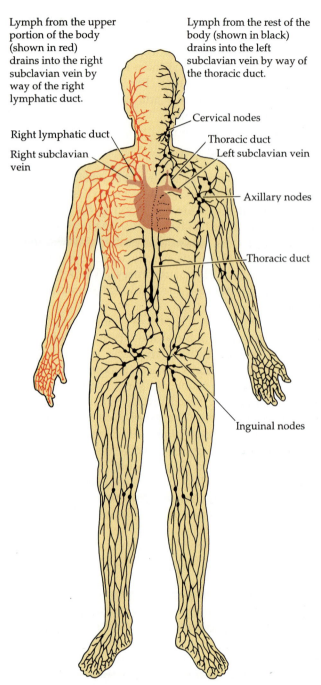

Lymph from the upper portion of the body (shown in red) drains into the right subclavian vein by way of the right lymphatic duct.

Lymph from the rest of the body (shown in black) drains into the left subclavian vein by way of the thoracic duct.

Cervical nodes

Right lymphatic duct

Thoracic duct

Right subclavian vein

Left subclavian vein

Axillary nodes

Thoracic duct

Inguinal nodes

16.3   The Lymphatic System
A network of ducts collects lymph from the body's tissues and carries it toward the heart, where it mixes with blood to be pumped back to the tissues. There are major lymph nodes in the neck, armpits, and groin. What we call "swollen glands" in the neck are cervical lymph nodes in which invading bacteria are trapped.

THE IMMUNE RESPONSE

Let us now consider the broad outlines of the complex processes mediated by the lymphocytes. Foreign organisms and substances that invade the body and escape the nonspecific phagocytes in the animal must

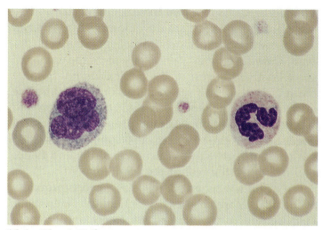

16.4   Blood Cells
The large, stained cell to the left is a leukocyte called a monocyte. The other stained cell is a leukocyte called a neutrophil. Notice that the numerous red blood cells in this blood smear do not have nuclei and hence are unstained.

come up against the immune system. The *humoral* immune system is based on highly specific protein molecules called **antibodies** that, among them, can react with virtually any conceivable foreign cell or biological macromolecule. Antibodies, secreted by those B cells that have differentiated into plasma cells, are also called **immunoglobulins** (Igs). Each *recognizes* and binds to a particular site—an **antigenic determinant**—found on one or more molecules foreign to the animal (Figure 16.5). The entire foreign body bearing the antigenic determinant, whether it is a microorganism or, say, a protein molecule, is called an **antigen**. An antigenic determinant is that part of an antigen to which an immunoglobulin (antibody) binds. Because an antigenic determinant is a specific chemical grouping, it may be found on many different molecules. A large antigen such as a whole cell may have many different antigenic determinants on its surface, each capable of eliciting and binding a specific antibody. Some immunoglobulin molecules travel free in the blood and lymph; others exist as integral membrane proteins on B cells.

The *cellular* immune system, in contrast to the humoral, does not use antibodies. Instead, its T cells have **T-cell receptors**, specific surface molecules that recognize and react to antigenic determinants on the surfaces of other cells. The T-cell receptors, which are glycoproteins, have specificities like those of the humoral immune system. They bind to specific antigenic determinants on nonself material, initiating the destructive activity of the T cell. The T cells recognize and help in the destruction of nonself cells or of any of the body's own cells that have been altered by virus infections. Because the T cells recognize and mobilize attacks on foreign material, they can bring about rejection of certain types of organ or tissue

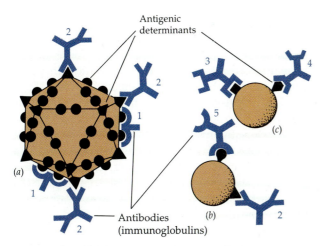

16.5 Each Antibody Recognizes a Unique Antigenic Determinant

Three foreign bodies are depicted in tan: a virus (a) and two distinct globular proteins (b and c). Each has on its surface antigenic determinants (black) that are recognized by specific antibodies (blue shapes). An antibody recognizes its antigenic determinant wherever it is; for example, antibody 2 locates its unique determinant both on the virus (a) and on protein (b).

transplants. The important difference between lymphocytes (both T and B) and phagocytes is that a particular T or B cell reacts specifically with only one specific antigenic determinant, whereas phagocytes react nonspecifically with any foreign matter they encounter.

One of the most striking features of the immune system is that an animal does not require a previous encounter with a particular antigen to mount an immune response. An invading foreign protein elicits the immune response even if it is the body's first contact with that protein. This observation prompted some biologists and chemists to propose that a specific immunoglobulin must be formed *after* the body encounters a specific antigen, perhaps by using the antigenic determinant as a template for the final three-dimensional folding of the immunoglobulin molecule. However, this idea had to be abandoned in the face of experimental results. It is now known that a person contains the means of producing *millions* of distinct immunoglobulins without foreign templates—even though that person may never have encountered the antigenic determinants corresponding to more than a small fraction of them. The problem of accounting for the origin of such a tremendous diversity of specific proteins will be considered later in this chapter.

Immunological Memory and Immunization

An important feature of the immune system is **immunological memory**. The first time a vertebrate animal is exposed to a particular antigen (for example,

the bacterium *Bordetella pertussis*, which causes whooping cough), there is a time lag (usually several days) before the number of antibody molecules and the number of T cells circulating in the bloodstream catch up with the number of invaders. But for years afterward, sometimes for life, the immune system "remembers" that particular antigen and remains capable of responding on shorter notice. Whereas the first exposure to the antigen results in a standard response, a second exposure causes a much greater, longer sustained, and more rapid production of antibodies and T cells (Figure 16.6).

The ability of the human body to remember a specific antigen explains why **immunization** has virtually wiped out such deadly diseases as smallpox, diphtheria, and polio in medically sophisticated countries. (Smallpox, in fact, has been eliminated worldwide from the spectrum of infectious diseases affecting humans, thanks to a concentrated international effort by the World Health Organization. As far as we know, the only remaining smallpox viruses on Earth are those kept in some laboratories.) In **vaccination** a small inoculum of virus or bacteria or their proteins (usually treated to make them harmless) is injected into the body. Later, if the same or very similar disease organisms should attack, prepared cells recognize the antigen and quickly overwhelm the invaders with a massive production of lymphocytes and immunoglobulins.

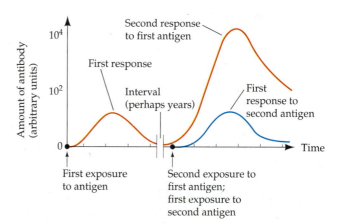

16.6 Immunological Memory

A first exposure to an antigen (red curve) stimulates the body to produce antibodies within a few days. A second exposure to the same antigen, perhaps years later, stimulates the body to respond almost immediately and to produce much larger amounts of antibody (note the logarithmic scale on the vertical axis). This is why vaccinations provide long-term protection against disease and why they sometimes provoke allergic reactions. However, a first exposure to a *second* antigen (blue curve) given at the same time as the second exposure to the first antigen, is not met by an enhanced response.

Clonal Selection and Its Consequences

Each person possesses an enormous number of different B cells and T cells, apparently capable of dealing with practically any antigen ever likely to be encountered. How does this diversity arise? Also, why do some of our immunoglobulins and T cells not attack and destroy the components of our own bodies? An individual can mount an immune response against another person's proteins, yet it rarely mounts one against its own. The immune system can distinguish self (one's own antigens) from nonself (those from outside the body). As we will now see, the versatility of immune responses, immunological memory, and the recognition of self can all be explained satisfactorily in terms of a particular theory of the origin of specific immunoglobulins.

In 1954 the Danish immunologist Niels K. Jerne proposed a new view of the relationship between antigen and antibody. His idea was that the antigen does not itself specify the structures of the antibodies formed against it; instead, in Jerne's view, those antibodies are already being produced in small quantity, and the antigen specifically stimulates the lymphocytes that are making those particular antibodies. There must be a population of different lymphocytes corresponding to each of the antigenic determinants to which the organism can respond. Jerne's model was improved and extended by the Australian immunologist MacFarlane Burnet, who named it the **clonal selection theory**. According to the theory, the individual animal contains an enormous variety of different B cells, each type able to produce only one kind of antibody (Box 16.A). The specificities exist prior to the organism's encountering the antigens. The cell-to-cell variation in antibody specificity results from variation in genotype. This variation in genotype is the exception, mentioned in Chapter 15, to the rule that all cells in an organism have identical genotypes.

After an antigen enters the body, it encounters B cells that can recognize its antigenic determinants. As a consequence of this meeting, each of these B cells begins to multiply, giving rise to a large clone of plasma cells, each of which secretes antibody of the same specificity (Figure 16.7).

The clonal selection theory accounts nicely for the body's ability to respond rapidly to any of a vast number of different antigens. In the extreme case, even a single B cell might be sufficient for an immunological response by the body, provided it encountered the antigen and then proliferated into a large enough clone rapidly enough to deal with the invasion. Clonal selection accounts for the proliferation of T cells as well as of B cells.

The clonal selection theory also explains two other phenomena. One, the recognition of self, is discussed in the next section. The other, immunological memory, was illustrated in Figure 16.6. According to the clonal selection theory, the activated lymphocyte produces *two* kinds of daughter cells. The ones that carry out the attack on the antigen are **effector cells**—either plasma cells that lead to antibody production or T cells that bind antigenic determinants. The other kind, called **memory cells**, are long-lived cells that

BOX 16.A

A B Cell Produces One Immunoglobulin

A fundamental postulate of the clonal selection theory is that a given B cell makes one, and only one, specific immunoglobulin, that is, an antibody directed against one antigenic determinant. How might we test this hypothesis? One approach would be to inoculate an animal simultaneously with two different antigens. After waiting to allow the immune response to develop, we could take a sample of blood, carefully isolate single plasma cells in individual microdrops, and test each to see whether it produced antibodies to either or both of the antigens. (Why do you suppose we inoculate the animal at the start of the experiment, rather than simply taking a control animal as the source of plasma cells? This question will be dealt with in the next paragraph.) Such an experiment was actually performed, with the following results. Most of the plasma cells did not produce antibody to either antigen; but of those that did, each produced only a *single* antibody type. No cell was found to produce both antibodies.

In principle, this experiment could have been done without first exposing the animal to the two antigens. However, the search for *any* plasma cell that responded to either antigen would then have been a needle-in-a-haystack problem. With perhaps millions of different, specific plasma cells present, one would probably never find the one that was desired. By inducing the animal to mount an immune response to one or more antigenic determinants, we cause an enormous increase in the fraction of plasma cells producing antibodies of the corresponding types.

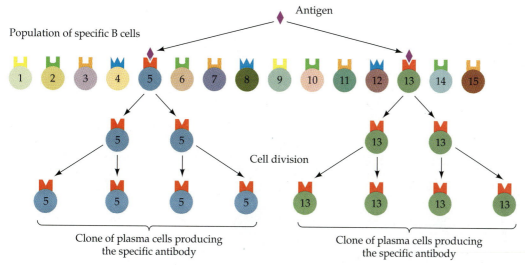

Population of specific B cells

Antigen

Cell division

Clone of plasma cells producing
the specific antibody

Clone of plasma cells producing
the specific antibody

16.7 Clonal Selection
An animal produces many kinds of B cells, each genetically unique and each
able to recognize a particular antigen. Of the B cells shown here, only two—5
and 13—recognize the antigen (red). The antigen stimulates these B cells to
proliferate, forming clones of plasma cells that produce the specific antibody
specified by the ancestral cell's genotype.

retain the ability to start dividing on short notice to
produce yet more effector and more memory cells.
Effector cells live only a few days, but memory cells
may survive for decades. When the body first encounters a particular antigen, one or more types of
lymphocytes become activated and divide to produce
clones of effector and memory cells. (Why more than
one type? Because the antigen may possess more
than one antigenic determinant.) The effector cells
destroy the invaders at hand, but one or more clones
of different memory cells have now been added to
the immune system. Thus if the animal encounters
the same antigen a second time it can respond more
rapidly and more massively, as indicated in Figure
16.6.

Self, Nonself, and Tolerance

In addition to explaining immunological memory, the
clonal selection theory may explain the recognition
of self. Given the great array of different lymphocytes
directed against particular antigens, how is it that a
healthy animal apparently does not produce self-de-
structive immune responses? Two possible mechanisms of this tolerance to self have been proposed;
there is evidence for both, and it is not yet known
whether one, the other, or both mechanisms are important in an animal. The older of the two theories
proposes a sensitivity to antigen on the part of lymphocytes that have not fully developed. "Anti-self"
lymphocytes would encounter corresponding self antigens before the differentiation of the lymphocytes
is complete, and this meeting may lead either to the

inactivation or to the complete elimination of the anti-
self lymphocytes. Thus no clones of anti-self lymphocytes would normally appear in the bloodstream.
This is the **clonal deletion theory**.

Much later, a class of lymphocytes known as **suppressor T cells** was discovered. These antigen-spe-
cific cells inhibit the activities of effector T and B cells.
It is likely that at least some aspects of self-tolerance
are the result of the inactivation of anti-self lymphocytes by specific suppressor T cells.

In 1945 Ray D. Owen (California Institute of Technology) observed that some *nonidentical* twin cattle
contained some of each other's red blood cells, even
though these were of differing types and might have
been expected to elicit immune responses resulting
in their elimination. Four years later Burnet suggested that the blood cells had passed between the
animals before the differentiation of immune specificities was complete—and were thus regarded as
"self" when the recognition of self developed. Burnet
further proposed that, if this were true, one should
be able to inject foreign antigen into an animal early
in its development and cause that animal henceforth
to recognize that antigen as "self." This induction of
immunological tolerance was demonstrated in 1953
by the English immunologist Peter B. Medawar, who
used strains of mice so highly inbred as to be almost
clones (Figure 16.8). He injected lymphoid cells from
adult mice of one strain into newborn mice of another
strain. Other newborn mice of the second strain
served as uninjected controls. Eight to ten weeks
later, he tested for tolerance in the treated and untreated mice by grafting skin onto them. The un-

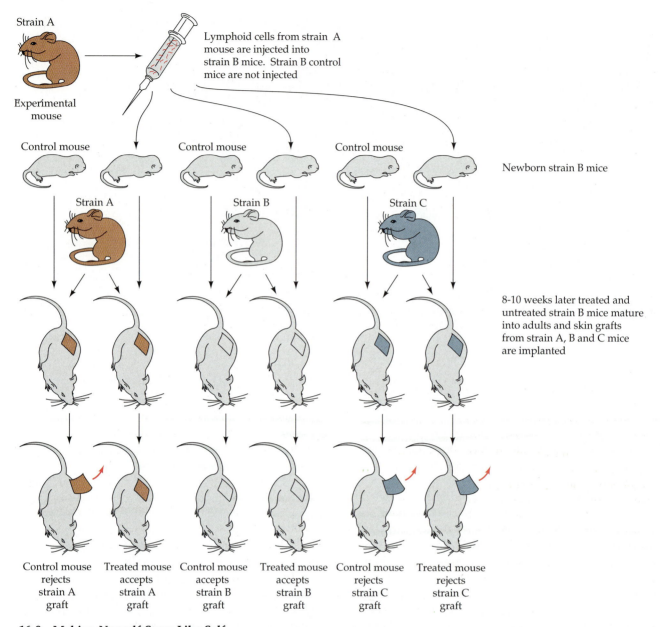

Strain A

Experimental mouse

Lymphoid cells from strain A mouse are injected into strain B mice. Strain B control mice are not injected

Control mouse   Control mouse   Control mouse   Newborn strain B mice

Strain A   Strain B   Strain C

8-10 weeks later treated and untreated strain B mice mature into adults and skin grafts from strain A, B and C mice are implanted

Control mouse rejects strain A graft

Treated mouse accepts strain A graft

Control mouse accepts strain B graft

Treated mouse accepts strain B graft

Control mouse rejects strain C graft

Treated mouse rejects strain C graft

16.8 Making Nonself Seem Like Self

The ability of adult mice to recognize and reject grafts of foreign skin can be overcome. Adult mice of strain B that were injected shortly after birth with lymphoid cells from strain A (upper left) tolerate grafts from strain A or strain B, but reject as foreign grafts from other strains, such as strain C. As shown at the bottom, adults of strain B raised from uninjected newborn mice accept grafts only from other strain B mice and reject skin from strain A as well as from strain C. What is recognized as "self" and "nonself" depends partly on when it is first encountered.

treated mice rejected grafts from the other strain, but the treated mice accepted them. Medawar thus discovered that immunological tolerance to an antigen can be induced by exposure to the antigen early in development.

The establishment of tolerance, whether by the production of appropriate suppressor T cells or by the process of clonal deletion, must be repeated throughout the life of the animal because lympho-

cytes are constantly produced. Continued exposure to self-antigen helps to maintain tolerance. If, for some reason, an animal stops producing a given protein—that is, a self-antigen—a clone of lymphocytes directed against that protein may become established. Then if the protein is synthesized again at a later time, it may elicit a full-fledged immune response, and an **autoimmune disease** (such as rheumatoid arthritis, in which an immune system attacks part of the body in which it resides) may result.

The role of continued exposure to self-antigen was nicely demonstrated in 1962 by Edward L. Triplett, of the University of California, Santa Barbara. He removed the pituitary glands of frog embryos and "stored" them under the skin of other very young embryos. Later, when the original owners had matured, Triplett returned their pituitaries to them—whereupon the glands were rejected. The frogs no longer recognized them as self, presumably because there had not been continuous exposure to them. As one of his controls, he removed only parts of some of the pituitaries and stored them as before in young embryos. When these pieces of pituitary were returned to their matured owners (which had grown up with partial pituitaries), they were not rejected. Research on the immune recognition of self continues, in part to resolve the uncertainty over its mechanism and in part because of the importance of finding cures to autoimmune diseases.

Development of Plasma Cells

Let us now turn to a more detailed consideration of the *mechanisms* of the immune response. When a B or T cell is activated by an antigen, it proliferates, with both effector and memory cells being produced. Recall that the effector cells that develop from B cells are called plasma cells. The plasma cells secrete antibodies with the same specificity as that of the antibody receptors on the surface of the parent B cell; that is, these new immunoglobulins are specific for the same antigenic determinant that triggered their formation.

The activation of a B cell begins with the binding of particular antigens to the particular antibodies carried on the B-cell surface. Activation requires the presence of a particular type of T cell, which also binds to the antigen. The effect of the T cell on the subsequent differentiation of the B cell has much in common with the phenomenon of *induction* discussed in Chapter 15. We will speak of these lymphocyte interactions again in this chapter.

Following the activation of the B cell, cellular division and differentiation lead to the formation of plasma cells and memory cells. As plasma cells develop, the number of ribosomes and the amount of endoplasmic reticulum in their cytoplasm increase greatly (Figure 16.9)—changes that one would expect in cells that will be devoted to the synthesis of large amounts of proteins (immunoglobulins) for secretion. All the plasma cells arising from a given B cell produce antibodies of specificity identical to that of the receptors on the parent B cell.

A problem of great interest for many years was that of obtaining cultures of plasma cells all producing the same antibody, so that large quantities of *pure* antibody could be prepared. That has been solved by

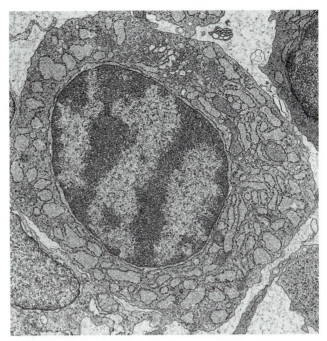

16.9 A Plasma Cell
Note the prominent nucleus (recognizable by the double membrane), the cytoplasm crowded with rough endoplasmic reticulum, and an extensive Golgi complex to the right of the center near the top of the cell—all structural features of a cell actively synthesizing and exporting proteins.

the discovery of methods for producing monoclonal antibodies.

Monoclonal Antibodies

Until recently, research on antibodies was complicated by the fact that most antigens carry many different antigenic determinants. Thus when an animal produced antibodies, it would produce a complex mixture; it was virtually impossible to separate the individual antibody types for chemical study. In the mid-1970s, however, Cesar Milstein (an Argentine biochemist living in Cambridge, England) and a colleague from Switzerland, Georges Köhler, made an important breakthrough. They knew that a single lymphocyte produces only a single species of antibody. In principle, all one needed to do was to cause a single lymphocyte to multiply in pure culture to get a large population of cells all dedicated to the production of the same antibody. However, the antibody-producing cells cannot be cultured. On the other hand, there are cancerous tumors of plasma cells called myelomas that grow rapidly in culture. The cells of a given tumor all produce the same immunoglobulin, but they produce far too little of the immunoglobulin to be useful sources. Milstein and Köhler made use of both cell types—normal lymphocytes and myeloma cells—to produce hybrid cells (**hybridomas**) that made specific normal antibodies in quantity and that, like the myeloma cells, could proliferate rapidly and indefinitely in culture.

Clones of hybrid cells are made as follows (Figure 16.10): An animal is inoculated with an antigen to trigger specific lymphocyte proliferation. Later, the spleen is dissected out and lymphocytes are collected from it. (The spleen, like the lymph nodes and certain other lymphoid tissues associated with the gut, is a site of lymphocyte accumulation and maturation.)

These lymphocytes are combined under appropriate conditions with myeloma cells from a single tumor. Cell fusion occurs between the lymphocytes and the myeloma cells, giving rise to hybridomas. The hybrid cells are grown in a suitable medium so that a clone forms from each. Individual clones are tested and the ones that produce the desired antibodies—specific for one antigenic determinant—are selected. These clones produce **monoclonal antibodies** (uniform an-

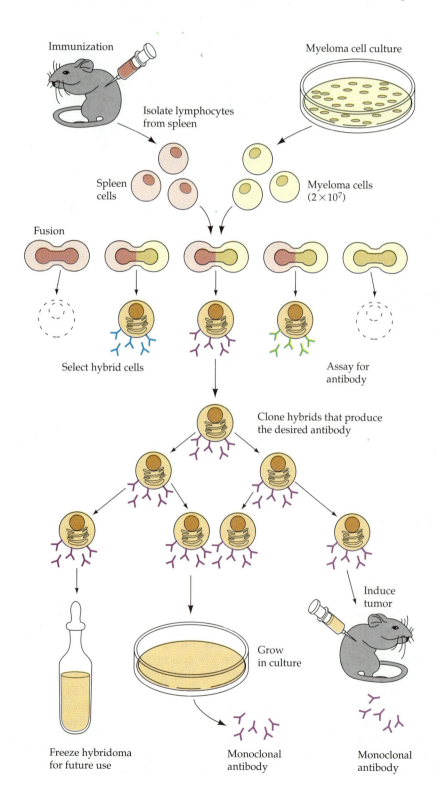

16.10 Preparation of Monoclonal Antibodies

First the antigen for which monoclonal antibodies are desired is injected into a mouse. The mouse's spleen is later removed and lymphocytes, some of which are specific for the injected antigen, are isolated from the spleen. The lymphocytes are mixed with mouse myeloma cells and treated to encourage fusion, yielding hybridomas. The cell mixture is next treated so as to destroy all nonhybrid cells, and surviving cells are tested one by one for their ability to produce the desired antibody. Positive-testing hybrids are cloned. One may then freeze the hybridoma for storage, amplify the clone by growing it in mass culture, or inject it into mice. Injected mice develop tumors that produce large amounts of monoclonal antibody.

tibodies from a single clone of cells) in large quantities, either from a mass culture or following transfer into an animal where they can grow as a tumor.

Monoclonal antibodies are ideal for the study of specific antibody chemistry, and they have been used to further our knowledge of cell membranes as well as for specialized laboratory procedures such as tissue typing for grafts and transplants. Medical uses are also being developed. One possibility is passive immunization—inoculation with specific antibody rather than with an antigen that would cause the patient to develop his or her own antibody. Monoclonal antibodies can also be used for such purposes as detecting specific cancers—diagnostic kits using monoclonal antibodies for colon cancers are already available. Antitumor monoclonal antibodies have been successfully employed in immunotherapy for tumors. In yet another use, poisons can be attached to specific monoclonal antibodies directed against tumors as a form of cancer treatment.

THE IMMUNOGLOBULINS

The chemical structure of the most common form of immunoglobulins was worked out by Gerald M. Edelman (Rockefeller University) and Rodney M. Porter (Oxford University). They found that the basic immunoglobulin molecule is a tetramer consisting of four polypeptides (Figure 16.11). The complex contains two identical "light" chains and two identical "heavy" chains, the chains being held together by disulfide bonds. Each chain consists of a **constant region** and a **variable region**. The constant regions of both light and heavy chains are similar from one immunoglobulin to another. On the other hand, the amino acid sequence of the variable region (considering the variable ends of the heavy and light chains together) is unique in each of the millions of different types of immunoglobulins in a person's humoral immune system. Thus the variable regions of a light and a heavy chain combine to form, on each of the immunoglobulin's "arms," a highly specific, three-dimensional structure that is not unlike the active site of an enzyme. It is this characteristic part of a particular immunoglobulin molecule that combines with a particular, unique antigenic determinant.

Although the variable regions are responsible for the *specificity* of an immunoglobulin, the constant regions are equally important, for it is the constant regions that determine the type of action to be taken in eliminating the antigen, as we will see in the next section. In particular, the constant regions of the heavy chains determine whether the antibody remains part of the cell's plasma membrane or is secreted into the bloodstream. The two halves of an antibody, each consisting of one light and one heavy chain, are identical, so each of the two arms can combine with an identical antigen, leading some-

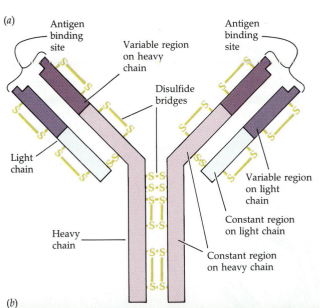

(a)

Antigen binding site

Variable region on heavy chain

Disulfide bridges

Light chain

Heavy chain

Antigen binding site

Variable region on light chain

Constant region on light chain

Constant region on heavy chain

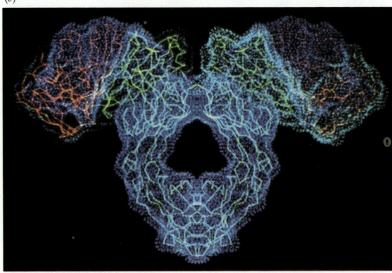

(b)

16.11 Structure of Immunoglobulins
(a) Disulfide bridges (yellow) hold the four polypeptide subunits of an immunoglobulin together. The variable regions recognize and interact with antigens. (b) An immunoglobulin molecule in roughly the same orientation as (a), drawn by a computer. Overall, the light chains are shown as green surfaces (the dot pattern) and the heavy chains as dark blue surfaces. The variable regions are shown as red skeletons (red lines); the constant regions are shown as yellow skeletons in the light chains and as light blue skeletons in the heavy chains.

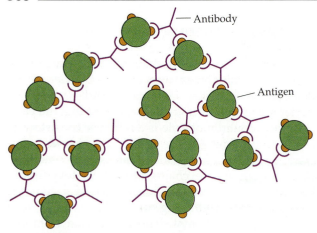

16.12 Antibody–Antigen Complex
Antibodies have two sites that can bind to different molecules of antigen, and more than one antibody may bind to the same antigen molecule. Large antibody–antigen complexes may precipitate from the blood; they play a role in many kinds of clinical laboratory tests.

times to the formation of a large complex of antigen and antibody molecules (Figure 16.12).

The Classes of Immunoglobulins

There are five immunoglobulin classes in all (Figure 16.13). One, called immunoglobulin M, or **IgM**, is always the first antibody product of a plasma cell. Four other classes arise by a process called class switching, which will be discussed later in this chapter. These four classes are called IgD, IgE, IgG, and IgA; each plays a specific role or roles in the immune system.

IgG molecules, which have the γ heavy-chain constant region, make up about 85 percent of the total immunoglobulin content of the bloodstream. They consist of a single immunoglobulin unit (two identical heavy chains and two identical light chains) and are produced in greatest quantity during a secondary immune response (Figure 16.6). IgG defends the body in several ways. For example, some IgG molecules that have bound antigens become attached by their heavy chains to phagocytes. This makes it easier for the phagocytes to ingest the antigens (Figure 16.14). Another major function of IgG is to activate the complement system, a potent nonspecific defense system discussed in the next section, which also enhances phagocytosis.

IgM, in contrast, constitutes the bulk of the antibodies produced at the beginning of a primary immune response. It differs from IgG in being composed of five immunoglobulin units (Figure 16.13). Because they have more binding sites, IgM molecules are more active than IgG molecules in activating the complement system and promoting the phagocytosis of antibody-coated cells. The major role of **IgD** is as membrane receptors on B cells (like IgM), but as soluble antibody IgD is found only at low levels.

IgE takes part in inflammation and allergic reactions. IgE helps kill worm pathogens such as those that cause the disease schistosomiasis, which affects some 200 million people in Africa and South America. Where inflammation occurs, IgE may participate in

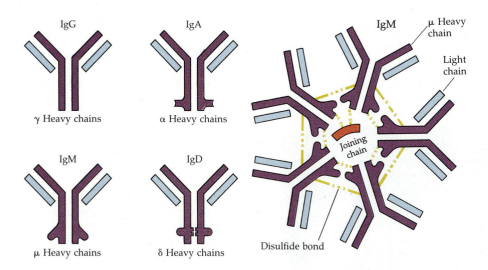

16.13 Classes of Immunoglobulins
The immunoglobulin classes differ in their heavy chains. IgM, unlike other antibodies, is made up of five immunoglobulin subunits. Two of the five units are bonded to a single joining chain.

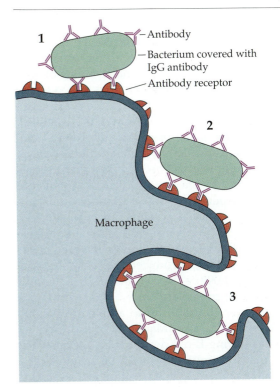

16.14 IgG Antibodies and Phagocytosis
Phagocytes called macrophages have receptors on their cell surfaces for part of the IgG molecule. Thus when a bacterium has reacted with IgG antibodies, the resulting complex binds readily to macrophages, activating phagocytosis.

bringing leukocytes, components of the complement system, and other factors into the inflamed region. For most of us, the effect of IgE is most apparent when we suffer allergies. IgE molecules bind to antigenic determinants on the substances—the allergens—that provoke the allergy, and they also bind to receptor sites on the surfaces of cells called mast cells. The mast cell–IgE–allergen complex stimulates the release of histamine and other compounds, leading in turn to inflammation. Hives, hay fever, eczema, and asthma are all common allergic reactions (Figure 16.15).

Body secretions such as saliva, tears, milk, and gastric secretions all contain immunoglobulins, specifically **IgA**. IgA molecules are transported across epithelial cells to join the secreted fluids. IgA exists as both monomers and dimers, but Figure 16.13 shows only the monomeric unit.

THE COMPLEMENT SYSTEM

A significant fraction of the total protein content of vertebrate blood consists of the about 20 different proteins that make up the **complement system**. These proteins, in different combinations, perform three types of defensive reactions. Their most impressive defense reaction is antibody-mediated lysis of foreign

cells—bacteria, for example. When antibodies bind to antigenic determinants on the surface of a foreign cell, this binding may bring about the binding of the first of the complement proteins to the cell surface. What follows is a cascade of reactions, with different complement proteins acting upon one another in succession. The final product of the complement cascade is a lytic complex—a doughnut-shaped structure in the foreign cell membrane that renders the membrane leaky and thus causes lysis (bursting) of the foreign cell (Figure 16.16).

Another activity of the complement system is to increase the effectiveness of endocytosis and the destruction of foreign microorganisms by phagocytes. Phagocytes can easily recognize foreign cells after complement proteins attach to the foreign cells. A third activity of the complement system is the attraction of phagocytes to sites of infection.

The complement system is *nonspecific* in its action—the specificity of recognition in antibody-mediated lysis is due to the action of the antibodies that bind specifically to cell surface antigens of the invading cells, not to the complement proteins themselves. Hereditary deficiencies in one or another of the complement proteins can cause characteristic diseases, mostly infections and hypersensitivity diseases.

THE ORIGIN OF ANTIBODY DIVERSITY

When a mammal is born, it possesses a full complement of genetic information for antibody synthesis. At each of the loci coding for the heavy and light chains it has an allele from the mother and one from the father. Throughout the animal's life, each of its cells begins with the same full complement. However, the genomes of B cells become modified during development in such a way that each cell eventually can produce one—and only one—specific type of an-

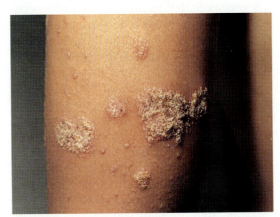

16.15 An Allergic Reaction
Eczema is a common allergic reaction characterized by itching, redness, and the appearance of crusted lesions. The condition is not contagious.

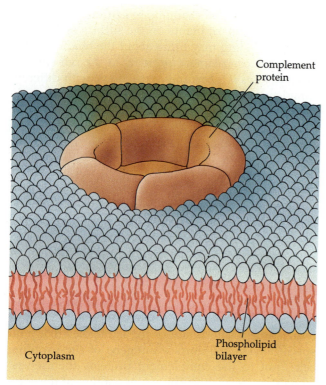

Complement protein

Cytoplasm

Phospholipid bilayer

16.16 Complement Action Produces a Lytic Complex
The end result of a cascade of reactions is a precisely arranged group of proteins extending through the phospholipid bilayer of the plasma membrane of a foreign cell. A pore in the complex makes the affected cell leaky; fluids rushing into the cell then cause it to lyse.

tibody. Different B cells develop different antibody specificities. A question that vexed immunologists and geneticists for decades was how a single organism could produce so many different specific immunoglobulins—perhaps a million, or even a billion antibody specificities. Research in recent years has effectively answered this question.

The most surprising part of the answer is that functional immunoglobulin genes are assembled from DNA segments that initially are spatially separate. Every cell has hundreds of DNA segments potentially capable of participating in the synthesis of the variable regions, the parts of the antibody molecule conferring immunological specificity. B-cell precursors differ from all other cells in that these segments are *rearranged* during B-cell development. Pieces of the DNA are deleted, and DNA segments formerly distant from one another are spliced together; thus a gene is assembled from randomly selected pieces. Each B-cell precursor in the animal assembles its own unique set of immunoglobulin genes. This remarkable process generates many diverse antibodies from the same starting genome. The assembly of immunoglobulin genes from spatially separate DNA segments was first demonstrated by Susumu Tonegawa, in Switzerland, in 1976.

In both humans and mice, the DNA segments coding for immunoglobulin heavy chains are on one chromosome and those for light chains are on others. The variable region of the light chain is coded for by two families of DNA segments, and the variable region of the heavy chain is coded for by three families. These families are assembled virtually at random; thus the diversity afforded by several hundred DNA segments is multiplied by the combination of different families. And, since light and heavy chains are synthesized independently of one another, the combination of light and heavy chains introduces further diversity. If, say, 1,000 different light-chain variable regions and perhaps 10,000 different heavy chain variable regions, could be produced, this would allow some $1,000 \times 10,000 = 10,000,000$ different immunoglobulins to be produced.

Another factor contributing to antibody diversity is *mutation*, which occurs frequently in the variable-region genes during B-cell development. It has been estimated that such mutations increase the total number of different antibody specificities by at least 10–100 times.

How a B Cell Produces a Particular Heavy Chain

To see how DNA rearrangement generates antibody diversity, let us consider how the heavy chain of IgM is produced. B cells produce this antibody, which then attaches to the plasma membrane of the B cells.

The locus governing heavy-chain synthesis is found on chromosome 12 of mice and on chromosome 14 of humans. In mice, the locus is laid out as shown in Figure 16.17, with a long stretch of DNA occupied by a family of 100 or more V segments (where V stands for "variable"). Humans have on the order of 300 V segments. In a given B cell, *one* of these segments is used to produce part of the variable region of the heavy chain, with the remaining V segments being discarded or rendered inactive. At a distance of many nucleotides from the V segments, there is a family of 10 or more D (for "diversity") segments. *One* of these, too, is used to produce part of the variable region of the heavy chain of a given B cell, as is *one* J (for "joining") segment from the family of four such segments (in mice) lying yet farther along the chromosome. This combination of V, D, and J segments forms a complete variable region for a functional gene. Still farther along, and separated from the suite of J segments, is a family of eight segments, one of which comes to code for the constant region of the heavy chain. Light chains are produced from similar families of DNA segments, but without D segments.

How does order emerge from this seeming chaos of DNA segments? It comes from two important steps. First, substantial chunks of DNA are deleted

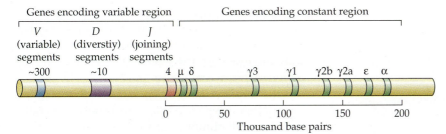

16.17 Heavy-Chain Genes
Immunoglobulin heavy chains are encoded by a locus that has many segments. The variable region for the heavy chain of a particular antibody is encoded by one *V* segment, one *D* segment, and one *J* segment. Each of these segments is taken from a pool of like segments. The constant region is selected from another pool of segments.

from the chromosome during the rearrangement of the segments. As a result of these deletions, a particular *D* segment is joined directly to a particular *J* segment, and then the *D* segment is joined to one of the *V* segments; thus a single "new" sequence, consisting of one *V*, one *D*, and one *J* segment, can now code for the variable region of the heavy chain. All the progeny of this cell constitute a clone having this same sequence for the variable region. In other B cells, DNA may be deleted in different places, leading to different variable-region sequences. In mice, with about 100 *V* segments, 10 or more *D* segments, and 4 *J* segments to choose from, there should be about $100 \times 10 \times 4 = 4,000$ different heavy-chain variable

regions produced in different B cells—before we allow for the further multiplication that is introduced by mutation.

The second step in organizing the synthesis of an immunoglobulin chain follows transcription. Splicing of the RNA transcript (Chapter 13) removes the product of an intron that includes any *J* segments lying between the selected *J* segment and the first constant-region segment. Splicing of the RNA transcript also removes the products of introns contained in both the *V* segment and the constant-region segment (Figure 16.18). The result is an mRNA that can be translated, directly yielding the heavy chain of the cell's specific antibody.

The Constant Region and Class Switching

Early in its life a plasma cell produces IgM molecules that are responsible for the specific recognition of a particular antigenic determinant. At this time, the constant region of the antibody's heavy chain is encoded by the first constant-region segment, the μ segment (Figure 16.17). Later in an immunological response—later in the life of the plasma cell—a further deletion may occur in the plasma cell's DNA, so that the heavy-chain variable-region gene (consisting of the same *V*, *D*, and *J* segments) is now juxtaposed with a constant-region segment farther down the original DNA, such as γ, ε, or α (Figure 16.19).

Such a deletion—called **class switching**—results in the production of an antibody with a different *function* but the same *antigen specificity*. The new antibody has the same variable regions of the light and heavy chains but a different constant region of the heavy chain. This antibody falls in one of the four other immunoglobulin classes (IgD, IgE, IgG, or IgA; Figure 16.13), depending on which of the constant-region segments is adjacent to the variable-region gene. Once class switching has occurred, the plasma cell cannot go back to making the previous immunoglobulin class—that part of the DNA has been deleted. On the other hand, it is possible that a fur-

16.18 Splicing Out Extra Joining Sequences
After *V* (variable), *D* (diversity), *J* (joining), and *C* (constant) DNA segments have been joined, the resulting functional gene for a heavy chain is transcribed. RNA splicing of the primary transcript removes the transcripts of any introns, along with transcripts of any extra *J* segments, yielding the mature mRNA.

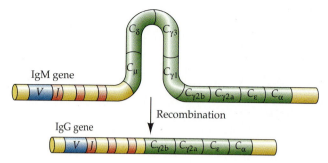

16.19 Class Switching
The functional gene produced by joining of *V*, *J*, and *C* (and, in heavy chains, *D*) segments may later be modified, so that a different *C* segment is transcribed. This is accomplished by deletion of part of the constant region, as shown in this diagram. Here we see class switching from an IgM gene (Figure 16.13) to an IgG gene.

ther class switch may occur, if further constant-region segments are still present.

THE CELLULAR IMMUNE SYSTEM

Thus far we have been concerned primarily with the humoral immune system, whose effector molecules are the antibodies secreted by plasma cells that develop from activated B cells. The cellular immune system, which guards against infections by fungi and some viruses, is mediated by various kinds of T cells and has both interesting parallels with and differences from the humoral system. The key similarity resides in the stunning *specificity* of both immune systems. In both systems, this is based upon the possession of extraordinarily specific receptors.

T-cell receptors were identified more recently than antibodies, and there is much to be learned about them. They are glycoproteins with molecular weights about half that of an IgG and are made up of two polypeptide chains (Figure 16.20). The genes that code for T-cell receptors are similar to those for immunoglobulins, suggesting that both are derived from a single, evolutionarily more ancient group of genes. Like the immunoglobulins, T-cell receptors include both variable and constant regions; and they are assembled by *V–D–J* joining. Once formed, they are bound to the plasma membrane of the T cell that produces them.

T-cell receptors do not recognize native, unprocessed antigen. Rather, they recognize *processed* antigen fragments bound to proteins of the major histocompatibility complex (MHC; see the next section). T-cell receptors thus have a dual specificity for processed antigen and MHC protein (Figure 16.21).

When T cells are activated by contact with a specific antigen, they develop and give rise to several distinct types of effector cells. One type is called cytotoxic T cells, or **T$_C$ cells**. These cells recognize virus-infected cells by their specific antigenic determinants and eliminate the infected cells directly by causing them to lyse (Figure 16.22). T$_C$ cells induce lysis in the target cells by releasing a toxic protein. T cells in another category participate as helpers in both the cellular and humoral immune systems. In fact, these are called helper T cells, or **T$_H$ cells**. As mentioned already, a T$_H$ cell of appropriate specificity must bind an antigen before a B cell can be activated by it. The activation and effector phases of antibody and T$_C$ cell responses are summarized in Figure 16.23.

Still other T cells differentiate into suppressor T cells, or **T$_S$ cells**. These regulatory cells inhibit the responses of both B and T cells to antigens. As noted already, they probably play an important role in immune tolerance and in the acceptance of one's own self antigens—the relative importances of T$_S$ cells and of clonal deletion are still not known.

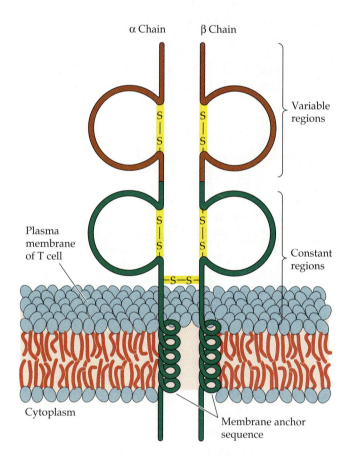

16.20 A T-Cell Receptor
A T-cell receptor consists of two polypeptide chains—an α chain and a β chain. Each polypeptide chain possesses a hydrophobic region (Chapter 5) that anchors the chain in the phospholipid bilayer of the T cell plasma membrane.

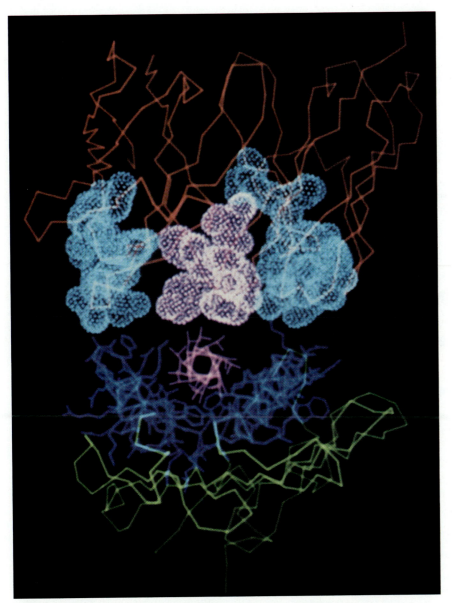

16.21 The T-Cell Receptor Binds Both Antigen and MHC Protein
An MHC protein (below) presents a processed fragment of an antigenic protein (pink pinwheel) to a T-cell receptor (above) in this computer graphic model. The purple region of the T-cell receptor recognizes antigen; the light blue regions of the T-cell receptor recognize the dark blue region of the MHC protein.

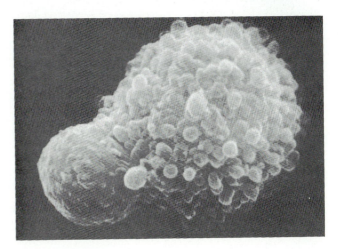

16.22 Death of a Tumor Cell
Cytotoxic T cells recognize foreign or abnormal antigens on cell surfaces, bind to the offending cells, and kill them by causing them to lyse. Here a T$_C$ cell (smaller sphere at lower left) has contacted a tumor cell and the tumor cell is lysing, as indicated by the blebs all over its surface.

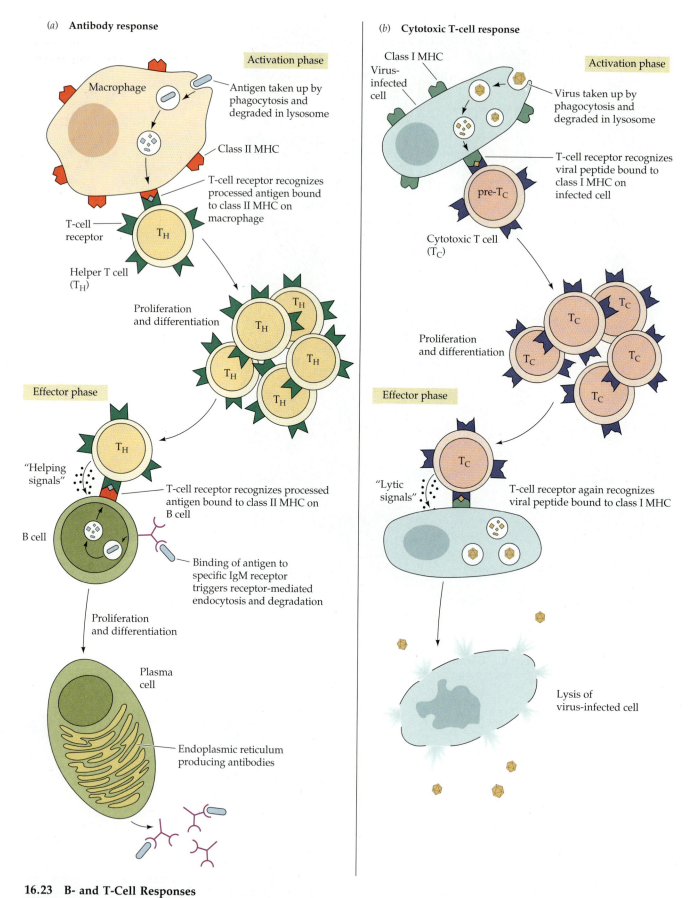

(a) Antibody response

Activation phase

Macrophage

Antigen taken up by phagocytosis and degraded in lysosome

Class II MHC

T-cell receptor recognizes processed antigen bound to class II MHC on macrophage

T-cell receptor

T_H

Helper T cell (T_H)

Proliferation and differentiation

T_H T_H T_H T_H T_H

Effector phase

T_H

"Helping signals"

T-cell receptor recognizes processed antigen bound to class II MHC on B cell

B cell

Binding of antigen to specific IgM receptor triggers receptor-mediated endocytosis and degradation

Proliferation and differentiation

Plasma cell

Endoplasmic reticulum producing antibodies

(b) Cytotoxic T-cell response

Class I MHC

Virus-infected cell

Activation phase

Virus taken up by phagocytosis and degraded in lysosome

T-cell receptor recognizes viral peptide bound to class I MHC on infected cell

pre-T_C

Cytotoxic T cell (T_C)

Proliferation and differentiation

T_C T_C T_C T_C T_C

Effector phase

T_C

"Lytic signals"

T-cell receptor again recognizes viral peptide bound to class I MHC

Lysis of virus-infected cell

16.23 B- and T-Cell Responses
The diagram summarizes the activation and effector phases of the B-cell antibody response and the cytotoxic T-cell (T_C) response.

THE MAJOR HISTOCOMPATIBILITY COMPLEX

The key to the interactions of B cells and the various classes of T cells lies in a tight cluster of loci called the major histocompatibility complex, or **MHC** (Figure 16.24). These loci code for specific proteins found on the surfaces of cells. Because of the number of MHC genes and the number of their alleles, different animals of the same species are highly likely to have different MHC genotypes—and that difference is what leads to the rejection of organ transplants.

Class I MHC loci code for proteins (antigens) that are present on the surface of every cell in the animal.

These proteins function in antiviral T-cell immunity (Figure 16.23b). The products of class II loci are found only on the surfaces of B cells, T cells, and certain phagocytes called macrophages. It is this class of MHC products that is primarily responsible for the interaction of T$_H$ cells, macrophages, and B cells in antibody responses (Figure 16.23a). In order to activate a T cell, its receptors must bind both to surface proteins coded for by MHC loci and to processed foreign antigen. Because T-cell receptors recognize self-MHC products, they rid the body of its own virus-infected cells and may also act against cancerous cells. Class III MHC loci code for some of the proteins

(a) Major histocompatibility complex of mouse

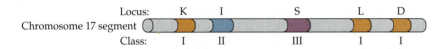

(b) Major histocompatibility complex of human

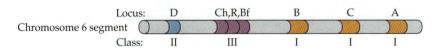

(c) Pattern of inheritance for human MHC alleles

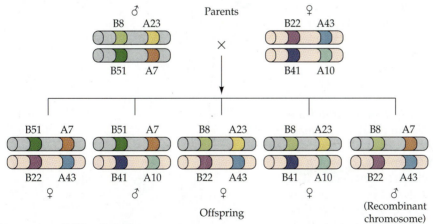

Offspring

16.24 Histocompatibility Complexes

Loci that code for antigens on mammalian cell surfaces are clustered into histocompatibility complexes, so named because they determine the compatibility of transplanted tissues. Loci of the histocompatibility complexes of (a) the mouse and (b) humans are classified by function. (c) The diversity of histocompatibility types derives from each of the loci's having many alleles. Two human parents are both heterozygous at the A and B loci of the complex and they have no alleles in common. Their offspring may fall into at least four histocompatibility types, more if recombinant chromosomes are formed by crossing over. Here, one child carries a recombinant chromosome from his father. The more than 20 known alleles at the A locus and

more than 40 known alleles at the B locus can generate nearly 1,000 unique haploid genotypes: A1B1, A1B2, . . . A20B40. Within a population having these 1,000 genotypes, the possible parental combinations can generate nearly a million unique histocompatibility genotypes just considering the A and B loci of chromosome 6. Multiple alleles at other histocompatibility loci multiply the possibilities to generate even greater diversity. Thus our "self" is defined by our MHC genotype as well as by our immune system. Given that a person can only accept a tissue transplant with a very similar histocompatibility genotype, it is not surprising that finding a compatible donor for a particular recipient can be difficult.

of the complement system, referred to earlier, that interact with antigen–antibody complexes and result in the lysis of foreign cells. Similarities in structure and base sequences between MHC genes and the genes coding for antibodies suggest the possibility that the MHC genes may be descended from the same ancestral genes as are those for antibodies and T-cell receptors. Major aspects of the immune systems may thus be woven together by a common evolutionary thread.

Transplants

A major side effect of the MHC antigens became important with the development of organ transplant surgery, with sometimes devastating effect. Because the proteins produced by the MHC are specific to each individual, they act as antigens if transplanted into another individual. If one attempts to transplant an organ or a piece of skin from one person to another, the transplant is recognized as nonself and soon provokes an immune response; the tissue is killed, or "rejected," by the cellular immune system. But if the transplant is done immediately after birth or if it comes from a genetically identical person (an identical twin), the material is recognized as self and is not rejected.

Physicians can overcome the rejection problem for a while by treating the patient with drugs (immunosuppressants) that reduce the activity and number of B and T cells. However, this technique leaves the patient virtually defenseless against bacteria and viruses, and the patient must then be protected from all such invaders by elaborate isolation procedures.

INTERLEUKINS

How do the cells of the immune system communicate with one another? What signals are passed from macrophages to T cells, and from T_H cells to B cells? The signals are proteins called **interleukins**, of which 10 or more are now known. T cells are activated by IL-1 (interleukin-1), which is released by macrophages (Figure 16.23). The activated T cells then produce both IL-2 and proteins that serve as receptors for IL-2. The binding of IL-2 to a T cell's IL-2 receptors causes the cell to divide. The result is the rapid growth of a clone of T cells (Figure 16.25). IL-2 produced by T_H cells helps B cells to start secreting antibodies, and it probably causes the B cells to divide after they are activated by antigen. IL-2 also activates natural killer cells.

The central role of IL-2 in the immune response is best illustrated by a medical example. When a tissue transplant is to be made, the immune system must be suppressed to reduce the danger of rejection of the transplanted tissue. The two drugs most com-

monly used for suppressing the immune system both function by inhibiting the production of IL-2.

DISORDERS OF THE IMMUNE SYSTEM

A common type of disease relating to the immune system arises when the system overreacts to a dose of antigen. Although the antigen itself may present no danger to the host, the immune response may, by the production of inflammation and other symp-

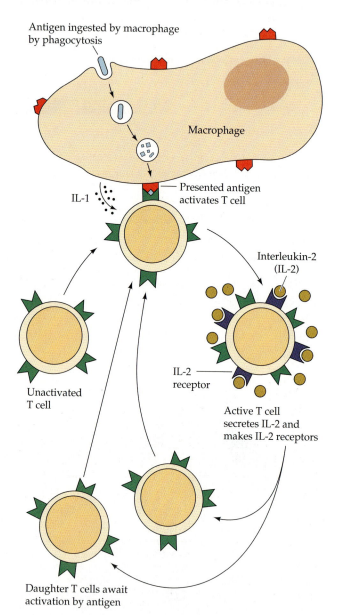

16.25 The Action of Interleukin-2
A T cell, activated by IL-1 from a macrophage, makes IL-2 receptors that become part of its own plasma membrane and secretes IL-2 that binds to these receptors and those of other activated T cells. The binding of the IL-2 causes the T cells to divide, and the daughter cells are ready to become activated in turn, leading to a clone of T cells with the same antigen specificity.

toms, cause serious illness or death. **Allergies** are the most familiar examples of such a problem. In extreme cases, an exposure to a particular antigen (such as the toxin of a bee sting) may lead to a fatal overreaction of the immune system—dilation of some blood vessels and constriction of others, causing shock and even death.

Sometimes the immune recognition of self fails, resulting in the appearance of one or more "forbidden clones" of B and T cells directed against self antigens. This failure does not always result in disease, but in some instances it can be disastrous. Among the **autoimmune diseases** of our species, in which components of the body are attacked by its own immune system, are rheumatic fever, rheumatoid arthritis (Figure 16.26), ulcerative colitis, myasthenia gravis, and several others. It is believed by many medical scientists that multiple sclerosis results from an abnormality in a type of T cell. The abnormal T cells mount an immune attack on the myelin sheath, an insulating material surrounding many nerve-cell processes (Chapter 36). The end result is a severely debilitating loss of nerve function, including blindness and loss of motor control.

Immune Deficiency Disorders

There are various **immune deficiency disorders**, such as those in which B cells are never formed and others in which B cells lose the ability to give rise to plasma cells. In either case, the person is unable to produce antibodies; thus one major line of defense against microbial pathogens is lost.

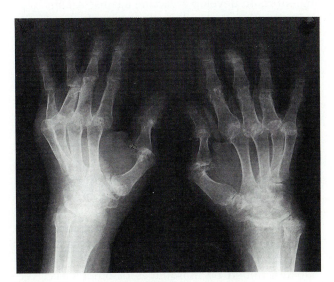

16.26 Rheumatoid Arthritis
This X ray shows the bones in the hands of a 50-year-old man with a 20-year history of rheumatoid arthritis. This autoimmune disease affects the joints, destroying the cartilage with an inflammatory reaction and causing pain and the severe deformities seen here.

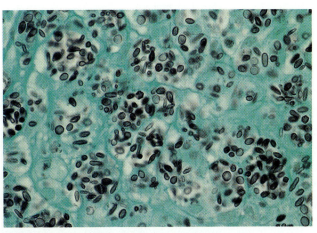

16.27 Bacterial Infection in an AIDS Patient
The cells of an AIDS patient are impaired in their ability to resist bacterial infection. As a result, this patient's cells are packed with spores (stained black) of the bacterium *Sporothrix schenckii*.

The T$_H$ cell, because of its essential roles in both antibody responses and cellular immune responses, is perhaps the most central of all the components of the immune system—the worst one to lose to an immune deficiency disorder. Unfortunately, there is a disease that homes in on the T$_H$ cells, leaving its victims with a much-impaired immune system (Figure 16.27). An epidemic of this disease, acquired immune deficiency syndrome, or **AIDS**, arose during the 1980s. AIDS is caused by the virus HIV-I (human immunodeficiency virus I), also known as HTLV-III (human T-cell lymphotrophic virus III). HIV-I is a retrovirus—a virus with RNA as its genetic material, capable of inserting its own genome into the genome of its animal host (Chapter 13).

The structure of HIV-I is shown in Figure 16.28. A central core, with a protein coat, contains two identical molecules of RNA as well as certain enzymes. An envelope, derived from the plasma membrane of the cell in which the virus was formed, surrounds the core. The envelope is studded with an envelope protein that enables the virus to infect its target T$_H$ cell.

HIV-I attacks host cells by way of a membrane protein (CD4) found only on T$_H$ cells. CD4 acts as the receptor for the viral envelope protein. When HIV-I infects a cell, the viral core is released into the cell. Among the enzymes in the core is **reverse transcriptase**, which catalyzes the formation of a double-stranded DNA molecule encoding the same information as the viral RNA (Chapters 13 and 14). Reverse transcriptase also catalyzes the destruction of the host's original RNA molecules. The DNA transcript enters the nucleus of the host cell and is spliced into a chromosome, much as bacteriophage DNA may become incorporated into a bacterial chromosome as a prophage (Chapter 12). Another HIV-I

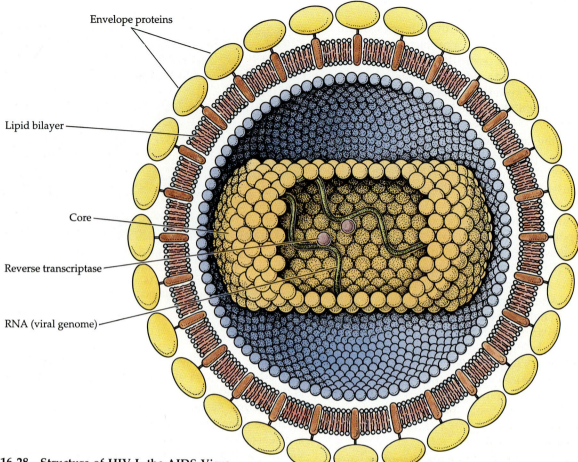

Envelope proteins

Lipid bilayer

Core

Reverse transcriptase

RNA (viral genome)

16.28 Structure of HIV-I, the AIDS Virus
HIV-I has a core containing RNA (the viral genome) and various proteins including the enzyme reverse transcriptase. Another specific protein bounds the core. The surface of the virus is complex—a phospholipid bilayer surrounds a layer of proteins and is studded with knobs of envelope protein.

enzyme, called **integrase**, catalyzes this splicing. The DNA transcript of the HIV-I RNA thus becomes a permanent part of the chromosome, replicating with it at each division of the host cell.

A DNA transcript of the HIV-I RNA, once incorporated into the genome of a T_H cell, may remain there latent for days, or even for a decade or more. The latent period ends if the infected T_H cell becomes activated. Then the viral DNA is transcribed, yielding many molecules of viral RNA, some of which are translated, forming the enzymes and structural proteins of a new generation of viruses. Other RNA molecules are incorporated directly into the new viruses as their genetic material. In controlled growth, formation of new viruses is slow; the viruses bud from the infected cell, surrounding themselves with modified plasma membrane from the host (see Figure 13.16). More rapid virus production leads to lysis of the host cell. Several viral genes control the rate of production of individual proteins and of whole viruses. One gene, responsible for the antigenic properties of the envelope protein, undergoes rapid mu-

tation, making HIV-I a rapidly moving target for what is left of the host's immune system—and complicating efforts to develop a vaccine against AIDS.

Lysis of infected cells leads to a gradual, selective depletion of the T_H cells, and the host's immune system becomes unable to function. Patients usually die of "opportunistic" infections, diseases caused by bacteria and fungi that are almost always eliminated by the immune systems of normal individuals. There is no cure for AIDS as of this writing, and more than half of all known AIDS patients have died. **AIDS-related complex**, a less severe form of the disease, has milder symptoms but appears to develop in most patients into full-blown AIDS.

AIDS has the highest incidence among drug addicts (from the use of shared, contaminated needles) and male homosexuals. It can be transmitted by transfusions, by sexual activity (either homosexual or heterosexual), and, sadly, from mother to fetus. It is not transmitted by mosquitoes or other insect vectors, or by kissing or casual contact. Hundreds of thousands of people will have died before the AIDS epidemic ends. In the meantime, medical and biological research on the subject is proceeding with great intensity.

CANCER

There is at least one case in which it is appropriate for an animal's defenses to mount an all-out assault against some of its own cells: when those cells are cancerous. The **transformation** of a normal cell to a cancer cell may begin with changes in the nucleus—perhaps by a series of mutations, perhaps by the conversion of a proto-oncogene to an oncogene (Chapter 13). Some cancers are attributable to infection with certain viruses that change nuclei. These tumor-inducing **oncogenic viruses** have genes that bring about little-understood metabolic and structural changes in their host cells that cause the infected cells to proliferate. The transformation is genetically stable, so that a single cell produces a clone of cancer cells.

The nuclear changes are followed by other changes in different parts of the cell, notably the plasma membrane. The membranes change so that their surface proteins no longer limit the cells' growth; thus cells continue to multiply, forming tumors (large masses of cells). Unlike **benign tumors**, cancers (**malignant tumors**) invade surrounding tissues and spread to other parts of the body. This spreading, called metastasis, proceeds in two stages: first, the cancer cells extend into surrounding tissues, then they enter either the bloodstream or the lymphatic system (Figure 16.29). Cancer cells metastasizing by way of the lymph are slowed by lymph nodes, where they must pause before proceeding to the next node. The removal of a series of lymph nodes and the ducts between them can often end the disease in a patient (mastectomy to stop breast cancer, for example, often

includes the removal of lymph nodes). Metastasis through the bloodstream is another matter—less common than metastasis through the lymphatic system, it is rapid and very commonly fatal.

In 1908 one of the founders of the science of immunology, the German microbiologist Paul Ehrlich, suggested that cancerous cells might differ sufficiently from normal cells so that they could be recognized as *nonself* and hence be attacked by the immune system. He also noted the possibility that cancer cells might appear rather often, perhaps several times a day, usually being disposed of by the immune system. It was later suggested that the cellular immune system might have evolved principally as a weapon—an "immune surveillance system"—against cancers. More recently, this idea has fallen into disrepute because of experimental findings that have not been reconciled with the theory. There continues to be interest, however, in the apparently nonspecific natural killer cells referred to earlier in this chapter.

The exact nature of the cellular transformation to the cancerous state is not yet known. As already mentioned, three possible causes of the transformation are accumulated mutations, the conversion of proto-oncogenes to oncogenes, and induction by an oncogenic virus. It is also known that cancer can be caused by certain physical and chemical agents. Ultraviolet radiation and X rays (which are potent mutagens) can cause the conversion of normal cells to cancerous ones. Excessive exposure to sunlight, with its substantial content of ultraviolet rays, can lead to various cancers of the skin. This is particularly true of fair-skinned persons and albinos, as well as persons with defective enzymes for the repair of damage to DNA. Other cancers can be induced by chemical **carcinogens** (cancer-causers); polluted air and tobacco smoke contain carcinogens. Many cancers are diseases of old age. When the body gets old, more cancers may be initiated, and any defense system (natural killer cells?) may become less effective. Thus cancerous cells may elude the older body's defenses and multiply unchecked.

Once a cancer has been detected and diagnosed, there are three principal lines of treatment. One is surgery that removes the affected tissues and organs. For this approach to be successful, the surgeon must know the exact locations of all cancerous tissues. Also, the tissue to be removed must obviously not be irreplaceable. **Radiotherapy** (exposure to massive doses of X or gamma rays) is a possible line of treatment for cancerous tissues that are irreplaceable. Treatment of cancer of the larynx by radiotherapy may leave the vocal cords intact, whereas surgery would remove them. Tissues exposed to massive doses of radiation suffer extensive chromosomal breakage; some cells not scheduled to divide again remain alive after the chromosomes are broken, but

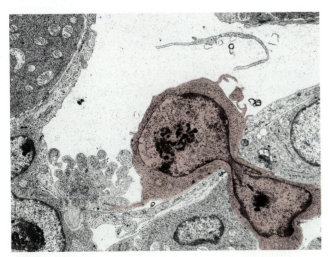

16.29 The Spread of Cancer
A tumor cell squeezes into the lumen of a small vein through a cell in the vein's wall. Cancer spreads through animal bodies as tumors invade tissues and malignant cells get into the bloodstream. The blood then transports the cancer cells throughout the body, starting new tumors in various organs.

dividing cells such as cancerous cells and the cells of the immune system die when they attempt to undergo mitosis. Because radiation is also harmful to normal cells, the radiation must be restricted to localized areas of the body; thus the radiologist, like the surgeon, must know the extent and location of the cancer. Another danger of radiotherapy is that the immune system may be seriously damaged by such treatment, leaving the patient virtually defenseless against bacterial, viral, or other infections.

The third major line of cancer treatment is **chemotherapy**, treatment with drugs that preferentially affect rapidly dividing cells. These drugs generally act by interfering with the metabolism of the nucleotides of DNA. In chemotherapy there is no need for precise knowledge of the locations of the cancerous tissues. However, there are effects on normal cells as well—especially those of the immune system (remember that the immune response requires cell division for the rapid proliferation of lymphocytes and their products). Thus chemotherapy has side effects that are often severe. In any case, it is now possible to achieve many cures by using one or more of these approaches. There is hope, too, that immunological techniques, such as vaccination against the few known oncogenic viruses, may reduce the incidence of cancer. And we can help ourselves by taking care of what we put on our skin and in our stomachs and lungs.

The danger of infection in cancer patients receiving chemotherapy or radiotherapy, and in AIDS patients, serves to emphasize an important aspect of development. Some organs, such as kidneys and hearts, are established early in life. However, others, notably the immune system, need constant replenishment—their development continues as long as the animal lives.

SUMMARY

Nonspecific defenses against pathogens include the "normal flora," the physical integrity of the skin, and localized secretions such as lysozyme. Phagocytes engulf pathogens. Other nonspecific defenses include inflammation, the complement system, natural killer cells, and interferon.

The immune system recognizes, selectively eliminates, and remembers nonself cells and macromolecules. The humoral immune response relies on antibodies (immunoglobulins) produced by plasma cells that develop from B cells. Each antibody recognizes and binds one specific antigenic determinant. The cellular immune response relies on T-cell receptors. Effector T cells include cytotoxic, helper, and suppressor T cells.

The immune system can act against virtually any antigenic challenge, even though most have never been encountered before. The immune response includes the formation of both effector and memory cells, the latter allowing a larger and more rapid response to a second exposure to the antigen. The immune system distinguishes between self and nonself materials.

The animal begins with a great variety of B and T cells. Specific ones become activated upon exposure to specific antigenic determinants, and they multiply and differentiate to give rise to clones of cells of identical specificity. Continued exposure to self-antigens beginning at or before birth results in the deletion or inactivation (by suppressor T cells) of cells that would otherwise produce anti-self antibodies.

The immunoglobulin molecule is a tetramer of two identical light chain polypeptides and two identical heavy chain polypeptides. Both light and heavy chains contain constant and variable regions. Immunoglobulin specificity is determined by the variable chains. Different heavy chains determine the differing types of effects produced by the different antibody classes. Antibody diversity results from the chance selection of different regions of a chromosome in constructing a single gene coding for the immunoglobulin molecule, as well as from frequent mutation of some of the genes.

Lymphocyte interactions are based on surface antigens, which are coded for by the major histocompatibility complex, that function in binding and presenting degraded fragments of foreign antigens to T cells. Histocompatibility antigens also are the basis for the rejection of nonself tissue transplants. Cells of the immune system communicate by releasing interleukins.

AIDS is caused by HIV-I, a retrovirus that attacks T_H cells. Reverse transcriptase, carried by the virus, catalyzes the formation of a DNA transcript of the viral RNA. The DNA transcript is spliced into the host genome and may remain latent for years before being expressed and producing new viruses, eventually crippling the host's immune system.

Normal cells may be transformed into cancer cells by mechanisms that are not yet understood. The proliferation of cancer cells is not limited as is that of normal cells, in part because they have differing surface proteins.

SELF-QUIZ

1. Which statement is *not* true of phagocytes?
   a. Some travel in the circulatory system.
   b. They ingest microorganisms by endocytosis.
   c. A single one can ingest 5–25 bacteria before it dies.
   d. Although they are important, an animal can do perfectly well without them.
   e. Lysosomes play an important role in phagocyte function.

2. Immunoglobulins:
   a. help antibodies do their job.
   b. recognize and bind antigenic determinants.
   c. are among the most important genes in an animal.
   d. are the chief participants in nonspecific defense mechanisms.
   e. are a specialized class of white blood cells.

3. Which statement is *not* true of an antigenic determinant?
   a. It is a specific chemical grouping.
   b. It may be part of many different molecules.
   c. It is the part of an antigen to which an antibody binds.
   d. It may be part of a cell.
   e. A single antigen has only a single antigenic determinant on its surface.

4. T-cell receptors:
   a. are the primary receptors for the humoral immune system.
   b. are carbohydrates.
   c. cannot function unless the animal has previously encountered the antigen.
   d. are produced by plasma cells.
   e. are important in combatting viral infections.

5. According to the clonal selection theory:
   a. an antibody changes its shape according to which antigen it meets.
   b. an individual animal contains only one kind of B cell.
   c. the animal contains many kinds of B cells, each producing one kind of antibody.
   d. each B cell produces many kinds of antibody.
   e. no clones of anti-self lymphocytes appear in the bloodstream.

6. Immunological tolerance:
   a. depends on repeated exposure throughout the life of the animal.
   b. develops late in life and is usually life-threatening.
   c. disappears at birth.
   d. results from the activities of the complement system.
   e. results from DNA splicing.

7. The extraordinary diversity of antibodies results in part from:
   a. the action of monoclonal antibodies.
   b. the splicing of RNA molecules.
   c. the action of suppressor T cells.
   d. the splicing of gene segments.
   e. their remarkable nonspecificity.

8. Which of the following play *no* role in the antibody response?
   a. Helper T cells.
   b. Interleukins.
   c. Macrophages.
   d. Phytoalexins.
   e. Products of class II MHC loci.

9. The major histocompatibility complex:
   a. codes for specific proteins found on the surfaces of cells.
   b. plays no role in T-cell immunity.
   c. plays no role in antibody responses.
   d. plays no role in graft rejection.
   e. is coded for by a single locus with multiple alleles.

10. Which of the following plays *no* role in AIDS?
    a. Integrase.
    b. Reverse transcriptase.
    c. Transcription.
    d. Translation.
    e. Transfection.

FOR STUDY

1. Describe the part of an antibody molecule that interacts with an antigenic determinant. How is it similar to the active site of an enzyme? How does it differ from the active site of an enzyme?

2. Contrast immunoglobulins and T-cell receptors, with respect to both structure and function.

3. Discuss the diversity of antibody specificities in an individual in relation to the diversity of enzymes. Does every cell in an animal contain genetic information for all the organism's enzymes? Does every cell contain genetic information for all the organism's immunoglobulins?

4. Describe and contrast two ways in which DNA splicing plays roles in the immune response.

5. Discuss the roles of monoclonal antibodies in medicine and in biological research.

READINGS

Ada, G. L. and G. Nossal. 1987. "The Clonal-Selection Theory." *Scientific American*, August. Fascinating historical account of the development of the central concept of immunology.

Atkinson, M. A. and N. K. Maclaren. 1990. "What Causes Diabetes?" *Scientific American*, July. An explanation of the origin of diabetes, in terms of an autoimmune response. It may lead to the development of preventive therapies for insulin-dependent diabetes.

Cohen, I. R. 1988. "The Self, the World, and Autoimmunity." *Scientific American*, April. The nature of autoimmune diseases, and an approach to their prevention.

Edelman, G. M. 1970. "The Structure and Function of Antibodies." *Scientific American*, August. The amino acid sequence of an antibody dictates its unique characteristics and determines its ability to interact with an antigen.

Feldman, M. and L. Eisenbach. 1988. "What Makes a Tumor Cell Metastatic?" *Scientific American*, November. Some tumor cells bear MHC molecules that enable them to evade the immune system.

Golub, E. S. and D. R. Green. 1991. *Immunology: A Synthesis*, 2nd Edition. Sinauer Associates, Sunderland, MA. An outstanding textbook of immunology, with special attention to the experimental basis for what we know. Also reflects on the history of the discipline.

Lerner, R. A. and A. Tramontano. 1988. "Catalytic Antibodies." *Scientific American*, March. A powerful new tool for biotechnology, combining the talents of enzymes and antibodies. Perhaps useful, as well, in augmenting the capabilities of the immune system.

Marrach, P. and J. Kappler. 1986. "The T Cell and Its Receptor." *Scientific American*, February. A detailed consideration of the key actors in the cellular immune system.

Prescott, D. M. and A. S. Flexer. 1986. *Cancer: The Misguided Cell*, 2nd Edition. Sinauer Associates, Sunderland, MA. A comprehensive but clear treatment of biological, medical, and personal aspects of cancer. In paperback.

Smith, K. A. 1990. "Interleukin-2." *Scientific American*, March. A clear description of the role of interleukin-2 in the expansion of a clone of T cells.

Tonegawa, S. 1985. "The Molecules of the Immune System." *Scientific American*, October. Beautifully illustrated account of the structures of antibodies and T-cell receptors and of how they are formed.

PART THREE
Evolutionary Processes

17

Origins

PREVIEW: Although the universe is between 10 and 20 billion years old, our solar system formed less than 5 billion years ago. Microscopic life probably appeared within a billion years after Earth was formed and was abundant 3.5 billion years ago. Experiments that mimic presumed ancient environmental conditions suggest several plausible mechanisms for the origin of life. The evolution of life involved the development of both metabolic processes and mechanisms of replication, but the order in which these events occurred is as yet unknown.

Comparative study of the metabolism of living organisms, particularly bacteria, helps us identify some of the metabolic pathways that may have been used by the earliest living organisms. Evidence of past life is preserved in Earth's rocks.

This chapter deals with the history of Earth and how it is dated, spontaneous generation, Earth's primitive atmosphere, and the evolution of metabolism, genetic replication, and photosynthesis.

At some time between 10 and 20 billion years ago, a mighty explosion is thought to have occurred. The matter of the universe, which had been highly concentrated, began to spread apart rapidly. The universe is still expanding today. The noise of the original explosion—the "big bang"— can still be "heard" today in the form of background radiation that permeates all of space, traces of which can be detected by radio telescopes. The big bang sent gases hurtling in all directions—perhaps to expand until the end of time. Some time after the bang, clouds of gases formed. Eventually these clouds collapsed upon themselves through gravitational attraction, forming the galaxies, which are great clusters of hundreds of billions of stars (Figure 17.1). Among those billions of galaxies is our own, the Milky Way.

Somewhat less than 5 billion years ago, toward the outer edge of the Milky Way, our solar system (the sun, Earth, and our sister planets) took form. Most of the planets probably formed by gravitational attraction and the aggregation of cold dust particles. As Earth slowly grew by this process, the weight of the outer layers compressed the interior of the planet. The resulting pressures and the energy from radioactive decay heated the interior until it melted. Within this viscous liquid, the settling of the heavier elements produced a fluid iron and nickel core with a radius of approximately 3,700 kilometers. Around the core lies a 3,000-km thick mantle of dense silicate materials. Over the mantle is a lighter crust, more than 40 km thick under the continents but thinning to 5 km in some places under the oceans.

Earth's atmosphere originally consisted largely of hydrogen. This atmosphere was thin because its hydrogen and other light gases were too light to be held by Earth's gravitational field and they continually escaped to outer space. However, heavier gases, such as carbon dioxide and nitrogen released from the mantle and crust, were held by Earth's gravitational field, and gradually formed a new atmosphere. Water vapor escaping from inside the planet condensed into seas. Lightning and other energy sources converted atmospheric gases into simple organic molecules, and these dissolved in the seas.

One of the most exciting challenges of modern biology is to understand the beginning of life itself. How did the first living inhabitants of Earth evolve from nonliving matter? Did they arise quickly, or only over hundreds of millions of years as chemicals began to react and the reactions were refined? How we view subsequent evolutionary processes and products depends in part on our answers to these fundamental questions.

SPONTANEOUS GENERATION: OLD IDEAS

Only in the last century or two did serious doubts arise about the origin of life. Before then, practical people "knew" that new life appeared regularly: flies and maggots arose from rotting meat and barnyard manure, lice from sweat, glowworms from rotting logs, eels and fish from sea mud, and frogs and mice from moist earth. No less an authority than Aristotle

17.1 The Birthplace of Stars
Places like the central portion of the great nebula in the constellation Orion, where the density of gases is very high, are thought to be "nurseries" for new stars.

vouched for such commonsense deductions. For more than 2,000 years, **spontaneous generation**—the formation of living organisms from nonliving matter—was accepted by most people as an obvious fact of nature.

Early Experiments on Spontaneous Generation

Until 1668, no one had performed an experiment to determine, for example, whether maggots really arose spontaneously from decaying meat. Then Francesco Redi, an Italian physician, demonstrated that maggots in meat are the larvae of flies and that if the meat is protected so that adult flies cannot lay their eggs on it, no maggots appear (Figure 17.2a). But when the Dutch lens grinder and microscope maker Anton van Leeuwenhoek discovered microorganisms in 1676, spontaneous generation received new support. Many people who were ready to concede that worms and maggots did not appear spontaneously from nonliving matter nonetheless thought that this was a plausible way to account for the appearance of the new creatures that Leeuwenhoek could find everywhere.

In the latter part of the eighteenth century, the Italian biologist Lazzaro Spallanzani showed that if broths are placed in sealed containers after being (as we say today) adequately sterilized, they remain devoid of life (Figure 17.2b). However, he failed to convince his contemporaries, partly because others performed the same experiments with less care and obtained different results. Also, some people argued that Spallanzani's techniques not only killed the microorganisms already present but also rendered the air unfit for the generation and growth of new ones. Experimental methods were not yet good enough to rule out spontaneous generation for those who wanted to believe in it.

Pasteur's Experiments

In 1862 the great French scientist Louis Pasteur obtained results that finally convinced most people that spontaneous generation does not occur. Pasteur performed a series of meticulous experiments showing that microorganisms came only from other microorganisms and that a genuinely sterile broth or solution remained sterile indefinitely unless contaminated by living creatures. His most elegant experiment relied on swan-necked flasks that were open to the air (ruling out the "spoiled air" objection raised against Spallanzani). Pasteur filled the flasks with nutrient medium, heated them to kill any microorganisms present, then cooled them slowly. The shape of the necks kept any new organisms from falling into the medium. No new growth appeared in the flasks (Figure 17.2c). As a result of these experiments, the aphorism *omne vivum e vivo* ("all life from life") became widely accepted.

SPONTANEOUS GENERATION REVISITED

Pasteur answered an old question, but his results posed a new and more fundamental question: If all life comes from preexisting life, where did the *first* life come from? In spite of its importance, this question did not attract serious scientific attention for another 60 years.

In 1924, the Russian biologist Alexander I. Oparin published a short monograph entitled *The Origin of Life*. Although it was never translated from Russian and had no impact on scientific thought at the time, it laid out a reasonable scenario of events and conditions leading to the beginnings of life on Earth. Five years later, the British biologist J. B. S. Haldane independently arrived at similar ideas and published them in *The Rationalist Annual*, again to little effect. Not until after Oparin expanded his ideas into a book—also entitled *The Origin of Life*, it was published in 1936 and translated into other languages—was the problem of the appearance of life on Earth studied experimentally.

(a) Redi's experiment

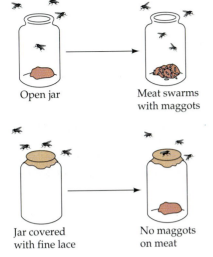

Open jar

Meat swarms with maggots

Jar covered with fine lace

No maggots on meat

(b) Spallanzani's experiment

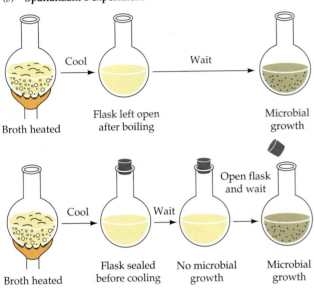

Broth heated — Cool → Flask left open after boiling — Wait → Microbial growth

Broth heated — Cool → Flask sealed before cooling — Wait → No microbial growth — Open flask and wait → Microbial growth

(c) Pasteur's experiment

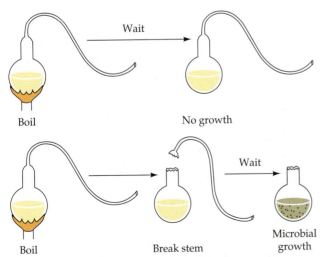

Boil — Wait → No growth

Boil — → Break stem — Wait → Microbial growth

17.2 Tests of Spontaneous Generation
Experiments by several scientists were needed to rule out all plausible theories of the spontaneous generation of life.

Today there is a substantial body of information on which to base plausible theories of the origins of life on Earth. The preserved remains of many early organisms have been found and described, and a number of laboratory experiments have studied chemical reactions under conditions similar to those believed to have prevailed on the early Earth. We know that all organisms share the same genetic machinery and have very similar basic cellular metabolic pathways. This strongly suggests that all living organisms are descendants of a single common ancestor. This does not imply that life originated only once, but it does suggest that only one of those origins led to all successful lineages of organisms. We will first examine the evidence, and then explore its implications for the ways life may have originated.

THE EARLY EARTH

The abundances of chemical elements on Earth and in its atmosphere differ strikingly from the cosmic abundances of elements. For example, the noble gases (neon, xenon) are proportionally at least a million times more abundant cosmically than they are on Earth. Therefore, most scientists believe that Earth formed initially from solid material that had already lost most of its hydrogen and noble gases. According to this view, the early atmosphere was formed by the release of volatiles trapped in the crust and mantle of Earth. If so, this early atmosphere would have been mainly carbon dioxide and water vapor, with lesser amounts of hydrogen, nitrogen, ammonia, hydrogen sulfide, carbon monoxide, and methane. Such an atmosphere could have led to a weak greenhouse effect, making Earth too hot for oceans to exist on its surface for perhaps a billion years.

Laboratory Simulation of Earth's Early Atmosphere

Experimental attempts to study chemical reactions under conditions similar to those believed to have prevailed on Earth when life evolved began in the early 1950s. Some scientists, such as Harold Urey of the University of Chicago, believed that life evolved in an atmosphere of hydrogen, methane, and ammonia. The first such experiments—performed by Stanley Miller, one of Urey's graduate students—demonstrated that organic compounds, including amino acids, are readily synthesized under these conditions. Miller set up a recirculating system of hydrogen, ammonia, and methane gases, and water vapor. He passed these gases over a spark to simulate lightning (Figure 17.3). Within a few hours the system contained numerous amino acids, simple acids, and other compounds. The electric discharge provided the energy for the formation of the compounds

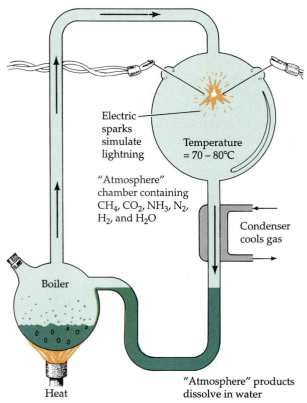

17.3 Simulation of Abiotic Synthesis on Early Earth
The apparatus shown here is similar to that used by Stanley Miller to test Harold Urey's theory of the origin of life. Samples were withdrawn from both the fluid at the bottom of the apparatus and from the boiler.

shown in Figure 17.4a. These compounds then reacted in water to yield all the substances in Figure 17.4b.

Further experiments by Miller and many others have shown that comparable results can be obtained using different proportions of those gases, and even using different gases, provided that free oxygen is absent. In fact, organic compounds are spontaneously synthesized under a wide variety of conditions. Scientists must now determine which of the many possible conditions actually were the ones present on the early Earth.

ENERGY SOURCES FOR LIFE'S ORIGINS

All living systems require an energy source to drive their chemical reactions. The Miller–Urey experiments, and many similar ones, used ultraviolet light or electric discharges as an energy source. Both these sources, however, are destructive of life and are not used today by any biological systems. Indeed, because the early Earth lacked a thick atmosphere, Earth's surface was exposed to high levels of ultraviolet light, which is believed to have prevented or-

ganisms from evolving in exposed locations. Both electric discharges and ultraviolet light act mostly in the atmosphere, whereas life almost certainly first arose in water or on protected moist surfaces on land. Other energy sources must have powered the origins of life.

One possible source of sufficiently high concentrations of energy is the hot solution of molecules such as hydrogen sulfide and carbon dioxide emitted from sea-floor volcanic vents. Today these vents serve as energy sources for rich and interesting ecosystems (Figure 17.5). Such vents, not necessarily restricted to deep-sea locations on the early Earth, might well have been focal points for the evolution of life.

(a) Gaseous products

$H-C \equiv N$ **Hydrogen cyanide**	$\overset{\displaystyle O}{\overset{\displaystyle \|}{H-C-H}}$ **Formaldehyde**
$N \equiv C - C \equiv N$ **Cyanogen**	$\overset{\displaystyle O}{\overset{\displaystyle \|}{CH_3 - C - H}}$ **Acetaldehyde**
$H-C \equiv C - C \equiv N$ **Cyanoacetylene**	$\overset{\displaystyle O}{\overset{\displaystyle \|}{CH_3CH_2 - C - H}}$ **Propionaldehyde**

(b) Products in solution

H_2N-CH_2-COOH **Glycine**	$\overset{CH_3}{\overset{\|}{HN-CH_2-COOH}}$ **Sarcosine**	$HO-CH_2-COOH$ **Glycolic acid**
$\overset{CH_3}{\overset{\|}{H_2N-CH-COOH}}$ **Alanine**	$\overset{H_3C \quad CH_3}{\overset{\backslash \quad /}{HN-CH-COOH}}$ **N-Methylalanine**	$\overset{CH_3}{\overset{\|}{HO-CH-COOH}}$ **Lactic acid**
$\overset{CH_3}{\overset{\|}{\underset{H_2N-CH-COOH}{CH_2}}}$ **α-Aminobutyric acid**	$\overset{CH_3}{\overset{\|}{\underset{CH_3}{H_2N-C-COOH}}}$ **α-Aminoisobutyric acid**	$\overset{COOH}{\overset{\|}{\underset{H_2N-CH-COOH}{CH_2}}}$ **Aspartic acid**
CH_3COOH **Acetic acid**	CH_3CH_2COOH **Propionic acid**	$\overset{COOH}{\overset{\|}{\underset{\underset{H_2N-CH-COOH}{CH_2}}{CH_2}}}$ **Glutamic acid**
$HN\overset{\nearrow CH_2COOH}{\searrow CH_2CH_2COOH}$ **Iminoaceticpropionic acid**	$\overset{O}{\overset{\|}{H_2N-C-NH_2}}$ **Urea**	

$HOOC-CH_2-CH_2-COOH$ **Succinic acid**	$HCOOH$ **Formic acid**

17.4 Many Molecules Were Synthesized in Miller's Experiments
Different small organic molecules formed in the "atmospheric" (a) and "oceanic" (b) compartments of Miller's apparatus. Most of these organic and amino acids are important components of contemporary living organisms.

17.5 A Deep-Sea Vent Ecosystem
Volcanic vents emit hydrogen sulfide and carbon dioxide, which are a source of energy for ecosystems on the floor of the deep sea; seen here are several species of tubeworms (*Riftia*) and brachyuran crabs.

THEORIES ABOUT THE ORIGIN OF LIFE

Any theory about the origin of life from nonlife must explain the origin of the two key features of living systems: metabolism and replication (see Chapter 1). However, there is no reason to believe that these features necessarily evolved simultaneously. Both are complex and precisely regulated processes that have been refined over millions of years of biological evolution. When metabolism and replication first appeared, they probably lacked most of the characteristics they exhibit today. Some theories about the origin of life suggest that metabolism evolved first, followed by the evolution of replication; other theories reverse the order. Determining the actual order of events will be difficult because, unlike the geophysical and chemical processes that have left traces in Earth's rocks and atmosphere, the transition from disordered to ordered chemical systems has left no direct evidence. We must use a variety of indirect clues, relying on theories to guide our guesses and interpretations.

Oparin's Theory

Oparin proposed that the sequence of events in the origin of life was cells first, metabolism second, and replication third. He spent much of his career studying complex solutions. If we place some olive oil in water and shake the mixture, the oil temporarily breaks up into tiny droplets. If instead of olive oil you use a large protein, such as gelatin, and a polysaccharide, such as gum arabic, shaking the mixture forms drops that are divided into two "phases." Their interiors, which are primarily protein and polysaccharide with some water, are surrounded by an aqueous solution with low concentrations of proteins and polysaccharides. These drops, known as **coacervates**, are quite stable and will form in solutions of many different kinds of polymers.

Coacervates have several properties relevant to the origin of life. Many substances, when added to a coacervate preparation, are preferentially concentrated within the drops. Lipids can coat the boundaries of drops with membranelike structures that strengthen the drops. As mentioned above, drops made from complex solutions may contain two phases, one inside the other.

Coacervate drops that contain enzyme molecules can absorb substrates, catalyze reactions, and let the products diffuse back out into the solution. Drops containing phosphorylase, for example, will absorb glucose 1-phosphate from the surrounding medium and polymerize it into starch. If a second enzyme, amylase, is also present in the solution, it will break up the starch into maltose, which escapes into the solution. These coacervate drops are, in effect, small factories that convert a monosaccharide into a disaccharide using the energy of the high-energy phosphate in glucose 1-phosphate (Figure 17.6). Oparin even succeeded in making chlorophyll-containing coacervate drops that absorbed an oxidized dye from the solution, used light energy to reduce it, and returned the reduced dye to the medium.

Oparin regarded complex coacervate drops as possible precursors to the cells that provide the physical framework within which metabolic reactions can take place. He believed that the first coacervate drops contained only relatively simple molecules. Because drops in which chemical reactions were better controlled would have survived longer than drops with

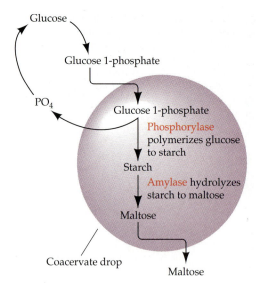

17.6 "Metabolism" of Coacervate Drops
In this hypothetical coacervate drop the enzyme amylase hydrolyzes starch to maltose.

more poorly controlled reactions, refinements of metabolic processes by the use of enzymes would have evolved. Much later, Oparin believed, genes appeared. Oparin's hypothesis was widely accepted by biologists because it was the only plausible theory to have been proposed. As might be expected, many scientists switched their allegiances when alternative theories were developed.

Eigen's Theory

About 30 years ago, Manfred Eigen proposed an alternative theory with the reverse order of events—genes first, followed by metabolism, and still later by cells. Molecular biologists were attracted to this approach because RNA is much simpler than proteins and so might have evolved more readily, and because replication appeared to them to be the key process in the evolution of life. Eigen's theory was stimulated by experiments in which RNA was stripped of all its accompanying cellular baggage and studied in isolation in simple solutions. These experiments showed that a solution of nucleotide monomers can give rise to a nucleic acid polymer molecule that can replicate and mutate. However, to produce this result, the test-tube environment must contain a polymerase enzyme—a catalyst that must be extracted from a bacteriophage. This catalyst would not have been available to the first organisms. Other experiments, carried out by Leslie Orgel, showed that nucleotide monomers will, under appropriate conditions, polymerize to form RNA, even in the absence of a polymerase enzyme, if an RNA template is present.

Thus there are experiments in which RNA can be synthesized using an enzyme but no template, and other experiments in which RNA can be made using a template but no enzyme. All living cells use both templates and enzymes. However, if RNA was the original molecule of life, it must have appeared without either templates or enzymes. To date, no experiments have succeeded in producing RNA without at least one of these substrates, but future experiments may succeed in accomplishing this.

Cairns-Smith's Theory

Another theory of the origin of life postulates an alternative order of events: enzymes first, cells second, and genes third. This theory, developed by A. G. Cairns-Smith in the early 1980s, suggests that naturally occurring microscopic mineral crystals in clays might have served as the basis for replication until the time when nucleic acids evolved and took over the function of replication. Thus, we emend our list to clays first, enzymes second, cells third, and genes fourth. Clay microcrystals consist of flat plates of silicate lattices with regular arrays of ionic sites occupied by various metals. When such a crystal is contained in a droplet of water, the metal ions form irregular patterns of electrostatic potential that can attract particular molecules to the surfaces of the lattice and catalyze chemical reactions. Which reactions are catalyzed depends upon the precise arrangement of the metal ions. Molecules synthesized in this manner could be released back to the water. Because a crystal grows by incorporating silicate and metal ions from the surrounding water, the new materials are similar in composition to the original parts of the crystal that generated them. Thus crystals could, in principle, both replicate information and transfer it to other molecules. What is uncertain is whether they could have done so with sufficient precision to serve as a basis for the evolution of life.

According to this theory, the clay lattice first directed the synthesis of primitive enzymes. These enzymes, in turn, catalyzed the formation of membranes that surrounded droplets, perhaps similar to those surrounding Oparin's coacervate drops. For a long time the clay crystals functioned as primitive genetic material, but at some point, by as yet unspecified mechanisms, RNA evolved and took over the role of replicating and transferring information. Once RNA appeared, it was so much better as a genetic material that clay-based life was quickly outcompeted by RNA-based life and driven to extinction. This scenario is plausible, but critical experiments to test some of its key assumptions have not yet been performed.

The fact that we cannot clearly choose among such strikingly different theories indicates how much is yet to be learned. On the other hand, the fact that there are such theories and that we can test at least parts of them shows we have made progress in our attempts to understand the origins of life.

THE EVOLUTION OF CATALYTIC ACTIVITY

Although the sequence of events in the origin of life is not yet agreed upon, there is considerable evidence to suggest that metabolism, whether it originated through coacervate drops or clay crystals, preceded the evolution of genes. In the following discussion, we explore in more detail how genes may have originated and how metabolism might have been refined, basing our discussion on recent advances in molecular biology and comparative biochemistry.

Evolution of Enzymes

The control of metabolic processes in all living organisms depends upon enzymes (Chapter 6). Enzymes do two major things: they speed up reactions, and they determine which of the myriad of chemical reactions that *could* occur actually *do* occur at rates high enough to be biologically important. The specificity of enzymes need not have evolved at the same time as their catalytic functions; indeed, because specificity is a more complex function, it might well have evolved after primitive enzymes had acquired the ability to speed up reactions.

There is a problem, however: Enzymes are proteins, and proteins are produced by the translation of genetic messages. How could enzymes have evolved before there were genes? A possible answer is suggested by the fact that many enzymes consist of a protein plus some nonprotein cofactor (Chapter 6). Interestingly, most enzymatic reactions associated with membranes require cofactors. One model for the origin of enzymes, then, is that the primitive control of metabolism began as cofactor-mediated processes occurring in association with membranes. Gradually, over evolutionary time, enzymes might have become associated with these cofactors, thereby making the control of reactions more precise.

According to this view, some of the functions now performed by enzymes were carried out by cofactors alone before the evolution of the genetic code and mechanisms for its translation. This protometabolism would have been imprecise, but at those early times there were no precisely controlled systems competing with the imprecise systems we are imagining. Such imprecise systems probably could not survive today, but they might well have done so some billions of years ago.

Pre-Enzymatic Control of Chemical Reactions

Another piece of evidence suggesting that control of metabolic functions may have developed before enzymes is the recent discovery that the RNA precursor of ribosomal RNA (Chapter 11) can catalyze the splic-

ing of its own intron, without the involvement of any enzyme. Thus, before the evolution of proteins and DNA and mechanisms for its translation, nucleic acids may have existed and influenced which chemical reactions occurred and at what rates.

This possibility is supported by the fact that AMP (adenosine monophosphate) and ATP (adenosine triphosphate) are structurally very similar despite their strikingly different functions (Figure 17.7). ATP is the universal energy carrier in cells, whereas AMP is one of the nucleotides that make up RNA. To convert ATP to AMP, it is necessary only to remove two phosphate groups. Primitive cells lacking a genetic apparatus but having AMP could have, in the presence of enzymes, produced RNA molecules by processes very similar to those observed in Eigen's experiments in which RNA was synthesized with an enzyme but no template. This RNA would have been, in effect, a primitive form of parasite that infected enzyme-based cells. Many infected cells were probably killed, but some would have adapted to the RNA parasite, eventually benefitting from its ability to produce more precise replication than was possible in protein-based living systems. There is no direct evidence of such a transformation, but in Miller's experiments and others like them, amino acids were readily synthesized. Nucleotides, however, are much more difficult to synthesize and are much less stable than amino acids. Accumulations of amino acids in the prebiotic aqueous solutions on Earth were much more likely than accumulations of nucleotides.

17.7 AMP and ATP
AMP and ATP are similar in structure but have very diffent functions.

THE ORIGIN OF GENES

All available evidence suggests that nucleic acids evolved long before there were genes. Somehow, by processes as yet unknown, genes evolved, together with the mechanisms for their translation. It is unlikely that the early stages of this process closely resembled the mechanisms we observe today. The current DNA–RNA–protein system is highly complex and very precise, the result of a long period of evolution under the influence of natural selection and other processes.

Because of their simple, monotonous, and uniform structure, nucleic acids are easy to replicate. However, this simple structure makes them unsuitable for catalyzing metabolic reactions. Therefore, whereas DNA is an excellent replicator and hence good for transmitting information between generations, it is ill-suited for controlling chemical reactions. What does provide such control is the translation of the base sequences of DNA by RNA into proteins, whose physical properties are determined by their specific sequences of amino acids. A key problem in understanding the evolution of life as we know it is to determine how the process of transcribing DNA into mRNA and translating mRNA into proteins evolved. Our current guesses are crude, but recent findings about ribosomal RNA are providing important clues about this process.

COMPARATIVE METABOLISM

The oldest remains of living things discovered so far are 3.5 billion years old. Such remains—**fossils**—are central to the study of how life evolved on Earth. How fossils are formed and how they are studied will be discussed in Chapter 27. The chemical reactions used by these early organisms cannot be studied because they have left no identifiable traces. However, clues about these processes are provided by the study of metabolic processes used by various living organisms today.

The central metabolic processes of all modern eukaryotes are essentially the same. As discussed in Chapter 7, these organisms extract energy from food by glycolysis, followed by either fermentation or aerobic respiration. Plants, as well as some protists, also have the ability to synthesize food from simple inorganic compounds by the process of photosynthesis (Chapter 8). Many other metabolic reactions are identical or closely similar in all eukaryotes. Where they differ, they are often variations on the same basic theme.

Of all the possible energy-yielding reactions involving oxidation, eukaryotes use only the one that reduces oxygen to water. In contrast, prokaryotes,

which evolved before eukaryotes, display a staggering variety of mechanisms for energy storage and food synthesis. Oxygen gas is not the only possible oxidizing agent for cellular respiration among prokaryotes; many bacteria use oxygen, just as plants and animals do, but others use sulfur-containing compounds (Figure 17.8). *Desulfovibrio desulfuricans* can use sulfuric acid for respiration, a process that yields hydrogen sulfide (H_2S) instead of water. Still other bacteria conduct reactions in which carbon dioxide is the oxidizing agent; these organisms liberate methane gas or acetic acid rather than water. Denitrifying bacteria oxidize their food with nitrate ions (NO_3^-) and give off nitrous oxide (N_2O), nitrogen gas (N_2), or ammonia (NH_3).

Many bacteria function with only the energy they receive from fermentation (Chapter 7). Moreover, a number of plants and fungi, as well as a few protists, can function both anaerobically and aerobically, depending on environmental conditions. Other bacteria obtain energy from chemical reactions. A few species of *Pseudomonas* use the energy of the reactions between hydrogen and oxygen to synthesize carbohydrates from carbon dioxide and hydrogen gas. Many sulfur bacteria are not photosynthetic but obtain their energy by oxidizing hydrogen sulfide to sulfur, or sulfur or thiosulfate to sulfate ions. Some *Thiobacillus* species oxidize ferrous ions (Fe^{2+}) to insoluble ferric hydroxide. Some nitrifying bacteria oxidize ammonia to nitrite (NO_2^-), and others oxidize nitrite to nitrate (NO_3^-). These basic metabolic differences among the bacteria provide clues about the early evolution of life.

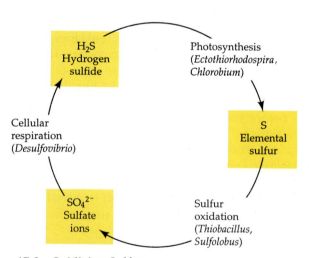

17.8 Oxidizing Sulfur
Several species of bacteria that use sulfur-containing compounds as oxidizing agents and hydrogen donors may function together to carry out sulfur cycles, one of which is shown here. Representative genera able to conduct the processes are given in parentheses. Many such cycles are found in nature.

Anaerobic Microorganisms

Because Earth's atmosphere lacked oxygen at the time when life first appeared, we can learn a great deal about early life from the bacteria that survive today in environments lacking oxygen. These bacteria are **obligate anaerobes**—organisms that obtain their energy without using oxygen. They must remain in an oxygen-free environment because oxygen is poisonous to them. The earliest organisms must have been obligate anaerobes. The first photosynthetizers were also anaerobic, and their photosynthetic processes did not generate oxygen.

The first organisms to give off oxygen were probably cyanobacteria. As cyanobacteria proliferated more than 2.5 billion years ago, the amount of oxygen in Earth's atmosphere greatly increased. This free oxygen began to oxidize minerals dissolved in Earth's oceans, all of which were present in reduced forms. Thus we know that rocks containing oxidized forms of minerals are not much older than 2.5 billion years. After the reduced minerals had all been oxidized, oxygen began to accumulate in the atmosphere.

Aerobic Microorganisms

Because of its toxicity to obligate anaerobes, oxygen was probably the first major pollutant produced by living organisms. To this day it is the most important pollutant produced by life because it has dramatically changed Earth's atmosphere and climate. The upsurge of oxygen created conditions favoring any mutant organisms capable of tolerating the presence of free oxygen. Even more important, it favored those able to use oxygen in respiration. Because aerobic respiration is much more efficient than fermentation (Chapter 7), aerobes outcompeted anaerobes and soon (on a macroevolutionary time scale) replaced them in all environments where free oxygen was present in significant quantities.

THE EVOLUTION OF PHOTOSYNTHESIS

We cannot study the evolution of photosynthesis directly because no traces of early photosynthetic pathways are preserved. However, we can make reasoned guesses about stages in the evolution of photosynthesis by studying the photosynthetic systems of bacteria living today. Three major types of anaerobic photosynthetic bacteria—green sulfur bacteria, purple sulfur bacteria, and purple nonsulfur bacteria—live today in sediments that lack oxygen. The green sulfur bacteria contain a type of chlorophyll called chlorobium, whereas the other two types contain bacteriochlorophyll. Anaerobic photosynthetic bacteria also contain red and yellow carotenoids, which absorb light of wavelengths that are not absorbed by chlorophyll, and pass the energy along to chlorophyll. The photosynthetic system of these bacteria is embedded in membrane complexes called thylakoids (Chapter 4). They possess the electron transport chains by which captured solar energy is passed along and used to generate ATP (Chapter 8). Photosynthetic bacteria were so abundant about 3.4 billion years ago that their partly decomposed remains formed extensive deposits of carbon resembling the coal deposits produced by vascular plants 3 billion years later.

To reduce carbon dioxide (CO_2), a photosynthetic cell requires a source of hydrogen atoms. Green plants use water as their hydrogen source, and thus their photosynthesis liberates oxygen. Like plants, many bacteria use light energy to generate ATP and $NADPH + H^+$, but their photosynthesis does not liberate oxygen (Chapter 8) because they use different sources of hydrogen atoms. The green and purple sulfur bacteria obtain their hydrogen atoms from hydrogen sulfide and generate sulfur as a waste product (Figure 17.9). The purple nonsulfur bacteria obtain hydrogen atoms from organic compounds such as ethanol, lactic acid, or pyruvic acid, or directly from hydrogen gas. In some environments today, the hydrogen is provided by other bacteria as the end products of their fermentations. Under the anaerobic conditions of the early Earth, hydrogen sulfide and other compounds containing hydrogen would have been more abundant than they are today, because now atmospheric oxygen quickly oxidizes them into water, carbon dioxide, and sulfur oxides.

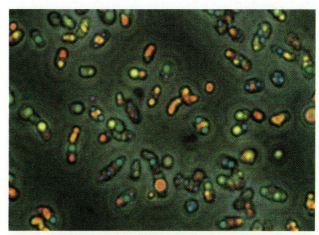

17.9 Sulfur as a By-Product of Photosynthesis
The purple sulfur bacteria in this photo obtain the hydrogen atoms they need for photosynthesis from hydrogen sulfide—a substance that was more plentiful in the anaerobic environment of the early Earth than it is today. The bright yellow granules in the photo are sulfur, formed as a by-product of this bacterial photosynthesis.

The most recent step in the evolution of photosynthesis is the evolution of the ability to use water as the source of hydrogen. This step first appeared in certain sulfur bacteria, which evolved into cyanobacteria. This was the first group of organisms to evolve the ability to split water, which doubtless was the cause of their extraordinary success. The oxygen liberated by this process opened the way for the evolution of oxidation reactions as the energy source for the synthesis of ATP, and thus for the evolution of the full respiratory chain of reactions now carried out by all aerobic cells.

DATING THE PAST

The great age of Earth was not recognized until the middle of the last century. Even when it became evident that Earth was very old, scientists had only imprecise estimates of its age and of when different events had taken place. The sequence of past events was established before scientists developed methods for accurate dating of those events.

Relative Dating

Organisms whose remains were found in sedimentary rocks initially provided the best evidence of the relative ages of those rocks. Indeed, geologists using such evidence had established the relative chronology of geological periods long before Darwin advanced his theory of evolution. The guiding concept was the **principle of superposition**, the fact that as rocks form by the piling up of sediments, younger rocks are deposited on top of older ones (Figure 17.10). Remains of organisms that lived when the sediments were accumulating were preserved within the rocks. Geologists could see slight differences among similar organisms as they compared the layers, or **strata**, of older and younger rocks in the same place. They also found remains of some kinds of organisms at widely separated locations. By assuming that rocks containing a particular type of fossil were likely to be of approximately the same age, they inferred that two widely separated sedimentary deposits that each contained the same type of fossil had been laid down at approximately the same time. By making such comparisons among many locations, and always considering the superposition of strata, geologists had determined the general order of events in the history of life long before they knew the actual times of their occurrences.

Absolute Dating

As rocks form, radioactive isotopes of uranium, thorium, rubidium, and potassium are incorporated into them in proportion to the isotopes' presence in the environment. Each type of radioactive isotope then begins to lose electrons at its own constant rate, becoming by this decay a stable isotope. These rates of decay can serve as **radiometric clocks** because the ages of rocks can be calculated from the proportions of radioactive and stable isotopes present. Uranium-238, for example, spontaneously decays into lead at a slow but precisely known rate. By comparing the amount of ^{238}U still present in a rock with the amount of lead derived from its decay, geochemists can estimate the age of the rock with less than a 5 percent error. Those radioisotopes that decay rapidly are useful for dating the more recent past; those that decay slowly can be used to date the more distant past.

17.10 Younger Rocks Lie on Top of Older Ones
The positions of the rocks in the Grand Canyon of the Colorado River reveal their *relative* ages. However, radiometric clocks were needed to estimate the *absolute* ages given in the figure.

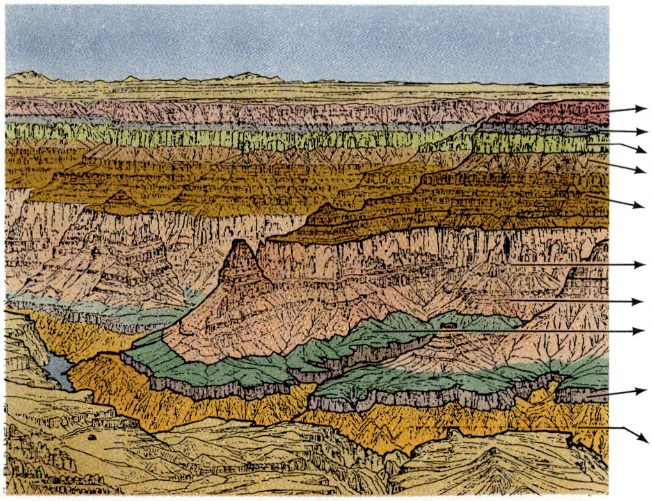

Rock formation	Approximate age in millions of years	Era
Kaibab limestone	250	
Toroweap limestone	255	
Coconino sandstone	260	
Hermit shale	265	
Supai sandstone	285	
Redwall limestone	335	Paleozoic
Mauv limestone	515	
Bright Angel shale	530	
Tapeats sandstone	545	
Zoroaster granite and Vishnu schist	1700-2000	Precambrian

TABLE 17.1
Earth's Geological History

EON	ERA	PERIOD	BEGAN[a]	MAJOR EVENTS IN THE HISTORY OF LIFE
Hadean			4.5 bya	
Archean			3.8 bya	Origin of life; prokaryotes flourish.
Proterozoic			2.5 bya	Eukaryotes evolve; several animal phyla appear.
Phanerozoic			600 mya	
	Paleozoic	Cambrian	600 mya	Most animal phyla present; diverse algae.
		Ordovician	500 mya	Diversification of many animal phyla; first jawless fishes. **Mass extinction** at end of period.
		Silurian	440 mya	Diversification of jawless fishes; first bony fishes; invasion of land by plants and animals.
		Devonian	400 mya	Diversification of fishes; first insects and amphibians. **Mass extinction** late in period.
		Carboniferous	345 mya	Extensive forests; first reptiles; insects radiate.
		Permian	290 mya	Continents aggregate into Pangaea; reptiles radiate; amphibians decline; many types of insects. **Mass extinction,** especially of marine forms, at end of period.
	Mesozoic	Triassic	245 mya	Continents begin to drift; early dinosaurs; first mammals; diversification of marine invertebrates. **Mass extinction** near end of period.
		Jurassic	195 mya	Continents drifting; diverse dinosaurs; first birds
		Cretaceous	138 mya	Most continents widely separated; continued dinosaur radiation; flowering plants and mammals diversify. **Mass extinction** at end of period.
	Cenozoic	Tertiary	66 mya	Continents nearing current positions; radiations of birds, mammals, flowering plants, and pollinating insects.
		Quaternary	2 mya	Repeated glaciations; people evolve; extinctions of large mammals.

[a]bya, billion years ago; mya, million years ago

Geologists divide Earth's history into four **eons**, as shown in Table 17.1: the Hadean eon, the Archean eon, the Proterozoic eon, and the Phanerozoic eon. The Phanerozoic eon is subdivided into **eras**, and the eras are further subdivided into **periods**. The boundaries between these units were originally based on major differences in the fossils contained in successive strata. When radioactivity was discovered and understood early in the present century, scientists were able to determine the dates given in the table. This dated record of the patterns of evolution of life on Earth will be examined more closely in Chapter 27.

Numbers in the billions are so large they have little meaning for most people. To convey a better sense of the ages of fossil organisms, it is useful to scale down the history of Earth to a hypothetical 30-day month, as shown in Figure 17.11. Each "day" on this geological calendar represents approximately 150 million years. On the calendar, the Hadean, Archean, and Proterozoic eons stretch across the first 26 days. The Cambrian period, which opened the Paleozoic era 600 million years ago, marks a great divide in the fossil record. By the dawn of the Cambrian, on day 27, the ancient seas teemed with life. Representatives of most modern groups of organisms had appeared, as had species from groups that left no survivors.

The absolute dating made possible by the use of radiometric clocks greatly increased the estimates of the age of Earth. More recently, detailed studies of ancient rocks demonstrated that cellular life evolved on Earth much earlier than previously suspected. Many scientists now believe that life appeared within the first half billion years of Earth's existence—the first four days or so on our calendar. A concise summary of the probable chronicle of these events is given in Box 17.A.

The evolution of life irrevocably changed the nature of our planet. Life not only created the oxygen of our atmosphere, it also removed most of the carbon dioxide from the atmosphere by transferring it into sediments. On the early Earth, volcanoes poured large quantities of hydrogen, nitrogen, carbon dioxide, methane, and hydrogen sulfide into the atmosphere. In the waters, compounds such as ammonium, nitrates, carbon dioxide, sulfates, and phosphates also circulated. Under these conditions, a variety of anaerobic bacteria evolved and thrived. The cycles of some of these elements cannot be completed even today without the involvement of anaer-

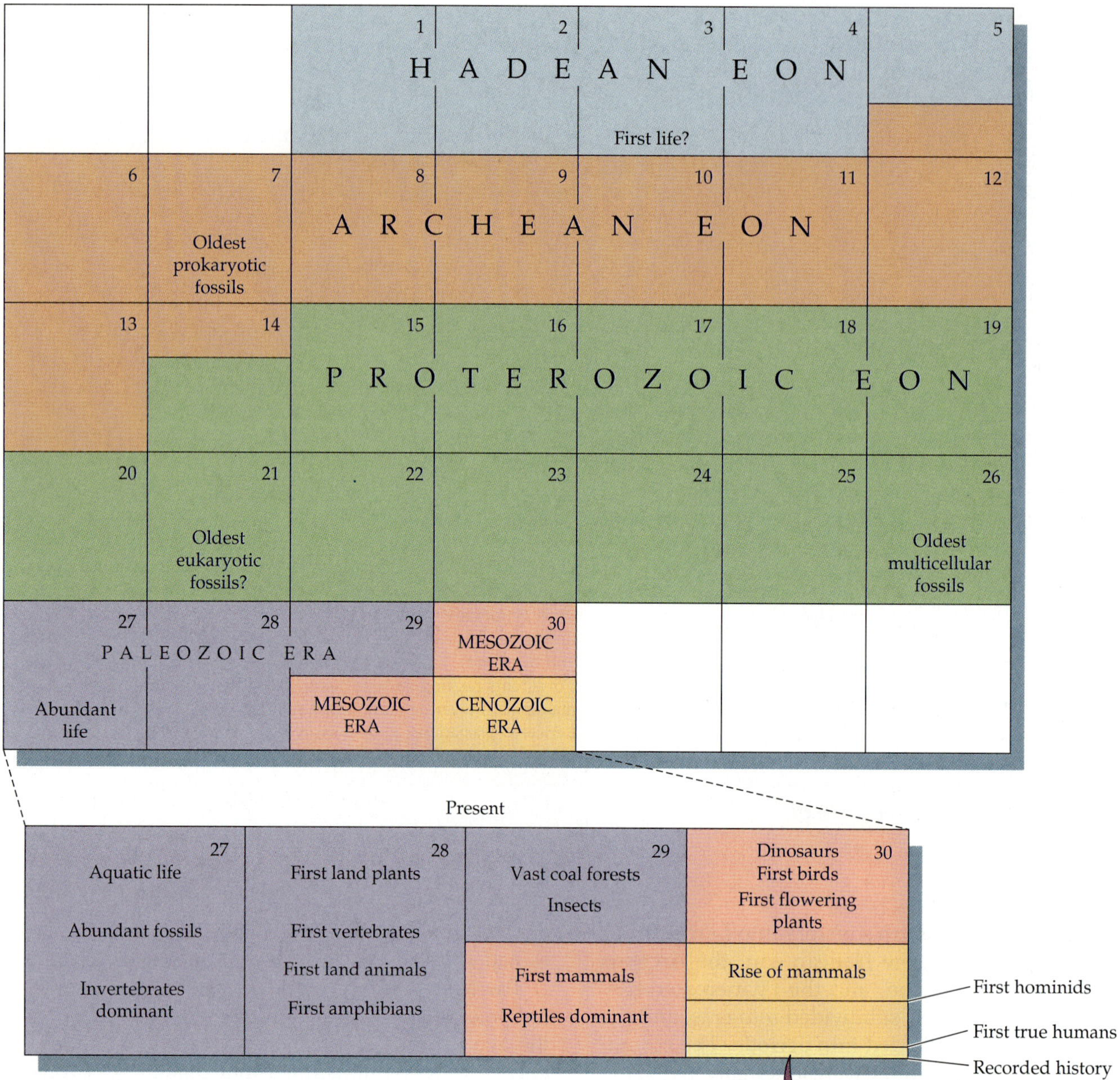

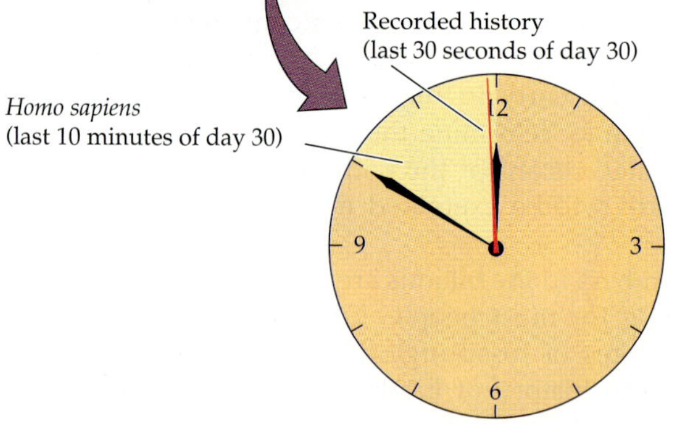

		1	2	3	4	5
		HADEAN EON		First life?		
6	7 Oldest prokaryotic fossils	8	9 ARCHEAN EON	10	11	12
13	14	15	16 PROTEROZOIC EON	17	18	19
20	21 Oldest eukaryotic fossils?	22	23	24	25	26 Oldest multicellular fossils
27 PALEOZOIC ERA Abundant life	28	29 MESOZOIC ERA	30 MESOZOIC ERA / CENOZOIC ERA			

17.11 Life's Calendar
Major periods and events in the history of life on Earth are represented in this calendar, on which a "day" lasts about 150 million years. On this scale, *Homo sapiens* evolved in the last 10 minutes of day 30, and recorded history is confined to the final 30 seconds.

Present

27 Aquatic life — Abundant fossils — Invertebrates dominant	28 First land plants — First vertebrates — First land animals — First amphibians	29 Vast coal forests — Insects — First mammals — Reptiles dominant	30 Dinosaurs — First birds — First flowering plants — Rise of mammals

— First hominids
— First true humans
— Recorded history

Recorded history (last 30 seconds of day 30)

Homo sapiens (last 10 minutes of day 30)

obic bacteria (see Chapter 47). Thus, even though anaerobic bacteria survive only in environments lacking oxygen, life as we know it would not have evolved without their metabolic activities, and it still depends on them today.

Is new life still being assembled from nonliving matter on today's Earth? We don't know with certainty, but probably it is not. Any simple biological molecules released into today's environment are quickly consumed by already living things. Also, the Earth's atmosphere is now so rich in oxygen that such molecules can no longer accumulate. Generation of life from nonlife on Earth did happen, but it is an event of the past.

BOX 17.A

A Concise Scenario for the Origin and Early Evolution of Life

1. The solar system condenses from a dust cloud; Earth and other planets form.

2. Earth's early atmosphere (mostly hydrogen) is lost. Heat from radioactive decay and gravitational compression melts Earth's interior.

3. Gases escape from the hot interior, producing a reducing atmosphere of nitrogen, ammonia, water vapor, carbon monoxide, methane, and other reduced forms of carbon. There is no oxygen yet.

4. Water vapor condenses into seas. Ultraviolet light, radioactive decay, volcanic heat, and lightning provide the energy to produce organic compounds from atmospheric gases.

5. In places such as sea-floor volcanic vents, high concentrations of energy, hydrogen sulfide, and carbon dioxide provide the initial conditions for protobiological chemical interactions. Naturally occurring ATP provides an energy source for synthesis.

6. Proto-living systems evolve primitive enzymes to catalyze metabolic processes. Reproduction occurs but it is poorly controlled.

7. Living systems evolve the ability to use RNA as a template, first for catalyzing reactions, later for protein synthesis. Replication becomes more precise.

8. Proto-living systems evolve DNA as the genetic material and slowly improve the precision with which it is copied and used as a template for protein synthesis. **Life has evolved.** Copying errors (mutations) yield variations upon which natural selection acts.

1 BILLION YEARS have elapsed

9. Simple prokaryotic organisms evolve varied synthetic and respiratory pathways, using a variety of molecules as substrates for their energy. One of these pathways uses glucose as the energy source for the synthesis of ATP. This pathway is later used by all eukaryotes. Several photosynthetic pathways evolve, but none of them liberates oxygen gas.

2 BILLION YEARS have elapsed

10. Some cyanobacteria evolve photosynthetic pathways that liberate large quantities of oxygen.

11. Oxygen, a potent biological poison, accumulates in the atmosphere. Some prokaryotes evolve the ability to combine it with their metabolic products—aerobic respiration begins.

12. As oxygen builds up, some is converted to ozone, which accumulates in the upper atmosphere. This ozone shields Earth from ultraviolet radiation, allowing life to invade shallow waters and the land.

13. Respiring organisms, using efficient oxidation reactions as their energy source for ATP synthesis, evolve the cytochromes of the terminal respiratory chain.

3 BILLION YEARS have elapsed

14. Eukaryotic cells evolve. Endosymbiotic cyanobacteria become chloroplasts. Other endosymbiotic bacteria become mitochondria. Multicellular organisms evolve from these efficient eukaryotic cells.

SUMMARY

The experiments of Redi, Spallanzani, Pasteur, and others demonstrated that life does not arise from nonliving matter on Earth today. However, life *did* arise by chemical evolution—a form of spontaneous generation—under conditions much different from today's. The initial energy source for life's evolution is not known, but volcanic vents were probably important sites for the evolution of protolife. The early atmosphere did not contain significant quantities of oxygen, and it probably had much more carbon dioxide than does the current atmosphere.

A key problem in understanding the origin of life is to determine whether metabolism preceded replication or vice versa. A likely scenario for the evolution of the genetic code is that RNAs similar to ribosomal RNAs were the first genetic materials, and that more complicated mechanisms for replication and translation of the genetic code evolved slowly over long time periods. Life evolved more than 3.5 billion years ago. By 3.4 billion years ago, organisms were already so plentiful that their partly decomposed remains formed extensive carbon-rich seams in rocks. The earliest organisms lived in an oxygen-free environment, but the great success of the early cyanobacteria generated the oxygen-rich atmosphere that has characterized Earth ever since. Control over metabolism evolved gradually, leading to the complex processes of cellular metabolism that characterize all living organisms today. Anaerobic bacteria were of paramount importance in the evolution of life, and they are vital even today, because without them the great biogeochemical systems of elements upon which all life depends would be disrupted.

SELF-QUIZ

1. The atmosphere of early Earth consisted largely of:
   a. water vapor.
   b. hydrogen.
   c. carbon dioxide.
   d. helium.
   e. nitrogen.

2. Pasteur's experiments convinced most people that spontaneous generation of life did not happen because:
   a. Pasteur was extremely meticulous.
   b. Pasteur used very fine mesh screens to cover his flasks.
   c. Pasteur did not boil his flasks for a long time.
   d. Pasteur's swan-necked flasks ruled out the "spoiled air" objection to Spallanzani's experiments.
   e. Leeuwenhoek's microscopic observations of tiny organisms made spontaneous generation less probable.

3. To test Harold Urey's theory of the origin of life, Stanley Miller used an apparatus with an atmosphere of:
   a. oxygen, hydrogen, and nitrogen.
   b. oxygen, hydrogen, ammonia, and water vapor.
   c. oxygen, hydrogen, and methane.
   d. hydrogen, methane, and carbon dioxide.
   e. hydrogen, ammonia, methane, and water vapor.

4. The sequence of events in the origin of life proposed by Oparin was:

   a. cells first, metabolism second, and replication third.
   b. metabolism first, cells second, and replication third.
   c. replication first, cells second, and metabolism third.
   d. cells first, replication second, and metabolism third.
   e. metabolism first, replication second, and cells third.

5. The sequence of events in the evolution of life postulated by Eigen was:
   a. metabolism first, genes second, and cells third.
   b. genes first, metabolism second, and cells third.
   c. genes first, cells second, and metabolism third.
   d. metabolism first, cells second, and genes third.
   e. cells first, genes second, and metabolism third.

6. Some biologists think that enzymes might have evolved before there were genes because:
   a. amino acids were produced in Stanley Smith's apparatus.
   b. when life evolved, proteins might have been synthesized without the use of genes.
   c. early cells could have used RNA to catalyze chemical reactions.
   d. control of metabolism was achieved by carbohydrates to which enzymes were later attached.
   e. No biologist actually believes this.

7. Biologists believe that the current DNA–RNA–protein system is the result of a long period of evolution because:

   a. the transcription of DNA to mRNA and translation of mRNA into proteins consists of many steps.
   b. DNA replication is complicated but relatively error-free.
   c. the current system is very complex and precise.
   d. DNA is an excellent replicator but is ill-suited for controlling chemical reactions.
   e. All of the above.

8. The first organisms to give off oxygen were probably:
   a. bacteria that use sulfuric acid for respiration.
   b. iron-oxidizing bacteria.
   c. cyanobacteria.
   d. green algae.
   e. colonial flagellates.

9. To reduce carbon dioxide, a photosynthetic cell needs a source of:
   a. hydrogen atoms.
   b. water.
   c. sulfur.
   d. hydrogen sulfide.
   e. bacteriochlorophyll.

10. The principle of superposition states that:
    a. lighter elements rest on top of more dense ones.
    b. lighter rocks rest on top of more dense ones.
    c. organisms are preserved in rocks if they land on top of hard places.
    d. younger rocks lie on top of older ones.
    e. as rocks age, they are thrust upward by earthquakes.

FOR STUDY

1. In comparison with the kingdom Monera, eukaryotes use a very limited set of reactions to conduct their energy metabolism. A very sketchy explanation of this situation is offered in the text. Starting from this version, develop a more complete and explicit one. Your explanation should deal both with the environment in which the first eukaryotes appeared and with their evolutionary origins.

2. Describe the evidence supporting

the assertion that prokaryotes were already abundant 3.5 billion years ago.

3. On the one hand, we are confident that life no longer arises from nonliving matter. On the other hand, most biologists believe that life did arise on this planet, billions of years ago, from nonliving matter. Account for this apparent discrepancy.

4. Laboratory studies of the origin of life have been able to pinpoint nei-

ther the exact composition of Earth's early atmosphere nor the energy source that powered the chemical reactions leading to early life. In view of this, why have biologists not abandoned Oparin's thesis?

5. How might each of the following have been involved in the evolution of coacervate drops?
   a. coating of drop boundaries with lipids
   b. wave action in bodies of water
   c. catalysts within the drops

READINGS

Bernal, J. D. 1967. *The Origin of Life.* World Publishing Company, Cleveland, OH. A clear—but dated—consideration of technical and philosophical issues. Includes reprints of original papers by Oparin and Haldane, as appendices.

Cairns-Smith, A. G. 1982. *Genetic Takeover and the Mineral Origins of Life.* Cambridge University Press, New York. An account of the clay matrix theory of the origin of life.

Dyson, F. J. 1985. *The Origins of Life.* Cambridge University Press, New York. A concise argument in favor of multiple origins of life. Favors the primacy of metabolism over replication.

Eigen, M., W. Gardiner, P. Schuster, and R. Winkler-Oswatitch. 1981. "The Origin of Genetic Information." *Scientific American*, April. A molecular biology view of the origin of genes.

Hawking, S. W. 1988. *A Brief History of Time from the Big Bang to Black Holes.* Bantam, New York. A readable account, written for informed laypersons, of current concepts of time and the origin of the universe.

Life: Origin and Evolution. 1979. W. H. Freeman, San Francisco. (A Scientific American Book.) A collection of articles on all aspects of the origin of life.

Loomis, W. F. 1988. *Four Billion Years.* Sinauer Associates, Sunderland, MA. A very readable book on the evolution of genes and organisms, concentrating on the first billion years.

Margulis, L. 1984. *Early Life.* Jones and Bartlett, Boston. An engaging account of the earliest organisms and how they evolved. Good treatment of the symbiotic theory of the origin of eukaryotes by its principal exponent.

Woese, C. R. 1984. "The Origin of Life." Carolina Biology Readers, Carolina Biological Supply Company, Burlington, NC. Discusses why old theories of the origins of life may be incorrect.

18

The Mechanisms
of Evolution

PREVIEW: Evolution is the accumulation of heritable changes within populations over time. We study the patterns of those changes and conduct experimental manipulations of populations in order to discover the underlying evolutionary mechanisms. The agents of evolution include nonrandom mating, mutation, genetic drift, migration, and natural selection. Of these, only natural selection adapts populations to their environments; the other agents produce new opportunities for natural selection to work. Genetic variation is the raw material of evolutionary change, and nearly all species are genetically highly variable, both locally and geographically. Many detailed studies have demonstrated the action of natural selection and other agents of evolution over short time frames.

This chapter deals with evolutionary change, the genetic structure of populations, the Hardy–Weinberg rule, agents that change genotype and allele frequencies, forms of natural selection, geographic variation, and micro- and macroevolution.

Evolutionary processes operating since life first appeared on Earth more than 3 billion years ago have given rise to the millions of species currently living, as well as to the much larger number of species that lived in the past but no longer survive today. The extensive record of the past reveals many patterns in the evolution of life, but that record provides little information about the factors that led to the evolutionary changes. To identify the agents of evolution and demonstrate their modes of action, we must carry out short-term observational and experimental studies in the field and in the laboratory. In this chapter we examine the agents of evolution and the short-term studies designed to investigate them. In later chapters we consider how to use this information to explain the broad features of the evolutionary record.

Because the events that led to the appearance of life and guided its evolution over several billion years cannot be observed directly, much speculation has accompanied attempts to understand evolution. Studies of simple organisms living today provide important clues about early evolutionary processes, but answering key questions about the processes guiding evolutionary changes requires hypotheses that can be tested observationally and experimentally. In Chapter 1, we pointed out that ideas about evolution have been advanced for centuries. What was lacking until this century, however, were testable hypotheses about the causes of evolutionary change. The essential ingredients of these hypotheses were the mechanisms of evolution that Charles Darwin first outlined and the underlying genetic basis provided by the rediscovered work of Gregor Mendel. Mendel's discovery of how traits are inherited in families was extended to explain how gene frequencies are determined in populations.

WHAT IS EVOLUTION?

Biological evolution is a change in the genetic composition of a population over time. Such change requires variation among the population's individuals in traits that are **heritable**—that is, traits that can be passed on to offspring. The genetic constitution governing a heritable trait is called its **genotype**, as we saw in Chapter 10. A population changes genetically—it evolves—when individuals having different genotypes survive or reproduce at different rates. This may come about because the heritable traits influence the ability of individuals to obtain mates, find food, or avoid hazards. However, which genotypes survive and reproduce can also be determined by events very different from those normally affecting survival and reproductive success.

The physical expression of a genotype, its **phenotype**, is what the agents of evolutionary change

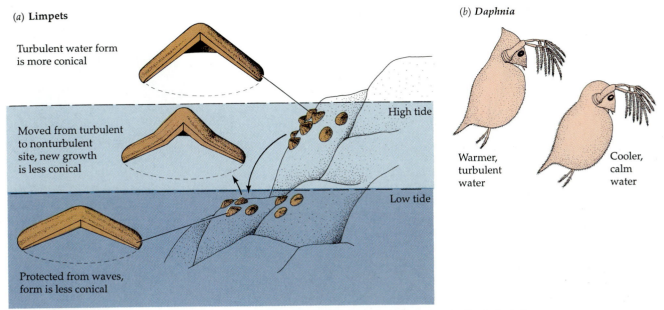

(a) **Limpets**

Turbulent water form
is more conical

High tide

Moved from turbulent
to nonturbulent
site, new growth
is less conical

Low tide

Protected from waves,
form is less conical

(b) *Daphnia*

Warmer,
turbulent
water

Cooler,
calm
water

18.1 Variation Can Be Environmentally Induced
Some organisms adjust their shapes to the different environments they encounter during their lives. *(a)* The way limpets grow is partly determined by
the turbulence of the water in which they live. *(b)* Head shape in water fleas
(*Daphnia*) depends on both water temperature and turbulence.

act upon. But not all phenotypic variation is governed
by genotype. Some of the variation observed within
populations of living organisms is genetically determined but some of it is not.

Environmentally Induced Variation

The shapes and sizes of many marine animals that
live attached to a substrate depend not only on their
genotypes, but also on water temperature, concentration of nutrients (particularly calcium), competition with neighbors, and the turbulence of the water.
For example, limpets growing high in the intertidal
zone, where they experience heavy wave action, are
more cone-shaped than limpets of the same species
growing in the subtidal zone, where they are protected from wave action. This difference is not genetic; limpets taken from high in the intertidal zone
and transplanted to the subtidal zone will add new
growth to their shells to produce a flatter, subtidal
shape (Figure 18.1a). If a water flea, *Daphnia cucullata*,
grows in cool or calm water, it develops a rounded
head, but if it is moved to warm or turbulent water,
it develops a pointed "helmet" (Figure 18.1b). This
variation cannot be genetically determined, because
during the periods over which the shape changes,
the populations consist entirely of females producing
genetically identical daughters by means of a mitosis-
like oogenesis. Environmental variation is reponsible
for the phenotypic variation in these examples.

The cells of the leaves on a tree or shrub are normally genetically identical. Yet leaves on the same
tree often differ in shape and size. Leaves closer to

the top of an oak tree, where they receive more wind
and sunlight, are more deeply lobed than leaves
lower down in the same tree (Figure 18.2). However,
these within-plant variations are not passed on to
offspring. What *is* passed on is the ability to form
various types of leaves from the same genotype in
response to different environmental conditions.

Genetically Based Variation

Although environmentally induced variation is common, high levels of genetic variation characterize

Grown in sun

Grown in shade

18.2 Leaf Shape Depends on Light
The sun leaves of white oaks have more edge per unit of
surface area than shade leaves on the same tree do; thus
they dissipate heat more rapidly, and also allow light to
pass to the shade leaves growing lower down on the
tree.

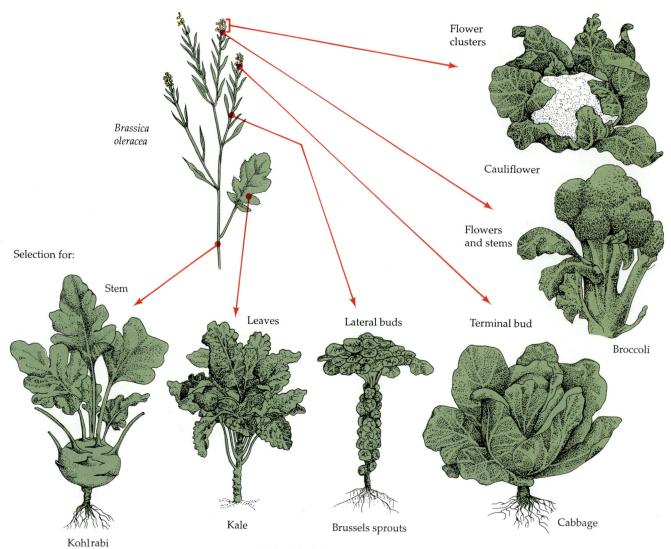

18.3 The Domestication of Mustard
There was enough genetic variation in the common wild mustard (*Brassica oleracea*) for European agriculturalists to select for and produce all the divergent crop plants shown here.

nearly all natural populations. The first indication that this is true came from the success of people attempting to produce economically desirable traits in plants and animals. For example, artificial selection for different traits in a common European wild mustard produced many important crop plants (Figure 18.3). Such results by plant and animal breeders can be achieved only if the original population is genetically variable. Therefore, the almost universal success of breeders indicates that genetic variation is common.

To understand evolution we need to know how genetic variability is maintained and expressed in populations, and how populations are distributed in space and over time. We also need to know the agents that change the genetic variation in populations, how they act, and their relative importance in affecting the direction of evolutionary changes.

THE STRUCTURE OF POPULATIONS

The appropriate unit for defining and measuring genetic variation is a **population**, a group of organisms of the same species occupying a particular geographic region. The size of the geographic region we recognize depends on the objectives of a particular investigation. Figure 18.4 shows the distribution pattern of the shrub *Clematis fremontii* in Missouri at several different scales, ranging from local aggregations to the entire range of the species. The individuals in a single aggregation may be a useful population for the study of pollen movement. The populations in different glades would be appropriate subjects for studies of adaptation to local conditions. An investigation to determine why *Clematis fremontii* has such a limited geographic range would need information from populations throughout the range of the species.

For some purposes, a population needs to be precisely defined as the group of individuals in a particular place that mate with each other. A population in this sense may be part of a larger population—a geographic population—that extends over so great an area that its members in some places are unlikely to mate with those in others. A locally interbreeding group within a geographic population may be called a subpopulation, a deme, or a Mendelian population. Populations within which members interbreed are often the subjects of evolutionary studies. They will be our concern throughout this chapter.

Genetic variation is a necessary ingredient of evolutionary change. The evolutionary potential of a population depends in part on the amount of genetic variation present in it. That is because the greater the number of variable genetic loci in a population and the more alleles present at each locus, the more likely it is that which individuals survive will be determined by their genotypes rather than by chance. Therefore, the study of genetic variation and its causes is a central concern of evolutionary biologists.

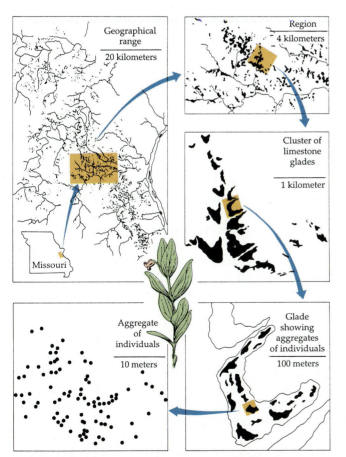

18.4 Populations Are Structured Hierarchically
Clematis fremontii is a shrub that grows only in Missouri, in glades with limestone outcrops that have very specific conditions of moisture, soil nutrients, and soil structure. At all spatial scales, *Clematis* individuals aggregate in suitable sites.

How can we determine how much genetic variation a particular population has? The **gene pool** is the sum total of genetic information present in a population at any given moment. **Allele frequencies** are central to the study of genetic variation in populations. If we could count every allele at every locus in every organism in a population, we could measure the genetic variation in that population; that is, we could determine the number of alleles for each locus and their relative proportions, or frequencies, in the population. In fact we cannot count alleles, but it is possible to make useful estimates using the methods that follow. Once you understand how to calculate allele frequencies, you will find it easy to think about the importance of the amount of genetic variation in populations.

Assume that there are only two alleles at a given locus. For convenience, we will label one allele *A* and the other *a*. (Whether allele *A* is dominant over *a* does not matter at this point.) As you know, two different alleles can combine to form three different genotypes: *AA*, *Aa*, and *aa*. The relative frequencies of *A* and *a* can be calculated as follows: In a population consisting of *N* diploid individuals, let *X* be the number of individuals that are homozygous for the *A* allele (*AA*); let *Y* be the number that are heterozygous (*Aa*); and let *Z* be the number homozygous for the *a* allele (*aa*). Note that $X + Y + Z = N$, the total number of individuals in the population, and that the number of alleles is $2N$. Each *AA* individual has two *A* alleles and each *Aa* individual has one *A* allele. Therefore, the total number of *A* alleles in the population is $2X + Y$, and the total number of *a* alleles in the population is $2Z + Y$. If *p* represents the frequency of the *A* allele and *q* the frequency of *a*, then

$$p = \frac{2X + Y}{2N} = \frac{X + 0.5Y}{N}$$

and

$$q = \frac{2Z + Y}{2N} = \frac{Z + 0.5Y}{N}$$

To see how this works, let us calculate the allele frequencies in two populations, each consisting of 200 diploid individuals. Population 1 has mostly homozygotes—90 *AA*, 40 *Aa*, and 70 *aa*—while population 2 has mostly heterozygotes—45 *AA*, 130 *Aa*, and 25 *aa*. In population 1, where $X = 90$, $Y = 40$, and $Z = 70$,

$$p = \frac{X + 0.5Y}{N} = \frac{90 + (0.5)(40)}{200} = 0.55$$

and

$$q = \frac{2Z + 0.5Y}{N} = \frac{70 + (0.5)(40)}{200} = 0.45$$

In population 2, where $X = 45$, $Y = 130$, and $Z = 25$,

$$p = \frac{X + 0.5Y}{N} = \frac{45 + (0.5)(130)}{200} = 0.55$$

and

$$q = \frac{Z + 0.5Y}{N} = \frac{25 + (0.5)(130)}{200} = 0.45$$

These calculations demonstrate two important points. First, notice that for each population $p + q = 1$. Frequencies are measures that range from 0 to 1; the sum of any set of frequencies is thus 1. If there is only one allele in a population, its frequency is 1. If an allele is missing from a population, its frequency is 0, and the locus in that population is represented by one or more other alleles. Because $p + q = 1$, $q = 1 - p$, which means that when there are two alleles at a locus in a population, we can calculate the frequency of one allele and then easily obtain the second frequency by subtraction.

The second thing to notice in these calculations is that these two populations—one consisting mostly of homozygotes and the other mostly of heterozygotes—have exactly the same allele frequencies for A and a. Therefore, they have the same gene pool for this locus. However, the alleles in the gene pool are distributed differently among genotypes. These distribution patterns, together with information about allele frequencies, describe the **genetic structure** of a population.

When studying the genetic structure of populations, evolutionary biologists begin with numbers of genotypes, as we have in this section. For many purposes, genotypes, like alleles, are best thought of as frequencies. The genotype frequencies of population 1 are 0.45 AA, 0.20 Aa, and 0.35 aa. What are the genotype frequencies of population 2?

The Hardy–Weinberg Rule

We have noted that evolution is change in the genetic composition of a population over time. A population that is not changing, that has the same allele and genotype frequencies from generation to generation, is said to be at **equilibrium**. How can we tell whether a population is changing or is at equilibrium? The major method is a statistical result known as the Hardy–Weinberg rule, named after the British mathematician G. H. Hardy and the German biologist W. Weinberg, who each derived it independently in 1908. The rule specifies the conditions a population must meet to be at equilibrium and the genotype frequencies that will be found in such a population. By comparing the frequencies we observe in real populations with those specified by the Hardy–Weinberg rule, we can detect changes and direct our attention to the most likely causes of a change.

The conditions for Hardy–Weinberg equilibrium, which are precise, are considered in detail in the

sections that follow. To meet the conditions, a population must be very large and be made up of sexually reproducing diploid individuals. All individuals in the population must survive and reproduce equally well, and mating must combine genotypes at random. The rule says that if all these conditions hold, the frequencies of alleles at a locus remain constant from generation to generation. It further says that the frequencies of the genotypes, which also remain constant, are related to the allele frequencies. When there are two alleles at the same locus, allele and genotype frequences are related as follows:

$$p^2_{(AA)} + 2pq_{(Aa)} + q^2_{(aa)} = 1$$

To see why this is so, we will consider an example in which the frequency p of A alleles is 0.6. Because individuals select mates without regard to genotype, gametes carrying A and a combine at random—that is, as predicted by the frequencies p and q. The probability that any given sperm or egg bears an A allele rather than an a allele is 0.6. Six out of 10 random selections of a sperm or an egg will bring up an A allele. Since $q = 1 - p$, the probability of drawing an a allele is $1 - 0.6 = 0.4$. To obtain the probability of two A-bearing gametes coming together at fertilization, we must multiply the two independent probabilities of drawing them, $p \times p = p^2 = 0.36$. Therefore, 0.36, or 36 percent, of the offspring in the next generation will have AA genotypes. Similarly, the probability of bringing together two a-bearing gametes is $q \times q = q^2 = 0.16$, so there will be 16 percent aa genotypes in the next generation. As Figure 18.5 shows, there are two ways of producing a heterozygote: an A sperm may combine with an a egg, the probability of which is $p \times q$; or an a sperm may combine with an A egg, the probability of which is also $p \times q$. Consequently, the overall probability of obtaining a heterozygote is $2pq$. What percentage of the next generation will be heterozygotes?

It is easy now to show that the allele frequencies p and q remain constant each generation. Notice that total of $p^2 + pq$ represents the total of the A alleles. The fraction that this constitutes of all alleles is

$$\frac{p^2 + pq}{p^2 + 2pq + q^2} = p^2 + pq = p^2 + p(1 - p) = p$$

By a parallel procedure we can show that the frequency of a in the next generation will be q. Thus the original allele frequencies are preserved.

The most important message of the Hardy–Weinberg rule is that allele frequencies remain the same unless some agent acts to change them. Also, the rule shows us exactly what distribution of genotypes to expect for a population at equilibrium at any value of p and q. Figure 18.6 shows the Hardy–Weinberg frequencies for the three genotypes, AA, Aa, and aa, at all values of p and q. The values we used in Figure 18.5 are highlighted with a dashed line.

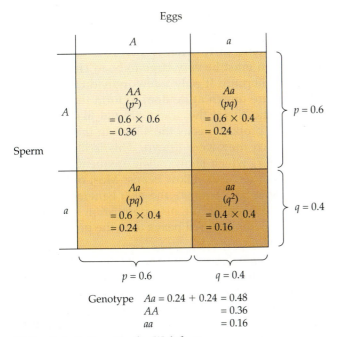

18.5 Calculating Hardy–Weinberg Genotype Frequencies
The areas within the squares are proportional to the expected frequencies of possible matings if mating is random with respect to genotype and the frequency (p) of A alleles is 0.6 and the frequency (q) of a alleles is 0.4.

Populations in nature rarely—perhaps never—meet the conditions of the Hardy–Weinberg rule. Why, then, is the rule considered so important for the study of evolution? The answer is that without it we cannot tell whether evolutionary agents are operating. If individuals in a population mate randomly and no other agents are operating to change allele frequencies, then genotype frequencies will approximate those calculated from the Hardy–Weinberg formula. However, if the frequencies of genotypes deviate significantly from the expected Hardy–Weinberg values, that fact is evidence either of nonrandom mating, or of the action of some other agent of evolution. In other words, if the genotype frequencies in a population do not fit Hardy–Weinberg frequencies, we know that something of evolutionary interest is influencing the population. Without the rule, there is no way of knowing whether observed genotype frequencies are interesting or even surprising.

AN AGENT THAT CHANGES GENOTYPE FREQUENCIES

Evolutionary agents act to change the genetic structure of a population. They are violations of the conditions for Hardy–Weinberg equilibrium. Evolutionary agents can change genotype frequencies in a population while leaving allele frequencies unperturbed or they can also change allele frequencies. Because a population evolves when its allele frequencies change, it might seem that only agents that change allele frequencies should be called evolutionary agents. In fact, as we will see, a major change in genotype frequencies can have a tremendous evolutionary effect. We first consider an agent that influences genetic variation in populations without changing allele frequencies. Then we will turn to those agents that do change allele frequencies.

Nonrandom Mating

One Hardy–Weinberg condition specifies that individuals not choose mates on the basis of their genotypes. Often, however, individuals with certain genotypes (or phenotypes) do mate more often with individuals of the same (or different) genotypes (or phenotypes) than would be expected on a random basis. When such **assortative mating** takes place, homozygous genotypes are overrepresented, in comparison with Hardy–Weinberg expectations, and heterozygous genotyes are underrepresented in the next generation of the population. Assortative mating and self-fertilization are two forms of nonrandom mating. The latter, which is common in many groups of organisms, especially plants, tends to reduce the frequencies of heterozygous individuals in populations. Among vertebrates, assortative mating is common. For example, in the United States, tall women tend to marry tall men and short women tend to marry short men. Nonrandom mating alters the frequencies of genotypes but not the frequencies of the alleles themselves.

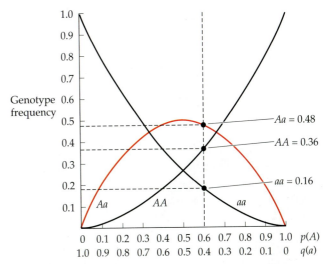

18.6 Hardy–Weinberg Genotype Frequencies
Hardy–Weinberg frequencies are shown for the three genotypes AA, Aa, and aa at all values of p and q. Notice that the proportion of heterozygotes increases as the values of p and q approach 0.5.

AGENTS THAT CHANGE ALLELE FREQUENCIES

Mutation

The origin of genetic variation is **mutation** (Chapter 10). Most mutations are harmful or neutral to their bearers, but if the environment changes, previously neutral or deleterious alleles may become advantageous. In addition to creating the raw material that makes evolution possible, mutations can also influence the direction of evolution. Mutation rates are very low for most loci that have been studied. Rates as high as one mutation in a thousand zygotes per generation are rare; one in a million is more typical. Nonetheless, over long time spans these rates are sufficient to create considerable genetic variation. In addition, mutations can restore to a population alleles that other evolutionary agents remove. Thus, mutations both create and help to maintain variation within populations. One condition for Hardy–Weinberg equilibrium is that there be no mutation. Although this condition is never strictly met, the rate at which mutations arise at single loci is so low that mutations will result in only very small deviations from Hardy–Weinberg expectations. If large deviations are found, it is appropriate to dismiss mutation as the cause and to look for evidence of the operation of other evolutionary agents.

Genetic Drift

As you know, on average half the times a fair coin is tossed it will come up heads, but for any particular toss it is equally likely to come up heads or tails. A ratio of 8 heads to 2 tails is not uncommon in 10 tosses of a coin; a ratio of 80 heads to 20 tails is much less likely. To demonstrate the principle of probability that guides the outcome of coin tossing, you need to toss a coin many times. Otherwise you would be demonstrating the effects of chance.

Just as chance determines the outcome of tossing a coin just once or a few times, chance alters allele frequencies in small populations. Such alteration is called **genetic drift**. This is why we noted that a population must be very large to be in Hardy–Weinberg equilibrium. If only a few individuals or a few gametes are drawn at random to form the next generation, the alleles they carry are not likely to be in the same proportions as the alleles in the gene pool from which they were drawn. This is illustrated in Figure 18.7, which shows allele frequencies as proportions of red and yellow beans. Most of the beans that survive to germinate the next generation in this example are red, so the new population has a much higher frequency of red beans than the previous generation had.

In very small populations, genetic drift may be strong enough to change allele frequencies even when other evolutionary agents are pushing the frequencies in different directions. Deleterious alleles, for example, may increase because of genetic drift, and rare advantageous alleles may be lost. Two important causes of genetic drift are population bottlenecks and the founder effect.

BOTTLENECKS. Even organisms that normally have large populations may pass through occasional periods when only a small number of individuals survive. During these population **bottlenecks**, genetic variation can be lost by chance. For example, suppose we have performed a cross of $Aa \times Aa$ individuals of a species of *Drosophila* to produce an offspring population in which $p = q = 0.5$ and in which the genotype frequencies are 0.25 AA, 0.50 Aa, and 0.25 aa. If we randomly select four individuals from among the offspring to form the next generation, the allele frequencies in this small sample may differ markedly from $p = q = 0.5$. If, for example, we happen to draw two AA homozygotes and two heterozygotes, in this "surviving population" $p = 0.75$ and $q = 0.25$. If we begin a large evolutionary study by replicating this experiment 1,000 times, one of the two alleles will be missing entirely from about eight of the 1,000 "surviving populations."

Populations in nature pass through bottlenecks for many different reasons. Climatic changes may reduce the range of a species to a very small area. For example, during the last Ice Age, the southward move-

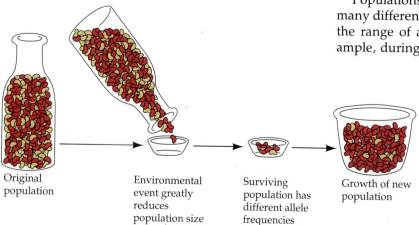

Original population

Environmental event greatly reduces population size

Surviving population has different allele frequencies

Growth of new population

18.7 Genetic Drift
As this imaginary population passes through a bottleneck, the frequencies of the yellow and red alleles change because of random sampling.

18.8 Northern Elephant Seals Lack Genetic Variability
A large male in control of this beach trumpets amid females who are here to have pups and then mate. Because a few such males in the breeding colony sire most of the offspring, the effective population size is much smaller than the actual population size.

ment of glaciers in North America and Europe greatly compressed the ranges of many plant species and reduced populations to very low levels. Predation may also reduce populations to very small sizes. During the 1890s, hunting by humans reduced the number of northern elephant seals to about 20 animals in a single population on the coast of Mexico. The actual breeding population may have been even smaller because elephant seals are highly polygynous, and only a few males inseminate all the females in any generation (Figure 18.8). Northern elephant seals have less genetic variation than any other seal that has been studied. This was established by analyzing small samples of tissue collected from populations of these seals on the California coast. Using a laboratory procedure called electrophoresis, which is explained later in this chapter, the investigators studied 24 proteins from each animal, looking for evidence of genetic variation among the seals at the loci encoding the proteins. They found no evidence of variation among the individual animals in those 24 proteins. The southern elephant seal, which had not been severely reduced in numbers by hunting, is much more variable.

THE FOUNDER EFFECT. When a species expands into new regions, populations may be started by a small number of pioneering individuals. The pioneers are not likely to have all the alleles found in their source population. Even if they do, the allele frequencies are likely to differ from those in the source population. The situation is equivalent to that for a large population reduced by a bottleneck, but rather than a small surviving population, there is a small founding population. This type of genetic drift is called the **founder effect**. Because many plant species often reproduce sexually by self-fertilization, a new population may be started by a single seed, an extreme example of a founder effect.

Migration

Few populations are completely isolated from other populations of the same species, so it is likely that some migration between populations will take place. When we speak of **migration**, we mean *movement followed by breeding in the new location*. This is sometimes called gene flow. It ranges from extremely low to very high, depending on the number of migrating individuals and their genotypes. Immigrants may add new alleles to the pool of a population or may change the frequencies of alleles already present. Emigrants may completely remove alleles or may reduce the frequencies of alleles when they leave a population. If genotype frequencies do not differ greatly among the partly isolated subpopulations of a large population, the migration of a very few individuals suffices to keep allele frequencies among subpopulations from diverging due to genetic drift. For a population to be in Hardy–Weinberg equilibrium, there must be no migration between it and other subpopulations.

The edible blue mussel, *Mytilus edulis*, has genetic variation for an enzyme that helps to maintain cells in osmotic equilibrium with the surrounding water by metabolizing proteins into free amino acids. The allele governing one form of the enzyme, which has a high activity rate, is at very high frequencies in populations growing in high salinity seawater. However, mussels with this form of the enzyme do not survive well in low-salinity estuaries, where the osmotic environment is different. In estuaries, the frequency of the allele for this form of the enzyme is reduced within each generation by the higher mortality of its possessors. Nonetheless, each spring its frequency rises again when large numbers of larvae, produced by adults living in high-salinity environments, invade and settle in the estuaries (Figure 18.9). Because estuarine populations are a small fraction of the total, they are continually swamped by the migrating offspring from individuals growing in marine environments.

The way that evolutionary agents change the frequency of this allele in estuary populations of mussels illustrates that the operation of one evolutionary agent, such as natural selection, may oppose the effects of another evolutionary agent, such as migration. However, unless the opposing effects are exactly balanced, a population will not be in Hardy–Weinberg equilibrium.

Natural Selection

Not all individuals survive and reproduce equally well in a given environment. Therefore, some individuals contribute more offspring to the next generation, on a per capita basis, than do other individuals. Individuals vary for heritable traits that determine the success of their reproductive efforts. The differential contribution of offspring resulting from different heritable traits was called **natural selection** by Charles Darwin because of its similarity to the process of artificial selection as practiced by animal and plant breeders. When reproductive success differs among genotypes, natural selection is operating as an evolutionary agent. One condition for Hardy–Weinberg equilibrium is that there must be no natural selection. Biologists investigate the action of natural selection by comparing allele (or phenotype) frequencies between generations and attempting to determine the reasons why some genotypes (or phenotypes) are better represented than others in subsequent generations.

You have probably noticed that up to this point we have been considering how evolution changes single loci in a population. When we consider more than a single locus, the useful tool of the Hardy–Weinberg rule does not apply. Natural selection usually operates simultaneously on many traits or on traits jointly governed by more than one locus. Sexual recombination continually enhances selective opportunities at this level. In asexually reproducing organisms, the daughter cells resulting from the mitotic division of a single cell normally contain identical genotypes at all loci. In a population of organisms producing new individuals in this asexual way—without any combination of genetic material from different individuals—every new individual is genetically identical with its parent, unless there has been a mutation. The vast majority of organisms, however, engage in sexual exchange of genetic material when reproducing. The offspring differ from their parents because chromosomes assort randomly during meiosis and because the act of fertilization brings together material from two different cells.

Sexual recombination generates an endless variety of genotypic combinations that increases the *evolutionary potential* of populations. Because it increases the variation among the offspring produced by an individual, it improves the chance that some of them will be successful in the varying and often unpredictable environments they will encounter. However, sexual recombination does not influence the frequency of different alleles or the rate of evolutionary change. Rather, it generates new combinations of genetic material upon which selection can act.

Depending on which traits are being favored in a population, natural selection can produce one or another of several quite different results. In the example in Figure 18.10, the variable trait is size, but many other traits would serve just as well. If many different factors—genetic and environmental—contribute to size, then the actual distribution of sizes in a population should approximate the bell-shaped curve shown in the left parts of the figure. If the smallest and the largest individuals contribute relatively fewer offspring to the next generation than those closer to the center, **stabilizing selection** is operating. Natural selection frequently acts in this way, and by doing so may counter increases in variation brought about by mutation or migration. We know from the fossil rec-

18.9 Migration Influences Genotype Frequencies
The white organisms are young blue mussels; they are adapted to the full-strength seawater in which they were spawned, but they have migrated and are now settled in an estuary with lower-salinity water. They will not survive or reproduce well there. The few large adults are survivors of an earlier larval settling.

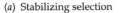

(a) Stabilizing selection

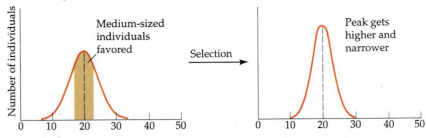

(b) Directional selection

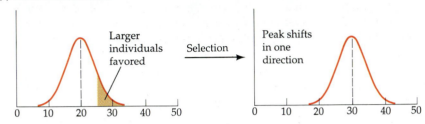

(c) Disruptive selection

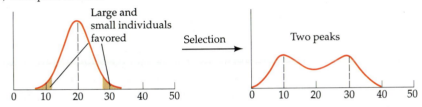

18.10 Natural Selection Alters Population Variability
Phenotypes favored by natural selection are shown in color under the bell-shaped curves that plot the distributions of phenotypes in the entire population before and after selection.

ord that most populations do not evolve rapidly most of the time. This may be largely an effect of stabilizing selection.

If individuals toward only one extreme of the size distribution—say, the larger ones—reproduce better than other individuals do, then **directional selection** is operating. If directional selection operates over many generations, then an evolutionary trend within the population results. However, what is favored often differs over time, so evolutionary trends may be reversed. **Disruptive selection**, selection simultaneously favoring the survival of individuals toward both extremes of the distribution, is apparently a much rarer phenomenon. When it operates, it tends to produce two peaks in the distribution of a trait.

Natural selection is the only agent of evolution that can adapt populations to their environments. The changes in allele frequencies produced by natural selection derive from the abilities of individuals to survive and reproduce in the environments in which they live. Therefore, the study of natural selection is central to evolutionary investigations.

During the 130 years since the publication of *The*

Origin of Species, biologists have studied many examples of the evolutionary effects of natural selection. Here we describe two, selected because they illustrate how environmental changes alter the way that natural selection influences populations.

INDUSTRIAL MELANISM. Two centuries ago, before the Industrial Revolution, a luxuriant growth of pale gray lichens covered the trunks of trees throughout Britain. Many insect species, among them moths, rested on these lichens, protected from visually hunting predators by their close resemblance to the color of the lichen-coated trunks. But lichens, being very sensitive to airborne pollutants, are unable to survive on tree trunks near major industrial centers. In the late 1840s, the proportion of very dark (melanic) individuals began to increase in moth populations near cities, where pollution killed the lichens and soot darkened the tree trunks. Because of its association with industrial centers, this phenomenon was called **industrial melanism**.

Among peppered moths (*Biston betularia*), the melanic form had been present at very low frequencies

(a)

(b)

18.11 Industrial Melanism
(a) Melanic peppered moths are conspicuous on lichen-covered bark. (b) Light individuals are conspicuous on soot-blackened bark. (c) The proportion of melanic individuals in moth populations around industrial areas early in this century is shown by the proportion of each circle that is black.

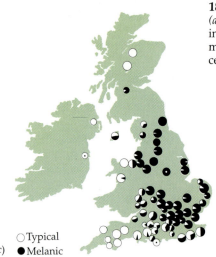

○ Typical
(c) ● Melanic

for centuries, but during the last century it increased until 98 percent of the moth population around the city of Manchester was melanic. The dark color of this moth is under the control of a single dominant allele that arises spontaneously by mutation from the allele determining light color. The selective agent that caused this nearly complete allele substitution within 50 years was not the pollution itself, but birds. Insect-eating birds obtain part of their food by searching for insects on tree trunks. Insects whose coloration blends in with the background are more difficult to see and are thus more likely to escape predation (Figure 18.11a,b). In the less polluted regions of Britain, where tree trunks were still covered with light-colored lichens, the light form of the peppered moth continued to predominate, but in urban areas, where the tree trunks were dark, melanics increased relative to light individuals (Figure 18.11c). To determine that birds were the major selective agents, investigators released marked melanic and nonmelanic moths in polluted and nonpolluted woods and estimated how well they survived. These experiments demonstrated

that the form most closely resembling the background color of tree trunks survived best (Table 18.1). By watching from blinds, observers were able to see birds capturing a higher proportion of the moths that did not match their backgrounds.

These experiments and observations demonstrate several things. First, directional selection in nature can be strong enough to lead to nearly complete allele substitution within 50 generations (peppered moths have a 1-year life cycle). Second, the direction of selection depends on local environmental conditions and so can result in a complex pattern of geographical variation. Third, changes at a single locus can produce phenotypic changes that markedly affect sur-

---

TABLE 18.1
Survival Rates of Moths that Match and Contrast with the Tree Trunks on Which They Rest

	LIGHT MOTHS	DARK MOTHS
Dorset, England: Unpolluted woodland (light background)		
Released by investigators	496	473
Recaptured later by investigators	62	30
Percent recaptured	12.5	6.3
Birmingham, England: Polluted woodland (dark background)		
Released by investigators	137	447
Recaptured later by investigators	18	123
Percent recaptured	13.1	27.5

vival. It is also worth noting that, because of pollution controls, lichens are now increasing in many areas of Britain. As a result, melanic moths are decreasing relative to nonmelanic ones in these areas.

GALAPAGOS FINCHES. Extensive field studies have been made of the finches found on the Galapagos Islands off the coast of Ecuador (see also Chapter 19). Low elevations in the Galapagos Islands are very arid. The dry season is long and severe, and some years no rain falls even during the normal wet season. During drought periods, seeds are the most abundant foods available to the finches. The only insects that remain are hidden within the tissues of plants or in the ground. Many finches, especially younger individuals, die during long dry periods.

Natural selection acts strongly on Galapagos finches nearly every year, but its direction changes seasonally as different food types become abundant or scarce. Sometimes this selection is disruptive. The large cactus finch (*Geospiza conirostris*) has been studied intensively on Isla Genovesa, where it exhibits an unusual amount of variation in bill size. Conditions there were good for finches from 1972 through 1976, but there was a prolonged drought in 1977, during which about 70 percent of the cactus finches died. Interestingly, individuals with more extreme bill sizes survived better during the drought than individuals with average bill sizes (Figure 18.12). This difference is related to the four ways cactus finches feed during the dry season. They are (1) bark stripping to expose insects, (2) cracking the seeds of cacti, (3) extracting seeds from cactus fruits and eating the fleshy rewards attached to them, and (4) tearing open cactus pads to obtain insects living within them. Birds that strip bark have significantly deeper bills than those that do not. Birds that crack hard seeds have significantly larger bills than those not observed to do so. Birds that open fruits have significantly longer bills than those seen feeding only on already opened fruits. As a result, during droughts, birds with unusually long, large, or deep bills survive better than those with less extreme bills because they have available to them a vital food source that other birds cannot use. Selection is disruptive because there are so few food types available during the long dry season and because each one of those sources is more readily exploited by birds with extreme bill sizes and shapes. In contrast, during wetter times, many insects and small seeds are available. These foods are more efficiently harvested by birds with average-sized bills.

Fitness

A central concept in the study of natural selection is **fitness**. The fitness of a genotype or phenotype is its reproductive contribution to subsequent generations *relative* to the contribution of other genotypes or phenotypes. The word *relative* is critical because the absolute number of offspring produced by an individual does not influence allele frequencies. Changes in *absolute* numbers of offspring are responsible for increases and decreases in the *size* of a population, but it is the relative success among genotypes within a population that leads to changes in allele frequencies, that is, to evolution.

BEHAVIOR AND FITNESS. By its activities, an individual may influence its fitness in two different ways. First, it may produce its own offspring, contributing to its **individual fitness**. Second, it may help the survival of relatives that bear the same alleles by descent from a common ancestor. This is called **kin selection.** Together, individual fitness and kin selection determine the **inclusive fitness** of the individual. Among species that are either solitary or live in groups no larger than a pair and its offspring, individual fitness strongly dominates inclusive fitness. Among highly social

Bills favored during normal years

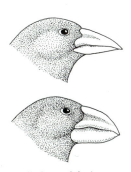

Bills favored during drought years

18.12 Natural Selection Alters Bill Size
Bill size is highly variable among large cactus finches. Individuals with unusually long, large, or deep bills survive better during droughts than do individuals with average-size bills.

BOX 18.A

Why Do Daughter Wasps Stay Home and Help to Rear Younger Siblings?

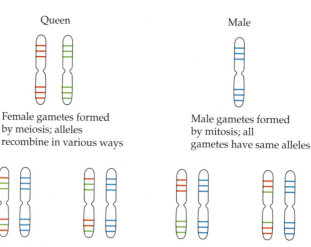

Queen

Female gametes formed by meiosis; alleles recombine in various ways

Male

Male gametes formed by mitosis; all gametes have same alleles

Female offspring:
    Inherit   50% of alleles from male
    Share   50% of these alleles with sisters
    Inherit   50% of alleles from female (queen)
    Share   25% of these alleles with sisters

Total shared alleles: 75%

A female wasp can raise more offspring if her daughters stay home and help her. But why should the daughters stay? Their behavior may be explained by the distinctive kinship situation found among social hymenopterans—ants, bees, and wasps.

Male hymenopterans are haploid and produce gametes by mitosis rather than by meiosis (see Figure 10.13). If the queen wasp mates only once, all the alleles contributed by the male to her female offspring are identical; therefore, all the queen's daughters inherit an identical set of alleles—50 percent of their genotype —from the male. The queen wasp is diploid; her gametes are produced by meiosis. Thus, each daughter shares with her sisters approximately one-half (25 percent) of the 50 percent of the alleles she inherits from the queen. So the percentage of alleles

Because of haplodiploid mating, offspring of a queen wasp share approximately 75 percent of their alleles. In most other animals, including humans, both males and females produce their gametes by meiosis; given this fact, can you calculate the approximate percentage of alleles you share with a sibling?

shared by a female wasp and any subsequent offspring the queen produces from the same mating is 75 percent (see figure). However, if the daughters leave, mate, and reproduce on their own, they produce offspring that share only 50 percent of their alleles. Thus, by staying home, female wasps help raise individuals to whom they are more closely related than they would be to the indi-

viduals they would produce if they bred on their own.

Of course, offspring do not consciously make these decisions. Rather, they have been molded by natural selection to behave "as if" they did. The haplodiploid sexual system of hymenopterans "predisposes" offspring to remain with their mothers as helpers. Other factors that influence whether offspring remain at home and assist their mothers are the number of times queens mate, whether the daughters also occasionally lay eggs, and the importance of colony size for reproductive success and survival.

species, such as social insects and many primates, kin selection also may be very important. An example of how kin selection can help explain otherwise puzzling behavior is provided in Box 18.A.

GENETIC VARIATION AND EVOLUTION

By the 1930s, evolutionists brought the contributions of natural selection and the other evolutionary agents together with the Hardy–Weinberg rule to create what became known as the modern synthetic view

of evolution. This synthesis helped to explain much about evolution and the results of many evolutionary studies. However, evolutionsts were unable to obtain the precise data on genetic variation needed for short-term studies of natural populations until the 1960s.

The recent development of molecular techniques now enables us to measure genetic variation with the needed precision. The DNA of eukaryotes is transcribed into RNA, which is, in turn, translated into proteins. DNA fragments and some proteins can be distinguished on the basis of their rate of migration through a gel in an electric field, in a process called

TABLE 18.2
Genetic Variation among Animals and Plants

	AVERAGE NUMBER OF SPECIES STUDIED	AVERAGE NUMBER OF LOCI STUDIED	PROPORTION OF POLYMORPHIC LOCI PER POPULATION	PROPORTION OF HETEROZYGOUS LOCI PER POPULATION
Insects				
Drosophila	28	24	0.529	0.150
Wasps	6	15	0.243	0.062
Others	4	18	0.531	0.151
Marine invertebrates	9	26	0.587	0.147
Marine snails	5	17	0.175	0.083
Land snails	5	18	0.437	0.150
Fish	14	21	0.306	0.078
Amphibians	11	22	0.336	0.082
Reptiles	9	21	0.231	0.047
Birds	4	19	0.145	0.042
Rodents	26	26	0.202	0.054
Large mammals	4	40	0.233	0.037
Plants	8	8	0.464	0.170

gel electrophoresis, which was explained in detail in Chapter 14. Proteins are dissolved, placed on the edge of a sheet of jellylike material, and exposed to an electric field. Their rate of migration through the gel is determined by both the charge on the molecules and by their size: smaller molecules move faster than larger ones. DNA fragments containing different numbers of base pairs can be separated clearly from one another with this technique. Results from the hundreds of species whose protein variations have been analyzed electrophoretically clearly show how universal genetic variation is (Table 18.2). Analyses of restriction sites and DNA sequences also show that variation in the fine structure of DNA itself is extensive in most populations. Indeed, few populations have so little variation that there is nothing for evolutionary agents to act upon.

As you learned in Chapter 10, if an individual has two different alleles at a given locus, it is said to be heterozygous for that locus. If a population has two or more genotypes for a given locus, all of them at levels above 1 percent, it is said to be **polymorphic** for that locus. Transitional polymorphism exists when one allele is replacing another, as occurred during the evolution of industrial melanism among peppered moths. If the alleles arrive at a stable intermediate frequency, the population is said to be in balanced polymorphism.

We have just defined polymorphism in genetic terms, but the genotype of an organism determines its phenotype. If we know the relationship between the two, genotype frequencies can be determined by measuring phenotypes, which is usually much easier. Also, because natural selection acts on phenotypes and not directly on genotypes, measuring the extent of phenotypic polymorphism reveals how natural selection is acting. Phenotypic traits include the visible features of an organism (its color, the number of hairs on its legs) and also its enzymes and the reactions and products they catalyze.

GEOGRAPHIC VARIATION

In addition to the variation found within local populations, populations usually vary geographically in both their phenotypes and the underlying genotypes. Some of this variability is immediately obvious, especially when it is in our own species (Figure 18.13), but some of it is not. For example, the variability among plants in the chemicals they synthesize to defend themselves against herbivores is not apparent from external morphology. Some individuals of *Trifolium repens*, a European clover, produce cyanide. Cyanogenic individuals are less palatable to herbivores—particularly mice and slugs—than acyanogenic individuals. However, plants with cyanide are more susceptible to being killed by frost because freezing damages cell membranes, releasing the toxic cyanide into the plant's own tissues. In populations of this clover, the frequency of cyanide-producing individuals increases gradually from north to south and from east to west across Europe (Figure 18.14). Cyanogenic plants typify clover populations only in those areas where winters are mild. In order to distinguish among cyanogenic and acyanogenic individuals we must perform elaborate chemical tests in the laboratory.

(a)

(b)

18.13 Humans Are Highly Variable
Members of the Yali tribe of Irian Jaya (a) differ among
themselves; however, they are all distinct from the
Ja'aliyin people (b) who live along the Nile River in
northern Sudan.

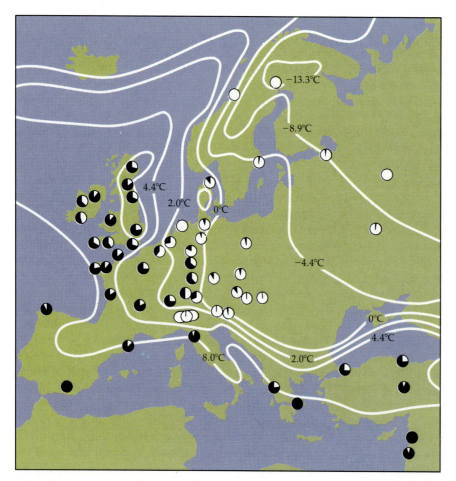

**18.14 Geographic Variation in
Cyanogenic Clovers**
The frequencies of cyanide-producing
individuals in populations of white
clover (*Trifolium repens*) are represented by the proportion of each circle that is blackened. The isotherms
(lines connecting points with equal
temperatures) plot January mean
temperatures.

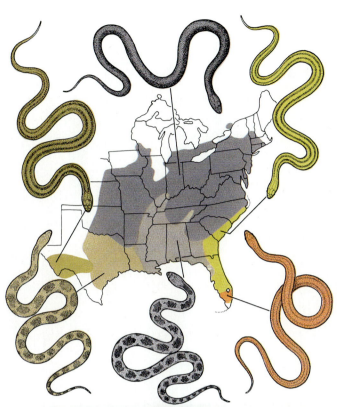

18.15 Step Clines in the Rat Snake
Notice how abruptly the snakes' color patterns change. There are no obvious environmental changes at these locations.

Gradual geographic changes in frequencies of phenotypes and genotypes, as illustrated by cyanide production among clovers, are called **clines**. Clines are widespread among most groups of organisms. In some regions, however, frequencies of certain traits change abruptly, creating **step clines**. A good example is provided by color patterns in the rat snake, *Elaphe obsoleta* (Figure 18.15). In this species the color differences are complex and striking. Nonetheless, a single color pattern dominates an extensive region, and the prevailing color changes rather abruptly where different forms come into contact. There are no obvious environmental changes in the regions where the color patterns change dramatically.

HOW MUCH OF GENETIC VARIATION IS ADAPTIVE?

If we know the amount of genetic variation in a population very precisely, we can determine how natural selection acts on externally visible phenotypic traits and can also determine how the fine structure of molecules has been molded by evolutionary agents. Consider the case of a well-studied protein, cytochrome *c*. Cytochrome *c* is one component of the electron transport chain of mitochondria (see Chapter

7). Together with other proteins of the citric acid cycle and respiratory chain, it is part of the common heritage of all eukaryotic organisms. Cytochrome *c* is relatively easy to study because it is not bound tightly to the mitochondrial membrane, because it is a small polymer with about 104 amino acids and a molecular weight of 12,400, and because it is highly stable.

Cytochrome *c* amino acid sequences are known for nearly 100 species of organisms ranging from yeasts, bread molds, sunflowers, and wheat to fruit flies, horses, and people. In addition, the three-dimensional structure of cytochrome *c* has been determined for several species, enabling us to match primary sequences to molecular structure. The amino acid sequences for cytochrome *c* from 33 species are compared in Figure 18.16. The great sequence similarity among all species indicates that all cytochromes *c* are variations on a common theme. Thirty-five positions along the chain are occupied by the same amino acid in every species, suggesting that there has been no change at that position throughout the entire evolutionary history of eukaryotes. Because there have undoubtedly been many mutations in the DNA from which these amino acids are read, we may assume that all mutations at those positions were deleterious and were weeded out by natural selection. The side chains of the invariant amino acids at positions 14, 17, 18, and 80 (marked by arrows on the figure) all interact with the heme. Evidently any changes in these amino acids would adversely affect the functioning of the vital heme group. Most of the remaining positions show little variation, but there are eight positions each with six or more different amino acids, indicating that the functional requirements of those positions are rather nonspecific.

The figure shows that the amino acid, but not the charge, may vary at some positions. For instance, region 80–85 is always hydrophobic, or at least free of charged groups. Region 86–93 is mostly charged, with a predominant separation of positive and negative charges. Region 94–98 is never charged, and 99–104 is rarely hydrophobic. These charge consistencies suggest that all of these cytochrome *c* molecules are folded in the same way and that conserving charge is necessary to preserve the three-dimensional structure of the molecules. This interpretation is confirmed by the three-dimensional structure of rice and tuna cytochromes *c*, which are folded in exactly the same way (Figure 18.17). Most of the glycine molecules in cytochrome *c* are in places where the molecule makes sharp bends. Glycine (blue in Figure 18.16) is the only amino acid lacking a side chain, making it the best one for these positions.

This detailed look at cytochrome *c* brings several important points into focus. First, the reasons for the lack of variation in certain positions would have been difficult to determine without detailed knowledge of the molecule's functioning and how its three-dimen-

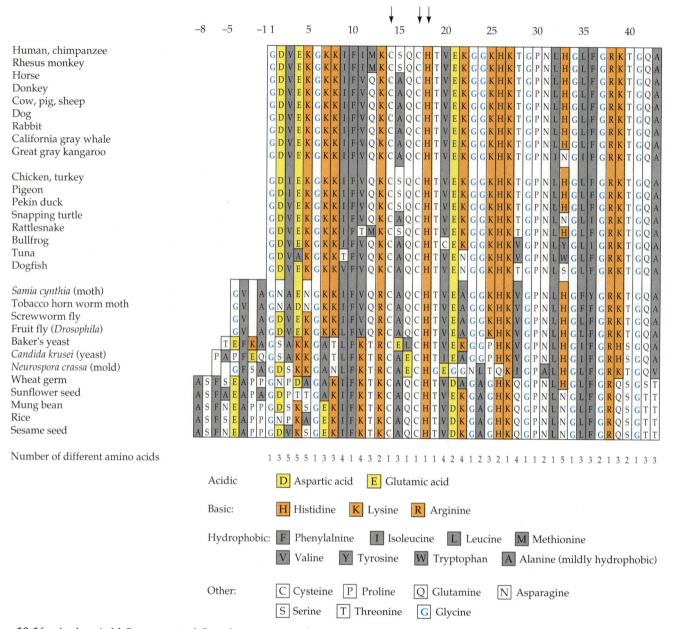

18.16 Amino Acid Sequences of Cytochromes _c_
The number of amino acids that appear in each position for cytochrome _c_ in 33 different species is listed across the bottom of the figure; 35 positions do not vary among the species. The pattern of vertical bands (gray, white, and color) indicates how the chemical nature of side chains on the amino acids was preserved during evolution.

sional structure is related to that functioning. For a poorly known protein, the same data on sequences would be difficult to interpret. Molecular biologists now detect functionally important regions of molecules by searching for and then studying positions that lack variation because this usually indicates that the action of natural selection has been strong and stabilizing.

Determining the significance of the amino acids that have changed is more complicated. The changes could represent important adaptations in the cyto-

chrome _c_ molecules to the cellular physiology of the organisms in which they are found. Alternatively, they could indicate that those amino acid substitutions did not affect the functioning of the molecules and, hence, they slowly accumulated over time uninfluenced by natural selection. Substitutions that are functionally neutral should accumulate slowly but at a constant rate. This has been the case for the cytochromes _c_ of eukaryotes, which have diverged from one another at a rate of about three amino acid substitutions per 100 million years (Figure 18.18).

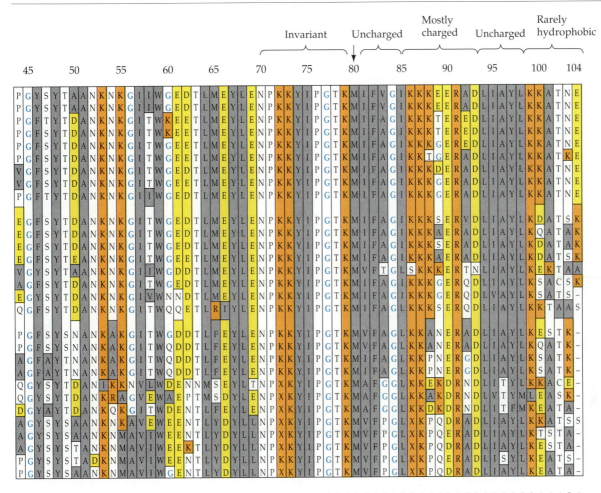

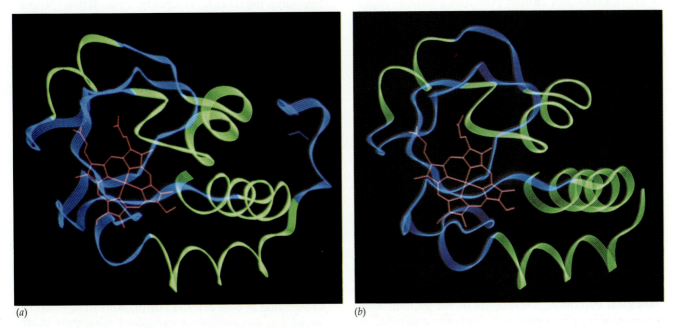

(a)　(b)

18.17 Cytochrome c Molecules Fold in the Same Way
Computer-generated models of cytochrome c from rice (left) and tuna (right). Both molecules fold to produce the same three-dimensional structure, even though one is from a plant and the other is from an animal. Analogous portions of the molecules are shown in similar colors. The heme group is shown in pink.

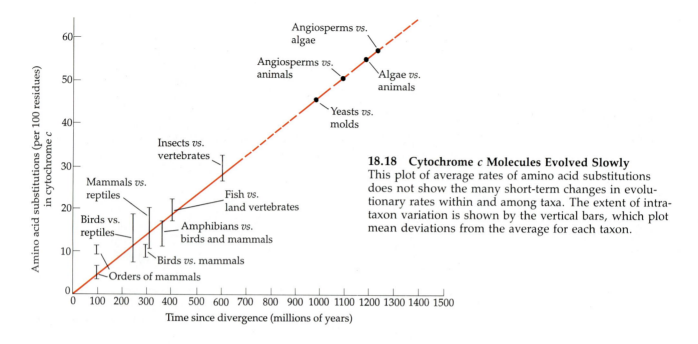

18.18 Cytochrome *c* Molecules Evolved Slowly
This plot of average rates of amino acid substitutions does not show the many short-term changes in evolutionary rates within and among taxa. The extent of intrataxon variation is shown by the vertical bars, which plot mean deviations from the average for each taxon.

MICROEVOLUTION AND MACROEVOLUTION

The short-term changes in a population's gene pool that we have been discussing in this chapter are often called **microevolution**. Microevolutionary studies are an important part of evolutionary biology because short-term changes can be observed directly and subjected to experimental manipulations. Studies of short-term changes reveal much about evolution, but by themselves they cannot provide a complete explanation of the long-term changes that are often called **macroevolution**. Macroevolutionary changes can be strongly influenced by events that occur so infrequently that they are unlikely to be observed during microevolutionary studies. Also, because the way evolutionary agents act changes over time, we cannot interpret the past simply by extending today's results backward in time. Additional types of evidence must be gathered if we wish to understand the course of evolution over more than a billion years. Parts of this evidence will be presented in the chapters that follow. Finally, in Chapter 27, we will synthesize this information to provide a picture of patterns of macroevolution and their causes.

SUMMARY

Biological evolution is a change in the genetic composition of a population over time. For a population to evolve, it must have genetic variation. Genetic variation among individuals within populations is the raw material upon which agents of evolution act. Some variation in populations is genetically determined, but some is caused by environmental conditions. Only the former can be passed on to offspring. Many techniques are used to measure genetic variation. They reveal that nearly all living species are highly variable, even those that have changed very little in appearance for millions of years.

We can often tell that some agent of evolution is acting on a population by examining frequencies of genotypes within it. We begin by analyzing genotypes governed by a single locus with two alleles and applying the Hardy–Weinberg rule. If the proportions of different genotypes are close to Hardy–Weinberg equilibrium values, we have no evidence that evolutionary agents are affecting the population. However, if genotype proportions deviate from Hardy–Weinberg expectations, we know that some agent—or agents—is operating. The nature of the deviation may provide clues about which agent it is. Changes in variation within populations are caused by mutation, genetic drift, migration, nonrandom mating, and natural selection. Of these agents, only natural selection adapts populations to local environmental conditions. However, some other evolutionary agents may play important roles in creating new combinations of genes upon which natural selection can act.

Well-studied cases of the action of agents of evolution, such as the northern elephant seal's loss of genetic variation, the cyclic changes in frequency of an enzyme in blue mussel populations in low-salinity estuaries, the spread of industrial melanism in European moths, and changes in the sizes and shapes of bills in Galapagos finches, reveal that evolution may be driven both by changes in the external environment and by interactions among individuals within populations.

SELF-QUIZ

1. The phenotype of an organism is:
   a. the type specimen of its species in a museum.
   b. its genetic constitution which governs its traits.
   c. the temporal expression of its genes.
   d. the physical expression of its genotype.
   e. the morphology it achieves as an adult.

2. The appropriate unit for defining and measuring genetic variation is:
   a. the cell.
   b. the individual.
   c. the population.
   d. the community.
   e. the ecosystem.

3. Which of the following is *not* true of allele frequencies?
   a. The sum of any set of allele frequencies is always 1.
   b. If there are two alleles at a locus and we know the frequency of one of them, we can obtain the frequency of the other by subtraction.
   c. If an allele is missing from a population, its frequency is 0.
   d. If two populations have the same gene pool for a locus, they will have the same proportion of homozygotes at that locus.
   e. If there is only one allele at a locus, its frequency is 1.

4. In a population at Hardy–Weinberg equilibrium in which the frequency, p, of A alleles is 0.3, the expected frequency of Aa individuals is:
   a. 0.21
   b. 0.42
   c. 0.63
   d. 0.18
   e. 0.36

5. Which of the following agents does *not* change allele frequencies?
   a. Nonrandom mating
   b. Mutation
   c. Genetic drift
   d. Migration
   e. Natural selection

6. Natural selection that preserves existing allele frequencies is called:
   a. unidirectional selection.
   b. bidirectional selection.
   c. prevalent selection.
   d. stabilizing selection.
   e. preserving selection.

7. The evolutionary agent that caused an increase in the frequency of melanism among some moths around cities in industrialized countries was:
   a. blackened tree trunks.
   b. soot in the leaves eaten by the caterpillars.
   c. deposition of soot on the wings of the moths.
   d. selective predation by birds on resting moths.
   e. selective predation by birds on caterpillars.

8. A population is said to be polymorphic for a locus if:
   a. it has at least three different alleles at that locus.
   b. it has at least two different alleles at that locus.
   c. it has at least two genotypes for that locus.
   d. it has at least three genotypes for that locus.
   e. it has at least two genotypes for that locus, all of them present at a frequency greater than 1 percent.

9. A cline is defined as:
   a. the distribution of an organism across a slope.
   b. an abrupt change in frequencies of certain traits over time.
   c. a gradual change in frequencies of certain traits over time.
   d. an abrupt change in frequencies of certain traits over space.
   e. a gradual change in frequencies of certain traits over space.

10. Certain positions in the cytochrome c molecule are occupied by the same amino acids in all species because:
   a. there have been no mutations at those positions.
   b. there have been no mutations in hydrophobic regions.
   c. there have been no mutations in charged regions.
   d. all mutations at those positions were deleterious and were weeded out by natural selection.
   e. there has been no selection affecting those positions.

FOR STUDY

1. During the past 45 years, over 200 species of insects that attack crop plants have become highly resistant to DDT and other pesticides. Using your recently acquired knowledge of evolutionary processes, explain the rapid and widespread evolution of resistance. Propose ways of using pesticides that would slow down the rate of evolution of resistance. Now that DDT has been banned in the United States, what do you expect to happen to levels of resistance to DDT among insect populations? Justify your answer.

2. In nature, there is never truly random mating among individuals in a population, and natural selection is seldom totally absent. Why, then, does it make sense to use the Hardy–Weinberg model, which is based on assumptions known to be generally false? Can you think of other models in science that also are based on false assumptions? How are such models used?

3. An investigator is studying populations of house mice living in barns and sheds on a large farm. Each building has a population of between 25 and 50 mice. Populations in different buildings have strikingly different frequencies of alleles determining coat color and tail length. By marking most individuals, the investigator determines that mice only rarely move between buildings. He interprets his study as providing evidence for random genetic drift. Could other agents of evolution plausibly account for the pattern? If so, how could they be distinguished?

4. As far as we know, natural selection cannot adapt organisms to future events. Yet many organisms exhibit responses in advance of natural events. For example, many mammals go into hibernation while it is still quite warm. Similarly, many birds leave the temperate zone for their tropical wintering grounds long before winter has arrived. How can these "anticipatory" behaviors evolve?

5. Many people believe that species, like individual organisms, have life cycles. They believe that species are born by some process of speciation, undergrow growth and expansion, and inevitably die out as a result of "species old age." Is there any agent of evolution that could cause such a species life cycle? If not, how do you explain the high rates of extinction of species in nature?

READINGS

Crow, J. F. 1986. *Basic Concepts in Population, Quantitative, and Evolutionary Genetics*. Freeman, New York. An excellent introduction to all aspects of modern population genetics.

Dickerson, R. E. 1972. "Cytochrome *c*: The Structure and History of an Ancient Protein." *Scientific American*, April. Discussion of the evidence for the 1.3-billion-year history of cytochrome *c*, the best-known protein.

Endler, J. A. 1986. *Natural Selection in the Wild*. Princeton University Press, Princeton, NJ. A thorough review of the problems and successes in measuring natural selection in nature.

Futuyma, D. J. 1986. *Evolutionary Biology*, 2nd Edition. Sinauer Associates, Sunderland, MA. A comprehensive review of all aspects of evolutionary biology.

Grant, P. R. 1986. *Ecology and Evolution of Darwin's Finches*. Princeton University Press, Princeton, NJ. Summarizes all the studies of the evolution of these fascinating birds that have played such an important role in our understanding of evolution.

Wilson, E. O. and W. H. Bossert. 1971. *A Primer of Population Biology*. Sinauer Associates, Sunderland, MA. A brief, self-teaching textbook designed to form a bridge between elementary textbooks and more advanced treatments of population genetics, evolutionary theory, and ecology.

19

The Multiplication of Species

PREVIEW: Speciation—the process by which one evolving population or group of populations splits into two such units—has produced the millions of species of organisms that are alive today and that have lived in the past. Species may form following the division of one population into two populations by a geographic barrier or when any kind of strong barrier prevents gene flow in the absence of geographic isolation. Speciation is reinforced by the development of reproductive isolating mechanisms that reduce the frequency of hybridization between populations. Little genetic or phenotypic change need accompany speciation. Speciation may result in an evolutionary radiation within a lineage into many species, each exploiting different environmental resources.

This chapter deals with species concepts; geographic, parapatric, and sympatric speciation; the maintenance of separate gene pools; genetic changes accompanying speciation; time required for speciation, evolutionary radiations, and the significance of speciation.

The organisms that live in the tops of tropical trees are virtually unknown. Most of the individuals now being collected from these trees (Figure 19.1) belong to undescribed species. Because so many species are still being found, some biologists estimate that there may be as many as 30–50 million species of organisms living on Earth today, even though only 1.5 million have so far been described and named. All these species, plus all those that lived in the past, are believed to have descended from a single ancestor. Therefore, many millions of new species must have developed over evolutionary time by the division of ancestral species into two or more daughter species. But what are species, and when and how were all these species formed?

WHAT ARE SPECIES?

Species means "kind." Kinds of organisms were originally recognized by their appearances—by the types of differences pictured in the field guides that we use to help us distinguish red oaks from white oaks and robins from wood thrushes. Among organisms in which males and females look very different, the two sexes were sometimes identified as different species. But as soon as such individuals were discovered to be males and females of the same population, they were placed together in the same species. The fact that this was done shows that biologists have long recognized that interbreeding is important for how organisms evolved and how they should be named.

A rigorous definition of species that incorporated the concept of shared reproduction was given by Ernst Mayr in 1940. He stated that "Species are groups of actually or potentially interbreeding natural populations which are reproductively isolated from other such groups." All parts of this definition identify important aspects of the **biological species concept**, as it is now called. The "groups" in this definition are the local populations or demes we discussed in Chapter 18. The words "actually or potentially" acknowledge that even if some members of a species are isolated from other members and hence are unable to mate with them, they should not be placed in separate species if they would be likely to interbreed if they should cross the barrier. The word "natural" is a vital part of the definition because, although individuals of two species may be able to interbreed in captivity, this does not affect evolutionary processes if they do not interbreed in nature (Figure 19.2).

The biological species concept recognizes the fact that species are key evolutionary units, united by sharing alleles inherited from common ancestors, and evolving independently from other such units. Species are also taxonomic units; that is, they are units in classification systems in which biologists name

421

19.1 Tropical Organisms Are Poorly Known
Most animals (primarily insects) that live in tropical tree canopies are unknown to us. (*a*) A biologist fogs the canopy to collect specimens of the animals living there. (*b*) Tiny leaf beetles from tropical trees. These five species of the family Chrysomelidae were unknown before recent research on forest canopy insects.

(*a*)

(*b*)

organisms and group them into categories designed to reflect certain kinds of relationships (Chapter 20). Usually, but not always, taxonomic species are equivalent to biological species; but for many groups of organisms, taxonomic species designations are still made on the basis of appearances rather than reproductive behavior. In most of these cases there is no direct information about their reproductive behavior.

HOW NEW SPECIES ARISE

The process by which one evolutionary unit splits into two such units that thereafter evolve as distinct lineages is known as **speciation**, or **cladogenesis**. Interestingly, although Charles Darwin entitled his book *The Origin of Species*, he did not really write about the origin of species. Darwin did not discuss how a single species splits into two or more daughter species. Rather, he was concerned with demonstrating that species are altered by natural selection over time.

Speciation is a temporal process, and species are its products. Because speciation is a process, populations exist at various stages of this process. For scientists interested in speciation processes, the various intermediate stages provide excellent objects for study. However, for those interested in classifying or otherwise using species as frames of reference, populations at intermediate stages in the process of speciation are a nuisance. It is difficult to decide how to name them and where to recognize boundaries.

The speciation events that gave rise to the millions of current species took place in the past. We will never know exactly what happened in most of those cases. However, biologists have learned much about speciation processes by studying existing populations and by applying knowledge of current processes to past events. These kinds of studies have shown that speciation occurs in several different ways.

Geographic Speciation

Often one species evolves into two daughter species after a physical barrier to movement develops within its range. Speciation that begins in this manner is known as **geographic speciation**, or **allopatric speciation** (*allo* = "different"; *patria* = "country"). Once such a barrier is present, the two separated populations may diverge genetically from each other. You will recall from Chapter 18 that natural selection operates to adapt populations to their environments. If the environments on the two sides of the physical barrier differ, different selective pressures may cause the populations to diverge genetically. Genetic drift may also bring about genetic changes. The two populations may also have started with different mixtures of alleles. Differences that accumulate while the

19.2 Captivity Lowers the Barriers to Interbreeding
Although horses and zebras do not interbreed in the wild, they have done so on this ranch in Kenya. The resulting hybrids, known as "zebroids," display traits of both parental species.

barrier is in place may become so large that the populations remain distinct if the barrier breaks down some time later.

Because geographic speciation is a slow process, it cannot be observed directly. The best evidence for it comes from groups of populations that differ in how much they are geographically and reproductively isolated from each other. For example, many different populations of platyfish live in the rivers of eastern Mexico. These populations differ in the extent to which they are physically and reproductively isolated from one another. The northernmost species, *Xiphophorus couchianus*, is restricted to a single river system where it is the only species of platyfish. It is sterile when crossed with the other species in the laboratory. *Xiphophorus variatus* has three distinct populations, each in a different river, but they produce fertile offspring when crossed in the laboratory. The southernmost species, *Xiphophorus maculatus*, is widely distributed in many stream systems from Veracruz, Mexico, to Belize. It is highly variable in the appearance of its spots (Figure 19.3), but individuals from all *Xiphophorus maculatus* populations readily interbreed with those from other populations. At some time in the past, an ancestral platyfish colonized all these streams. The subpopulations, isolated in different streams, have been diverging since then. Some of them, such as *Xiphophorus couchianus*, have diverged so much that they could not interbreed and

produce viable offspring with individuals of the other species even if the other species were to colonize the stream in which *Xiphophorus couchianus* lives. Others, such as *Xiphophorus maculatus*, would certainly interbreed with immigrants from other stream systems.

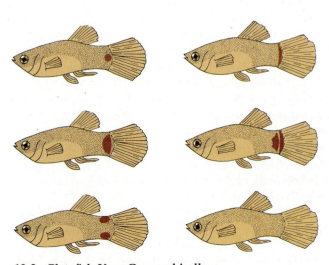

19.3 Platyfish Vary Geographically
Six of the many patterns of tail spotting found in *Xiphophorus maculatus* in Mexico are shown here. Fish with any of these patterns can interbreed with any of the others and produce fertile hybrids.

Thus, populations of platyfishes exist at various stages in the process of speciation, some being well isolated reproductively, others less so.

The importance of geographic isolation is also demonstrated by the unusual finches of the Galapagos Islands that we considered in Chapter 18. Darwin's finches (as they are usually called, because Darwin was the first scientist to study them) arose on the Galapagos Islands by speciation from a single South American species that colonized the islands, which are 1,000 kilometers off the coast of Ecuador. Today there are 14 species of Galapagos finches, all of which differ strikingly from the probable mainland ancestor (Figure 19.4). There would probably be only one species of finch in the Galapagos today if there had been only one Galapagos island. Biologists hold that view because on Cocos Island, well to the north of the Galapagos, there is only one species of finch. Finches must have reached Cocos Island from the Galapagos a long time ago, because the Cocos finch is very different in appearance from the other Galapagos finches. Consistent with the idea that geographic speciation requires two populations of a species to be separated by a physical barrier, this species has not divided into two daughter species. In the main Galapagos archipelago, however, some of the islands are sufficiently distant from one another that there is little migration between those islands. Therefore, populations of finches on different islands have differentiated enough that when occasional immigrants do arrive from other islands, they either do not breed with the residents or, if they do, their offspring do not survive as well as those produced by pairs composed of island residents. Therefore, the genetic distinctiveness of the different populations is maintained.

How wide a barrier must be to prevent gene flow depends on the size and mobility of the species. What is a firm barrier for a terrestrial snail may be no barrier at all to a butterfly or a bird. Populations of wind-pollinated plants are totally isolated at the maximum distance pollen may be blown by the wind, but individual plants are effectively isolated at distances much shorter than that. For animal-pollinated plants, the width of the barrier is the distance pollinators can travel while carrying pollen. Even animals with great powers of dispersal are often reluctant to cross narrow strips of unsuitable habitat. For animals that cannot swim or fly, narrow water-filled gaps may be effective barriers.

Studies of island populations of organisms indicate that the smallest area within which a single species has split into two daughter species varies greatly among types of organisms (Figure 19.5). Birds apparently have not speciated on islands smaller than Madagascar (600,000 km^2), whereas small mammals have speciated on Luzon and Cuba (approximately 120,000 km^2). Fishes, reptiles, and amphibians—all much less mobile than birds—have speciated in even smaller areas. Among terrestrial snails, species have formed in single mountain valleys isolated from adjacent valleys by high ridges where snails cannot live. As shown in the figure, even the largest parks and wildlife preserves are too small for speciation of large organisms within their boundaries.

Populations of organisms with poor dispersal abilities can diverge genetically very rapidly. For example, the land snail *Helix aspera* was introduced into California from France in 1859. It has subsequently spread over much of the southwestern United States. *Helix* lives primarily in gardens, and its habitat is divided into patches by the grid of roads in residential areas of cities. *Helix* was introduced into Texas from California in the 1930s; already populations within city blocks are locally differentiated. Figure 19.6 shows variations in the frequencies of two alleles of the enzyme malate dehydrogenase-1 among populations of snails in two adjacent city blocks in Bryan, Texas. This differentiation into locally distinct subpopulations has happened within 60 years. If all gene flow across city streets were to be prevented for a long time, sufficient differences might accumulate, either through genetic drift or natural selection, so that snails on different sides of roads might fail to interbreed even when individuals moved across the roads.

THE EVOLUTION OF REPRODUCTIVE ISOLATION

What happens when populations that have been evolving in isolation reestablish contact as a result of climatic changes, changes in sea level, shifting courses of rivers, habitat alterations, and so on? One possibility is that the populations may have become so different that they do not interbreed. Individuals may not find members of the other population attractive as mating partners. The two populations may have evolved different breeding seasons, so that their members are never reproductively active at the same time. The populations may have become so different physiologically that matings between their members do not produce any viable offspring. If any of these changes has happened, we can say that the process of speciation was completed during the period of isolation. The populations have become reproductively isolated from each other and will henceforth evolve as distinct units.

At the other extreme, so few differences may have accumulated that the individuals freely interbreed with members of the other population and produce offspring that are as successful as those resulting from within-population matings. The offspring of ge-

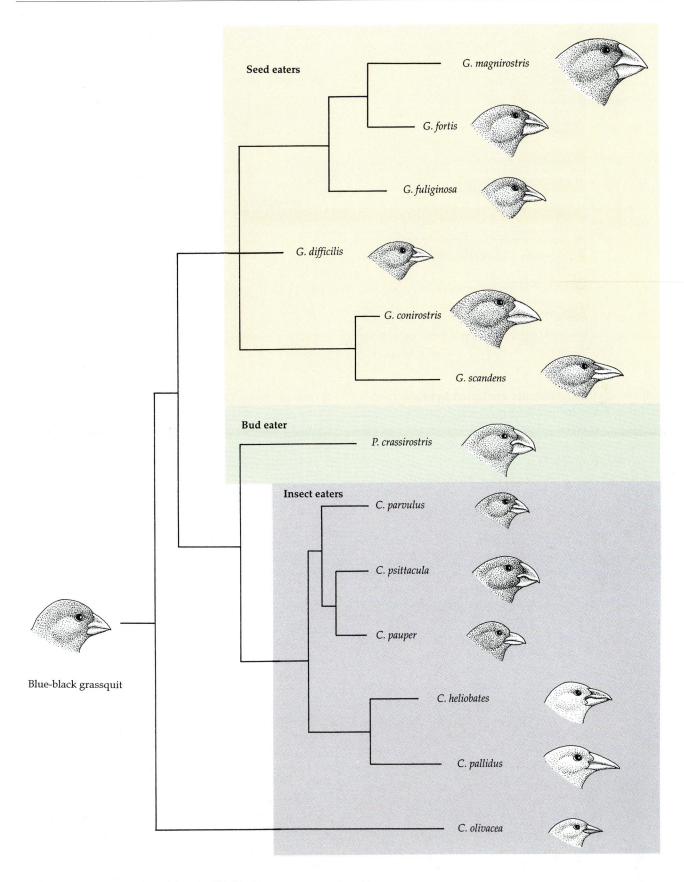

19.4 Evolution among Galapagos Finches
The descendants of the blue-black grassquits that colonized the Galapagos Islands several million years ago evolved into 14 species whose members are variously adapted to feed on seeds, buds, or insects.

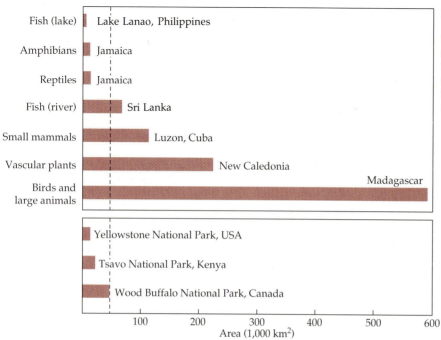

19.5 Speciation Requires Large Areas
The smallest islands on which a single species appears to have given rise to two species are shown for a variety of organisms. Compare these areas with the sizes of three of the world's largest national parks.

netically dissimilar parents are called **hybrids**. If successful hybrids spread through both populations and reproduce with the other individuals, the gene pools quickly amalgamate and no new species result from the period of isolation.

Alternatively, there may be limited interbreeding where two populations come into contact, resulting in a hybrid zone. For example, the ranges of carrion crows and hooded crows do not overlap, but meet in a line extending across western Europe. Members of the two species hybridize extensively where their ranges come into contact (Figure 19.7). This zone has been in the same position for more than a century, and the frequency of hybridization between the two

forms of crows is not diminishing. There is no obvious environmental change at the location of the hybrid zone. Biolologists believe that ancestral crow populations were divided into two groups—one in Spain and Portugal and another in the eastern Mediterranean—during the last glacial period, when northern and central Europe were too cold for crows. As the climate warmed up, the two forms moved northward and met approximately where they now hybridize. This particular hybrid zone is apparently stable. But the frequency of hybridization often decreases after contact is established between two previously separated populations because isolating mechanisms evolve.

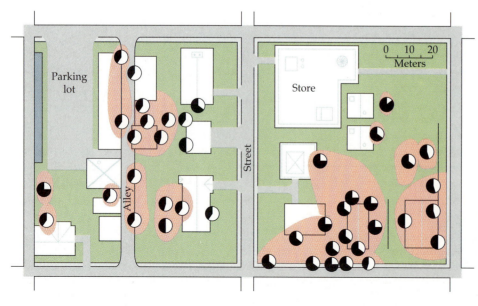

19.6 Local Variation in an Introduced Snail
Each circle represents a small colony of snails, *Helix aspera*, living in a patch of vegetation adjacent to buildings. Between-block variation is greater than within-block variation at this and other loci. Local demes are shown in pink.

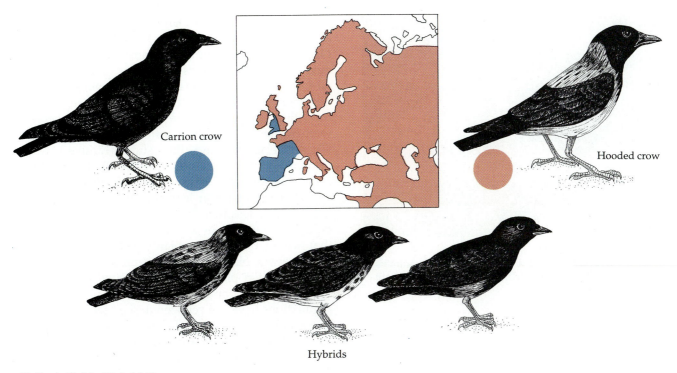

Carrion crow

Hooded crow

Hybrids

19.7 A Stable Hybrid Zone
The zone of contact between the breeding ranges of hooded and carrion crows is narrow and stable. Because the plumages of the two species are so distinctive, hybrids are easily identified.

Isolating Mechanisms

Suppose now that individuals from two different populations come into contact. What influences the possibility that these individuals will actually interbreed? Factors that reduce this possibility are called **isolating mechanisms**. They fall into two categories: **prezygotic mechanisms**, which lower the probability that hybrid zygotes will be formed, and **postzygotic mechanisms**, which reduce the viability or fertility of hybrids. We have already mentioned several types of prezygotic mechanisms, such as temporal and behavioral isolation and physiological differences. Differences in mating behavior provide many prezygotic isolating mechanisms among animals. Mating behavior is stimulated in many species by courtship pheromones, or scents (discussed in Chapter 44). Individual and geographic variation in such pheromones is known among larch budworm moths, pine beetles, and garter snakes. If such differences develop between the populations while they are isolated, individuals from the different populations may be less attracted to one another as mates.

Postzygotic mechanisms reduce the number of offspring produced by between-population matings. The zygotes formed may be abnormal; or the hybrids may mature normally but be infertile when they in turn attempt to reproduce. For example, the offspring of matings between horses and donkeys—mules—

are vigorous, but because mules are sterile, they leave no descendants (Figure 19.8). Individuals producing hybrid zygotes have wasted their time and energy engaging in courtship and matings that leave fewer surviving offspring than would have resulted had they chosen mating partners from their own population. Prezygotic isolating mechanisms prevent such waste of effort; therefore, mechanisms that cause rejection of individuals from other populations as mating partners often evolve.

To determine what happens when formerly separate populations come together, studies would ideally begin when contact is first established. Such an opportunity to observe a hybrid zone being formed is provided by blue and snow geese (Figure 19.9). These geese breed in arctic North America and spend winter in the southern United States. White-plumaged birds (snow geese) dominate breeding populations in the West; dark-plumaged birds (blue geese) dominate in the East. Historical evidence shows that the two color forms were almost completely separated geographically until the third decade of this century. The recent hybrid zone is due to a change in the winter feeding ranges of the birds. Birds of both types now winter in large flocks in the rice-growing regions of inland Texas and Louisiana. The geese select mates while on the wintering grounds, and pairs migrate to nest on the breeding ground from which the fe-

19.8 Sturdy but Sterile
Mules are hybrids that are widely used as pack animals because of their stamina. For that purpose, their infertility is unimportant.

male came. Interbreeding is common between the two forms, and a hybrid zone is developing in a small region of the Canadian Arctic. Biologists are monitoring its spread and trying to determine whether isolating mechanisms are developing.

Parapatric Speciation

Parapatric speciation (*para* = "near") is the name given to the development of reproductive isolation among members of a continuous population in the absence of a geographic barrier. This rarely happens, but occasionally a species boundary forms where there is an important environmental discontinuity, as

19.9   A New Hybrid Zone Forms
Blue and snow geese are forming mixed pairs because the species now winter together in Louisiana rice fields.

between populations of plants growing at the boundaries of different soil types. There are natural soil discontinuities, but unusually striking ones are created by mining activities that leave rubble (tailings) with high concentrations of heavy metals that are detrimental to plant growth. The soils developing on the tailings at the Goginian lead mine near Aberystwyth, Wales, are highly contaminated with lead, but where the tailings end, they suddenly give way to normal rich pastureland (Figure 19.10). The pasture grass *Agrostis tenuis* is common to all sites, but there is a sharp gradient in lead tolerance among plants separated by less than 20 meters. Plants on the mine tailings grow well in soil with lead concentrations that would be lethal to plants growing just a few meters away. Nearly complete reproductive isolation exists between plants on contaminated and normal soil because they flower at different times. These two populations have not yet been designated as separate species, but reproductive isolation between them has already evolved, demonstrating that gene flow can stop even in the absence of a distinct physical barrier.

Sympatric Speciation

In **sympatric speciation** (*sym* = "same"), a gene pool becomes subdivided even though members of the daughter species overlap in their range during the speciation process. A common means by which this comes about is **polyploidy**, a multiplication of the number of chromosomes (Chapter 9). Polyploidy can arise by the duplication of the chromosomes of a single species, which is called **autopolyploidy**, or by the combination of chromosomes from two different species, called **allopolyploidy**. Polyploid individuals of either type usually cannot interbreed successfully with members of the parent populations because the polyploid individuals have twice as many chromosomes as their parents. Therefore, chromosomes cannot pair properly during the first metaphase of meiosis (Chapter 9). As a result, the zygotes usually fail to develop properly. Thus the matings either result in no offspring, or, if offspring are produced, they die before they mature.

Polyploidy can create a new species quickly, provided the polyploid individuals can mate among themselves or self-fertilize. This is much more easily accomplished among plants than among animals because individual plants of many normally outcrossing species can also be self-fertile. Also, if the polyploidy arises in a number of the offspring of a single parent, the siblings can fertilize one another. Animals that have speciated by polyploidy either are parthenogenetic—that is, they produce young from unfertilized eggs—or they probably survived through the initial generations by means of matings among siblings.

19.10 A Grass Speciates Parapatrically
The grasses growing near this Welsh lead mine—even those of the same species—vary greatly in their tolerance of the soil's heavy-metal contaminants. This variation has led to reproductive isolation between strains of *Agrostis tenuis* growing on or adjacent to contaminated soil.

Polyploidy has been very important in the evolution of flowering plants. Approximately 47 percent of all species of flowering plants are polyploid. Most of these are allopolyploids that have arisen from hybridization between two species, followed by self-pollination. The red-tubular flowered gilias of western North America illustrate these processes. One group of species, originally considered to be members of a single species, lives in the Mojave Desert of California. The group is now known to contain five species: three diploids and two tetraploids (which have four, rather than two, sets of chromosomes). *Gilia transmontana* arose from hybridization between the diploids *Gilia minor* and *Gilia clokeyi*, and the other tetrapoid, *Gilia malior*, from hybridization between *Gilia minor* and *Gilia aliquanta* (Figure 19.11). These five species, although similar in appearance, are sterile in all interspecific mating combinations.

Among animals, sympatric speciation can result from the way species exploit resources. For example, many parasites and insect herbivores specialize on one or a few host species. Many insects find their mates on their host plants and subsequently mate and lay their eggs there. Individuals that colonize a new type of host plant will thus mate with each other rather than with individuals still feeding on the original host. Developmental rates often vary on different hosts, so individuals feeding on different plant species may come into reproductive condition at different times. In that case, even if they move from one plant species to another they are unlikely to find a suitable mate there. Reproductive isolation can arise in this manner.

Some evolutionists are convinced that insects often speciate sympatrically by such host-plant preferences, but it is difficult to obtain critical evidence.

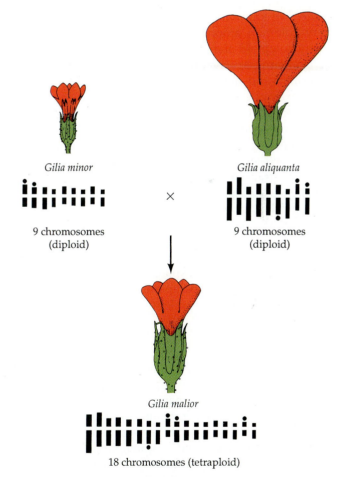

Gilia minor

9 chromosomes (diploid)

×

Gilia aliquanta

9 chromosomes (diploid)

Gilia malior

18 chromosomes (tetraploid)

19.11 Speciation by Polyploidy
Plants formerly assigned to a single species, *Gilia transmontana*, actually belong to five distinct species—three diploid and two tetraploid. Their ancestry is revealed by their chromosomes, whose relative sizes are shown here; *G. malior*, a tetraploid, arose from hybridization between the diploids *G. minor* and *G. aliquanta*.

Why it is difficult to do so is illustrated by two species of Japanese ladybird beetles. *Epilachnea niponica* feeds on thistles and *Epilachnea yasutomii* feeds on a number of other plants that grow near thistles. Adults of both species feed and mate on their own host plant. In the laboratory, individuals of the two species readily hybridize, but hybrids have never been discovered in nature. Fossils show that these two beetle species have maintained their differences for at least 20,000 years. Nevertheless, although reproductive isolation is currently maintained by host-plant preference, it is impossible to be certain that this is a case of sympatric speciation because host-plant specialization could have evolved at a time in the past when the species were geographically separated. Without historical evidence, we cannot tell which process produced today's result.

How Important Is Each Speciation Mode?

The three modes of speciation—geographic, parapatric, and sympatric—differ in their relative importance. We know that sympatric speciation has been very common among plants because there are so many polyploid plant species. To date, few examples of sympatric speciation among animals have been identified and not all biologists agree that they are good examples.

Among larger animals, the evidence strongly indicates that geographic separation has been the dominant mode of speciation. Also, because these species do not self-fertilize or form polyploid populations, they lack traits conducive to providing reproductive isolation in sympatry. Nonetheless, it is difficult to distinguish between geographic and parapatric speciation after the fact. Contact zones with hybridization between two species, a common pattern among animals, could result from either geographic speciation followed by range expansion that puts the species in contact, or from parapatric speciation. In many cases, we may never be able to determine the actual course of past events.

THE GENETICS OF SPECIATION

Sympatric species that maintain their distinctness need not have diverged from one another very much genetically. Small genetic differences can cause substantial morphological and physiological differences. These differences may influence the success of hybrids or may prevent individuals from being acceptable as mates to members of the other species. Also, evolutionists suspect that the genetic changes accompanying speciation may be small because of the existence of **sibling species**, species that are so similar morphologically that they are difficult to distinguish by appearance alone. For example, many sibling species of *Drosophila* are very difficult to identify from their external appearances, and share nearly all of their alleles. The prevalent allele at most loci in one species is nearly always found at a frequency of at least 0.01 in other closely related species. Most morphological differences *among* the species are based upon variability already present *within* each of the species. Speciation in *Drosophila* has not involved major reorganization of the genome. All of the several hundred species of this genus that have evolved in Hawaii during the past 40 million years, even though they have diverged morphologically (Figure 19.12), are relatively similar genetically. Among other *Drosophila* groups, sibling species may be much more different genetically.

Research confirms that the differences among species generally are similar in kind to the differences found within species. There is no compelling reason to think that evolution of the sorts of differences that separate most closely related species requires any mechanisms or processes other than those known to operate within species.

THE COHESION OF SPECIES

A biologist who is knowledgeable about a particular group of organisms usually has little difficulty in identifying all members of that group and distinguishing them from members of other groups. The patterns of morphological similarities among organisms that allow us to do this, although they are so familiar to us that they seem "natural," are really quite surprising. It is rather remarkable that we can, for example, recognize as members of the same species both red-winged blackbirds from New England and red-winged blackbirds from California, even though there is no gene exchange between individuals in California and New England (Figure 19.13). The standard field guides to birds, mammals, insects, and flowers are possible only because most species change very little in appearance over large distances. The morphological similarity exhibited by individuals of a species spread over a wide area is especially surprising when their ranges are broken up into well-separated areas. Most individuals remain within the areas where they were born, and environmental conditions may differ strikingly among areas.

Some organisms reproduce without any sexual recombination. Surprisingly, widespread populations of these organisms also are composed of individuals that are very similar to one another. Species within which self-fertilization is the rule, even though some sexual recombination occurs, also show cohesion. Although many plant species have this type of reproductive system, the individuals in their populations are all very similar in appearance—they have not diverged genetically.

(a)

(b)

19.12 Morphologically Different, Genetically Similar
Although these fruit flies, a small sample of the hundreds of species found only in the Hawaiian Islands, are extremely variable in appearance, they are genetically almost identical. (a) *Drosophila sylvestris;* (b) *D. conspicua;* (c) *D. balioptera.*

(c)

Many species maintain their distinctiveness even though they regularly exchange genes with other species. For example, wolves and coyotes occasionally hybridize in nature, but the fossil record indicates that these two carnivores have been evolving as separate and distinct lineages for perhaps as long as 2 million years. They differ strikingly in their social organization, hunting methods, and size and shape (Figure 19.14). Similarly, cottonwoods and balsam poplars have been hybridizing with one another for at least 12 million years. Yet they have maintained their differences so that individuals of the two species are easily identifiable. It is not understood why these species, among which there is considerable gene exchange, remain distinct whereas others, with similar amounts of gene exchange, blend into one variable population.

(b)

(a)

19.13 A Redwing Is a Redwing
Nobody would have difficulty telling that both of these birds are red-winged blackbirds, even though one (a) comes from California and the other (b) comes from New England.

(a)   (b)

19.14 Distinct Despite Occasional Hybridization
Coyotes (left) are small and slender; they are easily distinguished from the larger, more robust timber wolf shown on the right, even though they can be found in the same regions and occasionally hybridize.

Factors Promoting Cohesion

An important factor promoting cohesion of speces is **gene flow**—the movement of individuals so that they reproduce in places some distance away from where they were born. Gene flow may be critical in maintaining the cohesion of most species because even a very low rate of gene flow among populations prevents them from differentiating as a result of genetic drift. The effects of gene flow are reinforced if natural selection operates in the same direction in different places, because then migrating individuals will not differ greatly genetically from individuals in the populations they join.

HOW LONG DOES SPECIATION TAKE?

There is no general rule about how much time speciation requires. A population may speciate by polyploidy within one breeding season. At the other extreme, some populations that have been isolated for millions of years remain reproductively compatible. American and European sycamores have been isolated from one another for at least 20 million years. They are, nonetheless, morphologically very similar and they form fertile hybrids (Figure 19.15). Other examples cover virtually the entire range between these two extremes. For example, a strain of *Drosophila paulistorum* could interbreed with other strains when it was first collected. But after being isolated in the laboratory for only a few years, its matings with other strains were sterile. Several reproductively isolated species of cichlid fishes are found only in Lake Nabugabo in Africa, which has been separated from Lake Victoria for only 4,000 years. One reason for such differences in speciation rates is that organisms differ greatly in their generation times. *Drosophila* and tropical cichlid fishes may have many generations per year, whereas larger organisms may have generation times as long as decades. The shorter its generation time, the faster a population can potentially evolve.

Behaviorally complex animals appear to speciate more quickly than simpler ones. An important characteristic of behaviorally complex organisms is that they make sophisticated discriminations among potential mating partners. They not only discriminate members of their own species from members of other species, but they also make subtle within-species discriminations on the basis of size, shape, appearance, and behavior. These discriminations relate to both the quality of the genes of the potential partner and to the quality of parental care likely to be given. Such behavioral discrimination can greatly amplify differences in fitness already caused by the interactions of the pool of potential partners with the physical environment, prey, or predators. Therefore, mate selection is probably a major cause of rapid evolution in general and of the development of reproductive isolation between species.

THE SIGNIFICANCE OF SPECIATION

Speciation has resulted in millions of species and in rich ecological communities. However, most speciation events are byproducts of selection acting on populations in isolation. Where and how often barriers appear is unrelated to any ecological "needs" of the populations. Also, many differences between diverging populations may be caused initially by genetic drift. The fringes of species' ranges are more likely to become fragmented than the central areas. Frag-

(a)   (b)

19.15 Geographically Separated but Similar
Despite the fact that they occur on different continents and have been separated for at least 20 million years, (a) American and (b) European sycamores (*Platanus occidentalis* and *P. hispanica*) are similar in appearance.

mented ranges are also typical of areas where environmental conditions change dramatically over short distances. The deer mouse *Peromyscus maniculatus*, the most widely distributed small mammal in North America, varies greatly geographically, especially in coat color (which matches the background color of soils, providing camouflage) and in tail and foot length (which determines how readily the mice can climb). However, over large areas where there is little topographic and vegetational change, deer mice are relatively uniform. In mountainous areas, where environmental conditions change dramatically from place to place, and on islands, where populations are isolated from one another by water barriers, differentiation is much more pronounced (Figure 19.16). Some of these local populations appear so different that they are sometimes regarded as distinct species.

Some biologists believe that most evolutionary change takes place at the time of speciation. They suggest that the isolation of small populations, together with random changes in allele frequences, disrupts the integrated functioning of the genome and allows for rapid responses to agents of evolution. According to this view, once the speciation process has been completed, the better integrated new genotypes resist modifications, leading to long periods of stasis that are interrupted only by the next round of speciation. Unfortunately, the fossil record is not well enough understood to allow scientists to determine whether evolution typically is more rapid during speciation than at other times.

LONG-TERM EFFECTS OF SPECIATION

Although many studies have been made of evolutionary changes, we still know little about how different evolutionary agents interact with one another, both in the short term and in the long term. Nonetheless, as pointed out by G. Evelyn Hutchinson, the evolutionary play is performed in an ecological theater, and much can be learned by investigating the ecological context of evolutionary changes.

Ecology and the Rate of Speciation

The fossil record shows that some lineages give rise to many more species than do others. How evolutionary agents may influence the rate of accumulation of differences that affect speciation rates is shown by the large hoofed mammals of Africa. These mammals are abundant on African savannas today, and they have good fossil records going back millions of years. A great deal is known about feeding behavior, habitat specificity, population sizes, and social organizations of the living species.

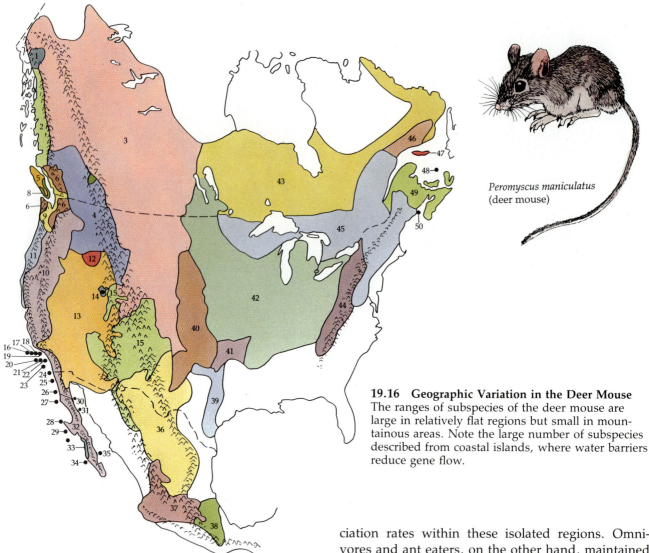

Peromyscus maniculatus
(deer mouse)

19.16 Geographic Variation in the Deer Mouse
The ranges of subspecies of the deer mouse are large in relatively flat regions but small in mountainous areas. Note the large number of subspecies described from coastal islands, where water barriers reduce gene flow.

The fossil record reveals that different lineages varied markedly in their rate of speciation. Speciation rates could possibly be correlated with birth rates, because animals with higher birth rates would produce more offspring per generation upon which natural selection can act. However, there is no pattern among the birth rates of current members of these lineages that supports this hypothesis. Nor can such a pattern be found in comparisons of birth rates of living species and the likely birth rates of their ancestors. However, speciation rates *are* correlated with the diets of the mammals, being higher among grazers (grass eaters) and browsers (eaters of branches and leaves of woody plants) than among omnivores and ant eaters (Figure 19.17). The grazers and browsers require large expanses of open grasslands and woodlands, respectively. In Africa, these resources disappeared from and reappeared in large areas during periods of climatic change, isolating populations and causing both high extinction rates and high spe-

ciation rates within these isolated regions. Omnivores and ant eaters, on the other hand, maintained more continuous populations during those climate changes and did not speciate as often.

This evolutionary pattern is consistent with the action of natural selection within lineages, but it could not be proposed as an explanation from short term studies of living populations alone. It is only by using a combination of information about the ecology of living species and about the fossil record that the patterns of speciation and the long-term evolution of these mammals can be understood. This example also shows how some groups may show stasis, long periods during which they change little, while other lineages in the same environment evolve rapidly.

Evolutionary Radiations

The fossil record and current distributions of organisms reveal that some species or groups have given rise to a large number of daughter species, a phenomenon called **evolutionary radiation**. There have been such radiations on all continents, but they are much more obvious and easier to study on islands, where there are many fewer species. In addition,

1	Wildebeest
2	Topi
3	Warthog
4	Kob
5	Duiker
6	Patas monkey
7	Kudu
8	Giraffe
9	Springbok
10	African buffalo
11	Klipspringer
12	Impala
13	Spring hare
14	Eland
15	African bush pig
16	Duiker
17	Hamadryas baboon
18	Aardvark

19.17 Speciation Rates in Some African Mammals
Grazers (who eat grass and other herbaceous plants) and browsers (who eat leaves and other parts of woody plants) have speciated more rapidly than omnivores and anteaters. (Animals belonging to two different genera are both commonly referred to as "duikers"—a phenomenon discussed further in Chapter 20; thus duikers appear as both browsers and omnivores.)

because many organisms disperse poorly over large water-filled gaps, islands lack many plant and animal groups found on the mainland. This creates ecological opportunities that often result in rapid evolutionary changes.

The islands on which the most remarkable evolutionary radiations have occurred are those of the Hawaiian archipelago. The Hawaiian Islands are especially useful as a natural laboratory for evolutionary studies because the archipelago is the most isolated one in the world. It lies 4,000 km from the nearest major land mass and 1,600 km from the nearest group of islands. Also, the islands are arranged in a line of decreasing age, with the youngest islands to the southeast and the oldest to the northwest. The biota of the Hawaiian Islands includes 1,000 species of flowering plants, 8,000 species of insects, 1,000 land snails, and more than 100 birds. However, there are no amphibians, no terrestrial reptiles, and only one native mammal—a bat.

More than 90 percent of all plant species on the Hawaiian Islands are endemic, that is, they are found nowhere else. Several groups of flowering plants are more diverse in their forms and life histories on the islands than their mainland counterparts, and they live in a wider variety of habitats. An outstanding example is the group of sunflowers called silverswords and tarweeds (*Argyroxiphium*, *Dubautia*, and *Wilkesia*). The 28 species in the silversword group are believed to be derived from a single species of tarweed from the Pacific coast of North America. Variation among silverswords in Hawaii is far greater than among mainland tarweeds. Hawaiian species include prostrate herbs, upright herbs, rosettes, shrubs, trees, and vines (Figure 19.18). They occupy nearly all the islands' habitats from sea level to above the timberline in the mountains.

In silverswords, the extraordinary diversification is produced by a very few genes. There are a few key loci whose alleles control the developmental processes that result in widely differing morphologies. In the laboratory, hybrids are readily produced among many of the species that appear very different. This indicates that they differ by only a small number of genes.

Silversword radiation illustrates some important points about evolution. First, major changes in organisms can be produced by changes in a small number of genes. Second, the fact that island represen-

19.18 Rapid Evolution among Hawaiian Plants
Silverswords and tarweeds in three closely related genera of the sunflower family (Asteracea) are believed to have descended from a single, herbaceous ancestor that colonized Hawaii from the Pacific coast of of North America.

Illustrated here are (a) *Argyroxiphium sandwicense*; (b) *Wilkesia gymnoxiphium*; and (c) *Dubautia laxa*. Their rapid evolution makes them appear more distantly related than they actually are.

tatives of these plants are much more diverse in their sizes and shapes than the mainland relatives from which they are derived is clearly due to the fact that the colonizers arrived on islands that had very few plant species. In particular, trees and shrubs tend to have large seeds that are dispersed only short distances; therefore, such plants only rarely migrate to oceanic islands. Many island trees and shrubs have evolved from nonwoody ancestors. On the mainland, however, tarweeds lived in ecological communities that contained tree and shrub lineages older than their own. Therefore, opportunities to exploit the tree way of life were already preempted.

Exploiting a New Feeding Technique: Gaping

Although evolutionary radiations are striking on islands, they do occur in species-rich mainland communities as well. Sometimes this is associated with evolution of the ability to gather food in a different way. An example is the American blackbird subfamily, a group of 94 species that obtains a variety of foods by means of **gaping**—opening the bill forcibly against the object it wants to move. This exposes insects and other small animals not available to a bird that can pick prey only off the surface. The blackbirds, unlike most birds, have powerful muscles for opening their bills (Figure 19.19a). They gape into

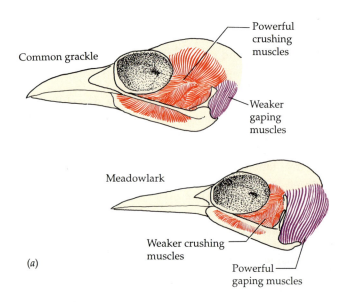

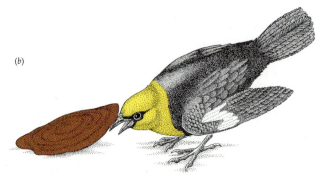

19.19 Gaping Exposes New Food
(a) Meadowlarks gape into soil to expose prey. They have much larger muscles for opening their bills than do grackles, which crush their food. (b) Gaping is also used to turn over objects on the ground to expose prey, as this yellow-headed blackbird is doing.

wood, fruits, leaf clusters, and stems of nonwoody plants; under sticks, stones, and animal droppings on the ground; and into the soil (Figure 19.19b). With this feeding method, they have come to occupy nearly all habitat types in North and South America, and they are among the most abundant birds throughout that region.

SUMMARY

Speciation—the splitting of one species into two separate species— is a central process of evolution. As defined by the biological species concept, a species is a group whose members can mate to produce fertile offspring but which are isolated reproductively from other groups. Species are also taxonomic units recognized on the basis of morphological traits, usually in the absence of information on reproductive interactions. In most cases, both definitions apply to the same group, but there are many exceptions.

Geographic separation is the most important means by which a single species can be split, but the width of the barrier needed to accomplish the separation varies strikingly. For small, relatively sedentary organisms, barriers as narrow as a city street may be effective. Differentiation can occur within continuous populations, especially if there are abrupt environmental changes. Species may be formed quickly by a multiplication of chromosome numbers (polyploidy), which makes offspring sterile in crosses with members of the parent species. Polyploidy has been a major factor in plant speciation, but it has been rare among animals.

Once a barrier is established, two populations may diverge in response to local conditions or as a result of genetic drift. If at a later time they come together again, they may have diverged so much that they no longer interbreed. However, divergence may not have been great enough to prevent rapid amalgamation of the two units through hybridization. Reproductive isolation may be due to either prezygotic or postzygotic isolating mechanisms, or both. Hybrid zones may be established when two formerly separated populations come together. Some of these hybrid zones persist for long periods of time in the same place.

Speciation may be accompanied by relatively little genetic change. The differences separating species involve the same types of traits that vary within species. As a result of speciation, new lineages are created that henceforth evolve in genetic independence of one another. Speciation is often simply the incidental byproduct of adaptations among populations to local conditions. However, speciation may open up new avenues of adaptation resulting in many subseqent speciation events as members of a lineage exploit new environmental resources.

SELF-QUIZ

1. A biological species is:
   a. a group of actually interbreeding natural populations that is reproductively isolated from other such groups.
   b. a group of potentially interbreeding natural populations that is reproductively isolated from other such groups.
   c. a group of actually or potentially interbreeding natural populations that is reproductively isolated from other such groups.
   d. a group of actually or potentially interbreeding natural populations that is reproductively connected to other such groups.
   e. a group of actually interbreeding natural populations that is reproductively connected to other such groups.

2. Taxonomic species are not always equivalent to biological species because:
   a. taxonomic species are often described on the basis of appearances rather than reproductive behavior.
   b. taxonomic designations often ignore reproductive behavior.
   c. taxonomic species often refer to artificial rather than natural populations.
   d. Taxonomic species *are* always equivalent to biological species.
   e. taxonomic species are not reproductively isolated from other species.

3. Cladogenesis is another name for:
   a. the formation of glades.
   b. the formation of two species by the splitting of one evolutionary lineage.
   c. the formation of a new species by the coming together of two evolutionary lineages.
   d. the reduction of two lineages by the extinction of one of them.
   e. the formation of new species by a taxonomic revision of a group.

4. Geographic speciation may happen when:
   a. continents drift apart, separating previously connected lineages.
   b. a mountain range separates formerly connected populations.
   c. genetic drift causes evolutionary changes on two sides of a barrier.
   d. the range of a species is separated by loss of intermediate habitat.
   e. All of the above.

5. Finches speciated on the Galapagos Islands because:
   a. the Galapagos Islands are a long way from the mainland.
   b. the Galapagos Islands are very arid.
   c. the Galapagos Islands are small.
   d. the islands in the Galapagos archipelago are sufficiently isolated from one another that there is little interisland migration.
   e. the islands in the Galapagos archipelago are close enough to one another that there is considerable interisland migration.

6. Which of the following is *not* a potential prezygotic isolating mechanism?
   a. Temporal segregation of breeding seasons
   b. Differences in mating pheromones
   c. Sterility of hybrids
   d. Spatial segregation of mating sites
   e. Inviability of sperm in female reproductive tracts

7. A common means by which sympatric speciation comes about is:
   a. polyploidy.
   b. hybrid sterility.
   c. temporal segregation of breeding seasons.
   d. spatial segregation of mating sites.
   e. imposition of a geographical barrier.

8. The existence of sibling species tells us that:
   a. brother and sister matings are often successful.
   b. genetic changes accompanying speciation are often small.
   c. genetic changes accompanying speciation are usually large.
   d. speciation usually requires major reorganization of the genome.
   e. species differ from one another in traits other than those that differ among individuals within species.

9. Which of the following is *not* true of speciation?
   a. Speciation always takes thousands of years.
   b. Speciation often takes thousands of years but may happen within a single generation.

   c. Among animals, speciation usually requires some physical barrier.
   d. Among plants, speciation often happens as a result of polyploidy.
   e. Speciation has produced the millions of species living today.

10. Evolutionary radiations:
   a. often happen on continents but rarely on island archipelagos.
   b. characterize birds and plants but not other taxonomic groups.
   c. have happened on all continents and on islands as well.
   d. require major reorganizations of the genome.
   e. never happen in species-rich environments.

FOR STUDY

1. Gene exchange between populations is prevented by geographic isolation, by behavioral responses before mating (e.g., females reject courting males of other species), and by mechanisms that function after mating has occurred (e.g., hybrid sterility). All of these are commonly called "isolating mechanisms." In what ways are these types very different? If you were to apply different names to them, which one would you call "isolating mechanisms?" Why?

2. The blue goose of North America occurs in two very distinct color types—a blue one and a white one. As we have seen, they are interfertile, and matings between the two color types are common. However, on their breeding grounds in northern Canada, blue individuals mate with blue individuals much more frequently than would be expected by chance. Suppose that 75 percent of all mated pairs consisted of two individuals of the same morph. What would you conclude about speciation processes in these geese? If 95 percent of pairs were of the same morph? If 100 percent of pairs were of the same morph?

3. Many species of butterflies are divided into local populations among which there is little gene flow. Yet many of these butterflies show relatively little geographic variation. Describe the studies you would carry out to determine what maintains this morphological similarity.

4. Distinguish among the following terms: allopatric speciation, parapatric speciation, and sympatric speciation. For each of the three statements below, indicate which type of speciation is implied:
   a. This process occurs most commonly in nature as a result of auto- or allopolyploidy.
   b. The present sizes of national parks and wildlife preserves may be too small to allow this type of speciation among organisms restricted to those areas.
   c. Generally occurs in species inhabiting areas where sharp environmental discontinuities exist.

5. Evolutionary radiations are common and easily studied on oceanic islands. In what types of mainland situations would you expect to find major evolutionary radiations? Why?

READINGS

Atchley, W. R., and D. S. Woodruff, Editors. 1981. *Evolution and Speciation: Essays in Honor of M. D. J. White.* Cambridge University Press, New York. A rich collection of essays on many aspects of speciation.

Bush, G. L. 1975. "Modes of Animal Speciation." *Annual Review of Ecology and Systematics*, vol. 6, pages 339–364. A thorough review of the types of speciation and the evidence for them by a strong proponent of sympatric speciation. Has an extensive bibliography.

Endler, J. T. 1977. *Geographic Variation, Species, and Clines.* Princeton University Press, Princeton, NJ. A theoretical analysis of the ways in which sympatric and parapatric speciation might occur.

Mayr, E. 1970. *Populations, Species, and Evolution.* Harvard University Press, Cambridge, MA. An abridged version of the most thorough work on speciation theory as applied to animals.

Otte, D. and J. A. Endler, Editors. 1989. *Speciation and its Consequences.* Sinauer Associates, Sunderland, MA. A comprehensive collection of essays on concepts, methods, and consequences of speciation. Includes both general treatments and analyses of specific cases.

Stebbins, G. L. 1950. *Variation and Evolution in Plants.* Columbia University Press, New York. Although now dated, this is one of the great classics on speciation, full of many examples about speciation among plants.

20

Systematics and Phylogeny

PREVIEW: Systematics is the study of biological diversity and its evolution. Taxonomy, a subdivision of systematics, is the science of biological classification. Classification systems help to clarify relationships among organisms; they help us remember organisms and their traits; they enable us to communicate clearly the identity of organisms being studied; they improve our predictive powers; and they provide stable names. Taxonomic systems used by biologists are hierarchical, that is, each higher group contains all the groups below it. A major goal of biological classification systems is to reflect evolutionary relationships among organisms. Many different traits are used to classify organisms because no one type of information is always the most appropriate. Knowing phylogenetic relationships is essential for interpreting the evolution of traits of organisms.

This chapter deals with the schools of biological classification, the construction of phylogenies, taxonomic characters and how they are used, the use of biochemistry in systematics, and taxonomic keys.

From the simple beginnings discussed in Chapter 17, life has evolved into the almost overwhelming richness of species found today. People have long been interested in the richness of the living world and have attempted to understand its origin and maintenance. At first, the study of biological diversity was motivated by purely practical reasons—to determine which plants and animals might be useful sources of food, medicine, and other products. During the seventeenth and eighteenth centuries, the study of nature was strongly stimulated in Western culture by the desire to reveal the hidden order and harmony as thought to have been provided by God. Because God was assumed to have had a plan in mind, scientists of that period believed that the diversity of living organisms must obey some general laws, and that these laws could be revealed, in part, by the way organisms were classified. These early attempts at classification led to the complex systems we use today. In this chapter we will consider the goals and methods of modern taxonomy and show how biologists use classification systems to express different kinds of relationships among both living and fossil organisms.

THE IMPORTANCE OF BIOLOGICAL CLASSIFICATION

Dealing with a complex world requires an ability to recognize similarities and differences among objects. Classification systems serve four very important roles. First, they are an aid to memory. It is impossible to remember the characteristics of a large number of different things unless we can group them into categories, whose members share many characteristics. Second, classification systems greatly improve our predictive powers. If, for example, we know that females of all known mammalian species have mammary glands with which they produce milk for their offspring, we can be quite certain that a newly discovered animal with other typical mammalian traits, such as hair and a constant, high body temperature, will also have this method of provisioning its offspring, even if the first individuals we happen to find are males, and hence lack functional mammary glands.

Third, classification systems improve our ability to explain relationships among things. For biologists, this is especially important when we attempt to reconstruct the evolutionary pathways that have produced the diversity of organisms living today. Finally, taxonomic systems provide relatively stable, unique, and unequivocal names for organisms. If those names are changed, the systems provide means of tracing the changes. Common names, even if they exist (most organisms do not have any common names at all), are unreliable and often confusing. For example, there are plants called "bluebells" in England, Scotland, the eastern United States, and the Rocky Mountains—but none of the bluebells in any of those places is closely related to the bluebells in any of the other places (Figure 20.1). Fish called pickerel are prized for eating in central Ontario, Canada, whereas to the south, in the Great Lakes region, pickerels are regarded as undesirable for the table. The inconsistency is due to the fact that around the Great Lakes the name "pickerel" is applied to a fish species that is called the "grass pike" in central Ontario. In neither location is that species regarded as good eating. A different species is called "pickerel" in central Ontario. These cases illustrate the need for formal, unique names for organisms.

TAXONOMIC HIERARCHIES

Recognizing and interpreting similarities and differences among organisms is easier if the organisms are assigned to groups that are ordered and ranked—that is, if the organisms are classified. Any group of organisms treated as a unit in a classification system is called a **taxon** (plural, taxa). **Taxonomy** is the theory and practice of classifying organisms. **Systematics** is the scientific study of the diversity of organisms. Its goal is to assess evolutionary relationships among organisms and to express these relationships as taxonomic systems.

The biological classification system used today is based on the work of the great Swedish biologist Carolus Linnaeus (1707–1778). In the Linnaean system, each species is assigned two names, one identifying the species itself and the other the genus to which the species belongs (Figure 20.2). A **genus** is a group of closely related species (the plural of genus is genera, and its adjectival form is generic). In many cases the name of the taxonomist who first proposed the species name is added at the end. Thus, *Homo*

(a)

(b)

(c)

20.1 Many Different Plants Are Called Bluebells
(a) These North Dakotan flowers, a species of *Campanula*, family Campanulaceae, are usually called bluebells. (b) In Virginia, these *Mertensia*, family Boraginaceae, are also called bluebells. (c) This English bluebell, *Endymion nonscriptus*, is a member of the family Liliaceae.

The facsimile page reads:

86   AVES ACCIPITRES. *Vultur.*

I. ACCIPITRES.

Roſtrum e Mandibula ſuperiore denticulum utrinque exſerens.

40. **VULTUR.** *Roſtrum* rectum, apice aduncum.
Caput impenne, antice nuda cute.
Lingua bifida.

Gryphus. 1. V. maximus, caruncula verticali longitudine capitis. †
Vultur Gryps Gryphus. *Klein. av.* 45.
Cuntur. Raj. av. 11.
Habitare fertur Chili.
Rara avis in terris; mihi ignota; videſis Kleinium.

Harpyja. 2. V. occipite ſubcriſtato. †
Yzquauhtli *Hern. mex. pp.* 34.
Aquilæ criſtatæ genus. *Raj. av.* 161.
Habitat in Mexico.
Magnitudo arietis. Collum, Dorſum, Cauda, Criſta
ſurrecta nigra; Subtus *nigro candidoque fulvo mixta. Homines etiam adoritur. Hernand.*
Cfr. *Aquila mexicana coronata. Roſtrum Vulturis. O-culi membrana nictitante. Sub* ingluvie *pennæ albæ, quas iratus dimittit usque ad pedum digitos. Alæ Caudaque ſubtus albo nigroque punctatæ colore Tigridis. Erectus conſidet. Pennas occipitis ſæpius erigit in formam coronæ. Fertur unico ictu cranium hominis iratus findere. Viſus Madriti in Vivario Regio a Z. Hallman.*

Papa. 3. V. naribus carunculatis, vertice colloque denudato.
Vultur elegans. *Edv. av.* 2 t. 2.
Vultur. *Alb. av.* 4. p. 4. t. 4.
Habitat in India occidentali.
Obſ. Vultur. *Alb. av.* 3. p. 1. t. 1. *an femina hujus?*
Caput & Collum *quaſi excoriata retrahere poteſt intra vaginam cutis plumoſæ colli inferioris.*

Aura. 4. V. fuſcogriſeus, remigibus nigris, roſtro albo.
Tzopilotle ſ. Aura. *Hernand. mex.* 331.

Uru-

20.2 Part of Linnaeus' Classification of Vultures
This facsimile of a page from the tenth edition of *Systema Naturae* (1758) shows the descriptions Linnaeus provided for the species he named.

sapiens Linnaeus is the name of the modern human species. *Homo* is the genus to which the species belongs, *sapiens* identifies the species, and Linnaeus proposed the species name *sapiens*. You can think of the generic name *Homo* as being equivalent to your surname and the specific name *sapiens* as being equivalent to your first name. This two-name system, referred to as **binomial nomenclature**, is universally employed throughout biology. The generic name is always capitalized; the species name is not. As we have done in this paragraph, both the generic and specific names are always italicized, whereas common names are not. A reference to more than one species in a genus is expressed with the abbreviation spp. after the generic name (for example, *Drosophila* spp.); the abbreviation sp. is used after a genus name when the identity of the species is uncertain. Rather than repeating a genus name when it is used several times in the same paragraph, it is spelled out once and abbreviated to the initial letter thereafter.

Most classification systems group smaller units into successively larger ones. The number of features shared by members of the larger units are fewer than the number shared by members of smaller units. For instance, ostriches, owls, hummingbirds, and sparrows are quite different animals, but they share more features with one another than they do with whales, cats, mice, or deer. By calling the former birds and the latter mammals, and placing them in separate but equivalent categories, we direct attention to certain significant features that serve to unite the members of each taxon and distinguish them from members of other taxa.

In the Linnaean system, the species are grouped into higher taxonomic categories. The number and limits of these categories are somewhat arbitrary, but there are some guiding rules. One is purely practical: If every species were put into its own genus, or conversely, all species were lumped into one genus, the genus would not carry any information not already present in the designation of the species. A second consideration is the relative amount of similarity or dissimilarity among the organisms. A higher taxonomic category may have a single species in it if that species is very different from all other species. Some genera, on the other hand, contain hundreds of species. In a taxonomic system designed to reflect evolutionary relationships, genera and higher taxonomic categories are based on the length of time since the taxa last shared a common ancestor. For example, genera might be separated by, say, 15 million years of independent evolution; higher taxonomic categories by still longer times.

The category above the genus in the Linnaean system is the **family**. The names of animal families end in -idae. Thus, Formicidae is the family that contains all ant species, whereas Hominidae contains humans and a few of our fossil relatives. Family names are based on the name of a member genus. Formicidae is based on *Formica* (remove the -a and add -idae); Hominidae is based on *Homo*. Plant classification follows the same procedure except that the ending -aceae is used instead of -idae. Thus Rosaceae is the family that includes the genus of roses (*Rosa*) and its immediate relatives. Unlike the generic and species names, family names are not italicized, but they are capitalized.

Families are, in turn, grouped into **orders**, and orders into **classes**. Classes of animals and protists are grouped into **phyla** (singular, phylum), and the equivalent category in plants, bacteria, and fungi is **division**. Phyla are grouped into **kingdoms**. The hierarchical units into which the blackburnian warbler (*Dendroica fusca*) and the moss rose (*Rosa gallica*) are classified are shown in Figure 20.3.

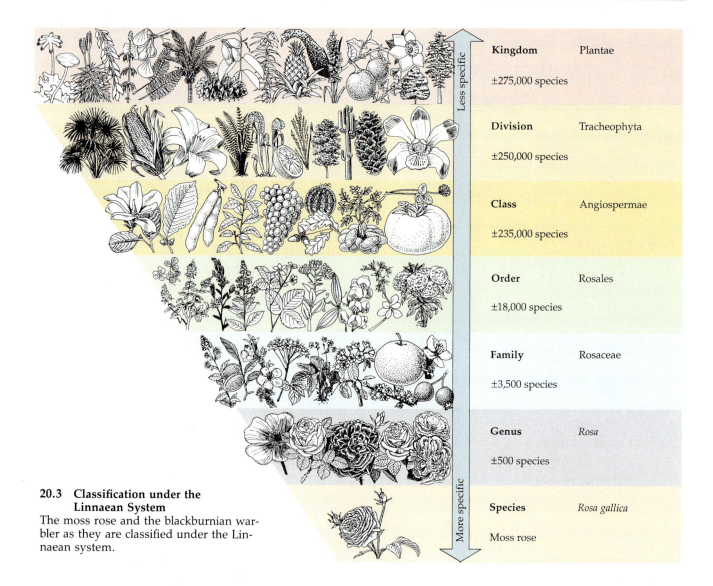

Kingdom	Plantae
±275,000 species	
Division	Tracheophyta
±250,000 species	
Class	Angiospermae
±235,000 species	
Order	Rosales
±18,000 species	
Family	Rosaceae
±3,500 species	
Genus	*Rosa*
±500 species	
Species	*Rosa gallica*
	Moss rose

Less specific → More specific

20.3 Classification under the Linnaean System
The moss rose and the blackburnian warbler as they are classified under the Linnaean system.

THE GOALS OF CLASSIFICATION SYSTEMS

The different biological classification systems all are designed to express relationships among organisms, but they differ with respect to the kinds of relationships they attempt to express. Classifications are based on features selected according to the goals of the system. If, for instance, we were interested in a system that would help us decide what plants and animals were desirable as food, we might erect a classification system based on palatability, ease of capture, and the edible parts each organism possessed. Early Hindu classifications of plants were designed according to such criteria. Hindu classifications of animals were made from several points of view, such as method of reproduction, habitat, mode of life, usefulness to people, and number of senses. One ancient system divided animals into (1) those with placentas, (2) those formed from eggs, (3) those that generated spontaneously, and (4) those born of vegetable matter. Another system divided animals

into (1) those born of moisture and heat, (2) those that bear live young, (3) those that lay eggs, and (4) those that burst forth from the ground.

We do not use such systems today, but they served the needs of the people who developed them. It is inappropriate to ask whether those classifications, or any others, including contemporary ones, are right or wrong. Classification systems can be judged only in terms of their utility and consistency with their stated goals. To evaluate any classification system we must first ask: What is it trying to accomplish? Then we can ask: How well does it accomplish those objectives?

The major goal of the biological classification systems in regular use today is to reflect evolutionary relationships. Evolutionary history has two important components. One is **phylogeny**, the pattern of genetic linkages between ancestors and their descendants. The other is the rate of evolution of traits among groups of organisms. Because the rates of change in the structures of organisms differ among

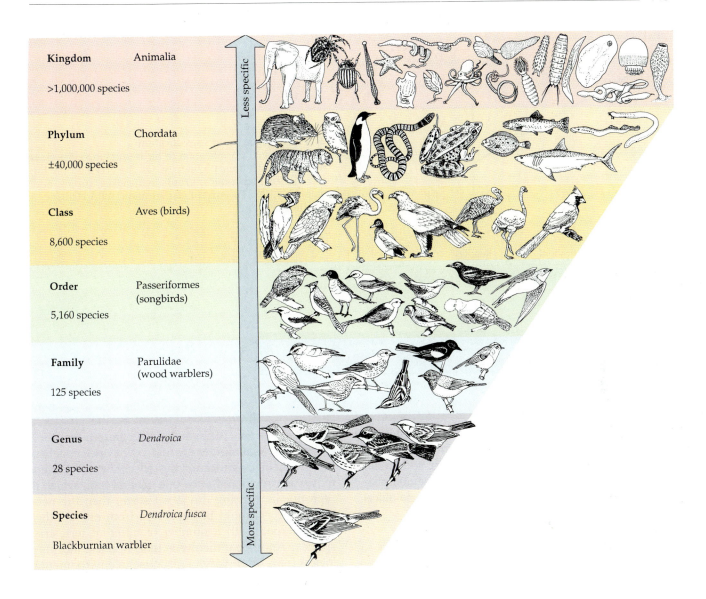

Kingdom	Animalia
>1,000,000 species	
Phylum	Chordata
±40,000 species	
Class	Aves (birds)
8,600 species	
Order	Passeriformes (songbirds)
5,160 species	
Family	Parulidae (wood warblers)
125 species	
Genus	*Dendroica*
28 species	
Species	*Dendroica fusca*
Blackburnian warbler	

Less specific → More specific

lineages, some organisms that are very similar to one another have been evolving independently within their lineages much longer than organisms in other lineages that are structurally much more different from one another. Classification systems that combined both phylogeny and rate of morphological change preceded in time those that emphasize only one of these factors.

Orthodox Systematics

The Linnaean classification system was developed before biologists were aware of the length of evolutionary time and when only a few methods for comparing organisms were available. Consequently, organisms were grouped primarily on the basis of their gross morphological similarities and differences. Later, when biologists accepted the fact of evolution, existing classifications were modified to reflect evolutionary relationships. Therefore, most currently used classifications reflect both evolutionary relation-

ships and the amount of morphological difference among organisms. As evolutionary relationships have become better known, taxonomies have been modified to more closely reflect phylogenies.

In general, current classifications deviate from phylogeny when a descendant is so different from its ancestor than it is more similar to organisms in other lineages that it is to other descendants of the same ancestor. For example, we know that birds and crocodiles share a more recent common ancestor than crocodiles do with snakes and lizards. However, the commonly used classification places crocodiles with snakes, lizards, and turtles, and separates birds into their own group. This classification emphasizes the similarity of crocodiles to snakes, lizards, and turtles, and the great differences between birds and those animals, but is not in accord with evolutionary relationships. In recent classification systems designed to show only evolutionary relationships, birds are grouped with crocodiles and their ancestors into a single taxon separate from snakes, lizards, and turtles

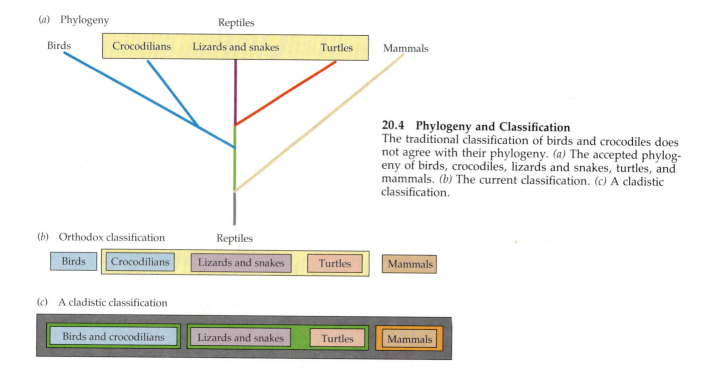

(a) Phylogeny

Birds   Reptiles   Mammals

Crocodilians   Lizards and snakes   Turtles

20.4 Phylogeny and Classification
The traditional classification of birds and crocodiles does not agree with their phylogeny. (a) The accepted phylogeny of birds, crocodiles, lizards and snakes, turtles, and mammals. (b) The current classification. (c) A cladistic classification.

(b) Orthodox classification

Reptiles

Birds   Crocodilians   Lizards and snakes   Turtles   Mammals

(c) A cladistic classification

Birds and crocodilians   Lizards and snakes   Turtles   Mammals

(Figure 20.4). Many taxa that reflect both evolutionary relationships and morphological similarity are in use today. They persist because they are so familiar to workers in the field and also because they emphasize some important morphological and physiological differences.

Phenetic Systematics

Phenetic systematics, a school of systematics that arose in the 1950s and 1960s, attempts to erect classification systems strictly on the basis of overall similarity among organisms. Systematists of this school measure as many traits as possible and use the measurements to estimate overall similarity. They argue that because no information on evolutionary history is available for most organisms, it is a mistake to attempt to reflect that history in classifications.

It might seem to be a straightforward task to measure a large number of traits of organisms and then assess the degree of similarity among them. In practice, however, this is not so simple, because decisions must be made about whether some traits are more important than others, and about whether a group of traits that are all direct responses to a single selective pressure should be given the same weight as traits influenced by different selective pressures. For example, most salamanders metamorphose from an aquatic larval form to a terrestrial adult before becoming reproductively mature. However, some salamanders reproduce while they are still in an aquatic larval form, a pattern known as **neoteny** (Figure 20.5). Many traits are associated with this reproductive pat-

tern. If they were all treated as separate traits, then all neotenic salamanders would be grouped together in a phenetic classification system. However, if neoteny was considered to be one general trait, which is reproduction during the larval stage, then the neotenic species would be separated from one another and grouped with a number of different terrestrial species. Another difficulty is that, because morphological traits often vary in response to environmental conditions, environmentally induced variation may be confused with genetically based differences.

20.5 A Neotenous Salamander
This mud puppy (*Necturus maculosus*) of eastern North America still has gills, even though it is reproductively an adult.

Cladistic Systematics

Some recent classifications attempt to show only evolutionary relationships among organisms, ignoring their degree of morphological similarity or difference. The objective of **cladistic systematics** is to determine the evolutionary histories of organisms and then to express those relationships in phylogenetic trees. A **clade** is the entire portion of a phylogeny that is descended from a single ancestral species. The closeness of organisms on a **cladogram** indicates the presumed time since they diverged from their most recent common ancestor. Because the goal is to show phylogenies, taxa in a cladistic classification are clades and are **monophyletic**; that is each taxon is a single lineage that includes all—and only—the descendants of a single ancestor.

Traits shared due to descent from a common ancestor are called **ancestral traits**. But how can ancestral traits be recognized? One important way is to study the traces of organisms that lived in the past. A **fossil** is any recognizable structure from an organism, or any impression from such a structure, that has been preserved. Every living organism is a potential fossil, but few actually become fossils. After they die, most organisms are destroyed quickly by biological degradation (decay), mechanical breakage, or chemical dissolution. Despite these difficulties, we do have an extensive supply of fossils available to us. A good fossil record helps reveal ancestral traits. For example, the excellent fossil record of horses shows that modern horses, which have one toe on each foot, evolved from ancestors that had multiple toes. A trait, such as the modern horse's single toe, that differs from the ancestral trait in the lineage is called a **derived trait**.

Even in the absence of a fossil record, however, reasonable inferences about ancestral traits can sometimes be made. For example, among butterflies, species in two families, the brush-footed butterflies (Nymphalidae) and the monarchs (Danaidae) have four functional and two very small legs, whereas the swallowtails (Papilionidae) and the sulfurs (Pieridae) have six functional legs (Figure 20.6). Given that moths and all other orders of insects have six functional legs, having six legs is probably ancestral. By inference, then, the four-legged trait in monarchs and brush-footed butterflies is probably a derived trait of butterflies descended from six-legged ancestors.

(a)

(b)

(c)

(d)

20.6 Six Legs is the Ancestral Number
Having four rather than six functional legs is a derived trait among butterflies. As you can tell from these photographs (a) Mourning cloaks (*Nymphalis antiopa*) and (b) gulf fritillaries (*Agraulis vanillae*) have only four functional legs. The six functional legs of (c) the anise swallowtail (*Papilio zelicaon*) and (d) Harford's sulfur (*Colias harfordii*) are also readily seen.

Deciphering ancestry is often difficult because over evolutionary time, a character may be lost from a lineage and later regained. Also, the same trait may appear independently in several lineages. To erect classification systems that accurately reflect phylogenies, it is necessary to be able to distinguish ancestral from derived traits. Therefore, cladists devote much effort to gathering and interpreting evidence to determine which traits are really ancestral and which are derived in different groups of organisms.

HOMOLOGY. Any two structures derived from a common ancestral structure are said to be **homologous**. Although homologous structures have a common ancestor, they may not have similar appearances or functions because, over time, such structures can diverge until they are very different. Nevertheless, homologous structures usually retain certain basic features that betray their common ancestry. Consider, for example, the leaves of plants. Several lines of evidence, especially details of their structure, indicate that all leaves are homologous, but they have been modified in many ways to become not only light-trapping devices but also protective spines, tendrils, and brightly colored lures that attract pollinators (Figure 20.7). Because all these structures are modified leaves, they are all homologues of one another.

HOMOPLASY. Not all resemblances are products of a common ancestry. If a structure evolves in different lineages, so that it is possessed by several species although it was not found in their most recent common ancestor, it exhibits **homoplasy**. Homoplasy can result from **convergent evolution, reverse evolution, or parallel evolution**. Under convergent evolution, structures that were formerly very different come to resemble one another because they have undergone selection to perform similar functions. For example, the structures that aid plants in climbing over other plants have evolved from stipules, leaflets, leaves, and inflorescences (Figure 20.8).

Under parallel evolution, the same character evolves in different lineages, often from a common basis. This is clearly seen among butterflies and moths where similar patterns of banding in the wings have evolved a number of times as a result of similar modifications of a common basic banding pattern shared by most species (Figure 20.9). Homoplasy is strongly suspected when organisms that are known to have been separated evolutionarily for a long time share some derived trait that appears to have evolved independently in each lineage after the two lineages separated.

Comparison of Taxonomic Methods

For all classification systems, the more information that is available, the better. Therefore, new knowledge from all areas of biological inquiry is rapidly incorporated into modern taxonomies. However, as we have just seen, that information is used very differently by orthodox, phenetic, and cladistic taxonomists. The similarities and differences among these schools are summarized in Table 20.1.

(a)

(b)

20.7 Homologous Structures Derived from Leaves
(a) The orange spines of the barrel cactus (*Ferrocactus acanthodes*) and (b) the colorful bracts of *Heliconia rostrata* are both modified leaves.

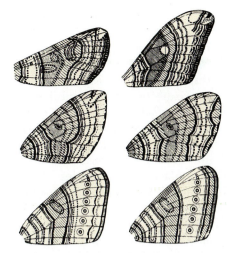

20.9 Parallel Evolution of Banding in Moth Wings
The similar banding patterns in the wings of these moths, which belong to five different families, were probably inherited from a common ancestor.

Passiflora rubra

Tendril derived from stipule

Normal stipule

Bignonia capreolata

Tendrils derived from 3-lobed leaves

Tendrils derived from leaf

Hooks derived from stems of flower heads

Clematis afoliata

Uncaria gambir

20.8 Convergent Evolution of Climbing Structures
The similarity of the tendrils of these plants, all of which help them climb over other plants, is due to convergent evolution.

CONSTRUCTING PHYLOGENIES

Biologists are interested in the phylogenetic relationships among organisms for many reasons. To understand the evolution of structures they need to know which traits are ancestral and which are derived. To determine how fast various traits have evolved in different lineages they need a good phylogeny. Phylogenetic information is essential for the study of nearly all aspects of adaptation. Indeed, it is difficult to think of a problem in biology that is not made easier to solve by the availability of a good phylogeny of the organisms being studied.

For several reasons, many different types of data are used in developing phylogenetic classifications. Traits that evolve slowly are useful for determining evolutionary relationships at the level of phyla or classes in the taxonomic hierarchy. Traits that evolve rapidly are useful at the level of families and genera. The traits that can be measured in flowering plants, for example, differ from those that can be measured in mammals. Some traits are readily preserved in fossils, whereas others, such as behavior and molecular structure, rarely survive fossilization processes. Therefore taxonomists use many different traits and use them differently when constructing phylogenies.

TABLE 20.1
Classification Methods Compared

ATTRIBUTE	PHENETICS	CLADISTICS	ORTHODOX SYSTEMATICS
Relationship shown	Overall similarity or difference	Genealogy	Genealogy + similarity
Characters used	All kinds	Shared derived traits	All kinds
Homology	Not considered	Primary importance	Important
Fossils	Not used	Important	Important
Ecological data	Not used	Rarely used	Sometimes important
Rates of evolution	Not considered	Not considered	Very important

Structure and Behavior

An important source of taxonomic information is **gross morphology**, that is, sizes and shapes of body parts. Living organisms have been measured for centuries, so we have a wealth of morphological data. This is also the kind of information that is most readily available from fossils. Sophisticated methods are now available for measuring morphology and for estimating the amount of morphological variability among individuals, populations, and species.

The early developmental stages of many organisms reveal similarities with other organisms that are lost by the time adulthood is reached. For example, the larvae of some marine creatures called sea squirts have a dorsal supporting rod in their backs that disappears as they develop into adults. Many other animals—all the animals called "vertebrates"—also have such a structure. Because larval sea squirts share this and other structures with vertebrates, they are known to be more closely related to vertebrates than would be suspected by examining adult sea squirts (Figure 20.10). Larval morphology does not, however, always reflect evolutionary ancestry. Many larvae have been highly modified by evolution for their particular existence. Butterflies are not closely related to animals that resemble their caterpillars. Therefore, care must be taken when using different stages of organisms to infer their phylogenies.

Living organisms often reveal their close affinities by similarities in their behavior. This information is most useful for detecting relationships among rather closely related organisms. For example, the German ethologist Konrad Lorenz showed that similarities in courtship behavior patterns support other evidence that several species of ducks with quite different plumages are nonetheless very closely related (Figure 20.11). Many of these ducks, despite these substantial differences in plumage, can mate and produce fertile hybrid offspring, showing that they are genetically similar.

Biochemical Traits

The molecules of organisms constitute their **micromorphology** just as shapes of body parts constitute their gross morphology. Among the most important biochemical traits of organisms are their nucleic acids—DNA and RNA—and the proteins whose synthesis the nucleic acids direct. Because they are so important for determining phylogenies, especially for groups with poor fossil records, we will discuss these biochemical traits in some detail.

PROTEIN STRUCTURE. It is possible to estimate similarities and differences among proteins without knowing the details of their structures by measuring the **immunological distance** between them. If a small amount of blood serum from an animal is injected into a test animal, such as a rat, the foreign blood acts as an antigen, causing the rat to produce antibodies that combine with the antigen and destroy it (Chapter 16). If the immunized rat is then reinjected with blood from the same species, large antigen–antibody aggregates form and will precipitate from solution. Antigen–antibody reactions are highly spe-

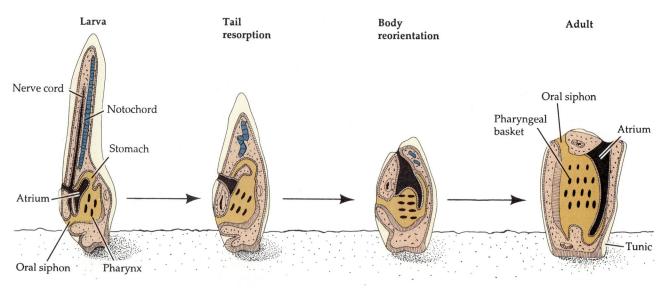

20.10 Larvae Reveal Evolutionary Relationships
Sea squirt larvae have a well-developed notochord (blue) that reveals their relationship with other chordates. As the larvae metamorphose they lose the notochord, and only the pharyngeal basket of the adults shows their chordate affinities.

(a)   (b)

20.11   Homologous Displays among Ducks
(a) The North American ruddy duck (*Oxyura jamaicensis*) and (b) the Asian white-headed duck (*O. leucocephala*), though geographically isolated and possessing different plumages, are in fact closely related, as the similar courtship displays engaged in by males of both species indicate. The raised-tail posture attracts the brown-feathered female.

cific; an antibody that precipitates the blood of one species does not react as strongly with the blood of other species. Nonetheless, there are interspecific reactions that decrease as the proteins become less similar. The strength of such immunological reactions can be used to estimate the similarity of proteins from different species. For example, a rat can be immunized with the blood of a salamander. Serum from the rat can then be divided among several test tubes, to which serum from other species of salamanders is added. The amount of precipitate formed is proportional to the similarities in amino acid sequences in the serum proteins of the different species (Figure 20.12).

This procedure has been used to study lineages among the plethodontines, a group of three genera of terrestrial North American salamanders—*Plethodon, Ensatina,* and *Aneides* (Figure 20.13). Based on studies of morphology, the distributions of living species, and fossils, investigators many years ago concluded that *Plethodon* resembles the ancestor of the group; that *Ensatina* is derived from that ancient stock; and that *Aneides* diverged from *Plethodon* by specializing for arboreal life. Immunological data confirm these conclusions and suggest that *Aneides* separated from *Plethodon* about 50 million years ago. The immunological data indicate that there was a burst of change among the species of *Aneides,* all of which apparently evolved at about the same time. *Plethodon,* on the other hand, has changed very little over the past 50 million years. In a purely cladistic system, the *Aneides* species would be included in the *Plethodon* genus even though the *Aneides* species are very different from their ancestors. Maintaining *Aneides* as a separate genus emphasizes the substantial differ-

ences between those species and the living species of *Plethodon.* To do so, however, requires that *Plethodon* violate the cladistic criterion that taxa should include all the species within their lineages. However, whatever system is chosen, the true phylogeny of this group has been established. It was correctly deduced from morphological data, and it has been confirmed by immunological data.

Electrophoresis, a method described in Chapter 14, can be used on both whole proteins and protein fragments. Because proteins of similar sizes move through an electrophoretic gel at similar rates, an estimate of the difference between proteins can be made from the difference in their mobility. Some proteins, such as collagens of bone and skin, keratins, and seed proteins, are often well preserved over geological time, allowing the method to be applied to fossil as well as living materials.

More precise information about phylogenies can be obtained by comparing the microstructure of proteins. The **amino acid sequences** of proteins can be determined easily by a process that sequentially removes amino acids from the amino terminus of polypeptides. The cleaved amino acids are then identified by gas–liquid or thin-layer chromatography, or by mass spectroscopy. A direct measure of genetic differences between two taxa is given by obtaining homologous proteins from the two taxa and determining the number of amino acids that have changed since the lineages of the taxa diverged from a common ancestor. This was the method employed to determine the sequences of amino acids in the cytochromes *c* of the organisms discussed in Chapter 18. As we saw, that information revealed a great deal about how natural selection influenced the evolution

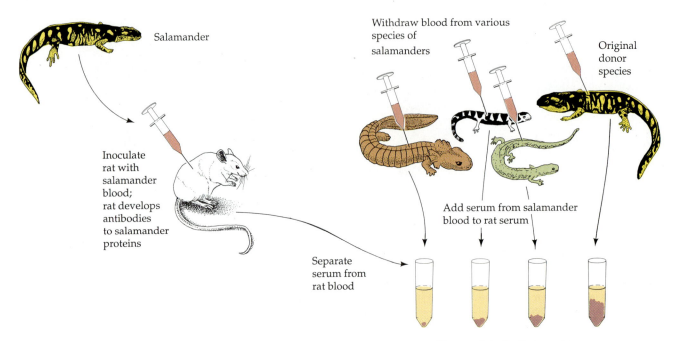

20.12 Immunological Distance
Reading from right to left, the three other salamander species, each in a different genus, are increasingly less closely related to the donor species (yellow). The degree of relatedness is estimated by measuring the extent of reaction between the serum proteins of a species and antibodies from the donor species. One could learn still more about the relationships among these species by using different species as the original blood donor.

(a)

20.13 Anatomy and Immunology Agree
The phylogenetic relationships suggested by the anatomy of these salamanders have been confirmed by immunological data. (a) *Plethodon jordani* is similar to the animals that gave rise to (b) *Ensatina eschscholtzi*, a terrestrial species, and (c) *Aneides flavipunctatus*, an arboreal salamander.

(b)

(c)

of cytochrome *c*. It was also used to estimate the approximate times that the lineages of those organisms separated.

NUCLEIC ACID STRUCTURE. The structure of the genes themselves—their **base sequences**—provides the most direct evidence of evolutionary relationships among organisms. DNAs can be compared, even if the precise sequences of their bases are not known, by a process called **nucleic acid hybridization**. As you may recall from Chapter 13, the two strands of the DNA double helix can be separated by heating them, but they will reanneal when cooled. If DNAs from two species are mixed and heated, they form interspecific double helices when cooled. However, because of differences in the base sequences of the two DNAs, they do not match up well. Less heat is required to separate these hybrid helices than helices composed of DNA from a single species. The degree of mismatching of the DNA is related to its thermal stability in a very consistent way. About 1 percent base-pair mismatching lowers by 1 percent the temperature at which 50 percent of the helices dissociate. DNA sequences differing by more than about 20 percent of their base pairs do not form stable duplexes, so this method can be used to compare only relatively similar species.

Nucleic acid hybridization has already yielded some surprising results. For instance, the DNAs of humans and chimpanzees are much more similar than would be expected given the considerable morphological differences between the two species (Table 20.2). This indicates that humans and chimps diverged from a common ancestor more recently than previously thought. A long-standing debate among biologists over whether the giant panda of China was related to bears or to racoons was resolved by DNA hybridization data, which clearly indicate that the

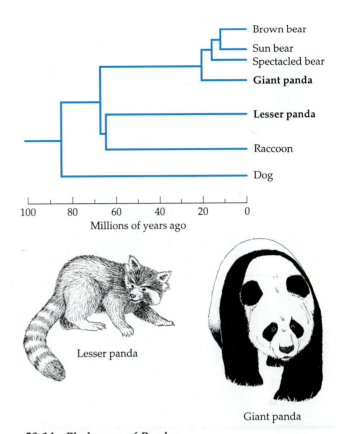

20.14 Phylogeny of Pandas
DNA hybridization data indicate that the giant panda of Asia is a bear specialized for eating bamboo. The lesser panda, also of Asia, was long thought to be closely related to the giant panda but is actually more closely related to raccoons and their allies of the Americas.

giant panda is a bear (Figure 20.14). The unusual features of the giant panda are recent adaptations to its specialized diet—bamboo.

Attempts to resolve avian phylogenies using only morphological data never produced a consensus; different morphological traits gave different phylogenies. As a result of DNA hybridization studies, avian phylogenies are being extensively revised. For example, New World vultures were revealed to be closely related to storks, whereas the Old World vultures are closely related to hawks and eagles (Figure 20.15). Convergent evolution for life as a scavenger —rather than descent from a common ancestor— produced the obvious similarities between Old and New World vultures.

Base sequences are determined in several ways, among which is the cleavage of DNA and RNA into short sections by the use of enzymes that recognize specific base sequences. These cleaved pieces can then be separated electrophoretically and their sequences are determined. (You may want to review the sections on cleaving and sequencing DNA in Chapter 14.)

Electrophoretic analysis has been carried out on

TABLE 20.2
Genetic Similarities among Some Vertebrates as Estimated by DNA Hybridization

TAXA COMPARED	PERCENTAGE DIFFERENCE IN DNA SEQUENCES
Human–chimpanzee	1.6
Human–gibbon	3.5
Human–rhesus monkey	5.5
Human–galago (nocturnal primate)	28.0
House mouse–Norway rat	20.0
Cow–sheep	7.5
Cow–pig	20.0

(a)

(b)

20.15   Vultures Have Converged
DNA hybridization results show that New and Old
World vultures look alike because they have converged
toward one another from very different ancestors. *(a)* The
turkey vulture is the most common and widespread New
World vulture. *(b)* White-backed and Ruppell's vultures
are scavengers of large mammal carcasses in African
grasslands.

three regions of the chicken's nuclear genome: those
containing structural genes for lysosome *c*, those for
three different "α-like" globins, and those for four
"β-like" globins. Each of these regions is located on
a different chromosome. These parts of the genome
have also been measured in other members of the
group called phasianoid birds: jungle fowl (the ancestor of modern chickens), chukar, turkeys, pheasants,
peafowl, and guinea fowl. The genealogy suggested
by these data differs significantly from the one based
on traditional morphological criteria (Figure 20.16).
Taxonomists were misled by considerable convergent
evolution among phasianoid birds, just as they were
by convergent evolution between New World and
Old World vultures.

THE FUTURE OF SYSTEMATICS

Biochemical methods will certainly continue to increase in importance. Nonetheless, information from
many sources will always be valuable for determining
lineages of organisms. The fossil record, which reveals when lineages diverged and began their independent evolutionary histories, is necessary to pro-

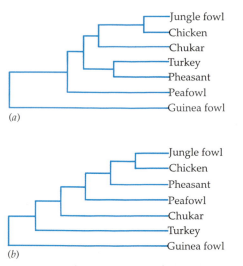

(a)

(b)

(c)

20.16   Phylogeny of Phasianoid Birds
Because of morphological convergence, the phylogeny
suggested by nuclear gene maps *(a)* and the phylogeny
suggested by morphological studies of phasianoid birds
(b) differ. Both phylogenies agree, however, that the
guinea fowl *(c)* is evolutionarily distant from the other
phasianoid birds.

Morphological traits

(a) Cladistic analysis

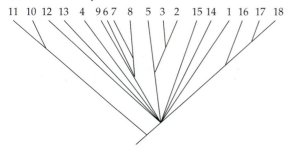

(b) Traditional analysis

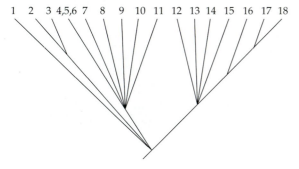

Molecular traits

(c) Protein sequence data

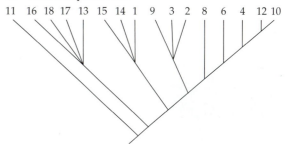

(d) Cladistic analysis of proteins

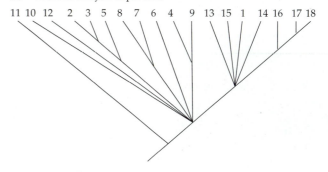

1 = Whales, dolphins
2 = Rodents
3 = Rabbits, hares, pikas
4 = Shrews, moles
5 = Elephant shrews
6 = Tree shrews
7 = Flying foxes
8 = Bats
9 = Primates
10 = Pangolins
11 = Sloths, anteaters, armadillos
12 = Carnivores (cats, dogs, hyenas, weasels)
13 = Aardvark
14 = Horses, tapirs, rhinoceroses (odd-toed hoofed animals)
15 = Pigs, hippos, deer, antelope, cattle (even-toed hoofed animals)
16 = Hyraxes
17 = Dugongs, manatees
18 = Elephants

20.17 Mammalian Phylogenies Are Still Unknown
Because different methods give different results, evolutionary relationships among mammals are still unresolved. These four proposed phylogenies differ in many ways. The phylogenies are based on (a) cladistic analysis of morphological traits; (b) traditional analysis of morphological data; (c) protein sequence data; and (d) cladistic analysis of proteins.

vide absolute timing for evolutionary events. Also, no one method is suitable for all time frames and for all kinds of organisms. Therefore, the range of data used in classification is likely to increase rather than decrease in the future. For this reason, systematics integrates activities from many different biological disciplines. A systematist needs to have a command of both molecular techniques and natural history.

Many kinds of information are needed because molecular data do not always resolve taxonomic

problems. Mammals are among the best known groups of animals. Their fossil record is extensive, and nearly all extant species have been named and described. The fossil record reveals that there was an explosive radiation of mammals about 60 million years ago. At that time all recent orders of mammals, as well as a number of groups not very similar to any of the surviving lineages, evolved. However, the mammalian groups radiated over such a short period that the phylogenies could not be reconstructed from the fossil discoveries. Molecular data, including data on myoglobin, hemoglobin, lens proteins, fibrinopeptides, cytochrome c, and ribonuclease, have been gathered for mammals for more than 20 years. By using this information, researchers have erected cladograms for the mammalian lineages (Figure 20.17). These attempts have been only partly successful owing to gaps in the data and to the fact that different molecules do not suggest the same phylogenies. Also, the computer programs used to generate phylogenies from the molecular and morphological data assume that there has been very little parallel evolution or reverse evolution, both of which probably occurred rather frequently. In addition, extensive work with fossils had already resolved those aspects of the phylogenies that were most readily dealt with by molecular techniques. Thus, despite two decades of extensive work, mammalian phylogenies are almost as uncertain as they were when molecular techniques were first employed. More research, using an even wider variety of data, will be necessary to resolve mammalian phylogenies.

1A Front and hind wings similar in size and shape. When at rest, wings folded together and parallel to body.

DAMSELFLIES see **2**

1B Hind wings wider than front wings near base. When at rest, wings extended on either side at right angles to the body.

DRAGONFLIES see **3**

2A Wings with 5 or more cross veins. Wings not narrowed to form a stalk at base.

FAMILY AGRIONIDAE
(Broad-winged damselflies)

3A Triangles in fore and hind wings similar in shape and position.

FAMILY AESCHNIDAE

(c)

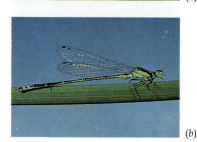

(a)

2B Wings with only 2 or 3 cross veins. Wings narrow to form a stalk at base.

FAMILY COENAGRIONIDAE
(Narrow-winged damselflies)

3B Triangles in fore and hind wings unlike in shape and position.

FAMILY LIBELLULIDAE

(b)

(d)

20.18   A Key to Odonates
Damselflies fold their wings over their backs, whereas dragonflies sit with their wings held horizontally. Further definition among the species is provided by the venation patterns in the wings, which a taxonomist would study with magnification. The representatives of the four families described in this key are *(a) Agrion splendens*, family Agrionidae; *(b) Enallagma* sp., family Coenagrionidae; *(c) Aeschna multicolor*, family Aeschnidae; and *(d) Sympetrum corruptum*, family Libellulidae.

TAXONOMIC KEYS

Taxonomic systems are used not only to indicate relationships among organisms, but also to help identify organisms, whether we wish to do so purely for pleasure or for research purposes. Taxonomists publish written descriptions of species being characterized and named for the first time. In addition, a description of a group of organisms is usually accompanied by a **taxonomic key**, which taxonomists design to help identify specimens belonging to the group in question. One of the most useful forms of key is **dichotomous**: At each step, species are divided into two groups on the basis of the presence, absence, or degree of development of some characters. As an example, consider the key in Figure 20.18 for the identification of insects belonging to the order Odonata. This simple key allows identification only to the family level. Other keys must be used to identify species. Note that the characters chosen for use in the key are simple and easy to see. They are not necessarily the characters that reveal the most about evolutionary relationships among dragonflies and damselflies.

It is not necessary to know anything about evolutionary relationships within a group to make a useful taxonomic key. All you need are characters that clearly separate the organisms and that can be used on most specimens. For example, it is often important to be able to identify plants that are not in flower. To do so requires the use of keys based on leaf and stem features. However, evolutionary relationships among plants are revealed much more clearly by their reproductive structures (flowers) because those structures evolve more slowly than leaves and stems do.

SUMMARY

Classification is a basic and vital activity. By means of it we order our environment to make it easier to respond appropriately to complex inputs. All biological classification systems are designed to serve particular purposes and must be judged with reference to their specific goals. Such systems can reflect similarities in the measurable traits of organisms (phenetic classifications), their phylogenetic relationships (cladistic classifications), or some combination of the two (orthodox classifications). Mixed systems are common because often people wish to accommodate more than one goal within a system.

In biology we give organisms two names, a generic name and a specific name. We group species into higher taxonomic categories such as genera, families, orders, classes, phyla, and kingdoms. The size and number of these higher categories is somewhat arbitrary.

Classification systems are erected from a wide variety of data—mostly from gross morphology, embryology, behavior, sequences of amino acids in proteins, and base sequences in DNA. Molecular

techniques are increasingly being used. In some cases they are causing major changes in previously hypothesized phylogenies. In other cases, molecular data confirm earlier classifications. In still other cases, all available data in combination do not resolve the uncertainties.

Taxonomic keys serve in identifying specimens. They may be based either on characters that reflect actual evolutionary relationships or on characters that are simply convenient. Knowing evolutionary relationships among organisms is important for almost all types of biological investigations.

SELF-QUIZ

1. Which of the following is *not* a major role of a classification system?
   a. A classification system is an aid to memory.
   b. A classification system improves predictive powers.
   c. A classification system helps explain relationships among things.
   d. A classification system provides relatively stable names for things.
   e. A classification system helps us design identification keys.

2. Any group of organisms treated as a unit in a classification system is:
   a. a species.
   b. a genus.
   c. a taxon.
   d. a clade.
   e. a phylogen.

3. A genus is:
   a. a group of closely related species.
   b. a group of genera.
   c. a group of similar genotypes.
   d. a taxonomic unit larger than a family.
   e. a taxonomic unit smaller than a species.

4. The two important components of evolutionary history are:
   a. phylogeny and ontogeny.
   b. phylogeny and cladistics.
   c. phylogeny and rate of evolution of traits.

   d. rate of evolution of traits and cladistics.
   e. rate of evolution of traits and ontogeny.

5. Orthodox systematics establishes classifications by considering:
   a. only the phylogenies of organisms.
   b. only the amount of similarity or dissimilarity among organisms.
   c. both the amount of similarity among organisms and their phylogenies.
   d. the similarities among organisms using a large number of traits.
   e. neotenous traits as more important than all other traits.

6. Which of the following *cannot* result in homoplasy?
   a. Convergent evolution
   b. Reverse evolution
   c. Parallel evolution
   d. Divergent evolution
   e. Independent evolution of the same trait

7. Traits that evolve very slowly are useful for determining relationships at the level of:
   a. phyla.
   b. genera.
   c. orders.
   d. families.
   e. species.

8. The immunological distance between two proteins is measured by:
   a. the difference in the rate at which they migrate in a gel.
   b. the amount of precipitate formed in an antigen-antibody reaction.
   c. sequencing their amino acids.
   d. determining the base sequences in the genes that code them.
   e. None of the above.

9. Molecular methods have not resolved mammalian taxonomic problems because:
   a. molecular methods have not yet been applied to mammals.
   b. mammals have a very poor fossil record.
   c. most mammalian lineages separated at about the same time 60 million years ago.
   d. molecular methods are difficult to use on mammalian cells and tissues.
   e. many mammalian lineages became extinct.

10. Taxonomic keys are used to:
    a. help identify unknown organisms.
    b. show how evolutionarily conservative traits are distributed.
    c. show how evolutionarily labile traits are distributed.
    d. reveal evolutionary relationships among organisms.
    e. establish clades.

FOR STUDY

1. The great blue heron (*Ardea herodias*) is found over most of North America. The very similar gray heron (*Ardea cinerea*) ranges over most of Europe and Asia. They are currently treated as different species. A colleague argues that they should be combined into one species. What facts should you consider in evaluating your colleague's suggestion?

2. Describe the goals of the classification systems erected by cladists and pheneticists. Could these goals be combined into a single classification system?

3. How would you classify crocodiles if you were a cladist? A pheneticist? How do you personally think crocodiles should be classified? Why?

4. A student of the evolution of frogs has performed DNA hybridization experiments on about 25 percent of all frog species. As a result of these experiments she proposes a new classification of frogs that differs strikingly from the traditionally accepted one. Should frog taxonomists immediately accept this new classification? Why or why not?

5. Linnaeus developed his system of classification before Darwin proposed his theory of evolution by natural selection, and most classifications of organisms were developed by people who were not evolutionists. Yet most of those classifications are still used today, with minor modifications, by evolutionary biologists. Why has this happened?

READINGS

Eldredge, N. and J. Cracraft. 1980. *Phylogenetic Patterns and the Evolutionary Process*. Columbia University Press, New York. A good sampling of various perspectives on concepts and practices in systematics.

Ferguson, A. 1980. *Biochemical Systematics and Evolution*. Blackie & Son, Glasgow. A concise introduction to the use of biochemical techniques at all levels of taxonomy.

Hillis, D. M., and C. Moritz, Editors. 1990. *Molecular Systematics*. Sinauer Associates, Sunderland, MA. An introduction to the experimental techniques of molecular systematics, this book also demonstrates the importance of such work to the study of evolution.

Mayr, E. 1982. *The Growth of Biological Thought*. Belknap Press, Cambridge, MA. A good treatment of the history of thinking about systematics, biological diversity, and evolution.

Ridley, M. 1986. *Evolution and Classification: The Reform of Cladism*. Longmans, Harlow, Essex. A critical evaluation of current schools of thought in systematics that provides clear statements about the goals and methods of all approaches.

Sneath, P. H. A., and R. R. Sokal. 1973. *Numerical Taxonomy: The Principles and Practice of Numerical Classifications*. W. H. Freeman, San Francisco. The most complete treatment of quantitative methods in taxonomy and why such methods are important.

Wiley, E. O. 1981. *Phylogenetics: The Theory and Practice of Phylogenetic Systematics*. Wiley-Interscience, New York. A clear overview of the phylogenetic approach to evolution and classification.

PART FOUR
The Evolution of Diversity

21

Viruses and Monera

PREVIEW: The most abundant organisms on Earth are so small that we cannot see them with the naked eye. However, they play numerous critical roles in the biosphere, interacting in one way or another with every living thing. Some bacteria perform key steps in the cycling of nitrogen, sulfur, and carbon; others trap energy from the sun or from inorganic chemical sources. Certain viruses and bacteria are agents of disease in multicellular organisms.

This chapter deals with diversity among the viruses and in the kingdom Monera.

The microscopic members of the kingdom Monera are the most successful of all creatures on Earth—if success is measured by numbers of individuals. The bacteria in one person's mouth outnumber all the humans who have ever lived. Despite their minute sizes (Table 21.1), the monerans outdo all other groups in metabolic diversity, and they occupy more—and more extreme—habitats than any other group. Their effects on our environment are manifold and profound.

Viruses are even smaller than the monerans. Although small and structurally simple, they are not "primitive" or evolutionarily ancient. Rather, they are believed to have arisen from the plant, animal, and bacterial groups that they infect. We have already considered viruses in some detail (Chapters 4 and 12), but here we will discuss the diversity of viruses.

VIRUSES

Discovery of the Viruses

Viruses have become well understood only within the last half century, but the first step on this path of discovery was taken by the Russian botanist Dmitri Ivanovsky in 1892. Ivanovsky was studying tobacco mosaic disease and tried to isolate the causal agent of the disease by passing an infectious extract of diseased tobacco leaves through a fine porcelain filter, a technique that had been used previously to isolate disease-causing bacteria. To his surprise, the liquid that passed through the filter still caused tobacco mosaic disease. Other workers soon showed that similar filterable agents, so tiny that they cannot be seen under the light microscope, cause several plant and animal diseases. They discovered that alcohol, which kills cultures of bacteria, does not destroy the viruses' ability to cause disease.

Wendell Stanley, of what is now Rockefeller University, first succeeded in crystallizing viruses in 1935. The crystalline viral preparation became infectious again when it was dissolved. It was soon shown that crystallized viral preparations consist primarily of protein, with a significant amount of nucleic acid. Finally, direct observation of viruses with electron microscopes clearly showed how greatly they differ from bacteria and other organisms.

Viral Structure

Unlike the organisms making up the five taxonomic kingdoms of the living world, the viruses are **acellular**; that is, they are not cells and do not consist of cells. Unlike the cellular creatures, they do not metabolize energy—they neither produce ATP nor do they conduct fermentation, cellular respiration, or photosynthesis.

Whole viruses never arise directly from preexisting viruses. They are *obligate intracellular parasites*; that is, they develop and reproduce only within the cells of specific hosts. The cells of animals, plants, fungi, protists, and bacteria serve as hosts to viruses. Viruses outside host cells exist as individual particles called **virions**. The virion, the basic unit of a virus, consists of a central core of either DNA or RNA (but not both) surrounded by a **capsid**, or coat, which is composed of one or at most a few kinds of proteins (Figure 21.1). These proteins are so assembled as to give the virion a characteristic shape. Many animal viruses also acquire a membrane consisting of lipids

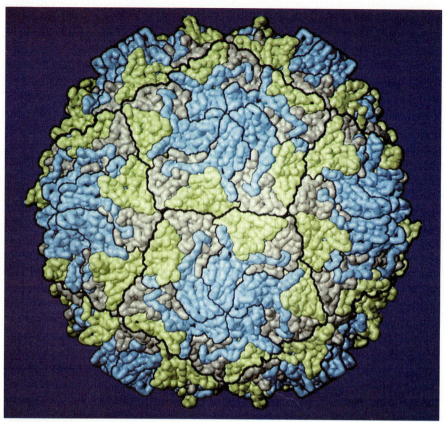

21.1   A Virion
The capsid of a poliovirus as drawn by computer. There are three major proteins in the capsid, each shown in a different color. The proteins are organized into building blocks, outlined by the wavy black lines.

and proteins as they bud through host cell membranes in the course of viral reproduction, and many bacterial viruses have specialized "tails" made of protein. The complex architecture of HIV-I, the AIDS virus, is shown in Figure 16.28.

TABLE 21.1 Common Sizes of Microorganisms		
MICROORGANISM	**TYPE**	**TYPICAL SIZE RANGE (CUBIC MICROMETERS)**
Protists	Eukaryote	5000–50,000
Photosynthetic bacteria	Prokaryote	5–50
Spirochetes	Prokaryote	0.1–2
Mycoplasmas	Prokaryote	0.01–0.1
Pox viruses	Virus	0.01
Influenza virus	Virus	0.0005
Polio virus	Virus	0.00001

Modified from R. Y. Stanier, E. A. Adelberg, and J. Ingraham. 1976. *The Microbial World,* 4th Edition.

Reproduction of Viruses

Viruses reproduce by taking over their host cell's metabolism; the viral nucleic acid directs the production of new viruses from host materials. Animal viruses begin the process by attaching to the plasma membrane of the host cell. They are then taken up by endocytosis (Chapter 5), which leaves them trapped within a membranous vesicle inside the cell. The host membrane breaks down, then the host cell digests the protein capsid. At this point, the viral nucleic acid takes charge. The host cell replicates the viral nucleic acid and synthesizes new capsid protein as directed by the viral nucleic acid. New capsids and new viral nucleic acid combine spontaneously, and, in due course, the host cell releases the new virions. Animal viruses usually escape from the host cell by budding through virus-modified areas of the plasma membrane (or sometimes the nuclear envelope). During this process the completed virions pick up a membrane somewhat similar to that of the host cell (Figure 21.2).

Plant viruses and bacterial viruses (bacteriophages) must get through a cell wall as well as through the host plasma membrane. Infection of a

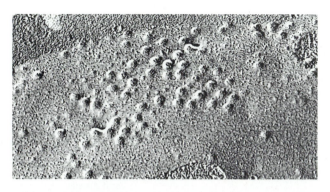

21.2 Buds of an Animal Virus
The numerous small bumps on the curved surface of a cell are buds of Sindbis virus in the plasma membrane. At this stage the capsids are acquiring a membrane envelope that completely surrounds each virion when the process is complete. These membrane envelopes make the first contact when these virions infect new host cells.

plant usually results from attack by a virion-laden insect vector—intermediate carriers of disease from one organism to another are called **vectors**. The insect uses its proboscis to penetrate the cell wall, and the virions then move from the insect into the plant. Plant viruses, such as tobacco mosaic virus, can be artificially introduced without the assistance of insect carriers if a leaf or other part is bruised mechanically before a suspension of virions is applied. Bacteriophages are often equipped with tail assemblies that inject their nucleic acid through the cell wall into the host bacterium. Virions escape from plant or bacterial cells by lysing the host cell, rather than by budding. Some bacteriophages have lytic life cycles (cycles that result in rapid lysis of host cells); others have lysogenic life cycles, in which the viral and host nucleic

acids replicate at the same time, and the virus may be present as a "silent" provirus for many bacterial cell generations. Bacteriophage life cycles were described in detail in Chapter 12.

How does a virion recognize a suitable host? Some bacteriophages use a specific interaction between the proteins of the bacteriophage tail and of the host cell wall. Membrane-surrounded animal viruses probably depend on the membrane to recognize suitable cells: because the membrane was obtained from the previous host, it can readily fuse with the plasma membrane of a new host cell. It is not known how other virions recognize their host cells.

Classification of Viruses

A common way to classify viruses separates them first by whether they have DNA or RNA and then by whether their nucleic acid is single- or double-stranded (Table 21.2). As noted in the table, some of the RNA viruses have more than one molecule of RNA, and the DNA of one virus family is circular. Further levels of classification depend upon such factors as the overall shape of the virus and the symmetry of the capsid. As illustrated in Figures 21.3 and 4.31, most capsids may be categorized as **helical** (being coiled like a spring), **icosahedral** (being a regular solid with 20 faces), or **binal** (having a polyhedral, or many-faced, head with a helical tail). Another level of classification is based on the presence or absence of a membranous envelope around the virion; still further subdivision relies on capsid size. In Table 21.2 we show only the major levels and some examples; technical names of the taxonomic groups are omitted.

The distribution of viruses in terms of host organ-

   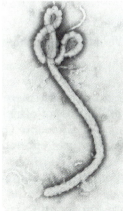

21.3 The Shapes of Viruses
Virions come in different shapes. (a) Tobacco mosaic virus consists of an inner helix of RNA covered with a helical array of protein molecules. This computer model corresponds to about one-seventh of the entire, long virus. (b) Adenoviruses have an icosahedral capsid as an outer shell. Inside this 20-faced structure is a spherical mass of other proteins and DNA. (c) These T2 bacteriophages illustrate the binal form of capsid. (d) Not all virions are regular in shape. Wormlike virions of Ebola virus infect humans, causing hemorrhages.

TABLE 21.2
A Classification Scheme for Some Animal Viruses

VIRUS GROUP	NUCLEIC ACID MOL. WT. (millions)	KIND	STRANDS	VIRION SHAPE	SIZE (nm)	NOTES
Families of viruses affecting both vertebrates and other hosts						
Poxviridae	160–200	DNA	2	Brick-shaped	300×240×100	Pox viruses
Parvoviridae	1.2–1.8	DNA	1	Icosahedral	20	Hosts include rats and insects
Reoviridae	15	RNA	2	Icosahedral	50–80	Vertebrate, insect and plant hosts
Rhabdoviridae	4	RNA	1	Bullet-shaped	175 × 70	Rabies, vesicular stomatitis
Families of viruses of vertebrates						
Herpetoviridae	100–200	DNA	2	Icosahedral	150	Herpes
Adenoviridae	20–29	DNA	2	Icosahedral	70–80	Adenovirus
Papovaviridae	3–5	DNA	2	Icosahedral	45–55	Papillomas
Retroviridae	10–12	RNA	1	Spherical	100–200	Tumor viruses
Paramyxoviridae	7	RNA	1	Spherical	100–300	Measles, Newcastle disease
Orthomyxoviridae	5	RNA	1	Spherical	80–120	Influenza
Togaviridae	4	RNA	1	Spherical	40–60	Rubella, hog cholera, arboviruses
Coronaviridae	?	RNA	1	Spherical	80–120	
Arenaviridae	3.5	RNA	1	Spherical	85–120	
Picornaviridae	2.6–2.8	RNA	1	Icosahedral	20–30	Digestive and respiratory diseases
Bunyaviridae	6	RNA	1		90–100	

isms is puzzling. Viral diseases of flowering plants are very common, but they are rare in the cone-bearing seed plants, ferns, algae, and fungi. Almost all vertebrates are susceptible to viral infection, but among invertebrates such infections are common only in arthropods. A group of viruses called **arboviruses** (short for arthropod-borne viruses) causes serious diseases such as encephalitis in humans and other mammals. They are transmitted to the mammalian host through a bite (certain arboviruses are carried by mosquitoes, for example). Though carried within the arthropod vector's cells, arboviruses apparently do not affect the insect host severely—just the bitten and infected mammal.

Viroids: RNA without a Capsid

Pure viral nucleic acids can produce viral infections under laboratory conditions, although only inefficiently. Might there be infectious agents in nature that consist of nucleic acid without a protein capsid? In 1971, Theodore Diener of the U. S. Department of Agriculture reported the isolation of agents of this type, called **viroids**. Viroids are single-stranded RNA molecules consisting of 270–380 nucleotides. They are one-thousandth the size of the smallest viruses. All the viroids thus far studied have substantial regions of internal complementarity, so that they fold into double-stranded rods. These rods are most abundant in the nuclei of infected cells.

Viroids have been found only in plant cells. There they produce a variety of diseases. Viroids are transmitted from plant to plant by one of two mechanisms. If two plants (one infected and one healthy) are mechanically injured and their wounded surfaces come into contact, viroids may be transmitted from the infected plant to the healthy one. The other mechanism of transfer operates from generation to generation—if a pollen grain or an ovule produced by an infected plant contains viroids, these will infect the plant produced by subsequent fertilization.

There is no evidence that viroids are translated to synthesize proteins, and it is not known how they cause disease. Viroids are replicated by the enzymes of their plant hosts. Similarities in base sequences between viroids and transposable genetic elements strongly suggest that viroids evolved from transposable elements (which were discussed in Chapters 12 and 13).

Scrapie-Associated Fibrils: Infectious Proteins?

A class of protein fibrils, called **scrapie-associated fibrils**, or prions, has been associated with certain degenerative diseases of mammalian central nervous systems. These fibrils consist entirely of protein, with no evident nucleic acid component. The fibrils are associated with scrapie, a disease of sheep and goats, and may be the infective agent of the disease. Such fibrils have also been identified in connection with two similar diseases of the human central nervous system, kuru and Creutzfeld–Jakob disease. One investigator of scrapie-associated fibrils, Stanley Prusiner of the University of California, San Francisco Medical Center, has suggested a possible relationship between the fibrils and Alzheimer's disease, a severe dementia commonest in the elderly.

The mechanism of action of scrapie-associated fibrils is unknown. Genes coding for the known parts of their amino acid sequences exist in the chromosomes of both infected and healthy animals. It is possible that these genes are, in fact, proviruses (Chapter 12); further research will surely produce a detailed picture of the relationship between these genes and scrapie-associated fibrils.

Now we will stop our "descent" to simpler and simpler nonliving infectious agents. The remainder of this chapter will be devoted to the properties and diversity of the most ancient *living* things, the members of the kingdom Monera.

GENERAL BIOLOGY OF THE MONERA

Bacteria and Disease

The late nineteenth century was one of the most productive eras in the history of medicine, a time during which bacteriologists, chemists, and physicians established the fact that many diseases are caused by microbial agents. The German physician Robert Koch laid down a set of rules for testing the relationship between a disease and a microorganism. According to Koch, the disease in question could be laid to a particular microorganism if (1) the microorganism could always be found in diseased individuals, (2) the microorganism taken from the host could be grown in pure culture, (3) a sample of the culture produced the disease when injected into a new, healthy host, and (4) the newly infected host yielded a new, pure culture of microorganisms identical to that obtained in step 2. These rules—called **Koch's postulates**—are still important among modern procedures for the investigation of new diseases.

In Chapter 16 we considered the immune system and other modes of protection against diseases of microbial origin. We also examined, briefly, the problems faced by a pathogenic bacterium in establishing itself in a host. The consequences of an infection for the host depend on a number of factors. One is the **invasiveness** of the pathogen: its ability to multiply within the body of the host. Another is its **toxigenicity**: its ability to produce chemical substances injurious to the tissues of the host. Often bacterial toxins can kill the cells of a host that has not previously been exposed to them. *Corynebacterium diphtheriae*, the agent of diphtheria, has low invasiveness and multiplies only in the throat, but its toxigenicity is so great that virtually the entire body is affected. In contrast, *Bacillus anthracis*, which causes anthrax (a disease primarily of cattle and sheep), has low toxigenicity but an invasiveness so great that the entire bloodstream ultimately teems with the bacteria.

Keep in mind that in spite of our frequent mention of human pathogens, only a small minority of the known bacterial species are agents of disease. Far more species play positive roles in our lives and in the biosphere—participating in the digestive processes of animals (Chapter 41), in the processing of nitrogen and sulfur in soils (Chapter 30), as decomposers in all ecosystems (Chapter 47), and as key participants in many of our own industrial and agricultural processes.

The Kingdom Monera

The kingdom Monera, which includes all organisms whose cells are prokaryotic, has the most ancient origins of any group still present today. Its earliest fossil representatives date back at least 3.5 billion years, and, as noted in Chapter 17, these ancient traces indicate that there was considerable diversity among the prokaryotes even during the Archean Eon. The prokaryotes reigned supreme on an otherwise sterile earth for more than 2 billion years, adapting to new environments and to changes in existing ones. The Monera have spread to every conceivable habitat on the planet, including the insides of other organisms. By any standard, they must be judged enormously successful creatures.

The Monera of today are astoundingly diverse (Figure 21.4), and it is likely that they represent the current products of many independent evolutionary lines that have been separate for hundreds of millions of years. That is, while having a common prokaryotic heritage, they have been following their separate evolutionary paths for most of the history of life on the planet. In addition, each of these evolutionary lines has spread over the surface of Earth and adapted to many or most of the possible environmental challenges, so that diversity is great within each of the lines, even as compared with the diversity within the other kingdoms. The result has been a taxonomist's nightmare.

There are at least three possible motivations for setting up schemes of classification. One, favored by many biologists, is to group organisms by evolutionary affinity, providing where possible a natural classification. A second one, followed by the great classic *Bergey's Manual of Systematic Bacteriology*, is to group species in such a way as to facilitate the identification

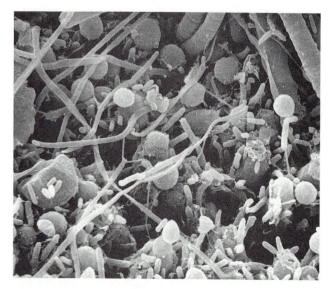

21.4 Bacterial Diversity
These bacteria—large, small, long, short, spherical, spiral —all grew on the surface of a granite rock placed for 42 days in a Canadian river.

of unknown organisms. Yet another is to set up groupings that display the diversity of the living world, without in the long run being excessively concerned about incorporating all the newest evidence.

We follow *Bergey's Manual* for the most part in this chapter. We will divide the kingdom Monera into four **divisions**, based on the nature of the cell wall. (You will recall from Chapter 20 that bacteriologists use the term "division" rather than "phylum.") The four divisions are the **gram-negative bacteria**, the **gram-positive bacteria**, the **mycoplasmas**—bacteria lacking cell walls—and the **Archaebacteria**, a group whose cell walls are chemically unrelated to those of the gram-negative and gram-positive bacteria (Table 21.3). The cell walls of the gram-negative and the gram-positive bacteria will be discussed later in this chapter.

Carl Woese, of the University of Illinois, defines two separate *kingdoms* of bacteria, the Archaebacteria and the Eubacteria. As we will see, the Archaebacteria differ strongly from all other bacteria, and it is possible that Woese's definition will gain favor. Two of the competitors to the currently popular five-kingdom scheme of living things are a *three*-kingdom scheme—Archaebacteria, Eubacteria, and Eukaryotes—favored by Woese and a *seven*-kingdom classification scheme—Archaebacteria, Eubacteria, Protista, Slime Molds, Fungi, Plantae, and Animalia—favored by some other biologists. A *thirteen*-kingdom scheme was proposed in 1989. We will adhere to the five-kingdom scheme, in which Archaebacteria and three other bacterial divisions are subsumed within a single kingdom Monera that includes all

prokaryotes. In this chapter we will describe several kinds of bacteria. Many aspects of the genetics, growth, and biochemistry of the Monera have been considered elsewhere in the book, as part of our unified study of biological principles.

Prokaryotes versus Eukaryotes

The architectures of prokaryotic and eukaryotic cells were compared in Chapter 4, and you may wish to review Figures 4.3–4.7 as well as Box A in that chapter. The basic unit of the Monera is the prokaryotic cell, which contains a full complement of genetic and protein-synthesizing systems, including DNA, RNA, and all the enzymes needed to transcribe and translate the genetic information into protein (Chapter 11). The cell also contains at least one system for generating the ATP it needs (Chapter 7).

The prokaryotic cell differs from the eukaryotic cell in three important ways, as discussed in Chapter 4. First, the organization and replication of the genetic material differs. The DNA of the prokaryotic cell is not organized within a membrane-bounded nucleus, and it is not complexed with histones to form chromatin, as in eukaryotes (Chapter 9). DNA molecules in prokaryotes are circular (Chapter 12). The elaborate mechanism of mitosis (Chapter 9) is missing; prokaryotic cells divide by their own elaborate method, fission, after replicating their DNA. Second,

---

TABLE 21.3
Selected Groups within the Kingdom Monera

GROUP	REPRESENTATIVE GENERA
Archaebacteria	
Thermoacidophiles	*Sulfolobus*
Methanogens	*Methanopyrus*
Strict halophiles	*Halobacterium*
Gram-negative bacteria	
Gliding bacteria	*Beggiatoa, Thiothrix*
Spirochetes	*Borrelia, Treponema*
Curved and spiral bacteria	*Bdellovibrio, Spirillum*
Gram-negative rods	*Escherichia, Rhizobium, Salmonella*
Gram-negative cocci	*Neisseria, Nitrosococcus*
Rickettsias and chlamydias	*Chlamydia, Rickettsia*
Cyanobacteria	*Anabaena, Anacystis, Oscillatoria*
Gram-positive bacteria	
Gram-positive rods	*Bacillus, Clostridium*
Gram-positive cocci	*Staphylococcus, Streptococcus*
Actinomycetes	*Actinomyces, Corynebacterium, Streptomyces*
Mycoplasmas	*Mycoplasma*

prokaryotes have none of the membrane-bounded cytoplasmic organelles that eukaryotes have—mitochondria, chloroplasts, Golgi apparatus, endoplasmic reticulum—but the cytoplasm of a prokaryotic cell may contain a variety of mesosomes and photosynthetic membrane systems not found in eukaryotes. Third, the cell walls of most prokaryotes (but not the Archaebacteria) contain peptidoglycan, a substance whose composition is unique to the kingdom Monera (Figure 4.3).

The absence of characteristic eukaryotic organelles should not be construed as a total absence of internal membranes and other internal structures from the prokaryotic cell. Membranous mesosomes frequently associate with new cell walls during cell division, and a bacterial cell's DNA is often seen in electron micrographs to be attached to a mesosome (Figure 21.5). Many aerobic bacteria have respiratory enzymes that are bound to elaborate internal membrane systems, and photosynthetic bacteria have extensive and highly organized internal membranes laden with photosynthetic pigment systems (see Figure 4.4 and Figure 21.18).

Though many prokaryotes are not motile, some can move by means of flagella—whiplike filaments that extend singly or in tufts from one or both ends of the cell, or all around it (Figure 21.6). A bacterial flagellum consists of a single fibril made of the protein flagellin, in contrast to the flagellum of eukaryotes, which usually contains a circle of nine pairs of microtubules surrounding two central microtubules, all made of the protein tubulin (Chapter 4). The bacterial flagellum (Figure 4.6) rotates about its base, rather than beating like a eukaryotic flagellum or cilium.

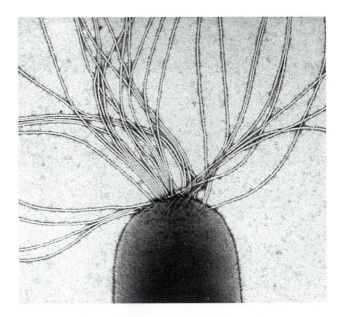

21.6 Bacterial Flagella
A tuft of flagella at one end of a *Spirillum graniferum*; the other end bears a similar tuft.

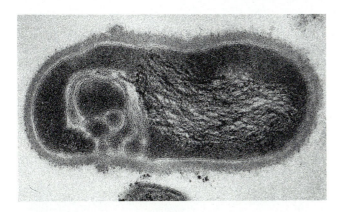

21.5 Membranes in Bacteria
A large mesosome is continuous with the plasma membrane on the left in this cell of *Corynebacterium parvum*. A large, fibrous mass of DNA, attached to the mesosome, fills most of the remainder of the cell. The mesosome is not a separate, membrane-bounded compartment like a eukaryotic organelle.

Metabolic Diversity in the Kingdom Monera

Organisms capable of being either anaerobic or aerobic are called **facultative anaerobes**. Although many types of prokaryotes can obtain energy either by fermentation or by cellular respiration (Chapter 7) and are therefore facultative anaerobes, others can live only by fermentation and are, in fact, poisoned by oxygen gas. These oxygen-sensitive fermenters are called **obligate anaerobes**. Some facultative anaerobes can conduct only fermentation but are not damaged by oxygen when it is present. At the other extreme, some Monera are **obligate aerobes**, unable to survive for extended periods in the absence of oxygen.

Some of the Monera carry out respiratory electron transport (Chapter 7) without using oxygen as an electron acceptor. These forms use oxidized inorganic ions such as nitrate, nitrite, or sulfate as electron acceptors. Among these organisms are the denitrifiers, bacteria that return nitrogen to the atmosphere, completing the cycle of nitrogen in nature (Chapter 30).

Biologists recognize four broad nutritional categories in the kingdom Monera. Organisms of the first category, the **photoautotrophs**, are photosynthetic bacteria. Photoautotrophs use light as their source of energy and carbon dioxide as their source of carbon. One group, the cyanobacteria, like the photosynthetic eukaryotes, performs photosynthesis with chlorophyll *a* as the key pigment and produces oxygen as a by-product of noncyclic photophosphorylation (Chapter 8).

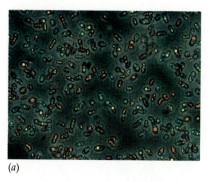

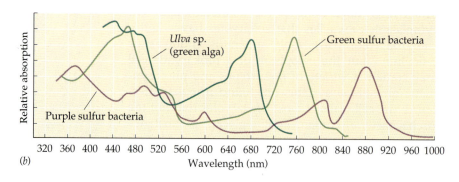

(a)

(b)

21.7 Photosynthetic Bacteria
(a) The light spots inside some of these cells of *Chromatium*, a purple sulfur bacterium, are globules of elemental sulfur that were produced when hydrogen sulfide (H_2S) was used as an electron donor for photosynthesis. (b) Notice that the alga does not absorb any wavelengths longer than 750 nm. Consequently, the bacteria can conduct photosynthesis beneath the alga, using the longer wavelengths.

The other photosynthetic bacteria, in contrast, use bacteriochlorophyll as their key photosynthetic pigment, and they do not release oxygen gas. They produce particles of pure sulfur instead because they use hydrogen sulfide (H_2S) rather than H_2O as an electron donor for photophosphorylation (Figure 21.7a). Bacteriochlorophyll absorbs light of longer wavelength than the chlorophyll used by all other photosynthesizing organisms does. As a result, bacteria using this pigment can grow in water beneath fairly dense layers of algae, because light of the wavelengths they can use is not appreciably absorbed by the algae (Figure 21.7b).

The second nutritional category is the **photoheterotrophs**. These bacteria use light as their source of energy but must obtain their carbon atoms from organic compounds made by other organisms. The photoheterotrophs are the purple **nonsulfur** bacteria. They use such compounds as carbohydrates, fatty acids, and alcohols as their organic "food."

Chemoautotrophs, organisms of the third category, obtain their energy by oxidizing inorganic substances, and they use some of that energy to fix carbon dioxide in reactions analogous to those of the photosynthetic carbon reduction cycle. The chemoautotrophs include the nitrifiers (Chapter 30), which oxidize ammonia or nitrite ions to form nitrate ions that are taken up by plants, as well as other bacteria that oxidize hydrogen gas, hydrogen sulfide, sulfur, and other materials. Scientists exploring the ocean bottom near the Galapagos Islands in 1977 discovered a spectacular example of chemoautotrophy. They found an entire ecosystem based on chemoautotrophic bacteria (Figure 21.8) that are eaten by a large community of crabs, mollusks, and giant worms (pogonophorans, Chapter 26)—all at a depth of 2,500 meters, far below any hint of light from the

sun but in the immediate neighborhood of volcanic vents in the ocean floor (see Figure 17.5). The bacteria obtain energy by oxidizing hydrogen sulfide released from the vents.

Finally, there are the **chemoheterotrophs**, which typically obtain both energy and carbon atoms from one or more organic compounds. The great majority of bacteria are chemoheterotrophs—as are all animals, fungi, and many protists.

Nitrogen fixation, in which atmospheric nitrogen gas is converted into chemical forms usable by bacteria and other living things, is another important metabolic adaptation. This vital process is carried out by a wide variety of prokaryotes, including cyanobacteria, actinomycetes (Table 21.3), some other phototrophic bacteria, and numerous others. The only nitrogen fixers in nature are members of the kingdom Monera.

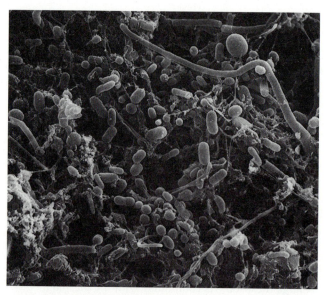

21.8 Chemoautotrophic Bacteria
All of these bacteria are chemoautotrophs that live near a hot water vent along the Galapagos Rift in the eastern Pacific, where their fixing of carbon supports an entire community of organisms that thrives in total darkness.

Structural Characteristics of Bacteria and Bacterial Associations

Most bacteria have a thick and relatively stiff cell wall. If the wall contains a great deal of peptidoglycan, it takes on a blue to purple color when treated with the Gram stain (Box 21.A); bacteria with such walls are called **gram-positive**. **Gram-negative** bacteria contain less peptidoglycan and appear pink to red following Gram staining. This difference is useful in classifying bacteria, but it should be noted that a variety of cell wall structures are found among the Monera. Some bacteria, the mycoplasmas, have no cell wall at all; and the Archaebacteria produce walls, but their walls contain no peptidoglycan.

Three shapes are particularly common among the bacteria, as shown in Figure 4.7 and in various figures in this chapter. These are spheres, rods, and curved or spiral forms. A spherical bacterium is a **coccus** (plural: cocci); a rod-shaped one is a **bacillus** (plural: bacilli). Cocci may live singly or may associate in two- or three-dimensional arrays as chains, plates, or blocks of cells. Bacilli and spiral forms may be single or may form chains. Note that associations such as chains do not signify multicellularity, because each cell is fully viable and independent. Most bacteria reproduce by the fission of one cell into two. Associations arise as cells adhere to one another after fission.

Some bacteria have other structural features. Some

BOX 21.A

The Gram Stain and Bacterial Cell Walls

In 1884 Christian Gram, a Danish physician, developed an uncomplicated staining process that has lasted into our high-technology era as the single most common tool in the study of bacteria. The **Gram stain** separates most kinds of bacteria into two distinct groups: gram-positive and gram-negative. A smear of bacteria on a microscope slide is soaked in a violet dye and treated with iodine; it is then washed with alcohol and counterstained with safranine. Gram-positive bacteria retain the violet dye and appear blue to purple, as shown in the illustration. The alcohol washes the violet stain off gram-negative bacteria; these bacteria then pick up the safranine counterstain and appear pink to red, as shown below. Gram-staining characteristics are a crucial consideration in grouping some kinds of bacteria and are important in determining the identity of bacteria in an unknown sample.

It is likely that the different staining reactions relate to differences in the amount and accessibility of peptidoglycan in the cell walls of bacteria. The electron micrographs and associated sketches show a thick layer of peptidoglycan exterior to the plasma membrane on gram-positive cells (top); the gram-negative cell wall typically has only one-fifth as much peptidoglycan, and outside the peptidoglycan layer the cell is surrounded by a second, outer membrane quite distinct in chemical makeup from the plasma membrane (bottom). The consequences of these different features are endless, and relate to the disease-causing characteristics of some bacteria. Indeed, the bacterial cell wall is a favorite target in medical combat against diseases that are caused by bacteria because it has no counterpart in eukaryotic cells. Antibiotics and other agents that specifically destroy bacterial cell walls tend to have little, if any, effect on the cells of humans and other eukaryotes.

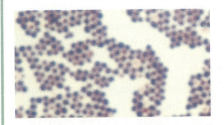

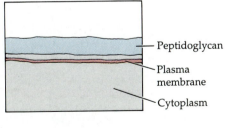

Peptidoglycan
Plasma membrane
Cytoplasm

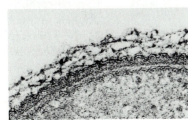

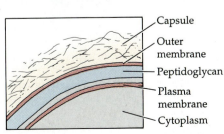

Capsule
Outer membrane
Peptidoglycan
Plasma membrane
Cytoplasm

attach to their substrate by stalks that may be an extension of the cell wall or an extracellularly secreted product. Some bacteria associate in chains that become enclosed within delicate tubular sheaths. These associations are called filaments. All the cells of a filament divide simultaneously.

Reproduction and Resting

Most bacteria reproduce by fission, an asexual, or **vegetative**, process. But as you will recall, there are also sexual processes—transformation, conjugation, and transduction—that allow the exchange of genetic information between some bacteria (Chapters 11 and 12).

Some other bacteria produce **endospores** (Figure 21.9). These spores are *resting* structures, not reproductive ones. Harsh environmental conditions—such as extreme heat or cold, or lack of water—cause endospore production. A cell replicates its DNA and encapsulates one copy, along with some of its cytoplasm, in a tough cell wall. The parent cell then breaks down, releasing the endospore. This is not a reproductive process; the endospore merely replaces the parent cell. The endospore can survive the harsh environmental conditions. Later, if it encounters fa-

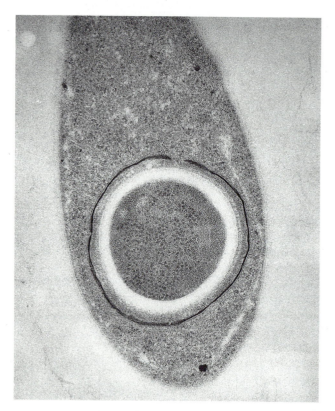

21.9   A Bacterial Endospore
Clostridium tetani, the causal agent of tetanus, produces endospores as resistant resting structures. This endospore—the round, thick-walled structure—lies near one end of the cell.

vorable conditions, it germinates—it becomes metabolically active and divides, forming new cells like the original parent. Endospores have germinated after more than a thousand years of dormancy.

MAJOR GROUPS OF THE BACTERIA

Archaebacteria

The division Archaebacteria consists of a few prokaryotic genera that live in habitats notable for such characteristics as extreme salinity, low oxygen concentration, high temperature, or high or low pH (Figure 21.10a). On the face of it, the Archaebacteria seem to be an ill-assorted group. However, they share a number of characteristics, such as the definitive lack of peptidoglycan in their walls and the possession of lipids of distinctive composition. Also, the base sequences of their ribosomal RNAs show great similarities. By the criterion of rRNA base sequence, the Archaebacteria differ as thoroughly from the other bacteria as either group does from the eukaryotic kingdoms. Biologists had classified the members of the Archaebacteria in widely separated groups of bacteria until the late 1970s, when Woese and others, using techniques of molecular biology, recognized their close relationship and ancient origin and grouped them together as the Archaebacteria (from a Greek root meaning "ancient"). The division can be clearly divided into three major subgroups: the thermoacidophiles, the methanogens, and the strict halophiles.

THERMOACIDOPHILES. Some of the Archaebacteria are both thermophilic (heat-loving) and acidophilic (acidloving). *Sulfolobus* is a typical genus of such thermoacidophiles. Bacteria of this genus live in hot sulfur springs at temperatures of 70–75°C. They die of "cold" at 55°C (131°F). Hot sulfur springs are also extremely acidic. *Sulfolobus* grows best in the range from pH 2 to pH 3, but it readily tolerates pH values as low as 0.9. Those thermoacidophiles that have been tested maintain an internal pH near 7 in spite of the acidity of their environment. Thermoacidophiles thrive where very few other organisms can even survive.

METHANOGENS. Ten species of prokaryotes, previously assigned to unrelated bacterial groups, share the property of producing methane (CH_4) by the reduction of carbon dioxide. All these methanogens are obligate anaerobes, and they use methane production as the key step in their energy metabolism. Woese established, by comparing rRNA base sequences, that all methanogens are closely related to one another. Methanogens release approximately 2 billion tons of methane gas into Earth's atmosphere each year, accounting for all the methane in our air, in-

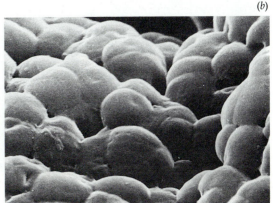

21.10 Archaebacteria
(a) Masses of archaebacteria form a pink mat around a hot spring (upper right) at Yellowstone National Park. (b) Other archaebacteria, such as these methane-producing bacteria, live anaerobically in the digestive tracts of animals.

cluding that associated with mammalian flatulence. Approximately one-third of the methane production comes from methanogens in the guts of grazing herbivores such as cows (Figure 21.10b; Chapter 41). Another methanogen, *Methanopyrus*, lives on the ocean bottom near blazing volcanic vents. *Methanopyrus* survives and grows at 110°C—it is the current record-holder for temperature tolerance. It grows best at 98°C and does not grow at all at temperatures below 84°C.

STRICT HALOPHILES. The strict halophiles live exclusively in extremely salty environments. They contain pink carotenoids (Chapter 3) and are thus easily seen under some circumstances. They grow in the Dead Sea and also in brines of all types—pickled fish may sometimes show reddish-pink spots that are colonies of halophilic bacteria. Photographs of salt flats taken from orbiting satellites show a distinct pink tinge resulting from the presence of vast numbers of *Halobacterium* and its relatives. Few other organisms can live in the saltiest of the homes of the strict halophiles—other organisms would "dry" to death, losing too much water by osmosis to the hypertonic environment (Chapter 5). Strict halophiles have been found in lakes with pH values as high as 11.5—the most alkaline environment used by living organisms, and almost as alkaline as household ammonia.

Gram-Negative Bacteria

The great majority of bacteria are gram-negative—this division takes up more than three-quarters of *Bergey's Manual*. The members of this division are highly diverse in form and metabolism. We will describe only a few of the groups of gram-negative bacteria.

GLIDING BACTERIA. The gliding bacteria are rods that "glide" from place to place. Members of the genus *Beggiatoa* (Figure 21.11a) are an example of gliding bacteria. The physical basis of the gliding movement is not yet known. The cells may use a secreted slime in their locomotion—they leave slimy trails behind them as they move over soil or dead organic matter.

Gliding bacteria form remarkable structures called fruiting bodies. A group of cells aggregates to make one of these structures. The fruiting bodies of some species are simple globes more than 1 mm in diameter; those of other species are somewhat more complex, branched structures (Figure 21.11b). Within the fruiting bodies of some species, single cells transform themselves into thick-walled spores that can resist harsh environmental conditions; within those of other species, whole clusters of cells form a cyst that is resistant to drying. Both spores and cysts can germinate under favorable conditions to yield the typical gliding cells.

SPIROCHETES. The spirochetes are characterized by unique structures called **axial filaments**, composed of flagella, running along the cell body between a thin, flexible cell wall and an outer envelope (Figure 21.12a). The cell body is a long cylinder coiled spirally. The flagella constituting the axial filaments begin at either end of the cell and overlap in the middle. The axial filaments are thought to be responsible for the motility of these organisms, and there are typical basal rings where the flagella are attached to the cell wall, but the precise mechanism by which they work is unknown.

Many spirochetes are parasites in humans, including *Treponema pallidum*, the organism that causes the venereal disease syphilis (Figure 21.12b). Others live free in mud or water.

(a)

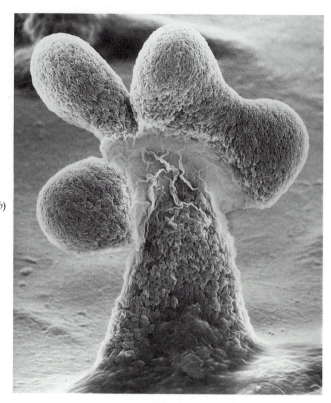

(b)

21.11 Gliding Bacteria
(a) This filament of *Beggiatoa* was isolated from ocean mud. (b) Individual bacteria (*Stigmatella aurantiaca*) aggregated to make up the stalk and knobs of this fruiting body. Within the knobs some bacteria have become individually surrounded by a hard layer of slime that provides resistance to heat, dehydration, and ultraviolet radiation. These cells—the cysts—are destined to become the next generation. When conditions are right, each cyst will germinate to become a gliding cell of this species.

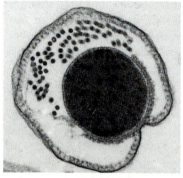

(a)

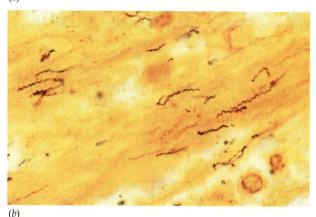

(b)

21.12 Spirochetes
(a) A spirochete from the gut of a termite, seen in cross section. The axial filaments are seen as dots inside the outer envelope but outside the dense cell body. (b) Corkscrew-shaped cells of *Treponema pallidum* in fetal tissue. *T. pallidum*, transmitted by sexual contact—and, as here, from mother to fetus—causes syphillis.

CURVED AND SPIRAL BACTERIA. The curved and spiral bacteria are diverse in properties, sharing little more than a curved form and a gram-negative reaction (Box 21.A). The spiral bacteria in this group differ from the spirochetes in not having an axial filament. Some, notably members of the genus *Spirillum*, live free in fresh or salt water, whereas others are parasitic. Some species cause diseases in animals. Several strains of *Campylobacter fetus* are increasingly recognized as causes of intestinal inflammation and other damage to humans (Figure 21.13). *Bdellovibrio* is a particularly interesting genus in this group. Some species of *Bdellovibrio* penetrate and reproduce within various other bacteria—one bacterium parasitizing another.

GRAM-NEGATIVE RODS. The gram-negative rods, as a group, demonstrate both the advantage and weakness of the classification scheme employed here. The advantage is that such groups are readily recognized in the laboratory by their shape and staining properties. The disadvantage is that the group is so large and diverse that it can scarcely be a "natural" group with close evolutionary ties among its members. Some gram-negative rods are aerobic, others are facultative anaerobes, and still others are obligate anaerobes. Nitrogen-fixing genera such as *Rhizobium* (Figure 30.11) are included, as are *Nitrobacter*, *Thiobacterium*, and other anaerobes that use nitrogen or sulfur compounds instead of oxygen for respiration.

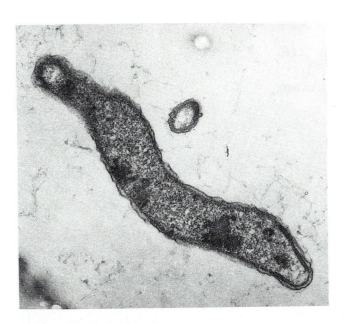

21.13 A Spiral Bacterium
Various strains of the spiral *Campylobacter fetus* cause abortions in sheep and cattle and intestinal inflammation in humans.

Escherichia coli, probably the most studied organism of all, is a gram-negative rod (see Box A in Chapter 4). So, too, are many of the most famous human pathogens, such as *Yersinia pestis* (the cause of plague), *Shigella dysenteriae* (dysentery), *Vibrio cholerae* (cholera), and *Salmonella typhimurium* (a common agent of food poisoning in humans). A bacterium from this group is shown in Figure 21.14.

Certain gram-negative rods invade animal cells, where they survive and cause diseases. For example, *Yersinia pseudotuberculosis*, the agent of guinea-pig plague, invades intestinal cells of guinea pigs and other mammals. Its ability to invade mammalian cells results from the possession of a single gene, called *inv*, that codes for a single large protein. Biologists using recombinant DNA techniques have successfully transferred this gene from *Y. pseudotuberculosis* to *E. coli*, with the result that the recipient *E. coli* cells were able to invade mammalian cells.

Most plant diseases are caused by fungi, and viruses cause others. However, about 200 plant diseases are of bacterial origin. **Crown gall**, with its characteristic tumors (Figure 21.15), is one of the most striking. The causal agent of crown gall is *Agrobacterium tumefaciens*, a gram-negative rod. *A. tumefaciens* harbors a plasmid containing the genes responsible for the crown gall disease. The plasmid is used in recombinant DNA studies as a vehicle for inserting genes into new plant hosts (Chapter 14).

GRAM-NEGATIVE COCCI. Cocci are spherical bacterial cells. There are both gram-negative and gram-positive cocci, providing a structural basis for grouping species. Some of the gram-negative cocci use oxides of nitrogen or sulfur as terminal electron acceptors for cellular respiration; among them are the nitrifiers (Chapter 30). Because of the human infections it causes, *Neisseria gonorrhoeae* (Figure 21.16), the cause of the common venereal disease gonorrhea, gets the most publicity. *N. gonorrhoeae* changes its antigenic determinants frequently by means of transposable elements, making it an elusive target for the human

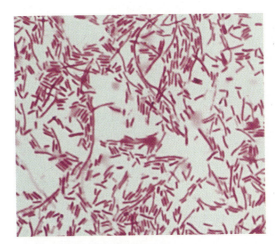

21.14 Gram-Negative Rods
Salmonella typhi, the causal agent of typhoid fever, is a gram-negative rod. The pink color is the "negative" response to the Gram stain.

21.15 Crown Gall
This massive growth on a white oak trunk is crown gall, a plant disease caused by *Agrobacterium tumefaciens*.

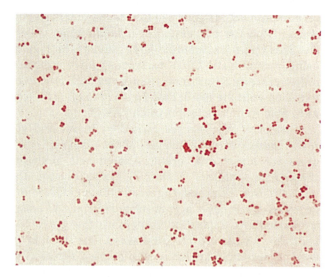

21.16 Gram-Negative Cocci
Neisseria gonorrhoeae, which causes gonorrhea, often aggregates in groups of two to four cells.

immune system. The mechanism is similar to the cassette mechanism used by yeasts to change their mating type (Box 13.A). *N. meningitidis* causes a frequently fatal infection of the linings of the nervous system. Other genera of gram-negative cocci, such as *Acinetobacter* and *Moraxella*, also cause infections.

RICKETTSIAS AND CHLAMYDIAS. The rickettsias and related organisms were once grouped with viruses because of their size and because they reproduce only within the cells of other organisms. The rickettsias,

extremely small intracellular parasites approximately 1 μm in length and 0.3 μm in diameter, have cell walls and chemical characteristics similar to those of some gram-negative bacteria. With one exception, rickettsias have never been grown outside living cells. Rickettsias are agents of several serious diseases in humans, notably Rocky Mountain spotted fever (which is actually more common in the southeastern United States than in the Rocky Mountains) and typhus (Figure 21.17a). Rickettsias are frequently carried by arthropods, particularly fleas and ticks, but they do not seem to cause any disease symptoms in their arthropod hosts.

Chlamydias are often lumped with the rickettsias because of their size (0.2–1.5 μm in diameter) and because they too are obligate intracellular parasites. These tiny spheres are unique prokaryotes because of their complex reproductive cycle, in which two different types of cells are seen (Figure 21.17b). In humans, various strains of chlamydias cause eye infections (especially trachoma), venereal disease, and some forms of pneumonia.

CYANOBACTERIA. The cyanobacteria (blue-green bacteria) are very independent nutritionally. They perform photosynthesis by using chlorophyll *a* and liberating oxygen gas, they carry out fermentation under anaerobic conditions, and many fix nitrogen. They require only water, nitrogen gas, oxygen, a few mineral elements, light, and carbon dioxide. Despite their ability to do the kind of photosynthesis otherwise characteristic only of eukaryotic photosynthesizers, they are true prokaryotes. They contain none of the

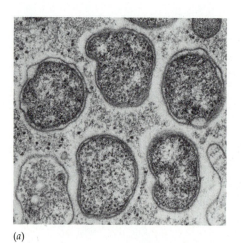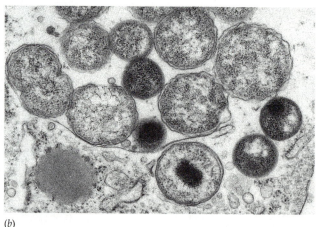

(a) *(b)*

21.17 Rickettsias and Chlamydias
(a) A cluster of tiny rickettsias in the cytoplasm of an infected animal cell. The species shown here is *Rickettsia prowazekii*, the cause of typhus; it is transmitted from person to person by lice. *(b)* Like rickettsias, chlamydias are small, intracellular parasites. This electron micrograph shows both the small, dense elementary bodies and the larger, thin-walled initial bodies that are the two major

phases of the life cycle of *Chlamydia psittaci*. The elementary bodies are taken into a cell by phagocytosis and then develop into initial bodies; the initial bodies, which cannot infect other cells, grow and divide (note the dividing initial body at the left). Finally, initial bodies reorganize into elementary bodies, which are liberated by rupture of the host cell.

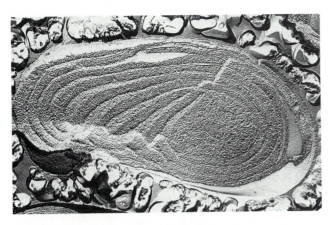

21.18 Thylakoids in Cyanobacteria
A cell of a cyanobacterium prepared by the freeze-etch method (Chapter 5) to emphasize the extensive system of internal membranes. These photosynthetic membranes are seen through most of the cytoplasm.

familiar membrane-bounded organelles of eukaryotic cells, they do not have a discrete nucleus, their chromosomes lack histones, and their cell walls contain peptidoglycan. Even so, the cyanobacteria contain elaborate and highly organized internal membrane systems, the photosynthetic lamellae, or thylakoids

(Figure 21.18). They are also the only prokaryotic photosynthesizers that contain chlorophyll *a*, with the exception of one other cyanobacteriumlike genus, *Prochloron*.

The cyanobacteria form a closely related, homogeneous, logical grouping. They are placed in a group by themselves because of their mode of photosynthesis; they also are distinguishable on the basis of rRNA sequence. However, lest we forget the complexity of prokaryote taxonomy, it should be mentioned that a few leading microbiologists classify the cyanobacteria as a subgroup of the gliding bacteria because mobile cyanobacteria use the gliding type of locomotion.

Cyanobacteria can associate in colonies or can live free as single cells (Figure 21.19). Depending on the species and on growth conditions, colonies of cyanobacteria may range from flat sheets one cell thick to spherical balls of cells. Some filamentous colonies show differentiation into at least three different cell types: vegetative cells, spores, and **heterocysts** (Figure 21.19*b*). Heterocysts are important structures for two reasons. First, when filaments break apart in order to reproduce, the heterocyst may serve as a breaking point. More important, heterocysts contain an enzyme called nitrogenase and thus are special-

(*a*)

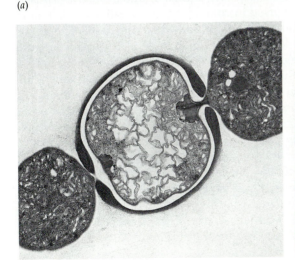

(*b*)

(*c*)

21.19 Cyanobacteria
(*a*) *Anabaena* is a genus of filamentous cyanobacteria. The filament shown here has a resting spore (the elongated, thick cell), and a heterocyst (the thick-walled, circular cell near the other end). (*b*) The spherical cell in the center of this electron micrograph—attached to each of the other cells by a thin neck—is a heterocyst. A thick wall sepa-

rates the cytoplasm of the nitrogen-fixing heterocyst from the surrounding environment. (*c*) Cyanobacteria appear in enormous numbers in some environments. Here we see a tidal pool on an atoll in the Indian Ocean; it is low tide and the green mat is made up of various unicellular cyanobacteria.

ized for nitrogen fixation. All the known cyanobacteria with heterocysts fix nitrogen.

Although a form of sexual reproduction is found in some bacteria, it is absent in the cyanobacteria. The cyanobacteria reproduce by fission. There are viruses that can infect cyanobacteria and then transfer genetic material from one organism to another by transduction (Chapter 12), but true sexuality has never been observed.

Gram-Positive Bacteria

There are far fewer gram-positive than gram-negative bacteria, but the gram-positive bacteria are a diverse division.

GRAM-POSITIVE RODS. There are two principal subgroups of gram-positive rods. One produces endospores—highly resistant structures containing a copy of the bacterium's DNA, some ribosomes and other cytoplasmic constituents, a thick peptidoglycan coat, and an outer spore coat (Figure 21.9). Members of this group include the many species of *Bacillus* (Fig-

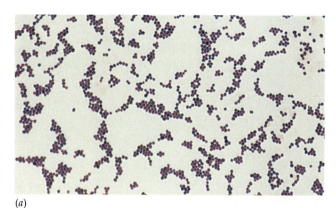

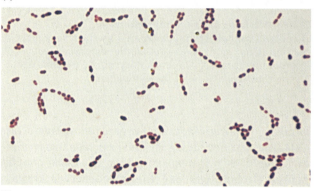

(a)

(b)

21.21 Gram-Positive Cocci
(a) "Grape clusters" are the usual arrangement of gram-positive staphylococci such as these *Staphylococcus aureus* cells. (b) Streptococci usually aggregate in chains of two or more cells, as shown here by *Streptococcus pyogenes*.

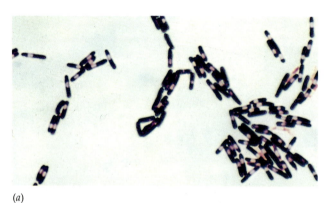

(a)

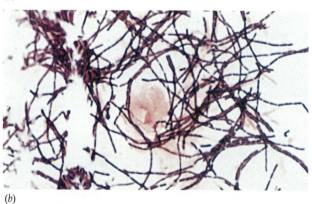

(b)

21.20 Gram-Positive Rods
(a) These deeply stained Gram-positive *Bacillus subtilis* have formed endospores (see Figure 21.9) in response to unfavorable changes in their environment. Most of the bacterial population makes the spores at the same time. Developing endospores appear here as clear areas in otherwise deeply stained cells. (b) Gram-positive rods of *Lactobacillus* do not form endospores.

ure 21.20a), including *B. anthracis*, the agent of anthrax in sheep and humans, and *B. thuringiensis*, a species used in commercial preparations to kill moth and butterfly larvae that are plant pests. The genus *Clostridium*, whose members also produce endospores, includes *C. denitrificans*, a free-living bacterium that plays an important role in this planet's nitrogen cycle; two producers of potent toxins, *C. botulinum* (the agent of various forms of botulism) and *C. tetani* (tetanus); and *C. perfringens* (food poisoning, gas gangrene). The toxins produced by *C. botulinum* are among the most poisonous ever discovered—the lethal dose for humans is about one-millionth of a gram (1 μg).

Other gram-positive rods do not form endospores. These include bacteria of the genera *Lactobacillus* (a lactic acid producer; Figure 21.20b), *Listeria*, and others.

GRAM-POSITIVE COCCI. The gram-positive cocci are numerous. Species of the genus *Staphylococcus* (Figure 21.21a) are about 1 μm in diameter and are called staphylococci. They are abundant on the human body surface and are responsible for boils and many other skin problems. *S. aureus* is the best-known human pathogen; it is found in 20–40 percent of normal adults (and in 50–70 percent of hospitalized adults) and can cause respiratory, intestinal, and wound infections, in addition to skin diseases. Staphylococci produce toxins that are a major cause of food poisoning and are the cause of toxic shock syndrome.

Streptococcus is another important genus of gram-positive cocci. Cells of this group are known as streptococci. These cells divide along a single axis and stay together after fission, forming chains (Figure 21.21*b*). *S. mutans* is an acid-producing oral species that can erode tooth enamel. There is no major organ system in the human body that is not subject to one form of streptococcal infection or another. The first demonstration that DNA is the hereditary material was performed by Avery, MacLeod, and McCarty, using cultures of *S. pneumoniae,* formerly called *Diplococcus pneumoniae* (see Chapter 11).

ACTINOMYCETES. The actinomycetes develop an elaborately branched system of filaments called a **mycelium** (Figure 21.22). They closely resemble the filamentous bodies of fungi and were, in fact, once classified as fungi.

Some actinomycetes reproduce by forming chains of spores at the tips of the filaments. Nevertheless, the actinomycetes are true prokaryotes. In the species that do not form spores, the branched, filamentous growth ceases and the structure breaks up into typical cocci or rods.

The actinomycetes include several medically important members, such as *Actinomyces israelii,* a cause of infections in the oral cavity and elsewhere; *Mycobacterium tuberculosis,* the species that causes tuberculosis; and *Streptomyces,* the genus that produces streptomycin as well as several other antibiotics. We derive most of our antibiotics from members of the actinomycetes.

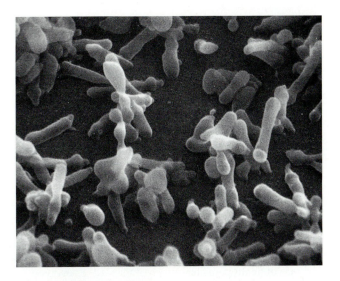

21.23 The Smallest Monera
Cells of *Mycoplasma gallisepticum.* Lacking cell walls and containing only about half as much DNA as other bacterial cells, mycoplasmas are the smallest known members of the kingdom Monera.

Mycoplasmas

The fourth division of the Monera consists of bacteria that lack cell walls. These bacteria, the mycoplasmas, are the smallest cellular creatures ever discovered—they are even smaller than rickettsias and chlamydias. Mycoplasmas have diameters between 0.1 and 0.2 μm (Figure 21.23). The mycoplasmas are "small" in another crucial sense: They have less than half as much DNA as do the other prokaryotes. It has been speculated that the amount of DNA in a mycoplasma may be about the minimum amount required to code for the absolutely essential properties of a cell.

Most mycoplasmas are parasites found within the cells of animals and plants. As intracellular parasites, they are not subjected to the osmotic challenges faced by free-living bacteria, so they need no cell walls. As a result, they take on irregular shapes; and they cannot be killed with penicillin, which kills other bacteria by interfering with wall synthesis (Chapter 12).

SUMMARY

Viruses do not metabolize energy, and they are incapable of reproducing on their own. They are acellular, and they are obligate internal parasites. Some have bacterial hosts; others parasitize animals and plants. Host specificity is expressed by the capsid of a virus, but this and all other information relating to the virus is determined ultimately by its nucleic acid core. Viruses are classified by whether their nucleic acid core is DNA or RNA, whether the nucleic acid is single- or double-stranded, the overall shape of the capsid, the symmetry of the capsid, and other factors.

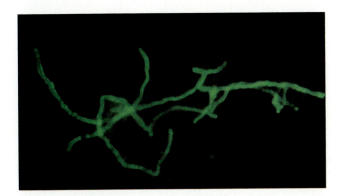

21.22 Filaments of an Actinomycete
Branching filaments of *Actinomyces israelii,* visualized with a fluorescent stain. This species is part of the normal flora in the human tonsils, mouth, intestinal tract, and lungs but will invade body tissues and cause severe abscesses when afforded the opportunity.

Viroids are infectious RNA molecules, and scrapie-associated fibrils (sometimes called prions) may be infectious protein molecules. Both are recognized only when they are found in diseased organisms.

The Monera are notable for their great abundance, extraordinary diversity, and success in adapting to every conceivable habitat. There are many forms of metabolism among the bacteria. The four main categories of energy and carbon metabolism are photoautotrophy, photoheterotrophy, chemoautotrophy, and chemoheterotrophy.

Taxonomic classification of bacteria continues to be a difficult task. The kingdom Monera comprises all the prokaryotes, and its members—all bacteria—may be assigned among four divisions: Archaebacteria, gram-negative bacteria, gram-positive bacteria, and bacteria without cell walls (Table 21.3). It has been suggested that the Archaebacteria merit the status of a separate kingdom.

SELF-QUIZ

1. Viruses:
   a. are cellular in structure.
   b. produce ATP.
   c. are obligate parasites.
   d. all have DNA as their genetic material.
   e. undergo mitosis.

2. In viral reproduction,
   a. capsid protein structure is encoded in the host's DNA.
   b. new viruses are produced from host materials.
   c. the host cell must die before new viruses can be made.
   d. viral core nucleic acid is encoded in the host's DNA.
   e. new capsids are synthesized around new viral nucleic acid molecules.

3. Which statement is *not* true of viroids?
   a. They are smaller than viruses.
   b. They consist solely of RNA.
   c. They produce diseases in plants.
   d. They are translated to synthesize proteins.
   e. Their nucleic acid folds into double-stranded rods.

4. Most bacteria:
   a. are agents of disease.
   b. lack ribosomes.
   c. evolved from the most ancient protists.
   d. lack a cell wall.
   e. are chemoheterotrophs.

5. All photosynthetic bacteria:
   a. use chlorophyll *a* as their photosynthetic pigment.
   b. use bacteriochlorophyll as their photosynthetic pigment.
   c. release oxygen gas.
   d. produce particles of sulfur.
   e. are photoautotrophs.

6. Gram-negative bacteria:
   a. appear blue to purple following Gram staining.
   b. are the most abundant of the bacterial groups.
   c. are all either rods or cocci.
   d. contain no peptidoglycan in their walls.
   e. are all photosynthetic.

7. Endospores:
   a. are produced by viruses.
   b. are reproductive structures.
   c. are very delicate and easily killed.
   d. are resting structures.
   e. lack cell walls.

8. The Archaebacteria:
   a. are Gram-negative.
   b. lack peptidoglycan in their cell walls.
   c. survive only at moderate temperatures and near neutrality.
   d. all produce methane.
   e. show little similarity among their subgroups.

9. The actinomycetes:
   a. are important producers of antibiotics.
   b. belong to the kingdom Fungi.
   c. are never pathogenic to humans.
   d. are Gram-negative.
   e. are the smallest known bacteria.

10. Which statement is *not* true of mycoplasmas?
    a. They lack cell walls.
    b. They are the smallest known cellular organisms.
    c. They contain the same amount of DNA as other prokaryotes.
    d. They cannot be killed with penicillin.
    e. Most live within plant or animal cells.

FOR STUDY

1. Contrast the biology of animal viruses, plant viruses, and bacteriophages.

2. Differentiate among the members of the following sets of related terms:
   a. prokaryotic/eukaryotic
   b. obligate anaerobe/facultative anaerobe/obligate aerobe
   c. photoautotroph/photoheterotroph/chemoautotroph/chemoheterotroph
   d. Gram-positive/Gram-negative

3. For each of the types of organism listed below, give a single characteristic that may be used to differentiate it from the related organism(s) in parentheses:
   a. spirochetes (spiral bacteria)
   b. *Bacillus* (*Lactobacillus*)
   c. mycoplasmas (free-living bacteria)
   d. cyanobacteria (other photoautotrophic bacteria)

4. Until fairly recently, the cyanobacteria were called blue-green *algae* and not grouped with the bacteria. Suggest several reasons for this (abandoned) tendency to separate the bacteria and cyanobacteria. Why *are* the cyanobacteria now grouped with the other bacteria?

5. The rickettsias were once grouped with the viruses. Why? Now they are grouped with the other bacteria. Why?

READINGS

Balows, A., H. G. Trüper, M. Dworkin, W. Harder and K.-H. Schleifer, Editors. 1991. *The Prokaryotes* (3 volumes). Springer-Verlag, New York, 1991. The ultimate reference on the bacteria: ecophysiology, isolation, identification, applications.

Brock, T. D. and M. T. Madigan. 1991. *Biology of the Microorganisms*, 6th Edition. Prentice-Hall, Englewood Cliffs, NJ. An excellent general textbook, including a chapter on viruses.

Fischett, V. A. 1991. "Streptococcal M Protein." *Scientific American*, June. How rheumatic fever and strep throat bacteria evade the body's defenses.

Hirsch, M. S. and J. C. Kaplan. 1987. "Antiviral Therapy." *Scientific American*, April. Approaches to killing viruses without damaging their host cells.

Hogle, J. M., M. Chow and D. J. Filman. 1987. "The Structure of Poliovirus." *Scientific American*, March. The relationship of viral structure and function is considered. The authors also speculate on viral evolution.

Koch, A. L. 1990. "Growth and Form of the Bacterial Cell Wall." American Scientist, vol. 78, pages 327–341. How does this cell wall, a single peptidoglycan molecule, allow the cell to grow but keep it from bursting?

McEvedy, C. 1988. "The Bubonic Plague." *Scientific American*, February. It shaped world history, and it hasn't disappeared yet.

Shapiro, J. A. 1988. "Bacteria as Multicellular Organisms." *Scientific American*, June. The behavior of highly regular colonies.

Stanier, R. Y., J. L. Ingraham, M. L. Wheelis and P. R. Painter. 1986. *The Microbial World*, 5th Edition. Prentice-Hall, Englewood Cliffs, NJ. A wide-ranging, authoritative text.

Tiollais, P. and M.-A. Buendia. 1991. "Hepatitis B Virus." *Scientific American*, April. Virus structure, epidemiology, vaccines, genetics, and replication.

Woese, C. 1981. "Archaebacteria." *Scientific American*, June. An early treatment of the subject by their leading student.

22

Protista

PREVIEW: The kingdom Protista is extraordinarily diverse, in part because it is defined largely by exclusion —it contains those eukaryotic organisms that do not fit comfortably into the fungal, animal, or plant kingdoms. Its members include major pathogens, as well as decomposers and most of the major producers of aquatic ecosystems. The kingdoms of multicellular eukaryotes all evolved, independently, from the protists.

This chapter deals with diversity among the protists, including the protozoa, the algae, and the funguslike protists.

The kingdom Protista is a great evolutionary grab bag. All its members are *eukaryotic*, and all evolved from the Monera. The remaining three kingdoms— Plantae, Fungi, Animalia—evolved from protists. The most precise definition of the protists, as we are using the term, is "all eukaryotes that are not included among the plants, fungi, or animals."

Most protists are unicellular. Some biologists reserve the term Protista for unicellular eukaryotes and nothing else. The organisms that we call here multicellular protists are, in such a classification, assigned variously to the other three eukaryotic kingdoms. But we find that a classification based on all-unicellular Protista creates more problems than it solves. Remember: All such classification schemes are for the convenience of biologists; they have no significance to the organisms themselves. There are many multicellular protists. The giant kelps are plantlike marine protists that are among the longest organisms in existence.

Protists are strikingly diverse in their metabolism, perhaps second only to the monerans (Chapter 21). Nutritionally, some are autotrophs, whereas others are absorptive heterotrophs, and still others are ingestive heterotrophs. Some switch with ease between the autotrophic and heterotrophic modes of nutrition. One phylum consists entirely of nonmotile organisms, but the others include cells that move by amoeboid motion, by ciliary action, or by means of flagella (Table 22.1).

Most protists are aquatic. Some live in marine environments, others in freshwater, and still others in the body fluids of other organisms. The slime molds inhabit damp soil and the moist, decaying bark of rotting trees.

PROTISTA AND THE OTHER EUKARYOTIC KINGDOMS

Part of the difficulty in placing certain organisms in the appropriate kingdom is a natural consequence of the evolutionary origin of the other eukaryotic kingdoms from the kingdom Protista. The other eukaryotic kingdoms—the fungi, plants, and animals— arose from protists in various ways. There are real difficulties in deciding just where to draw the lines between the Protista and the other eukaryotic kingdoms. Different protists tend to resemble animals, plants, or fungi to one extent or another. The animallike protists are referred to as **protozoa**. Plantlike protists are referred to as **algae**. There are also **funguslike protists**—the slime molds, chytrids, and water molds.

In this book, we tend to assign eukaryotic organisms to the kingdom Protista if they are unicellular or colonial, although this is not a hard-and-fast rule. The distinction between Fungi and Protista depends upon the presence of flagella (the kingdom Fungi has no members with flagella), the nature of the gametes (some Protista have visibly different male and female gametes, Fungi do not), the possession or lack of a dikaryotic stage (Fungi have such a stage; Chapter 23), and the chemistry of the cell wall (all members of the kingdom Fungi have cell walls consisting primarily of chitin). Plantae and algae differ in that plants develop from embryos protected by tissues of the parent plant. The separation between Protista and Animalia is relatively easy—we call an organism an animal if it is a multicellular heterotroph with ingestive metabolism. The status of the sponges will be considered in Chapter 25.

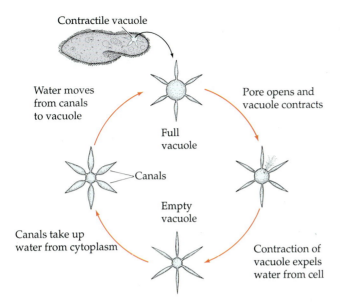

Contractile vacuole

Water moves
from canals
to vacuole

Pore opens and
vacuole contracts

Full
vacuole

Canals

Empty
vacuole

Canals take up
water from cytoplasm

Contraction of
vacuole expels
water from cell

22.1 Contractile Vacuoles
A tiny pore in the contractile vacuole near a protist's surface opens when the vacuole is full, connecting it with
the outside world; then the vacuole quickly contracts, expelling its contents.

SPECIAL FEATURES OF SOME PROTISTS

Contractile Vacuoles

Members of several of the protist phyla have **contractile vacuoles** that help them cope with their hypotonic environments. These organisms have a more
negative osmotic potential than their fresh water environment does and, hence, constantly take in water
by osmosis (Chapter 5). Excess water collects in the

contractile vacuole and is then pushed out, as shown
in Figure 22.1.

A beautifully simple experiment confirms that bail-
ing out water is the principal function of the contractile vacuole. One observes some cells under the microscope and notes the rate at which the vacuoles are
contracting—they look like little eyes winking. Then
one takes other cells of the same type and places
them in solutions of differing osmotic potential
(Chapter 5). The less negative the osmotic potential
of the surrounding solution, the more hypertonic are
the cells, so water rushes into them faster and
faster—and the contractile vacuoles pump more rapidly. Conversely, the contractile vacuoles stop pumping if the solute concentration of the medium is increased so that it is isotonic with the cells.

Endosymbiosis

In Chapter 4, we introduced the concept of endosymbiosis (organisms living together, one inside the
other). As one of the most bizarre examples, we selected the protist *Myxotricha paradoxa*, which has a
variety of bacteria living inside it and on its surface.
Endosymbiosis is very common among the protists,
and, in some instances, the endosymbionts are themselves protists. All radiolarians (protists in the phylum Sarcodina), for example, harbor photosynthetic
protists as endosymbionts (Figure 22.2). As a result,
the radiolarians appear greenish or yellowish, depending on the type of endosymbiont they harbor.
This arrangement is beneficial to the radiolarian, for
it can make use of the food produced by its photosynthesizing guest. The guest presumably profits
from metabolites made by the host or simply by being
given physical protection.

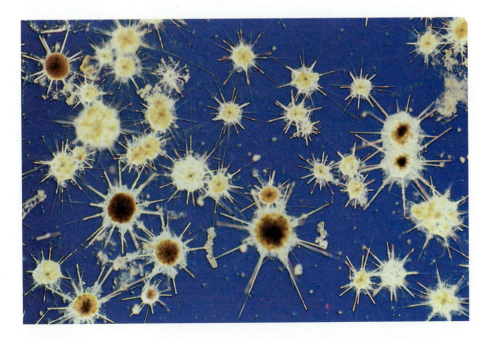

22.2 Radiolarians
Photosynthetic algae living as endosymbionts within these radiolarians provide food for the radiolarians and also part of the
pigmentation seen through the
glassy skeletons. Both the algae
and the radiolarians are protists.

TABLE 22.1
Classification of Protists

PHYLUM	COMMON NAME	FORM	LOCOMOTION	PHOTOSYNTHETIC PIGMENTS	STORAGE PRODUCT[a]
Protozoa					
Mastigophora	Flagellates	Unicellular, some colonial	One or more flagella	Some have chlorophylls a and b, β-carotene	Some have starch
Sarcodina	Amoebas and relatives	Unicellular, no definite shape	Pseudopods	None	
Sporozoa	Amoeboid parasites	Unicellular	None	None	
Ciliophora	Ciliates	Unicellular	Cilia	None	
Funguslike protists					
Gymnomycota	Slime molds (cellular and acellular)	Single cells and aggregates or coenocytes	Amoeboid	None	
Protomycota	Chytrids and hypochytrids	Unicellular or mycelium	None	None	
Oomycota	Water molds and downy mildews	Coenocytic mycelium	None	None	
Algae					
Pyrrophyta	Dinoflagellates	Unicellular	Two flagella	Chlorophylls a and c, β-carotene	Starch, fat, oils
Chrysophyta	Diatoms	Usually unicellular	Usually none	Chlorophylls a and c, β-carotene, and fucoxanthin	Chrysolaminarin, oils
Phaeophyta	Brown algae	Multicellular	Two flagella on reproductive cells	Chlorophylls a and c, β-carotene, and fucoxanthin	Chrysolaminarin, oils, mannitol
Rhodophyta	Red algae	Multicellular or unicellular	None	Chlorophylls a and d, β-carotene, phycoerythrin, and phycocyanin	Floridean starch
Chlorophyta	Green algae	Unicellular, colonial, or multicellular	Most have flagella at some stage	Chlorophylls a and b, β-carotene	Starch

[a]Shown only for photosynthetic forms
[b]Marine forms only

Alternation of Generations

Some algae and funguslike protists demonstrate the phenomenon of **alternation of generations**, in which a multicellular diploid, spore-producing organism gives rise to a multicellular haploid, gamete-producing organism. The fusion of two gametes once again creates a diploid organism (Figure 22.3). Either the haploid organism, the diploid organism, or both may also reproduce asexually. The two organisms differ genetically, in that one has haploid cells and the other has diploid cells, but they may or may not differ morphologically.

Gametes are not generally produced directly by meiosis in multicellular protists, fungi, or plants. Instead, specialized cells of the diploid organism, called **sporocytes**, divide meiotically to produce four spores. The spores may eventually germinate and divide mitotically to produce multicellular haploid organisms. These haploid organisms constitute the **gametophyte** generation, so called because it is the haploid organisms that produce gametes—by *mitosis* and cytoki-

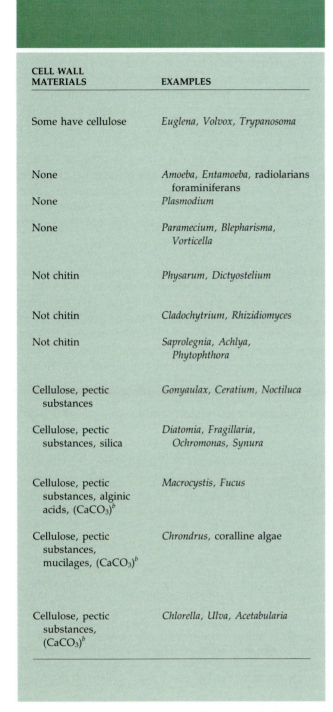

CELL WALL MATERIALS	EXAMPLES
Some have cellulose	Euglena, Volvox, Trypanosoma
None	Amoeba, Entamoeba, radiolarians foraminiferans
None	Plasmodium
None	Paramecium, Blepharisma, Vorticella
Not chitin	Physarum, Dictyostelium
Not chitin	Cladochytrium, Rhizidiomyces
Not chitin	Saprolegnia, Achlya, Phytophthora
Cellulose, pectic substances	Gonyaulax, Ceratium, Noctiluca
Cellulose, pectic substances, silica	Diatomia, Fragillaria, Ochromonas, Synura
Cellulose, pectic substances, alginic acids, $(CaCO_3)^b$	Macrocystis, Fucus
Cellulose, pectic substances, mucilages, $(CaCO_3)^b$	Chrondrus, coralline algae
Cellulose, pectic substances, $(CaCO_3)^b$	Chlorella, Ulva, Acetabularia

of the sporophyte generation undergo meiosis at some point and produce haploid spores, starting the cycle anew.

PROTOZOA

The Protozoa are all unicellular. Most ingest their food by endocytosis.

Phylum Mastigophora

All members of the Mastigophora, which includes both protozoa and algae, possess a flagellum (or more than one) and are thus called flagellates. This is the largest and by far the most heterogeneous protist phylum, numbering far more than 10,000 species. Figure 22.4 depicts a cell of the genus *Euglena*. Like most other members of the phylum, this common freshwater organism has a complex cell plan, including a well-formed nucleus. It propels itself through the water with one of its two flagella, which sometimes doubles as an anchor to hold the organism in place. The flagellum provides power by means of a wavy motion that spreads from base to tip. *Euglena* reproduces vegetatively by mitosis and cytokinesis—the simplest and most direct way possible. It has very flexible nutritional requirements. In sunlight it is fully autotrophic, using its chloroplasts to synthesize organic compounds through photosynthesis. When kept in the dark, the organism loses its photosynthetic pigment and begins to feed exclusively on dead organic material floating in the water around it. Such a "bleached" cell of *Euglena* resynthesizes its photosynthetic pigment when returned to the light and, hence, becomes autotrophic again. *Euglena* cells treated with certain antibiotics or mutagens lose their photosynthetic pigment completely; neither they nor their descendants are ever autotrophs again. However, those descendants function perfectly well as heterotrophs.

Because of their great diversity in nutrition, the Mastigophora are sometimes said to bridge the gap

nesis. Gametes, unlike spores, can only produce new organisms by fusing with other gametes. Fusion of two gametes produces a diploid **zygote**, which then undergoes mitotic divisions to produce a diploid organism: the **sporophyte** generation. The sporocytes

22.3 Alternation of Generations
A diploid generation that produces spores alternates with a haploid generation that produces gametes in some algae and funguslike protists.

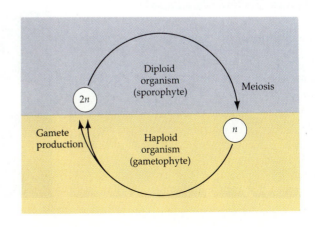

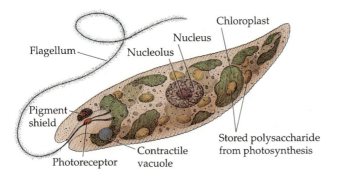

Labels on figure: Flagellum, Pigment shield, Photoreceptor, Nucleolus, Nucleus, Chloroplast, Contractile vacuole, Stored polysaccharide from photosynthesis

22.4 A Photosynthetic Flagellate
Several species of *Euglena* are among the best-known flagellates. Photosynthetic chloroplasts are among the most prominent features in a typical cell of *Euglena*.

between plants and animals at the unicellular level. Some relatively large, green, colonial flagellates appear quite similar to green algae, and other flagellates resemble the motile aquatic stages of some terrestrial plants. Still other flagellates, in contrast, strongly resemble tiny animals. Some species related to *Euglena* are devoid of chloroplasts and make their living by preying on other protists, including *Euglena*. An impressive variety of other "zooflagellates" (animallike flagellates), so named to distinguish them from the plantlike "phytoflagellates," live as internal parasites on larger animals, including humans. Within the guts of certain wood-eating roaches and termites live an array of huge zooflagellates possessing some of the most bizarre and complicated body forms found anywhere among the protists (Chapter 4).

Some idea of the diversity of this phylum is conveyed by Figure 22.5. The genera *Gonium* and *Volvox* are both photosynthetic and might well be treated as green algae, whereas *Trypanosoma* and the cellulose-digesting flagellate from the cockroach *Cryptocercus* are sometimes considered animals. *Euglena* and its relatives, some 800 species, all photosynthetic flagellates, are sometimes treated as a separate phylum, Euglenophyta.

One group of flagellates, the Choanoflagellida, is thought to be ancestral to the sponges, the most ancient of the surviving phyla of animals (Chapter 25). Sponges are colonial, rather than truly multicellular, and the Choanoflagellida bear a striking resemblance to the most characteristic type of cell found in the sponges (see Figure 25.7).

The Mastigophora also span the gap between single-celled and many-celled organisms. Surprisingly large and well-formed colonies of cells are found in such freshwater groups as the genus *Volvox* (Figure 22.5c). The cells are not differentiated into tissues and organs as in the plants and animals, but the colonies show vividly how the preliminary step of this great evolutionary development might have been taken. In addition, the intermediate stages between the one-celled state of *Euglena* and the extreme colonial state of *Volvox* are preserved in such loosely colonial forms as *Gonium* and *Pandorina*.

Some of the Mastigophora are human pathogens. Sleeping sickness is one of the most dreaded diseases of Africa. The vector (intermediate host) for sleeping sickness is an insect, the tsetse fly. Massive efforts have been made to exteriminate the tsetse fly by the

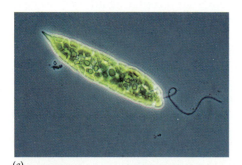

(a)

22.5 Flagellate Diversity
Cellular features vary widely in the Mastigophora. (a) You can compare this photograph of an individual *Euglena* with Figure 22.4, noting the chloroplasts. (b) *Trypanosoma gambiense*, a parasitic flagellate. In trypanosomes the flagellum runs along one edge of the cell body as part of a structure called the undulating membrane. (c) Colonies of *Volvox*, showing the precise spacing of cells and containing a number of daughter colonies. This colonial, photosynthetic organism is sometimes classified with the green algae—another group of protists—or even with the plants rather than with the flagellate protists.

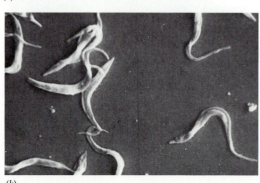

(b)

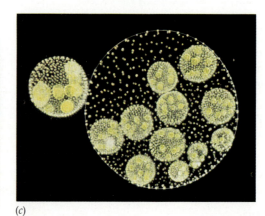

(c)

use of DDT and other insecticides, but these have met with only partial success. The tsetse fly has diverse tastes and bites livestock and wild animals as well as humans, infecting all of them with parasitic flagellates of the genus *Trypanosoma* (Figure 22.5*b*). The trypanosomes multiply in the mammalian bloodstream and produce toxic substances. When these parasites invade the nervous system, the symptoms of sleeping sickness appear—and are followed by death. Other disease-causing flagellates include *Giardia lamblia*, which contaminates water supplies and causes the intestinal disease giardiasis, and *Trichomonas vaginalis*, which causes a common but usually mild venereal disease.

Phylum Sarcodina

The Sarcodina—amoebas and their relatives—are protists that form pseudopods, extensions of their constantly changing body mass. Amoebas have often been portrayed in popular writing as simple blobs— the simplest form of "animal" life imaginable. A superficial examination of a typical amoeba (see Figure 22.7*a*) shows how such an impression might have been obtained. An amoeba consists of a single cell with no definite shape. It feeds on small organisms and particles of organic matter by phagocytosis, engulfing them with pseudopods. Particles of food are sealed off in food vacuoles within the cytoplasm of the amoeba. The material is then slowly digested and assimilated into the main body of the organism. The pseudopods are also the organs of locomotion. The mechanism of amoeboid motion will be discussed in Chapter 38.

The amoeba is probably not a primeval organism, despite its apparently "primitive" characteristics. Compelling evidence points to the conclusion that its simplicity is a secondarily derived condition in evolution. The phylum Sarcodina apparently originated from ancestors within the phylum Mastigophora. Some intermediate forms still exist. The example shown in Figure 22.6, *Mastigamoeba aspera*, is such an exactly intermediate link that it could equally well be placed in either phylum, the Mastigophora or the Sarcodina. Amoebas of the free-living genus *Naegleria*, some of which can enter humans and cause a fatal disease of the nervous system, have a two-stage life cycle, one stage having amoeboid cells and the other flagellated cells.

Amoebas are, in fact, rather advanced forms of protists. Many are specialized for life on the bottoms of lakes, ponds, and other bodies of water. Their creeping locomotion and their manner of engulfing food particles fit them for life close to a relatively rich supply of sedentary organisms or organic particles (Figure 22.7*a*). Some other Sarcodina are even more specialized. All are animallike, existing as predators, parasites, or scavengers. None is photosynthetic.

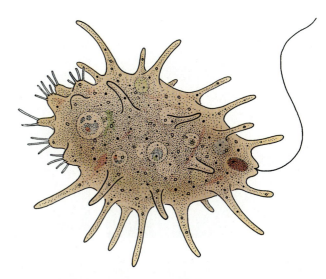

22.6 An "Intermediate" Amoeba
A cell of *Mastigamoeba aspera* displays both pseudopods and a flagellum; this species is usually considered to be an intermediate form between the phyla Mastigophora and Sarcodina.

There are shelled amoebas that live in casings of sand grains glued together or in spiny or scaly shells secreted by the organism itself.

Foraminiferans are marine creatures that secrete shells of calcium carbonate (Figure 22.7*b*). Their pseudopods are long, threadlike, and branched. The pseudopods reach out through numerous microscopic pores in the shells and interconnect with one another to create a sticky net, which the foraminiferan uses to catch smaller plankton (free-floating microscopic organisms). After foraminiferans reproduce, by mitosis and cytokinesis, the daughter cells abandon the parent shell and make new shells of their own. The discarded skeletons of infinite legions of ancient foraminiferans make up extensive limestone deposits in various parts of the world, forming a covering hundreds to thousands of meters deep over millions of square kilometers of ocean bottom. Foraminiferan skeletons also make up the sand of some beaches. A gram of such sand may contain as many as 50,000 foraminiferan shells.

The shells of individual foraminiferan species have distinctive shapes, and they are easily preserved as fossils in marine sediments. Each geological period had its own distinctive foraminiferan species. For this reason, plus the fact that they are so abundant, foraminiferan remains are especially valuable as indicators in the classification and dating of sedimentary rocks and also serve as indicators in oil prospecting.

Heliozoans are sarcodines surrounded by a bristling array of long pseudopods (Figure 22.7*c*). Like the foraminiferans, they drift in the water and use their pseudopods to trap smaller organisms. Most heliozoans live in freshwater.

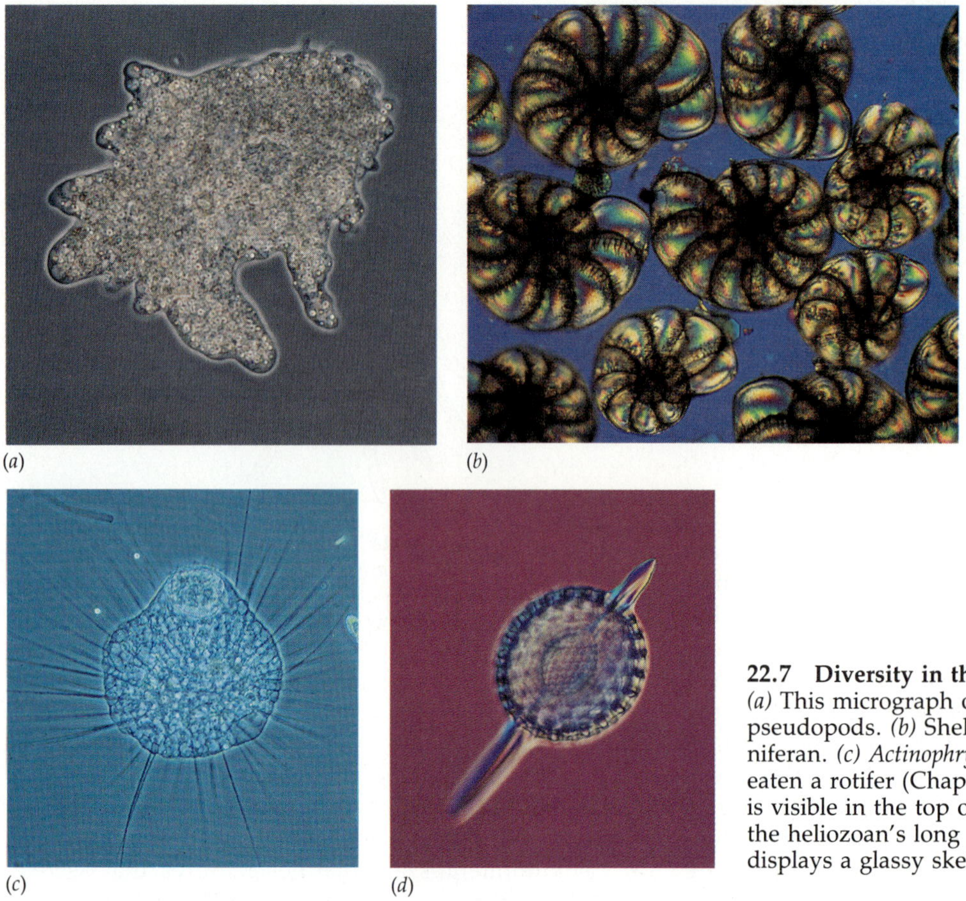

(a)

(b)

(c)

(d)

22.7 Diversity in the Sarcodina
(a) This micrograph of *Amoeba proteus* shows several pseudopods. (b) Shells of *Elphidium*, a fossil foraminiferan. (c) *Actinophrys sol*, a heliozoan, has just eaten a rotifer (Chapter 25); the microscopic animal is visible in the top of the heliozoan's body. Note the heliozoan's long pseudopods. (d) A radiolarian displays a glassy skeleton of delicate intricacy.

The radiolarians are a third group whose feeding technique relies on drifting and trapping. These exclusively marine protists are perhaps the most beautiful of all microorganisms (Figure 22.7d). Almost all radiolarian species secrete glassy skeletons from which needlelike pseudopods project. The skeletons of the different species are as varied as snowflakes, and many have elaborate geometric designs. A few of the radiolarians are among the largest of the protists, with skeletons measuring several millimeters across. Uncountable numbers of radiolarian skeletons, some as old as 700 million years, form the sediment under some seas in the tropics.

Phylum Sporozoa

The sporozoans are exclusively parasitic protozoa that derive their name from the fact that some of them produce sporelike infective stages. Sporozoans differ from many protists in lacking contractile vacuoles. Because their rigid cell walls limit expansion, they do not take in excess water. Sporozoans generally have an amoeboid body form, but this in no way indicates a relationship to the Sarcodina. Rather, this body form has evolved over and over again in parasitic protists. This mark of "degeneracy" has appeared, for example, even in parasitic dinoflagellates,

which some biologists assign to the plant kingdom (although we and others include the dinoflagellates in the protists). Sporozoans, like many animal obligate parasites, have elaborate life cycles featuring asexual and sexual reproduction by a series of very dissimilar life stages. Often these stages are associated with two kinds of host organisms.

The best known sporozoans are the malaria parasites of the genus *Plasmodium*, a highly specialized group that spends part of its life cycle within human red blood cells (Figure 22.8). Malaria continues to be a major problem in some tropical countries although it has been almost eliminated from the United States; indeed, malaria is one of the world's most serious diseases in terms of number of people infected. Female mosquitoes of the genus *Anopheles* transmit *Plasmodium* to humans. *Plasmodium* enters the human circulatory system when a hungry, infected *Anopheles* penetrates the human skin in search of blood. The parasite cells find their way to the liver and the lymphatic system, change their form, multiply, and reenter the bloodstream, attacking red blood cells. The attackers multiply in a red blood cell for approximately two days, producing up to 36 new *Plasmodium* cells each. The victimized cell then bursts, releasing a new swarm of parasites to attack other red blood cells.

If another *Anopheles* bites the victim, some of the parasitic *Plasmodium* cells are taken into the mosquito along with the blood, thus infecting the mosquito. The parasites attack cells of the mosquito's gut, reproduce, and move into the salivary glands, thence to be passed to another human host. The *Plasmodium* life cycle that spreads malaria is best broken by the removal of stagnant water, in which mosquitoes breed. The use of insecticides to reduce the *Anopheles* population can be effective but possible ecological, economic, and health risks should be considered.

Phylum Ciliophora

The Ciliophora are characterized by the possession of hairlike cilia (appendages described in Chapter 4), and they thus have the common name ciliates. This protozoan phylum ranks with the Mastigophora in diversity and ecological importance (Figure 22.9). Ciliates are all heterotrophic, and are much more specialized in body form than most flagellates and other protists. The Ciliophora are also characterized by the possession of two types of nuclei, a large **macronu-**

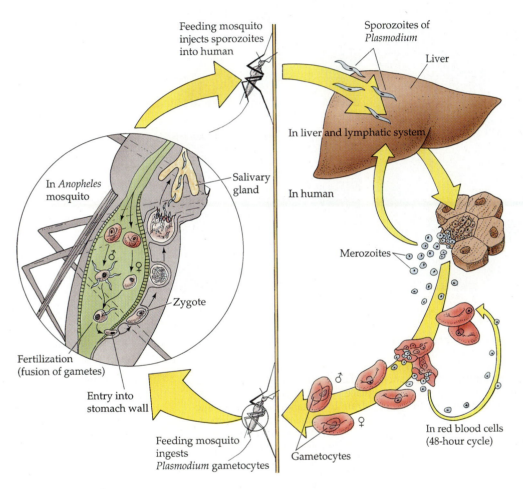

22.8 Life Cycle of *Plasmodium*
Malaria-causing *Plasmodium* species spend part of their life in humans and part in mosquitoes. In the gut of the mosquito, male and female gametocytes obtained from a human during a blood meal develop into male and female gametes, which then fuse to form a zygote. The zygote develops, enters the gut wall, and produces a cyst, which eventually divides into numerous slender cells called sporozoites. The sporozoites invade the salivary gland of the mosquito and are injected into a human's bloodstream during another blood meal. In the human, the sporozoites penetrate cells of the liver and lymphatic system, where they divide and develop into cells of another stage, called merozoites. Merozoites may, in turn, invade fresh cells of the human liver or lymphatic system, where they divide and develop into another generation of merozoites; or they may enter red blood cells, where they may also divide, grow, lyse cells, and reinvade fresh red blood cells on a 48-hour cycle. Eventually some merozoites inside red blood cells develop into male and female gametocytes, ready to be picked up by a hungry mosquito to start the life cycle again.

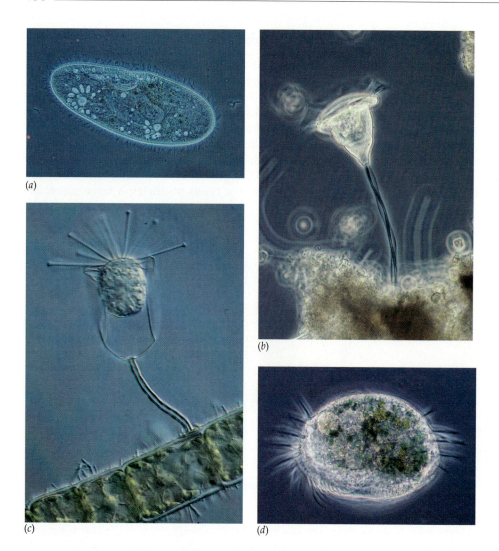

22.9 Ciliates
(a) *Paramecium caudatum*, a free-swimming organism, belongs to the ciliophoran subgroup called holotrichs, which are characterized by having many cilia of uniform length. (b) Members of the peritrich subgroup of Ciliophora have cilia on their mouthparts and usually attach to their substrate, as demonstrated by this *Vorticella*. (c) Tentacles replace cilia as suctorians, another subgroup, develop. Attachment to the substrate by a stalk characterizes suctorians, such as this *Paracineta*. (d) The organelles of *Euplotes* are developed for specific uses. This ciliate "walks" on fused cilia called cirri that project from its body. Other cilia in *Euplotes* are fused into flat sheets that sweep in food particles; this individual has just fed on green algae.

cleus and, within the same cell, from one to as many as 80 **micronuclei**. The micronuclei, which are typical eukaryotic nuclei, are essential for genetic recombination. The macronucleus contains many copies of the genetic information, packaged in units containing very few loci each; the macronuclear DNA is transcribed and translated to control the life of the cell.

Paramecium, a frequently studied genus, exemplifies the complex structure and behavior of ciliates (Figure 22.10). The slipper-shaped cell is covered by an elaborate **pellicle**, a structure composed principally of an outer membrane and an inner layer of closely packed, kidney-shaped structures (called alveoli) that embrace the cilia. Defensive organelles called trichocysts are also present as a layer of the pellicle. A microscopic explosion expels the trichocysts in a few milliseconds, and they emerge as sharp darts, driven forward at the tip of a long, expanding shaft.

The cilia provide a form of locomotion that is generally more precise than that made possible by flagella or pseudopods. A paramecium can direct the beat of its cilia to propel itself either forward or backward in a spiraling manner (Figure 22.11). A paramecium can back off swiftly when it encounters a barrier or a negative stimulus. Some of the large ciliates hold

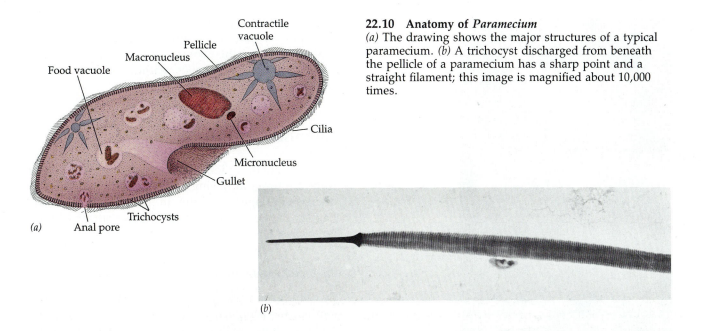

22.10 Anatomy of *Paramecium*
(a) The drawing shows the major structures of a typical paramecium. *(b)* A trichocyst discharged from beneath the pellicle of a paramecium has a sharp point and a straight filament; this image is magnified about 10,000 times.

the speed record for the kingdom Protista—faster than 2 millimeters per second. A few of the cilia of a paramecium are sensory in function, and they are somehow able to transmit stimuli back through the remainder of the cytoplasm, thereby coordinating the rapid movements of the entire organism.

Paramecia reproduce by cell division. The micronuclei divide mitotically, but the macronucleus simply pinches apart to give two daughter macronuclei.

Paramecia also have an elaborate sexual behavior called **conjugation** (Figure 22.12). Two paramecia line up tightly against each other and fuse in the oral region of the body. There is an extensive reorganization and exchange of nuclear material during the next several hours, as follows. In each cell, all the micronuclei but one degenerate. The remaining micronucleus in each cell divides meiotically, a process that reduces the chromosome number from diploid to haploid (Chapter 9). Each of the four meiotic products in each cell divides mitotically. Then all but one of the eight resulting haploid nuclei break down, and the last one divides mitotically and produces two haploid nuclei. The macronuclei break up and dis-

appear. One of the two haploid nuclei in each cell remains in its "home" cell and the other migrates to the partner cell, where it fuses with its counterpart. The exchange is fully reciprocal—each of the two paramecia gives and receives an equal amount of DNA. The two organisms now separate and go their separate ways, each genetically "refreshed" by the recombination that occurred during conjugation. The new, recombined diploid nucleus of each cell divides mitotically, producing two diploid nuclei, one that is destined to be the new macronucleus and a second that gives rise to the appropriate number of micronuclei by further mitotic divisions.

Conjugation in *Paramecium* is a *sexual* process of genetic recombination, but it is not a *reproductive* process. Two cells began the process, and the same two cells are still there at the end. As a rule, each clone of paramecia must periodically go through the process of conjugation. Laborious experimentation has shown that if some species are not permitted to conjugate, the asexual clones can live through no more than approximately 350 cell divisions before they die out.

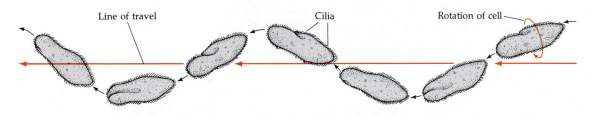

22.11 Locomotion of *Paramecium*
Beating its cilia in coordinated waves that progress from one end of the cell to the other, a paramecium can move in either direction with respect to the long axis of the cell. The cell itself rotates in a spiraling line of travel.

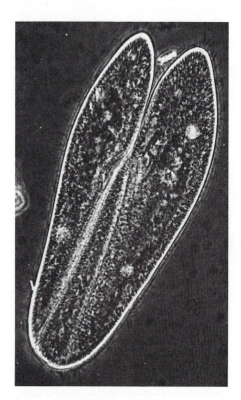

22.12 Conjugating Paramecia
The two *Paramecium* individuals shown here are conjugating. Their cells fuse and exchange micronuclei, thereby permitting genetic recombination. After conjugation, the cells separate and continue their lives as two individuals.

CYTOPLASMIC ORGANIZATION IN THE CILIATES. Most ciliates possess all the traits just described for Paramecium. Some, however, are notable for the exceptional degree of development of their individual organelle systems. Certain ciliates, for example, have the equivalent of legs. Fused cilia called **cirri** move in an independent, but coordinated, fashion and enable the organism to walk over surfaces (Figure 22.9*d*). Nervelike **neurofibrils** leading to individual cirri coordinate this locomotion. The coordination is lost if the neurofibrils are experimentally cut. Many kinds of ciliates possess **myonemes**: musclelike fibers within the cytoplasm. Contraction of myonemes causes a rapid retraction of the stalk in forms such as *Vorticella* when the organism is disturbed.

Possibly the ultimate in cytoplasmic organization is displayed by highly specialized ciliates that live in the digestive tracts of cows and many other hoofed mammals. They possess not only myonemes, neurofibrils, and elaborately fused cilia, but also a cytoplasmic "skeleton" and a "gut" complete with "mouth," "esophagus," and "anus" (Figure 22.13).

When examining the intricate structures of these organisms and many other protists, one must pause and remember that one is looking at only one cell. Structural complexity in multicellular organisms—

fungi, animals, and plants—is based on the diversity and coordination of cell types. Most protists, on the other hand, owe their complexity to the diversity and coordination of organelles within a single cell or within an acellular body.

FUNGUSLIKE PROTISTA

As mentioned earlier, the funguslike protists differ from the fungi in having flagella and in *not* having a dikaryotic stage, or chitin in their cell walls. The flagella may characterize only certain stages of their life cycles.

Phylum Gymnomycota

Two important groups of slime molds together constitute the funguslike phylum Gymnomycota: the acellular slime molds (class Myxomycetes), also called the plasmodial slime molds, and the cellular slime molds (class Acrasiomycetes). The two groups of slime molds share a number of characteristics. Both are motile, both ingest particulate food by endocytosis, and both form spores on erect fruiting bodies. They undergo striking changes in organization during their life cycles, and one stage consists of isolated cells engaging in absorptive nutrition.

Some slime molds may attain areas of 1 m or more in diameter while in their less–aggregated stage. Such

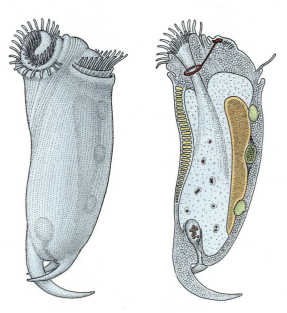

22.13 An Exceptional Ciliate
Surface and cutaway views of the rumen ciliate *Diplodinium dentatum*. The "mouth" is at the top, the "esophagus" (surrounded by a contractile ring that can close it) is just below, and the "anus" is toward the bottom. The plates comprising the "skeleton" are stacked on the left.

22.14 Acellular Slime Molds
(a) Plasmodia of the yellow slime mold *Physarum* cover a rock in Nova Scotia. (b) Sporangiophore and sporangia of *Physarum polycephalum*.

a large slime mold may weigh more than 50 g. Slime molds of both classes favor cool, moist habitats, primarily in forests. They range from colorless to brilliant yellows and oranges.

CLASS MYXOMYCETES. If the nucleus of an amoeba began rapid mitotic division accompanied by a tremendous increase in cytoplasm and organelles, the resulting organism might resemble the vegetative phase of the *acellular* slime molds. During most of its life history, an acellular slime mold exists as a wallless mass of cytoplasm with numerous diploid nuclei; this mass streams over its substrate in a remarkable network of strands called a **plasmodium** (Figure 22.14*a*). A myxomycete plasmodium is an example of a **coenocyte**—a body in which many nuclei are enclosed in a single plasma membrane. The outer cytoplasm of the plasmodium, which is normally in a less fluid state than the interior, provides a degree of structural rigidity.

Myxomycetes such as *Physarum* (a popular research subject) provide a dramatic example of cytoplasmic streaming. The outer cytoplasmic region becomes more fluid in places, and there is a rush of cytoplasm into those areas, stretching the plasmodium in one direction or another. This streaming somehow reverses its direction every few minutes as cytoplasm rushes into a new area and drains away from an older one. As the plasmodium spreads over its substrate, it engulfs food particles—predominantly bacteria, yeasts, spores of fungi, and other small organisms, as well as decaying animal and plant remains. Sometimes an entire wave of plasmodium moves across the substrate, leaving strands behind. A contractile protein called myxomyosin participates in the

streaming mechanism. Minute fibers—microfilaments—mediate the movement in conjunction with a myosinlike molecule.

An acellular slime mold can grow almost indefinitely in its plasmodial stage, as long as the food supply is adequate and other conditions, such as moisture and pH, are favorable. However, if conditions become unsuitable, one of two things can happen. The plasmodium can form a resistant structure, an irregular mass of hardened cell-like components called a **sclerotium**, which rapidly becomes a plasmodium again upon restoration of favorable conditions; or the plasmodium can transform itself into spore-bearing fruiting structures (Figure 22.14*b*). Rising from heaped masses of plasmodium, these stalked or branched fruiting structures—called **sporangiophores**—derive their rigidity from the deposition of cellulose or chitin at the surfaces of their component cells.

The nuclei of the plasmodium are diploid, and they divide by meiosis during the development of the sporangiophore. One or more knobs—variously colored and shaped **sporangia**—develop on the end of the stalk; within a sporangium, haploid nuclei become surrounded by walls and form spores. Eventually, as the sporangiophore dries, it sheds its spores. The spores germinate into wall-less, flagellated, haploid cells called **swarm cells**, which can either divide mitotically to produce more haploid swarm cells or function as gametes. Swarm cells can manage on their own and can become walled and resistant cysts when conditions are unfavorable. Upon the return of more favorable conditions, the cysts release flagellated swarm cells. Two swarmers can fuse to form a diploid zygote, which divides by mitosis but without wall formation between the nuclei, and thus forms a new, coenocytic plasmodium.

CLASS ACRASIOMYCETES. The other principal group of Gymnomycota encompasses the cellular slime molds. Whereas the plasmodium is the basic vegetative unit of acellular slime molds, an amoeboid cell is the vegetative unit of the cellular slime molds. Large numbers of cells called **myxamoebas**, which have single haploid nuclei, engulf bacteria and other food particles and reproduce by mitosis and fission. This simple developmental stage can persist indefinitely as swarms of independent, isolated cells, as long as food and moisture are available.

When conditions become unfavorable, however, cellular slime molds aggregate and then form fruiting structures, as do their acellular counterparts. The apparently independent myxamoebas aggregate into an irregular mass called a **pseudoplasmodium** (Figure 22.15a). Unlike the true plasmodium of the acellular slime molds, this structure is not simply a giant sheet of cytoplasm, for the individual myxamoebas retain their plasma membranes and, therefore, their identity. The pseudoplasmodium is not a coenocyte. The chemical signal that causes the myxamoebas to aggregate into a pseudoplasmodium is 3′,5′-cyclic adenosine monophosphate (cAMP), a "messenger" that plays many important roles in chemical signaling in animals (Chapter 34). A pseudoplasmodium may migrate over the substrate for several hours before ultimately constructing a delicate, stalked fruiting structure (Figure 22.15b). Cells at the top of the fruiting structure develop into thick-walled spores. The spores are released; later, under favorable conditions, they germinate, releasing myxamoebas. The cycle from myxamoebas through a pseudoplasmodium and spores to new myxamoebas is asexual. There is also a sexual cycle, in which two myxamoebas (possibly of different mating types; see Chapter 23) fuse. The product of this fusion develops into a spherical structure that ultimately germinates, releasing new myxamoebas.

Phylum Protomycota

Chytrids and hypochytrids, the two groups that constitute the phylum Protomycota, are aquatic microorganisms sometimes classified as fungi. We place them among the protists because they possess flagellated cells, but they resemble fungi in that their cell walls consist primarily of chitin.

All chytrids and hypochytrids are either parasitic (on organisms such as algae, mosquito larvae, and nematodes) or **saprobic**, which means that they obtain nutrients by breaking down dead organic matter. They live in freshwater habitats or in moist soil. Some of the Protomycota are unicellular, while others have **mycelia**—masses of filaments—made up of branching chains of cells. Hypochytrids and chytrids differ in some structural and metabolic details. Chytrids reproduce both sexually and asexually, but no sexual stages have been observed in the hypochytrids.

Allomyces is a well-studied genus of the chytrids. *Allomyces* has an **isomorphic alternation of generations**, that is, it has haploid and diploid phases of the life cycle that are indistinguishable except on the bases of their chromosome number and of their reproductive products. (Later in this chapter we will examine the isomorphic life cycle of a marine alga, the sea lettuce *Ulva lactuca*.)

The diploid generation of *Allomyces* produces numerous diploid flagellate spores, called **zoospores**, by mitosis and cytokinesis. (Many funguslike protists and algae reproduce vegetatively from zoospores, produced by the haploid and the diploid phases of the life cycle.) Haploid zoospores are also produced, by meiotic divisions and cytokinesis, in discrete sporangia (spore cases), which are cell-like units delimited by distinct cross walls. Diploid zoospores simply

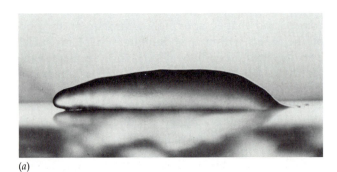

(a)

22.15 A Cellular Slime Mold
(a) A pseudoplasmodium of *Dictyostelium discoideum* migrating over the substrate. (b) Fruiting structures of *D. discoideum* in various stages of development.

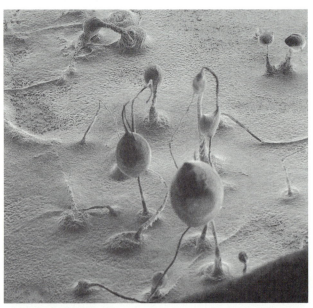

(b)

produce new diploid organisms, whereas haploid zoospores produce haploid organisms. The haploid organism produces gametes at the tips of **hyphae** (filaments). A single haploid organism can produce both male and female gametes, with the male gametes at the very ends of the hyphae and the female gametes just below, in specialized, walled-off structures known as **gametangia** (gamete cases) (Figure 22.16). Both female and male gametes have flagella. The motile female gamete produces a chemical attractant (pheromone) that attracts the swimming male gamete.

One of the hypochytrids, *Rhizidiomyces apophysatus*, parasitizes funguslike protists of the phylum Oomycota (water molds), to be considered next. Upon contacting water mold cells, the zoospores of *R. apophysatus* lose their flagella and develop germination tubes that extend into the host tissue. Ultimately a branching mass of **rhizoids** (rootlike structures) ramifies throughout the infected water mold. The hypochytrid forms a sporangium, and a tube develops, through which a mass of the hypochytrid's cytoplasm and nuclei is released. The mass cleaves, freeing a large number of zoospores. The zoospores swim away, some coming in contact with water molds and repeating the life cycle.

Phylum Oomycota

The phylum Oomycota consists in large part of the water molds and their funguslike terrestrial relatives, such as the downy mildews. If you have seen a whitish, cottony mold growing on dead fish or dead insects in water, it was probably the mycelium of a water mold of the common genus *Saprolegnia* (Figure 17). *Saprolegnia* itself is a common target of parasitism by the hypochytrid *Rhizidiomyces* just mentioned.

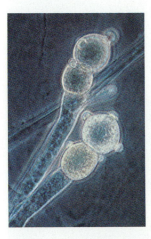

22.16 Reproductive Structures of a Chytrid
Gametangia of the water mold *Allomyces*. The rounded structures at the ends of the wall-less hyphae are male gametes.

22.17 *Saprolegnia*
The mycelium of a water mold radiates from the carcass of a salamander.

These funguslike protists are coenocytic—they have hyphae with no cross walls to divide them into discrete cells. Their cytoplasm is continuous throughout the mycelium, and there is no single structural unit with a single nucleus, except in certain reproductive stages. A distinguishing feature of the Oomycota is that they have flagellated reproductive cells. They are diploid throughout most of their life cycle and have cellulose in their cell walls.

The water molds such as *Saprolegnia* are all aquatic and saprobic. Some other members of the phylum Oomycota are terrestrial. Although most terrestrial members of the phylum are harmless or helpful saprobes, a few are serious plant parasites attacking crops such as avocados, grapes, and potatoes. The mold *Phytophthora infestans* is the causal agent of "late blight" of potatoes, which brought about the great Irish potato famine of 1845–1847. *P. infestans* destroyed the entire Irish potato crop in a matter of days in 1846. Among the consequences of the famine were a million deaths from starvation and the emigration of about 2 million people, mostly to the United States. *Albugo*, another member of the Oomycota, is a well-known parasitic genus causing a mealy blight on sweet potato leaves, morning glories, and numerous other plants. It is an obligate parasite and has never been grown on any medium other than its plant host.

ALGAE

The algae are photosynthetic. They probably carry out 50–60 percent of all the photosynthesis on the planet, with the kingdom Plantae accounting for most of the rest. The overall contribution of cyanobacteria and other photosynthetic bacteria is smaller, although it is locally important in some aquatic ecosystems. The algae differ from the plants in that the zygote of an alga is on its own—the parent gives the zygote no protection, whereas a plant zygote grows into a multicellular embryo that is protected by parental tissue.

The algae exhibit a remarkable range of growth forms. Some are simply unicellular; others are filaments composed either of distinct cells or of multinuclear structures without cross walls (coenocytes). Others—including the algae commonly known as seaweeds—are multicellular and intricately branched or arranged in multicellular, leaflike extensions. The bodies of a few types of algae are even subdivided into tissues and organs. Certain algal phyla—for example, phylum Chlorophyta, the green algae—include representatives of almost all these growth forms.

Algal life cycles also show extreme variation, but all except members of the phylum Rhodophyta (red algae) have forms with flagellated motile cells in at least one stage of their life cycle. Some are unicellular and motile throughout most of their existence (for example, the dinoflagellates—the protist phylum Pyrrophyta).

Table 22.1, shown earlier in this chapter, shows some of the principal biochemical characteristics of the algae. Except for mannitol (a sugar alcohol), fats, and oils, all the storage compounds listed are polymers of glucose. These polymers differ among themselves in the kind of chemical linkage between their adjacent glucose molecules, in the degree of branching of their polysaccharide chains, and in their chain size (Chapter 3). The pectic substances found in algal cell walls are all variously modified polymers of a sugar acid, galacturonic acid.

Phylum Pyrrophyta

The Pyrrophyta are unicellular, and a distinctive mixture of photosynthetic and accessory pigments gives their chloroplasts a golden-brown color. The dinoflagellates, the major group within the Pyrrophyta, are of great ecological, evolutionary, and morphological interest. Dinoflagellates are probably second in importance only to diatoms (members of the algal phylum Chrysophyta; see the following section) as primary photosynthetic producers of organic matter in the oceans. Many dinoflagellates are endosymbionts, living within the cells of other organisms, including various invertebrates and even other marine protists. Dinoflagellates are particularly common endosymbionts in corals, to whose growth they contribute mightily by photosynthesis. Some dinoflagellates are nonphotosynthetic and live as parasites within other marine organisms.

The dinoflagellates are notably peculiar cells (Figure 22.18a). They have two flagella, with one in an equatorial groove around the cell and the other starting at the same point as the first and passing down

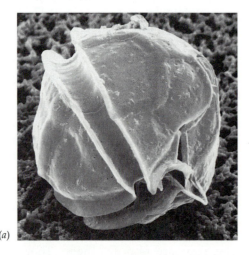

22.18 Dinoflagellates
(a) A single *Gonyaulax tamarensis*. Each groove in the side of the cell normally contains a flagellum, but the flagella are not seen here. (b) In astronomical numbers, this species causes a toxic "red tide," as seen here off the coast of Baja California.

(a)

(b)

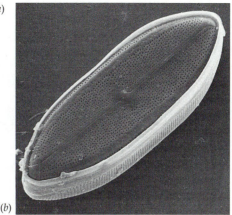

(a)

(b)

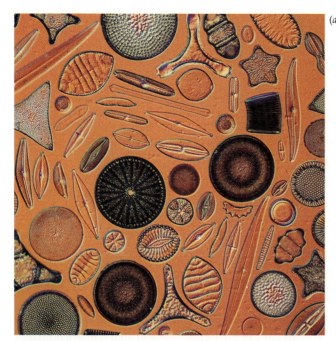

22.19 Diatoms
(a) Diatoms exhibit a splendid variety of species-specific forms. (b) *Neidium iridus.* The dark and light areas of this scanning electron micrograph emphasize the distinct two-piece construction of the cell wall.

a longitudinal groove before extending free into the surrounding medium. Most dinoflagellates are marine organisms. Some dinoflagellates reproduce in enormous numbers in warm and somewhat stagnant waters. The result can be a "red tide," so called because of the reddish cast of the sea that results from fluorescence of the chlorophyll in the dinoflagellates (Figure 22.18b). During a red tide, the concentration of dinoflagellates may reach 60 million cells per liter of ocean water. Certain species produce a potent nerve toxin; a red tide made up of these species can kill great numbers of fish. A particularly severe red tide in the Gulf of Mexico in the summer of 1971 killed tons of fish along the west coast of Florida. Likewise, the genus *Gonyaulax* produces a potent toxin that can accumulate in shellfish in amounts that, although not fatal to the shellfish, may kill a person who eats it.

Many dinoflagellates are bioluminescent (Box 7.A). In complete darkness, cultures of these organisms emit a faint glow. If one suddenly disturbs a culture physically, by stirring it or bubbling air through it, the organisms each emit a number of bright flashes, perhaps a thousandfold brighter than the dim glow of a quiet culture. The flashing then rapidly subsides. A ship passing through a tropical ocean containing a rich growth of these species produces a bow wave and a wake that glow eerily as billions of these dinoflagellates discharge their light systems.

Phylum Chrysophyta

Diatoms and their relatives make up the Chrysophyta. Some species are single-celled; others are filamentous. Many have sufficient carotenoids (Chapter 3) in their chloroplasts to give them a yellow or brownish color; all make chrysolaminarin (a carbohydrate) and oils as photosynthetic storage products. Many diatoms deposit silicon in their cell walls. The cell wall of some species is constructed in two pieces, with the walls of the top overlapping the walls of the bottom like the top and bottom of a petri plate. An organism such as this that has a "top" and a "bottom"—but no front or rear—is said to be radially symmetrical. (Most organisms that are not radially symmetrical are bilaterally symmetrical; that is, they have right and left sides that are mirror images of each other.)

Architectural magnificence on a microscopic scale is the hallmark of the diatoms (Figure 22.19). The silicon-impregnated walls produced by these marine or freshwater organisms have intricate patterns; in fact, the taxonomy of the diatoms is entirely based on their wall patterns. Looking down at the top, we see one of two basic shapes: A diatom is either radially symmetrical or bilaterally symmetrical.

Diatoms reproduce both sexually and asexually. Asexual reproduction is by cell division and is somewhat constrained by the silica-containing cell wall. Both the top and the bottom of the "petri plate" become tops of new "plates" without changing appreciably in size, and, as a result, the new cells made from former bottoms are smaller than the parent cells. If the process continued indefinitely, one cell line would simply vanish. Sexual reproduction largely solves the problem. Gametes form, shed their cell walls, and fuse. The zygote is sometimes called an auxospore, a name whose Greek origin (*auxein,* "increase") refers to the fact that it increases substantially in size before a new wall is laid down (Figure 22.20).

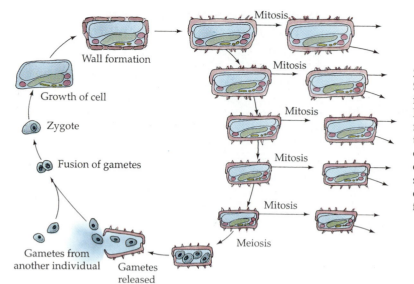

22.20 Diatom Reproduction
Silicon-impregnated cell walls, shown edge-on in this diagram, are two-part "petri dishes." In asexual reproduction by mitosis and cytokinesis (vertical sequence on right-hand side), the parts separate and each becomes the top of a new dish. In the process, the offspring cells within the cell wall become progressively smaller. Zygotes produced by sexual reproduction (cycle at left) grow and lay down new full-size cell skeletons.

Diatoms are ubiquitous in the ocean and are frequently present in great numbers. They are also common in fresh water. Because the walls of dead cells resist decomposition, certain sedimentary rocks are composed almost entirely of these silica-containing skeletons that sank to the sea floor. Diatomaceous earth, which is obtained from such rocks, has many industrial uses, from insulation and filtration to metal polishing.

Phylum Phaeophyta

The Phaeophyta, which are called brown algae, are always multicellular, composed either of branched filaments or of leaflike growths called **thalli** (Figure 22.21). The brown algae obtain their namesake color from the carotenoid fucoxanthin, present in large amount in the plastids. The combination of this yellow-orange pigment with the green of chlorophylls a and c yields a dirty brown color.

The Phaeophyta include the largest of the algae. Giant kelps, such as those of the genus *Macrocystis*, may be up to 60 meters long (see Figure 29.2). The brown algae are almost exclusively marine. Some float in the open ocean—the most famous example is the genus *Sargassum*, which forms dense mats of vegetation in the Sargasso Sea in the mid-Atlantic. Most brown algae, however, are attached to rocks

(a)

(b)

22.21 Brown Algae
(a) This species of *Hornosira* is an intertidal brown alga growing in Australia. (b) A species of the filamentous *Ectocarpus* as seen through a light microscope.

near the shore. A few thrive only where they are regularly exposed to heavy surf—a notable example is the sea palm *Postelsia palmaeformis* of the Pacific coast. The attached forms all develop a specialized structure, called a **holdfast**, that literally glues them to the rocks (Box 22.A). Some brown algae may differentiate extensively into stemlike stalks and leaflike blades, and some develop gas-filled cavities or bladders. For biochemical reasons that are only poorly understood, these gas cavities often contain as much as 5 percent carbon monoxide—a concentration high enough to kill a human.

In addition to organ differentiation, the larger brown algae also exhibit considerable tissue differentiation. Most of the giant kelps have photosynthetic filaments only in the outermost regions of the various organs. Inside the photosynthetic region lie filaments of long cells that very much resemble the food-conducting tissue of plants (Chapter 29). Called trumpet cells because they have flaring ends, they are tubes that rapidly conduct the products of photosynthesis (mostly mannitol) through the body of the alga.

The brown algae exemplify the extraordinary diversity found among the algae. One example of a genus of simple brown algae is *Ectocarpus* (Figure 22.21*b*). Its branched filaments, a few centimeters in length, commonly grow on shells and stones. The gametophyte and sporophyte phases of its alternation of generations can be distinguished only by chromosome number or reproductive products (zoospores or gametes). Thus the generations are isomorphic. In contrast, some kelps of the genus *Laminaria* and some other brown algae show a more complex **heteromorphic alternation of generations**. The large and obvious alga of these species is the

sporophyte. Meiosis in special fertile regions of the leaflike fronds produces haploid zoospores. These germinate to form a tiny, filamentous gametophyte. The gametophytes produce either eggs or sperm. The genus *Fucus* carries gametophyte reduction still further: It has no multicellular haploid phase, only a multi*nuclear* haploid phase. The gametes themselves are formed directly by meiosis.

The cell walls of brown algae may contain as much as 25 percent alginic acid, a gummy polymer of sugar acids. The substance serves to cement cells and filaments together and also provides good holdfast glue. It is used commercially as an emulsifier in ice cream, cosmetics, and other products.

Phylum Rhodophyta

The red algae—the Rhodophyta, Figure 22.22—are almost all multicellular. The characteristic color of the red algae is a result of the pigment phycoerythrin, which is found in relatively large amounts in the chloroplasts of many species. In addition to phycoerythrin, red algae contain phycocyanin, carotenoids, and chlorophyll. The red algae include species that grow in the shallowest tide pools and also the algae found deepest in the ocean (as deep as 170 meters where nutrient conditions are right and the water is clear enough to permit the penetration of light). Very few red algae inhabit fresh water. Most grow attached to some substrate by a holdfast.

In a sense the red algae, along with several other groups of algae, are misnamed. They have the capacity to change the relative amounts of their various photosynthetic pigments depending upon the light conditions where they are growing. Thus the leaflike *Chondrus crispus*, a common North Atlantic red alga,

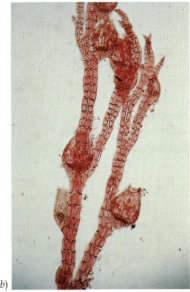

22.22 Red Algae
(a) Dulce is an edible red alga. (b) A species of *Polysiphonia*, with both vegetative and reproductive structures, seen under the light microscope. Both these algae get their red color largely from the pigment phycoerythrin.

BOX 22.A

Algae in a Turbulent Environment

One of the most interesting environments, because of both its animal and its algal life, is the intertidal zone of a rocky beach such as that shown in the figure. Here the resident organisms are subjected to frequent, dramatic changes in water supply, availability of oxygen, temperature, and other factors, not the least of which is an intermittent pounding by surf. Multicellular algae that grow in this environment are equipped with firm holdfasts that anchor them against the surging water. Although seaweeds are common in the rocky intertidal zone, they rarely grow on sandy beaches, where there are no stable attachment sites.

Two algal growth forms are evident in a rocky intertidal environment, one featuring upright growth and the other a prostrate, fleshy, crustlike growth. The more upright algae may be more subject to grazing by animals, but they can seize a photosynthetic advantage by overgrowing and shading the prostrate forms. The prostrate algae, on the other hand, have the reciprocal advantage and disadvantage—they may be shaded by the upright algae, but they are less subject to grazing. A number of algal species alternate between these two general forms, a phenomenon known as **heteromorphy**. Where grazing pressure varies substantially and predictably, there is a seasonal alternation between the two patterns of growth (upright and prostrate) in the same species, with the prostrate form predominating at times of high grazing pressure. Other

Plants growing in the intertidal zone on an exposed rocky beach take a tremendous pounding by the surf. (a) Sea palm (*Postelsia palmaeformis*), a brown alga growing along the California coast. (b) A tough, branched holdfast that anchors a brown alga dominates this photograph.

heteromorphic algae, subject to variable but unpredictable grazing pressure, produce both forms at all times. The population as a whole may profit under such unpredictable conditions by always having at least one well-adapted growth form available.

One might think of this violent environment as one where only a few hardy organisms eke out an existence, clinging to the rocks. In fact, some of the rocky intertidal areas where wave action is heaviest are the most productive communities on Earth, yielding from two to ten times as much photosynthetic product as the most productive tropical rain forests. Shrubby kelp in such areas carry up to two and one-half times as much photosynthetic surface (fronds) as do the plant communities of tropical rain forests. What factors make

this environment supportive of such intense production? The heavy pounding of the surf contributes in at least four ways. Wave action constantly replaces the nutrient-laden water in contact with the algae, so that the supply of mineral elements is always at a maximum. Heavy wave action keeps certain important consumers, such as sea urchins, from attacking and eating the algae. Similarly, heavy wave action dislodges some of the less tenacious organisms from the rocks, making it possible for shrubby kelps, sea palms, and other well-adapted algae to take their places and form dense "forests." Finally, wave action keeps the fronds of the algae in constant motion, so that no fronds are shaded for long—all are lit, maximizing photosynthesis.

may appear bright green when it is growing at or near the surface of the water, and deep red when growing at greater depths. The pigmentation—the ratio of pigments present—depends to a remarkable degree upon the wavelengths of the light that reaches the alga. In deep water, the photosynthetically effective light penetrating the ocean is mostly in the blue-green part of the spectrum, and the algae there accumulate large amounts of phycoerythrin, the pigment that absorbs light in that region (Figure 22.23). (If you do not recall what the color of a pigment tells us about the wavelengths it absorbs, review Figures 8.8 and 8.9 and the accompanying text.) The algae in deeper water have as much chlorophyll as the green ones near the surface, but the accumulated phycoerythrin makes them look red. Variation in pigmentation depending on light wavelength is known as **chromatic adaptation**.

In addition to being the only algae with phycoerythrin and phycocyanin among their pigments, the red algae have two other unique characteristics. They store the products of photosynthesis in the form of floridean starch, which is composed of very small, branched chains of approximately 15 glucose units; and they produce no motile, flagellated cells at any stage in their life cycle. The male gametes are naked and slightly amoeboid, whereas the female gametes are completely immobile.

Some red algal species enhance the formation of coral reefs. They share with the coral animals the biochemical machinery for depositing calcium carbonate as a precipitate both in and around their cell walls. After the death of the algal cells, the calcium carbonate persists, sometimes forming substantial rocky masses.

Some red algae also produce large amounts of mucilaginous polysaccharide substances, which are based mostly on the sugar galactose with a sulfate group attached. This material readily forms solid gels and is the source of agar, a substance widely used in the laboratory for making a solid aqueous medium on which tissue cultures and many microorganisms may be grown.

Some marine red algae are parasitic upon other red algae. The hosts are photosynthetic, but the parasites are often colorless and nonphotosynthetic, deriving their nutrition from the host. Lynda Goff and Annette Coleman of Brown University discovered that the parasitic red alga *Choreocolax* inserts its nuclei into cells of its host red alga, *Polysiphonia*. This is the first example found of regular introduction of parasite nuclei into living host cells. Apparently parasite genes are expressed in the host cytoplasm, diverting the host's metabolism.

Phylum Chlorophyta

The Chlorophyta, commonly known as green algae, and the photosynthetic Mastigophora such as *Euglena* and *Volvox*, which many biologists include with the green algae, are the only protist groups that contain the full complement of photosynthetic pigments also characteristic of the kingdom Plantae. Chlorophyll a predominates, and a major pigment is chlorophyll b, which none of the other algae have. The carotenoids found in these groups, predominantly β-carotene and certain xanthophylls (carotenoids with one or more hydroxyl groups), are likewise those characteristic of plants. The principal photosynthetic storage product, like that of the plant kingdom, is long, straight, or branched chains of glucose that together make up starch (Chapter 3). Because of these similarities, and for other reasons, it is thought that the plant kingdom evolved from one or more representatives of the phylum Chlorophyta.

Uniformity of pigmentation and photosynthetic storage product is combined in the green algae with an incredible variety in shape and construction of the algal body. *Chlorella* is an example of the simplest type: unicellular and flagellated. *Oedogonium* is filamentous, with each cell having only one nucleus. *Cladophora* is multicellular, but each cell is multinucleate. *Bryopsis* is tubular and coenocytic, forming cross walls only with the formation of reproductive structures. *Acetabularia* (see Figure 4.13) is a single, giant uninucleate cell with remarkable morphology,

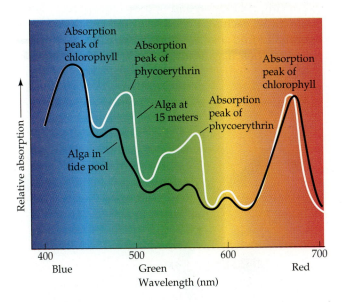

22.23 Light Absorption by a Red Alga
When the red alga *Chondrus crispus* grows under 15 meters of seawater, it accumulates phycoerythrin; with large amounts of this pigment in its cells, it absorbs blue-green wavelengths strongly. When this alga grows at the surface, it contains little phycoerythrin and, using its molecules of chlorophyll, absorbs blue and red wavelengths most strongly.

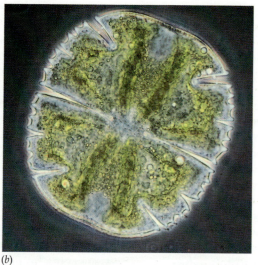

(a)   *(b)*

22.24   Green Algae
(a) A stand of sea lettuce, *Ulva lactuca*, submerged in a tide pool. *(b)* A micro-scopic desmid of the genus *Micrasterias*. A narrow, nucleus-housing isthmus joins two elaborate semicells; a single, large, ornate chloroplast fills much of the volume of each semicell.

becoming multinucleate only at the end of the repro-ductive stage. *Ulva lactuca* is a membranous sheet two cells thick; the unusual appearance of *U. lactuca* jus-tifies its common name of sea lettuce (Figure 22.24*a*). Finally, there are the remarkable unicellular desmids with their elaborately sculptured cell walls (Figure 22.24*b*).

LIFE CYCLES IN THE CHLOROPHYTA.   There is also great diversity of life cycles among the green algae. We will

examine two of these in detail, beginning with the life cycle of the sea lettuce *Ulva lactuca*, which is summarized in Figure 22.25. The diploid sporophyte of this common seashore alga is a "leaf" a few cen-timeters in diameter. Specialized cells (sporocytes) differentiate and undergo meiosis and cytokinesis, producing motile haploid spores (zoospores). These swim away, each propelled by four flagella, and some eventually find a suitable place to settle. The spores then lose their flagella and begin to divide mitotically,

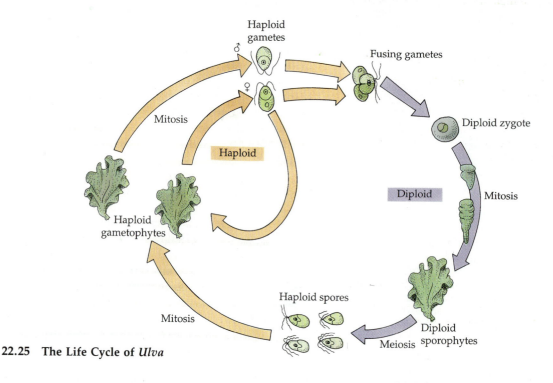

22.25   The Life Cycle of *Ulva*

producing a thin filament that develops into a broad sheet only two cells thick. The haploid gametophytes thus produced look just like the diploid sporophytes.

Each spore contains genetic information for just one mating type, and a given gametophyte can produce only male or female gametes, never both. The gametes arise mitotically within single cells (gametangia), rather than within any specialized multicellular structure as in mosses and vascular plants (Chapter 24). Both types of gametes bear two flagella (in contrast to the four flagella of a haploid spore) and hence are motile. However, the female gamete is distinctly larger, so *Ulva* is **anisogamous**—having gametes of differing appearance. (This is in contrast to some other algae and funguslike protists that are **isogamous**, that is, whose gametes are indistinguishable structurally.) Sperm and egg come together and unite, losing their flagella as the zygote forms and settles. After resting briefly, the zygote begins mitotic division, forming a new sporophyte. Any gametes that fail to find partners can settle down on a favorable substrate, lose their flagella, undergo mitosis, and produce new gametophytes directly; in other words, the gametes can also function as zoospores. Motile gametes that can also function as zoospores, as in *Ulva*, are rare.

A life cycle like that of *Ulva* is isomorphic: Sporophyte and gametophyte generations are identical in structure. In the funguslike protists and many other algae, the generations are heteromorphic; that is, the sporophyte differs from the gametophyte in form. In one variation of the heteromorphic cycle—the **haplontic** life cycle—multicellular gametophytes produce gametes that fuse to form a zygote. The zygote functions directly as a sporocyte, undergoing meiosis to produce spores, which in turn produce new gametophytes. In the entire haplontic life cycle, only one cell—the zygote—is diploid. The filamentous green algae of the genus *Ulothrix* are examples of haplontic algae (Figure 22.26).

Other algae go through a **diplontic** life cycle like that of many animals. Meiosis of sporocytes produces gametes directly; these fuse, and the zygote divides mitotically to form a new multicellular sporophyte. In such organisms, every cell except the gametes is diploid. Between these two extremes, one finds algae whose gametophyte and sporophyte generations are both multicellular, but having one phase (usually the sporophyte) that is much larger and more prominent than the other.

Among the algae, a number of evolutionary trends have been discerned. For example, there is a trend toward multicellularity in some groups. Another trend leads from the mass release of large numbers of tiny, seemingly undifferentiated gametes (isogamy) toward increased protection of a single, large egg. For example, the female gamete of the green alga *Oedogonium* is large and immobile and the male

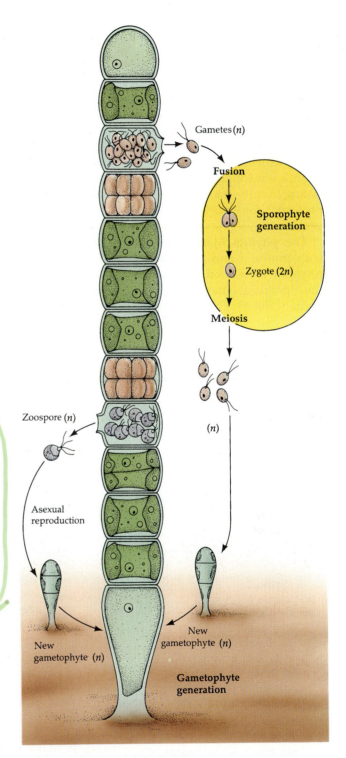

22.26 A Haplontic Life Cycle
In the life cycle of *Ulothrix*, a filamentous, multicellular gametophyte generation alternates with a sporophyte generation consisting of a single cell. Some gametophyte cells divide mitotically under certain conditions, forming zoospores that develop into other multicellular gametophytes. Sexual reproduction is initiated when some cells of a gametophyte divide mitotically to form gametes that fuse in pairs to form zygotes. A zygote undergoes meiosis, producing haploid cells that develop into the next gametophyte generation.

gamete is free-swimming and flagellated—an extreme case of anisogamy known as **oogamy**.

SUMMARY

All members of the Protista are eukaryotic, and all evolved from the Monera. Most—but not all—protists are unicellular. The kingdom is more heterogeneous than the other kingdoms—its members are often regarded simply as those that biologists do not accept as true fungi, animals, or plants. As a consequence, the kingdom Protista may be viewed as consisting of three major subgroupings, the (animallike) protozoa, the (plantlike) algae, and the funguslike protists.

The phylum Mastigophora (flagellates) includes both protozoa and algae. The remaining protozoan phyla are the Sarcodina (amoebas and their relatives), the Sporozoa (amoeboid parasites), and the Ciliophora (ciliates). Funguslike protist phyla include the Gymnomycota (slime molds), the Protomycota (chytrids and hypochytrids) and the Oomycota (water molds). The unicellular algal phyla of the kingdom Protista include the Pyrrophyta (dinoflagellates and their relatives) and the Chrysophyta (diatoms and their relatives). There are three phyla of multicellular algae, the Phaeophyta (brown algae, some of which are quite large), the Rhodophyta (red algae), and the Chlorophyta (green algae, a highly diverse group from which the plant kingdom evolved). The algal phyla are defined on the bases of the chemistry of their photosynthetic pigments and products and of their life cycles.

Photosynthetic protists play a major role in the energy balance of the living world, and saprobic protists are among the important decomposers. Some parasitic protists are agents of disease in plants and animals. Many protists have highly differentiated bodies even though they consist of but a single cell. It is believed that the multicellular kingdoms all evolved from protists.

SELF-QUIZ

1. The members of the Mastigophora:
   a. all possess flagella.
   b. are all algae.
   c. are all protozoa.
   d. are never colonial.
   e. are never pathogenic.

2. Which statement is *not* true of the Sarcodina?
   a. They apparently evolved from the Mastigophora.
   b. They use amoeboid movement.
   c. They include the foraminiferans.
   d. They include the radiolarians.
   e. They possess flagella.

3. The Sporozoa:
   a. possess flagella.
   b. being amoeboid, are closely related to the Sarcodina.
   c. are all parasitic.
   d. are among the funguslike protists.
   e. include the trypanosomes that cause sleeping sickness.

4. The Ciliophora:
   a. move by means of flagella.
   b. use amoeboid movement.
   c. include *Plasmodium*, the agent of malaria.
   d. possess both a macronucleus and micronuclei.
   e. are autotrophic.

5. The Myxomycetes:
   a. are also called the acellular slime molds.
   b. lack fruiting bodies.
   c. consist of large numbers of myxamoebas.
   d. consist at times of a mass called a pseudoplasmodium.
   e. possess flagella.

6. The Acrasiomycetes:
   a. are also called the acellular slime molds.
   b. lack fruiting bodies.
   c. form a plasmodium that is a coenocyte.
   d. use cAMP is a "messenger" to signal aggregation.
   e. possess flagella.

7. Which statement is *not* true of the algae?
   a. They differ from plants in lacking protected embryos.
   b. They are photosynthetic autotrophs.
   c. They all use the same compounds for storage.
   d. They include both unicellular and multicellular forms.
   e. Their life cycles show extreme variation.

8. Which statement is *not* true of the Phaeophyta?
   a. They are all multicellular.
   b. They use the same photosynthetic pigments as do plants.
   c. They are almost exclusively marine.
   d. Many have alginic acid in their cell walls.
   e. Some have extensive tissue differentiation.

9. The members of the phylum Rhodophyta:
   a. are mostly unicellular.
   b. are mostly marine.
   c. owe their red color to a special form of chlorophyll.
   d. have flagella on their gametes.

10. Which statement is *not* true of the phylum Chlorophyta?
    a. These algae use the same photosynthetic pigments as do plants.
    b. Some members of the phylum are unicellular.
    c. Some members are multicellular.
    d. All its members are microscopic in size.
    e. There is great diversity of life cycles among the members.

FOR STUDY

1. For each type of organism given below, give a single characteristic that may be used to differentiate it from the other, related organism(s) in parentheses.
   a. Foraminifera (Radiolaria)
   b. Vorticella (Paramecium)
   c. Euglena (Volvox)
   d. Trypanosoma (Giardia)
   e. amoeba (foraminiferan)
   f. Physarum (Dictyostelium)

2. For each of the groups listed below, give at least two characteristics used to delimit the group from others.
   a. Ciliophora
   b. Sporozoa
   c. Sarcodina
   d. Mastigophora

3. What is a major role in the world played by the photosynthetic protists? What is a major role in the world played by the saprobic protists?

4. Giant seaweed (mostly brown algae) have "floats" that aid in keeping their fronds suspended at or near the surface of the water. Why is it important that the fronds be suspended?

5. Justify the placement of *Euglena* in the Mastigophora. Why might it just as well be placed in the Chlorophyta?

READINGS

Donelson, J. E. and M. J. Turner. 1985. "How the Trypanosome Changes Its Coat." *Scientific American*, February. Trypanosomes evade the host's immune system by constantly switching on new genes that code for different surface antigens.

Farmer, J. N. 1980. *The Protozoa*. Mosby, St. Louis. A full treatment of the animal-like protists.

Godson, G. N. 1985. "Molecular Approaches to Malaria Vaccines." *Scientific American*, May. The life cycle and molecular biology of another problem pathogen.

Saffo, M. B. 1987. "New Light on Seaweeds." *BioScience*, October. Experimental observations on vertical distribution of algae.

Vidal, G. 1984. "The Oldest Eukaryotic Cells." *Scientific American*, February. Ancient marine protists.

23

Fungi

PREVIEW: Some fungi are parasitic on animals, plants, protists, or other fungi. Others are saprobic and return carbon and other elements to the environment for further cycling. Still others are mutualistic, associating with photosynthetic organisms to form mycorrhizae or lichens. Many fungi are economically useful. The fungi exploit virtually every habitat on Earth.

This chapter deals with diversity among the lichens and the fungi, including the division Eumycota and its classes, the Zygomycetes, the Ascomycetes, the Basidiomycetes, and the Deuteromycetes; and with mycelia, hyphae, rhizoids, haustoria, dikaryons, and some fungal diseases.

The fungi play a crucial role in the biosphere: They are the most important degraders of dead organic matter. Members of the kingdom Fungi are found in all environments, and parasitic fungi attack virtually all eukaryotes, including other fungi (Figure 23.1). Many fungi are exquisitely constructed, and their life cycles are extremely complex. The great diversity in anatomy and life cycles of fungi and certain protists has caused some taxonomic difficulty. However, most problems are resolved if we view the kingdom Fungi as encompassing *heterotrophic organisms with absorptive nutrition.* The fungi, as we define them, are **saprobes** (organisms living on dead matter), parasites, or **mutualists**—organisms living in mutually beneficial symbiosis with other organisms. They form spores, and none of their cells ever possess flagella. Sexual reproduction in the fungi is by a process called **conjugation**, in which filaments of different mating types fuse. The cell walls of all fungi consist of the polysaccharide **chitin**, which is also found in the cell walls of certain funguslike protists as well as in the skeletons of arthropods (Chapter 3). Most fungi are multicellular.

These criteria enable us to distinguish between the fungi and certain other groups that we have presented as phyla of the kingdom Protista (Chapter 22). The slime molds (protist phylum Gymnomycota) consist of one class (Acrasiomycetes) that takes up food by phagocytosis rather than by absorption and another (Myxomycetes) that has cells with flagella. The protist phyla Protomycota and Oomycota also have flagellated cells, and the Oomycota have cellulose, rather than chitin, in their cell walls.

We consider the kingdom Fungi to consist of a single division, the **Eumycota,** or true fungi (Table 23.1). We distinguish the classes of the Eumycota on the basis of the methods and structures they use for sexual reproduction and, to a lesser extent, on criteria such as the presence or absence of cross walls separating their cells. The Eumycota are divided into four classes: the Zygomycetes (conjugating fungi); Ascomycetes (sac fungi); Basidiomycetes (club fungi); and the Deuteromycetes (imperfect fungi), which do not form sexual structures by which they might be identified as members of one of the other three classes.

GENERAL BIOLOGY OF THE FUNGI

The Fungal Body

The body of a multicellular fungus is called a **mycelium.** Its cells are organized into rapidly growing individual filaments called **hyphae.** Certain hyphae, the **rhizoids,** anchor saprobic fungi to their substrate. Parasitic fungi may have modified hyphae, called **haustoria,** that penetrate the cells of the host organism. The total hyphal growth of a mycelium (not the growth of an individual hypha) may exceed 1 kilometer per day. The hyphae may be highly dispersed or may be clumped into a cottony mass. Sometimes the mycelium becomes organized into elaborate fruiting bodies such as mushrooms or puffballs.

The attack of a parasitic fungus on a plant illustrates the roles of some fungal structures (Figure 23.2). The hyphae of a fungus invade a leaf through pores in the leaf called stomata (Chapters 28, 29) or wounds. Once inside, the hyphae form a mycelium. Haustoria grow into living plant cells, absorbing the

(a)

(b)

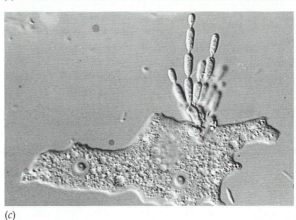

(c)

23.1 Parasitic Fungi

(a) The gray masses on this ear of corn are the parasitic fungus commonly called corn smut (*Ustilago maydis*). (b) The tropical fungus whose fruiting body is growing out of the carcass of this ant has developed from a spore ingested by the ant. The spores of this fungus must be ingested by insects before they will germinate and develop into a mycelium. The mycelium absorbs organic and inorganic nutrients from the ant's body, eventually killing it, after which the fruiting body produces a new crop of spores. (c) An amoeba (below) being parasitized by a fungus of the genus *Amoebophilus* ("the amoeba-lover").

nutrients within the cells. Eventually fruiting bodies form, either within the plant body or on its surface. Later in this chapter we describe in more detail a complex fungal life cycle—that of the black stem rust of wheat.

Fungi and Their Environment

The tubular hyphae of a fungus give it a unique relationship with its physical environment: The fungal mycelium has an enormous surface-to-volume ratio compared with most other large multicellular organisms. This large surface-to-volume ratio is a marvelous adaptation for absorptive nutrition, in which nutrients are absorbed across the cell surfaces. Throughout the mycelium, except in fruiting bodies, all the cells are very close to their environmental food source.

Another characteristic of fungi is their tolerance for highly hypertonic environments (ones with more negative osmotic potentials; Chapter 5). Many fungi

TABLE 23.1
Classification of Fungi

DIVISION	CLASS	COMMON NAME	FEATURES	EXAMPLES
Eumycota (80,000 species)	Zygomycetes	Conjugating fungi	No hyphal cross walls, usually no fleshy fruiting body	*Rhizopus*
	Ascomycetes	Sac fungi	Ascus, perforated cross walls	*Neurospora*, yeast
	Basidiomycetes	Club fungi	Basidium, complete cross walls	*Puccinia*, mushrooms
	Deuteromycetes	Imperfect fungi	No known sexual stages	*Arthrobotrys* (traps nematodes)
(20,000 species)		Lichens	Mutualistic association of a true fungus and a photosynthetic microorganism	*Cladonia, Letharia*

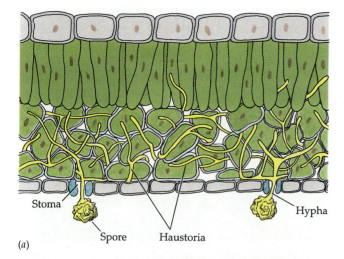

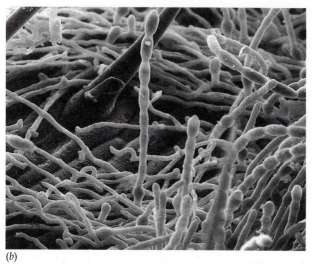

23.2 A Fungus Attacks a Leaf
(a) Fungal spores germinate on the surface of the leaf. Elongating hyphae pass through stomata into the interior of the leaf, elongate further, and branch. Some hyphae develop into haustoria that penetrate cells within the leaf. (b) The surface of the grass *Stenotaphrum secundatum* infected with the fungus *Erysiphe grammis*; the plant cells are darker and fungal cells lighter in shade.

are more resistant than bacteria to damage in hypertonic surroundings—for example, jelly in the refrigerator will not become a growth medium for bacteria, but it may eventually harbor mold colonies. The refrigerator example illustrates a further trait of fungi: many fungi tolerate low temperatures, 5 or 6 degrees below freezing, and some tolerate high temperatures of 50°C or more. You have probably seen the green mold *Penicillium* growing on oranges in the refrigerator (Figure 23.3).

Nutrition of Fungi

The fungi are all heterotrophs that obtain their food by absorption. Many of them are saprobes, obtaining their energy, carbon, and nitrogen directly from dead

organic matter by the action of enzymes. Many other fungi are parasites. Still other fungi form mutualistic associations with other organisms. Saprobic fungi, along with the bacteria, are the major decomposers of the biosphere, contributing to decay and thus to the recycling of the various elements used by living things (Figure 23.4).

Saprobic fungi can be grown on artificial media, so their exact nutritional requirements can be determined. Sugars are their favored sources of carbon, and most fungi obtain their nitrogen in the form of proteins or protein breakdown products. Many, but not all, fungi can use nitrate or ammonium ions as their sole source of nitrogen. No known fungus can use nitrogen gas directly, as can some bacteria and plant–bacteria associations (Chapter 30). Vitamins, which are defined and discussed in Chapter 41, play some role in fungal nutrition. Most fungi are unable to synthesize their own thiamin (vitamin B_1) or biotin and hence must absorb them from their environment. Other compounds that are vitamins for animals can be made by the fungi. All fungi, like all other organisms, require a number of mineral elements.

The parasitic fungi are particularly interesting from a nutritional point of view. **Facultative** parasites are those that can grow parasitically but can also be grown by themselves on defined artificial media. Biologists can work out the exact nutritional requirements of a facultative parasite by varying the composition of the growth medium. The **obligate** parasites—those that cannot be grown on any available defined medium—are perhaps more interesting. These fungi, including various mildews and rusts, can grow only on their specific hosts—usually plants (Figure 23.1a). It is obvious that they have unusual nutritional requirements, and it will be of great importance to learn what they are—which biologists will do eventually with new approaches.

Some fungi have adaptations allowing them to function as active predators, trapping microscopic

23.3 A Kitchen Nuisance
The green mold *Penicillium digitatum* on an orange.

23.4 A Saprobic Fungus
The invisible mycelium producing these showy structures of the turkey-tail, *Trametes versicolor*, is obtaining nutrients from a fallen tree branch, thus taking part in the decomposition of the branch.

protists or animals in their vicinity. The most common approach is to secrete sticky substances from their hyphae, so that passing organisms stick tightly to them. Fungal haustoria then quickly invade the prey, growing and branching within it, spreading through its body, absorbing nutrients, and eventually killing it. A more dramatic method of predation is the **constricting ring** formed by some species of *Arthrobotrys*, *Dactylaria*, and *Dactylella* (Figure 23.5). All these fungi grow in soil; when nematodes (tiny roundworms) are present, the fungi form three-celled rings with a diameter that can just accommodate a nematode. A nematode crawling through a ring stimulates it, causing the cells of the ring to swell and trap the worm. Fungal hyphae quickly invade and digest the unlucky victim. In 1984, R. G. Thorn and G. L. Barron of the University of Guelph reported that 11 of 27 species of wood-decaying mushrooms tested also have the ability to supplement their nitrogen supply by trapping nematodes.

Certain highly specific associations between fungi and other organisms have nutritional consequences for the fungal partner. **Lichens** are mutualistic associations of a fungus with either a unicellular alga or a cyanobacterium; the combination functions effectively as a "plant." In the union, the fungus protects

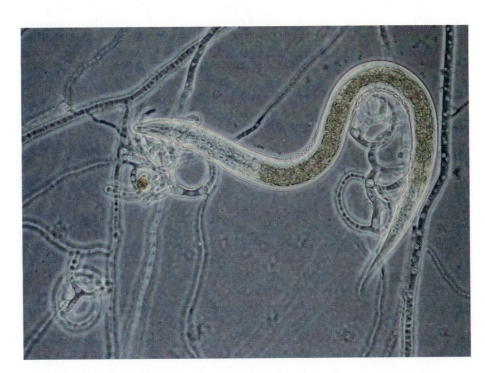

23.5 Predaceous Fungi
A nematode (roundworm) trapped in a sticky loop of the soil-dwelling fungus *Arthrobotrys conoides*; an empty loop is seen at lower left.

23.6 Ants Farming a Fungus
Workers of the Costa Rican ant species *Atta cephalotes* add a cut piece of leaf to their fungal garden. The fungus is the white material. In response to this care, the fungus grows and serves as food for the ants.

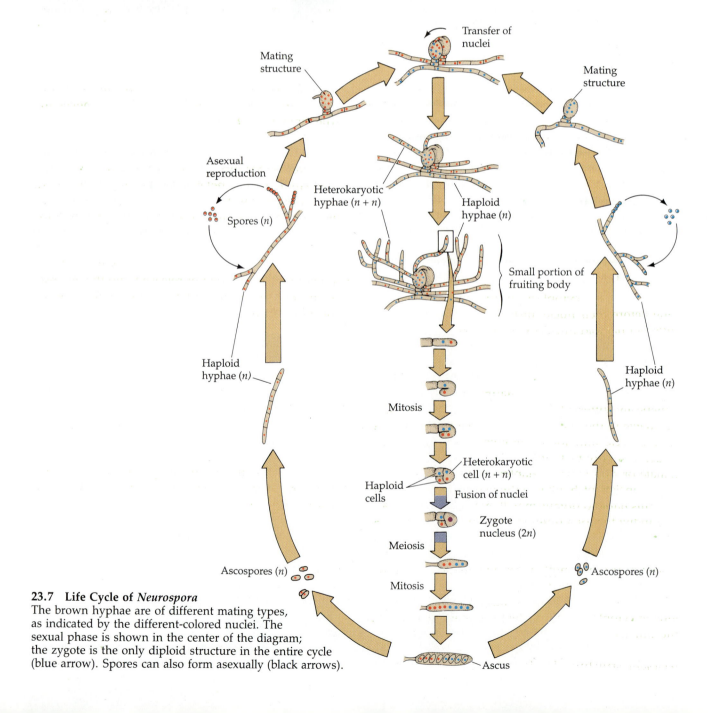

23.7 Life Cycle of *Neurospora*
The brown hyphae are of different mating types, as indicated by the different-colored nuclei. The sexual phase is shown in the center of the diagram; the zygote is the only diploid structure in the entire cycle (blue arrow). Spores can also form asexually (black arrows).

the photosynthetic bacterium or alga while simultaneously drawing nutrition from it. In Box 28.A we will describe **mycorrhizae**: associations between fungi and the roots of certain plants. In such associations, the fungus is fed by the plant but provides minerals to the root (primarily phosphorus), so that the plant's nutrition is promoted, too. Seed germination in most orchid species depends upon the presence of a specific mutualistic fungus, which itself derives nutrients from the seed and seedling.

Perhaps the most striking fungal associations, though, are with insects. Some leaf-cutting ants farm fungi, feeding the fungi and later harvesting and eating them (Figure 23.6). The ants collect leaves and flower petals, chew them into small bits, and "plant" bits of fungal mycelium on their surfaces. They even "weed" these gardens by removing other fungal species. The species of fungus cultivated by the ants are found nowhere else than in these gardens. Scale insects—which parasitize some shrubs and trees by attaching to them permanently, sinking their proboscises into them, and sucking sap—live in association with the fungus *Septobasidium*. The fungus spreads over a colony of scales, infecting some of them and thus parasitizing them—without killing them. As new scales hatch from their eggs, some of them become infected by the fungus and take it along as they establish new colonies. The fungus protects the scale colony against drying and against some predators but also draws its nutrition from some of the insects.

Dikaryon Formation in the Fungi

Sexual reproduction begins in some fungi in an unusual way: The cytoplasm of two individuals fuses long before their nuclei fuse, so that two genetically different haploid nuclei exist within the same cell. To see how this comes about, we will take the bread mold *Neurospora crassa* as a representative example (Figure 23.7). *Neurospora*, like many fungi, protists, and bacteria, has various **mating types**, which are genetically different from one another but indistinguishable to the eye. *Neurospora* is haploid for part of its life cycle and forms different male and female structures. Any mating type can function either as a female or as a male. A haploid individual of *Neurospora* consists of hyphae entwined into a mycelium, and this haploid organism will mate only with a haploid individual of a different, but compatible, mating type.

Hyphae from individuals of different mating types grow toward one another, forming mating structures that eventually touch (top of Figure 23.7). At the points of contact, enzymes digest the cell walls, allowing cytoplasm and nuclei from one cell to invade the other. The recipient mating structure then contains two haploid nuclei, one from each parent. The recipient structure develops a hypha consisting of

cells that each have two nuclei—one of each mating type. This hypha is called a **dikaryon** (having *two* nuclei). Because the two nuclei in each cell of a dikaryon differ genetically, the hypha is also called a **heterokaryon** (having *different* nuclei). The heterokaryotic hypha develops into a heterokaryotic mycelium consisting of both undifferentiated regions and fruiting structures. Within the fruiting structures, many pod-shaped asci are produced, each containing two dissimilar nuclei—one from each parent. Ultimately, these pairs of nuclei fuse, giving rise to zygotes long after the original "mating" of the *Neurospora* hyphae. The zygote—the only diploid structure in the entire life cycle of *Neurospora*—undergoes meiosis, producing four haploid nuclei. Each of the four products of meiosis divides mitotically; then cytokinesis results in a total of eight spores. The spores are shed and germinate to form new haploid mycelia, each of a specific mating type. As indicated in Figure 23.7, spores may also form asexually.

The reproduction of *Neurospora* displays several unusual features. First, there are no gamete cells as such, only gamete nuclei. Second, there is never any true diploid tissue, although for a long period during development the genes of both parents are present in the dikaryon and can be expressed. In effect, the cells are neither diploid (2n) nor haploid (n); rather, they are (n + n). A deleterious recessive mutation in one nucleus may be compensated for by a normal allele on the same chromosome in the other nucleus. Dikaryosis is perhaps the most significant of the genetic peculiarities of the fungi. Finally, although these organisms grow in moist places, the gamete nuclei are not motile and are not released into the environment; therefore water in liquid form is not required for fertilization. (As we will see, some members of the plant kingdom are dependent upon liquid water for the meeting of gametes, as are a number of aquatic animals.)

Multiple Hosts in a Fungal Life Cycle

Some fungi are very specific about what host organism they use as a source of nutrition, and some even use different hosts for different stages of their life cycles. (A **host** is an organism that harbors a symbiotic organism—generally a parasite—and provides it with nutrition.) One of the most striking examples of having two different hosts is the complicated life cycle of the black stem rust of wheat, *Puccinia graminis*, which causes a major agricultural disease. In the epidemic of 1916, *P. graminis* caused a crop loss in Canada and the United States amounting to 280 million bushels of wheat, and approximately one-quarter of the North American crop was lost in the epidemic of 1935.

We present the life cycle of *P. graminis* to illustrate three principal points: the utilization of two different hosts, the extent of dikaryosis, and the sheer complexity of some fungal life cycles (Figure 23.8). During the summer, dikaryotic (n + n) hyphae of *P. graminis* proliferate in the stem and leaf tissues of wheat plants, drawing their nutrition from the wheat plants and thereby damaging them severely. The hyphae also produce extensive amounts of summer spores—**uredospores**. These one-celled, orange spores are scattered by the wind and infect other wheat plants, spreading the disease. Like the hyphae, the uredospores are dikaryotic. Dark brown winter spores—**teliospores**—begin to appear on the hyphae in the late summer. Each teliospore consists of two dikaryotic cells. Both cells are thick-walled and resistant to winter freezing. The teliospores remain dormant until spring. Then the two haploid nuclei in each cell

finally fuse, so that the two cells each become diploid (2n). These are the only truly diploid cells in the entire life cycle. The teliospore germinates once the nuclei have fused. Each of the two cells develops into a reproductive structure, and each of the reproductive structures divides meiotically to produce four haploid **basidiospores** (discussed later in this chapter, under the class Basidiomycetes). The basidiospores are of two different mating types, called + and −. They are carried by the wind and, if they land on a leaf of the common barberry plant, they germinate and produce haploid hyphae (+ or −, depending upon the mating type of the germinating basidiospore). These hyphae invade the barberry leaf. The hyphae form flask-shaped structures on the upper surfaces of the leaf, within which some of the hyphae pinch off tiny, colorless haploid **pycniospores** from their tips. The hyphae also form another type of structure, called an **aecium**, near the lower surface

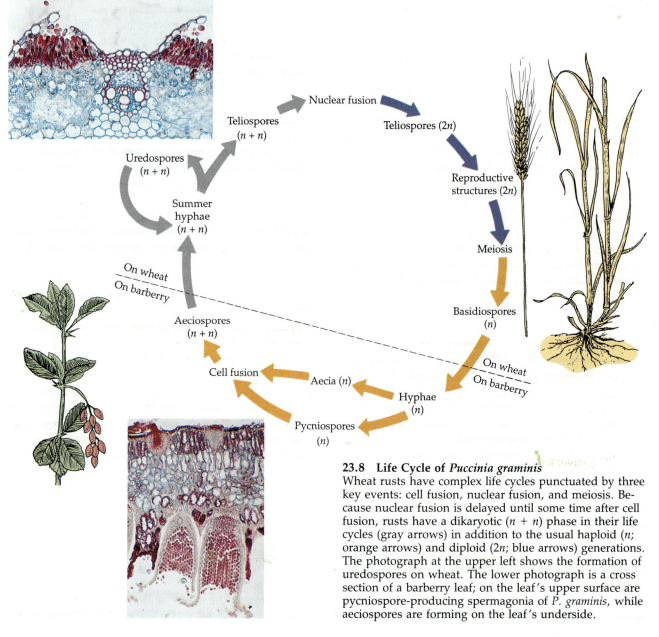

23.8 Life Cycle of *Puccinia graminis*
Wheat rusts have complex life cycles punctuated by three key events: cell fusion, nuclear fusion, and meiosis. Because nuclear fusion is delayed until some time after cell fusion, rusts have a dikaryotic (n + n) phase in their life cycles (gray arrows) in addition to the usual haploid (n; orange arrows) and diploid (2n; blue arrows) generations. The photograph at the upper left shows the formation of uredospores on wheat. The lower photograph is a cross section of a barberry leaf; on the leaf's upper surface are pycniospore-producing spermagonia of *P. graminis*, while aeciospores are forming on the leaf's underside.

of the barberry leaf. Thus the leaf contains "flasks" on the upper surface and aecia near the lower one, and these are connected by the hyphae. Insects are attracted to a sweetish liquid produced in the flasks, and they carry pycniospores from one flask to another. A pycniospore of one mating type fuses with a receptive hypha of the other type within a flask. The nucleus of the pycniospore repeatedly divides mitotically, and the products move down through the hyphae into immature aecia, where they produce dikaryotic cells with two haploid nuclei, one of each mating type. These cells develop into **aeciospores**, each containing two unlike nuclei. The aeciospores are scattered by the wind, and some germinate on wheat plants, continuing the life cycle. When the dikaryotic aeciospores germinate, they produce the summer hyphae with two nuclei in each cell. In all stages of its life cycle from the aeciospore to the teliospore, *P. graminis* is dikaryotic, that is, the individual cells contain nuclei from both "parents." Overwintered teliospores, at first dikaryotic, become diploid. Basidiospores are produced by meiosis and cytokinesis, so they are haploid—and each one is either + or − for mating type.

The different types of spores produced during the life cycle of *P. graminis* play very different roles. Wind-borne uredospores are the primary agents for spreading the rust from wheat plant to wheat plant and to other fields. Resistant teliospores allow the rust to survive the harsh winter but contribute little to the spreading of the rust. Basidiospores spread the rust from wheat to barberry plants. Pycniospores and receptive hyphae initiate the sexual cycle of the rust. Finally, aeciospores spread the rust from barberry to wheat plants.

The principal means of combatting *P. graminis* in wheat country used to be removing barberry from the area. Without this obligate (necessary) intermediate host, *P. graminis* cannot infect a new generation of wheat. Modern control of the disease focuses on the development of resistant strains of wheat—a continual race against the rapid evolution of the rust. A new wheat variety carrying new resistance genes has to be released every year.

DIVISION EUMYCOTA

Class Zygomycetes

The conjugating fungi, or zygomycetes, have hyphae without cross walls. They produce no motile cells of any kind, and only one diploid cell, the zygote, appears in the entire life cycle. No fleshy fruiting body is formed; rather, the hyphae spread in an apparently aimless fashion, with occasional stalked sporangia reaching up into the air (Figure 23.9). In all these regards, the zygomycetes resemble the actinomycetes, a group of bacteria discussed in Chapter 21

23.9 A Zygomycete
This small forest of filamentous structures is made up of sporangiophores of the fungus *Phycomyces*. The stalks end in tiny, rounded sporangia.

(see Figure 21.22). The similarity between these two types of organism is an example of convergent evolution, in which the two types have analogous but not homologous structures (Chapter 20).

About 600 species of zygomycetes have been described. A zygomycete you have probably seen at one time or another is *Rhizopus stolonifer*, the black bread mold. The mycelium of a zygomycete spreads over its substrate, growing forward by means of specialized hyphae. In vegetative reproduction, large numbers of stalked sporangiophores are produced, each bearing a single sporangium containing many hundreds of minute spores. As in other filamentous fungi, the spore-forming structure is separated from the rest of the hypha by a wall.

The zygomycetes reproduce sexually by conjugation and are thus called conjugating fungi. Adjacent hyphae of two different mating types grow together, fuse, and form zygotes (Figure 23.10). Zygotes develop into thick-walled, highly resistant **zygospores** that may remain dormant for months before germinating and undergoing meiosis. The products of meiosis form haploid hyphae and later sporangia that release haploid spores, and the spores develop into a new generation of hyphae. Pheromones participate in the conjugation process, as in gamete attraction in *Allomyces* (Chapter 22). Because *Rhizopus* is terrestrial, it is perhaps not surprising that at least two of the pheromones are passed as gases from one organism to another.

Class Ascomycetes

The ascomycetes, or sac fungi, are a large and diverse group of fungi distinguished by a unique saclike structure called the **ascus** (plural, asci; Figure 23.11).

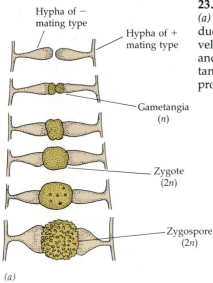

Hypha of − mating type

Hypha of + mating type

Gametangia (*n*)

Zygote (2*n*)

Zygospore (2*n*)

(a)

23.10 Conjugation in a Zygomycete
(a) When two hyphae of differing mating type grow side by side they produce branches that grow toward each other. The tips of these branches develop into gametangia, structures that produce gametes. The gametangia—and then the gametes—fuse, and the resulting zygote develops into a resistant zygospore. (b) Conjugation in the black bread mold *Rhizopus stolonifer* produced these zygospores.

(b)

Both nuclear fusion and subsequent meiosis take place within individual asci. The meiotic products form **ascospores** that are ultimately shed to begin the new gametophyte generation. Ascomycete hyphae are segmented by cross walls (unlike zygomycete hyphae), but a pore in each cross wall permits extensive movement of cytoplasm and organelles (including the nuclei) from one cell to the next.

There are about 30,000 known species of ascomycetes. They can be divided into two broad groups, depending on whether or not the asci are contained within a specialized fruiting structure. Those species having a fruiting structure, the **perithecium**, are collectively called **euascomycetes** (true ascomycetes), whereas those without perithecia are called **hemiascomycetes** (half-ascomycetes). The sexual cycle of euascomycetes includes the formation of a dikaryon stage (Figure 23.7). When the hyphae of two compatible mating types of a particular euascomycete fuse as though they were conjugating, a heterokaryon is formed. Nuclei are passed from one mycelium to the other, but then the introduced nuclei proceed to divide simultaneously with the host nuclei for a considerable period of time. Only with the formation of asci do the nuclei finally fuse. Nuclear fusion of this sort is called **karyogamy**.

Ascus formation itself is a peculiar process (Figure 23.7). A hook (crozier) forms at the tip of the heterokaryotic hypha, and the paired nuclei come to lie on either side of it. Both nuclei then undergo mitosis simultaneously, with their mitotic axes parallel to the axis of the hypha. New walls are laid down as shown in the figure, and then the two nuclei in the top of the hook finally fuse. Meiosis begins ascospore formation.

SUBGROUPS OF THE ASCOMYCETES. The hemiascomycetes are very small in general, and many species are unicellular. Perhaps the best known are the yeasts, especially baker's or brewer's yeast (*Saccharomyces cerevisiae*; Figure 23.12a). Of all the fungi, the yeasts are among the most important domesticated ones. *S. cerevisiae* metabolizes glucose, obtained from the yeast's environment, to ethanol and carbon dioxide. Carbon dioxide bubbles form in bread dough and give the final product its light texture. The ethanol and carbon dioxide are both retained in beer, but both are baked away in bread making. Other yeasts live on fruits such as figs and grapes and play an important role in the making of wine.

Yeasts multiply either by fission or, in the better-known genera, by a process of budding (the outgrowth of a new cell from the surface of an old one).

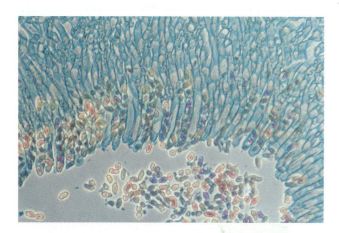

23.11 Asci
Asci and ascospores of the black morel *Morchella elata*. Each ascus contains eight ascospores—the products of meiosis followed by a single mitotic division and cytokinesis.

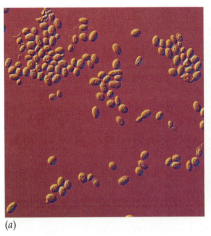

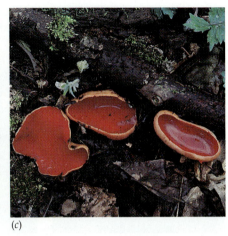

(a)   (b)   (c)

23.12 Some Ascomycetes
(a) A number of cells of baker's yeast (*Saccharomyces cerevisiae*). Some of the cells are budding. These hemiascomycetes are facultative anaerobes—organisms that can grow in either the presence or the absence of free oxygen. (b)

Two yellow morels, *Morchella esculenta*. Morels are characterized by their netlike caps and subtle flavor. (c) These scarlet cups, *Sarcoscypha coccinea*, are representative cup fungi. Morels and cup fungi are euascomycetes.

The single cells are haploid. Conjugation takes place only occasionally between two adjacent, compatible cells; nuclear fusion is followed immediately by meiosis and a single mitosis, so that the entire structure becomes an ascus. Yeasts have no dikaryon stage.

Among the euascomycetes—the other major group of sac fungi—are several common genera of mold, including *Neurospora*, the pink molds. *Neurospora* are found everywhere, frequently occurring on old bread. *N. crassa* will be familiar to you as one of the important organisms used for experimental genetics (Chapter 10).

A large number of euascomycetes are serious parasites on higher plants. Chestnut blight and Dutch elm disease are caused by euascomycetes. The powdery mildews are a group that infects cereal grains, lilacs, and roses, to name but a few host plants. They may also be a serious problem to grape growers, and a great deal of research has been done on ways to control these serious agricultural pests.

Two particularly delicious euascomycete fruiting structures are the morels (Figure 23.12b) and the truffles. Truffles grow in a mutualistic symbiotic association with the roots of some species of oaks. People once used pigs to find truffles—some truffles secrete a substance with an odor similar to a pig's sex-attraction substance. Unfortunately, pigs also eat truffles, so dogs are now the usual truffle hunters.

The euascomycetes also include the cup fungi (Figure 23.12c). In these organisms, the fruiting structures are small cups that can be as large as several centimeters across. The inner surfaces of the cups are covered with a mixture of sterile filaments and asci, and they produce huge numbers of spores. Although these fleshy structures appear to be composed of

distinct tissue layers, microscopic examination shows that their basic organization is still filamentous—a tightly woven mycelium. Such fruiting structures are formed only by the dikaryotic mycelium.

The euascomycetes reproduce asexually by means of **conidiospores** that form at the tips of specialized hyphae called **conidiophores** (from Greek *konis*, "dust"; Figure 23.13). These small chains of conidiospores are produced literally by the millions and are sufficiently resistant to survive for weeks in nature. It is the conidiospores that give molds their characteristic colors.

Class Basidiomycetes

About 25,000 species of the club fungi, or basidiomycetes, have been described. Some basidiomycetes

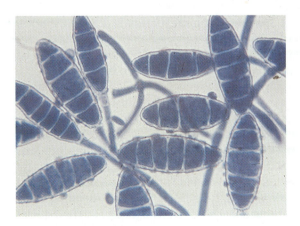

23.13 Conidiospores
These macroconidia of *Microsporum gypseum* are a type of conidiospore.

(a)   (b)   (c)

23.14 Basidiomycete Fruiting Structures
Although some only persist a few days or weeks, the fruiting structures of
the class Basidiomycetes are probably the most familiar structures produced
by fungi. (a) *Lycoperdon perlatum*, the gem-studded puffball. (b) A highly poisonous species of *Amanita*, a mushroom. (c) A slug is eating this bracket fungus, *Merulius incarnatus*, which in turn is parasitizing a tree.

produce the most spectacular fruiting structures
found anywhere among the fungi (Figure 23.14).
These are the puffballs (which may be over half a
meter in diameter), the mushrooms of all kinds (more
than 3,250 species, including the familiar *Agaricus
campestris* and the poisonous toadstools), and the
giant bracket fungi often encountered on trees and
fallen logs in a damp forest. Bracket fungi do great
damage to cut lumber and to stands of timber. Some
basidiomycetes are among the most damaging plant
pathogens, including wheat rust (*Puccinia graminis*;
Figure 23.8) and the smut fungi (Figure 23.1a) that
parasitize cereal grains. In sharp contrast, other basidiomycetes contribute to the well-being of plants as
fungal partners in mycorrhizae.

Basidiomycete hyphae are characterized by being
completely septate (walled off). (Recall that the zygomycetes have hyphae without walls, and that the
ascomycetes have walls with large pores.)

The **basidium** (plural, basidia), a swollen cell at
the tip of a hypha, is the site of nuclear fusion and
meiosis in the basidiomycetes (Figure 23.15). The basidium plays the same role in the basidiomycetes as
the ascus does in the ascomycetes. After nuclear fusion takes place in the basidium, the resulting diploid
cell undergoes meiosis, and the four products are
extruded from the basidium, forming haploid **basidiospores** on tiny stalks. The basidiospores are scattered and then germinate, giving rise to haploid hyphae. As these hyphae grow, haploid hyphae of
different mating types meet and fuse, forming dikaryotic hyphae, each cell of which contains two nuclei,
one from each parental hypha. The dikaryotic mycelium grows and it eventually produces fruiting
structures.

The elaborate fruiting structure of a fleshy
basidiomycete—such as the gill mushroom in Figure
23.16—is topped by a cap, or pileus, which has gills
on its underside. Basidia develop in enormous numbers between the gills. The basidia discharge their
spores into the air spaces between adjacent gills, and

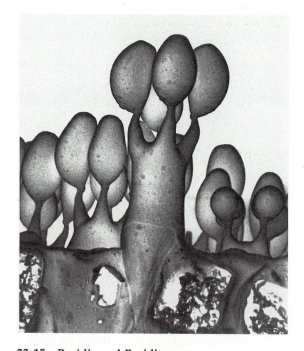

23.15 Basidia and Basidiospores
Two nuclei fuse in a basidium; meiosis and further development lead to the production of four or eight narrow
projections through which nuclei and other organelles
squeeze, forming basidiospores.

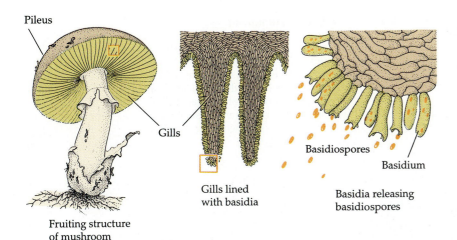

Pileus

Gills

Fruiting structure
of mushroom

Gills lined
with basidia

Basidiospores

Basidia releasing
basidiospores

Basidium

23.16 Anatomy of a Mushroom
The gills are on the underside of a gill
mushroom's cap. The gills are lined
with basidia (see Figure 23.15), and basidiospores are shed into the air from
the basidia.

the spores sift down into air currents for dispersal
and germination as new haploid mycelia. The exact
pattern of the gills and the spore color are used in
determining mushroom species. One simply places a
mature cap, gill side down, on a piece of paper for a
few hours in a quiet place. The ejected basidiospores
settle from between the gills, leaving on the paper
an elegant replica of the gill pattern and also clearly
revealing spore color.

Class Deuteromycetes

Their mechanisms of sexual reproduction readily distinguish the three preceding classes of the Eumycota
(Zygomycetes, Ascomycetes, and Basidiomycetes).
However, a large number of fungi, both saprobes and
parasites, lack sexual stages entirely—presumably
these stages have been lost in evolution. It thus becomes difficult to classify these fungi with any of the
three major classes. True fungi that cannot be classified taxonomically are simply dumped into the or-
phanage known as the imperfect fungi class, or deuteromycetes (Figure 23.17). At this time, about 25,000
species are classified as imperfect fungi. Among them
are the pathogens that cause athlete's foot and ring-
worm.

From time to time, sexual structures are found on
a fungus classified as a deuteromycete. When that
happens, the fungus is reassigned to the appropriate
class. That happened, for example, with a fungus
that produces plant growth substances called gibberellins (Chapter 32). This fungus was originally classified as the deuteromycete *Fusarium moniliforme*; but
it was later found to produce asci, whereupon it was
renamed *Gibberella fujikuroi* and transferred to the
class Ascomycetes.

Penicillium (Figure 23.3) is a deuteromycete genus
of green molds that produce the antibiotic penicillin,
presumably for defense against competing bacteria.
Two species, *P. camemberti* and *P. roqueforti*, are the
organisms responsible for the characteristic flavors of

Camembert and Roquefort cheeses. Not surprisingly,
people who are hypersensitive to penicillin may react
violently if they eat one of these cheeses. Another
deuteromycete genus of importance in some diets is
Aspergillus, the genus of brown molds. *A. tamarii* acts
on soybeans in the production of soy sauce, and *A.
oryzae* is used in brewing the Japanese alcoholic bev-
erage sake.

LICHENS

A lichen is not an organism—it is a meshwork of
two radically different organisms: a fungus and a
photosynthetic microorganism. Together the organisms constituting a lichen survive some of the harsh-
est environments on earth (Figure 23.18). Paradoxi-

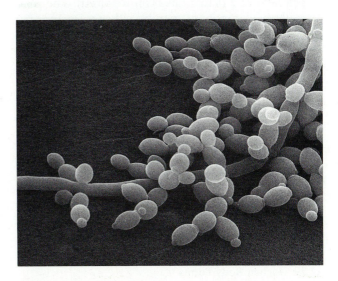

23.17 A Deuteromycete
Candida albicans represents the imperfect fungi in this
scanning electron micrograph; both a long hypha and
oval, budding yeast cells are shown. This species causes
the disease candidiasis in humans and can be life-threatening in persons with defective immune systems.

(a)

(b)

23.18 Lichens in Frigid Environments

(a) This fractured sandstone rock from the cold desert of Antarctica shows the black, white, and green layers of a lichen that grows around crystals under the rock's hard surface crust. The black layer consists of fungal hyphae and algal cells; the white layer is mostly colorless fungal hyphae; and the green layer consists of hyphae and algae. No signs of life are visible on the outer surface of this environment. (b) Many types of lichens growing on the Alaskan tundra.

(a)

(b)

(c)

cally, lichens are very sensitive to air pollution, because they are unable to excrete any toxic substances that they absorb. Hence they are not commonly found in industrial cities. Lichens are good biological indicators of air pollution because of their sensitivity. The fungal components of most lichens are ascomycetes, but some are basidiomycetes or imperfect fungi (one zygomycete serving as the fungal component of a lichen has been reported). The photosynthetic component may be either a cyanobacterium or a green alga. Relatively little experimental work has been done with lichens, perhaps because they grow so slowly—typically less than 1 centimeter per year. Thus it is only recently that workers have been able to culture the fungal and photosynthetic partners separately and then to reconstruct a lichen from the two.

There are about 20,000 "species" of lichens. They are found in all sorts of exposed habitats: tree bark, open soil, or bare rock. Reindeer "moss," actually not a moss at all but the lichen *Cladina*, covers vast

23.19 Lichen Body Forms

Lichens fall into three principal classes with respect to body form. (a) Crustose lichens such as the orange, white, and gray species in this photograph often grow on otherwise bare rock, as shown here, or on tree bark. (b) A wet foliose lichen growing on a dead twig. (c) A fruticose lichen growing on a wooden fence.

areas in arctic, subarctic, and boreal regions, where it serves as an important dietary item for reindeer and other large mammals. Lichens come in various forms and colors. Crustose (crustlike) lichens look like colored powder dusted over their substrate, while foliose (leafy) and fruticose (shrubby) lichens may appear quite complex (Figure 23.19).

The most widely held interpretation of the lichen relationship is that it represents a type of mutually beneficial symbiosis. Hyphae of the fungal mycelium are tightly pressed against the photosynthetic cells (Figure 23.20a) and sometimes even invade them. The bacterial or algal cells not only survive these indignities but continue their growth and photosynthesis. In fact, algal cells in a lichen "leak" photosynthetic products at a greater rate than do similar cells growing on their own.

A cross section of a typical foliose lichen is shown in Figure 23.20b. There is a tight upper region of fungal hyphae alone, a layer of cyanobacteria or algae, a looser hyphal fungal layer, and finally hyphae (rhizoids) that attach the whole structure to its substrate. The meshwork has properties that enable it to hold water fairly tenaciously. Certain nutrients for the photosynthetic cells arrive in part through the fungal hyphae, the meshwork provides a suitably moist environment for the photosynthetic cells, and the fungi derive fixed carbon from the photosynthesis of the algal or cyanobacterial cells.

Lichens can reproduce simply by fragmentation of the vegetative body, which is called the **thallus**, or else by specialized structures called **soredia**. The soredia consist of one or a few photosynthetic cells surrounded by fungal hyphae (Figure 23.20b). The soredia become detached, move in air currents, and on arriving at a favorable location, develop into a new lichen. If the fungal partner is an ascomycete or a basidiomycete, it may go through its sexual cycle, producing either ascospores or basidiospores. When these are discharged, however, they leave the lichen unaccompanied by the photosynthetic partner and thus may not be capable of reestablishing the lichen association. Nevertheless, many lichens produce characteristic fruiting structures in which the asci or basidia are located.

Lichens are often the first colonists when new areas of bare rock appear. Lichens satisfy the great majority of their wants from the air and from rainwater, augmented by the absorption of some minerals from their rocky substrate. A lichen begins to grow shortly after a rainfall, as it begins to dry. As it grows, the lichen acidifies its environment slightly, and this acid contributes to the slow breakdown of rocks, an early step in soil formation. When the lichen dries still further, its photosynthesis ceases. The water content of the lichen may drop to less than 10 percent of its dry weight, at which point the lichen becomes highly insensitive to extremes of temperature. The flora of Antarctica features more than 100 times as many species of lichens as of plants.

23.20 Lichen Anatomy
(a) Soredia of a fruticose lichen. Each soredium consists of a photosynthetic cell surrounded by fungal hyphae. Soredia detach readily from the parent lichen and travel in air currents, founding new lichens when they settle in a suitable environment. (b) Layers of a typical lichen, seen in a cross section. Algal cells are shown as round, green objects and fungal hyphae are tinted orange.

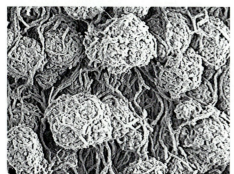
(a)

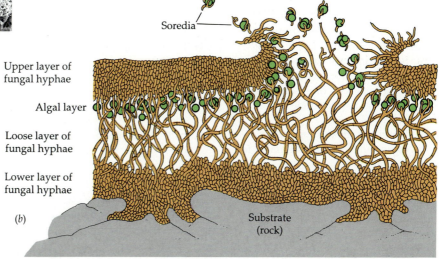
Soredia
Upper layer of fungal hyphae
Algal layer
Loose layer of fungal hyphae
Lower layer of fungal hyphae
(b)
Substrate (rock)

SUMMARY

Fungi are heterotrophic, usually multicellular organisms with absorptive nutrition. They are saprobes, mutualists, or parasites. Of the parasites, some are facultative and some obligate. A very few species are active predators on microscopic animals. Some fungi form mutualistic symbiotic associations with algae or bacteria (in lichens) or plant roots (in mycorrhizae), and a few species are even farmed by leaf-cutting ants. Other symbiotic associations are formed with insects such as scales.

The fungal body is composed of hyphae, massed to form a mycelium. Specialized hyphae include the rhizoids of zygomycetes and the haustoria of many parasites. Other characteristic features of the fungi are sexual reproduction by conjugation, the possession of chitinous cell walls, and the absence of flagellated cells. Fungi demonstrate a third nuclear condition in addition to the haploid and diploid states:

the dikaryotic, or ($n + n$) state. Fungal life cycles are often complex, involving various sources of nutrition and different kinds of spores.

Some fungi are important pathogens. Others are used in a variety of processes such as baking, brewing, and the production of various foods and antibiotics.

The kingdom Fungi consists of the division Eumycota (true fungi). The classes of the division Eumycota are the Zygomycetes (conjugating fungi), Ascomycetes (sac fungi), Basidiomycetes (club fungi), and Deuteromycetes (imperfect fungi, which lack known sexual reproductive stages). The classes differ with respect to their reproductive structures, spore formation, and, less importantly, the cross walls (if any) of their hyphae.

Lichens are mutualistic combinations of a fungus, usually an ascomycete, with a green alga or a cyanobacterium. Lichens are found in some of the seemingly most inhospitable environments on the planet.

SELF-QUIZ

1. Which statement is *not* true of the fungi?
   a. A multicellular fungus has a body called a mycelium.
   b. Hyphae are composed of individual mycelia.
   c. Many fungi tolerate highly hypertonic environments.
   d. Many of them tolerate low temperatures.
   e. They are anchored to their substrate by rhizoids.

2. The absorptive nutrition of fungi is aided by:
   a. heterokaryon formation.
   b. spore formation.
   c. the fact that they are all parasites.
   d. their large surface-to-volume ratio.
   e. their possession of chloroplasts.

3. Which statement about fungal nutrition is *not* true?
   a. Some fungi are active predators.
   b. Some fungi form mutualistic associations with other organisms.
   c. All fungi require mineral nutrients.
   d. Fungi can make some of the compounds that are vitamins for animals.
   e. Facultative parasites can grow only on their specific hosts.

4. Which statement is *not* true of heterokaryosis?
   a. The cytoplasm of two cells fuses before their nuclei fuse.
   b. The two haploid nuclei are genetically different.
   c. The two nuclei are of the same mating type.
   d. The heterokaryotic stage ends when the two nuclei fuse.
   e. Not all fungi have a heterokaryotic stage.

5. Reproductive structures consisting of a photosynthetic cell surrounded by fungal hyphae are called:
   a. ascospores.
   b. basidiospores.
   c. conidiospores.
   d. soredia.
   e. gametes.

6. The zygomycetes:
   a. have hyphae without cross walls.
   b. produce motile gametes.
   c. form fleshy fruiting bodies.
   d. are haploid throughout their life cycle.
   e. have structures homologous to those of the actinomycetes.

7. Which statement is *not* true of the ascomycetes?
   a. They include the yeasts.
   b. They form reproductive structures called asci.

   c. Their hyphae are segmented by cross walls.
   d. Many of their species have a heterokaryotic stage.
   e. All of them have fruiting structures called perithecia.

8. The basidiomycetes:
   a. often produce fleshy fruiting structures.
   b. have hyphae without walls.
   c. have no sexual stage.
   d. never produce large fruiting structures.
   e. form diploid basidiospores.

9. The deuteromycetes:
   a. have distinctive sexual stages.
   b. are all parasitic.
   c. include some commercially important species.
   d. include the sac fungi.
   e. are never components of lichens.

10. Which statement is *not* true of the lichens?
   a. They can reproduce by fragmentation of their vegetative body.
   b. They are often the first colonists in a new area.
   c. They render their environment more basic (alkaline).
   d. They contribute to soil formation.
   e. They may contain less than 10 percent water by weight.

FOR STUDY

1. You are shown an object that looks superficially like a pale green mushroom. Describe at least three criteria (including anatomical and chemical ones) that would enable you to tell whether the object was a piece of a plant or a piece of a fungus.

2. Differentiate between the members of each of the following pairs of related terms:
   a. hypha/mycelium
   b. euascomycete/hemiascomycete
   c. ascus/basidium
   d. rhizoids/haustoria

3. For each of the types of organism listed below, give a single characteristic that may be used to differentiate it from the other, related organism(s) in parentheses:
   a. Zygomycetes (Ascomycetes)
   b. Basidiomycetes (Deuteromycetes)
   c. Ascomycetes (Basidiomycetes)
   d. Baker's yeast (*Neurospora crassa*)

4. Many fungi are dikaryotic during part of their life cycle. Why are dikaryons described as ($n + n$) instead of $2n$?

5. If all the fungi on Earth were suddenly to die, how would the surviving organisms be affected? Be thorough and specific in your answer.

READINGS

Alexopoulos, C. J. and C. W. Mims. 1979. *Introductory Mycology*, 3rd Edition. John Wiley & Sons, New York. Still a leading textbook on the biology of fungi.

Moore-Landecker, E. 1991. *Fundamentals of Fungi.* Prentice-Hall, Englewood Cliffs, NJ. The most up-to-date introduction to the kingdom.

Newhouse, J. R. 1990. "Chestnut Blight." *Scientific American*, July. A new biological method for controlling the fungus responsible for a disease that virtually eliminated an important tree species.

Raven, P. H., R. F. Evert and S. Eichhorn. 1986. *Biology of Plants*, 4th Edition. Worth, New York. Includes an excellent discussion of the fungi.

24

Plants

PREVIEW: Over much of the Earth's surface, the most visible organisms are plants. They render their surroundings green, brown, or gray, depending on the season and place. Both major groups of plants (bryophytes and vascular plants) appeared on Earth more than 350 million years ago. The vascular plants have undergone remarkable and diverse evolutionary development, with the flowering plants now the dominant vegetation in most ecosystems.

This chapter deals with bryophytes, psilopsids, lycopsids, sphenopsids, ferns, gymnosperms, and angiosperms; and evolution, microphylls, megaphylls, homospory, heterospory, cones, flowers, fruits, and double fertilization.

What a thrill it would be to *see* the face of Earth at different stages of its evolution. Photographs from space would show the drifting of the continents. Up close, we would see scenes scarcely imaginable to us today. For more than four billion years, the terrestrial environment was basically a mass of rock. Its appearance changed relatively little until plants colonized the land and slowly spread over its surface.

Earth did not take on a green tint until less than half a billion years ago, long after the ancestors of today's plants had invaded the land at some time in the Paleozoic era (Table 17.1). The earliest land plants were small, but their metabolic activities helped convert native rock to soil that could support the needs of their successors. Evolution led rapidly (in geological terms) to larger and larger plants, and in the Carboniferous period (345–290 million years ago) great forests were widespread. These forests would be unrecognizable to us, for they were composed of trees mostly unlike any to be seen today (Figure 24.1). Over many tens of millions of years, these trees were replaced by others more familiar to us—pinelike conifers and their relatives. It is only during the last 100 million years that today's vegetation gradually became dominant. Today the Earth is a patchwork of widely differing environments, each supporting an appropriately adapted community of plants, and each plant community determining the characteristics of its associated community of animals. What do we see when we look at any of these environments? We see its *plants*—or, in a very harsh environment, its relative *lack* of plants.

THE PLANT KINGDOM

A broad, loose definition of the word **plant** is: *a multicellular photosynthetic eukaryote.* The definition is loose in that a few organisms that can only be regarded as plants are not photosynthetic. These parasitic species are clearly related to photosynthetic plants and possess adaptations that afford them alternative modes of nutrition. The definition is broad in that it casts an enormous net over a wide range of organisms that, while all multicellular and photosynthetic, differ in size, cellular organization, photosynthetic and associated pigments, and cell wall chemistry. We prefer to narrow the definition of a plant.

According to the narrower definition used in this book, plants are those multicellular, photosynthetic eukaryotes that have the following additional properties. They develop from embryos protected by tissues of the parent plant. (This characteristic is definitive; an older classification scheme used the term Embryophyta to refer to precisely those organisms that we include here in the kingdom Plantae.) Their cells have walls that contain cellulose as the major strengthening polysaccharide. Their chloroplasts contain chlorophylls a and b and a limited array of specific carotenoids (Chapter 8). Their storage carbohydrate is starch.

Alternation of Generations in Plants

A universal feature of the life cycles of plants is the alternation of generations (Chapters 9, 22). If we con-

24.1 An Ancient Forest
A little over 300 million years ago, a forest grew in a setting similar to tropical river delta habitats of today. A dense forest of lycopsids 10–20 meters high grew along the low natural levee forming the edge of the river. A few tree ferns approximately 10 m high were present. A small fern grew on the trunk of a tree fern. On the ground, scattered clumps of weak-stemmed seed ferns grew, as did tangled clumps of lycopsids. Beyond the levee a swampy area supported growths of lycopsids, seed ferns, and sphenopsids. Visible in the distance, groups of sphenopsid trees about 10 m high grew by the shores of lakes. Yet farther in the distance giant gymnosperms—up to 40 m tall—towered over the forest.

sider the plant life cycle to begin with a single cell, a haploid spore, then the first phase of the cycle features the formation, by mitosis and cytokinesis, of a haploid plant (Figure 22.3). This multicellular, haploid plant, the **gametophyte**, produces haploid gametes. (Gametophyte means "gamete plant.") The fusion of two gametes (syngamy, or fertilization; Chapter 9) results in the formation of a diploid cell, the zygote. Successive mitotic divisions of the zygote give rise to a multicellular embryo and eventually the mature **sporophyte** ("spore plant"). Cells contained in **sporangia** (singular, sporangium, "spore reservoir") on the sporophyte undergo meiosis to produce haploid, unicellular spores, and the cycle begins again. The sporophyte generation extends from the zygote through the adult, multicellular diploid plant; the gametophyte generation extends from the spore through the adult, multicellular haploid plant, to the gamete. The transitions between the phases are ac-

complished by the processes of fertilization and of meiosis.

Alternation of generations in plants is **heteromorphic**. That is, the gametophyte and sporophyte of any plant are totally different in appearance, unlike the situation in some algae such as *Ulva* in which the generations are indistinguishable to the eye (Figure 22.25). In both heteromorphic and homomorphic alternation of generations, the sporophyte and gametophyte differ genetically, in that one has diploid cells and the other has haploid cells.

Classification of Plants

We may divide the kingdom Plantae into smaller, more closely related groups in several ways. Many botanists prefer the term **division** to the term phylum, which is commonly used by zoologists. Accepting this preference, we must next consider which groups have division status. One currently popular approach is to divide the plant kingdom into ten divisions of equal taxonomic status. Another approach is a more hierarchical one, dividing the kingdom into just two divisions, one of which is broken into a further hierarchy. Since adherents of the ten-division scheme often group certain of the divisions together for convenience, thus approximating the second scheme, we have decided to follow the second, more hierarchical scheme (Table 24.1). We believe that this emphasizes the evolutionary trends that you will be learning.

PLANT COLONIZERS OF THE LAND

It was on land that the plant kingdom first appeared and evolved. The terrestrial environment differs dramatically from the aquatic environment in several ways. Water buoys up the organisms living in a pond or ocean, supporting them against gravity. A multicellular plant on land must either have some type of system to provide support against gravity or else sprawl unsupported on the ground. It also must either have some way to transport water and minerals from the soil to its aerial (raised) parts or else live where water is so abundant that all of its parts are bathed fairly regularly.

Origins of the Plant Kingdom

It is widely agreed that the plant kingdom arose from the green algae (Phylum Chlorophyta; Chapter 22). It is less clear which particular kind of green alga was ancestral to the plants, and it is possible that the two divisions of the plant kingdom arose independently from different green algal ancestors. The characteristics of the green algae that make them the most likely source of the plant kingdom are these: First, the photosynthetic and accessory pigments in their plastids are closely similar to those of plants. Second, both the green algae and the plants produce starch as their principal storage carbohydrate. Third, cellulose (Chapter 3) is the principal component of the cell walls of plants and green algae.

Sexual reproduction in the plant kingdom features a large, stationary egg. Not all green algae show this oogamous type of reproduction (Chapter 22), but some of them do. Like the plants, some—but far from all—green algae have a life cycle with both a multicellular gametophyte and a multicellular sporophyte generation. A few green algae have bulky, three-dimensional bodies like plants, although most green algae are unicellular, filamentous, or two-dimensional in form. Thylakoids are arranged into grana in the chloroplasts of plants and many green algae. Some green algae produce new cell walls after following cell division by a mechanism similar to that employed by plants. No other phylum of algal protists shares so many traits with the kingdom Plantae.

We do not know what the specific ancestors of the plants looked like. However, those ancestral green algae probably lived at the margins of ponds or marshes, ringing them with a green mat.

THE TWO DIVISIONS OF THE PLANT KINGDOM

The two distinct lines of terrestrial plants that evolved from the algae are represented by the division Bryophyta and the division Tracheophyta (Figure 24.2). The two divisions took differing approaches to the challenges of the terrestrial environment.

The bryophytes never evolved into large plants. Bryophytes—mosses, liverworts, and hornworts—have little or no water-transporting tissue, yet they are found in some dry environments. They grow in dense masses, and water can move through the mass by capillary action. The "leafy" structure of bry-

---

TABLE 24.1
Classification of Plants

DIVISION	SUBDIVISION	CLASS	COMMON NAME	CHARACTERISTICS
Bryophyta				No vascular tissues
		Musci	Mosses	Filamentous stage
		Hepaticae	Liverworts	No filamentous stage
		Anthocerotae	Hornworts	Embedded archegonia
Tracheophyta				Vascular tissues
	Psilopsida			Microphylls, no roots (known only as fossils)
	Lycopsida		Club mosses	Microphylls in spirals
	Sphenopsida		Horsetails	Microphylls in whorls
	Pteropsida			Megaphylls
		Filicinae	Ferns	No seeds
		Gymnospermae	Gymnosperms	Naked seeds
		Angiospermae	Flowering plants	Double fertilization, seeds in fruit

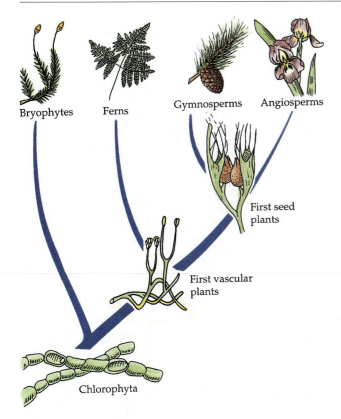

24.2 Bryophytes and Tracheophytes
The green algae (Chlorophyta) gave rise to two distinct lines of terrestrial plants—the bryophytes (mosses, liverworts, and hornworts) and the vascular plants of the division Tracheophyta.

ophytes readily catches and holds water that splashes onto them. The bryophytes are small enough so that minerals can be distributed by diffusion. They lack the leaves, stems, and roots that characterize other plants, although they have structures analogous to each.

Tracheophytes, or vascular plants, are the ferns, conifers, and flowering plants. Tracheophytes differ from bryophytes in crucial ways, one of which is the possession of a **vascular system** consisting of specialized tissues for the transport of materials from one part of the plant to another. One tissue, the **xylem**, conducts water and minerals from the soil to aerial parts of the plant; the other, the **phloem**, conducts the products of photosynthesis from sites of production to sites of utilization or storage (Chapters 28, 29). In plants whose stems and roots grow in diameter, another specialized tissue, **cambium**, produces new xylem and phloem, the tissues that contribute to the widening.

The tracheophytes appeared before the bryophytes in the course of evolution. The oldest tracheophyte fossils date back more than 410 million years, whereas the oldest bryophyte fossils are some 350 million years old, dating from a time when tracheophytes were already widely distributed.

Most of the characteristics that distinguish bryophytes and tracheophytes from algae are evolutionary adaptations to life on land. Both bryophytes and tracheophytes have protective coverings that prevent their drying out, and both have means of taking up water from the soil (tracheophytes have roots and bryophytes have rhizoids). Support against gravity is provided by the turgor of bryophyte and tracheophyte cells, by a woody stem in some tracheophytes, and by thickened cell walls in bryophytes. The bryophytes and tracheophytes, unlike the algae, form embryos, young sporophytes contained within a protective structure. We will examine the adaptations of the vascular plants presently, but let us concentrate first upon the bryophytes.

DIVISION BRYOPHYTA

Most bryophytes are small and grow in dense mats in moist habitats (Figure 24.3). The largest bryophytes are about 20 centimeters tall, and most are only a few centimeters tall or long. Why have larger

24.3 A Mat of Bryophytes
Moss colonizing an old lava bed in Iceland.

bryophytes never evolved? The probable answer is that they never evolved an efficient system for conducting water and minerals from the soil to distant parts of the plant body. A second likely size-limiting factor is that bryophyte sperm must swim through water to reach and fertilize eggs. Rain or dew on a small moss plant enables newly released sperm to make their way to a nearby part of the plant, but would such travel be efficient enough on a large plant?

Most bryophytes live on the soil or on other plants, but others grow on bare rock, dead and fallen tree trunks, and even on the buildings in which we live and work. It is often said that the bryophytes are incompletely adapted to life on land because of their dependence upon standing water for fertilization. Realize, though, that these small plants have been here many times longer than we and perhaps a quarter of a billion years longer than the flowering plants; they are widely distributed over six of the seven continents (all except Antarctica). These *are* successful plants, nicely adapted to the niches in which they appear. Their niches are virtually all terrestrial, although a few bryophytes live in fresh water—these aquatic forms are descended from terrestrial ones. There are no marine bryophytes. All bryophytes *do* have a waxy covering that retards water loss, and their embryos are protected within layers of maternal tissue.

Plasticity of Form in the Bryophytes

Many bryophytes take on different growth habits in different environments. Depending on how much water is available in the environment, these plants may form matted cushions of intertwining gametophytes or carpets of varying depths with the gametophytes perfectly erect; or other forms may be assumed. Gametophytes of one species may, in some extreme environments, take on the appearance of other species. Thus it is difficult or impossible to identify some bryophytes without seeing the sporophytes, which show less variation within a species than do the gametophytes.

The Bryophyte Life Cycle

The life cycles of bryophytes differ sharply from those of other plants and of their algal ancestors. The bryophyte gametophyte is the conspicuous generation, the green plant we recognize. The gametophyte is photosynthetic and thus nutritionally independent. In contrast, the sporophyte is dependent upon the gametophyte and remains permanently attached to it.

The bryophyte sporophyte produces unicellular, haploid spores as products of meiosis. A spore germinates, giving rise to a multicellular haploid game-

tophyte whose cells contain chloroplasts and are thus photosynthetic. Eventually, gametes form within specialized sex organs. The **archegonium** is a multicellular, flask-shaped female sex organ with a long neck and a swollen base (Figure 24.4). The base contains a single egg. The **antheridium** is a male sex organ in which sperm, each bearing two flagella, are produced in large numbers. Once released, the sperm must swim to a nearby archegonium on the same or a neighboring plant. The sperm are aided in this by chemical attractants released by the egg or archegonium. Before sperm can enter the archegonium, certain of the cells in the neck of the archegonium must break down, leaving a water-filled canal through which the sperm swim to complete their journey. Note the dependence on liquid water for all of these events.

Mitotic divisions of the zygote produce a multicellular, diploid sporophyte embryo. The base of the archegonium grows during the early growth of the embryo, so that the archegonium continues to protect the embryo. Eventually the developing sporophyte elongates sufficiently to break free of the archegonium, but the sporophyte has a foot that is embedded in the gametophyte and absorbs water and nutrients from it—the sporophyte remains attached to the gametophyte throughout its life. The sporophyte produces a **capsule**, or sporangium, within which

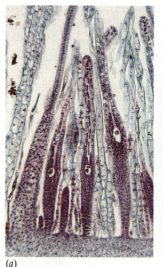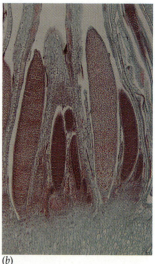

(a)   (b)

24.4 Sex Organs of a Moss
(a) Archegonia of the moss *Mnium* (division Bryophyta). The large egg cells are in the centers of the archegonia and look like eyes. These structures develop at the tip of a gametophyte. It is in the archegonium that the egg will be fertilized and begin development into a sporophyte. (b) Antheridia, also located at the tip of a gametophyte of *Mnium*. These male organs contain a large number of sperm. When released, the sperm must locate an archegonium and swim down its neck to the egg.

meiotic divisions give rise to spores and, thus, to the next gametophyte generation.

The structure and pattern of elongation of the sporophyte differ among the three classes of bryophytes—the liverworts, hornworts, and mosses. The evolutionary relationship of the three classes is uncertain.

Class Hepaticae: Liverworts

The gametophytes of some liverworts are "leafy" and prostrate. However, the simplest liverwort gametophytes are nothing but flat plates of cells perhaps a centimeter or so in length; they produce antheridia or archegonia on their upper surfaces and rhizoids on the lower. Liverwort sporophytes are shorter than those of mosses and hornworts, rarely exceeding a few millimeters in length. The sporophyte has a stalk that connects capsule and foot. The stalk elongates and thus raises the capsule above ground level, favoring dispersal of spores when they are released. The elongation of the sporophyte is distributed broadly over the length of the stalk, which has no specific growing zone. There are no stomata (pores allowing gas exchange between the atmosphere and the plant's interior) on the liverwort sporophyte. The capsules of liverworts are simple—a globular capsule wall surrounding a mass of spores. In some species of liverworts, spores are not released by the sporophyte until the surrounding capsule wall rots. In other liverworts, however, the spores are disseminated by structures called **elaters**, which are located within the capsule. Elaters are long cells with a helical thickening of the cell wall (Figure 24.5c). As an elater

loses water, the whole cell shrinks longitudinally to a fraction of its former length, thus compressing the helical thickening like a spring. When the stress becomes sufficient, the compressed "spring" snaps back to its resting position, throwing spores in all directions.

Among the most familiar liverworts are species of the genus *Marchantia* (Figure 24.5). *Marchantia* is easily recognized by the characteristic structures on which its male and female gametophytes bear their antheridia and archegonia. Like most liverworts, *Marchantia* reproduces vegetatively by simple fragmentation of the gametophyte. *Marchantia* and some other liverworts and mosses also reproduce vegetatively by means of **gemmae** (singular, gemma), lens-shaped clumps of cells loosely held in structures called gemma cups.

Class Anthocerotae: Hornworts

The hornworts—so named because their sporophytes look like little horns—appear at first glance to be liverworts with very simple gametophytes (Figure 24.6a). These gametophytes are flat plates of cells, a few cells thick. However, hornworts have two characteristics found in no other bryophytes. First, the archegonia are embedded in the gametophytic tissue instead of being borne on stalks. Second, of all the bryophytic sporophytes, those of the Anthocerotae come closest to being capable of indefinite growth. Consider the stalk of either the liverwort or moss sporophyte—its growth terminates when a capsule forms, so that the elongation of the sporophyte is strictly limited. In a plant such as the hornwort *An-*

(a)

(b)

(c)

24.5 *Marchantia*, a Liverwort
Marchantia is a representative of the division Bryophyta, class Hepaticae. (a) Gametophytes. (b) Cups filled with gemmae—small, lens-shaped outgrowths of the body, each of which is capable of developing into a new plant. (c) The disk-headed structures bear antheridia; the finger-headed structures bear archegonia.

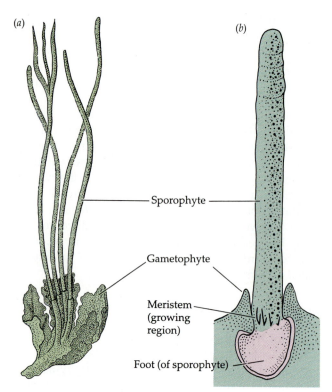

(a)

(b)

Sporophyte

Gametophyte

Meristem (growing region)

Foot (of sporophyte)

24.6 Hornworts and Their Growth
(a) The hornwort *Anthoceros* (division Bryophyta, class Anthocerotae) drawn slightly larger than actual size. *(b)* The sporophyte of a hornwort grows from its basal end.

thoceros, however, there is no stalk, but a basal region of the capsule remains capable of indefinite cell division, continuously producing new spore-bearing tissue above (Figure 24.6*b*). Sporophytes of some hornworts growing in mild and continuously moist conditions can become as tall as 20 centimeters, making them the tallest known bryophytes.

Hornworts differ both from other bryophytes and from tracheophytes in having a single, large chloroplast in each cell. Photosynthetic cells in other plants have many small chloroplasts. Hornworts have internal cavities filled with a mucilage, often populated by cyanobacteria that convert atmospheric nitrogen gas into a form usable by the host plant (Chapter 30).

Class Musci: Mosses

The most familiar bryophytes are the mosses. Let us consider the reproductive cycle of a moss (Figure 24.7). The gametophyte that develops following spore germination is a branched, filamentous plant, or **protonema** (plural, protonemata) that looks much like a filamentous green alga and is unique to this class. Some of the filaments contain chloroplasts and are photosynthetic; other filaments are nonphotosynthetic and anchor the protonema to the substrate. The nonphotosynthetic filaments, called rhizoids, are

the bryophyte counterpart of the root hairs of vascular plants. After a period of growth, cells close to the tips of the photosynthetic filaments commence rapid cell division in three dimensions to form buds. The buds eventually differentiate a distinct apex and produce the familiar leafy moss plant with the "leaves" spirally arranged. These leafy **gametophores**, as they are called, produce sex organs at their tips. Sometimes a gametophore may produce both antheridia and archegonia, sometimes only one or the other. The antheridia release sperm that travel through liquid water to the archegonia, where they fertilize the eggs. Sporophyte development in most mosses follows a precise pattern, resulting ultimately in the formation of an absorptive foot, a stalk, and, at the tip, a swollen capsule. The moss sporophyte grows at its apical end, in contrast to the situation in the hornworts. Cells at the tip of the stalk divide, supporting elongation of the structure and giving rise to the capsule. The moss sporophyte often possesses stomata, allowing gas exchange with the atmosphere, unlike the sporophyte of liverworts.

The archegonial tissue also grows rapidly as the stalk elongates, and for a time it keeps pace with the rapidly expanding sporophyte. Finally, however, the archegonium loses out, and it is split apart around its middle. The top portion of the archegonium frequently persists on the top of the rapidly elevating capsule as a little pointed cap, the **calyptra**. The top of the capsule is ultimately shed, after meiosis has led to the creation of numerous mature spores within. Groups of cells just below the lid form a series of teeth surrounding the opening. Highly responsive to humidity, the teeth arch into the mass of spores when the atmosphere is dry and then out again as it becomes moist. The spores are thus dispersed when the surrounding air is moist, that is, under the best conditions for their subsequent germination.

Only a few mosses depart from this pattern of capsule development. A familiar exception is *Sphagnum*, which has a very simple capsule with an air chamber in it. In some way, air pressure builds up in this chamber, so that eventually the capsule pops its lid open, dispersing the spores with an audible explosion. In terms of biomass, *Sphagnum* probably outweighs all other mosses put together, being found in tremendous quantities in northern bogs and tundra, and extending well into the Arctic (Figure 24.8).

DIVISION TRACHEOPHYTA

The Tracheophyta, or vascular plants, are an extraordinarily large and diverse group, yet they can be said to have been launched by a single evolutionary event. Sometime during the Paleozoic era, probably well before the Silurian period, the sporophyte generation

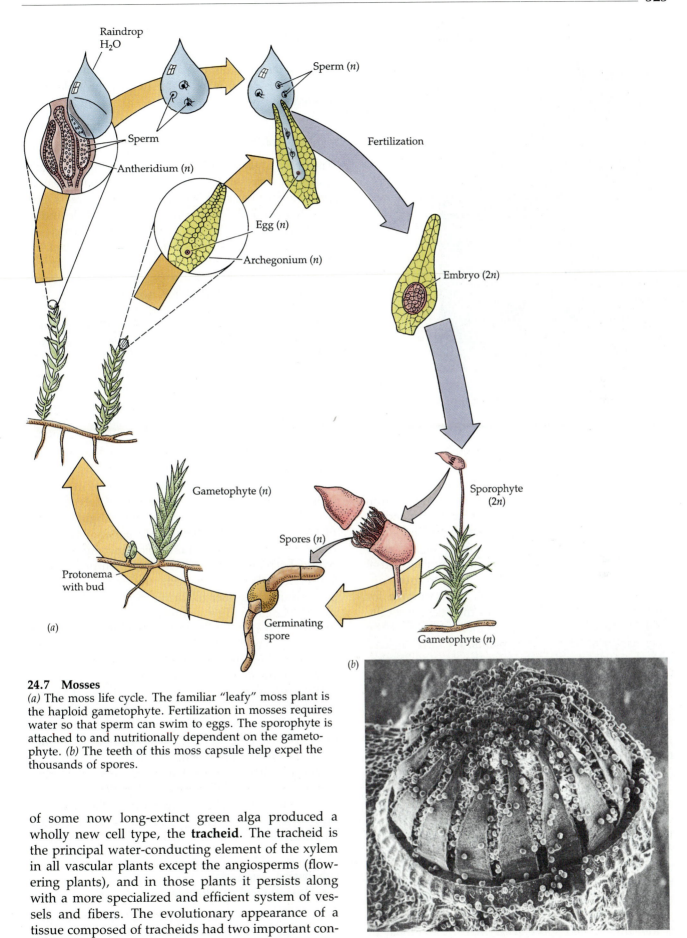

Raindrop
H₂O

Sperm (n)

Sperm

Antheridium (n)

Fertilization

Egg (n)

Archegonium (n)

Embryo (2n)

Gametophyte (n)

Sporophyte
(2n)

Spores (n)

Protonema
with bud

Germinating
spore

Gametophyte (n)

(a)

(b)

24.7 Mosses
(a) The moss life cycle. The familiar "leafy" moss plant is
the haploid gametophyte. Fertilization in mosses requires
water so that sperm can swim to eggs. The sporophyte is
attached to and nutritionally dependent on the gametophyte. (b) The teeth of this moss capsule help expel the
thousands of spores.

of some now long-extinct green alga produced a
wholly new cell type, the **tracheid**. The tracheid is
the principal water-conducting element of the xylem
in all vascular plants except the angiosperms (flowering plants), and in those plants it persists along
with a more specialized and efficient system of vessels and fibers. The evolutionary appearance of a
tissue composed of tracheids had two important con-

24.8   A Peat Bog
The moss *Sphagnum* is responsible for the formation of peat, which is used for fuel. This peat bog in New Brunswick, Canada is being harvested for commercial use.

sequences. First, it provided a pathway for long-distance transport of water and mineral nutrients from a source of supply to regions of need. Second, it provided something almost completely lacking—and unnecessary—in the largely aquatic algae: rigid structural support. This is important in a terrestrial environment because land plants tend to grow upward (thus requiring support) as they compete for sunlight for photosynthesis. Thus the tracheid set the stage for the complete and permanent invasion of the land masses of the world by plants. We have decided to include all vascular plants in the division Tracheophyta on the grounds that members of every single group within it possess this one particular type of cell, the tracheid, in the xylem.

The life cycle of those vascular plants that lack seeds is remarkably uniform. Alternation of generations is uniformly heteromorphic, with both haploid and diploid phases being free-living and independent at maturity. The tracheophyte sporophyte is the large and obvious plant that one normally notices in nature, in contrast to the bryophyte sporophyte, which is attached to and completely dependent upon the gametophyte. Tracheophyte gametophytes are rarely more than a centimeter or two in length. By contrast, the sporophyte of a tree fern may be 15 or 20 meters tall and may live for years. The most prominent resting stage in the life cycle of a seedless vascular plant is the single-celled spore. This life cycle is similar to those of the fungi, the algae, and the bryophytes—

but not, as we will presently see, to that of the seed plants. The seedless vascular plants must have an aqueous environment at one stage of their life cycle, because syngamy is accomplished by a motile, flagellated sperm.

Evolution of the Tracheophytes

The green algal ancestors of the plant kingdom made their first successful attempts to invade the terrestrial environment between 400 and 500 million years ago. The evolution of a water-impermeable cuticle and of protective layers for the gamete-bearing structures helped to make the invasion permanent, as did the initial absence of herbivores. By the late Silurian period vascular plants—early psilopsids—were being preserved as fossils that we can study today. The Devonian period, 400 to 345 million years ago, saw several remarkable developments. Three groups of non-seed-producing tracheophytes that are still among us made their first appearances during that period: the lycopsids (club mosses), sphenopsids (horsetails), and ferns. The proliferation of these plants made the terrestrial environment ever more hospitable to animals, with amphibians and insects arriving soon after they became established. Fossil remains about 360 million years old provide the first evidence of seed plants.

Trees of various kinds, which appeared in the Devonian, came into their own in the Carboniferous period (345 to 290 million years ago). Mighty forests of lycopsids up to 40 meters tall, sphenopsids, and tree ferns flourished in the tropical swamps of what would become North America and Europe. In the subsequent Permian period, the continents came ponderously together to form a single, gigantic land mass, Pangaea. The continental interior become warmer and dryer, but late in the period glaciation was extensive. The 200-million-year reign of the lycopsid-fern forests came to an end, to be replaced by gymnosperm forests that ruled throughout the Triassic and Jurassic periods. The gymnosperm forests changed with time as the gymnosperm groups evolved. The gymnosperm forests dominated during the era in which the continents drifted apart and dinosaurs strode the Earth.

The oldest evidence of angiosperms—flowering plants—dates into the Cretaceous period, about 120 million years ago. The angiosperms radiated almost explosively over a period of about 55 million years and became the dominant plant life of the planet.

Subdivision Psilopsida:
The Most Ancient Tracheophytes

Psilopsids, the first vascular plants, belong to a now-extinct subdivision (Psilopsida). The psilopsids appear to have been the only vascular plants in the

Silurian period of the Paleozoic era, and they dominated the landscape. Early versions of the structural features of all the other groups of the Tracheophyta appeared in the psilopsid plants of that time, and these shared features strengthen the case for a common origin of all vascular plants from some common green algal ancestor.

In 1917 the British paleobotanists Robert Kidston and William H. Lang first reported some well-preserved fossils of vascular plants embedded in Devonian rocks near Rhynie, Scotland. Their preservation was remarkable, considering the age of the rocks (over 395 million years). These fossil plants all had a simple vascular system consisting of tracheids and phloem. Furthermore, flattened scales on the stems of some of the plants entirely lacked vascular tissue and thus were not comparable with the true leaves of any other vascular plants. The plants were without roots, apparently anchored in the soil by horizontal portions of stem (**rhizomes**) that bore rhizoids. These rhizomes also bore aerial branches, and sporangia—homologous with the bryophyte capsule—were found at the tips of the stems. Branching was dichotomous, that is, the shoot apex divided to produce two equivalent new branches, each pair diverging at approximately the same angle from the original stem, as shown in Figure 24.9a for the genus *Rhynia*. Scattered fragments of such plants—psilopsids—had been found earlier, but never in such profusion or so well preserved as those discovered near Rhynie by Kidston and Lang.

The presence of tracheids clearly indicated that plants such as *Rhynia* were vascular plants. But were they sporophytes or gametophytes? Close inspection of thin sections of fossil sporangia revealed that the spores were in groups of four. In virtually all the living seedless vascular plants (with no evidence to the contrary from fossil forms), the four products of a meiotic division and cytokinesis remain attached to one another during their development into spores. The spores finally separate only when they are mature, and even after separation, their walls reveal the exact geometry of the way in which they were attached. Thus a group of four closely packed spores is inevitably found only immediately after meiosis, and a plant that produces such a group of four must be a diploid sporophyte.

There is still some disagreement as to whether psilopsids are entirely extinct. The confusion arises because of the existence today of two genera of rootless and spore-bearing plants, *Psilotum* and *Tmesipteris* (Figure 24.9b). Are they the living relics of the psilopsids? There are good arguments, based on anatomy, morphology, and development, that these two genera are probably ferns rather than psilopsids. There are indeed no roots, but a few ferns such as *Stromatopteris* form roots only quite late in development. Although *Psilotum nudum* has no true leaves, only minute scales, plants of the genus *Tmesipteris*

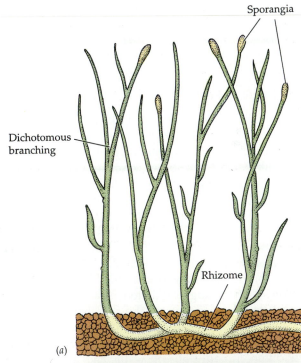

(a)

(b)

24.9 Ancient Psilopsid and Modern Look-Alike
(a) A drawing of an extinct psilopsid (division Tracheophyta, subdivision Psilopsida) in the genus *Rhynia*, named after the discovery of fossil remains near Rhynie, Scotland. The horizontal underground stem is called a rhizome; the aerial shoots were less than 50 cm in height, and several drawn here are topped by sporangia. (b) Aerial branches of *Psilotum nudum*, a plant considered by some to be a surviving psilopsid and by others to be a fern (subdivision Pteropsida, class Filicinae).

have respectable flattened photosynthetic organs with well-developed vascular tissue. Finally, there is an enormous hole in the geologic record to explain: No psilopsid fossils appear anywhere after the Devonian period.

FURTHER DEVELOPMENTS IN THE ANCIENT TRACHEOPHYTES

Within a few tens of millions of years, during the Devonian period, three new subdivisions of the Tracheophyta appeared on the scene, arising from psilopsid ancestors. These new groups featured various advances over the psilopsids, including one or more of the following: true roots, true leaves, and a differentiation between two types of spores. Let us consider these advances, one at a time.

The Origin of Roots

Rhynia and its close relatives lacked true roots. They had only rhizoids, arising from a prostrate rhizome, with which to gather water and minerals. How, then, did subsequent groups of vascular plants come to have the complex roots we see today?

A French botanist, E. A. O. Lignier, proposed an attractive hypothesis in 1903 that is still widely accepted today. Lignier argued that the green algal ancestors of the first vascular plants were dichotomously branching. This accounts for the dichotomous branching observed in the psilopsids themselves. Lignier suggested, too, that such a branch could bend and penetrate the soil, branching there (Figure 24.10). The underground portion could anchor the plant firmly, and, even in this primitive condition, it could absorb water and minerals. The subsequent discovery of numerous fossil plants from the Devonian period, all having horizontal stems with both underground and aerial branches, supported Lignier's hypothesis.

The underground branches, being in an environment sharply different from that above the ground, were subjected to very different selection pressures over succeeding millions of years. Thus, the two parts of the plant axis (the shoot and root systems) diverged in structure, so that they came to have distinct internal and external anatomies (Chapter 28). However, we believe that the root and shoot systems of vascular plants were once homologous—in fact, part of the same organ.

The Origin of True Leaves

Thus far, we have used the term *leaf* rather loosely. We spoke of "leafy" mosses; we also commented on the absence of "true leaves" in psilopsids. In the strictest sense, a **leaf** is a flattened photosynthetic

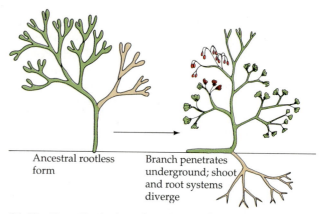

24.10 Root Evolution: One Hypothesis
According to Lignier's hypothesis, branches from ancestral rootless plants could have penetrated the soil, where they gradually evolved into a root system.

structure emerging laterally from a main axis or stem and possessing true vascular tissue. This tight definition allows a closer look at true leaves in the Tracheophyta, and we then see that there are two different types of leaves, probably of different evolutionary origins.

The first type of leaf, called a **microphyll** ("small leaf"), is usually small and only rarely has more than a single vascular strand, at least in plants alive today. Plants in two subdivisions have microphylls: the Lycopsida (club mosses) and the Sphenopsida (horsetails, or scouring rushes), of which only a few genera still survive. The evolutionary origin of this kind of leaf is thought to be the progressive development of vascular tissue within small, scalelike outgrowths of the stem (Figure 24.11a). The principal characteristic of a microphyll is that its vascular strand departs from the vascular system of the stem in such a way that there is scarcely any perturbation in the conformation of the stem's vascular system. This was true even in the fossil lycopsid and sphenopsid trees of the Carboniferous period, many of which had microphylls several centimeters long.

The other type of leaf, called a **megaphyll** ("large leaf"), is encountered only in the subdivision Pteropsida (ferns and seed plants). The megaphyll is thought to have arisen from the flattening of a dichotomously branching stem system, with the development of extensive photosynthetic tissue between the branch members (Figure 24.11b). Another feature of the megaphyll leaf type is that its vascular system creates a major alteration in the architecture of the stem vascular system where it departs for the leaf base (Figure 24.11c).

Homospory and Heterospory

In the vascular plants of most ancient origin, the gametophyte and sporophyte are both free-living, photosynthetic, and independent. Spores produced

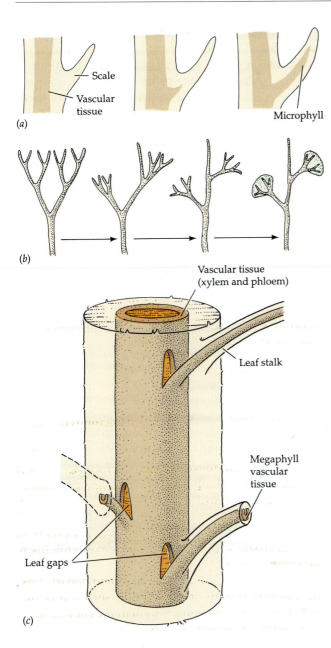

(a)

(b)

(c)

24.11 Evolution of Leaves
(a) Microphylls might have evolved from scales such as those on the stems of psilopsids. The diagram shows progression from a scale without vascular tissue (left) to one with some vascular tissue (center), to a true microphyll (right). All three types appear in fossil plants. (b) Megaphylls may have originated as a branching stem system (left) and become progressively reduced (left center) and flattened (right center). Flat plates of photosynthetic tissue developed between small end branches (right). The end branches evolved into the veins of leaves. (c) Where vascular tissue departs from a stem into a megaphyll, there is a gap in that tissue in the stem immediately over the junction.

by the sporophytes are of a single type, and they develop into a single type of gametophyte, bearing both female and male reproductive organs. Such plants, bearing a single type of spore, are said to be **homosporous** (Figure 24.12a). The sex organs on the gametophytes of homosporous plants are of two types. The female organ is a multicellular archegonium, typically containing a single egg. The male organ is an antheridium, containing many sperm.

A different system, with two distinct types of spores, evolved somewhat later. Plants of this second type are said to be **heterosporous** (Figure 24.12b). They function as follows. One type of spore, the **megaspore**, develops into a larger, specifically female gametophyte (megagametophyte) that produces only eggs. The other type, the **microspore**, develops into a smaller, male microgametophyte that produces only sperm. Megaspores are produced in small numbers in megasporangia on the sporophyte, and microspores in large numbers in microsporangia.

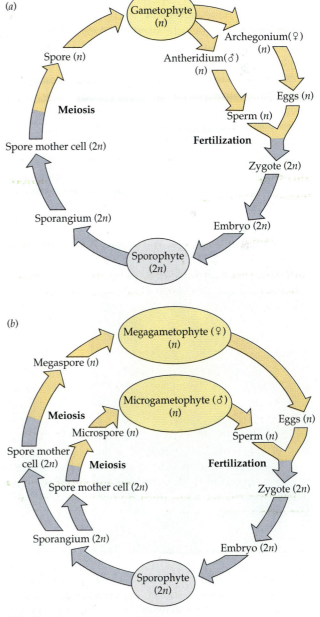

(a)

(b)

24.12 Homospory and Heterospory

The psilopsids were all homosporous. Heterospory evidently evolved a number of times, independently, in the early evolution of the vascular plants descended from the psilopsids. The fact that heterospory evolved repeatedly suggests that it affords selective advantages. We will see, in fact, that subsequent evolution in the plant kingdom featured ever greater specialization of the heterosporous condition.

SURVIVING SEEDLESS TRACHEOPHYTES

Subdivisions Lycopsida and Sphenopsida: Club Mosses and Horsetails

The Lycopsida, or club mosses, and the Sphenopsida, or horsetails (also called scouring rushes because silica deposits found in the cell walls made them useful for this purpose), are represented by relatively few current species. Both lycopsids and sphenopsids have true roots, and both bear only microphylls. Both include homosporous species and heterosporous species. Like all seedless tracheophytes, both have a heteromorphic alternation of generations with a large, independent sporophyte and a small, independent gametophyte. Here, however, the resemblance ends. The leaves are arranged spirally on the stem in the Lycopsida (Figure 24.13), whereas they form distinct whorls in the Sphenopsida (Figure 24.14). The sporangia in lycopsids appear in conelike structures called strobili (singular, strobilus) and are tucked in the upper angle between a specialized microphyll and the stem, whereas the sporangia of the sphenopsids are recurved toward the stem on the ends of short stalks (sporangiophores). Growth in lycopsids comes entirely from groups of dividing cells at the tips of the stems, whereas growth in sphenopsids comes to a large extent from discs of dividing cells just above each whorl of leaves, so that each segment of the stem grows from its base.

Though only minor elements of the vegetation of today, these two subdivisions appear to have been the dominant vegetation during the Carboniferous period. One kind of coal, called cannel coal, is formed almost entirely from fossilized spores of a tree lycopsid named *Lepidodendron*; this finding is an indication of the importance of this genus in the forests of that time.

Subdivision Pteropsida: Plants with Megaphylls

The subdivision Pteropsida consists of both seedless plants—the ferns—and seed plants—the gymnosperms and the angiosperms. The sporophytes of the Pteropsida have true roots, stems, and leaves. Their leaves are typically large, although some species have small leaves as a result of evolutionary reduction. The pteropsid leaf is a megaphyll. The sporangia of the Pteropsida are on the lower surfaces of leaves or modified leaves or, more rarely, on the margins of leaves. Pteropsid sporophytes are independent and much larger than the gametophytes. The gametophytes of ferns are independent, but those of the gymnosperms and angiosperms are dependent upon the sporophytes.

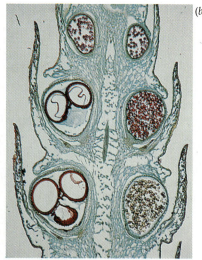

24.13 Lycopsids
(a) *Lycopodium annotinum*, a club moss (division Tracheophyta, subdivision Lycopsida), with conelike strobili. Club mosses have small leaves (microphylls) arranged spirally on their stems. (b) Thin section through a strobilus of *Selaginella*, another club moss.

24.14 Sphenopsids
(a) Vegetative and sterile shoots of the horsetail *Equisetum telmateia* (division Tracheophyta, subdivision Sphenopsida). Leaves form in spaced whorls at nodes on the stems of the vegetative shoots; a few of the candle-shaped fertile shoots are seen toward the lower right. *(b)* Sporangia and sporangiophores of the horsetail *Equisetum arvense*.

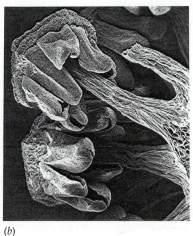

(a)   *(b)*

Class Filicinae: Ferns

The true ferns constitute the class Filicinae, which first appeared during the Devonian period and today consists of about 12,000 species (Figure 24.15). Ferns are characterized by the possession of large leaves (fronds), by the absence of seeds, and by a requirement for water as a medium for the transfer of the male gametes.

(b)

(a)

24.15 Fern Fronds
(a) Fronds of the cinnamon fern (division Tracheophyta, subdivision Pteropsida, class Filicinae). *(b)* Tiny fronds of a species of the water fern *Marsilea*. *(c)* "Fiddleheads" (developing fronds) of a Christmas fern; these structures will unfurl and expand to give rise to the complex adult frond.

(c)

24.16 Fern Sori
Sori, each with many spore-producing sporangia, on the underside of a frond of the western sword fern.

Most ferns inhabit shaded, moist woodlands and swamps. Some, the tree ferns, reach heights of up to 20 meters. Tree ferns lack the rigidity of woody plants, and thus do not grow in sites exposed directly to strong winds but rather in ravines or beneath trees in forests.

The fern frond is a megaphyll. During its development, the fern frond unfurls from a tightly coiled "fiddlehead" (Figure 24.15c). Some fern leaves become climbing organs and may grow to be as much as 30 meters in length. The sporangia are found on the undersurfaces of the leaves, sometimes covering the whole undersurfaces and sometimes at the edges; in some species the sporangia are clustered in groups called sori (singular, sorus; Figure 24.16).

Devonian fossil beds have yielded ferns with some characteristics that are psilopsid and some that resemble those of other subdivisions of the Tracheophyta. The genus *Protopteridium* had flattened branch systems with extensive photosynthetic tissue between the branches—like a pteropsid—but bore terminal sporangia—like a psilopsid (Figure 24.17). This plant—like a psilopsid—evidently lacked true roots. During late Paleozoic times, the ferns underwent considerable evolutionary experimentation in the structure of their leaves, and particularly in the arrangement of their vascular tissue.

The Fern Life Cycle

The undersides or edges of fern fronds carry sporangia in which cells undergo meiosis to form haploid spores (Figure 24.18). Once shed, spores often travel great distances and eventually germinate to form small, independent gametophytes. These gametophytes produce antheridia and archegonia, although not necessarily at the same time or on the same gametophyte. Sperm swim through water to archegonia, often on other gametophytes, and the result-

ing zygote develops into a new sporophyte embryo. The young sporophyte sprouts a root and can thus grow independently of the gametophyte. In the alternating generations of a fern, the gametophyte is small, delicate, and short-lived, but the sporophytes can be very large and can sometimes survive for hundreds of years.

Most ferns are homosporous. However, two orders of aquatic ferns, the Marsileales and Salviniales, have evolved heterospory. Male and female spores of these plants (which germinate to produce male and female gametophytes, respectively) are produced in different sporangia, and the male spores are always much smaller and greater in number.

A few genera of ferns produce a tuberous, fleshy gametophyte instead of the characteristic, flattened photosynthetic structure described above. These tuberous gametophytes depend upon a mutalistic fungus for nutrition; in some genera (for example, *Psilotum*), even the sporophyte embryo must become associated with the fungus before any extensive development will proceed.

THE SEED PLANTS

The most recent group to appear in the evolution of the plant kingdom is the seed plants, consisting of the gymnosperms (such as pines and their relatives)

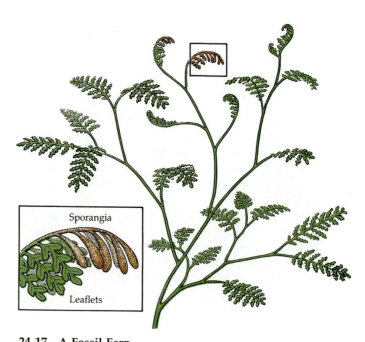

24.17 A Fossil Fern
Ferns exhibited endless and sometimes strange combinations of structures during their evolutionary history. The fossil fern shown here (genus *Protopteridium*, from the Devonian period) lacked true roots and had branches with both leaflets and terminal sporangia.

24.18 Life Cycle of a Fern

Unlike the mosses, the ferns have a life cycle in which the most familiar stage is the mature, diploid sporophyte. In sporangia on the undersides of fern fronds, cells called sporocytes undergo meiosis to produce haploid spores. When spores germinate, they form the small, heart-shaped gametophytes. Eggs in archegonia on gametophytes are fertilized by swimming sperm from antheridia. The embryo that develops from the diploid zygote in the archegonium eventually sends out roots and develops into a sporophyte.

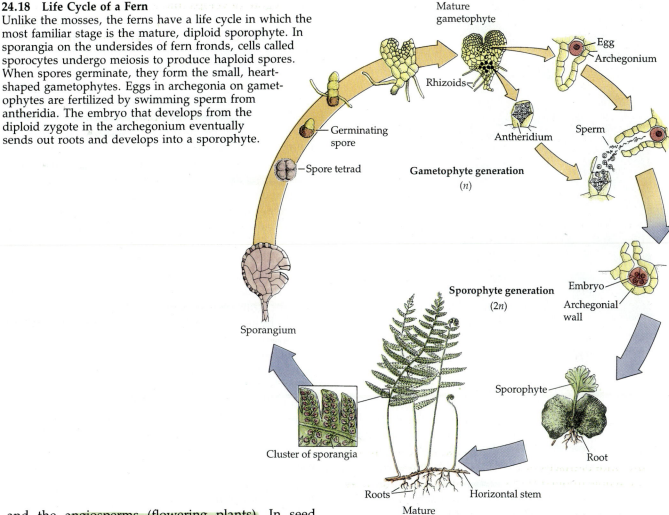

and the angiosperms (flowering plants). In seed plants, the gametophyte generation is reduced even further than in ferns. The haploid gametophyte develops partly or entirely while attached to—and nutritionally dependent upon—the diploid sporophyte. Only the earliest types of gymnosperms to evolve have swimming sperm and therefore require environmental water for the meeting of gametes. All other gymnosperms and all the angiosperms have evolved other means of bringing gametes together. This marks the culmination of one of the most striking evolutionary trends in the plant kingdom—independence from liquid water for the purposes of reproduction.

Seed plants form separate megasporangia and microsporangia—female and male sporangia, respectively—on modified leaves called sporophylls ("spore leaves"). In most seed plants, sporophylls are grouped on short axes to form strobili such as the cones of some gymnosperms and the flowers of angiosperms.

Spores are produced by meiosis within the sporangia of seed plants, but the spores are not shed. Instead, the gametophytes develop within the sporangia and depend upon them for food and water.

In most species only one of the products of a meiotic division in a megasporangium survives. The surviving haploid nucleus divides mitotically, and those products divide again to produce a small multicellular female gametophyte. In the angiosperms, female gametophytes do not normally include more than eight nuclei in all. The female gametophyte of a seed plant is retained within the megasporangium even at maturity—and during the early development of the next sporophyte generation.

Meanwhile, within the microsporangium, meiotic divisions produce microspores that undergo one or a few cell divisions to form the male gametophytes, which are the familiar pollen grains (Figure 24.19; see also Figure 31.2). Distributed by wind, an insect, a bird, or a plant breeder, a pollen grain that reaches the appropriate surface of a sporophyte, near the female gametophyte, develops further. It develops a slender pollen tube that grows and digests its way through the sporophyte tissue toward the female gametophyte.

24.19 Pollen
The wind carries pollen grains from a male (above) to a female strobilus.

When the tip of the male gametophyte's pollen tube reaches the female gametophyte, either sperm nuclei or a sperm cell is released from the tube, and fertilization occurs. The resulting diploid zygote divides repeatedly, forming a young sporophyte that develops to some embryonic state; then the entire system becomes dormant. The end product at this stage is a **seed**. *A seed may contain tissues from three generations.* The seed coat develops from tissue of the diploid parent sporophyte. Within the seed coat is a layer of haploid female gametophyte tissue from the next generation (this tissue is fairly extensive in most gymnosperm seeds, but its place is taken by a tissue called endosperm, to be discussed presently, in angiosperm seeds). In the center of the seed package, the third generation is found in the form of the embryo of the new sporophyte.

The embryos of seedless plants develop directly into sporophytes, which either survive or die, depending on environmental conditions at the time. There is no resting stage in the life cycle. In contrast, the multicellular seed of a gymnosperm or angiosperm is a well-protected resting stage. Layers of cells enclose the dormant embryo, and the seeds of some species may remain viable for many years, germinating when conditions are favorable for the growth of the sporophyte. When the young sporophyte begins to grow, it draws on food reserves present in the seed. The possession of seeds is a major reason for the enormous evolutionary success of seed plants, which are the prominent elements of Earth's modern land flora in most areas.

CLASS GYMNOSPERMAE

Although there are probably fewer than 750 species of living gymnosperms (class Gymnospermae), these plants are second only to the angiosperms (flowering plants) in their dominance of the land masses. There are four orders of gymnosperms living today (Figure 24.20). The cycads are palmlike plants of the tropics, growing as tall as 20 meters. Ginkgos, which were common during the Mesozoic era, are represented today by a single genus, the maidenhair tree. The order Gnetales consists of three disparate genera that share a characteristic type of cell (the vessel element) in their xylem tissue found in no other group of plants except the angiosperms. One member of the order is *Welwitschia*, a long-lived desert plant with just two straplike leaves that sprawl on the sand and can become as long as 3 meters. Far and away the most abundant of the gymnosperms are the conifers, cone-bearing plants such as pines.

All living gymnosperms have active secondary growth (growth in diameter of the axis, produced by cambium; Chapter 28), and all but the Gnetales have only tracheids as water-conducting and support cells in the xylem. Despite this apparently suboptimal design of their water transport and support system (as compared with the cell types found in angiosperms), the gymnosperms include the tallest trees known. The coastal redwoods of California are the record holders—the largest are well over 100 meters tall. Xylem produced by gymnosperms is the principal resource of the lumber industry.

Fossil Gymnosperms

The earliest fossil evidence of gymnosperms is found in Devonian rocks. The early gymnosperm story is an especially interesting one, illustrating the difficulties under which students of fossil plants frequently labor. Many years ago, a relatively rare and poorly preserved plant fossil, *Archaeopteris*, was described. The fossil appeared to many workers to be an ancient heterosporous fernlike plant similar to *Protopteridium* (Figure 24.17). However, another fairly common Devonian fossil, *Callixylon*, remained a puzzle. It consisted only of well-petrified logs, some wider than a meter in diameter and longer than 20 meters. Among the delicate and herbaceous psilopsids it seemed entirely out of place. Its wood was composed of pitted tracheids, and it obviously had possessed a highly effective mechanism for secondary growth. Charles Beck of the University of Michigan resolved the puzzle when, in 1960, he found fronds of *Archaeopteris* clearly attached to and part of *Callixylon* logs. Because the name *Archaeopteris* had been published first, the name *Callixylon* was dropped, and the "rare" *Archaeopteris* was suddenly recognized as being a common Devonian plant. It had both psilopsid and fern-

24.20 Gymnosperms
Division Tracheophyta, subdivision Pteropsida, class Gymnospermae. (a) This "sago palm" belongs to a group, the cycads, that are the most ancient order of gymnosperms. Cycads often have growth forms resembling both ferns and palms. (b) Characteristic broad leaves of the maidenhair tree, *Ginkgo biloba*; the leaves turn brilliant gold in the autumn. (c) *Welwitschia mirabilis* growing in the Namib Desert of Africa. Two huge, straplike leaves grow throughout the life of the plant, breaking and splitting as they grow. Seeds are produced on the cone-bearing branches. *Welwitschia* belongs to the Gnetales, the gymnosperm order most closely related to the flowering plants. (d) A dramatic conifer, the giant sequoia, *Sequoiadendron giganteum*, growing in Yosemite National Park, California.

like characteristics, but its woody tissue, based on tracheids, was clearly that of a gymnosperm.

Several new lines of gymnosperms had evolved by the Carboniferous period, including one group, the seed ferns, that possessed fernlike foliage but had characteristic gymnosperm seeds attached to the leaf margins. The first true conifers appeared at approximately the same time. Either they were not dominant trees or else they did not grow where conditions were right for fossilization. However, during the Permian period the conifers and cycads came into their own. Gymnosperms dominated the forests until less than 80 million years ago; in fact, they dominate some present-day forests.

Order Coniferales: Conifers

The great Douglas fir and cedar forests of the northwestern United States and the massive forests of pine, fir, and spruce that clothe the northern continental regions and upper slopes of mountain ranges rank among the great vegetation formations of the world (the "boreal forest;" see Figure 48.13). All these trees belong to one particular order of gymnosperms, Coniferales—the conifers, or cone-bearers. The sporophylls of the conifers are borne in apical cones that are specialized strobili. All conifers are heterosporous. Male and female sporophylls are found in separate male and female cones.

The Gymnosperm Life Cycle

We will use the life cycle of a pine to illustrate reproduction in gymnosperms (Figure 24.21). The production of male gametophytes as pollen grains frees the plant once and for all from a need for liquid water for fertilization. The wind assists conifer pollen grains

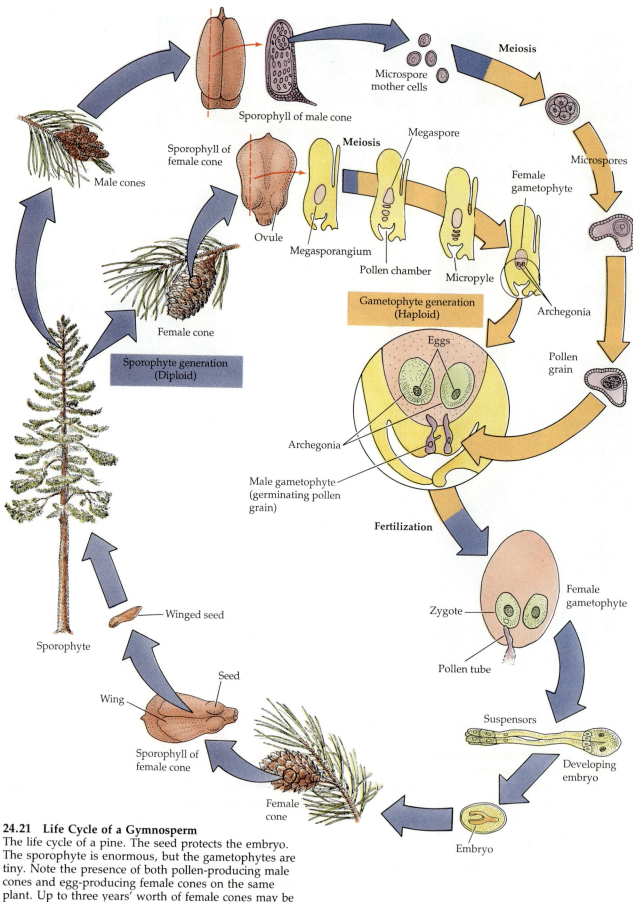

Microspore mother cells

Meiosis

Microspores

Sporophyll of male cone

Sporophyll of female cone

Meiosis

Megaspore

Female gametophyte

Ovule

Male cones

Megasporangium

Pollen chamber

Micropyle

Archegonia

Female cone

Gametophyte generation (Haploid)

Pollen grain

Eggs

Sporophyte generation (Diploid)

Archegonia

Male gametophyte (germinating pollen grain)

Fertilization

Sporophyte

Winged seed

Zygote

Female gametophyte

Pollen tube

Wing

Seed

Suspensors

Sporophyll of female cone

Developing embryo

Female cone

Embryo

24.21 Life Cycle of a Gymnosperm
The life cycle of a pine. The seed protects the embryo.
The sporophyte is enormous, but the gametophytes are
tiny. Note the presence of both pollen-producing male
cones and egg-producing female cones on the same
plant. Up to three years' worth of female cones may be
present at the same time.

in their first stage of travel to the female gameto-phyte. The pollen tube provides the means for the last stage of travel, by growing and digesting its way through maternal sporophytic tissue and eventually releasing a sperm nucleus near the egg. The mega-sporangium is enclosed in a special layer of sporo-phytic tissue, called the **integument**, that is destined eventually to develop into the seed coat. The integ-ument, together with its enclosed megasporangium and the tissue attaching it to the maternal sporo-phyte, forms the **ovule**. The small opening in the integument at the apex of the ovule, through which the pollen grain travels, is called the **micropyle**. The gymnosperm female gametophyte produces eggs en-closed in archegonia surrounded by the integument.

The word *gymnosperm* means, literally, "naked-seeded." Conifer ovules (which develop into seeds upon fertilization) are borne exposed on the upper surfaces of sporophylls without fruit tissue to protect the seeds. Any protection from the environment de-rives merely from the fact that the sporophylls are tightly pressed against each other within the cone. Indeed, some pines have such tightly closed female cones that normally only fire suffices to split them open and release the seeds. One example is the lodgepole pine, which covers vast fire-ravaged areas in the Rocky Mountains and elsewhere.

CLASS ANGIOSPERMAE: FLOWERING PLANTS

The flowering plants constitute the class Angiosper-mae. This highly diverse class includes about 275,000 different species. In other chapters, when "plants" are mentioned in discussing processes such as long-distance transport in the xylem and phloem, or the chemical regulation of development, we generally mean this very particular group of plants, the an-giosperms. With the angiosperms, we arrive at the culmination of an evolutionary trend that runs throughout the plant kingdom, in which the *sporo-phyte* generation becomes *larger* and *more independent* of the gametophyte, while the *gametophyte* generation becomes *smaller* and *more dependent* upon the sporo-phyte (Figure 24.22).

Angiosperms differ from other plants in several ways, although exceptions exist. **Double fertilization** is the single most reliable distinguishing character-istic of the angiosperms. *Two* male gametes, con-tained within a single male gametophyte, participate in fertilization events within the female gametophyte of an angiosperm. One sperm nucleus combines with the egg to produce a diploid zygote, the first cell of the sporophyte generation. The other sperm nucleus combines with *two* other haploid nuclei of the female gametophyte. The result of this second fertilization

Sporophyte generation (2*n*)

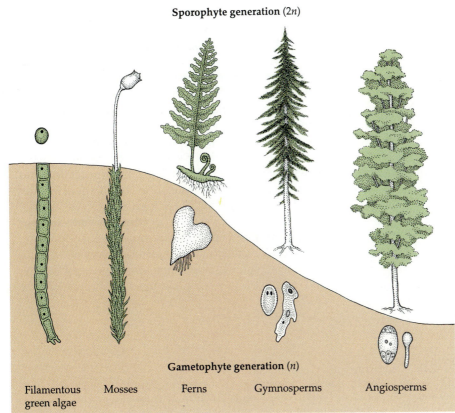

Gametophyte generation (*n*)

| Filamentous green algae | Mosses | Ferns | Gymnosperms | Angiosperms |

24.22   Shifting to a Dominant Sporophyte
In most filamentous green algae, the gametophyte (the filament) is much larger than the sporophyte (repre-sented by the zygote). Subsequent evolution has gradually shifted the emphasis to the sporophyte genera-tion. In mosses, the two generations are of comparable size, but the spo-rophyte is attached to and partially dependent upon the gametophyte. The trend to relatively smaller game-tophytes and larger sporophytes con-tinues through the ferns, gymno-sperms, and angiosperms. In the seed plants the male gametophytes are microscopically small, and the fe-male gametophytes are totally depen-dent on the sporophyte tissue.

is a triploid (3*n*) nucleus. Divisions of the triploid nucleus give rise to a triploid tissue, the **endosperm**, that nourishes the embryonic sporophyte during its early development. All present-day angiosperms use double fertilization, but no other plant does—with one exception, *Ephedra* (a member of the gymnosperm order Gnetales), discovered in 1990 to have double fertilization. We cannot be sure when and how double fertilization evolved, since there is no fossil evidence on this point.

A second diagnostic characteristic of angiosperms is the possession of specialized water-transporting cells called **vessel elements** in the xylem, but these are also found in a few gymnosperms and ferns, and even in a lycopsid and a sphenopsid. Another distinctive cell found in angiosperm xylem is the **fiber**, which plays an important role in supporting the plant body. For another diagnostic characteristic, the very name *angiosperm* refers to the fact that the seeds of these plants are enclosed in a modified leaf called a carpel. Still another diagnostic feature of flowering plants is the possession of, well, flowers.

The Flower

The reproductive organ of angiosperms is the **flower**, a form of strobilus. A generalized flower (for which there is no exact counterpart in nature) is shown in Figure 24.23. In the flower, those sporophylls bearing microsporangia are called **stamens**, and those bearing megasporangia are called **carpels**. In addition, there often are a number of specialized sterile (non-spore-bearing) leaves below the sporophylls, the upper being called **petals** (collectively, the **corolla**), and the lower **sepals** (collectively, the **calyx**). The corolla and calyx often play roles in attracting animal pollinators to the flower. From base to apex, the sepals, petals, stamens, and carpels are arranged in whorls and attached to a central receptacle.

Each stamen is composed of a **filament** ending in two **anthers**, each composed of sporangia in which pollen is produced. A structure composed of one or more fused carpels is called a **pistil.** The swollen base of the pistil, containing one or more ovules, is called an **ovary**; the apical stalk of the pistil is called a **style**; and the terminal surface that receives the pollen is called a **stigma.**

Because the flower shown in Figure 24.23 produces both megasporangia and microsporangia, it is said to be **perfect**, meaning that it contains both female and male parts. Many angiosperms produce two types of flowers on the same plant, one type with only megasporangia and the other with only microsporangia; consequently, either the stamens or the carpels are nonfunctional or absent in a given flower. Species having both female and male flowers on the same plant are said to be **monoecious** (one-housed—but, it must be added, one house with separate rooms). The sexes are completely separated in some other species of angiosperms; in these species, a given plant produces either male or female sporophylls, but never both. Such species are said to be **dioecious** (two-housed). In other words, there are truly female plants and truly male plants.

In the ideal flower of Figure 24.23, we also illustrated distinct petals and sepals arranged in distinct whorls, whereas in nature sometimes the petals and sepals are arranged in a continuous spiral and there is no distinguishing between the two. Such indistinguishable appendages are called **tepals**. Sometimes appendages of any sort—petals, sepals, or tepals—are completely absent. Would you expect to find a flower that had neither stamens nor carpels? Such a flower would have no biological purpose. It could play no reproductive role, nor would there be any point in attracting insects or birds to it.

Flowers may be single or clustered together to form an **inflorescence**. Different families of flowering plants have their own, characteristic types of inflorescences, such as the umbels of the carrot family, the heads of the aster family, and others (Figure 24.24).

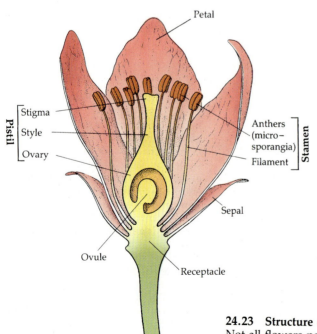

24.23   Structure of a Generalized Flower
Not all flowers possess all the structures shown here, but they must possess stamens, pistils, or both in order to play their role in sexual reproduction. This flower, possessing both, is a perfect flower.

24.24 Inflorescences
(a) This poison hemlock's inflorescences are umbels. Each umbel bears flowers on stalks that arise from a common center. (b) *Inula helenium* is a member of the aster family; its inflorescences are heads. In a head, each of the long, petallike structures is a ray flower; the central portion of the head consists of dozens to hundreds of disc flowers. Flowers of *I. helenium* contain a cough suppressant.

Evolution of the Flower

Botanists disagree about which type of flower is the *most* ancient, from an evolutionary point of view. However, one of the earliest types has a large number of tepals (or sepals and petals), carpels, and stamens, all spirally arranged (Figure 24.25a). Evolutionary change within the angiosperms included a number of striking modifications from this early condition: reduction in the number of each type of organ, differentiation of petals from sepals, stabilization of each type of organ to a fixed number, arrangement in whorls, and finally, change in symmetry from radial (as in a lily) to bilateral (as in a sweet pea or orchid), often accompanied by an extensive fusion of parts (Figure 24.25b). A great variety of corolla types have emerged in the course of evolution, as you will realize if you think of some of the flowers you recognize.

The first carpels to evolve were clearly modified leaves, appearing as folded but incompletely closed sporophylls—really intermediate between those of the gymnosperms and those of the angiosperms that evolved later. In the groups of angiosperms that evolved later, the carpels fused and then became progressively more and more buried in receptacle tissue (Figure 24.26); in the flowers of the latest groups to evolve, the other flower parts are attached at the very top of the ovary rather than at the bottom. The stamens of the most ancient flowers were also leaflike, little resembling those of the "ideal" flower.

Why do so many flowers have pistils with long styles and anthers with long filaments? Natural selection has probably favored length in both of these structures because it increases the likelihood of successful pollination. Long filaments may bring the anthers in contact with insect bodies, or they may put the anthers where they catch the wind better. Similar

24.25 Flower Form and Evolution
(a) A flower of *Magnolia grandiflora*, showing major features of early flowers: radial symmetry with the individual tepals, carpels, and stamens separate, numerous, and attached at their bases in a spiral arrangement. (b) The orchid *Pophiopedilum villosum* represents a bilaterally symmetrical type of structure that evolved much later than the form of the magnolia flower in (a). One of the three petals evolved into the complex lower "lip." Inside, the stamen and pistil are fused, and there is a single anther.

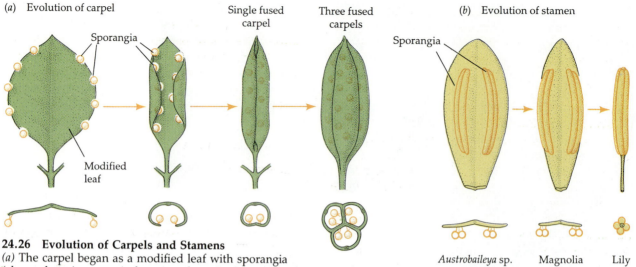

(a) Evolution of carpel

Sporangia

Modified leaf

Single fused carpel

Three fused carpels

(b) Evolution of stamen

Sporangia

Austrobaileya sp.   Magnolia   Lily

24.26 Evolution of Carpels and Stamens
(a) The carpel began as a modified leaf with sporangia (shown here in orange) along its edges. In the course of evolution, leaf edges curled inward and finally fused. At the end of the sequence, three carpels have fused to form a three-chambered ovary. (Cross sections of the several stages are seen across the bottom.) (b) Three modern plants are used to show the major stages in stamen evolution. The leaflike portion of the structure was progressively reduced until only the microsporangia (orange) remained.

arguments may apply to long styles. A long style may serve another purpose as well. If several pollen grains land on one stigma, a pollen tube will start growing from each grain toward the ovary. If there are more pollen grains than ovules, there is a "race" for the ovules. If there is a genetic correlation be-

(a)

(b)

(c)

(d)

24.27 Fruits
(a) A simple fruit: sour cherry. (b) An aggregate fruit: blackberry. (c) A multiple fruit: pineapple. (d) An accessory fruit: apple; the fleshy part external to the core consists of modified stem tissue.

tween the growth rate of the pollen tube and the overall success of the plant, the race down the style can be viewed as "mate selection" by the plant holding that style.

The Fruit

The ovary of a flowering plant (together with its seeds) develops into a fruit after fertilization. Because fruits arise only from flowering parts, they can be produced only by flowering plants—that is, by angiosperms. A fruit may consist only of the mature ovary and its seeds, or it may include other parts of the flower or structures closely related to it. A **simple fruit** such as a cherry (Figure 24.27) is one that develops from a single ovary. A raspberry is an example of an **aggregate fruit**—one developing from several carpels of a single flower. Pineapples and figs are examples of **multiple fruits**, formed from a cluster of flowers (an inflorescence). Fruits derived from parts in addition to the ovary and seeds are called **accessory fruits** and are exemplified by apples, squash, and bananas. The development and ripening of fruits will be considered in Chapter 31.

The Angiosperm Life Cycle

The life cycle of the angiosperms will also be considered in detail in Chapter 31. As illustrated in Figure 24.28, it has points of similarity with and difference from that of the gymnosperms (Figure 24.21). The angiosperms are heterosporous, like the gymnosperms. The female gametophyte is even more reduced than in the gymnosperms. The ovules are con-

24.28 Life Cycle of an Angiosperm
A haploid female gametophyte develops in the ovule (colored arrows), and haploid male gametophytes (pollen grains) develop in the anthers. When pollen reaches the stigma of a pistil, a pollen tube grows through the pistil until it reaches the female gametophyte.

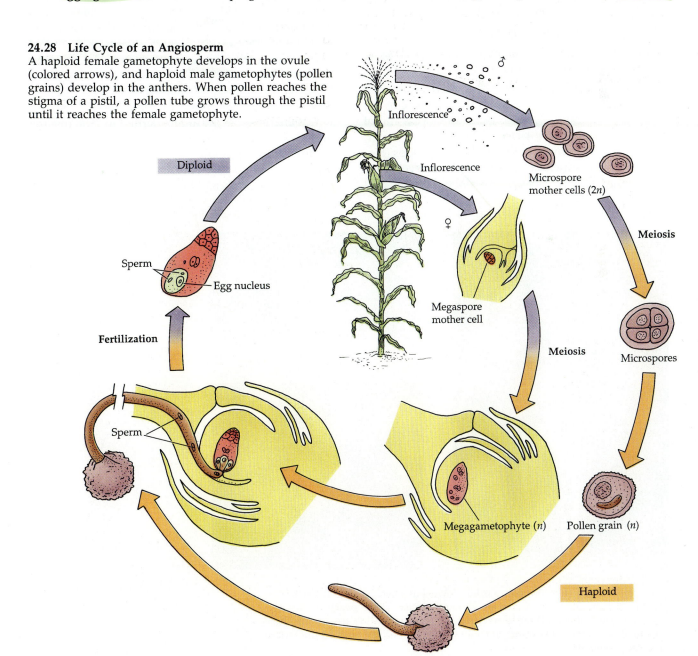

tained within fused carpels, rather than being exposed on the surfaces of sporophylls as in the gymnosperms. The male gametophytes are, again, pollen grains.

The ovule develops into a seed, containing the products of the double fertilization that characterizes the angiosperms. The triploid endosperm serves as a storage tissue, containing starch or lipid reserves, storage proteins, and other reserve substances. The diploid zygote develops into an embryo consisting of an embryonic axis and one or two **cotyledons**. The cotyledons, also called seed leaves, have different fates in different plants. In many, they serve as absorptive organs that take up and digest the endosperm. In others, they enlarge and become photosynthetic upon the germination of the seed. Cotyledons often play both of these roles.

Subclasses of the Angiospermae

The angiosperms are divided into the subclass Monocotyledonae (the monocots; Figure 24.29) and the subclass Dicotyledonae (the dicots; Figure 24.30). The names derive, respectively, from the existence of but a single embryonic cotyledon in the monocots and of two cotyledons in the dicots. There are, however, other major differences between the two groups (Figure 28.6). These include differences in leaf vein patterns, arrangement of vascular tissue in the stem and root, numbers of flower parts, and presence or absence of secondary growth (produced by a cambium). The cotyledons of some, but not all, dicots store the reserves originally present in the endosperm. Some of the differences between the two subclasses are illustrated in Figures 24.29 and 24.30.

The monocots include among others grasses, cattails, lilies, orchids, and palm trees. The dicots include the vast number of familiar seed plants: most of the herbs, weeds, vines, trees, and shrubs. Among them are oaks, willows, violets, sunflowers, and chrysanthemums.

Origin and Evolution of the Angiosperms

How did the angiosperms arise? Modern cladistic analyses (Chapter 20) have settled this once vexing question. It is widely agreed that the angiosperms and two groups of gymnosperms, the Gnetales and the long-extinct cycadeoids, a subgroup of the cycads, arose from a single ancestral species that gave rise to no other groups. A close relationship with the Gnetales was long suspected, primarily on the grounds that some Gnetales have vessel elements, which characterize the angiosperms. In 1990 this theory was strengthened by the electron-microscopic confirmation of double fertilization in *Ephedra*, a member of the Gnetales. The cycadeoids, which became extinct at about the same time as did the di-

(a)   (b)

24.29 Monocots
Monocots include such popular garden flowers as lilies, as well as orchids (Figure 24.25b). Other monocots are shown here. (a) These peach palms are known as pāyē bāyē in their West African homeland; palms are among the few monocot trees. (b) Grasses, such as this field of wheat seen in the summer, are monocots.

(a)

(b)

(c)

24.30 Dicots
(a) The cactus family is a large group of dicots, with about 1,500 species in the Americas. This one is called a fishhook cactus because of the shape of its spines. (b) This mountain ash from West Virginia is a member of the family Rosaceae, as are the familiar roses from your local florist. (c) This "African tulip tree" was photographed in Hawaii. Other dicots are seen in Figures 24.24 and 24.25a.

nosaurs, shared several important characteristics with the Gnetales and the angiosperms. The reproductive organs of one of the cycadeoids, although clearly a gymnosperm structure with naked seeds, had suggestive similarities to the flower of *Magnolia*.

The next great area of controversy is likely to be this: Which were the first angiosperms? The leading candidates at this point are the magnolia family (Figure 24.25a) and another family, the Chloranthaceae, whose flowers are much simpler in anatomy than those of the magnolias. One thing that complicates the search for the earliest fossil angiosperms is the unlikelihood of our being able to tell whether an ancient fossil plant practiced double fertilization.

The first angiosperms, according to fossil evidence, were probably trees. The gymnosperms from which they evolved must also have been trees; in fact, the history of vascular plant evolution up to that point featured a progression from herbaceous (nonwoody) early tracheophytes to larger, woody forms. During angiosperm evolution, herbaceous forms appeared once again. Today the predominant ground cover consists of herbaceous angiosperms. How do we explain the evolution of small, nonwoody plants from tall, woody ones?

Where Herbaceous Plants Predominate

When a piece of ground becomes available for new colonization—perhaps as a result of fire, or the appearance of a new sand bar, or clearing by humans—the first colonizers are generally herbs. Their seeds may have been present, dormant, in the soil, or they may have been brought in by wind or animals. The rapidly-growing herbs produce seeds that germinate in turn and contribute to the growth of the herb population. Before the land was cleared, the foliage of shrubs and trees may have shaded the ground so that little light was available for photosynthesis by small herbs, but the herbs have their day in the sun once the taller plants are removed. Later, as the land is modified by these herbaceous plants, other, larger forms appear (or reappear) and generally take over as succession proceeds (Chapter 47).

Although not dominating the scene, some herbaceous angiosperms do well in forests. They succeed through good timing, doing their growth, photosynthesis, and reproduction early in the season before the foliage develops fully on the trees above. Once the leafy canopy of the forest becomes dense, the shoots of these herbs die back, leaving underground organs (rhizomes or tubers) with stored food for the beginning of the following year's growth.

Herbs predominate in at least one extreme environment: tundras (Figure 24.31). Here the ability of some herbs to store photosynthate in underground organs stands them in particularly good stead, because they leave no above-ground parts to be damaged during the long cold season. The cold is so intense that shrubs and trees are unable to survive, but the dormant underground parts of herbs succeed.

SUMMARY

Plants are multicellular, photosynthetic eukaryotes that develop from embryos protected by parental tissue. They evolved from unknown green algal ancestors. All plants have a heteromorphic alternation of generations. Among the major trends in the evolution of the plant kingdom are: growing independence of liquid water for reproduction, increasing size and

24.31 Herbs and Tundra
Here in the tundra of Denali National Park, Alaska, the landscape is dominated by herbs.

independence in the sporophyte generation, increasing plant size (a trend reversed in the angiosperms, with a substantial return in some members to the herbaceous habit), and a shift from homospory to heterospory.

The division Bryophyta is composed of the classes Hepaticae (liverworts), Anthocerotae (hornworts), and Musci (mosses). Bryophytes lack vascular tissues and have no true leaves or roots. Their sporophyte generation is smaller than the gametophyte generation, and it is dependent upon the gametophyte for water and nutrition.

The vascular plants constitute the division Tracheophyta. All have tracheids and other cells associated with vascular tissue. The vascular tissues are xylem, for conduction of water and minerals, and phloem, for conduction of foods.

The tracheophyte subdivision Psilopsida is extinct. Its members lacked leaves and roots and consisted only of stems with sporangia. The subdivisions Lycopsida and Sphenopsida are the club mosses and horsetails, respectively. Now minor groups, they once were part of the planet's dominant vegetation.

The Filicinae (ferns), Gymnospermae (conifers and their relatives), and Angiospermae (flowering plants) are classes of the subdivision Pteropsida. Pteropsids are characterized by having megaphylls rather than microphylls as the other tracheophytes do.

Gymnosperms and angiosperms differ from ferns in possessing seeds. They produce pollen, so liquid water is not needed for the transfer of male gametes.

The Gymnospermae include pines, firs, cycads, and other large plants, some of which constitute the dominant vegetation of great forests. The largest gymnosperms are trees that are more than 100 meters tall.

Angiosperms are distinguished by the possession of flowers. Also, their vascular tissues contain cell types (such as vessel elements) rarely found elsewhere in the plant kingdom. Double fertilization, producing a zygote and a triploid endosperm, is unique to flowering plants and one member of the Gnetales. The angiosperms have two subclasses, Monocotyledonae and Dicotyledonae, distinguishable on such bases as vein patterns in leaves, number of cotyledons, and numbers of flower parts.

SELF-QUIZ

1. Plants differ from algae in that:
   a. only plants are photosynthetic.
   b. only plants are multicellular.
   c. only plants possess chlorophyll.
   d. only plants have multicellular embryos protected by the parent.
   e. only plants are eukaryotic.

2. Which statement is *not* true of the alternation of generations in plants?

   a. It is heteromorphic.
   b. Meiosis occurs in sporangia.
   c. Gametes are always produced by meiosis.
   d. The zygote is the first cell of the sporophyte generation.
   e. The gametophyte and sporophyte differ genetically.

3. Which statement is is *not* evidence for the origin of plants from the green algae?

   a. Some green algae have multicellular sporophytes and multicellular gametophytes.
   b. Both plants and green algae have cellulose in their cell walls.
   c. The two groups have the same photosynthetic and accessory pigments.
   d. Both produce starch as their principal storage carbohydrate.
   e. All green algae produce large, stationary eggs.

4. The bryophytes:
   a. gave rise to the tracheophytes.
   b. grow in dense masses, allowing capillary movement of water.
   c. possess xylem and phloem.
   d. possess microphylls and megaphylls.
   e. possess true roots.

5. The psilopsids:
   a. possessed vessel elements.
   b. possessed true roots.
   c. possessed sporangia at the tips of stems.
   d. possessed microphylls.
   e. lacked branching stems.

6. Lycopsids and sphenopsids:
   a. possess microphylls.
   b. possess megaphylls.
   c. are represented today primarily by trees.
   d. have never been a dominant part of the vegetation.
   e. produce only simple fruits.

7. Which statement is *not* true of ferns?
   a. The sporophyte is larger than the gametophyte.
   b. Most ferns are heterosporous.
   c. The young sporophyte can grow independently of the gametophyte.
   d. The frond is a megaphyll.
   e. The gametophytes produce archegonia and antheridia.

8. The gymnosperms:
   a. dominate all the land masses today.
   b. have never dominated the land masses.
   c. have active secondary growth.
   d. all have vessel elements.
   e. lack sporophylls.

9. Which statement is *not* true of flowers?
   a. Pollen is produced in the anthers.

b. Pollen is received on the stigma.
   c. An inflorescence is a cluster of flowers.
   d. A species having female and male flowers on the same plant is dioecious.
   e. A flower with both mega- and microsporangia is said to be perfect.

10. Which statement is *not* true of fruits?
    a. They develop from ovaries.
    b. They may include other parts of the flower.
    c. A multiple fruit develops from several carpels of a single flower.
    d. Fruits are produced only by angiosperms.
    e. A cherry is a simple fruit.

FOR STUDY

1. Mosses and ferns share a common trait that makes water droplets a necessity for sexual reproduction. What is this trait?

2. Ferns display a dominant sporophyte stage (with large fronds). Describe the major advance in anatomy that enables most ferns to grow much larger than mosses.

3. What features distinguish lycopsids from sphenopsids? What features distinguish these groups from psilopsids? from ferns?

4. Suggest an explanation for the great success of the angiosperms in occupying terrestrial habitats.

5. Contrast microphylls with megaphylls, in terms of structure, evolutionary origin, and occurrence among plants. And what are sporophylls?

6. In many locales, large gymnosperms predominate over large angiosperms. Under what conditions might gymnosperms have the advantage, and why?

READINGS

Burnham, C. R. 1988. "The Restoration of the American Chestnut." *American Scientist*, vol. 76, pages 478–487. If you read the article on chestnut blight recommended in the previous chapter, it may encourage you to see that there are methods, based on Mendelian genetics, to rebuild the native population of this important tree species.

Crosson, P. R. and N. J. Rosenberg. 1989. "Strategies for Agriculture." *Scientific American*, September. Discusses various approaches to increasing yields for an expanding population, but points out that some social and economic changes will also be required.

Graham, L. E. 1985. "The Origin of the Life Cycle of Land Plants." *American Scientist*, vol. 73, pages 178–186. How plants made it to the terrestrial environment.

Heyler, D. and C. M. Poplin. 1988. "The Fossils of Montceau-les-Mines." *Scientific American*, September. Plants and animals of the Carboniferous period, discovered in a rich fossil lode in central France. Presents a detailed picture of life in a time long past.

Hinman, C. W. 1986. "Potential New Crops." *Scientific American*, July. Prospects for food and other materials from such plants as jojoba, buffalo gourd, and others.

Niklas, K. J. 1986. "Computer-Simulated Plant Evolution." *Scientific American*, March. An examination of some hypotheses concerning plant evolution. The hypotheses were modeled on computers, generating testable suggestions. The article presumes no knowledge of computers.

Niklas, K. J. 1987. "Aerodynamics of Wind Pollination." *Scientific American*, July. Do the mechanics of pollination seem improbable to you? Wind-pollinated plants have many adaptations that favor successful pollination.

Raven, P. H., R. F. Evert and S. Eichhorn. 1986. *Biology of Plants*, 4th Edition. Worth, New York. An excellent general botany textbook.

25

Sponges and Protostomate Animals

PREVIEW: The millions of animal species living on Earth all require some form of complex organic molecules as sources of energy. Animals live in nearly every environment that supports life, and they are adapted for procuring energy in diverse ways. The evolution of animals may have been strongly influenced by the types of food available on the early Earth and by the selective pressures exerted by predators. The earliest animals fed on algal mats and filtered small prey from the water. With improved hydrostatic skeletons, animals were able to move faster and capture larger prey. Eventually, some animals developed the ability to maintain their positions in the water column. Other lineages invaded the land.

This chapter deals with the features of the major phyla of the protostomate lineages of animals, their characteristics and their adaptations, and outlines the major evolutionary trends in the kingdom Animalia.

Members of the kingdom **Animalia** are among the most conspicuous living things in the world around us. As members of this kingdom, we have a special interest in its other members, especially those that are evolutionarily closely related to us. Animals are also highly diverse. Most of the 30 million or so species of living organisms are animals. Because there are so many species, we will devote two chapters to the kingdom Animalia, dividing animals into two major groups believed to represent major lineages that have been evolving separately since the Cambrian period. The lineages differ fundamentally in their early embryological development. The words *protostomate*, as used in this chapter title, and *deuterostomate*, as used in the next chapter title, identify these developmental types. Sponges are animals, but their developmental patterns differ from those of all other animals. The developmental patterns of sponges, protostomes, and deuterostomes will be described shortly.

Why are there so many more species of animals than plants? The reasons are to be found in differences in their ways of life. The plant way of life can be characterized as the utilization of simple organic molecules from the environment by an immobile, attached organism that powers its metabolic machinery using the direct energy of sunlight. In contrast, animals require a variety of complex organic molecules as sources of energy, and they obtain these molecules by active expenditure of energy. This energy is used either to move the animals through their environment or to cause the environment and the food it contains to move to the animals. The food of animals is highly varied and includes most other members of the animal kindom as well as members of all the other kingdoms. Much of the diversity of animal sizes and shapes is the result of adaptation to the kinds of foods animals eat.

Lacking rigid cell walls, animals cannot utilize high osmotic pressures to control exchange of materials with their environments. Instead they have evolved other, more elaborate mechanisms for regulating their internal composition. The need to find food has given a strong selective advantage to structures that provide detailed information about the environment and structures able to receive and coordinate this information. Consequently, most animals are behaviorally much more complex than plants. A real appreciation of animal structure and functioning can be achieved only through first-hand experience in the field and laboratory. However, the accounts in this chapter and the one following it can serve as an orientation to the major groups of animals, their similarities and differences, and the evolutionary pathways that resulted in the current numbers and variety of animal species.

Animals accomplish their diverse activities within the constraints of basic structural plans or designs. The German word *Bauplan* (plural, *Baupläne*) is used to refer to both the body plan of an animal and the functional interrelationship of its parts. A *Bauplan* includes the entire animal, its organ systems, and the integrated functioning of its parts. *Baupläne* reflect, and thus provide clues to, the evolutionary history of animal lineages. Consequently we use them as a way to organize our treatment of animal groups in this and the next chapter. As you will see, animals in many lineages have evolved greater body complexity, but many simple animals are successful today even in a world with many complex species. An overview of the *Baupläne* of animals is given in Table 25.1. The terms used in the table will be explained when we discuss the animal lineages.

HOW ARE ANIMALS CLASSIFIED?

Classifications of animals attempt to reflect their evolutionary relationships. Clues to these relationships are found in the fossil record, in the patterns of embryological development of animals, and in the comparative morphology and physiology—*Baupläne*—of living and fossil animals. As animals diversified, natural selection often acted more strongly on later developmental stages than on earlier ones. Therefore, early developmental stages often evolved more slowly than later stages, so they are said to be **evolutionarily conservative**. These early developmental stages often reveal relationships that are no longer evident among adults.

Body Symmetry

A fundamental aspect of an animal's *Bauplan* is its overall shape, that is, its **symmetry**. A symmetrical animal can be divided along at least one plane into similar halves. Animals that have no plane of sym-

TABLE 25.1
Grades of Body Complexity among Animal Phyla

GRADE	PHYLA
Without true tissues	Porifera
With true tissues	
Two embryonic tissue layers	Cnidaria, Ctenophora
Three embryonic tissue layers	
Acoelomate (no body cavity)	Platyhelminthes, Nemertea
Pseudocoelomate (blastocoel persists as body cavity)	Nematoda, Rotifera
Coelomate (mesodermal body cavity develops)	Pogonophora, Annelida, Arthropoda, Mollusca, Echinodermata, Chordata

metry are said to be **asymmetrical**. Many sponges are asymmetrical (Figure 25.1*a*). However, most animals have some kind of symmetry. The simplest form is **spherical symmetry**, in which body parts radiate out from a central point (Figure 25.1*b*). An infinite number of planes passing through the central point can divide a spherically symmetrical animal into similar halves. Spherical symmetry is widespread among protists, but most animals possess other forms of symmetry. One common type is **radial symmetry**, a cylindrical form with one main axis around which body parts are arranged (Figure 25.1*c*). A perfectly radially symmetrical animal can be divided into similar halves by any plane that passes through the main axis. Some simple sponges and a few other animals

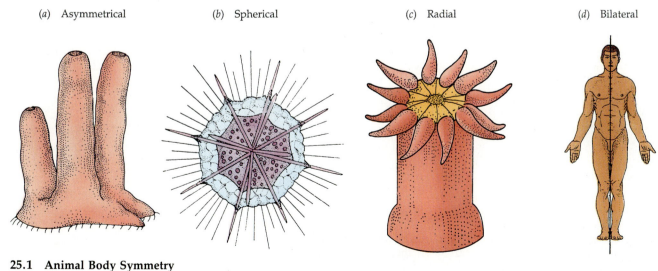

(a) Asymmetrical   *(b)* Spherical   *(c)* Radial   *(d)* Bilateral

25.1 Animal Body Symmetry
The types of body symmetry are illustrated by *(a)* an asymmetrical sponge, *(b)* a spherical radiolarian, *(c)* a radial sea anemone, and *(d)* a bilateral human.

have such symmetry, but most radially symmetrical animals are so modified that only two planes can divide them into similar halves. These animals are said to have **biradial symmetry**. We will describe three phyla of radially symmetrical animals.

The bodies of **bilaterally symmetrical** animals can be divided into mirror images only by a cut through the midline of the body from its front (anterior) to its back (posterior) end (Figure 25.1d). The upper side of a bilaterally symmetrical animal is its dorsal surface; the bottom side is its ventral surface. Bilateral symmetry is especially characteristic of animals that move through their environments. It is strongly correlated with the development of sense organs and central nervous tissues at the anterior end of the animal, a process known as **cephalization**. Cephalization may have been selected for among motile animals because the anterior end is the part that encounters new environments first.

Developmental Pattern

Because early embryological development is often evolutionarily conservative, it reveals a great deal about evolutionary relationships among animals. Consequently, early developmental patterns have been extensively used to identify major lineages of animals. During development from a single-celled zygote to a multicellular adult, animals form a number of distinct cell layers (see Chapter 15). These layers behave as distinct units during early embryological development and give rise to different tissues and organs in the adult animal. **Diploblastic** animals have only two embryonic cell layers—the outer ectoderm and the inner endoderm. **Triploblastic** animals have, in addition to ectoderm and endoderm, a third layer. This layer, the mesoderm, derives in most species from the endoderm; in a few species, it derives from the ectoderm. The mesoderm lies between the ectoderm and the endoderm. Most protostomes and all deuterostomes are triploblastic. We will describe the diploblastic phyla first.

On the basis of differences in their early developmental patterns, all animals other than sponges can be divided into two major lineages. In one lineage, called the **protostomes**, cleavage of the fertilized egg is **determinate**; that is, if the egg is allowed to undergo a few cell divisions, and the cells are then separated, each cell will develop only into a partial embryo, the outcome depending on the cell's original position in the blastula. Cleavage of the fertilized egg in the other lineage, the **deuterostomes**, typically is **indeterminate**; that is, cells separated after several cell divisions can still develop into complete embryos. Another basic difference is that the cleavage pattern of deuterostomes is radial: Cells divide along a plane either parallel or at right angles to the long axis of the fertilized egg (Figure 25.2). Among protostomes, the cleavage pattern is spiral: The plane of division is oblique to the long axis of the egg, causing the

Radial cleavage

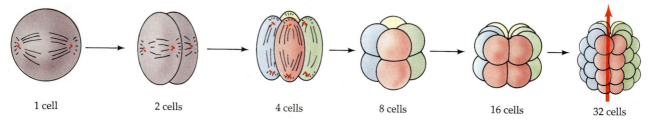

| 1 cell | 2 cells | 4 cells | 8 cells | 16 cells | 32 cells |

Spiral cleavage

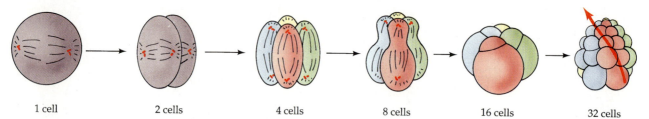

| 1 cell | 2 cells | 4 cells | 8 cells | 16 cells | 32 cells |

25.2 Egg Cleavage Patterns
The cells derived from the four cells produced by the first two divisions of the zygote are color-coded so that you can follow their positions. Radial cleavage produces equal-sized cells that lie directly above one another. In spiral cleavage, the plane of division (red arrow) is oblique and the cells are arranged in a spiral. Spiral cleavage is unequal, so that some cells are larger than others.

25.3 Animal Body Cavities

The three major types of animal body cavities differ in the types of cells lining them. Tissues derived from ectoderm are colored blue, those from mesoderm are pink, and those from endoderm are yellow.

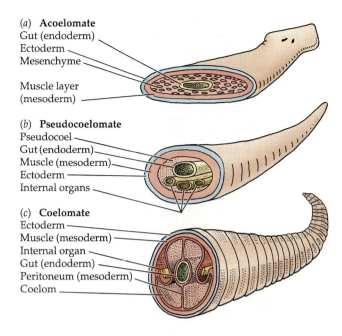

(a) **Acoelomate**
Gut (endoderm)
Ectoderm
Mesenchyme
Muscle layer
(mesoderm)

(b) **Pseudocoelomate**
Pseudocoel
Gut (endoderm)
Muscle (mesoderm)
Ectoderm
Internal organs

(c) **Coelomate**
Ectoderm
Muscle (mesoderm)
Internal organ
Gut (endoderm)
Peritoneum (mesoderm)
Coelom

cells to be arranged in a spiralling pattern. Finally, the mouth of the deuterostome embryo originates at some distance from the blastopore, which becomes the anus, whereas the mouth of the protostome embryo arises from or next to the blastopore.

Body Cavities

Protostomes are divided into several groups that differ in the type of body cavity they have and how it develops. Body cavity formation is another important example of how embryology gives clues about evolutionary relationships among animals. The **acoelomates** lack an internal body cavity. The space between the gut and the body wall is filled with masses of cells. The **pseudocoelomates** have a body cavity, the **pseudocoel**, derived directly from the blastocoel (Figure 25.3), the first cavity formed inside the proliferating ball of embryonic cells. The pseudocoel provides a liquid-filled space in which many of the body organs float. **Coelomate** animals have a **coelom**, a body cavity that develops within the embryonic mesoderm and is lined with a special mesodermal lining called the **peritoneum**. The internal organs of coelo-

mate animals may hang down into the coelom, but they are slung in pouches of the peritoneum rather than floating within the cavity. All deuterostomes are coelomate animals.

The coelom forms by different means in deuterostomes and coelomate protostomes (Figure 25.4). Among deuterostomes, the coelom arises by an outpocketing of the embryonic gut (enterocoelous development). These pockets ultimately pinch off and

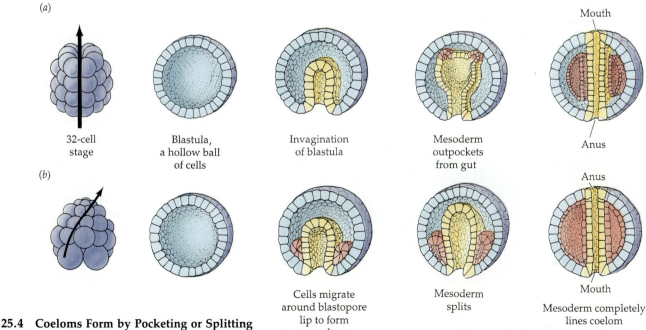

(a)

32-cell stage
Blastula, a hollow ball of cells
Invagination of blastula
Mesoderm outpockets from gut
Mouth
Anus

(b)

Cells migrate around blastopore lip to form mesoderm
Mesoderm splits
Anus
Mouth
Mesoderm completely lines coelom

25.4 Coeloms Form by Pocketing or Splitting
These sequences continue the developmental processes shown in Figure 25.2 to the stage of coelom formation. (a) Among deuterostomes, the coelom arises by outpocketing. (b) Among coelomate protostomes, the coelom arises from a "split" in the mesoderm.

come to lie in the blastocoel. Among coelomate protostomes, mesoderm develops within the embryo from a single cell, often near the blastopore, which migrates into the blastocoel. A split then forms in the mesoderm, creating the coelomic cavities (schizocoelous development). The early embryological differences between protostomes and deuterostomes, summarized in Table 25.2, indicate they are lineages that separated early in the evolution of animals.

Body cavities are of great functional significance to animals, as we will soon see. For the moment we note that the fundamental developmental patterns that produce different types of cavities are useful, together with other traits, for comparing the *Baupläne* of animal phyla. We also use these traits to help us understand the ways of life of animal groups and patterns of evolution within the phyla. We will also discuss how the evolution of animals was influenced by the general ecological conditions on Earth over evolutionary time.

THE ORIGINS OF ANIMALS

Animals probably arose from ancestral colonial protists as a result of division of labor among their aggregated cells. Division of labor probably evolved because an undifferentiated mass of cells exchanges materials with its environment relatively slowly and because some functions are best performed by specialized cells. Within ancestral colonies of cells—perhaps similar to those still existing in *Volvox* and other colonial flagellated protists (see Figure 22.5*b*)—some cells became specialized for movement, others for nutrition, and still others differentiated into gametes. Once division of labor began to evolve, the units continued to differentiate, all the while improving their coordination with other working groups of cells. These coordinated groups of cells evolved into the larger and more complex organisms that we now call animals.

Multicellular animals may have arisen from the protists three times. The sponges (phylum Porifera), cnidarians and ctenophores (phyla Cnidaria and Ctenophora), and flatworms (phylum Platyhelminthes) may represent three separate evolutionary lines. The other 28–30 animal phyla probably evolved from a flatworm or flatwormlike ancestor. Possible evolutionary relationships among protostomes are shown in Figure 25.5. New information will certainly modify and refine our understanding of these relationships.

More than a million living species of animals have been described and named. Estimates of the actual number of species living today range up to 30 million or more. These species have arisen over more than a billion years of evolution, much of which has left no fossil traces that we have discovered. However, the fossil record is good enough to provide us with a general idea of the physical environments in which animals lived and to characterize the ecological communities within which they interacted. Using this information, we can make educated guesses as to why animals evolved as they did.

Oxygen and the Evolution of Animals

The stage was probably set for the appearance of animals by the generation of atmospheric oxygen by mats of cyanobacteria (see Chapter 17). All animals must take up oxygen from the environment and distribute it to their cells, and they must get rid of carbon dioxide; this process is called **gas exchange**. Very small unicellular aquatic organisms that are adapted to stagnant water can obtain enough oxygen by simple diffusion even when oxygen concentrations are very low. Larger unicellular organisms, however, have lower surface-to-volume ratios. They can obtain enough oxygen by simple diffusion only if concentrations of oxygen are higher than the concentrations that support small prokaryotic cells. For example, large eukaryotic cells require oxygen levels of at least 2 to 3 percent of current atmospheric concentrations, whereas bacteria can thrive on only 1 percent of atmospheric levels. Small multicellular animals with a high-surface-to-volume ratio can exchange gases through body surfaces; larger animals have evolved specialized structures, such as gills and lungs, to exchange gases, and circulatory systems to transport oxygen and carbon dioxide to and from their cells (see Chapters 39 and 40).

About 1,500 mya (million years ago), oxygen concentration became high enough for large eukaryote cells to flourish and diversify (Figure 25.6). By 700 mya, protist communities were rich in species and ecological types. Photosynthesizing species were

TABLE 25.2 Developmental Differences between Protostomes and Deuterostomes	
PROTOSTOMES	**DEUTEROSTOMES**
Spiral cleavage	Radial cleavage
Blastopore becomes mouth	Blastopore becomes anus
Mesoderm derives from cells on lip of blastopore	Mesoderm derives from walls of developing gut
Mesoderm splits to form coelom (schizocoelous development)	Mesoderm outpockets to form coelom (enterocoelous development)

25.5 Animal Lineages
This is the evolutionary tree of animals we will use in this chapter and the next one. Note the multiple origins of animals and the two major animal lineages—protostomes, on the left, and deuterostomes, on the right.

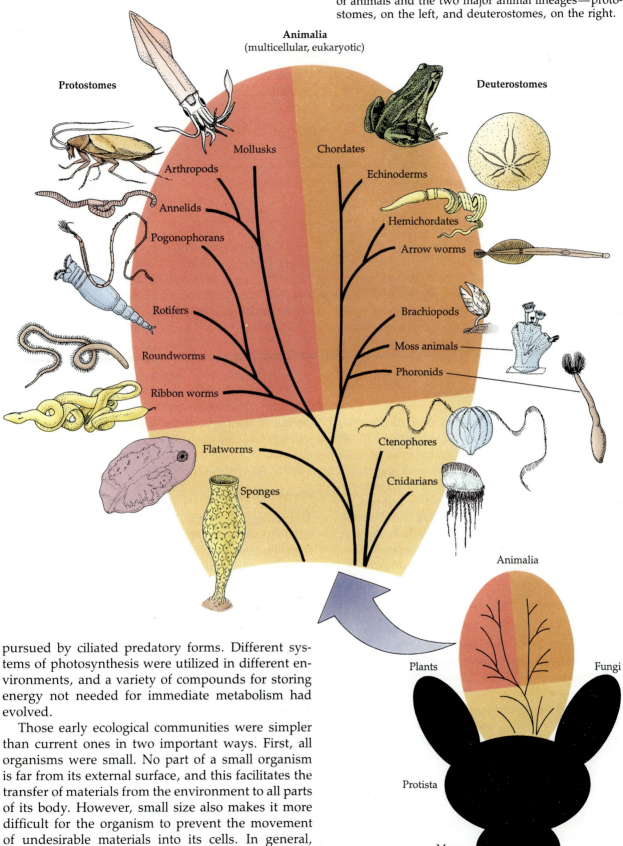

Animalia
(multicellular, eukaryotic)

Protostomes

Deuterostomes

Mollusks

Chordates

Arthropods

Echinoderms

Annelids

Hemichordates

Pogonophorans

Arrow worms

Rotifers

Brachiopods

Roundworms

Moss animals

Ribbon worms

Phoronids

Flatworms

Ctenophores

Sponges

Cnidarians

Animalia

Plants

Fungi

Protista

Monera

pursued by ciliated predatory forms. Different systems of photosynthesis were utilized in different environments, and a variety of compounds for storing energy not needed for immediate metabolism had evolved.

Those early ecological communities were simpler than current ones in two important ways. First, all organisms were small. No part of a small organism is far from its external surface, and this facilitates the transfer of materials from the environment to all parts of its body. However, small size also makes it more difficult for the organism to prevent the movement of undesirable materials into its cells. In general, small organisms are more strongly influenced by the

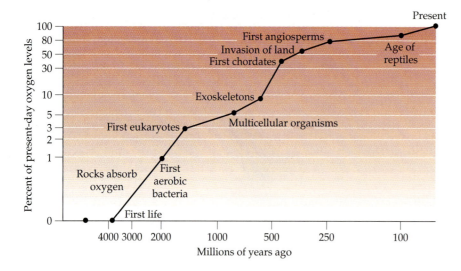

25.6 Large Cells Need More Oxygen
Hundreds of millions of years of evolution of life elapsed before atmospheric oxygen levels became high enough to support large eukaryotes and multicellular organisms.

physical environment than are large animals. Small organisms are also highly vulnerable to larger predators. It is likely that a combination of changes in the physical environment, especially oxygen concentrations, and interactions between predators and their prey favored increases in the sizes of organisms during the Vendian period (700–570 mya). The steady increase in atmospheric concentrations of oxygen made it possible for larger animals to evolve because they could obtain enough oxygen to maintain the metabolism of all their cells.

Early communities probably also differed from today's in the absence of sexual reproduction. Most modern protists lack sex, or reproduce sexually only rarely, but sex is almost universal among larger organisms. Why sexual reproduction evolved during the Vendian and why it is more advantageous to larger organisms than to smaller ones cannot be considered here, but its development greatly increased potential rates of evolution because of the many new allele combinations it generated (Chapter 18).

Early Animal Evolution

When the first animals evolved, the mats of cyanobacteria that previously had been the primary photosynthesizers were being replaced by algae. Therefore, the primary food supplies available to the earliest animals were algae floating in the water (**phytoplankton**) and the extensive algal mats that covered the shallow sea bottoms. Later animals fed on free-floating protists, tiny animals, and animal larvae, collectively called **zooplankton**.

The earliest animals were probably colonies of flagellated cells that fed on phytoplankton. Some of these animals developed specialized cells with stinging tentacles that allowed the capture of larger plankton. Others evolved morphological and behavioral adaptations for grazing in the algal mats. Both of these changes favored the evolution of larger animals that

could move about. Not surprisingly, the presence of larger grazing animals created opportunities for still larger animals that fed on them. The presence of predators, in turn, may well have led to selection for shells and other protection, burrowing and the use of safe refuges such as caves and crevices, and better locomotor abilities. These themes in the evolution of animals continue today. For the moment, however, let us examine how they operated during the early evolution of the invertebrates, remembering that all life was confined to the seas during this extensive period of the evolution of life on Earth.

SIMPLE AGGREGATIONS

Phylum Porifera

The sponges (phylum **Porifera**, Latin for "pore bearers") are sedentary animals that move water through their bodies and filter food from the water. The sponges were one of the earliest groups of animals to evolve from protists. Sponges move water by the beating of flagella on cells lining the body chamber. These unique feeding cells (**choanocytes**) have collars consisting of modified cilia that surround the flagella. Similar cells are also found among flagellated protists of the order Choanoflagellida, suggesting an evolutionary relationship between the two groups. Changes in *Baupläne* during sponge evolution led to better capture of suspended food from the water circulating through the body cavity. Sponges are so distinct that many specialists place them in a subkingdom of their own, the Parazoa.

All sponges, even large ones, have a very simple *Bauplan*. The body of a sponge consists of a loose aggregation of cells built around a water canal system. A sponge has no mouth or true digestive cavity, no muscles, and no nervous system. In fact, there are no organs at all in the usual sense of the word. A sponge is so loosely organized that even if it is

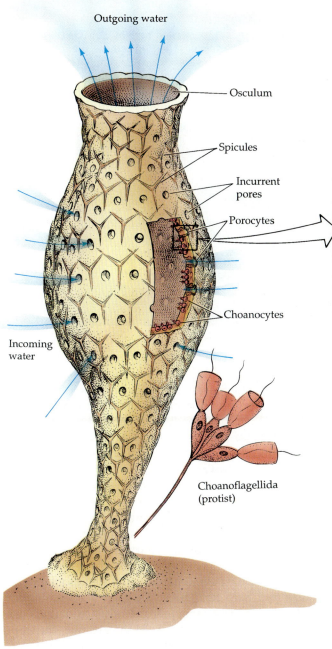

Outgoing water

Osculum

Spicules

Incurrent pores

Porocytes

Choanocytes

Incoming water

Choanoflagellida (protist)

25.7 Sponge *Bauplan*
The flow of water through the sponge is shown by blue arrows. The enlargements show the detailed structure of the body wall and the similarity between sponge choanocytes and the choanoflagellate protists (Chapter 22).

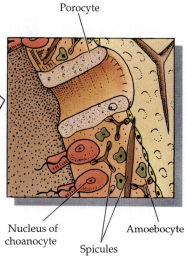

Porocyte

Nucleus of choanocyte

Spicules

Amoebocyte

completely disassociated by straining it through filter paper its cells can reassociate into a new sponge.

Throughout most of its life an individual sponge is immobile, attached to the substratum. It feeds by drawing water into itself and capturing the small organisms and nutrient particles that flow past the walls of its inner cavity (Figure 25.7). Water currents are set up by the flagellated choanocytes lining the inside of the body cavity. Water flows into the animal either by way of small **incurrent pores** that perforate special epidermal cells (in simple sponges) or through intercellular pores (in complex sponges). The water passes into a chamber of the body where food particles are captured by the choanocytes. It then exits through one or more larger openings called **oscula** (singular, osculum). Between the thin epidermis and

the choanocytes lies a layer of cells containing **amoebocytes**, wandering amoebalike cells that are responsible in part for whatever low levels of communication occur among the cells. Also present are spicules—thin, spined structures produced by the cells of the sponge that stiffen and support its body. Among larger sponges the spicular skeleton may be very complex.

The sponge *Bauplan* differs from the *Baupläne* of all other animals in that sponges have no distinct body tissues, no cavity between the layers of cells, and no recognizable body symmetry. Nonetheless, sponges come in a wide variety of sizes and shapes (Figure 25.8). The different ways water moves in different environments have probably influenced the evolution of sponge morphologies, as have the advantages accruing to sponges if they do not repeatedly filter the same water. Sponges living in intertidal or shallow subtidal environments, where they are subjected to strong wave action, hug the substrate. They spread laterally and have multiple oscula scattered over the body surface. Many sponges living in calm waters are simple, having a single large osculum situated on top of the body. Water is taken in through pores on the sides of the body and expelled upward through the osculum. Sponges living in flowing water do not need to exert much energy to move water through their bodies. Most of them are flattened and are oriented at right angles to the direction of current flow, so they intercept water as it flows past them.

Sponges reproduce both sexually and asexually. In most species, a single individual produces both eggs and sperm. Water currents carry sperm from one individual to another. Asexual reproduction is by bud-

25.8 Sponges Differ in Size and Shape
(a) Glass sponges (class Hexactinella), illustrated here by *Euplectella aspergillum*, are named after their glasslike spicules. (b) You can see the forms of spicules clearly in this close-up of *Leucosolenia* (class Calcarea). (c) The brown volcano sponge (class Demospongiae) is typical of many simple marine sponges. The feathery structures among the sponges are the feeding tentacles of annelid worms.

(a)

(b)

(c)

ding and other processes that produce fragments able to develop into new sponges. Most of the 10,000 species of sponges are marine animals, but a few are found in fresh water.

THE EVOLUTION OF DIPLOBLASTIC ANIMALS

Animals in all phyla other than Porifera have distinct cell layers and symmetrical bodies. There are two animal phyla whose members have only two embryonic cell layers; that is, they are diploblastic. They have no body cavity and are radially symmetrical. These animals are believed to represent a lineage evolved from protists that is separate from all other animals.

Phylum Cnidaria

The key innovation in the evolution of cnidarians was the evolution of specialized cells called cnidocytes, which contain stinging structures called **nematocysts** that can be discharged to capture prey (Figure 25.9). One hypothesis is that cnidocytes evolved through a symbiotic relationship with protists that lived within the bodies of the ancestors of cnidarians. Cnidocytes allowed the early cnidarians to capture large prey, an ability that was beneficial both in the algal mats and in the plankton. Once cnidocytes had evolved, cnidarians radiated into many different species. In fact, they may have constituted more than half of the early Cambrian animal species. Today's cnidarian species —jellyfish, sea anemones, corals, and hydroids— number roughly 10,000. All but a few species are marine.

Cnidarians possess tentacles on which the cnidocytes are borne; epithelial cells with muscle fibers whose contractions enable cnidarians to move; and nerve nets for the integration of body activities. They have a mouth connected to a dead-end digestive sac called the **gastrovascular cavity**. The same opening thus serves as both mouth and anus in these animals. The digestive apparatus and the row of tentacles armed with nematocysts that normally surrounds the mouth enable the animal to capture and swallow a much wider range of food particles than is available to sponges. After food is captured, it is transported to the mouth by the tentacles. The nematocysts paralyze and help hold prey. They are responsible for the sting that some jellyfish and other cnidarians can inflict on human swimmers. At the extreme, the tropical Pacific sea wasp (genus *Chironex*) and the Portuguese man-of-war (genus *Physalia*) can cause fatal injuries.

ALTERNATION OF BODY FORMS IN CNIDARIANS. Most cnidarian species pass through an alternation of body forms and reproductive types during their life cycles (Figure 25.10). The first form is the **polyp**, in which the body consists of a cylindrical stalk with mouth and tentacles at the opposite end from a site of attachment to the substrate. An animal that lives attached to a substrate and does not move about is said to be **sessile**. This is an asexual stage, but individual polyps may reproduce by budding, thereby forming a colony. The **medusa** (plural, medusae) is the familiar, free-swimming, sexual stage shaped like a bell or an umbrella. It floats with its mouth and tentacles facing downward. Medusae produce eggs and sperm and release them into the water. When an egg is

Portuguese man-of-war

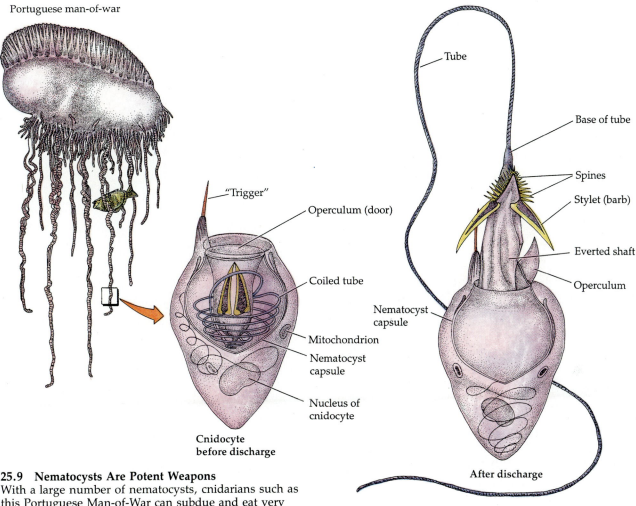

"Trigger"

Operculum (door)

Coiled tube

Mitochondrion

Nematocyst capsule

Nucleus of cnidocyte

Cnidocyte before discharge

Tube

Base of tube

Spines

Stylet (barb)

Everted shaft

Operculum

Nematocyst capsule

After discharge

25.9 Nematocysts Are Potent Weapons
With a large number of nematocysts, cnidarians such as this Portuguese Man-of-War can subdue and eat very large prey. The styles and spines are used to inject toxins into the prey.

fertilized, it develops into a free-swimming, ciliated larva called a **planula** (plural, planulae) that eventually settles to the bottom and transforms into a polyp.

The cnidarian *Bauplan* is characterized by radial symmetry and by only two embryonic cell layers. A middle cell layer eventually develops from the ectoderm, but it never produces complex internal organs such as those found in the triploblastic phyla. Although the polyp and medusa stages are very different in appearance, they actually share a similar *Bauplan*. A medusa is essentially a polyp without a stalk. Conversely, a polyp is a medusa with a stalk, with which it attaches itself to the substrate. Most of the outward differences between these stages are due to the development of the **mesoglea**, the middle body layer. The mesoglea of polyps is usually thin, whereas in medusae it is very thick and makes up the bulk of the animal.

HYDROZOANS. Among the class **Hydrozoa**—the group containing the only freshwater representatives of the phylum—the polyp usually dominates the life cycle. A few species have solitary polyps, but most hydrozoans are colonial: A single planula eventually giving rise to a large number of polyps, all of which are interconnected and share a continuous gastrovascular cavity (Figure 25.11). There is often differentiation among the individuals in a single colony. Some individuals have tentacles with many nematocysts, and these individuals capture prey for the colony. Others, which lack tentacles and are unable to feed, are specialized for the production of medusae. Still others have fingerlike projections and are adapted for defense. All of these individuals, however, are ultimately derived from a single, sexually produced planula.

The **siphonophores** are free-floating hydrozoans in which medusae and polyps are combined to create a complex colony. Some of the individual medusae are modified to function as gas-filled floats. Others are modified to move the colony through the water by jet propulsion. Still others are modified for defense. And, of course, there are also feeding and reproductive indivduals. All siphonophores are **carnivores** (eaters of other animals) that subdue prey by stunning them with their highly toxic nematocysts.

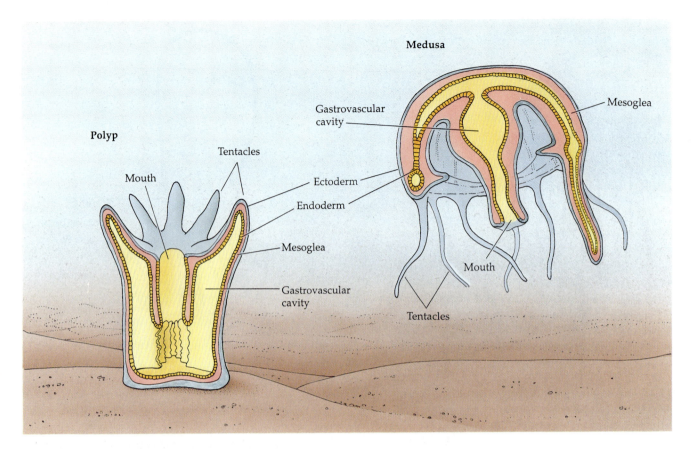

25.10   Cnidarians Have Two Body Forms
During the life cycle of a cnidarian, the usually sessile, asexual polyp alternates with the free-swimming, sexual medusa. As the positions of the mouth and tentacles indicate, the medusa is "upside down" from the polyp—or vice versa.

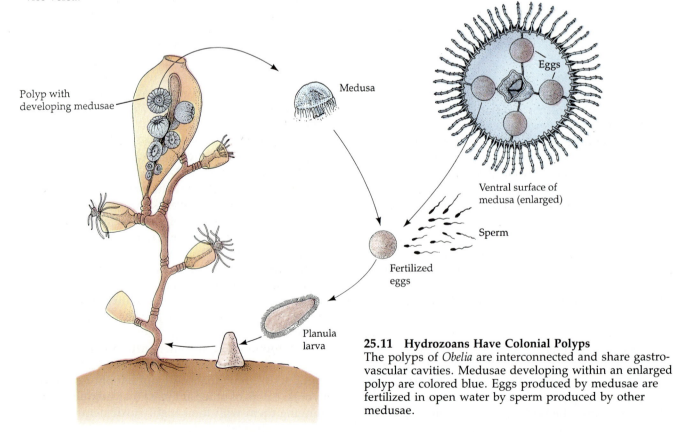

25.11   Hydrozoans Have Colonial Polyps
The polyps of *Obelia* are interconnected and share gastrovascular cavities. Medusae developing within an enlarged polyp are colored blue. Eggs produced by medusae are fertilized in open water by sperm produced by other medusae.

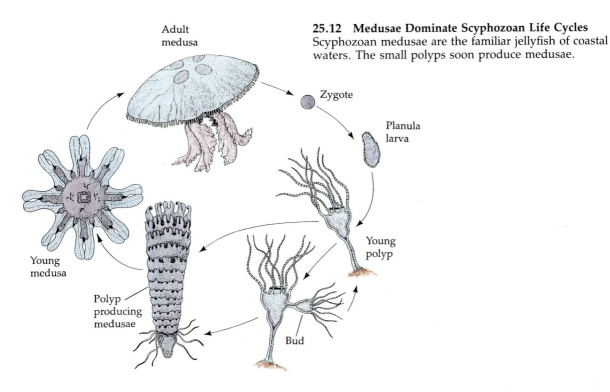

25.12 Medusae Dominate Scyphozoan Life Cycles
Scyphozoan medusae are the familiar jellyfish of coastal waters. The small polyps soon produce medusae.

SCYPHOZOANS. The several hundred species of the class **Scyphozoa** are all marine. Some are as large as half a meter in diameter. The mesoglea of their medusae is very thick and firm, giving rise to their common name of jellyfish. The medusa typically has the form of an inverted cup, and the tentacles with their nematocysts extend downward from the margin of the cup. Active swimming is achieved by the contraction of muscles ringing the cup margin, which expels water from the cup. When the muscles relax, the cup expands and again fills with water. Food captured by the tentacles is passed to the mouth and then distributed to one of four gastric pouches, where enzymes begin the process of digesting the food.

The life cycle of scyphozoans is dominated by the medusa rather than by the polyp. Gonads (sex organs) develop in the gastrodermal tissue close to the gastric pouches. An individual medusa is either male or female, and eggs and sperm are usually released into the open sea. The fertilized egg develops into a small, heavily ciliated planula that quickly settles on a substrate and changes into a small polyp. This polyp feeds and grows and may produce additional polyps by budding. After a period of growth, the polyp begins to bud off small, medusalike individuals by transverse division of its body column (Figure 25.12). These small individuals feed, grow, and transform themselves into adult medusae. Thus a polyp that grows from a single fertilized egg is capable of producing many genetically identical medusae that will eventually engage in sexual reproduction.

ANTHOZOANS. The roughly 6,000 species of sea anemones and corals that constitute the class **Anthozoa** are all marine animals. They differ from other cnidarians in that the medusa stage has been completely eliminated from the life cycle. Instead, gametes (eggs and sperm) are produced by the polyp, and the fertilized egg develops into a planula that metamorphoses directly into another polyp. Many species can also reproduce asexually by budding or fission. Anthozoans, like all other cnidarians, are carnivores that capture prey with nematocyst-studded tentacles. However, the digestive cavity of anthozoans is more complex than that of other cnidarians, being partitioned by numerous sheets (mesenteries) that increase the surface area available for the secretion of digestive enzymes and absorption of nutrients (Figure 25.13). Gonads also develop on these sheets.

Sea anemones are solitary anthozoans that lack specialized protective coverings. They are widespread in both warm and cold ocean waters. Many of them are able to crawl slowly on their pedal discs and a few species actually can swim. The corals, in contrast, are usually sessile and colonial. Each individual polyp secretes an organic matrix upon which calcium carbonate, the eventual skeleton of the colony, is deposited. The forms of coral skeletons are species-specific and highly diverse (Figure 25.14a). The common names of the coral groups—horn corals, brain corals, staghorn corals, organ-pipe corals, sea fans, and sea whips, among others—accurately convey the impression of their appearances. As a

Mouth

Wall

Incomplete mesentery

Longitudinal muscles (on complete mesentery)

Pedal disc

Complete mesentery

Pharynx

Gonads

Transverse muscles

Basilar muscle

25.13 Anthozoans Have Complex Polyps
This diagram shows the muscular structures of the sea anemone and some of the numerous sheets (mesenteries) that partition the body cavity. The nematocysts on the tentacles are not shown here.

colony grows, old individuals die, leaving their calcareous skeletons intact. The living members form a layer on top of a growing reef of skeletal remains. Reef-building corals are restricted to clear, warm waters. They are especially abundant in the Indo-Pacific region, where they form chains of islands and reefs. The Great Barrier Reef along the northeast coast of Australia is more than 200 kilometers long and up to 150 kilometers wide. A continuous coral reef hundreds of miles long in the Red Sea has been calculated to contain more material than all the buildings in the major cities of North America!

Corals flourish in nutrient-poor, clear, tropical waters. For a long time scientists wondered how they captured enough zooplankton to grow as rapidly as they do. The answer is that highly modified dinoflagellates live symbiotically within the bodies of the corals and, through their photosynthesis, provide carbohydrates to their hosts and help with calcium deposition (Figure 25.14b). This is why corals are restricted to surface waters with high light levels. The dinoflagellates benefit by being protected from predators within the skeletons of the corals.

KEYS TO CNIDARIAN SUCCESS. The cnidarians became the dominant marine organisms early in the Cambrian period, 600 mya, and they remain important components of marine ecological communities today (Figure 25.15). Their success lies in having a *Bauplan* that combines a very low metabolic rate with an ability to

(a)

(b)

25.14 Corals
(a) Many different species of corals grow together on this reef in Honduras. (b) The ends of the branches of the elkhorn coral (*Acropora palmata*) are spread so that they maximize the amount of sunlight the photosynthetic dinoflagellates within their cells receive.

(a)   (b)

(c)

25.15 Diversity among the Cnidarian Classes

(a) The structure of the polyps of this North Atlantic coastal hydrozoan, *Gonothyraea loveni*, is clearly seen in this photo. (b) The sea nettle jellyfish *Chrysaora melanaster* illustrates the complexity of some scyphozoan medusae. (c) The nematocyst-studded tentacles of the stubby rose anemone *Tealia coriacia* (class Anthozoa) are poised to capture large prey that may happen to be carried to the animal by water movement.

capture relatively large prey. The bulk of the body of medusae and many polyps is made up of the largely inert mesoglea. As a result, even a large cnidarian, such as a sea anemone, requires relatively little food and can fast for weeks or months. Nematocysts allow cnidarians to subdue prey that are much more active and structurally complex than the cnidarians themselves. Nonetheless, many cnidarians, such as corals, eat microscopic prey. Corals have added a symbiotic association with dinoflagellates to further their ability to grow where prey is relatively scarce. These traits allow cnidarians to survive in environments where encounter rates with prey are much lower than would be required to sustain animals with higher metabolic needs.

Phylum Ctenophora

The comb jellies, also known as sea walnuts or sea gooseberries, constitute the diploblastic phylum

Ctenophora (comb bearers). The *Baupläne* of ctenophores and cnidarians are superficially similar (Figure 25.16), but there are substantial differences. Both ctenophores and cnidarians have two cell layers separated by a thick, gelatinous mesoglea, and both have radial symmetry and feeding tentacles. However, ctenophores have a gut that opens through two anal pores opposite the mouth; thus material moves in one direction through the gut. The ctenophores have eight rows of ciliated plates called ctenes; the animal moves by beating these cilia rather than by muscular contractions. Ctenophoran tentacles are solid and lack nematocysts; instead, the tentacles are covered with sticky filaments to which prey stick. The tentacles are then retracted and brought into contact with the mouth. In some species, the entire surface of the body is coated with a sticky mucus and serves as the prey-capturing structure. All of the 100 known species of ctenophores are carnivorous marine animals.

Ctenophores cannot capture prey as large relative to their own bodies as can some cnidarians, but their sticky tentacles, which dangle in the water, accumulate large amounts of planktonic prey. Ctenophores have low metabolic rates because they, like cnidarians, are composed primarily of inert mesoglea. Therefore, they are successful predators in open seas where prey abundances are often very low.

The life cycles of ctenophores are simple. From gonads located on the walls of the gastrovascular cavity, gametes are liberated into the cavity and then discharged through the mouth or through pores. Fertilization takes place in open seawater. In nearly all species the fertilized egg develops directly into a miniature ctenophore that gradually grows into an adult.

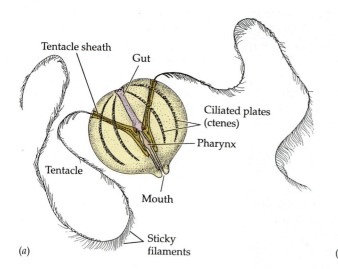

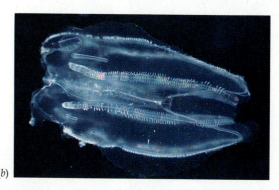

25.16 Comb Jellies Feed with Tentacles
(a) The sticky filaments to which prey adhere are clearly seen on the long tentacles of this comb jelly, or ctenophore. (b) *Leucothea* is a comb jelly with much shorter tentacles.

THE EVOLUTION OF BILATERAL SYMMETRY

Bilateral symmetry probably first arose in simple organisms, consisting of flattened masses of cells, that crawled over a substrate, feeding as they went. Some surviving organisms are no more complex than the earliest of these simple animals. One of these is *Trichoplax adhaerens*, an animal that looks like a large amoeba but is composed of more than 1,000 cells formed into a flattened disc. It can crawl in any direction. To feed, it first secretes enzymes into the space between its ventral surface and the substrate and then absorbs food that has been digested by the enzymes. *Trichoplax* has no distinct tissues or organs. It uses its cilia to move over the substratum, and it is small enough that oxygen and food digested outside the body can simply diffuse into it.

Phylum Platyhelminthes

An early lineage that evolved from flattened masses of cells is the phylum **Platyhelminthes**, the flatworms. Flatworms are triploblastic, bilaterally symmetrical animals that have more complex internal organs than cnidarians and ctenophores do (Figure 25.17). They have no body cavity and they lack any organs for excreting metabolic wastes or for transporting oxygen to internal tissues. This *Bauplan* dictates that each cell must be near a body surface in order to respire, a need facilitated by the flattened body form. The digestive tract of a flatworm, if there is one, consists simply of a mouth opening into a dead-end sac. However, the sac (intestine) is often highly branched, forming intricate geometric patterns that increase the surface area usable for absorption of nutrients.

The flatworms believed to be most similar to the ancestral forms are the turbellarians (class **Turbellaria**; Figure 25.17*a*)—small, free-living marine and freshwater worms (a few live in moist terrestrial habitats). The simplest marine species lack a digestive cavity, excretory organs, and distinct reproductive organs. Freshwater turbellarians of the genus *Dugesia*, better known by the popular name planaria, are the most familiar species of flatworms. They have a head at the front end with a chemoreceptor organ, two very simple eyes called ocelli (singular, ocellus), and a tiny brain composed of anterior thickenings of the longitudinal nerve cords.

All living flatworms feed on animal tissues, some as predators and parasites, others as scavengers. The earliest flatworms, however, may have been herbivores (plant eaters), later adapting to feed on small animals found in the algal mats, and, finally, to feed on or within the bodies of much larger animals. Flatworms that move around a great deal use broad layers of cilia to glide over surfaces. This form of movement is very slow, but it is sufficient for small scavenging animals.

Although the earliest flatworms were free-living, the flatworm *Bauplan* was readily adapted to a parasitic existence. Inside a host, nutrients can simply be absorbed through the body surface, much as they are by *Trichoplax*. A likely evolutionary progression was from feeding on carrion, to invading the bodies of dying hosts, to invading and consuming parts of living, healthy hosts. Today the most familiar members of the phylum—the tapeworms (class **Cestoda**) and flukes (class **Trematoda**; Figure 25.17*c*)—are parasitic. These worms inhabit the bodies of many vertebrates and cause some serious human diseases. Other flukes (class **Monogenea**) are important external parasites of fishes and other aquatic vertebrates.

Parasites live in nutrient-rich environments and do not need to search for food, but they face other challenges. To complete their life cycles, parasites must overcome the defenses of their hosts. They must also colonize new hosts before they actually kill their current host, at which time they will also die. Some parasitic flatworms simply void their eggs with the host's feces; later, these eggs are ingested directly by other host individuals. However, most parasitic species, such as the broad fish tapeworm, have complex life cycles involving two or more hosts and several larval stages (Figure 25.18). Such complex life cycles may have evolved because they increase the probability that the life cycle will be completed.

THE DEVELOPMENT OF BODY CAVITIES

Rapid movement is advantageous for both prey and the predators that pursue them. Fast-moving prey and predators evolved in the early Cambrian period (600 mya). Fluid-filled body cavities probably first evolved because they helped animals to move more rapidly than the organisms we have discussed so far.

Such cavities function as skeletons that can transfer forces generated by the contraction of muscles from one part of the body to another (see Chapter 38).

Fluids are good **hydrostatic skeletons** because they are incompressible. When muscles around some part of a fluid-filled body cavity contract, the fluid must move to another part of the cavity. If the body tissues around the cavity are flexible, fluids moving from one region will cause the expansion of some other region. Fluids can be used to move specific body parts, or to move the whole animal, provided that temporary attachments can be made to the substrate. In the remaining phyla we will encounter many variations on the theme of fluid-filled body cavities that are used to change the shapes of organisms and move them around. Indeed, the types and numbers of body cavities provide a major key to the lives of these animals and the degree to which they can control their shapes and movements.

Phylum Nemertea

Ribbon worms (phylum **Nemertea**) are triploblastic, dorsoventrally flattened animals with nervous and excretory systems similar to those of flatworms. Unlike flatworms, however, they have a complete digestive tract, with a mouth at one end and an anus at the other. The body of most ribbon worms is solid over most of its length, but in some species a fluid-filled cavity called the rhynchocoel, within which floats a hollow, muscular proboscis, extends much of the length of the worm. Contraction of the muscles

(a)

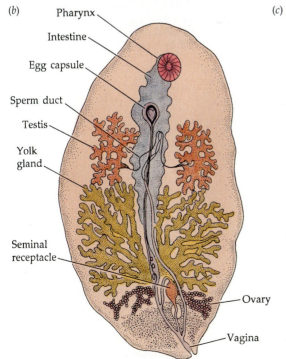

(b)

Pharynx
Intestine
Egg capsule
Sperm duct
Testis
Yolk gland
Seminal receptacle
Ovary
Vagina

(c)

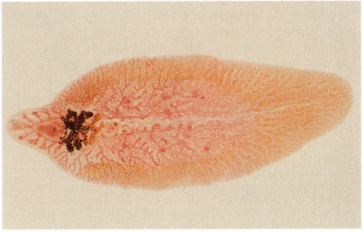

25.17 Flatworms Live Freely and Parasitically
(a) Prosthecereaus bellostriatus is a free-living marine flatworm of the Pacific Coast of North America. *(b)* This flatworm, *Syndesmis*, lives parasitically in the gut of sea urchins. As is typical of internal parasites, its body is filled primarily with sexual organs. *(c)* The parasitic sheep liver fluke *Fasciola hepatica* also has the simple morphology of many internal parasites.

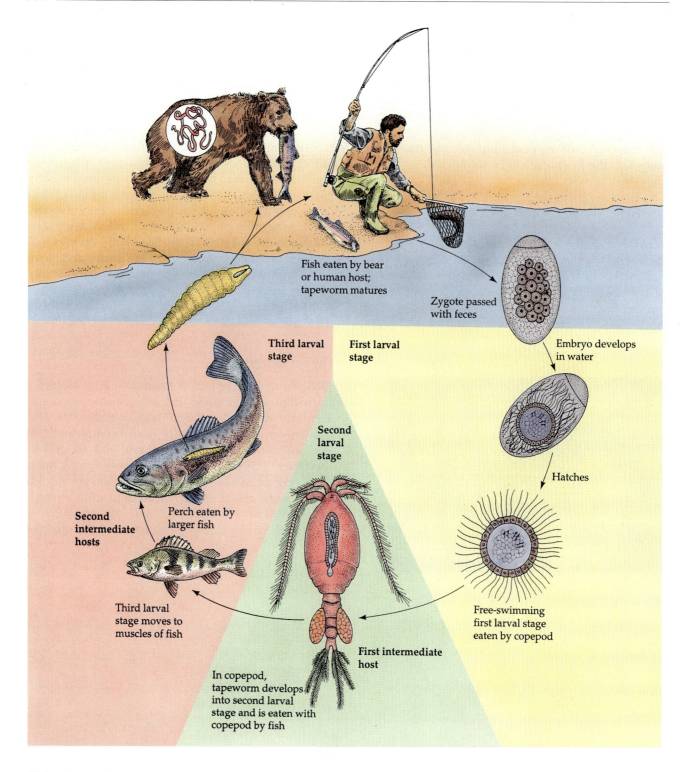

Fish eaten by bear or human host; tapeworm matures

Zygote passed with feces

Embryo develops in water

Hatches

Third larval stage

First larval stage

Second larval stage

Second intermediate hosts

Perch eaten by larger fish

Third larval stage moves to muscles of fish

First intermediate host

In copepod, tapeworm develops into second larval stage and is eaten with copepod by fish

Free-swimming first larval stage eaten by copepod

25.18 Returning to a Host By a Complex Route
The broad fish tapeworm *Diphyllobothrium latum* must pass through both a copepod and a fish before it can reinfect its primiary mammalian host. Such complex life cycles assist recolonization of hosts, but they also offer opportunities for humans to break the cycle with hygienic measures.

surrounding the rhynchocoel causes the proboscis to be ejected explosively through an anterior opening (Figure 25.19). Thus the rhynchocoel transmits forces that move the proboscis rapidly without moving the rest of the animal. Small ribbon worms move by beating their cilia. Larger ones employ waves of contraction of body muscles to move on the surface of sediments or to burrow. Movement by both of these methods is slow.

The proboscis of most ribbon worms is armed with

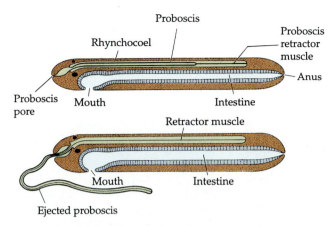

25.19 The Proboscis Is the Feeding Organ of Ribbon Worms
The proboscis (green) can be moved rapidly, but the worm moves only slowly. Notice that the entrance to the rhyncocoel and digestive tracts are separate.

a stylet that pierces prey. Paralytic toxins produced by the proboscis are discharged into the wound made by the stylet. Some species lack a stylet and capture prey by wrapping the muscular proboscis around it. The proboscis is then withdrawn into the rhynchocoel by means of a retractor muscle, and the ribbon worm takes the prey into its mouth. In addition to capturing prey, the rhynchocoel also helps the ribbon worm to burrow. The proboscis is shot into the substratum and when it contracts and fattens, the worm is pulled forward. Similar burrowing mechanisms are found in other phyla.

Because ribbon worms have a complete digestive tract with an anterior mouth and a posterior anus, food items move in one direction through the tract and are acted upon by a series of digestive enzymes —a more efficient system than one in which food remains must be ejected through the mouth.

The approximately 900 species of ribbon worms are nearly all marine, but some are found in fresh water and a few live in moist terrestrial tropical environments. Ribbon worms are all carnivores, feeding mostly upon arthropods and small worms in several different phyla.

PSEUDOCOELOMATE ANIMALS

As we have seen (Chapter 15), a blastocoel develops within the blastula early in the development of most animals. In some animals the blastocoel persists into adulthood as a **pseudocoelom**, a fluid-filled cavity in which body organs float. A pseudocoelom can function as a hydrostatic skeleton, but in some pseudocoelomate animals the cavity is partly invaded by masses of cells. Of the many phyla of pseudocoelomate animals, we will discuss only the two that have many species.

Phylum Nematoda

The increased speed of movement of particular body parts made possible by the possession of even a small fluid-filled cavity at the front end resulted in selection favoring the extension of such cavities to the full length of the body. One group that evolved as a result of this change was the roundworms (phylum **Nematoda**). The internal organs of a roundworm float freely in the pseudocoel. The shape of the body is provided by a thick, multilayered cuticle secreted by the underlying epidermis (Figure 25.20*a*). As a roundworm grows, it sheds and resecretes its cuticle four times.

Because the cross-sectional shape of their bodies is round, roundworms have a relatively small amount of body surface available for the exchange of oxygen and other materials with the environment. Exchange does take place through the cuticle, but it also takes place through the intestine, which is only one cell layer thick. Materials are moved through the gut despite the high hydrostatic pressures in the pseudocoel by rhythmic contractions of a highly muscular pharynx at the anterior end of the worm's gut.

Roundworms are one of the most abundant and universally distributed of all animal groups (Figure 25.20*b*). We unintentionally eat and drink enormous numbers of roundworms in our lifetimes. A single rotting apple from the ground of an orchard was found to contain 90,000 roundworms. One square meter of mud off the coast of The Netherlands yielded 4,420,000 individuals. The topsoil of rich farmland has up to 3 billion individuals per acre. Countless roundworms live as scavengers in the upper layers of the soil, in the bottoms of lakes and streams, and as parasites in the bodies of most kinds of plants and animals. Only about 20,000 species have been described, but the actual number of living species may be in excess of 1 million!

The diets of roundworms are as varied as their habitats. Many roundworms live parasitically within their hosts. Many are predators, preying on protists and other small animals (including other roundworms). Much study of roundworms has been stimulated by the desire to control parasitic species. Roundworms are parasites of people, cats, dogs, cows, sheep, and most economically important plants. The largest known roundworm, which reaches a length of nine meters, is found in the placenta of female sperm whales. The structure of parasitic roundworms does not differ much from that of free-living species, but the life cycles of parasitic species have evolved to enable the transfer of individuals among hosts. A relatively simple method is found in *Trichinella spiralis*, the causal agent of the disease trichinosis. The larvae of *Trichinella* form calcium-containing cysts in the muscles of their mammalian hosts (Figure 25.20*c*); if present in great numbers,

25.20 Gut and Gonad Fill the Body

(a) The large gut (blue) and testis (red) fill most of the body of a male *Trichinella spiralis*, a roundworm that causes trichinosis. (b) This free-living roundworm moves through marine sediments in an undulating pattern based on alternating relaxation and contraction of its longitudinal muscles. (c) A cyst of *Trichinella spiralis* infects its host's muscle tissue.

Nerve ganglion

Cuticle

Pharynx

Dorsal nerve

Ventral nerve

Excretory tube

Pseudocoelom

Intestine

Testis

Anus

(a)

(b)

(c)

these cysts cause severe pain or death. *Trichinella* is transmitted when another mammal eats the flesh of the infected individual. The larvae bore through the intestinal wall of the new host and are carried in the bloodstream to the muscles, where they feed and again form cysts. Thus the intermediate host is likely to be another mammal—usually a pig in the case of human infections. There is no special stage in the life history of *Trichinella* that lives in an alternative host. Other nematode life cycles are as complex as that of the broad fish tapeworm (Figure 25.18).

Phylum Rotifera

The most abundant and widespread of the other groups of pseudocoelomate animals are the rotifers (phylum **Rotifera**). Rotifers are triploblastic, bilateral, unsegmented animals. Most rotifers are tiny (50–500 μm long), not much larger than ciliate protists, but they have a highly developed organ structure (Figure 25.21). There is a complete gut that passes from an anterior mouth to a posterior anus. In the rotifer *Bauplan*, the pseudocoel is greatly reduced and does

not function as the primary hydrostatic skeleton. Most rotifers are very active animals, but they propel themselves through the water by means of rapidly beating cilia rather than by muscular contraction. Rapid movement by this means is possible because rotifers are so small.

The most distinctive organs of rotifers are those used to collect and process food. The **corona** is a conspicuous ciliated organ surmounting the head of many species. Coordinated beating of cilia provides the locomotor force and also sweeps particles of organic matter from the water into the mouth and down to the **mastax**, a complicated skeletal structure bearing teeth that grind the food. A few rotifer species that prey on protists and small animals can evert the mastax through the mouth to seize small objects. This is accomplished by the contraction of muscles surrounding the pseudocoel.

A few rotifer species are marine, but most of the 1,800 known species live in fresh water. A small number, loosely referred to as terrestrial, actually occupy an unusual aquatic habitat: They rest on the surfaces of mosses and lichens in a desiccated, inactive state until it rains. When rain falls, they become active, swimming about and feeding in the films of water that temporarily cover the plants. Most rotifers are short-lived; their life spans are typically between one and two weeks.

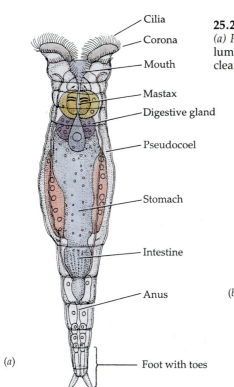

Cilia
Corona
Mouth
Mastax
Digestive gland
Pseudocoel
Stomach
Intestine
Anus
Foot with toes

(a)

25.21 A Rotifer
(a) *Philodina roseola*, a free-living rotifer, is similar to many species in this phylum. (b) The internal anatomy of this small rotifer (*Epiphanes senta*) shows clearly under a phase contrast microscope.

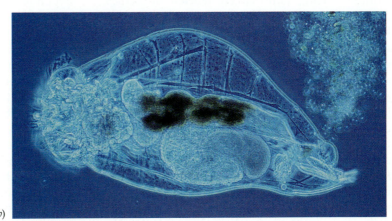

(b)

COELOMATE ANIMALS

Although the evolution of body cavities provided many new movement capabilities, control over body shape is rather crude if the cavity has muscles only on its exterior surface, as a pseudocoel does. Because a coelom is surrounded by muscles, better control over the movements of the fluids it contains is possible. Even with a coelom, however, control over movement is limited if an animal has only a single, large body cavity. But if the cavity is separated into compartments, localized changes in shape, produced by contractions of circular and longitudinal muscles in individual segments, are possible. Thus the animal can change the shape of each segment independently of the others. Segmentation of the coelom evolved a number of different times among both protostomes and deuterostomes.

Phylum Pogonophora: Losing the Gut

Many early protostomes were small animals with thin body coverings. Much of the exchange of gases and waste products took place through the body wall. Marine waters and sediments contain abundant food particles in the form of bacteria as well as dissolved organic matter that also can be absorbed directly through the body wall. One lineage of protostomes evolved into burrowing forms with a crown of tentacles through which gas exchange occurs, and they entirely lost their digestive tracts (Figure 25.22). These animals, the pogonophores (phylum **Pogonophora**) were not discovered until this century because they live only in deep water, often many thousands of meters below the surface. However, pogonophores are extremely abundant in deep oceanic sediments, reaching densities of many thousands per square meter.

25.22 Pogonophores
The red bodies of these deep-sea vent pogonophores project from the long white tubes in which they live.

The coelom of a pogonophore consists of an anterior compartment into which the tentacles can be withdrawn and a long, subdivided cavity extending much of the length of its body. As demonstrated by experiments using radioactively labeled molecules, pogonophores take up dissolved organic matter at high rates, even against concentration gradients. However, because they live buried in sediments where dissolved organic matter accumulates to high concentrations, they normally do not work against concentration gradients to take up food.

The largest and most remarkable pogonophores, reaching up to 2 meters in length, lie near deep-ocean hydrothermal vents—openings in the sea floor through which hot, sulfur-rich water pours. The large body cavity of these species is filled with bacteria that fix carbon by the oxidation of hydrogen sulfide gas. The pogonophores either consume these bacteria directly or live on their metabolic by-products. Hydrothermal vent ecosystems, which are not based upon light as their energy source, have attracted considerable interest because they may be the kind of environment in which life on Earth actually originated (see Chapter 17).

Phylum Annelida: Many Subdivisions of the Coelom

You will recall that the locomotion of roundworms depends on alternating contractions of longitudinal muscles on opposite sides of the body. Because the forces generated by the contracting muscles are transmitted through the entire undivided body cavity, roundworms can achieve only poorly controlled movements. If the body cavity is subdivided, an animal can alter the shape of its body in complex ways and thus it can control its movements much more precisely.

The potential of a *Bauplan* based on a highly subdivided body cavity is illustrated by the annelids (phylum **Annelida**), a very diverse group of worms. The approximately 15,000 species of annelids live in marine, freshwater, and terrestrial environments. Each body segment of a typical annelid has a pair of cavities. Segmentation is essential to the crawling movement of an annelid. The waves of bulges that move up and down the length of the body are made possible by its segmentation. There are separate nerve centers controlling each segment, but the cen-

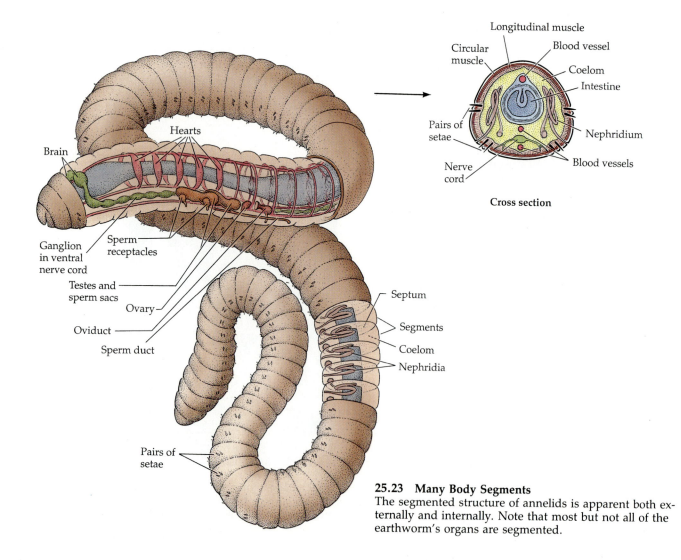

25.23 Many Body Segments
The segmented structure of annelids is apparent both externally and internally. Note that most but not all of the earthworm's organs are segmented.

ters are also connected by nerve cords that coordinate their functioning. The coelom in each segment is isolated from those in other segments (Figure 25.23). Because most annelids lack a rigid external protective covering, the flexible outer body wall plays a key role in locomotion. The thin body wall also serves as a general surface for gas exchange in most species. However, this thin, permeable body surface restricts annelids to moist environments because they lose body water rapidly.

POLYCHAETES. More than half of all annelid species are placed in the class **Polychaeta**. Nearly all polychaetes are marine animals. Most have one or more pairs of eyes and one or more pairs of tentacles at the anterior end of the body. The body wall in most segments extends laterally as a series of thin outgrowths, called parapodia, that have many blood vessels and function in both gas exchange and locomotion. Stiff bristles called setae protrude from each parapodium. They form temporary attachments with the substrate and prevent the animal from slipping backward when its muscles contract. Many species live in burrows in soft sediments and capture prey from surrounding water with elaborate feathery tentacles (Figure 25.24a). Most polychaetes are **dioecious**, which means the sexes are separate. Gametes are often released into the water, where fertilization occurs. The fertilized egg develops into a ciliated larva

known as a **trochophore** (Figure 25.24b). As a trochophore feeds, it forms body segments at its rear end; it eventually metamorphoses into a small adult worm.

OLIGOCHAETES. More than 90 percent of the approximately 3,000 described species of the class **Oligochaeta** live in freshwater or terrestrial habitats. Oligochaetes have no parapodia, eyes, and anterior tentacles, and they have relatively few setae. Earthworms are the most familiar oligochaetes. They are scavengers and ingestors of soil, from which they extract food particles. Unlike polychaetes, all oligochaetes are **hermaphroditic**, that is, each individual is both male and female (Chapter 35). Sperm are exchanged simultaneously between two copulating individuals (Figure 25.24c). Eggs are laid in a cocoon outside the adult's body. Later the cocoon is shed and, when development is complete, miniature worms emerge and begin independent life.

LEECHES. The leeches (class **Hirudinea**; Figure 25.24d) are believed to have evolved from oligochaetelike ancestors. Most species live in freshwater or terrestrial habitats, and, like oligochaetes, they lack parapodia and tentacles. Leeches are hermaphroditic; each individual serves as a sperm donor and a sperm recipient during copulation. Unlike those of other annelids, the body segments of leeches are not divided into compartments, and the coelomic space is largely filled with mesenchyme tissue. Therefore, leeches move in a very different manner than other annelids. Groups of segments at each end of a leech are modified to form suckers which serve as tempo-

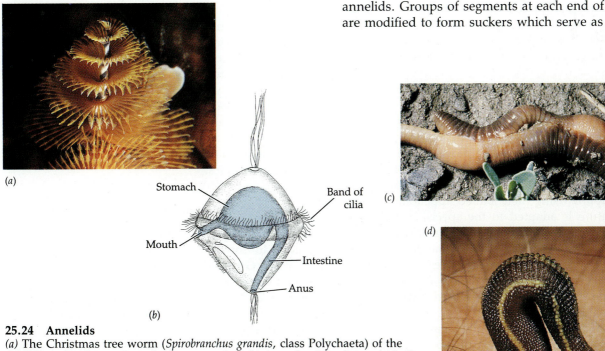

(a)

Stomach

Band of cilia

(c)

Mouth

Intestine

Anus

(b)

(d)

25.24 Annelids
(a) The Christmas tree worm (*Spirobranchus grandis*, class Polychaeta) of the Caribbean Sea has striking feeding tentacles. (b) A trochophore larva. (c) Individual earthworms (*Lumbricus* sp., class Oligochaeta) are hermaphroditic. When they copulate, as the pair in this photo are doing, each individual both donates and receives sperm. (d) Notice the conspicuous anterior and posterior suckers of this freshwater leech.

rary anchors. With its posterior anchor attached to the substrate, the leech extends its body by contraction of its circular muscles. The anterior sucker is then attached, the posterior one detached, and the leech shortens itself by contracting its longitudinal muscles.

Most leeches are external parasites of other animals. The mouth has three toothed jaws with which the leech makes an incision in its host. An anticoagulant secreted into the wound keeps the blood flowing. Leeches used to be widely employed in medicine for blood-letting. Even today they are used to reduce fluid pressures in tissues damaged by, for example, a snake bite, and to eliminate pools of coagulated blood.

THE SHIFT TO AN EXTERNAL SKELETON

Among most of the animals we have discussed, skeletal support is provided by fluid-filled body cavities acted upon by muscles surrounding them. Variations on this *Bauplan* have involved the size and extent of those cavities, how they are lined, and whether they are subdivided. The external coverings of most of these animals are relatively thin and flexible, serving as gas-exchange surfaces in many species. Some, however, such as roundworms, have tough cuticles. In Precambrian times, evolution took a different direction in one of the annelidlike lineages. In those animals, the body covering became thickened by the incorporation of layers of protein and a strong, flexible, modified polysaccharide called **chitin**. Following this change, which was initially probably protective in function, the body covering acquired support and locomotor functions—it became an **exoskeleton**.

Phylum Arthropoda

How can an animal with a rigid exoskeleton and no cilia move? One possibility is to have appendages with flexible joints, which are moved by muscles. A lineage in which this method evolved led to the arthropods (phylum **Arthropoda**). Most arthropods can move rapidly using their jointed appendages. Their musculature is adapted to the special needs of armored existence. Arthropod muscles are specialized to operate particular segments of the body and the appendages attached to them (Figure 25.25*a*). Within most of the evolutionary lineages of arthropods, certain of the appendages were modified from their original walking or swimming functions to aid in eating and reproduction. Other appendages acquired sensory functions. This division of labor among body regions afforded a versatility to each individual arthropod species not found in its annelid ancestors (Figure 25.25*b*).

A rigid exoskeleton affects functioning in other ways. First, the coelom is rendered useless as a hydrostatic skeleton. During arthropod evolution, the coelom became filled with blood, which directly bathes the animal's organs. Second, the exoskeleton cannot serve as a gas-exchange surface, so arthropods have evolved other means of taking in oxygen and releasing carbon dioxide. For example, many arthropods have evolved **gills**, structures specialized for gas exchange, which are often greatly branched or folded to increase their surface area. Third, an animal with a rigid, nonliving exoskeleton cannot grow by gradually increasing in size. Growth among arthropods is accomplished by **molting**, a periodic shedding of the exoskeleton followed by the rapid hardening of a new and larger exoskeleton that has formed under the old one. Therefore, arthropod growth takes place in spurts immediately following each molt, while the new exoskeleton expands and hardens. While this process is under way, movement is difficult or impossible and the animals are highly vulnerable to predators. The soft-shelled crabs of commerce are individuals captured just after they have shed their old exoskeletons.

Diversification of the Arthropods

The evolution of the exoskeleton and its accessories had yet another curious and profound influence on arthropod evolution. Encasement within armor does more than just protect an animal from predators. It also provides support for walking on dry land, and, of equal importance, it keeps the animal from drying out quickly in air. Arthropods were, in short, the best candidates to invade the land. This they did repeatedly, with spectacular success. Many different groups of arthropods colonized the land, but all of the other groups are completely overshadowed in numbers and diversity by the insects. The great majority, not only of arthropods but of all animal species, are insects. Nearly one million species of insects have been named and described. The actual number of living species may be as high as 30 million, most of which live in the tropics.

Ancient Arthropods and Modern Descendants

A small group of inconspicuous tropical animals placed in their own phylum, the onychophorans (phylum **Onychophora**), are similar to what many people think were ancestors of the earliest arthropods (Figure 25.26*a*). In fact, the onychophorans are so intermediate between annelids and arthropods that they are sometimes included in one of those phyla. Their internal structure is closer to that of annelids, but their circulatory system, in which the blood bathes the internal organs, resembles that of arthropods. The body is covered by a thin, flexible cuticle that contains chitin, but it is not divided into

(a)

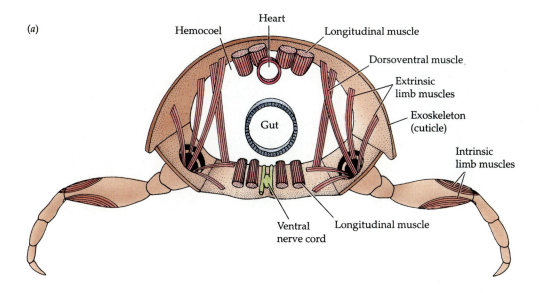

(b)

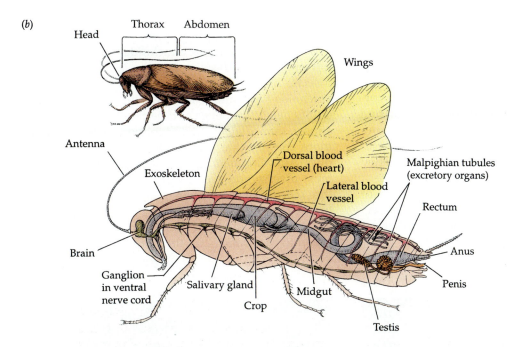

25.25 External and Jointed
(a) You can best see the structure of an arthropod by "looking through" a segment of a generalized individual. (b) The *Bauplan* of this insect differs in many details from that of other arthropods, but the basic theme of a segmented body with modified, highly functional appendages is general to the entire phylum.

segments.

The trilobites (subphylum **Trilobita**) were arthropods that flourished in the seas of the Cambrian and Ordovician periods but were extinct by the close of the Paleozoic era. Trilobites were heavily armored and their body segmentation and appendages followed a relatively simple, repetitive plan, again reminiscent of annelids (Figure 25.26b). Why trilobites declined in abundance and eventually became extinct is unknown.

Living arthropods are classified into three large

subphyla: the **Chelicerata**, the **Crustacea**, and the **Uniramia**.

Subphylum Chelicerata

The bodies of all chelicerates are divided into two major regions, the anterior of which bears two pairs of appendages modified to form mouthparts and four pairs of walking legs. The 65,000 described species are placed in three classes, only one of which contains many species.

25.26 Annelid–Arthropod Linkages
(a) Annelid or arthropod? *Peripatus* has traits of both phyla. (b) The relatively simple, repetitive segments of the now-extinct trilobites—illustrated here by fossils of *Dalamites limulurus*, a species that lived during the Silurian period—are similar to those of modern annelids, but they have some of the differentiation characteristic of arthropods.

The pycnogonids (class **Pycnogonida**) or sea spiders are a small group of marine species that are seldom seen except by marine biologists (Figure 25.27a). The class **Merostomata** contains a single order, the Xiphosura or horseshoe crabs. These marine animals, which have a long fossil history during which they changed very little, have a large horseshoe-shaped covering over most of the body. They are common in shallow waters along the eastern coasts of North America and Southeast Asia, where they scavenge and prey on bottom-dwelling invertebrates. Periodically they crawl into the intertidal zone to mate and lay eggs (Figure 25.27b).

The arachnids (class **Arachnida**) are relatives of horseshoe crabs that invaded the land and, like the insects, achieved early and lasting success. Most arachnids have simple life cycles in which miniature adults hatch from eggs and begin independent lives almost immediately. Some species, however, have more complex life cycles. Others retain their eggs during development, giving birth to live young. The most diverse and ecologically important groups are the scorpions, harvestmen, spiders, and acarines (mites and ticks) (Figure 25.28). Spiders have evolved into effective predators. Some make use of excellent vision to chase and seize their prey. Others use elab-

25.27 Minor Chelicerates
(a) Although they are not spiders, it is easy to see why sea spiders (class Pycnogonida) were given their common name. (b) This spawning aggregation of horseshoe crabs (class Merostomata) was photographed on a beach in Delaware.

25.28 Arachnid Diversity
(a) Scorpions, such as this *Uroctonus mondax* from California, are nocturnal predators. (b) Wolf spiders are active diurnal predators with large eyes and good vision. (c) Harvestmen, often called daddy longlegs, are generalized scavengers. (d) Ticks are the most bothersome of arachnids. This wood tick, *Ixodes ricinus*, is piercing the skin of its human host.

orate silken webs to snare prey. The webs constructed by different groups of spiders are strikingly varied and enable spiders to position their snares in many different environments. Spiders also use silk to construct safety lines during climbing, and for homes, mating structures, protection for developing young, and dispersal. The silk is produced by modified abdominal appendages, called spinnerets, that are connected to internal glands that secrete the proteins of which the silk is constructed.

Subphylum Crustacea

The crustaceans (subphylum **Crustacea**) are the dominant arthropods of the oceans. One group alone, the copepods, is so numerous in plankton communities that they may well be the most abundant of all animals. Most crustaceans have a body that is divided into three regions: head, thorax, and abdomen. The head bears five pairs of appendages, and each thoracic and abdominal segment usually bears one pair. The segments of the head are fused together, and, in many species, a fold of the exoskeleton, the **carapace**, extends dorsally and laterally back from the head to cover and protect some of the other segments (Figure 25.29a).

Sexes are separate in nearly all crustaceans and individuals come together to copulate. The fertilized eggs of most crustacean species are attached to the outside of the female's body, where they are brooded during their early development. At hatching, the young of some species are released as larvae; those of other species are released as juveniles similar in form to the adults. Some species, however, release fertilized eggs that float in the water or attach the eggs to some object in the environment. The typical crustacean larva, called a **nauplius**, has three pairs of appendages and a median compound eye (Figure 25.29b).

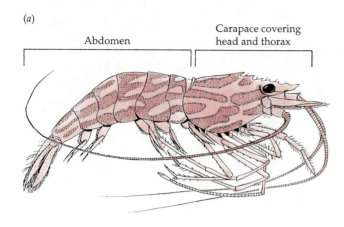

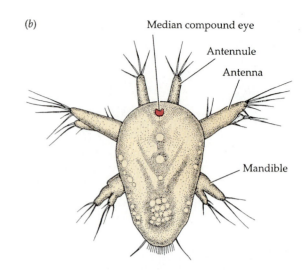

25.29 Crustacean Structure
(*a*) The bodies of crustaceans are divided into three regions, each of which bears appendages. (*b*) A nauplius larva has one compound eye and three pairs of appendages.

25.30 Crustacean Diversity
(*a*) Decapod crustaceans are represented here by a pair of California spiny lobsters, *Panulirus interruptus*. (*b*) *Ligia occidentalis* is an intertidal isopod found on the California coast. (*c*) *Eucheda* is a typical representative of the abundant planktonic copepods. (*d*) The appendages of this gooseneck barnacle (*Lepas anatifera*) are protruded from its shell and in the feeding position.

The most familiar crustaceans are shrimps, lobsters, crayfish, crabs, sowbugs, and sand fleas (Figure 25.30*a,b*). There is also a wide variety of other small species, many of which superficially resemble shrimps; the abundant copepods mentioned previously are members of one of these groups (Figure 25.30*c*). The barnacles (class **Cirripedia**) are unusual crustaceans that are hermaphroditic and sessile as

adults (Figure 25.30d). With their unique calcareous shells, they superficially resemble mollusks, but, as the zoologist Louis Agassiz remarked over a century ago, a barnacle is "nothing more than a little shrimp-like animal, standing on its head in a limestone house and kicking food into its mouth."

Subphylum Uniramia

The *Bauplan* of unirames is based on a body divided into two or three regions. The anterior regions have few segments, but the posterior region, the abdomen, has many segments. The subphylum Uniramia contains two classes of primarily terrestrial animals, both of which have elaborate systems of channels to bring oxygen to the cells of the internal organs.

MYRIAPOD UNIRAMES. Animals in the class **Myriapoda**—centipedes, millipedes, and two other groups of inconspicuous animals—have two body regions, a head and an abdomen. Centipedes and millipedes have well-formed heads and long, flexible, segmented abdomens that bear many pairs of legs (Figure 25.31). Centipedes are predators of insects and other small animals. Millipedes are scavengers and plant eaters. More than 3,000 species of centipedes and 7,500 species of millipedes have been described; many species remain unknown. Most of them are less than a few centimeters long, but some tropical species are ten times that long.

INSECTS. Insects (class **Insecta**) have three basic body parts (head, thorax, abdomen), a single pair of antennae on the head, and three pairs of legs attached to the thorax. Gas exchange is accomplished by means of air sacs and tubular channels called **tracheae** (singular, trachea) that extend from external openings inward to tissues throughout the body. The adults of those insects able to fly have two pairs of stiff-membranous wings attached to the thorax—except for flies, which have only one pair. The one million species of insects that have been described are believed to be only a small fraction of the total number living on Earth today. Insects are found in nearly all terrestrial and freshwater habitats, and they utilize as food nearly all species of plants and many species of animals. Some are internal parasites of animals, and others are external suckers of blood or consumers of superficial body tissues. Because of their ability to fly, insects effectively transmit many viral, bacterial, and protist diseases among plants and animals.

Very few species of insects live in the oceans, and those that do are found only at the shoreline. In freshwater environments, on the other hand, insects are the dominant animals, burrowing through substrata or extracting suspended prey from water columns. However, rather than using tentacles or other modifications of the anterior part of the body, as do most suspension feeders in other phyla, insects extract prey from water primarily by constructing silken webs.

Most insects have the full number of adult segments when they hatch from their eggs, but they differ strikingly in their state of maturity at hatching and the processes by which they achieve adulthood. Individuals of the groups of wingless insects believed to be most similar to insect ancestors have **simple development**. They hatch from the egg as juveniles very similar to the adults, and they mature mostly by increasing in size. Among the winged insects, development is more complex. The hatchlings are not similar in form to adults, and they undergo substantial changes at each molt in the process of becoming an adult. The intermolt immature stages of insects are called **instars**. If changes between these instars are gradual, an insect is said to have **incomplete metamorphosis** (see Figure 15.11). Grasshoppers and katydids are two examples of insects that undergo incomplete metamorphosis. If there are dramatic changes between one or more instars, an insect is

(a)

(b)

25.31 Myriapods
(a) Centipedes, represented here by *Scolopendra heros*, have powerful jaws for capturing active prey. (b) Millipedes, which are scavengers and plant-eaters, have smaller jaws and legs.

(a)

(b)

(c)

(d)

(e)

(f)

(g)

(h)

25.32 A Diversity of Insects

(a) The apterygotes are illustrated here by a firebrat (*Thermobia domestica*), a typical thysanuran. (b) This mayfly is a paleopterous insect. It lives a few hours, mates, and dies. The other paleopterous insects—dragonflies and damselflies—were shown in Figure 20.18. Most insects belong to one of the neopterous orders; shown here are representatives of some of the largest orders. (c) A broad-winged katydid (order Orthoptera). (d) A mating pair of harlequin bugs (order Hemiptera). (e) A predaceous diving beetle (order Coleoptera). (f) A robberfly (order Diptera). (g) A buckeye butterfly (order Lepidoptera). (h) A bumblebee (order Hymenoptera).

said to have **complete metamorphosis** (Figure 15.9). Complete metamorphosis is associated with a specialization of different life stages for living in different environments and utilizing different food sources. Among many species, the larvae are adapted for feeding and growing and the adults are specialized for reproduction and dispersal. The adults of some species do not feed at all, living only long enough to mate, disperse, and lay eggs. Adults of many other species are active feeders, usually using different food resources than the larvae do. The most familiar example of complete metamorphosis is the "caterpillar-into-butterfly" phenomenon, but other insects, including many species of beetles, also experience it.

Entomologists, students of insects, divide insects into about 26 different orders; Figure 25.32 illustrates several orders. One can make sense out of this bewildering variety by recognizing the existence of three major adaptive types: apterygotes, paleopterous pterygotes, and neopterous pterygotes. The **apterygotes** are entirely wingless. They include springtails (order Colembola) and silverfish (order Thysanura). Both are small insects that mostly live hidden in the soil and humus, beneath rocks, in rotting logs, in the crevices of tree trunks, and among old books.

There are two major evolutionary lineages of winged insects, or **pterygotes**. One lineage has wings that cannot be folded back against the body. They are often excellent flyers, but they require a great deal of open space in which to maneuver. The only surviving members of these **paleopterous** groups are the orders Odonata (dragonflies and damselflies) and Ephemeroptera (mayflies). These insects all have aquatic larvae called nymphs that metamorphose into flying adults after they crawl out of the water. Dragonflies and damselflies are active predators as adults, but adult mayflies lack functional digestive tracts and live only long enough to mate and lay eggs.

The other evolutionary lineage led to the **neopterous** orders: Orthoptera (grasshoppers, crickets, roaches, mantids, walking sticks), Isoptera (termites), Plecoptera (stone flies), Dermaptera (earwigs), Thysanoptera (thrips), Hemiptera (true bugs), Homoptera (aphids, cicadas, leafhoppers), Neuroptera (lacewings and their relatives), Coleoptera (beetles), Trichoptera (caddisflies), Lepidoptera (butterflies and moths), Diptera (flies), Hymenoptera (sawflies, bees, ants, wasps), and other less familiar orders.

Because they can fold their wings over their backs when not in use, neopterous insects are able to fly from one place to another and then, upon landing, tuck the wings out of the way and crawl into crevices and other tight places as wingless insects do. Several neopterous orders, including the Phthiraptera (lice) and Siphonaptera (fleas) are parasitic and, although descended from flying insects, have lost the ability to fly.

Why have the insects undergone such incredible evolutionary diversification? Insects may have originated from a centipedelike ancestor at least as far back as the Devonian period, more than 350 mya. This early start apparently gave them an advantage in exploiting the newly formed forests and other ancient forms of land vegetation. By Carboniferous times, a great diversity of types already swarmed over the land, and winged insects had appeared—the first animals to fly. The terrestrial environments penetrated by insects were like a new planet, an ecological new world comparable in size and superior in complexity to the surrounding seas, but one with relatively few species of competing terrestrial animals. Viewed in this light, it is not surprising that insects have become richer in species than all other animals put together.

CALCIFIED PROTECTION

In several lineages, the selective pressures brought on by predators during the early Cambrian period led to the evolution of calcified body coverings. Such protection evolved among unicellular protists, algae, and many animals, some of which we have already discussed. However, the most spectacular calcified shells evolved in the phylum we will treat next.

Phylum Mollusca

The mollusks (phylum **Mollusca**) are a group that underwent one of the most remarkable animal evolutionary radiations. The earliest known mollusks were wormlike animals with calcified spicules on their dorsal surface, a rasping feeding structure, known as the **radula**, at the anterior end, and gills at the posterior end. They are believed to have been grazers on algae, much like some of their modern descendants. Selective pressures from predators probably favored further development of the protective spicules into a series of plates or a single, caplike shell, both of which provide better protection (Figure 25.33).

The pattern of embryological development is very similar among all mollusks. Also, their remarkable radiation was based upon a *Bauplan* with three structural components: the foot, the mantle, and the visceral mass. The bodies of such different appearing animals as snails, clams, and squids are all built from these components. For these reasons, invertebrate zoologists place all 100,000 species of mollusks in a single phylum.

The **foot** is a large, muscular structure that originally was the molluscan organ of locomotion as well as the support for the internal organs. The foot has undergone modification in the lineage leading to squids and octopuses to form tentacles borne upon

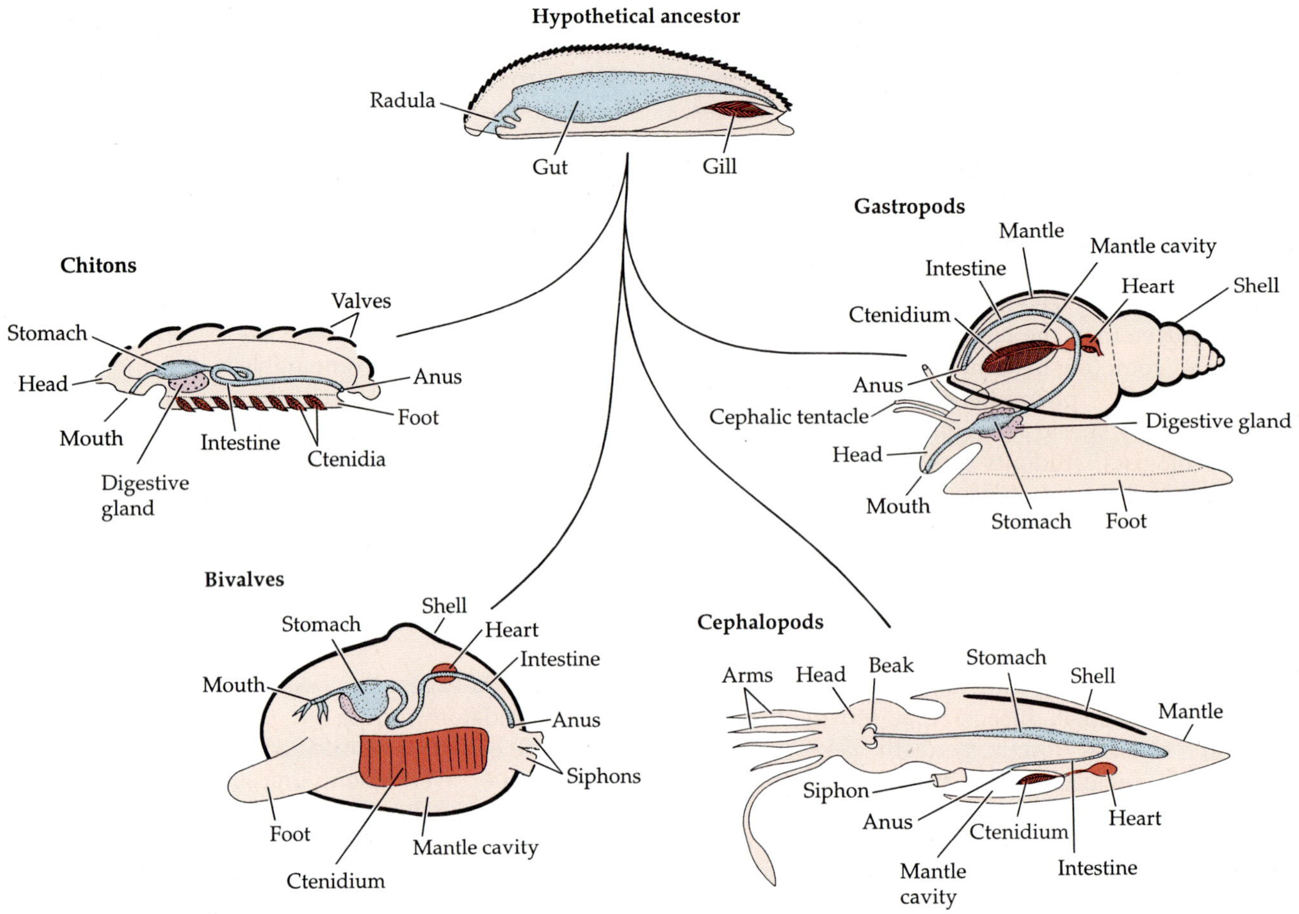

25.33 Evolution among Mollusks
Modifications of the hypothetical ancestor yielded a diverse array of molluscan *Baupläne*.

a head with complicated sense organs. In other groups it has undergone modifications into a burrowing organ. In some lineages it is greatly reduced.

Mollusks also possess a **mantle**, a sheet of specialized tissue that covers the internal organs like a body wall. The mantle secretes the shell, which provides external protection in most molluscan groups, but which has been modified into an internalized support in slugs and squids and lost altogether in octopuses. The featherlike gills of mollusks are located between the mantle and the **visceral mass**, the major internal organs. Contractions of cilia on the mantle create a flow of fresh, oxygenated water over the ctenidia (gills). The actual body cavity of mollusks is much reduced, but the open circulatory system has large fluid-filled cavities that serve as major components of their hydrostatic skeleton. The radula was originally an organ for scraping algae from rocks, a function it still has in many living mollusks. However, the radula has become modified to perform a number of different functions—in some species it serves as a drill, in others as a poison dart. Mollusks range in size from snails only 1 millimeter high to giant squids more than 18 meters long—the largest known protostomes.

CHITONS. The lineage in which the spicules became united to form plates gave rise to the chitons (class **Polyplacophora**). Chitons are the living mollusks most similar to probable molluscan ancestors (Figure 25.34*a*). The body is symmetrical, and the internal organs, particularly the digestive and nervous systems, are relatively simple. Development proceeds through a trochophore larva almost indistinguishable from that of annelids. Chitons are marine herbivores that scrape algae from rocks with their sharp radulae. An adult chiton spends most of its life clamped tightly to rock surfaces by its large, muscular foot. It can move slowly by means of rippling waves of muscular contractions in its foot.

MONOPLACOPHORANS. The lineage in which the spicules became united into a single caplike shell gave rise to the monoplacophorans (class **Monoplacophora**). The evolution of the shell, although it provided protection for the gills and other body parts, tended to isolate the gills from oxygen-bearing water. This favored gill enlargement and the development of a cavity under the shell through which water could readily circulate. The monoplacophorans were the most abundant mollusks during the Cambrian pe-

(a)

(b)

(c)

(d)

(e)

(f)

(g)

25.34 Mollusk Diversity
(a) Tonicella lineata is a chiton (class Polyplacophora) common in the intertidal zone of the Pacific Coast of North America. *(b)* Scallops like this bay scallop (*Aequipecten irradians*) are unusual among bivalves in being able to swim by rapidly opening and closing their valves. *(c)* Most gastropods, such as the chestnut cowry *Cypraea spadicea*, have a coiled shell; however, terrestrial and marine slugs *(d)* have evolutionarily lost their shells. This sea slug, *Dirona albolineata*, is brightly colored. *(e)* The land snail *Monadenia infumata* is typical of many terrestrial gastropods. *(f)* Octopuses are active predators. This one is searching for mollusks on a coral head off the coast of Mexico. *(g) Nautilus pompilius* is one of a few surviving species of a rich evolutionary radiation of nautiloids.

riod, 600 mya. They were thought to have become extinct many millions of years ago, but in 1952, in the deep Pacific off the coast of Costa Rica, an oceanographic vessel dredged up 10 specimens of an unusual little mollusk with a cap-shaped shell. The discovery of these animals, which were placed in the genus *Neopilina*, created a sensation because they turned out to be living monoplacophorans. *Neopilina* are unusual because, unlike other living mollusks, they have multiple gills, muscles, and excretory structures that are repeated down the length of the body.

BIVALVES AND GASTROPODS. One lineage of monoplacophorans developed shells that extended over the sides of the body as well as the top, giving rise to the bivalves (class **Bivalvia**), the familiar clams, oysters, scallops, mussels, and other important edible shellfish, together with a host of similar, less known forms (Figure 25.34*b*). In addition to becoming larger, the shells of these animals evolved into two major pieces connected by a flexible hinge, hence the name bivalves. The larger shells allowed enlargement of the mantle cavity and the gills, but they also made locomotion more difficult. As a result, bivalves are largely sedentary and have greatly reduced heads. The foot is compressed and is used by many clams for burrowing into mud and sand. Food is removed from the water by the enlarged gills that today serve as the main sites of gas exchange and as feeding devices.

Another lineage of monoplacophorans evolved spiralled shells and became the gastropods (class **Gastropoda**). Gastropods are motile, using their large foot to move slowly across the substrate or to burrow through it. The shell and visceral mass of a gastropod larva undergo a 180° counterclockwise turn relative to the foot, termed **torsion**, during development. The result is that the digestive tract and nervous system are twisted into a U shape. The anus, the opening of the mantle, and the gills are moved to the front of the body, just behind and over the head. The gills of gastropods are the primary sites of gas exchange. In some species, they also are feeding devices.

Gastropods are the most diverse and widely distributed of the molluscan classes (Figure 25.34*c–e*). They are the only mollusks that live in terrestrial environments. Among the terrestrial species, the mantle cavity is modified into a highly vascularized lung that is folded into the body. The crawling forms include a rich variety of snails, whelks, limpets, slugs, abalones, and the often brilliantly ornamented nudibranchs. Still other gastropods, the sea butterflies and pteropods, have modified the foot into a swimming organ with which they move through the open waters of the sea.

CEPHALOPODS. Another lineage of monoplacophorans

evolved an exit tube on the shell for currents escaping from the mantle cavity. At first, this functioned to improve the flow of water over the gills, but subsequently it became modified to allow the early cephalopods (class **Cephalopoda**) to regulate mixtures of air and water in the shell cavity. By regulating these mixtures, the animals could control their buoyancy and, hence, their position in the water column. Cephalopods were the first animals able to move vertically in the ocean. This ability, combined with the modification of the exit tube into a device for forcibly ejecting water from the cavity, enabled cephalopods to move rapidly through the water. With this greatly enhanced mobility, some cephalopods such as squids were able to invade the open ocean and become the major predators there (Figure 25.34*f*). Today they continue to be important pelagic predators. In their evolutionary radiation they developed many of the same methods of locomotion utilized by fish, which were later invaders of the same environment. As active predators, cephalopods have complicated sense organs, especially their eyes, which are comparable to those of vertebrates in their ability to resolve images (see Chapter 37).

Cephalopods appeared relatively late in molluscan evolution—at the end of the Cambrian period. By the mid-Devonian period nearly all types had appeared. Their subsequent evolution primarily has involved increases in size in many lineages and reductions in external hard parts. Structurally, cephalopods are mollusks in which the foot is modified into a head with tentacles, a siphon, and a large brain. The mantle is large and muscular, the gills hang within the mantle cavity, and there is a beak that enables the animals to capture and subdue a variety of prey.

The earliest cephalopods developed shells partitioned into chambers connected by tubes through which liquids can be moved from one chamber to another. As fluid is moved out of a chamber, gas diffuses into it. Changing the gas content of the chambers allows the animal to adjust its buoyancy so that it can maintain any position it desires in the water column. Of these early cephalopods only the nautiloids (genus *Nautilus*) survive (Figure 25.34*g*).

Nautiloids effectively control their buoyancy, but the bulky, chambered shell is an impediment to rapid locomotion. Also, nautiloids cannot live at great depths because the high water pressures there would collapse the shell unless it were so thick as to virtually prevent locomotion. The *Bauplan* thus changed in the subsequent evolution of cephalopods: The shell was reduced and other mechanisms of controlling buoyancy were substituted. Among some of the pelagic squids, a few low-pressure internal gas spaces substitute for the chambered shell. Other species employ mechanisms of "chemical lift," secreting body fluids less dense than sea water into coelomic spaces or into vacuoles in the muscles of the tentacles. Still other cephalopods have abandoned control of their buoyancy entirely. They either maintain themselves in the water column by swimming actively or they live on the bottom, as octopuses do.

The lifestyles of cephalopods are similar to those of many fishes. Like fishes, some deep-sea species have luminous organs. In shallow waters, some squids lie buried in the sand during the day and emerge at night to forage as do some fishes. Some species form large schools high in the water column and undertake long-distance migrations.

THEMES IN PROTOSTOME EVOLUTION

Most protostome evolution took place in the oceans. Because water provides good support for animal bodies, early animals used simple body fluids as the basis for their support. Fluid-filled body cavities evolved into hydrostatic skeletons, which, when acted upon by surrounding circular and longitudinal muscles, enabled the animals to move. Subdivisions of the body cavities allowed better control of movement and permitted different parts of the body to be moved somewhat independently of one another. Thus protostomes gradually evolved abilities to change their shapes in complex ways and to move with great speed through the substrate and in the water. Not until the appearance of the cephalopods, however, were any protostomes able to move rapidly in water and control their position in the water column.

Predation may have been the major selective pressure for the development of hard, external body coverings. External body covers evolved in many invertebrate phyla. These coverings, originally protective in function, became key elements in the development of new systems of locomotion. Locomotory abilities permitted prey to escape more readily from predators but also allowed predators to pursue their prey more effectively. Thus the evolution of protostomes has been and continues to be a complex arms race among predators and prey.

Over much of the time of protostome evolution, the only food available in the water was dissolved organic matter and very small organisms. Consequently, many different lineages of animals evolved feeding structures designed to extract small prey from water and structures for moving water through or over their prey-collecting devices. Because water flows readily, bringing food with it, sessile lifestyles evolved repeatedly during protostome evolution. Most protostome phyla today have at least some sessile members. A sessile animal gains access to local resources but forfeits access to more distant resources. A sessile animal is also exposed to physical agents and predators from which it cannot escape by movement.

Sessile animals cannot get together for mating purposes and must instead rely on the extrusion of gametes into the water where fertilization can occur. Some species eject both eggs and sperm into the water. Others retain their eggs within their bodies and extrude only their sperm which are carried by the water to other individuals. Many species whose adults are sessile have motile larvae, many of which have complicated mechanisms for locating suitable sites for settling. In addition, many colonial protostomes are able to grow in the direction of better resources or into sites offering better protection.

Another consequence of a sessile existence is that there is often intense competition for space. Competition for space, which provides access to light, is intense among plants in most terrestrial environments. In the sea, where there are large numbers of sessile species, animals also compete directly for space. They have evolved mechanisms for overgrowing one another and for engaging in toxic warfare where they come into contact (Figure 25.35). Coloniality also enables animals to compete more effectively for space because colonies are better at overgrowing neighbors than are single individuals.

All of the phyla of protostomate animals had evolved by the Cambrian period, 600 mya, but diversification within those lineages continues to the present. As many as 30 million species of animals may be living today, most of which are arthropods, especially insects—the first animals to evolve the ability to fly. The characteristics of the protostome phyla are summarized in Table 25.3.

Many of the evolutionary trends demonstrated by protostomes also dominated the evolution of deuterostomes, the lineage that led to the chordates, the group to which we belong. Hard external body coverings evolved in and were later abandoned by many lineages. Consideration of evolution in deutero-

25.35 Toxic Warfare Between Sea Anemones
When two sea anemones come into contact they attack one another by developing special tentacles that produce toxins.

TABLE 25.3
General Characteristics of Major Protostomate Animal Phyla

PHYLUM	SYMMETRY	BODY CAVITY	DIGESTIVE TRACT	CIRCULATORY SYSTEM
Porifera	Asymmetrical, radial	None	None	None
Cnidaria	Radial, biradial	None	Dead-end sac	None
Ctenophora	Biradial	None	Complete	None
Platyhelminthes	Bilateral	None	Dead-end sac	None
Nemertea	Bilateral	Rhynchocoel	Complete	None
Nematoda	Bilateral	Pseudocoelom	Complete	None
Rotifera	Bilateral	Pseudocoelom	Complete	None
Pogonophora	Bilateral	Coelom	None	None
Annelida	Bilateral	Coelom	Complete	Closed or open
Arthropoda	Bilateral	Coelom	Complete	Closed or open
Mollusca	Bilateral	Reduced coelom	Complete	Open except in cephalopods

stomes, the subject of the next chapter, will help to enrich our understanding of the major factors influencing animal evolution and the patterns they have produced.

SUMMARY

Members of the kingdom Animalia are multicellular, heterotrophic eukaryotes with ingestive nutrition. They are believed to have arisen independently from protistan flagellates. The millions of animal species are all built on two basic development patterns and a small number of *Baupläne*. Slight modifications of those *Baupläne* have yielded an amazing variety of forms. For example, animals with long, cylindrical, wormlike *Baupläne* include some species that burrow through soil, species that burrow into marine substrates, and species that live inside the bodies of plants and other animals. The tripartite (foot, mantle, visceral mass) *Bauplan* of mollusks has been modified

to produce such diverse animals as chitons, snails, clams, squids, and octopuses, which exploit many different marine, fresh water, and terrestrial environments.

The first multicellular animals were simple organisms. As evolution proceeded, animals in many lineages became larger and more complex. Some components of complexity are: the presence of many tissue types, body cavities (initially simple and subsequently subdivided), locomotory appendages, and the ability to swim and fly. These traits enabled animals to protect themselves better from predators and from their physical environment, and to exploit a wide variety of food resources. At the same time, the evolution of larger and more complex animals provided new food supplies for still other animals, resulting in an evolutionary arms race that continues to this day. Despite the evolution of many species of complex animals, many species of very small, simple animals survive today.

SELF-QUIZ

1. The *Bauplan* of an animal is:
   a. its body plan.
   b. the functional interrelationship of its parts.
   c. its body plan and the functional interrelationship of its parts.
   d. its body plan and its evolutionary history.
   e. the functional interrelationship of its parts and its evolutionary history.

2. A bilaterally symmetrical animal can be divided into mirror images by:
   a. any cut through the midline of its body.
   b. any cut from its anterior to its posterior end.
   c. any cut from its dorsal to its ventral surface.
   d. only a cut through the midline of its body from its anterior to its posterior end.
   e. only a cut through the midline of its body from its dorsal to its ventral surface.

3. Among protostomes, cleavage of the fertilized egg is:
   a. delayed while the egg continues to mature.
   b. determinate; that is, cells separated after a few divisions develop into only partial embryos.
   c. indeterminate; that is, cells separated after a few divisions develop into complete embryos.
   d. triploblastic.
   e. diploblastic.

4. In which of the following ways did early communities differ from modern ones?
   a. All organisms were small and engaged in sexual reproduction.
   b. Some organisms were large but none of them engaged in sexual reproduction.
   c. Some organisms were large but all of them lived on the surface of the substrate.

d. All organisms were small; they lived both on the surface and below it.

e. All organisms were small and lacked sexual reproduction.

5. The sponge *Bauplan* is characterized by:
   a. a mouth and digestive cavity but no muscles or nerves.
   b. muscles and nerves but no mouth or digestive cavity.
   c. a mouth, digestive cavity, and spicules.
   d. muscles and spicules but no digestive cavity or nerves.
   e. no mouth, digestive cavity, muscles, or nerves.

6. The phyla of diploblastic animals are:
   a. Porifera and Cnidaria.
   b. Cnidaria and Ctenophora.
   c. Cnidaria and Platyhelminthes.
   d. Ctenophora and Platyhelminthes.
   e. Porifera and Cnidaria.

7. The success of cnidarians is believed to be due to:
   a. their ability to live in both salt and fresh water.

b. their ability to move rapidly in the water column.
   c. their ability to capture and consume a large number of small prey.
   d. their low metabolic rates and their ability to capture large prey.
   e. their ability to capture large prey and to move rapidly.

8. Which of the following are coelomate protostome phyla?
   a. Rotifera, Pogonophora, Annelida, and Arthropoda
   b. Pogonophora, Annelida, Arthropoda, and Nematoda
   c. Pogonophora, Annelida, Arthropoda, and Mollusca
   d. Rotifera, Annelida, Arthropoda, and Mollusca
   e. Nematoda, Rotifera, Pogonophora, and Arthropoda

9. Insects that hatch from eggs into juveniles similar to the adults are said to have:
   a. instars.
   b. neopterous development.
   c. simple development.
   d. incomplete metamorphosis.
   e. complete metamorphosis.

10. Which of the following is *not* part of the molluscan *Bauplan*?
    a. Mantle
    b. Foot
    c. Visceral mass
    d. Radula
    e. Jointed skeleton

11. Many lineages of protostomes evolved feeding structures designed to extract small prey from water because:
    a. over much of the time of protostome evolution, the only food available was dissolved organic matter and very small organisms.
    b. over much of the time of protostome evolution, small animals were more abundant than large animals.
    c. large animals were available as food but they were difficult to capture.
    d. to be successful in competition for space, protostomes had to feed on small prey.
    e. water flows naturally over their feeding structures, so early protostomes did not have to work to get food.

FOR STUDY

1. Differentiate among the members of each of the following sets of related terms.
   a. radial symmetry/bilateral symmetry
   b. protostome/deuterostome
   c. indeterminate cleavage/determinate cleavage
   d. spiral cleavage/radial cleavage
   e. enterocoelous/schizocoelous
   f. coelomate/pseudocoelomate/acoelomate

2. For each of the types of organisms listed below, give a single trait that may be used to distinguish them

from the organisms in parentheses.
   a. cnidarians (sponges)
   b. gastropods (all other mollusks)
   c. polychaetes (other annelids)
   d. ribbon worms (roundworms)

3. Segmentation has arisen a number of times during protostomate evolution. What advantages does segmentation provide? Given these advantages, why do so many unsegmented animals survive?

4. Many animals extract food from the surrounding medium. What protostomate phyla contain animals that extract suspended food from the water column?

5. Name the phyla in which animals form large colonies of attached individuals. What advantages does coloniality provide to animals?

6. Discuss the structures animals have evolved to enable them to capture large prey.

7. A major factor influencing the evolution of animals was predation. What major animal features appear to have evolved in response to predation? How do they help their bearers avoid becoming a meal for some other animal?

READINGS

Barnes, R. D. 1987. *Invertebrate Zoology*, 5th Edition. Saunders, Philadelphia. One of the best general textbooks, covering all groups from the animal-like protists to the invertebrate chordates.

Borror, D. J., C. A. Triplehorn, and N. F. Johnson. 1989. *An Introduction to the Study of Insects*, 6th Edition. Saunders, Philadelphia. A clearly written introduction to the richness of the insects.

Brusca, R. C. and G. J. Brusca. 1990. *Invertebrates*. Sinauer Associates, Sunderland, MA. A thorough account of the invertebrates that provides detailed treatments of *Baupläne* and includes excellent discussions of phylogenies.

Cloudsley-Thompson, J. L. 1976. *Insects and History*. St. Martins Press, New York. A vivid account of instances in which insects have affected human so-

ciety; special emphasis on insects as carriers of disease organisms.

Noble, E. R. and G. A. Noble. 1982. *Parasitology: The Biology of Animal Parasites*, 5th Edition. Lea & Febiger, Philadelphia. A general text covering the major parasitic protists and worms that affect animals, including people.

26

Deuterostomate Animals

PREVIEW: The deuterostome animal lineage, stemming perhaps from ancestral flatworms, led to the echinoderms, chordates, and their relatives. In this lineage, which probably was initially dependent primarily upon plankton as a source of food, a coelom evolved and became divided into several compartments. Most of the groups maintained a relatively sedentary existence, but one offshoot, the chaetognaths, became motile predators. The sedentary lineages gave rise to the echinoderms, marine animals with radial symmetry and calcified skeletons, and the chordates, in which swimming adults evolved from a suspension-feeding larval form. This lineage later colonized the land, giving rise to the tetrapod vertebrates, the most conspicuous land animals today. A recent mammalian lineage gave rise to humans, now the dominant animal on Earth.

This chapter deals with the deuterostome lineages, the evolution of the lophophore and the pharynx as devices for filtering prey from the water, internal skeletons, the evolution of aquatic and terrestrial vertebrates, and the recent evolution of *Homo sapiens*.

As we saw in Chapter 25, early embryological development patterns suggest that members of the kingdom Animalia separated into two major lineages early in the evolution of animals from protist ancestors. The deuterostome lineage, which we will consider in this chapter, is characterized by indeterminate cleavage in the early embryo, formation of the mesoderm from an outpocketing of the embryonic gut, and a blastopore that becomes the anus (see Table 25.2). The ancestry of the deuterostomes is not clearly established because by the time the first deuterostomes appear in the fossil record they had already split into several identifiable lineages. Some of these followed a sessile way of life, extracting suspended prey from seawater. All deuterostomes are quite similar to one another in their basic body structure: They are all triploblastic and have well developed body cavities (Table 26.1). However, those cavities typically do not function as hydrostatic skeletons as they do in so many protostome phyla. There are not nearly as many species of animals in the deuterostome phyla as there are among protostomes, but the deuterostome phyla are more variable in their *Baupläne* than are the phyla of protostomes.

TRIPARTITE DEUTEROSTOMES

One group of animals that clearly belongs to the deuterostome lineage nonetheless has many resemblances to some protostomate animals. Generally known as lophophorate animals (Greek, "crest bearers"), they have a tripartite *Bauplan* in which the body is divided into three parts: an anterior **prosome**, a middle **mesosome**, and a posterior **metasome**. In most species, each of these regions has a separate coelomic compartment, the **protocoel**, **mesocoel**, and **metacoel**, respectively. Typically these animals secrete some variety of tough outer body covering. All lophophorate animals have a U-shaped gut, with the anus located close to the mouth but outside the tentacles.

The most conspicuous feature shared by these animals is the **lophophore**, a circular or U-shaped ridge around the mouth that bears either one or two rows of ciliated, hollow tentacles (Figure 26.1). This large and complex structure, which is both a food-collecting organ and a surface for the exchange of gases, is held in position and moved by the contraction of muscles surrounding the mesocoel. Adult animals of all species in these groups are sessile and use the cilia of the lophophore to capture phytoplankton and zooplankton.

Nearly all of the members of the three lophophorate phyla are marine; a few live in fresh water. About 4,500 living species are known, but many times that number of species existed during the Paleozoic and Mesozoic eras. Fossils of these animals provide important records of past ecological communities and the evolution of life.

582

TABLE 26.1
General Characteristics of Deuterostomate Animal Phyla

PHYLUM	SYMMETRY	BODY CAVITY	DIGESTIVE TRACT	CIRCULATORY SYSTEM
Lophophorate phyla	Bilateral	Coelom	Complete	None in most
Chaetognatha	Bilateral	Coelom	Complete	None
Echinodermata	Biradial	Coelom	Complete	Open or none
Hemichordata	Bilateral	Coelom	Complete	Closed
Chordata	Bilateral	Coelom	Complete	Closed

Phylum Phoronida

There are only about a dozen species of phoronids (phylum **Phoronida**), worms that live in muddy or sandy sediments or attached to rocky substrates. They are found from intertidal waters to depths of about 400 meters. Phoronids live in chitinous tubes they secrete, and their most conspicuous external feature is the lophophore (Figure 26.2*a*). Cilia drive water into the top of the lophophore. Water exits through the narrow spaces between the tentacles.

Suspended food particles adhere to the mucus on the cilia and are transported, also by ciliary action, to the food groove and into the mouth. At the opposite end of the body is an inflated area that contains the stomach and helps to anchor the animal to the substrate.

Phylum Ectoprocta

Moss animals (phylum **Ectoprocta**) are colonial lophophorate animals that live in a "house" secreted

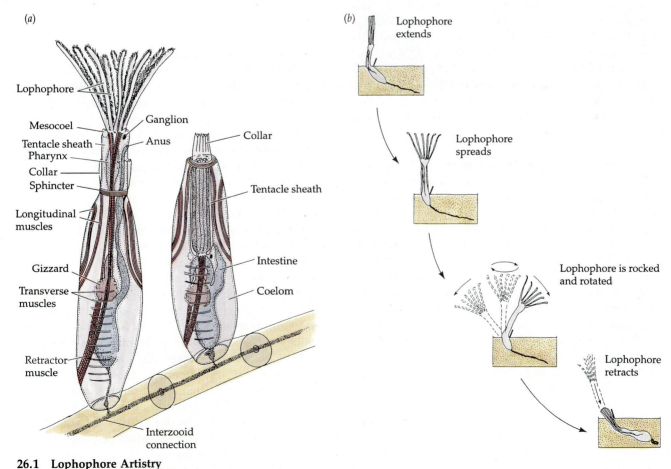

26.1 Lophophore Artistry
(*a*) Expanded or retracted, the lophophore dominates the anatomy of an ectoproct moss animal. (*b*) The lophophore in action. To increase contact with prey, moss animals rock and rotate the lophophore.

(a)

(c)

(b)

26.2 Lophophorate Animals
(a) These *Phoronis* sp. from California have their lophophores extended in the feeding position. (b) A colony of freshwater moss animals, *Cristatella mucedo*. Notice the fine structure of the lophophores. (c) The stalk of this North Pacific brachiopod, *Laqueus* sp., is barely visible to the left of the shell. You can see the lophophore between the valves of the shell.

by the body wall. They are called moss animals because of the plantlike appearance of their colonies (Figure 26.2b). Most are marine animals but a few live in fresh water. An individual moss animal and its house are referred to as a **zooid**. An entire colony of zooids is called a **zoarium**. The body cavities of zooids are connected to one another by cords of tissues along which materials can be moved. Moss animals are the only lophophorates able to protract and retract their lophophores. A moss animal can rock and rotate the extended lophophore as it feeds to increase contact with prey in the water (Figure 26.1).

A zoarium is created by the asexual reproduction of its founding member. Colonies of different species vary in shape, and one colony may contain up to 2 million zooids. In some species, individual colony members are specialized in anatomy and function, some being concerned solely with feeding and others with reproduction, defense, or support. Sperm are released into the external environment and are collected from the surrounding water by other individuals. Eggs are fertilized internally, and developing embryos are brooded for a while before exiting as larvae to seek suitable sites for attachment.

Phylum Brachiopoda

Brachiopods, or lamp shells (phylum **Brachiopoda**), are solitary marine lophophorate animals that superficially resemble clams (Figure 26.2c). Brachiopods possess a mantle and a calcareous shell with two opposing valves, which can be pulled shut to protect the soft body inside. The shell differs from that of clams in that the two halves move dorsoventrally rather than laterally. The lophophore of a brachiopod is extended into two arms. Unlike the lophophores of phoronids and moss animals, it is positioned within the protection of the shell. Water is drawn into the slightly opened shell by the beating of cilia on the lophophore. Food is trapped in the lophophore and directed to a ridge along which it is transferred to the mouth.

Nearly all brachiopods are permanently attached to a solid substrate or are imbedded in soft sediments. Most species are attached by means of a long, flexible stalk that holds the animal above the substrate. Gas exchange takes place across nonspecialized body surfaces, especially the tentacles of the lophophore. Most brachiopods release their gametes into the water, where fertilization takes place. The larvae, which resemble the adults, remain in the plankton for only a few days before settling and metamorphosing into adults.

Brachiopods reached their peak in Paleozoic and Mesozoic times. More than 12,000 fossil species have been described. Only about 350 species remain today, but they are common in some marine environments. The decline of the brachiopods is sometimes attributed to competition from mollusks, animals that exploit similar prey. This is, however, speculation; we may never know what caused the precipitous decline of brachiopods at the end of the Mesozoic.

MODIFYING THE LOPHOPHORE

Evolution among one lineage of deuterostomes resulted in several different modifications of the lophophore and the coelomic cavity surrounding it. Living representatives of this lineage are all wormlike animals that live buried in marine sands or muds, under rocks, or attached to algae.

Phylum Hemichordata

The hemichordates (phylum **Hemichordata**) all have a tripartite *Bauplan*. The three regions of the body, called the proboscis, collar, and trunk, are believed to be homologous to the prosome, mesosome, and metasome of lophophorate animals. The animals in the two highly divergent groups within this phylum are adapted to different ways of capturing food.

The pterobranchs (class **Pterobranchia**) are believed to have changed relatively little from the ancestors of their lineage. They are colonial worms that live in tubes secreted by their proboscises. Surrounding the proboscis is a collar with 1–9 pairs of arms bearing long tentacles that serve as both prey-capturing and gas-exchange structures. The digestive tract is U-shaped, with the anus emptying next to the tentacles. The proboscis is surrounded by a coelomic cavity that has a pair of openings to the exterior through which undigestible particles ingested during feeding can be extruded.

In the other hemichordate lineage, the lophophore was reduced and eventually lost and the proboscis grew larger and became a digging organ. The survivors of this lineage are the acorn worms (class **Enteropneusta**). These animals live in burrows in muddy or sandy sediments (Figure 26.3). The enlarged proboscis is coated with a sticky mucus that traps prey items and is then conveyed by ciliary action to the mouth. In the esophagus, the food-laden mucus is compacted into a ropelike mass that is moved through the digestive tract by ciliary action. Behind the mouth is a pharynx that opens to the outside through a number of slits, more than 100 pairs in some species. These openings probably originally functioned as exits for sand, but they are expanded greatly among acorn worms and are used as a gas-exchange apparatus. Thus an acorn worm breathes with the anterior end of its gut by pumping water into its mouth and out through the gill slits.

26.3 An Acorn Worm
This *Saccoglossus kowaleskii* has been extracted from its burrow so that you can see it; its proboscis is at the upper right.

ADAPTATIONS TO PREDATORY LIFE

Adult lophophorate animals attach to the substrate and extract prey from the water either by creating currents that bring prey to the lophophore or by extending the lophophore and moving it through the water. However, in one lineage of deuterostomes, the tripartite *Bauplan* characteristic of lophophorate animals was modified for active pursuit of prey in the open water.

Phylum Chaetognatha

Arrow worms (phylum **Chaetognatha**) are tripartite, streamlined, bilaterally symmetrical animals. Most of them swim in the open sea, but a few live on the sea floor. Arrow worms have a long fossil history. Their abundance as fossils indicates that they were already common more than 500 mya (million years ago). The 100 or so known species of arrow worms are small marine carnivores all less than 10 centimeters long (Figure 26.4). They are so small that their gas-exchange and excretion requirements can be met by diffusion through the body surface. Arrow worms lack a circulatory system. Wastes and nutrients are moved around the body in the coelomic fluid, which is propelled by cilia that line the coelom. The *Bauplan* is based on a coelom that is divided into head, trunk, and tail compartments. There is no distinct larval stage. Miniature adults hatch directly from eggs that are released into the water.

Arrow worms, which are nearly transparent, typically lie motionless in the water until movement of the water signals the approach of a prey item. At that time the arrow worm darts forward and grasps the prey with stiff spines adjacent to its mouth. Arrow worms are stabilized in the water by means of one or two pairs of lateral fins and a taillike, or caudal, fin. Arrow worms are not powerful enough to swim against strong water currents, but a number of species undertake diurnal vertical migrations of up

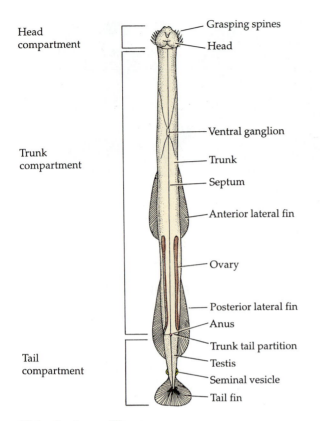

Head compartment

Trunk compartment

Tail compartment

Grasping spines

Head

Ventral ganglion

Trunk

Septum

Anterior lateral fin

Ovary

Posterior lateral fin

Anus

Trunk tail partition

Testis

Seminal vesicle

Tail fin

26.4 An Arrow Worm
Notice the tripartite *Bauplan*. The fins and grasping spines are powerful adaptations for a predatory life.

to several hundred meters, moving to deeper water during the day and back to the surface at night. (Similar vertical migrations are undertaken by many other small oceanic animals, probably because surface waters are dangerous during the day when large visually hunting predators are active.)

CALCIFYING THE SKELETON

Bilateral symmetry, as we have seen, is characteristic of animals that move about actively. Radial symmetry is found primarily among sessile or slow-moving animals. In several of the deuterostome phyla we have been discussing, planktonic larvae settle on substrates and metamorphose into sessile adults whose symmetry is roughly biradial. In another lineage, which originated with a Precambrian burrowing deuterostome, radial symmetry evolved in association with a complex *Bauplan*. The result was one of the most successful adaptive radiations in animal evolution, which, surprisingly, combines radial symmetry with locomotion.

Phylum Echinodermata

This remarkable lineage gave rise to the echinoderms (phylum **Echinodermata**), among the most unusual

of living organisms and the most structurally complex of all radially symmetrical animals. One of the two main changes during the evolution of the echinoderms was the calcification of an internal skeleton, which gave the animals protection against predators. The calcified plates of the early ancestors became enlarged and thickened until they fused inside the entire body. The plates are covered by thin layers of skin and some muscles.

The second major change was the evolution of the **water vascular system**, a series of seawater channels and spaces derived by the enlargement and extension of one of the three coelomic cavities of the ancestral forms—in most species, the mesocoel. The water vascular system is a network of hydraulic canals leading to extensions called **tube feet** that function in gas exchange, locomotion, and feeding. Seawater enters the water vascular system through a sievelike pore, the **madreporite**. In most species a calcified canal, the stone canal, leads from the madreporite to another canal, the ring canal, which rings the esophagus. From the ring canal, other canals radiate out, extending through the arms of the species that have arms and connecting with the tube feet (Figure 26.5*a*). Water is moved through the water vascular system by the contraction of muscles, especially those around the tube feet.

Echinoderms have an extensive fossil record. About 23 classes have been described, of which only six survive today. There are about 6,000 species of modern echinoderms, but an additional 13,000 species, probably only a small fraction of those that actually lived, have been described from their fossil remains. Nearly all living species have a bilaterally symmetrical, ciliated larva (Figure 26.5*b*) that feeds for a while as a planktonic organism before settling and transforming into a radially symmetrical adult. Living echinoderms are divisible into two lineages: subphylum Pelmatozoa and subphylum Eleutherozoa (Figure 26.6). The two groups differ in the forms of their water vascular systems and in the number of arms they have.

PELMATOZOANS. In the only surviving pelmatozoan lineage (class **Crinoidea**), the lophophore became calcified along with the rest of the body. The two groups within the lineage are sea lilies and feather stars. Sea lilies were abundant 300–500 mya, but only about 80 species survive today. Sea lilies attach to the substrate by means of a flexible stalk consisting of a stack of calcareous discs. The rest of the animal is attached to the stalk by a cup-shaped structure containing a tubular digestive system. From five to several hundred arms, usually in multiples of five, extend outward from the cup. Because the calcareous plates of the arms are jointed, the arms can bend. A groove runs down the center of each arm. On both sides of the grooves are tube feet covered with mucus-secreting

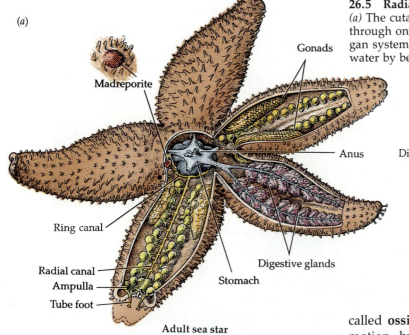

Madreporite

Gonads

Anus

Ring canal

Radial canal

Digestive glands

Ampulla

Stomach

Tube foot

Adult sea star

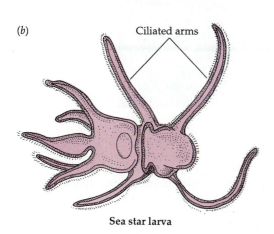

Ciliated arms

Sea star larva

26.5 Radial but Locomotory
(*a*) The cutaway view of a sea star and a cross section through one of its arms reveal its internal organs and organ systems. (*b*) The sea star larva moves through the water by beating its cilia.

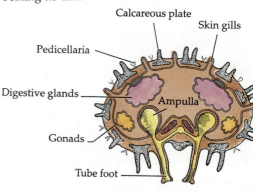

Calcareous plate

Skin gills

Pedicellaria

Digestive glands

Ampulla

Gonads

Tube foot

Cross section through arm

glands. A sea lily feeds by moving its arms around by muscular contraction. Food particles stick to the tube feet, are transferred to the groove, and are carried to the mouth by ciliary action. The tube feet of sea lilies are also used for gas exchange and elimination of nitrogenous wastes.

Feather stars (Figure 26.6*a*) are similar to sea lilies except that they have flexible appendages with which they grasp the substrate while they are feeding and resting. Feather stars can walk on the tips of their arms or swim by rhymically beating their arms. Feather stars feed in much the same manner as sea lilies. About 600 species of feather stars have been described.

ELEUTHEROZOANS. Most surviving echinoderms are members of the eleutherozoan lineage. Some, such as the brittle stars (class **Ophiuroidea**), are similar in structure to crinoids. The flexible arms of brittle stars (Figure 26.6*b*) are composed of jointed calcified plates

called **ossicles**. The arms are thrashed during locomotion, but the tube feet play some locomotor role in burrowing forms and in young individuals of most species. Brittle stars generally have five arms, but each arm may branch a number of times. Unlike most other members of the phylum, brittle stars have only one opening to the digestive tract. Most of the 2,000 species of brittle stars are deposit feeders. They ingest sediments, assimilate the organic material from them, and eject the remainder through their mouths. Some species remove suspended food particles from the water, and still others capture small animals.

The most familiar echinoderms are the sea stars (class **Asteroidea**; Figure 26.6*c*). Sea stars are similar in basic form to brittle stars but their arms are not as flexible. Many sea stars are predators on polychaetes, gastropods, bivalves, and fishes. Each tube foot of a sea star is a little adhesive organ consisting of an internal bulb connected by a muscular tube to an external sucker. The tube foot creates suction by hydraulic expansion and contraction. When the bulb contracts, the tube foot elongates. When it touches a surface, the sucker draws back slightly, creating a small vacuum. Adhesion is simultaneously increased by the secretion of a sticky substance around the sucker. With hundreds of tube feet acting simultaneously, a sea star can exert enormous force. It can grasp a clam in its arms, anchor the arms with tube feet, and, by steady contraction of the muscles in the arms, gradually exhaust the clam's muscles and pull the shell apart. Tube feet serve as organs of locomotion, and because of their thin walls, they also are important sites for gas exchange.

Sea star species that feed on bivalves are able to push their stomachs out through their mouths and then through the narrow spaces between the shells of a bivalve. Digestive enzymes are secreted into the soft parts of the bivalve, and the animal is digested

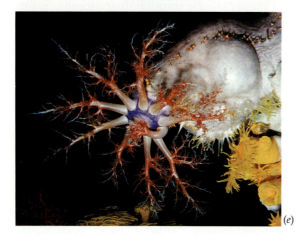

26.6 A Diversity of Echinoderms
(a) You can clearly see the flexible arms of these feather stars. (b) This brittle star, *Ophiothrix swensonii*, is resting on a pink sponge. (c) The rainbow sea star *Orthasterias koehleri* is typical of many sea stars; some species, however, have more than five arms. (d) Purple sea urchins (*Strongylocentrotus purpuratus*) are important grazers of algae in the intertidal zone of the Pacific Coast of North America. (e) This tropical sea cucumber, *Cucumaria* sp., is actively retracting its feeding tentacles so that it can digest the food particles that have adhered to them.

and consumed. Other species feed on smaller prey or suspended particles and do not extrude their stomachs. Sea stars are important predators in many marine environments, such as coral reefs and rocky intertidal zones, where they strongly influence the species composition of animal communities.

OTHER ECHINODERMS. The remaining three classes of echinoderms lack arms. The sea daisies (class **Concentricycloidea**) were only recently discovered, and little is known about them. They have round bodies with a ring of marginal spines, but no arms. Their tube feet, which lack suckers, are arranged in a ring near the margin of the body. Sea daisies are the only echinoderms in which the water vascular system has two ring canals and in which the tube feet are arranged in a peripheral circle rather than along grooves radiating from the center. Sea daisies are found on rotting wood in deep ocean waters, and apparently feed on bacteria, which they absorb either through a membrane that covers the oral surface or by means of a shallow, saclike stomach.

Sea urchins (class **Echinoidea**; Figure 26.6d) are armless, hemispherical animals that are covered with spines attached to the underlying skeleton via ball-and-socket joints. They resemble sea stars with their "arms" folded and fused over their backs. The spines come in varied sizes and shapes, and some produce highly toxic substances. The skeleton of a sea urchin is totally fused together. Many sea urchins are vegetarians, scraping algae from rocks with a complex rasping structure. Others feed on small organic debris collected by their tube feet or spines.

Sea cucumbers (class **Holothuroidea**; Figure 26.6e) resemble stretched-out, flexible sea urchins lacking spines and with greatly reduced skeletal plates. Tube feet are located on either side of five grooves along the body of the animal. They are used primarily for attaching to the substrate rather than for moving. Some sessile species lack tube feet entirely. The tube feet at the anterior end of a sea cucumber are modified into large, feathery tentacles that can be protruded from the mouth. The tentacles are coated with a sticky substance to which prey or the surrounding substrate adhere. Periodically, the sea cucumber retracts its tentacles into the body and then digests the adhered material.

EVOLUTION OF THE PHARYNX

Another deuterostome lineage evolved a different way of exploiting the abundant food provided by marine phytoplankton and zooplankton. The structures that enlarged in this lineage to become a device for removing plankton from the water are the **pharyngeal gill slits** which originally functioned simply as sites for gas exchange. The requirement for effective gas exchange—a large surface area—also serves well for prey capture; indeed, the lophophore serves this dual function in many lophophorate animals. In the lineage we will now discuss, enlargement of the gill slits eventually led to remarkable evolutionary developments that produced animals quite unlike members of any other animal phyla.

Phylum Chordata

The loss of the lophophore, combined with the enlargement of the pharyngeal gill slits as a substitute feeding device, was the direction taken in the evolutionary lineage leading to the chordates (phylum **Chordata**). Chordates are bilaterally symmetrical animals that have gill slits at some stage in their development. The main features of their *Bauplan* are an internal skeleton, a dorsal, hollow nervous system, and a ventral heart. Skeletons vary among chordate groups, but all species have a dorsal supporting rod, the **notochord**, at some stage during their development. In some species, the notochord is lost during metamorphosis to the adult stage. In other species, it is replaced by other skeletal structures having the same function.

The tunicates (subphylum **Urochordata**) may be related to the ancestry of the chordates. The 2,500 species of tunicates are all marine animals, most of which are attached to the substrate as adults. Swimming tadpolelike larvae are the dispersal stage of the tunicate life cycle. These larvae reveal the close evolutionary relationships between the tunicates and other chordates. In addition to its pharyngeal gill slits, a tunicate larva has a dorsal, hollow nerve cord and a notochord. The notochord serves as a relatively rigid support and as an attachment for muscles. After a short time in the plankton, the larva settles on the bottom and transforms into a sessile adult that feeds by extracting plankton from the water with its pharynx, which is enlarged into a pharyngeal basket (see Figure 20.10).

More than 90 percent of known species of tunicates are sea squirts (class **Ascidiacea**). The baglike bodies of the adults are surrounded by a tough tunic, composed of protein and a complex polysaccharide, secreted by the epidermal cells. Much of the body is occupied by the large pharyngeal basket, which is lined with cilia whose beating moves water through the animal. These cilia also move the thin layer of mucus that lines the basket and to which the food particles adhere. Water enters the body through an anterior opening (the buccal siphon), passes through the pharyngeal basket into a chamber, called the atrium, that is enclosed by the tunic, and flows out through another opening (the atrial siphon), well removed from the site where the water entered. Some sea squirts are solitary (Figure 26.7), but others produce large colonies by asexual budding from a single founder. In some colonial species the individuals each have their own buccal siphon, but they share a single atrial siphon.

In another lineage of tunicates, the larvaceans (class **Appendicularia**), individuals become reproductively mature and complete their life cycles in the plankton. They never settle on the bottom, but swim in the water, filtering prey with their large pharyn-

26.7 Tunicates: Pharyngeal Basket Specialists
The large pharyngeal basket of the sea squirt *Ciona intestinalis* occupies most of its body cavity.

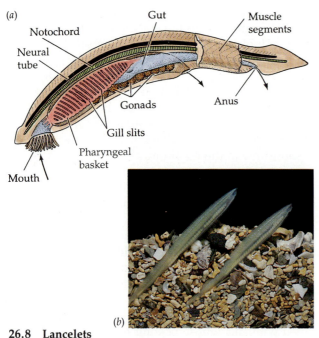

(a)

Notochord

Neural tube

Gut

Muscle segments

Gonads

Anus

Gill slits

Pharyngeal basket

Mouth

(b)

26.8 Lancelets
(a) Cutaway diagram showing internal structure. Note the large pharyngeal basket (shown in red) with gill slits.
(b) The anterior ends of two adult lancelets (*Branchiostoma lanceolatum*) protrude from shell gravel.

geal baskets. There are only a few species of larvaceans, but they are widespread in the world's oceans.

The lancelets (subphylum **Cephalochordata**) are small, fishlike animals that live partly buried in soft marine sediments and extract small prey from the water with their pharyngeal baskets (Figure 26.8). The notochord of a lancelet extends the entire length of the body throughout its life.

SUCKING MUD: THE RISE OF THE VERTEBRATES

The pharyngeal gill basket is a very efficient means of extracting prey from surrounding water. Because of the many exit openings through the basket, the system also is effective in very muddy situations where many inedible particles are inevitably ingested together with the food. In the late Cambrian, over 500 mya, early chordates evolved improved abilities to extract food from mud and sand. Their descendants evolved a jointed, dorsal **vertebral column** that functionally replaced the notochord as the primary support. These chordates—the **vertebrates**—also evolved external armor that enabled them to live above the substrate, where predators, principally arthropods at that time, were abundant, rather than having to burrow.

A key to the *Bauplan* of most vertebrates (Figure 26.9) is the vertebral column, which typically supports a skull in front and two pairs of appendages. These structures evolved after the vertebrates first

appeared. Vertebrates have a large coelom in which the body organs are slung, but it serves neither as a hydrostatic skeleton nor as a gas-exchange structure. Oxygen is carried to internal organs through a well-developed circulatory system, driven by the contractions of a ventral heart.

The early vertebrates swam over the bottom, sucking mud as they went. They were the jawless fishes (class **Agnatha**), for which a fossil record exists. These early vertebrates, called ostracoderms, were small animals typically between 6 and 30 centimeters in length. The name means "shell-skinned," referring to their bony external armor. With their heavy armor, these fishes could swim only slowly, but swimming above the substrate was easier than having to burrow through it as all previous sediment feeders had done. This new mobility may have been the major breakthrough that enabled vertebrates to exploit their environments in a number of new ways. One of those ways was to attach to carrion and use the gills to create suction to pull fluids and partly decomposed tissues into the mouth.

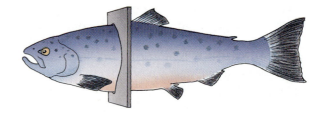

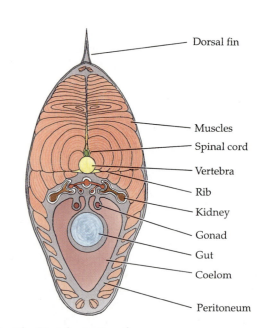

Dorsal fin

Muscles

Spinal cord

Vertebra

Rib

Kidney

Gonad

Gut

Coelom

Peritoneum

26.9 The Vertebrate *Bauplan*
The key elements in the vertebrate *Bauplan*—internal skeleton, dorsal nervous system (spinal cord), organs slung in the coelom, and segmented muscles—are diagrammed for a bony fish.

26.10 Agnaths
These sea lampreys (*Petromyzon marinus*) have attached to a carp with their large, sucking mouths and are rasping away at its flesh.

Modern fishes that feed this way are the lampreys and hagfishes (Figure 26.10), the only jawless fishes to survive beyond the Devonian period. These fishes lack an external armor, having a tough scaly skin instead. Hagfishes ingest the tissues of dead animals whereas lampreys suck the blood of living fishes or eat the flesh of dying fishes. The round mouth is a sucking organ with which the animals attach to their prey and rasp at the flesh. Hagfishes are entirely marine, but lampreys live in both fresh and salt water.

Jaws: A Key Evolutionary Novelty

In the Devonian period, often referred to as the Age of Fishes (even though fishes are still the most numerous and species-rich group of vertebrates today), an immense variety of new kinds of fishes evolved in the seas and fresh waters. Many of these were jawless, but in one lineage, jaws evolved from some of the cartilaginous or bony hyoid arches that supported the gill region (Figure 26.11). The advantage of a jaw is that it allows a living prey item to be grasped while its tissues are being consumed. Further development of the jaws and teeth led to the ability to chew both soft and hard body parts of prey. Although many intermediates must have existed between jawless fishes and the fully jawed ancestors of modern fishes, it is not difficult to imagine how each stage would have functioned better than those that preceded it.

One group representing the early jawed fishes were the heavily armored placoderms (class **Placodermi**). Some of them evolved elaborate fins and sleek body forms that must have improved their maneu-

verability in open water. A few attained huge size and were probably, together with squids, the most important predators in the Devonian oceans. Despite this early success, however, most placoderms disappeared by the end of the Devonian period, and none survived to the end of the Paleozoic era.

Fins and Mobility

Two other groups of fishes that still survive today became numerically important during the Devonian period. Members of one group, the sharks, skates and rays, and chimaeras—the cartilaginous fishes (class **Chondrichthyes**; Figure 26.12)—have a skeleton composed entirely of a firm but pliable material called **cartilage**. Their skin, unlike that of many early jawless fishes and placoderms, is not armored. For the most part it is flexible and leathery, sometimes bearing bristly projections that give it the consistency of sandpaper. The loss of external armor was favored

(a) Jawless fishes
(agnaths)

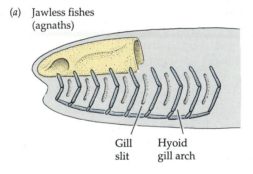

(b) Early jawed fishes
(placoderms)

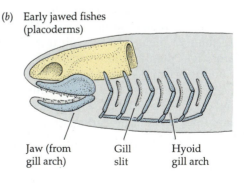

(c) Modern jawed fishes
(cartilaginous and bony fishes)

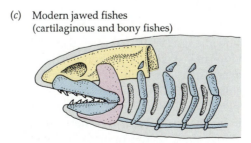

26.11 Jaws from Gill Arches
A possible stage in the evolution of jaws from the anterior gill arches is shown in (b).

(a)

(b)

26.12 Cartilaginous Fishes
(a) Most sharks, such as the seven-gill shark *Notorhyncus cepedianus*, are active predators living in open waters. (b) Skates and rays, represented here by the bullseye electric ray *Diplobatis ommata*, feed on the ocean bottom and have their mouths on the ventral surface of their body.

because it increased mobility and, hence, ability to escape from predators; indeed, reduction in armor and rapid swimming evolved together. Most sharks and their relatives are rapid swimmers. Control of swimming was improved by the evolution of two pairs of appendages, a pair of pectoral fins just behind the gill slits, and a pair of pelvic fins just in front of the anal region that help the fish balance itself in the water as it moves. These are the two pairs of limbs that are another feature of the vertebrate *Bauplan* (Figure 26.9). Forward movement in sharks is provided by the tail and its fins. Among skates and rays, however, the lateral fins are greatly enlarged, and these animals swim with undulating movements of those fins.

Most sharks are predators, but others evolved to be strainers of plankton. Indeed, the world's largest fish, the whale shark (*Rhincodon typhus*), which may grow to more than 15 meters and weigh more than 9,000 kilograms, is a plankton strainer. Most skates and rays live on the ocean floor and feed on mollusks and other invertebrates buried in the sediments. The chimaeras feed on mollusks, whose shells they crack with their hard, flat teeth. The cartilaginous fishes originated in the sea, and they only rarely penetrate fresh water.

Mobility and Buoyancy

Most of the early evolution of fishes took place in the oceans, but one lineage evolved in fresh water. These fishes have internal skeletons of bone rather than cartilage, hence their common name, bony fishes (class **Osteichthyes**). In fresh water, oxygen is often in short supply. The gill slits of bony fishes open into a single chamber covered by a hard flap called the **operculum**. Movement of the operculum improves the flow of water over the gills and brings more oxygen in contact with the gas-exchange surfaces.

Early bony fishes also evolved lunglike sacs that supplemented the gills in respiration. As we will see in the next section, this evolutionary step was important for another lineage of fishes, which invaded the land. However, in most fishes, the lungs evolved into **swim bladders** which serve as organs of buoyancy that help keep the fish suspended in water. By adjusting the amount of gas in its swim bladder, a fish can control the depth in the water column at which it is stable. Today only a small group of species, the lungfishes, still uses the lungs for their original purpose of respiration. The external armor of bony fishes is greatly reduced, but most species are covered with flat, smooth, thin scales that provide some protection while being light in weight. With their light skeletons and their swim bladders, some bony fishes reinvaded the seas to become major players in marine ecological communities.

Among the more than 20,000 species of bony fishes living today, there is a remarkable diversity of sizes and shapes and ways of making a living (Figure 26.13). The smallest bony fish is a goby that is just over 1 centimeter long as an adult. The largest are ocean sunfishes that weigh up to 900 kilograms. Fishes are adapted to exploit nearly all types of food sources available in fresh and salt water. In the oceans they filter plankton from the water, rasp algae from rocks, eat corals and other colonial invertebrates, dig invertebrates from soft sediments, and prey upon virtually all other vertebrates in the oceans except large whales and dolphins. In fresh water they also eat plankton, devour insects of all aquatic orders, harvest fruits that fall into the water in flooded forests, and prey upon other aquatic vertebrates. Many live buried in soft sediments where they grab passing prey or from which they emerge at night to feed in the water column above. Many are solitary, but others form large schools in open water.

With their fins and swim bladders, fishes can read-

(a)

(b)

(c)

(d)

(e)

(f)

26.13 Bony Fishes

(a) The lungish *Protopterus aethiopicus* survives the frequent East African droughts by breathing air. *(b) Latimeria* is the sole survivor of a lineage thought to have been extinct. *(c)* The Volga sturgeon *Huso huso*, the major source of Russian caviar, is a survivor of an ancient lineage of bony fishes. *(d)* Salmonids, illustrated here by the rainbow trout *Salmo gairdneri*, are commercially important fishes of the north temperate zone. *(e)* Angelfishes like the emperor angelfish *Pomocanthus imperator* are dominant fishes on tropical coral reefs. *(f)* Several dozen species of surfperches are found in marine waters. The Pacific white surfperch *Phanerodon furcatus* is typical of these species.

ily control their positions in the open water. Their eggs, however, tend to sink. Most fishes attach their eggs to plants or to the substrate. Some species with very small eggs discharge them directly into surface waters where they are bouyant enough to complete their development before they sink very far. Most fishes, however, move to shallow waters to spawn, which is why coastal waters and estuaries are so important in the life cycles of many species. Some fishes, such as salmon, actually abandon salt water for breeding, ascending rivers to spawn in streams and freshwater lakes. In contrast, other species, such as eels, that live most of their lives in fresh water migrate to the sea to spawn there.

BREATHING AIR AND EXPLORING THE LAND

The evolution of lunglike sacs by the early bony fishes set the stage for the invasion of the land by some of their descendants. Early bony fishes probably used their lungs to supplement the gills at times when oxygen levels in the water were low. This ability would also have allowed them to leave the water temporarily when pursued by predators unable to breath air. However, with their rather simple, unjointed fins, bony fishes were unable to do more than flop around on land, as most fishes do today if placed out of water. Evolution of joints in the fins was necessary before fishes were able to move over land to find new bodies of water when those in which they lived dried up or became overpopulated. Later, these fishes began to use terrestrial food sources and became more fully adapted to life on land.

The bony fishes ancestral to land vertebrates were the crossopterygians, or lobe-finned fishes (subclass **Crossopterygii**). As the name suggests, these fishes evolved jointed limbs that enabled them to locomote, albeit clumsily, on land. Lobe-fins flourished from the Devonian period into Mesozoic times. They were thought to have become extinct about 25 million years ago. However, in 1939 a crossopterygian was caught by a commercial fisherman in deep waters off the east coast of Africa. Since that time several dozen specimens of this extraordinary fish, which was given the name *Latimeria chalumnae*, have been taken. *Latimeria*, a predator on other fishes, grows to about 1.5 meters long and weighs up to 82 kilograms (Figure 26.13b). Its skeleton is mostly composed of cartilage and its swim bladder contains fat rather than gases. Its survival provided an unexpected opportunity for anatomists. Because soft organs are rarely preserved as fossils, it was assumed that we could not learn about the details of these organs in lobe-fins. The discovery of *Latimeria* changed all that, and biologists seized the chance for studying lobe-fin anatomy. As a deep-sea fish, *Latimeria* is clearly not in the lineage of lobe-fins that invaded land, but it does reveal the ways in which the limbs of early fishes became jointed.

In and Out of the Water: The Amphibians

During the Devonian period, the amphibians (class **Amphibia**) arose from ancestors that were similar to crossopterygians. The ancestral bony fishes already had lungs with which they could breath air, but amphibians evolved thin skins that they also use for respiration. In addition, the stubby, lobed fins of the crossopterygians gradually evolved into the walking legs of amphibians (Figure 26.14). The design of these legs has remained largely unchanged throughout the evolution of terrestrial vertebrates. Devonian crossopterygians were probably able to crawl from one pond or stream to another by pulling themselves along on their fins, as do some modern species of catfishes and other fishes. The earliest amphibians, which were very fishlike in most of their body structure, merely evolved modifications of this locomotor ability.

About 4,000 species of amphibians live on Earth today, many fewer than the number known only from fossils. Living amphibians belong to three classes (Figure 26.15): the wormlike, tropical, burrowing caecilians (order **Gymnophiona**); frogs and toads (order **Anura** = tailless); and salamanders (order **Urodela** or **Caudata** = tailed). Most species of frogs and toads live in tropical and warm temperate regions, although a few species are found at very high latitudes. Salamanders, in contrast, are more diverse in temperate regions, but many species are found in the tropics, particularly in the mountains where cool, moist conditions prevail.

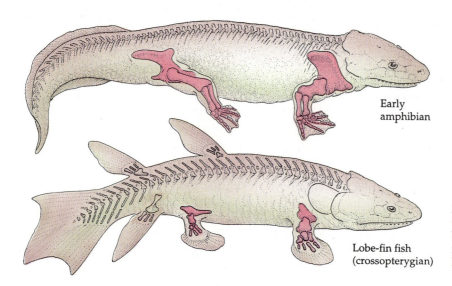

Early amphibian

Lobe-fin fish (crossopterygian)

26.14 Legs from Fins
Only a modest enlargement of the bones of the fins of a lobe-fin fish is needed to convert them into walking legs.

(a)

(c)

(b)

26.15 Amphibians
(a) *Gymnophis mexicana*, a burrowing Central American caecilian, looks more like a worm than an amphibian. (b) Male and female golden toads (*Bufo periglenes*) mate. The population of this amphibian in the cloud forests of Costa Rica has recently declined for unknown reasons. (c) The bright colors of *Tylototriton verrucosus*, a Chinese salamander, indicate to visually hunting predators that it is poisonous.

Most amphibians must live in water at some time in their lives. In the typical life cycle, part or all of the adult life is spent on land, usually in a moist habitat, but adults return to fresh water to lay their eggs (Figure 26.16). An amphibian egg must remain moist in order to develop because it is surrounded by a delicate envelope through which it will lose water readily if the surroundings are dry. The egg of most species gives rise to an aquatic larva that lives in water for some time before metamorphosing into a terrestrial adult.

There are, however, interesting variations on this life cycle. Some amphibians are entirely aquatic, never leaving the water at any stage in their lives. Others are entirely terrestrial, laying their eggs in moist places on land. In many lungless salamanders that live in rotting logs or in the soil, gas exchange takes place entirely through the skin and the lining of the mouth. However, all terrestrial species are confined to moist environments because amphibian skins cannot prevent water loss when exposed to dry air. Some toads have tough skins that enable them to live for periods of time in dry places, but they, too, must return periodically to water.

Attention has recently been drawn to amphibians because populations of many species throughout the world are declining rapidly. For example, the golden toad is disappearing from the Monteverde Cloud Forest Reserve in Costa Rica, a reserve established primarily to protect this rare species. The reasons for the declines are not known, but biologists are monitoring amphibian populations closely in order to learn more about the causes of their difficulties and to determine the implications of amphibian declines for other organisms.

Conquest of the Land

Two morphological changes allowed vertebrates to exploit the full range of terrestrial habitats: The first was a tough skin impermeable to water. The second was an egg whose shell is relatively impermeable to water but is permeable to gases; such an egg can be deposited in relatively dry places. Reptiles (class **Reptilia**) evolved both of these traits and, as a result, were the first vertebrates to become common over much of the terrestrial surface of Earth. They arose from early amphibians in the Carboniferous period, some 300 mya. About 6,000 species live today.

The reptilian egg has a leathery or brittle calcium-impregnated shell that retards the evaporation of the fluids inside. Within the shell and surrounding the embryo are three membranes—the **amnion**, the **chorion**, and the **allantois**—that give protection from desiccation and assist the embryo in excretion and respiration. The embryo is also supplied with large quantities of food—yolk—that permit it to attain a relatively advanced state of development before it hatches and must feed itself (Figure 26.17). Most reptiles do not care for their offspring after the eggs

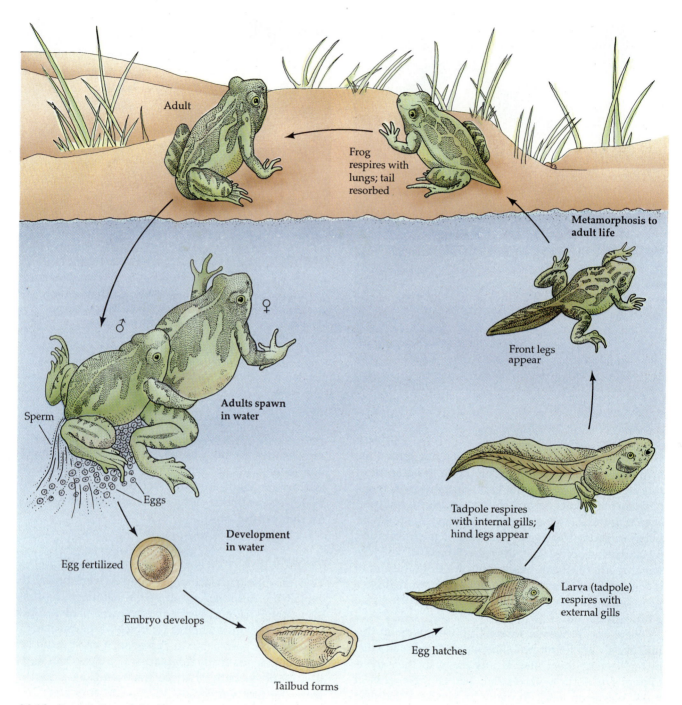

26.16 In and Out of the Water
Most events in the life cycle of a frog take place in the water. Only adults can live on land.

hatch; in fact, most of them desert the eggs once they are laid. In some species, eggs are retained inside the female's body until hatching. Still others have placentae that nourish the developing embryos.

The skin of a reptile is covered with horny keratinized scales that greatly reduce loss of water from the body surface. Thus the skin is unavailable as an organ of gas exchange. That function is performed almost entirely by the lungs, which are much larger than those of amphibians. Air is forced into and out of the lungs by bellowslike movements of the ribs. The reptilian heart is divided into chambers, which separate oxygenated from unoxygenated blood and permit oxygenated blood to be pumped to needy tissues more efficiently. Thus reptiles can sustain higher levels of muscular activity than can amphibians, although they tire much more rapidly than birds or mammals do.

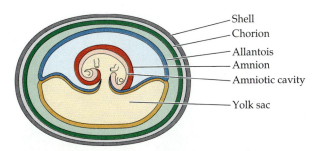

26.17 Reptilian Egg
The evolution of the reptilian egg, with its shell, extraembryonic membranes (amnion, chorion, and allantois), and embryo-nourishing yolk sac, was a major step in the conquest of the terrestrial environment.

Shell
Chorion
Allantois
Amnion
Amniotic cavity
Yolk sac

Modern Reptiles

There are three subclasses of modern reptiles (Figure 26.18). Turtles and tortoises (subclass **Testudinata**) have an armor of bony plates, both dorsally and ventrally, forming a shell into which the head and limbs can be withdrawn. Most turtles live in lakes and ponds, but tortoises are completely terrestrial, and sea turtles spend their entire lives at sea except when they come ashore to lay eggs. Most turtles and tortoises are vegetarians, eating a variety of aquatic and terrestrial plants, but some species are strongly carnivorous.

There are two other reptilian lineages. Only one species of the subclass **Rhynchocephalia** survives—the tuatara (*Sphenodon punctatus*), a lizard-like animal that lives only on a few islands off the coast of New Zealand. The other lineage, the subclass **Squamata**, includes lizards, snakes, and amphisbaenians. The most abundant and familiar reptiles belong to this

(a)

(b)

(c)

(d)

26.18 Reptilian Diversity
(a) This green sea turtle, *Chelonia mydas*, is coming ashore on a beach in Baja California, Mexico, to lay its eggs in the sand just above the high-tide mark. (b) The tuatara *Sphenodon punctatus* looks like a typical lizard, but it is the sole survivor of a lineage that separated from lizards long ago. (c) The collared lizard *Crotaphytus collaris* preys on smaller lizards in the deserts of the southwestern United States. (d) The ringneck snake *Diadophus punctatus* is a nonpoisonous species found in many parts of western North America.

group. Most lizards are insectivores, but some are herbivores, and still others prey on vertebrates. The monitors of the East Indies are the largest lizards, growing as long as 4 meters. Most lizards walk on four limbs, but some are totally limbless and others have small, virtually functionless legs. All snakes are legless, probably having evolved from burrowing lizards. All snakes are carnivores, adapted for swallowing objects much larger than their own diameter. Three groups of snakes have evolved poison glands and inject venom into their prey with their teeth. Snakes range in size up to pythons more than 10 meters long.

Dinosaurs and their Descendants

During the Mesozoic era, one lineage (class **Archosauria**) split from the reptiles and underwent an extraordinary diversification, thus becoming the dominant large animals on land. The Ornithischia and Saurischia—popularly called dinosaurs—were the

26.20 An Early Bird
An artist's recreation of *Archaeopteryx* shows features of both its reptilian relatives and of the modern birds that would come later.

(a)

(b)

26.19 Dinosaurs and Their Descendants
(a) This "terrible claw" dinosaur (*Deinonychus*) lived in Montana during the Cretaceous period. A running predator, it stood about 15 feet tall. (b) Most crocodilians are tropical, but alligators live in warm temperate environments in Asia and, like this *Alligator mississippiensis*, in the southeastern United States.

prevalent large terrestrial reptiles. Plesiosaurs and ichthyosaurs flourished in the seas, and membrane-winged pterosaurs capable of flight took to the skies.

The dinosaurs that capture the imagination of children and adults became extinct. However, their lineage survives as crocodiles, caimans, gharials, alligators, and birds. Crocodilians (subclass **Crocodilia**; Figure 25.19) are confined to tropical and warm temperate environments. They spend much of their time in water, but they build nests on land or on floating piles of vegetation. Eggs are not incubated, but they may be tended by the female until they hatch. Crocodilians are all carnivorous and prey upon vertebrates of all classes, including large mammals.

Zoologists sometimes lightly refer to birds (subclass **Aves**) as "feathered dinosaurs." There is an important truth embodied in that phrase. The early birds that evolved from dinosaurs in the Mesozoic era, such as *Archaeopteryx* (Figure 26.20), were intermediate between ancestral dinosaurs and modern birds in many ways. Although *Archaeopteryx* was covered with feathers and had well-developed wings, the fingers of its forearms were not much reduced, and it had a long tail. It may have been a weak flapping flier, relying much on gliding. Its brain case and breastbone were not significantly modified from the reptilian condition.

The single most characteristic feature of birds is their feathers, which are highly modified versions of reptilian scales (Figure 26.21*a*). The flying surface of the wing is created by large quills that arise from the forearm and from the reduced, stubby fingers (Figure 26.21*b*). Other strong feathers sprout like a fan from the shortened tail and serve as stabilizers during flight. Still other feathers, the contour feathers and down feathers, that arise from well-defined tracts (Figure 26.21*c*), cover the body like a garment and provide insulation to control loss of body heat.

The body skeleton of birds is adapted for flight. Because avian bones are hollow and have internal struts, they are light but strong. The sternum (breastbone) forms a large, vertical keel to which the breast muscles are attached. These muscles pull the wings downward during the main propulsive movement in flight. Flight is metabolically expensive, and a flying bird consumes energy at a very high rate. The lungs have a flow-through pattern that allows a more complete exchange of respiratory gases than does the pattern of the mammalian lung (Chapter 39).

Because birds have high metabolic rates, they also have high body temperatures. Heat is generated by the contraction of muscles. Its rate of loss is controlled by the feathers, which may be appressed to the body or elevated to alter the amount of trapped air they contain and, hence, the amount of insulation they provide. Indeed, avian metabolic rates are so high that a bird uses about eight times the amount of energy per day as a lizard of the same weight! The avian brain is relatively large in proportion to body size, the enlargement being primarily in the cerebellum, the center of sight and muscular coordination. The beaks of modern birds completely lack teeth (truly nothing is scarcer than hens' teeth).

Most birds lay their eggs in a nest where they are incubated by the body heat of the adult. The high body temperatures of birds result in relatively short incubation periods, less than two weeks in many small species. The megapodes of Australia and New Guinea incubate their eggs in a pile of rotting organic matter, using the heat of decomposition to keep the eggs warm. The nestlings of some groups of birds hatch at a relatively helpless stage and are fed for some time by their parents. Such nestlings are termed **altricial**. In other groups, the nestlings are **precocial**; that is, they can run about and feed themselves shortly after hatching. Adults attend most precocial offspring for some time, however, warning them of and protecting them from predators, guiding them to good foraging places, protecting them from bad weather, and, in some cases, feeding them as well. Because birds invest heavily in parental care by laying large, yolk-laden eggs, incubating the eggs, and guarding and feeding the nestlings, they have lower reproductive rates than nearly all groups of animals we have discussed so far.

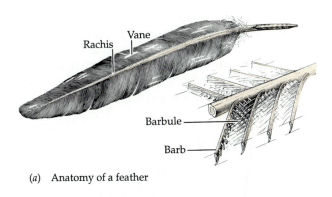

(a) Anatomy of a feather

(b) Arrangement of feathers in wing

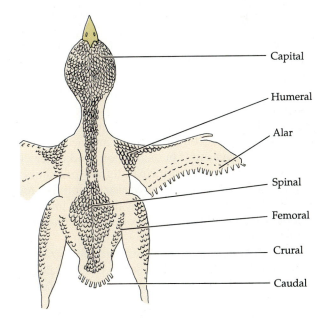

(c) Feather tracts

26.21 Feathers and Wings
(a) A feather is sturdy and light, with a complicated anatomy. (b) Large feathers create flying surfaces. (c) The distinct feather tracts on the dorsal surface of a bird.

(a)   *(b)*   *(c)*   *(d)*

26.22 Birds: Feathered Dinosaurs
(a) Most penguins, like this emperor penguin, live in harsh environments of the Antarctic, although there is one species that breeds as far north as the equator. *(b)* Parrots are most abundant and diverse in Australia and New Guinea. The crimson rosella, a common bird in temperate Australia, feeds primarily on the seed capsules of *Eucalyptus*. *(c)* Most species of birds are found in the order Passeriformes, known as the perching birds or songbirds. The robin (*Turdus migratorius*) is a well-known example. *(d)* The superb starling of East Africa (*Spreo supurbus*), also a passerine songbird, is actually a rather poor songster.

Birds as a group eat almost all types of animal and plant material. A few aquatic species have bills modified for filtering small food particles from the water. In terrestrial environments, insects are the most important dietary item for birds. Indeed, about 60 percent of all species of birds are found in the order **Passeriformes** (the perching birds), the majority of which are primarily or exclusively insectivorous. In addition, however, birds eat fruits and seeds, nectar and pollen, leaves and buds, carrion, and other vertebrates. Birds are the principal diurnal predators of flying insects, and some species also exploit that food source at night. Also, birds are the most important eaters of fruits and dispersers of seeds in many terrestrial ecosystems.

As adults, birds range in size from the 2-gram bee hummingbird of the West Indies to the 150 kilogram ostrich. Some birds of Madagascar and New Zealand known from fossils were even larger, but they were exterminated by early people when they first reached those islands. Although there are more than 8,600 species of living birds, more species than in any other vertebrate group except fishes, birds are less diverse structurally than are other vertebrates, probably be-

cause of the constraints imposed by flying (Figure 26.22). Also, because the avian fossil record is poor, patterns of evolution in this group are not well established. As we saw in Chapter 20, the use of modern molecular methods is resulting in major changes in our ideas about avian phylogenies.

THE ORIGINS OF MAMMALS

Approximately 63 mya, the Mesozoic era—the Age of Reptiles—gave way to the Cenozoic era—the Age of Mammals. However, mammals did not simply arrive on the scene and displace reptiles. On the contrary, mammals (class **Mammalia**), appeared in the early part of the Mesozoic era, branching from the now-extinct reptilian order Therapsida. Small mammals coexisted with the reptiles for 150–200 million years. When the giant reptiles and dinosaurs disappeared at the close of the Mesozoic era, mammals increased dramatically in numbers, diversity, and size.

Skeletal simplification accompanied the evolution of mammals from their therapsid ancestors. As mammals evolved, most lower-jaw bones migrated to the middle ear, leaving a single bone in the lower jaw. At the same time, the skull evolved a simpler form. The bulk of the limbs and the bony girdles from which they are slung was reduced, and the limbs were oriented beneath the body rather than poking out to the side and then down as in reptiles. Skeletons of later therapsids have most of these features, showing that the early mammals represented only a continuation of changes that had been under way for a long time (Figure 26.23). Indeed, the boundary between therapsids and mammals is an arbitrary one.

The skeletal features we have been discussing are readily fossilized. However, the important soft parts of mammals were seldom preserved in the fossil record. Such key mammalian features as mammary glands, sweat glands, hair, and a four-chambered heart may have evolved among the later therapsids, but the existing record does not tell us when this happened. Mammals are unique among animals in suckling their young with a nutritive fluid (milk) secreted by mammary glands. Mammalian eggs are fertilized within the body of the female, and the embryos undergo some development within the uterus prior to being born. Mammals have a protective and insulating covering called hair, which is extremely luxuriant in some species but almost entirely absent

in other species, such as whales and dolphins. The latter have, instead, thick layers of fat (blubber) under their skins for insulation.

Mammals have far fewer, but more highly differentiated, teeth than reptiles do. Dentition differences among mammals reflect their varied diets. By understanding the relationships between dentition and diet among living mammals, it is possible to infer most features of the diets of extinct groups. Mammals range in size from tiny shrews weighing only about 2 grams to the blue whale, which measures up to 31 meters long and weighs up to 160,000 kilograms, the largest animal ever to live on Earth. The approximately 4,000 species of living mammals are placed into three major groups.

The subclass Prototheria contains a single order, the **Monotremata**, represented by two families and a total of three species in Australia and New Guinea. These mammals, the duck-billed platypus and the spiny anteaters or echidnas, differ from other mammals in laying eggs and in possessing a number of reptilelike anatomical features (Figure 26.24a). Their young are nursed on milk but there are no nipples on the mammary glands; rather, the milk oozes out and is lapped off the fur by the offspring.

The other two groups of mammals are in the subclass Theria. In one group, the **marsupials**, contain-

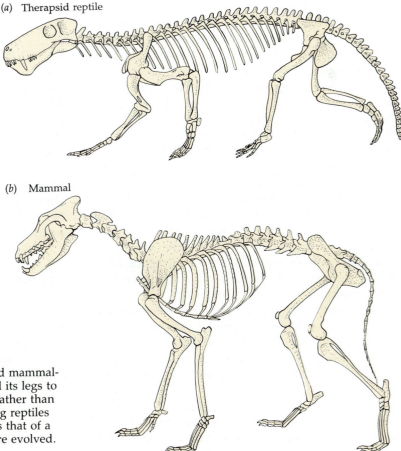

(a) Therapsid reptile

(b) Mammal

26.23 Tucking the Legs Under the Body
(a) *Lycaenops*, a Permian therapsid reptile, had mammallike thoracic and pelvic girdles that permitted its legs to be positioned directly underneath the body rather than extending out to the sides as the legs of living reptiles do. (b) Modern mammalian skeletons such as that of a wolf reveal how the trend in skeletal structure evolved.

26.24   Monotremes and Marsupials
(a) The Australian echidna, or spiny anteater, is one of three surviving species of monotremes. It is a fairly common and widespread Australian mammal. (b) Australia's kangaroos are perhaps the most familiar marsupials. Here a female red kangaroo (*Macropus rufus*) carries her offspring in the distinctive pouch. (c) This mouse opossum (*Marmosa* sp.) from the Amazon basin is an example of a South American marsupial.

ing about 240 species, females have a ventral pouch or folds in which the young are carried and fed (Figure 26.24b). Gestation in marsupials is short, and the young are born very tiny but with well-developed forelimbs with which they climb to the pouch. Once the offspring has left the uterus, female marsupials may become sexually receptive again. They can then carry a fertilized egg capable of initiating development and of replacing the offspring in the pouch should something happen to it. The marsupial mode of reproduction is adapted to relatively harsh and uncertain conditions where adults must travel long distances to find food and where droughts may cause loss of already born offspring.

At one time, marsupials were widely distributed on the southern continents, but today the majority of species are restricted to the Australian region, with a modest representation in South America (Figure 26.24c) and only one species in North America. Marsupials radiated into virtually all mammalian lifestyles except marine and flying ones. The largest living marsupial is the red kangaroo of Australia, which weighs up to 90 kilograms, but much larger marsupials existed in Australia until quite recent times. These animals were probably exterminated by people soon after they reached Australia about 40,000 years ago.

Most living mammals are **eutherians** (sometimes they are called placentals, but this is not a good name because some marsupials also have placentas).

Young eutherians are more highly developed at birth than marsupials, and there is no external pouch in which they are housed after birth. The nearly 4,000 species of eutherians are placed into 16 orders (Figure 26.25), the largest of which is the Rodentia, with about 1,700 species. The next largest order, the Chiroptera (bats), has about 850 species, followed by the order Insectivora (moles and shrews) with just over 400 species. The largest mammals are marine, but some terrestrial mammals, such as elephants and rhinoceroses, weigh up to several thousand kilograms. Eutherians are extremely varied in form and ecology. They are, together with insects, the most important grazers in terrestrial ecosystems. They have exerted strong selective pressures on the evolution of a number of features of terrestrial plants, such as spines, tough leaves, and growth forms.

HUMAN EVOLUTION

The family **Hominidae** (humans and their relatives) separated approximately 5 million years ago from a common ancestor shared with chimpanzees. Because we are especially interested in our own evolution, a great deal of effort has been expended to gather evidence about our ancestors, how they lived, and the pathways by which we evolved. Important new hominid fossil finds are made nearly every year.

In 1974, a fossil skeleton of a hominid approxi-

(a)

26.25 Eutherian Diversity

The golden-mantled ground squirrel (order Rodentia) is one of many species of small, diurnal rodents of the western North American mountains. (b) With their powers of echolocation (see Chapter 37), many bats can locate and capture prey even in complete darkness. This big brown bat *Eptesicus fuscus* (order Chiroptera) is about to capture a large moth. (c) The African chimpanzee (order Primates) is the closest living human relative. (d) Cats (order Carnivora) ambush their prey and capture them by short chases. The massive legs of the African lion are not suited to long-distance running. (e) Large hoofed mammals (order Artiodactyla) are important herbivores over much of Earth. Sheep, such as these bighorns, are primarily animals of mountainous regions in the northern hemisphere.

(b)

(c)

(d)

(e)

mately 3.6 million years old was discovered in Ethiopia. Although it is the oldest known hominid fossil, it is also one of the best preserved and most complete hominid skeletons ever discovered. That individual, a young female known to the world as "Lucy," attracted a great deal of attention and controversy. Lucy has been assigned to the species *Australopithecus afarensis*, now regarded as the most likely ancestor of later hominids.

From *Australopithecus afarensis* ancestors, a number of species of australopithecines evolved. Several million years ago, two distinct types of australopithecines lived at the same time over much of eastern Africa. The more robust type (about 40 kilograms) is represented by at least two species. One, *A. boysei*, is common in fossil sites in Ethiopia, at Lake Turkana in Kenya, and in Olduvai Gorge in Tanzania. The other, *A. robustus*, is known from South Africa. Both of these species died out suddenly about 1.5 million years ago. The smaller (25–30 kilograms), more slender *A. africanus* (Figure 26.26), is just as old, but it is much rarer as a fossil, suggesting that it was less common than the other species.

Being less agile, members of the robust species

26.26 Australopithecines
From skeletal remains archaeologists can reconstruct the approximate appearance of australopithecines such as *Australopithecus africanus.*

prey. Both these advantages were probably important for early australopithecines. The evolution of the hand as a precise grasping instrument was valuable for carefully selecting high-quality plant food items, which in savannas are mostly the meristematic growing regions of grasses close to or under the ground. The more precise the control of hand movement, the better an animal is able to pull on grasses in such a way that the meristem is harvested rather than being left behind because the stem breaks off above it. Also, bipedal australopithecines could carry large food items and infants long distances.

Over many generations, the feet of *A. africanus* evolved to become better running appendages and the hands to become better grasping instruments, both of which increased its ability to exploit the dry savanna environments of Africa. During this stage, its diet was composed primarily of plant materials, but animal foods—such as bird eggs and nestlings, insects, and newborn mammals—were probably eaten when they were available.

The Rise of *Homo*

Many experts believe that a population of *Australopithecus africanus* or a similar species gave rise to the genus *Homo* about 2.5 mya; then early members of *Homo* lived contemporaneously with australopithecines for perhaps half a million years. Two major changes accompanied the evolution of *Homo* from *Australopithecus*: an increase in body size and a striking increase in brain size to about double that of the late australopithecines (Figure 26.27). The oldest fossil remains of members of the genus *Homo, Homo*

probably stayed relatively close to trees to which they retreated at night and when predators were near. Members of the smaller *A. africanus* were able to run faster, and (because of their smaller size) needed less food per day to survive. They probably lived in more open, drier savannas where food was less abundant than in the moister areas inhabited by the more robust species. Their small size and greater agility enabled them to exploit these more dangerous and less productive areas.

The australopithecines had distinct morphological adaptations for **bipedalism**—locomotion in which the body is held erect and moved exclusively by the movements of the hind legs beneath it. *A. africanus,* which had a broad, bowl-shaped pelvis similar to that of modern humans, was presumably already well adapted for walking upright. Bipedal locomotion is slower and energetically more costly than walking on all four legs, but it has two important advantages: First, it frees the hands to manipulate objects, and second, it elevates the eyes, enabling the animal to see over tall vegetation to spot both predators and

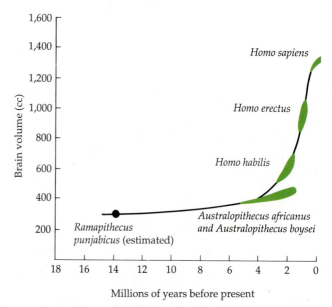

26.27 Hominid Brains Enlarged Rapidly
Notice how rapidly, in evolutionary terms, the brain tripled in volume.

habilis, discovered in the Olduvai Gorge, Tanzania (Figure 26.28*a*), have been dated as 2 million years old. Tools used by these early hominids to obtain food were found with the fossils. Other fossils of *H. habilis* have been found in Kenya, Ethiopia, and South Africa, indicating that the species had a wide range in Africa.

H. habilis lived in relatively dry savannas where, for much of the year, the main food reserves are subterranean roots, bulbs, and tubers. Exploitation of these food resources requires an ability to dig into the hard, dry savanna soils, something that cannot be done with the unaided hand. Moreover, although these underground storage organs of plants are good sources of carbohydrates, they are much lower in protein than are leaves, so *H. habilis* would have needed to supplement its diet with animal foods. Roots can be harvested in large quantities in a relatively short time by an individual with a simple digging tool, and *H. habilis* women could have harvested them this way even while carrying infants. Therefore, male *H. habilis* may have had more time for cooperative hunting of large, dangerous animal prey than if their diet had been dominated by leafy foods.

The only other known extinct species of our genus, *Homo erectus*, evolved in Africa about 1.6 mya. Soon thereafter it had spread as far as eastern Asia. Members of this species were as large as modern people, but their bones were somewhat heavier. *H. erectus* used fire for cooking and for hunting large animals, and they made characteristic stone tools that have been found in many parts of the Old World (Figure 26.28*b*). These tools were probably used for a variety of purposes, including digging, capturing animals, cleaning and cutting meat, scraping hides, and cutting wood. Although *H. erectus* survived in Eurasia until about 250,000 years ago, it was replaced in Africa by our species, *Homo sapiens*, about one-half million years ago.

Homo sapiens Evolves

The trends we have observed in the transition from *Australopithecus* to *Homo erectus* continued with the evolution of our own species. The earliest humans had larger brains and smaller teeth than members of the earlier species of *Homo*. These changes were probably favored by an increasingly complex social life. The ability of group members to communicate with one another was valuable in cooperative hunting, for sharing information about the location and exploitation of food sources, and for improving one's status in the complex social interactions that must have characterized those societies just as they do ours today.

Several types of *H. sapiens* existed during the mid-Pleistocene epoch. All were skilled hunters of large mammals, even though plants continued to be important components of their diets. This period also witnessed the emergence of another distinctly human trait: religious practice and a concept of life after death. These beliefs are revealed by finding that deceased individuals were buried and were provided with tools and clothing, in their graves, presumably for their existence in the next world.

One type of *H. sapiens*, generally known as Neanderthal because it was first discovered in the Neander Valley in Germany, was widespread in Europe and Asia between about 75,000 and 30,000 years ago. Neanderthals were short, stocky, and powerfully built people whose massive skulls housed brains somewhat larger than our own. They manufactured a variety of tools and were highly efficient at hunting

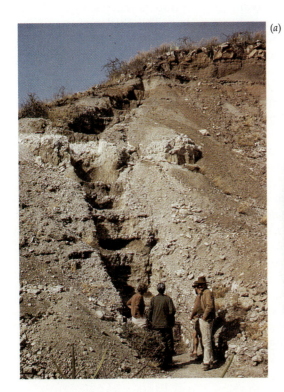

(a)

(b)

26.28 *Homo* Is Excavated
(a) Excavations at Olduvai Gorge, Tanzania, have revealed much about hominid evolution. (b) Such excavations have unearthed stone tools used by *Homo erectus*.

26.29 Cro-Magnon Cave Painting
Cro-Magnons often painted animals with arrows and spears flying toward them, suggesting that these paintings may have served as part of religious rituals designed to increase success during hunts.

large mammals, which were ambushed and subdued in close combat by several individuals attacking together. For a short time, their range overlapped that of Cro-Magnon people, a more modern form of *H. sapiens*, but then they abruptly disappeared. Many scientists believe that they were exterminated by the Cro-Magnons, just as *H. habilis* may have been exterminated by *H. erectus*. Cro-Magnons are believed to have migrated from Africa to Europe where fossils having their skeletal features have been dated at more than 100,000 years of age.

Cro-Magnon people made and used a variety of sophisticated tools. They made the remarkable paintings of large mammals, many of them showing scenes of hunting, that have been discovered in caves in various parts of Europe (Figure 26.29). The animals depicted were characteristic of the cold steppes and open savannas that occupied much of Europe during periods of glacial expansion. Cro-Magnon people spread across Asia, reaching North America perhaps as long as 20,000 years ago, although the date of their arrival in the New World is still uncertain. As they rapidly spread southward through North and South America, they exterminated, by overhunting, populations of many species of large mammals that had been abundant on those continents.

The Evolution of Language and Culture

Increases in brain size among our ancestors caused a dramatic increase in certain behavioral capabilities, especially the capacity for language. Most animal communication consists of a limited number of signals, nearly all which pertain to immediate circumstances and are associated with changed emotional states induced by those circumstances. The language of honeybees is unusual in that it contains a symbolic component referring to events distant in both space and time (Chapter 43). Human language is far richer in its symbolic character than any animal utterances. Our words can refer to past and future times and to distant places. We are capable of learning thousands of words, many of them referring to abstract concepts, and of arranging those words to form sentences with complex meanings.

The expanded mental abilities of humans are largely reponsible for the development of **culture**, the process by which knowledge and traditions are passed along from one generation to another by teaching and observation. Culture can change rapidly because genetic changes are not necessary for a cultural trait to spread through a population. The primary disadvantage of culture is that each generation must be taught its norms. The challenge is to pass on the valuable norms of a culture without inhibiting creative changes.

Cultural traditions are clearly revealed by the designs of the tools and other implements associated with human fossils and in the cave paintings these people created. The rapid spread of the domestication of plants and animals, and the resultant conversion of most human societies from ones in which food was obtained by hunting and gathering to ones in which pastoralism (herding large animals) and agriculture dominated, was greatly facilitated by cultural learning. The agricultural revolution, in turn, led to an increasingly sedentary life, the development of cities, greatly expanded food supplies, and the rapid growth of the human population (Figure 26.30). Twenty-five thousand years ago only 3 million people lived on Earth. Ten thousand years ago Earth still housed only about 5 million people. However, as agriculture spread during the subsequent 5,000 years, the human population increased rapidly to about 100 million. Rapid increase in the human population has continued unabated until today.

Agriculture developed in the Middle East approximately 11,000 years ago, and from there it spread rapidly northwestward across Europe, finally reaching the British Isles about 4,000 years ago. The plants and animals domesticated by these people were cereal grains such as wheat and barley; beans, lentils, and peas; and woody plant crops such as grapes and olives. Others, such as rye, cabbage, celery, and carrots, were domesticated as agriculturalists spread across Europe. Cattle, sheep, goats, horses, pigs, dogs, cats, and chickens were their most important domesticated animals.

Agriculture developed independently in east Asia, contributing to our modern diets such plants as soybeans, rice, citrus fruits, and mangoes. There was some exchange, even at early times, between agricultural centers in the Old World, but when people crossed the cold and barren Bering land bridge into the New World, they apparently brought no domesticated plants with them. These people eventually

developed such important crops as corn, tomatoes, kidney and lima beans, peanuts, potatoes, chili peppers, and squashes.

The Human Lineage

When and how people spread around Earth from Africa, the home of hominid evolution, has been the subject of much controversy. Resolution of the debate has been achieved by the study of mitochondrial DNA from contemporary human populations in Africa, Asia, Europe, Australia, and New Guinea. Mitochondrial DNA is particularly useful for determining lineages over short time spans because it accumulates random changes about ten times as fast as chromosomal DNA. Also, because all mitochondria are of maternal origin, changes in nucleotide sequences result only from the ticking of the molecular clock, not from sexual recombination.

26.30 Pastoralism and Agriculture (a) This Fulani herder occupies dry season temporary quarters in Burkina Faso, Africa. The bicycle is a bit of modern technology incorporated into an otherwise ancient system. (b) Agricultural development has totally transformed the landscape in these hills above Port-au-Prince, Haiti.

TABLE 26.2
Summary of the Kingdom Animalia

PHYLUM	NUMBER OF LIVING SPECIES	SUBGROUPS
Porifera: Sponges	10,000	
Cnidaria: Coelenterates	10,000	Hydrozoa: Hydras and hydroids Scyphozoa: Jellyfish Anthozoa: Corals, sea anemones
Ctenophora: Comb jellies	100	
Platyhelminthes: Flatworms	25,000	Turbellaria: Free-living flatworms Trematoda: Flukes (all parasitic) Cestoda: Tapeworms (all parasitic)
Nemertea: Ribbon worms	900	
Nematoda: Roundworms	20,000	
Rotifera: Rotifers	1,800	
Pogonophora: Pogonophores	145	
Annelida: Segmented worms	15,000	Polychaeta: Polychaetes (all marine) Oligochaeta: Earthworms, freshwater worms Hirudinea: Leeches
Arthropoda: Arthropods	>1,500,000	Trilobita: Trilobites (extinct) Onychophora: Onychophorans Merostomata: Horseshoe crabs Arachnida: Scorpions, harvestmen, spiders, acarines Crustacea: Crabs, shrimp, lobsters, barnacles, copepods Chilopods: Centipedes Diplopoda: Millipedes Insecta: Insects
Mollusca: Mollusks	100,000	Polyplacophora: Chitons Monoplacophora: Monoplacophorans Bivalvia: Clams, oysters, mussels Gastropoda: Snails, slugs, limpets Cephalopoda: Squids, octopuses, nautiloids
Phoronida: Phoronids	12	
Ectoprocta: Moss animals	4,000	
Brachiopoda: Lamp shells	350	More than 12,000 fossil species described!
Hemichordata: Acorn worms	100	
Chaetognatha: Arrow worms	70	
Echinodermata: Echinoderms	6,000	Crinoidea: Sea lilies, feather stars Ophiuroidea: Brittle stars Asteroidea: Sea stars Concentricycloidea: Sea daisies Echinoidea: Sea urchins Holothuroidea: Sea cucumbers
Chordata: Chordates	40,000	Urochordata: Sea squirts Cephalochordata: Lancelets Agnatha: Lampreys, hagfishes Placodermi: Placoderms (extinct) Chondrichthyes: Cartilaginous fishes (sharks, rays) Osteichthyes: Bony fishes Amphibia: Amphibians Reptilia: Reptiles Dinosauria: Dinosaurs, crocodilians, birds Mammalia: Mammals

The most surprising result of these investigations is the discovery that the human evolutionary tree apparently has a single ancestor, often called "mitochondrial Eve." Despite this name, the data do not suggest that all modern people are descended from a single female. Rather, they show that the ancestral population was small enough that, perhaps as a result of genetic drift, only one set of mitochondrial genes was transmitted to later generations. Also, the mitochondrial clock suggests that this ancestral group lived only about 200,000 years ago, and that human populations left Africa for other continents only a little more than 100,000 years ago. Clearly, we are a much more closely related species than anybody had previously suspected.

THEMES IN DEUTEROSTOME EVOLUTION

A summary of all phyla in the kingdom Animalia is provided in Table 26.2. In several important ways, deuterostome evolution paralleled protostome evolution. Both lineages exploited the abundant food supplies buried in soft marine substrates, attached to rocks, or suspended in the water column. Because of the ease with which water can be moved, many groups of both lineages developed elaborate structures for extracting prey from water. In both groups, a coelomic cavity evolved and subsequently became divided into compartments that allowed better control of body shape and movement. Both groups evolved locomotor abilities, and some members of both groups evolved mechanisms for controlling their buoyancy in water, using gas-filled internal spaces whose contents can be adjusted to control the depth at which the animal is stable.

Both protostomes and deuterostomes invaded the land, but with very different consequences. The jointed external skeletons of arthropods, although they provide excellent support and protection in air, are not suitable for large animals. In addition, an arthropod must shed its skin and become temporarily vulnerable in order to grow. The internal, jointed skeletons of vertebrates, however, permit growth to a large size without any temporary vulnerable stages. Consequently, although arthropods are abundant and diverse on land, vertebrates are the only very large terrestrial animals to have appeared on Earth.

Terrestrial lineages of vertebates reinvaded aquatic environments a number of times. Interestingly, suspension feeding re-evolved in several of these lineages. For example, the largest living mammals, the baleen whales (the toothless whales, including blue whales, humpback whales, and right whales), feed upon relatively small prey they extract from the water with large plates in their mouths. Prey are available in the air, but their concentrations are so low that remaining in one place and screening them from the air is not a viable way of life. Predators, such as bats and some birds, that feed on aerial prey are highly mobile and usually forage by pursuing individual prey items.

Unlike the oceans, where the dominant photosynthesizers are unicellular algae, most photosynthesis on land is carried out by vascular plants. The dominant plants in most terrestrial environments are large and complex organisms that provide the primary physical structure of those ecological communities. Although plants are suitable sources of energy for animals, most plant tissues are difficult to digest. Herbivores must ingest large quantities of fibers and defensive chemicals along with the energy-rich molecules they need. Because larger animals can exist on poorer quality food than can small animals, a common pattern in herbivore evolution is a steady increase in body size. This pattern is striking in the evolution of reptiles and in the later evolution of mammalian herbivores. The evolution of large herbivores, in turn, favored the evolution of larger carnivores able to attack and overpower them. This evolutionary trend may have come to a temporary halt because of the invention of weapons by a moderately sized, omnivorous primate.

SUMMARY

Deuterostome evolution began early in the history of life on Earth. These animals evolved ways of utilizing all major food types in marine environments. Locomotion was first based upon a divided coelomic cavity, but internal skeletons developed early in deuterostome evolution. The echinoderms and chordates became animals that moved about on the substrate, finding and exploiting food buried in sediments or attached to rocks. The echinoderm *Bauplan* is based on radial symmetry, and echinoderms remained slow-moving animals. However, some chordates, with their jointed internal skeletons, evolved the ability to swim rapidly. They achieved control over their buoyancy with swim bladders, structures that later enabled them to invade the land and breathe air. The first dominant terrestrial vertebrates were amphibians, followed by reptiles and, finally, by dinosaurs and their descendants (crocodilians and birds) and by mammals. Terrestrial environments are now strongly controlled by a single species of mammal—*Homo sapiens.*

SELF-QUIZ

1. Which of the following are deuterostomate phyla with a tripartite
body plan?
   a. Rotifera, Phoronida, Ectoprocta,
   and Brachiopoda
   b. Phoronida, Ectoprocta, Brachiopoda, and Hemichordata
   c. Phoronida, Ectoprocta, Hemichordata, and Chordata
   d. Echinodermata, Ectoprocta, Brachiopoda, and Chordata
   e. Phoronida, Ectoprocta, Hemichordata, and Echinodermata

2. The structure used by brachiopods
to capture food is a:
   a. pharyngeal gill basket.
   b. proboscis.
   c. lophophore.
   d. mucus net.
   e. radula.

3. The water vascular system of echinoderms is a:
   a. series of seawater channels derived by enlargement and extension of a coelomic cavity.
   b. series of seawater channels derived by enlargement and extension of the pharyngeal cavity.
   c. series of channels derived by
   enlargement and extension of a
   coelomic cavity and filled with
   coelomic fluid.
   d. series of channels derived by
   enlargement and extension of a
   coelomic cavity and filled with
   fresh water.
   e. series of channels that can be
   filled to different degrees with
   water to enable the animal to
   control its buoyancy.

4. The pharyngeal gill slits of chordates originally functioned as sites
for:
   a. uptake of oxygen only.
   b. release of carbon dioxide only.
   c. both uptake of oxygen and release of carbon dioxide.

   d. removing small prey from the
   water.
   e. for forcibly expelling water to
   move the animal.

5. The key to the vertebrate *Bauplan*
is:
   a. a pharyngeal gill basket.
   b. a vertebral column to which internal organs are attached.
   c. a vertebral column to which
   two pairs of appendages are
   attached.
   d. a vertebral column to which a
   pharyngeal gill basket is
   attached.
   e. a pharyngeal gill basket and two
   pairs of appendages.

6. Which of the following fishes do
not have a cartilaginous skeleton?
   a. Chimeras
   b. Lungfishes
   c. Sharks
   d. Skates
   e. Rays

7. In most fishes, lunglike sacs
evolved into:
   a. pharyngeal gill slits.
   b. true lungs.
   c. coelomic cavities.
   d. swim bladders.
   e. none of the above.

8. Most amphibians return to water
to lay their eggs because:
   a. water is isotonic to egg fluids.
   b. adults must be in water while
   they guard their eggs.
   c. there are fewer predators in
   water than on land.
   d. amphibians need water to produce their eggs.
   e. amphibian eggs quickly lose
   water and desiccate if their surroundings are dry.

9. The horny scales that cover the
skin of reptiles prevents them
from:

   a. using their skin as an organ of
   gas exchange.
   b. sustaining high levels of metabolic activity.
   c. laying their eggs in water.
   d. flying.
   e. crawling into small places.

10. Which of the following is *not* true
of the feathers of birds?
   a. They are highly modified reptilian scales.
   b. They provide insulation for the
   body.
   c. They arise from well-defined
   tracts.
   d. They help birds fly.
   e. They are important sites of gas
   exchange.

11. Monotremes differ from other
mammals by:
   a. not producing milk.
   b. lacking body hairs.
   c. laying eggs.
   d. living in Australia.
   e. having a pouch in which the
   young are raised.

12. Bipedalism is believed to have
evolved in the human lineage
because:
   a. bipedal locomotion is more efficient than quadrupedal
   locomotion.
   b. bipedal locomotion is more efficient than quadrupedal locomotion and it frees the hands to
   manipulate objects.
   c. bipedal locomotion is less efficient than quadrupedal locomotion but it frees the hands to
   manipulate objects.
   d. bipedal locomotion is less efficient than quadrupedal locomotion but bipedal animals can run
   faster.
   e. bipedal locomotion is less efficient than quadrupedal locomotion but natural selection does
   not act to improve efficiency.

FOR STUDY

1. In what animal phyla has the ability to fly evolved? How do structures used for flying differ among
these animals?

2. Extracting suspended food from
the water column is a common
mode of foraging among animals.

Which groups contain species that
extract prey from the air? Why is
this mode of obtaining food so
much less common than extracting
prey from water?

3. Compare the buoyancy systems of
cephalopods and fishes.

4. Why does possession of an external skeleton limit the size of a terrestrial animal more than possession of an internal skeleton?

5. Large size both confers benefits
and poses certain risks. What are
these risks and benefits?

READINGS

Alexander, R. M. 1975. *The Chordates*. Cambridge University Press, New York. A comprehensive and readable account of the biology of members of the phylum Chordata.

Bakker, R. T. 1975. "Dinosaur Renaissance." *Scientific American*, April. Discusses the relationships between birds and dinosaurs; presents evidence that dinosaurs were warm-blooded.

Bond, C. E. 1979. *Biology of Fishes*, Saunders, Philadelphia. A leading text for courses on ichthyology.

Carroll, R. C. 1987. *Vertebrate Paleontology and Evolution*. W. H. Freeman, San Francisco. A thorough account of the fascinating evolutionary history of the vertebrates.

Colbert, E. H. 1980. *Evolution of the Vertebrates: A History of the Backboned Animals Through Time*, 3rd Edition. Wiley-Interscience, New York. A thoughtful discussion of the origins and evolutionary radiations of the vertebrate groups.

Langston, W., Jr. 1981. "Pterosaurs." *Scientific American*, February. An account of the largest animals ever to fly. Notes on evolutionary relationships among birds and reptiles.

Gill, F. B. 1990. *Ornithology*. W. H. Freeman, New York. A technically accurate and readable introduction to bird biology for students at any level.

Pough, F. H., J. B. Heiser and W. N. McFarland. 1989. *Vertebrate Life*. Macmillan, New York. An excellent treatment of the natural history of the vertebrates.

Vaughan, T. A. 1978. *Mammalogy*, 2nd Edition. Saunders, Philadelphia. The leading textbook on mammals. Offers good coverage of both the orders of mammals and general aspects of mammalian biology.

Willson, M. F. 1984. *Vertebrate Natural History*. Saunders, Philadelphia. A thorough treatment of all aspects of the lives of vertebrates.

27

Patterns in the Evolution of Life

PREVIEW: Life evolved early in the history of Earth. Since that time, organisms have increased in size, complexity, and number of species, but there are fewer body plans among living animals than there were among the animals living 600 million years ago. Although there have been six major extinction periods since life first evolved, the number of species has increased overall. The possible causes of extinctions include movements of the continents, which resulted in mountain building, changes in climates and sea levels, and mixing and separation of biotas; volcanic activity; and the collision of Earth with asteroids. Life invaded the land after a long period of being restricted to the oceans. The pace of evolution has been very irregular. Periods of relatively rapid change have alternated with periods of little change, but those periods have not coincided for all groups of organisms. The fossil record reveals a great deal about the phylogeny of life, especially of those species with hard skeletons.

This chapter deals with the turbulent history of Earth, the fossil record and the patterns of evolution it reveals, rates of evolutionary change, and the rise and fall of the major evolutionary lineages.

The development of a comprehensive picture of the evolution of life is one of the most exciting parts of modern biology. Information from many sciences, including the physical sciences, is being integrated to show how life evolved. In previous chapters we discussed the origins of life, mechanisms of evolution, modes of speciation, and the diversity of life that resulted from evolutionary processes. In this chapter we draw this information together to provide a general picture of the broad features of biological evolution, and of the origins, radiations, and extinctions of lineages. As this picture unfolds, it is important to keep in mind the millions of years over which evolution has taken place and how much Earth and its living inhabitants have changed during that time.

HOW EARTH HAS CHANGED

The past is, to us, a foreign land. The Earth of the distant past is a foreign planet on which continents were not where they are today. Knowing how the positions of continents changed is vital to understanding the evolution of life on Earth. Earth's crust consists of solid plates approximately 40 kilometers thick that float on a liquid mantle. The plates move because the sea floor is spreading along ocean ridges

where material from the mantle rises and pushes the plates aside. Where plates come together, they either move laterally past one another along fault lines, or one plate slides under the other, creating mountain ranges. The movement of the plates and the continents they contain, has had enormous effects on climates, sea levels, and the distributions of organisms.

Table 17.1 showed the time units and dates of the geological history of Earth. In the late Cambrian period, there were six continents, all located at equatorial latitudes. All of them were united into a single large continent, known as Pangaea, during Permian times (Figure 27.1a). During the Paleozoic era there were several periods when sea levels rose and shallow seas spread over parts of the continents. There were also several periods when sea levels dropped and parts of the continental shelves were exposed. During the Permian period, Pangaea began to break up by a process known as **continental drift**. By the late Triassic period, North America and Eurasia formed a single large continent, Laurasia. The southern land masses were united into another continent, Gondwanaland. These two continents were separated by a gradually widening oceanic channel. By the mid-Cretaceous period, Gondwanaland had begun to break up into Africa, South America, and a third land mass consisting of what are now Australia,

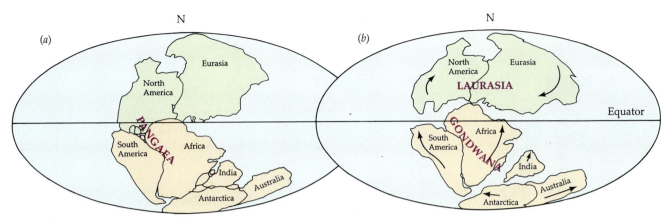

27.1 The Continents Drift
(a) The Paleozoic era, about 300 million years ago. The supercontinent Pangaea is alone in a great ocean. (b) The Mesozoic era, about 150 million years ago. Pangaea has split into the northern and southern land masses of Laurasia and Gondwanaland. During the Mesozoic era Gondwanaland began to break up into several land masses.

Antarctica, and India (Figure 27.1b). Some organisms still have a distribution encompassing the southern continents once united as Gondwanaland. By the late Cretaceous period, India had broken free and begun its northward drift, eventually colliding with Asia. The Himalaya Mountains formed as a result of this impact. South America and Africa continued to drift apart, and Australia separated from Antarctica and slowly moved to its present position much closer to the equator. North America and Eurasia also drifted apart, but they were later rejoined at the present Bering Sea. About 3 mya (million years ago), the Isthmus of Panama arose, connecting North and South America for the first time in more than 250 million years.

Over much of its history, the climate of Earth was considerably warmer than it is today, and temperatures declined more slowly toward the poles. At other times, Earth was much colder than it is today. Large areas were covered with glaciers during the late Precambrian, the Carboniferous, the Permian, and the Pleistocene, but these cold periods were separated by very long periods of more equable climates. We live in one of the colder periods in the history of Earth, a fact that makes it difficult for us to imagine the mild climates characteristic of high latitudes during much of the history of life. Usually climates change slowly, but there have been major climate shifts over periods as short as 5,000–10,000 years, primarily as a result of changes in Earth's orbit around the sun. A few shifts appear to have been even more rapid than that. For example, the ice cover of Antarctic seas changed from a glacial to an interglacial state in less than 100 years. Therefore, climates sometimes have changed rapidly enough that extinctions due to them could appear "instantaneous" in the fossil record.

THE FOSSIL RECORD

Much of what we know about past life comes from a rather small number of sedimentary rocks in which organisms are especially well preserved. The richest fossil beds are composed both of organisms that lived in or on the sediments and others that were transported there after they died. That is because conditions for preservation are best in environments lacking oxygen, whereas ecological communities in oxygenated environments are much richer in species. Thus most fossil assemblages are collections of organisms that were transported by wind or water to sites without oxygen. Occasionally, however, organisms are preserved where they lived. In such cases we can reconstruct the nature of past ecological communities, especially those of cool, anaerobic swamps where conditions for preservation were excellent.

The Completeness of the Fossil Record

About 300,000 species of fossil organisms have been described, and the number is steadily growing. However, this is only a tiny fraction of the species that have ever lived. We do not know how many species really did live in the past, but there are ways of making reasonable estimates. Of the present-day **biota**—the species in all groups (monerans, protists, fungi, plants, and animals) living today—approximately 1.5 million species have already been described. The actual number of extant species is probably at least 10 million (possibly as high as 50 million) because most species of insects, the richest animal group, have not yet been described. Thus the number of known fossil species is less than 2 percent of the probable total of living species. Yet, because there has been life on Earth for 3.5 billion years, and be-

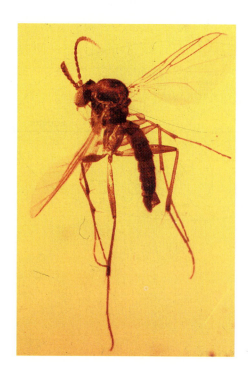

27.2 A Fossil Fly
This fly was trapped and exquisitely preserved in amber fossilized from the sap of a tree in an ancient Baltic forest.

denly appears in the Asian fossil record as a major new form of organism. If we lacked fossil evidence of horse evolution in North America, we might conclude that horses evolved very rapidly somewhere in Asia.

Patterns in the Fossil Record

The general diversity of evolving lineages of organisms in three of the five kingdoms over geological time is shown in Figure 27.3. The thickness of the pathways indicates the number of species in the different phyla, but the figure does not show the many different lineages that arose, prospered, and died out within the major phyla.

The dates of these events have been determined by the use of naturally occurring radioactive isotopes as clocks. In successive, equal periods of time, the same *fraction* of the remaining radioactive material of any radioisotope decays. For example, in 14.3 days, one-half of any sample of phosphorus-32 (^{32}P), a radioactive isotope of phosphorus, decays. In the next 14.3 days, one-half of the remaining half decays, leaving only one-quarter of the original sample of ^{32}P. After 42.9 days, three **half-lives** have passed, so only 1/8 (that is, 1/2 × 1/2 × 1/2) of the original radioactive material remains, and so forth. Each radioisotope has a characteristic half-life. Tritium, for example, has a half-life of 12.3 years, and carbon-14 (^{14}C) has a half-life of about 5,700 years.

Because of this regularity of radioactive decay, we can use ^{14}C to determine the time of death of anything that has died within the last 15,000 years or so and has left carbon-containing remains. The ratio of radioactive ^{14}C to nonradioactive ^{12}C in a living creature is always the same as that in the atmosphere in which it lives because plants and animals are constantly exchanging carbon with the atmosphere. The production of new ^{14}C in the upper atmosphere (by the reaction of neutrons with ^{14}N) just balances the natural radioactive decay of ^{14}C, and so a steady state exists.

However, as soon as a tree or any other living thing dies, it ceases to equilibrate its carbon compounds with the rest of the world. Its decaying ^{14}C is not replenished from outside, so the ratio of ^{14}C to ^{12}C falls. By measuring what fraction of the total carbon in a specimen is ^{14}C, we can easily calculate how much time has elapsed since it died. Some radiocarbon dates of archaeological objects are shown in Figure 27.4. Note that sometimes, as in point 9 in the figure, we have confirming evidence from cultural or other sources.

cause species survive, on average, less than 10 million years, the number of species that lived in the past must have been many times the number that are alive today. There have been many replacements of the biota during geological history. Even if at any moment in the past the number of species was no greater than at present (or even if it was substantially less), the total number of species over evolutionary time would be much greater than the current biota.

The sample of fossils, though small in relation to the total number of extinct species, is not uniformly poor. On the one hand, insects rarely fossilize. Although past insect communities were probably very rich in species, only about 10,000 species have been described from fossils (Figure 27.2). In contrast, the record is especially good for the phyla of marine animals that have hard skeletons. Among the nine major phyla that have hard-shelled members, approximately 200,000 species have been described from fossils, roughly twice the number of living marine species in these same groups. Paleontologists lean heavily on these groups in their interpretations of the evolution of life in the past.

Unfortunately, most described fossils come from a relatively small number of sites. Therefore, an organism may appear suddenly at a site, giving the impression that it evolved rapidly from one of the species previously found there when, in fact, it had evolved slowly elsewhere and moved to the site. For example, horses evolved slowly over millions of years in North America. Many different lineages arose and died out. Ancestors of horses crossed the Bering Land Bridge into Asia several times, the last one only several million years ago. Evidence of each crossing sud-

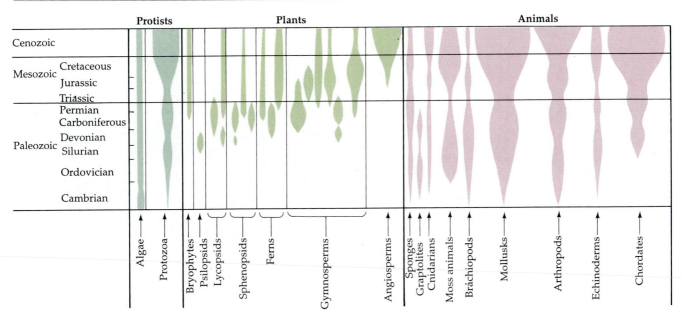

27.3 Broad Features of the Fossil Record
The times of origin of the major phyla of protists, plants, and animals are shown in this diagram. The approximate number of species in each group is indicated by the width of the pathways.

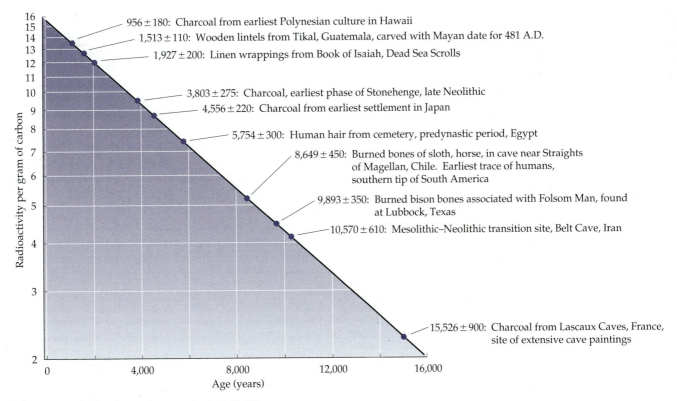

27.4 The Radioactive Carbon Clock Tells Time
Sometimes, as with the wooden lintels from Tikal (second from top), radioactively determined dates are confirmed by evidence from cultural or other sources.

(a)

(b)

27.5 Ediacaran Animals
Fossils excavated at Ediacara in southern Australia are 600 million years old and illustrate the diversity of life that evolved in the Precambrian era. *(a) Spriggina floundersi. (b) Mawsonites,* an Ediacaran metazoan.

Precambrian Life

Most kingdoms of eukaryotic organisms—protists, fungi, and animals, but not plants—evolved prior to the Cambrian period, but the known fossil record for that ancient time is still very fragmentary. The record is best for animals. Significant fossil finds, dated about 640 mya, have been made in Australia, southwestern Africa, and the English Midlands. The animals at those sites, all soft-bodied invertebrates, are known as the Ediacaran fauna, named after the Australian site where they were first discovered. (The term **fauna** refers to all of the animals living in a particular area; the corresponding term for plants is **flora**.) Among these animals are forms believed by some biologists to represent early annelids, arthropods, echinoderms, and cnidarians, but they are very different from later members of those phyla (Figure 27.5). They may instead represent a separate radiation of animal groups that have no living descendants. About two-thirds of the species so far identified in the Ediacaran fauna are thought to be cnidarians. Whatever the final judgment concerning the relationships of these animals, it is clear that a rich variety of animals had already evolved more than 600 mya.

The Paleozoic Era

By the mid-Cambrian period, about 530 mya, all extant animal phyla were already present, as revealed by the excellent remains in the Burgess Shale in British Columbia. More significantly, there are many species belonging to about 10 extinct phyla known only

from this fossil bed (Figure 27.6). Thus animals with many different body plans evolved in the Cambrian period, but most of them failed to survive past Cambrian times. The evolution of hard skeletons among representatives of so many phyla between 600 and 530 mya suggests that predation pressure may have been a major factor influencing the evolution of early animals.

During the Ordovician period (505–438 mya) many animal phyla radiated, creating a great profusion of classes and orders. Sponges were abundant and not much different from those living today. Cnidarians built large reefs, but the builders were different from modern corals, which first appeared in the mid-Ordovician period. There was a great increase in animals that filtered small prey from the water, such as bivalve mollusks and echinoderms. Floating graptolites, members of a now extinct phylum, were abundant (Figure 27.7). Trilobites were the dominant arthropods (see Figure 25.26b), but they also became extinct. On land, ancestors of club mosses and horsetails appeared, but they were still relatively small. At the end of the Ordovician, when sea levels dropped considerably, numerous families and orders of animals became extinct.

During the Silurian period (438–408 mya), graptolites declined, corals proliferated, and the colonization of land continued. Nevertheless, late Silurian marine communities probably looked little different from those of the Ordovician period. Early vertebrates diversified during this period and the first terrestrial arthropods—scorpions and millipedes—appeared. Among many groups of organisms, evolutionary change greatly accelerated during the Devonian period (408–360 mya). There was great adaptive radiation of corals, trilobites, and shelled squidlike cephalopods (Figure 27.8). Fishes became common, radiated, and succeeded one another over time. Jawed fishes replaced jawless ones, and the heavy armor characteristic of most earlier fishes gave

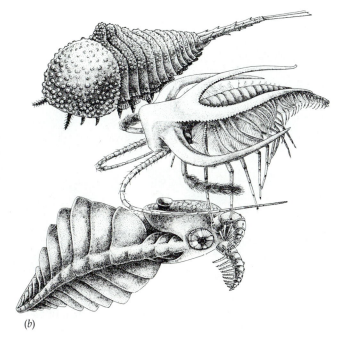

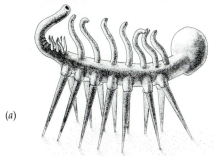

(a)

(b)

27.6 Extinct Phyla in the Burgess Shale
(a) The actual fossil of *Hallucigenia*, and an artist's conception of its probable appearance. (b) An artist's conception of several other organisms of the many represented in the Burgess Shale.

way to the less rigid outer coverings of modern fishes. Terrestrial communities also changed markedly during the Devonian period. Land plants became common, and some reached the size of trees. They were mostly lycopsids and sphenopsids, along with some ferns. Toward the end of the period the first gymnosperms appeared. A springtail from this period is the first known fossil of an insect. The first fishlike amphibians began to occupy the land.

Extensive forests grew during the Carboniferous period (326–320 mya), as was shown in Figure 24.1. The remains of those trees are the coal beds we now mine for energy. Coal is the compressed remains of plants that grew in swampy forests where trees fell into deep, anaerobic mud, which preserved them from biological degradation. Carboniferous beds are rich in fossils, many of which retain traces of the fine details of their structure (Figure 27.9). The diversity of terrestrial animals increased greatly in the Carboniferous period. Snails, scorpions, centipedes, and insects were present in great abundance and variety. Amphibians became better adapted to terrestrial existence. Some of them were large animals more than 5 meters long, quite unlike any surviving today. From one amphibian stock, the first reptiles evolved late in the period.

Deposits from the Permian period (286–248 mya)

contain representatives of most orders of insects, including giant dragonflies, the largest insects that ever lived. By the end of the period, reptiles greatly outnumbered amphibians. These reptiles included a variety of terrestrial forms as well as species that were major predators in aquatic environments. In fresh waters, the Permian period was a time of extensive radiation of bony fishes. Ammonites (cephalopod mollusks) proliferated in the oceans, but their radiation was abruptly terminated at the end of the period when the greatest of all mass extinctions occurred. At the end of the Permian period, about 95 percent of all invertebrate species became extinct. Trilobites were extinguished completely, and the numbers of species in such groups as anthozoan corals, ammonites, ostracods, crinoids, and brachiopods declined

27.7 Graptolites
Graptolites were abundant in the oceans of the Ordovician period; these were fossilized in black shale.

27.8 A Devonian Marine Community
During the Devonian period what is now the state of New York was an underwater reef. This reconstruction shows how part of it may have appeared; notice the trilobites and the unusual corals and cephalopods.

sharply. On land, most of the species of trees that dominated the great coal-forming forests became extinct. They were replaced by cycadeoids, cycads, ginkgos, and conifers.

The Mesozoic Era

During the Mesozoic era, as Pangaea separated into individual continents, the world biota, which had until that time been relatively homogeneous, became increasingly **provincialized**; that is, individual continents acquired distinctive terrestrial floras and faunas, and the faunas of continental shelves also diverged regionally. The provincialization that began during the Mesozoic still influences the geography of life today.

In the Triassic period (248–213 mya), the ammonites underwent a second great radiation, modern corals proliferated, and some invertebrate groups, such as bivalve mollusks, increased in diversity. Many invertebrate groups that previously had been restricted to living on surfaces of bottom sediments evolved burrowing forms. On land, gymnosperms and seed ferns became the dominant woody plants. The great radiation of reptiles that gives the Mesozoic its popular name, the "Age of Reptiles," began (Figure 27.10). Late in the Triassic period a lineage of reptiles gave rise to the first mammals.

The beginning of the Jurassic period (213–144 mya) is marked by a mass extinction that eliminated most of the ammonites and many other marine invertebrates. The ammonites underwent yet another radiation, and many invertebrate groups evolved sturdier shells, probably because of increased predation pressures from crabs, gastropods, and fishes. Bony fishes began the great radiation that culminated in their dominance of the seas, a position they have held to this day. Frogs, salamanders, and lizards first appeared during the Jurassic. Flying reptiles evolved, and dinosaurs radiated into bipedal predators and large quadrupedal herbivores. One lineage of reptiles gave rise to the first birds in the mid-Jurassic period. A number of groups of mammals, derived from another reptilian lineage, evolved during the period.

During the Cretaceous period (144–65 mya), marine invertebrates increased in variety and number of

27.9 A Carboniferous Fossil
A fossilized tree fern excavated in France. Like the fossil fuels we burn today, this fossil is a remnant of the massive forests of the Carboniferous period.

27.10   The "Age of Reptiles"
The remarkable dinosaur fauna of the Mesozoic era still captures our imagination. This painting illustrates some of the large species from the Jurassic and Cretaceous periods.

species, and terrestrial biotas became increasingly provincialized. Dinosaurs continued to diversify. The first snakes appeared during the Cretaceous, but their lineage did not radiate until much later. By the end of the period, most groups of modern mammals had evolved, but these animals were generally rather small. Early in the Cretaceous period, possibly somewhat earlier, flowering plants evolved from gymnosperm ancestors and began the radiation that led to their current dominance on land. The earliest angiosperms were pollinated by insects. Indeed, the diversification of angiosperms in the Cretaceous coincides with major increases in the diversity of insect groups, such as moths, butterflies, wasps, bees, ants, and flies, whose ecology is closely bound up with plants, especially with their flowers and fruits.

The end of the Cretaceous period, a time when sea levels dropped substantially, saw another great extinction in the history of life on Earth. Ammonites and dinosaurs, which already had been declining for some time, became extinct, except for one dinosaur lineage that gave rise to crocodiles and birds. On land, all vertebrates larger than about 25 kilograms in body weight apparently became extinct. Mass extinction was most prevalent among marine planktonic organisms and bottom-dwelling invertebrates. Small terrestrial vertebrates survived the end of the period much better, and there was almost no extinction among fishes and terrestrial plants.

The Cenozoic Era

The Cenozoic era (65 mya–present) is often called the "Age of Mammals" because of the extensive radiation of that group, but other taxa were undergoing important changes as well. Because of the richness of its fossil record, paleontologists divide the Cenozoic era more finely than they do the earlier eras by subdividing the periods into epochs.

In the Tertiary period (65–2 mya) the angiosperms diversified extensively and came to dominate world forests except in cool regions. In the middle of the Tertiary, when the climate became considerably drier and cooler, many lineages of angiosperms evolved herbaceous (nonwoody) forms, and parklands and grasslands spread over much of Earth. By the beginning of the Cenozoic era, invertebrate faunas were essentially modern in most respects. It is among the vertebrates that evolutionary change was most rapid during the last 65 million years. Living groups of reptiles, such as snakes and lizards, underwent extensive radiations during this period. In fact, snakes have undergone speciation as rapidly as any taxon during the past few million years.

The main radiation of birds took place in the early Tertiary period. By the Eocene epoch (55 mya), all modern orders were present. By the Miocene epoch (25 mya), birds differed little from their modern relatives. Mammals were common at the beginning of the Tertiary period, but the largest were no bigger than a small dog. The differentiation of most orders of mammals took place early in the Tertiary. The class Condylartha, generalized herbivores of the early Tertiary, gave rise to many orders of hooved mammals, including four orders of South American mammals that became extinct before the end of the period. Grazing and browsing mammals proliferated during those periods when Earth was drier and herbaceous vegetation dominated large expanses of the continents, as it does today. The groups of mammals that have undergone the most extensive diversification in recent times are the bats, rodents, hooved mammals, and primates.

The most recent geological period, the Quaternary, began with the Pleistocene epoch about 2 mya. The Pleistocene was a time of drastic cooling and climatic fluctuations. There were four major and several minor glacial episodes during which Earth became much cooler and distributions of animals and plants moved toward the equator. The last glaciers retreated from temperate latitudes less than 10,000 years ago. Organisms are still adjusting to these changes. Many high-latitude ecological communities have occupied their current locations for no more than a few thousand years. Interestingly, the Pleistocene climatic fluctuations resulted in few extinctions of species. Instead, most species shifted their ranges. The not-

able exception is the extinction of large birds and mammals in North and South America and Australia, coincident with the arrival of people on those continents. These were the first of the human-caused extinctions. How our species is affecting biological diversity today will be discussed in the last chapter of this book.

PATTERNS IN THE EVOLUTION OF LIFE

Despite the incompleteness of the fossil record, especially for early periods in the history of life, some general patterns in the origins of lineages, extinctions, and changes in size are evident.

Origins of Lineages

Three times during the history of life, many new evolutionary lineages originated (Figure 27.11). The

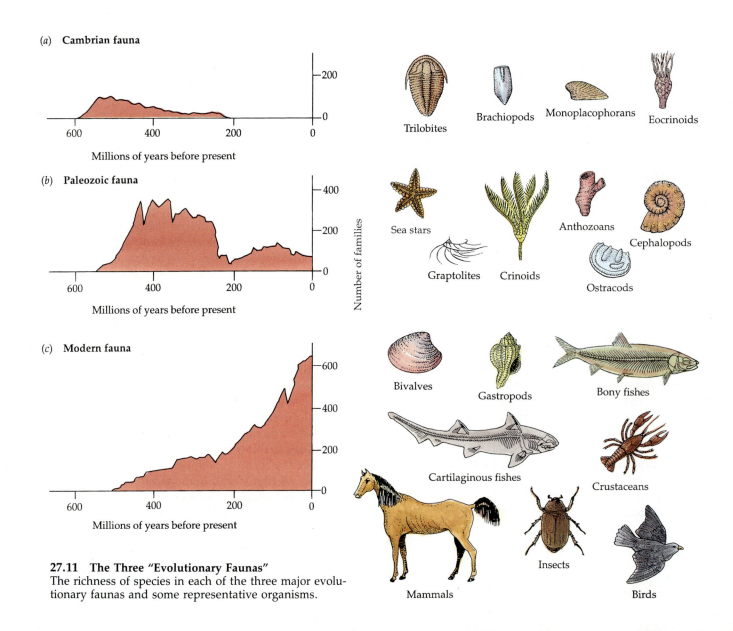

(a) **Cambrian fauna**

Millions of years before present

(b) **Paleozoic fauna**

Millions of years before present

Number of families

(c) **Modern fauna**

Millions of years before present

Trilobites Brachiopods Monoplacophorans Eocrinoids

Sea stars Graptolites Crinoids Anthozoans Ostracods Cephalopods

Bivalves Gastropods Bony fishes

Cartilaginous fishes Crustaceans

Mammals Insects Birds

27.11 The Three "Evolutionary Faunas"
The richness of species in each of the three major evolutionary faunas and some representative organisms.

first such event, known as the Cambrian explosion, was about half a billion years ago. The second, about 60 million years later, resulted in the Paleozoic fauna. The great Permian extinctions 300 million years later were followed by the third event, the Triassic explosion, which led to our modern fauna. These explosions all resulted in the evolution of many new species, but qualitatively they were very different. During the Cambrian explosion, virtually all the major groups of living organisms appeared, along with a number of phyla that subsequently became extinct. As many as 100 different phyla of organisms may have evolved in the Cambrian, of which only 30 survive today. The Paleozoic and Triassic explosions greatly increased the number of families, genera, and species, but no new phyla of organisms evolved. These later explosions resulted in many new species of organisms, but all those species had modifications of the *Baupläne* already present when the great biological diversifications began.

Evolutionists have long puzzled over these striking differences. The most commonly accepted theory is that because the Cambrian explosion took place in a world that contained few species of organisms, all of them small, the ecological setting was favorable for the evolution of many new *Baupläne* and different ways of life. Many types of organisms were able to survive initially in this world, but as competition intensified and new types of predators evolved, many forms were unable to persist. The post-Cambrian world was also relatively poor in species at the time of both later evolutionary explosions, but the species that were already present included a wide array of *Baupläne* and ways of life. Therefore, major new innovations were less likely to evolve in this world than in the Cambrian period.

The fossil record for marine invertebrates with hard skeletons is good enough to provide a general picture of the number of lineages, as estimated by the number of families in those phyla, over the course of evolutionary history. From the Cambrian period, the number of families steadily increased through the Ordovician period, but then remained roughly constant, except for a temporary drop at the end of the Devonian period, followed by a return to the former number. Following the Permian extinctions, diversity has increased steadily throughout the Mesozoic and Cenozoic eras to a maximum today. Approximately 40,000 species of marine invertebrates lived in the Paleozoic and Mesozoic eras, a number that increased to about 250,000 in the late Cenozoic era.

Two principal factors contributed to this rise in diversity. As we already mentioned, one is increasing provinciality resulting from the breakup of Earth's ocean by the drifting continents. Much of the Cenozoic era increase in numbers of species appears to be due to provincialization. The second major factor is the increasing number of species within ecological communities, that is, groups of interdependent organisms of various species. For example, late Cenozoic era communities appear to have about twice as many species in them as earlier communities did, primarily because organisms in later communities are more diverse in their ways of living. In the earlier communities, most organisms lived on or near the surface of bottom sediments. Later there were many more burrowing forms, species able to move actively around on the surface, and species able to swim in the water column.

On land, the numbers of species of vascular plants increased from the Devonian period to the Permian period, decreased at the end of the Permian period, and then rose again to a plateau that was maintained until angiosperms began to diversify in the late Cretaceous period. There are many more species of angiosperms living today than at any time previously. Because the fossil record of insects is so poor, we lack good estimates of insect species numbers in the past. However, because so many insect species feed on only a single species of plant or a few closely related species, it is certain that the number of insect species increased along with that of angiosperms. Thus the assemblage of insects today is richer than ever before. The number of genera of mammals also increased throughout the Cenozoic era.

Trends in Size

The earliest organisms were small. Large organisms did not evolve until Cambrian times. The maximum sizes of living organisms have increased, irregularly to be sure, during the course of evolution. The largest plants and animals that ever lived are alive today (Figure 27.12). E. D. Cope first noted that sizes often increase within lineages over time. Cope observed this trend among vertebrates, but it has since been verified among foraminiferans, arthropods, echinoderms, brachiopods, ammonites, and various plant groups. The most striking exception to this trend is among insects, which have remained small throughout their evolutionary history. Nonetheless, because sizes of organisms sometimes decrease within lineages, Cope's generalization is only true on average. Size in lineages can decrease, even when there is selection *within species* for larger sizes, if species composed of larger organisms have higher extinction rates than other species in the same lineage whose members are smaller. In fact, the record suggests that this is true (Figure 27.13). Decreases in size are usually followed either by another period of evolution of increasing size within the same lineage or by the replacement of the lineage by another group that also undergoes prolonged increases in size.

Short-term studies of the action of natural selection on the bills of Darwin's finches, as discussed in Chapter 18, have revealed the importance of food

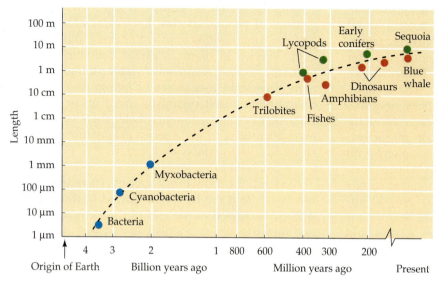

27.12 The Largest Organisms Live Today
Only the largest species living at each time are plotted.

supplies and competition in this process. Other studies have shown that mate selection and competition among males for access to females are important factors favoring large sizes among males of many vertebrate species. Two types of studies are revealing. One uses a comparative approach in which the extent of size differences between males and females in many species is correlated with such factors as habitat, diet, and mating system. For example, among the American blackbirds, size differences between

males and females is greatest among species in which one male mates with a number of females, a mating pattern called **polygyny** (Table 27.1). These results suggest that competition among males for breeding opportunities is the key factor leading to larger males. This interpretation of the general pattern is reinforced by detailed studies of individual species. For example, among Montezuma oropendolas, a polygynous blackbird species, heavier males are better fighters than lighter males, and they achieve most of the copulations.

Evolutionary studies of size changes among mammals colonizing islands are also revealing. Interestingly, small mammals tend to increase in size on islands, whereas large mammals tend to decrease in size. A likely reason is that on small islands there is less for large mammals to eat, so small size is favored among large species. However, for small mammals, the absence of predators from which they must hide may favor larger sizes and the more efficient exploitation of plant resources that larger size confers. Thus whether mammals become larger or smaller on islands depends on the initial size of the colonizers.

Rates of Evolutionary Change

One of the most interesting features of the history of life is that, for most lineages, there are long periods when rates of morphological evolution appear to be very slow and other periods when rates of morphological evolution are much faster. For example, the average rate of change of the height of molars of horses during the Tertiary was 4 percent per million years. In laboratory experiments under strong artificial selection, rates of evolution hundreds of times that fast have been reported for short time periods.

27.13 Larger Species Live a Shorter Time
The bars indicate the average duration of species within the given animal groups plotted against the average size of species within a group. Larger species tend to have shorter evolutionary lives than smaller species, although many small species were also short-lived.

TABLE 27.1
Sexual Size Dimorphism and Mating Systems among American Blackbirds

Group	Mating system	Spacing pattern	% Size difference between sexes (mean and range)	♂	♀
Oropendolas	Polygynous	Colonial	25 (15–35)		
Caciques	Polygynous	Colonial	22 (21–23)		
Caciques	Monogamous	Territorial	12 (10–15)		
Orioles	Monogamous	Territorial	6 (0–14)		
Marsh nesting species	Polygynous	Territorial	16 (12–20)		
Marsh nesting species	Monogamous	Territorial	7 (6–14)		
Grackles	Monogamous	Territorial	13 (11–14)		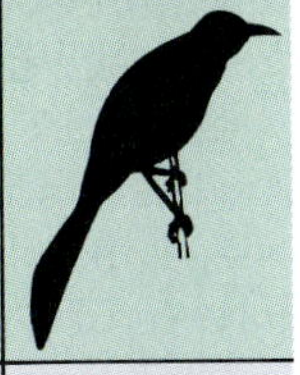
Grackles	Polygynous	Territorial	21 (20–22)		

Higher rates of evolution have also been measured in nature. For example, skeletal proportions among house sparrows introduced into North America have changed 5–30 percent during the past century. Thus, even during periods of "rapid" evolutionary change that are known from the fossil record, rates of evolution are much slower than those that are known to be possible.

Many patterns in the fossil record suggest that there are frequently long periods during which rates of morphological evolutionary change are extremely slow (**stasis**), interspersed with periods when they are much faster, a pattern known as **punctuated equilibrium**. Evolutionary rates are seldom constant within lineages. Also, many organisms have changed very little over millions of years—at least as revealed by their skeletal remains—whereas others have undergone remarkable morphological evolution during the same time period.

Are such patterns consistent with the idea that evolution is influenced primarily by natural selection? Are they predictable from theories of evolution by natural selection? These are two very different questions because an inability to predict rates of future change does not mean that the patterns are inconsistent with the theory.

A useful way of viewing this problem is from the perspective of the **Red Queen effect**. Recall that the Red Queen explained to Alice in Wonderland that "it takes all the running you can do, to keep in the same place. If you want to get somewhere else, you must run at least twice as fast as that!" The evolutionary interpretation of the Red Queen effect is that any evolutionary change in a species is experienced by coexisting species as a change in their environment. In addition, most changes in coexisting species are likely to be detrimental from the point of view of their associates. Predators and parasites become better at exploiting their prey; thus prey evolve better defenses against their predators, and so forth. On the other hand, evolution is not that simple. Plants do not evolve to be totally unpalatable. Predators reach some sort of equilibrium with their prey. That is because there are costs to changes, and a change that improves performance at one task often reduces performance at other tasks. For example, slender legs that enable a horse to run faster also break more readily. Wild horses living in environments with predators actually run more slowly than race horses bred for speed. However, in the wild, on rough terrain, race horses probably would not fare as well as slower but sturdier horses.

Therefore, communities of organisms may reach a stage in which no change in any of the species is favored by natural selection. This stage could persist for long periods of time if the environment does not change. However, a change in the environment might mean that mutants that were not previously favored were now well adapted to the new environment. An example of such a process is provided by the mollusks of Lake Turkana in Kenya, for which there is an excellent fossil record extending from the Pliocene epoch to recent times. Most of the lineages are still living, and details of evolutionary pathways are available for 19 of them. Major phenotypic changes occurred at about the same time in all the lineages, and these were associated with major drops in the level of the lake (Figure 27.14) that isolated populations in small basins. Later, when lake levels rose again and populations were reunited, many of the more recently evolved forms died out and the original species regained their dominance. Thus periods of rapid evolution were triggered by important changes in both the physical environment and the kinds and numbers of species with which each mollusk interacted.

EXTINCTION RATES

More than 99 percent of the species that have ever lived are extinct. However, we know relatively little about the causes of extinction, except for species that have become extinct in historical times. We do know, however, that rates of extinction have not been constant. There have been extinctions at all times throughout the history of life, but rates have changed dramatically. In addition, some groups suffered high extinction rates while others were proliferating. Paleontologists distinguish between "normal" or "background" extinction rates and rates characterizing **mass extinctions**, periods when rates of extinction are much higher than during intervening times.

There were six mass extinctions during the history of life (Figure 27.15). The earliest of these, at the end of the Cambrian period, destroyed about half of the known animal families. At the end of the Devonian period, 345 million years ago, another 30 percent of animal families became extinct. The catastrophe at the close of the Permian period affected both marine and terrestrial organisms. In the oceans, brachiopods declined almost to extinction and trilobites became totally extinct. On land, the trees that formed the great coal forests became extinct, as did most groups of amphibians. At the end of the Triassic period, about 180 million years ago, nearly all ammonoids and approximately 80 percent of reptile species vanished. Sixty-five million years ago, at the end of the Cretaceous period, dinosaurs, large reptiles, and a large part of marine life disappeared.

Mass extinctions changed the flora and fauna of the next period by eliminating some types of organisms and thus increasing the relative abundance of others. For example, among planktonic Foraminifera, important marine protists, the only survivors of mass extinction periods were relatively simple species with broad geographic ranges. Among the mollusks of the Atlantic coastal plain of North America, *species* with broad geographic ranges were less likely to go extinct during background extinction periods than were species with small geographic ranges. However, during the mass extinction of the late Cretaceous period, *groups of species* with large geographic ranges survived better than those with small ranges, even if the *individual* species in them had small ranges. Similar patterns are found in other molluscan groups elsewhere, suggesting that traits favoring long-term survival during "background" times are often different from those that favor survival during times of mass extinctions.

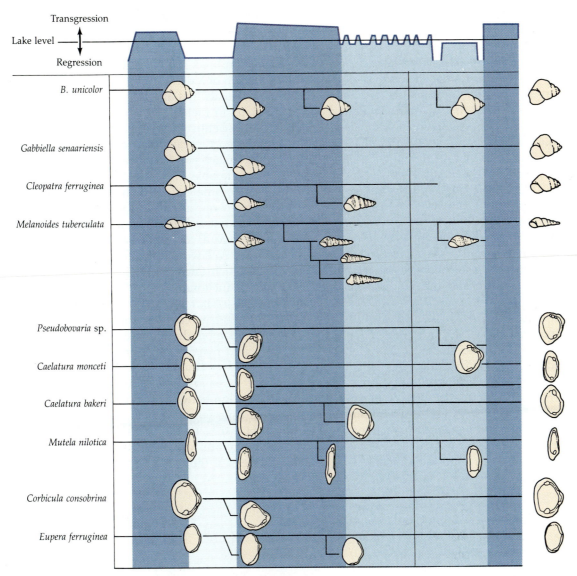

27.14 Environmental Changes Stimulate Evolution
Notice how these ten lineages of mollusks evolved rapidly at times when the lake level changed dramatically.

On land, at the end of the Cretaceous period, extinction rates were much higher among large vertebrates than among small ones. The same was true in the Pleistocene when extinction rates were high only among large mammals and birds. In addition, as we have seen, during some mass extinctions, marine organisms were heavily hit while terrestrial organisms survived well. Other extinctions affected organisms in both environments.

Causes of Extinctions

It is difficult to determine the causes of extinctions by studying fossils alone. It is helpful to have systems in which the changes documented in the fossil record are continuing and can be studied today. A good

example is provided by the three-spined stickleback (*Gasterosteus aculeatus*), a tiny (75 mm or less), widespread marine fish that has repeatedly invaded fresh water throughout its long evolutionary history. Sticklebacks in all marine and most freshwater populations have well-developed pelvic girdles with prominent spines that make it difficult for other fishes to swallow them. Large predatory insects, however, can readily grasp the stickleback's spines, and these insects prey selectively on the stickleback individuals with the largest pelvic girdles. When sticklebacks invade freshwater habitats where predatory fish are absent but predatory insects are present, the lineages rapidly evolve smaller spines. Populations with reduced spines are found primarily in young lakes that were covered by ice during the last glaciation (Figure

27.16). The extensive fossil record of sticklebacks shows that spine reduction evolved repeatedly over millions of years in populations that invaded fresh water. In addition, molecular data reveal that freshwater populations are all closely related to adjacent marine populations. Therefore, spine reduction has evolved many times in different places. The fossil record also shows that lineages with smaller spines have had much higher extinction rates than lineages with well-developed spines. That has happened because the lakes in which spine reduction evolves are ephemeral habitats from which it is difficult to migrate to other habitats. The recently glaciated lakes where sticklebacks with small spines live today are no more than a few thousand years old; as they fill in with sediments, those populations will also go extinct. Newly formed lakes elsewhere will be colonized by marine sticklebacks, and the process will repeat itself.

The differences among mass extinctions in severity and in which organisms became extinct suggest that there is no single cause for mass extinctions. Some theories advanced to explain mass extinctions invoke extraordinary events with an external origin; others propose that normal events on Earth were responsible. The most widely accepted external catastrophe theory suggests that a large meteorite or asteroid collided with Earth at the end of the Cretaceous period and threw a cloud of dust into the atmosphere large enough to darken skies and lower temperatures worldwide. The impact site of the presumed meteorite tentatively has been identified in the Caribbean Sea. There is indirect geological evidence for the impact in the form of a widespread layer of iridium and other heavy metals that are typically in much higher concentrations in meteorites than on the surface of Earth. Not all scientists are convinced of the external catastrophe theory because it requires that mass extinctions take place within a small number of years. There is considerable evidence that many groups of animals that became extinct at the end of the Cretaceous had been declining for some time. Also, extinctions were not synchronous in different regions of Earth. Therefore, although evidence that a large foreign object collided with Earth is quite strong, there is still much uncertainty concerning its role in causing the extinctions.

In addition to extraterrestrial causes of mass extinctions, normal geological processes and events, such as mountain building, massive volcanic activity, glaciations, and major regressions in sea levels, are also likely to have profoundly influenced extinction rates. There is, as we have already pointed out, a strong association between periods when these processes were especially active and times of mass extinctions, but this temporal association does not prove that they are related causally. Because of the incompleteness of the fossil record, and because no

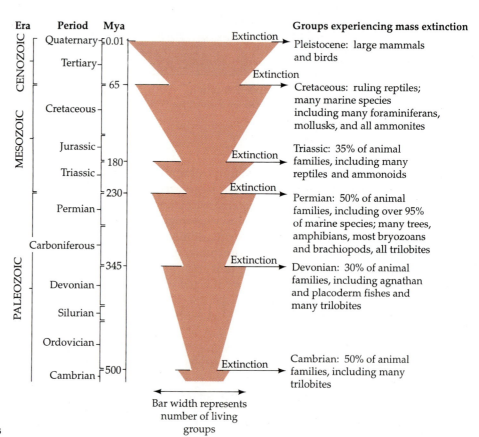

27.15   The Six Mass Extinctions

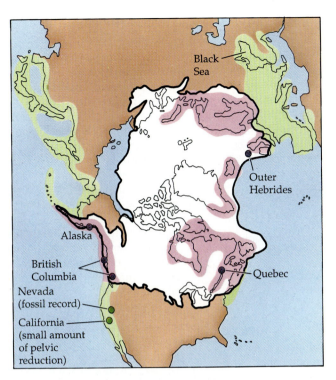

27.16 Natural Selection Acts on Spines
The region covered by Pleistocene glaciers is outlined in black. The current range of sticklebacks includes formerly glaciated areas (shown in lavender) and unglaciated areas (yellow). Orange regions lie outside the range of the species. Places where sticklebacks are known to have evolved reduced spines are indicated by circles.

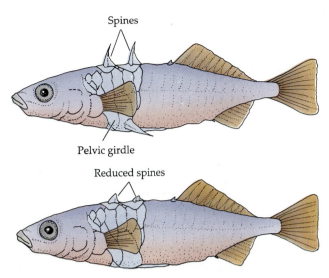

single cause is likely to explain all six mass extinctions, controversy over the causes of particular mass extinctions is likely to continue for a long time.

HOW PREDICTABLE IS EVOLUTION?

The early evolution of life is the result of physico-chemical events in the unique environment present on the early Earth. No other planet in our solar system provided conditions suitable for anything closely resembling life as we know it on Earth. But could other forms of life also have arisen? Was the emergence of life as we know it inevitable, given the environments of those remote times? Once life evolved, were the phyla as we know them almost certain to have evolved? Biologists hold very different views on these important questions.

We may never know how predictable evolution is, but there are reasons to believe that the world might look very different today if the continents had drifted apart at different times and at different rates than they actually did. If asteroids had not collided with Earth or mass eruptions of volcanoes had not taken place, dinosaurs might still dominate our planet. There might never have been an Age of Mammals.

The massive extinctions of the present century are being caused by a single species whose appearance was by no means certain.

For these reasons, we probably cannot predict the future course of evolution. However, this does not mean that the course of evolution has been random or chaotic. The evolutionary radiations that followed major perturbations and extinctions were probably responses to the new conditions created by those changes. By careful study we may be able to explain those patterns. We may also be able to explain why certain groups of organisms became extinct but others survived when the environment changed. Therefore, the course of evolution may be *understandable* in scientific terms, even though it may be too strongly affected by unpredictable events to be *predictable*.

SUMMARY

Life appeared more than 3 billion years ago. The length of time over which life has been evolving is very difficult to grasp. Nonetheless, the broad patterns of the evolution of life on Earth are evident. Most of what we know about past life has been discovered through the study of fossils, but knowledge of the important physical changes on Earth, particularly the drifting of continents, is necessary if we are to understand the patterns revealed by fossils. During the Cambrian period, all surviving phyla of organisms, plus some that no longer survive, evolved. Since then, evolution has modified the *Baupläne* that first appeared in the Cambrian period. The rapid diversification of organisms following several periods of mass extinction has not produced any new phyla for more than 500 million years. However, the number of species living on Earth has increased since the end of the Permian period, and more species live today than at any time in the past.

SELF-QUIZ

1. The number of species of fossil organisms that has been described is:
   a. 50,000
   b. 100,000
   c. 200,000
   d. 300,000
   e. 500,000

2. Radioactive carbon can be used to date the age of fossil organisms because:
   a. all organisms contain many carbon compounds.
   b. radioactive carbon has a regular rate of decay to nonradioactive carbon.
   c. the ratio of radioactive to nonradioactive carbon in living organisms is always the same as that in the atmosphere.
   d. the production of new radioactive carbon in the atmosphere just balances the natural radioactive decay of ^{14}C.
   e. all of the above.

3. About two-thirds of the species so far identified in the Ediacaran fauna are thought to be:
   a. algae.
   b. protists.
   c. echinoderms.
   d. cnidarians.
   e. ferns.

4. The coal beds we now mine for energy are the remains of:
   a. trees that grew in swamps during the Carboniferous period.
   b. trees that grew in swamps during the Devonian period.
   c. trees that grew in swamps during the Permian period.
   d. herbaceous plants that grew in swamps during the Carboniferous period.
   e. none of the above.

5. The Mesozoic period is commonly called the:
   a. Age of Fishes.
   b. Age of Reptiles.
   c. Age of Mammals.
   d. Age of Gymnosperms.
   e. Age of Seed Ferns.

6. The times during the history of life when many new evolutionary lineages appeared were:
   a. Precambrian, Cambrian, and Triassic.
   b. Precambrian, Cambrian, and Tertiary.
   c. Cambrian, Paleozoic, and Triassic.
   d. Cambrian, Triassic, and Devonian.
   e. Paleozoic, Triassic, and Tertiary.

7. The largest organisms that evolved during the history of life lived:
   a. during the Cambrian period.
   b. during the Carboniferous period.
   c. during the Cretaceous period.
   d. during the Tertiary period.
   e. during the Quaternary period and today.

8. We know that organisms can evolve rapidly because:
   a. the fossil record reveals periods of very rapid evolutionary change.
   b. theoretical models of evolutionary change show that rapid change can be produced by natural selection.
   c. rapid evolutionary changes have been produced under artificial selection.
   d. rapid evolutionary changes have been measured in natural populations of organisms during the past century.
   e. all of the above.

9. In which of the following periods was there *no* mass extinction of living organisms?
   a. The end of the Cambrian period
   b. The end of the Devonian period
   c. The end of the Permian period
   d. The end of the Triassic period
   e. The end of the Silurian period

10. We cannot predict the future course of evolution because:
    a. prediction of future events is always difficult.
    b. the course of evolution has been random.
    c. the course of evolution has been chaotic.
    d. we cannot predict what we do not understand.
    e. the course of evolution has been influenced by major events whose occurrence is highly unpredictable.

READINGS

Bonner, J. T. 1988. *The Evolution of Complexity by Means of Natural Selection.* Princeton University Press, Princeton, NJ. An excellent treatment of broad patterns in the evolutionary record and the developmental and physiological bases for them.

Futuyma, D. J. 1986. *Evolutionary Biology,* 2nd Edition. Sinauer Associates, Sunderland, MA. The best general treatment of evolution and its mechanisms.

Gould, S. J. 1989. *Wonderful Life: The Burgess Shale and the Nature of History.* W. W. Norton, New York. An engaging account of the remarkable fauna, containing representatives of many phyla of animals that left no modern descendants, found in the Burgess Shale. Explores the implications of this fauna for our view of the history of life.

Hawking, S. W. 1988. *A Brief History of Time: From the Big Bang to Black Holes.* Bantam Books, New York. A concise account, written for a lay audience willing to tackle difficult concepts, of current theories of the evolution of the universe.

Loomis, W. F. 1988. *Four Billion Years.* Sinauer Associates, Sunderland, MA. A very readable essay on the evolution of genes and organisms and how they are related.

Morris, S. C., and H. B. Whittington. 1979. "The Animals of the Burgess Shale." *Scientific American,* July. A short account of the finds in one of these rich and important fossil beds.

Raup, D. M., and S. M. Stanley. 1978. *Principles of Paleontology,* 2nd Edition. W. H. Freeman, San Francisco. An excellent account of the history of life and how it is studied.

Russell, D. A. 1982. "The Mass Extinctions of the Late Mesozoic." *Scientific American,* January. An interpretation of the causes of one of the great extinctions.

Simpson, G. G. 1944. *Tempo and Mode in Evolution.* Columbia University Press, New York. A dated but classic book providing clear pictures of the long-term patterns of evolutionary change.

Stanley, S. M. 1979. *Macroevolution: Pattern and Process.* W. H. Freeman, San Francisco. A text that argues that macroevolutionary mechanisms are different from microevolutionary mechanisms.

White, M. E. 1986. *The Greening of Gondwana.* Reed Books, Sydney. A beautifully illustrated account of the geological history of Australia and the unusual plants that evolved there.

FOR STUDY

1. Some groups of organisms have radiated into a large number of species while others only developed into a small number of species. Is it meaningful to consider the former groups as more "successful" than the latter? What does the word "success" really mean in evolution? How does your answer influence your thinking about *Homo sapiens*, the only surviving representative of the Hominidae—a family that never had many species in it?

2. If extinction rates in groups of organisms are relatively constant over long time spans, as they sometimes are, then recently evolved species should be no better adapted to their environments than older species. Does this observation contradict the belief that natural selection adapts organisms to their environments?

3. Why is it useful to be able to date past events absolutely as well as relatively?

4. In a study performed a number of years ago, lungfish fossils and modern specimens were assigned scores based on the number of advanced versus ancestral characters they displayed. On this scale, 0 represents the inferred ancestral condition and 100 the most modern forms. The pace of evolution in this group is plotted in two graphs below. Is this a fairly typical or an atypical pattern? What does it suggest about adaptive radiation among lungfishes?

5. What factors favor increases in body size? How could average body size among species in a lineage decrease while natural selection is favoring large body sizes in most of the species in the lineage?

6. Does the uneven pace of evolution negate the belief that natural selection is the principal evolutionary agent? What does the theory of natural selection, by itself, predict about the rates of evolutionary change?

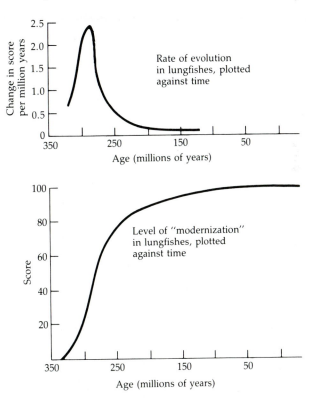

PART FIVE

The Biology of Flowering Plants

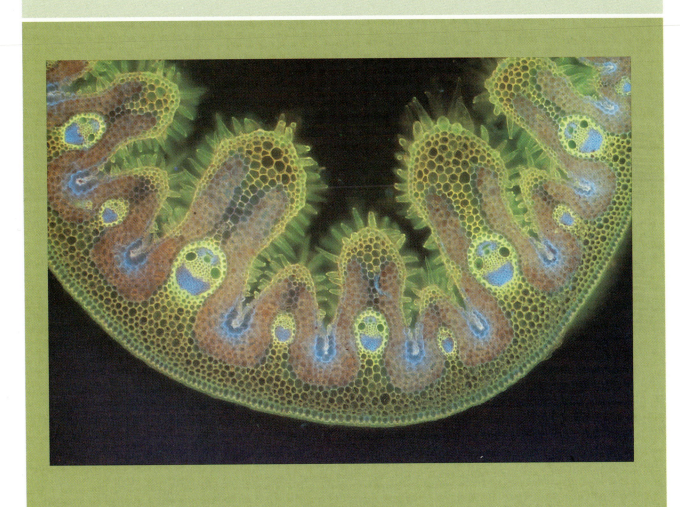

28

The Flowering Plant Body

PREVIEW: The flowering plants—monocots and dicots—consist of a few important organs whose functions constitute the life of the plant. The functions of these organs can be understood in terms of their structure, including the microscopic structure of their component cells. The cells are grouped into tissues, and the tissues are grouped into organs. The plant body is highly modular, but it functions as a coherent unit.

This chapter deals with roots, stems, leaves, meristems, parenchyma, sclerenchyma, collenchyma, xylem, phloem, vascular bundles, primary growth, and secondary growth.

More than three centuries ago Robert Hooke became the first person to observe cells under a magnifying lens. He was examining tissue from the stem of a flowering plant—specifically, cork. In fact, what he described was the cell walls of dead tissue. Improvements in microscopy over succeeding centuries gradually gave us a detailed knowledge of the anatomy of plants. Compare Hooke's drawing of cork (Figure 1.16) with a modern micrograph of the same material (Figure 28.1).

As we saw in Chapter 24, flowering plants dominate the scene on many of Earth's land masses. You will recall that flowering plants are vascular plants that are distinguished by double fertilization and by seeds enclosed in modified leaves; their xylem contains vessel elements and fibers. If you have not been thinking about plants for a while, you might want to review the sections entitled "The Two Divisions of the Plant Kingdom" and "Class Angiospermae: Flowering Plants" in Chapter 24 before reading the rest of Chapter 28.

In this chapter we will present some basic anatomical features common to many flowering plants. As always in biology, it is important to remember that there *are* differences from group to group of organisms, and even from species to species. Let us begin, then, by looking at four important or familiar species: coconut palm, red maple, rice, and soybean.

FOUR EXAMPLES

Coconut Palm

In some cultures the coconut palm (*Cocos nucifera*, Figure 28.2) is called the Tree of Life, because every above-ground part of the plant has value to humans. Where the coconut palm grows, in coastal lowlands throughout the tropical world, people use its stem—the trunk—as lumber. They dry the sap from its trunk to yield a sugar, or they ferment the sap and drink it as "toddy." They use the leaves as thatching for cottages and also for making hats and baskets. They eat the apical bud at the top of the trunk as "hearts of palm" in salads. The coconut fruit serves in many ways. The hard shell can be used as a utensil or burnt as fuel, while the fibrous middle layer, or coir, of the fruit wall can be made into mats and rope. The seed of the coconut palm has both a liquid endosperm—"coconut milk"—and a solid endosperm—"coconut meat." The refreshing and delicious milk contains no bacteria or other pathogens and is thus a particularly important drink in locales where the water is unpotable. Millions of people get most of their protein from coconut meat. Much coconut meat is dried and marketed as copra, from which coconut oil is pressed. Coconut oil is the most widely used vegetable oil in the world; it is used in the manufacture of products from hydraulic brake

633

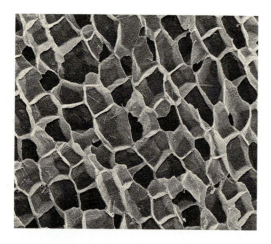

28.1 Cork
Commercial cork from the cork oak is seen in this scanning electron micrograph.

fluid to synthetic rubber and, although nutritionally poor, as food. Ground copra serves as fertilizer and as food for livestock.

The trunk of a coconut palm differs in three basic ways from the trunks of many other familiar trees. The first and most striking difference is that the trunk bears no branches, and all the leaves are borne in a cluster at the top of the trunk. The second difference is that the coconut trunk tapers little from the base

of the tree to the top—even the youngest part of the trunk is essentially as thick as the base. We will discuss this phenomenon later in this chapter. The third difference is that when the trunk is cut through, it reveals no annual rings.

Each coconut palm tree has separate male and female flowers; both are small and inconspicuous. The male flowers have six stamens. The fronds—leaves—of the coconut palm are large and made up of numerous long, narrow leaflets, each having veins running parallel to one another.

Red Maple

One of the most familiar native trees in the eastern United States is the red, or scarlet, maple (*Acer rubrum*, Figure 28.3). Unlike the coconut palm, the red maple does not provide us with a panoply of useful products. However, it enriches us by its beauty—not only is it abundant in forests, but we admire it in parks and as a street tree growing to 10–30 meters in height. We use its wood as lumber, although the sugar maple is a more important commercial source of maple wood.

Microscopic examination of a young maple stem reveals a number of vascular bundles of water- and food-conducting cells, arranged in a cylindrical pattern. The mature trunk is thick and massive; examination of a cut stump shows that the wood is made

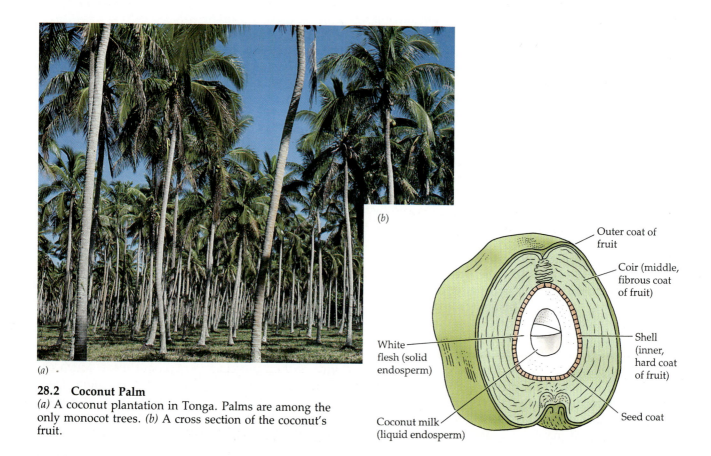

28.2 Coconut Palm
(*a*) A coconut plantation in Tonga. Palms are among the only monocot trees. (*b*) A cross section of the coconut's fruit.

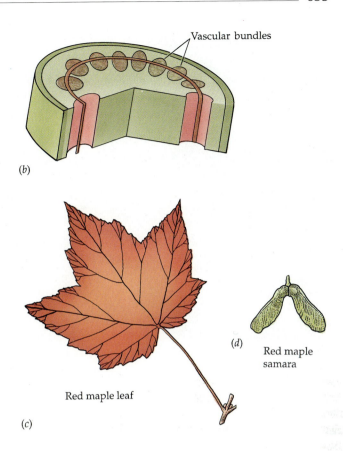

Vascular bundles

(a)

(b)

Red maple leaf

(c)

(d) Red maple samara

28.3 Red Maple
(a) A red maple tree, *Acer rubrum*, in autumn. *(b)* Vascular bundles are visible in this diagram of a young maple stem in cross section. *(c)* A leaf of red maple. *(d)* The characteristic winged fruit of the maple family.

up of many annual rings. The roots, too, are woody. The red maple leaf consists of a single blade with three to five lobes, and with veins that radiate out from a single focal point. The scarlet flowers have four sepals, four petals, eight stamens, and one pistil. The distinctive, winged fruit of the maple family contains two seeds.

Rice

More than half of the world's human population derives the bulk of its food energy from the seeds of a single plant, rice (*Oryza sativa*, Figure 28.4). Rice is particularly important in the diets of people in the Far East, where it has been cultivated for nearly 5,000 years. People use rice straw in many ways, such as thatching for roofs, food and bedding for livestock, and clothing. Rice hulls have many uses ranging from fertilizer to fuel.

Some rice is fed to livestock, but most is eaten by humans. Rice, when milled for human consumption, is an incomplete food because milling removes the bran that contains B vitamins. Even unmilled rice is a poor source of protein; thus, it should be eaten with other, supplementary foods such as soybeans

or fish. Most rice varieties are grown submerged in water for the bulk of the growing season. Fish are often raised in rice paddies, where they serve as supplemental food for humans, add fertilizer to the paddy, and control the mosquito population.

The rice plant looks much like other cereal grain plants. The leaves are long, narrow, flat, and more than half a meter in length, with veins running parallel to one another along the length of the leaf. Rice stems do not thicken and become woody as do the stems of trees and shrubs. Rice flowers have six stamens and one ovary. The vascular bundles in the rice stem are scattered, rather than lying in a ring as in the red maple stem.

Soybean

Soybeans (*Glycine max*, Figure 28.5) were first grown in China thousands of years ago, but today the United States is the largest single producer. Soybeans are featured in many foods and sauces. They also yield a commercially important oil, used in the manufacture of adhesives, paints, inks, and plastics. After oil has been squeezed from the seeds, the residue may be fed to livestock or made into soy flour. Soybean stems may be used for straw.

The soybean plant stands from less than a meter to more than 2 meters in height. Soybean leaves are three-lobed, with veins radiating out in a netlike pattern. The vascular bundles of the young soybean

28.4 Rice
(a) Terraced rice paddies in Bali, Indonesia. (b) A rice plant.

stem, like those of the red maple, are arranged in a cylindrical pattern. The flowers are small, either white or blue, and consist of five sepals, five petals, ten stamens, and one pistil. Soybean plants tend to be drought-resistant owing to their richly branching root systems, which often extend more than 1.5 meters below the soil surface.

SUBCLASSES OF FLOWERING PLANTS

Comparison of the features of these four plants— coconut palm, red maple, rice, and soybean—sug- gests at least two ways in which they may be classi- fied. We may divide them, according to their growth plan, into trees (coconut palm, red maple) and herbs (rice, soybean). Alternatively, we may divide them, according to several important anatomical characters illustrated in Figure 28.6, into monocots (coconut palm, rice) and dicots (red maple, soybean). Mono- cots are generally narrow-leaved flowering plants such as grasses (including rice), lilies, orchids, and palms. Dicots are broad-leaved flowering plants such as soybeans, roses, sunflowers, and maples. The monocots and dicots are the two subclasses, Mono- cotyledonae and Dicotyledonae, that make up the

28.5 Soybean
Mature and young soybeans are growing here in adjacent fields.

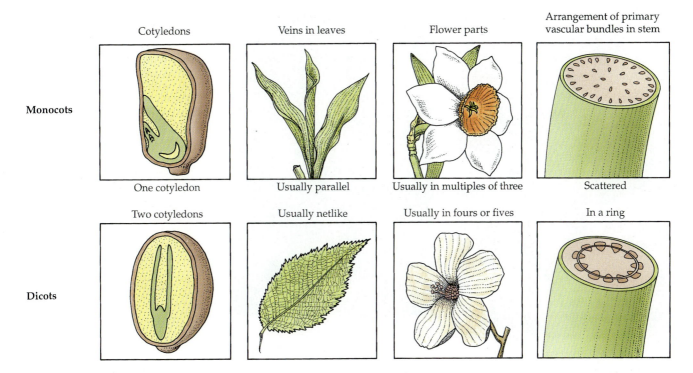

	Cotyledons	Veins in leaves	Flower parts	Arrangement of primary vascular bundles in stem
Monocots	One cotyledon	Usually parallel	Usually in multiples of three	Scattered
Dicots	Two cotyledons	Usually netlike	Usually in fours or fives	In a ring

28.6 Monocots versus Dicots

class Angiospermae—flowering plants. They get their names from the possession of, respectively, one or two **cotyledons** or "seed leaves."

ORGANS OF THE PLANT BODY

The body of most vascular plants is divided into three principal organs: the **leaves**, the **stem**, and the **root system** (Figure 28.7). A stem and its leaves, taken together, are called a shoot. The **shoot system** of a plant consists of all stems and all leaves. Broadly speaking, the leaves are the chief organs of photosynthesis, whereas the roots anchor the plant in place and absorb water and mineral nutrient elements from the soil. The stem holds and displays the leaves to the sun, maximizing the photosynthetic yield, and provides transport connections between the roots and leaves. The points where leaves attach to the stem are called **nodes**, and the stem regions between nodes are **internodes** (Figure 28.7).

Each of the principal organs can best be understood in terms of its function and its structure. By structure we mean both gross form and microscopic anatomy—the component tissues as well as their arrangement.

Roots

Water and minerals usually enter the plant through the root system. There are two principal types of root

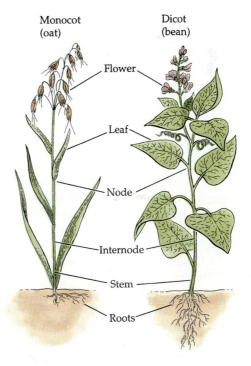

28.7 Body Plans of Monocots and Dicots
Both monocots and dicots absorb water through a root system that anchors and provides nutrients for a shoot system made of stems and leaves in which photosynthesis takes place. Flowers, made up of specialized leaves, are specialized for sexual reproduction.

(a)

(b)

28.8 Root Systems
The fleshy taproot system of a cactus (a) contrasts with the fibrous root system of grasses (b). Fibrous root systems are diverse (c–e), with forms adapted to the different environments in which they grow.

(c)　　　　　　　　　　　(d)　　　　　　(e)

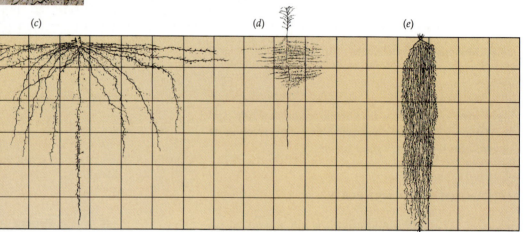

systems. Many dicots have a **taproot system**, in which a single, large, deep-growing root is accompanied by less prominent secondary roots (Figure 28.8a). The taproot itself may function as a food-storing organ, as in carrots and radishes. In contrast, monocots and some dicots have a **fibrous root system**, which is composed of numerous thin roots roughly equal in diameter (Figure 28.8b). Fibrous root systems often have a tremendous surface area for the absorption of water and minerals. A fibrous root system holds soil very well, giving grasses with such systems a protective role on steep hillsides where runoff from rain could cause erosion. The structural diversity of root systems is indeed striking (Figure 28.8c–e).

At the tip of each root there is a **root cap** (Figure 28.9) that protects the delicate growing region of the

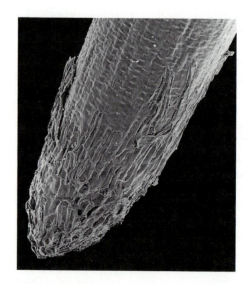

28.9 A Root Cap
The root cap protects the delicate underlying tissue of this wheat root tip. As the root grows through the soil, the root cap wears off and is replaced.

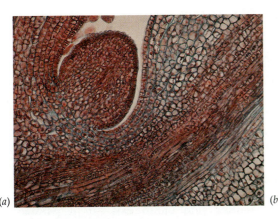

28.10 A Selection of Stems
(a) A lateral bud developing at the junction between leaf and stem of a lilac; the bud contains vascular tissue and may develop into a branch. (b) A potato is a modified stem called a tuber; the sprouts that grow from its eyes are branches. (c) Runners (red) of beach strawberry are horizontal stems; roots are produced at intervals, providing a local water supply and allowing rooted portions to live independently if the runner is cut.

root as it pushes through the soil. Cells of the root cap are often damaged or scraped away and must therefore be replaced constantly. The root cap is also the structure that detects the pull of gravity and thus causes the root to grow downward.

A tissue composed of rapidly dividing cells is located at the tip of the root proper, just behind the root cap. This tissue is the **root apical meristem**, which produces all the cells that contribute to growth in the length of the root. (Tissues responsible for growth in other parts of the plants are also called meristems.) Some of the daughter cells from the root apical meristem are contributed to the root cap, but most—those that are produced at the other end of the meristem—elongate. Following elongation, these cells differentiate, giving rise to the various tissues of the mature root.

Stems

Unlike most roots, a stem may be green and capable of photosynthesis. A stem bears leaves at its nodes, and where each leaf meets the stem there is a **lateral bud**, which develops into a branch if it becomes active (Figure 28.10a). A branch is also a stem. The branching patterns of plants are highly variable, de-

pending upon the species, environmental conditions, and a gardener's pruning activities.

Some stems are highly modified. The potato **tuber**, for example, is actually a portion of the stem. Its eyes (Figure 28.10b) contain lateral buds, and a sprouting potato is just a branching stem. The "runners" of strawberry plants and Bermuda grass are horizontal stems, giving rise to roots at frequent intervals (Figure 28.10c). Runners allow a form of asexual reproduction (Chapter 31). If a runner is broken, independent plants can develop from each side of the break.

Stems bear buds—embryonic shoots—of various types. We have already mentioned the lateral buds, which give rise to branches. At the tip of each stem or branch is an **apical bud**. The apical bud contains a **shoot apical meristem**, which produces the cells for the growth and development of the stem (Figure 28.11). **Leaf primordia**, which will expand into mature leaves, are also contained within the apical bud. The stem, at a point or points varying from species to species, also produces flowers.

Shoot systems have various forms, with branches taking on different relationships to the plant as a whole (Figure 28.12). Some plants branch underground, and their branches emerge from the soil looking like separate plants.

28.11   The Shoot Apex
Just below the dome at the tip are the dividing cells of the shoot apical meristem. Leaf primordia extend upward on either side. (See also Figure 28.19.)

Leaves

In most plants, the leaves are responsible for most of the photosynthesis, producing food for the plant and releasing oxygen gas. Leaves also carry out metabolic reactions that make nitrogen available to the plant for the synthesis of proteins and nucleic acids (Chapter 30). Leaves are important food-storage organs in some species; in others—the succulents—the leaves store water. A less obvious but often crucial function of leaves is to shade neighboring plants. Plants, like all organisms, compete; if a plant can reduce the photosynthetic capability of its neighbors by intercepting sunlight, it can obtain a greater share of the available water and mineral nutrients. Finally, as we will see in Chapter 31, the "timer" by which some plants measure the length of the night is located in the leaves.

Leaves are marvelously adapted to serve as light-gathering, photosynthetic organs. Typically, the **blade** of a leaf is flat, and during the daytime it is held by its stalk, or **petiole** at an angle almost perpendicular to the rays of the sun. Some leaves "track" the sun, moving so that they constantly face it. If leaves were thicker, the outer layers of cells would

absorb so much of the light that the interior layers of cells would be too dark, and thus would be unable to photosynthesize.

The different leaves of a single plant may have quite different shapes—the form of a leaf results from a combination of genetic, environmental, and developmental influences. Still, a given species tends to bear leaves of some broadly defined type (Figure 28.13). The leaf is either **simple**, consisting of a single blade, or **compound**, consisting of more than one blade, or leaflet. A compound leaf may bear its leaflets in either a **pinnate** fashion—featherlike, with leaflets on either side of an axis—or a **palmate** fashion—like a palm, or hand, with leaflets radiating from a central point. In a simple leaf, or in a leaflet of a compound leaf, the veins may be parallel to one another or in a netlike arrangement. Leaves are classified by their outlines, margins, tips, and bases and also by how they are arranged on the plant. At least some of the forms in Figure 28.13 probably look familiar to you. Plant taxonomists recognize and have named more than 20 types of leaf margins and almost as many different forms or outlines. Similarly, they recognize about a dozen shapes of leaf tips and another dozen of leaf bases. The general development of a specific pattern is programmed in the individual's genes and is expressed by differential growth of the leaf veins and of the tissue between the veins.

LEVELS OF ORGANIZATION IN THE PLANT BODY

Newly formed cells expand to their final size, and differentiate, that is, become structurally or chemically specialized for particular functions. A **tissue** is an organized group of cells, working together as a functional unit. Simple tissues are composed of a single type of cell, while compound tissues are composed of several cell types. Plant tissues, in turn, are organized into three tissue systems that extend throughout the body of the plant, from organ to organ. These three systems are the vascular tissue

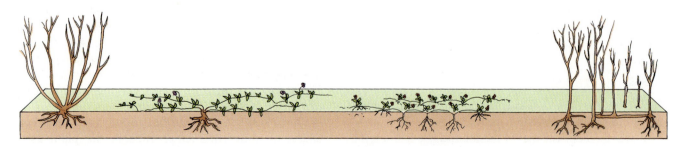

28.12   Types of Shoot Systems
Do you recognize some of these stem types? You can probably find most of them among the weeds on your campus.

Leaf shapes

Margins

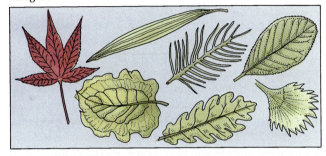

Apices and bases

Arrangements on stem

Parts and types

28.13 The Diversity of Leaves
There are many kinds of leaves. Do you recognize some of these?

system—composed of xylem and phloem—which conducts materials from one part of the body to another; the dermal tissue system, which protects the body surface; and the ground tissue system, which plays many roles, including producing and storing food materials.

To understand the structures and functions of the tissue systems, we must know the nature of their building blocks. Some cells are alive when functional, while others function only after they die and disintegrate. Some cells develop chemical capabilities not demonstrated by other cells. Several cell types differ most dramatically in the structure of their cell walls. We will first consider the various types of cells that make up the plant body and then see how aggregations of cells form both functioning tissues and tissue systems.

PLANT CELLS

Plant Cell Walls

The division of a plant cell is completed when cell walls form, separating the daughter cells. In the middle of this newly formed barrier is the **middle lamella**, a layer that cements the daughter cells to one another (Figure 28.14a). Each daughter cell, as it expands to its final size, secretes polysaccharides, such as cellulose, that form the **primary wall**.

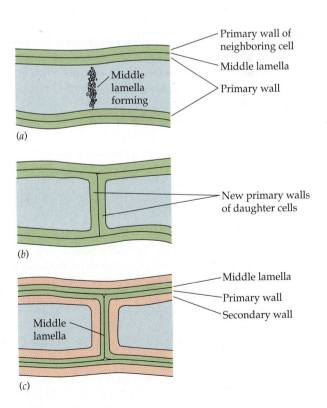

28.14 Cell Wall Formation
The middle lamella is the first wall layer to form. New layers appear as the wall develops.

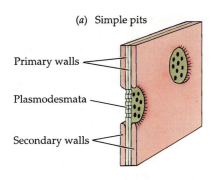

(a) Simple pits

Primary walls

Plasmodesmata

Secondary walls

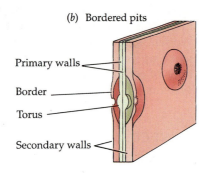

(b) Bordered pits

Primary walls

Border

Torus

Secondary walls

28.15 Pits
(a) At a simple pit, there is no secondary wall and the primary wall is thin. (b) Separation of the secondary from the primary wall produces a bordered pit. The border is an extension of the secondary wall. The primary walls in the bordered pits of some gymnosperms thicken, forming a torus that blocks the hole in the border when pressure imbalances force the torus toward one cell or the other.

Once cell expansion stops, a plant cell may deposit more polysaccharides and other materials such as lignin, characteristic of wood, or suberin, characteristic of cork, in one or more layers internal to the primary wall (Figure 28.14b,c). These layers collectively form the **secondary wall**, which often serves supporting or waterproofing roles.

The cell wall lies outside the plasma membrane of the cell, but it is not a chemically inactive region. Chemical reactions in the wall play an important role in cell expansion. Cell walls may thicken or be sculptured or perforated as part of the differentiation of various cell types.

The cell wall does not completely seal a cell off from the cell next to it. Water and dissolved materials can move readily from cell to cell because plant cells have structures called **pit pairs** connected by strands of cytoplasm called plasmodesmata (Chapter 4). A pit is a thinning in the primary wall of a cell at a place where the secondary wall is either absent or separated from the primary wall by a space. Where there is a pit in one cell, there is usually a corresponding pit in the adjacent cell—together they are a pit pair. Plasmodesmata pass through pit pairs and the middle lamella between them, allowing molecules to pass freely from one cell to the other (Figure 28.15).

Parenchyma Cells

The most numerous cells in the young plant body are the **parenchyma** cells (Figure 28.16a). Parenchyma cells are alive when they perform their functions in the plant—some other cell types become functional only after they die. Parenchyma cells usually have thin walls, consisting only of a primary wall and the shared middle lamella. Many parenchyma cells have shapes similar to those of soap bubbles crowded into a limited space—figures with 14 faces. They are not elongated or otherwise asymmetrical. Most have large central vacuoles.

Many parenchyma cells function as storage depots for various substances, such as starch or lipids. In the cytoplasm of these cells, starch is often stored in specialized plastids called leucoplasts (Chapter 4). Lipids may be stored as oil droplets, also in the cytoplasm. Other parenchyma cells appear to serve simply as "stuffing," helping to support the stem. Leaves have a particularly important type of parenchyma cell that is specialized for photosynthesis and is equipped with abundant chloroplasts. Some other parenchyma cells—but not these photosynthetic cells—retain the capacity to divide and hence may give rise to new meristems, as when a branch root forms within a region of parenchyma cells inside a taproot.

Sclerenchyma Cells

Whereas parenchyma cells function when alive, **sclerenchyma** cells function when *dead*. A heavily thickened secondary wall performs their function: support. There are two types of sclerenchyma cells: elongated **fibers** and variously shaped **sclereids**. Fibers, often organized into bundles, provide a relatively rigid support both in wood and in other parts of the plant (Figure 28.16b). The bark of trees owes much of its mechanical strength to long fibers. Sclereids are not elongated like fibers but may pack together very densely, as in a nut's shell or other types of seed coats (Figure 28.16c). There are isolated clumps of sclereids, called stone cells, in pears and some other fruits, which give them their characteristic gritty texture.

Collenchyma Cells

Another type of supporting cell, the **collenchyma** cell, remains alive even after laying down rather thick cell walls (Figure 28.16d). Collenchyma cells are generally elongated. In these cells, the primary wall thickens and no secondary wall forms. Collenchyma provides support to petioles, nonwoody shoots, and growing organs. Tissue made of collenchyma cells, although resistant to bending, is more flexible than sclerenchyma; stems and leaf petioles strengthened

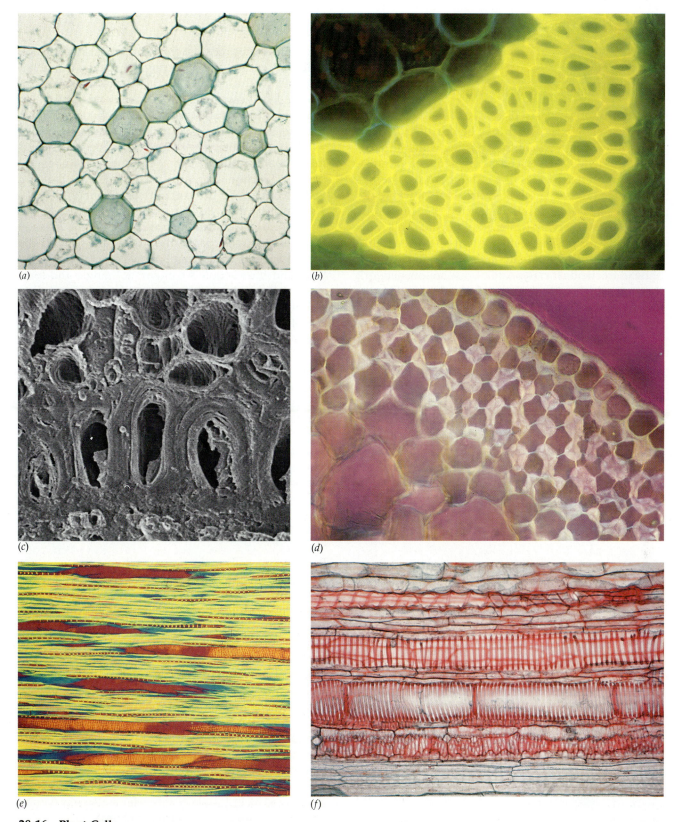

28.16 Plant Cells

(a) Parenchyma cells in the pith of a sunflower stem; note the uniformly thin walls. (b) Sclerenchyma: Fibers in a broad bean pod. A stain causes the heavily thickened walls to fluoresce a brilliant gold. (c) Sclerenchyma: Sclereids in the seed coat of a squash; the extremely thick secondary cell walls are laid down in layers. (d) Collenchyma cells make up the five outer cell layers of this spinach leaf vein. They are recognizable by the very thick primary walls at the corners of the cells and by the thinness of the remainder of the wall. (e) Tracheids appear a brilliant red in this fluorescence micrograph of pine wood; note the complexity of the cell walls. (f) Vessel elements in the stem of a squash. The secondary walls are stained red; note the different patterns of thickening, including rings and spirals. Which cells are alive and which dead at maturity?

by collenchyma are able to sway in the wind without snapping as they might if they were strengthened by sclerenchyma.

Water-Conducting Cells of the Xylem

The xylem of vascular plants contains cells called tracheary elements, which die before they assume their ultimate function of transporting water and dissolved minerals. The tracheary elements of gymnosperms and angiosperms differ significantly. The tracheary elements of gymnosperms are **tracheids**—spindle-shaped cells interconnected by numerous pits in their otherwise water-impermeable cell walls (Figure 28.16*e*). The cell contents—nucleus and cytoplasm—disintegrate upon death, so that a group of dead tracheids forms a continuous hollow network through which water can readily be drawn. Flowering plants evolved a still better-adapted type of water-conducting system made up of vessels (Figure 28.16*f*). The individual cells, called **vessel elements**, also die before they become functional. Vessel elements are generally larger in diameter than tracheids; they are laid down end-to-end and lose all or part of their end walls, so that each vessel is a continuous hollow tube consisting of many vessel elements and providing a clear pipeline for water conduction. In the course of angiosperm evolution, vessel elements have become shorter, and their end walls have become less and less obliquely oriented and less obstructed (Figure 28.17).

28.17 Conducting Cells of Vascular Systems
The xylem of angiosperms contains vessels that conduct water and minerals. These four drawings represent different stages in the evolution of the vessel element; the one at the left is the most ancient type, and the fourth is the most recently evolved type.

Sieve Tube Elements

The phloem, in contrast to xylem, consists primarily of living cells. In the flowering plants, the characteristic cell of the phloem is the **sieve tube element**. These, like vessel elements, meet end-to-end and form long sieve tubes, which transport foods from their sources to tissues that do not photosynthesize—tissues in roots, for example. As sieve tube elements mature during their development, a chemical "drilling" action expands small holes in the end walls, connecting the contents of neighboring cells. The result is that the end walls look like sieves and are called **sieve plates** (Figure 28.18). As the holes in the sieve plates expand, the membrane around the central vacuole disappears, allowing some of the cytosol and the vacuole's contents to mingle and form a single fluid; this mixture can be forced from cell to cell along the sieve tube. The nucleus and some of the other organelles in the sieve tube element also break down and thus do not clog the holes of the sieve. However, a "fixed," stationary layer of cytoplasm remains, lining the cell wall and confining the remaining organelles. In some, but not all, flowering plants, the sieve tube elements have adjacent **companion cells**. A parent cell divides, thereby producing a sieve tube element and its companion cell. Companion cells retain all their organelles and may, through the activities of their nuclei, regulate the performance of the sieve tube elements.

PLANT TISSUES AND TISSUE SYSTEMS

Parenchyma cells make up parenchyma tissue, a simple tissue composed of only one type of cell. Sclerenchyma and collenchyma are other simple tissues. But cells of various types also combine to form tissues. Both xylem and phloem are complex tissues, composed of more than one type of cell. All xylem contains parenchyma cells, which store food. The xylem of angiosperms contains vessel elements, as well as thick-walled sclerenchyma fibers that provide considerable mechanical strength to the xylem. In most gymnosperms, tracheids serve both in water conduction and in support because vessels and fibers are absent. As a result of its cellular complexity, xylem can perform a variety of functions, including transport, support, and storage. The phloem of angiosperms includes sieve tube elements, companion cells, fibers, sclereids, and parenchyma cells.

The **vascular tissue system**, which includes the xylem and phloem, serves as the conductive, or "plumbing," system of the plant. All living cells require a source of energy and chemical building blocks. As already mentioned, the phloem transports food from the sites of production—sources—commonly the leaves, to sites of utilization or storage—

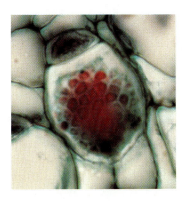

28.18 A Sieve Plate
The perforated end of a sieve tube element. Phloem sap passes through the holes in sieve plates from one sieve tube element to the next.

sinks—elsewhere in the plant. The xylem distributes water and mineral ions taken up by the roots to the stem and leaves.

The **dermal tissue system** is the outer covering of the plant. All parts of the young plant body are covered by an **epidermis**, either a single layer of cells or several layers. The shoot epidermis secretes a layer of wax, the **cuticle**, that helps retard water loss. The protective covering of the stems and roots of older woody plants is the **periderm**, which is composed of cork and other tissues that will be discussed later in this chapter.

The **ground tissue system** makes up the rest of a plant and consists primarily of parenchyma tissue, often supplemented by collenchyma or sclerenchyma tissue. The ground tissues function primarily in storage, support, and photosynthesis. Let us now see how the tissue systems are organized in the different organs of a flowering plant, as well as how this organization develops as the plant grows.

GROWTH AND MERISTEMS

Animals grow all over, that is, virtually all parts of the body grow as the individual develops from embryo to adult. Plants, in contrast, grow in specific regions of active cell division. These regions of cell division are called meristems. At the tip of each shoot or branch is a shoot apical meristem, and at the tip of each root is a root apical meristem. Tissues produced by **primary growth**—that is, by the activity of root or shoot apical meristems—are called primary tissues. Taken together, they are referred to as the primary plant body (Figure 28.19).

Many plants develop what we commonly refer to as wood and bark. These complex tissues are derived from other meristems. One, called a **vascular cambium**, is a cylindrical tissue consisting primarily of

vertically elongated cells that divide frequently, producing derivative cells both to the inside of the vascular cambium layer, forming new xylem, and to the outside, forming new phloem. As trees grow in diameter, the outermost layers of the stem are sloughed off. This would expose the tree to damage, including excessive water loss or invasion by microorganisms, but such problems are forestalled by the activity of **cork cambia**, which, in a tree, form continuously in the bark. Each cork cambium produces new cells, primarily in the outward direction. The walls of these cells become impregnated with the waxy substance suberin, and thus augment the dermal tissue system. Growth in the diameter of stems and roots, produced by the vascular and cork cambia, is called **secondary growth**. The wood and bark it produces are called the secondary tissues and constitute the secondary plant body.

In some plants, meristems may remain active for years—even centuries. Such plants grow in size, or at least in diameter, throughout their lifetimes. This phenomenon is known as indeterminate growth, in contrast to the determinate growth characteristic of most animals, which do not grow through their whole lifetimes but stop at some point. The life cycles of plants fall into three categories: annual, biennial, and perennial. **Annuals,** such as many food crops, live less than a year. **Biennials,** such as carrots and cabbage, grow for all or part of one year and live on into a second year during which they flower, set seed, and die. **Perennials** are plants, such as oak trees, that live for a few to many years.

THE MERISTEMS AND THEIR PRODUCTS

The Young Root

Cell divisions in the root apical meristem produce both the protective root cap and the other primary tissues of the growing root (Figure 28.20). When a meristematic cell divides, the products initially take up no more volume between them than did the dividing cell. One of the products of each cell division develops into another meristematic cell the size of its parent, while the other product develops differently. The products above the apical meristem—away from the root cap—constitute three cylindrical primary meristems that give rise to the three tissue systems of the root. The innermost primary meristem, the **procambium**, gives rise to the vascular tissue system; the **ground meristem** gives rise to the ground tissue system; and the outermost, the **protoderm**, gives rise to the dermal tissue system. The apical and primary meristems together constitute the root's zone of cell division, the source of all the cells of the root's primary tissues. Just above this zone, where the cells are somewhat older, is the zone of cell elongation, in which cells are elongating and thus causing the root

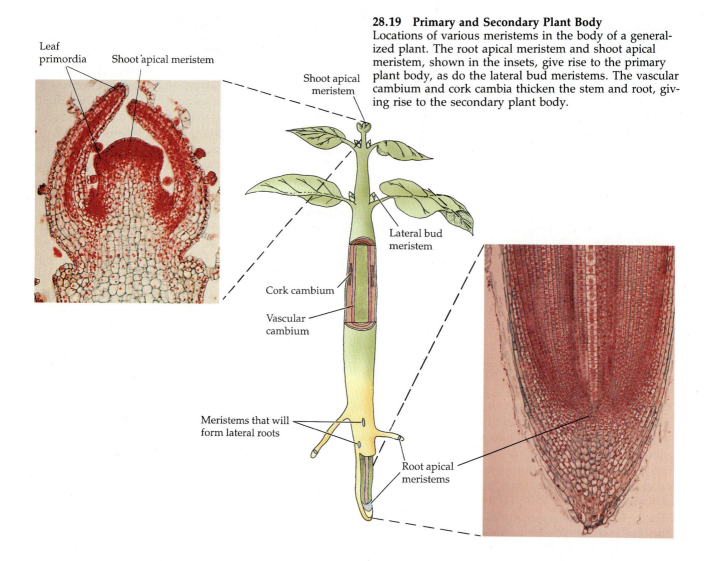

Leaf primordia

Shoot apical meristem

Shoot apical meristem

Lateral bud meristem

Cork cambium

Vascular cambium

Meristems that will form lateral roots

Root apical meristems

28.19   Primary and Secondary Plant Body
Locations of various meristems in the body of a generalized plant. The root apical meristem and shoot apical meristem, shown in the insets, give rise to the primary plant body, as do the lateral bud meristems. The vascular cambium and cork cambia thicken the stem and root, giving rise to the secondary plant body.

itself to reach ever farther into the soil. Where the cells are older yet, and where they are differentiating—taking on specialized forms and functions—is the zone of cell differentiation. These three zones grade imperceptibly into one another—there is extensive cell division even as far up as the zone of cell differentiation, and some cells differentiate even in the zone of cell division.

The protoderm gives rise to the outer layer of cells of the root, the epidermis, which is adapted for protection and for the absorption of mineral ions and water (Figure 28.21). Epidermal cells are flattened, and many of them produce amazingly long, delicate **root hairs** that vastly increase the surface area of the root (Figure 28.21d). It has been estimated that a

28.20   Tissues and Regions of the Root Tip
Cells divide in the root apical meristem. Some of the daughter cells become part of the root cap, which is constantly being eroded away, but most daughter cells develop on the side away from the tip and differentiate into the primary tissues of the root.

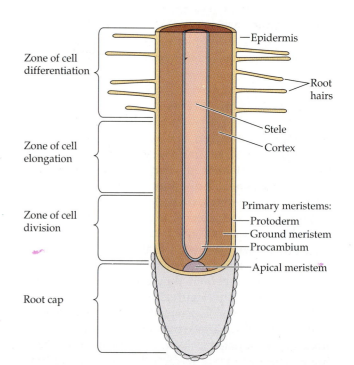

Zone of cell differentiation

Epidermis

Root hairs

Zone of cell elongation

Stele

Cortex

Zone of cell division

Primary meristems:
— Protoderm
— Ground meristem
— Procambium

Apical meristem

Root cap

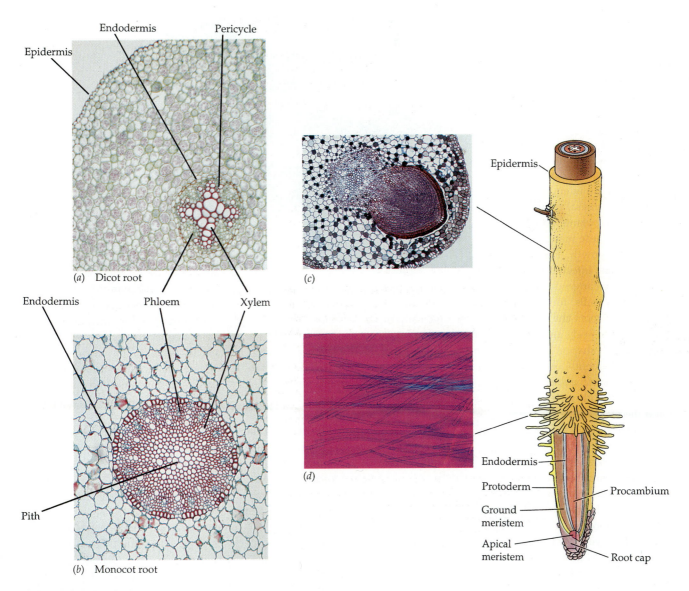

28.21 Root Anatomy

The drawing shows a generalized root structure. In the photographs are the primary root tissues of *(a)* a young dicot and *(b)* a monocot. The monocot (an orchid) has a central pith region; the dicot (ranunculus) does not. *(c)* A branching root tip. Cells in the pericycle divide and the products differentiate, forming the tissues of a branch root. *(d)* Root hairs, viewed under polarized light.

mature rye plant has a total root surface of more than 1,500 square kilometers (600 square miles)—all contained within about 6 liters of soil. Root hairs grow out among the soil particles, probing nooks and crannies and taking up water and minerals.

Internal to the root's epidermis is a region of ground tissue many cells in thickness, called the **cortex**. The cells of the cortex are relatively unspecialized and often function in food storage. In many plants, but especially in trees, epidermal and sometimes cortical cells form an association with a fungus. This association, called a **mycorrhiza**, increases the absorption of minerals and water by the plant (Box 28.A). There are some species that have poorly de-

veloped root hairs, or no root hairs at all. Plants of these species do not survive unless they develop mycorrhizae that help in mineral absorption.

Proceeding inward, we come to the **endodermis** of the root, a single layer of cells that is the innermost cell layer of the cortex. Endodermal cells differ markedly in structure from the rest of the cortical cells; parts of their walls contain suberin, a waxy substance that forms a waterproof seal. The endodermal cells control the access of water and dissolved substances to the inner, vascular tissues (see Figure 29.6). Elsewhere in the root, water can pass freely through cell walls and between cells.

Once past the endodermis, we enter the domain

BOX 28.A

Roots and Fungi

Many plants, including almost all tree species, depend on a mutually beneficial symbiotic association with fungi to get an adequate supply of water and mineral elements. The root hairs of such plants do not absorb enough of these materials to sustain maximum growth. The roots, however, become infected with a fungus, which wraps around them and may invade their cells, forming a mycorrhiza. Such roots characteristically branch extensively and become swollen and club-shaped (see figure). The hyphae (filaments) of the fungus increase the surface area for the absorption of both water and minerals, and the mass of the mycorrhiza, like a sponge, holds water efficiently in the neighborhood of the root.

Some flowering plants in most families form mycorrhizae, as do liverworts, ferns, club mosses, and gymnosperms. Fossils of mycorrhizal structures more than 300 million years old have been found. Certain plants that live in nitrogen-poor habitats, such as cranberry bushes and orchids, invariably have mycorrhizae. Orchid seeds will not germinate in

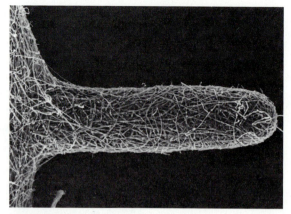

Hyphae of the fungus *Pisolithus tinctorius* cover this eucalyptus root, forming a mycorrhiza.

nature unless they are already infected by the fungus that will form their mycorrhizae. Plants without chlorophyll, such as the parasite dodder (Figure 30.24), always have mycorrhizae, which are often shared with the roots of green, photosynthetic plants.

The symbiotic fungus–plant association of the mycorrhiza is important to both partners. The fungus obtains important organic compounds such as sugars and amino acids from the plant. In return, the fungus greatly increases the absorption of water and minerals (especially phosphorus) by the plant. Also, the fungus often provides certain growth substances (including auxin and cytokinins, which will be discussed in Chapter 32), and it provides some protection

against attack by microorganisms. Plants with active mycorrhizae typically are a deeper green and may resist drought and temperature extremes better than plants of the same species with little mycorrhizal development. Attempts to introduce plant species to new areas have often failed until a bit of soil from the native area was added. The failure to grow without that soil presumably resulted from an inability to form mycorrhizae. The partnership with the fungus results in a plant better adapted for life on land. K. A. Pirozynzki and D. W. Malloch, of the Canadian Department of Agriculture, have suggested that the evolution of this symbiotic association was the single most important step leading to the colonization of the terrestrial environment by living things.

---

produced by the procambium. This domain, the vascular cylinder or **stele** (Figure 28.22), consists of three tissues: the pericycle, the xylem, and the phloem. The **pericycle** consists of one or more layers of relatively undifferentiated cells. It is the tissue within which branch roots arise (Figure 28.21c); the pericycle also provides a few of the dividing cells that enable the root to grow in diameter. The xylem—seen in cross section as a star with a variable number of points—lies at the very center of the root of a dicot. Between the points of the xylem star are bundles of phloem. In a monocot root, a region of parenchyma cells, the **pith**, lies internal to the xylem. It is useful to try picturing these structures in three dimensions, as in Figure 28.22), rather than attempting to understand their functions solely on the basis of two-dimensional cross sections.

Primary Tissues of the Stem

The shoot apical meristem, like the root apical meristem, forms three primary meristems: the procambium, ground meristem, and protoderm, which in turn give rise to the three tissue systems. Leaves arise from leaf primordia that form as cells divide on the sides of shoot apical meristems. The growing stem has no cap analogous to the root cap, but the leaf primordia can act as a protective covering. Dicot stems grow in a region of elongation below the shoot apical meristem. Grasses and some other monocots, however, elongate at the *bases* of internodes and leaves, where there is some meristematic tissue. Lawn and range grasses can grow back after mowing or grazing because they grow from basal meristems close to the soil surface.

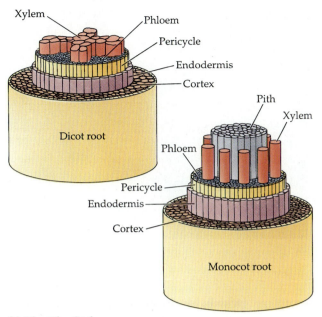

Xylem
Phloem
Pericycle
Endodermis
Cortex

Dicot root

Pith
Xylem
Phloem
Pericycle
Endodermis
Cortex

Monocot root

28.22 The Stele
The distribution of tissues in the stele—the region internal to the endodermis—differs in the roots of dicots and monocots.

The plumbing of angiosperm stems differs from that of roots. In roots, the vascular tissue lies in the middle, with the xylem at or near the very center. In contrast, the vascular tissue of a young stem is divided into a number of discrete **vascular bundles**, which generally form a cylinder in the dicots (Figure 28.23*a*) but are seemingly scattered throughout the cross section of the stem in the monocots (Figures 28.23*b* and *c*). Internal to the vascular bundles of dicots is a storage tissue, pith, and to the outside lies a similar storage tissue, the cortex. The cortex may contain strengthening collenchyma cells with thickened walls (Figure 28.24*a*). In many monocots the pith is hollowed out (as in Figure 28.24*b*). The pith, the cortex, and the regions between the vascular bundles in dicots—pith rays—constitute the ground tissue system of the stem. The outermost cell layer of the young stem is the epidermis, which functions primarily to minimize the loss of water from the cells within.

Secondary Growth of Stems and Roots

Whereas some stems and roots show little or no growth in diameter, remaining slender, many others undergo considerable thickening. This thickening—limited to some dicots—is of great importance and interest to us because it gives rise to wood and bark, as well as making the support of large trees possible.

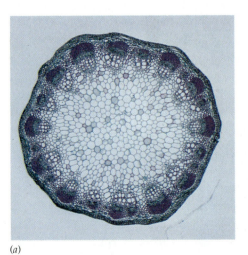

(*a*)

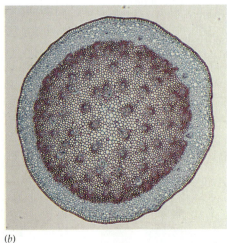

(*b*)

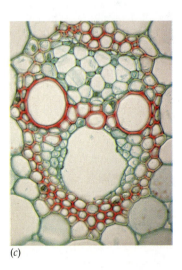

(*c*)

28.23 Vascular Bundles in Stems
The vascular tissues in stems are organized into bundles. (*a*) In dicots the vascular bundles are arranged in a circle with pith in the center and cortex outside the ring, as in this young sunflower stem. (*b*) This cross section of a stem of *Dracaena fragrens* shows the seemingly scattered array of bundles typical of monocot stems. In both monocots and dicots, the bundles are oriented so that xylem is toward the center of the stem and phloem is to the outside. (*c*) A single vascular bundle from a corn stem. The large (red-walled) cells are xylem and the cluster of green cells forming the "forehead" of the "face" are phloem. The single large opening below the vessels in the xylem is a channel that was left as embryonic "protoxylem" was stretched and destroyed.

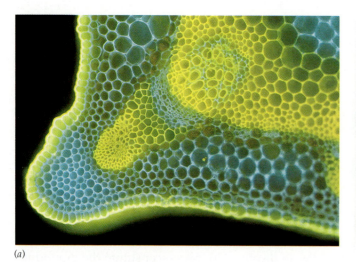

(a)

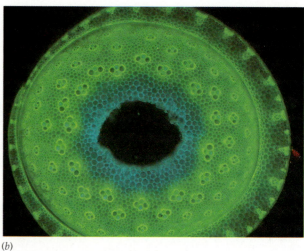

(b)

28.24 Other Stem Structures
(a) The stem of a broad bean resists bending but is not brittle. Collenchyma, the tissue in bright blue in the projection to the lower left of this stem cross section, provides flexible support. The next tissue toward the interior, shown in gold, consists of phloem fibers that are stiffer than the collenchyma. (b) This bamboo looks like a "typical" monocot, except for its hollowed-out pith.

28.25 Vascular Cambium Thickens Stems and Roots
Stems and roots grow thicker because a thin layer of cells, the vascular cambium, remains meristematic—capable of dividing. (a) When a vascular cambial cell (gray) divides, it produces either a new xylem cell toward the inside of the stem or root, or a new phloem cell toward the outside. Older xylem and phloem cells are pushed farther from the cambium with each division of the cambium. (b) This longitudinal section of a woody stem shows the vascular cambium thickening the stem by producing secondary xylem and secondary phloem.

Secondary growth results from the activity of two meristematic tissues, vascular cambium and cork cambium. Vascular cambia consist of cells that divide to produce new—secondary—xylem and phloem cells, while cork cambia produce mainly waxy-walled cork cells.

Initially, the vascular cambium is a single layer of cells between the primary xylem and the primary phloem. If the root or stem is going to increase in diameter, cells of the vascular cambium begin to divide, producing secondary xylem cells toward the inside of the organ and secondary phloem cells toward the outside (Figure 28.25a). In the stem, cells of the pith rays between the vascular bundles also divide, forming a continuous cylinder of vascular cambium running the length of the stem. This cyl-

(a)

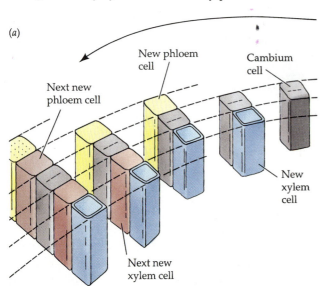

New phloem cell

Cambium cell

Next new phloem cell

New xylem cell

Next new xylem cell

(b)

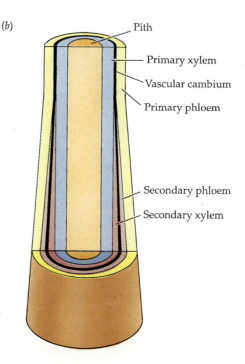

Pith

Primary xylem

Vascular cambium

Primary phloem

Secondary phloem

Secondary xylem

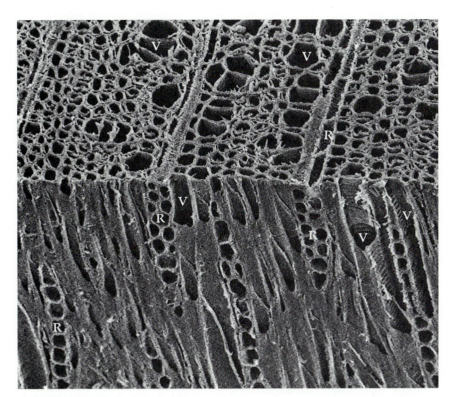

28.26 Vascular Rays
Wood of tulip poplar, showing that the orientation of xylem vessels (V) is perpendicular to that of vascular rays (R). The longitudinal section of the stem shows the xylem vessels as long vertical tubes and the ray cells as circles. The cross section in the upper part of the micrograph, which is perpendicular to the longitudinal section, shows the xylem vessels as circles and the rays as tubes. Rays transport food horizontally from the phloem to storage tissues; xylem vessels conduct water vertically.

inder, in turn, gives rise to complete cylinders of secondary xylem—wood—and secondary phloem—bark (Figure 28.25*b*).

As the vascular cambium produces secondary xylem and phloem, its principal products are vessel elements and supportive fibers in the xylem, and sieve tube elements, companion cells, and fibers in the phloem. Not all xylem and phloem cells are adapted for transport or support—some serve for the storage of materials in the stem or root. Living cells such as these storage cells must have some connection with the sieve tubes of the phloem, or they would starve to death. The connections are provided by **vascular rays,** which are composed of cells derived from the vascular cambium. The rays are laid down progressively as the cambium divides, resulting in rows of living parenchyma cells running perpendicular to the xylem vessels and phloem sieve tubes (Figure 28.26). As the root or stem continues to increase in diameter, new vascular rays are initiated so that this storage and transport tissue continues to meet the needs of the bark and of the living cells in the xylem. The cambium itself increases in circumference with the growth of the root or stem—if it did not, it would split. The vascular cambium grows by the division of some of its cells in a plane at right angles to the plane that gives rise to secondary xylem and phloem. The products of each of these divisions both lie within the vascular cambium itself.

Many dicots have vascular cambia and cork cambia and thus undergo secondary growth. In the rare cases in which monocots form thickened stems—palm trees, for example—a greater girth is achieved by quite a different mechanism (Box 28.B).

Wood

Most trees from temperate zone forests have **annual rings** (Figure 28.27), which result from changing environmental conditions during the growing season. In the springtime, when water is relatively plentiful, the tracheids or vessel elements produced by the vascular cambium tend to be large in diameter and thin-walled. As water becomes less available during the summer, narrower cells with thicker walls are produced; we see this summer wood as darker and perhaps more dense. Thus each year is usually recorded in a tree trunk by a clearly visible annual ring consisting of one light and one dark layer. Trees in the tropics do not lay down regular rings.

The difference between old and new regions also contributes to the appearance of wood. As a tree grows in diameter, the xylem toward the center becomes clogged with resins and ceases to conduct water and minerals. This heartwood is darker; the sapwood, which is that portion that is actively conducting all water and minerals in the tree, is lighter and more porous. Knots—which we find attractive in knotty pine, but regard as a defect in structural timbers—are branches: As a trunk grows, a branch extending out of it becomes buried in new wood and appears as a knot when the trunk is cut lengthwise.

BOX 28.B

An Unusual Tree Trunk

If palm trees have no vascular cambium and hence no secondary growth, how can they develop such thick stems? Why are their trunks cylindrical (of the same diameter at top and base) rather than conical like those of other trees? The thickening of the primary body of a palm tree is achieved with the aid of a tissue

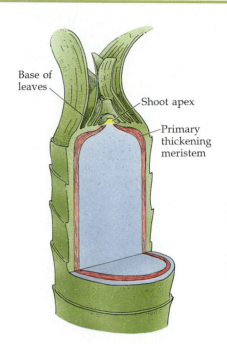

Base of leaves

Shoot apex

Primary thickening meristem

called the primary thickening meristem. Most of the primary body is not produced directly by the apical meristem. The apical meristem, located in the shoot apex, gives rise to the central part of the stem and to leaf primordia. The primary thickening meristem, whose cellular products elongate laterally and produce the flat-topped structure shown in the figure, lies immediately below the leaf primordia. Internodes elongate between the new leaves, and the whole structure grows upward as cell divisions continue in the apical and primary thickening meristems.

Secondary Growth: Periderm

Obviously, as secondary growth continues, something has to give. Expansion of the vascular tissues stretches and breaks the epidermis and cortex, which are ultimately lost. Derivatives of the phloem then become the outermost tissue of the stem. Woody roots behave similarly. Because the epidermis is specialized in part for the retention of water, how does the plant cope if this tissue is shed? Before layers of

epidermal cells are broken away, cells lying near the surface begin to divide and produce layers of cork, a tissue composed of cells with thickened, waterproof walls (Figure 28.1). The dividing cells, derived from the phloem, constitute a cork cambium. Sometimes cells are also produced to the inside by the cork cambium; these cells constitute the phelloderm. Cork is waterproofed by suberin. The cork soon becomes the outermost tissue of the stem or root. Cork, cork cambium, and phelloderm—if present—make up the periderm of the secondary body. As the vascular cambium continues to produce secondary vascular tissue, the corky layers are in turn lost, but a similar process of cell division in the underlying phloem gives rise to new corky layers.

As periderm forms, there is still a need for gas exchange with the environment—carbon dioxide must be released and oxygen must be taken up for cellular respiration. Lenticels are spongy regions that allow such gas exchange (Figure 28.28).

Leaf Anatomy

A cross section of a typical dicot leaf is shown in Figure 28.29. Generally a leaf has two zones of photosynthesizing tissues referred to as mesophyll, meaning "middle of the leaf." The upper layer or layers of mesophyll consist of roughly cylindrical cells. This zone is referred to as palisade mesophyll. The lower layer or layers consist of irregularly shaped cells called spongy mesophyll. Within the mesophyll there is a great deal of air space through which carbon

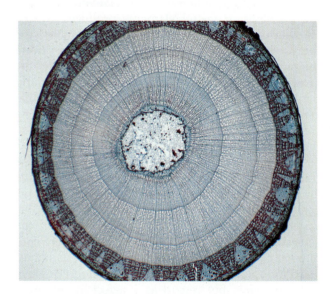

28.27 Annual Rings
Rings of xylem vessels are the most noticeable feature of this cross section from a three-year-old basswood stem.

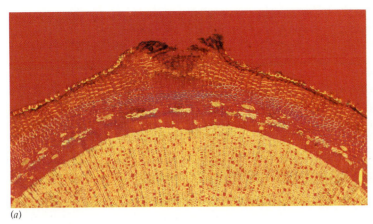

(a)

(b)

28.28 Lenticels
Lenticels allow gas exchange through the periderm. (a) Micrograph of a lenticel in a year-old elder twig; note the spongy tissue that constitutes the lenticel. (b) Gigantic lenticels on the trunk of a Chinese plum tree.

dioxide can diffuse to surround all photosynthesizing cells. Vascular tissue branches extensively in the leaf, forming a network of **veins.** These veins extend to within a few cell diameters of all the cells of the leaf, ensuring that the mesophyll cells are well supplied with water. The products of photosynthesis are loaded into the phloem of the veins for export to the rest of the plant (Figure 28.30a).

Covering the entire leaf is a layer of nonphotosynthetic cells constituting the epidermis. To retard water loss, the epidermal cells and their overlying waxy cuticle must be highly impermeable, but this impermeability poses a problem: While keeping water within the leaf, the epidermis keeps carbon dioxide, the raw material of photosynthesis, out. The problem of balancing water retention and carbon

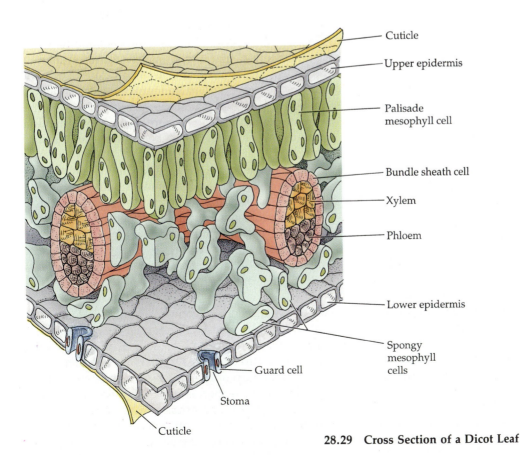

Cuticle

Upper epidermis

Palisade mesophyll cell

Bundle sheath cell

Xylem

Phloem

Lower epidermis

Spongy mesophyll cells

Guard cell

Stoma

Cuticle

28.29 Cross Section of a Dicot Leaf

(a)

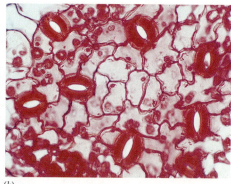

(b)

28.30 Structures of a Dicot Leaf

(a) The network of fine veins in this Japanese maple leaf carries water to the mesophyll cells and carries photosynthetic products away from them. (b) The lower epidermis of a dicot leaf, stained. The small, heavily stained, paired cells are guard cells; the gaps between them are stomata, through which carbon dioxide enters the leaf.

dioxide availability is solved by an elegant regulatory system that will be discussed in more detail in Chapter 29. This system is based on pairs of **guard cells**, modified epidermal cells that change shape, thereby opening or closing pores, called **stomata** (singular, stoma), between the guard cells (Figure 28.30b; see also Figure 29.10). When the stomata are open, carbon dioxide can enter, but water can be lost.

In Chapter 8 we described the C_4 plants, which can fix carbon dioxide efficiently even when the carbon dioxide supply falls to a level at which the photosynthesis of C_3 plants is inefficient. One adaptation that helps C_4 plants do this is a modified leaf anatomy, as shown in Figure 8.27. Here the photosynthetic cells are grouped around the veins in concentric layers: an outer mesophyll layer and an inner **bundle sheath**. These layers each contain different types of chloroplasts, leading to the biochemical division of labor described in Chapter 8.

MODULAR CONSTRUCTION OF THE PLANT BODY

One of the most striking features of the plant body is that it is *modular*. Each branch may be thought of as a module, in many ways independent of the other branches. A branch of a plant does not bear the same relationship to the remainder of the body as does an arm to the remainder of the human body. Each branch lives out its own history, and branches grow independently, exploring different parts of the surrounding environment. Branches may respond differently to gravity, some growing more or less vertically and others horizontally. Leaves represent modules of another sort, being produced in fresh batches to take over the constant function of feeding

the plant. Often the shapes of different leaves of the same plant differ in accordance with their differing local environments. Leaves are much shorter-lived modules than are branches. Branch roots, too, are semi-autonomous modules.

The modularity of plants results in a decentralization of control systems. Branches experiencing different local environments send differing reports to the rest of the plant. If a plant has more than one stem, different parts of the shoot system may receive different reports as to the availability of water and minerals, because they are served by different parts of the root system. In spite of this decentralization, the plant functions as a coherent unit.

SUMMARY

Most flowering plants consist of a shoot system—the leaves and stem—and a root system. The root system anchors the plant and absorbs water and minerals, the leaves conduct photosynthesis, and the stem displays the leaves to the sun. Roots have a protective root cap. Stems bear leaves and lateral buds.

Some flowering plant cells, such as collenchyma cells and fibers, support the plant body and its organs. Vessel elements conduct water and minerals, and sieve tube elements conduct food materials. Parenchyma cells play various roles—for example, some are photosynthetic and others store reserve foods. Meristematic cells can divide to produce other cells with various developmental fates.

Flowering plants possess simple tissues, such as collenchyma, that are composed of a single type of cell; they also possess complex tissues, such as xylem, that are composed of several types of cells. Plants have three tissue systems: the vascular tissue system (made up of the water- and mineral-conducting xylem and the food-conducting phloem), the dermal tissue system (epidermis and periderm), and the ground tissue system (composed of parenchyma, collenchyma, and sclerenchyma cells).

Plant growth results from the activities of meristems: the apical meristems of stems and roots, and the vascular and cork cambia. These tissues retain the capacity for cell division. The apical meristems give rise to the primary plant body, and the vascular cambium and cork cambia give rise to the secondary plant body.

The parts of an individual plant are more independent than are those of many animals; that is, plants are highly modular.

SELF-QUIZ

1. Which of the following is *not* a difference between monocots and dicots?
   a. Dicots more frequently have broad leaves.
   b. Monocots commonly have flower parts in multiples of three.
   c. Monocot stems do not generally undergo secondary thickening.
   d. The vascular bundles of monocots are commonly arranged as a cylinder.
   e. Dicot embryos commonly have two cotyledons.

2. Roots:
   a. always form a fibrous root system that holds the soil.
   b. possess a root cap at their tip.
   c. form branches from lateral buds.
   d. are commonly photosynthetic.
   e. do not show secondary growth.

3. The plant cell wall:
   a. lies immediately inside the plasma membrane.
   b. is an impenetrable barrier between cells.
   c. is always waterproofed with either lignin or suberin.
   d. consists of a primary wall and secondary wall, separated by a middle lamella.
   e. contains cellulose and other polysaccharides.

4. Which statement about parenchyma cells is *not* true?
   a. They are alive when they perform their functions.
   b. They typically lack a secondary wall.
   c. They often function as storage depots.
   d. They are the most numerous cells in the primary plant body.
   e. They are found only in stems and roots.

5. Tracheids and vessel elements:
   a. die before they become functional.
   b. are important constituents of all bryophytes and tracheophytes.
   c. have walls consisting of middle lamella and primary wall.
   d. are always accompanied by companion cells.
   e. are found only in the secondary plant body.

6. Which statement is *not* true of sieve tube elements?
   a. Their end walls are called sieve plates.
   b. They die before they become functional.
   c. They link end-to-end, forming sieve tubes.
   d. They form the system for translocation of foods.
   e. They lose the membrane that surrounds their central vacuole.

7. The pericycle:
   a. separates the stele from the cortex.
   b. is the tissue within which branch roots arise.
   c. consists of highly differentiated cells.
   d. forms a star-shaped structure at the very center of the root.
   e. is waterproofed by Casparian strips.

8. Secondary growth of stems and roots:
   a. is brought about by the apical meristems.
   b. is common in both monocots and dicots.
   c. is brought about by vascular cambia and cork cambia.
   d. produces only xylem and phloem.
   e. is brought about by vascular rays.

9. Periderm:
   a. contains lenticels that allow for gas exchange.
   b. is produced during primary growth.
   c. is permanent; once formed it lasts as long as the plant does.
   d. is the innermost part of the plant.
   e. contains vascular bundles.

10. Which statement about leaf anatomy is *not* true?
    a. Stomata are controlled by paired guard cells.
    b. The cuticle is secreted by the epidermis.
    c. The veins contain xylem and phloem.
    d. The cells of the mesophyll are packed together, minimizing air space.
    e. C_3 and C_4 plants differ in leaf anatomy.

FOR STUDY

1. When a young oak was 5 meters tall, a thoughtless person carved his initials in its trunk at a height of 1.5 meters above the ground. Today that tree is 10 meters tall. How high above the ground are those initials? Explain your answer in terms of the manner of plant growth.

2. Consider a newly formed sieve tube element in the secondary phloem of an oak tree. What kind of cell divided to produce the sieve tube element? What kind of cell divided to produce that parent cell? Keep on tracing back in this manner until you arrive at a cell in the apical meristem.

3. Distinguish between sclerenchyma cells and collenchyma cells in terms of structure and of function.

4. Distinguish between primary and secondary growth. Do all angiosperms undergo secondary growth? Explain.

5. What anatomical features make it possible for a plant to retain water as it grows? Describe the tissues and how and when they form.

READINGS

Esau, K. 1977. *Anatomy of Seed Plants*, 2nd Edition. John Wiley, New York. A comprehensive treatment; particularly good on secondary growth.

Feldman, L. J. 1988. "The Habits of Roots." *BioScience*, vol. 38, pages 612–618. Considers many aspects of the biology of roots, including structure, competition, associations with soil microorganisms, and others.

Mangelsdorf, P. C. 1986. "The Origin of Corn." *Scientific American*, August. What was the ancestry of this popular vegetable? The gross anatomy of some possible ancestors is compared.

Raven, P. H., R. F. Evert and S. Eichhorn. 1986. *Biology of Plants*, 4th Edition. Worth, New York. An excellent general botany textbook.

Swaminathan, M. S. 1984. "Rice." *Scientific American*, January. This article deals primarily with ways to increase the yield of rice, but it also describes the structure of the plant.

29

Transport, Support, and Protection in Plants

PREVIEW: Terrestrial vascular plants show many adaptations to life on land: They can support themselves, they can protect themselves from predators, and they can transport and conserve water. Water and minerals are transported as sap in a complex tissue, the xylem, and food and some minerals are transported in another complex tissue, the phloem. Plants must have openings to let in carbon dioxide from the atmosphere without losing too much water through the same openings. Certain plants can successfully occupy habitats in which little water is available and others can survive in exceptionally salty soils because they are structurally and physiologically adapted to these uninviting habitats. Plants use chemical defenses to deter herbivores and fungi.

This chapter deals with vascular tissues—xylem and phloem—the evaporation–cohesion–tension model, the pressure flow model, stomata, crassulacean acid metabolism, reaction wood, xerophytes, halophytes, and secondary plant products.

Vascular plants, like all organisms, face myriad environmental challenges. They meet these challenges with the help of many adaptations for support of the plant body, transport of water and nutrients, and protection against other organisms and toxic elements in the surroundings. These adaptations arose and have been refined through hundreds of millions of years of natural selection. What we see today is one frame of a long movie of adaptations to life on Earth; some of the adaptations visible in this single frame are being challenged and perhaps further refined. Let us begin by looking at the current status of a deadly "game" between a plant and one of its insect enemies.

Milkweeds are latex-producing, or laticiferous, plants. When damaged, the milkweed releases copious amounts of a white, rubbery, toxic liquid, **latex**, from tubes called laticifers. It has long been suspected that the latex deters insects from eating the plant, because laticiferous plants are not attacked by insects that feed on neighboring plants of other species. This is consistent with, but does not prove, the hypothesis that the latex keeps the insects at bay. Stronger support for the hypothesis was afforded in 1987 by David Dussourd, now of the University of Maryland, and Thomas Eisner of Cornell University, who studied field populations of *Labidomera clivicollis*, a beetle that is one of the few insects that do feed on *Asclepias syriaca*, the field milkweed.

The two zoologists observed a remarkable prefeeding behavior, in which the beetles cut a few veins in the leaves before settling down to dine (Figure 29.1). The latex is kept under pressure in the undamaged plant, so cutting the veins, with their adjacent laticifers, causes massive leakage and depressurizes the system. By cutting a few veins, the beetles interrupt the latex supply to a downstream portion of the leaf. The beetles then move to the relatively latex-free portion and eat their fill. Some other insects that do not feed on undamaged milkweeds will eat parts of leaves that have had the latex supply cut off. When presented simultaneously with leaves on undamaged plants and leaf parts that have had their laticifers cut, *L. clivicollis* and other insects that share this vein-cutting behavior select the relatively latex-free leaf parts.

Does this behavior of the beetles negate the adaptational value of latex protection? Not at all—there are still great numbers of potential insect pests that are effectively deterred by the latex. And this is just one frame of the movie—it may be that, over time, milkweed plants producing higher concentrations of toxins will be selected by virtue of their ability to kill beetles that cut the laticifers.

(a)

(b)

29.1 Demolishing a Plant Defense
Meal preparation precedes dining among certain beetles; *Labidomera clivicollis* is shown here. In (a) the beetle cuts the veins that carry latex in the milkweed. (b) The beetle begins to dine. Note that no latex is dripping from the edges of this cut, although white drops are visible at the vein cuts in (a). We can see the original and some subsequent cuts in the veins as well as the extent of the meal in (c). Larvae of this species as well as the adults cut veins before eating.

(c)

CHALLENGES OF THE TERRESTRIAL ENVIRONMENT

Life first arose and flourished in the oceans, as we have seen. What was needed for multicellular life to colonize the land? Animals and fungi could not spread widely over dry land that contained no food. Plants thus took the crucial evolutionary step for the spread of life by invading the terrestrial environment. Because they use photosynthesis, sunlight, and inorganic materials to produce food—both for themselves and for animals and fungi—plants were the first eukaryotes to face the challenges of life out of water.

The challenges of terrestrial life are many. Living things require water—the water within cells is the solvent for virtually all biochemical reactions. Therefore, terrestrial organisms must gather water from the environment and retain it within their bodies. For taking in and conserving water, they need mechanisms different from those that serve for aquatic organisms, and they need different mechanisms for supporting their bodies as well. In a watery environ-

ment, a giant kelp—a marine algal protist—can spread out like an enormous tree because of the buoying action of the surrounding water (Figure 29.2). On land, large organisms can resist the pull of gravity only with the help of rigid materials such as wood or bone. The pressure of water in the tissues provides much of the support for nonwoody terrestrial plants, which wilt in the absence of water.

Sexual reproduction and the dispersal of progeny in the terrestrial environment also call for different mechanisms from those used in lakes and seas. All in all, it is not surprising that most land plants look quite different from their cousins in watery environments. Reproduction and dispersal will be dealt with in Chapter 31. Here we will concentrate on the matters of *transport*, *support*, and *protection* in terrestrial plant bodies.

Multicellularity in the kingdom Plantae has led to considerable specialization of cell and tissue types (Chapter 28). In the vascular plants—notably the ferns, gymnosperms, and flowering plants—we find cells and tissues specialized for absorbing water and minerals, for transporting water and minerals, for conducting photosynthetically produced food, for supporting against gravity, for strengthening against the force of winds, for preventing or limiting water loss, for producing branches, for storing reserve food, and even for forming new tissues. At least five of these functions meet the special challenges of the terrestrial environment.

UPTAKE AND TRANSPORT OF WATER AND MINERALS

Terrestrial plants obtain both water and mineral nutrients from the soil, usually by way of their roots. What are the mechanisms by which water and min-

(a)   (b)

29.2   Support for Plants
(a) Supported by the surrounding water, the flexible bodies of kelps can reach great heights. (b) Large terrestrial plants withstand the pull of gravity with the aid of rigid tissues such as wood. These California coast redwoods belong to the species that includes the world's tallest trees.

erals enter the plant body through the dermal tissue system of the root, pass through the ground tissue system, enter the stele, and ascend as sap in the xylem of the vascular tissue system? Because neither water nor minerals can enter the vascular tissue system without first crossing at least one plasma membrane, we will first consider two membrane phenomena, osmosis and the uptake of minerals.

Osmosis

Osmosis, the movement of a solvent through a membrane in accordance with the laws of diffusion, was discussed in Chapter 5. Recall that the **osmotic potential** of a solution results from the presence of dissolved solutes. The greater the solute concentration, the more negative the osmotic potential and, hence, the greater the tendency of water to move into that solution from another solution of lower concentration. The two solutions must be separated by a differentially permeable membrane (permeable to water but impermeable to the solute). Solutions—or cells—with identical osmotic potentials are isotonic; if two solutions differ in osmotic potential, the one with the less negative osmotic potential is hypotonic to the other. Recall, too, that osmosis is a passive process—ATP is not required.

Unlike animal cells, plant cells are surrounded by

a relatively rigid cell wall. After a certain amount of water enters a plant cell, the entry of more water is resisted by an opposed **pressure potential**, sometimes called turgor pressure, owing to the rigidity of the wall. As more and more water enters, the pressure potential becomes greater and greater. The pressure potential is analogous to the air pressure in an automobile tire; it is a real pressure that can be measured with a pressure gauge. Cells with walls do not burst when placed in pure water because of the rigidity of the walls. Water enters by osmosis until the pressure potential exactly balances the osmotic potential. At this point, the cell is quite turgid, that is, it has a high pressure potential.

The overall tendency of a solution to take up water from pure water is called the **water potential** (Figure 29.3). The water potential is simply the sum of the negative osmotic potential and the positive pressure potential. For pure water under no applied pressure, all three of these potentials are defined as equal to zero. In all cases in which water moves between two cells, or between a cell and its environment, or between two solutions separated by a membrane, the following rule of osmosis applies: *Water always moves toward the region of more negative water potential.*

Osmotic phenomena are of great importance. Most plants depend upon the pressure potentials of their cells to maintain the turgidity of their bodies—if the

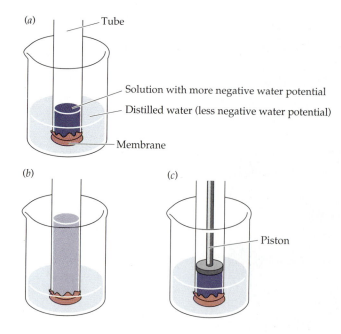

(a)
Tube
Solution with more negative water potential
Distilled water (less negative water potential)
Membrane

(b)

(c)
Piston

29.3 Water Potential and Pressure Potential

(a) The solution in the tube has a negative water potential owing to the presence of dissolved solutes; the beaker contains distilled water (water potential = 0). A differentially permeable membrane separates the solution in the tube from the water in the beaker. Water moves from the beaker to the tube (toward the more negative water potential) by osmosis until equilibrium is reached. (b) At osmotic equilibrium, there is no further net movement of water. The solution in the tube still has a more negative *osmotic* potential than does the water in the beaker, but the pressure resulting from the height of the column of solution in the tube offsets the difference in osmotic potential. (c) A piston was placed in the tube to prevent the entry of water into the tube; because of the difference in osmotic potential, a pressure potential forms. That pressure potential is the same as the pressure resulting from the column height of solution in (b). In both (b) and (c) the final pressure potential in the tube is equal but opposite to the osmotic potential in the tube, giving the solution in the tube a water potential of zero. The water in the beaker contains no solutes and is not under pressure, so its water potential is also zero, and the systems in both (b) and (c) are at osmotic equilibrium.

pressure potential is lost, the plant wilts (Figure 29.4). The movement of water within a plant follows a gradient of water potential, and, as we will see, the flow of phloem sap through the sieve tubes is driven by a gradient in pressure potential.

Uptake of Minerals

Mineral nutrient ions are taken up across plasma membranes with the help of proteins. You may wish to review the section on diffusion in Chapter 5. Some of these proteins are carriers for the facilitated diffusion of particular ions. Facilitated diffusion does not require ATP. The concentrations of some ions in the soil solution are lower than those required inside the plant, however. Thus, the plant must take up these ions against a concentration gradient. Such active transport is an energy-requiring process, and it depends upon cellular respiration (Chapter 7) as a source of ATP. Active transport, too, requires specific carrier proteins.

Water and Ion Movement from Soil to Xylem

Water moves along a gradient of water potential, toward ever more negative regions. Water moves into the stele of the root because the water potential is more negative within the stele than in the cortex. The

(a)

(b)

29.4 Turgor in Plants

(a) This tomato plant remains turgid as long as the pressure potential of its cells is high. (b) When cells lose too much water, their pressure potential drops and the plant wilts.

cortex, in turn, has a more negative water potential than does the soil solution. Minerals enter and move in plants in various ways. Where water is actually flowing, dissolved minerals are carried along. Where water is moving more slowly, minerals diffuse. At certain points, where plasma membranes are being crossed, some minerals are sped along by active transport.

Water and minerals from the soil pass through the dermal and ground tissue systems to the stele through either of two major compartments. Plant cells are surrounded by cell walls that lie outside the plasma membrane, and intercellular spaces are common in many tissues. Water and dissolved substances have ready access to the walls and intercellular spaces, which together constitute a compartment called the **apoplast** (Greek, "away from living material"). The apoplast is a continuous meshwork through which material can flow or diffuse without ever having to cross a membrane (Figure 29.5). Movement of materials through the apoplast is thus unregulated. The remainder of the plant body is the compartment called the **symplast** ("together with living material"), another continuous meshwork consisting of the living cells, connected by plasmodesmata (Chapter 4). The selectively permeable plasma membranes of the cells control access to the symplast.

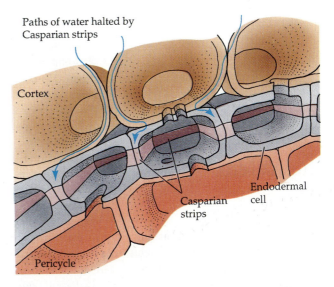

29.6   Casparian Strips
The walls of endodermal cells include a region impregnated with water-repellent suberin, constituting the Casparian strips. These waxy "gaskets" are so placed that water cannot pass between the endodermal cells and into the stele. Water must first enter the living endodermal cells.

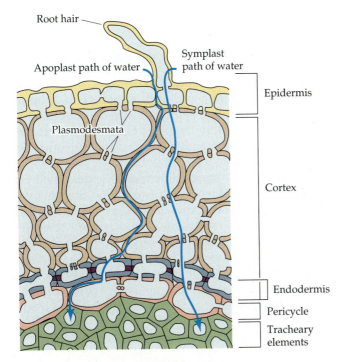

29.5   Apoplast and Symplast
Water and minerals diffuse freely through the apoplast, which consists of cell walls and intercellular spaces. At some point, water and minerals must enter the symplast —the remainder of the plant body—or they will be unable to pass from the cortex to the stele.

Water and minerals can pass from the soil solution through the apoplast to the inner border of the cortex. As you may recall from Chapter 28, water and minerals must then pass through the inner cell layer of the cortex, called the **endodermis** (Figure 28.22) to enter the stele. What distinguishes the endodermis from the rest of the ground tissues is the presence of **Casparian strips**. These waxy, suberin-containing structures line the endodermal cells at their tops, bottoms, and sides, acting as a "gasket" that prevents water and ions from moving between them (Figure 29.6). The endodermis thus completely separates the apoplast of the cortex from the apoplast of the stele. The Casparian strips do not obstruct the outer or inner faces of the endodermal cells. Accordingly, water and ions can enter the stele only by way of the symplast—that is, by entering and passing through the cytoplasm of the endodermal cells. Thus, membrane proteins determine which minerals pass, and at what rates.

Once they have passed the endodermal barrier, water and minerals leave the symplast. Parenchyma cells in the pericycle or xylem help minerals move back into the apoplast. Some of these parenchyma cells, called **transfer cells**, are structurally modified for transporting mineral ions from their cytoplasm—part of the symplast—into their cell walls—part of the apoplast. The wall that receives the transported ions has many knobby growths extending into the transfer cell, increasing the surface area of the plasma

BOX 29.A

There Are No Pressure Pumps in the Xylem

Eduard Strasburger was the leading plant cytologist of the late nineteenth century. He was one of those who established that the nucleus is the carrier of hereditary information, and he is best remembered today for pioneering work that led to the discovery of meiosis. He also performed some of the first important experiments that led to our current understanding of water movement in the xylem. His contemporaries generally believed that living cells in the xylem played a key role, probably by acting as "pressure pumps" pushing the sap upwards.

Strasburger worked with trees about 20 meters in height. He sawed them through at their bases and plunged the cut ends into buckets containing solutions of poisons such as picric acid. The solutions rose through the trunks, as was readily evident from the progressive death of the bark higher and higher up. When the solutions reached the leaves, the leaves died, too, and the solutions were no longer transported—the liquid levels in the buckets stopped dropping at that point. This simple experiment established three important points: (1) Living, "pumping" cells are not responsible for the upward movement of the solutions, for the solutions themselves killed all living cells with which they came in contact. (2) The leaves play a crucial role in causing the transport. As long as they were alive, the solutions continued to be transported upward; when the leaves died, transport ceased. (3) Transport in these experiments, which covered distances of 20 meters and more, was not caused by root pressure, for the trunks had been completely separated from the roots. It is interesting, if disappointing, that theories of xylem transport based on root pressure were entertained for some years following Strasburger's definitive experiments.

---

membrane, the number of transport proteins, and, thus, the rate of transport. Transfer cells also have many mitochondria that produce the ATP needed to power the active transport of mineral ions. As mineral ions move into the solution in the walls, the water potential of the wall solution becomes more negative; thus water moves out of the cells into the wall solution by osmosis. Active transport of ions moves the ions directly, and water follows passively. The end result is that water and minerals find themselves in the xylem, constituting the sap.

Ascent of Sap in Xylem: The Magnitude of the Problem

The water and minerals in the xylem must be transported to the entire shoot system, all the way to the highest leaves and apical buds. Before we consider the mechanisms underlying this transport, we should know what needs to be explained: How much sap is transported, and how high must it be taken in the tallest trees?

A single 15-meter maple tree was estimated to have some 177,000 leaves, with a total leaf surface area of 675 square meters. On a summer day, that tree lost 220 liters of water *per hour* to the atmosphere by evaporation from the leaves. To prevent wilting, xylem transport in that tree needed to provide 220 liters of water to the leaves every hour.

How high must the xylem sap be transported? The question may be rephrased: How tall are the tallest trees? The tallest gymnosperms, the coast redwoods —*Sequoia sempervirens*—exceed 110 meters in height, as do the tallest angiosperms, the Australian *Eucalyptus regnans*. If we add in the length of the roots, we see that any successful explanation of transport in the xylem must account for transport over vertical distances of 120 meters and more.

Early Models of Transport in the Xylem

Some of the earliest models to explain the rise of sap in the xylem were based on a hypothetical "pumping" action by living cells in the stem. Experiments published in 1893 by the great German botanist Eduard Strasburger definitively ruled out such models (Box 29.A).

Another early suggestion was a model based on capillary action, the rising of watery solutions in very thin tubes or in woven materials like paper. At first blush this seems reasonable. The diameters of vessel elements and tracheids are tiny. The narrower the tube, the higher water will rise by capillary action. However, the actual diameters of tracheids are only small enough to support a capillary column of about 40 centimeters, shorter than many shrubs, let alone a giant eucalyptus towering over us to a height greater than the length of a football field.

Root Pressure

After the capillary model was questioned, some plant physiologists turned to a model based on root pressure—a pressure exerted by the root tissues that would force liquid up the xylem. The basis for root pressure is a higher solute concentration, and accordingly a more negative water potential, in the xylem sap than in the soil solution. This draws water into the stele; once there, the water has nowhere to go but up.

There is, in fact, good evidence for root pressure, as for example in the phenomenon of **guttation**, in which liquid water is forced out through openings in leaves (Figure 29.7). Plants guttate only under conditions of high atmospheric humidity and plentiful water in the soil. Root pressure is also the source of the sap that oozes from the cut stumps of some plants such as *Coleus* when their tops are removed. However, root pressure cannot account for the ascent of sap in trees. Root pressures seldom exceed one or two times atmospheric pressure, and they are actually less at times when transport in the xylem is most rapid. If root pressure were driving sap up the xylem, we should observe a positive pressure potential in the xylem at all times. In fact, as we are about to see, the xylem sap is under a tension—a *negative pressure potential*—when it is ascending. Furthermore, as Strasburger had already shown, materials can be transported upward in the xylem even when the roots have been removed (Box 29.A).

The Evaporation–Cohesion–Tension Mechanism

To understand how sap actually rises in the xylem, even to the tops of the tallest trees, we must begin by looking at the final step in the process of water movement from soil to root to leaf and out to the atmosphere. At the end of the line, water evaporates from the moist walls of mesophyll cells, diffuses through the air spaces of the leaf, and finally leaves as water vapor through the open stomata.

The evaporation of water from mesophyll cells makes their water potential more negative—effec-

tively, it increases the solute concentration of the cells—so more water enters osmotically from the nearest tiny vein. The removal of water from the xylem of the veins establishes a tension, or pull, on the entire column of water contained within the xylem, so the column is drawn upward all the way from the roots. The ability of water to be pulled upward through a tiny tube results from the remarkable cohesiveness of water—the tendency of water molecules to adhere to one another through hydrogen bonding (Chapter 2). The narrower the tube, the greater the tension the water column can withstand without breaking. As the water column in the xylem is pulled upward, more water enters the xylem in the root by osmosis from surrounding cells.

In summary, the key elements of water transport in the xylem are *evaporation* from the moist cells in the leaves and a resulting *tension* in the remainder of the xylem's water owing to its *cohesion*, so that water is pulled up to replace that which has been lost (Figure 29.8). All this requires no work on the part of the plant. At each step between soil and atmosphere, water moves passively to a region with a more strongly negative water potential. Dry air has the most negative water potential, and the soil solution has the least negative water potential; xylem sap has a water potential more negative than that of cells in the root but less negative than that of mesophyll cells in the leaf.

Mineral ions contained in the xylem sap rise passively as the solvent, water, ascends from root to leaf. In this way the nutritional needs of the shoot are met. Some of the mineral elements brought to the leaves are subsequently redistributed to other parts of the plant by way of the phloem, but the initial delivery from the roots is through the xylem.

The evaporative loss of water from the shoot is called **transpiration**. In addition to promoting the transport of minerals, transpiration contributes to temperature regulation. As water evaporates from mesophyll cells, heat is taken up from the cells, and the leaf temperature drops. This cooling effect may be important in enabling plants to live in certain environments.

The Xylem Sap Is Under Tension

The evaporation–cohesion–tension model can be true only if, in fact, the column of solution in the xylem *is* under tension. The most elegant demonstrations of this tension, and of its adequacy to account for the ascent of sap in the tallest trees, were performed in the early 1960s by Per Scholander of the Scripps Institution of Oceanography in La Jolla, California. Scholander measured tension in stems with a device called a pressure bomb.

The principle of the pressure bomb is as follows: Consider a stem in which the xylem sap is under

29.7 Guttation
Root pressure forces water through openings in the tips of this strawberry leaf.

tension. If the stem is cut, the sap pulls away from the cut, into the stem. Now the tissue is placed in a cylinder—the bomb—in which the pressure may be raised. The cut surface remains outside the bomb. As pressure is applied, the xylem sap is forced back to the cut surface. When the sap first becomes visible again at the cut surface, the pressure in the bomb is recorded. This pressure is the same as the tension that was originally present in the xylem (Figure 29.9).

Scholander used the pressure bomb to study dozens of plant species, from diverse habitats, growing under a variety of conditions. In all cases in which the xylem sap was ascending, it was found to be under tension. The tension disappeared in some of the plants at night. In developing vines, the xylem sap was not under tension until leaves formed. Once leaves appeared, transport in the xylem began, and tensions were recorded.

Suppose you wanted to measure tensions in the xylem at various heights in a large tree, such as a

Douglas fir more than 80 meters tall, to confirm that the tensions were sufficient to account for the rate at which sap was moving up the trunk. How would you get stem samples for measurement? Scholander surveyed a tree to determine the heights of particular twigs and then had a sharpshooter with a high-powered rifle shoot the twigs from the tree. As quickly as the twigs fell to the ground, they were inserted in the pressure bomb and their xylem tensions recorded. Scholander had twigs shot from a tree at heights of 27 and 79 meters at four different times of day. At each hour the difference in tensions was great enough to keep the xylem sap ascending.

TRANSPIRATION AND THE STOMATA

The epidermis of leaves and stems minimizes transpirational water loss by secreting a waxy cuticle, which is quite impermeable to water. However, the

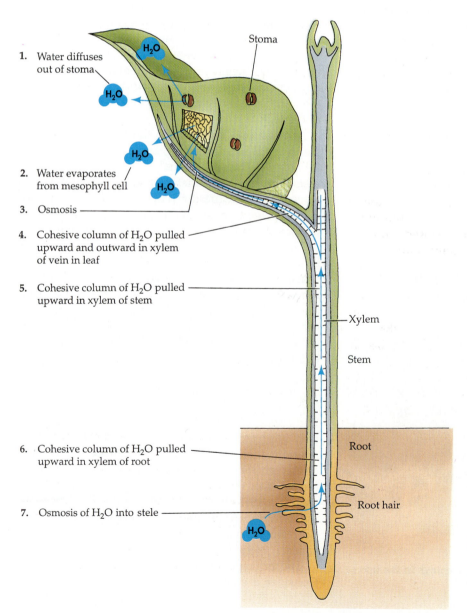

1. Water diffuses out of stoma
2. Water evaporates from mesophyll cell
3. Osmosis
4. Cohesive column of H$_2$O pulled upward and outward in xylem of vein in leaf
5. Cohesive column of H$_2$O pulled upward in xylem of stem
6. Cohesive column of H$_2$O pulled upward in xylem of root
7. Osmosis of H$_2$O into stele

Stoma
Xylem
Stem
Root
Root hair

29.8 Water Transport in Plants
Evaporation, cohesion, and tension account for the movement of water from the soil to the atmosphere.

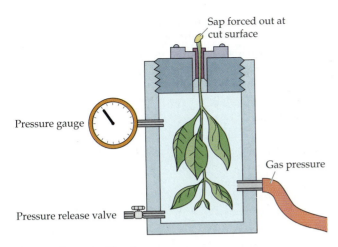

29.9 A Pressure Bomb
By applying just enough pressure so that xylem sap is pushed back to the cut surface of a plant sample, a scientist determines the tension on the sap in the living plant.

cuticle is also impermeable to carbon dioxide, posing a problem: How is the leaf to balance its need to retain water with its need to obtain carbon dioxide for photosynthesis? An elegant compromise has evolved, in the form of stomata (singular: stoma). A **stoma** is a gap in the epidermis; its opening and closing is controlled by a pair of specialized epidermal cells called **guard cells** (Figure 29.10). When the stomata are open, carbon dioxide can enter the leaf by diffusion—but water vapor may also be lost in the same way. Closed stomata prevent water loss—but also exclude carbon dioxide from the leaf. Most plants compromise by opening the stomata only when the light intensity is sufficient to maintain a good rate of photosynthesis. At night, when darkness precludes photosynthesis, the stomata remain closed—no carbon dioxide is needed at this time, and water is conserved. Even during the daytime, the stomata close if water is being lost at too great a rate.

The mechanism by which stomata open and close is now beginning to be understood. The guard cells control the size of the stomatal opening. When the stomata are about to open, potassium ions are actively transported into the guard cells from the surrounding epidermis. (The redistribution of potassium can be visualized by means of an instrument called an electron microprobe.) The accumulation of potassium ions makes the water potential of the guard cells more negative. To maintain osmotic balance, water then enters the guard cells, making them more turgid and stretching them in such a way that a gap, the stoma, appears between them. To close the stoma, this process is reversed: Potassium ions diffuse passively *out* of the guard cells, water follows by osmosis, turgidity is lost, and the guard cells collapse together and seal off the stoma. Negatively charged chloride and organic ions also move along with the potassium ions, maintaining electrical bal-

ance and contributing to the change in osmotic potential of the guard cells.

The term **stomatal complex** refers to the stoma and its associated guard cells. The complex shown in Figure 29.10 is typical of dicots; it contrasts sharply with the more ornate stomatal complexes of monocots, which include further accessory epidermal cells in addition to the stomata and guard cells. The principle of operation is the same for both monocot and dicot stomata.

What controls the movement of potassium into and out of guard cells? The control system is complex, with more than one type of sensor system. For one thing, the level of carbon dioxide in the spaces inside the leaf is monitored; a low level favors opening of the stomata, thus allowing an increased carbon dioxide level and an enhanced rate of photosynthesis. On the other hand, certain cells monitor their own water potentials. If they are too dry—that is, if their water potential is too negative—they release a substance called abscisic acid (Chapter 32). According to one hypothesis, abscisic acid then acts on the guard cells, causing them to release potassium ions, thus closing the stomata and preventing further drying of the leaf. (Some scientists think that abscisic acid serves only to keep the stomata closed, rather than causing the closure; further experiments on the timing of abscisic acid production are needed to resolve this doubt.)

Crassulacean Acid Metabolism

The daily changes in stomatal aperture of most plants are shown in the black curve of Figure 29.11. The stomata are typically open for much of the day and closed at night (they may, however, close during very

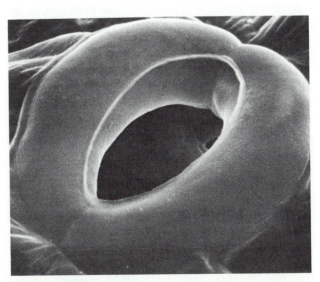

29.10 A Gaping Stoma
Mesophyll cells and air space inside this dicot leaf are seen through the stoma between the two sausage-shaped guard cells.

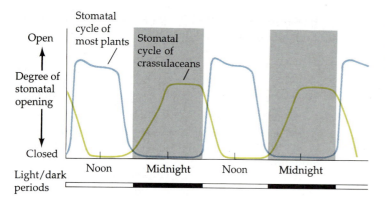

29.11 Stomatal Cycles
Most plants open their stomata during the daytime. Plants of the family Crassulaceae have evolved means to reverse this stomatal cycle. Crassulacean stomata open during the night.

hot days). But not all plants follow this pattern. Many **succulent plants**—fleshy plants that live in dry areas or near the ocean—are members of the flowering plant family Crassulaceae (Figure 8.29). The crassulaceans share a few unusual biochemical and behavioral features. One that was particularly surprising to its discoverers is their "backward" stomatal cycle: Their stomata are open at night and closed by day (colored curve in Figure 29.11). It was then discovered that crassulacean leaf tissues become quite acidic by night and more neutral in the daytime.

The mystery was resolved when the following facts were discovered. At night, while the stomata are open, carbon dioxide diffuses freely into the leaf and reacts in the mesophyll cells with phosphoenolpyruvic acid to produce organic acids such as malic acid and aspartic acid (Chapter 8). These acids accumulate to high concentrations in the vacuoles of the mesophyll cells. With daybreak, the stomata *close*, thus preventing water loss. Throughout the day, the organic acids are broken down to release the carbon dioxide once again—and it is released behind closed stomata. Because the carbon dioxide cannot diffuse out of the plant, it is available for photosynthesis. This set of chemical reactions is referred to as **crassulacean acid metabolism, or CAM**. CAM and the accompanying stomatal behavior were subsequently observed in species of many other plant families besides the Crassulaceae. Biochemically, the CAM reactions are similar to those used by C_4 plants in photosynthesis (Chapter 8). In C_4 photosynthesis, acid formation and sugar formation are separated in space, taking place in different chloroplasts; in CAM, acid and sugar formation are separated in time.

Notice that the formation of organic acids is absolutely essential to the functioning of the reversed stomatal pattern of CAM plants. Without the acid formation, carbon dioxide could still be admitted to the leaf at night and saved for daytime. However, it could only build up in the intercellular spaces of the leaf to the same level—0.03 percent of the atmosphere—as in the surrounding air. This amount would be used up by the Calvin–Benson cycle of photosynthesis in a matter of minutes in the daytime.

Instead, during the night a CAM plant makes the carbon dioxide into organic acids as fast as it comes in, thus allowing yet more carbon dioxide to enter. Acid formation is, in effect, a pump that fills the leaf with carbon dioxide. CAM is well adapted to environments where water is scarce—a leaf with its stomata open at night loses much less water, because of the lower temperature, than does a leaf with its stomata open by day.

TRANSLOCATION OF SUBSTANCES IN THE PHLOEM

How substances in the phloem are moved from sources, such as leaves, to sinks, such as the root system, remains a topic of interest in plant physiology. Sugars, amino acids, some minerals, and a variety of other substances are translocated in the phloem. Any model to explain this translocation must account for a few important facts: Translocation stops if the phloem tissue is killed by heating or other methods; thus, the mechanism must be different from that of transport in the xylem. Translocation often proceeds in both directions—up and down the stem or petiole—simultaneously. This may be explained in terms of neighboring sieve tubes conducting in opposite directions, with each sieve tube transporting all its contents in a single direction. Translocation is inhibited by compounds that inhibit cellular respiration and thus limit the ATP supply.

To answer some of the most pressing questions about translocation, plant physiologists needed to obtain samples of pure phloem sap from individual sieve tube elements. This task was simplified by the recognition that a common garden pest, the aphid, feeds by drilling into a sieve tube. An aphid inserts its stylet, or feeding organ, into a stem until the stylet enters a sieve tube. Within the sieve tubes, the pressure is much greater than in the surrounding plant tissues, so phloem sap is forced up the stylet and into the aphid's digestive tract. So great is the pressure that a sugary drop of liquid is forced out the insect's anus (Figure 29.12). At times, ants collect this

29.12 Phloem Sap Gets Around
Aphids—the white organisms with "sculpted" abdomens
—are drilling into a plant to obtain phloem sap. A drop
of the sap, or "honeydew," has formed at the anus of
one of the aphids, and an ant is about to collect it.

sugary discharge as food, and some species of ants
actually "farm" colonies of aphids, moving them
from place to place and protecting them from enemies. Plant physiologists use the aphid somewhat
differently. When liquid appears on the aphid's abdomen, indicating that it has struck a sieve tube, the
physiologist freezes the aphid and cuts its body away
from the stylet. Phloem sap continues to exude from
the cut stylet, where it may be collected for analysis.

The Pressure Flow Model

Most phloem transport can be accounted for by a
bulk flow of phloem sap, under pressure, through
the sieve tubes (Figure 29.13). Two important features of this model are the active, ATP-requiring
transport of sugars and other solutes *into* the sieve
tubes in source areas and the *removal* of the solutes
by active transport where the sieve tubes enter sinks.
According to this model, the solute concentration at

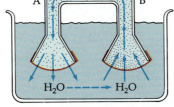

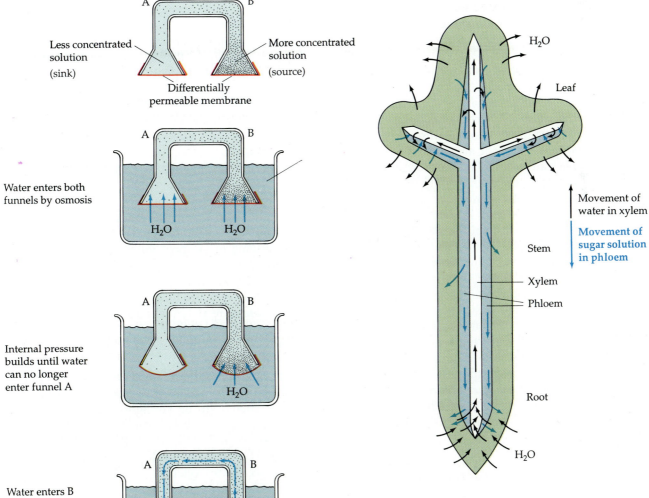

29.13 Pressure Flow in a Sieve Tube
The pressure flow model depends on a difference in solute concentrations at the source and sink ends of the
sieve tube.

the source end of the sieve tube is higher, so that water has a greater tendency to enter the sieve tube by osmosis at that end. In turn, this entry of water causes a greater pressure potential at the source end, so that the entire fluid content of the sieve tube is in effect squeezed toward the sink end of the tube. This mechanism, called the **pressure flow model**, was first proposed more than half a century ago, but some of its features are still debated.

Other mechanisms have been proposed to account for translocation in sieve tubes. Some have been disproved, and none of the rest have been supported by a weight of evidence comparable to that for the pressure flow model.

Loading and Unloading of the Phloem

The pressure flow model depends entirely on the existence of mechanisms for loading sugars and other solutes into the phloem in source regions and for unloading them in sink regions. Phloem loading has been amply demonstrated in a number of plants. Sugars and other solutes to be transported are passed from cell to cell through the symplast in the mesophyll. When these substances reach cells adjacent to the ends of leaf veins, they leave the mesophyll cells and enter the apoplast, sometimes with the help of transfer cells. Then specific sugars, amino acids, some mineral elements, and a few other compounds are actively transported into cells of the phloem, thus reentering the symplast. Passage through the apoplast is important in that it allows selectivity as to what substances are accumulated for translocation—substances can enter the phloem only upon passing through a differentially permeable membrane. In many plants, the cells thus loaded are companion cells (Chapter 28), which then transfer the solutes to the adjacent sieve tube elements. Loading of the phloem with solutes results in a very negative water potential in the sieve tubes; thus, water enters by osmosis from the surrounding tissue and maintains a high pressure potential within the sieve tubes.

In sink regions, the transported solutes are actively transported out of the sieve tubes and into the surrounding tissues. This unloading serves two purposes: It helps to maintain the gradient of osmotic potential and hence of pressure potential in the sieve tubes, and it also promotes the buildup of sugars to high concentrations in storage regions such as developing fruits.

An Alternative Form of Sucrose Transport

The xylem, as well as the phloem, may transport sugars. In sugar maple trees and many other deciduous trees and shrubs of the temperate zones, excess photosynthate produced in late summer and early fall is stored as starch in living xylem cells of the

29.14 Collecting Maple Sap
Sap from several taps flows through tubes to a central location for collection. Nobody is perfect—here, sap leaks from two connections.

trunk and twigs. Later, in early spring, the starch is digested into sugars that are transported initially in the xylem sap, which people in those regions collect and concentrate into syrup (Figure 29.14). The activities of plants may vary with the time of year, and patterns of transport may change accordingly.

SUPPORT IN A TERRESTRIAL ENVIRONMENT

Water buoys up aquatic plants, but terrestrial plants must either sprawl on the ground or have systems to support them against gravity. There are two principal systems of support for terrestrial plants. One support system is the pressure potential of the cells in the body. A small plant can maintain an erect posture if its cells are turgid, but it collapses—wilts—if the pressure potential falls too low. Support by the pressure potential is often augmented by the presence of strengthening tissues such as collenchyma (Chapter 28).

The most important support system found in many plants is wood—a mass of secondary xylem. Wood is such a strong yet lightweight material that we have used it in buildings, furniture, and other structures for millennia. Not all wood is the same, however. Let us next consider some special adaptations of secondary xylem.

Reaction Wood

Growing trees would seem to face a problem: As their branches grow longer and heavier, why don't they simply sag to the ground? This problem is averted by means of a gravity-induced asymmetry in wood structure: Specialized **reaction wood**, differing from normal wood, keeps the limb straight. Angiosperms and gymnosperms have different kinds of reaction wood, and in different places. In gymnosperms, **compression wood** forms on the lower side of a branch. It is prestressed—it is laid down under compressive stress—and *expands*, thus tending to push the branch upward (Figure 29.15*a*). Compression wood contains thicker and shorter tracheids, with more lignin and less cellulose in their walls than normal wood. In contrast, the reaction wood of angiosperms, called **tension wood**, is formed on the upper side of the branch. It is laid down under tension and *shrinks*, thus tending to pull the branch upward, or at least to resist downward bending (Figure 29.15*b*). In tension wood the fibers have more heavily thickened walls, containing less lignin and more cellulose; and there are fewer and smaller vessels than in normal wood.

That a gravitational stimulus, not the sagging of the branch, determines reaction wood deposition is illustrated by an experiment performed by the plant physiologist A. B. Wardrop (Figure 29.15*c*). He bent the trunk of a young angiosperm sapling into a circle and allowed reaction wood to develop. Reaction wood formed on the top side of the top of the loop, where it was under tension, and also on the top side of the bottom of the loop, where it was under compression.

Trees grown indoors tend to be much more spindly than their outdoor counterparts, apparently because indoor trees are not subjected to buffeting by wind. They develop a firmer trunk if they are simply shaken or pounded with a padded mallet from time to time. The change in wood deposition caused by such treatments may be akin to reaction wood formation.

Quonset Huts and Marble Palaces

People put up quonset huts for cheap, relatively short-term shelter. They build palaces of marble if they want to create a monument for the ages. We find similar contrasts if we compare the wood of trees growing under very different conditions. Consider balsa and mahogany—one is extremely light and soft, the other dark and hard. Balsa, like most species with very soft woods, is a fast-growing tree, frequently found in areas recently burnt or cut to the ground. Balsa wood has cells with relatively large diameters, and the wood fibers have very thin walls. For a given volume of wood, the amount of structural material laid down is slight; thus the rapidly growing balsa plant can display its foliage to the sun without a great commitment of resources to structural support. In short, balsa is the botanical equivalent of a quonset hut.

In contrast, mahogany is extremely slow growing. It is very sturdy. Its fine-textured wood has tiny cells, and the wood fibers have thick walls. In contrast to balsa, mahogany wood is "expensive" to form—it has much more dry weight per volume. As a result of its hardness, it can support the plant as a long-lived tree in rain forests. Mahogany contains impregnating materials that darken the wood and help render it resistant to fungal attack. It is the botanical equivalent of a palace.

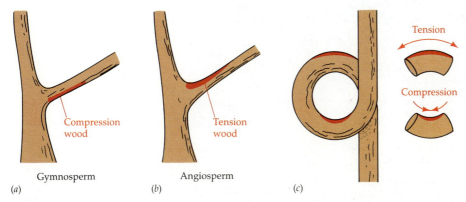

29.15 Reaction Wood
Reaction wood reduces the tendency of branches to sag. (*a*) Compression wood, the reaction wood of gymnosperms, is a heavily lignified wood that forms on the lower sides of branches. (*b*) Tension wood, the reaction wood of angiosperms, forms on the upper sides of branches. (*c*) An angiosperm sapling was bent into a loop and tied in place. Tension wood formed on the tops of both horizontal regions; as shown in the cutouts, one region of tension wood was under tension, but the other was actually under compressive stress. This result indicates that the stimulus to tension wood formation is gravitational (tension wood forms on the upper side) rather than a response to the stress itself.

CHALLENGES POSED BY DRY AND WET ENVIRONMENTS

Some environments impose severe demands upon plants, many of which have evolved adaptations enabling them better to cope with the problem. As we have seen, the terrestrial environment in general confronts plants and other organisms with a shortage of water. Some terrestrial habitats, such as deserts, intensify this challenge, and many plants that inhabit particularly dry areas have one or more structural adaptations that allow them to conserve water. Plants with special adaptations to dry environments are called **xerophytes**.

Some desert plants have, in fact, no special *structural* adaptations for water conservation other than those found in almost all flowering plants. Instead they have an alternative strategy. Simply put, these plants carry out their entire life cycle—from seed to seed—during a brief period in which the surrounding desert soil is sufficiently moist (Figure 29.16). Through the long dry periods that intervene, only the seeds remain alive, until enough moisture is present to trigger the next life cycle. These desert annuals simply evade the periods of drought. Plants that remain active during the dry periods must have special adaptations that enable them to survive.

Special Adaptations of Leaves to Dry and Wet Environments

The secretion of a heavier layer of cuticle over the leaf epidermis to retard water loss is a common ad-

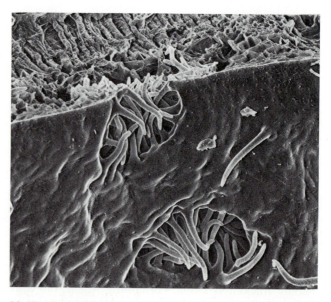

29.17 Stomatal Crypts
In leaves that must conserve water, stomata are often located in sunken pits called stomatal crypts. The hairs covering these two crypts presumably trap moist air. (A section of the leaf's interior is seen toward the top.)

aptation to dry environments. An even more common adaptation is a dense covering of epidermal hairs. In another adaptation to avoid drying air currents, some species have stomata only in sunken cavities below the leaf surface (Figure 29.17); often these cavities contain hairs as well. Ice plants and their relatives have fleshy leaves in which water may be stored. Others, such as ocotillo, produce leaves

29.16 Desert Annuals
Seeds of desert plants often lie dormant for long periods awaiting conditions appropriate for germination. When they do germinate, they grow and reproduce rapidly before the short season passes. They cover the desert landscape with color for only a few weeks, water being inadequate at other times.

(a)   *(b)*

29.18 Opportune Leaf Production
Plants in hot, dry environments lose great amounts of water through their leaves. One adaptation to counter the problem is to have leaves only when water is available in the soil. The ocotillo, which lives in the lower deserts of the southwestern United States and Northern Mexico, does this. *(a)* During dry periods, the thorny, leafless stems of an ocotillo appear almost dead; but when water is on hand, leaves develop rapidly *(b)* and provide the plant with photosynthetic products.

29.19 Shade at Midday
Eucalyptus leaves hang vertically; thus their flat surfaces are not presented directly to the midday sun. This adaptation minimizes heating and water loss.

only when water is abundant, shedding them as the soil dries out (Figure 29.18). Cacti and similar plants have spines rather than typical leaves, and photosynthesis is confined to the fleshy stems. The spines may serve to reflect incident radiation or perhaps to dissipate heat. Corn and a number of related grasses have leaves that roll up during dry periods, thus reducing the leaf surface area through which water is lost. Some trees that grow in arid regions have leaves that hang vertically at all times, thus evading the midday sun. Characteristic examples are found among the eucalyptuses (Figure 29.19).

Xerophytic adaptations of leaves minimize water loss by the plant. However, such adaptations simultaneously minimize the uptake of carbon dioxide and thus limit photosynthesis. In consequence, most xerophytes grow slowly, but they utilize water more efficiently than do other plants—that is, they fix more grams of carbon by photosynthesis per gram of water lost to transpiration than other plants do. The slow growth of xerophytes is not a problem where there are no other plants to compete with them.

Other Adaptations to a Limited Water Supply

We have described some adaptations to desert life, emphasizing those adaptations relating to leaves. Roots, too, may contribute to the adaptation of a species to environments low in water. The desert of Atacama in northern Chile often goes several years without receiving any measurable rainfall. The landscape there is almost barren save for a substantial number of surprisingly large mesquite trees of the genus *Prosopis* (Figure 29.20). How do these trees obtain water? They have taproots that grow to very great depths, sufficient to reach underground water supplies.

A more common adaptation of desert plants is to have root systems that grow each rainy season but die back during dry periods. Cacti, on the other hand, have shallow but extensive fibrous root systems that effectively trap water at the surface of the soil following even light rains.

Xerophytes and other plants receiving inadequate water may accumulate the amino acid proline to substantial concentrations in their vacuoles. As a consequence, the osmotic potential and water potential of the cells become more negative, thus tending to extract more water from the soil.

Floating Plants

Floating aquatic angiosperms, such as water lilies and duckweeds, face a different problem. Whereas the stomata of most flowering plants are most abundant on the lower epidermis of the leaf, this would be of no use to a floating plant. On them, the stomata are

29.20 Mining Water with Deep Taproots
This California desert is not as arid as the Chilean Atacama, but *Prosopis juliflora*, a mesquite, still reaches far down in the sand dunes for its water supply.

confined to the upper epidermis, which is the only one exposed to the atmosphere.

Where Water Is Plentiful and Oxygen Is Scarce

When soils become waterlogged, the availability of oxygen from the soil declines. Most plants cannot tolerate this situation for long. However, a number of species are adapted to life in such a swampy habitat. Their roots grow slowly and, hence, do not penetrate deeply. With an oxygen level too low to support aerobic respiration, the roots carry on alcoholic fermentation (Chapter 7), which provides a supply of ATP for the activities of the root system.

In some plants in swampy environments, the root systems give rise to **pneumatophores**, extensions that grow out of the water and up into the air (Figure 29.21). Oxygen diffusing into pneumatophores aerates the submerged parts of the root system.

Submerged or partially submerged aquatic plants often have large air spaces in the leaf parenchyma and in the petioles. Tissue with such air spaces is called **aerenchyma** (Figure 29.22). Aerenchyma stores and permits the diffusion of oxygen, and it provides buoyancy.

SALINE ENVIRONMENTS

No toxic substance plays a more significant role on a world scale in restricting plant growth than does salt. Saline—salty—habitats support, at best, a sparse vegetation. Plants growing in such habitats come from a wide variety of flowering plant groups and

are referred to as **halophytes**. Saline environments themselves are diverse, ranging from hot, dry, salty deserts to moist, cool, salty marshes. Along the seashore are salty environments created by ocean spray. The ocean itself is a saline environment, as are river estuaries, where fresh and salt water meet and mingle. The salinization of agricultural land is an ever-increasing world problem—where crops are irri-

29.21 Coming Up for Air
The mangroves in this tidal swamp have given rise to pneumatophores—the peglike extensions seen punctuating the sand, under which the remainder of the root system is buried.

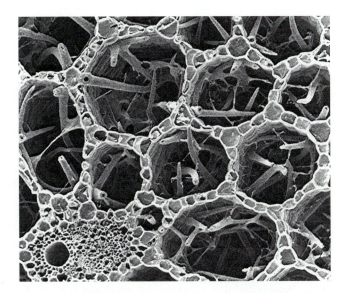

29.22 Aerenchyma
Cross section of a petiole of the yellow water lily, showing a vascular bundle and a number of open channels that constitute the aerenchyma. This scanning electron micrograph shows that the channels are lined by cells, including branched ones that send projections into the channels. Aerenchyma has far fewer cells than does a comparable volume of most petioles; hence less respiratory metabolism is carried on and the need for oxygen is much reduced.

gated, sodium ions in the water accumulate in the soil to ever greater concentrations. Biologists in Israel and elsewhere have had some success in breeding crops that can be watered with seawater or diluted seawater.

Saline environments pose an osmotic problem. Because of a high salt concentration, the environment has an unusually large negative water potential. To obtain water from such an environment, resident plants must have an even more negative water potential than that of plants in nonsaline environments; otherwise, the plants lose water and wilt. A second problem related to the saline environment is the potential toxicity of high concentrations of certain ions, notably sodium. Chloride ions may also be toxic at high concentrations. In some saline environments the concentrations of other ions, such as lithium, sulfate, borate, and bicarbonate, are dangerously high.

Salt Accumulation and Salt Glands

How can halophytes cope with a highly saline environment, while nonhalophytes cannot? Most halophytes share one specific adaptation: They accumulate sodium and, usually, chloride ions and they transport these ions to the leaves. Nonhalophytes accumulate relatively little sodium, even when placed in a saline environment; of the sodium that is absorbed by their roots, very little is transported to the

shoot. The increased salt concentration in halophytes makes their water potential more negative, so that they can take up water from the saline environment. We still do not know how halophytes are able to tolerate such high internal sodium and chloride concentrations without being poisoned.

A number of halophytes have other adaptations to life in saline environments. For example, some have salt glands in their leaves. These glands excrete salt, which collects on the leaf surface until it is removed by rain or wind (Figure 29.23). This kind of adaptation, which reduces the danger of poisoning due to excessive levels of accumulated salt, is found both in some desert plants, such as tamarisk, and in some mangroves growing in seawater in the tropics.

Salt glands can play multiple roles, as in the arid-zone shrub *Atriplex halimus*. In *Atriplex*, the salt secreted by the glands into small bladders on the leaves contributes to the ability of the leaves to obtain water from the roots by increasing the gradient in water potential. At the same time, by making the water potential of the leaves more negative, the salt reduces the transpirational loss of water to the atmosphere.

Adaptations Common to Halophytes and Xerophytes

Many, but not all, halophytes accumulate the amino acid proline in their cell vacuoles. Proline, unlike sodium, is relatively nontoxic. As in xerophytes, the accumulated proline makes the water potential more negative.

Succulence—the possession of fleshy, gummy leaves—has been mentioned as an adaptation to dry environments. The same adaptation is very common

29.23 Salt Secretion
This salty mangrove has used special glands to secrete salt; the salt now appears as crystals on the leaves.

among halophytes, as might be expected, considering that saline environments also make it difficult for plants to take up water. Succulence characterizes many halophytes occupying salt marshes. There the salt concentration in the soil solution may change throughout the day—while the tide is out, evaporation increases the salt concentration. Succulence may offer a reserve supply of water for the plant during the period of maximum salinity; when the salinity drops upon reversal of the tide, the leaf's store of water is replenished.

Other general adaptations to a saline environment are of the same sorts observed in xerophytes. These include high root-to-shoot ratios, sunken stomata, a reduced leaf area, and thick cuticles.

A Versatile Halophyte

The halophyte *Triglochin maritima* is unusual in being able to adjust to a wide range of environmental salinities. Recent work in Toronto, Canada, has established that *T. maritima* plants change in many ways when they are shifted to environments with lower or higher salt concentrations. Researchers watered some plants with seawater and some with diluted seawater. The plants watered with undiluted seawater produced much smaller leaves, with smaller cells, than those watered with diluted seawater; but they retained their leaves longer. The *rate* of leaf production is the same regardless of salinity, but when the environment is changed, the *pattern* of leaf production changes accordingly.

These morphological differences in *T. maritima* leaves are accompanied by physiological changes as well. The leaf cells of the plants grown on undiluted seawater contained much higher concentrations of sodium and chloride ions as well as of proline. The rates of photosynthesis in the leaves of the plants watered with undiluted seawater were higher.

PLANTS AND HERBIVORES

Herbivores—animals that eat plants—depend on plants as their source of energy and nutrients. Plants have defense mechanisms to protect them against herbivores, as we will see; but first let us consider cases in which herbivores actually have a positive effect on the plants that they eat.

Grazing and Plant Productivity

Consider the phenomenon of grazing, the predation of plants by animals. What are the consequences of grazing? Is it detrimental to the plants, or are they somehow adapted to their place in the food chain of nature? In fact, certain plants and their predators evolved together, each acting as the agent of natural

selection on the other. Because of this, grazing increases photosynthetic production in certain plant species.

The removal of some leaves from a plant typically increases the rate of photosynthesis of the remaining leaves. This probably is the result of a number of factors. For one thing, nitrogen obtained from the soil by the roots no longer needs to be divided among so many leaves. For another, the transport of sugars and other photosynthetic products from the leaves may be enhanced, because the demand for those products in sinks—such as roots—is undiminished, while the sources—leaves—have been decreased. A particularly significant factor, especially in grasses, is an increase in the availability of light to the younger, more active leaves or leaf parts. If a grazer removes older, dying—or even dead—leaves and leaf parts, this decreases the shading of younger leaves. In grasses, growth is from the base of the shoot and leaf, in contrast with the situation in most other plants, which grow from their shoot and leaf tips.

Some grazed plants continue to grow until much later in the season than ungrazed but otherwise similar plants. This results in part because the removal of apical buds by the grazers stimulates lateral buds to become active, thus producing a more heavily branched plant (Chapter 32). Also, leaves on ungrazed plants may die earlier in the growing season than leaves on grazed plants.

A clear case of grazing increasing productivity was reported in 1987 by Ken Paige of the University of Utah and Thomas Whitham of Northern Arizona University. Mule deer and elk graze on many plants, including one called scarlet gilia. The grazing removes about 95 percent of the above-ground part of each scarlet gilia. However, each plant quickly regrows not one but four replacement stems. The cropped plants produce three times as many fruits by the end of the growing season as do ungrazed plants (Figure 29.24). Paige and Whitham cropped some scarlet gilia in the laboratory; these plants, too, produced more fruits than did uncropped plants. Not only does their productivity increase, but we may also conclude that grazing by herbivores *increases the fitness* of scarlet gilia, as the cropped plants pass their genes on to more surviving offspring than do uncropped plants.

Chemical Defenses against Herbivores

Some plants, like scarlet gilia, profit from moderate herbivory. Also, it is to a plant's advantage to attract animals that spread the plant's pollen or that eat its fruit and thus distribute its seeds to other places. However, it is also to the plant's advantage to resist attacks by fungi and herbivorous animals and to inhibit the growth of neighboring plants. These attractions, resistances, and inhibitions are often brought

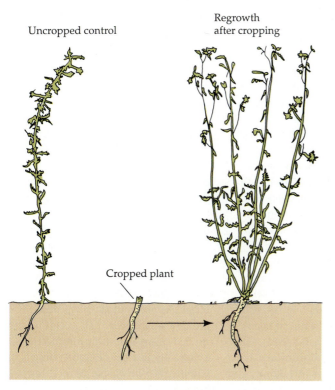

Uncropped control

Regrowth after cropping

Cropped plant

29.24 Overcompensation for Being Eaten
The scarlet gilia on the right, originally single-stemmed, was cropped to the point indicated. It then grew four new stems and produced almost three times as many off-spring as did uncropped plants like the one on the left.

about by special chemicals, known as **secondary products**, produced by the plants. (Primary products are substances such as proteins, nucleic acids, carbohydrates, and lipids, that are produced and used by all living things.) Although different kinds of organisms share a biochemical heritage of primary products, they differ as radically in chemical content as in external appearance. Animals and fungi, for example, need various enzymes to digest their food, a need not shared by the vast majority of plants. The plant kingdom is noteworthy for its profusion of special-purpose secondary products. These compounds help plants compensate for being unable to move. Although a plant cannot flee its herbivorous enemies, it may be able to defend itself chemically.

The effects of defensive secondary products on animals are diverse. Some act upon the nervous systems of herbivorous insects, mollusks, or mammals. Others mimic the natural hormones of animals, causing some insect larvae to fail to develop into adults. Still others damage the digestive tracts of herbivores. Some are toxic to fungal pests.

There are more than ten thousand secondary plant products, ranging in molecular weight from about 70 to more than 400,000 daltons; most, however, are of low molecular weight. Some are produced by only a single species, while others are characteristic of an

entire genus or even family. Their roles are diverse —and in most cases unknown. However, many function in the roles mentioned above, such as attracting pollinators and protecting against herbivores. The major classes of secondary plant products are listed in Table 29.1. A few proteins and amino acids also protect plants against herbivores (Box 29.B).

Most secondary plant products are stored in vacuoles, isolated from the cytoplasmic machinery that they might otherwise damage or destroy. When plant cells are damaged physically by herbivores or chemically by fungi, defensive secondary products are released from the vacuoles and thus freed to act upon the attacking organism. Some other secondary products are stored in laticifers (Figure 29.1). A specific example of an insecticidal secondary product will be considered next.

A Versatile Secondary Product

Some plants produce canavanine, an amino acid that is not found in proteins but that is closely similar to the amino acid arginine, which is found in almost all proteins (Figure 29.25). Canavanine has recently been found to have two important roles in those plants that produce it in significant quantity. The first role is as a nitrogen-storing compound in seeds. The second role is a defensive one and is based on the similarity of canavanine to the protein amino acid arginine. Many insect larvae that consume canavanine-containing plant tissue end up poisoned—the canavanine gets mistakenly incorporated into the insect's proteins in some of the places where the DNA has coded for arginine. Canavanine is different enough in structure from arginine that some of the proteins end up with modified tertiary structures and, hence,

TABLE 29.1
Secondary Plant Products

CLASS	SOME ROLES
Alkaloids	Affect nervous systems of herbivores
Other N, S compounds	Carcinogenesis, nerve degradation in herbivores
Phenolics	Affect nervous systems of herbivores; obnoxious taste; fungicidal action
Quinones	Inhibit growth of competing plants
Terpenes	Fungicidal and insecticidal actions; attract pollinators
Steroids	Mimic animal hormones; prevent normal development of insect herbivores
Flavonoids	Attract pollinators and animals that disperse seeds

BOX 29.B

A Protein for Defense against Insects

A group of scientists at the ARCO Plant Cell Research Institute and the University of Wisconsin recently studied the abilities of wild and domesticated common beans (*Phaseolus vulgaris*) to resist attack by two species of bean weevils. The investigation began with the observation that some wild bean seeds show high resistance to the weevils, whereas no cultivated beans show such resistance. The scientists discovered that all weevil-resistant beans contain a specific seed protein, arcelin. This protein has never been found in cultivated beans. Therefore, the scientists speculated that arcelin is responsible for the resistance of some seeds to predation by the weevils.

As other differences between wild and cultivated beans might have been responsible, two series of experiments were performed to test the relationship between resistance and the possession of arcelin. In one series, cultivated and wild beans were crossed. In the progeny of such crosses, there was an absolute correlation between the presence of arcelin and resistance to weevils. In the other series of experiments, the scientists worked with "artificial bean seeds" made by removing seed coats of cultivated beans and grinding the remainder of the seeds into flour. Different concentrations of arcelin were added to different batches, and the flour was molded into artificial seeds. Bean weevils were then allowed to attack the artificial seeds. The more arcelin the artificial seeds contained, the more resistant they were to weevils.

The scientists then proceeded to prepare, clone, and sequence a cDNA (complementary DNA, Chapter 14) for arcelin, so that they could compare the structure of arcelin with that of other, possibly related proteins that may confer insect resistance on seeds of beans and other legumes. The goal of the work is to introduce genes for arcelin or other resistance-conferring proteins into agriculturally important crops such as beans. In preliminary tests, this group also showed that arcelin in cooked beans is not harmful to rats— a first step toward demonstrating whether arcelin is safe in food for humans.

reduced biological activities. The defects in protein structure and function lead, in turn, to developmental abnormalities in the insect.

A few insect larvae are able to eat canavanine-containing plant tissue and still develop normally. How can this be? In these larvae the enzyme that charges the tRNA specific for arginine discriminates accurately between arginine and canavanine. The canavanine they ingest is thus not incorporated into the proteins they form. The corresponding enzyme in other larvae discriminates much less effectively between those two amino acids, so that canavanine is frequently substituted for arginine.

SUMMARY

Most plants have structural adaptations to their environments for support, protection, and the transport of water and solutes. Support for the erect plant is provided by the pressure potential of the individual cells, by specialized strengthening tissues such as sclerenchyma and collenchyma, and in trees and shrubs by wood.

Water and minerals are transported by tracheids or vessel elements. Transport in the xylem depends on evaporation from moist-walled cells in the leaf,

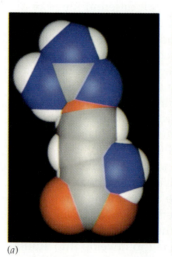 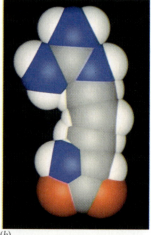

(a)   (b)

29.25 A Toxic Secondary Product and Its Analogue (a) Computer model of canavanine, a nitrogen-storing compound produced by some plants. (b) The amino acid arginine. Because of the similarity of these structures, canavanine is toxic to insect larvae that consume it and then incorporate canavanine in place of arginine in their proteins.

lowering the water potential of those cells and thus pulling water—held together by its cohesiveness—up through the xylem from the root. Products of photosynthesis, and some minerals, are translocated through the phloem by a bulk flow of phloem sap through the sieve tubes. The difference in solute concentration between sources and sinks allows a difference in pressure potential, resulting in bulk flow.

A compromise between water retention and carbon dioxide uptake by leaves is allowed by stomata, whose opening is controlled by osmotically regulated movements of guard cells. In most plants, stomata are open for most of the daylight hours but closed at night. Most succulent plants have a reversed stomatal cycle. Such plants perform crassulacean acid metabolism: They form organic acids during the night and use that carbon for photosynthesis during the day.

Any part of a plant may show specialized adaptations. Stems may grow rapidly and with little development of fiber cell walls, as in balsa, or slowly with heavily developed xylem, as in mahogany. Some plants drop their leaves during periods of drought. Others have sunken stomata in pits in the leaf or roll up their leaves when it is dry; still others store water in fleshy stems or leaves. Leaves of some xerophytes and halophytes are specialized in various ways, including succulence and proline accumulation. Root adaptations include the pneumatophores found in some mangroves. Desert annuals evade drought entirely, spending most of their time as seeds.

Plants produce thousands of secondary products, some of which are directed against herbivores. Others serve to attract animal pollinators.

SELF-QUIZ

1. Water potential:
   a. is the difference between the osmotic potential and the pressure potential.
   b. is analogous to the air pressure in an automobile tire.
   c. is the movement of a solvent through a membrane.
   d. determines the direction of water movement between cells.
   e. is defined as 1.0 for pure water under no applied pressure.

2. Which of the following statements is *not* true?
   a. The symplast is a meshwork consisting of the (connected) living cells.
   b. Water can enter the stele without entering the symplast.
   c. The Casparian strips prevent water from moving between endodermal cells.
   d. The endodermis is a cell layer in the cortex.
   e. Water can move freely in the apoplast without entering cells.

3. Which of the following is *not* part of the transpiration–cohesion–tension model?
   a. Water evaporates from the walls of mesophyll cells.
   b. Removal of water from the xylem exerts a pull on the water column.
   c. Water is remarkably cohesive.
   d. The wider a tube, the greater the tension its water column can withstand.

   e. At each step, water moves to a region with a more strongly negative water potential.

4. Stomata:
   a. control the opening of guard cells.
   b. release less water to the environment than do other parts of the epidermis.
   c. are usually most abundant on the upper epidermis of a leaf.
   d. are covered by a waxy cuticle.
   e. close when water is being lost at too great a rate.

5. Plants that perform Crassulacean acid metabolism:
   a. incorporate carbon dioxide into organic acids.
   b. have leaves that become more acidic during the daylight hours.
   c. close their stomata at night.
   d. are also called C_4 plants.
   e. must live in environments where water is plentiful.

6. Which statement is *not* true of phloem transport?
   a. It takes place in sieve tubes.
   b. It depends upon mechanisms for loading solutes into the phloem in sources.
   c. It stops if the phloem is killed by heat.
   d. In sinks, solutes are actively transported into sieve tube elements.
   e. A high pressure potential is maintained in the sieve tubes.

7. Compression wood:
   a. forms in angiosperms.
   b. forms on the upper side of a branch.
   c. contains less lignin and more more cellulose than ordinary wood.
   d. contains thinner and longer tracheids than ordinary wood.
   e. expands to push a branch upward.

8. Which of the following is *not* an adaptation to dry environments?
   a. A less negative osmotic potential in the vacuoles
   b. Hairy leaves
   c. A heavier cuticle over the leaf epidermis
   d. Sunken stomata
   e. Root systems that grow each rainy season and die back when it is dry

9. Halophytes:
   a. all accumulate proline in their vacuoles.
   b. have less negative osmotic potentials than other plants.
   c. are often succulent.
   d. have low root-to-shoot ratios.
   e. rarely accumulate sodium.

10. Which statement is *not* true of secondary plant products?
    a. Some attract pollinators.
    b. Some are poisonous to herbivores.
    c. Most are proteins or nucleic acids.
    d. Most are stored in vacuoles.
    e. Some mimic the hormones of animals.

FOR STUDY

1. Epidermal cells are described as protective against excess water loss. How do they perform this function?

2. Phloem transports material from "sources" to "sinks." What is meant by a source and a sink? Give examples of each.

3. Contrast the transport of organic substances through the phloem with the transport of water and minerals through the xylem.

Touch on mechanisms and on overall direction.

4. The common oleander (*Nerium oleander*) has its stomata sunk in crypts in its leaves. Whether or not you know what an oleander is, you should be able to describe an important feature of its natural habitat.

5. Transpiration exerts a powerful pulling force on the water column in the xylem. When would you ex-

pect transpiration to proceed most rapidly? Why? Describe the source of the pulling force.

6. Plants that can perform crassulacean acid metabolism (CAM) are thereby adapted to environments in which water is available in limited supply—they open their stomata only at night. Could a non-CAM plant, such as a pea plant, enjoy a similar advantage if it opened its stomata only at night? Explain.

READINGS

Barrett, S. C. H. 1987. "Mimicry in Plants." *Scientific American*, September. Some plants use camouflage to avoid predation. Some weeds survive by mimicking crops, so that humans will select them.

Dussourd, D. E. and T. Eisner. 1987. "Vein-Cutting Behavior: Insect Counterploy to the Latex Defense of Plants." *Science*, vol. 237, pages 898–901. Describes the phenomenon cited at the beginning of this chapter. Clear, readable account of field observations and sound experimentation.

Lewin, R. 1987. "On the Benefits of Being Eaten." *Science*, vol. 236, pages 519–520. Describes the work on scarlet gilia cited (under "Grazing") in this chapter. Discusses other examples of cropping.

Niklas, K. J. 1989. "The Cellular Mechanics of Plants." *American Scientist*, vol. 77, pages 344–349. This fine article, subtitled "How Plants Stand Up," details how cell walls and other aspects of stem architecture enable terrestrial plants to stand erect.

Raven, P. H., R. F. Evert and S. Eichhorn. 1986. *Biology of Plants*, 4th Edition. Worth, New York. A sound general botany textbook.

Rosenthal, G. A. 1986. "The Chemical Defenses of Higher Plants." *Scientific American*, January. Plants employ many chemicals that repel or poison herbivores or that retard the growth of herbivorous insects; some herbivores use these plant-derived compounds for their own defense.

Salisbury, F. B. and C. W. Ross. 1985. *Plant Physiology*, 3rd Edition. Wadsworth, Belmont, CA. An authoritative textbook with excellent chapters on transport and translocation.

Shigo, A. L. 1985. "Compartmentalization of Decay in Trees." *Scientific American*, April. Trees cannot flee from threats, nor do their injuries heal; rather, trees defend themselves by walling off the damage done to them.

Wilson, B. F. and R. R. Archer. 1979. "Tree Design: Some Biological Solutions to Mechanical Problems." *BioScience*, vol. 29, pages 293–298. As trees grow, they constantly redesign themselves. This article examines aspects of this process, such as reaction wood formation, from an engineering viewpoint.

30

Plant Nutrition

PREVIEW: Plants feed themselves with only a little help from other organisms. They take in materials from the atmosphere and the soil and use them in photosynthesis and other activities. Plants depend on the soil for mineral elements essential to their growth; they also modify the soil in which they grow. Soils may contain harmful minerals in addition to essential mineral elements; some plants have mechanisms to cope with these poisons. The soil also contains bacteria that supply nitrogen and sulfur to plants. Plants have many adaptations to improve their ability to capture light for photosynthesis. There are a few species of carnivorous plants.

This chapter deals with essential elements, soils, fertilizers, toxic environments, nitrogen fixation, nitrification, denitrification, nitrate reduction, sulfur metabolism, foliage adaptations, epiphytes, and heterotropic plants.

One year, the older leaves of some crop plants turned yellow instead of remaining green as usual. The yellow leaves on some of the plants became tan in color and then dropped to the ground. A consulting plant physiologist suggested that the soil in which the plants grew contained too little nitrogen and that ammonium sulfate, $(NH_4)_2SO_4$, should be added to increase the nitrogen in the soil.

Why do plants need nitrogen? The answer is simple, if we recall the chemical structures of amino acids—and, hence, proteins—and nucleic acids. These vital components of all living things contain nitrogen, as do chlorophyll and many other important biochemical compounds. If a plant cannot get enough nitrogen, it cannot synthesize these compounds at a rate adequate to keep itself healthy.

How did the physiologist know that the plants were suffering a nitrogen deficiency? For two principal reasons: First, it was a good guess, because a nitrogen deficiency is the most common mineral deficiency of plants; second, because different deficiencies produce characteristic **deficiency symptoms** in a given plant species. Nitrogen deficiency commonly causes yellowing, or chlorosis, of leaves because chlorophyll contains nitrogen, and without chlorophyll, a leaf is not green. Inadequate iron in the soil can also cause chlorosis—iron is not contained in the chlorophyll molecule, but it is required for chlorophyll synthesis. However, iron deficiency commonly causes chlorosis of the youngest leaves, while nitrogen deficiency causes chlorosis of the oldest leaves. The reason for this difference is that nitrogen is readily translocated in the plant and can be redistributed from older tissues to younger, growing tissues to favor their growth; iron is contained in compounds that do not get translocated and hence cannot be redistributed to the young tissues.

NUTRIENTS

Every living thing needs raw materials from its environment. These **nutrients** include the ingredients of macromolecules: carbon, hydrogen, oxygen, and nitrogen. Carbon and oxygen enter the living world through photosynthesis carried out by plants and by some monerans and protists; these organisms obtain carbon and oxygen from atmospheric carbon dioxide. The principal source of hydrogen is water, usually taken up from the soil solution by plants—for hydrogen, too, photosynthesis is the gateway to the living world. Nitrogen, which constitutes about four-fifths of the atmosphere, exists as the virtually inert gas N_2, dinitrogen. A large amount of energy is required to break the triple covalent bond linking the two nitrogen atoms and to obtain a reasonably reactive form from which amino acids and other nitrogen-containing organic compounds may be synthesized. Dinitrogen is initially processed—fixed and oxidized—by some highly specialized monerans in the soil, yielding materials that can be taken up by plants. The plants, in turn, provide organic nitrogen and carbon to animals, fungi, and many microorganisms.

In addition to containing carbon, nitrogen, oxy-

679

gen, and hydrogen, the proteins of organisms contain sulfur, and their nucleic acids contain phosphorus. There is magnesium in chlorophyll, and iron in many important compounds such as the cytochromes. Within the soil, minerals dissolve in water, forming a solution that contacts the roots of plants. Plants take up most of these **mineral nutrients** from the soil solution in ionic form.

Autotrophs and Heterotrophs

The plant kingdom plays a key role in the provision of carbon, oxygen, hydrogen, and nitrogen to the rest of the living world. Plants and some protists and monerans are autotrophs; that is, they make their own organic food from simple inorganic nutrients—carbon dioxide, water, nitrate or ammonium ions containing nitrogen, and a few soluble minerals (Figure 30.1). Organisms that require at least one of their raw materials in the form of organic compounds are called heterotrophs; herbivores depend directly and carnivores indirectly upon autotrophs as their source of nutrition.

Most autotrophs are photosynthetic, using light as the source of energy for synthesizing organic compounds from inorganic raw materials. Some autotrophs, however, are **chemosynthetic**, deriving their energy not from light but from reduced inorganic substances such as hydrogen sulfide (H_2S) in their environment. All chemosynthesizers are monerans and thus, seemingly, not something to discuss in this chapter. However, as we will see, the activities of

certain chemosynthetic bacteria are vital to the nutrition of plants, for these bacteria carry on one of the key steps in the processing of nitrogen.

How Does a Sessile Organism Find Nutrients?

To be sessile—stationary—an organism must exploit energy that is somehow brought to it. Sessile animals depend primarily upon the movement of water to bring energy in the form of food to them, but a plant's supply of energy arrives at the speed of light! A plant's supply of essential materials, however, is strictly local, and it may deplete its local environment of water and minerals as it develops. How do plants cope with such a problem? One answer is to *grow* into new resources—growth is a plant's version of locomotion. Roots must grow to reach new sources of minerals, and a more elaborate root system can more effectively obtain water. Growth also helps a plant secure light and carbon dioxide—to do this, it grows leaves. A plant may compete with other plants for light by outgrowing them, not only capturing more light for itself but also aggressively shading its neighbors.

Ingestion versus Controlled Uptake of Nutrients

Animals ingest their meals, taking in unneeded and sometimes toxic materials along with needed nutrients; what an animal ingests is not determined by its actual needs. Part of what animals ingest must be disposed of as waste products such as urea. Plants, in contrast, do not urinate or produce wastes in other obvious ways. Instead, they control their uptake of most substances, matching the uptake rates to their biochemical needs. The major waste products released to the environment by plants are carbon dioxide or, during active photosynthesis, oxygen gas.

This is not to say that plants do not take up toxic substances from the soil. We have seen examples of such a problem in Chapter 29 and will see more in this chapter. However, plants exert a more systematic control over what can enter their bodies than do animals.

MINERAL NUTRIENTS

The essential mineral elements required by higher plants are described in Figure 30.2 and Table 30.1. They all come from the soil solution and derive ultimately from rock. Four other elements needed by all plants are carbon, oxygen, hydrogen, and nitrogen. The criteria for calling something an **essential element** are the following: (1) The element must be necessary for normal growth and reproduction. (2) The element cannot be replaceable by another element. (3) The requirement must be direct, that is, not

30.1 Meeting Nutrient Requirements
Autotrophic plants require only light and simple inorganic nutrients such as carbon dioxide, water, nitrate or ammonium ions, and several essential minerals. These Boston lettuce plants are growing on nothing more than a solution containing these ingredients.

Normal

Nitrogen deficiency

Calcium deficiency

Iron deficiency

30.2 Mineral Deficiency Symptoms
The tobacco plant at the left lacks no mineral elements and appears normal. In plants deficient in nitrogen, older leaves are yellow and the oldest are yellow and dried.

Plants deficient in calcium have distorted young leaves. Plants deficient in iron have young leaves that are yellow or almost white but with green veins.

the result of some indirect effect, such as the need to relieve toxicity caused by some other substance.

Several essential elements fulfill multiple roles, some of which are structural and others catalytic. Magnesium, as we have mentioned, is a constituent of the chlorophyll molecule and hence essential to photosynthesis. It is also required as a cofactor by numerous enzymes in cellular respiration and other metabolic pathways. Iron is a constituent of many molecules, including some proteins, that participate in oxidation-reduction reactions. Phosphorus, usu-

ally in phosphate groups, is found in many compounds, particularly in pathways of energy metabolism such as photosynthesis and glycolysis. The transfer of phosphate groups is important in many energy-storing and energy-releasing reactions, notably those that use or produce ATP. Other roles of phosphate groups include the activation and inactivation of various enzymes.

Plant tissues contain high concentrations of potassium, which serves in moving water from cell to cell. There are no "pumps" for the active transport of

TABLE 30.1
Elements Required by Higher Plants

ELEMENT	SOURCE	ABSORBED FORM	MAJOR FUNCTIONS
Nonmineral elements			
Carbon (C)	Atmosphere	CO_2	In all organic molecules
Oxygen (O)	Atmosphere	CO_2	In most organic molecules
Hydrogen (H)	Soil	H_2O	In most organic molecules
Nitrogen (N)	Soil	NH_4^+ and NO_3^-	In proteins, nucleic acids, etc.
Mineral nutrients			
Macronutrients			
Phosphorus (P)	Soil	$H_2PO_4^-$	In nucleic acids, ATP, phospholipids, etc.
Potassium (K)	Soil	K^+	Enzyme activation; water balance, etc. (Poorly understood)
Sulfur (S)	Soil	SO_4^{2-}	In proteins, coenzymes
Calcium (Ca)	Soil	Ca^{2+}	Affects cell walls, membranes, and many enzymes
Magnesium (Mg)	Soil	Mg^{2+}	In chlorophyll; required by many enzymes; stabilizes ribosomes
Micronutrients			
Iron (Fe)	Soil	Fe^{3+}	In active site of many redox enzymes and electron carriers; needed for chlorophyll synthesis
Chlorine (Cl)	Soil	Cl^-	Photosynthesis; ionic balance
Manganese (Mn)	Soil	Mn^{2+}	Activates many enzymes
Boron (B)	Soil	$H_2BO_3^-$, HBO_3^{2-}	May be needed for carbohydrate transport (Poorly understood)
Zinc (Zn)	Soil	Zn^{2+}	Enzyme activation; auxin synthesis
Copper (Cu)	Soil	Cu^{2+}	In active site of many redox enzymes and electron carriers
Molybdenum (Mo)	Soil	MoO_4^{3-}	Nitrogen fixation; nitrate reduction

water, yet water must be moved from place to place as, for example, into the guard cells surrounding stomata. Plants and animals solve this problem by actively transporting potassion ions (K^+) from one cell to another. Chloride ions follow the K^+ passively, maintaining electrical balance. Movement of these ions changes the water potential of the cells, and water then moves passively to maintain osmotic balance. Thus, to move water from one cell to another, the plant pumps K^+—and the water follows.

Before plants deficient in an essential element die, they usually produce characteristic deficiency symptoms (Figure 30.2). Such symptoms—for example, yellow leaf tissue between the green veins of younger leaves, typical of iron-deficient plants of many species—help horticulturists diagnose a deficiency in plants that suffer from inadequate nutrition.

Table 30.1 gives the essential minerals as two categories: the macronutrients and the micronutrients. Plant tissues need **macronutrients** in concentrations of at least 1 milligram per gram of their dry matter, and they need **micronutrients** in concentrations of less than 100 micrograms per gram of their dry matter. (Dry matter, or dry weight, is what remains after all the water has been removed from a tissue sample.) As we saw in the introduction to this chapter, some of these elements, such as nitrogen, may move around within the plant, while others, such as iron, do not. *All* these elements are essential to the life of all plants. As we are about to see, it may be that there are other elements essential to some plants—and perhaps to all.

Discovering Which Minerals Are Essential

Plant physiologists discovered most of the essential elements by the technique outlined in Figure 30.3. A seedling's not growing, flowering, and producing viable seed in the absence of a particular element gives evidence that the element is essential. The technique is limited by the possibility that some elements thought to be excluded from the solutions are actually present. Some of the chemicals used in early experiments were so impure that they provided micronutrients not suspected by the first investigators. Some minerals are required in such tiny amounts that there may be enough in a seed coat to feed the embryo and the resultant plant throughout its entire lifetime *and* leave enough in the next seed coat to get the next generation well started. There was enough chloride on dust particles and water droplets in the *air* in Berkeley, California (where some of the work on essential elements was performed) to provide the infinitesimal amounts needed to keep experimental plants growing. The essentiality of chlorine was not established until 1954, after special air filters had been installed in the laboratory! Simply touching a plant may give it a significant dose of chlorine in the form of chloride ions from sweat.

Only rarely are new essential elements reported now—either the list is virtually complete or, more likely, we will need more sophisticated techniques to find others. Some minerals are essential for certain plants but apparently not for others. It appears that the next new essential element to be confirmed—the first since chlorine—may be nickel, which was shown in 1984 to be essential for legumes and may soon be demonstrated to be generally required.

30.3   Discovering Essential Plant Nutrients
The plant physiologist transplants a seedling to a solution lacking only one of the ingredients thought to be essential for growth, substance "A" in this example. If the plant grows and reproduces normally after being transplanted, the missing ingredient is assumed to be nonessential. The experimental environment must be rigorously controlled because some essential nutrients are needed in tiny amounts that may be present as contaminants of other materials or on objects.

SOILS

Soils are of great importance to plants, and plant–soil interactions are complex. Plants obtain their mineral nutrients from the soil or the water in which they grow. Water for terrestrial plants also comes from the soil, as does the supply of oxygen for the roots; soil also provides mechanical support for plants on land. Soil harbors bacteria that perform chemical reactions leading to products required for plant growth; on the other hand, soil may also contain organisms harmful to plants.

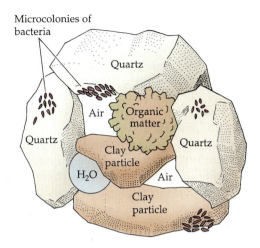

30.4 The Complexity of Soil
Soil consists of more than inorganic particles. It contains living organisms. Air and water are present in pores in soil crumbs like this one. Fungi and roots are also important components of soil.

Soils consist of both living and nonliving components. The living components include plant roots as well as populations of many bacterial and fungal species (Figure 30.4). The nonliving portion of the soil includes rock fragments ranging in size from large boulders to the finest of **clay** particles (2 μm and less in diameter, Table 30.2). Soils also contain minerals, water, gases, and organic matter from animals, plants, fungi, and bacteria. Soils change constantly because of both nonhuman natural causes—such as rain, high and low temperatures, and the activities of plants and animals—and human activities, farming in particular. Soils from different parts of the world differ dramatically in their chemical composition and physical structure, owing to the different circumstances of temperature, water supply, and other factors during their formation.

The structure of any soil changes with depth, revealing a soil profile. Soils differ so greatly that it is difficult to generalize, but virtually all soils consist of two or more **horizons**—recognizable horizontal layers—lying on top of one another in predictable patterns (Figure 30.5). Mineral nutrients tend to be leached from the top horizon, that is, dissolved in rain or irrigation water, and carried to deeper horizons; other processes also cause movement of materials up and down in the soil. Soil scientists recognize three major zones in the profile of a typical soil. The A horizon is the zone from which minerals have been depleted by leaching. Most of the organic matter in the soil is in the A horizon, as are most roots, earthworms, soil insects, nematodes, and soil protists. Successful agriculture depends upon the presence of a suitable A horizon. The B horizon is the zone of infiltration and accumulation of materials leached from above, and the C horizon is the original parent material from which the soil is derived. Some deep-growing roots extend into the B horizon, but roots rarely enter the C horizon.

Soils and Plant Nutrition

The supply of minerals to plants depends upon the presence of clay particles, which have a net negative charge. The minerals that are important for plant nutrition, such as potassium, magnesium, and calcium, are found in soil as positive ions chemically attached to clay particles. The ions are detached from the clay by reaction with carbonic acid (H_2CO_3). Car-

TABLE 30.2 Soil Particles	
SOIL TYPE	**PARTICLE SIZE (mm)**
Coarse sand	0.2–2.0
Fine sand	0.02–0.2
Silt	0.002–0.02
Clay	<0.002

30.5 A Soil's Profile
The A, B, and C horizons can sometimes be seen in road cuts such as this one in Australia. The upper layers developed from the bedrock. The dark upper layer is home to most of the living organisms in the soil.

bonic acid is almost universally present in soils because it is formed whenever carbon dioxide (CO_2) from the atmosphere dissolves in water according to the reaction $CO_2 + H_2O \rightarrow H_2CO_3$. The permanently attached or "fixed" charged groups on clay particles are predominantly negatively charged, and the positively charged ions in solution may associate reversibly with these negative ions. Protons—hydrogen ions (H^+)—released from roots or obtained from the ionization of carbonic acid trade places with ions such as potassium (K^+) and calcium (Ca^{2+}) on the clay particles, thus bringing those nutrients into the soil solution. This trading of places is called **ion exchange** (Figure 30.6). The fertility of a soil is determined primarily by its ability to provide such nutrients as potassium, magnesium, and calcium in this manner.

The harvesting of crops, which contain mineral nutrients obtained from the soil during growth, and leaching often create a need for the addition of fertilizers to agricultural soils. Crop yields are reduced if too much of any element is removed. Minerals may be restored by adding such organic fertilizers as rotted manure or inorganic fertilizers of various types. The three elements most commonly added to agricultural soils are nitrogen, phosphorus, and potassium. The ratios of these elements vary among fertilizers, which are often characterized by their N–P–K percentages. A 5–10–5 fertilizer, for example, contains 5 percent nitrogen, 10 percent phosphate, and 5 percent potassium. Both organic and inorganic fertilizers can provide the necessary minerals. Organic fertilizers contain materials that improve the physical properties of the soil, providing air pockets for gases, root growth, and drainage. Inorganic fertilizers, on the other hand, provide a rapid increase in soil nutrients and can be formulated to meet the requirements of a particular soil and a particular crop.

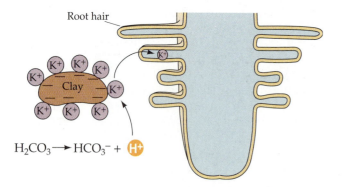

30.6 Ion Exchange
A clay particle binds cations such as the potassium ions shown here. The potassium is exchanged for hydrogen ions obtained from carbonic acid or from the plant itself. If ion exchange is too extensive, the soil will be depleted of potassium and thus require fertilizer.

Soil Formation

The type of soil that forms in a given area depends on the rock from which it is formed, the climate, the topography, the organisms living there, and the length of time soil-forming processes have been acting. Rocks are broken down in part by **mechanical weathering**, the physical breakdown—without any accompanying chemical changes—of materials by wetting, drying, and freezing. The most important parts of soil formation, however, include **chemical weathering**, the chemical alteration of at least some of the materials in the rocks. The key process is the formation of clay. Both the physical and the chemical properties of soils depend on the amount and kind of clay particles they contain. Just grinding up rocks does not produce a clay that swells and shrinks and is chemically active. The rock must be chemically changed as well. The initial step in the chemical weathering of most soil minerals is hydrolysis, as illustrated for feldspar, a common soil mineral, in the following formula:

$$\langle Si, Al, O \rangle K^+ + H^+OH^- \rightarrow \langle Si, Al, O \rangle H^+ + K^+OH^-$$

FELDSPAR   WATER   HYDROLYZED FELDSPAR   POTASSIUM HYDROXIDE

Two examples illustrate the diversity of soil-forming processes. In wet tropical regions, where rainfall and temperatures are high and the soils are usually moist, water moves rapidly downward through the soil. Silica and soluble nutrients are quickly leached, leaving insoluble iron and aluminum compounds in the A horizon. These are often oxidized and give bright reddish colors to those soils. This type of soil-formation process is known as **laterization**. The resulting soil is very poor in nutrients. In semiarid regions, where water from rainfall evaporates rapidly, there is no net movement of water downward through the soil. Instead, water penetrates for a distance, stops, and then is drawn back up by the roots of plants and by evaporation from the soil surface. Under these conditions the soil remains rich in mineral nutrients, and a hard layer of calcium carbonate often forms in the B horizon. The B horizon may be so hard that it prevents deeper penetration of the soil by plant roots. These soils are, however, very fertile when supplied with additional water and nitrogen. Much of the success of irrigated agriculture depends on the high nutrient content of arid-zone soils.

Effects of Plants on Soils

How soil forms in a particular place also depends on the types of plants growing there. Plant litter is a major source of carbon-rich materials that are broken down to form **humus**—dark-colored organic material, each particle of which is too small to be recognizable with the naked eye.

Soils rich in exchangeable positive ions of mineral nutrients tend to support plants that extract large quantities of nutrients for incorporation into their tissues. These tissues produce a rich, alkaline humus called a mull when they die and decompose. Plants growing on nutrient-poor soils extract fewer nutrients and form tissues that yield a poor, acidic humus known as mor. Conifers, in particular, produce a mor humus that is resistant to decay and may accumulate in thick layers on the surface of the soil.

Clay particles slowly decompose, chemically, as a soil weathers. After hundreds of thousands of years, all the exchangeable nutrient ions are leached out of the mineralized soil, and its fertility is greatly reduced. A general picture of these changes over a period of 1 million years is shown in Figure 30.7. Changes in plant biomass—the total weight of living plants (green curve)—are also shown in the figure to demonstrate how plant productivity also declines over time.

Biologists were slow to recognize the importance of the long-term changes in soils because the soils of the north temperate zone, where most soil scientists live and work, are nearly all very young, dating from the last glacial period only a few thousand years ago. In large areas of the tropics and subtropics, however, especially away from areas of recent mountain formation, soils are ancient and have few remaining

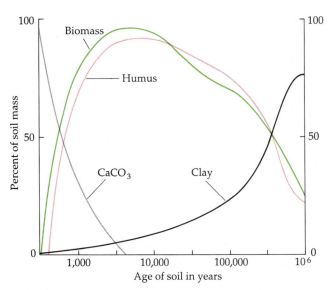

30.7 Aging of Soils
Over many thousands of years, weathering processes gradually remove most exchangeable nutrient ions from a soil, leaving an infertile residue. Many tropical and subtropical soils have ages near the right margin of this figure. Typical changes in humus, clay, calcium carbonate, and quantity of supported vegetation (biomass) in a hypothetical soil over a period of one million years are shown here. The behavior of calcium is typical of many soil nutrients.

nutrients. These soils are largely useless for agriculture unless heavily fertilized, but they do support lush vegetation if rainfall is high (see Chapter 48).

NUTRITIONAL ADAPTATIONS TO DIFFERING SOILS

Because terrestrial plants cannot move from place to place, they must settle for whatever is available where they begin life. The range of soil conditions is vast, with great possible variation in acidity, water availability, air supply, microbial content, and salinity. Which mineral nutrients are present, and in what proportions, also varies widely, as does the concentration of potentially toxic minerals.

Even an individual plant, or a single root, must deal with environmental diversity. Animal droppings give high local concentrations of nitrogen. A particle of calcium carbonate in the soil may make a tiny area alkaline, while dead organic matter may make a nearby area quite acidic.

Habitats Laden with Heavy Metals

High concentrations of heavy metal, such as copper, lead, nickel, and zinc, poison most plants, even though plants require some heavy metals at low concentrations. Some sites are naturally rich in heavy metals as a result of normal geological processes. Human activities, notably the mining of metallic ores, leave localized areas—known as tailings—with substantial heavy-metal concentrations and low nutrient concentrations. Such sites are hostile to most plants, and seeds falling on them generally do not produce adult plants.

However, mine tailings rich in heavy metals generally are not barren. Instead, they support plant populations that differ genetically from populations of the same species on the surrounding normal soils (Figure 30.8). Within some species, a few individuals may have genotypes that allow them to survive in soils rich in heavy metals. Those individuals may grow poorly on such soils compared with their potential for growth on more normal soils, but they survive. In fact, because few other plants can survive in such habitats, competition is sharply reduced, so the few plants growing in them tend to thrive.

It was initially thought that some plants tolerate heavy metals by excluding them: By not taking up the metal ions, the plant could avoid being poisoned. However, measurements have shown that tolerant plants growing on mine tailings do take up the heavy metals, accumulating them to concentrations that would kill most plants. Thus the tolerant plants must have some mechanism for dealing with the heavy metals taken up.

The British biologist D. Jowett made an interest-

30.8 Plant Life on a Mine Tailing
This barren scene is a tailing of a copper mine at Copper-hill, Tennessee. Although high concentrations of copper kill most plants, Bermuda grass is colonizing the hillock in the foreground. How can it do this?

30.9 Serpentine Barrens
The sparsely vegetated serpentine soil in the foreground contrasts with the forest on the nonserpentine soil in the background at a place in Baltimore County, Maryland.

ing discovery about heavy-metal-tolerant plants. In Wales and Scotland, bent grass (*Agrostis*) is found growing near many mines (see Figure 19.10). From mine to mine, the heavy metals in the soil differ. Jowett obtained bent grass samples from several such sites and tested their ability to grow in various solutions, each containing a single one of the heavy metals. In general, the plants tolerated a particular heavy metal—the one most abundant in their habitat—but were sensitive to other heavy metals. That is, their tolerance was for one or two heavy metals only, rather than for the heavy metals as a group.

Tolerant populations can evolve and colonize an area surprisingly rapidly. The bent grass population around a particular copper mine in Wales is copper-resistant and relatively abundant, yet the copper-rich soil dates from mining done late in the nineteenth century, only a century ago.

Plants on Serpentine Soils

One unproductive soil type that is found in many parts of the world is derived from rock called **serpentine**. Calcium is in short supply in serpentine soils, as are some other macronutrients; magnesium is present in greater concentration than calcium; and chromium, nickel, and certain other heavy metals may be abundant. These factors make serpentine soils inhospitable to many plants. Indeed, the vegetation on most serpentine soils differs dramatically from that on immediately adjacent nonserpentine soils, with the serpentine vegetation being more sparse and less diverse (Figure 30.9).

It seems probable that the shortage of calcium and

the high magnesium concentration are the principal challenges facing potential colonizers of a serpentine soil. The challenges are successfully met by a number of species that have suitable physiological adaptations. Figure 30.10 illustrates a striking difference between species in response to environmental calcium supply. Biologists divided some serpentine soil into several samples, adjusted the calcium level of each, and grew jewel flower—*Streptanthus glandulosis*, which grows on serpentine—and tomato—a crop plant intolerant of serpentine—on each sample. The growth of the tomato plants was sharply dependent

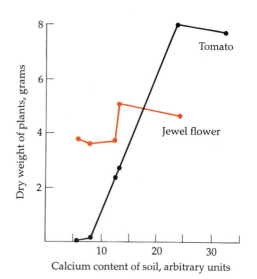

30.10 Differing Responses to Calcium Supply
Adding calcium to serpentine soil permits growth of a crop plant but does not increase growth of a serpentine plant.

on calcium concentration, while that of the jewel flower was remarkably insensitive to it. Serpentine plants such as jewel flower can absorb calcium efficiently even from soils highly deficient in that element. They may also be able either to exclude excess magnesium or to tolerate high internal magnesium concentrations.

In our discussion of plants and soils, we have seen that some plants have special adaptations that allow them to thrive, or at least survive, on soils, such as serpentine, that are inhospitable to other plants. Next we will consider the roles that certain soil bacteria play in rendering soils more hospitable to plants.

NITROGEN FIXATION

Plants cannot use dinitrogen gas (N_2) directly as a nutrient. N_2 is a highly unreactive substance, and only a few species of monerans are capable of converting it into something more generally useful. These organisms—the **nitrogen fixers**—convert N_2 to ammonia (NH_3). All nitrogen fixers are prokaryotes, but some of them must live in intimate association with specific eukaryotes before they develop functional nitrogen-fixing machinery. Not only are there relatively few kinds of nitrogen fixers, but what few there are have a small biomass relative to the mass of other organisms on Earth. However, without the nitrogen fixers, no other known organisms would survive! This elite group of prokaryotes is just as essential in the biosphere as are the photosynthetic autotrophs.

Nitrogen-fixing species are widely distributed in the kingdom Monera. One group of microorganisms fixes nitrogen only in close association with the roots of certain seed plants; the best known of these microorganisms belong to the bacterial genus *Rhizobium*. Species of *Rhizobium* are found living free in the soil, where they do not fix nitrogen. However, *Rhizobium* living in nodules on the roots of plants in the legume family—which includes peas, soybeans, alfalfa, and a large number of tropical shrubs and trees—do fix nitrogen (Figure 30.11). Some cyanobacteria fix nitrogen in association with fungi in lichens or with ferns, cycads, or bryophytes. Finally, the filamentous bacteria called actinomycetes fix nitrogen in association with root nodules on shrub species such as alder.

The ancient Chinese, Greeks, and Romans, and probably members of other early civilizations, recognized that plants such as clover, alfalfa, and peas improve the soil in which they are grown. Two German chemists, Hellriegel and Wilfarth, first showed in 1888 that the root nodules on these plants are caused by bacteria and are in fact sites of nitrogen fixation. These particular plant-infecting bacteria all belong to the genus *Rhizobium*, and the various species of *Rhizobium* show a fairly high specificity for the species of legume they nodulate.

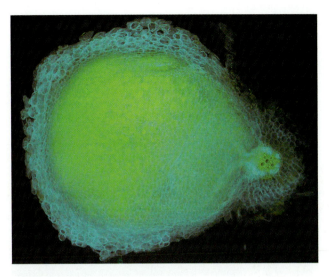

30.11 Root Nodules
This large, round, tumorlike nodule is developing from a broad bean root—the smaller structure to the right. The nodule houses bacteria of the genus *Rhizobium*, which are emitting yellow fluorescence in this stained nodule. The vascular tissues of the root are also yellow; parenchyma cells of the nodule and root are blue.

In the oceans, various photosynthetic bacteria, including cyanobacteria, fix nitrogen; in fresh water, cyanobacteria are the principal nitrogen fixers. On land, free-living soil bacteria make some contribution, but it is the root nodules of plants that produce most of the fixed nitrogen. Unlike various free-living prokaryotic nitrogen fixers that fix what they need for their own uses and release the fixed nitrogen only upon their deaths, root nodules release up to 90 percent of the nitrogen they fix to the rest of the plant and actually excrete some amino acids into the soil, making nitrogen immediately available to other organisms. Some farmers alternate their crops, planting clover or alfalfa occasionally to increase the useful nitrogen content of the soil.

The pioneering role—the ability to occupy environments having few or no other plants—of some of the nonleguminous plants with nodules is an extremely important one (Figure 30.12; see also Figure 46.24). Shrubs such as alder thrive in mountainous areas, with their roots grasping chunks of rock in the talus (debris) slopes below the cliffs. The western mountain lilac *Ceanothus* grows well in extremely gravelly soils that have little or no organic matter and hence no fixed nitrogen. Eastern sweet gale (*Myrica*) flourishes on almost pure sand; their growth is dependent on their nodules. These plants, with their bacterial partners, play a vital role in making otherwise barren habitats available to other plants and to the animals that depend on them.

Organisms fix approximately 90 million tons of atmospheric dinitrogen per year. Tens of millions of

tains molybdenum and several iron atoms. Nitrogen fixation also requires a strong reducing agent to transfer hydrogen atoms to dinitrogen and the intermediate products, and the reactions require a great deal of energy which is supplied by ATP. Depending upon the species of nitrogen fixer, either respiration or photosynthetic metabolism may provide the necessary ATP and reducing agent. The activity of nitrogenase in living nodules may be monitored as described in Box 30.A.

Nitrogenase is extremely sensitive to oxygen—so much so that its discovery was delayed because investigators had not thought to seek it under anaerobic conditions, which are inconvenient to establish in the laboratory. Because of the great sensitivity of nitrogenase to oxygen, it is not surprising that many nitrogen fixers are anaerobes, living in the absence of oxygen. Aerobic nitrogen fixers must decrease their internal oxygen levels drastically. One means for doing this—the production of a special type of hemoglobin—will be described in the next section.

tons are also fixed industrially, by a method to be described later. A smaller amount of nitrogen is fixed in the atmosphere by nonbiological means such as lightning, volcanic eruption, and forest fires; the products thus formed are brought down by rainwater. But by far the greatest share of total world nitrogen fixation is that performed biologically by nitrogen-fixing organisms.

Chemistry of Nitrogen Fixation

Nitrogen fixation progressively reduces the dinitrogen molecule by the addition of pairs of hydrogen atoms, and finally the last of the three bonds between the nitrogen atoms is cleaved:

$$N \equiv N \xrightarrow{2H} HN = NH \xrightarrow{2H} H_2N - NH_2 \xrightarrow{2H} 2\ NH_3$$

DINITROGEN   AMMONIA

In each of these reactions, the reactants are firmly bound to the surface of a single enzyme called **nitrogenase** (Figure 30.13). The nitrogenase molecule con-

Symbiotic Nitrogen Fixation

The legume nodule represents an excellent example of symbiosis, a situation in which two different organisms live in physical contact and, in association, do things that neither organism can do separately. In the form of symbiosis called mutualism, both organisms benefit from the relationship. Neither free-living *Rhizobium* species nor uninfected legumes can fix ni-

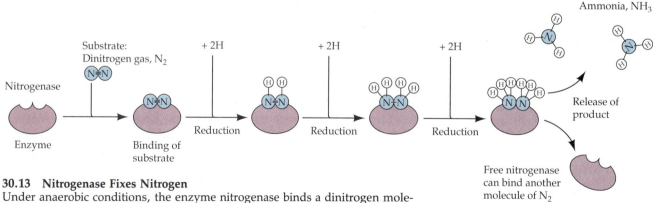

30.13 Nitrogenase Fixes Nitrogen
Under anaerobic conditions, the enzyme nitrogenase binds a dinitrogen molecule while the nitrogen is reduced by the addition of three successive pairs of hydrogen atoms. The final products—two molecules of ammonia—are released, freeing the nitrogenase to bind another dinitrogen molecule.

BOX 30.A

Biotechnology in a Plant

When should nitrogen-containing fertilizer be added to a crop? If too much fertilizer is added during the growing season, the farmer is wasting money. If too little fertilizer is added, the crop's yield—and the farmer's income—is low. The best judge of a plant's nitrogen status is the plant itself—and one crop plant has been "taught" how to report its nitrogen status to scientists.

Aladar Szalay of the University of Alberta and Thomas Baldwin of Texas A&M University isolated genes from *Vibrio harveyi*, a luminescent marine bacterium, that code for the production of luciferase, the enzyme that catalyzes the light-producing reaction. These scientists then inserted the luciferase genes at a key locus in the chromosome of *Rhizobium japonicum*, a bacterium that participates in the nitrogen-fixing root

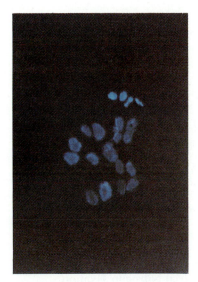

The photo on the left shows soybean root nodules in normal light. The right-hand photo is the same scene, shot in the dark in the presence of decanal fumes.

nodules of soybean plants. The locus is between the gene coding for nitrogenase production and the promoter of the nitrogenase gene.

The nitrogenase genes in soybean root nodules are inactive when the plants are getting enough nitrogen from the soil. If more nitrogen is needed, the nitrogenase promoter activates the nitrogenase gene, allowing the nodules to fix their own ni-

trogen directly. The insertion of *V. harveyi* luciferase genes between the *R. japonicum* nitrogenase gene and its promoter has the following result: When more nitrogen is needed, the nitrogenase promoter activates the neighboring genes—including those that code for luciferase. One can tell when luciferase is present by adding a suitable substrate, in this case a volatile substance known as decanal. Decanal vapor is absorbed directly by cells. If decanal and luciferase are both present, light is emitted.

trogen. Only when the two are closely associated in root nodules does the reaction take place.

The establishment of this symbiosis between *Rhizobium* and a legume requires a complex series of steps with active contributions by both the bacterium and the root. The first to act are root hairs, projecting from the epidermal cells of very young roots. As indicated at the beginning of Figure 30.14, the root hairs release a substance that attracts the *Rhizobium* and stimulates the bacteria to multiply; the bacteria, in turn, produce one or more growth substances that cause changes in the root hair so that its cell wall invaginates—folds inward. The invagination proceeds inward through several cells as an infection thread; the bacteria in the thread continue to divide, although slowly. At this stage the bacteria are still *outside* the plant in a sense, for the thread is lined with cellulose and other cell-wall materials. The thread grows into the cortex tissue of the root until

it encounters one or more cells that, by accident, are tetraploid—$4n$—these are common in older tissues of many plant species. The tetraploid cells begin to divide rapidly, and the infection thread bursts, releasing the bacteria into the cytoplasm of these host cells. The bacteria now undergo a remarkable transformation, increasing about tenfold in size, developing an outside membranous envelope, and forming an elaborately folded internal membrane. At this stage the infecting bacteria are called bacteroids (Figure 30.15).

As the final step before the fixation of nitrogen can begin, the plant produces hemoglobin, which surrounds the bacteroids. Hemoglobin is an oxygen-carrying pigment that one seldom associates with plants, but some nodules contain enough of it to be bright pink when viewed in cross section. The hemoglobin traps oxygen, keeping the bacteroids and their nitrogenase anaerobic.

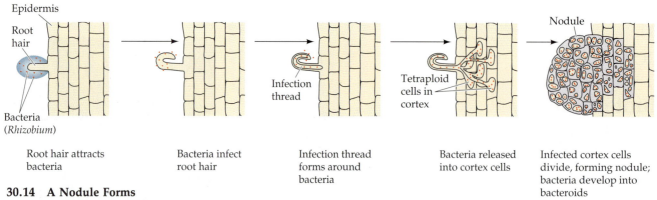

Epidermis

Root hair

Bacteria (Rhizobium)

Infection thread

Tetraploid cells in cortex

Nodule

Root hair attracts bacteria

Bacteria infect root hair

Infection thread forms around bacteria

Bacteria released into cortex cells

Infected cortex cells divide, forming nodule; bacteria develop into bacteroids

30.14 A Nodule Forms
Rhizobium develops the ability to fix nitrogen only after entering a legume root. The drawings show the sequence of events in nodule formation.

Industrial Nitrogen Fixation

Bacterial nitrogen fixation does not suffice to support the needs of agriculture. Native Americans used to plant dead fish along with corn, so that the decaying fish would release fixed nitrogen usable by the developing corn seedlings. Industrial nitrogen fixation is becoming ever more important to world agriculture because of the degradation of soils and the need to feed a rapidly expanding population. Research on biological nitrogen fixation is being vigorously pursued, with commercial applications very much in mind. At present, obtaining a supply of nitrogen-containing fertilizer takes more energy than any other aspect of crop production in the United States. The most common industrial method for producing nitrogen fertilizer is the so-called Haber process, in which dinitrogen gas and hydrogen gas combine to make ammonia. It takes the energy obtained from 1,000 kilograms (1.1 tons) of coal to produce 2.5 kilograms

of ammonia by the Haber process—nitrogen fixation is very expensive in terms of energy.

There is an urgent need to develop an alternative to the Haber process because of the cost of energy and other economic factors. One line of investigation centers on recombinant DNA technology as a means of "teaching" new plants to produce nitrogenase. Workers in many industrial and academic laboratories are working on the insertion of bacterial genes coding for nitrogenase into plasmids, and on the incorporation of such plasmids into the cells of angiosperms, particularly crop plants. However, developing crops that can fix their own nitrogen will take more than just the insertion of genes for nitrogenase, because there must also be provisions for excluding free oxygen and for obtaining strong reducing agents. Biological nitrogen fixation, like industrial nitrogen fixation, is extremely expensive in terms of energy. Looking to nature for evidence of this, we find that legumes compete successfully with grasses only where there is a real shortage of nitrogen in the soil. Ultimately, the need for ATP represents a greater

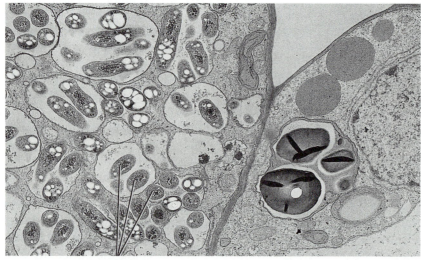

Bacteroids

Uninfected cell

30.15 *Rhizobium* in a Legume
Bacteroids of *Rhizobium japonicum* in vacuoles within a soybean root cell. A portion of an uninfected root cell is seen to the right.

technical challenge than the insertion of nitrogenase genes. However, the stakes—especially the financial ones—are great, and a great amount of effort is being invested in research along these lines. This is, in fact, one of the busiest areas in the burgeoning field of biotechnology.

DENITRIFICATION

Because nitrogen fixation acts to decrease the amount of dinitrogen gas in the atmosphere, we should mention, if only briefly, the opposite process, called **denitrification**. There are a number of normally aerobic bacteria, mostly species of the genera *Bacillus* and *Pseudomonas*, that can use nitrate (NO_3^-) as a terminal electron acceptor in place of oxygen if they are kept under anaerobic conditions:

$$2NO_3^- + 10e^- + 12H^+ \rightarrow N_2 + 6H_2O$$

These bacteria are extremely common and, as can be seen in the equation, return dinitrogen to the atmosphere. Nature's nitrogen cycle is discussed in Chapter 47.

NITRIFICATION

When nitrogen fixers release fixed nitrogen into the soil, it is primarily in the form of ammonia and ammonium ions (NH_4^+). Although ammonia is toxic to plants, ammonium ions can be taken up safely. However, most plants grow better with nitrate than with ammonium ions as a source of nitrogen. There is also some dependence on soil pH, with nitrate ions being taken up preferentially under more acidic conditions and ammonium ions under more basic ones.

Where do nitrate ions in the soil come from? Here again, plants depend upon specific soil bacteria, which among them accomplish nitrification (the oxidation of ammonia to nitrate). Bacteria of two genera, *Nitrosomonas* and *Nitrosococcus*, are capable of converting ammonia to nitrite ions (NO_2^-), and *Nitrobacter* bacteria oxidize nitrite to nitrate. Thus these three genera of prokaryotes constitute a critical ecological link, taking the products of other crucial bacteria, the nitrogen fixers, and converting them into a form more available to plants and, hence, to the rest of the biosphere.

What is in it for these bacteria to be so ecologically helpful? Actually, they carry on nitrification for their own selfish ends. These three genera are chemosynthetic autotrophs; that is, their chemosynthesis is powered by the energy released in oxidizing ammonia or nitrite. For example, by passing the electrons from nitrite through an electron transport chain, *Nitrobacter* can make ATP, and, using some of this ATP, it can also make NADH. With the ATP and NADH,

the bacterium can convert carbon dioxide and water to glucose and other foods. In short, the nitrifiers base their entire biochemistry—their entire lives—on the oxidation of ammonia or nitrite ions. *Nitrobacter* can convert 6 molecules of carbon dioxide to 1 molecule of glucose for every 78 nitrite ions that get oxidized—not terribly efficient, but it is efficient enough to keep *Nitrobacter* living, growing, and reproducing.

NITRATE REDUCTION

So far we have seen dinitrogen gas *reduced* to ammonia in nitrogen fixation and ammonia *oxidized* to nitrate in nitrification. We will now see that plants proceed to *reduce* the nitrate all the way back to ammonia before using it further to manufacture amino acids (Figure 30.16). The reactions of **nitrate reduction** are all carried on by the plant's own enzymes. The later steps, from nitrite to ammonia, take place in the chloroplasts—these organelles, we are gradually learning, have other crucial functions in addition to photosynthesis. The final products of nitrate reduction are amino acids, from which the plant's proteins and all its other nitrogen-containing compounds are formed.

Nitrogen metabolism, in bacteria and in plants, is complex. It is also of great importance. Nitrogen atoms constitute approximately 1–5 percent of the dry weight of a leaf, and nitrogen-containing compounds constitute 5–30 percent of the plant's total dry

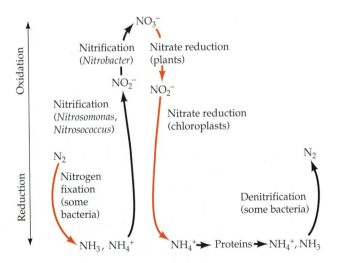

30.16 Bacteria, Plants, and Nitrogen
The path of nitrogen from the atmosphere to plant proteins is complex. Bacteria are responsible for nitrogen fixation and nitrification. Plants reduce nitrates back to ammonia, the form in which nitrogen is actually incorporated into proteins. Finally, some denitrifying bacteria can oxidize ammonia back to dinitrogen gas, which returns to the atmosphere.

weight. The nitrogen content of animals is even higher, and all the nitrogen in the animal world gets there by way of the plant kingdom.

SULFUR METABOLISM

All living things require sulfur, which is a constituent of two amino acids, cysteine and methionine, and hence of almost all proteins. Sulfur is also a component of other biologically crucial compounds, such as coenzyme A (Chapter 7). Animals must obtain their cysteine and methionine from plants; but plants can start with sulfate ions (SO_4^{2-}) obtained from the soil or from a liquid environment. It is interesting that the most abundant elements in plants are all taken up from the environment in their most oxidized forms—sulfur as sulfate, carbon as carbon dioxide, nitrogen as nitrate, phosphorus as phosphate, and hydrogen as water. In plants, sulfate is reduced and incorporated into cysteine; from this all the other sulfur-containing compounds in the plant are made. These important processes—sulfate reduction and the utilization of cysteine—are closely analogous to the reduction of nitrate to ammonia, and the subsequent utilization of ammonia, by plants.

Numerous bacteria base their metabolism on the modification of sulfur-containing ions and compounds in their environment. One group, for example, performs reactions analogous to nitrification. Just as the nitrifiers oxidize ammonia to nitrate, the chemosynthetic sulfur bacteria oxidize hydrogen sulfide (H_2S) to sulfate, using the energy thus released to make ATP and fix carbon dioxide.

ADAPTATION TO AVAILABLE SUNLIGHT

As you know, plants require light, as well as chemical nutrients, for photosynthesis. Plants have evolved many adaptations to help them exploit the available sunlight, especially in the face of crowding by other plants or by their own leaves. A great variety of evolutionary modifications—of leaf structure, chlorophyll content, and leaf arrangement—make particular plants photosynthetically efficient in particular environments. Plants of the same species may differ structurally and metabolically when grown in environments with differing levels of available light. In other words, an individual plant can develop so as to fit its particular environment.

Even in a single plant, individual leaves may differ if some are shaded during the day and others are not. Leaves adapted to the lower light intensities in the shade of other leaves are called shade leaves. Shade leaves typically have lower light compensation points; that is, the light intensity at which their rate of photosynthesis balances their rate of cellular res-

piration is lower than that of sun leaves. The respiratory rate of shade leaves is lower. Shade leaves are usually greener, with their greater concentration of chlorophyll ensuring the capture of whatever light is available. The total activity of RuBP carboxylase—the enzyme that catalyzes the initial CO_2-fixing reaction in photosynthesis—is greater in sun leaves, as is the photosynthetic rate on a per area basis.

Leaves exposed to more light become hotter and thus lose more water by transpiration for each unit of carbon fixed in photosynthesis. This tendency may be balanced by various adaptations. Such exposed leaves may, for example, be thicker than their shaded neighbors, reducing their surface-to-volume ratio and thus decreasing evaporational water loss. Another adaptation that minimizes leaf overheating is the possession of finely dissected or needlelike leaves—their pointed shapes maximize the rate of heat loss to the environment (Figure 30.17).

Leaf Orientation

On a single plant, each leaf may differ from the others in various respects: size, shape, surface appearance, and intensity of color. Variations in color reflect a difference in the proportions of chloroplast pigments. As ecologists Harold Mooney of Stanford University and Nona Chiariello of the University of Utah have expressed it, "Each leaf on a plant has a unique history and lives in a distinct microclimate." Recall, too, what we said at the end of Chapter 28 about the modularity of the plant body.

Leaves in different microclimates may be held at different angles, enhancing the overall photosyn-

30.17 Leaves as Radiators of Heat
Besides affording protection against potential grazers, the spines of this cholla cactus serve as efficient radiators of heat to the surrounding air. Loss of heat is especially rapid at the points of the spines.

thesis of the plant. Those usually shaded by others tend to be held at angles such that their upper surfaces aim toward the midday sun, taking maximum advantage of the limited amount of direct light they receive. On the other hand, those least shaded tend to be oriented with the upper surface directed more toward the horizon than straight up. Thus they can photosynthesize efficiently in the (usually cooler) early morning when the sun is low in the sky. When the sun is overhead, and the risk of water loss is greatest, leaves so oriented perform less photosynthesis but also heat up less and lose less water. Toward midday, these unshaded leaves may be held parallel to the incoming light—with their edges aimed at it—thus intercepting less of it. This can result in improved light utilization by a plant, or by a population of plants. When the sun is at its brightest, leaves in the upper layers can be held parallel to the light and still photosynthesize at a maximum rate. Having the upper leaves in such an orientation also exposes leaves in the lower layers to light; the lower leaves then photosynthesize at a high rate too.

Most flowering plants have leaves that move, more or less, during the day, tracking the sun across the sky and maximizing their exposure to light. In some species this tracking is virtually perfect, so that the leaves are held perpendicular with their broad surfaces to the light at all times of day.

Leaf Arrangement

The leaves on a plant are arranged so as to permit efficient use of sunlight in the plant's habitat. Leaves are "expensive" to produce, but they pay back the investment through photosynthesis. A plant that positions leaves so that they shade one another as little

30.18 Mosaic Leaf Arrangement
The leaves forming the canopy of this *Cecropia peltata* are arranged in a mosaic pattern. The leaves shade each other only minimally, allowing a maximum of photosynthesis by each leaf.

30.19 Maximizing Light Absorption
The leaves of these water lilies are arranged so that there is minimal overlap. Differential growth of the petioles accounts for this efficient arrangement.

as possible has a greater net energy gain than a plant with leaves that interfere with one another. Plants growing in deep shade often have a "mosaic" arrangement of leaves as viewed from above, with the leaves arranged so as to maximize light absorption by each leaf and minimize shading of one leaf by another (Figure 30.18).

Similarly, the leaves of water lilies grow in such a way as to avoid being shaded by each other (Figure 30.19). As new leaves form and grow, they do not permanently submerge and shade the older leaves— if an older leaf is submerged by a newer one lying on top of it, the petiole of the older leaf elongates until that leaf is no longer shaded. Overcrowding can interfere with this mechanism.

Some plants have a climbing habit; part of their foliage is brought to more brightly lit areas by growth up tree trunks, walls, and other vertical surfaces. This growth is assisted by various anatomical adaptations such as coiled stems, tendrils, or even adhesive pads (Figure 30.20). Tendrils often coil faster in the light, thereby anchoring the plant in regions where light is available.

Epiphytes in a Tropical Rain Forest

The dense foliage of a tropical rain forest permits little light to reach the forest floor. As a result, only a few herbaceous—small, nonwoody—plants are found on the forest floor, except in clearings. However, there is a profusion of such plants growing high up in the forest, attached to the branches of trees. Plants that grow on others without being parasites

30.20 An Adaptation for Climbing
Plants achieve a climbing habit, elevating their foliage and intercepting more light, by a variety of means. This passion vine can climb using its tendrils, which actively coil around objects upon contact. Other climbing adaptations were shown in Figure 20.8.

are called **epiphytes** (Figure 30.21). They generally obtain water and even minerals from the very humid air around the forest foliage. By being situated well up in the canopy, epiphytes have access to a higher light intensity than that available on the floor.

The epiphytic habit of these plants is, in itself, an adaptation, but many epiphytes have other adaptations as well. For example, most epiphytes are CAM plants (Chapter 29). Some orchids have an extreme adaptation: A thickened *root* epidermis is the only site of photosynthesis in the plant. These roots hang down in the air as do those of many other epiphytes.

Another dramatic adaptation in an epiphyte is found in *Dischidia rafflesiana*. As shown in Figure 30.22, some of the leaves of *D. rafflesiana* form structures that serve as homes for ant colonies. The ant colonies accumulate organic matter, forming humus. Higher parts of the plant produce roots that grow downward, into the ant-housing structures, and take up nutrients from the humus provided by the ants —the ants feed the plant.

HETEROTROPHIC SEED PLANTS

Not all plants are photosynthetic autotrophs. A few have, in the course of their evolution, lost the ability to feed themselves by photosynthesis. A few are **parasites**—they obtain their food directly from the

30.21 Epiphytes
One way for small plants to obtain light in a dense forest is to grow as epiphytes on other, taller plants. This tree in Trinidad is supporting at least five different species of epiphytes at various points on its trunk.

(a)   (b)

30.22 A Plant That Collects Rent from Animals
Dischidia rafflesiana, an epiphyte, harbors ant colonies in modified leaves. (a) The "house," a tubular leaf, contains the ants and the humus they produce. (b) Roots growing into the house (which has been cut open) absorb nutrients.

(a)   (b)

30.23 Parasitic Plants
(a) Orange-brown tendrils of parasitic dodder wrap around other plants. The dodder gets water, sugars, and other nutrients from its host through tiny, rootlike protuberances that penetrate the surface of the host. (b) Indian pipe also does not conduct photosynthesis; it gets its nutrients from another plant by way of a "fungal bridge" in the soil.

living bodies of other plants (Figure 30.23). Perhaps the most familiar parasitic plants are the mistletoes and dodders. Mistletoes are green and actually carry on some photosynthesis, but they parasitize other plants for water and mineral nutrients and may derive photosynthetic products from them as well. Another parasitic plant, the Indian pipe, once was thought to obtain its food from dead organic matter; however, it is now known to derive all its nutrition from nearby actively photosynthesizing plants by way of a "fungal bridge" connecting the roots of the two plants. Hence it too is a parasite.

Some other heterotrophic plants are the 450 or so **carnivorous** species—those that augment their nitrogen and phosphorus supply by capturing and digesting flies and other insects (Figure 30.24). The best-known plant carnivores are Venus's-flytrap (genus *Dionaea*), sundews (genus *Drosera*), and pitcher plants (genus *Sarracenia*). These plants are normally found in boggy regions where the soil is acidic. Most decay-causing organisms require a more neutral pH, so relatively little available nitrogen is recycled into these acidic soils; accordingly, the carnivorous plants have adaptations allowing them to augment their supply of nitrogen.

Sarracenia produces pitcher-shaped leaves that collect small amounts of rainwater. Insects are attracted into the pitchers either by bright colors or by nectar, and are prevented from getting out again by stiff, downward-pointing hairs. The insects eventually die and are digested by a combination of enzymes and bacteria in the water. Sundews have leaves covered with hairs that secrete a clear, sticky liquid high in sugar. An insect touching one of these hairs becomes stuck, and more hairs curve over the insect and stick to it as well. Digestion is accomplished by the secretion of enzymes, and the carbon- and nitrogen-containing products of digestion are absorbed by the sundew. The Venus's-flytrap springs a mechanical trap, triggered by three hairs in the center of a partially closed leaf lobe. When the trap is triggered by the movement of an insect, the two halves of the leaves close, and spiny outgrowths at the margins of the leaves interlock to make a prison. Again, digestion is accomplished by the secretion of enzymes onto the trapped insects. None of the carnivorous plants *must* feed on insects. They grow well without insects; but in nature, they grow faster and are a darker green when insects are available to them. The extra supply of nitrogen is used to make more proteins, chlorophyll, and other nitrogen-containing compounds.

SUMMARY

Nearly all plants are photosynthetic autotrophs. A very few species are parasitic heterotrophs that feed on other plants. Carnivorous species are autotrophs but supplement their nitrogen supply by capturing and digesting insects. The only nutrients required by most plants are carbon dioxide, water, nitrate or ammonium ions, and several mineral salts. The essential

(a)

(b)

30.24 Carnivorous Plants
(a) A Scottish sundew has trapped an insect on its sticky hairs. Secreted enzymes will accomplish external digestion. (b) The glistening hairs at the tops of these pitcher plant leaves point downward and help trap insects. The leaves contain rainwater, along with bacteria and protists that participate in digesting trapped insects.

elements—both macronutrients and micronutrients—generally perform multiple roles in the plant.

The habitats of plants differ in soil composition, including mineral content. Some soils are poor in essential nutrients, while others contain toxic minerals at levels sufficient to poison many species. Other species are adapted to life on such soils.

An adequate supply of nitrogen becomes available to plants only through the activities of a few species of soil bacteria. Atmospheric dinitrogen cannot be used directly by any eukaryotic organism. Nitrogen fixation, the conversion of N_2 to usable nitrogen compounds, is performed by bacteria, some living free in the soil and others, such as the species of *Rhizobium*, living symbiotically within the roots of plants. Ammonium ions are the main product of nitrogen fixa-

tion in the soil, yet most plants preferentially take up nitrate ions under most conditions. Nitrate ions are available because other species of soil bacteria carry out the process of nitrification; the nitrifying bacteria are chemosynthetic autotrophs. Once in the plant, nitrate is reduced to ammonia, which is incorporated into organic compounds. Similarly, sulfate ions are produced in the soil by bacteria that oxidize hydrogen sulfide; and the sulfate ions are, like nitrate ions, reduced within the plant.

Plants have various adaptations to maximize the efficiency with which their leaves capture light for photosynthesis. These include adaptations of leaf structure, biochemistry, orientation, and arrangement, as well as adaptations for climbing or for life on top of other plants.

SELF-QUIZ

1. Which of the following is *not* an essential mineral element for plants?
   a. Potassium
   b. Magnesium
   c. Calcium
   d. Lead
   e. Phosphorus

2. Fertilizers:
   a. are often characterized by their N–P–O percentages.
   b. are not required if crops are re-

   moved frequently enough.
   c. restore needed mineral nutrients to the soil.
   d. are needed to provide carbon, hydrogen, and oxygen to plants.
   e. are needed to destroy soil pests.

3. Which of the following is *not* an important step in soil formation?
   a. The removal of bacteria.
   b. Mechanical weathering.
   c. Chemical weathering.

   d. Clay formation.
   e. Hydrolysis of soil minerals.

4. Laterization:
   a. results in a very productive soil.
   b. often takes place in mine tailings.
   c. produces a soil rich in copper, lead, nickel, and zinc.
   d. produces a soil rich in chromium and poor in calcium.
   e. produces a soil rich in insoluble iron and aluminum compounds.

5. Nitrogen fixation:
   a. is performed only by plants.
   b. is the oxidation of nitrogen gas.
   c. is catalyzed by the enzyme nitrogenase.
   d. is a single-step chemical reaction.
   e. is possible because N_2 is a highly reactive substance.

6. Nitrification:
   a. is performed only by plants.
   b. is the reduction of ammonium ions to nitrite and nitrate ions.
   c. is the reduction of nitrate ions to nitrogen gas.
   d. is catalyzed by the enzyme nitrogenase.
   e. is performed by certain bacteria in the soil.

7. Nitrate reduction:
   a. is performed by plants.
   b. takes place in mitochondria.
   c. is catalyzed by the enzyme nitrogenase.
   d. includes the reduction of nitrite ions to nitrate ions.
   e. is known as the Haber process.

8. Which of the following statements about sulfur is *not* true?
   a. All living things require it.
   b. It is a component of DNA and RNA.
   c. It is a constituent of two amino acids.
   d. Its metabolism is similar to the metabolism of nitrogen.
   e. Many bacteria base their metabolism on reactions of sulfur-containing ions.

9. Which of the following statements is *not* true of epiphytes?
   a. They grow on other plants.
   b. They are parasitic.
   c. Many are CAM plants.
   d. They receive more light than similar-sized plants on the forest floor.
   e. They generally obtain their water supply from the air.

10. The heterotrophic seed plants:
   a. are all parasites.
   b. are all carnivores.
   c. are all incapable of photosynthesis.
   d. all derive their nutrition from animals.
   e. all develop from multicellular embryos.

FOR STUDY

1. Methods for determining whether a particular element is essential have been known for over a century. Since the methods are so old and well established, why is it that the essentiality of several elements has been discovered only recently?

2. If a Venus's flytrap were to be deprived of soil sulfates and hence made unable to synthesize the amino acids cysteine and methionine, would it die from lack of proteins? Discuss.

3. Soils are dynamic systems. What changes might result when land is subjected to heavy irrigation for agriculture, after previously having been relatively dry for many years? What changes in the soil might result when a virgin, deciduous forest is replaced by crops that are harvested each year?

4. What is the significance of nitrification to plants? to the bacteria that carry on nitrification?

READINGS

Brill, W. J. 1981. "Agricultural Microbiology." *Scientific American*, September. Using microbes to support plant growth.

Epstein, E. 1984. "Rhizostats: Controlling the Ionic Environment of Roots." *BioScience*, November. Computers, roots, and plant nutrition.

Power, J. F. and R. F. Follett. 1987. "Monoculture." *Scientific American*, March. There is an increasing tendency to grow the same crop year after year. How does this affect soils? Is it a good practice?

Raven, P. H., R. F. Evert and S. Eichhorn. 1986. *Biology of Plants*, 4th Edition, New York, Worth, 1986. An excellent general botany textbook.

Salisbury, F. B. and C. W. Ross. *Plant Physiology*, 3rd Edition, Belmont, CA, Wadsworth, 1985. An authoritative textbook with good chapters on mineral nutrition and related topics.

31

Reproduction and Development in Flowering Plants

PREVIEW: Sexual reproduction in angiosperms requires the transfer of a male gametophyte—a pollen grain—from one flower part to another part of the same flower or to a second flower. The unique feature of angiosperm reproduction is a process of double fertilization, in which two sperm nuclei combine with nuclei of the female gametophyte, giving rise to both an embryo and a distinctive, triploid nutritive tissue, the endosperm. Sexual reproduction promotes genetic diversity; asexual reproduction allows a well-adapted clone to spread rapidly in a stable environment.

This chapter deals with asexual and sexual reproduction, alternation of generations, flowers and flowering, pollination, fruit growth, seed dormancy, and seed germination.

Plants have many ways of reproducing themselves—and with humans helping, there are even more ways. As flowers are sex organs, it is no surprise that virtually all flowering plants reproduce sexually. But many reproduce asexually as well; some reproduce asexually most of the time. Which is the better way to reproduce? As we will see, most of the answers to this question relate to genetic diversity or genetic recombination.

The details of sexual reproduction differ among different species of flowering plants. In our discussion of Mendel's work (Chapter 10), we saw that some plants can reproduce sexually either by cross-pollinating or self-pollinating. This is possible because, as explained in Chapter 24, in many species each single individual has both male and female sex organs.

Both sexual and asexual reproduction are important in agriculture. Annual crops, including wheat, rice, millet, and corn—the great grain crops, all of which are grasses—as well as plants in other families such as soybeans and safflower are grown from seed, that is, sexually. Other crops begin asexually with slips, grafts, or other means. Orange trees, which have been under cultivation for centuries, are grown from seed except for one type, the navel orange. This plant has apparently arisen but once in history. Early in the nineteenth century, on a plantation on the Brazilian coast, one seed gave rise to one tree that had aberrant flowers. Parts of the flowers aborted, and seedless fruits were formed. Every navel orange in the world comes from a navel orange tree derived asexually from another, which came from another, and another, and so on back to that original Brazilian tree. Navel oranges *must* be propagated asexually. Strawberries need not be, because they are perfectly capable of setting seed. Nonetheless, growing strawberries asexually is common because vast numbers of plants that are genetically identical to one particularly desirable plant can be produced from strawberry runners.

According to an old saying, "A plant is a seed's way of making another seed." In keeping with this, we trace the life cycle of flowering plants to conclude this chapter.

SEXUAL REPRODUCTION

Sexual reproduction provides for genetic diversity through *recombination*. Meiosis and mating shuffle genes into new combinations, giving a population a variety of genotypes in each generation. This genetic diversity serves well as the environment changes or as the population expands into new environments. This is why sexual reproduction is often thought of as being "better" than asexual reproduction.

As you know, an angiosperm's organ of sexual reproduction is the flower. The structure of a "typical" flower was described in Chapter 24; you might find it useful to refer back to pages 538–540. A complete flower consists of four groups of modified leaves: the carpels, stamens, petals, and sepals. To review briefly: The female structures, bearing megasporangia, are the carpels. Enclosed within the carpels are the ovules. Each ovule consists of the megasporangium it encloses, its protective layers of surrounding integuments, and its stalk. There is an opening in the integuments, the micropyle, through which the pollen tube grows on its way to the megasporangium. A pistil, consisting of one or more fused carpels, contains one or more ovules in the ovary at its base, and bears a pollen-receptive stigma at the tip of its elongated style. The male structures, bearing microsporangia, are the stamens. Each stamen consists of a filament bearing two anthers, and each anther consists of two microsporangia.

In addition to these sexual parts, many flowers have sterile parts, which do not participate directly in reproduction, arranged in whorls around the spore-bearing carpels and stamens. The petals, often colored, constitute the corolla. Below them the sepals, often green like true leaves, constitute the calyx. All the parts of a flower are borne on a stem tip, the receptacle.

Before continuing with the next section, you may want to review the section entitled "Alternation of Generations" at the beginning of Chapter 24.

Gametophytes

The female gametophyte, the gamete-producing generation, located within the ovule, is very simple (Figure 31.1). The megasporocyte, a cell within the ovule's megasporangium, divides meiotically to produce four haploid megaspores. All but one of the megaspores then degenerate. The surviving megaspore undergoes mitotic divisions, usually producing eight nuclei, all initially contained within a single cell—three nuclei at one end, three at the other, and two in the middle. Subsequent cell wall formation leads to an elliptical, seven-celled megagametophyte with a total of eight nuclei. At the end of the megagametophyte nearest the micropyle are three tiny cells: the **egg** and two **synergids.** At the opposite end are three antipodal cells, and in the large central cell are two **polar nuclei.** The entire seven-celled, eight-nucleate structure is called the **embryo sac** and constitutes the female gametophyte.

The male gametophyte consists of fewer cells than the female (Figure 31.1). Meiosis of microsporocytes takes place within the four microsporangia, fused into two anthers, sometimes very early in development. Each microspore normally undergoes one mitotic division within the spore wall before the anthers open and shed the pollen. Further development is delayed until the pollen arrives at the stigma of a pistil. The pollen grain is the male gametophyte.

Pollination

The gymnosperms and angiosperms evolved independence from water as a medium for gamete travel and fertilization—a freedom not shared by other plant groups. Their sperm nuclei travel within a pollen grain (Figure 31.2), and the pollen tube provides a route to the ovary. But how does pollen travel from an anther to a stigma? The mechanisms that have evolved for pollen transport are many and various. In one, pollination is accomplished before the flower bud opens, as in peas and their relatives—resulting in the self-fertilization described in Chapter 10. In self-pollination, pollen is transferred by the direct contact of anther and stigma within the same flower—very efficient, but self-fertilization negates the advantage of sexual reproduction that comes with cross-fertilization: the opportunity for genetic recombination.

Wind is the vehicle for pollen transport in many species, especially grasses (Figure 31.3). Wind-pollinated flowers have sticky or featherlike stigmas, and they produce pollen grains in great numbers. Some aquatic angiosperms are pollinated by water action, with water carrying pollen grains from plant to plant. Animals are important pollinators; the mutually beneficial aspects of such plant–animal associations are discussed at length in Chapter 46.

Double Fertilization

When a pollen grain lands on the stigma of a pistil, a pollen tube develops from the pollen grain and either grows downward on the inner surface of the style or digests its way down the spongy tissue of this female organ, growing millimeters or even centimeters in the process. The pollen tube follows a chemical gradient of calcium ions or other substances in the style until it reaches the micropyle. Of the two nuclei in the pollen grain, one is the **tube nucleus,** close to the tip of the pollen tube, and the other is the **generative nucleus** (Figure 31.4). The pollen tube eventually digests its way through megasporangial tissue and reaches the female gametophyte. The generative nucleus meanwhile has undergone one mitotic division to produce two **sperm nuclei,** *both* of which are released into the cytoplasm of one of the synergids.

From this synergid, which degenerates, the sperm

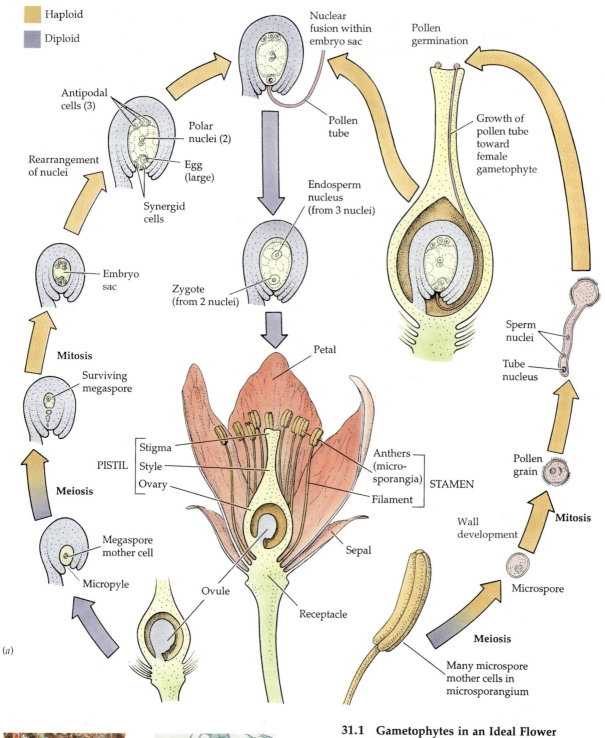

Haploid
Diploid

Antipodal cells (3)

Polar nuclei (2)

Egg (large)

Rearrangement of nuclei

Synergid cells

Nuclear fusion within embryo sac

Pollen germination

Pollen tube

Growth of pollen tube toward female gametophyte

Endosperm nucleus (from 3 nuclei)

Embryo sac

Zygote (from 2 nuclei)

Mitosis

Surviving megaspore

Petal

Sperm nuclei

Tube nucleus

Stigma
Style
Ovary

PISTIL

Anthers (microsporangia)

STAMEN

Filament

Meiosis

Megaspore mother cell

Micropyle

Sepal

Pollen grain

Wall development

Mitosis

Ovule

Receptacle

Microspore

Meiosis

Many microspore mother cells in microsporangium

(a)

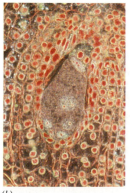

(b)

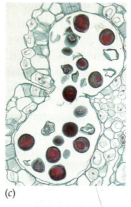

(c)

31.1 Gametophytes in an Ideal Flower
(a) Gametophytes in context: the angiosperm life cycle. (b) A mature female gametophyte surrounded by smaller ovule cells in the ovary. The gametophyte is oriented as at the upper left in (a); locate and identify its nuclei. (c) Male gametophytes—pollen grains—in the anthers.

nuclei enter two different cells. One sperm nucleus enters the egg cell and fuses with its nucleus, producing the diploid zygote. The other sperm nucleus enters the central cell of the embryo sac and fuses with the two polar nuclei to form a triploid (3n) nucleus. While the zygote nucleus begins division to

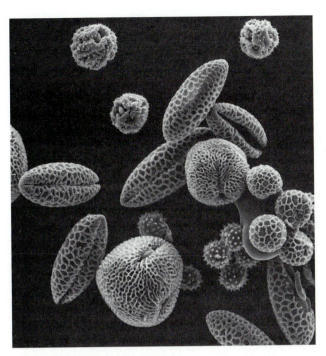

31.2 A Pollen Grain Sampler
Pollen grains of geranium, tiger lily, phlox, marigold, and dandelion. Each species has a characteristic size, shape, and pattern of wall sculpturing.

form the new sporophyte embryo, the triploid nucleus undergoes rapid mitosis to form a specialized nutritive tissue, the **endosperm.** The female antipodal cells and synergids simply degenerate.

Shortly after fertilization, a highly coordinated growth and development of embryo, endosperm, integuments, and carpel ensues. As large amounts of nutrients are moved in from other parts of the plant, the endosperm begins accumulating starch, lipids, and proteins. The integuments develop into a double-layered seed coat, sometimes fleshy, and sometimes heavily lignified and hard. The carpel ultimately becomes the wall of the fruit that encloses the seed.

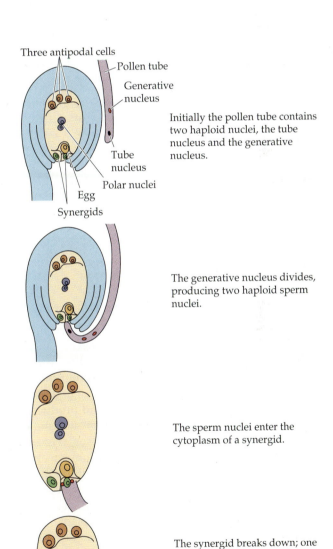

Initially the pollen tube contains two haploid nuclei, the tube nucleus and the generative nucleus.

The generative nucleus divides, producing two haploid sperm nuclei.

The sperm nuclei enter the cytoplasm of a synergid.

The synergid breaks down; one sperm nucleus fertilizes the egg, forming the zygote, the first cell of the 2n sporophyte generation. The other unites with the two polar nuclei, forming the first cell of the 3n endosperm.

31.3 Wind Pollination
A flowering inflorescence of a grass. The numerous anthers all point away from the stalk and stand free of the plant, promoting dispersal of the pollen by the wind.

31.4 Pollen Nuclei and Double Fertilization
The sperm nuclei contribute to the formation of the diploid zygote and the triploid endosperm. Double fertilization is a characteristic feature of angiosperm reproduction.

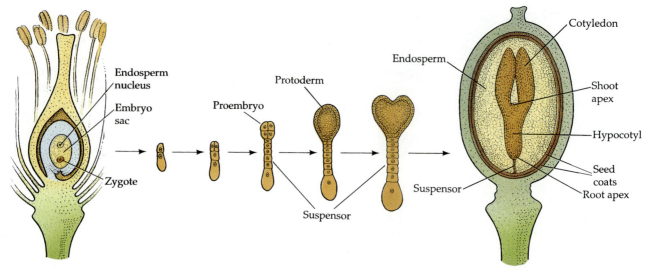

31.5 Early Development of a Dicot
The zygote nucleus divides mitotically, one daughter cell giving rise to the embryo proper and the other to the suspensor. The embryo develops through intermediate stages to form the torpedo stage embryo. The tissues surrounding the embryo sac develop into seed coats.

Of all the characteristic traits of the angiosperms, only one trait, the double fertilization mechanism just described, is found in *all* angiosperms and *only* in angiosperms—and in a tiny gymnosperm group. The origin of this process in geological time is wholly unknown.

Embryo Formation in Flowering Plants

The first step in the formation of the embryo is a mitotic division of the zygote, the fertilized egg in the embryo sac, giving rise to two daughter cells. Even at this stage, the two cells face different fates. There is an asymmetric—uneven—distribution of contents within the zygote so that one end produces the embryo proper and the other end produces an early supporting structure, the **suspensor** (Figure 31.5). A filamentous suspensor and a globular embryo are distinguishable after just four mitotic divisions. Soon thereafter the embryo develops a covering layer just one cell thick—the protoderm—that will become the epidermis of the new plant. The suspensor then ceases to elongate, and, as development continues, the first organs take form within the embryo.

In dicots—monocots are somewhat different—the embryo soon takes on a characteristic heart-shaped form as the cotyledons start to grow. Further elongation of the cotyledons and of the main axis of the embryo gives rise to what is called the torpedo stage, during which some of the internal tissues begin to differentiate. The elongated region below the cotyledons is called the hypocotyl. At the top of the

hypocotyl, between the cotyledons, is the shoot apex; at the other end is a root apex. Each of these apical regions contains an apical meristem whose dividing cells give rise to the organs of the mature plant. In many species, such as peas and peanuts, the cotyledons absorb the reserve foods from the surrounding endosperm and end up very large in relation to the rest of the embryo (Figure 31.6). In others, including the castor bean, the cotyledons remain thin, drawing on the reserves in the endosperm as needed when the seeds germinate. In either case, the endosperm is, at the outset, the maternal plant's contribution to the nutrition of the next generation (Box 31.A).

Fruit

After fertilization, the ovary wall of a flowering plant—together with its seeds—develops into a

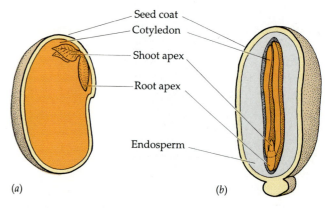

31.6 Variety in Dicot Seeds
In some dicots, the cotyledons absorb much of the endosperm and fill most of the seed (a). In others, the endosperm remains separate and the cotyledons remain thin (b).

BOX 31.A

Maternal "Care" in Plants

The mammalian mother bears a special relationship to her embryonic and newborn offspring. The mammalian embryo receives its nourishment by way of the maternal bloodstream, and it is subjected to hormonal influences from the mother, but not from the father. Maternal *plants* also influence their offspring in important ways. Of course, the offspring receive half their genetic endowment by way of the egg, but the maternal parent has other, specific effects, especially ones relating to the size of seeds produced.

The seeds develop from and within tissues of the maternal flower.

Barbara A. Schaal, of Washington University, St. Louis, has studied maternal effects in the Texas bluebonnet, *Lupinus texensis*. She demonstrated that variation in seed weight in these plants was attributable to responses of the maternal plant to environmental variation. Differences in seed weight, in turn, affected the subsequent performance of the offspring, as shown in the figure. A higher percentage of larger seeds germinated, and they germinated earlier. Larger seeds gave rise to larger seedlings. The seedlings from larger seeds grew more leaves and survived better during the first month of life after germination. By six weeks of age, differences in size and survivorship between seedlings from larger and smaller seeds were no longer evident. However, between the ages of 80 and 130 days,

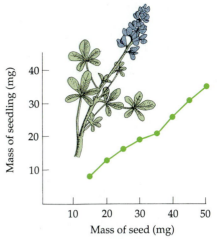

the period of most intense reproduction in the bluebonnet, plants from larger seeds produced more ovules than did those from smaller seeds. Because the size of the seed is determined more by the maternal parent than by the genotype of the embryo, Schaal concluded that maternal effects were important influences on the development and reproduction of the offspring.

---

fruit. A fruit may consist of only the mature ovary and its seeds, or it may include other parts of the flower or structures that are closely related to it. The major variations on this theme are illustrated in Figure 24.27.

Some fruits play a major role in reproduction because they help disperse seeds over substantial distances. A number of trees, including ash, elm, maple, and tree of heaven, produce a dry, winged fruit called a samara (see Figure 28.3). A samara spins like a helicopter blade and, while whirling downward, holds the fruit aloft long enough for it to be blown some distance from the parent tree. The dandelion fruit is another that is marvelously adapted for dispersal by wind. Water disperses some fruits; coconuts have been spread in this way from island to island in the Pacific (Figure 31.7a). Still other fruits travel by hitching rides with animals—either inside or outside them (Figure 31.7b). Burdocks, for example, have hooks that adhere to animal fur, and many other plants have prickled, barbed, hairy, or sticky fruits. Fleshy fruits such as berries provide food for mammals or birds; their seeds travel safely through the animal's digestive tract or are regurgitated, in either case being deposited some distance from the parent plant.

Having discussed mechanisms for dispersing

seeds and pollen, we should make it clear that most seeds, and most pollen grains, end up close to their sources. Long-range dispersal is the exception, not the rule. In those cases in which dispersal is extensive, paternal genes usually travel farther than maternal genes—both maternal and paternal genes travel with the seed, but only paternal genes travel with the pollen grain.

ASEXUAL REPRODUCTION

As mentioned earlier in this chapter, when a plant reproduces sexually by means of self-fertilization, there are fewer opportunities for genetic recombination than there are with cross-fertilization. With self-fertilization, the only genetic variability that can be arranged into new combinations is that possessed by the single parental plant. A plant that is heterozygous for a locus can, for example, produce among its progeny both kinds of homozygotes for that locus plus the heterozygote. But it cannot produce any progeny that carry any alleles that it does not itself possess. Asexual reproduction goes a step further: It eliminates genetic recombination. When a plant reproduces asexually, it produces a clone of progeny all having genotypes identical to its own.

(a)

(b)

31.7 Dispersal of Fruits
(a) Coconuts are germinating where they washed ashore on a Tahitian beach.
(b) Many sticky fruits are hitching a ride on this buffalo's fur, thus traveling far from their parent plants.

If a clone is highly adapted to its environment, asexual reproduction allows many more copies of that genotype to form, spreading throughout that environment. This ability to exploit a stable environment is a major advantage of asexual reproduction, providing another answer to the question asked at the beginning of this chapter: In some circumstances, asexual reproduction is better.

Asexual Reproduction in Nature

We call the basic organs of the plant body—stems, leaves, and roots—vegetative parts, distinguishing them from the reproductive parts. The modification of a vegetative part of a plant is what makes **vegetative reproduction** possible. The stem is the part modified in many cases. Strawberries and some grasses produce **stolons,** horizontal stems that form roots at intervals and establish potentially independent daughter plants (see Figure 28.10). Other stolons are produced as branches that sag to the ground and put out roots. The rapid multiplication of water hyacinths demonstrates the effectiveness of stolons for vegetative reproduction. Some plants, such as potatoes, form **tubers,** the fleshy tips of underground stems. **Rhizomes** are underground stems that can give rise to new shoots. Bamboo is a particularly striking example of a plant that reproduces vegetatively by means of rhizomes. A single bamboo plant can give rise to a stand—even a forest—of plants constituting a single, physically connected entity. Whereas stolons and rhizomes are horizontal stems,

bulbs and corms are short, vertical, underground stems. Lilies and onions form **bulbs** (Figure 31.8a), short stems with many fleshy, modified leaves, such as the familiar "scales" of onions. The leaves make up most of the bulb. Bulbs are, thus, large buds that function in food storage and can later give rise to new plants. Crocuses, gladioli, and many other plants produce **corms,** which are underground stems that function very much like bulbs. Corms are conical and consist primarily of stem tissue, lacking the fleshy scales characteristic of bulbs.

Not all vegetative reproduction is accomplished by means of stems. Leaves may also be the source of new plantlets, as in the succulent plants of the genus *Kalanchoe* (Figure 31.8b). Many kinds of angiosperms, ranging from grasses to trees such as aspens and poplars, form interconnected, genetically homogeneous populations by means of **root suckers**—horizontal roots. What appears to be a whole stand of aspen trees, for example, is a clone derived from a single tree by root suckers (see Figure 45.6).

Vegetatively reproducing plants often grow in unstable environments. Plants with stolons or rhizomes, such as beach grasses, rushes, and sand verbena, are common pioneers on coastal sand dunes. Rapid vegetative reproduction enables the plants, once introduced, not only to multiply but also to survive burial by the shifting sand; in turn, the dunes are stabilized by the extensive network of rhizomes or stolons.

Dandelions and some other plants reproduce by **apomixis,** the asexual production of seeds. As you

31.8 Vegetative Reproduction
(a) The short stem is visible at the very bottom of this sectioned daffodil bulb. White storage leaves grow from the stem; the yellow parts contain flower buds. (b) The plantlets forming on the margin of this *Kalanchoe* leaf will fall to the ground and start independent lives.

know, meiosis reduces the number of chromosomes in gametes and fertilization restores the sporophytic number of chromosomes in the zygote. Consider that a plant might somehow skip over *both* meiosis and fertilization while producing seeds. Apomixis produces seeds within the female gametophyte without the mingling and segregation of chromosomes and without the union of gametes. The ovule develops into a seed and the ovary wall develops into a fruit. An apomictic embryo has the sporophytic number of chromosomes. The result of apomixis is a fruit with seeds genetically identical to the parent plant.

Interestingly, apomixis sometimes requires pollination. In some apomictic species a sperm nucleus must combine with the polar nuclei in order for the endosperm to form. In other apomictic species, the pollen provides the signals for embryo and endosperm formation, although neither sperm nucleus participates in fertilization. Pollination and fertilization are *not* the same thing.

Asexual Reproduction in Agriculture

Farmers take advantage of some of these natural forms of vegetative reproduction. Farmers and scientists have also added new types of asexual reproduction by manipulating plants. One of the oldest methods of vegetative reproduction used in agriculture consists simply of making cuttings, or **slips,** of stems, inserting them in soil, and waiting for them to form roots and thus become autonomous plants. Rooting is sometimes hastened by treating the slips

with a plant growth substance, auxin, as described in Chapter 32.

Agriculturists reproduce many woody plants by **grafting.** A person makes a graft by taking a piece of one plant and attaching it to the stem or root of another plant. The part of the resulting plant that comes from the root-bearing "host" is called the **stock;** the part grafted on is called the **scion.** Three types of grafts are illustrated in Figure 31.9. In order for a graft to "take," the cambium of the scion must become associated with the cambium of the stock (Chapter 28). The cambia of scion and stock both form masses of wound tissue. If the two masses meet and fuse, the resulting continuous cambium can produce xylem and phloem, allowing transport of water and minerals to the scion and of photosynthate to the stock. Grafts are most often successful when the stock and scion belong to the same or closely related species. Grafting techniques are of great importance in agriculture, with most fruit grown for the market in the United States being produced on trees grown from grafts.

There are many reasons for grafting plants for fruit production, the most common being the desire to combine a hardy root system with a shoot system that produces the best-tasting fruit. This is illustrated by the story of the wine grape, *Vitis vinifera.* In 1863 plant lice of the genus *Phylloxera* inflicted great damage in French vineyards. The roots of vines on more than 2.5 million acres were destroyed. The problem was solved by importing great numbers of *V. vinifera* plants, which have *Phylloxera*-resistant root systems,

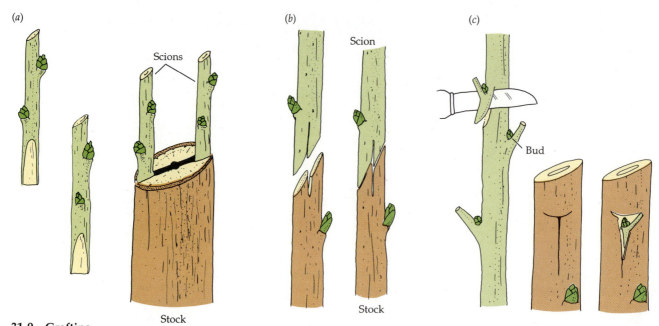

31.9 Grafting
The scions are shown in green and the stocks in brown. *(a)* Cleft grafting; the scions are placed so that their vascular cambia are aligned with the vascular cambium in the stock. *(b)* Whip grafting. *(c)* Budding.

from California. These plants were used as stocks to which French vines were grafted as scions. Thus the fine French grapes could be grown using roots resistant to the lice.

Scientists in universities and industrial laboratories have been developing new ways to produce valuable plant materials. For example, gene splicing can provide plants with capabilities they previously lacked (Chapter 14). By causing cells of different sorts to fuse, one can obtain plants with exciting new combinations of properties. By cloning—making genetically identical copies of—small bits of tissue, one can obtain large numbers of equally desirable plants. A problem remains: How can one efficiently take such small, delicate materials and get them to grow in the field? Plants in nature solved this problem long ago. The product of sexual and apomictic reproduction in flowering plants is a compact package, protectively wrapped, containing an embryonic member of the next generation along with a supply of the nutrients it needs to begin its independent existence. This package is the seed. What is needed as a tool of plant biotechnology, and what is actually being developed, is an artificial seed, containing the product of laboratory invention (Figure 31.10).

Artificial seeds contain a multicellular "somatic embryo"—not a sexually produced embryo, but an embryolike product of mitotic divisions in tissue culture. So that the "embryo" does not dry out, and so that it may be stored and transported before planting, it is embedded in a water-soluble gel; then the combined embryo and gel are encapsulated in a protec-

tive plastic coat. The coat and gel dissolve away after the artificial seed is planted. Other materials may be added to the gel, among them suitable inorganic nutrients, fungicides, and pesticides. Such scientifically designed artificial seeds should come into increasingly common use as the remaining problems are solved and methods are perfected for the mass production of these tiny packages.

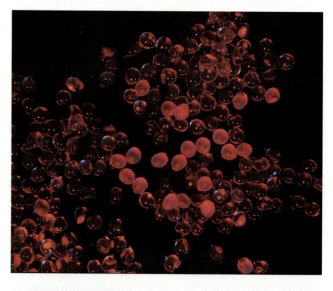

31.10 Artificial Seeds
These water-soluble capsules, developed by Plant Genetics, Inc., house somatic embryos along with nutrients and other chemicals.

FROM SEED TO FLOWER

The completion of the plant life cycle requires germination—sprouting—of the seed and then a period of vegetative growth of the plant that arises from the embryo. Eventually, flowers appear on the plant. New gametophytes are formed, and the cycle begins again.

Seed Dormancy

Some seeds are, in effect, "instant plants," in that all they need for germination is water. Yet many other species have seeds that do not germinate so readily; that is, they are initially **dormant.** Seed dormancy may last for weeks, months, years, or even decades! The mechanisms of dormancy are numerous and diverse. One is the presence of an impermeable seed coat, which prevents the entry of water or, sometimes, of oxygen. The breaking of dormancy in such seeds depends on the abrasion of the seed coat, perhaps by the seed's tumbling across the ground or through creek beds, by its passing through the digestive tracts of birds, or by other means. Such modification of the seed coat is called **scarification.**

The seed coat may also impose dormancy by the simple fact of mechanically restraining the embryo—if the embryo cannot expand, the seed cannot germinate. In the laboratory, we can promote germination of such a seed by simply cutting away part of the coat or by partially dissolving it with strong acid. In nature, soil microorganisms probably play a major role in softening seed coats of this type, although the action of digestive enzymes in the guts of birds or other animals is also important. Another agent of scarification to release mechanical restraint is fire, which affords significant control of germination in some natural habitats. Fire can also melt wax in seed coats, making water available to the embryo (Figure 31.11; see also Figure 45.19).

The presence of chemical inhibitors of germination is another mechanism of seed dormancy. As long as the concentration of inhibitor is high, the seed remains dormant. One means of reducing the level of inhibitor is by leaching, that is, prolonged exposure to water. Another is the scorching of the seeds by fire. Usually inhibitors of germination are already present in the dry seed, but in a few cases they are produced only after the seed has begun to take up water. The most common chemical inhibitor of seed germination is **abscisic acid.** In some seeds the level of abscisic acid or other inhibitors does not decline during germination; rather, the effect of the inhibitor is overcome by gradually increasing the concentrations of growth *promoters.* Abscisic acid and growth substances will be considered in Chapter 32.

There are still other mechanisms for breaking dormancy. Some seeds, such as those of the tomato and

31.11 Fire and Seed Germination
Fireweed germinated and flourished after a great fire in Yellowstone National Park.

lima bean, remain dormant until they have dried extensively. Temperature can be an environmental cue initiating germination. Even a brief exposure to temperatures near freezing may stimulate germination, but more commonly a period of many days or weeks of low temperature is required to end dormancy. In agriculture, it is common practice to refrigerate seeds such as those of apples to hasten germination. This refrigeration procedure is known as **stratification;** typically, it consists of a month or two at 5°C. The effects of cold treatment vary from species to species, but one result may be a gradual decrease in the content of germination inhibitors such as abscisic acid.

Roles of Seed Dormancy

What is the adaptive advantage of seed dormancy? For one thing, it may result in germination at a favorable time. Seeds that require a long cold period for germination commonly germinate in the spring, when water is usually abundant—germination in the dry days of late summer could be risky. Some other seeds need a period of **afterripening**—they simply will not germinate until a certain amount of time has passed, regardless of how they are treated. This prevents germination while the seed of a cereal grain, for example, is still attached to the parent plant, and it tends to favor dispersal of the seed.

The mechanism of dormancy increases the likelihood of a seed germinating in the right place, as well. Some cypress trees grow in standing water (Figure 31.12), and their seeds germinate only if leached extensively by water. Many weeds must have their seed coats damaged before they will germinate, and other

31.12 Leaching of Germination Inhibitors
The seeds of *Nyssa aquatica* germinate only after being leached by water, insuring that seeds germinate in a situation suitable for growth.

weed seeds will not germinate unless they have been exposed to light. Either type of weed germinates best in disturbed soils. You may have noticed how a freshly cultivated patch of soil quickly teems with weeds that are then free from competition with other plants. Seeds that must be scorched by fire in order to germinate also avoid competition—they germinate only when the area has been largely cleared by fire. Light-requiring seeds tend to germinate only at or near the surface of the soil; these tend to be tiny seeds with few food reserves. Other seeds whose germination is *inhibited* by light germinate only when well buried. Light-inhibited seeds are generally large and well stocked with nutrients.

Seed dormancy helps annual plants counter the effects of year-to-year variations in the environment. Some seeds remain dormant throughout an unfavorable year, and other seeds germinate at different times during the year. Seed dormancy can also contribute to the dispersal of a plant species. Those seeds that remain dormant until they have passed through the guts of birds or other animals will likely be carried some distance before they are deposited. Seeds carried by birds in their digestive tracts can give rise to the first plants to grow on newly formed volcanic islands, for example.

Seed Germination

Imbibition—the uptake of water—is the first step in the germination of a seed. Because, typically, the seed is dry prior to the start of germination, its water potential (Chapter 29) is very negative and water can be taken up readily if the seed coat allows it. The

magnitude of the water potential may be demonstrated by observing the force exerted by seeds expanding in water. Imbibing cocklebur seeds can exert a pressure of up to 1,000 atmospheres against a restraining force. Once a seed has taken up enough water, it undergoes metabolic changes: Certain preformed enzymes become activated, RNA and then new enzymes are synthesized, the rate of cellular respiration increases, and other metabolic pathways become activated. Interestingly, there is no DNA synthesis during these early stages of seed germination. DNA is synthesized only after the radicle—the embryonic root—begins to grow and poke out beyond the seed coat (Figure 31.13). As there is no DNA synthesis, we can see that the early stages of germination do not include cell division.

Mobilizing Food Reserves

Until the **seedling,** the young plant developing from a seed, becomes able to carry on photosynthesis, it must depend on built-in reserves from the seed to meet its needs for energy and materials. For seeds of some species, the principal reserve of energy and carbon-containing building blocks is the carbohydrate starch, which is a polymer of glucose, the starting point for glycolysis and cellular respiration (Chapter 7). More species, however, store lipids—fats or oils —as reserves in their seeds. This allows more energy to be stored in a smaller space because lipids contain more calories per unit of weight than does starch. Even in species that store lipids in their seeds, starch is often stored in the roots or tubers of the plants themselves, where space is not a concern. The lipids can be digested to release glycerol and fatty acids, both yielding energy through cellular respiration. Glycerol and fatty acids can also be converted to

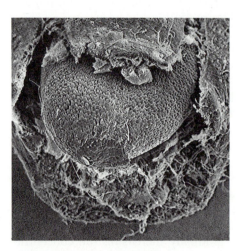

31.13 The Radicle Emerges
The tip of this barley seed's radicle has just broken through its protective sheath. The appearance of the radicle is one of the first externally visible events in seed germination.

glucose, so that the fat-storing families of plants, too, can make all the needed building blocks for seedling growth. Typically, the endosperm of the seed contains a reserve of amino acids in the form of enormous polymers—proteins—rather than as free amino acids. The growing embryo can break down the polymers to obtain the amino acids it needs to assemble into its own myriad proteins.

Giant molecules such as proteins, lipids, and starch must be digested into monomers before they can enter the cells of the embryo to be used as building blocks and energy sources. Germinating barley and other cereal seeds digest proteins and starch as follows: As the embryo becomes active, it secretes substances called **gibberellins.** These diffuse through the endosperm to a surrounding tissue called the **aleurone layer,** which lies inside the seed coat. The gibberellins trigger a crucial series of events in the aleurone layer. First, protein-containing bodies called aleurone grains break down, releasing amino acids. The aleurone layer then uses the amino acids in the assembly of digestive enzymes, including amylases (starch-degrading enzymes), proteases (protein-degrading enzymes), and ribonucleases (RNA-degrading enzymes). These enzymes, along with certain others already present in the aleurone layer, are next secreted into the endosperm, where they catalyze the

digestion of the reserve polymers. As a result, the necessary sugar and amino acid monomers become available to the growing embryo (Figure 31.14). Other activities of the gibberellins are described in Chapter 32; also, gibberellins are the growth promoters that antagonize abscisic acid to break the dormancy of some seeds.

Vegetative Growth

Repeated mitotic divisions and expansion of the cells make the new seedling grow. The root and shoot systems expand, and new leaves continue to appear. Older leaves may wither and fall from the growing plant. The shoot may branch, depending on the species of the plant and on local environmental conditions. Such vegetative growth may continue for days or for centuries. Throughout this time, all aspects of development are regulated by the environment and by chemical adjustments within the plant. These regulatory aspects are the subject of the next chapter.

Transition to the Flowering State

Vegetative growth may be terminated, or repeatedly interrupted, or accompanied, by flowering. The transition to the flowering state often marks the end of

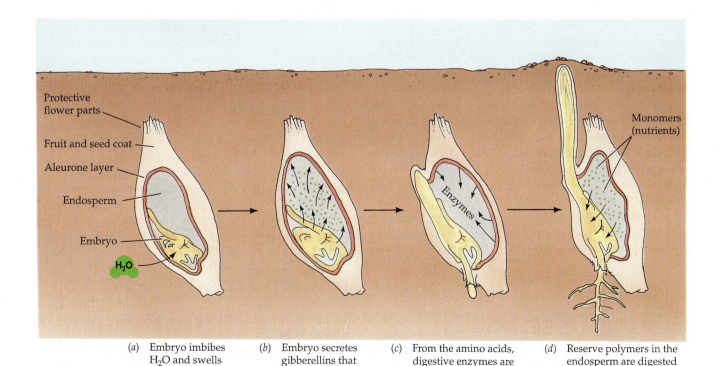

(a) Embryo imbibes H₂O and swells

(b) Embryo secretes gibberellins that diffuse into the aleurone layer, where they trigger the digestion of proteins to amino acids

(c) From the amino acids, digestive enzymes are synthesized; and the enzymes, along with other enzymes from the aleurone, attack the endosperm

(d) Reserve polymers in the endosperm are digested by the enzymes, releasing small nutrient molecules from which the embryo synthesizes new cells

31.14 Embryos Mobilize Polymer Reserves
Seed germination in cereal grasses consists of a cascade of processes. Gibberellin signals the conversion of reserve polymers into monomers usable by the developing embryo.

BOX 31.B

Reproductive versus Vegetative Growth

Legions of corn plants senesce in unison.

Flowering is the beginning of the end for plants of many species. Plants subject to **monocarpic senescence** are ones that die after setting fruit only once (see figure), while plants of other species produce fruit many times, over many years. In some species, there is a causal relationship between the formation of flowers or fruit and the subsequent senescence and death of the plant. One can keep such plants alive and growing vigorously by removing their flower buds as quickly as they form. Tobacco farmers remove flowers and fruits from their plants in order to increase the production of leaves.

The observation that senescence can be retarded by removing flower buds suggested that there may be a simple competition for nutrients between the reproductive and vegetative organs of the plant. Further support for this idea came from the observation that in many species senescence immediately follows completion of seed and fruit development, processes that represent a nutrient drain on the maternal plant. This competition for nutrients ex-

tends to competition among the fruits on the same plant. Removal of some of the flower buds on a tomato plant will cause the remaining flowers to produce larger fruit. In some species, treatments that promote vegetative growth inhibit flowering, and, sometimes, treatments that inhibit vegetative growth—heavy pruning, treatment with chemical growth retardants—promote flowering.

Life is not simple, and neither is senescence. A simple model of competition for nutrients between reproductive and vegetative tissues fails to explain most cases of monocarpic senescence. Several lines of evidence support this conclusion. For one, removal of flower buds from some plants, such as cocklebur, does *not* retard senescence. For another, one might expect that supplying extra

nutrients as fertilizer would enable a plant to set fruit and still have enough resources to continue to grow; however, augmented nutrition does not usually retard senescence. Yet another contrary indication comes from studies of spinach, a monocarpic plant that has separate male and female individuals. The simple competition model predicts that female plants, because they are the ones that produce seeds and fruit, should senesce and die sooner than male plants, which do not face the same nutrient drain. In fact, male and female spinach plants senesce at the same rate and time.

Our understanding of senescence is still limited. However, various plant growth substances are known to play roles in regulating senescence. These will be discussed in the next chapter.

vegetative growth for the plant (Box 31.B). If we view a plant as something produced by a seed for the purpose of bearing more seeds, then the act of flowering is one of the supreme events in a plant's life.

Flowering is triggered in very different ways in different plant species. As we will see in the next chapter, environmental cues include seasonal change in night length and seasonal temperature change. Many plants use their leaves to measure the length of the night, some flowering when the nights are

longer than some critical length, others flowering when the nights are shorter than some critical length, and still others following more complicated rules. Most plants, however, do not use night length to control their flowering. For those plants that do, we still do not know how information about night length is transmitted from the leaves to the meristems that will be transformed into flower buds. The physiology of flowering continues to pose some of the most difficult questions in biology.

SUMMARY

Flowers are organs of sexual reproduction. In the flower, female gametophytes develop within specialized structures on the maternal sporophyte. The mature megagametophyte typically contains eight nuclei in a total of seven cells. The entire male gametophyte travels as a pollen grain. Sperm nuclei travel through growing pollen tubes to the ovaries. Once in the embryo sac, one sperm nucleus unites with the egg, and the other unites with the two polar nuclei to produce the first cell of the endosperm; this is double fertilization. The zygote develops into an embryo with its attached suspensor, then a seedling, and ultimately a mature plant.

Both pollen and fruit must be dispersed. There are many mechanisms of pollination using wind or a variety of animals, but water is required for pollination only in certain aquatic plants. Flowers develop into seed-containing fruits, which often play important roles in the dispersal of the species.

Some flowering plants reproduce asexually. Stolons, tubers, rhizomes, bulbs, corms, or root suckers are means by which plants may reproduce vegetatively. Humans may reproduce plants vegetatively from other tissues. Some species produce seeds by apomixis.

Seeds may be kept dormant by some mechanism, such as chemical inhibition or a rigid or impermeable seed coat. Dormancy may be broken by various means, including leaching of inhibitors by water, drying, abrasion of the seed coat, fire, low temperature, and light. Among the events in seed germination are the imbibition of water and the mobilization of reserve foods.

The plant forms flowers at one or several times in its life, beginning another round of sexual reproduction. In some species, individuals senesce and die after producing seeds.

SELF-QUIZ

1. Which of the following does *not* participate in asexual reproduction?
   a. Stolons
   b. Rhizomes
   c. Fertilization
   d. Tubers
   e. Apomixis

2. Apomixis includes:
   a. sexual reproduction.
   b. meiosis.
   c. fertilization.
   d. a diploid embryo.
   e. no production of a seed.

3. Sexual reproduction in angiosperms:
   a. is by way of apomixis.
   b. requires the presence of petals.
   c. can be accomplished by grafting.
   d. gives rise to genetically diverse offspring.
   e. cannot result from self-pollination.

4. The typical angiosperm female gametophyte:
   a. is called a megaspore.
   b. has 8 nuclei.
   c. has 8 cells.
   d. is called a pollen grain.
   e. is carried to the male gametophyte by wind or animals.

5. Pollination in angiosperms:
   a. never requires water.
   b. never occurs within a single flower.
   c. always requires help by animal pollinators.
   d. is also called fertilization.
   e. makes most angiosperms independent of water for reproduction.

6. Which statement is *not* true of double fertilization?
   a. It is found in all angiosperms.
   b. It is found in no other plant.
   c. One of its products is a triploid nucleus.
   d. One sperm nucleus fuses with the egg nucleus.
   e. One sperm nucleus fuses with two polar nuclei.

7. The suspensor:
   a. gives rise to the embryo.
   b. is heart-shaped in dicots.
   c. separates the two cotyledons of dicots.
   d. ceases to elongate early in embryo development.
   e. is larger than the embryo.

8. Which of the following is *not* an advantage of seed dormancy?
   a. It makes the seed more likely to be digested by birds that disperse it.
   b. It counters the effects of year-to-year variations in the environment.
   c. It increases the likelihood of a seed germinating in the right place.
   d. It favors dispersal of the seed.
   e. It may result in germination at a favorable time of year.

9. Which of the following does *not* participate in seed germination?
   a. Imbibition of water
   b. Metabolic changes
   c. Growth of the radicle
   d. Mobilization of food reserves
   e. Extensive mitotic divisions

10. To mobilize its food reserves, a germinating barley seed:
    a. becomes dormant.
    b. undergoes apomixis.
    c. secretes gibberellins into its endosperm.
    d. converts glycerol and fatty acids into lipids.
    e. lets its embryo take up proteins from the endosperm.

FOR STUDY

1. For a crop plant that reproduces both sexually and asexually, which method of reproduction might be preferable from the viewpoint of the farmer?

2. Thompson seedless grapes are produced by vines that are triploid. Think about the consequences of this chromosomal condition for meiosis in the flowers. Why are these grapes seedless? Describe the role played by the flower in fruit formation when no seeds are being formed. How do you suppose Thompson seedless grape plants are propagated?

3. How may it be advantageous for some species to have seeds whose dormancy is broken by fire?

4. Cocklebur fruits contain two seeds each, and the two seeds are kept dormant by two different mechanisms. How may this be advantageous to cockleburs?

5. Discuss the roles of the various parts of a seed. Which of these parts should be present in an artificial seed?

READINGS

Barrett, S. C. H. 1987. "Mimicry in Plants." *Scientific American*, September. Some plants mimic insects, encouraging the mimicked insects to act as pollinators.

Handel, S. N. and A. J. Beattie. 1990. "Seed Dispersal by Ants." *Scientific American*, August. Some plants produce seeds bearing specialized fat bodies that are eaten by ants after the ants carry the seeds to their nests. This and other mechanisms are discussed.

Niklas, K. J. 1987. "Aerodynamics of Wind Pollination." *Scientific American*, July. Do the mechanics of pollination seem improbable to you? Wind-pollinated plants have many adaptations that favor successful pollination.

Raven, P. H., R. F. Evert and S. Eichhorn. 1986. *Biology of Plants*, 4th Edition. Worth, New York. A well-balanced general botany textbook.

Salisbury, F. B. and C. W. Ross. 1985. *Plant Physiology*, 3rd Edition. Wadsworth, Belmont, CA. An authoritative textbook with good chapters relating to reproductive development.

Regulation of Plant Development

PREVIEW: An adult plant develops from a zygote by a series of steps controlled by internal factors, including a variety of chemical growth substances, as well as by external factors such as light. Each of the growth substances has multiple effects on the growth and development of the plant, and particular developmental processes typically are regulated by more than one growth substance.

This chapter deals with auxin, gibberellins, cytokinins, abscisic acid, ethylene, and phytochrome, and with etiolation, seedling growth, bud dormancy, leaf abscission, photoperiodism, flowering, and senescence.

As a leaf develops from a leaf primordium into its mature form, new tissues must be laid down, and each cell type must be equipped with its appropriate complement of enzymes and structural substances. Consider, as an example, leaf development in a plant, such as corn, that performs C_4 photosynthesis (Chapter 8). Two types of photosynthetic tissue surround the vascular bundles in C_4 leaves (Figure 8.27). The outer photosynthetic layer, mesophyll tissue, has chloroplasts that use the enzyme PEP carboxylase to capture carbon dioxide. The inner photosynthetic layer, bundle sheath tissue, has chloroplasts that use another enzyme, rubisco, to transfer the carbon dioxide to a different acceptor molecule. Different genes are transcribed in each of the two types of photosynthetic tissue (Figure 32.1). The leaf primordium does not contain specific cells predestined to become bundle sheath cells or mesophyll cells. How is the fate of a leaf primordium cell determined? Research now in progress suggests that distance from the vascular bundles determines the fate of a cell, but we have yet to learn the chemical nature of the signal from the vascular bundles.

We showed in Chapter 28 that the plant body is *modular* to an extent unknown in organisms such as humans. Each leaf is a module with its own microenvironment and its own life history. Different parts of the shoot system explore different parts of the environment, as do different parts of the root system. The development of each part of the plant body must be regulated in accordance with its own environment, and yet the whole body must be so regulated as to function as an integrated whole. Both levels of developmental regulation—local and global—are accomplished by the chemical and physical systems to be described in this chapter.

AN OVERVIEW OF DEVELOPMENT

In preceding chapters, especially Chapter 31, we have considered the life of a flowering plant. Let us now review this life history, from seed to death, this time focusing on how various events are regulated.

From Seed to Seedling

To begin, consider a seed. It may be dormant. Seed dormancy may be broken by one of the physical mechanisms—scarification, fire, leaching of inhibitors by water—described in Chapter 31. Some seeds will not germinate until they have been exposed to light. Exposure to a tiny amount of light energy suffices. This light is absorbed by phytochrome, a pigment that plays many developmental roles, as we will see.

As the seed germinates, the growing embryo must obtain building blocks—monomers—by digesting the polymeric reserve foods in the endosperm. The embryo secretes gibberellins, growth substances that direct the mobilization of the reserves (Figure 31.14). Gibberellins, like phytochrome, play many roles in the life of the plant.

If the seed germinates underground, it must elongate rapidly and cope with life in the absence of light. Phytochrome controls this stage and ends it when

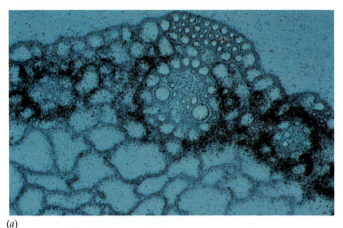 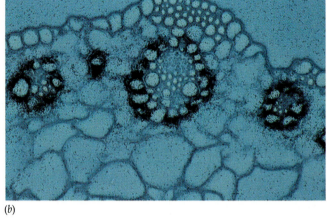

(a) (b)

32.1 Differing Development in Neighboring Cells
Autoradiography reveals which genes are being transcribed in which cells in this corn leaf. (a) The radioactive probe for PEP carboxylase mRNA accumulates only on mesophyll cells, marking them with black silver grains. (b) The radioactive probe for rubisco mRNA accumulates only on bundle sheath cells. The two cell types are producing different enzymes for photosynthesis.

the shoot reaches the light. The growth of the seedling, both in darkness and light, is controlled by other chemical substances as well. These substances include auxin, a growth substance produced by the shoot tip and by leaves, and gibberellin. Auxin and gibberellin help determine the growth rate of the shoot (Figure 32.2).

As the seedling develops, it not only elongates but it also forms tissues and organs. Auxin regulates tissue and organ formation, too. Other information regulating these phenomena comes from the root system in the form of the cytokinins, another group of regulators.

Reproductive Development

Eventually, the plant flowers. This may happen when the plant simply reaches an appropriate age or size. However, some plant species flower at particular times of the year, meaning that the plant must sense the appropriate date. These plants are photoperiodic—they measure the length of the night (shorter in the summer, longer in the winter) with great precision. We know quite a lot about photoperiodism— but not how it works. We know that the plant measures the length of darkness, and we know that there is a biological clock—itself a mystery—and we know that light absorbed by phytochrome can affect the time-measuring process.

Once a leaf has determined, by measuring the night length, that it is time for the plant to flower, that information must be transported to the places where flowers will form. But how this comes about remains a mystery. It seems likely that a "flowering hormone"—named florigen, but never discovered—

32.2 From Seed to Seedling
Controlling factors in the first stages of plant growth.

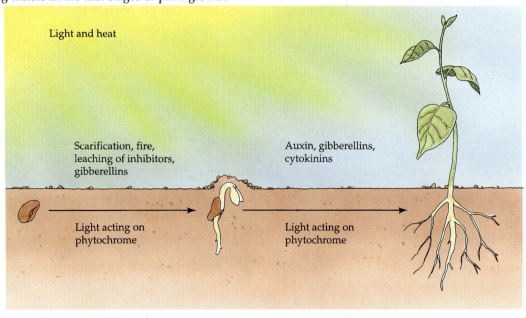

Light and heat

Scarification, fire, leaching of inhibitors, gibberellins

Auxin, gibberellins, cytokinins

Light acting on phytochrome

Light acting on phytochrome

32.3 Flowering and Fruit Formation
Controlling factors in plant reproduction.

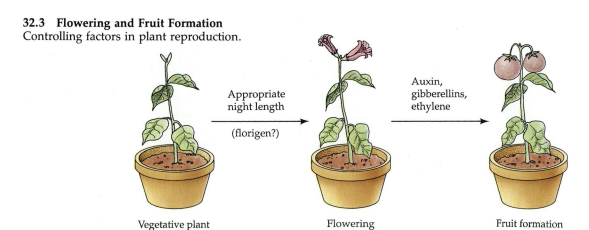

Vegetative plant   Appropriate night length (florigen?) → Flowering   Auxin, gibberellins, ethylene → Fruit formation

travels from the leaf to the point of flower formation. Perhaps information travels in some other form, but this form has never been discovered, either.

After flowers have formed, regulators including auxin and gibberellin play further roles. Chemicals control the growth of a pollen tube down the style of a pistil. Following fertilization, a fruit develops—again under the control of chemicals such as gibberellin and auxin (Figure 32.3). The ripening of the fruit is also under the control of chemicals, commonly the gas ethylene.

Dormancy and Senescence

Some perennials have buds that enter a state of winter dormancy during the cold season. The state of dormancy may be maintained by a number of factors, including the growth substance abscisic acid. Abscisic acid, like all the growth regulators, plays multiple roles in the plant. Another of its roles is in the regulation of the degree of opening of stomata (Chapter 29).

Finally, the plant dies—a phenomenon that may be under environmental control. Death follows senescent changes that are controlled by regulators such as ethylene (Figure 32.4).

All the elements of this hasty overview will be considered in detail in the pages that follow. However, it should be clear by this point that a plant is subject to regulation throughout its life.

PLANT GROWTH SUBSTANCES

Several types of **plant growth substances**—naturally occurring compounds that regulate development—were mentioned in the preceding paragraphs. The roles of the plant growth substances in development are summarized in Table 32.1. Generally speaking, each kind of growth substance is produced in one or more specific parts of the plant's body and travels to other parts where it has its effects. Each of the known plant growth substances plays multiple roles in the plant. We will now consider each of the major plant growth substances, beginning with the gibberellins.

GIBBERELLINS

We first encountered the gibberellins in Chapter 31, as we described the mechanism by which germinating barley and other cereal grass seeds converted their reserve proteins and starch into soluble monomers (Figure 31.14). This is an elegant sequential mechanism: The embryo produces a signal (gibberellin), the aleurone layer responds by synthesizing and secreting digestive enzymes into the endosperm, and the enzymes digest the storage compounds, giving the embryo access to the nutrients it needs. Gibberellins produce a wide variety of effects on plant development in addition to this triggering of digestive enzyme synthesis.

32.4 Senescence and Death
Controlling factors in the final stages of plant growth.

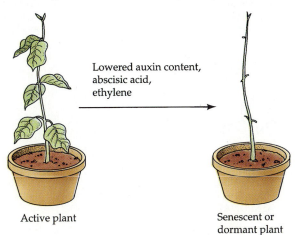

Active plant   Lowered auxin content, abscisic acid, ethylene → Senescent or dormant plant

TABLE 32.1
Growth Substance Activities during the Life of a Plant

	GIBBERELLINS	AUXIN	CYTOKININS	ETHYLENE	ABSCISIC ACID
Site of Production	Embryo, young leaves, root and shoot apices	Embryo, young leaves, shoot apical meristem	Roots	Ripening fruit, senescing tissue, stem nodes	Root cap, older leaves, stem
Activity					
Seed dormancy	Breaks				Imposes
Seed germination	Promotes				Inhibits
Seedling growth	Promotes cell division and expansion	Promotes cell expansion	Promotes cell division		
Apical dominance		Inhibits lateral buds	Promotes lateral buds		
Leaf abscission		Inhibits	Inhibits	Promotes	
Winter dormancy	Breaks				Imposes
Flowering	Stimulates in some plants				
Fruit development	Promotes	Promotes		Promotes ripening	

Discovery of the Gibberellins

The gibberellins are a large family of closely related compounds (Figure 32.5*a*, *b*), some found in higher plants and others in a pathogenic (disease-causing) fungus, where they were first discovered. The discovery of the gibberellins followed a crooked path, beginning with a book dictated in 1809 by a Japanese farmer named Konishi. In it, he described the symptoms of *ine bakanae-byo*, the "foolish seedling" disease of rice. Seedlings affected by the *bakanae* disease grow more rapidly than their healthy neighbors, but the rapid growth gives rise to spindly plants that die before producing seed. The disease has been of considerable economic importance in several parts of the world. In 1898 it was learned that the *bakanae* disease, including both its growth-promoting and toxic effects, was caused by a fungus now known as *Gibberella fujikuroi*.

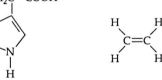

(a) Gibberellin A$_1$ (Important in stem growth)

(b) Gibberellin A$_3$ (Commercially available)

(c) Auxin (Indoleacetic acid)

(d) Ethylene (The "senescence hormone")

(e) Kinetin (A cytokinin discovered in aged DNA)

(f) Zeatin (A naturally occuring cytokinin in plants)

(g) Abscisic acid (The "stress hormone")

32.5 Plant Growth Substances
Chemical structures of some representative compounds.

In 1925 the Japanese biologist Eiichi Kurosawa undertook a study of how *G. fujikuroi* caused the excessive spindly growth characteristic of the *bakanae* disease. He grew the fungus on a liquid medium and then separated the fungus from the medium by filtering. He heated the medium to kill any remaining fungus. The resulting heat-treated filtrate still stimulated the growth of uninfected rice seedlings. He found no such effects using medium that had never contained the fungus (Figure 32.6). Thus, he established that *G. fujikuroi* produces a chemical substance with growth-promoting properties. In the late 1930s it became clear that there was more than one gibberellin, as the new growth substance was called. It was also shown that the toxic effects of the fungus were caused by another, inhibitory substance.

Were the gibberellins simply exotic products of an obscure fungus, or did they play a more general role in the growth of plants? Bernard O. Phinney of the University of California, Los Angeles, partially answered this question in 1956, when he reported the spectacular growth-promoting effect of gibberellins on certain dwarf mutant strains of corn. These plants were known to be genetic dwarfs—each phenotype was produced when a particular recessive allele was present in the homozygous condition (see Chapter 10 if you need to review these genetic terms). Gibberellin applied to normal—tall—corn seedlings had virtually no effect, but gibberellin applied to the dwarfs caused them to grow as tall as their normal relatives. (A comparable effect of gibberellin on dwarf tomatoes is shown in Figure 32.7.) This suggested to Phinney that (1) gibberellins are normal constituents of corn and perhaps of all plants and (2) the mutants are short because they cannot produce their own gibberellin. According to these hypotheses, normally tall plants manufacture enough gibberellins to promote their full growth. Phinney and other scientists tested extracts from numerous plant species to see if they promoted growth in dwarf corn; it was found that, indeed, many such extracts promoted growth. These findings provided direct evidence that normal, healthy plants contain gibberellinlike substances.

Why So Many Gibberellins?

We do not yet know how many different gibberellins exist. Each year brings reports of more, with several dozen now having been characterized. Some gibberellins are produced in the root system, and others are produced in young leaves. For many years, plant physiologists were puzzled by the existence of such a great number of different gibberellins. However, recent work has led to the conclusion that only one gibberellin, gibberellin A_1, actually controls stem elongation; the other gibberellins found in stems are simply intermediates in the production of gibberellin A_1. As discussed next, gibberellins affect processes

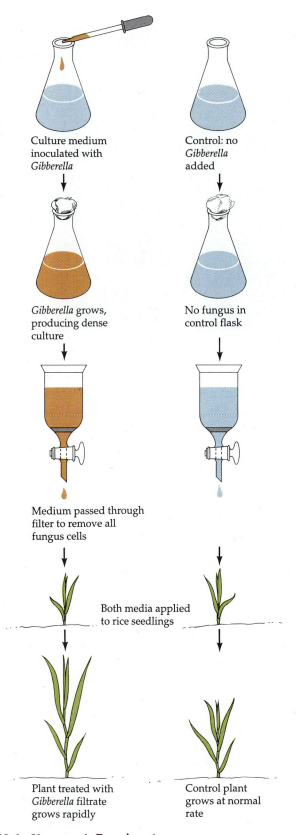

Culture medium inoculated with *Gibberella*

Control: no *Gibberella* added

Gibberella grows, producing dense culture

No fungus in control flask

Medium passed through filter to remove all fungus cells

Both media applied to rice seedlings

Plant treated with *Gibberella* filtrate grows rapidly

Control plant grows at normal rate

32.6 Kurosawa's Experiment

Kurosawa demonstrated that *bakanae* disease is caused by substances produced by *Gibberella fujikuroi*. Fungus-free medium previously used to culture the fungus made rice seedlings grow rapidly, as if they had the *bakanae* disease. As a control, Kurosawa determined that medium in which the fungus had never been cultured did not cause disease symptoms.

32.7   Effect of Gibberellin on Dwarf Tomatoes
A tiny amount of gibberellin—just 4 micrograms—was
enough to induce normal growth in the plant on the
right. On the left is an untreated control plant.

other than stem elongation, but we do not yet know
which gibberellin has any other particular effect.

Other Activities of Gibberellins

Gibberellins and other growth substances regulate
the growth of fruits. It has long been known that
seedless varieties of grapes form smaller fruit than
their seeded relatives. In one experiment, removal of
seeds from very young seeded grapes prevented their
normal growth, suggesting that the seeds are sources
of a fruit growth regulator. It was then shown that
spraying young seedless grapes with a gibberellin
solution causes them to grow as large as seeded varieties. Subsequent biochemical studies showed that
the developing seeds produce gibberellin, which diffuses out into the immature fruit tissue.

Some biennial plants—plants that grow vegetatively in their first year and flower and die in their
second year—respond dramatically to an increased
level of gibberellin. In their second year, in response
either to the increasing length of days or to the winter
cold period, the apical meristems of these biennials
produce elongated shoots that eventually bear flowers. This elongation is called **bolting**. When the plant
senses the appropriate environmental cue—longer
days, or a sufficient winter chilling—it produces
more gibberellins, raising the gibberellin concentration to a level that causes the shoot to bolt. Plants
that have not experienced the cue will bolt if sprayed
with a gibberellin solution (Figure 32.8).

Gibberellins also cause fruit to grow from unfertilized flowers; they promote seed germination in lettuce and some other species; and in the spring they
help bring buds out of their winter dormancy.
Growth substances, such as the gibberellins, usually
have multiple effects within the plant. Growth substances often interact with one another in regulating
developmental processes. In controlling stem elongation, for example, gibberellins interact with another growth substance, auxin.

AUXIN

If you pinch off the apical bud at the top of a bean
plant, lateral buds that were once quiescent become
active. Similarly, pruning a shrub causes an increase
in branching. If you excise the blade of a leaf but
leave its petiole attached to the plant, the petiole
drops off sooner than it would if the leaf were intact.

No gibberellin   With gibberellin

32.8   Bolting
Spraying with gibberellin causes cabbage and some other
plants to bolt. While untreated control plants retain their
compact leafy heads, the internodes of treated plants
elongate dramatically, resulting in shoots that tower over
the person who sprayed them.

If a plant is kept indoors, its shoot system grows toward a bright window. What these diverse responses of shoot systems have in common is that they are mediated by a plant growth substance called **auxin**—or **indoleacetic acid** in chemical terms (Figure 32.5*c*).

Discovery of Auxin

The discovery of auxin and its numerous physiological activities traces back to work done in the 1880s by Charles Darwin and his son Francis, who were interested in plant movements. One type of movement they studied was **phototropism,** the growth of plant structures toward light (as in shoots) or away from it (as in roots).

An obvious question they could ask was: What part of the plant senses the light—where is the "eye" of the plant, as it were? Much of their work on this question was done with canary grass seedlings grown in the dark. The dark-grown grass seedling has a coleoptile, or leaf sheath, covering the immature shoot. To find the light-receptive region of the coleoptile, the Darwins tried "blindfolding" it in various places and then illuminating it from one side (Figure 32.9). The coleoptile grew toward the light as long as its very tip was exposed. If, however, the top millimeter or more of the coleoptile was covered, there was no phototropic response. Thus, the tip is the photoreceptor. However, the actual bending

takes place in a growing region a few millimeters below the tip. Therefore, the Darwins reasoned, some type of "message" must travel within the coleoptile from the tip to the growing region.

Others later demonstrated that the message is a chemical substance; it cannot pass through a barrier impermeable to chemicals but does pass through certain nonliving materials such as gelatin. The tip of the coleoptile produces a growth substance that moves down the coleoptile to the growing region. If the tip is removed, the growth of the coleoptile is sharply inhibited; if the tip is then carefully replaced, growth resumes, even if the tip and base are separated by a thin layer of gelatin. The growth substance moves down from the tip but does not move from one side of the coleoptile to the other. If the tip of an oat coleoptile is cut off and replaced so that it covers only one side of the cut end of the shoot, the coleoptile curves as the cells on the side below the replaced tip grow more rapidly than those on the other side.

With this information as a beginning, the Dutch botanist Frits W. Went managed to isolate the growth substance from oat coleoptiles. (Previous attempts to extract the material had invariably failed.) Went removed coleoptile tips and placed their cut surfaces on a block of gelatin, hoping the growth substance would diffuse into the gelatin. Then he placed pieces of the gelatin block on decapitated coleoptiles—positioned to cover only one side, just as coleoptile tips had been placed in some of the earlier experiments. As they grew, the coleoptiles curved toward the side away from the gelatin. This curvature demonstrated that the growth substance had indeed diffused into the gelatin block from the isolated coleoptile tips (Figure 32.10). The growth substance had at last been isolated from the plant. It was later named auxin; still later, it was shown to be indoleacetic acid. Went's historic experiment was performed in 1926—the very year Kurosawa published his classic account of the isolation of a growth substance from the fungus *Gibberella fujikuroi*, an accomplishment closely analogous to Went's.

Auxin Transport

Auxin could be studied in a number of ways once it had been isolated from the plant. Early experiments showed that its movement through certain tissues is strictly polar, that is, unidirectional along a line from apex to base. Figure 32.11 outlines how these experiments were conducted and the results obtained. By inverting the setups in half of the experiments, the scientists determined that the apex-to-base direction of auxin movement has nothing to do with gravity; the polarity of this movement is a totally biological matter. Many plant parts show at least partial polarity of auxin transport. Auxin moves in leaf petioles from the blade end toward the stem end.

All coleoptiles exposed to light from same side.

Coleoptile Blindfold Light

Coleoptile

Grass seedling

Primary root

Phototropism occurs only in coleoptiles that had *tips* exposed to light.

32.9 The Darwins' Experiment
The top drawings show some of the ways in which Charles and Francis Darwin "blindfolded" dark-grown grass seedlings; the lower drawings show what they observed in each case. Coleoptiles responded to light only when the top millimeter or so was exposed; they responded as they grew by bending toward the light a few millimeters below the tip. These observations suggested that the plant's "eye" is in the tip, and that it sends a message from the tip to the region of bending.

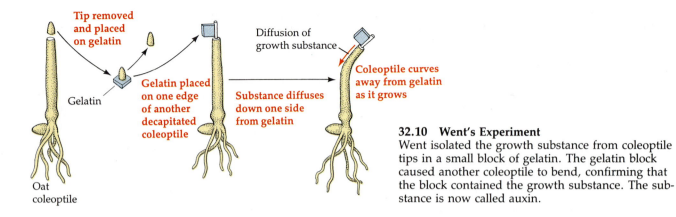

Oat coleoptile

32.10 Went's Experiment
Went isolated the growth substance from coleoptile tips in a small block of gelatin. The gelatin block caused another coleoptile to bend, confirming that the block contained the growth substance. The substance is now called auxin.

Phototropism and Gravitropism

It is the lateral movement of auxin in the apex that affects the direction of plant growth. When light strikes a coleoptile from one side, auxin at the tip moves laterally toward the shaded side. The imbalance thus established is maintained down the coleoptile, so that below in the growing region there is more auxin on the shaded side, causing the unequal growth that results in a coleoptile bent toward the light—phototropism (Figure 32.12). Similarly, but even in the dark, auxin moves to the lower side of a shoot that has been tipped over, causing more rapid growth in the lower side and, hence, an upward bend of the shoot. This phenomenon is **gravitropism,** the growth of a plant part in a direction determined by gravity. The gravitropism of shoots is defined as negative; the gravitropism of roots—which bend downward—is positive.

Auxin and Vegetative Development

Auxin, like the gibberellins, plays a multiplicity of roles in the plant. Cuttings from the shoots of some plants can produce roots and thus grow into entire new plants. For this to happen, certain undifferentiated cells in the shoot, originally destined to function only in food storage, must set off on an entirely new mission: to differentiate and become organized into the growing zone of a root. Cuttings of many species can be stimulated to grow roots profusely by dipping the cut surfaces into an auxin solution. Quite different is the effect of auxin on leaf **abscission,** the separation of leaves from stems, as happens when leaves fall from trees before winter. If the blade of the leaf is excised, the petiole abscises more rapidly than if the leaf had remained intact (Figure 32.13). If, however, the cut surface is treated with an auxin solution, the petiole remains attached to the plant, often longer than an intact leaf would have. It appears that the time of abscission of leaves in nature is determined in part by a decrease in the movement of auxin, produced in the blade, through the petiole.

Auxin maintains **apical dominance,** the tendency of some plants to grow a single main stem with minimal branching. This phenomenon can be shown by an experiment with dark-grown pea seedlings. If the plant remains intact, the stem elongates and the lateral buds remain quiescent. Removal of the apical bud—the site of auxin production—causes the lateral buds to grow out vigorously, but this growth is prevented if the cut surface of the stem is treated with an auxin solution (Figure 32.14). Branches themselves show apical dominance, with their own lateral buds being inactive unless they are pruned off. Note

32.11 Auxin Moves from Apex to Base
To reveal the direction in which auxin moves, experimenters cut bits of tissue away from the tip and the base of coleoptiles. They put the cut ends of the pieces against agar blocks. When auxin was applied to the apical end, it moved through the tissue—even against gravity. Auxin did not move through the tissue from the basal end.

Tip

Apical end

Excised segment of oat coleoptile

Base

Basal end

Coleoptile

Auxin transmitted from donor agar block to receiver block

Donor

Receiver

Receiver

Donor

Auxin not transmitted from donor agar block to receiver block

Receiver

Donor

Donor

Receiver

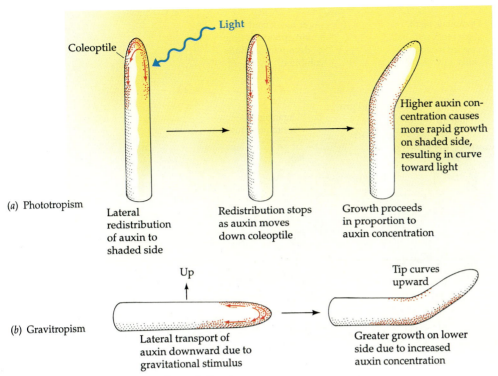

Light

Coleoptile

Higher auxin con-
centration causes
more rapid growth
on shaded side,
resulting in curve
toward light

(a) Phototropism

Lateral
redistribution
of auxin to
shaded side

Redistribution stops
as auxin moves
down coleoptile

Growth proceeds
in proportion to
auxin concentration

Up

Tip curves
upward

(b) Gravitropism

Lateral transport of
auxin downward due to
gravitational stimulus

Greater growth on lower
side due to increased
auxin concentration

32.12 Plants Respond to Light and Gravity
Auxin in the coleoptile tip moves toward the shaded
side, beginning the phototropic response. Auxin accumulates on the lower side of a horizontal coleoptile, beginning the gravitropic response.

Petiole

Blade

(a) Blade removed   No auxin   Petiole abscises

Auxin in
lanolin

(b) Blade removed   Auxin added   Petiole remains
                                          on plant

32.13 Auxin and Leaf Abscission
Auxin delays leaf abscission. The leaf blade is a source of
auxin throughout the growing season.

that in the experiments on leaves and stems that we
have discussed, removal of a particular part of the
plant produces an effect—abscission or loss of apical
dominance—and that the effect is prevented by treatment with auxin. These results are consistent with
other data showing that the excised part of the leaf
or stem is an auxin source and that auxin in the intact
plant helps maintain apical dominance and delays
the abscission of leaves.

Many synthetic auxins—chemical analogues of indoleacetic acid—have been produced and studied.
One of the synthetic auxins, 2,4-dichlorophenoxy-
acetic acid (2,4-D), has the striking property of being
lethal to dicots at concentrations that are harmless to
monocots. This property made 2,4-D a widely used
selective herbicide that could be sprayed on a lawn
or a cereal crop to kill the dicots, which eliminates
most of the weeds. Because 2,4-D takes a long time
to break down, it pollutes the environment. Now that
this has been recognized, scientists are seeking new
approaches to selective weed killing (Box 32.A).

Auxin and Fruit Development

Although fruit development normally depends on
prior fertilization of the egg, in many species treatment of an unfertilized ovary with auxin or gibberellin causes **parthenocarpy**—fruit formation without
fertilization of the egg. Parthenocarpic fruit form
spontaneously in some plants, including dandelions and the cultivated varieties of pineapple and
banana—and seedless grapes.

The strawberry is an unusual "fruit." What we

(a) In intact plants lateral buds are inhibited by apical bud

(b) Decapitate plant. Agar block (no auxin) applied to stump. Buds grow out

(c) Agar block containing auxin applied at time of decapitation. Buds inhibited

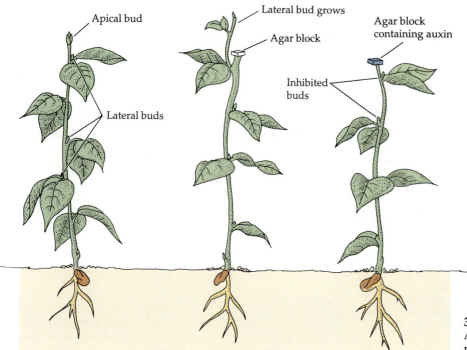

Apical bud

Lateral buds

Lateral bud grows

Agar block

Agar block containing auxin

Inhibited buds

Inhibited buds

32.14 Auxin and Apical Dominance
Apical dominance results from auxin produced by the apical bud, as suggested by these experiment results.

BOX 32.A

Strategies for Weed Killing

The first commercial herbicides—weedkillers—were substances that killed any plant they touched. Today we still use such general herbicides when it is not necessary to save other vegetation; for example, products containing 3-aminotriazole (3AT) are used to kill stands of poison oak or poison ivy. We cannot use such an all-out, nonspecific approach if we wish to kill certain plants but retain others.

The next approach was the use of selective herbicides such as 2,4-D. Although 2,4-D is an effective herbicide, its use is controversial.

The advent of recombinant DNA technology opened entirely new avenues of herbicide research. One of these avenues is the transfer of genes conferring resistance to herbicides. The concept is simple and attractive: First, develop a highly effective herbicide that will kill all the plants in a field; second, genetically engineer the chosen crop plants so that they are resistant to that herbicide; finally, spray the field and watch the weeds die while the crop, protected by its inserted gene, prospers. Biotechnology companies view this as a promising method for weed control, although some scientists fear its environmental and health consequences.

Scientists at Calgene, a biotechnology company in Davis, California, recently reported the results of such an experiment. Their goal was to render selected plants insensitive to bromoxynil, a potent inhibitor of photosynthesis. The Calgene scientists discovered that a particular soil bacterium, *Klebsiella ozaenae*, converts bromoxynil to an inactive product. From this bacterium they isolated the gene that codes for the enzyme that inactivates bromoxynil. They then inserted the *Klebsiella* gene into tobacco plants. To make sure the gene would be expressed in the tobacco plants, they inserted it along with, and under the control of, a light-activated promoter. The promoter is one that normally controls the gene that codes for rubisco, the enzyme that fixes carbon dioxide in photosynthesis. The experiment was a success: Transgenic plants containing the bacterial gene grew vigorously after being sprayed with bromoxynil solutions that killed control plants.

commonly call the fruit is actually a modified stem, or receptacle, with the tiny, dry "seeds" being the true fruits, called achenes. The achenes produce auxin, and the auxin induces the growth of the fleshy receptacle, as demonstrated by the French botanist Jean-Pierre Nitsch (Figure 32.15). When he removed all the achenes within three weeks after pollination, the stem tissue did not develop into a "strawberry." If he pollinated only one to three of the many pistils and kept the others virgin, he observed localized receptacle growth in the area of pollination. He could induce normal expansion if, after removing all of the achenes, he spread an auxin-containing paste over the receptacle. These three results are all consistent with the hypothesis that the achenes cause the growth of the receptacle by producing auxin.

Growth substances control other aspects of fruit physiology—and of the development and senescence of flower parts as well. This activity of the growth substances illustrates yet again the great diversity of important roles that are played by these compounds.

Plant Cell Walls

Each of the known plant growth substances causes a variety of responses that are often seemingly unrelated, thus raising one of the major questions of plant physiology: Do all the effects of a particular growth substance arise from a single mechanism—a common master chemical reaction? We know so little about how the growth substances act at the molecular

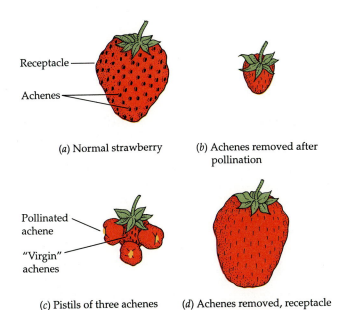

Receptacle
Achenes

(a) Normal strawberry

(b) Achenes removed after pollination

Pollinated achene
"Virgin" achenes

(c) Pistils of three achenes pollinated

(d) Achenes removed, receptacle treated with auxin

32.15 Auxin and Strawberry Development
Auxin produced by the achenes is responsible for the normal growth of the strawberry "fruit."

level that this question cannot yet be answered. For one growth substance—auxin—we are beginning to gain some insight as to what one of its central mechanisms may be, if indeed there is more than one molecular mechanism. To appreciate this mechanism, we must first briefly consider the architecture of the plant cell wall.

The principal strengthening component of the plant cell wall is cellulose, a large polymer of glucose (Chapter 3). In the wall, cellulose molecules tend to associate with one another, forming crystalline regions called micelles. Individual cellulose molecules may extend from one micelle across relatively noncrystalline regions to other micelles (Figure 32.16a). Bundles of approximately 250 cellulose molecules, including many micelles, constitute microfibrils visible with an electron microscope. What makes the cell wall rigid is a network of cellulose microfibrils embedded in a jellylike matrix of other, smaller polysaccharides (Figure 32.16b). Peter Albersheim and his colleagues at the University of Colorado proposed a model for the molecular architecture of the cell wall, showing how the other polysaccharides interconnect the cellulose microfibrils (Figure 32.16c).

Cell Walls and Growth

The growth of a plant cell is driven primarily by the uptake of water, which enters the cytoplasm of the cell and its vacuole (see Figure 4.25). As the vacuole expands, the cell grows rapidly, with the vacuole often making up more than 90 percent of the volume of a mature cell. As the vacuole expands, it presses the cytoplasm against the cell wall, and the wall resists this pressure potential. For the cell to grow, its wall must loosen and be stretched. As the wall stretches, it should get thinner; however, cell wall thickness is maintained because new polysaccharides are deposited throughout the wall and new cellulose microfibrils are deposited at the inner surface. Thus, the cellulose microfibrils in the outermost part of the wall are the oldest, and those in the innermost part the youngest.

The wall plays a key role in controlling the growth rate of a plant cell. How does the plant determine the behavior of its cell walls?

Auxin and the Cell Wall

Auxin can loosen cell walls—make them more stretchable—as first demonstrated half a century ago by the Dutch physiologist A. J. N. Heyn. Heyn took segments of oat coleoptiles and made them plasmolyze (go limp) by placing them in highly concentrated—hypertonic—solutions (Chapter 5). He reasoned that the only thing keeping the segments from collapsing was the network of cell walls—turgor played no role under these circumstances. He hung

(a)

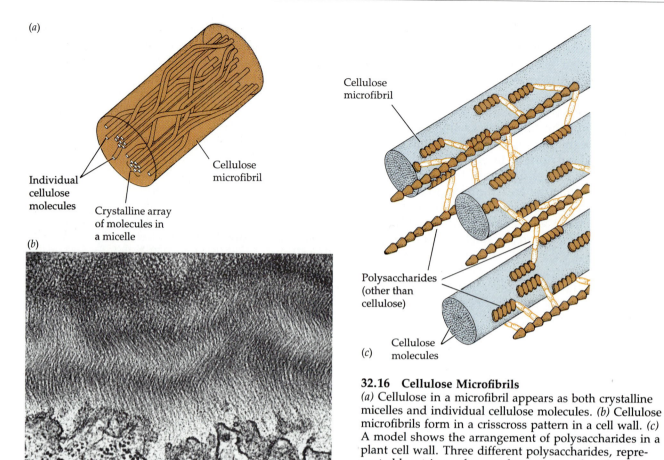

Individual cellulose molecules

Crystalline array of molecules in a micelle

Cellulose microfibril

(b)

Cellulose microfibril

Polysaccharides (other than cellulose)

Cellulose molecules

(c)

32.16 Cellulose Microfibrils

(a) Cellulose in a microfibril appears as both crystalline micelles and individual cellulose molecules. (b) Cellulose microfibrils form in a crisscross pattern in a cell wall. (c) A model shows the arrangement of polysaccharides in a plant cell wall. Three different polysaccharides, represented by strings of cones, hexagons, and ovals, help organize cellulose microfibrils in the wall.

the segments on pins, as shown in Figure 32.17, and hung weights on their ends, causing the segments to bend. He next removed the weights, allowing the segments to bend back. This recovery was incomplete, and he called the reversible bending **elasticity** and the irreversible bending **plasticity.** Pretreatment of the coleoptile segments with auxin significantly increased their plasticity—it loosened the wall. This suggested that auxin-induced cell expansion might result from just such a loosening effect.

It was later shown that auxin itself does not loosen cell walls upon contact; rather, there must be an intervening step—auxin may cause the release of a "wall-loosening factor" from the cytoplasm. Work in the 1970s in the United States and in Europe indicated that the wall-loosening factor was simply hydrogen ions (protons, H^+). Simple acidification of the growth medium (that is, addition of H^+) causes segments of stems or coleoptiles to grow as rapidly as do others treated with auxin, and treatment of coleoptile segments with auxin causes acidification of the medium. Treatments that block acidification by auxin also block auxin-induced growth. It has been proposed that hydrogen ions, secreted into the cell wall as a result of auxin action, activate some enzyme in the wall. This enzyme may digest specific linkages

connecting the cellulose microfibrils, or it may alter bonds in the matrix in which the microfibrils reside. The end result in either case would be a temporary loosening of the wall.

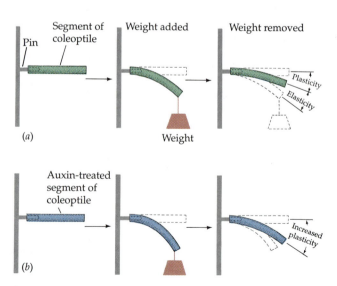

32.17 Auxin Affects Cell Walls

Auxin increases the plasticity, but not the elasticity, of cell walls.

The cell wall is an important site for the major activities regulating plant development. Auxin is not the only agent that affects the properties of the cell wall. Gibberellins, too, can loosen the wall, which is not surprising in view of their growth-promoting activities. Other growth substances also modify the cell wall.

Auxin Receptors

We will see in Chapter 34 that animal hormones begin to act only after binding with specific receptor proteins. Does the action of plant growth substances also require their recognition by receptor proteins? Over the past several years it has been shown that several proteins can bind various plant growth substances. However, it has not been shown that these proteins actually function in the living plant as receptors that mediate the effects of the regulators.

In 1989 Glenn Hicks, David Rayle, and Terri Lomax, at Oregon State University, provided the first solid evidence for a connection between such apparent receptor proteins and an auxin-related plant response. They were working with the diageotropica (*dgt*) mutation of tomato. Plants homozygous for the *dgt* mutation fail to show normal gravitropism—instead of growing upright, they simply sprawl on the ground. They also show other symptoms indicating that they are unable to respond to their own auxin. The Oregon State workers demonstrated that stems of the *dgt* homozygotes lack a pair of auxin-receptor proteins that are present in normally gravitropic tomatoes and in many other plant species. It seems, then, that these proteins participate in normal auxin responses, and that the absence of these proteins in *dgt* homozygotes accounts for their aberrant growth.

Differentiation and Organ Formation

What, within a plant, signals the different types of cells and organs to form? Much of the research on these questions has been done with cultured tissues. One tissue that is easily grown in culture is pith, the spongy, innermost tissue of a stem. Pith tissue cultures proliferate rapidly but show no differentiation; all the cells are similar, and similarly unspecialized. If a notch is cut in the cultured tissue and a stem tip inserted into the notch, then cells differentiate below the inserted tip. Some of the cells differentiate to form water-conducting cells of the sort found in xylem (Chapter 28). Differentiation also begins if, instead of a stem tip, a mixture of auxin and coconut milk is placed in the notch—coconut milk is a rich source of growth substances. A similar effect can be observed in intact plants. If notches are cut in the stems of *Coleus blumei* plants, interrupting some of the strands of conducting tissues, the strands gradually regenerate from the top side of the cut to the lower—recall

that auxin moves from the tip to the base of a stem. If the leaves above the cut are removed, regeneration is slowed; if, however, the leaves are replaced with an auxin solution, then new conductive tissue regenerates. Auxin and other growth substances signal the formation of specific cell types.

Other work with cultured tissues has helped to clarify which growth substances control organ formation. Undifferentiated cultures of tobacco pith tissue form roots when treated with an appropriate concentration of auxin. Another group of growth substances—the **cytokinins**—causes buds and then shoots to form in such cultures. The pattern of organ formation depends upon the ratio of auxin to cytokinin in the medium. A high proportion of auxin favors roots and a high proportion of cytokinins favors buds, but both processes are most active when both growth substances are present.

CYTOKININS

The cytokinins, which have a variety of effects besides stimulating bud formation, promote cell division in cultured tissues, an activity that led to their discovery.

Discovery of the Cytokinins

For a number of years Folke Skoog, at the University of Wisconsin, had been studying cell division. He and his associate Carlos Miller reasoned that there should be some growth substance that regulated cell division, by analogy with the role of auxin, which regulates cell expansion. They suspected that the hypothetical growth substance should regulate the metabolism of nucleic acids—because cell division requires replication of DNA—and that it might even consist of DNA itself. In 1955 they and other members of Skoog's group were studying the effects of various compounds on the rate of cell division in cultures of carrot root tissue. They took an old bottle of herring sperm DNA (the only DNA on hand) off the shelf and added a bit to the medium in which some of the cultures were growing. They were gratified to observe that these cultures began to proliferate much more rapidly than the controls. Determined to discover what component of the DNA preparation was the active material, they purchased a supply of fresh herring sperm DNA—and were disappointed to find that it was quite inactive. Deciding that the only difference between the two samples of DNA was age, they tried to "age" the new DNA by heating it in a sterilizer. This worked! The heated DNA preparation caused rapid cell division in the carrot cultures. By careful chemical work, they found that a single substance in the old or sterilized DNA preparations was the active material. They named this substance kinetin (Figure 32.5e) and suggested that it might be just one of a family of compounds, which are now called cytokinins.

For several years they and other investigators tried in vain to find kinetin in plant tissues. However, two closely related compounds called zeatin (Figure 32.5f) and isopentenyl adenine *were* found. These two are naturally occurring cytokinins, whereas kinetin may be considered to be a synthetic one.

Other Activities of the Cytokinins

Cytokinins are believed to form primarily in the roots and to move to other parts of the plant. Rapid growth of the tissues is obtained by providing an appropriate combination of auxin and cytokinin in the medium. Cytokinins can cause the germination of certain light-requiring seeds when the seeds are kept in constant darkness. Cytokinins usually inhibit the elongation of stems, but they cause lateral swelling of stems and roots—the fleshy roots of radishes are an extreme example. Cytokinins stimulate lateral buds to grow into branches; thus, the balance between auxin and cytokinin levels controls the "bushiness" of a plant. Cytokinins increase the expansion of cut pieces of leaf tissue, so they may regulate normal leaf expansion. Cytokinins also delay the senescence of leaves. If leaf blades are detached from a plant and floated on water or a nutrient solution, they quickly turn yellow and show other signs of senescence. If instead they are floated on a solution containing a cytokinin, they remain green and senesce much more slowly. Cytokinins apparently regulate the redistribution of biologically active materials from one part of a plant to another. When one of a pair of leaves opposite each other on the stem of a bean plant is treated with a cytokinin, the treated leaf remains dark green and healthy. The untreated leaf opposite it, on the other hand, turns completely yellow and senesces rapidly as a result of its loss of nutrients to the treated leaf.

ETHYLENE

Whereas the cytokinins oppose or delay senescence, there is another growth substance that promotes it. This is the gas **ethylene**, $H_2C{=}CH_2$, which is sometimes called the "senescence hormone" (Figure 32.5d). Ethylene is produced by all parts of the plant and, like all plant growth substances, it exerts a number of effects. Back when streets were lit by gas rather than by electricity, leaves on trees near street lamps abscised earlier than those on trees farther from the lamps. We now know that it was ethylene, a combustion product of the illuminating gas, that caused the abscission. Auxin delays leaf abscission, but ethylene strongly promotes it; thus, a balance of auxin and ethylene controls abscission (Figure 32.18).

32.18 Leaf Abscission
The petiole to the left is about to abscise from the stem. The abscission layer—the black band—will break, allowing the leaf to fall. There is a bud in the axil of the leaf.

As another senescence-related activity, ethylene makes fruit ripen. "One rotten apple spoils the barrel," as they used to say. That rotten apple acts as a rich source of ethylene, which triggers the ripening and subsequent rotting of the others in the barrel. As fruit ripens, it loses chlorophyll and its cell walls break down. Both processes are promoted by ethylene produced in the fruit tissue. Ethylene also causes its own production to be increased. Thus, once ripening begins, more and more ethylene is formed, and, because it is a gas, ethylene diffuses readily throughout the fruit and even to neighboring fruit on the same or other plants.

Another gas, carbon dioxide, antagonizes the effects of ethylene. In consequence, commercial shippers and storers of fruit can precisely control the ripening of their wares. They hasten ripening by adding ethylene to the storage chambers; they postpone ripening by adding carbon dioxide. In fact, this use of ethylene is the single most important use of a plant growth substance in agriculture and commerce.

Ethylene is active at other stages of plant development as well. The stems of young dicot seedlings that have not yet seen light often form an apical hook (Figure 32.19). The apical hook is maintained through an asymmetric production of ethylene gas, which inhibits the elongation of cells on the inner surface of the hook. Ethylene inhibits stem elongation in general, promotes lateral swelling of stems (as do the cytokinins), and causes stems to lose their sensitivity to gravitropic stimulation.

32.19 The Apical Hook of a Dicot
Asymmetric production of ethylene is responsible for the apical hook of this young pea seedling, which was grown in the dark.

Senescence

In some places, autumn is a time of striking change in the colors of the leaves of deciduous trees and shrubs. The display of colors is followed by the falling of the leaves, which have, in effect, passed through "old age" and died (Figure 32.20). Equally dramatic, in a different way, is the aging and death of entire plants, especially when they grow in great fields. Many crop plants grow vigorously throughout most of the year but then, after flowering and setting fruit, die together by the thousands. In both of these cases, we are dealing with the phenomenon of senescence, that is, irreversible, deteriorative changes with aging.

Is senescence simply an undesirable, but unavoidable, fact of life, or does it play a useful role? Leaf senescence and the subsequent abscission are of real importance for the survival of the plant. Leaves senesce and abscise at the end of the growing season, shortly before the onset of the severe conditions of winter. Many species of plants have delicate leaves that could be a liability during a typical winter in the temperate zone—it would be too cold for efficient photosynthesis (and the ground might be frozen, making water unavailable), yet water could still be lost from the stomata in the leaves. Damage from freezing would render the leaf unable to function normally during the next growing season in any case. Before the leaves die and are shed, their proteins are hydrolyzed to yield amino acids, which are then exported from the leaves to the stems—an important form of resource conservation. Thus, controlled leaf abscission costs the plant little and benefits it greatly. In other parts of the world, plants shed their leaves during the harsh dry season and grow them during the wet periods, which are more favorable for growth.

What about the senescence and death of the entire plant that follows flowering and seed setting in some species? This appears to be an adaptation for producing more offspring—by pumping so much energy (food) and so many nutrients into the seeds that the parent essentially starves itself to death.

ABSCISIC ACID

Abscisic acid is another plant growth substance with multiple effects in the living plant (Figure 32.5g). This compound inhibits stem elongation and is generally present in high concentrations in dormant buds and some dormant seeds. As we saw in Chapter 29, it also regulates gas and water vapor exchange between leaves and the atmosphere, through its effects on the stomata in the leaf surface. Abscisic acid is sometimes referred to as the "stress hormone" of plants because of its possible role in maintaining winter dormancy

32.20 Leaf Senescence
Where leaves would be a liability under winter conditions, they senesce and die in autumn. Leaves are senescing in this forest in Michigan; only a skeleton of branches will remain to face the elements.

32.21 Winter Dormancy
As winter approaches, many deciduous plants cease growth, cover their buds with scales, and shed their leaves—all changes that aid survival in harsh conditions. Here a winter-dormant twig is coated with ice.

of buds and because it accumulates when plants are deprived of water.

Bud Dormancy and Abscisic Acid

In the temperate zones, the shoots of perennial plants—those that grow for a number of years—do not grow constantly and in all seasons. At some time in the year, the terminal buds of temperate-zone perennials become inactive, and growth ceases until the next spring. This minimizes damage to the plant during a harsh winter (Figure 32.21). Buds on a typical deciduous tree of the northern temperate zone undergo a number of changes well in advance of winter. These changes include the formation of thickened, overlapping **bud scales** that are covered with wax, which helps waterproof the bud contents—the leaf primordia and the growing point of the stem. The winter bud often contains an insulating material consisting of modified, cottony leaves.

Elsewhere in the plant, there are other changes. Leaves abscise, and the scars produced where leaves were formerly attached to the stems are sealed with a corky material. Lateral growth of the trunk ceases, and the solute concentrations in the transport systems increase, lowering the freezing point of the sap. These are several of the changes that constitute **winter dormancy**. In at least some species, some of these changes appear to be associated with an increased concentration of abscisic acid in the buds. Both the onset and termination of winter dormancy are precisely controlled.

ENVIRONMENTAL CUES

An environmental cue, the length of the night, determines the onset of winter dormancy. As summer wears on, the days get shorter—that is, the nights

become longer. Leaves have a mechanism for measuring the length of the night, as we will see presently. This is a marvelous way to determine the season of the year. If a plant determined the season by the temperature, it could be fooled by a winter warm spell or by unseasonable cold weather in the summer. The length of the night, on the other hand, is determined by Earth's rotation around the sun, and does not vary; plants use this accurate indicator to time several aspects of growth and development.

Length of night is one of several environmental cues detected by plants, or by individual, modular plant parts such as leaves. Light—its presence or absence, its intensity, and its duration—provides various cues. Temperature, too, provides important environmental cues, both by its value at any particular time and by the distribution of warmer and colder stretches over a period of time. The plant "reads" an environmental cue and then "interprets" it, often by stepping up or decreasing production of growth substances.

PHYTOCHROME

Light regulates many aspects of plant development. For example, some seeds will not germinate in darkness but do so readily after even a brief exposure to light. Studies showed that blue and red light are highly effective in promoting germination, whereas green light is not. Far-red light *reverses* the effect of a prior exposure to red light! Far-red is a very deep red, bordering on the limit of human vision and centered upon a wavelength of 730 nm; red wavelengths are around 660 nm. If alternating, brief exposures to red and far-red light are given in close succession, the seeds respond only to the final exposure: If red, they germinate; if far-red, they remain dormant (Figure 32.22). Such red, far-red reversibility is observed in many other aspects of plant development.

The basis for the red and far-red effects resides in a bluish pigment—a protein called **phytochrome**—which is found in plants in two interconvertible forms. The pigment's bluish color results from its absorption of red or far-red light (Figure 32.23). Red light converts phytochrome into one form; far-red converts that form back into the other (Figure 32.24). Light drives the interconversion of the two forms of phytochrome, both in the test tube and in the living plant. The form that absorbs principally red light is called P_r. Upon absorption of a photon of red light, a molecule of P_r is converted into P_{fr}, the far-red absorbing form. P_r is stable in the darkness, but P_{fr} gradually disappears once it has been formed. Some of the P_{fr} converts spontaneously, in the dark, to P_r, but another fraction is destroyed. P_{fr} has a number of important biological effects—as we have just seen, one of them is the initiation of germination in certain

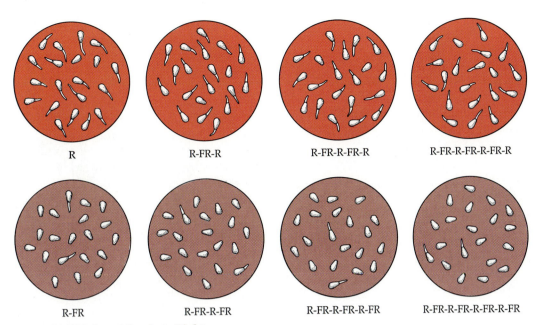

32.22 Sensitivity of Seeds to Light
Lettuce seeds were exposed to alternating periods of red light for 1 minute and far-red light for 4 minutes. Seeds germinated if the final period was red and remained dormant if the final exposure was to far-red. In each case, the final exposure reversed the effect of the preceding exposure to the other wavelength of light.

seeds. We will see examples of other activities of P_{fr} in a moment.

We do not yet know how P_{fr} produces its many effects. Some of a plant's phytochrome may be included in membranes, where it may regulate how ions move into and out of cells and organelles. P_{fr} may function as a channel through which ions move until the conversion to P_r closes the channel. This and other ideas are currently undergoing active investigation.

Phytochrome and Early Seedling Growth

The radicle, the embryonic root, is the first portion of the seedling to escape the seed coat. The shoot emerges later. Frequently, seeds germinate below the soil surface, so the young plant is **etiolated**—kept in darkness. It must reach the surface before its food reserves are expended—it must begin photosynthesis or it will starve and die. Plants have evolved

32.23 Absorption Spectra of Phytochrome
P_r absorbs most strongly at 660 nm, in the red region of the spectrum, whereas P_{fr} absorbs most strongly at 730 nm, in the far-red.

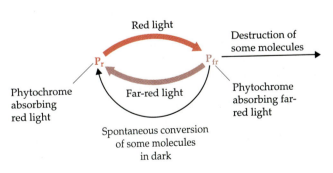

32.24 Behavior of Phytochrome
Phytochrome can exist either as P_r, which absorbs red light, or as P_{fr}, which absorbs far-red light. Each form is converted to the other by light of the appropriate color. In the living plant, some P_{fr} is spontaneously converted to P_r and some P_{fr} is destroyed.

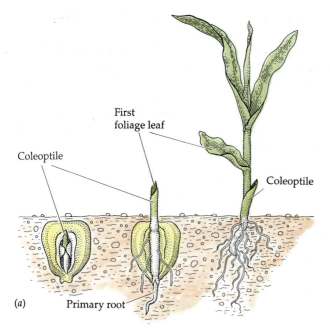

(a)

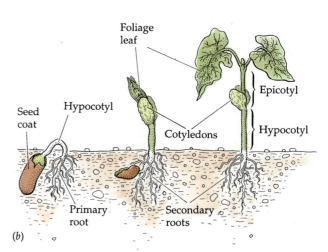

(b)

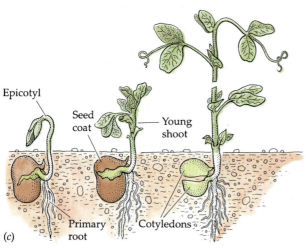

(c)

32.25 Patterns of Early Shoot Development

Shoots are protected as they penetrate the soil during germination. (a) A coleoptile covers the early shoot of corn and other monocots. After the shoot emerges from the soil, it pierces the coleoptile and grows out. (b) The shoot apex of most dicots, such as the bean shown here, is protected by the cotyledons as the upper part of the plant is pulled through the soil by the elongating hypocotyl. When the epicotyl elongates, the first foliage leaves emerge. (c) The cotyledons of other dicots such as peas remain in the soil. The shoot apex is pulled up as the bent epicotyl elongates.

a variety of ways to cope with this problem. For one thing, etiolated flowering plants do not form chlorophyll. They turn green only upon exposure to light. This conserves precious resources, for chlorophyll would be of no use in the darkness. Only when light is available does it "pay" to expend metabolic energy on the production of chlorophyll. An etiolated shoot elongates rapidly. The rapid growth brings the buried shoot to the soil surface, where photosynthesis quickly begins.

Early shoot development varies among the flowering plants (Figure 32.25). In some monocots, such as grasses, the shoot is initially protected by a leaf sheath, the coleoptile. The developing shoot later grows out through the coleoptile. Dicots lack this protective structure. In most dicots, the hypocotyl elongates, and the cotyledons are carried to the surface, where they become the first important photo-

synthetic structures. In other dicots, such as peas, the cotyledons remain below the soil surface, and tissue above them, the epicotyl, grows up through the soil. In all three cases, elongation—whether of the coleoptile, the hypocotyl, or the epicotyl—proceeds much more rapidly in the dark than in the light.

The shoot of etiolated dicot seedlings curves back into a hook at the apex during part of their development, in response to ethylene production (Figure 32.19). The hook serves a protective function, ensuring that the tender apical bud is not forced directly upward to batter its way through the soil. The leaves of an etiolated seedling do not expand. An underground leaf cannot photosynthesize, and it would be a real problem to drag an expanded leaf through the soil.

All these etiolation phenomena are adaptive, and they are regulated by the pigment phytochrome. In a seedling that has never been exposed to light, all the phytochrome is in the red-absorbing (P_r) form. Exposure to light converts P_r to P_{fr}, and the P_{fr} initiates the reversal of the etiolation phenomena we have just described: chlorophyll synthesis begins, shoot elongation slows, the hook at the apex opens, and the leaves start to expand.

PHOTOPERIODIC CONTROL OF FLOWERING

In 1920 W. W. Garner and H. A. Allard of the U.S. Department of Agriculture studied the behavior of a newly discovered mutant tobacco plant. The mutant was named Maryland Mammoth because of its large leaves and exceptional height (and where it was found). The other plants in the field flowered, but the Maryland Mammoth continued to grow. Garner and Allard took cuttings of the Maryland Mammoth into their greenhouse, where they finally flowered in December. Garner and Allard also noticed that some soybean plants all flowered at about the same time, in late summer, even though they had been planted at different times in the spring.

Garner and Allard guessed that both of these observations had something to do with the seasons. They tested a number of likely seasonal variables, such as temperature, but the key proved to be the length of day. By artificially varying the length of day—moving plants between light and dark rooms at different times—they established a direct link be-

tween flowering and day length. The **critical day length** for Maryland Mammoth tobacco proved to be 14 hours (Figure 32.26). The plants did not flower if the light period was longer than 14 hours each day, but flowering commenced after the days became shorter than that. Both soybeans and Maryland Mammoth tobacco are **short-day plants** (SDPs). Others, such as spinach and clover, are **long-day plants** (LDPs), flowering only when the day is *longer* than some critical minimum. Generally LDPs are triggered to flower in midsummer and SDPs in late summer, or sometimes in the spring.

Other Patterns of Photoperiodism

Some plants require more complex signals in order to flower. One group, the short-long-day plants, must first experience short days and then long ones. Accordingly, because they pass first through the short days of early spring and then through ever longer ones, they flower during the long days before midsummer. Another group, the long-short-day plants, cannot flower until the long days of summer have been followed by shorter ones, so that they come into bloom only in the fall. Long-short-day plants will not bloom in the spring, in spite of its short days, nor will a short-long-day plant flower in late summer.

This phenomenon of day-length-controlled processes is called **photoperiodism.** Other effects besides flowering are also under photoperiodic control. We have learned, for example, that the onset of winter dormancy is triggered by short days. (Animals, too, show a variety of photoperiodic behaviors; in aphids, for example, long days favor the development of sexually reproducing females, whereas females that reproduce asexually develop when days are short.) It is important to note that the flowering of some angiosperms is not photoperiodic. In fact, there are more **day-neutral plants** than short-day and long-day plants. Some plants are photoperiodically sensitive only when young and become day-neutral as they grow older. Others require specific combinations of day length and other factors to flower.

Importance of Night Length

It is a historical accident that we use the terms short-day plant and long-day plant—the SDPs could as well have been called long-night plants and the LDPs short-night plants, because the natural day has a fixed length of 24 hours. But the terms SDP and LDP became entrenched before it was learned that plants actually measure the length of *night*, or of darkness. This was demonstrated by Karl Hamner of the University of California at Los Angeles and James Bonner of the California Institute of Technology. Working with cocklebur, an SDP, they ran a series of experi-

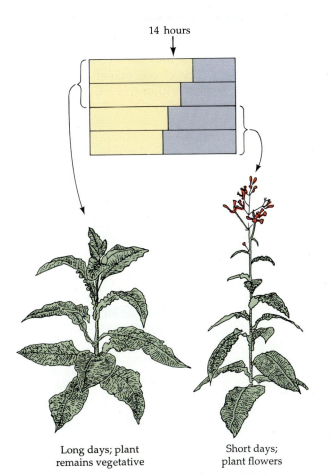

14 hours

Long days; plant remains vegetative

Short days; plant flowers

32.26 Flowering of Maryland Mammoth Tobacco
By varying the length of day, Garner and Allard showed that Maryland Mammoth tobacco flowers only with day lengths shorter than 14 hours—that is, the critical day length is 14 hours.

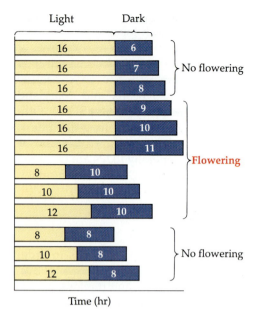

32.27 Night Length and Flowering
In the experiments symbolized by the six upper bars, plants were exposed to 16 hours of light, followed by various dark periods. Only plants given 9 or more hours of dark flowered. In the 6 experiments indicated at the bottom, plants were exposed to various light periods followed by 8 or 10 hours of dark. Only plants given 10 hours of dark flowered. Experiments like these showed that short-day plants are actually long-night plants.

ments in which either (1) the light period was kept constant, either shorter or longer than the critical day length, and the dark period was varied or (2) the dark period was kept constant and the light period was varied (Figure 32.27). The plants flowered under all treatments in which the dark period exceeded 9 hours, regardless of the length of the light period. Thus, it is the night length that matters; for cocklebur, the critical night length is about 9 hours.

In cocklebur, a single long night is enough of a photoperiodic stimulus to trigger full flowering some days later, even if the intervening nights are short ones. Most plants, less sensitive than the cocklebur, require from two to many nights of appropriate length to induce flowering. Plants of some species must experience an appropriate night length *every* night before they can flower—a single shorter night, even one day before flowering would have commenced, inhibits flowering.

Hamner and Bonner showed the role of night measurement by plants in yet another way (Figure 32.28a). Plants, both SDPs and LDPs, were grown on a variety of light regimes. In some regimes, the dark period was interrupted by a brief exposure to light; in others, the light period was interrupted briefly by darkness. Interruptions of the light period by darkness had no effect on the flowering of either short-day or long-day plants. However, even very brief interruptions of the dark period completely nullified

the effect of a long night. An SDP flowered only if the long nights were uninterrupted. An LDP experiencing long nights flowered if these were broken by exposure to light. Thus, there must be some timing mechanism that measures the length of a continuous dark period and uses the result to trigger flowering or to keep the plant vegetative. The nature of the timing mechanism is still unknown in spite of much study.

Phytochrome seems to participate somehow in the photoperiodic timing mechanism. In the interrupted-night experiments, the most effective wavelengths of light were red (Figure 32.28b). The effect of a red-light interruption of the night was fully reversed by a subsequent exposure to far-red light. It was once thought that the timing mechanism might simply be the slow conversion of phytochrome during the night from the P_{fr} form—produced during the light hours —to the P_r form. However, this suggestion is inconsistent with most of the experimental observations and must be wrong. Phytochrome must play a different role. Another important role is played by a biological clock.

Circadian Rhythms and the Biological Clock

It is abundantly clear that organisms have a way of measuring time and that they are well adapted to the 24-hour day–night cycle of our planet. Some sort of biological clock resides within the cells of all eukar-

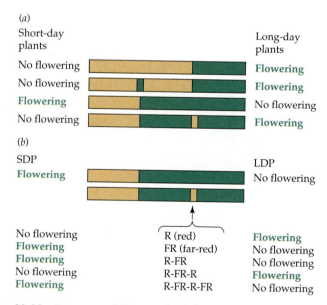

32.28 Interrupted Days and Nights
(a) Short-day plants require long, uninterrupted nights to flower. Long-day plants flower when the night is short; a long night can be shortened by being interrupted by a brief period of light. (b) When red and far-red light are given in alternation, the last treatment determines the effect of the light interruption, suggesting that phytochrome participates in photoperiodic responses.

yotes, and the major outward manifestations of this clock are known as **circadian rhythms** (Latin *circa*, about; *dies*, a day). Plants provide innumerable examples of approximately 24-hour cycles. The leaflets of a plant such as clover or the tropical tree *Albizzia* normally hang down and fold at night and rise and expand during the day. Flowers of a number of plants show similar "sleep movements," closing at night and opening during the day. Under dim light they continue to open and close on an approximately 24-hour cycle (Figure 32.29).

Circadian rhythms of protists, animals, fungi, and plants share some important characteristics. First, the **period** is remarkably insensitive to temperature, although the **amplitude** of the fluctuation may be drastically reduced by lowering the temperature (Figure 32.30 explains these terms). Second, circadian rhythms are highly persistent. They continue even in an environment in which there is no alternation of light and dark. Third, circadian rhythms can be **entrained,** within limits, by light–dark cycles that differ from 24 hours. That is, the period an organism expresses can be made to coincide with that of the light–dark regime. The period in nature is *approximately 24 hours.* If an *Albizzia* tree, for example, were to be placed under artificial light on a day–night cycle totaling *exactly* 24 hours, the rhythm expressed would show a period of exactly 24 hours. If, however, an experimenter used a day–night cycle of, say, 22 hours, then the rhythm would be entrained to a 22-hour period.

If light–dark cycles can entrain circadian rhythms, it follows that light should also be able to shift the

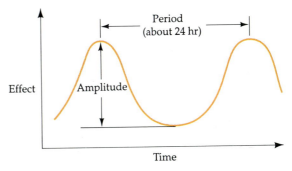

32.30 Features of Circadian Rhythms
Circadian rhythms are characterized on the basis of time, measured in periods of about 24 hours, and on the basis of the magnitude of the rhythmic effect, measured by the cycle's amplitude.

rhythm. If an organism is maintained under constant darkness, with its circadian rhythm expressed on the approximately 24-hour period, a brief exposure to light can make the next peak of activity appear either later or earlier than one would have predicted, depending on when the stimulus is applied. Moreover, the organism does not then return to its old schedule if kept in darkness. If the first peak is delayed by 6 hours, the subsequent peaks are all 6 hours late. Such phase shifts are permanent—unless the system is again upset by further stimuli.

Important questions about circadian rhythms remain to be answered. We do not know, for example, how light acts to reset the biological clock. In fact, we still do not know what the clock is, that is, what its biochemical or biophysical basis may be.

There is now ample evidence that photoperiodic behavior is based on the interaction of night length with the biological clock. However, the manner of coupling of the clock with flowering is unclear. Experiments with the small lawn weed *Chenopodium rubrum* provided one sort of evidence. As a short-day plant, *Chenopodium* will flower in response to a single long night. It shows at least some flowering response to a night as long as 96 hours. A single red flash during this extremely long night can either enhance or inhibit this response, depending upon precisely when it is administered (Figure 32.31). The effect of the red flash oscillates on an approximately daily basis (like the clock), but just how the underlying daily rhythm relates to flowering remains to be learned.

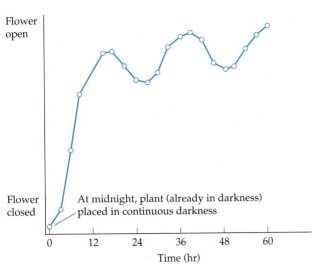

32.29 Sleep Movements of Flowers
Kalanchoe flowers close at night. In the middle of the night (time = 0), when the flowers were completely closed, biologists transferred the plant to a dark box and kept it there for another 60 hours. During that time the flowers continued to open and partially close on a 24-hour cycle.

A Flowering Hormone?

Is the timing device for flowering located in a particular part of an angiosperm, or are all parts able to sense the length of night? As in the Darwins' study of the light receptor for phototropism, this question was resolved by "blindfolding" different parts of the plant. It quickly became apparent that each leaf is

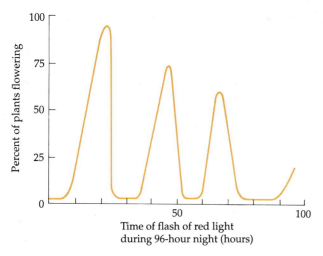

32.31 A Flowering Rhythm
The short-day plant *Chenopodium rubrum* flowers in response to a single 96-hour "night." If single flashes of red light are given during such a 96-hour dark period, they either enhance or inhibit the flowering response. The peaks of enhancement are on about a 24-hour cycle initiated by the light-to-dark transition that started the night.

capable of timing the night. If a short-day plant is kept under a regime of short nights and long days, but one or more leaves are covered so as to give those leaves long nights, the plant will flower (Figure 32.32*a*). This type of experiment works best if only one leaf is left on the plant. In fact, if one leaf is

given a photoperiodic treatment conducive to flowering—an inductive treatment—other leaves kept under noninductive conditions will tend to inhibit flowering.

Although it is the leaves that sense an inductive night period, the flowers form elsewhere on the plant. Thus, some message must be sent from the leaf to the site of flower formation. Three lines of evidence suggest that this message is a chemical substance—a flowering hormone. First, if a photoperiodically induced leaf is removed from the plant shortly after the inductive night period, the plant does not flower. If, however, the induced leaf remains attached for several hours, the plant flowers. This suggests that something—the hypothetical hormone—must be synthesized in the leaf in response to the inductive night, then move out of the leaf to induce flowering.

The second line of evidence for the existence of a flowering hormone comes from grafting experiments. If two cocklebur plants are grafted together, and if one plant is given inductive long nights and its graft partner is given noninductive, short nights, *both* plants flower (Figure 32.32*b*). Grafting experiments also provided the third line of evidence for a flowering hormone. Jan A. D. Zeevaart, a plant physiologist now at Michigan State University, induced a single leaf of the SDP *Perilla* and detached it from the plant. This leaf could then be grafted onto another,

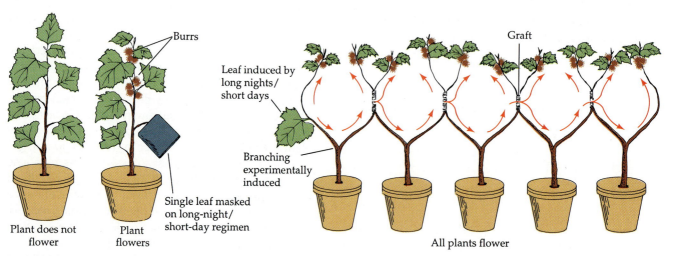

(a) Cockleburs on short nights/long days

(b) Grafted cockleburs on short nights/long days

32.32 Evidence for a Flowering Hormone
(a) Cocklebur, a short-day plant, will not flower if kept under long days and short nights. If even one leaf is masked for part of the day, thus shifting that leaf to short days and long nights, the plant will flower—note the burrs. Because the flowers are formed far from the induced leaf, it is likely that some substance carries the flowering message from the leaf. (b) Five cocklebur plants grafted together and kept under long days and short nights, with most leaves removed. If a leaf on a plant at one end of the chain is subjected to long nights, all of the plants will flower. Arrows indicate the routes of the hypothetical flowering hormone from the induced leaf.

noninduced, *Perilla* plant and cause it to flower. The same leaf could even be grafted onto successive hosts and cause each of them to flower in turn. As long as three months after the induction of the leaf, it could still cause plants to flower.

The Search for Florigen

From experiments such as Zeevaart's, it certainly appears that the photoperiodic induction of a leaf causes a more or less permanent change in it, inducing it to start and to continue producing a flowering hormone that is transported to other parts of the plant, switching those target parts to the reproductive state. So reasonable is this idea that biologists have named the hormone, even though there is yet to be a convincing isolation and characterization of such a compound after decades of active searching. It is called **florigen.** The direct demonstration of florigen activity remains a cherished goal of plant physiologists. Gibberellins regulate the flowering of many species, especially of long-day plants, but these growth substances do not have the properties of florigen.

As a final teaser, we will describe an experiment that suggests that the florigen of short-day plants is identical with that of long-day plants, even though SDPs produce it only under long nights and LDPs only under short nights. An SDP and an LDP were grafted together, and *both* flowered, as long as the photoperiodic conditions were inductive for *one* of the partners. Either the SDP or the LDP could be the one induced, but both would always flower—suggesting the transfer of a flowering-inducing substance, the elusive florigen, from one plant to the other.

VERNALIZATION AND FLOWERING

In both wheat and rye, we distinguish two categories of flowering behavior. Spring wheat, for example, is sown in spring and flowers the same year. It is an annual plant. Winter wheat, on the other hand, is biennial, and must be sown in the fall and go through a winter before flowering in the second year. If winter wheat does not experience a winter cold exposure, it will not flower normally the next year. This was of great agricultural interest in the Soviet Union because winter wheat is a better producer than spring wheat, but the winters in parts of the USSR are too cold for its survival. A number of studies were performed in Russia during the first two decades of this century. These studies demonstrated that if seeds of winter wheat were premoistened and prechilled, they would develop and flower normally when sown in the

spring. Thus, high-yielding winter wheat could be grown even in previously hostile regions. This phenomenon—the induction of flowering by wintry temperatures—is called **vernalization.**

Vernalization may require as many as 50 days of low temperature (in the range from about -2 to $+12°$ C). Some plant species require both vernalization and long days to flower. There is a long wait from the cold days of winter to the long days of summer, but as the vernalized state easily lasts 200 days and more, these plants do flower once the appropriate night length is experienced. Thus, vernalization, once accomplished, is a stable condition.

SUMMARY

The pigment phytochrome and several growth substances help control plant development. Each of the growth substances produces a variety of responses. Abscisic acid has been dubbed the "stress hormone" and ethylene the "senescence hormone." Cytokinins appear to be important from seed germination to senescence. Auxin and the gibberellins have effects so diverse that one wonders whether a single molecular mechanism can account for all the effects of either. Phytochrome exists in two forms, P_r and P_{fr}, that are interconverted by red and far-red light. P_{fr} triggers many developmental events. Normal development depends on a proper dynamic balance of all of these substances.

If a seed germinates underground, the seedling remains etiolated and grows rapidly until it reaches the surface. A growing plant orients itself with respect to light and gravity by the processes of phototropism and gravitropism. Some species show strong apical dominance, by which the main shoot apex prevents the growth of lateral shoots.

Some perennial plants undergo winter dormancy, triggered by the increasing length of the night, sensed by the leaves, and perhaps mediated by abscisic acid. The breaking of dormancy generally requires changes in growth substance levels following winter chilling of individual buds. Leaf abscission, a type of senescence, protects the plant, removing potential sites of water loss or other damage.

Some plants are photoperiodic in their flowering behavior, being short-day or long-day plants or sometimes having more complex requirements, but most are day-neutral. The mechanism of photoperiodic control appears to include a biological clock and phytochrome. Plants measure the nightly dark period. In some species, vernalization is required for the plants to flower. There is evidence for a flowering hormone, florigen; but that substance has yet to be convincingly isolated from any plant.

SELF-QUIZ

1. The gibberellins:
   a. are responsible for phototropism and gravitropism.
   b. are gases at room temperature.
   c. are produced only by fungi.
   d. cause bolting in some biennial plants.
   e. inhibit the synthesis of digestive enzymes by barley seeds.

2. In coleoptile tissue, the transport of auxin:
   a. is from base to tip.
   b. is from tip to base.
   c. is by simple diffusion, with no preferred direction.
   d. depends on the orientation of the coleoptile with respect to gravity.
   e. does not occur, because auxin is used where it is made.

3. Which process is *not* directly affected by auxin?
   a. Apical dominance
   b. Leaf abscission
   c. Synthesis of digestive enzymes by barley seeds
   d. Root initiation
   e. Parthenocarpic fruit development

4. Plant cell walls:
   a. are strengthened primarily by proteins.
   b. often make up over 90 percent of the total volume of an expanded cell.
   c. can be loosened by an increase in pH.
   d. gets thinner and thinner as the cell grows longer and longer.

   e. are made more plastic by treatment with auxin.

5. Which statement is *not* true of the cytokinins?
   a. They promote bud formation in tissue cultures.
   b. They delay the senescence of leaves.
   c. They usually promote the elongation of stems.
   d. They cause certain light-requiring seeds to germinate in the dark.
   e. They stimulate the development of branches from lateral buds.

6. Ethylene:
   a. is antagonized by carbon dioxide.
   b. is liquid at room temperature.
   c. delays the ripening of fruits.
   d. generally promotes stem elongation.
   e. inhibits the swelling of stems, in opposition to cytokinin effects.

7. Phytochrome:
   a. is a nucleic acid.
   b. exists in two forms interconvertible by light.
   c. is a red or far-red colored pigment.
   d. is sometimes called the "stress hormone."
   e. is the photoreceptor for phototropism.

8. Which statement is *not* true of photoperiodism?
   a. It is related somehow to the biological clock.

   b. Phytochrome plays some role in the timing process.
   c. It is based on measurement of the length of the night.
   d. Most plant species are day-neutral.
   e. It is limited to the plant kingdom.

9. Although florigen has never been isolated, we think that it may exist because:
   a. night length is measured in the leaves, but flowering occurs elsewhere.
   b. it is produced in the roots and transported to the shoot system.
   c. it is produced in the coleoptile tip and transported to the base.
   d. we think that gibberellin and florigen are the same compound.
   e. it may be activated by prolonged (more than a month) chilling.

10. Which statement is *not* true of vernalization?
    a. It may require more than a month of low temperature.
    b. The vernalized state generally lasts for about a week.
    c. Vernalization makes it possible to get two winter wheat crops each year.
    d. Is accomplished by subjecting moistened seeds to chilling.
    e. It was of interest to Russian scientists because of the climate in the USSR.

FOR STUDY

1. Corn stunt virus causes a great reduction in the growth rate of infected corn plants, so the diseased plants take on a dwarfed form. Since their appearance reminds you of the genetically dwarfed corn studied by Phinney, you suspect that the virus may inhibit the synthesis of gibberellins by the corn plants. Describe two experiments you might conduct to test this hypothesis, only one of which should require chemical measurement.

2. While relatively low concentrations of auxin promote the elongation of segments cut from young plant

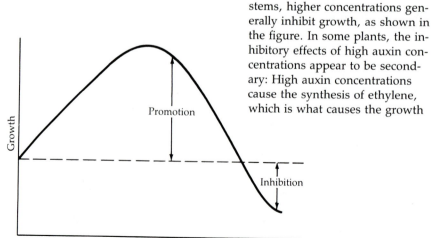

stems, higher concentrations generally inhibit growth, as shown in the figure. In some plants, the inhibitory effects of high auxin concentrations appear to be secondary: High auxin concentrations cause the synthesis of ethylene, which is what causes the growth inhibition. Cobalt ions inhibit ethylene synthesis. If cobalt ions were added to the solutions in which the stem segments grew, how do you think that would affect the appearance of the graph?

3. When carbon dioxide is applied to plants, it has effects opposite to those of ethylene. Carbon dioxide is sometimes added to the atmosphere around fruit being shipped to other parts of the country. What do you suppose is the purpose of this procedure?

4. Poinsettias are popular ornamental plants that typically bloom just before Christmas. Their flowering is photoperiodically controlled. Are they long-day or short-day plants? Explain.

5. You plan to induce the flowering of a crop of long-day plants in the field by using artificial light. Is it necessary to keep the lights on continuously from sundown until the critical night length is reached? Explain.

READINGS

Cleland, C. E. 1978. "The Flowering Enigma." *BioScience*, April. Florigen eluded physiologists when this article was published, and it still does.

Evans, M. L., R. Moore and K.-H. Hasenstein. 1986. "How Roots Respond to Gravity." *Scientific American*, December. Classical and modern experimentation on the mechanisms of gravitropism.

Mandoli, D. F. and W. R. Briggs. 1984. "Fiber Optics in Plants." *Scientific American*, August. How plants guide light to regions of high phytochrome concentration. Includes an excellent description of the light environment in a wheat field.

Moses, P. B. and N.-H. Chua. 1988. "Light Switches for Plant Genes." *Scientific American*, April 1988. How is light absorption by phytochrome transduced into developmental effects? Some stretches of DNA respond to phytochrome by turning on specific genes.

Raven, P. H., R. F. Evert and S. Eichhorn. 1986. *Biology of Plants*, 4th Edition. Worth, New York. A well-balanced general botany textbook.

Salisbury, F. B. and C. W. Ross. 1985. *Plant Physiology*, 3rd Edition. Wadsworth, Belmont, CA. An authoritative textbook with excellent chapters on growth substances and development.

PART SIX

The Biology of Animals

33

Physiology, Homeostasis, and the Regulation of Body Temperature

PREVIEW: Physiology is the study of how animals work and how organ systems maintain a relatively constant internal environment. An important focus in physiology is the way organ systems control and regulate conditions in the body. The regulation of body temperature is an example of this process. Temperature has strong effects on biological functions. Animals can either compensate for the effects of temperature or regulate body temperature at a constant level. Animals control their body temperature by altering heat exchange with the environment and by altering metabolic heat production. A part of the vertebrate brain acts as a thermostat. Regulated body temperature can be raised to produce fever or lowered to conserve energy, as in hibernation.

This chapter deals with organs and organ systems, homeostasis, body temperature, metabolic rate, metabolic compensation, homeothermy, poikilothermy, endothermy, ectothermy, thermoregulation, heat production, heat loss, fever, torpor, and hibernation.

Many of the places animals live and many of the things that animals do are truly amazing. Animals live in the most extreme environments on our planet. Consider emperor penguins that walk over 100 kilometers across Antarctic ice in the winter season to breed in just about the coldest and most inhospitable place on Earth (Figure 33.1). After the female lays an egg, she walks back to the sea to feed; the male incubates the egg, then protects and feeds the chick until she returns with her body fat replenished. The female then takes over the feeding of the chick, and the male walks back to the sea, having fasted for over 4 months. You will learn about many incredible adaptations of animals in this section; for example, those that allow birds to fly at altitudes at which humans cannot exist without technological assistance, seals to remain underwater for over half an hour, mice to live without water in the hottest deserts, and bats to capture flying insects in total darkness. The explanations of how animals achieve such feats are found in physiology, the study of how animals work—the subject of this section. Unusual adaptations are not just the spice of physiology; they are extensions of basic physiological mechanisms and therefore help us to understand the principles of normal physiological functions in humans and other animals.

In earlier sections of this book we learned how cells work. Animals, of course, consist of cells, so many considerations of how cells function, including their energy metabolism, their needs for nutrients, their production of waste products such as carbon dioxide, and their osmotic balance can be extended to whole animals. For the simplest animals, sponges and cnidarians, this extension is fairly straightforward, since most of those species live in the sea and their bodies are only two cell layers thick. Seawater contains nutrients, it has a suitable composition of salts, and it provides a stable physical environment. Each cell of a sponge or a jellyfish is in direct contact with the environment and practically functions as an autonomous unit. It receives its nutrients directly from and releases its wastes directly into the seawater. The support of cell functions in aquatic animals more complex than sponges and cnidarians is complicated by the fact that most of their cells are not in direct contact with the external environment. The cells of terrestrial animals cannot be exposed directly to their external environment, air, because it would

33.1 A Stroll Across the Ice
These emperor penguins are migrating across Antarctic sea ice to their breeding colony. After they lay their eggs, the females walk back to the sea to feed. The males care for the chicks and fast for about four months, until they are relieved by their mates. Then the males walk back to the sea to feed. The distance may be over 100 kilometers.

dry them out and kill them. For those reasons, most cells of most animals are served by an internal environment consisting of extracellular fluids. That internal environment provides appropriate physical conditions for the cells of the animal, supplies all the nutrients they need, and removes all their wastes.

HOMEOSTASIS

Most of physiology focuses on how the internal environment is maintained in a condition that enables it to serve the needs of the cells. The internal environment is obviously not as vast as the sea. Its nutrient content can be rapidly exhausted, and its physical conditions are altered by the metabolic activities of the very cells it serves. Organs and organ systems of animals function to keep various aspects of the internal environment at a steady state, a condition called **homeostasis**. Gas-exchange organs provide oxygen to and remove carbon dioxide from the extracellular fluids, digestive organs supply nutrients, and excretory systems eliminate wastes.

Homeostasis is an essential feature of complex animals, and it has made it possible for animals to adapt to nearly every environment on Earth. If an organ fails to function properly, homeostasis of the internal environment is lost, and as a result cells get sick and die. The afflicted cells are not just those of the organ

that functions improperly, but the cells of all other organs as well. Loss of homeostasis is therefore a serious problem that makes itself worse. To avoid loss of homeostasis, the activities of organs must be controlled and regulated in response to changes in the internal environment. Control and regulation require information; hence, the organ systems of information—the nervous system and the endocrine system—must be included in our discussions of every physiological function. For that reason, we treat the endocrine and nervous systems at the beginning of this section of the book. Subsequent chapters deal with the organ systems that are responsible for homeostasis of various aspects of the internal environment.

ORGANS AND ORGAN SYSTEMS

The diversity of adaptations that enable animals to live in just about any environment creates a bewildering number of details that can make the study of physiology seem daunting rather than fascinating. Therefore, it is useful to begin our study with a road map of the organs, organ systems, and physiological functions of at least one species. That species might as well be *Homo sapiens*, but the road map generated will also apply to most other vertebrates and occasionally to invertebrates as well.

The Structure of Organs

Organs are made of tissues, and a tissue consists of cells with similar structure and function. Biologists who study cells and tissues recognize many types of cells, but group them into only four general types of tissues: epithelial, connective, muscle, and nervous. Epithelial tissues are usually sheets of tightly connected cells like those that cover the body surface and those that line various hollow organs of the body such as the digestive tract or the lungs. Some epithelial cells have secretory functions, for example, those that secrete mucus, digestive enzymes, or sweat. Other epithelial cells have cilia and assist in the movement of substances over surfaces or through tubes. Since epithelial cells create boundaries between the inside and the outside of the body and between body compartments, they frequently have absorptive and transport functions. An epithelium can be stratified, like the skin, which consists of many layers of cells, or it can be simple, like the lining of the gut which consists of a single layer of cells. We'll encounter epithelial tissues in our discussions of the linings and tubules of reproductive systems (Chapter 35), the linings of gas-exchange systems (Chapter 39), the linings of digestive tracts (Chapter 41), and the tubules of excretory systems (Chapter 42).

Connective tissues support and reinforce other tissues. Unlike epithelial tissues, which consist of densely packed, tightly connected populations of cells, most connective tissues consist of a dispersed population of cells embedded in an extracellular matrix. The properties of the matrix differ in different types of connective tissues. The connective tissue in skin contains lots of elastic fibers that can be stretched and then return to their original position. The connective tissues that connect muscles to bone and bones to one another have lots of collagen fibers with high tensile strength. Bone is a connective tissue in which the extracellular matrix has been hardened by mineral deposition. We will learn about bone and associated connective tissues in Chapter 38. Two other major types of connective tissue are adipose tissue (fat cells) and the cellular components of the blood, which we will discuss in Chapter 40.

Muscle tissue consists of cells that can contract and therefore cause movements of organs, limbs, or just about any part of the body. Muscles are the most important effectors of the body; they enable it to do things. We'll discuss the three types of muscles, their mechanisms of contraction, and their control by the nervous system in Chapter 38. But since muscle tissues play important roles in most organs and organ systems, we'll encounter them in many places in this section.

Nervous tissue enables animals to deal with information. The cells of nervous tissue, neurons, are extremely diverse. Some respond to specific types of stimuli, such as light, sound, pressure, or certain molecules, by generating electrical signals in their membranes. These electrical signals can be conducted via long cell processes to other parts of the body, where they are passed on to other neurons, muscle cells, or secretory cells. Nervous tissue is the subject of Chapters 36, 37, and 38, but since neurons are involved in controlling the activities of most organ systems, they will be mentioned in other chapters as well.

Organs are usually made up of more than one tissue type, and most organs include all tissue types. For example, a section of the gut, the organ of digestion and absorption of nutrients, is lined with a single layer of columnar epithelial cells (Figure 33.2). Some secrete mucus or enzymes and others function mainly to absorb nutrients. Beneath the gut lining is connective tissue, within which there are glands and blood vessels. Concentric layers of muscle tissue are responsible for moving food through the gut and mixing it with the secretions of the epithelial cells. Neurons run between the layers of other tissues to control both the secretions and the movements of the gut.

An individual organ such as the gut is frequently organized with other organs that have complementary functions into an organ system. The major organ systems of the body are outlined in Figures 33.3 through 33.8.

The Organ Systems for Information and Control

The principal organ system that processes information and uses that information to control the physiology and behavior of the animal is the nervous system (Figure 33.3a). It consists of the brain and spinal cord (the central nervous system) along with peripheral nerves that conduct electrical signals from sensors to the central nervous system and conduct signals from the central nervous system to effectors, which are either muscle tissue or secretory tissue. The sensors of the nervous system are quite diverse, but include eyes, ears, organs of taste and smell, and cells sensitive to temperature, touch, pressure, stretch, and pain.

The endocrine system (Figure 33.3b) also processes information and controls the functions of organs, but its messages are distributed in the blood to the entire body as chemical signals called hormones. This is in contrast to the electrical signals of the nervous system, which are routed to specific targets. The principal organs of the endocrine system are ductless glands that secrete specific hormones into the blood. In addition, many other tissues contain individual cells that secrete hormones. There are strong interactions between the nervous system and the endo-

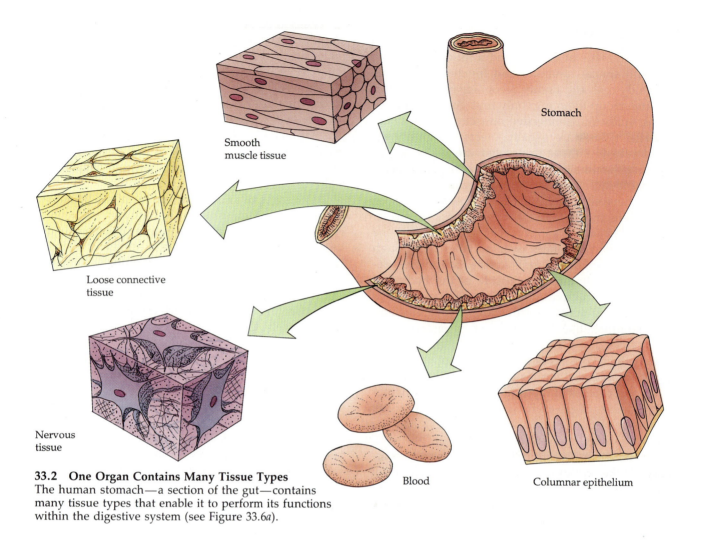

33.2 One Organ Contains Many Tissue Types
The human stomach—a section of the gut—contains many tissue types that enable it to perform its functions within the digestive system (see Figure 33.6*a*).

crine system. Cells in the brain produce hormones that control parts of the endocrine system. In turn, there are cells in the brain that respond to the chemical messages produced by endocrine glands.

The Organ Systems for Protection, Support, and Movement

The largest organ of the body is the skin, along with its special elaborations—hair and nails (Figure 33.4*a*). The skin protects the body from organisms that carry disease, from the physical environment, and from excessive loss of water. The skin contains nervous tissue that is sensitive to various stimuli, so it is a major sense organ. The skin is also an effector organ in that it helps to control body temperature by being a route of heat exchange with the environment.

The skeletal system supports and protects the body (Figure 33.4*b*). In addition, the skeleton is an important effector in that it forms the supports and the levers that muscles pull on to cause behavior.

The muscle system (Figure 33.4*c*) includes the skeletal muscles that are under our conscious control and

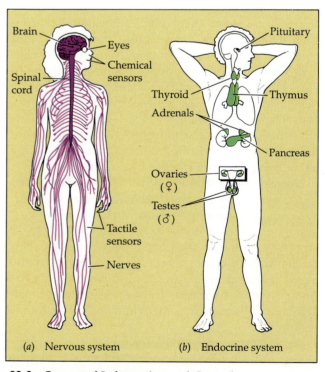

33.3 Organs of Information and Control

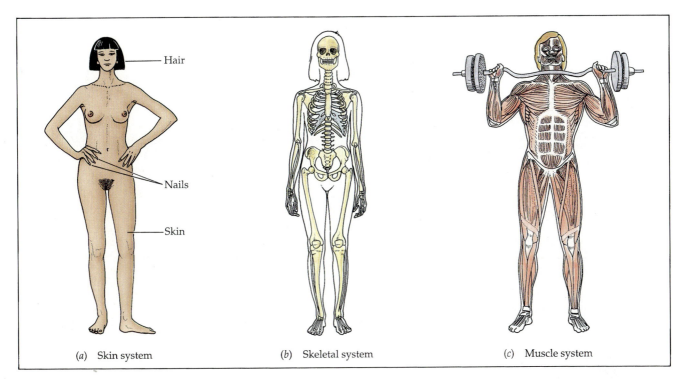

(a) Skin system (b) Skeletal system (c) Muscle system

33.4 Organs of Protection, Support, and Movement

cause all voluntary movements, the muscles of the internal organ systems that are not under our conscious control, and the muscles that constitute the heart.

The Organs of Reproduction

The male and female reproductive systems consist of gonads (testes and ovaries), which produce sex cells. In addition, there are organs for delivering the sex cells to the site where fertilization takes place (Figure 33.5). The female reproductive system includes the uterus, the organ that supports the development of the embryo. The female's mammary glands provide nutrients for the infant. In the gonads and uterus are tissues that secrete hormones that play roles in sexual development and reproduction.

The Organs of Nutrition

The digestive system is largely a continuous tubular structure that extends from mouth to anus (Figure 33.6a). This tube, also called the gut, is divided into different segments that serve different functions in the processing and digestion of food and the absorption of nutrients. Glands associated with the gut deliver into it digestive enzymes and other molecules that play roles in the breakdown of complex food molecules. The lower gut stores and eliminates solid wastes.

The gas-exchange system, also called the respiratory system, provides oxygen, which, in addition to fuel molecules from food is essential for cellular respiration (Figure 33.6b). Carbon dioxide, a waste product of cellular respiration, is eliminated by the gas-

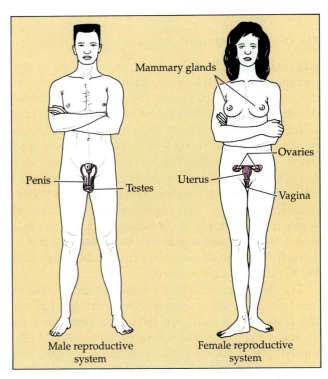

Male reproductive system Female reproductive system

33.5 Organs of Reproduction

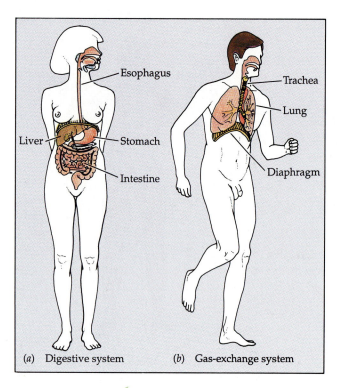

(a) Digestive system   (b) Gas-exchange system

33.6 Organs of Nutrition

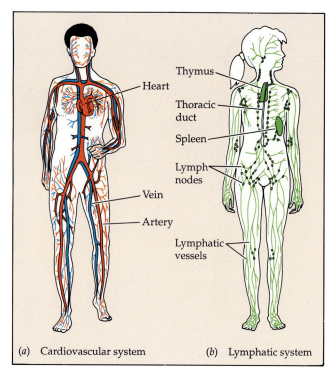

(a) Cardiovascular system   (b) Lymphatic system

33.7 Organs of Transport

exchange system. The gas-exchange organs of humans are lungs, which consist of a system of progressively dividing airways leading to tiny but numerous membranous gas-exchange surfaces that have a very large combined surface area. The muscles used to move air into and out of the lungs are another component of the gas-exchange system.

The Organs of Transport

Gas exchange in the lungs must be supported by the transport of oxygen from the gas-exchange organs to the tissues of the body and the transport of carbon dioxide from the tissues to the lungs. This is accomplished by the circulatory system, which includes a pump (the heart), a system of blood vessels, and blood (Figure 33.7a). The circulatory system also transports nutrients from the gut, delivers nitrogenous wastes to the excretory system, transports hormones, transports heat, and generates mechanical forces. Blood is made up of cellular components in a liquid medium called plasma. The blood plasma is virtually continuous with the extracellular fluids that are the internal environment.

The lymphatic system is another transport system consisting of a set of vessels that ramify throughout the body, but it does not include a pump and its vessels do not form a complete circuit (Figure 33.7b). The lymphatic system picks up extracellular fluid and eventually delivers it back into the blood circulatory system.

The Organs of Excretion

The kidneys are the site of urine formation. Urine includes nitrogenous wastes from the metabolism of proteins and nucleic acids as well as excess salts and some other substances that the body excretes. The kidneys play crucial roles in maintaining the correct water content of the body and the correct salt composition of the extracellular fluids. Urine is delivered to a bladder for storage until it is released to the exterior through the urethra (Figure 33.8).

In the chapters that follow we will explore the various physiological functions listed above and the organ systems that accomplish them. In the remainder of this chapter, however, we will discuss general principles of homeostasis that can be applied to considerations of all of animal physiology. We'll illustrate those principles by treating the example of the regulation of body temperature in detail. Temperature is an important physical parameter of the internal environment. It can be perturbed by the activities of cells and by changes in the outside environment. Animals have evolved a number of adaptations for dealing with changes in temperature.

GENERAL PRINCIPLES OF HOMEOSTASIS

Homeostasis is the word used to describe a constant state of internal conditions in the body. Homeostasis depends on the functions of the various organs and

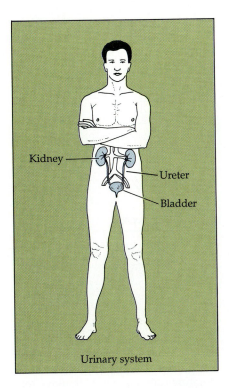

33.8 Organs of Excretion and Water/Salt Balance

organ systems of the body; these functions must be controlled and regulated to achieve a relatively constant internal environment. The terms *control* and *regulation* might seem interchangeable, but their meanings differ. Control implies the ability to *change* the rate of a reaction or process. Regulation is the more sophisticated and more specific physiological concept; it refers to *maintaining* a variable within specific levels or limits. An analogy is that you control the speed of your car by using the accelerator and brake, but you regulate the speed of your car when you use the accelerator and brake to maintain a particular speed.

Set Points and Feedback

Regulation requires, in addition to control mechanisms, the ability to obtain and use information. You can regulate the speed of a car only if you know the speed at which you are traveling and the speed you wish to maintain (Figure 33.9). The desired speed is a **set point** and the reading on your speedometer is feedback information. If the set point and the feedback information are compared, any difference is an

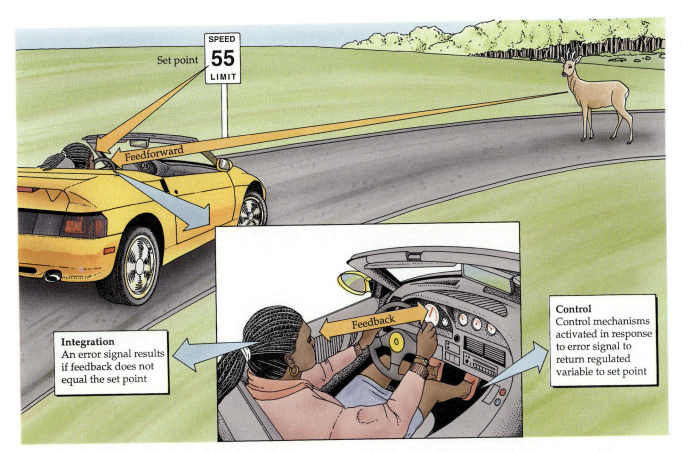

33.9 Control, Regulation, and Feedback
As you drive a car, the posted speed limit is your set point and the speedometer gives you feedback information. Comparing the speed limit to the speedometer reading gives you error signals that you convert into corrective actions by using the brakes and the accelerator to regulate the car's speed. The sight of a deer in the road ahead is feedforward information.

error signal. Error signals suggest corrective actions, which you make by using the accelerator or brake.

Understanding physiological regulation requires knowledge not only of the mechanisms of action of the molecules, cells, tissues, organs, and organ systems—**the controlled systems**—but also knowledge of how relevant information is obtained, processed, integrated, and converted into commands by the regulatory systems. A fundamental way to analyze a regulatory system is to identify its source of feedback information. Feedback information is most often used in regulatory systems as **negative feedback**, so called because it is used to reduce or reverse change. In our car analogy, recognition that you are over the speed limit is negative feedback if it causes you to slow down. Conversely, if you recognize that you are slowing down while going up a hill, that information is negative feedback if it causes you to step harder on the accelerator. Another analogy will illustrate this point.

Regulatory Systems

The classic analogy for regulatory systems is a thermostat—a relevant analogy in that attention will be paid in this chapter to the biological thermostats of vertebrate animals. The heating–cooling system thermostat on the wall of a room is a regulatory system. It has set points that you can adjust, and it receives feedback information from a sensor. The circuitry of the thermostat converts any differences between the set points and the sensor into signals that activate the controlled systems—the furnace and the air conditioner. When room temperature rises above the high set point, the air conditioner is activated, thus reducing room temperature below the set point. When room temperature falls below the low set point, the furnace is activated, thus raising room temperature up toward that set point. Hence the sensor of room temperature is providing information that is used as negative feedback (Figure 33.10).

Negative feedback makes good sense for physiological regulatory systems, so you may wonder if there is any such thing as **positive feedback** in physiology. It is not as common as negative feedback, but it does exist. Rather than returning a system to a set point, it is used to amplify a response. One example is sexual behavior, in which a little stimulation can cause more behavior, which causes more stimulation, and so on. Positive feedback is not used by regulatory systems that maintain stability!

Feedforward information is another feature of regulatory systems. The function of feedforward information is to change the set point. If you wanted to regulate the temperature of your house at a lower level at night than during the day, you could add a clock to the thermostat to provide feedforward information about time of day. These general considera-

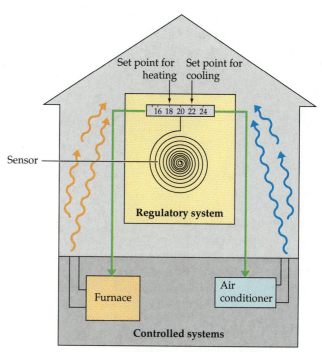

33.10   A Thermostat Regulates House Temperature
Change in room temperature causes a sensor to move relative to set points on the thermostat and activates the furnace or the air conditioner. Room temperature, as detected by the sensor, is feedback to the regulatory system.

tions about control and regulation and regulatory systems help to organize our thinking about physiological systems, but the physiological systems can be far more complex than the thermostat analogy. In some systems we do not even know the nature of the feedback information. For most people, body weight is regulated, although it might not be at the level we prefer. Without consciously counting calories, the brain controls hunger so that food intake matches energy expenditure. We do not understand what information the brain uses to achieve this remarkable feat of regulation, but there are some interesting hypotheses and active research on the problem.

One regulatory system that we do understand well is the system that regulates body temperature. Let's begin by asking, Why do organisms need to thermoregulate? What is the effect of temperature on living systems?

THE EFFECTS OF TEMPERATURE ON LIVING SYSTEMS

Over the face of the Earth, temperatures vary enormously. Think of boiling hot springs and the frigid Antarctic plateau. Heat always moves from a warmer object to a cooler object, so any change in the temperature of the environment will cause a change in the temperature of an organism in that environment unless the organism does something to regulate its temperature. Living cells are restricted to a rather narrow range of temperatures. If cells cool to below 0°C, they are in danger of having ice crystals form within them; this can fatally damage their structures. There are adaptations that prevent freezing and others that permit cells to survive freezing, but generally cells must remain above 0°C to be viable. The upper temperature limit is around 45°C for most cells. Some very specialized algae can grow in hot springs at 70°C, and some bacteria can live at near 100°C, but in general, proteins begin to denature as temperatures rise above 45°C. As proteins denature, they lose their functional properties. Cellular functions are mostly limited to the range of temperatures between 0 and 45°C, which can be taken as the thermal limits for life.

The Q_{10} Concept

Even within the range of 0 to 45°C, changes in temperature can create problems for living systems. Most physiological processes, like the biochemical reactions of which they are made up, are temperature sensitive, going faster at higher temperatures (see Figure 6.29). A concept that describes the temperature sensitivity of a reaction or process is the Q_{10}, which is simply a quotient calculated by dividing the rate of a process or reaction at a certain temperature, R_T, by the rate of that process or reaction at a temperature 10° lower, R_{T-10}.

$$Q_{10} = R_T/R_{T-10}$$

The Q_{10} can be measured for a simple enzymatic reaction or for a complex physiological phenomenon such as gill ventilation in a fish. Most biological Q_{10}'s are between 2 and 3, which means the reaction rates double or triple as the temperature increases by 10°C (Figure 33.11). If a reaction or process is not temperature sensitive, it has a Q_{10} of 1.

Changes in temperature can be particularly disruptive to living systems because all the component reactions in a living system do not have exactly the same Q_{10}. When individual reactions with different Q_{10}'s are linked together in the complex networks that carry out physiological processes, changes in temperature disrupt the balance and integration of the functioning of the organism. To maintain homeostasis, organisms must be able to compensate for or prevent changes in temperature.

Metabolic Compensation

The body temperatures of some animals are tightly coupled to the temperature of the environment. Think of a fish in a pond in a highly seasonal envi-

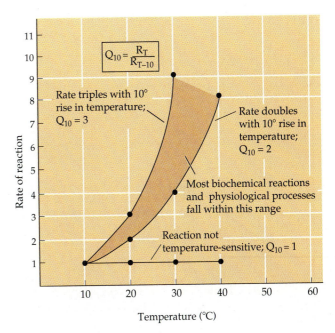

33.11 Q_{10} and Reaction Rate
The larger the Q_{10}, the faster the reaction rate rises as the temperature rises.

ronment. As the temperature of the water in that pond changes from 4°C in midwinter to 24°C in midsummer, the body temperature of the fish will do the same. If we bring such a fish into the laboratory in the summer and measure its metabolic rate (the sum total of the energy turnover of its cells) at different water temperatures, we might calculate a Q_{10} of 2 and plot our data as shown by the red line in Figure 33.12. We predict from our graph that in the winter, when the temperature is 4°C, the fish's metabolic rate will be only one-fourth of what it was in the summer. We then return the fish to its pond. When we bring the fish back to the laboratory in the winter and repeat the measurements, we find, as plotted by the blue line on the graph, that its metabolic rate at 4°C is not as low as we predicted, but is almost the same as it was at 24°C in the summer. In fact, if we repeat the measurement over a range of temperatures, we now find that the fish's metabolic rate is always higher than the rate we measured at the same temperature in the summer-acclimatized fish. **Acclimatization** is the process of physiological and biochemical change an animal undergoes in response to seasonal changes in climate.

The reason for the difference between our prediction from the summer data and our measurements on the winter fish is that seasonal acclimatization in the fish has produced **metabolic compensation**. Metabolic compensation readjusts the biochemical machinery to counter the effects of temperature. What might account for such a change? Look back again at Figure 6.29. Its caption suggests a hypothesis that might be investigated. If fish of the species we are

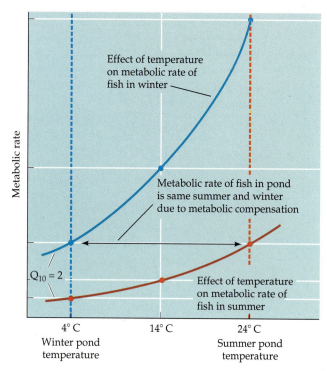

33.12 Metabolic Compensation for Seasonal Differences

When the metabolic rate of a fish is measured both in summer and in winter, it is shown to be temperature-sensitive. Yet at normal environmental temperatures in nature, the fish may have the same metabolic rate summer and winter. Metabolic compensation acclimatizes a fish to its changing environment.

studying have duplicate enzymes with differing optimal temperatures, metabolic compensation might be brought about by their catalyzing reactions with one set of enzymes in the summer and with another set in the winter. The end result of such readjustment is that metabolic functions can be much less sensitive to seasonal changes in temperature than they are to shorter-term thermal fluctuations.

THERMOREGULATORY ADAPTATIONS

Organisms have evolved numerous behavioral and physiological mechanisms for maintaining optimal body temperatures. A discussion of these thermoregulatory adaptations is best organized by a thermoregulatory classification of the animals that possess them. Animals are commonly described as cold-blooded or warm-blooded. Taken at face value, these terms could lead to some rather ridiculous errors. Someone might classify desert reptiles as warm-blooded during the day and cold-blooded at night, many insects might be classified warm-blooded during flight and cold-blooded at rest, and hibernating mammals would be very cold-blooded during most of the winter.

Biologists prefer a different set of terms. **Homeotherm** refers to an animal that regulates its body temperature at a rather constant level, and **poikilotherm** to an animal whose temperature changes. This system of classification says something about the biology of the animals, but it also presents problems. Should a fish in the deep ocean where the temperature changes very little be called a homeotherm, and should a hibernating mammal that allows its body temperature to drop to nearly the temperature of its environment be called a poikilotherm? The problem posed by the hibernator has been set aside by coining a third category, the **heterotherm**—an animal that regulates its body temperature at a constant level some of the time. Homeotherm, poikilotherm, and heterotherm are useful descriptive terms.

Another set of terms classifies animals on the basis of thermoregulatory mechanisms. **Ectotherms** are animals that largely depend on external sources of heat, such as solar radiation, to maintain their body temperatures above the environmental temperature. **Endotherms** are animals that can mobilize substantial metabolic sources of heat production as well as active mechanisms of heat loss to regulate body temperature. Mammals and birds behave as endotherms, whereas animals of all other species behave as ectotherms most of the time.

Laboratory Studies of Ectotherms and Endotherms

Let us choose a small lizard to represent ectotherms and a mouse of the same body size as the lizard to represent endotherms. In the laboratory we put each animal in a small metabolism chamber that enables us to measure the body temperature of the animal and its metabolic rate as we change the temperature of the chamber from 0 to 35°C. The results obtained from the two species are very different (Figure 33.13). As graph a shows, the body temperature of the lizard always equilibrates with that of the chamber, whereas the body temperature of the mouse remains at 37°C. We see in graph b that the metabolic rate of the lizard increases with temperature. Below about 28°C the metabolic rate of the mouse increases as chamber temperature decreases (notice that you must read the graph right to left to see this). It seems the lizard cannot regulate its body temperature or metabolism independently of environmental temperature. The mouse, however, regulates its body temperature by altering its rate of metabolic heat production.

Field Study of an Ectotherm

A logical next step would be to test in nature our laboratory conclusion that the lizard cannot regulate its body temperature. This can be done by implanting a capsule containing a radio telemeter in the lizard's

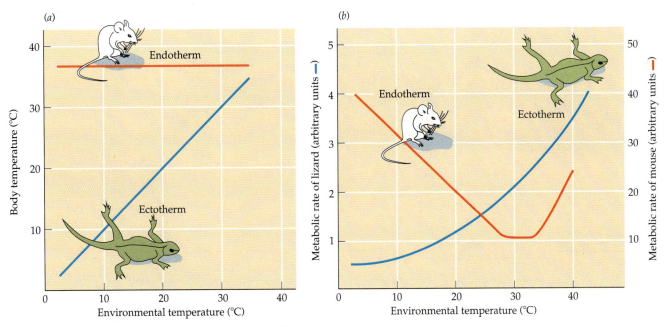

33.13 Effects of Environmental Temperature Differ
(a) The body temperatures of a representative ectotherm (a lizard; blue line) and endotherm (a mouse; red line) respond differently to changes in environmental temperature. (b) The reason for the difference is that the rate of metabolic heat production in the lizard decreases as the environmental temperature falls, whereas the mouse increases its metabolic heat production to compensate for declines in ambient temperature.

body and then releasing the lizard in its desert environment. The radio telemeter measures the lizard's body temperature and then converts it to a radio signal that can be heard through a portable radio. We can thus observe the body temperature of the lizard as it goes about its normal behavior. Our prediction is that the temperature of the lizard should follow the temperature of the environment, which can change more than 40°C in a few hours. The results of the experiment differ strikingly from this prediction (Figure 33.14). At night the temperature in the desert may drop close to freezing, but the temperature of the lizard remains absolutely stable at 16°C. This is not difficult to explain; the lizard spends the night in a burrow where the soil temperature is a constant 16°C. Early in the morning, soon after sunrise, the lizard emerges from its burrow. The air temperature is still quite cool, but the body temperature of the lizard rises to 35°C in less than 30 minutes. The lizard achieves this by basking on a rock with maximum exposure to the sun. As its dark skin absorbs solar radiation, its body temperature rises considerably above the surrounding air temperature. By altering its exposure to the sun, the lizard maintains its body temperature at around 35°C all morning as it seeks food, avoids predators, and interacts with potential mates or competitors. By noon the air temperature near the surface of the desert has risen to 50°C, but the lizard's body temperature remains

around 35°C. It is now staying mostly in shade, frequently up in bushes where there is a cooling breeze. As afternoon progresses, air temperature declines, and the lizard again spends more of its time in the sun and on hot rocks so that its body temperature still remains around 35°C. It returns to its burrow just before sunset, and its body temperature rapidly drops to 16°C.

This field experiment shows that the lizard can regulate its body temperature quite well by behavioral mechanisms rather than by metabolic mechanisms. The deficiency in our laboratory experiment was that the lizard in the chamber could not use its thermoregulatory behavior. If we give a lizard access to a thermal gradient in the laboratory, it is capable of regulating its body temperature by selecting the right place on the gradient. If only a hot place and a cold place are available, it will shuttle back and forth. It will maintain a different body temperature during the night than during the day, and if it is infected with pathogenic bacteria it will give itself a fever by selecting higher temperatures on the gradient (see Box 33.A). The lesson learned from the discrepancies between the results from the laboratory and field experiments are encapsulated by a quotation from the famous German embryologist Hans Spemann: "An experiment is like a conversation with an animal, but the animal must be permitted to answer in its own language."

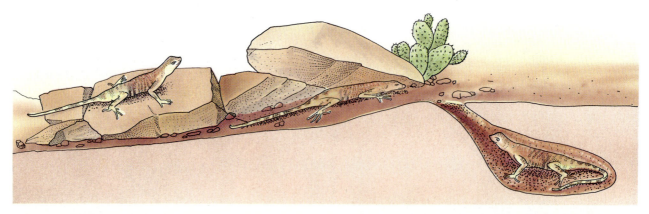

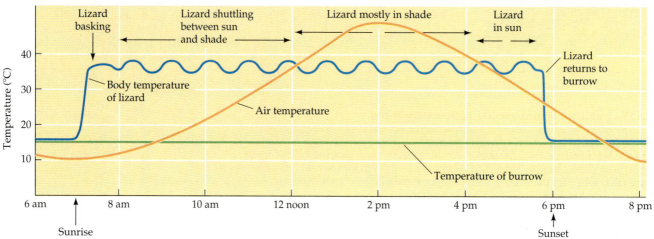

33.14 A Lizard Regulates Its Body Temperature Behaviorally
In natural environments, ectotherms such as lizards regulate their body temperatures by behavior. In the desert environment shown here, the lizard maintains its body temperature by basking to absorb solar radiation in the cool morning, but at midday it avoids the sun and maximizes its heat loss to the cooler air. At night it burrows to avoid the near-freezing desert temperature.

Behavioral Thermoregulation

Behavioral thermoregulation is not the exclusive domain of ectotherms. It is also the first line of defense for endotherms. When the option is available, most animals select thermal microenvironments that are best for them. They may change their posture, orient to the sun, move between sun and shade, and move between still air and moving air, as demonstrated by the lizard in our field experiment. Examples of more complex thermoregulatory behavior are nest construction and social behavior such as huddling. In humans, the selection of clothing is quite important. Behavioral thermoregulation is widespread in the animal kingdom (Figure 33.15).

Control of Blood Flow to the Skin

Physiological thermoregulation is not the exclusive domain of endotherms. Various physiological thermoregulatory adaptations are found in ectotherms. Both ectotherms and endotherms can alter the rate of heat exchange between their bodies and their environments by controlling the flow of blood to the skin. For example, when a person's body temperature rises as a result of exercise, blood flow to the skin increases, and the skin surface gets quite warm. The extra heat brought from the body core to the skin by the blood is lost to the environment, thus tending to bring body temperature back to normal. In contrast, when a person is exposed to cold, the blood vessels supplying the skin constrict, decreasing blood flow and heat transport to the skin, thus reducing heat loss to the environment.

The control of blood flow to the skin is an important adaptation for ectotherms like the marine iguana of the Galapagos Islands. The Galapagos are volcanic islands on the equator, bathed by cold oceanic currents. Marine iguanas are reptiles that bask on black lava rocks near the ocean and swim in the sea where they feed on submarine algae. When the iguanas cool to the temperature of the sea, they are slower and more vulnerable to predators, and probably incapable

BOX 33.A

Fevers and "Feeling Crummy"

You respond to a large variety of infectious illnesses by getting a fever and feeling crummy. You lose your appetite, you have no energy, your joints and muscles ache, you get the chills, and you just feel like putting on flannel pajamas and getting into bed. Are these well-known nonspecific symptoms of illness simply unfortunate side effects? To the contrary, scientists are beginning to think that getting a fever and feeling crummy are adaptive responses that help us fight diseases. The immediate causes of these responses are chemical messages from the immune system. When an infectious virus or bacterium invades the body, it is grabbed by scavenger cells called macrophages. One of the things mac-

rophages do is release chemicals called interleukins that sound the alarm to other cells of the immune system throughout the body. Interleukins cause many other responses as well. They make neurons transmitting pain more sensitive, and therefore we ache. They make us sleepy. They stimulate a part of the brain called the hypothalamus to release corticotropin-releasing hormone, which initiates the stress responses of the body. Interleukins also cause a rise in the hypothalamic set point for thermoregulatory responses. Intracellular messengers activated by the interleukins include prostaglandins. A potent inhibitor of prostaglandin synthesis is aspirin, thus explaining how this drug can reduce fever and make us feel better. But, is it a good idea to reduce fever and feel less crummy if these are adaptive responses to infection?

The first convincing evidence that fever is an adaptive response to infection came from experiments on lizards by Matthew Kluger at the University of Michigan. Lizards with access to a heat lamp shuttle in and

out of the light during the day and maintain their body temperatures at about 38°C. When injected with pathogenic bacteria, the lizards spent more time under the light and raised their body temperatures to between 40 and 42°C—they developed fevers. Does a fever help the lizard fight infection? To answer this question, groups of lizards receiving equal inoculations of bacteria were kept in incubators at 34, 36, 38, 40, and 42°C. All of the lizards at 34 and 36°C died, about 25 percent at 38°C survived, and about 75 percent at 40 and 42°C survived. Fevers do help.

Even though it seems clear that fever is an adaptive response to infection, high fevers can still be quite dangerous and even lethal. Even modest fevers can be dangerous to people with weakened hearts and people who are chronically ill. A fetus can be endangered when a pregnant woman has a high fever. Antipyretics such as aspirin are valuable drugs in such cases, but perhaps they should not be taken by most people at the first sign of aches or chills.

(a)

(b)

33.15 Endotherms Use Behavior to Thermoregulate
Humans and other endotherms adjust their behavior to the environmental temperature in many ways. (a) In this situation of extreme Antarctic cold, two people have put on many layers of insulating clothing. The ice hut they are building will provide them with a shelter against even more extreme conditions. (b) This elephant uses a cool shower of water to bring relief from the heat in Kenya, Africa.

of efficient digestion. They therefore intersperse feeding forays into the sea with basking. It is advantageous for iguanas to retain body heat as long as possible while swimming and to warm up as fast as possible when basking. They adjust their cooling and heating rates by changing the flow of blood to the skin. Blood vessels feeding the skin constrict when an iguana is in the ocean and dilate when it is basking. Also, an iguana's heart rate is slower when it is swimming than when it is basking. Slowed heart rate and constricted vessels when swimming mean that less blood is being pumped through the skin, and therefore less heat is being transported from deep in the iguana's body to its skin to be lost to the water. Faster heart rate and dilated vessels when the iguana is basking increase the transport of heat from the skin to the rest of its body. Of course, basking on black rocks under the equatorial sun can be too much of a good thing. When an iguana reaches an optimal body temperature, it lifts its body off the rocks and orients itself to minimize its surface area that is directly exposed to solar radiation. Thus, the marine iguana uses both physiological and behavioral mechanisms to regulate its body temperature.

Metabolic Heat Production

The use of metabolic heat production to maintain a body temperature above that of the environment is surprisingly common among ectotherms. For example, the powerful flight muscles of many insects, such as dragonflies, moths, bees, and beetles, must reach a fairly high temperature (35 to 40°C) before the insects can fly, and they must maintain these high temperatures during flight, even at air temperatures around 0°C. Such insects use flight muscles themselves to produce the required heat. These muscles are about 20 percent efficient—about 20 percent of the energy they consume goes into useful work and 80 percent is lost as heat. So during flight there is an enormous production of heat, which keeps body temperature elevated. To reach flight temperature from resting temperature, the insects contract their flight muscles isometrically—that is, the muscles that contract alternately during flight to produce the wingbeats contract simultaneously during warm-up. There is virtually no wing movement even though the muscles are contracting and producing heat (Figure 33.16). During warm-up, some bees and moths appear to be shivering because the wings show small amplitude movements.

The heat-producing ability of insects can be quite remarkable. It enables moths to fly at night when air temperatures are low and solar basking is not possible. The heat-producing ability of a species of scarab beetle that lives in the mountains north of Los Angeles, California, has made it possible for these beetles to have a rather unusual mating behavior. The

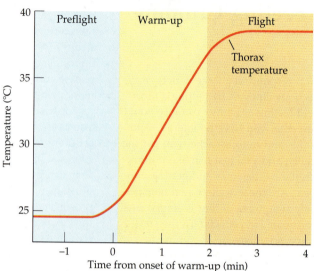

33.16 Preflight Warm-Up in a Moth
Prior to "takeoff," insects such as the sphinx moth can contract the flight muscles in their thoraxes to generate heat and bring the temperature of the muscles up to the level required for flight.

beetles spend most of their life cycle in the soil, except for mating, at which time females and males emerge from the soil and the males undertake flights in search of females. What is unusual is that they engage in this behavior in winter, at night, during snowstorms. The drop in barometric pressure associated with a storm probably triggers the emergence. These beetles were long considered to be very rare because very few entomologists look for beetles in the mountains, in winter, at night, during snowstorms. Presumably the same is true for potential predators!

Honeybees engage in colonial temperature regulation. They live in large colonies consisting mostly of female worker bees that maintain the hive and rear young hatched from eggs laid by the single queen bee in the colony. During winter, honeybee workers combine their individual heat-producing abilities to regulate the temperature of the brood. They cluster in the area of the hive where the brood is located and adjust their joint metabolic heat production and density of clustering so that brood temperature remains remarkably constant at about 34°C even as outside air temperature drops below freezing.

Some reptiles use metabolic heat production to raise body temperature above air temperature. The female Indian python protects her eggs by coiling her body around them. If air temperature falls, she uses isometric contractions of her body wall muscles to generate heat. This adaptation of the python, like the use of flight muscles by insects, is analogous to shivering in mammals. The python is able to maintain the temperature of her body—and therefore that of her eggs—considerably above air temperature.

Biological Heat Exchangers

The heat that active muscles produce can be used to raise body temperature above the temperature of the surrounding air or water if that heat is not rapidly lost to the environment. It is particularly difficult for fish to slow the loss of body heat to the environment because blood pumped from the heart comes into close contact with water flowing over the thin gill membranes before that blood is distributed throughout the body. Therefore, any heat transferred to the blood from active muscles is lost rapidly to the environment. It is surprising therefore to find that some large, rapidly swimming fishes, such as bluefin tuna

and great white and mako sharks, can maintain temperature differences between their bodies and the surrounding water as great as 10 to 15°C (Figure 33.17). The heat comes from their powerful swimming muscles, of course, but the ability to conserve that heat is due to remarkable arrangements of the blood vessels.

In the usual fish circulatory system, oxygenated blood from the gills collects in a large, dorsal vessel, the aorta, which travels through the center of the fish, distributing blood to all organs and muscles. "Hot fish" such as bluefin tuna have smaller central dorsal aortas. Most of their oxygenated blood is transported in large vessels just under the skin (Figure 33.18). Hence, the cold blood from the gills is kept close to the surface of the fish. Smaller vessels transporting this cold blood into the muscle mass run in parallel with vessels transporting warm blood from the muscle mass back toward the heart. Because of the close contact between vessels carrying cold blood into the muscle and ones carrying warm blood away, there is a flow of heat from the warm to the cold blood. Because heat is exchanged between blood vessels carrying blood in opposite directions, this adaptation is called a countercurrent heat exchanger. It keeps the heat within the muscle mass and therefore makes it possible for the fish to have a deep body temperature considerably above water temperature. Why is it advantageous for the fish to be warm? Each ten-degree rise in muscle temperature increases its sustainable power output almost threefold!

THERMOREGULATION IN ENDOTHERMS

An endotherm responds to changes in the temperature of its environment by changing its metabolic rate. Within a narrow range of environmental temperatures called the **thermoneutral zone**, the metabolic rate of the endotherm is low and independent of temperature. The metabolic rate of a resting animal at a temperature within the thermoneutral zone is called the **basal metabolic rate.** It is usually measured on animals that are quiet but awake and that are not using energy in the digestive processes or for reproduction. A resting animal consumes energy at the basal metabolic rate just to carry out all of its metabolic functions other than thermoregulation.

It is interesting to note at this point that the basal metabolic rate of an endotherm is about six times greater than the metabolic rate of a similarly sized ectotherm at rest and at the same body temperature as the endotherm. You can see this difference in Figure 33.13b. A gram of mouse tissue consumes energy at a much higher rate than does a gram of lizard tissue when both tissues are at 37°C. This difference is due to a basic change in cell metabolism that accompanied the evolution of endotherms from

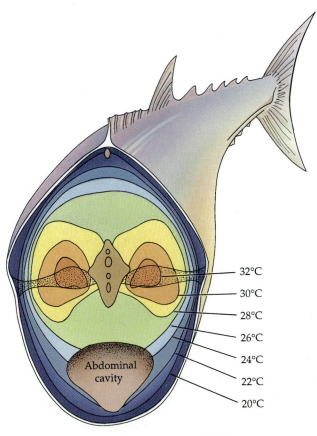

32°C
30°C
28°C
26°C
24°C
22°C
20°C

Abdominal cavity

Water temperature 20°C

33.17 Cold Water, Warm Muscles
In an actively swimming fish such as the bluefin tuna, the muscles that power swimming generate heat that keeps the fish's internal body temperature much higher than that of the surrounding water.

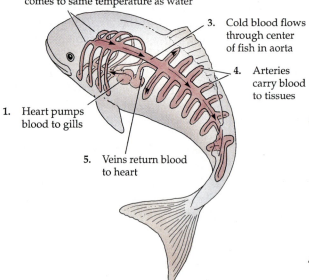

"Cold" fish

2. In gills, blood is oxygenated and comes to same temperature as water

3. Cold blood flows through center of fish in aorta

4. Arteries carry blood to tissues

1. Heart pumps blood to gills

5. Veins return blood to heart

"Hot" fish

2. In gills, blood is oxygenated and comes to same temperature as water

3. Cold blood flows from gills to body in arteries just under the skin

1. Heart pumps blood to gills

5. Veins return blood to heart

4. Blood flowing into muscles in arteries is warmed by blood flowing out of muscles in veins; heat remains in muscles

33.18 Hot and Cold Fishes
(a) The circulatory systems of most fish conduct the oxygenated blood from the gills to the organs of the fish through a large dorsal aorta. Because the blood comes into equilibrium with water temperature in the gills, the blood the aorta carrys through the interior of the fish's body is at water temperature. (b) "Hot" fish species such as the bluefin tuna (Figure 33.17) retain the heat produced by their muscles because the anatomy of their blood vessels allows for heat exchange between the warm blood leaving the muscle and the cold blood entering the muscle.

their ectothermic ancestors. The higher level of heat production by endotherms makes it easier for them to maintain a temperature difference between the body and the environment.

Active Heat Production and Heat Loss

The thermoneutral zone is bounded by a lower critical temperature and an upper critical temperature. Below the lower critical temperature, an endotherm's metabolic rate increases as environmental temperature declines because the animal must produce more and more heat to maintain a constant body temperature as heat loss to the environment increases. As the environment gets colder, eventually the animal reaches its **summit metabolism**, or maximum possible thermoregulatory heat production. If the environmental temperature falls still lower, the animal's body temperature will begin to drop. When the environmental temperature goes above the upper critical temperature, the animal pants or sweats. Since these active heat loss responses require an increased expenditure of energy, the metabolic rate rises. The thermoregulatory responses of endotherms can be

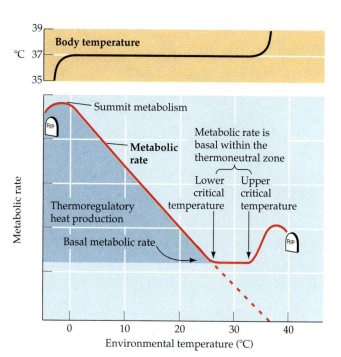

33.19 Environmental Temperature and Mammalian Metabolic Rates
Within a mammal's thermoneutral zone its metabolic rate is low and constant; the animal thermoregulates behaviorally and by changing its thermal insulation. Above the upper critical temperature the animal must expend energy to lose heat by panting or sweating. Below the lower critical temperature the animal increases metabolic heat production to compensate for increased heat loss to the environment.

illustrated by graphs, such as Figure 33.19, that depict metabolic rate as a function of environmental temperature.

Heat for thermoregulation is produced by two mechanisms in mammals, shivering and nonshivering heat production. Birds use only shivering heat production. Shivering uses the contractile machinery of skeletal muscles to consume ATP without causing overt behavior. The muscles pull against each other so that little movement other than a tremor results. All of the energy from the conversion of ATP to ADP in this process is released as heat. Most nonshivering heat production occurs in specialized tissue called **brown fat**. Its brown appearance comes from abundant mitochondria and a rich blood supply (Figure 33.20). In brown fat cells a unique protein called **thermogenin** uncouples oxidative phosphorylation. This means that metabolic fuels are consumed to produce heat without the production of ATP. Brown fat is especially abundant in newborn infants of many mammalian species, in some adult mammals that are highly acclimatized to cold, and in mammals that hibernate.

Living in the Cold

The coldest habitats on Earth are found in the Arctic, the Antarctic, and at the tops of high mountains. Many birds and mammals can be found in these cold places, but no reptiles or amphibians. Clearly, the ability to produce a substantial amount of heat metabolically has made it possible for endotherms to exploit these formidable, frigid environments. However, most tropical species of birds and mammals would not fare well in those environments. What adaptations besides endothermy alone characterize species that live in the cold?

The most important adaptations of endotherms to cold environments are those that reduce their heat loss to the environment. Since most heat loss is from the body surface, many cold-climate species have smaller surface areas than their warm-climate cousins, even when their body masses are the same. Rounder body shapes and shorter appendages reduce the surface area-to-volume ratios of some cold-climate species; compare the desert jackrabbit and the arctic hare shown in Figure 33.21. Another means of decreasing heat loss is to increase thermal insulation. Fur, feathers, and layers of fat decrease the loss of heat from the body of an endotherm to the environment. You can experience the effectiveness of thermal insulation by comparing what it feels like to sit on a cold stone or metal bench wearing just cotton shorts with what it feels like after you put a feather pillow or a wool blanket between you and the bench. Arctic and alpine animals, and animals adapted to cold winter conditions, have much thicker layers of fur, feathers, or fat than do their warm-climate equivalents. The fur of an arctic fox or a northern sled dog provides such good thermal insulation that those animals don't even begin to shiver until air temperature reaches as low as −20 to −30°C.

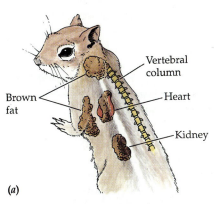

(a)

33.20   Brown Fat: A Heat-Producing Tissue
In many mammals, brown fat produces heat. (a) In a ground squirrel, brown fat occurs in specific anatomical locations. (b) A light microscope view of white fat. Each cell is filled with a globule of lipid and has few organelles. The tissue has few blood vessels. (c) A light microscope view of brown fat at the same magnification reveals cells with many intracellular structures and multiple droplets of lipid. Numerous capillaries run through the tissue. (d) An electron micrograph of brown fat shows the tight packing of mitochondria in a brown fat cell. A portion of a lipid droplet can be seen in the upper right; in the upper left is part of a blood capillary.

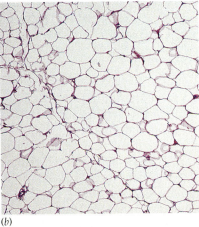

(b)

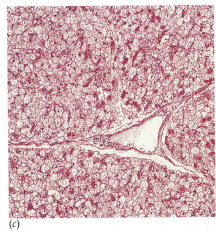

(c)

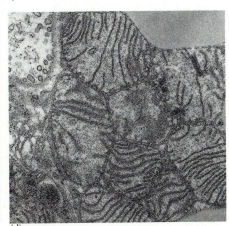

(d)

33.21 Adaptations to Hot and Cold Climates
(a) The desert hare has a large surface area for its body mass, largely due to its long extremities. The large ears serve as heat exchangers, passing heat from the hare's blood to the surrounding air. (b) The arctic hare has shorter extremities and therefore a smaller surface area for its body size. The fur of the arctic hare is longer and thicker than that of the desert hare and provides good insulation.

Changing Thermal Insulation

Humans change their thermal insulation by putting on or taking off clothes. How do animals do it? We have already discussed one example, the marine iguana. By changing the blood flow to its skin, the marine iguana increases or decreases the exchange of heat between the environment and its body—in other words, the iguana changes its thermal insulation. Increasing or decreasing blood flow to the skin is an important thermoregulatory adaptation for endotherms as well. In a hot environment, your skin feels hot because of the high rate of blood flow through it, but when you are sitting in an overly air-conditioned theater, your hands, feet, and other body surfaces feel cold as blood flow to those areas decreases. The wolf has an elegant mechanism for decreasing heat loss from its feet without the risk of freezing them. As long as the wolf's foot temperature is more than a few degrees above freezing, certain blood vessels in its foot are constricted and blood flow to the foot is minimal. As foot temperature gets closer to 0°C, these vessels open and allow more blood to flow through the foot, thus keeping it from freezing.

For highly insulated arctic animals and for many large mammals from all climates, getting rid of excess heat can be a serious problem, especially during exercise. Arctic species usually have some part of the body surface such as the abdomen that is thinly furred and can act as a window for heat loss. Large mammals like elephants, rhinoceroses, and water buffalo have little or no fur and seek places where they can wallow in water when the air temperature is too hot. Having water in contact with the skin greatly increases heat loss because water has a much greater capacity for absorbing heat than does air.

Evaporative Water Loss

The evaporation of water is a very effective means of dissipating heat. A gram of water absorbs about 580 calories when it evaporates. But water is heavy, animals do not carry a great excess supply of it, and hot environments tend to be arid environments where water is a scarce resource. Therefore, evaporation of water by sweating or panting is usually a last resort for animals adapted to hot environments. Sweating or panting are active processes and require the expenditure of metabolic energy. That is why the graph of metabolic rate versus environmental temperature goes up above the upper critical temperature (see Figure 33.19). A sweating or panting animal is producing heat in the process of dissipating heat. This can be a losing battle. Animals can survive in environments that are below their lower critical temperature much better than they can in those above their upper critical temperature.

THE VERTEBRATE THERMOSTAT

The various thermoregulatory mechanisms and adaptations discussed above are the controlled systems for the regulation of body temperature. These controlled systems must receive commands from a regulatory system that integrates information relevant to the regulation of body temperature. A convenient word for the regulatory system in this case is the thermostat. All animals that thermoregulate, both vertebrate and invertebrate, must have regulatory systems, but discussion here will focus on the vertebrate thermostat.

Where is the vertebrate thermostat? The major integrative center is at the bottom of the brain in a structure called the **hypothalamus**. If you slide your tongue back as far as possible along the roof of your mouth, it will be just a few centimeters below your hypothalamus. The hypothalamus is a part of many regulatory systems, so we will refer to it many times in the chapters to come. If the hypothalamus of a mammal's brain is damaged, the animal loses its ability to regulate its body temperature, which then rises in a warm environment and falls in a cold one.

Set Points and Feedback

What information does the thermostat use? In many species the temperature of the hypothalamus itself is a major source of feedback information to the thermostat. Cooling the hypothalamus causes fishes and reptiles to seek a warmer environment, and heating the hypothalamus causes them to seek a cooler environment. In mammals, cooling of the hypothalamus can stimulate constriction of blood vessels to the skin and increase metabolic heat production. Because

tion of temperature change. Hence, hypothalamic temperature serves as a negative feedback signal.

There is not a single set point for body temperature in an animal, but separate set points for the activation of different thermoregulatory responses. If the hypothalamus of a mammal is heated and cooled, the vessels supplying blood to the skin constrict at a specific hypothalamic temperature. A hypothalamic temperature slightly lower than that will initiate shivering, and a hypothalamic temperature two or three degrees higher will initiate panting. The characteristics of hypothalamic control of each response can be described. For example, if we measure metabolic heat production while heating and cooling the hypothalamus as was shown in Figure 33.22, we can describe the results graphically, as shown by either one of the curves in Figure 33.23. Over a certain range of hypothalamic temperatures, metabolic heat production remains low and constant, but cooling the hypothalamus below a certain level, the set point, stimulates increased metabolic heat production. The increase in heat production is proportional to how far the hypothalamus is cooled below the set point. This is much more sophisticated than a simple on–off thermostat like the one in a house.

The thermostat integrates other sources of information in addition to hypothalamic temperature.

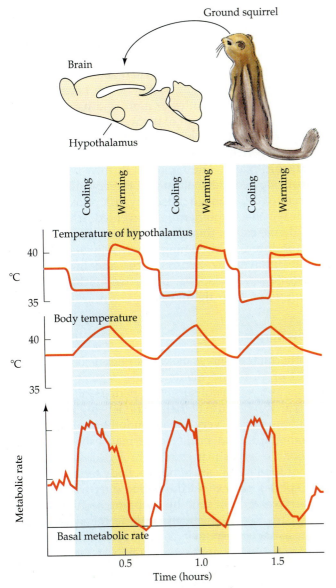

33.22 The Hypothalamic Thermostat
In this experiment, a ground squirrel was maintained at cold environmental temperatures so that its metabolic heat production was initially high. Cooling the hypothalamus increased metabolic heat production even further and the animal's body temperature rose. Heating the hypothalamus reduced metabolic rate and the animal's body temperature fell.

of the activation of these thermoregulatory responses, body temperature rises when the hypothalamus is cooled. Conversely, warming of the hypothalamus stimulates dilation of blood vessels to the skin and sweating or panting, and the overall body temperature falls when the hypothalamus is warmed (Figure 33.22). The hypothalamus appears to generate a set point like a setting on a thermostat. When actual hypothalamic temperature goes above or below that set point, thermoregulatory responses (the controlled system) are activated to reverse the direc-

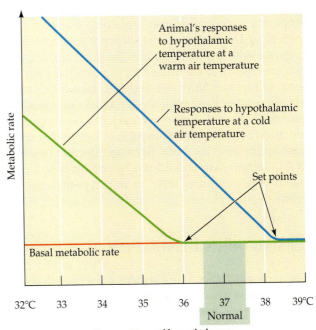

33.23 Adjustable Set Points
The set point for the metabolic heat production response to hypothalamic temperature changes at different environmental temperatures. Other factors, such as being asleep or awake, the time of day, or the presence of a fever can also affect the set point.

One example is information about the temperature of the environment as registered by temperature sensors in the skin. Changes in skin temperature shift the hypothalamic set points for responses. As shown in Figure 33.23, in a warm environment you might have to cool the hypothalamus of a mammal to stimulate it to shiver, but in a cold environment you would have to warm the hypothalamus of the same animal to stop its shivering. The set point for the metabolic heat production response is higher when the skin is cold and lower when the skin is warm. Information from the skin is feedforward information that adjusts the hypothalamic set point. Many other factors also shift hypothalamic set points for responses. Set points are higher during wakefulness than during sleep. They are higher during the active part of the daily cycle than during the inactive part, even if the animal is awake at both times.

Fever is another example of a shift in set points. Fevers are rises in body temperature in response to substances called **pyrogens** derived from bacteria or viruses that invade the body. Injections of the killed bacteria or even the purified cell walls of the killed bacteria can also cause fever. The presence of the pyrogen in the body causes a rise in the hypothalamic set point for the heat-production response. If you are the unlucky person developing a fever, you may feel unbearably cold (the chills) even though your body temperature is normal. You shiver, put on more clothes, and turn up your room temperature or electric blanket. As a result your body temperature rises until it matches the new set point. At the higher body temperature you no longer feel cold, and you may not feel as if you are hot, but someone touching your forehead will say that you are "burning up." If you take an aspirin, it has the effect of reducing your set point to normal. Now you feel hot, take off clothes, and even sweat until your elevated body temperature returns to normal. Extreme fevers can be dangerous and must be reduced, but there is evidence that moderate fevers help the body fight an infection (see Box 33.A). Perhaps we should not be too hasty in using medication to counter shifts in our hypothalamic set points.

TURNING DOWN THE THERMOSTAT

If fever results from a turning up of the hypothalamic thermostat, are there cases of turning down the thermostat so that body temperature is regulated at a lower level? The answer is yes, but not all decreases in body temperature are regulated. Hypothermia is the condition in which body temperature is below normal. It can result from a natural turning down of the thermostat, or it can result from traumatic events such as starvation (lack of fuel), exposure, serious illness, or anesthesia. Due to Q_{10} effects, hypothermia slows metabolism, slows the heart, weakens muscle contractions (including those of the heart), decreases nerve conduction, and causes unconsciousness. This is not a happy state of affairs for most endotherms and can lead to death, but it can also be somewhat protective. There are a number of cases in which drowning victims have been under water for 10 or 15 minutes or more and have shown no pulse when pulled out of the water. Nevertheless, some—mostly small children drowned in cold water—were revived by paramedics and recovered to a remarkable extent. The rapid fall in their body temperature slowed metabolism and thereby slowed the progress of cell damage caused by lack of oxygen. In this way, the hypothermia prevented irreversible brain damage even though it was a pathological condition induced by drowning. Some animals can anticipate unfavorable circumstances and induce hypothermia as an adaptive, protective mechanism by turning down their thermostats.

Shallow Torpor

Hypothermia conserves metabolic energy. Numerous species of birds and mammals use regulated hypothermia as a means of surviving periods of cold and food scarcity. Because of their extremely high surface-to-volume ratios, very small endotherms such as hummingbirds and pocket mice may exhaust their metabolic reserves just getting through a single day without food if they are at normal body temperature. Such species can extend the period over which they can survive without food by dropping body temperature. This adaptive hypothermia is called shallow torpor or daily torpor; it usually occurs on a daily basis, with body temperature falling at the time of day the animal normally becomes inactive. Body temperature can drop 10 to 15°C during shallow torpor, resulting in an enormous saving of metabolic energy.

A small bird, the willow tit, provides an example of shallow torpor that shows how well regulated this process can be. Willow tits live through the winter above the Arctic Circle. In spite of their good thermal insulation, these tiny birds must become hypothermic to survive the long, cold arctic nights. Each evening the bird lowers its metabolic rate to a level that it maintains all night, and its body temperature falls as a result of the decreased heat production. How low the bird's metabolic rate drops is different on different nights (Figure 33.24). Randi Reinertsen at the University of Trondheim has shown that the decrease in its metabolism depends on air temperature, on season (hence, on length of night), and on the bird's fat reserves at roosting time. Every morning the bird has depleted its fat reserves and must immediately feed on seeds it has cached nearby. This is living on the razor's edge! The brain of this small animal integrates all the relevant information, resulting in just the right resetting of its thermostat to get it through the night.

Hibernation

Regulated hypothermia can also last for days or even weeks, with drops to very low temperatures; this phenomenon is called **hibernation**. Many diverse species of mammals hibernate, but only one species of bird—the poorwill—has been shown definitely to hibernate. For the deep sleep of hibernation, the body's thermostat is turned down to an extremely low level to maximize energy conservation. Many hibernators maintain body temperatures around 2 to 4°C during hibernation. The metabolic rate needed to sustain an animal after this incredible drop in body temperature is only ⅓₀ to ⅟₅₀ of basal metabolic rate —an enormous saving of metabolic energy. Animals hibernate when temperatures are low and food is scarce. Some hibernators cache a food supply in their well-insulated nests. Individual bouts of hibernation may last from less than a day to over a week (Figure 33.25). A bout terminates spontaneously when the hibernator's body temperature returns to normal. The animal may remain at its normal temperature for a few hours to a day or more before entering another bout of hibernation.

The hibernation season is controlled by an endogenous biological clock (or calendar), which continues to run with a periodicity of about a year even when animals are kept under constant conditions in the laboratory. This is called a **circannual rhythm** (circa = about; annual = a year). A typical circannual cycle for a hibernator such as a ground squirrel includes an active season, during which it cannot hibernate even if exposed to cold temperatures and deprived of food. During the active season, usually spring through fall, the animals breed, raise their young, prepare their nests for winter, fatten their bodies, and store food. During the hibernation season, as Figure 33.25 shows, animals hibernate in recurrent bouts. They progressively lose body weight even if food is available. Toward the end of the hibernation season the reproductive organs grow and become functional. Body temperature falls during hibernation because the hypothalamic set point drops, and arousal from a bout of hibernation is due to a return of the set point to a normal mammalian level. The ability of hibernators to reduce the set point so dramatically probably evolved as an extension of the set point decrease that accompanies sleep even in nonhibernating species of mammals and birds.

SUMMARY

A multicellular animal must maintain a suitable environment for all the cells of its body. Organs and organ systems function to maintain constant conditions in the internal environment—a state of homeo-

Willow tit

33.24 Hypothermia Deepens with Cold and Dark
The curves show how a willow tit's body temperature changes on long nights at different environmental temperatures. The colder the air, the deeper is the bird's hypothermia. Notice that the depth of hypothermia is set early in the night, and must therefore be a result of information available at that time— not simply a consequence of running out of fuel reserves faster during colder nights. If nights are made shorter (which can be done in the laboratory), the birds maintain higher body temperatures at these same air temperatures.

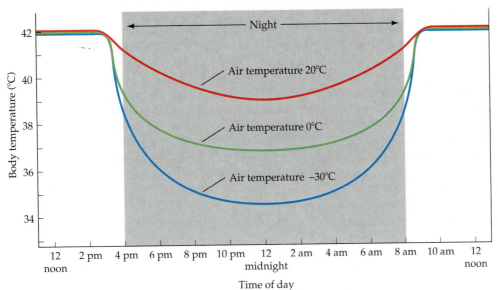

Body temperature (°C) vs Time of day. Night spans approximately 4 pm to 8 am.

Air temperature 20°C
Air temperature 0°C
Air temperature −30°C

Time of day: 12 noon, 2 pm, 4 pm, 6 pm, 8 pm, 10 pm, 12 midnight, 2 am, 4 am, 6 am, 8 am, 10 am, 12 noon

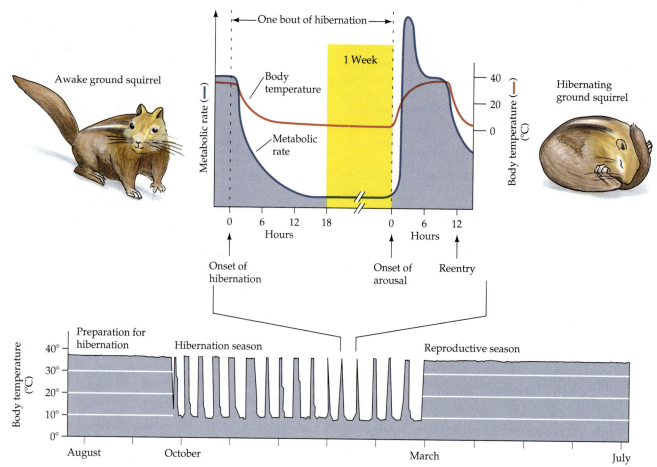

Awake ground squirrel

Hibernating ground squirrel

33.25 Hibernation of a Mammal Occurs in Bouts
During the hibernation season, ground squirrels hibernate in bouts. A bout of hibernation ends when the animal's body temperature returns to normal mammalian levels; it may remain there for a few hours or even a day or more before another bout of hibernation begins.

stasis. The activities of organs and organ systems are controlled by regulatory systems that respond to feedback information from the internal environment.

An example of physiological regulation is the regulation of body temperature. Temperature is a physical variable that can have strong influences on living systems. Within temperature limits compatible with life, the Q_{10} calculation quantifies the effect of temperature on biochemical reactions and physiological processes. Animals can avoid the effects of changes in temperature by regulating body temperature independently of environmental temperature. Homeotherms maintain body temperatures that are rather constant, whereas poikilotherms may have variable body temperatures.

Unlike ectotherms, endotherms have well-developed metabolic capacities for producing heat internally as a means of maintaining a constant body temperature above that of the environment. Endotherms have adapted to a wide variety of thermal environments through differences in thermal insulation. Evaporative water loss through sweating and panting are effective mechanisms for heat loss.

Thermoregulatory mechanisms of vertebrates are controlled by a thermostat that resides in the hypothalamus. Its major source of negative feedback information is hypothalamic temperature, and it uses feedforward information from the temperature of the skin to generate set points for thermoregulatory responses such as metabolic heat production. Thermoregulatory set points vary with such variables as fever, time of day, and sleep–wake transitions. Some species conserve metabolic energy by turning down their thermostats. This is done on a daily basis by many species of very small endotherms, and for longer periods by hibernators.

SELF-QUIZ

1. If the Q_{10} of the metabolic rate of an animal is 2, then:
   a. that animal is better acclimatized to a cold environment than if its Q_{10} were 3.
   b. the animal is an ectotherm.
   c. the animal consumes half as much oxygen per hour at 20°C as it does at 30°C.
   d. the animal's metabolic rate is not at basal levels.
   e. the animal produces twice as much heat at 20°C than at 30°C.

2. Which of the following statements is *true* of brown fat?
   a. It produces heat without producing ATP.
   b. It insulates animals acclimatized to cold.
   c. It is a major source of heat production for birds.
   d. It is only found in hibernators.
   e. It provides fuel for muscle cells responsible for shivering.

3. What is the most important and most general difference between mammals and birds adapted to cold climates in comparison to species adapted to warm climates?
   a. Higher basal metabolic rates
   b. Higher Q_{10}'s
   c. Brown fat
   d. Greater insulation
   e. Ability to hibernate

4. Which of the following would cause a decrease in the hypothalamic temperature set point for metabolic heat production?
   a. Going into a cold environment
   b. Taking an aspirin when you have a fever

   c. Arousing from hibernation
   d. Getting an infection that causes a fever
   e. Cooling the hypothalamus

5. Mammalian hibernation:
   a. occurs when animals run out of metabolic fuel.
   b. is a regulated decrease in body temperature.
   c. is less common than hibernation in birds.
   d. can occur at any time of year.
   e. lasts for a period of several months during which body temperature remains close to environmental temperature.

6. Which of the following is an important difference between an ectotherm and an endotherm of similar body size?
   a. Ectotherms have higher Q_{10}'s.
   b. Only ectotherms use behavioral thermoregulation.
   c. Only endotherms can constrict and dilate the blood vessels to the skin to alter heat flow.
   d. Only endotherms can get fevers.
   e. At body temperatures of 37°C, the ectotherm has a lower metabolic rate than the endotherm.

7. The function of the countercurrent heat exchanger in "hot fish" is:
   a. to trap heat in the muscles.
   b. to produce heat.
   c. to heat the blood returning to the heart.
   d. to dissipate excess heat generated by powerful swimming muscles.
   e. to cool the skin.

8. What is the difference between a winter- and a summer-acclimatized fish that is termed "metabolic compensation"?
   a. The winter acclimatized fish has a higher Q_{10}.
   b. The winter acclimatized fish develops greater insulation.
   c. The winter acclimatized fish hibernates.
   d. The summer acclimatized fish has a countercurrent heat exchanger.
   e. The summer acclimatized fish has a lower metabolic rate at any given water temperature than does the winter fish.

9. Which of the following is an important characteristic of epithelial cells?
   a. They generate electrical signals.
   b. They contract.
   c. They have an extensive extracellular matrix.
   d. They have secretory functions.
   e. They are only found on the surface of the body.

10. Negative feedback:
    a. works in opposition to positive feedback to achieve homeostasis of a physiological variable.
    b. always turns off a process.
    c. acts to reduce an error signal in a regulatory system.
    d. is responsible for metabolic compensation.
    e. is a feature of the thermoregulatory systems of endotherms but not of ectotherms.

FOR STUDY

1. Make a table that lists all of the properties of the internal environment that you think are critical to keep the cells of the body healthy. Next to each property list the organs or organ system responsible for maintaining it.

2. What are the major differences between ectotherms and endotherms; compare and contrast their major thermoregulatory adaptations?

3. Why is an environment that is above the upper critical temperature of an endotherm more dangerous for that animal than is an environment that is below its lower critical temperature?

4. Why is it difficult for a fish to be endothermic, and how do "hot" fish overcome these difficulties?

5. If the temperature of the hypothalamus of a mammal is the feedback information for its thermostat, why is it that hypothalamic temperature hardly changes at all when that animal moves between environments hot enough and cold enough to stimulate the animal to pant and to shiver?

READINGS

Crawshaw, L. I., B. P. Moffitt, D. E. Lemons and J. A. Downey. 1981. "The Evolutionary Development of Vertebrate Thermoregulation." *American Scientist*, vol. 69, pages 543–550. All vertebrates thermoregulate, and the nervous system mechanisms involved appear to have a common origin even though the effector mechanisms may differ.

French, A. R. 1986. "The Patterns of Mammalian Hibernation." *American Scientist*, vol. 76, pages 569–575. Body size has important consequences for energy metabolism; therefore, a variety of patterns of hibernation have evolved as discussed in this article.

Heinrich, B. 1981. "The Regulation of Temperature in the Honeybee Swarm." *Scientific American*, June. When honeybees leave their hive in a swarm, they thermoregulate, as described in this article.

Heller, H. C., L. I. Crawshaw and H. T. Hammel. 1978. "The Thermostat of Vertebrate Animals." *Scientific American*, August. This article describes research on and properties of the brain mechanisms responsible for thermoregulation in vertebrates and the adaptations in those mechanisms that make hibernation possible.

Schmidt-Nielsen, K. 1981. "Countercurrent Systems in Animals." *Scientific American*, May. Countercurrent exchanges are the basis for a variety of physiological adaptations, some of which are presented in this article. Developed in special detail is the case of water conservation in the camel's nose.

Schmidt-Nielsen, K. 1990. *Animal Physiology: Adaptation and Environment*, 4th Edition. Cambridge University Press, New York. This is an excellent advanced textbook on comparative animal physiology. Chapter 8, "Temperature Regulation," expands on many of the topics presented in this chapter.

34

Animal Hormones

PREVIEW: The control and coordination of the activities of different cells, tissues, and organs in the body require communication between cells. One of the two major types of communication between cells in animals is hormonal communication. Hormones are chemical messages produced by cells, released into the extracellular fluids, and circulated around the body. A cell responds to a hormone if it has receptor molecules that bind that hormone. Receptors for hormones that do not cross the plasma membrane are located on the cell surface. When a hormone binds to a receptor on the cell surface, it triggers the production of a second messenger molecule within the cell. The second messenger

then initiates the response of the cell. Some hormones do cross the plasma membranes of receiving cells and bind to receptors in the cytoplasm, thus initiating the response of the cell. Different types of cells may show different responses to the same hormone. In vertebrates, the brain exerts control over many endocrine glands.

This chapter deals with local and circulating hormones of invertebrates and vertebrates, including their chemical nature, the cells or endocrine glands that produce them, mechanisms by which the brain controls endocrine glands, modes of action of hormones, and the responses of the target cells.

Hormones, the chemical messengers of the body, receive much popular attention. We blame "hormone storms" for a lot of adolescent behavior. We associate surges of the hormone adrenaline with the "fight-or-flight" response to frightening situations. Farmers use insect hormones in their wars against pests. Hormones have been added to livestock feed to make chicken and beef tender and juicy. There is great concern over the abuse of hormones by some athletes and body builders who take steroids to increase the mass of their muscles. Many hormones have important medical applications. In the United States alone, there are more than 1.5 million diabetics who could not survive without their regular doses of the hormone insulin. Treatment with hormones can decrease the problem of brittle bones in older women and may alleviate some of the other physical effects of aging in both men and women. In this chapter we will examine the science of **endocrinology**—the study of hormones and their actions.

CHEMICAL COMMUNICATION

A **hormone** is a substance that serves as a chemical message between cells of a multicellular organism. A chemical communication system that uses a hormone is made up of at least two cells: one cell produces and releases the hormone—the message—and a sec-

ond cell with appropriate **receptors** receives the message. We call the receiving cell the **target cell**. The receipt of the message activates mechanisms within the target cell that interpret the message and respond to it. The response may be developmental, physiological, or behavioral.

A simple way to classify chemical communication systems is according to the distance over which the messages operate: Are their effects local or are they distributed throughout the body? (The effects of some chemical messages are even exerted on other organisms in the environment. We will learn about those substances—pheromones—in Chapter 44).

Local Hormones

Some chemical messages are released by cells into the extracellular fluids nearby and exert their effects locally (Figure 34.1*a*). These hormones are inactivated so rapidly by local enzymes or taken up so completely by cells in the immediate vicinity that they usually do not get into the blood in sufficient quantities to exert effects on distant cells.

An excellent example of a local hormone is **histamine**, one of the mediators of inflammation. Histamine is released by specialized cells in the blood when they come into contact with foreign objects such as bacteria. The release of histamine is one of the causes of inflammation, an immune system re-

(a)   **Local hormone**

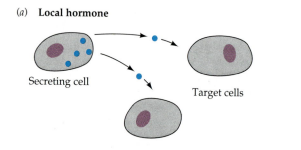

(b)   **Circulating hormone**

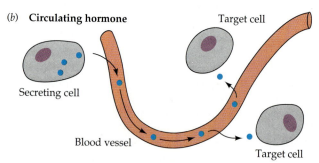

34.1   Chemical Signaling Systems
(a) Many cells in the body secrete chemical messages that influence nearby cells and even themselves. (b) Some cells secrete chemical messages into the bloodstream, which carries them to target cells elsewhere in the body.

sponse that protects the body from invasion by foreign organisms or materials. When the skin is cut by a dirty object, the area around the cut becomes "inflamed"—red, hot, and swollen. The tissue becomes red and hot because local blood vessels expand (dilate) and the blood flow to the area increases. It becomes swollen because the blood vessels in the area become more permeable—leaky—allowing blood plasma, together with protective blood proteins and white blood cells, to move into the infected tissue.

Local responses to histamine are clearly protective, but widely spread responses to histamine can cause problems, such as the symptoms of hay fever. When a person sensitized to a kind of pollen inhales that pollen, it causes certain cells in the respiratory passages to release histamine. The histamine causes the tissues of the passages to swell and to increase their secretions of mucus, leading to congestion, coughing, sneezing, and a runny nose. Such allergic reactions are unpleasant but rarely dangerous. In a person who is extremely allergic, however, or in a person who has a blood-borne infection, histamine and other mediators of the inflammation response may be released in such large amounts that they enter the blood and circulate around the body. The resulting expansion of blood vessels and leakage of fluid from the circulatory system can cause blood pressure to drop severely. Fluid may leak into the lungs and the airway may become severely congested. The result-

ing failure of the circulatory and respiratory systems, termed anaphylactic shock, can be lethal. In some highly sensitive people, a single bee sting, an injection of an antibiotic to which they are allergic, or ingestion of a food to which they are allergic can quickly lead to anaphylactic shock.

Circulating Hormones

The classical concept of a hormone is that of a chemical message secreted by cells and distributed throughout the body by the circulatory system (Figure 34.1b). Wherever such a hormone encounters a cell with a receptor to which it can bind, it triggers a response. The nature of the response depends on the responding cell. The same hormone can cause different responses in different types of cells. As an example, consider the hormone adrenaline. If a lion creeps up behind you and roars, your brain sends signals through your nervous system to adrenaline-containing cells, which immediately release adrenaline. The hormone diffuses into the blood and rapidly circulates around your body. What does it do for you? Adrenaline activates receptors in the heart to make the heart beat faster and pump more blood. Adrenaline activates receptors in the vessels supplying blood to your digestive tract, causing those vessels to constrict—you can digest lunch later. Your heart is pumping more blood, and a greater percentage of that blood is going to the muscles needed for your escape. In the liver, adrenaline stimulates the breakdown of glycogen into glucose for a quick energy supply. In fatty tissue, adrenaline stimulates the breakdown of fats as another source of energy. These are some of the many actions triggered by this one hormone. They all contribute to increasing your chances of escaping the lion. Whether a cell in your body responds to the surge of adrenaline depends on whether it has adrenaline receptors. The specific response of each cell with adrenaline receptors depends on the type of cell it is.

Glands and Hormones

Many hormones are secreted by cells that are clumped together to form secretory organs called **glands**. However, glands in the body are of two types. Some, such as sweat glands and salivary glands, release secretions that are not hormones into ducts that open outside the body. Sweat gland ducts open onto the surface of the skin, and salivary gland ducts open into the mouth. Such glands are called **exocrine glands** because they secrete their products to the outside of the body (exo = Greek, "outside of"). The glands that secrete hormones do not have ducts; they are called **endocrine glands** because they secrete their products into the blood, which is inside the body. Endocrine glands store hormones until

34.2   The Endocrine Glands of Mammals

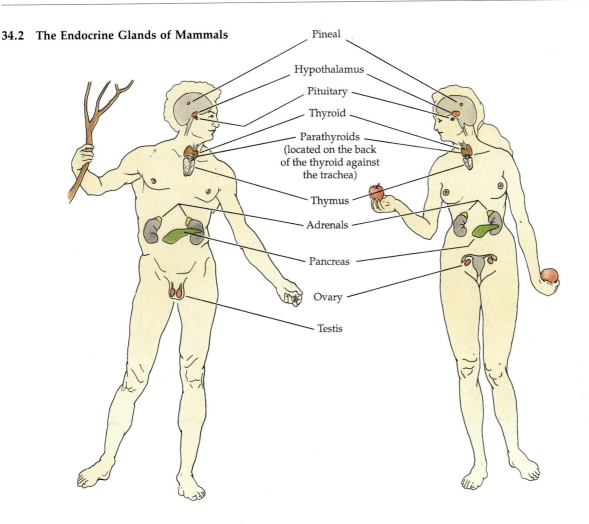

they are needed. Collectively, they make up the endocrine system. The endocrine glands of the human body are shown in Figure 34.2.

Not all hormones are secreted by discrete endocrine glands. Some are secreted by cells that are dispersed among other cells rather than being clumped together to form glands. For example, many hormones of the digestive tract are produced and secreted by cells in the lining of the tract. These hormones enter the blood, and, like adrenaline, circulate throughout the body and activate cells that have the appropriate receptors. Many hormones are secreted by cells in the nervous system. These hormones, called neurohormones, can act locally, but some are taken up by the circulatory system and act on cells at some distance from their site of release.

INVERTEBRATE HORMONES

Although we will focus most of our attention in this chapter on the hormones of mammals, a few invertebrate hormones will receive our attention because they have been so well studied and have made major contributions to our knowledge of how endocrine

systems function. Many invertebrate hormones have molecular structures similar or identical to hormones of vertebrates but their functions may be completely different.

Hormonal Control of Molting in Insects

The British physiologist Sir Vincent Wigglesworth was a pioneer in the study of hormonal control of growth and development in insects. He conducted experiments on the blood-sucking bug *Rhodnius*. Upon hatching from the egg, *Rhodnius* is nearly a miniature version of the adult, but it lacks some adult features. Because insects have rigid exoskeletons, their growth is episodic, punctuated with molts (shedding) of the exoskeleton. Each growth stage between two molts is called an instar. A blood meal triggers each episode of molting and growth in *Rhodnius*, which goes through five molts before developing into a complete adult. *Rhodnius* is a hardy experimental animal; it can even live a long time after it is decapitated. If decapitated soon after it has a blood meal, a *Rhodnius* may live up to a year, but it does not molt. If decapitated a week after its blood meal, it will molt. These observations led to the hypothesis

(a)

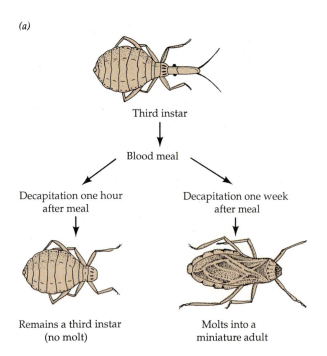

(b)

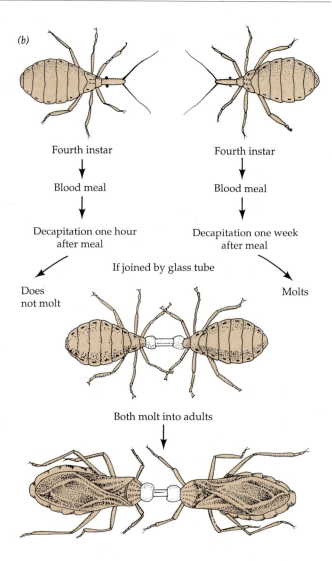

34.3 Hormonal Control of Molt
The blood-sucking bug *Rhodnius* grows and molts after a blood meal. *(a)* If it is decapitated just after that meal, it will live but not molt. If it is decapitated one week after a blood meal, it will molt. *(b)* Apparently a diffusable factor from the head region is necessary for molt; this is demonstrated by connecting two bugs—decapitated at different times—with a glass tube.

that something diffusing slowly from the region of the head controls the molt (Figure 34.3*a*).

The proof that one or more diffusing substances cause the molt came from a clever experiment in which Wigglesworth decapitated two *Rhodnius*: one that had just had its blood meal and another that had had its blood meal one week earlier. The two decapitated bodies were connected with a short piece of glass tubing—and they both molted (Figure 34.3*b*). Thus, one or more substances from the bug fed earlier crossed through the glass tube and stimulated the molting process in the other bug.

We now know that two hormones regulate the molt process. Cells in the brain produce a substance, simply called brain hormone, that is transported to a pair of neuroendocrine structures attached to the brain called the **corpora cardiaca** (singular, corpus cardiacum). After appropriate stimulation—which for *Rhodnius* is a blood meal—the corpora cardiaca release brain hormone, which diffuses to an endocrine gland, the **prothoracic gland**. The prothoracic gland then produces and releases a hormone called ecdysone, which directly stimulates the molt. The control of molting by brain hormone and ecdysone is a general mechanism in insects.

Hormonal Control of Development in Insects

The *Rhodnius* decapitation experiments described above produced a curious result: Regardless of the instar used for the experiment, the decapitated bug always molted directly into an adult form. Additional experiments by Wigglesworth demonstrated that a hormone other than those responsible for molting must determine whether the bug would molt into another juvenile instar or into an adult. The head of *Rhodnius* is long, so it was possible to remove just the front part of the head which contains the neuroendocrine cells that secrete and release brain hormone, but spare the rear part, which contains two other endocrine structures called the **corpora allata**. When Wigglesworth did the glass-tube experiment with an unfed, completely decapitated, fifth-instar bug connected to a fourth-instar bug that had been fed but had had only the front part of its head removed, both bugs molted into juvenile forms (Figure 34.4). Some substance coming from the rear part of the head of the fourth-instar bug prevented the ex-

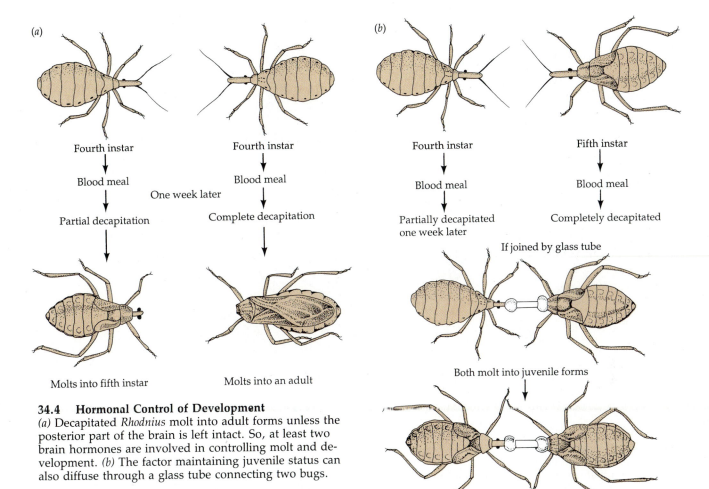

(a) Fourth instar

Blood meal

Partial decapitation

One week later

Molts into fifth instar

(b) Fourth instar

Blood meal

Partially decapitated one week later

Fourth instar

Blood meal

Complete decapitation

Molts into an adult

Fifth instar

Blood meal

Completely decapitated

If joined by glass tube

Both molt into juvenile forms

34.4 Hormonal Control of Development
(a) Decapitated *Rhodnius* molt into adult forms unless the posterior part of the brain is left intact. So, at least two brain hormones are involved in controlling molt and development. *(b)* The factor maintaining juvenile status can also diffuse through a glass tube connecting two bugs.

pected result that both bugs would molt into adult forms. We now know that substance is juvenile hormone and it comes from the corpora allata. As long as juvenile hormone is present, *Rhodnius* molts into another juvenile instar. The corpora allata normally stop producing juvenile hormone during the fifth instar. If juvenile hormone is absent, the bug molts into the adult form.

The role of juvenile hormone is more complex in insects that undergo complete metamorphosis. Complete metamorphosis includes a pupal stage that metamorphoses into the adult (see Chapter 15). An excellent example is the silkworm moth, *Hyalophora cecropia* (Figure 34.5). As long as juvenile hormone is present in high concentrations, larvae molt into larvae. When levels of juvenile hormone fall, larvae molt into pupae. No juvenile hormone is produced in the pupae, so they molt into adults. In our perpetual war against insects, juvenile hormone is a new weapon. Synthetic forms of juvenile hormone can be distributed in the environment to prevent the development of adult insects capable of reproduction.

The critical experiments identifying insect hormones and their functions preceded by many years the actual chemical identification of the hormones.

That is not surprising when you consider the tiny amounts of certain hormones that exist in an organism. In one of the earliest studies of the prothoracic gland hormone ecdysone, biochemists produced 250 milligrams of pure ecdysone (about one-fourth the weight of an apple seed), but they started with 4 tons of silkworms!

VERTEBRATE HORMONES

The evolution of endocrine systems has been a process of using the same chemical messages but coupling them to new physiological responses. In many cases the same chemical structure serves as a hormone in widely divergent species but has completely different actions. Many vertebrate hormones have molecular structures similar or identical to those of invertebrate hormones, but once again their functions are different. The hormone thyroxine, for example, is found in animal species ranging from tunicates (sea squirts) to humans, but thyroxine's function differs greatly among these species. In mammals, it elevates cellular metabolic rate; in frogs, it is

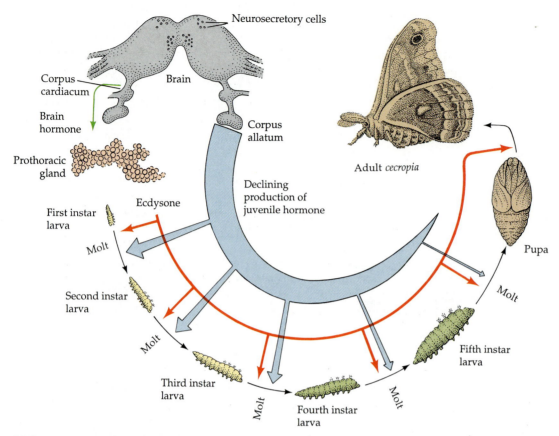

34.5 Three Hormones Control Molt and Metamorphosis
Neurosecretory cells in the brain of the *cecropia* moth produce brain hormone
(green arrow) that is stored in and released from the corpora cardiaca. Brain
hormone stimulates the prothoracic gland to secrete ecdysone (red pathway).
The corpora allata produce juvenile hormone (blue crescent). As long as juvenile hormone is abundant, the larva molts into a larger larva in response to
ecdysone. As juvenile hormone wanes the larva molts into a pupa. The pupa
does not produce juvenile hormone, so it metamorphoses into an adult. The
release of ecdysone is episodic; each red arrow indicates a molt.

essential for metamorphosis from tadpole to adult.
Another example is prolactin, which stimulates milk
production in female mammals after they give birth.
In pigeons and doves, prolactin stimulates the production of crop milk for nourishment of the young.
Crop milk is really not a milk secretion at all, but a
sloughing off of cells lining the upper digestive tract.
In amphibians, prolactin causes the animals to prepare for reproduction by seeking water, and in fishes,
such as salmon, that migrate between salt and fresh
water, prolactin regulates the mechanisms that maintain osmotic balance with the changing environment.

The endocrine systems of vertebrates are quite varied and complex. We recognize at least nine endocrine glands (see Figure 34.2), most of which produce
and release more than one hormone. In addition,
cells in many other organs produce and release hormones. The list of chemical messages in the body is
long and growing longer (Table 34.1).

In this section we will focus on the hormones of
mammals. The same hormones may exist in other
vertebrates, but as just noted, they may serve different functions than they do in mammals. We begin
our survey with an examination of the "master
gland," the **pituitary**, which produces a number of
hormones, some of which have as their targets other
hormone-secreting cells elsewhere in the body.

The pituitary gland sits in a depression at the bottom of the skull just over the back of the roof of the
mouth (Figure 34.6). The posterior pituitary originates as an outpocketing of the developing brain,
whereas the anterior pituitary originates as an outpocketing of the upper region of the embryonic
digestive tract. Thus the posterior pituitary derives
from nervous system tissue; the hormones it stores
and releases are called neurohormones. The anterior
pituitary consists of epithelial tissue that develops
endocrine functions.

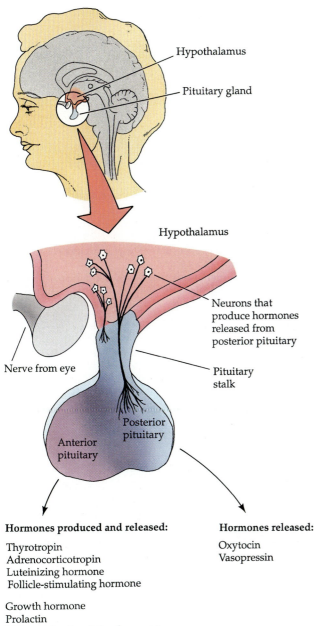

34.6 A Powerful Gland
The human pituitary gland is only the size of a blueberry, but it secretes many hormones. The posterior pituitary secretes two peptides produced in the hypothalamus. The anterior pituitary produces and secretes four tropic hormones and a number of other hormones.

Labels in figure:
Hypothalamus
Pituitary gland
Hypothalamus
Neurons that produce hormones released from posterior pituitary
Nerve from eye
Pituitary stalk
Posterior pituitary
Anterior pituitary

Hormones produced and released:
Thyrotropin
Adrenocorticotropin
Luteinizing hormone
Follicle-stimulating hormone

Growth hormone
Prolactin
Melanocyte-stimulating hormone
Endorphins
Enkephalins

Hormones released:
Oxytocin
Vasopressin

Neurohormones of the Posterior Pituitary

The posterior pituitary releases two neurohormones, vasopressin (also called antidiuretic hormone or ADH) and oxytocin. These are small peptides synthesized in nerve cells in the hypothalamus. Vasopressin and oxytocin move down long extensions of these nerve cells that course down the pituitary stalk into the posterior pituitary, where the hormones are stored in the nerve endings (Figure 34.6).

The posterior pituitary increases its release of vasopressin whenever blood pressure falls or the blood becomes too salty. The main action of vasopressin is to increase the amount of water conserved by the kidneys. When vasopressin secretion is high, the kidneys reabsorb more water and produce a small volume of highly concentrated urine. When vasopressin secretion is low, the kidneys produce a larger volume of dilute urine. We will come back to the mechanism of vasopressin action in Chapter 42.

When a woman is about to give birth, her posterior pituitary releases oxytocin, which stimulates the contractions of the muscles that push the baby out of her body. Oxytocin also brings about the letdown of milk from her breasts. The baby's suckling stimulates nerve cells in the mother, causing secretion of oxytocin. Even the sight and sounds of her baby can cause a nursing mother to secrete oxytocin and release milk from her breasts.

Hormones of the Anterior Pituitary

The anterior pituitary produces and secretes a diversity of peptide hormones. Each of these hormones is produced by a different type of pituitary cell. Four of these hormones, called **tropic hormones,** control the activity of other endocrine glands. The tropic hormones are thyrotropin, adrenocorticotropin, luteinizing hormone, and follicle-stimulating hormone. We will say more about these tropic hormones when we describe their target glands (thyroid, adrenal cortex, testes, and ovaries). The other hormones produced by the anterior pituitary influence tissues that are not endocrine glands. These hormones are growth hormone, prolactin, melanocyte-stimulating hormone, and endorphins and enkephalins.

Growth hormone consists of about 200 amino acids and acts on a wide variety of tissues to promote growth directly and indirectly. One of its important direct effects is to stimulate cells to take up amino acids. Growth hormone promotes growth indirectly by stimulating the liver to produce growth-regulating factors called somatomedins, which circulate in the blood and stimulate the growth of bone and cartilage. Overproduction of growth hormone in children causes gigantism, and underproduction causes dwarfism (Figure 34.7a,b). High levels of growth hormone in adults cannot cause increased height, because the shafts and the growth plates of the long bones have fused. Rather, abnormally high levels of growth hormone in adults cause thickening of the hands, feet, jaw, nose, and ears, creating a condition known as acromegaly.

Beginning in the late 1950s, children diagnosed as having a serious deficiency of growth hormone, and

TABLE 34.1
Principal Hormones of Humans

SECRETING TISSUE OR GLAND	HORMONE	CHEMICAL NATURE	TARGETS	IMPORTANT PROPERTIES OR ACTIONS
Hypothalamus	Releasing and release-inhibiting hormones (see Table 34.2)	Peptides	Anterior pituitary	Control secretion of hormones of anterior pituitary
	Oxytocin, vasopressin	Peptides	(See Posterior pituitary)	Stored and released by posterior pituitary
Anterior pituitary: Tropic hormones	Thyrotropin	Glycoprotein	Thyroid gland	Stimulates synthesis and secretion of thyroxine
	Adrenocorticotropin	Polypeptide	Adrenal cortex	Stimulates release of hormones from adrenal cortex
	Luteinizing hormone	Glycoprotein	Gonads	Stimulates secretion of sex hormones from ovaries and testes
	Follicle-stimulating hormone	Glycoprotein	Gonads	Stimulates growth and maturation of ova in females; stimulates sperm production in males
Anterior pituitary: Other hormones	Growth hormone	Protein	Bones, liver, muscle	Stimulates protein synthesis and growth
	Prolactin	Protein	Mammary glands	Stimulates milk production
	Melanocyte-stimulating hormone	Peptide	Melanocytes	Controls skin pigmentation
	Endorphins and enkephalins	Peptides	Spinal cord neurons	Decreases painful sensations
Posterior pituitary	Oxytocin	Peptide	Uterus, breasts	Induces birth by stimulating labor contractions; causes milk flow
	Vasopressin (antidiuretic hormone)	Peptide	Kidneys	Stimulates water reabsorption
Thyroid	Thyroxine	Iodinated amino acid derivative	Many tissues	Stimulates and maintains metabolism necessary for normal development and growth
	Calcitonin	Peptide	Bone	Stimulates bone formation; lowers blood calcium
Parathyroids	Parathormone	Protein	Bone	Absorbs bone; raises blood calcium
Thymus	Thymosins	Peptides	Immune system	Activate immune responses of T cells in the lymphatic system
Pancreas	Insulin	Protein	Muscle, liver, fat, other tissues	Stimulates uptake and metabolism of glucose; increases conversion of glucose to glycogen and fat
	Glucagon	Protein	Liver	Stimulates breakdown of glycogen and raises blood sugar

therefore destined to become hypopituitary dwarfs, were treated with human growth hormone extracted from human pituitaries taken from cadavers. The treatment was successful in stimulating substantial growth, but it was extremely costly; a year's supply of human growth hormone for one individual required up to 50 pituitaries. In the mid-1980s scientists using genetic engineering technology isolated the gene for human growth hormone and introduced it into bacteria, which produced enough of the hormone to make it commercially available.

It is now feasible and affordable to prevent hypopituitary dwarfism, but the availability of growth hormone raises new questions. Should any child at the lower end of the height charts be treated? Should a normal child whose parents are interested in basketball be given growth hormone? These types of questions are impossible to answer with scientific data

TABLE 34.1
Principal Hormones of Humans

SECRETING TISSUE OR GLAND	HORMONE	CHEMICAL NATURE	TARGETS	IMPORTANT PROPERTIES OR ACTIONS
Pancreas	Somatostatin	Peptide	Digestive tract; other cells of the pancreas	Inhibits insulin and glucagon release; also decreases secretion, motility, and absorption in the digestive tract
Adrenal medulla	Adrenaline, noradrenaline	Modified amino acids	Heart, blood vessels, liver, fat cells	Stimulate "fight-or-flight" reactions: increase heart rate, redistribute blood to muscles, raise blood sugar
Adrenal cortex	Glucocorticoids (cortisol)	Steroids	Muscle, immune system, various other tissues	Mediate response to stress; reduce metabolism of glucose, increase metabolism of proteins and fats; reduce inflammation and immune responses
	Mineralcorticoids (aldosterone)	Steroids	Kidneys	Stimulates excretion of potassium ions and reabsorption of sodium ions
Stomach lining	Gastrin	Peptide	Stomach	Promotes digestion of food by stimulating release of digestive juices; stimulates stomach movements that mix food and digestive juices
Lining of small intestine	Secretin	Peptide	Pancreas	Stimulates secretion of bicarbonate solution by ducts of pancreas
	Cholecystokinin	Peptide	Pancreas, liver, and gall bladder	Stimulates secretion of digestive enzymes by pancreas and other digestive juices from liver; stimulates contractions of gall bladder and ducts
	Enterogastrone	Polypeptide	Stomach	Inhibits digestive activities in the stomach
Pineal	Melatonin	Modified amino acid	Hypothalamus	Involved in biological rhythms
Ovaries	Estrogens	Steroids	Breasts, uterus, other tissues	Stimulate development and maintenance of female characteristics and sexual behavior
	Progesterone	Steroid	Uterus	Sustains pregnancy; helps to maintain secondary female sexual characteristics
Testes	Androgens	Steroids	Various tissues	Stimulate development and maintenance of male sexual behavior and secondary male sexual characteristics; stimulate spermatogenesis
Most cells	Prostaglandins	Modified fatty acids	Various tissues	Many diverse actions
Heart	Atrial natriuretic hormone	Peptide	Kidneys	Increases sodium ion excretion

alone. The controversy around growth hormone has become even more complex as a result of a recent study suggesting that when growth hormone is administered to older persons, it reverses some of the effects of aging. In comparison to a control group not receiving growth hormone, a group of elderly persons receiving growth hormone decreased their body fat, increased their muscle mass, and reported feeling more energetic. It is not known whether these changes will last beyond the period of treatment, or whether the treatment has side effects.

Prolactin is another hormone produced by the anterior pituitary. Earlier in the chapter, we described the evolutionary diversity of the functions of prolactin. In human females the major function of prolactin is to stimulate the production and secretion of milk. In some mammals prolactin also functions as an important hormone during pregnancy. In human males

(a)

34.7 Effects of Abnormal Amounts of Growth Hormone

(a) Gigantism results from the overproduction of growth hormone in childhood. In this news photo from 1939, a young man and his father visit Chicago on business; the son (left) is over 8 feet tall, whereas his father is a very average 5 feet, 11 inches. (b) When the anterior pituitary does not produce enough growth hormone during childhood, pituitary dwarfism results. The man on the left is P. T. Barnum, circus entrepreneur. With him is Charles Stratton, who appeared in Barnum's circus under the name General Tom Thumb.

(b)

prolactin plays a role along with other pituitary hormones in controlling the endocrine function of the testes.

Melanocyte-stimulating hormone is produced in very low amounts by the human anterior pituitary, and its functions in humans are not well understood. Melanocytes are cells containing melanin, a black pigment. In fishes, amphibians, and reptiles that can change their color, melanocyte-stimulating hormone changes the way melanin is distributed in the melanocytes, thereby darkening the tissue containing them.

The remaining hormones of the anterior pituitary are endorphins and enkephalins, which are sometimes referred to as the body's "natural opiates." These molecules help to control pain. What is interesting about the production of endorphins and enkephalins in the pituitary is that they are encoded by the same gene as are two other pituitary hormones.

The gene codes for a large parent molecule called pro-opiomelanocortin. This large molecule is cleaved to produce several peptides, some of which have hormonal functions. Adrenocorticotropin, melanocyte-stimulating hormone, endorphins, and enkephalins all result from the cleavage of pro-opiomelanocortin.

Hypothalamic Releasing Neurohormones

The idea of the pituitary as the "master gland" received quite a blow with the discovery that it is really a "middleman" controlled by the hypothalamus. The hypothalamus is a part of the brain, and the hormones it releases are called neurohormones. The hypothalamus receives information about conditions in the body and in the external environment through the nervous system. If the connection between the

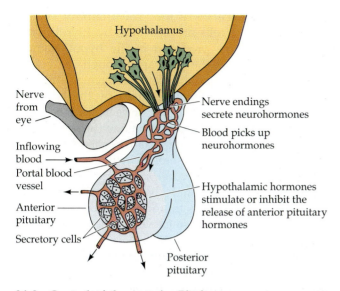

34.8 Control of the Anterior Pituitary
A system of blood vessels runs between the hypothalamus and the anterior pituitary. Peptide hormones produced by hypothalamic cells enter these blood vessels in the median eminence and are transported to the anterior pituitary, where they control the activity of the pituitary cells that synthesize and release hormones.

hypothalamus and the pituitary is cut, the release of pituitary hormones no longer responds to changes in the environment or in the body. If pituitary cells are maintained in culture, extracts of hypothalamic tissue stimulate some of those cells to release their hormones into the culture medium. Therefore, scientists hypothesized that secretions of the hypothalamic cells control the activities of the cells of the anterior pituitary. The route by which such chemical messages could reach the anterior pituitary was known —namely, a special set of blood vessels called **portal blood vessels** that run between the hypothalamus and the anterior pituitary (Figure 34.8). It was proposed that secretions from nerve endings in the hy-

pothalamus are absorbed into the blood and are conducted down the portal vessels to the anterior pituitary where they cause the release of anterior pituitary hormones.

In the 1960s two large teams of scientists, led by Roger Guillemin and Andrew Schally, initiated the search for the hypothalamic releasing neurohormones. Because the amounts of releasing hormones in any individual mammal are tiny, "bucket biochemistry" was required for their isolation. The scientists set up teams in slaughterhouses to collect massive numbers of hypothalami from pigs and sheep. The resulting *tons* of tissue were shipped to laboratories in refrigerated trucks. One effort began with the hypothalami from 270,000 sheep and yielded only 1 milligram of purified thyrotropin-releasing hormone. Biochemical analysis of this pure sample revealed that thyrotropin-releasing hormone contains only three amino acids; it is a tripeptide. Closely following the discovery of thyrotropin-releasing hormone, the scientists identified gonadotropin-releasing hormone, which controls the release of follicle-stimulating hormone and luteinizing hormone from the anterior pituitary. The 1972 Nobel prize in medicine was awarded to Guillemin and Schally for these discoveries. Isolation techniques have been improved enormously so that now we need only milligrams of tissue to isolate and characterize a peptide. Many more hypothalamic neurohormones are now known, and they include both releasing hormones and release-inhibiting hormones (see Table 34.2).

Hormones of the Thyroid Gland

The thyroid gland consists of two lobes, one on either side of the windpipe or trachea, connected by a strip of thyroid tissue that wraps around the front of the trachea. If you gently place your thumb and forefinger on either side of your trachea just below your Adam's apple and swallow, you will feel your thyroid

TABLE 34.2
Releasing and Release-Inhibiting Neurohormones of the Hypothalamus

NEUROHORMONE	ACTION
Thyrotropin-releasing hormone	Stimulates thyrotropin release
Gonadotropin-releasing hormone	Stimulates release of follicle-stimulating hormone and luteinizing hormone
Prolactin release-inhibiting hormone	Inhibits prolactin release
Prolactin-releasing hormone	Stimulates prolactin release
Somatostatin (growth hormone release-inhibiting hormone)	Inhibits growth hormone release; interferes with thyrotropin release
Growth hormone-releasing hormone	Stimulates growth hormone release
Adrenocorticotropin-releasing hormone	Stimulates adrenocorticotropin release
Melanocyte-stimulating hormone release–inhibiting hormone	Inhibits melanocyte-stimulating hormone release

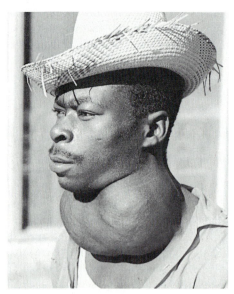

34.9 Goiter
A greatly enlarged thyroid gland is a symptom of the condition goiter. Worldwide, goiter affects about 5 percent of the population. The addition of iodine to table salt has greatly reduced the incidence of the condition in industrialized nations, but many cases still occur in the less developed countries of the world.

gland move up and down under your fingertips. The thyroid gland produces the hormones thyroxine and calcitonin. The thyroxine molecule consists of two molecules of the amino acid tyrosine and four atoms of iodine. Another form of thyroid hormone has only three iodine atoms, but we will refer to both as thyroxine for convenience.

Thyroxine in mammals plays many roles in regulating cell metabolism. It elevates the metabolic rates of most cells and tissues and promotes the use of carbohydrates rather than fats as fuel. Exposure to cold over a period of days leads to an increased release of thyroxine and an increase in basal metabolic rate. Thyroxine is especially crucial during development and growth. It promotes amino acid uptake and protein synthesis by cells. Insufficient thyroxine in a human fetus or growing child greatly retards physical and mental growth, resulting in a condition known as cretinism.

Malfunction of the thyroid gland causes a condition called goiter in which the thyroid gland gets very large and appears as a swelling on the front of the neck (Figure 34.9). Goiter can be associated with either **hyperthyroidism** (very high levels of thyroxine) or **hypothyroidism** (very low levels of thyroxine). To understand how that can be so, we must consider the control of thyroid activity by the pituitary and by the hypothalamus.

The activity of the thyroid gland is determined by thyrotropin, which is secreted into the blood by the pituitary. Thyrotropin activates the thyroid gland

cells that produce thyroxine. The thyrotropin-producing pituitary cells are activated by thyrotropin-releasing hormone produced in the hypothalamus and transported to the pituitary through the portal vessels. Environmental information such as ambient temperature or day length is used by the brain to determine whether to increase or decrease the secretion of thyrotropin-releasing hormone (Figure 34.10). There is at least one very important negative feedback loop in this sequence of steps—circulating thyroxine inhibits the response of the pituitary cells to thyrotropin-releasing hormone. Therefore, when thyroxine levels are high, less thyrotropin is released, and conversely, when thyroxine levels are low, more thyrotropin is released.

Hyperthyroid goiter results when the pituitary cells are not turned off by thyroxine. Thyrotropin levels remain high and the thyroid gland is activated so much that it grows bigger. Production of thyroxine by the thyroid stays abnormally high. Hyperthyroid patients have high metabolic rates, are jumpy and nervous, usually feel hot, and may have a buildup of fat behind the eyeballs, causing their eyes to bulge.

Hypothyroid goiter results when there is not enough circulating thyroxine to turn off thyrotropin production. Its most common cause is a deficiency of dietary iodide, without which the thyroid gland cannot make *functional* thyroxine. Without thyroxine, thyrotropin levels remain high, so the thyroid continues to produce large amounts of nonfunctional thyroxine and gets very large. The symptoms of hypothyroidism are low metabolism, intolerance of cold, and general physical and mental sluggishness.

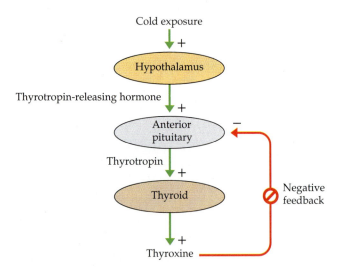

34.10 Regulation of Thyroid Function
Chronic exposure to cold results in the hypothalamus producing more thyrotropin-releasing hormone, which stimulates certain pituitary cells to produce more thyrotropin. Thyrotropin stimulates increased activity in thyroid cells resulting in increased production of thyroxine. Thyroxine exerts negative feedback on the pituitary, decreasing the release of thyrotropin.

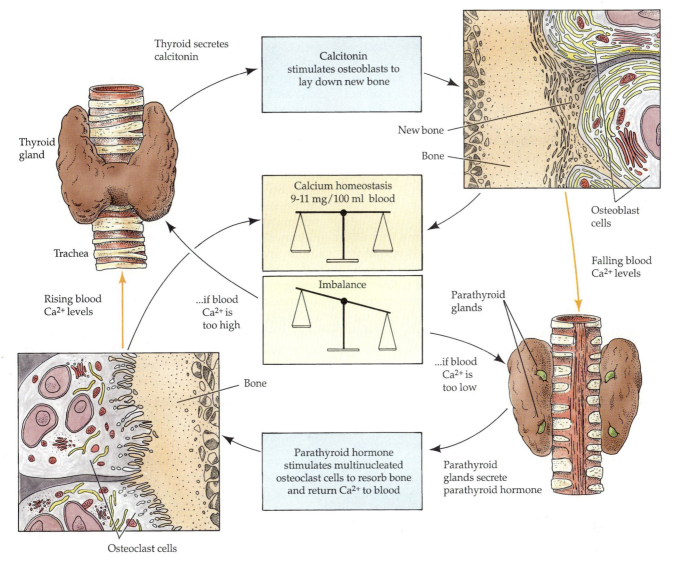

34.11 Calcium Balance in Bone
The calcium levels of the blood are regulated in part by calcitonin and parathyroid hormone. Bone can be a source of calcium or a sink for excess calcium. Osteoclasts break down bone and release calcium, whereas osteoblasts build new bone using calcium from the blood.

Hypothyroid goiter used to be extremely common in mountainous areas and regions far from the oceans where there is little iodide in the soil or water. The addition of iodide to table salt has greatly reduced the incidence of the disease.

Another hormone of the mammalian thyroid gland is calcitonin. It is not produced by the same cells that produce thyroxine. Calcitonin helps to regulate the levels of calcium circulating in the blood (Figure 34.11). Bone is a huge repository of calcium in the body, and bone is continually being remodeled. Cells called osteoclasts break down bone and release calcium, and cells called osteoblasts use circulating calcium to deposit new bone. Calcitonin decreases the activity of osteoclasts and stimulates the activity of osteoblasts; thus, it shifts the balance from adding

calcium ions to the blood to taking calcium ions from the blood. The regulation of blood calcium levels is more strongly influenced by the hormone to be considered in the next section than it is by calcitonin, but calcitonin plays an important role in protecting the bones of women during pregnancy.

Parathyroid Hormone

The **parathyroid glands** are four tiny structures superficially embedded on the surface of the thyroid gland. Their single hormone product is parathyroid hormone, or parathormone, a critical control element in the regulation of blood calcium levels. Bone growth and remodeling require calcium; so do many cellular processes, such as nerve and muscle functions, which are very sensitive to changes in calcium concentration. Muscle contraction and nerve function are severely impaired if the blood calcium level rises or falls by as little as 30 percent of normal values. A fall in blood calcium triggers the release of parathormone, which in turn stimulates a number of actions that add calcium to the blood (Figure 34.11). Parathormone stimulates osteoclasts to dissolve bone and release calcium. It also helps the digestive tract to absorb calcium from food and helps the kidneys reabsorb calcium before excreting wastes.

Hormones of the Pancreas

Prior to the 1920s, severe diabetes mellitus was an invariably fatal disease characterized by weakness, lethargy, and body wasting. The disease was known to be connected with a gland located just below the stomach, the **pancreas**, and with abnormal glucose metabolism, but the link was not clear. Today we know that diabetes mellitus is caused by a lack of the hormone insulin. Insulin replacement therapy makes it possible for more than 1.5 million diabetics in the United States to lead almost normal lives.

Most cells in an untreated diabetic's body cannot use glucose in the blood for metabolic fuel. As a result, the blood glucose level rises until glucose is lost in the urine. High blood glucose causes water to move from cells into the blood by osmosis, and the kidneys increase urine output to excrete the excess fluid volume in the blood. The name *diabetes* refers to copious production of urine, and *mellitus* (derived from the Greek word for honey) reflects the fact that the urine of the untreated diabetic is sweet. Since the cells of the body cannot use blood glucose for fuel, they must burn fat and protein. As a result, the body of the untreated diabetic wastes away and critical tissues and organs are damaged. The change in the prognosis for diabetics came almost overnight in 1921, when medical doctor Frederick Banting and medical student Charles Best of the University of Toronto discovered that they could reduce the symptoms of diabetes with an extract they prepared from pancreatic tissue. The work of Banting and Best led to enormous relief of human suffering.

The active component of the extract Banting and Best prepared was a small protein hormone, insulin, consisting of 51 amino acids. Insulin is produced in clusters of cells in the pancreas, called **islets of Langerhans** after the German medical student who discovered them. Other cells in the islets produce two additional hormones, glucagon and somatostatin. The rest of the pancreas produces enzymes and secretions that are delivered through ducts to the intestine, where they play roles in digestion. Thus, the pancreas is both an endocrine and an exocrine gland.

Following a meal, the concentration of glucose in the blood rises as glucose is absorbed from the gut.

This rise in glucose concentration stimulates the pancreas to release insulin. Insulin stimulates cells to use glucose as fuel and to convert it into storage products such as glycogen and fat. When there is no longer food in the gut, the blood's glucose concentration falls, and the pancreas stops releasing insulin. As a result, most of the cells of the body shift to using glycogen and fat rather than glucose for fuel. If the concentration of glucose in the blood falls below normal, cells in the islets release the hormone glucagon, which stimulates the liver to convert glycogen back to glucose to resupply the blood. These relationships will be discussed in greater detail in Chapter 41.

Somatostatin is a curious hormone. Released from the pancreas in response to rapid rises of glucose and amino acids in the blood, it inhibits the release of both insulin and glucagon. Somatostatin also slows the digestive activities of the gut. The function of this hormone may be to extend the period of time over which nutrients are absorbed from the gut and used by the cells of the body. This substance was first discovered as a hypothalamic neurohormone that *inhibits* the release of growth hormone by the pituitary. It was called growth hormone release-inhibiting hormone, but somatostatin is a more convenient name.

The Adrenal Hormones

An adrenal gland sits over each kidney. Functionally and anatomically an adrenal gland is actually a gland within a gland (Figure 34.12). The core, called the

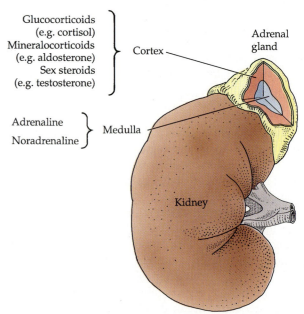

Hormones released into blood:

Glucocorticoids (e.g. cortisol)
Mineralocorticoids (e.g. aldosterone)
Sex steroids (e.g. testosterone) } Cortex

Adrenaline
Noradrenaline } Medulla

Adrenal gland

Kidney

34.12 The Adrenals: Two Glands in One
The adrenal medulla and the adrenal cortex produce different hormones. Together they form an adrenal gland, one of which is situated on top of each kidney.

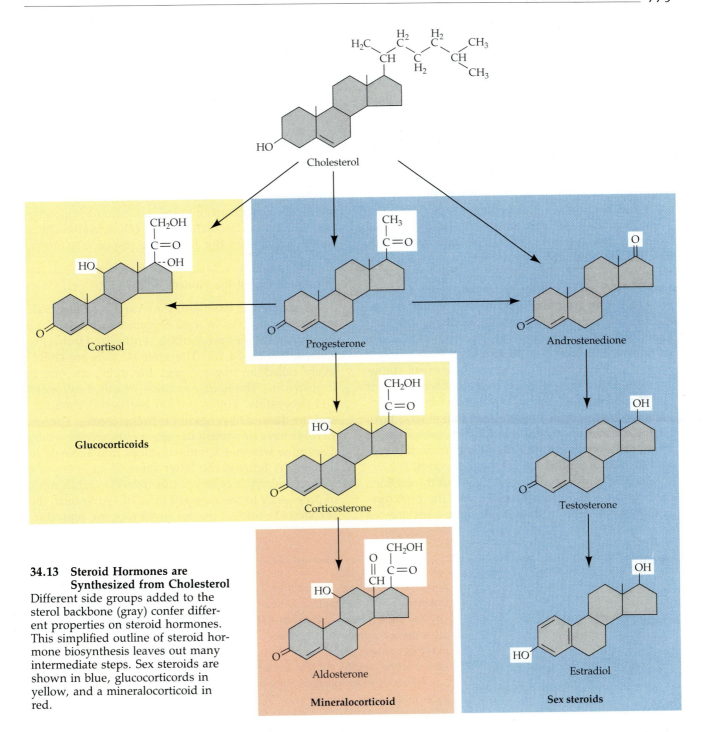

34.13 Steroid Hormones are Synthesized from Cholesterol
Different side groups added to the sterol backbone (gray) confer different properties on steroid hormones. This simplified outline of steroid hormone biosynthesis leaves out many intermediate steps. Sex steroids are shown in blue, glucocorticords in yellow, and a mineralocorticoid in red.

adrenal medulla, produces the hormones adrenaline and, to a lesser degree, noradrenaline. Surrounding the medulla (as an apricot surrounds its pit) is the **adrenal cortex**, which produces a number of hormones. The medulla develops from nervous system tissue and is under the control of the nervous system; the cortex is under hormonal control, largely by adrenocorticotropin from the anterior pituitary.

The adrenal medulla gives rise to the fight-or-flight reactions to stressful situations. Adrenaline arouses the body to action. As we saw earlier in the chapter, it increases heart rate, breathing rate, and blood pres-

sure, and it diverts blood flow to active skeletal muscles and away from the gut.

The hormones produced by the adrenal cortex are all steroids synthesized from cholesterol (Figure 34.13). In general they are called the corticosteroids, and they are divided into three functional classes. The glucocorticoids influence blood glucose concentrations as well as other aspects of fat, protein, and carbohydrate metabolism. The mineralocorticoids influence the ionic balance of the extracellular fluids. The sex steroids stimulate sexual development and reproductive activity. Sex steroids are secreted in only small amounts by the adrenal cortex and will be discussed further in the section on the gonads. Of the 30 or so different steroids produced by the adrenal cortex, only two are of great importance in human physiological functions under normal conditions. Those are the mineralocorticoid aldosterone (which helps to regulate salt concentration in the blood, as will be discussed in Chapter 42) and the glucocorticoid cortisol.

Cortisol plays important roles in helping the body to respond to short-term stress. Recall the lion that roared at you earlier in the chapter— now imagine it chasing you into high grass and stalking you for hours as you try to get to safety. Your immediate reaction is stimulated by your nervous system and by the release of adrenaline. Your heart is beating faster, you are breathing faster, and your running muscles are getting maximal supplies of oxygen and glucose. This is not a sustainable situation, so within about 5 minutes, cortisol levels rise and help you to sustain your escape. You need a high level of blood glucose for your brain to function, so cortisol stimulates the other cells of your body to decrease their use of glucose and start to metabolize fats and proteins for energy. This is not a time to feel sick, have allergic reactions, or heal wounds, so cortisol blocks immune system reactions. This is why cortisol is useful for reducing inflammations and allergies.

The effects of adrenaline and cortisol in reactions to short-term stress are beneficial, but the prolongation of these effects by the chronic, long-term stresses of modern life can be very damaging to the body. High blood pressure, poor gastrointestinal function, inhibition of protein synthesis, fat mobilization, and inhibition of the immune system are not healthy responses over long periods of time. Additional details on the interactions of stress, aging, and the cortisol response are discussed in Box 34.A.

Cortisol release is controlled by the pituitary hormone adrenocorticotropin, which in turn is controlled by the hypothalamic adrenocorticotropin-releasing hormone. Because the cortisol response to stress has this chain of steps, each involving secretion, diffusion, circulation, and cell activation, it is much slower than the adrenaline response to stress.

The Hormones of the Gonads

The testes of the male and the ovaries of the female produce hormones as well as gametes. Most of the gonadal hormones are steroids synthesized from cholesterol (Figure 34.13). The male steroids are collectively called androgens, and the dominant one is testosterone. The female steroids are called estrogens and progestins. The dominant estrogen is estradiol, and the dominant progestin is progesterone. The sex steroids have important developmental effects—they determine whether a fetus develops into a female or a male. (A fetus is the latter stage of an embryo; a human embryo is called a fetus from the eighth week of pregnancy to the moment of birth.) After birth the sex steroids control the maturation of the reproduc-

BOX 34.A

The Rat Race

We usually assume that excessive stress leads to premature aging. Stress reactions *increase* blood pressure and fat metabolism while they *inhibit* digestion and immune system function. When prolonged, the effects of stress can contribute to cardiovascular disease, strokes, ulcers, and susceptibility to diseases. Therefore, assuming that stress accelerates aging is not unreasonable. The exact physiological interactions between stress and aging have only recently been elucidated, however, and they are quite interesting. Research by Robert Sapolsky at Stanford University has shown that old rats can initiate a stress response just as effectively as young rats, but they cannot turn it off as rapidly. All the deleterious effects of stress persist longer in older rats than in younger ones. Why should this be so?

The answer involves the mechanism of negative feedback in the control of stress responses. In a region of the brain called the hippocampus there are cells with receptors for cortisol. When a rat experiences stress, cortisol levels rise in the blood. The cortisol activates these hippocampal cells, which inhibit the secretion of adrenocorticotropin-releasing hormone by the hypothalamus. Decreased adrenocorticotropin-releasing hormone secretion causes a decline in adrenocorticotropin release from the pituitary, so the adrenal cortex stops producing cortisol. As rats age, they lose the hippocampal cells of this negative feedback loop. Apparently cortisol contributes to the demise of these cells. So, the more stress experienced, the more of these crucial hippocampal cells are lost. The result is a premature loss of the negative feedback mechanism that protects the body from the deleterious effects of sustained stress responses. The increased incidence of stress responses leads to many disorders usually associated with aging— strokes, cardiovascular disease, digestive system malfunctions, and impaired immune system function that increases susceptibility to cancers and other diseases. It is believed that these relationships between stress and aging also pertain to humans.

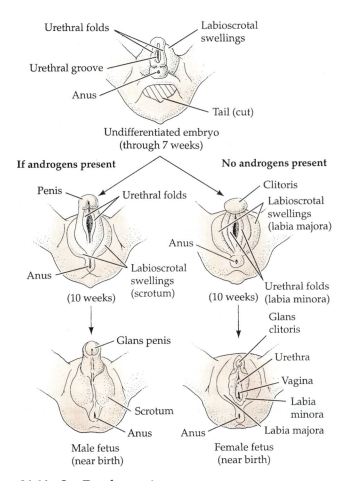

34.14 Sex Development
Hormones direct sex development along one pathway or its alternative as human external sex organs form. The sex organs of early embryos are similar (top). The testes of genetic males begin to secrete androgens at about 7 weeks after fertilization. Under the influence of androgens, a penis and scrotum form. Without the influence of androgens, female external organs develop.

tive organs and the development and maintenance of secondary sexual characteristics such as breasts and facial hair.

The developmental effects of the sex steroids begin at about the seventh week of gestation in the human embryo. Up until that time, the embryo could develop into either sex. The ultimate instructions for sex determination reside in the genes. Individuals receiving two X chromosomes normally become females, and individuals receiving an X and a Y chromosome normally become males. These instructions, however, are carried out through the production and action of the sex steroids, and the potential for error exists. The presence of the Y chromosome normally causes the embryonic, undifferentiated gonads to begin producing androgens in the seventh week. In response to the androgens, the reproductive system develops into that of a male (Figure 34.14). However, if the androgens are not produced at that time, or the androgen receptors do not function, then the

fetus develops female reproductive structures, even if it is a male genetically. In humans, female development is the default, or neutral, course; a fetus will develop female characterists unless androgens are present to trigger male development. The opposite situation exists in some other vertebrates—male development is the default condition, which is switched to female development if estrogens are present.

Error in the hormonal control of sexual development results in the occasional production of intersex individuals. The most extreme (but rare) case is a true **hermaphrodite** who has both testes and ovaries. **Pseudohermaphrodites** have the gonads of one sex and the external sex organs of the other. For example, an XY individual can develop testes, but if his tissues are insensitive to the androgens produced by those testes, the external sex organs and the secondary sexual characteristics of a female can develop (Figure 34.15).

The effects of the sex steroids are dramatic at the time of puberty. Throughout childhood the production of sex steroids by the gonads is extremely low. As puberty approaches, the hypothalamus begins to produce and secrete gonadotropin-releasing hormone, which causes the pituitary to produce the gonadotropins luteinizing hormone and follicle-stimulating hormone. In the preadolescent male, the increased level of luteinizing hormone stimulates groups of cells in the testes to synthesize androgens, which in turn initiate the profound physiological, anatomical, and psychological changes associated with adolescence. The voice deepens, hair begins to grow on the face and body, the testes and the penis grow, and skeletal muscles enlarge. Even an active program of weight lifting will not lead to massive muscle development in preadolescent boys or in women because the increase in muscle bulk requires a level of androgens not normally found in individuals other than males past puberty. The taking of synthetic androgens to enhance muscle development is a dangerous practice and a serious type of substance abuse (Box 34.B).

Puberty in the female is also ushered in by an increased release of gonadotropin-releasing hormone by the hypothalamus. The levels of luteinizing hormone and follicle-stimulating hormone rise, stimulating the ovaries to mature and produce the female sex hormones, estrogens and progestins. The increased circulating levels of the sex steroids initiate the development of the traits characteristic of a sexually mature woman: enlarged breasts, vagina, and uterus, a broad pelvis, increased subcutaneous fat, pubic hair, and the initiation of the menstrual cycle.

Other Hormones

We have discussed all of the major endocrine glands and "classical" hormones in this chapter, but there

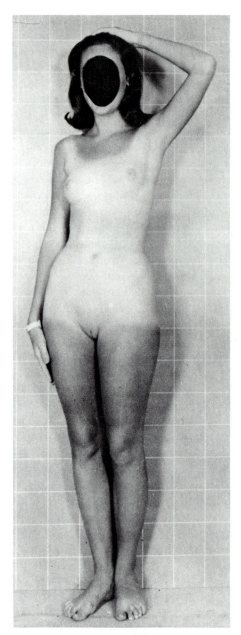

34.15 Testicular Feminization
This person, who is genetically a male with an XY geno-type, carries a mutation that leaves body cells unrespon-sive to male sex hormones. During early fetal life, the developing testes produced normal amounts of testoster-one, but because of the mutation, the cells forming the sex organs could not respond. Development thus fol-lowed the female pattern. Fully functional testes devel-oped within the body, but external female organs devel-oped. The person does not have a uterus or ovaries, however, and the vagina ends blindly.

are many hormones we have not mentioned. Exam-ples include the numerous hormones produced in the digestive tract that help organize the way the gut processes food. Even the heart has endocrine func-tions. When blood pressure rises and causes the walls of the heart to become more stretched, certain cells in the walls of the heart release atrial natriuretic hor-

mone. This hormone increases the excretion of so-dium ions and water by the kidneys, thereby low-ering blood volume and blood pressure. As we discuss the physiology of the organ systems of the body in subsequent chapters, we will frequently mention hormones those organs produce.

MECHANISMS OF HORMONE ACTION

We have discussed the functions of many hormones in this chapter, but we have not yet addressed the question of how these chemical messages are read by the cells that receive them. How do hormones induce their actions at the cellular and molecular lev-els? Two general mechanisms have been discovered, one for water-soluble hormones such as the peptide and protein hormones, and one for the lipid soluble hormones—the steroids and thyroxine.

Water-Soluble Hormones and Second Messengers

Water-soluble hormones act by way of a **second mes-senger**, the hormone itself being the first messenger. In this metaphor, the first messenger (the hormone) arrives at the plasma membrane of the target cell, but because water-soluble hormones cannot cross a plasma membrane, a second messenger is needed to "notify" the interior of the cell that the first messen-ger has arrived (Figure 34.16).

A well-characterized second messenger for some water-soluble hormones is **cAMP** (cyclic adenosine 3,5 monophosphate). The receptors that extend out-ward from the plasma membrane to bind hormones are proteins (Figure 34.16). For each specific hor-mone, there is a specific protein that acts as its re-ceptor. When a receptor binds a hormone molecule, the shape (the tertiary structure, as explained in Chapter 3) of the receptor changes; in its new form, the receptor can interact with a second protein in the membrane, enabling the second protein (G-protein) to bind a molecule of guanosine triphosphate (GTP). The activated G-protein splits off a subunit that in turn activates a third membrane protein, the enzyme **adenylate cyclase,** which extends to the inside of the membrane. Adenylate cyclase, when activated, con-verts ATP to cAMP within the cell.

The main points of the cAMP story were first dis-covered by E. W. Sutherland of Washington Univer-sity, who began this work in the 1950s. He was in-vestigating how adrenaline stimulates liver cells to break down storage molecules and liberate glucose. Several extremely important discoveries came out of this research. It became evident early on that there were a number of steps between the hormone–recep-tor interaction and the liberation of glucose. Suther-land was able to show that one process involved was

BOX 34.B

Muscle-Building Anabolic Steroids

The androgen testosterone helps skeletal muscles grow, especially when they are exercised regularly. The bulging biceps, triceps, pectorals, and deltoids of body builders are extreme examples of the skeletal muscle growth that occurs in every male past puberty. Natural muscular development can be exaggerated by both men and women who want to increase their maximum strength in athletic competition if they take synthetic androgens—anabolic steroids. Unfortunately, you can have too much of a good thing; anabolic steroids have serious negative side effects. In women, the use of artificial androgens causes the breasts and uterus to shrink, the clitoris to enlarge, menstruation to become irregular, facial and body hair to grow, and the voice to deepen. In men, the testes shrink, hair loss increases, the breasts enlarge, and sterility can result. What is even more serious, taking anabolic steroids greatly increases the risk of heart disease, certain cancers, kidney damage, and personality disorders such as depression, mania, psychoses, and extreme aggression. Most official athletic organizations, including the International Olympic Committee, ban anabolic steroid use. Frequently the blood of competing athletes is tested to determine whether they are in compliance. A gold-medal winner in track in the 1988 Olympic Games was required to forfeit his medal and records because he had used anabolic steroids.

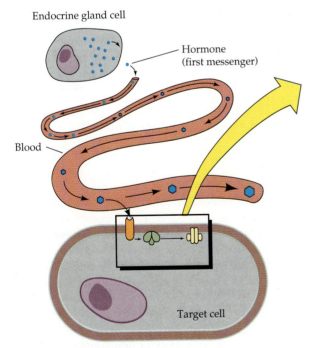

34.16 Second Messengers
Water-soluble hormones bind to receptors on the cell surface and activate second messengers such as cAMP.

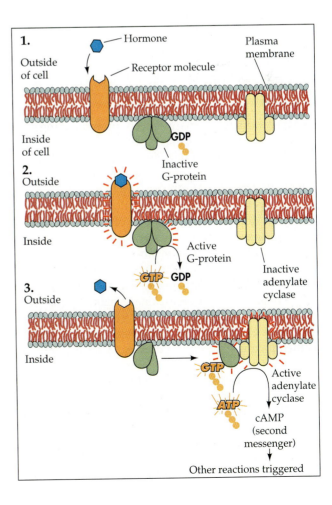

the control of enzyme activity through phosphorylation. This was the first demonstration of what is now known to be a common mechanism of regulation of enzyme function. Sutherland then demonstrated that adrenaline could stimulate fragmented liver cells to release glucose as long as pieces of their plasma membranes were present. This demonstration of hormone action in a cell-free system was also a major landmark in biochemistry. The third major discovery in this research program was that the interaction of the hormone with the membrane fragments produced a small molecule that could then stimulate the phosphorylation of enzymes in a preparation of liver cell fragments that did not contain any membranes. This molecule was identified as cAMP.

In the years after Sutherland's work, the list of systems activated by cAMP grew rapidly. Many hormones in vertebrate tissues act by means of this second messenger. These systems include adrenaline's stimulation of the breakdown of stored carbohydrates and fats, adrenocorticotropin's stimulation of the production of glucocorticoids in the adrenals, luteinizing hormone's stimulation of androgen synthesis, and many more. Different target cells have different, specific secondary targets within them that are activated by the second messenger. These secondary targets can activate different responses in a cell. The specificity of hormone action resides not only in the receptors that determine *which* cells respond to a given hormone, but also in the *way* a given cell responds, and that depends on what responding mechanisms it has.

The Targets of cAMP

cAMP produces its many effects by interacting with a variety of **protein kinases**, enzymes that catalyze the transfer of phosphate from ATP to specific proteins, and with **phosphoprotein phosphatases**, enzymes that remove these phosphate groups from the phosphorylated proteins. The action of adrenaline on liver cells demonstrates the roles of cAMP-dependent protein kinases (Figure 34.17). When adrenaline binds to its receptor on the plasma membrane, adenylate cyclase is activated, and cAMP is formed. cAMP, the second messenger, activates a specific protein kinase. That protein kinase acts on *two* other enzymes, adding a phosphate group from ATP to each one. One of the newly phosphorylated proteins is the enzyme **glycogen synthase**. This enzyme catalyzes the joining of glucose molecules to synthesize the energy-storing molecule glycogen, but it is *inactivated* by the addition of the phosphate group. The other enzyme phosphorylated by the cAMP-activated protein kinase is **phosphorylase kinase**, which is *activated* by the addition of the phosphate group. Phosphorylase kinase, itself a protein kinase, catalyzes the

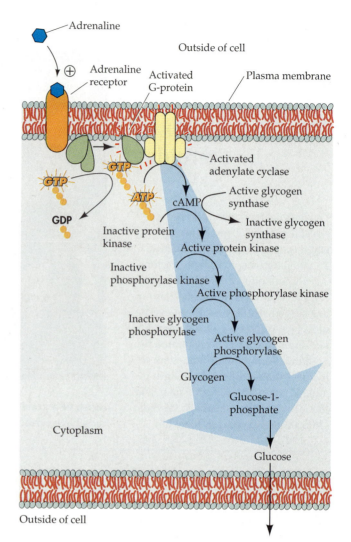

34.17 Adrenaline and cAMP
Adrenaline stimulates production of the second messenger cAMP in liver cells. cAMP in liver cells triggers a cascade of events resulting in stimulating breakdown of glycogen and inhibition of glycogen synthesis.

phosphorylation of the enzyme **glycogen phosphorylase**. Glycogen phosphorylase participates in the breakdown of glycogen to glucose. Thus cAMP, through its effects on two protein kinases, inhibits the storage of glucose as glycogen and promotes the release of glucose through glycogen breakdown. Both these effects increase glucose levels in liver cells and hence in the blood as well.

This cascade of regulatory steps results in a remarkable amplification of the effect of a single hormone molecule. A molecule of adrenaline binds to a single receptor molecule, but the activated receptor activates many molecules (let us say 10) of the G-protein. Each activated G-protein activates one molecule of adenylate cyclase, but adenylate cyclase is an enzyme, and it can catalyze the production of perhaps 100 molecules of cAMP. Each molecule of cAMP activates only one protein kinase molecule, but each protein kinase may activate, perhaps, 100 phosphorylase kinase molecules, and each phosphorylase kinase may activate, perhaps, 100 glycogen phosphorylase molecules, and each of *them* catalyzes the production of perhaps 100 molecules of glucose from glycogen. Thus an amplification of $10 \times 100 \times 100 \times 100 \times 100$ is achieved—that is, each molecule of adrenaline could cause the production of about one billion molecules of glucose.

Unless there is a continuing supply of a hormone, it diffuses away from the receptor or is enzymatically degraded, allowing the receptor to revert to its inactive tertiary structure. In turn, the concentration of active adenylate cyclase decreases as the enzyme self-inactivates by breaking down the GTP, and cAMP is no longer formed. The cAMP still present is quickly removed by the action of specific **phosphodiesterases**, enzymes that catalyze the conversion of cAMP to an inactive product. Recall that cAMP inactivates phosphoprotein phosphatase; that enzyme is thus no longer inhibited when cAMP levels fall. Phosphoprotein phosphatase is thus free to dephosphorylate the three enzymes of glycogen metabolism; glycogen synthase is activated, and phosphorylase kinase and glycogen phosphorylase are inactivated. Thus the entire sequence of hormone-mediated activations is reversed when the hormone is no longer present.

Other Second Messengers

There are other second messengers besides cAMP in animal cells. Two second messengers, **inositol triphosphate** and **diacylglycerol**, are produced from a single class of membrane phospholipids, the **polyphosphoinositides**. When certain water-soluble hormones bind to their specific receptor proteins, polyphosphoinositides are cleaved to form inositol triphosphate and diacylglycerol. Diacylglycerol stimulates a protein kinase, and inositol triphosphate increases the permeability of membranes to calcium ions. Calcium ions are, in effect, a *third* messenger; their effects are discussed below.

Yet another second messenger, active in some target cells, is **cGMP** (cyclic guanosine monophosphate), a close chemical analogue of cAMP. The effects of insulin on some target cells are mediated by cGMP; that is, insulin is the first messenger and cGMP is the second messenger.

Calcium Ions

Calcium ions (Ca^{2+}) mediate many responses in many kinds of cells. As we will see in Chapters 36, 37, and 38, Ca^{2+} ions play crucial roles in both nerve cell and muscle cell functions. To be effective, Ca^{2+} ions must combine with a calcium-binding protein, of which the most widely distributed is a specific protein, **calmodulin**. Another calcium-binding protein, troponin, regulates a key reaction in the contraction of the skeletal muscles of vertebrates.

Ca^{2+} ions were once viewed as second messengers, because various first messengers cause an influx of Ca^{2+}. However, it now appears that the influx of Ca^{2+} in response to a hormone is caused by the intermediary action of a true second messenger such as inositol triphosphate.

Lipid-Soluble Hormones

Steroid hormones, such as estrogen, progesterone, and the hormones of the adrenal cortex, as well as thyroxine, generally do not react with receptors on the target-cell surface (although it is now known that there are *some* steroid receptors bound to plasma membranes). These hormones are all lipid-soluble, and as you will recall from Chapter 5, this means that they pass readily through the lipid-rich plasma membrane. They act by stimulating the synthesis of new kinds of proteins through gene activation rather than by altering the activity of proteins already present in the target cells. Once inside a cell, a lipid-soluble hormone binds to a receptor protein in the cytoplasm (Figure 34.18). The presence of a receptor protein is what distinguishes a responsive cell from a nonresponsive one. A receptor protein is specific for a particular hormone, and it changes shape when it binds its hormone. The hormone–receptor complex quickly associates with acidic chromosomal proteins, and thus with the DNA of the chromosomes. The receptor protein itself cannot bind the acidic chromosomal proteins unless it has already bound a hormone molecule and undergone the necessary change in structure. Once associated with the acidic chromosomal proteins, the hormone activates the transcription of certain genes into messenger RNAs, which are exported to the cytoplasm and translated into specific proteins.

HORMONES AND BEHAVIOR

Just as there are anatomical and physiological differences between male and female animals, there are also differences in the behavior of the two sexes. In many cases, differences in sexual behavior have been shown to be due to the actions of the sex steroids on the developing brain and on the mature brain. We will consider two examples.

Sexual Behavior in Rats

Rats, like most other animals, behave sexually in accord with **fixed action patterns**—stereotyped, spe-

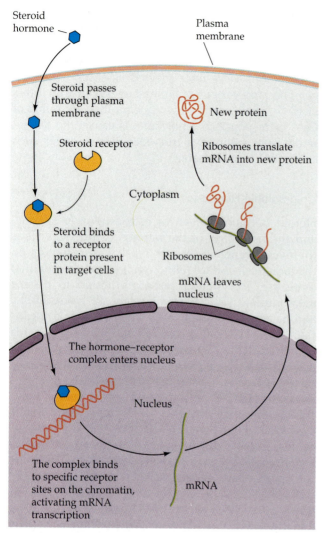

Steroid hormone

Plasma membrane

Steroid passes through plasma membrane

New protein

Steroid receptor

Ribosomes translate mRNA into new protein

Cytoplasm

Steroid binds to a receptor protein present in target cells

Ribosomes

mRNA leaves nucleus

The hormone–receptor complex enters nucleus

Nucleus

The complex binds to specific receptor sites on the chromatin, activating mRNA transcription

mRNA

34.18   Action of Lipid-Soluble Hormones
Lipid-soluble hormones have cytoplasmic receptors and activate gene transcription.

cies-specific behaviors that do not have to be learned. These behavior patterns are built into the nervous system. When a female rat is in heat (receptive to males), she will respond to a tactile stimulus of her hindquarters with a stereotyped posture called **lordosis**. She lowers her front legs, extends her hind legs, arches her back, and deflects her tail to one side. When a male encounters a female in heat, he copulates with her with the following sequence of behaviors: he mounts her from the rear, clasps her hindquarters, inserts his penis in her vagina, and thrusts. What are the roles of the genotypes and the sex hormones in the development of the fixed action patterns of lordosis and copulation? This question is answered by a series of experiments shown in Figure 34.19.

If a female rat has her ovaries removed, either as a newborn or as an adult, she will not show lordosis. However, if such a spayed female is injected with female sex steroids, she will show lordosis. The hor-

mones are necessary for the expression of the female sexual behavior. If a spayed adult female is injected with testosterone, she does not show male sexual behavior. There is a surprising variation on this experiment, however. If a *newborn* female rat is spayed *and* injected with testosterone, she will not show lordosis if treated with female sex hormones when she becomes an adult. If, however, this genetically female adult rat is treated with testosterone, she will mount other females in heat and show the male fixed action patterns associated with copulation. That is, the presence of testosterone in the newborn spayed female masculinizes her developing nervous system, so that when she reaches adulthood, her nervous system responds to male rather than to female steroids and generates male fixed action patterns.

Similar experiments on genetic males do not yield entirely reciprocal results. Castration (removal of the testes) of an adult male does not alter its response to treatment with sex steroids. Such a castrated male does not show lordosis when treated with female hormones, and it does show male sexual behavior when injected with testosterone. If a *newborn* male is castrated, however, it *will* show lordosis when injected with female hormones as an adult, but it will *not* show male sexual behavior when injected with testosterone as an adult. If the newborn male is castrated *and* injected with testosterone, it will not show lordosis when it is subsequently injected with female hormones as an adult. It will, however, show normal male sexual behavior in response to testosterone.

These results indicate that the nervous systems of both genetic males and genetic females develop female fixed action patterns if testosterone is not present at an early stage. Testosterone in the newborn triggers the development of male fixed action patterns whether the animal is a genetic male or a genetic female.

Bird Brains and Bird Song

Only the males of most species of songbirds sing, and they do so only during the season of reproduction. With their songs they claim territory, compete with other males, declare dominance, and attract females. Bird song has some characteristics of a genetically determined behavior—it is highly species-specific and stereotypical, being expressed in exactly the same form every time. Yet most songbirds must learn their species-specific songs. When young birds are still in the nest, they hear their fathers sing but they do not sing. After leaving the nest and their parents, they may migrate and flock with other species, but they do not sing, and they may not hear their species' song until the following spring. As that spring approaches and days get longer, the young male's testes begin to grow and mature. As his testosterone level rises, he begins to try to sing. Even if he is isolated

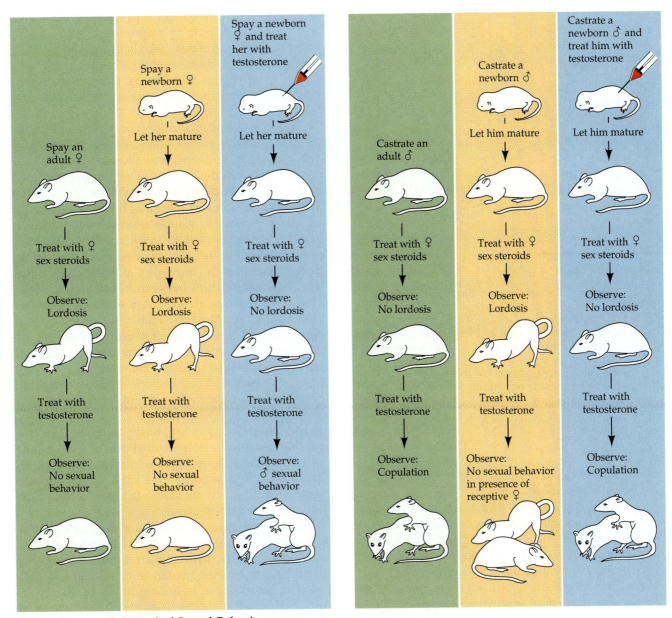

34.19 Hormonal Control of Sexual Behavior
From a series of experiments with rats we see that the sex steroids are involved in both the development and expression of sexual behavior. The presence of testosterone in newborn, gonadectomized rats of both sexes establishes male behavior patterns, and its absence establishes female patterns. Injections of sex steroids in gonadectomized adult rats stimulate expression of the sexual behavior pattern that developed in response to genotype and early steroid exposure.

from all other males of his species, his song will gradually improve until it is a proper rendition of his species-specific song. At that point the song is **crystallized**—that is, the bird expresses it in that form every spring thereafter. The juvenile bird's brain learned the pattern of the song by hearing the father. During the subsequent spring, under the influence of testosterone, the bird learns to express that song —a behavior that then becomes rigidly fixed in its nervous system.

Why don't the females sing? Can't they learn the patterns of their species-specific song? Don't they have the muscular or nervous system capabilities necessary to sing? Or, do they simply lack the hormonal stimulus for developing the behavior? To answer these questions, female songbirds were injected with testosterone in the spring. In response to the injections, they developed their species-specific song and sang just as the males do. Apparently, females learn the song pattern of their species and have the

34.20 Effects of Testosterone on Bird Brains
Testosterone induces growth in regions of a songbird's brain that are responsible for song. During the nonbreeding season, the brains of male and female zebra finches are similar, but in spring rising testosterone levels in the male cause the song regions of its brain to develop. The sizes of the circles are proportional to the volume of the brain occupied by that region; dashed circles indicate estimated volumes. The area labeled "X" is not found in the brains of female finches.

capabilities to express it, but they normally lack the hormonal stimulation.

What does testosterone do to the brain of the songbird? A remarkable discovery is that testosterone causes the parts of the brain necessary for learning and expressing song to grow larger (Figure 34.20). Each spring certain regions of the males' brains grow. The individual cells increase in size, they grow longer extensions and—most surprisingly—the *numbers* of brain cells in those regions of the bird brain also increase. Prior to this work on bird brains, it was assumed that newborn vertebrates have their full complement of brain cells, and that they lose these cells progressively throughout life without replacing them. Thus, research on the neurobiology of bird song has revealed that hormones can influence brain structure as well as brain function on both a developmental and seasonal basis.

SUMMARY

Hormones are chemical messages from one cell to another. Hormonal communication may be between neighboring cells or between cells in different parts of the body. To reach distant parts, hormones circulate in the bloodstream. Any hormonal communication system requires not only a specific chemical message but also specific receptors to receive the message and induce intracellular responses. Receptors for water-soluble hormones, which cannot enter cells, are on the cell surface; receptors for fat-soluble hormones generally are within cells. Cell-surface receptors usually activate second messengers within the cell.

The growth and maturation of insects are under the control of several hormones. Brain hormone controls timing of molt, ecdysone induces molt, and juvenile hormone determines whether the molt includes a change in developmental state.

The production and release of hormones by many vertebrate endocrine glands are controlled by tropic hormones from the anterior pituitary. Releasing and release-inhibiting neurohormones from the hypothalamus control the hormones of the anterior pituitary. The relationships between hypothalamic, pituitary, and endocrine gland secretions are regulated by information from the brain and by the negative feedback effects of hormones and neurohormones on the pituitary or on hypothalamic secretory cells. The anterior pituitary also produces hormones that are not tropic hormones. They include growth hormone, prolactin, melanocyte-stimulating hormone, and peptides with opiatelike actions.

The posterior pituitary stores and releases two peptide neurohormones synthesized by cells in the hypothalamus. Vasopressin is important in controlling water conservation by the kidney, and oxytocin stimulates labor contractions and the secretion of milk from mammary glands.

Other vertebrate hormones are thyroxine, which influences metabolic rate; parathyroid hormone and calcitonin, which control calcium metabolism; the pancreatic hormones insulin and glucagon, which regulate blood sugar; hormones of the adrenal medulla, which stimulate "fight-or-flight" reactions; and the steroid hormones of the adrenal cortex, which control stress responses and salt balance.

The hormones of the gonads, the estrogens and the androgens, control the development of primary and secondary sexual characteristics as well as sexual behavior and physiology.

Some hormones can have major effects on behavior. In rats, species-specific sexual behavior in adulthood is dependent on the presence of sex steroids, but whether a rat displays male or female sexual behavior as an adult depends on whether it was exposed to androgens or estrogens as a newborn. Singing in songbirds is also dependent on sex steroids. Testosterone stimulates the growth and development of areas of the brain necessary for song expression. Normally only male songbirds sing, and only when their testosterone levels rise in spring, but females can also sing their species-specific song if treated with testosterone.

SELF-QUIZ

1. Which of the following statements is true for *all* hormones?
   a. They are secreted by glands.
   b. They have receptors on cell surfaces.
   c. They may stimulate different responses in different cells.
   d. They effect target cells distant from their site of release.
   e. When the same hormone occurs in different species, it has the same action.

2. The hormone ecdysone:
   a. is released from the posterior pituitary.
   b. stimulates molt and metamorphosis in insects.
   c. maintains an insect in larval stages unless brain hormone is present.
   d. stimulates secretion of juvenile hormone from the prothoracic glands.
   e. Keeps the insect exoskeleton flexible to permit growth.

3. The posterior pituitary:
   a. produces oxytocin.
   b. is under the control of hypothalamic releasing neurohormones.
   c. secretes tropic hormones.
   d. secretes neurohormones.
   e. is under feedback control by thyroxine.

4. Growth hormone:
   a. can cause adults to grow taller.
   b. stimulates protein synthesis.
   c. is released by the hypothalamus.
   d. can only be obtained from cadavers.
   e. is a steroid.

5. Both adrenaline and cortisol are secreted in response to stress. What else is true for both of these hormones?
   a. They act to increase blood glucose.
   b. The receptors are on the surfaces of target cells.
   c. They are secreted by the adrenal cortex.
   d. Their secretion is stimulated by adrenocorticotropin.
   e. They are both secreted into the blood within seconds of the onset of stress.

6. Prior to puberty:
   a. the pituitary is secreting luteinizing hormone and follicle-stimulating hormone, but the gonads are unresponsive.
   b. the hypothalamus does not secrete much gonadotropin-releasing hormone.
   c. males can stimulate massive muscle development through a vigorous training program.
   d. testosterone plays no role in development of the male sex organs.
   e. genetic females will develop male genitals unless estrogen is present.

7. Which of the following is *not* true of cyclic AMP?
   a. It is broken down by adenylate cyclase.
   b. It is involved in the chain of events whereby adrenaline stimulates liver cells to break down glycogen.
   c. It is a second messenger mediating intracellular responses to many hormones.
   d. Many of its effects are mediated by protein kinases.
   e. A molecule of cAMP activates only a single protein kinase molecule.

8. Steroid hormones:
   a. are all produced by the adrenal cortex.
   b. have only cell surface receptors.
   c. are lipophobic.
   d. act through alteration of the activity of proteins in the target cell.
   e. act through stimulating production of new proteins in the target cell.

9. Which is *not* true of a genetic female rat spayed at birth?
   a. When she becomes an adult, she will be able to show lordosis if injected with estrogen.
   b. When she becomes an adult, she will show male sexual behavior without hormone treatment.
   c. If treated with testosterone as a newborn, she will show male sexual behavior if injected with testosterone as an adult.
   d. When she becomes an adult, she will display no sexual behavior without hormone treatment.
   e. As an adult, she will have high circulating levels of gonadotropins.

10. Which of the following statements about bird song is true?
    a. Young birds can learn their species song from either their mother or their father.
    b. After song has crystallized, it is not dependent on hormones for expression.
    c. Even if female songbirds are treated with testosterone, they do not sing.
    d. Parts of the bird brain responsible for singing behavior grow in response to testosterone.
    e. Male songbirds sing all year long.

FOR STUDY

1. Compare the mechanisms of action of peptide and steroid hormones.

2. Explain how both hyperthyroidism and hypothyroidism can cause goiter. Include the roles of the hypothalamus and the pituitary in your answers.

3. Explain the developmental abnormalities that can produce a genetic male with female secondary sexual characteristics. Describe the gonads of such an individual.

4. How did Sutherland's experiments demonstrate that the result of adrenaline combining with a membrane-bound receptor is the production of a second messenger?

5. How can cAMP working through a protein kinase activate one enzyme while inactivating another enzyme in the same cell?

READINGS

Atkinson, M. A. and N. K. MacLaren. 1990. "What Causes Diabetes?" *Scientific American*, July. Diabetes is a major disease that involves a hormone deficiency. This paper reveals how malfunctions of the immune system cause insulin-dependent diabetes.

Berridge, M. J. 1985. "The Molecular Basis of Communication Within the Cell." *Scientific American*, October. An authoritative account of second messengers and their roles in a variety of biological phenomena.

Bloom, F. E. 1981. "Neuropeptides." *Scientific American*, October. A description of the discovery, synthesis, distribution, and actions of peptides that serve as chemical messengers in the nervous system and as hormones in the body.

Focuses on vasopressin, oxytocin, endorphins, enkephalins, and a few others.

Cantin, M., and J. Genest. 1986. "The Heart as an Endocrine Gland." *Scientific American*, February. Interesting account of the discovery and characterization of a hormone half a century after its existence had been predicted.

Carafoli, E., and J. T. Penniston. 1985. "The Calcium Signal." *Scientific American*, November. Calcium as a second messenger; the roles of calcium-binding proteins.

Carmichael, S. W., and H. Winkler. 1985. "The Adrenal Chromaffin Cell." *Scientific American*, August. How a cell synthesizes, stores, and secretes hormones.

Eckert, R., D. Randall and G. Augustine. 1988. *Animal Physiology*, 3rd Edition. W. H. Freeman, San Francisco. An excellent textbook, particularly useful with respect to second messengers and regulatory physiology.

Snyder, S. H. 1985. "The Molecular Basis of Communication Between Cells." *Scientific American*, October. An overview of the relationships between the nervous and endocrine systems; the focus is on chemical messengers and their molecular biology.

Vander, A. J., J. H. Sherman, and D. S. Luciano. 1990. *Human Physiology: The Mechanisms of Body Function*, 5th Edition. McGraw-Hill, New York. Chapter 10 deals specifically with hormonal regulation, but material on hormones appears throughout this fine text.

35

Animal Reproduction

PREVIEW: Sex has some advantages, but it isn't necessary. Animals reproduce asexually and sexually. Asexual reproduction can produce large numbers of exact copies of the parent, whereas sexual reproduction allows for genetic recombination and produces genotypic diversity. In sexual reproduction, meiosis produces haploid gametes that are united through external or internal fertilization. Male and female reproductive systems are adapted to produce and deliver gametes. The female reproductive systems of mammals and some other species provide for the protection and nurture of the embryo during development. The events of reproduction proceed under the control of many hormonal mechanisms.

This chapter deals with asexual reproduction, the male and female reproductive systems, gametogenesis, fertilization, human sexual responses, nurture of embryos, pregnancy, childbirth, and birth control.

Asexually reproducing organisms seem to be practically immortal. For example, an amoeba reproduces itself almost exactly by splitting in two. The resulting cells appear to be identical rather than being a parent and an offspring; thus, the parent never ceases to exist. Some multicellular animals reproduce asexually. Many cells in the bodies of these animals have the ability to produce whole new organisms. The genetic lineage of an asexually reproducing animal is continuous and unchanging except for the effects of mutations. Sexually reproducing animals, in contrast, produce sex cells, or **gametes**, that recombine the genetic material in each generation. The subsequent union of pairs of gametes creates new individuals that are genetically distinct from their parents.

We are sexually reproducing animals. We think of our gametes and our reproductive functions as the means of reproducing ourselves. But try taking another view of this. Very early in the life of a new embryo, during the first cell divisions, a population of cells arises that wanders in the body until the sex organs form. These nomadic cells then migrate to the sex organs; eventually they give rise to the gametes that will produce the next generation. It is as if we as organisms are the devices produced by our gametes to reproduce themselves. The gametes themselves are part of an almost continuous lineage of cells called the germ line, which is punctuated by meiosis and recombination. This view of the relationship between the body and the gametes points out the centrality of reproductive processes in the lives of animals.

ASEXUAL REPRODUCTION

Asexual reproduction produces offspring genetically identical to one another. This mode of reproduction is highly efficient because all the individuals in a population can convert resources into offspring, and the population can grow as rapidly as resources permit. However, asexual reproduction does not provide for the generation of genotypic diversity in a population. Thus an asexually reproducing population does not have a wide variety of genotypes on which natural selection can act as the environment changes. Nevertheless, some animals reproduce asexually.

Budding

A common mode of asexual reproduction in simple multicellular animals is for a new individual to arise as an outgrowth of an older one—a process called **budding** (Figure 35.1a). Some sponges form buds of undifferentiated cells on the outsides of their bodies. These buds grow through mitotic cell division, and the cells undergo differentiation before the buds break away from the parents and become independent sponges. Many freshwater sponges produce internal buds, or gemmules. A gemmule consists of several undifferentiated cells. Eventually the gemmules escape from the parent and become free-living individuals, genetically identical to the parent. Budding is part of the life cycle of some cnidarians. The bud resembles the parent and may grow as large as the parent before it becomes an independent organism.

791

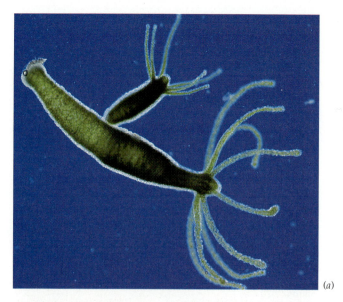

(a)

(b)

(c)

35.1 Asexual Reproduction in Animals
(a) Budding. A new individual buds from an adult *Hydra*. *(b)* Regeneration. A single amputated arm from the sea star *Asterias rubens* develops into a new animal. *(c)* Parthenogenesis. Greenflies hatch from unfertilized eggs. In all these forms of asexual reproduction, the offspring are genetically identical to the parent.

Regeneration

The cells that initiate budding are **totipotent**; that is, they have the ability to give rise to new, complete organisms. That is not true of most of the cells of most animals (see the discussion of determination and differentiation in Chapter 15). There are notable exceptions, however, where pieces of animals can develop into whole animals (Figure 35.1*b*). A dramatic example of such **regeneration** was unwittingly produced by a group of public officials who tried to protect oyster beds by instituting a search-and-destroy mission aimed at sea stars that were preying on the oysters. They "killed" these echinoderm predators by cutting them into pieces and dumping the pieces back into the sea. However, sea stars have remarkable abilities of regeneration. If they lose arms, they regenerate new ones—and if a severed arm includes a portion of the central disk of the animal's body, it can regenerate into a complete sea star. In their attempt to eliminate the sea stars, the public officials created large numbers of pieces that had the capability of regeneration, and the predator population increased enormously.

Regeneration is a form of asexual reproduction that usually follows an animal's being broken by an outside force, but in some species the breakage is a normal event initiated by the animal itself. Certain species of segmented worms (annelids) develop seg-ments with rudimentary heads bearing sensory organs, then break apart. Each fragmented segment forms a new worm.

Parthenogenesis

Another type of asexual reproduction is **parthenogenesis**, the development of offspring from unfertilized eggs (Figure 35.1*c*). Many species, especially arthropods, reproduce parthenogenetically, as do some species of fish, amphibians, and reptiles. Most species that reproduce parthenogenetically also engage in sexual reproduction or sexual behavior. Aphids, for example, are parthenogenetic in the spring and summer, multiplying rapidly while conditions are favorable. Some of the unfertilized eggs laid in spring and summer develop into male aphids, others develop into females. As conditions become less favorable, the aphids mate and the females lay fertilized eggs. These eggs do not hatch until the following spring, and they yield only females. Species capable of parthenogenesis frequently switch from asexual to sexual reproduction when environmental conditions change. Parthenogenesis is used when conditions are stable and favorable; sexual reproduction introduces genotypic variability when conditions are changing, stressful, or unpredictable.

Parthenogenetic reproduction in some species requires a sex act even though the ovum is diploid and fertilization is not necessary. The eggs of parthenogenetically reproducing ticks and mites develop only after the animals have mated, but the eggs remain unfertilized. Some species of beetles have no males at all, and parthenogenesis is their only means of reproduction. Yet their eggs require sperm to trigger their development. These beetles therefore mate with males of closely related, but different, species.

SEXUAL REPRODUCTIVE SYSTEMS OF ANIMALS

The enormous genotypic diversity in most sexually reproducing species derives from the independent assortment of chromosomes and the recombination of alleles on those chromosomes. As you know, a sexually reproducing animal packages single copies of its genes into reproductive cells called gametes. Because an animal has two alleles for each gene, each gamete, while containing a complete set of genes, contains a unique assortment of alleles. Sexual reproduction requires **fertilization**—the fusion of two gametes (almost always from different individuals) to form a **zygote**. The zygote receives half of its alleles from each parent, and therefore has a new, unique genotype. Natural selection acts on the genotypic diversity produced by this process. Those individuals having genotypes best suited to environmental conditions are the most likely to survive and produce the largest number of offspring.

Both sexes, female (♀) and male (♂), produce haploid gametes from germ cells. The tiny gametes of males are called **sperm**; they move by beating their flagella (Figure 35.2). The much larger female gametes are called **eggs**, or **ova** (singular, ovum) and are nonmotile. Sperm and eggs are produced in the primary sex organs, the **gonads**. Male gonads are **testes** (singular, testis), and female gonads are **ovaries**. In addition to primary sex organs, most animals (except the sponges and cnidarians) have accessory sex organs, including ducts, glands, and structures that deliver and receive gametes. The primary and accessory sex organs of an animal constitute its reproductive system.

Gametogenesis

In virtually all animals except sponges, gametes are produced within the gonads, but, as described in the introduction to this chapter, the gametes derive from a special lineage of cells called the germ line. Those cells are not produced by the gonads; they come to reside in the gonads only after the gonads have formed in the embryo. The germ cells are diploid, and they proliferate by mitosis. The cells resulting from the mitotic proliferation of germ cells in the

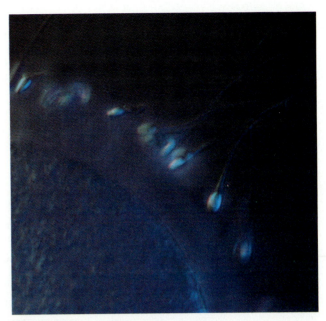

35.2 Gametes Differ in Size
Male gametes—sperm—are small and motile; a large number of them appear from the upper right. Female gametes—eggs—are large and well provisioned to nourish the early stages of development. Only a portion of a single egg is seen in this micrograph.

gonads of females are called **oogonia** (singular, oogonium) and those in the gonads of males, **spermatogonia** (singular, spermatogonium). Meiosis, the next step in gametogenesis, reduces the chromosomes to the haploid number, and the haploid cells mature into sperm and ova. We will consider next how spermatogonia become sperm and after that how oogonia become ova. Although there are considerable differences in these two processes, meiosis is central to both. Because this is so, you may want to review the discussion of meiosis in Chapter 9 before reading more about gametogenesis.

Spermatogenesis is the process by which sperm form from germ cells (Figure 35.3). Sperm form in seminiferous tubules within the mammalian testes. The process begins when the diploid spermatogonia near the wall of a seminiferous tubule increase in size and divide by mitosis to become **primary spermatocytes**. The primary spermatocytes undergo the first meiotic division to form **secondary spermatocytes**, which are haploid. (As you will recall, the first meiotic division halves the number of chromosomes.) These cells remain connected by bridges of cytoplasm after each division. The second meiotic division produces four haploid **spermatids** for each primary spermatocyte that entered meiosis. Throughout spermatogenesis, the germ cells are intimately associated with the Sertoli cells that line the seminiferous tubule. The Sertoli cells provide an appropriate environment for spermatogenesis and transfer nutrients to the developing spermatocytes and spermatids.

35.3 Spermatogenesis

The left panel shows the formation of haploid spermatids from diploid spermatogonia. Spermatids, all of which are different genetically, will differentiate into sperm. A mature mammalian sperm cell is drawn at the right. Spermatogenesis is prolific, producing thousands of sperm cells; the micrograph shows a cluster of human sperm.

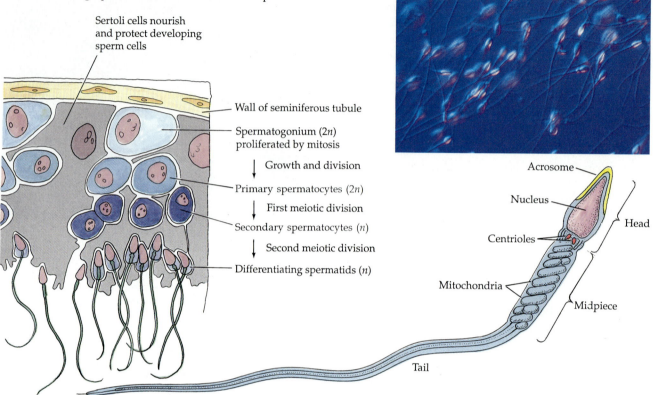

Sertoli cells nourish and protect developing sperm cells

Wall of seminiferous tubule

Spermatogonium (2n) proliferated by mitosis

↓ Growth and division

Primary spermatocytes (2n)

↓ First meiotic division

Secondary spermatocytes (n)

↓ Second meiotic division

Differentiating spermatids (n)

Acrosome

Nucleus

Centrioles

Head

Mitochondria

Midpiece

Tail

Spermatids differ from one another genetically because the random orientation of chromosomes at the first meiotic metaphase shuffles the parental genomes. A given spermatid contains some chromosomes the male inherited from his father and others inherited from his mother, with the particular combination being a matter of chance. As you know, crossing over during the first meiotic division also contributes to the genetic differences among spermatids.

Just after being produced by meiosis, a spermatid bears little resemblance to sperm. As it differentiates into a sperm, its nucleus becomes compact, its motile flagellum develops into a tail, and most of its cytoplasm is lost. As the head of the sperm forms, it is capped by an **acrosome** which contains enzymes that will enable the sperm to digest its way into an egg. Between the head and tail of the mature sperm is a midpiece containing two centrioles and mitochondria to provide energy for locomotion. The microtubules that extend from the centrioles into the flagellum have the same pattern as in all typical eukaryotic flagella, the standard "9 + 2" arrangement described in Chapter 4.

As the spermatocytes develop into spermatids, and the spermatids develop into sperm, they move progressively from the outermost region of the seminiferous tubule toward the center. The fully differentiated sperm are finally released from the Sertoli cells. The entire process takes about 10 weeks. Each day a human male produces about 30 million sperm.

Oogenesis is the process of meiosis and differentiation of the oogonia into eggs (Figure 35.4). Some oogonia develop into **primary oocytes,** which enter the first meiotic division but arrest in prophase I. During this arrest, the primary oocytes enlarge, gaining yolk, ribosomes, cytoplasmic organelles, and energy stores. They accumulate rRNA, mRNA, and tRNA, as well as materials from follicle cells and from the blood. Many of the lipids and proteins stored by vertebrate ova, for example, are made in the liver and transported to the ovaries in the bloodstream. The primary oocyte acquires all of the energy, raw materials, and RNA that the egg needs to get through its first cell divisions after fertilization.

Each month during a human female's fertile years, at least one primary oocyte comes out of its resting stage and matures into an egg. As this primary oocyte resumes meiosis, the nucleus completes its first meiotic division near the surface of the cell. The

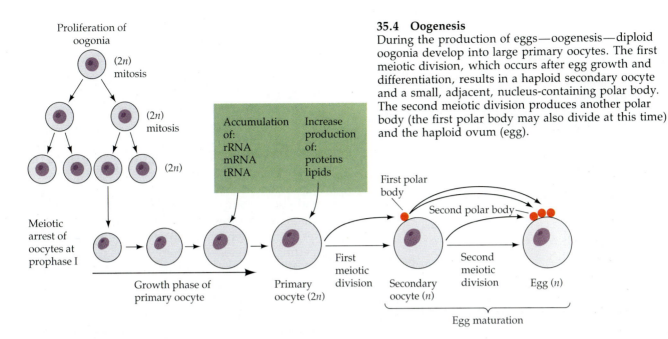

Proliferation of
oogonia

Meiotic
arrest of
oocytes at
prophase I

Accumulation of: rRNA mRNA tRNA

Increase production of: proteins lipids

First polar body

Second polar body

Growth phase of primary oocyte

Primary oocyte (2n)

First meiotic division

Secondary oocyte (n)

Second meiotic division

Egg (n)

Egg maturation

35.4 Oogenesis

During the production of eggs—oogenesis—diploid oogonia develop into large primary oocytes. The first meiotic division, which occurs after egg growth and differentiation, results in a haploid secondary oocyte and a small, adjacent, nucleus-containing polar body. The second meiotic division produces another polar body (the first polar body may also divide at this time) and the haploid ovum (egg).

daughter cells of this division receive a grossly unequal share of the cytoplasm of the primary oocyte. One receives almost all of the cytoplasm and becomes the **secondary oocyte**, and the other receives almost none and forms the **first polar body**. The second meiotic division of the large secondary oocyte is, once again, accompanied by an asymmetric division of the cytoplasm. One daughter cell forms the large, haploid **ootid**, and the other forms the **second polar body**. The polar bodies degenerate, so the end result of oogenesis is one very large, haploid ovum that is well provisioned for the rapid divisions of the cleavage stage of development.

Sex Types

Most animals are a distinct sex type—male or female. Species having male and female members are called **dioecious** (Greek, "two houses"). In contrast, a single individual of some other species may possess both female and male reproductive systems. Such species are called **monoecious** ("one house") or **hermaphroditic** (from the name of a male, Hermaphroditus, whose body, according to a Greek myth, was joined with that of a nymph). Almost all invertebrate groups have some hermaphroditic species; the common earthworm is an example (Figure 35.5; see Figure 25.24c). Some hermaphroditic species have members that are both male and female at the same time (**simultaneous hermaphrodites**); the members of other species are **sequential hermaphrodites**, being male and female at different times in their life cycles. As we learned in the last chapter, the term hermaphroditism can also be used to describe developmental abnormalities in normally dioecious species that give rise to individuals with both male and female sex organs.

35.5 A Hermaphrodite

The common earthworm, *Lumbricus terrestris*, is a good example of the hermaphroditic sex type. (a) Each earthworm has both male and female sex organs. (b) Two individuals exchange gametes.

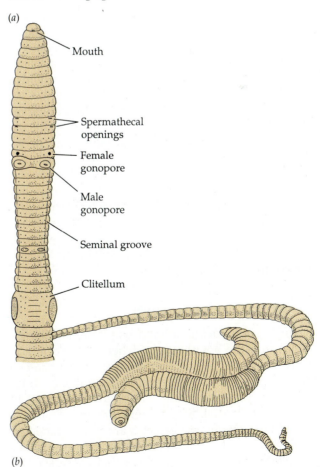

(a)

Mouth

Spermathecal openings

Female gonopore

Male gonopore

Seminal groove

Clitellum

(b)

35.6 Hermaphroditic Copulation
Although some hermaphroditic species can fertilize themselves, most must mate with at least one other individual; the sea slug *Aplysia* often copulates *en masse*. Here a group of sea slugs forms a mating chain in which each animal is functioning as a female for the animal behind and as a male for the animal in front.

Some simultaneous hermaphrodites have a low probability of meeting a potential mate. An example is the tapeworm, an intestinal parasite that may be quite large and cause lots of trouble, but may still be the only tapeworm in your intestine. Such species can usually fertilize themselves. Most simultaneous hermaphrodites must, however, mate with another individual (Figure 35.6).

Sequential hermaphroditism confers different advantages on different species. It can reduce the possibility of inbreeding among siblings by making them all the same sex at the same time and therefore incapable of mating with one another. Sequential hermaphroditism can also maximize the reproductive success of all the individuals in a polygamous species in which a few dominant males have most of the reproductive success. Sequential hermaphroditism makes it possible for an individual to reproduce as a female until the opportunity arises to be a dominant male. An excellent example is the tropical Pacific fish *Labroides dimidiatus*, a wrasse. All individuals of this species are born female. The population consists of social groups described as harems; each harem consists of a number of females controlled by one dominant male. The male defends the territory of the group from intruders. If the male dies or is removed, the largest, most dominant female in the group changes sex and becomes a functional male, assuming control of the group.

Getting Eggs and Sperm Together

Sexually reproducing animals may release their gametes into the environment, where the meeting of gametes results in fertilization; or the male gametes may be inserted into the female's reproductive tract, where fertilization occurs. Thus fertilization can be external or internal. External fertilization is limited to aquatic habitats where gametes are not in danger of drying out, and is the more common pattern among simpler animals, especially those that are sessile. External fertilization places a premium on adaptations that increase the probability that male and female gametes will actually meet. One simple adaptation is the production of huge numbers of gametes. A female oyster, for example, may produce 100 million eggs per year, and the number of sperm produced by a male oyster is truly astronomical. Numbers alone do not guarantee that gametes will meet, however, without mechanisms of timing that synchronize the reproductive activities of the males and the females of a population. Seasonal breeders may use photoperiod cues, changes in temperature, or changes in weather to time their production and release of gametes. Sexual behavior also plays an important role in getting gametes together. Many species travel great distances to congregate with potential mates and release their gametes at the same time in a suitable environment. An excellent example is the remarkable migration of salmon. These fish hatch and go through juvenile stages in fresh water. They then migrate to the ocean, where they live and grow for 3 to 5 years. When they are finally ready to breed, they migrate back to the stream in which they were hatched, where they spawn and die.

Because gametes released into a dry environment die quickly, internal fertilization is a major adaptation for terrestrial life. Many aquatic species also practice internal fertilization, however. A great advantage of internal fertilization is the protection it provides for the early developmental stages. Animals have evolved an incredible diversity of sexual behaviors and accessory sex organs that facilitate internal fertilization. In general, a tubular structure, the **penis**, enables the male to deposit sperm in the female's accessory sex organ, the **vagina**, or in some species, the **cloaca** (a cavity common to the digestive, urinary, and reproductive systems).

Copulation is an act that permits sperm to move directly from the male's reproductive system into the female's reproductive system. The transfer of sperm can also be indirect. In some species of mites and scorpions (among the arthropods) and salamanders (among the vertebrates), the males secrete containers called **spermatophores** in which they deposit sperm. When a female mite finds a spermatophore, she straddles it and opens a pair of plates in her abdomen so that the tip of the spermatophore enters her reproductive tract and allows the sperm to enter. Some female salamanders scoop up the portion of the gelatinous spermatophore containing the sperm with the lips of their cloacae.

Male squids and spiders play a more active role in spermatophore transfer. The male spider secretes a drop containing sperm into a bit of web; then, with a special structure on a foreleg, he picks up the sperm-containing web and inserts it through the female's genital opening. The male squid uses one special tentacle to pick up a spermatophore and insert it

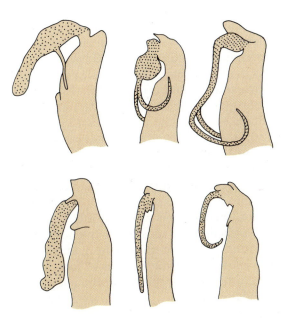

35.7 Insect Penises Show Morphological Diversity
The penises of six species of *Arfia*, a genus of damselflies, are quite different. Each fits into the corresponding female organ as a key in a lock to facilitate a tight union during the prolonged copulation process. The penis may also have an elaborated structure to scoop sperm deposited by other males out of the female's reproductive tract.

into the female's genital opening. In the process, the tip of his tentacle may break off and remain in the female's body along with the sperm.

Most male insects copulate and transfer spermatophores to the female's vagina through a tubular penis. The genitalia (external parts of the sex organs) of insects often have species-specific shapes that

match in a lock-and-key fashion. This assures a tight, secure fit between the mating pair during the prolonged period of sperm transfer (Figure 35.7).

This is a mere introduction to the fascinating diversity of animal reproductive systems. We will now look at the mammalian reproductive system in greater depth, using the human as the model animal. Many of the details, of course, are held in common with other vertebrates.

REPRODUCTIVE SYSTEMS IN HUMANS AND OTHER MAMMALS

The Male

The testes of mammals, except those of bats, elephants, and aquatic mammals, are lodged outside the body cavity in a pouch of skin, the **scrotum** (Figure 35.8). In most mammals spermatogenesis can take place only at a temperature slightly lower than normal body temperature. The scrotum keeps the testes at a temperature optimal for spermatogenesis. Muscles in the scrotum contract in a cold environment, bringing the testes closer to the warmth of the body; in a hot environment they relax, and the testes are suspended farther from the body.

A testis consists of tightly coiled seminiferous tubules encased in connective tissue (Figure 35.8b). Between the tubules are the cells that produce the male sex hormones. As we have discussed, sperm are produced in the seminiferous tubules (Figure 35.3). From the lumen of the seminiferous tubules, sperm move into a storage structure called the **epididymis**, where they mature and become motile. The epididymis connects to the **urethra** by a tube called the **vas deferens**.

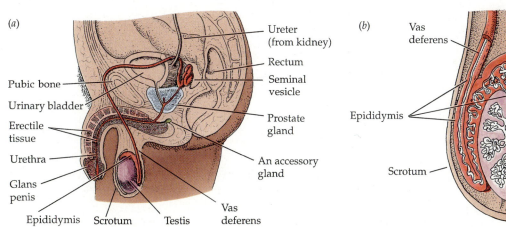
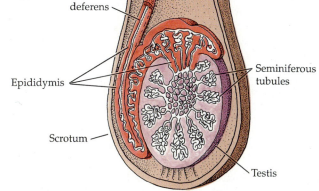

35.8 Reproductive Organs of the Human Male
(a) Cross section of the male reproductive system, as seen from the left side of the body. Sperm produced in the testes are stored in the epididymis. Sperm pass via the vas deferens to the seminal vesicle and from there to the urethra at the base of the penis. (b) Cross section of a testis; human males have a pair of testes.

The urethra comes from the bladder, runs through the penis, and opens to the outside of the body at the tip of the penis. It is the common duct for the urinary and reproductive systems.

The shaft of the penis is covered with normal skin, but the tip, or **glans penis,** is covered with thinner, more sensitive skin that is especially responsive to sexual stimulation. A fold of skin called the foreskin covers the glans of the human penis. The practice of circumcision removes a portion of the foreskin. There is no rationale or justification for circumcision based on health, yet it remains a cultural or religious tradition for many people.

The penis becomes hard and erect during sexual arousal because blood fills shafts of spongy tissue that run the length of the penis (see Figure 35.8a). The presence of this blood creates pressure that closes off the vessels that normally drain the penis. Thus, the penis becomes engorged with blood, facilitating insertion into the vagina. Some species of mammals, but not humans, have a bone in the penis; those species still depend on erectile tissue, however, for copulation.

The culmination of the male sex act propels sperm through the vas deferens and the urethra. This process of sperm movement has two steps, **emission** and **ejaculation.** During emission, sperm and the secretions of several accessory glands move into the urethra at the base of the penis. Together, the sperm and these secretions constitute **semen,** the fluid that is ejaculated into the female's vagina. About 60 percent of the volume of the semen is seminal fluid, which comes from the **seminal vesicles.** Seminal fluid is thick because it contains mucus and protein. It also contains fructose, which serves as an energy reserve for the sperm, and modified fatty acids called prostaglandins that stimulate contractions in the female reproductive tract. Another source of secretions is the **prostate gland,** which produces a thin, milky, alkaline fluid. The prostate fluid helps to neutralize the acidity of the urethra and the female reproductive tract to create a favorable environment for the sperm. Prostate fluid also contains a clotting enzyme that works on the protein in the seminal fluid to convert the semen into a gelatinous mass.

Ejaculation follows emission, and it is caused by wavelike contractions of muscles at the base of the penis surrounding the urethra. The rigidity of the erect penis allows these contractions to force the semen through the urethra and out of the body.

The Female

The female gonads, the ovaries, are paired structures in the body cavity (Figure 35.9). At birth, a baby girl has about a million primary oocytes in each ovary. By the time she reaches puberty (sexual maturity), she has only about 200,000 primary oocytes in each ovary—the rest have degenerated. Over the course of a woman's fertile years, only about 450 of these oocytes will mature completely into eggs and be released. When she is about 50 years old, she reaches **menopause,** the end of fertility. Only a few oocytes are then left in each ovary. Throughout a woman's life, oocytes are degenerating, and no new ones are being produced.

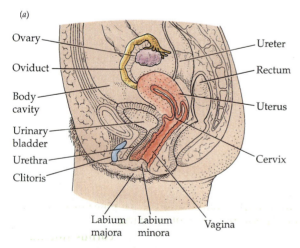

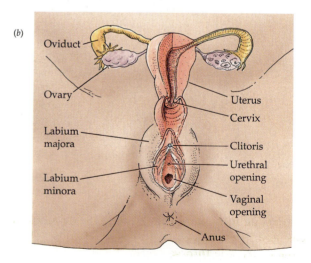

35.9 Reproductive Organs of the Human Female
(a) This cross section shows a single ovary and oviduct. It shows only half of the uterus, cervix, and vagina since they are single, midline structures. After ovulation occurs, the egg enters an oviduct that leads to the uterus. If sperm are present in the upper regions of the oviduct, the egg might be fertilized and begin development on its way to the uterus. (b) Frontal view of the female reproductive tract. The right-hand side of the drawing shows the internal structures in a cutaway view. External structures appear in the lower half of the drawing.

35.10 The Ovarian Follicle Develops

The progressive stages of development of the ovarian follicle, ovulation, and growth and degeneration of the corpus luteum are shown in this idealized cross section of the ovary. The developing follicle is seen in sequence, starting at the upper right and moving clockwise. The micrograph at the upper left shows a mature mammalian follicle; the oocyte is in the center.

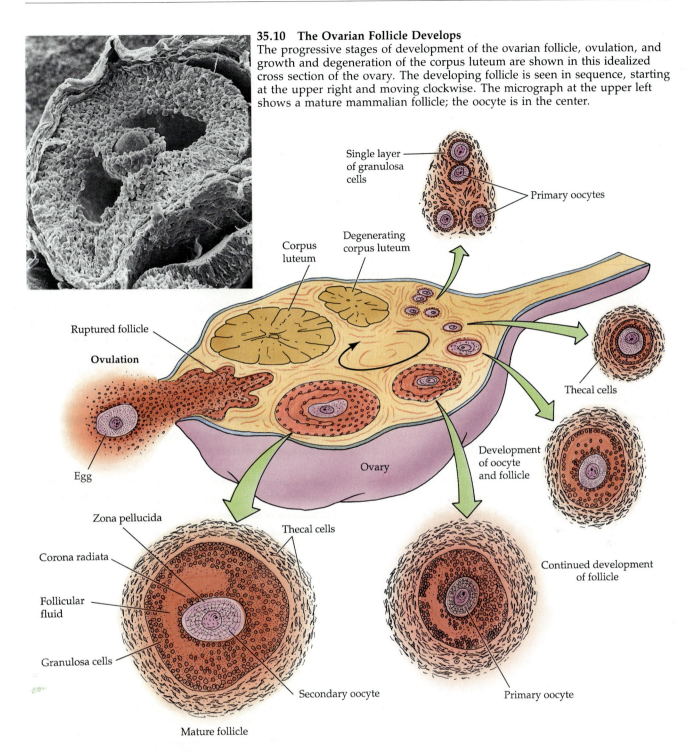

Single layer of granulosa cells

Primary oocytes

Degenerating corpus luteum

Corpus luteum

Thecal cells

Ruptured follicle

Ovulation

Development of oocyte and follicle

Ovary

Egg

Continued development of follicle

Zona pellucida

Thecal cells

Corona radiata

Follicular fluid

Granulosa cells

Secondary oocyte

Primary oocyte

Mature follicle

Each egg in the ovary is surrounded by a layer of cells, and together they constitute the functional unit of the ovary, the **follicle** (Figure 35.10). Between the time of puberty and that of menopause, 6 to 12 follicles mature within the ovaries of a human female each month. In each of these follicles, the egg enlarges and the surrounding cells proliferate. Usually only one of these follicles matures fully each month to the point that it ruptures and releases an egg. The process of releasing an egg from the ovary is called **ovulation**. Following ovulation, the follicular cells continue to proliferate and form a mass of endocrine tissue about the size of a marble. This structure, which remains in the ovary, is the **corpus luteum**. It functions as an endocrine gland, producing estrogen and progesterone for about 2 weeks. It then degenerates unless the egg meets a sperm and is fertilized. We will return to the corpus luteum in the discussion of the menstrual cycle.

At the time of ovulation, the egg is released into the body cavity. But before it can get lost in the body cavity, the egg is swallowed by the fringed end of one of the paired tubes called **oviducts** (also called fallopian tubes). Cilia lining the oviduct propel the egg slowly through the tube. If a sperm reaches the oviduct and fertilizes the egg, the resulting zygote undergoes its first cell divisions, becoming a **blastocyst** as it continues to move down the oviduct. Several days later, the blastocyst reaches the **uterus**, or womb, a muscular, thick-walled cavity. During the week before ovulation, the lining of the uterus, the **endometrium**, has grown and has become laced with blood vessels. The blastocyst implants itself in the endometrium, and a **placenta** develops. The placenta exchanges nutrients and waste products between the mother and the developing human, which is termed an **embryo** for the first 8 weeks of development and a **fetus** thereafter. The uterus, in which the fetus completes its development, is shaped like an upside-down pear, and its neck ends in a muscular opening called the **cervix**.

The cervix leads into the vagina, in which sperm are deposited during copulation and through which the baby passes at birth. The vagina opens to the outside through a chamber that also contains an opening for the excretion of urine. The chamber is protected by two sets of skin folds, the delicate **labia minora** and the thicker, fatty **labia majora**. At the anterior tip of the labia minora is the **clitoris**, a small bulb of erectile tissue that is the anatomical homologue of the penis. The clitoris is highly sensitive and plays an important role in sexual response. The labia minora and the clitoris consist of erectile tissue and become engorged during sexual excitation. The opening of an infant female's vagina is covered by a thin membrane, the hymen, which has no known function. It is eventually ruptured by vigorous physical activity or first intercourse, but it can make first intercourse difficult or painful for the female.

The Menstrual Cycle

Ovulation is part of the human female's regular reproductive cycle, called the menstrual cycle because it ends conspicuously with **menstruation**, the sloughing off of the endometrium, or uterine lining. This sloughed-off tissue and blood from the uterine wall are lost through the vagina. The menstrual cycle really consists of two coordinated cycles, one in the ovary, which results in the release of an egg each month, and one in the uterus, which prepares the endometrium to receive a blastocyst. The human reproductive cycle has a period of about 28 days or one month; a synonym for menstruation is menses, the Latin word for months. Some mammals have shorter ovarian cycles and others have longer ones. Rats and mice have ovarian cycles of about 4 days, and many species have only one cycle per year.

Most mammals do not end their cycles with men-

struation; instead, the uterine lining is reabsorbed rather than being sloughed off. In these species the reproductive cycle is called the **estrous cycle** because its most striking event is the sexual receptivity of the female at the time of ovulation, called estrus, or "heat." When the female comes into estrus, she actively solicits male attention and may be aggressive to other females. She attracts males by releasing chemical signals as well as by her behavior. The human female is unusual among mammals in that she is potentially sexually receptive throughout her reproductive cycle and at all seasons of the year.

Hormonal Control of the Human Female Reproductive Cycle

The ovarian and menstrual cycles of human females are coordinated and timed by hormonal controls. Gonadotropins secreted by the anterior pituitary are the central elements in this control process. Prior to puberty, the secretion of gonadotropins is low, and the ovaries are inactive. At the time of puberty, the hypothalamus increases its release of gonadotropin-releasing hormone, thus stimulating the anterior pituitary to secrete follicle-stimulating hormone and luteinizing hormone. In response to these two gonadotropins, ovarian tissue grows and produces estrogen, and the follicles go through early stages of maturation. The rise in estrogen causes the development of secondary sexual characteristics including the maturation of the uterus. Between puberty and menopause (when menstrual cycles cease), the interactions of gonadotropin-releasing hormone, the gonadotropins, and the sex steroids control the reproductive cycle.

Menstruation marks the beginning of the menstrual or ovarian cycle (Figure 35.11). A few days before menstruation begins, the anterior pituitary begins to increase its secretion of follicle-stimulating hormone and luteinizing hormone. In response to these gonadotropins, follicles begin to mature in the ovaries and estrogen levels begin to rise slowly. After about a week of growth, all but one of these follicles wither away. The one still-growing follicle secretes increasing amounts of estrogen, stimulating the endometrium to grow. Estrogen exerts a negative feedback effect on gonadotropin release by the pituitary during the first 12 days of the cycle. Then, on about day 12 of the cycle, estrogen exerts a positive, rather than a negative, feedback effect on the pituitary. As a result, there is a great surge of luteinizing hormone, and to a lesser extent, follicle-stimulating hormone. The luteinizing hormone surge triggers the mature follicle to rupture and release the egg. The luteinizing hormone surge also stimulates the follicle cells to develop into the corpus luteum and to secrete estrogen and progesterone.

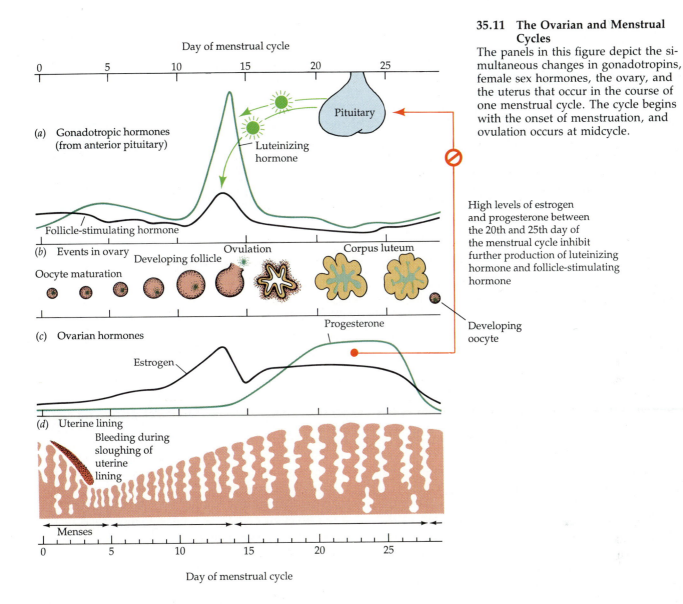

35.11 The Ovarian and Menstrual Cycles

The panels in this figure depict the simultaneous changes in gonadotropins, female sex hormones, the ovary, and the uterus that occur in the course of one menstrual cycle. The cycle begins with the onset of menstruation, and ovulation occurs at midcycle.

(a) Gonadotropic hormones (from anterior pituitary)

Luteinizing hormone

Follicle-stimulating hormone

(b) Events in ovary

Ovulation

Developing follicle

Corpus luteum

Oocyte maturation

High levels of estrogen and progesterone between the 20th and 25th day of the menstrual cycle inhibit further production of luteinizing hormone and follicle-stimulating hormone

Developing oocyte

(c) Ovarian hormones

Progesterone

Estrogen

(d) Uterine lining

Bleeding during sloughing of uterine lining

Menses

Day of menstrual cycle

Estrogen and especially progesterone secreted by the corpus luteum following ovulation are crucial to the continued development and maintenance of the endometrium. In addition, these sex steroids have negative feedback on the pituitary, inhibiting gonadotropin release, so new follicles do not begin to mature. If the egg is not fertilized, the corpus luteum degenerates on about the twenty-sixth day of the cycle. Without the production of steroids by the corpus luteum, the endometrium sloughs off, and new follicles begin to develop. The cycle begins again.

If the egg is fertilized, a zygote is created. The zygote undergoes numerous cell divisions becoming a blastocyst as it travels down the oviduct. When the blastocyst arrives in the uterus and implants in the endometrium, a new hormone comes into play. A layer of cells covering the blastocyst begins to secrete human chorionic gonadotropin, which keeps the corpus luteum functional. These same tissues also produce large amounts of estrogen and progesterone.

Eventually these tissues derived from the blastocyst take over for the corpus luteum. Continued high levels of estrogen and progesterone prevent the pituitary from secreting gonadotropins, and therefore the ovarian cycle ceases for the duration of the pregnancy. This same mechanism is exploited by birth control pills, which contain synthetic hormones resembling estrogen and progesterone and therefore prevent the ovarian cycle through negative feedback to the pituitary.

Human Sexual Responses

The sexual responses of both women and men consist of four phases: excitement, plateau, orgasm, and resolution. As sexual excitement begins in a woman, her heart rate and blood pressure rise, muscular tension increases, her breasts swell, and her nipples become erect. Her external genitals, including the sensitive clitoris, swell as they become filled with

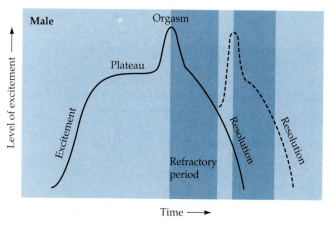

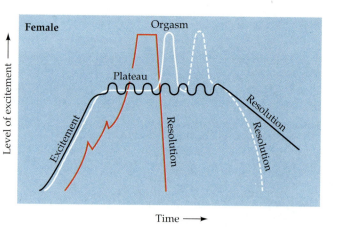

35.12 Human Sexual Responses
The dashed lines show that both males and females may have repeated orgasms, but in the male they are separated by refractory periods during which sexual excitement cannot be maintained. Females have a greater diversity of response cycles, as shown by the three sets of lines. The cycle most similar to that of the male is shown in white. Alternatively, a female may experience sustained multiple orgasms (shown in black) or may omit the plateau phase in a surge toward a very intense orgasm (red curve). Females do not have refractory periods.

blood, and the walls of the vagina secrete lubricating fluid that facilitates copulation.

As a woman's sexual excitement increases, she enters the plateau phase. Her blood pressure and heart rate rise further, her breathing becomes rapid, and the glans and shaft of the clitoris begin to retract—the greater the excitement, the greater the retraction. The sensitivity that once focused in the clitoris spreads over the external genitals, and the clitoris itself becomes even more sensitive.

Orgasm begins with a 2–4 second contraction of the outer third of the vagina, followed by shorter contractions approximately 1 second apart. Orgasm may last as long as a few minutes, and, unlike men, some women can experience several orgasms in rapid succession. During the resolution phase, blood drains from the genitals and body physiology returns to close to normal. The resolution phase lasts for approximately 5–10 minutes after orgasm; if she does not experience orgasm, a woman's resolution phase may take 30 minutes or longer.

The cycle of a man's sexual responses is very similar to that of the woman. The excitement phase is marked by an increase in blood pressure, heart rate, and muscle tension. The penis fills with blood and becomes hard and erect. In the plateau phase, breathing becomes rapid, the diameter of the glans increases, and a clear lubricating fluid oozes from the penis. The testes also swell and the scrotum tightens. Pressure and friction against the nerve endings in the glans and in the skin along the shaft of the penis eventually trigger orgasm. Massive spasms of the muscles in the genital area and contractions in the accessory reproductive organs result in ejaculation.

Within a few minutes after ejaculation, the penis shrinks to its normal size, and body physiology returns to resting conditions.

A difference between the male and female cycles is the presence of a refractory period in males immediately after orgasm. During this period, which may last 20 minutes or longer, a man cannot achieve a full erection or another orgasm, regardless of the intensity of sexual stimulation. Figure 35.12 shows the male and female response cycles.

FERTILIZATION

As you know, the union of sperm and egg, or fertilization of the egg, results in a diploid zygote and initiates the development of the embryo, which was described in Chapter 15. Fertilization is not a single event, but a complex series of processes. It begins with the juxtaposition of sperm and egg, accomplished in most species by sexual behavior. The final distance between sperm and egg must be bridged by the motility of sperm because eggs are universally nonmotile. When egg and sperm finally meet, several events take place in sequence: the sperm is activated, the sperm gains access to the plasma membrane of the egg, sperm and egg membranes fuse, and the egg is activated. Egg activation sets up blocks to entry by additional sperm, stimulates the final meiotic division of the egg nucleus, and initiates the first stages of development. The last event of fertilization is the actual fusion of the egg and sperm nuclei to create the diploid nucleus of the zygote. We will look at each of these steps.

Sperm Activation

Mammalian sperm face a formidable task after they are ejaculated into the female's reproductive tract. They must swim up from the vagina, through the uterus, and into the oviducts, where they might find an egg. They are aided in their journey by waves of muscular contractions of the vagina that are part of the female response to sexual stimulation and are also stimulated by the prostaglandins in the semen. Sperm can reach the upper ends of the oviducts within 10 minutes of ejaculation. The mammalian egg, like any other cell, is bounded by a plasma membrane. Immediately surrounding the plasma membrane is a glycoprotein envelope called the **zona pellucida**. Surrounding all of that is a layer called the **cumulus** consisting of follicle cells in a jelly matrix (Figure 35.13). When sperm are first deposited in the vagina, they are not capable of getting through all of these barriers to fertilize the egg. In the uterine environment, the sperm undergo **capacitation**; that is, they become capable of interacting with the egg and its barriers. The response of a capacitated sperm to an egg is mediated by the acrosome of the sperm, so it is called the **acrosomal reaction**.

There are species differences as to where and when the acrosomal reaction is initiated, but in all cases the first step is the breakdown of the membranes bounding the sperm head and the acrosome. This releases the enzymes contained in the acrosome. One of those enzymes, **hyaluronidase**, helps to disperse the cumulus cells surrounding the egg. Hyaluronidase digests the hyaluronic acid in the extracellular matrix that binds the cumulus cells together. Other enzymes released from the acrosome also help to disrupt the cumulus. Even though only one sperm fuses with the egg, a large number of sperm releasing their acrosomal enzymes make the plasma membrane of the egg more accessible.

Once the sperm gets through the cumulus layer, an enzymatic reaction occurs between the sperm head and the zona pellucida. The surface of the sperm head contains enzyme molecules and the zona pellucida contains substrate molecules. The lock-and-key binding of enzyme and substrate links the sperm to the egg. Acrosomal enzymes then digest a path through the zona pellucida so that the sperm can come into contact with, and eventually fuse with, the plasma membrane of the egg.

Egg Activation and Blocks to Polyspermy

The unfertilized egg is metabolically sluggish, conserving its resources for the early stages of development. The binding of the sperm to the plasma membrane of the egg and the entry of the sperm into the egg activates the egg and initiates a programmed sequence of events. The very first responses to fer-

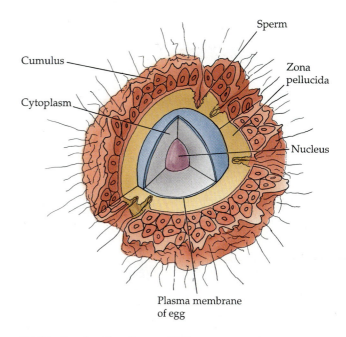

35.13 Barriers to a Sperm Cell
The human egg has several layers that the sperm cell must penetrate to reach the nucleus and fertilize the egg. The cumulus and zona pellucida must both be penetrated before the sperm can come into contact with, and eventually fuse with, the egg's plasma membrane.

tilization are **blocks to polyspermy**, that is, mechanisms that prevent more than one sperm from entering the egg. If more than one sperm enters the egg, the resulting embryo will probably not survive.

Blocks to polyspermy have been studied intensively in sea urchins. Because sea urchins have large eggs that can be fertilized in dishes of seawater, they are excellent experimental subjects for studying fertilization. Within a tenth of a second after the first sperm enters a sea urchin egg, the egg takes in sodium ions. This increase in sodium concentration within the egg changes the electrical potential across the egg's plasma membrane. This change prevents the entry of additional sperm and is called the fast block to polyspermy.

There is also a slow block to polyspermy that takes 20 to 30 seconds (Figure 35.14). A sea urchin egg has a membranous structure called a **vitelline envelope**, rather than a zona pellucida, surrounding its plasma membrane. The vitelline envelope is bonded to the plasma membrane and has sperm-binding receptors on its surface. Just under the plasma membrane are cortical vesicles filled with enzymes. The sea urchin egg, like all animal eggs, contains calcium sequestered in organelles within the cell. When a sperm enters, the egg releases calcium into its own cytoplasm. The increase in calcium causes the cortical vesicles to bond with the plasma membrane and release their enzymes, which break the bonds between the vitelline envelope and the plasma membrane.

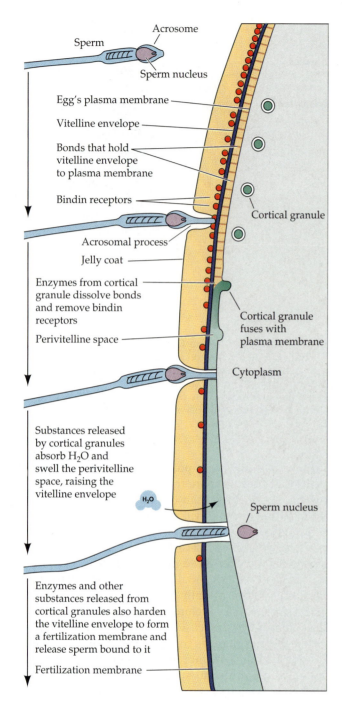

Acrosome

Sperm

Sperm nucleus

Egg's plasma membrane

Vitelline envelope

Bonds that hold
vitelline envelope
to plasma membrane

Bindin receptors

Acrosomal process

Jelly coat

Enzymes from cortical
granule dissolve bonds
and remove bindin
receptors

Perivitelline space

Cortical granule

Cortical granule
fuses with
plasma membrane

Cytoplasm

Substances released
by cortical granules
absorb H₂O and
swell the perivitelline
space, raising the
vitelline envelope

H₂O

Sperm nucleus

Enzymes and other
substances released from
cortical granules also harden
the vitelline envelope to form
a fertilization membrane and
release sperm bound to it

Fertilization membrane

35.14 Slow Block to Polyspermy
Immediately following sperm entry, the electrical polarity
of the egg's plasma membrane changes, creating a fast
block to entry by additional sperm. Within seconds the
slow block, depicted in this figure for a sea urchin egg,
is initiated. The first step is fusion of cortical granules
with the egg plasma membrane, releasing their contents
into the space between the plasma membrane and the
vitelline envelope. The raised vitelline envelope loses
its bindin receptors and becomes the fertilization
membrane.

Water then flows by osmosis into the space between
the vitelline envelope and the plasma membrane.
This raises the vitelline envelope to form the **fertilization membrane**. The enzymes also remove the

sperm-binding receptors from the surface of the fertilization membrane and cause it to harden, preventing the passage of additional sperm through it.

The release of calcium ions within the egg following fertilization activates the egg metabolically. The
pH of the cytoplasm increases, oxygen consumption
rises, and protein synthesis increases. The actual fusion of sperm and egg nuclei does not take place
until some time after fertilization—about 1 hour in
sea urchins and about 12 hours in mammals. The egg
nucleus must complete its second meiotic division
before egg and sperm nuclei unite.

Most methods of birth control are focused on
events surrounding fertilization. Physical barriers
and behavioral changes are used to prevent the meeting of sperm and egg. Hormonal manipulations are
employed to disrupt the ovarian cycle and prevent
ovulation. Most recently, a chemical means of preventing implantation of the fertilized egg has been
developed. These various birth control methodolo-
gies and their relative effectiveness are discussed in
Box 35.A.

CARE AND NURTURE OF THE EMBRYO

Once development begins, the embryo requires access to oxygen, removal of carbon dioxide, a continuous source of nutrients, and a suitable physical environment. Two general patterns of care and nurture
of the embryo have evolved in animals: oviparity and
viviparity.

Oviparity

Oviparous animals lay eggs in the environment,
which means that their offspring go through the embryonic stages outside the body of the mother. We
have noted that the egg is always much larger than
the sperm. Its large size is mostly due to its stored
nutrients, or yolk, on which the entire course of development of an oviparous animal depends. Oviparous terrestrial animals, such as reptiles, birds, and
insects, coat their eggs with tough, waterproof membranes or shells to keep them from drying out and
to protect against predators. The protective coverings
of terrestrial eggs must, however, be permeable to
oxygen and carbon dioxide. Oviparous animals may
engage in various forms of protective parental behavior focused on their eggs—nest construction and
incubation being good examples—but until the eggs
hatch, the embryos are entirely dependent on the
nutrients originally stored in the egg at the time of
fertilization. Once the offspring leaves the protective
coverings of the egg, it may receive continuing parental care as it completes its development into a
mature organism. Among the mammals, only the
monotremes—the spiny anteater, and the duck-
billed platypus—are oviparous.

Viviparity

Viviparous animals retain the embryo within the mother's body for part of its development, during which time it depends on nutrients supplied by the mother, not on nutrients originally stored in the egg. Viviparous animals are said to give birth to "live offspring"—a curious choice of words, because the offspring of oviparous animals are certainly not dead. Most viviparous animals are mammals, and most mammals are viviparous. Viviparous mammals have an enlarged and thickened portion of the female reproductive tract that holds the developing embryo—as you know, this structure is called the uterus. In marsupials, the order of mammals that includes kangaroos and opossums, the uterus simply holds the embryo and does not have special adaptations to supply it with nutrients. Marsupials are very immature when born. They crawl into a pouch called a marsupium on the mother's belly, attach firmly to the nipple of a mammary gland, and complete their development outside of the mother's body. Mammals other than monotremes and marsupials are called **eutherian mammals**. A distinguishing feature of eutherian mammals is the intimate association of blood supplies of mother and embryo in the placenta. Nutrients pass from mother to embryo and wastes pass from embryo to mother through the placenta. We will discuss its structure in the next section.

To complicate developmental classification, the eggs of some fishes, amphibians, and reptiles are fertilized internally and then retained within the body of the female until they hatch. The young then leave the mother's body. In such cases, however, the developing embryos still receive all of their nutrition from the yolk stored in the eggs. This reproductive pattern is called **ovoviviparity**.

35.15 Extraembryonic Membranes
Specialized membranes enclose the embryo inside the amniotic egg of a chicken. These membranes—the amnion, the allantois, and the chorion—protect the embryo and serve other vital functions. Reptiles, birds, and mammals all have these three extraembryonic membranes.

The Extraembryonic Membranes and the Beginning of Development

The embryos of reptiles, birds, and mammals are surrounded by a series of membranes that play major roles in mediating the exchange of nutrients between the embryo and its environment. Figure 35.15 uses the chicken egg to demonstrate how these extraembryonic membranes form as outgrowths of the basic tissue layers of the embryo. The bird embryo starts out as a disk of cells sitting on top of an enormous body of yolk. The **yolk sac** is the first extraembryonic membrane to form. It forms from the layer of embryonic cells immediately on top of the yolk. These cells are continuous with the cells that will develop into the digestive system. Thus, the yolk sac can be viewed as an extension of the gut. Other membranes extend out from the embryo to form cavities. The **amnion** forms a fluid-filled cavity that immediately surrounds the embryo. The **allantois** forms a cavity that receives wastes. The **chorion** is the outermost membrane that lines the inside surface of the egg.

These extraembryonic membranes are the basic features of the **amniotic egg**, which was a major step in the evolution of reptiles from amphibian ancestors about 300 million years ago. The amniotic egg was the adaptation that freed terrestrial vertebrates from dependence on an aquatic environment for reproduction. Fish or amphibian eggs rapidly dry out if they are exposed to air, but the amniotic egg provides an aqueous environment within which the embryo can develop.

The same extraembryonic membranes found inside birds' eggs also form in mammals. The mammalian blastocyst is a hollow ball of cells that has a central fluid-filled cavity (Figure 35.16). The first membrane to appear is the chorion; it is apparent by the fifth cell division after fertilization and completely surrounds the blastocyst. It takes more than 3 days for the human blastocyst to travel down the oviducts to the uterus, where it lives free for the next 2 to 3 days. On about day 6 following fertilization, the blastocyst attaches to the lining of the uterus. The chorion plays an important role in the implantation process

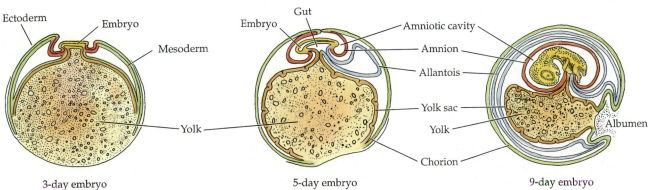

3-day embryo   5-day embryo   9-day embryo

BOX 35.A

The Technology of Birth Control

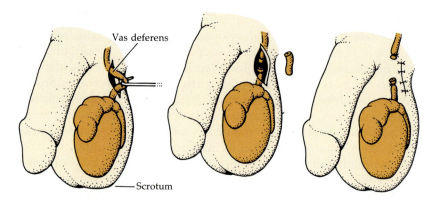

Vas deferens

Scrotum

A vasectomy is a minor operation in which each vas deferens is cut and the cut ends are tied closed. The pathway from the testes to the penis is thus interrupted, and the man's ejaculate will not contain sperm.

There are many methods of contraception (birth control) that people use to help them control the number of their children and the spacing between their births. Some of these methods are used by the woman, and others are used by the man. Here we review some of the most commonly employed contraception methods and their relative failure rates. Birth-control techniques have become more effective over time, and research continues, with the goals of still-greater effectiveness, safety, and convenience.

One of the oldest and simplest methods of contraception is coitus interruptus, the withdrawal of the penis from the vagina before ejaculation. The failure rate of this method can run almost as high as 40 percent because a few drops of fluid are released by the penis during arousal, and these drops may carry enough sperm to bring about fertilization. Sometimes ejaculation near the vagina allows some sperm to find their way into it, or withdrawal may not be soon enough.

The only certain methods of contraception are (at present) virtually irreversible ones—the sterilization of either the man or the woman. Male sterilization by vasectomy is a simple

operation done under a local anesthetic in a doctor's office. As shown in the figure, each vas deferens is cut and the cut ends are then tied off. After this minor surgery, sperm cannot pass through the vas deferens after leaving the epididymis, so the man's ejaculate no longer contains sperm. The operation does not affect his hormone levels or his sexual responses.

The most common method of female sterilization is tubal ligation ("having the tubes tied"). A small piece is removed from each oviduct, and the ends of the oviduct are tied off. Alternatively, the oviducts may be burned (cauterized) to seal them off, a process called endoscopy. With the oviducts blocked, the egg cannot reach the uterus, and sperm cannot reach the egg.

The birth control pill is the most effective method of contraception other than sterilization. It works by preventing ovulation, so there is no egg to be fertilized. The most commonly used pills contain high levels of estrogen and a synthetic proges-

terone. These hormones exert negative feedback effects on the pituitary, so gonadotropins are not released in amounts sufficient to permit development of the ovum and the follicle, and the ovarian cycle is suspended. On about day 5 of her menstrual cycle, the woman starts taking a pill each day, thus raising her hormone levels. After 20 or 21 days, the pill is discontinued, the lining of the uterus disintegrates, and slight menstrual bleeding occurs. The pill is no longer recommended for use by women over 40 years of age. It causes a small but significant increase in the tendency to suffer blood clotting disorders, which can be fatal in extreme cases; but the risk of death from using the pill is less than that associated with a full-term pregnancy.

Another highly effective method of contraception (with a failure rate varying from 1 percent to about 5 percent) is the intrauterine device (IUD). The IUD is a small piece of plastic or copper that is inserted in the uterus. The IUD probably works by preventing implantation of the

by inducing numerous responses in the endometrium. As the blastocyst invades it, the endometrium proliferates and develops more blood vessels. The interaction of the chorion with the wall of the uterus is the beginning of the placenta, which will grow and become the site of exchange of nutrients and wastes between mother and embryo.

A compact inner mass of cells within the blastocyst forms the embryo (Figure 35.16). As we saw in the

bird embryo, the amnion surrounds the embryo, creating a fluid-filled cavity within which the developing embryo floats. The allantois forms a stalk or cord that connects the embryo with the chorion at the location where the placenta will form. This allantoic stalk becomes the **umbilical cord**. Blood vessels from the embryo grow down the umbilical cord and are the conduits carrying nutrients from and wastes to the placenta. You can now appreciate why astronauts

Common Methods of Birth Control

METHOD	MODE OF ACTION	FAILURE RATE (PREGNANCIES PER 100 WOMEN PER YEAR)
Coitus interruptus	Withdrawal of penis before ejaculation	10–40
Vasectomy	Prevents release of sperm	0.0–0.15
Tubal ligation	Prevents egg from entering uterus	0.0–0.05
"The Pill"	Prevents ovulation	0–3
Intrauterine device (IUD)	Prevents implantation of fertilized egg	1–5
Condom	Prevents sperm from entering vagina	3–20
Diaphragm/jelly	Prevents sperm from entering uterus; kills sperm	3–25
Vaginal jelly or foam	Kills sperm; blocks sperm movement	3–30
Rhythm method	Abstinence near time of ovulation	15–35
Douche	Supposedly flushes sperm from vagina	80
(Unprotected)	(No form of birth control)	(80)

fertilized egg. Complications that can arise from its use, including uterine infections that can cause sterility, have led many women to consider other options. Lawsuits against IUD manufacturers and subsequent insurance considerations have resulted in a decline in its use and manufacture in the United States, although it is still available in a number of other countries.

Two primary mechanical methods of contraception have been in use for over a century. The condom ("rubber," or "prophylactic") is a sheath made of latex or of lamb intestinal material that can be fitted over the erect penis. A condom traps the ejaculate so that sperm do not enter the vagina. Latex condoms are also of value in preventing the transmission of sexually transmitted diseases such as AIDS, syphilis, and gonorrhea. In theory, the use of a condom can be highly effective, with a failure rate near zero; in practice, the failure rate is about 15 percent, usually because of faulty technique.

The diaphragm is a dome-shaped piece of rubber with a firm rim that fits over the woman's cervix and thus blocks sperm from entering the uterus. It is treated first with contraceptive jelly or cream and then inserted in the vagina before intercourse. Failure rates are about the same as for condoms. A device simpler than the diaphragm is the contraceptive vaginal sponge. It is a donut-shaped, highly absorbent, polyurethane sponge permeated with a spermicide. Placed in the upper region of the vagina, it blocks, absorbs, and kills sperm. The sponge stays effective for about a day. It is easier to use than the diaphragm and has about the same failure rate.

Used alone, spermicidal foams, jellies, and creams have a failure rate of 25 percent or more. About all that can be said for them is that they are more effective than total avoidance of contraception. In spite of popular belief, douching (flushing the vagina with liquid after intercourse) is useless as a method of birth control.

Some people attempt to avoid pregnancy by one or another form of the rhythm method. The couple avoids sex from day 10 to day 17 of the menstrual cycle, when the woman is most likely to release an egg. The use of a calendar to track the cycle may be supplemented by the basal body temperature method, which is based on the observation that a woman's body temperature drops on the day of ovulation and rises sharply on the day afterward. Other methods of predicting the time of ovulation are under development. Significant improvements must be made if the rhythm method's failure rate (between 15 and 35 percent) is to be reduced.

A recent addition to birth control technology is a drug, RU 486, developed in France. RU 486 is not a contraceptive pill, but a *contragestational* pill. It blocks implantation and causes the expulsion of the blastocyst. RU 486 opposes the actions of the progesterone produced by the corpus luteum, which is essential for maintenance of the uterine lining. If RU 486 is taken at the time of the first missed menses after fertilization it causes the uterine lining to be sloughed off, along with the embryo, which is in very early stages of development and implantation.

taking space walks refer to the cables and the air hoses attaching them to the spacecraft as their umbilical cords.

Cells slough off the embryo and float in the amniotic fluid that bathes the embryo. If a small sample of the amniotic fluid is withdrawn with a needle—a process called **amniocentesis**—some of these cells can be cultured and used for biochemical and genetic analyses that can reveal the sex of the embryo as well as genetic markers for diseases such as cystic fibrosis, Tay–Sachs disease, and Down syndrome (Figure 35.17).

Amniocentesis usually is not performed until after the fourteenth week of pregnancy, and the tests require 2 weeks. Should abnormalities in the fetus be detected, termination of the pregnancy at that stage by therapeutic abortion would put the woman's health at greater risk than would an abortion per-

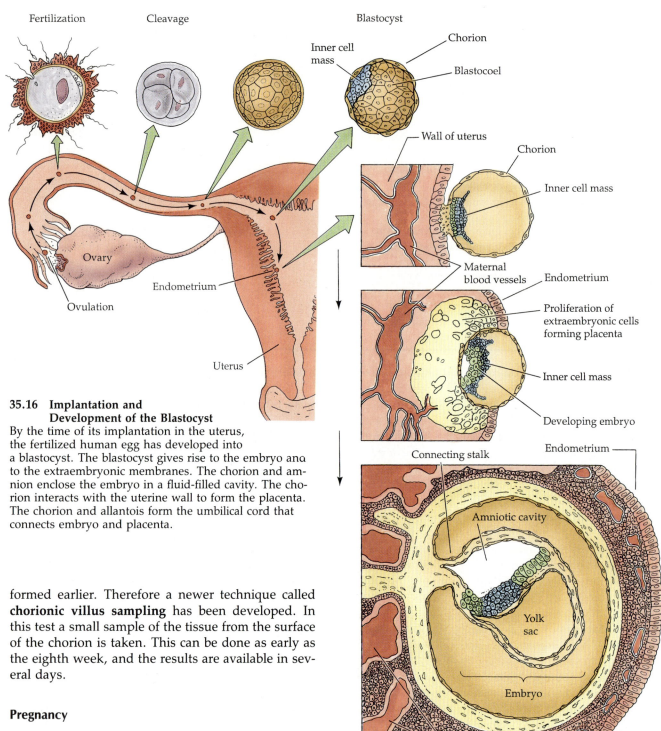

35.16 Implantation and Development of the Blastocyst
By the time of its implantation in the uterus, the fertilized human egg has developed into a blastocyst. The blastocyst gives rise to the embryo and to the extraembryonic membranes. The chorion and amnion enclose the embryo in a fluid-filled cavity. The chorion interacts with the uterine wall to form the placenta. The chorion and allantois form the umbilical cord that connects embryo and placenta.

formed earlier. Therefore a newer technique called **chorionic villus sampling** has been developed. In this test a small sample of the tissue from the surface of the chorion is taken. This can be done as early as the eighth week, and the results are available in several days.

Pregnancy

Gestation, or pregnancy, is the period from conception (fertilization of egg by sperm) to birth, during which time the embryo develops in the uterus. In general, the duration of pregnancy in mammals correlates positively with body size; in mice it is about 21 days, in cats and dogs about 60 days, in humans about 266 days, in horses about 330 days, and in elephants about 600 days. In discussing the events of human pregnancy, we divide it into three trimesters of about 3 months each.

The first trimester begins with fertilization. The blastocyst, as we know, goes through a series of rapid cell divisions, known as **cleavage**, before implanta-

tion. After implantation, the differentiation of tissues and organs we discussed in Chapter 15 begins. The first trimester is the main period of **organogenesis**. The heart begins to beat in week 4; limbs form by week 8; and most organs are present in at least rudimentary form by the end of the first trimester. At that time the embryo appears to be a miniature version of the adult and is called a fetus (Figure 35.18). Because the first trimester is a time of rapid cell di-

Withdrawal of
amniotic fluid

Centrifugation

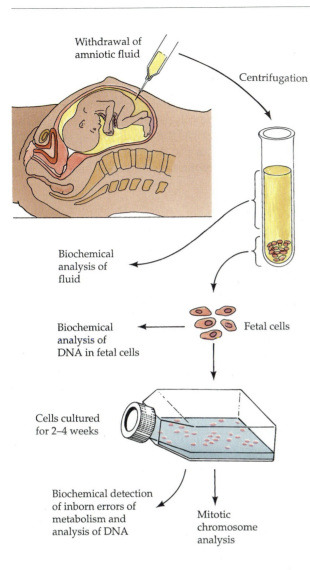

Biochemical
analysis of
fluid

Biochemical
analysis of
DNA in fetal cells

Fetal cells

Cells cultured
for 2–4 weeks

Biochemical detection
of inborn errors of
metabolism and
analysis of DNA

Mitotic
chromosome
analysis

35.17 Amniocentesis
Genetic information, including the sex of the fetus, can be gained by amniocentesis—the withdrawal and analysis of a small amount of amniotic fluid. The procedure is usually performed late in the third or early in the fourth month of pregnancy.

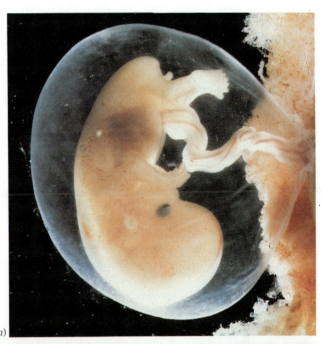

(a)

(b)

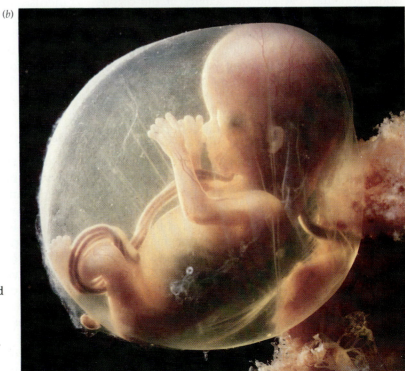

35.18 A Human Embryo
(a) The first trimester of pregnancy is a period of rapid cell division and differentiation; the organs and body structures of this six-week-old embryo are forming rapidly. (b) At four months the fetus moves freely within its protective amniotic membrane. The fingers and toes are fully formed.

vision and differentiation, it is the period during which the embryo is most sensitive to radiation, drugs, and chemicals that can cause birth defects. An embryo can be damaged before the mother even knows she is pregnant.

Hormonal changes cause major and noticeable responses in the mother during the first trimester, even though the fetus at the end of that time is still so small that it would fit into a teaspoon. Soon after the blastocyst implants, it begins to secrete human chorionic gonadotropin, the hormone that stimulates the corpus luteum to continue producing estrogen and progesterone. The high levels of these steroids prevent menstruation, which would abort the embryo. The high estrogen and progesterone levels also exert negative feedback on the pituitary, inhibiting the release of follicle-stimulating hormone and luteinizing hormone and preventing a new round of ovulation. Side effects of these hormonal shifts are the well-known "symptoms" of pregnancy: morning sickness, mood swings, changes in the senses of taste and smell, and swelling of the breasts.

During the second trimester the fetus grows rapidly to about 600 grams, and the mother's abdomen enlarges considerably. The limbs of the fetus elongate, and the fingers, toes, and facial features become well formed (Figure 35.18b). Fetal movements are first felt by the mother early in the second trimester, and they get progressively stronger and more coordinated. By the end of the second trimester, the fetus may suck its thumb.

The production of estrogen and progesterone by the placenta increases during the second trimester. As these hormones increase, the level of human chorionic gonadotropin and the activity of the corpus luteum decrease. The corpus luteum degenerates by the second trimester, but ovulation and menstruation are still inhibited by the high levels of steroids secreted by the placenta. Along with these hormonal changes, the unpleasant symptoms of early pregnancy usually disappear.

The fetus and the mother continue to grow rapidly during the third trimester. As the fetus approaches its full size, pressure on the mother's internal organs can cause indigestion, constipation, frequent urination, shortness of breath, and swelling of the legs and ankles. Throughout pregnancy the circulatory system of the fetus has been functioning, and as the third trimester approaches its end, other internal organs mature. The digestive system begins to function, the liver stores glycogen, the kidneys produce urine, and the brain undergoes cycles of sleep and waking.

BIRTH

Throughout pregnancy the uterus periodically undergoes slow, weak, rhythmic contractions called Braxton–Hicks contractions. These contractions become gradually stronger during the third trimester and are sometimes called false labor contractions. True labor contractions usually mark the beginning of childbirth, or **parturition**. In some women, however, the first signs of labor are the discharge of the mucous plug that blocks the uterus during pregnancy ("a bloody show") or the rupture of the amnion and the loss of the amniotic fluid ("waters breaking").

Labor

Numerous factors contribute to the onset of labor. Hormonal and mechanical stimuli increase the contractility of the uterus. Progesterone inhibits and estrogen stimulates contractions of uterine muscle. Toward the end of the third trimester the estrogen–progesterone ratio shifts in favor of estrogen. Oxytocin stimulates uterine contraction; its secretion by the pituitaries of both mother and fetus increases at the time of labor. Mechanical stimuli come from the stretching of the uterus by the fully grown fetus and the pressure of the fetal head on the cervix. These mechanical stimuli act through reflexes to increase pituitary release of oxytocin, which in turn increases the activity of the uterine muscle (Figure 35.19). A positive feedback hypothesis proposes that when the sensitivity of the uterus reaches a certain threshold, the weak, slow, rhythmic contractions of the uterus change into stronger labor contractions. These stronger contractions push the head of the fetus against the cervix, which stimulates, in turn, the reflexes that cause further contractions of the uterus.

In the early stage of labor, the contractions of the uterus are 15 to 20 minutes apart, and each lasts 45 to 60 seconds. During this time the contractions pull the cervix open until it is large enough to allow the baby to pass through. This stage of labor lasts an average of 12 to 15 hours in a first pregnancy and 8 hours or less in subsequent ones. Gradually the contractions become more frequent and more intense.

Delivery

In the second stage of labor, which begins when the cervix is fully dilated, the baby's head moves into the vagina and becomes visible from the outside. The usual head-down position of the baby at the time of delivery is set up by a shift in the orientation of the fetus during the seventh month. If the fetus fails to reorient head down, a different part of the fetus enters the vagina first, and the birth is more difficult. Passage of the fetus through the vagina is assisted by the woman's bearing down with her abdominal and other muscles to help push the baby along. Once the head and shoulders of the baby clear the cervix, the rest of its body eases out rapidly, but it is still connected to the placenta in the mother by the umbilical cord. This second stage of labor may take as

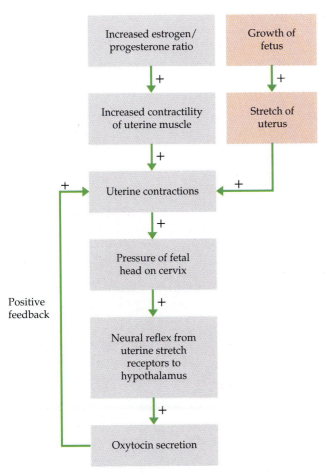

35.19   Oxytocin Reflex and Birth
Release of oxytocin by the pituitary gland increases uterine contractions during labor and birth. The diagram illustrates a positive feedback loop.

little as a minute or up to half an hour or more in a first pregnancy.

As soon as the baby clears the birth canal, it can start breathing and become independent of its mother's circulation. The umbilical cord may then be clamped and cut. The segment still attached to the baby dries up and sloughs off in a few days, leaving behind its distinctive signature, the belly button—more properly called the umbilicus. The detachment and expulsion of the placenta and fetal membranes takes from a few minutes to an hour, and may be accompanied by a number of uterine contractions.

A caesarian section is the surgical extraction of the baby from the uterus (Julius Caesar was supposedly born this way, hence the term). It may be necessary for a number of reasons: if the fetus is large and the mother's pelvis small, if the first stage of labor lasts too long, if the cervix fails to dilate sufficiently, or if there is any sudden threat to the health of the baby or the mother.

Lactation

Throughout pregnancy, the high circulating levels of estrogen and progesterone cause the mammary glands of the breasts to develop in preparation for lactation (the secretion of milk). Prolactin secretion also increases progressively during pregnancy, but its effect is countered by estrogen and progesterone, which inhibit the production of milk. Just before birth, the breasts may secrete a few milliliters of fluid each day. This fluid, called **colostrum**, contains little fat, and its rate of production is very low. With expulsion of the placenta, the estrogen and progesterone levels in the mother's bloodstream fall rapidly, and within a few days the well-developed mammary glands are producing milk. This milk does not flow readily into the ducts of the breasts, however. If it did, it would dribble out continuously, rather than just flowing out when the baby suckled. Oxytocin plays an important role in controlling lactation (Figure 35.20). When the baby suckles, a neural signal from the breast causes the release of oxytocin from the posterior pituitary. In about 30 seconds, this oxytocin reaches the breasts and stimulates contraction of the muscle cells that surround the milk-secreting cells. The milk is thereby "let down," or ejected, into the ducts of the breasts.

Oxytocin also has a role just after the delivery of the baby. If the newborn baby is placed at the mother's breast, even though the breast cannot deliver more than colostrum at this time, the suckling of the infant causes oxytocin release, which stimulates continued uterine contractions that help to expel the placenta and inhibit bleeding of the uterine wall.

SUMMARY

Some simple animals reproduce asexually (by budding, regeneration, or parthenogenesis), but most animals reproduce sexually. Sex produces genotypic diversity through recombination. Male and female reproductive structures are in separate individuals in dioecious species, and in single hermaphroditic individuals in monoecious species. Fertilization is external in some species that release their gametes into aqueous environments, and is internal in other species. In both cases behavior plays an important role in bringing eggs and sperm together.

The reproductive organs of the male and female produce, process, and deliver mature gametes to the site where fertilization takes place. In the case of internal fertilization, the complementary structures of the male and female external reproductive organs are adapted for the act of copulation which involves a suite of behavioral and physiological sexual responses. The male testes and the female ovaries are the internal reproductive organs where gametogenesis occurs. Both spermatogenesis and oogenesis require mitotic proliferation of primary germ cells, meiosis, and maturation of gametes. Sperm production in the seminiferous tubules of the testes is continuous throughout the reproductive life of the hu-

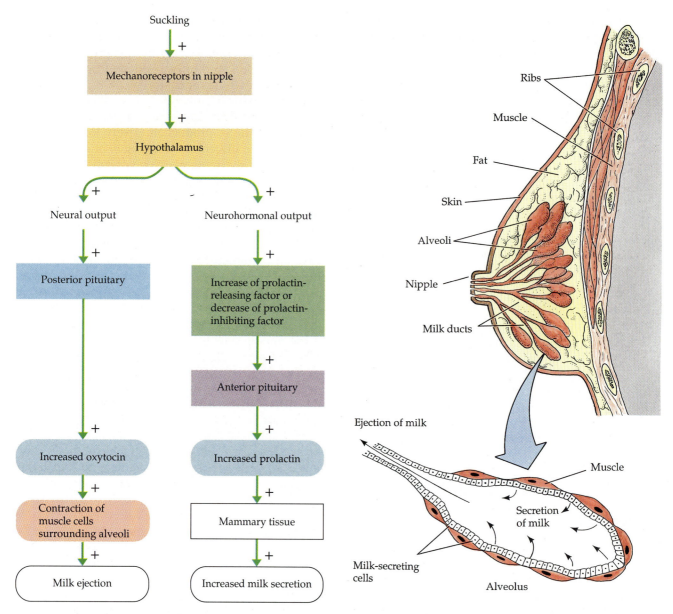

35.20   Hormonal Control of Lactation
The role of hormones in lactation is diagrammed at left. The mammary gland is shown in cross section at the right; the alveoli are reservoirs for milk.

man male. In contrast, the human female has her lifetime supply of primary oocytes at birth, and each month during her reproductive years at least one oocyte develops into a mature egg. This monthly ovarian cycle is closely tied to the menstrual cycle through which the uterus is prepared each month to receive an embryo. If an embryo (blastocyst) does not arrive and implant itself in the wall of the uterus, the lining of the uterus sloughs off, causing menstrual bleeding and discharge.

Gametogenesis and the female ovarian and menstrual cycles are under hormonal control. The hypothalamus produces releasing factors that control the production and release of gonadotropins from the anterior pituitary. Gonadotropins stimulate testes to mature and to produce sperm and male sex steroids. Gonadotropins are under feedback control from the sex steroids. Gonadotropin release is cyclical in the female, thus controlling the monthly cycle of follicle maturation, ovulation, and production of sex steroids that in turn control the menstrual cycle.

Fertilization involves a complex series of interactions between sperm and egg. Sperm are motile and must swim up the oviducts to fertilize an egg. The egg is surrounded by a glycoprotein envelope and follicle cells. Passage of the sperm through these barriers is made possible by enzymes released from the acrosome on the head of the sperm. Final contact and binding of sperm with the plasma membrane of the egg is facilitated by a molecular recognition mechanism. Fusion of sperm and egg initiates several responses of the egg that include blocks to polyspermy, metabolic activation, and the final step of meiosis.

The mammalian female reproductive tract provides protection for the developing embryo. The placenta, which includes tissues from the mother and from the embryo, supplies the embryo with oxygen and nutrients and removes its waste products throughout the period of gestation. In humans, childbirth involves a positive feedback hormonal mechanism in which stretching of the uterine wall stimulates oxytocin release, which in turn causes uterine muscles to contract. Oxytocin and prolactin are involved in controlling lactation following childbirth.

SELF-QUIZ

1. Match each of the following modes of asexual reproduction with the statement or description that characterizes it. (Each letter may be used more than once, and more than one letter may apply to each statement.)
   a. Budding
   b. Regeneration
   c. Parthenogenesis
       (i) A form of asexual reproduction that usually follows an animal being broken by an external force, but it can also be initiated by the animal itself.
       (ii) Many freshwater sponges produce clusters of undifferentiated cells which eventually "escape" the parent and become free-living organisms genetically identical to the parent.
       (iii) Offspring develop from unfertilized eggs.
       (iv) The process requires totipotent cells.
       (v) Species that reproduce this way may also engage in sexual reproduction.

2. A species in which the individual possesses both male and female reproductive systems is termed (choose *all* that apply):
   a. dioecious.
   b. parthenogenic.
   c. hermaphroditic.
   d. diploid.
   e. monoecious.

3. The major advantage of internal fertilization is that:
   a. it ensures paternity.
   b. it permits the fertilization of a large number of gametes.
   c. it reduces the incidence of destructive competitive interactions between the members of a group.
   d. it results in the formation of a stable pair-bond between mates.
   e. it allows the developing organism to enjoy a greater degree of protection during the early phases of development.

4. Which one of the following statements is *true*?
   a. At the time of birth, the human female has produced all of the oocytes she will ever produce.
   b. At the onset of puberty, ovarian follicles produce new oocytes in response to hormonal stimulation.
   c. At the onset of menopause, the human female stops producing oocytes.
   d. Oocytes are produced by the human female throughout adolescence.
   e. The oocytes produced by the female are kept "in storage" in the seminiferous tubules.

5. Spermatogenesis and oogenesis differ in that:
   a. spermatogenesis produces gametes with greater energy stores than those produced by oogenesis.
   b. spermatogenesis produces four equally functional diploid cells per meiotic event and oogenesis does not.
   c. oogenesis produces four equally functional haploid cells per meiotic event and spermatogenesis does not.
   d. spermatogenesis produces many gametes with meager energy reserves, whereas oogenesis produces relatively few well-provisioned gametes.
   e. in humans, spermatogenesis begins before birth, whereas oogenesis does not start until the onset of puberty.

6. The acrosome of the sperm:
   a. carries genetic information.
   b. provides energy for movement.
   c. carries the enzymes that facilitate fertilization.
   d. induces ovulation.
   e. prevents polyspermy.

7. During oogenesis in mammals, the second meiotic division occurs:
   a. after capacitation.
   b. after implantation.
   c. before ovulation.
   d. before the acrosomal reaction.
   e. after a sperm enters the egg.

8. One of the major differences between the sexual response cycles in human males and females is:
   a. the increase in blood pressure in males.
   b. the increase in heart rate in females.
   c. the presence of a refractory period in females after orgasm.
   d. the presence of a refractory period in males after orgasm.
   e. the increase in muscle tension in males.

9. Which of the following membranes is part of the embryonic contribution to placenta formation?
   a. Amnion
   b. Chorion
   c. Uterine membrane
   d. Corona radiata
   e. Zona pellucida

10. Contractions of muscles in the uterine wall and in the breasts are stimulated by:
   a. progesterone.
   b. estrogen.
   c. prolactin.
   d. oxytocin.
   e. human chorionic gonadotropin.

FOR STUDY

1. Compare and contrast spermatogenesis and oogenesis in terms of the products of each process and the timetable over which each process proceeds.

2. Describe how the events occurring during sperm activation and egg activation lead to successful fertilization.

3. Describe the hormonal and neural regulatory mechanisms operating in lactation.

4. Ovarian and uterine events in the month following ovulation differ depending on whether or not fertilization occurs. Describe those differences and explain their endocrine controls.

5. Explain how positive feedback plays a role in birth.

READINGS

Beaconsfield, P., G. Budwood and R. Beaconsfield. 1980. "The Placenta." *Scientific American*, August. This article describes the process of implantation and development of the organ that is the intermediary between fetus and mother. The many functions of the placenta are described.

Epel, D. 1977. "The Program of Fertilization." *Scientific American*, November. 1977. The initial events in the sperm–ovum interaction.

Gilbert, S. F. 1991. *Developmental Biology*, 3rd Edition. Sinauer Associates, Sunderland, MA. This excellent text on animal development includes chapters on fertilization as well as on the germ line and gametogenesis.

Johnson, M. H. and B. J. Everitt. 1988. *Essential Reproduction*, 3rd Edition. Blackwell Scientific, Oxford. A concise and comprehensive technical account of the biology of gametogenesis, fertilization, and pregnancy.

Katchadourian, H. A. 1989. *Fundamentals of Human Sexuality*, 5th Edition. Saunders, Philadelphia. An introductory text that covers the anatomy and physiology of sex and reproduction in the first two chapters, and then goes on to treat developmental, behavioral, and social aspects of sex.

Short, R. V. 1984. "Breast Feeding." *Scientific American*, April. Breast feeding has hormonal consequences that have contraceptive effects. Trends toward bottle feeding in many developing nations may be causing rises in their rates of population growth.

Wassarman, P. M. 1988. "Fertilization in Mammals." *Scientific American*, December. This article examines the molecular and cellular events that surround the fusion of sperm and egg.

36

Neurons and the Nervous System

PREVIEW: We saw in Chapter 34 that hormones are one means by which cells communicate with one another. The second major means of communication among cells is the nervous system. Nervous systems process and integrate information about events in the body and in the external environment. They control the organ systems and the muscles of the body, orchestrate behavior, and give us the ability to learn and remember. Nerve impulses are the language of nervous systems. These electrical events are generated and conducted throughout the body by the membranes of neurons. Ion pumps and ion channels in the membranes give neurons their electrical properties. Neurons are organized in complex, interconnected networks and circuits. They communicate with one another and with other cells mostly by means of neurotransmitters. The major focus of this chapter is the anatomy and function of the human nervous system.

This chapter deals with neurons, glial cells, dendrites, axons, ion channels, synapses, neurotransmitters, resting potentials, action potentials, structures of the vertebrate brain, the autonomic nervous system, reflexes, sleep, learning, memory, and language.

The human brain weighs about one and a half kilograms, is mostly water, and is the consistency and color of vanilla custard (Figure 36.1). Yet the complexity of this small mass of tissue exceeds that of any other known matter. The work of the brain is to process and store information and to control the physiology and behavior of the body. The brain is constantly receiving information from all of the senses, integrating and interpreting that information, and generating commands to the muscles and organs of the body. The brain senses the need to act, decides on the appropriate action, orchestrates it, initiates and coordinates it, monitors it, and remembers it. The essence of individuality and personality resides in the brain. You could imagine remaining yourself after replacing any organ of your body except your brain.

Some of the actions controlled by the brain are conscious, or voluntary; others, such as the functions of the heart, lungs, and gut are involuntary, or autonomic. What is remarkable is that a myriad of voluntary and involuntary actions are all going on simultaneously. Every second of its life, the brain processes thousands of bits of information. But the brain cannot function alone; it is part of a nervous system that receives information from **sensors**, such as the eyes and ears, and communicates information to **effectors**, such as muscles and glands. The brain and spinal cord make up the **central nervous system**, and networks of nerves constitute the **peripheral nervous system**, which carries information to and from the central nervous system. The information that flows through the nervous system consists of both electrical and chemical messages. The electrical messages are nerve impulses generated in individual nerve cells by ions moving through channels in their membranes. The chemical messages are released by individual nerve cells and influence the activities of neighboring nerve, muscle, and secretory cells.

Throughout the animal kingdom, nervous systems vary in complexity, ranging from that of humans to the simple nerve nets of cnidarians, which seem to do little more than detect food or danger and cause retraction of tentacles and constriction of the body (Figure 36.2). In all cases, however, nervous systems control behavior and the functions of the body. In general, the more complex the behavior and physiological capabilities of a species, the larger its nervous system. Sometimes small size belies incredible capacity, however. Consider, for example, the nervous systems of small spiders that have programmed within them the thousands of precise movements necessary to construct a beautiful web without any prior experience. One of the greatest challenges of biology is to understand how the human brain functions, but it would be a major breakthrough even to understand a much simpler nervous system. Much progress in neurobiology has come from research on simple ner-

36.1 The Human Brain
The most complex structure in the known universe.

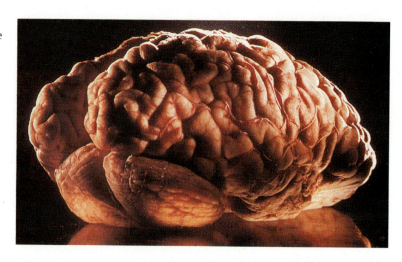

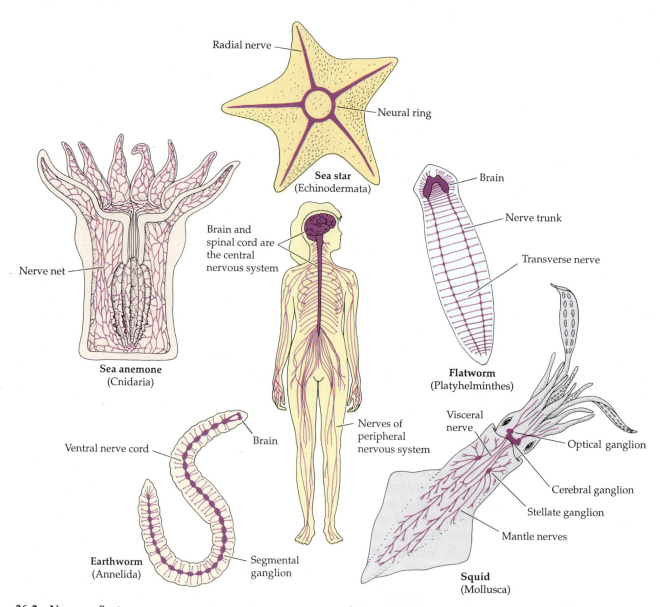

Radial nerve

Neural ring

Sea star
(Echinodermata)

Brain

Nerve trunk

Transverse nerve

Brain and spinal cord are the central nervous system

Nerve net

Sea anemone
(Cnidaria)

Flatworm
(Platyhelminthes)

Visceral nerve

Optical ganglion

Cerebral ganglion

Stellate ganglion

Mantle nerves

Nerves of peripheral nervous system

Ventral nerve cord

Brain

Earthworm
(Annelida)

Segmental ganglion

Squid
(Mollusca)

36.2 Nervous Systems

As animals become more complex in both sensory and behavioral abilities, information processing is increasingly centralized in ganglia (collections of nerve cells) or in a brain. The brain and spinal cord in the human constitute the central nervous system; the central nervous system communicates with the body through nerves that make up the peripheral nervous system.

vous systems. In this chapter we will focus on the human nervous system, but you should keep in mind that much of our knowledge comes from research on a long list of animal species. The way neurons function is almost identical in animals as different as squids and humans.

CELLS OF THE NERVOUS SYSTEM

The functions of the brain depend on the properties of its cells. **Neurons** are the cells that make it possible for the brain and the entire nervous system to transmit and integrate information. The important property of neurons is that their plasma membranes can generate electrical signals called nerve impulses or **action potentials**. Their plasma membranes can also conduct these electrical signals rapidly from one location on a cell to the most distant reaches of that cell—a distance that can be over a meter for some neurons, as we'll see. Where a neuron contacts another neuron or a muscle or gland cell, special adaptations enable the message carried by the action potential to be delivered to that cell. Much of neurobiology focuses on the structure and function of neurons, but they are not the only type of cell in the nervous system. In fact, there are more **glial cells** than neurons in the brain. Glial cells, however, do not generate or conduct action potentials; they serve various supporting roles for neurons.

Neurons

Most neurons have four regions: a cell body, dendrites, an axon, and axon terminals (Figure 36.3), but the variation between different types of neurons is considerable. The **cell body** contains the nucleus and most of the cell's organelles. Many nerve-cell processes may sprout from the cell body. (A process is any part extending from a cell, an organ, or an organism.) Most nerve-cell processes are bushlike **dendrites** (from the Greek word for tree, dendron) that bring information from other neurons or sensory cells to the neuron's cell body. In most neurons, one nerve-cell process is much longer than the others and is called the **axon**. Axons usually carry information away from the cell body. The length of the axon varies greatly in different types of neurons—as does the degree of branching of the dendrites. The axons of some neurons are remarkably long. The neuron that causes your little toe to flex has its cell body in the spinal cord in the middle of your back, and its axon goes all the way down your leg to the muscles in the little toe. Axons are the "telephone lines" of the nervous system. Information received by the dendrites can influence the cell body to generate an action potential that is then conducted along the axon to the cell that is its target. At the target cell, the

axon divides, like the frayed end of a rope, into a spray of fine processes. Each of these tiny processes ends in a swelling called an **axon terminal**, which is the site of storage and release of the chemical messages that will transmit the information to the next cell.

A neuron communicates with another cell at a **synapse**, the place where its axon terminals come very close to the plasma membrane of another cell, with a gap or cleft of about 25 nanometers between the membranes of the two cells. The neuron's axon terminals release a chemical message that diffuses across the synaptic cleft (the space between the cells) to receptors on the membrane of the target cell. Most individual neurons make and receive thousands of synapses. The essence of integration of information by neurons is the summation of synaptic inputs.

Glia

Glial cells, like neurons, come in many forms. Some glial cells act like a scaffolding to support and orient the neurons and help them to make the right contacts during their embryonic development. Another class of glial cells creates a form of "insulation" for some axons, as will be described later in this chapter. Some glial cells supply neurons with nutrients; still others act as phagocytes that consume foreign particles and cell debris. Glial cells help to maintain the proper ionic environment around neurons. Glial cells do not have axons, nor do they generate or conduct nerve impulses, but some communicate with one another electrically through a special kind of synapse called a **gap junction**, which we described in Chapter 5 as a connection that enables electrical charges to flow between cells.

Glial cells called **astrocytes** (because they look like stars) create the **blood–brain barrier**. Blood vessels throughout the body are very permeable to a great variety of chemicals, some of which are toxic. The brain is better protected from toxic substances than most tissues of the body because of the blood–brain barrier. Processes of the astrocytes form this barrier by covering the smallest and most permeable blood vessels in the brain. Protection of the brain is crucial because, unlike other tissues of the body, the brain cannot recover from damage by generating new cells. Shortly after birth, neurons in the brain cease cell division, and at that time we have the greatest number of neurons we will ever have in our lives. Throughout the rest of life there is a progressive loss as neurons die. Without the blood–brain barrier, the rate of neuron loss could be much greater. The barrier is not perfect, however. Because it consists mostly of cell membranes, it is most permeable to fat-soluble substances. Anesthetics are fat-soluble chemicals and so is alcohol; both have well-known effects on the brain.

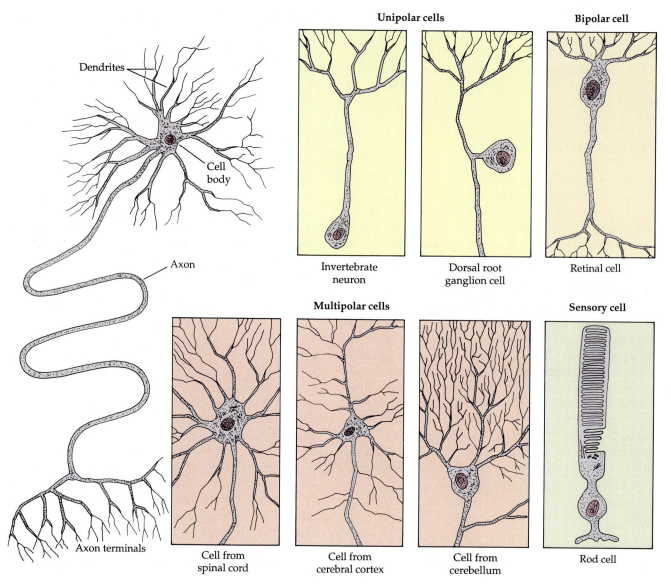

Unipolar cells

Bipolar cell

Dendrites

Cell body

Axon

Invertebrate neuron

Dorsal root ganglion cell

Retinal cell

Multipolar cells

Sensory cell

Axon terminals

Cell from spinal cord

Cell from cerebral cortex

Cell from cerebellum

Rod cell

36.3 Neurons: Building Blocks of Nervous Systems
A generalized neuron (far left) includes a cell body, dendrites that collect input, an axon that conducts action potentials, and axon terminals that make synapses with target cells. The general neuronal "plan" is modified in many different types of specialized neurons.

Cells in Circuits

The human brain contains about 100 billion neurons, and each neuron may make synapses with 1,000 or more other neurons. So there may be as many as a million billion synapses in the human brain. Therein lies the incredible ability of the brain to process and store information. The thousands of circuits in the nervous system serve many different functions; specific regions of the nervous system contain the circuits for particular functions.

ANATOMY OF THE NERVOUS SYSTEM

Major Functional Divisions

In vertebrates, the brain and spinal cord make up the central nervous system. Cranial nerves and spinal nerves conduct information between the central nervous system and the other parts of the body; these nerves constitute the peripheral nervous system. Each **nerve** is a bundle of axons (Figure 36.4). A nerve carries information about many things simultaneously. Some axons in a nerve may be carrying information to the central nervous system while other axons in the same nerve are carrying information from the central nervous system to the organs of the body. The peripheral nervous system reaches to every tissue of the body.

Conceptually, in terms of function, the peripheral

(a)

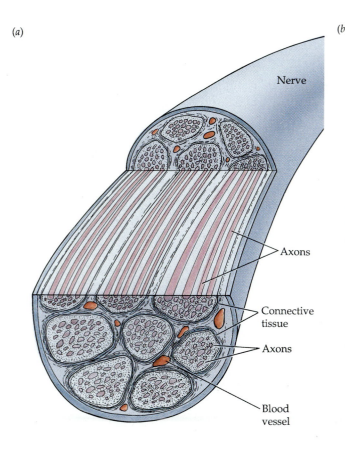

Nerve

Axons

Connective tissue

Axons

Blood vessel

(b)

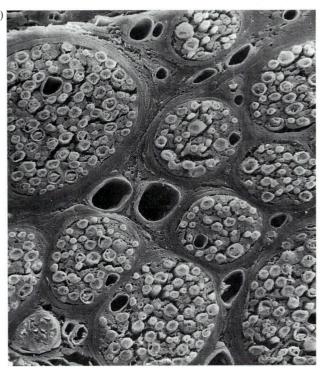

36.4 Many Axons Make Up a Nerve

Some of the axons in a nerve conduct information toward the central nervous system and others conduct information away from the central nervous system. Some axons are wrapped by glial cell membranes (myelinated) and others are not (unmyelinated). The scanning electron micrograph shows a cross section of part of a nerve.

nervous system can be divided into an **afferent** side that brings information to the central nervous system, and an **efferent** side that carries information from the central nervous system (Figure 36.5). Afferent and efferent axons may travel together in the same nerve. The afferent side has a division that carries information from our conscious senses and a division that carries information of which we are unaware. We are consciously aware of vision, hearing, touch, taste, pain, balance, and the position of the limbs of our body. We are not consciously aware of most physiological conditions, such as our blood pressure, body temperature, blood sugar level, and oxygen supply. The efferent side of the peripheral nervous system also has two divisions: a voluntary division that executes our conscious movements and an autonomic division that controls involuntary physiological functions.

Development of the Vertebrate Nervous System

As a vertebrate embryo develops, its nervous system begins as a hollow tube of neural tissue. The tube runs the length of the embryo on its dorsal side. At the head end of the embryo, this neural tube forms three swellings that give rise to the basic divisions of the brain: the hindbrain, the midbrain, and the forebrain. The rest of the neural tube forms the spinal

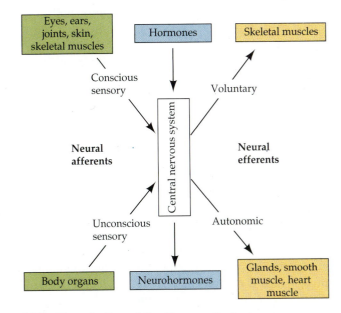

36.5 Organization of the Nervous System

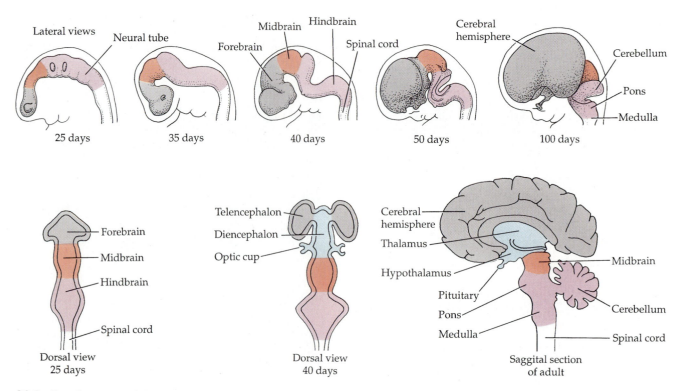

36.6 Development of the Human Nervous System
A hollow neural tube forms in the early embryo. Three swellings at the anterior end of the neural tube develop into the parts of the adult brain. Cranial and spinal nerves sprout from this neural axis to form the peripheral nervous system.

cord. The cranial and spinal nerves sprout from the neural tube and grow throughout the embryo.

Each of the three basic brain regions develops into a number of structures in the adult brain. The hindbrain gives rise to the **medulla**, the **pons**, and the **cerebellum**. The medulla is continuous with the spinal cord, the pons is above it, and the cerebellum sits above the pons. The medulla and pons control a number of basic physiological functions such as breathing and circulation. The cerebellum orchestrates and refines behavior patterns. The midbrain forms structures that process some aspects of visual and auditory information.

The forebrain forms the **diencephalon** and the surrounding **telencephalon**, which consists of two **cerebral hemispheres**. The telencephalon increases in size, complexity, and importance as we go up the phylogenetic scale. In mammals it plays major roles in sensory perception, learning, memory, and conscious behavior. The diencephalon forms the **thalamus**—the final relay station for sensory information going to the telencephalon—and the **hypothalamus**, which is responsible for the regulation of many physiological functions and biological drives (Chapters 33 and 34).

Figure 36.6 shows how the hollow neural tube of the vertebrate embryo develops into these hindbrain, midbrain, and forebrain structures. It is important to keep in mind the somewhat linear arrangement of the structures of the adult brain as revealed by their development from the embryonic neural tube. Information moves up and down this linear neural axis. A communication from the spinal cord to the telencephalon travels through the medulla, pons, midbrain, and diencephalon before reaching the telencephalon. These structures are referred to as the **brain stem**. As information passes through the brain stem it is frequently modified and relayed to other subsystems. In general, more primitive and autonomic functions are localized farther down the neural axis, and more complex and evolutionarily advanced functions are found farther forward on the neural axis. As we go up the phylogenetic scale from fish to mammal, we observe increasing relative size and dominance of the forebrain. Damage to the telencephalon in mammals results in severe impairment or even coma, but a shark swims almost normally with its telencephalon removed.

The Spinal Cord

The spinal cord conducts information between the brain and the organs of the body, and it processes and integrates information. A cross section of the spinal cord reveals a central area of gray matter in the shape of a butterfly surrounded by an area of white matter (Figure 36.7). The gray matter contains the cell bodies of the spinal neurons, and the white matter contains the axons that conduct information up and down the spinal cord. Spinal nerves leave the spinal cord at regular intervals on each side. Each spinal nerve has two roots, one connecting with the **dorsal horn** of the gray area, and one connecting with the **ventral horn** of the gray area. Each spinal nerve carries both afferent information from the sense organs and efferent information to the muscles and glands of the body. The afferent fibers enter the spinal cord through the dorsal roots, and the efferent fibers leave the spinal cord through the ventral roots.

Information entering the dorsal horn can be transmitted to neurons that will carry it to the brain, or it can be relayed to efferent neurons or to other neurons in the central gray called **interneurons**. Interneurons can make connections with efferent neurons in the ventral horns, and they can communicate with other spinal neurons up and down the spinal axis to amplify the response and to generate more complex motor patterns. The spinal cord does a considerable amount of information processing and integration. For example, when you step on a tack, spinal circuits coordinate the rapid pulling back that is carried out by many muscles on both sides of your body. Spinal circuits can also generate repetitive motor patterns such as those of walking.

The Reticular System

A structure called the **reticular system** extends throughout the core of the medulla, pons, and midbrain (Figure 36.8). Information coming up the neural axis passes through the reticular system, where many connections are made to various neurons. The reticular system controls sleep and wakefulness. High levels of activity in the reticular system maintain the brain in a waking condition, hence this structure is sometimes called the reticular activating system. If the brain stem is damaged at midbrain or higher levels, the individual is likely to remain in a coma, but if the damage is below the reticular system, the individual will have normal patterns of sleeping and waking.

The Limbic System

Immediately surrounding the diencephalon are some forebrain structures that are the only forebrain structures in more primitive vertebrates such as fishes, amphibians, and reptiles. In birds and mammals,

36.7 The Spinal Cord
The gray matter is rich in cell bodies and is surrounded by white matter made up of ascending and descending axons. Sensory information (afferent) enters through the dorsal horns, and motor output (efferent) leaves via the ventral horns. A considerable amount of information processing is done by interneurons in the gray matter.

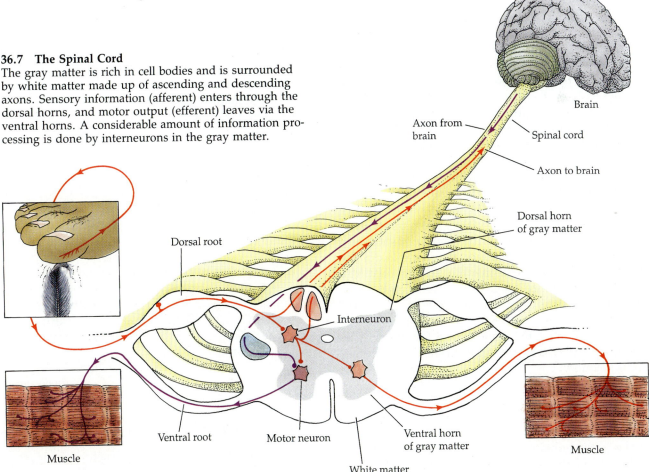

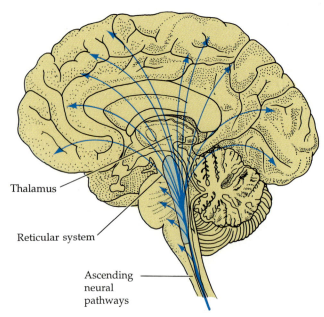

36.8 The Reticular System
A network of neurons and groups of neurons extending throughout the core of the brain stem controls levels of alertness and sleep.

these primitive forebrain structures are completely covered by the evolutionarily more recent elaboration of the telencephalon, the cerebral cortex. The primitive forebrain still has important functions in birds and mammals, and is referred to as the **limbic system** (Figure 36.9). The limbic system is responsible for basic physiological drives, instincts, and emotions. Within the limbic system are areas that when stimulated with small electrical currents can cause intense sensations of pleasure, pain, or rage. If a rat is given the opportunity to stimulate its own pleasure centers by pressing a switch, it will ignore food, water, and even sex, pushing the switch until it is exhausted. Pleasure and pain centers in the limbic system are believed to play major roles in learning and in physiological drives.

One part of the limbic system, the **hippocampus**, is necessary in humans for the transfer of short-term memory to long-term memory. If you are told a new telephone number, you may be able to hold it in short-term memory for a few minutes, but within half an hour it is forgotten unless you make a real effort to remember it. The phenomenon of remembering something for more than a few minutes requires the process of transfer from short-term to long-term memory.

The Cerebrum

The cerebral hemispheres are the dominant structures in the human brain. They are so large that they cover all of the other parts of the brain except the cerebellum (Figure 36.10). A sheet of gray matter (tissue rich in neuronal cell bodies) called the **cortex** covers each cerebral hemisphere. The cortex is about 4 millimeters thick and covers a total surface area over both hemispheres of one square meter. Since it would be rather inconvenient to have flat structures a meter square on top of our heads, the cortex is folded into ridges called convolutions and valleys called sulci so that it fits into the skull. Under the cortex is white matter, made up of the axons that interconnect the cell bodies in the cortex with each other and with other areas of the brain. The human cerebral cortex contains about 80 percent of all of the nerve cell bodies in the entire nervous system.

Specific regions of the cerebral cortex govern certain functions, so it helps to have an anatomical road map. As viewed from the left side, a left cerebral hemisphere looks like a boxing glove for the right hand with the fingers pointing forward, the thumb pointing out, and the wrist at the rear (Figure 36.10). The "thumb" area is the **temporal cortex**, which processes auditory information and the use of language. The area on the underside of the temporal cortex is specialized for the recognition of faces. If that area is damaged, names of people are remembered, but they cannot be associated with the correct faces. The part of the glove covering the fingers is the **frontal cortex**, the part of the glove covering the back of the hand is the **parietal cortex**, and the wrist area is the **occipital cortex**. The occipital cortex processes visual information.

The frontal and the parietal cortices are separated by a deep valley called the **central sulcus**. The strip of parietal cortex just behind the central sulcus is the **primary somatosensory cortex**. This area receives information through the thalamus about touch and

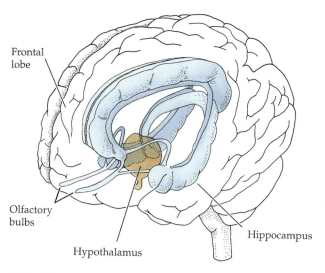

36.9 The Limbic System
Structures surrounding the hypothalamus and bordering the cerebral cortex control aspects of motivation, drives, emotions, and memory.

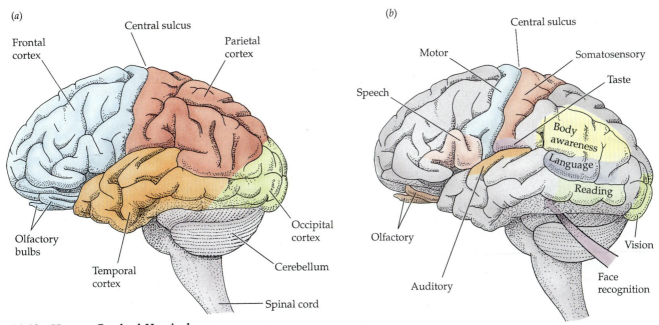

(a)

Frontal cortex
Central sulcus
Parietal cortex
Olfactory bulbs
Temporal cortex
Occipital cortex
Cerebellum
Spinal cord

(b)

Motor
Central sulcus
Speech
Somatosensory
Taste
Body awareness
Language
Reading
Olfactory
Auditory
Vision
Face recognition

36.10 Human Cerebral Hemispheres
These highly convoluted structures cover most of the other structures of the brain. The cerebral cortex is a thin layer of cells on the surfaces of the cerebral hemispheres. (a) The cerebrum is divided into four lobes. (b) Different functions are localized in the cerebral lobes.

pressure sensations. The whole body surface is represented in this strip of cortex, with the head being at the bottom and the legs at the top (Figure 36.11). Areas of the body that have lots of sensory neurons and are capable of making fine distinctions in touch (such as the lips and the fingers) have disproportionately large representation. If the somatosensory cortex is stimulated electrically, the subject reports feel-

ing specific sensations, such as touch, from a very localized part of the body.

A strip of the frontal cortex just in front of the central sulcus is the **primary motor cortex**. The cells in this region send axons to muscles in specific parts of the body. Once again, parts of the body with fine motor control have the greatest representation with the head being represented at the top. If a stimulating

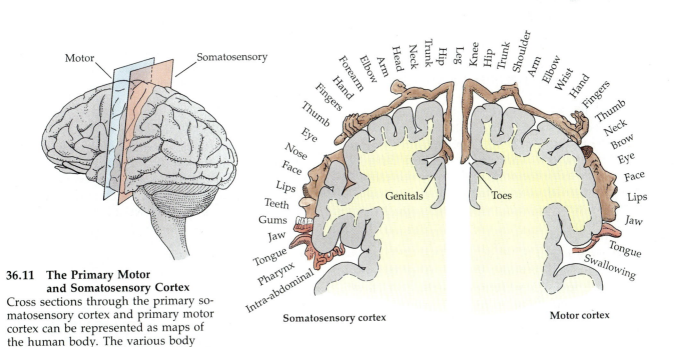

36.11 The Primary Motor and Somatosensory Cortex
Cross sections through the primary somatosensory cortex and primary motor cortex can be represented as maps of the human body. The various body parts are shown in relation to the brain area devoted to those parts.

Somatosensory cortex

Motor cortex

electrode is used in this region of the frontal cortex, the response is the twitch of a muscle, but not a coordinated, complex behavior. Large areas of the cerebral cortex are designated as the **association cortex**. Cells in the association cortex receive input from multiple sensory areas and send output to multiple motor areas.

As we go along the phylogenetic tree from fish to mammal, the size of the telencephalon increases very substantially. Even when we consider only mammals, the cerebral cortex increases in size and complexity as we compare animals such as rodents, whose behavioral repertoires are relatively simple, with animals such as primates that have much more complex behavior. The most dramatic increase in the size of the cerebral cortex took place during the last several million years of human evolution. The incredible intellectual capacities of *Homo sapiens* are the result of this enlargement of the cerebral cortex. Humans do not have the largest brains in the animal kingdom; elephants, whales, and porpoises have larger brains in terms of mass. If we compare brain size to body size, however, humans and dolphins top the list. Humans have the largest ratio of brain size to body size, and they have the most highly developed cerebral cortices. Another feature of the cerebral cortex that reflects increasing behavioral and intellectual capabilities is the ratio of association cortex to primary sensory and motor cortices. Humans have the largest relative amount of association cortex.

Now that we have some appreciation for the whats and the wheres of brain functions, it is time to turn to the hows. We will examine the properties of neurons that enable them to respond to stimulation, create action potentials, conduct action potentials, and to process and integrate information.

THE ELECTRICAL PROPERTIES OF NEURONS

Neurons, like all cells, have a slight excess of negative electrical charges within them. The inside of a neuron is usually about 70 millivolts more negative than the outside of the cell. This electrical charge difference across the plasma membrane of a neuron is called its **resting potential**. The resting potential provides a means for neurons to be responsive to specific stimuli. A neuron is sensitive to any chemical or physical factor that causes a change in the resting potential across a portion of its plasma membrane. The most extreme change in membrane potential is the electrical event known as an **action potential**, which is a sudden and rapidly reversed flip-flop in the charge across a portion of the membrane. For a brief moment, only one or two milliseconds, the inside of that part of the membrane becomes more positive than the outside. An action potential can move along a

membrane from one part of a neuron to its farthest extensions. This conduction of action potentials along the membranes of neurons is the way the nervous system transmits information.

To understand how resting potentials are created, how they are perturbed, and how action potentials are generated and conducted along membranes, it is necessary to know a little about electricity, ions, and the special ion channel proteins in the membranes of neurons. Voltage is the tendency for electrons to move between two points. Voltage is to the flow of electrons what pressure is to the flow of water. If the negative and the positive poles of a battery are connected by a copper wire, electrons flow from negative to positive. This flow of electrons can be used to do work. As you may recall from Chapter 2, electrical charges cross cell membranes not as electrons but as charged ions. The major ions that carry electrical charges across the membranes of neurons are sodium (Na^+), chloride (Cl^-), potassium (K^+), and calcium (Ca^{2+}). With these basics of bioelectricity in mind, we can ask how the resting potential of the membrane is maintained, and how the flows of ions through membrane channels are turned on and off to generate action potentials.

Pumps and Channels

Like those of all cells, the membranes of neurons are lipid bilayers that are rather impermeable to ions. But the membranes of neurons contain three classes of proteins that give neurons their special electrical properties. These proteins act as pumps, channels, and receptors.

Membrane pumps use energy to move ions or other molecules against their concentration gradients. The major pump in neuronal membranes is the sodium–potassium pump that we learned about in Chapter 5. The action of this pump expels Na^+ ions from the inside of the cell, exchanging them for K^+ ions from outside the cell (Figure 36.12). The pump expels about three Na^+ ions for every two K^+ ions it brings in. The sodium–potassium pump keeps the concentration of K^+ inside the cell greater than that of the external medium, and keeps the concentration of Na^+ inside the cell less than that of the external medium. The concentration differences established by the pump mean that K^+ would diffuse out of the cell and Na^+ would diffuse into the cell if the ions could cross the lipid bilayer. By itself, the unequal distribution of K^+ and Na^+ ions on the two sides of the plasma membrane does not create the resting potential of the cell. To explain the resting potential, we must introduce **ion channels**.

Ion channels are pores formed by proteins in the lipid bilayer. These water-filled pores allow ions to pass through. They are selective in that they may allow only one kind of ion to pass through; thus there

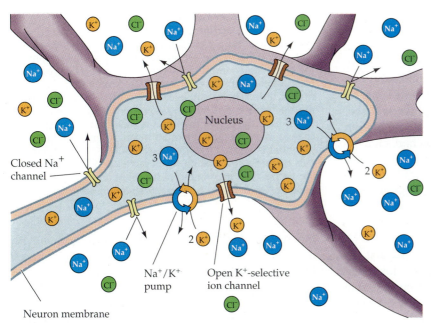

36.12 Ion Pumps, Ion Channels, and Resting Potential
The sodium–potassium pumps of a neuron establish ion concentration gradients across its plasma membrane. Under resting conditions more K^+ channels are open in the membrane than are Na^+ channels. Because of the diffusion of K^+ ions out of the cell, the membrane has an excess of positive charges on the outside and negative charges on the inside. This is the resting potential.

are potassium channels, sodium channels, chloride channels, and calcium channels. Ions can move in either direction through a channel. Because most ion channels open or close by changing their shapes, they are said to be gated. Voltage differences across the membrane change the shapes of certain channel proteins, so those channels are **voltage-gated**. The shapes of other channel proteins change when a chemical binds to a receptor associated with the channel; such channels are **chemically gated** channels. Both voltage-gated and chemically gated channels play important roles in neuronal functions, as we will see. But first, we will see how nongated, or open, channels are responsible for maintaining the resting potential in neurons and other cells.

The Resting Potential

Potassium channels are the commonest type of open channel in the plasma membranes of neurons. These K^+ channels make neurons much more permeable to K^+ than to any other ions. As Figure 36.12 shows, this is the fact that explains the resting potential. Because the neuron's plasma membrane is permeable to K^+, and because the sodium–potassium pump keeps the concentration of K^+ inside the cell much higher than that outside the cell, K^+ tends to diffuse out of the cell. If a K^+ ion leaves the cell, it leaves behind an unmatched negative charge. The tendency of the K^+ ions to diffuse out of the cell through the open channels thus causes the inside of the cell to become negatively charged in comparison to the external medium. Negative charges attract positive charges, such as those carried by K^+ ions. Eventually, a balance is established between the tendency for K^+ ions to diffuse down their concentration gradient to

the outside of the cell and the attraction of the unmatched negative charges to pull them back inside the cell. The resting potential is the voltage difference that creates that balance.

Changes in Membrane Potentials

Because of the resting potential, the plasma membrane of the neuron is polarized—that is, it separates regions of unequal electric charge. If a stimulus perturbs the resting potential, and the inside of the cell becomes *more* negative, the membrane is **hyperpolarized**. If the inside of the cell becomes *less* negative, the membrane is **depolarized**. Changes in the various gated channels can cause a membrane to hyperpolarize or depolarize. Think, for example, what would happen if some sodium channels in the cell membrane opened. Na^+ ions would diffuse into the cell because of their higher concentration on the outside of the membrane, and they would also be attracted into the cell by the excess negative charges. As a result of the entry of Na^+ ions, the membrane would become depolarized in comparison to its resting condition. In addition to the permanently open potassium channels, there can also be gated potassium channels in the same membrane. What would happen to membrane polarity if some of those opened? The opening and closing of ion channels resulting in changes in the polarity of the membrane is the basic mechanism by which neurons respond to electrical, chemical, or other stimuli.

What good does it do a neuron to undergo a change in its resting membrane potential at one particular location? Can that information be passed on to other parts of the cell? A local perturbation of membrane potential causes electrical currents to flow, which results in a spreading of the change in membrane potential. The nerve cell is a poor conductor of electricity, however, and the change in membrane polarity diminishes and disappears before it gets very far from the site of stimulation. Communication of a stimulus by the flow of an electrical current is useful only over very short distances. But, as we will see, it is an important part of action potential generation and synaptic interactions. It is also a mechanism used by many types of sensory neurons to transduce stimuli such as light, sound, or chemical stimuli into the language of the nervous system—action potentials.

Action Potentials

Action potentials are remarkable phenomena that enable neurons to convey information over long distances with no loss of the signal. An action potential is a sudden and major change in membrane potential that lasts for only one or two milliseconds. It is conducted along the axon of a neuron at speeds up to 100 meters per second, which is equivalent to running the length of a football field in one second. If we place the tip of an electrode in a resting axon and measure the voltage difference across its membrane, it is about -70 mV (Figure 36.13). If this electrode is exposed to an action potential traveling down the axon it registers a rapid change in membrane potential which rises from -70 mV to about $+40$ mV. The membrane potential rapidly returns to its resting level of -70 mV as the action potential passes by. At any location along this axon we could insert another electrode and record the same action potential. The height of the action potential does not change as it travels along the axon. The action potential is an all-or-nothing, self-regenerating event.

Voltage-gated sodium channels are primarily responsible for the action potential (see Figure 36.13). These channels are normally closed, but when the membrane potential reaches a certain threshold, they suddenly flip open briefly—for less than a millisecond. The threshold at which these sodium channels open is less negative (more positive) than the resting potential, so any stimulus that depolarizes the membrane brings the membrane potential toward that threshold. Due to the action of the sodium–potassium pump, Na+ concentration is much higher outside the axon than inside, so whenever the sodium channels open in a part of the membrane, Na^+ ions from the outside enter the cell at that location. The entering Na^+ ions make the inside of the membrane positive.

The opening of the Na^+ channels causes the rise of the action potential—what neurobiologists call the "spike." What causes the drop back to resting potential? The main reason for the drop is that, after opening briefly, the sodium channels close and remain inactive for a few milliseconds. This is long enough for the membrane to return to resting potential. In some axons there are also voltage-gated potassium channels. They open more slowly than the sodium channels and stay open longer, so they help return the voltage across the membrane to its resting level by allowing K^+ ions to carry excess positive charges out of the cell.

The difference in concentration of Na^+ ions across the plasma membrane of neurons is the "battery" that drives the action potential. How rapidly does the battery run down? It might seem that a substantial number of Na^+ and K^+ ions would have to cross the membrane for the membrane potential to go from -70 mV to $+40$ mV, and back to -70 mV again. In fact, only about one Na^+ (or K^+) ion in 10 million actually passes through the channels during the passage of an action potential. Thus the effect of a single action potential on the concentration ratios of Na^+ or K^+ is minute. Even hundreds of action potentials barely change the concentration differences of Na^+ and K^+ on the two sides of the membrane. So it is not difficult for the sodium–potassium pump to keep the "battery" charged, even when the cell is generating many action potentials every second.

Once a membrane has generated an action potential, it cannot immediately fire another. It has a **refractory period** that lasts for a few milliseconds because the sodium channels are inactivated and unable to reopen. When the sodium channels can again respond to depolarization of the membrane by opening and allowing Na^+ to pass, another action potential can be fired.

Propagation of the Action Potential

How does an action potential move over long distances? When one part of an axon fires an action potential, the adjacent regions of membrane also become depolarized because of local electrical current spread. Such movements of ions, as shown in Figure

36.13 The Action Potential Can Be Visualized on an Oscilloscope ▶

An electrode in an axon detects the voltage change in the membrane that is the action potential. The signal from the electrode is amplified and fed into an oscilloscope. A beam of electrons sweeps across the screen in a set period of time. That beam is deflected up if the signal from the electrode is positive and down if the signal is negative. Thus, the action potential is seen on the screen as a change in membrane potential through time. The action potential is created by voltage-gated Na^+ and K^+ channels opening and closing, as depicted at the bottom of the figure. The membrane potential at any given time depends on which and how many channels are open.

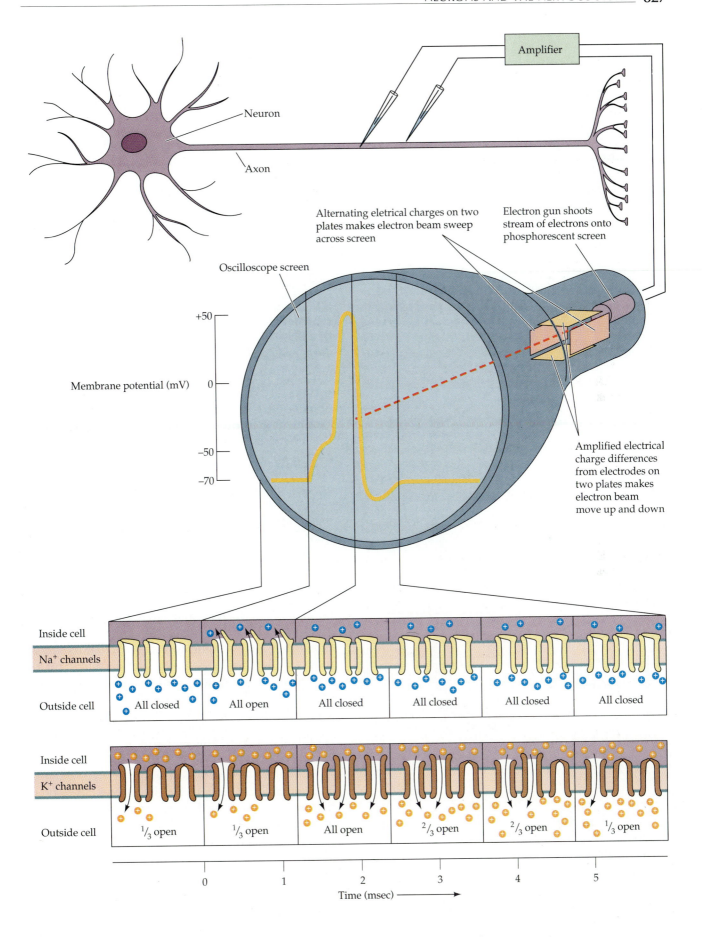

Neuron

Axon

Amplifier

Alternating eletrical charges on two plates makes electron beam sweep across screen

Electron gun shoots stream of electrons onto phosphorescent screen

Oscilloscope screen

Membrane potential (mV)

+50

0

−50

−70

Amplified electrical charge differences from electrodes on two plates makes electron beam move up and down

Inside cell

Na^+ channels

Outside cell

All closed | All open | All closed | All closed | All closed | All closed

Inside cell

K^+ channels

Outside cell

$^{1}/_{3}$ open | $^{1}/_{3}$ open | All open | $^{2}/_{3}$ open | $^{2}/_{3}$ open | $^{1}/_{3}$ open

0   1   2   3   4   5

Time (msec)

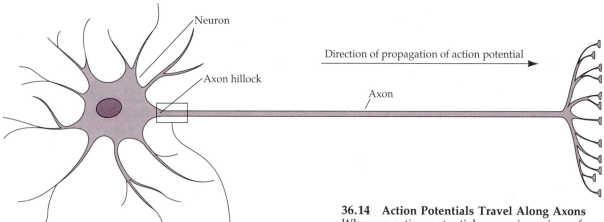

Neuron

Axon hillock

Direction of propagation of action potential

Axon

1. Resting conditions
   ∏ Na⁺ channels
   ∏ K⁺ channels

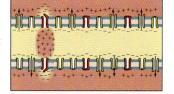

2. Action potential initiated at axon hillock by opening of Na⁺ channels

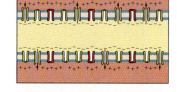

3. Current spreads down membrane and causes neighboring Na⁺ channels to open

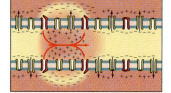

4. Channels recently opened become inactivated and K⁺ channels open causing membrane to repolarize

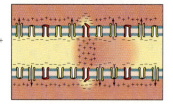

5. Action potential continues to be regenerated along axon as steps **2, 3,** and **4** are repeated

36.14 Action Potentials Travel Along Axons
When an action potential occurs in a piece of membrane, electrical currents flow to adjacent areas of membrane and depolarize them. As voltage-gated channels in these areas reach threshold, they create an action potential. In this way, an action potential continuously regenerates itself along the axon.

36.14, depolarize the region of the membrane adjacent to that experiencing the action potential. Outside the axon, positive ions flow rapidly toward the depolarized region, attracted by the negative charge balance there. On the inside of the cell, positive charges move *away* from the depolarized region, where they are more abundant, toward the adjacent, more negative regions. The net result is a tendency to repolarize the membrane at the point of the exist-

ing action potential and to depolarize the adjacent regions. If the depolarization of an adjacent region of membrane brings it to the threshold of the sodium channels, they open and create an action potential. Because an action potential always brings the area of membrane adjacent to it to threshold, it is self-regenerating and propagates itself along the axon. The action potential propagates itself in only one direction—it cannot reverse itself because the part of the membrane it came from is undergoing its refractory period.

The action potential does not travel along all axons at the same speed. Action potentials travel faster in large-diameter axons than in small-diameter axons. Among invertebrates, the axon diameter determines the rate of conduction, and axons that transmit messages involved in escape behavior are very thick. The axons that enable squid to escape predators are almost a millimeter in diameter. These squid giant axons were the most important experimental material used in the classic studies in neurophysiology that produced basic discoveries about action potentials and their conduction. The British physiologists A. L. Hodgkin and A. F. Huxley performed most of those studies in the 1940s and 1950s.

In nervous systems more complex than those of invertebrates, increasing the speed of action potentials by increasing the diameter of axons would result in enormous nerves. The optic nerve from each of our eyes contains about one million axons. If we used plain, simple axons to build a nerve that conducts information from the eye to the brain as fast as ours does, each optic nerve would have to be about the diameter of the eyeball itself. Other groups of axons in the brain, the spinal cord, and the peripheral nervous system would be equally unwieldy. Vertebrates have evolved an adaptation that maximizes propagation velocity without having to use huge axons. Their axons are insulated with membranous wrappings produced by specialized glial cells called **Schwann cells** (Figure 36.15). The membranous wrapping is called myelin. It is the myelin that gives the light, shiny appearance to "white matter," which is any nervous system tissue containing mostly axons. The myelination of axons is not continuous. At regular intervals called **nodes of Ranvier** the bare axon is exposed.

How do the nodes of Ranvier and the myelin insulation of the axon increase propagation velocity? At the nodes of Ranvier an axon can fire action potentials. But in the adjacent regions of the axon that are insulated with myelin, electrical charges cannot accumulate or cross the membrane. Therefore, the local depolarization caused by the action potential causes electrical current to flow to the next place in the membrane where charges can accumulate and cross—the next node of Ranvier. When that node reaches threshold, it fires an action potential, and so forth, with the action potentials jumping from node to node down the axon. This form of impulse propagation is called **saltatory** (jumping) **conduction** and

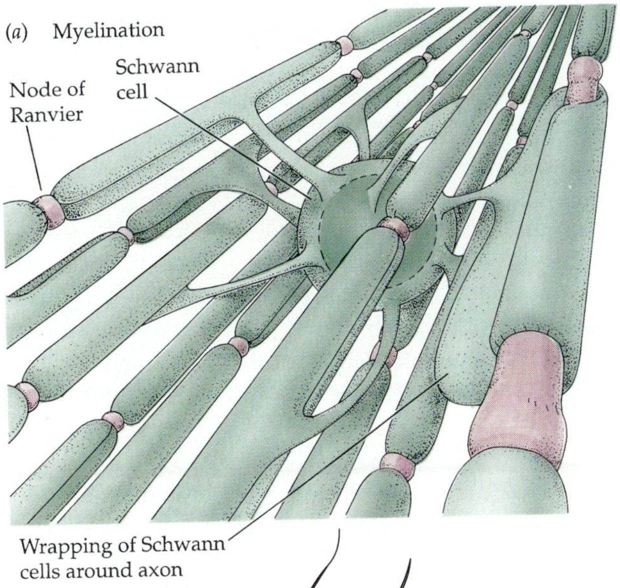

(a) Myelination

Node of Ranvier

Schwann cell

Wrapping of Schwann cells around axon

(b)

36.15 Jumping Action Potentials
(a) Some axons are myelinated by the wrappings of Schwann cell membranes. (b) Action potentials can occur only at gaps in the myelin wrap (nodes of Ranvier), but electrical currents created by an action potential can flow to adjacent nodes and depolarize them. As a result, the action potential is conducted down the axon by jumping from node to node.

Myelin sheath

Neuron

Axon

Myelin sheath

Node of Ranvier

Axon

Plasma membrane

Direction of action potential propagation

Na⁺ channels

K⁺ channels

Na⁺ channels inactivated; K⁺ channels open; membrane repolarizes

Na⁺ channels open; membrane depolarizes

Action potential jumps from node to node

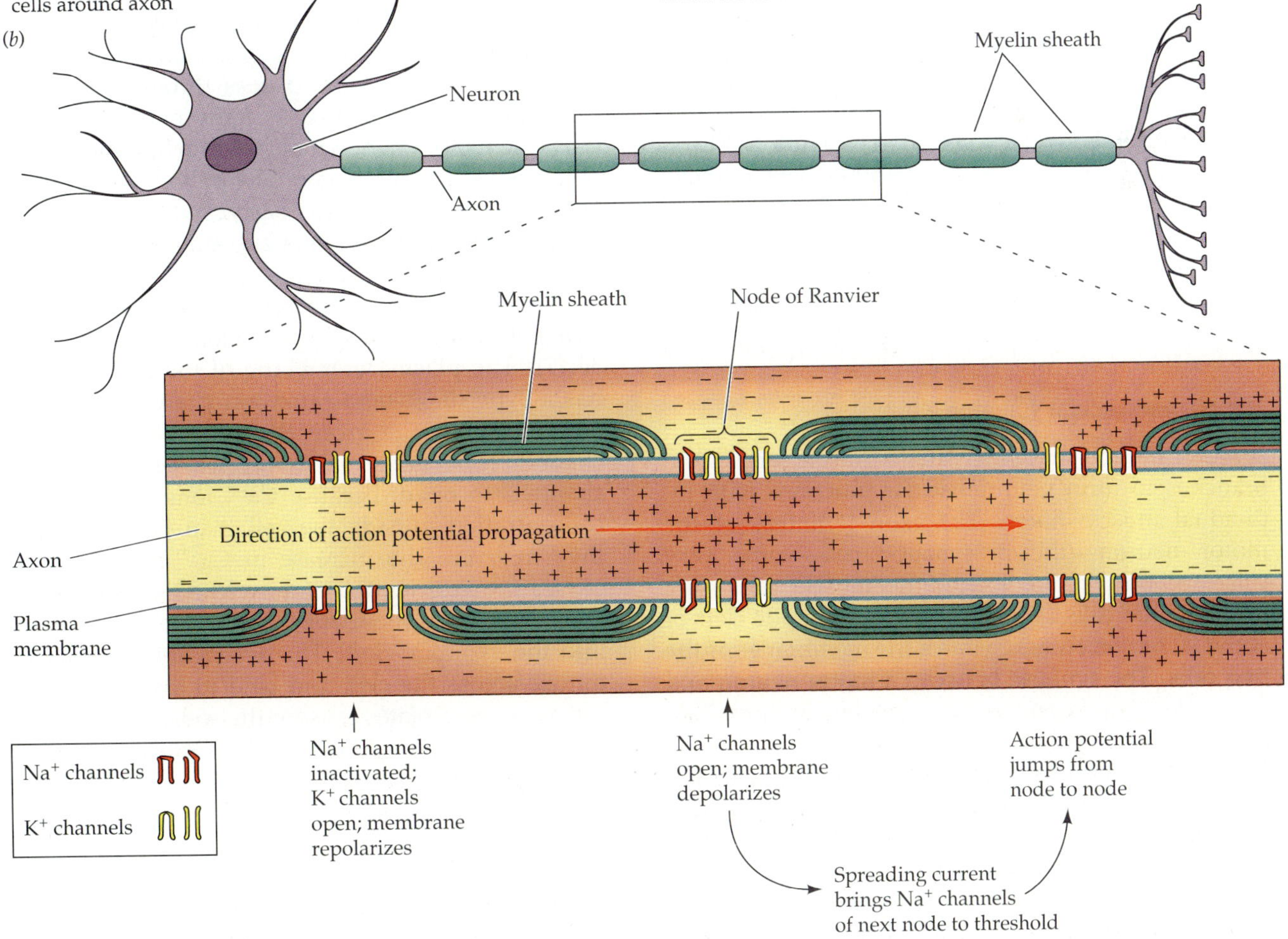

Spreading current brings Na⁺ channels of next node to threshold

is much quicker than continuous impulse conduction down an unmyelinated axon.

You have probably experienced the difference in the velocity at which action potentials travel down myelinated and unmyelinated axons. If you touch a very hot or very cold object, you experience a sharp pain before you sense that the object is hot or cold. With the sensing of the temperature also comes a burning pain different from the first sharp pain. Sensory axons carrying sharp-pain sensation are myelinated, but most axons carrying information about temperature, and axons carrying burning, aching pain, are unmyelinated. As a result, you know that something is wrong before you know what it is or how bad it is. The unmyelinated temperature and burning-pain axons conduct impulses at only 1 to 2 meters per second, whereas the myelinated sharp-pain axons conduct impulses at velocities up to 5 or 6 meters per second. The largest myelinated axons in the human nervous system conduct impulses at velocities up to 120 meters per second.

SYNAPTIC TRANSMISSION

The most remarkable abilities of nervous systems stem from the interactions of neurons. These interactions process and integrate information. Our nervous systems can orchestrate complex behaviors, deal with complex concepts, and learn and remember because large numbers of neurons interact with one another. The mechanisms for those interactions lie in the synapses between cells. Synapses are locations where one cell influences another cell directly through the transfer of a chemical or an electrical message. Most synapses are chemical, and we will focus first on those. Information crosses chemical synapses in one direction only, from the **presynaptic cell** to the **postsynaptic cell**.

The Neuromuscular Junction

A motor neuron that innervates a muscle sends axon branches to form neuromuscular junctions with individual muscle fibers. The axon terminals of the motor neuron—the presynaptic cell—come very close to, but do not touch, the muscle cells—the postsynaptic cells (Figure 36.16). A synaptic cleft of about 20 nanometers separates the membranes of the two cells. The synapse between a motor neuron and a muscle fiber is an excellent model for how fast, excitatory chemical synapses work.

The axon terminal contains many spherical vesicles filled with chemical messenger molecules called **neurotransmitters**. All motor neurons innervating vertebrate skeletal muscles have acetylcholine as their neurotransmitter. The postsynaptic membrane is part of the muscle cell's plasma membrane, but it is

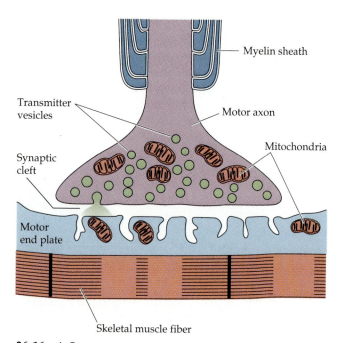

36.16 A Synapse
Chemical communication between neurons and muscle cells occurs at synapses called neuromuscular junctions.

slightly modified in the area of the synapse and is called a **motor end plate**. The modification that makes a patch of membrane a motor end plate is the presence of acetylcholine receptor molecules. The receptors function as chemically gated channels that allow both Na^+ and K^+ ions to pass through. Since the resting membrane is already fairly permeable to K^+ ions, the major change that occurs when these channels open is the movement of Na^+ ions into the cell. When a receptor binds acetylcholine, a channel opens, and Na^+ ions move into the cell, making the cell more positive inside.

Chemical synaptic transmission begins with an action potential arriving at the axon terminal (Figure 36.17). The plasma membrane of the axon terminal has a type of voltage-gated ion channel found nowhere else on the axon: the voltage-gated calcium channel. The action potential causes the calcium channels to open; because Ca^{2+} ions are in greater concentration outside the cell than inside the cell, they rush in. The increase in Ca^{2+} inside the cell causes the vesicles full of acetylcholine to fuse with the presynaptic membrane and eject their contents into the synaptic cleft. The acetylcholine molecules diffuse across the cleft and bind to the receptors on the motor end plate, causing the sodium channels to open briefly and depolarize the postsynaptic cell membrane.

Events in the Postsynaptic Membrane

The electrical properties of postsynaptic membranes are different from those of axon membranes in an

1. Action potential arrives at axon terminal (Na⁺ channels open, depolarizing terminal membrane)

2. Depolarization of terminal membrane causes Ca^{2+} channels to open. Ca^{2+} enters cell and triggers fusion of transmitter vesicles with presynaptic membrane

3. Neurotransmitter molecules diffuse across synaptic cleft and bind to receptors on postsynaptic membrane. Activated receptors open Na⁺ channels and depolarize postsynaptic membrane

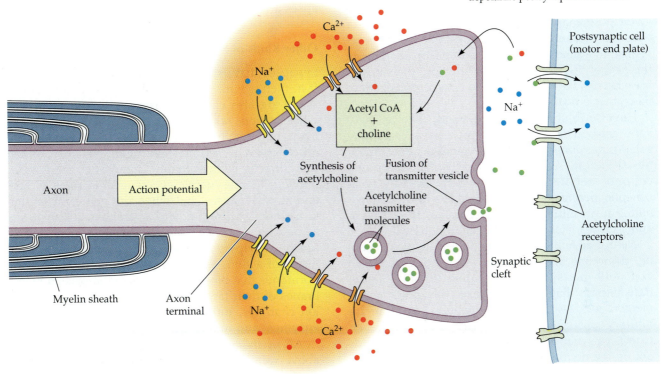

36.17 Synaptic Transmission Begins with the Arrival of an Action Potential

Voltage-gated Ca^{2+} channels at the nerve ending open in response to an action potential. Ca^{2+} ions cause vesicles of neurotransmitter to fuse with the presynaptic membrane and dump neurotransmitter into the synaptic cleft. Molecules of neurotransmitter bind to receptors on the postsynaptic membrane and cause chemically gated ion channels to open.

important way: Because motor end plates have very few voltage-gated sodium channels, they do not fire action potentials. This is true of motor end plates, it is true of dendrites, and it is true of most regions of nerve cell bodies. What the binding of neurotransmitter to receptors and the opening of chemically gated ion channels does is to perturb the resting potential of the postsynaptic membrane. This local change in membrane potential spreads to neighboring regions of the plasma membrane of the postsynaptic cell. Eventually, the spreading depolarization may reach an area of membrane that does contain voltage-gated channels. The entire plasma membrane of a skeletal muscle fiber, except for the motor end plates, has voltage-gated sodium channels. If a presynaptic axon terminal releases sufficient neurotransmitter to depolarize a motor end plate enough to bring its neighboring membrane to threshold, action potentials are fired. These action potentials are then conducted throughout the muscle fiber's system of membranes and cause it to contract. (We'll learn

about the coupling of muscle membrane action potentials and contraction of muscle fibers in Chapter 38.)

How much neurotransmitter is enough? Neither a single acetylcholine molecule nor the contents of an entire vesicle (about 10,000 acetylcholine molecules) is enough to bring the plasma membrane of a muscle cell to threshold. But a single action potential in an axon terminal releases about 100 vesicles, and that *is* enough to fire an action potential in the muscle fiber and cause it to twitch.

Excitatory and Inhibitory Synapses

The synapses between neurons and skeletal muscle are always excitatory; that is, motor end plates always respond to acetylcholine by depolarizing. There are several different kinds of synapses between neurons, however. Recall that a given neuron may have many dendrites. Axon terminals from many other neurons may make synapses with those dendrites and with the cell body as well. The axon terminals of different presynaptic neurons may store and release different neurotransmitters, and membranes of the dendrites and cell body of a postsynaptic neuron may have receptors to a variety of neurotransmitters. Thus a given postsynaptic neuron can receive a variety of chemical messages. If the postsynaptic neuron's response to a neurotransmitter is depolarization, as in the case of the neuromuscular junction, the synapse is excitatory, but if the response is hyperpolarization, the synapse is inhibitory.

How do inhibitory synapses work? The postsynaptic cells in inhibitory synapses have chemically gated potassium or chloride channels as their receptors. When these channels are activated by binding with a neurotransmitter, they hyperpolarize the postsynaptic membrane. Thus, the release of neurotransmitter at an inhibitory synapse makes the postsynaptic cell *less* likely to fire an action potential.

Neurotransmitters that depolarize the postsynaptic membrane are excitatory; they bring about an **excitatory postsynaptic potential (EPSP)**. Neurotransmitters that hyperpolarize the postsynaptic membrane are inhibitory; they bring about an **inhibitory postsynaptic potential (IPSP)**. However, whether a synapse is excitatory or inhibitory depends not on the neurotransmitter but on the postsynaptic receptors—on what kind of ion channels the postsynaptic cell has. The same neurotransmitter can be excitatory at some synapses and inhibitory at others.

Summation

Individual neurons make "decisions" about whether or not to fire action potentials by summing excitatory and inhibitory postsynaptic potentials. The ability of neurons to sum excitatory and inhibitory postsynaptic potentials is the major mechanism by which the nervous system integrates information. Each neuron may receive 10,000 or more synaptic inputs, yet it has only one output—action potentials. All the information contained in the thousands of inputs a neuron receives is reduced to the rate at which that neuron generates action potentials. For most neurons the critical area for "decision making" is the **axon hillock**, the region of the cell body at the base of the axon. The plasma membrane of the axon hillock is not insulated by glia and it has many voltage-gated channels. Excitatory and inhibitory postsynaptic potentials from anywhere on the dendrites or the cell body spread to the axon hillock. If the resulting potential depolarizes this area of membrane to threshold, it fires an action potential. Because postsynaptic potentials decrease as they spread from the site of the synapse, all postsynaptic potentials do not have equal influences on the axon hillock. A synapse at the end of a dendrite has less influence than a synapse on the cell body near the axon hillock.

Excitatory and inhibitory postsynaptic potentials can be summed over space or over time (Figure 36.18). Spatial summation adds up the simultaneous influences of synapses from different sites on the postsynaptic cell. Temporal summation adds up synapses arriving at the same site in a rapid sequence.

Other Kinds of Synapses

Synapses that use chemically gated ion channels are fast; their actions happen within a few milliseconds. There are also chemically mediated slow synapses, whose actions take hundreds of milliseconds or even many minutes. Neurotransmitters at these slow synapses activate second-messenger systems, rather than directly controlling ion channels in the postsynaptic cell. Presynaptic events are the same in fast and slow synapses, but when the neurotransmitter of a slow synapse binds to a receptor, it activates a second messenger such as cAMP. The mechanisms of slow synapses are therefore similar to the mechanisms of action, explained in Chapter 34, of certain hormones that bind to receptors in the plasma membranes of their target cells. Slow synapses may open ion channels, influence membrane pumps, activate enzymes, and induce gene expression.

All of the neuron-to-neuron synapses discussed above were between the axon terminals of a presynaptic cell and the cell body or dendrites of a postsynaptic cell. Synapses can also be formed between the axon terminals of one cell and the axon of another cell. Such a synapse can modulate how much neurotransmitter the second cell releases in response to action potentials traveling down its axon. We refer to this mechanism of regulating synaptic strength as **presynaptic excitation** or **presynaptic inhibition**.

Electrical synapses, or gap junctions, are completely different from chemical synapses because they directly couple neurons electrically. In gap junctions, the presynaptic and postsynaptic cell membranes are separated by 2 to 3 nanometers, but the membrane proteins of the two neurons form connexons—molecular tunnels actually bridging the two cells, through which electric currents, ions, and small molecules can readily pass. Electric transmission across gap junctions is very fast and can proceed in either direction; that is, stimulation of either neuron can result in the appearance of an action potential in the other. Gap junctions are less common in the complex nervous systems of vertebrates than they are in the simple nervous systems of invertebrates. There are two reasons for this. First, electrical continuity between neurons does not enable the process of summation, which is the essence of the ability of complex nervous systems to integrate information. Second, an effective electrical synapse requires a large area of contact between the presynaptic and postsynaptic cells. This rules out the possibility of a neuron's having thousands of synaptic inputs—which is the rule in complex nervous systems.

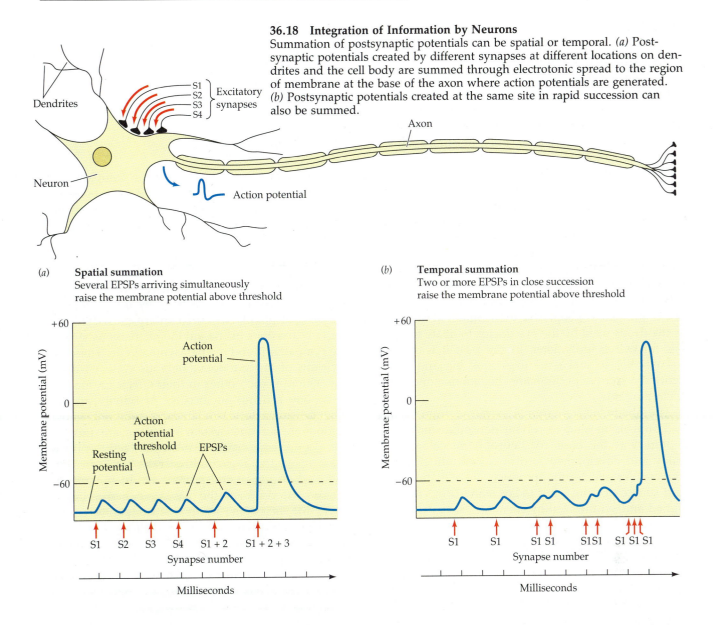

36.18 Integration of Information by Neurons
Summation of postsynaptic potentials can be spatial or temporal. (a) Postsynaptic potentials created by different synapses at different locations on dendrites and the cell body are summed through electrotonic spread to the region of membrane at the base of the axon where action potentials are generated. (b) Postsynaptic potentials created at the same site in rapid succession can also be summed.

Neurotransmitters and Receptors

At present a couple of dozen neurotransmitters are recognized, and more will surely be discovered. No others are as thoroughly understood as acetylcholine, the neurotransmitter at all synapses between motor neurons and skeletal muscles. In addition to their role in voluntary movement, acetylcholine and noradrenaline serve as neurotransmitters in the other division of the peripheral nervous system, the autonomic. These two neurotransmitters also play roles in the central nervous system, but they constitute only a small percent of the neurotransmitter content of the brain. The workhorse neurotransmitters of the brain are simple amino acids. Glutamic acid and aspartic acid are excitatory, whereas glycine and gamma-aminobutyric acid are inhibitory. Another important group of neurotransmitters in the central nervous system is the monoamines, which are derivatives of amino acids. They include dopamine and noradrenaline (derivatives of tyrosine) and serotonin

(a derivative of tryptophan). A number of peptides also function as neurotransmitters.

A neurotransmitter may have several different types of receptors in different tissues and may induce different actions. For example, acetylcholine has two well-known receptor types, which are called **muscarinic** and **nicotinic** because of other compounds that also bind to them. Nicotine, the active ingredient in tobacco, binds to acetylcholine receptors in the skeletal muscles, but not to those in heart muscle or in the autonomic nervous system. In contrast, muscarine, a compound found in poisonous mushrooms called *Amanita muscaria*, binds to the acetylcholine receptors in heart muscle and in the autonomic nervous system, but not to those in skeletal muscle. Both types of acetylcholine receptors are found in the central nervous system, where nicotinic receptors tend to be excitatory and muscarinic tend to be inhibitory. These acetylcholine receptors are the reason why smoking tobacco has behavioral and physiological effects and is addictive and why a number of ethnic cultures around the world have used *Amanita* mushrooms as hallucinogenic drugs.

The drug **curare**, extracted from the bark of a South American plant and used by native peoples to make poisoned darts and arrows, binds to nicotinic receptors but does not activate them. Therefore, skeletal muscles in an animal poisoned by curare cannot respond to motor neuron activation. The animal goes into flaccid (relaxed) paralysis and dies because it stops breathing. Curare is used medically to treat severe muscle spasms and to prevent muscle contractions that would interfere with surgery. Another compound called **atropine**, extracted from the plant *Atropa belladonna*, binds with muscarinic receptors and prevents acetylcholine from activating them. It is used medically to increase heart rate, decrease secretions of digestive juices, and decrease spasms of the gut. Most people have encountered atropine as the drops used by the eye doctor to dilate the pupils for eye examinations. In the past it was used cosmetically to make the eyes look big and dark, hence the plant's species name, belladonna, meaning "beautiful lady."

The ability of compounds extracted from plants and animals to bind to certain neurotransmitter receptors was the basis for neuropharmacology, the study and development of drugs that influence the nervous system. Natural products are still an important source of drugs, but today many drugs are designed and synthesized by chemists.

Clearing the Synapse of Neurotransmitter

It is as important to turn off the action of neurotransmitter as it is to turn it on. If released neurotransmitter molecules simply remained in the synaptic cleft, the postsynaptic membrane would become saturated with neurotransmitter, and its receptors would be constantly bound. As a result, the postsynaptic neuron would remain hyperpolarized or depolarized and would be unresponsive to short-term changes in the presynaptic neuron. Thus, neurotransmitter must be cleared from the synaptic cleft shortly after it is released by the axon terminal.

There are several means by which neurotransmitter action is terminated. First, the neurotransmitter may be destroyed by enzymes. For example, acetylcholine is rapidly destroyed by the enzyme **acetylcholinesterase**, which is present in the synaptic cleft in close association with the acetylcholine receptors on the postsynaptic membrane. Some of the most deadly nerve gases developed for chemical warfare are inhibitors of acetylcholinesterase. They cause acetylcholine to linger in the synaptic clefts, causing the victim to die of spastic muscle paralysis. Agricultural insecticides such as malathion also inhibit acetylcholinesterase. Second, neurotransmitter may simply diffuse away from the cleft. Third, neurotransmitter

may be taken up via active transport by nearby cell membranes. Each of these mechanisms—enzymatic destruction, diffusion, and active transport—can clear the synaptic cleft so that a new, discrete signal can pass through the synapse.

NEURONS IN CIRCUITS

Because neurons can interact in the complex ways just discussed, networks of neurons can process and integrate information. Next we will examine networks in different parts of the nervous system.

The Autonomic Nervous System

The autonomic nervous system controls the organs and organ systems of the body by influencing the activities of glands and involuntary muscles. There are two divisions of the autonomic nervous system, the **sympathetic** and the **parasympathetic**. These two divisions of the autonomic nervous system work in opposition to each other in their effects on most organs, one causing an increase in activity and the other causing a decrease. The most commonly known autonomic nervous system functions are those of the sympathetic division called the "fight-or-flight" mechanisms that increase heart rate, blood pressure, and cardiac output and prepare the body for emergencies. In contrast, the parasympathetic division slows the heart and lowers blood pressure. It is tempting to think of the sympathetic system as the one that speeds things up and the parasympathetic system as the one that slows things down, but that is not always a correct distinction. The sympathetic system slows the digestive system down and the parasympathetic system speeds it up. The two divisions of the autonomic nervous system are easily distinguished from each other by their anatomy, their neurotransmitters, and their actions (Figure 36.19).

Both divisions of the autonomic nervous system are efferent pathways of the central nervous system. Each autonomic efferent begins with a neuron that has its cell body in the brain stem or spinal cord and uses acetylcholine as its neurotransmitter. These cells are called **preganglionic neurons** because the second neuron in each autonomic output pathway resides in a ganglion (collection of neuron cell bodies) that is outside of the central nervous system (Figure 36.19). The second neuron in an autonomic nervous system output pathway is a **postganglionic neuron** because its axon extends out from the ganglion. The axons of the postganglionic cells end on the cells of the target organs.

The postganglionic neurons of the sympathetic division use noradrenaline as their neurotransmitter and those of the parasympathetic division use acetylcholine as their neurotransmitter. In organs that receive both sympathetic and parasympathetic input,

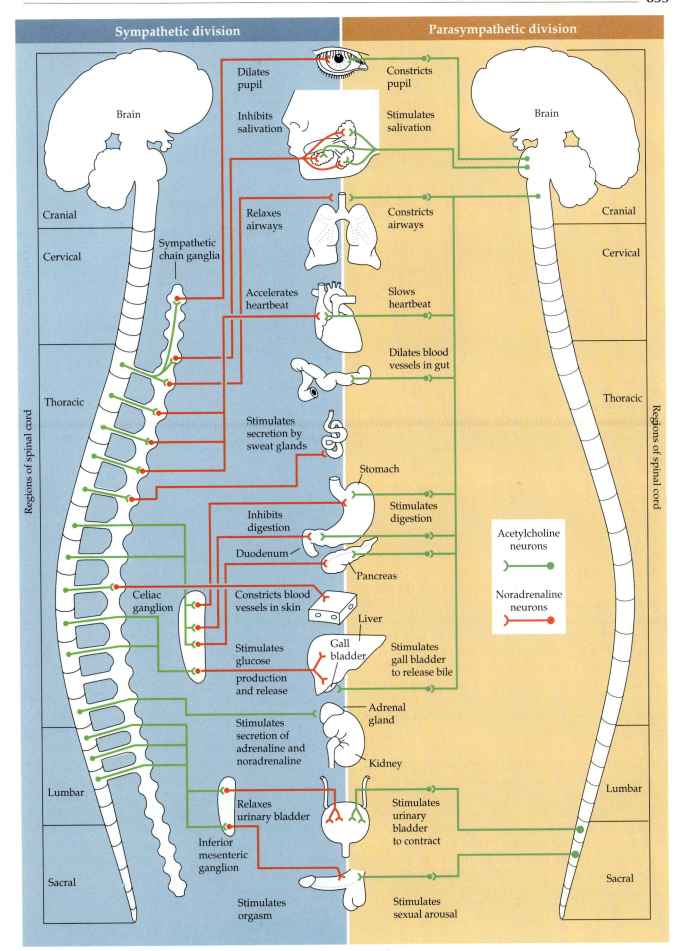

36.19 Organization of the Autonomic Nervous System

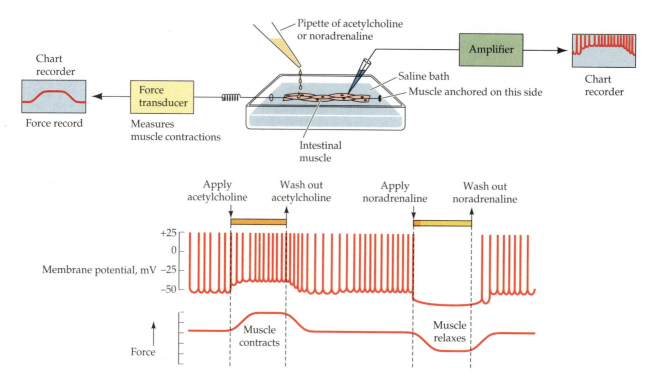

36.20   Responses to Postganglionic Transmitter
In this experiment, a strip of intestinal muscle is mounted in an incubation bath so that the force of its contractions can be measured. An electrode records action potentials in a muscle cell. When acetylcholine is dripped onto the muscle, the cells depolarize, fire action potentials more rapidly, and increase their force of contraction. Noradrenaline causes the cells to hyperpolarize, decrease their rate of firing, and decrease their force of contraction.

the target cells respond in opposite ways to noradrenaline and acetylcholine. A region of the heart called the pacemaker which generates the heartbeat provides an example. Noradrenaline depolarizes pacemaker cells and causes the heart to beat faster. Acetylcholine hyperpolarizes cardiac pacemaker cells and causes the heart to beat slower. In contrast, noradrenaline causes muscle cells in the digestive tract to hyperpolarize and slows digestion. Acetylcholine depolarizes muscle cells in the gut and accelerates digestion (Figure 36.20).

A distinction between the sympathetic and parasympathetic divisions of the autonomic nervous system is in their anatomical organization (Figure 36.19). The preganglionic neurons of the parasympathetic division come from the brainstem and the last segment of the spinal cord. The preganglionic neurons of the sympathetic division come from the upper regions of the spinal cord below the neck—the thoracic and lumbar regions. The ganglia of the sympathetic nervous system are mostly lined up in two chains, one on either side of the spinal cord. The parasympathetic ganglia are close to, sometimes sitting on, the target organs.

Monosynaptic Reflexes

Much information is processed through neural circuits within the spinal cord. The simplest example of a spinal neural circuit controlling behavior is the **monosynaptic reflex loop**. It is quite common in a visit to a physician to have your reflexes tested by being struck below the kneecap with a rubber mallet. The response is a rapid, involuntary extension of the leg—the knee-jerk reflex. Similar reflexes can be elicited by sharp blows to tendons of the wrist, elbow, ankle, and other joints. These reflexes depend on a neural circuit made up of a sensory neuron and a motor neuron with just one synapse between them, hence the term monosynaptic reflex (Figure 36.21).

The sensory neuron in this circuit detects the stretching of a muscle. The mallet blow on the tendon causes a quick stretch of the muscle attached to that tendon. Within the muscle are modified muscle fibers wrapped in connective tissue. These **muscle spindles** are stretch receptors that activate sensory neurons when they are stretched. The number of nerve impulses per second carried by the sensory neuron signals the degree of stretch. The sensory neuron from the muscle spindle has its cell body in a ganglion on the dorsal root of the spinal cord, but the fiber of this neuron continues all the way to the gray matter of the ventral horn of the spinal cord. There the sensory fiber branches and makes synapses with motor neurons for the same muscle from which the sensory neuron originated. Each motor neuron sends impulses along its axon which leaves the spinal cord through the ventral root. That axon synapses on the

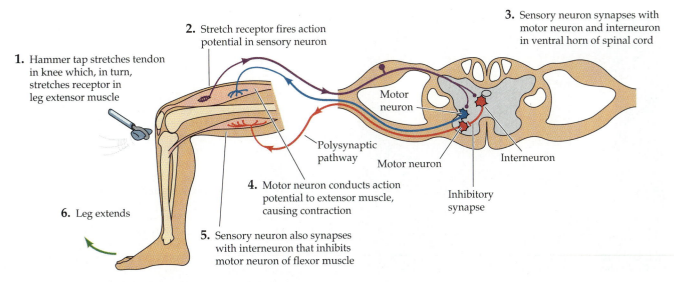

1. Hammer tap stretches tendon in knee which, in turn, stretches receptor in leg extensor muscle

2. Stretch receptor fires action potential in sensory neuron

3. Sensory neuron synapses with motor neuron and interneuron in ventral horn of spinal cord

Motor neuron

Polysynaptic pathway

Motor neuron

Interneuron

Inhibitory synapse

4. Motor neuron conducts action potential to extensor muscle, causing contraction

6. Leg extends

5. Sensory neuron also synapses with interneuron that inhibits motor neuron of flexor muscle

36.21 Monosynaptic and Polysynaptic Spinal Reflexes
When the muscle is stretched, the stretch receptor fires action potentials in the sensory neuron. Transmitter released by the sensory neuron onto the motor neuron causes it to fire action potentials and stimulates constriction in the muscle that was stretched. Because muscles work in antagonistic pairs, one of the pair must relax when the other contracts. This is accomplished by a parallel reflex that involves spinal interneurons and therefore more than one synapse.

stretched muscle, causing it to contract. The function of this reflex loop is to adjust the contraction in the muscle to changing loads. An increased load on the limb stretches the muscle, and the stretch reflex returns it to the desired position.

Even though the knee-jerk reflex is involuntary, you are still aware of being struck on the knee. This means the information has to get to your brain. Branches of the sensory neuron make synapses with interneurons in the dorsal horn of the spinal cord. These interneurons send axons up the dorsal white matter tracts to the thalamus and on to the cerebral cortex. Motor commands from the cerebral cortex descend the spinal cord in other white matter tracts to make synapses with the same motor neurons involved in the reflex loop. Thus, the same muscle can be controlled both by involuntary reflexes and by conscious commands.

Polysynaptic Reflexes

Most involuntary reflex circuits include more than one synapse. A number of interneurons in the central nervous system are necessary for more complex responses. For example, a limb can move in opposite directions because muscles work in pairs. One muscle of a pair is an extensor and the other is a flexor. At the same time a flexor contracts, the extensor must relax, or the limb can not move. A polysynaptic reflex is added on to the monosynaptic circuit controlling the stretch reflex to cause relaxation of the opposing muscle (Figure 36.21).

Much more complex polysynaptic reflexes are responsible for coordinated escape movements. If you step on a tack, many muscles in your foot and leg work together to produce a coordinated withdrawal response of that limb while muscles on the opposite side of the body cause extension of your other leg to support your body weight and maintain your balance. The large number of muscle contractions and relaxations required for this sequence of movements are initially orchestrated by interneurons in the spinal cord.

HIGHER BRAIN FUNCTIONS

Very few functions of the nervous system have been worked out to the point of identifying the neural circuits that underlie them. Brain processes responsible for phenomena such as thought, perception, memory, and learning are extremely complex. Nevertheless, neurobiologists using a wide range of techniques are making considerable progress in understanding some of the neural mechanisms involved in such higher functions of the nervous system. There have been rapid advances in understanding how the nervous system processes visual information. That will be discussed in the next chapter. The following discussion presents several complex aspects of brain and behavior that present challenges. Neurobiologists want to learn how individual neurons function and how neurons interact in circuits to produce these behaviors.

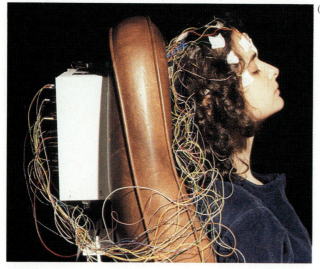

(a)

36.22 Patterns of Electrical Activity in the Cortex Characterize Stages of Sleep

(a) Electrical activity of the cerebral cortex is detected by electrodes placed on the scalp and recorded on moving chartpaper by a polygraph. The resulting record is the electroencephalogram. Muscle activity and eye movements are usually recorded at the same time. (b) The pattern of the electroencephalogram is different in wakefulness and in different stages of sleep. (c) During a night we cycle through the different stages of sleep. REM sleep, or dreaming sleep, usually occurs in four episodes. Deepest slow-wave sleep, or nonREM sleep, occurs in the first half of the night. It is common for brief awakenings to occur after REM sleep.

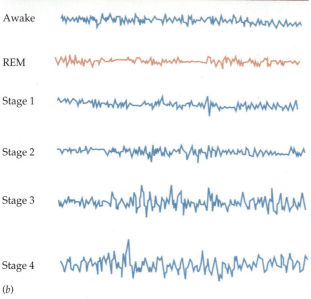

(b)

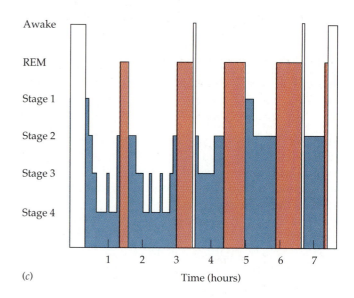

(c)

Sleeping and Dreaming

A dominant feature of our behavior is the daily cycle of being asleep and awake. All birds and mammals sleep, and probably all other vertebrates sleep as well, but it has not been proven that sleep of a fish or reptile is the same as that of a mammal. We spend one-third of our lives sleeping, yet we do not know why or how. We do know that we need to sleep, however. Loss of sleep impairs alertness and performance. Most people in our society—certainly most college students—are chronically sleep-deprived. Large numbers of accidents and serious mistakes that endanger lives can be attributed to impaired alertness due to sleep loss. Yet, insomnia (difficulty in falling asleep) is one of the most common medical complaints. It is important to learn more about the neural control of sleep.

A common tool of sleep researchers is the **electroencephalogram** (EEG). The electroencephalogram detects regional electrical activity over the cerebral

cortex and represents that activity as wavy lines drawn by pens on a moving chart (Figure 36.22). Usually sleep researchers also record the electrical activity of one or more skeletal muscles as an **electromyogram** (EMG) on the same moving chart along with the EEG. EEG and EMG patterns reveal the transition from being awake to being asleep; they also reveal that there are different stages of sleep. EEGs record sleep onset in a mammal as a transition from the very fast, low-amplitude wave pattern of wakefulness to slower and higher amplitude wave patterns. This stage of sleep in mammals other than humans is called **slow-wave sleep** because of the big, slow, wave tracings made by the pens recording the EEG. About 80 percent of sleep time is spent in slow-wave sleep, and the remaining 20 percent of sleep time is spent in a state known as **rapid-eye-movement sleep**. The electroencephalogram recorded during rapid-eye-movement sleep is similar to that recorded during periods of wakefulness: a fast, low-amplitude pattern. In humans, rapid-eye-movement sleep is when we have vivid dreams and nightmares. Rapid-eye-movement sleep gets its name from jerky movements of the eyes, but its most remarkable feature is the fact that descending commands from the brain almost completely paralyze the skeletal muscles. The waves of an electromyogram recorded during rapid-eye-movement sleep are flat. Occasional muscle twitches break through the paralysis as in a dog that appears to be trying to run in its sleep. If you look closely at a sleeping dog when its legs and paws are twitching, you will be able to see the rapid eye movements as well. Humans usually have about four episodes of rapid-eye-movement sleep each night, which vary from about 10 to 45 minutes in length (Figure 36.22). Because rapid-eye-movement sleep is called REM sleep, the rest of sleep time is called nonREM sleep. Human nonREM sleep is divided into four stages of progressively deeper sleep.

What is the function of sleep? Some believe that it is a restorative process, but we do not know what is restored. Similarly, we do not know why we dream. It has been hypothesized that dreaming facilitates memory consolidation and it has also been hypothesized that dreaming facilitates forgetting. The concept is the same in both cases—housekeeping in the circuits of the memory banks. We do know more about the neural control of sleep than we know about its function. The neural control of sleep resides in the brain stem. If the brain stem is damaged above the level of the pons, long-lasting or permanent coma results. If the brain stem is damaged just above the spinal cord, the subject may be paralyzed but have normal sleep–wake cycles. A number of specific groups of neurons in the brain stem play crucial roles in sleep processes.

Learning and Memory

Learning is the modification of behavior by experience, and memory is the ability of the nervous system to retain what is learned and what is experienced. Even very simple animals can learn and remember, but these two abilities are most highly developed in humans. They make possible language, culture, artistic creativity, and scientific progress. Consider the amount of information associated with learning a language. The capacity of human memory and the rate at which items can be retrieved are remarkable features of the nervous system. Is it possible to understand these phenomena in terms of the cells and molecules that make up the brain?

Habituation and sensitization are simple forms of learning that can be studied in all nervous systems from the nerve nets of cnidarians to the human central nervous system. **Habituation** is learning to ignore a repeated stimulus that conveys little information. **Sensitization** is learning to be especially aware of a

stimulus that conveys important information. For example, humans can habituate to noisy, busy, crowded environments, ignoring most of the barrage of sensory information they receive from those environments. Yet the small sound of a baby's cry amidst the turmoil will immediately get the attention of its sensitized mother or father. As complex as this example is, habituation and sensitization can be understood as processes that rely on individual synapses.

The synaptic basis of habituation and sensitization has been extensively studied by Eric Kandel and his colleagues at Columbia University. The animal they use in their experiments is a marine mollusk called a sea slug or a sea hare (*Aplysia californica*). Although the sea slug is a mollusk, it does not have a shell, so it is very vulnerable to predators. When undisturbed, its siphon is extended to take in water to ventilate its gill membranes. If the siphon is touched, the animal withdraws it. However, if the siphon is touched repeatedly, the animal habituates to the stimulus and no longer withdraws it. The researchers found that the siphon-withdrawal reflex depends on sensory neurons and motor neurons with only one synapse between them. By studying the characteristics of that synapse, they learned that habituation was due to a decrease in the amount of neurotransmitter released by the axon terminals of the presynaptic cells. This reduced release of neurotransmitter could last a considerable time, and was not due simply to fatigue or depletion of neurotransmitter.

The researchers also studied sensitization. To sensitize the sea slug, they applied mild electric stimulation to its tail just before they gently touched its siphon. Now the animal responded to the touch of its siphon with a much more vigorous withdrawal. In studying the synapse in the withdrawal reflex pathway, they discovered that sensitization was due to the action of a third neuron that made synapses onto the axon terminals of the sensory cell. This sensitizing neuron released the transmitter serotonin onto the presynaptic cell. Serotonin activates a second-messenger system that creates a series of intracellular changes, the end result of which is that calcium channels stay open longer in response to each action potential that arrives at the axon terminal. Consequently, the axon terminal releases more transmitter in response to each action potential (Figure 36.23).

Another form of learning that is widespread among animal species is **associative learning**, in which two unrelated stimuli are linked to the same response. The simplest example of associative learning is the **conditioned reflex** discovered by the Russian physiologist Ivan Pavlov. Pavlov was studying control of digestive functions in dogs and observed that a dog salivates at the sight or smell of food—a simple autonomic reflex. He discovered that if he

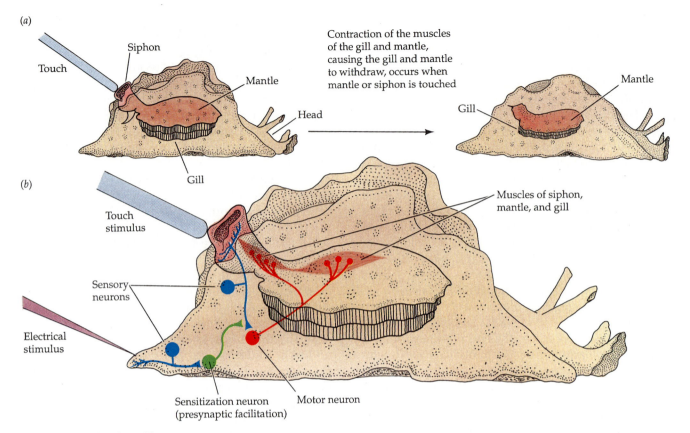

(a)

Touch

Siphon

Mantle

Head

Gill

Contraction of the muscles of the gill and mantle, causing the gill and mantle to withdraw, occurs when mantle or siphon is touched

Mantle

Gill

(b)

Touch stimulus

Muscles of siphon, mantle, and gill

Sensory neurons

Electrical stimulus

Sensitization neuron (presynaptic facilitation)

Motor neuron

36.23 Learning in a Slug

The siphon-withdrawal reflex of the sea slug *Aplysia*, shown in (a), can be overcome by habituation—after repeated stimuli, the animal no longer withdraws its siphon. (b) The opposite effect, sensitization, can also be achieved: Applying a mild electrical shock to the animal's tail just prior to touching the siphon intensifies the siphon-withdrawal reflex. Habituation and sensitization can be understood in terms of the underlying synaptic events in *Aplysia*.

rang a bell just before the food was presented to the dog, after a few trials the dog would salivate at the sound of the bell, even if no food followed. The salivation reflex was conditioned to be associated with the sound of a bell, which normally is quite unrelated to feeding and digestion. This simple form of learning has been studied extensively in efforts to understand its underlying neural mechanisms.

Attempts to treat human diseases have sometimes led to increases in scientific understanding. Epilepsy is a disorder characterized by uncontrollable increases in neural activity in specific parts of the brain. The resulting "epileptic fits" can endanger the afflicted individual. Serious cases of epilepsy were sometimes treated by destroying the part of the brain from which the surge of activity originated. To find the right area, the surgery was done under local anesthesia and different regions of the brain were electrically stimulated with fine electrodes while the patient reported on the resulting sensations. When some regions of association cortex were stimulated, patients sometimes reported vivid memories that were like reliving events from the past. Such obser-

vations were the first evidence that memories have anatomical locations in the brain and exist as properties of neurons and networks of neurons. Yet the destruction of a small area does not completely erase a memory, so it is postulated that memory is a function distributed over many brain regions and that a memory may be stimulated via many different routes.

You can recognize several forms of memory from your own experience. There is immediate memory for events that are happening now. Immediate memory is almost perfectly photographic, but it lasts only seconds. Short-term memory contains less information, but it lasts longer—on the order of 10 to 15 minutes. If you are introduced to a group of new people, you may remember most of their names for 5 or 10 minutes, but will have forgotten them in an hour or so if you have not repeated them, written them down, or used them in a conversation lasting longer than the round of introductions. Repetition, use, or reinforcement by something that gets your attention (such as the title President or Queen) facilitates the transfer of short-term memory to long-term memory, which can last for days, months, or years. Knowledge about neural mechanisms for the transfer of short-term memory to long-term memory has come from observations of patients who have lost parts of the limbic system, notably the hippocampus. A famous case is that of H. M., who had his hippocampus removed on both the left and right sides of his brain in an effort to control severe epilepsy. Following his surgery, H. M. could not transfer any information to long-term memory. If someone were introduced to him, had a conversation with him, and then left the room for an hour, when that person returned, he or she was unknown to H. M., and it was as if the previous conversation had never taken place. H. M. had normal memory for events that happened before his surgery, but he could only remember post-surgery events for 10 or 15 minutes. Interestingly, such patients learn and retain motor skills quite normally, but they cannot remember learning them. The hippocampus is essential for the formation of conscious, long-term memories.

Language, Lateralization, and Human Intellect

No aspect of brain function is as integrally related to consciousness and intellect as is language. Therefore, studies of the brain mechanisms underlying the acquisition and use of language are extremely interesting to neuroscientists. A curious fact of language abilities is that they are mostly located in one cerebral hemisphere—which is the left hemisphere in 97 percent of all people. This phenomenon is referred to as the **lateralization** of language functions. Some of the most fascinating research on this subject has been done by Roger Sperry and his colleagues at the California Institute of Technology; Sperry received the Nobel prize in Medicine for this work. The two cerebral hemispheres are connected by a white-matter tract called the **corpus callosum**. In one severe form of epilepsy, bursts of action potentials travel from hemisphere to hemisphere across the corpus callosum. Cutting the tract eliminates the problem, and patients function quite normally following the surgery. Experiments, however, revealed interesting deficits in the language abilities of these "split-brain" patients. With the connections between the two hemispheres cut, the knowledge or experience of the right hemisphere could no longer be expressed in language, nor could language be used to communicate with the right hemisphere.

Sensory input from the right hand goes to the left cerebral hemisphere, and sensory input from the left hand goes to the right cerebral hemisphere. If a split-brain patient is blindfolded and a familiar tool is placed in his or her right hand, the patient can identify the tool and describe its use. If the tool is placed in the left hand, however, the patient can use the tool correctly, but cannot name it or describe its use.

In split-brain individuals, the right hemisphere has lost access to the language abilities that reside predominantly in the left hemisphere. It is said that language is *lateralized* to the left hemisphere.

The brain mechanisms of language in the left hemisphere have been the focus of much research. Once again, the experimental subjects are persons who have suffered damage to some region of the left hemisphere and are left with one of many forms of **aphasias**, deficits in the abilities to use or understand words. The known language areas of the left hemisphere are shown in Figure 36.24a. Broca's area is in the frontal lobe just in front of the motor cortex, and it is essential for speech. Damage to Broca's area results in halting, slow, poorly articulated speech or even complete loss of speech, but the patient can still read and understand language. In the temporal lobe, close to its border with the occipital lobe, is Wernicke's area, which is more involved with sensory than motor aspects of language. Damage to Wernicke's area can cause a person to lose the ability to speak sensibly while retaining the abilities to form the sounds of normal speech and to imitate its cadence. Moreover, such a patient cannot understand spoken or written language. Near Wernicke's area is the angular gyrus, which is believed to be essential for integration of spoken and written language.

Normal language ability depends on the flow of information between various areas of the left cerebral cortex (Figure 36.24b). Input from spoken language travels from the primary auditory cortex to Wernicke's area. Input from reading language travels from the primary visual cortex to the angular gyrus to Wernicke's area. Commands to speak are formulated in Wernicke's area and travel to Broca's area and from there to the primary motor cortex. Damage to any one of those areas or the pathways between them can result in aphasia.

Young children can recover remarkably from even severe damage to the left cerebral hemisphere because lateralization of language abilities to the left hemisphere has not fully developed, and the right hemisphere can take over all of the language-related functions. One such patient in Sperry's split-brain studies produced provocative results with respect to the relationship between language and intellect. This patient had language functions in both hemispheres. Due to left-brain damage in childhood, his right hemisphere developed language functions, and over time, his left hemisphere recovered. Then, because of his epilepsy, his corpus callosum was cut. Afterwards, it was possible to communicate with this individual through either his left or his right cerebral hemisphere. Each had a separate personality with its individual likes and dislikes. Each responded differently to evaluating events and projecting plans into the future. It was as if there were two persons housed in one brain.

(a) Speaking a heard word

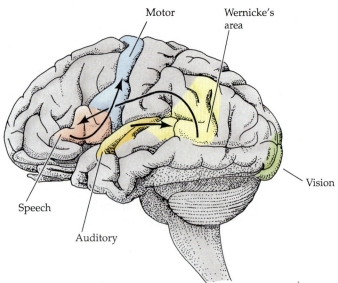

(b) Speaking a written word

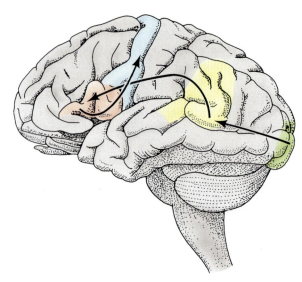

36.24 Language Areas
Different regions of the left cerebral cortex participate in
the use of spoken (a) and written (b) language.

SUMMARY

Nervous systems sense, process, and integrate information and generate motor commands to effector organs. The cells directly responsible for these properties are neurons. Neurons come in many forms, but usually have a cell body with many extensions. Extensions called dendrites bring information into the cell body, and other extensions, called axons, conduct information away from the cell body. Neurons communicate with each other and with other cells at synapses, where the membranes of two communicating cells come into proximity. There are electrical and chemical synapses.

An action potential is a sudden change in the resting electrical potential across the cell membrane. Resting and action potentials result from asymmetric distribution of ions on the two sides of the plasma membrane. The asymmetry results from the action of Na^+–K^+ pumps. The action potential is generated by selective opening and closing of specific ion channels—sodium ions move into the axon, followed by the departure of potassium ions. At synapses, there are chemically gated ion channels in the postsynaptic membrane that open in response to neurotransmitters. These changes in the ion channels can result in depolarizing or hyperpolarizing the membrane and can make it more or less likely that the cell will fire an action potential; synapses can therefore be either excitatory or inhibitory.

Neurons are organized into nervous systems. The vertebrate nervous system is divided into the central division, consisting of the brain and spinal cord, and the peripheral division, consisting of nerves that bring information into or conduct information away from the central nervous system. The central nervous system develops from a hollow neural tube that forms three swellings at its anterior end early in embryogenesis. These swellings give rise to the major divisions of the brain—the forebrain, the midbrain, and the hindbrain.

The functions of the nervous system can be divided into afferent (sensory) and efferent (motor) functions, and into voluntary and involuntary functions. Voluntary functions depend on sensory information of which we are consciously aware and motor outputs under conscious control. Involuntary functions depend on sensory information of which we are not always consciously aware and on the autonomic control of organs and organ systems.

Neurons are involved in different functions and are organized into circuits. The simplest neural circuit is the monosynaptic reflex loop. This reflex depends on a stretch receptor in a muscle that generates action potentials in a sensory nerve in proportion to the degree to which the muscle is stretched. Through this reflex loop, stretching the muscle causes it to contract and oppose the stretch. Other reflex loops are more complex, with more synapses, with interneurons, and with cross-innervation activating complementary muscles and causing antagonistic muscles to relax. Information from spinal inputs is also transmitted to the brain, and spinal circuits are activated or modified by information descending from the brain.

Attempts are being made to understand higher functions of the nervous system, such as learning, memory, and language, in terms of cellular mechanisms and neural circuits.

SELF-QUIZ

1. In the nervous system, the *most* abundant cell type is the:
   a. motor neuron.
   b. sensory neuron.
   c. preganglionic parasympathetic neuron.
   d. glial cell.
   e. preganglionic sympathetic neuron.

2. Within the nerve cell, information moves from:
   a. dendrite to cell body to axon.
   b. axon to cell body to dendrite.
   c. cell body to axon to dendrite.
   d. axon to dendrite to cell body.
   e. dendrite to axon to cell body.

3. Which of the following statements is *false*?
   a. Sensory afferents carry information of which we are consciously aware.
   b. Visceral afferents carry information about physiological functions of which we are not consciously aware.
   c. The voluntary motor division of the efferent side of the peripheral nervous system executes conscious movements.
   d. The cranial nerves and spinal nerves are parts of the peripheral nervous system.
   e. Afferent and efferent axons never travel in the same nerve.

4. Which of the following statements is *false*?
   a. In the spinal cord, the white matter contains the axons conducting information up and down the spinal cord.

   b. The limbic system is involved in basic physiological drives, instincts, and emotions.
   c. The limbic system consists of primitive forebrain structures.
   d. The vast majority of the nerve cell bodies in the human nervous system are contained within the limbic system.
   e. In humans, a part of the limbic system is necessary for the transfer of short-term memory to long-term memory.

5. Which of the following statements accurately describes an action potential?
   a. The magnitude of the action potential *increases* along the axon.
   b. The magnitude of the action potential *decreases* along the axon.
   c. All action potentials in a single neuron are of the same magnitude.
   d. During an action potential, the transmembrane potential of a neuron remains constant.
   e. An action potential permanently shifts a neuron's transmembrane potential away from its resting value.

6. A neuron that has just fired an action potential cannot be immediately restimulated to fire a second action potential. The short interval of time during which restimulation is not possible is called:
   a. hyperpolarization.
   b. the resting potential.
   c. depolarization.
   d. repolarization.
   e. the refractory period.

7. The rate of propagation of an action potential depends on:
   a. whether or not the axon is myelinated.
   b. the axon's diameter.
   c. whether or not the axon is insulated by glial cells.
   d. the cross-sectional area of the axon.
   e. all of the above.

8. The binding of neurotransmitter to the postsynaptic receptors in an inhibitory synapse results in:
   a. depolarization of the transmembrane potential.
   b. generation of an action potential.
   c. hyperpolarization of the transmembrane potential.
   d. increased permeability of the membrane to sodium ions.
   e. increased permeability of the membrane to calcium ions.

9. Whether or not a synapse is excitatory or inhibitory depends on:
   a. the type of neurotransmitter.
   b. the presynaptic terminal.
   c. the size of the synapse.
   d. the nature of the postsynaptic neurotransmitter receptors.
   e. the concentration of neurotransmitter in the synaptic space.

10. The part of the brain that differs the most in complexity between mammals and amphibians is:
    a. the midbrain.
    b. the forebrain.
    c. the cerebellum.
    d. the limbic system.
    e. the hippocampus.

FOR STUDY

1. Compare and contrast the two divisions of the autonomic nervous system. Emphasize distinctions with respect to their anatomical organization, neurotransmitters used, and general effects on the functions of specific organ systems.

2. Outline the development of the vertebrate nervous system. Where on the neural axis are the more evolutionarily primitive and advanced functions localized?

3. Describe the electrochemical and structural elements involved in the establishment and maintenance of the neuron's transmembrane resting potential.

4. Describe the processes and structures involved in (a) the initiation and propagation of an action potential, and (b) synaptic transmission.

5. Define and describe the synaptic basis for habituation and sensitization in *Aplysia*.

READINGS

Camhi, J. M. 1984. *Neuroethology: Nerve Cells and the Natural Behavior of Animals*. Sinauer Associates, Sunderland, MA. This advanced text covers the properties and functions of neurons but emphasizes aspects relevant to sensory abilities of animals and the neural control of behavior. Examples are taken from a wide variety of vertebrates and invertebrates.

Kuffler, S. W., J. G. Nicholls and A. R. Martin. 1984. *From Neuron to Brain*, 2nd Edition. Sinauer Associates, Sunderland, MA. An advanced text on cellular neurobiology.

Llinás, R. R. 1988. *The Biology of the Brain: From Neurons to Networks*. 1990. *The Workings of the Brain: Development, Memory, and Perception*. W. H. Freeman, New York. These two volumes are a rich collection of articles on various aspects of neuroscience that have been published in *Scientific American* over the past 15 years. They provide a wonderfully broad yet selectively indepth survey of modern neurobiology.

Thompson, R. F. 1985. *The Brain: An Introduction to Neuroscience*. W. H. Freeman, New York. A well written and easy to understand introductory text on neuroscience, covering topics from membrane events to the neural basis of behavior.

37

Sensory Systems

PREVIEW: An animal's view of the world is dependent on its sensory systems. All of the information an animal has about its external and internal worlds comes from sensory cells that convert stimuli into neural signals that are processed by the nervous system. Sensory organs such as eyes and ears gather environmental stimuli and focus them on sensory cells. Although sensory cells are specialized to respond to specific stimuli, their general response is a change in membrane resting potential. This change alters the rate at which the sensory cells release neurotransmitters or the rate at which they fire action potentials. All animals, including humans, are sensitive to only a portion of the information available in the environment. Some animals can sense environmental stimuli that humans cannot.

This chapter deals with mechanisms of smell, taste, touch, balance, hearing, and vision, especially as they function in humans. This chapter also describes some ways that animals use other kinds of sensory information that humans cannot, including ultraviolet light, high frequency sound, electrical fields, and magnetism.

"Angela, you *look* great. I *hear* that you and Carl are going to the Khyber Pass for dinner."

"Yes, you raved so much about the wonderful *aromas* and complex, spicy *flavors* that we couldn't resist. After all, you have good *taste*, my friend."

"You'll also like the *feel* of the place. The tables are low and you sit on *soft* cushions. The walls are covered with *brightly colored* hangings. Stay in *touch*, I'd like to know how you liked it."

"I just hope the food isn't too *hot*, or you'll *hear* from me sooner than you think."

It is impossible to discuss anything for very long without using words that refer to our senses. Our senses are the window through which we view the world, and the world is what our senses tell us it is. Different species look through different sensory windows, so their views of the world differ. Dogs do not see color, but they have keener senses of hearing and smell than humans do. As you gaze at a beautiful sunset, your dog may be sniffing around at your feet and pricking up its ears as it hears small animals moving in the underbrush. Bees can see patterns on flowers that reflect ultraviolet light; we cannot. In environments that are totally dark to us, some snakes can "see" the infrared radiation emitted by bodies warmer than the environment. Bats can use reflected sound to avoid obstacles and catch small insects. The sounds bats emit are extremely intense, but they are beyond our range of hearing. In murky waters, the duck-billed platypus uses its sensitive bill to feel and taste food items, and electric fishes detect other fishes by the electric fields they create. How the environ-ment "looks" to any animal depends on what information that animal receives from its sensors.

What is a sensor? A sensor is a cell that transduces (converts) a physical or chemical stimulus into action potentials that are conducted to other parts of the nervous system for processing and interpretation. Most sensors are modified neurons, but some are other types of cells closely associated with a neuron. A sensor is specialized to respond to a specific type of stimulus, such as light or pressure. In general, the stimulus causes a change in the flow of ions across the plasma membrane of the sensor, resulting in a change in the membrane potential of the cell. The membrane potential of a sensor is called a **receptor potential**. Changes in receptor potential can cause the sensor to generate action potentials or to change the rate at which it releases transmitter onto a neigh-boring neuron that generates action potentials. Re-gardless of the type of stimulus that activates a par-ticular sensor, the message sent to the nervous system is always a train of action potentials. The intensity of the stimulus affects the frequency of the action potentials.

How action potentials from sensors are interpreted depends on where in the nervous system the mes-sages are sent. In a small patch of skin on your arm there are sensors that respond to warming by increas-ing the rate at which they fire action potentials. Other sensors in the same patch of skin respond when they are cooled, and still others are activated by touch, movements of hairs, or painful stimuli such as cuts

845

or burns. As we learned in Chapter 36, these sensors all transmit their messages through axons that enter the central nervous system through the dorsal horns of the spinal cord. The synapses made by those axons in the dorsal horns, and the subsequent pathways of transmission in the nervous system, determine whether the stimulation of the patch of skin on your arm is perceived by you as warmth, cold, pain, touch, itch, or tickle.

An important characteristic of sensors is that they can stop being excited by a stimulus that initially caused them to be active. The process of becoming insensitive to a continuing source of stimulation is known as **adaptation**. Adaptation enables an animal to ignore background or unchanging conditions while remaining sensitive to changes or to new information. For example, when you dress, you feel each item of clothing touch your skin, but the sensation of clothes touching your skin is not constantly on your mind throughout the day. You are immediately aware, however, when a seam rips, your shoe comes untied, or someone touches your back ever so lightly. Our ability to discriminate between important and unimportant stimuli is partly due to the fact that some sensors adapt; it is also due to information processing by the central nervous system. Some sensors adapt very little or do so slowly—examples are pain receptors and sensors responsible for balance.

Some sensory systems include organs, such as eyes and ears, that gather environmental stimuli and direct them onto sensors. In this chapter, we will learn how sensory systems gather stimuli, transduce stimuli to action potentials, transmit action potentials to the central nervous system, and how the central nervous system processes that information to yield perceptions about the internal and external worlds. Sensors are also very important components of autonomic regulatory mechanisms, as we will see in the subsequent chapters on physiological systems.

CHEMOSENSORS

Animals receive information about chemical stimuli through **chemosensors** that respond to specific molecules in the environment. Chemosensors are responsible for smell, taste, and the monitoring of aspects of the internal environment such as the level of carbon dioxide in the bloodstream. Chemosensitivity is universal among animals. A colony of corals responds to a small amount of meat extract in the seawater around it by extending bodies and tentacles and searching for food. A single amino acid can stimulate this response. If, however, an extract from an injured individual of the colony is released into the water, the colony members retract their tentacles and bodies to avoid danger. Similar reactions to chemical stimuli can be observed in humans. Upon smelling

freshly baked bread, we salivate and feel hungry, but we gag and retch when we smell diamines from rotting meat. Information from chemosensors can cause powerful behavioral and autonomic responses.

Chemosensation in Arthropods

Arthropods use chemical signals to attract mates. These signals, called pheromones, demonstrate the sensitivity of chemosensory systems. The female silkworm moth releases a pheromone called bombykol from a gland at the tip of her abdomen. The male silkworm moth has sensors for this molecule on his antennae (Figure 37.1). Each feathery antenna carries about 10,000 bombykol-sensitive hairs, and each hair has a dendrite of a sensor cell at its core. A single molecule of bombykol is sufficient to activate a dendrite and generate action potentials in the antennal nerve that transmits the signal to the central nervous

(a)

(b)

37.1 Chemosensors of the Silkworm Moth.
Mating in silkworms of the genus *Bombyx* is coordinated by a chemical attractant that the female moth releases from a gland at the tip of her abdomen (a). A male moth detects this chemical attractant in the air passing over his antennae, which are covered with chemosensitive hairs (b).

37.2 Tasting with the Feet
Using chemosensors on their feet, flies such as this *Drosophila melanogaster* can tell the nature of a potential food source by stepping in it. Chemoreceptors in the sensory hairs distinguish among sugars, amino acids, salts, and other substances.

system. When approximately 200 hairs per second are activated, the male flies upwind in search of the female. It is likely that the threshold stimulation of 200 hairs per second will be reached when there is only 1 molecule of bombykol for every 10^{15} other molecules in the air. Because of the males' extremely high degree of sensitivity, the sexual message of a female moth is likely to reach any male that happens to be within a downwind area stretching over several kilometers. Because the rate of firing in the male's sensory nerves is proportional to bombykol concentrations in the air, the male can follow a concentration gradient and home in on the emitting female.

Many arthropods have chemosensory hairs, with each hair containing one or more types of specific sensor cells. For example, crabs and flies have chemosensory hairs on their feet, which they use to taste potential food by stepping in it. These hairs have receptors for sugars, amino acids, salts, and other molecules (Figure 37.2). After a fly steps in a drop of sugar water and tastes it, its proboscis extends and it feeds. Potential food items will stimulate extension of the proboscis; other substances will not.

Olfaction

The sense of smell is called **olfaction**, and it is another form of chemosensation. In vertebrates, the smell sensors are neurons embedded in a layer of epithelial cells at the back of the nasal cavity (Figure 37.3). Their axons project to the olfactory bulb of the brain, and their dendrites end in olfactory hairs that project through to the surface of the epithelium. A protective layer of mucus covers the epithelium. Molecules

must diffuse through this mucus to get to the receptors on the olfactory hairs. When you have a cold or an attack of hay fever, the amount of mucus increases and the epithelium swells. With this in mind, study Figure 37.3 and you will easily understand why you lose your sense of smell at those times. A dog has up to 40 million nerve endings per square centimeter of nasal epithelium, many more than humans do. Humans do have a very sensitive olfactory system, but we are unusual among mammals in depending more on vision than on olfaction (we tend to join bird-watching societies more often than mammal-smelling societies). Whales and porpoises have no olfactory sensors and hence no sense of smell.

How does an olfactory sensor transduce the structure of a molecule into action potentials? A molecule that triggers an olfactory sensor is called an odorant molecule. Odorant molecules bind to receptor molecules on the olfactory hairs of the sensors. Olfactory receptors are specific for particular odorant molecules, and the way they work is like a lock-and-key mechanism. If a "key" (an odorant molecule) fits the "lock" (the receptor), then a G protein is activated, which in turn activates an enzyme that causes an increase of a second messenger in the cytoplasm of the sensor cell. The second messenger binds with sodium channel proteins in the sensor's plasma membrane and opens the channels, causing an influx of Na^+. The sensor thus depolarizes to threshold and fires action potentials (Figure 37.4).

How does the sensor signal the intensity of a smell? It responds in a graded fashion to the concentration of odorant molecules; the more odorant molecules that bind to receptors, the more action poten-

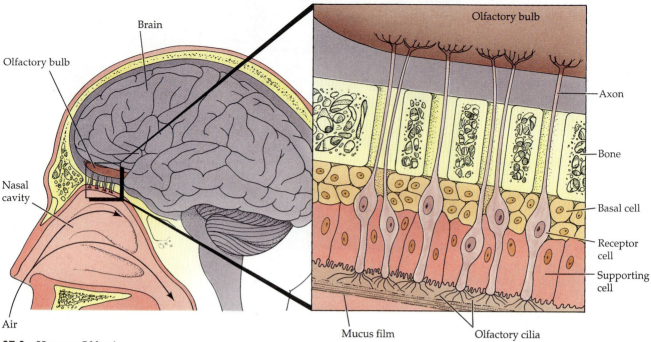

Olfactory bulb

Brain

Olfactory bulb

Nasal cavity

Air

Axon

Bone

Basal cell

Receptor cell

Supporting cell

Mucus film   Olfactory cilia

37.3 Human Olfaction

The sensors of the human olfactory system are embedded in the tissue lining the nasal cavity. Receptor molecules are on olfactory cilia of receptor cells. These cells send their axons to the olfactory bulb of the brain.

37.4 Molecular Events in Olfaction

When we "smell" something, odorant molecules bind to receptor molecules on cilia of receptor cells. There is a large family of receptor molecules expressed in different olfactory cells. Binding of odorant molecules to receptors initiates molecular events that culminate in the transmission of action potentials and the sensation of smell.

1. Odorant molecule binds to specific receptor in the plasma membrane of chemoreceptor

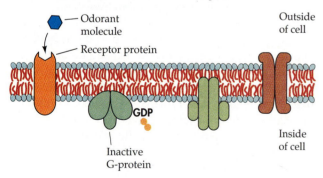

Odorant molecule

Receptor protein

Outside of cell

GDP

Inside of cell

Inactive G-protein

2. Receptor odorant complex activates G-protein, which combines with a molecule of GTP displacing GDP

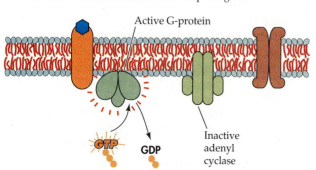

Active G-protein

GTP   GDP

Inactive adenyl cyclase

3. G-protein subunit dissociates and activates adenyl cyclase, which produces cAMP

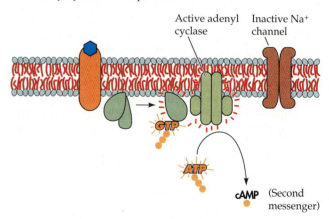

Active adenyl cyclase

Inactive Na⁺ channel

GTP

ATP

cAMP   (Second messenger)

4. cAMP (the second messenger) binds calcium channel, opens it, and Na⁺ enters the cell.

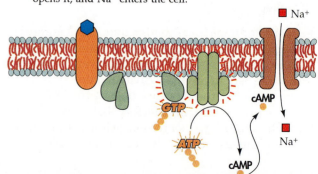

Na⁺

cAMP

GTP

ATP

cAMP

Na⁺

tials are generated and the greater the intensity of the perceived smell. The activation of different receptors on different sensors or combinations of sensors enables animals to perceive a wide variety of different smells.

Gustation

The sense of taste in humans and other vertebrates depends on clusters of sensors called **taste buds**. The taste buds of terrestrial vertebrates are confined to the mouth cavity, but some fishes have taste buds in the skin that enhance the fish's ability to taste its environment. Some fishes living in murky water are very sensitive to small amounts of amino acids in the water around them and can find food without the use of vision. The duck-billed platypus, a monotreme mammal, has similar talents due to taste buds on the sensitive skin of its bill. What is a taste bud and how does it work?

A taste bud is a cluster of many taste sensors. The 10,000 or so taste buds on our tongues provide good examples. The taste buds are embedded in the epithelium of the tongue, and many are found on the raised papillae of the tongue. Look at your tongue in a mirror—the papillae make it look fuzzy. There are many taste buds on each papilla. On its outer surface, a taste bud has a pore that exposes the tips of the taste sensors (Figure 37.5). Microvilli (tiny hairlike projections) increase the surface areas of the sensors where their tips converge at the taste pore. Taste sensors, unlike olfactory sensors, are not neurons. At their bases, taste sensors form synapses with dendrites of sensory neurons.

Gustation begins at receptors in the membranes of the microvilli. As with olfactory transduction, receptor molecules on the sensor cells bind molecules, and the binding causes changes in the membrane polarity of the taste sensors. Because the taste sensors are not neurons, however, they do not fire action potentials. Instead, they release neurotransmitter onto the dendrites of the sensory neurons. The sensory neurons respond to the neurotransmitter by firing action potentials that are conducted to the central nervous system. The tongue does a lot of hard work, so its epithelium is shed and replaced at a rapid rate. Individual taste buds last only a few days before they are replaced, but the sensory neurons associated with them live on, always forming new synapses as new taste buds form.

It is popular to claim that humans can perceive only four tastes: sweet, salt, sour, and bitter. Particular regions of the tongue have taste buds responsible for these general categories of taste, but the regions overlap to a large extent (Figure 37.6). You can map your own tongue by dipping toothpicks in dif-

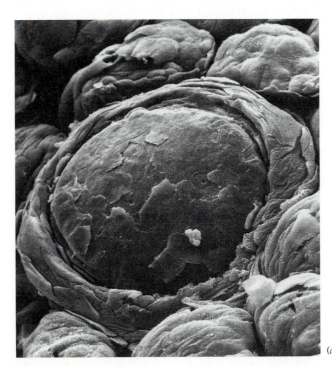

(a)

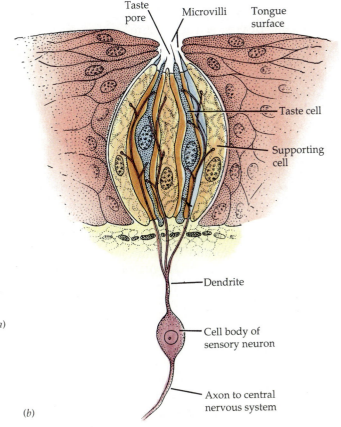

(b)

37.5 Taste Buds
(a) Scanning electron micrograph of a human taste bud. Taste buds are clusters of taste-sensing cells embedded in the papillae of the tongue. (b) Drawing of a taste bud shown in cross section. Unlike the olfactory system, the primary receptors are not neurons.

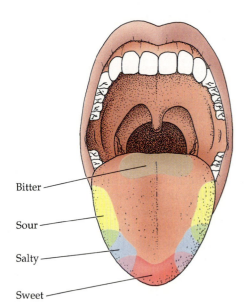

37.6 Regions of Taste
The sensors for sweet, salt, sour, and bitter are found in specific regions of the human tongue; the regions overlap in some areas.

ferent solutions and then touching the toothpicks to different regions of the surface of your tongue. In actuality, taste buds can distinguish between a variety of sweet-tasting molecules and a variety of bitter-tasting molecules. The full complexity of the chemosensitivity that enables us to enjoy subtle flavors of delicious foods comes from the combined activation of gustatory and olfactory sensors; hence, you lose some of your sense of "taste" when you have a cold.

What about a snake continually sampling the air with its forked tongue darting in and out? If the snake, like us, only tasted sweet, salt, sour, and bitter with its tongue, it would not get much useful information by tasting the air. The forks of the snake's tongue fit into cavities in the roof of its mouth that are richly endowed with olfactory receptors. The tongue samples the air and presents the sample directly to olfactory receptors. So the snake is really using its tongue to smell its environment, not to taste it. Why doesn't the snake simply use the flow of air to and from its lungs as we do to smell the environment? Air flow to and from the lungs is relatively slow in reptiles, but the tongue can dart in and out many times in a second. It is a quick source of olfactory information.

MECHANOSENSORS

Mechanosensors are specialized cells that are sensitive to mechanical forces that distort their membranes. A variety of mechanosensors in the skin are responsible for our perception of touch, pressure, and tickle. Stretch receptors in muscles, tendons, and joints give us information about the position of the parts of the body in space and the forces acting on them. Stretch receptors in the walls of blood vessels signal blood pressure. "Hair" cells with processes that are sensitive to being bent are incorporated into mechanisms for signaling the body's position with respect to gravity as well as mechanisms for hearing.

Physical distortion of a mechanoreceptor's membrane causes ion channels to open and alters the resting potential of the cell. The magnitude of this receptor potential depends on the intensity of the mechanical stimulus. When receptor potential rises above threshold, action potentials are generated and transmitted to the central nervous system via a sensory nerve. The farther above threshold the receptor potential rises, the faster action potentials are generated. The rates of action potentials in the sensory nerves tell the central nervous system the strengths of the stimuli exciting the mechanoreceptors.

Touch and Pressure

Objects touching the skin generate a wide variety of sensations due to the richness and diversity of mechanosensors found in the skin (Figure 37.7). The outer layers of skin, especially hairless skin such as lips and fingertips, contain many whorls of nerve endings enclosed in connective tissue capsules. These very sensitive mechanoreceptors are called **Meissner's corpuscles**, and they permit us to sense objects that touch the skin even lightly. Meissner's corpuscles adapt very rapidly, however. That is one reason why you roll a small object between your fingers, rather than holding it still, to discern its shape and texture. As you roll it, you continue to stimulate sensors anew. When you hold an object still, the sensors originally activated adapt. Try it. Also in the upper regions of the skin are **expanded-tip tactile receptors** of various kinds. They differ from Meissner's corpuscles in adapting only partially and slowly. They are useful for providing steady-state information about objects that continue to touch the skin.

The density of the various tactile receptors differs across the surface of the body. A two-point discrimination test demonstrates this fact. If you lightly touch someone's back with two pins or two pencil tips, you can determine how far apart the two stimuli have to be before the person can distinguish whether he or she was touched with one or two points. The same test applied to the person's lips or fingertips reveals a finer spatial discrimination; that is, the person can identify as separate two stimuli that are closer together.

Deep in the skin, processes of neurons wrap around hair follicles. When the hairs are displaced,

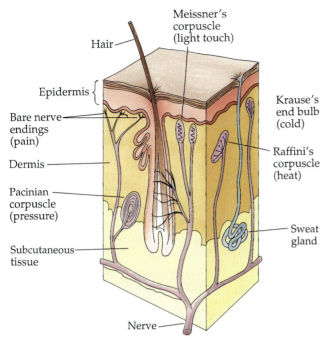

37.7 The Skin Feels Many Sensations
Heat, cold, pain, and pressure are all perceived through receptors in the skin. Even a very small patch of skin contains a diversity of receptors that send information to the brain. Some of these receptors have specialized structures associated with their nerve endings; others simply have bare nerve endings.

those neurons are stimulated. Also deep within the skin is another type of mechanoreceptor, the **Pacinian corpuscles**. These look like onions because they are made up of concentric layers of connective tissue cells encapsulating a process of a sensory neuron. Pacinian corpuscles respond especially well to vibrations applied to the skin, but they adapt rapidly to steady pressure. The connective tissue capsule is important in the adaptation of these sensors. An initial pressure distorts the corpuscle and the membrane of the neuron at its core, but the layers of the capsule rapidly rearrange to redistribute the force, thus eliminating the distortion.

Stretch Receptors

A variety of **stretch receptors** informs an animal about the position of its limbs and the stresses on its muscles and joints. These mechanosensors are activated by being stretched. They continuously feed information to the central nervous system, and that information is essential for coordinating movements. We encountered one important type of stretch receptor, the muscle spindle, when we discussed the monosynaptic reflex loop in Chapter 36. Muscle spindles are embedded in connective tissue within skeletal muscles. They consist of modified muscle fibers that are innervated in the center with processes of

sensory neurons (Figure 37.8). Whenever the muscle is stretched, the spindle cells are also stretched, and the neurons transmit action potentials to the central nervous system. In the monosynaptic reflex loop, this information causes the skeletal muscle to contract, counteracting the stretch. The monosynaptic reflex is good for keeping a limb in a constant position as load changes, but if the limb is to move, the sensitivity of the muscle spindle must be changed. Motor neurons from the central nervous system innervate the muscle spindles. When those motor neurons fire action potentials, they cause the spindle fibers to contract. Because a spindle fiber is attached to connective tissue at both ends, it stretches its own central region when it contracts, activating its sensory neuron. This mechanism adjusts the sensitivity of the muscle spindle, thus permitting the muscle to assume different positions.

Another stretch receptor is found in tendons and ligaments. It is called the **Golgi tendon organ**. Its role is to provide information about the force generated by a contracting muscle. When a contraction becomes too forceful, the information from the Golgi tendon organ feeds into the spinal cord, inhibits the motor neuron, and causes the contracting muscle to relax.

Hair Cells

Hair cells are mechanosensors that are not neurons. From one surface they have a number of projections called **stereocilia**, which look like a set of organ pipes. When these stereocilia are bent, they alter the receptor potential of the hair cell membrane. In some hair cells, when the stereocilia are bent in one direction the receptor potential becomes more negative, and when they are bent in the opposite direction, it becomes more positive. When hair cells become more positive, they release neurotransmitter to the sensory neurons associated with them, and the sensory neurons send nerve impulses to the brain.

Hair cells are found in the **lateral line** sensory system of fishes. The lateral line consists of a canal just under the surface of the skin that runs down each side of the fish. The canal has numerous openings to the external environment. Many structures called cupulae project into the lateral line canal. Each cupula contains hair cells whose stereocilia are embedded in gelatinous (jellylike) material. Movements of water in the lateral line canal move the cupulae and stimulate the hair cells (Figure 37.9). Thus the lateral line provides information about the movements of the fish through the water as well as information about other moving objects, such as predators or prey, that cause pressure waves in the surrounding water.

Many invertebrates have equilibrium organs called **statocysts** that use hair cells to signal the position of the animal with respect to gravity. A typical statocyst

(a)

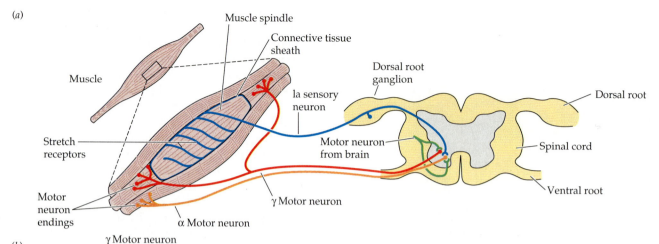

Muscle spindle

Connective tissue sheath

Dorsal root ganglion

Dorsal root

Muscle

Ia sensory neuron

Motor neuron from brain

Spinal cord

Stretch receptors

γ Motor neuron

Ventral root

Motor neuron endings

α Motor neuron

(b)

γ Motor neuron

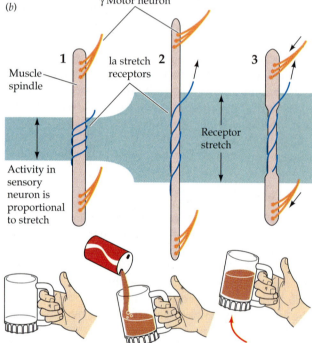

Muscle spindle

Ia stretch receptors

Receptor stretch

Activity in sensory neuron is proportional to stretch

1. Muscle at rest

2. Added load causes muscle and spindle to stretch, activity in reflex loop increases, arm returns to original position

3. Increased activity in γ motor neurons due to commands from brain stretches spindle, increases reflex loop activity, and raises arm

37.8 Stretch Receptors and Muscle Contraction

(a) Embedded in a skeletal muscle are bundles of modified muscle fibers that serve as stretch receptors. These muscle spindles communicate to the spinal cord through sensory neurons and receive motor neurons from the spinal cord. Motor neurons to muscle spindles are separate from the motor neurons to the surrounding muscle. (b) Activity in the sensory neuron from a muscle spindle is proportional to its degree of stretch. Loading a limb stretches a muscle and its spindles, initiating the reflex that increases the strength of contraction in the muscle to compensate for the load. The limb can be moved by firing action potentials in the motor neuron to the muscle spindle. This causes fibers in the spindle to contract and stretch their own sensitive region. This activates their sensory neuron, which in turn activates motor neurons to muscle.

37.9 The Lateral Line Sensory System

Hair cells in the lateral line organs of a fish detect movement of the water around the animal. This gives the fish information about its own movements and the movements of other objects nearby.

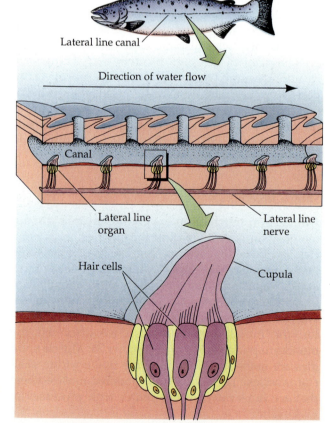

Lateral line canal

Direction of water flow

Canal

Lateral line organ

Lateral line nerve

Hair cells

Cupula

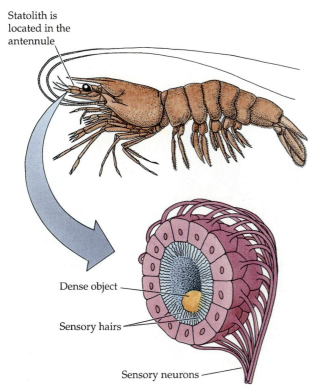

Statolith is located in the antennule

Dense object

Sensory hairs

Sensory neurons

37.10 Invertebrate Organs of Equilibrium
The statocyst is a sense organ found in many invertebrates. The force of gravity acting on statoliths within the hair-cell-lined statocyst gives the animal information about its position.

is a chamber lined with hair cells whose stereocilia project into the center of the chamber (Figure 37.10). Within the chamber are a number of statoliths, which may be grains of sand or other dense granules. Gravity pulls the statoliths down, so at any particular moment the statoliths are stimulating the hair cells that are lowest, as determined by the position of the animal. In an ingenious experiment, a scientist replaced the statoliths in the statocysts of lobsters with iron filings. When he held a magnet over the lobsters, they swam upside down, and when he held the magnet to their sides, they swam on their sides.

The vertebrate equilibrium organ, the **vestibular apparatus**, also uses hair cells to detect the position of the body with respect to gravity. In the next section we will examine the structure of the ear. For the moment it is enough to know that the inner ear contains three **semicircular canals** at right angles to one another (Figure 37.11). Each semicircular canal has a swelling called an ampulla, which contains a group of hair cells with their stereocilia embedded in a gelatinous cupula. The canals are filled with fluid. As an animal's head changes position, the fluid in its semicircular canals moves, puts pressure on the cupulas, and bends the stereocilia of the hair cells. In addition to the semicircular canals, the vestibular apparatus also has two chambers whose function is like that of the statocysts of invertebrates. Hair cells line

the floor of the chambers; their stereocilia are embedded in a layer of gelatinous material. On top of this layer are many otoliths (literally, ear stones), which are granules of calcium carbonate. As the head moves, gravity pulls on the dense otoliths, which bend the stereocilia of the hair cells.

Auditory Systems

Auditory systems use mechanosensors to transduce sound pressure waves into action potentials. These systems usually include special adaptations to gather sound, direct it to the sensors, and amplify it. Human hearing provides a good example of these various aspects of auditory systems. The organs of hearing are the ears. The two prominent structures on the sides of our heads usually thought of as ears are the **ear pinnae**. The pinna of an ear collects sound waves and directs them into the auditory canal leading to the real hearing apparatus in the middle ear and the inner ear.

As shown in Figure 37.12, the eardrum, or **tympanic membrane**, covers the end of the auditory canal. The tympanic membrane vibrates in response to sound waves traveling down the auditory canal. The chamber of the middle ear, an air-filled cavity, lies on the other side of the tympanic membrane. The middle ear is open to the throat at the back of the mouth through the **eustachian tube**. Because air flows through the eustachian tube, pressure equilibrates between the middle ear and the outside world. When you have a cold or hay fever, the tube becomes blocked by mucus or by tissue swelling, and you have difficulty "clearing your ears," or equilibrating the pressure in the middle ear with the outside air pressure. Then, the flexible tympanic membrane bulges in or out and may cause an earache.

The middle ear contains a set of three delicate bones called the **ear ossicles**, individually named the **malleus** (hammer), **incus** (anvil), and **stapes** (stirrup). The ossicles transmit the vibrations of the tympanic membrane to the fluid-filled inner ear, where they will be transduced into action potentials. The lever-like action of the ossicles amplifies the vibrations about 20-fold. The malleus is attached to the center of the tympanic membrane, and at the other end of the chain of ossicles, the stapes is attached to a smaller membrane called the **oval window**, which covers an opening into the inner ear. The incus serves as a fulcrum. When the tympanic membrane moves in, the lever action of the ossicles pushes the stapes, and the oval window bulges into the inner ear. When the tympanic membrane moves out, the stapes and the oval window are also pulled out. In this way, pressure waves in the auditory canal are converted into pressure waves in the fluid-filled inner ear.

Pressure waves are transduced into action potentials in the inner ear. The inner ear is a long, narrow, coiled chamber, hence its name the **cochlea,** from

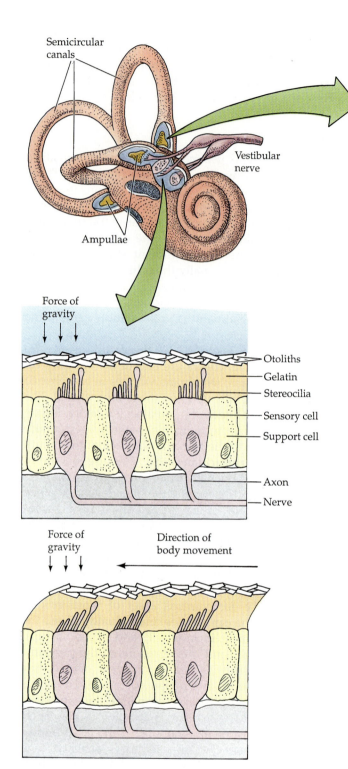

Semicircular canals

Vestibular nerve

Ampullae

Force of gravity

Otoliths
Gelatin
Stereocilia
Sensory cell
Support cell

Axon
Nerve

Force of gravity

Direction of body movement

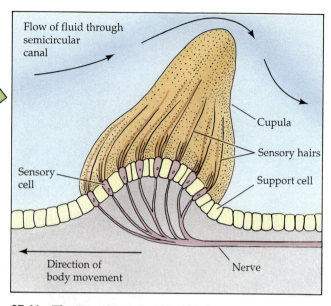

Flow of fluid through semicircular canal

Cupula

Sensory hairs

Sensory cell

Support cell

Direction of body movement

Nerve

37.11   The Two Organs of Equilibrium in the Mammalian Inner Ear
The bony inner ear has three parts: the snail-shaped cochlea, the vestibule, and the semicircular canals. The vestibule and semicircular canals house organs of equilibrium. In the vestibule, layers of otoliths are moved by gravity and by angular momentum when the head changes position. In the ampullae of the semicircular canals, the gelatinous cupulae are pushed one way or the other when movement of the head causes the fluid in the canals to shift.

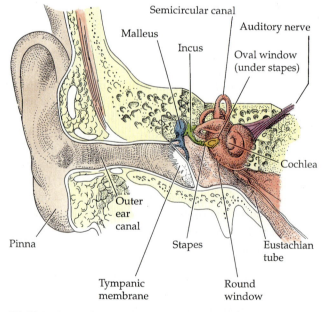

Semicircular canal

Malleus

Incus

Auditory nerve

Oval window (under stapes)

Cochlea

Outer ear canal

Pinna

Tympanic membrane

Stapes

Round window

Eustachian tube

Latin and Greek words for snail or shell. A cross section of this chamber reveals that it is really composed of three parallel canals separated by two membranes; Reissner's membrane and the basilar membrane (Figure 37.13a). Sitting on the basilar membrane is the **organ of Corti**, the apparatus that transduces pressure waves into action potentials in the auditory nerve that conveys information from the

37.12   The Human Auditory System
The small bones of the middle ear convert movements of the eardrum to pressure waves in the fluid-filled inner ear, or cochlea (Figure 37.13).

37.13 The Inner Ear
The inner ear converts pressure waves into action potentials. (a) The sensory cells are part of the organ of Corti, which sits on the basilar membrane. (b) Pressure waves of different frequencies flex the basilar membrane at different locations.

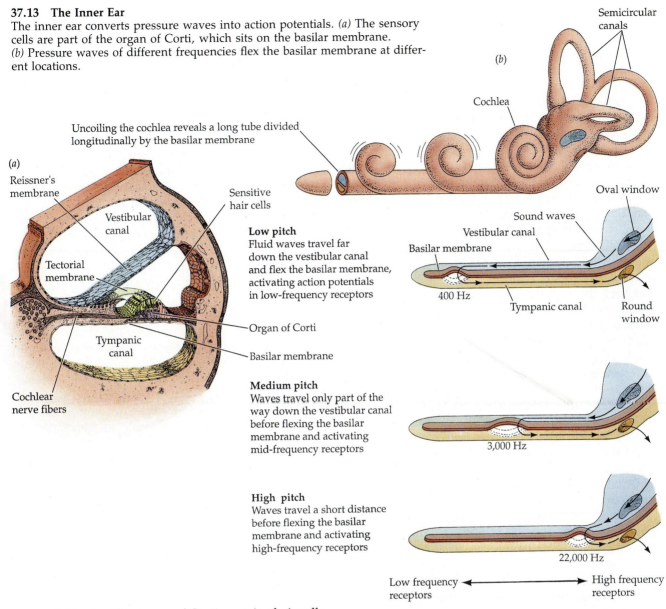

Uncoiling the cochlea reveals a long tube divided longitudinally by the basilar membrane

(a)

Reissner's membrane

Vestibular canal

Tectorial membrane

Tympanic canal

Cochlear nerve fibers

Sensitive hair cells

Organ of Corti

Basilar membrane

Low pitch
Fluid waves travel far down the vestibular canal and flex the basilar membrane, activating action potentials in low-frequency receptors

Medium pitch
Waves travel only part of the way down the vestibular canal before flexing the basilar membrane and activating mid-frequency receptors

High pitch
Waves travel a short distance before flexing the basilar membrane and activating high-frequency receptors

(b)

Semicircular canals

Cochlea

Oval window

Sound waves

Vestibular canal

Basilar membrane

400 Hz

Tympanic canal

Round window

3,000 Hz

22,000 Hz

Low frequency receptors ⟷ High frequency receptors

ear to the brain. The organ of Corti contains hair cells whose stereocilia are in contact with an overhanging, rigid shelf called the tectorial membrane. Whenever the basilar membrane flexes, the tectorial membrane bends the hair cell stereocilia. As a consequence, the hair cells depolarize or hyperpolarize and alter the rate of action potentials transmitted to the brain by their associated sensory neurons. What causes the basilar membrane to flex, and how does this mechanism distinguish sounds of different frequencies? In Figure 37.13*b*, the cochlea is uncoiled to make it easier to understand its structure and function. Also, to simplify matters, we have left out Reissner's membrane, thus combining the upper and the middle chambers into one functional chamber. The purpose of Reissner's membrane is to contain a specific aqueous environment for the organ of Corti separate from the aqueous environment in the rest of the cochlea. This is important for the nutrition of the sensitive organ of Corti, but it does not play a role

in the transduction of sound waves. The simplified model of the cochlea shown in Figure 37.13*b* reveals two additional features that are important for its function. First, the upper and lower chambers separated by the basilar membrane are joined at the distal end of the cochlea, making one continuous chamber that folds back on itself. Second, just as the oval window is a flexible membrane at the beginning of the cochlea, the **round window** is a flexible membrane at the end of the long cochlear chamber.

Air is highly compressible, but fluids are not. Therefore, a sound pressure wave can travel through air without much displacement of the air, but a sound pressure wave in fluid causes displacement of the fluid. Imagine someone holding a screen-door spring slightly stretched between her two hands. You could grab the spring in the center and move it back and forth without moving its ends—it is compressible. Now imagine the person holding a broomstick in the same way. If you grab its middle and move it back and forth, obviously its ends will move too—the broomstick is incompressible.

How does this comparison of springs and broomsticks relate to the inner ear? When the stapes pushes the oval window in, the fluid in the upper chamber of the cochlea is displaced. Think about what happens if the oval window moves in very slowly. The cochlear fluid displacement travels down the upper chamber, round the bend, and back through the lower chamber. At the end of the lower chamber the displacement is absorbed by the round window membrane bulging outward. Now what happens if the oval window vibrates in and out rapidly? The waves of fluid displacement do not have enough time to travel all the way to the end of the upper chamber and back through the lower chamber. Instead, they take a shortcut by crossing the basilar membrane, causing it to flex. The more rapid the vibration, the closer to the oval and round windows the wave of displacement will flex the basilar membrane. Thus, different pitches of sound will flex the basilar membrane at different locations and activate different sets of hair cells. The ability of the basilar membrane to respond to vibrations of different frequencies is enhanced by its structure. Near the oval and round windows, at the proximal end, it is narrow and stiff, but it gradually gets wider and more flexible toward its distal end. So it is easier for the proximal basilar membrane to resonate with high frequencies and for the distal basilar membrane to resonate with lower frequencies. A complex sound made up of many frequencies will distort the basilar membrane at many places simultaneously and activate a unique subset of hair cells.

Action potentials generated by the mechanosensors at different places along the organ of Corti travel to the brain stem along the auditory nerve. The auditory pathways make several synapses in the brain stem and send off collateral fibers to the reticular activating system. That is why sudden or loud noises wake us up or get our attention. Eventually the auditory pathways reach the temporal lobes of the cerebral cortex. In the primary auditory cortex are tone maps analogous to the maps of the body found in the primary somatosensory and the primary motor cortices. High-frequency sounds are represented at one end of this patch of cortex and low-frequency sounds are represented at the other. Surrounding the primary auditory cortex are the areas of association cortex that process auditory input, interpret it, and integrate the information with inputs from other senses and from memory.

Deafness is the loss of the sense of hearing, and it has two general causes. One type of deafness, known as conduction deafness, is due to the loss of

function of the tympanic membrane and the ossicles of the middle ear. Repeated infections of the middle ear can cause scarring of the tympanic membrane and stiffening of the connections between the ossicles. The consequence is less efficient conduction of sound waves from the tympanic membrane to the oval window. With increasing age there is a normal progressive stiffening of the ossicles that results in a gradual loss of the ability to hear high-frequency sounds. The other type of deafness is caused by damage to the inner ear or the auditory pathways; it is therefore termed nerve deafness. A common cause of nerve deafness is damage to the hair cells of the delicate organ of Corti by exposure to loud sounds such as jet engines, pneumatic drills, or highly amplified rock music. This damage is cumulative and permanent.

PHOTOSENSORS AND VISUAL SYSTEMS

Sensitivity to light (photosensitivity) confers upon the simplest animals the ability to orient to the sun and sky and gives more complex animals instantaneous and extremely detailed information about objects in the environment. It is not surprising that simple and complex animals can sense and respond to light. What is remarkable is that evolution has conserved the same basis for photosensitivity across the whole range of animal species—the molecule rhodopsin. In this section we will learn how rhodopsin responds when stimulated by light energy, how that response is transduced into neural signals, and how the brains of vertebrates process those signals to result in the sense of vision. We will also examine the structures of eyes, the organs that gather and focus light energy onto photosensitive cells. Last, we will learn how the brain uses action potentials from the retina to create our mental image of the visual world.

Rhodopsin

Photosensitivity depends on the ability of a molecule to absorb photons of light and to respond by changing its conformation. The molecule that does this in the eyes of all animals is **rhodopsin** (Figure 37.14). Rhodopsin consists of a protein, **opsin**, which by itself does not absorb light, and a light-absorbing group, **11-*cis* retinal**. The light-absorbing group is cradled in the center of the opsin, and the entire rhodopsin molecule sits in the plasma membrane of a photosensor cell. When the 11-*cis* retinal absorbs a photon of light energy, its shape changes so that it becomes a different isomer of retinal—all-*trans* retinal. This conformational change puts a strain on the bonds between retinal and opsin, and the two components break apart. This disassociation of retinal

(a)   Rhodopsin molecule

Disk membrane
of rod cell

11-*cis* retinal
group

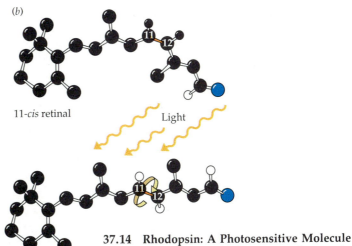

(b)

11-*cis* retinal

Light

All-*trans* retinal

37.14   Rhodopsin: A Photosensitive Molecule
Rhodopsin is a transmembrane protein—opsin —containing a light-responsive group, 11-*cis* retinal. When 11-*cis* retinal absorbs a photon of light energy, it changes shape, becoming all-*trans* retinal, which is not responsive to light. The molecule will return spontaneously to the 11-*cis* conformation.

and opsin is referred to as bleaching, and it results in the molecule losing its photosensitivity. The retinal spontaneously returns to its 11-*cis* isomer and recombines with opsin to become, once again, photosensitive rhodopsin.

The rhodopsin molecule sits in the membrane of a photosensitive cell. How does this molecule communicate to the cell that it has absorbed a photon? And how does the photosensor cell communicate to the nervous system that its rhodopsin molecules are receiving light? To answer these questions, we must see how photosensitive cells respond to light. A good example of a photosensitive cell is a vertebrate photoreceptor, a **rod cell**, which is a modified neuron (Figure 37.15). A dense layer of photoreceptor calls at the back of the eye forms the **retina**, which, as we will see below, is the structure that transduces the visual world into the language of the nervous system.

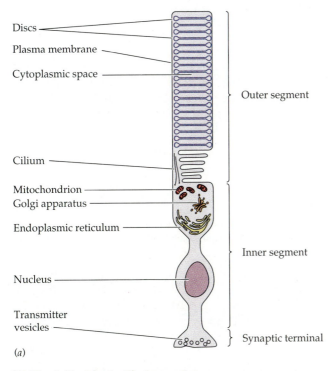

Discs
Plasma membrane
Cytoplasmic space

Outer segment

Cilium

Mitochondrion
Golgi apparatus

Endoplasmic reticulum

Inner segment

Nucleus

Transmitter
vesicles

Synaptic terminal

(a)

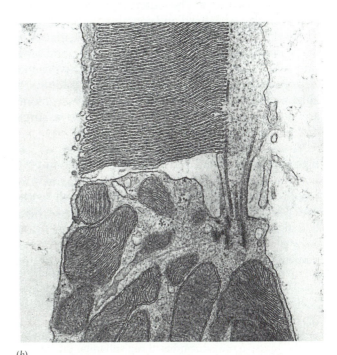

(b)

37.15   A Vertebrate Photoreceptor
(a) The rod cell of the vertebrate retina is a neuron modified for photosensitivity. The membranes of a rod cell's disks are densely packed with rhodopsin. (b) A transmission electron micrograph of a section through the cone of a squirrel's eye shows some of the cell's fine structure drawn in (a).

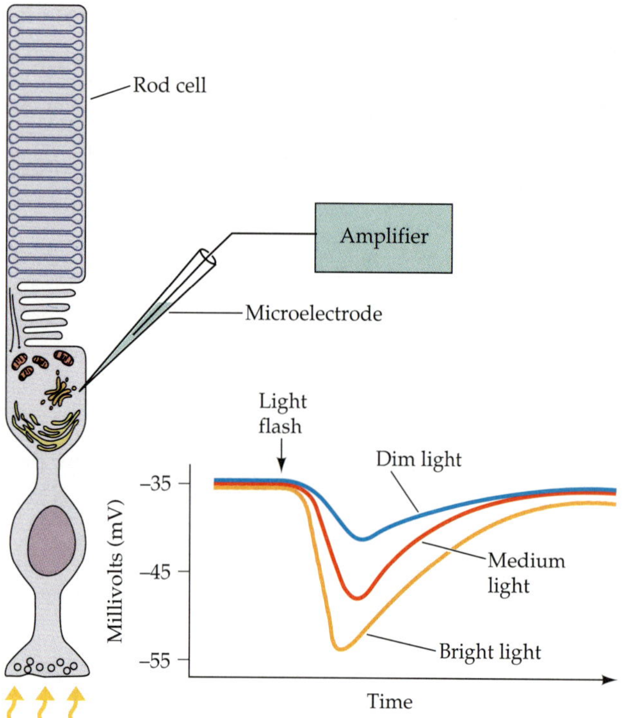

Rod cell

Amplifier

Microelectrode

Light flash

Dim light

Medium light

Bright light

Time

37.16 A Rod Cell Responds to Light
The receptor potential of a rod cell hyperpolarizes—becomes more negative—in response to a flash of light.

Light

A single rod cell from the retina has an inner segment that contains the usual organelles of a cell and has a synaptic terminal at its base where it communicates with other neurons. The outer segment of the rod cell is highly specialized and contains a stack of discs. These discs form at the base of the outer segment by invagination and pinching off of the cell membrane. The disc membranes are densely packed with rhodopsin and their function is to capture photons of light passing through the rod cell. Outer segments are periodically shed and regenerated, so a single disc has a life of about a month.

To see how the rod cell responds to light, we can take a piece of retina and penetrate a single rod cell with an electrode. Through this electrode, we can record the receptor potential of the rod cell in the dark and in the light (Figure 37.16). The first surprise is that the resting potential of a rod cell kept in the dark is not very negative in comparison with other neurons. In fact, the plasma membrane of the rod

37.17 Molecular Events of Photoreception
A key player in these events is a G protein called transducin. Inactive transducin has three subunits, one of which binds GDP. Activated rhodopsin causes this GDP to be replaced by GTP, and the transducin molecule splits. The subunit with GTP can activate a phosphodiesterase molecule. Active phosphodiesterase degrades cGMP and causes Na$^+$ channels to close.

1. Rhodopsin in rod cell disc is excited by light.

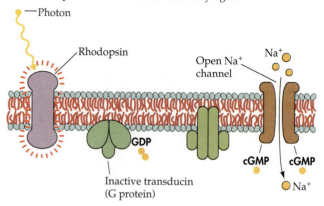

2. Excited rhodopsin activates ~500 transducin molecules.

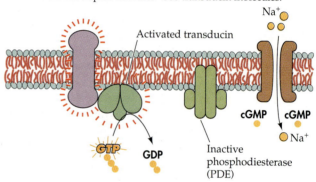

3. Activated transducin activates PDE which begins hydrolyzing cGMP.

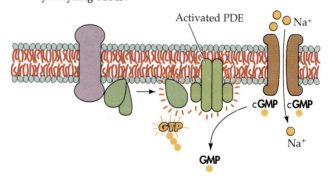

4. PDE hydrolizes up to 4000 molecules of cGMP. Na$^+$ channels close when they no longer bind cGMP.

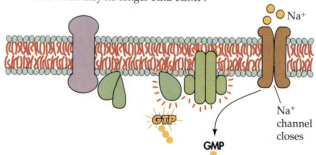

One photon of light can result in closing of up to **one million** Na$^+$ channels!

cell is fairly permeable to Na$^+$ ions, so these positive charges are continually entering the cell. The second surprise is that when a light is flashed onto this piece of retina, the receptor potential of the rod cell becomes more negative—it hyperpolarizes. The rod cell itself does not generate action potentials.

How does the absorption of light by rhodopsin hyperpolarize the rod cell? When rhodopsin is excited by light, it initiates a cascade of events (Figure 37.17). The photoexcited rhodopsin combines with and activates another protein, a G protein called transducin. Activated transducin in turn activates a phosphodiesterase. Active phosphodiesterase converts cyclic GMP to 5'-GMP. What was that cyclic GMP doing before it was converted to 5'-GMP? It was holding open the sodium channels and keeping the cell depolarized. As cyclic GMP is destroyed, the sodium channels close, and the cell hyperpolarizes. This may seem like a roundabout way of doing business, but its significance is its enormous amplification ability. Each molecule of photoexcited rhodopsin can activate about 500 transducin molecules, thus activating about 500 phosphodiesterase molecules. The catalytic prowess of a molecule of phosphodiesterase is great; it can hydrolyze over 4,000 molecules of cyclic GMP per second. The bottom line is that a single photon of light can result in the closing of over a million sodium channels resulting in a change in the rod cell's receptor potential. Now let us see how photosensors work in animals.

Visual Systems of Invertebrates

Planarians, simple multicellular animals, obtain directional information about light from photosensitive cells organized into eye cups (Figure 37.18). The eye cups are bilateral structures, and each is partially shielded from light by a layer of pigmented cells lining the cup. Since the openings of the eye cups face in opposite directions, the photoreceptors on the two side of the animal are unequally stimulated unless the animal is facing directly toward or away from a light source. The planarian uses directional information about light sources to move away from light, thereby becoming less visible to predators that hunt visually.

Arthropods (such as crustaceans, spiders, and insects) have evolved **compound eyes** that provide them with information about patterns or images in the environment. Each compound eye consists of many optical units called **ommatidia** (Figure 37.19). The number of ommatidia in a compound eye varies from only a few in some ants, to 800 in fruit flies, to 10,000 in some dragonflies. Each ommatidium has a lens structure that directs light onto photosensitive cells called retinula cells. Fruit flies have seven elongated retinula cells in each ommatidium. The inner borders of the retinula cells are covered with microvilli that contain rhodopsin and constitute a light trap. Since the microvilli of the different retinula cells overlap, they appear to form a central rod, called a rhabdom, down the center of the ommatidium. A single nerve fiber leaves each ommatidium. Since each ommatidium of a compound eye is directed at a slightly different part of the visual world, only a crude or perhaps a broken-up image of the visual field can be communicated from the compound eye to the central nervous system.

Vertebrate and Cephalopod Eyes

Vertebrates and cephalopod mollusks both have evolved eyes with exceptional abilities to form images of the visual world. These eyes operate like cameras, and considering that they evolved independently of each other, their high degree of analogy is remarkable (Figure 37.20). The vertebrate eye is a spherical, fluid-filled structure bounded by a tough connective tissue layer called the sclera. At the front of the eyeball, the sclera forms the transparent **cornea** through which light enters the eye. Just inside the cornea is the **iris**

37.18 A Simple Photosensory System Although planarians do not "see" as we understand it, this flatworm's eye cups enable it to move away from a light source to an area where it may be less visible to predators.

Planarian responds to light by moving directly away from the source toward darkness

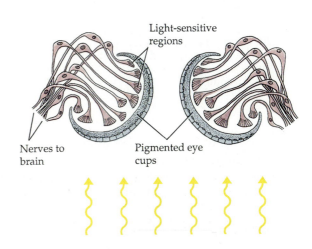

Light-sensitive regions

Nerves to brain

Pigmented eye cups

(a)

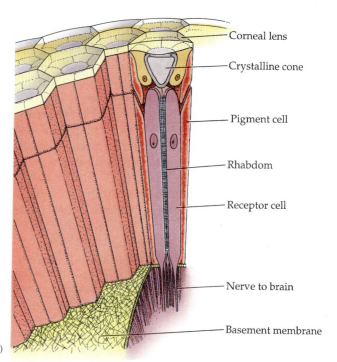

Corneal lens

Crystalline cone

Pigment cell

Rhabdom

Receptor cell

Nerve to brain

Basement membrane

(b)

37.19 Eye of an Insect
(a) The compound eye of this hoverfly contains hundreds of ommatidia. (b) Each ommatidium focuses light on a rhabdom consisting of interdigitating, light-sensitive membranes of a small number of receptor cells.

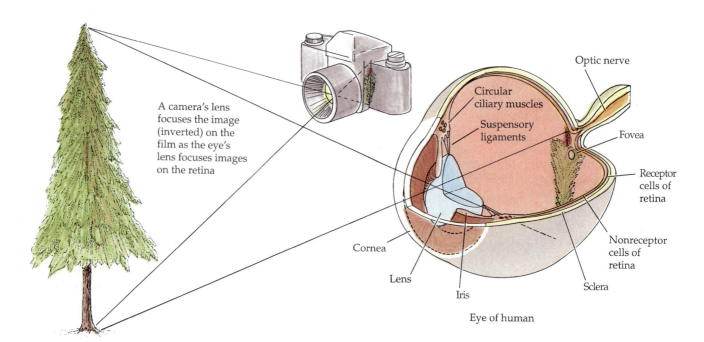

A camera's lens focuses the image (inverted) on the film as the eye's lens focuses images on the retina

Optic nerve

Circular ciliary muscles

Suspensory ligaments

Fovea

Receptor cells of retina

Nonreceptor cells of retina

Sclera

Cornea

Lens

Iris

Eye of human

37.20 Eyes Like Cameras
Lenses of cephalopod and vertebrate eyes focus images on layers of receptor cells, just as a camera lens focuses images on film.

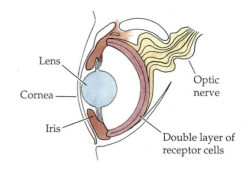

Lens

Cornea

Iris

Optic nerve

Double layer of receptor cells

Eye of squid

with its central aperture, the **pupil**. The iris is pigmented and gives the eye its color, but its important function is to control the amount of light reaching the photosensitive tissue at the back of the eyeball. The iris is under control of the autonomic nervous system. In bright light the iris constricts and the pupil

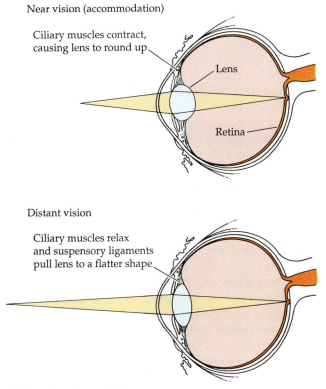

Near vision (accommodation)

Ciliary muscles contract, causing lens to round up

Lens

Retina

Distant vision

Ciliary muscles relax and suspensory ligaments pull lens to a flatter shape

37.21 Coming into Focus
Humans focus their eyes on close objects by changing the shape of the lens.

is very small, but as light levels fall, the iris relaxes and the pupil enlarges.

Behind the iris is the crystalline lens, which helps to focus images on the photosensitive layer, the retina, at the back of the eye. The cornea and the fluids of the eye chambers also help to focus light on the retina, but the lens is responsible for the ability to accommodate—to focus on objects at various locations in the near visual field. To focus a camera on objects close at hand, you must adjust the distance between the lens and the film. Fishes, amphibians, and reptiles accommodate in a similar manner, moving the lenses of their eyes closer to or farther from their retinas. Mammals and birds use a different method; they alter the shape of the lens. The lens is contained in a connective tissue sheath that tends to keep it in a spherical shape. The lens is suspended by suspensory ligaments, however, that pull it into a flatter shape. Circular muscles called the ciliary muscles counteract the pull of the suspensory ligaments and permit the lens to round up. With the ciliary muscles at rest, the flatter lens has the correct optical properties to focus distant images on the retina, but not close images. Contracting the ciliary muscles rounds up the lens, changing its light-bending properties to bring close images into focus (Figure 37.21). As we get older, our lenses become less elastic, and we lose the ability to focus on objects close at hand without the help of corrective lenses. Pro-

longed concentration on small, close objects (such as the type on these pages) tires and strains the eyes by overworking the ciliary muscles.

The Vertebrate Retina

The retina is really an extension of the brain. During development, neural tissue grows out from the brain to form the retina. In addition to a layer of photoreceptor cells, the retina includes other layers of cells that process the visual information from the photoreceptors and transmit it to the brain in the form of action potentials in the optic nerves. We will examine these functions of the retina in some detail, but first let us describe its features in general.

The density of photoreceptors is not the same across the entire retina. Light coming from the center of the field of vision falls on an area of the retina called the **fovea,** where the density of receptor cells is the highest. The human fovea has about 160,000 receptor cells per square millimeter. A hawk has about 1,000,000 receptor cells per square millimeter of fovea, giving it a visual acuity about eight times greater than ours. In addition, the hawk has two foveas in each eye. One fovea receives light from straight ahead and the other receives light from below. Thus, while the hawk is flying, it sees both its projected flight path and the ground below where it might detect a mouse scurrying in the grass. The horse has an interesting fovea. It is a vertical patch of retina with high receptor density. The horse's lens is not good at accommodation, but it focuses distant objects that are straight ahead on one part of this long fovea and close objects that are below the head on another part of the fovea. When horses are startled by an object close at hand, they pull their heads back and rear up to bring the object into focus on the close-vision fovea.

Where blood vessels and the bundle of axons going to the brain pass through the back of the eye, there is a blind spot on the retina. You are normally not aware of your blind spot, but you can find it. Stare straight ahead, holding a pencil in your outstretched hand so that the eraser is in the center of your field of vision. While continuing to stare straight ahead, slowly move the pencil to the side until the eraser disappears. When this happens, the light from the eraser is focused directly on your blind spot.

Until now we have referred to only one kind of photoreceptor, the rod cell. There are, however, two major kinds of photoreceptors, both named for their shapes—rod cells and **cone cells** (Figure 37.22). A human retina has about 3 million cones and about 100 million rods. Rod cells have a higher sensitivity to light, but do not contribute to color vision. Cones are responsible for color vision, but they are less sensitive to light. Cones are also responsible for our highest acuity of vision. Even though there are many

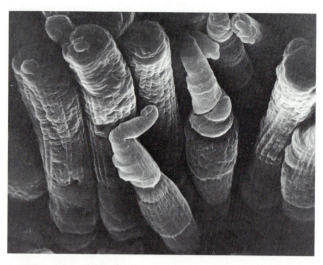

37.22 Rods and Cones
This scanning electron micrograph of photoreceptor cells in the retina of a mudpuppy, an amphibian, shows cylindrical rods and tapered cones.

more rods than cones in human retinas, our foveas contain mostly cones. Thus, cones give us our greatest visual acuity, but because they have low sensitivity to light, they are of no use at night. Our visual acuity is low at night, and we see only in black and white. You may have trouble seeing a small object such as a keyhole at night when you are looking straight at it—that is, when its image is falling on your fovea. If you look a little to the side, so that the image falls on a rod-rich area of retina, you can see the object better. Astronomers looking for faint objects in the sky learned this trick a long time ago. Animals that are nocturnal (such as flying squirrels) may have only rods in their retinas and have no color vision. In contrast, some animals that are strictly diurnal (such as chipmunks and ground squirrels) have only cones in their retinas.

How do cone cells enable us to see color? There are at least three kinds of cone cells, each possessing slightly different types of opsin molecules. Because different cone cells have different opsin molecules, they differ in the wavelengths of light they absorb best. Although the retinal group is the light absorber, its molecular interactions with opsin tune its spectral sensitivity. Some opsins cause retinal to absorb most efficiently in the blue region, some in the green, and some in the red (Figure 37.23). Intermediate wavelengths of light excite these classes of cones in different proportions. Recently the genes for the different opsins of humans were identified: one for blue-sensitive opsin, one for red-sensitive opsin, and several for green-sensitive opsin.

The human retina is organized into five layers of neurons that process visual information before it is sent to the brain (Figure 37.24). Strangely enough, the layer of photoreceptors is all the way at the back

of the retina, so that light has to traverse all of the other layers before reaching the rods and cones. The outer segments of the rods and cones are partly buried in a layer of pigmented epithelium that absorbs photons not captured by rhodopsin and prevents any backscattering of light that might decrease visual acuity. In contrast, nocturnal animals such as cats have a highly reflective layer, called the tapetum, behind the photoreceptors. Photons not captured on their first pass through the photoreceptors are reflected back, thus increasing visual sensitivity (but not acuity) in low-light conditions. The reflective tapetum is what makes cats' eyes appear to glow in the dark.

A first step in investigating how the human retina processes visual information is to study how the five layers of neurons in the retina are interconnected and how they influence one another. The cells in the outermost layer are photoreceptors. As we know, the photoreceptors hyperpolarize in response to light, but they do not generate action potentials. The cells in the innermost layer are ganglion cells. Ganglion cells do fire action potentials, and their axons form the optic nerves that travel to the brain. The photoreceptors and ganglion cells are connected by bipolar cells. As the membrane potentials of rods and cones alter in response to light, the photoreceptor cells alter the rate at which they release neurotransmitter at their synapses with the bipolar cells. Like rods and cones, bipolar cells do not fire action potentials. In response to neurotransmitter from the photoreceptors, the membrane potentials of bipolar cells change, which alters the rate at which they release neurotransmitter onto ganglion cells. The ganglion cells generate action potentials, and the rate of neurotransmitter release from the bipolar cells determines the rate at which they fire action potentials. So, the direct flow of information in the retina is from pho-

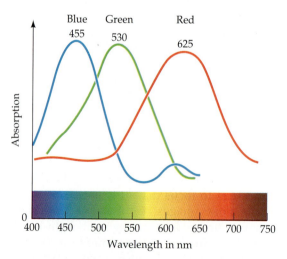

37.23 Absorption Spectra of Cone Cells
Human color vision is based on three kinds of cone cells with different spectral properties.

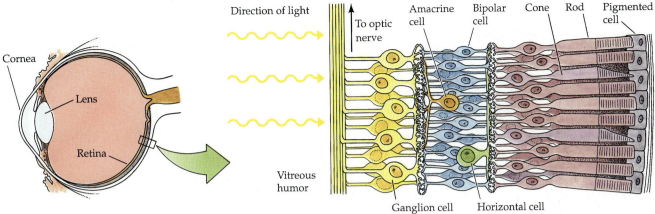

37.24 The Retina
The retina processes visual information through a network of five layers of neurons: ganglion cells, amacrine cells, horizontal cells, bipolar cells, and photoreceptors (rods and cones).

toreceptor to bipolar cell to the ganglion cell. Ganglion cells send the information to the brain.

What do the other two layers, the horizontal cells and the amacrine cells, do? They communicate laterally across the retina. Horizontal cells form connections between neighboring photoreceptor–bipolar cell combinations. Thus the communication between a photoreceptor and its bipolar cell can be influenced by neighboring photoreceptors. One reason for this lateral flow of information is to sharpen the perception of contrast between light and dark patterns falling on the retina. Amacrine cells form connections between neighboring bipolar cell–ganglion cell pairs. One role of amacrine cells is to adjust the sensitivity of the eyes according to the overall level of light falling on the retina.

Knowing the paths of information flow through the retina still doesn't tell us how that information is processed. What does the eye tell the brain in response to a pattern of light falling on the retina? One aspect of information processing in the retina is information reduction. There are over 100 million photoreceptor cells in each retina, but only about 1 million ganglion cells sending messages to the brain. By what rules is the information from all those photoreceptors reduced to the messages sent to the brain by the ganglion cells? This question was addressed through some elegant, classic experiments in which scientists used electrodes to record the activity of single ganglion cells in living animals while their retinas were stimulated with spots of light (Figure 37.25). They found that each ganglion cell has a well-defined **receptive field**—a specific group of photoreceptors that, when stimulated with light, has a strong influence on the activity of that ganglion cell. The receptive fields of ganglion cells are all circular. However, the way a spot of light influences the activity of the ganglion cell depends on where in the

receptive field it falls. Each ganglion cell's receptive field is divided into two concentric areas, called a center and a surround. If stimulation of the center excites the ganglion cell, stimulation of the surround inhibits it. This kind of receptive field is called an on-center receptive field. In an off-center receptive field, the situation is reversed; stimulation of the center inhibits the ganglion cell, and stimulation of the surround excites it. Center effects are always stronger than surround effects. The response of a ganglion cell to stimulation of the center of its receptive field depends on how much of the surround area is also stimulated. A small dot of light directly on the center has the maximal effect, a bar of light hitting the center and pieces of the surround has less of an effect, and a large, uniform patch of light falling equally on center and surround has no effect. Ganglion cells communicate information about light–dark contrasts falling on their receptive fields to the brain.

How are receptive fields related to the connections between the cells of the retina? The photoreceptors in the center of a ganglion cell's receptive field are connected to that ganglion cell by bipolar cells. The photoreceptors in the surround send information to the center photoreceptors and thus to the ganglion cell through the lateral connections of horizontal cells. Thus, the receptive field of a ganglion cell is due to a pattern of synapses between photoreceptors, horizontal cells, bipolar cells, and ganglion cells. The receptive fields of neighboring ganglion cells can overlap greatly; a given photoreceptor can be connected to several ganglion cells.

The eye sends the brain simple messages about the pattern of light intensities falling on small, circular patches of retina. How does the brain create a mental image of the visual world from this information? David Hubel and Torsten Wiesel of Harvard University tackled this question in the 1960s. Information from the retina is transmitted through the optic nerves to the brain's visual processing area, in the occipital cortex at the back of the cerebral hemisphere. Hubel and Weisel recorded the activities of

(a) **Experimental design**

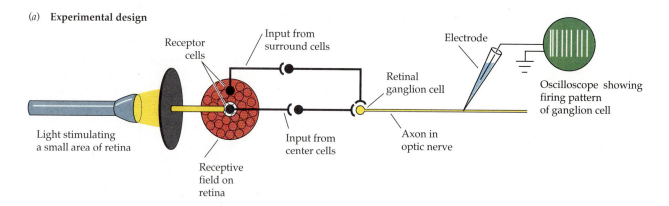

(b) **Results**

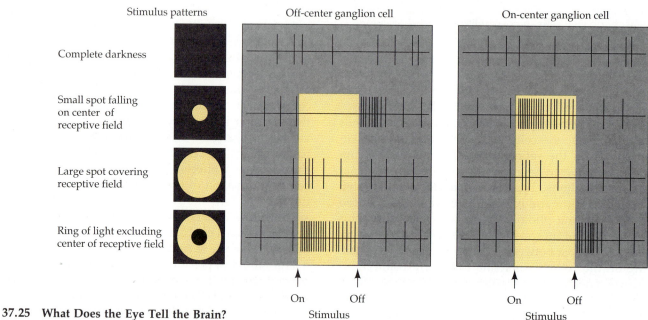

37.25 What Does the Eye Tell the Brain?
When the retina is stimulated with dots and rings of light, individual ganglion cells show different response properties. Each ganglion cell is spontaneously active and responds to light shining on a small circular area of retina in its receptive field. Some ganglion cells are stimulated and some are inhibited by a spot of light falling on the receptive field centers. A ganglion cell with an "on-center" is inhibited by a ring of light falling on the peripheral area of its receptive field. The opposite is true of ganglion cells with "off-centers." Cells may show a transient response when the stimulus is turned off.

single cells in the visual processing areas of the brains of living animals while they stimulated the animals' retinas with spots and bars of light. They found that cells in the visual cortex, like ganglion cells, had receptive fields—specific areas of the retina that when stimulated by light would influence the rate at which the cells fired action potentials (Figure 37.26). Cells in the visual cortex showed different types of receptive fields. One type of cells, called **simple cells**, was maximally stimulated by a bar of light with a certain orientation falling on one place on the retina. So, simple cells probably receive input that comes from a number of ganglion cells that have their circular receptive fields lined up in a row. Another type of cortical cells, called **complex cells**, was also maxi-

mally stimulated by a bar of light with a particular orientation, but the bar could fall anywhere on a large area of retina described as that cell's receptive field. Complex cells seem to receive input from a number of simple cells that share a certain stimulus orientation but that have receptive fields in different places on the retina. Some complex cells responded best when the bar of light was moving in a particular direction. This could possibly be due to certain combinations of on-center and off-center receptive fields.

The concept that emerges from these experiments is that the brain assembles a mental image of the visual world by analyzing edges of patterns of light falling on the retina. This analysis is conducted in a massively parallel fashion. Each retina sends 1 mil-

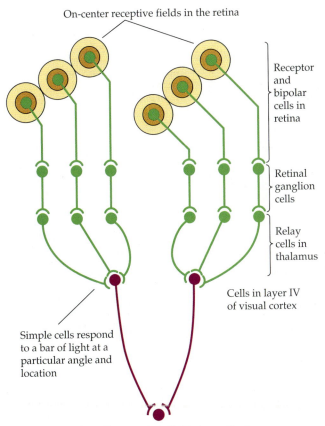

On-center receptive fields in the retina

Receptor and bipolar cells in retina

Retinal ganglion cells

Relay cells in thalamus

Cells in layer IV of visual cortex

Simple cells respond to a bar of light at a particular angle and location

Complex cell responds to light at a particular angle at any location on the retina, or to a bar of light moving across the retina

37.26 Receptive Fields
Cells in the visual cortex respond to specific patterns of light falling on the retina. Ganglion cells project information about circular receptive fields to simple cells in the cortex in such a way that simple cells have linear receptive fields. Simple cells project to complex cells in such a way that complex cells respond to a linear stimulus falling on different areas of the retina.

lion axons to the brain, but there are at least 200 million neurons in the visual cortex. Each bit of information from a retinal ganglion cell is received by hundreds of cortical cells, each responsive to a different combination of orientation, position, and even movement of contrasting lines in the pattern of light falling on the retina.

Binocular Vision

How do we see objects in three dimensions? The quick answer is because our two eyes see overlapping, yet slightly different, fields. Turn a typical conical flowerpot upside down and look at it so that the bottom of the pot is exactly in the center of your overall field of vision. You see the bottom of the pot, and you see equal amounts of the sides and rim of the pot as concentric circles around the bottom. Now close one eye and then the other. With your left eye

closed you see more of the right side and right rim of the pot. With your right eye closed you see more of the left side and left rim of the pot. The discrepancies in the information coming from your two eyes are interpreted by the brain to provide information about the depth and the three-dimensional shape of the flower pot. If you are blind in one eye, you have great difficulty in discriminating distances. Animals whose eyes are on the sides of their heads have nonoverlapping fields of vision and, as a result, poor depth vision.

The story of how the brain integrates information from two eyes begins with the paths of the optic nerves. If you look at the underside of the brain, the optic nerves from the two eyes would appear to join together just under the hypothalamus and then separate again. The place where they join is called the optic chiasm. Axons from the half of each retina closest to your nose cross in the optic chiasm and go to the opposite side of your brain. This means that all visual information from your left visual field (everything left of straight ahead) goes to the right side of your brain, as shown in red in Figure 37.27. All visual information from your right visual field goes to the left side of your brain, as indicated in green in the figure. Both eyes transmit information about a specific spot in your right visual field, for example, to the same place in the left visual cortex. How are the two sources of information integrated?

Cells in the visual cortex are organized in columns. These columns alternate: left eye, right eye, left eye, right eye, and so on. Cells closest to the border between two columns receive input from both eyes and are therefore **binocular cells**. Binocular cells interpret distance by measuring the disparity between where the same stimulus falls on the two retinas. What is disparity? Hold your finger out in front of you and look at it, closing one eye and then the other. Your finger appears to jump back and forth because its image falls on a different position on each retina. Repeat the exercise with an object at a distance. It doesn't appear to jump back and forth as much because there is less disparity in the positions of the image on the two retinas. Certain binocular cells respond optimally to a stimulus falling on both retinas with a particular disparity. Which set of binocular cells is stimulated depends on how far away the stimulus is.

When we look at something we can detect shape, color, depth, and movement. Where does all this information come together? Is there a single cell that only fires when a red sports car drives by? Probably not. A specific visual experience most likely comes from simultaneous activity in a large, but not random, collection of cells. Its complexity is even greater in that a visual experience is not strictly visual—it is enhanced by information from the other senses and from memory as well.

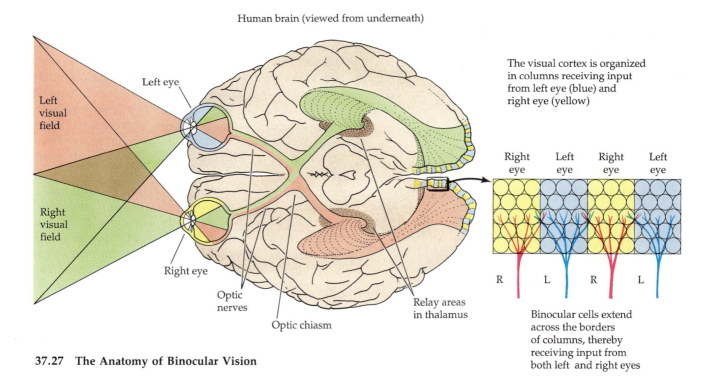

Human brain (viewed from underneath)

The visual cortex is organized in columns receiving input from left eye (blue) and right eye (yellow)

Left eye

Left visual field

Right visual field

Right eye

Optic nerves

Optic chiasm

Relay areas in thalamus

Right eye | Left eye | Right eye | Left eye

R   L   R   L

Binocular cells extend across the borders of columns, thereby receiving input from both left and right eyes

37.27   The Anatomy of Binocular Vision

OTHER SENSORY WORLDS

After emphasizing the incredible neural complexity of sensory integration, we must recognize that humans make use of only a subset of the information available to us in the environment. Other animals have sensory systems that enable them to use different subsets and different types of information.

Infrared and Ultraviolet Detection

When discussing vision, we use the term "the visible spectrum," but what we really mean is light visible to us. The visible spectrum is a very narrow region of the entire, continuous range of electromagnetic radiation in the environment (see Figure 8.6). For example, we cannot see ultraviolet radiation, but many insects can. One of the seven photosensitive cells in each ommatidium of a fruit fly is sensitive to ultraviolet light. The visual sensitivity of many pollinating insects includes the ultraviolet part of the spectrum. Some flowers have patterns that are invisible to us but that show up if we photograph them with film that is sensitive to ultraviolet light (see Figure 46.22). Those patterns provide information to prospective pollinators, but humans are not equipped to receive that information.

At the other end of the spectrum is infrared radiation, which we sense as heat. Other animals extract much more information from the infrared radiation —especially infrared radiation emitted by potential

prey. Pit vipers such as rattlesnakes have bilateral pit organs just in front of their eyes that can sense infrared radiation (Figure 37.28). In total darkness these snakes can locate a prey item such as a mouse, orient to it, and strike it with great accuracy based on the directional information they derive from the warmth of the mouse's body.

37.28   Pit Organs in a Rattlesnake
The western rattlesnake, *Crotalus atrox*, is a pit viper. The "holes" that appear in front of its eyes are pit organs that sense infrared radiation. Such snakes can locate prey in total darkness based on directional information they receive through these organs.

37.29 In the Dark
Using echolocation this bat avoids an obstacle course of thin wires while flying in complete darkness.

Echolocation

A number of species emit intense sounds and create images of their environments from the echoes of those sounds. Bats, porpoises, dolphins, and, to a lesser extent, whales, are accomplished echolocators. Some species of bats have elaborate modifications of their noses to direct the sounds they emit, and they also have impressive ear pinnae to collect the returning echoes. The sounds they emit as pulses (about 20 to 80 per second) are above our range of hearing, but they are extremely loud in contrast to their faint echoes that bounce off small insects. The situation of the echolocating bat is similar to that of a construction worker trying to overhear a whispered conversation on a street corner while using a pneumatic drill. To avoid deafening themselves, bats use muscles in their middle ears to dampen their sensitivity while they emit sounds, then relax them quickly enough to hear the echoes. The ability of bats to use echolocation to "see" their environment is so good that in a totally dark room strung with fine wires, bats can capture tiny flying insects while navigating around the wires (Figure 37.29).

Detection of Electric Fields

We have discussed the mechanoreceptors in the lateral lines of fishes. The lateral lines of some species, especially ones such as catfish that live in murky waters, also contain electroreceptors. These receptors enable the fish to detect weak electrical fields generated by potential prey. The use of electroreceptors is even more sophisticated in species called electric fishes. These animals have evolved electric organs in their tails that generate a continuous series of electrical pulses, creating a weak electrical field around the fish (Figure 37.30). Any objects in the environ-

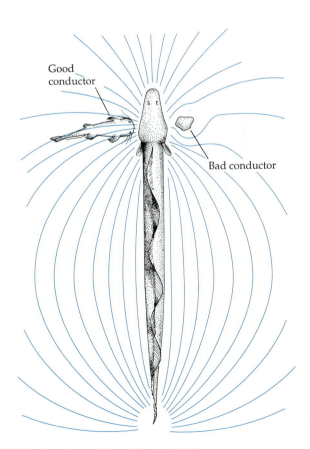

37.30 Sensing with an Electric Field
Fish such as the electric eel generate personal electric fields and sense perturbations within their fields. Both animate and inanimate objects disrupt the electric field and are detected by the eel.

ment, such as rocks, logs, plants, or other fish, disrupt the electric fish's electrical field, and the electroreceptors of the lateral line detect those disruptions.

Magnetic Sense

As astonishing as this may seem, the magnetic lines of force of the planet provide directional information for some animals. From experiments with homing pigeons, biologists have accumulated evidence that this is so. Normally, homing pigeons released from locations distant from their homes can orient and find their way home, even if they are fitted with translucent contact lenses that eliminate all directional information from the sun. However, pigeons that are released on cloudy days with tiny magnets glued to their heads become disoriented and cannot navigate. The magnetic sense is poorly understood at present, but single magnetosensory neurons have been discovered.

SUMMARY

Sensory systems gather information about an animal's external and internal environments and transmit it to the nervous system. Sensory cells and organs transduce a variety of stimuli into changes in receptor potentials, which subsequently alter the rates at which action potentials are transmitted through sensory nerves to the central nervous system. Most sensory systems can adapt to constant levels of stimulation.

Chemosensors have receptor molecules that enable them to respond to the presence of specific substances in the environment. The binding of a receptor molecule with a stimulating molecule leads to intracellular events that alter ion channels in the cell membrane, resulting in changes in receptor potential. Olfaction, taste, and responses to pheromones are examples of chemosensitivity.

The skin contains several types of mechanoreceptors that respond to distortion of their membranes by generating action potentials. Muscle spindles contain stretch receptors that provide information about the position and movements of the body. The sensitivity of these stretch receptors is controlled by the central nervous system through motor neurons. Hair cells have stereocilia that alter the receptor potential of the cell when they are bent. The lateral lines of fishes, equilibrium organs, and auditory systems use hair cells to transduce pressure into action potentials.

Photosensitivity throughout the animal kingdom depends on rhodopsin's capturing photons of light. In photoreceptor cells, changes in rhodopsin molecules are translated into changes in receptor potential. Vertebrate photoreceptors are rod cells, which are responsible for dim light and black-and-white vision, and cone cells, which are responsible for color vision.

The vertebrate eye focuses light on the retina, which transduces photons of light into action potentials sent by ganglion cells to the visual cortex of the brain. Ganglion cells have circular receptive fields with concentric centers and surrounds. Cells in the visual cortex integrate the input derived from retinal ganglion cells into a mental image by analyzing patterns of contrasting and moving edges of light.

Many animals have sensory worlds that we do not share. For example, bats echolocate, insects see ultraviolet radiation, pit vipers sense infrared radiation, fish feel electrical fields, and homing pigeons can orient by sensing Earth's magnetic fields.

SELF-QUIZ

1. Which of the following statements is *false*?
   a. Sensory transduction in vertebrate sensory systems involves the conversion (direct or indirect) of a physical or chemical stimulus into nerve impulses.
   b. In general, a stimulus causes a change in the flow of ions across the plasma membrane of a sensory receptor.
   c. The term adaptation is given to the process by which a sensory system becomes insensitive to a continuing source of stimulation.
   d. The more intense a stimulus, the greater the magnitude of each action potential fired by a receptor.
   e. Sensory adaptation plays a role in the ability of organisms to discriminate between important and unimportant information.

2. The female silkworm moth releases a chemical called bombykol from a gland at the tip of her abdomen. Bombykol is:
   a. a sex hormone.
   b. detected by the male only when present in large quantities.
   c. not species-specific.
   d. detected by hairs on the antennae of male silkworm moths.
   e. a chemical basic to the taste process in arthropods.

3. Which of the following statements is *false*?
   a. Dogs are unusual among mammals in depending more upon olfaction than vision as their dominant sensory modality.
   b. Olfactory stimuli are recognized by the interaction between the stimulus and a specific macromolecule on olfactory hairs.
   c. The more odorant molecules binding to receptors, the more action potentials are generated.
   d. The greater the number of action potentials generated by an olfactory receptor, the greater the intensity of the perceived smell.
   e. The perception of different smells results from the activation of different combinations of olfactory receptors.

4. The touch receptors that are located very close to the skin surface are:
   a. relatively insensitive to light touch.
   b. very quick to adapt to stimuli.
   c. uniformly distributed throughout the surface of the body.
   d. called Pacinian corpuscles.
   e. adapt slowly and only partially to stimuli.

5. The membrane that gives us the ability to discriminate different pitches of sound is the:
   a. round window.
   b. oval window.
   c. tympanic membrane.
   d. tectorial membrane.
   e. basilar membrane.

6. Which of the following statements is *false*?
   a. A rod cell's transmembrane potential becomes more negative when the rod cell is exposed to light after a period of darkness.
   b. A photoreceptor releases the most neurotransmitter (per unit time) when in total darkness.
   c. Whereas in vision the intensity of a stimulus is encoded by the degree of hyperpolarization of photoreceptor cells, in audition the intensity of a stimulus is encoded by changes in firing rates of sensory cells.
   d. Stiffening of the ossicles in the middle ear can lead to deafness.
   e. The interaction between hammer (malleus), anvil (incus), and stirrup (stapes) conducts sound waves across the fluid-filled middle ear.

7. In primates, the region of the retina where the central part of the visual field falls is called the:
   a. central ganglion cell.
   b. fovea.
   c. optic nerve.
   d. cornea.
   e. pupil.

8. The region of the vertebrate eye where the optic nerve passes out of the retina is called the:
   a. fovea.
   b. iris.
   c. blind spot.
   d. pupil.
   e. optic chiasm.

9. Which of the following statements about the cones in a human eye is *false*?
   a. They are responsible for high acuity vision.
   b. They encode for color vision.
   c. They are more sensitive to light than rods.
   d. They are fewer in number than rods.
   e. They exist in high numbers at the fovea.

10. The color in vision results from the:
    a. ability of each cone to absorb all wavelengths of light equally.
    b. lens of the eye acting like a prism and separating the different wavelengths by light.
    c. different absorption of wavelengths of light by different classes of rods.
    d. three different isomers of retinal in different classes of cone cells.
    e. absorption of different wavelengths of light by amacrine and horizontal cells.

FOR STUDY

1. Drawing upon your knowledge of the structure and function of the human auditory system, describe how the brain is able to distinguish sounds of different frequencies.

2. Describe the molecular mechanisms whereby a single photon is able to cause hyperpolarization of a rod cell.

3. Compare and contrast the functioning of olfactory receptors and photoreceptors. What is the basis whereby each system discriminates between an apple and an orange?

4. Describe and contrast two sensory systems that enable animals to "see" in the dark. What problems or limitations are inherent in these systems in comparison with vision?

5. Describe what is meant by a receptor potential and how it functions to encode intensity of stimulus. Use a specific sensor as your example.

READINGS

Camhi, J. M. 1984. *Neuroethology: Nerve Cells and the Natural Behavior of Animals.* Sinauer Associates, Sunderland, MA. This is a particularly good text for putting basic neurophysiology in the context of whole animals and their behaviors. Part II deals with the sensory worlds of animals.

Hubel, D. H. 1988. *Eye, Brain, and Vision.* Scientific American Library Series No. 22. W. H. Freeman, New York. A comprehensive and beautifully illustrated book about the neurophysiology and neuroanatomy of vision. It is very readable for the nonexpert, yet it presents the depth and breadth of knowledge and experience of someone who has been a major contributor to this area of research.

Hudspeth, A. J. 1983. "The Hair Cells of the Inner Ear." *Scientific American,* January. The inner ear transduces mechanical forces of pressure waves into action potentials transmitted to the brain. This paper describes the cells that accomplish the transduction and explain how they do it.

Knudsen, E. I. 1981. "The Hearing of the Barn Owl." *Scientific American,* 1981. The barn owl can use its remarkably precise and sensitive auditory system to locate prey in complete darkness. This paper explains the neurophysiological basis for its extreme accuracy.

Newman, E. A. and P. H. Hartline. 1982. "The Infrared 'Vision' of Snakes." *Scientific American,* March. Some snakes are able to use infrared radiation emitted by objects in their environment to construct a sensory world.

Stryer, L. 1987. "The Molecules of Visual Excitation." *Scientific American,* July. This paper describes the molecular chain of events that transduce photons of light falling on the retina into action potentials that are transmitted to the brain.

Suga, N. 1990. "Biosonar and Neural Computation in Bats." *Scientific American,* June 1990. The use of echolocation by bats to construct a view of their environment requires complex neural processing of information by their auditory systems.

38

Effectors

PREVIEW: Effectors enable animals to respond to sensory information in various ways: by moving their bodies, or by emitting chemicals, sound, light, or electrical energy. Within cells, organelles called microtubules and microfilaments use the energy of ATP to generate forces needed for movement. Muscle cells contract by means of a system in which molecules of two proteins, actin and myosin, slide past each other. Skeletons provide protection, support, and systems of levers necessary to translate the contractions of muscle cells into movements of the animal and its appendages.

This chapter deals with cilia and flagella, actin, myosin, tropomyosin, troponin, muscle cells, hydrostatic skeletons, exoskeletons, endoskeletons, cartilage, bone, joints, ligaments, and tendons.

Information obtained from sensors is not of much value to an animal unless the animal can do something in response. The mechanisms an animal uses to respond—internally or externally—are its **effectors**. This broad definition includes the internal organs and organ systems that the animal uses to control its internal environment; these effectors are the subjects of subsequent chapters. In this chapter we will focus on the mechanisms of creating mechanical forces and using those forces to change shape and to move—which are the basis for just about all of animal behavior. A fish swims, an earthworm crawls, a mosquito flies, and a kangaroo jumps because cells move.

Two cellular structures, microtubules and microfilaments (introduced in Chapter 4), create cell movement. Both of these structures cause movement by the sliding of long protein molecules past one another. Microtubules generate the small movements of cilia and flagella. The microfilament system reaches its highest level of organization in muscle cells, which can generate large-scale movements.

Muscles come in a wide variety of types with different characteristics. The catch muscles of clams and the sphincter muscles of our excretory systems can generate long, continuous, sustained contractions. The flight muscles of insects and hummingbirds can cycle through contraction and relaxation hundreds of times a second. In this chapter we will examine the structure, function, and neural control of different muscle types. Muscles alone, however, would be just a quivering, amorphous mass of tissue without the rigid support of skeletal systems. Skeletons enable animals to use the contractile forces of muscles to do specific tasks.

CILIA, FLAGELLA, AND MICROTUBULES

Ciliated Cells

Most animal species have some cells with tiny, hairlike appendages called **cilia**. Each cilium is tiny, about 0.25 μm in diameter, but they occur in dense patches. Most animals use ciliated cells to move liquids and particles over cell surfaces. Many invertebrates use ciliated cells to obtain food and oxygen. Cilia circulate a current of water across the gill surfaces of some mollusks, for example. Oxygen diffuses across the gill membranes, and food, consisting of tiny organisms and detritus, is filtered from the water by the netlike gills and ingested (Figure 38.1). Cilia around the mouths of rotifers simply sweep microorganisms and detritus directly into the gut.

The airways of many animals are lined with and cleaned by ciliated cells (Figure 38.2). The cilia continuously sweep a layer of mucus from deep down in our lungs, up through the windpipe, and into the throat. The mucus carries particles of dirt and dead cells. We can then either swallow or spit out the mucus, and with it the trapped detritus. Ciliated cells lining the female reproductive tract create currents that sweep eggs from the ovaries into the oviducts and all the way down to the uterus, as you may remember from Chapter 35.

A cilium pushes against a liquid with the same basic motion as a swimmer's arms during the breast-

(a)

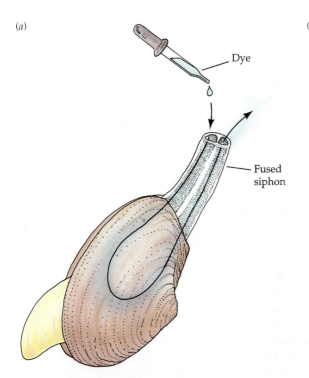

Dye

Fused
siphon

(b)

38.1 Cilia Create Water Currents
(a) In burrowing mollusks such as this clam, cilia lining
the siphons maintain a unidirectional flow of water: in
one siphon, over the gills, and out the other siphon. The
gills extract oxygen and food from this flow of water. (b)
Only the clam's siphons protrude in this photograph; the
rest of the animal is completely buried in the substrate.

stroke (Figure 38.3a). During the **power stroke**, the
cilium projects stiffly outward and moves backward
through the liquid, propelling the cell forward (or the
liquid backward). During the **recovery stroke**, the
cilium folds as it returns to its original position. The
power stroke is fast and the recovery stroke is slow.
Because the resistance encountered by an object moving through a fluid is proportional to the square of
its velocity, the resistance of the medium to the recovery stroke is slight compared with its resistance
to the power stroke. Fluids exposed to the beating of

38.2 Cilia Line Respiratory Passages
A scanning electron micrograph of a rabbit's airway.

(a) Movement of cilium

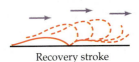

Power stroke   Recovery stroke

(b) Movement of flagellum

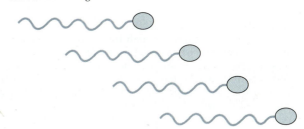

38.3 Cilia and Flagella Move Differently
(a) A cilium moves in a pattern similar to the arms of a
swimmer doing the breaststroke. (b) Flagella are much
longer than cilia. A flagellum moves in an undulating,
whiplike pattern.

cilia are thus propelled in the direction of the power stroke. Cilia typically beat in coordinated waves. At any particular moment, some cilia of a cell are moving through the power stroke and others are recovering.

Cells with Flagella

The **flagella** of eukaryotes are essentially identical to cilia except that they are longer and occur singly or in groups of only a few on any one cell. Flagellated cells maintain a flow of water through the bodies of sponges, bringing in food and oxygen and removing carbon dioxide and wastes. Flagella power the movement of the sperm of most species. Because of their greater length, flagella have a whiplike stroke pattern rather than the swimming-like stroke pattern of cilia (Figure 38.3b).

How Cilia and Flagella Move

The central structure of a cilium or a flagellum is called the **axoneme**. It contains a ring of nine pairs of microtubules, as we learned in Chapter 4. In the center of the ring may be one additional pair of microtubules, a single microtubule, or no microtubule (see Figure 4.27). Microtubules are hollow tubes formed from polymerization of the globular polypeptide **tubulin**. Other proteins in the axoneme form spokes, sidearms, and crosslinks (Figure 38.4a). Sidearms composed of the protein **dynein** are responsible for generating force. Dynein is a **mechanoenzyme** that catalyzes the hydrolysis of ATP and uses the released energy to change its orientation, thereby generating mechanical force. When the dynein arms on one microtubule pair contact a neighboring microtubule pair and bind to it, ATP is broken down, and the resulting conformational changes in the dynein molecules cause the arms to point downward, toward the base of the axoneme. This action pushes the first microtubule pair upward in relation to its neighbor. The dynein arms then detach from the neighboring pair and reorient to their starting horizontal position.

38.4 Microtubules and Motion
Cilia and flagella move through the actions of microtubules in a central structure called the axoneme. (a) A cross section of an axoneme. The microtubules occur in doublets that run the length of the axoneme. (b) Dynein arms of the doublets generate force by making and breaking crosslinks with the neighboring microtubule. (c) When microtubule pairs try to slide past each other, the axoneme bends because the microtubules are anchored together at the bottom. (d) If all links between microtubules except the dynein arms are eliminated, as in this experimental preparation, ATP causes the microtubules to slide past each other and the axoneme elongates enormously.

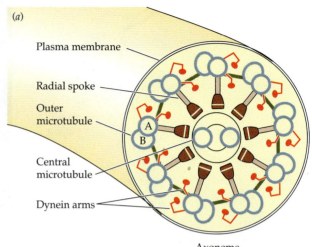

(a)

Plasma membrane

Radial spoke

Outer microtubule

A

B

Central microtubule

Dynein arms

Axoneme

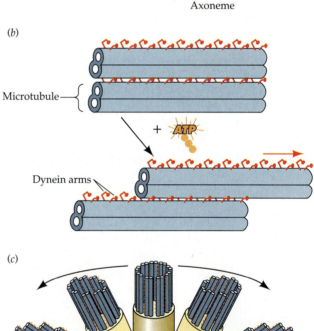

(b)

Microtubule

+ ATP

Dynein arms

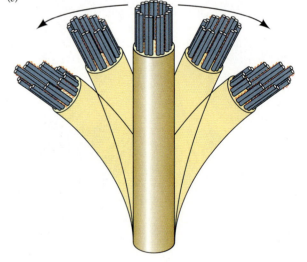

(c)

(d)

Telescoping

As the cycle is repeated, adjacent microtubule pairs try to "row" past each other, with the dynein side-arms acting as "oars" (Figure 38.4b). But since the microtubules are anchored at the bottom, the axoneme bends, instead of elongating, as the microtubule pairs slide past one another (Figure 38.4c). In ways not fully understood, the central microtubule and the other proteins that bind the axoneme together control the dynein action so all the microtubule pairs are not "rowing" at the same time.

Scientists have investigated the motile mechanisms of the axoneme by selectively removing its proteins. If axonemes are severed from cells, they will continue to flex in a normal pattern if exposed to Ca^{2+} and ATP, demonstrating that the motile mechanism is intrinsic to the axoneme itself. If the axoneme is gently treated with proteolytic enzymes, it is possible to disrupt the spokes and crosslinks and leave only the microtubules and dynein arms intact. If they are then exposed to Ca^{2+} and ATP, the microtubules row past one another and the whole structure elongates manyfold (Figure 38.4d), demonstrating that the forces moving microtubule pairs along one another are the basis for the bending of the intact axoneme. It is also possible to selectively extract the dynein from isolated axonemes. When this is done, they lose their ATPase activity and their motility. Putting purified dynein back restores ATPase activity and motility.

Microtubules as Intracellular Effectors

Microtubules play important roles in cell movements. As components of the cytoskeleton, microtubules contribute to the cell's shape. Cells can change shape and move by polymerizing and depolymerizing the tubulin in their microtubules. During mitosis, the spindle that moves chromosomes to the mitotic poles at anaphase forms by the polymerization of tubulin. Another example of microtubule involvement in cell movement is the growth of the processes of neurons in the developing nervous system. Neurons find and make their appropriate connections by sending out long processes that search for the correct contact cells. If polymerization of tubulin is chemically inhibited, the neurons do not extend processes. Microtubules are important intracellular effectors for changing cell shape, moving organelles, and enabling cells to respond to the environment.

MICROFILAMENTS AND CELL MOVEMENT

Protein **microfilaments**, like microtubules, are another basic mechanism for changing cell shape and causing cell movements. The dominant microfilament in cells is the protein **actin**, and bundles of cross-linked actin strands form important structural components of cells. For example, the microvilli that increase the absorptive surface area of the cells lining the gut are stiffened by internal bundles of actin (Figure 38.5a), and the stereocilia of the sensory hair

(a)

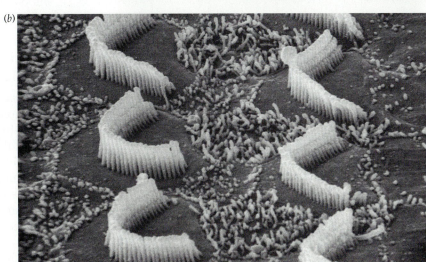
(b)

38.5 Microfilaments Support Cell Projections
(a) The cells lining the gut have numerous fingerlike projections. (b) Some mechanoreceptors, such as these auditory receptors in the organ of Corti, have stereocilia. Both these types of cell projections are stiffened by microfilaments.

cells described in Chapter 37 are stiffened by actin microfilaments (Figure 38.5 *b*). Like microtubules, actin microfilaments can change the shape of a cell simply by polymerizing and depolymerizing. An example of this action is the projections sent out by phagocytic cells, as was shown in Figure 16.2.

The most ubiquitous role of actin microfilaments is their generation, together with the protein myosin, of the contractile forces responsible for many aspects of cell locomotion and changes in cell shape. For example, the contractile ring that divides a mitosing cell into two daughter cells is composed of actin microfilaments in association with myosin. Another example is the mechanisms many cells employ to engulf materials—phagocytosis and pinocytosis, which were discussed in Chapter 4. Nets of actin and myosin beneath the cell membrane are responsible for the changes in cell shape involved in phagocytosis and pinocytosis, as well as in the locomotion of the whole cell. A third example is the role of actin and myosin microfilaments in the contractile mechanism of muscles.

Amoeboid Movement

Amoeboid movement occurs in a variety of cell types in multicellular animals. During development many cells migrate by amoeboid movement, and throughout life phagocytic cells (Chapter 16) circulate in the blood, squeeze through the walls of the blood vessels, and wander through the tissues by amoeboid movement. The mechanisms of amoeboid movement have been studied extensively in the protist for which this type of movement has been named: the amoeba, which lives in freshwater streams and ponds.

The amoeba moves by extending lobe-shaped projections called pseudopods and then seemingly squeezing itself into them (Figure 38.6). The cytoplasm in the core of the amoeba is relatively liquid and is called **plasmasol**, but just beneath its plasma membrane the cytoplasm is much thicker and is called **plasmagel**. To form a pseudopod, the thick plasmagel in a certain area of the cell thins, allowing a bulge to form. Just under the cell surface, in the plasmagel, there is a network of actin microfilaments that interacts with myosin to squeeze plasmasol into the bulge, thus forming a pseudopod. As the plasmagel continues to contract, cytoplasm streams in the direction of the pseudopod. Eventually the cytoplasm at the leading edge of the pseudopod converts to gel and the pseudopod stops forming. Thus the basis for amoeboid motion is the ability of the cytoplasm to cycle through sol and gel states and the ability of the microfilament network under the cell membrane to contract and cause the cytoplasmic streaming that pushes out a pseudopod.

MUSCLES

Muscle contraction is the most important effector mechanism animals have for responding to their environments. All behavioral and most physiological responses depend on muscle cells. Muscle cells are specialized for contraction, and they have high densities of microfilaments consisting mostly of actin and myosin. Such cells are found throughout the animal kingdom. They account for the thrashing movements of nematodes, the expansion–contraction movements of earthworms, the pulsating movements of jellyfish, and the limb movements of arthropods and vertebrates. Muscle cells are found in the walls of blood vessels, guts, bladders, and hearts. Wherever contraction of whole tissues takes place in animals, muscle cells are responsible. In all cases, the molecular mechanism of contraction is the same, but there are a large number of specializations of muscle cells fitting them to the wide variety of functions they serve. We begin our study of muscle by looking at the three types of muscle cells found in vertebrates (Figure 38.7): smooth muscle, skeletal muscle, and cardiac (heart) muscle.

Smooth Muscle

Smooth muscle provides the contractile forces for most of our internal organs that are under the control of the autonomic nervous system. Smooth muscle moves food through the digestive tract, controls the flow of blood through blood vessels, and empties the urinary bladder. Smooth muscle cells are the simplest muscle cells in structure. They are usually long and spindle-shaped, and each cell has a single nucleus. The microfilaments of smooth muscle are not as regularly arranged as those we will see in the other muscle types, so the contractile machinery is not obvious when the cells are viewed under the light microscope (Figure 38.7a). If we study smooth muscle

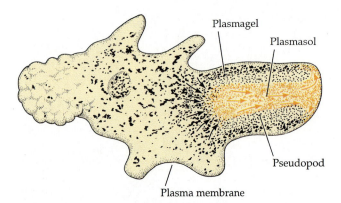

38.6 Amoeboid Movement
Cells can move by squeezing themselves into pseudopods. An amoeba (see Chapter 22) best illustrates this type of movement, which is used by many cells.

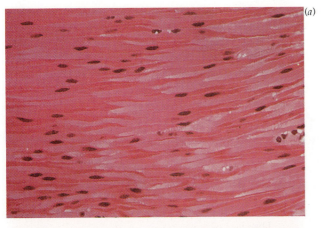

(a)

(b)

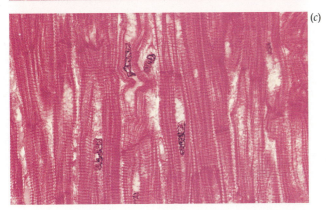

(c)

38.7 Muscle Tissue

(a) Smooth muscle cells are arranged in sheets; the dark structures are nuclei. Smooth muscle contracts when it is stretched; how does this adapt it for its role in lining the digestive tract? (b) Skeletal, or striated, muscle is characterized by a highly regular arrangement of microfilaments that results in the appearance of stripes, or bands (discussed at length in Figure 38.8). (c) Vertebrate cardiac muscle is also striated, but cardiac muscle fibers create a meshwork that resists tearing or breaking.

tion that is initiated. How membrane depolarization triggers contraction will be discussed later in this chapter.

Skeletal Muscle

Skeletal muscle is responsible for all voluntary movements, such as running or playing a piano, and it is also responsible for breathing. Skeletal muscle is called **striated muscle** because the highly regular arrangement of its contractile microfilaments gives it a striped appearance (Figure 38.7b). Skeletal muscle cells, or **muscle fibers**, are quite large, and they have many nuclei because they develop through the fusion of many individual cells. A muscle such as your biceps (which bends your arm) is composed of many muscle fibers.

What is the relation between a muscle fiber and the microfilaments responsible for contraction? Each muscle fiber is composed of bundles of microfilaments, called **myofibrils** (Figure 38.8). Within each myofibril, the microfilaments of actin and myosin are organized in an extremely regular way. First, if we look at a cross section of a myofibril, we can see that there are thick filaments—the myosin—and thin filaments—the actin. If we cut the myofibril at certain locations, we see only thick filaments, in other locations only thin filaments, but in most regions of the myofibril, each thick myosin filament is surrounded by six thin actin filaments.

If we next look at a longitudinal section of a myofibril, we can see clearly the striated appearance of skeletal muscle. The band pattern of the myofibril is due to repeating units called **sarcomeres**, which are the actual units of contraction (Figure 38.8a). Each sarcomere is made of overlapping filaments of actin and myosin. As the muscle contracts, the sarcomeres shorten, and the appearance of the band pattern changes.

The observation that the widths of the bands in the sarcomeres change when a muscle contracts led two British biologists, Hugh Huxley and Andrew Huxley, to propose the first formulation of the molecular mechanism of muscle contraction. Let us look at that band pattern in detail (Figure 38.8e). Each sarcomere is bounded by Z lines that serve as anchors

from a particular organ, such as the walls of the digestive tract, we find it has interesting properties, however. The cells are arranged in sheets, and individual cells in the sheets are in electrical contact with one another through gap junctions. As a result, an action potential generated in the membrane of one smooth muscle cell can spread to all of the cells in the sheet of tissue. Another interesting property of a smooth muscle cell is that the resting potential of its membrane is sensitive to being stretched. If the wall of the digestive tract is stretched in one location (such as by receiving a mouthful of food), the membranes of the stretched cells depolarize, reach threshold, and fire action potentials that cause the cells to contract. Thus, smooth muscle contracts when stretched, and the harder it is stretched, the stronger the contrac-

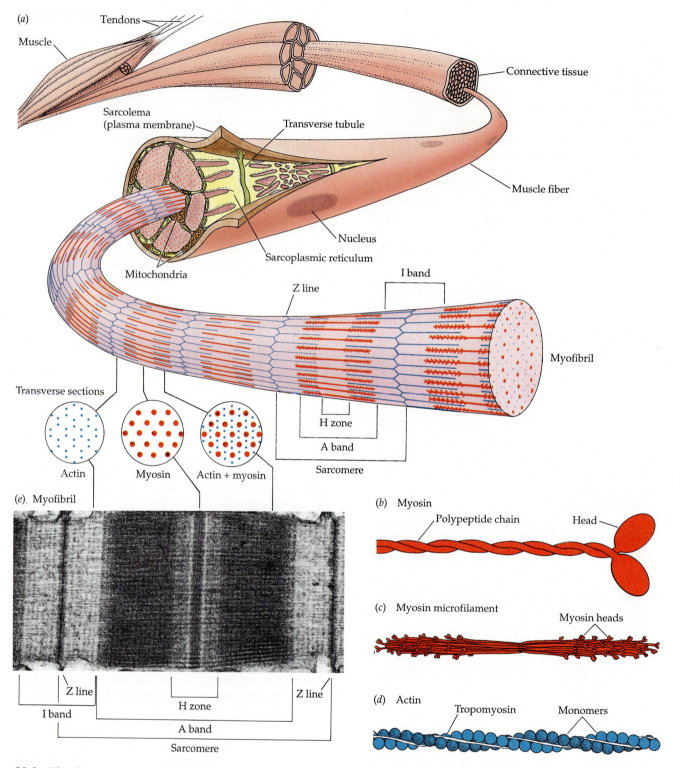

38.8 The Structure of Skeletal Muscle from Molecules to Tissue

(a) A skeletal muscle is made up of bundles of muscle fibers. Each fiber contains numerous myofibrils made up of microfilaments. Each muscle fiber is a multinucleate cell. The plasma membrane of the muscle cell is continuous with an internal system of transverse tubules. Within each muscle fiber the nuclei, the mitochondria, and the sarcoplasmic reticulum surround myofibrils, which are highly ordered assemblages of microfilaments. The two types of microfilaments are the thick ones, myosin, and the thin ones, actin. They partially overlap in the myofibril, and where they do, they surround each other.

Where there are only actin microfilaments, the myofibril appears light, and where they overlap, the myofibril appears dark. (b,c) Myosin microfilaments consist of bundles of molecules with long polypeptide tails and globular heads. (d) Actin microfilaments consist of two chains of monomers twisting around each other. Two polypeptide chains of tropomyosin twist around the actin chains. The formation of crossbridges between the globular myosin heads and the actin chains is the basis for muscle contraction. (e) The structure of the myofibrils gives muscle fibers their characteristic striated appearance.

for the thin actin filaments. In the center of the sarcomere is the band that corresponds to the myosin filaments. The H zone and the I band are light-appearing regions where actin and myosin filaments do not overlap in the relaxed muscle. When the muscle contracts, however, the sarcomere shortens. The H zone and I band become much narrower, and the A band (the myosin filaments) approaches the Z lines. This observation led Huxley and Huxley to propose the **sliding-filament theory** of muscle contraction: Actin and myosin filaments slide past each other as the muscle contracts.

To understand what makes the filaments slide, we must examine the structure of actin and myosin. The myosin molecule consists of two long polypeptide chains coiled together, each ending in a globular head (Figure 38.8b). The myosin filament is made up of many myosin molecules arranged in parallel with their heads projecting laterally from one or the other end of the filament (Figure 38.8c). The actin filament (Figure 38.8d) consists of a helical arrangement of two chains of monomers like two strands of pearls twisted together. The myosin heads have sites that can bind to actin and thereby form bridges between the myosin and the actin filaments. The myosin heads also have ATPase activity; they bind and hydrolyze ATP. The energy released changes the orientation of the myosin head.

Putting these details together explains the cycle of events that cause the actin and myosin filaments to slide past one another and shorten the sarcomere. A myosin head binds to an actin filament. Upon binding, the head changes its orientation with respect to the myosin filament, thus exerting a force causing the filaments to slide about 5 to 10 nanometers. Next the myosin head binds a molecule of ATP, which causes it to release the actin. When the ATP is hydrolyzed, the energy released causes the myosin head to return to its original conformation in which it can again bind to actin. An analogy is that the hydrolysis of the ATP is like cocking the hammer of a pistol, and binding with actin pulls the trigger.

An interesting aspect of this contractile mechanism is that ATP is needed to break the actin–myosin bonds, but not to form them. Thus muscles require ATP to stop contracting. This fact explains why muscles stiffen when animals die, a condition known as rigor mortis. After death, replenishment of the ATP stores of muscle cells stops, so the myosin–actin bridges cannot be broken, and the muscles stiffen. Eventually the proteins begin to lose their integrity and the muscles begin to soften. Because these events have regular time courses that differ somewhat for different regions of the body, an examination of the stiffness of the muscles of a corpse sometimes enables a coroner to estimate the time of death.

Controlling the Actin–Myosin Interaction

Muscle contractions are initiated by nerve action potentials arriving at the neuromuscular junction. Motor neurons are generally highly branched and can innervate up to 100 muscle fibers. All of the fibers innervated by a single motor neuron are a **motor unit** and contract simultaneously in response to that motor neuron firing action potentials. To understand the fine control the nervous system has over the sliding of actin and myosin filaments, we must examine the membrane system of the muscle fiber and some additional protein components of the actin microfilaments.

Vertebrate skeletal muscle fibers are excitable cells; when they are depolarized to a threshold that opens their voltage-gated sodium channels, their plasma membranes generate action potentials, just as the membranes of axons do. The initial depolarization that spreads across the muscle cell membrane is generated at the neuromuscular junction—the synapse between the motor neuron and the muscle cell membrane. As discussed in Chapter 36, neurotransmitter from the motor neuron binds to receptors in the postsynaptic membrane, causing ion channels to open. The depolarization of the postsynaptic membrane spreads to the surrounding plasma membrane of the muscle cell which contains voltage-gated ion channels. When threshold is reached, the plasma membrane fires an action potential that is rapidly conducted to all points on the surface of the cell.

The plasma membrane of the muscle fiber is continuous with a system of tubules that descends into and ramifies throughout the cytoplasm of the muscle fiber—also called the sarcoplasm (see Figure 38.8). These are the **T-tubules**, and they communicate with a network of membranes, called the **sarcoplasmic reticulum**, that surrounds every myofibril. The wave of depolarization that spreads over the plasma membrane of the muscle fiber also spreads throughout the T-tubule system. Junctions between the T-tubules and the sarcoplasmic reticulum are called **triads**, and they contain specialized protein molecules that open calcium channels in the sarcoplasmic reticulum. Ordinarily, calcium pumps in the membranes of the sarcoplasmic reticulum cause this membrane-bound compartment of the cell to take up and sequester Ca^{2+} ions. When a wave of depolarization reaches the triads, the calcium channels in the sarcoplasmic reticulum open, resulting in a massive diffusion of Ca^{2+} ions out of the sarcoplasmic reticulum and into the sarcoplasm surrounding the microfilaments. The Ca^{2+} stimulates the interaction of actin and myosin and the sliding of the filaments. How does it do so?

Remember that an actin microfilament is a helical arrangement of two strands of actin monomers.

Lying in the grooves between the two actin strands is another two-stranded protein, **tropomyosin** (Figure 38.9). At regular intervals the microfilament also includes another globular protein, **troponin**. The troponin molecule has three subunits; one binds actin, one binds tropomyosin, and one binds Ca^{2+}. When Ca^{2+} is sequestered in the sarcoplasmic reticulum, the tropomyosin strands block the sites where myosin heads can bind to the actin. When the T-tubule system depolarizes, Ca^{2+} is released into the sarcoplasm, where it binds to the troponin, changing the shape of the troponin molecule. Because the troponin is also bound to the tropomyosin, this conformational change of the troponin twists the tropomyosin enough to expose the actin–myosin binding

sites. This initiates the cycle of making and breaking of actin–myosin bridges; the microfilaments are pulled past one another, and the muscle fiber contracts. When the T-tubule system repolarizes, the calcium pumps remove the Ca^{2+} ions from the sarcoplasm, causing the tropomyosin to return to the position in which it blocks the binding of the myosin heads to the actin strands, and the muscle fiber returns to its resting condition.

Twitches, Graded Contractions, and Tonus

In vertebrate skeletal muscle, an action potential arriving at the neuromuscular junction is sufficient to cause an action potential in the muscle fiber. The

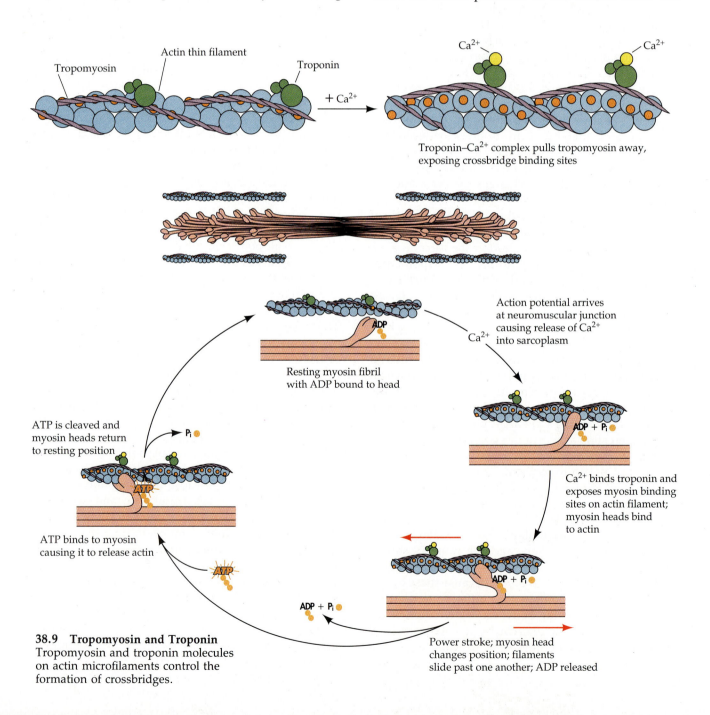

38.9 Tropomyosin and Troponin
Tropomyosin and troponin molecules on actin microfilaments control the formation of crossbridges.

action potential spreading throughout the membrane system of the muscle fiber will cause a minimum unit of contraction called a **twitch**. A twitch can be measured in terms of the tension it generates (Figure 38.10). If action potentials in the muscle fiber are adequately separated in time, each twitch is seen as a discrete, all-or-none phenomenon. If action potentials are fired at a more rapid rate, however, new twitches are triggered before the microfilaments have had a chance to return fully to their resting condition. As a result, the twitches sum and the tension generated by the fiber increases and becomes more continuous. Thus, the individual muscle fiber can show a graded response to increased levels of stimulation by its motor neuron. At high levels of stimulation, the calcium pumps in the sarcoplasmic reticulum can no longer remove Ca^{2+} ions from the sarcoplasm between action potentials, and the contractile machinery generates maximum tension—a condition known as **tetanus**. (This condition must not be confused with the disease tetanus, caused by a bacterial toxin and characterized by spastic contractions of skeletal muscles.) How long a muscle fiber can maintain a tetanic contraction depends on its supply of ATP. Eventually it will fatigue. This may seem paradoxical since binding with ATP breaks the actin–myosin bonds. But remember that the energy released from the hydrolysis of ATP "recocks" the myosin heads, allowing them to recycle through another power stroke. Think of the analogy of rowing a boat upstream. You cannot maintain your position relative to the stream bank by just holding the oars out against the current, you have to keep rowing.

The graded ability of a whole muscle to generate tension depends also on how many muscle fibers in that muscle are activated. Thus whether a contraction of a muscle is strong or weak depends both on how many motor neurons to that muscle are firing and on the rate at which those neurons are firing. Many muscles of the body maintain a low level of tension called **tonus** even when we are at rest. For example, the muscles of our neck, trunk, and limbs that maintain our posture against the pull of gravity are always working, even when we are standing or sitting still. Muscle tonus comes from the activity of a small but changing number of motor units in a muscle, so that at any one time some of the muscle's fibers are contracting and others are relaxed. Tonus is constantly being readjusted by the nervous system.

Fast and Slow Twitch Fibers

Not all skeletal muscle fibers are alike in their twitch characteristics, and in one muscle there may be more than one type of fiber. There are two major types of skeletal muscle fibers, called **slow twitch fibers** and **fast twitch fibers** (Figure 38.11a). Slow twitch fibers are also called red muscle because they have lots of

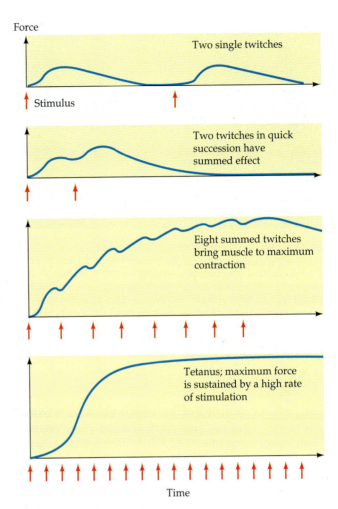

38.10 Twitches and Tetanus
A twitch is a minimum unit of contraction of a muscle. It can be elicited by a single, brief electrical pulse to the nerve innervating the muscle. Twitches that occur in rapid succession have a summed effect. Tetanus is the maximum state of tension that can be achieved in a muscle.

the oxygen-binding molecule myoglobin, they have lots of mitochondria, and they are well supplied with blood vessels. The tension developed from a single twitch in a slow twitch fiber is low, and its maximum tension is low and slow to develop, but it is highly resistant to fatigue. Slow twitch fibers have substantial reserves of glycogen and fat, so their abundant mitochondria can maintain a steady, prolonged production of ATP if oxygen is available. Muscles with high proportions of slow twitch fibers are good for long-term, aerobic work (that is, work that requires lots of oxygen). Champion long-distance runners, cross-country skiers, swimmers, and bicyclists have leg and arm muscles consisting mostly of slow twitch fibers (Figure 38.11b).

Fast twitch skeletal muscle fibers are also called white muscle because they have fewer mitochondria, little or no myoglobin, and fewer blood vessels than slow twitch fibers do. The white meat of domestic

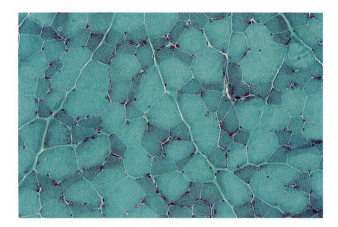

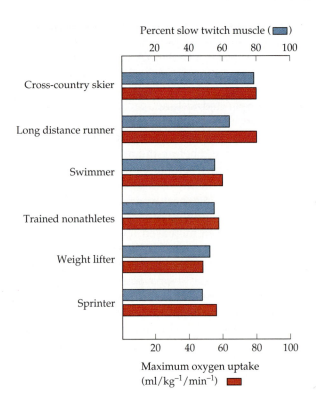

38.11 Two Types of Fibers
(a) Skeletal muscle consists of fast and slow twitch fibers. In this stained micrograph, which is a cross section of a skeletal muscle, slow twitch fibers appear dark and fast twitch fibers appear light. (b) World-class athletes in different sports have different distributions of fiber types. Slow twitch fibers are best adapted for sustained aerobic activity. Fast twitch fibers can generate maximum tension quickly, but they also fatigue quickly.

chickens is composed of fast twitch fibers. Fast twitch fibers can develop maximum tension more rapidly than slow twitch fibers, and that maximum is greater, but they fatigue rapidly. Their myosin has high ATPase activity, so they can put the energy of ATP to work very rapidly, but they cannot replenish it rapidly enough to sustain contraction for a long time. Fast twitch fibers are especially good for short-term work that requires maximum strength. Champion weight lifters and sprinters have leg and arm muscles with high proportions of fast twitch fibers.

What determines the proportion of fast and slow twitch fibers in your muscles? The most important factor is your genetic heritage, so there is some truth to the statement that champions are born, not made. To a certain extent, however, you can alter the properties of your muscle fibers through training. With aerobic training, the oxidative capacity of fast twitch fibers can improve substantially. But a person born with a high proportion of fast twitch fibers will never become a champion marathon runner, and a person born with a high proportion of slow twitch fibers will never become a champion high jumper.

Cardiac Muscle

The cardiac (heart) muscle of vertebrates is striated, but it differs from skeletal muscle in several ways. Cardiac muscle fibers branch to create a meshwork of contractile elements (see Figure 38.7c). Unlike skeletal muscles, which can easily be torn along the length of the muscle fibers, the meshwork of the cardiac muscle cannot be separated in this way. Therefore the heart walls can withstand high pres-

sures without danger of tearing or forming leaks. Unlike skeletal muscle fibers, cardiac muscle fibers are made up of individual, uninucleate cells. These cells are joined to one another by structures called **intercalated disks** which provide strong mechanical adhesion, and they also have gap junctions that present low resistance to ions or electrical currents. As a result, the cardiac muscle cells are in electrical continuity with one another, and a depolarization spreads rapidly throughout the walls of the heart.

The origin of the depolarization that triggers heart contraction is another interesting feature of cardiac muscle. Some heart muscle fibers are specialized for pacemaking function; they initiate the rhythmic contraction of the heart. This pacemaking function is due to a unique class of potassium ion channels found in cardiac muscle fibers. These channels tend to remain somewhat open following an action potential, but they gradually close. As they close, the cell becomes less negative and eventually reaches threshold, thereby initiating the next action potential (Figure 38.12). Because of the pacemaking property of cardiac muscle fibers, a heart removed from an animal continues to beat with no input from the nervous system; we say that the heartbeat is **myogenic**—generated by the muscle itself. The autonomic nervous system modifies the rate of the pacemaker cells, but is not essential for their continued, rhythmic function. The myogenic nature of the heartbeat is a major factor in making heart transplants possible, because the implanted heart does not depend on neural connections to beat.

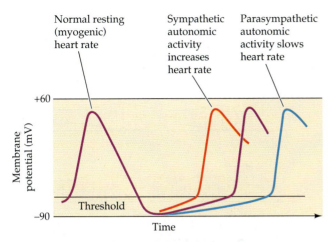

Normal resting (myogenic) heart rate

Sympathetic autonomic activity increases heart rate

Parasympathetic autonomic activity slows heart rate

+60

Membrane potential (mV)

Threshold

−90

Time

38.12 The Membrane Potentials of Pacemaker Cells
The resting potentials of pacemaker cells spontaneously depolarize to threshold and fire action potentials. The rate of spontaneous depolarization determines heart rate. Sympathetic stimulation increases the rate of depolarization and parasympathetic stimulation decreases the rate of depolarization of pacemaker cell membranes.

SKELETAL SYSTEMS

Muscles can only contract and relax, so without something rigid to pull against, muscles can do little more than lie in a formless mass that twitches and perhaps changes shape. **Skeletal systems** provide rigid supports against which muscles can pull and therefore create directed movements. Next, we will examine the three types of skeletal systems found in animals.

Hydrostatic Skeletons

The simplest type of skeleton is the **hydrostatic skeleton** of cnidarians, annelids, and many other soft-bodied invertebrates. It simply consists of a volume of incompressible fluid (water) enclosed in a body cavity surrounded by muscle. When muscles oriented in a certain direction contract, the fluid-filled body cavity bulges out in the opposite direction. An example of a hydrostatic skeleton is that of the sea anemone. Its body cavity is filled with seawater. To extend its body and its tentacles, it closes its mouth and constricts muscle fibers that are arranged in circles around its body. As the circular muscles contract, the liquid in the body cavity is put under pressure, and that pressure forces the body and tentacles to extend. If the anemone is alarmed, it retracts its tentacles and body by contracting muscle fibers that are arranged longitudinally in the body wall and in the long dimension of the tentacles.

The hydrostatic skeletons of some animals have become adapted for locomotion. An annelid such as the earthworm uses its hydrostatic skeleton to crawl.

The earthworm's body cavity is divided into many separate segments. The body wall has a muscle layer in which the muscle fibers are arranged in circles around the body cavity, and another muscle layer in which the muscle fibers run lengthwise. A closed compartment in each segment of the worm is filled with fluid. If the circular muscles in a segment contract, the compartment in that segment gets narrower and elongates. If the lengthwise, or longitudinal, muscles of a segment contract, the compartment gets shorter and bulges outward. Alternating contractions of the the circular and longitudinal muscles create waves of constriction and widening, lengthening and shortening, that travel down the body of the earthworm. The bulging, short segments serve as anchors as the narrowing, expanding segments project forward and longitudinal contractions pull other segments forward. Bristles help the widest parts of the body to hold firm against the substrate. By passing alternating waves of contraction and extension along its body, the earthworm can make fairly rapid progress through or over the soil (Figure 38.13).

An interesting adaptation of the hydrostatic skeleton for locomotion is the jet propulsion used by the squid and the octopus. Muscles surrounding a water-filled cavity in these cephalopods contract, putting the water under pressure and expelling it from the animal's body. As the water shoots out under pressure, it propels the animal in the opposite direction.

Exoskeletons

An **exoskeleton** is a hardened outer surface to which internal muscles can be attached. Contractions of those muscles can cause jointed segments of the exoskeleton to move relative to each other. The simplest example of an exoskeleton is the shell of a mollusk, which generally consists of just one or two pieces. Some marine bivalves and snails have shells composed of protein strengthened by crystals of calcium carbonate (a rock-hard material). These shells can be quite massive, affording significant protection against predators. The shells of land snails generally lack the hard mineral component and are much lighter. Molluscan shells can grow as the animal grows, and growth rings are usually quite apparent on the shells. The soft parts of the molluscan body have a hydrostatic skeleton as well. It is this second, hydrostatic skeleton that is used in locomotion, while the exoskeleton mainly provides protection. Some scallops, however, swim by opening their shells and snapping them shut—another version of jet propulsion.

The most complex exoskeletons are found in the phylum Arthropoda. Plates of exoskeleton cover all the outer surfaces of the arthropod's body and all its appendages. The plates are secreted by a layer of cells just below the exoskeleton. A continuous, layered waxy covering, the **cuticle**, is laid down over

(a)

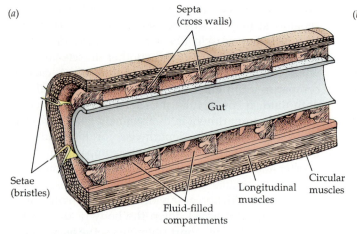

Septa
(cross walls)

Gut

Setae
(bristles)

Circular
muscles

Longitudinal
muscles

Fluid-filled
compartments

38.13 A Hydrostatic Skeleton and Locomotion
(a) The earthworm's hydrostatic skeleton consists of fluid-filled compartments separated by septa. (b) Contractions of circular muscles cause compartments (or segments) to elongate, and contractions of longitudinal muscles cause compartments (or segments) to shorten. Alternating waves of elongation and contraction move the earthworm through the soil. Bristles prevent parts of the worm from moving backward as waves of contractions pass by.

(b)

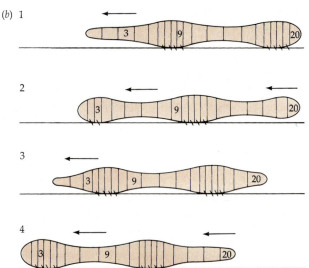

the entire body. Then stiffening materials are added everywhere but at the joints where flexibility must be retained. The layers of cuticle include a thin, waxy epicuticle that protects the bodies of terrestrial arthropods from drying out, and a thicker inner layer, the endocuticle, that forms the bulk of the structure. The endocuticle is a tough, pliable material found

only in arthropods. It consists of a complex of protein and chitin, a nitrogen-containing polysaccharide. In marine crustaceans, the endocuticle is further toughened by the addition of insoluble calcium salts. The thickness of the exoskeleton varies, so that a very efficient armor is formed. Muscles attached to the inner surfaces of the arthropod exoskeleton move its various parts around the joints (Figure 38.14).

An exoskeleton has the advantage of protecting all the soft tissues of the animal, but it is itself subject to damage such as abrasion and crushing. The greatest drawback of the arthropod exoskeleton is that it cannot grow. Therefore, if the animal is to become

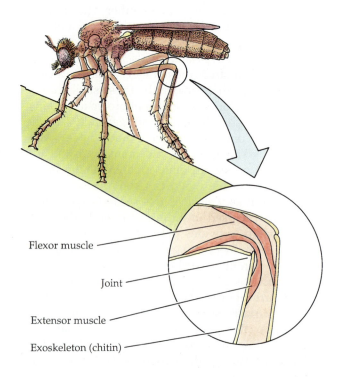

Flexor muscle

Joint

Extensor muscle

Exoskeleton (chitin)

38.14 An Insect's Exoskeleton
Muscles attached to the exoskeleton move parts around flexible joints. The insect on the right is a katydid.

larger, it must **molt**, shedding its exoskeleton and forming a new, larger one. This leads to a period of great vulnerability because the new exoskeleton takes time to harden. During this period the animal's body is temporarily unprotected, and it is unable to move rapidly, not having the firm exoskeleton against which its muscles can exert maximum tension. Soft-shelled crabs, a gourmet delicacy, are crabs caught at the time of year when molting occurs.

Endoskeletons of Vertebrates

The **endoskeleton** of vertebrates is an internal scaffolding to which the muscles attach. It is composed of rodlike, platelike, and tubelike bones, which are connected to each other at a variety of joints that allow a wide range of movements. The human skel-

eton consists of 206 bones (not all of which are shown in Figure 38.15), divided into an **axial skeleton**, which includes the skull, vertebral column, and ribs, and an **appendicular skeleton**, which includes the **pectoral girdle**, the **pelvic girdle**, and the bones of the arms, legs, hands, and feet. Endoskeletons do not provide the protection that exoskeletons do, but their real advantage is that bones can continue to grow. Because the bones are on the inside of the body, the body can enlarge without shedding its skeleton.

The endoskeleton consists of two kinds of connective tissue: cartilage and bone. Connective tissue cells produce large amounts of extracellular matrix material. The matrix material produced by cartilage cells is a rubbery mixture of proteins and polysaccharides. The principal protein in the matrix is collagen. Col-

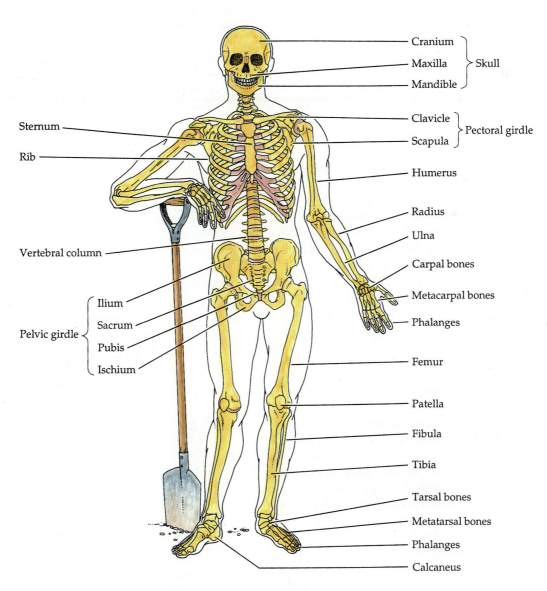

38.15 The Human Endoskeleton

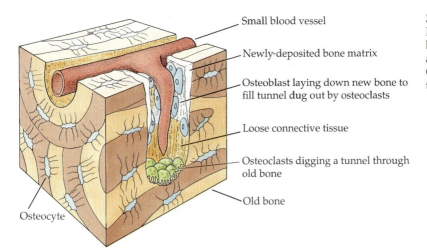

Small blood vessel

Newly-deposited bone matrix

Osteoblast laying down new bone to fill tunnel dug out by osteoclasts

Loose connective tissue

Osteoclasts digging a tunnel through old bone

Old bone

Osteocyte

38.16 Renovating the Substructure
Bones are constantly being remodeled by osteoblasts, which lay down bone, and osteoclasts, which dissolve bone. Osteocytes are osteoblasts that become trapped by their own handiwork.

lagen fibers run in all directions through the gel-like matrix and give it the well-known strength and resiliency of "gristle." Cartilage is found in parts of the endoskeleton where both stiffness and resiliency are required, such as on the articulating surfaces of joints where bones move against each other. Cartilage is also the supportive tissue in stiff but flexible structures such as the larynx (voice box), the nose, and the ears. The skeletons of sharks and rays are composed entirely of cartilage and therefore those animals are called the cartilaginous fishes. In all other vertebrates, cartilage is the principal component of the embryonic skeleton, but it is gradually replaced by bone over the course of development.

Bone consists mostly of extracellular matrix material that contains collagen fibers as well as crystals of insoluble calcium phosphate, which give bone its rigidity and hardness. The skeleton, in fact, serves as a reservoir of calcium for the rest of the body, and it is in dynamic equilibrium with soluble calcium in the extracellular fluids of the body. This equilibrium is under hormonal control by calcitonin and parathyroid hormone, as was described in Chapter 34. If too much calcium is taken from the skeleton, the bones are seriously weakened.

The living cells of bone are responsible for the dynamic remodeling of bone structure that is constantly under way. Osteoblasts lay down new matrix on bone surfaces. These cells gradually become surrounded by matrix and eventually find themselves enclosed within the bone. In this situation they cease laying down matrix, but they continue to exist within small lacunae (cavities) in the bone and are now called osteocytes. In spite of the vast amounts of matrix between them, osteocytes remain in contact with one another through long cellular processes that run through tiny channels in the bone. Communication between osteocytes is believed to be important in controlling the activities of the cells that are laying down new bone or eroding it away.

The cells that erode or reabsorb bone are the osteoclasts, and they are derived from the same cell lineage that produces the white blood cells. Osteoclasts burrow into bone, forming cavities and tunnels. Osteoblasts follow behind and deposit new bone (Figure 38.16). Thus, the interplay of osteoblasts and osteoclasts constantly replaces and remodels the bones. How the activities of these cells are coordinated is still not understood, but stress placed on bones provides information used in the process. A remarkable finding in studies of astronauts spending long periods in zero gravity was that their bones decalcified. Conversely, certain bones of athletes can become considerably thicker than they were prior to training or than the same bones in nonathletes. Both thickening and thinning of bones are experienced by someone who has a leg in a cast for a long time. The bones of the uninjured leg, which carries the person's weight, thicken, but the bones of the inactive leg in

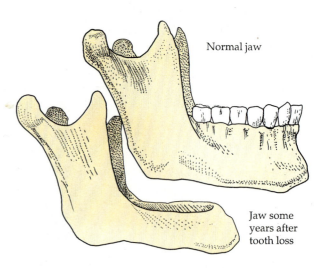

Normal jaw

Jaw some years after tooth loss

38.17 Tooth Loss and Jawbone Structure
Human jawbone is resorbed after tooth loss because of lack of compressional forces.

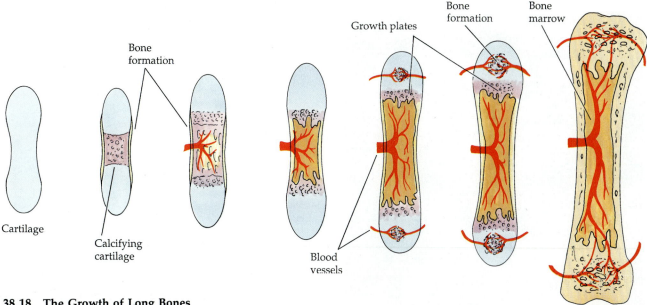

Cartilage

Bone formation

Calcifying cartilage

Growth plates

Bone formation

Bone marrow

Blood vessels

Mature bone

38.18 The Growth of Long Bones

Long bones develop in the embryo as structures made of cartilage. Calcification begins in the shaft and secondary sites of calcification form at the ends. The regions between the calcified parts are where growth can occur. Eventually the areas of calcification fuse and elongation of the bone ceases.

the cast thin. The jawbones of people who lose their teeth experience less compressional force during chewing, and become considerably remodeled (Figure 38.17).

Types of Bone

Bones are divided into two types, based on how they develop. Dermal bone forms directly as bony plates in the skin, whereas cartilage bone forms first as cartilaginous structures and is gradually ossified to become bone. The outer bones of the skull are examples of dermal bones, and the bones of the limbs are examples of cartilage bones. Cartilage bones can grow throughout the process of ossification. In the long bones of the legs and arms, for example, ossification occurs first at the centers and later at either end (Figure 38.18). Growth can continue until these areas of ossification join. The dermal bones forming the skull cap grow until their edges meet. The fontanel, or "soft spot," on the top of a baby's head is where the skull bones have not yet joined.

The composition of bone may be **compact**—solid and hard— or **cancellous**— having numerous internal cavities that make it appear spongy, even though it is quite rigid. The architecture of a specific bone depends on its position and function, and it usually has both compact and cancellous regions. The shafts of the long bones of the limbs, for example, consist of cylinders of compact bone surrounding marrow-filled central cavities, but the ends of those bones consist of cancellous bone (Figure 38.19a). The cancellous bone is light in weight because of its numerous cavities, but it is also strong, because its internal meshwork constitutes a system of supporting struts and ties. It can withstand considerable forces of compression. The rigid, tubelike structure of the shaft can withstand compression as well as bending forces. Architects and nature alike use hollow tubes as lightweight structural elements. In a solid rod that is subjected to a bending force, one side of the rod is compressed while the other side is stretched (Figure 38.19b), and both help to resist the force. The center of a solid rod contributes very little to its ability to resist bending, so hollowing out a rod reduces its weight but not its strength.

Most of the compact bone in mammals is called Haversian bone because it is composed of structural units called Haversian systems (Figure 38.20). Each system is a set of thin, concentric bony cylinders, between which are the osteocytes in their lacunae. Through the center of each Haversian system runs a narrow canal containing blood vessels. The osteocytes in one Haversian system connect only with osteocytes in the same system, as no channels cross the boundaries (called glue lines) between systems. An important feature of Haversian bone is its resistance to fracturing. If a crack forms in one Haversian system, it tends to stop at the nearest glue line.

Joints and Levers

Muscles and bones work together around **joints** where two or more bones come together. Since mus-

(a)

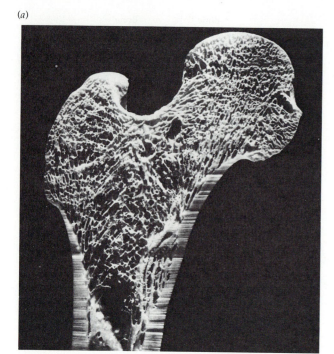

(b)   Solid rod subjected to bending force

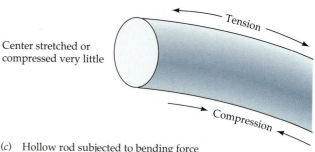

Center stretched or
compressed very little

Tension

Compression

(c)   Hollow rod subjected to bending force

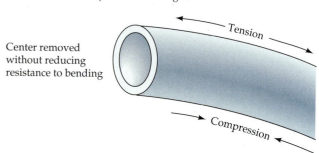

Center removed
without reducing
resistance to bending

Tension

Compression

38.19   Internal Architecture of Bone
(a) Bone may have spongy and compact regions. The
ends of long bones are spongy and the shafts are hollow
tubes of compact bone. *(b)* A solid rod resists bending
because peripheral areas resist forces of tension and
compression. *(c)* A hollow rod also resists forces of tension and compression, but is lighter.

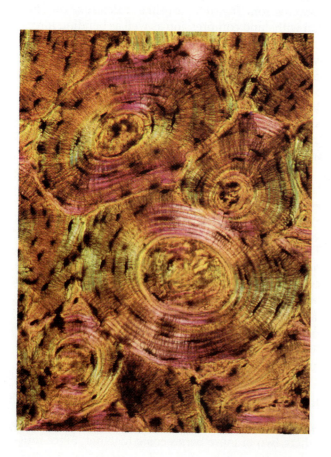

cles can only contract, they create movement around
joints by working in antagonistic pairs; when one
contracts, the other relaxes. With respect to a particular joint, such as the knee, we can refer to the
muscle that flexes the joint as the **flexor** and the
muscle that extends the joint as the **extensor** (Figure
38.21). The bones that articulate at the joint are held
in place by **ligaments**, which are flexible bands of
connective tissue. Other straps of connective tissue,
tendons, attach the muscles to the bones. In many
kinds of joints, it is only the tendon that spans the
joint, sometimes moving over the surfaces of the
bone like a rope over a pulley. It is the tendon of the
quadriceps muscle traveling over the knee-joint that
is tapped to elicit the knee jerk reflex. Ligaments can
also hold tendons in place and change the direction
of the force they exert. For example, many of the
muscles that extend your toes are actually in your
lower leg. The tendons from these muscles travel
over the front of your ankle, over the upper surfaces
of the foot bones, and attach to the bones of your
toes. Straps of ligaments over the ankle hold these
tendons in place and allow them to bend at a right
angle at the ankle (Figure 38.21).

The human skeleton has a wide variety of joints

38.20   Haversian Systems in Bone
Osteoblasts lay down bone in layers. In long bones these
layers form concentric tubes parallel to the long axis of
the bone. At the center of the tube is a canal containing
blood vessels and nerves. This micrograph is colored because the bone cross section was illuminated with polarized light.

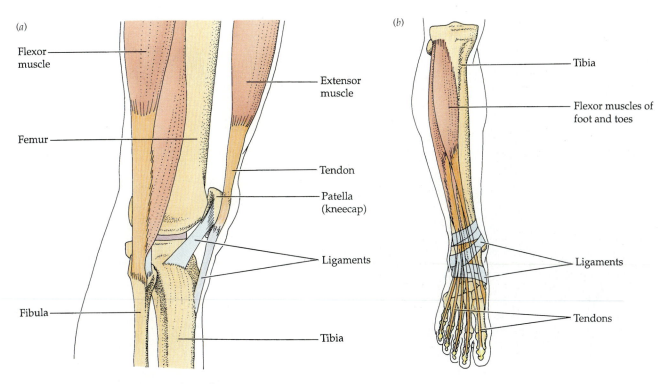

38.21 Joints, Ligaments, and Tendons
(a) This side view of the human knee shows tendons that attach muscle to bone and ligaments that attach bone to bone. (b) Tendons connect muscles in the front of the leg to foot and toe bones. These tendons, which pass under strapping ligaments at the front of the ankle, elevate the foot and toes.

with different ranges of movement. The knee joint is a simple hinge that has almost no rotational movement, and can flex in one direction only. At the shoulders and hips are ball-and-socket joints that allow movement in just about any direction. A pivotal joint between the two bones of the forearm where they meet at the elbow allows the smaller bone, the radius, to rotate when the wrist is twisted from side to side. Several kinds of joints permit some rotation, but not in all directions as do the ball-and-socket joints. Examples of these joints are found in the bones of the hands, and they give the hands a very wide range of possible movements (Figure 38.22).

Bones and the muscles that work with them around joints can be thought of as levers. A lever has a power arm and a load arm that work around a fulcrum. If the power arm is long and the load arm is short, the lever can exert a lot of force over a short distance. If the load arm is long and the power arm is short, strength is sacrificed for speed and distance. Compare now the joint of the jaw and the knee joint (Figure 38.23). The power arm of the jaw is almost as long as the load arm, and the jaw can apply great pressures over a small distance, such as when you crack a nut with your teeth. The power arm of the lower leg is short and the load arm is long, so you can run fast, jump high, and deliver swift kicks, but you can't apply nearly the pressure with a leg that you can with your jaws.

OTHER EFFECTORS

Unlike muscles, which are an almost universal effector, any other effectors are more specialized and are not shared by large numbers of animal species. Some specialized effectors are used for defense, some for communication, some for capture of prey or avoidance of predators. It would take an entire volume to discuss all of the effectors animals use, but we can briefly mention a few here to give a sampling of their evolutionary diversity.

Some animals possess highly specialized organs that are fired like miniature missiles to capture prey and repel enemies. **Nematocysts** are elaborate cellular structures produced only by hydras, jellyfish, and other members of the phylum Cnidaria. They are concentrated in huge numbers on the outer surface of the tentacles of the animal. As was shown in Figure 25.9, each nematocyst is made up of a slender thread coiled tightly within a capsule, which is armed with a spinelike trigger projecting to the outside. When a potential prey organism brushes the trigger, the nematocyst fires, turning the thread inside out

38.22 Types of Joints

The designs of joints are similar to mechanical counterparts and enable a variety of movements.

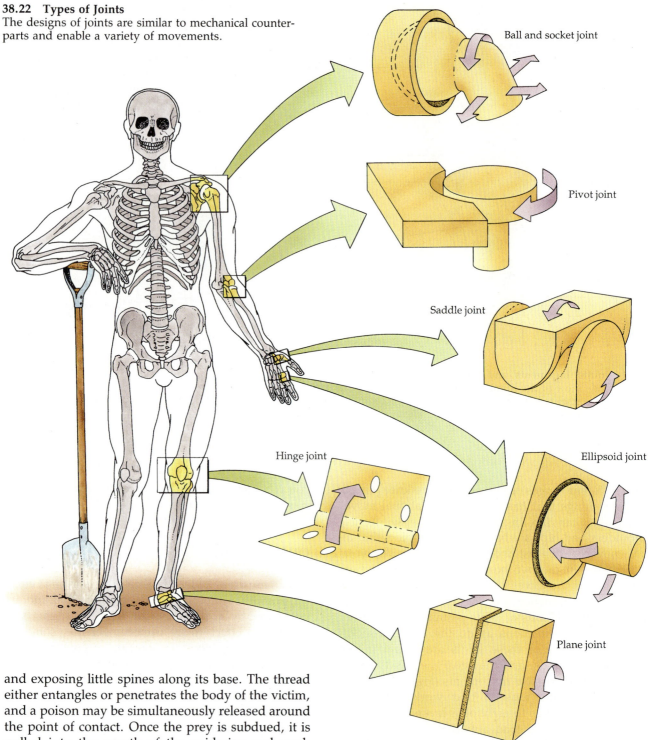

Ball and socket joint

Pivot joint

Saddle joint

Ellipsoid joint

Hinge joint

Plane joint

and exposing little spines along its base. The thread either entangles or penetrates the body of the victim, and a poison may be simultaneously released around the point of contact. Once the prey is subdued, it is pulled into the mouth of the cnidarian and swallowed. A jellyfish called the Portuguese man-of-war has tentacles that can be several meters long. They can capture, subdue, and devour full-grown mackerel, and the poison of their nematocysts is so potent that it can kill a human who gets tangled in the tentacles.

Chromatophores

A change in body color is an effector response that some animals use to camouflage themselves in a par-

ticular environment or to communicate with other animals. **Chromatophores** are pigment-containing cells in the skin that can change the color of the animal. Chromatophores are under nervous or hormonal control, or both; in most cases, they can effect a change within minutes or even seconds. In squids, soles, flounders, the famous chameleons (a group of African lizards), and a few other animals, chromatophores enable the animal to blend in with the back-

undergo shape changes as a result of the action of muscle fibers radiating outward from the cell. When the muscles are relaxed, the chromatophores are small and compact, and the animal is pale. To darken the animal, the muscles contract and spread the chromatophores over more of the surface. Chromatophores with different pigments enable animals to assume different hues or to become mottled to match the background more precisely.

Glands

Glands are effector organs that produce and release chemicals. We saw in Chapter 34 that some glands produce chemicals that are responsible for communication among animal cells. Other glands produce chemicals that are used defensively or to capture prey. Look ahead at Figure 46.8b to see a fish that uses venomous secretions to protect itself. Certain snakes, frogs, salamanders, spiders, mollusks, and fish have poison glands. Many of these poisons have

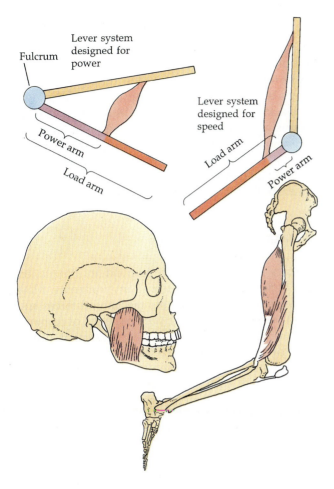

38.23 Bones and Joints as Systems of Levers
A lever works around a fulcrum and has a power arm and a load arm. If the ratio of the load arm to the power arm is low, the lever can generate much force over a small distance. An example is the human jaw. If the ratio of the load arm to the power arm is high, the lever can rapidly move small weights over a long distance. An example is the human leg.

ground on which it is resting and thus be more likely to escape discovery by predators (Figure 38.24). In other kinds of fishes and lizards, a color change is used as a signal to communicate with potential mates and territorial rivals of the same species.

There are three principal types of chromatophore cells. The most common type has fixed cell boundaries, within which pigmented granules may be moved about by microfilaments. When the pigment is concentrated in the center of each chromatophore, the animal is pale. The animal turns darker when the pigment is dispersed throughout the cell. Some other chromatophores are capable of amoeboid movement. They can mold themselves into shapes with a minimal surface area, leaving the tissue relatively pale; or they can flatten out to make the tissue appear darker. Finally, cephalopods have chromatophores that can

(a)

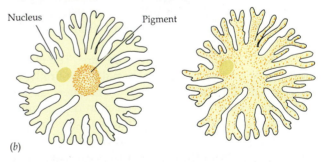

(b)

38.24 Chromatophores and Camouflage
Chromatophores are pigment-containing cells that bring about color and pattern changes in their bearers. (a) This octopus has adapted its chromatophores so that it is almost indistinguishable from its background of coral, sponges, and other animals. The octopus is the largest element in the photograph; its eyes are partially closed and its head fills much of the lower left quarter of the scene. (b) When pigment is concentrated in the center of a chromatophore (left), the chromatophore is pale; when the pigment is dispersed throughout the cell (right), it appears dark.

proven to be of practical use to humans. The poison dendrotoxin, which certain tribes of the Amazonian jungles use on the tips of their arrows, comes from the skin of a frog, and it blocks certain potassium channels. The snake venom bungarotoxin inactivates the neuromuscular acetylcholine receptors. The pufferfish poison tetrodotoxin blocks voltage-gated sodium channels. A poison from a mollusk, conotoxin, blocks calcium channels. There are many such examples, and as you can imagine, they are very useful research tools for neurobiologists.

Defensive secretions are not all poisonous. A well-known example is the odoriferous chemical called mercaptan sprayed by skunks. Human olfaction is more sensitive to mercaptan than to any other compound. See Figure 46.8a for another spectacular example of an animal that releases an irritating, defensive secretion.

Another category of effectors are the glands that produce and release pheromones, chemicals used in communication among animals, which we mentioned when we discussed chemical sensors in Chapter 37, and which will be discussed further in Chapter 44. Still other effector glands produce secretions necessary to facilitate physiological functions. Examples are salivary glands, whose secretions aid digestion, and sweat glands, whose secretions are an effective means of heat loss.

Sound and Light Producers

The ability to produce sound is an extremely important effector for humans, as speech is one of the most distinctive features of our species. Mammals, birds, and amphibians have evolved a variety of organs that create sound by passing streams of air over structures that vibrate. Insects such as cicadas produce sounds by rubbing together rough surfaces on their appendages. Sound production is by no means a universal effector, however. Most species of animals do not produce sounds.

Production of light is even more rare. The classic example of a light-producing animal is the firefly. In an organ at the tip of the firefly's abdomen, an enzyme, luciferase, catalyzes the reaction of a protein, luciferin, with ATP, releasing light energy. The primary function of the light is to attract mates. Bioluminescence is used by a number of plant and animal species (see Box 7.A), and in many cases its function is not known.

Electric Organs

A number of fishes can generate electricity, including the electric eel, the knife fish, the torpedo (a kind of ray), and the electric catfish. The electrical fields they generate are used for sensing the environment (as was shown in Figure 37.30), for communication, and for stunning potential predators or prey. The electric organs of these animals evolved from muscle, and they produce electric potentials in the same general way as nerves and muscles do. Electric organs consist of very large, disc-shaped cells arranged in long rows like stacks of coins. When the cells discharge simultaneously, the electric organ can generate far more current than can nerve or muscle. Electric eels, for example, can produce up to 600 volts with an output of approximately 100 watts—enough to light a row of light bulbs, or temporarily stun a person.

SUMMARY

Effectors enable animals to respond to stimulation from their internal and external environments. Most effector mechanisms generate mechanical forces and use those forces to change the shapes of or to move cells or whole animals. Cellular movement comes from two structures: microtubules and microfilaments. Both depend on long protein molecules that can slide past each other. Microtubules move cilia and flagella; microfilaments are responsible for amoeboid movement and for muscle cell contractions.

Muscle is classified as smooth, striated, or cardiac. The contractile mechanism of each is a system of microfilaments consisting mostly of actin and myosin. In a striated muscle cell, the microfilaments are organized into myofibrils. The myofibrils are made up of units called sarcomeres that contain overlapping myosin and actin filaments. Myosin has globular heads that bind with actin, change their orientation, and release. This cycle of events causes the filaments to slide past each other and contract the muscle fiber. Action potentials cause the sacroplasmic reticulum to release Ca^{2+} ions into the cytoplasm of the muscle fiber. Ca^{2+} binds to troponin molecules on the actin filaments and thereby exposes the sites to which myosin can bind. In striated muscle, a single action potential causes a minimum unit of contraction called a twitch. Fast twitch muscle fibers can generate maximum tension quickly but also fatigue rapidly. Slow twitch fibers generate less tension, and do so more slowly, but are resistant to fatigue.

Skeletal systems provide rigid structures against which muscles can pull. Skeletons may be hydrostatic, as in earthworms; exoskeletons, as in arthropods; or endoskeletons, as in vertebrates. Exoskeletons must be shed for growth to occur, but endoskeletons grow and are remodeled continuously.

Bones are connected to one another by ligaments and by muscles. Muscles are attached to bones by tendons. Muscles act in antagonistic pairs to move bones around joints. Different kinds of joints permit different movements. The combination of muscles and bones around a joint constitutes a lever system. Depending where the muscles are attached to the bones, the lever can apply great force over a short distance or it can create fast movements over long distances.

Other effectors include nematocysts, which are used by cnidarians to capture prey or to repel enemies, and chromatophores, which enable animals to change color. Various glands produce and release chemicals for communication, defense, or capture of prey. Other specialized effector organs emit sound, light, or electrical energy.

SELF-QUIZ

1. The movement of cilia and flagella is due to:
   a. polymerization and depolymerization of tubulin.
   b. making and breaking of cross-bridges between actin and myosin.
   c. contractions of microtubules.
   d. changes in conformations of dynein molecules.
   e. the spokes of the axoneme using energy of ATP to contract.

2. Smooth muscle differs from both cardiac and skeletal muscle in that:
   a. it can act as a pacemaker for rhythmic contractions.
   b. contractions of smooth muscle are not due to interactions between neighboring microfilaments.
   c. neighboring cells can be in electrical continuity through gap junctions.
   d. neighboring cells are tightly coupled by intercalated disks.
   e. the membranes of smooth muscle cells are depolarized by stretching.

3. Fast twitch fibers differ from slow twitch fibers in that:
   a. they are more common in the leg muscles of champion sprinters.
   b. they have more mitochondria.
   c. they fatigue less rapidly.
   d. their abundance is more a product of genetics than training.
   e. they are more common in the leg muscles of champion cross-country skiers.

4. The role of Ca^{2+} in the control of muscle contraction is:
   a. to cause depolarization of the t-tubule system.
   b. to change the conformation of troponin thus exposing myosin binding sites.
   c. to change conformation of myosin heads causing microfilaments to slide past each other.
   d. to bind to tropomyosin and break actin–myosin cross-bridges.
   e. to block the ATP binding site on myosin heads enabling muscle to relax.

5. Which of the following statements about muscle contractions is *false*?
   a. A single action potential at the neuromuscular junction is sufficient to cause a muscle to twitch.
   b. Once maximum muscle tension is achieved, no ATP is required to maintain that level of tension.
   c. An action potential in the muscle cell activates contraction by releasing Ca^{2+} into the sarcoplasm.
   d. Summation of twitches leads to a graded increase in the tension that can be generated by a single muscle fiber.
   e. The tension generated by a muscle can be varied by controlling how many of its motor units are active.

6. Which of the following statements about the structure of skeletal muscle is *true*?
   a. The bright bands of the sarcomere are the regions where actin and myosin filaments overlap.
   b. When a muscle contracts, the A bands of the sarcomere (dark regions) get longer.
   c. The myosin filaments are anchored in the Z lines.
   d. When a muscle contracts, the H bands of the sarcomere (light regions) get shorter.
   e. The sarcoplasm of the muscle cell is contained within the sarcoplasmic reticulum.

7. The long bones of our arms and legs are strong and can resist both compressional and bending forces because:
   a. they are solid rods of compact bone.
   b. their extracellular matrix contains crystals of calcium carbonate.
   c. their extracellular matrix is mostly collagen and polysaccharides.
   d. they have a very high density of osteoclasts.
   e. they consist of lightweight cancellous bone with an internal meshwork of supporting elements.

8. If we compare the jaw joint with the knee joint as lever systems:
   a. the jaw joint can apply greater compressional forces.
   b. their power arm to load arm ratios are about the same.
   c. the knee joint has greater rotational abilities.
   d. the knee joint has a greater power arm to load arm ratio.
   e. only the jaw is a hinge joint.

9. Which of the following statements about skeletons is *true*?
   a. A skeleton can consist only of cartilage.
   b. Hydrostatic skeletons can only be used for amoeboid locomotion.
   c. An advantage of exoskeletons is that they can continue to grow throughout the life of the animal.
   d. External skeletons must remain flexible, so they never include calcium carbonate crystals as do bones.
   e. Internal skeletons consist of four different types of bones: compact, cancellous, dermal, and cartilage.

10. Chemicals used by neurophysiologists to block voltage-gated sodium channels have come from:
   a. chromatophores.
   b. nematocytes.
   c. electric eels.
   d. luciferase.
   e. poison glands of fish.

FOR STUDY

1. Describe in outline form all of the events that occur between the arrival of an action potential at a motor nerve terminal and the contraction of a muscle fiber.

2. How do we know that the basis for the movement of cilia resides in the dynein components of the axoneme?

3. Wombats are powerful digging animals and kangaroos are powerful jumping animals. How do you think the structure of their legs would compare in terms of their designs as lever systems?

4. Maria and Margaret are identical twin sisters. Their mother was an olympic marathon runner and their father was on the varsity rowing team in college. Maria has become a serious cross-country skier and Margaret has joined the track team as a sprinter. Which one do you think will have the greatest chance of becoming a champion in her sport, and why?

5. If an adolescent breaks a leg bone close to the ankle joint, after the break heals that leg may not grow as long as the other one. Explain why, and in your explanation explain why the leg grows at all.

READINGS

Cameron, J. N. 1985. "Molting in the Blue Crab." *Scientific American*, May. How an arthropod deals with its exoskeleton in order to grow.

Caplan, A. I. "Cartilage." 1984. *Scientific American*, October. This tissue, a component of the vertebrate skeletal system, plays a surprisingly diverse group of roles in the developing and mature animal.

Carafoli, E. and J. T. Penniston. 1985. "The Calcium Signal." *Scientific American*, November. Calcium as a second messenger; calcium in muscle contraction.

Cohen, C. 1975. "The Protein Switch of Muscle Contraction." *Scientific American*, November. Interaction of muscle proteins and calcium ions.

Eckert, R., D. Randall and G. Augustin. 1988. *Animal Physiology*, 3rd Edition. W. H. Freeman, New York. An excellent advanced textbook. Chapter 10 deals with muscle and Chapter 11 with cell motility.

Gans, C. 1974. *Biomechanics: An Approach to Vertebrate Biology*. University of Michigan Press, Ann Arbor. A small classic on the architecture of animals and how their structure is adapted to their environment and lifestyle.

Hadley, N. F. 1986. "The Arthropod Cuticle." *Scientific American*, July. Describes the properties of the exoskeleton of the most abundant and diversified phylum of animals.

Lazarides, E. and J. P. Revel. 1979. "The Molecular Basis of Cell Movement." *Scientific American*, May. Microtubules and microfilaments in action.

Schmidt-Nielsen, K. 1990. *Animal Physiology: Adaptation and Environment*, 4th Edition. Cambridge University Press, New York. Chapter 11 gives a marvelous treatment of skeletons, muscles, and other effectors.

Vander, A. J., J. H. Sherman and D. S. Luciano. 1985. *Human Physiology: The Mechanisms of Body Function*, 4th Edition. McGraw-Hill, New York. Chapter 10 deals with the structure and function of human muscle.

39

Gas Exchange in Animals

PREVIEW: Animal cells require oxygen from the environment to produce the ATP that fuels their metabolism. They must also get rid of the carbon dioxide produced by metabolism. These problems are solved by the exchange of the two gases across membranes by diffusion. Gas exchange between an animal and the environment is influenced by concentration gradients of the gases and by whether the animal lives in water or in air. Adaptations that enhance gas exchange include anatomical specializations of gas-exchange surfaces, mechanisms for ventilating the gas-exchange surfaces with water or air, and mechanisms for transporting gases between the gas-exchange surfaces and the other cells of the body.

This chapter deals with fish gills, bird lungs, mammalian lungs, the gas-exchange system of insects, diffusion, partial pressures of gases, altitude, countercurrent exchange, tidal ventilation, surface tension, surfactants, hemoglobin, myoglobin, and the neural regulation of gas exchange.

The wail that heralds the birth of an infant and brings joy to its parents also initiates breathing, a process that must continue throughout every minute of its life. Whether awake or asleep, exercising or resting, talking or eating, we must breathe (Figure 39.1). For brief moments, such as when swimming underwater, we can hold our breath, but the urge to breathe mounts rapidly and soon becomes overwhelming. Ignoring the demands of a petulant child threatening to hold its breath until it gets its way will not endanger its life, but any involuntary event, such as drowning or choking, that prevents breathing can lead to death in a short time, and death by any other means is marked by the cessation of breathing.

The reason for the absolute coupling of breathing and life was introduced in Chapter 7. Cells need a constant supply of energy to carry out their functions. That energy comes in the form of ATP produced through the oxidation of nutrient molecules. Thus the production of ATP requires oxygen gas (O_2). Some cells, such as muscle cells, can survive short periods without O_2 by deriving ATP from glycolysis only and incurring an O_2 debt that has to be paid back at a later time. Brain cells, however, have little capacity to function in the absence of O_2. In the short run, lack of O_2 causes loss of consciousness. As the time without O_2 increases, cell functions grind to a halt, cells die, and tissues and organs suffer irreversible damage. Breathing provides the body with the

O_2 required to support the energy metabolism of all of its cells. Breathing also is the means of eliminating one of the waste products of cell metabolism, carbon dioxide (CO_2).

For humans and for all other large animals, breathing facilitates the exchange of these two gases between the body and the environment. The actual exchange of gases, however, takes place by the diffusion of molecules across membranes that separate the internal environment of the body from the external environment. Some animals, especially small and inactive ones, accomplish gas exchange without breathing. The contact between their gas-exchange membranes and the environment is adequate to support diffusion of O_2 in and diffusion of CO_2 out without their expending energy to move the environment over those membranes, which is what breathing really does. A word commonly used for gas exchange is respiration, as in artificial respiration, but whole-animal gas exchange should not be confused with cellular respiration, even though those two processes are tightly linked.

Animals differ greatly in the rates of gas exchange necessary to support their energy metabolism. At room temperature, a frog consumes about 0.01 liters of O_2 per hour, and a resting human consumes about 15 liters of O_2 per hour. If the resting human is a marathon runner, he or she will consume about 150 liters of O_2 per hour during a race. The runner, how-

39.1   Portable Life Support
In space there is no oxygen to breathe. Because we cannot live without it, astronauts who leave their spaceship take their own atmosphere to support respiration.

ever, can be outdistanced and outlasted by fish breathing water and by birds flying at very high altitudes where there is little O_2. We'll learn how they do it in this chapter.

Gas Exchange Is By Diffusion

The respiratory gases are always exchanged by **diffusion** only. There are no active transport mechanisms for respiratory gases. This general fact unifies the diversity in the structure and function of gas-exchange systems. It will be easier for you to understand the various adaptations of gas-exchange systems if you review the discussion of diffusion in Chapter 5 beginning on page 99. Because diffusion is strictly a physical phenomenon, it is limited by a number of physical factors such as whether animals breathe air or water.

Breathing Air or Water

Whether an animal lives in air or water, its physical environment places constraints on the design of its gas-exchange system. It is much easier to achieve high levels of gas exchange in air than in water for several reasons. First, the oxygen content of water is much lower than the oxygen content of an equal volume of air. The maximum O_2 content of a rapidly flowing stream splashing over rocks and tumbling over waterfalls is less than 10 ml of O_2 per liter of water. The O_2 content of fresh air, however, is about

200 ml of O_2 per liter of air. Next, O_2 diffuses much more slowly in water than in air. In a still pond, the O_2 content of the water can fall to zero only a few millimeters below the surface if the water is not stirred. Finally, if an animal breathes, it does work to ventilate its gas-exchange surfaces with air or water. It takes more energy to move water than to move air because water has a much higher density and viscosity than air does.

The slow rate of diffusion of O_2 molecules in water is not only a problem for water-breathing animals; it is also a constraint in the gas-exchange systems of animals that live in air. Eukaryotic cells respire in their mitochondria, which are in the cytoplasm—an aqueous medium. Cells are bathed with extracellular fluid—also an aqueous medium. The fact that O_2 diffuses slowly in water is a factor limiting the efficiency of O_2 distribution from gas-exchange membranes to the sites of cellular respiration in both air-breathing and water-breathing animals. In water, an adequate supply of O_2 to support the metabolism of a typical animal cell can be obtained only if the diffusion path length is no greater than about 1 millimeter. Therefore, an animal without an internal system for transporting gases must be built so that none of its cells is more than about 1 millimeter from the outside world. This is a rather severe size limit, but one way to accommodate it and still grow bigger is to have a flat, leaflike body plan, which is common among the lower invertebrates (Figure 39.2a). Another way is to have a very thin body built around a central cavity through which water circulates (Figure 39.2b). Otherwise, specialized structures are required to provide an increased surface area for diffusion, and an internal circulatory system is needed to carry gases to and from these exchange structures (Figure 39.2c).

Effects of Temperature on Water Breathers

Temperature is a crucial factor influencing gas exchange in animals that depend on or live in water. As described in Chapter 33, animals that breathe water are virtually all ectotherms. The body temperatures of aquatic ectotherms are closely tied to the temperature of the water around them. As the temperature of the water rises, so does body temperature, and because of Q_{10} effects (Chapter 33), energy expenditure and oxygen demand rise exponentially. But warm water holds less gas than cold water. (Just think of what happens when you open a warm bottle of beer or soda.) So aquatic ectotherms are in a double bind; as the temperature of their environment goes up, their demand for O_2 goes up, but the availability of O_2 in their environment goes down (Figure 39.3). If the animal does work to ventilate its gas-exchange surfaces (as fish do, for example), the energy the animal must expend increases as water temperature

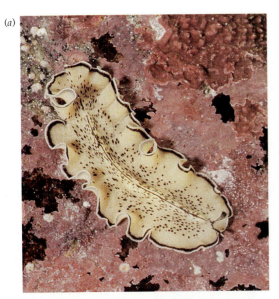

(a)

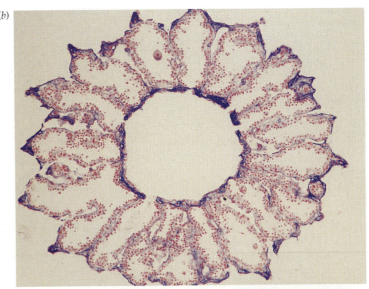

(b)

39.2 Keeping in Touch with the Medium

(a) The leaflike structure of this marine flatworm means that no cell in its body is more than a millimeter away from seawater. (b) The same is true of sponges; they have body walls perforated by many channels lined with flagellated cells. These channels communicate with the outside world and with a central cavity. The flagella maintain currents of water through the channels, through the central cavity, and out of the animal. Every cell in the sponge is very close to the respiratory medium. (c) The gills of a newt provide a large surface area for gas exchange. Blood circulating through the gills comes into close contact with the respiratory medium.

(c)

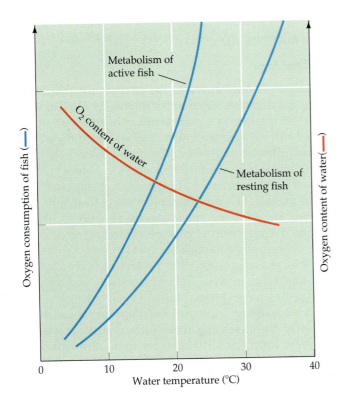

rises. Thus, as water temperature goes up, the water breather must extract more O_2 from the environment, or it must decrease its energy expenditures for activities other than breathing.

Altitude

Just as rising temperature makes it more difficult for aquatic animals to get an adequate supply of O_2, increasing altitude makes it more difficult for air breathers to get enough O_2. The amount of O_2 in the atmosphere decreases with increasing altitude. One way of expressing the concentration of gases in air and in water is by their **partial pressures**. At sea level, the pressure exerted by the atmosphere is the equiv-

39.3 The Double Bind of Water Breathers

As water temperature increases so do the body temperatures of water-breathing ectotherms, and therefore their oxygen needs increase. However, warmer water carries less oxygen in solution than does colder water.

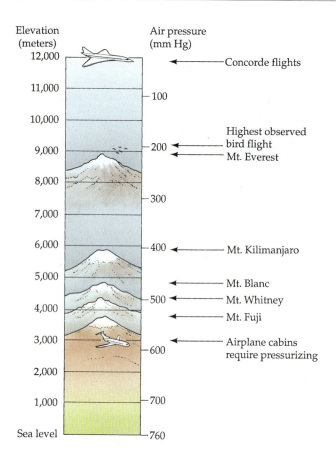

39.4 Scaling Heights
The oxygen content of the atmosphere decreases with altitude. Therefore, airplane cabins must be pressurized and mountain climbers must carry pressurized containers of oxygen. Birds, however, have been observed flying over even the highest peaks.

alent to that produced by a column of mercury 760 millimeters high. We therefore say that the **barometric pressure** is 760 mm of mercury (Hg). Because dry air is 20.9 percent O_2, the partial pressure of oxygen (pO_2) at sea level is 20.9 percent of 760 mm Hg, or 159 mm Hg. As you go higher in elevation, there is less and less air above you, so barometric pressure declines. At an altitude of 5,300 meters, barometric pressure is only half as much as it is at sea level, so the pO_2 at that altitude is only 80 mm Hg. At the summit of Mount Everest (8,848 meters) the pO_2 is only about 50 mm Hg or roughly one-third what it is at sea level. Remember that diffusion of O_2 into the body is dependent on O_2 concentration differences between the air and the body fluids, so the drastically reduced O_2 concentration in the air at a high altitude constrains O_2 uptake. This is why mountain climbers that venture to the heights of Mount Everest breathe O_2 from pressurized bottles that they carry with them (Figure 39.4).

Carbon Dioxide Exchange with the Environment

Respiratory gas exchange is a two-way street. CO_2 diffuses out of the body as O_2 diffuses into the body. In both air and water, CO_2 molecules diffuse at about the same rate as do O_2 molecules. However, the concentration of CO_2 in the atmosphere is so low, and

its solubility in water is so high, that diffusion of CO_2 from the animal is usually not as great a problem as is the diffusion of O_2 into the animal.

RESPIRATORY ADAPTATIONS

Surface Area

There is great diversity in the adaptations that facilitate gas exchange in animals (Figure 39.5). The simplest adaptation is a maximization of the specialized body surface area over which gases can diffuse. **External gills** are highly branched and folded elaborations of the body surface that provide a large surface area for gas exchange. They consist of thin and delicate membranes that minimize the path length traversed by diffusing molecules of O_2 and CO_2 (see Figure 39.2c). External gills are vulnerable to damage and are tempting morsels for carnivorous organisms, so it is not surprising that protective body cavities for gills have evolved. Many mollusks, arthropods, and fishes have **internal gills** in such cavities.

Air-breathing animals, like aquatic ones, have adapted by increasing their surface area for gas exchange, but their structures are quite different from gills. First, gas-exchange surfaces in air breathers must be in moist internal cavities or they will be in

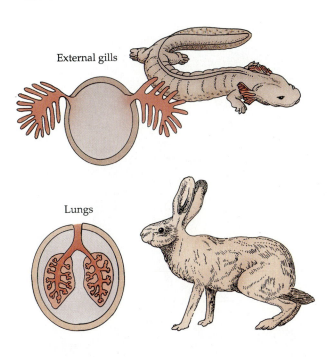

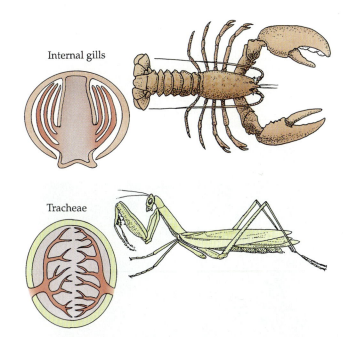

39.5 The Designs of Gas-Exchange Organs
Increased surface area for the diffusion of respiratory gases is a common feature. Gills are adaptations for breathing water; lungs and tracheae are adaptations for breathing air.

danger of drying out. Second, surface elaborations such as gills work only in water because without water for support they collapse and stick together like the pages of a wet magazine, and thus lose effective surface area. That is why a fish suffocates in air in spite of the much higher O_2 concentration. The gas-exchange structures, or **lungs**, of most air-breathing vertebrates are highly divided, elastic **air sacs**, described in greater detail below. The gas-exchange structures of insects are highly branched systems of air-filled tubes that ramify through all the tissues of the insect's body. In both cases the surface areas for gas exchange are greatly enhanced by their division into many small units.

Ventilation

Internal gas-exchange surfaces, both lungs and gills, must be ventilated. The medium containing the respiratory gases must be moved over those surfaces. Actually, two ventilation systems are required; one for ventilating the external side of the exchange surface with the respiratory medium (air or water) and one for ventilating the internal side of the exchange surface with the blood that transports the respiratory gases to and from the cells and tissues of the body. An animal's gas-exchange system is made up of its gas-exchange surfaces and the mechanisms it uses to ventilate those surfaces. In the following sections four gas-exchange systems will be described, two of which are remarkable for their efficiency—fish gills and bird lungs—and another that is relatively inefficient—mammalian lungs. But first we will discuss gas exchange in insects, which is unique.

Insect Respiration

Respiratory gases diffuse through air most of the way to and from every cell of the insect's body. This is achieved through a system of air tubes, or **tracheae**, that open to the outside environment through holes called **spiracles** in the sides of the abdomen (Figure 39.6). The tracheae branch into even finer tubes until they end in tiny **air capillaries**. In highly active insect tissue such as flight muscle, no mitochondrion is more than a few micrometers away from an air capillary. The diffusion rate of oxygen is about 300,000 times higher in air than in water, so air capillaries enable insects to supply oxygen to their cells at high rates. Many insects metabolize at extremely high rates, but this relatively simple gas-exchange system is well able to provide them with the oxygen they need.

An interesting variation on the diffusion theme is used by some species of bugs that dive and stay under water for long periods. These bugs carry with them a bubble of air. A small bubble may not seem like a very large reservoir of oxygen, yet these bugs can stay underwater almost indefinitely with their small air tanks. The secret is that when the bug dives, the air bubble contains about 80 percent nitrogen and 20 percent O_2. As the insect consumes the O_2 in its bubble, the bubble shrinks, but its nitrogen concentration increases and its O_2 concentration decreases. When the partial pressure of O_2 in the bubble falls below the partial pressure of O_2 in the surrounding

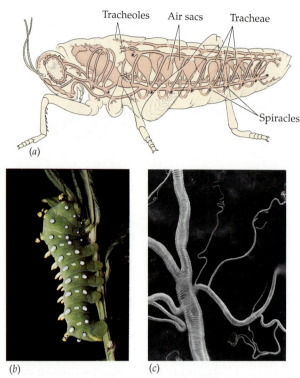

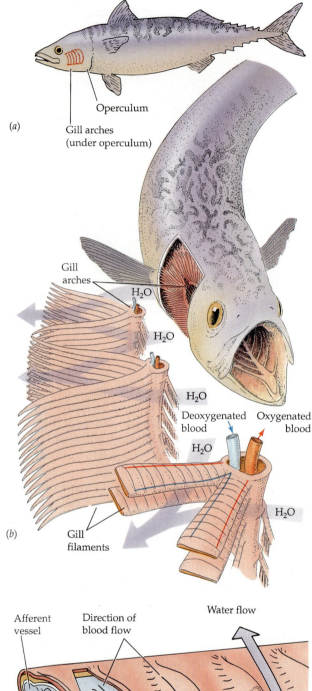

39.6 The Tracheal Gas-Exchange System of Insects
(a) The tracheal system extends throughout the body and opens to the exterior through spiracles. (b) The spiracles of the larva of a silk moth look like golden eyes down the side of the animal. (c) A scanning electron micrograph of part of the tracheal system shows the air capillaries.

water, O_2 diffuses into the bubble. For many of these small bugs, the rate of O_2 diffusion into the bubble is enough to meet the O_2 demand of the animal while it is under water.

Fish Gills

The internal gills of fishes are marvelously adapted for maximum possible rates of gas exchange. They offer a large surface area for gas exchange between blood and water. They are supported by five or six bony **gill arches** on either side of the fish between the mouth cavity and the protective **opercular flaps** (Figure 39.7a). Each gill arch is lined with hundreds of leaf-shaped **gill filaments** arranged in two rows. These rows of gill filaments point toward the opercular opening, which is the direction of water flow, and the tips of the filaments of adjacent arches interdigitate. The upper and lower flat surfaces of each gill filament have rows of evenly spaced folds, or **lamellae**, which greatly increase the gill surface area (Figure 39.7b). The enormous surface area of the lamellae is the site of gas exchange. The result of this interdigitating network of gill filaments and lamellae is that practically all water that passes across the gills comes into close contact with the gas-exchange surfaces.

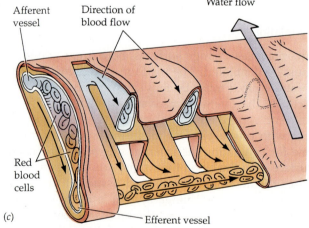

39.7 The Internal Gills of Fish
(a) Opercular flaps protect and help to ventilate the gills. (b) Each gill arch supports two rows of gill filaments; each filament is folded into many thin, flat lamellae that are the actual gas-exchange surfaces. (c) Blood flow through the lamellae is countercurrent to the flow of water over the lamellae.

(a) Mouth open, operculum closed, opercular cavity expanding

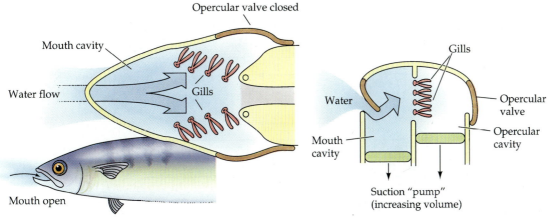

(b) Mouth closed, mouth cavity contracting, operculum opening

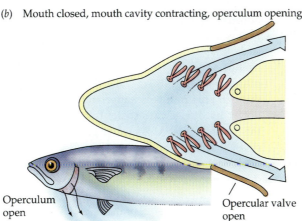

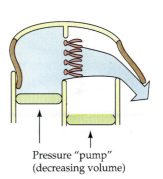

39.8 Pumping Together
Two pumps work together to ventilate fish gills. One pump is the mouth cavity (a positive-pressure pump) and the other is the opercular cavity (a negative-pressure pump).

The very delicate structure of the lamellae minimizes the path length for diffusion of gases between blood and water. Blood travels in blood vessels through the gill arches and the gill filaments, but in the lamellae the blood flows between the two surfaces of the lamellae as a sheet not much more than one red blood cell thick. The surfaces of the lamellae consist of highly flattened epithelial cells with practically no cytoplasm, so the water and the red blood cells are separated by little more than 1 or 2 μm.

Besides a large surface area and a short diffusion path length, what more can be done to maximize the rate of diffusion? It is still possible to maximize the concentration difference of O_2 between water and blood. Fish accomplish this by ventilating both the external and the internal surfaces of the lamellae. A constant flow of water moving over the gills maximizes the O_2 concentration on the external surfaces. On the internal side, the concentration of O_2 is minimized as the circulation of blood sweeps the O_2 away as rapidly as it diffuses across.

Most fishes ventilate the external surfaces of their gills by means of a two-pump mechanism that maintains a unidirectional and constant flow of water over the gills. The closing and contracting of the mouth cavity acts as a **positive-pressure pump**, pushing water over the gills. The opening and closing of the opercular flaps acts as a **negative-pressure pump**, pulling water over the gills. Because these pumps are slightly out of phase with each other, they maintain an almost continual flow of water across the gills. Hence, moving water is always bathing the gill surfaces (Figure 39.8).

The flow of blood over the inner surfaces of the lamellae is also unidirectional because of the arrangement of the **afferent blood vessels**, which bring blood to the gills, and the **efferent blood vessels**, which take blood away from the gills (Figure 39.7c). Blood flow through the lamellae is in the opposite direction to the water flow over the lamellae. Such **countercurrent flow** makes gas exchange much more efficient than parallel flow, as is illustrated in Box 39.A. Countercurrent exchange is an important principle in a number of different physiological systems.

In summary, the fish is able to extract an adequate supply of O_2 from meager environmental sources by

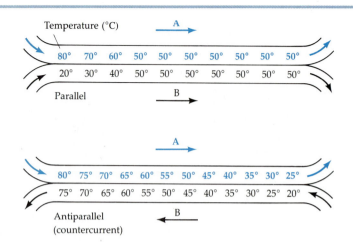

BOX 39.A

Countercurrent Exchangers

Consider two pipes, side by side, containing parallel streams of water flowing in the same direction. In pipe A, the water that enters has a temperature of 80° C; in B, the water that enters has a temperature of 20° C. If we assume that heat may be exchanged between the two pipes but may not be lost to the environment, then it is clear what happens: Heat is transferred from A to B until the temperatures in the two pipes are identical—approximately 50° C in this example. The second law of thermodynamics (Chapter 6) says, among other things, that heat cannot flow spontaneously from a cooler to a warmer system; therefore, once both pipes reach 50° no further heat can be transferred from A to B.

Suppose now that we make one seemingly innocuous change; we have water flow in opposite directions (antiparallel) in pipes A and B. As before, heat flows from A to B;

however, in this case the transfer of heat is much more complete—instead of approximately one-half the heat being transferred, almost all of it passes from A to B. As water flows through pipe B, it gets hotter and hotter, but it is always cooler than the water in pipe A. As water gets closer to the end of pipe A, it is relatively cool, but it is still warmer than that in pipe B, so it continues to give up heat. Thus heat transfer occurs along the entire region of overlap of the two pipes. The only difference between these two examples is that in the first case the flow is parallel, and in the second case antiparallel. The antiparallel system is usually called a **countercurrent exchanger**.

Countercurrent systems are legion in the animal kingdom. In fish gills,

the efficiency of the oxygenation of blood is maximized by having blood flow in the direction opposite to that of the oxygen-bearing water (Figure 39.8c). When the water leaves the gill, it has lost much of its oxygen, and the blood has become maximally oxygenated. If blood flowed in the same direction as the water, oxygen exchange would be much less complete. In Chapter 42 we will see the operation of a countercurrent exchange system in the vertebrate kidney—in a structure called the loop of Henle—allowing the formation of a steep salt concentration gradient within the kidney. Yet another example is heat exchangers as we saw in the case of "hot fish" and in the extremities of arctic animals in Chapter 33.

maximizing the surface area for diffusion, minimizing the path length for diffusion, and maximizing oxygen extraction efficiency by constant, unidirectional, countercurrent ventilation of the exchange surfaces with blood and water.

Bird Lungs

For a long time, avian lungs presented a puzzle. Birds can sustain extremely high levels of activity for much longer periods than mammals can, and they can do so at very high altitudes where mammals cannot even survive. Yet the lungs of a bird are smaller than those of a similar-sized mammal. Another unusual feature of birds is that in addition to lungs they have **air sacs** at several locations in their bodies. The air sacs connect with one another and with the lungs (Figure 39.9). Even though the air sacs as well as the lungs

receive inhaled air, the air sacs are not gas-exchange surfaces. If a sample of air or pure oxygen is tied off in an air sac, its composition does not change rapidly, as it would if the O_2 were diffusing into the blood and CO_2 were diffusing into the air sac.

The microscopic anatomy of the bird lung is unique among air-breathing vertebrates. The common lung design of other air-breathing vertebrates is a system of branching airways that get finer and finer until they end in clusters of microscopic air sacs where gas exchange takes place. Air flow in such a system is tidal—it enters and leaves by the same route. The avian lung, in contrast, does not have blind air sacs, but has tubelike **parabronchi** that permit unidirectional air flow through the lungs (Figure 39.10). Air capillaries off the parabronchi increase the surface area for gas exchange. Another unusual feature of avian lungs is that in comparison to mam-

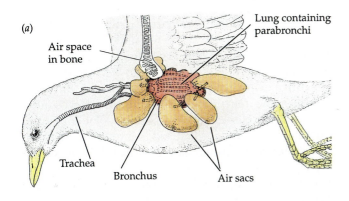

(a)

Air space in bone

Lung containing parabronchi

Trachea

Bronchus

Air sacs

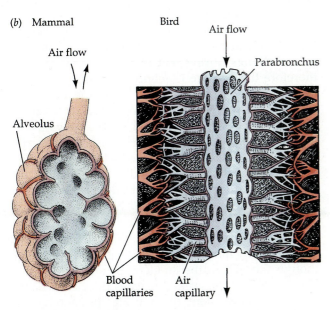

(b) Mammal

Bird

Air flow

Air flow

Air flow

Parabronchus

Alveolus

Blood capillaries

Air capillary

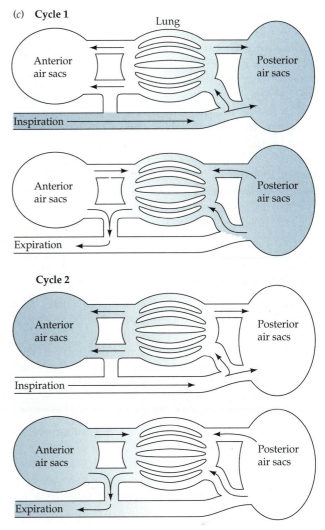

(c) **Cycle 1**

Lung

Anterior air sacs

Posterior air sacs

Inspiration

Anterior air sacs

Posterior air sacs

Expiration

Cycle 2

Anterior air sacs

Posterior air sacs

Inspiration

Anterior air sacs

Posterior air sacs

Expiration

39.9 Air Flow Through Bird Lungs Is Constant and Unidirectional
(a) The respiratory system of a bird. The air sacs are unique to bird anatomy. (b) The gas-exchange surfaces of mammals are alveoli, which are blind sacs, so air flow must be tidal. The gas-exchange surfaces of birds are air capillaries branching off the parabronchi that run through the lungs. (c) The unidirectional flow of air during avian respiration.

malian lungs they expand and contract relatively little during a breathing cycle.

The puzzle of how birds breathe was finally solved when researchers discovered that a single breath of air remains in the bird's gas-exchange system for two inhalation–exhalation cycles. On the first inhalation, the inhaled air goes primarily to the posterior air sacs. When the bird exhales, the air in the posterior air sacs flows through the parabronchi of the lungs. Upon the next inhalation, the air in the parabronchi flows into the anterior air sacs, as fresh air fills the posterior air sacs. Finally, the air in the anterior air sacs leaves the bird on the second exhalation after it entered. The two sets of air sacs act as bellows, maintaining a constant, unidirectional flow of air through the lungs.

The advantages of this unique gas-exchange system are similar to those of fish gills. Because air from

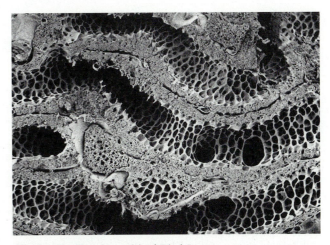

39.10 The Parabronchi of Bird Lungs
This scanning electron micrograph shows avian parabronchi in oblique section and shows the air capillaries between them.

the outside flows unidirectionally and practically continuously over the gas-exchange surfaces, the concentration of O_2 on the environmental side of those surfaces is maximized. Furthermore, the unidirectional flow of air through the system makes possible a pattern of blood flow to minimize the O_2 concentration on the internal side of the exchange surfaces. However, in birds, the flow appears to be **crosscurrent** (at right angles) rather than countercurrent to the airflow. It is now clear how birds can fly over Mount Everest. A bird is able to supply its gas-exchange surfaces with a continuous flow of fresh air that has an oxygen concentration close to that of the ambient air. Next we will see why mammals could not fly over Mount Everest, even if they could fly!

Breathing in Mammals

Vertebrate lungs have their evolutionary origins in outpocketings of the digestive tract (Figure 39.11). The lungs began as blind sacs, and they remain so today in all air-breathing vertebrates other than birds. Because of this feature, ventilation cannot be constant and unidirectional; it must be tidal. Air comes in and then flows out by the same route. The characteristics of lung ventilation for humans are shown in Figure 39.12. When we are at rest, the amount of air that our normal breathing cycle moves per breath is called the **tidal volume** (about 500 ml for an average human adult). We can breathe much more deeply and inhale more air than our resting tidal volume, and the additional volume of air we can take in above

normal tidal volume is our **inspiratory reserve volume**. Conversely, we can forcefully exhale more air than we normally do during a resting exhalation. This additional amount of air that can be forced out of the lungs is the **expiratory reserve volume**. But there is still some remaining air in the lungs even after the most extreme exhalation possible. The lungs and airways cannot be collapsed completely; they always contain a **residual volume**. The **total lung capacity** is the sum of the residual volume, expiratory reserve volume, tidal volume, and inspiratory reserve volume.

Tidal breathing places severe limits on the concentration difference available to drive the diffusion of O_2 from air into the blood. Fresh air is not moving into the lungs during half of the respiratory cycle; therefore the average O_2 concentration of air in the lungs is considerably less than it is in the air outside of the lungs. Also, the incoming air mixes with the stale air that was not expelled by the previous exhalation. **Dead space** is the term used to describe lung volume that is not ventilated with fresh air. This dead space consists of the residual volume and, depending on the depth of breathing, some or all of the expiratory reserve volume. Notice the scale in Figure 39.12. It tells us that a tidal volume of 500 ml of fresh air mixes with up to 2,000 ml of stale moist air before reaching the gas-exchange surfaces. Even though the pO_2 in the ambient air may be 150 mm Hg, the pO_2 of the air that reaches the gas-exchange surfaces is only about 100 mm Hg. In contrast, the pO_2 in the water bathing the lamellae of the fish gills

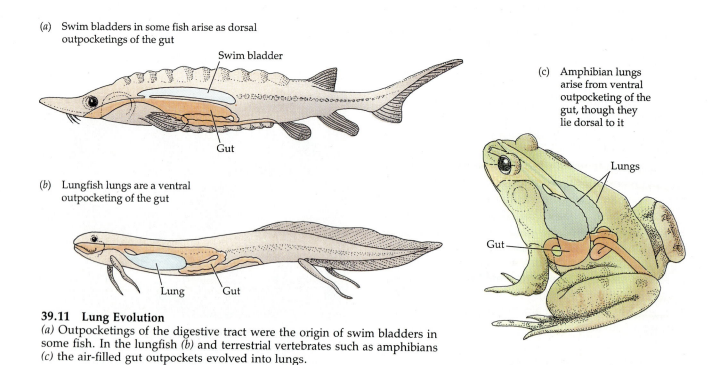

(a) Swim bladders in some fish arise as dorsal outpocketings of the gut

Swim bladder

Gut

(b) Lungfish lungs are a ventral outpocketing of the gut

Lung   Gut

(c) Amphibian lungs arise from ventral outpocketing of the gut, though they lie dorsal to it

Lungs

Gut

39.11 Lung Evolution
(a) Outpocketings of the digestive tract were the origin of swim bladders in some fish. In the lungfish (b) and terrestrial vertebrates such as amphibians (c) the air-filled gut outpockets evolved into lungs.

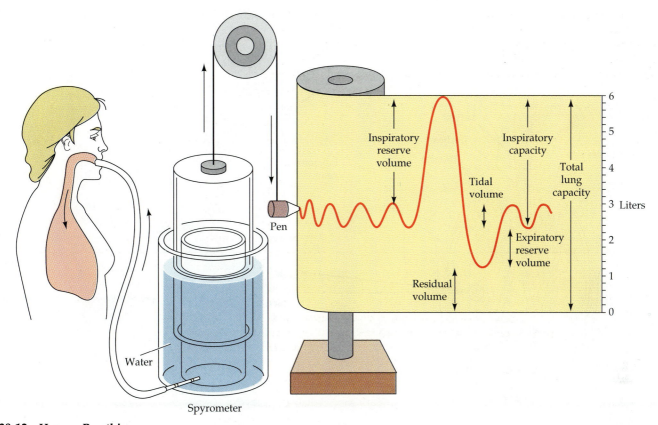

39.12 Human Breathing
The characteristics of breathing can be demonstrated by breathing from a closed reservoir of air and measuring the changes in the volume of that reservoir.

or in the air flowing through the air capillaries of the bird lung is essentially the same as the pO_2 in the outside water or air.

As well as reducing the concentration difference, tidal breathing detracts from the efficiency of gas exchange in another way. It precludes the possibility of countercurrent gas exchange between air and blood. Because the air enters and leaves the gas exchange structures by the same route, there is no anatomical way that blood can flow parallel to and countercurrent to the air flow. Consequently, the pO_2 of the blood leaving the lung can never exceed the pO_2 of the exhaled air. In contrast, the pO_2 of the blood leaving the bird lung or fish gill can be almost as high as the pO_2 of the outside air or water.

Lest we sell the mammalian lungs short, we must take a closer look at their anatomy. They do possess some interesting, important design features that maximize the rate of gas exchange: an enormous surface area and a very short path length for diffusion. Mammalian lungs serve the respiratory needs of mammals quite well, considering their ecologies and lifestyles. There are environmental factors other than low O_2 concentration that make it difficult for mammals to live on top of Mount Everest!

Air enters the lungs through the oral cavity or nasal passages, which join in the **pharynx,** as shown in Figure 39.13. The pharynx gives rise to the esophagus, through which food reaches the stomach, and to the airways. At the beginning of the airways is the **larynx,** or "voice box", which houses the vocal cords. The larynx is the "Adam's apple" that you can see and feel on the front of your neck. The larynx opens into the major airway, the **trachea,** which is about the diameter of a garden hose. The trachea has thin walls, but they are supported by rings of cartilage that prevent the trachea from collapsing as air pressure changes during the breathing cycle. If you run your fingers down the front of your neck just below your larynx, you can feel a couple of these rings of cartilage.

The trachea branches into two slightly smaller airways called **bronchi,** one leading to each lung. These bronchi branch repeatedly to generate a treelike structure of progressively smaller airways going to all regions of the lungs. Each of these bronchi is a smaller version of the trachea in terms of structure; they all have supporting cartilage rings. As the branching of the bronchial tree continues to give rise to still smaller airways, the cartilage supports even-

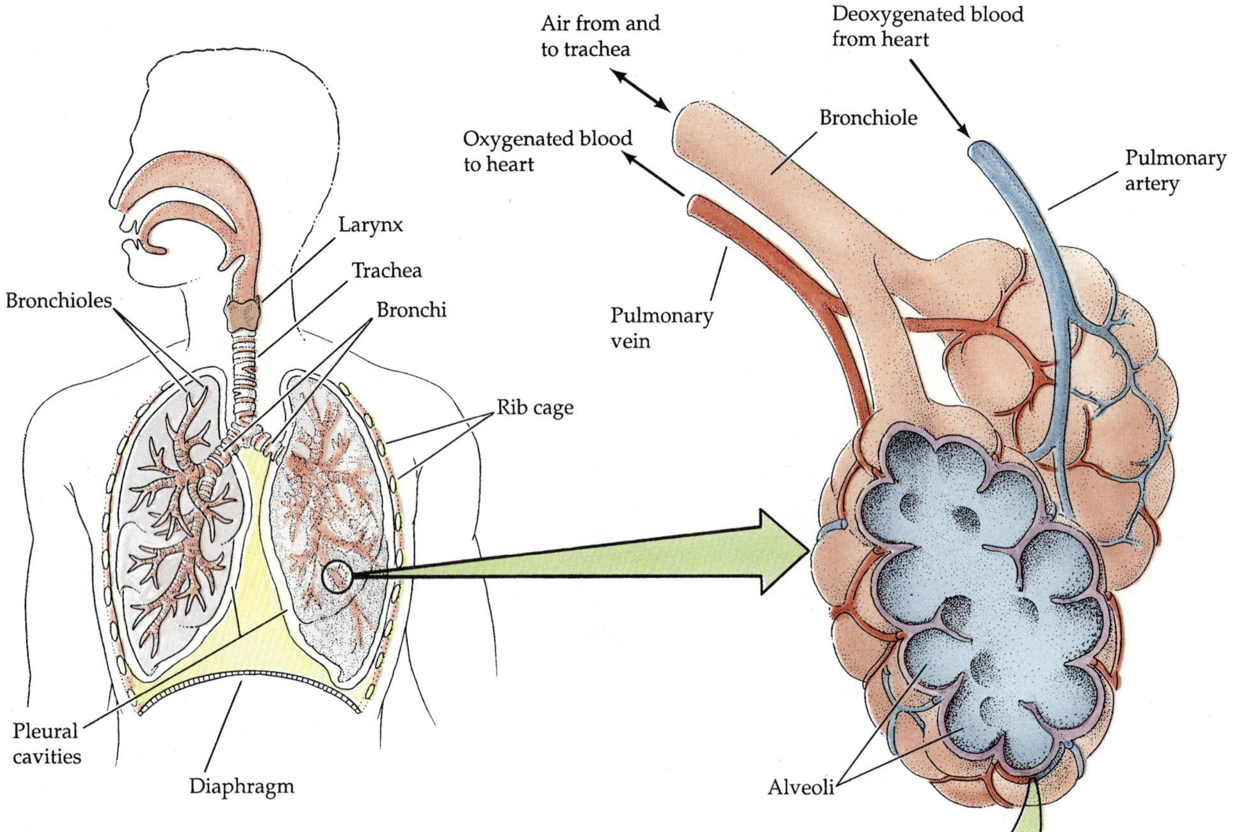

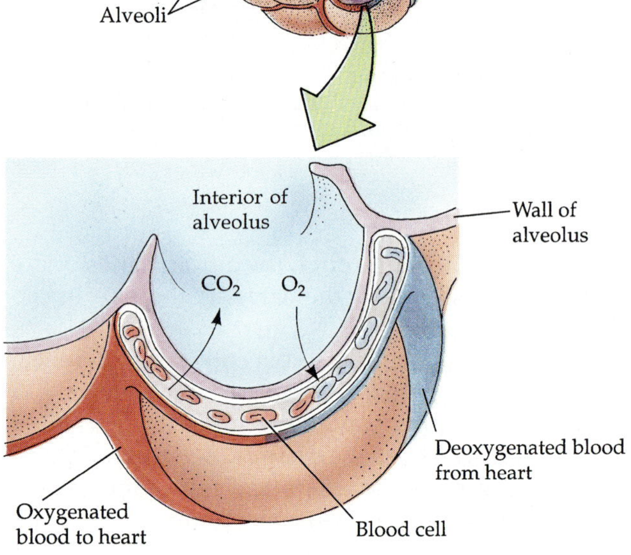

39.13 The Human Respiratory System
The lungs lie within the thoracic cavity, which is bounded by the ribs and the diaphragm. Pleural membranes line the part of the thoracic cavity containing the lungs, so the lungs are actually in the pleural cavity. Air enters the lungs from the oral-nasal passages via the trachea and bronchi, and eventually reaches the alveoli. The air in the alveoli is in intimate contact with the blood flowing through the networks of fine blood vessels surrounding the alveoli.

tually disappear, marking the transition to **bronchioles**. The branching continues until the bronchioles are less than the diameter of a pencil lead, at which point the tiny, thin-walled air sacs called **alveoli** begin. Alveoli resemble clusters of grapes on a system of stems. The "stems" are the bronchioles, which go through about six more branch points. Finally, terminal bronchioles end in alveoli. The alveoli are the sites of gas exchange. The airways only conduct the air to and from the alveoli, and therefore their volume is physiological dead space. If you trace an airway from the primary bronchus leaving the trachea down to the very last terminal bronchiole, you pass about 23 branching points. Thus, there are 2^{23} terminal bronchioles—a very large number. The number of alveoli is even larger, about 300 million in the human lungs. Even though each alveolus is very small, the combined surface area for diffusion of respiratory gases is about 70 square meters, or the size of a badminton court.

Each alveolus consists of very thin cells. Between and surrounding the alveoli are networks of the smallest of blood vessels, also made up of exceedingly thin cells. Where blood vessel meets alveolus there is very little space between them, so the diffusion path length between the air and the blood is only 2 μm. In comparison, even the diameter of a red blood cell is much greater—about 7 μm.

Surfactant and Mucus

Two other fascinating adaptive features of mammalian lungs bear mentioning even though they do not directly influence gas-exchange properties. They are the production of mucus and **surfactant**. Surfactant is a substance that reduces the **surface tension** of the liquid lining the insides of the alveoli.

What is surface tension and why need we consider it as we study lung function? Surface tension arises from the attractive (cohesive) forces between the molecules of a liquid. At the surface of the liquid, these cohesive forces are unbalanced and give the surface the properties of an elastic membrane. This is why certain insects called water striders can walk on the surface of water (see Figure 2.17) and why a carefully placed razor blade can float. A surfactant is a substance that interferes with the cohesive forces that create surface tension. Detergent is a surfactant and when added to water it can cause the floating razor blade to sink and can make things difficult for the water strider.

The thin aqueous layer lining the alveoli has surface tension, which can make it very difficult to inflate the lungs. Surface tension in the alveoli is normally reduced by surfactant molecules produced by certain cells in the alveoli. If a baby is born a month or more prematurely, its alveoli may not be producing surfactant. It is very difficult for such a baby to breathe because an enormous inhalation effort is required to stretch the alveoli against the surface tension. This condition, called respiratory distress syndrome, may cause a baby to die from exhaustion and suffocation. The common treatment is to put the baby on a respirator to assist its breathing and to treat it with hormones to speed lung development. A new approach in which surfactant is applied to the lungs in an aerosol is very promising.

Many cells lining the airways produce a sticky mucus that captures bits of dirt and microorganisms as they are inhaled. The mucus must be continually cleared from the airways, however, which is done by the beating of cilia lining the airways, as was shown in Figure 38.2. The cilia move the mucus with its trapped debris up toward the pharynx, where it is swallowed. This phenomenon, called the **mucus escalator**, can be adversely affected by inhaled pollutants. The smoking of one cigarette can immobilize the cilia of the airways for hours. Hacking, or smoker's cough, results from the need to clear the obstructing mucus from the airways when the mucus escalator is out of order.

Mechanics of Ventilation

As was shown in Figure 39.13, the lungs are suspended in the **thoracic cavity**, which is bounded on the top by the shoulder girdle, on the sides by the rib cage, and on the bottom by a domed sheet of muscle, the **diaphragm**. The thoracic cavity is lined on the inside by the **pleural membranes**, which divide it into right and left **pleural cavities**. The pleural cavities are closed spaces, so any effort to increase their volume will create negative pressure—suction—within them. Negative pressure within the pleural cavities causes the lungs to expand as air flows into them from the outside. This is the mechanism of inhalation. The diaphragm contracts to begin an inhalation. This contraction pulls the diaphragm down, increasing the volume of the thoracic and pleural cavities (Figure 38.14). As pressure in the pleural cavities becomes more negative, air enters the lungs. Exhaling begins when the contraction of the diaphragm ceases, the diaphragm relaxes, and the elastic recoil of the lungs pushes air out through the airways.

The diaphragm is not the only muscle that increases the volume of the thoracic cavity. Intercostal muscles between the ribs can expand the thoracic cavity by lifting the ribs up and outwards. When heavy demands are placed on the respiratory system, such as during strenuous exercise, the intercostal muscles and the diaphragm contract together and increase the volume of air inhaled.

Inhalation is always an active process, with muscles contracting, and exhalation is usually a passive process, with muscles relaxing. There are, however, sets of muscles between the ribs that can be called into play for forceful exhalations. Place your hands on your ribs and on your abdomen while breathing first shallowly and then deeply. Feel which muscles are active during inhalation and which are active during exhalation.

With the diaphragm at rest between breaths, the pressure in the pleural cavities is still slightly negative. This slight suction keeps the alveoli partially

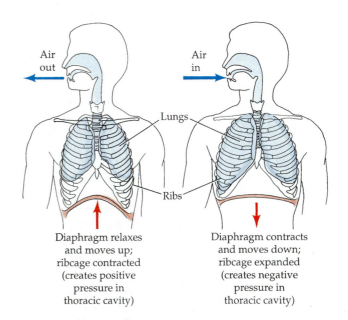

Air out

Air in

Lungs

Ribs

Diaphragm relaxes and moves up; ribcage contracted (creates positive pressure in thoracic cavity)

Diaphragm contracts and moves down; ribcage expanded (creates negative pressure in thoracic cavity)

39.14 Getting Air into the Lungs
Inhalation is an active process. Contraction of the diaphragm and raising of the ribs expands the thoracic cavity, sucking air into the lungs.

inflated. If the thoracic wall is punctured, by a knife wound for example, air leaks into the pleural cavity, causing the lung to collapse. If the hole in the thoracic wall is not sealed, the breathing movements of diaphragm and intercostal muscles pull air into the pleural cavity rather than the lung, and gas exchange across the walls of the alveoli in that lung ceases.

TRANSPORT OF RESPIRATORY GASES BY THE BLOOD

The circulatory system uses a pump, the heart, and a network of blood vessels to transport blood and the substances it carries around the body. The circulatory system is the subject of the next chapter, but since two of the substances it transports are the respiratory gases, we must mention aspects of it here. The circulatory system sweeps O_2 away from the inside of the gas-exchange surfaces as it diffuses across. This internal ventilation of the gas-exchange surfaces minimizes the concentration of oxygen on the internal side and promotes the diffusion of O_2 across the surface at the highest possible rate. The blood then delivers the O_2 it picks up to the cells and tissues of the body.

The ability of the blood to pick up and transport O_2 would be quite limited if the O_2 were carried only in solution in the liquid part of the blood, the **blood plasma**. Blood plasma can carry about 0.3 ml oxygen per 100 ml, so to support the O_2 needs of a person at *rest*, the heart would have to pump about 5,000 liters of blood plasma per hour (enough to fill the gas tanks of about 100 cars!). Fortunately, the blood also contains **red blood cells** (Figure 39.15), which are red because they are loaded with an oxygen-binding pigment called **hemoglobin**. Hemoglobin increases the capacity of the blood to transport oxygen by about 60-fold.

Hemoglobin

Red blood cells contain enormous numbers of hemoglobin molecules. Hemoglobin is a protein consisting of four polypeptide subunits. Each of these polypeptides surrounds a heme group—an iron-containing ring structure that can reversibly bind a molecule of O_2. As O_2 diffuses into the red blood cells, it binds to hemoglobin. Once O_2 is bound, it cannot diffuse back across the cell membrane. By mopping up O_2 molecules as they enter the red blood cells, hemoglobin maximizes the concentration difference driving the diffusion of O_2 into the red blood cells. In addition, hemoglobin enables the red blood cells to carry a large amount of O_2 for use by the tissues of the body.

The ability of hemoglobin to pick up or release oxygen depends on the concentration or partial pres-

39.15 Red Blood Cells
Seen here from several angles, the "flattened jelly doughnut" shape of red blood cells maximizes their surface area for oxygen exchange.

sure of oxygen in its environment. When the pO_2 of blood plasma is high, as it usually is in the lungs, each molecule of hemoglobin can carry its maximum load of four molecules of O_2. As the blood circulates through capillary beds elsewhere in the body, it encounters lower pO_2's, and the hemoglobin releases some of the O_2 it is carrying. The lower the tissue or plasma pO_2, the more O_2 is released from hemoglobin to diffuse out of the red blood cells and into the tissues. This relationship between pO_2 and the amount of O_2 bound to hemoglobin—an important property of hemoglobin—is not linear, but is described by a sigmoid (S-shaped) hemoglobin/oxygen dissociation curve (Figure 39.16).

Remember that a hemoglobin molecule consists of four subunits, each of which can bind one molecule of O_2. At low pO_2's, one subunit will bind one O_2 molecule; as a result, the shape of this subunit changes, causing an alteration in the quarternary structure of the entire hemoglobin molecule (see Chapter 3). This structural change makes it easier for the other subunits to bind O_2—their O_2 affinity is increased. So even a small increase in pO_2 results in most hemoglobin molecules picking up a second, and then a third, molecule of O_2. This influence of the binding of O_2 by one subunit on the binding affinity of the other subunits is called positive cooperativity: the first molecule to bind makes it easier for the second and third. After the third O_2 molecule is bound, however, it takes quite a large increase in pO_2 to load the fourth subunits of all the hemoglobin molecules, because fewer and fewer binding sites are available as more and more hemoglobin molecules become fully saturated with oxygen.

The significance of the sigmoid shape of the he-

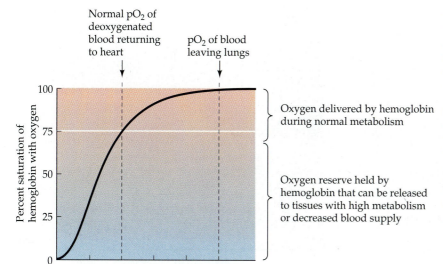

Normal pO₂ of deoxygenated blood returning to heart

pO₂ of blood leaving lungs

Oxygen delivered by hemoglobin during normal metabolism

Oxygen reserve held by hemoglobin that can be released to tissues with high metabolism or decreased blood supply

39.16 Oxygen Binds to Hemoglobin
The binding of oxygen to hemoglobin depends on the concentration of oxygen in the blood plasma. As the blood flows through the lungs, the partial pressure of oxygen in the plasma reaches about 100 mm Hg. At this pO_2, each hemoglobin molecule can bind four molecules of oxygen and is 100 percent saturated with oxygen. In normal tissues, the pO_2 in the plasma falls to about 40 mm Hg. At this lower oxygen concentration, each molecule of hemoglobin can only bind three molecules of oxygen, so the hemoglobin is only 75 percent saturated.

moglobin/oxygen dissociation curve is best appreciated by considering the dynamics of unloading the O_2 in the tissues. The pO_2 that normally exists in the alveoli of the lungs is about 100 mm Hg; at this pO_2 the hemoglobin is 100 percent saturated with oxygen. The pO_2 in mixed venous blood is usually about 40 mm Hg. From the curve in Figure 39.16 you can see that the hemoglobin returning to the heart from the systemic circulation is still about 75 percent saturated, which means most hemoglobin molecules drop only one of their four O_2 molecules as they circulate through the body. This may seem to be a very inefficient system for the delivery of oxygen to the tissues, but it is actually extremely adaptive. When a tissue becomes oxygen-starved and its local pO_2 falls below 40 mm Hg, the hemoglobin flowing through that tissue will drop much more of its oxygen load—the steep portion of the sigmoid dissociation curve comes into play when tissue pO_2 falls below the normal 40 mm Hg. The cooperative oxygen-binding property of hemoglobin is very effective in making O_2 available to the tissues precisely when and where it is most needed.

Myoglobin

Muscle cells have their own oxygen-binding molecule, called **myoglobin**. Myoglobin consists of just one polypeptide chain associated with an iron-containing ring structure that can bind one molecule of oxygen (see Figure 3.21). Myoglobin has a higher affinity for O_2 than hemoglobin does (Figure 39.17), so it picks up and holds oxygen at pO_2's at which hemoglobin is releasing its bound O_2. Myoglobin provides a reserve of oxygen for the muscle cells for times when metabolic demands are high and blood flow is interrupted as contracting muscles constrict blood vessels. When hemoglobin has no more O_2 to

give up, and tissue pO_2 falls even lower, myoglobin releases its bound O_2. Diving mammals such as seals that can remain active underwater for many minutes have high concentrations of myoglobin in their muscles. Muscles called upon for extended periods of work frequently have more myoglobin than muscles that are used for short, intermittent periods. As we mentioned in Chapter 38, this is one of the reasons for the difference in appearance of the "white" and "dark" meat of chickens and turkeys. These birds are not long-distance fliers, and their flight muscles (the white meat) have little myoglobin. But ducks and geese come from distinguished lineages of long-distance fliers. Their flight muscles have much myoglobin, as well as more mitochondria and more blood vessels; thus they appear dark.

Regulation of Hemoglobin Function

Various factors influence the oxygen-binding properties of hemoglobin and thereby have important influences on oxygen delivery to tissues. For example, there are variations in the chemical composition of the polypeptide chains that form the hemoglobin molecule. The normal hemoglobin of adult humans has two each of two kinds of polypeptide chains—two α chains and two β chains—and has the oxygen-binding characteristics shown in Figure 39.16. Before birth, the human fetus has a different form of hemoglobin consisting of two α chains and two γ chains. The chemical composition of fetal hemoglobin enables fetal blood to pick up O_2 from maternal blood when both are at the same pO_2. Fetal hemoglobin thus has an oxygen-dissociation curve that plots to the left of the adult curve. This difference between maternal and fetal hemoglobin greatly facilitates the transfer of O_2 from the mother's blood to the blood of the fetus in the placenta (Figure 39.17).

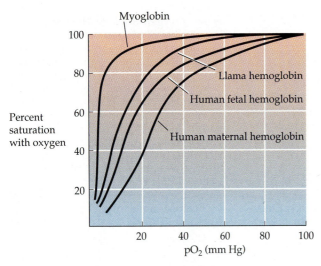

39.17 Oxygen-Binding Adaptations
The oxygen-binding properties of myoglobin and different hemoglobins reveal important adaptations. Myoglobin can serve as an oxygen reservoir because it remains 100 percent saturated until the partial pressure of oxygen falls so low that hemoglobin has given up most of its oxygen. The higher oxygen affinity of fetal hemoglobin in comparison to maternal hemoglobin enables fetal blood to pick up oxygen from maternal blood when both are at the same pO_2. Llama hemoglobin has such a high affinity for oxygen it is 100 percent saturated even at the low pO_2's found in the high Andes Mountains; they are used as pack animals because they are so well adapted to high altitudes (photograph).

Llamas and vicunas are mammals native to high altitudes in the Andes mountains of South America. Like those of the fetus discussed above, the hemoglobins of these animals must pick up O_2 in an environment with a low pO_2. In the animal's natural habitat, over 5,000 meters above sea level, the pO_2 is below 85 mm Hg, and the pO_2 in their lungs is about 50 mm Hg. The hemoglobins of the llama and the vicuna have oxygen-dissociation curves much to the left of the curves of hemoglobins of most other mammals—in other words, they can become saturated with O_2 at lower pO_2's than those of other animals can. These curves reflect the important adaptations of their hemoglobin for life at high altitude (Figure 39.17).

The oxygen-binding properties of normal adult hemoglobin are influenced by a number of physiological conditions. The influence of pH on the function of hemoglobin has been well studied and is known as the **Bohr effect**. As the pH of the blood plasma falls, the oxygen-dissociation curve shifts to the right (Figure 39.18). This means that the hemoglobin will release more O_2 to the tissues. Where does hemoglobin encounter a decreased pH as it circulates through the body? In tissues with very high metabolic rates the pH is reduced by the release of acidic metabolites such as lactic acid, fatty acids, and CO_2, which combines with water to form carbonic acid. Because of the Bohr effect, hemoglobin releases more of its bound oxygen in these tissues—another way that O_2 is supplied where and when it is most needed.

Diphosphoglyceric acid is a normal intermediary metabolite that plays an important role in regulating hemoglobin function. The mature mammalian red blood cell is a rather simple cell. It is little more than a sac of hemoglobin, but it has a very high content of diphosphoglyceric acid. The concentration of diphosphoglyceric acid in red blood cells increases in response to exercise and during acclimation to high altitude. Diphosphoglyceric acid reversibly combines with deoxygenated hemoglobin and changes its shape so that it has a lower affinity for O_2. The result is that at any particular pO_2, hemoglobin releases more of its bound O_2 than it otherwise would. In other words, diphosphoglyceric acid shifts the oxygen-dissociation curve of mammalian hemoglobin to the right.

It is interesting that the llama and the human employ opposite adaptations of hemoglobin function for life at high altitudes. The llama's hemoglobin has a left-shifted oxygen-dissociation curve, which means that it can become 100 percent saturated with O_2 at the low pO_2's at high altitude. As a consequence, the llama's tissues must operate at a lower pO_2 (Figure 39.17). In contrast, human hemoglobin acquires, through acclimation, a right-shifted oxygen-dissociation curve. The result is that human hemoglobin never becomes fully saturated with O_2 at high altitude. However, more of the O_2 carried by that hemoglobin is released to the tissues.

Transport of Carbon Dioxide from the Tissues

Delivering O_2 to the tissues is only half of the respiratory function of the blood. It also must take metabolic waste products away from the tissues. Because

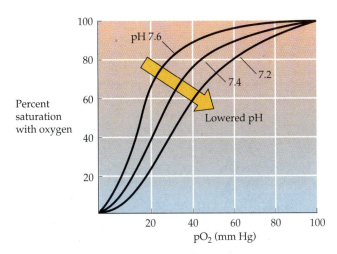

39.18 The Oxygen-Binding Properties of Hemoglobin Can Change
Factors such as pH can affect the oxygen-binding capacity of hemoglobin. Lowered pH shifts the dissociation curve to the right—more oxygen is being released to the tissues.

we are concerned with respiratory gases in this chapter, the metabolic waste product we will consider is carbon dioxide (CO_2). CO_2 is highly soluble and readily diffuses through cell membranes. It diffuses from its site of production in a cell across the plasma membrane into the blood, where its concentration is lower. Very little CO_2 is transported by the blood in this dissolved form, however. Most CO_2 produced by tissues is transported to the lungs in the form of

the **bicarbonate ion**, HCO_3^-. How and where the CO_2 becomes HCO_3, is transported, and then is converted back to CO_2 is an interesting story.

When CO_2 dissolves in water, some of it slowly reacts with the water molecules to form carbonic acid (H_2CO_3), some of which then dissociates into a proton (H^+) and a bicarbonate ion (HCO_3^-). This sequence of events is expressed:

$$CO_2 + H_2O \rightleftharpoons H_2CO_3 \rightleftharpoons H^+ + HCO_3^-$$

If this reaction proceeded rapidly, converting CO_2 to HCO_3^- as it entered the blood, it would facilitate removal of CO_2 from the tissues. Why? By converting CO_2 to HCO_3^-, the concentration of CO_2 in the blood would remain low. This would favor diffusion of CO_2 from the tissues into the blood. Unfortunately, the reaction between CO_2 and H_2O in the blood plasma goes too slowly to have much of an effect. It is a different story, however, in the red blood cells, where the enzyme **carbonic anhydrase** speeds up the conversion of CO_2 to HCO_3^-. Therefore, as shown in Figure 39.19, the normal path of CO_2 is to diffuse from tissues into the plasma and from there into the red blood cells. In the red blood cells carbonic anhydrase catalyzes the formation of carbonic acid. When this carbonic acid dissociates, the resulting bicarbonate diffuses back out into the plasma. Some CO_2 is also carried in chemical combination with deoxygenated hemoglobin as **carboxyhemoglobin**.

In the alveoli, the reactions involving CO_2 and bicarbonate are reversed. CO_2 diffuses from the blood plasma into the air in the alveoli, and is ex-

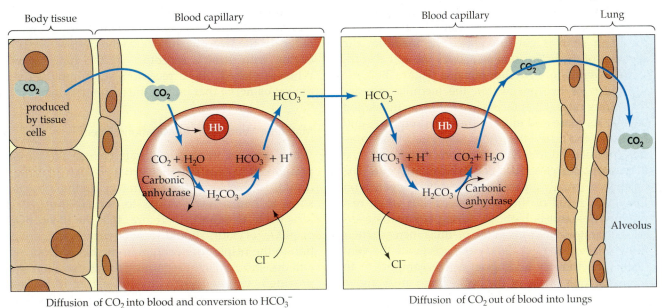

Diffusion of CO_2 into blood and conversion to HCO_3^- | Diffusion of CO_2 out of blood into lungs

39.19 Carbon Dioxide Is Transported as Bicarbonate Ions
In tissues, CO_2 diffuses from cells into plasma and into the red blood cells. In red blood cells, CO_2 is converted to bicarbonate ions because carbonic anhydrase is present. Bicarbonate ions leave red blood cells in exchange for chloride ions. In the lungs, these processes are reversed.

haled. This creates a concentration difference so that CO_2 leaves the red blood cells and enters the plasma. The loss of CO_2 from the red blood cells shifts the equilibrium between CO_2 and bicarbonate. As the HCO_3^- in the red blood cells gets converted back to CO_2, more HCO_3^- moves into the red blood cells from the plasma.

REGULATION OF VENTILATION

At the beginning of this chapter, we noted that we must breathe every minute of our lives, yet we don't worry about it or even think about it very often. Breathing is an autonomic function of the nervous system. The breathing pattern can be adjusted around other activities such as speech and eating. Most impressively, our breathing rates are adjusted to match the changing metabolic demands of our bodies. Now we will learn how the regular respiratory cycle is generated and how it is controlled so that we get the oxygen we need and eliminate the carbon dioxide we produce as our levels of activity change.

The Ventilatory Rhythm

Breathing is an involuntary function. The complex, coordinated movements of the diaphragm and other muscles do not require conscious thought. Automatically, the central nervous system maintains a ventilatory rhythm and modifies its depth and frequency to meet the demands of the body for O_2 supply and CO_2 elimination. The ventilatory rhythm ceases if the spinal cord is severed in the neck region, which indicates that the rhythm is generated in the brain. If the brain stem is cut just above the **medulla**, the segment of the brain stem just above the spinal cord, a crude ventilatory rhythm remains (Figure 39.20).

Groups of neurons within the medulla increase their firing rates (Chapter 36) just before an inhalation begins. As more and more of these neurons fire, and fire faster and faster, the inspiratory (inhalation) muscles contract. Suddenly the neurons stop firing, the inspiratory muscles relax, and exhalation begins. Exhalation is usually a passive process that depends on the elastic recoil of the lung tissues. But when respiratory demand is high, as during strenuous exercise, expiratory neurons in the medulla increase their firing rates during exhalation and accelerate the ventilatory rhythm by adding an active component to the exhalation phase of the cycle. Brain areas above the medulla can also modify the ventilatory rhythm. The rhythm is modified to accommodate speech, ingestion of food, coughing, and emotional states.

As respiratory demands increase, the activities of the inspiratory neurons also increase, thus contributing to greater depth of inspiration. There is an

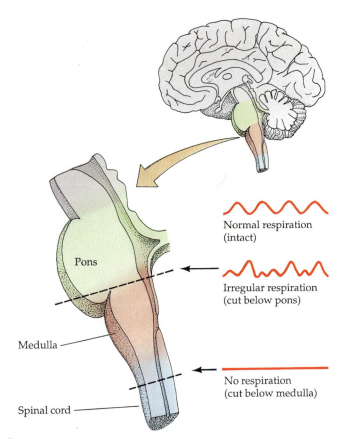

39.20 The Brainstem Generates and Controls Breathing Rhythm
Transections of the brainstem at different levels reveal that the basic breathing rhythm is generated in the medulla and modified by neurons in or above the pons.

override reflex, however, that prevents the ventilatory muscles from overdistending and damaging the lung tissue. This reflex is named the **Hering–Breuer reflex** for the two physiologists who discovered it. It begins with stretch receptors in the lung tissue. When stretched, these receptors send impulses via the **vagus nerve** that inhibit the inspiratory neurons.

Matching Ventilation to Metabolic Needs

When changes in metabolic rate change the partial pressure of oxygen and the partial pressure of carbon dioxide in the blood, the respiratory rhythm changes so as to return these values to normal levels. It is reasonable to expect, therefore, that the blood pO_2, or pCO_2, or both, should provide feedback information to the respiratory centers in the brain. Dramatic and disastrous insight regarding this expectation was provided by three French physiologists in 1875. They wanted to investigate the physiological effects of breathing low concentrations of O_2. Sophisticated gas pumps and pressure chambers did not exist in 1875, so the three decided to go up in a balloon to very high altitudes and observe the effects of the rarefied atmosphere on one another. They noted no ill effects in their notebooks, and continued to throw out ballast, going higher than 8,000 meters. Then all three became unconscious. The balloon finally descended on its own, and one of the physiologists regained consciousness to find his two colleagues dead. This infamous flight of the balloon *Zenith* is tragic proof that the human body is not very good at sensing its own need for O_2.

Humans and other mammals are very sensitive, however, to increases in the pCO_2 of the blood, whether it is influenced by energy demands or by the composition of the air breathed. If you rebreathe a small volume of air so that the pCO_2 of that reservoir gradually rises, your breathing becomes deeper and more rapid, and you become anxious and agitated. You react the same way even if pure O_2 is constantly released into the reservoir to keep its pO_2 constant. Typical ventilatory responses to changes in blood pO_2 and pCO_2 are shown in Figure 39.21.

It makes sense that CO_2 rather than O_2 should be the dominant feedback stimulus for ventilation. As we have seen, animals have evolved respiratory systems and hemoglobin properties that work to keep the blood leaving the gas-exchange surfaces fully saturated with O_2 over a broad range of alveolar pO_2's and metabolic rates. Normal fluctuations in metabolism and ventilation have very little effect on the maximum amount of O_2 carried by the blood. In contrast, small changes in metabolism and alveolar pCO_2 do influence the concentration of CO_2 in the blood. Consequently, changes in blood pCO_2 are a

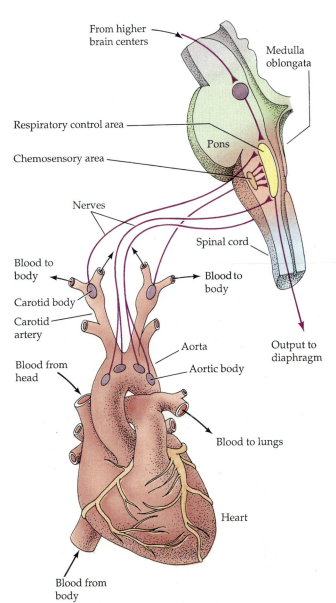

39.22 Chemosensors Provide Information
Chemosensors on large blood vessels leaving the heart are sensitive to the partial pressure of oxygen in the blood; other chemosensors on the surface of the medulla are sensitive to the partial pressure of carbon dioxide in the blood. The body uses information from these chemosensors to match breathing rate to metabolic demand.

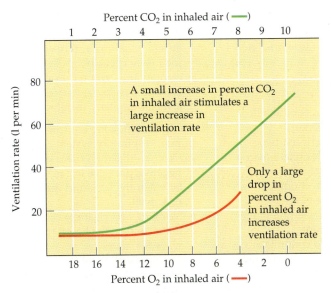

39.21 Carbon Dioxide Affects Breathing
Breathing is more sensitive to increased carbon dioxide than it is to decreased oxygen. Even a small change in the partial pressure of carbon dioxide in the blood causes a large change in breathing, whereas a large change in the blood's partial pressure of oxygen causes only a small change in breathing.

much finer index of energy demands and respiratory performance than is the O_2 content of the blood.

Where are gas concentrations in the blood sensed? The major site of CO_2 sensitivity is an area on the ventral surface of the medulla, not far from the groups of neurons that generate the ventilatory rhythm. Sensitivity to the O_2 concentration of the blood resides in nodes of tissue on the large blood vessels leaving the heart, the aorta and the carotid arteries (Figure 39.22). These carotid and aortic bodies receive enormous supplies of blood, and they contain chemosensory nerve endings. If the blood supply to these structures decreases, or if the pO_2 of the blood falls dramatically, then the chemosensors are activated and send impulses to the respiratory centers. Although we are not very sensitive to changes in blood pO_2, the carotid and aortic bodies can stimulate increases in ventilation during exposure to very high altitude or when blood volume or blood pressure are very low.

SUMMARY

The respiratory gas exchange of animals involves the uptake of O_2 from the environment and the release of CO_2 to the environment. These exchanges take place by diffusion only, but the rates of diffusion can be maximized by adaptations that ventilate both the internal and external surfaces of the respiratory gas-exchange surfaces. An aqueous environment, high altitude, and high metabolic rates place heavy demands on respiratory systems. Fishes and birds have respiratory systems that are very effective in extracting oxygen from the environment; both have evolved ventilatory systems that provide continuous, unidirectional flows of respiratory media over the gas-exchange surfaces. Because of their tidal ventilation, mammalian lungs can extract a lower percentage of the oxygen content of the respiratory medium than do fish gills or bird lungs. Nevertheless, the mammalian lung, with its millions of alveoli, has a very large surface area for diffusion. A unique system for gas exchange is found in insects, which have a network of tracheae throughout their bodies. The tracheae enable gases to diffuse rapidly between individual cells and the external environment.

Transport of O_2 from the lungs or gills of vertebrates is enhanced by hemoglobin contained within the red blood cells. Hemoglobin binds with O_2 in the lungs, where pO_2 is high, and releases O_2 in the tissues, where pO_2 is lower. The affinity of hemoglobin for oxygen is influenced by a number of factors, including the structure of hemoglobin, temperature, pH, and an effector molecule, diphosphoglyceric acid. Another oxygen-binding molecule, myoglobin, is found in muscle cells. Myoglobin acts as a reservoir of O_2, which it releases when the pO_2 of the muscle falls to very low levels.

The rate of ventilation, or breathing, is matched to the metabolic demands of the body by the nervous system. The pO_2 of the blood is sensed by chemoreceptors on blood vessels near the heart, and the pCO_2 of the blood is sensed by chemoreceptors in the brain. Ventilatory rate is much more sensitive to changes in blood pCO_2 than it is to changes in blood pO_2.

SELF-QUIZ

1. Which of the following statements is *false*?
   a. In mammalian respiratory systems, respiratory gases are exchanged by diffusion only.
   b. Oxygen has a lower rate of diffusion in water than in air.
   c. The oxygen content of water falls as the temperature of water rises, all other things being equal.
   d. The amount of oxygen in the atmosphere decreases with increasing altitude.
   e. Because of the relative inefficiency of their breathing apparatus, mammals have evolved active transport mechanisms to augment their respiratory gas exchange.

2. Which of the following statements about the respiratory system of birds is *false*?
   a. Respiratory gas exchange does not occur in the air sacs.
   b. The respiratory system of birds can achieve more complete exchange of O_2 from air to blood than that of humans.
   c. Air passes through birds' lungs in only one direction.
   d. The gas-exchange surfaces in bird lungs are the alveoli.
   e. A breath of air remains in the bird respiratory system for two breathing cycles.

3. In a countercurrent exchange system:
   a. the fluids in two conduits flow in opposite directions.
   b. gas exchange between two streams is less complete than if they flowed in parallel.
   c. the fluids in two conduits flow in the same direction.
   d. the materials in solution move against the current.
   e. the fluids in two conduits are funneled into a final common vessel.

4. In the human respiratory system:
   a. the lungs and airways are completely collapsed after a forceful exhalation.
   b. the average O_2 concentration of air inside the lungs is always less than it is in the air outside of the lungs.
   c. the pO_2 of the blood leaving the lungs is greater than the pO_2 of the exhaled air.
   d. the amount of air that is moved per breath during normal, at-rest breathing is termed the total lung capacity.
   e. oxygen and carbon dioxide are actively transported across the alveolar–capillary membranes.

5. All of the following are true of the human respiratory system *except*:
   a. During inhalation there is a negative pressure in the space between the lung and the thoracic well.
   b. Smoking one cigarette can immobilize the cilia lining the airways for hours.
   c. The respiratory control center in the medulla responds more strongly to changes in arterial O_2 concentration than to changes in arterial CO_2 concentrations.
   d. Without surfactant, the work of breathing is greatly increased.
   e. The diaphragm contracts during inhalation and relaxes during exhalation.

6. The hemoglobin of a human fetus:
   a. is the same as that of an adult.
   b. has a higher affinity for O_2 than that of an adult.
   c. has only two protein subunits instead of four.
   d. is supplied by the mother's red blood cells.
   e. has a lower affinity for O_2 than that of the adult.

7. The amount of oxygen carried by hemoglobin depends upon the partial pressure of oxygen in the blood. Hemoglobin in active muscles:
   a. becomes saturated with oxygen.
   b. takes up only a small amount of oxygen.
   c. readily unloads oxygen.

   d. tends to decrease the partial pressure of oxygen in the muscle tissues.
   e. is denatured.

8. Most carbon dioxide is carried in the blood:
   a. in red blood cell cytoplasm.
   b. dissolved in the plasma.
   c. in the plasma as bicarbonate ions.
   d. bound to plasma proteins.
   e. in red blood cells bound to hemoglobin.

9. Myoglobin:
   a. binds O_2 at pO_2's at which hemoglobin is releasing its bound O_2.
   b. has a lower affinity for O_2 than hemoglobin.

   c. consists of four polypeptide chains like hemoglobin.
   d. provides an immediate source of O_2 for muscle cells at the onset of activity.
   e. can bind four O_2 molecules at once.

10. Carbon dioxide, a product of cellular respiration, is carried in the bloodstream. When the level of CO_2 in the blood becomes greater than the set operating range:
   a. the rate of respiration decreases.
   b. the pH of the blood rises.
   c. the respiratory centers become dormant.
   d. the rate of respiration increases.
   e. the blood becomes more alkaline.

FOR STUDY

1. Compare and contrast the respiratory systems of birds, fish, and humans. Why can birds and fish outperform mammals in environments where the concentration of O_2 is low?

2. What does the following chemical equation represent and how does it relate to gas exchange in human lungs and in active tissues?

$$CO_2 + H_2O \rightleftharpoons H_2CO_3 \rightleftharpoons H^+ + HCO_3^-$$

3. Describe and contrast the adaptations of llamas and humans for gas exchange at high altitude.

4. Describe, in terms of the structure of the human respiratory system, how air is brought into and then expelled from the lungs.

5. Workers A and B must inspect two large gas storage tanks. Tank 1 contains 100 percent N_2; tank 2 contains 100 percent CO_2. The tanks are *not* flushed out before inspection. Worker A goes into tank 1, becomes unconscious and dies. Worker B goes into tank 2, feels strangely short of breath, and leaves the tank while feeling somewhat dizzy.

   Explain why worker A died whereas worker B lived.

READINGS

Eckert, R., D. Randall and G. Augustine. 1988. *Animal Physiology: Mechanisms and Adaptations*, 3rd Edition. W. H. Freeman, New York. An outstanding textbook of animal physiology. Chapter 14 covers gas exchange.

Feder, M. E. and W. W. Burggren. 1985. "Skin Breathing in Vertebrates." *Scientific American*, November. Not all breathing involves lungs or gills. This article discusses adaptations possessed by many vertebrates for gas exchange through the skin.

Perutz, M. F. 1978. "Hemoglobin Structure and Respiration." *Scientific American*, December. An authoritative article on hemoglobin structure and function by the man who received the Nobel prize for his work on this subject.

Schmidt-Nielsen, K. 1971. "How Birds Breathe." *Scientific American*, December. Describes the complex adaptations of the avian respiratory system.

Schmidt-Nielsen, K. 1990. *Animal Physiology: Adaptation and Environment*, 4th Edition. Cambridge University Press, New York. An outstanding textbook that emphasizes the comparative approach.

Vander, A. J., J. H. Sherman and D. S. Luciano. 1990. *Human Physiology: The Mechanisms of Body Function*, 5th Edition. McGraw-Hill, New York. Chapter 14 deals with human respiration.

40

Internal Transport and Cardiovascular Systems

PREVIEW: All but the smallest of animals must be able to transport within their bodies nutrients, respiratory gases, hormones, metabolic products and wastes, and elements of the immune system. This transport is carried out by cardiovascular systems, and blood is the vehicle that transports material. The cardiovascular systems of many invertebrates and all vertebrates have a muscular pump, or heart, that moves blood through a closed network of vessels. In the smallest of these vessels, exchanges of materials between the blood and the cells of the body take place. The cardiovascular system of verte-brates is controlled by the autonomic nervous system, which alters the output of the heart and controls the amount of blood flowing to different regions of the body. Hormones play important roles in regulating cardiovascular function.

This chapter deals with open and closed circulatory systems, the evolution of the vertebrate heart, atria and ventricles, pulmonary and systemic circuits, arteries, veins, and lymph vessels, pacemaker activity, clotting factors, and the regulation of cardiovascular functions.

Sweating, severe chest pain, fainting, 911, paramedics, flashing lights, emergency room, intensive care—heart attack. (Figure 40.1). We all will experience this sequence of events directly or indirectly. More than one-third of the deaths in the United States are due to heart failure, and many such deaths are those of a sudden nature that we call heart attacks. Why is it so serious when the heart fails? The answer is that all organs of the body depend on the heart.

Remember from Chapter 33 that the needs of all the cells in the body of an animal are served by the internal fluid environment. Cells take up nutrients from the fluid that bathes them, and they release their waste products into that fluid. As a result, the activities of cells change the internal fluid environment in ways that are not healthy for the cells themselves. The individual organs function to return different aspects of the internal environment to optimal levels. The lungs take in oxygen and eliminate carbon dioxide, the digestive tract takes in nutrients, the liver controls nutrient levels and eliminates toxic compounds, and the kidneys control salt concentrations and eliminate toxic wastes. The cells of each organ depend on the activities of all of the other organs, and it is only through a system of transport, or circulation, that the activities of each organ can influence the internal environment of the entire body. So, when the heart stops, transport stops, the internal environment deteriorates, and cells get sick and die.

A heart is not all there is to a transport system; it is just the pump. In addition, there must be a vehicle to transport materials through the system—the blood—and a series of conduits through which the materials can be pumped around the body—the blood vessels. Heart, blood, and vessels together comprise a cardiovascular system (*kardia* is the Greek word for heart, and *vasculum* is the Latin word for small vessel). In this chapter we mainly explore the structure and function of vertebrate cardiovascular systems, which reach the highest level of performance in birds and mammals, but we will also examine the simpler transport systems of some invertebrates.

TRANSPORT SYSTEMS OF VARIOUS ANIMAL GROUPS

Although this chapter focuses on cardiovascular systems, the simplest transport systems are not

914

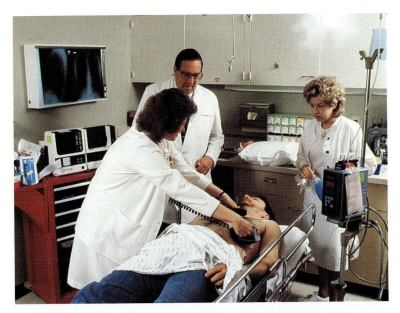

40.1 Heart Attack!
Hospital emergency rooms deal with the aftermath of sudden heart attacks. Here an emergency team attempts to stimulate a failed heart to resume beating.

vascular—they do not use vessels. Some do not even use a pump to move things around. But they carry out the essential task of transport systems: transporting substances to and away from every cell of the bodies of animals.

Gastrovascular Cavities

A circulatory system is unnecessary if all the cells of an organism are close enough to the external environment so that nutrients, respiratory gases, and wastes can diffuse directly between the cells and the outside environment. Among small aquatic invertebrates, various structures and shapes permit such direct diffusional exchange. The hydra, a cnidarian, provides a good example (see Figures 25.11 and 25.15). This aquatic animal is cylindrical and only two cell layers thick. Each of the hydra's cells contacts water either surrounding the animal or circulating through its **gastrovascular cavity**, a dead-end sac that serves both for digestion (gastro) and transport (vascular) (Figure 40.2a). The cells of the bodies of some other invertebrates are served by diffusion from highly branched gastrovascular systems. In addition, flattened body shapes minimize the diffusion path length—the distance that molecules have to diffuse between cells and the external environment (Figure 40.2b). A central gastrovascular system cannot, however, serve the needs of larger animals with many layers of cells. Such animals must have a transport or circulatory system. Also, for terrestrial animals, interstitial fluid, rather than an external medium, bathes all the cells of the body; a circulatory system is needed to maintain the composition of this interstitial fluid.

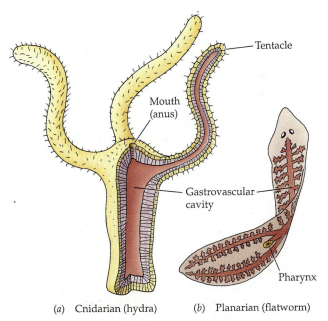

(a) Cnidarian (hydra)   (b) Planarian (flatworm)

40.2 Gastrovascular Cavities
Gastrovascular cavities in animals without circulatory systems serve the metabolic needs of the innermost cells of the body. (a) The gastrovascular cavity of a cnidarian, *Hydra*, even extends into the tentacles. No cell of the hydra is more than one cell removed from either the gastrovascular cavity or the external medium. (b) The gastrovascular cavity of a flatworm, *Planaria*, extends into all regions of the animal's flattened body.

(a) Insect

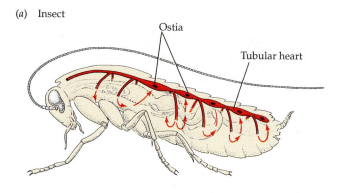

(b) Mollusk

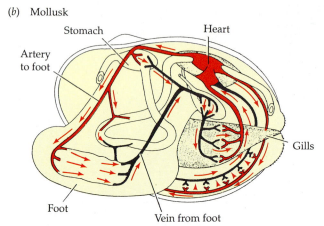

40.3 Open Circulatory Systems
The open circulatory systems of an arthropod (a) and a mollusk (b). Blood is pumped by a tubular heart through vessels to all parts of the body, where it trickles through interstitial spaces. Eventually the blood, or hemolymph, reenters the heart through the ostia.

Open Circulatory Systems

In the simplest circulatory systems, the interstitial fluid is simply squeezed through intercellular spaces as the animal moves. In these **open circulatory systems**, there is no distinction between interstitial fluid and blood. Usually a muscular pump, or heart, assists the distribution of the fluid throughout the tissues. The contractions of the heart propel the interstitial fluid through vessels leading to different regions of the body, but the fluid leaves those vessels to trickle through the tissues and eventually to return to the heart. The fluid enters the heart through simple holes called **ostia** when the heart relaxes. Examples of open circulatory systems can be seen in mollusks and arthropods (Figure 40.3).

Closed Circulatory Systems

In a **closed circulatory system** the blood never leaves the vessels. Blood circulates through the vascular system, pumped by one or more muscular hearts. The system keeps the circulating blood separate from the interstitial fluid.

A simple example of a closed circulatory system is that of the common earthworm, an annelid (Figure 25.23). One large blood vessel on the ventral side of the earthworm carries blood from its anterior end to its posterior end. In each segment of the worm, smaller vessels branch off and transport the blood to even smaller vessels in the tissues of that segment. Here, respiratory gases, nutrients, and metabolic wastes diffuse between the blood and the interstitial fluids. The blood then flows on into larger and larger vessels that eventually lead into a single large vessel on the dorsal side of the worm. The dorsal vessel carries the blood from the posterior to the anterior end. Five pairs of vessels connect the large dorsal and ventral vessels in the anterior end, thus completing the circuit (Figure 40.4). Contractions of the dorsal vessel and of the five pairs of connecting vessels keep the blood circulating, and therefore those vessels serve as hearts for the earthworm.

There are several advantages of closed circulatory systems over open ones. First, closed systems can deliver oxygen and nutrients to the tissues and carry away metabolic wastes more rapidly than open systems can. Second, blood can be directed to specific tissues. Third, cellular elements and large molecules that function within the vascular system can be kept within it; examples are red blood cells and large molecules that help in the distribution of hormones and nutrients. Overall, closed circulatory systems are superior in metabolic capacity. How then do highly active insect species achieve high levels of metabolic output with their open circulatory systems? The key to answering this question is to remember something you learned in the previous chapter: Insects do not depend on their circulatory systems for respiratory gas exchange (see Figure 39.6).

All vertebrates have closed circulatory systems and chambered hearts. Unlike the contractile vessels of the earthworm, a chambered heart has valves that prevent the backflow of blood when the heart contracts. From fishes to amphibians to reptiles to birds and mammals, there is an increasing complexity of

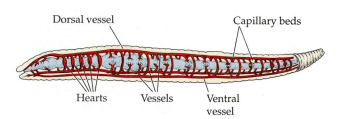

40.4 A Closed Circulatory System
In a closed circulatory system, blood is confined to the blood vessels, kept separate from the interstitial fluid, and is pumped by one or more muscular hearts. The earthworm exemplifies this system, with large dorsal and ventral blood vessels and a branching network of smaller capillaries.

the heart and an increase in the number of chambers. An important consequence of this increased complexity is the separation of the circulation into two circuits, one to the lungs and one to the body. In fishes, blood is pumped from the heart to the gills and then to the tissues of the body and back to the heart. In birds and mammals, blood is pumped from the heart to the lungs and back to the heart in the **pulmonary circuit**, and from the heart to the rest of the body and back to the heart in the **systemic circuit**. We will see how the separation of the circulation into two circuits improves the efficiency and capacity of the cardiovascular system.

The vascular system includes **arteries** that carry blood away from the heart, and **veins** that carry blood back to the heart. **Arterioles** are small arteries, and **venules** are small veins. **Capillaries** are very small, thin-walled vessels that connect arterioles and venules. Exchanges between the blood and the interstitial fluid take place only across capillary walls.

Circulatory Systems of Fishes

Fishes have two-chambered hearts. A less muscular chamber called the **atrium** receives blood from the body and pumps it into a more muscular chamber, the **ventricle**, which then pumps the blood to the gills, where gas exchange takes place (Figure 40.5a). Blood leaving the gills collects in a large dorsal artery, the **aorta**, which distributes it to smaller arteries and arterioles leading to all the organs and tissues of the body. In the tissues the blood flows through capillary beds, then collects in venules and veins, and is eventually returned to the heart. Most of the pressure imparted to the blood by the contraction of the ventricle is dissipated by the high resistance of the many tiny, narrow spaces the blood flows through in the gills. As a result, the blood entering the aorta of the fish is not under very high pressure, and this limits

40.5 Vertebrate Circulatory Systems
All vertebrates have closed circulatory systems. (a) Fishes have a two-chambered heart—a single atrium and a single ventricle. Blood is pumped first through the gills and then to the other tissues of the body. (b) In the lungfish, some of the blood that flows through the gills then flows through the lungs and back to the heart. The lungfish heart has two atria, one receiving oxygenated blood from the lung and one deoxygenated blood from the body. (c) In adult amphibians, there is partial separation of the pulmonary and systemic circulations. The heart is three-chambered, with separate atria to receive oxygenated blood from the lungs and deoxygenated blood from the body. Even though there is only one ventricle, its anatomy tends to direct blood from the right atrium to the lungs and blood from the left atrium to the aorta. (d) In reptiles, the ventricle of the three-chambered heart is partially divided by a septum to further direct the flow of oxygenated blood to the body and deoxygenated blood to

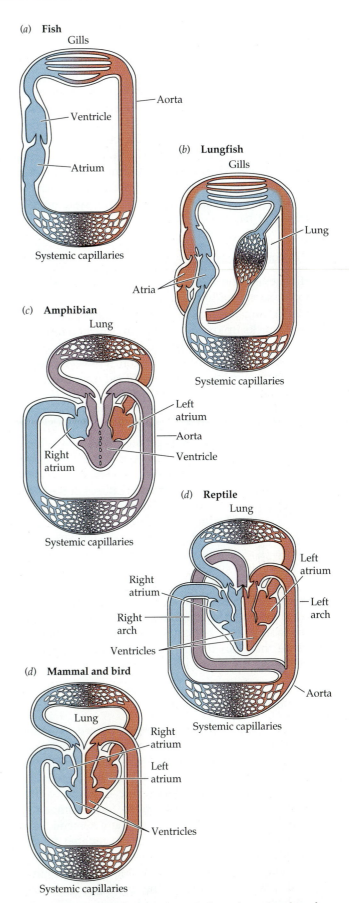

(a) **Fish**

(b) **Lungfish**

(c) **Amphibian**

(d) **Reptile**

(d) **Mammal and bird**

the lungs. (e) Birds and mammals have four-chambered hearts. Their pulmonary and systemic circulatory systems are totally separate.

the ability of the fish circulatory system to supply the tissues with oxygen and nutrients.

An important evolutionary step is suggested by the circulatory systems of lungfish. The lungfish has a simple outpocketing of its gut that serves as an air bladder (Figure 39.11b). Some of the blood that would normally flow through the gills is routed to the air bladder. This blood returns directly to the heart rather than joining the arterial blood in the aorta (Figure 40.5b). This simple anatomical feature in a modern-day species provides a clue about how the evolution of separate pulmonary and systemic circuits might have taken place.

Circulatory Systems of Amphibians

Partial separation of pulmonary and systemic circulation is seen in adult amphibians such as frogs and toads. A single ventricle pumps blood to the lungs, where it picks up oxygen and dumps carbon dioxide, and also to the rest of the body, where it picks up carbon dioxide and dumps oxygen. However, separate atria receive the oxygenated blood from the lungs and the deoxygenated blood from the body (Figure 40.5c). Because both of these atria deliver blood to the same ventricle, there is a potential for mixing of the oxygenated and deoxygenated blood in which case the blood going to the tissues would not carry a full load of oxygen. In reality, however, mixing is rather limited because anatomical features of the ventricle tend to direct the flow of deoxygenated blood from the right atrium to the pulmonary circuit and the flow of oxygenated blood from the left atrium to the aorta. The advantage of this partial separation of pulmonary and systemic circulation is that the high resistance of the capillary beds of the gas-exchange organ no longer lies between the heart and the tissues. Therefore, the amphibian heart delivers blood to the aorta, and hence to the body, at a higher pressure than can the fish heart, which pumps the blood through the gills first.

Circulatory Systems of Reptiles

Most reptiles have three-chambered hearts, but the ventricle is partially divided by a septum that directs the deoxygenated blood to the pulmonary circulation and the oxygenated blood to the systemic circulation (Figure 40.5d). In crocodilians, the division of the ventricle is complete, producing a four-chambered heart that separates the pulmonary and systemic circulation. The result is that fully oxygenated blood can be delivered under high pressure to the tissues of the body, and deoxygenated blood can be delivered at lower pressure to the pulmonary circuit, which has lower resistance than the systemic circuit.

Circulatory Systems of Birds and Mammals

Birds and mammals have four-chambered hearts and fully separated pulmonary and systemic circulations (Figure 40.4e). Several advantages arise from this design. First, as noted for the crocodilians, the systemic arterial blood can have the highest possible oxygen content. Second, respiratory gas exchange is maximized by sending blood with the lowest oxygen concentration to the lungs. Third, because the two circulatory systems are separate, they can operate at different pressures. The pulmonary circuit has lower resistance to blood flow than the systemic, so it can operate at a lower pressure. A systemic circuit at high pressure efficiently supplies oxygen and nutrients to the tissues of the body.

THE HUMAN HEART

Structure and Function

Like all mammalian hearts, the human heart has four chambers, two atria and two ventricles (Figure 40.6). The atrium and ventricle on the right side of your body are called the right atrium and right ventricle. The atrium and ventricle on the left side of your body are called the left atrium and left ventricle. Each atrium pumps blood into its respective ventricle, and the ventricles pump blood into arteries. The right ventricle pumps blood through the pulmonary circuit, and the left ventricle pumps blood through the systemic circuit. Valves between the atria and ventricles, the **atrioventricular valves**, prevent backflow of blood into the atria when the ventricles contract. The **pulmonary** and **aortic valves** positioned between the ventricles and the arteries prevent the backflow of blood into the ventricles.

The right atrium receives blood from the **superior** and **inferior venae cavae**, which are large veins that collect blood from the upper and lower body, respectively. The veins of the heart itself also drain into the right atrium. The right ventricle pumps blood into the **pulmonary artery**, which transports it to the lungs. The **pulmonary vein** returns the oxygenated blood from the lungs to the left atrium, which pumps it into the left ventricle. The walls of the left ventricle are powerful muscles that contract around the blood with a wringing motion starting from the bottom. When pressure in the left ventricle is high enough to push open the **aortic valve**, the blood rushes into the aorta to begin its circulation throughout the body. Because there are many more arterioles and capillaries in the systemic circuit than in the pulmonary circuit, the systemic resistance is much higher than the pulmonary resistance. Because the left ventricle must squeeze the blood with greater force than does

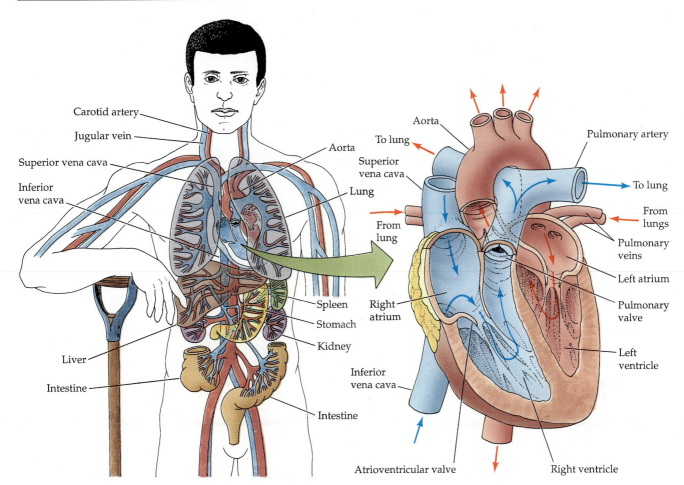

40.6   The Human Heart and Circulation
Deoxygenated blood from the tissues of the body returns to the right atrium and flows through an atrioventricular valve into the right ventricle. The right ventricle pumps the blood into the pulmonary circuit, from which it returns to the left atrium and flows into the left ventricle through an atrioventricular valve. The left ventricle pumps blood into the systemic circuit. The atrioventricular valves prevent blood from flowing back into the atria when the ventricles contract. Pulmonary and aortic valves prevent blood from flowing back into ventricles from the arteries when the ventricles relax.

the right ventricle, the muscular wall of the left ventricle is about six times thicker than that of the right ventricle.

The pumping of the heart—contraction of the two atria followed by contraction of the two ventricles—is the **cardiac cycle** (Figure 40.7). Contraction of the ventricles is called **systole** and relaxation of the ventricles is called **diastole**. When a doctor measures the blood pressure in a large artery such as the one in your upper arm, the pressure rises to a maximum during systole and falls to a minimum during diastole (Figure 40.8). In a conventional blood-pressure reading, the systolic value is placed over the diastolic value. Normal values for a young adult might be 120 mm of mercury (Hg) during systole and 80 mm Hg during diastole, or 120/80.

The sounds of the cardiac cycle, the "lub-dub" heard through a stethoscope placed on the chest, are created by the slamming shut of the heart valves. As the ventricles begin to contract, the atrioventricular valves close (lub), and when the ventricles begin to relax, the pressure in the aorta and pulmonary artery causes the aortic and pulmonary valves to bang shut (dub). Defective valves produce the sounds of heart murmurs. For example, if an atrioventricular valve is defective, blood will flow back into the atria with a "whoosh" sound following the "lub."

Cardiac Muscle and the Heartbeat

The cardiac muscle has some unique properties, as we saw in Chapter 38, that help it function as an effective pump. One of those properties is the fact that cardiac muscle cells are in electrical continuity with one another. Special junctions called gap junctions enable action potentials to spread rapidly from cell to cell. (Recall what we learned about gap junctions in Chapters 5 and 36). Because a spreading

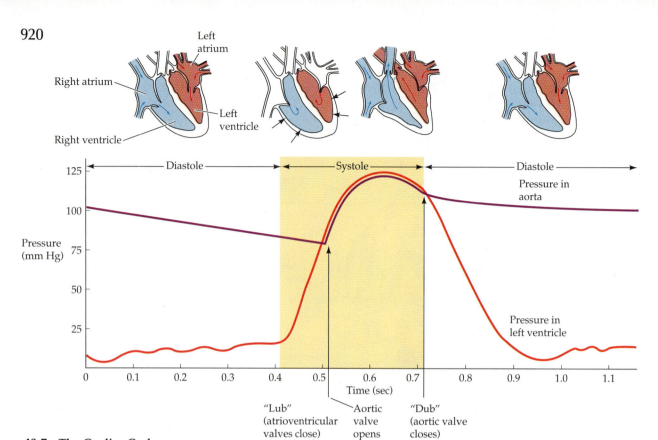

Left atrium

Right atrium

Left ventricle

Right ventricle

"Lub"
(atrioventricular valves close)

Aortic valve opens

"Dub"
(aortic valve closes)

40.7 The Cardiac Cycle

The rhythmic contraction and relaxation of the atria and ventricles is called the cardiac cycle. The cardiac cycle can be felt in the pulse and heard through a stethoscope. The heart sounds are caused by the heart valves slamming shut. Systole refers to the contraction of the ventricles; their relaxation is called diastole.

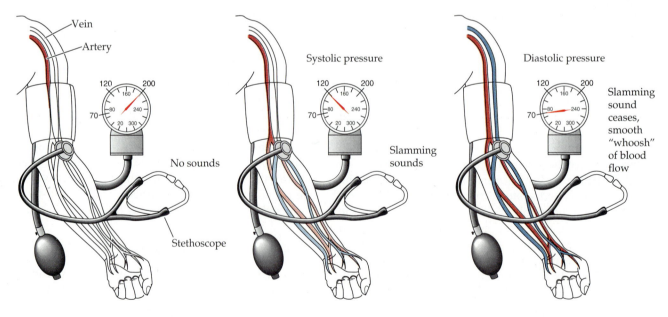

(a) Pressure in cuff is increased to close both arteries and veins. No sound is audible

(b) Pressure in cuff is gradually lowered until sound of artery opening and closing during systole is heard. At this time pressure in cuff is just below systolic pressure in artery

(c) Pressure is further lowered and sound ceases when artery remains open for entire heart cycle. Cuff is just below the diastolic pressure in artery

Blood pressure in this person is 120/70

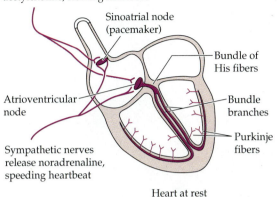

Parasympathetic nerves release acetylcholine, slowing heartbeat

Sinoatrial node (pacemaker)

Bundle of His fibers

Atrioventricular node

Bundle branches

Purkinje fibers

Sympathetic nerves release noradrenaline, speeding heartbeat

Heart at rest

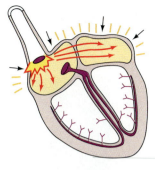

Sinoatrial node fires, action potentials spread through atria which contract

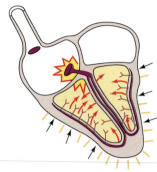

Atrioventricular node fires, sending impulses along conducting fibers; ventricles contract

action potential stimulates contraction, large groups of cardiac muscle cells contract in unison, which is important for pumping blood. The massive electrical events associated with the heartbeat can be measured on the body surface as an electrocardiogram or EKG (see Box 40.A).

Another important property of cardiac muscle cells is that some of them have the ability to initiate action potentials and therefore can contract spontaneously without stimulation from the nervous system. These cells stimulate neighboring cells to contract, thereby acting as pacemakers. The important characteristic of a pacemaker cell is that its resting membrane potential gradually becomes less negative until it reaches the threshold voltage for initiating an action potential. The nervous system controls the heartbeat (speeds it up or slows it down) by influencing the

◄ 40.8 Blood Pressure Measured in an Artery of the Arm

(a) Blood pressure in a major artery can be measured with an inflatable pressure cuff called a sphygmomanometer. When the inflation pressure of the cuff exceeds the blood pressure in the artery, blood flow in the artery stops. *(b)* If the pressure in the cuff is gradually released, a point is reached when blood pressure at the peak of systole is greater than the pressure in the cuff. When this occurs, a little blood will squeeze through the occluded artery and the artery will slam shut, producing a sound that can be heard through a stethoscope applied to the arm. The pressure at which these slamming sounds are heard first is the systolic pressure. *(c)* As the pressure in the cuff is released even more, the slamming sound decreases and becomes more of a periodic "whoosh" sound. This is because the artery still occludes whenever the blood pressure falls below the cuff pressure at some time during the cardiac cycle. Eventually a cuff pressure is reached at which the sound of blood flow is continuous; this is the diastolic pressure. When cuff pressure is below diastolic pressure, the artery does not occlude at any time during the cardiac cycle.

40.9 The Heartbeat

Pacemaker cells in the sinoatrial node initiate action potentials that spread through the walls of the atria, causing them to contract. Because the walls of the ventricles are not in electrical continuity with the atrial muscle tissue, the action potentials initiated in the pacemaker must pass through the atrioventricular node. When cells of the atrioventricular node fire action potentials, they spread rapidly through the bundle of His and Purkinje fibers to all regions of the ventricular muscle, causing it to contract. Parasympathetic nerves slow the pacemaker; sympathetic nerves speed up the pacemaker.

rate at which pacemaker cells undergo their gradual depolarization between action potentials.

Under normal circumstances the pacemaker activity of the heart originates from modified cardiac muscle cells located at the junction of the superior vena cava and right atrium, in the **sinoatrial node** (Figure 40.9). An action potential spreads from the sinoatrial node across the atrial walls, causing the two atria to contract in unison. There are no gap junctions, however, between the atria and the ventricles. The action potential initiated in the atria passes to the ventricles through another node of modified cardiac muscle cells called the **atrioventricular node**. The atrioventricular node passes the action potential on to the ventricles via modified muscle fibers called the **bundle of His**. The bundle of His divides into right and left bundle branches, which connect with **Purkinje fibers** that branch throughout the ventricular muscle (Figure 40.9).

The timing of the spread of the action potential from atria to ventricles is important. The atrioventricular node imposes a short delay in the spread of the action potential from atria to ventricles. Then, the action potential spreads very rapidly throughout the ventricles, causing them to contract. Thus the atria contract before the ventricles do, so that the blood passes progressively from the atria to the ventricles to the arteries.

BOX 40.A

The Electrocardiogram

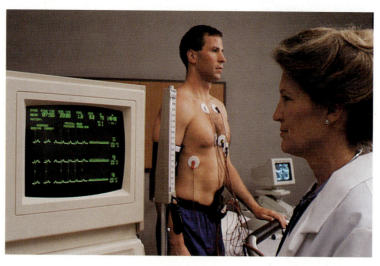

The electrocardiogram is used to monitor heart function during an exercise tolerance test.

A normal electrocardiogram (EKG)

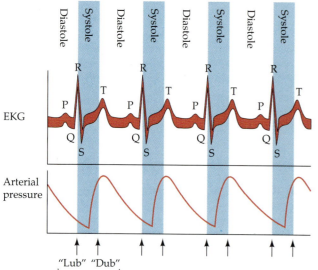

Diastole | Systole | Diastole | Systole | Diastole | Systole | Diastole | Systole

EKG

Arterial pressure

"Lub" "Dub"

The sounds heard through a stethoscope occur at the beginning and end of systole

Some abnormal EKGs

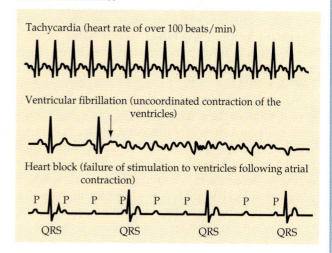

Tachycardia (heart rate of over 100 beats/min)

Ventricular fibrillation (uncoordinated contraction of the ventricles)

Heart block (failure of stimulation to ventricles following atrial contraction)

P   P   P   P   P   P   P   P

QRS   QRS   QRS   QRS

Besides detecting rhythmic irregularities in heart beat (arrhythmias), EKGs can detect damage to the heart muscle (infarctions) or decreased blood supply to the heart muscle (ischemias) by changes in the size and shape of the EKG curves

During the cardiac cycle, there are electrical events in the cardiac muscle that can be recorded by placing electrodes on the surface of the body. The recording is called an **electrocardiogram**, or EKG (remember that the Greek word for heart is *kardia*, hence EKG, but ECG is also used). The EKG is an important tool for diagnosing heart problems. The action potentials that sweep through the muscles of the atria and the ventricles prior to their contraction are such massive, localized electrical events that they cause electrical currents to flow outward from the heart to all parts of the body. Electrodes placed on the surface of the body at different locations—usually on the

wrists and ankles—detect those electrical currents at different times because the heart is positioned asymmetrically in the chest cavity. The appearance of the EKG depends on the exact placement of the electrodes used for the recording. Placing them on the right wrist and left ankle produced the EKG shown.

The waves of the EKG are designated as P, Q, R, S, and T. P corresponds to the depolarization and contraction of the atrial muscles, QRS together correspond to the depolarization of the ventricles, and T corresponds to the relaxation and repolarization of the ventricles. Where is the electrical event corresponding to the repolarization of the atria? Repolari-

zation of the atria occurs at the same time as the massive depolarization of the ventricles, so the QRS complex completely masks the smaller electrical event resulting from atrial repolarization. Below the EKG tracing are drawn the corresponding pressures in the aorta as well as the timing of the heart sounds.

From EKGs recorded after a person has a heart attack, cardiologists (heart specialists) can determine which regions of the heart were damaged. To obtain such an EKG, electrodes are positioned around the heart on the chest wall. Comparing EKGs from the different electrodes tells the cardiologist which region of the heart is behaving abnormally.

Control of the Heartbeat

The activity of pacemaker cells, and therefore the heartbeat rate, can be altered by the autonomic nervous system. Parasympathetic nerves can release acetylcholine at the sites of the sinoatrial and atrioventricular nodes. Acetylcholine slows the pace of action potential generation, thereby leading to a slowing of the heartbeat. Overactivity of the parasympathetic system can even lead to fainting, which is referred to as a **vagal reaction** because parasympathetic fibers reach the heart from the brain via a nerve called the vagus nerve. A vagal reaction can be stimulated by deeply felt grief or by having blood withdrawn from a vein. In contrast, sympathetic nerves can release noradrenaline onto the cells of the sinoatrial and atrioventricular nodes, speeding up the heartbeat. An increase in sympathetic nervous system activity also elevates the level of adrenaline in the blood, which contributes to the excitatory effect on the heart. Noradrenaline and adrenaline also strengthen the contractions of the cardiac muscle cells.

THE VASCULAR SYSTEM

Arteries and Arterioles

Blood pressure is highest in the vessels that carry blood away from the heart—the arteries and arterioles—and their structure reflects this fact (Figure

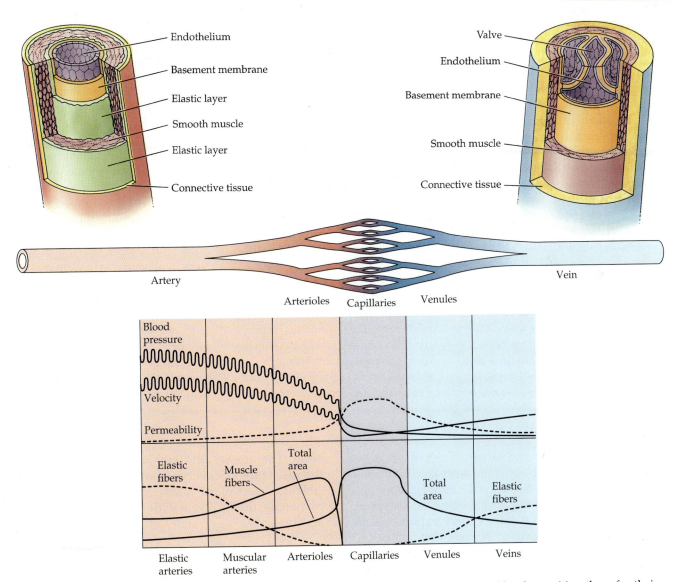

40.10 Anatomy of Blood Vessels
The anatomical characteristics of blood vessels match their functions. Arteries have lots of elastic fibers and muscle fibers. They must withstand high pressures. Arterioles have muscle fibers that control blood flow to different capillary beds. The total cross sectional area of capillaries is larger than for any other class of vessels and they are more permeable, thus suiting them for their function of exchange of nutrients and wastes with the extracellular fluids. Because they operate under low pressure, veins have valves to prevent backflow of blood. As veins get larger they have more elastic fibers, and that gives them the ability to accommodate changing volumes of blood. At rest, most blood is in the large veins.

BOX 40.B

Cardiovascular Disease

Cardiovascular disease is by far the largest single killer in the developed western world; it is responsible for about half the deaths each year. The immediate cause of most of these deaths is heart attack or stroke, but those events are the end result of a disease process called **atherosclerosis**, which begins many years before symptoms are detected. Hence, atherosclerosis is called the silent killer. What is atherosclerosis, and how can it be prevented?

Healthy arteries have a smooth internal lining of endothelial cells. This lining can be damaged by chronic high blood pressure, smoking, a high-fat diet, and other causes. Fatty deposits called **plaque** begin to form at sites of endothelial damage. Platelets, which are discussed later in this chapter, stick to the plaque, and fibrous connective tissue cells form a cap over it, but the plaque continues to grow and narrow the artery. Calcium deposits in the plaque cause the arteries to become less elastic, a condition known as **arteriosclerosis**, or hardening of the arteries. If the cap

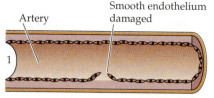

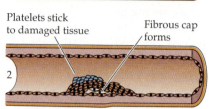

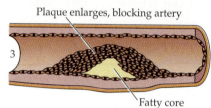

The development of atherosclerotic plaque.

over the plaque breaks and circulating blood-clotting factors come into contact with the fatty core of the broken-off part, a blood clot or **thrombus** forms, further blocking the artery. The blood supply to the heart itself flows through the **coronary arteries**. These arteries are highly susceptible to the processes of atherosclerosis; as they narrow, blood flow to the heart muscles decreases. Chest pains and shortness of breath during mild exertion are symptoms of this condition. A person with atherosclerosis is at high risk of forming a thrombus in a coronary artery. Such a **coronary thrombosis** can totally

block the vessel, causing a heart attack, or **coronary infarction**. A piece of a thrombus breaking loose, called an **embolus**, is likely to travel to and become lodged in a vessel of smaller diameter, blocking its flow (an **embolism**). Arteries already narrowed by plaque formation are likely places for an embolus to lodge. If there is an embolism in an artery in the brain, the cells fed by that artery die. This is called a stroke. The specific damage resulting from a stroke, such as memory loss, speech impairment, or paralysis, depends on the location of the blocked artery.

The most important solution to cardiovascular disease is prevention, not treatment. The risk factors for developing atherosclerosis are: high-fat and high-cholesterol diet, smoking, a sedentary lifestyle, high blood pressure or **hypertension**, obesity, certain medical conditions such as diabetes, and genetic predisposition. There is not much you can do about the genes you inherit or about some forms of diabetes, but the other risk factors can be avoided, thus decreasing the significance of genetic predisposition. It is never too early to take steps to prevent atherosclerosis. Many American children are overweight and up to 25 percent may have cholesterol levels that are too high. Many teenagers already have well-developed plaques in their arteries. Changes in diet and behavior can prevent and reverse these trends and help to fend off the silent killer.

40.10). The walls of the large arteries have many elastic fibers that enable them to withstand high pressures. These elastic fibers have another important function as well. During systole they are stretched, and thereby store some of the energy imparted to the blood by the heart. During diastole they return this energy by squeezing the blood and pushing it forward. As a result, the flow of blood through the arterial system is smoother than it would be through a system of rigid pipes.

Smooth muscle fibers are abundant in the arteries and arterioles. They alter the diameter of those vessels, and hence their resistance, thereby controlling the distribution of blood to different tissues of the

body. The arteries and arterioles are referred to as the **resistance vessels** because their resistance is variable. Diseases of the arteries cause about half of the human deaths each year in developed countries (Box 40.B).

Capillaries

Beds of capillaries connect arterioles to venules. No cell of the body is more than a couple of cell diameters away from a capillary. The needs of cells are served by the exchange of materials between blood and interstitial fluid. This exchange takes place across the capillary walls. It is possible because capillaries have

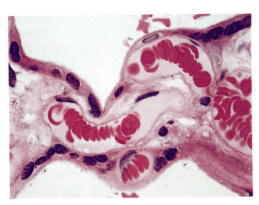

40.11 A Narrow Lane
Red blood cells pass through capillaries slowly and in single file.

thin, permeable walls and because blood flows through them slowly under very low pressure (Figure 40.10). To anyone who has played with a garden hose, it may seem strange that big arteries have high pressure and fast flow, but when the blood flows into the small capillaries the pressure and flow decrease. When you restrict the diameter of the garden hose by placing your thumb over the opening, the pressure in the hose increases, which in turn increases the velocity of the water spraying out of the hose. This puzzle is resolved by one more piece of information. Arterioles branch into so many capillaries that the total cross-sectional area of capillaries is much greater than that of any other class of vessels. Even though each capillary is so small that the red blood cells pass through in single file (Figure 40.11), each arteriole gives rise to such a large number of capillaries that together they have a much greater capacity for blood than do the arterioles. An analogy would be a fast-flowing river dividing up into many small rivulets flowing across a flat, broad delta. Each rivulet may be small and its flow sluggish, yet all together they accommodate all of the water poured into the delta by the river.

Exchange in Capillary Beds

The walls of capillaries are permeable to water and small molecules, but not to large molecules such as proteins. Blood pressure therefore tends to squeeze water and small molecules out of the capillaries and into the surrounding interstitial spaces. This is the process of filtration. The large molecules that cannot cross the capillary wall create an osmotic potential that tends to draw water back into the capillary (Figure 40.12).

Blood pressure is highest on the arterial side of a capillary bed and steadily decreases as the blood flows to the venous side. Therefore, more water is squeezed out of the capillaries on the arterial side of

the bed. The osmotic potential pulling water back into the capillary gradually becomes the dominant force as the blood flows toward the venous side of the bed. The interactions of the two opposing forces—the blood pressure versus the osmotic potential—determines the net flow of water.

The balance between blood pressure and osmotic potential changes if the blood pressure in the arterioles and the permeability of the capillary walls change. An example of such a change is associated with the inflammation that accompanies injuries to the skin or allergic reactions. The inflamed area becomes hot and red due to increased blood flow to the area. The inflamed tissue also becomes swollen. The major cause of these events is a chemical called **histamine** that is released in the inflamed tissue. Histamine makes blood vessels expand, which increases

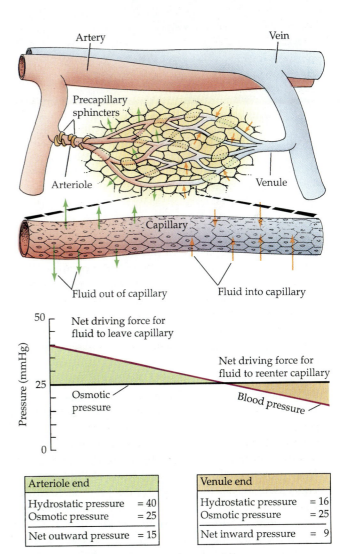

Fluid out of capillary   Fluid into capillary

Arteriole end	
Hydrostatic pressure	= 40
Osmotic pressure	= 25
Net outward pressure	= 15

Venule end	
Hydrostatic pressure	= 16
Osmotic pressure	= 25
Net inward pressure	= 9

40.12 A Balance of Forces Controls Extracellular Fluid Volume
Fluids are squeezed out of capillaries by blood pressure and pulled back in by osmotic potential created by large molecules that cannot leave the capillaries.

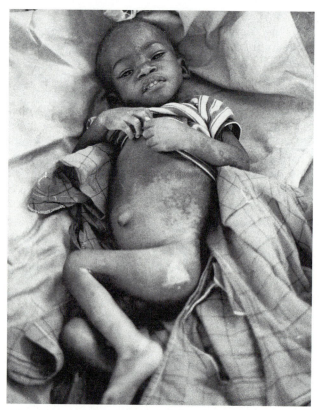

40.13 Kwashiorkor, "The Rejected One"
In people on marginal diets, the symptoms of protein malnutrition frequently show up soon after breast feeding ceases. Swollen abdomen, face, hands, and feet due to edema and spindly limbs due to muscle atrophy are hallmarks of serious protein starvation.

terstitial fluid. Capillary walls are membranous, and as we learned in Chapter 5, lipid-soluble substances pass freely across a membrane from the area of higher concentration to that of lower concentration. Oxygen, which is in high concentration in the blood coming from the arteriole, but is depleted from active skeletal muscle tissues, readily moves from the blood into the muscle. At the same time, carbon dioxide, which is in high concentration in the working muscle but low in the blood, rapidly moves into the blood. The concentrations of these gases in the blood thus change rapidly as the blood travels through the capillary beds.

Small molecules in the blood can generally pass through the capillary walls, but the capillaries in different tissues are differentially selective to the sizes of molecules they will allow to pass from blood to interstitial fluids. In all capillaries, O_2, CO_2, glucose, lactate, and small ions such as Na^+ and Cl^- can cross. In the capillaries of the brain, not much else can cross unless it is a lipid-soluble substance such as alcohol; we therefore speak of the **blood–brain barrier** to characterize the high selectivity of brain capillaries. In other tissues the capillaries are much less selective and have actual pores to permit the passage of large molecules. Such capillaries are found in the digestive tract, where nutrients are absorbed, and in the kidneys, where wastes are filtered. Some capillaries have large gaps that permit the movement of even larger substances. These are found in the bone marrow, spleen, and liver. Substances are moved across many capillary walls by **endocytosis** (Chapter 4).

The Lymphatic System

The fluid that accumulates outside the capillaries contains water and small molecules, but no red blood cells and less protein than is found in the blood. This interstitial fluid is called **lymph**. A separate system of vessels—the **lymphatic system**—returns the lymph to the blood. These vessels begin as fine lymphoid capillaries, merge progressively into larger and larger vessels, and end in a major vessel—the **thoracic duct**—which empties into the superior vena cava returning blood to the heart (see Figure 16.3). Lymphatic vessels have one-way valves that keep the lymph flowing toward the thoracic duct. The propelling force moving the lymph is pressure on the lymphatic vessels from the contractions of nearby skeletal muscles.

Mammals and birds have lymph nodes along the major lymphatic vessels. Lymph nodes are an important component of the defensive machinery of the body. They are a major site of lymphocyte production and of the phagocytic action that removes microorganisms and other foreign materials from the circulation. The lymph nodes also act as mechanical filters. Particles become trapped there and are digested by

blood flow to the area and increases pressure in the capillaries. Histamine also increases the permeability of the capillaries and venules. Therefore, more water leaves the capillaries and venules, and the tissue swells due to the accumulation of interstitial fluids, a condition known as **edema**. The use of drugs called **antihistamines** can alleviate inflammation and allergic reactions.

The loss of water from the capillaries increases if the osmotic potential of the blood decreases, as is seen in the disease kwashiorkor. This disease is caused by severe protein starvation. When the body has no amino acids available for the synthesis of essential proteins, it begins to break down its own blood proteins. This means that there are fewer molecules available in the blood to maintain the osmotic potential that pulls water back into the capillaries. Consequently, interstitial fluids build up, creating swelling or edema which is primarily seen in the abdomen and the extremities (Figure 40.13).

Whether specific small molecules cross a capillary wall depends on the architecture of the capillary, the type of substance under consideration, and the concentration difference between the blood and the in-the phagocytes that are abundant in the nodes. Lymph nodes swell during infection. Some of them, particularly those on the side of the neck or in the armpit, become noticeable at such times. The nodes also trap metastasized cancer cells, that is, those that have broken free of the original tumor. Because such cells may start additional tumors, surgeons often remove the neighboring lymph nodes when they excise a malignant tumor.

Venous Return

Blood flows back to the heart through the veins, but what propels it? The pressure of the blood flowing from capillaries to venules is extremely low, so it cannot be the beating of the heart that propels blood through the veins. If the veins are above the level of the heart, gravity will help; below the level of the heart, blood must be moved against the pull of gravity. In actuality, blood does tend to accumulate in veins, and the walls of veins are more expandable than the walls of arteries. As much as 80 percent of the total blood volume may be in the veins at any one time. Veins are called **capacitance vessels** because of their high capacity to store blood.

Blood must be returned from the veins so that circulation can continue. If too much blood remains in the veins, then too little blood returns to the heart, and thus too little blood is pumped to the brain; a person may faint as a result. Fainting is self-correcting because a fainting person falls, thereby changing from the position in which gravity caused blood to accumulate in the lower body. There are means other than fainting, however, by which blood is moved from the tissues back to the heart.

The most important of the forces that propel venous and lymphatic return from the regions of the body below the heart is the milking action of skeletal muscle contraction around the vessels. As muscles contract, the vessels are squeezed and the blood moves forward. Blood flow might be temporarily obstructed as a muscle contraction is held, but with relaxation of the muscles the blood is free to move again. Within the veins are valves that prevent backflow of blood (Figure 40.14). Thus, whenever a vein is squeezed, blood is propelled forward because the valves prevent it from flowing backward. In this way blood is gradually pushed toward the heart. As we noted above, similar valves are found in the lymphatic vessels.

The role of muscle contractions in venous return is appreciated by people who must stand still for prolonged periods, thereby accumulating blood in the veins of the lower body. The guards at Buckingham Palace in England, for instance, are well aware that they must shift their weight and contract their leg muscles periodically to prevent the blood flow to the heart from dropping to the extremely low

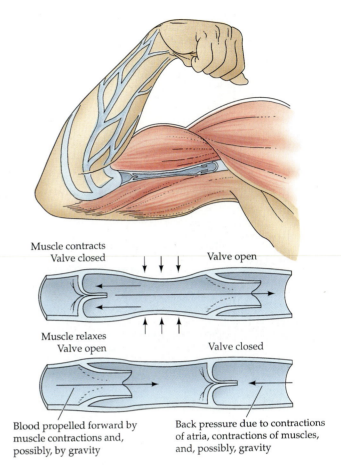

Muscle contracts
Valve closed   Valve open

Muscle relaxes
Valve open   Valve closed

Blood propelled forward by muscle contractions and, possibly, by gravity

Back pressure due to contractions of atria, contractions of muscles, and, possibly, gravity

40.14 One-Way Flow
Contractions of skeletal muscles squeeze blood toward the heart because of valves in veins.

level that causes fainting. Gravity causes edema as well as blood accumulation in veins. The back pressure that builds up in the capillaries when blood accumulates in the veins shifts the balance between blood pressure and osmotic potential so that there is a net movement of fluid into the interstitial spaces. This is why you have trouble putting your shoes back on after you sit for a long time with your shoes off, such as on an airline flight. In persons with very expandable veins, the veins may become so stretched that the valves can no longer prevent backflow. This condition produces varicose veins. Draining these veins is highly desirable and can be aided by wearing support hose and periodically elevating the legs above the level of the heart.

During exercise, the milking action of muscles ensures that the heart receives an adequate volume of blood to pump to the lungs and then to the respiring tissues. As an animal runs, its legs act as auxiliary vascular pumps returning blood to the heart from the veins of the lower body. As more blood is returned to the heart, the heart contracts more forcefully, and its pumping action is more effective. This strengthening of the heart beat is due to an intrinsic property

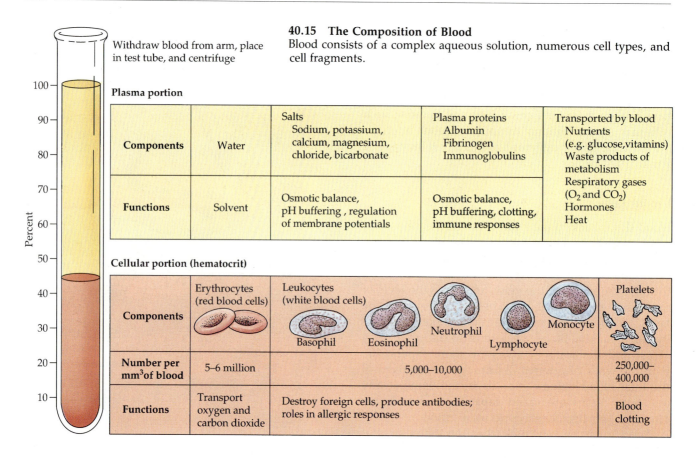

40.15 The Composition of Blood
Blood consists of a complex aqueous solution, numerous cell types, and cell fragments.

Withdraw blood from arm, place in test tube, and centrifuge

Plasma portion

Components	Water	Salts Sodium, potassium, calcium, magnesium, chloride, bicarbonate	Plasma proteins Albumin Fibrinogen Immunoglobulins	Transported by blood Nutrients (e.g. glucose, vitamins) Waste products of metabolism Respiratory gases (O_2 and CO_2) Hormones Heat
Functions	Solvent	Osmotic balance, pH buffering, regulation of membrane potentials	Osmotic balance, pH buffering, clotting, immune responses	

Cellular portion (hematocrit)

Components	Erythrocytes (red blood cells)	Leukocytes (white blood cells) Basophil Eosinophil Neutrophil Lymphocyte Monocyte	Platelets
Number per mm³ of blood	5–6 million	5,000–10,000	250,000–400,000
Functions	Transport oxygen and carbon dioxide	Destroy foreign cells, produce antibodies; roles in allergic responses	Blood clotting

of cardiac muscle fibers referred to as the **Frank–Starling law**: If the fibers are stretched, as they are when the volume of returning blood increases, they contract more forcefully. This principle holds (within a certain range) whenever venous return increases, by any mechanism.

The actions of breathing also help to return venous blood to the heart. The **"ventilatory pump"** works because the suction created by the ventilatory muscles that pulls air into the lungs (Chapter 39) also pulls blood and lymph toward the chest, increasing venous return to the right atrium. Some smooth muscle in the walls of the veins also moves venous blood back to the heart by constricting the veins and moving the blood forward. These muscles are rare in most of the veins, and totally absent from lymphatic vessels in humans. They do not play a major role in venous return. However, in the largest veins closest to the heart, smooth muscle contraction at the onset of exercise can suddenly increase venous return and stimulate the heart in accord with the Frank–Starling law, thus increasing cardiac output.

THE BLOOD

We have considered the circulation of the blood in detail without looking closely at the blood itself. Blood is a tissue; it has cellular elements suspended in an aqueous medium of specific, yet complex, composition. The cells of the blood can be separated from the aqueous medium called **plasma** by centrifugation (Figure 40.15). If we take a 100-ml sample of blood and spin it in a centrifuge, all of the cells move to the bottom of the tube, leaving the straw-colored, clear plasma on top. The **packed cell volume**, or the percentage of the blood volume made up by cells, is about 40 percent in normal humans, and is also referred to as the **hematocrit**. Normal hematocrit is about 38 for women and 42 for men, but the values can vary considerably. They are usually higher, for example, in people living and doing heavy work at high altitude. We will consider next the three classes of cellular elements in blood: the red blood cells, or erythrocytes; the white blood cells, or leukocytes; and the platelets, which are not really cells, but pinched-off fragments of cells.

Red Blood Cells

Most of the cells in the blood are **erythrocytes**, or red blood cells. The function of red blood cells is to transport the respiratory gases, as we saw in Chapter 39. There are about five million red blood cells per milliliter of blood whereas there are only five to ten thousand white blood cells in that same volume. Red blood cells form from special cells in the bone marrow called **stem cells**, particularly in the ribs, breastbone,

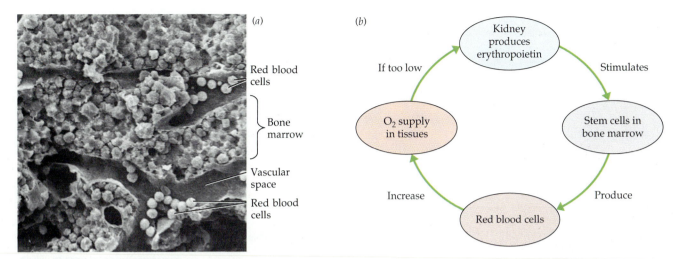

40.16 Red Blood Cells Form in the Bone Marrow
(a) In this scanning electron micrograph of a section of bone, the marrow is surrounded by vascular spaces. As new red blood cells mature, they squeeze through the endothelium lining the vascular spaces and enter the blood. (b) Erythropoietin stimulates stem cells in the bone marrow to produce red blood cells.

pelvis, and vertebrae (Figure 40.16a). Red blood cell production is controlled by a hormone, **erythropoietin**, which is released by cells in the kidney in response to insufficient oxygen (Figure 40.16b).

Erythropoietin stimulates stem cells to produce red blood cells. Under normal conditions your bone marrow is producing about two million red blood cells every second. The developing red blood cells go through many cell divisions while still in the bone marrow, and during this time they are producing hemoglobin. When the hemoglobin content of a red blood cell approaches about 30 percent, its nucleus, endoplasmic reticulum, Golgi apparatus, and mitochondria begin to break down. This process is almost complete when the new red blood cell squeezes through pores in the capillary walls and enters the circulation. Each red blood cell circulates for about 120 days and is then broken down. The iron from its hemoglobin molecules is recycled back to the bone marrow. Mature red blood cells are biconcave flexible disks, packed with hemoglobin. Their shape gives them a large surface area for gas exchange, and their flexibility enables them to squeeze through the capillaries (see Figures 39.15 and 40.11).

White Blood Cells

Leukocytes, or white blood cells, defend the body against infection, as discussed in Chapter 16. Some search for and destroy foreign cells; some are phagocytes that consume bacteria, debris, and even dead or damaged cells from our own bodies; and some manufacture antibodies. Leukocytes squeeze through capillary walls and spend a great deal of time outside the vascular system. They move about by amoeboid motion and follow cues from chemicals released by dead or sick cells to sites of infection and cell damage.

Platelets and Blood Clotting

Platelets bud off from large cells in the bone marrow. A platelet is just a tiny fragment of a cell with no nucleus, but it is packed with enzymes and chemicals necessary for its function of sealing leaks in the blood vessels. When a vessel is damaged, collagen fibers are exposed. When a platelet encounters collagen fibers, it is activated. It swells, becomes irregularly shaped and sticky, and releases a number of chemicals that activate other platelets and initiate the clotting of blood. The sticky platelets form a plug at the damaged site, and the subsequent clotting forms a stronger patch on the vessel.

The clotting of blood involves many steps and many **clotting factors**. The absence of any of these factors can cause excessive bleeding and thus can be lethal. Because the liver produces most of the clotting factors, liver diseases such as hepatitis and cirrhosis can result in excessive bleeding. The well-known sex-linked trait hemophilia is an example of a genetic inability to produce one of the clotting factors. The various clotting factors are involved in a cascade of steps that activate other substances circulating in the blood. The cascade begins with cell damage and platelet activation and ends with the conversion of an inactive circulating enzyme, **prothrombin**, to its active form, **thrombin** (Figure 40.17). Thrombin causes circulating protein molecules called **fibrinogen** to polymerize and form **fibrin** threads. The fibrin

Injury to the lining of a blood vessel exposes collagen fibers; platelets adhere and get sticky

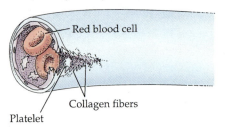

Red blood cell

Collagen fibers

Platelet

Platelets release substances that cause the vessel to contract. Sticky platelets form a plug and initiate formation of a fibrin clot

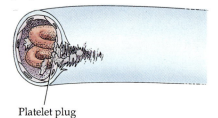

Platelet plug

The fibrin clot seals the wound until the vessel wall heals

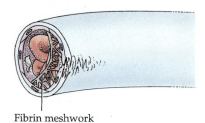

Fibrin meshwork

Clotting factors
1. Released from platelets and injured tissues
2. Plasma proteins synthesized in liver and circulating in inactive form

Prothrombin circulating in plasma → Thrombin

Fibrinogen circulating in plasma → Fibrin

(a)

(b)

40.17 Blood Clotting
(a) Damage to a blood vessel initiates a cascade of events that produces a blood clot. (b) As a blood clot forms, red blood cells become enmeshed in a network of fibrin threads as shown in this scanning electron micrograph.

threads form the network that stabilizes the blood clot, seals the vessel, and provides a scaffold for the formation of scar tissue.

Plasma

Plasma is a complex solution of gases, ions, nutrient molecules, and proteins, but this complex composition is regulated within narrow limits. Most of the ions are Na^+ and Cl^- (hence the salty taste of blood), but a large number of other ions are also present. Nutrient molecules in plasma include glucose, amino acids, lipids, cholesterol, and lactic acid. The circulating proteins have many functions. We have just noted those that function in blood clotting; others of interest include albumin, which is largely responsible for the osmotic potential in capillaries that prevents a massive loss of water from plasma to interstitial spaces; antibodies (the immunoglobulins); hormones; and various carrier molecules such as **transferrin**,

which carries iron from the gut to where it is stored or used. Plasma is very similar to interstitial fluid in composition, and most of its components move readily between these two fluid compartments of the body. The main difference between them is the higher concentration of proteins in the plasma.

CONTROL AND REGULATION OF THE CIRCULATION

Control and regulation in the cardiovascular system operates at many levels. Every tissue requires an adequate supply of blood saturated with O_2, carrying essential nutrients, and relatively low in waste products. The nervous system cannot monitor and control every capillary bed in the body. Instead, each bed regulates its own blood flow through **autoregulatory mechanisms** that cause the arterioles supplying the bed to constrict or dilate.

The oxygen content, waste production, and blood flows in all of the capillary beds are eventually integrated through their effects on the composition and pressure of the arterial blood leaving the heart. If resistance in the capillaries falls due to high demand for blood, then arterial pressure will fall. If metabolic demands exceed the capacity of the pulmonary circuit for gas exchange, then O_2 and CO_2 concentrations in arterial blood, as well as its pH, will change. Changes in all of these parameters provide information used by the nervous and endocrine systems to control breathing, heart rate, and blood flow, and to match them to the metabolic needs of the body.

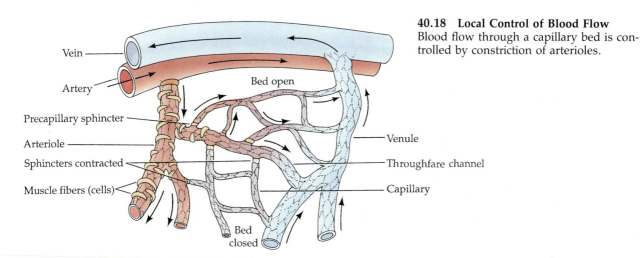

40.18 Local Control of Blood Flow
Blood flow through a capillary bed is controlled by constriction of arterioles.

Autoregulation

The autoregulatory mechanisms that adjust the flow of blood to a tissue are intrinsic to the local tissue itself, but they can be influenced by the nervous system and certain hormones as well. They control the flow of blood through a capillary bed by controlling the degree of contraction of the smooth muscles in the arteries and arterioles feeding that bed. The flow of blood in a typical capillary bed is diagrammed in Figure 40.18. Blood flows into the bed from an arteriole. Smooth muscle "cuffs," or precapillary sphincters, on the arteriole can completely shut off the supply of blood to the capillary bed. When the arteriole is open, the arterial blood pressure pushes blood into the capillaries. Autoregulation depends on the sensitivity of the smooth muscle to the composition of its chemical environment. Low O_2 concentrations and high CO_2 concentrations cause the smooth muscle to relax, thus increasing the blood supply, which brings in more O_2 and carries away the CO_2. Increases in the concentration of products of metabolism other than CO_2, such as lactate, hydrogen ions, potassium, and adenosine, also promote increased blood flow by this mechanism. Hence, activities that increase the metabolism of a tissue also increase the blood flow to that tissue.

Systemic Control

The same smooth muscles of arteries and arterioles that respond to autoregulatory stimuli also respond to a variety of central nervous system and endocrine signals. Most arteries and arterioles are innervated by the autonomic nervous system, particularly the sympathetic division. Most sympathetic neurons release noradrenaline, which causes the smooth muscle fibers to contract, thus constricting the vessel and increasing its resistance to blood flow. An exception is found in skeletal muscle, where some specialized sympathetic neurons release acetylcholine and cause the smooth muscles of the arterioles to relax and the vessels to dilate.

Hormones also can cause arterioles to constrict. Adrenalin has actions similar to those of noradrenalin. It is released from the adrenal medulla during massive sympathetic activation—the fight-or-flight response. Angiotensin, produced when blood pressure to the kidneys falls, causes arterioles to constrict. Vasopressin, released by the posterior pituitary, has similar effects (Figure 40.19). These hormones influence arterioles located for the most part in peripheral tissues (extremities) or in tissues whose functions need not be maintained continuously (such as the gut). By reducing blood flow in those arterioles, the hormones increase the central blood pressure and blood flow to essential organs such as heart, brain, and kidneys.

Central Nervous System Regulation of Blood Pressure and Composition

The autonomic nervous system activity that controls constriction of blood vessels and heart rate originates in cardiovascular centers in the medulla of the brain stem. Many inputs converge on this central integrative network and influence the commands it issues via parasympathetic and sympathetic fibers (Figure 40.20). Of special importance is information about changes in blood pressure from stretch receptors in the walls of the arteries leading to the brain—the aorta and the carotid arteries. Increased activity in the stretch receptors indicates rising blood pressure and inhibits sympathetic nervous system output. As a result, the heart slows and arterioles in peripheral tissues dilate. If pressure in the arteries falls, the activity of the stretch receptors goes down, stimulating sympathetic output. Increased sympathetic output causes the heart to beat faster and the arterioles in peripheral tissues to constrict. When arterial pressure falls, the change in stretch-receptor activity also

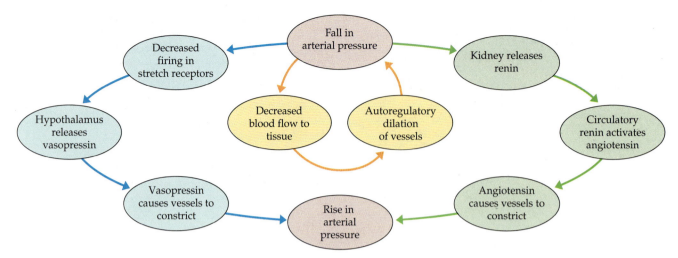

40.19 Controlling Blood Pressure through Vascular Resistance
A fall in blood pressure triggers hormonal controls that work to increase
blood pressure.

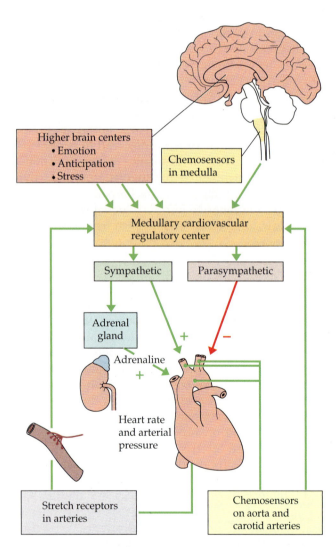

40.20 Controlling Blood Pressure through Heart Rate
The autonomic nervous system regulates heart rate in response to information about blood pressure and blood
composition.

causes the hypothalamus to release vasopressin,
which helps to increase blood pressure by stimulating
peripheral arterioles to constrict.

Other information that causes the medullary cardiovascular control system to increase heart rate and
blood pressure comes from the carotid and aortic
bodies (see Figure 39.22). These nodules of modified
smooth muscle tissue are chemosensors that respond
to inadequate O_2 supply. If arterial blood flow falls
or the O_2 content of the arterial blood falls drastically,
these sensors are activated and send signals to the
cardiovascular control center.

The cardiovascular control center also receives input from other brain areas. Emotions and the antic-
ipation of intense activity, such as at the start of a
race, can cause the center to increase heart rate and
blood pressure. A reflex that slows the heart is the
so-called diving reflex, which is highly developed in
marine mammals (Figure 40.21). Humans also have
a diving reflex that causes the heart to slow when
the face is immersed in water.

A question to ask about any physiological system
is, "What parameter is being regulated?" For the respiratory system, as discussed in Chapter 39, the primary answer to this question is that the CO_2 concentration of the blood is being regulated, with
regulation of O_2 concentration playing a lesser role.
The answer to what is being regulated in the cardiovascular system is more complex and has at least
two parts. First, the blood flow to individual tissues
is regulated by local, autoregulatory mechanisms.
Second, the pressure, O_2 content, and CO_2 content
of the arterial blood is regulated directly or indirectly
by central nervous system mechanisms that interact
to match cardiovascular function to the regional and
overall needs of the body.

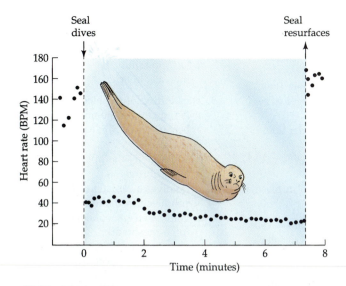

40.21 Master Divers
When a marine mammal dives, its heart slows. In addition, arteries to most organs constrict so that almost all blood flow and available oxygen goes to the animal's heart and brain. These adaptations enable some seals to remain under water for up to an hour.

SUMMARY

The metabolic needs of the individual cells of very small animals can be met by direct exchange of materials with the external medium bathing the animal and possibly filling its gastrovascular cavity as well. In larger and more complex animals, the needs of cells are met by a circulatory system that transports nutrients, respiratory gases, and metabolic wastes to and from the cells. Circulatory systems can be open, with the blood leaving vessels to percolate through tissues, or closed, with the blood contained in a system of vessels.

The circulatory systems of vertebrates consist of a pump (or heart) and a closed system of vessels. Arteries and arterioles carry blood from the heart. Cap-illaries are the site of exchange between blood and interstitial fluids. Venules and veins carry blood back to the heart. The heart has evolved from two chambers in fishes to four chambers in birds and mammals, resulting in greater separation of the pulmonary and systemic circuits.

The contractions of the human heart are coordinated by electrical continuity between individual cardiac muscle cells. A pacemaker initiates the contraction of the atria and the resulting electrical signal is transmitted with a slight delay to the ventricles by a conducting system of fibers. The autonomic nervous system modulates the activity of the pacemaker.

The high but selective permeabilities of capillaries determine the exchange of substances between blood and interstitial fluid. Pressure in the capillaries forces fluid out of capillaries, and the osmotic potential of molecules left behind attracts water back in. Fluid lost to the interstitial spaces can cause edema, but interstitial fluid is usually transported back to the venous system through the lymphatic vessels.

Blood is a tissue consisting of cellular elements in an aqueous medium. The cellular elements include red blood cells, which transport oxygen, and white blood cells, which serve as body defenses. Both are made in the bone marrow. Cell fragments called platelets and circulating proteins participate in clotting responses.

The cardiovascular system is controlled and regulated at many levels. Autoregulatory mechanisms control the blood flow in individual capillary beds. Constriction of arterioles decreases blood flow in capillary beds and tends to increase arterial blood pressure. Dilation of arterioles has opposite effects. The central nervous system, acting through the autonomic division, responds to changes in blood pressure and composition of the blood by altering heart rate and degree of constriction of arterioles. Hormonal mechanisms that cause the constriction of arterioles in peripheral tissue also respond to decreases in blood pressure.

SELF-QUIZ

1. An open circulatory system is characterized by:
   a. absence of a heart.
   b. absence of blood vessels.
   c. blood with a composition different from interstitial fluid.
   d. absence of capillaries.
   e. a higher pressure circuit through gills than to other organs.

2. Which of the following statements about vertebrate circulatory systems is *not* true?
   a. In fish, oxygenated blood from the gills returns to the heart through the left atrium.
   b. In mammals, deoxygenated blood leaves the heart through the pulmonary artery.
   c. In amphibians, deoxygenated blood enters the heart through the right atrium.
   d. In reptiles, the blood in the pulmonary artery has a lower oxygen content than the blood in the aorta.
   e. In birds, the pressure in the aorta is higher than the pressure in the pulmonary artery.

3. Which of the following statements about the human heart is *true*?
   a. The walls of the right ventricle are thicker than the walls of the left ventricle.
   b. Blood flowing through atrioventricular valves is always deoxygenated blood.
   c. The second heart sound is due to the aortic valve closing.
   d. Blood returns to the heart from the lungs in the vena cava.
   e. During systole the aortic valve is open and the pulmonary valve is closed.

4. Pacemaker actions of cardiac muscle:
   a. are due to opposing actions of noradrenaline and acetylcholine.
   b. are localized in the bundle of His.
   c. depend on the gap junctions between cells making up the atria with those making up the ventricles.
   d. are due to spontaneous depolarization of the plasma membranes of some cardiac muscle cells.
   e. result from hyperpolarization of cells in the sinoatrial node.

5. Blood flow through capillaries is slow because:
   a. lots of blood volume is lost from the capillaries.
   b. pressure in venules is high.
   c. total cross sectional area of capillaries is larger than that of arterioles.
   d. the osmotic pressure in capillaries is very high.
   e. red blood cells are bigger than capillaries and must squeeze through.

6. How are lymphatic vessels like veins?
   a. Both have nodes where they join together into larger common vessels.
   b. Both carry blood under low pressure.
   c. Both are capacitance vessels.
   d. Both have valves.
   e. Both carry fluids rich in plasma proteins.

7. The production of red blood cells:
   a. ceases if the hematocrit falls below normal.
   b. is stimulated by erythropoietin.
   c. is about equal to the production of white blood cells.
   d. is inhibited by prothrombin.
   e. occurs in bone marrow before birth and in lymph nodes after birth.

8. All of the following increase blood flow through a capillary bed *except*:
   a. high CO_2 concentration.
   b. high lactate and hydrogen ion concentrations.
   c. histamine.
   d. vasopressin.
   e. increase in arterial pressure.

9. The clotting of the blood:
   a. is impaired in hemophiliacs because they don't produce platelets.
   b. is initiated when platelets release fibrinogen.
   c. involves a cascade of factors produced in the liver.
   d. is initiated by leukocytes forming a meshwork.
   e. requires conversion of angiotensinogen to angiotensin.

10. Autoregulation of blood flow to a tissue is due to:
    a. sympathetic innervation.
    b. release of vasopressin by the hypothalamus.
    c. increased activity of baroreceptors.
    d. chemosensors in carotid and aortic bodies.
    e. effect of local environment on arterioles.

FOR STUDY

1. How is cardiac output increased at the beginning of a race? Involve the Frank–Starling Law in your answer.

2. The final stages of alcoholism involve loss of liver function and accumulation of fluids in extremities and abdominal cavity. Explain how these two consequences of alcoholism are related.

3. A sudden and massive loss of blood results in a fall in blood pressure. Describe several mechanisms that act to return blood pressure to normal.

4. You can describe the cycle of events in a ventricle of the heart by a graph that plots the pressure in the ventricle on the *y* axis and the volume of blood in the ventricle on the *x* axis. What do you think such a graph would look like? Where would the heart sounds occur on this graph? How would the graph differ for the left and the right ventricles?

5. Why doesn't diastolic blood pressure fall to zero between heartbeats? Why does systolic blood pressure increase with (a) sympathetic activity, (b) increased venous return, and (c) age?

READINGS

Eckert, R., D. Randall and G. Augustine. 1988. *Animal Physiology: Mechanisms and Adaptations*, 3rd Edition. W. H. Freeman, New York. An outstanding textbook of animal physiology, with excellent coverage of the circulatory system.

Golde, D. W. and J. C. Gasson. 1988. "Hormones that Stimulate the Growth of Blood Cells." *Scientific American*, July.

Robinson, T. F., S. M. Factor and E. H. Sonnenblick. 1986. "The Heart as a Suction Pump." *Scientific American*, July. A new proposal concerning the filling of the heart. This article also gives interesting general information on cardiac muscle and the connective tissues of the heart.

Scholander, P. F. 1963. "The Master Switch of Life." *Scientific American*, December. Delightful, classic description of the discovery of the diving adaptations of marine mammals.

Vander, A. J., J. H. Sherman and D. S. Luciano. 1990. *Human Physiology: The Mechanisms of Body Function*, 5th Edition. McGraw-Hill, New York. Chapter 13 deals with circulation.

Zapol, W. M. 1987. "Diving Adaptations of the Weddell Seal." *Scientific American*, June.

Zucker, M. 1980. "The Functioning of Blood Platelets." *Scientific American*, June. On the role of platelets in blood clotting.

41

Animal Nutrition

PREVIEW: Unlike green plants, animals cannot use solar energy to synthesize organic molecules from carbon dioxide and water. Animals depend on carbon skeletons made by plants as building blocks for their own molecules, and they depend on the energy put into chemical bonds by plants to suppy their needs for metabolic energy. They obtain this energy by eating plants, or by eating other animals that have eaten plants. Animals show an amazing variety of adaptations for obtaining and processing food. However, the molecules obtained from food are too large and complex to be used by the bodies of animals. Animals must break down these complex molecules into simple subunits that their cells can absorb and metabolize. Animals' diets must also include small quantities of certain minerals and complex molecules called vitamins that they cannot synthesize themselves. The processes of food acquisition and digestion involve many fascinating neural and hormonal control mechanisms, and hormones direct the internal traffic of molecules used for metabolic energy.

This chapter deals with nutrient requirements, vitamins, deficiency diseases, adaptations for acquisition and processing of food, digestive systems, enzymatic digestion, absorption of nutrients, and the control and regulation of digestion and fuel metabolism.

Animals must eat to stay alive, and in a sense, they are what they eat. Animals are **heterotrophs**; they derive both their energy and their structural molecules from their food. This is in contrast to the **autotrophs** (most plants, some monerans, and some protists) that can trap solar energy through photosynthesis and use it to synthesize all of their structures from inorganic materials. Almost all life runs on solar energy, but it comes to the heterotrophs indirectly through the autotrophs. A dramatic exception to this rule is found in the ecosystem around deep sea vents, where sunlight never reaches (see Figure 17.5). The bacteria that provide the nutritional foundation for that unusual ecosystem use the energy in the bonds of hydrogen sulfide that flows from the vents to synthesize organic molecules. Photosynthetic autotrophs provide the nutritional foundation for all other known ecosystems, and heterotrophs have evolved an enormous diversity of adaptations for exploiting that wellspring of life (Figure 41.1).

Heterotrophic nutritional lifestyles span a great range. At one end are the **saprophytes** such as fungi that absorb organic molecules from dead organisms, and at the other end are large predators such as lions. Some animals feed only on the remains of dead organisms. For example, **detritivores** such as earthworms process environmental deposits containing organic matter for their nutritional needs. All other animals that feed on other living organisms can be considered **predators**. **Herbivores** are predators that prey on plants; **carnivores** prey on other animals. **Omnivores**, including humans, prey on both plants and animals. **Filter feeders** such as clams and blue whales prey on small organisms by filtering them out of the environmental medium. We are only too familiar with examples of **fluid feeders** such as mosquitos, aphids, and leeches. In this chapter we will examine some of the anatomical adaptations for this diversity of nutritional lifestyles.

The statement "we are what we eat" is true only in a limited sense. Although some parasitic animals can simply absorb all of the basic nutrients they need from their environment, most animals must break down the complex molecules in their food into simple units such as amino acids, fatty acids, and sugars. These simplest nutrient units can then be used for synthesis of new molecules or metabolized as an energy source. The breakdown of food molecules is the process of **digestion**. The cells of some animals, such as sponges, engulf particles of food and digest them intracellularly. Most animals process their food through **extracellular digestion** in a digestive cavity called a gut. The gut of simple animals such as flatworms and jellyfish is a saclike structure with only one opening to the environment. More complex animals have a tubular gut with a separate entrance and exit for food. Such animals **ingest** food through a mouth, where it may be broken up. The food then

(a)   *(b)*

(c)   *(d)*

41.1 A Focus on the Consumers

Heterotrophs have evolved an amazing range of adaptations for exploiting sources of energy. *(a)* The giraffe is a herbivore whose source of food is out of reach for most other ground-dwelling animals. *(b)* The long bill and tongue of the hummingbird and its hovering flight enable it to harvest the tiny amounts of nectar in individual flowers. *(c)* The red angler fish, cleverly disguised as a sponge, lures its prey within gulping distance. *(d)* A constrictor, such as this corn snake, squeezes its prey to death and, by unhinging its jaws, swallows the victim in one piece.

passes through the tubular digestive tract, where digestive enzymes break it down. Throughout this process the food is really outside the body because it has not crossed any cell membranes. Digestion occurs *outside* the cells. The products of digestion are then absorbed into the body and taken up by the cells.

In this chapter we will first consider what nutrients are required and why. Then, how are they procured? Once ingested, how are they processed, digested, and absorbed? Finally we will learn how the body regulates its traffic in molecules used for metabolic fuel.

NUTRIENT REQUIREMENTS

Nutrients as Fuel

In Chapters 6 and 7 we learned that energy in the chemical bonds of food molecules is transferred to the high-energy phosphate bonds of ATP. That ATP then provides energy for active transport, biosynthesis of molecules, degradation of molecules, muscle contraction, and other work. Heat is a by-product of these energy conversions, since they are never 100 percent efficient. Eventually all of the energy that is transferred to ATP from the chemical bonds of food molecules is released to the environment as heat. It is convenient, therefore, to talk about the energy requirements of animals and the energy content of food in terms of a measure of heat energy, the **calorie**. A calorie is the amount of heat necessary to raise the temperature of one gram of water one degree centigrade. As this is a tiny amount of energy in comparison to the energy requirements of many animals, the physiologist commonly uses the **kilocalorie** (kcal) as a unit (1000 calories = 1 kcal). Nutritionists also use the kilocalorie as a standard unit of energy, but they traditionally refer to it as the **Calorie**, always spelled with a capital C to distinguish it from the single calorie. When a person is on a 1,000 Calorie per day diet, it means 1000 kcal/day. Such confusion of terms is unfortunate, but we live with it.

The metabolic rate of an animal (see Chapter 33) is a measure of the overall energy needs that that animal must meet by the ingestion and digestion of food. The components of food that provide energy are fats, carbohydrates, and proteins. Fat yields 9.5 kcal/gram when it is metabolically oxidized, carbohydrate yields 4.2 kcal/gram, and protein yields about 4.1 kcal/gram. The basal metabolic rate of a human is about 1,300 to 1,500 kcal/day for an adult female and 1,600 to 1,800 kcal/day for an adult male. Any physical activity adds to this basal energy requirement. Some equivalencies between food, energy, and exercise are shown in Figure 41.2. (It should be noted that more and more, physiologists are abandoning the calorie as an energy unit as they switch to the System International units. The replacement for the calorie in the System International is the joule; one calorie equals 4.184 joules.)

The cells of the body use energy continuously, but most animals do not eat continuously. Humans generally eat several meals a day, a lion may eat once in several days, a boa constrictor may eat only once a month, and hibernators may go 5–6 months without eating. Therefore animals must store fuel molecules that can be released as needed when the animal is between meals. Carbohydrate is stored in liver and muscle cells as **glycogen**, but the total glycogen stores are usually not more than the equivalent of a day's energy requirements. Fat is the most important form of stored energy in the bodies of animals. Not only does fat have the highest energy content per gram, but it can be stored with little associated water, making it more compact. If migrating birds had to store energy as glycogen rather than fat to fuel long flights, they would be too heavy to fly! Protein is not used to store energy, although body protein can be metabolized as an energy source as a last resort.

If an animal takes in too little food to meet its needs for metabolic energy, it is **undernourished** and must make up the shortfall by metabolizing some of the molecules of its own body. Consumption of self for fuel begins with the storage compounds glycogen and fat. If starvation continues after those reserves are gone, the body begins to use its own proteins for fuel. Breakdown of body proteins causes impairment and loss of function, and eventually leads to death. Blood proteins are among the first to go, resulting in loss of fluid to the interstitial spaces (edema; see Chapter 40). Muscles atrophy, and eventually even brain protein is lost. Figure 41.3 shows the course of

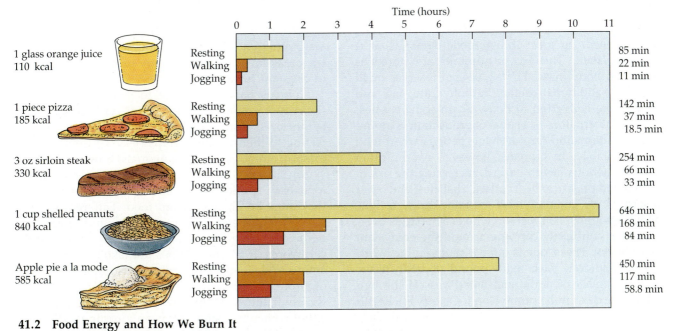

41.2 Food Energy and How We Burn It
The energy in kilocalories for several common food items is shown on the left. The graph indicates about how long it would take a person with a basal metabolic rate of about 1,800 kcal/day to utilize the equivalent amount of energy while involved in various activities.

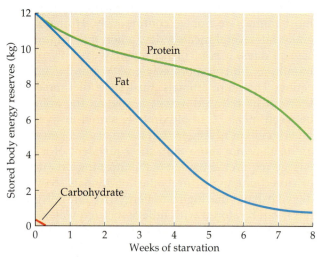

41.3 Depletion of Body Energy Reserves during Starvation

The carbohydrate reserves of our bodies are meager and are depleted by only a single day without food intake. Our major energy reserve is fat, and even a person of average body weight has enough fat to survive four or five weeks without food. When most body fat has been exhausted, the only remaining fuel is protein, and it begins to be lost at an accelerating rate, with serious consequences and eventual death.

starvation. Undernourishment is rampant among people in underdeveloped and war-torn nations, and a billion people—one-fifth of the world's population—are undernourished. Ironically, a serious cause of undernourishment in Western, developed nations is a self-imposed starvation called **anorexia nervosa** that results from a psychological aversion to body fat.

When an animal consistently takes in more food than it needs to meet its energetic demands, it is **overnourished**. The excess nutrients are stored as increased body mass. First, glycogen reserves build up; then additional dietary carbohydrate, fat, and protein are converted to body fat. In some species, such as hibernators, seasonal overnutrition is an important adaptation for surviving periods when food is unavailable. In humans, however, overnutrition can be a serious health hazard, increasing the risk of high blood pressure, heart attack, diabetes, and other disorders. A common clay building brick weighs about 5 pounds, so a person who is 50 pounds overweight is constantly carrying around the equivalent of 10 bricks. That alone is quite a strain on the heart, but in addition, each extra pound of body tissue includes miles of additional blood vessels through which the heart must pump blood. Obesity is a health hazard—but so are poorly planned fad or crash diets that can lead to malnutrition (discussed below). People spend billions of dollars every year on schemes to lose weight, even though one need only follow a simple rule: take in fewer calories than your body burns, but maintain a balanced diet.

Nutrients as Building Blocks

Green plants can synthesize for themselves all the organic molecules they need as long as they have sunlight, water, carbon dioxide, and sources of nitrogen and certain mineral nutrients. This is not true of animals. Every animal requires certain basic organic molecules (carbon skeletons) that it cannot synthesize for itself, but from which it can build the more complex organic molecules it needs. As an example of a required carbon skeleton, consider the acetyl group (Figure 41.4). Animals cannot make acetyl groups from carbon, oxygen, and hydrogen, but they can readily obtain them by metabolizing either carbohydrates or fats. From these acquired acetyl groups, animals make a wealth of other needed compounds, including fatty acids, steroid hormones, electron carriers for cellular respiration, certain amino acids, and, indirectly, legions of other compounds. It is important to recognize that the three major classes of nutrients—the carbohydrates, the fats, and the proteins—provide both energy and carbon skeletons for biosynthesis.

The acetyl group can be derived from the metabolism of any carbohydrate or fat, so it is unlikely ever to be in short supply for an animal with adequate food. Other carbon skeletons, however, are derived from more limited sources, and an animal can suffer a deficiency of these materials even if its caloric intake is adequate. This state of deficiency is called **malnutrition**. A good example of such substances is the amino acids that serve as the building blocks for protein. Humans obtain amino acids by digesting dietary protein and absorbing the resulting amino acids. The body then synthesizes its own protein molecules, as specified by its DNA, from these dietary amino acids. Another source of amino acids is the breakdown of existing body proteins, as they are in constant turnover.

Animals have the ability to synthesize some of their own amino acids by taking carbon skeletons synthesized from acetyl or other groups and transferring to them amino groups (—NH₂) from already existing amino acids. Most animals, however, cannot synthesize all of the amino acids they need. Therefore, each species has certain **essential amino acids** that must be obtained from food. Different species have different essential amino acids, and in general, herbivores have fewer essential amino acids than do carnivores. If an animal does not take in an essential amino acid, its protein synthesis is impaired. Think of protein synthesis as writing a story on a typewriter. If the typewriter is missing a key, the story either comes to a stop or has an error in it wherever the letter represented by that key is needed. In protein synthesis, the story usually comes to a stop and a functional protein is not produced.

Humans require eight essential amino acids in the diet: isoleucine, leucine, lysine, methionine, phen-

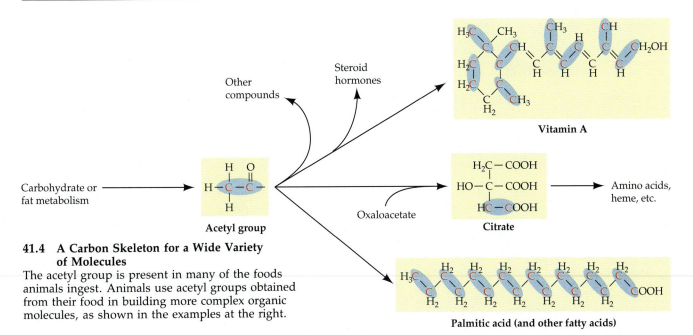

41.4   A Carbon Skeleton for a Wide Variety of Molecules
The acetyl group is present in many of the foods animals ingest. Animals use acetyl groups obtained from their food in building more complex organic molecules, as shown in the examples at the right.

ylalanine, threonine, tryptophan, and valine. All eight are available together in milk, eggs, and meat, but no plant food contains all eight. A strict vegetarian can run the risk of protein malnutrition. However, an appropriate dietary mixture of plant foods supplies all eight essential amino acids (Figure 41.5). Wheat, corn, rice, and other grains are deficient in lysine but are well stocked with most of the others. Beans, lentils, and other legumes have lots of lysine but are low in methionine. Eating only grains or only beans would lead to a serious deficiency of one or more essential amino acids. If, however, one eats grains and beans together, then one gets enough lysine *and* enough methionine and the rest. Beans and grains complement each other, each providing one or more of the essential amino acids lacking in the other. In general, grains are complemented by legumes or by milk products; legumes are complemented by grains, and also by seeds and nuts. Long before the chemical basis for this complementarity was understood, societies with little access to meat learned appropriate dietary practices through trial and error. Thus many Central and South American peoples ate beans with corn, and the native peoples of North America complemented their beans with squash. It is important to remember that we do not retain great stores of free amino acids in our bodies, yet we synthesize proteins continuously. Therefore, it makes little nutritional sense to eat grains one day and beans the next; they must be eaten together for proper amino acid balance. Excess amino acids are burned for fuel, converted to fat, or excreted.

Why are dietary proteins completely digested to their constituent amino acids before being absorbed into the body? Wouldn't it be more energy-efficient to reuse some dietary proteins directly? There are several reasons why ingested proteins are not reused.

First, macromolecules such as proteins are not readily taken up through plasma membranes, but when these macromolecules are broken down into their constituent monomers (such as amino acids), the monomers are readily transported. Second, protein structure and function are highly species-specific. A protein that functions optimally in one species might not function well in another species. Third, foreign proteins entering the body directly from the gut would be recognized as invaders and be attacked by the immune system. Most animals avoid these problems by digesting food proteins extracellularly and then absorbing the amino acids into the body. The

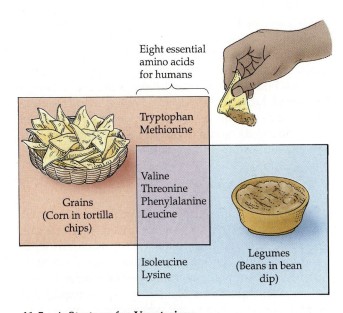

41.5   A Strategy for Vegetarians
By combining cereal grains and legumes, a vegetarian can obtain all of the essential amino acids.

TABLE 41.1
Mineral Elements Required by Animals

ELEMENT	SOURCE IN HUMAN DIET	MAJOR FUNCTIONS
Macronutrients		
Calcium (Ca)	Dairy foods, eggs, green leafy vegetables, whole grains, legumes, nuts	In bones and teeth; blood clotting; nerve and muscle action; enzyme activation
Chloride (Cl)	Table salt (NaCl), meat	Water balance; digestion (as HCl); principal negative ion in fluid around cells
Magnesium (Mg)	Green vegetables, meat, whole grains, nuts, milk, legumes	Required by many enzymes; found in bones and teeth
Phosphorus (P)	Dairy foods, eggs, meat, whole grains, legumes, nuts	In nucleic acids, ATP, and phospholipids; bone formation; buffers; metabolism of sugars
Potassium (K)	Meat, whole grains, fruits, vegetables, legumes	Nerve and muscle action; protein synthesis; principal positive ion in cells
Sodium (Na)	Table salt, dairy foods, meat, eggs, vegetables	Nerve and muscle action; water balance; principal positive ion in fluid around cells
Sulfur (S)	Meat, eggs, dairy foods, nuts, legumes	In proteins and coenzymes; detoxification of harmful substances
Micronutrients		
Chromium (Cr)	Meat, dairy foods, whole grains, dried beans, peanuts, brewers' yeast	Glucose metabolism
Cobalt (Co)	Meat, tap water	Vitamin B_{12}; formation of erythrocytes
Copper (Cu)	Liver, meat, fish, shellfish, legumes, whole grains, nuts	In active site of many redox enzymes and electron carriers; production of hemoglobin; bone formation
Fluoride (F)	Most water supplies	Improves resistance to tooth decay
Iodide (I)	Fish, shellfish, iodized salt	In thyroid hormones
Iron (Fe)	Liver, meat, green vegetables, eggs, whole grains, legumes, nuts	In active site of many redox enzymes and electron carriers; hemoglobin; myoglobin
Manganese (Mn)	Organ meats, whole grains, legumes, nuts, tea, coffee	Activates many enzymes
Molybdenum (Mo)	Organ meats, dairy foods, whole grains, green vegetables, legumes	Required by some enzymes
Selenium (Se)	Meat, seafood, whole grains, eggs, chicken, milk, garlic	Involved in metabolism of fats
Zinc (Zn)	Liver, fish, shellfish, and many other foods	Required by some enzymes; involved in physiology of insulin

new proteins formed from these amino acids are recognized as "self" by the immune system.

From acetyl units obtained from carbohydrates or fats, we can synthesize almost all the lipids required by the body, but we must have a dietary source of essential **fatty acids**, notably linoleic acid. Essential fatty acids are necessary components of membrane phospholipids, and a deficiency can lead to problems such as infertility and impaired lactation.

Mineral Nutrients

The principal **mineral elements** required by animals are listed in Table 41.1. Animals need large amounts of some of these elements, called macronutrients, and only tiny amounts of others called micronutrients. We know that certain species require various other elements. Some elements may be required in such minute amounts that deficiencies are never observed, but nevertheless, the elements are essential.

Animals need calcium and phosphorus in great quantity for many uses. Calcium phosphate is the principal structural material in bones and teeth. Muscle contraction, nerve function, and many other intracellular functions in animals require calcium. Phosphorus is an integral component of nucleic acids. We learned in Chapter 7 about the role of phosphate groups in biological energy transfers. Sulfur is part of the structure of two amino acids and is therefore found in almost all proteins. Sulfur is also found in a number of other essential compounds. Iron is the oxygen-binding atom in both hemoglobin and myoglobin, which are the oxygen-carrying proteins in vertebrate blood and muscle (Chapter 39). In addition, iron undergoes redox reactions in some of the electron-carrying proteins of cellular respiration (Chapter 7). A number of mineral nutrients act as cofactors for enzymes—among these are magnesium, manganese, zinc, and cobalt. Potassium, sodium, and chloride ions are particularly important in

BOX 41.A

Beriberi and the Vitamin Concept

Beriberi is a Singhalese word meaning "extreme weakness." This disease is found wherever unbalanced diets are common. It became particularly prevalent in Asia in the nineteenth century, when it became standard practice to mill rice to a high, white polish and discard the hulls that are present in brown rice. There are several forms of beriberi, with different symptoms, but the heart is generally adversely affected.

In 1897 Christian Eijkman, working in what is now Indonesia, discovered that chickens developed beriberi-like symptoms when fed a diet of polished rice. During the next decade, others found that people in Malaysia on a polished rice diet developed beriberi, whereas those who ate brown rice did not. Finally, in 1912, Polish-born Casimir Funk showed that pigeons with beriberi could be cured of their symptoms by feeding them a concentrate of rice polishings —the hulls that were discarded to make the rice more "appealing." He went on to suggest that beriberi and a number of other diseases are dietary in origin, and that they result from deficiencies in specific substances, for which he coined the term *vitamines* because he mistakenly thought those substances were all amines. In 1926 thiamin (vitamin B_1), which is the substance lost in the rice milling process, was the first vitamin to be isolated in pure form; in 1936 its structure was determined, and it was synthesized for the first time.

The work on birds by Eijkman and Funk was the first in a series of experiments on animals that established that diseases can result from dietary deficiencies. Until this work was done, it had been thought that all diseases were caused by microorganisms.

---

the osmotic balance of tissues and in the electrical properties of membranes, including resting potentials and action potentials (Chapter 36).

Animals require large amounts of both sodium and chloride ions. Because plants contain few of those ions, herbivores may travel considerable distances to natural salt licks. Ranchers and game wardens frequently supply salt licks for animals that do not have access to natural sources.

Specific requirements for individual elements vary considerably from species to species. In vertebrates, copper is essential in trace amounts for certain enzymes to function properly. For example, hemoglobin synthesis requires copper, even though copper is not part of the hemoglobin molecule. In numerous invertebrate species, however, copper is part of the respiratory pigment, hemocyanin, and those animals require more copper than vertebrates do.

Vitamins

Another group of essential nutrients is the **vitamins**. Like essential amino acids and fatty acids, vitamins are organic compounds that an animal cannot make for itself but that are absolutely required for its normal growth and metabolism (Box 41.A). Vitamins function mostly as coenzymes or parts of coenzymes and are required in very small amounts compared with essential amino acids and fatty acids that have structural roles. The list of vitamins needed varies from species to species. For example, ascorbic acid (vitamin C) can be made by most mammals, so it is not a vitamin for them; however, primates cannot synthesize ascorbic acid. If we do not get it in our diet, we develop the disease known as **scurvy** (Box 41.B). There are 13 such compounds that humans cannot synthesize in sufficient quantities (Table 41.2). They are divided into two groups, the **water-soluble vitamins** and the **fat-soluble vitamins**.

Water-soluble vitamins (the B complex and vitamin C) play roles in both vertebrates and invertebrates. The B vitamins are coenzymes or parts of coenzymes. The B vitamin niacin, for example, we have encountered already under its other name, nicotinamide. It is the portion of NAD (nicotinamide adenine dinucleotide) and NADP (Chapter 7) that undergoes oxidation and reduction in the respiratory chain and in other key redox systems in all living things. Riboflavin (vitamin B_2), similarly, is the site of oxidation and reduction in the respiratory chain intermediates FAD (flavin adenine dinucleotide) and FMN (flavin mononucleotide). Vitamin C (ascorbic acid) has a number of functions, among them an essential role in the formation of the structural protein collagen. That fibrous protein is a major constituent of bone, cartilage, tendons, ligaments, and skin. The water-soluble compounds that are vitamins for humans are essential to all animals. However, some species can make some of those compounds in sufficient quantity so as not to require them in the diet.

Fat-soluble vitamins have diverse functions. Vitamin A (retinol) is a precursor of retinal, the visual pigment in our eyes. Vitamin D (calciferol) regulates the absorption and metabolism of calcium. Although vitamin D may be obtained in the diet, it can also be produced in human skin by the action of ultraviolet wavelengths of sunlight on certain lipids already present in the body. Therefore, it is technically only

BOX 41.B

Scurvy and Vitamin C

In 1498 Vasco da Gama sailed around the Cape of Good Hope. In so doing, he lost all but 60 of his 160-man crew to a disease that was becoming all-too-familiar on ocean voyages of some months' duration. The symptoms of **scurvy** include general debility, hemorrhaging and decay of skin and flesh, bleeding gums and loss of teeth, and finally, death.

Over a century later, the British physician James Lind found that shipboard scurvy could be prevented by having sailors eat citrus fruit or sauerkraut. The effects were dramatic. In 1760 one British naval hospital treated 1754 cases of scurvy. Beginning in 1795, the Royal Navy required that all sailors take a daily ration of lemon juice— and the same hospital treated exactly one case of scurvy in 1806. The navy switched from lemons to limes in 1865, and since that time British sailors have been referred to as "limeys."

In 1907 researchers induced scurvy in guinea pigs by giving them a diet of dried hay and oats, with no fresh plant material. By supplementing that diet with various foods, the researchers were able to determine which foods prevent scurvy. At last, in 1932, the anti-scurvy factor was isolated from lemon juice. The anti-scurvy factor is ascorbic acid, a compound identified in plant extracts some 6 years earlier by Albert Szent-Györgyi and now commonly called vitamin C.

It is noteworthy that vitamin deficiency diseases are not usually found among primitive societies living according to long-established tradition. Rather, they occur when new habits are thrust upon people by civilization or technology. Scurvy and other deficiency diseases may also arise when people are cut off from their normal modes of living by such things as ocean voyages or imprisonment, or when armies are on campaign or cities are under siege.

a vitamin for individuals with inadequate exposure to the sun, such as people living at high latitudes where clothing usually covers the body and the sun may not shine for long periods of time. Human races that evolved in the equatorial and lower latitudes acquired or retained dark skin pigmentation as a protection against the damaging ultraviolet rays of the sun. The loss of dark skin pigmentation in races that became adapted to higher latitudes presumably facilitated ultaviolet absorption by limited skin areas exposed briefly to sunlight. Exceptions to the correlation between light skin and adaptation to high latitudes are the Eskimo peoples of the Arctic region, who obtain abundant vitamin D from the large amounts of fish oils and other animal foods in their diets.

Vitamin E is still poorly understood. Its principal function may be to protect unsaturated fatty acids from oxidation and inactivation in cellular membranes. Vitamin K functions in blood clotting following an injury and hence plays a crucial role in the protection of the body. The fat-soluble vitamins are generally required by vertebrates but are not needed by invertebrates.

When water-soluble vitamins are ingested in excess of bodily needs, they are simply eliminated in the urine. (This is the fate of much of the vitamin C many people take in excessive doses.) The fat-soluble vitamins, however, accumulate in body fat and may build up to life-threatening levels if taken in excess. Eskimos generally do not eat polar bear liver, which is exceptionally rich in vitamin A.

Nutritional Deficiency Diseases in Humans

When a person experiences chronic shortage of a nutrient, a characteristic deficiency disease results; if the deficiency is not remedied, death may follow. An example is the condition known as kwashiorkor. As we learned in Chapter 40, kwashiorkor results from protein deficiency, which causes swelling of the extremities, distension of the abdomen (see Figure 40.12), immune system breakdown, degeneration of the liver, mental retardation, and other problems.

Shortage of any of the vitamins results in specific deficiency symptoms, which are listed in Table 41.2. Two deficiency diseases, beriberi and scurvy, were discussed in Boxes 41.A and 41.B. **Pellagra**, which results from a deficiency of the B vitamin nicotinamide, is a common and severe disease in many poor areas. It also occurs frequently in conjunction with chronic alcoholism. Its symptoms include diarrhea, itching skin, abdominal pain, and other problems. Vitamin D deficiency decreases calcium absorption and use and thus leads to a softening of the bones and a distortion of the skeleton. This deficiency disease is known as **rickets**. Vitamin B_{12} (cobalamin) is produced by microorganisms that live in our intestines and use the cobalt in our diet. Cobalamin is present in all foods of animal origin. Plants neither use nor produce vitamin B_{12}, and a strictly vegetarian diet (not supplemented by vitamin pills) can lead to **pernicious anemia**, the B_{12} deficiency disease.

Inadequate mineral nutrition can also lead to deficiency diseases. Iodine, for example, is a constituent

TABLE 41.2
Vitamins in the Human Diet

VITAMIN	SOURCE	FUNCTION	DEFICIENCY SYMPTOMS
Water-Soluble			
B₁, thiamin	Liver, legumes, whole grains, yeast	Coenzyme in cellular respiration	Beriberi, loss of appetite, fatigue
B₂, riboflavin	Dairy foods, organ meats, eggs, green leafy vegetables	Coenzyme in cellular respiration (in FAD and FMN)	Lesions in corners of mouth, eye irritation, skin disorders
Niacin (nicotinamide, nicotinic acid)	Meat, fowl, liver, yeast	Coenzyme in cellular metabolism (in NAD and NADP)	Pellagra, skin disorders, diarrhea, mental disorders
B₆, pyridoxine	Liver, whole grains, dairy foods	Coenzyme in amino acid metabolism	Anemia, slow growth, skin problems, convulsions
Pantothenic acid	Liver, eggs, yeast	In coenzyme A	Adrenal problems, reproductive problems
Biotin	Liver, yeast, bacteria in gut	In coenzymes	Skin problems, loss of hair
B₁₂ cobalamin	Liver, meat, dairy foods, eggs	Coenzyme in formation of nucleic acids and proteins, in red blood cell formation	Pernicious anemia
Folic acid	Vegetables, eggs, liver, whole grains	Coenzyme in formation of heme and nucleotides	Anemia
C, ascorbic acid	Citrus fruits, tomatoes, potatoes	Aids formation of connective tissues, prevents oxidation of cellular constituents	Scurvy, slow healing, poor bone growth
Fat-Soluble			
A, retinol	Fruits, vegetables, liver, dairy foods	In visual pigments	Night blindness, damage to mucous membranes
D, calciferol	Fortified milk, fish oils, sunshine	Absorption of calcium and phosphorus	Rickets
E, tocopherol	Meat, dairy foods, whole grains	Muscle maintenance, prevents oxidation of cellular components	Anemia
K, menadione	Intestinal bacteria, liver	Blood clotting	Blood-clotting problems (in the newborn)

of the hormone thyroxin (Chapter 34), which is produced in the thyroid gland. If insufficient iodine is obtained in the diet, the thyroid gland grows larger in an attempt to compensate for the inadequate production of thyroxin. The swelling that results is called a **simple goiter** (Figure 34.10). Such goiters are common in mountain areas such as the Andes of South America because of low iodine levels in the soil and hence in the crops grown there. Goiters were once common in Switzerland and in the Great Lakes area of the United States, but the problem was largely solved by the addition of small amounts of iodine to table salt or to drinking water.

ADAPTATIONS FOR FEEDING

The ways an animal acquires its nutrients and its adaptations for doing so are frequently its most distinguishing characteristics. The role that a species plays in nature is described as its ecological niche,

and its feeding specializations and adaptations are major dimensions of that ecological niche (Chapter 47). The crucial adaptations that enable a species to exploit a particular source of nutrition are frequently physiological and biochemical. Think, for example, of the Australian koala, which eats nothing but leaves of eucalyptus trees. Eucalyptus leaves are tough, low in nutrient content, and loaded with pungent, toxic compounds that evolved to protect the trees from predators. The gut of the koala can digest and detoxify the leaves and absorb all of the nutrients the animal needs from this highly specialized, formidable diet. The feeding adaptations that are most obvious to us, however, are the anatomical and behavioral features that animals use to acquire and ingest their food.

Food Acquisition by Carnivores

The predatory behaviors of many carnivores are legendary. One need only call to mind the hunting skills

41.6   Inside-Out Digestion
This sea star is eating two whelks. It is holding them with its arms as the tissue from its everted stomach digests them.

of hawks, wolves, and lions, or the chills down the spine inspired by images of the great white shark. Carnivores have evolved stealth, speed, power, large jaws, sharp teeth, and strong gripping appendages. A cheetah, for instance, first stalks its prey stealthily from downwind, aided by its natural camouflage. When close enough, it dashes after the prey at speeds as fast as 110 kilometers per hour. It then brings the prey down with its sharp, powerful claws and teeth. Carnivores also have evolved remarkable means of detecting prey. As we learned in Chapter 37, bats use echolocation, pit vipers sense infrared radiation from the warm bodies of their prey, and certain fishes detect electric fields created in the water by their prey.

Adaptations for killing and ingesting prey are diverse and highly specialized. These adaptations can be especially important when the prey species are capable of inflicting damage on the predators. Many species of snakes take relatively large prey that are well equipped with sharp teeth and claws. A snake may strike with poisonous fangs and immobilize its prey before ingesting it. A boa or python immobilizes and kills its prey by squeezing it with coils of its powerful body. To swallow large prey, a snake's lower jaw disengages from its joint with the skull. The tentacles of jellyfish, corals, squid, and octopus, the long, sticky tongues of frogs and chameleons, and the webs of spiders are other examples of fascinating adaptations for capturing and immobilizing prey. Some prey can be impossible for the predator to ingest, so some digestion is accomplished externally. Sea stars evert their stomachs to digest their molluscan prey while they are still in their shells (Figure 41.6). Similarly, the prey of spiders are usually

other insects with indigestible exoskeletons. The spider can inject its prey with digestive enzymes and then suck out the liquified contents, leaving behind the empty exoskeletons that are frequently seen in old spider webs.

Food Acquisition by Herbivores

Herbivores obtain food less dramatically than predators do. Cows or sheep graze in grassy meadows, and caterpillars munch steadily on leaves. Some herbivores have striking adaptations for feeding, such as the trunk (a flexible, gripping nose) of the elephant, the long neck of the giraffe, or the tiny wings that are responsible for the hovering flight that allows the hummingbird to gather nectar swiftly from large numbers of flowers.

Behavior that can almost be described as agricultural is an unusual adaptation of some herbivores. Some African termites and tropical leaf-cutter ants prepare and tend subterranean fungus gardens (Figure 41.7). The reason for this mutually beneficial relationship is that the fungi produce enzymes that can break down the cellulose of the woody materials the termites ingest. Without the fungi, the termites can derive no nutrition from the wood. Other species of ants and termites depend on protozoa in their guts to provide the same benefit.

Food Acquisition by Filter Feeders

Filter feeders are animals that strain out particles or small organisms suspended in water. Most sessile aquatic animals, such as sponges, corals, barnacles, and bivalve mollusks, feed in this way, and there are

41.7   Fungus Gardens
These rain forest termites harvest plant matter, which they cannot digest, and spread it on the walls of their underground chambers. The fungus that grows in these carefully prepared beds is food for the termites.

also mobile filter feeders, such as baleen whales, tadpoles, mosquito larvae, hundreds of fishes (such as herring, sardines, and menhaden), and some birds. All filter feeders have some device for passing great volumes of water through some filter-like part at the front end of the gut. Flamingos sift through water and mud with their grooved bills, capturing insect larvae, worms, seeds, bacteria, and other matter. Many stationary filter feeders employ mucus to extract particles from water. Oysters, clams, and other bivalves draw water over their gills, where sticky mucus traps particulate matter. Their gills are densely covered with cilia that convey the mucus and its trapped particles toward the mouth, where coarse matter is rejected and the rest allowed to enter.

The most dramatic case of filter feeding is found among the baleen whales, the largest of which (in fact, the largest animal ever to live on this planet) is the blue whale (*Balaenoptera musculus*). A blue whale may grow to 100 feet long—equivalent to several school buses end to end; its tongue is the size of an elephant, and its heart is the size of a small automobile. Yet the blue whale eats tiny crustaceans called krill, which it filters from seawater. If you run your tongue along the roof of your mouth you will feel ridges. These ridges are greatly enlarged in size in baleen whales, and form **baleen plates**, which have fringed edges (Figure 41.8). The whale approaches a swarm of krill and gulps it in, along with thousands of liters of water. As it closes its mouth, its tongue forces the seawater out through the fringe

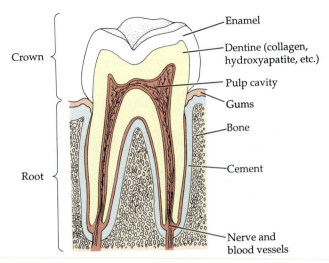

41.9 A Mammalian Tooth Has Several Layers
A section through a mammalian tooth shows that the crown above the gums is covered with hard enamel. Below the enamel is a thick dentine layer that extends into the skull to form the root, and inside the dentine is the pulp cavity containing the tooth's supply of blood vessels and nerves. The tooth is held in it bony socket in the skull by a fairly soft "cement." The teeth of other vertebrates, such as lizards, are not entrenched in bony sockets and consequently must be replaced frequently.

of the baleen plates, leaving the krill behind. Krill are very abundant in cold, nutrient-rich waters and make it possible for the whale to support its metabolic rate of about a million kilocalories per day.

Vertebrate Teeth

Many vertebrate species have distinctive teeth. Teeth are adapted for the acquisition and initial processing of specific types of foods, and because they are one of the hardest structures of the body, they may remain in the environment long after the animals die. Paleontologists use teeth to identify animals that lived in the distant past and to understand their behavior.

All mammalian teeth have a general structure consisting of three layers (Figure 41.9). An extremely hard material called **enamel**, composed principally of calcium phosphate, covers the crown of the tooth. Both the crown and the root contain a layer of a bony material called **dentine**, within which is a pulp cavity containing blood vessels, nerves, and the cells that produce the dentine. The shapes and organization of mammalian teeth, however, can be very different, as they are adapted to specific diets (Figure 41.10). In general, incisors are teeth used for cutting, chopping, or gnawing; canines are teeth used for stabbing, ripping, and shredding; and cheek teeth (molars and premolars) are used for shearing, crushing, and

Baleen sieve plates

41.8 The Largest Eating the Smallest
A feeding right whale shows the baleen plates that hang from the roof of its mouth and filter small animals from the huge volumes of water that pass through them.

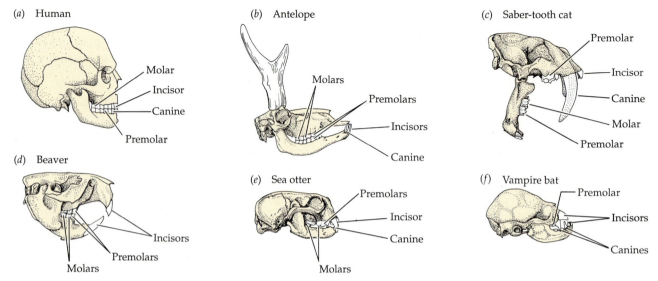

(a) Human — Molar, Incisor, Canine, Premolar

(b) Antelope — Molars, Premolars, Incisors, Canine

(c) Saber-tooth cat — Premolar, Incisor, Canine, Molar, Premolar

(d) Beaver — Incisors, Premolars, Molars

(e) Sea otter — Premolars, Incisor, Canine, Molars

(f) Vampire bat — Premolar, Incisors, Canines

41.10 Mammalian Teeth Are Specialized for Different Diets
(a) A human is an omnivore and has a generalized set of teeth. (b) The pronghorn antelope grazes on tough grasses, herbs, and shrubs. It has no incisors or canines in its upper jaw, and in the lower jaw they are far forward and used to tear the leaves off plants. The antelope's cheek teeth are highly adapted for grinding coarse vegetation. (c) Carnivores such as the extinct saber-tooth cat have greatly enlarged canine teeth for gripping, killing, and tearing their prey. The incisors are used for scraping muscle off of bone, and the cheek teeth are used for shearing flesh and crushing bones. (d) Rodents such as the beaver have enlarged incisors that they use for gnawing. The beaver cuts down and strips the bark from trees with its incisors. The incisor teeth of rodents grow continuously to compensate for wear and tear. The beaver has no canines, but its cheek teeth are well adapted for grinding fibrous plant material. (e) The sea otter has well-developed cheek teeth for crushing the shells of marine invertebrates. (f) The upper incisors and canines of the vampire bat are triangular and extremely sharp. They are used to make incisions in the skin of large mammals so the bat can drink their blood.

grinding. The highly varied diet of humans is dealt with by a generalized, multipurpose set of teeth, as is common among omnivores. We have four upper and four lower incisors for biting, and two upper and two lower canines for tearing. Behind the canines are premolars (two on each side of each jaw) and molars (three on each side of each jaw) for crushing and grinding. In all, there are 32 teeth in the adult mouth, before the dentist starts extracting them. A child develops a first set of 20 "milk teeth," which are lost and replaced by the permanent teeth.

DIGESTION

Most animals digest food extracellularly. Animals take food into a body cavity that is continuous with the outside environment, then secrete digestive enzymes into that cavity. The enzymes act on the food, reducing it to nutrient molecules that can be absorbed by the cells lining the cavity. Only after they are absorbed by the cells are the nutrients within the body of the animal. The simplest digestive system is a gastrovascular cavity that connects to the outside world through a single opening. After a cnidarian captures a prey with its stinging nematocysts (see Figure 25.9), its tentacles cram the prey into the gastrovascular cavity (Figure 40.1) where digestive enzymes partially digest it. Extracellular digestion in cnidarians is supplemented by a certain amount of intracellular digestion—some small food particles are taken by endocytosis into the cells lining the gut.

The gastrovascular cavity of flatworms is somewhat more complex than that of cnidarians, but it still has only one opening through which food enters and waste products exit. Where the gastrovascular cavity meets the mouth, it narrows to form a tubular **pharynx**. This muscular structure can be pushed out through the mouth during feeding. In addition, extensive branching of the gastrovascular cavity increases the efficiency of absorption of nutrient molecules; the branches increase the surface area through which absorption can occur (Figure 41.11).

Some multicellular animals have no digestive systems at all. Many of these are internal parasites such as tapeworms. They live in an environment so rich in already digested nutrient molecules that they can just absorb them directly into their cells.

Tubular Guts

The guts of all animal groups other than sponges, cnidarians, and flatworms are tubular, with an opening at either end. A mouth takes in food; solid diges-

nutrition from the food passing through the host's gut while they pay rent of a sort by contributing to the digestive processes of the host. Members of the leech genus *Hirudo* produce no enzymes that can digest the proteins in the blood they suck from vertebrates; however, a colony of gut bacteria produces the enzymes necessary to break down those proteins into amino acids, which are subsequently used by both leeches and bacteria. Many animals obtain some vitamins from bacteria in their hindguts, and herbi-

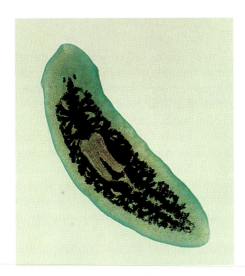

41.11 The Planarian Digestive Tract
The gastrovascular system of this flatworm has been stained to reveal all of its branches. This system communicates to the outside through the muscular pharynx visible in the center of the animal.

tive wastes, or **feces**, are excreted through an anus. Different regions in such a tubular gut are specialized for particular functions (Figure 41.12), and we will now look at those compartments. But remember that all locations within the tubular gut are really outside the body of the animal. Only by crossing the membranes lining the gut does a nutrient molecule enter the body.

At the anterior end of the gut are the mouth (the opening itself) and **buccal cavity** (mouth cavity). Food may be broken up by teeth (in some vertebrates) or by mandibles (in insects), or somewhat further along the gut by structures such as the **gizzards** of birds and earthworms, where muscular contractions of the gut grind the food together with small stones. Some animals simply ingest large chunks of food with little or no fragmentation. **Stomachs** and **crops** are mainly storage chambers, enabling animals to ingest relatively large amounts of food and digest it at leisure. Digestion and absorption may or may not occur in such a storage chamber, depending on the species. Food delivered into the next section of gut, the **midgut** or **intestine**, is well minced and well mixed. Most digestion and absorption occurs here. Specialized glands secrete digestive enzymes into the intestine, and the gut wall itself secretes other digestive enzymes. The **hindgut**, which often includes a muscular **rectum** near the anus, recovers water and ions and retains feces.

Found within the hindguts of many species are colonies of bacteria that live in cooperation with their hosts. The association of animal and bacteria is an example of organisms living together to their mutual benefit, a topic that will be considered in detail in the ecology chapters. The bacteria obtain their own

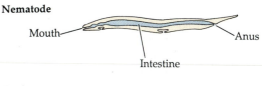

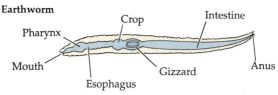

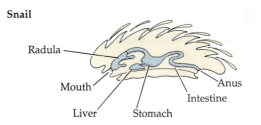

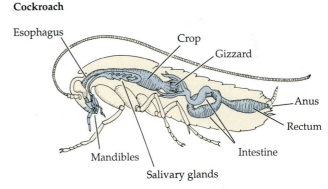

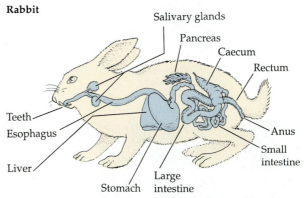

41.12 Compartments Specialized for Digestion and Absorption

vores such as cattle, termites, and cockroaches depend on microorganisms in various parts of their guts for the digestion of cellulose.

Those parts of the gut that absorb nutrients have, in many animals, evolved extensive surface areas for absorption, as we have already seen in flatworms. In the earthworm, a long dorsal infolding of the intestine, called the typhlosole (Figure 41.13a), provides extra absorptive surface area. The shark's intestine has a spiral valve, forcing food to take a longer path and thus encounter more absorptive surface (Figure 41.13b). In many vertebrates, the wall of the gut is richly folded, with the individual folds bearing legions of tiny fingerlike projections called **villi** (Figure 41.13c). The villi in turn have microscopic projections, which are called microvilli, on the cells that compose their surfaces (Figure 41.13d). All of the surface area of all the microvilli throughout the long vertebrate intestine absorbs nutrients.

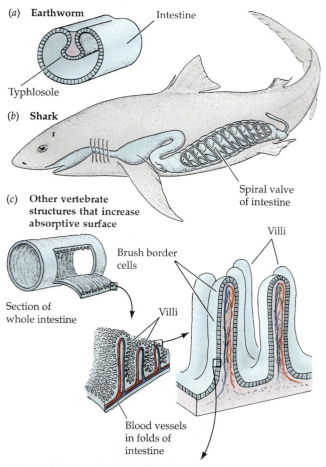

(a) **Earthworm**   Intestine

Typhlosole

(b) **Shark**

Spiral valve of intestine

(c) **Other vertebrate structures that increase absorptive surface**

Section of whole intestine

Brush border cells

Villi

Villi

Blood vessels in folds of intestine

(d)

Digestive Enzymes

The breakdown of protein, carbohydrate, and fat macromolecules into their simplest monomeric units is accomplished by digestive enzymes. All of these enzymes cleave the chemical bonds of macromolecules through a reaction that adds a water molecule at the site of cleavage, hence they are generally called hydrolytic enzymes. Examples of hydrolytic cleavage are the breaking of the bonds between adjacent amino acids of a protein or peptide, as was shown in Figure 6.16, and the breaking of the bonds between adjacent glucose units in starch, as was shown in Figure 6.8.

Digestive enzymes are classified according to the substances that they hydrolyze: carbohydrases hydrolyze carbohydrates; proteases, proteins; peptidases, peptides; lipases, fats; and nucleases, the nucleic acids. The prefixes *exo-* (outside) and *endo-* (within) indicate where the enzyme cleaves the molecule. Thus an endoprotease hydrolyzes a protein at an internal site along the polypeptide chain, whereas an exoprotease snips away amino acids at the ends of the molecule (see Figures 6.16 and 6.17).

How can an organism produce enzymes to digest biological macromolecules without digesting itself? The answer is that the digesting, as you know, is always done *outside* the animal. The gut is simply a tunnel through the animal; food in the gut is outside the body and hence can receive treatment (such as high acidity or potent enzymes) that would be intolerable within a cell or a tissue. (When cnidarians digest small prey intracellularly, the hydrolytic reactions occur within food vacuoles.) Many digestive enzymes are produced in an inactive form, known as a **zymogen**, and become active only when they reach the outside of the cell. When secreted into the gut, the zymogens are activated, sometimes by exposure to a different pH but more often by the action of another enzyme. For example, the walls of the vertebrate intestine secrete an enzyme called **enterokinase**, which activates proteases that are secreted in an inactive form from the pancreas into the intestine.

The gut itself is not digested by activated enzymes because it is protected by a covering of mucus, a slimy material secreted by special cells in the lining of the gut. The mucus also lubricates the gut and protects it from abrasion. If mucus production is inadequate, digestive enzymes or stomach acid can act

41.13 Greater Intestinal Surface Area Means More Absorption
(a) In earthworms, the adaptation is a typhlosole—a simple longitudinal infolding of the intestinal wall. (b) Sharks have evolved a spiral valve that increases the surface area of the intestine. (c) In most vertebrates an enormous absorptive surface is achieved by the sheer length of the tubular small intestine. The inner intestinal wall is folded. (d) The thin, fingerlike microvilli that cover the villi increase absorptive surface area enormously.

upon the gut, producing sores called **ulcers**. Insects rely on a different trick to prevent digestion of the gut lining. Within the gut they secrete a thin tube of chitin, a modified polysaccharide (see Figure 3.13) that is also found in the insect's protective exoskeleton. The chitin tube encloses the food and enzymes and protects the gut from abrasion and self-digestion.

STRUCTURE AND FUNCTION OF THE VERTEBRATE GUT

The Tissue Layers

The separate compartments that have specific functions in the digestive tracts of vertebrates are all part of a continuous tube that runs from mouth to anus. The cellular architecture of that tube follows a common plan throughout. Four major layers of different cell types form the wall of the tube (Figure 41.14). These layers differ from compartment to compartment, but they are always present. Starting in the cavity, or **lumen**, of the gut, the first tissue layer is the **mucosa**. Nutrients are absorbed across the membranes of the mucosal cells; in some regions of the gut, those membranes have many folds that increase their surface area. Mucosal cells also have secretory functions. Some secrete mucus that lubricates the food and protects the walls of the gut. Others secrete digestive enzymes, and still others in the stomach secrete hydrochloric acid (HCl). At the base of the mucosa are some smooth muscle cells. Just outside the mucosa is the layer called the **submucosa**. Here we find the blood and lymph vessels that carry absorbed nutrients to the rest of the body. The submucosa also contains a network of nerves; those neurons are both sensory (responsible for stomach aches!) and regulatory—they control the various secretory functions of the gut.

External to the submucosa are two layers of smooth muscle cells responsible for the movements of the gut. The innermost layer has its cells oriented around the gut and is therefore called the circular layer. It constricts the lumen. The outermost layer of muscle, the longitudinal layer, has its cells arranged along the length of the gut. When it contracts, the gut shortens. Between these two layers of muscle is a network of nerves that controls the movements of the gut, coordinating the different regions with one another.

Surrounding the gut is a fibrous coat called the **serosa**. Like other abdominal organs, the gut is also covered and supported by a tissue, the **peritoneum**.

Movement of Food in the Gut

Food entering the mouths of most vertebrates is chewed and mixed with the secretions of salivary glands. The muscular tongue then pushes the

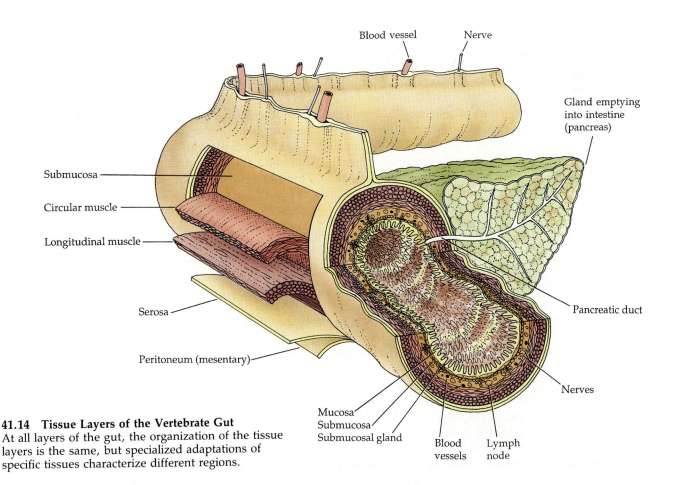

41.14 Tissue Layers of the Vertebrate Gut
At all layers of the gut, the organization of the tissue layers is the same, but specialized adaptations of specific tissues characterize different regions.

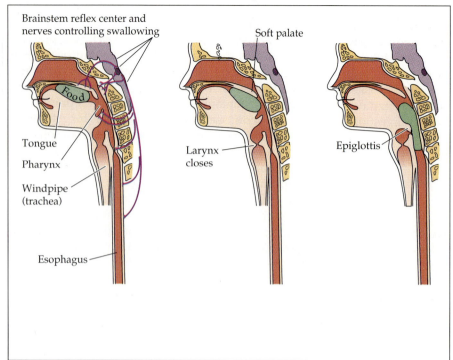

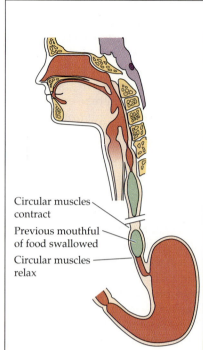

1. Food is chewed and tongue pushes bolus of food to back of mouth. Sensory nerves initiate the swallowing reflex

2. Soft palate is pulled up, vocal cords close larynx

3. Larynx pulled up and forward and covered by epiglottis; bolus of food enters esophagus

4. Peristaltic contractions propel food to stomach

41.15   The Act of Swallowing
Food pushed to the back of the mouth triggers the swallowing reflexes. Once food enters the esophagus, peristalsis propels it to the stomach.

mouthful, or bolus, of food toward the back of the mouth cavity. By making contact with the soft tissue at the back of the mouth, the food initiates a complex series of neural reflex actions commonly known as swallowing. Stand in front of a mirror and gently touch this tissue at the back of your mouth with the eraser of your pencil or with a cotton swab. You may gag slightly, but you will also experience an uncontrollable urge to swallow. The act of swallowing involves many muscles doing a variety of jobs that propel the food through the **pharynx** and into the **esophagus** without allowing any of it to enter your windpipe (trachea) or nasal passages (Figure 41.15).

Once the food is in the esophagus, **peristalsis** takes over and pushes the food toward the stomach. Peristalsis is a wave of smooth muscle contraction that moves progressively down the gut from pharynx to anus. The smooth muscle of the gut contracts in response to being stretched. The act of swallowing a bolus of food stretches the upper end of the esophagus, and this initiates a wave of contraction that slowly progresses down the entire gut, pushing the contents of the gut toward the anus. Peristalsis tends to move down the gut in the direction of the anus, but it can run in the opposite direction as well. When your stomach is very full, pressure on your abdomen or a sudden movement can push some stomach con-

tents into the lower end of your esophagus. This can initiate peristaltic movements that bring the acidic, partially digested food into your mouth. When you vomit, contractions of the abdominal muscles explosively force stomach contents out through the esophagus. Prior to vomiting, waves of reverse peristalsis can bring the contents of the upper regions of the intestine into the stomach.

The movement of food from the stomach into the esophagus is normally prevented by a thick ring of circular smooth muscle at the esophageal–stomach junction. This ring of muscle, called a **sphincter**, is normally constricted. Waves of peristalsis cause it to relax enough to let food pass through. Sphincter muscles are found elsewhere in the digestive tract as well. The **pyloric sphincter** governs the passage of stomach contents into the intestine. Another important sphincter surrounds the anus.

Digestion in the Mouth and Stomach

In addition to physically disrupting food, the mouth initiates the digestion of carbohydrates through the action of the enzyme **amylase**, which is secreted with the saliva and mixed with the food as it is chewed. Amylase is a carbohydrase; it hydrolyzes the bonds between the 6-carbon sugar units that make up the long-chain starch molecules. The action of amylase is what makes a piece of bread or cracker taste sweet if you chew it long enough.

A large volume of food can be consumed rapidly

by most vertebrates, but digestion is a long, slow process. The stomach stores the food devoured in the course of a meal and continues its physical breakdown. The secretions of the stomach help kill microorganisms taken in with the food and begin the digestion of proteins. The major enzyme produced by the stomach is an endopeptidase called **pepsin**. Pepsin is secreted as the zymogen **pepsinogen** by cells in the **gastric pits**—deep folds in the stomach lining. Other cells in the gastric pits produce hydrochloric acid, and still others near the openings of the gastric pits and throughout the stomach mucosa secrete mucus.

Hydrochloric acid maintains the stomach fluid (the gastric juice) at a pH of 1 to 3, which is important for several reasons. First, the low pH activates the conversion of pepsinogen to pepsin. This conversion is amplified as the newly formed pepsin activates other pepsinogen molecules. Hydrochloric acid also provides the right pH for the enzymatic action of pepsin. Also, the low pH helps dissolve the intercellular substances holding the cellular structure of the food together. This breakdown of the ingested tissues exposes more food surface area to the action of digestive enzymes. Mucus secreted by the stomach mucosa coats and protects the walls of the stomach from being eroded and digested by the HCl and pepsin.

Contractions of the muscles in the walls of the stomach churn its contents, thoroughly mixing them with the stomach secretions. The acidic, fluid mixture of digestive juices and partially digested food in the stomach is called **chyme**. A few substances can be absorbed from the chyme across the stomach wall, including alcohol (hence its rapid effects), aspirin, and caffeine, but even these are absorbed in rather small quantities in the stomach. The chyme is pushed toward the bottom end of the stomach by peristaltic contractions of the stomach walls. These waves of peristalsis cause the pyloric sphincter to relax briefly so that little squirts of the chyme can enter the first region of the intestine, where digestion continues and absorption of nutrients begins. The human stomach empties itself gradually, over a period of approximately 4 hours. This enables the intestine to work on a little material at a time and prolongs the digestive and absorptive processes throughout much of the time between meals.

The Small Intestine, Gall Bladder, and Pancreas

The **small intestine** takes its name from its diameter. However, it is really a very large organ and is the place where the major events of digestion and absorption take place. The small intestine of an adult human is more than 6 meters long; its coils fill much of the lower abdominal cavity (Figure 41.16). As a consequence of its length, and because of the folds, villi, and microvilli of its lining, its inner surface area is enormous—about 550 square meters, or roughly

the size of a tennis court. Across this surface it absorbs all of the nutrient molecules derived from food. The small intestine has three sections. The short initial section—the **duodenum**—is where most digestion occurs. Two additional sections—the **jejunum** and the **ileum**—carry out 90 percent of the absorption of nutrients.

Digestion requires many specialized enzymes as well as several other secretions. Two accessory organs, not part of the digestive tract, the liver and the pancreas, provide many of these enzymes and secretions. The liver synthesizes a substance called **bile** from cholesterol. Bile emulsifies fats just as soap emulsifies grease on your clothes or hands. Bile secreted from the liver flows through a duct that leads to the small intestine. A side branch of that duct delivers the bile to the gall bladder, where it is stored until it is needed to assist in fat digestion (Figure 41.17). When undigested fats enter the duodenum, the gall bladder releases bile, which flows down the bile duct and enters the duodenum.

To understand the role of bile in fat digestion, think of the oil in salad dressing; it is not soluble in water (it is **hydrophobic**) and tends to aggregate together in large globules. The enzymes that digest fat, the **lipases**, are water-soluble and must do their work in an aqueous medium. The interface between the aqueous digestive juices and large globules of fat would be very small. Bile solves this problem by stabilizing tiny droplets of fat so that they cannot aggregate into large globules. Bile molecules have one end that is soluble in fat (**lipophilic** or hydrophobic) and one end that is water soluble (**hydrophilic** or **lipophobic**). Bile molecules, therefore, tend to bury their lipophilic ends in fat droplets, leaving their lipophobic ends sticking out. As a result, they prevent the fat droplets from coalescing with each other. Keeping the fat droplets very small maximizes the droplet surface area exposed to lipases.

The **pancreas** is a large gland that lies just beneath the stomach. It has both endocrine and exocrine functions. Here we will consider just the exocrine products that it delivers to the duodenum through the pancreatic duct. As Figure 41.17 shows, the pancreatic duct joins the bile duct before entering the gut. The pancreas produces a host of digestive enzymes, listed in Table 41.3. The enzymes are released as zymogens; otherwise they would digest the pancreas and its ducts before they ever reached the gut. Once in the gut, one of these inactive enzymes, **trypsinogen**, is activated by enterokinase, produced by cells lining the duodenum (Figure 41.18). Active **trypsin** can cleave other trypsinogen molecules to release even more active trypsin. Similarly, trypsin acts on the other zymogens secreted by the pancreas and releases their active enzymes. The mixture of zymogens produced by the pancreas can be very dangerous if the pancreatic duct is blocked or if the pancreas is injured by an infection or a severe blow to

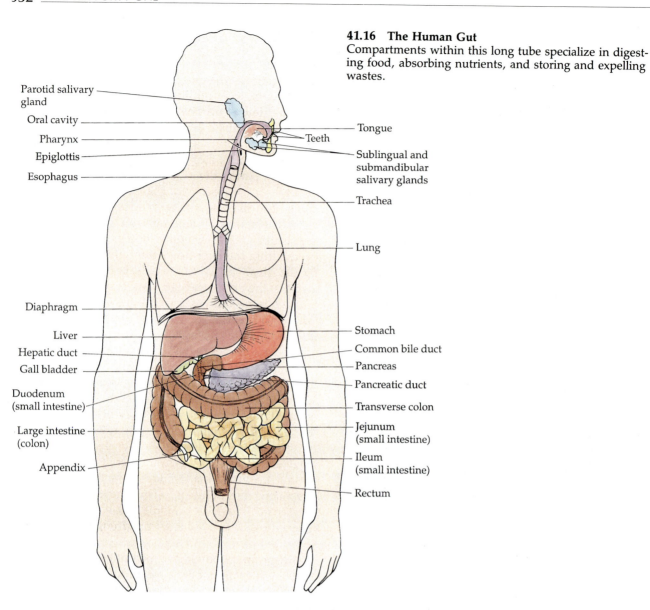

41.16 The Human Gut
Compartments within this long tube specialize in digesting food, absorbing nutrients, and storing and expelling wastes.

Parotid salivary gland
Oral cavity
Pharynx
Epiglottis
Esophagus
Tongue
Teeth
Sublingual and submandibular salivary glands
Trachea
Lung
Diaphragm
Liver
Hepatic duct
Gall bladder
Duodenum (small intestine)
Large intestine (colon)
Appendix
Stomach
Common bile duct
Pancreas
Pancreatic duct
Transverse colon
Jejunum (small intestine)
Ileum (small intestine)
Rectum

TABLE 41.3
Sources and Functions of the Major Digestive Enzymes of Humans

ENZYME	SOURCE	ACTION	SITE OF ACTION
Salivary amylase	Salivary glands	Starch → Maltose	Mouth
Pepsin	Stomach	Proteins → Peptides	Stomach
Pancreatic amylase	Pancreas	Starch → Maltose	Small intestine
Lipase	Pancreas	Fats → Fatty acids and glycerol	Small intestine
Nuclease	Pancreas	Nucleic acids → Nucleotides	Small intestine
Trypsin	Pancreas	Proteins → Peptides	Small intestine
Chymotrypsin	Pancreas	Proteins → Peptides	Small intestine
Carboxypeptidase	Pancreas	Peptides → Peptides and amino acids	Small intestine
Aminopeptidase	Small intestine	Peptides → Peptides and amino acids	Small intestine
Dipeptidase	Small intestine	Dipeptides → Amino acids	Small intestine
Enterokinase	Small intestine	Trypsinogen → Trypsin	Small intestine
Nuclease	Small intestine	Nucleic acids → Nucleotides	Small intestine
Maltase	Small intestine	Maltose → Glucose	Small intestine
Lactase	Small intestine	Lactose → Galactose and glucose	Small intestine
Sucrase	Small intestine	Sucrose → Fructose and glucose	Small intestine

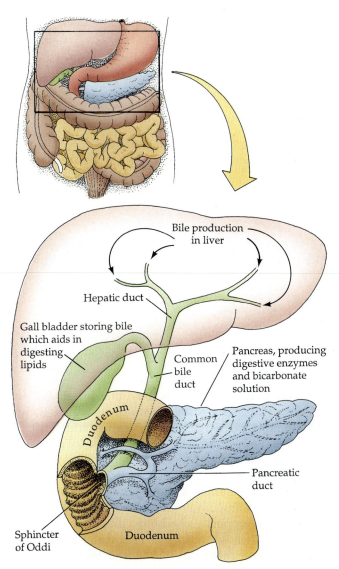

41.17 The Ducts of the Gall Bladder and Pancreas
Bile produced in the liver leaves the liver via the hepatic duct. Branching off this duct is the gall bladder, which stores bile. Below the gall bladder, the hepatic duct is called the common bile duct and is joined by the pancreatic duct before entering the duodenum.

the abdomen. A few trypsinogen molecules spontaneously converting to trypsin can initiate a chain reaction of enzyme activation that digests the pancreas in a very short period of time. Such an event would, of course, destroy both the endocrine as well as the exocrine function of the pancreas. The pancreas produces, in addition to digestive enzymes, a secretion rich in bicarbonate ions (HCO_3^-). Bicarbonate ions neutralize the pH of the chyme that enters the duodenum from the stomach. This is essential because intestinal enzymes function best at a neutral or slightly alkaline pH.

Absorption in the Small Intestine

Only the smallest products of digestion pass through the mucosa of the small intestine and into the blood

and lymphatic vessels that lie in the submucosa. The final digestion of proteins and carbohydrates takes place right among the microvilli. The mucosal cells with microvilli produce dipeptidase, which cleaves dipeptides into individual amino acids that the cells can absorb. These cells also produce the enzymes maltase, lactase, and sucrase, which cleave the common disaccharides into their constituent, absorbable monosaccharides—glucose, galactose, and fructose. Many humans stop producing the enzyme lactase at about the age of 4 years and thereafter have difficulty digesting lactose, which is the sugar in milk. Lactose is a disaccharide and cannot be absorbed without being cleaved into its constituent units glucose and galactose. If a substantial amount of lactose is unabsorbed and passes into the large intestine, its metabolism by bacteria in the large intestine causes abdominal cramps, gas, and diarrhea.

The mechanisms by which the cells lining the intestine absorb nutrient molecules and inorganic ions are diverse and not completely understood. Many inorganic ions are actively transported into the cells. For example, active transporters exist for sodium, calcium, and iron. Transporters also exist for certain classes of amino acids and for glucose and galactose, but, curiously, their activity is much reduced if active sodium transport is blocked. Sodium diffuses from the gut contents into the mucosal cells, and is then actively transported from the mucosal cells into the submucosa. To diffuse into a mucosal cell, a sodium ion binds to a carrier molecule in the mucosal cell membrane, which also binds a nutrient molecule such as glucose or an amino acid. The diffusion of the sodium ion, driven by a concentration difference, therefore drives the absorption of the nutrient molecule. This mechanism is called **sodium cotransport** (see Figure 5.16).

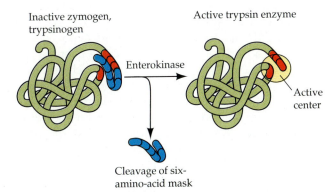

41.18 Zymogen Activation by Enterokinase
Powerful digestive enzymes often exist as inactive precursors called zymogens until they reach a location where their catalytic activity is required. Trypsin, an enzyme that cleaves polypeptide chains, is secreted from the pancreas as the inactive zymogen trypsinogen. The inactivity is due to a nullifying "mask" of six amino acids that is removed by the enzyme enterokinase when the zymogen reaches the lumen of the small intestine.

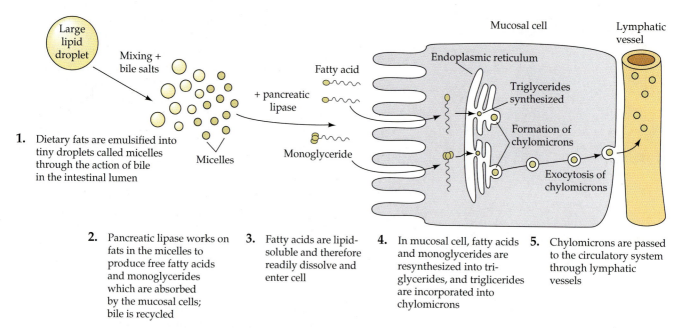

1. Dietary fats are emulsified into tiny droplets called micelles through the action of bile in the intestinal lumen

2. Pancreatic lipase works on fats in the micelles to produce free fatty acids and monoglycerides which are absorbed by the mucosal cells; bile is recycled

3. Fatty acids are lipid-soluble and therefore readily dissolve and enter cell

4. In mucosal cell, fatty acids and monoglycerides are resynthesized into triglycerides, and triglicerides are incorporated into chylomicrons

5. Chylomicrons are passed to the circulatory system through lymphatic vessels

41.19   The Digestion and Absorption of Fats

The absorption of the products of fat digestion is much simpler (Figure 41.19). Lipases break fats down into fatty acids and monoglycerides, which are lipid-soluble and are thus able to dissolve in the membranes of the microvilli and diffuse into the mucosal cells. Once in the cells, they are resynthesized into triglycerides, combined with cholesterol and phospholipids, and coated with protein to form water-soluble **chylomicrons**, which are really little particles of fat. The chylomicrons pass into the lymphatic vessels in the submucosa and into the bloodstream through the thoracic duct. Following a meal rich in fats, the chylomicrons can be so abundant in the blood that they give it a milky appearance.

The bile that emulsifies the fats is not absorbed along with the monoglycerides and the fatty acids, but is recycled back and forth between the gut contents and the microvilli. Finally, in the ileum, bile is actively reabsorbed and returned to the liver via the bloodstream. Bile is synthesized from cholesterol, and the loss of unreabsorbed bile from the digestive tract is the major means by which we remove cholesterol from the body. One rationale for including certain kinds of fiber in our diets is that the fiber binds the bile and decreases its reabsorption in the ileum.

The Large Intestine

Peristalsis gradually pushes the contents of the small intestine into the large intestine, or **colon**. The rate of peristalsis is controlled so that food passes through the small intestine slowly enough for digestion and absorption to be complete, but quickly enough to insure an adequate supply of nutrients for the body. The material that enters the colon has had most of its nutrients removed, but it contains a lot of water and inorganic ions. The colon reabsorbs water and ions, producing semisolid **feces** from the slurry of indigestable materials it receives from the small intestine. If too much water is reabsorbed, constipation results. The opposite condition, diarrhea, results if too little water is reabsorbed or if water is secreted into the colon (both can be induced by toxins from microorganisms). Feces are stored in the last segment of the colon and periodically excreted.

Immense populations of bacteria live within the colon. One of the resident species is *Escherichia coli*, the bacterium so popular with researchers in biochemistry, genetics, and molecular biology (see Box 4.B). This inhabitant of the colon subsists on matter indigestible to humans and produces some products useful to the host. For example, vitamin K and biotin are synthesized by the bacteria and absorbed across the wall of the colon. Many species of mammals maximize the nutritional benefits from such bacterial activity by reingesting their own feces, a behavior called **coprophagy**. Excessive or prolonged intake of antibiotics can lead to vitamin deficiency because the antibiotics kill the normal intestinal bacteria at the same time they are killing the disease-causing organisms for which they are intended. The intestinal bacteria produce gases such as methane and hydrogen sulfide as byproducts of their largely anaerobic metabolism. Humans expel gas after eating beans because the beans are rich in carbohydrates that bacteria—but not humans—can break down.

The large intestine of humans has a small, finger-like pouch called the **appendix** that is best known for the trouble it causes when it becomes infected. The appendix of humans plays no essential role in digestion, but it does contribute to immune system function. It can be surgically removed without serious consequences. The part of the gut that forms the appendix in humans forms the much larger **caecum** in herbivores, where it functions in cellulose digestion. As our primate ancestors evolved to exploit diets less rich in indigestible cellulose, the caecum no longer served an essential function and gradually became **vestigial**, like the nonfunctional eyes of cave fish or the dewclaws of dogs and cats.

Digestion by Herbivores

Cellulose is the principal organic compound eaten by herbivores. However, most herbivores cannot produce **cellulases**, the enzymes that hydrolyze cellulose. Exceptions include silverfish (well known for eating books and stored papers), earthworms, and shipworms. Other herbivores, from termites to cattle, rely on microorganisms living in their digestive tracts to digest cellulose for them. These microorganisms inhabit various parts of the gut, where they may be present by the billions. Most are bacteria, but some are fungi or protists.

The digestive tracts of **ruminants** such as cattle, goats, and sheep are specialized to maximize benefits from microorganisms. In place of the usual mammalian stomach, ruminants have a large, four-chambered organ (Figure 41.20). The first and largest of these chambers is the **rumen**, the second is the **reticulum**, and both are packed with anaerobic microorganisms that break down cellulose. These two chambers serve as fermentation vats for the digestion of cellulose. The ruminant periodically regurgitates the contents of the rumen (the cud) into the mouth for rechewing. When the more thoroughly ground-up vegetable fibers are swallowed again, they present more surface area to the microorganisms for their digestive actions.

The microorganisms in the rumen and reticulum metabolize cellulose and other nutrients to simple fatty acids. Ruminants produce and swallow large quantitites of alkaline saliva to buffer this acid production. Although fatty acids are the major nutrients the host derives from its microorganisms, the microorganisms themselves provide an important source of protein. A cow can derive more than 100 grams of protein per day from digestion of its own microorganisms. The plant materials ingested by a ruminant are a poor source of protein, but they contain inorganic nitrogen that the microorganisms use to synthesize their own amino acids.

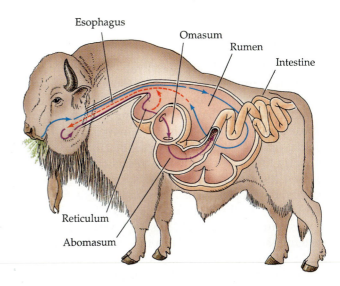

41.20 The Ruminant Stomach
Specialized compartments enable ruminants to digest and subsist on protein-poor plant material.

The byproducts of the fermentation of cellulose are carbon dioxide and methane, which the animal belches. A single cow can produce 400 liters of methane a day. Methane is the second most abundant "greenhouse gas," whose concentration in the atmosphere is increasing, and domesticated ruminants are second only to industry as a source of methane emitted into the atmosphere.

The food leaving the rumen carries with it enormous numbers of the cellulose-fermenting microorganisms. This mixture passes through the **omasum** where it is concentrated by water reabsorption. It then enters the true stomach, or **abomasum**, which secretes hydrochloric acid and proteases. The microorganisms are killed by the acid, digested by the proteases, and passed on to the small intestine for further digestion and absorption. The rate of multiplication of microorganisms in the rumen is great enough to offset their loss, so a well-balanced, mutually beneficial relationship is maintained.

Mammalian herbivores other than ruminants have microbial farms and cellulose fermentation vats in a branch off the large intestine called the caecum. Rabbits and hares are good examples (see Figure 41.12). There is no small intestine after the caecum, however, so how are the nutrients produced by the microorganisms absorbed? The answer, as described in the last section, is coprophagy. These animals frequently produce two kinds of feces, one of pure waste, which they discard, and one consisting of mostly caecal material, which they ingest directly from the anus. By passing again through the stomach and the small intestine, the contents of the caecum are digested and absorbed.

CONTROL AND REGULATION OF DIGESTION

Neural Reflexes

Everyone has experienced salivation stimulated by the sight or smell of food. That response is a neural reflex, as is the act of swallowing following tactile stimulation at the back of the mouth. Many neural reflexes operate to coordinate activities in different regions of the digestive tract so that it works in a properly timed, assembly-line manner. For example, loading the stomach with food stimulates increased activity in the colon, which can lead to a bowel movement. This is the **gastrocolic reflex**. The digestive tract is unusual in that it has its own **intrinsic nervous system**. Neural messages can travel from one region of the digestive tract to another without being processed by the central nervous system.

About one hundred years ago, a Russian physiologist, Ivan Pavlov, was making an effort to explain all regulation and control of the digestive tract in terms of neural reflexes. One result of his research was the discovery of a very basic form of learning, the **conditioned reflex**. When a dog is presented with food, it salivates—an unconditioned reflex. If a bell is rung whenever the dog is presented with food, then after a number of trials, the dog will salivate whenever the bell is rung, even if no food is present—a conditioned reflex. In the course of his experiments, Pavlov tried to explain the control of the secretory activity of the pancreas in terms of a neural reflex. He showed that the presence in the duodenum of acidic chyme from the stomach stimulated the pancreas to secrete digestive juices. Even hydrochloric acid alone applied to the duodenum stimulated the pancreas to secrete. When Pavlov tried to discover the path of the neural reflex controlling this response, however, he ran into trouble. The response could not be eliminated by destroying neural connections. Other researchers even removed a section of duodenum from the digestive tract of an animal and sutured it into a closed loop with absolutely no neural connections. They placed the loop back in the abdominal cavity, and when acid was injected into it, the pancreas was stimulated to secrete.

You have probably guessed the explanation of Pavlov's dilemma, but the answer was first demonstrated by two British physiologists, Bayliss and Starling, who were working at the same time as Pavlov. They removed a section of small intestine from an animal and scraped off its mucosal lining. They ground the mucosa with sand, filtered it, and injected the extract into the bloodstream of an animal. The recipient animal secreted pancreatic juices. This was the first demonstration of a chemical message traveling through the circulatory system and having an effect on a specific tissue. They named their newly discovered chemical message **secretin** and also coined the

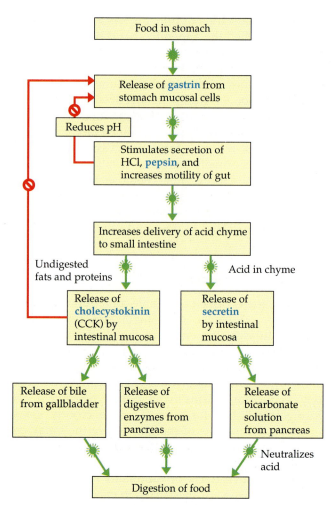

41.21 Controlling Digestion
Many hormones are involved in controlling the sequential processing of food in the digestive tract.

general term **hormone**. The moral of the story is that control of the functions of the digestive tract is not by neural reflexes alone; it is also by hormonal mechanisms.

Hormonal Controls

The activities of the digestive tract and its accessory organs are under the control of several hormones (Figure 41.21). Bayliss and Starling's mucosal cell extract probably contained multiple hormones, but the dominant one, secretin, is primarily responsible for stimulating the pancreas to secrete a solution rich in bicarbonate ions. In response to fats and proteins in the chyme, the mucosa of the small intestine also secretes **cholecystokinin**, which stimulates the gall bladder to release bile and the pancreas to release digestive enzymes. Cholecystokinin also slows down the movements of the stomach, which slows the delivery of chyme into the small intestine.

The stomach also secretes a hormone; it is called

gastrin. Cells in the lower region of the stomach release gastrin when they are stimulated by the presence of food. Gastrin circulates in the blood until it reaches cells in the upper areas of the stomach wall, where it stimulates the secretions and movements of the stomach. Gastrin release is inhibited when the stomach contents become too acidic—another example of negative feedback.

CONTROL AND REGULATION OF FUEL METABOLISM

The Role of The Liver

The liver directs the traffic of nutrient molecules used in energy metabolism. When nutrients are abundant in the circulatory system, the liver can store them in the forms of glycogen (animal starch) and fat. The liver also synthesizes plasma proteins from circulating amino acids. When the availability of fuel molecules in the bloodstream declines, the liver delivers glucose and fats back to the blood. The liver has an enormous capacity to interconvert fuel molecules. For liver cells, the monosaccharides are interchangeable and can be converted to either glycogen or fat. Certain amino acids and some other molecules such as pyruvate and lactate can be converted into glucose —a process called **gluconeogenesis**. The liver is also the major controller of fat metabolism through its production of lipoproteins (Box 41.C).

Hormonal Control of Fuel Metabolism

The **absorptive period** refers to the time that food is in the gut and nutrients are being absorbed and circulated in the blood. During this time the liver converts glucose to glycogen and fat, the body fat tissues convert glucose and fatty acids to stored fat, and the cells of the body preferentially use glucose for metabolic fuel. When there is no longer food in the gut —the **postabsorptive period**—these processes reverse. The liver breaks down glycogen to supply glucose to the blood, the liver and the fat tissues supply fatty acids to the blood, and most of the cells of the body preferentially use fatty acids for metabolic fuel. The major exception to this rule is the cells of the nervous system, which require a constant supply of glucose for their energetic needs. Although the nervous system can use other fuels to a limited extent, its overall dependence on glucose is the reason why it is so important for other cells of the body to shift to fat metabolism during the postabsorptive period. This shift preserves the available glucose and glycogen stores for the nervous system for as long as possible.

What directs the traffic in fuel molecules? Two hormones produced and released by the pancreas, **insulin** and **glucagon**, are largely responsible for controlling the metabolic directions fuel molecules will take (Figure 41.22). The most important of these hormones is insulin, which is produced by cells in the

---

BOX 41.C

Lipoproteins: The Good, the Bad, and the Ugly

In the intestine, bile solves the problem of processing hydrophobic fats in an aqueous medium. The transportation of fats in the circulatory system presents the same problem, but the solution is lipoproteins. A lipoprotein is a particle made up of a core of fat and cholesterol and a covering of protein that makes it hydrophilic. The largest lipoprotein particles are the chylomicrons produced by the cells lining the intestine to transport dietary fat and cholesterol into the circulation. As lipoproteins circulate through the liver and the fat tissues around the body, receptors recognize the protein coat, then lipases begin to hydrolyze the fats which are then absorbed into fat or liver cells. Thus the protein coat of the lipoprotein serves to make it water-soluble, and also serves as an "address" that targets the lipoprotein to a specific tissue.

Other lipoproteins originate in the liver, and they are classified according to their density. As you know, fat has a low density (it floats in water), so the more fat a lipoprotein contains, the lower its density. Very-low-density lipoprotein (called VLDL) produced by the liver contains mostly triglyceride fats that are being transported to fat cells in tissues around the body. Low-density lipoproteins (LDL) contain mostly cholesterol, which they transport to tissues around the body for use in biosynthesis and for deposition. High-density lipoproteins serve as acceptors of cholesterol and are believed to remove cholesterol from tissues and return it to the liver where it can be used to synthesize bile. Because of their differing functions in cholesterol regulation, LDL is sometimes called "bad cholesterol" and HDL "good cholesterol," but those designations are somewhat controversial at present. We do know, however, that a high ratio of LDL to HDL in a person's blood is a risk factor for atherosclerotic heart disease. Cigarette smoking lowers your HDL levels, and regular exercise increases your HDL levels.

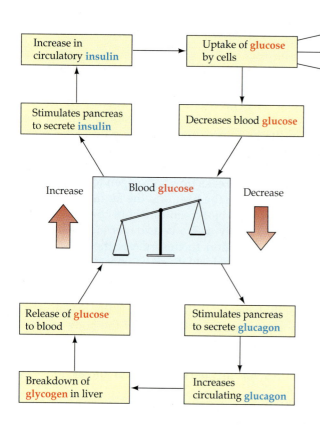

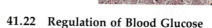

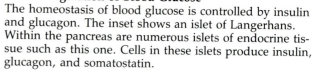

41.22 Regulation of Blood Glucose
The homeostasis of blood glucose is controlled by insulin and glucagon. The inset shows an islet of Langerhans. Within the pancreas are numerous islets of endocrine tissue such as this one. Cells in these islets produce insulin, glucagon, and somatostatin.

cose levels during the postabsorptive period are adrenaline and the glucocorticoid cortisol. Low blood glucose is a stress that triggers glucose-sensitive cells in the hypothalamus to signal to the adrenal medulla, through the sympathetic nervous system, to secrete adrenaline. Adrenaline has an effect on liver cells similar to that of glucagon. Through the cAMP second-messenger cascade, it increases the breakdown of glycogen and increases gluconeogenesis. The stress of low blood glucose also stimulates the adrenal cortex to secrete cortisol. Cortisol inhibits glucose metabolism in many cells while promoting metabolism of fats and proteins.

The traffic of fuel molecules during the absorptive and postabsorptive periods is summarized in Figure 41.23. The steps that are controlled by insulin and glucagon are indicated. Note that during the absorptive period, the direction of traffic in all fuel molecules is toward storage, and glucose is the preferred energy source for all cells. During the postabsorptive period, most cells switch to metabolizing fat so that blood glucose reserves are saved for the nervous system. The level of circulating glucose is maintained through glycogen breakdown and gluconeogenesis.

SUMMARY

Animals depend on organic molecules in their diets, both as sources of metabolic energy and for carbon skeletons from which to build their own organic molecules. The major dietary sources of both energy and carbon skeletons are carbohydrates, fats, and proteins. An animal with insufficient caloric intake is undernourished and must metabolize its own carbohydrate, fat, and, finally, its own protein, for energy. The need for some carbon skeletons can be met from a wide variety of sources; for example, the acetyl group can be derived from any carbohydrate or fat

islets of Langerhans in the pancreas. Insulin is released into the circulatory system by those cells when blood glucose rises above the normal postabsorptive level of about 90 mg of glucose per 100 ml of plasma. Insulin facilitates the entry of glucose into most cells of the body. So when insulin is present, most cells burn glucose as their metabolic fuel, fat cells use glucose to make fat, and liver cells convert glucose to glycogen and fat. As soon as blood glucose falls back to postabsorptive levels, insulin release diminishes rapidly, and the entry of glucose into cells other than those of the nervous system is inhibited. Without a supply of glucose, cells switch to using glycogen and fat as their metabolic fuels. In the absence of insulin, the liver and fat cells stop synthesizing glycogen and fat and begin breaking them down. As a result, the liver supplies glucose to the blood rather than taking it from the blood, and both the liver and the fat tissues supply fatty acids to the blood.

Glucagon is released from the islets of Langerhans when the blood glucose concentration falls below the normal postabsorptive level. It has the opposite effect of insulin; it stimulates liver cells to break down glycogen and carry out gluconeogenesis, thus releasing glucose into the blood. The major hormonal control of fuel metabolism during the postabsorptive period, however, is *the lack of insulin*; glucagon plays a secondary role.

Two other hormones that help preserve blood glu-

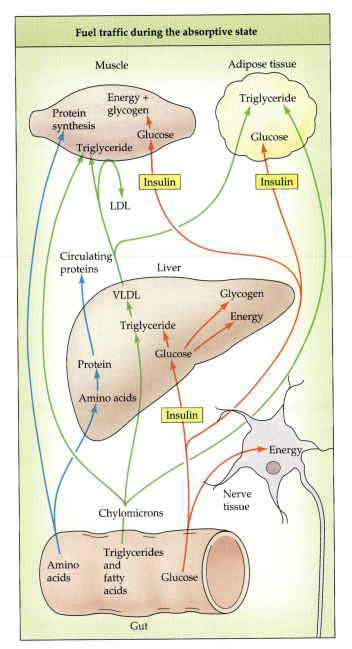

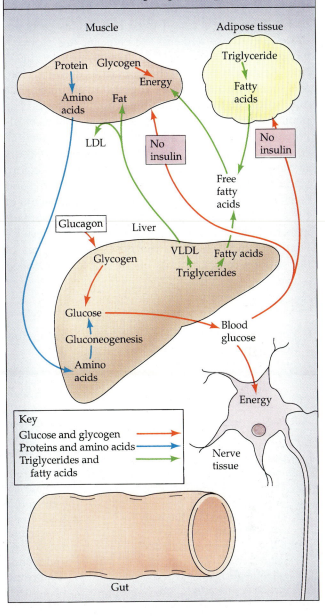

41.23 Fuel Molecule Traffic during the Absorptive and Postabsorptive Periods

and used to synthesize a wide variety of molecules. Other requirements are very specific, such as those for essential amino acids, essential fatty acids, vitamins, and certain mineral elements. Inadequate supplies of required nutrients result in malnutrition and may lead to specific deficiency diseases.

Animals have evolved a great variety of adaptations for exploiting sources of food. Predation consists of the acts of capturing and ingesting other organisms. Animals that eat other animals are carnivores, those that eat plants are herbivores, and those that eat both are omnivores. Detritivores eat the decomposition products of dead organisms. Some animals are deposit feeders, some others are fluid feeders, and still others are filter feeders. Teeth are a general but highly variable feeding adaptation of vertebrates.

Guts are cavities or tubes where digestion and absorption occur. The mouth is often specialized for acquisition and initial processing of a particular food. Tubular guts have the same basic tissue structure throughout, but they are usually divided into separate compartments with specialized functions. In vertebrate guts, food passes from the mouth through the esophagus to the stomach. The stomach stores food and initiates some aspects of digestion. Digestive enzymes reduce carbohydrate, fat, and protein molecules to their simplest monomeric units that can be absorbed by intestinal cells and circulated to the body. Amylase, secreted in saliva, begins carbohydrate digestion. Pepsin, secreted in the stomach along with hydrochloric acid, begins protein digestion. Acidic chyme from the stomach is delivered into the small intestine, where digestion continues. Bile emulsifies fat to facilitate the action of lipase. The pancreas secretes digestive enzymes as well as bicarbonate ions that neutralize the pH of the intestinal contents. Most of the absorption of nutrients occurs in the small intestine. In the large intestine, or colon, water and ions are reabsorbed into the body, and feces are formed and stored. Some compartments of the gut may have large populations of microorganisms that aid in digesting molecules which otherwise would be indigestible to the host. The activities of the various compartments of the gut are controlled

and coordinated by neural reflexes and by hormones. The liver manages the traffic in fuel molecules by absorbing the excess during the absorptive period, interconverting the different types of fuel molecules, and storing them as glycogen and fat. During the postabsorptive period, the liver breaks down its glycogen and fat to supply the blood with glucose and fatty acids. The breakdown of fats in fat tissue also provides fatty acids to the blood. Fuel metabolism is regulated mainly by the pancreatic hormones insulin and glucagon. Insulin is secreted when blood glucose rises; it promotes uptake and storage of glucose by many cells of the body. Lack of insulin promotes use of fat as the major fuel of the body thus preserving glucose for the nervous system. Glucagon is secreted when blood glucose falls; it promotes gluconeogenesis and breakdown of glycogen. With prolonged starvation, adrenaline and cortisol also play roles in maintaining blood glucose levels.

SELF-QUIZ

1. Most of the metabolic energy a bird requires for a long-distance migratory flight is stored as:
a. glycogen.
b. fat.
c. protein.
d. carbohydrate.
e. ATP.

2. Which statement about essential amino acids is *true*?
a. They are not found in vegetarian diets.
b. They are stored by the body for when they are needed.
c. Without them one is undernourished.
d. All animals require the same essential amino acids.
e. Humans can acquire all their essential amino acids by eating milk, eggs, and meat.

3. Which statement about vitamins is *true*?
a. They are essential inorganic nutrients.
b. They are required in larger amounts than are essential amino acids.
c. Many serve as coenzymes.
d. Vitamin D can only be acquired by eating meat or dairy products.
e. When vitamin C is eaten in large quantities, the excess is stored in fat for later use.

4. The digestive enzymes of the small intestine:

a. do not function best at a low pH.
b. are produced and released in response to circulating secretin.
c. are produced and released under neural control.
d. are all secreted by the pancreas.
e. are all activated by an acidic environment.

5. Which of the following is true about nutrient absorption across the gut epithelium?
a. Carbohydrates are absorbed as disaccharides.
b. Fats are absorbed as fatty acids and monoglycerides.
c. Amino acids only move across by diffusion.
d. Bile salts transport fats across.
e. It mostly occurs in the duodenum.

6. Chylomicrons are like the tiny particles of dietary fat found in the lumen of the small intestine in that:
a. both are coated with bile salts.
b. both are lipid-soluble.
c. both travel in lacteals.
d. both contain triglyceride.
e. both are coated with lipoproteins.

7. Microbial fermentation in the guts of cattle:
a. produces fatty acids as a major nutrient for the cattle.
b. occurs in specialized regions of the small intestine.

c. occurs in the caecum from which food is regurgitated to be chewed again and swallowed into the true stomach.
d. produces methane as a major nutrient.
e. is possible because the stomach wall does not secrete HCL.

8. Which of the following is stimulated by cholecystokinin?
a. Stomach motility
b. Release of bile
c. Secretion of HCL
d. Secretion of bicarbonate ions
e. Secretion of mucus

9. During the absorptive period:
a. breakdown of glycogen supplies glucose to blood.
b. glucagon secretion is high.
c. circulating lipoproteins are low.
d. glucose is the major metabolic fuel.
e. synthesis of fats and glycogen in muscle is inhibited.

10. During the postabsorptive period:
a. glucose is the major metabolic fuel.
b. glucagon stimulates the liver to produce glycogen.
c. insulin facilitates uptake of glucose by brain cells.
d. the major metabolic fuel is fatty acids.
e. liver functions slow down because of low insulin levels.

FOR STUDY

1. From what you have learned about nutrition in this chapter, discuss some of the problems with "crash" or "fad" diets. What should one take into account when considering or planning a diet aimed at weight reduction?

2. The digestive tract must move food slowly enough to enable digestion and absorption but fast enough to supply the animal's energetic needs. Describe controls that speed up and slow down the activities of the digestive tract.

3. Describe the role of the liver in the homeostasis of blood glucose. What are the controlling factors?

4. Why is obstruction of the common bile duct so serious? Consider in your answer the multiple functions of the pancreas and the way digestive enzymes are processed.

5. Trace the history of a fatty acid molecule from being on a piece of buttered toast to being in a plaque on a coronary artery. What possible forms and structures could it have passed through in the body? Describe a direct and an indirect route it could have taken.

READINGS

Atkinson, M. A. and Noel K. MacLaren. 1990. "What Causes Diabetes?" *Scientific American*, July. The body's own immune system can cause this serious disease that destroys the ability to regulate fuel metabolism.

Brown, M. S. and J. L. Goldstein. 1984. "How LDL Receptors Influence Cholesterol and Atherosclerosis." *Scientific American*, November. A detailed discussion of what does and does not happen to the cholesterol and fatty acids we consume.

Davenport, H. 1972. "Why the Stomach Does Not Digest Itself." *Scientific American*, January. The stomach contains millimolar hydrochloric acid but does not digest its own lining . . . usually.

Degabriele, R. 1980. "The Physiology of the Koala." *Scientific American*, July. Amazing adaptation to an unusual and limited diet.

Moog, F. 1981. "The Lining of the Small Intestine." *Scientific American*, November. How nutrients are absorbed.

Schmidt-Nielsen, K. 1990. *Animal Physiology: Adaptation and Environment*, 4th Edition. Cambridge University Press, New York. An outstanding textbook, emphasizing the comparative approach.

Scrimshaw, N. S. 1991. "Iron Deficiency." *Scientific American*, October. This article discusses one of the most common dietary deficiencies in humans and its consequences.

Uvnas-Moberg, K. 1989. "The Gastrointestinal Tract in Growth and Reproduction." *Scientific American*, July. This article provides a concise treatment of hormonal controls of digestive tract function and how they change to accommodate special needs of pregnancy and lactation.

Vander, A. J., J. H. Sherman and D. S. Luciano. 1990. *Human Physiology: The Mechanisms of Body Function*, 5th Edition. McGraw-Hill, New York. Chapters 16 and 17 provide a readily understandable treatment of digestion and absorption of food.

42

Salt and Water Balance and Nitrogen Excretion

PREVIEW: The salt and water content of cells and the specific concentrations of individual salts and inorganic ions are crucial parameters on which life depends. Cells of multicellular animals exchange ions and water with the fluid that surrounds them. Excretory organs and systems regulate the salt and water balance of that extracellular fluid environment. Excretory systems in terrestrial animals also eliminate the nitrogenous waste products of metabolism from the animal's body. The basic unit of structure of an excretory organ is a tubule that receives extracellular fluid. The cells of the tubule alter the composition of this fluid by reabsorption of some molecules and secretion of others, and thereby produce urine, which differs in composition from the extracellular fluid. The specific functions of the excretory tubules, and hence the excretory organ, of a species depend on its external environment—fresh water, salt water, or terrestrial.

This chapter deals with osmoregulation, volume regulation, nitrogen excretion, protonephridia, metanephridia, nephrons, kidneys, filtration, secretion, reabsorption, the countercurrent multiplier, and hormonal control of salt and water balance.

Blood, sweat, and tears all taste salty, like seawater. Life evolved in the seas, and if a complex animal is to exist elsewhere, it must carry its own internal sea to bathe the cells of its body. This statement is taken too literally sometimes to mean that the mineral ion, or salt, composition of our blood is similar to that of the ancient seas. That conclusion is wrong. Though all animals on Earth share the same distant origins, there is great diversity in the composition of their body fluids, and no animal can be taken to have an internal environment just like the environment in which life evolved. However, all animals require both water and salts (Figure 42.1), and most animals must regulate the salt and water contents of their bodies. This chapter focuses on how they do it.

As noted in Chapter 33, the extracellular fluids that we carry with us service all the cells of our bodies. In addition to supplying cells with oxygen and nutrients and carrying away waste products, these extracellular fluids also determine the water balance of the cells. To understand what is meant by water balance, recall that cell membranes are permeable to water, and that the movement of water across membranes depends on differences in osmotic potential. (You may find it useful to review the discussion of osmosis in Chapter 5.) If the osmotic potential of the extracellular fluid is less negative than that of the intracellular fluids, water moves into the cells, swelling and bursting them. If the osmotic potential of the extracellular fluid is more negative than that of the intracellular fluids, the cells lose water and shrink. The osmotic potential of the extracellular environment determines both the volume and osmotic potential of the intracellular environment.

Excretory systems regulate the osmotic potential and the volume of the extracellular fluids. In addition, excretory systems regulate the composition of the extracellular fluids by excreting molecules that are in excess (such as NaCl when we eat lots of salty popcorn) and conserving those that are valuable or in short supply (such as glucose and amino acids). In terrestrial organisms, excretory systems also serve to eliminate the toxic waste products of nitrogen metabolism.

Exactly what the excretory system of a particular species must do depends on the environment in

962

42.1 Water Is Essential for Life
During the dry season on the African savanna, animals congregate around scarce sources of water. Sometimes even predators and prey must come together at a water hole.

which that species lives. In this chapter we will examine excretory systems that accomplish salt and water balance and nitrogen excretion in marine, freshwater, and terrestrial habitats. In spite of the evolutionary diversity of the anatomical and physiological details, these systems all obey a common rule and employ common mechanisms. The common rule is that there is no active transport of water—water must be moved either by pressure or by a difference in osmotic potential. The common mechanisms derive from the fact that most excretory organs consist of systems of tubules that receive extracellular fluid and alter its composition to produce **urine**—the fluid waste product that is excreted. The extracellular fluid enters the excretory tubules by a process of **filtration**, and its composition is changed by processes of **active secretion** and **reabsorption** of specific molecules by the cells of the tubules. As we will see, these same three mechanisms are used both in systems that excrete water and conserve salts and in systems that do the opposite, that conserve water and excrete salts.

WATER, SALTS, AND THE ENVIRONMENT

We think of marine and freshwater environments as being distinctly different, one salty and the other not. In reality, aqueous environments grade continuously from fresh to extremely salty. Consider a place where a river enters the sea through a bay or a marsh. Aqueous environments within that bay or marsh range in **salinity** (salt content) from the fresh water of the river to the open sea. Evaporating tide pools can become much saltier than the sea. Animals live in all of those environments.

Most marine invertebrates can acclimate to a wide range of environmental salinities by allowing their body fluids to have the same osmotic potential as the environment; they thus avoid the risk of being burst or shrunk by osmotic movement of water. Such animals are called **osmoconformers**. There are limits to osmoconformity, however. No animal could come into osmoconformity with fresh water and survive; nor could animals survive with internal salt concentrations as high as those that may be reached in an evaporating tide pool. Such concentrations cause proteins to denature. Animals that maintain an osmotic potential of their internal fluids different from that of their environment are called **osmoregulators**. Even animals that osmoconform over a wide range of osmotic potentials must osmoregulate at the extremes of environmental salinity.

To osmoregulate in fresh water, animals must continuously excrete the water that invades their bodies by osmosis, but while doing so they must conserve salts. Hence, they produce large amounts of dilute urine. To osmoregulate in salt water, animals must conserve water and excrete salts. They tend to produce small amounts of urine.

(a)

(b)

42.2   Evaporating Salt Ponds Vary in Salinity
(a) Impoundments of water around San Francisco Bay yield salt when they evaporate. Different species of bacteria and algae grow best at different salinities, so the colors of the salt ponds change as they evaporate. (b) One animal that can live in these ponds at all salinities is a tiny crustacean, the brine shrimp.

The brine shrimp, *Artemia*, is an example of an animal adapted to live in environments of almost any salinity. *Artemia* are found in huge numbers in the most saline environments possible, such as the Great Salt Lake in Utah or coastal evaporation ponds where salt is obtained for commercial purposes. The salinity of such water reaches 300g/liter—in comparison, normal seawater contains about 35g/liter. *Artemia* are harvested from these environments and sold directly or processed for fish food (Figure 42.2). *Artemia* cannot survive for long in fresh water, but they can live in very dilute seawater, in which they maintain the osmotic potential of their body fluids above the osmotic potential of the environment. Under these conditions, *Artemia* is a **hypertonic osmoregulator**, meaning that it regulates the concentration (the osmolarity) of its body fluids so they are more concentrated than (hypertonic to) the environment (Figure 42.3). At high environmental salinities, *Artemia* is an exceptionally effective **hypotonic osmoregulator**, keeping the osmotic potential of its body fluids well below that of the water in which it is living. Very few organisms can survive in the crystallizing brine in which *Artemia* thrive. The main mechanism used by this small crustacean for osmoregulation is the active transport of NaCl across its gill membranes.

An osmoconformer also can be an **ionic conformer**, which means that the ionic composition as well as the osmolarity of its body fluids matches that of its environment. Most osmoconformers, however, are **ionic regulators** to some degree. That means that they employ active transport mechanisms to maintain specific ions in their body fluids at concentrations different from those in the environment.

The terrestrial environment presents entirely different sets of problems for salt and water balance. The terrestrial environment is extremely desiccating

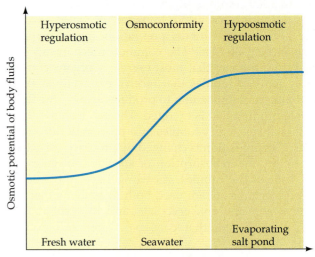

42.3   Osmoregulators and Osmoconformers
Over a broad range of environmental salinities, many marine invertebrates are osmoconformers. Animals that live at the extremes of environmental salinities, however, can display osmoregulatory abilities. They become hyperosmotic regulators in very dilute water, or hypoosmotic regulators in very saline water.

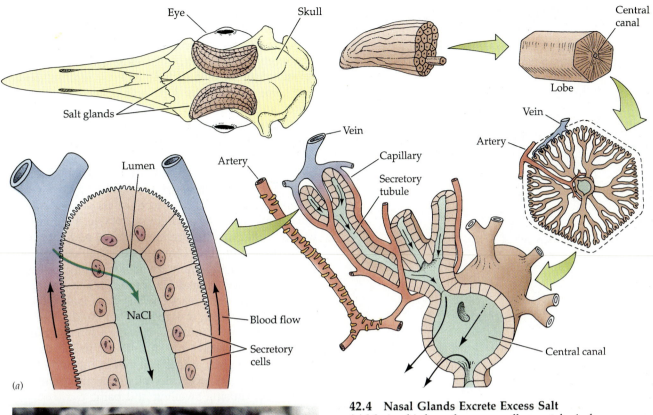

42.4 Nasal Glands Excrete Excess Salt
(a) Marine birds such as sea gulls can take in large amounts of salt in their diets. One adaptation for excreting excess salt is the nasal salt gland. These glands can produce extremely salty secretions that the bird then clears from its nostrils by shaking its head or sneezing. (b) This Adélie penguin has returned from a feeding trip at sea. As it rests, its nasal salt gland is excreting the excess salt it consumed. Note the patch of evaporated salt on the rock below the bird's beak.

(drying), so most terrestrial animals must conserve water. Exceptions are animals such as muskrats and beavers that practically live in water. Terrestrial animals get their salts from their foods. Plants generally have low concentrations of sodium, so most herbivores must be very effective in conserving sodium ions. As noted in Chapter 41, some terrestrial herbivores travel long distances to naturally occurring salt licks to supplement their dietary intake of sodium. In contrast, birds that feed on marine animals must excrete the large excess of sodium they ingest with their food. They do this by means of **nasal salt glands**, which excrete a concentrated solution of sodium chloride via a duct that empties into the nasal cavity. Birds such as sea gulls and penguins that have

nasal salt glands can be seen frequently sneezing or shaking their heads to get rid of the very salty droplets that form (Figure 42.4).

THE EXCRETION OF NITROGENOUS WASTES

The end products of the metabolism of carbohydrates and fats are simply water (H_2O) and carbon dioxide (CO_2); they present no problems for excretion. Proteins and nucleic acids, however, contain nitrogen in addition to carbon, hydrogen, and oxygen; when they are metabolized, there is a nitrogenous waste product in addition to H_2O and CO_2. That waste product is mostly **ammonia** (NH_3). Ammonia is highly toxic and either must be excreted continuously so concentrations do not build up, or detoxified by conversion into other molecules for excretion. Those molecules are principally **urea** and **uric acid** (Figure 42.5).

Ammonia excretion is relatively simple for aquatic animals. Ammonia diffuses and is highly soluble in

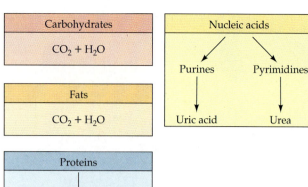

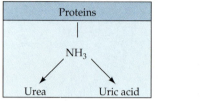

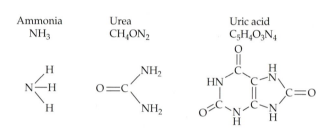

42.5 Waste Products of Metabolism
Ammonia, urea, and uric acid are the nitrogenous waste products of the metabolism of proteins and nucleic acids.

water. Animals that breathe water continuously lose ammonia from their blood to the environment by diffusion across their gill membranes. Animals that excrete nitrogen as ammonia are called **ammonotelic,** and they include aquatic invertebrates and bony fishes. Crocodiles and amphibian tadpoles are also ammonotelic.

Ammonia is a more dangerous metabolite for terrestrial animals that have limited access to water. In mammals, ammonia is lethal when it reaches only 5 mg/100 ml of blood. Therefore, terrestrial (and some aquatic) animals convert ammonia into either urea or uric acid. **Ureotelic** animals, such as mammals, amphibians, and cartilaginous fishes (sharks and rays), excrete urea as their principal nitrogenous waste product. Urea is quite soluble in water, but excretion of urea solutions at low concentrations could result in a large loss of water that many terrestrial animals could ill afford. Later in the chapter we will see that mammals have evolved excretory systems that produce urine that is hypertonic to their body fluids, thereby conserving water while excreting urea. The cartilaginous fishes are a story unto themselves. These marine species actually maintain their body fluids hypertonic to the marine environment by retaining high concentrations of urea. Water moves

into their bodies by osmosis and must be excreted, so water conservation is not a problem for them.

Some groups of terrestrial animals conserve water by excreting nitrogenous wastes as uric acid; they are called **uricotelic** and include insects, reptiles, birds, and some amphibians. Uric acid is very insoluble in water and is excreted as a semisolid—the whitish material in bird droppings. Therefore, the uricotelic animal loses very little water as it disposes of its nitrogenous waste.

Although we can classify animals on the basis of their major nitrogenous waste product as being ammonotelic, ureotelic, or uricotelic, most species produce more than one nitrogenous waste. Humans are ureotelic, yet we also excrete uric acid and ammonia, as anyone who has changed diapers knows. (Actually, most of the ammonia in diapers is produced by the bacterial breakdown of urea. The bacterium that performs this reaction was first isolated by a microbiologist from a diaper of his child.) The uric acid in human urine comes largely from the metabolism of nucleic acids and of caffeine. In the classical disease called gout, uric acid levels in the body fluids increase, and uric acid precipitates in the joints and elsewhere, causing swelling and pain.

In some species, different developmental forms live in quite different habitats and have different forms of nitrogen excretion. Tadpoles of frogs and toads, for example, excrete ammonia across their gill membranes, but when they develop into adult frogs or toads they generally excrete urea. Some adult frogs and toads that live in arid habitats excrete uric acid. Thus, there is considerable evolutionary flexibility in how nitrogenous wastes are excreted.

INVERTEBRATE EXCRETORY SYSTEMS

Marine invertebrates are mostly osmoconformers, so they have few adaptations for salt and water balance other than active transport of specific ions. For nitrogen excretion they can passively lose ammonia by diffusion to the seawater. Freshwater and terrestrial invertebrates, however, display a variety of fascinating adaptations for salt and water balance and nitrogen excretion. Even though these adaptations are quite diverse, they are all based on the same basic principles of filtration of body fluids and active secretion and reabsorption of specific ions.

Flame Cells of Flatworms

Many flatworms, such as *Planaria* (Figure 37.18), live in fresh water and excrete water through an elaborate network of tubules running throughout their bodies. The tubules end in **flame cells,** so called because each flame cell has a tuft of cilia beating inside the tubule, giving the appearance of a flickering flame (Figure

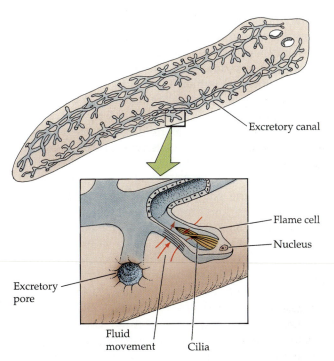

Excretory canal

Flame cell

Nucleus

Excretory pore

Fluid movement

Cilia

42.6 Excretion in Flatworms
The excretory system of the flatworm *Planaria* consists of a network of tubules ending in specialized cells called flame cells. A flame cell and its associated tubules are called a protonephridium. Body fluids enter the space enclosed by the flame cell and are driven down the tubules by the beating of the cilia in the flame cells. The tubule cells modify the composition of this fluid.

42.6). Flame cells and tubules together are called **protonephridia** (*proto* is the Greek word for before and *nephros* is the Greek word for kidney). The beating of the cilia causes fluid in the tubules to flow toward the exterior of the animal. This fluid, as it leaves the planarian, is hypotonic to the animal's internal body fluids.

How does the hypotonic fluid get into the tubules?

Two possibilities exist. The first is that the beating of the cilia create a slight negative pressure in the tubule. Water from the surrounding tissue could then filter into the tubule because of the pressure difference between the tissues and the excretory tubules. The fluid that filtered into the tubule would be the same osmolarity as the tissue fluids, but as it flowed down the tubules, the cells of the tubules could actively reabsorb ions from this tubular fluid or urine. The second possibility is that ions are actively transported into the excretory tubules at their upper ends. Water would follow because of the increased osmotic potential of the tubular fluid. Then, as the fluid moved down the tubules, active reabsorption of ions could occur. In either case, the urine leaving the protonephridia is hypotonic to the internal fluids of the planarian.

Metanephridia

Filtration of the body fluids and tubular processing of the urine are highly developed in the **metanephridia** of annelid worms, such as the earthworm (meta is from a Greek word meaning akin to). Recall that annelids have fluid-filled body cavities called coeloms and closed circulatory systems through which blood is pumped under pressure (see Figure 40.4). The pressure causes the blood to be filtered across the thin, permeable capillary walls into the coelom. This process is called filtration because the cells and large protein molecules of the blood stay behind in the capillaries while water and small molecules leave the capillaries and enter the coelom. Where does this coelomic fluid go?

The metanephridia of earthworms are paired tubules; each segment of the worm contains one pair. The metanephridia begin in ciliated funnel-like openings in the coelom called **nephrostomes** and end in pores called nephridiopores that open to the outside of the animal (Figure 42.7). Coelomic fluid enters the

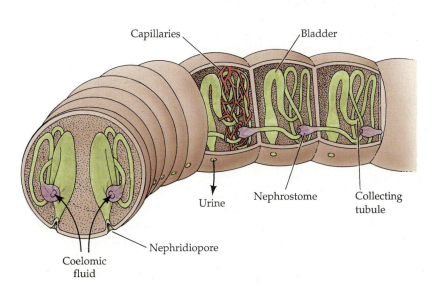

Capillaries

Bladder

Urine

Nephrostome

Collecting tubule

Nephridiopore

Coelomic fluid

42.7 Earthworm Metanephridia
The metanephridia of earthworms are arranged segmentally. Coelomic fluid enters the metanephridia through nephrostomes. The tubule cells of the metanephridia alter the composition of this fluid, producing a dilute urine that is excreted through the nephridiopores.

metanephridia through the nephrostomes. As the coelomic fluid passes through the tubules, the cells of the tubules actively reabsorb certain molecules from it and actively secrete other molecules into it. What leaves the animal through the nephridiopores is a hypotonic urine containing nitrogenous wastes, among other solutes.

Why should earthworms excrete a hypotonic urine? Aren't they terrestrial animals? Earthworms live in the soil, an environment that has a 100 percent relative humidity, and they are in constant contact with films of water covering the soil particles. So earthworms, like freshwater aquatic animals, must excrete the excess water that is always entering their bodies by osmosis.

In the metanephridium we see all of the basic processes used in the excretory systems of vertebrates to be discussed later in this chapter: filtration of the body fluids, and tubular processing of the filtrate by active secretion and reabsorption of ions and molecules.

Malpighian Tubules

Insects and other terrestrial arthropods have a remarkable system for the excretion of nitrogenous wastes with very little loss of water. As a result, these animals can live in the driest habitats on earth (as well as the wettest). The insect excretory system consists of blind tubules (from 2 to over 100) attached to the gut and hanging into the fluid-filled coelom (Figure 42.8). The cells of these **Malpighian tubules** ac-

tively transport uric acid, potassium ions, and sodium ions from the coelomic fluid into the tubules. As solutes are secreted into the tubules, water follows because of the difference in osmotic potential. The walls of the Malpighian tubules have muscle fibers whose contractions move the contents of the tubules toward the gut.

In the gut, the tubular fluid continues to change in composition. The gut contents are more acidic than the tubular fluids, and as a result, the uric acid becomes less soluble and precipitates out of solution. The cells of the gut walls actively transport sodium and potassium ions from the gut contents back into the coelom. Because the uric acid molecules have precipitated out of solution, water is free to follow the reabsorbed salts back into the coelom through osmosis. Left in the gut are crystals of uric acid mixed with undigested food, and this dry matter is all that the insect eliminates. The Malpighian tubule system is a highly effective mechanism for excreting nitrogenous wastes and some salts without giving up any significant fraction of the animal's precious water supply.

Crustacean Green Glands

Malpighian tubules are found in some arthropods, such as insects and spiders, but not in others, such as the crustaceans (crabs, lobsters, crayfish, and their relatives). In the crustaceans, the excretory system consists of an **end sac** connected by a **nephridial canal** to a bladder that empties to the outside by way

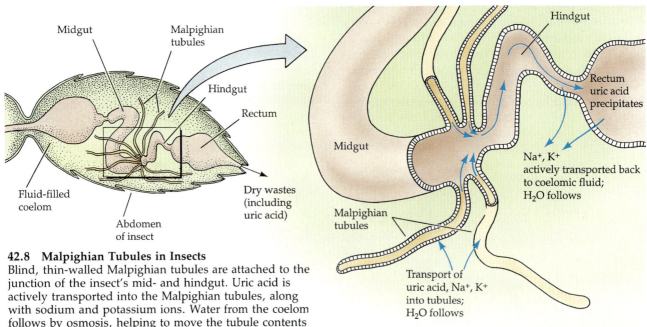

42.8 Malpighian Tubules in Insects
Blind, thin-walled Malpighian tubules are attached to the junction of the insect's mid- and hindgut. Uric acid is actively transported into the Malpighian tubules, along with sodium and potassium ions. Water from the coelom follows by osmosis, helping to move the tubule contents to the hindgut. Sodium and potassium ions are actively transported out of the hindgut and are returned to the coelomic fluid, and water follows osmotically. Uric acid is excreted as nearly dry waste, while precious water and salts are conserved.

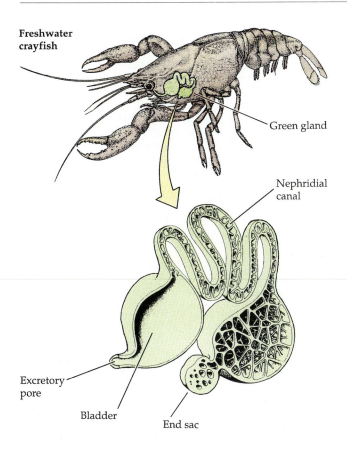

Freshwater crayfish

Green gland

Nephridial canal

Excretory pore

Bladder

End sac

42.9 The Green Gland

Lobsters and crayfish illustrate a type of crustacean "kidney" called a green gland. The pair of green glands are located anteriorly in the body and empty to the exterior through excretory pores located below each antenna. The entire green gland is bathed in hemolymph. A hemolymph filtrate is forced into the bulbous end sac, then passes through a nephridial canal to the bladder and out the excretory pore.

of an excretory pore (Figure 42.9). The whole assembly in a crayfish or lobster is called a **green gland**, regarded by many as a gourmet treat. A key feature of this excretory system is that coelomic fluid (hemolymph) is filtered into the end sac by the high pressure within the coelom. The filtrate is processed by the cells of the canal, and the resulting urine is stored in the bladder until it is excreted through the excretory pore.

There is an interesting structural and functional difference between the green glands of freshwater and marine crayfish. Freshwater crayfish must excrete large amounts of hypotonic urine and conserve salts. The cells of the nephridial canal actively transport salts back to the coelom, and at the same time actively secrete molecules to be eliminated into the canal. For marine crayfish, salt conservation is not a major problem, and those species have a much shorter nephridial canal.

VERTEBRATE EXCRETORY SYSTEMS

The major vertebrate organ for salt and water balance and nitrogen excretion is the **kidney**. The functional unit of the kidney is the **nephron**. In each human kidney there are about one million nephrons. To understand how the kidney works, it is essential to understand the structure and function of the nephron. A remarkable fact about the kidneys of ver-

tebrates is that in different species the same basic functional unit serves different needs. The kidneys of freshwater fishes excrete water, whereas the kidneys of most mammals conserve water. To understand how this can be so, once again we need to look at the nephron.

The Structure and Functions of the Nephron

Each nephron has a vascular and a tubular component (Figure 40.10). The vascular component is unusual in consisting of two capillary beds between the arteriole that supplies it and the venule that drains it. The first capillary bed is a dense knot of very permeable vessels called a **glomerulus** (Figure 42.11a). Blood enters the glomerulus through an **afferent arteriole** and leaves through an **efferent arteriole**. The efferent arteriole gives rise to the second set of capillaries, the **peritubular capillaries** that surround the tubular component of the nephron. The tubular component begins with **Bowman's capsule**, which encloses the glomerulus. Together, the glomerulus and its surrounding Bowman's capsule are called the **renal corpuscle**. The cells of the capsule that come into direct contact with the glomerular capillaries are called **podocytes**. These highly specialized cells have numerous armlike extensions, each with hundreds of fine, fingerlike projections. The podocytes wrap around the capillaries so that their fingerlike projections cover the capillaries completely (Figure 42.11b and c).

The function of the glomerulus is to filter the blood to produce a tubular fluid without cells and large molecules. Both the cells of the capillaries and the podocyctes of the Bowman's capsule participate in filtration. The walls of the capillaries have pores that allow water and small molecules to leave the capillary but are too small to permit red blood cells and very large protein molecules to pass. Even smaller than the pores in the capillaries are the narrow slits between the fingerlike projections of the podocytes. The result is that water and small molecules pass from the capillary blood and enter the tubule of the nephron (Figure 42.11d), but red blood cells and proteins remain in the capillaries.

The force that drives the process of filtration in the glomerulus is the pressure of the arterial blood. As in every capillary bed, the pressure of the blood en-

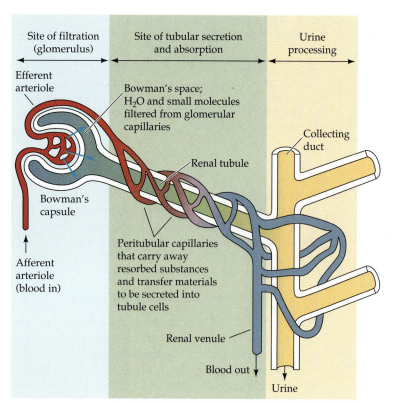

Site of filtration (glomerulus)

Site of tubular secretion and absorption

Urine processing

Efferent arteriole

Bowman's space; H₂O and small molecules filtered from glomerular capillaries

Collecting duct

Renal tubule

Bowman's capsule

Afferent arteriole (blood in)

Peritubular capillaries that carry away resorbed substances and transfer materials to be secreted into tubule cells

Renal venule

Blood out

Urine

42.10   The Vertebrate Nephron
The nephron is a system of tubules closely associated with a system of blood vessels. An afferent renal arteriole supplies a knot of capillaries called a glomerulus. The glomerulus is the site of blood filtration. It is drained by an efferent arteriole, which gives rise to another set of capillaries that surrounds the tubules of the nephron. A renal venule drains the peritubular capillaries. The end of the renal tubule system envelops the glomerulus so that the filtrate from the glomerular capillaries enters the tubules. The products of individual nephrons enter collecting ducts.

tering the capillary causes the filtration of water and small molecules until the osmotic potential created by the large molecules remaining in the blood is sufficient to counter the outward flow of water (see Chapter 40). The glomerular filtration rate is high because afferent arterioles in the kidney are larger than those in other tissues, so that blood pressure in the glomerular capillaries is unusually high, and also because the capillaries of the glomerulus, along with their covering of podocytes, are much more permeable than other capillary beds in the body.

The composition of the fluid that is filtered into the tubule is similar to that of the blood plasma and different from that of urine. It contains glucose, amino acids, ions, and nitrogenous wastes at the same concentrations as in the blood plasma. The filtrate lacks the plasma proteins, however. As this fluid passes down the tubule, its composition changes as the cells of the tubule actively reabsorb certain molecules from the tubular fluid and secrete other molecules into it. When the tubular fluid leaves the kidney as urine, its composition is very different from that of the original filtrate. The function of the **renal tubules**—the tubules of the nephrons—is to determine the composition of the urine through processes of active secretion and reabsorption of specific molecules. The peritubular capillaries serve the needs of the renal tubules by bringing to them the molecules to be secreted and carrying away the molecules that are reabsorbed.

The Evolution of the Nephron

It is believed that the earliest vertebrates lived in fresh water, and that the nephron evolved as a structure that excreted the excess water constantly entering those animals by osmosis. Does a study of the evolution of the vertebrate nephron support this view? Those earliest vertebrates are long extinct, and the soft tissues of kidneys are not preserved in the fossil record, so we must employ indirect means to reveal the evolution of the nephron. We can study the kidneys of present-day members of the oldest of the vertebrate groups with the expectation that they may still retain some primitive features of structure and organization. Another approach is to study the embryonic stages of development in such species, because those developmental stages frequently reveal primitive structure and organization not seen in the adult organism. The embryonic development of the kidneys of fishes and amphibians offers suggestions about the sequence of evolutionary stages of the vertebrate nephron.

The earliest nephrons to appear in embryonic fishes and amphibians are tubules that open directly into the coelomic cavity through a nephrostome (a ciliated opening) (Figure 42.12a). This is very similar to the organization of the annelid metanephridium (see Figure 42.7). Near the nephrostome is a knot of capillaries that protrudes right up under the membrane lining the coelomic cavity, where it can filter

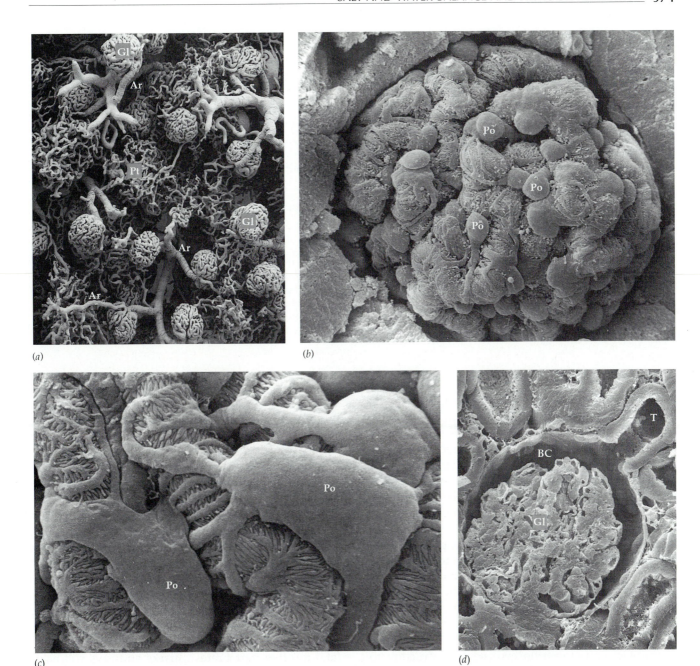

(a)

(b)

(c)

(d)

42.11 An SEM Tour of the Nephron
Scanning electron micrographs show the anatomical bases for kidney function. *(a)* Blood vessels in the kidney show the knots of capillaries that are the glomeruli (Gl). Each glomerulus has an afferent and an efferent arteriole (Ar). Peritubular capillaries (Pt) are looser networks surrounding the tubules of the nephron. *(b)* The capillaries of the glomeruli are tightly wrapped by specialized tubule cells called podocytes (Po). *(c)* Each podocyte has hundreds of tiny, fingerlike projections that create filtration slits between them. Anything passing from the glomerular capillaries into the tubule of the nephron must pass through these slits. *(d)* Bowman's capsule (BC) surrounds the glomerulus, collects the filtrate, and funnels it into the tubule (T) of the nephron.

the blood directly into the coelom. The resulting coelomic fluid flows through the nephrostome into the tubules that process the fluid to form urine.

In a slightly more advanced stage of kidney development, the knot of capillaries does not protrude up under the coleomic lining, but instead is encap-

sulated by an elaboration of the renal tubule as it leaves the nephrostome (Figure 42.12*b*). In the most advanced stage of nephron development, the nephrostome is lost and all of the tubular fluid is derived directly from filtration in the glomerulus (Figure 42. 12*c*).

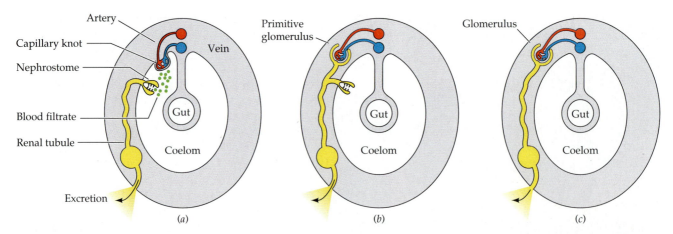

Artery
Capillary knot
Nephrostome
Vein
Gut
Coelom
Blood filtrate
Renal tubule
Excretion
(a)

Primitive glomerulus
Gut
Coelom
(b)

Glomerulus
Gut
Coelom
(c)

42.12 Evolution of the Nephron
A model of the evolution of the vertebrate nephron is based on studies of the kidneys of present-day descendants of the oldest vertebrate groups, and especially of embryological stages of those kidneys. *(a)* The most primitive nephron had a tubule with the nephrostome open- ing into the coelom; a filtrate of the blood entered the coelom from knots of capillaries along its borders. *(b)* The next evolutionary stage probably involved a specialization of the tubule to enable filtration directly into the tubule, but the nephrostome remained. *(c)* The final stage eliminates the nephrostome entirely.

It appears that the function of the earliest vertebrate nephron was to eliminate coelomic fluid while conserving important molecules. Subsequent stages in the evolution of the nephron enhanced its ability to handle a large volume of filtrate derived directly from the blood. Thus the original function of the nephron was most likely to bail excess water out of animals while conserving valuable molecules. This conclusion supports the notion that the earliest vertebrates lived in fresh water and that some of their descendents had to evolve secondary adaptations to enable them to live in habitats where it was necessary to conserve water.

Water Conservation in Vertebrates

If the vertebrate nephron evolved as a structure to excrete water while conserving salts and essential small molecules, how then have vertebrates adapted to environments where water must be conserved and salts excreted? The answer to this question is different for each vertebrate group. Even among marine fishes, the bony fishes have a different set of adaptations than do the cartilaginous fishes.

Marine bony fishes cannot produce urine more concentrated than their body fluids, and their body fluids are only one-fourth to one-third the osmotic potential of seawater. They prevent excess loss of water by producing very little urine. Their urine production is low because their kidneys have fewer glomeruli than do the kidneys of freshwater fishes. The kidneys of some species of marine bony fishes have no glomeruli at all! Even though the glomeruli are reduced or absent, the renal tubules are retained for purposes of active excretion of ions and certain molecules. Marine bony fishes meet their water needs by drinking seawater, but this results in a large salt load. They handle this salt load by active excretion of ions from the gill membranes and from the renal tubules. Nitrogenous wastes are lost as ammonia from the gill membranes.

The cartilaginous fishes are osmoconformers, but not ionconformers. Unlike marine bony fishes, the cartilaginous fishes convert nitrogenous waste to urea and then retain large amounts of that urea in their bodies so that their body fluids have the same osmotic concentration as seawater. In some cases they are even slightly hypertonic to the sea so that water moves into their bodies by osmosis. Such a concentration of urea in the body fluids would prove fatal to other vertebrates. Sharks and rays still have the problem of excreting the large amounts of salts they take in with their food. They have several sites of active secretion of NaCl; the major one is a salt-secreting **rectal gland**.

Most amphibians live in or near freshwater and are limited to humid habitats when they venture from the water. Typical species of amphibians, like freshwater fishes, produce large amounts of dilute urine and conserve salts. Some amphibians, however, have adapted to habitats that require water conservation, and their adaptations are diverse.

There is at least one species of saltwater amphibian, the crab-eating frog of Southeast Asia. For marine fishes there are two completely different evolutionary solutions to the osmotic problems of living in saltwater; which is employed by the crab-eating frog? The answer is the cartilaginous fishes solution; the

42.13 Waxy Frogs
Phyllomedusa is an arboreal frog that lives in a seasonally dry and hot habitat. It reduces its evaporative water loss by secreting waxes and fats from skin glands and spreading these secretions all over its body.

crab-eating frog retains urea in its body fluids so that it is slightly hypertonic to the seawater in the mangrove swamps where it lives. This adaptation is only seen in adults, however, so the species requires fresh water for reproduction.

Amphibians from very dry terrestrial environments have been studied by Vaughn Shoemaker at the University of California. An important adaptation in these species is a reduction in the water permeability of their skins. Some secrete a waxy substance which they spread over the skin to make it waterproof (Figure 42.13). Remarkable adaptations are seen in several species of frogs that live in arid regions of Australia. These animals burrow deep in the ground and estivate during long dry periods. **Estivation** is a state of very low metabolic activity. When it rains, frogs come out of estivation, feed, and reproduce. Their most interesting adaptation, however, is that they have enormous urinary bladders. Prior to entering estivation, they fill their bladders with dilute urine, which may make up one-third of their entire body weight. This dilute urine serves as a water reservoir that the frog gradually uses during the long period of estivation. Australian aborigines have learned to use these estivating frogs as an emergency source of drinking water.

Reptiles occupy habitats ranging from aquatic en-vironments to extremely hot, dry deserts. Three major adaptations have freed the reptiles from maintaining the close association with water that is necessary for amphibians. First, they do not need fresh water to reproduce. They employ internal fertilization and lay eggs with shells that retard evaporative water loss. Second, they have scaly, dry skins that are much less permeable to water than is amphibian skin. Third, they excrete nitrogenous wastes as uric acid solids and therefore lose little water in the process.

Birds employ the same adaptations for water conservation as do reptiles: internal fertilization, shelled eggs, skin that retards water loss, and uric acid as the nitrogenous waste product. In addition, some birds can produce urine that is hypertonic to their body fluids. This ability is much more highly developed in the mammals.

STRUCTURE AND FUNCTION OF THE MAMMALIAN KIDNEY

The ability of mammals and birds to produce urine that is hypertonic to their body fluids represents a major step in kidney evolution. In these species, we see for the first time the kidney playing the major role in water conservation. A structure that originally evolved to excrete water has been converted to a structure to do the opposite, conserve water. To understand how this evolutionary switch has occurred, we must examine the structure and function of the whole kidney.

Anatomy

We can focus on humans to provide an example of the mammalian excretory system. Humans have two kidneys located at the rear of the abdominal cavity (Figure 42.14). Each kidney releases the urine it produces into a tube (the **ureter**) that leads to the **urinary bladder**, where the urine is stored until it is excreted through the **urethra**. The urethra is a short tube opening to the outside at the end of the penis in males or just anterior to the vagina in females. Two **sphincter muscles** surrounding the base of the urethra control the timing of urination. One of these sphincters is a smooth muscle and is controlled by the autonomic nervous system. When the bladder is full, a spinal reflex acts to relax this sphincter. This is the only control of urination in infants, but the reflex gradually comes under the influence of higher centers in the nervous system as a child grows older. The other sphincter is a skeletal muscle, and it is controlled by the voluntary or conscious nervous system. That is why, when the bladder is *very* full, only serious concentration prevents urination.

The kidney is shaped like a kidney bean; when cut down its long axis and split open as a bean splits open, its important anatomical features are revealed (Figure 42.14). The ureter and the **renal artery** and **vein** enter the kidney on its concave (i.e., punched-in) side. The ureter divides into several branches, the ends of which envelop projections of kidney tissue called **renal pyramids**. These renal pyramids make up the internal core or **medulla** of the kidney. The medulla is capped by a distinctly different tissue called the **cortex**. The renal artery and vein give rise to many arterioles and venules in the region between the cortex and the medulla.

The secret of the ability of the mammalian kidney to produce concentrated urine is in the relationship between the structures of the nephron and the anatomy of the kidney. There are about one million nephrons in each human kidney, and their organization within the kidney is very regular (Figure 42.14*b*). All of the glomeruli are located in the cortex. The initial segment of the tubule of a nephron is called the **proximal convoluted tubule**—proximal because it is closest to the glomerulus, and convoluted because it is twisted. The proximal convoluted tubules are also all located in the cortex. At a certain point, the proximal tubule takes a dive directly down into the medulla, giving rise to the portion of the tubule called the **loop of Henle**. It is called a loop because it runs straight down into the medulla, makes a hairpin turn, and comes straight back to the cortex. This ascending limb of the loop of Henle becomes the **distal convoluted tubule** in the cortex. The distal convoluted tubules of many nephrons join a common **collecting duct** in the cortex. The collecting ducts then run in parallel with the loops of Henle down through the medulla and empty into the ureter at the tips of the renal pyramids.

The blood vessels of the kidney are organized in a manner similar to that of the nephrons (Figure 42.14*c*). The renal arterioles branch from the renal arteries and radiate into the cortex, supplying each glomerulus with an afferent arteriole. Draining each glomerulus is an efferent arteriole that gives rise to the peritubular capillaries, which mostly surround the cortical portions of the tubules. A small number of the peritubular capillaries run into the medulla in parallel with the loops of Henle and the collecting ducts. These vessels are the **vasa recta**. All of the

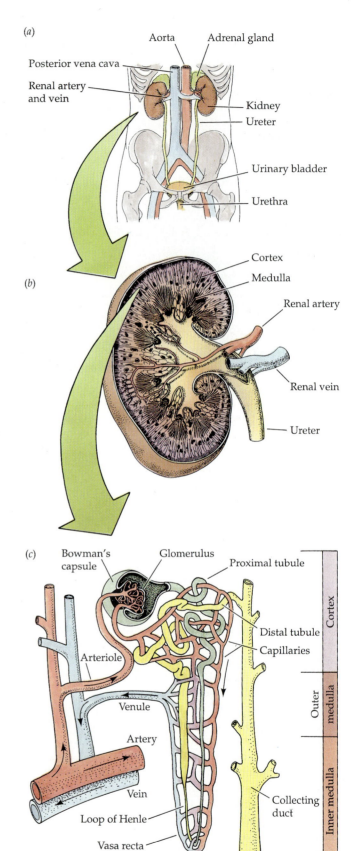

42.14 The Human Excretory System
(*a*) The human excretory system consists of two kidneys positioned in the upper rear of the abdominal cavity. The urine they produce is conducted to the urinary bladder through the ureters. The urethra drains the bladder. (*b*) A longitudinal section of the kidney reveals an internal structure that includes a cortex and, beneath it, a medulla. Urine leaves the kidney from the inner surface of the medulla and is collected in branches of the ureter.

(*c*) A closer look at the cortex and medulla reveals that the glomeruli, the proximal tubules, and the distal tubules are all in the cortex. The loops of Henle and the vasa recta are all in the medulla. Collecting ducts run from the cortex down to the tips of the medulla.

peritubular capillaries from a nephron join back together into a venule that joins with venules from other nephrons and eventually leads to the renal vein, which takes blood from the kidney. Remember that anything coming into the kidney comes through the renal artery, and everything that comes into the kidney must leave either through the renal vein or the ureter (there is also some drainage of lymph from the kidney, but it is minor). Only a small, selective percentage of everything that is filtered leaves the kidney in the urine. To understand kidney function, you must understand how most of the substances and water filtered from the blood in the glomerulus get back into the venous blood draining the kidney.

Glomerular Filtration Rate

Most of the water and solutes filtered in the glomerulus are reabsorbed and do not appear in the urine. This is emphasized when we appreciate the sheer quantity of the filtrate as compared to the volume of urine produced each day. The kidneys receive about 20 percent of the blood pumped into arteries by the heart. The cardiac output of a human at rest is about 5 liters per minute, so the kidney processes over 1,400 liters of blood per day—an enormous volume. How much of this huge volume is filtered? The answer is about 12 percent. This is still a large number—180 liters per day! Since we normally only urinate on the order of 2 to 3 liters per day, about 98 to 99 percent of the fluid volume that is filtered in the glomerulus is being reabsorbed into the blood.

Tubular Reabsorption

Where and how is this enormous fluid volume reabsorbed from the renal tubules back into the blood? Most of the water and solutes in the glomerular filtrate are reabsorbed in the proximal convoluted tubule. The cells of this section of the renal tubule are cuboidal, and their surfaces facing into the tubule have thousands of **microvilli**, which increase the surface area for reabsorption (Figure 42.15). There are lots of mitochondria in these cells—an indication that they are biochemically active. The work they do is to transport NaCl and other solutes like glucose and amino acids out of the tubular fluid. Virtually all glucose molecules and amino acid molecules that are filtered from the blood are actively reabsorbed by the cells of the proximal convoluted tubules. This movement of solutes into the tubular cells makes the cells hypertonic to the tubular fluid, and water flows into the cells in response to this difference. The water and the solutes moved out of the tubular fluid by this process are taken up by the peritubular capillaries and thereby returned to the venous blood leaving the kidney.

In spite of the large volume of reabsorption of

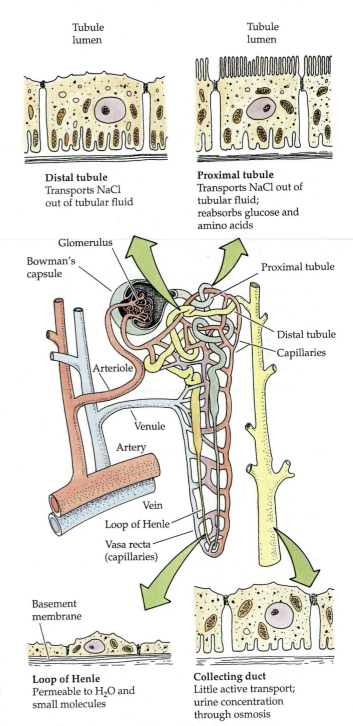

Distal tubule
Transports NaCl out of tubular fluid

Proximal tubule
Transports NaCl out of tubular fluid; reabsorbs glucose and amino acids

Loop of Henle
Permeable to H$_2$O and small molecules

Collecting duct
Little active transport; urine concentration through osmosis

42.15 Cells of the Renal Tubules
The structures of the cells of the renal tubules reflect the functions of the different tubule segments. The cells of the proximal tubule have many mitochondria; a well-developed "brush border" lines the inside of the tubule to increase the surface area available for the absorption of substances from the urine. Intercellular spaces and indentations at the basal end of the cells increase the area of cell contact with interstitial fluids. Distal tubule cells also have many mitochondria and extensively folded basal surfaces. Collecting duct cells are less adapted for active secretion and reabsorption, and cells in the thin regions of the loop of Henle are flat, with few mitochondria or surface indentations.

water and solutes by the proximal convoluted tubule, the overall concentration, or osmotic potential, of the fluid that enters the loop of Henle is not different from that of the blood plasma, even though their compositions are quite different.

The Countercurrent Multiplier

How does the structure of the mammalian kidney enable it to produce a urine that is more concentrated than the blood plasma? Humans can produce a urine that is 4 times more concentrated than their plasma, but some mammals such as kangaroo rats that live in very dry deserts are able to conserve water so well that their urine may be 12 to 15 times more concentrated than their blood plasma. This remarkable ability is due to the loops of Henle, which function as a **countercurrent multiplier system**. Countercurrent refers to the direction of urine flow in the descending versus the ascending limbs of the loop, and multiplier refers to the ability of this system to create a concentration gradient in the renal medulla.

The cells of the descending limb of the loop of Henle, and the initial cells of the ascending limb, are rather unspecialized (Figure 42.15). They are flat cells with no microvilli and few mitochondria. The part of the tubule made up of these cells is permeable to water and small molecules. Partway up the ascending limb, however, the cells become specialized for transport again. They are cuboidal, have lots of mitochondria, and have some microvilli on their surfaces facing into the tubule (Figure 42.15). The portion of the ascending limb made up of these cells is impermeable to water, but the cells actively transport NaCl out of the tubular fluid. As a result, the urine becomes more dilute as it flows toward the distal convoluted tubule (Figure 42.16). Where does the NaCl that is transported out of the ascending limb go? It goes into the interstitial fluid in the renal medulla, from which it can diffuse back into the descending limbs of the loops of Henle. The NaCl that enters the descending limbs flows up the ascending limbs, where it is transported out of the urine once again. As a result of this process, NaCl concentrations build up in the interstitial fluid of the renal medulla, with the highest concentrations near the tips of the renal pyramids.

How does this action of the loop of Henle concentrate the urine? We can see that the urine is less concentrated when it leaves the loop than when it entered. The active secretion of substances to be excreted and reabsorption of substances to be conserved continues in the distal convoluted tubules. It is in the collecting duct that the urine is actually concentrated. The collecting duct runs from the cortex, where it receives filtrate from the distal tubules, down through the medulla, to the tip of the renal pyramids, where it discharges into the ureter. Over this distance it is surrounded by increasingly concen-

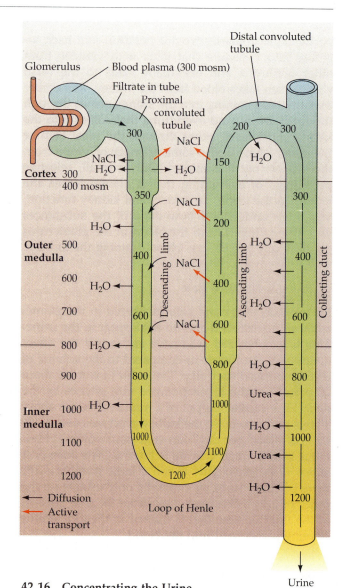

42.16 Concentrating the Urine
The loop of Henle acts as a countercurrent multiplier to establish a concentration gradient in the renal medulla. The numbers in the left column indicate the osmolarity (osm) of the interstitial fluids in the renal medulla. The basis for the multiplier is that the thick segment of the ascending limb pumps NaCl out of the urine and into the interstitial fluid, but H_2O cannot follow this movement of NaCl because this region of the tubule is impermeable to water. However, the NaCl in the interstitial fluid can diffuse back into the descending loop of Henle. If it does, it is pumped *out* of the urine again when it flows up the ascending limb. In this way the solute concentration in the renal medulla increases. The urine entering the collecting duct is much less concentrated than the interstitial fluid of the medulla, so as it passes down the collecting duct it loses water to the interstitial fluid and becomes more and more concentrated.

trated interstitial fluid. The collecting duct is permeable to water, but not to mineral ions, so the osmotic potential of the interstitial environment draws water from the fluid in the collecting duct and leaves behind an increasingly concentrated solution. The urine that leaves the collecting duct at the tip of a renal pyramid can be almost as concentrated as the highest interstitial concentration established by the countercurrent multiplier system.

How does the water reabsorbed from the collecting duct get out of the renal medulla? It leaves in the vasa recta, which are highly permeable to salts and water.

In summary, the mammalian kidney works in the following manner: the glomeruli filter large volumes of blood plasma. The proximal convoluted tubules reabsorb most of this volume along with valuable molecules such as glucose and amino acids. They do so by actively transporting NaCl and other solutes from the tubular fluid; water follows because of the local difference in osmotic potential created by the transport of the solutes. The loops of Henle create a concentration gradient in the medulla of the kidney. As the urine flows in the collecting ducts through this concentration gradient, water is reabsorbed, thus creating a urine hypertonic to the blood plasma.

REGULATION OF KIDNEY FUNCTIONS

The function of kidneys is to regulate the volume, the osmolarity, and the chemical composition of the extracellular fluids. If the fluids do not have the right composition to meet the needs of the body cells, the cells cannot survive. Thus, if the kidneys fail, death will ensue—unless the victim has access to an **artificial kidney** machine, which can cleanse the blood through the process of **dialysis** (Box 42.A). There are multiple systems involved in the control and regulation of kidney functions. Although we discuss them separately, they are always working together in an integrated fashion to match kidney function to the needs of the body.

Autoregulation of Glomerular Filtration Rate

If the kidneys cease to filter the blood, they cannot accomplish any of their functions. Therefore, there are mechanisms that act to keep the blood filtering through the glomeruli at a constant high rate, regardless of what is happening elsewhere in the body. Because these adaptations of the kidney are to maintain its own functions, these mechanisms are called autoregulatory. The glomerular filtration rate (GFR) depends on an adequate blood supply to the kidneys at an adequate blood pressure. The autoregulatory mechanisms act to compensate for decreases in cardiac output or decreases in blood pressure so that the GFR remains high (Figure 42.17).

One autoregulatory mechanism is the dilation of the afferent renal arterioles when blood pressure falls. This decreases the resistance in the arterioles and helps to maintain blood pressure in the glomerular capillaries. If that response does not keep the

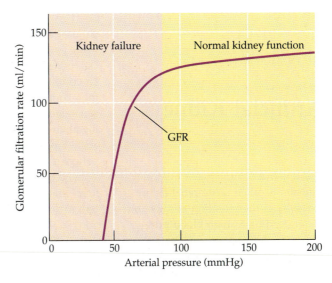

42.17 Glomerular Filtration Rate
Glomerular filtration is driven by arterial pressure, yet because of autoregulatory mechanisms the glomerular filtration rate is independent of arterial pressure over a wide range. When arterial pressure falls too low, however, the kidney fails to produce urine.

GFR from falling, then the kidney releases an enzyme, **renin**, into the blood. Renin acts on a circulating protein to begin a process that converts this protein into an active hormone called **angiotensin**. Angiotensin has a number of effects that help to restore the GFR to normal. First, angiotensin causes the efferent renal arteriole to constrict, which elevates blood pressure in the glomerular capillaries. Second, angiotensin causes peripheral blood vessels to constrict—an action that elevates central blood pressure. Third, angiotensin stimulates the adrenal cortex to release the hormone **aldosterone**. Aldosterone stimulates sodium reabsorption, which makes the reabsorption of water more effective. Enhancing water reabsorption helps to maintain blood volume and therefore central blood pressure. Finally, angiotensin acts on structures in the brain to stimulate thirst. Increased water intake increases blood volume and blood pressure.

Regulation of Blood Volume

When you lose blood, your blood pressure tends to fall. Besides activating the kidney autoregulatory mechanisms described above, a fall in blood pressure decreases the activity of stretch receptors in the walls of the large arteries such as the aorta and the carotids. These stretch receptors provide information to cells in the hypothalamus that produce **antidiuretic hormone** (also called vasopressin) and send it down their axons to the posterior pituitary gland (Chapter 34). As stretch receptor activity falls, the production and release of this hormone increases (Figure 42.18). An-

BOX 42.A

Artificial Kidneys

Sudden and complete loss of kidney function is called acute renal failure. It results in the retention of salts and water, leading to high blood pressure, and it also results in the retention of urea and metabolic acids. A patient with acute renal failure will die in one to two weeks if not treated. It is now possible to compensate for renal failure and even

surgical removal of the kidneys by using "artificial kidneys." In an artificial kidney, or dialysis unit, the blood of the patient and a dialyzing fluid come into very close contact, separated only by a semipermeable membrane. This membrane allows small molecules to diffuse from the patient's blood into the dialysis fluid. As we know, molecules and ions diffuse from an area of high concentration to an area of lower concentration, so the composition of the dialysis fluid is crucial. For molecules or ions we want to conserve, such as glucose or Na, their concentration in the dialysis fluid must be the same as their concentration in the plasma. For molecules and ions we want to clear from the plasma, such as urea and

sulfate, their concentration in the dialysis fluid must be zero. The total osmotic potential of the dialysis fluid must equal the osmotic potential of the plasma.

A schematic drawing of a dialysis machine is shown in the figure. Arterial blood flows between semipermeable membranes, which are surrounded with dialysis fluid at body temperature. The "cleansed" blood is returned to the body through a vein and the used dialysis fluid is discarded. At any one time only about 500 ml of blood are in the dialysis unit, and it processes several hundred ml/min. A patient with no kidney function must be on the dialysis machine for 4 to 6 hours three times a week.

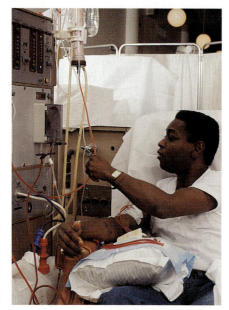

The man in the photograph is monitoring the flow of his blood through a dialysis unit that eliminates metabolic waste products normally removed from the blood by the kidneys. The mechanisms of the dialysis unit are illustrated in the diagram.

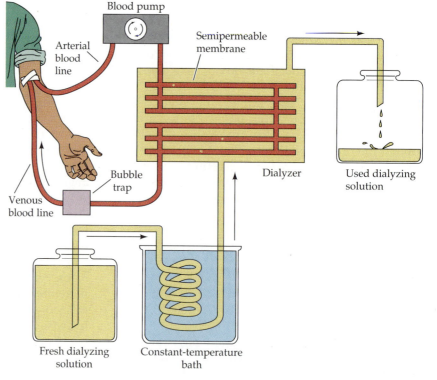

tidiuretic hormone acts on the collecting ducts of the kidney by increasing their permeability to water. When there is a high circulating level of antidiuretic hormone, the collecting ducts are very permeable to water, more water is reabsorbed, and we produce small quantities of concentrated urine, thus conserving blood volume and blood pressure. Without antidiuretic hormone, water cannot be reabsorbed from the collecting ducts, and we produce lots of very

dilute urine. Diabetes insipidus is a disease resulting from lack of antidiuretic hormone. The word *insipidus* derives from the dilute (tasteless) character of the urine. Diabetes mellitus, caused by an inability of cells to take up glucose from the blood, also causes copious urine production, but the urine tastes sweet (mellitus means honeylike). Caffeine and alcohol inhibit the actions of antidiuretic hormone and increase the production of urine. The resulting dehydrating

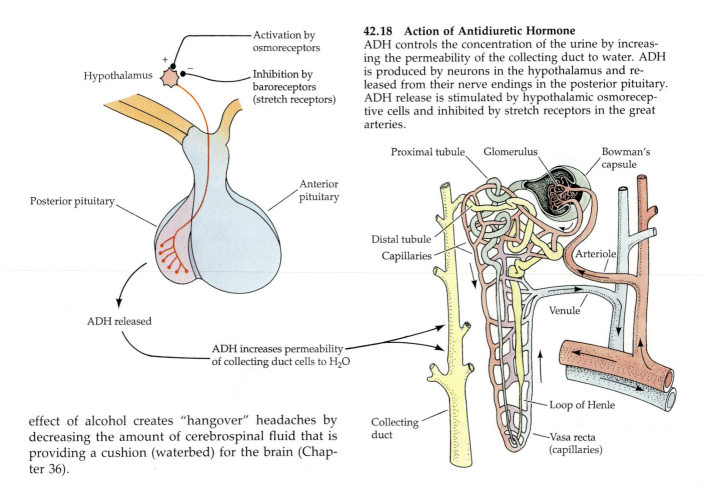

42.18 Action of Antidiuretic Hormone

ADH controls the concentration of the urine by increasing the permeability of the collecting duct to water. ADH is produced by neurons in the hypothalamus and released from their nerve endings in the posterior pituitary. ADH release is stimulated by hypothalamic osmoreceptive cells and inhibited by stretch receptors in the great arteries.

effect of alcohol creates "hangover" headaches by decreasing the amount of cerebrospinal fluid that is providing a cushion (waterbed) for the brain (Chapter 36).

Regulation of Blood Osmolarity

There are cells in the hypothalamus that sense the osmotic potential of the blood. If blood osmolarity increases, these **osmoreceptors** stimulate increased release of antidiuretic hormone that enhance water reabsorption from the kidney. The osmoreceptors also stimulate thirst. Increased water intake dilutes the blood as it expands blood volume. It is difficult and probably artificial to separate osmotic regulation from volume regulation because changes in plasma osmolarity cause osmotic movement of water between the extracellular and intracellular compartments, resulting in changes in the volume of extracellular fluid.

Regulatory Flexibility

The ability of the mammalian kidney to produce a concentrated urine has made it possible for mammals to inhabit some of the most arid habitats on Earth. Some of these animals, such as the desert gerbil, have such extremely long loops of Henle that their renal pyramids stick way out of the concave surface of the kidney (Figure 42.19). These animals are so effective in conserving water that they can survive on the water released by the metabolism of their rather dry food; they do not need to drink! The concentrating ability of the mammalian kidney, coupled with the

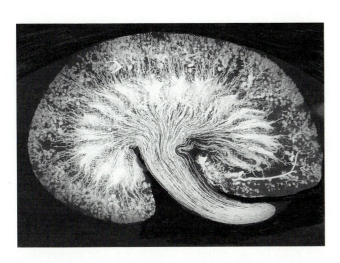

42.19 Concentration Ability

The ability of the mammalian kidney to concentrate the urine depends on the lengths of its loops of Henle relative to the overall size of the kidney. Some desert rodents have single renal pyramids that are so long they protrude out of the kidney and into the ureter.

BOX 42.B

Water Balance in the Vampire Bat

The vampire bat (*Desmodus*) is a small mammal that feeds at night on the blood of sleeping large mammals, such as cattle. Blood is a very liquid, high-protein diet. In order to process this diet, the renal system of the vampire bat must shift from drought conditions to flood conditions and then back to drought conditions in minutes. At sunset, when the bat has not had a meal for many hours, it is producing a highly concentrated urine at a low rate to conserve its precious body water. If it is successful in finding prey, it is important for the bat to process as much blood in as short a time as possible, before the victim wakes up. To maximize its nutrient intake, the bat concentrates its blood meal by rapidly excreting water content. Accordingly, within minutes the bat produces copious amounts of very dilute urine. The warm fluid running down the victim's neck and waking it is not blood!

As soon as the meal is ended—usually abruptly—the bat begins to digest the concentrated blood in its gut. Because the blood is mostly protein, a large amount of nitrogenous waste is produced, and this must be excreted as urea in solution. But now water is in short supply. The bat must limit its water loss for it may be a long time until the next meal. Consequently, the bat's kidneys produce small amounts of an extremely concentrated urine. This urine can be more than 20 times the concentration of the bat's plasma. Humans, in comparison, can produce a urine only about 4 times as concentrated as their plasma. In this way the remarkable regulatory abilities of the vampire bat kidney enable the animal to process its unusual diet.

The graph shows the changes in a vampire bat's urine concentration and urine flow rate before and after its meal of blood. In the photograph, two vampire bats roost on the ceiling of their cave during the day.

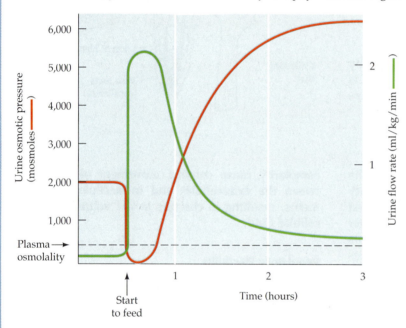

remarkable flexibility of its regulatory systems, enables it to adapt to rapidly changing conditions. An interesting case of this regulatory flexibility is that of the vampire bat, described in Box 42.B. The vampire bat can display the full range of extremes of mammalian salt and water balance mechanisms in a matter of minutes.

SUMMARY

The problems of salt and water balance and nitrogen excretion that animals face depend on their environments. In marine environments animals can be osmoconformers or osmoregulators. Marine osmoregulators must conserve water and excrete salts. Freshwater animals must continually excrete water and conserve salts. Virtually all animals are ionic regulators to some degree. On land, water conservation is essential; whether salts must be conserved or excreted depends on the diet. Aquatic animals can eliminate nitrogenous wastes such as ammonia by diffusion across their gill membranes. Terrestrial animals detoxify ammonia by converting it to urea or uric acid for excretion.

In excretory systems body fluids are filtered into a system of tubules. Filtration is driven by hydrostatic pressure. The composition of the filtrate is altered by the cells of the tubules, which actively secrete and reabsorb solutes. The protonephridia of flatworms consist of flame cells and excretory tubules that open to the environment through nephridiopores. The metanephridia of segmented worms take in coelomic fluid and alter its composition through active secretion and active reabsorption of solutes by tubule cells. Arthropod excretory organs include the green glands of crustaceans and the Malpighian tubules of insects.

The vertebrate nephron originally evolved as an adaptation for excreting water and conserving solutes, but the kidneys of marine vertebrates and terrestrial vertebrates have evolved a variety of adaptations for conserving water. Mammals and birds are the only vertebrates that have evolved the capacity for producing urine hypertonic to their body fluids as a means of water conservation. The organization of the renal tubules in the mammalian kidney is the basis for its ability to produce concentrated urine. Glomeruli in the cortex of the kidney filter water and solutes from the blood. The proximal convoluted tubule reabsorbs most of this water and many of the solutes. The loop of Henle creates a concentration gradient in the tissues of the renal medulla. The distal convoluted tubule alters the composition of the tubular fluid by active transport. The collecting duct concentrates the urine by allowing the osmotic loss of water to the surrounding interstitial fluid of the medulla.

Regulation of kidney function in mammals involves autoregulatory mechanisms for maintaining a constant high glomerular filtration rate even if blood pressure varies. An important autoregulatory mechanism is the release of renin by the kidney when blood pressure falls. Renin activates angiotensin, which causes peripheral vasoconstriction, causes release of aldosterone, and stimulates thirst. Changes in blood pressure and blood osmolarity influence the release of antidiuretic hormone from the posterior pituitary. Antidiuretic hormone controls the permeability of the collecting duct to water and therefore the ultimate concentration of the urine.

SELF-QUIZ

1. Which statement about osmoregulators is *true*?
   a. Most marine invertebrates are osmoregulators.
   b. All freshwater invertebrates are hypertonic osmoregulators.
   c. Cartilaginous fish are hypotonic osmoregulators.
   d. Bony marine fish are hypertonic osmoregulators.
   e. Mammals are hypotonic osmoregulators.

2. The excretion of nitrogenous wastes:
   a. by humans can be as urea and uric acid.
   b. by mammals is never in the form of uric acid.
   c. in marine fish is in the form of urea.
   d. does not contribute to the osmotic potential of the urine.
   e. requires more water if the waste product is the rather insoluble uric acid.

3. How are earthworm metanephridia like mammalian nephrons?
   a. Both process coelomic fluid.
   b. Both take in fluid through a ciliated opening.
   c. Both produce hypertonic urine.
   d. Both employ tubular secretion and reabsorption to control urine composition.
   e. Both deliver urine to a urinary bladder.

4. What is the role of renal podocytes?

   a. They control the glomerular filtration rate by changing resistances of renal arterioles.
   b. They reabsorb most of the glucose that is filtered from the plasma.
   c. They prevent red blood cells and large molecules from entering the renal tubules.
   d. They provide a large surface area for tubular secretion and reabsorption.
   e. They release renin when glomerular filtration rate falls.

5. Which of the following is not found in a renal pyramid?
   a. Collecting ducts
   b. Vasa recta
   c. Peritubular capillaries
   d. Convoluted tubules
   e. Loops of Henle

6. Which part of the nephron is mostly responsible for the difference in mammals between the glomerular filtration rate and the urine production rate?
   a. The glomerulus
   b. The proximal convoluted tubule
   c. The loop of Henle
   d. The distal convoluted tubule
   e. The collecting duct

7. For mammals of the same size, what feature of their excretory systems would give them the greatest ability to produce a hypertonic urine?
   a. High glomerular filtration rate

   b. Longer convoluted tubules
   c. Increased number of nephrons
   d. Highly permeable collecting ducts
   e. Longer loops of Henle

8. Which of the following would not be a response stimulated by a large fall in blood pressure?
   a. Constriction of afferent renal arteriole
   b. Increased release of renin
   c. Increased release of antidiuretic hormone
   d. Increased thirst
   e. Constriction of efferent renal arteriole

9. Which of the following statements is *true* for angiotensin?
   a. It is secreted by the kidney when GFR falls.
   b. It is released by the posterior pituitary when blood pressure falls.
   c. It stimulates thirst.
   d. It increases permeability of the collecting ducts to water.
   e. It acts to decrease glomerular filtration rate when blood pressure rises.

10. Birds that feed on marine animals ingest a lot of salt, but excrete most of it by means of:
   a. Malpighian tubules.
   b. rectal salt glands.
   c. green glands.
   d. hypertonic urine.
   e. nasal salt glands.

FOR STUDY

1. What do marine fish, reptiles, mammals, and insects have in common with respect to water balance? Compare their physiological adaptations for dealing with their common problem.

2. What are the relative advantages and disadvantages of ammonia, urea, and uric acid as nitrogenous waste products of animals?

3. Explain how the kidney is able to maintain a constant glomerular filtration rate over a wide range of arterial blood pressures. Also, referring back to what you learned about regulation of cardiovascular function in Chapter 40, how can a fall in glomerular filtration rate cause an increase in cardiac output?

4. Inulin is a molecule that is filtered in the glomerulus, but it is not secreted or reabsorbed by the renal tubules. If you injected inulin into a subject and after a brief time measured the concentration of inulin in the blood and the urine of the subject, how could you determine the subject's glomerular filtration rate? Assume that the rate of urine production is 1 milliliter per minute.

5. Explain the roles of the loop of Henle and the collecting duct in producing a hypertonic urine in mammals. How is this mechanism controlled in response to changes in osmolarity of the blood and blood pressure?

READINGS

Cantin, M. and J. Genest. 1986. "The Heart as an Endocrine Gland." *Scientific American*, February. Heart tissue secretes a hormone that helps control salt and water balance.

Eckert, R., D. Randall and G. Augustine. 1988. *Animal Physiology: Mechanisms and Adaptations*, 3rd Edition. W. H. Freeman, San Francisco. An outstanding textbook of animal physiology. Chapter 12 covers water and salt balance and excretion.

Heatwole, H. 1978. "Adaptations of Marine Snakes." *American Scientist*, vol. 66, pages 594–604. A variety of adaptations, including means of maintaining salt and water balance, allows several groups of snakes to exploit the marine environment.

Schmidt-Nielsen, K. 1990. *Animal Physiology: Adaptation and Environment*, 4th Edition. Cambridge University Press, New York. An excellent textbook, emphasizing the comparative approach.

Smith, H. W. 1961. *From Fish to Philosopher*. Doubleday, Garden City, NJ. A classic, using salt balance and excretory physiology as the organizing principle for a survey of vertebrate evolution.

Stricker, E. M. and J. G. Verbalis. 1988. "Hormones and Behavior: The Biology of Thirst and Sodium Appetite." *American Scientist*, vol. 76, page 261. The control of water and salt intake is an important part of osmoregulation.

Vander, A. J., J. H. Sherman and D. S. Luciano. 1985. *Human Physiology: The Mechanisms of Body Function*, 4th Edition. McGraw-Hill, New York. Chapter 13 deals with the regulation of water and salt balance.

43

Animal Behavior

PREVIEW: Behavioral adaptations are essential for survival and reproductive success, and are shaped by natural selection. Behaviors range from those that are encoded in the genome to those that are mostly the result of experience. Even with learned behaviors, however, natural selection shapes what can be learned and when it can be learned. Natural selection favors genetically determined behaviors in situations where learning is not possible, where it is possible to learn the wrong behavior, and where mistakes would be extremely costly or dangerous. Learning makes it possible for the nervous system to acquire complex and changing sets of information. Mechanistic studies of behavior reveal the abilities of animals and their underlying neural mechanisms.

This chapter deals with fixed action patterns, releasers, deprivation experiments, the genetics of behavior, imprinting, the evolution of aggressive behavior, territoriality, circadian rhythms, and navigation.

Behavior is an important part of an animal's biology. In the previous ten chapters we have focused on the physiology of animals largely from the standpoint of the functions of cells, tissues, organs, and organ systems, but in some cases we have extended our discussions into behavior. For example, we described thermoregulatory behavior, the effects of hormones on behavior, sexual behavior, feeding behavior, and some cases of migratory behavior. Behavior is a crucial aspect of the physiology of most animals; it is essential in finding a place to live, finding food, finding mates, caring for young, and surviving threats from the environment. Like molecular, cellular, and morphological adaptations, behaviors have been shaped by natural selection. We can study behavior in terms of mechanisms, in terms of adaptive significance, in terms of genetics, and in terms of evolutionary relationships. Let us consider one interesting behavior, web spinning by spiders.

Spider webs are objects of beauty and marvels of engineering. Classic webs have the obvious function of capturing insect prey, but web structures serve other functions as well: they play roles in mating, they provide nests, and they protect offspring. Some spiders spin cocoons or simple sheets or funnels of silk. The design of a web structure and its construction are so regular that they identify the species of the builder. Because there are thousands of species of spiders, there are thousands of web designs; all are products of natural selection. One example is the orb web of the garden spider, *Araneus*, immortalized in the children's story *Charlotte's Web* by E. B. White. The garden spider spins a new web every day in the early morning hours before dawn (Figure 43.1). From an initial attachment point, it strings a horizontal thread. From the middle of that thread it drops a vertical thread to a lower attachment point. The fork of the resulting Y will be the hub of the finished orb. Next a few more radial supports and a few surrounding "framing" threads are added. All of the radial spokes are then filled in according to a regular set of rules. Finally, a spiral of sticky threads is laid down with regular spacing and attachment points to the radial spokes. This remarkable feat of engineering takes only half an hour, but it requires thousands of specific movements done in just the right sequence. Where is the blueprint for Charlotte's web? How does she learn the construction skills needed to build that web? How did natural selection produce this complex behavior?

The answer to those questions is that the blueprint is coded in the genes and built into the spider's central nervous system as a motor score. In this case, learning plays no role in the expression of that complex blueprint. Newly hatched spiders disperse to new locations and usually spin their first webs without ever having experienced a web built by an adult of their species. Nevertheless, they build perfect webs the first time; each of the thousands of movements comes in just the right sequence. It is remarkable that the genetic code and the simple nervous system of a spider could contain and express such a complex score for such a detailed sequence of precise behaviors.

Because many animal behaviors are, like the spinning of spider webs, genetically determined, we can

983

43.1 An Orb-Weaving Spider and Its Web
Where is the blueprint?

study them and their mechanisms as direct products of evolution. Other behaviors are more complex, however, in that they involve learning. Nevertheless, the mechanisms and parameters of learning in a species are also evolutionary adaptations. What an animal can learn, and when it can learn it, can be adaptations that reflect the species' ecology and life history. The study of the ecology and evolution of animal behavior is the field of **ethology**, the subject of this chapter. The field of psychology also deals with animal behavior, but focuses more on questions of how behavior is modified by experience through the processes of cognition, learning, and memory— properties of the nervous system most highly developed in primates and especially in humans. Overlapping the study of the evolution of behavior and the study of how behavior is modified by experience are investigations of the neurophysiological mechanisms of behavior. We touched upon neurobiology of learning and memory in Chapter 36, and in this chapter we will examine some more examples of mechanistic studies of behavior. Let us begin with a detailed examination of behaviors that appear to be rather direct expressions of the genetic code.

THE STUDY OF INSTINCT

The spinning of a web by a spider is an example of **instinctive behavior**—behavior that is genetically determined rather than learned. Instinctive behaviors are highly stereotypic, that is, they are performed the same way every time. They are also species-specific in that there is very little variation in the way different individuals of the same species perform the behavior. A term ethologists use for such a genetically determined behavior is **fixed action pattern**.

Fixed action patterns do not require learning or prior experience for their expression, and they are generally not modified by learning. Another spider example illustrates this point. A certain cocoon-spinning spider makes its egg cocoon by spinning a base plate, building up the walls, laying its eggs inside, and spinning a lid to close the cocoon. An ethologist determined that this behavior requires about 6,400 individual movements. It is performed exactly the same way every time and is not modifiable by experience. If the spider is moved to a new location after she finishes the base plate, she will continue to spin the sides of the cocoon, lay her eggs (which fall out the bottom), and spin the lid. If the next time she is ready to begin a cocoon, she is placed on her previously completed base plate, she will nevertheless start by spinning a new base plate right over the old one as if it were not there. If she is nutritionally deprived and runs out of silk in the midst of spinning a cocoon, she will nevertheless continue all of the 6,400 movements in a pantomime of cocoon building. Once started, the cocoon-building motor score runs from beginning to end, and it can only be started at the beginning.

The ability to spin a perfect web or a perfect cocoon without having to learn the design and the construction skills is definitely advantageous for a short-lived animal with little or no exposure to parents or other members of the species. On the other hand, there are behaviors, such as finding food or interacting with other members of the population, that are highly dependent on learning. There is a continuous spectrum of animal behaviors ranging from those with completely closed programs not modifiable by experience to those with programs that are completely open and acquired only through learning. An important evolutionary question is: Under what conditions does natural selection favor open versus closed behavior patterns? To answer this question for a species and for a behavior, it is necessary to discover whether or not the behavior is genetically determined and to what extent it is modifiable by experience.

Deprivation Experiments

Just because a behavior is highly stereotypic and species-specific is not sufficient evidence for it to be called a fixed action pattern. The songs of most species of birds are stereotypic and identify the species that is singing. Yet, learning is an essential step in the acquisition of song in many of those species. If the eggs of white-crowned sparrows are hatched in an incubator and the young male birds are reared in isolation so that they do not hear the song of their species, when they reach sexual maturity their songs will be an unusual assemblage of sounds (Figure 43.2a,b). This species cannot express its species-specific song if it does not experience that song as a nestling. For a behavior to be described as a fixed action pattern, it must be expressed without prior experience, and therefore without learning.

The **deprivation experiment**, rearing an animal so that it experiences neither its parents, other members of its species, or its natural environment, is a method for demonstrating that a behavior is a fixed action pattern. The example of the white-crowned sparrow mentioned above is a case where the deprivation experiment revealed that a stereotypic, species-specific behavior was not a fixed action pattern. The opposite result was obtained in a deprivation experiment in which an ethologist hand-reared a tree squirrel in isolation, on a liquid diet, and in a cage where there was no soil or particulate substrate. When the young squirrel was given a nut, it put the nut in its mouth and ran around the cage. Eventually it oriented toward a corner of the cage and went through stereotypic digging movements, placed the nut in the corner, went through movements of refilling the imaginary hole, and ended by tamping the nonexistent soil with its nose. The squirrel had never handled a food object and had never experienced soil, yet the fixed action patterns involved in burying its nut were fully expressed.

There are naturally occurring deprivation experiments. Many species, especially among insects living in seasonal environments, have life cycles of one year with non-overlapping generations. This means that the adults lay eggs and die before the eggs hatch or the young mature into adults. In such species learning from adults of the parental generation is impossible, and complex behaviors necessary for survival and reproductive success must be genetically programmed. Web spinning in the spiders discussed above is an example of complex behavior in species

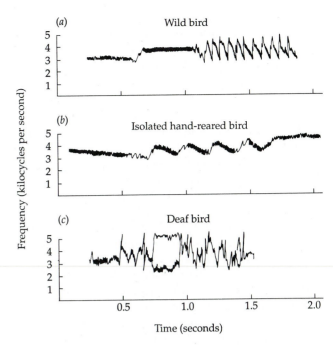

43.2 Song Learning in White-Crowned Sparrows
These sonograms visually record sound frequencies and plot them over time. (a) Sonogram of the species-specific song of a male white-crowned sparrow in its natural state. (b) In a deprivation experiment, a bird hatched in an incubator and reared in isolation from other birds does not learn the species-specific song. (c) In a variation on the experiment, a young bird that heard the correct song but was deafened before it reached maturity could not reproduce the song.

that have no opportunity to learn from other members of their species. That example can be extended to include spider courtship behavior. A male spider must approach a female in her web. If he simply blundered into the web, he would give the same signals as a prey item caught in the web, and the female would kill him and eat him. To avoid having his reproductive effort cut short, he is genetically programmed to approach an anchor strand of the web and pluck it in just the right way to send a courtship message to the female. If the message is correct and the female is receptive, he can enter the web and mate with the female rather than be her dinner. In some species, she eats him anyway after they mate, but at least he has achieved reproductive success rather than simply becoming nutriment for some other male's progeny.

Triggering Fixed Action Patterns

If a behavior is not expressed during a deprivation experiment, that is not proof that the behavior is not genetically programmed. It may just mean that the right conditions are not available to stimulate the behavior. Just as the nut stimulated the digging and burying behaviors of the squirrel, specific stimuli are usually required to elicit the expression of most fixed action patterns. These stimuli are called **sign stimuli** or **releasers** by ethologists. Two great ethologists, Konrad Lorenz and Niko Tinbergen, did classic studies of releasers and fixed action patterns. Their work provided a number of insights into the properties of releasers.

Releasers are usually a simple subset of all of the sensory information available to an animal. Male European robins have red feathers on their breasts. During the breeding season, the sight of a male robin stimulates another male robin to sing, give aggressive displays, and attack the intruder if he does not heed the warnings. An immature male robin, whose feathers are all brown, will not elicit aggressive behavior. However, an unformed tuft of red feathers on a pole will elicit an attack (Figure 43.3). A patch of red in certain locations is a sufficient releaser for male aggressive behavior in robins. Just as the motor score of the fixed action pattern is genetically programmed, so is the information that enables the animal to recognize the releaser for that fixed action pattern. It is more feasible to evolve a genetic mechanism to respond to a simple stimulus than to recognize a complex set of stimuli. The simplicity of most releasers has resulted in ethologists making some curious discoveries.

Tinbergen and A. C. Perdeck dissected in great detail the releasers and fixed action patterns involved in the interactions between herring gulls and their chicks during feeding. The adult gull has a red dot at the end of its bill (Figure 43.4). When it returns to its chicks, they peck at the red dot, and that stimulates the adult to regurgitate food for the chicks to eat. Tinbergen and Perdeck asked what the essential characteristics of the parent gull were that released food solicitation behavior in the chicks. They made

43.4 The Dot Marks the Spot
This gull is incubating eggs. When the young hatch they will peck at the parent's red bill spot, and that act will stimulate the parent to regurgitate food.

paper cutout models of gull heads and bills but varied the colors and the shapes. Then they scored each model according to how many pecks it received from chicks. The surprising results were that the shape or color of the head didn't matter at all. In fact, a head was not even necessary; the chicks responded just as well to models of bills alone. The color of the bill and the dot also were not critical as long as there was a contrast between the two. The most amazing result was that the most effective releaser for chick pecking was a long, thin object with a dark tip that had no resemblance to an adult herring gull at all (Figure 43.5).

The simplicity of the properties of releasers makes it possible for a **supernormal releaser** to exist that is more effective in eliciting a fixed action pattern than is the natural condition. In the case of a bird called the oystercatcher, the sight of its clutch of eggs releases incubation behavior. But, if you give an oystercatcher the choice of its own clutch of two eggs or a clutch of three artificial eggs, it will sit on the larger clutch of artificial eggs. If you give the oystercatcher the choice of its own clutch of two eggs or one very large artificial egg, it will try to incubate the large egg, even if it can hardly straddle it. Of course, these choices would not occur in nature, and therefore there has not been counterselection to prevent the evolution of such maladaptive behavior.

The production of supernormal releasers by ethologists is a curiosity, but exploitation of the simplicity of releasers by natural selection has produced some dramatic results as well. Many of the elaborate behaviors and physical attributes used by species in courtship displays have arisen through natural selection favoring more effective releasers. In one group, the bowerbirds, males use colorful objects collected from the environment to enhance their courtship dis-

43.3 Triggering Aggressive Behavior
A mounted immature male robin (right) with no red feathers does not stimulate aggression from a territorial adult male. The formless clump of red feathers on the left, however, *does* trigger aggressive territoriality.

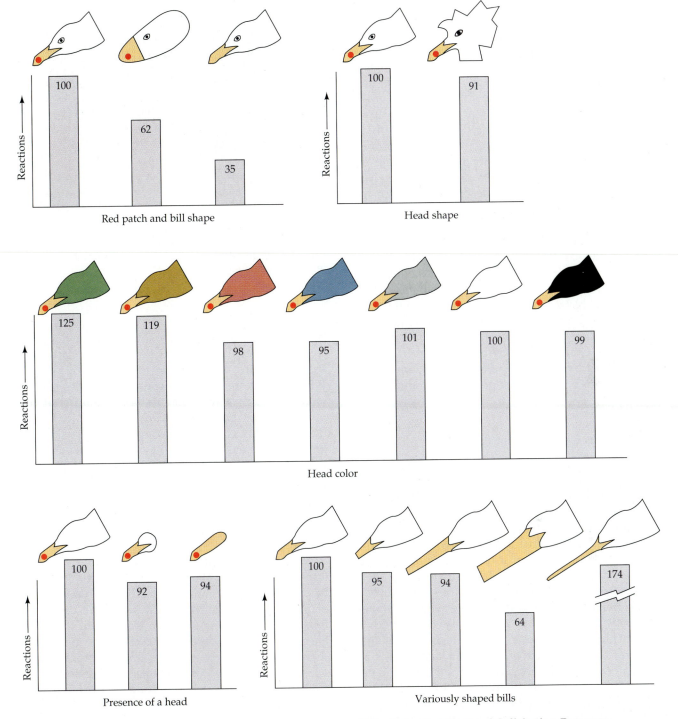

43.5 Releasing the Food Solicitation Response
A series of experiments scored the pecking response of herring gull chicks. The presence of the red dot and the shape of the bill seem to be releasers; the shape and color of the adult's head have little or no influence on the response.

plays by providing supernormal releasers. We'll examine the evolution of courtship displays in the next chapter.

Motivation

Another reason that a fixed action pattern may not be expressed in a deprivation experiment is that the animal is not in the appropriate developmental or physiological state. Juvenile animals do not show courtship behavior even if the appropriate releasers are present. An adult animal may not engage in aggressive display or courtship display when it is not in reproductive condition. The same animals that may be highly aggressive to one another during the reproductive season may ignore one another at other times of the year. Some species that are solitary and aggressive during the reproductive season gather together in flocks or herds for mutual protection or greater feeding efficiency at other times of the year.

The internal conditions of an animal determine its motivational state, and the motivational state determines which fixed action pattern is expressed at any particular time. After all, the total behavior of an animal is not simply a random sequence of fixed action patterns depending on what releasers it happens to encounter. Depending on its motivational state, an animal may search for the appropriate releaser and ignore many others. Ethologists use the term **appetitive behavior** to describe the search for an appropriate releaser. The associated term for the fixed action pattern after the releaser is encountered is **consummatory behavior**. It is easy to think of these terms in the context of a hungry predator that is searching for prey. The appetitive behavior—the search—may depend heavily on previous experience. Where has it found prey before? When the prey is spotted, the predator orients to it and engages in the fixed action patterns of seize and kill. Appetitive behavior is characterized by being modified by experience and by being quite flexible. Consummatory behavior may be genetically programmed and highly stereotypic.

Closed versus Open Behavior Patterns

The ability to learn and to modify behavior as a result of experience is often highly adaptive. Most of human behavior is the result of learning. Why then are so many behavior patterns in so many species of animals closed, genetically determined, and not modifiable? We've already considered some of the answers to that question. If role models and opportunities to learn are not available, there is no alternative to programming behavior in the genes. Closed behavior patterns are also adaptive when mistakes are costly or dangerous. If a female mates with a male of the wrong species, that is a costly mistake. The function of much of courtship behavior is to guarantee correct species recognition. If courtship behavior is learned in an environment in which incorrect as well as correct models exist, it would be possible to learn the wrong pattern. As we will see later on in this chapter, fixed action patterns governing mating behavior can prevent such mistakes. In behaviors such as predator avoidance or capture of dangerous prey, there is no room for mistakes. If the behavior is not performed promptly and accurately the first time, there may be no second chance. Whether a rattlesnake is predator or prey, it is dangerous. Some rattlesnakes prey on kangaroo rats. A kangaroo rat that has never encountered a rattlesnake can avoid it in total darkness because the sound of the snake moving through the air to strike releases the powerful escape jump of the kangaroo rat (Figure 43.6). A kangaroo rat does not have the luxury of

43.6   An Escape Response
The sound of a striking rattlesnake triggers an automatic escape jump in a kangaroo rat; the rat does not have to learn this behavior.

learning what a rattlesnake sounds like when it is striking. On the other hand, king snakes eat rattlesnakes but are not immune to rattlesnake venom. When a king snake first encounters a rattlesnake, it strikes in such a way that its jaws clamp shut the mouth of the rattlesnake, and from there it begins the long process of swallowing. A king snake that grabs a rattlesnake at any other place on its body will not have the opportunity to learn by trial and error. So, closed behavior patterns are highly adaptive for species that have little opportunity to learn complex behaviors, for species that might learn the wrong behavior, and in situations where mistakes are costly or dangerous.

Many behaviors are intricate interactions of genetically programmed elements and elements modified by experience. One example that has been the subject of elegant experiments is bird song. We discussed bird song in Chapter 34 as an example of the influence of hormones on brain function. As mentioned there, song is largely a species-specific characteristic used by adult males in territorial display and courtship. A few species, such as song sparrows, express their species-specific song during a deprivation experiment, but others, such as white-crowned sparrows, do not (Figure 43.2b). The white-crowned sparrow must hear the song of its own species during its nestling period. It does not sing as a juvenile, but it apparently forms a template in its nervous system which it then matches through trial and error when it reaches sexual maturity the following spring. If a bird that has heard its correct song as a juvenile is deafened before it begins to express its song, it will never develop its species-specific song (Figure 43.2c) It must be able to hear itself to match the template stored in its nervous system. If it is deafened after it expresses its correct song, it will continue to sing like a normal bird. So, two periods of learning are essential: the first occurs in the nestling stage and the second at the onset of sexual maturity.

Studies of what birds can learn and when they can learn, however, reveal strong genetic limits to the modifiability of their behavior through experience. In the case of the white-crowned sparrow, it must hear its species' song within a rather narrow **critical period** during its development. Once this critical period has passed, the bird cannot learn to sing its species-specific song, regardless of how many role models it experiences. What a bird can learn during its critical period is also severely limited, as revealed by experiments on hand-reared chaffinches that were played various tape recordings of bird song during their critical periods. If exposed to the songs of other species, the chaffinches did not learn them. They also did not learn a chaffinch song played backwards or with the elements scrambled. Even if they heard a chaffinch song played in pure tones, they did not form a template. But, if they heard a normal chaffinch song along with all of these other sounds, they developed templates and learned to sing the proper song the following spring. Thus, the chaffinch is genetically programmed to recognize the appropriate song to learn and when to learn it.

What advantage are genetic limits on what a bird can learn and when? The acoustic environment can be quite complex. Many species of songbirds can be singing in the same garden. The critical period limits learning to the period of time when there is the most intimate contact between the young bird and its parents so that the father's song is the one experienced most intensively. Further limits on what song the young bird is sensitive to help to guarantee that the template it forms is not contaminated with other sounds it hears.

The learning of a song template by a nestling bird is an example of the phenomenon of **imprinting**. We mentioned above that releasers are generally simple subsets of available information because there are limits to what can be programmed genetically. Imprinting makes it possible to encode complex information in the nervous system rather than in the genes. Offspring can imprint on their parents and parents on their offspring to insure individual recognition, even in a crowded situation such as a colony or a herd. If a mother goat does not nuzzle and lick her newborn within 5 to 10 minutes after birth, she will not recognize it as her own. In this case imprinting depends on olfactory cues, and the critical period is determined by the high levels of circulating oxytocin at the time of birth.

THE GENETICS OF BEHAVIOR

To say that behavior is genetically programmed might be taken to suggest that there are specific genes for behavior. That is almost certainly not true. Genes code for proteins, and it is a long way from a gene or a protein to a behavior. Specific proteins play critical roles in the development of patterns in the nervous system and in the functions of the nervous system. Yet there are many steps between the expression of a gene and the expression of a behavior, and in no case are all of those steps known. Nevertheless, genetic experiments can tell us a great deal about the characteristics of heritability of behavior. Genetics employs a number of traditional approaches that can be applied to fixed action patterns. They include hybridization, artificial selection, crossing of selected strains, and molecular analysis of genes and gene products.

Hybridization Experiments

The material for genetic analysis is variability, and variability is most pronounced among species. Closely related species frequently show large differences in fixed action patterns, and if such species can be hybridized, the offspring reveal interesting disruptions of their behavior. A classic case is nest building in lovebirds of the genus *Agapornis*. One species, *A. roseicollis*, carries nesting material tucked under its tail feathers. Another species, *A. fischeri*, carries nesting material in its beak (Figure 43.7). Are these simple behaviors learned or genetically programmed? When the two species are crossed, the hybrid offspring display a maladaptive combination of the two carrying methods. The hybrid picks up nesting material and tucks it into its tail feathers, but does not release the object immediately. As a result, the hybrid inevitably pulls the nesting material out of its tail feathers and drops it. With years of experience, the hybrids learn to carry material in their beaks, but they always make the intention movement toward their tail feathers whenever they pick up nesting material. This hybridization study indicates that the ways in which birds of the two species carry nesting materials are genetically determined.

Konrad Lorenz conducted a series of hybridization experiments on ducks to investigate the genetic determinants of their elaborate courtship behaviors. Dabbling ducks such as mallards, teals, pintails, and gadwalls are closely related and interfertile, but they rarely interbreed in nature because of the specificity of their courtship displays. Each male duck performs a carefully choreographed water ballet (Figure 43.8). It is unlikely that the female will accept his advances if the entire display is not successfully completed. When Lorenz studied the displays of the various species, he found that they consisted of about 20 components. The display of each species included a subset of these components put together in a certain sequence. When he crossbred the species, he found that the hybrids expressed some components of the

(a)

(b)

43.7 A Behavior Influenced by Genes
(a) Peach-faced lovebirds carry nest-building materials tucked in their back feathers. (b) Fischer's lovebirds carry the objects in their bills. Hybrid offspring of the two species display a confused combination of the two behaviors, indicating the behaviors are genetically programmed.

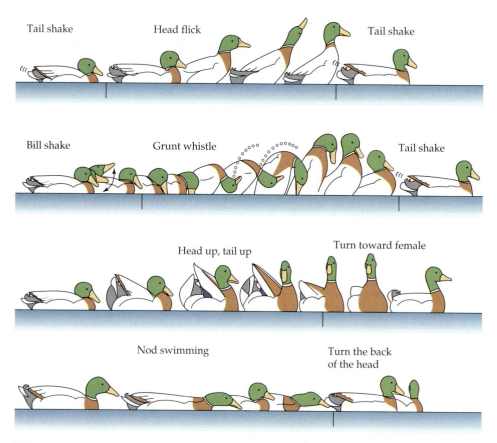

43.8 Courtship of the Mallard Duck
There are about ten elements in the courtship display of the male mallard duck. Closely related duck species may display some of the same ten elements, but they will have other elements not displayed by mallards. The elements of the courtship display and their sequence are species-specific and act to prevent hybridization.

display of each parent put together in new combinations. Most interesting, the hybrids sometimes showed display components that were not in the repertoire of either parent, but were characteristic of the displays of other species. This series of hybridization studies clearly demonstrated that the motor patterns of the courtship displays were genetically programmed. Females were not interested in males showing the hybrid displays, thus demonstrating the adaptive significance of the species-specific fixed action patterns.

Selection and Crossing of Selected Strains

Domesticated animals provide abundant evidence that artificial selection of mating pairs on the basis of their behavior can result in strains with distinct behavioral as well as anatomical characteristics. Among dogs, consider retrievers, pointers, and shepherds. Each has a particular behavioral tendency that can be honed to a fine degree by training, whereas other strains cannot be so trained. Dogs and other large animals, however, are not the best subjects for genetic studies. Most controlled selection experiments in behavioral genetics have been done on more convenient laboratory animals with short life cycle times and high numbers of offspring. A favorite subject for such studies has been the fruit fly, genus *Drosophila*. Artificial selection has been successful in shaping a variety of behaviors in fruit flies, especially aspects of their courtship and mating behaviors. Crossing of selected strains reveals that such behavioral differences produced by artificial selection are usually due to multiple genes that probably influence the behavior indirectly by altering general properties of the nervous system.

There are very few behavioral genetic studies that reveal simple Mendelian segregation of behavioral traits. One such case is the genetics of hygienic behavior in honeybees. There is a bacterium that infects and kills the larvae of honeybees. One strain of honeybees that is rather resistant to this disease practices hygienic behavior—when a larva dies, workers uncap its brood cell and remove the carcass from the hive. Another strain of honeybees did not show hygienic behavior and was, therefore, more susceptible to the spread of the disease. When these two strains were crossed, the interesting results indicated that the hygienic behavior was controlled by two recessive genes (Figure 43.9). First, the F_1 generation were all nonhygienic, indicating that the behavior was controlled by recessive genes. Backcrossing the F_1 with the pure hygienic strain produced the typical 3:1 ratio expected for a two-gene trait. The behavior of the nonhygienic individuals was very interesting. One-third of them showed no hygienic behavior at all, one-third would uncap the cells of dead larvae but not remove them, and one-third would not uncap

cells but would remove carcasses if the cells were open. Even though these results appear to indicate a gene for uncapping and a gene for removal, these behaviors are each quite complex. They involve sensory mechanisms, orientation movements, and motor patterns. Each of those categories depends on many properties of many cells. The genetic deficits of nonhygienic bees could simply influence very small, specific, yet critical properties of some cells. Lacking one critical property, such as a specific synapse or a particular sensory receptor, the whole behavior would not be expressed. The responsible gene is not a gene for the behavior.

Molecular Genetics of Behavior

The powerful techniques of molecular genetics enable the investigation of the functions of specific genes, and those techniques can be applied to genes that influence behavior. For example, in the marine mollusk *Aplysia*, egg laying involves a sequence of fixed action patterns. The eggs are extruded from the animal in long strings by contractions of the muscles of the reproductive duct. The animal stops whatever it is doing (usually eating or crawling), takes the egg string in its mouth, and with a series of stereotypic head movements, it pulls the egg string from the duct and coils it into a mass glued together by secretions from its mouth. Finally, with a strong head movement, it affixes the entire mass of eggs to a solid substrate. *Aplysia* has a very simple nervous system, and it was discovered that certain cells in its nervous system produce a peptide that can elicit certain aspects of egg-laying behavior, but not all. This peptide is called egg-laying hormone. When the amino acid sequence of egg-laying hormone was determined, it was possible to use molecular genetic techniques to find the gene that coded for it. The surprising discovery was that the gene codes for a precursor molecule that has almost 300 amino acids, whereas egg-laying hormone has only 36 amino acids. The precursor molecule also contains a number of other peptides that function as neural signals controlling aspects of egg-laying behavior. Thus one gene could code for a set of neural signals necessary to elicit the coordinated motor scores involved in egg-laying behavior. This example is about as close as we can get to making connections between a specific gene and a specific behavior. But the connections depend on the existence of a highly organized nervous system, which of course is a product of many other genes.

THE EVOLUTION OF BEHAVIOR

To the extent that behavior is genetically determined, it is subject to natural selection. The basic tenet of natural selection is that a trait is favored if it confers

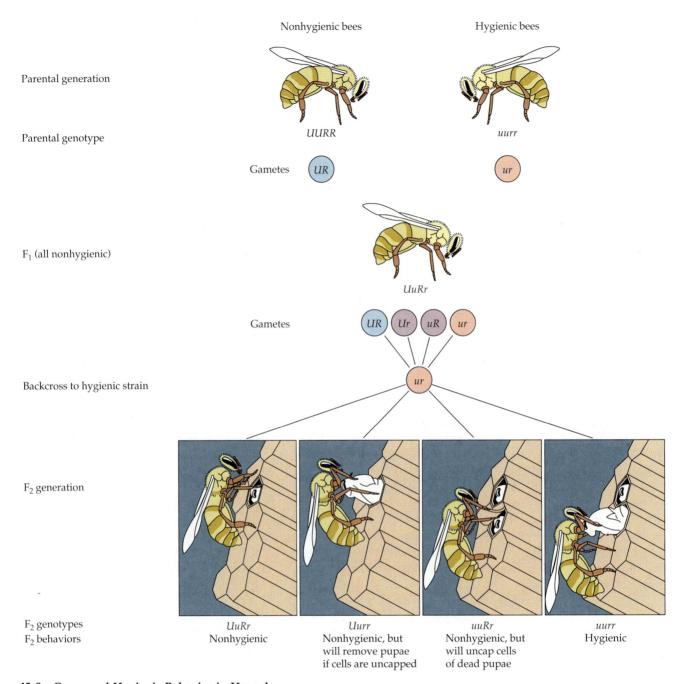

Nonhygienic bees

Hygienic bees

Parental generation

Parental genotype

UURR

uurr

Gametes

UR

ur

F₁ (all nonhygienic)

UuRr

Gametes

UR   *Ur*   *uR*   *ur*

Backcross to hygienic strain

ur

F₂ generation

F₂ genotypes   *UuRr*   *Uurr*   *uuRr*   *uurr*
F₂ behaviors   Nonhygienic   Nonhygienic, but Nonhygienic, but Hygienic
                              will remove pupae will uncap cells
                              if cells are uncapped of dead pupae

43.9 Genes and Hygienic Behavior in Honeybees
The females of some honeybee strains make a practice of removing the car-casses of dead larvae from their nests. This behavior seems to have two com-ponents—uncapping the larval cell (*u*), and removing the carcass (*r*)—each of which is under the control of a recessive gene.

on the individual a net gain in reproductive success relative to other members of the population. To study the evolution of a behavior, it is necessary to identify the selective pressures operating on it. A cost is as-sociated with the performance of any behavior, so if that behavior is to be favored by natural selection, the benefit it bestows on an animal must be greater than its cost. A cost–benefit analysis can be done for any behavior. For example, the costs of elaborate courtship behaviors, such as the water ballets of dab-bling ducks described above, include time and energy that could be spent on other activities, and increased vulnerability to predators. The benefit is that the in-dividuals do not waste their limited reproductive in-vestments (time, eggs, sperm, and parental care) on sterile offspring. In the next chapter, we will consider in cost–benefit terms the evolution of other forms of social behavior. Here we will explore the selective pressures operating on a particularly ubiquitous be-havior, aggression.

Selection For and Against Aggressive Behavior

The costs of aggressive behavior are quite clear. It is dangerous; animals engaged in aggressive encounters are likely to be injured. Even if they are not, aggressive behavior takes much time and energy, and because it is frequently done in the open, it exposes the participants to predators. Aggressive behavior can cause stress which can suppress reproductive physiology. More directly, it can interrupt reproductive behavior and take time away from parental care once young are produced. A challenge is to identify what resources an animal gains by aggressive behavior and to demonstrate that those resources result in an increase in the animal's reproductive success. Figure 43.10 presents a model for thinking about this problem. Competition between members of a population can result in differential reproductive success. A hypothesis for the evolution of aggressive behavior states that if competition exists, and if resources are economically defensible, then aggressive behavior should be favored by selection.

Many examples can be cited to show that animals are aggressive over resources that are defensible but are not aggressive over resources that cannot be defended, even if they are requisites for reproduction. Marine birds that are highly aggressive over nest sites on the cliffs of islands are not aggressive over patches of ocean where they feed. Similarly, some species of tropical hummingbirds are highly aggressive when feeding on flowers but are not aggressive when feeding on insects. Aggression can guarantee exclusive access to resources fixed in space like nest sites or patches of flowers, but cannot guarantee priority of access to moving resources such as schools of fish or flying insects.

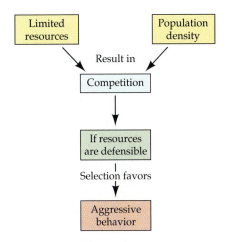

43.10 A Model for the Evolution of Aggression

Territoriality

Aggressive behavior in defense of space is called **territoriality**. Since this is the most common type of aggressive behavior, we can ask whether or not, and how, the possession of territory influences reproductive success. Ethologists recognize four different types of territories, which are simply labelled types

43.11 Type A Territory Provides Everything Needed for Reproduction and the Rearing of Young
The territories of male song sparrows on a brushy flatland in Ohio were mapped over many years by ornithologist Margaret Nice. Each number identifies a bird; the underlines indicate previous years on the same territory. In years when the song sparrow population is high, all available space is occupied; when the population is low, much habitat is unoccupied. In some species the size of the territory increases in poor habitat or in years with poor food supply, but in these song sparrows it remains about 4,000 square meters.

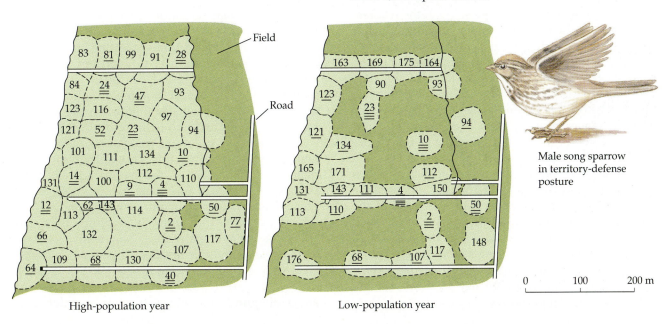

High-population year

Low-population year

Male song sparrow in territory-defense posture

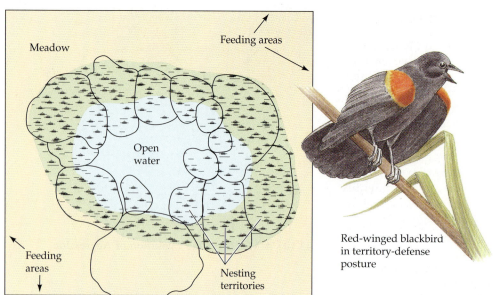

Red-winged blackbird
in territory-defense
posture

43.12 Type B Territory Provides a Nesting Site and Some Food

Each enclosed area is a territory of a male red-winged blackbird. In preferred habitat—where reeds grow over water—the territories are small because of intense competition. In the marsh periphery the territories are less contested and thus are larger. Females prefer males with central territories. The birds feed in meadows surrounding the marsh.

A, B, C, and D. Type A territories are all-purpose and are the type held by most songbirds. They provide a mating site, a nesting site, and all of the food necessary to rear offspring (Figure 43.11). Studies have supported the hypothesis that the reproductive success of species that defend type A territories is limited by food supply and that the possession of a territory guarantees an adequate food supply for raising young. Individuals that do not acquire territories are unlikely to breed, and have a higher mortality rate than do holders of territories. Thus, the relationship between holding a territory and reproductive success is clear in the case of type A territories.

Type B territories include a large breeding and nesting area but do not supply most of the food necessary to rear young. One animal that defends type B territories is the red-winged blackbird, which nests in emergent vegetation in marshes but gets most of its food from areas around the marsh (Figure 43.12). Nesting in reeds over the water is excellent protection from land predators, so for such a species optimal nest sites are more limiting for reproduction than is food supply. Evidence for the value of the nest site comes from the fact that female red-winged blackbirds preferentially choose as mates males with territories in the emergent vegetation. As a result, a male in the center of the marsh may acquire two or three mates while males on dry land around the marsh may still be bachelors.

Type C territories are strictly for nesting and for rearing young. As mentioned above, many marine birds such as gulls are highly aggressive over very small nest sites in crowded colonies on islands. The advantage of such a site is the protection it provides from predators that are found on the mainland but not on the island. In other cases the colonies may cling to narrow cliffs on the mainland, but still their inaccessibility to predators is a crucial factor. Another example of type C territories are those held by colonies of penguins (Figure 43.13). Here the value of the real estate may be that it is clear of vegetation or snow at the right time of year and provides access to the sea. Also, nesting in colonies offers protection from aerial predators.

Type D territories are the most curious of all. They are small pieces of land used only for courtship display and mating. Typical of species defending type D territories is an antelope called the kob. On the seemingly endless plains of their African habitat, male kobs gather for the breeding season on small traditional display grounds. Males battle intensely for possession of central territories on this display ground. They may go without eating for weeks to defend and display on these tiny territories. Females come to the display grounds to choose a mate, and they usually select a male holding a central territory (Figure 43.14). The females then leave the area and raise their young with no help from the male.

What selective pressures could work on male and female kobs to produce this strange form of territoriality? Its origins are in the mating system called polygamy, in which one male has multiple mates. The evolution of polygamous mating systems is discussed in the next chapter. It is sufficient here to know that males of many species aggressively defend their access to multiple females. This situation creates great variability in the reproductive success of males. Males that are dominant have several mates and many offspring, and males that are not dominant

43.13 Type C Territories Can Be Small
The spacing of individual breeding territories of king penguins is determined by how far the penguins can reach to peck their neighbors. There are no nests; the egg is incubated under a flap of skin. Every time a penguin walks through the colony to relieve its mate or to go to the sea to feed, it creates a commotion and receives many pecks.

have few or no offspring. Females of the species do not have such variability in their potential for offspring. But females can have a huge variability in their potential for passing on their genes to grandchildren depending on whether or not their male offspring become dominant, successful males. To the extent that male attributes leading to dominance are inherited, females that mate with dominant males will have dominant male offspring and hence more grandchildren. The type D territory is the extreme result of the sexual selection arising from a polygamous mating system. The male that gains the central display territory will have enormous reproductive success and other males will have little or no reproductive success. A female that mates with the most dominant male rather than with a peripheral male is more likely to have a male offspring that one day will be king of the display ground and spread her genes to many, many grandchildren.

THE TIMING OF BEHAVIOR: BIOLOGICAL RHYTHMS

From a study of the evolution of behavior, we turn for the rest of this chapter to considerations of me-

43.14 Type D Territories Are Display Grounds
This male Uganda kob has attracted several potential mates.

chanistic studies of behavior. Many behaviors have been studied quite extensively, but in very few cases are the underlying neurophysiological mechanisms understood. An almost ubiquitous behavior is daily rhythmicity, and in many animals, including mammals and birds, the brain mechanisms of daily rhythms of behavior are becoming well understood.

Circadian Rhythms

Our planet turns on its axis once every 24 hours, creating a cycle of environmental conditions that has existed throughout the evolution of life. Many organisms evolved rhythmicity—Chapter 32 discussed the phenomenon in some plants—but daily cycles are a characteristic of just about all animals. What is surprising, however, is that daily rhythmicity does not depend on the 24-hour cycle of light and dark. If animals are kept under absolutely constant environ-

mental conditions, such as constant dark and constant temperature, with food and water available all of the time, they still demonstrate a daily cycle in activity, sleeping, eating, drinking, and just about anything else you can measure, suggesting that the animal has an internal (endogenous) clock. Without time cues from the environment, however, these daily cycles are not exactly 24 hours. They are therefore called **circadian rhythms** (*circa* means about, *dies* means a day).

Some terminology is necessary for discussing rhythms. A rhythm can be thought of as a series of cycles, and the length of one of those cycles is the **period** of the rhythm. Any point on the cycle is a **phase** of that cycle. Hence, when two rhythms completely match, they are in phase, and if a rhythm is shifted (like resetting a clock), it is phase-advanced or phase-delayed. Since the period of a circadian rhythm is not exactly 24 hours, it has to be phase-

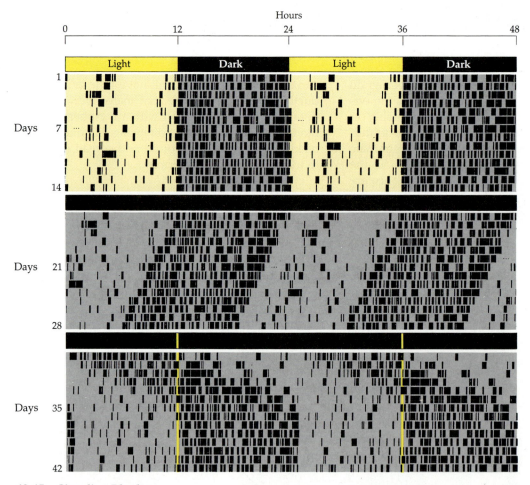

43.15   Circadian Rhythms
The black marks indicate that a mouse is running on an activity wheel. The time of day runs across the chart, and sequential days run down the chart. Notice, however, that two days are recorded on each line. This means that the data for each day are plotted twice, once on the right side of a line and again on the left side of the next line; this double plotting makes patterns easier to see.
Changes in the schedule of light and dark are indicated

by the solid bars running across the figure. First the mouse sees 12 hours of light and 12 hours of dark every day, then it is placed in total darkness, and finally it is given a 10-minute exposure to light each day. In constant darkness the circadian rhythm is free-running, but a 10-minute flash of light can entrain it. This figure is idealized, but represents results taken from real experiments.

advanced or phase-delayed every day to remain in phase with the daily cycle of the environment. This process of the resetting of the rhythm by environmental cues is called **entrainment**. An animal kept in constant conditions will not be entrained to the 24-hour cycle of the environment, and its circadian clock will free-run with its natural periodicity. If its period is less than 24 hours, the animal will begin its activity a little earlier each day (Figure 43.15). Animals showing free-running circadian rhythms can be used in experiments to investigate stimuli that phase-shift or entrain the clock. Under natural conditions, environmental cues such as the onset of light or dark entrain the free-running rhythm to the 24-hour cycle of the real world. In the laboratory it is possible to entrain circadian rhythms in animals held under constant conditions with short pulses of light or dark administered every 24 hours. Researchers can also entrain animals to light or dark pulses given at intervals not equal to 24 hours, as long as those intervals are not too short or too long. There are limits to the entrainability of the endogenous clock.

When you fly across a number of time zones, your circadian clock is out of phase with the real world at your destination. This is the cause of jet lag. Gradually your endogenous rhythm comes back into synchrony with the real world as it is reentrained every day by environmental cues. However, your internal rhythm cannot be shifted by more than 30 to 60 minutes each day, so it takes a number of days to reentrain your endogenous clock to real time in your new location. This period of reentrainment is when you have the symptoms of jet lag, since your endogenous rhythm is waking you up, making you sleepy, initiating activities in your digestive tract, and stimulating many other physiological functions at inappropriate times of day.

Where is the circadian clock? In mammals there is a master circadian clock that is located in two tiny groups of cells just over the optic chiasm, the place where the two optic nerves join. Hence these structures are called the suprachiasmatic nuclei. If these two little groups of cells are destroyed, the animal loses circadian rhythmicity. Under constant conditions it is equally likely to be active or asleep, eat or drink, at any time of day. Its activities are randomly distributed (Figure 43.16). Recent experiments have shown that circadian rhythmicity can be restored in an animal whose suprachiasmatic nuclei have been destroyed by transplanting those nuclei from another animal.

Because circadian rhythms are found in virtually every animal group, as well as in protists, plants, and fungi, the molecular mechanisms for generating

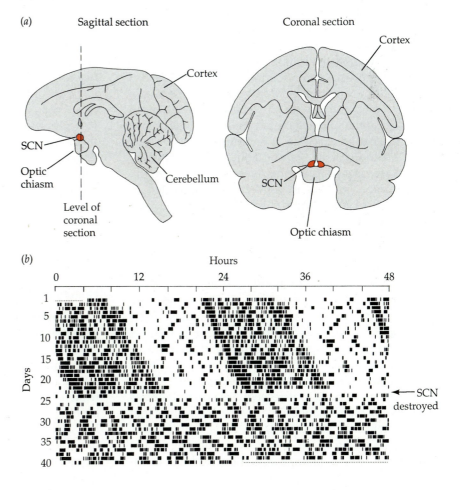

43.16 Where the Clock Is
(a) The circadian clock of mammals is in the suprachiasmatic nuclei of the brain. These nuclei are located just over the site on the bottom of the brain where the optic nerves join. (b) If its suprachiasmatic nuclei are destroyed, a mammal loses its circadian rhythm.

circadian rhythms must be very general properties of cells. We should expect, however, that a diversity of master clocks such as the suprachiasmatic nuclei have been produced by natural selection. After all, invertebrates do not have suprachiasmatic nuclei. Diversity of circadian control systems exists even within the vertebrates. The circadian clock of birds resides in the pineal gland. This small mass of tissue sticks up between the cerebral hemispheres and produces the hormone melatonin. If the pineal gland of a bird is removed, the bird will no longer have circadian rhythms. In mammals, light entrains the circadian clock via photoreceptors in the eyes, but in birds, the pineal gland itself is sensitive to light and is sometimes called the third eye. If a small amount of black ink is injected under the skin on the top of a bird's head so that it blackens the skull over the pineal gland, the bird will not entrain to the cycle of light and dark, but will show a free-running circadian rhythm.

Circannual Rhythms

Our planet also revolves around the sun once every 365 days. Since Earth is tilted on its axis, its revolution around the sun results in seasonal changes in day length, temperature, humidity, weather, and other variables at all locations except at the equator. The behavior of animals must adapt to these seasonal changes, and therefore it is important that animals anticipate the seasons and adjust their behavior accordingly. Most animals should not come into reproductive condition and mate just prior to winter, for example, because their offspring would then be born during a time of little food and harsh weather conditions. For many species, the change in day length is an excellent and absolutely reliable indicator of seasonal changes to come. If photoperiod has a direct effect on the physiology and behavior of a species, that species is said to be photoperiodic. For example, if male deer are held in captivity and subjected to two cycles of day-length change in one year, they will grow and drop their antlers twice during that year. Many species of birds kept in captivity come into or out of reproductive condition depending on day length.

For some animals, changing day length is not a reliable cue. Hibernators, for example, spend long months in dark burrows underground but have to be physiologically prepared to breed almost as soon as they emerge in the spring. The timing of their breeding is important because their young must have time to grow and fatten before the next winter. Other examples of animals that receive little or ambiguous information from changes in day length are birds that overwinter near the equator or birds that migrate across the equator. For a bird overwintering in the tropics, there is no change in photoperiod that it can

use to time its migration north to the breeding grounds. A bird that crosses the equator must fly south as day length decreases at one time of year but fly north when day length decreases at another time of year.

Hibernators and equatorial migrants have endogenous annual rhythms which are called circannual rhythms. In other words, their nervous systems have built-in calendars. Just as circadian rhythms are not exactly 24 hours, circannual rhythms are not exactly 365 days. The circannual rhythm of an animal under constant conditions may be 360 days, or 345 days. Rarely is it longer than 365 days, because being late for an annual event such as breeding would be a very costly mistake.

ORIENTATION AND NAVIGATION

Orientation is a highly modifiable behavior with reference to a place or a stimulus. The place or stimulus can be a predator or prey, a mate or offspring, or a nest or feeding ground in the immediate surroundings of the animal. What if the place or stimulus is at some distance; how does the animal orient to it and find its way? In most cases that answer is quite simple: the animal knows and remembers the structure of its environment. It uses landmarks to find its nest, a safe hiding place, or a food source. Orienting by means of landmarks is called **piloting**. Even long-distance migrations of animals can be achieved by piloting, and it need not depend on specific landmarks. For example, the gray whales that spend the summer feeding in the Gulf of Alaska and the Bering Sea and migrate south in the winter to Scammon's Lagoon on the coast of Baja California can find their way by following two simple rules (Figure 43.17): Keep the land to the left in the fall and to the right in the spring. By following the west coast of North America, they can travel from summer to winter areas and back again by piloting. Coastlines, mountain chains, rivers, water currents, and wind patterns serve as piloting cues for many species. Yet there are remarkable cases of long-distance orientation and movement that cannot be explained on the basis of piloting by landmarks.

Homing

The ability of an animal to return to a nest site, burrow, or any other specific location is **homing**. In most cases homing is just piloting in a known environment, but animals are capable of much more sophisticated feats of navigation to find the way home from a distant point. People who breed and race homing pigeons take the pigeons from their home loft and release them at a remote site where they have never been. The first pigeon home wins. Their departure directions, as well as their known flying

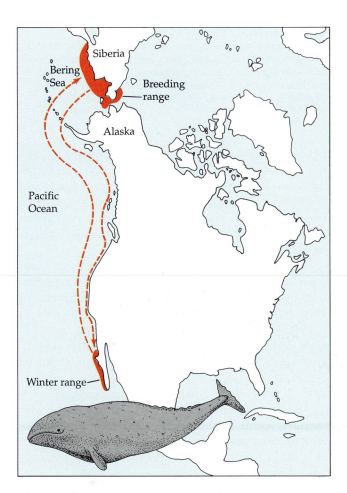

43.17 Piloting
Gray whales migrate south in winter, from the Bering Sea to the coast of Baja California. They follow a landmark—the west coast of North America. Such navigation is called piloting.

speed and the distance they have to travel, show that the pigeons fly fairly directly from the point of release to home; they do not randomly search until they encounter familiar territory. Scientists have used homing pigeons to investigate the mechanisms of animal navigation. In one series of experiments the pigeons were fitted with frosted contact lenses so they could see no details other than light and dark. These pigeons still homed and fluttered down to the ground in the vicinity of their loft. They were able to navigate without visual images of the landscape.

Marine birds provide many dramatic cases of homing over great distances in an environment where landmarks are rare. In daily feeding trips, many marine birds fly over hundreds of miles of featureless ocean and then return directly to a nest site on a tiny island. Remarkable feats of homing are demonstrated by albatrosses. When a young albatross leaves its nest on an oceanic island, it flies widely over the southern oceans for eight or nine years before it reaches reproductive maturity. At that time it flies back to the island where it was raised to select a mate and build a nest (Figure 43.18). After that first mating season

the pair separates, and each bird resumes its solitary wanderings over the oceans. The next year they return to the same nest site at the same time, reestablish their pair bond, and breed. Thereafter they return to the nest to breed every other year, spending many months in between at sea. These long-distance, synchronous homing trips are amazing feats of navigation and timing.

Migration

Ever since humans inhabited temperate and subpolar latitudes, they must have been aware of the fact that whole populations of animals, especially birds, disappear and reappear seasonally. It was not until the early 19th century, however, that patterns of migration were established by marking individual birds with identifying bands around their legs. Being able to identify individual birds in a population made it possible to demonstrate that the same birds and their offspring returned to the same breeding grounds year after year, and that these same birds were found during the nonbreeding season at distant locations hundreds or even thousands of kilometers from the breeding grounds.

How do migrants find their way over such great distances? A reasonable hypothesis might be that young birds on their first migration follow experienced birds and learn the landmarks by which they could pilot in subsequent years. However, adult birds of many species leave the breeding grounds before the young finish fattening and are ready to begin their first migration. So, naive birds must be able to navigate on their own, and there is little room for mistakes. Some species of small songbirds that breed in eastern North America migrate south by first moving to the coast, fattening there, and then taking off

43.18 Coming Home
A pair of black-browed albatrosses engage in courtship display over their partially completed mud nest. Many albatrosses return to the site of their own birth to find a mate, and will return to that site year after year.

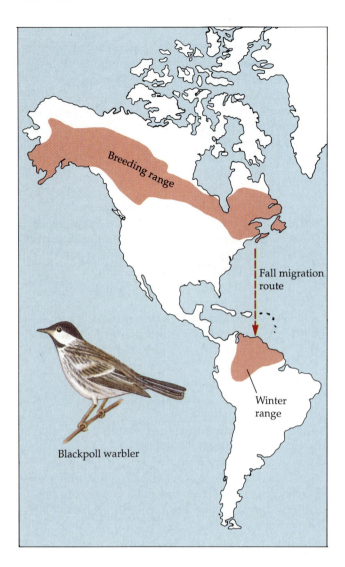

Breeding range

Fall migration route

Winter range

Blackpoll warbler

43.19 Songbirds Migrate over the Atlantic
The blackpoll warbler is one of many species that breeds over the northern United States and Canada and winters in South America. Its fall migration is first to the northeast coast of North America, where it feeds in preparation for the nonstop overwater flight to South America.

have gone, you can navigate. Bicoordinate navigation, called true navigation, means that you know the latitude and longitude (the map coordinates) of where you are and of where you want to go. From that information you can plot a route to your destination. Do animals have these sophisticated abilities to navigate?

Researchers conducted an experiment with European starlings to see which method of navigation they use. This population of starlings did not have a long migration. Their breeding grounds were in northern Germany, and their wintering grounds were to the southwest in southern England and western France (Figure 43.20). The birds were captured before the fall migration and transported to Switzerland. If they were capable of true navigation, they should have flown northwest to their traditional wintering ground. But they did not—they flew southwest as they normally did, and therefore ended up in Spain. The researchers concluded that the starlings used distance and direction for navigation.

How do animals tell distance and direction? In many instances, distance is not a problem as long as the animal recognizes its destination. Homing animals pick up landmarks and can pilot once they reach familiar areas. There is evidence, however, that biological rhythms play a role in determining migration distances. Birds kept in captivity display increased and oriented activity at the time of year when they would normally migrate (Figure 43.21). This is called **migratory restlessness**, and it has a definite duration. Since distance is determined by how long an animal moves in a given direction, the programming of duration of migratory restlessness sets the distance for its migration.

There are two obvious candidates for the means of telling direction: the sun and the stars. During the day, the sun is an excellent compass as long as you know what time it is. The sun rises in the east, sets in the west, and in the northern hemisphere, it points south at noon. We learned above that animals can tell the time of day by means of their circadian clocks. Clock-shifting experiments demonstrate that animals use circadian clocks to determine direction from the position of the sun. Researchers placed birds in a circular cage that enabled them to see the sun and sky, but not other visual cues. Food bins were arranged around the sides of the cage, and the birds were trained to expect food in the bin in one particular direction, let us say south as an example. After

over the North Atlantic on a direct route to South America (Figure 43.19). They cannot land on water and their fuel reserves are limited by their small size, so given the distance, their flight speed, and their metabolic rate, they must be extremely efficient and accurate in navigating to their landfall on the coast of South America.

Navigation

Homing and migrating animals find their way by several mechanisms of navigation. Piloting is a type of navigation, but it cannot explain the abilities of many species to take direct routes to their destinations through areas they have never experienced. Humans use two systems of navigation that differ in complexity: distance and direction navigation and bicoordinate navigation. Distance and direction means that you know the direction you must go in to reach your destination and you know how far away that destination is. So if you have a compass to tell you direction, and a means of measuring how far you

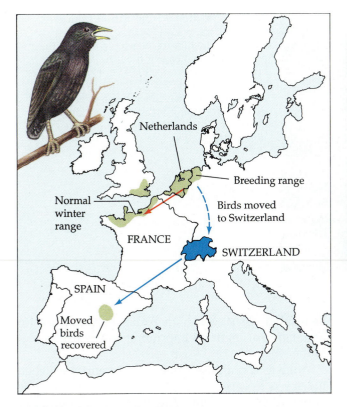

43.20 Navigation with a Compass
European starlings normally make a short winter migration in a southwesterly direction, from the Netherlands to coastal France and southern England (red arrow). Experimental populations of starlings moved to a site in Switzerland did not fly northwest to their traditional grounds, but followed the same southwesterly route (blue arrow), which took them to Spain.

43.21 Raring to Go
A bird ready to migrate shows migratory restlessness in a circular cage. The cage is lined with a paper funnel, and on the floor is an ink pad. The bird's feet mark the orientation of its activity.

training, no matter when they were fed, and even with the cage rotated between feedings, they always went to the food bin at the southern end of the cage for food (Figure 43.22). Next, the birds were placed in a room with a controlled light cycle and their circadian rhythms were phase-shifted. As an example, let us say that in the controlled light room the lights went on at midnight. After a couple of weeks the circadian clocks of the birds were phase-advanced by 6 hours. Now the birds were returned to the circular cage under natural light conditions with sunrise at 6 A.M. Because of the shift in their circadian rhythms, their endogenous clocks were indicating noon at the time the sun came up. Where should they look for food at noon if food was always in the south? In the direction of the sun, of course. So, at sunrise, they looked for food in the east food bin rather than the south food bin. The 6-hour phase shift in their circadian clocks resulted in a 90° error in their orientation. These types of experiments on many species have shown that animals can orient by means of a time-compensated solar compass.

(a) Pigeon placed in a circular cage from which it can see the sky (but not the horizon) can be trained to seek food in one direction, even when cage is rotated between trials

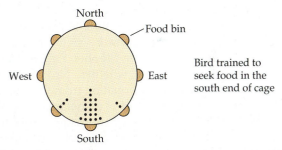

Bird trained to seek food in the south end of cage

(b) Pigeon placed on altered light–dark cycle and its circadian rhythm phase-advanced by 6 hours. Bird is then returned to training cage under natural sky

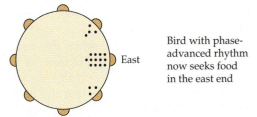

Bird with phase-advanced rhythm now seeks food in the east end

43.22 The Time-Compensated Solar Compass
(a) In the circular cage experiments, pigeons were trained to search for food in the south by filling only the southernmost food bin. Each dot represents a peck in search of food. (b) Birds whose circadian rhythms were phase-shifted forward by 6 hours oriented as though the dawn sun was at its noon position, searching for the south food bin in the east.

43.23 Directional Information from the Stars
As the Earth rotates over the course of the night, the positions of the stars appear to change, as shown in this long-exposure photograph taken at Kitt Peak Observatory in Arizona. In the northern hemisphere, there is one star—the North Star—that does not change position and therefore provides directional information.

Many animals are active at night, and many species of birds that are usually day-active migrate at night; how do they tell direction? Two sources of information about direction are available from the stars. If you look at the sky at different times of night, the constellations change position because Earth is rotating (Figure 43.23). So if you have a star map and a clock, you can tell direction from any constellation. There is one point in the sky, however, that does not change position throughout the night, and that is the point directly over the axis on which the Earth turns. In the northern hemisphere, a star called Polaris or the North Star is in that position. So, we can always tell which direction is north by sighting the North Star. Clever experiments by Stephen Emlen at Cornell University investigated whether birds use these sources of directional information from the stars. Emlen raised young birds in a planetarium—a large, domed room in which star patterns are projected on the ceiling. The star patterns in the planetarium could be slowly rotated to simulate the rotation of Earth. When the star patterns were not rotated, birds caught in the wild could still orient perfectly well in the planetarium, but birds raised in the planetarium under a still sky could not. If the star patterns in the planetarium were rotated each night as the young birds matured, then they were able to orient in the planetarium. This showed that birds can learn to use star patterns for orientation if the sky rotates. There was no evidence, however, that the birds used their circadian clocks to derive directional information from the star patterns. Experienced birds were not confused by a still sky or a sky that rotated faster than normal. The birds were orienting to the fixed point in the sky, the North Star. If young birds were raised under a sky that rotated around some other star, they would imprint on that star and orient to it as if it were the North Star (Figure 43.24). These studies showed that birds raised in the northern

hemisphere learn a star map that they can use for orientation at night by imprinting on the fixed point in the sky.

Animals cannot use sun and star compasses when the sky is overcast, but they still home and migrate under such conditions. Are there other sources of information they can use for orientation? In fact, there appears to be considerable redundancy in animals' abilities to sense direction. Pigeons home perfectly well on overcast days, but this ability is severely impaired by attaching small magnets to their heads. These experiments and subsequent ones with more sophisticated ways of disrupting the magnetic field around the bird have demonstrated a magnetic sense. Cells have been found that contain small particles of the magnetic mineral magnetite, but the neurophysiology of the magnetic sense is largely unknown. There are still other cues that animals can use. Many are sensitive to the planes of polarization of light, which can give directional information even under heavy cloud cover. Very low frequencies of sound can give information about coastlines and mountain chains. Weather patterns can also provide considerable directional information.

Much less is known about the mechanisms of bicoordinate navigation than about distance and direction navigation, but some animals definitely have the ability to sense their geographical locations and where they should go. In other words, they have a map sense. Distance and direction capabilities do you little good without a map. Information about longitude and latitude is available from natural cues, but the evidence that animals can or do use those sources of information is meager. Longitude can be determined by position of the sun and time of day: If the sun comes up earlier than expected, you must be east of home, and if the sun comes up later than expected, you must be west of home. Time and sun position also give information about latitude: At a given time

43.24 Star Patterns Can Be Altered in a Planetarium
A scientist has placed birds in orientation cages (see Figure 43.21) in a planetarium. By changing the positions or movements of the stars projected on the planetarium ceiling, he can investigate what information the birds use to orient their migratory restlessness.

of day (in the northern hemisphere), if the sun is higher in the sky than expected, you must be south of home, and if it is lower in the sky than expected, you must be north of home. Other sources of information about longitude and latitude can come from sensing Earth's magnetic lines of force and from the positions of the stars. To pinpoint home using these sources of information, an animal would have to have extremely precise and accurate sensory capabilities that have yet to be demonstrated. Perhaps, however, pinpointing home is not required of an animal's bicoordinate navigational abilities. If an animal gets anywhere near home, piloting can take over, and piloting can use a variety of long-distance cues such as coastlines, smells, and low-frequency sounds that have greater ranges than specific visual landmarks.

HUMAN BEHAVIOR

The behavior of an animal is a mixture of some components that are genetically programmed and others that can be molded by learning. Even learned behaviors, however, may have genetic determinants in terms of what can be learned and when it can be learned. Thus the behavior of a species, like its physiology and morphology, is shaped by natural selection. There are situations in which natural selection favors closed behavior programs, others in which open behavior programs are favored, and still others

in which a mixture of fixed and learned behavioral components is the optimal adaptation. Given these considerations, how would we characterize human behavior?

No one would question the fact that an important characteristic of human behavior is the extent to which it is modifiable by experience. That is the basis for human culture, the hallmark of our species. Nevertheless, the structure and many functions of our brain are coded in our genome, and therefore, so are some aspects of our behavior, including drives, limits to and propensities for learning, and even some motor patterns. Biological drives such as hunger, thirst, sex, and sleepiness are inherent to our nervous systems. Is it reasonable, therefore, to expect that emotions such as anger, aggression, fear, love, hate, and jealousy are solely the consequences of learning? Just as our sensory systems enable us to make use of certain subsets of information from the environment, so the structure of our nervous system makes it more or less possible to process certain types of information. Consider, for example, how basic and simple it is for an infant to learn spoken language, yet how many years that same child must struggle to master reading and writing. Finally, there is evidence that some motor patterns are programmed into our nervous systems. Studies of diverse human cultures from around the world reveal basic similarities of facial expressions and body language in human populations that have had little or no contact with one another. Infants born blind smile, frown, and show other facial expressions at appropriate times even though they have never observed such expressions in others. It in no way detracts from the value we place on the learning abilities of humans to also acknowledge that our behavior has been shaped through evolution, even though that shaping may be more in terms of broad outlines than fine detail.

SUMMARY

Behavioral adaptations are essential for survival and reproductive success, and are shaped by natural selection. Many behaviors of many species are genetically determined and expressed without prior experience. Such a behavior is called a fixed action pattern. Deprivation experiments are a means of demonstrating whether or not a given behavior is a fixed action pattern. A fixed action pattern may not be expressed in a deprivation experiment, however, if the animal is not in the appropriate stage of development or motivation and if required stimuli are not present. A releaser is a stimulus that elicits a fixed action pattern, and the parameters of a releaser may be a simple subset of the information available. Supernormal releasers can be produced by exaggerating the essential features of a releaser. Fixed action patterns are adaptive in situations where there are no opportunities to learn, where it is possible to learn the wrong behavior, and where mistakes are costly and dangerous. So, fixed action patterns are common in species with short lives and non-overlapping generations, in courtship behavior, in predator avoidance, and in dealing with dangerous prey. Genetic experiments on animals with fixed action patterns demonstrate the heritability of behaviors, the fact that artificial selection can change behaviors, and some possible molecular mechanisms connecting a gene to a behavior.

Learned behavior may be shaped by natural selection in terms of what can be learned and when it can be learned. Bird song provides examples that show a continuum from genetically determined to learned behaviors, with most species showing interactions between genetic determinants and learning. Many species must hear their species-specific song during a genetically determined critical period while they are nestlings if they are to express that song when they become sexually mature. They must also be able to hear themselves sing during the time they are developing song expression.

The evolution of a behavior can be studied in terms of a cost–benefit analysis. Every behavior has a cost and the benefits in terms of reproductive success must outweigh the costs if it is to be favored by selection. Aggression is a costly behavior and is favored only if it can secure limited resources at a reasonable cost. Territoriality is aggressive behavior in defense of space, and different types of territorial behavior reflect what resources are limiting and are economically defensible.

Mechanistic studies reveal the parameters of the behavioral abilities of animals. Underlying daily rhythms of behavior are endogenous circadian rhythms that continue under constant conditions with free-running periods that are not exactly 24 hours. Photoperiodism and jet lag are phenomena associated with circadian rhythms. In mammals and birds, circadian clock mechanisms have been localized to specific brain structures. Species that migrate to or through the tropics and species that hibernate also have endogenous circannual rhythms that give them the ability to anticipate seasonal events.

Long-distance movements of animals require abilities to navigate, and such abilities are highly developed in some species. Piloting is the simplest form of navigation, but distance and direction navigation and bicoordinate navigational systems are also used by animals. Time-compensated solar compasses and star maps are used for directional information. Endogenous rhythms may be used for distance information. Animals are clearly capable of bicoordinate navigation, but the mechanisms are not known.

Although most human behavior is due to or influenced strongly by learning, its broader outlines are shaped by natural selection. Behavioral drives, aspects of emotions, and some motor patterns, such as those for basic facial expressions, may be genetically determined and built into the human nervous system, as are our propensities and abilities to learn certain types of information.

SELF-QUIZ

1. The building of a web by a spider is an example of:
   a. a fixed action pattern.
   b. a releaser.
   c. an open behavior program.
   d. imprinting.
   e. a learned behavior.

2. If courtship behavior is not seen in a deprivation experiment, you can conclude:
   a. the animal is not sexually mature.
   b. the animal has low sexual drive.
   c. it is the wrong time of year.
   d. the appropriate releaser is not present.
   e. None of the above

3. Which of the following statements about releasers is *true*?
   a. The appropriate releaser always triggers a fixed action pattern.
   b. A releaser is a simple subset of sensory cues available to the animal.

   c. Releasers are learned through imprinting.
   d. A releaser triggers appetitive behavior.
   e. An animal only responds to a releaser when it is sexually mature.

4. Which of the following statements about the genetics of behavior is *true*?
   a. About 20 genes control the courtship displays of male dabbling ducks.
   b. One gene can code for several neural signals involved in controlling a behavior.
   c. Genes for retrieving, pointing, and herding have been described in dogs.
   d. A single gene causes love birds to carry nesting material tucked in their tail feathers.
   e. Hygienic behavior in bees has been shown to be due to two dominant genes.

5. Which of the types of territorial behavior maximizes the reproductive success of the females of the species that displays that type of territoriality?
   a. Type A
   b. Type B
   c. Type C
   d. Type D
   e. All of the above

6. If the sun were to come up earlier than expected on the basis of a circadian rhythm:
   a. it could cause symptoms of jet lag.
   b. it could phase advance the circadian rhythm.
   c. the animal could be east of home.
   d. it could entrain the circadian rhythm.
   e. All of the above

7. To have the ability to pilot, an animal must:

a. have a time-compensated solar compass.
b. orient to a fixed point in the night sky.
c. be able to know distance between two points.
d. know landmarks.
e. know its longitude and latitude.

8. Birds that migrate at night:
a. inherit a star map.
b. determine direction by knowing the time and the position in the sky of a star constellation.
c. orient to the fixed point in the sky.

d. imprint on one or more key constellations.
e. determine distance, but not direction, from the stars.

9. The most likely explanation for the observation that humans from entirely different societies smile when they greet a friend is:
a. they share a common culture.
b. they have imprinted on smiling faces when they were infants.
c. they have learned that smiling does not stimulate aggression.
d. smiling is a fixed action pattern.

e. smiling is an open behavior pattern.

10. If (1) a bird is trained to seek food on the western side of a cage open to the sky; (2) the bird is removed from the cage and its circadian rhythm is phase-delayed by 6 hours; and (3) the bird is returned to the open cage at noon real time, in which direction will it seek food?
a. North
b. South
c. East
d. West

FOR STUDY

1. Critique this statement: Hygienic behavior of bees is controlled by two genes as demonstrated by hybridization and backcrossing experiments.

2. Photoperiod (daylength) can provide information about season (time of year), so why do some birds have circannual rhythms?

3. If you raised a songbird in a deprivation experiment and it did not sing the song of its species the following fall, what possible hypotheses could you formulate about this result and how could you test them?

4. In early spring male prairie grouse congregate in a small area of wide open grasslands and aggressively defend small territories where they strut around, display their feather patterns, and make booming sounds. What would you expect to be the behavior of female prairie grouse, and why? Discuss both mating and rearing of young.

5. Pick an animal (other than a human) that you think would have mostly open behavior patterns and another animal that you think would have mostly closed behavior patterns. What differences in their biological characteristics could account for the differences in their behavioral repertoires?

READINGS

Alcock, J. 1989. *Animal Behavior*, 4th Edition. Sinauer Associates, Sunderland, MA. A balanced textbook, recommended to readers searching for a good next step into the subject.

Emlen, S. 1975. "The Stellar-Orientation System of a Migratory Bird." *Scientific American*, August. Experiments on stellar-orientation mechanisms of birds done in a planetarium.

Gould, J. L. 1981. *Ethology: The Mechanisms and Evolution of Behavior*. W. W. Norton, New York. A detailed treatment of behavior from a physiological point of view.

Gould, J. L. and P. Marler. 1987. "Learning by Instinct." *Scientific American*, January. An article on the interactions of learning and instinct, focusing on bees and on bird song.

Gwinner, P. 1986. "Internal Rhythms in Bird Migration." *Scientific American*, April. Circannual rhythms play critical roles in long-distance migration.

Lorenz, K. 1958. "The Evolution of Behavior." *Scientific American*, December. An essay on the evolution of releasers.

Moore-Ede, M. C., F. M. Sulzman and C. A. Fuller. 1982. *The Clocks that Time Us*. Harvard University Press, Cambridge, MA. A well-written book on circadian rhythmicity that covers just about all aspects of the subject.

Scheller, R. H. and R. Axel. 1984. "How Genes Control an Innate Behavior." *Scientific American*, March. One gene codes for a number of neural signals.

Tinbergen, N. 1960. *The Herring Gull's World*. Doubleday, Garden City, NJ. A delightful account of the behavior of one species from the pen of one of the founders of modern ethology.

Tinbergen, N. 1952. "The Curious Behavior of the Stickleback." *Scientific American*, December. A classic study of releasers and fixed action patterns.

PART SEVEN

Ecology and Biogeography

44

Behavioral Ecology

PREVIEW: Ecology is the study of relationships between organisms and their environments. Interactions with conspecifics form a major part of these relationships. Animals may improve their survival and reproductive success by joining together in groups if they thereby gain better foraging success and better protection against predators. However, group living also exposes them to diseases, competition, and social interference. These costs and benefits of social living vary with age, sex, experience, and the physical condition of individuals. Social groups are maintained by chemical, visual, auditory, and tactile signals. Complex social systems have evolved from offspring remaining with parents to help rear future broods and from cooperation among adults of the same generation. The type of social system an animal has is influenced by its diet and by the habitat in which it lives.

This chapter deals with the effects of group living on survival and reproductive success, the roles of the sexes, types of communication signals and their evolution, choice of associates, and the evolution of animal societies.

Ecology, the subject of Part Seven, is the branch of biology that investigates the interactions of organisms with one another and with their environments. The term **environment**, as used by ecologists, includes all factors, both physical and biological, that influence the life of an individual. Physical factors include water, nutrients, sunlight, temperature, and wind. Biological factors are other organisms—both those of the same species (conspecifics) and those belonging to other species—with which the individual interacts. Note that plants are part of an animal's environment, and vice versa. Because species differ in their characteristics and live in many different environments, their interactions with their physical and biological environments differ greatly. An environmental factor that exerts a strong influence on members of one species may not have any influence on another species.

Interactions between organisms and their environments are two-way processes: Organisms both influence and are influenced by their environments. Indeed, managing environmental changes caused by our own species is one of the major problems of the modern world. For this reason, ecologists are often asked to help analyze causes of environmental problems and to assist in finding solutions for them. However, it is important not to confuse the *science* of ecology with the term "ecology," which is often used in popular writing to refer to the functioning of nature. Pollution does not destroy the "ecology" of an area, but it may well destroy or damage important ecological processes.

Ecologists study patterns of distribution and abundances of organisms to determine how those patterns are established and maintained and how they change over short and long time periods. From its roots in descriptive natural history, ecology has developed into a complex field of inquiry dealing with levels of organization ranging from the relationship between an individual organism and its physical and biological environments to the structure of communities and ecosystems. This complex subject can be approached in many different ways. We begin by discussing how animals are organized socially to improve their survival and reproductive success. Then we turn to the structure and dynamics of populations whose members all belong to the same species. In later chapters we add levels of complexity and discuss the functioning of all the species living in a region and their interactions with the physical environment.

All sexually reproducing organisms, including solitary ones, interact to exchange gametes. Among many sessile marine organisms, the gametes simply associate in the water, but most mobile animals get together to copulate, even if they part immediately thereafter. From these brief and simple interactions, social organization evolved, becoming more elaborate in some lineages and culminating in the complex societies of vertebrates and social insects. Within social systems, individuals compete for opportunities

1009

to reproduce, but they may also cooperate to defend one another from predators.

COSTS AND BENEFITS

To find out how animals are organized socially, ecologists study the behaviors of individuals. Ecologists often analyze their observations of animal behavior in terms of "costs and benefits." Such analyses are based on the principle that an animal has only a finite amount of time and resources to apportion among different kinds of acts. A behavioral act may be costly to the animal performing it, but it may also confer benefits on the animal. The **energetic cost** of a behavior is the difference between the energy the animal would have expended had it rested and the energy expended in performing the behavior. The **risk cost** of a behavior is the increased chance of being injured or killed as a result of performing it, compared with resting. The **opportunity cost** is the sum of the benefits the animal forfeits by not being able to perform other behaviors during the same time interval; for instance, to defend its mate or territory, an animal must stop feeding.

An animal is not expected to perform any behavior when the sum of these costs is greater than the sum of the **benefits**—the improvements in survival and reproductive success that the animal achieves by performing the behavior. It is very difficult to measure these costs and benefits directly, but a great deal can be learned about them by observing how behavior changes when environmental conditions change. Ecologists interested in animal behavior observe the actions of animals to determine which costs and benefits are most important to them. Ecologists do not believe the animals are making conscious calculations of costs and benefits, but rather that, over many generations, natural selection has molded the behavior in accordance with costs and benefits.

Social behavior evolves when individuals that join with other conspecific individuals in groups survive better and produce more offspring than solitary individuals. The key to the evolution of social behavior is *differential reproductive success*, so throughout this chapter we will compare the reproductive successes of individuals who engage in different types of social behavior. However, determining the effects of group living on survival and reproduction is not easy. A given behavior may be advantageous for individuals of one species but disadvantageous for those of another species, or even for the same individuals at a different time or place.

Some Benefits of Group Living

Some animals can improve their chances of obtaining food by foraging (seeking food) in a group. For example, by hunting together, animals of some species are able to capture prey that would be too large for any one of them to handle. Other animals cooperate to maneuver prey into places where they are easier to catch. Cooperative hunting was a key component of the evolution of human sociality. By hunting in groups, our ancestors were able to kill large mammals they could not have subdued as individual hunters, and to defend their prey and themselves from other carnivores (Figure 44.1).

44.1 The Earliest True People
Having killed their prey, a band of *Homo habilis* drives rival predators—spotted hyenas and sabertooth cats—from a fallen dinothere (an extinct relative of modern elephants).

44.2 Defensive Postures of Musk Oxen
This compact circle of formidable adults encloses young
oxen.

Individuals of many species are better protected
from predators if they live in groups. Predators may
be able to find a group of animals more easily than
they could a solitary animal, but a group may defend
itself better once it is found. When attacked by
wolves, musk-oxen form a circle with the young animals inside (Figure 44.2). It is difficult for wolves to
penetrate the barrier of large heads and massive
horns of the adult animals. Many small birds, when
attacked by a hawk, form tight flocks; the hawk,
meanwhile, attempts to isolate one group member
from the rest. Clumping deters the hawk because it
risks injury if, while attempting to penetrate the tight
group, its wing were to hit one of the prey. Also,
single individuals are difficult to follow in a rapidly
moving group. To investigate the importance of
flocking as an antipredator adaptation, scientists released a trained goshawk near flocks of wood pigeons
in England and measured the percentage of attacks
that were successful. They found that the hawk was
most successful when it attacked solitary pigeons and
was increasingly less successful in capturing a pigeon
as the number of pigeons in the flock increased. The
main reason was that larger flocks reacted sooner to
the hawk's approach (Figure 44.3).

Some Costs of Group Living

An almost universal cost associated with group living
is higher exposure to diseases and parasites. Long
before the causes of diseases were known, people
sensed that association with sick persons increased
their chances of contracting illnesses. Quarantine has
been used as a means of combating the spread of
illness for as long as we have had written records.
The diseases of wild animals are not well known, but
most of those that have been studied are spread by
close contact. Nests of bank swallows in large,
crowded colonies are more heavily infested with fleas
than nests in smaller colonies.

Other costs of group living, like the benefits, depend on the circumstances. Grouped individuals may
compete for food, interfere with one another's foraging, injure one another's offspring, or inhibit one
another's reproduction. The effects of group living
on the survival and reproductive success of an individual also depend on its age, sex, size, and physical
condition. A large, dominant male that controls access to all the females in a group greatly increases
his reproductive success by living in the group; in
contrast, a small, subordinate male that cannot mate
at all may survive better in the group, but he has no
reproductive success. An infant who would quickly
die if it were separated from its mother benefits differently from group living than a young animal capable of surviving on its own. *Social groups are organized collections of individuals whose survival and
reproductive success are influenced very differently by
group living.*

TYPES OF SOCIAL ACTS

Individuals living in social groups perform many different types of acts. These acts can be grouped into
four categories according to their effects on individuals (Table 44.1). An **altruistic act** confers a benefit

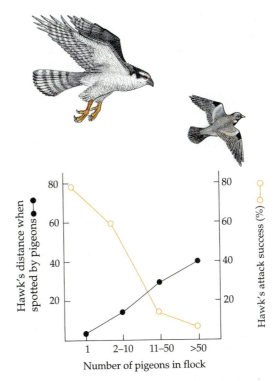

44.3 Flocking Gives Protection from Predators
Goshawks are less successful when they attack wood pigeons in flocks than when they attack solitary pigeons.

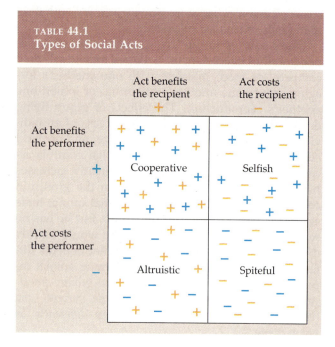

TABLE 44.1
Types of Social Acts

on some other individual at a cost to the performer. A **selfish act** benefits the performer but inflicts a cost on some other individual. A **cooperative act** benefits both individuals, and a **spiteful act** inflicts costs on both. These terms are purely descriptive; they do not imply any conscious motivation or awareness on the part of the performer.

The relative costs and benefits determine the conditions under which behaviors can evolve. Many current studies of the social behavior of animals attempt to measure the relative costs and benefits of social acts, how the effects of the acts are distributed among the individuals of the group, and how the individuals are related genetically. To see how cooperative and selfish acts could evolve into patterns of behavior, we need only recall from Chapter 18 that an individual's **fitness** is its reproductive contribution to successive generations relative to the reproductive contributions of others in its population. If there is a genetic basis for a cooperative or a selfish act, and if performing it increases the fitness of the performer, the genes governing the act will increase in frequency in the lineage—that is, cooperative or selfish behaviors will evolve.

It is not simple to see how altruistic behaviors could evolve. In fact, this has been the subject of a lively debate, ongoing since the 1970s, among biologists interested in animal behavior. How could an act that *lowers* the performer's chances for survival evolve into a behavior? The key lies in genetic relatedness: Altruistic behaviors can evolve when performers and recipients are genetically related. As was pointed out in Chapter 18, an individual may influ-

ence its fitness in two different ways. First, it may produce its own offspring, contributing to its **individual fitness**. Second, it may help the survival of relatives that bear the same alleles due to their descent from a common ancestor (see Box 18.A). This process is called **kin selection**. Together, individual fitness and kin selection determine the **inclusive fitness** of the individual. Altruistic acts may evolve into altruistic behaviors when the benefits of increasing the reproductive success of related individuals exceed the cost of decreasing the altruist's own reproductive success; that is, when they increase the altruist's inclusive fitness. Box 44.A shows how the relatedness of an individual is determined.

As we have seen in Chapter 18, cooperative behaviors can evolve among related individuals. But unrelated animals—even members of different species—have been observed giving warnings of danger, sharing food, and grooming each other. How can the evolution of these cooperative behaviors be explained? The model that has been proposed as an answer is called **reciprocal altruism**. According to the model, acts of reciprocal altruism evolve if the performer is, in turn, the recipient of beneficial acts from the individuals it has helped. If there is a genetic basis for the acts, natural selection may increase the frequency of the alleles governing the cooperative behavior.

Individual Adjustments to Group Living

Males and females generally pay different costs and derive different benefits from group living. Males of most species increase their reproductive success by mating with a large number of females, whereas females are unlikely to increase their reproductive success by increasing the number of males with which they mate. One male provides a female with enough sperm for a very large number of offspring—usually many more than the number of eggs the female can produce, or the number of young she can nourish. Therefore, males of many species attempt to increase the number of females with which they copulate, and they fight with one another for access to females. Females, on the other hand, usually are more discriminating in their choice of mates, and they resist most attempts by males to copulate with them.

Individuals may be larger or smaller than average for their age and sex. Variation in skills, competitive abilities, and attractiveness to potential mates is often associated with these size differences. The largest males of *Centris pallida*, a solitary bee of the American Southwest, are three times the size of the smallest males (Figure 44.4). These size differences, although they are environmentally determined, are fixed for an individual's life. Large males search for females about to emerge from their buried pupae. When they find a female, they dig her up and copulate with her,

BOX 44.A

Calculating the Coefficient of Relatedness

Suppose that two cousins in your family share some trait such as eye color not only with each other but also with the grandfather through whom they are related. The cousins may both have copies of an eye-color allele inherited from their grandfather through their parents. Such alleles are said to be identical by descent.

The **coefficient of relatedness** r is the probability that an allele in one individual is an identical copy, by descent, of an allele in another individual. To calculate r we construct a diagram showing the individuals concerned and their common ancestors, linked across generations by arrows. At each generation link, meiosis takes place so that the probability that a copy of an allele gets passed on is 0.5. For k generation links, the probability is $(0.5)^k$. To calculate r we sum this value for all possible pathways between the two individuals; that is, $r = \Sigma(0.5)^k$.

Some specific examples are diagramed here. Values of r are calculated between two individuals, represented by solid circles. Other relatives are indicated by open circles. The generation links used in the calculations are represented by solid lines. The dashed lines represent other links in the pedigree. In the cases of full siblings and of cousins, identical genes can be inherited by two pathways, from mother or father.

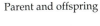
Parent and offspring

$r = (0.5)^1$
$= 0.5$

Grandparent and grandchild

k_1

k_2

$r = (0.5)^2$
$= 0.25$

Sibling

$k_1 \quad k_2 \quad k_3 \quad k_4$

$r = (0.5)^2 + (0.5)^2$
$= 0.5$

Full cousins

$k_1 \quad k_2 \quad k_5 \quad k_6$

$k_3 \quad k_7 \quad k_4 \quad k_8$

$r = 2(0.5)^4$
$= 0.125$

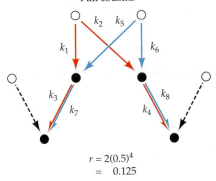
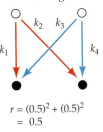
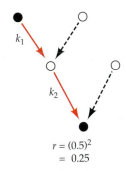

but the digging takes several minutes, and it often attracts other males. These newly arrived males fight with the original one, and the largest male usually wins. If a small male searched and dug for buried females, he would probably just serve as a female-finder for larger males. Instead, smaller males patrol potential pupation sites and wait for females that emerge without being discovered by large males. Intermediate-sized males sometimes dig and sometimes patrol. Large males inseminate more females than small males do, but the small males probably get more matings by patrolling than they would by digging.

Individuals may adjust their social behavior when there are changes in the environment. For example, small birds gain protection from predators by forming flocks, but individuals in flocks interfere with one another's foraging. Individuals of some species change their flocking patterns when the weather or the risk of predation changes. For example, yellow-eyed juncos, small seed-eating finches, fight with one another over food. Dominant individuals usually defend an area from which they attempt to exclude all

44.4 Size Variation in Male Bees
The small male on the left is waiting for females to emerge rather than attempting to dig them from the soil.

other juncos. Juncos also spend time watching for predators, principally by looking up and scanning. When they are fighting or scanning, they cannot look for food. When they are feeding or fighting, they are vulnerable to surprise attack by a hawk. The best way to avoid predators—watching all the time—would lead to starvation. The risk of surprise attack decreases as the number of juncos in a flock increases because, even if each individual spends only a small amount of time scanning, the chance that at least one bird is scanning at any moment increases. The larger the flock, the less each bird looks up, and the more time each bird spends feeding. But the larger the flock, the more time each bird also spends fighting.

On cold days, the juncos need to eat more than they do on warm days, so they have less time to defend space and scan for predators. On cold days, wintering junco flocks in Arizona contained an average of 7 birds, whereas on warm days they contained an average of only 2 birds. When a trained hawk was flown over the canyon in which the juncos were feeding, mean flock size increased from 3.9 birds before the hawk flew by to 7.3 birds afterward. That is, the birds tolerated more interference with foraging when the need for protection from predation was higher.

ROLES OF THE SEXES

Parental investment is an act of parental care that increases the chances of survival of an offspring but reduces the ability of the parent to produce additional offspring. It may also lower the chances of survival of the parent itself, because the parent could have used the time spent in parental investment to engage in other activities that would improve its own status.

Males and females differ in the kinds and amounts of parental investment they can make. Males produce tiny sperm; females produce much larger eggs. Males and females also differ in the type of care they are able to give to fertilized eggs and to offspring. In turn, these differences in the kinds of parental investment males and females can make strongly influence the types of social systems that evolve in different animal groups. Females of most species invest more energy in the production of each gamete than do males which is why females are usually the choosier sex. However, one sex does not always invest more than the other.

The reproductive roles of males and females are different in different animal groups. For example, only female mammals have functional mammary glands; male mammals cannot produce milk and generally contribute nothing to the feeding of offspring. However, birds do not produce milk; among birds there is no part of reproduction, except for laying eggs, that cannot be performed equally well by males

44.5 Male Damselfish Guarding Young
A male damselfish (*Acanthochromis acanthus*) guards a brood of silvery young, which may number up to 300. The young are vigorously defended by both parents for as long as 48 days.

and females. As you might expect from this fact, biparental care is prevalent among birds. Sex role specialization in birds, when it exists, is due to competition among individuals for reproductive opportunities rather than to the inability of individuals to perform other behaviors.

Sex roles among fishes differ from those in birds and mammals because most species of fishes do not feed their young. Parental investment consists primarily of guarding eggs and young from predators (Figure 44.5). In many fish species, males are the primary guarders. A male can guard a clutch of eggs while attracting additional females to lay eggs in his nest. On the other hand, a female can produce another clutch of eggs sooner if she resumes foraging immediately after mating, rather than spending time guarding eggs.

Among species with parental care, which parent provides it is also influenced by the effectiveness of care by one versus two parents, how exhausted the female is at the end of egg laying, and the male's likelihood of having fathered the offspring. In animals with internal fertilization, whether a certain male's sperm fertilized a certain female's eggs depends on when they copulated and whether she copulated with other males during her fertile period. In animals with external fertilization, only those sperm released at the same time the female's eggs are laid can fertilize them; consequently, a male attending a female when she spawns has a high probability of having fathered her offspring. Male parental behavior in fishes and amphibians—groups that include species with both internal and external fertilization—illustrates the influence of reliability of paternity on the evolution of male parental behavior. Male

TABLE 44.2
Relationship between Mode of Fertilization and Parental Care in Fishes and Amphibians

| | NUMBER OF FAMILIES WITH MALE PARENTAL CARE | | NUMBER OF FAMILIES WITH FEMALE PARENTAL CARE | |
	INTERNAL FERTILIZATION	EXTERNAL FERTILIZATION	INTERNAL FERTILIZATION	EXTERNAL FERTILIZATION
TAXON				
Fishes	0	48	15	21
Amphibians	2	13	11	7
Total	2	61	26	28

parental care is much more prevalent in species with external fertilization than in species with internal fertilization (Table 44.2). If there is unequal parental investment by the two sexes, it is nearly always the male who does more among fishes, but the female who does more among mammals and birds.

While an individual is investing in one group of offspring, it may not be able to begin investing in another group. An individual is expected to stop investing in one group of offspring when the costs to that group of its doing so are less than the benefits likely by investing in the next group of offspring.

(a)

(b)

44.6 Some Female Birds are Brighter than Males
Female red phalaropes (*Phalaropus fulicarius*) are larger and more brightly colored than males. (a) A dull-plumaged male has taken a break from incubation in order to feed himself. (b) The more brightly colored female, once she has laid her eggs, abandons the male.

Among most birds, it is difficult for one parent to rear a brood of nestlings successfully. If either parent deserts, it is unlikely that parent will improve its fitness because the survival of offspring may be much lower if only one parent attends them, and opportunities for additional matings may be poor for individuals of both sexes.

However, special conditions can allow an individual bird to increase its fitness with little or no parental investment. The conditions for the evolution of desertion by females are especially stringent, but they have been met in a few species of sandpipers with reversed sex roles. The females of these species are brighter than the males, unlike those of most bird species, in which, if there is a plumage difference, males are brighter than females (Figure 44.6). Females of these sandpiper species compete with one another for males, lay eggs, and then leave the incubation of the eggs and the care of the young to the males. These species lay no more than three or four large eggs, which hatch into precocial young that are already able to walk around and feed themselves. One parent can guard these few offspring and lead them to good foraging sites. Therefore, females improve their reproductive success by terminating parental care early and gathering more energy to lay additional clutches of eggs.

COMMUNICATION: MAINTENANCE OF SOCIAL GROUPS

The functioning of social groups depends on mutual adjustments of the members' behavior. Sometimes animals simply adjust their behavior by observing and responding to the behavior of others in their group. However, communication usually improves group cohesion. A **display**, or **signal**, is a behavior that has evolved to influence the actions of other individuals. We know that scanning for predators by juncos in flocks is not a display, because solitary juncos look up in the same way. The behavior has evolved because it affords protection, not because it

TABLE 44.3
Characteristics of Different Modes of Communication

| | MODE OF COMMUNICATION | | | |
CHARACTERISTIC	CHEMICAL	AUDITORY	VISUAL	TACTILE
Range	Medium to long	Medium to long	Medium to long	Short
Rate of change of signal	Slow	Fast	Fast	Fast
Flows around barrier	Yes	Yes	No	Yes
Locatability	Variable	Medium	High	High
Useful at night	Yes	Yes	No	Yes
Energetic cost	Low	High	Low	Low
Risk of performing	Low to medium	Medium	High	Low

communicates to the group; the looking up has not been exaggerated into a display. The head-forward posture juncos use when fighting over food, however, is a display. Solitary juncos never use it. The distinction is important because to explain displays, we need to understand why the exaggeration evolved.

Information is sent and received by means of one or more sensory modes—chemical, visual, auditory, and tactile (see Chapter 37). The properties of each of these modes and the ways they are used in communication are summarized in Table 44.3.

Chemical Communication

The two main types of chemical communication are contact communication, known as **taste**, and distance communication, known as **smell**. You may wish to review the material on taste and smell in Chapter 37. Smells, but not tastes, originate at some distance from the receiving individual, but in practice they may be difficult to distinguish. Foods cannot be "tasted" as strongly if we have a cold and our nostrils are clogged, indicating that much of our taste sensation is really a combination of taste and smell. Chemical communication is the most widespread and most ancient form of communication. However, it is harder for us to study chemical communication than visual or auditory communication because humans have such a poor sense of smell. Many animals are sensitive to lower concentrations of chemicals than we can detect with our best machines.

Chemical signals among individuals of the same species are called **pheromones**. The communication between individuals by pheromones is analogous to the communication between cells by hormones. The chemical molecules used in communication are relatively small and are energetically inexpensive to produce. A few molecules of a chemical signal can influence a great deal of behavior, and the variety of signals that can be produced is large. Even with our poor olfactory abilities, humans can recognize over 10,000 odors. Because it takes a long time for pheromones to diffuse until they can no longer be detected, chemical signals can be broadcast over great distances, for very little cost, and can last a long time. Animals can locate signal sources by detecting concentration gradients, but the long persistence of chemicals makes them a poor means of communicating rapid changes in position.

Chemical signals are used in trail marking (by ants and stingless bees), in territory marking (Figure 44.7a), and in sexual stimulation. Many female insects attract males from long distances with chemical sex signals (Figure 44.7b). Males of some insects, such as bumblebees, deposit scent on vegetation and then patrol the area for females attracted by the deposits. Some bee species recognize hive members by their odors and attack bees that have the wrong odors.

Visual and Auditory Communication

Many animals use visual communication. Visual signals are easy to produce, an endless variety of signals is possible, and the position of the signaler is automatically indicated. Because visual signals can be changed very rapidly, they are excellent for indicating rapid shifts in position or mood. Most animals can receive visual information, although the sharpness of vision varies widely among animals. Among fishes, gills and fins are especially important for visual signaling, but among terrestrial vertebrates, including humans, the head and face are most important. Relatively minor alterations in these regions influence the responses of conspecifics more than major alterations elsewhere.

Some animals communicate by means of auditory signals. Animals use vocalizations to identify species and individuals, to maintain contact among members of groups, to communicate the emotional states of the callers, and to warn of the presence of predators. However, they may be very costly to produce. Some

(a)

44.7 Communicating Chemically

Many animals communicate with other members of their own species using chemicals, called pheromones. *(a)* To mark his territory, this male cheetah is spraying pheromonal secretions from a scent gland in his hindquarters onto a tree. Other cheetahs passing the spot will know that the area is "claimed." *(b)* A female gypsy moth secretes the pheromone gyplure to attract males; the males sense the presence of gyplure via receptors in their antennae.

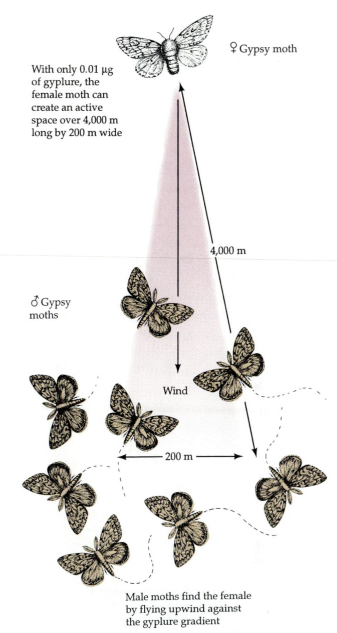

With only 0.01 µg of gyplure, the female moth can create an active space over 4,000 m long by 200 m wide

♀ Gypsy moth

4,000 m

♂ Gypsy moths

Wind

200 m

Male moths find the female by flying upwind against the gyplure gradient

(b)

frogs expend as much energy calling as they do hopping. Mammals and birds produce most of their sounds by forced movement of air across internal vocal cords or other structures whose properties are controlled by a simple set of muscles. Fishes produce sounds mainly with their swim bladders or by vibrating one part of their skeleton against another, such as grinding their teeth or vibrating their vertebrae. Only a few insects produce sounds that serve as signals. Cicadas have a complex drum that they vibrate with special sets of muscles, and crickets and grasshoppers produce sounds by rubbing one part of their body against another. However, the majority of animal species are mute and do not communicate by means of sound.

To study the vocalizations of animals, scientists use a sound spectrograph, a machine originally invented to assist deaf people in learning to speak, to help them analyze auditory signals. The sound spec-

trograms produced by these machines are "pictures" of sounds showing the pitch, loudness, and duration of the components of the sounds (Figure 44.8).

Tactile Communication

Tactile signaling—communication by touch—is especially important for animals with poor vision or for those living in places where conditions are unsuitable for visual communication. Many deep-sea fishes, living in perpetual darkness, have highly elaborate "feelers" used for both prey detection and sexual signaling (Figure 44.9). Many spiders, particularly web-building species that have poor vision, use tactile mating signals. A male spider usually seeks out a female on her web. It is dangerous for him to

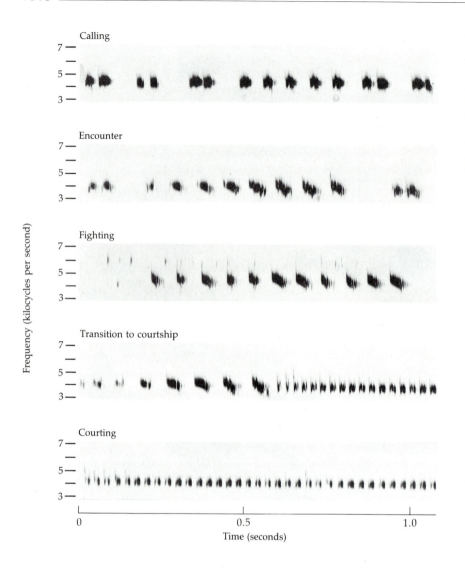

Calling

Encounter

Fighting

Transition to courtship

Courting

Frequency (kilocycles per second)

Time (seconds)

44.8 Calls of a Cricket
By changing the intensity and rate of repetition of its calls, a cricket can change from a pattern typical of general calling (top) to that typical of an encounter, or to the even more rapid repetition of calls accompanying fighting. Should a male cricket discover that the individual he has encountered is a female, he may switch to the rapidly trilled courtship call.

approach her, however, because she normally detects prey in her web by the vibrations they produce. Approaching males might be mistaken for prey, and might become meals rather than mates. This risk has favored the evolution of elaborate plucking rituals during which the male produces vibrations distinct from those caused by struggling prey in the web.

The most highly evolved communication signals with an important tactile component are the dances of honeybees, carefully studied by Karl von Frisch and his associates in Germany. Dancing bees make sounds and carry odors on their bodies, but they convey a great deal of information by movement, which is monitored by other bees that follow and touch the dancer. When a foraging bee finds food, she returns to the hive and communicates her dis-

44.9 Feeling in the Dark
This anglerfish ambushes its prey with a lightening-fast gulp. By extending its rays, the fish exposes sensory filaments that cover its body and provide the fish with a great sensitivity to low-frequency vibrations.

covery by dancing in the dark on the vertical surface of the honeycomb. If the food is less than 80–100 meters from the hive, she performs a **round dance,** running rapidly in a circle, reversing her direction after each circumference. The odor on her body indicates the flower to be looked for, but the dance contains no information about the direction in which to go. If the food source is farther than 80 meters, she performs a **waggle dance,** which conveys information about both the direction and the distance of the food source. In the waggle dance, a bee repeatedly traces out a figure-eight pattern as she runs on the vertical surface. She runs in a short straight line while vigorously wagging her abdomen, then alternately makes half circles to the left and the right to begin the waggle run again. The angle of the straight

line indicates the direction of the food source relative to the direction of the sun, as Figure 44.10 shows. The speed of the dancing indicates the distance to the food source: the farther away it is, the slower the waggle run. The dances of honeybees are unusual because they are based on an arbitrary convention: straight down could just as well indicate the direction of the sun as straight up. Arbitrary, symbolic conventions like this have been developed to an extreme degree in human language.

ORIGINS OF COMMUNICATION SIGNALS

Charles Darwin was the first person to give serious attention to the problem of the origin and evolution

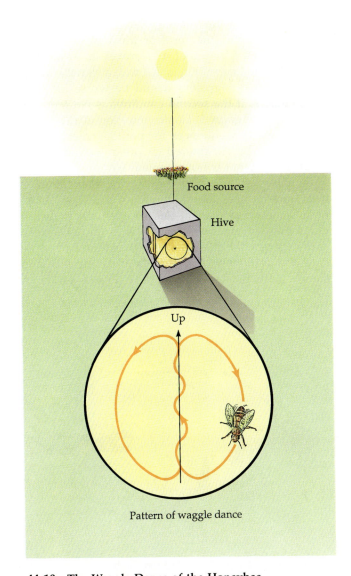
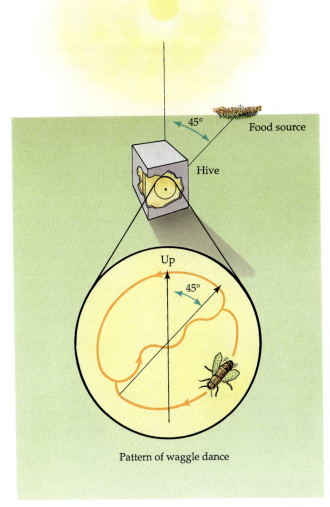

44.10 The Waggle Dance of the Honeybee
By running straight *up,* a honeybee tells her hive mates that there is a food source in the direction of the sun and at least 80 meters from the hive. If she knew of a food source in the opposite direction from the sun, she would begin each phase of her dance by running straight *down.* When her waggle dance is run at an angle, the other bees know that the same angle separates the direction of the food source from the direction of the sun.

of communication signals. In 1872 he published the results of his detailed studies in a book entitled *The Expression of the Emotions in Man and Animals*, but unlike his earlier books, *The Voyage of the Beagle* and *The Origin of Species*, this one attracted little attention. Nonetheless, Darwin identified several important means by which communication signals originate.

Intention Movements

An important source of communication signals pointed out by Darwin is preparation for future behavior. A bird about to take off usually performs certain preparatory behaviors known as **intention movements**: it sleeks its feathers, flexes its legs to lower its body, and raises its wings slightly at the shoulders. These movements prepare it for takeoff, but they also inform other birds what the individual is about to do. These intention movements are not signals, but they may become elaborated and modified by natural selection into communication signals. The head-forward threat displays of juncos and other birds, in which an individual faces its opponent and slightly raises its wings at the shoulders, are believed to have evolved from intention movements of flight. Many visual displays are thought to have arisen from such intention movements.

Autonomic Responses

A second source of signals comes from **autonomic nervous system responses** (Chapter 36). The action of the autonomic nervous system during moments of excitement produces many effects that are visible to another individual. For example, an animal confronted with a dangerous situation stops digesting the food it has eaten. Its heart speeds up, blood rushes to its brain and skeletal muscles, its blood pressure increases, its breathing becomes quicker and deeper, and its temperature-regulating mechanisms are strongly activated. Those changes, in turn, result in readily visible responses, such as defecation, urination, perspiration, and the erection of fur and feathers, which can evolve into communication signals. Many mammals, for example, urinate during territorial boundary disputes. The erection of fur and feathers that are almost universal components of social signals among both birds and mammals probably arose from such autonomic nervous system responses.

Displacement Activities

A third type of behavior that evolves into displays is **displacement activities.** Many animals in situations of emotional conflict perform what appear to be irrelevant activities. People scratch their heads, bite their fingernails, light up cigarettes, or attack unin-

volved bystanders. In the spring, a male three-spined stickleback defends a small territory within which he constructs a small nest. At the boundary of his territory, his tendencies to attack a neighbor or to escape from him are equally strong. During boundary fights, males often perform a head-down threat display that closely resembles "nest building," but no nest building takes place. Motivational conflict and thwarted intentions provided the behavior that led to the evolution of this display.

EVOLUTION OF COMMUNICATION SIGNALS

Signals evolve for reasons that differ from their origins. Whatever its origin, if a movement is to evolve into a signal under the influence of natural selection, two important conditions must be met. First, a receiver must not be able to respond so as to make conditions worse for the signaler than if it had not signaled. Second, the receiver must be able to respond in such a way that it improves its own situation. If the first condition is not met, a signal will not evolve because signalers would be worse off than nonsignalers. If the second condition is not met, the signal will be ignored. Communication is a two-way process; the consequences for both signaler and receiver must be considered in order to understand what is and what is not signaled.

Signals evolve in part because the elaboration of behaviors reduces the ambiguity of messages. Elaborate signals may make it easier for females to distinguish males of various species. On tropical beaches, many species of fiddler crabs live together. Each male crab digs a burrow, then stands at its entrance and signals with his large front claw (Figure 44.11). A

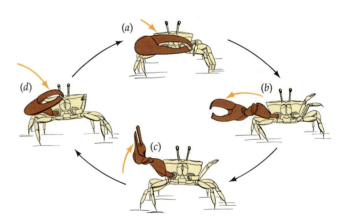

44.11 Visual Display of a Fiddler Crab
A lateral wave of his large right claw dominates the species-specific display of *Uca lactea*, a fiddler crab from Fiji. (*a*) The display begins with the claw flexed in front. (*b*) The claw is then unflexed outward, (*c*) raised upward, and (*d*) brought back to its original position.

female is attracted, enters the burrow, and copulates with him underground. The name "fiddler crab" derives from the movements of the large claws, which give the impression of bowing a violin. Each species of crab has its own pattern of claw movements. Experienced observers can recognize the species to which a male belongs at some distance by his movement patterns—and more importantly, so can female crabs.

A display also may signal clearly that the signaler is (or is not) about to attack, that it is (or is not) ready to mate, or that it is about to move to a different site. However, an individual does not always benefit by signaling its intentions clearly. It may be to the advantage of a signaler to communicate accurately its *ability* to attack without indicating *whether* it will do so because attacking and fighting are dangerous. For example, almost all red deer stags are injured at some time during their lives by fighting with other males; 20–30 percent of them become permanently disabled by being blinded in one eye or by having a leg broken; and they are unable to reproduce thereafter. Therefore, if a male can use a threat display to induce a challenger to retreat without a fight, he may improve his chances of survival and his reproductive success. An opponent is likely to retreat if he believes that if he fights, he will lose. Initially, both individuals may be uncertain about the outcome of a potential fight. Size and skill play a part, but so do chance events: one of the animals may slip or just happen to gain an advantageous position. But if one animal accurately signals its uncertainty, it may invite an attack. Therefore, threat displays may not communicate accurate information about an individual's real emotional state or future behavior (Figure 44.12), but they may signal something about its physiological potential to sustain a fight should one ensue.

CHOOSING ASSOCIATES

Social systems are complex, but they are based on a rather small set of choices by their members. Individuals choose which other individuals to associate with, how to interact with them, and when to leave them. Variations in how those decisions are made can result in an amazing variety of social systems. The most widespread choice of associates, made by members of most sexually reproducing species, is the choice of mating partner. An individual chooses a mate based on the other individual's traits, on the resources it holds, or on a combination of the two. Among those species in which individuals do not control any resources, the traits of the potential partner are the only criteria for mate selection.

Sexual Selection

Traits may evolve among individuals of one sex as a result of **sexual selection**, the spread of traits that confer advantages to their bearers in competition for mates or resources or during courtship. Individuals of one sex, usually males, may fight, display, or otherwise compete for control of resources, such as food or nesting sites. The winners then have exclusive access to the females that are attracted to those resources. Traits that improve success in courtship evolve as a result of the mating preferences of individuals of the opposite sex.

Elephant seals illustrate the results of intrasexual competition for breeding resources. Females come ashore on beaches where they give birth to their calves, nurse them to independence, and become pregnant again. Females congregate on a small number of high-quality pupping beaches, and males fight for possession of those beaches. Males that control

44.12 Roaring Threat Display of a Red Deer Stag
This red deer stag defends a harem of females by roaring at an approaching challenger.

44.13 A Male Demonstrates His Foraging Prowess
The male hanging fly on the left has just presented a moth to his mate, thus demonstrating his foraging skills. She feeds on the moth while they copulate.

good pupping beaches achieve most of the copulations. Fewer than 10 percent of male elephant seals ever copulate with females; of those that do, some inseminate more than 100 females during their lifetime. In contrast, nearly all females that survive to adulthood breed, and it is rare for a female to wean as many as ten pups in her lifetime. Because larger males generally defeat smaller ones, natural selection has favored large size in males, and thus the average male today is several times as large as the average female. The males have large canine teeth with which they fight. Their odd elephant-like snouts are resonating chambers that help exaggerate their threat displays, which can cause an opponent to back down without a fight.

Females of many species choose their mates on the basis of the resources they hold or can acquire. A female hanging fly will mate with a male only if he provides her with a morsel of food. The bigger the food item, the longer she copulates with him, and the more of her eggs he fertilizes (Figure 44.13). A female gains from this behavior because she obtains a better supply of energy for egg production. Many birds also engage in courtship feeding, in which males bring food to their potential mates. In some species there is a considerable amount of "breaking off of engagements" at these early stages. A female common tern is more likely to leave a male who provides infrequent meals for her than a male who feeds her more regularly. Males who bring food often during courtship will be better at providing food for young later in the season, and are, therefore, more desirable mates.

Males of some species do not bring food to females, but control areas where food or other important resources are available. The best-known cases involve **territorial behavior**, the defense of an area that contains food, nesting sites, or other resources. If a male controls sufficient resources, he may attract more than one female, a phenomenon known as **resource defense polygamy**. The number of females attracted to a male red-winged blackbird's territory depends very little on the traits of the male holding the area (Figure 44.14). Rather, the females respond to the quality of nest sites and the quantity of food on and close to his territory. The probability that a female will return to a particular territory for the next breeding season is just as high when the owner has changed as when her previous mate is still present, and a female continues her nesting activities even if the territorial male changes within a breeding season.

Among species in which males do not control resources, females choose mates primarily by their traits, and males of these species often evolve highly elaborate structures and displays. These structures and displays increase the males' reproductive success. Males of the African long-tailed widowbird have remarkable tails that are longer than their heads and bodies combined. In an experimental study, the tails of some males were shortened while the tails of other males were lengthened. All the manipulated males successfully defended their display sites, indicating that the long tail does not confer an advantage in male–male competition. However, males with artificially elongated tails attracted about four times as many females as males with shortened tails did (Figure 44.15).

44.14 A Male Red-Winged Blackbird Defends His Territory
This display, which is always accompanied by a song, is called a song-spread. It warns other males to stay away, but it attracts females searching for breeding sites.

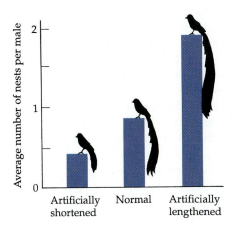

44.15 The Longer the Tail, the Better the Male
Male long-tailed widowbirds whose tails were artificially lengthened by gluing on extra feathers attracted more females and had greater reproductive success than males with normal or shortened tails.

Males of the satin bowerbird of Australia build and decorate elaborate structures called bowers that are used only for courting and mating with females. Males do not associate with females after mating, and they provide no parental care. Bowers are often close enough to one another that males are able to steal decorations from one another's bowers (Figure 44.16). The ability of a male to steal from his neighbors is positively correlated with his dominance at feeding sites. Therefore, females can use the number of objects a male has accumulated in his bower as an indication of his competitive ability. Males have a strong preference for blue objects. Because blue flowers are rare in Australian rain forests, the number of blue flowers a male has may be a better indicator of his raiding abilities than the number of objects of more common colors. The larger the number of ob-

jects present at his bower, the more matings a male obtains. Females that mate with males who have elaborately decorated bowers produce sons who will, in turn, build and control attractive bowers.

Choosing Nonmating Associates

An important criterion for choosing an associate other than a mate is that of genetic relatedness. Relatives that live together are more likely to cooperate with one another than with nonrelatives because relatives share alleles they have received from common ancestors. As we saw at the beginning of the chapter, an individual can increase its fitness—the frequency in future generations of the alleles it possesses—not only by producing its own offspring but also by increasing the survival and reproductive success of relatives that share those alleles. An individual can improve its fitness by lowering its own reproductive success while helping relatives, that is, by being altruistic, provided that the fitness costs imposed on itself are compensated for by improvements in the reproductive success of the relatives. The relatives need not do anything in return, nor do the individuals need to know who their relatives are.

Nonetheless, individuals of some species *do* know who their relatives are and adjust their behavior accordingly. White-fronted bee-eaters are colonial, cooperatively breeding African birds in which most nesting pairs are assisted by nonbreeding adults who help incubate eggs and feed nestlings (Figure 44.17). Males and females are equally likely to help at a nest, and individuals whose own nests fail may help at other nests later in the same breeding season. However, about half the nonbreeding individuals who could help do not do so. Individuals that do not help usually do not have any close relatives in their social group. Nearly all individuals that do help assist close relatives. When helpers have a choice of two nests at which to help, about 95% of the time they choose the nest with the young most closely related to them. Helping among white-fronted bee-eaters is altruistic because individuals that help are not more successful as breeders than those who do not help, nor do they appear to gain any other advantage from helping. Helping nonrelatives does not improve fitness, and white-fronted bee-eaters rarely help nonrelatives.

THE EVOLUTION OF ANIMAL SOCIETIES

The social systems found among animals today are the result of a long period of evolution, but there are very few records of the social systems of the past. Behavior leaves very few traces in the fossil record, so biologists must infer possible routes of the evolution of social systems from current patterns of social organization. Fortunately, many stages of social com-

44.16 Satin Bowerbird at His Bower
The male satin bowerbird in this photo has a predeliction for blue objects, some of which he has stolen from neighbors.

44.17 White-Fronted Bee-Eaters Help Close Relatives
These bee-eaters are perched at and near the entrances to their nests in a river bank. The colored tags on the birds enable investigators to identify individuals.

plexity exist among species, and the simpler systems provide clues about the stages through which the more complex ones may have passed. Sociality has arisen in two rather different ways: the familial route and the parasocial route.

The **familial route** to sociality begins with solitarily breeding females or pairs that give little or no care to their offspring. Then parental care behavior evolves. If natural selection then favors a more extended period of parental care, or if the breeding season is longer than the time it takes for the young to mature, the adults may still be caring for younger offspring when older offspring reach the age at which they could help their parents. If the older offspring assist their parents in nest building, in foraging, and in warning and defense against predators, sociality has evolved. This appears to be the route of evolution of bee-eater societies.

The **parasocial route** also begins with solitarily breeding adults. The first step toward sociality is the aggregation of unrelated adults of the same generation into nonsocial groups around clumped resources. Within these aggregates, cooperative behav-

ior, such as defense of the resources, may be advantageous. If such cooperation becomes better developed, division of labor may evolve; for example, some individuals may remain at the nesting area to guard while others leave to forage. Eventually, colonies may persist long enough for generations to overlap and for individuals of two or more generations to share responsibilities. For both of these routes, what is critical is that individuals begin to cooperate. The main difference is that in the familial route the cooperating individuals are close relatives, whereas in the parasocial route they are mostly unrelated individuals.

Insect Social Systems

Biologists have found patterns all the way from solitary breeding to sociality among bees and wasps. Most of the complex social systems found among wasps and bees evolved by the familial route. Solitary species are either parasitoids—they lay their eggs in the nests or on the larvae of other organisms—or species in which a nest is built and provisioned by a single female. The egg chamber of most nest-building species is sealed before the young emerge from their pupae, and there is no association across generations. In some species, however, the female remains with the nest longer and reopens the chamber to provide more food to the larvae. This may continue long enough that the female is still present when the young emerge from their pupae. The newly emerged females may remain with their mothers to help rear additional broods. Among some social insects this evolutionary trend has culminated in the production of sterile offspring that assist the fertile individuals. To refresh your memory of why these sterile individuals remain to help their mothers rather than attempting to reproduce on their own, reread Box 18.A.

Among some groups of bees and a few wasps, sociality evolved via the parasocial route. In species with simple societies, unrelated females of the same generation aggregate their nests in the same general location (Figure 44.18). The next evolutionary stage is cooperative building of a large nest with a single entrance. Within this nest each female lays her eggs in cells she herself has built, feeds her own larvae, and seals her own cells. This step probably evolved because females had greater reproductive success when they raised their offspring in a large nest with many defenders. For reasons that are not clear, cooperative rearing of the brood then evolved in some species. In a cooperative system, however, advantages accrue to those individuals able to dominate others. The result was the evolution of a dominant female, the queen, who lays most of the eggs, and subordinate females who do most of the work.

(a)

(b)

44.18 Social Wasps and Ants
Two possible paths to sociality are demonstrated in these social insects. (a) The parasocial route. The communal nests of this night-flying wasp (*Aploca pallida*) from Costa Rica are built by several females, most of which are not closely related to one another. (b) The familial route. These army ants, *Eciton burchelli*, are in a raiding column that searches the forest floor for prey. By hunting cooperatively, these small ants, which are sisters, are able to capture relatively large prey items.

Vertebrate Social Systems

As illustrated by white-fronted bee-eaters, most bird social systems probably evolved via the familial route, which results in offspring of one brood remaining with their parents and helping to rear additional broods. Florida scrub jays live all year on territories, each of which has a breeding pair and up to six helpers (Figure 44.19). Approximately three-fourths of the helpers are offspring from the previous breeding season. If both parents have survived, these individuals are full siblings of the young they are

helping. If one parent has died and the living partner has remated, they are half siblings. Helpers assist their parents by defending the nest, by giving alarm calls, and by feeding the nestlings. As a result, nests with helpers fledge more young than nests without helpers. The helpers could do better by breeding on

44.19 Cooperation among Scrub Jays
Florida scrub jay helpers, most of whom are offspring from the previous breeding season, are helping to feed nestlings. They also defend the nest against predators such as the approaching snake.

44.20   Jackals and Their Helpers
An older male, who has been helping his parents by bringing food to his younger siblings, is being groomed by the mated pair.

communal nest in which all the females lay eggs. All of the adults incubate the eggs, feed the young, and defend the nest against predators. But the females do not all begin to lay eggs at the same time. During the time before she begins to lay eggs, a female throws eggs laid by the other females out of the nest. A female usually stops throwing eggs out when she starts to lay her own. The dominant female of the group lays last and, accordingly, more of her eggs remain in the nest than those of other females. The dominant female, who has the most offspring in the nest, performs less parental care than the subordinate females do. Subordinate females, although they do worse than dominant females, nonetheless do better than they would as solitary breeders in the same habitat because groups compete more successfully for good breeding areas than do solitary pairs.

Most mammals evolved sociality via the familial route. Some present-day mammals represent the first stage, in which solitary females or male–female pairs care for their young. Many mammalian lineages have evolved a long period of parental care; in such lineages earlier young are still present when the next generation is born, and they often help rear their younger siblings (Figure 44.20). The general pattern among mammals is that female offspring remain in their **natal group**—the group in which they were born—but males tend to leave, or are driven out, and must seek other social groups. Therefore, among mammals, most helpers are females. In some species of bats, stable colonies of unrelated females defend communal foraging areas, suggesting that they evolved sociality via the parasocial route. In a few of these species, large numbers of females place their young in a central creche, or nursery. Each female gives birth to a single offspring, but females often suckle more than one infant, suggesting that at least some communal care of offspring has evolved.

their own if territories were available to them, but they remain with their parents because all suitable areas are occupied and they must wait for a vacancy before they become breeders. Most helpers spend 1–3 years with their parents and then leave to establish their own territories. Young males remain longer with their parents and help more than young females do. Males can gain more from helping because they may take over the parental territory if their father dies or they may help to enlarge the territory and claim a portion for themselves.

The parasocial route among birds is illustrated by the groove-billed ani, a black, ground-dwelling cuckoo of the American tropics. Anis live in groups of one to six monogamous breeding pairs whose members are usually unrelated to one another. Group members cooperate to defend a territory and build a

44.21   Communal Nesting in Weaverbirds
Many African weaverbirds nest in colonies in isolated trees. Although these nests are highly conspicuous, it is difficult for most avian, mammalian, and reptilian predators to get to them. The weaverbirds build these nests at the tips of branches, where thorns deter climbing predators.

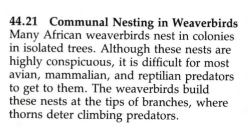

TABLE 44.4
Social Organization and Feeding Ecology in African Hoofed Mammals

SPECIES	BODY WEIGHT (kg)	FEEDING[a]	GROUP SIZE	SOCIAL UNIT
Dikdik, duiker	3–60	Selective browsing and grazing	1 or 2	Pair
Reedbuck, gerenuk	20–60	Selective browsing and grazing	2–12	Male with small harem
Impala, gazelles	20–250	Grazing and browsing	2–100	Territory-defending male with harem
Wildebeest, hartebeest	90–270	Grazing	Up to thousands	Herd in which males defend females
Eland, buffalo	300–600	Unselective grazing	Up to thousands	Herd with male dominance hierarchy

[a] "Browsing" means eating the leaves and other portions of woody plants; "grazing" means eating grasses and other herbaceous plants.

ECOLOGY AND SOCIAL ORGANIZATION

The type of social organization a species evolves is strongly related to the environment in which it lives. For example, among the African weaverbirds, the species that live in forests eat insects, feed solitarily, and build well-hidden nests. Most of these species are monogamous, and males and females have identical plumage. In marked contrast, weaverbirds that live on the tree-studded grasslands called savannas eat primarily seeds, feed in large flocks, and nest in colonies, usually in isolated trees, where their nests are large and conspicuous (Figure 44.21). Most of these species are polygamous, with brightly colored males and dull females. These striking differences probably evolved because nest sites and insects in forests, being widely dispersed, can best be exploited by solitary pairs rather than by groups. In savannas, the *Acacia* trees are few and widely scattered, so good nesting sites are scarce and are highly clumped. Also, it would be difficult to hide nests in the isolated trees. Indeed, many species build bulky nests that provide insulation from the hot sun to which they are exposed most of the day. Males compete for these limited nest sites, and the males that hold the better sites attract more females. Males spend their time attempting to attract additional mates rather than helping to rear the offspring they already have, which explains the evolution of brighter plumage among males.

All species of African hoofed mammals eat plant materials; their social organization and feeding ecology are correlated with the size of the animal (Table 44.4). Because of their higher metabolic demands per unit of body weight (see Chapter 23), smaller hoofed mammal species are very selective in what they eat, feeding preferentially on high-protein foods such as buds, young leaves, and fruits. These foods are widely scattered in forested environments where it is possible to hide from predators. Hiding is a tactic that is most effective when animals live solitarily. The largest hoofed mammal species are able to eat lower-quality food, but they must process great quantities of it each day. They feed in grasslands with high standing crops of herbaceous vegetation, follow the rains to areas where grass growth is best, and live in large herds (Figure 44.22). Living in herds makes it possible for males to compete among themselves for the control of females, and dominant males defend medium to large harems. In these open environments it is impossible to hide, but it is difficult for predators to approach the herds undetected. Species of intermediate sizes have feeding ecologies and social systems intermediate between those of the smaller and larger species.

Differences in primate social systems are also correlated with differences in environments. Nocturnal forest-dwelling insectivores, such as lorises, some lemurs, and the owl monkey, live in pairs and usually forage solitarily. Many diurnal species take insects and other animal food when they are available, but most of them depend primarily on fruits, seeds, and leaves for their energy. In Africa and Asia, troop sizes are smallest among arboreal forest-dwelling species, whatever their diets, and largest among the ground-dwelling savanna species, such as baboons (Figure 44.23).

Female primates, like other female mammals, usually remain with their natal troop, whereas young males are driven away by older males before they become reproductively active. In troops with more than one male, there are strong dominance hierarchies among the males, and just a few of the males do most of the copulating. Females may also have dominance relationships, and young females may assume the status of their mothers when they mature. Among some species, such as vervets, females frequently interact cooperatively, grooming and defending one another.

Defense of Space

Some of the species we have described defend a space containing their food and nesting resources;

44.22 Living in Large Herds
East African wildebeest live in large herds that move from place to place to obtain the fresh, green grass on which they feed. Their major predators—lions—live on permanent territories and often have little to eat when the wildebeest herds are far away.

others do not. The key factor determining this difference is whether the resource is economically defensible—that is, whether the costs of defense, such as the energy expenditure, the risk of injury, and the exposure to predation while defending, are less than the benefits conferred by having control of the resources. This general idea can be made much more precise. Excellent studies have been made with birds that exploit floral nectar, such as hummingbirds in the New World and sunbirds in the Old World. Ecologists have measured defensibility by estimating the concentrations and quantities of nectar in flowers, the energetic costs of flying, sitting, and fighting, and the increase in the amount of energy a bird can ingest per day as a result of defending a territory. These measurements yield a complete accounting of the costs and benefits of defending a food resource. An example of such a calculation, given in Box 44.B, shows that defense of a territory is profitable for the golden-winged sunbird.

SOCIAL BEHAVIOR AND NATURAL HISTORY

We have described only a few animal social systems, but the sample reveals some important concepts. First, social systems are best studied not by asking why they benefit the species, but by asking how the individuals that join together to form a social system benefit by the association. Second, social systems are dynamic; individuals are constantly communicating with one another and adjusting their relationships. Third, how individuals relate to one another depends in part on the degree of their genetic relatedness. Certain types of helping behavior evolve much more readily among close relatives than among more distantly related individuals. And finally, social systems evolve in relation to the animals' size, diet, and the environment in which they live. The great variability of social organization among animals and the highly dynamic nature of these systems demand that students of social systems be good natural historians.

44.23 Baboons in Groups
Baboons, which forage in open grasslands, travel in large groups. Typically, adult males are on the outer edges of the group while females and infants are toward the center. If a predator approaches, the formidable males cooperate in defending the group.

BOX 44.B

Economics of Territory Defense in the Golden-Winged Sunbird

Golden-winged sunbirds of East Africa feed on the nectar of flowers, a resource that can be depleted very rapidly but renews itself daily.

The golden-winged sunbird engages in three major activities, which have the following energy costs in calories per hour (as measured in the laboratory):

Foraging for nectar	1,000 cal/hour
Sitting on perch	400 cal/hour
Defending territory	3,000 cal/hour

The birds get their energy from nectar, but the amount of time a bird needs to forage is less when each flower contains more nectar:

	1	2	3
Average amount of nectar per flower (microliters)			
Foraging time required to meet daily energy needs (hours)	8	4	2.7

By defending a territory and excluding other nectar consumers, a bird increases the average amount of nectar available per flower. It therefore needs to spend less time foraging. Instead, it sits on a perch more of the time, an activity that is less costly. For example, if holding a territory increases the average amount of nectar per flower from 2 microliters to 3 microliters, the bird saves 1.3 hours of foraging time. Thereby it saves

$$(1,000 \times 1.3) - (400 \times 1.3) = 780 \text{ calories}$$

To achieve this saving a bird must spend, on average, 0.28 hours each day defending its territory—time it could otherwise spend sitting. The cost of defense is therefore

$$(3,000 \times 0.28) - (400 \times 0.28) = 728 \text{ calories}$$

Thus if territory defense can raise the average nectar content from 2 microliters to 3 microliters per flower, territories are economically defensible. Field measurements of nectar content on defended and nondefended areas have shown that defense results in just enough increase in nectar content per flower to make territoriality beneficial. If, however, the gains in nectar were just a little less, or the amount of time required to defend the territory a bit higher, it would no longer pay the birds to defend those resources.

They must know a great deal about the lives of the species they study and must be good field observers.

SUMMARY

Some animals survive and reproduce better in groups than they do solitarily. The advantages of group living may include enhanced foraging success, better protection against predators, and better ability to defend rich foraging areas. Disadvantages of group living include higher exposure to disease, competition for resources, and direct social interference. Social groups consist of different individuals that experience different benefits and costs of group living.

Males usually increase their fitness by maximizing the number of copulations they achieve. Males and females often differ in the amount of parental care they provide. Males of species with social systems in which they are more likely to be the actual fathers of the offspring of their mates provide more parental care than males of species in which there is greater uncertainty of paternity.

Social groups are maintained by communication among group members. Chemical signals are diverse, energetically inexpensive, and can act at low concentrations. However, they have long fade-out times, their sources are difficult to locate, and they are not suitable for communicating rapid changes. Visual signals are easy to produce and locate and can be changed rapidly. Auditory signals are also easy to produce and can be changed rapidly, but they are energetically expensive and their source is more difficult to locate. Tactile communication is practiced by spiders and bees, usually supplemented by odors and sounds. Communication signals have arisen from intention movements, as by-products of the action of the autonomic nervous system, and from the apparently irrelevant activities animals perform in situations of conflict. These behaviors become elaborated as signals if performing them in more conspicuous and stereotyped ways benefits the signaler.

Animals choose associates on the basis of their traits, the resources they hold, and their relatedness to the chooser. Close relatives share alleles with one another and can increase the frequency of these alleles in future generations by helping one another even if the aid is not returned. Among nonrelatives, helping behavior can evolve only if individuals reciprocate.

Complex social systems have evolved by two routes. The familial route begins when offspring remain with their parents and help to rear future broods. The parasocial route begins when adults of the same generation associate around clumped resources, and cooperation evolves among them. This cooperation may later be extended to include other generations. Among insects, social systems of most wasps and some bees evolved by the familial route, whereas most bee social systems evolved via the parasocial route. The familial route has been the predominant one in birds and mammals. Social systems of animals evolve in relation to diet and the type of habitat in which they live.

SELF-QUIZ

1. Which of the following is *not* a component of the cost of performing a behavioral act?
   a. Its energetic cost.
   b. The risk of being injured.
   c. Its opportunity cost.
   d. The risk of being attacked by a predator.
   e. Its information cost.

2. An almost universal cost associated with group living is:
   a. increased risk of predation.
   b. interference with foraging.
   c. higher exposure to disease and parasites.
   d. poorer access to mates.
   e. poorer access to sleeping sites.

3. An act is said to be altruistic if it:
   a. confers a benefit to the performer by inflicting some cost on some other individual.
   b. confers a benefit both to the performer and to some other individual.
   c. confers a cost both to the performer and to some other individual.
   d. confers a benefit on another individual at some cost to the performer.
   e. imposes a cost on a performer without benefitting any other individual.

4. Which of the following statements about male and female roles in social systems is *not* correct?
   a. Females invest more in gamete production but they may invest more or less than males in care of offspring.
   b. Biparental care is prevalent among birds.
   c. Males of most mammal species help feed offspring.
   d. Males with a high probability of parentage invest more in parental care than males that are less certainly related to the offspring of their mates.
   e. Among fishes, if there is unequal parental care by individuals of the two sexes, it is nearly always the male who does more.

5. A display or signal is a behavior that:
   a. has evolved to influence the actions of other individuals.
   b. stimulates one or more sensory modes.
   c. stimulates the endocrine or reproductive systems of other individuals.
   d. began as an intention movement.
   e. began as a displacement activity.

6. Choice of mating partner may be based on:
   a. the inherent qualities of a potential mate.
   b. the resources held by a potential mate.
   c. both the inherent qualities of a potential mate and the resources it holds.
   d. the success of individuals of the opposite sex in courtship.
   e. all of the above.

7. A male that holds a territory in which more than one female breeds is an example of:
   a. an individual with a high coefficient of relatedness.
   b. an individual with a high inclusive fitness.
   c. resource defense polygamy.
   d. the parasocial route of social evolution.

   e. the familial route of social evolution.

8. The first step in the parasocial route to sociality is:
   a. pairs providing care for their offspring.
   b. an aggregation of unrelated adults of the same generation.
   c. an aggregation of unrelated adults of different generations.
   d. offspring remaining with their parents to help raise siblings.
   e. a division of labor among adults of the same generation.

9. Among social mammals most helpers are females because:
   a. only females have functional mammary glands.
   b. males mature too slowly to help.
   c. females form closer ties with their mothers than males do.
   d. young males leave their social group or are driven out by their fathers.
   e. young males are too aggressive to help.

10. Smaller African hoofed mammals are usually solitary because:
    a. they feed on scattered high-quality foods in forested environments.
    b. the low quality of their food does not permit them to assemble in groups.
    c. they are too small to defend themselves against predators.
    d. they are too small to follow the rains to areas where grass growth is best.
    e. they are usually driven from their natal groups.

FOR STUDY

1. Most hawks are solitary hunters. Swallows often hunt in groups. What are some plausible explanations for this difference? How could you test your ideas?

2. Among birds, males of promiscuous species that display on communal grounds are usually much larger and more brightly colored than females, whereas in monogamous species, males are usually similar in size to females, whether or not they are more brightly colored. What hypotheses can be advanced to explain this difference?

3. For many years ethologists believed that communication signals evolved to enable the signaler to communicate more clearly to others just what it intended to do. Why has this view been abandoned? What common knowledge about human behavior should have created strong initial skepticism about that view?

4. Water striders are semiaquatic insects that skate on the surface of ponds and streams. They feed on other insects that fall onto the water. Prey are located both visually and by means of special receptors in the legs that detect ripples in the water. In certain water strider species, the male uses the latter mechanism to attract a mate: he creates ripples in the water and when a female approaches to investigate he copulates with her. Is this a case of intraspecific deceit? Adult water striders often eat immature ones. If immature individuals, as well as adult females, were attracted to displaying males, would this be a case of intraspecific deceit? Male water striders generate ripples with specific frequency patterns; prey generate ripples at frequencies that depend on their size and how they struggle in the water. What factors might have favored the evolution of such specific reproductive signals? How would such signals benefit the males that give them? The females that respond to them?

5. Among frogs, a male clasps a gravid female behind her front legs and stays with her until she lays her eggs, at which time he fertilizes them. In most species of frogs, the male remains clasped to the female for a short time, usually no longer than a few hours. However, in some species, pairs may remain together for up to several weeks. In view of the facts that a male cannot court or mate with any other female while clasping one female and that a female lays only a single clutch of eggs, why is it advantageous for males to behave that way? What can you guess about the breeding ecology of frogs that remain clasped for long periods? Why should females permit males to clasp them for so long? (Females do *not* struggle!)

READINGS

Alcock, J. 1989. *Animal Behavior: An Evolutionary Approach*, 4th Edition. Sinauer Associates, Sunderland, MA. A highly readable textbook of behavior; recommended as the next step in reading about the subject.

Brown, J. L. 1987. *Helping and Communal Breeding in Birds: Ecology and Evolution*. Princeton University Press, Princeton, NJ. A review of available data on helping and communally breeding birds, together with a thorough review of underlying theory.

Gould, J. L. and C. G. Gould. 1989. *Sexual Selection*. Scientific American Library, New York. A beautifully illustrated treatment of sexual behaviors and how they may have influenced the course of evolution.

Krebs, J. R. and N. B. Davies. 1986. *An Introduction to Behavioral Ecology*, 2nd Edition. Blackwell Scientific Publications, Oxford. An excellent account of the methods and results of modern studies of behavioral ecology.

Lack, D. 1968. *Ecological Adaptations for Breeding in Birds*. Methuen, London. Includes a definitive account of social behavior in birds.

Rubenstein, D. I. and R. W. Wrangham, Editors. 1986. *Ecological Aspects of Social Evolution*. Princeton University Press, Princeton, NJ. The results of long-term field studies of 18 species of birds and mammals are treated in an evolutionary ecological context. Rich in natural history details.

Trivers, R. 1985. *Social Evolution*. Benjamin/Cummings, Menlo Park, CA. A good overview of social behavior, with a great deal to say about human social behavior.

45

Structure and Dynamics of Populations

PREVIEW: Some species are common; others are rare. Some species have broad ranges; others have narrow ones. Within its range, a species is more common in some places than in others, and its density and distribution change over time. Population ecology is the science that studies these patterns and attempts to explain them. Populations usually grow rapidly in environments with few individuals and abundant resources, but growth slows down as the population approaches the environmental carrying capacity. Many populations fluctuate within narrow density ranges, usually because birth and death rates change as population density changes. Other populations fluctuate widely in abundance because disturbances which repeatedly reduce them to densities well below the environmental carrying capacity are followed by rapid population growth. People use information on how birth and death rates change to manage populations of desirable species so as to provide maximum sustainable yields of individuals or products and to keep populations of undesirable species at low levels.

This chapter deals with population structure, population dynamics, life histories, disturbance, population regulation, population management, and human population growth.

One obvious pattern in nature is that some species are much more common than others. Also, some species seem to be just about as abundant one year as another, whereas other species are numerous some years but scarce other years. These observations stimulate a variety of questions: What causes a species to be common or rare? Why does the size of some populations fluctuate yearly and seasonally? Why is a species common in some parts of its geographic range, rare in others, and absent outside its range? To answer such questions, we must describe populations in ways that capture the essential details of the processes that produce these results. An individual, of course, is born only once and dies only once, and occupies only one space at a time. But when we put together information about a number of individuals, we can see patterns in the population. Population ecologists are interested in finding out what these patterns are—whether a population is increasing or decreasing in number, and how its members are distributed in space. When the patterns are known, ecologists can investigate what causes them.

Populations of most species are divided into a number of subpopulations that are at least partly isolated from one another. Different agents often affect subpopulations uniquely in different parts of the range. Amphibians that breed in ponds scattered in forests are good examples of species with fragmented population structure.

The red-spotted newt is a colorful salamander that breeds in small ponds throughout eastern North America. Farm ponds in rural areas and wildlife management ponds in national forests are soon populated by newts, showing that they are good colonizers of new habitats. In most parts of its range, the red-spotted newt has a complex life cycle. Adults feed in ponds and lay their eggs there. The eggs hatch into larvae that grow and develop in the ponds for several months. Eventually the larvae change into efts, immature newts that leave the water to live terrestrial lives for 4 to 9 years. The red efts are highly toxic to their natural predators. Most efts return to the ponds in which they were born but some travel long distances and colonize new ponds (Figure 45.1).

In some ponds in some years, survival of eggs and larvae is good and many efts are produced, but most ponds produce very few efts most years. That is because diseases and attacks by leeches are intense most years. Survival of larvae to become efts is high

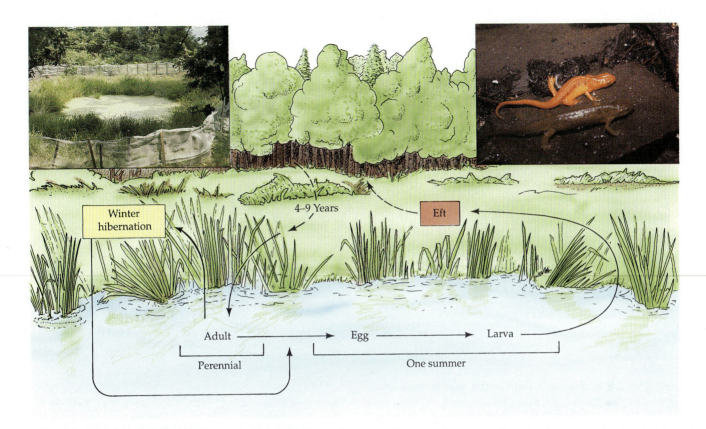

45.1 Red-Spotted Newts Colonize Young Ponds
Red-spotted newts breed in small ponds in the Appalachian Mountains. The poisonous red efts (the newt's juvenile form) live for 4–9 years on land. Most efts return as adults to the pond in which they were born, although some migrate to new ponds. Adult newts spend winters hibernating near their breeding ponds and return to the same pond each spring until they die.

only in those ponds where disease-causing organisms and leeches are not well established. An ecologist needed to study many different ponds to understand the structure and ecology of the red-spotted newt. If only a few ponds had been studied, the pattern of widespread reproductive failure across subpopulations could not have been known. Recruitment of adults to ponds would have been attributed primarily to immigration rather than to homing of efts that lived for many years on land.

POPULATION STRUCTURE

In Chapter 18 we discussed the genetic structure of populations and its evolutionary importance. **Population structure** is an ecological concept as well as a genetic one. When geneticists look at population structure, they are interested in the distribution of genotypes, whereas ecologists are more interested in the distribution of the members of a species in space than in their genetic similarities or differences. Ecologists and geneticists both need to consider whether groups of individuals in a species are partly or completely isolated from one another. To find out about

the population structure of a species, an ecologist asks, about the members of the species, "How many are there? Where are they? How old are they?" Ecologists study population structure at scales ranging from local subpopulations to the entire species.

Population Density

The number of individuals of a species per unit of area (or volume) is called the **population density**. Ecologists are interested in population densities because denser populations exert stronger influences on populations of other species. Also, people often wish to manage species so as to raise their densities (crop plants, aesthetically attractive species) or reduce their densities (agricultural pests, disease organisms). To manipulate densities, we must know what factors make populations grow and shrink and how these factors work.

Because organisms and their environments differ, densities are measured in more than one way. Ecologists usually measure the density of organisms in terrestrial environments by number of individuals per unit of area, but number per unit of volume is generally a more useful measure for organisms living

45.2 Even Spacing
The spacing between the nests of these king cormorants is such that the incubating birds cannot peck their neighbors without rising from their eggs.

in water. For species whose members differ markedly in size, as is the case with most plants and some animals, such as mollusks, fishes, and reptiles, the total mass of individuals—their **biomass**—is sometimes the most useful measure. Sometimes individuals can be counted directly without missing any of them or counting any of them twice. But this is usually impossible or too laborious, so densities are commonly estimated by sampling the population in representative areas and extrapolating to the number in the whole area.

Spacing

Ecologists studying population structure look at the way the individuals in a population are spaced. Ecologists are interested in spacing because it reveals why individuals settled and survived where they did. Individuals in a population may be tightly aggregated into clumps, evenly spaced, or randomly scattered. Distributions can become clumped when young individuals settle close to their birthplaces, when suitable environments are "islands" separated by unsuitable areas, or by chance. Plants often have evenly spaced distributions, produced by competition for light, water, and soil nutrients, and by their rubbing against one another when moved by wind or water

45.3 Age Distributions Reveal Birth and Death Patterns
(a) The human population of India in 1970. (b) The human population of Sweden in 1977. The length of the bars shows the percent of the population in each age class.

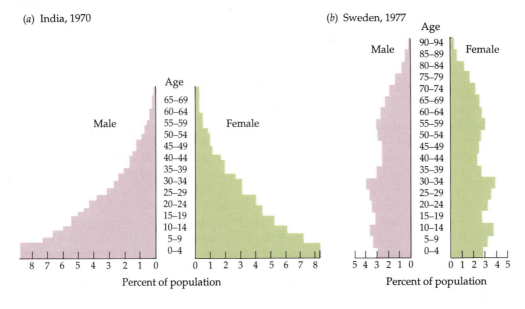

(a) India, 1970

(b) Sweden, 1977

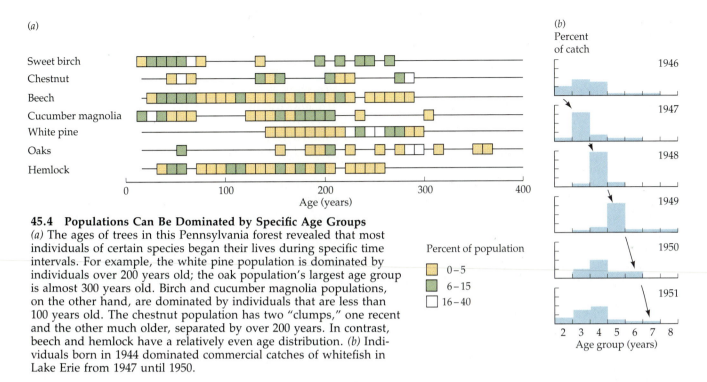

45.4 Populations Can Be Dominated by Specific Age Groups
(a) The ages of trees in this Pennsylvania forest revealed that most individuals of certain species began their lives during specific time intervals. For example, the white pine population is dominated by individuals over 200 years old; the oak population's largest age group is almost 300 years old. Birch and cucumber magnolia populations, on the other hand, are dominated by individuals that are less than 100 years old. The chestnut population has two "clumps," one recent and the other much older, separated by over 200 years. In contrast, beech and hemlock have a relatively even age distribution. (b) Individuals born in 1944 dominated commercial catches of whitefish in Lake Erie from 1947 until 1950.

currents. Among animals, defense of space is the most common cause of even distributions (Figure 45.2). Random distributions may result when many factors interact to influence where individuals settle and survive.

AGE DISTRIBUTION

Populations are composed of individuals ranging from newborns to postreproductive adults. The proportions of individuals in each age group in a population make up its **age distribution**. The timing and rates of births and deaths determine age distributions. If birth rates and death rates are both high, a population is dominated by young individuals, as illustrated by the human population of India in 1970 (Figure 45.3a). If birth rates and death rates are low, there is a relatively even distribution of individuals of different ages, as illustrated by the human population of Sweden in 1977 (Figure 45.3b). Thus, examination of the age distribution of a population can reveal much about the recent history of births and deaths in the population.

The timing of births and deaths may influence age distributions for many years in populations of long-lived species. A striking example is provided by the age distributions of trees in a Pennsylvania forest. Most of the white pines and oaks in the forest began their lives over 200 years ago, whereas most of the birches and magnolias germinated within the last 100 years. The populations of all these species are dominated by individuals that germinated and survived during a few particularly favorable years (Fig-

ure 45.4a). Fish populations also are often dominated by individuals of one age group. In Lake Erie, 1944 was such an excellent year for reproduction and survival of whitefish that individuals of that age group dominated whitefish catches in the lake for several years, as the blue bars in Figure 45.4b show.

The age distributions of populations reveal much about their history, their potential for future growth, and their relationships to other species. By comparing the age distribution of trees in the Pennsylvania forest, we can tell that the tree species differ in their germination requirements, and we can infer that some species are being replaced by others. Age distributions also influence how populations grow and use environmental resources. Young individuals may depend on their parents for resources. Because of their small size, juveniles consume less per day than adults do, whether they are fed by their parents or forage for themselves. Prereproductive and postreproductive individuals do not produce offspring. Therefore, the rate at which a population grows and the amount of resources it consumes depends on its age distribution.

POPULATION DYNAMICS

An individual organism, if it completes its entire life cycle, is born, grows to maturity, reproduces, and dies. During that time it ingests nutrients or food, grows, interacts with other individuals of the same and other species, and usually moves or is moved so that it does not die exactly where it was born. Such activities of the individuals making up a population determine the dynamics of the population—determine, that is, whether the population as a whole is increasing, decreasing, or staying the same, and whether its structure is changing.

Life History

The stages an individual organism goes through during its life constitute its **life history**. Life histories are very diverse, but they all have a period during which the individuals grow—no organism begins its life at its adult size. Among unicellular organisms that reproduce by cellular fission, new individuals are half the size of adults, but in multicellular organisms, newborns are relatively much smaller. Every newborn individual starts life with some source of energy from its maternal parent. Orchid seeds receive very little maternal energy, whereas coconuts, birds, and newborn placental mammals receive large amounts. During their growth periods, individuals of many species are completely independent of their parents, obtaining all their own energy and providing for their own protection. In some animal species, however, the parents provide additional care and protection during the growth stage that may extend, as it does in many birds and mammals, until the individual has reached adult size.

Another important part of the life history of all organisms is a dispersal stage, when individuals move or are moved from their birthplaces to the places where they will live as adults. Some organisms, such as plants and sessile animals, disperse as spores or seeds before much growth has taken place. Other organisms, such as insects and birds, disperse primarily as adults. Still others may disperse during several different stages. Individuals of some species can change their locations many times during their lives in response to environmental changes. Others must remain in the first place they settle.

All life cycles include a reproductive stage. Many species do not begin to reproduce until they have reached full size, but others start while they are still relatively small and continue to reproduce as they grow. Most plants begin to reproduce before they reach full size. Among continuously growing animals, such as many mollusks, fishes, and reptiles, growth and reproduction overlap extensively during the lives of individuals. Nonetheless, reproduction usually reduces growth because these two processes compete for the limited amount of energy an individual has at its disposal.

For at least part of their lives, all organisms are in an energy-gathering stage. This stage may encompass the entire life history, but in some species energy harvesting is confined to a particular stage. For example, most moths feed only when they are larvae. The adults lack mouth parts and digestive tracts.

They live on energy gathered as larvae and survive only long enough to disperse, mate, and lay eggs.

Many organisms have resting or reorganization stages in their life cycles. During resting stages, such as spores, seeds, and many eggs, metabolic rates are low and the individuals are usually highly resistant to changes in the physical environment. Reorganization stages are found primarily among insects, such as beetles, flies, moths, butterflies, and bees, which undergo radical changes from their larval to their adult forms. During these reconstructions, which take place in a form called a pupa, larval tissues and organs are broken down and adult tissues and organs are constructed from the larval material.

Life history stages may overlap. An organism may be doing more than one of the things mentioned in the five preceding paragraphs at any one time (Figure 45.5). Nonetheless, understanding how different types of organisms express these stages, and how an individual's performing one activity constrains its ability to perform others, provides a useful perspective for thinking about the lives of organisms and the dynamics of their populations. The life history patterns in Figure 45.5 illllustrate the great variability found in nature. The patterns include ones in which reproduction terminates the life of the individual (annual plant, *Cecropia* moth, salmon), one in which growth and reproduction overlap broadly (tree), one in which growth is limited to a short period but reproduction continues intermittently for a long time after the individual has reached full size (marmot), and one in which both growth and reproduction are interspersed throughout the life of the individual (*Paramecium*).

Births and Deaths

A surprising amount can be learned about a population simply by knowing when individuals are born and die. This was the first aspect of population dynamics people studied because of its great practical importance. When a person dies, some expenses, such as the cost of burial and support for surviving relatives, usually are borne by the society at large. Societies have, not surprisingly, been interested in the probability that individuals of different ages will die, because that determines when, and how much, money must be spent. The cost of modern life insurance—really death insurance, but we all prefer the pleasant euphemism—is based on elaborate and accurate estimates of death schedules, but the practice has ancient roots. Roman societies provided funeral insurance for their members, and they developed and used the first tables showing ages of death of their citizens. These tables were based on rather inaccurate knowledge, but the fact that they were constructed at all indicates the importance death schedules had 2,000 years ago.

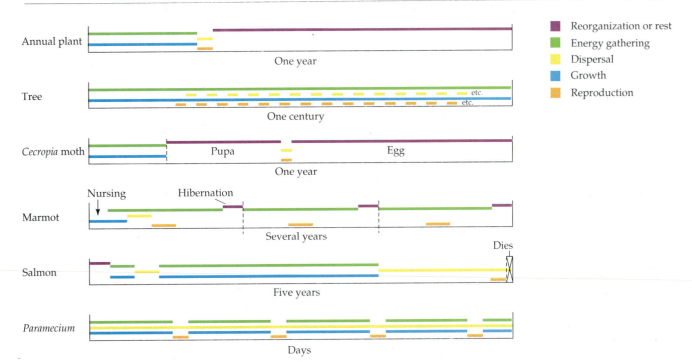

Legend:
- ■ Reorganization or rest
- ■ Energy gathering
- □ Dispersal
- ■ Growth
- ■ Reproduction

Annual plant — One year

Tree — One century

Cecropia moth — Pupa / Egg — One year

Marmot — Nursing / Hibernation — Several years / Dies

Salmon — Five years

Paramecium — Days

45.5 Life Histories Differ
All life histories have growth, dispersal, reproductive, energy-gathering, and resting stages, but when these stages happen and how much they overlap varies dramatically.

These six patterns are some of the common ones. To make comparisons among them easier, all six are started at the beginning of a growth period.

Births, deaths, immigration, and emigration are **demographic processes** that determine the numbers of individuals in a population. Ecologists measure the rates at which those events take place. The rates are influenced both by environmental factors and by the properties of the species in question. The number of individuals present in a population is equal to the number present at some time in the past, plus the number born between then and now, minus the number that died, plus the number that immigrated, minus the number that emigrated.

Analyses of population dynamics are usually discussed, as we have done, in terms of the number of individuals. For many kinds of organisms, called **unitary organisms**, individuals are easy to distinguish, and most adult members of a population are similar to one another in size and shape. For **modular organisms**, on the other hand, individuals are not easy to distinguish, and the members of a population have highly variable sizes and shapes. The fertilized egg of a modular organism develops into a unit of construction—a module—that then produces additional modules much like itself (Fig. 45.6). Usually these modules remain attached to one another, and the aggregate is usually immobile. Most plants are modular, and there are many important groups of modular protists, fungi, and animals—sponges, corals, moss animals, colonial tunicates.

The limits of modular individuals are often difficult to determine. Even more important for the study of

45.6 Modular Organisms May Look Like Populations
Each of these quaking aspen groups consists of only one genetic individual that has spread by underground roots and has sent up many stems, each of which appears to be a separate tree.

ecology, the impact of modular organisms on their environment often depends less on the number of genetically distinct individuals than on the number and size of modules. A large tree with many branches is a very different organism than a small sapling with just a few branches. Also, the modules of a single organism may differ markedly in size and age. Therefore, students of modular organisms are often concerned primarily with the number, size, and shape of modules rather than with the number of genetically distinct individuals.

Life Tables

A useful way to visualize patterns of births and deaths in a population is to construct a **life table**, which indicates for a group of individuals born at the same time—a **cohort**—the number still alive at later times and the number of offspring they produced during each of the time intervals. A life table for a cohort of the short-lived grass *Poa annua* is shown in Table 45.1. As you can see from the table, members of this cohort began producing seeds some time after they were 3 months old and continued to produce seeds throughout the rest of their lives. By the end of 2 years, all members of the cohort were dead. Note that the life table includes numbers observed—such as 527 plants alive at the beginning of the 6–9 month interval and 211 plants dying during this interval—as well as rates calculated from those numbers. We can calculate the death rate during the 6–9 month age interval by dividing the number dying by the number alive at the beginning: 211/527 = 0.4.

Ecologists use graphs to highlight the most important changes in birth and death rates in populations. For example, if we plot the seed-production data in Table 45.1 as a graph, we can see clearly that the number of seeds produced per individual peaks

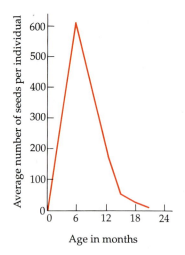

45.7   Age-Specific Reproductive Rate of *Poa annua*
The average number of seeds produced per individual peaks at 6 months of age. By the time they reach 15 months of age, individuals of *Poa annua* produce very few additional seeds.

at 6 months of age and that individuals produce very few seeds after they are a year old (Figure 45.7).

Graphs of survivorship—the mirror image of death rate—in relation to age show clearly when individuals survive well and when they do not. To interpret survivorship data, ecologists have found it useful to compare real data with several hypothetical curves that illustrate a range of possibilities. A useful type of graph plots the proportion of individuals of a cohort that are still alive at different times during their total potential life span (Figure 45.8a). At one extreme—hypothetical curve I—nearly all individuals survive for their entire potential life span and die almost simultaneously. An intermediate possibility—hypothetical curve II—is that survivorship remains

---

TABLE 45.1
A Life Table for a Cohort of 843 Individuals of *Poa annua*, a Short-Lived Grass

AGE INTERVAL[a]	NUMBER ALIVE AT BEGINNING	PROPORTION ALIVE AT BEGINNING	NUMBER DYING	DEATH RATE	NUMBER OF SEEDS PRODUCED
0–3	843	1.000	121	0.144	0
3–6	722	0.857	195	0.270	300
6–9	527	0.625	211	0.400	620
9–12	316	0.375	172	0.544	430
12–15	144	0.171	95	0.625	210
15–18	54	0.064	39	0.722	60
18–21	15	0.018	12	0.800	30
21–24	3	0.004	3	1.000	10
24	0	—	—	—	—

[a] Period between two exact ages (in months).

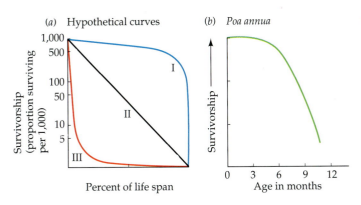

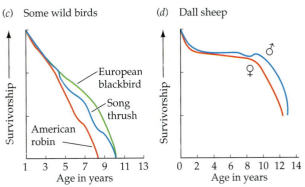

45.8 Survivorship Curves

Survivorship curves show the proportion of individuals of a cohort still surviving at different times in their lives. (a) Hypothetical survivorship curves showing three possible patterns. (b) The survivorship curve of *Poa annua* resembles the Type I pattern. (c) Survivorship curves of three species of thrushes show the Type II form. (d) Survivorship curves of dall sheep combine elements of Types I and III.

the same throughout the life span. At the other extreme—hypothetical curve III—the survivorship of young individuals is very low but then survivorship is high for the remainder of the life span.

Survivorship data from real populations are often very similar to one of these hypothetical curves. For example, survivorship of *Poa annua* seedlings is very high for the first 6 months, but then becomes much poorer among older individuals (Figure 45.8b). Most wild birds have survivorship curves similar to hypothetical curve II, that is, the probability of surviving is about the same throughout the life span of the individuals once they are a few months old (Figure 45.8c). A more common type of survivorship pattern, especially among organisms that produce large numbers of offspring, each of which receives little energy and no subsequent parental care, is one with low survivorship of young individuals, followed by high survivorship during the middle part of the life span, and then low survivorship toward the end of the life span. The dall sheep, although it does not have a high birth rate, has such a survivorship curve (Figure 45.8d), one that combines the first part of the type III curve with the middle and late parts of the type I curve.

Survivorship may differ between males and females of a single species. For example, survivorship of male red deer differs from that of females. The males engage in heavy combat during the breeding season; many are injured in fights over females. The females engage in no such combat. Their survival is nearly constant once they reach reproductive age (2 to 3 years old), whereas death rates of males rise sharply when they reach their breeding age of 7 to 8 years (Table 45.2).

POPULATION GROWTH WHEN RESOURCES ARE ABUNDANT

Imagine that we could select a single bacterium at random from the surface of this book and could endow it and all its descendants with the power to grow and reproduce without any restriction. In a month this bacterial colony would weigh more than the visible universe and would be expanding outward at the speed of light. Similarly, a single pair of Atlantic cod and their descendants reproducing without hindrance would in six years fill the Atlantic Ocean with their bodies. People mature and breed very slowly, but if the existing human population could achieve the impossible feat of continuing to increase at its present rate (which is less than its maximum potential), in 2,000 years the human population would weigh as much as the entire Earth. And 4,000 years later, it would weigh as much as the visible universe and be expanding at the speed of light!

All populations have this potential for explosive growth because, as the number of individuals in the

TABLE 45.2
Death Rates of Male and Female Red Deer on Rhum Island, Scotland[a]

AGE	DEATH RATE FEMALES	DEATH RATE MALES
0–6 months	12.4	12.4
6–12 months	15.0	20.8
Yearlings	7.4	13.0
2 years	1.1	1.8
3–4 years	3.6	1.7
5–6 years	3.8	2.2
7–8 years	2.3	6.1
9–10 years	2.8	16.3
11–12 years	8.7	37.0

[a] Percentage of those in each age class alive at the beginning of a particular year that died during that year.

(a) Exponential (unrestricted) growth

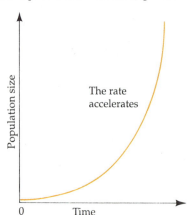

(b) Logistic (restricted) growth

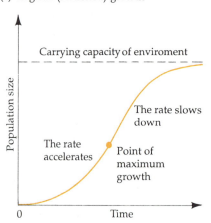

45.9 Population Growth Curves
(a) Theoretically, a population in an imaginary environment with unlimited resources could grow like this.
(b) Typically, a population in an environment with limited resources grows like this.

population increases, the rate of unrestricted growth accelerates (Figure 45.9a). The larger the population, the faster its potential for growth. This form of increase is called **exponential growth** and is expressed mathematically in the following way:

$$\frac{\text{Rate of}}{\substack{\text{increase in} \\ \text{number of} \\ \text{individuals}}} = \left(\substack{\text{Average} \\ \text{birth rate} \\ \text{per} \\ \text{individual}} - \substack{\text{Average} \\ \text{death rate} \\ \text{per} \\ \text{individual}}\right) \times \substack{\text{Number} \\ \text{of} \\ \text{individuals}}$$

or, more concisely,

$$G = (b - d)N$$

When conditions are optimal for the population, the difference between the average birth rate b and the average death rate d is called r_{max}—the **intrinsic rate of increase**. It has a characteristic value for each species. Therefore, the rate of growth of a population under optimal conditions is $G = r_{max}N$.

POPULATION GROWTH WHEN RESOURCES ARE LIMITED

No population can maintain exponential growth for very long because environmental limitations cause birth rates to drop and death rates to rise. In fact, over long time periods, the size of most populations fluctuates around a relatively constant number. The simplest way to picture the limits imposed by the environment is to recognize that an environment can support no more than a certain number of individuals of any particular species. This number is called the environmental **carrying capacity**, and it is determined by the availability of resources—food, nest sites, shelter—as well as by disease, predators, and perhaps social interactions. As a result, the increase of a population soon departs from exponential growth and instead, when graphed, follows a curve that bends to the right as the population approaches the carrying capacity (Figure 45.9b).

The S-shaped growth pattern, which is characteristic of many populations growing in environments with limited resources, can be treated mathematically by adding some environmental resistance to the exponential growth equation. The simplest such equation is one for **logistic growth** in which each additional individual depresses future population growth equally. Logistic growth has the following form:

$$G = (b - d) \times \left(\frac{K - N}{K}\right) N$$

where K is the carrying capacity and the other symbols are the same as in the equation for exponential growth. The biological assumption in this equation is that each additional individual makes things slightly worse for the others because it competes with them for available resources, or for other reasons. Population growth finally stops when $N = K$. Then $(K - N) = 0$, so that

$$\left(\frac{K - N}{K}\right) = 0$$

and

$$G = 0$$

Logistic growth can be readily demonstrated in the laboratory by introducing a few individuals into an environment with abundant resources. When this is done, populations often grow rapidly, then slow down, and eventually begin to maintain a size that fluctuates around K, whose value can be varied by altering the amount or availability of resources.

The logistic growth equation contains some important simplifications that are not characteristic of most populations. The most critical assumptions are that each individual exerts its effects immediately at birth and that all individuals produce equal effects. However, in nature, organisms grow during their lives, and their effects on others normally increase with age, so there may be a delay between the birth of an individual and the time at which it begins to

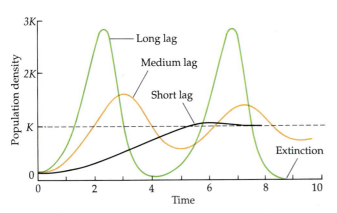

45.10 Time Lags Produce Oscillations
The longer the delay, the greater the oscillations.

affect the other members of the population. Also, the number of offspring produced by an individual usually does not depend on the resources available at the time the offspring are born but on the resources available when they were being provisioned prior to birth. As a result, if there is a long delay between fertilization and birth, a population may shoot past the carrying capacity when the offspring are born. The way populations approach and fluctuate around K depends on the durations of such delays (Figure 45.10). A population that exceeds the carrying capacity will later drop down below it. If the delay is very long, the oscillations may be so great that the population may crash all the way to extinction.

Changes in Birth Rates

The birth rate characteristic of a species is determined by the timing of its reproduction and by the number of offspring an individual produces in a given batch, known as a **clutch** or **litter** in animals or a **seed crop** in vascular plants. The basic limit to the number of offspring produced is the amount of energy and nutrients parents have available for their production. As populations adapt to different types of environments, their birth and death rates may change. As an animal population approaches the carrying capacity of its environment, the amount of energy each female can acquire above what she needs to maintain herself usually gets smaller, and the number of offspring she can produce thus declines. A dominant individual plant, however, by extracting resources from a large area, may continue to produce large numbers of seeds even when the population is at K.

Some organisms reproduce only once and then die. A bacterium that divides to form two daughter cells may be considered to have died when it split. Annual plants consume so much of their energy for seed production that they do not survive long after reproducing. Some longer-lived organisms also reproduce once in their lifetimes and die very soon

afterward. Pacific salmon of the genus *Onchorhynchus* hatch in fresh water, spend a number of years at sea, return to fresh water, spawn, and die. Most agaves (century plants) of the American Southwest likewise store up energy for many years, produce a large flowering stalk, form many seeds, and then die (Figure 45.11). Yucca plants, which grow in the same environments, appear similar, but they invest less in each reproduction and live to reproduce many times.

Organisms that reproduce only once in their lifetimes are called **semelparous**. Those that reproduce more than once are called **iteroparous**. A semelparous individual can produce more offspring in a single clutch than an iteroparous individual that has the same amount of available energy can because the semelparous individual reserves no energy for its future survival. Data from many species, including plants, insects, shrimp, snails, fishes, birds, and mammals, show that engaging in reproduction reduces adult survival rates. The more an adult invests in reproduction, the lower the probability that it will survive and, if it is still growing, the more slowly it will grow after reproducing. In the extreme case, death always follows reproduction, and the organism is semelparous.

The timing of reproduction is crucial to its success.

45.11 Agaves Reproduce Once and Die
The plant shown here has mobilized the energy stored during its long life to produce a large flowering stalk with hundreds of flowers, literally reproducing itself to death.

A useful analogy compares the production of off-spring to earning interest on money deposited in a bank. It pays to deposit money in the bank as soon as possible so that it can begin earning interest. Off-spring that are produced early likewise "yield interest" sooner because they themselves begin to reproduce sooner. However, if juvenile survival is very poor and reproduction greatly reduces the chances for parental survival, then natural selection may favor the parents' delaying reproduction until they are older. Among birds, for example, most individuals begin to reproduce when they are a year old, but gulls, penguins, and albatrosses do not breed until they are 3 to 9 years old. All these birds reach full size within a year, but young individuals of these species cannot compete effectively with older ones for food, mates, or breeding sites. Therefore, if they attempted to breed, they would be forced to use poor sites, and they and their offspring would thus suffer higher death rates than if they waited until they gained more experience.

There is a tight tradeoff between the number of offspring produced and their size because it "costs more" to produce a large offspring than a small one. If we hold constant the amount of energy available to a parent for reproduction, doubling the cost of raising an individual offspring reduces by half the number that can be raised. Natural selection favors fewer but larger offspring if large offspring survive so much better than small offspring that more of them survive to reproductive maturity. This pattern is common in highly competitive environments, where offspring do not survive well if they become independent from their parents when still at a small size. In fluctuating environments, where disturbances create unoccupied sites, offspring survival depends mostly on finding an opening. Most species that colonize disturbed areas produce many small offspring. These patterns are illustrated in Table 45.3, which gives the sizes of seeds of plants in four environments arranged in order from disturbed areas to dense woodlands.

Changes in Death Rates

As we have just seen, the demands of reproduction often cause death rates to be higher among reproducing than among nonreproducing individuals. Death rates are also influenced by reproductive rates through what is known as **reproductive value**. Reproductive value is a measure of the contribution an individual alive at age x can make to the growth rate of the population. Many newborn individuals die before they have a chance to reproduce. The older a prereproductive individual is, the greater the probability that it will survive to reproduce. The reproductive value of an individual thus steadily increases

TABLE 45.3
Average Seed Size in Relation to Environment in British Plants

ENVIRONMENT	NUMBER OF SPECIES	AVERAGE SEED WEIGHT (MILLIGRAMS)
Open, disturbed	70	1.19
Semiclosed	22	2.21
Scrub and wood margins	32	4.44
Woodland (ground plants)	27	13.69
Woodland (shrubs)	21	93.70

until it begins to reproduce. Once reproduction begins, however, an individual has already transferred some of its reproductive value to its offspring. Therefore, its reproductive value declines, and reaches zero when it has finished reproducing, at which point it can make no further contribution to the growth rate of the population.

Because reproductive value declines with age, the power of natural selection acting on alleles that produce their phenotypic effects only at older ages is increasingly weaker. After the last offspring have been born and raised, natural selection is powerless to influence phenotypic traits, even those that are highly detrimental to survival. However, alleles that delay the phenotypic expression of deleterious alleles *are* favored by natural selection. The result is a pushing back in the time of expression of many deleterious alleles. These alleles increasingly express themselves phenotypically as individuals age, causing increased mortality, especially after reproduction has ceased.

The fact that undesirable phenotypic effects of alleles are often expressed late in life poses serious social problems for people in modern industrial societies. As a result of improved hygiene and nutrition, most people in these societies are now spared the serious childhood infections that cause death rates to be high in nonindustrial societies. Most people live to the time when the so-called "genetic diseases of old age" begin to afflict them. Cancer and heart diseases are the main killers in industrialized societies. They are much more difficult to deal with than the infectious diseases that were the former causes of death. The social costs of extending life for a few more years for persons six decades old and older are great. Therefore, despite the expenditure of enormous resources to extend the lives of old people, the average age at death in the United States has changed very little during the past 30 years. As one of the killers is eliminated, another takes its place (Figure 45.12). There appears to be no escape from this situation.

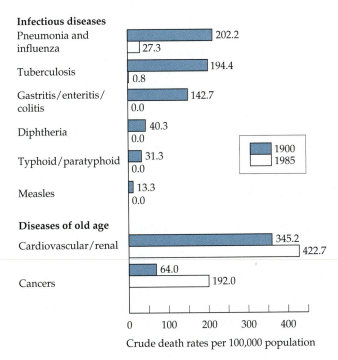

Infectious diseases

Pneumonia and influenza: 202.2 / 27.3

Tuberculosis: 194.4 / 0.8

Gastritis/enteritis/colitis: 142.7 / 0.0

Diphtheria: 40.3 / 0.0

Typhoid/paratyphoid: 31.3 / 0.0

Measles: 13.3 / 0.0

Diseases of old age

Cardiovascular/renal: 345.2 / 422.7

Cancers: 64.0 / 192.0

Legend: 1900 / 1985

Crude death rates per 100,000 population

45.12 The Causes of Human Death Change
During the past century the major causes of human deaths in the United States have shifted from contagious diseases to the diseases of old age.

POPULATIONS IN PATCHY ENVIRONMENTS

The environment in which an organism normally lives is called its **habitat**. Everything that the organism needs must be provided by its habitat. Over a geographic area, suitable habitat may be broken into a number of patches, some of which are large enough to support one or more individuals for their lifetimes, or at least for a complete breeding season. In most areas, patches of suitable habitat are separated by spaces in which the environment is not favorable for the species. To understand the dynamics of most populations, we need to examine the problems that patchiness causes for individuals. Also, although each patch of suitable habitat provides the basic needs of an organism, patches may differ from one another in ways that influence how the organisms use them.

Habitat Selection

Some organisms, such as plants and many marine organisms, do not choose their habitats. Their spores, seeds, or larvae are carried by external forces, and once they are deposited, the developing organisms cannot move. However, most animals choose habitats, and in patchy environments, they make two basic choices: The first is where to settle. The second is when to leave their current location for another.

Some organisms make both these choices many times in their lives, whereas others make only the first choice and make it only once, after which they must live where they are until they die. Barnacle larvae settling out of the water can move for short distances, assessing the densities of already settled individuals and the substrate, but once they metamorphose, they are permanently attached.

TRADEOFFS IN HABITAT SELECTION. An organism seeking a suitable habitat is unlikely to encounter a top-quality one at the beginning of its search. Should it accept a suboptimal habitat it has just encountered or reject that habitat and search for a better one? The consequences associated with either choice depend upon a number of factors. How much poorer than the best possible habitat is the one just found? What is the probability that a better habitat will be found in a given amount of searching? How risky is it to keep on searching? How easy is it to search? How much time is available to search? How fast are habitats filling up? Organisms that can search easily and at low risk should be less willing to accept a poor habitat than organisms that search with higher risk and greater difficulty.

Barnacle larve have little energy to devote to habitat selection, but many birds can fly long distances quickly and efficiently. They are at relatively low risk while they search and can assess many different habitats in a short time. Jaegers are seabirds that feed their young on lemmings—rodents whose abundance fluctuates dramatically from year to year. Because of their mobility, jaegers can search widely for places where lemmings are abundant (Figure 45.13). A given pair may move its nesting location hundreds of kilometers from one breeding season to the next.

At the other extreme, seeds of plants rarely move

45.13 A Long-Tailed Jaeger Searches for Lemming Outbreaks
Dense populations of lemmings necessary to support successful breeding by jaegers are patchy in space and time, so the birds must spend a great deal of time searching for suitable breeding areas.

once they have landed. However, seeds can wait for changes in patch quality. Many seeds are viable for decades and may lie in the soil waiting for a change in conditions to break their dormancy and trigger germination. Seeds of some desert plants do not germinate until the seed coat is scarified, which normally happens when there has been a flash flood and the seed has been scraped by rocks and boulders. This means that they germinate only when the soil is full of moisture and thus favorable for seedling growth. Other seeds wait for a rain of a particular magnitude, not germinating until certain compounds have been leached out of the seed coat.

POPULATION DENSITY AND HABITAT SELECTION. In general, as habitats fill up with members of a population, conditions deteriorate for each individual. Therefore, habitat selection involves assessment of both the quality of the site and the density of conspecifics (other individuals of the same species) already living there. The first members of a population to settle in an area select the best available sites, but as densities of individuals in the best sites increase, survival and reproductive success are lowered, and those sites become less suitable. Therefore, poorer but empty sites may afford survival and reproductive rates equal to those in better but more crowded sites.

If settling individuals are able to assess both the quality of sites and the densities of conspecifics in them, they should settle in the best available situation, occupying increasingly poorer sites as better ones fill up. If their assessments are reasonably accurate, the average success will be the same for individuals in all habitats, because the higher densities in better habitats will exactly offset the higher quality of those habitats. Ecologists call the distribution of organisms that would result from such assessment an **ideal free distribution** because the organisms are assumed to possess complete knowledge about the qualities of the habitats and the densities of conspecifics in them (the *ideal* part), and to be *free* to settle wherever they wish. As you probably recognize, the ideal free distribution is what scientists call a model. It predicts what the distribution of organisms and their reproductive success will be if those two assumptions are met.

However, the two assumptions of this model are rarely met in nature. The "free" assumption is violated when the individuals already living at a site block newcomers' attempts to settle there. The most widespread form of such behavior is **territoriality** — the defense of a space from which all conspecifics except one's mate and recent offspring are excluded. The defense of a territory reserves its resources for the exclusive use of its occupants. Its cost is measured in the time and energy expended and the risks that may be associated with acting conspicuously in an environment with many predators (see Box 44.B).

Where there is territoriality or other defense of space, additional individuals may be unable to settle even if it would be advantageous for them to do so. They have no choice but to settle in a poorer place where their success will not be as good. Territorial species often have a **dominance distribution**, which differs from an ideal free distribution in that the individuals that breed in better habitats have better reproductive success than individuals that breed in poorer habitats.

In monogamous species, both males and females have the same distribution, but among polygamous species the sexes may have very different distributions. Male red-winged blackbirds defend territories within which up to a dozen females settle. These females sometimes return to the same territory in consecutive years, but they often move among territories, changing locations and mates, especially if their nests fail. Their distribution is close to an ideal free one, whereas the males have a very marked dominance distribution (Figure 45.14).

The "ideal" assumption is also violated because individuals do not usually have complete information about habitat quality and densities of already settled individuals. Female red-winged blackbirds do not defend territories, and they can easily gather fairly complete information about potential habitats because they breed in marshes with low vegetation whose quality can be assessed quickly and within which

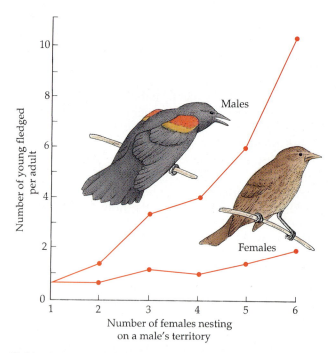

45.14 Reproductive Success of Blackbirds Fits Two Distribution Models
Because they are territorial (see Figure 44.14), male red-winged blackbirds have a dominance distribution, whereas females have an ideal free distribution.

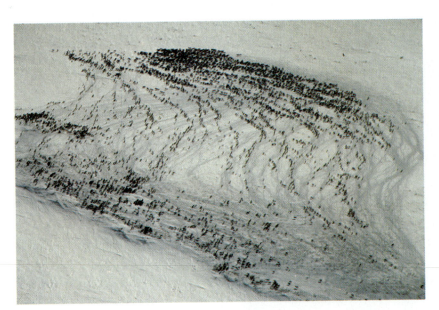

45.15 Animals Migrate to Remain in Suitable Environments
These caribou in the American Arctic are migrating from the open tundra to winter feeding grounds at the edge of the boreal forest where food, some of it in the form of lichens on the branches of trees, is more readily available when the ground is covered with snow.

other birds are easily seen. This is probably why their distributions approximate ideal free ones.

When individuals select habitats with less than complete information, their distributions frequently deviate from an ideal free one. However, in such a case, their reproductive success is not usually higher in the better habitats, nor is it the same in all habitats. In other words, their distributions do not fit either the ideal free distribution model or the dominance distribution model well. This tells us that the "ideal" assumption has been violated, and that we need to find out why the environmental assessments of the individuals are inaccurate.

Dispersal

Up to this point we have been considering patchiness that breaks up space. Environments can also be patchy over time. Such temporal patchiness provokes many kinds of responses from organisms. A favorable habitat can change temporally into an unfavorable place because of environmental changes, such as a drought or the coming of winter. At some point, habitat quality may decline so much that individuals can improve their chances of survival and reproduction by going elsewhere. If regularly repeated seasons are the prime cause of temporal changes in a habitat, life cycles adjust in equally regular ways so that organisms appear to anticipate the changes.

One of the most spectacular responses to seasonal changes in habitat quality is **migration,** the regular seasonal movement of animals from one place to another. This behavior is most widespread among birds, but some insects and mammals also migrate (Figure 45.15). Migrations of birds, mammals, and insects serve primarily to keep them in good foraging areas at all times of the year. Wildebeests, large an-

telopes of East Africa, follow the rains to places where there is fresh growth of grasses. Most insectivorous birds leave high latitudes in the winter for more favorable wintering grounds.

Movements of animals that are less regular than annual migrations may be caused by **irruptions,** which are buildups of large populations when food supplies are favorable. Irruptions are often followed by mass dispersal when supplies deteriorate. The most famous case is that of "migratory" locusts in Africa that irrupt in certain centers and spread out over areas where they do not normally live, devouring nearly everything in their path (Figure 45.16). Lemmings in the Arctic irrupt, deplete their habitat, and then move, searching for areas with better food supplies.

Movement to a different place is not the only way organisms cope with habitat change. Many organisms have a life cycle with a resting stage in which they survive unfavorable periods. In plants this is the seed or spore stage. Many plants, such as lichens, mosses, and some ferns, can dry up, then rehydrate when moist conditions return. Vascular plants may die back to the ground and then regrow during the next favorable season, or they may lose their leaves during an unfavorable period and regrow them again.

Animals also have a variety of ways of coping with unfavorable periods without moving to another place. Many invertebrates pass unfavorable periods as eggs and pupae, which have very low metabolic rates. If the unfavorable period is cold, the adults themselves may crawl into a safe place for the winter. As you will recall from Chapter 33, many mammals **hibernate** in winter; that is, their metabolic rates drop to very low levels to conserve energy. If the unfavorable period is dry, some animals seek moist sites

45.16 A Swarm of Migratory Locusts
A swarm like this one over a cotton field in Ethiopia can totally defoliate a field within minutes. These insects are also captured and eaten by people, and are considered a delicacy in some areas.

or bury themselves in the soil and then secrete an impermeable membrane that prevents water from escaping. Some other animals store food for times of shortage.

Because habitats for most organisms are patchy in space and over time, a map of the distribution of individuals of a species will show places where the species is not found, as we saw in Figure 18.4. In those areas where the species is found, the number of individuals per unit of area (population density) will reflect the current and past quality of the habitat. The distribution map will change over time, and the changes reveal much about the factors that influence the distribution and abundance of individuals in the population. Therefore, even though the study of patchiness and how organisms respond to it is much more difficult than the study of a population in a single area, the effort is essential in order to understand the dynamical behavior of populations.

POPULATION REGULATION

Some species are more common than others, and the common ones tend to stay common and the rare ones rare. This suggests that the sizes of populations in nature are regulated in some way. There are, of course, substantial fluctuations in abundances, some examples of which we have just discussed. But many populations fail to attract our attention precisely because they do not fluctuate very much. Although we have just discussed the importance of patchiness for population ecology, we begin our discussion of population regulation by ignoring patchiness. Once the basic ideas have been explained, it will become clear how they are influenced by habitat patchiness.

Density Dependence

A population may be more likely to decrease in density when its members are common and to increase in density when they are rare. Population regulation that works in this **density-dependent** way can be caused by birth or death rates changing in response to density. We can think of several situations in which death or birth rates might change in a density-dependent manner. As a species increases in abundance, it may deplete its food supply so that each individual gets less. This decreased nutrition may increase death rates and lower birth rates. Predators may be attracted to regions where the numbers of their prey have increased. If the predators are then able to capture a larger proportion of the prey than they did when the prey were scarce, the death rate of the prey goes up. Diseases, which may increase death rates, spread more easily in dense than in sparse populations.

If the death rate in a population is unrelated to the population's density, the resulting change in population size is said to be **density-independent**. For example, a very cold spell in winter may kill a large proportion of individuals of a species regardless of the population's density. Nonetheless, seemingly density-independent environmental factors often act together with population density in changing population size. Cold weather may not kill organisms directly, but it may increase the amount of food they need to eat each day, so that individuals pushed by population density into poorer foraging areas are more likely to die than those with better foraging areas. Or the probability of dying may be related to the quality of sleeping places for the night. If the density of the population is high, a larger fraction of

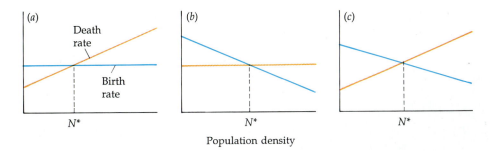

45.17 Density-Dependent Population Regulation
If either birth or death rates are density-dependent, a population tends to
fluctuate around an equilibrium density, designated here as N*. (a) Density-
independent birth and density-dependent death rates. (b) Density-dependent
birth and density-independent death rates. (c) Both birth and death rates are
density-dependent.

the individuals may be sleeping in poorer places
where they will experience greater chilling.

The concept of density dependence is important
for understanding population regulation. Population
regulation can result from various combinations of
density-dependent and density-independent processes, as shown in Figure 45.17. The graphs in this
figure show how birth and death rates could change
in relation to population density. If either birth or
death rates, or both, are density-dependent, there is
an equilibrium, shown by N* in the figure, to which
the population density tends to return if it moves
either above or below that density. If neither rate is
density-dependent, there is no equilibrium, but, as
we have just discussed, some of the agents influencing birth and death rates are likely to act in a density-
dependent manner. The actual abundance of a species is determined by the combined effects of all the
factors and processes, density-dependent and density-independent, impinging upon its subpopulations.

Populations differ markedly in the magnitude and
frequency of their fluctuations in density. Some species or local populations, such as most insects in the
temperate zone, are recovering from past crashes
much of the time. Others, such as many bird species,
appear to be close to the environmental carrying capacity much of the time. How changes in densities
are related to the proportion of time that populations
are close to environmental carrying capacity is illustrated in Figure 45.18.

Disturbance

The processes of growth, death, immigration, and
emigration guarantee that populations are always in
dynamic flux. In addition, populations are exposed
to disturbances. A **disturbance** is a short-term event
that disrupts populations, communities, or ecosystems by changing their environment. Disturbances
differ in spatial distribution, frequency, predictability,
and severity. Common physical disturbances are
fires, hurricanes, ice storms, floods, landslides, and
lava flows. Biological disturbances include tree falls,
diseases, and the burrowing and trampling activities
of animals.

Responses of organisms depend upon the frequency and severity of the disturbances. Organisms
respond behaviorally and physiologically to disturbances that occur regularly and repeatedly during their

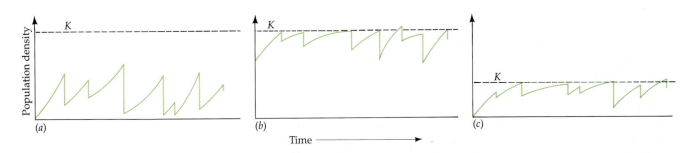

45.18 Population Dynamics and Disturbance
(a) Dynamics of a population dominated by phases of population growth after
repeated disasters. The population density is well below K most of the time.
(b) Dynamics of a population dominated by limitations on environmental
carrying capacity. The population density is close to K most of the time.
(c) Same as in (b), but with a much lower carrying capacity.

45.19 Opened by Fire
These capsules of *Eucalyptus pyriformis* from Western Australia were opened by the heat of a bush fire. They scattered their seeds onto the bed of ashes, where some of the seeds germinated.

ground parts are destroyed by fire. Some plants that are adapted to fire-prone environments mature their seeds in tough capsules or cones that remain closed until heated to very high temperatures. The heat of a passing fire causes the capsules to open and release the seeds within a few days. The seeds then fall and germinate in the bed of ashes (Figure 45.19).

Organisms also influence the frequency of disturbances. Immediately after a fire, there is not enough fuel to carry another fire. However, as vegetation grows back, fuel accumulates, gradually increasing the chances of another fire. Thus, the frequency of fires may be proportional to the rate at which vegetation accumulates as fuel. As many trees age, their roots become weakened by fungal infections. Old, large trees are thus susceptible to being toppled by high winds. Therefore, the likelihood of a major blowdown increases with forest age. A striking example is provided by the waves of blowdowns that move across fir forests in northern New England, generating a complex pattern of forests at different ages of regrowth (Figure 45.20).

lifetimes. Animals seek shelter in storms and go into hibernation in winter. Trees drop their leaves in winter and change physiologically so that they can tolerate frosts. However, if a particular disturbance is unusually severe, the tolerances of individuals may be exceeded and heavy mortality may result. An organism is likely to have adaptations for tolerating a particular kind of disturbance only if the disturbance is frequent relative to the organism's life span, because only when that is the case can natural selection be effective.

Adaptations to disturbance are widespread among organisms. Many plants resprout after their above-

POPULATION REGULATION AND MANAGEMENT

For many centuries, humans have tried to decrease populations of species they consider undesirable and to increase populations of desirable species. This has motivated many studies of natural populations. The desirability of species changes as new uses for their products are discovered or old products lose their markets. For example, guayule (*Parthenium argentatum*), a shrub of the deserts of the southwestern United States and Mexico, produces a milky sap with many of the properties of rubber. Guayule is now being cultivated, and there is intensive research on

45.20 Waves of Mortality among Fir Trees
Several waves of mortality can be seen in this forest of Balsam firs in northern New England.

the properties of its rubberlike sap. Thus, a very obscure shrub of no apparent value has rapidly become an object of considerable interest. There is, however, some stability in our perceptions of desirability and undesirability of species. Animals that eat crop plants have long been considered undesirable, whereas many kinds of fishes, game birds, and mammals are consistently desirable. Understanding how populations grow and are regulated provides the basis for control and management policies.

A general principle derived from population studies, shown in Figure 45.9*b*, is that the total number of births and the growth rates of individuals tend to be highest when a population is well below its carrying capacity. Therefore, if we wish to maximize the number of individuals that can be harvested, we should manage populations so that they are far enough below carrying capacity to have high birth and growth rates. Hunting seasons for birds and mammals are set with this objective in mind.

However, in populations, such as many species of fishes, in which individual females lay thousands or millions of eggs and in which growth is density-dependent and extends over long periods of time, both the *size* and the number of individuals harvested affect the sustainable yield. If prereproductive individuals are harvested at a high rate, the growth rates of the remaining individuals normally increase. Most fish populations can be harvested very heavily on a sustained basis because only a small number of females must survive to reproductive age to produce the eggs needed to maintain high rates of entry of young fish into the population.

Fish can, of course, be overharvested. Indeed, many populations have been greatly reduced because so many individuals were harvested that too few survived to reproductive age to maintain the population. The Georges Bank off the coast of New England—a source of cod, halibut, and other prime food fishes—is a case in point. As fish populations were reduced, more effective equipment had to be used for their harvest, more effort had to be expended to get the same yield, income to fishermen dropped, and the price of fish to the consumer increased. Another example of excessive harvesting is provided by the whaling industry. The blue whale, Earth's largest animal, was the first whale species to be hunted nearly to extinction. The fishery then turned successively to smaller and smaller species of whales, the only ones still numerous enough to support commercially viable whaling operations (Figure 45.21). Management of whale populations is difficult for two reasons. The first is that whales have very low reproductive rates, so that a large number of whales is needed to produce even a small number of offspring. The second is that whales are distributed widely throughout Earth's oceans—thus they are an international resource whose conservation and wise

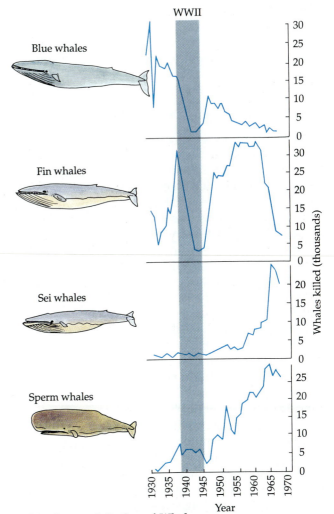

45.21 Overexploitation of Whales
These curves show the number of whales of four species killed each year between 1930 and 1968. The graph for the largest species is at the top; the one for the smallest species is at the bottom.

management depends upon cooperative action by *all* whaling nations. This has proved very difficult to achieve.

The same principles apply if we wish to reduce the size of populations of undesirable species and keep them at low densities. At densities well below carrying capacity, populations have high birth rates and can therefore withstand higher death rates than they could at densities close to carrying capacity. Taking away a species' resources reduces the carrying capacity of its environment—a permanent alteration—whereas killing part of the population is a temporary measure which the animals, by their reproduction, usually counteract swiftly. It is better to get rid of rats around dumps and in cities by making garbage unavailable, that is, by lowering carrying capacity, rather than attempting to poison the rats. Similarly, if we wish to preserve a rare species, the most important step is to provide it with suitable habitat. If habitat is available, the species will usually reproduce at rates sufficient to maintain the population. If there is insufficient habitat, preserving the species usually requires expensive and continuing intervention, such as providing extra food.

Management of the Human Population

Management of our own population has become an urgent ecological issue because human population growth is responsible for most environmental problems, from pollution to extinctions of other species. For thousands of years, Earth's carrying capacity for human populations was set by food, water supplies, and disease—that is, by the technology available to garner resources and to combat diseases. Domestication of plants and animals and cultivation of land enabled our ancestors to increase dramatically the resources at their disposal. These developments stimulated rapid population growth up to the next carrying capacity limit, which was set by the agricultural productivity possible with human- and animal-powered tools. Agricultural machines and artificial fertilizers came with the tapping of fossil fuels, greatly increasing agricultural productivity and thereby raising Earth's carrying capacity for people. The development of modern medicine made diseases lose their effectiveness as limiting factors on human populations. In combination with hygiene, medicine has allowed people to live in large numbers in areas where diseases formerly kept numbers very low. This raised global carrying capacity still further (Figure 45.22).

We have now arrived at the stage where carrying capacity is set by Earth's ability to absorb the by-products of our enormous consumption of fossil fuel energy and by whether we are willing to cause the extinction of millions of species to accommodate our increasing use of environmental resources. How many people Earth can and should support has become the most complex and urgent debate of our times. We will explore some of the consequences of high human population densities for the survival of other species in Chapter 49.

SUMMARY

The density of the members of a population, their spacing, and their age distribution constitute its population structure. Population dynamics are changes in population structure that result from the reproduction, growth, energy gathering, dispersal, and death of members of the population. All individuals, if they survive, go through stages during which they carry out these activities at some time during their life history. Ecologists can study how births and deaths affect population dynamics by constructing a

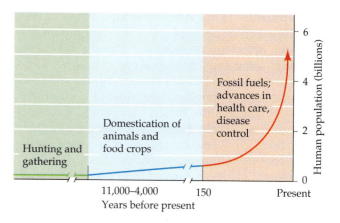

45.22 Human Population Growth
For thousands of years the human population was relatively stable. It increased somewhat with domestication of plants and animals. Recently the human population has grown dramatically as a result of increases in Earth's carrying capacity for humans brought about by modern medicine and the use of energy from fossil fuels.

life table for a cohort. Populations in resource-rich environments increase exponentially, but population growth slows and ceases when the carrying capacity of the environment is reached. Birth rates are affected by the amount of resources available to the parents. Death rates of young individuals are usually high among species that produce large numbers of offspring and provide them with little energy or care. Parental care reduces juvenile death rates, allowing more individuals to survive to postreproductive age, when death rates again become very high.

Most plants and many sessile animals do not select habitats, but most mobile animals do. Because increasing population densities usually reduce environmental quality, individuals often find it profitable to prevent others from settling. If its habitat changes from favorable to unfavorable on a regular seasonal basis, a population may migrate. Members of some species pass times of resource shortage in resting stages.

Most populations fluctuate in density less than might be expected from their potential birth and death rates because a higher proportion of individuals die when densities are high than when densities are low. Regulation of population numbers is usually achieved by several different factors acting together.

Populations are often managed by people to increase numbers of desirable species and decrease numbers of undesirable ones. Desirable species are best encouraged by providing them with suitable habitats. Under favorable conditions many animals and perennial plants can withstand high rates of harvesting, but many commercially important species have been overexploited. Management of the human population has become the most urgent environmental and social problem of the present time.

SELF-QUIZ

1. The number of individuals of a species per unit of area is known as its:
   a. population size.
   b. population density.
   c. population structure.
   d. subpopulation.
   e. biomass.

2. The age distribution of a population is determined by:
   a. the timing of births.
   b. the timing of deaths.
   c. the timing of both births and deaths.
   d. the rate at which the population is growing.
   e. the rate at which the population is decreasing.

3. Which of the following is *not* a stage in the life history of all organisms?
   a. Growth stage
   b. Dispersal stage
   c. Reproductive stage
   d. Reorganization stage
   e. Energy-gathering stage

4. Which of the following is *not* a demographic process?
   a. Growth
   b. Births
   c. Deaths
   d. Immigration
   e. Emigration

5. A group of individuals born at the same time is known as a:
   a. deme.
   b. subpopulation.
   c. Mendelian population.
   d. cohort.
   e. taxon.

6. A population grows at its intrinsic rate of increase when:
   a. its birth rates are the highest.
   b. its death rates are the lowest.
   c. conditions are optimal.
   d. it is close to the environmental carrying capacity.
   e. it is well below the environmental carrying capacity.

7. Organisms that reproduce only once in their lifetime are said to be:
   a. semelparous.
   b. annuals.
   c. perennials.
   d. iteroparous.
   e. semiparous.

8. Which of the following is *not* true of reproductive value?
   a. Reproductive value is a measure of the contribution an individual of age x can make to the growth rate of the population.
   b. The reproductive value of an individual increases until it begins to reproduce.
   c. Reproductive value reaches its maximum when an individual completes reproduction.
   d. Reproductive value reaches its maximum when an individual begins to reproduce.
   e. Reproductive value declines during the reproductive life of an individual.

9. To achieve an ideal free distribution organisms must have:
   a. complete knowledge of qualities of habitats and the densities of conspecifics in them.
   b. complete knowledge of the qualities of habitats and the ability to settle wherever they wish.
   c. complete knowledge of qualities of habitats, the densities of conspecifics in them, and the ability to settle wherever they wish.
   d. the ability to settle in all ideal habitats.
   e. the ability to defend a territory in an ideal habitat.

10. Density-dependent population regulation results when:
    a. only birth rates change in response to density.
    b. there is little immigration.
    c. diseases spread in dense populations.
    d. birth or death rates or both change in response to density.
    e. population densities fluctuate very little.

11. The best way to reduce the population of an undesirable species is to:
    a. reduce the carrying capacity of the environment for the species.
    b. selectively kill reproducing adults.
    c. selectively kill prereproductive individuals.
    d. attempt to kill individuals of all ages.
    e. sterilize individuals.

FOR STUDY

1. Huntington's chorea is a severe disorder of the human nervous system that generally results in death. It is caused by a dominant allele that does not usually express itself phenotypically until its bearer is 35 to 40 years old. How fast is the gene causing Huntington's chorea likely to be eliminated from the human population? How would your answer change if the gene expressed itself when its bearer was 20 years old? 10 years old?

2. Many people have improperly formed wisdom teeth and must spend considerable sums of money to have them removed. Assuming, as is probably the case, that the presence or absence of wisdom teeth and their mode of development are partly under genetic control, will we gradually lose our wisdom teeth by evolutionary processes?

3. Some organisms, such as oysters and elm trees, produce vast quantities of offspring, nearly all of which die before they reach adulthood. What fraction of such deaths are likely to be selective, that is, dependent on the geno-types of the individuals dying? If in fact most such deaths are non-selective, what does that imply for the rates of evolution of oysters and elms?

4. Most organisms whose populations we wish to manage for higher densities are long-lived and have low reproductive rates, whereas most organisms whose populations we attempt to reduce are short-lived but have high reproductive rates. What is the significance of this difference for management strategies and effectiveness of management practices?

5. In the mid-nineteenth century, the human population of Ireland was largely dependent upon a single food crop, the potato. When a disease caused the potato crop to fail, the Irish population declined drastically for three reasons: (1) a large percentage of the population emigrated to the United States and other countries; (2) the average age of a woman at marriage increased from about 20 to about 30 years; and (3) many families starved to death. None of these social changes was planned at the national level, yet all contributed to adjusting population size to the new carrying capacity. Discuss the ecological strategies involved, using examples from other species. What would you have done had you been in charge of the national population policy for Ireland?

6. From a purely ecological standpoint, can the problem of world hunger ever be overcome by improved agriculture alone? What other components must a hunger-control policy include?

READINGS

Begon, M. and M. Mortimer. 1986. *Population Ecology: A Unified Study of Animals and Plants*, Second Edition. Blackwell Scientific Publications, Oxford. An introduction to population dynamics that stresses the differences between plants and animals.

Begon, M., J. L. Harper and C. R. Townsend. 1990. *Ecology: Individuals, Populations and Communities*, 2nd Edition. Blackwell Scientific Publications, Oxford. A basic text for all aspects of contemporary ecology.

Harper, J. L. 1977. *The Population Biology of Plants*. Academic Press, New York. The most complete review of the literature on plant populations; an excellent advanced reference for most topics in plant population biology.

Hutchinson, G. E. 1978. *An Introduction to Population Ecology*. Yale University Press, New Haven, CT. An advanced text with a particularly good historical account of the development of modern ideas.

Lack D. 1954. *The Natural Regulation of Animal Numbers*. Clarendon Press, Oxford. A readable classic work on population ecology.

Ricklefs, R. E. 1990. *Ecology*, Third Edition. W. H. Freeman and Company, New York. An excellent text that covers both dynamical and evolutionary aspects of ecology.

Wilson, E. O. and W. H. Bossert. 1971. *A Primer of Population Biology*, Sinauer Associates, Sunderland, MA. A self-teaching book that covers the basic principles of population ecology.

46

Interactions within Biological Communities

PREVIEW: The distributions and abundances of species are determined by the physical environment and by interactions with other species. Organisms interact by eating one another, by reducing one another's food supplies or limiting their access to other resources, by benefiting one another, by helping without being helped, or by harming without being harmed. Predators and prey influence one another's abundances and also influence one another's evolution. Competing organisms exclude one another from some habitats, but many species with similar resource requirements live together. Mutualisms connect dissimilar organisms, some so different as to be in different kingdoms. Mutualisms are especially widespread between plants and other organisms because microorganisms help plants obtain nutrition and because animals transport many plant gametes and seeds. Species influence the communities in which they live by changing local climates, by providing structure, and by altering community composition. The numbers of species living together in biological communities are determined by many factors, both local and regional, acting together.

This chapter deals with resources, ecological niches, foraging theory, predator–prey interactions, host–parasite interactions, competition, commensalism, mutualism, coevolution, how species affect biological communities, and patterns of species richness.

Reproduction by most species of fig trees is intimately linked to the reproduction of certain wasps. Most wasp species visit only one fig species and most fig species are visited by only one wasp species. A fig tree begins reproduction by producing a large number of closed inflorescences called syconia. Inside each syconium many female flowers form. Female fig wasps pollinate these flowers. A female fig wasp, bearing both fertilized eggs and fig pollen, enters a syconium through a small hole that soon seals. She pollinates receptive female flowers, lays her eggs in the ovaries of some of the flowers, and then dies. Each wasp larva develops within, and eats, one seed. As the syconium ripens, pollen-bearing male flowers mature. Just before the fruit ripens, wingless male fig wasps chew their way out of the seeds in which they developed and crawl around inside the syconium searching for seeds housing female wasps. The males chew open these seeds and mate with the females inside them. The females then emerge from their seeds and collect pollen from male flowers. Females of most species have specialized structures on their thoraxes to hold the pollen. The females leave the syconium through a hole cut in its wall by the males, fly to another fig tree, and begin the reproductive cycle again (Figure 46.1). The syconium then undergoes final ripening, becoming a soft, sweet fig that may eventually be consumed by some bird or mammal.

Fig wasps depend completely on fig trees for the completion of their life cycles. To produce offspring, a female fig wasp must carry pollen to a receptive fruit and pollinate the flowers. Otherwise the ovaries of the flowers will not develop into seeds that can be eaten by wasp larvae. In some fig species, only one female wasp normally enters a syconiuim. If she is not carrying pollen, none of her offspring survive.

Although a fig requires fig wasps in order to produce viable seeds, wasp larvae consume many potential fig seeds. A fig thus pays a considerable price in seeds to get its flowers pollinated. Seeds eaten by fig wasps are part of the investment a fig makes in order to be a pollen recipient and pollen donor. If a syconium is not visited by a fig wasp bearing pollen, no seeds mature and the fruit is aborted.

Some other species of wasps do not pollinate figs. Instead, they penetrate the fruit walls with their ovipositors and lay their eggs either on developing seeds

46.1 Figs and Fig Wasps
Tetrapus costaricensis wasps are emerging from seeds of a fig. The black wasps are winged females; the brown individuals are wingless males. Dark seeds contain wasps that have not yet emerged. The large, cream-colored seeds are viable and intact fig seeds. The white structures are pollen-bearing fig flowers.

or on fig wasp larvae. These wasps either kill fig seeds directly or kill fig wasp larvae without providing any benefit to the fig.

The interaction between figs and fig wasps shows that populations of organisms interact with one another in complex and surprising ways. These interactions can be grouped into a few major types (Table 46.1). If one organism uses another as its source of energy, the eater is called a predator or parasite and the eaten is its prey or host. Their interaction is known as a **predator–prey** or **host–parasite interaction**. If two organisms use the same resources and those resources are insufficient to supply their combined needs, the organisms are competitors, and their interactions constitute **competition**. If both participants benefit, we call them mutualists, and their interaction constitutes a **mutualism**. If one participant benefits but the other is unaffected, the interaction is a **commensalism**. If one participant is harmed but the other is unaffected, the interaction is an **amensalism**.

As illustrated by the fig–fig wasp interaction, these types of interactions are not clear-cut, both because

the strengths of interactions vary and because many cases do not fit the categories neatly. Both figs and fig wasps are completely dependent on their mutualism, but the fig wasp is also a predator on fig seeds. Some of the birds and mammals that eat figs kill fig seeds, whereas others pass the seeds unharmed through their digestive tracts, thus dispersing them to new locations. Nonetheless, many interactions fit well within the categories in the simple scheme shown in Table 46.1, so we will use them as a guide for exploring interactions among species.

Organisms living together in a small areas form a biological community. Figs and fig wasps, for example, are part of a community that includes other plants and animals, fungi, protists, and monerans. Each of the species interacts in unique ways with other species in its community. The study of such interactions, and how they determine which and how many species live in a place, is the focus of community ecology.

RESOURCES AND CONSUMERS

Many interactions within communities center on resources and consumers. A **resource** is any substance directly consumed by an organism that can potentially lead to the growth of its population. The two key properties of a resource are that its amount or availability is reduced by being consumed and that it is used by an organism for its own maintenance and growth. We usually think first of resources that can be consumed by being eaten, but organisms also consume space, hiding places, and nest sites by occupying them. Factors such as temperature, humidity, salinity, and pH, even though they may strongly affect population growth, are not resources because they are not consumed. Some resources, such as space, are not altered by being used. Occupied space may be unavailable to other organisms, but it immediately becomes available for occupancy when the user leaves. Other resources must regenerate after being reduced in quantity by consumers if they are to be available to the community in the future.

TABLE 46.1
Types of Ecological Interactions

		EFFECTS ON ORGANISM 2		
		BENEFIT	HARM	NO EFFECT
EFFECTS ON ORGANISM I	BENEFIT	Mutualism	Predation or parasitism	Commensalism
	HARM	Predation or parasitism	Competition	Amensalism
	NO EFFECT	Commensalism	Amensalism	—

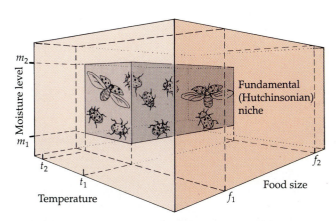

46.2 Three Dimensions of a Fundamental Niche
Only within particular environmental ranges of temperature, food-item size, and moisture can a ladybird beetle survive

Ecological Niches

Ecologists say that each species has an **ecological niche**, an abstraction that sums up the ecological position of the species within the community and the range of physical conditions and resources required by the species. However, this general definition serves only to indicate what we are talking about. Precise thinking about niches requires a more formal definition. The most useful one was proposed by G. E. Hutchinson. He suggested that niches could be characterized by determining the critical factors that directly influence a species and the range of values of those factors under which the species can survive. For example, individuals of a certain species might be able to grow and reproduce well only if the temperature lies between two values we will call t_1 and t_2; if moisture lies between levels m_1 and m_2; if habitat structure lies between h_1 and h_2; and if the individuals have food items between sizes f_1 and f_2. The **fundamental niche** of the organism can then be defined as the space within a hypervolume—a volume with more than three dimensions—determined by the intersection of all those ranges of critical factors. It is difficult to visualize a hypervolume, but a representation of three dimensions of a Hutchinsonian niche might look like Figure 46.2.

A fundamental niche represents the range of conditions under which a species could survive if there were no competitors, predators, or other negative influences. If some other species were added, however, the new species might reduce the resources used by the first species, or the new species might prey upon the first species so successfully in certain places that the first species could not survive there. In the presence of other species, therefore, the fundamental niche of a species is reduced to its **realized niche**.

A study of two species of barnacles, *Balanus bal-*

anoides and *Chthamalus stellatus*, provides a clear demonstration of the differences between fundamental and realized niches. These two species live in the intertidal zone of rocky North Atlantic shores. Adult *Chthamalus* generally live higher in the intertidal zone than do adult *Balanus*, but young *Chthamalus* settle in large numbers in the *Balanus* zone. In the absence of *Balanus*, young *Chthamalus* survive and grow well, but if *Balanus* are present the *Chthamalus* in the *Balanus* zone become eliminated by being smothered, crushed, or undercut by the larger, more rapidly growing species. Young *Balanus* may settle in the higher part of the intertidal zone, but they grow poorly because they lose water rapidly when exposed to air; therefore *Chthamalus* compete successfully with them there. The result is intertidal zonation, with *Chthamalus* growing above *Balanus* (Figure 46.3). The realized niche of *Chthamalus* is less than its fundamental niche because of competition with *Balanus* in the lower region, and the realized niche of *Balanus* is less than its fundamental niche because of competition with *Chthamalus* in the higher region. Experiments have shown that the ranges of both barnacles are greater if the other species is removed. Figure 46.3 shows the differences between fundamental and realized niches of the barnacles for only one dimension: height in the intertidal zone. Other important niche dimensions for barnacles include the type of substrate, the amount of wave action, and the water temperature.

One value of the Hutchinsonian niche concept is that it directs attention toward the environmental requirements that differ most among coexisting species. Even though a species requires a particular resource, that resource may not be significant for understanding the species' functioning in the community. For example, most terrestrial animals have strict but similar requirements for a certain minimum level of oxygen. However, studying the use of oxygen reveals very little about the structure of terrestrial communities because oxygen is nearly always present above that level; animals seldom deplete the supply. Interactions that influence distributions and abundances of terrestrial species are primarily over resources, such as food, that are depletable and renew slowly. In aquatic environments, however, where the oxygen supply is highly variable, organisms regularly deplete dissolved oxygen. Aquatic ecologists, unlike terrestrial ecologists, pay careful attention to oxygen as a key niche variable.

Foraging Theory

Obtaining food is the major activity driving both competitive and predator–prey interactions. Competing individuals attempt to obtain the same food. In predator–prey interactions, one animal attempts to eat the other. Therefore, ecologists want to know

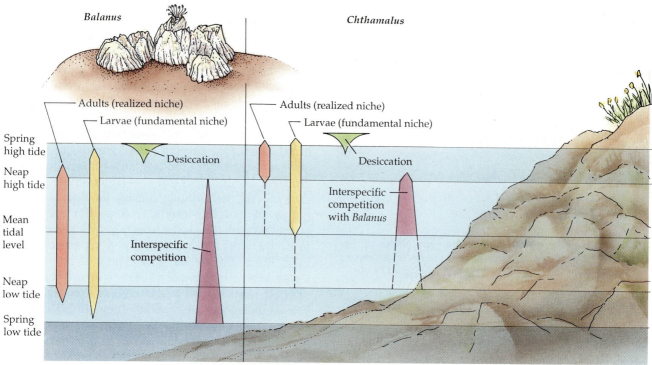

46.3 Fundamental and Realized Niches of Barnacles
The single dimension of vertical distribution tells much about the niches of these two species of intertidal barnacles. The width of the red and orange bars is proportional to the density of the populations. The width of the green and purple wedges indicates the importance of that factor in limiting population density. The lower end of the distribution of adult *Chthamalus* is strongly curtailed by interspecific competition with *Balanus*; the lower end of the distribution of adult *Balanus* contracts slightly because of interactions with other species.

how animals find and choose their food. Foraging theory was developed to help discover how predators decide where to forage and which prey to choose from among the potential prey they encounter.

Let us analyze the act of predation. First, a predator searches for prey. When a potential prey item is encountered, the predator decides whether to pursue it. If it decides to, a pursuit ensues. If the prey is captured, it must be "handled"—that is, subdued and eaten. Finally, there may be a postconsumption digestive pause if the predator cannot resume foraging immediately after eating (as when a snake swallows a very large prey). The time and energy required for each of these different activities must be accounted for if we wish to understand why predators forage as they do.

Ecologists who tackled this problem already knew that most animals eat fewer types of prey than they encounter and could capture. The ecologists also made some assumptions about how natural selection might operate. If one predator in a population, the ecologists reasoned, managed to obtain food more efficiently than other members of the population, it would have more time for other activities, such as reproduction, than the others. It might therefore leave more offspring. If there were a genetic basis for the superior foraging ability, the ability could, over evolutionary time, be selected for and become the characteristic foraging behavior of the species. The

ecologists then went on to model such foraging as follows.

If a predator is choosing prey so as to maximize its energy intake rate, then whether or not a prey item is included in the predator's diet does not depend upon the abundance of that prey type, but depends only on the abundances of other prey types that yield more energy per unit of time spent capturing and eating them. If the top-ranked prey type—the one yielding the most energy per unit of time invested—is sufficiently abundant, a predator captures the most energy per unit of time spent foraging by taking only that prey type and ignoring all others. As the abundance of the top-ranked prey type decreases, an energy-maximizing predator adds lower-ranked prey items to its diet in order of the energy per unit of time that they yield.

Ecologists tested this model using bluegill sunfish. Laboratory experiments with bluegills enabled investigators to measure energy content of different prey, handling time for prey, energy spent searching for and handling prey, and actual encounter rates with prey under different prey densities. Using these values, the investigators predicted the diets of bluegills—that is, they predicted the proportions of large, medium, and small prey bluegills would take in each of three environments stocked with different densities and proportions of the prey. For example, they predicted that in an environment stocked with a low

(a)

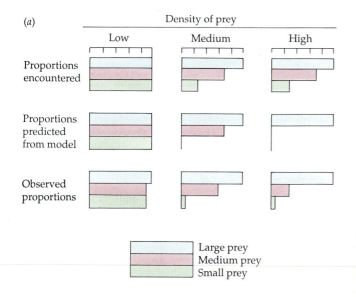

Density of prey

Low   Medium   High

Proportions encountered

Proportions predicted from model

Observed proportions

Large prey
Medium prey
Small prey

(b)

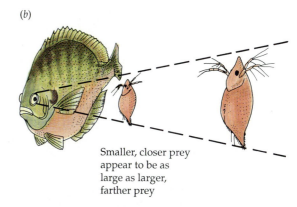

Smaller, closer prey appear to be as large as larger, farther prey

46.4 Bluegills Are Energy Maximizers
(a) Widths of the bars showing observed proportions in the diet are very similar to those showing predicted proportions, supporting the hypothesis that bluegills select prey so as to maximize energy intake. (b) How an energy maximizer can make a mistake.

prey density and equal proportions of the three sizes of prey, the bluegills would take equal proportions of all three sizes. The investigators put the bluegills in the three environments and observed the proportions of the prey they actually took. Figure 46.4a provides the predictions and the observed results for all three environments.

The proportions of large, medium, and small prey taken by the fish were very close to those predicted by the model. The major difference between predicted and observed results was that the fish ate some small prey that, according to the model, they should have ignored. The most likely reason for this is that the fish could not estimate accurately the true sizes of some prey. Determining sizes of objects is particularly difficult for a predator in open water where there are few clues about the distance to a prey. A smaller prey close to the predator and a larger prey farther away make images of the same size on the predator's retina (Figure 46.4b). Predators may thus attack and eat small prey that happen to be very close to them when first seen. This is in fact what the bluegills did.

PREDATOR–PREY INTERACTIONS

As we saw in the previous section, the sizes of predators and prey strongly influence their interactions. The relative sizes of a predator and its prey determine how the predator captures and handles the prey. If the predator is much larger than its prey, the prey are handled in bulk. The world's largest predators, baleen whales, feed on very small prey, which they filter from the water. Predators that are only moderately larger than their prey usually pursue and capture their prey one at a time. Many predators are as small as or even smaller than their prey and feed

upon them internally or after attaching to them externally. Such predators are usually called **parasites**. Parasites that are about the same size as their prey and are able singly or in a small group totally to consume an individual prey are known as **parasitoids**. Parasitoid wasps and flies prey on many species of insects.

Prey types also differ in whether and how vigorously they defend themselves against predators. Sessile prey may defend themselves chemically (by possessing chemicals that are toxic to their predators), physically (by having hard shells, spines, thorns, and so on), or by their behavior. Motile prey depend on being able to escape from predators. Prey also defend themselves by camouflage (making themselves difficult for predators to find) and by resembling inedible or dangerous objects.

In some environments, populations of prey are replenished immediately after being reduced by predators. On rocky shorelines, each wave brings a new and undiminished supply of planktonic food to suspension-feeding animals such as barnacles. The numbers of many prey, however, increase only as fast as they can reproduce. If their populations are reduced, their numbers remain small until the next breeding season. How consistently prey are available determines the kinds of behavior predators can employ. Sessile predators, such as web-spinning spiders, mussels, and barnacles, live primarily where their prey are consistently available for long time periods.

Suspension feeders, as we saw in Chapters 25 and 26, are found in many animal phyla (Brachiopoda, Mollusca, Annelida, Arthropoda, Phoronida, Ectoprocta, Echinodermata, Chordata). They feed on prey much smaller than themselves that are suspended in water or air. Every suspension feeder has an apparatus with which it extracts prey from the medium.

The structure of this apparatus determines the upper and lower size limits of prey that can be captured. Prey that are too small pass through the mesh of the structure, while prey that are too large bounce off it. Most suspension feeders depend on the movement of the surrounding medium through their filtering apparatus. This can be accomplished either by movement of the medium or of the animal. An important feature of suspension feeding is that most of the work of feeding is expended *before* the prey contact the animal. The predator cannot save much energy by choosing prey; in fact, it must usually expend energy to reject unwanted prey. For this reason most suspension feeders eat virtually all prey that are retained by their collecting devices. A striking exception is spiders, whose prey—flying insects—bring themselves into contact with the spiders' webs. Spiders reject many prey items that land in their webs, even cutting some of them out and dropping them.

Because parasitoids and parasites are the same size as or smaller than their prey, many of them are able to complete their life cycles by feeding on a single prey item (Figure 46.5). They choose foods only once or a few times during a life cycle. Usually the egg-laying females make the choice. Parasites lay their eggs in food-rich environments, but hosts have defenses against parasites and may be able to destroy the eggs. The ability of a host to defend itself depends on its condition and on the number and kind of parasites attacking it. If a host is already weakened by stresses imposed by the physical environment or shortage of food, parasites are more likely to gain a

foothold. Bark beetles attack pine trees by tunneling into the nutritive layers just below the outer bark. A tree defends itself by exuding the sticky pitch for which pines are famous. Weakened trees exude less pitch than healthy trees. Also, the more beetles that attack, the greater the probability that they will overcome the tree's defenses. Each colonizing beetle releases a powerful pheromone that attracts other individuals to the same tree. If there are enough beetles in the general area, a mass attack results that may kill the tree (Figure 46.6).

Herbivores eat the tissues of plants. Plant tissues are abundant, but many of them are of poor quality as nutrition for other organisms. Wood is primarily cellulose, itself very difficult to break down, impregnated with lignins, which are even more difficult to digest. Because wood is hard to digest, few organisms attack branches and trunks unless the plant is already weak or dead. If this were not the case, the world's forests would not be as tall and lush as they are. Plants also produce roots, leaves, flowers, nectar, pollen, fruits, and seeds, which differ in chemistry, size, structure, and pattern of production. The organisms specialized for eating these different plant tissues are many and diverse.

Typical predators, many of which are **carnivores** —eaters of animals—are the animals we usually think of when predation comes to mind. They are generally moderately larger than their prey, but some, such as wolves and lions, are smaller than the largest prey they attack. Typical predators pursue and capture their prey one by one. They eat many

(a)　　　　　　　　　　　　　　(b)

46.5 Parasites and Parasitoids
(a) This Caribbean soldierfish is host to a parasitic isopod that feeds on its body tissues. The fish has no way to remove the isopod even when aware of its presence. (b) This wasp (*Ephialtes*) is laying an egg on a beetle larva deep within a log. Her ovipositor has extremely sensitive chemoreceptors that help her locate the larva and tell whether it is already parasitized. A parasitoid larva will hatch from the egg.

46.6 Pine Attacked by Bark Beetles
Masses of bark beetles have attacked this tree. The galleries under the bark were created by the developing larvae burrowing through the tissues, eating as they went.

individual prey items during their lives. The number of prey consumed may be very large indeed. A small bird in a temperate forest during winter must eat an average-sized insect or seed every few seconds to maintain itself.

Short-Term Predator–Prey Dynamics

When a typical predator captures and eats a prey, it reduces the size of the prey population by one. To understand how numbers of predators and prey fluctuate, we will first consider a single predator species eating a single prey species in a homogeneous environment. The predators can find enough to eat if the rate at which they encounter prey is above a certain threshold value. Below that threshold they will lose weight and eventually starve. Nevertheless, the predators may continue to eat prey even when they are scarce, reducing the prey population to an even lower level. Eventually the starvation or emigration of predators may allow the prey population to increase its numbers. This increase of prey may, in turn, permit the predators to increase. You can easily see why the recovery of a predator population often lags behind the recovery of its prey population. Thus predator–prey interactions often change the population densities of both species, producing oscillations —decreases and increases that appear wavelike when graphed.

Many laboratory and field studies have demonstrated that population densities of predators and prey oscillate. Figure 46.7*a* shows oscillations of a laboratory population of the bean weevil (*Callosobruchus*) and its parasitoid, a wasp (*Heterospilus prosopidis*). Note that the oscillations were quite regular for the first 54 generations but then disappeared for a while before being reestablished.

Population density oscillations of small mammals living at high latitudes and their predators are the best-known examples from nature. Populations of Canada lynxes have a 9- to 11-year cycle (one decrease followed by an increase) that lags behind a similar oscillation in the populations of their principal prey, snowshoe hares. These oscillations, as estimated by the number of pelts received by the Hudson Bay Company in Canada, are shown in Figure 46.7*b*. Both predation by lynxes and changes in the quality of the plants on which the hares feed contribute to the oscillations of hare populations. Populations of Arctic lemmings and their chief predators—snowy owls, jaegers, and Arctic foxes—have oscillations with a 3- to 4-year cycle.

Predator–prey interactions do not always generate oscillations in population densities. Environmental heterogeneity (spatial patchiness), for example, may stabilize the densities of prey and predator populations over a wide area. In heterogeneous environments, prey may be more vulnerable to predators in some patches than in others. In the absence of predation, prey are distributed over all the patches in accordance with the availability of food and other resources. As predation increases, prey are removed selectively from those patches where the predators can harvest them most efficiently. A striking example is provided by the interaction between St. John's wort (*Hypericum perforatum*) and a beetle (*Chrysolina quadrigemina*) that eats it. St. John's wort, an annual plant introduced into California from Europe, increased rapidly and soon dominated vast expanses of rangeland. Because it is much less desirable as a forage species for cattle than the grasses it replaced, people attempted to control St. John's wort by introducing *Chrysolina*. The experiment was a spectacular success. The beetle nearly eliminated St. John's wort from sunny patches. The plant is still common in shadier areas, where the beetle does not do well.

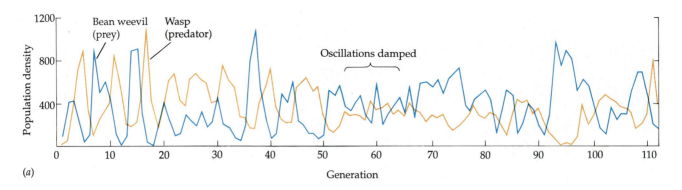

(a)

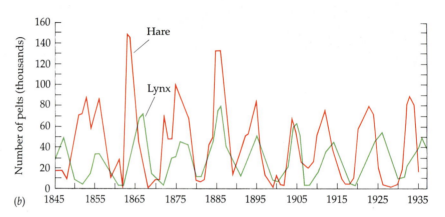

46.7 Predator–Prey Oscillations
(a) Oscillations in population sizes of laboratory populations of the azuki bean weevil and its predator, a parasitoid wasp. (b) An example from nature: the 9- to 11-year population cycles of snowshoe hares and lynx in Canada. Highest densities of lynx populations regularly were reached soon after hare populations reached their highest densities.

(b)

Stability can also result from temporal patchiness. The cactus *Opuntia*, introduced into Australia from North America, also spread rapidly and became common over vast expanses of valuable sheep-grazing land. It was controlled by introducing a moth species (*Cactoblastis cactorum*) that completely destroys patches of *Opuntia* when the egg-laying females find them. However, new patches of cactus arise in other places from seeds dispersed by birds. These new patches flourish until they are found and destroyed by moths. Today, over a large region, the numbers of both *Opuntia* and *Cactoblastis* are fairly constant and low, but in the local areas that make up the whole, there are vigorous oscillations that result in the extermination of first prey and then predator. As we saw in the example of the red-spotted newt in Chapter 45, a study of a single local population of *Opuntia* could not reveal the regional pattern of stability of population density or what produced it.

Predator–Prey Interactions over the Long Term

So far we have considered the short-term dynamics of predator–prey systems. As might be expected, predators are agents of evolution as well as agents of mortality. Prey have evolved a rich variety of responses to predation. Prey generally evolve to become more difficult to capture, subdue, and eat. By evolving, the prey themselves become agents of selection too. Predators, in turn, evolve to be more effective in overcoming prey defenses. Animals have evolved some spectacular defenses, such as the ejection of noxious chemicals, toxic hairs and bristles, tough spines, and mimicry of inedible objects or of larger or dangerous organisms (Figure 46.8). Plants too have evolved spectacular defenses, as we will see in a later section.

MIMICRY. Among the most interesting and best studied of evolved responses to predation is **mimicry**: taking on the appearance of some inedible or less desirable food object. In **Batesian mimicry** a palatable species mimics a noxious or harmful one. Examples are the mimicry of ants by spiders, of bees and wasps by many different insects (Figure 46.9), of poisonous coral snakes by a number of species of harmless snakes, and of toxic salamanders by palatable ones. Batesian mimicry works because a predator that captures an individual of the unpalatable model species learns to avoid any prey that appears similar. However, if a predator captures a palatable mimic, it is rewarded, and it learns to associate palatability with prey of that appearance. As a result, models are attacked more often than they would be if there were no mimics. Because models that differ from their mimics more than the average are less likely to be attacked by predators that have eaten a mimic, directional selection causes models to evolve away from mimics. Batesian mimicry systems are stable only if a mimic evolves toward a model faster than the model evolves away from it.

(a) (b)

46.8 Defenses of Animal Prey
(a) A bombardier beetle, *Brachinus*, ejects a noxious spray at the temperature of boiling water in the direction of an antagonist. The spray is ejected in high-speed pulses more than 20 times in succession. (b) The Indo-Pacific lionfish *Pterois volitans* is among the most toxic of all reef fishes. Glands at the base of its spines can inject poison into an attacker; its bright markings are thought to warn potential predators of this capability.

The convergence over evolutionary time in the appearance of two or more unpalatable species is called **Müllerian mimicry**. In a Müllerian mimicry system, all species, including the predator, benefit when inexperienced predators eat individuals of any of the species, because the predators learn more rapidly that all similar species are unpalatable. Some of the most spectacular tropical butterflies are members of Müllerian mimicry systems (Figure 46.10), as are many kinds of bees and wasps.

PARASITE EVOLUTION. The ability of parasites to infect new hosts depends on the infectiousness of the parasite, the rate of parasite population growth within a host, and the length of time a host survives. If the parasites kill their hosts quickly, they can transfer to new hosts for only a short time, unless they can continue to do so after the host dies. However, a parasite that grows slowly within its host may be outcompeted by faster-growing individuals of its own or other species. The outcome of host–parasite evolution depends on the relative importance of these effects.

Parasites transmitted to new hosts by biological vectors benefit greatly if their hosts live longer. Parasite species of this sort thus may evolve in ways that make them less deadly to their hosts. In contrast, parasites that are transmitted by nonliving vehicles, such as water, may not be affected adversely if their host dies quickly. Such species seldom evolve in ways that reduce their deadliness because more rapidly growing—and thus more deadly—parasite genotypes that quickly dominate populations within hosts are still transmitted readily after their hosts die. It is thus not surprising that the most severe human contagious diseases, such as dysentery, cholera, and hepatitis, are waterborne.

Parasites that are transmitted to vertebrates by arthropods grow and reproduce only in vertebrates; the arthropods are simply dispersal agents. Because of their large sizes, vertebrates offer large amounts of tissues that can be converted to parasite growth and reproduction without stopping the functioning of the host. In contrast, the small arthropod vectors would be destroyed quickly if the parasites grew within their bodies. Thus, parasites are usually more benign in arthropod vectors than in their vertebrate hosts. The malaria-causing *Plasmodium*, for example, has little effect on the mosquitoes that transmit it.

46.9 A Batesian Mimic
This mantispid fly is an effective mimic of a wasp.

46.10 Müllerian Mimics
In this array of Costa Rican butterflies and moths, the fourth individual down in the first vertical column, the third and fifth of the second column, the third and sixth in the third column, and the third in the fourth column are members of highly unpalatable species; the sixth down in the first column, fourth in the second column, and the fifth in the fourth column are members of moderately unpalatable species; and the fifth in the first column, sixth in the second column, fourth in the third column, and sixth in the fourth column belong to highly palatable species. Palatabilities of other species have not yet been tested with birds.

PLANT DEFENSES. Leaves are the prey taken by many predators. Herbivores that eat leaves of herbaceous (nonwoody) plants are called **grazers**. Herbivores that eat leaves of woody plants are called **browsers**. Many leaves defend themselves physically by being tough or by having hairs or spines. Most leaves also contain chemicals that have negative effects on predators. These substances act in two principal ways. One group of defensive chemicals, the acute toxins, interferes with herbivore metabolic functions. Some of these toxins, such as nicotine, interfere with transmission of nerve impulses to muscles. Others imitate insect hormones and thereby block insect metamorphosis. Still other toxins are unusual amino acids that become incorporated into herbivore proteins and interfere with their functioning.

Defensive chemicals of the second type make leaves extremely difficult to digest, thus reducing their suitability as food for herbivores. The most common of these digestibility-reducing substances are tannins, which are present in the leaves of some herbaceous and most woody species. As most leaves age, their concentrations of tannins increase, and the leaves also become tougher. Tannins may be present in such large quantities that waters draining from areas dominated by tanniferous plants are tea-colored. The most famous of such "blackwater rivers" is the Río Negro in Brazil.

Some plants respond to being grazed or browsed by increasing the levels of defenses in their leaves. Mountain birches in northern Finland are eaten by caterpillars of the moth *Oporinia autumnata*. When caterpillars attack a birch, the tree responds by increasing the concentration of defensive chemicals in its leaves, sometimes within a few days of the initial attack. The larval periods of *Oporinia*, of another moth, and of two sawflies—all of which feed on birches—are longer when larvae are fed on leaves growing close to leaves heavily damaged earlier in the *same* growing season. Caterpillars that were experimentally fed leaves from birch trees that had been severely damaged the previous year grew more slowly than caterpillars fed leaves from birches only lightly damaged the previous year.

COMPETITION

The Sonoran Desert of the southwestern United States and adjacent parts of Mexico is a region with mild winters and very hot summers. Throughout the area, trees grow only along rivers, where their roots can reach subsurface water. Evergreen shrubs, well separated from each other because their roots compete for water, dominate the vegetation (Figure 46.11). Rain falls primarily in winter in the western

46.11 Shrubs Dominate Sonoran Desert Communities
In the dry Sonoran Desert, trees grow only along the washes because else-where soil moisture is insufficient to support trees.

part of the desert, in summer in the eastern part, and in both summer and winter in the center. Spring and autumn are dry throughout the desert. Following the first heavy rains of a wet period, the shrubs produce new leaves and increase their photosynthetic rates. Seeds of annual plants also germinate. The annuals grow rapidly, flower, and produce large quantities of seeds, which fall to the soil at the end of the rainy season.

Many animals in the Sonoran Desert eat the vegetative tissues of the perennial and annual plants. Some species visit flowers for rewards of nectar and pollen, whereas others eat the flowers themselves. Ants, rodents, and birds harvest seeds after they are shed. Birds primarily exploit local dense patches of seeds and take a relatively small proportion of the total seeds produced. Ants and rodents, on the other hand, consume large quantities of seeds.

Ecologists interested in how species interact studied the ants and rodents in the Sonoran Desert. They found, as might be expected, that the ants, on the average, eat somewhat smaller seeds than the rodents do, but that there is much overlap in the sizes of seeds taken by the two kinds of animals (Figure 46.12). The ecologists removed ants from some sites, rodents from other sites, and both ants and rodents from a third set of sites. If either ants or rodents were removed, population densities of the other group increased (Table 46.2). These experimental results demonstrate that competition for food links the ants

and the rodents. The experiments also showed that ants and rodents greatly reduce seed densities.

Individuals of the same species as well as individuals of different species may compete for resources. **Intraspecific competition,** competition among individuals of the same species, may result in reduced growth and reproductive rates for some individuals, may exclude some individuals from better habitats, and may cause the death of others. Few of the shrubs

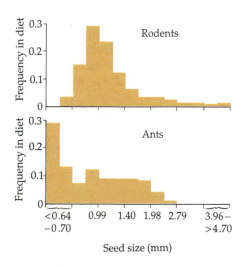

46.12 Ants and Rodents Harvest Different-Size Seeds
Frequency histograms show the proportions of seeds of various sizes harvested by individuals of the two groups of animals.

TABLE 46.2
Experiments Show How Ants and Rodents Interact with Their Food Supply

	RODENTS REMOVED	ANTS REMOVED	RODENTS AND ANTS REMOVED	CONTROL
Number of ant colonies	543	0	0	318
Number of rodents	0	144	0	122
Density of seeds relative to control	1.0	1.0	5.5	1.0

in Figure 46.11 are very close to others because intraspecific competition for water has thinned the population. **Interspecific competition**, competition among different species, affects individuals in the same way, but in addition an entire species may be excluded from habitats where it cannot compete successfully. In extreme cases, a competitor may cause the extinction of another species.

Individuals and species can use the same resources without competing if the quantity of resources is large enough that the presence of one individual or species does not reduce the other's access to the resources. If, however, the resources are reduced to a great degree by the users—as seeds in the Sonoran Desert are by ants and rodents—the interaction is called **exploitation competition**. When an individual, by its behavior, directly prevents other individuals from using a resource, the interaction is called **interference competition**. Behaviorally complex animals often interfere with one another, whereas plants and simpler animals usually compete by reducing the supply of resources.

As Charles Darwin first pointed out, competition is often most intense among individuals of the same species because they are so similar to one another in size, shape, and requirements. Nonetheless, organisms of different species may overlap greatly in the resources they use, so interspecific competition often is also intense. Interspecific competition is especially strong among plants because most species require the same mineral nutrients and all are powered by sunlight.

The first interspecific competition experiments were performed by the Russian ecologist G. F. Gause and reported in an influential book, *The Struggle for Existence*, published in 1934. Gause began by conducting experiments with microorganisms in test tubes within which the environment was homogeneous. In every experiment, the winner completely excluded the loser, an outcome called **competitive exclusion**. Which of the two species was the competitive winner depended on the conditions in the test tube. The results of one of Gause's experiments with *Paramecium*, in which *P. caudatum* was the winner, are shown in Figure 46.13. In simple laboratory

environments such as these, competitive exclusion is common; it is not unusual for a species to be a loser in an environment in which it would survive well if alone.

In later experiments, Gause was able to prevent competitive exclusion by providing some environmental heterogeneity, so that there were places where one species did better and other places where its competitor did better. For example, *Paramecium bursaria* and *P. aurelia* exclude one another in homo-

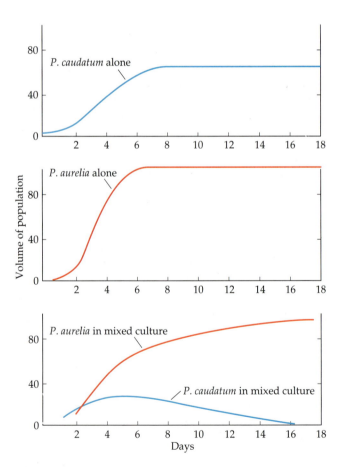

46.13 Competition between Two Protists
Although *Paramecium caudatum* (blue curve) and *P. aurelia* (red curve) both flourish alone, when they are grown together in the same container, one species soon eliminates the other.

geneous environments, but they form stable mixtures if there is deoxygenated water at the bottom of the containers. This happens because *P. bursaria*, which has mutualistic algae within its cell, can feed in deoxygenated water, whereas *P. aurelia* cannot.

One way to detect interspecific competition and determine its effects is to remove individuals of one or more potentially competing species and observe the responses of the remaining species. This is how ecologists identified the exploitation competition between ants and rodents in the Sonoran Desert. Other researchers found rapid responses to the removal of competitors among three species of small rodents: the common meadow vole (*Microtus pennsylvanicus*), primarily a grassland species; and the red-backed vole (*Clethrionomys gapperi*) and the deer mouse (*Peromyscus maniculatus*), both of which are normally woodland species. Researchers confined the animals in enclosures 0.2 hectares in size, each of which was half woodland and half grassland. When any one of the three species was alone in an enclosure, it occupied both habitat types. The presence of *Microtus* caused restriction of the other two species to the woodland and the immediately adjacent grassland, but if *Microtus* was removed, the other species expanded back into the grassland. These changes were not due to exploitation but to direct aggressive interference by the behaviorally dominant *Microtus*.

Plant competition is easy to study because investigators can manipulate the densities of the plants and availability of light, water, and nutrients, and can measure growth and reproduction. Ecologists studied competition among two species of weedy wild oats (*Avena fatua* and *A. ludoviciana*) and a species of domestic oats (*A. sativa*). They sowed a set of plots all containing the crop species at the same density. In each plot they also sowed one of the weeds at either high or low density. They also sowed plots with only the two weeds. They harvested some of the plants when they were young and others when they were mature. They dried and weighed the plants and compared the weights from the different plots. They found *A. fatua* to be the most vigorous interspecific competitor of the three species under the experimental conditions. *A. fatua* depressed the weight of stems, leaves, and flower heads of the other two species more than the other species depressed it. Ranked in order of their competitive success, the three species are: *A. fatua* > *A. sativa* > *A. ludoviciana*.

We have discussed predation and competition as if they were very different processes, but they are intimately related. Competitors are actually rival predators that are trying to capture the same "prey," that is, the same resources. Interactions among competitors may also affect their relationships with their own predators. For example, a competitor may force an animal to forage in more dangerous places,

46.14 A Sea Star Determines Rocky Intertidal Zonation This sea star, *Pisaster ochraceous*, is resting on rocks from which it has harvested all the mussels. Other organisms, including the algae visible in the photograph, will colonize the site.

thereby exposing it to predators. Or, a predator, by preying upon a dominant competitor, may create openings for species that would otherwise be competitively excluded. The sea star *Pisaster ochraceous* is an abundant predator in rocky intertidal communities on the Pacific coast of North America. Its preferred prey are mussels (*Mytilus californianus*), which in the absence of sea star predation dominate a broad belt of the intertidal zone and push out other competitors. *Pisaster*, by heavily cropping the mussels, creates bare spaces that are occupied by a variety of other species (Figure 46.14). Competition, predation, and physical factors such as desiccation and wave action—all acting jointly—determine which species live in these rocky intertidal communities.

OTHER INTERSPECIFIC INTERACTIONS

During predator–prey and competitive interactions, either one or both partners in the interaction are harmed. As we saw in Table 46.1, amensalism is a kind of interaction that causes harm to one of the partners without affecting the other. In the other two kinds of interspecific interactions—commensalism and mutualism—neither partner is harmed, but one or both may benefit.

Amensalism and Commensalism

An individual may harm another organism without benefitting by doing so. Mammals, by trampling, create bare spaces around waterholes. They benefit by

46.15 Commensalisms
(a) Cattle egrets, such as these individuals foraging around Cape buffalo in East Africa, catch more insects with less work than do egrets foraging away from the larger beasts. (b) This tropical tree supports a heavy load of ferns, orchids, bromeliads, and mosses.

drinking water, but not by trampling the plants they kill. Leaves and branches falling from trees damage smaller plants beneath them. The trees benefit by dropping old structures, whether or not they damage other plants. Such interactions—amensalisms—are widespread and important. Herbs, shrubs, and small trees in tropical forests often are damaged more by falling objects than they are by herbivores.

Commensalism benefits one partner but has no effect on the other. Determining that one of the partners in an interaction is neither aided nor harmed can be difficult. Nonetheless, there are cases, such as the relationship between cattle egrets and grazing mammals, where one partner clearly appears to be unaffected. Cattle egrets are found throughout the tropics and subtropics. In the New World they range from the central United States to northern Argentina. They are most often found foraging on the ground around cattle (Figure 46.15a). They concentrate their attention near the heads of cattle, catching insects flushed by the cows' feet and mouths. Cattle egrets

foraging close to cows capture more food for less effort than egrets foraging away from cows, so the benefit to them is clear. The cattle neither gain nor lose.

Other commensalisms include the relationship between epiphytes such as mosses, bromeliads, and orchids, and the trees on whose trunks and branches they grow (Figure 46.15b). Epiphytes do not intercept much light, and unless they grow so large that they actually break a tree's branches, they probably have little effect on the trees. The epiphytes obviously benefit, however, because the trees provide attachment sites with access to light and also offer food in the form of nutrients running down the surfaces of the trunks and branches.

Mutualism

Mutualisms—interactions that benefit both participants—are widespread and important, and they are found in virtually all plant and animal groups. Mutualistic interactions are found between plants and microorganisms, protists and fungi, plants and insects, and among plants. Animals also have mutualistic interactions with protists and with one another. Indeed, the evolution of eukaryotic organisms is believed to be the result of mutualistic interactions between previously free-living mitochondria and chloroplasts and the cells they originally infected (see Chapters 4 and 17).

Some of the most complex and ecologically important mutualisms are between members of different kingdoms. Nitrogen-fixing bacteria of the genus *Rhizobium* receive protection and nutrients from their host plant and provide their host with nitrates which are often in short supply in terrestrial environments (see Figure 30.12). Many plants have mutualistic associations with fungi attached to their roots (see Box 28.A).

As you know from Chapter 23, lichens are compound organisms consisting of highly modified fungi that harbor either cyanobacteria or green algae among their hyphae. The fungus absorbs water and nutrients and provides a supporting structure; the microorganism conducts photosynthesis. This mutualistic combination is especially successful at occupying inhospitable habitats such as rock surfaces, tree bark, and bare, hard ground.

Animals also have important mutualistic interactions with protists. Corals and some tunicates gain most of their energy from photosynthetic protists that live within their tissues. In exchange, they provide the protists with nutrients from small animals they capture. Termites have protists in their guts that help them digest cellulose in the wood they eat. Young termites must acquire their protists by eating feces of other termites; if they are prevented from doing so, they soon die. The protists are given a suitable environment in which to live and an abundant supply of wood.

Animal–Animal Mutualisms

Many species of ants have mutualistic relationships with aphids. Ants "milk" these small, plant-sucking insects by stroking them with their forelegs and antennae. The aphids respond by secreting droplets of partly digested plant sap that has passed through their guts. The ants protect the aphids from predatory wasps, beetles, and other natural enemies. The aphids lose nothing, because plant sap has an excess of sugar relative to the amino acids aphids need.

Some coral reef fishes and shrimps obtain their energy by eating parasites from the scales and gills of larger fish (Figure 46.16). In Africa a species of wading bird removes parasites from among crocodiles' teeth. These mutualisms are particularly interesting because the cleaners are potentially suitable prey that actually enter the mouths of dangerous predators. The predators refrain from attacking the cleaners, but such restraint could not have been present when the interactions first began to evolve. Biologists suspect that such cleaning began with the cleaners removing parasites from less dangerous locations.

A fascinating mutualism exists in parts of Africa between small birds called honeyguides and the mammals they lead to bees' nests. Honeyguides are unusual among birds in having enzymes for digesting beeswax. They cannot, however, get at the wax from intact bees' nests, most of which are in hollow trees. A honeyguide leads a mammal to a bees' nest by flying toward the nest, perching conspicuously and calling loudly, and waiting until the mammal

46.16 A Cleaning Mutualism
This prawn (*Lysmata grabhami*) is cleaning a coral reef fish known as sweetlips.

catches up. It then flies farther, repeating the performance. When it is close to the nest, it perches lower and gives a special call. The mammal then digs out the nest and the bird feeds on scraps of wax and honey. Honeyguides usually lead ratels (honey badgers) to bees' nests, but humans also follow honeyguides (Figure 46.17). People who find a nest in this way leave pieces of wax in a conspicuous place to ensure that the honeyguide has enough to eat. The complexity of this mutualism indicates that the birds have been interacting with mammals for a very long time.

46.17 Honeyguides Lead People to Bees' Nests
A lesser honeyguide (*Indicator minor*) pauses while the person it is guiding to a bees' nest catches up.

(a)

(b)

(c)

46.18 Ant–Acacia Mutualisms
Acacia trees have large, swollen, hollow thorns (a) that house ants. The ants patrol the trees, eating the eggs and larvae of herbivorous insects and cutting away tips of vines and branches from neighboring plants that would otherwise smother the acacia. In an experiment, small acacia trees were cut down and ants were allowed to recolonize some trees but not others. Those with ant colonies (b) grew back quickly, but those without ants (c) were heavily attacked by other insects and refoliated very slowly.

Plant–Animal Mutualisms

Terrestrial plants have many mutualisic interactions with animals in which a variety of benefits accrue to both the plants and the animals. A complex mutualism between trees and ants that live in Central America illustrates several such benefits. Trees of the species *Acacia cornigera* have large, hollow thorns in which ants of the genus *Pseudomyrmex* construct their nests and raise their young (Figure 46.18a). These ants live only on acacias. The trees have special nutritive bodies on their leaves upon which the ants feed, as they do on nectar produced at the bases of the leaf petioles. The ants attack and drive off leaf-eating insects; they even bite and sting browsing mammals. They also cut down other plants, particularly vines, that grow over their host tree. An ecologist experimentally removed ants from some acacias. He allowed ants to recolonize some of them but prevented ants from recolonizing others. Those plants that did not regain their ants were quickly overgrown by other plants (Figure 46.18c) and soon died. In this mutualism the animals get room and board and the plants get protection against both predators and competitors.

Many flowering plants depend on animals to move their pollen and seeds (Table 46.3). In pollination, the plants benefit by having their pollen carried to other plants and by receiving pollen to fertilize their ovules. The animals benefit by obtaining food in the form of

TABLE 46.3 Mutualistic Relationships of Plants with Animals		
BENEFIT TO PLANTS	**BENEFIT TO ANIMALS**	**SOME EXAMPLES**
Animals disperse pollen	Animals feed on pollen or nectar	Most plants with brightly colored flowers
Animals disperse seeds	Animals feed on fleshy rewards surrounding or attached to seeds	Conifers such as junipers and yews
Animals disperse both pollen and seeds	Animals feed on both floral and fruit rewards	Most tropical trees, shrubs at all latitudes

nectar and pollen. Animals are not the only means by which pollen is transferred among plants, but it has been favored by natural selection in many instances because it is efficient. Animal pollination is generally a "payment on delivery" system. Plants provide animals with attractive rewards, and movement to another flower of the same species is encouraged by the existence of similar rewards there. As a result, the animals transfer the pollen efficiently to the stigmas of conspecific plants. But there is a price: The energy and materials the plants spend to produce rewards for the animals cannot be used for growth or seed production.

Animals are induced to move seeds, too, by the presence of nutritive rewards attached to or surrounding them. Many seeds are surrounded by fleshy fruits that are eaten by animals, which either regurgitate or defecate the seeds some time later away from the plant (Figure 46.19a). A nutritive body is attached to many seeds that are dispersed by ants. The ants carry this body, with the seed attached to it, back to their nests for later consumption (Figure 46.19b). Although ants carry seeds only short distances they often bury them in good germination sites where they are protected from fires and other predators.

Interactions between plants and their pollinators and seed dispersers are clearly mutualistic, but they are not without conflict. As we saw at the beginning of this chapter, fig wasps are both pollinators and seed predators. Many animals visit flowers without transferring any pollen, sometimes cutting holes in them to get at the nectaries at the base of the corolla. Some plants, in their turn, attract insects without providing any rewards. The flowers of certain orchids mimic female insects, enticing the male insects to copulate with the flowers (Figure 46.20). The male insects neither sire any offspring nor obtain any reward, but they do pollinate the plant. Many seed dispersers are also seed predators that destroy some of the seeds they remove from plants.

INTERACTION AND COEVOLUTION

The richness and ubiquitousness of the interactions among organisms that we have been discussing in this chapter demonstrate that traits of all species have been influenced by interactions with other coexisting species. That is, species have **coevolved** with one another. The traits of predators have influenced those of their prey. Parasites coevolve with their hosts, and mutualists coevolve with one another. What is much less clear, however, is the extent to which particular traits of species are the result of interactions with only one other species. Such species-specific coevolution has been clearly demonstrated in only a few cases. One such example is the coevolution between figs and fig wasps we discussed at the beginning of this chapter. Another is the relationship between some species of yucca and the moths of the genus *Tegeticula* that pollinate them. A female moth enters a yucca flower and lays 1 to 5 eggs on the ovary. When the eggs hatch, the larvae burrow into the ovary and feed upon the developing seeds. This is typical predation—but after she has laid her eggs, the female moth scrapes pollen from the anthers in

(a)

(b)

46.19 Fruits Attract Different Frugivores
(a) Bright red fruits are attractive to many birds, such as this redwing, a European thrush. (b) A *Formica* ant removes a ripe seed from a pod of a golden smoke plant in the Colorado Rockies. The ant-attracting elaiosome is the white tissue wrapped around the black seed.

46.20 Some Orchids Mimic Female Insects
Flowers of some orchids so closely resemble female wasps that males are fooled into attempting to copulate with the flowers, as this male wasp is trying to do.

the flower, rolls it into a small ball, flies to another yucca plant, and places the pollen ball on the stigma of the flower before laying another batch of eggs. *Yucca* has no other pollinators, *Tegeticula* larvae eat no other food, and each yucca species has a specific moth species associated with it (Figure 46.21). The relationship shows other signs of close coevolution. The moth refrains from laying more than a few eggs on any one ovary, and the yucca pollen is unusually sticky and readily formed into a ball.

Although species-specific coevolution is relatively rare, **diffuse coevolution**, in which traits of a species are influenced by interactions with a wide variety of predators, parasites, prey, and mutualists, is widespread. Most flowers are pollinated by a number of pollinators, and most pollinators visit many species of flowers. Most flowers adapted for bird pollination are red, a color that attracts most birds, not just a few species. Many flowers adapted for insect pollination have honey guides, contrasting colors that lead to the entrance to the flowers. These lines, which are conspicuous when viewed under ultraviolet light (Figure 46.22), are visible to bees and butterflies but not to birds. Bat-pollinated flowers open at night and have wide openings into which the head of a bat can enter, but the floral rewards are accessible to many species of bats.

Diffuse coevolution also accounts for the traits of the fleshy fruits that surround many seeds. Most bird

(a)

(b)

46.21 Yucca–Yucca Moth Coevolution
(a) The Joshua tree, *Yucca brevifolia*, is pollinated only by (b) the yucca moth *Tegeticula yuccasella*, shown here on a yucca flower.

(a)

(b)

46.22 Bees See Flowers by Reflected Ultraviolet Light
(a) Under normal sunlight these black-eyed Susans appear familiar to us.
(b) Bees, however, see the same flowers more as they appear in this ultraviolet photograph, in which the central flowers in the heads are more conspicuous than under normal light.

dispersed fruits are red or some combination of red and another color. Mammal-dispersed fruits are typically purple. But very few fruits are adapted for dispersal by only a few species of birds or mammals.

The traits of flowers and fruits are the result of diffuse coevolution between plants and animals because most flower visitors and fruit dispersers must exploit many different plant species to survive throughout the year. Most plant species produce flowers and fruits for only a few weeks or months. Animals must travel to wherever flowers and fruits are available and must switch to feeding on whichever plant species are flowering or bearing fruit.

HOW SPECIES AFFECT BIOLOGICAL COMMUNITIES

Species influence the communities in which they live in many different ways. They may alter microclimate, soil structure, and water movement. Such changes in the physical environment change its suitability as habitat for other organisms. As we have just seen, species also change the amount and distribution of resources, and they consume one another.

The Role of Plants

In terrestrial communities, plants are the major modifiers of physical environments. They also provide the pathway through which energy enters communities. Because of the variety and complexity of plants and the sizes of some of them, plants also form most of the structural environment for other organisms. Anyone who has walked into the shade of a tree on

a hot, sunny day knows that the climate near the ground is strongly influenced by vascular plants. Except where temperatures are very cold or moisture is scarce, trees grow large enough to shade smaller plants beneath them and the surface of the soil. Temperatures fluctuate less between day and night underneath the trees than they do in the open and light levels are much lower there. Also, the leaves of trees intercept and evaporate much of the rain that falls on them so that less reaches the ground than in open areas. However, the rain that does reach the ground evaporates more slowly inside a forest than in the open because temperatures are lower, humidities are higher, and it is less windy inside forests.

In addition to modifying climates, vascular plants are the major structural elements of terrestrial environments. A typical forest is a mixture of trees, shrubs, and herbaceous plants (Figure 46.23). The crowns of the trees form the **canopy** of the forest. The canopy, which is exposed to sun, wind, and precipitation, is the major modifier of the local climate. Beneath the canopy is the **understory**, a mixture of smaller trees, shrubs, and herbs. Some of these plants are young trees that may eventually become members of the canopy, but many of them belong to species that never attain such large sizes. These plants live in a climate modified by the trees above them and they are often damaged by leaves, branches, and fruits falling from the trees.

The plants of a forest differ in many ways in addition to size. They differ in their bark, the number of their branches and the angles at which they are held, the sizes and shapes of their leaves, the kinds of flowers and fruits they produce, and how much they die back during unfavorable seasons. It is in

46.23 Forests Have Complex Structure
This tropical evergreen forest on Barro Colorado Island in Panama has trees of different ages and sizes, vines, and many shrubs.

such structurally complex plant communities that members of the other kingdoms live. The types of food available to them, when the food is available, and the ways in which it can be found and captured depend on the structure of the plant community. This is why ecologists spend much time measuring the structure of plant communities and observing how animals move and find their food in them.

Succession

A complex forest of the kind shown in Figure 46.23 does not develop quickly. Some of the large trees may be several hundred years old. Other individuals of the same and different species may have occupied the site before today's plants grew there. Long ago people noticed that following the destruction of most of the plants on a site by a disturbance such as a fire or a landslide, the first invading plants differed from those that colonized the site later. The gradual process by which the species composition of a community changes is called **ecological succession.** Patterns

of succession differ according to the climate and soils of the area, but all change the soils of the sites and the conditions under which later-arriving plants must grow.

Succession at sites that at the beginning support no organisms is known as **primary succession.** Consider, for example, the succession of plants in Glacier Bay, Alaska, following the rapid retreat of the glaciers there during the last 200 years. A retreating glacier leaves a series of moraines—gravel deposits formed where the glacial front was stationary for a number of years. The ages of moraines can be determined approximately by counting growth rings on the oldest trees present on each of them. The pattern of succession of plants and changes in soil nitrogen content at Glacier Bay are shown in Figure 46.24. The glacier retreated about 100 kilometers during this period (see Figure 30.12).

Each boulder-strewn moraine was first invaded by pioneer plants such as lichens, mosses, and a few species of shallow-rooted herbs. These plants usually are small and simple, and they can grow on the nutrient-poor moraines where they are exposed to the full force of the climate. These small plants were followed by shrubby willows, then by alders, and eventually by conifers. Primary succession is determined in part by changes in the soil caused by the plants themselves. Alder trees have nitrogen-fixing bacteria associated with their roots. Because nitrogen is virtually absent from glacial moraines, it required a century for alders to fix enough nitrogen in the soil to allow good growth of conifers. The conifers then outcompeted and displaced the alders.

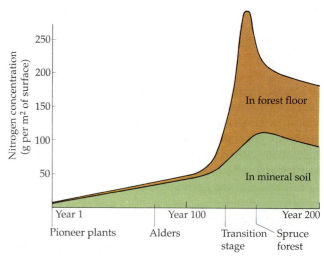

46.24 Soil Properties Change during Vegetation Succession
As the plant community occupying a moraine changed from pioneering plants to a spruce forest, nitrogen accumulated in both the forest floor and in the mineral soil. Alder trees fixed nitrogen in the soil, paving the way for the coniferous forest.

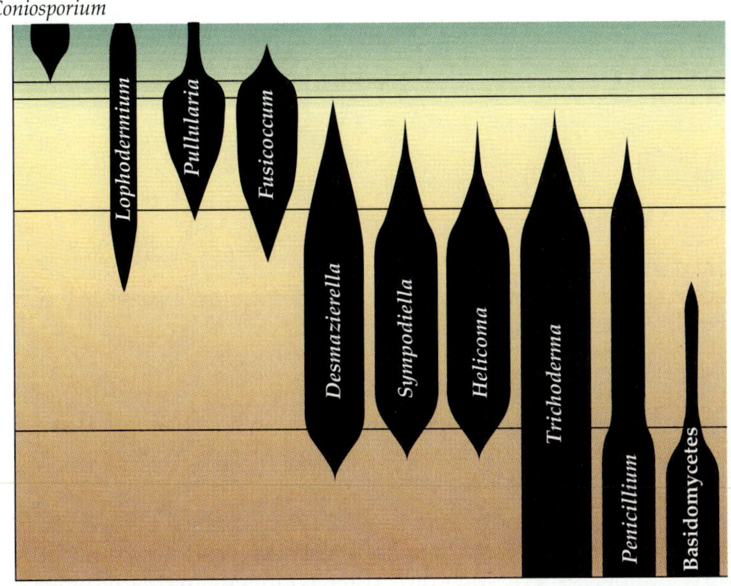

Coniosporium

46.25 Degradative Succession on Pine Needles
The abundances of ten types of fungi in pine litter change over time from the
layer of living needles on the surface to highly decomposed needles below.
This degradative succession lasts about 7 years.

Succession initiated after disturbances, such as fires and logging, that remove only parts of the community is called **secondary succession**. Unlike primary succession, which is driven for a long time by changes in the soil, secondary succession is most strongly influenced throughout the successional changes by interactions among the plants for resources, especially light, soil nutrients, and water. The first invaders of disturbed sites are usually fast-growing plants that photosynthesize well in full sunlight. However, these plants create shade in which their own seedlings grow more poorly than seedlings of shade-adapted plants. Thus, plants that grow well under full sun are replaced by plants that germinate and grow better in shade.

Another type of succession, known as **degradative succession**, takes place when all or part of the dead body of some plant or animal decomposes. Degradative succession is driven by activities of early species that remove some nutrients and change others, making available new resources for later species. Degradative succession terminates when the resource is completely consumed.

The succession of fungal species in decomposing pine needles in litter beneath Scots pines (*Pinus sylvestris*) is shown in Figure 46.25. New litter is continuously deposited under pines, so that the surface layer is young and deeper layers of litter are progressively older. Degradative succession begins when the first group of organisms starts consuming the needles as soon as they fall, and continues over a period of about seven years, after which the last group of organisms—basidiomycetes—have decom-

posed the remaining cellulose and lignin, and the remains are no longer recognizable as pine needles.

The Role of Animals

The effects of animals on terrestrial communities are not as profound as those of plants, but their influences, especially those of large mammalian herbivores, are considerable. The alterations of ecological communities by moose have been studied for more than four decades on Isle Royale in Lake Superior. Between 1948 and 1950 ecologists began an experiment to evaluate the long-term effects of moose browsing. They built fences around 100-square-meter plots of land to keep the moose out. Then they laid out, but did not fence, control plots outside the exclosures. Moose preferentially feed on early successional deciduous plants, such as mountain ash, mountain maple, aspen, and birch. They rarely eat white spruce and balsam fir, species that replace deciduous trees during succession, because the foliage of these conifers has high concentrations of indigestible resins and low concentrations of nitrogen. In the control plots, deciduous species were so heavily eaten by moose that spruce and fir were the only plants that grew above the height at which moose feed. Inside the exclosures, on the other hand, deciduous trees remained abundant. Thus, by reducing the abundances of deciduous trees, the moose accelerate the successional change from deciduous trees to conifers.

Beavers strongly influence ecological communities by cutting trees and building dams. In the Kabeto-

1940

1961

1986

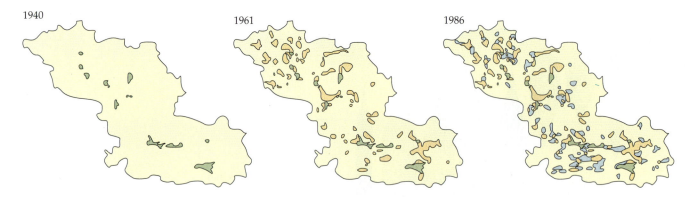

46.26 Beavers Change Vegetation
Aquatic ecological communities increased on this 45-square-kilometer watershed in Minnesota between 1940 and 1986 because of dam building by beavers. Aquatic communities present in 1940 are shown in green. Those added betwen 1940 and 1961 are in orange; those added between 1961 and 1986 are blue.

gama Peninsula in northern Minnesota, logging and fires in the late 1930s caused large increases in aspen trees, which grow well in cleared, disturbed areas. Aspen trees are the favorite food of beavers, so the number of beaver dams in the area increased from 71 in 1940 to 835 in 1986. Less than 1 percent of the peninsula was covered by beaver ponds in 1940, but this increased to 13 percent in 1986 (Figure 46.26). As a result, the amount of open water, marsh, bog, standing dead trees, and seasonally flooded lands increased greatly. Beavers also modify vegetation by cutting trees in the vicinity of their dams. A beaver family typically cuts about a metric ton of wood within 100 meters of its pond each year. In northern Minnesota, beavers often virtually clear-cut stands of aspen, favoring the succession to alders and hazels and, eventually, to white spruce and balsam fir.

PATTERNS OF SPECIES RICHNESS

All the interactions among all species living together in an ecological community could theoretically be described in detail. However, because there are many thousands of species in most communities, such a description would be as complex as nature itself. Even if all the necessary information were available, such descriptions would not help us identify the most important interactions. To build comprehensible pictures of complex ecological communities we need terms that describe some aggregate feature of groups of organisms. When we choose our aggregate terms wisely, they capture important information about ecological communities.

An important aggregate feature of any community is **species richness**—most simply defined as the number of species living in it. But often we want to know both the number of species and their relative abundance or rarity. Other times we want to know about the relative sizes of members of the different

species. For such purposes, we need a weighted measure of species richness, sometimes called **species diversity**. Species are usually weighted by their abundance, biomass, or energy production and consumption, because these traits influence community dynamics. For example, consider two imaginary forest stands, each of which has 100 trees belonging to 10 species. In one stand, there are 10 individuals of each species, whereas in the other stand there are 91 individuals of one species and one of each of the others. The number of species is the same (10 species) in the two stands, but the stand in which most individuals are members of one species is less diverse.

Ecological communities differ dramatically in the number of species that live in them. Within a region, forests typically have more species than grasslands do, and tropical communities have more species than temperate communities do. The number of species found in a community depends on factors acting on different time and spatial scales. Competition, predation, disease, and mutualisms are processes that often produce their effects very quickly. They operate primarily at local levels. Immigration, emigration, and habitat fragmentation usually produce their effects more slowly, and on regional scales. Evolutionary changes are even slower. In this chapter we concentrate on local and regional factors affecting patterns of species richness. The role of long-term historical factors will be explored in Chapter 48.

Local Species Richness

What determines how many species can fit into a community? Some visual modeling can help us think about the possibilities. From Figure 46.27 you can see that to answer this question we need to know about the resources available to organisms in a community, the range of resources used by each species, and the amount of overlap between species in resource use. You can see that the number of species present in a

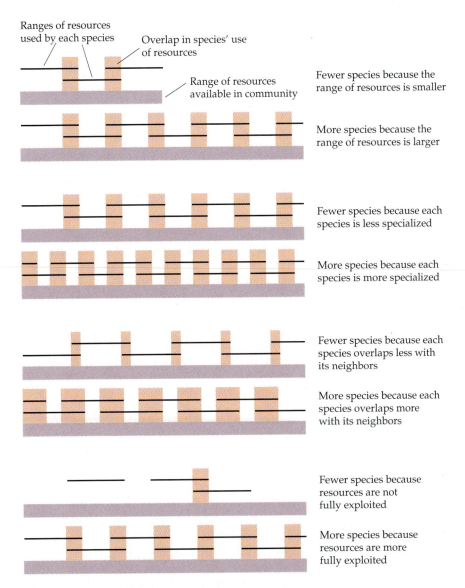

Ranges of resources used by each species

Overlap in species' use of resources

Range of resources available in community

Fewer species because the range of resources is smaller

More species because the range of resources is larger

Fewer species because each species is less specialized

More species because each species is more specialized

Fewer species because each species overlaps less with its neighbors

More species because each species overlaps more with its neighbors

Fewer species because resources are not fully exploited

More species because resources are more fully exploited

46.27 A Simple Model of Local Species Richness
These comparisons show us how three factors that are important for influencing the richness of communities can interact.

community may increase if the variety of resources available increases. More species may be fit in if each one uses a relatively narrow portion of the resources. More species may be fit in if they overlap extensively in their use of resources. Finally, more species may be fit in if the range of resources is fully used than if parts of it are unexploited.

STRUCTURAL COMPLEXITY AND SPECIES RICHNESS. Communities dominated by structurally complex plants support more species of freshwater fishes, birds, and insects than communities dominated by structurally simple plants. Ecologists studied the richness of fish species in 18 freshwater lakes in Wisconsin and found it to be positively correlated with the structural complexity of submerged vegetation. More complex vegetation offers fish more hiding places from predators

and more places to seek food. Similarly, terrestrial communities with more structurally complex plants offer birds a greater range of resources and a larger number of ways they can be exploited. Trees have trunks, branches of various sizes, and leaves from which prey can be captured, and there are many ways to move through trees. Grasses provide only a few kinds of food, and there are only a few ways for animals to move through grasses.

Structurally complex terrestrial communities have more species of insects because they provide both a greater range of food resources and a greater diversity of hiding places. Insects are highly vulnerable to predators if they sit conspicuously on leaves or branches. Many insects mimic leaves or twigs, hide in crevices in bark, or resemble their backgrounds in ways that make them harder to see. Structurally complex plants provide more different kinds of hiding places than simple plants do; therefore more kinds of insects can and do hide on them.

DISTURBANCE AND SPECIES RICHNESS. Disturbance, if it is not too severe, often increases species richness in ecological communities. Grazing may increase species richness. When rabbits were removed from an area at the Rothamsted Experimental Station in England, there was a slow but steady reduction in the number of plant species because the more vigorous competitors, no longer subjected to heavy grazing, eliminated less competitive species. Plants may create their own disturbances that allow more species to persist. Among the most important of these are tree falls, which create gaps in the canopy within which many tree species germinate and grow. Many forest trees cannot reach adult sizes unless they grow in such gaps, where there are more light and nutrients than in the closed forest.

Regional Species Richness Influences Local Species Richness

Much has been learned about factors influencing species richness from studies of small areas of a hectare or less. However, if species richness is determined only by local events, then habitats that have similar structures, rates of photosynthesis, and are subject to similar disturbances should support biological communities with similar species richness. Numerous studies demonstrate that this is not the case. The nature of the differences shows that regional processes also affect local species richness.

Local species richness is often correlated with the number of species found regionally. For example, ecologists have found that the number of gall-wasp species found locally on a particular species of oak in the state of California is positively correlated with the total number of gall-wasp species known to feed on the oak throughout its entire range (Figure 47.28a). Thus the number of wasp species on an oak tree within a small area is determined by the tree's structure and also by the number of gall-wasp species living on that oak species in a larger region. A similar pattern has been revealed by studies of songbirds on Caribbean islands: The more species found on an island, the greater the number of species found in a single habitat on the island (Figure 46.28b).

Ecologists have also studied the influence of regional processes on species richness by comparing the sizes of animal species found in areas of different sizes. Figure 46.29 shows the frequency of different sizes (by average weight) of North American mammalian species for the entire continent, for the Sonoran Desert, and for a small patch of relatively uniform habitat within the Sonoran Desert. In the continental sample, small mammals (those weighing about 100 grams) outnumber all others, but their dominance decreases as the size of the area sampled decreases. In local areas there are roughly equal numbers of mammalian species in all size classes. This pattern, which holds true for mammals in all major habitat types in North America cannot be explained entirely by local processes.

The presence of only a few species of small mammals at local sites, despite the large number of such species living on the entire continent, suggests that competition within local habitats prevents coexistence of similar-sized species. Large species, because of their high daily energy requirements, exist at lower population densities and usually have much broader geographic ranges than smaller species do. Even though individuals of large species require more food per day than individuals of smaller species, large animals can cover much larger areas, ingest more food, retain material for a longer time in their guts, extract a greater fraction of the energy and nutrients from their food, and go longer without eating than can small animals. Therefore, larger animals can feed on lower-quality foods and can include more kinds of food in their diets. Also, small species, with their smaller geographic ranges, have higher extinction rates and species replace one another frequently be-

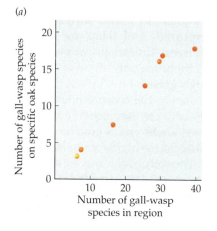

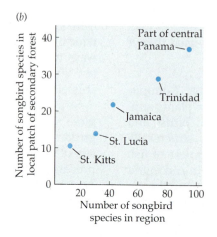

46.28 Local Richness Is Increased by Regional Richness
The number of species present in a local environment correlates positively with the number of species present in the larger region. (a) The number of gall-wasp species on different species of California oak trees. (b) The number of songbird species in patches of secondary forest on some Caribbean islands and on the Panamanian mainland. The smaller the area, the more even the distribution.

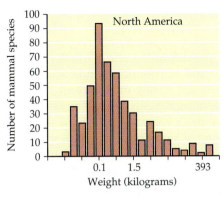
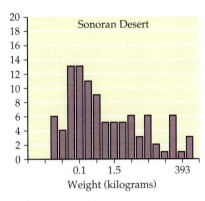
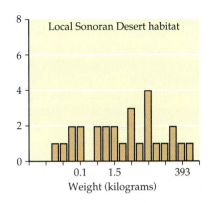

46.29 Distribution of Mammal Weights in Communities
The numbers of mammal species, grouped according to their weights, are plotted on a logarithmic scale for North America, the Sonoran Desert, and for a local habitat within the Sonoran Desert.

cause they are confined to habitats where food is abundant enough to meet their more stringent energy and nutrient requirements. A combination of these factors—high nutrient requirements of small species, local competitive exclusion among similar-sized small species, and higher extinction rates of small species with small geographic ranges, can explain why local mammalian communities have relatively fewer small species than we would expect from species lists of larger areas.

HABITAT EXTENT AND SPECIES RICHNESS. The more widely distributed and extensive a habitat, the more likely organisms are to evolve adaptations to that habitat. There is great variation in salinity between seawater and fresh water, but the salinity of seawater is relatively uniform worldwide, as is the salinity of most fresh water. Waters of intermediate salinity and waters that are more saline than the oceans are rel-

atively rare. The species richness of aquatic invertebrates throughout the world parallels this pattern (Figure 46.30).

A similar pattern is found among plants growing on outcrops of serpentine rocks. Soils derived from serpentine are highly deficient in several essential plant nutrients and are often regarded as highly unfavorable for plants. Indeed, in western North America, where serpentine outcrops are small and scattered among other rock types, serpentine areas support relatively few plant species. However, in parts of South Africa, where serpentine is the most common rock type, plant species richness is higher on serpentine than on soil derived from other rock types. Therefore, to understand many aspects of species richness, we need to know the geographic extent of different habitats.

SUMMARY

Organisms interact with one another in varied and complex ways. During competitive interactions both of the participants are harmed. During predator–prey interactions, one participant, the prey, is harmed, whereas the other, the predator, benefits. Both participants benefit in mutualistic interactions, one benefits while the other is unaffected in commensal interactions, and one is harmed while the other is unaffected in amensal interactions. The resources a species uses and its interactions with its predators and competitors define its niche. Most species could exist in a wider range of environments than they do if competitors and predators were not present.

Predators and prey influence one another's abundances over both the short and long terms. Predator–prey interactions often lead to oscillations in abundances, but these oscillations are reduced if the environment is heterogeneous. Over the long term, predators influence the evolution of their prey, lead-

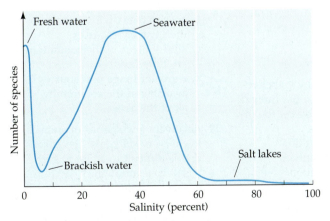

46.30 Salinity and Species Richness
Many species of invertebrates live in fresh water, and many live in seawater. Brackish water and the highly saline salt lakes—rare habitats compared to fresh and seawater—are species-poor.

ing, for example, to mimicry and the development of chemical and mechanical defenses.

Species that overlap in their use of resources may compete with one another if resources are in short supply or if the species interfere with one another behaviorally. Competitive exclusion of one species by another is common in laboratory experiments, but in more complex natural situations, competing species ordinarily eliminate one another from only specific habitats.

Mutualisms are found between nearly all groups of organisms. They are especially common between plants and other organisms. Plant–animal mutualisms are widespread because of the importance of animals as transporters of gametes and seeds. Fungi and protists interact mutually to produce lichens. There are also mutualistic interactions between animals, such as those between fishes and cleaner shrimps and between mammals and honeyguides.

An ecological community consists of all the species living together in a given area. Species change the communities in which they live by altering the physical environment, providing structure, modifying resource abundances and distributions, and consuming one another. Plants are major modifiers of climate and providers of community structure. The structure of plant communities develops slowly by a process of succession during which species replace one another over time. Animals, especially large mammals, also exert major effects on biological communities.

The number of species found living together in ecological communities depends on local and regional processes. At the local level, species richness is often correlated with community structure and the amount of disturbance. Immigration, emigration, and total regional species richness interact with local processes to produce species richness patterns that differ from those that would result if only local processes were acting.

SELF-QUIZ

1. Two organisms that use the same resources when those resources are in short supply are said to be:
   a. predators.
   b. competitors.
   c. mutualists.
   d. commensalists.
   e. amensalists.

2. Which of the following is *not* a resource?
   a. Food
   b. Space
   c. Hiding places
   d. Nest sites
   e. Temperature

3. A fundamental niche represents the range of conditions under which a species could survive if:
   a. there were no predators or competitors.
   b. there were no predators or other negative influences.
   c. there were no competitors, predators, or other negative influences.
   d. environmental conditions were ideal.
   e. the size of its hypervolume were the greatest.

4. An energy-maximizing predator adds prey types to its diet in order of:
   a. their abundance.
   b. their size.
   c. the ease with which they can be captured.

d. the energy per unit of time they yield.
   e. none of the above.

5. Which of the following factors tends to stabilize populations of predators and their prey?
   a. A high birth rate of the prey.
   b. A high birth rate of the predator.
   c. The ability of predators to further reduce prey when they are scarce.
   d. The ability of predators to search widely for prey.
   e. Environmental heterogeneity.

6. The convergence over evolutionary time in the appearance of two or more unpalatable species is called:
   a. cladism.
   b. mutual adaptation.
   c. Müllerian mimicry.
   d. Batesian mimicry.
   e. convergent mimicry.

7. When an individual, by its behavior, directly prevents other individuals from using a resource, the interaction is called:
   a. interference competition.
   b. exploitation competition.
   c. behavioral competition.
   d. intraspecific competition.
   e. competitive exclusion.

8. Damage caused to shrubs by branches falling from overhead trees is an example of:
   a. interference competition.

b. partial predation.
   c. amensalism.
   d. commensalism.
   e. diffuse coevolution.

9. Ecological succession is:
   a. the changes in species over time.
   b. the gradual process by which the species composition of a community changes.
   c. the changes in a forest as the trees grow larger.
   d. the process by which a species becomes abundant.
   e. the build-up of soil nutrients.

10. More species may be fit into a community if:
    a. each species uses a narrow range of resources.
    b. species overlap extensively in their use of resources.
    c. all resources in the community are fully used.
    d. a greater range of resources is available in the community.
    e. all of the above.

11. Relatively few species of animals live in very salty lakes because:
    a. it is difficult to adapt to high salt concentration.
    b. there is little to eat in very salty lakes.
    c. organisms are less likely to adapt to rare habitats.
    d. organisms evolved in the ocean.
    e. organisms evolved in fresh water.

FOR STUDY

1. A general rule commonly accepted by ecologists states that a "jack of all trades is master of none." Yet most ecological communities are mixtures of "jacks" and "masters," that is, generalists and specialists. Under what conditions would you expect jacks to be more successful? Masters? Why?

2. What features of predator–prey interactions tend to generate instabilities that lead to fluctuations in densities of both species? Given that instabilities are expected, what keeps populations of either predator or prey from fluctuating to extinction?

3. Pests and pathogens usually have generation times much shorter than those of their hosts. Consequently, they should be able to evolve faster. What prevents them from evolving so fast that they completely overcome the resistances of their hosts and exterminate them?

4. If wind pollination is inefficient, why are there so many wind-pollinated plants? If seeds that land close to the parent plant survive less well than those that are carried farther away, why do so many plants produce seeds lacking dispersal devices?

5. On the eastern side of the Sierra Nevada in California, four species of chipmunks occupy adjacent habitats from which they exclude each other by direct aggressive interference. In the San Jacinto Mountains of southern California, three other chipmunk species similarly occupy adjacent habitats but no interspecific aggression is observed. Each species simply remains in its own habitat. Which of these two assemblages do you think is the older one? Why?

6. Some direct interactions between two species benefit only one of those species. Give examples of such "one-way" benefits in each of the following cases:
   a. between two species of plants (give one example of energetic and another example of physical support).
   b. between a woody plant and a browser.
   c. between a predator and its prey.

7. Wood is an abundant food source that has been available for millions of years. Why have so few animals evolved to be able to eat wood?

8. In the text we discussed several cases showing that large mammals modify the communities in which they live. Give some examples in which smaller animals modify their communities.

READINGS

Barbour, M. G., H. J. Burk and W. D. Pitts. 1980. *Terrestrial Plant Ecology*. Benjamin Cummings, Menlo Park, CA. A general text on the interactions between plants and their surroundings, including several good chapters on communities and vegetation types.

Begon, M., J. L. Harper and C. R. Townsend. 1990. *Ecology: Individuals, Populations and Communities*, 2nd Edition. Blackwell Scientific Publications, Oxford. A comprehensive treatment of all aspects of ecology.

Futuyma, D. J. and M. Slatkin, Editors. 1983. *Coevolution*. Sinauer Associates, Sunderland, MA. An excellent collection of essays that summarize current knowledge of coevolutionary relationships among all types of living organisms.

Harborne, J. B. 1982. *Introduction to Ecological Biochemistry*, 2nd Edition. Academic Press, New York. A presentation of the types of chemicals produced by living organisms and their roles in ecological interactions.

Ricklefs, R. E. 1990. *Ecology*, 3rd Edition. W. H. Freeman, New York. An excellent textbook covering all aspects of ecology.

Stephens, D. W. and J. R. Krebs. 1986. *Foraging Theory*. Princeton University Press, Princeton, NJ. An in-depth review of the development of foraging theory and its application to a variety of ecological problems.

Thompson, J. N. 1982. *Interaction and Coevolution*. Wiley, New York. A review of patterns in and conditions favoring the evolution of close interactions among species.

Wickler, W. 1968. *Mimicry in Plants and Animals*. Wiedenfeld and Nicholson, London. A general and easily followed presentation of the evolution of close resemblances among organisms that are not closely related to one another.

47

Ecosystems

PREVIEW: The sun, the principal energy source for living organisms, provides the heat that drives global patterns of air and ocean circulation and precipitation. The elements of which living organisms are composed cycle through the oceans, atmosphere, fresh waters, and land. Living organisms exert strong influences on the amounts of those elements that cycle and their rates of movement. Energy flows through ecosystems when predators eat prey, but because not all prey are eaten, and because both predators and prey must expend energy to maintain themselves, the amount of available energy drops rapidly with each trophic level. Because agricultural ecosystems extract materials and release them elsewhere, rather than cycling them, agricultural productivity depends on replenishing these materials from sources outside the system.

This chapter deals with climates, global atmospheric and oceanic circulation, biogeochemical cycles, human modification of biogeochemical cycles, energy flow in ecosystems, trophic levels, and agricultural ecosystems.

Life exists in a thin layer on and near the surface of Earth. No organisms are found higher than a few tens of kilometers above sea level. In the soil, life becomes scarce a few meters down (although recent studies have revealed the existence of communities of monerans and protists several hundred meters below the surface). There are living organisms in the deepest trenches in the oceans, but those trenches are only about 10 kilometers deep. Marine organisms, like their terrestrial counterparts, do not burrow deeply into the substrata.

Nonetheless, living organisms are influenced by physical and chemical processes that occur far away from where they actually live. Communities of living organisms together with the physical environment with which they interact are known as **ecosystems**. To understand how ecosystems function, we need to understand the dynamics of the physical environment with which they interact; that is, we need a picture of climates, of global circulation patterns of air and water, and of the great biogeochemical cycles associated with them.

CLIMATES ON EARTH

The sun is the source of energy that warms Earth and makes life possible. The sun's energy drives the global circulation patterns of air and ocean waters. This warming and cooling of moving masses of air and water, in turn, explains much of Earth's climatic patterns. There is great climatic variability on Earth, primarily because the amount of solar energy received, and when it arrives, differs strikingly between the equator and the poles.

Solar Energy Fluxes

Every place on Earth gets the same total number of hours of sunlight each year—an average of 12 hours per day—but not the same amount of *heat*. The rate at which heat arrives per unit of ground area, known as the solar energy flux, depends primarily on the angle of the sunlight. If the sun is low in the sky, a given amount of solar energy is spread over a larger area than if the sun is directly overhead. In addition, when the sun is low in the sky, the sunlight must pass through more atmosphere. Thus, more energy is absorbed and reflected before the sunlight reaches the ground.

Because day length, as well as the angle at which solar energy arrives, varies more at higher latitudes, temperature differences between seasons increase with latitude (Figure 47.1). At high latitudes there is a negative energy balance during part of the winter; that is, less energy arrives than is lost to space by radiation from Earth's surface and atmosphere. On average, the mean annual air temperature decreases about 0.4°C for every degree of latitude (about 110 kilometers) at sea level.

Air temperature also decreases with elevation. The reason for this lies in the properties of gases. As a parcel of air rises, it expands, its pressure drops, and energy is expended in pushing neighboring molecules aside. With that loss of energy, its temperature drops. When the parcel of air descends, it is com-

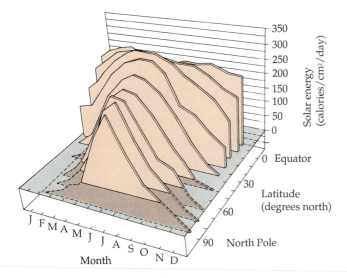

47.1 Solar Energy Fluxes at Earth's Surface
Energy flux changes throughout the year at nine different latitudes north of the equator as plotted. Notice that the energy balance is negative (less than 0 cal/cm²) for part of the year at latitudes higher than 40 degrees; that the total amount of solar energy received during the year decreases with latitude; and that tropical regions receive about the same amount of solar energy throughout the year.

pressed, its pressure rises, and the same amount of energy is recovered and its temperature increases.

When wind patterns bring air into contact with a mountain range, the air must rise to pass over the mountains, and it cools as it does so. Cool air cannot hold as much moisture as warm air can. Thus, unless the air is very dry, clouds form, and moisture is released as rain or snow. For this reason, the windward side of a mountain range generally has higher rainfall than the leeward side does. On the leeward side, the air descends and warms, and the dry air picks up rather than releases moisture. A place on the leeward side of a mountain range that recieves little rainfall for this reason is said to be in a **rain shadow** (Figure 47.2).

Global Atmospheric Circulation

Earth's climates are strongly influenced by global air circulation patterns that result from the variation in solar energy fluxes and the properties of air that we have just discussed. Air rises not only when it crosses mountains, but also when it is heated by the sun. The height to which it rises, which is proportional to the amount the air is heated by the sun, is greatest in the tropics. When air rises at the equator, winds blow toward the equator from the north and south to take its place. That air is, in turn, replaced by air from aloft that descends again after having traveled away from the equator at great heights. At roughly 30° north and south latitude, air that cooled and lost its moisture when it rose at the equator descends and warms. Many of Earth's deserts, such as the Sahara and the Australian deserts, are located at those latitudes. Deserts at higher latitudes are mostly in rain shadows. At about 60° north and south latitude, air rises again, for reasons that are not well understood. Cold, dense air descends at the poles, where there is little solar energy input. The black arrows in Figure 47.3 show these vertical patterns.

The spinning of Earth on its axis also influences surface winds because Earth's velocity varies from very rapid at the equator to very slow close to the poles. An air mass at a particular latitude has the same velocity as Earth at that latitude. If an air mass moves toward or away from the equator, it is deflected by the changing velocity of Earth beneath it —to the right in the northern hemisphere and to the left in the southern hemisphere. Therefore, the winds blowing toward the equator from the north and south veer to become the northeast and southeast **trade winds**, respectively. Winds blowing away from the equator also veer and become the **westerlies** that prevail at mid-latitudes. These surface winds are shown by the brown arrows in Figure 47.3.

Because Earth's axis is tilted, the location of greatest solar energy flux likewise shifts seasonally as

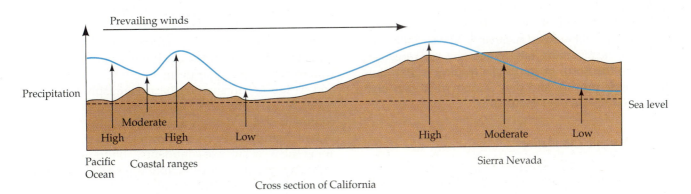

47.2 California Rain Shadows
Average annual rainfall is lower in the lee of California's coastal ranges, and at the highest elevations and eastern slopes of the Sierra Nevada.

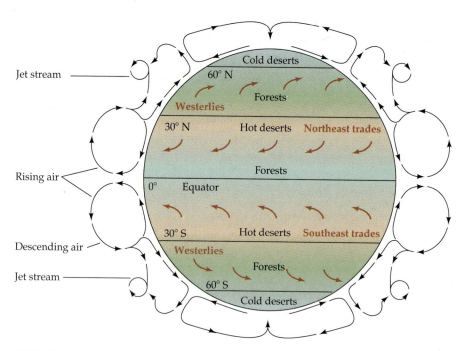

Jet stream

Rising air

Descending air

Jet stream

Cold deserts
60° N
Forests
Westerlies
30° N Hot deserts **Northeast trades**
Forests
0° Equator
30° S Hot deserts **Southeast trades**
Westerlies
Forests
60° S
Cold deserts

47.3 Circulation of Earth's Atmosphere
If we could stand outside Earth and observe the movement of the air, we
would observe vertical movements like those indicated by the black arrows
and surface winds like those shown by the brown arrows.

Earth orbits around the sun. The flux of solar energy is greatest during the time of year when the sun is directly overhead at noon, but the region where the sun is directly overhead at noon shifts north and south seasonally. As a result, the **intertropical convergence zone**, the area of greatest solar energy flux, where the trade winds converge and air rises most strongly, also shifts. The intertropical convergence zone shifts to the north during the northern summer (southern winter) and to the south during the northern winter (southern summer), as far as the Tropic of Cancer and the Tropic of Capricorn, respectively. However, the intertropical convergence zone lags behind the overhead passage of the sun by a bit more than a month because it takes that long for the sun to heat the surface mass of Earth. Seasonal changes in climate in the tropics and subtropics are associated with the movement of the intertropical convergence zone because whenever an area is in the zone, air rises and heavy rains fall. When the convergence zone is to the north or south of a tropical region, the trade winds prevail, and they seldom yield rain unless forced to rise over mountains.

Global Oceanic Circulation

The oceans play an important role in world climates because their waters move long distances and because water has a high specific heat. The **specific heat** of a substance is the amount of energy required to raise the temperature of 1 gram of the substance 1°C.

For water, this value is 1 cal/g. Similarly, 1 gram of water cooling 1°C gives off 1 cal/g of energy. Air and land surfaces have a much lower specific heat than water does. Consequently, in comparison with continents, oceans warm up more slowly in summer—because it takes more heat to raise their temperature—and cool off more slowly in winter—because more heat must be released to cool them.

The global pattern of wind circulation, driven by solar energy input on a rotating Earth, drives the circulation of ocean waters. Ocean water generally moves in the direction of the prevailing winds (Figure 47.4). Winds blowing toward the equator from the northeast and southeast cause water to converge at the equator and move westward until it is blocked by a continental land mass. At that point the water splits, some of it moving north and some of it moving south along continental shores. This poleward movement of ocean water is a major mechanism of heat transfer to high latitudes. As it moves poleward, the water veers to the right in the Northern Hemisphere and to the left in the Southern Hemisphere. Thus, water turns eastward until it is blocked by another continent and is deflected laterally along its shores. In both hemispheres, water flows toward the equator along the west sides of continents, continuing to veer right and left, respectively, until it meets at the equator and flows westward again.

The typical circulation pattern of ocean water, clockwise in the northern hemisphere and counterclockwise in the southern hemisphere, is modified

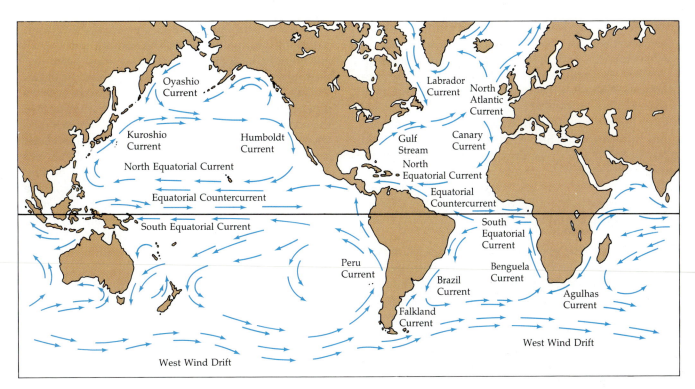

47.4   Global Oceanic Circulation
To see that ocean currents are driven primarily by the wind, compare the
currents shown here with the average surface winds shown in Figure 47.3.
Note also how the continents deflect ocean currents.

by the positions of continents. In the North Pacific,
for example, movement of water into the Arctic
Ocean is mostly blocked by land and very shallow
continental shelves. Most water turns south along
the west coast of North America. However, in the
north Atlantic, there is a wide gap between Iceland
and the British Isles, through which large amounts
of warm water flow poleward. At latitude 60° south,
where no continental land masses impede water
flow, there is a powerful eastward-flowing ocean current, driven by the strong westerly winds that blow
most of the time in that region.

Ocean water on the west sides of continents is
cool, both because of the equatorward flow of ocean
currents and because, as water veers away from the
continental shores, warm surface water is replaced
by cooler water upwelling from below. At higher
latitudes, west coast currents are warmer than adjacent land. Air passing over the oceans picks up large
quantities of moisture, which is released when the
air moves over the cooler land. Farther toward the
equator, however, ocean waters are cooler than the
adjacent land most of the year. Westerly winds are
warmed as they blow over land, so they pick up
rather than release water. Therefore, along the west
coasts of continents, precipitation falls mostly in winter when the oceans are warmer than the land. Annual precipitation is much greater in coastal British

Columbia than in Baja California, and in southern
than in northern Chile, for example, because the period of time when the land is cooler than the ocean
becomes shorter and shorter toward the equator.

At high latitudes, the temperature of the interior
of large continents fluctuates greatly with the seasons, becoming very cold in winter and hot in summer, giving these regions a pattern called a **continental climate**. The coasts of continents, particularly
those on west sides at middle latitudes where the
prevailing winds blow from ocean to land, have **maritime climates** with smaller differences between winter and summer temperatures. Seasonal temperatures change the most on the largest land mass—
Asia—where strong winter high pressure (descending air) over Siberia causes winds to blow out of the
continent toward the coasts. In the summer, however, strong low pressure (rising air) over Siberia
draws great quantities of moist air over the land from
the Indian Ocean, producing the great summer
monsoons—rainy seasons—characteristic of southern Asia.

Because the latitudinal gradient of solar energy
flux, the global circulation patterns of air and ocean
waters, and the warming and cooling of moving air
masses explain so much of Earth's climatic patterns,
we can predict fairly accurately the climate of a site
if we know its latitude, its continental location, and

its position with respect to mountain ranges. It would be a mistake, however, to conclude that Earth's climates are due entirely to physical factors. The properties of the atmosphere, and hence climate, are profoundly influenced by living organisms, and they have been since life first released into the atmosphere large amounts of oxygen, which was toxic to most of the early forms of life (Chapter 17). To understand how living organisms influence climate, we will now discuss the global ecosystem.

THE GLOBAL ECOSYSTEM

Carbon, nitrogen, phosphorus, calcium, sodium, sulfur, hydrogen, and oxygen, together with smaller amounts of other elements, are the prime materials of which living systems are constructed. They are vital to the functioning of life today, as they were during its evolutionary past, and will be in its future. These elements, necessary for life as we know it, move through living organisms and the physical environment. This movement depends in large part upon the activities of living organisms themselves. Thus, the quantities of these elements available to living organisms are strongly influenced by how living organisms get them, how long they hold onto them, and what they do with them while they have them.

It is convenient to divide the global ecosystem into four compartments—oceanic, fresh water, atmospheric, and terrestrial—because the nature of the physical environments in each of those four compartments and the types of organisms living there are very different. The amounts of elements found in the different compartments, what happens to those elements, and the rates at which they enter and leave the compartments differ strikingly. After we have described these compartments, we will consider them together to present a picture of the cycling of the elements among the compartments.

The Oceans

The oceans are very deep in many places, but they exchange materials with the atmosphere only at their surface. Oceans receive materials from land as runoff from rivers. Over time scales of hundreds to thousands of years, oceans are also the ultimate repository of most materials produced by human activity, even though the immediate receivers are often other compartments of the global ecosystem. Oceans, because of their huge size, and because they exchange materials with the atmosphere only at their surface, respond very slowly to outside disturbances. Also, except on continental shelves, ocean waters mix very slowly and are strongly stratified.

Elements that enter the oceans from other compartments gradually sink to the bottom, unless they are brought back to the surface by cool water that rises near continental west coasts. Such zones of upwelling are rich in nutrients, and many of the world's great fisheries are concentrated there. Over most of the oceans, nutrient concentrations are very low. Oxygen is usually present at all depths, because even slow mixing suffices to replenish the small amount of oxygen consumed by the respiration and decomposition of the few organisms that can live in the nutrient-poor water. Most of the elements that enter the oceans settle to the bottom and remain there until the bottom sediments are elevated above sea level by movements of Earth's crust. Many millions of years may elapse before this happens.

Fresh Waters

Much less water is found in lakes and rivers than in the oceans, but because these bodies of water are relatively small, most elements entering them are not buried in bottom sediments for long periods of time. Some elements enter fresh water in rainfall, but most of them are released by the weathering of rocks and are carried to lakes and rivers via **groundwater**, the water that resides in the soil and rocks, or by surface flow.

Once elements enter rivers, they are usually carried rapidly to the oceans. In lakes, however, they are taken up by living organisms and are incorporated into their cells. These organisms eventually die and sink to the bottom. Decomposition of their tissues uses up the oxygen in deeper waters. Consequently, surface waters of lakes quickly become depleted of nutrients, and oxygen is depleted in deeper waters. But the decrease in photosynthesis rates that would result from nutrient depletion is countered by vertical movements of water that bring nutrients back to the surface and oxygen to deeper water. In shallow lakes, wind is an important mixing agent. In deeper lakes, however, wind usually mixes only surface waters. In temperate regions, a very important mixing process is an overturning, which occurs twice each year (Figure 47.5). This process is driven by wind, but it depends on the peculiar behavior of water in relation to temperature, as we will now explain.

In spring in the temperate zone, the sun warms the surface layer of a lake. Initially, the warm layer is very shallow, but as spring and summer progress, its depth gradually increases. However, there is still a well-defined zone—the **thermocline**—where temperature drops abruptly to about 4°C. Only if a lake is shallow enough to warm all the way to the bottom does the temperature of the deepest water rise above 4°C. Water is most dense at 4°C; above and below

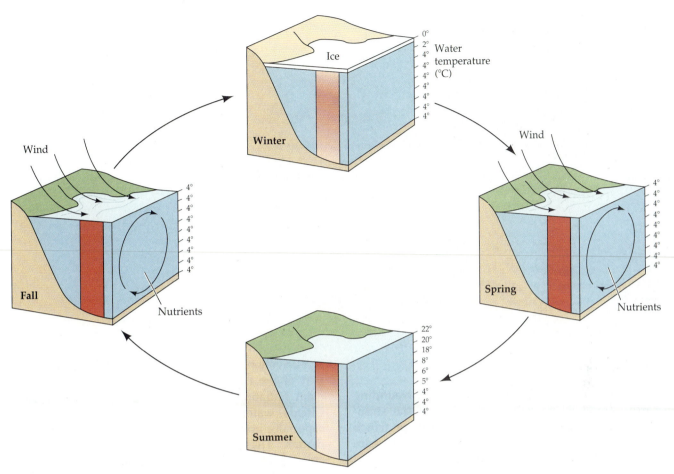

47.5 Annual Temperature and Oxygen Cycles in a Temperate Lake

These vertical temperature profiles are typical of temperate zone lakes that freeze in winter. Turnovers that occur in spring and fall allow nutrients and oxygen to become evenly distributed in the water column. How much surface waters warm during summer varies with the size and depth of the lake and with the local climate. Oxygen concentrations are shown by the intensity of red in the vertical columns.

that temperature, it expands. As the surface of a lake cools in autumn, the cooler surface water, being more dense than the warmer water below it, sinks, and is replaced by warmer water from below. This process continues until the entire water column has reached 4°C. At this point, the density of the water is uniform throughout the lake, and even modest winds readily mix the entire water column. Colder weather then cools the surface water below 4°C. This water becomes less dense than the 4°C water below it and floats at the top. Another turnover occurs in spring, when surface layers above the thermocline warm to 4°C, and the water column, again being of uniform density throughout, is easily mixed by wind.

Deep tropical and subtropical lakes may become permanently stratified because they never become cool enough to have uniformly dense water. Their

bottom waters lack oxygen because decomposition quickly depletes any oxygen reaching them. Many tropical lakes, however, are overturned at least periodically by strong winds, so that their deeper waters are occasionally oxygenated.

The Hydrological Cycle

The movement of water from the land to the oceans is obvious. Its return to the land from the oceans is not as evident, even though the amount moving in each direction is the same. The cycling of water between the oceans, atmosphere, and land, known as the **hydrological cycle**, begins with the evaporation of water, most of it from ocean surfaces. Some water returns as precipitation over the oceans themselves, but much less falls back on the ocean than is evaporated from it. The rest of the water is carried by winds over the land, where it falls as rain or snow. Water also evaporates from the soil, from freshwater lakes and rivers, and from the leaves of plants, but the total amount that evaporates from the land is less than the amount that falls on it as precipitation. The excess water eventually returns to the oceans via rivers, coastal runoff, and subterranean discharge (Figure 47.6).

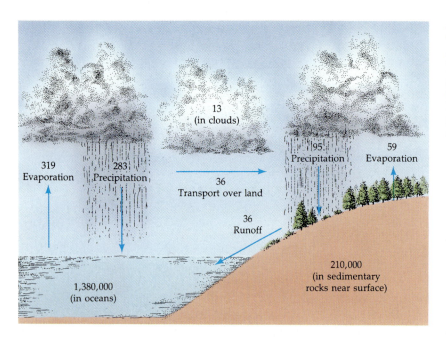

47.6 The Global Hydrological Cycle
The numbers give the relative amounts of water (expressed as units of 10^{18} g) held or exchanged. Notice that the greatest exchanges take place at the ocean surface. Although rocks contain large quantities of water, this "locked-in" water plays very little role in the hydrological cycle.

The Atmosphere

The atmosphere is a thin sphere of gases surrounding Earth. About 80 percent of the mass of the atmosphere lies in its lowest layer, the **troposphere**, which extends upward from Earth's surface about 17 kilometers in the tropics and subtropics, decreasing to about 10 km at higher latitudes (Figure 47.7). Most global air circulation takes place within the troposphere, and virtually all atmospheric water vapor is located there. Above the troposphere, the **stratosphere,** extending upward to about 50 kilometers above Earth's surface, contains about 99 percent of the remaining atmospheric mass but is extremely dry. Materials enter the stratosphere at the intertropical convergence zone and tend to remain there for a relatively long time because stratospheric air circulation is horizontal. The ozone (O_3) in the stratosphere absorbs most short wavelengths of ultraviolet radiation.

The atmosphere is 78.08 percent nitrogen, 20.95 percent oxygen, 0.93 percent argon, and 0.03 percent carbon dioxide, plus traces of hydrogen gas, neon, helium, krypton, xenon, ozone, and methane. The atmosphere contains Earth's biggest pool of nitrogen and large supplies of oxygen. Although carbon dioxide constitutes a very small fraction of the atmosphere, it is the source of the carbon used by terrestrial photosynthetic organisms. Concentrations of atmospheric water vapor are highly variable in space and time.

The atmosphere plays a decisive role in regulating temperatures at and close to Earth's surface. Without an atmosphere, the average surface temperature on Earth would be about −18°C rather than its actual +17°C. This difference exists because the atmosphere is relatively transparent to sunlight but it traps a large part of the outgoing infrared radiation (heat)—the main radiation emitted by a cool body like Earth. Clouds, water vapor, carbon dioxide, and ozone are especially important in this trapping of infrared radiation. That is why, as we will discuss later, increased concentrations of atmospheric carbon dioxide may lead to important climatic changes.

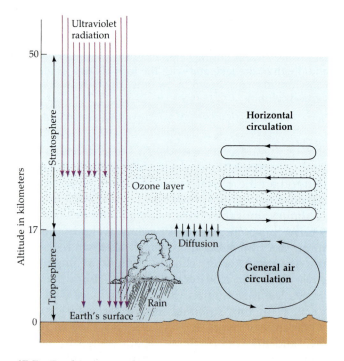

47.7 Earth's Atmosphere
The two lowest layers, the troposphere and the stratosphere, differ in their circulation, the amount of moisture they contain, and the amount of ultraviolet radiation they receive.

The atmosphere is a transport medium for many gases, as well as for airborne particles containing carbon, nitrogen, sulfur, phosphorus, and other nutrient elements. Many of these compounds are oxidized in the atmosphere by photochemical reactions involving the —OH radical, which forms when ozone is bombarded by solar ultraviolet radiation in the presence of water vapor. Carbon, nitrogen, and sulfur are returned to the terrestrial part of the ecosystem either dissolved in precipitation or by direct gas transfer.

Scientists have measured the fluxes of atmospheric gases (the rates at which they arrive on and are generated on Earth) and have estimated what those fluxes would be in the absence of life. If monerans on Earth were not constantly replenishing them, gases such as methane (CH_4) and hydrogen would vanish within a few decades. Nitrogen, the most abundant element in Earth's atmosphere, is not stable in the presence of oxygen or carbon dioxide on a watery planet. The stable compound of nitrogen on a lifeless Earth would be nitrate ion in the oceans. The probable atmosphere on a lifeless Earth would be similar to those of Mars and Venus, which are 95 and 96.6 percent carbon dioxide, and 2.7 and 3.2 percent nitrogen gas, respectively. Earth would be very much hotter if it had such an atmosphere.

The Land

About one-fourth of Earth's surface, most of it in the northern hemisphere, is currently above sea level. Most of this land is covered by a layer of soil of varying depths, weathered from parent rocks beneath it or carried to its present location by erosional agents. Unlike air and water, soils and rocks are solids that tend to remain where they are. Therefore, elements on land move much more slowly than they do in air and water, and they usually move only short distances. For this reason, we will emphasize local rather than global ecosystems in our discussion of the land compartment.

The land is connected to the other ecosystem compartments because living organisms take elements from and release elements to the air. Also, elements in soils are carried in solution into the groundwater and eventually into rivers and oceans, where they are lost to living organisms until an episode of geological uplifting raises marine sediments and a new cycle of erosion and weathering begins.

As you may recall from Chapter 30, the type of soil that forms in an area depends on the underlying rock, as well as on climate, topography, the organisms living there, and the length of time soil-forming processes have been acting. As a soil weathers, its clay particles slowly decompose chemically. After hundreds of thousands of years, most nutrients needed by plants have weathered out and have been carried by groundwater to the oceans. Therefore, very old soils are much less fertile than young soils. A general picture of these changes over a period of 1 million years was shown in Figure 30.7. Thus, even though the global supply of nutrients is constant, regional and local variations strongly affect ecosystem processes on the land.

Plants extract nutrients from the soil and incorporate them into their tissues. If the soil is fertile, those tissues are rich in nutrients. When those tissues then fall and decompose, they produce a dark, rich, alkaline, organic material—**humus**—called **mull**. Plants growing on infertile soils form nutrient-poor tissues that decompose into an acidic humus known as **mor**. Conifers in particular produce a mor humus that is resistant to decay and may accumulate in thick layers on the soil surface. An example of this is the decay of pine needles in England discussed in Chapter 46 (see Figure 46.25). Because needle decomposition takes seven years, there is a permanent thick layer of partly decomposed needles on the soil surface under pines.

There is an intricate relationship between plants and the microorganisms with which they share the soils. Plants allocate a high proportion of photosynthate to their roots, much of which is diverted to the mycorrhizal fungi living there. Ignoring the importance and fragility of the interconnections between plants and microorganisms has sometimes led to failure of ecosystem regeneration. For example, a 15-hectare plot in the Klamath Mountains of southern Oregon, clear-cut in 1968, has been replanted four times. All the plantings were failures, even though forests in this area regenerate readily after wildfires (Figure 47.8). The reason for the failures is that the site was both burned and treated with herbicides to open it up for better growth of conifer seedlings. However, the early successional deciduous trees and shrubs that were killed by the herbicide treatment are the species that support soil organisms and ameliorate temperatures and moisture, thereby maintaining the populations of soil microorganisms. When those plants were eliminated, most soil microorganisms died, and conifers then were unable to grow. Many such nonregenerating clear-cuts dot high mountains in this region.

BIOGEOCHEMICAL CYCLES

As we saw in Chapter 30, living organisms need large quantities of carbon, hydrogen, oxygen, nitrogen, phosphorus, and sulfur. Earth is essentially a closed system with respect to these elements. They cycle through living organisms to the physical environment, and back again through organisms, without inputs from external sources. The carbon and nitrogen atoms of which life is composed today are the

47.8 Nonregenerating Clear-Cuts in Oregon
No conifers are growing in the 15- to 20-year-old clear-cuts in this photo, despite the fact that each of them has been planted with seedlings three to five times since the trees were cut.

same atoms that were in dinosaurs, insects, and trees in the Mesozoic Era. Some of these elements circulate constantly, but large quantities of other elements are temporarily lost from circulation through deposition in deep sea sediments. The pattern of movement of an element through living organisms and through different reservoirs in the physical environment is called its **biogeochemical cycle**.

Each element has its own distinctive biogeochemical cycle whose properties depend on the physical and chemical nature of that element and how it is used by organisms. The elements all cycle quickly through organisms because no individual, even of the longest-lived species, lives very long. Elements, such as carbon and nitrogen, that exist in the atmosphere as a gas cycle faster than elements that are not found in a gaseous state. We will illustrate the properties of biogeochemical cycles by discussing those elements that are needed by living organisms in large amounts.

The Carbon Cycle

Living organisms are triumphs of carbon chemistry. To survive, living organisms must have access to carbon atoms. As we have seen, nearly all the carbon that enters organisms does so as carbon dioxide from the atmosphere, which enters the leaves of plants. There it is incorporated into organic molecules by photosynthesis. All organisms in other kingdoms get their carbon by eating either other organisms or their remains.

Although atmospheric carbon dioxide is the immediate source of carbon for living organisms, only a small proportion of Earth's carbon is found in the atmosphere. Most of it is found as nongaseous dissolved carbon—bicarbonate and carbonate ions—in oceans, and as carbonate minerals in rocks. Sedimentary rocks hold most of Earth's carbon, but movement of carbon between rocks and other reservoirs of carbon is very slow. The quantities of carbon in each ecosystem compartment and the yearly carbon fluxes between compartments are shown in Figure 47.9. Notice that although marine organisms have very little carbon, they have a profound influence on the distribution of carbon in the seas: They convert soluble carbonate ions from seawater into insoluble ocean sediments by depositing carbon in their shells and skeletons, which eventually sink to the bottom. Biological processes redistribute carbon between atmospheric and terrestrial reservoirs, removing it from the atmosphere during photosynthesis and returning it to the atmosphere during respiration. Growing plants at middle and high latitudes in the Northern Hemisphere incorporate so much carbon into their bodies during the summer that they reduce the concentration of atmospheric carbon dioxide from about 350 parts per million in winter to 335 parts per million midsummer (see Figure 47.14). This carbon is released back into the atmosphere by decomposition in the autumn.

At times in the remote past, large quantities of carbon were removed from the global carbon cycle when organisms died in large numbers in environ-

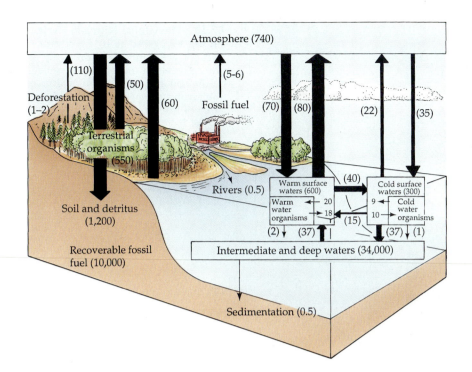

47.9 The Global Carbon Cycle
The two large reservoirs of carbon are dissolved carbon in the oceans and carbonate minerals in rocks (not shown). The numbers show the quantities in the reservoirs and the amounts that move between the various compartments. The widths of the arrows are proportional to the sizes of the fluxes. Units are billions of metric tons (10^{15} g).

ments without oxygen. In such environments, decay organisms do not reduce organic carbon to carbon dioxide. Instead, organic molecules accumulate, and eventually become oil, natural gas, coal, or peat. Humans have discovered and used these deposits, known as fossil fuels, at ever-increasing rates during the past 150 years. As a result, carbon dioxide, the final product of the burning of those fuels, is being released into the atmosphere faster than it is being removed by living organisms and transferred to the oceans. This buildup of atmospheric carbon dioxide will warm the global climate during the next century if current trends in the use of fossil fuels continue. Carbon dioxide is one of the so-called greenhouse gases; that is, it is transparent to sunlight but opaque to heat radiation. Carbon dioxide thus permits sunlight to strike and warm Earth, it but traps some of the heat Earth radiates back to space. This addition to the carbon cycle is also shown in Figure 47.9. Enough carbon is released by burning fossil fuels to alter the heat balance of Earth even though the absolute quantity is small relative to other components of the carbon cycle.

The Nitrogen Cycle

Nitrogen is an essential component of many organic molecules, such as nucleic acids and proteins. Nitrogen is an abundant, chemically inert gas that makes up 78 percent of the atmosphere. But, unlike carbon dioxide, which is readily used directly by photosynthetic plants, most organisms cannot use nitrogen in its gaseous form. Nitrogen can be converted into biologically useful forms only by a few species of

bacteria and cyanobacteria (Chapter 22). Therefore, despite its abundance, nitrogen is often in short supply in ecosystems. This is why nearly all commercial fertilizers contain biologically useful compounds of nitrogen.

Just as organisms cannot take up nitrogen gas directly from the atmosphere, they do not respire nitrogen back to the atmosphere. Instead, organic molecules containing nitrogen are converted to inorganic molecules in several stages by different organisms. Most of these nitrogen-containing compounds, such as nitrates or ammonia, are again taken up by plants. This movement of nitrogen among organisms accounts for about 95 percent of all nitrogen fluxes on Earth (Figure 47.10).

The quantity of nitric oxides and nitrate that humans produce by burning fossil fuels and by manufacturing and using fertilizers is about half the amount the rest of the living world produces naturally. The larger amount of nitrogen being cycled has greatly increased agricultural productivity but, as we will soon discuss, it also contributes to the problem of acid precipitation.

The Phosphorus Cycle

The phosphorus cycle differs from those of carbon and nitrogen—as well as those of sulfur, oxygen, and hydrogen—in that it lacks a gaseous phase. Some phosphorus is transported on dust particles, but in general the atmosphere plays a very minor role in the phosphorus cycle. Phosphorus exists mostly as phosphate (PO_4^{3}) or similar compounds. Most phosphate deposits are of marine origin. De-

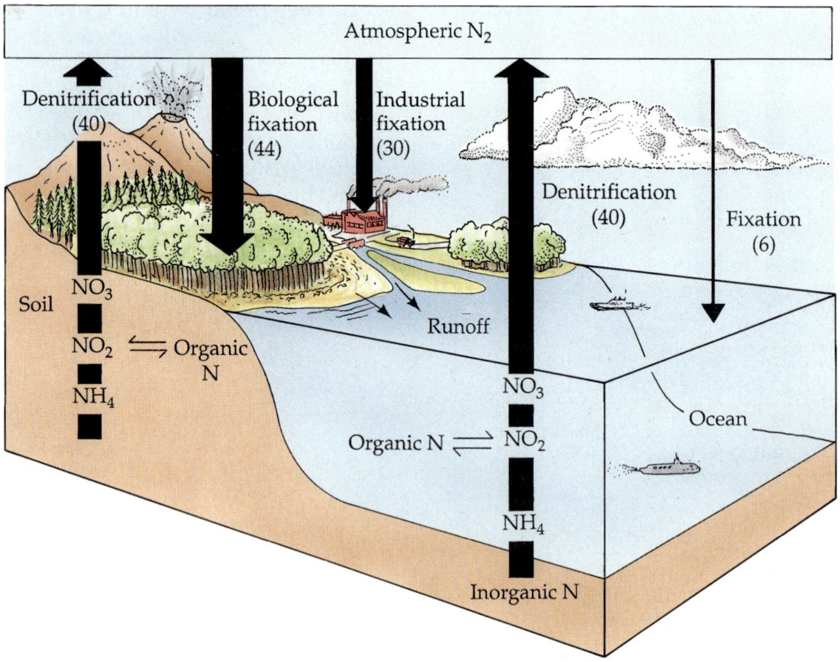

47.10 The Global Nitrogen Cycle
Fluxes, expressed as 10^9 kg of nitrogen per year, are driven primarily by the activities of unicellular organisms in the soil and oceans. The widths of the arrows are proportional to the sizes of the fluxes.

composing biological materials on the sea floor release phosphate, which enters sediment pores and is eventually incorporated into a type of rock known as apatite. Such deposits are forming today on continental margins underlying regions of upwelling that support large concentrations of organisms.

We have seen in our discussions of carbon and nitrogen that the reservoirs of elements and fluxes between reservoirs tell us a lot about how the elements cycle. Now you can learn how phosphorus cycles without seeing a figure—by studying the numbers in Table 47.1. Notice how much more rapidly phosphorus is cycled through marine organisms than through terrestrial organisms. Phosphorus also moves readily between the surface and the bottom

of the oceans. On average, a phosphorus atom is cycled about 50 times between deep waters and surface waters before being removed to the sediments. Each time a phosphorus atom reaches surface waters, it is cycled between the oceanic biota and the dissolved phosphates in the water about 25 times before it returns to deep water. As a result, the average phosphorus atom is incorporated into marine organisms about 1,250 times during its stay in the ocean! The overall phosphorus cycle is fairly constant, but phosphorus is often a limiting nutrient in soils and lakes. This is why phosphate is a component of most fertilizers and why additions of phosphate to lakes causes marked increases in their biological productivity.

TABLE 47.1
Reservoirs, Fluxes, and Residence Times of Phosphorus in the Global Ecosystem

RESERVOIR	AMOUNT IN RESERVOIR (10^6 METRIC TONS)	FLUX (10^6 METRIC TONS/YEAR)	RESIDENCE TIME (YEARS)
Atmosphere	0.0028	4.5	0.0006 (53 hrs)
Land biota	3000	63.5	47.2
Land	2,000,000	88–100	2000.0
Shallow ocean	2710	1058	2.56
Ocean biota	138	1040	0.1327 (48 days)
Deep ocean	87,100	60	1452
Sediments	4×10^9	214	1.87×10^8
Total ocean ecosystem	89,810	1.9	47,270

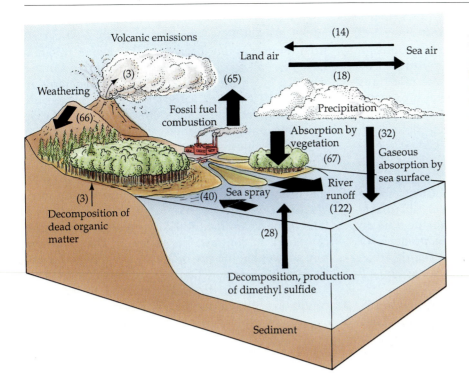

47.11 The Global Sulfur Cycle
The transfer rates here, expressed as 10^9 kg of sulfur per year, include both natural and human-caused fluxes. The total fluxes are now more than twice what they were a century ago, primarily because of fossil fuel combustion.

The Sulfur Cycle

Volcanoes and fumaroles emit sulfur dioxide and hydrogen sulfide. This is the only significant natural nonbiological flux of sulfur. These emissions release, on average, between 10 and 20 percent of the total natural flux of gaseous sulfur to the atmosphere, but they vary greatly in time and space. Large eruptions spread large quantities of sulfur over broad areas, but they are rare events. Volatile sulfur compounds are also emitted by both terrestrial and marine organisms. Certain marine algae produce large amounts of dimethyl sulfide. Why they do so is not certain, but because coastal algae produce the largest amounts, researchers believe that dimethyl sulfide protects the algae from desiccation and excess salinity. Dimethyl sulfide release by algae accounts for about half of the sulfur in the biotic part of the sulfur cycle. The other half is produced by terrestrial organisms. On land, the breakdown of organosulfur compounds during fermentation is the most important mechanism of sulfur release (Figure 47.11).

Sulfur is apparently always abundant enough to meet the needs of living organisms. It also plays an important role in global climate. Even if air is moist, clouds do not form readily unless there are nuclei around which water can condense. Dimethyl sulfide is the major source of such nuclei. Therefore, increases or decreases in sulfur emissions can change cloud cover and, hence, climate. Sulfur emitted by human activities now equals natural sulfur fluxes, and the increased sulfur cycle is already influencing global climate. The increase in atmospheric concentrations of carbon dioxide has not warmed Earth as much during the past century as expected, because cloud cover has increased along with increases in atmospheric concentrations of dimethyl sulfide.

HUMAN ALTERATIONS OF BIOGEOCHEMICAL CYCLES

The elements circulating in the biogeochemical cycles we have just discussed are essential for the metabolism of living organisms. The interaction between organisms and elemental cycles is a reciprocal one. Organisms influence the rates of cycling of elements, and the rates of cycling, in turn, influence the rates of metabolic processes.

Human activity has greatly modified the quantities of elements being cycled and where they enter and leave ecosystems. These changes can increase metabolic rates if they increase the availability of nutrients. They can also decrease metabolic rates if levels of elements become high enough to be toxic to organisms or if high levels cause other detrimental environmental changes. We will now consider several examples of consequences of human modifications of biogeochemical cycles. These consequences range from local to global, and they include both increases and decreases of metabolic rates.

Lake Eutrophication: A Local Effect

The most striking and best studied example of local effects of altered biogeochemical cycles is **eutrophication**—the addition of nutrients, especially phosphorus, to fresh water. Human activity, which tends to be concentrated around water, dumps large quantities of materials directly and indirectly into lakes and rivers. Most of this is domestic and industrial sewage, but much of it also comes from the leaching of fertilizers and pesticides from agricultural lands that drain into rivers and lakes. Some nutrients arrive in precipitation.

In fresh water, photosynthesis is most often limited by supplies of phosphorus. In eutrophic—enriched—lakes, the extra phosphorus provided by fertilizers and detergents allows algae and monerans to multiply rapidly, forming **blooms** that turn water green. Usually spherical green algae are replaced by filamentous blue-green bacteria as eutrophication proceeds. The decomposition of the dead cells produced by this increased biological activity uses up all the oxygen in the lake until the water column overturns again. Therefore, anaerobic organisms come to dominate the sediments. These organisms are unable to break down organic compounds all the way to carbon dioxide, and many of the end products of their activities have unpleasant odors. The decrease in oxygen in the water also harms some aquatic animals.

The changes accompanying eutrophication are illustrated by Lake Erie. Two hundred years ago Lake Erie had moderate levels of photosynthesis and clear, oxygenated water. Today over 14 million people live in the Lake Erie basin. Nearby cities pour over 250 billion liters of domestic and industrial wastes into the lake annually. The entire basin is also intensely farmed and heavily fertilized. In the early part of this century, nutrients in the lake increased greatly and algae proliferated. At the water filtration plant at Cleveland, for example, algae increased from 81 per milliliter in 1929 to 2,423 per milliliter in 1962. Algal blooms and populations of bacteria also increased. The numbers of *Escherichia coli* increased enough to cause the closing of many beaches on the south shore of the western end of the lake because of health hazards.

As the oxygen level dropped in deeper lake waters, many native species that thrive only in oxygenated water declined. They were replaced by species of clams, snails, and midges that can survive in relatively anaerobic environments. For example, nymphs of the mayfly genus *Hexagenia* were replaced by worms as the dominant organisms of the lake bottom (Figure 47.12). The fish fauna also changed. Prior to the turn of the century, dominant fishes in Lake Erie were lake herring, blue pike, carp, yellow perch, sauger, whitefish, and walleye. Lake trout were common in deeper waters. By 1925, herring had become too scarce to support the herring industry. After 1945, blue pike, sauger, and whitefish became very scarce, and lake trout disappeared. Currently the fishing industry is dependent upon yellow perch, smelt, sheepshead, white bass, carp, catfish, and walleye, most of which are less valuable commercially than the species that declined.

In 1972 and in 1978 the United States and Canada developed plans to clean up the Great Lakes. The most important part of those plans was the building of sewage treatment plants to reduce phosphorus discharges. As a result, the amount of phosphates added to Lake Erie decreased more than 80 percent from the maximum discharge. The deeper waters of Lake Erie still become poor in oxygen during summer months, but if the goal of reducing phosphorus inputs to 11,000 metric tons per year is achieved, the lake should return to a state similar to that of 50 years ago. This example, together with those of other lakes adjacent to cities, shows that lakes recover if nutrient inputs to their waters are greatly reduced.

The rate at which a lake recovers depends on the rate of turnover in its waters. Because it takes many years for the water of Lake Erie to be replaced, it will take many years for the lake to return to its former condition. In contrast, the waters of Lake Washington, adjacent to the city of Seattle, are replaced within 3 years. Therefore, when sewage was diverted from Lake Washington, the lake returned to its former condition within a decade.

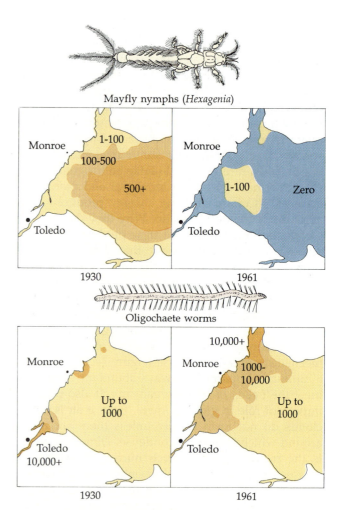

Mayfly nymphs (*Hexagenia*)

Oligochaete worms

47.12 Eutrophication Changes Lake Erie
As Lake Erie became more polluted during the middle part of the present century, the original mayfly population decreased (upper maps) but oligochaete worms increased (lower maps). The numbers and colors indicate the number of organisms per square meter.

Acid Precipitation: A Regional Effect

An important *regional* effect of the human alteration of two major biogeochemical cycles is **acid precipitation**—rain or snow whose pH is reduced by the presence of sulfuric and nitric acids, derived in large part from the burning of fossil fuels. Acid precipitation was first detected in Scandinavia, where it acidified lakes. In Norway, populations of brown trout dropped to half their previous levels by 1978, and populations declined another 40 percent by 1983. Other species of fish were similarly affected. Acid precipitation is now a phenomenon of all major industrial countries and is particularly widespread in eastern North America (Figure 47.13). The normal pH of precipitation in this region is about 5.6, but today precipitation in New England averages about pH 4.1, and there are occasional storms with a pH as low as 3.0. These extreme values cause direct damage to the leaves of plants. In Central Europe, 15–20 percent of the growing stock of harvestable forest has suffered moderate to severe damage or death, due in part to acid precipitation.

Ecologists in Canada studied the effects of acid precipitation by adding enough sulfuric acid to reduce the pH of two lakes from about 6.6 to 5.2. In both lakes, nitrifying bacteria failed to adapt to these moderately acidic conditions, with the result that the nitrogen cycle was blocked and ammonium accumulated in the water. When the ecologists stopped adding acid to one lake, pH increased to 5.4 and nitrification resumed after a lag of about 1 year. These experiments show that lakes are very sensitive to acidification but that they can recover quickly when pH returns to normal values.

Oxides of nitrogen and sulfur entering the atmosphere from installations burning fossil fuels may travel hundreds of kilometers before they settle to

Earth in precipitation or as dry particles. The source of acid precipitation in New England is primarily the Ohio Valley; that of Scandinavia is primarily the industrial areas of England and Germany. Because the pollution originates in one area but causes problems in another, the solution of this problem is politically very difficult. Moreover, acid precipitation is caused by the generation of the energy upon which modern societies depend. Oxides of nitrogen and sulfur can be removed from the smokestack gases of large installations, but costs rise sharply as the percentage removed rises above 90 percent. The number of sources emitting these oxides is now so great that almost complete removal will be necessary to correct the problem, even if no new sources are added.

Alterations of the Carbon Cycle: A Global Effect

The carbon cycle is the biogeochemical cycle most seriously perturbed globally by human activity. Climatologists have made measurements of atmospheric concentrations of carbon dioxide on top of Mauna Loa in Hawaii since 1958. These measurements reveal a slow but steady increase in atmospheric carbon dioxide concentrations (Figure 47.14). From a variety of calculations, atmospheric scientists believe that 150 years ago, before the Industrial Revolution, the concentration of atmospheric carbon dioxide was probably about 265 parts per million. Today it is 350 parts per million. This increase has been caused primarily by combustion of fossil fuels and secondarily by the burning of forests. If current trends in both these activities continue, atmospheric carbon dioxide is expected to double its 1900 value of 290 parts per million by the middle of the next century. This carbon dioxide will eventually be transferred to the oceans and deposited in sediments as calcium carbonate ($CaCO_3$), but the rate at which this happens is much

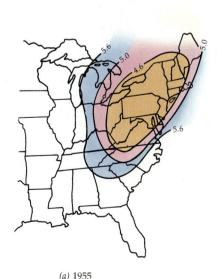

(a) 1955

(b) 1980

47.13 Increases in Acid Precipitation in Eastern North America
These maps show the average annual pH of precipitation in eastern North America. The oxides of nitrogen and sulfur—the principal contributors to acid precipitation—travel far enough from their sources that the effects of many sources blend together to produce the pattern shown here.

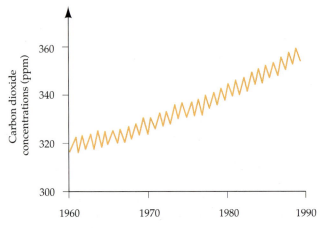

47.14 Atmospheric Carbon Dioxide Concentrations Are Increasing

These concentrations, expressed as parts per million by volume of dry air, were recorded on top of Mauna Loa, Hawaii. Each year CO_2 concentrations rise during the winter, when respiration exceeds photosynthesis, and fall during the summer, when photosynthesis exceeds respiration. However, the steady rise in overall concentrations is very apparent.

slower than the rate at which humans are introducing carbon dioxide into the atmosphere.

Climatologists have analyzed air trapped in the Antarctic and Greenland ice caps and found that concentrations of carbon dioxide in the atmosphere have varied considerably in the past. During the last Ice Age—between 15,000 and 30,000 years ago—the concentration was as low as 200 parts per million, whereas during a warm interval 5,000 years ago it may have been slightly higher than it is today. Global

climate models developed by climatologists can predict the likely consequences of further increases in concentrations of atmospheric carbon dioxide. Predictions from these models are imprecise and uncertain because nature is so much more complicated than the models. Nonetheless, if the concentration of atmospheric carbon dioxide doubles from its current level, the mean temperature of Earth is expected to increase 3 to 5°C, especially at higher latitudes. A carbon dioxide doubling would also shift climatic patterns latitudinally, causing droughts in the central regions of continents, and would increase precipitation in coastal areas. Global warming might also cause melting of the Arctic and Antarctic ice caps, raising sea levels and flooding coastal cities and agricultural lands.

Because carbon dioxide is carried thousands of kilometers away from where it enters the atmosphere by air movements, the problem is global. It is very difficult for societies to decide how much they should invest today to avert potential future climatic problems. Nonetheless, many nations have now committed themselves to undertake efforts to reduce their emissions of carbon dioxide. Doing so will not be easy, because carbon dioxide is the inevitable end product of fossil fuel combustion, and modern societies are powered by fossil fuels. The cleanest possible power plant burning fossil fuel emits as much carbon dioxide as a very dirty one. Nonetheless, because so much energy is currently wasted, many steps can be taken to increase energy-use efficiency so that we get more valuable services for the same amount of fuel burned. Also, we can substitute energy sources that do not contain carbon (solar, geothermal, nuclear) for fossil fuels.

TABLE 47.2
Areas, Biomass of Plants, and Net Primary Production of Earth's Major Vegetation Zones

	AREA		MASS OF PLANTS		PRIMARY PRODUCTION	
VEGETATION ZONE	10^6 KM^2	PERCENT	10^9 TONS	PERCENT	10^9 TONS	PERCENT
Polar	8.05	1.6	13.77	0.6	1.33	0.6
Conifer forest	23.20	4.5	439.06	18.3	15.17	6.5
Temperate	22.53	4.5	278.67	11.5	17.97	7.7
Subtropical	24.26	4.8	323.90	13.5	34.55	14.8
Tropical	55.85	10.8	1347.10	56.1	102.53	44.2
Total land	133.4	26.2	2402.5	100	171.54	73.8
Glaciers	13.9	2.7	0	0	0	0
Lakes and rivers	2.0	0.4	0.04	<0.01	1.0	0.4
All continents	149.3	29.3	2402.54	100	172.54	74.2
Ocean	361.0	70.7	0.17	<0.001	60.0	25.8
Earth total	510.3	100	2402.71	100	232.54	100

ENERGY FLOW THROUGH ECOSYSTEMS

Almost all energy utilized by living organisms comes from the sun. Even the fossil fuels—coal, oil, and natural gas—upon which the economy of modern civilization is based are reserves of captured solar energy locked up in the remains of organisms that lived millions of years ago.

Much of the solar energy that arrives on Earth is reflected back into the atmosphere as heat, especially in places where the surface is bare because there is too little water to support plant growth. Even where Earth's surface is covered with green leaves, much solar energy goes to evaporate water from the plants. This evaporation, called transpiration, creates a flow of water through the plant that carries nutrients from the soil to the leaves where photosynthesis takes place. Only a small portion of the arriving solar energy is captured by the process of photosynthesis.

Net Primary Production

Plants use most of the energy they capture to maintain themselves. This energy is eventually converted to heat during respiration, but some of the energy is used to produce new tissues that are available to be eaten by herbivores, or, after the plant dies, by organisms that eat their remains. The total amount of energy plants assimilate by photosynthesis is called **gross primary production**. Because plants use much of this energy to power their own metabolism, they always contain much less energy than the total amount they assimilated. The amount of energy that remains after we subtract the energy the plants use for maintenance and biosynthesis is known as **net primary production**. The total net primary production on Earth in different climatic regions is summarized in Table 47.2. The global distribution of this production is shown in Figure 47.15.

Net primary production correlates strongly with the quantity of water available because, as just mentioned, water must move through plants as they photosynthesize. A tropical lowland forest can transpire about 16 cubic centimeters of water per square centimeter of ground area per month. If more than this falls as rain, it runs off into streams and rivers. At higher latitudes, less water can be transpired much of the year because the evaporative power of air is less at lower temperatures. In many areas, transpiration is limited not by temperature but by available

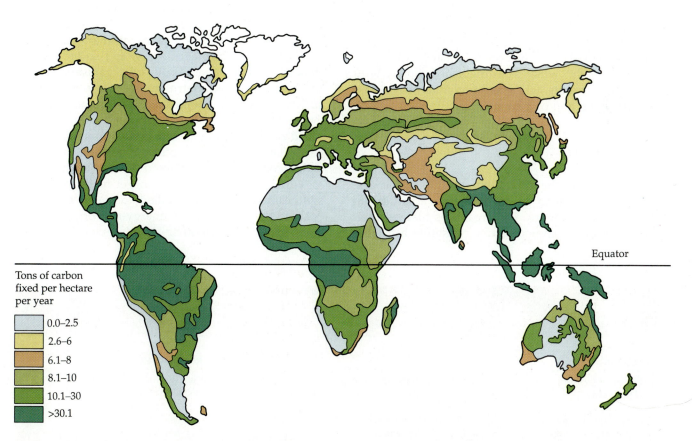

Tons of carbon fixed per hectare per year

- 0.0–2.5
- 2.6–6
- 6.1–8
- 8.1–10
- 10.1–30
- >30.1

Equator

47.15 Biological Production of Terrestrial Ecosystems
Areas of high annual production are in wet tropical and subtropical regions and the wetter parts of temperate latitudes. Low production characterizes the hot subtropical deserts (where moisture is limiting) and high latitudes (where cool temperatures lower photosynthetic rates). This map plots primary production in tons of carbon fixed per hectare per year.

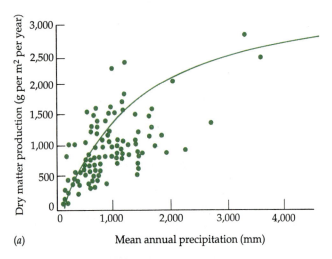

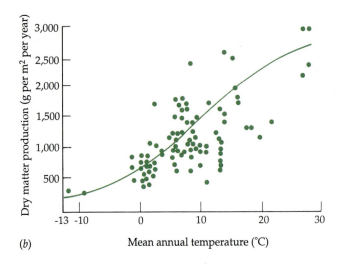

(a) Mean annual precipitation (mm)

(b) Mean annual temperature (°C)

47.16 Ecosystem Production, Temperature, and Precipitation

(a) Net primary production increases rapidly with precipitation but reaches a plateau because a forest can transpire only so much water no matter how much falls. (b) Net primary production increases steadily with increasing mean annual temperature but also tends to plateau be-cause plant respiration increases almost exponentially with temperature, whereas the rate of photosynthesis increases linearly with temperature. The data points are scattered around the curves because other factors also influence production.

soil moisture and by soil fertility (Figure 47.16). Moisture limitation is especially widespread in arid regions, where net primary production rises linearly with increases in precipitation. When precipitation increases over about 2,000 millimeters per year, production is limited by available soil moisture only during dry periods. Production in aquatic systems is limited by light, which is extinguished rapidly with depth; by nutrients, which sink and must be replaced by upwelling; and by temperature.

TROPHIC LEVELS

When we study the flow of energy through ecosystems, it is useful to group organisms according to their source of energy. The organisms that get their energy from a common source constitute a **trophic level**. Photosynthetic plants get their energy directly from sunlight. Collectively, they constitute the trophic level called photosynthesizers. Organisms that eat plants constitute the trophic level called herbivores. Organisms that eat herbivores are called primary carnivores. Those that eat primary carnivores are called secondary carnivores, and so on. Organisms that eat the dead bodies of organisms or their remains are called detritivores. Many organisms, such as ourselves, obtain their food from more than one trophic level below them. Such organisms are called omnivores (Table 47.3). Thus, organisms in a particular trophic level occupy a position in an ecological community determined by the number of energy-transfer steps required to get to them. Photosynthesizers are often called primary producers because they produce the energy-rich organic molecules upon which all other organisms feed. All other

TABLE 47.3
The Major Trophic Levels

TROPHIC LEVEL	EXAMPLES	SOURCE OF ENERGY
Photosynthesizers	Green plants, photosynthetic monerans and protists	Solar energy
Herbivores	Termites, grasshoppers, water fleas, anchovies, deer, geese	Tissues of photosynthesizers
Primary carnivores	Spiders, warblers, wolves, copepods	Herbivores
Secondary carnivores	Tuna, falcons, killer whales	Primary carnivores
Omnivores	Humans, opossums, crabs, robins	Several trophic levels
Detritivores	Fungi, many bacteria, worms, millipedes, vultures	Dead bodies and waste products of other organisms

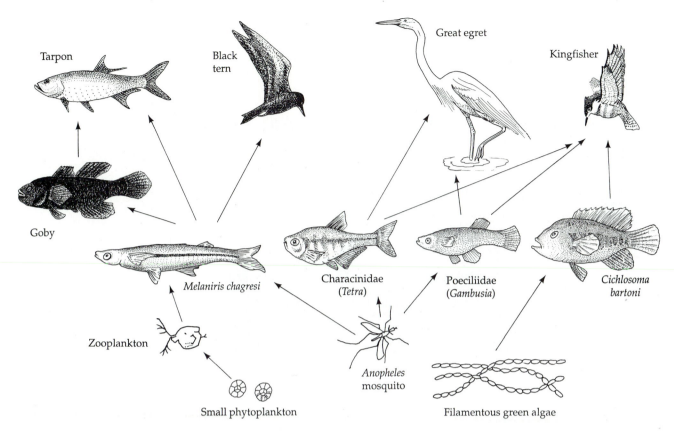

47.17 Food Web of Gatun Lake, Panama
This diagram groups together many species of phytoplankton and zooplankton, but the essential features of this food web are little influenced by this lumping.

organisms are often called consumers because they consume, either directly or indirectly, energy-rich organic molecules produced by green plants.

A set of linkages in which a plant is eaten by an herbivore, which is in turn eaten by a primary carnivore, and so on, is called a **food chain**. Food chains are usually interconnected to make a **food web**. A simplified food web for Gatun Lake, Panama, is shown in Figure 47.17. The arrows in this food web show who eats whom. Food webs provide a useful summary of predator–prey interactions within a community. Because most biological communities have many species, similar species are often combined in food webs, as they are in the diagram of the Gatun Lake food web.

The Maintenance of Organisms and Energy Flow

The energy organisms use to maintain themselves is dissipated as heat, a form of energy that cannot be transferred to other organisms. An organism's net production—its growth plus reproduction—is the amount of energy available to organisms at the next level. The efficiency of energy transfer through food webs depends on the fraction of net production at one trophic level that is eaten by organisms at the next level, and how those organisms apportion the ingested energy between production and maintenance (respiration). The values shown in Table 47.4 for different animal taxa, expressed as average pro-

TABLE 47.4
Average Production Efficiencies, $P/(P + R)$, for Various Groups of Animals

GROUP	$P/(P + R)$ (PERCENT)
Insectivores (mammals)	0.9
Birds	1.3
Small-mammal communities	1.5
Other mammals	3.1
Fishes and social insects	10.0
Invertebrates other than insects	
Herbivores	21
Carnivores	28
Detritivores	36
Nonsocial insects	
Herbivores	39
Carnivores	56
Detritivores	47

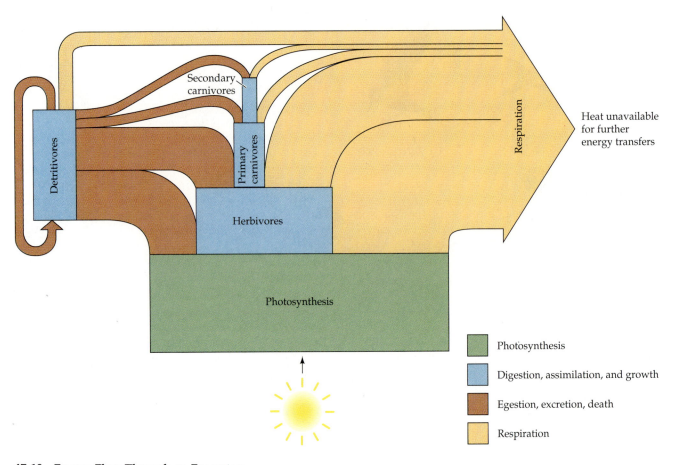

47.18 Energy Flow Through an Ecosystem
The quantities of energy flowing through an ecosystem can be visualized using a diagram like this one, in which the width of the channels is roughly proportional to the amount of energy flowing through them. Trophic levels are indicated by blocks of blue or green, energy channels are in tan or brown, and the direction of energy flow is shown by arrows.

duction efficiency [production/(production + respiration)], show two interesting patterns. First, birds and mammals have very low efficiencies because they expend so much energy maintaining constant high body temperatures. Second, herbivores are less efficient than carnivores because plant tissues are generally more difficult to digest than animal tissues.

Even when production efficiencies and consumption rates are both high, seldom is as much as 20 percent of the energy assimilated by a trophic level converted to production at the next trophic level. The amount of energy reaching an upper trophic level is determined by the net primary production and the efficiencies with which food energy is converted to biomass—the total weight of organisms—in the trophic levels below it (Figure 47.18). To show how energy decreases in moving between trophic levels, ecologists construct a graph in which the trophic levels are piled on top of one another (Figure 47.19a). Such plots have the form of **pyramids of energy** because the quantities of energy captured per time period are much greater in lower trophic levels than they are in higher ones. A plot of a **pyramid of**

biomass, which shows the standing biomass of organisms at different trophic levels, illustrates the amount of biomass energy that is available for organisms of the next trophic level (Figure 47.19b).

Usually pyramids of energy and biomass for an ecosystem have similar shapes, but sometimes they do not. Differences are caused by the structures of organisms and how they allocate their energies. On land, dominant photosynthetic plants store energy for moderately long periods, and they dominate both energy flow and standing biomass. In aquatic communities, on the other hand, the dominant photosynthesizers are monerans and protists. They have such high rates of cell division that a small standing biomass can feed a much larger biomass of herbivores that grow and reproduce much more slowly. This can produce an inverted pyramid of biomass, as shown in Figure 47.19b, even though the pyramid of energy for the same ecosystem has the typical shape.

GRAZING RATES IN GRASSLANDS. Although most terrestrial ecosystems are dominated by large plants, they differ strikingly in pattern of energy flow. We can see

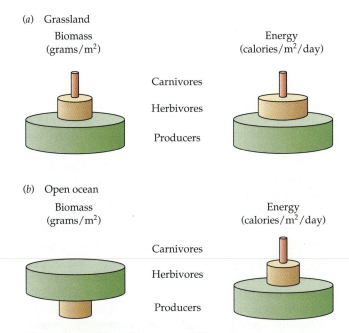

(a) Grassland

Biomass (grams/m²) Energy (calories/m²/day)

Carnivores

Herbivores

Producers

(b) Open ocean

Biomass (grams/m²) Energy (calories/m²/day)

Carnivores

Herbivores

Producers

47.19 Pyramids of Energy and Biomass
Ecosystems can be compared in terms of the amount of material present in organisms at different trophic levels, and in terms of energy flow. *(a)* A grassland is typical of most terrestrial ecosystems. Most of the biomass is found in the green plants, and most of the energy flows through them. *(b)* An inverted pyramid of biomass from a marine community. The producers here are unicellular algae, which divide so rapidly that a small biomass can support a much larger biomass of herbivores.

this by comparing grasslands and forests. Grassland plants produce few woody tissues and can support a great load of herbivores. Mammals—wild or domestic—may consume 30–40 percent of the aboveground biomass, that is, 30–40 percent of the annual aboveground net primary production. Insects may consume an additional 5–15 percent. Soil animals, primarily nematodes, may consume 6–40 percent of the belowground biomass.

Consumption rates in local areas may be even higher. For example, in certain prairies, grazers in the areas around prairie dog colonies often harvest 60–80 percent of the biomass. The prairie dogs take most of the harvest, but bison, elk, and pronghorn that graze within the prairie dog colonies part of the year take some as well (Figure 47.20). Typically, the adjacent areas are only lightly grazed. As a result, aboveground plant biomass within colonies is much less than it is outside prairie dog colonies. This intense grazing has little effect on net primary production within the colonies, which remains very similar to that of the adjacent areas. But the grazing by herbivores changes the patterns of nutrient cycling and competition between soil microorganisms and plants. Plants in prairie dog colonies allocate most of their energy to regrowth of aboveground tissues and relatively little to roots. Also, because soil is more exposed within colonies, surface soil temperatures are several degrees higher in summer than in adjacent

47.20 Mammals Graze in Prairie Dog Colonies
Bison prefer to graze within this prairie dog colony because the plants there have higher protein levels than plants outside the colonies. Several prairie dogs are visible at the right center of the photo; notice how short the grass is within the colony.

areas, so conversion of organic nitrogen-containing compounds to inorganic compounds proceeds faster. However, because plants allocate less energy to roots, soil microorganisms have less energy for growth, and they compete less well with plants for available nitrogen. As a result, plant tissues in prairie dog colonies have higher nitrogen levels than elsewhere—a desirable outcome from the perspective of the prairie dogs.

BROWSING RATES IN FORESTS. In contrast to those in grasslands, the dominant plants in forests allocate substantial energy to forming wood, which accumulates at high rates in growing forests. As we have seen, wood is heavily defended by the plants and is rarely eaten unless a plant is diseased or otherwise weakened. Also, in most forests, leaves fall to the ground relatively undamaged at the end of the growing season. There are outbreaks of defoliating insects in forests, but in general, browsing rates are so low that forest ecologists ignore losses to herbivores when calculating forest production.

DETRITIVORES. As shown in diagrams of energy flow in ecosystems, much of the energy ingested by organisms is converted to biomass that is eventually consumed by detritivores. Detritivores reduce the remains of organisms to mineral nutrients that can be taken up by plants again. If there were no detritivores, most nutrients would eventually be tied up in dead bodies, where they would be unavailable to plants. Therefore, continued ecosystem productivity depends on rapid decompositon of detritus. Under the warm, wet conditions found in tropical forests, detritus is decomposed within a few weeks or months, and no litter accumulates on the soil surface. Rates of decomposition are slower under colder and drier conditions. On cold mountains at high latitudes, decomposition of leaf litter may take decades; decomposition of trunks may take over a century.

AGRICULTURE AND ECOSYSTEM PRODUCTIVITY

Agriculture is a system by which humans exploit ecosystems by replacing species of low value to them with ones of high value. We do this by helping some species compete better and by manipulating the system to increase the production that is useful to us. Agriculture has several intricately intertwined components. We can eliminate competition with other plants—weeds—by cultivating and by applying herbicides. We eliminate pests, usually by applying toxic chemicals. We augment photosynthesis by fertilizing and irrigating. And we develop special high-yielding strains of plants that respond to additional fertilizer by increasing their growth rates. All these

components must work together because "miracle" strains do not actually yield more than other strains unless they are provided with fertilizers and protected from competitors and herbivores. Agriculture also depends on energy from outside the system for cultivating and harvesting. In modern agriculture this energy comes from fossil fuels. In preindustrial agricultural systems it is supplied in smaller amounts, primarily by animal muscle power.

Although human manipulations of agricultural systems have spectacularly increased food production per hectare, they have also created problems. Insecticides and herbicides pollute lakes, rivers, and groundwaters in most industrialized countries. Many agricultural pests have evolved resistances to pesticides; in response, we increase the use of pesticides, creating even more severe pollution problems. Agriculturalists are combatting these problems by developing new methods of pest control in agriculture. Generally known as **integrated pest management**, these methods combine chemicals with cultural strategies, such as crop rotation, mixed plantings of crop plants, and mechanical tillage of the soil, and biological methods, such as development of pest-resistant strains of crops, use of predators and parasites rather than pesticides, and use of chemical attractants. Under integrated pest management, farmers use chemicals sparingly enough to avoid most pollution problems. As a result, pests do not evolve resistance to chemicals, or do so very slowly.

Organisms and materials move into and out of all natural ecosystems. Some ecosystems, such as rivers and the deep sea, are powered primarily by organic compounds and materials imported from elsewhere. But in most ecosystems, losses of materials through one component are made up via gains through another. Agricultural ecosystems differ from natural ecosystems because humans *extract* as many energy-rich compounds as possible from the systems and consume them elsewhere. Preindustrial agricultural people lived near their fields and returned most wastes, including their own excretory products, to the fields. The flow of nutrients changed dramatically when most people moved to cities. Today, most humans eat their food far from where it was produced, and we release our wastes in large quantities in small areas. Modern agricultural systems require heavy applications of fertilizers to replace those elements removed when crops are harvested and transported elsewhere.

Agriculture developed in temperate regions, and temperate methods often work poorly in the tropics. Lush forests grow even on ancient, impoverished tropical soils. Their total photosynthesis per unit of ground area is often not much less than that of tropical forests on much better soils. This is possible because nutrients in the tropical forest system are very tightly cycled. Trees produce long-lived leaves that they shed asynchronously. The plants remove many minerals from their leaves before they drop them, and microorganisms and tree roots quickly break down litter and capture mineral nutrients before they leach down through the soil. Minerals in tropical forests are concentrated within the plants themselves. Deeply weathered tropical soils thus are exhausted if they are cultivated for only a few years. Larger concentrations of minerals reside in temperate forest soils, which is why these soils provide good agricultural yields for many years.

SUMMARY

The climate of Earth is determined primarily by Earth's atmosphere, which lets in sunlight but traps heat emanating from Earth's surface; by the pattern of solar energy input at different latitudes; and by Earth's rotation on its axis. These factors create a pattern of surface winds, which drive global oceanic circulation in turn.

The main elements of living organisms—carbon, nitrogen, phosphorus, sulfur, hydrogen, and oxygen—cycle among the four compartments of the global ecosystem: the oceans, fresh waters, the atmosphere, and the land. Each element has its own biogeochemical cycle, its pattern and rate of movement among living organisms, and different reservoirs in the physical environment. Human activity greatly

modifies cycles of these elements on local, regional, and global scales. The most serious local problems are the pollution of lakes and rivers. A serious regional problem is increasing acidification of precipitation, due in large part to the burning of fossil fuels and wood. Increasing concentrations of atmospheric carbon dioxide are expected to cause warming of the global climate by the early part of the next century.

Ecosystems are powered by solar energy. Most energy enters living organisms via photosynthesis. On land, green plants are the largest and longest-lived organisms, but in aquatic environments, most photosynthesizers are very small. Much of the energy taken in by an organism is used for maintenance and is eventually dissipated as heat, so the efficiency of energy transfer to higher trophic levels is usually very low. In terrestrial ecosystems, transfer rates are higher in grasslands, where plants produce few woody tissues, than in forests, where much net primary production is allocated to relatively indigestible wood.

Humans design agricultural ecosystems to maximize production of useful products. We accomplish this goal by eliminating competitors and predators, adding nutrients and water, and genetically altering domesticated plants and animals. Agriculture extracts materials at high rates from ecosystems; these materials must be replaced by fertilizers. Modern agriculture is heavily dependent upon external energy sources to carry out all these activities.

SELF-QUIZ

1. Which of the following is *true* about the amount of sunlight and heat arriving on Earth?
   a. Every place on Earth gets the same annual number of hours of sunlight and the same amount of heat.
   b. Every place on Earth gets the same annual number of hours of sunlight but not the same amount of heat.
   c. Every place on Earth gets the same annual amount of heat but not the same number of hours of sunlight.
   d. Both the annual amount of sunlight and the amount of heat received vary over the surface of Earth.
   e. None of the above.

2. When an area is within the intertropical convergence zone:
   a. the northeast trade winds blow steadily.
   b. the southeast trade winds blow steadily.

   c. air is descending and it seldom rains.
   d. air is rising and heavy rains fall frequently.
   e. westerly winds blow steadily.

3. Zones of marine upwelling are important because:
   a. they help scientists measure the chemistry of deep ocean water.
   b. they bring to the surface organisms that are difficult to observe elsewhere.
   c. ships can sail faster in these zones.
   d. they increase marine productivity by bringing nutrients back to surface ocean waters.
   e. they bring oxygenated water to the surface.

4. Which of the following is *not* true of the troposphere?
   a. It contains nearly all atmospheric water vapor.
   b. Materials enter it primarily at the intertropical convergence zone.

   c. It is about 17 kilometers deep in the tropics.
   d. Most global atmospheric circulation takes place there.
   e. It contains about 80 percent of the mass of the atmosphere.

5. The humus formed from decomposition of nutrient-rich leaves is called:
   a. the thermocline.
   b. mor.
   c. mull.
   d. eutrophication.
   e. dirt.

6. Carbon dioxide is called a greenhouse gas because:
   a. it is used in greenhouses to increase plant growth rates.
   b. it is transparent to heat radiation but opaque to sunlight.
   c. it is transparent to sunlight but opaque to heat radiation.
   d. it is transparent to both sunlight and to heat radiation.
   e. it is opaque to both sunlight and heat radiation.

7. The phosphorous cycle differs from those of carbon and nitrogen in that:
   a. it lacks a gaseous phase.
   b. it lacks a liquid phase.
   c. only phosphorous is cycled through marine organisms.
   d. living organisms do not need phosphorous.
   e. the phosphorous cycle does not differ importantly from the carbon and nitrogen cycles.

8. Acid precipitation results from human modifications of:
   a. the carbon and nitrogen cycles.
   b. the carbon and sulfur cycles.
   c. the carbon and phosphorous cycles.
   d. the nitrogen and sulfur cycles.
   e. the nitrogen and phosphorous cycles.

9. The total amount of energy plants assimilate by photosynthesis is called:
   a. gross primary production.
   b. net primary production.
   c. biomass.
   d. a pyramid of energy.
   e. eutrophication.

10. The amount of energy reaching an upper trophic level is determined by:
    a. net primary production.
    b. net primary production and the efficiencies with which food energy is converted to biomass.
    c. gross primary production.
    d. gross primary production and the efficiencies with which food energy is converted to biomass.
    e. gross primary production and net primary production.

11. Which of the following is *not* a component of integrated pest management?
    a. Use of cultural strategies such as crop rotation and mixed plantings.
    b. Use of pest-resistant strains of crops.
    c. Use of predators and parasites of crop pests.
    d. Use of chemical attractants.
    e. Use of chemical pesticides whenever pests are discovered.

FOR STUDY

1. How would you expect temperature and oxygen profiles to appear in a broad, shallow tropical lake? In a very deep tropical lake? Why?

2. The waters of Lake Washington, adjacent to the city of Seattle, rapidly returned to their preindustrial levels of purity when sewage was diverted from the lake to Puget Sound, an arm of the Pacific Ocean. Would all lakes being polluted with sewage clean up as rapidly as Lake Washington if pollutant input were stopped? What characteristics of a lake are most important to its rate of recovery following removal of pollutant inputs? What is the diverted sewage likely to do to Puget Sound?

3. Tropical forests are being cut at a very rapid rate at the present time. Does this necessarily mean that deforestation is a major source of input of carbon dioxide to the atmosphere? If not, why not?

4. The two drawings below represent biomass pyramids for (a) an old field in Georgia, and (b) the English Channel. Explain the significance of the inversion of the second pyramid compared with the first.

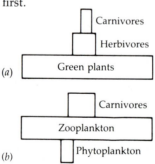

5. The amount of energy flowing through a food chain declines more or less rapidly depending upon the nature of the organisms in the chain. Which of the following simplified food chains is likely to be more efficient? Why? What criteria of efficiency are you using?
   a. phytoplankton → zooplankton → herring
   b. shrubs → deer → wolf

6. A government official authorizes the construction of a large power plant in a former wilderness area. Its smokestacks discharge into the air great quantities of waste resulting from the combustion of coal. List and describe all *likely* ecological results, at local, regional, and global levels. Suppose the wastes were thoroughly scrubbed from the stack gases. Which of the ecological results you have just outlined would still happen?

READINGS

Anderson, J. M. 1981. *Ecology for the Environmental Sciences*. Wiley, New York. A good discussion of relationships between ecosystem dynamics and current environmental problems.

Jordan, C. F. 1985. *Nutrient Cycling in Tropical Forest Ecosystems*. Wiley, New York. A useful summary of nutrient cycling patterns in tropical forests and how they differ from those in temperate forests.

Likens, G. E., F. H. Bormann, R. S. Pierce, J. S. Eaton and N. M. Johnson. 1977. *Biogeochemistry of a Forested Eco-*system. Springer-Verlag, New York. An account of intensive studies of a watershed in New England; clearly demonstrates the value of long-term studies for understanding ecosystem processes.

Rambler, M. B., L. Margulis, and R. Fester, Editors. 1989. *Global Ecology: Towards a Science of the Biosphere*. Academic Press, San Diego. A collection of essays covering a wide variety of the interactions between organisms, global biogeochemical cycles, and global climate.

Southwick, C. H. 1985. *Global Ecology*. Sinauer Associates, Sunderland, MA. A collection of readings about ecosystem processes and human impacts on them.

Whittaker, R. H. 1975. *Communities and Ecosystems*, 2nd Edition. Macmillan, New York. A good short textbook with excellent coverage of theoretical topics; an excellent follow-up to materials in this chapter.

48

Biogeography

PREVIEW: Historical and ecological biogeographers study past and present distributions of organisms to determine why species live where they do. Species distributions have been strongly influenced by the former positions of the continents as well as by present-day barriers to dispersal. The number of species in a region is the result of a balance between rates of arrival of new species and extinction of species already present. The tropics contain more species of most taxa than temperate and arctic regions, and islands are much poorer in species than continents. The major biomes of Earth result from particular types of climates, which influence biological productivity, plant life forms, and species interactions.

This chapter deals with historical and ecological biogeography, continental drift, dispersal, habitat islands, community convergence, and biomes.

The plants and animals of Australia differ from those of other continents. The first Europeans to visit Australia were perplexed by eucalyptus trees, strange flowers pollinated by parrots and other unusual birds, and mammals that hopped around on their hind legs, carrying their young in pouches. The first Europeans to visit North America, on the other hand, did not encounter the same degree of strangeness; the plants and animals of North America are similar to those of Europe. To explain these differences in species distributions, we must understand processes operating over longer time frames and broader spatial frames than those we considered in Chapter 46.

Biogeography is the study of the diversity of organisms over space and time. A basic question asked by biogeographers is: Why does taxon X occupy a certain area? Two answers to the question are possible: Either the taxon evolved there, or it evolved elsewhere and dispersed to the area. A parallel question is: Why *doesn't* taxon X occupy a certain area? The possible answers are that the taxon evolved elsewhere and never dispersed to the area, or that it was once present in the area but no longer lives there. If you were a biogeographer, your key problem would be to determine which distribution patterns can be explained by where the taxa evolved and which ones result from dispersal. Because millions of years of changes in both the physical environment of Earth and the organisms themselves have determined distributions of organisms, obtaining answers to these questions is difficult.

THE HISTORY OF BIOGEOGRAPHY

Biogeography, a historical science, itself has an interesting history. The field began when eighteenth-century travelers first noted intercontinental differences in distributions of organisms. As the patterns became better known, biogeographers attempted to identify the areas of origin and pathways of dispersal of taxa (species or groups of species). Early investigators generally believed that taxa originated in rather small areas and thereafter dispersed more widely. Therefore, they thought that most patterns had to be explained by dispersal. Most early studies focused on only a single taxon. At that time, most scientists believed that the continents had always been where they are now. As you will recall from Chapter 27, this belief is now known to be false.

In the next stage in the history of biogeography, ecologists began to seek alternatives to untestable historical assumptions about the distributions of particular taxa. Those ecologists concentrated on how present-day distributions are influenced by interactions among species and between species and their physical environments. Because local habitats always contain fewer species than the number living in the general region, ecologists believed that interactions of the types we discussed in Chapter 46 were important in explaining the distributions of organisms.

In the third stage, which biogeography has just entered, biogeographers are attempting to synthesize and extend both of the previous approaches by combining them with new methods, especially improved cladistic methods that allow the reconstruction of phylogenies in rigorous, standardized ways (see Chapter 20). These methods, combined with more complete morphological and genetic data, are providing the first reliable phylogenies of taxa with poor fossil records. At the same time, better geological knowledge is providing evidence of the positions of the continents at different times in Earth's history.

Assume that you are a biogeographer studying the distribution of related groups of organisms living on different continents. You can use cladistic methods and fossil data to estimate how long ago the groups shared their most recent common ancestor. Then you can compare this estimate with the time of separation of the continents to determine whether the ancestor lived before or after the continents separated. If it lived before, no long-distance dispersal need be assumed to account for the distribution. If the ancestor lived only after the continents separated, you must assume that some members of the group moved from one continent to another.

Because distributions of organisms are influenced by many historical and contemporary factors, individual biogeographers usually concentrate on one of two major approaches to the field. **Historical biogeography** concerns itself primarily with evolutionary histories of groups of organisms: Where and when did they originate? How did they spread? What does their present-day distribution tell us about their past histories? **Ecological biogeography** concentrates on current interactions of organisms with the physical environment and with one another to understand how ecological relationships influence where species and higher taxa are found today. The integration of

these subdisciplines of biogeography is essential to a full understanding of the geographic distributions of organisms.

The names of these two subdisciplines are somewhat misleading. All biogeography is historical. Ecological biogeography concentrates on recent history, current interactions, and changes within the past few thousand years. It also concentrates on patterns of distributions within local areas and regions. Historical biogeography examines longer time periods and larger spatial scales. Because time and space are continuous variables, these two subdisciplines blend into other another (Figure 48.1). Recognizing this, we will, for the sake of simplicity, discuss the approaches separately.

THE SHAPING OF THE MODERN WORLD

To understand why particular organisms are found in certain places today, we need to remember that continental positions have changed strikingly and that Earth has not always looked as it does today. We must imagine things being very different from what we observe today.

One of the first people to suggest that the continents had very different positions in the past was a German meteorologist, Alfred Wegener. He based his ideas on the shapes of the continents, which seemed to have once fit together, and on distributions of plants and animals that were hard to explain if one assumed that the continents had always been where they are now. When Wegener proposed his ideas in 1912, however, few scientists took them seriously, primarily because there were no known mechanisms

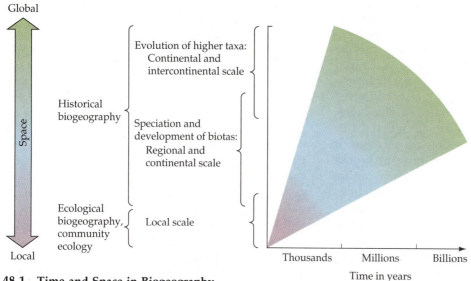

48.1 Time and Space in Biogeography
Biogeographical processes are studied at local and global spatial scales and at temporal scales ranging up to billions of years.

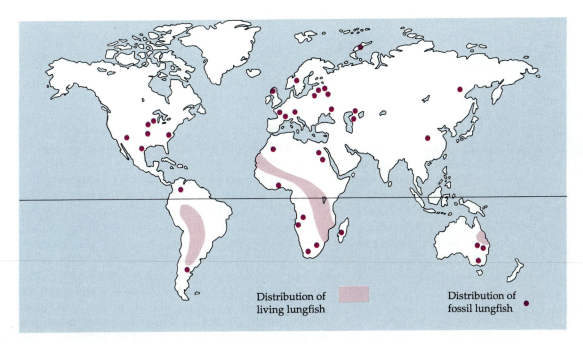

Distribution of
living lungfish

Distribution of
fossil lungfish

48.2 Lungfish Distributions
Although lungfish live today only in South America, Africa, and Australia,
their worldwide fossil record suggests that they evolved before the breakup of
Pangaea.

by which continents could have moved, but also because no convincing geological evidence of such movements existed. As we learned in Chapter 27, geological evidence and plausible mechanisms are both available today, and the broad pattern of continental movements is now clear.

In the early Mesozoic era, about 245 million years ago, many groups of nonmarine organisms, including insects, freshwater fishes, and frogs, had already evolved. All the continents were still very close to one another and there were no marine barriers blocking the dispersal of terrestrial and freshwater organisms. Some organisms that are found on most continents today may have been present on those land masses when they were all part of Pangaea (Figure 48.2). However, by the close of the Mesozoic era, about 65 million years ago, the continents were breaking apart and drifting away from one another (see Figure 27.1).

HISTORICAL BIOGEOGRAPHY

Historical biogeographers use several types of data. A key type is phylogenies, which both indicate relationships among organisms and give approximate times of the separation of lineages based on molecular clocks. A second type of data is fossils, which reveal where members of lineages lived in the past. Fossils may show how long a taxon has been present in an area and whether its members formerly lived in areas where they are no longer found. Fossils are usually helpful, but the fossil record is always incomplete. The first and last members of a taxon that lived in an area are extremely unlikely to have become fossils that are discovered and described. A third type of data is the distribution of living species. Much more complete and extensive information can be gathered on such distributions than will ever be available from fossils. Much can be learned by examining the distribution patterns of *many different groups* of living organisms to determine how similar they are.

Biogeographic study of the flightless weevil *Lyperobius huttoni* in New Zealand demonstrates the value of both geological and comparative distributional data from many taxa. New Zealand comprises two main islands separated by a channel called Cook Strait. The weevil is widely distributed in the mountains of South Island and also lives on sea cliffs at the extreme southwest corner of North Island (Figure 48.3). Considering this distribution without any other information, you might guess that *L. huttoni* had somehow managed to cross Cook Strait, even though it cannot fly. However, more than 60 other animal and plant species, including other species of flightless insects, share this disjunct distribution. The assumption that all of them made the same ocean crossing seems less likely. In fact, that assumption is unnecessary, because recent geological evidence, summarized by the Pliocene map in Figure 48.3, indicates that the southwest tip of North Island was once part of South Island and was separated by a water gap

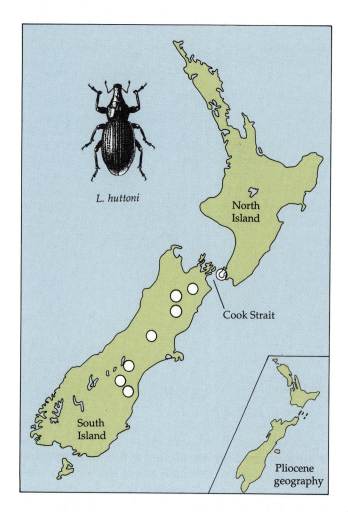

now separated by a barrier because it was in those areas before the barrier was imposed, it is said to have a **vicariant distribution**. If a species is in both areas because it crossed the barrier, it is said to have a **dispersal distribution**. In principle, vicariant and dispersal distributions are distinct, but in fact, they blur, because some barriers are harder to cross than others. Cook Strait is a modest barrier compared to the distance between Australia and New Zealand or between New Zealand and South America. Also, because all organisms disperse some distance from their birthplaces to where they eventually live, the question, "How far must an organism move before its distribution should be attributed to dispersal?" has no simple answer.

Species that live all over the Earth are of little use in reconstructing the history of life. Species with restricted distributions are much more informative. Species, genera, and families found only in one region are said to be **endemic** to that location. As far as we know, all species are endemic to Earth. Some species are endemic to one continent. Still others are restricted to very small areas, such as tiny islands or single mountaintops. Endemism must be evaluated carefully when reconstructing biogeographic history. Endemic taxa can be either very old ones that are becoming extinct or very young taxa that have recently evolved in a restricted area.

The longer an area has been isolated from other areas by a vicariant event, such as continental drift, the more endemic taxa it is likely to have, because there has been more time for evolutionary divergence to take place. Australia, which has been separated the longest from the other continents, about 65 million years, has the most distinct biota. South America has the next most distinct biota, having been isolated from other continents for nearly 60 million years. North America and Eurasia, which were joined together for much of Earth's history, have very similar biotas. That is why the Europeans felt more at home in North America than in Australia.

We will now discuss an example that illustrates how biogeographers use modern methods to reconstruct the biogeographic history of taxa. Midges are a family of flies (Chironomidae) with aquatic larvae. Ten genera within the family have species in both South America and Australia, all living in cool, temperate areas. But each species is found on only one of the continents. In nine of the ten genera, the most closely related species are found only on the other continent. Figure 48.4 shows a cladogram of five of

from North Island. Therefore, none of those 60 species need have made a water crossing.

This example illustrates an important method commonly used in biogeographic analysis. Although organisms *do* cross major oceanic and terrestrial barriers, biogeographers quite often apply the rule of **parsimony** when interpreting distribution patterns. A parsimonious interpretation of a distribution pattern is one that requires the smallest number of dispersal events to account for it. You do not need to postulate dispersal events for *L. huttoni* if you assume that it and the other species with that currently disjunct distribution already lived on the southwest tip of North Island when it was part of South Island. Of course, some of those species, such as birds and flying insects, may indeed have dispersed across Cook Strait, but when there is no evidence indicating that they did, most biogeographers favor the more parsimonious interpretation.

Reconstructing Biogeographic History

We have noted that a species can be found in an area either because it evolved there or because it dispersed to the area. If a species lives in two distinct areas

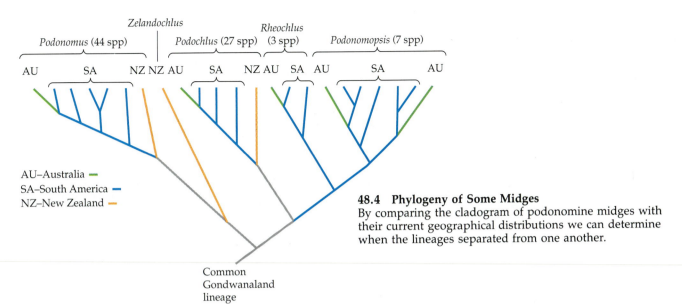

AU–Australia —
SA–South America —
NZ–New Zealand —

48.4 Phylogeny of Some Midges
By comparing the cladogram of podonomine midges with their current geographical distributions we can determine when the lineages separated from one another.

Common
Gondwanaland
lineage

the genera, together with the distributions of their species in Australia, South America, and New Zealand (you may want to review the section on cladistic systematics in Chapter 20). You will notice that the New Zealand species are relatively older than the Australian ones and that they are more closely related to South American than to Australian midges. Surprisingly, the Australian species are more closely related to the South American ones than they are to the New Zealand species. Two of the genera, *Rheochlus* and *Podonomopsis*, are not found in New Zealand, and *Zelandochlus* is found only there.

The cladogram, combined with the current geographical distributions of these midges, strongly suggests that the clade originated before the breakup of Gondwanaland (see Figure 27.1) and that the Australian species of *Podonomus*, *Podochlus*, and *Podonomopsis* evolved from ancestors that dispersed from South America some time after the breakup. The cladograms and current distributions of the midges do not fit with an interpretation that there have been recent exchanges of midges between New Zealand and Australia.

No fossil midges have been found. If such fossils are found, the interpretation that appears most probable today might have to be modified. We might learn, for example, that genera currently not found in one of these geographic areas did live there in the past. But for now, the combination of phylogenetic and geographic information available makes this the strongest interpretation of the biogeographic histories of these insects.

Major Biogeographic Regions

To some degree, the continents have been separated from one another long enough to have evolved dis-

tinctive biotas. The differences among continental biotas, first recognized more than a century ago, formed the basis for dividing Earth into the major biogeographic regions shown in Figure 48.5. Notice that, with the exception of Australia, these regions today are not completely separated from each other by water, although they were in the past. Their biological distinctness is maintained today by mountain and desert barriers to dispersal.

Over most of the history of life, the seas have been in fairly continuous contact. Moreover, all the phyla of marine animals evolved long before the continents separated in the late Paleozoic to early Mesozoic eras, so all of them have a nearly worldwide distribution today. Little long-range dispersal is needed to explain current distributions. Some taxa are restricted to warm waters, others to colder waters, but biogeographic regions of the oceans are not as well defined as those on land. Subtidal and intertidal coastal biotas are more distinct from one another than open ocean biotas.

ECOLOGICAL BIOGEOGRAPHY

Ecological biogeographers use the wealth of information on current distributions of organisms to test their theories. They can also use experiments to test hypotheses, something that historical biogeographers cannot do. As an example, we will discuss a model that attempts to account for the species richness of an area, and we will look at some experiments that have been conducted to test this model.

An Equilibrium Model of Species Richness

What determines the species richness of an area? It is obvious that species immigrating into an area increase its richness and that species becoming extinct

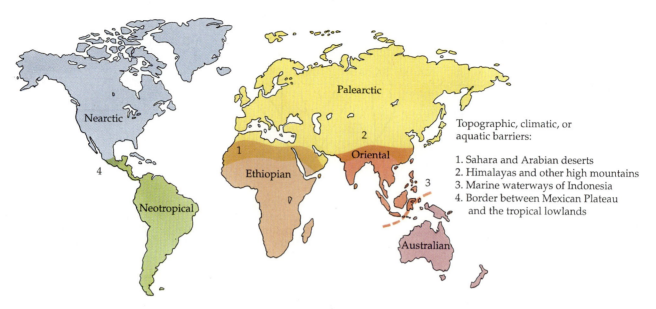

48.5 Major Biogeographic Regions
Barriers separating the biogeographic are shown as broad bands rather than thin lines because floras and faunas change gradually rather than abruptly from one region to another.

decrease it. How might biogeographers consider the effects of these two processes so as to account for species richness? It is easiest to visualize the effects if we consider, as did Robert MacArthur of Princeton University and E. O. Wilson of Harvard University, oceanic islands that initially have no species at all. Consider a newly formed oceanic island that receives colonists from a mainland area. Let us call the list of species present on the mainland that might possibly invade the island the species pool. The first colonists to arrive on the island are all "new" species because there are no species there already. As the number of species on the island increases, a larger fraction of immigrants are members of species already present, so that even if the same number of species arrives as before, the rate of arrival of *new* species decreases, until it reaches zero when the island has all the species in the pool.

Now let us think about extinction rates. First there will be only a few species on the island, and their populations may grow large. As more species arrive and their numbers increase, the resources of the island will have to be divided among more species. We therefore expect the average population size to become smaller as the number of species becomes larger. The smaller a population, the more likely it is to go extinct. Also, the number of species available to become extinct increases as species accumulate on an island. New arrivals to an island may include pathogens and predators that increase the probability of extinction of other species, further increasing the number of species becoming extinct per unit of time.

Therefore, because the rate of arrival of new spe-

cies decreases while the extinction rate increases with the number of species present, eventually the number of species should reach an equilibrium. These processes are pictured as curves in Figure 48.6*a*. If there are more species than the equilibrium number, extinctions should exceed arrivals. If there are fewer species than the equilibrium number, arrivals should exceed extinctions. Such an equilibrium is dynamic, because even if species richness remains relatively constant, species composition may change as different species replace those that become extinct. The model does not predict which species will arrive and which will become extinct. It predicts only the equilibrium number of species if arrival and extinction rates are known and are constant. If either rate fluctuates very much, there will be no equilibrium number of species.

The same ideas can be used to predict how species richness should differ among islands of different sizes and different distances from the mainland. We expect smaller islands to have lower rates of arrival of new species because fewer species would find small islands. Similarly, we expect fewer colonizers to reach islands more distant from the mainland. Figure 48.6*b* gives relative species richness equilibria for islands of different sizes and distances from the mainland.

Tests of the Species Richness Equilibrium Model

As we have just seen, the equilibrium model predicts that the number of species should be positively correlated with island size and negatively correlated

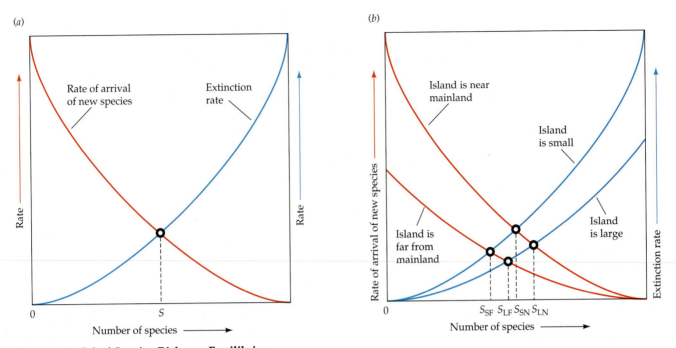

(a)

(b)

48.6 A Model of Species Richness Equilibrium
Rates of arrival of new species and extinction of species already present determine the equilibrium number of species. (a) Equilibrium (S) occurs when the number of species arriving equals the number of species going extinct. (b) Small islands far from the mainland (S_{SF}) have the fewest species; large islands near the mainland (S_{LN}) have the largest number of species at equilibrium.

with distance from the mainland. This pattern is shown by the numbers of bird species found on some islands in the Pacific Ocean (Figure 48.7). New Guinea serves as the mainland for most of these islands. Similar patterns are known for plants, insects, lizards, and mammals.

Natural disturbances sometimes permit tests of this equilibrium model of biogeography. A striking example is provided by Krakatoa, a volcanic island located between Sumatra and Java. In August 1883, Krakatoa erupted in the greatest volcanic explosion in recent history, one that covered the island with red-hot lava and destroyed all life on its surface. After the lava cooled, Krakatoa was colonized rapidly by plants and animals from nearby islands. By 1933, Krakatoa was again covered with a tropical evergreen forest, and 271 species of plants and 31 species of birds were found there. Birds had reached an equilibrium number of species by 1920, after which the number of new arrivals was approximately equal to the number of extinctions (Table 48.1), but the number of plant species is still increasing. Repeated censuses of Krakatoa will yield direct measures of arrival and extinction rates of species and will show when different taxa have reached equilibrium.

An experimental test of the equilibrium model of island biogeography was carried out off the southern tip of Florida, a region dotted by thousands of small islands consisting entirely of red mangrove trees

rooted in shallow water. Islands that are about 12 meters in diameter each contain 20 to 45 species of arthropods. Smaller islands have fewer species, larger ones more. Six islands were fumigated with methyl bromide, which destroyed all arthropods on them (Figure 48.8). Methyl bromide decomposes rapidly and does not inhibit recolonization. The rates of recolonization of the islands were very high. Within

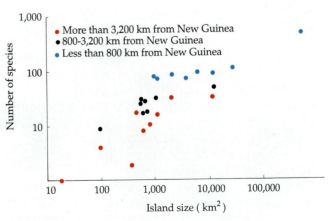

48.7 Isolated Islands Have Fewer Bird Species
The dots give numbers of land and freshwater bird species on islands of different sizes in the Moluccas, Melanesia, Micronesia, and Polynesia. These islands have been divided into three groups according to their distance from the source area, New Guinea.

TABLE 48.1
Number of Species of Nonmarine Birds on Krakatoa

	NUMBER OF SPECIES				
PERIOD	MIGRANTS	NONMIGRANTS	TOTAL	EXTINCTIONS	COLONIZATIONS
1908	0	13	13		
1908–1919				2	20
1919–1921	4	27	31		
1921–1932				5	4
1932–1954	3	27	30		

a year the fumigated islands had about their original number of species, indicating that an equilibrium had been reached.

The models and examples we have discussed pertain to oceanic islands, where water provides barriers to dispersal. Mainland areas are full of **habitat islands**, that is, patches of habitat separated from other similar patches by different types of habitats. For many species living in habitat islands, the intervening areas may be just as unsuitable to live in as if they were covered with water. We can apply the island model of species richness to these habitat islands, recognizing that some intervening areas, though unsuitable for permanent occupancy, may nonetheless permit a brief stopover. Therefore, arrival rates are higher for most habitat islands than they for similarly sized oceanic islands. We will discuss the importance of habitat islands for conservation of species richness in the next chapter.

SIMILAR ENVIRONMENTS, SIMILAR SPECIES

The first European botanists to travel around the Earth in the middle of the nineteenth century noticed that areas with similar climates in different parts of the Earth have similar vegetation. This is why we apply the name "Mediterranean vegetation" to communities dominated by evergreen shrubs with tough leaves, both in the Mediterranean region and in the other four regions of the world that have mild, wet winters and hot, dry summers. What impressed early travelers most was that these communities, despite their similarities, were composed of plants in very different taxa. The similarities are due to **convergent evolution**, that is, the development of similarities among organisms that were originally very different from one another.

The evolution of convergence, as opposed to the **parallel evolution** of organisms that were originally

48.8 Experimental Island Biogeography
Scaffolding is erected by scientists to enclose a small mangrove island in the Florida Keys. Methyl bromide introduced into the enclosure killed all the arthropods inside it. When the enclosure was removed, arthropods quickly recolonized the island.

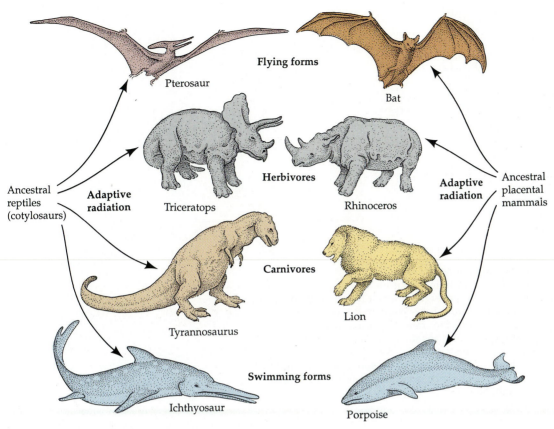

48.9 Convergent Body Forms of Reptiles and Mammals
Reptiles and mammals both radiated into species that were somewhat convergent to one another even though the reptiles are clearly recognizable as reptiles and the mammals as mammals.

very similar and have evolved in the same direction, depends on three things. First, the organisms must have been unlike one another at some previous time. Second, their similar environments must favor similar adaptations. Third, there must have been enough time for convergence to take place. Evolutionary changes, as we have seen, are usually slow. Convergence takes millions of years. Similar ecological opportunities in separate places in the remote past led to the convergence of certain animals. The adaptive radiation of reptiles in the Mesozoic era produced some animals similar to some of the animals that evolved when mammals radiated millions of years later (Figure 48.9).

Plants in different areas with Mediterranean climates converged because of the similarity of the climates. Because the winters are cool and wet in such places, whereas the summers are hot and dry, soil moisture levels are most favorable when temperatures are lowest. When temperatures rise in the spring, the soils dry out. Under this regime, natural selection favors the ability to photosynthesize in winter, when leaves may be exposed to occasional freezing weather, and the ability to extract water from relatively dry soils. Shrubs with tough, drought-re-

sistant leaves are well adapted to this regime, and they dominate most Mediterranean vegetation. Another type of plant common in all Mediterranean climates is perennials that grow in winter and early spring, but die back to underground storage organs during the long, hot, dry summer.

BIOMES

Ecologists group the many types of terrestrial communities into a few **biomes**, each consisting of the organisms that occupy one of the major climatic regions of Earth; that is, biomes are Earth's major terrestrial ecosystems. Biomes are usually identified and named by their characteristic vegetation, sometimes supplemented by their location or climate. Thus the names of the major biomes—tundra, boreal forest, temperate deciduous forest, grassland, cold desert, hot desert, Mediterranean, thorn forest, tropical savanna, tropical deciduous forest, tropical evergreen forest, and tropical montane forest—tell us the dominant vegetation in them and, sometimes, where we can find those biomes.

Because climate plays a key role in determining

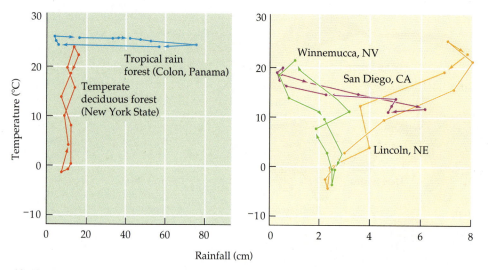

48.10   Climates of Some Major Biomes
The graphs indicate seasonal changes in temperature and precipitation for several biomes. On the left, a tropical evergreen forest biome (little fluctuation in temperature, heavy rainfall most months) is compared with a temperate deciduous forest biome (relatively constant rainfall but marked temperature variations). The right-hand graph shows the data for Lincoln, Nebraska, in a temperate grassland biome (sharp temperature variation; rainfall peaking in summer); a warm desert biome in Nevada (sharp seasonal temperature differences and low rainfall); and the Mediterranean biome of San Diego (moderate temperature range, winter rainfall peak).

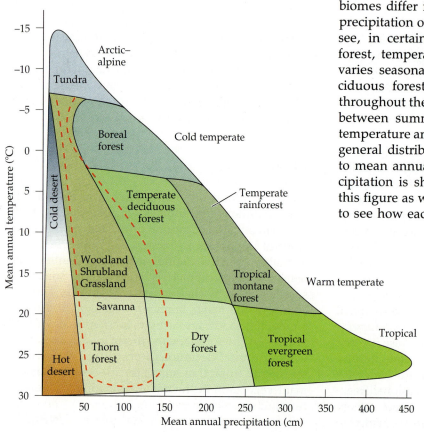

which types of plants are in a given environment, the distribution of biomes on Earth is strongly influenced by annual temperature ranges and rainfall amounts. One way to see clearly how climates of the biomes differ is to plot changes in temperature and precipitation over the year (Figure 48.10). As you can see, in certain biomes, such as tropical evergreen forest, temperatures are nearly constant but rainfall varies seasonally. In others, such as temperate deciduous forest, precipitation is relatively constant throughout the year but temperature varies strikingly between summer and winter. In still others, both temperature and precipitation change seasonally. The general distribution of terrestrial biomes in relation to mean annual temperature and mean annual precipitation is shown in Figure 48.11. By referring to this figure as we discuss the biomes, you will be able to see how each one fits into global climatic types.

48.11   Biome and Climate
Here biome distributions are plotted on axes of mean annual temperature and mean annual precipitation. Within the region indicated by dashed lines, other factors—such as seasonality of drought, fire, and grazing—strongly affect which biome is present.

(a)

48.12 Arctic and Alpine Tundra Biomes
(a) By early September, when this picture was taken in Denali National Park, Alaska, autumn is setting in and the tundra will soon be covered with snow for 8 months or more. (b) This alpine area in the Teleki Valley on Mount Kenya is dominated by giant groundsells (*Senecio* spp.) whose thick leaves and densely pubescent trunks help to keep them from freezing at night.

(b)

Tundra

The tundra biome is found in the Arctic and high in mountains at all latitudes. Because the climate in these areas is too cold for trees, tundra vegetation is dominated by low-growing perennial plants (Figure 48.12). Arctic tundra localities have short, but often warm, summers during which the sun shines 24 hours every day, whereas alpine tundras in the tropics have 12-hour days and 12-hour nights all year. It is never very warm at these high altitudes, and it freezes on most clear nights. In some tropical alpine plants, old leaves remain on the stem and provide an insulating cover that prevents water in the stem from freezing at night. Plants photosynthesize slowly all year in tropical alpine communities, and most animals are resident there all year. Arctic tundra plants, in contrast, photosynthesize only during the short summer. Most Arctic tundra animals either migrate into the biome for the summer and go elsewhere in the winter, or they are dormant most of the year. Arctic tundra is underlain by permanently frozen ground called **permafrost**. Therefore most Arctic tundra is very wet even though precipitation is low, because water cannot drain down through the soil.

Boreal Forest

Equatorward from the Arctic tundra, and below the alpine tundra on temperate-zone mountains, is the boreal forest biome, which is dominated by coniferous evergreen trees (Figure 48.13a). Over most of the boreal forest biome, winters are long and very cold and summers short but often warm. Boreal forests also grow along west coasts of continents at middle to high latitudes, where winters are mild but very wet and summers are cool and relatively dry. These forests have Earth's tallest trees, and they support the highest standing biomasses of wood of all ecological communities. Southern hemisphere boreal forests are dominated by southern beech trees, which are angiosperms that have very small leaves and appear remarkably similar to conifers (Figure 48.13b).

Boreal forests have only a few tree species, nearly all of which are wind pollinated and have wind-dispersed seeds. The dominant animals, such as insects, moose, and hares, eat leaves. Because there are so few species, many prey have a single major predator species, and population densities of both predators and prey fluctuate markedly, as you saw in Chapter 46. The seeds in the cones of the conifers are eaten both by species that store seeds, such as squirrels, jays, and nutcrackers, and by those, such as crossbills, that travel long distances to find areas with many cones.

48.13 Boreal Forest Biome
(a) In mountainous areas boreal forests grow up to the timberline, the upper limit of tree growth. This northern-hemisphere forest is in Jasper National Park, Alberta, Canada. (b) A southern boreal forest dominated by southern beeches (*Nothofagus*) along the banks of the Arrayanes River in the Andes Mountains of Patagonia, Argentina.

Temperate Deciduous Forest

The temperate deciduous forest biome is found on the eastern sides of North America and Asia and in parts of Western Europe. These are regions where temperatures fluctuate dramatically between summer and winter and ample precipitation falls throughout the year. Deciduous trees, which lose their leaves during the cold winter but produce leaves that photosynthesize rapidly during the warm, moist summers, dominate these environments (Figure 48.14). There are many more tree species in deciduous forests than in boreal forests, and many trees have animal-dispersed pollen and fruits. The forests richest in species are in the southern Appalachian Mountains of the United States and in eastern China and

Japan, areas that were little disturbed by Pleistocene glaciation.

Seasonal cycles of activity are striking in deciduous forests, the most conspicuous being the changes in the leaves themselves. Many birds migrate into this biome in summer when insects are abundant. Understory plants often flower and fruit in early spring before the canopy above them has leafed out. Most trees and shrubs produce their fruits in autumn at the end of the growing season.

Grassland

The grassland biome is found in many climates, but all of them are relatively dry much of the year. In some grasslands, most precipitation falls in winter;

(a) *(b)*

48.14 Temperate Deciduous Forest Biome
(a) The trees in these Rhode Island woods are mostly oaks and maples. This photo catches them in July, with a full complement of leaves. *(b)* The same scene six months later, in early January. The trees have shed their leaves and are in a condition of winter dormancy.

in others, most of it falls in summer. Tropical grasslands are hot all year, but winters are very cold in most temperate grasslands. Grasslands are found in many different climates because grasses are able to survive disturbances. Because they store much of their energy underground, grasses quickly resprout after they are heavily grazed or burned. Trees and shrubs store more of their energy above ground, so they are more easily killed by fires and heavy brows-ing. As we saw in Chapter 47, grasslands typically support large populations of grazing mammals, and fires are common in those dry regions.

Grasslands are structurally simple, but they are rich in species of grasses, sedges, and **forbs**—broad-leaved herbaceous species that live together with grasses and sedges. Many forbs have showy flowers, and grasslands are often riots of color in late summer when forbs are in full bloom (Figure 48.15).

48.15 Temperate Grassland Biome
Late in summer, the broad-leaved plants growing among the tall grasses of the Great Plains of Kansas begin to flower, creating patches of gold and other colors among the green grass.

48.16 Cold Desert Biome
This cold desert in California is on the eastern slopes of the Sierra Nevada. It is dominated by a few species of low-growing shrubs and large bunchgrasses with small grasses and forbs growing among them.

Cold Desert

You will find the cold desert biome in dry regions at middle to high latitudes, especially in the interiors of large continents. Many cold deserts are in the rain shadows of mountain ranges, where seasonal changes in temperature are great, and where most of the little rain that falls does so in winter. Cold deserts are dominated by a few species of low-growing shrubs (Figure 48.16). The surface soil layers are recharged with moisture in the winter, and plant growth is concentrated in the spring. By early summer these deserts are often rather barren; so little rain falls that plants cannot photosynthesize much during the hot summers. Cold deserts are relatively poor in species in most taxonomic groups, but be-

cause the plants produce many seeds, seed-eating birds, ants, and rodents are common.

Hot Desert

The hot desert biome is found in a belt centered on 30° north and 30° south latitude, where air from aloft descends, warms, and picks up moisture (see Figure 47.3). Hot deserts receive most of their rainfall in summer, when the intertropical convergence zone moves poleward, but they also receive winter rains from storms that form over middle-latitude oceans. The driest regions are where neither summer nor winter rains penetrate regularly, such as the center of Australia and the middle of the Sahara Desert. Hot deserts have a richer and structurally more diverse

48.17 Hot Desert Biome
In this New Mexican desert large organpipe cacti share a springtime scene with the ephemeral blooming of several species of shrubs and a few herbs. Note the structural richness of this desert.

48.18 Mediterranean Biome
Scrubby evergreen *Eucalyptus* trees several meters high dominate this winter wet–summer dry vegetation in the Stirling Range, Western Australia. Many species of shrubs and herbs—photosynthetically active primarily in winter and spring—grow among the trees.

vegetation than cold deserts, except in the driest areas. Succulent plants that store large quantities of water in their expandable stems and photosynthesize primarily with their stems rather than with leaves are conspicuous in many hot deserts (Figure 48.17). Annual plants are abundant during rainy periods. Animal pollination and animal dispersal of fruits are common, and great quantities of seeds are produced. Population densities of rodents and ants are often remarkably high, and lizards and snakes are typically very common.

Mediterranean Biome

The Mediterranean biome is found on the west sides of continents at moderate latitudes where winters are cool and wet and summers are hot and dry. Such climates are found in California, central Chile, southwestern Africa, and southwestern Australia as well as in the Mediterranean region. The Mediterranean biome is dominated by low-growing shrubs with tough evergreen leaves (Figure 48.18). The shrubs carry out most of their growth and photosynthesis in early spring, which is when insects are active and birds breed.

Many shrubs of northern hemisphere Mediterranean vegetation produce bird-dispersed fruits that ripen in late fall or winter, when large numbers of migrant birds arrive from the north. One such fruit—the olive—has played an important role in human history, providing a rich food source for people at a period of otherwise low food availability. Annual plants grow abundantly in Mediterranean climates, producing seeds that store well during the hot, dry summers. Therefore, this biome also supports large populations of small rodents, most of which store seeds in underground burrows.

Thorn Forest and Tropical Savanna

The remaining biomes are found at low latitudes where seasonal temperature fluctuations are small and annual cycles are dominated by wet and dry seasons. In general, the length of time that a region is close to the intertropical convergence zone, and hence receives rainfall, increases toward the equator.

The thorn forest biome is found where no rain may fall for 8 or 9 months of the year. Typically, thorn forests are found on the equatorial sides of hot deserts, and they contain many plants similar to those found in the deserts (Figure 48.19a). Animals are especially active during the short but intense summer rainy season.

Dry, tropical regions of Africa, South America, and Australia have extensive areas of savanna biome. Savannas are dominated by grasses and grasslike plants with scattered trees. The savannas of Africa support large populations of mammals that change the vegetation by their grazing and browsing (Figure 48.19b). Many of the herbaceous plants of African savannas are well adapted to heavy grazing pressure and persist only because of it. Many of these areas revert to dense thorn forest if they are protected from fire or grazing.

Tropical Deciduous Forest

Where the rainy season lengthens toward the equator, thorn forest is replaced by the tropical deciduous forest biome. Tropical deciduous forests are taller, have fewer succulent plants, and are much richer in species than are thorn forests. Most of the trees, except those growing along rivers, lose their leaves during the dry season, but many of them flower then (Figure 48.20). Most trees are pollinated by animals

48.19   Thorn Forest and Savanna Biomes
(a) As this photograph from Sonora, Mexico shows, thorn forests are dominated by low-stature trees, many of which are leafless during the long dry season. (b) African savannas still support large herds of grazing and browsing mammals, represented here by impalas.

(a)

(b)

48.20   Tropical Deciduous Forest Biome
Only a few trees, growing in the better-watered sites, still have leaves during the late dry season (March) in this Costa Rican forest. As is typical for this time of year in northwestern Costa Rica, no significant rain had fallen since December.

48.21 Tropical Evergreen Forest Biome
An example of the most productive biome on Earth is seen from a low-flying plane in Mt. Spec National Park, Australia. You can tell that this forest is rich in tree species by looking at the great variety of canopy shapes.

and many have animal-dispersed fruits, but wind-dispersed fruits are also common. The community is very rich in species of both plants and animals. The long dry season is very hot and often quite windy.

Tropical deciduous forest soils are less leached of nutrients than soils in wetter areas. Thus they are some of the best tropical soils for agriculture. As a result, most tropical deciduous forests have been cleared by humans for cattle raising and crops. Where dry-season water is available, farmers can irrigate their fields to grow both a dry-season and a rainy-season crop.

Tropical Evergreen Forest

The tropical evergreen forest biome is found where total rainfall exceeds 250 centimeters of rain annually and the dry season lasts no more than a few months. This is the richest of all biomes in species of both plants and animals, and it has the highest overall energy flow of any ecological community (Figure 48.21). As we saw in Chapter 47, however, this productivity is based on very tight cycling of mineral nutrients, most of which are tied up in vegetation. The soils, except where they are very young, are deeply weathered, and usually cannot support agriculture without massive applications of fertilizers.

Tropical wet forests may have up to 300 species of trees per square kilometer. Most of these species are rare, and nearly all of them rely on animals to transport their pollen and disperse their fruits. Food webs are extremely complex. Most species of invertebrates living in these forests have not yet been named or described by scientists. Human activity currently is destroying this biome at a very high rate.

Tropical Montane Forest

Temperatures steadily drop with altitude on the slopes of tropical mountains. Trees in the tropical montane forest biome are shorter than lowland tropical trees. Leaf sizes are also smaller, and epiphytes are more luxuriant, especially at those elevations where clouds form regularly, bathing the forest in moisture on most days. Photosynthesis in tropical montane forests is depressed by low temperatures and because the leaves are wet most of the time.

AQUATIC ECOSYSTEMS

In aquatic ecosystems, the communities are built upon the same predator–prey, competitive, and mutualistic interactions among their component species as are terrestrial communities, but the differences between water and air have caused different outcomes. First, water, compared with air, is a very dense medium, and it provides much more support for organisms living in it. At the same time, moving water creates forces much greater than those of moving air. Submerged plants depend on water to support them and are flexible so that they move with water currents rather than resisting them. Second, movement through water is more difficult than movement through air. Third, because sunlight penetrates only short distances through water, communities at moderate depths in lakes and oceans receive little light. Fourth, oxygen is often in short supply in aquatic environments, whereas most terrestrial environments have abundant oxygen.

A number of other features of aquatic communities

are related to these basic properties of water. Except in shallow shoreline waters, most aquatic photosynthesizers are very small, and all their tissues are eaten readily by animals. Photosynthesizers, most of which are algae, provide relatively little physical structure to aquatic communities. Water currents carry large amounts of food to organisms on the margins of oceans, leading to communities in which animals compete directly with one another and with algae and plants for space. In terrestrial communities animals rarely become abundant enough to fill up space. In many aquatic communities, animals, rather than plants, are the largest and longest-lived organisms.

Ecosystems in Rivers and Lakes

Green algae and filamentous cyanobacteria conduct most of the photosynthesis in lakes. Vascular plants grow only in shallow waters near shores. Rotifers, cladocera, insects, and fishes live in lakes. Insects in particular occupy a wide variety of ecological niches. There are also freshwater clams, cnidarians, various worms, crustaceans, and sponges, but, except for crustaceans, they do not assume the prominence that their relatives achieve in the sea. Many groups of insects, such as dragonflies, damselflies, mayflies, stone flies, caddis flies, and true flies, have aquatic larvae that become terrestrial adults. Their emergence and metamorphosis link aquatic and terrestrial communities.

Rivers and streams are dynamic bodies of water whose properties change dramatically with seasons as well as between their sources in highlands and their coastal floodplains. They are small and usually clear near their sources and become larger, warmer, and murkier toward their mouths. The large quantities of solids carried in suspension in many rivers reduce the depth to which sunlight penetrates, which limits possibilities for photosynthesis. In addition, because of currents, only plants that are firmly attached to the bottom can retain a fixed position. As a result, plant densities are low and there is little photosynthesis in most streams and rivers. The organisms living in them depend on food that falls into the water from terrestrial communities. There are attached plants in shallow, slow-moving water that support grazing animals, but most stream animals eat food drifting downstream (Figure 48.22).

Marine Ecosystems

Oceans cover about 71 percent of Earth's surface, but most of that area is not very productive (see Chapter 47). Attached multicellular plants are limited to a very thin strip along the edges. Most of the photosynthesis of the oceans is carried out by single-celled organisms, particularly bacteria, diatoms, and dinoflagellates (Chapter 23). These are most abundant in zones of coastal upwelling, where they are maintained in surface waters by upward-moving currents, and where nutrient and light levels are high. The most important grazers of phytoplankton in the sea are crustaceans of species different from those that dominate fresh waters.

Different parts of the oceans are given names for convenient reference (Figure 48.23). At all depths,

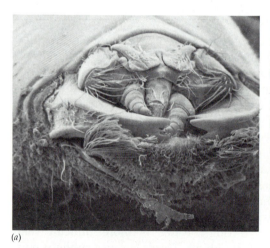

(a)

(b)

48.22 Hunting with a Net in the River
Caddisfly larvae live in tubes they construct and attach to rocks in streams. They capture prey by means of finely-constructed nets that are suspended in the water. (a) This scanning electron micrograph of the head of a caddisfly shows both the mouthparts (above), with which the net and its attached prey are manipulated into the digestive tract, and the net-spinning structures (below). (b) Magnified about 30 times, the net is seen as a fine lattice of silk to which diatoms, green algae, and fine particles of debris adhere.

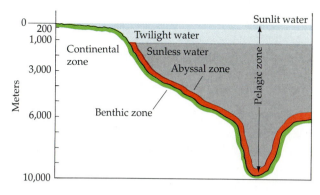

48.23 Zones of the Ocean
Oceanic zones are shown schematically in relation to depth and sunlight penetration. Abyssal and benthic zones coincide below the penetration of light. The vertical scale is much exaggerated.

the bottom of the ocean is called the **benthic zone**, and the open water column, the **pelagic zone**. The benthic zone below the level of sunlight penetration is called the **abyssal zone**. Animals living in the abyssal zone (except for those living around thermal vents) depend, directly or indirectly, on the remains of organisms drifting down from sunlit parts of the ocean. Food is distributed very sparsely in the abyssal zone, and organisms living there have some remarkable adaptations enabling them to devour prey items that are very large compared with themselves (Figure 48.24a).

Pelagic communities are relatively simple, consisting of zooplankton and phytoplankton, supplemented by planktonic larvae of benthic organisms during brief periods of the year. These small organisms are eaten by planktivores ranging in size from small fishes to giant whales. Most pelagic fishes live in large schools and are streamlined for rapid movement through the open water (Figure 48.24b).

Communities of the **littoral zone**—the coastal zone from the uppermost limits of tidal action down to the depth where the water is thoroughly stirred by wave action—are richer in species and more complex in structure than pelagic communities. Communities on and in mud and sand consist mostly of burrowing animals. Some feed by extracting prey from the water column, others construct nets on the mud surface, and still others burrow through the substrate, extracting very small prey or seeking out larger suspension-feeding organisms (Figure 48.25).

(a)

(b)

48.24 Abyssal and Pelagic Fishes
(a) The anglerfish, with its very large mouth and small eyes, is typical of many deep-sea fishes. The bulb at the tip of the projection from its forehead contains luminous bacteria; the light from the bacteria lures prey that are captured by the anglerfish. With its huge mouth, an anglerfish is able to swallow prey as large as itself. (b) In the open ocean there are no hiding places; fishes that live there must be able to swim rapidly to escape predators; the streamlined body shape is an adaptation for swift movement.

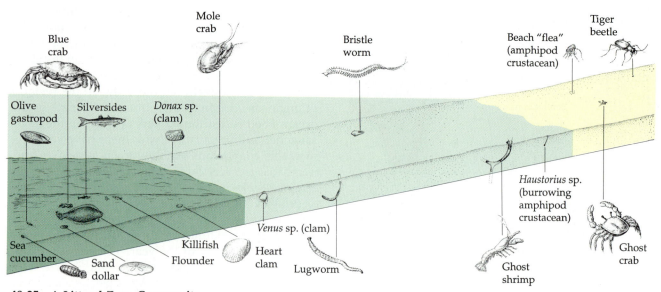

48.25   A Littoral-Zone Community
In this depiction of bottom-dwelling species in an Atlantic beach littoral zone, the yellow area is always above water. The light green area is the intertidal; the darker green region is permanently under water but is subject to turbulence from wave action.

48.26   A Crowded Rocky Intertidal Community
This community on the California coast is dominated by mussels (dark) and barnacles (light), among which clumps of algae are growing. Many of the small acorn barnacles are growing directly on the shells of the mussels. The goose-necked barnacles (exoskeletons with many small plates) are attached to the rocks and are projected above the mussels by their long stalks. Many smaller organisms not visible in this photograph live in the crevices among the mussels.

Communities on rocky substrates in the littoral zone are sites of intense competition for space because waves carry such a plentiful supply of food to the animals living there. Among competitors for space in a single small area may be algae from several different phyla, plus animals from diverse phyla—sponges, cnidarians, mollusks, annelids, moss animals, echinoderms, and chordates (Figure 48.26).

Production by intertidal communities exposed to wave action is extremely high. Intertidal algae on the coast of Washington State produce over 6 kilograms of dry matter per square meter of substrate per year, twice as much as rainforests do. Such high production is possible because waves, in addition to providing planktonic food continuously, also wash away the layer of used water next to the algal fronds, keeping them in constant contact with nutrient-rich water. Also, by stirring the fronds, waves ensure that no frond is either always in sun or always in shade. Intertidal mussel beds are as productive as rainforests, even though mussels are one trophic level above plants.

Muddy intertidal shores in tropical regions are dominated by low-stature forests composed of a number of different tree species collectively called mangroves. Mangroves are among the few trees able to grow in salt water, and, by trapping sediments among their roots and branches, they contribute to the building up of tropical shorelines. Many animal species live permanently among mangroves, and many pelagic species come there to spawn.

The richest and most structurally complex of all marine communities are coral reefs, which fringe coasts of tropical continents and islands. Like the trees of terrestrial communities, corals provide struc-

48.27 Coral Reef Fishes
Yellowtail surgeons and rainbow wrasses surround this coral reef in the Sea of Cortez, Mexico.

tures within which many different ways of life are possible. The richness of coral reef species, which may include over 100 species of corals alone, can only be hinted at even in the best of illustrations. Fishes, in particular, assume a much wider variety of shapes than in the pelagic zone, in part because of the variety of prey on which they feed and in part because refuges from predators are always close at hand (Figure 48.27).

Continental shores are important as breeding grounds for many pelagic species. All pelagic species that lay large eggs move to coastal waters, particularly sheltered bays and estuaries, to breed. The vitality of many populations of marine organisms, particularly fishes and arthropods, depends on the existence of these spawning grounds. Unfortunately, these areas are often excellent ones for commercial and residential developments, and they also receive pollutants from the rivers that flow into them. If estuaries are not managed wisely, many marine species may become extinct.

SUMMARY

Biogeographers analyze and explain the past and present distributions of organisms on Earth. Histor-ical biogeographers study the evolutionary histories of groups of organisms, where they originated, and if and how they spread. Ecological biogeographers study present-day interactions among organisms to determine how these relationships influence where species are found. Many groups of organisms are found on all the continents because they evolved before the continents separated. The continents have been isolated long enough, however, for each one to develop its own distinctive organisms.

The number of species in an island area (oceanic or habitat) is the result of a dynamic balance between extinctions and arrivals. Arrival rates depend upon the number of species in the source area, the distance between the source and the island , and the number of species already present. Patterns of recolonization of islands that have been naturally or artificially deprived of their species shows that equilibria may be established very rapidly.

Because similar climates favor similar species, different regions on Earth with similar climates have similar ecological communities. The major biome types are identified by their characteristic vegetation. A particular biome is found where the climate is suitable for that type of vegetation, but the species of organisms in a particular biome differ on different continents.

SELF-QUIZ

1. Biogeography as a science began when:
   a. eighteenth-century travelers first noted intercontinental differences in distributions of organisms.
   b. Europeans went to the the Middle East during the Crusades.
   c. cladistic methods were developed.
   d. the fact of continental drift was accepted.
   e. Charles Darwin proposed the theory of natural selection.

2. Historical and ecological biogeography differ in that:
   a. only historical biogeography is concerned with history.
   b. historical biogeography is concerned with longer time periods and larger space scales.
   c. both are concerned with the same time scales but historical biogeography deals with larger space scales.
   d. both are concerned with the same space scales but historical biogeography deals with longer time scales.
   e. historical biogeography is not concerned with the current distributions of organisms.

3. Marine biogeographic regions are less distinct than terrestrial ones because:
   a. the ocean biota is more poorly known than the terrestrial one.
   b. there are currently few barriers to dispersal of marine organisms.
   c. most marine families and higher taxa evolved before the oceans were separated by continental drift.

   d. we know less about distributions of marine organisms.
   e. oceanic circulation is faster than atmospheric circulation.

4. A parsimonious interpretation of a distribution pattern is one that:
   a. requires the smallest number of vicariant events.
   b. requires the smallest number of dispersal events.
   c. requires the smallest total number of vicariant plus dispersal events.
   d. accords with the cladogram of a group.
   e. accounts for centers of endemism.

5. The only major biogeographic region that today is isolated by water from other regions is:
   a. Greenland.
   b. Africa.
   c. South America.
   d. Australia.
   e. None of the major regions is isolated by water today.

6. Equilibrium species richness is reached in the MacArthur-Wilson model when:
   a. immigration rates of new species and extinction rates of species are equal.
   b. immigration rates of all species and extinction rates of species are equal.
   c. the rate of vicariant events equals the rate of dispersal events.
   d. the rate of island formation equals the rate of island loss.
   e. there is no equilibrium number of species in that model.

7. Mediterranean vegetation is dominated by:
   a. deciduous trees.
   b. evergreen trees.
   c. deciduous shrubs.
   d. evergreen shrubs.
   e. grasses.

8. Which of the following is *not* true of tropical evergreen forests?
   a. They have large numbers of species of trees.
   b. Most plant species are animal pollinated.
   c. Most plant species have animal-dispersed fruits.
   d. Biological energy flow is very high.
   e. Productivity depends on a rich supply of soil nutrients.

9. Which of the following is *not* true of river ecosystems?
   a. Only plants that are attached to the bottom can remain in a single location.
   b. Sunlight does not penetrate very far into river water.
   c. There is little photosynthesis and animals depend on imported food.
   d. Animal communities are dominated by mollusks and cnidarians.
   e. Stream properties change dramatically between the headwaters and lowlands.

10. At all depths, the bottom of the ocean is known as the:
    a. benthic zone.
    b. abyssal zone.
    c. pelagic zone.
    d. interoceanic convergence zone.
    e. subtidal zone.

FOR STUDY

1. Horses evolved in North America but subsequently became extinct there. They survived to modern times only in Africa and Asia. In the absence of a fossil record we would probably infer an Old World origin for horses. Today, by far the greatest richness of species of fruit flies (*Drosophila*) is in the Hawaiian Islands. Would you conclude that the genus *Drosophila* originally evolved in Hawaii and spread to other regions? Under what circumstances do you think it is safe to conclude that a group of organisms evolved close to where the greatest number of species live today?

2. For nearly every ecological community, the number of species present is much fewer than the number potentially available to colonize it. Is this evidence for species equilibrium? Is there anything that you consider stronger evidence for species equilibrium?

3. A well-known legend states that Saint Patrick drove the snakes out of Ireland. Give some alternative explanations, based on sound biogeographic principles, for the absence of indigenous snakes in that country.

4. What are some significant present-day human concerns whose solutions involve biogeographic considerations? What kinds of biogeographic knowledge are most important for each one?

5. Most of the world's flightless birds are either nocturnal and secretive (e.g., the kagu of New Caledonia) or large, swift, and well-armed (e.g., the ostrich of Africa). The exceptions are found primarily on islands, and many of these island species have become extinct with the arrival of humans and their domestic animals. What special biogeographic conditions on islands might permit the survival of flightless birds? Why has human colonization so often resulted in the extinction of such birds? The power of flight has been lost secondarily in representatives of many groups of birds; what are some possible evolutionary advantages of flightlessness that might offset its obvious disadvantages?

READINGS

Brown, J. H. and A. C. Gibson. 1983. *Biogeography*. Mosby, St. Louis. A comprehensive, up-to-date treatment of both ecological and historical biogeography.

Humphries, C. J., and L. R. Parenti. 1986. *Cladistic Biogeography*. Clarendon Press, Oxford. A concise treatment of the ways in which cladistic methods are used to determine the causes of current distributions of organisms.

MacArthur, R. H. 1972. *Geographical Ecology: Patterns in the Distribution of Species*. Harper and Row, New York. The best introduction to quantitative theories of ecological biogeography.

MacArthur, R. H., and E. O. Wilson. 1967. *The Theory of Island Biogeography*. Princeton University Press, Princeton, NJ. A classic book that launched modern investigations of ecological biogeography.

Myers, A. A. and P. S. Giller. 1988. *Analytical Biogeography: An Integrated Approach to the Study of Animal and Plant Distributions*. Chapman & Hall, London. Contains chapters by different authors on many aspects of ecological and historical biogeography, including discussions of modern cladistic methods and their significance.

49

Conservation Biology

PREVIEW: The impending extinction of a large fraction of the world's biota during the coming century has spawned a new discipline—conservation biology—that focuses on those species, ecological communities, and ecosystems being negatively affected, directly or indirectly, by human activities. Conservation biology provides the biological concepts and tools for preserving biological diversity and ecosystem functioning. The extinction of species is driven by climate modification, over-exploitation, introduced species, and habitat destruction. Conservation biology research is concentrated on those regions with high species richness and endemism, and on threatened species and communities. The preservation of biodiversity requires plans for conserving threatened individual species, development and management of large reserves, restoration of degraded habitats, and fundamental changes in human attitudes towards other species.

This chapter deals with patterns of species richness and endemism, the biology of rarity, climate modification, habitat fragmentation, and restoration and landscape ecology.

It is both remarkable and sobering that the future of so many species depends on the activities of just one species: *Homo sapiens.* Other organisms alter environments for their associates, but no other species dominates the rest of the living world as humans do. Despite our dominance, we depend on other species in many ways for our quality of life. We depend on other organisms for essential ecosystem services, such as the purification of the air we breathe and the water we drink, the maintenance of the soil that grows our food, and the safe disposal of much of our waste. Over half the medical prescriptions written in the United States contain some natural plant or animal product, even though the search for and exploitation of such products from the living world has barely begun (Figure 49.1). As fossil fuels—from which we manufacture most artificial substitutes for natural products—become more expensive, natural plant and animal materials may increase in importance. We also derive enormous aesthetic pleasure from interacting with other organisms. Most people would see a world with far fewer species as a less desirable one in which to live. Finally, many people feel it is morally objectionable for humans to "play God" by causing the extinction of other species.

Despite the value of other organisms to the quality of human life, our activities are directly and indirectly causing the extinctions of many species and the loss of many ecosystems. We do not know how many extinctions will happen during the next 50 years. Some scientists believe that more than 25 percent of the world's biota may be extirpated within the next century if corrective measures are not taken. The number of extinctions will depend both on what we do and on unexpected events. To develop sound strategies for preserving both ecosystem and species diversity, we need to understand current trends and the causes of current extinctions.

Conservation biology is a relatively new discipline that has developed in response to the biodiversity crisis. It studies the causes of species richness and the means by which genes, species, and communities can be preserved. The science of conservation biology draws heavily on concepts and knowledge from population genetics, evolution, ecology, and wildlife management, and the needs of conservation are stimulating new research in those fields in turn. As a result, the basic sciences of ecology and evolutionary biology are being enriched while they are also helping to solve important conservation problems. Concern about the overall implications of the biodiversity crisis and associated global climate changes is resulting in a better understanding of the services Nature provides and a critical reassessment of their true economic value to humans.

CAUSES OF EXTINCTIONS

Human activities have caused extinctions of species for thousands of years, but today we have more powerful tools than our ancestors had for decimating

1126

49.1 A Source of Life-Saving Medicine
An important drug for combating leukemia was derived from an obscure plant, the Madagascar rosy periwinkle. Many more medicinal products from wild species remain to be discovered.

species, and there are billions more of us to use them. The major causes of extinctions are overexploitation, habitat destruction, and the introduction of predators and diseases. In the future, climate modification may be added to this list.

Overexploitation

Until recently, humans caused extinctions primarily by overhunting. Some of the most striking losses of species were on islands, where animals had evolved in relatively predator-free conditions, and thus had few defenses against human hunters or any other predators. For example, when Polynesian people settled in Hawaii, they quickly exterminated, probably by overhunting, at least 39 species of endemic land birds, including 7 species of geese, 2 species of flightless ibises, a sea eagle, a small hawk, 7 flightless rails, 3 species of owls, 2 large crows, a honeyeater, and at least 15 finches (Figure 49.2). When the Maori colonized New Zealand, they exterminated a number of species of large flightless birds, including the moas, which were larger than ostriches. Birds were more severely affected on islands than other vertebrates were, both because birds could reach oceanic islands and colonize them more readily, and because they repeatedly evolved flightless forms on predator-free islands.

On the continents, large mammals have suffered most from human activities. When humans arrived in North America over the Bering Land Bridge, about

20,000 years ago, they encountered a rich fauna of large mammals. The great majority of those species were exterminated within a few thousand years. A similar extermination of large animals followed the human colonization of Australia. About 30,000 years ago Australia had 15 genera of marsupials larger than 50 kilograms, a genus of gigantic lizards, and a genus of heavy flightless birds. All the species in 13 of those 15 genera had become extinct by 18,000 years ago.

Some North American species did not survive the European immigration of the last century. The passenger pigeon, the most numerous bird in North America in the early 1800s, became extinct by 1914, largely due to overhunting. Russian whalers exterminated the unusual Steller's sea cow of the North Pacific late in the last century. The American bison was on the brink of extinction at the turn of the

49.2 Many Hawaiian Birds Are Now Extinct
The goose, flightless ibis, and flightless rail pictured here were among the many species of Hawaiian birds quickly exterminated by the Polynesians when they colonized the islands.

century, and might well be extinct today if its hunting had not been outlawed. Loss of species through over-hunting continues today. Elephants and rhinoceroses are threatened in Africa because poachers kill them for their valuable tusks and horns. At present rates of killing, these two species will disappear from most of their already reduced range in 20 years.

Habitat Destruction

Today's huge human population is fed, clothed, and housed by agricultural and forestry industries that convert natural ecological communities containing many species into highly modified communities dominated by one or a few species of plants. Within these communities, humans discourage the presence of other species by applying biocides that kill competing plants, bacteria, fungi, nematodes, insects and other arthropods, and vertebrates. Agricultural ecosystems have reduced species richness for thousands of years, but traditional agroecosystems cultivate many more economically valuable species together than modern, high-energy input systems do, and the traditional systems support many other species incidentally (Figure 49.3). When ecosystems, such as agricultural lands and plantation forests, are managed so as to divert most of their primary production to certain species intended for human use, we say that their production is **coopted**. Agriculture and forestry today are so extensive that more than 30 percent of all terrestrial production is coopted for human use (Table 49.1), and the percentage is rising. Thus, all other species have only two-thirds of the total global

HABITAT CATEGORY	NET PRIMARY PRODUCTION COOPTED[a]
Cultivated land	15.0
Grazing land	11.6
Forest land	13.6
Human-occupied areas	0.4
Total	40.6
Total net primary production	132.1
Percent coopted	30.7

TABLE 49.1.
Cooption of Net Primary Production by Human Manipulations of Ecosystems

[a] Values are in petagrams (one petagram = 10^{15} grams).

terrestrial production available for their use, and the fraction is steadily decreasing.

Because of increasing habitat modification and the cooption of biological production, habitat loss is certain to be the most important cause of species extinctions during the next century. The habitats required by some species are being completely destroyed. Other habitats, particularly old-growth forests, natural grasslands, and estuaries, are being reduced to small, widely separated patches that may be thought of as habitat islands.

Introduced Pests, Predators, and Competitors

Deliberately or accidentally, people move many species of organisms from one continent to another. Pheasants and partridges were introduced into North America for hunting. European settlers brought their crops and domesticated animals with them to Australia. Other species, such as rabbits and foxes, were introduced there for sport. Weed seeds were carried as ballast in sailing ships or as contaminants in sacks of crop seeds. Despite quarantines, disease organisms spread rapidly, carried by infected plants, animals, and people.

As you know, a species that has evolved over time in a community with certain predators and competitors may be vulnerable to a newly introduced predator or competitor. Introduced species have caused the extinctions of thousands of native species worldwide. Nearly half of the small marsupials and rodents of Australia have been extirpated during the last 100 years by a combination of competiton with rabbits for forage and predation by foxes. Black rats carried to remote oceanic islands on ships are especially destructive predators. Native rice rats survive in the Galapagos archipelago only on islands not invaded by black rats. On some Galapagos islands, introduced

49.3 Vegetables and Fruits Grow Together
Beans and papayas dominate the visible part of this tropical garden in Trinidad, but many other species are growing among them. This traditional form of agriculture allows diversity to flourish.

49.4 Foreign Disease Kills Chestnuts
These American chestnut trees (*Castanea dentata*) have sprouted from the stumps of trees that were attacked by blight, but they are already dying. Today no American chestnuts survive long enough to produce seeds.

pigs and rats regularly excavate all the nests of the giant Galapagos tortoises and devour the eggs. Populations of some tortoises are maintained today only by humans who remove eggs and rear the young tortoises in captivity until they are large enough to defend themselves against pigs and rats.

Among the most striking consequences of cross-continental introductions by humans are outbreaks of pests in new environments. Forest trees in eastern North America have been attacked by several European diseases. The chestnut blight, a fungus, virtually eliminated the American chestnut, once a dominant tree in forests of the Appalachian Mountains (Figure 49.4). Some individuals still resprout, but sprouts are soon found by the blight and killed. Nearly all American elms over large areas of the East and Midwest have been killed by Dutch elm disease, caused by the fungus *Ceratocystis ulmi*. The disease is thought to have originated in Asia, and it was first recorded in western Europe about 1920; the first infestation in North America was reported in 1930. Ecologists suspect that intercontinental movement of disease organisms caused extinctions in the past, but evidence of disease outbreaks is not usually preserved in the fossil record.

Climate Modification

No species is known to have been extirpated by the current global warming, but past climatic changes, particularly cooling, resulted in many extinctions of species. Europe has fewer tree species than eastern Asia and North America because the area of Euro-

pean forests was extremely compressed during the height of the last glacial period. Trees could not extend their ranges southward because they were blocked by the Mediterranean Sea. To the east, aridity formed a similar barrier. European forests were much richer in species before the Pleistocene than they are now.

STUDIES OF INDIVIDUAL SPECIES

Studies undertaken by conservation biologists range from those focused on preserving a single species to those with the goal of preserving entire communities and ecosystems. Often single species of special economic, aesthetic, or ecological value serve as surrogates for entire communities. The protection of species that require large areas of suitable habitat, such as elephants and eagles, may ensure the survival of most or all other species living in that ecosystem.

The Probability of Survival

Local populations can be reduced to very small numbers by the loss of habitat, so that remaining patches can accommodate only a few individuals, and by deterioration of habitat quality, so that the remaining habitat supports a lower density of individuals. Several factors increase the probability that a population will become extinct if its numbers are reduced to low levels. Populations with only a small number of individuals are highly susceptible to local deleterious effects such as fires, unusual weather, disease, and predators. Estimating the risks faced by small populations is an important component of preservation analyses. The development of the concept of a **minimum viable population** (MVP), now widely used to estimate a population's risk of extinction, was stimulated by the National Forest Management Act of 1976, which requires the U.S. Forest Service to maintain viable populations of all native vertebrate species in each national forest. A minimum viable population is the estimated density or number of individuals necessary for the species to maintain or increase its numbers in a region. No sharp threshold exists above which populations are viable and below which they are not, and population viability estimates depend on the time period over which survival is measured. Nonetheless, an MVP analysis can estimate a population's risk of extinction over decades and centuries, time frames that are appropriate for management plans.

A **population vulnerability analysis** (PVA) is carried out to estimate how the size of a population influences its risk of going extinct in some time period. A PVA is based on knowledge of the interactions between the genetic variability, morphology, physiology, and behavior of a population, and its environment, both physical and biological. These interactions combine to influence the spatial structure, age structure, size, sex ratio, maximum density, and growth rates of a population. Such analyses must be made specifically for each species and each environment, but a number of general rules and concepts are used in making those estimates.

One component of a PVA is estimation of the extent and significance of **demographic stochasticity**, that is, the amount of variation in birth and death rates. In a small population, extinction is likely when, due to random variation, high death rates coincide with low birth rates. Estimates of the sizes of local populations at high risk of immediate extinction due to demographic stochasticity range from 10 individuals among microorganisms reproducing by fission to about 50 for larger, sexually reproducing animals with lengthy prereproductive periods. Larger populations also may be at high risk because the same environmental conditions that cause low birth rates are likely to cause high death rates.

Another component of a PVA is analysis of **genetic stochasticity**, the amount of heterozygosity and genetic variance. Of major concern is the depression of individual fitness that can result from the increased inbreeding that occurs in small populations. The inferiority of homozygotes is a common cause of inbreeding depression. For example, in the sulfur butterfly, *Colias philodice* (Figure 49.5), individuals heterozygous at the locus coding for the enzyme glucosephosphate isomerase (GPI) can produce energy faster, and thus have greater endurance and are able to fly during a broader array of environmental conditions than are homozygotes. But as the number of individuals in the population decreases, inbreeding increases, and this leads to an ever-increasing percentage of the less fit homozygous butterflies in the population.

49.6   Southern African Cheetahs Are Genetically Homogenous
High juvenile mortality among captive cheetahs sparked studies that revealed the lack of within-population genetic variability. This mother and her offspring not only look alike, they are almost identical genetically.

The threat of inbreeding depression may influence management strategies. Cheetahs have unusually high juvenile mortality in captivity and are extremely sensitive to diseases. These statistics prompted a genetic survey of over 200 structural loci among cheetahs of southern Africa which revealed almost no genetic variation among them (Figure 49.6). In fact, all the cheetahs in the southern African population are so similar that, unlike most vertebrates, they can accept skin grafts from one another. This indicates that there may be no genetic variation at the major histocompatibility complex (MHC), a normally highly polymorphic genetic region associated with disease resistance as well as with graft rejection (see Chapter 16). For these reasons, conservation biologists are especially concerned about the vulnerability of wild cheetah populations and are planning to crossbreed cheetahs from east Africa with southern African cheetahs to increase the genetic heterozygosity of the southern populations.

Another genetic difficulty faced by small populations is that the action of natural selection is weak relative to that of genetic drift (see Chapter 18). As a result, small local populations may be less able to evolve in response to environmental changes than larger populations would be. Thus, small populations may be more vulnerable to environmental variation than are larger populations.

49.5   A Precarious Position
Mint nectar nourishes the common sulfur butterly, *Colias philodice*. Decreased populations have led to inbreeding and the loss of valuable heterozygosity among these butterflies.

(a)

(b)

49.7 Both Glades and Lizards Are Endangered
(a) Open glades exist as small patches on south-facing slopes in the Ozark Mountains. (b) Collared lizards, which in the Ozarks live only in these open glades, are genetically different from those elsewhere in the range of the species.

Knowledge of population genetics is also important in planning introductions of individuals from captivity or from other geographical regions to reestablish populations in areas from which they have been extirpated. For example, when ibexes (*Capra ibex ibex*) were extirpated in Czechoslovakia by overhunting, the species was successfully reestablished using ibexes from nearby Austria. Some years later, however, ibexes from Turkey (*C. ibex aegagrus*) and from the Sinai (*C. ibex nubiana*) were also introduced to the herd. Animals from all these populations are interfertile, but hybrid females came into rut in early fall instead of winter and gave birth to kids in midwinter when they could not survive. The reproductive failures were so severe that the entire population died out.

An attempt to avoid this type of genetically caused disaster is being carried out in the Ozark Mountains of southern Missouri. Prairie plant communities from the southwestern United States became established in these mountains about 8,000 years ago during an unusually hot, dry period. When the climate became cooler and moister again about 4,000 years ago, oak forests developed in the Ozarks. The prairie vegetation survived only in isolated, fire-prone, open glades on south-facing slopes with shallow soils. Animals such as the collared lizard (Figure 49.7) that depend on these plant communities were reduced to local, isolated populations, and their habitat has been further reduced by agriculture and fire prevention. The Missouri Conservation Commission has instituted a program of glade recovery that includes cutting trees, regular burning, and reintroducing locally extinct animals.

Collared lizards had been extirpated from all the glades in the area where the reintroduction experiments are taking place, so individuals had to be imported from other places. Collared lizards are abundant in the southwestern United States, but the Ozark populations, which have been isolated for about 2,000 lizard generations, are genetically distinct from those in the Southwest. Therefore, a decision was made to use lizards from other Missouri glades. Genetic analyses revealed that the lizards from a single glade are all genetically identical, the result of 2,000 generations of genetic drift in populations whose sizes are nearly always below 50 individuals. Founding lizards could have been taken from a single glade, but a minimum of 10 mature lizards was considered necessary to achieve a high probability of success, and donor populations would have been threatened by removing such a large fraction of their members at one time. Therefore, lizards from at least five different glades were released together. Donor populations were selected to have distinct, maternally inherited mitochondrial DNA markers so that, by sampling offspring over the years, investigators will be able to measure the reproductive success of the released lizards and their descendants.

Captive Propagation

Species being threatened by overexploitation, loss of habitat, or environmental degradation through pollution can be preserved in captivity while the external threats to their existence are reduced or removed. Success at this venture requires research on nutrition and the preparation of suitable diets, on the use of

49.8 Adjusting to a New Home
These two young peregrine falcons are resting on their release site in downtown Denver, Colorado. They will soon be released to forage on their own.

vaccinations and antibiotics, and on the control and enhancement of reproduction by both behavioral and technical means (artificial insemination, embryo transfers). Such research is necessary because information about these topics existed previously for only a few domesticated animals, such as cattle, sheep, goats, pigs, and chickens. Results from these species are usually not transferable to other species without substantial modifications.

Captive propagation is only a temporary measure that helps buy time. There is simply not enough space in existing zoos and botanical gardens to maintain adequate populations of more than a small fraction of rare and endangered species. In addition, a species maintained in captivity can no longer evolve together with the other species in its ecological community. Nonetheless, captive propagation plays an important role in maintaining species during critical periods and in providing a source of individuals to be reintroduced into the wild. Captive propagation projects in zoos also have been very influential in raising public awareness of the biodiversity crisis.

A successful example of the use of captive propagation is the reintroduction of the peregrine falcon (Figure 49.8) to parts of its range from which it had been extirpated, primarily due the widespread use of organochlorine pesticides, such as DDT and dieldrin. These pesticides break down very slowly, so they gradually accumulate in animals that are at the top of the food chain like the falcon. High concentrations of the pesticides interfered with the deposition of calcium in eggshells, causing the falcons' eggs to break easily and resulting in a high rate of reproductive failure. Much of the habitat from which peregrines had been eliminated again became suitable for these birds when some nations greatly restricted the use of organochloride pesticides. Where sufficient individuals survived, peregrine populations recovered unassisted. In Great Britain, for example, the population had declined, but about 350 pairs still survived in 1963. Today, there are over 1,000 pairs in

Britain, probably more than at any other time in this century. In areas such as eastern United States where the species was extirpated, breeding populations have been restored by releasing captive-reared birds.

In 1942, about 350 pairs of peregrines bred in the United States east of the Mississippi River. This breeding population entirely disappeared by 1960, and no peregrines are known to have reproduced in this region during the next 20 years. Captive breeding of peregrines began at Cornell University in 1970, and by 1974 there were enough young being raised to begin experimental releases. By the end of 1986 more than 850 captive-reared peregrines had been released in 13 eastern states, with spectacular success (Figure 49.9). This success depended on the development of suitable methods for rearing chicks in captivity and for maintaining young birds at release sites long enough for them to become attached to those sites and remain there to breed when they matured.

Captive propagation efforts are expensive. The Peregrine Fund at Cornell has spent nearly three million dollars, and the expenses of other cooperating agencies add at least another half million to the total. However, these amounts are small compared with the prices of other forms of human activities. It is estimated that the work needed to restore all of the world's threatened birds of prey could be accomplished with five million dollars per year, the approximate cost of one armored tank.

BIOLOGY OF RARE SPECIES

Much attention is paid to rare and endangered species because they are likely to become extinct unless protective action is taken. A species may be considered to be rare if it has a small geographic range or

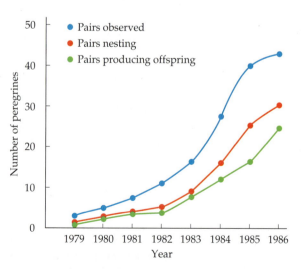

49.9 Peregrine Populations Have Been Reestablished
Throughout eastern North America, many pairs now attempt to reproduce, most of them successfully.

if it never achieves high population densities. There is nothing mysterious about rarity. A new species that forms as a result of the geographical isolation of a small population (Chapter 19) is rare at the beginning of its existence. Species have low population densities if their preferred foods are rare or only periodically available. Predators at the top of the food chain are usually rare because they require large foraging areas.

Interestingly, rarity is an inevitable consequence of high species richness. A typical tropical forest contains several hundred species of trees, whereas a temperate one may have only a few dozen species. However, the number of individual trees per hectare in tropical forests is about the same as that in temperate forests. Therefore, the same number of individuals is divided among a much larger number of species, many of which must be rare. Similar arguments apply for most taxa of organisms. Thus, the regions of the world with the greatest species richness inevitably have the largest number of rare species. For the same reason, species-rich genera tend to have more rare species than do species-poor genera. The 113 threatened Australian species of *Acacia*, the 11 threatened species of *Eucalyptus*, and the 57 threatened species of *Grevillea*, all speciose genera, make up more than twice the proportion of species in these genera than the proportional representation of these genera in the Australian flora.

In addition to having fewer individuals, rare species may be qualitatively different from their more common relatives. A plant with a restricted range and low population densities interacts with other plants and animals differently than a more common plant does (Table 49.2). Because most of a rare plant's neighbors belong to other species, its success in interspecific competition affects its survival more than its success in intraspecific competition does. Herbivores and pollinators are less likely to evolve specialized relationships with a rare plant than with a common one. Therefore, rare species are more likely to defend themselves against generalized herbivores

49.10 Orchid Bee Carries a Pollinium
An Ecuadorian orchid has "packaged" its pollen into a neat little sac—the yellow structure—that an orchid bee (*Eulaema* sp.) now carries on its back.

and to have adaptations for placing pollen precisely on the bodies of generalized pollinators. Extreme adaptations for precision of pollen placement are found in orchids, where the unusual flower shapes restrict pollinator access to a single route. The pollen of orchids is packaged in compact sacs (pollinia) that are placed precisely on the bodies of the pollinators so that the grains are less likely to brush off when the pollinators visit flowers of other species (Figure 49.10). It is not unusual for individual orchid bees, the major pollinators of many tropical orchids, to carry pollinia of as many as half a dozen species of orchids.

Differences between rare and common species have important implications for conservation practices. Species that have been rare for a long time may evolve traits that enable them to reproduce even though they are widely separated from conspecific individuals. However, species that have been common for a long time may be especially vulnerable if their population densities or ranges are suddenly re-

TABLE 49.2.
Probable Results of Natural Selection Acting on Rare and Common Plants

TRAIT	POPULATION CHARACTERISTICS OF PLANT		
	SMALL RANGE, LOCALLY DENSE	LARGE RANGE, LOCALLY RARE	LARGE RANGE, LOCALLY COMMON
Competitive abilities	Intraspecific > interspecific	Interspecific > intraspecific	Intraspecific > interspecific
Target of chemical defenses	Generalist herbivores	Generalist herbivores	Specialists and generalists
Flower longevity	Short	Long	Short
Reward to pollinators per flower	Moderate	Large or none	Moderate
Number and placement of stamens	Many, scattered	Few, precisely placed	Number and location variable

duced. Therefore, newly rare species may require special management if they are to persist at densities much lower than those under which they evolved.

CONSERVATION AND CLIMATE CHANGE

As a result of global warming, average temperatures in North America will probably increase 2–5°C within the next century. Conservation biologists are attempting to predict the effects of this warming trend on North American deciduous forest trees. Each 1° rise translates into a range shift northward of about 150 kilometers—that is, as the climate warms, the average temperature found at a certain location will instead be found 150 kilometers north of that location, and an organism that survives best at that average temperature would have to move that distance. Therefore, trees would have to shift their ranges as much as 500–800 kilometers in a single century. Most deciduous forest trees have long prereproductive periods, and their seeds move only very short distances. The American beech appears to be especially vulnerable. If temperatures were to increase, adult beeches would begin to produce fewer seeds. Within a few decades, seedlings would no longer survive in the forest understory. Adult trees would survive much longer, giving the appearance of a healthy population long after reproduction had already ceased.

The beech, whose seeds are dispersed primarily by jays, advanced at the frontiers of its range only about 20 kilometers per century during past climatic warming. Thus, beeches would have to migrate 40 times faster than they did in the past to keep up with the anticipated rate of climatic change. Therefore, even though areas of suitable climate might remain, beeches probably could not reach them without human assistance (Figure 49.11). We may need to intervene by moving seeds and possibly by assisting seedling establishment if beech forests are going to be maintained.

Forest models project difficulties for the Kirtland's warbler, an endangered species that nests only in young stands of jack pine on sandy soils in Michigan (Figure 49.12). The current population of warblers is less than 1,000 individuals. The stands of jack pines depend on periodic forest fires for their persistence. The cones of jack pines remain closed on the branches until they are heated by a hot fire. They then open and release their seeds, which germinate in the ash on the floor of the burned forest. Kirtland's warblers nest only in jack pine forests that are 8–18 years old. Current management of the warbler population involves controlled burns of forest areas, which ensures the establishment of new stands of the trees. In today's climate, jack pine grows rapidly between fires, but growth rates would decrease if the climate were to warm. Because central Michigan is at

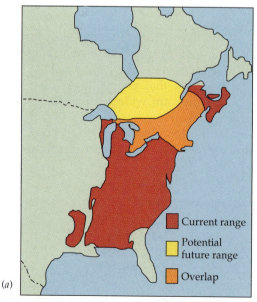

(a)

Current range

Potential future range

Overlap

(b)

49.11 Global Warming Threatens Beech Trees
(a) There is little overlap between the current range of beech and its likely future range if the climate of eastern North America warms according to current predictions. (b) Seedlings and saplings abound in this healthy beech forest; however, reproductive failure is likely to accompany climate warming.

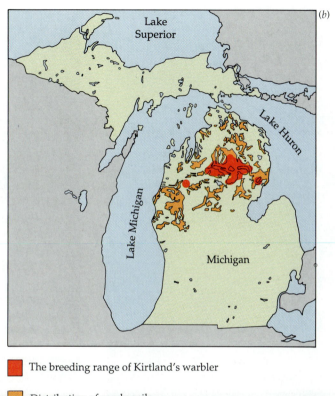

49.12 Kirtland's Warbler Is In Extreme Danger
(a) A male Kirtland's warbler sings in a young jack pine. (b) Although jack pines grow in other areas of the Great Lakes states, the warblers breed only in a few areas in the northern part of lower Michigan, where pine barrens are restricted to very sandy soils and frequent fires maintain conditions suitable for germination of the pines.

🟥 The breeding range of Kirtland's warbler

🟧 Distribution of sandy soils

the southern boundary of the range of jack pine, growth rates are currently depressed during warm summers. During a period of steady climatic warming, jack pine would first be replaced by white pine, red pine, and sugar maple, trees able to tolerate warmer climates. However, if the climate continued to warm, those trees also would decline. Models of forest dynamics, based on physiological responses of tree species of the region, project that by the middle of the next century many areas currently forested by jack pine would become treeless shrublands. Northward movement of the jack pine forests would be difficult because a broad intervening region lacks the coarse, sandy soils that both jack pines and warblers need.

The rates of past global climatic changes were much slower than the rate predicted for the coming century. In addition, because of habitat fragmentation, migration routes may be blocked by extensive areas of unsuitable habitats. And, of course, climatic zones would not simply shift northward if the globe warms; instead, new climates would develop and some existing ones would disappear. New climates are certain to develop in tropical lowlands. All climate models predict that warming will be less in tropical regions than at high latitudes, but even a warming of 2°C would result in lowland climates hotter than those found anywhere in the humid tropics today.

COMMUNITY-LEVEL CONSERVATION

Because many species can survive only in the ecological communities in which they evolved, conservation biologists are as much concerned about preserving complete ecological communities as they are about maintaining individual species. Although the types of research conducted for these two purposes overlap, there are distinct themes in community-level conservation research.

Endemism

Tropical ecosystems are generally richer in species than ecosystems at higher latitudes are (Chapter 48). Therefore, loss of tropical habitats threatens more species than loss of comparable areas of temperate habitats. The number of species that become extinct as a result of habitat destruction also depends on how many local species are **endemic**—are found nowhere else. For example, nearly all the mammals and birds of Madagascar are found only on that island. Therefore, if the small fragments of tropical forests remaining on Madagascar are destroyed, the species dependent on them are certain to be extirpated in the wild (Figure 49.13).

Endemism is especially marked on islands, but there are also mainland regions with high degrees of

49.13 Madagascar Abounds with Endemic Species
Among the many species found only on the island of Madagascar are (a) the *Euphorbia* and *Alluaudia* species of the dry forest; (b) an entire primate group, the lemurs, exemplified here by the ring-tailed lemur (*Lemur catta*); and (c) this insectivorous mammal, a streaked tenrec.

endemism. For example, the Rift Valley lakes of Africa harbor over 1,000 fish species, most of which live in only one lake. The Atlantic coastal forests of southeastern Brazil are another center of endemism. Because only about 1 percent of the original extent of those forests remains, many species there have become extinct or are in danger of immediate extinction. Mountainous regions have many endemic species because temperature and rainfall change rapidly with elevation, creating many distinct habitats within a small area.

Centers of endemism are not the same for all groups of organisms. The Cape region at the southern tip of Africa has a flora of 8,500 species, 80 percent of which are endemic, but only 4 of the 187 species of birds found there are endemic (Table 49.3). The reason is that the Cape region, an area of only 90,000 square kilometers, is too small for in situ speciation to occur among birds, but plants readily speciate in areas of that size.

Keystone Species

Keystone species are species that influence the structure and functioning of an ecological community to a much greater extent than would be expected simply from their abundance. The first keystone species to be recognized in nature were predators whose prey would dominate their competitors if the predators were absent. The sea star *Pisaster ochraceous*, discussed in Chapter 46, is one such keystone species. Since then, important keystone relationships have also been detected among species with mutualistic interactions. For example, in Peruvian forests, only a dozen species of figs and palms support an entire community of large frugivorous (fruit-eating) birds and mammals during the period of the year when fruits are least available. Loss of those few tree species would probably eliminate most of the frugivores even if hundreds of other tree species remained. In turn, loss of the frugivores might seriously impair

TABLE 49.3.
African Centers of Endemism[a]

UNIT	AREA (1000 km²)	PLANTS NUMBER OF SPECIES	PLANTS PERCENT ENDEMIC	MAMMALS NUMBER OF SPECIES	MAMMALS PERCENT ENDEMIC	BIRDS NUMBER OF SPECIES	BIRDS PERCENT ENDEMIC
Guinea–Congo	2,815	8,000	**80**	58	**45**	655	36
Zambezian	3,939	8,500	**54**	55	4	650	15
Sudanian	3,565	2,750	33	46	2	319	8
Somali–Masai	1,990	2,500	**50**	50	14	345	32
Cape	90	8,500	**80**	14	0	187	4
Karoo–Namib	692	3,500	**50**	13	0	112	9
Montane	647	3,000	**75**	50	4	220	**65**

[a] Percentages of unusually high endemism are in boldface type.

the dispersal of the seeds of many other species of trees. Thus, a keystone mutualistic relationship is probably maintained by a few tree species that constitute only a small fraction of the 2,000 species of trees in the forest.

Other keystone species include herbivores that alter habitat structure and change vegetation succession (termites and large mammals, such as moose and elephants), species that maintain particular landscape features (beavers), and parasites and pathogenic microorganisms. Because of the importance of keystone species, we need to identify them quickly and take action to preserve them, because their extinction could result in the extinction of many other species in their communities.

Habitat Fragmentation

Human population growth and the cooption of a large fraction of global terrestrial production has resulted in extreme fragmentation of natural habitats. Small habitat patches differ from larger patches in ways that affect the survival of species. Small habitat patches are influenced by **size effects**, that is, they are qualitatively different from larger patches of the same habitat. Small patches cannot support populations of species that require large areas, and they can harbor only small populations of many of the species that can survive there. Also, the fraction of a patch that is influenced by **edge effects**—phenomena that occur where one kind of habitat meets another—rapidly increases as patch size decreases (Figure 49.14). In forest patches, winds are stronger, temperatures are higher, humidities are lower, and light levels are higher close to the edges than they are farther inside the forest. Also, species typical of surrounding habitats can invade the edges of patches to compete with or prey upon the species living there.

Patch size and the size of animal home ranges interact to influence the likelihood that a species can persist in a habitat fragment of a certain size. At one extreme, if the home range of an animal is larger than the size of an available patch of suitable habitat, the animal either disappears from that patch or it greatly expands its home range to include more than one patch. In such a situation, animals must repeatedly cross areas of unsuitable habitat where they may be considered pests or where they may be at great risk of predation or other dangers. For example, the number of black bears killed by automobiles on Florida

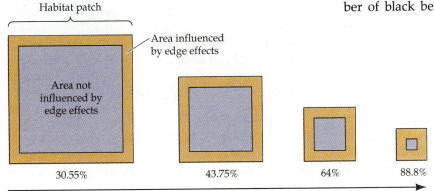

Habitat patch

Area influenced by edge effects

Area not influenced by edge effects

30.55%   43.75%   64%   88.8%

Increasing percentage of patch influenced by edge effects

49.14 Edge Effects
These diagrams show the proportions of square areas influenced by edge effects if such effects penetrate equal distances in patches of all sizes. Under such conditions, edge effects are more important in smaller habitat patches.

(b)

(a)

49.15 Old-Growth Forests and Owls
(a) Old-growth coniferous forests of the Pacific Northwest have trees of all ages and there are many large logs on the forest floor. (b) The northern spotted owl reproduces successfully only in these forests.

highways as the animals search widely for small patches of suitable habitat, is higher than the number of bears born, so the population is declining.

In the northwestern United States over 70 percent of the old-growth coniferous forests of the region have been cut. Most of the stands that remain are at middle and high elevations where conditions are less suitable for lowland species. Because current forestry practices are based on clear-cutting stands every 50–80 years, no second-growth forests become old enough to acquire key characteristics of old growth, such as having trees of all ages (some of these conifers live over 500 years), having many dead trees and snags, and having large logs on the forest floor (Figure 49.15a). Among the species that require old growth to maintain viable populations are salamanders, which live inside rotting logs, the only microhabitat that retains moisture during the dry summers. Another is the spotted owl, a species that hunts for rodents, primarily flying squirrels, that live in mature forests (Figure 49.15b). As the fraction of the area remaining in old growth is reduced, the home ranges of owls must increase; hunting success and,

hence, reproductive success, decreases, and juvenile mortality during dispersal increases. As a result, the proportion of remaining suitable habitat occupied by spotted owls decreases. The species faces a high risk of extinction in most of Washington and Oregon within the next 50 years if forestry practices are not changed.

Information on the distribution and abundances of organisms in a landscape before its habitats become fragmented is usually not available. However, near Manaus, Brazil, a major research project was launched before logging took place (Figure 49.16). Land owners agreed to preserve forest patches of certain sizes and locations and censuses of those

49.16 Species Loss Is Studied in Brazilian Forest Fragments
Research indicated that isolated patches lost species much more quickly than patches that remained connected to the main forest, even if the isolated patches were larger than the connected ones.

patches were conducted before logging began, while the areas were still a part of continuous forest. Within a few years of clearing the surrounding areas, species were already disappearing from the isolated patches. The first species to be eliminated were monkeys with large home ranges, such as the black spider monkey, the tufted capuchin, and the bearded saki. The birds that follow army ant swarms to capture insects flushed by the ants (Figure 49.17) also disappeared quickly from small patches. A particular army ant colony is a useful resource for the birds only when it is actually raiding, about 27 days of the 35-day period between colony moves. Therefore, the birds must have access to a number of army ant colonies to be guaranteed of always having one in the raiding stage. Smaller patches have so few ant colonies that there are periods when none are raiding.

49.17 Extinctions in Patches
The white-plumed antbird, a common species in Brazilian forests, has become extinct in isolated forest plots.

LANDSCAPE MANAGEMENT

An important component of conservation biology is the establishment of parks, sanctuaries, and reserves whose primary function is to maintain species and ecosystems relatively free of human disturbance. The National Park System of the United States plays a vital role, not only in species preservation, but also in providing opportunities for people to enjoy natural environments and develop an appreciation for them. More parks are being added to systems in many countries, but it is unlikely that their size and number will ever be equal to the task of ecosystem and species preservation. There will never be enough parks, and they are, and will be, too small to maintain all species and to permit evolutionary adaptations to continue within their boundaries. Moreover, most parks have traditionally been established around areas of monumental geological features, not areas of high species richness.

The United States model of national parks must be modified before it is exported to most other countries. Parks in the United States were established primarily in areas where Europeans had not yet settled and where the indigenous people had already been exterminated. In most countries, parks must be established in already heavily settled areas. The people living there cannot be evicted, nor is it possible, in most cases, to prevent other hungry people from settling in or hunting in the parks. The high rates of population growth in most tropical countries guarantee that pressures on parks from agricultural settlers will increase rather than decrease.

For these reasons, lands currently exploited for food, medicines, and fiber must also play an important role in management for species preservation. These lands are far more extensive than parks and reserves, and they include climates and ecosystems not represented in the parks. Fortunately, many patterns of economic exploitation of lands are compatible with the preservation of most species of organisms. Only a few species, such as predators on people and domestic animals, or large, destructive herbivores, are incompatible with human uses of landscapes.

Megareserves

A key development in conservation practice is the concept of **megareserves**. A megareserve is a large area of land that includes, at its core, an undisturbed natural area. Surrounding that core are buffer areas in which economic activity that does not destroy the ecosystem is permitted. This may include sustainable harvesting of animal populations and plant products, such as rubber, fruits, nuts, and wood. On the edges of the megareserve is a zone in which more intensive land use, such as agriculture or plantation forestry, is permitted (Figure 49.18). Costa Rica has pioneered the development of megareserves. It is consolidating its parks and reserves into eight megareserves that will, it is hoped, maintain about 80 percent of the country's biodiversity (Figure 49.19). Each megareserve includes natural areas and areas managed for economically valuable products. Some of them remain the homes of indigenous people who will continue to use the environment in their traditional ways.

The largest of the Costa Rican megareserves is La Amistad Biosphere Reserve, a mosaic of over 500,000 hectares that includes three national parks, a large biological reserve, five Indian reservations, and two large forest reserves. Altitudes within the reserve range from 100 to 3,819 meters. It contains the largest

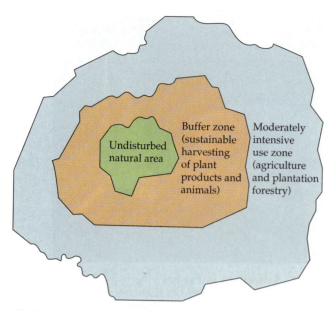

49.18 Design of a Megareserve
This idealized version of a Biosphere Reserve has been achieved on only a small fraction of existing reserves.

tract of highland vegetation in Central America, and it has considerable hydroelectric generating potential. All the native large predators still survive there. However, the reserve faces threats from surrounding agriculturists, overhunting, and logging.

A strategy for the conservation and development of La Amistad has been completed and is being implemented. If managed properly, the reserve can provide drinking water and electricity, forest products, nature tourism, and protection of indigenous cultures, as well as species preservation. In combination, these benefits outweigh what could be gained by logging the steep slopes and converting them to low-productivity agricultural systems.

An unusual reserve, created with the help of local farmers and landowners, is the Community Baboon Sanctuary in Belize. The original purpose of the reserve was the protection of the black howler monkey, known locally as "baboon," a species with a restricted range in Mexico, Belize, and Guatemala (Figure 49.20). The sanctuary was planned in consultation with people living in the area, and private landowners have agreed to use their land in accordance with standards established by the reserve. The formal plan grew out of extensive discussions among local people. Talks were given at local schools, and a booklet, *Baboons of Belize*, was distributed. Publicity for the program locally, nationally, and internationally increased local pride and helped generate tourism, an important goal. The sanctuary, which has been slowly expanding, now occupies 47 square kilometers of land, including seven villages along 32 kilometers of riverine forest and an 0.8-kilometer strip on either side of the river. In the future it will link up with two other sanctuaries to form a megareserve. A reserve

manager has been hired and a system of administration has been established.

Forest Reserves

Forest reserves in which economically valuable products are harvested can combine species preservation and economic development. Such a plan was proposed at the first annual meeting of rubber tappers in Brasilia, Brazil, in 1985. It is being implemented in Acre, the state where the rubber tappers are strongest and best organized. In Acre, the value of wild rubber, Brazil nuts, and several other forest products harvested in 1980 was estimated at over 26 million dollars. The estimated current value of the sustainable harvest of such products is less than the potential value of products of cattle ranching and agriculture, but agriculture is not sustainable in those regions, whereas forests can yield harvested products indefinitely. Whether such forest reserves can survive politically is uncertain. Officially recognized title to the land is difficult to obtain and maintain. Also, the tappers are caught between their debts to middlemen and increasing threats to their livelihood from forest clearing for timber and ranching. The latter activities are currently highly subsidized by the Brazilian government to the benefit of wealthy absentee landlords. Political organization on the part of the tappers and other extractive users, cooperation from the Brazilian federal government, and foreign assistance are vital ingredients for the success of forest reserves. If successful, Amazonian forest reserves may serve as models for other regions. Forest reserves could play a major role in conservation efforts worldwide.

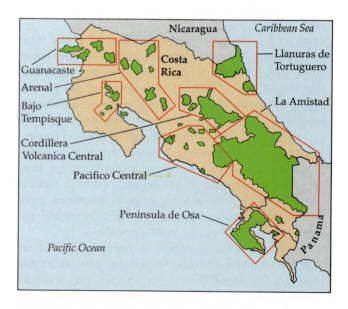

49.19 The Megareserves of Costa Rica
The dark lines enclose regions that are being managed for both biodiversity and economic activities.

49.20 Black Howler Monkey
The rare black howler monkey is now protected in community reserves in Belize, Central America.

Restoration Ecology

Many areas that could be incorporated into megareserves have been highly altered by human activities. Some of these must be restored to their original state if they are to play their intended roles in biodiversity conservation. In addition, many small reserves need to be restored and managed so that species that formerly lived there, and have for one reason or another already disappeared, can return. For these reasons, a subdiscipline of conservation biology, known as **restoration ecology**, is growing rapidly. Research on methods of restoring species and ecosystems is needed because many ecological communities will not recover, or will do so very slowly, if there is no creative intervention in the recovery process.

The world's largest restoration project is under way in Guanacaste National Park in northwestern Costa Rica (Figure 49.21). The goal is to restore a large area of tropical dry forest, the most threatened ecosystem in Central America, from small fragments that remain in an area converted primarily to cattle grazing. One method of restoration would be to exclude fire and domestic livestock from the park and let nature take its course. However, although grass patches of less than 120 hectares would be clothed by woody vegetation within 20 years, large expanses

of pasture would require 50–200 years to become completely forest-covered because of the slow rates of movement of tree seeds into large areas. Reforestation can be speeded up by habitat manipulation, which is what Daniel Janzen, architect of the restoration project, has chosen to do. Restoration requires basic ecological information about the abilities of different plant species to germinate and grow in the degraded pastures that surround most of the forest fragments. Information is also needed on the ability of existing seed dispersers to move seeds from the forest into the abandoned pastures, and on the populations of pastureland species that are serious predators on seeds and seedlings of forest trees.

The single most important threat to Guanacaste National Park during the coming decades is fire. A few fires in the region are due to natural causes, but most fires are started by people. Moreover, the dense stands of introduced grasses, if not grazed, produce highly flammable swards which carry hot fires that penetrate far into surrounding forests. Domestic livestock keep these grasses under control by grazing, and they are also important dispersers of the seeds of some native trees that are good at invading pastures and in whose shade many other species germinate. Therefore, the restoration program will encourage grazing by domestic livestock in the park until such time as woody succession in an area has progressed to the point where grass no longer poses serious competition to the woody species and is no longer sufficiently dense to carry hot fires.

The importance of basic ecological information for successful restoration is also illustrated by the design of a restoration project for abandoned mine tailings in Derbyshire, England (Figure 49.22). Many thousands of hectares of derelict lands exist in Britain, most of them the result of mining activities. Because many of these areas are close to or in scenically attractive areas, there is considerable public pressure to restore attractive, species-rich vegetation on them. Traditionally, abandoned mine tailings were sown with a mixture of seeds of pasture grasses and heavily fertilized. However, after an initial period of good growth, plant growth seriously declined in most such stands on these nutrient-poor soils. Repeated fertilization is expensive and is generally not feasible on those sites. Therefore, research in Derbyshire was directed toward determining the best mixtures of seeds to sow to produce a self-sustaining, species-rich plant community on poor soils. This required studies of the germination requirements of a variety of plant species and studies of competition during the establishment period of young plants. Restoration ecologists based some of their experiments on the theory that herbaceous plant communities on rich soils are species-poor because a few dominant plants outcompete many other species under those excellent growing conditions. Indeed, they discovered that by not fertilizing the lands to be reclaimed and by sow-

49.21 Restoring a Tropical Dry Forest
(a) Most of the tropical dry forests in Central America have been cut, burned, and converted to cattle pasture. (b) Turning the situation to use in Guanacaste National Park, the cattle grazing under the trees will be used to disperse the seeds of trees and to reduce grass cover during the forest restoration process.

(a)

(b)

ing seeds of species adapted for slow growth on nutrient-poor soils, they were able to establish communities whose species richness remained high for long periods of time. These results are being employed in reclamation projects elsewhere in Britain.

ECOSYSTEM SERVICES

Natural ecosystems provide many essential services for humans without which we could not survive or which we would otherwise have to provide for ourselves, at great cost, through artificial means. Ecosystems absorb carbon dioxide and other gases, emit oxygen, cleanse water, regulate stream flows, and provide recreational and aesthetic services. The economic value of ecosystem services is hard to quantify, but some estimates have been made. The cost of building a system that would duplicate the treatment of wastewater and the fish production now provided by a hectare of Louisiana wetland is estimated to be

about $200,000. And this figure does not include the value of the site for sulfate reduction, carbon dioxide fixation, oxygen release, or waterfowl production. The storage and purification of water, binding of soil, and fertilization of woodland provided by one hectare of streamside vegetation in Georgia was calculated to be worth $2,000 per year.

The cost of substituting technologies for lost ecosystem services has been aptly summarized by ecologist F. H. Bormann:

We must find replacements for wood products, build erosion control works, enlarge reservoirs, upgrade air pollution control technology, install flood control works, improve water purification plants, increase air conditioning, and provide new recreational facililties. These substitutes represent an enormous tax burden, a drain on the world's supply of natural resources, and increased stress on the natural system that remains. Clearly, the diminution of solar-powered natural systems and the expansion of fossil-powered human systems are currently locked in a positive feedback cycle. Increased consumption of fossil energy means increased

49.22 Vegetation Can Be Restored on Mine Tailings
(a) The ugly surface of this freshly abandoned mine consists of rocky, nutrient-poor soil. (b) Seven and a half years later, the tailings are covered with species-rich and aesthetically attractive vegetation.

(a)

(b)

stress on natural systems, which in turn means still more consumption of fossil energy to replace lost natural functions if the quality of life is to be maintained.

Some ecosystem services, such as aesthetic benefits, cannot be replaced with technological inventions. One of the largest sources of foreign income in Kenya is nature tourism. Probably the loss of a single species would not reduce the flow of tourists to Kenya, but if elephants, rhinoceroses, lions, leopards, and buffalo were all to disappear, fewer people would be likely to pay the high price of a Kenyan vacation. Populations of these species can be maintained only if large tracts of the ecosystems in which they live are preserved.

The preservation of biological diversity and ecosystem services is one of the greatest challenges facing humankind. Many of the scientific tools needed for the task are already available, but to implement them effectively, important changes in people's attitudes toward other species must take place. If species are valued only because they are economically useful to us, increased losses of species are inevitable because other uses of natural habitats are likely to be seen as more profitable, at least in the short run.

Even though a wetland has great value, some people usually can make enough money in the short run by destroying the wetland to be motived to do so. Only when we value high biological diversity and ecosystem functioning as the heritage of all humankind, a heritage to be passed along to our descendants as rich and full as possible, will the current alarming rate of ecosystem destruction and extirpation of species be reduced.

SUMMARY

Conservation biology is the study of the causes of and the preservation of biological diversity. Its rise in importance as a biological discipline has been stimulated by the high rates of habitat destruction currently under way on Earth and the resulting extinction "crisis." Humans cause extinctions of species by overexploitation, habitat destruction, and introductions of new species.

The preservation of a species involves determining its risk of extinction. Population vulnerability analysis takes into account the demographics and population genetics of a species and the ways it interacts with its environment. This knowledge may help biologists succeed in preserving a species in captivity or in reintroducing it to its former habitat. Rare species are particularly vulnerable to extinction, and require special management.

Global climatic warming is likely to result in climates and communities different from those found today. Rapid climatic shifts may result in extinctions of species unable to move with the changing climate.

Species can best be preserved by preserving the communities and habitats in which they live. Biologists focus on communities with high rates of endemism and on preserving the keystone species on which communities depend. The small patches of habitat left over when habitats are fragmented cannot support species as well as larger patches can. National parks have played an important role in habitat preservation, but larger areas are needed to preserve most species. Megareserves can protect natural habitats and accommodate many human activities as well. Some damaged ecosystems can be restored with human assistance.

Natural ecosystems provide essential services to humans that would be impossible, or extremely costly, to replace. The biodiversity crisis has forced us to reassess the value of species and ecosystems to the quality of human life.

SELF-QUIZ

1. Which of the following is *not* currently a major cause of species extinctions?
   a. Habitat destruction
   b. Climate change
   c. Overexploitation
   d. Introduction of predators
   e. Introduction of diseases

2. When ecosystems are managed to favor strongly those species intended for human use we say that their production is:
   a. modified.
   b. diverted.
   c. coopted.
   d. channeled.
   e. managed.

3. A minimum viable population is:
   a. the estimated number of individuals for the species to maintain genetic diversity.
   b. the estimated number of individuals for the species to persist in all U.S. national forests.
   c. the estimated number of individuals for the species to survive for several decades.
   d. the estimated number of individuals necessary for a species to maintain or increase its numbers in a region.
   e. the minimum density required for individuals to find mating partners.

4. Which of the following is *not* a component of a population vulnerability analysis?
   a. Spatial structure of a population.
   b. Sex ratio within a population.
   c. Amount of variation in birth and death rates.

   d. Amount of heterozygosity and genetic variance.
   e. Captive propagation of individuals.

5. Orchids cross-pollinate successfully even though they usually are very rare because:
   a. their unusual shapes restrict pollinator access to one route.
   b. the pollen of orchids is packaged in sacs called pollinia.
   c. pollen is placed very precisely on the bodies of visitors.
   d. pollinia seldom brush off on heterospecific stigmas.
   e. all of the above.

6. Conservation biologists are concerned about global warming because:
   a. the rate of change in climate is projected to be faster than the rate at which ranges of many species can move.
   b. it is already too hot in the tropics.
   c. climates have been so stable for thousands of years that many species lack abilities to tolerate variable temperatures.
   d. climate change will be especially harmful to rare species.
   e. none of the above.

7. A species that is found only in a particular region is said to be:
   a. an indicator species for that region.
   b. a restricted species.
   c. a vulnerable species.
   d. endemic to that region.
   e. demographically constrained.

8. A keystone species is one that:
   a. preys heavily on a particular species.
   b. is especially vulnerable to extinction.
   c. is restricted to a small geographic area.
   d. experiences considerable demographic stochasticity.
   e. strongly influences the structure and functioning of its ecological community.

9. As a habitat patch gets smaller it:
   a. cannot support populations of species that require large areas.
   b. supports only small populations of many species.
   c. is influenced to an increasing degree by edge effects.
   d. is invaded by species from surrounding habitats.
   e. all of the above.

10. Restoration ecology is an important discipline because:
    a. many areas being incorporated into megareserves have been highly degraded.
    b. many areas being incorporated into megareserves are vulnerable to global climate change.
    c. many species suffer from demographic stochasticity.
    d. many species are genetically depauperate.
    e. fire is a threat to many reserves.

11. Which of the following is *not* an ecosystem service?
    a. Producing carbon dioxide
    b. Flood control
    c. Water purification
    d. Air purification
    e. Preservation of biological diversity

FOR STUDY

1. Most species driven to extinction by people in the past were large vertebrates. Do you expect this pattern to persist into the future? If not, why not?

2. Species endangered as a result of global climatic warming might be preserved if we moved individuals from areas that are becoming unsuitable to those likely to be better for them in the future. What are the major difficulties associated with such interventions? For what types of species would they work well? poorly?

3. Conservation biologists have debated extensively whether it is better to set up many small reserves or a few large ones. What biological processes should be evaluated in making judgments about sizes and locations of reserves? To what extent should we be concerned with preserving the largest number of species rather than those species judged to be of unusual importance?

4. During World War I, French doctors adopted a "triage" system of dealing with wounded soldiers. The wounded were divided into three categories: those almost certain to die no matter what was done to help them, those likely to recover even if not assisted, and those whose probability of survival was greatly increased if they were given medical attention. The limited resources available to the doctors were directed primarily at the third category. What would be the implications of adopting a similar attitude toward species preservation?

5. Utilitarian arguments dominate discussions about the importance of preserving the biological richness of the planet. In your opinion, what role should moral arguments play?

READINGS

Defenders of Wildlife. 1989. *Preserving Communities & Corridors*. Washington, D.C. A short collection of essays exploring the roles of corridors in preserving wildlife and how important conservation legislation can best be implemented.

Gradwohl, J. and R. Greenberg. 1988. *Saving the Tropical Forests*. Island Press, Washington, D.C. A good account of both the causes of tropical forest destruction and of successful projects throughout the world where local communities have averted forest destruction while reaping social and financial benefits.

Janzen, D. H. 1986. *Guanacaste National Park: Tropical Ecological and Cultural Restoration*. Universidad Estatal a Distancia, Costa Rica. Describes the current efforts to restore a large area of tropical dry forest from small fragments and the scientific basis for the management plan.

National Academy of Sciences/National Research Council. 1987. *Ecological Knowledge and Environmental Problem-Solving*. National Academy Press, Washington, D.C. A thorough review of ecological knowledge and how it has been useful in helping to solve environmental problems. Includes descriptions and analyses of 13 successful projects.

Reid, W., J. N. Barnes and J. Blackwelder. 1988. *Bankrolling Successes*. Washington, D.C., Environmental Policy Institute and National Wildlife Federation, Washington, D.C. A careful analysis of successful projects in sustainable development, especially in tropical regions.

Soulé, Michael E. (Editor). 1986. *Conservation Biology: The Science of Scarcity and Diversity*. Sinauer Associates, Sunderland, MA. A thorough review, by several dozen contributors, of the status of knowledge and concepts for in situ and ex situ preservation of species.

Western, D. and M. Pearl (Editors). 1989. *Conservation for the Twenty-first Century*. Oxford University Press, New York. A rich set of essays by conservationists, governmental decision-makers, and wildlife managers that identifies gaps in knowledge and proposes agendas for conservation action worldwide.

Wilson. E. O. (Editor). 1988. *Biodiversity*. National Academy of Sciences Press, Washington, D.C. The published results of the National Forum on Biodiversity held in Washington, D. C. in September, 1986. Contributions from more than 60 leading biologists, economists, agricultural experts, philosophers, agency representatives, and other professionals cover a broad range of topics concerning the preservation of biodiversity.

GLOSSARY

Abdomen (ab' duh mun) [L.: belly] In arthropods, the posterior portion of the body; in mammals, the part of the body containing the intestines and most other internal organs, posterior to the thorax.

Abomasum (ab' oh may' sum) The true stomach of ruminants (animals such as cattle, sheep, and goats).

Abscisic acid (ab sighs' ik) [L. *abscissio*: breaking off] A plant growth substance having growth-inhibiting action. Causes stomata to close.

Abscission (ab sizh' un) [L. *abscissio*: breaking off] The process by which leaves, petals, and fruits separate from a plant.

Absolute temperature scale A temperature scale in which the degree is the same size as in the Celsius (centigrade) scale, and zero is the state of no molecular motion. Absolute zero is –273° on the Celsius scale.

Absorption (1) Of light: complete retention, without reflection or transmission. (2) Of liquids: soaking up (taking in through pores or cracks).

Absorption spectrum A graph of light absorption versus wavelength of light; shows how much light is absorbed at each wavelength.

Abyssal zone (uh biss' ul) [Gr. *abyssos*: bottomless] That portion of the deep ocean where no light penetrates.

Abzyme An immunoglobulin (antibody) with catalytic activity.

Accessory fruit A fruit derived from parts in addition to the ovary and seeds. (Contrast with simple fruit, aggregate fruit, multiple fruit.)

Accessory pigments Pigments that absorb light and transfer energy to chlorophylls for photosynthesis.

Acclimation (a cli ma' shun) A change in an organism which improves its ability to tolerate a changed environmental situation.

Acellular Not composed of cells.

Acetylcholine A neurotransmitter substance that carries information across vertebrate neuromuscular junctions and some other synapses.

Acetyl coenzyme A Compound that reacts with oxaloacetate to produce citrate at the beginning of the citric acid cycle; a key metabolic intermediate in the formation of many compounds.

Acid [L. *acidus*: sharp, sour] A substance that can release a proton. (Contrast with base.)

Acid precipitation Precipitation that has a lower pH than normal as a result of acid-forming precursors introduced into the atmosphere by human activities.

Acidic Having a pH of less than 7.0 (a hydrogen ion concentration greater than 10^{-7} molar).

Acoelomate Lacking a coelom.

Acquired Immune Deficiency Syndrome See AIDS.

Acrosome (a' krow soam) [Gr. *akros*: highest or outermost + *soma*: body] The structure at the forward tip of an animal sperm which is the first to fuse with the egg membrane and enter the egg cell.

ACTH (adrenocorticotrophic hormone) A pituitary hormone that stimulates the adrenal cortex.

Actin [Gr. *aktis*: a ray] One of the two major proteins of muscle; it makes up the thin filaments. Forms the microfilaments found in most eukaryotic cells.

Action potential An impulse in a neuron, propagated without change in magnitude, taking the form of a wave of depolarization or reverse polarization imposed on a polarized cell surface.

Action spectrum A graph of biological activity versus wavelength of light. It compares the effectiveness of light of different wavelengths.

Activating enzyme An enzyme that couples a low-energy compound with ATP to yield a high-energy derivative. As used in this book, an enzyme that condenses ATP with a specific amino acid to give enzyme-bound AMP-amino acid; it then transfers the amino acid to the appropriate tRNA, yielding a charged tRNA for use in translation.

Activation energy The energy barrier that blocks the tendency for a set of chemical substances to react. A reaction is speeded up if this energy barrier is surmounted by adding heat energy, or if the barrier is lowered by providing a different reaction pathway with the aid of a catalyst. Designated by the symbol E_a.

Active site The region on the surface of an enzyme where the substrate binds, and where catalysis occurs.

Active transport The transport of a substance across a biological membrane against a concentration gradient—that is, from a region of low concentration (of that substance) to a region of high concentration. Active transport requires the expenditure of energy and is a saturable process. (Contrast with facilitated diffusion, free diffusion.)

Acute toxins Chemicals present in prey organisms that directly poison the metabolic machinery of their predators. Usually produce significant effects even though present in very small amounts.

Adaptation (a dap tay' shun) In evolutionary biology, a particular structure, physiological process, or behavior that makes an organism better able to survive and reproduce. Also, the evolutionary process that leads to the development or persistance of such a trait.

Adenosine triphosphate See ATP.

Adenylate cyclase Enzyme catalyzing the formation of cyclic AMP from ATP.

Adrenal (a dree' nal) [L. *ad-*: toward + *renes*: kidneys] An endocrine gland located near the kidneys of vertebrates, consisting of two glandular parts, the cortex and medulla.

Adrenaline See epinephrine.

Adsorption Binding of a gas or a solute to the surface of a solid.

Aerenchyma (air eng' kyma) [Gr. *aer*: air + *enchyma*: infusion] Modified parenchyma tissue, with many air spaces, found in shoots of some aquatic plants. (See parenchyma.)

Aerobic (air oh' bic) [Gr. *aer*: air + *bios*: life] In the presence of oxygen, or requiring oxygen.

Afferents (af' ur unts) [L. *ad*: to + *ferre*: to bear] Neurons that carry impulses to the central nervous system. (Contrast with efferents.)

Age distribution The proportion of individuals in a population belonging to each of the age categories into which the population has been divided. The number of divisions is arbitrary.

Aggregate fruit A fruit developing from several carpels of a single flower. (Contrast with simple fruit, accessory fruit, multiple fruit.)

AIDS (Acquired immune deficiency syndrome) Condition in which the body's helper T lymphocytes are destroyed, leaving the victim subject to opportunistic diseases. Caused by the HIV-I virus.

Alcohol An organic compound with one or more hydroxyl (—OH) groups.

Aldehyde (al' duh hide) A compound with a —CHO functional group. Many sugars are aldehydes.

Aldosterone (al dahs' ter own) A steroid hormone produced in the adrenal cortex of mammals. Promotes secretion of potassium and reabsorption of sodium in the kidney.

Aleurone layer (al' yur own) [Gr. *aleuron*: wheat flour] In grass seeds, a specialized cell layer just between the seed coat and the endosperm, synthesizing hydrolytic enzymes under the influence of gibberellin, and thus helping mobilize reserves for the developing embryo.

Alga (al' gah) (plural: algae) [L.: seaweed] Any one of a wide diversity of protists belonging to the phyla Pyrrophyta, Chrysophyta, Phaeophyta, Rhodophyta, and Chlorophyta (and, formerly, Cyanophyta—"blue-green algae"). Most live in the water, where they are the dominant autotrophs; most are unicellular, but a minority are multicellular ("seaweeds" and similar protists).

Allantois (uh lan' toe iss) [Gr. *allantoeides*: sausage-shaped] In vertebrate development, one of the embryonic membranes; forms part of mammalian umbilical cord.

Allele (a leel') [Gr. *allos*: other] The alternate forms of a genetic character found at a given locus on a chromosome.

Allergy [Ger. *allergie*: altered reaction] An overreaction to an antigen in amounts that do not affect most people; often involves IgE antibodies.

Allometric growth A pattern of growth in which some parts of the body of an organism grow faster than others, resulting in a change in body proportions as the organism grows.

Allopatric (al' lo pat' rick) [Gr. *allos*: other + *patria*: fatherland] Pertaining to populations that occur in different places.

Allopatric speciation See geographical speciation.

Allopolyploid (al' lo pol' lee ploid) [Gr. *allos*: other + *poly*: many + *ploos*: fold] A polyploid constructed of sets of chromosomes that originated from two or more species.

Allostery (al' lo steer' y) [Gr. *allos*: other + *stereos*: structure] Regulation of the activity of an enzyme by binding, at a site other than the catalytic active site, of an effector molecule that does not have the same structure as any of the enzyme's substrates.

Alpha helix A type of protein secondary structure; a right-handed spiral.

Alternation of generations The succession of haploid and diploid phases in a sexually reproducing organism. In most animals (male wasps and honey bees are notable exceptions), the haploid phase consists only of the gametes. In fungi, algae, and plants, however, the haploid phase may be the more prominent phase (as in fungi and mosses) or may be as prominent as the diploid phase (see the life cycle of *Ulva*, for example). In vascular plants, the diploid phase is more prominent.

Altricial (al trish' ul) [L. *altor*: one who nourishes] Being born in a poorly developed condition that requires parental care for survival. Typically applied to birds and mammals.

Altruistic act A behavior whose performance harms the actor but benefits other individuals.

Alveolus (al ve' o lus) (plural: alveoli) [L. *alveus*: cavity] A small, baglike cavity, especially the blind sacs of the lung.

Amensalism (a men' sul ism) Interaction in which one animal is harmed and the other is unaffected. (Contrast with commensalism, mutualism.)

Amine An organic compound with an amino group (see Amino acid).

Amino acid An organic compound of the general formula H_2N—CHR—$COOH$, where R can be one of 20 or more different side groups. An amino acid is so named because it has both a basic amine group, —NH_2, and an acidic carboxyl group, —$COOH$. Proteins are polymers of amino acids.

Ammonotelic (am moan' o teel' ic) [Gr. *telos*: end] Describes an organism in which the final product of breakdown of nitrogen-containing compounds (primarily proteins) is ammonia. (Contrast with ureotelic, uricotelic.)

Amnion (am' nee un) [Gr.: sac] A membranous sac containing a watery fluid in which the embryo is suspended (in reptiles, birds, and mammals).

Amoeba (a mee' bah) [Gr. *amoibe*: change] Any one of a large number of different kinds of unicellular animals belonging to the phylum Sarcodina, characterized among other features by its ability to change shape frequently through the protrusion and retraction of cytoplasmic extensions called pseudopods.

Amoeboid (a mee' boid) Like an amoeba; constantly changing shape by the protrusion and retraction of pseudopodia.

Amphi- [Gr.: both] Prefix used to denote a character or kind of organism that occupies two or more states. For example, amphibian (an animal that lives both on the land and in the water).

Amphibian (am fib' ee an) A member of the vertebrate class Amphibia, such as a frog, toad, or salamander.

Amphipathic (am' fi path' ic) [Gr. *amphi*: both + *pathos*: emotion] Of a molecule, having both hydrophilic and hydrophobic regions.

amu (atomic mass unit, or dalton) The basic unit of mass on an atomic scale, defined as $1/12$ the mass of a carbon-12 atom. There are 6.023×10^{23} amu in one gram. This number is known as Avogadro's number.

Amylase (am' ill ase) Any of a group of enzymes that digest starch.

Anabolism (an ab' uh liz' em) [Gr. *ana*: up, throughout + *ballein*: to throw] Synthetic reactions of metabolism, in which complex molecules are formed from simpler ones. (Contrast with catabolism.)

Anaerobic (an ur row' bic) [Gr. *an*: not + *aer*: air + *bios*: life] Occurring without the use of molecular oxygen, O_2.

Analogy (a nal' o jee) [Gr. *analogia*: resembling] A resemblance in function, and often appearance as well, between two structures which is due to convergence in evolution rather than to common ancestry. (Contrast with homology.)

Anaphase (an' a phase) [Gr. *ana*: indicating upward progress] The stage in nuclear division at which the first separation of sister chromatids (or, in the first meiotic division, of paired homologues) occurs. Anaphase lasts from the moment of first separation to the time at which the moving chromosomes converge at the poles of the spindle.

Ancestral trait Trait shared by a group of organisms as a result of descent from a common ancestor.

Aneuploid (an' you ploy dee) A condition in which one or more chromosomes or pieces of chromosomes are either lacking or present in excess.

Angiosperm (an' jee oh spurm) [Gr. *angion*: vessel + *sperma*: seed] One of the flowering plants; literally, one whose seed is carried in a "vessel," which is the fruit. (See fruit.)

Angiotensin (an' jee oh ten' sin) A peptide hormone that raises blood pressure by causing peripheral vessels to constrict; maintains glomerular filtration by causing constriction of efferent glomerular arterioles; stimulates thirst; and stimulates the release of aldosterone. Circulates in blood in its inactive form, angiotensinogen.

Animal [L. *animus*: breath, soul] A member of the kingdom Animalia. In general, a multicellular eukaryote that obtains its food by ingestion.

Animal pole In some eggs, zygotes, and embryos, the pole away from the bulk of the yolk. (Contrast with vegetal pole.)

Anion (an' eye on) An ion with one or more negative charges. (Contrast with cation.)

Anisogamy (an' eye sog' a mee) [Gr. *aniso*: unequal + *gamos*: marriage] The existence of two dissimilar gametes (egg and sperm).

Annelid (an' el id) A member of the phylum Annelida; one of the segmented worms, such as an earthworm or leech.

Annual Referring to a plant whose life cycle is completed in one growing season. (Contrast with biennial, perennial.)

Annual ring In the wood of some gymnosperms and angiosperms, a figure resulting from the formation of large vessels in the spring and small vessels in the summer.

Anterior Toward the front.

Anther (an' thur) [Gr. *anthos*: flower] A pollen-bearing portion of the stamen of a flower.

Antheridium (an' thur id' ee um) (plural: antheridia) [Gr. *antheros*: blooming] The multicellular structure that produces the sperm in bryophytes and ferns.

Antibody One of millions of blood proteins, produced by the immune system, that specifically recognizes a foreign substance and initiates its removal from the body.

Anticodon A "triplet" of three nucleotides in transfer RNA that is able to pair with a complementary triplet (a codon) in messenger RNA, thus aligning the transfer RNA on the proper place on the messenger. The codon (and, reciprocally, the anticodon) codes for a specific amino acid.

Antigen (an' ti jun) Any substance that stimulates the production of an antibody or antibodies upon introduction into the body of a vertebrate.

Antigenic determinant A specific region of an antigen, which is recognized by and binds to a specific antibody.

Antiparallel Parallel, but running in opposite directions. The two strands of DNA are antiparallel.

Antipodals (an tip' o dulls) [Gr. *anti*: against + *podus*: foot] Cells (usually three) of the mature embryo sac of a flowering plant, located at the end opposite the egg (and micropyle).

Antiport A membrane transport protein that carries one substance in one direction and another in the opposite direction. (Contrast with symport.)

Anus (a' nus) Opening through which digestive wastes are expelled, located at the posterior end of the gut.

Aorta (a or' tuh) [Gr. *aorte*: aorta] The main trunk of the arteries leading to the systemic (as opposed to the pulmonary) circulation.

Apex (a' pecks) The tip or highest point of a structure, as the apex of a growing stem or root.

Apical (a' pi kul) Pertaining to the apex, as the apical meristem, which is the actively growing tissue at the tip of a stem or root.

Apomixis (ap oh mix' is) [Gr. *apo*: away from + *mixis*: sexual intercourse] The asexual production of seeds.

Apoplast (ap' oh plast) In plants, the continuous meshwork of cell walls and extracellular spaces through which material can pass without crossing a plasma membrane. (Contrast with symplast.)

Apterous Lacking wings. (Contrast with alate: having wings.)

Aquatic [L. *aqua*: water] Living in or on water, or taking place in or on water. (Contrast with marine, terrestrial.)

Aqueous [L. *aqua*: water] Containing water, or dissolved in water.

Archaebacteria (ark' ee bacteria) [Gr. *archaios*: ancient] A distinctive division of bacteria of ancient origin; they possess distinctive lipids and lack peptidoglycan. Some biologists assign the archaebacteria to their own, separate kingdom.

Archegonium (ar' ke go' nee um) [Gr. *archegonos*: first of a kind] The multicellular structure that produces eggs in bryophytes, ferns, and gymnosperms.

Archenteron (ark en' ter on) [Gr. *archos*: beginning + *enteron*: bowel] The earliest primordial animal digestive tract.

Arteriole One of the branches of an artery.

Artery A muscular blood vessel carrying oxygenated blood away from the heart to other parts of the body. (Contrast with vein.)

Artifact [L. *ars, artis*: art + *facere*: to make] Something made by human effort or intervention. In biology, something that was not present in the living cell or organism, but was unintentionally produced by an experimental procedure.

Ascospore (ass' ko spor) A fungus spore produced within an ascus.

Ascus (ass' cuss) [Gr. *askos*: bladder] In fungi belonging to the class Ascomycetes (sac fungi), the club-shaped sporangium within which spores are produced by meiosis.

Asexual Without sex.

Associative learning "Pavlovian" learning, in which an animal comes to associate a previously neutral stimulus (such as the ringing of a bell) with a particular reward or punishment.

Assortative mating A breeding system under which mates are selected on the basis of a particular trait or group of traits. Results in more pairs of individuals sharing traits than would be the case if mating were random.

Assortment (genetic) The random separation during meiosis of nonhomologous chromosomes and of genes carried on nonhomologous chromosomes. For example, if genes *A* and *B* are borne on nonhomologous chromosomes, meiosis of diploid cells of genotype *AaBb* will produce haploid cells of the following types in equal numbers: *AB*, *Ab*, *aB*, and *ab*.

Asymmetric carbon atom In a molecule, a carbon atom to which four different atoms or groups are bound.

Atherosclerosis (ath' er oh sklair oh' sis) A disease of the lining of the arteries characterized by fatty, cholesterol-rich deposits in the walls of the arteries. When fibroblasts infiltrate these deposits and calcium precipitates in them, the disease becomes arteriosclerosis, or "hardening of the arteries."

Atmosphere The gaseous mass surrounding our planet. Also: a unit of pressure, equal to the normal pressure of air at sea level.

Atom [Gr. *atomos*: indivisible] The smallest unit of a chemical element. Consists of a nucleus and one or more electrons.

Atomic mass unit See amu.

Atomic number The number of protons in the nucleus of an atom, also equal to the number of electrons around the neutral atom. Determines the chemical properties of the atom.

Atomic weight The average weight of an atom of an element on the amu scale. (The average depends upon the relative amounts of different isotopes of an element on Earth.)

ATP (adenosine triphosphate) A compound containing adenine, ribose, and three phosphate groups. When it is formed, useful energy is stored; when it is broken down (to ADP or AMP), energy is released to drive endergonic reactions. ATP is a universal energy storage compound.

Atrium (a' tree um) A body cavity, as in the hearts of vertebrates. The thin-walled chamber(s) entered by blood on its way to the ventricle(s). Also, the outer ear.

Autogenic succession Ecological succession on newly opened areas. Usually driven by competitive interactions among plants. (See also degradative succession.)

Autoimmune disease A disorder in which the immune system attacks the animal's own body.

Autonomic nervous system The system (which in vertebrates comprises sympathetic and parasympathetic subsystems) that controls such involuntary functions as those of guts and glands.

Autopolyploid (au' tow pol' lee ploid) [Gr. *auto*: self + *poly*: many + *ploos*: fold] A polyploid constructed entirely of sets of chromosomes that originated from the same species.

Autoradiography The detection of a radioactive substance in a cell or organism by putting it in contact with a photographic emulsion and allowing the material to "take its own picture." The emulsion is developed, and the location of the radioactivity in the cell is seen by the presence of silver grains in the emulsion.

Autosome Any chromosome (in a eukaryote) other than a sex chromosome.

Autotroph (au' tow trow' fik) [Gr. *autos*: self + *trophe*: food] An organisms that is capable of living exclusively on inorganic materials, water, and some energy source such as sunlight or chemically reduced matter. (Contrast with heterotroph.)

Auxin (awk' sin) [Gr. *auxein*: increase] In plants, a substance (indoleacetic acid) that regulates growth and various aspects of development.

Auxotroph (awks' o trofe) [Gr. *auxanein*: to grow + *trophe*: food] A mutant form of an organism that requires a nutrient or nutrients not required by the wild-type, or reference, form of the organism. (Contrast with prototroph.)

Avogadro's number The conversion factor between atomic mass units and grams. More usefully, the number of atoms in that quantity of anelement which, expressed in grams, is numerically equal to the atomic weight in amu; 6.023×10^{23} atoms. (See mole.)

Axon [Gr.: axle] Fiber of a neuron which can carry action potentials. Carries impulses away from the cell body of the neuron; releases a neurotransmitter substance.

Axoneme (ax' oh neem) The complex of microtubules and their crossbridges that forms the motile apparatus of a cilium.

Bacillus (buh sil' us) [L.: little rod] Any of various rod-shaped bacteria.

Bacteriophage (bak teer' ee o fayj) [Gr. *bakterion*: little rod + *phagein*: to eat] One of a group of viruses that infect bacteria and ultimately cause their disintegration.

Bacterium (bak teer' ee um) (plural: bacteria) [Gr. *bakterion*: little rod] A prokaryote. An organism with chromosomes not contained in nuclear envelopes.

Balanced polymorphism [Gr. *polymorphos*: having many forms] The maintenance of more than one form, or the maintenance at a given locus of more than one allele, at frequencies of greater than one percent in a population. Often results when heterozygotes are superior to both homozygotes.

Baroreceptor [Gr. *baros*: weight] A pressure-sensing cell or organ.

Barr body In mammals, an inactivated X chromosome.

Basal body Centriole found at the base of a eukaryotic flagellum or cilium.

Base A substance which can accept a proton (H⁺). (Contrast with acid.) In nucleic acids, a nitrogen-containing base (purine or pyrimidine) is attached to each sugar in the backbone.

Base pairing See complementary base pairing.

Basic having a pH greater than 7.0 (having a hydrogen ion concentration lower than 10^{-7} molar).

Basidium (bass id' ee yum) In fungi of the class Basidiomycetes, the characteristic sporangium in which four spores are formed by meiosis and then borne externally before being shed.

Batesian mimicry Mimicry by a relatively harmless kind of organism of a more dangerous one, by which the mimic enjoys protection from predators that mistake it for the dangerous model. (Contrast with Müllerian mimicry.)

B cell A type of lymphocyte involved in the humoral immune response of vertebrates. Upon recognizing an antigenic determinant, a B cell develops into a plasma cell, which secretes an antibody. (Contrast with a T cell.)

Benefit An improvement in survival and reproductive success resulting from a behavior. (Contrast with cost.)

Benthic zone [Gr. *benthos*: bottom of the sea] The bottom of the ocean. (Contrast with pelagic zone.)

Beta-pleated sheet A type of protein secondary structure resulting from hydrogen bonding between polypeptide regions running antiparallel to one another.

Biennial Referring to a plant whose life cycle includes vegetative growth in the first year and flowering and senescence in the second year. (Contrast with annual and perennial.)

Bilateral symmetry The condition in which only the right and left sides of an organism, divided exactly down the back, are mirror images of each other. (Contrast with radial symmetry.)

Bilateria Animal phyla with bilateral symmetry during their development and throughout their evolutionary history. (Contrast with radiata.)

Binomial (bye nome' ee al) Consisting of two names; for example, the binomial nomenclature of biology which gives the name of the genus followed by the name of the species.

Biogenesis [Gr. *bios*: life + *genesis*: source] The origin of living things from other living things.

Biogeochemical cycles Movement of elements through living organisms and the physical environment.

Biogeography The scientific study of the geographic distribution of organisms. Ecological biogeography is concerned with the habitats in which organisms live, historical biogeography with the complete geographic ranges of organisms and the historical circumstances that determine the ranges.

Biological species concept The view that a species is most usefully defined as a population or series of populations within which there is a significant amount of gene flow under natural conditions, but which is genetically isolated from other populations.

Biology [Gr. *bios*: life + *logos*: discourse] The scientific study of life in all its forms.

Bioluminescence The production of light by biochemical processes in an organism.

Biomass The total weight of all the living organisms, or some designated group of living organisms, in a given area.

Biome (bye' ome) A major division of the ecological communities of Earth; characterized by distinctive vegetation.

Biome type One of a broad category of biomes, such as all of the grasslands of the world taken together.

Biota (bye oh' tah) All of the organisms, including animals, plants, fungi, and microorganisms, found in a given area.

Biotic (bye ah' tik) Pertaining to any aspect of life, especially to characteristics of entire populations or ecosystems.

Bipedal locomotion (by ped' ul) [L. *bipes*: two-footed] Walking on two feet.

Bivalent (biv' uh lent) In meiosis, a synapsed pair of homologous chromosomes.

Blastocoel (blass' toe seal) [Br. *blastos*: sprout + *koilos*: hollow] The central, hollow cavity of a blastula.

Blastodisc (blass' toe disk) A disk of cells forming on the surface of a large yolk mass, comparable to a blastula, but occurring in forms in which the massive yolk restricts cleavage to one side of the egg only.

Blastomere A cell produced by the division of a fertilized egg.

Blastopore The opening from the archenteron to the exterior of a gastrula.

Blastula (blass' chu luh) [Gr. *blastos*: sprout] An early stage in animal embryology; in many species, a hollow sphere of cells surrounding a central cavity.

Bloom A sudden increase in the density of phytolankton, especially in a freshwater lake.

Bohr effect (boar) The reduction in affinity of hemoglobin for oxygen caused by acidic conditions, usually as a result of increased CO_2.

Bolting In rosetted angiosperms, a dramatic elongation of the stem, usually followed by flowering.

Bottleneck A combination of environmental conditions that causes a serious reduction in the size of a population.

Bowman's capsule An elaboration of kidney tubule cells that surrounds a knot of capillaries (the glomerulus). Blood is filtered across the walls of these capillaries and the filtrate is collected into Bowman's capsule.

Brachiation (bray' kee ay' shun) [L. *brachium*: arm] Movement by grasping branches above the head and swinging along beneath them.

Browser An animal that feeds on the vegetative tissues of woody plants.

Bryophyte (bri' uh fite') [Gr. *bruon*: moss + *phyton*: plant] Any member of the division Bryophyta, including mosses, liverworts, and hornworts.

Bud primordium [L. *primordium*: the beginning] In plants, a small mass of potentially meristematic tissue found in the angle between the leaf stalk and the shoot apex. Will give rise to a lateral branch under appropriate conditions.

Budding Asexual reproduction in which a more or less complete new organism simply grows from the body of the parent organism and eventually detaches itself.

Buffering A process by which a system resists change—particularly in pH, in which case added acid or base is partially converted to another form.

Bulb In plants, an underground storage organ composed principally of enlarged and fleshy leaf bases.

Bundle sheath In C_4 plants, a layer of photosynthetic cells between the mesophyll and a vascular bundle of a leaf.

C_3 photosynthesis The form of photosynthesis in which 3-phosphoglycerate is the first stable product, and ribulose bisphosphate is the CO_2 receptor.

C_4 photosynthesis The form of photosynthesis in which oxaloacetate is the first stable product, and phosphoenolpyruvate is the CO_2 acceptor. C_4 plants also perform the reactions of C_3 photosynthesis.

Caecum (see' cum) [L. *caecus*: blind] A blind branch off the large intestine. In many nonruminant mammals, the caecum contains a colony of microorganisms that contribute to the digestion of food.

Callus [L. *calleo*: thick-skinned] In plants, wound tissue, of relatively undifferentiated proliferating cell mass, frequently maintained in cell culture.

Calmodulin (cal mod' joo lin) A calcium-binding protein found in all animal and plant cells; mediates many calcium-regulated processes.

Calorie [L. *calor*: heat] The amount of heat required to raise the temperature of one gram of water by one degree Celsius (1°C) from 14.5°C to 15.5°C. In nutrition studies, "calorie" refers to the kilocalorie (1 kcal = 1000 cal), the amount of heat required to raise the temperature of one kilogram of water by 1°C.

Calvin–Benson cycle The stage of photosynthesis in which CO_2 reacts with RuBP to form 3PG, 3PG is reduced to a sugar, and RuBP is regenerated, while other products are released to the rest of the plant.

Calyptra (kuh lip' tra) [Gr. *kalyptra*: covering for the head] A hood or cap found partially covering the apex of the sporophyte capsule in many moss species, formed from the expanded wall and neck of the archegonium.

Calyx (kay' licks) [Gr. *kalyx*: cup] All of the sepals of a flower, collectively.

CAM See crassulacean acid metabolism.

Cambium (kam' bee um) [L. *cambiare*: to exchange] A meristem that gives rise to radial rows of cells in stem and root, increasing them in girth; commonly applied to the vascular cambium which produces wood and phloem, and the cork cambium, which produces bark.

cAMP (cyclic AMP) A compound, formed from ATP, that mediates the effects of numerous animal hormones. Also needed for the transcription of catabolite-repressible operons in bacteria. Used for communication by cellular slime molds.

Canopy The leaf-bearing part of a tree. Collectively the aggregate of the leaves and branches of the larger woody plants of an ecological community.

Capillaries [L. *capillaris*: hair] Very small tubes, especially the smallest blood-carrying vessels of animals between the termination of the arteries and the beginnings of the veins.

Capping In RNA processing in eukaryotes, the addition of a modified G at the 5' end of the molecule.

Capsid The protein coat of a virus.

Capsule In bryophytes, the spore case. In some bacteria, a gelatinous layer exterior to the cell wall.

Carbohydrates Organic compounds with the general formula $C_nH_{2m}O_m$. Common examples are sugars, starch, and cellulose.

Carbon budget The amount of atmospheric carbon (from carbon dioxide) incorporated into organic molecules by a plant.

Carboxylic acid (kar box sill' ik) An organic acid that contains the carboxyl group, —COOH, which dissociates to the carboxylate ion, —COO⁻.

Carcinogen (car sin' oh jen) A substance that causes cancer.

Cardiac (kar' dee ak) [Gr. *kardia*: heart] Pertaining to the heart and its functions.

Carnivore [L. *carn*: flesh + *vovare*: to devour] An organism that feeds on animal tissue. (Contrast with detritivore, herbivore, omnivore.)

Carotenoid (ka rah' tee noid) [L. *carota*: carrot] A yellow, orange, or red lipid pigment commonly found as an accessory pigment in photosynthesis; also found in fungi.

Carpel (kar' pel) [Gr. *karpos*: fruit] The organ of the flower that contains one or more ovules.

Carrier In facilitated diffusion, a membrane protein that binds a specific molecule and transports it through the membrane. In genetics, a person heterozygous for a recessive trait. In respiratory and photosynthetic electron transport, a participating substance such as NAD that exists in both oxidized and reduced forms.

Carrying capacity In ecology, the largest number of organisms of a particular species that can be maintained indefinitely in a given part of the environment.

Cartilage In vertebrates, a tough connective tissue found in joints, the outer ear, and elsewhere.

Casparian strip A band of cell wall containing suberin and lignin, found in the endodermis. Restricts the movement of water across the endodermis.

Caste In social insects, any set of individuals of a particular anatomical type, or age group, or both, that performs specialized labor in the colony.

Catabolism [Ge. *kata*: down + *ballein*: to throw] Degradational reactions of metabolism, in which complex molecules are broken down. (Contrast with anabolism.)

Catabolite repression The decreased synthesis of many enzymes that tend to provide glucose for a cell; caused by the presence of excellent carbon sources, particularly glucose.

Catalyst (cat' a list) [Gr. *kata*-, implying the breaking down of a compound] A chemical substance that accelerates a reaction without itself being consumed in the overall course of the reaction. Catalysts lower the activation energy of a reaction. Enzymes are biological catalysts.

Cation (cat' eye on) An ion with one or more positive charges. (Contrast with anion.)

Caudal [L. *cauda*: tail] Pertaining to the tail, or to the posterior part of the body.

cDNA See complementary DNA.

Cell adhesion molecules Molecules on animal cell surfaces that affect the selective association of cells during development of the embryo.

Cell cycle The stages through which a cell passes between one division and the next. Includes all stages of interphase and mitosis.

Cell theory The theory, well established, that organisms consist of cells, and that all cells come from preexisting cells.

Cell wall A relatively rigid structure that encloses cells of plants, fungi, many protists, and most bacteria. The cell wall gives these cells their shape and limits their expansion in hypotonic media.

Cellular immune system That part of the immune system that is based on the activities of T cells. Directed against parasites, fungi, intracellular viruses, and foreign tissues (grafts). (Contrast with humoral immune system.)

Cellular respiration See respiration.

Cellulose (sell' you lowss) A straight-chain polymer of glucose molecules, used by plants as a structural supporting material.

Central dogma of molecular biology The statement that information flows from DNA to RNA to polypeptide (in retroviruses, information also flows from RNA to cDNA).

Central nervous system That part of the nervous system which is condensed and centrally located, e.g., the brain and spinal cord of vertebrates; the chain of cerebral, thoracic and abdominal ganglia of arthropods.

Centrifuge [L. *fugere*: to flee] A device in which a sample can be spun around a central axis at high speed, creating a centrifugal force that mimics a very strong gravitational force. Used to separate mixtures of suspended materials.

Centriole (sen' tree ole) A paired organelle that helps organize the microtubules in animal and protist cells during nuclear division; composed of nine sets of three fused microtubules.

Centromere (sen' tro meer) [Gr. *centron*: center + *meros*: part] The region where sister chromatids join.

Cephalization (sef' uh luh zay' shun) [Gr. *kephale*: head] The evolutionary trend toward increasing concentration of brain and sensory organs at the anterior end of the animal.

Cephalopod (sef' a low pod) A member of the mollusk class Cephalopoda, such as a squid or an octopus.

Cerebellum (sair' uh bell' um) [L.: diminutive of *cerebrum*: brain] The brain region that controls muscular coordination; located at the anterior end of the hindbrain.

Cerebrum (su ree' brum) [L.: brain] The dorsal anterior portion of the forebrain, making up the largest part of the brain of mammals. In mammals, the chief coordination center of the nervous system.

Channel A membrane protein that forms an aqueous passageway though which specific solutes may pass by simple diffusion; some channels are gated: they open and close in response to binding of specific molecules.

Character In taxonomy, any trait of an organism used in creating a classification system.

Character displacement An alteration of the traits of a species as a result of competition or other interactions with associated species.

Chemical bond An attractive force stably linking two atoms.

Chemiosmotic mechanism According to this model, ATP formation in mitochondria and chloroplasts results from a pumping of protons across a membrane (against a gradient of electrical charge and of pH), followed by the return of the protons through a protein channel with ATPase activity.

Chemoautotroph An organism that uses carbon dioxide as a carbon source and obtains energy by oxidizing inorganic substances from its environment. (Contrast with chemoheterotroph, photoautotroph, photoheterotroph.)

Chemoheterotroph An organism that must obtain both carbon and energy from organic substances. (Contrast with chemoautotroph, photoautotroph, photoheterotroph.)

Chemoreceptor A cell or tissue that senses specific substances in its environment.

Chemosynthesis Synthesis of food substances, using the oxidation of reduced materials from the environment as a source of energy.

Chiasma (kie az' muh) (plural: chiasmata) [Gr.: cross] An "x"-shaped connection between paired homologous chromosomes in prophase I of meiosis. A chiasma is the visible manifestation of crossing-over between homologous chromosomes.

Chitin (kye' tin) [Gr. *chiton*: tunic] The characteristic tough but also flexible organic component of the exoskeleton of arthropods, consisting of a complex, nitrogen-containing polysaccharide. Also found in cell walls of fungi.

Chlorophyll (klor' o fill') [Gr. *chloros*: green + *phyllon*: leaf] Any of a few green pigments associated with chloroplasts or with certain bacterial membranes; responsible for trapping light energy for photosynthesis.

Chloroplast [Gr. *chloros*: green + *plast*: a particle] An organelle bounded by a double membrane containing the enzymes and pigments that perform photosynthesis. Chloroplasts occur only in eukaryotes.

Cholecystokinin (ko' lee sis to kai nin) A hormone produced and released by the lining of the duodenum when it is stimulated by undigested fats and proteins. It stimulates the gall bladder to release bile and slows stomach activity.

Chorion (kor' ee on) [Gr. *khorion*: afterbirth] The outermost of the membranes protecting mammal, bird, and reptile embryos; in mammals it forms part of the placenta.

Chromatid (kro' ma tid) Each of a pair of new sister chromosomes from the time at which the molecular duplication occurs until the time at which the centromeres separate at the anaphase of nuclear division.

Chromatin The nucleic acid–protein complex found in eukaryotic chromosomes.

Chromatography Any one of several techniques for the separation of chemical substances, based on differing relative tendencies of the substances to associate with a mobile phase or a stationary phase.

Chromatophore (krow mat' o for) [Gr. *chroma*: color + *phoreus*: carrier] A pigment-bearing cell that expands or contracts to change the color of the organism.

Chromoplast [Gr. *chroma-*: color + *plast*: a particle] An organelle, derived from a chloroplast, containing carotenoid pigments but not chlorophyll; it is inactive in photosynthesis.

Chromosomal aberration Any large change in the structure of a chromosome, including duplication or loss of chromosomes or parts thereof, usually gross enough to be detected with the light microscope.

Chromosome (krome' o sowm) [Gr. *chroma*: color = *soma*: body] In monera and viruses, the DNA molecule that contains most or all of the genetic information of the cell or virus. In eukaryotes, a structure composed of DNA and proteins that bears part of the genetic information of the cell.

Chromosome walking A technique, based on recognition of overlapping fragments, used as a step in DNA sequencing.

Chyme (kime) [Gr. *chymus*: juice] Created in the stomach; a mixture of ingested food with the digestive juices secreted by salivary glands and the lining of the stomach.

Ciliate (sil' ee ate) A member of the protist phylum Ciliophora, unicellular organisms that propel themselves by means of cilia.

Cilium (sil' ee um) (plural: cilia) [L. *cilium*: eyelash] Hairlike organelle used for locomotion by many unicellular organisms and for moving water and mucus by many multicellular organisms. Generally shorter than a flagellum.

Circadian rhythm (sir kade' ee an) [L. *circa*: approximately + *dies*: day] A rhythm in behavior, growth, or some other activity that recurs about every 24 hours under constant conditions.

Citric acid cycle A set of chemical reactions in cellular respiration, in which acetyl coenzyme A reacts with oxaloacetate to form citric acid, and oxaloacetate is regenerated. Acetyl-coA is oxidized to carbon dioxide, and hydrogen atoms are stored as NADH and $FADH_2$.

Clade (clayd) [Gr. *klados*: branch] All of the organisms, both living and fossil, descended from a particular common ancestor.

Cladistic systematics Systematics based entirely on the phylogenetic relationships among organisms. (Contrast with phenetic and evolutionary systematics.)

Cladogenesis (clay doh jen' e sis) [Gr. *klados*: branch + *gignesthi*: to be born] The formation of new species by the splitting of an evolutionary lineage.

Cladogram Graphic representation of a cladistic relationship.

Class In taxonomy, the category below the phylum and above the order; a group of related, similar orders.

Clathrin A fibrous protein, on the inner surface of animal cell membranes, that strengthens coated vesicles and thus participates in receptor-mediated endocytosis.

Clay A soil constituent composed of particles smaller than 2 micrometers in diameter.

Cleavages First divisions of the fertilized egg of an animal.

Climax In ecology, a community that terminates a succession and which tends to replace itself unless it is further disturbed or the physical environment changes.

Climograph (clime' o graf) Graph relating temperature and precipitation with time of year.

Cline A gradual change in the traits of a species over a geographical gradient.

Cloaca (klo ay' kuh) [L. *cloaca*: sewer] In some invertebrates, the posterior part of the gut; in many vertebrates, a cavity receiving material from the digestive, reproductive, and excretory systems.

Clonal deletion In immunology, the inactivation or destruction of lymphocyte clones that would produce immune reactions against the animal's own body.

Clonal selection The mechanism by which exposure to antigen results in the activation of selected T-cell or B-cell clones, resulting in an immune response.

Clone [Gr. *klon*: twig, shoot] Genetically identical cells or organisms produced from a common ancestor by asexual means.

Clutch The number of offspring produced in a given batch.

Coacervate (ko as' er vate) [L. *coacervare*: to heap up] An aggregate of colloidal particles in suspension.

Coacervate drop Drops formed when a mixture of large proteins and polysaccharides is shaken in water. The interiors of these drops, which are often very stable, contain most of the proteins and polysaccharides.

Coated vesicle Vesicle, sometimes formed from a coated pit, with characteristic "bristly" surface; its membrane contains distinctive proteins, including clathrin.

Coccus (kock' us) [Gr. *kokkos*: berry, pit] Any of various spherical or spheroidal bacteria.

Cochlea (kock' lee uh) [Gr. *kokhlos*: a land snail] A spiral tube in the inner ear of vertebrates; it contains the sensory cells involved in hearing.

Coding strand In a stretch of double-stranded DNA, that strand that is not transcribed.

Codominance A condition in which two alleles at a locus produce different phenotypic effects and both effects appear in heterozygotes.

Codon A "triplet" of three nucleotides in messenger RNA that directs the placement of a particular amino acid into a polypeptide chain. (Contrast with anticodon.)

Coefficient of relatedness The probability that an allele in one individual is an identical copy, by descent, of an allele in another individual.

Coelom (see' lum) [Gr. *koiloma*: cavity] The body cavity of certain animals, which is lined with cells of mesodermal origin.

Coenocyte (seen' a sight) [Gr.: common cell] A "cell" bounded by a single plasma membrane, but containing many nuclei.

Coenzyme A nonprotein molecule that play a role in catalysis by an enzyme. The coenzyme may be part of the enzyme molecule or free in solution. Some coenzymes are oxidizing or reducing agents, others play different roles.

Coevolved relationship Relationship between species that have influenced one another's evolution.

Cohort (co' hort) [L. *cohors*: company of soldiers] A group of similar-age organisms, considered as it passes through time.

Coitus (koe' i tus) [L. *coitus*: a coming together] The act of sexual intercourse.

Coleoptile (koe' lee op' til) [Gr. *koleos*: sheath + *ptilon*: feather] A pointed sheath covering the shoot of grass seedlings.

Collagen [Gr. *kolla*: glue] A fibrous protein found extensively in bone and connective tissue.

Collenchyma (cull eng' kyma) [Gr. *kolla*: glue + *enchyma*: infusion] A type of plant cell, living at functional maturity, which lends flexible support by virtue of primary cell walls thickened at the corners. (Contrast with parenchyma, sclerenchyma.)

Colon [Gr. *kolon*: large intestine] The large intestine.

Colostrum (koh los' trum) The secretion of the mammary glands around the time of an infant's birth. It contains protein and lactose but little fat, and its rate of production is much less than the rate of milk production two or three days after birth.

Commensalism The form of symbiosis in which one species benefits from the association, while the other is neither harmed nor benefited.

Communication Action on the part of one organism (or cell) that alters the pattern of behavior in another organism (or cell) in an adaptive fashion.

Community Any ecologically integrated group of species of microorganisms, plants, and animals inhabiting a given area.

Companion cell Specialized cell found adjacent to a sieve tube element in some flowering plants.

Comparative analysis An approach to studying evolution in which hypotheses are tested by measuring the distribution of states among a large number of species.

Compensation point The light intensity at which the rates of photosynthesis and of cellular respiration are equal.

Competition In ecology, use of the same resource by two or more species, when the resource is present in insufficient supply for the combined needs of the species.

Competitive exclusion A result of competition between species for a limiting resource in which one species completely eliminates the other.

Competitive inhibitor A substance, similar in structure to an enzyme's substrate, that binds the active site and thus inhibits a reaction.

Complement system A group of eleven proteins that play a role in some reactions of the immune system. The complement proteins are not immunoglobulins.

Complementary base pairing The A–T (or A–U), T–A (or U–A), C–G and G–C pairing of bases in double-stranded DNA, in transcription, and between tRNA and mRNA.

Complementary DNA (cDNA) DNA formed by reverse transcriptase acting with an RNA template; essential intermediate in the reproduction of retroviruses; used as a tool in recombinant DNA technology; lacks introns.

Complete metamorphosis A change of state during the life cycle of an organism in which the body is almost completely rebuilt to produce an individual with a completely different body form. Characteristic of insects such as butterflies, moths, beetles, ants, wasps, and flies.

Compound (1) A substance made up of atoms of more than one element. (2) Made up of many units, as the compound eyes of arthropods (as opposed to the simple eyes of the same group of organisms).

Compression wood See reaction wood.

Condensation reaction A reaction in which two molecules become connected by a covalent bond, and a molecule of water is released. ($AH + BOH \rightarrow AB + H_2O$.)

Cones (1) In the vertebrate retina: photoreceptors responsible for color vision. (2) In gymnosperms: reproductive structures consisting of many sporophylls packed relatively tightly.

Conidium (ko nid' ee um) [Gr. *konis*: dust] An asexual fungus spore borne singly or in chains either apically or laterally on a hypha.

Conifer (kahn' e fer) [Gr. *konos*: cone + *phero*: carry] One of the cone-bearing gymnosperms, mostly trees, such as pines and firs.

Conjugation (kahn' jew gay' shun) [L. *conjugare*: yoke together] The close approximation of two cells during which they exchange genetic material, as in *Paramecium* and other ciliates, or during which DNA passes from one to the other through a tube, as in bacteria.

Connective tissue An animal tissue that connects or surrounds other tissues; its cells are embedded in a collagen-containing matrix.

Connexon In a gap junction, a protein channel linking adjacent animal cells.

Consensus sequences Short stretches of DNA that appear, with little variation, in many different genes.

Constitutive enzyme An enzyme that is present in approximately constant amounts in a system, whether its substrates are present or absent. (Contrast with inducible enzyme.)

Continental climate A pattern, typical of the interiors of large continents at high latitudes, in which bitterly cold winters alternate with hot summers. (Contrast with maritime climate.)

Continental drift The gradual drifting apart of the world's continents that has occurred over a period of billions of years.

Continental islands Islands separated by rising waters from continents to which they were once attached.

Contractile vacuole An organelle, often found in protists, which pumps excess water out of the cell and keeps it from being "flooded" in hypotonic environments.

Convergent evolution Evolutionary process in which initially dissimilar organisms become more similar with respect to a particular structure.

Cooperation Behavior in which two or more individuals interact to their mutual benefit. No conscious awareness by the actors of the effects of their behavior is implied.

Cooption The act of capturing something for a particular use. In ecology refers to the diversion of ecological production for human use. Such production is said to be coopted.

Corepressor A low molecular weight compound that unites with a protein (the repressor) to prevent transcription in a repressible operon.

Cork A waterproofing tissue in plants, with suberin-containing cell walls. Produced by a cork cambium.

Corm A conical, underground stem that gives rise to a new plant. (Contrast with bulb.)

Corolla (ko role' lah) [L.: diminutive of *corona*: wreath, crown] All of the petals of a flower, collectively.

Coronary (kor' oh nair ee) Referring to the blood vessels of the heart.

Corpus luteum (kor' pus loo' tee um) [L. *corpus*: body + *luteum*: yellow] A structure formed from a follicle after ovulation; it produces hormones important to the maintenance of pregnancy.

Cortex [L.: bark or rind] (1) In plants: the tissue between the epidermis and the vascular tissue of a stem or root. (2) In animals: the outer tissue of certain organs, such as the adrenal cortex and cerebral cortex.

Cost See energetic cost, opportunity cost, risk cost.

Cotyledon (kot' ul lee' dun) [Gr. *kotyledon*: a hollow space] A "seed leaf." An embryonic organ which stores and digests reserve materials; may expand when seed germinates.

Covalent bond A chemical bond that arises from the sharing of electrons between two atoms. Usually a strong bond.

Crassulacean acid metabolism (CAM) A metabolic pathway enabling the plants that possess it to store carbon dioxide at night and then perform photosynthesis during the day with stomata closed.

Crista (plural: cristae) A small, shelflike projection of the inner membrane of a mitochondrion; the site of oxidative phosphorylation.

Critical night length In the photoperiodic flowering response of short-day plants, the length of night above which flowering occurs and below which the plant remains vegetative. (The reverse applies in the case of long-day plants.)

Critical period The age during which some particular type of learning must take place or during which it occurs much more easily than at other times. Typical of song learning among birds.

Cross-pollination The pollination of one plant by pollen from another plant. (Contrast with self-pollination.)

Cross (transverse) section A section taken perpendicular to the longest axis of a structure.

Crossing over The mechanism by which linked markers undergo recombination. In general, the term refers to the reciprocal exchange of corresponding segments between two homologous chromatids. However, the reciprocity of crossing-over is problematical in prokaryotes and viruses; and even in eukaryotes, very closely linked markers often recombine by a nonreciprocal mechanism.

Crown gall A plant disease, caused by *Agrobacterium tumefaciens*, in which tumors are formed.

CRP The cAMP receptor protein that interacts with the promoter to enhance transcription; a lowered cAMP concentration results in catabolite repression.

Crustacean (crus tay' see an) A member of the arthropod class Crustacea, such as a crab, shrimp, or sowbug.

Cryptic appearance The resemblance of an animal to some part of its environment, which helps it to escape detection by predators.

Culture A laboratory association of organisms under controlled conditions. Also the collection of knowledge, tools, values, and rules that characterize a human society.

Cuticle A waxy layer on the outer surface of a plant or an insect, tending to retard water loss.

Cutin (cue' tin) [L. *cutis*: skin] A mixture of long, straight-chain hydrocarbons and waxes secreted by the plant epidermis, providing a water-impermeable coating on aerial plant parts.

Cyanobacteria (sigh an' o bacteria) [Gr. *kuanos*: the color blue] A division of photosynthetic bacteria, formerly referred to as blue-green algae; they lack sexual reproduction, and they use chlorophyll *a* in their photosynthesis.

Cyclic AMP See cAMP.

Cyst (sist) [Gr. *kystis*: bladder, touch] (1) A resistant, thick-walled cell formed by some protists and other organisms. (2) An abnormal sac, containing a liquid or semisolid substance, produced in response to injury or illness.

Cytochromes (sy' toe chromes) [Gr. *kytos*: container + *chroma*: color] Iron-containing red proteins, components of the electron-transfer chains in photophosphorylation and respiration.

Cytokinesis (sy' toe kine ee' sis) [Gr. *kytos*: container + *kinein*: to move] The division of the cytoplasm of a dividing cell. (Contrast with mitosis.)

Cytokinin (sy' toe kine' in) [Gr. *kytos*: container + *kinein*: to move] A member of a class of plant growth substances playing roles in senescence, cell division, and other phenomena.

Cytoplasm The contents of the cell, excluding the nucleus.

Cytoplasmic determinants In animal development, gene products whose spatial distribution may determine such things as embryonic axes.

Cytoskeleton The network of microtubules and microfilaments that gives a eukaryotic cell its shape and its capacity to arrange its organelles and to move.

Cytosol The fluid portion of the cytoplasm, excluding organelles and other solids.

Cytotoxic T cells Cells of the cellular immune system that recognize and directly eliminate virus-infected cells. (Contrast with helper T cells, suppressor T cells.)

Day-neutral plants Plants in which flowering is not photoperiodically determined. More species are day-neutral than are short-day or long-day plants.

Deciduous (de sid' you us) [L. *decidere*: fall off] Referring to a plant that sheds its leaves at certain seasons. (Contrast with evergreen.)

Decomposer See detrivore.

Degeneracy The situation in which a single amino acid may be represented by any of two or more different codons in messenger RNA. Most of the amino acids can be represented by more than one codon.

Degradative succession Ecological succession occuring on the dead remains of the bodies of plants and animals, as when leaves or animal bodies rot.

Deletion A mutation resulting from the loss of a continuous segment of a gene or chromosome. Such mutations never revert to wildtype. (Contrast with duplication, point mutation.)

Deme (deem) [Gr. *demos*: common people] Any local population of individuals belonging to the same species and among which mating is random.

Demographic stochasticity Random variations in the factors influencing the size, density, and distribution of a population.

Demography The study of dynamical changes in the sizes, densities, and distributions of populations.

Denaturation Loss of activity of an enzyme or nucleic acid molecule as a result of structural changes induced by heat or other means.

Dendrite [Gr. *dendron*: a tree] A fiber of a neuron which often cannot carry action potentials. Usually much branched and relatively short compared with the axon, and commonly carries information to the cell body of the neuron.

Denitrification Metabolic activity by which inorganic nitrogen-containing ions are reduced to form nitrogen gas and other products; carried on by certain soil bacteria.

Density dependence Change in the severity of action of agents affecting birth and death rates within populations. Such changes may be directly or inversely related to population density.

Density independence The state where the severity of action of agents affecting birth and death rates within a population does not change with the density of the population.

Deoxyribonucleic acid See DNA.

Dependency guild A group of plant species that provide one another with support or nutrition.

Depolarization A shift toward zero in membrane potential.

Derived trait A trait found among members of a lineage that was not present in the ancestors of that lineage.

Dermal tissue system The outer covering of a plant, consisting of epidermis in the young plant and periderm in a plant with extensive secondary growth. (Contrast with ground tissue system and vascular tissue system.)

Desmosome (dez' mo sowm) [Gr. *desmos*: bond + *soma*: body] An adhering junction between animal cells.

Determinate cleavage A pattern of early embryological development in which the potential of cells is determined very early such that separated cells develop only into partial embryos. (Contrast with indeterminate cleavage.)

Determination Process whereby an embryonic cell or group of cells becomes fixed into a predictable developmental pathway.

Detritivore (di try' ti vore) [L. *detritus*: worn away + *vorare*: to devour] An organism that eats the dead remains of other organisms.

Deuterium An isotope of hydrogen, possessing one neutron in its nucleus; deuterium oxide is called "heavy water."

Deuterostome One of two major lines of evolution in animals, characterized by radial cleavage, enterocoelous development, and other traits.

Development Progressive change, as in structure or metabolism; in most kinds of organisms, development continues throughout the life of the organism.

Dialysis (dye ahl' uh sis) [Gr. *dialyein*: separation] The removal of ions or small molecules from a solution by their diffusion across a semipermeable membrane to a solvent where their concentration is lower.

Diastole (dye ahs' toll ee) [Gr. *diastole*: dilation] The portion of the cardiac cycle when the heart muscle is relaxing. (Contrast with systole.)

Dichotomous (dye kot' a mus) [Gr. *dikhotomos*: divided] Dividing at each step into two parts or classifications.

Dicot (short for dicotyledon) [Gr. *dis*: two + *kotyledon*: a cup-shaped hollow] Any member of the angiosperm class Dicotyledonae, flowering plants in which the embryo produces two cotyledons prior to germination. Leaves of most dicots have major veins arranged in a branched or reticulate pattern.

Differentiation Process whereby originally similar cells follow different developmental pathways. The actual expression of determination.

Diffuse coevolution The situation in which the evolution of a lineage is influenced by its interactions with a number of species, most of which exert only a small influence on the evolution of the focal lineage.

Diffusion Random movement of molecules or other particles, resulting in even distribution of the particles when no barriers are present.

Digestibility-reducing substances Chemicals present in organisms that combine with proteins to make their tissues less nutritious and more difficult to digest when ingested by predators. Unlike acute toxins, these chemicals do not poison any metabolic systems in the predator.

Digestion Enzyme-catalyzed process by which large, usually insoluble, molecules (foods) are hydrolyzed to form smaller molecules of soluble substances.

Dihybrid cross A mating in which the parents differ with respect to the alleles of two loci of interest.

Dikaryon (di care' ee ahn) [Gr. *dis*: two + *karyon*: kernel] A cell or organism carrying two genetically distinguishable nuclei. Common in fungi.

Dioecious (die eesh' us) [Gr.: two houses] Organisms in which the two sexes are "housed" in two different individuals, so that eggs and sperm are not produced in the same individuals. Examples: humans, fruit flies, oak trees, date palms. (Contrast with monoecious.)

Diploblastic Having two cell layers. (Contrast with triploblastic.)

Diploid (dip' loid) [Gr. *diploos*: double] Having a chromosome complement consisting of two copies (homologues) of each chromosome. A diploid individual (or cell) usually arises as a result of the fusion of two gametes, each with just one copy of each chromosome. Thus, the two homologues in each chromosome pair in a diploid cell are of separate origin, one derived from the female parent and one from the male parent.

Diplontic life cycle A life cycle in which every cell except the gametes is diploid. (Contrast with haplontic life cycle.)

Disaccharide A carbohydrate made up of two monosaccharides (simple sugars).

Dispersal stage Stage in its life history at which an organism moves from its birthplace to where it will live as an adult.

Displacement activity Apparently irrelevant behavior performed by an animal under conflict situations, especially when tendencies to attack and escape are closely balanced.

Display A behavior that has evolved to influence the actions of other individuals.

Disruptive selection Selection in which extreme phenotypes are favored. (Contrast with stabilizing selection.)

Distal Away from the point of attachment or other reference point. (Contrast with proximal.)

Diverticulum (di ver tic' u lum) [L. *divertere*: turn away] A small cavity or tube that connects to a major cavity or tube.

Division A term used by botanists and some microbiologists, corresponding to the term phylum used by zoologists.

DNA (deoxyribonucleic acid) The fundamental hereditary material of all living organisms. In eukaryotes, stored primarily in the cell nucleus. A nucleic acid using deoxyribose rather than ribose.

DNA ligase Enzyme that unites Okazaki fragments of the lagging strand during DNA replication; also mends breaks in DNA strands. It connects pieces of a DNA strand and is used in recombinant DNA technology.

DNA methylation Addition of methyl groups to DNA; plays role in regulation of gene expression; protects a bacterium's DNA against its restriction endonucleases.

DNA polymerase Any of a group of enzymes that catalyze the formation of DNA strands from a DNA template.

Dominance In genetic terminology, the ability of one allelic form of a gene to determine the phenotype of a heterozygous individual, in which the homologous chromosome carries both it and a different allele. For example, if *A* and *a* are two allelic forms of a gene, *A* is said to be dominant to *a* if *AA* diploids and *Aa* diploids are phenotypically identical and are distinguishable from *aa* diploids. The *a* allele is said to be recessive.

Dominance distribution Distribution of organisms resulting from territoriality or other defense of space, such that success is better for individuals with territories in better habitats. (Contrast with ideal free distribution.)

Dominance hierarchy The set of relationships within a group of animals, usually established and maintained by aggression, in which one individual has precedence over all others in eating, mating, and other activities; a second individual has precedence over all but the highest-ranking individual, and so on down the line.

Dormancy A condition in which normal activity is suspended, as in some seeds and buds.

Dorsal [L. *dorsum*: back] Pertaining to the back or upper surface. (Contrast with ventral.)

Double fertilization Process virtually unique to angiosperms in which one sperm nucleus combines with the egg to produce a zygote, and the other sperm nucleus combines with the two polar nuclei to produce the first cell of the triploid endosperm.

Double helix Of DNA: molecular structure in which two complementary polynucleotide strands, antiparallel to each other, form a right-handed spiral.

Duodenum (doo' uh dee' num) The beginning portion of the vertebrate small intestine. (Contrast with ileum, jejunum.)

Duplication (genetic) A mutation resulting from the introduction into the genome of an extra copy of a segment of a gene or chromosome. (Contrast with deletion, point mutation.)

Dynein [Gr. *dunamis*: power] A protein that undergoes conformational changes and thus plays a part in the movement of eukaryotic flagella and cilia.

Ecdysone (eck die' sone) [Gr. *ek*: out of + *dyo*: to clothe] In insects, a pair of hormones inducing molting.

Echinoderm (e kine' oh durm) A member of the phylum Echinodermata, such as a seastar or sea urchin.

Ecological biogeography The study of the distributions of organisms from an ecological perspective, usually concentrating on migration, dispersal, and species interactions.

Ecological community The species living together at a particular site.

Ecological niche (nitch) [L. *nidus*: nest] The functioning of a species in relation to other species and its physical environment.

Ecological time The time required for significant changes in the distribution and density of a population. Typically a small to moderate number of generations.

Ecology [Gr. *oikos*: house + *logos*: discourse, study] The scientific study of the interaction of organisms with their environment, including both the physical environment and the other organisms that live in it.

Ecosystem (eek' oh sis tum) The organisms of a particular habitat, such as a pond or forest, together with the physical environment in which they live.

Ecto- (eck' toh) [Gr.: outer, outside] A prefix used to designate a structure on the outer surface of the body. For example, ectoderm. (Contrast with endo- and meso-.)

Ectoderm [Gr. *ektos*: outside + *derma*: skin] The outermost of the three embryonic tissue layers first delineated during gastrulation. Gives rise to the skin, sense organs, nervous system, etc.

Ectotherm [Gr. *ektos*: outside + *thermos*: heat] An animal unable to control its body temperature. (Contrast with endotherm.)

Edge effect The changes in ecological processes in a community caused by physical and biological factors originating in an adjacent community.

Effector Any organ, cell, or organelle that moves the organism through the environment or else alters the environment to the organism's advantage. Examples include muscle, bone, and a wide variety of exocrine glands.

Efferents [L. *ex*: out + *ferre*: to bear] Neurons that carry impulses from the central nervous system. (Contrast with afferents.)

Egg In all sexually reproducing organisms, the female gamete; in birds, reptiles, and some other vertebrates, a structure witin which early embryonic development occurs.

Elasticity The property of returning quickly to a former state after a disturbance.

Electron (e lek' tron) [L. *electrum*: amber (associated with static electricity), from Gr. *slektor*: bright sun (color of amber)] One of the three most important fundamental particles of matter, with mass approximately 0.00055 amu and charge –1.

Electron microscope An instrument that uses an electron beam to form images of minute structures; the transmission electron microscope is useful for thinly-sliced material, and the scanning electron microscope gives surface views of cells and organisms.

Electrophoresis (e lek' tro fo ree' sis) [L. *electrum*: amber + Gr. *phorein*: to bear] A separation technique in which substances are separated from one another on the basis of their electric charges and molecular weights.

Electrotonic potential In neurons, a hyperpolarization or small depolarization induced by the application of a small electrical current. (Contrast with resting potential, action potential.)

Elemental substance A substance composed of only one type of atom.

Embolus (em' buh lus) [Gr. *embolos*: inserted object; stopper] A circulating blood clot. Blockage of a blood vessel by an embolus or by a bubble of gas is referred to as an **embolism**. (Contrast with thrombus.)

Embryo [Gr. *en-*: in + *bryein*: to grow] A young animal, or young plant sporophyte, while it is still contained within a protective structure such as a seed, egg, or uterus.

Embryo sac In angiosperms, the female gametophyte. Found within the ovule, it consists of eight or fewer cells, membrane bounded, but without cellulose walls between them.

Emergent property A property of a complex system that is not exhibited by its individual component parts.

Emigration The deliberate and usually oriented departure of an organism from the habitat in which it has been living.

Endemic (en demm' ik) [Gr. *endemos*: dwelling in a place] Confined to a particular region, thus often having a comparatively restricted distribution.

Endergonic reaction One for which energy must be supplied. (Contrast with exergonic reaction.)

Endo- [Gr.: within, inside] A prefix used to designate an innermost structure. For example, endoderm, endocrine. (Contrast with ecto-, meso-.)

Endocrine gland (en' doh krin) [Gr. *endon*: inside + *krinein*: to separate] Any gland, such as the adrenal or pituitary gland of vertebrates, that secretes certain substances, especially hormones, into the body through the blood.

Endocytosis A process by which liquids or solid particles are taken up by a cell through invagination of the plasma membrane. (Contrast with exocytosis.)

Endoderm [Gr. *endon*: within + *derma*: skin] The innermost of the three embryonic tissue layers first delineated during gastrulation. Gives rise to the digestive and respiratory tracts and structures associated with them.

Endodermis [Gr. *endon*: within + *derma*: skin] In plants, a specialized cell layer marking the inside of the cortex in roots and some stems. Frequently a barrier to free diffusion of solutes.

Endomembrane system Endoplasmic reticulum plus Golgi apparatus plus, when present, lysosomes; thus, a system of membranes that exchange material with one another.

Endometrium (en do mee' tree um) [Gr. *endon*: inside + *metrios*: womb] The epithelial cells lining the uterus of mammals.

Endoplasmic reticulum [Gr. *endon*: within + L. *plasma*: form; L. *reticulum*: little net] A system of membrane-bounded tubes and flattened sacs, often continuous with the nuclear envelope, found in the cytoplasm of eukaryotes. Exists as rough ER, studded with ribosomes, and smooth ER, lacking ribosomes.

Endorphins Naturally-occurring, opiate-like substances in the mammalian brain.

Endoskeleton A skeleton covered by other, soft body tissues. (Contrast with exoskeleton.)

Endosperm [Gr. *endon*: within + *sperma*: seed] A specialized triploid seed tissue found only in angiosperms; contains stored food for the developing embryo.

Endosymbiosis [Gr. *endon*: within + *syn*: together + *bios*: life] The living together of two species, with one living inside the body (or even the cells) of the other.

Endosymbiotic theory Theory that the eukaryotic cell evolved from a prokaryote that contained other, endosymbiotic prokaryotes.

Endotherm [Gr. *endon*: within + *thermos*: hot] An animal that can control its body temperature by the expenditure of its own metabolic energy. (Contrast with ectotherm.)

Energy The capacity to do work.

Energetic cost The difference between the energy an animal would have expended had it rested, and that expended in performing a behavior.

Enhancer In eukaryotes, a DNA sequence, lying on either side of the gene it regulates, that stimulates a specific promoter.

Enkephalins [Gr. *en-*: in + *kephale*: head] Two of the endorphins. (See endorphin.)

Enterocoelous development A pattern of development in which the coelum is formed by an outpocketing of the embryonic gut (enteron).

Enterokinase (ent uh row kine' ase) An enzyme secreted by the muscosa of the duodenum. It activates the zymogen trypsinogen to create the active pancreatic enzyme trypsin.

Entropy (en' tro pee) [Gr. *en*: in + *tropein*: to change] A measure of the degree of disorder in any system. A perfectly ordered system has zero entropy; increasing disorder is measured by positive entropy. Spontaneous reactions in a closed system are always accompanied by an increase in disorder and entropy. Designated by the symbol S.

Environment An organism's surroundings, both living and nonliving; includes temperature, light intensity, and all other species that influence the focal organism.

Enzyme (en' zime) [Gr. *en*: in + *zyme*: yeast] A protein, on the surface of which are chemical groups so arranged as to make the enzyme a catalyst for a chemical reaction.

Eon The largest division of geological time.

Epi- [Gr.: upon, over] A prefix used to designate a structure located on top of another; for example: epidermis, epiphyte.

Epicotyl (epp' i kot' il) [Gr. *epi*: upon + *kotyle*: something hollow] That part of a plant embryo or seedling that is above the cotyledons.

Epidermis [Gr. *epi*: upon + *derma*: skin] In plants and animals, the outermost cell layers. (Only one cell layer thick in plants.)

Epididymus (ep uh did' uh mus) [Gr. *epi*: upon + *didymos*: testicle] Coiled tubules in the testes that store sperm and conduct sperm from the seminiferous tubules to the vas deferens.

Epiphyte (ep' e fyte) [Gr. *epi*: upon + *phyton*: plant] A specialized plant that grows on the surface of other plants but does not parasitize them.

Episome A plasmid that may exist either free or integrated into a chromosome. (See plasmid.)

Epistasis An interaction between genes, in which the presence of a particular allele of one gene determines whether another gene will be expressed.

Epithelium In animals, a layer of cells covering or lining an external surface or a cavity.

Equilibrium, chemical A state in which forward and reverse reactions are proceeding at counterbalancing rates, so there is no observable change in the concentrations of reactants and products.

Equilibrium constant A particular kind of ratio of the concentrations of products to reactants when equilibrium has been reached in a chemical reaction. Designated by the symbol, K_{eq}.

Era The second largest division of geological time.

Erythrocyte (ur rith' row sight') [Gr. *erythros*: red + *kytos*: hollow vessel] A red blood cell.

Esophagus (i soff' i gus) [Gr. *oisophagos*: gullet] That part of the gut between the pharynx and the stomach.

Essential element An irreplaceable mineral element without which normal growth and reproduction cannot proceed.

Estrogen Any of several steroid sex hormones, produced chiefly by the ovaries in mammals.

Estrous cycle The cyclical changes in reproductive physiology and behavior in female mammals (other than some primates), culminating in estrus.

Estrus (es' truss) [L. *oestrus*: frenzy] The period of heat, or maximum sexual receptivity, in some female mammals. Ordinarily, the estrus is also the time of release of eggs in the female.

Ethology (ee thol' o jee) [Gr. *ethos*: habit, custom + *logos*: discourse] The study of whole patterns of animal behavior in natural environments, stressing the analysis of adaptation and evolution of the patterns.

Ethylene One of the plant growth substances, the gas $H_2C\!=\!CH_2$.

Etiolation Growth in the absence of light.

Euchromatin Chromatin that is diffuse and non-staining during interphase; may be transcribed. (Contrast with heterochromatin.)

Eukaryotes (you car' ry otes) [Gr. *eu*: true + *karyon*: kernel or nucleus] Organisms whose cells contain their genetic material inside a nucleus. Includes all life except the viruses and the kingdom Monera.

Eusocial Term applied to insects, such as termites, ants, and many bees and wasps, in which individuals cooperate in the care of offspring, there are sterile castes, and generations overlap.

Eustacian tubes (yew stay' shen) Tubes in each ear that connect the middle ear cavities with the pharynx. Necessary to equilibrate pressure in the middle ear with barometric pressure.

Eutherian (yew thir' ee an) A major division of mammals. Often called the placental mammals because their embryos are nourished by a placenta attached to the lining of the uterus.

Eutrophication (yoo trofe' ik ay' shun) [Gr. *eu-*: well + *trephein*: to flourish] The addition of nutrient materials to water. Especially in lakes, the subsequent flourishing of algae and microorganisms can result in oxygen depletion and the eventual stifling of life in the water.

Evergreen A plant that retains its leaves through all seasons. (Contrast with deciduous.)

Evolution Any gradual change. Organic evolution, often referred to as evolution, is any genetic and resulting phenotypic change in organisms from generation to generation.

Evolutionary agent Any factor that influences the direction and rate of evolutionary changes.

Evolutionary radiation The proliferation of species within a single evolutionary lineage.

Evolutionary systematics Systematics based on information from both cladistic and phenetic studies.

Excretion Release of metabolic wastes by an organism.

Exergonic reaction A reaction in which free energy is released. (Contrast with endergonic reaction.)

Exo- (eks' oh) Same as ecto-.

Exocrine gland (eks' oh krin) [Gr. *exo*: outside + *krinein*: to separate] Any gland, such as a salivary gland, that secretes to the outside of the body or into the gut.

Exocytosis A process by which a vesicle within a cell fuses with the plasma membrane and releases its contents to the outside. (Contrast with endocytosis.)

Exon A portion of a DNA molecule, in eukaryotes, that codes for part of a polypeptide. (Contrast with intron.)

Exoskeleton (eks' oh skel' e ton) A hard covering on the outside of the body; the exoskeleton of insects and other arthropods has many of the same functions as the bony internal skeleton of vertebrates.

Experiment A scientific method in which particular factors are manipulated while other factors are held constant so that the potential influences of the manipulated factors can be determined.

Exploitation competition Competition that occurs because resources are depleted. (Contrast with interference competition.)

Exponential growth Growth, especially in the number of organisms in a population, which is a simple function of the size of the growing entity: the larger the entity, the faster it grows. (Contrast with logistic growth.)

Expressivity The degree to which a genotype is expressed in the phenotype—may be affected by the environment.

Extinction The termination of a lineage of organisms.

Extrinsic protein A membrane protein found only on the surface of the membrane. (Contrast with intrinsic protein.)

F_1 generation The immediate progeny of a mating; the first filial generation.

F_2 generation The immediate progeny of a mating between members of the F_1 generation.

F-duction Transfer of genes from one bacterium to another, using the F-factor as a vehicle.

F-factor In some bacteria, the fertility factor; a plasmid conferring "maleness" on the cell that contains it.

Facilitated diffusion Passive movement through a membrane involving a specific carrier protein; does not proceed against a concentration gradient. (Contrast with active transport, free diffusion.)

Facultative Capable of occurring or not occurring, as in facultative aerobes. (Contrast with obligate.)

Familial route Origin of sociality when an older generation of offspring assists the parents in caring for young. (Contrast with parasocial route.)

Family In taxonomy, the category below the order and above the genus; a group of related, similar genera.

Fat A triglyceride that is solid at room temperature.

Fatty acid A molecule with a long hydrocarbon tail and a carboxyl group at the other end. Found in many lipids.

Fauna (faw' nah) All of the animals found in a given area. (Contrast with flora.)

Faunal dominance The ability of species originating in one part of the world to spread their range to other parts of the world and displace other species already living there.

Feces [L. *faeces*: dregs] Waste excreted from the digestive system.

Feedback control Control of a particular step of a multistep process, induced by the presence or absence of a product of one of the later steps. A thermostat regulating the flow of heating oil to a furnace in a home is a negative feedback control device.

Fermentation (fur men tay' shun) [L. *fermentum*: yeast] The degradation of a substance such as glucose to smaller molecules with the extraction of energy, without the use of oxygen (i.e., anaerobically). Involves the glycolytic pathway.

Fertilization Union of gametes. Also known as syngamy.

Fertilization membrane A membrane that surrounds an animal egg which becomes rapidly raised above the egg surface within seconds after fertilization, serving to prevent entry of a second sperm.

Fetus The latter stages of an embryo that is still contained in an egg or uterus; in humans, the unborn young from the eighth week of pregnancy to the moment of birth.

Fiber An elongated and tapering cell of vascular plants, usually with a thick cell wall. Serves a support function.

Fibrin A protein that polymerizes to form long threads that provide the structure for a blood clot.

Filter feeder An organism that feeds upon much smaller organisms, that are suspended in water or air, by means of a straining device.

Filtration In the excretory physiology of some animals, the process by which the initial urine is formed; water and most solutes are transferred into the excretory tract, while proteins are retained in the blood or hemolymph.

First law of thermodynamics Energy can be neither created nor destroyed.

Fission Reproduction of a prokaryote by division of a cell into two comparable progeny cells.

Fitness The contribution of a genotype or phenotype to the composition of subsequent generations, relative to the contribution of other genotypes or phenotypes. (See inclusive fitness.)

Flagellate (flaj' el late) A member of the phylum Mastigophora, unicellular eukaryotes that propel themselves by flagella.

Flagellin (fla jell' in) The protein from which prokaryotic (but not eukaryotic) flagella are constructed.

Flagellum (fla jell' um) (plural: flagella) [L. *flagellum*: whip] Long, whiplike appendage that propels cells. Longer than a cilium, but has a similar internal structure of microtubules. Prokaryotic flagella differ sharply from those found in eukaryotes, such as the flagella in sperm cells.

Flight distance Distance at which a prey organism initiates escape from a predator.

Flora (flore' ah) All of the plants found in a given area. (Contrast with fauna.)

Florigen A plant hormone (not yet isolated) involved in the conversion of a vegetative shoot apex to a flower.

Flower The total reproductive structure of an angiosperm; its basic parts include the calyx, corolla, stamens, and carpels.

Fluorescence The emission of a photon of visible light by an excited atom or molecule.

Follicle [L. *folliculus*: little bag] In female mammals, an immature egg surrounded by nutritive cells.

Follicle-stimulating hormone A gonadotropic hormone produced by the anterior pituitary.

Food chain A portion of a food web, most commonly a simple sequence of prey species and the predators that consume them.

Food subweb A group of organisms united as a result of being food for the same or similar sets of predators.

Food web The complete set of food links between species in a community; a diagram indicating which ones are the eaters and which are consumed.

Forb Any broad-leaved (dicotyledonous), herbaceous plant. Especially applied to such plants growing in grasslands.

Fossil Any recognizable structure originating from an organism, or any impression from such a structure, that has been preserved over geological time.

Founder effect Random changes in allele frequencies resulting from establishment of a population by a very small number of individuals.

Fovea [L. *fovea*; a small pit] The area, in the vertebrate retina, of most distinct vision.

Frame-shift mutation A mutation resulting from the addition or deletion of a single base pair in the DNA sequence of a gene. As a result of this, mRNA transcribed from such a gene is translated normally until the ribosome reaches the point at which the mutation has occurred. From that point on, codons are read out of proper register and the amino acid sequence bears no resemblance to the normal sequence.

Free diffusion Diffusion derectly across a membrane without the involvement of carrier molecules. Free diffusion is not saturable and cannot cause the net transport from a region of low concentration to a region of higher concentration. (Contrast with facilitated diffusion and active transport.)

Free energy That energy which is available for doing useful work, after allowance has been made for the increase or decrease of disorder. Designated by the symbol G (for Gibbs free energy), and defined by: $G = H - TS$, where H = heat, S = entropy, and T = absolute (Kelvin) temperature.

Frequency-dependent selection Selection that changes in intensity when the proportion of individuals under selection increases or decreases.

Fruit In angiosperms, a ripened and mature ovary (or group of ovaries) containing the seeds. Sometimes applied to reproductive structures of other groups of plants, and includes any adjacent parts which may be fused with the reproductive structures.

Fruiting body A structure that bears spores.

Fundamental niche The range of condition under which an organism could survive if it were the only one in the environment. (Contrast with realized niche.)

Fungus (fung' gus) A member of the kingdom Fungi, a (usually) multicellular eukaryote with absorptive nutrition.

G$_1$ phase In the cell cycle, the gap between the end of mitosis and the onset of the S phase.

G$_2$ phase In the cell cycle, the gap between the S (synthesis) phase and the onset of mitosis.

Gametangium (gam i tan' gee um) [Gr. *gamos*: marriage + *angeion*: vessel or reservoir] Any plant or fungal structure within which a gamete is formed.

Gamete (gam' eet) [Gr. *gamete*: wife, *gametes*: husband] The mature sexual reproductive cell: the egg or the sperm.

Gametocyte (ga meet' oh site) [Gr. *gamete*: wife, *gametes*: husband + *kytos*: cell] The cell that gives rise to sex cells, either the eggs or the sperm. (See oocyte and spermatocyte.)

Gametogenesis (ga meet' oh jen' e sis) [Gr. *gamete*: wife, *gametes*: husband + *gignomai*: be born] The specialized series of cellular divisions that leads to the production of sex cells (gametes). (Contrast with oogenesis and spermatogenesis.)

Gametophyte (ga meet' oh fyte) In plants with alternation of generations, the haploid phase that produces the gametes. (Contrast with sporophyte.)

Ganglion (gang' glee un) [Gr.: tumor] A group or concentration of neuron cell bodies.

Gap junction A 2.7-nanometer gap between plasma membranes of two animal cells, spanned by protein channels. Gap junctions allow chemical substances or electrical signals to pass from cell to cell.

Gaping A method of feeding used by some birds in which the bill is opened forcibly against some substrate to expose otherwise hidden food.

Gastrocolic reflex A neural reflex that increases activity of the colon when the oral cavity and stomach are stimulated by food.

Gastropod (gas' troh pod) A member of the molluskan class Gastropoda, such as a snail or conch.

Gastrovascular cavity Serving for both digestion (gastro) and circulation (vascular); in particular, the central cavity of the body of jellyfish and other cnidarians.

Gastrula (gas' true luh) [Gr. *gaster*: stomach] An embryo forming the characteristic three cell layers (ectoderm, endoderm, and mesoderm) which will give rise to all of the major tissue systems of the adult animal.

Gastrulation Development of a blastula into a gastrula.

Gated channel A channel (membrane protein) that opens and closes in response to binding of specific molecules or to changes in membrane potential.

Gene [Gr. *gen*: to produce] A unit of heredity. Used here as the unit of genetic function which carries the information for a single polypeptide.

Gene amplification Creation of multiple copies of a particular gene, allowing the production of large amounts of the RNA transcript (as in rRNA synthesis in oocytes).

Gene cloning Formation of a clone of bacteria or yeast cells containing a particular foreign gene.

Gene family A set of identical, or once-identical, genes, derived from a single parent gene; need not be on the same chromosomes; classic example is the globin family in vertebrates.

Gene flow The exchange of genes between different species (an extreme case referred to as hybridization) or between different populations of the same species. A principal driving force of evolution.

Gene frequency The proportion of a given allele in a population.

Gene pool All of the genes in a population.

Generative nucleus In a pollen tube, a haploid nucleus that undergoes mitosis to produce the two sperm nuclei that participate in double fertilization. (Contrast with tube nucleus.)

Genet The genetic individual of a plant that is composed of a number of nearly identical but repeated units.

Genetic drift Changes in gene frequencies from generation to generation in a small population as a result of random processes.

Genetics The study of heredity.

Genetic stochasticity Variation in the frequencies of alleles and genotypes in a population over time.

Genetic structure The frequencies of alleles and genotypes in a population.

Genome (jee' nome) The genes in a complete haploid set of chromosomes.

Genome project An effort to map and sequence the entire genome of a species.

Genotype (jean' oh type) [Gr. *gen*: to produce + *typos*: impression] An exact description of the genetic constitution of an individual, either with respect to a single trait or with respect to a larger set of traits. (Contrast with phenotype.)

Genus (jean' us) (plural: genera) [Gr. *genos*: stock, kind] A group of related, similar species.

Geographical (allopatric) speciation Formation of two species from one by the interposition of a physical barrier. (Contrast with parapatric, sympatric speciation.)

Geotropism See gravitropism.

Germ cell A reproductive cell or gamete of a multicellular organism.

Germination The sprouting of a seed or spore.

Gibberellin (jib er el' lin) [L. *gibberella*: hunchback (refers to shape of a reproductive structure of a fungus that produces gibberellins)] One of a class of plant growth substances playing roles in stem elongation, seed germination, flowering of certain plants, etc. Named for the fungus *Gibberella*.

Gill An organ for gas exchange in aquatic organisms.

Gizzard (giz' erd) [L. *gigeria*: cooked chicken parts] A very muscular port of the stomach of birds that grinds up food, sometimes with the aid of fragments of stone.

Gland An organ of secretion or excretion.

Glia (glee' uh) [Gr.: glue] Cells, found only in the nervous system, which do not conduct action potentials.

Glomerulus (glo mare' yew lus) (plural: glomeruli) [L. *glomus*: ball] Sites in the kidney where blood filtration takes place. Each glomerulus consists of a knot of capillaries served by afferent and efferent arterioles.

Glucagon A hormone produced and released by cells in the islets of Langerhans of the pancreas. It stimulates glycogenolysis in liver cells.

Gluconeogenesis The biochemical synthesis of glucose from other substances such as amino acids, lactate, and glycerol.

Glucose (glue' kose) [Gr. *gleukos*: sweet wine mash for fermentation] The most common sugar, one of several monosaccharides with the formula $C_6H_{12}O_6$.

Glycogen (gly' ko jen) A branched-chain polymer of glucose, similar to starch (which is less branched and may be of lower molecular weight). Exists mostly in liver and muscle; the principal storage carbohydrate of most animals and fungi.

Glycolysis (gly kol' li sis) [from glucose + Gr. *lysis*: loosening] The enzymatic breakdown of glucose to pyruvic acid. One of the oldest energy-yielding mechanisms in living organisms.

Glyoxysome (gly ox' ee soam) A type of microbody, found in plants, in which stored lipids are converted to carbohydrates.

Golgi apparatus (goal' jee) A system of concentrically folded membranes found in the cytoplasm of eukaryotic cells. Plays a role in the production and release of secretory materials such as the digestive enzymes manufactured in the pancreas. First described by Camillo Golgi (1844–1926).

Gonad (go' nad) [Gr. *gone*: seed, that which produces seed] An organ that produces sex cells in animals: either an ovary (female gonad) or a testis (male gonad).

Gonadotropin A hormone that stimulates the gonads.

Gram stain A differential stain useful in characterizing bacteria.

Granum Within a chloroplast, a stack of thylakoids.

Gravitropism A directed plant growth response to gravity.

Grazer An animal that eats the vegetative tissues of herbaceous plants.

Gross morphology The sizes and shapes of the major body parts of a plant or animal.

Gross primary production The total energy captured by plants growing in a particular area.

Ground meristem That part of an apical meristem that gives rise to the ground tissue system of the primary plant body.

Ground tissue system Those parts of the plant body not included in the dermal or vascular tissue systems. Ground tissues function in storage, photosynthesis, and support.

Groundwater Water present deep in soils and rocks; may be stationary or flow slowly eventually to discharge into lakes, rivers, or oceans.

Group transfer The exchange of atoms between molecules.

Growth Irreversible increase in volume (probably the most accurate definition, but at best a dangerous oversimplification).

Growth factors A group of proteins that circulate in the blood and trigger the normal growth of cells. Each growth factor acts only on certain target cells.

Growth stage That stage in the life history of an organism in which it grows to its adult size.

Guard cells In plants, paired epidermal cells which surround and control the opening of a stoma (pore).

Guild A group of organisms that share a common food resource. Guild members have strong interactions with one another but weak interactions with members of other guilds.

Gut Digestive tract.

Guttation The extrusion of liquid water through openings in leaves, caused by root pressure.

Gymnosperm (jim' no sperm) [Gr. *gymnos*: naked + *sperma*: seed] A plant, such as a pine or other conifer, whose seeds do not develop within an ovary (hence, the seeds are "naked").

Habit The form or pattern of growth characteristic of an organism.

Habitat The environment in which an organism lives.

Habitat island A patch of habitat, separated from similar patches by other habitats.

Habituation (ha bich' oo ay shun) The simplest form of learning, in which an animal presented with a stimulus without reward or punishment eventually ceases to respond.

Hair cell A type of mechanoreceptor in animals.

Half-life The time required for half of a sample of a radioactive isotope to decay to its stable, nonradioactive form.

Halophyte (hal' oh fyte) [Gr. *halos*: salt + *phyton*: plant] A plant that grows in a saline (salty) environment.

Haploid (hap' loid) [Gr. *haploeides*: single] Having a chromosome complement consisting of just one copy of each chromosome. This is the normal "ploidy" of gametes or of asexual spores produced by meiosis or of organisms (such as the gametophyte generation of plants) that grow from such spores without fertilization.

Haplontic life cycle A life cycle in which the zygote is the only diploid cell. (Contrast with diplontic life cycle.)

Hardy–Weinberg Law The law that the basic processes of Mendelian heredity (meiosis and recombination) do not alter either the frequencies of genes or their diploid combinations. The Law also states how the percentages of diploid combinations can be predicted from a knowledge of the proportions of alleles in the population.

Haustorium (haw stor' ee um) [L. *haustus*: draw up] A specialized hypha or other structure by which fungi and some parasitic plants draw food from a host plant.

Helper T cells T cells that participate in the activation of B cells and of other T cells; targets of the HIV-I virus, the agent of AIDS. (Contrast with cytotoxic T cells, suppressor T cells.)

Hemoglobin (hee' mo glow' bin) [Gr. *haima*: blood + L. *globus*: globe] The colored protein of vertebrate blood (and blood of some invertebrates) which transports oxygen.

Hepatic (heh pat' ik) [Gr. *hepar*: liver] Pertaining to the liver.

Herbicide (ur' bis ide) A chemical substance that kills plants.

Herbivore [L. *herba*: plant + *vorare*: to devour] An animal which eats the tissues of plants. (Contrast with carnivore, detritivore, omnivore.)

Hering–Breuer reflex A neural reflex that inhibits inhalation when lung tissues are stretched.

Heritable Able to be inherited; in biology usually refers to genetically determined traits.

Hermaphroditism (her maf' row dite' ism) [Gr. *hermaphroditos*: a person with both male and female traits] The coexistence of both female and male sex organs in the same organism.

Hertz (abbreviated as Hz) Cycles per second.

Hetero- [Gr.: other, different] A prefix used in biology to mean that two or more different conditions are involved; for example, heterotroph, heterozygous.

Heterochromatic Chromatin that retains its coiling during interphase; generally not transcribed. (Contrast with euchromatin.)

Heterocyst A large, thick-walled cell in the filaments of certain cyanobacteria; performs nitrogen fixation.

Heterogeneous nuclear RNA (hnRNA) The product of transcription of a eukaryotic gene, including transcripts of introns.

Heterokaryon (het' er oh care' ee ahn) [Gr. *heteros*: different + *karyon*: kernel] A cell or organism carrying a mixture of genetically distinguishable nuclei. A heterokaryon is usually the result of the fusion of two cells without fusion of their nuclei.

Heteromorphic (het' er oh more' fik) [Gr. *heteros*: different + *morphe*: form] having a different form or appearance, as two heteromorphic life stages of a plant. (Contrast with isomorphic.)

Heterosporous (het' er os' por us) Producing two types of spores, one of which gives rise to a female megaspore and the other to a male microspore. Heterosporous plants produce distinct female and male gametophytes. (Contrast with homosporous.)

Heterotroph (het' er oh trof) [Gr. *heteros*: different + *trophe*: food] An organism that requires preformed organic molecules as food. (Contrast with autotroph.)

Heterozygous (het' er oh zie' gus) [Gr. *heteros*: different + *zygotos*: joined] Of a diploid organism having different alleles of a given gene on the pair of homologues carrying that gene. (Contrast with homozygous.)

Hexose A six-carbon sugar, such as glucose or fructose.

Hfr (for "high frequency of recombination") Donor bacterium in which the F-factor has been integrated into the chromosome. This produces a bacterium that transfers its chromosomal markers at a very high frequency to recipient (F⁻) cells.

Hibernation [L. *hibernus*: winter] The state of inactivity of some animals during winter; often marked by a drop in body temperature and metabolic rate.

Histamine (hiss' tah meen) A substance released within a damaged tissue by a type of white blood cell. Histamines are responsible for aspects of allergic reactions, including the increased vascular permeability that leads to edema (swelling).

Histology The study of tissues.

Histone Any one of a group of basic proteins forming the core of a nucleosome, the structural unit of a eukaryotic chromosome. (See nucleosome.)

Historical biogeography The study of the distributions of organisms from a long-term, historical perspective.

hnRNA See heterogeneous nuclear RNA.

Holdfast In many large attached algae, specialized tissue attaching the plant to its substratum.

Homeo box A segment of DNA, found in a few genes, perhaps regulating the expression of other genes and thus controlling large-scale developmental processes.

Homeostasis (home' ee o sta' sis) [Gr. *homos*: same + *stasis*: position] The maintenance of a steady state, such as a constant temperature or a stable social structure, by means of physiological or behavioral feedback responses.

Homeotherm (home' ee o therm) [Gr. *homos*: same + *therme*: heat] An animal which maintains a constant body temperature by virtue of its own heating and cooling mechanisms. (Contrast with poikilotherm.)

Homeotic genes (home' ee ott' ic) Genes that determine what entire segments of an animal become.

Homeotic mutation A drastic mutation causing the transformation of body parts in *Drosophila* metamorphosis. Examples include the *Antennapedia* and *ophthalmoptera* mutants.

Homologue (home' o log') [Gr. *homos*: same + *logos*: word] One of a pair, or larger set, of chromosomes having the same overall genetic composition and sequence. In diploid organisms, each chromosome inherited from one parent is matched by an identical (except for mutational changes) chromosome—its homologue—from the other parent.

Homology (ho mol' o jee) [Gr. *homologi(a)*: agreement] A similarity between two structures that is due to inheritance from a common ancestor. The structures are said to be homologous. (Contrast with analogy.)

Homoplasy (home' uh play zee) [Gr. *homos*: same + *plastikos*: to mold] The presence in several species of a trait not present in their most common ancestor. Can result from convergent evolution, reverse evolution, or parallel evolution.

Homosporous Producing a single type of spore that gives rise to a single type of gametophyte, bearing both female and male reproductive organs. (Contrast with heterosporous.)

Homozygous (home' o zie' gus) [Gr. *homos*: same + *zygotos*: joined] Of a diploid organism having identical alleles of a given gene on both homologous chromosomes. An organism may be a "homozygote" with respect to one gene and, at the same time, a "heterozygote" with respect to another. (Contrast with heterozygous.)

Hormone (hore' mone) [Gr. *hormon*: excite, stimulate] A substance produced in one part of a multicellular organism and transported to another part where it exerts its specific effect on the physiology or biochemistry of the target cells.

Host An organism that harbors a parasite and provides it with nourishment.

Host race A subpopulation of an organism adapted to feeding only upon a particular species of plant even though, as a whole, the species feeds upon a number of different plants. Widespread among insects.

Humoral immune system The part of the immune system mediated by B cells; it is mediated by circulating antibodies and is active against extracellular bacterial and viral infections.

Humus The partly decomposed remains of plants and animals on the surface of a soil. Its characteristics depend primarily upon climate and the species of plants growing on the site.

Hyaluronidase (hill yu ron' uh dase) An enzyme that digests proteoglycans. Found in sperm cells, it helps digest the coatings surrounding an egg so the sperm can enter the egg cell membrane.

Hybrid (high' brid) [L. *hybrida*: mongrel] The offspring of genetically dissimilar parents.

Hybridoma A cell produced by the fusion of an antibody-producing cell with a myeloma cell; it produces monoclonal antibodies.

Hydrocarbon A compound containing only carbon and hydrogen atoms.

Hydrogen bond A chemical bond which arises from the attraction between the slight positive charge on a hydrogen atom and a slight negative charge on a nearby fluorine, oxygen, or nitrogen atom. Weak bonds, but found in great quantities in proteins, nucleic acids, and other biological macromolecules.

Hydrological cycle The sum total of movement of water from the oceans to the atmosphere, to the soil, and back to the oceans. Some water is cycled many times within compartments of the system before completing one full circuit.

Hydrolyze (hi' dro lize) [Gr. *hydro*: water + *lysis*: cleavage] To break a chemical bond, as in a peptide linkage, with the insertion of the components of water, —H and —OH, at the cleaved ends of a chain. The digestion of proteins is a hydrolysis.

Hydrophilic (Gr. *hydro*: water + *philia*: love] Having an affinity for water. (Contrast with hydrophobic.)

Hydrophobic [Gr. *hydro*: water + *phobia*: fear] Molecules and amino acid side chains, which are mainly hydrocarbons (compounds of C and H with no charged groups or polar groups), have a lower energy when they are clustered together than when they are distributed through an aqueous solution. Because of their attraction for one another and their reluctance to mix with water they are called "hydrophobic." Oil is a hydrophobic substance; phenylalanine is a hydrophobic animo acid in a protein.

Hydrophobic interaction A weak attraction between highly nonpolar molecules or parts of molecules suspended in water.

Hydrostatic skeleton The incompressible internal liquids of some animals that transfer forces from one part of the body to another when acted upon by the surrounding muscles.

Hydroxyl group The —OH group, characteristic of alcohols.

Hymenopteran (high' man op' ter an) A member of the insect order Hymenoptera, such as a wasp, bee, or ant.

Hypertension High blood pressure.

Hypertonic [Gr.: higher tension] Having a more negative osmotic potential, as a result of having a higher concentration of osmotically active particles. Said of one solution as compared with another. (Contrast with hypotonic, isotonic.)

Hypha (high' fuh) (plural: hyphae) [Gr. *hyphe*: web] In the fungi, any single filament. May be multinucleate (zygomycetes, ascomycetes) or multicellular (basidiomycetes).

Hypocotyl That part of the embryonic or seedling plant shoot that is below the cotyldons.

Hypothalamus The part of the brain lying below the thalamus; it coordinates water balance, reproduction, temperature regulation, and metabolism.

Hypothetico-deductive method A method of science in which hypotheses are erected, predictions are made from them, and experiments and observations are performed to test the predictions. The process may be repeated many times in the course of answering a question.

Hypotonic [Gr.: lower tension] Having a less negative osmotic potential, as a result of having a lower concentration of osmotically active particles. Said of one solution as compared with another. (Contrast with hypertonic, isotonic.)

Ideal free distribution Distribution of organisms resulting from a process in which each newly arriving individual selects the habitat that is best for its success. (Contrast with dominance distribution.)

Ileum (ill' ee um) [L. *ileum*: viscera] The last division of the small intestine. (See duodenum, jejunum.)

Imaginal disc In insect larvae, groups of cells that develop into specific adult organs.

Imbibition [L. *imbibo*: to drink] The binding of a solvent to another molecule. Dry starch and protein will imbibe water.

Immunoglobulins A class of proteins, with a characteristic structure, active as receptors and effectors in the immune system.

Immunological distance The amount of difference between two proteins as measured by the strength of the antigen-antibody reaction between them.

Immunological tolerance A mechanism by which an animal does not mount an immune response to the antigenic determinants of its own macromolecules.

Imprinting A rapid form of learning, in which an animal comes to make a particular response, which is maintained for life, to some object or other organism.

Inclusive fitness The sum of an individual's own fitness (the effect of producing its own offspring: the individual selection component) plus its influence on fitness in relatives other than direct descendants (the kin selection component).

Incomplete dominance Condition in which the heterozygous phenotype is intermediate between the two homozygous phenotypes.

Incus (in' kus) [L. *incus*: anvil] The middle of the three bones that conduct movements of th eardrum to the oval window of the inner ear. (Contrast with malleus, stapes.)

Indeterminate cleavage A pattern of development in which individual cells retain the potential to develop into complete organisms if separated from one another well into development.

Individual distance The fixed distance an animal strives to keep between itself and other members of the same species.

Individual selection component That component of inclusive fitness that results from an organism producing its own offspring. (Contrast with kin selection component.)

Indoleacetic acid See auxin.

Induced fit A change in the tertiary structures of some enzymes, caused by binding of substrate to the active site.

Inducer In enzyme systems, a small molecule which, when added to a growth medium, causes a large increase in the level of some enzyme. Generally it acts by binding to repressor and changing its conformation so that the repressor does not bind to the operator. In embryology, a substance that causes a group of target cells to differentiate in a particular way.

Inducible enzyme An enzyme that is present in much larger amounts when a particular compound (the inducer) has been added to the system. (Contrast with constitutive enzyme.)

Industrial melanism The spread of dark (melanic) forms of many species of butterflies and moths in regions where trunks and leaves of plants have been blackened by industrial pollution.

Inflammation A nonspecific defense against pathogens; characterized by redness, swelling, pain, and increased temperature.

Inflorescence A structure composed of several flowers.

Ingestion Taking in of food by swallowing.

Inhibitor A substance which binds to the surface of an enzyme and interferes with its action on its substrates.

Initiation complex Combination of a ribosomal light subunit, an mRNA molecule, and the tRNA charged with the first amino acid coded for by the mRNA; formed at the onset of translation.

Insertion sequence A large piece of DNA that can give rise to copies at other loci; a type of transposable genetic element.

Instar (in' star) [L.: image, form] An insect or arthropod between molts.

Instinct Behavior that is relatively highly steroetyped and self-differentiating, that develops in individuals unable to observe other individuals performing the behavior or to practice the behavior in the presence of the objects toward which it is usually directed.

Insulin (in' su lin) [L. *insula*: island] An animal hormone, synthesized in islet cells of the pancreas, which promotes the conversion of glucose to the storage material, glycogen.

Integrase An enzyme that integrates retroviral cDNA into the genome of the host cell.

Integrated pest management A method of control of pests in which natural predators and parasites are used in conjunction with sparing use of chemical methods to achieve control of a pest without causing serious adverse environmental side effects.

Integument [L. *integumentum*: covering] A protective surface structure. In gymnosperms and angiosperms, a layer of tissue around the ovule which will become the seed coat. Gymnosperm ovules have one integument, angiosperm ovules two.

Intention movement The preparatory motions that animals go through prior to a complete behavior response; for example, the crouch before flying, the snarl before biting, etc.

Intercalated disks (in turk' a lated) Junctions between cardiac muscle cells where the cells join end to end. The cell membranes making up a disk are fused and offer low resistance to electrical currents.

Intercalary meristem A meristematic region in plants which occurs not apically, but between two regions of mature tissue. Intercalary meristems occur in the nodes of grass stems, for example.

Interference competition Competition resulting from direct behavioral interactions between organisms. (Contrast with exploitation competition.)

Interferon A glycoprotein produced by virus-infected animal cells; increases the resistance of neighboring cells to the virus.

Interleukins Regulatory proteins, produced by macrophages and lymphocytes, that act upon other lymphocytes and direct their development.

Intermediate filaments Fibrous proteins that stabilize cell structure and resist tension.

Internode Section between two nodes of a plant stem.

Interphase The period between successive nuclear divisions during which the chromosomes are diffuse and the nuclear envelope is intact. It is during this period that the cell is most active in transcribing and translating genetic information.

Interspecific competition Competition between members of two or more species.

Intertropical convergence zone The tropical region where the air rises most strongly; moves north and south with the passage of the sun overhead.

Intraspecific competition Competition among members of a single species.

Intrinsic protein A membrane protein that is embedded in the phospholipid bilayer of the membrane. (Contrast with extrinsic protein.)

Intrinsic rate of increase The rate at which a population can grow when its density is low and environmental conditions are highly favorable.

Intron A portion of a DNA molecule that, because of RNA splicing, is not involved in coding for part of a polypeptide molecule. (Contrast with exon.)

Invagination An infolding.

Invasiveness Ability of a bacterium to multiply within the body of a host.

Inversion (genetic) A rare mutational event that leads to the reversal of the order of genes within a segment of a chromosome, as if that segment had been removed from the chromosome, turned 180°, and then reattached.

Invertebrate Any animal that is not a vertebrate, that is, whose nerve cord is not enclosed in a backbone of bony segments.

In vitro [L.: in glass] In a test tube, rather than in a living organism. (Contrast with in vivo.)

In vivo [L.: in the living state] In a living organism. Many processes that occur in vivo can be reproduced in vitro with the right selection of cellular components. (Contrast with in vitro.)

Ion (eye' on) [Gr.: wanderer] An atom or group of atoms with electrons added or removed, giving it a negative or positive electrical charge.

Ionic bond A chemical bond which arises from the electrostatic attraction between positively and negatively charged ions. Usually a strong bond.

Iris (eye' ris) [Gr. *iris*, rainbow] The round, pigmented membrane that surrounds the pupil of the eye and adjusts its aperature to regulate the amount of light entering the eye.

Irruption A rapid increase in the density of a population. Often followed by massive emigration.

Islets of Langerhans Clusters of hormone-producing cells in the pancreas.

Isogamy (eye sog' ah mee) [Gr. *isos*: equal + *gamos*: marriage] A kind of sexual reproduction in which the gametes (or gametangia) are not distinguishable on the basis of size or morphology.

Isolating mechanism Geographical, physiological, ecological, or behavioral mechanisms that lead to a reduction in the frequency of hybrid matings.

Isomers Molecules consisting of the same numbers and kinds of atoms, but differing in the way in which the atoms are combined.

Isomorphic (eye' so more' fik) [Gr. *isos*: equal + *morphe*: form] having the same form or appearance, as two isomorphic life stages. (Contrast with heteromorphic.)

Isotonic [Gr.: same tension] Having the same osmotic potential. Said of two solutions. (Contrast with hypertonic, hypotonic.)

Isotope (eye' so tope) [Gr. *isos*: equal + *topos*: place] Two isotopes of the same chemical element have the same number of protons in their nuclei, but differ in the number of neutrons.

Isozymes Chemically different enzymes that catalyze the same reaction.

Iteroparous organism An organism that reproduces more than once in its lifetime. (Contrast with semelparous.)

Jejunum (jih jew' num) The middle division of the small intestine, where most absorption of nutrients occurs. (See duodenum, ileum.)

Joule (jool, or jowl) A unit of energy, equal to 0.24 calories.

Juvenile hormone In insects, a hormone maintaining larval growth.

Karyogamy (care' ee og' uh me) [Gr. *karyon*: kernel, nut + *gamos*: marriage] Fusion of gamete nuclei.

Karyotype The number, forms, and types of chromosomes in a cell.

Keratin (ker' a tin) [Gr. *keras*: horn] A protein which contains sulfur and is part of such hard tissues as horn, nail, and the outermost cells of the skin.

Key In taxonomy, a device for quickly identifying a specimen down to the species (or higher category) to which it belongs. The key consists of a series of choices made according to whether the speciment possesses one trait as opposed to another.

Keystone species A species that exerts a major influence on the composition and dynamics of the community in which it lives.

Kidneys A pair of excretory organs in vertebrates.

Kin selection component The component of inclusive fitness resulting from helping the survival of relatives containing the same alleles by descent from a common ancestor.

Kinesis (ki nee' sis) [Gr.: movement] Orientation behavior in which the organism does not move in a particular direction with reference to a stimulus but instead simply moves at an increasing or decreasing rate until it ends up farther from the object or closer to it. (Contrast with taxis.)

Kinetochore (kin net' oh core) [Gr. *kinetos*: moving + *khorein*: to move] Specialized structure on a centromere to which microtubules attach.

Kingdom The highest taxonomic category in the Linnaean system.

Lactic acid The end product of fermentation in vertebrate muscle and some microorganisms.

Lagging strand In DNA replication, the daughter strand that is synthesized discontinuously.

Lamarckism (Lah mark' iz um) The theory of evolution by acquired characteristics, as propounded by Jean Baptiste de Lamarck.

Lamella Layer.

Larva (plural: **Larvae**) [L.: ghost, early stage] An immature stage of any invertebrate animal that differs dramatically in appearance from the adult.

Latent learning Learning that is not put to immediate use but is "stored" for possible future use.

Lateral Pertaining to the side.

Lateral inhibition In visual information processing in the arthropod eye, the mutual inhibition of optic nerve cells; results in enhanced detection of edges.

Laterization (lat' ur iz ay shun) The formation of a nutrient-poor soil that is rich in insoluble iron and aluminum compounds.

Leader sequence A sequence of amino acids at the N-terminal end of a newly synthesized protein, determining where the protein will be placed in the cell.

Leading strand In DNA replication, the daughter strand that is synthesized continuously.

Leaf axil The upper angle between a leaf and the stem, site of lateral buds which under appropriate circumstances become activated to form lateral branches.

Leaf primordium [L.: the beginning] A small mound of cells on the flank of a shoot apical meristem that will give rise to a leaf.

Lek A traditional courtship display ground, where males display to females.

Lenticel Spongy region in a plant's periderm, allowing gas exchange.

Leukocyte (loo' ko sight) [Gr. *leukos*: clear + *kutos*: hollow vessel] A white blood cell.

Lichen (lie' kun) [Gr. *leikhen*: licker] An organism resulting from the symbiotic association of a true fungus and either a cyanobacterium or a unicellular alga.

Life cycle The entire span of the life of an organism from the moment of fertilization (or asexual generation) to the time it reproduces in turn.

Life history The stages an individual goes through during its life.

Life table A table showing, for a group of equal-aged individuals, the proportion still alive at different times in the future and the number of offspring they produce during each time interval.

Ligament A band of connective tissue linking two bones in a joint.

Ligand (lig' and) A molecule that binds to a receptor site of another molecule.

Lignin The principal noncarbohydrate component of wood, a polymer that binds together cellulose fibrils in some plant cell walls.

Linkage (genetic) Association between markers on the same chromosome such that they do not show random assortment. Linked markers recombine with one another at frequencies less than 0.5; the closer the markers on the chromosome, the lower the frequency of recombination.

Lipase (lye' pase) An enzyme that digests fats.

Lipids (lip' ids) [Gr. *lipos*: fat] Substances in a cell which are easily extracted by organic solvents; fats, oils, waxes, steroids, and other large organic molecules, including those which, with proteins, make up the cell membranes. (See phospholipids.)

Litter The partly decomposed remains of plants on the surface and in the upper layers of the soil.

Littoral zone The coastal zone from the upper limits of tidal action down to the depths where the water is thoroughly stirred by wave action.

Liver A large digestive gland. In vertebrates, it secretes bile and is involved in the formation of blood.

Locus In genetics, a specific location on a chromosome. May be considered to be synonymous with "gene."

Logistic growth Growth, especially in the size of an organism or in the number of organisms that constitute a population, which slows steadily as the entity approaches its maximum size. (Contrast with exponential growth.)

Long-day plant A photoperiodically sensitive plant that flowers when exposed to short nights.

Loop of Henle (hen' lee) Long hairpin loop of the mammalian renal tubule that runs from the cortex, down into the medulla, and back to the cortex. Creates a concentration gradient in the interstitial fluids in the medulla.

Lophophore A U-shaped fold of the body wall with hollow, ciliated tentacles that encircles the mouth of animals in several different phyla. Used for filtering prey from the surrounding water.

Lumen (loo' men) [L.: light] The cavity inside any tubular part of an organ, such as a piece of gut or a kidney tubule.

Lungs A pair of saclike chambers within the bodies of some animals, functioning in gas exchange.

Luteinizing hormone A gonadotropin produced by the anterior pituitary. It stimulates the gonads to produce sex hormones.

Lymph [L. *lympha*: water] A clear, watery fluid that is formed as a filtrate of blood; it contains white blood cells; it collects in a series of special vessels and is returned to the bloodstream.

Lymphocyte A major class of white blood cells. Includes T cells, B cells, and other cell types important in the immune response.

Lysis (lie' sis) [Gr.: a loosening] Bursting of a cell.

Lysogenic The condition of a bacterium that carries the genome of a virus in a relatively stable form. (Contrast with lytic.)

Lysosome (lie' so soam) [Gr. *lysis*: a loosening + *soma*: body] A membrane-bounded inclusion found in eukaryotic cells (other than plants). Lysosomes contain a mixture of enzymes that can digest most of the macromolecules found in the rest of the cell.

Lysozyme (lie' so zyme) An enzyme in saliva, tears, and nasal secretions that attacks bacterial cell walls, as one of the body's nonspecific defense mechanisms.

Lytic Condition in which a bacterium lyses shortly after infection by a virus; the viral genome does not become stabilized within the bacterial cell. (Contrast with lysogenic.)

Macro- (mack' roh) [Gr. *makros*: large, long] A prefix commonly used to denote something large. (Contrast with micro-.)

Macrodescriptor An index or term describing some aggregate feature of groups of organisms.

Macroevolution Evolutionary changes occurring over long time spans and usually involving changes in many traits. (Contrast with microevolution.)

Macroevolutionary time The time required for macroveolutionary changes in a lineage.

Macromolecule A giant polymeric molecule. The macromolecules are proteins, polysaccharides, and nucleic acids.

Macronutrient A mineral element required by plant tissues in concentrations of at least 1 milligram per gram of their dry matter.

Madreporite The seive plate which connects the internal water vascular system of echinoderms with the exterior.

Major histocompatibility complex A complex of linked genes, with multiple alleles, that control a number of immunological phenomena; it is important in graft rejection.

Malleus (mal' ee us) [L. *malleus*: hammer] The first of the three bones that conduct movements of th eardrum to the oval window of the inner ear. (Contrast with incus; stapes.)

Malpighian tubule (mal pee' gy un) A type of protonephridium found in insects.

Mammal [L. *mamma*: breast, teat] Any animal of the class Mammalia, characterized by the production of milk by the female mammary glands and the possession of hair for body covering.

Mantle A sheet of specialized tissues that covers most of the viscera of mollusks; provides protection to internal organs and secretes the shell.

Mapping In genetics, determining the order of genes on a chromosome and the distances between them.

Marine [L. *mare*: sea, ocean] Pertaining to or living in the ocean. (Contrast with aquatic, terrestrial.)

Maritime climate Weather pattern typical of coasts of continents, particularly those on the western sides at mid latitudes, in which the difference between summer and winter is relatively small. (Contrast with continental climate.)

Marsupial (mar soo' pee al) A mammal belonging to the subclass Metatheria, such as opossums and kangaroos. Most have a pouch (marsupium) that contains the milk glands and serves as a receptacle for the young.

Maternal inheritance (cytoplasmic inheritance) Inheritance in which the phenotype of the offspring depends on factors, such as mitochondria or chloroplasts, that are inherited from the female parent through the cytoplasm of the female gamete.

Mating type In some bacteria, fungi, and protists, sexual reproduction can occur only between partners of different mating type. "Mating type" is not the same as "sex," since some species have as many as eight mating types; also, as in *Paramecium aurelia*, mating may be between hermaphroditic partners of opposite mating type, with partners acting as both "male" and "female" in terms of donating and receiving genetic information.

Maturation The automatic development of a pattern of behavior, which becomes increasingly complex or precise as the animal matures. Unlike learning, the development does not require experience to occur.

Medulla (meh dull' luh) [L.: narrow] The inner part of an organ.

Medusa (meh doo' suh) The tentacle-bearing, jellyfish-like, free-swimming sexual stage in the life cycle of a cnidarian.

Mega- [Gr. *megas*: large, great] A prefix often used to denote something large. (Contrast with micro-.)

Megaphyll (meg' ah fill) [Gr. *megas*: large + *phyllon*: leaf] In vascular plants, a leaf thought to be derived from a flattened branch system, its vascular tissue forming a leaf gap where it attaches to the stem vascular tissue.

Megareserve A large park or reserve; usually has associated buffer areas in which human use of the environment is restricted to activities that do not destroy the functioning of the ecosystem.

Megasporangium The special structure (sporangium) that produces the megaspores.

Megaspore [Gr. *megas*: large + *spora*:seed] In plants, a haploid spore that produces a female gametophyte. In many cases the megaspore is larger than the male-producing microspore.

Meiosis (my oh' sis) [Gr.: diminution] Division of a diploid nucleus to produce four haploid daughter cells. The process consists of two successive nuclear divisions with only one cycle of chromosome replication.

Memory cells Long-lived lymphocytes that retain the ability to start dividing and produce more effector and more memory cells. Responsible for immunological memory.

Mendelian population A local population of individuals belonging to the same species and exchanging genes with one another.

Meristem [Gr. *meristos*: divided] Plant tissue made up of actively dividing cells.

Mesenchyme (mez' en kyme) [Gr. *mesos*: middle + *enchyma*: infusion] Embryonic or unspecialized cells derived from the mesoderm.

Meso- (mez' oh) [Gr.: middle] A prefix often used to designate a structure located in the middle, or a stage that appears at some intermediate time. For example, mesoderm, Mesozoic.

Mesocoel (mez' uh seal) [Gr. *mesos*: mid + *koilos*: hollow] The middle of the three body cavities found in many animals with enterocoelous development.

Mesoderm [Gr. *mesos*: middle + *derma*: skin] The middle of the three embryonic tissue layers first delineated during gastrulation. Gives rise to skeleton, circulatory system, muscles, excretory system, and most of the reproductive system.

Mesoglea The jelly-like middle layer that constitutes the bulk of the bodies of the medusae of many cnidarians; not a true cell layer.

Mesophyll (mez' a fill) [Gr. *mesos*: middle + *phyllon*: leaf] Chloroplast-containing, photosynthetic cells in the interior of leaves.

Mesosome (mez' o soam') [Gr. *mesos*: middle + *soma*: body] A localized infolding of the plasma membrane of a bacterium.

Messenger RNA (mRNA) A transcript of one of the strands of DNA, it carries information (as a sequence of codons) for the synthesis of one or more proteins.

Meta- [Gr.: between, along with, beyond] A prefix used in biology to denote a change or a shift to a new form or level; for example, as used in metamorphosis.

Metabolic pathway A series of enzyme-catalyzed reactions so arranged that the product of one reaction is the substrate of the next.

Metabolism (meh tab' a lizm) [Gr. *metabole*: to change] The sum total of the chemical reactions that occur in an organism, or some subset of that total (as in "respiratory metabolism").

Metacoel (met' uh seal) [GR. *meta*: after + *koilos*: hollow] The posterior of the three body cavities found in many animals with enterocoelous development.

Metamorphosis (met' a mor' fo sis) [Gr. *meta*: between + *morphe*: form, shape] A radical change occurring between one developmental stage and another, as for example from a tadpole to a frog or an insect larva to the adult.

Metanephridia The excretory organs of annelids.

Metaphase (met' a phase) [Gr. *meta*: between] The stage in nuclear division at which the centromeres of the highly supercoiled chromosomes are all lying on a plane (the metaphase plane or plate) perpendicular to a line connecting the division poles.

Metasome (met' uh some) [Gr. *meta*: after + *soma*: body[The posterior of the three body divisions of lophophorate animals.

Metastasis (meh tass' tuh sis) The spread of cancer cells from their original site to other parts of the body.

Methanogen Any member of a group of Archaebacteria that release methane as a metabolic product. This group is considered to be an extremely ancient one.

Micro- (mike' roh) [Gr. *mikros*: small] A prefix often used to denote something small. (Contrast with macro-, mega-.)

Microbiology [Gr. *mikros*: small + *bios*: life + *logos*: discourse] The scientific study of microscopic organisms, particularly bacteria, unicellular algae, protistans, and viruses.

Microbody A small organelle, bounded by a single membrane and possessing a granular interior. Peroxisomes and glyoxysomes are types of microbodies.

Microevolution The small evolutionary changes typically occurring over short time spans; generally involving a small number of traits and minor genetic changes. (Contrast with macroevolution.)

Microevolutionary time The time required for microevolutionary changes within a lineage of organisms.

Microfilament Minute fibrous structure generally composed of actin found in the cytoplasm of eukaryotic cells. They play a role in the motion of cells.

Micromorphology The structure of the macromolecules of an organism.

Micronutrient A mineral element required by plant tissues in concentrations of less than 100 micrograms per gram of their dry matter.

Microorganism Any microscopic organism, such as a bacterium or one-celled alga.

Microphyll (mike' roh fill) [Gr. *mikros*: small + *phyllon*: leaf] A leaf possibly derived from a scalelike outgrowth of the stem. Its vascular tissue does not form a leaf gap where it attaches to the stem bascular tissue.

Micropyle (mike' roh pile) [Gr. *mikros*: small + *pyle*: gate] Opening in the integument(s) of a seed plant ovule through which pollen grows to reach the female gametophyte within.

Microsporangium The special structure (sporangium) that produces the microspores.

Microspores [Gr. *mikros*: small + *spora*: seed] In plants, a haploid spore that produces a male gametophyte. In many cases the microspore is smaller than the female-producing megaspore.

Microtubules Minute tubular structures found in centrioles, spindle apparatus, cilia, flagella, and other places in the cytoplasm of eukaryotic cells. These tubules play roles in the motion and maintenance of shape of eukaryotic cells.

Microvilli (singular: microvillus) The projections of epithelial cells, such as the cells lining the small intestine, that increase their surface area.

Middle lamella A layer of derivative polysaccharides that separates plant cells; a common middle lamella lies outside the primary walls of the two cells.

Migration The regular, seasonal movements of animals between breeding and nonbreeding ranges.

Mimicry (mim' ik ree) The resemblance of one kind of organism to another, or to some inanimate object; serves the function of making the organism difficult to find, of discouraging potential enemies or of attracting potential prey. (See Batesian mimicry and Müllerian mimicry.)

Mineral An inorganic substance other than water.

Minimal medium A medium for the growth of bacteria, fungi, or tissue cultures, containing only those nutrients absolutely required for the growth of wild-type cells.

Minimum viable population. The smallest number of individuals required for a population to persist in a region.

Mitochondrial matrix The fluid interior of the mitochondrion, enclosed by the inner mitochondrial membrane.

Mitochondrion (my' toe kon' dree un) (plural: mitochondria) [Gr. *mitos*: thread + *chondros*: cartilage, or grain] An organelle that occurs in eukaryotic cells and contains the enzymes of the ctric acid cycle, the respiratory chain, and oxidative phosphorylation. A mitochondrion is bounded by a double membrane.

Mitosis (my toe' sis) [Gr. *mitos*: thread] Nuclear division in eukaryotes leading to the formation of two daughter nuclei each with a chromosome complement identical to that of the original nucleus.

Mitotic center Cellular region that organizes the microtubules for mitosis. In animals a centrosome serves as the mitotic center.

Missense mutation A point mutation resulting in coding for a different amino acid. (Contrast with nonsense mutation, frameshift mutation.)

Mobbing Gathering of calling animals around a predator; their calls and the confusion they create reduce the probability that the predator can hunt successfully in the area.

Modular organism An organism which grows by producing additional units of body construction that are very similar to the units of which it is already composed.

Mole A quantity of a compound whose weight in grams is numerically equal to its molecular weight expressed in atomic mass units. Avogadro's number of molecules: 6.023×10^{23} molecules.

Molecular clock See radiometric clock.

Molecular formula A representation that shows how many atoms of each element are present in a molecule.

Molecular weight The sum of the atomic weights of the atoms in a molecule.

Molecule A particle made up of two or more atoms joined by covalent bonds or ionic attractions.

Mollusk (mol' lusk) A member of the phylum Mollusca, such as a snail, clam, or octopus.

Molting The process of shedding part or all of an outer covering, as the shedding of feathers by birds or of the entire exoskeleton by arthropods.

Moneran (moh neer' un) A member of the kingdom Monera—a bacterium.

Monoclonal antibody Antibody produced in the laboratory from a clone of hybridoma cells, each of which produces the same specific antibody.

Monocot (short for monocotyledon) [Gr. *monos*: one + *kotyledon*: a cup-shaped hollow] Any member of the angiosperm class Monocotyledonae, plants in which the embryo produces but a single cotyledon (seed leaf). Leaves of most monocots have their major veins arranged parallel to each other.

Monecious (mo nee' shus) [Gr.: one house] Organisms in which both sexes are "housed" in a single individual, which produces both eggs and sperm. (In some plants, these are found in different flowers within the same plant.) Examples: corn, peas, earthworms, hydras. (Contrast with dioecious, perfect flower.)

Monohybrid cross A mating in which the parents differ with respect to the alleles of only one locus of interest.

Monomer A small molecule, two or more of which can be combined to form oligomers (consisting of a few monomers) or polymers (consisting of many monomers).

Monophyletic (mon' oh fih leht' ik) [Gk. *monos*: single + *phylon*: tribe] Being descended from a single ancestral stock.

Monosaccharide A simple sugar. Oligosaccharides and polysaccharides are made up of monosaccharides.

Monotreme (mon' oh treem) An egg-laying mammal, in particular a platypus or an anteater, belonging to the subclass Prototheria.

Mor An acidic humus formed on poor soils, especially under conifers. Mor is resistant to decay and may accumulate to considerable depths.

Morphogenesis (more' fo jen' e sis) [Gr. *morphe*: form + *genesis*: origin] The development of form. Morphogenesis is the overall consequence of determination, differentiation, and growth.

Morphology (more fol' o jee) [Gr. *morphe*: form + *logos*: discourse] The scientific study of organic form, including both its development and function.

Mosaic development Pattern of animal embryonic development in which each blastomere contributes a specific part of the adult body. (Contrast with regulative development.)

Motor neuron A neuron carrying information from the central nervous system to an effector such as a muscle fiber.

Motor unit A motor neuron and the set of muscle fibers it controls.

mRNA (See messenger RNA.)

Mucosa (mew koh' sah) An epithelial membrane containing cells that secrete mucus. The inner cell layers of the digestive and respiratory tracts. (Contrast with submucosa.)

Mull A basic humus that forms on rich soils, especially under broad-leaved plants; decays more readily than mor.

Müllerian mimicry The resemblance of two or more unpleasant or dangerous kinds of organisms to each other; the mimicry gives each added protection because potential enemies that learn to avoid members of one group tend to avoid members of the others even though they lack prior experience with them.

Multicellular [L. *multus*: much + *cella*: chamber] Consisting of more than one cell, as for example a multicellular organism. (Contrast with unicellular.)

Multiple fruit A fruit formed from an inflorescence. (Contrast with accessory fruit, aggregate fruit, simple fruit.)

Mutagen (mute' ah jen) [L. *mutare*: change + Gr. *gignomai*: causing] An agent, especially a chemical, that increases the mutation rate.

Mutation In the broad sense, any discontinuous change in the genetic constitution of an organism. In the narrow sense, the word usually refers to a "point mutation," a change along a very narrow portion of the nucleic acid sequence.

Mutation pressure Evolution (change in gene proportions) by different mutation rates alone.

Mutualism The type of symbiosis, such as that exhibited by fungi and algae or cyanobacteria in forming lichens, in which both species profit from the association.

Mycelium (my seel' ee yum) [Gr. *mykes*: fungus] In the fungi, a mass of hyphae.

Mycorrhiza (my' ka rye' za) [Gr. *mykes*: fungus + *rhiza*: root] An association of the root of a plant with the mycelium of a fungus.

Myelin (my' a lin) A material forming a sheath around some axons. It is formed by Schwann cells that wrap themselves about the axon. It serves to insulate the axon electrically and to increase the rate of transmission of a nervous impulse.

Myofibril (my' oh fy' bril) A row of sarcomeres (contractile units) in a muscle cell.

Myogenic (my uh jen' ik) Originating in muscle.

Myoglobin (my' oh globe' in) [Gr. *mys*: muscle + L. *globus*: sphere] The special protein found in muscle, of lower molecular weight and carrying less oxygen than hemoglobin.

Myosin [Gr. *mys*: muscle] One of the two major proteins of muscle, it makes up the thick filaments. (Contrast with actin.)

NAD (nicotinamide adenine dinucleotide) A compound found in all living cells, existing in two interconvertible forms: the oxidizing agent NAD$^+$ and the reducing agent NADH.

NADP (nicotinamide adenine dinucleotide phosphate) Like NAD, but possessing another phosphate group; plays similar roles but is used by different enzymes.

Natal group The group into which an individual was born.

Natural killer cell A small leukocyte that nonspecifically kills certain tumor cells and virus-infected cells in tissue cultures.

Natural selection The differential contribution of offspring to the next generation by various genetic types belonging to the same population. The mechanism of evolution proposed by Charles Darwin.

Negative control The situation in which a regulatory macromolecule (generally a repressor) functions to turn off transcription. In the absence of a regulatory macromolecule, the structural genes are turned on.

Negative feedback A pattern of regulation in which a change in a measured variable results in a correction that opposes the change.

Nekton [Gr. *nekhein*: to swim] Animals, such as fish, that can swim against currents of water. (Contrast with plankton.)

Nematocyst (ne mat' o sist) [Gr. *nema*: thread + *kystis*: cell] An elaborate, threadlike structure produced by cells of jellyfish and other cnidarians, used chiefly to paralyze and capture prey.

Nematode (nem' ah toad) A member of the phylum Nematoda; a roundworm.

Neoteny (nee ot' e nee) [Gr. *neo*: new + *tein*: stretch] Sexual maturation in a larval stage.

Nephridium (nef rid' ee um) [Gr. *nephros*: kidney] An organ which is involved in excretion, and often in water balance, involving a tube that opens to the exterior at one end.

Nephron (nef' ron) [Gr. *nephros*: kidney] The basic component of the kidney, which is made up of numerous nephrons. Its form varies in detail, but it always has at one end a device for receiving a filtrate of blood, and then a tubule that absorbs selected parts of the filtrate back into the bloodstream.

Neritic zone The shallow part of the ocean, situated over the continental shelf.

Nerve A bundle of axons or neurons.

Net primary production Total photosynthesis minus respiration by plants.

Neural plate A thickened strip of ectoderm along the dorsal side of the early vertebrate embryo; gives rise to the central nervous system.

Neurohormone A hormone produced and secreted by neurons.

Neuron (noor' on) [Gr. *neuron*: nerve, sinew] A nerve cell.

Neurotransmitter A substance, produced in and released by one neuron, that diffuses across a synapse and excites or inhibits the postsynaptic neuron.

Neurula (nure' you la) [Gr. *neuron*: nerve] Embryonic stage during formation of the dorsal nerve cord by two ectodermal ridges.

Neutral alleles Alleles that differ so slightly that the proteins for which they code function identically.

Neutron (new' tron) [E.: neutral] One of the three most fundamental particles of matter, with mass approximately 1 amu and no electrical charge.

Nicotinamide adenine dinucleotide (See NAD.)

Nicotinamide adenine dinucleotide phosphate (See NADP.)

Nitrification The oxidation of ammonia to nitrite and nitrate ions, performed by certain soil bacteria.

Nitrogenase In nitrogen-fixing organisms, an enzyme complex that mediates the stepwise reduction of atmospheric N_2 to ammonia.

Nitrogen fixation Conversion of nitrogen gas to ammonia, which makes nitrogen available to living things. Carried out by certain prokaryotes, some of them free-living and others living within plant roots.

Node [L. *nodus*: knob, knot] In plants, a (sometimes enlarged) point on a stem where a leaf or bud is or was attached.

Nomenclature The method of assigning names in the classification of organisms.

Noncompetitive inhibitor An inhibitor that binds the enzyme at a site other than the active site. (Contrast with competitive inhibitor.)

Nondisjunction Failure of sister chromatids to separate in meiosis II or mitosis, or failure of homologous chromosomes to separate in meiosis I. Results in aneuploidy.

Nonpolar molecule A molecule whose electric charge is evenly balanced from one end of the molecule to the other.

Nonsense (chain-terminating) mutation Mutations that change a codon for an amino acid to one of the codons (UAG, UAA, or UGA) that signal termination of translation. The resulting gene product is a shortened polypeptide that begins normally at the amino-terminal end and ends at the position of the altered codon.

Normal flora The bacteria and fungi that live on animal body surfaces without causing disease.

Notochord (no' tow kord) [Gr. *notos*: back + *chorde*: string] A flexible rod of gelatinous material serving as a support in the embryos of all chordates and in the adults of tunicates and lancelets.

Nuclear envelope The surface, consisting of two layers of membrane, that encloses the nucleus of eukaryotic cells.

Nucleic acid (new klay' ik) [E.: nucleus of a cell] A long-chain alternating polymer of deoxyribose or ribose and phosphate groups, with nitrogenous bases (adenine, thymine or uracil, guanine, cytosine) as side chains. DNA and RNA are nucleic acids.

Nucleic acid hybridization A method for estimating the amount of genetic difference between species by measuring the heat stability of DNA molecules composed of one strand from each of the two species.

Nucleoid (new' klee oid) The region that harbors the chromosomes of a prokaryotic cell. Unlike the eukaryotic nucleus, it is not bounded by a membrane.

Nucleolar organizer (new klee' o lar) A region on a chromosome that is associated with the formation of a new nucleolus following nuclear division. The site of the genes that code for ribosomal RNA.

Nucleolus (new klee' oh lus) [from L. diminutive of *nux*: little kernel or little nut] A small, generally spherical body found within the nucleus of eukaryotic cells. The site of synthesis of ribosomal RNA.

Nucleoplasm (new' klee o plazm) The fluid material within the nuclear envelope of a cell, as opposed to the chromosomes, nucleoli, and other particulate constituents.

Nucleosome A portion of a eukaryotic chromosome, consisting of part of the DNA molecule wrapped around a group of histone molecules, and held together by another type of histone molecule. The chromosome is made up of many nucleosomes.

Nucleotide The basic chemical unit (monomer) in a nucleic acid. A nucleotide in RNA consists of one of four nitrogenous bases linked to ribose, which in turn is linked to phosphate. In DNA, deoxyribose is present instead of ribose.

Nucleus (new' klee us) [from L. diminutive of *nux*: kernel or nut] (1) The dense central portion of an atom, made up of protons and neutrons, with a positive charge. Surrounded by a cloud of negatively-charged electrons. (2) Centrally located chamber of eukaryotic cells that is bounded by a double membrane and contains the chromosomes. The information center of the cell.

Nutrient A food substance; or, in the case of mineral nutrients, an inorganic element required for completion of the life cycle of an organism.

Obligate (ob' li gut) Necessary, as in obligate anaerobe. (Contrast with facultative.)

Obligate anaerobe An animal that can live only in oxygenated environments.

Observational analysis A scientific method in which data are gathered in unmanipulated situations to test hypotheses. Often employed in the field where experimental manipulations are difficult or impossible.

Oceanic zone The deeper ocean basins.

Okazaki fragments Newly formed DNA strands making up the lagging strand in DNA replication. DNA ligase links the Okazaki fragments to give a continuous strand.

Olfactory Having to do with the sense of smell.

Oligomer A compound molecule of intermediate size, made up of two to a few monomers. (Contrast with monomer, polymer.)

Omasum (oh may' sum) The third division of the ruminant stomach. Its function is mostly the absorption of wastes. (Contrast with abomasum, rumen.)

Ommatidium [Gr. *omma*: an eye] One of the units which, collected into groups of up to 20,000, make up the compound eye of arthropods.

Omnivore [L. *omnis*: all, everything + *vorare*: to devour] An organism that eats both animal and plant material. (Contrast with carnivore, detritivore, herbivore.)

Oncogenic (ong' co jen' ik) [Gr. *onkos*: mass, tumor + *genes*: born] Causing cancer.

Ontogeny (on toj' e nee) [Gr. *onto*: from "to be" + *gignomai*: be born, produce] The development of a single organism in the course of its life history. (Contrast with phylogeny.)

Oocyte (oh' eh site) [Gr. *oon*: egg + *kytos*: cell] The cell that gives rise to eggs in animals.

Oogenesis (oh' eh jen e sis) [Gr. *oon*: egg + *gignomai*: to be born] Female gametogenesis, leading to production of the egg.

Oogonium (oh' eh go' nee um) In some algae and fungi, a cell in which an egg is produced.

Operator The region of an operon that acts as the binding site for the repressor.

Operculum The hardened portion of the dorsal surface of the foot of many gastropod mollusks; forms a plug that blocks the opening to the shell when the animal is withdrawn inside it.

Operon A genetic unit of transcription, typically consisting of several structural genes that are transcribed together; the operon contains at least two control regions: the promoter and the operator.

Opportunity cost The sum of the benefits an animal forfeits by not being able to perform some other behavior during the time when it is performing a given behavior.

Opsin (op' sin) [Gr. *opsis*, sight] The protein portion of the visual pigment rhodopsin. (See rhodopsin.)

Optical isomers Isomers that differ in the configuration of the four different groups attached to a single carbon atom; so named because solutions of the two isomers rotate the plane of polarized light in opposite directions. The two isomers are mirror images of one another.

Optimal yield The largest rate of increase that a population can sustain in a given environment. There exists a particular population size, less than the carrying capacity, at which this yield is realized.

Order In taxonomy, the category below the class and above the family; a group of related, similar families.

Organ A body part, such as the heart, liver, brain, root, or leaf, composed of different tissues integrated to perform a distinct function for the body as a whole.

Organelles (or' gan els') [L.: little organ] Organized structures that are found in or on cells. Examples: ribosomes, nuclei, mitochondria, chloroplasts, cilia, and contractile vacuoles.

Organic Pertaining to any aspect of living matter, e.g., to its evolution, structure, or chemistry. The term is also applied to any chemical compound that contains carbon.

Organism Any living creature.

Organizer, embryonic A region of an embryo which directs the development of nearby regions. In amphibian early gastrulas, the dorsal lip of the blastopore.

Osmoregulation Regulation of the chemical composition of the body fluids of an organism.

Osmosis The movement of water from one region to another where the water potential is more negative. This is often a region in which the concentration of dissolved molecules or ions is higher, although the effect of dissolved substances may be offset by hydrostatic pressure in cells with semi-rigid walls.

Osmotic potential A property of any solution, resulting from its solute content; it may be zero or have a negative value. A negative osmotic potential tends to cause water to move into the solution; it may be offset by a positive pressure potential in the solution or by a more negative water potential in a neighboring solution. (Contrast with pressure potential.)

Ossicle (ah' sick ul) [L. *os*: bone] The calcified construction unit of echinoderm skeletons.

Osteoblasts (ohs' tee oh blasts) Cells that lay down the protein matrix of bone. (Contrast with osteoclasts.)

Osteoclasts Cells that dissolve bone. (Contrast with osteoblasts.)

Ovary (oh' var ee) Any female organ, in plants or animals, that produces an egg.

Oviduct [L. *ovum*: egg + *ducere*: to lead] In mammals, the tube serving to transport eggs to the uterus or to outside of the body.

Oviparous (oh vip' uh rus) Reproduction in which eggs are released by the female and development is external to the mother's body. (Contrast with viviparous.)

Ovulation The release of an egg from an ovary.

Ovule (oh' vule) [L. *ovulum*: little egg] In plants, an organ that contains a gametophyte and, within the gametophyte, an egg; when it matures, an ovule becomes a seed.

Ovum (oh' vum) [L.: egg] The egg, the female sex cell.

Oxidation (ox i day' shun) Relative loss of electrons in a chemical reaction; either outright removal to form an ion, or the sharing of electrons with substances having a greater affinity for them, such as oxygen. Most oxidation, including biological ones, are associated with the liberation of energy.

Oxidative phosphorylation ATP formation in the mitochondrion, associated with flow of electrons through the respiratory chain.

Oxidizing agent A substance that can accept electrons from another. The oxidizing agent becomes reduced; its partner becomes oxidized.

P generation The individuals that mate in a genetic cross. Their immediate offspring are the F_1 generation.

Pacemaker That part of the heart which undergoes most rapid spontaneous contraction, thus setting the pace for the beat of the entire heart. In mammals, the sinoatrial (SA) node. Also, an artificial device, implanted in the heart, that initiates rhythmic contraction of the organ.

Paleobiology The study of fossil evidence, the comparative biochemistry of living organisms, and conditions on the early Earth to determine the stages in the evolution of life.

Paleobotany The scientific study of fossil plants and all aspects of extinct plant life.

Paleontology (pale' ee on tol' oh jee) [Gr. *palaios*: ancient, old + *logos*: discourse] The scientific study of fossils and all aspects of extinct life.

Palisade parenchyma In leaves, one or several layers of tightly packed, columnar photosynthetic cells, frequently found just below the upper epidermis.

Pancreas (pan' cree us) A gland, located near the stomach of vertebrates, that secretes digestive enzymes into the small intestine and releases insulin into the bloodstream.

Pangaea (pan jee' uh) The single land mass formed when all the continents came together in the Permian period.

Panmyxia (pan mix' ee ya) The condition in which mating is random.

Parabronchi (para brong' kee) Passages in the lungs of birds through which air flows.

Paradigm A general framework within which some scientific discipline (or even the whole Earth) is viewed and within which questions are asked and hypotheses are developed. Scientific revolutions usually involve major paradigm changes.

Parallel evolution The evolution of organisms that were originally very similar and have both evolved in the same direction. (Contrast with convergent evolution.)

Parapatric speciation Development of reproductive isolation among members of a continuous population in the absence of a geographical barrier. (Contrast with geographic, sympatric speciation.)

Parasitism The attacking and consumption of a host organism by predators that are much smaller than it is; attackers may be internal or external. Sometimes but not always kills the host.

Parasitoid A parasite that is so large relative to its host that only one individual or at most a few individuals can live within a single host.

Parasocial route Development of sociality by the aggregation of adults of the same generation to form nonsocial groups around clumped resources. (Contrast with familial route.)

Parasympathetic nervous system A portion of the autonomic (involuntary) nervous system. Activity in the parasympathetic nervous system produces effects such as decreased blood pressure and decelerated heart beat. The neurotransmitter for this system is acetylcholine. (Contrast with sympathetic nervous system.)

Parathyroids A pair of glands involved in controlling calcium balance, among other activities.

Parenchyma (pair eng' kyma) [Gr. *para*: beside + *enchyma*: infusion] A plant tissue composed of relatively unspecialized cells without secondary walls.

Parental investment Investment in one offspring or group of offspring that reduces the ability of the parent to assist other offspring.

Parsimony The principle of preferring the simplest among a set of plausible explanations of a phenomenon. Commonly employed in evolutionary and biogeographic studies.

Parthenocarpy Formation of fruit from a flower without fertilization.

Parthenogenesis (par thu no jen' uh sis) [Gr. *parthenos*: virgin + *genesis*: birth] Development from an unfertilized egg.

Pasteur effect The sharp decrease in rate of glucose utilization when conditions become aerobic.

Pastoralism A nomadic form of human culture based on the tending of herds of domestic animals.

Pathogen (path' o jen) [Gr. *pathos*: suffering + *gignomai*: causing] An organism that causes disease.

Pattern formation In animal embryonic development, the organization of differentiated tissues into specific structures such as wings.

Pedigree The pattern of transmission of a genetic trait in a family.

Pelagic zone (puh ladj' ik) [Gr. *pelagos*: the sea] The open waters of the ocean.

Pellicle (pell' ik el) [L. *pellis*: skin] A thin, filmy covering.

Penetrance Of a genotype, the proportion of individuals with that genotype who show the expected phenotype.

Penis (pee' nis) [L.: tail, penis] The male organ inserted into the female during coitus (sexual intercourse).

PEP carboxylase The enzyme that combines carbon dioxide with PEP to form a 4-carbon dicarboxylic acid at the start of C_4 photosynthesis or of Crassulacean acid metabolism (CAM).

Pepsin [Gr. *pepsis*: digestion] An enzyme, in gastric juice, that digests protein.

Peptide linkage The connecting group in a protein chain, —CO—NH—, formed by removal of water during the linking of amino acids, —COOH to —NH₂. Also called an amide linkage.

Peptidoglycan The cell wall material of many prokaryotes, consisting of a single enormous molecule that surrounds the entire cell.

Perennial (per ren' ee al) [L. *per*: through + *annus*: a year] Referring to a plant that lives from year to year. (Contrast with annual, biennial.)

Perfect flower A flower with both stamens and carpels, therefore hermaphroditic.

Periderm The outer tissue of the secondary plant body, consisting primarily of cork.

Period A minor category in the geological time scale.

Pericycle [Gr. *peri*: around + *kyklos*: ring or circle] In plant roots, tissue just within the endodermis, but outside of the root vascular tissue. Meristematic activity of pericycle cells produces lateral root primordia.

Peristalsis (pair' i stall' sis) [Gr. *peri*: around + *stellein*: place] Wavelike muscular contractions proceeding along a tubular organ, propelling the contents along the tube.

Peritoneum The mesodermal lining of the coelom among coelomate animals.

Permease A protein in membranes that specifically transports a compound or family of compounds across the membrane.

Peroxisome A microbody that houses reactions in which toxic peroxides are formed. The peroxisome isolates these peroxides from the rest of the cell.

Petal In an angiosperm flower, a sterile modified leaf, nonphotosynthetic, frequently brightly colored, and often serving to attract pollinating insects.

Petiole (pet' ee ole) [L. *petiolus*: small foot] The stalk of a leaf.

pH The negative logarithm of the hydrogen ion concentration; a measure of the acidity of a solution. A solution with pH = 7 is said to be neutral; pH values higher than 7 characterize basic solutions, while acidic solutions have pH values less than 7.

Phage (fayj) Short for bacteriophage.

Phagocyte A white blood cell that ingests microorganisms by endocytosis.

Phagocytosis [Gr.: *phagein* to eat; cell-eating] A form of endocytosis, the uptake of a solid particle by forming a pocket of plasma membrane around the particle and pinching off the pocket to form an intracellular particle bounded by membrane. (Contrast with pinocytosis.)

Pharynx [Gr.: throat] The part of the gut between the mouth and the esophagus.

Phenetic systematics Systematics based strictly on phenotypic similarities among living organisms. (Contrast with cladistic systematics; evolutionary systematics.)

Phenogram Graphic representation of phenetic similarities.

Phenotype (fee' no type) [Gr. *phanein*: to show + *typos*: impression] The observable properties of an individual as they have developed under the combined influences of the genetic constitution of the individual and the effects of environmental factors. (Contrast with genotype.)

Pheromone (feer' o mone) [Gr. *phero*: carry + *hormon*: excite, arouse] A chemical substance used in communication between organisms of the same species.

Phloem (flo' um) [Gr. *phloos*: bark] In vascular plants, the food-conducting tissue. It consists of sieve cells or sieve tubes, fibers, and other specialized cells.

Phosphate group The functional group —OPO₃H₂; the transfer of energy from one compound to another is often accomplished by the transfer of a phosphate group.

3-Phosphoglycerate The first product of photosynthesis, produced by the reaction of ribulose bisphosphate with carbon dioxide.

Phospholipids Cellular materials that contain phosphorus and are soluble in organic solvents. An example is lecithin (phosphatidyl choline). Phospholipids are important constituents of cellular membranes. (See lipids.)

Phosphorylation The addition of a phosphate group.

Photoautotroph An organism that obtains energy from light and carbon from carbon dioxide. (Contrast with chemoautotroph, chemoheterotroph, photoheterotroph.)

Photoheterotroph An organism that obtains energy from light but must obtain its carbon from organic compounds. (Contrast with chemoautotroph, chemoheterotroph, photoautotroph.)

Photon (foe' tohn) [Gr. *photos*: light] A quantum of visible radiation; a "packet" of light energy.

Photoperiod (foe' tow peer' ee ud) The duration of a period of light, such as the length of time in a 24-hour cycle in which daylight is present.

Photoperiodism The regulation of processes such as flowering by the changing length of day (or of night).

Photophosphorylation Photosynthetic reactions in which light energy trapped by chlorophyll is used to produce ATP and, in noncyclic photophosphorylation, is used to reduce NADP⁺ to NADPH.

Photorespiration Light-driven uptake of oxygen and release of carbon dioxide, the carbon being derived from the early reactions of photosynthesis.

Photosynthesis (foe tow sin' the sis) [literally, "synthesis out of light"] Metabolic processes, carried out by green plants, by which visible light is trapped and the energy used to synthesize compounds such as ATP and glucose.

Phototropism [Gr. *photos*: light + *trope*: a turning] A directed plant growth response to light.

Phyletic evolution Change through time in a lineage of organisms which is genetically continuous from generation to generation.

Phylogenetic tree Graphic representation of lines of descent among organisms.

Phylogeny (fy loj' e nee) [Gr. *phylon*: tribe, race + *gignomai*: be born, produce] The evolutionary history of a particular group of organisms; also, the diagram of the "family tree" that shows which species may have given rise to others. (Contrast with ontogeny.)

Phylum [Gr. *phylon*: tribe, stock] In taxonomy, a high-level category just beneath kingdom and above the class; a group of related, similar classes.

Physiological time The time required for significant changes in the physiological processes or states within an organism.

Physiology (fiz' ee ol' o jee) [Gr. *physis*: natural form + *logos*: discourse, study] The scientific study of the functions of living organisms and the individual organs, tissues, and cells of which they are composed.

Phytoalexins Substances toxic to fungi, produced by plants in response to fungal infection.

Phytochrome (fy' tow krome) [Gr. *phyton*: plant + *chroma*: color] A plant pigment regulating a large number of developmental and other phenomena in plants; can exist in two different forms, one of which is active and the other is not. Different wavelengths of light can drive it from one form to the other.

Phytoplankton (fy' tow plangk' ton) [Gr. *phyton*: plant + *planktos*: wandering] The autotrophic portion of the plankton, consisting mostly of algae.

Pigment A substance that absorbs visible light.

Pilus (pill' us) [Lat. *pilus*: hair]   A surface appendage by which some bacteria adhere to one another during conjugation.

Pinocytosis [Gr.: drinking cell]   A form of endocytosis; the uptake of liquids by engulfing a sample of the external medium into a pocket of the plasma membrane followed by pinching off the pocket to form an intracellular vesicle. (Contrast with phagocytosis and endocytosis.)

Pistil [L. *pistillum*: pestle]   The female structure of an angiosperm flower, within which the ovules are borne. May consist of a single carpel, or of several carpel fused into a single structure. Usually differentiated into ovary, style, and stigma.

Pit   In botany, a small cavity in a cell wall that is not a complete perforation.

Pith   In plants, relatively unspecialized tissue found within a cylinder of vascular tissue.

Pituitary   A small gland attached to the base of the brain in vertebrates. Its hormones control the activities of other glands. Also known as the hypophysis.

Placenta (pla sen' ta) [Gr. *plax*: flat surface]   The organ, found in most mammals, that provides for the nourishment of the fetus and elimination of the fetal waste products. It is formed by the union of membranes of the mother's uterine lining with the membranes from the fetus.

Placental (pla sen' tal)   Pertaining to mammals of the subclass Eutheria, a group that is characterized by the presence of a placenta and that contains the majority of living species of mammals.

Plankton [Gr. *planktos*: wandering]   The free-floating organisms of the sea and fresh water that for the most part move passively with the water currents. Consisting mostly of microorganisms and small plants and animals. (Contrast with nektons.)

Plant   A member of the kingdom Plantae. Usually multicellular, gaining its nutrition by photosynthesis.

Planula (plan' yew la) [L. *planum*: something flat]   The free-swimming, ciliated larva of the cnidarians.

Plaque (plack) [Fr.: a metal plate or coin]   (1) A circular clearing in a turbid layer (lawn) of bacteria growing on the surface of a nutrient agar gel. Produced by successive rounds of infection initiated by a single bacteriophage. (2) An accumulation of prokaryotic organisms on tooth enamel. Acids produced by the metabolism of these microorganisms can cause tooth decay.

Plasma (plaz' muh) [Gr. *plassein*: to mold]   The liquid portion of blood, in which blood cells and other particulates are suspended.

Plasma cell   An antibody-secreting cell that developed from a B cell. The effector cell of the humoral immune system.

Plasma membrane   The membrane that surrounds the cell, regulating the entry and exit of molecules and ions. Every cell has a plasma membrane.

Plasmid   A DNA molecule distinct from the chromosome(s); that is, an extrachromosomal element. May replicate independently of the chromosome.

Plasmodesma (plural: plasmodesmata) [Gr. *plasma*: formed or molded + *desmos*: band]   A cytoplasmic strand connecting two adjacent plant cells.

Plasmodium   In the noncellular slime molds, a multinucleate mass of protoplasm surrounded by a membrane; characteristic of the vegetative feeding stage.

Plasmolysis (plaz mol' i sis)   Shrinking of the cytoplasm and plasma membrane away from the cell wall, resulting from the osmotic outflow of water. Occurs only in cells with rigid cell walls.

Plastid   Organelle in plants that serves for food manufacture (by photosynthesis) or food storage; bounded by a double membrane.

Platelet   A membrane-bounded body without a nucleus, arising as a fragment of a cell in the bone marrow of mammals. Important to blood-clotting action.

Pleiotropy (plee' a tro pee) [Gr. *pleion*: more]   The determination of more than one character by a single gene.

Poikilotherm (poy' kill o therm) [Gr. *poikilos*: varied + *therme*: heat]   An animal whose body temperature tends to vary with the surrounding environment. (Contrast with homeotherm.)

Point mutation   A mutation that results from a small, localized alteration in the chemical structure of a gene. Such mutations can give rise to wild-type revertants as a result of reverse mutation. In genetic crosses, a point mutation behaves as if it resided at a single point on the genetic map. (Contrast with deletion.)

Polar body   A nonfunctional nucleus produced by meiosis, accompanied by very little cytoplasm. The meiosis which produces the mammalian egg produces in addition three polar bodies.

Polar molecule   A molecule in which the electric charge is not distributed evenly in the covalent bonds.

Polar nucleus   One of two nuclei derived from each end of the angiosperm embryo sac, both of which become centrally located. They fuse with a male nucleus to form the primary triploid nucleus that will prduce the endosperm tissue of the angiosperm seed.

Pollen [L.: fine powder, dust]   The fertilizing element of seed plants, containing the male gametophyte and the gamete, at the stage in which it is shed.

Pollination   Process of transferring pollen from the anther to the receptive surface (stigma) of the ovary in plants.

Polygamy [Gr. *poly*: many + *gamos*: marriage]   A breeding system in which an individual acquires more than one mate. In polyandry, a female mates with more than one male, in polygyny, a male mates with more than one female.

Polymerase chain reaction (PCR)   A technique for the rapid production of millions of copies of a particular stretch of DNA.

Polygenes   Multiple loci whose alleles increase or decrease a continuously variable phenotypic trait.

Polymer   A large molecule made up of similar or identical subunits called monomers. (Contrast with monomer, oligomer.)

Polymorphism (pol' lee mor' fiz um) [Gr. *poly*: many + *morphe*: form, shape]   (1) In genetics, the coexistence in the same population of two distinct hereditary types based on different alleles. (2) In social organisms such as colonial cnidarians and social insects, the coexistence of two or more functionally different castes within the same colony.

Polyp   The sessile, asexual stage in the life cycle of most cnidarians.

Polypeptide   A large molecule made up of many amino acids joined by peptide linkages. Large polypeptides are called proteins.

Polyploid (pol' lee ploid)   A cell or an organism in which the number of complete sets of chromosomes is greater than two.

Polysaccharide   A macromolecule composed of many monosaccharides (simple sugars). Common examples are cellulose and starch.

Polysome   A complex consisting of a thread-like molecule of messenger RNA and several (or many) ribosomes. The ribosomes move along the mRNA, synthesizing polypeptide chains as they proceed.

Polyspermy   The entry of more than one sperm into an ovum at the time of fertilization.

Polytene (pol' lee teen) [Gr. *poly*: many + *taenia*: ribbon]   An adjective describing giant interphase chromosomes, such as those found in the salivary glands of fly larvae. The characteristic, reproducible pattern of bands and bulges seen on these chromosomes has provided a method for preparing detailed chromosome maps of several organisms.

Population   Any group of organisms that coexist at the same time and in the same place and are capable of interbreeding with one another.

Population density   The number of individuals (or modules) of a population in a unit of area or volume.

Population dynamics   The sum of the activities of the members of a population.

Population structure   The proportions of individuals in a population belonging to different age classes (age structure). Also, the distribution of the population in space and the amount of migration between subpopulations.

Population vulnerability analysis   A determination of the risk of extinction of a population given its current size and distribution.

Portal vein   A vein connecting two capillary beds, as in the hepatic portal system.

Positive control   The situation in which a regulatory macromolecule is needed to turn transcription of structural genes on. In its absence, transcription will not occur.

Posterior   Toward or pertaining to the rear.

Postzygotic isolating mechanism   Any factor that reduces the viability of zygotes resulting from matings between individuals of different species.

Precocial (pre koh' shul) [L. *praecox*: early ripening]   Being born at a sufficiently advanced state such that little parental care is required for survival. Most commonly applied to birds and mammals.

Predator An organism that kills and eats other organisms. Predation is usually thought of as involving the consumption of animals by animals, but in the broad usage of ecology it can also mean the eating of plants.

Premating isolating mechanism Any intrinsic isolating mechanism that operates prior to the completion of mating.

Pressure potential The actual physical (hydrostatic) pressure within a cell. (Contrast with osmotic potential, water potential.)

Prey [L. *praeda*: booty] An organism hunted or caught as an energy source.

Prezygotic isolating mechanism A mechanism that reduces the probability that individuals of different species will mate.

Primary growth In plants, growth produced by the apical meristems. (Contrast with secondary growth.)

Primary structure The specific sequence of amino acids in a protein.

Primary succession Succession that begins in an areas initially devoid of life, such as on recently exposed glacial till or lava flows.

Primary wall Cellulose-rich cell wall layers laid down by a growing plant cell.

Primate (pry' mate) A member of the order Primates, such as a lemur, monkey, ape, or human.

Primitive streak A line running axially along the blastodisc, the site of inward cell migration during formation of the three-layered embryo. Formed in the embryos of birds and fish.

Primordium [L. *primordium*: origin] The most rudimentary stage of an organ or other art.

Principle of superposition The generalization that younger rocks lie on top of older rocks unless Earth movements have altered their positions.

Prion See scrapie-associated fibril.

Pro- [Gr.: first, before] A prefix often used in biology to denote a developmental stage that comes first or an evolutionary form that appeared earlier than another. For example, prokaryote, prophase.

Procambium Primary meristem that produces the vascular tissue.

Progesterone A vertebrate female sex hormone.

Prokaryotes (pro kar' ry otes) [Gr. *pro*: before + *karyon*: kernel, nucleus] Organisms whose genetic material is not contained within a nucleus. The bacteria. Considered an earlier stage in the evolution of life than the eukaryotes.

Prometaphase The phase of nuclear division that begins with the disintegration of the nuclear envelope.

Promoter The region of an operon that acts as the initial binding site for RNA polymerase.

Prophage (pro' fayj) The noninfectious units that are linked with the chromosomes of the host bacteria and multiply with them but do not cause dissolution of the cell. Prophage can later enter into the lytic phase to complete the virus life cycle.

Prophase (pro' phase) The first stage of nuclear division, during which chromosomes condense from diffuse, threadlike material to discrete, compact bodies.

Proplastid [Gr. *pro*: before + *plastos*: molded] A plant cell organelle which under appropriate conditions will develop into a plastid, usually the photosynthetic chloroplast. If plants are kept in the dark, proplastids may become quite large and complex.

Proprioceptor [L. *proprius*: own] A receptor that provides information about the position or orientation or stress in an animal's own tissue.

Prosome (pro' soam) [(Gr. *pro*: before + *soma*: body] The anteriormost of the three body divisions of lophophorate animals.

Prostaglandin Any one of a group of specialized lipids with hormone-like functions. It is not clear that they act at any considerable distance from the site of their production.

Prosthetic group Any nonprotein portion of an enzyme.

Protease (pro' tee ase) A proteolytic enzyme.

Protein (pro' teen) [Gr. *protos*: first] One of the most fundamental building substances of living organisms. A long-chain polymer of amino acids with twenty different common side chains. Occurs with its polymer chain extended in fibrous proteins, or coiled into a compact macromolecule in enzymes and other globular proteins.

Proteolytic enzyme An enzyme whose main catalytic function is the digestion of a protein or polypeptide chain. The digestive enzymes trypsin, pepsin, and carboxypeptidase are all proteolytic enzymes.

Prothoracic gland In insects, a gland that produces and secretes ecdysone, the hormone that controls molting.

Protist A member of the kingdom Protista, which consists of those eukaryotes not included in the kingdoms Animalia, Fungi, or Plantae. Many protists are unicellular. The kingdom Protista includes protozoa, algae, and fungus-like protists.

Protocoel (pro' tuh seal) [(Gr. *pro*: before + *koilos*: hollow] The coelomic compartment of a prosome.

Protoderm Primary meristem that gives rise to epidermis.

Proton (pro' ton) [Gr. *protos*: first] One of the three most fundamental particles of matter, with mass approximately 1 amu and an electrical charge of +1.

Protonema (pro' tow nee' mah) [Gr. *protos*: first + *nema*: thread] The hairlike growth form that constitutes an early stage in the development of a moss gametophyte.

Protonephridia (singular, protonephridium) The excretory organs of flatworms.

Proto-oncogenes The normal alleles of genes possessing oncogenes (cancer-causing genes) as mutant alleles. Proto-oncogenes encode growth factors and receptor proteins.

Protoplast A cell which would normally have a cell wall, but from which the wall has been removed by enzymatic digestion or by special growth conditions.

Protostome One of two major lines of animal evolution, characterized by spiral, determinate cleavage of the egg, and by schizocoelous development. (Contrast with deuterostome.)

Prototroph (pro' tow trofe') [Gr. *protos*: first + *trophein*: to nourish] The nutritional wild-type, or reference form, of an organism. Any deviant form that requires growth nutrients not required by the prototrophic form is said to be a nutritional mutant, or auxotroph.

Protozoa A group of single-celled organisms classified by some biologists as a single phylum; includes the flagellates, amoebas, and ciliates. This textbook follows most modern classifications in elevating the protozoans to a distinct kingdom (Protista) and each of their major subgroups to the rank of phylum.

Provincialized A biogeographic term referring to the separation, by environmental barriers, of the biota into units with distinct species compositions.

Provirus See prophage.

Proximal Near the point of attachment or other reference point. (Contrast with distal.)

Pseudocoelom A body cavity not surrounded by a peritoneum. Characteristic of nematodes and rotifers.

Pseudogene A DNA segment that is homologous to a functional gene but contains a nucleotide change that prevents its expression.

Pseudoplasmodium [Gr. *pseudes*: false + *plasma*: mold or form] In the cellular slime molds such as *Dictyostelium*, an aggregation of single amoeboid cells. Occurs prior to formation of a fruiting structure.

Pseudopod (soo' do pod) [Gr. *pseudes*: false + *podos*: foot] A temporary, soft extension of the cell body that is used in location, attachment to surfaces, or engulfing particles.

Pulmonary Pertaining to the lungs.

Pupa (pew' pa) [L.: doll, puppet] In certain insects (the Holometabola), the encased developmental stage that intervenes between the larva and the adult.

Purine (pure' een) A type of nitrogenous base. The purines adenine and guanine are found in nucleic acids.

Purkinje fibers Specialized heart muscle cells that conduct excitation throughout the ventricular muscle.

Pyramid of biomass Graphical representation of the total masses at different trophic levels in an ecosystem.

Pyramid of energy Graphical representation of the total energy contents at different trophic levels in an ecosystem.

Pyrimidine (peer im' a deen) A type of nitrogenous base. The pyrimidines cytosine, thymine, and uracil are found in nucleic acids.

Pyrogen A substance that causes fever.

Pyruvate A three-carbon acid; the end product of glycolysis and the raw material for the citric acid cycle.

Quantum (kwon' tum) [L. *quantus*: how great] An indivisible unit of energy.

Quaternary structure Of aggregating proteins, the arrangement of polypeptide subunits.

R factor (resistance factor) A plasmid that contains one or more genes that encode resistance to antibiotics.

Radial symmetry The condition in which two halves of a body are mirror images of each other regardless of the angle of the cut, providing the cut is made along the center line. Thus, a cylinder cut lengthwise down its center displays this form of symmetry. (Contrast with bilateral symmetry.)

Radiata Animal phyla with radial symmetry throughout their development and their evolutionary history; the cnidarians and the ctenophores.

Radioisotope A radioactive isotope of an element. Examples are carbon-14 (^{14}C) and hydrogen-3, or tritium (^{3}H).

Radiometric clock The use of the regular, known rates of decay of radioisotopes of elements to determine dates of events in the distant past.

Radula The toothed feeding organ of many mollusks used to scrape prey from hard substrates.

Rain shadow A region of low precipitation on the leeward side of a mountain range.

Ramet The repeated morphological units of sessile, modular organisms. (Contrast with genet.)

Random drift Evolution (change in gene proportions) by chance processes alone.

Rate constant Of a particular chemical reaction, a constant which, when multiplied by the concentration(s) of reactant(s), gives the rate of the reaction.

Reaction, chemical A process in which atoms combine or change bonding partners.

Reaction wood Modified wood produced in branches in response to gravitational stimulation. Gymnosperms produce compression wood that tends to push the branch up; angiosperms produce tension wood that tends to pull the branch up.

Realized niche The actual niche occupied by an organism; it differs from the fundamental niche because of the presence of other species.

Receptacle [L. *receptaculum*: reservoir] In an angiosperm flower, the end of the stem to which all of the various flower parts are attached.

Receptive field Of a neuron, the area on the retina from which the activity of that neuron can be influenced.

Receptor-mediated endocytosis A form of endocytosis in which macromolecules in the environment bind specific receptor proteins in the plasma membrane and are brought into the cell interior in coated vesicles.

Recessive See dominance.

Reciprocal altruism The exchange of altruistic acts between two or more individuals. The acts may be separated considerably in time.

Recombinant An individual, meiotic product, or single chromosome in which genetic materials originally present in two individuals end up in the same haploid complement of genes. The reshuffling of genes can be either by independent segregation, or by crossing over between homologous chromosomes.

For example, a human may pass on genes from both parents in a single haploid gamete.

Recombinant DNA technology The application of genetic tools (restriction endonucleases, plasmids, and transformation) to the production of specific proteins by biological "factories" such as bacteria.

Rectum The terminal portion of the gut, ending at the anus.

Redirected activity The direction of some behavior, such as aggression, away from the primary target and toward another, less appropriate object.

Redox reaction A chemical reaction in which one reactant becomes oxidized and the other becomes reduced.

Reducing agent A substance that can donate electrons to another substance. The reducing agent becomes oxidized, and its partner becomes reduced.

Reduction (re duk' shun) Gain of electrons; the reverse of oxidation. Most reductions lead to the storage of chemical energy, which can be released later by an oxidation reaction. Energy storage compounds such as sugars and fats are highly reduced compounds.

Reflex An automatic action, involving only a few neurons (in vertebrates, often in the spinal cord), in which a motor response swiftly follows a sensory stimulus.

Refractory period Of a neuron, the time interval after an action potential, during which another action potential cannot be elicited.

Region In biogeography, a major division of the world distinguished by its peculiar animals or plants. For example, Africa south of the Sahara is recognized as constituting the Ethiopian region.

Regulative development A pattern of animal embryonic development in which the fates of the first blastomeres are not absolutely fixed. (Contrast with mosaic development.)

Regulator gene A gene that contains the information for making a regulatory macromolecule, often a repressor protein.

Releaser A sign stimulus used in communication. Often the term is used broadly to include any sign stimulus.

Releasing factor One of several hypothalamic hormones that stimulates the secretion of anterior pituitary hormone.

Renal [L. *renes*: kidneys] Relating to the kidneys.

Replication fork A point at which a DNA molecule is replicating. The fork forms by the unwinding of the parent molecule.

Repressible enzyme An enzyme whose synthesis can be decreased or prevented by the presence of a particular compound.

Repressor A protein coded by the regulator gene. The repressor can bind to a specific operator and prevent transcription of the operon.

Reproductive value The expected contribution of an individual of a particular age to the future growth of the population to which it belongs.

Resolving power Of an optical device such as a microscope, the smallest distance between two lines that allows the lines to be seen as separate from one another.

Resource Something in the environment required by an organism for its maintenance and growth that is consumed in the process of being used.

Resource defense polygamy A breeding system in which individuals of one sex (usually males) defend resources that are attractive to individuals of the other sex (usually females); individuals holding better resources attract more mates.

Respiration (res pi ra' shun) [L. *spirare*: to breathe] (1) Cellular respiration; the oxidation of the end products of glycolysis with the storage of much energy in ATP. The oxidant in the respiration of eukaryotes is oxygen gas. Some bacteria can use nitrate or sulfate instead of O_2. (2) Breathing.

Respiratory chain The terminal reactions of cellular respiration, in which electrons are passed from NAD or FAD, through a series of intermediate carriers, to molecular oxygen, with the concomitant production of ATP.

Resting potential Of any living cell, the difference in electrical potential between the inside and outside of the cell. In cells at rest, the interior is negative to the exterior. (Contrast with electrotonic potential, action potential.)

Restoration ecology The science and practice of restoring damaged or degraded ecosystems.

Restriction endonuclease Any one of several enzymes, produced by bacteria, that break foreign DNA molecules at very specific sites. Some produce "sticky ends." Extensively used in recombinant DNA technology.

Restriction map A partial genetic map of a DNA molecule, showing the points at which particular restriction endonuclease recognition sites reside.

Retina (rett' in uh) [L. *rete*: net] The light-sensitive layer of cells in the vertebrate or cephalopod eye.

Retrovirus An RNA virus that contains reverse transcriptase. Its RNA serves as a template for cDNA production, and the cDNA is integrated into a chromosome of the mammalian host cell.

Reverse transcriptase An enzyme that catalyzes the production of DNA (cDNA), using RNA as a template; essential to the reproduction of retroviruses.

Reversion (genetic) A mutational event that restores wild-type phenotype to a mutant.

Rhizoids (rye' zoids) [Gr. *rhiza*: root] Hair-like extensions of cells in mosses, liverworts, and a few vascular plants that serve the same function as roots and root hairs in vascular plants. The term is also applied to branched, rootlike extensions of some fungi and algae.

Rhizome (rye' zome) [Gr. *rhizoma*: mass of roots] A special underground stem (as opposed to root) that runs horizontally beneath the ground.

Rhodopsin A photopigment used in the visual process of transducing photons of light into changes in membrane potential of photoreceptive cells.

Rhodopsin A photopigment used in the visual process of transducing photons of light into changes in membrane potential of photoreceptive cells.

Ribonucleic acid See RNA.

Ribose (rye' bose) A sugar of chemical formula $C_5H_{10}O_5$, one of the building blocks of ribonucleic acids.

Ribosomal RNA (rRNA) Several species of RNA that are incorporated into the ribosome.

Ribosome A small organelle that is the site of protein synthesis.

Ribozyme An RNA molecule with catalytic activity.

Ribulose 1,5-bisphosphate (RuBP) The compound in chloroplasts which reacts with carbon dioxide in the first reaction of the Calvin–Benson cycle.

Risk cost The increased chance of being injured or killed as a result of performing a behavior, compared to resting.

RNA (ribonucleic acid) A nucleic acid using ribose. Various classes of RNA are involved in the transcription and translation of genetic information. RNA serves as the genetic storage material in some viruses.

RNA polymerase An enzyme that catalyzes the formation of RNA from a DNA template.

RNA splicing The last stage of RNA processing in eukaryotes, in which the transcripts of introns are excised through the action of small nuclear ribonucleoprotein particles (snRNP).

Rods Light-sensitive cells in the retina. (Contrast with cones.)

Root cap A thimble-shaped mass of cells, produced by the root apical meristem, that protects the meristem and that is the organ that perceives the gravitational stimulus in root gravitropism.

Root hair A specialized epidermal cell with a long, thin process that absorbs water and minerals from the soil solution.

Root pressure Pressure in the xylem resulting from active transport of minerals by transfer cells. Results in guttation. Inadequate to account for the long-distance movement of water up the stem.

Round dance The dance performed on the vertical surface of a honeycomb by a returning honeybee forager when she has discovered a food source less than 100 meters from the hive.

rRNA See ribosomal RNA.

Rubisco (RuBP carboxylase) Enzyme that combines carbon dioxide with ribulose bisphosphate to produce 3-phosphoglycerate, the first product of C_3 photosynthesis. The most abundant protein on Earth.

Rumen (rew' mun) The first division of the ruminant stomach. It mainly serves for the storage of food and the initiation of bacterial fermentation. Food is regurgitated from the rumen for further chewing. (Contrast with abomasum, omasum.)

Ruminant A herbivorous, cud-chewing mammal such as a cow, sheep, or deer, having a stomach consisting of four compartments.

S phase In the cell cycle, the stage of interphase during which DNA is replicated. (Contrast with G_1 phase, G_2 phase.)

Sap An aqueous solution of nutrients, minerals, and other substances that passes through the xylem of plants.

Saprobe [Gr. *sapros*: rotten + *bios*: life] An organism (usually a bacterium or fungus) that obtains its carbon and energy directly from dead organic matter.

Sarcomere (sark' o meer) [Gr. *sark*: flesh + *meros*: a part] The contractile unit of a skeletal muscle.

Sarcoplasm The cytoplasm of muscle cells.

Sarcoplasmic reticulum The endoplasmic reticulum of muscle cells. It acts to store and release calcium in the control of muscle contraction.

Saturated hydrocarbon A compound consisting only of carbon and hydrogen, with the hydrogen atoms connected by single bonds.

Schizocoelous development Formation of a coelom during embryological development by a splitting of mesodermal masses.

Schwann cell A glial cell that wraps around part of the axon of a peripheral neuron, creating a myelin sheath.

Sclereid A type of sclerenchyma cell, commonly found in nutshells, that is not elongated.

Sclerenchyma (skler eng' kyma) A plant tissue composed of cells with heavily thickened cell walls, dead at functional maturity. The principal types of sclerenchyma cells are fibers and sclereids.

Scrapie-associated fibril A type of protein fibril found in nervous tissues of mammals infected with certain diseases, notably scrapie, kuru, and Creutzfeld-Jacob disease. Little is known about these fibrils, including whether they are the causal agents of the diseases.

Secondary growth In plants, growth produced by vascular and cork cambia, contributing to an increase in girth. (Contrast with primary growth.)

Secondary succession Succession that begins following a disturbance that does not eliminate organisms from the site.

Secondary wall Wall layers laid down by a plant cell that has ceased growing; often impregnated with lignin or suberin.

Second law of thermodynamics States that in any real (irreversible) process, there is a decrease in free energy and an increase in entropy.

Second messenger A compound, such as cyclic AMP, that is released within a target cell after a hormone or other "first messenger" has bound to a surface receptor on a cell; the second messenger triggers further reactions within the cell.

Secondary structure Of a protein, localized regularities of structure, such as the α-helix and the β-pleated sheet.

Secretin (si kreet' in) A peptide hormone secreted by the upper region of the small intestine when acidic chyme is present. Stimulates the pancreatic ducts to secrete bicarbonate ions.

Secretion In the kidney, addition of a material to the urine after the glomerular filtration step.

Section A thin slice, usually for microscopy, as a cross section.

Sediment survival The volume of sedimentary rocks preserved and recognized per million years of geologic time.

Seed A fertilized, ripened ovule of a gymnosperm or angiosperm. Consists of the embryo, nutritive tissue, and a seed coat.

Seed crop The number of seeds produced by a plant during a particular bout of reproduction.

Seedling A young plant that has grown from a seed (rather than by grafting or by other means.)

Segmentation genes In insect larvae, genes that determine the number and polarity of larval segments.

Segregation (genetic) The separation of alleles, or of homologous chromosomes, from one another during meiosis so that each of the haploid daughter nuclei produced by meiosis contains one or the other member of the pair found in the diploid mother cell, but never both.

Selective permeability A characteristic of a membrane, allowing certain substances to pass through while other substances are excluded.

Self-differentiating Behavior that develops without experience with the normal objects toward which it is usually directed and without any practice. (See also instinct.)

Selfish act A behavioral act that benefits its performer but harms the recipients.

Self-pollination The fertilization of a plant by its own pollen. (Contrast with cross-pollination.)

Semelparous organism An organism that reproduces only once in its lifetime. (Contrast with iteroparous.)

Semen (see' men) [L.: seed] The thick, whitish liquid produced by the male reproductive organ in mammals, containing the sperm.

Semiconservative replication The common way in which DNA is synthesized. Each of the two partner strands in a double helix acts as a template for a new partner strand. Hence, after replication, each double helix consists of one old and one new strand.

Senescence [L. *senescere*: to grow old] Aging; deteriorative changes with aging.

Sensory neuron A neuron leading from a sensory cell to the central nervous system. (Contrast with motor neuron.)

Sepal (see' pul) One of the outermost structures of the flower, usually protective in function and enclosing the rest of the flower in the bud stage.

Septum [L.: partition] A membrane or wall between two cavities.

Serum That part of the blood plasma that remains after clots have formed and been removed.

Sessile (sess' ul) [L. *sedere*: to sit] Permanently attached; not moving.

Sex chromosome In organisms with a chromosomal mechanism of sex determination, one of the chromosomes involved in sex determination. One sex chromosome, the X chromosome, is present in two copies in one sex and only one copy in the other sex. The autosomes, as opposed to the sex chromosomes, are present in two copies in both sexes. In many organisms, there is a second sex chromosome, the Y chromosome, that is found in only one sex—the sex having only one copy of the X.

Sexduction See F-duction.

Sex linkage The pattern of inheritance characteristic of genes located on the sex chromosomes of organisms having a chromosomal mechanism for sex determination. The sex that is diploid with respect to sex chromosomes can assume three genotypes: homozygous wild-type, homozygous mutant, or heterozygous carrier. The other sex, haploid for sex chromosomes, is either hemizygous wild-type or hemizygous mutant.

Sexuality The ability, by any of a multitude of mechanisms, to bring together in one individual genes that were originally carried by two different individuals. The capacity for genetic recombination.

Sexual selection Selection by one sex of characteristics in individuals of the opposite sex. Also, the favoring of characteristics in one sex as a result of competition among individuals of that sex for mates.

Shoot The aerial part of a vascular plant, consisting of the leaves, stem(s), and flowers.

Short-day plant A photoperiodically-sensitive plant that flowers upon exposure to long night.

Sibling A brother or sister.

Sibling species Species that are so similar that they are difficult to distinguish from one another.

Sieve plate In sieve tubes, the highly specialized end walls in which are concentrated the clusters of pores through which the protoplasts of adjacent sieve tube elements are interconnected.

Sieve tube A column of specialized cells found in the phloem, specialized to conduct organic matter from sources (such as photosynthesizing leaves) to sinks (such as roots). Found principally in flowering plants.

Sieve tube element A single cell of a sieve tube, containing cytoplasm but relatively few organelles, with highly specialized perforated end walls leading to elements above and below.

Sign stimulus The single stimulus, or one out of a very few stimuli, by which an animal distinguishes key objects, such as an enemy, or a mate, or a place to nest, etc.

Simple fruit A fruit that develops from a single ovary. (Contrast with accessory fruit, aggregate fruit, multiple fruit.)

Sinoatrial node (sigh' no ay' tree al) The pacemaker of the mammalian heart.

Sinus (sigh' nus) [L. *sinus*: a bend, hollow] A cavity in a bone, a tissue space, or an enlargement in a blood vessel.

Small nuclear ribonucleoprotein particle (snRNP) A complex of an enzyme and a small nuclear RNA molecule, functioning in RNA splicing.

Social insect One of the kinds of insect that form colonies with reproductive castes and worker castes; in particular, the termites, ants, social bees, and social wasps.

Society A group of individuals belonging to the same species and organized in a cooperative manner; in the broadest sense, includes parents and their offspring.

Sociobiology The scientific study of animal societies and communication.

Sodium–potassium pump The complex protein in plasma membranes that is responsible for primary active transport; it pumps sodium ions out of the cell and potassium ions into the cell, both against their concentration gradients.

Solute A substance that is dissolved in a liquid (solvent).

Solution A liquid (solvent) and its dissolved solutes.

Solvent A liquid that has dissolved or can dissolve one or more solutes.

Somatic Pertaining to the body, or body cells (rather than to germ cells).

Somite (so' might) [Gr. *soma*: body] One of the segments into which an embryo becomes divided longitudinally, leading to the eventual segmentation of the animal as illustrated by the spinal column, ribs, and associated muscles.

Sonar The mode of orientation used by bats, porpoises, and a few other animals in which the positions of objects are estimated by emitting sounds and listening for the echoes that bounce back from them.

Spatial summation In the production or inhibition of action potentials in a postsynaptic neuron, the interaction of depolarizations and hyperpolarizations produced by several terminal boutons.

Spawning The direct release of sex cells into the water.

Speciation (spee' shee ay' shun) The process of splitting one population into two populations that are reproductively isolated from one another.

Species (spee' shees) [L.: kind] The basic lower unit of classification, consisting of a population or series of populations of closely related and similar organisms. The more narrowly defined "biological species" consists of individuals capable of interbreeding freely with each other but not with members of other species.

Species-area curve The graphical representation of the relation between the sizes of geographic areas and the number of species that inhabit them.

Species diversity A weighted representation of the species of organisms living in a region; large and common species are given greater weight than are small and rare ones. (Contrast with species richness.)

Species equilibrium A condition in which the number of species going extinct in a given area per unit time equals the rate at which new species are arriving; thus the number of species in the area remains constant.

Species pool All the species potentially available to colonize a particular habitat.

Species richness The number of species of organisms living in a region. (Contrast with species diversity.)

Specific heat The amount of energy that must be absorbed by a gram of a substance to raise its temperature by one degree centigrade. By convention, water is assigned a specific heat of one.

Sperm [Gr. *sperma*: seed] A male reproductive cell.

Spermatocyte (spur mat' oh site) [Gr. *sperma*: seed + *kytos*: cell] The cell that gives rise to the sperm in animals.

Spermatogenesis (spur mat' oh jen' e sis) [Gr. *sperma*: seed + *gignomai*: to be born] Male gametogenesis, leading to the production of sperm.

Spermatogonia Undifferentiated germ cells that give rise to primary spermatocytes and hence to sperm.

Sphincter (sfingk' ter) [Gr. *sphinkter*: that which binds tight] A ring of muscle that can close an orifice, for example, at the anus.

Spindle apparatus An array of microtubules stretching from pole to pole of a dividing nucleus and playing a role in the movement of chromosomes at nuclear division. Named for its shape.

Spiracle (spy' rih kel) [L. *spirare*: to breathe] An opening of the tracheal respiratory systems of terrestrial arthropods.

Spiteful act A behavioral act that harms both the actor and the recipient of the act.

Spongy parenchyma In leaves, a layer of loosely packed photosynthetic cells with extensive intercellular spaces for gas diffusion. Frequently found between the palisade parenchyma and the lower epidermis.

Spontaneous generation The idea that life is generated continually from nonliving matter. Usually distinguished from the current idea that life evolved from nonliving matter under primordial conditions at an early stage in the history of earth.

Spontaneous reaction A chemical reaction which will proceed on its own, without any outside influence. A spontaneous reaction need not be rapid.

Sporangiophore [Gr. *phore*: to bear] Any branch bearing one or more sporangia.

Sporangium (spor an' gee um) [Gr. *spora*: seed + *angeion*: vessel or reservoir] In plants and fungi, any specialized stucture within which one or more spores are formed.

Spore [Gr. *spora*: seed] Any asexual reproductive cell capable of developing into an adult plant without gametic fusion. Haploid spores develop into gametophytes, diploid spores into sporophytes. In prokaryotes, a resistant cell capable of surviving unfavorable periods.

Sporophyll (spor' o fill) [Gr. *spora*: seed + *phyllon*: leaf] Any leaf or leaflike structure that bears sporangia; refers to carpels and stamens of angiosperms and to sporangium-bearing leaves on ferns, for example.

Sporophyte (spor' o fyte) [Gr. *spora*: seed + *phyton*: plant] In plants with alternation of generations, the diploid phase that produces the spores. (Contrast with gametophyte.)

Stabilizing selection Selection against the extreme phenotypes in a population, so that the intermediate types are favored. (Contrast with disruptive selection.)

Stamen (stay' men) [L.: thread] A male (pollen-producing) unit of a flower, usually composed of an anther, which bears the pollen, and a filament, which is a stalk supporting the anther.

Stapes (stay' peez) [L. *stapia*: stirrup] The third of the middle ear bones of mammals. It is attached to the membrane of the oval window of the inner ear and causes the window to move in response to vibrations of the eardrum. (Contrast with incus, malleus.)

Starch [O.E. *stearc*: stiff, from Gr. *stereos*: a solid] An α-linked polymer of glucose; used by plants as a means of storing energy and carbon atoms.

Stasis Period during which little or no evolutionary change takes place within a lineage or groups of lineages.

Stele (steel) [Gr. *stele*: pillar] The central cylinder of vascular tissue in a plant stem.

Step cline A sudden change in one or more traits of a species along a geographical gradient.

Steroid Any of numerous lipids based on a 17-carbon atom ring system.

Sticky ends On a piece of two-stranded DNA, short, complementary, one-stranded regions produced by the action of a restriction endonuclease. Sticky ends allow the joining of segments of DNA from different sources.

Stigma [L.: mark, brand] The part of the pistil at the apex of the style, which is receptive to pollen, and on which pollen germinates.

Stolon A horizontal stem that forms roots at intervals.

Stoma (plural: stomata) [Gr. *stoma*: mouth] Small opening in the plant epidermis that permits gas exchange; bounded by a pair of guard cells whose osmotic status regulates the size of the opening.

Stratosphere The part of the atmosphere above the troposphere; extends upward to approximately 50 kilometers above the surface of the earth; contains very little water.

Stratum (plural **strata**) A layer or sedimentary rock laid down at a particular time in a past.

Strobilus (strobe' a lus) [Gr. *strobilos*: a cone] The cone, or characteristic multiple fruit, of the pine and other gymnosperms. Also, a cone-shaped mass of sprophylls found in club mosses.

Stroma The fluid contents of an organelle, such as a chloroplast.

Stromatolite A composite, flat-to-domed structure composed of successive mineral layers. Some are known to be produced by the action of bacteria in salt or fresh water, and some ancient ones are considered to be evidence for early life on Earth.

Structural formula A representation of the positions of atoms and bonds in a molecule.

Structural gene A gene that encodes the primary structure of a protein.

Style [Gr. *stylos*: pillar or column] In flowering plants, a column of tissue extending from the tip of the ovary, and bearing the stigma or receptive surface for pollen at its apex.

Sub- [L.: under] A prefix often used to designate a structure that lies beneath another or is less than another. For example, subcutaneous, subspecies.

Suberin A waxy material serving as a waterproofing agent in cork and in the Casparian strips of the endodermis in plants.

Submucosa (sub mue koe' sah) The tissue layer just under the epithelial lining of the lumen of the digestive tract. (Contrast with mucosa.)

Subspecies (sub' spee shees) A subdivision of a species. Usually defined more narrowly as a geographical race: a population or series of populations occupying a discrete range and differing genetically from other geographical races of the same species.

Substrate (sub' strayte) The molecule or molecules on which an enzyme exerts catalytic action.

Succession In ecology, the gradual, sequential series of changes in species composition of a community following a disturbance.

Supercoiling Coiling on coiling, as in DNA during prophase.

Supernormal stimulus Any stimulus, or any intensity of a variable stimulus, that is preferred by animals over the natural sign stimulus.

Suppressor T cells T cells that inhibit the responses of B cells and other T cells to antigens. (Contrast with cytotoxic T cells, helper T cells.)

Surface tension A measure of the cohesiveness of the surface of a liquid. As a result of hydrogen bonding, water has a very high surface tension, allowing some insects to walk on the water surface.

Surface-to-volume ratio For any cell, organism, or geometrical solid, the ratio of surface area to volume; this is an important factor in setting an upper limit on the size a cell or organism can attain.

Surfactant A substance that decreases the surface tension of a liquid. Lung surfactant, secreted by cells of the alveoli, is mostly phosopholipid; it decreases the amount of work necessary to inflate the lungs.

Survivorship curve A plot of the logarithm of the fraction of individuals still alive, as a function of time.

Suspensor In plants, a cell or group of cells derived from the zygote, but not actually part of the embryo proper, which in some seed plants pushes the young embryo deeper into nutritive gametophyte tissue or endosperm by its growth.

Swim bladder An internal gas-filled organ that helps fishes maintain their position in the water column; later evolved into an organ for gas exchange in some lineages.

Symbiosis (sim' bee oh' sis) [Gr.: to live together] The living together of two or more species in a prolonged and intimate ecological relationship. (See parasitism, commensalism, mutualism.)

Symmetry In biology, the property that two halves of an object are mirror images of each other. (See bilateral symmetry and radial symmetry.)

Sympathetic nervous system A division of the autonomic (involuntary) nervous system. Its activities include increasing blood pressure and acceleration of the heartbeat. The neurotransmitter at the sympathetic terminals is epinephrine or norepinephrine. (Contrast with parasympathetic nervous system.)

Sympatric (sim pat' rik) [Gr. *syn*: together + *patria*: homeland] Referring to populations whose geographic regions overlap at least in part.

Sympatric speciation Formation of new species even though members of the daughter species overlap in their distribution during the speciation process. (Contrast with geographic, parapatric speciation.)

Symplast The continuous meshwork of the interiors of living cells in the plant body, resulting from the presence of plasmodesmata. (Contrast with apoplast.)

Symport A membrane transport protein that carries two substances in the same direction across the membrane. (Contrast with antiport.)

Synapse (sin' aps) [Gr. *syn*: together + *haptein*: to fasten] The narrow gap between the terminal bouton of one neutron and the dendrite or cell body of another.

Synapsis (sin ap' sis) The highly specific parallel alignment (pairing) of homologous chromosomes during the first division of meiosis.

Synaptic vesicle A membrane-bounded vesicle, containing neurotransmitter, which is produced in and discharged by the presynaptic neuron.

Synergids (sin nur' jids) Two cells found close to the egg cell in the angiosperm embryo sac; they disappear shortly after fertilization.

Syngamy (sing' guh mee) [Gr. *sun-*: together + *gamos*: marriage] Union of gametes. Also known as fertilization.

Syrinx (sear' inks) [Gr.: pipe, cavity] A specialized structure at the junction of the trachea and the primary bronchi leading to the lungs. The vocal organ of birds.

Systematics The scientific study of the diversity of life.

Systemic circulation The part of the circulatory system serving those parts of the body other than the lungs or gills.

Systole (sis' tuh lee) [Gr.: contraction] Contraction of a chamber of the heart, driving blood forward in the circulatory system.

T cell A type of lymphocyte, involved in the cellular immune response. The final stages of its development occur in the thymus gland. (Contrast with B cell; see also cytotoxic T cell, helper T cell, suppressor T cell.)

T cell receptor A protein on the surface of a T cell that recognizes the antigenic determinant for which the cell is specific.

Target cell A cell which has the appropriate receptors to bind and respond to a particular hormone or other chemical mediator.

TATA box An eight-base-pair sequence, found about 25 base pairs before the starting point for transcription in many eukaryotic promoters, that binds a transcription factor and thus helps initiate transcription.

Taxis (tak' sis) [Gr. *taxis*: arrange, put in order] The movement of an organism in a particular direction with reference to a stimulus. A taxis usually involves the employment of one sense and a movement directly toward or away from the stimulus, or else the maintenance of a constant angle to it. Thus a positive phototaxis is movement toward a light source, negative geotaxis is movement upward (away from gravity), and so on.

Taxon A unit in a taxonomic system.

Taxonomic key A guide to the identification of an unknown specimen. The most useful keys are dichotomous. (See natural key; artificial key.)

Taxonomy (taks on' oh me) [Gr. *taxis*: arrange, classify] The science of classification of organisms.

Telophase (tee' lo phase) [Gr. *telos*: end] The final phase of mitosis or meiosis during which chromosomes became diffuse, nuclear envelopes reform, and nucleoli begin to reappear in the daughter nuclei.

Template In biochemistry, a molecule or surface upon which another molecule is synthesized in complementary fashion, as in the replication of DNA. In the brain, a pattern that responds to a normal input but not to incorrect inputs.

Temporal summation In the production or inhibition of action potentials in a postsynaptic neuron, the interaction of depolarizations or hyperpolarizations produced by rapidly repeated stimulation of a single point.

Tendon A collagen-containing band of tissue that connects a muscle with a bone.

Tension wood See reaction wood.

Tepal In an angiosperm flower, a sterile modified leaf. This term is used to refer to such flower parts when one is unable to distinguish between petals and sepals.

Terrestrial (ter res' tree al) [L. *terra*: earth] Pertaining to the land. (Contrast with aquatic, marine.)

Territory A fixed area from which an animal or group of animals excludes other members of the same species by aggressive behavior or display.

Tertiary structure In reference to a protein, the relative locations in three-dimensional space of all the atoms in the molecule. The overall shape of a protein. (Contrast with primary, secondary, and quaternary structures.)

Test cross A cross of a dominant-phenotype individual (which may be either heterozygous or homozygous) with a homozygous-recessive individual.

Testis (tes' tis) (plural: testes) [L.: witness] The male gonad; that is, the organ that produces the male sex cells.

Tetanus [Gr. *tetanos*: stretched] (1) In physiology, a state of sustained, maximal muscular contraction caused by rapidly repeated stimulation. (2) In medicine, an often-fatal disease ("lockjaw") caused by the bacterium *Clostridium tetani*.

Thalamus A region of the vertebrate forebrain; involved in integration of sensory input.

Thallus (thal' us) [Gr.: sprout] Any algal body which is not differentiated into root, stem, and leaf.

Thermocline Zone, in a body of water, where the temperatures change abruptly to about 4°C.

Thermogenin A protein in the mitochondria of brown fat that uncouples oxidative phosphorylation and enables this tissue to produce heat without producing large amounts of ATP.

Thermoreceptor A cell or structure that responds to changes in temperature.

Thorax In an insect, the middle region of the body, between the head and abdomen. In mammals, the part of the body between the neck and the diaphragm.

Thrombin An enzyme that converts fibrinogen to fibrin, thus triggering the formation of blood clots.

Thrombus (throm' bus) [Gr. *thrombos*, clot] A blood clot that forms within a blood vessel and remains attached to the wall of the vessel. (Contrast with embolus.)

Thylakoid A flattened sac within a chloroplast. The membranes of the numerous thylakoids contain all of the chlorophyll in a plant, in addition to the electron carriers of photophosphorylation. Thylakoids stack to form grana.

Thymus A ductless, glandular portion of the lymphoid system, involved in development of the immune system of vertebrates.

Thyroid [Gr. *thyreos*: door-shaped] A two-lobed gland in vertebrates. Produces the hormone thyroxin.

Thyrotropic hormone A hormone that is produced in the pituitary gland of amphibia such as frogs and transported in the bloodstream to the thyroid gland, inducing the thyroid gland to produce the thyroid hormone that regulates metamorphosis from tadpole to adult frog.

Tight junction A junction between epithelial cells, in which there is no gap whatever between the adjacent cells. Materials may get through a tight junction only by entering the epithelial cells themselves.

Tissue A group of similar cells organized into a functional unit and usually integrated with other tissues to form part of an organ such as a heart or leaf.

Tornaria (tor nare' e ah) [L. *tornus*: lathe] The free-swimming ciliated larva of certain echinoderms and hemichordates; its existence indicates the evolutionary relationship of these two groups.

Torsion The turning of the body and shell of gastropods through 180° that occurs suddenly during development.

Totipotency In a cell, the condition of possessing all the genetic information and other capacities necessary to form an entire individual.

Toxigenicity The ability of a bacterium to produce chemical substances injurious to the tissues of the host organism.

Trachea (tray' kee ah) [Gr. *trakhoia*: a small rough artery] A tube that carries air to the bronchi of the lungs of vertebrates, or to the cells of arthropods.

Tracheid (tray' kee id) A distinctive conducting and supporting cell found in the xylem of nearly all vascular plants, characterized by tapering ends and walls that are pitted but not perforated.

Tracheophyte A vascular plant.

Trade winds The winds that blow toward the intertropical convergence zone from the northeast and southeast.

Transcription The synthesis of RNA, using one strand of DNA as the template.

Transcription factors Proteins that assemble on a eukaryotic chromosome, allowing RNA polymerase II to perform transcription.

Transduction Transfer of genes from one bacterium to another with a bacterial virus acting as the carrier of the genes.

Transfection Uptake, incorporation, and expression of recombinant DNA.

Transfer cell A modified parenchyma cell that transports mineral ions from its cytoplasm into its cell wall, thus moving the ions from the symplast to the apoplast.

Transfer RNA (tRNA) A category of relatively small RNA molecules (about 75 nucleotides). Each kind of transfer RNA is able to accept a particular activated amino acid from its specific activating enzyme, after which the amino acid is added to a growing polypeptide chain.

Transformation Mechanism for transfer of genetic information in bacteria in which pure DNA extracted from bacteria of one genotype is taken in through the cell surface of bacteria of a different genotype and incorporated into the chromosome of the recipient cell. By extension, the term has come to be applied to phenomena in other organisms in which specific genetic alterations have been produced by treatment with purified DNA from genetically marked donors.

Transgenic Containing recombinant DNA incorporated into its genetic material.

Translation The synthesis of a protein (polypeptide). This occurs on ribosomes, using the information encoded in messenger RNA.

Translocation (1) In genetics, a rare mutational event that moves a portion of a chromosome to a new location, generally on a nonhomologous chromosome. (2) In vascular plants, movement of solutes in the phloem.

Transpiration [L. *spirare*: to breathe] The evaporation of water from plant leaves and stem, driven by heat from the sun, and providing the motive force to raise water (plus ions) from the roots.

Transposable element A segment of DNA that can move to, or give rise to copies at, another locus on the same or a different chromosome. May be a single insertion sequence or a more complex structure (transposon) consisting of two insertion sequences and one or more intervening genes.

Trichocyst (trick' o sist) [Gr. *trichos*: hair + *kystis*: cell] A threadlike organelle ejected from the surface of ciliates, used both as a weapon and as an anchoring device.

Triglyceride A simple lipid in which three fatty acids are combined with one molecule of glycerol.

Triplet See codon.

Trisomic Containing three, rather than two members of a chromosome pair.

tRNA See transfer RNA.

Trochophore (troke' o fore) [Gr. *trochos*: wheel + *phoreus*: bearer] The free-swimming larva of some annelids and mollusks, distinguished by a wheel-like band of cilia around the middle, and indicating an evolutionary relationship between these two groups.

Trophic level A group of organisms united by obtaining their energy from the same part of the food web of a biological community.

Tropism In plants, growth toward or away from a stimulus such as light (phototropism) or gravity (gravitropism).

Tropomyosin (troe poe mie' oh sin) A protein that, along with actin, constitutes the thin filaments of myofibrils. It controls the interactions of actin and myosin necessary for muscle contraction.

Troposphere The atmospheric zone reaching upward approximately 17 km in the tropics and subtropics but only to about 10 km at higher latitudes. The zone in which virtually all the water vapor in the atmosphere is located.

T-system A set of transverse tubes that penetrates skeletal muscle fibers and terminates in the sarcoplasmic reticulum. The T-system transmits impulses to the sacs, which then release CA^{2+} to initiate muscle contraction.

Tube foot In echinoderms, a part of the water vascular system. It grasps the substratum, prey, or other solid objects.

Tube nucleus In a pollen tube, the haploid nucleus that does not participate in double fertilization. (Contrast with generative nucleus.)

Tuber [L.: swelling] A short, fleshy underground stem, usually much enlarged, and serving a storage function, as in the case of the potato.

Tubulin A protein that polymerizes to form microtubles.

Tumor A disorganized mass of cells, often growing out of control. Malignant tumors spread to other parts of the body.

Turgor See pressure potential.

Tympanic membrane [Gr. *tympanum*: drum] The eardrum.

Typical predator See carnivore.

Understory The aggregate of smaller plants growing beneath the canopy of dominant plants in a forest.

Unicellular (yoon' e sell' yer ler) [L. *unus*: one + *cella*: chamber] Consisting of a single cell; as for example a unicellular organism. (Contrast with multicellular.)

Unitary organism An organism that consists of only one module.

Unsaturated hydrocarbon A compound containing only carbon and hydrogen atoms. One or more pairs of carbon atoms are connected by double bonds.

Upwelling The upward movement of nutrient-rich, cooler water from deeper layers of the ocean.

Urea A compound serving as the main excreted form of nitrogen by many animals, including mammals.

Ureotelic Describes an organism in which the final product of the breakdown of nitrogen-containing compounds (primarily proteins) is urea. (Contrast with ammonotelic, uricotelic.)

Ureter (your' uh tur) [Gr. *ourein*: to urinate] A long duct leading from the vertebrate kidney to the urinary bladder or the cloaca.

Urethra (you ree' thra) [Gr. *ourein*: to urinate] In most mammals, the canal through which urine is discharged from the bladder and which serves as the genital duct in males.

Uric acid A compound that serves as the main excreted form of nitrogen in some animals, particularly those which must conserve water, such as birds, insects, and reptiles.

Uricotelic Describes an organism in which the final product of the breakdown of nitrogen-containing compounds (primarily proteins) is uric acid. (Contrast with ammonotelic, ureotelic.)

Uterus (yoo' ter us) [L.: womb] The uterus or womb is a specialized portion of the female reproductive tract in certain mammals. It receives the fertilized egg and nurtures the embryo in its early development.

Vaccination Injection of virus or bacteria or their proteins into the body, to induce immunization. The injected material is usually attenuated (weakened) before injection.

Vacuole (vac' yew ole) [Fr.: small vacuum] A liquid-filled cavity in a cell, enclosed within a single membrane. Vacuoles play a wide variety of roles in cellular metabolism, some being digestive chambers, some storage chambers, some waste bins, and so forth.

Vagina (vuh jine' uh) [L.: sheath] In female mammals, the passage leading from the external genital orifice to the uterus; receives the copulatory organ of the male in mating.

Van der Waals interaction A weak attraction between atoms resulting from the interaction of the electrons of one atom with the nucleus of the other atom. This attraction is about one-fourth as strong as a hydrogen bond.

Vascular (vas' kew lar) Pertaining to organs and tissues that conduct fluid, such as blood vessels in animals and phloem and xylem in plants.

Vascular bundle In vascular plants, a strand of vascular tissue, including conducting cells of xylem and phloem as well as thick-walled fibers.

Vascular ray In vascular plants, radially oriented sheets of cells produced by the vascular cambium, carrying materials laterally between the wood and the phloem.

Vascular tissue system The conductive system of the plant, consisting primarily of xylem and phloem. (Contrast with dermal tissue system, ground tissue system.)

Vector (1) An agent, such as an insect, that carries a pathogen affecting another species. (2) A plasmid or virus that carries an inserted piece of DNA into a bacterium for cloning purposes in recombinant DNA technology.

Vegetal pole In some eggs, zygotes, and embryos, the pole near the bulk of the yolk. (Contrast with animal pole.)

Vegetative Nonreproductive, or nonflowering, or asexual.

Vein [L. *vena*: channel] A blood vessel that returns blood to the heart. (Contrast with artery.)

Vena cava [L.: hollow vein] One of a pair of large veins that carry blood from the systemic circulatory system into the heart.

Ventral [L. *venter*: belly, womb] Toward or pertaining to the belly or lower side. (Contrast with dorsal.)

Vernalization [L. *vernalis*: belonging to spring] Events occurring during a required chilling period, leading eventually to flowering. Vernalization may require many weeks of below-freezing temperatures.

Vertebrate An animal whose nerve cord is enclosed in a backbone of bony segments, called vertebrae. The principal groups of vertebrate animals are the fishes, amphibians, reptiles, birds, and mammals.

Vessel [L. *vasculum*: a small vessel] In botany, a tube-shaped portion of the xylem consisting of hollow cells (vessel elements) placed end to end and connected by perforations. Together with tracheids, vessel elements conduct water and minerals in the plant.

Vicariance (vye care' ee unce) [L. *vicus*: change] The splitting of the range of a taxon by the imposition of some barrier to dispersal of its members. May lead to cladogenesis.

Villus (vil' lus) (plural: villi) [L.: shaggy hair] A hairlike projection from a membrane; for example, from many gut walls.

Virion (veer' e on) The virus particle, the minimum unit capable of infecting a cell.

Viroid (vye' roid) An infectious agent consisting of a single-stranded RNA molecule with no protein coat; produces diseases in plants.

Virus [L.: poison, slimy liquid] Any of a group of ultramicroscopic infectious particles constructed of nucleic acid and protein (and, sometimes, lipid) that can reproduce only in living cells.

Vitamin [L. *vita*: life] Any one of several structurally unrelated organic compounds that an organism cannot synthesize itself, but nevertheless requires in small quantity for normal growth and metabolism.

Viviparous (vye vip' uh rus) [L. *vivus*: alive] Reproduction in which fertilization of the egg and development of the embryo occur inside the mother's body. (Contrast with oviparous.)

Waggle dance The running movement of a working honey bee on the hive, during which the worker traces out a repeated figure eight. The dance contains elements that transmit to other bees the location of the food.

Water potential In osmosis, the tendency for a system (a cell or solution) to take up water from pure water, through a differentially permeable membrane. Water flows toward the system with a more negative water potential. (Contrast with osmotic potential, pressure potential.)

Water-vascular system The array of canals and tubelike appendages that serves as the circulatory system, locomotory system, and food capturing system of many echinoderms; is in direct connection with the surrounding sea water.

Wavelength The distance between successive peaks of a wave train, such as electromagnetic radiation.

Wild-type Geneticists' term for standard or reference type. Deviants from this standard, even if the deviants are found in the wild, are said to be mutant.

Xanthophyll (zan' tho fill) [Gr. *xanthos*: yellowish-brown + *phyllon*: leaf] A yellow or orange pigment commonly found as an accessory pigment in photosynthesis, but found elsewhere as well. An oxygen-containing carotenoid.

X chromosome See sex chromosome.

Xerophyte (zee' row fyte) [Gr. *xerox*: dry + *phyton*: plant] A plant adapted to an environment with a limited water supply.

Xylem (zy' lum) [Gr. *xylon*: wood] In vascular plants, the woody tissue that conducts water and minerals; xylem consists, in various plants, of tracheids, vessel elements, fibers, and other highly specialized cells.

Y chromosome See sex chromosome.

Yolk The stored food material in animal eggs, usually rich in protein and lipid.

Z-DNA A form of DNA in which the molecule spirals to the left rather than to the right.

Zoarium (zoh air' ee um) A colony of moss animals (phylum Ectoprocta).

Zooid (zoh' oyd) [Gr. *zoon*: animal + *eidos*: appearance] An individual animal of a colony-forming species in several animal phyla.

Zooplankton (zoe' o plangk' ton) [Gr. *zoon*: animal + *planktos*: wandering] The animal portion of the plankton.

Zoospore (zoe' o spore) [Gr. *zoon*: animal + *spora*: seed] In algae and fungi, any swimming spore. May be diploid or haploid.

Zygote (zye' gote) [Gr. *zygotos*: yoked] The cell created by the union of two gametes, in which the gamete nuclei are also fused. The earliest stage of the diploid generation.

Zygospore A highly resistant type of fungal spore produced by the zygomycetes (conjugating fungi).

Zymogen An inactive precursor of a digestive enzyme secreted into the lumen of the gut, where a protease cleaves it to form the active enzyme. Zymogens make it unnecessary for some digestive enzymes to be formed inside cells, which active ezymes might digest.

ANSWERS TO SELF-QUIZZES

Chapter 2
1. b   6. a
2. e   7. c
3. c   8. b
4. c   9. e
5. d   10. d

Chapter 3
1. e   6. a
2. d   7. c
3. c   8. e
4. d   9. a
5. b   10. d

Chapter 4
1. a   6. e
2. e   7. a
3. c   8. d
4. e   9. b
5. c   10. d

Chapter 5
1. e   6. b
2. d   7. c
3. a   8. b
4. d   9. e
5. c   10. c

Chapter 6
1. c   6. a
2. e   7. e
3. b   8. b
4. c   9. d
5. c   10. e

Chapter 7
1. a   6. d
2. d   7. a
3. c   8. e
4. e   9. c
5. c   10. e

Chapter 8
1. c   6. d
2. b   7. c
3. d   8. d
4. b   9. b
5. e   10. b

Chapter 9
1. e   6. a
2. c   7. e
3. b   8. d
4. d   9. b
5. c   10. a

Chapter 10*
1. d   6. d
2. a   7. b
3. e   8. a
4. d   9. b
5. d   10. c

Chapter 11
1. c   6. d
2. a   7. b
3. c   8. d
4. b   9. d
5. e   10. a

Chapter 12
1. c   6. d
2. d   7. c
3. b   8. a
4. b   9. b
5. e   10. d

Chapter 13
1. d   6. b
2. c   7. c
3. a   8. d
4. b   9. e
5. e   10. a

Chapter 14
1. c   6. c
2. b   7. a
3. e   8. b
4. d   9. c
5. a   10. d

Chapter 15
1. d   6. c
2. c   7. d
3. e   8. b
4. a   9. a
5. b   10. b

Chapter 16
1. d   6. a
2. b   7. d
3. e   8. d
4. e   9. a
5. c   10. e

Chapter 17
1. b   6. c
2. d   7. e
3. e   8. c
4. a   9. a
5. b   10. d

Chapter 18
1. d   6. d
2. c   7. d
3. d   8. e
4. b   9. e
5. a   10. d

Chapter 19
1. c   6. c
2. a   7. a
3. b   8. b
4. e   9. a
5. d   10. c

Chapter 20
1. e   6. d
2. c   7. a
3. a   8. b
4. c   9. c
5. c   10. a

Chapter 21
1. c   6. b
2. b   7. d
3. d   8. b
4. e   9. a
5. e   10. c

Chapter 22
1. a   6. d
2. e   7. c
3. c   8. b
4. d   9. b
5. a   10. d

Chapter 23
1. b   6. a
2. d   7. e
3. e   8. a
4. c   9. c
5. d   10. c

Chapter 24
1. d   6. a
2. c   7. b
3. e   8. c
4. b   9. d
5. c   10. c

Chapter 25
1. c   7. d
2. d   8. c
3. b   9. c
4. e   10. e
5. e   11. a
6. b

*Answers to the Genetics "For Study" questions in Chapter 10 appear at the end of this section.

Chapter 26

1. b	7. d
2. c	8. e
3. a	9. a
4. c	10. e
5. c	11. c
6. b	12. c

Chapter 27

1. d	6. c
2. e	7. e
3. d	8. e
4. a	9. e
5. b	10. e

Chapter 28

1. d	6. b
2. b	7. b
3. e	8. c
4. e	9. a
5. a	10. d

Chapter 29

1. d	6. d
2. b	7. e
3. d	8. a
4. e	9. c
5. a	10. c

Chapter 30

1. d	6. e
2. c	7. a
3. a	8. b
4. e	9. b
5. c	10. e

Chapter 31

1. c	6. b
2. d	7. d
3. d	8. a
4. b	9. e
5. e	10. c

Chapter 32

1. d	6. a
2. b	7. b
3. c	8. e
4. e	9. a
5. c	10. b

Chapter 33

1. c	6. e
2. a	7. a
3. d	8. e
4. b	9. d
5. b	10. c

Chapter 34

1. c	6. b
2. b	7. a
3. d	8. e
4. b	9. b
5. a	10. d

Chapter 35

1. (i) b	4. a
(ii) a	5. d
(iii) c	6. c
(iv) a, b, c	7. e
(v) a, b, c	8. d
2. c, e	9. b
3. e	10. d

Chapter 36

1. d	6. e
2. a	7. e
3. e	8. c
4. d	9. d
5. c	10. b

Chapter 37

1. d	6. e
2. d	7. b
3. a	8. c
4. b	9. c
5. e	10. d

Chapter 38

1. d	6. d
2. e	7. b
3. a	8. a
4. b	9. a
5. b	10. e

Chapter 39

1. e	6. b
2. d	7. c
3. a	8. c
4. b	9. a
5. c	10. d

Chapter 40

1. d	6. d
2. a	7. b
3. c	8. d
4. d	9. c
5. c	10. e

Chapter 41

1. b	6. d
2. e	7. a
3. c	8. b
4. a	9. d
5. b	10. d

Chapter 42

1. b	6. b
2. a	7. e
3. d	8. a
4. c	9. c
5. d	10. e

Chapter 43

1. a	6. e
2. e	7. d
3. b	8. c
4. b	9. d
5. e	10. b

Chapter 44

1. e	6. e
2. c	7. c
3. d	8. b
4. c	9. d
5. a	10. a

Chapter 45

1. b	7. a
2. c	8. c
3. d	9. c
4. a	10. d
5. d	11. a
6. c	

Chapter 46

1. b	7. a
2. e	8. c
3. c	9. b
4. d	10. e
5. e	11. c
6. c	

Chapter 47

1. b	7. a
2. d	8. d
3. d	9. a
4. b	10. b
5. c	11. e
6. c	

Chapter 48

1. a	6. a
2. b	7. d
3. c	8. e
4. b	9. d
5. d	10. a

Chapter 49

1. b	7. d
2. c	8. e
3. d	9. e
4. e	10. a
5. e	11. a
6. a	

Answers to Genetics "For Study" Questions, Chapter 10

1. Yellow parent = $s^Y s^b$; offspring 3 yellow (s^Y–):1 black ($s^b s^b$). Black parent = $s^b s^b$; offspring all black ($s^b s^b$). Orange parent = $s^O s^b$; offspring 3 orange (s^O–):1 black ($s^b s^b$). Both s^O and s^Y are dominant to s^b.

2. Each of the eight boxes should contain Tt.

3. See Figure 10.4.

4. Autosomal. Mother $dp\,dp$, father $Dp\,dp$. If sex-linked, all daughters would be wild-type and sons would be *dumpy*.

5. All females wild-type; all males spotted.

6. F_1 all wild-type, $PpSwsw$; F_2 9:3:3:1 in phenotypes. See Figure 10.7 for analogous genotypes.

7a. Ratio of phenotypes in F_2 is 3:1 (double dominant to double recessive). See Figure 10.12.
7b. The F_1 are $Pby\,pBy$; they produce just two kinds of gamets (Pby and pBy). Combine them carefully and see the 1:2:1 phenotypic ratio fall out in the F_2.

7c. Pink-blistery.
7d. See Figures 9.14 and 9.16. Crossing over took place in the F_1 generation.

8. The genotypes are $PpSwsw$, $ppSwsw$, $Ppswsw$, and $ppswsw$ in a ratio of 1:1:1:1.

9a. 1 black:2 blue:1 splashed white.
9b. Always cross black with splashed white.

10a. $w^+ > w^e > w$.
10b. Parents $w^e w$ and $w^+ Y$. Progeny $w^+ w^e$, $w^+ w$, $w^e Y$, and wY.

11. All will have normal vision because they inherit Dad's wild-type X chromosome, but half of them will be carriers.

12. Agouti parent: $AaBb$. Albino offspring $aaBb$ and $aabb$; black offspring $Aabb$; agouti offspring $AaBb$.

13. The purples parent must be AaB–. If it were AAB–, there would be no white progeny—all the progeny contain B from the white parent. To be purple, that parent must contain at least one B; from the data given we cannot tell whether it is $AaBB$ or $AaBb$.

ILLUSTRATION CREDITS

Authors' Photographs

William K. Purves by Heidi D. Levin
Gordon H. Orians by Carl C. Hansen
H. Craig Heller by Meera C. Heller

Part-Opening Photographs

Part One: Robert Brons/BPS
Part Two: Andrew S. Bajer, Univ. of Oregon
Part Three: Barbara J. Miller/BPS
Part Four: © Bill Wood/NHPA
Part Five: Phil Gates, Univ. of Durham/BPS
Part Six: Gary J. James/BPS
Part Seven: Jon Stewart/BPS

Chapter 1 1.1: Frans Lanting/Minden Pictures. 1.2: © Stephen Krasemann/ NHPA. 1.3a: David E. Rowley/Planet Earth Pictures. 1.3b: R. L. Matthews/Planet Earth Pictures. 1.4: Grant Heilman Photography. 1.5a: Julian Hector/Planet Earth Pictures. 1.5b: Robert Tyrrell. 1.6: © M. I. Walker/ NHPA. 1.7: Ray F. Evert, Dept. of Botany, Univ. of Wisconsin. 1.9: © Stephen Dalton/ NHPA. 1.10: Chip Clark, National Museum of Natural History. 1.11a: © Stephen Dalton/NHPA. 1.11b: David Maitland/ Planet Earth Pictures. 1.12a: Donald Perry. 1.12b: Schering-Plough Inc. 1.14: Frans Lanting/Minden Pictures. 1.19: © Stephen Dalton/NHPA. 1.21a: T. J. Beveridge, Univ. of Guelph, and K. Jarrell, Queens Univ./ BPS. 1.21b: R. Rodewald, Univ. of Virginia/ BPS.

Chapter 2 2.1: Courtesy of IBM Corporation, Research Division, Almaden Research Center. 2.12: Bruce F. Molnia/TERRAPHO-TOGRAPHICS/BPS. 2.17: J. N. A. Lott, McMaster Univ./BPS. Box 2.A upper right: Lara Hartley. Box 2.A lower left: Courtesy of Walter Gehring.

Chapter 3 3.2: J. R. Kennedy, Univ. of Tennessee/BPS. 3.12b: W. F. Schadel, Small World Enterprises/BPS. 3.17a, 3.18a & b: D. Tronrud, Univ. of Oregon. 3.19: Jerome Gross. 3.20a: Richard Alexander, Univ. of Pennsylvania. 3.20b: C. Herd, using INSIGHT, BIOSYM Technologies. 3.20b & c: After B. Alberts et al., 1983, *Molecular Biology of the Cell*, Garland Publishing. 3.22: R. Feldmann, NIH. 3.26: N. L. Max, Lawrence Livermore Natl. Lab.

Chapter 4 4.1a: Paul W. Johnson/BPS. 4.1b: Leon J. Le Beau/BPS. 4.1c: Paul W. Johnson/BPS. 4.1d: Robert Brons/BPS. 4.1e: Jim Solliday/BPS. 4.1f: P. Gates, Univ. of Durham/BPS. 4.1g: G. W. Willis. Ochsner Medical Institution/BPS. 4.1h: M. G. Gabridge, cytoGraphics Inc./BPS. 4.2: J. N. A. Lott, McMaster Univ./BPS. 4.3: J. J. Cardamone, Jr., & B. K. Pugashetti, Univ. of Pittsburgh/BPS. 4.4: S. W. Watson, Woods Hole Oceanographic Inst. 4.5: S. C. Holt, Univ. of Texas Health Science Center, San Antonio/BPS. 4.6a: J. J. Cardamone, Jr., Univ. of Pittsburgh/BPS. 4.6c: S. Abraham & E. H. Beachey, V. A. Medical Center,

Memphis, TN. 4.7a & b: Z. Skobe, Forsyth Dental Center/BPS. 4.7c: N. S. Hayes et al., Univ. of North Carolina School of Medicine. 4.8a & b: Lara Hartley. 4.9 upper left & center: E. H. Newcomb, Univ. of Wisconsin/BPS. 4.9 lower left: E. H. Newcomb & W. P. Wergin, Univ. of Wisconsin/BPS. 4.9 lower right: J. J. Cardamone, Jr., Univ. of Pittsburgh/BPS. 4.10 upper left & center, lower right: B. F. King, Univ. of California, Davis, School of Medicine/BPS. 4.10 lower left: R. Rodewald, Univ. of Virginia/BPS. 4.10 lower center: Hilton Mollenhauer, U.S.D.A. Research Unit, College Station, Texas. 4.11: R. Rodewald, Univ. of Virginia/BPS. 4.12: Jim Solliday/BPS. 4.15: Hilton Mollenhauer, U.S.D.A. Research Unit, College Station, Texas. 4.16: P. Gates, Univ. of Durham/BPS. 4.17: E. H. Newcomb & W. P. Wergin, Univ. of Wisconsin/ BPS. 4.18: D. J. Wrobel, Monterey Bay Aquarium/BPS. 4.20: B. F. King, Univ. of California, Davis, School of Medicine/BPS. 4.21b: G. T. Cole, Univ. of Texas, Austin/ BPS. 4.22c: H. S. Pankratz, Michigan State Univ./BPS. 4.23b: R. Rodewald, Univ. of Virginia/BPS. 4.24: E. H. Newcomb & S. E. Frederick, Univ. of Wisconsin/BPS. 4.25: M. C. Ledbetter, Brookhaven National Laboratory. 4.26a: J. G. Izant, Fred Hutchinson Cancer Research Center. 4.26c: L. E. Roth, Univ. of Tennessee/BPS. 4.27a & b: W. L. Dentler, Univ. of Kansas/BPS. 4.28a: B. F. King, Univ. of California, Davis, School of Medicine/BPS. 4.29: Ned Lamb, Cold Spring Harbor Lab. 4.30a: J. R. Waaland, Univ. of Washington/BPS. 4.30b: E. H. Newcomb, Univ. of Wisconsin/BPS. 4.31a & c: Robley C. Williams, Univ. of California. 4.31b: Timothy S. Baker, Purdue Univ. Box 4.A (photographs): Jim Solliday/BPS. Box 4.A (drawing): After N. Campbell, 1990, *Biology*, 2nd Ed., Benjamin/Cummings Publishing Co. Box 4.B: G. W. Willis, Ochsner Medical Institution/BPS.

Chapter 5 5.2: After L. Stryer, 1981. *Biochemistry*, 2nd ed., W. H. Freeman. 5.4a & b: L. A. Staehelin, Univ. of Colorado/BPS. 5.7: J. D. Robertson, Duke Univ. 5.8c: G. T. Cole, Univ. of Texas, Austin/BPS. 5.9 top: D. S. Friend, Univ. of California, San Francisco. 5.9 center: D. E. Kelly, Univ. of Washington. 5.9 bottom: Courtesy of C. Peracchia. 5.20: M. M. Perry, J. Cell Sci. 39:266, 1979.

Chapter 6 6.1: Peter J. Bryant/BPS. 6.10: Richard Alexander, Univ. of Pennsylvania. 6.11*a* & *b*: After W. S. Bennett & T. A. Steitz, J. Mol. Biol. 140:211-230, 1980. 6.15: © Kay Chernush for Howard Hughes Medical Institute. 6.18*b*: Richard Alexander, Univ. of Pennsylvania. 6.19*b*: Richard Alexander, Univ. of Pennsylvania.

Chapter 7 7.9*a*: Hilton Mollenhauer, U.S.D.A. Research Unit, College Station, Texas. 7.9*b*: E. Racker, Cornell Univ. Box 7.A*a*: Bill Cibula, Picayune, MS. Box 7.A*b*: Ken Lucas/BPS. Box 7.A*c*: G. M. Thomas & G. Poinar, Univ. of California, Berkeley. Box 7.A*d*: K. V. Wood, courtesy of M. DeLuca, Univ. of California, San Diego.

Chapter 8 8.1 inset: J. H. Troughton & L. A. Donaldson. 8.19: J. R. Waaland, Univ. of Washington/BPS. 8.20*a* & *b*: J. A. Bassham, Lawrence Berkeley Lab., Univ. of California. 8.28*b*: E. H. Newcomb & S. E. Frederick, Univ. of Wisconsin/BPS. 8.29: Jon Stewart/BPS. 8.30*b*: E. H. Newcomb & S. E. Frederick, Univ. of Wisconsin/BPS.

Chapter 9 9.1*a*: P. Gates, Univ. of Durham/BPS. 9.1*b*: R. Rodewald, Univ. of Virginia/BPS. 9.2: G. F. Bahr, Armed Forces Inst. of Pathology. 9.3*b*: A. L. Olins, Univ. of Tennessee, Oak Ridge Grad. School of Biomedical Sciences. 9.7: Andrew S. Bajer, Univ. of Oregon. 9.8: J. B. Rattner & S. G. Phillips, J. Cell Biol. 57:359-372, 1973. 9.9*b* & *c*: C. L. Rieder, New York State Dept. of Health/BPS. 9.10*a*: T. E. Schroeder, Univ. of Washington/BPS. 9.10*b*: B. A. Palevitz & E. H. Newcomb, Univ. of Wisconsin/BPS. 9.11: G. T. Cole, Univ. of Texas, Austin/BPS. 9.13*a* & *b*: David Ward, Yale Univ. School of Medicine. 9.14 & 9.15: C. A. Hasenkampf, Univ. of Toronto/BPS. 9.18: B. Schuh, Monmouth Medical Center. 9.19: © Ruth Kavenoff, Designergenes Ltd., P. O. Box 100, Del Mar, CA 92014. 9.20: J. J. Cardamone, Jr., Univ. of Pittsburgh/BPS.

Chapter 10 10.1: Courtesy of Rayla G. & Howard M. Temin, Univ. of Wisconsin, Madison. 10.9 & 10.10: Courtesy of W. Ellis Davies. 10.11: Carl W. May/BPS. 10.16: Namboori B. Raju, Stanford Univ., Eur. J. Cell Biol. 23:208-223, 1980. 10.19: Peter J. Bryant/BPS. 10.22: After N. Campbell, 1990, *Biology*, 2nd Ed., Benjamin/Cummings Publishers. 10.23: © Walter Chandoha, 1991. *Study Question 1:* Barbara J. Miller/BPS.

Chapter 11 11.3: G. W. Willis, Ochsner Medical Institution/BPS. 11.6: M. H. F. Wilkins, Biophysics Dept., Kings College, London. 11.20*c*: N. L. Max, Lawrence Livermore Natl. Lab. 11.23*b*: Courtesy of J. E. Edstrom and EMBO J.

Chapter 12 12.1: Richard Humbert/BPS. 12.4: L. Caro & R. Curtiss. 12.18: Brian Matthews, Univ. of Oregon.

Chapter 13 13.8: Karen Dyer, Vivigen. 13.9: Joseph Gall, Carnegie Institution of Washington. 13.10: O. L. Miller, Jr., & B. R. Beatty.

Chapter 14 14.5: Pal Maliga, Advanced Genetic Sciences. 14.6: J. S. Yun & T. E. Wagner, Ohio Univ. 14.12*b*: Mike Tincher, courtesy of Agracetus. 14.14 bottom: M. L. Pardue & J. G. Gall, Chromosomes Today 3:47-52, 1972. 14.18*a*: P. Gates, Univ. of Durham/BPS. 14.19: Courtesy of Paul F. Umbeck, Agracetus, Inc., subsisiary of W. R. Grace & Co. 14.20: Advanced Genetic Sciences. 14.21: © Wide World Photos Inc., Rob Stapleton, 1989.

Chapter 15 15.1, urchin larva: George Watchmaker. 15.1, urchin: D. J. Wrobel, Monterey Bay Aquarium/BPS. 15.1, tadpole: Peter J. Bryant/BPS. 15.1, frog: Barbara J. Miller/BPS. 15.1, chick embryo: Jim Solliday/BPS. 15.1, chick: Leon J. Le Beau/BPS. 15.9: Peter J. Bryant/BPS. 15.11: Peter J. Bryant/BPS. 15.17: Susan Strome. 15.21: C. Rushlow and M. Levine. 15.22*a* & *b*: F. R. Turner, Indiana Univ. 15.24: E. B. Lewis. 15.25*a*: Kathy Tosney, Univ. of Michigan. 15.25*b*: From B. Alberts et al., 1983, *Molecular Biology of the Cell*, Garland Publishing, New York. 15.26, 15.27, 15.28, & 15.30: After B. Alberts et al., 1983, *Molecular Biology of the Cell*, Garland Publishing, New York. 15.29: After J. E. Sulston & H. R. Horvitz, Dev. Biol. 56:110-156, 1977.

Chapter 16 16.1: Z. Skobe, Forsyth Dental Center/BPS. 16.2: Courtesy of Lennart Nilsson; © Boehringer Ingelheim GmbH. 16.4: G. W. Willis, Ochsner Medical Institution/BPS. 16.9: R. Rodewald, Univ. of Virginia/BPS. 16.11*b*: Arthur J. Olson, Scripps Research Institute; used with permission. 16.15: L. Winograd, Stanford Univ. 16.21: Courtesy of Pamela Bjorkman. 16.22: A. Liepins, Sloan-Kettering Research Inst. 16.26: A. Calin, Stanford Univ. School of Medicine. 16.27: G. W. Willis, Ochsner Medical Institution. 16.28: After R. C. Gallo, "The AIDS Virus," © 1987 by Scientific American, Inc. 16.29: P. P. H. DeBruyn, Univ. of Chicago.

Chapter 17 17.1: Lick Observatory, Univ. of California. 17.5: Courtesy of Richard A. Lutz and Robert C. Vrijenhoek, Rutgers Shellfish Research Laboratory, Rutgers Univ. 17.9: Paul W. Johnson/BPS.

Chapter 18 18.4: After R. Ricklefs, 1989, *Ecology*, 3rd Ed., W. H. Freeman. 18.8: Frank S. Balthis, Nature's Design. 18.9: Harold W. Pratt/BPS. 18.11*a*: Kim Taylor/Bruce Coleman, Inc. 18.11*b*: M. W. F. Tweedie/Bruce Coleman, Inc. 18.12: Francisco Erize/Bruce Coleman, Inc. 18.13*a* & *b*: Gary J. James/BPS. 18.14 & 18.15: After D. Futuyma, 1987, *Evolutionary Biology*, 2nd Ed., Sinauer Assoc. 18.17*a* & *b*: Richard Alexander, Univ. of Pennsylvania.

Chapter 19 19.1*a*: Linda Sims, Smithsonian Institution. 19.1*b*: Raymond A. Mendez. 19.2: Des and Jen Bartlett/Bruce Coleman, Inc. 19.6: Courtesy of Robert Selander, Pennsylvania State Univ. 19.8: Gary J. James/BPS. 19.9: Paul A. Johnsgard, Univ. of Nebraska. 19.10: Anthony D. Bradshaw, Univ. of Liverpool. 19.12*a* & *c*:

Peter J. Bryant/BPS. 19.12*b*: Kenneth Y. Kaneshiro, Univ. of Hawaii at Manoa. 19.13*a*: Edward Ely/BPS. 19.13*b*: © John Shaw/NHPA. 19.14*a*: Ken Lucas/BPS. 19.14*b*: © William Munoz 19.15*a*: Larry Lefever/Grant Heilman Photography, Inc. 19.15*b*: © Heather Angel/BIOFOTOS. 19.18*a*: Gary J. James/BPS. 19.18*b* & *c*: © Jim Denny.

Chapter 20 20.1*a*: Helen E. Carr/BPS. 20.1*b*: Barbara J. Miller/BPS. 20.1*c*: © L. Campbell/NHPA. 20.5: Ken Lucas/BPS. 20.6: Peter J. Bryant/BPS. 20.7*a*: Jon Stewart/BPS. 20.7*b*: Barbara J. Miller/BPS. 20.11*a* & *b*: Paul A. Johnsgard. 20.13*a*: E. D. Brodie, Jr., Univ. of Texas, Arlington/BPS. 20.13*b* & *c*: Ken Lucas/BPS. 20.15*a*: Edward Ely/BPS. 20.15*b*: Barbara J. Miller/BPS. 20.16*c*: Elizabeth N. Orians. 20.18: Peter J. Bryant/BPS.

Chapter 21 21.1: Arthur J. Olson, Scripps Research Institute. 21.2: D. T. Brown et al., J. Virol. 10:524-536, 1972. 21.3*a*: D. L. D. Caspar, Brandeis Univ. 21.3*b*: D. S. Goodsell & A. J. Olson, Scripps Research Institute. 21.3*c*: S. C. Holt, Univ. of Texas Health Science Center, San Antonio/BPS. 21.3*d*: F. A. Murphy, Centers for Disease Control, Atlanta. 21.4: T. J. Beveridge & S. Schultze, Univ. of Guelph/BPS. 21.5 & 21.6: T. J. Beveridge, Univ. of Guelph/BPS. 21.7*a*: Paul W. Johnson/BPS. 21.8: H. W. Jannasch, Woods Hole Oceanographic Inst. 21.9: T. J. Beveridge, Univ. of Guelph/BPS. 21.10*a*: Barbara J. O'Donnell/BPS. 21.10*b*: C. Forsberg & T. J. Beveridge, Univ. of Guelph/BPS. 21.11*a*: Paul W. Johnson/BPS. 21.11*b*: K. Stephens, Stanford Univ./BPS. 21.12*a*: J. A. Breznak & H. S. Pankratz, Michigan State Univ./BPS. 21.12*b*: G. W. Willis, Ochsner Medical Institution/BPS. 21.13: T. J. Beveridge, Univ. of Guelph/BPS. 21.14: G. W. Willis, Ochsner Medical Institution/BPS. 21.15: D. A. Glawe, Univ. of Illinois/BPS. 21.16: G. W. Willis, Ochsner Medical Institution/BPS. 21.17*a*: W. Burgdorfer, Rocky Mountain Lab. 21.17*b*: Natl. Animal Disease Center, Ames, IA. 21.18: S. C. Holt, Univ. of Texas Health Science Center, San Antonio/BPS. 21.19*a*: Paul W. Johnson/BPS. 21.19*b*: H. S. Pankratz, Michigan State Univ./BPS. 21.19*c*: Brian A. Whitton/BPS. 21.20*a* & *b*: Leon J. Le Beau/BPS. 21.21*a* & *b*: G. W. Willis, Ochsner Medical Institution/BPS. 21.22: Centers for Disease Control, Atlanta. 21.23: M. G. Gabridge, cytoGraphics Inc./BPS. Box 21.A upper left: Centers for Disease Control, Atlanta. Box 21.A upper center: S. C. Holt, Univ. of Texas Health Science Center, San Antonio/BPS. Box 21.A lower left: Leon J. Le Beau/BPS. Box 21.A lower center: J. J. Cardamone, Jr., Univ. of Pittsburgh/BPS.

Chapter 22 22.2: © Manfred Kage, Peter Arnold, Inc. 22.5*a*: Paul W. Johnson/BPS. 22.5*b*: G. W. Willis, Ochsner Medical Institution/BPS. 22.5*c*: Dennis D. Kunkel/BPS. 22.7*a*: Paul W. Johnson/BPS. 22.7*b*: Robert Brons/BPS. 22.7*c* & *d*; 22.9*a*: Jim Solliday/BPS. 22.9*b*, *c*, & *d*: Paul W. Johnson/BPS. 22.10*b*: M. A. Jakus, NIH. 22.12: E. V. Gravé. 22.14*a*: Barbara J. Miller/

BPS. 22.14b: Henry Aldrich, Institute of Food and Agricultural Sciences, Univ. of Florida. 22.15a: D. W. Francis, Univ. of Delaware. 22.15b: © David Scharf, Peter Arnold, Inc. 22.16: J. R. Waaland, Univ. of Washington/BPS. 22.17: Dwight R. Kuhn. 22.18a: Paul W. Johnson/BPS. 22.18b: Gary J. James/BPS. 22.19a: Jim Solliday/BPS. 22.19b: V. Cassie. 22.21a: J. N. A. Lott, McMaster Univ./BPS. 22.21b: J. R. Waaland, Univ. of Washington/BPS. 22.22a: Maria Schefter/BPS. 22.22b: J. N. A. Lott, McMaster Univ./BPS. 22.24a: Harold W. Pratt/BPS. 22.24b: J. R. Waaland, Univ. of Washington/BPS. Box 22.Aa: Gerald Corsi/Tom Stack and Associates. Box 22.Ab: J. N. A. Lott, McMaster Univ./BPS.

Chapter 23 23.1a: D. A. Glawe, Univ. of Illinois/BPS. 23.1b: L. E. Gilbert, Univ. of Texas, Austin/BPS. 23.1c: G. L. Barron, Univ. of Guelph/BPS. 23.2b: G. T. Cole, Univ. of Texas, Austin/BPS. 23.3: D. A. Glawe, Univ. of Illinois/BPS. 23.4: Barbara J. Miller/BPS. 23.5: G. L. Barron, Univ. of Guelph/BPS. 23.6: Barbara J. Miller/BPS. 23.8 insets: W. F. Schadel, Small World Enterprises/BPS. 23.9: J. R. Waaland, Univ. of Washington/BPS. 23.10b: D. A. Glawe, Univ. of Illinois/BPS. 23.11: W. F. Schadel, Small World Enterprises/BPS. 23.12a: Jim Solliday/BPS. 23.12b: D. A. Glawe, Univ. of Illinois/BPS. 23.12c: Barbara J. Miller/BPS. 23.13: Centers for Disease Control, Atlanta. 23.14: Barbara J. Miller/BPS. 23.15: © Biophoto Associates. 23.17: G. T. Cole, Univ. of Texas, Austin/BPS. 23.18a: E. I. Friedmann, Florida State Univ. 23.18b: Barbara J. O'Donnell/BPS. 23.19a & c: Barbara J. Miller/BPS. 23.19b & 23.20a: J. N. A. Lott, McMaster Univ./BPS.

Chapter 24 24.1 Information provided by Professor Hermann Pfefferkorn, Dept. of Geology, Univ. of Pennsylvania. 24.3: Gary J. James/BPS. 24.4a & b: © Robert & Linda Mitchell. 24.5a: Lars Egede-Nissen/BPS. 24.5b: P. Gates, Univ. of Durham/BPS. 24.5c: Grant Heilman Photography. 24.7b: J. H. Troughton. 24.8: J. N. A. Lott, McMaster Univ./BPS. 24.9b: © Robert & Linda Mitchell. 24.13a: Barbara J. O'Donnell/BPS. 24.13b: J. N. A. Lott, McMaster Univ./BPS. 24.14a: Carl W. May/BPS. 24.14b: J. N. A. Lott, McMaster Univ./BPS. 24.15a & c: Barbara J. Miller/BPS. 24.15b & 24.16: J. N. A. Lott, McMaster Univ./BPS. 24.19: P. Gates, Univ. of Durham/BPS. 24.20a & c: Gary J. James/BPS. 24.20b: Ken Lucas/BPS. 24.20d: Carl W. May/BPS. 24.24a: Ken Lucas/BPS. 24.24b: J. N. A. Lott, McMaster Univ./BPS. 24.25a & b: Barbara J. Miller/BPS. 24.27a, b, & d: J. N. A. Lott, McMaster Univ./BPS. 24.27c: Barbara J. Miller/BPS. 24.29a: Catherine Pringle/BPS. 24.29b: Lara Hartley/TERRAPHOTOGRAPHICS/BPS. 24.30a: Jon Stewart/BPS. 24.30b & c: Barbara J. Miller/BPS. 24.31: Barbara J. O'Donnell/BPS.

Chapter 25 25.8a: Ken Lucas/BPS. 25.8b: Robert Brons/BPS. 25.8c: D. J. Wrobel, Monterey Bay Aquarium/BPS. 25.9, 25.10, 25.11, 25.12: Adapted from F. M. Bayer and H. B. Owre, 1968, *The Free-Living Lower Invertebrates*, Macmillan. 25.13: After G. &

R. Brusca, 1990, *Invertebrates*, Sinauer Associates. 25.14a & b: D. J. Wrobel, Monterey Bay Aquarium/BPS. 25.15a: Robert Brons/BPS. 25.15b & c; 25.16a; 25.17b: After G. & R. Brusca, 1990, *Invertebrates*, Sinauer Associates. 25.16b; 25.17a: D. J. Wrobel, Monterey Bay Aquarium/BPS. 25.17c & 25.20b: Robert Brons/BPS. 25.20c & 25.21b: Jim Solliday/BPS. 25.21a: After G. & R. Brusca, 1990, *Invertebrates*, Sinauer Associates. 25.22: R. R. Hessler, Scripps Inst. of Oceanography. 25.24a: S. K. Webster, Monterey Bay Aquarium/BPS. 25.24c: Roger K. Burnard/BPS. 25.24d: © Robert & Linda Mitchell. 25.26a: M. P. L. Fogden/Bruce Coleman, Inc. 25.26b: J. N. A. Lott, McMaster Univ./BPS. 25.27a: D. J. Wrobel, Monterey Bay Aquarium/BPS. 25.27b: Barbara J. Miller/BPS. 25.28a: Ken Lucas/BPS. 25.28b: Peter J. Bryant/BPS. 25.28c: L. E. Gilbert, Univ. of Texas, Austin/BPS. 25.28d: Robert Brons/BPS. 25.30a: Ken Lucas/BPS. 25.30b: Peter J. Bryant/BPS. 25.30c: D. J. Wrobel, Monterey Bay Aquarium/BPS. 25.30d: C. R. Wyttenbach, Univ. of Kansas/BPS. 25.31a: Ken Lucas/BPS. 25.31b: Roger K. Burnard/BPS. 25.32a: Richard Humbert/BPS. 25.32b through h: Peter J. Bryant/BPS. 25.34a, d, e, & g: Ken Lucas/BPS. 25.34b: Harold W. Pratt/BPS. 25.34c: D. J. Wrobel, Monterey Bay Aquarium/BPS. 25.34f: J. W. Porter, Univ. of Georgia/BPS. 25.35: D. J. Wrobel, Monterey Bay Aquarium/BPS.

Chapter 26 26.2a & c: D. J. Wrobel, Monterey Bay Aquarium/BPS. 26.2b: Robert Brons/BPS. 26.3: C. R. Wyttenbach, Univ. of Kansas/BPS. 26.6: D. J. Wrobel, Monterey Bay Aquarium/BPS. 26.7: Robert Brons/BPS. 26.8b: © Heather Angel/BIOFOTOS. 26.10: Tom Stack/Tom Stack and Associates. 26.12a & b: D. J. Wrobel, Monterey Bay Aquarium/BPS. 26.13a, c, d, e, & f: Ken Lucas/BPS. 26.13b: Peter Scoones/Planet Earth Pictures. 26.15a: Ken Lucas/BPS. 26.15b: M. P. L. Fogden/Bruce Coleman, Inc. 26.15c: E. D. Brodie, Jr., Univ. of Texas, Arlington/BPS. 26.18a, c, & d: Ken Lucas/BPS. 26.18b: Carl Gans, Univ. of Michigan/BPS. 26.19a: Ken Lucas/BPS. 26.19b: Larry Lipsky/Tom Stack and Associates. 26.20: Courtesy of Carnegie Museum of Natural History, Pittsburgh. 26.22a: B. F. Molnia/TERRAPHOTO-GRAPHICS/BPS. 26.22b: J. N. A. Lott, McMaster Univ./BPS. 26.22c: Timothy W. Ransom/BPS. 26.22d: Peter J. Bryant/BPS. 26.24a: John Cancalosi/Tom Stack and Associates. 26.24b: © Heather Angel/BIOFOTOS. 26.24c: M. P. L. Fogden/Bruce Coleman, Inc. 26.25a: J. N. A. Lott, McMaster Univ./BPS. 26.25b: Merlin D. Tuttle, Bat Conservation International. 26.25c: Ken Lucas/BPS. 26.25d: Gary J. James/BPS. 26.25e: Robert Stottlemyer/BPS. 26.26: From R. Leakey and R. Lewin, 1977, *Origins*, Dutton. Courtesy of Robert Harding Picture Library. 26.28a: © I. Devore/Anthro-Photo. 26.28b: J. Desmond Clark, Dept. of Anthropology, Univ. of California, Berkeley. 26.29: Edward S. Ross. 26.30a & b: Robert E. Ford/TERRAPHOTOGRAPHICS/BPS.

Chapter 27 27.2: W. B. Saunders/BPS. 27.5a: Ken Lucas/BPS. 27.5b: S. M. Awramik, Univ. of California/BPS. 27.6a

(photograph): S. Conway Morris. 27.6a & b (drawings): From S. J. Gould, 1989, *Wonderful Life*, © W. W. Norton. 27.7: Courtesy of the Natural History Museum of London. 27.8: Courtesy of the Smithsonian Institution. 27.9: Ken Lucas/BPS. 27.10: Painting by Rudolph Zallinger; courtesy of the Peabody Museum of Natural History, Yale Univ.

Chapter 28 28.1, 28.2: J. N. A. Lott, McMaster Univ./BPS. 28.3: © Ed Degginger. 28.4: Gary J. James/BPS. 28.5: Grant Heilman Photography. 28.8a: Kevin Schafer/Peter Arnold, Inc. 28.8b: © Ed Degginger. 28.9: Grant Heilman Photography. 28.10a: J. R. Waaland, Univ. of Washington/BPS. 28.10b: © Ed Degginger. 28.10c: Carl W. May/BPS. 28.11; 28.16a & f: J. R. Waaland, Univ. of Washington/BPS. 28.16b, d, & e: P. Gates, Univ. of Durham/BPS. 28.16c: J. N. A. Lott, McMaster Univ./BPS. 28.18; 28.19 insets: J. R. Waaland, Univ. of Washington/BPS. 28.21a: L. Elkin, Hayward State Univ./BPS. 28.21b & c: Jim Solliday/BPS. 28.21d: P. Gates, Univ. of Durham/BPS. 28.23a: Dwight R. Kuhn. 28.23b: © Ed Degginger. 28.23c: J. R. Waaland, Univ. of Washington/BPS. 28.24a & b: P. Gates, Univ. of Durham/BPS. 28.26: J. N. A. Lott, McMaster Univ./BPS. 28.27: Jim Solliday/BPS. 28.28a & b: P. Gates, Univ. of Durham/BPS. 28.30a: W. F. Schadel, Small World Enterprises/BPS. 28.30b: E. J. Cable/Tom Stack and Associates. Box 28.A: R. L. Peterson, Univ. of Guelph/BPS.

Chapter 29 29.1: Thomas Eisner, Cornell Univ. 29.2a: D. J. Wrobel, Monterey Bay Aquarium/BPS. 29.2b: Carl W. May/BPS. 29.4a & b: J. N. A. Lott, McMaster Univ./BPS. 29.7: P. Gates, Univ. of Durham/BPS. 29.10: J. H. Troughton & L. A. Donaldson. 29.12: Thomas Eisner, Cornell Univ. 29.14: J. N. A. Lott, McMaster Univ./BPS. 29.16: Robert E. Ford/TERRAPHOTOGRAPH-ICS/BPS. 29.17; 29.18a & b: J. N. A. Lott, McMaster Univ./BPS. 29.19: Carl W. May/BPS. 29.20: Gary J. James/BPS. 29.21: Barbara J. Miller/BPS. 29.22: J. N. A. Lott, McMaster Univ./BPS. 29.23: © Robert & Linda Mitchell. 29.25: Richard Alexander, Univ. of Pennsylvania.

Chapter 30 30.1: Grant Heilman Photography. 30.5: © William E. Ferguson. 30.8: J. Antonovics, Duke Univ. 30.9: Barbara J. Miller/BPS. 30.11: P. Gates, Univ. of Durham/BPS. 30.12: Barbara J. O'Donnell/BPS. 30.15: E. H. Newcomb & S. R. Tandon, Univ. of Wisconsin/BPS. 30.17 & 30.18: Barbara J. Miller/BPS. 30.19: J. N. A. Lott, McMaster Univ./BPS. 30.20: © William E. Ferguson. 30.21: Barbara J. Miller/BPS. 30.22a & b: Courtesy of D. Steingraeber, Colorado State Univ. 30.23a & b: Barbara J. Miller/BPS. 30.24a: Peter J. Bryant/BPS. 30.24b: Barbara J. Miller/BPS. Box 30.A: Aladar A. Szalay, Univ. of Alberta.

Chapter 31 31.1b: J. R. Waaland, Univ. of Washington/BPS. 31.1c: Jim Solliday/BPS. 31.2: Dennis D. Kunkel/BPS. 31.3: J. N. A. Lott, McMaster Univ./BPS. 31.7a: Gary J. James/BPS. 31.7b: Charles Palek/Tom Stack and Associates. 31.8*a*: J. N. A. Lott, McMaster Univ./BPS. 31.8*b*: P. Gates, Univ. of Durham/BPS. 31.10: Plant Genetics, Inc. 31.11: Jon Stewart/BPS. 31.12: Barbara J. Miller/BPS. 31.13: J. N. A. Lott, McMaster Univ./BPS. Box 31.B: Grant Heilman Photography.

Chapter 32 32.1*a* & *b*: Jane Langdale and Timothy Nelson, Yale Univ. 32.7: Jan Zeevaart, Michigan State Univ. 32.16*b*: B. A. Palevitz, Univ. of Georgia. 32.18: Grant Heilman Photography. 32.19: J. N. A. Lott, McMaster Univ./BPS. 32.20: Robert Stottlemyer/BPS. 32.21: J. N. A. Lott, McMaster Univ./BPS.

Chapter 33 33.1: Yvon LeMaho, Laboratory of Respiratory Physiology, C.N.R.S., Paris. 33.15*a*: B. F. Molnia TERRAPHOTO-GRAPHICS/BPS. 33.15*b*: Timothy W. Ransom/BPS. 33.20*b* & *c*: G. W. Willis, Ochsner Medical Institution/BPS. 33.20*d*: Fran Thomas, Stanford Univ. 33.21*a*: Jon Stewart/BPS. 33.21*b*: Jim Brandenberg/Minden Pictures.

Chapter 34 34.7*a*: AP/Wide World Photos. 34.7*b*: The Bettmann Archive, Inc. 34.9: S. H. Ingbar, Harvard Medical School. Box 34.B: James Sugar/Black Star.

Chapter 35 35.1*a*: © M. Walker/NHPA. 35.1*b*: J. Greenfield/Planet Earth Pictures. 35.1*c*: Geoff du Feu/Planet Earth Pictures. 35.2: © Lennart Nilsson, Bonnier Fakta. 35.3 (photo): M. I. Walker/Photo Researchers. 35.10 (photo): P. Bagavandoss. 35.18*a* & *b*: From *A Child Is Born*. © Lennart Nilsson, Bonnier Fakta.

Chapter 36 36.1: From *Behold Man*. © Lennart Nilsson, Bonnier Fakta. 36.4 (micrograph): From R. G. Kessel and R. H. Kardon, *Tissues and Organs*. © 1979 W. H. Freeman and Company. 36.22 (photo): Dan McCoy, Rainbow.

Chapter 37 37.1*a*: R. A. Steinbrecht. 37.1*b*: G. I. Bernard/Oxford Scientific Films/Animals, Animals. 37.2 (photo): Peter J. Bryant/BPS. 37.5 (micrograph): From R. G. Kessel and R. H. Kardon, *Tissues and Organs*. © 1979 W. H. Freeman and Company. 37.15: S. Fisher, Univ. of California, Santa Barbara. 37.21 (photo): © Anthony Bannister/NHPA. 37.22: E. R. Lewis, Y. Y. Zeevi, & F. S. Werblin, Univ. of California, Berkeley/BPS. 37.28: Alex Kerstich/Planet Earth Pictures. 37.29: © Stephen Dalton/NHPA.

Chapter 38 38.1*b*: D. J. Wrobel, Monterey Bay Aquarium/BPS. 38.2: CNRI/Science Photo Library/Photo Researchers. 38.5*a* & *b*: From R. G. Kessel and R. H. Kardon, *Tissues and Organs*. © 1979 W. H. Freeman and Company. 38.7*a*: M. I. Walker/Science Source/Photo Researchers. 38.7*b* & *c*: G. W.

Willis, Ochsner Medical Institution/BPS. 38.8*e*: F. A. Pepe, Univ. of Pennsylvania School of Medicine/BPS. 38.11: G. W. Willis, Ochsner Medical Institution/BPS. 38.14 (photo): John Dudak/Phototake. 38.19*a*: G. Mili. 38.20: Robert Brons/BPS. 38.24*a*: S. K. Webster, Monterey Bay Aquarium/BPS.

Chapter 39 39.1: Courtesy of NASA. 39.2*a*: © Sea Studios, Inc. 39.2*b*: Robert Brons/BPS. 39.2*c*: © Robert & Linda Mitchell. 39.4 right: © Eric Reynolds/Adventure Photo 1991. 39.6*b*: Peter J. Bryant/BPS. 39.6*c*: Thomas Eisner, Cornell Univ. 39.10: Walt Tyler, Univ. of California/Davis. 39.15: Dennis D. Kunkel/BPS. 39.17 right: Gary J. James/BPS.

Chapter 40 40.1: © SIU School of Medicine/Bruce Coleman, Inc. 40.11: © Ed Reschke. 40.13: UNICEF/Maggie Murray-Lee. 40.15: After N. Campbell, 1990, *Biology*, 2nd Ed., Benjamin/Cummings Publishing. 40.16*a*: From R. G. Kessel and R. H. Kardon, *Tissues and Organs*. © 1979 W. H. Freeman and Company. 40.17*b*: Dennis D. Kunkel/BPS. Box 40.A top: Jon Feingersh/Tom Stack and Associates.

Chapter 41 41.1*a*: Timothy W. Ransom/BPS. 41.1*b*: J. N. A. Lott, McMaster Univ./BPS. 41.1*c*: Ken Lucas/BPS. 41.1*d*: Barbara J. Miller/BPS. 41.6: D. J. Wrobel, Monterey Bay Aquarium/BPS. 41.7: © Robert & Linda Mitchell. 41.8 (photograph): © James D. Watt. 41.11: © Ed Reschke. 41.13*d*: E. S. Strauss. 41.22 inset: Dennis D. Kunkel/BPS.

Chapter 42 42.1: Frans Lanting/Minden Pictures. 42.2*a*: Helen E. Carr/BPS. 42.2*b*: © Ed Degginger. 42.4*b*: Marc Chappell, Univ. of California/Riverside. 42.11: From R. G. Kessel and R. H. Kardon, *Tissues and Organs*. © 1979 W. H. Freeman and Company. 42.13: Vaughan H. Shoemaker, Univ. of California, Riverside. 42.19: Lise Bankir, I.N.S.E.R.M. Unité 90, Hôpital Necker, Paris. Box 42.A left: © Custom Medical Stock Photos. Box 42.B right: © Robert & Linda Mitchell.

Chapter 43 43.1: Peter J. Bryant/BPS. 43.4: Marc Chappell, Univ. of California/Riverside. 43.7*a* & *b*: W. C. Dilger. 43.13: Animals Animals © 1991 G. L. Kooyman. 43.14: Animals Animals © 1991 Michael Dick. 43.18: H. C. Heller. 43.21: © Jonathan Blair/Woodfin Camp & Associates. 43.23: National Optical Astronomy Observatories. 43.24: © Jonathan Blair/Woodfin Camp & Associates.

Chapter 44 44.2: Jim Bradenburg/Minden Pictures. 44.4: John Alcock, Arizona State Univ. 44.5: Georgette Douwma/Planet Earth Pictures. 44.6*a* & *b*: J. Erckmann. 44.7*a*: Anup & Manos Shah/Planet Earth Pictures. 44.9: Norbert Wu. 44.12: T. H. Clutton-

Brock, Univ. of Cambridge. 44.13: John Alcock, Arizona State Univ. 44.14: Elizabeth N. Orians. 44.16: C. H. Greenewalt/VIREO. 44.17: Natalie J. Demong. 44.18*a*: © Anthony Bannister/NHPA. 44.18*b*: Doug Wechsler. 44.20: Patricia Moehlman. 44.21: Jim Brandenburg/Minden Pictures. 44.22: Jonathan Scott/Planet Earth Pictures. 44.23: Frans Lanting/Minden Pictures. Box 44.B: C. H. Greenewalt/VIREO.

Chapter 45 45.1: Douglas Gill, Univ. of Maryland, College Park. 45.2: © Brian Hawkes/NHPA. 45.6: Dennis Johns. 45.11: Frans Lanting/Minden Pictures. 45.13: D. Roby/VIREO. 45.15: Lowell Georgia/Photo Researchers. 45.16: G. Tortoli/Food and Agriculture Organization of the United Nations. 45.19: Elizabeth N. Orians. 45.20: Douglas Sprugel, Univ. of Washington.

Chapter 46 46.1: Alan Herre, Smithsonian Tropical Research Institution, Panama. 46.2: After G. E. Hutchinson, 1978, *An Introduction to Population Ecology*, Yale Univ. Press. 46.5*a*: S. K. Webster, Monterey Bay Aquarium/BPS. 46.5*b*: E. S. Ross. 46.6: Mark Mattock/Planet Earth Pictures. 46.8*a*: Thomas Eisner, Cornell Univ. 46.8*b*: Ken Lucas/BPS. 46.9: © James Carmichael/NHPA. 46.10: Thomas Eisner, Cornell Univ. 46.11: © Peter Kresan. 46.14: Charlie Ott/Photo Researchers. 46.15*a*: Jonathan Scott/Planet Earth Pictures. 46.15*b* © G. I. Bernard/NHPA. 46.16: A. Kerstitch/Planet Earth Pictures. 46.17: © Peter Johnson/NHPA. 46.18*a*: Larry E. Gilbert, Univ. of Texas/Austin. 46.18*c* & *d*: Daniel Janzen, Univ. of Pennsylvania. 46.19*a*: © Laurie Campbell/NHPA. 46.19*b* © Raymond A. Mendez. 46.20: John Alcock, Arizona State Univ. 46.21*a*: © Larry Ulrich. 46.21*b*: E. S. Ross. 46.22*a* & *b*: Thomas Eisner, Cornell Univ. 46.23: Elizabeth N. Orians. 46.25: After M. Begon, J. Harper, and C. Townsend, 1986, *Ecology*, Blackwell Scientific Publications.

Chapter 47 47.8: D. A. Perry, Oregon State Univ. 47.20: Robert Stottlemyer/BPS.

Chapter 48 48.8*a* & *b*: E. O. Wilson, Harvard Univ. 48.12*a*: Barbara J. O'Donnell/BPS. 48.12*b*: Gary J. James/BPS. 48.13*a*: J. N. A. Lott, McMaster Univ./BPS. 48.13*b*: M. Sutton/Tom Stack and Associates. 48.14*a* & *b*: Paul W. Johnson/BPS. 48.15: D. W. Kaufman, Kansas State Univ. 48.16: Edward Ely/BPS. 48.17: Robert Stottlemyer/BPS. 48.18, 48.19*a*: Elizabeth N. Orians. 48.19*b*: Frans Lanting/Minden Pictures. 48.20: Elizabeth N. Orians. 48.21: J. N. A. Lott, McMaster Univ./BPS. 48.22*a* & *b*: R. Apkarian, Univ. of Louisville. 48.24*a*: Bruce H. Robison. 48.24*b*: Larry Tackett/Tom Stack and Associates. 48.26: Carl W. May/BPS. 48.27: D. J. Wrobel, Monterey Bay Aquarium/BPS.

Chapter 49 49.1: Kevin Schafer/Tom Stack and Associates. 49.2: Paintings by H. Douglas Pratt. Courtesy of the Bernice P. Bishop Museum, Honolulu. 49.3: Barbara J. Miller/BPS. 49.4: Robert Stottlemyer/BPS. 49.5: John Gerlach/Tom Stack and Associates. 49.6: Joe McDonald/Tom Stack and Associates. 49.7*a* & *b*: Danny R. Billings, Missouri Dept. of Conservation. 49.8: W. Perry Conway. 49.10: Edward S. Ross. 49.11*b*: Robert Stottlemyer/BPS. 49.12*a*: Richard P. Smith/Tom Stack and Associates. 49.13*a*: © Walt Anderson. 49.13*b*: Warren & Genny Garst/Tom Stack and Associates. 49.13*c*: Frans Lanting/Minden Pictures. 49.15*a*: Thomas Kitchin/Tom Stack and Associates. 49.15*b*: Kevin Schafer/Tom Stack and Associates. 49.16 & 49.17: Courtesy of the Smithsonian Institution Office of Environmental Awareness; Richard Bierregaard, photographer. 49.20: Kevin Schafer/Tom Stack and Associates. 49.21*a* & *b*: © Bill Gabriel, BIOGRAPHICS. 49.22*a* & *b*: Peter Walhern, The University College of Wales.

INDEX

ABOUT THE BOOK

Typography: The text and captions are set in Palatino.
               Display elements are set in Optima.

Editor: Andrew D. Sinauer

Project Editor: Carol J. Wigg

Developmental Editor: Elmarie Hutchinson

Copy Editor: Norma S. Roche

Production Manager: Joseph J. Vesely

Book Production Coordinator: Janice Holabird

Art Editing and Illustration Program: J/B Woolsey Associates

Photo Research: Biological Photo Service; Travis Amos; Jane Potter

Book and Cover Design: Rodelinde Graphic Design

Composition: DEKR Corporation

Color Separations: Tru-Color Reproductions, Inc.

Prepress: R.R. Donnelley & Sons Company, Crawfordsville, IN

Cover Manufacture: New England Book Components, Inc.

Book Manufacture: R.R. Donnelley & Sons Company, Willard, OH